Metals Handbook® Ninth Edition

Volume 7
Powder Metallurgy

Prepared under the direction of
the ASM Handbook Committee

Planned, organized, and reviewed
by the ASM Powder Metallurgy Committee

Coordinator
Erhard Klar

William H. Cubberly, Director of Publications
Robert L. Stedfeld, Assistant Director of Publications
Kathleen Mills, Managing Editor
Joseph R. Davis, Technical Editor
Sunniva K. Refsnes, Technical Editor
Bonnie R. Sanders, Production Editor
Heather J. Frissell, Editorial Assistant
Diane M. Jenkins, Word Processing Specialist

Editorial Assistance
Robert T. Kiepura
Terri L. Weintraub

AMERICAN SOCIETY FOR METALS
METALS PARK, OHIO 44073

Metals Handbook is a collective effort involving thousands of technical specialists. It brings
together in one book a wealth of information from world-wide sources to help scientists, engineers, and technicians solve current and long-range problems.

Great care is taken in the compilation and production of this volume, but it should be made
clear that no warranties, express or implied, are given in connection with the accuracy or completeness of this publication, and no responsibility can be taken for any claims that may arise.

Nothing contained in the Metals Handbook shall be construed as a grant of any right of manufacture, sale, use, or reproduction, in connection with any method, process, apparatus, product,
composition, or system, whether or not covered by letters patent, copyright, or trademark, and
nothing contained in the Metals Handbook shall be construed as a defense against any alleged
infringement of letters patent, copyright, or trademark, or as a defense against any liability for
such infringement.

Comments, criticisms, and suggestions are invited, and should be forwarded to the American
Society for Metals.

Library of Congress Cataloging in Publication Data

American Society for Metals

Metals handbook.

Includes bibliographical references and indexes.
Contents: v. 1. Properties and selection—v. 2.
Properties and selection—nonferrous alloys and pure
metals—[etc.]—v. 7. Powder metallurgy.

1. Metals—Handbooks, manuals, etc. I. American
Society for Metals. Handbook Committee.
TA459.M43   1978   669   78-14934
ISBN 0-87170-013-1

SAN 204-7586

Printed in the United States of America

Foreword

Due to the great demand for the papers presented at the 1940 and 1941 Conferences on Powder Metallurgy held at the Massachusetts Institute of Technology, the American Society for Metals published its first book on metal powder technology in 1942. Many of the developments that led to substantial growth in the powder metallurgy industry during the 1940's were presented at these conferences. Of particular significance were advancements reported in the fabrication of parts made from iron powder, which contributed toward consideration of P/M processing as a viable alternative to conventional forming methods.

Volume 7 of the 9th edition of *Metals Handbook* represents ASM's recognition of the continued growth and sophistication of the metal powder industry. In view of the growing importance and acceptance of this technology, it is appropriate that *Powder Metallurgy* is the first completely new volume in the 9th edition, rather than an updated and expanded version of an 8th edition volume.

The ASM Handbook Committee and Powder Metallurgy Committee, the authors, and the reviewers, under the capable leadership of Dr. Erhard Klar, have collaborated to produce a book that meets the high technical standards established by previous Handbook volumes. In addition to in-depth studies on production of powders, testing and characterization of powders, and consolidation of powders, the text is supplemented by a comprehensive list of P/M terms and definitions. The extensive review of applications of metal powders that serves as the book's conclusion is designed to foster increased awareness of the current status and potential of the P/M industry. To all who contributed toward the completion of this task, we extend our sincere thanks.

Donald J. Blickwede
President

Edward L. Langer
Managing Director

The Ninth Edition of Metals Handbook
is dedicated to the memory of
TAYLOR LYMAN, A.B. (Eng.), S.M., Ph.D.
(1917–1973)
Editor, Metals Handbook, 1945–1973

Preface

In previous volumes of *Metals Handbook*, the subject of powder metallurgy has been dealt with in a few small chapters. The devotion, in the 9th edition, of an entire volume to powder metallurgy serves justice to both the increasing importance of this technology and the growing interest of ASM members in the subject. Although powder metallurgy is still a small industry (less than 1% of the total metals industry), P/M products are found in virtually all facets of life; without them, many achievements and conveniences of our civilization would be inconceivable.

The success of the P/M industry—it has enjoyed an above-average growth rate for the past four decades—derives from its ability to mass-produce complex structural parts with savings in labor, material, and/or energy. In recent years, an entirely new dimension has been added with the achievement of full density and improved control of purity and microstructure. This has resulted in the fabrication of high-performance materials, permitting extended service life or more efficient use. Examples of commercial uses include aerospace superalloys, low-alloy steels, dispersion-strengthened alloys, and tool steels. These developments have added considerable diversification to the conventional press-and-sinter technique and are opening up new markets beyond the traditional ones of the automotive, farm, lawn and garden, and office equipment industries.

After a brief overview of the history of powder metallurgy, this Volume, paralleling the segmentation of the industry itself, is divided into four major Sections: "Production of Metal Powders," "Characterization and Testing of Metal Powders," "Consolidation of Metal Powders" (followed by a series of articles on special and developing consolidation processes), and "Powder Systems and Applications." Major emphasis is placed on describing the present state of the art—that is, on materials and processes as they are currently used in industry.

The description of powder production processes is much more detailed than what can be found in the published literature. Similarly, the applications section is more detailed than what is taught in textbooks on powder metallurgy. Some of the more recent methods of powder characterization (particle measurement and surface analysis, for example) were given more space in the hope that they may contribute to a better appreciation of the often subtle but critical interdependence of powder properties and subsequent powder processing.

The space devoted to presses, tooling, and furnaces reflects progress made in production equipment as well as recent concerns regarding energy conservation. Because of their importance, considerable space is devoted to processes leading to full density, such as vacuum sintering, rolling, forging, extruding, and hot isostatic pressing.

The presentation of information is such that the book serves as an introduction to powder metallurgy for the technically trained individual and as a reference book for the experienced engineer. The materials design engineer will find the information necessary to judge whether P/M can offer a cost-effective solution or alternative for a given application and whether a P/M parts producer should be consulted for more details. Thus, this Volume represents an important contribution to the metallurgical, materials, and design engineering communities.

Planning and organization of this Volume were accomplished through the Metals Handbook and Powder Metallurgy Committees of ASM. The articles were written by a large number of P/M specialists from industry, private and governmental laboratories, universities, and consulting firms. ASM expresses appreciation and thanks to these experts. I would also like to thank the Metal Powder Industries Federation for their cooperation in this project.

Special thanks are due to the various chairmen and to the ASM editorial staff for their efforts in organizing and integrating individual manuscripts within broad subject areas, and to the many reviewers for their contributions. Finally, the dedication and general support by ASM editorial staff members made this book into a very readable document. Publication of this Volume would have been impossible without the contributions of all of these individuals and groups.

Dr. Erhard Klar
Coordinator
Powder Metallurgy Handbook

Policy on Units of Measure

By a resolution of its Board of Trustees, the American Society for Metals has adopted the practice of publishing data in both metric and customary U.S. units of measure. In preparing this Handbook, the editors have attempted to present data primarily in metric units based on Système International d'Unités (SI), with secondary mention of the corresponding values in customary U.S. units. The decision to use SI as the primary system of units was based on the aforementioned resolution of the Board of Trustees, the widespread use of metric units throughout the world, and the expectation that the use of metric units in the United States will increase substantially during the anticipated lifetime of this Handbook.

For the most part, numerical engineering data in the text and in tables are presented in SI-based units with the customary U.S. equivalents in parentheses (text) or adjoining columns (tables). For example, pressure, stress, and strength are shown both in SI units, which are pascals (Pa) with a suitable prefix (see the description of SI at the back of this Volume), and in customary U.S. units, which are pounds per square inch (psi). To save space, large values of psi have been converted to kips per square inch (ksi), where one kip equals 1000 pounds. Some strictly scientific data are presented in SI units only.

On graphs and charts, grids correspond to SI-based units, which appear along the left and bottom edges; where appropriate, corresponding customary U.S. units appear along the top and right edges.

Data pertaining to a specification published by a specification-writing group may be given in only the units used in that specification or in dual units, depending on the nature of the data. For example, the typical yield strength of aluminum sheet made to a specification written in customary U.S. units would be presented in dual units, but the thickness specified in that specification might be presented only in inches.

Conversions and rounding have been done in accordance with ASTM Standard E 380, with careful attention to the number of significant digits in the original data. For example, a sintering temperature of 2050 °F contains three significant digits. In this instance, the equivalent temperature would be given as 1120 °C; the exact conversion to 1121.11 °C would not be appropriate. For an invariant physical phenomenon that occurs at a precise temperature (such as the melting of pure silver), it would be appropriate to report the temperature as 961.93 °C or 1763.5 °F. In many instances (especially in tables and data compilations), temperature values in °C and °F are alternatives rather than conversions.

The policy on units of measure in this Handbook contains several exceptions to strict conformance to ASTM E 380; in each instance, the exception has been made to improve the clarity of the Handbook. The most notable exception is the use of g/cm^3 rather than kg/m^3 as the unit of measure for density (mass per unit volume). In addition, density measurements in either lb/ft^3 or $lb/in.^3$ have not been used in text. Conversion factors for these units of measure can be found in the Appendix "Metric and Conversion Data for Powder Metallurgy" in this Volume. Other examples of such exceptions are the use of "L," rather than "l," as the abbreviation for litre, and reporting stress intensity in MPa $\sqrt{m}$ rather than $MNm^{-3/2}$.

SI practice requires that only one virgule (diagonal) appear in units formed by combination of several basic units. Therefore, all of the units preceding the virgule are in the numerator and all units following the virgule are in the denominator of the expression; no parentheses are required to prevent ambiguity.

Handbook Committee, Officers and Trustees

Members of the ASM Handbook Committee
(1983-1984)

ix

ASM
Powder Metallurgy
Committee

x

Author/Reviewer Committees

Production of Metal Powders

Erhard Klar
Chairman
SCM Metal Products

J.C. Agarwal
AMAX Specialty Metals

Harry D. Ambs
SCM Metal Products

Vidhu Anand
Cabot Corp.

Donald H. Antonsen
The International Nickel Co., Inc.

Daniel R. Barch
Aluminum Co. of America

L. James Barnard
Kelsey-Hayes Co.

Robert P. Beliles
George Washington University

A. David Booz
Alcoa Laboratories

Robert J. Brandt
Bethlehem Steel Corp.

Graham Brown
Consultant

Robert S. Busk
International Magnesium
Consultants, Inc.

Joseph M. Capus
Quebec Metal Powders Ltd.

Richard F. Cheney
GTE Products Corp.

Richard R. Corle
Rockwell International

C. James Dahn
Safety Consulting Engineers, Inc.

Russell T. Dotter
Patterson-Kelley Co.
Harsco Corp.

Robert E. Droegkamp
Fansteel, Inc.

Edward J. Dulis
Crucible Research Center
Colt Industries

Peter C. Eloff
General Electric Co.

D. Eylon
Metcut-Materials Research Group

James W. Fesko
SCM Metal Products

Charles W. Fox
Homogeneous Metals, Inc.

Ira L. Friedman
Metallurgical Industries, Inc.

F.H. Froes
U.S. Air Force
AFWAL Materials Laboratory

Howard D. Glicksman
Engelhard Industries, Inc.

Gary Goller
Greenback Industries, Inc.

Per Ulf Gummeson
Hoeganaes Corp.

Harold T. Harrison
Pyron Corp.

Warren J. Haws
Brush Wellman, Inc.

C.D. Himmelblau
Lockheed Missiles & Space Co., Inc.

David L. Houck
GTE Products Corp.

Walter V. Knopp
P/M Engineering & Consulting
Co., Inc.

Edward J. Kosinski
Nuclear Metals, Inc.

William E. Kuhn
Dymatron, Inc.

Kishor M. Kulkarni
Cabot Corp.

John B. Lambert
Fansteel, Inc.

Alan Lawley
Drexel University

Fritz V. Lenel
Rensselaer Polytechnic Institute

Gordon R. Love
Sprague Electric Co.

James M. Marder
Brush Wellman Inc.

William D. Marks
SCM Metal Products

Albert J. Mastrangelo
SCM Metal Products

Paul E. Matthews
United States Bronze Powders, Inc.

Basil Meddings
Sherritt Gordon Mines Ltd.

Albert J. Neupaver
Pfizer, Inc.

M. Novotny
Engelhard Industries, Inc.

Robert J. Patterson II
Pratt & Whitney Aircraft Group

Otto P. Preuss
Brush Wellman, Inc.

Ranjan Ray
Marko Materials, Inc.

Clifford Reid
Reade Manufacturing Co., Inc.

Peter R. Roberts
Nuclear Metals, Inc.

Bruce D. Rowan
Handy & Harman

Louis G. Roy
Domfer Metal Powders Ltd.

Prasanna K. Samal
SCM Metal Products

Herbert E. Stokinger
Consultant in Industrial
Toxicology

Vyacheslav Styskin
Ervin Industries, Inc.

Wade Summers
Sweco, Inc.

Mark Svilar
SCM Metal Products

Arno Szegvari
Union Process Inc.

Pierre W. Taubenblat
AMAX Base Metals Research
& Development, Inc.

C.B. Thompson
SCM Metal Products

John H. Tundermann
Huntington Alloys, Inc.

William J. Ullrich
Alcan Ingot and Powders

James E. Williams, Jr.
Aluminum Co. of America

K.C. Yang
Handy & Harman

C.F. Yolton
Crucible Research Center
Colt Industries

Characterization and Testing of Metal Powders

Harry D. Ambs
Co-Chairman
SCM Metal Products

David H. Ro
Co-Chairman
TRW Inc.

Leo A. Adams
Metallurgical Industries, Inc.

Phillip J. Andersen
Zimmer, Inc.

Norbert A. Arnold
Keystone Carbon Co.

A. Ashurst
SCM Metal Products

Arthur B. Backensto
SCM Metal Products

J.K. Beddow
University of Iowa

Robert H. Berg
Particle Data, Inc.

Chi-Ren Chang
Industrial Vision Systems, Inc.

John E. Davidson
Quebec Metal Powders Ltd.

Manek R. Dustoor
Imperial Clevite, Inc.

Carl E. Evans
Pyron Corp.

H.E. Exner
Max-Planck-Institut für
   Metallforschung

Cynthia L. Freeby
The New Jersey Zinc Co.

Harold N. Frock
Leeds and Northrup Instruments

Warren J. Haws
Brush Wellman, Inc.

Peter J. Heinzer
Imperial Clevite, Inc.

Louise H. Hua
University of Iowa

James Lee Hubbard
Georgia Institute of Technology

A. Joshi
Lockheed Research Laboratory

Brian H. Kaye
Laurentian University

Erhard Klar
SCM Metal Products

Albert J. Neupaver
Pfizer, Inc.

Mary A. Pao
SCM Metal Products

Donald S. Parsons
GTE Products Corp.

Krishnakant B. Patel
Alcan Ingot and Powders

Willard E. Soper
Adamas Carbide Corp.

Athan Stosuy
Hoeganaes Corp.

C.B. Thompson
SCM Metal Products

Milton W. Toaz
Imperial Clevite, Inc.

William J. Ullrich
Alcan Ingot and Powders

Ervin E. Underwood
Georgia Institute of Technology

John C. Witsberger
Pfizer, Inc.

Consolidation of Metal Powders

S.W. McGee
Co-Chairman
Ervin Industries, Inc.

Peter W. Lee
Co-Chairman
The Timken Co.

Harry D. Ambs
SCM Metal Products

Maurice D. Ayers
Metal Innovations Inc.

Karl K. Breit
Autoclave Engineers, Inc.

Laird M. Brislen
A.C. Compacting Presses Inc.

Howard E. Boyer
Consultant

Robert Burns
Cincinnati Incorporated

Richard F. Cheney
GTE Products Corp.

Leonard L. Confer
National Carbide Die

A.P. Crease, Jr.
Drever Co.
(deceased)

William C. Diman
C.I. Hayes Inc.

William R. Duncan
Sherritt Gordon Mines Ltd.

Mark Eisenmann
Airco Industrial Gases
Division of Airco, Inc.

Peter C. Eloff
General Electric Co.

B. Lynn Ferguson
Deformation Control Technology

J.J. Fischer
Huntington Alloys, Inc.

Gary W. Gaines
Union Carbide Corp.

Randall M. German
Rensselaer Polytechnic Institute

David O. Gothard
Huntington Alloys, Inc.

Thomas E. Hale
General Electric Co.

Francis G. Hanejko
Hoeganaes Corp.

Hugh Hanes
ASEA Pressure Systems Inc.

Richard W. Heckel
Michigan Technological University

Arun D. Jatkar
NOVAMET Aluminum
Inco Alloy Products Co.

William Q. Judge
Engineered Sinterings & Plastics

Erhard Klar
SCM Metal Products

Roger A. Klein
Trexler Rubber Co.

Walter V. Knopp
P/M Engineering & Consulting
Co., Inc.

Steven P. Kohler
Massachusetts Institute of
Technology

Howard A. Kuhn
University of Pittsburgh

Fritz V. Lenel
Rensselaer Polytechnic Institute

Paul E. Matthews
United States Bronze Powders, Inc.

Prakash K. Mirchandani
Michigan Technological University

Alan J. Moses
Pfizer, Inc.

Thomas G. Mulcavage
Reese Metal Products

Harbhajan S. Nayar
Airco Industrial Gases
Division of Airco, Inc.

Leander F. Pease III
Powder-Tech Associates, Inc.

Manfred K. Piper
Drever Co.

Vernon C. Potter
Merriman Division
Litton Industrial Products

Peter E. Price
Industrial Materials Technology, Inc.

Howard I. Sanderow
Supermet
Division of Stanadyne Inc.

Fred A. Schaa, Jr.
A.C. Compacting Presses, Inc.

Darrell W. Smith
Michigan Technological University

Henry R. Stephenson
Reynolds Engineering Co.

Jerome P. Straub
Keystone Carbon Co.

Mark Svilar
SCM Metal Products

Milton W. Toaz
Imperial Clevite, Inc.

Victor A. Tracey
Inco Europe Ltd.

John H. Tundermann
Huntington Alloys, Inc.

Norman L. Ward
Burgess-Norton Manufacturing Co.

Fred S. Wheeler
Pentronix Inc.

Special and Developing Consolidation Processes

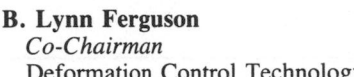

B. Lynn Ferguson
Co-Chairman
Deformation Control Technology

Claus G. Goetzel
Co-Chairman
Stanford University

Vidhu Anand
Cabot Corp.

Russell W. Burman
AMAX Specialty Metals Corp.

Arnold R. Erickson
Brunswick Corp.
Technetics/Witec

Peter Gebhard
Impco Inc.

Hugh Hanes
ASEA Pressure Systems Inc.

Alfred Hofstatter
Metal Alloys, Inc.

Clifford A. Kelto
Kelsey-Hayes Powder Technology Ctr.

William B. Kent
Universal-Cyclops Steel

Alan Lawley
Drexel University

Fritz V. Lenel
Rensselaer Polytechnic Institute

Paul E. Matthews
United States Bronze Powders, Inc.

S.W. McGee
Ervin Industries, Inc.

Stanley Mocarski
Ford Motor Co.

G.J. Perelli
Major Gauge & Tool Co.

Roger Pinney
ASEA Pressure Systems Inc.

S.J. Savage
U.S. Air Force
AFWAL Materials Laboratory

Oliver D. Smith
The Ceracon Corp.

John E. Smugeresky
Sandia National Laboratories

Jerry Toops
ASEA Pressure Systems Inc.

Raymond E. Wiech, Jr.
Geneva Materials Research, Ltd.

Powder Systems and Applications

Kenneth H. Moyer
Co-Chairman
Hoeganaes Corp.

Krishna M. Vedula
Co-Chairman
Case Western Reserve University

Vidhu Anand
Cabot Corp.

Phillip J. Andersen
Zimmer, Inc.

R.J. Arena
General Electric Co.

Norbert A. Arnold
Keystone Carbon Co.

Frank S. Babish
Sandvik Steel Co.

Louis W. Baum
Remington Arms Co., Inc.

Roy E. Beal
Amalgamated Technologies, Inc.

Kermit R. Beckman
Union Carbide Corp.

Ruth Bellows
Engineered Product Sales Corp.

John S. Benjamin
Inco Alloy Products Co.

Arnold Berlage
Oregon State University

M.J. Bila
Austin Powder Co.

Donald L. Bly
General Electric Co.

John A. Campbell
Carmet Co.

Joseph M. Capus
Quebec Metal Powders Ltd.

James L. Chevalier
U.S. Army Tank-Automotive
Command

Carl F. Cline
Lawrence Livermore Laboratory

Ron Collins
Deutsch Co.

Andrew Crowson
U.S. Army Research &
Development Center

E.H. deButts
Hercules Aerospace

Arthur S. Diamond
Diamond Research Corp.

John J. Doyle
Chrysler Corp.

Joseph Y. Dress
Roper Corp.

Robert E. Droegkamp
Fansteel, Inc.

Manek R. Dustoor
Imperial Clevite, Inc.

John T. English
Memorex D.I.C. Corp.

D. Eylon
Metcut-Materials Research Group

Charles P. Farago
Kirkwood Carbon Co.

Mae I. Fauth
Naval Surface Weapons Center

B. Lynn Ferguson
Deformation Control Technology

Gerald Friedman
TRW Inc.

F.H. Froes
U.S. Air Force
AFWAL Materials Laboratory

Randall M. German
Rensselaer Polytechnic Institute

O.M. Gigantino
Linde Division
Union Carbide Corp.

Paul S. Gilman
Inco Alloy Products Co.

Claus G. Goetzel
Stanford University

Nicholas J. Grant
Massachusetts Institute of
Technology

Mark S. Greenfield
Kennametal, Inc.

J.J. Grisik
General Electric Co.

Robert J. Hagenbach
Xerox Corp.

John R. Howell
Deere & Co.

Norman S. Hoyer
Westinghouse Electric Corp.

Saul Isserow
Army Materials & Mechanics
Research Center

Myron I. Jaffe
Sintered Metals Inc.

Warren R. Jensen
Raymark Corp.

William R. Johnson
Newmet Products Inc.

Lewis O. Jones
D.M. Steward Manufacturing Co.

Ihab L. Kamel
Drexel University

Sydney M. Kaufman
Ford Motor Co.

Gerald R. Kilp
Westinghouse Nuclear Fuel Division

Jon A. Kish
Rhenium Alloys, Inc.

Erhard Klar
SCM Metal Products

Charles S. Kortovich
TRW, Inc.

Kishor M. Kulkarni
Cabot Corp.

Robert P. Kusy
University of North Carolina

Chaman Lall
Remington Arms Co., Inc.

John B. Lambert
Fansteel, Inc.

Alan Lawley
Drexel University

Karl W. Lazar
Clad Metals Industries, Inc.

Sam Leber
General Electric Co.

Peter W. Lee
The Timken Co.

Fritz V. Lenel
Rensselaer Polytechnic Institute

Frank D. Leone
Pitney Bowes Inc.

James M. Marder
Brush Wellman, Inc.

William F. Marley Jr.
U.S. Army Foreign Science
and Technology Center

Paul E. Matthews
United States Bronze Powders, Inc.

John Mikurak
Sandvik Steel Co.

Stanley Mocarski
Ford Motor Co.

John H. Moll
Crucible Research Center
Colt Industries

L.N. Moskowitz
Standard Oil Co. (Indiana)

Lowell Mott
Hobart Brothers Co.

Gail G. Munk
Lovelock Seed Co.

Peter C. Murphy
Advanced Metallurgy Inc.

Anil V. Nadkarni
SCM Metal Products

James J. Oakes
Teledyne Firth Sterling

Donald T. Ostberg
U.S. Army Tank-Automotive
Command

George Otto
The Maytag Co.

Seyfi Ozsever
Raymark Corp.

John Patrick, Jr.
SCM Metal Products

Henry L. Patts
LECO Carbide

Robert L. Peaslee
Wall Colmonoy Corp.

Thomas W. Penrice
Teledyne Firth Sterling

Larry Pitchford
Reynolds Metals Co.

Harold L. Pope
Caterpillar Tractor Co.

Ranjan Ray
Marko Materials, Inc.

William E. Ray
Westinghouse Electric Corp.

Peter R. Roberts
Nuclear Metals, Inc.

Kempton H. Roll
Metal Powder Industries Federation
American Powder Metallurgy
Institute

Howard I. Sanderow
Supermet
Division of Stanadyne Inc.

Peter Schmey
United States Bronze Powders, Inc.

J. Thomas Schmidt
Magnaflux Corp.

Murray A. Schwartz
U.S. Bureau of Mines

Otto A. Siegwarth
Uddeholm Corp.

Michael S. Sierdzinski
Alloy Rods, Inc.

James J. Snyder
Gould, Inc.
Ocean Systems Division

F.B. Stern, Jr.
(retired)

Ralph W. Stevenson
Midland Ross Corp.

James E. Synk
SCM Metal Products

Russell S. Tarr
Thermet, Inc.

Edward L. Thellmann
Gould, Inc.
Ocean Systems Division

Barry Van Hoose
Alcan Ingot and Powders

Peter Vernia
General Motors Research Laboratories

Roger C. Waugh
Lockheed Georgia Co.

Charles A. Winters
GAF Corp.

James R. Woodruff
CMW Inc.

Contents

Terms and Definitions*

A

absolute density. See *density, absolute.*

absolute pore size. The maximum pore opening of a porous material, such as a filter, through which no large particle will pass.

acicular powder. A powder composed of needle or sliverlike particles.

activated sintering. The use of additives, such as chemical additions to the powder or additions to the sintering atmosphere, to improve the densification rate.

activation. The enlargement of the surface free energy or lattice binding energy of a solid. Also, the changing of a chemically passive surface of a metal to a chemically active state.

activation energy. Generally, the energy required for initiating a chemical reaction or physical process such as diffusion or plastic flow.

activator. The additive used in activated sintering, also called a dopant.

adhesion. The force of attraction between the atoms or molecules of two different phases.

agitator. A device to intensify mixing. Example: a high-speed stirrer or paddle in a blender or drum of a mill.

agglomerate (noun). An assembly of powder particles of one or more constituents clustered closely together.

agglomerate (verb). To develop an adherent cluster of particles.

aggregate (noun). A mass of particles.

aggregate (verb). To create a mass of particles; see *agglomerate.*

air classification. The separation of a powder into particle size ranges by means of an air stream of controlled velocity.

alloy powder, alloyed powder. A metal powder consisting of at least two constituents that are partially or completely alloyed with each other.

amorphous powder. A powder that consists of particles that are substantially noncrystalline in character.

angle of repose. The angular contour that a powder pile assumes.

annealed powder. A powder that is heat treated to render it soft and compactible.

antechamber. The entrance vestibule of a continuously operating sintering furnace.

aperture size. The opening of a mesh as in a sieve.

apparent density. The weight of a unit volume of powder, usually expressed as grams per cubic centimetre, determined by a specified method.

apparent hardness. The value obtained by testing a sintered object with standard indentation hardness equipment. Because the reading is a composite of pores and solid material, it is usually lower than that of a wrought or cast material of the same composition and condition. Not to be confused with *particle hardness.*

apparent pore volume. The total pore volume of a loose powder mass or a green compact. It may be calculated by subtracting the apparent density from the theoretical density of the substance.

arborescent powder. See preferred term *dendritic powder.*

atomization. The disintegration of a molten metal into particles by a rapidly moving gas or liquid stream or by other means.

attritor. A high-intensity ball mill whose drum is stationary and whose balls are agitated by rotating baffles, paddles, or rods at right angle to the drum axis.

attritor grinding. The intensive grinding or alloying in an attritor. Examples: milling of carbides and binder metal powders and mechanical alloying of hard dispersoid particles with softer metal or alloy powders. See also *mechanical alloying.*

automatic press. A self-acting machine for repeated compacting, sizing, or coining. See *press.*

average density. The density measured

on an entire body or on a major number of its parts whose measurements are then averaged.

axial loading. The application of pressure on a powder or compact in the direction of the press axis.

B

bake (verb). (1) To remove gases from a powder at low temperatures. (2) To heat treat a compacted powder mixture of a metal and polymer at the curing temperature.

ball mill. A machine in which powders are blended or mixed by ball milling.

ball milling. Grinding, blending, or mixing in a receptacle of rotational symmetry that contains balls of a metal or nonmetal harder than the material being milled.

barrel. See *die cavity.*

base plate. The plate that supports the press tool assembly.

batch. (1) The total output of one mixing, sometimes called a lot. (2) The tray or basket of compacts placed in a sintering furnace.

batch sintering. Presintering or sintering in such a manner that compacts are sintered and removed from the furnace before additional unsintered compacts are placed in the furnace.

bell. A jar-like enclosure for containing a vacuum or a controlled atmosphere in sintering equipment used for tungsten, tantalum, and other refractory metal bars.

bell-type furnace. A furnace for the sintering of large batches of small pieces under a controlled atmosphere.

billet. A compact, green or sintered, that will be further worked by forging, rolling, or extrusion; sometimes called an ingot.

binder. A substance added to the powder to (1) increase the strength of the compact and (2) cement together powder

*Compiled by Claus G. Goetzel, Consultant and Lecturer, Department of Materials Science and Engineering, Stanford University

particles that alone would not sinter into a strong object.

binder metal. A metal used as a binder.

binder phase. The soft metallic phase that cements the carbide particles in cemented carbides. More generally, a phase in a heterogeneous sintered material that gives solid coherence to the other phase(s) present.

blank. A pressed, presintered, or fully sintered compact, usually in the unfinished condition, to be machined or otherwise processed to final shape or condition.

bleedout. See preferred term *exudation*.

blend (noun). Thoroughly intermingled powders of the same nominal composition.

blend, blending (verb). The thorough intermingling of powder fractions of the same nominal composition to adjust physical characteristics.

blistering. The formation of surface bubbles on the compact during sintering caused by dynamic evolution of air or gases.

boat. A box or container used to hold the green compacts during passage through a continuous sintering furnace.

bonding. The joining of compacted or loose particles into a continuous mass under the influence of heat.

bottom punch. The part of the tool assembly that closes the die cavity at the bottom and transfers the pressure to the powder during compaction.

box furnace. A furnace used for batch sintering, normally utilizing a controlled atmosphere-containing sealed retort.

bridging. The formation of arched cavities or pores in a loose or compacted powder mass.

briquet(te). A self-sustaining mass of powder of defined shape; see preferred term *compact* (noun).

bubble. Specifically, small gas holes formed during sintering of doped tungsten.

buffer gas. A protective gas curtain at the charge or discharge end of a continuously operating sintering furnace.

bulk density. Powder in a container or bin expressed in mass per unit volume.

bulk volume. The volume of the powder fill in the die cavity.

burnoff. The removal of additives (binder or lubricant) by heating.

burr. An edge protrusion on a pressed compact or a coined part caused by plastic flow of metal into the clearance space between a punch and a die cavity. Synonymous with *flash*.

C

CIP. The acronym representing the words cold isostatic pressing.

cake. A coalesced mass of unpressed metal powder.

calibrate (verb). See preferred term *sizing*.

can. A sheathing of soft metal that encloses a sintered metal billet for the purpose of hot working (hot isostatic pressing, hot extrusion) without undue oxidation.

capillary attraction. The driving force for the infiltration of the pores of a sintered compact by a liquid.

carbide. A generic term applied to carbide compounds of refractory metals. Carbides are characteristically hard and wear resistant.

carbon tube furnace. An electric furnace that has a carbon retort for a resistor element and is especially suitable for the batch or continuous sintering of carbon-insensitive materials such as cemented carbides.

carbonyl powder. Powders prepared by the thermal decomposition of a metal carbonyl compound such as nickel tetracarbonyl $Ni(CO)_4$ or iron pentacarbonyl $Fe(CO)_5$. See *thermal decomposition*.

cascade separator. A special device to separate powder fractions of different particle size or specific gravity.

cemented carbide. A sintered material whose components are a carbide and a metallic binder matrix used mainly in tools, dies, and wear-resistant parts.

cermet. A material consisting of ceramic particles bonded with a metal.

chamber furnace. A batch sintering furnace usually equipped with a retort that can be sealed gastight.

charge. The powder fed into a die for compacting.

chemical decomposition. The separating of a compound into its constituents.

chemical deposition. The precipitation of a metal from a solution of its salt by the addition to the solution of another metal or a reagent.

chemical vapor deposition. The precipitation of a metal from a gaseous compound onto a solid or particulate substrate. Also known as CVD.

chemically precipitated powder. A metal powder that is produced as a fine precipitate by chemical displacement.

chute. A feeding trough for powder to pass from a fill hopper to the die cavity in an automatic press.

classification, classifying, classify. Separation of a powder into fractions according to particle size.

clearance. A gap between mating parts with particular reference to moving and stationary parts of a press tool assembly.

cloth. Metallic or nonmetallic screen or fabric used for screening or classifying powders.

closed pore. A pore completely surrounded by solid material and inaccessible from the surface of the body.

coarse fraction. The large particles in a powder spectrum.

coin, coining. The final pressing of a sintered compact to obtain a definite surface configuration (not to be confused with repressing or sizing).

cold compacting. See preferred term *cold pressing*.

cold pressing. The forming of a compact at or below room temperature.

cold welding. Cohesion between two surfaces of metal, generally under the influence of externally applied pressure at room temperature.

Coldstream process. A method of producing cleavage fractures in hard particles through particle impingements in a high-velocity cold gas stream. Also referred to as impact crushing.

collapsible tool. A press tool that can be easily disassembled.

comminution. See *pulverization*.

compact (noun). The object produced by compression of metal powder, generally while confined in a die.

compact, compacting, compaction (verb). The operation or process of producing a compact; sometimes called pressing.

compactibility. See *compressibility*.

compacting crack. A crack in a compact that is generated during the major phases of the pressing cycle, such as load application, load release, and ejection.

compacting force. The force that acts on the powder to be densified expressed in newtons or tons.

compacting pressure. The specific compacting force related to the area of contact with the press punch expressed in megapascals, meganewtons per square metre, or tons per square inch.

compacting tool set. See *die*.

component. (1) The metals or additives in a blend or mixture. (2) A sintered part of a specific configuration.

composite coating. A coating on a metal or nonmetal that consists of two or more components, one of which is often particulate in form. Example: a cermet coating.

composite compact. A metal powder compact consisting of two or more adhering layers, rings, or other shapes of different metals or alloys with each material retaining its original identity.

composite material. A heterogeneous, solid structural material consisting of two or more distinct components that are mechanically or metallurgically bonded together, such as a cermet or a wire of a high melting substance embedded in a metal or nonmetal matrix.

composite powder. A powder in which each particle consists of two or more different materials.

composite structure. The structure of a sintered composite compact.

compressibility. (1) The ability of a powder to be formed into a compact having well-defined contours and structural stability at a given temperature and pressure; a measure of the plasticity of powder particles. (2) A density ratio determined under definite testing conditions. Also referred to as compactibility.

compressibility curve. A plot of the green density of a compact with increasing pressure.

compressibility test. A test to determine the behavior of a powder under applied pressure. It tells of the degree of densification and cohesiveness of a compact as a function of the magnitude of the pressure.

compression crack. See *compacting crack*.

compression ratio. The ratio of the volume of the loose powder in a die to the volume of the compact made from it.

constituent. Any powder or additive in a powder mixture or compact.

contact area. The area involved as one particle touches another.

contact infiltration. The process of infiltration whereby the initially solid infiltrant is placed in direct contact with the compact and the pores are filled with the liquid phase by capillary force after the infiltrant has become molten. See also *infiltrant* and *infiltration*.

contact material. For heavy-duty applications, the material is often a composite of a hard, wear resistant, and arc transfer resistant refractory metal phase and a soft, electrical conductor metal phase. The proportion of each depends on the application, such as make and break contacts or welding electrodes.

continuous compaction. The production of relatively long compacts having a uniform cross section, such as sheet, rod, tube, etc., by direct extrusion or rolling of loose powder.

continuous furnace. Used for the uninterrupted sintering of compacts.

continuous sintering. Presintering or sintering in such a manner that the objects are advanced through a furnace at a fixed rate by manual or mechanical means; sometimes called stoking.

controlled atmosphere. Specifically, as applied to sintering, to prevent oxidation and destruction of the powder compacts.

core. A magnetic component for high frequency electronic applications made from carbonyl iron powder or ferrite powder.

core rod. A member of a die assembly used in molding a hole in a compact.

cored bar. A compact of bar shape heated by its own electrical resistance to a temperature high enough to melt its interior.

cracked gas. A generic term for a gas mixture obtained by thermal decomposition, with or without catalysis, of a gaseous compound. Examples: cracked ammonia (NH_3) is a mixture of nitrogen and hydrogen, and cracked natural gas hydrocarbons such as methane (CH_4) are a mixture of carbon and hydrogen. Cracked ammonia is also known as *dissociated ammonia*.

critical stress intensity factor. Specifically, a measure of toughness of a powder metallurgy part. Symbol: K_{Ic}.

crucible furnace. A furnace sometimes used for batch sintering.

crushing. A process of comminuting large pieces of metal or ore into rough size fractions prior to grinding into powder. A typical machine for this operation is a jaw crusher.

cyclone. A collector of fractions of a powder from air, water, or other gases or liquids; the device operates with the aid of centrifugal force that acts on the powder suspension in the fluid.

D

decomposition. Separation of a compound into its chemical elements or components.

deformation. The change of shape resulting from the application of a force.

degassing. Specifically, the removal of gases from a powder by a vacuum treatment at ambient or at elevated temperature.

delube. The removal of a lubricant from a powder compact, usually by burnout, or alternatively by treatment with a chemical solvent.

demixing. (1) The undesirable separation of one or more constituents of a powder mixture. (2) Segregation due to overmixing.

dendritic powder. Particles usually of electrolytic origin typically having the appearance of a pine tree.

density, absolute. (1) The ratio of the mass of a volume of solid material to the same volume of water. (2) The mass per unit volume of a solid material expressed in grams per cubic centimetre.

density, dry. The mass per unit volume of an unimpregnated sintered part.

density ratio. The ratio of the determined density of a compact to the absolute density of metal of the same composition, usually expressed as a percentage. Also referred to as percent theoretical density.

density, wet. The mass per unit volume of a sintered part impregnated with oil or other nonmetallic material.

dewaxing. The removal of wax from a powder compact by treatment with a chemical solvent or by burnout.

dewpoint. The temperature at which water vapor begins to condense. An index of water vapor content in a gas. Example: -40 °C (-40 °F) dewpoint contains 0.02% water vapor by volume.

dezincification. The evaporative loss of zinc from brass powder compacts during sintering.

diametrical strength. A property that is calculated from the load required to crush a cylindrical sintered test specimen in the direction perpendicular to the axis. See also *radial crushing strength*.

die. The part or parts making up the confining form in which a powder is pressed or a sintered compact is repressed or coined. The term is often used to mean a die assembly.

die barrel. A tubular liner for a die cavity.

die body. The stationary or fixed part of a die assembly.

die bolster. The external steel ring that is shrunkfit around the hard parts comprising the die barrel.

die breakthrough. The bursting of the die.

die cavity. That portion of the die body in which the powder is compacted or the sintered compact is repressed or coined.

die fill. A die cavity filled with powder.

die insert. A removable liner or part of a die body or punch.

die liner. A thin, usually hard and wear-resistant lining of the die cavity, such as produced by hard chromium plating. It is usually thinner than a die insert.

die lubricant. A lubricant applied to a

punch or the walls of a die cavity to minimize die wall friction and to facilitate pressing and ejection of the compact.

die opening. Entrance to the die cavity.

die plate. The base plate of a press into which the die is sunk.

die set. (1) The aligned mountings onto which the punch and die assemblies are secured. (2) The die system ready to install in the press.

die volume. See preferred term *fill volume*.

diffusion. The movement of atoms within a substance, usually from an area of high constituent concentration to an area of low constituent composition in order to achieve uniformity.

diffusion porosity. The porosity that is caused by the diffusion of one metal into another during sintering of an alloy. Also known as Kirkendall porosity.

dimensional change. Object shrinkage or growth resulting from sintering.

direct sintering. A method whereby the heat needed for sintering is generated in the body itself, such as by induction or resistance heating. Contrast with *indirect sintering*.

discontinuous sintering. Presintering or sintering of the objects in a furnace according to a specified cycle that is tailored to the charge. Examples: batch sintering, bell furnace sintering, box furnace sintering, induction furnace sintering.

dispersing agent. A substance that increases the stability of a suspension of particles in a liquid medium by deflocculation of the primary particles.

dispersion hardening. See *dispersion strengthening*.

dispersion-strengthened material. A metallic material that contains a fine dispersion of nonmetallic phase(s), such as Al_2O_3, MgO, SiO_2, ThO_2, Y_2O_3, or ZrO_2 singly or in combination, to increase the hot strength of the metallic matrix.

dispersion strengthening. The strengthening of a metal or alloy by incorporating chemically stable submicron size particles of a nonmetallic phase that impede dislocation movement at elevated temperature.

dissociated ammonia. A frequently used sintering atmosphere. See also *cracked gas*.

distribution contour. The shape of the particle size distribution curve.

dividing cone. A conical powder heap that is divided in quarters for the purpose of sample taking.

dopant. A substance added in small quan-

tity to a metallic powder to prevent or control recrystallization or grain growth during sintering or during use of the resultant sintered object. See also *activator*.

doping. The addition of a small amount of an activator to promote sintering.

double action press. A press that provides pressure from two sides, usually opposite each other, such as from top and bottom.

double cone mixer. A vessel in the shape of two cones abutting at their base that rotates on an axis through the base, and that provides thorough mixing or blending of a powder by cascading.

double pressing. A method whereby compaction is carried out in two steps. It may involve removal of the compact from the die after the first pressing for the purpose of storage, drying, baking, presintering, sintering, or other treatment, before reinserting into a die for the second pressing.

double sintering. A method consisting of two separate sintering operations with a shape change by machining or coining performed in between.

draft. An angle or taper on the surface of a punch or die.

drum test. A test of the green strength of compacts by tumbling them in a drum and examining the sharpness of the edges and corners.

dust. Specifically, a superfine metal powder having predominantly submicron size particles.

dwell time. The time period during which maximum pressure is applied to a compact in cold pressing or hot pressing.

dynamic hot pressing. Method of applying a vibrational load to the punches or die during hot pressing. See *hot pressing, static hot pressing*.

E

edge stability. An indicator of strength in a green compact, as may be determined by tumbling in a drum. See *drum test*.

edge strength. The resistance of the sharp edges of a compact against abrasion, as may be determined by tumbling in a drum. See *drum test*.

ejection. Removal of the compact after completion of pressing, whereby the compact is pushed through the die cavity by one of the punches. Also called *knockout*.

ejector punch. See *knockout punch*.

electrolytic deposition. The process of removing a metal from a solution of its

salt by the passing of an electric current through the solution.

electrolytic powder. Powder produced by electrolytic deposition or by pulverizing of an electrodeposit.

elutriation. A test for particle size in which the speed of a liquid or gas is used to suspend particles of a desired size, with larger sizes settling for removal and weighing, while smaller sizes are removed, collected, and weighed at certain time intervals.

endothermic atmosphere. A gas mixture produced by the partial combustion of a hydrocarbon gas with air in an endothermic reaction. Also known as endogas.

exfoliation. The spallation of a face layer of a compact, usually the result of air entrapment or faulty pressing technique.

exothermic atmosphere. A gas mixture produced by the partial combustion of a hydrocarbon gas with air in an exothermic reaction. Also known as exogas.

expansion. (1) An increase in size of a compact, usually related to an increase in temperature. A decrease in temperature produces an opposite effect. (2) Sometimes used to mean growth.

explosive compacting. See *high energy rate compacting*.

extrusion. Shaping metal powder into a chosen form by forcing it through an orifice or die of the appropriate shape. The powder may either be mixed with a plasticizer or contained in a can which is extruded with the contained powder.

exudation. The action by which all or a portion of the low melting constituent of a compact is forced to the surface during sintering; sometimes referred to as bleed out or sweating.

F

feed hopper. A container used for holding the powder prior to compacting in a press.

feed shoe. The part of the powder feed system that delivers the powder into the die cavity.

ferrite magnet. A magnet made of a mixture of ferric oxide and a strong basic oxide. Example: sodium ferrite $NaFeO_2$. The oxides are used as compacted powders for rectifiers or permanent magnets or as powders on memory or record tapes.

fiber metallurgy. The technology of producing solid bodies from fibers or chopped filaments, with or without a metal matrix. The fibers may consist of such nonmetals as graphite or aluminum oxide, or of such metals as tungsten or boron.

fiber reinforced composite. The term is generic for composite materials and applies to metal and nonmetal matrices and metal and nonmetal reinforcements. The metallic components may be made by powder metallurgy techniques.

fibrous structure. A microstructure with elongated grains in the direction of deformation, such as in hot pressed or hot extruded powder compacts, or in refractory metal rods and wires.

filament. A fine wire, such as made from tungsten for incandescent light bulbs. The filament may be a straight loop or coiled into a helix.

fill density. See preferred term *apparent density*.

fill depth. Synonymous with *fill height*.

fill factor. The quotient of the fill volume of a powder over the volume of the green compact after ejection from the die. It is the same as the quotient of the powder fill height over the height of the compact. Inverse parameter of *compression ratio*.

fill height. The distance between the lower punch face and the top plane of the die body in the fill position of the press tool.

fill position. The position of the press tool which enables the filling of the desired amount of powder into the die cavity.

fill ratio. See *compression ratio*.

fill shoe. See preferred term *feed shoe*.

fill volume. The volume that a powder fills after flowing loosely into a space that is open at the top, such as a die cavity or a measuring receptacle.

filter. Metal filters are porous products made either from wires and fibers or from sintered powders.

final density. The density of a sintered product.

fines. The portion of a powder composed of particles smaller than a specified size, usually 44 μm (−325 mesh).

flake powder. Flat or scalelike particles whose thickness is small compared to the other dimensions.

flame spraying. A coating process that uses a torch or plasma gun and metal powder as feed.

flash. Excess metal forced out between the punches and die cavity wall during compacting or coining. See also *burr*.

floating die. A die body that is suspended on springs or an air cushion which causes the die to move together with the upper punch over a stationary lower punch; the rate of die movement is lower than that of the upper punch and is a function of the friction coefficient of the powder in relation to the wall of the die cavity.

floating die pressing. The compaction of a powder in a floating die, resulting in densification at opposite ends of the compact. Analogous to double action pressing.

flow factor. See preferred term *flow rate*.

flow meter. A metal cylinder whose interior is funnel shaped and whose bottom has a calibrated orifice of standard dimensions to permit passage of a powder and the determination of the flow rate.

flow rate. The time required for a powder sample of standard weight to flow through an orifice in a standard instrument according to a specified procedure.

flow test. A standardized test to measure how readily a powder flows. See *flow rate*.

fluidized bed. A contained mass of a finely divided solid that behaves like a fluid when brought into suspension in a moving gas or liquid.

fluidized bed reduction. The finely divided solid is a powdered ore or reducible oxide, and the moving gas is reducing; the operation is carried out at elevated temperature in a furnace.

forging. (1) Reshaping a billet or ingot by hammering. (2) The process of placing a powder in a container, removing the air from the container, and sealing it; followed by conventional forging of the powder and container to the desired shape.

forming. A generic term in powder metallurgy describing the first step in changing a loose powder into a solid of specific configuration.

fraction. That portion of a powder sample which lies between two stated particle sizes.

fragmentation. The process of breaking a solid into finely divided pieces.

fragmented powder. A powder obtained by fragmentation and mechanical comminution into fine particles.

friction material. A sintered material exhibiting a high coefficient of friction designed for use where rubbing or frictional wear is encountered.

fugitive binder. An organic substance added to a metal powder to enhance the bond between the particles during compaction and thereby increase the green strength of the compact, and which decomposes during the early stages of the sintering cycle.

G

gall. To damage the surface of a compact or die parts, caused by adhesion of powder to the die cavity wall or a punch surface.

gap. The clearance between a moving punch and the die cavity.

gas classification. The separation of a powder into its particle size fractions by means of a gas stream of controlled velocity flowing counterstream to the gravity-induced fall of the particles. The method is used to classify submesh-size particles.

gas classifier. A device for gas classification; it may be of laboratory size for quality control testing or of industrial capacity to accommodate powder production requirements.

gaseous reduction. (1) The reaction of a metal compound with a reducing gas to produce the metal. (2) The conversion of metal compounds to metallic particles by the use of a reducing gas.

getter. A substance that is used in a sintering furnace for the purpose of absorbing or chemically binding elements or compounds from the sintering atmosphere that are damaging to the final product.

gettering box. A container for the getter substance that is readily accessible to the atmosphere and prevents contamination of the sintered product by direct contact.

grade. A specific, nominal chemical analysis powder identified by a code number. Example: cemented carbide manufacturers grade 74 M 60 FWC (74 is usage; M is equipment manufacturer; 60 is nominal HRC; FWC is fine cut tungsten carbide).

grain. An individual crystal within a polycrystalline metal or alloy. Sometimes used for *particle*, which is the preferred term.

grain boundary diffusion. One of the diffusion mechanisms in sintering. It is characterized by a very high diffusion rate because of an abundance of imperfections in the grain boundaries. See also *surface diffusion* and *volume diffusion*.

grain size. In sintered metals, a measure of the areas or volumes of the grains. Not to be mistaken for the particles of the original powder that have not yet dissolved in the structure.

graining. The process of vigorously stirring or agitating a partially solidified material to develop large grains having a thin oxide coating.

granular powder. A powder having equidimensional but nonspherical particles.

granulation, granulating. The production of coarse metal particles by pouring molten metal through a screen into water or by agitating the molten metal violently while it is solidifying.

green. Unsintered (not sintered).

green compact. An unsintered compact.

green density. The density of a green compact.

green strength. The ability of a green compact to maintain size and shape during handling and storage prior to sintering.

grit, grit size. The particle size of an abrasive powder, such as carborundum, corundum, silicon carbide, or diamond used in cutting and machining operations.

growth. An increase in compact or part size as a result of excessive pore formation during sintering.

H

HIP. The acronym representing the words hot isostatic pressing.

hard metal. A collective term that designates a sintered material with high hardness, strength, and wear resistance, and is characterized by a tough metallic binder phase and particles of carbides, borides, or nitrides of the refractory metals. The term is in general use abroad, while for the carbides the term *cemented carbides* is preferred in the U.S., and the boride and nitride materials are usually categorized as *cermets*.

heavy alloy, heavy metal. A sintered tungsten alloy with nickel, copper, and/or iron, the tungsten content being at least 10 wt% and the density being at least 16.5 g/cm³.

high energy rate compacting. Compacting of a powder at a very rapid rate by the use of explosives in a closed die.

hot densification. Rapid deformation of a heated powder preform in a die assembly for the purpose of reducing porosity. Metal is usually deformed in the direction of the punch travel. See *hot pressing*.

hot forging. The plastic deformation of a pressed and/or sintered compact in at least two directions at temperatures above the recrystallization temperature.

hot isostatic pressing. A process for simultaneously heating and forming a compact in which the powder is contained in a sealed flexible sheet metal or glass enclosure and the so-contained powder is subjected to equal pressure from all directions at a temperature high enough to permit plastic deformation and sintering to take place.

hot pressing. Simultaneous heating and forming of a compact. See also *pressure sintering*.

hydraulic press. A machine in which a compacting die is placed, the compacting force being provided by a hydraulic cylinder.

hydride powder. A powder produced by removal of the hydrogen from a metal hydride.

hydride process. The hydrogenation of such reactive metals as titanium and zirconium, followed by comminution of the brittle compound and vacuum treatment to remove the hydrogen from the powder.

hydrogen loss. The loss in weight of metal powder or a compact caused by heating a representative sample according to a specified procedure in a purified hydrogen atmosphere. Broadly, a measure of the oxygen content of the sample when applied to materials containing only such oxides as are reducible with hydrogen and no hydride-forming element.

hydrostatic compacting. See *hydrostatic pressing*.

hydrostatic mold. A sealed flexible mold made of rubber, a polymer, or pliable sheet made from a low melting metal such as aluminum.

hydrostatic pressing. A special case of isostatic pressing that uses a liquid such as water or oil as a pressure transducing medium and is therefore limited to near room temperature operation.

hysteresis loop. Characterizes the magnetization/demagnetization characteristics as a function of the applied magnetic field for ferromagnetic materials. Powder metallurgy products may differ from wrought products because of porosity or impurity traces.

I

impact sintering. An instantaneous sintering process during high energy rate compacting that causes localized heating, welding, or fusion at the particle contacts.

impregnation. The process of filling the pores of a sintered compact with a nonmetallic material such as oil, wax, or resin.

indirect sintering. A process whereby the heat needed for sintering is generated outside the body and transferred to the compact by conduction, convection, radiation, etc. Contrast with *direct sintering*.

induction sintering. Sintering in which the required heat is generated by subjecting the compact to electromagnetic induction. See also *direct sintering*.

infiltrant. Material used to infiltrate a porous sinter. The infiltrant as positioned on the compact is called a slug. See also *contact infiltration*.

infiltration. The process of filling the pores of a sintered or unsintered compact with a metal or alloy of lower melting temperature. See also *contact infiltration*.

ingot. See *billet*.

injection molding. A process similar to plastic injection molding using a plastic coated metal powder.

intercommunicating porosity. See preferred term *interconnected porosity*.

interconnected pore volume. The volume fraction of pores that are interconnected within the entire pore system of a compact or sintered product.

interconnected porosity. A network of connecting pores in a sintered object that permits a fluid or gas to pass through the object. Also referred to as interlocking or open porosity.

interface. A surface that forms the boundary between phases in a sintered compact.

interface activity. A measure of the chemical potential between the contacting surfaces of two particles in a compact or two grains in a sintered body.

intergranular. Between the grains of a metal. See *grain*.

internal oxidation. The preferential *in situ* oxidation of certain constituents or phases within the bulk of a solid alloy, accomplished by diffusion of oxygen into the body. The process is suitable for the production of dispersion-strengthened alloys, if the constituent to be oxidized forms a stable oxide and the major alloy component permits a high rate of oxygen diffusion.

irregular powder. Particles lacking symmetry.

isostatic mold. A sealed container of glass or sheet of carbon steel, stainless steel, or a nickel-based alloy. See *isostatic pressing*.

isostatic pressing. Cold or hot pressing of a powder using equal pressure from all directions.

J

jaw crusher. A machine for the primary disintegration of metal pieces or agglomerates into coarse powder. See *crushing*.

jet pulverizer. A machine that comminutes metal pieces or agglomerates by means of pressurized air or steam injected into a chamber.

K

keying. The deformation of metal particles during compacting to increase interlocking and bonding.

knockout (verb). Ejecting of a compact from a die cavity.

knockout punch. A punch used for ejecting compacts.

Kroll process. A process for the production of metallic titanium sponge by the reduction of titanium tetrachloride with a more active metal, such as magnesium or sodium. The sponge is further processed to granules or powder.

L

lamination. (1) A discontinuity, crack, or separation in a plane perpendicular to the axis of applied pressure that may be the result of air entrapment or misalignment of the pressing tools during compacting. (2) A thin compressed or rolled powder product with two or more layers.

lampblack. Fine soot used in the reduction and carburization of tungsten trioxide and titanium dioxide to produce tungsten carbide and titanium carbide powder, respectively.

lattice diffusion. See *volume diffusion*.

lift beam furnace. A continuously operating sintering furnace. The term in general use by the U.S. furnace industry is *walking beam furnace*.

lift rod, lifting rod. Part of the press tooling used for the raising or lifting of one or more punches.

linear shrinkage. The shrinkage in one dimension of a compact during sintering. Contrast with *volume shrinkage*.

liquid disintegration. The process of producing powders by pouring molten metal on a rotating surface.

liquid phase sintering. Sintering of a compact or loose powder aggregate under conditions where a liquid phase is present during part of the sintering cycle.

loading. Filling a die cavity with powder.

loading sheet. The part of a die assembly used as a container for a specific amount of powder to be fed into the die cavity. Sometimes it is part of the feed shoe.

loading weight. See preferred term *apparent density*.

loop classifier. A cyclone-type classifier, sometimes connected with a conical ball mill in an airtight system.

loose powder. Uncompacted powder.

loose powder sintering. Sintering of uncompacted powder using no external pressure.

lot. See *batch*.

lower punch. The lower member of a die assembly which forms the bottom of the die cavity. It may or may not move with respect to the die body.

lower ram. The part of a pneumatic or hydraulic press that is moving in a lower cylinder and transmits pressure to the lower punch.

lubricant. A substance mixed with a powder to facilitate compacting and subsequent mold ejection of compact; often a stearate or a proprietary wax. It may also be applied as a film to the surfaces of the punches or the die cavity wall, such as by spray coating.

lubricating. Mixing or incorporating a lubricant with a powder to facilitate compacting and ejecting of the compact from the die cavity; also, applying a lubricant to die walls and/or punch surfaces.

M

macropore. Pores in pressed or sintered compacts that are visible with the naked eye.

magnetic core. See *core*.

mandrel. A part of the press tool to produce a bore hole in the compact in the direction of the applied pressure.

master alloy powder. A prealloyed powder of high concentration of alloy content, designed to be diluted when mixed with a base powder to produce the desired composition.

matrix metal. The continuous phase of a polyphase alloy or mechanical mixture; the physically continuous metallic constituent in which separate particles of another constituent are embedded.

maximum pore size. See *absolute pore size*.

mechanical alloying. An alternate cold welding and shearing of particles of two or more species of greatly differing hardness. The operation is carried out in high-intensity ball mills, such as attritors, and is the preferred method of producing hard phase dispersed alloy powders. See also *attritor grinding* and *dispersion-strengthened material*.

mechanical disintegration. See preferred terms *comminution* and *pulverization*.

mechanical press. A machine in which a compacting die is placed, the compacting force being transmitted through a mechanical rather than a hydraulic system.

Menstruum method. A method of producing multicarbide powder, such as WC + TiC solid solution, by introducing the individual elements into a molten bath of a noncarbide-forming metal such as aluminum or nickel. The multicarbide is formed above 2100 °C (3800 °F), slowly cooled in the dispersed condition in the menstruum to room temperature, and finally won by chemical separation.

mesh, mesh number. (1) The number of screen openings per linear inch of screen; also called *mesh size*. (2) The screen number of the finest screen through which all of the particles of a given powder sample will pass.

mesh-belt conveyor furnace. A continuously operating sintering furnace that uses a conveyor belt for the transport of the charge.

mesh size. The width of the aperture in a cloth or wire screen.

metal powder. Elemental metals or alloy particles, usually in the size range of 0.1 to 1000 μm.

micromesh. A sieve with precisely square openings in the range of 10 to 120 μm produced by electroforming.

micromesh sizing. The process of sizing micromesh particles using an air or a liquid suspension process.

micron, micrometer. A dimension of 0.001 mm, written with the abbreviation μm.

micropore. The pores in a sintered product that can only be detected under a microscope.

micropulverizer. A machine that disintegrates powder agglomerates by strong impacts from small hammers fastened to a solid disk that rotates at very high velocity.

microshrinkage cavity. A fine void found microscopically in the low melting metal phase of infiltrated compacts due to contraction during solidification.

mill. A machine for preparing a powder. Common mills are ball, hammer, disk, roll, vortex, rod, and impact.

milling. The mechanical comminution of a metal powder or a metal powder mixture, usually in a ball mill, to alter the size or shape of the individual particles, to coat one component of a mixture with another, or to create uniform distribution of components.

milling fluid, milling liquid. An organic liquid, such as hexane, in which ball milling is carried out. The liquid serves to reduce the heat of friction and resulting surface oxidation of the particles during grinding, and to provide protection from other surface contamination.

mill scale powder. Pulverized iron oxide

scale that is a byproduct of hot rolling of steel. The material is readily reduced to a soft spongy iron powder free of mineral inclusions and other solid impurities.

minus mesh. The portion of a powder sample that passes through a screen of stated size. See preferred term *minus sieve*.

minus sieve. The portion of a powder sample that passes through a standard sieve of a specified number. See also *plus sieve*.

mix (noun). See preferred term *mixture*.

mix, mixing (verb). Causing the thorough intermingling of two or more powdered substances.

mixture. Two or more components thoroughly intermingled as by mixing.

mold. See preferred term *die*.

molding. See preferred terms *compacting* and *pressing*.

muffle furnace. A common type of sintering furnace for continuous or discontinuous operation. The muffle may be ceramic to support electric heating elements or resistor wire windings, or it may be a gastight metallic retort to retain the furnace atmosphere and support the work trays; both kinds may be used in the same furnace.

multiple die pressing (verb). The simultaneous compaction of powder into several identical parts with a press tool consisting of a number of components.

multiple punch press. A mechanical or hydraulic press that actuates several punches individually and independently of each other.

N

neck. The contact area between abutting particles in a compact undergoing sintering.

neck formation. The growth of interparticle contacts through diffusion processes during sintering.

needles. Elongated or rodlike particles with a high aspect ratio.

net shape. The shape of a sintered part that conforms closely to specified dimensions.

network structure. A structure in which one constituent occurs primarily at the grain boundaries, thus enveloping the grains of the other constituent(s). A desirable feature in cemented carbides, as in the system cobalt/tungsten carbide, where the cobalt phase forms a ductile network surrounding the brittle carbide grains.

nib. (1) A pressed preheated, shaped, sin-

tered, hot pressed, rough drilled, or finished compact. (2) A generic term for a piece of cemented carbide intended for use as a wire drawing die.

nodular powder. Irregular powder having knotted, rounded, or similar shapes.

nuclear cermet fuel. A sintered fuel rod composed of a fissile carbide or oxide constituent and a metallic matrix.

O

oil content. The amount of oil which an impregnated part, such as a self-lubricating bearing, retains.

oil flow rate. See *rate of oil flow*.

oil impregnation. The filling of a sintered skeleton body with oil by capillary attraction or under influence of an external pressure or a vacuum.

oil permeability. A measure of the capacity of the sintered bearing to allow the flow of an oil through its open pore system.

oilless bearing. See preferred term *self-lubricating bearing*.

open pore. A pore open to the surface of a compact. See *intercommunicating porosity*.

open porosity. See *interconnected porosity*.

overfill. The fill of a die cavity with an amount of powder in excess of specification.

overmix (verb). Mixing of a powder longer than necessary to produce adequate distribution of powder particles. Overmixing may cause particle size segregation.

oversinter (verb). The sintering of a compact at higher temperature or for longer time periods than necessary to obtain the desired microstructure or physical properties. It often leads to swelling due to excessive pore formation.

oversize powder. Powder particles larger than the maximum permitted by a particle size specification.

P

P/M. The acronym representing the words powder metallurgy.

packing, packing material. Any material in which compacts are embedded during the presintering or sintering operations. The material may act as a getter to protect the compacts from contamination. See *getter*.

packed density. See preferred term *tap density*.

packing density. See preferred term *apparent density*.

paddle mixer. A mixer that uses paddles mounted on a rotating shaft or disk to move and mix the powder.

particle. A minute portion of matter. A metal powder particle may consist of one or more crystals.

particle hardness. The hardness of an individual particle as measured by a Knoop or Vickers type microhardness indentation test.

particle morphology. The form and structure of an individual particle.

particle shape. The appearance of a metal particle, such as spherical, rounded, angular, acicular, dendritic, irregular, porous, fragmented, blocky, rod, flake, nodular, or plate.

particle size. The controlling linear dimension of an individual particle as determined by analysis with screens or other suitable instruments.

particle size analysis. See preferred term *sieve analysis*.

particle size classification. See preferred term *sieve classification*.

particle size distribution. The percentage by mass, by numbers, or by volume of each fraction into which a powder sample has been classified with respect to size.

particle size range. The limits between which a variation in particle size is allowed.

particle spacing. The distance between the surfaces of two or more adjacent particles in a loose powder or a compact.

paste compound. A metal powder mixed with a thin binder for use as a brush-on brazing filler metal.

pellet. Similar to a shotted particle. Generally, a small rounded or spherical solid body. See *shotting*.

percent theoretical density. See *density ratio*.

permanent magnet. Sintered or cast ferromagnetic alloy that has been strongly magnetized and retains its magnetism indefinitely. Example: Alnico alloys.

permeability. The ability of a compact to permit liquid or gas to flow through it as measured under specific conditions.

plasticizer. A substance added to a powder or powder mixture to render it more formable during cold pressing or extrusion.

plasma spray coating, plasma spraying. A thermal spraying process in which the coating material is melted with the heat of a plasma torch, and the molten melt droplets are propelled against a substrate by the hot, ionized gas of the torch.

platelets, plates. Flat particles of metal powder having considerable thickness.

The thickness, however, is smaller when compared to the length and width of the particles.

plus mesh. The powder sample retained on a screen of stated size, identified by the retaining mesh number.

plus sieve. The powder sample retained on a standard sieve of specified number.

pneumatic press. A press that uses air or a gas to deliver the pressure to the upper and lower rams.

pore. An inherent or induced cavity (void) within a particle or within an object.

pore area. The effective surface porosity of a sintered compact to determine the permeability to a test fluid.

pore channels. The connections between pores in a sintered body.

pore formation. The natural formation of pores during compaction and/or sintering. See *pore-forming material*.

pore-forming material. A substance included in a powder mixture that volatilizes during sintering and thereby produces a desired kind of porosity in a finished compact.

pore size. Width of a pore in a compacted and/or sintered powder or within a particle.

pore size distribution. Indicates the volume fractions of different pore size categories, which are determined metallographically.

pore size range. The limits between which a variation in pore size is allowed.

pore structure. Pattern of pores in a solid body indicating such characteristics as pore shape, pore size, pore size distribution.

pore wall. The interface between the pore and the solid.

porosimeter. A test apparatus to measure the interconnected porosity in a sintered compact by means of determining its permeability through use of a test fluid such as mercury, which either partially or completely fills the open pores.

porosity. The amount of pores (voids) expressed as a percentage of the total volume of the powder metallurgy part.

porous bearing. See *self-lubricating bearing*.

powder. An aggregate of discrete particles that are usually in the size range of 1 to 1000 μm.

powder designation. A code number identifying a specific powder.

powder fill. The filling of a die cavity with powder.

powder lubricant. An agent or component incorporated into a mixture to facilitate compacting and ejecting of the compact from its mold.

powder metallurgy. The technology and art of producing metal powders and of the utilization of metal powders for the production of massive materials and shaped objects.

powder metallurgy part. A shaped object that has been formed from metal powders and sintered by heating below the melting point of the major constituent. A structural or mechanical component made by the powder metallurgy process.

powder production. The process by which a powder is produced, such as machining, milling, atomization, condensation, reduction, oxide decomposition, carbonyl decomposition, electrolytic deposition, or precipitation from a solution.

powder rolling. See preferred term *roll compacting*.

powder technology. A broad term encompassing the production and utilization of both metal and nonmetal powders.

prealloyed powder. A metallic powder composed of two or more elements that are alloyed in the powder manufacturing process and in which the particles are of the same nominal composition throughout.

precipitation. The removing of a metal from a solution caused by the addition of a reagent by displacement; also, the removal of a metal from a gas by displacement.

precision part, precision sintered part. A powder metallurgy part that is compacted and sintered, closely conforming to specified dimensions without a need for substantial finishing.

preform. The initially pressed compact to be subjected to repressing.

preforming. The initial pressing of a metal powder to form a compact that is to be subjected to a subsequent pressing operation other than coining or sizing.

preheat. An early stage in the sintering procedure when, in a continuous furnace, lubricant or binder burnoff occurs without atmosphere protection prior to actual sintering in the protective atmosphere of the high heat chamber.

premix (noun). A uniform mixture of components prepared by a powder producer for direct use in compacting.

premix (verb). A term sometimes applied to the preparation of a premix; see preferred term *mixture*.

presintered blank. A compact sintered at a low temperature but at a long enough time to make it sufficiently strong for metal working. See *presintering*.

presintered density. The relative density of a presintered compact. See *presintering*.

presintering. Heating a compact to a temperature below the final sintering temperature, usually to increase the ease of handling or shaping of a compact or to remove a lubricant or binder (burnoff) prior to sintering.

press (noun). The machine used for compacting, sizing, or coining. Presses may be mechanical: eccentric, crank, cam, toggle, knuckle joint, rotary (table); or hydraulic: single action, multiple action, double action; or combination mechanical-hydraulic.

press (verb). See preferred term *compact*.

pressed bar. (1) A compact in the form of a bar. (2) A green rectangular compact.

pressed density. The weight per unit volume of an unsintered compact. Same as green density.

pressing (noun). See preferred term *compact*.

pressing (verb). See preferred term *compacting*.

pressing crack. A break in the compact caused by mass slippage of part of the compact. Also called slip crack.

pressing skin. The surface of a compact that is superficially more deformed than the interior due to a preferential alignment of the particles caused by contact with the die wall and punch faces.

pressing tool, press tool. The complete tool assembly consisting of the die, a die adaptor, the punches, and, when required, a core rod.

pressure bonding. Joining at elevated temperature of powder metallurgy parts or sections to one another or to wrought metal, sheet, or castings, without using liquid auxiliary metal, such as solder, brazing alloy, or weld metal.

pressure sintering. A hot pressing technique that usually employs low loads, high sintering temperatures, continuous or discontinuous sintering, and simple molds to contain the powder.

pressureless sintering. Sintering of loose powder.

protective atmosphere, protective gas. The atmosphere in the sintering furnace designed to protect the compacts from oxidation, nitridation, or other contamination from the environment.

puffed compact. A compact expanded by internal gas pressure.

pulverization. The process of reducing metal powder particle sizes by mechanical means; also called comminution or mechanical disintegration.

punch. The part of a die assembly that transmits pressure to the powder in the die cavity.

purge (verb). The removal of air from a sintering furnace chamber by replacing it with a vacuum or an inert gas prior to the introduction of the sintering atmosphere.

pusher furnace, pusher-type furnace. A continuous furnace widely used for sintering. Parts to be heated are periodically charged into the furnace in containers or trays, which are pushed along the hearth against a line of previously charged containers, thus advancing the containers toward the discharge end of the furnace, where they are removed.

pyrophoric powder. A powder whose particles self-ignite and burn upon exposure to oxygen or air. Example: fine zirconium powder.

pyrophoricity. The property of a substance with a large surface area to self-ignite and burn when exposed to oxygen or air.

Q

Q value. A quality factor of a magnetic core material, also called energy factor or coil magnification factor. It represents the ratio of the reactance of a coil to its series resistance. Its specific change with the frequency is a function of the type and composition of the magnetic powder used for the core.

quartering. A method of sampling a powder by dividing a cone-shaped heap into four parts, selecting one of them randomly, dividing this again into four parts, and repeating the procedure until the sample is small enough for analysis.

R

radial crushing strength. The relative capacity of a powder metallurgy sleeve specimen to resist fracture induced by a force applied between flat parallel plates in a direction perpendicular to the axis of the specimen.

ram. The moving member of a press or forge hammer to which the forming tool unit is fastened.

rare earth metal. One of the group of 15 chemically similar metals with atomic numbers 57 through 71, commonly referred to as the lanthanides. Some, such as samarium, are used in sintered cobalt/rare earth permanent magnets.

rate of oil flow. The rate at which a specified oil will pass through a porous

sintered compact under specific test conditions.

reaction sintering. The sintering of a powder mixture consisting of at least two components that chemically react during the treatment.

recrystallization. The formation, under the influence of heat, of new undistorted grains from the distorted grains resulting from compacting, milling, or from cold working after sintering.

reduced powder. Generic term for any metal or nonmetal powder produced by the reduction of an oxide, hydroxide, carbonate, oxalate, or other compound without melting.

reduction of oxide. The process of converting a metal oxide to metal by applying sufficient heat in the presence of a solid or gaseous material, such as hydrogen, having a greater attraction for the oxygen than does the metal.

reduction ratio. (1) The quotient of the reduced oxygen content into the total initial oxygen content of a powder. (2) The quotient of the reduced cross section into the original cross section in metal working such as extrusion, an indication of the degree of plastic deformation.

refractory metal. A metal characterized by its high melting temperature, generally above 2000 °C (3600 °F).

relative sintering temperature. The ratio of the sintering temperature to the melting temperature of the substance as expressed on the Kelvin scale.

repressing. The application of pressure to a sintered compact, usually for the purpose of improving a physical or a mechanical property or for dimensional accuracy.

resintering. (1) A second sintering operation. (2) Sintering a repressed compact.

restrike. Additional compacting of a sintered compact.

retracting die. Part of a method of stripping the compact from the die used extensively abroad. See *stripped die method.*

roll compacting, roll compaction. The progressive compacting of metal powders by the use of a rolling mill.

rolled compact. A compact made by passing metal powder between rollers so as to form a relatively long, sheetlike compact.

roller air analyzer. An air-elutriation apparatus suitable for the particle size determination of metal powders, especially in the subsieve range.

roller hearth furnace. A modification of the pusher-type continuous furnace that

provides for rolls in the hearth or muffle of the furnace whereby friction is greatly reduced and lightweight trays can be used repeatedly without risk of unacceptable distortion and damage to the work. See *pusher furnace.*

rotary press. A machine fitted with a rotating table carrying multiple die assemblies in which powder is compacted.

rotating electrode powder. An atomized powder consisting exclusively of solid spherical or near-spherical particles.

rotating electrode process. A consumable metal or alloy electrode is rotated during arc melting at very high speeds, and the molten droplets are propelled by centrifugal force toward the wall of a large collecting chamber while solidifying in flight.

RZ powder. The reduced iron powder made in Germany from the scale of pig iron.

S

sample splitter. A device to divide a powder pile for sampling.

sample thief. A pointed, hollow, tubular device to withdraw a representative powder sample from a shipping drum or other packing unit.

screen. The woven wire or fabric cloth, having square openings, used in a sieve for retaining particles greater than the particular mesh size. U.S. standard, ISO, or Tyler screen sizes are commonly used.

screen analysis. See *sieve analysis.*

screen classification. See *sieve classification.*

screening. Separation of a powder according to particle size by passing it through a screen having the desired mesh size.

secondary operation. Any operation performed on a sintered compact, such as sizing, coining, repressing, impregnation, infiltration, heat or steam treatment, machining, joining, plating, or other surface treatment.

sedimentation. The settling of particles suspended and dispersed in a liquid through the influence of an external force, such as gravity or centrifugal force.

segment die, segmented die. A die made of parts which can be separated for the easy removal of the compact; sometimes called a split die.

segregation. Separation of a blend or mixture into fine and coarse portions as a result of overmixing or vibration.

seize. To prevent a part from being ejected from a die as a result of galling.

self-lubricating bearing. A sintered product whose accessible pore volume is filled with a liquid lubricant which automatically produces a lubricating film on the bearing surface during running of the shaft. This is due to a pumping action of the shaft and frictional heat that lowers the viscosity of the oil. After completion of the running cycle, the oil is reabsorbed into the pore system of the bearing by capillary attraction.

shape accuracy. The shape of a sintered or sized product as it conforms to or deviates from specified dimensions and tolerances.

sheath. A sheet metal or glass covering of a sintered billet to protect it from oxidation or other environmental contamination during hot working. See also *can*.

shot. Small spherical particles of metal.

shotting. The production of shot by pouring molten metal in finely divided streams. Solidified spherical particles are formed during descent in a tank of water.

shrinkage. The decrease in dimensions of a compact occurring during sintering.

sieve. A powder separator using a set of graduated mesh size screens.

sieve analysis. Particle size distribution, usually expressed as the weight percentage retained on each of a series of standard screens of decreasing mesh size and the percentage passed by the screen of finest size; also called screen classification.

sieve classification. The separation of powder into particle size ranges by the use of a series of graded sieves. Also called screen analysis.

sieve fraction. That portion of a powder sample that passes through a sieve of specified number and is retained by some finer mesh sieve of specified number.

sieve shaker. A device for shaking, knocking, or vibrating a single sieve or a stack of sieves. It consists of a frame, a motorized knocker, shaker or vibrator, and fasteners for the sieve(s).

sieve underside. The underside of the mesh to which loose powder often adheres. To ensure accuracy of a sieve analysis, this material must be removed and included in the weight determination.

single action press. A press that provides pressure from one side.

sinter, sintering (verb). The bonding of adjacent particles in a powder mass or compact by heating to a temperature below the melting point of the main constituent.

sinter, sinterings (noun). See preferred term *powder metallurgy part*.

sintered density. The quotient of the mass (weight) over the volume of the sin-

tered body expressed in grams per cubic centimetre.

sintered density ratio. The ratio of the density of the sintered body to the solid, pore-free body of the same composition or theoretical density.

sintering atmosphere. See *protective atmosphere*.

sintering cycle. A predetermined and closely controlled time-temperature regime for sintering compacts, including the heating and cooling phases.

sintering temperature. The maximum temperature at which the compact is sintered. The temperature is either measured directly on the surface of the body by optical pyrometer, or indirectly by thermocouples installed in the furnace chamber.

sintering time. The time period during which the compact is at sintering temperature.

sintrate. Controlled heating so that a compact is sintered before the melting point of an infiltrating material is reached.

size fraction. A separated fraction of a powder whose particles lie between specified upper and lower size limits.

sizing. The pressing of a sintered compact to secure a desired dimension.

sizing die. A die used for the sizing of a sintered compact.

sizing knockout. An ejector punch used for ejecting a sintered compact from the sizing die.

sizing punch. A punch used for the pressing of a sintered compact during the sizing operation.

sizing stripper. A stripper punch used during the sizing operation.

skeleton. An unsintered or sintered porous compact with a large proportion of interconnected porosity that makes it suitable for infiltration.

slip casting. A process used in the production of refractory products. Ground material is mixed with water to form a creamy liquid, which is poured into plaster molds, where the surplus water is absorbed and a solid replica of the inside of the mold is obtained.

slip crack. See *pressing crack*.

slug. See *infiltrant*.

slurry. A free-flowing, pumpable suspension of fine solid material in a liquid.

soft magnetic alloy. A ferromagnetic alloy that becomes magnetized readily upon application of a field and returns to a nonmagnetic condition when the field is removed. Example: sintered pure iron or iron-3% silicon alloy.

solid density. See *density, absolute*.

solid state sintering. A sintering proce-

dure for compacts or loose powder aggregates during which no component melts. Contrast with *liquid phase sintering*.

spark sintering. A pressure sintering or hot pressing method that provides for the surface activation of the powder particles by electric discharges generated by a high alternating current applied during the early stage of the consolidation process.

specific gravity. The ratio of the density of a material to the density of some standard material, such as water at a specified temperature, or (for gases) air at standard conditions of pressure and temperature. Also referred to as relative density.

specific pressure. The pressure applied to a green or sintered compact per unit of area of punch cross section.

specific surface. The surface area of a powder expressed in square centimetres per gram of powder or square meters per kilogram of powder.

spherical powder. A powder consisting of ball-shaped particles.

spheroidal powder. A powder consisting of oval or rounded particles.

splat powder. A rapidly cooled or quenched powder whose particles have a flat shape and a small thickness compared to the other dimensions. Similar to *flake powder*.

splat quenching. The process of producing splat powder.

split die. See *segment die*.

split punch. A segmented punch or a set of punches that allow(s) a separate positioning for different powder fill heights and compact levels in dual-step and multistep parts. See *stepped compact*.

sponge iron. A coherent, porous mass of substantially pure iron produced by solid state reduction of iron oxide (mill scale or iron ore).

sponge iron powder. Ground and sized sponge iron that may have been purified or annealed or both.

sponge titanium powder. Ground and sized titanium sponge. See *Kroll process*.

spongy. A porous condition in metal powder particles usually observed in reduced oxides.

spray drying. The process of removing an organic liquid from a milled powder by the use of hot nitrogen.

spray nozzle. An orifice through which a molten metal passes to form a stream that can be further disintegrated by a gas, a liquid, or by mechanical means.

springback. The slight increase in compact size after compacting force is removed and the compact is removed from the die.

sputtering. The disintegration of a metal cathode in a vacuum tube due to bombardment by positive ions, which can be used to produce very fine powders or fine-grained coatings or films.

static hot pressing. A method of applying a static load uniaxially during hot pressing. Contrast with *dynamic hot pressing* and *isostatic hot pressing*. See *hot pressing*.

steam treatment. The treatment of a sintered ferrous part in steam at about 500 °C (930 °F) in order to produce a thin Fe$_2$O$_3$ layer on the exposed surface for the purpose of increasing hardness and wear resistance.

stepped compact. A compact with one (dual step) or more (multistep) abrupt cross-sectional changes, usually obtained by pressing with split punches, each section of which uses a different pressure and a different rate of compaction. See *split punch*.

stoking. See *continuous sintering*.

strip (noun). A roll-compacted metal powder product. See *roll compacting, rolled compact*.

strip (verb). Removal of the compact from the die. An alternative to ejecting or knockout. See *ejection, knockout*.

stripped die method. A specific method of removal of the compact after pressing, which keeps it in position between the punches while the die is retracted either upward or downward until the compact is fully exposed and freed by an upward withdrawal of the upper punch. The method is in general use abroad.

stripper punch. A punch which, in addition to forming the top or bottom of a die cavity, later moves further into the die to eject the compact.

submicron powder. Any powder whose particles are smaller than ~1 μm.

subsieve analysis. Size distribution of particles that will pass through a standard 325-mesh sieve having 44-μm openings.

subsieve fraction. Particles that will pass through a 44-μm (325-mesh) screen.

subsieve size. See preferred term *subsieve fraction*.

superfines. The portion of a powder composed of particles smaller than a specified size, currently less than 10 μm.

supplemental operation. See *secondary operation*.

surface contact points. The points at which abutting particles make contact during compacting and which grow into

necks during sintering. See *neck, neck formation*.

surface diffusion. One of the primary diffusion mechanisms during sintering. It is predominant for smaller particles and lower sintering temperatures as compared to other diffusion mechanisms, such as lattice or volume diffusion, which are prevalent for larger particles and higher temperatures. See *volume diffusion*.

surface film. Any continuous contamination on the surface of a powder particle.

swaging. A process for reducing the diameter or tapering of rods, wires, and tubes by forging, hammering, or squeezing.

swarf. An intimate mixture of fine machining or grinding chips of metal and extraneous matter such as resin or abrasive powder resulting from the operation.

sweating. See preferred term *exudation*.

T

tap density. The density of a powder when the volume receptacle is tapped or vibrated under specified conditions while being loaded.

theoretical density. The density of the same material in the wrought condition; see *density, absolute*.

thermal decomposition. The decomposition of a compound into its elemental species at elevated temperatures. A process whereby fine solid particles can be produced from a gaseous compound. See *carbonyl powder*.

thermal spray powder. A metal powder mixture designed for use with arc-spray welding equipment, flux core electrode wire, and acetylene welding overlay work.

thief. See *sample thief*.

throughput. Volume of charge passed in a time unit through a production sintering furnace.

tilt furnace. A furnace that has been used abroad for the infiltration of copper into porous sintered tungsten for heavy-duty contacts. The furnace is tilted to one position for the separate sintering of the tungsten and melting of the copper, then tilted to the opposite position to let the melt run to and contact or infiltrate, respectively, the tungsten pieces.

toggle press. A mechanical press used for compacting powders.

tool set. See *die*.

tooling. A generic term applying to compacting die assemblies and related items.

transition temperature. An arbitrarily defined temperature range in which the ductility changes rapidly with temperature. Examples: brittle-to-ductile transition temperature of tungsten is 200 to 300 °C (390 to 570 °F), that of molybdenum is near room temperature, and that of tantalum below −200 °C (−330 °F).

transverse-rupture strength. The stress, as calculated from the flexure formula, required to break a specimen as a simple beam supported near the ends while applying the load midway between the centerlines of the supports.

triple action press. A press that provides pressure from three sides, such as from top and bottom and from one side, either to impress a side indentation or recess, or to clamp a segmented die on top of a press table.

tube furnace. A furnace used for continuous or batch sintering utilizing a dense ceramic tube or a metallic retort to contain the controlled sintering atmosphere.

U

undersize powder. Powder particles smaller than the minimum permitted by a particle size specification.

uniaxial compacting. Compacting of powder along one axis, either in one direction or in two opposing directions. Contrast with *isostatic pressing*.

unidirectional compacting. Compacting of powder in one direction.

upper punch. The member of a die assembly or tool set that closes the die and forms the top of the part being produced.

upper ram. The part of a pneumatic or hydraulic press that is moving in an upper cylinder and transmits pressure to the upper punch or set of upper punches.

upset pressing. The pressing of a powder compact in several stages, which results in an increase in the cross-sectional area of the part prior to its ejection.

V

V-cone blender. A machine that has two cone-shaped containers arranged in a V and open to each other.

V-mixer. A machine that has two cylindrical containers arranged in the shape of a V and open to each other.

vacuum sintering. Sintering at subatmospheric pressure, such as in a technical vacuum or in a high vacuum.

vacuum sintering furnace. A furnace wherein sintering is conducted in a vacuum. The furnace may be of a design either for batch sintering or for continuous sintering. See *vacuum sintering.*

vibration compaction, vibratory compaction. A compacting process where vibration of the die assembly is used in addition to the usual pressure.

vibration density. The apparent density of a powder mass when the volume receptacle is vibrated under specified conditions while being loaded. Similar to *tap density.*

vibratory mill. A mill wherein comminution is aided by subjecting the balls or rods to a vibratory force.

void. See preferred term *pore.*

volatilize. The process of a substance becoming a gas. In powder metallurgy, lubricants and binders may be removed from the compact by heating. Also, in sintering brass powder compacts, the zinc content may be diminished by evaporation due to the high vapor pressure of the metal. See *dezincification.*

volume diffusion. One of the primary diffusion mechanisms during sintering. It is predominant for larger particles and higher temperatures, and its diffusion coefficient for the same conditions is smaller than that for grain boundary diffusion, and much smaller than that for surface diffusion. See *grain boundary diffusion, surface diffusion.*

volume filling. Filling the volume of a die cavity or receptacle with loose powder, and striking off any excess amount.

volume fraction. The volume percentage of a constituent or of porosity in a sintered body. Example: the amount of a refractory oxide phase in a dispersion-strengthened alloy.

volume ratio. The volume percentage of solid in the total volume of the sintered body.

volume shrinkage. The volumetric size reduction a compact undergoes during sintering. Contrast with *linear shrinkage.*

W

walking beam furnace. A continuous sintering furnace that transports the containers or trays by lifting them and swinging them in arcs toward the discharge end.

warpage. The distortion that occurs in a compact during sintering.

wax. Preferred lubricant for pressing cemented carbide powder mixtures.

wet, wettability. The property of a liquid, such as molten metal, to spread on a solid surface due to a low contact angle. This angle is a measure of the degree of wetting obtained in the solid-liquid system. The property is significant in liquid phase sintering and infiltration.

wetting agent. An additive that promotes wetting by conditioning the solid surface and thereby lowering the interfacial tension.

wet-bag tooling. A rubber or plastic sheet mold used in cold isostatic or hydrostatic pressing of powders.

whiskers. (1) Metallic filamentary growths, often microscopic in size, that attain very high strengths. (2) Oxide whiskers, such as sapphire, which because of their strength and inertness at high temperatures are used as reinforcements in metal matrix composites.

Z

zone sintering. Highly localized, progressive heating during sintering to produce a desired grain structure, such as grain orientation, and directional properties without subsequent working.

SOURCES

- *Glossary of Metallurgical Terms and Engineering Tables,* American Society for Metals, 1979
- *Definitions and Terms,* MPIF Standard 9-71, Metal Powder Industries Federation, Princeton, NJ, 1971
- *Standard Definitions of Terms Used in Powder Metallurgy,* ASTM Standard B 243-82, American Society for Testing and Materials, Philadelphia, 1982
- *Powder Metallurgical Materials and Products—Vocabulary,* ISO/DIS Standard 3252, International Organization for Standardization, Geneva, Switzerland, 1973 (available from American National Standards Institute, New York)
- *Glossary of Terms Relating to Powders,* British Standard 2955, British Standards Institution, 1958
- *International Powder Metallurgy Glossary,* Metal Powder Report, MPR Publishing Services, Shrewsbury, England, 1977-1983
- *Terminologie der Pulvermetallurgie—Ordnungsmerkmale, Begriffe,* DNA DIN-30900, Deutscher Normenausschuss, May 1972 (available from Beuth-Vertrieb GmbH, Cologne, Fed. Rep. Germany)
- *Métallurgie des Poudres—Terminologie,* Standard NF A 95-001, Association française de normalisation, Paris, 1971
- *Terms Used in Powder Metallurgy,* International Plansee Society for Powder Metallurgy, Reutte-Tirol, Austria, 1975

History of Powder Metallurgy*

By Kempton H. Roll
Executive Director
Metal Powder Industries Federation
American Powder Metallurgy Institute

POWDER METALLURGY has been called a lost art. Unlike clay and other ceramic materials, the art of molding and firing practical or decorative metallic objects was only occasionally applied during the early stages of recorded history. Sintering of metals was entirely forgotten during the succeeding centuries, only to be revived in Europe at the end of the 18th century, when various methods of platinum powder production were recorded.

Metal powders such as gold, copper, and bronze, and many powdered oxides (particularly iron oxide and other oxides used as pigments), were used for decorative purposes in ceramics, as bases for paints and inks, and in cosmetics since the beginnings of recorded history. Powdered gold was used to illustrate some of the earliest manuscripts. It is not known how these powders were produced, but it is possible that some of the powders were obtained by granulation after the metal was melted. Low melting points and resistance to oxidation (tarnishing) favored such procedures, especially in the case of gold powder. The use of these powders for pigments and ornamental purposes is not true powder metallurgy, because the essential features of the modern art are the production of powder and its consolidation into a solid form by the application of pressure and heat at a temperature below the melting point of the major constituent.

Early man learned by chance that particles of metal could be joined together by hammering, resulting in a solid metallic structure. In time, man learned how to build furnaces and develop temperatures high enough to melt and cast metals and to form lower melting alloys, such as copper and tin to make bronze.

Earliest Developments

Long before furnaces were developed that could approach the melting point of metal, P/M principles were used. About 3000 B.C., the Egyptians used a "sponge iron" for making tools. In this early process, iron oxide was heated in a charcoal and crushed shell fire, which was intensified by air blasts from bellows to reduce the oxide to a spongy metallic iron. The resulting hot sponge iron was then hammered to weld the particles together. Final shapes were obtained by simple forging procedures. Although the product often contained large amounts of nonmetallic impurities, some remarkably solid and sound structures have been discovered (Ref 1).

W.D. Jones (Ref 2) wrote of a process modification developed by African tribes. After reduction, the sponge was broken into powder particles, washed, and sorted by hand to remove as much of the slag and gangue as possible. The powder was then either compacted or sintered into a porous material, which was subsequently forged. Another example of ancient reduction of iron oxide was carried out in the fabrication of the Delhi Pillar, which weighs 5.9 metric tons (6.5 tons).

These crude forms of powder metallurgy ultimately led to the development of one of the commercial methods for producing iron powder. By grinding the sponge iron into fine particles, and heating in hydrogen to remove oxides and anneal or soften the particles, this process is today a viable technique for producing high-quality iron powder.

Powder metallurgy practices were used by the Incas and their predecessors in making platinum before Columbus made his voyage to the "New World" in 1492. The technique used was based on the cementing action of a lower melting binder, a technique similar to the present practice of making sintered carbides.

The technique consisted of cementing platinum grains (separated from the ore by washing and selection) by the addition of an oxidation-resistant gold-silver alloy of a fairly low melting point to wet the grains, drawing them together by surface tension and forming a raw ingot suitable for further handling (Ref 3).

A color change from the yellow of the sintered material to the whitish platinum of the final metal was caused by diffusion during heating prior to working. Heating is thought to have been accomplished by means of charcoal fires fanned by blowpipes. Analyses of these alloys vary considerably. The platinum content ranged from 26 to 72%, and the gold content ranged from 16 to 64%. Silver additions were found to vary from 3 to 15%, and amounts of copper up to 4% were traced.

Powder Metallurgy of Platinum

The metallurgy of platinum, as practiced in the 18th and 19th centuries in Europe, is considered to be one of the most important stages of development for modern powder metallurgy. For the first time, complete records were available that provided insight into the various methods of powder production and the processing of these powders into solid, useful implements.

Between 1750 and 1825, considerable attention was given to the manufacture of

*Portions of this article have been excerpted from Goetzel, C.G., *Treatise on Powder Metallurgy*, Vol 1, Interscience Publishers Inc., New York, 1949, and from Smith, C.S., The Early Development of Powder Metallurgy, in *Powder Metallurgy*, J. Wulff, Ed., American Society for Metals, 1942.

platinum. In 1755, Lewis (Ref 4) discovered that when a lead-platinum alloy was oxidized at high temperatures, a spongy, workable mass remained after lead oxide impurities had been volatilized. Scheffer (Ref 5) found that when platinum was heated with arsenic, the platinum showed signs of melting. This finding was confirmed in 1781 by Achard (Ref 6), who described the production of a fusible platinum-arsenic alloy, probably by forming the eutectic containing 87% Pt and melting at 600 °C (1110 °F). Achard formed solid platinum by hot hammering a sponge, welding the individual particles into a large solid. The sponge was obtained by high-temperature working of the platinum-arsenic alloy, which caused volatilization of the arsenic.

This procedure formed the basis for a method of producing platinum that was first used in about 1790 in commercially manufactured chemical vessels by Jannetty in Paris. Mercury was used later in a similar process by von Mussin-Puschkin (Ref 7). Other metals worked in this way include palladium, by using sulfur instead of arsenic, and iridium (using phosphorus). Ridolfi (Ref 8) made malleable platinum for chemical vessels using lead and sulfur.

In 1786, Rochon (Ref 9) successfully produced solid platinum without using arsenic by welding small pieces of scrap platinum. He produced malleable platinum by uniting purified platinum grains.

Knight (Ref 10) found that if chemically precipitated platinum powder was heated at high temperatures in a clay crucible, it softened and could be compressed and forged. Tilloch (Ref 11) put platinum powder into tubes made of rolled platinum sheet, which were then heated and forged to produce a compact mass. In 1813, Leithner (Ref 12) reported production of thin, malleable platinum sheets by drying out successive layers of powder suspended in turpentine and heating the resulting films at high temperatures without pressure.

In 1882, a French process was reported by Baruel (Ref 13), in which 14 kg (30 lb) of platinum powder was made into a solid ingot by a series of operations. Platinum was precipitated in powdered form, slightly compressed in a crucible, and heated to white heat. The powder was then put in a steel matrix and put under pressure with a screw coining press. The compact platinum was repeatedly reheated and re-pressed until a solid ingot was formed. The final heat treatments were made in a charcoal fire at lower temperatures. Because the platinum powder was placed in the steel die while hot, this process was based on the hot pressing technique.

In Russia in 1826, a high-temperature sintering operation was applied to previously compressed powder compacts on a commercial basis for the first time. This was in contrast to methods based on hot pressing. Sobolewskoy (Ref 14) described sifted platinum powder pressed into a cast iron cylinder that featured a steel punch actuated by a screw press. The resulting compacts were annealed for $1\frac{1}{2}$ days at high temperature in a porcelain firing kiln. The final product was highly workable, especially if the platinum powder had been well washed and was of high purity. Annealing, however, caused a decrease in volume; a cylinder 100 mm (4 in.) in diameter and 19 mm ($^3/_4$ in.) in height shrank 19 mm ($^3/_4$ in.) and 6 mm ($^1/_4$ in.) in these dimensions, respectively.

Another Russian method was reported by Marshall (Ref 15) in 1832. Platinum powder in a ring-shaped iron mold was pressed by a screw press, heated to a red heat, and re-pressed. After working in a rolling mill, the compacted discs were used as coins.

The Wollaston process of producing compact platinum from platinum sponge powder is generally considered the beginning of modern powder metallurgy. At least 16 years prior to his publication of 1829 (Ref 16), describing the manufacture of a product much superior to that of contemporary manufacturers, Wollaston devised the foundations for modern P/M technique. Wollaston was the first to realize all the difficulties connected with the production of solid platinum ingot from powdered metal, and thus concentrated on the preparation of the powder. He found that pressing the powder while wet into a hard cake (to be subsequently baked at red heat) was best done under considerable pressure. In addition, because available screw presses were not powerful enough, Wollaston developed a horizontal toggle press of the simple construction shown in Fig. 1. Wollaston used the following nine steps in the manufacture of compact platinum metal (Ref 17):

- Precipitating ammonium-platinum-chloride from diluted solutions
- Slowly decomposing the finely divided and carefully washed ammonium-platinum-chloride precipitate into loose sponge powder
- Grinding this sponge powder without applying pressure to the powder particles, thus avoiding any burnishing of the particles and preserving all the surface energy of the particles
- Sieving the sponge powder
- Washing the sponge powder with water to remove all remnants of volatile salts
- Separating fine particles from coarser particles through sedimentation (only the finest sponge particles were used)
- Pressing the wet mass containing the finest platinum particles into a cylindrical cake
- Drying the wet cake very slowly and then heating it to about 800 to 1000 °C (1475 to 1830 °F)
- Forging the cake while it was still hot

Fig. 1 Simple toggle press used by Wollaston for making platinum powder compacts

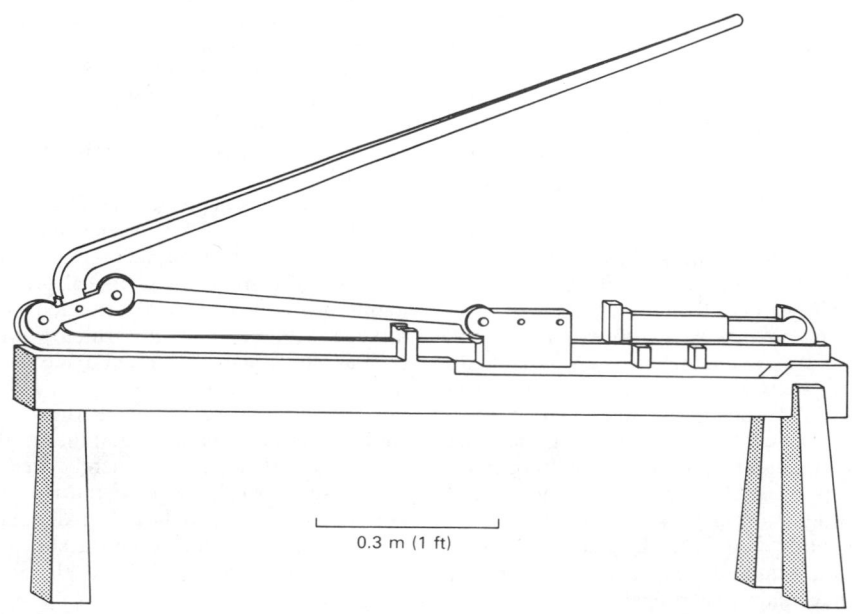

0.3 m (1 ft)

By applying these steps, Wollaston succeeded in producing compact platinum, which when rolled into thin sheet was practically free of gas blisters. Crucibles made from this sheet were the best quality platinum implements of their time. Wollaston's process was used for more than a generation and became obsolete only with the advent of the platinum fusion procedure developed by Sainte-Claire Deville and Debray in 1859 (Ref 17). They succeeded in producing a powerful flame with illuminating gas and oxygen, the oxygen being manufactured from manganese dioxide. However, the fused metal which they produced was superior to Wollaston platinum in quality and homogeneity, and the fusion procedure was also less expensive and quicker than the Wollaston method. Fusion, therefore, was soon adopted by every platinum refinery. It is still considered the superior method for manufacturing standard-quality platinum.

Further Developments

The use of P/M technology to form intricately shaped parts by pressing and sintering was introduced in the 19th century. In 1830, while determining the atomic weight of copper, Osann (Ref 18) found that the reduced metal could be sintered into a compact. Osann then developed a process for making impressions of coins from copper powder produced by the reduction of precipitated copper carbonate (Cu_2CO_3).

Osann found that reduction was best done at the lowest possible temperatures that could be used to produce a metal powder of the fineness known in platinum manufacture. High reduction temperatures resulted in granular masses that did not sinter well. Contamination of the powder by the atmosphere was eliminated by using the powder immediately after reduction or storing it in closed glass bottles. The powder was separated into three grades, determined by particle size, before use. To make an impression of a coin, fine powder was sprayed on the surface, followed by layers of coarser grades. The powder and a die were placed in a ring-shaped mold and compressed by the pressure of hammer blows on a punch or use of a knuckle press. Volume of the copper powder was reduced to one sixth of the original powder during compression. Sintering was done at temperatures close to the melting point of copper, after the compacts were placed in airtight copper packets sealed with clay. A nondistorted 20% shrinkage occurred, but the sintered copper was harder and stronger than cast copper.

Osann also produced medals of silver, lead, and copper by the same procedure. Although he considered his process especially suitable as an alternative to the electrotype method of reproducing coins and medallions, Osann advocated its use as an initial production method for these articles. He believed powder metallurgy could be used for producing printing type and for making convex and concave mirrors by pressing on glass. Osann thought that measurement of the shrinkage of copper compacts could be used to calculate temperature, as the shrinkage of clay cylinders was used in the Wedgewood pyrometer.

Among the advancements in the P/M industry during the second half of the 19th century were Gwynn's attempts to develop bearing materials from metal powders. Patents issued to Gwynn in 1870 (Ref 19) were the forerunners of a series of developments in the area of self-lubricating bearings. Gwynn employed a mixture of 99 parts of powdered tin, prepared by rasping or filing, and 1 part of petroleum-still residue. The two constituents were stirred while being heated. A solid form of desired shape was then produced by subjecting the mixture to extreme pressure while enclosing it in a mold. The patent specifically states that journal boxes made by this method or lined with material thus produced would permit shafts to run at high speeds without using any other lubrication.

Commercial Developments

The first commercial application of powder metallurgy occurred when carbon, and later osmium, zirconium, vanadium, tantalum, and tungsten, was used for incandescent lamp filaments. Methods were developed from 1878 to 1898 for making carbon filaments by the extrusion and subsequent sintering of carbonaceous materials.

Osmium filaments were used for a short time from 1898 to 1900. Auer von Welsbach (Ref 20) described the production of filaments of osmium by chemical precipitation of the powder and formation of a mixture with sugar syrup, which served both as binder and, if osmium oxide powder was used instead of the metal, as reducing agent as well. The mixture was squirted through fine dies, and the resulting fine threads were subsequently fired in protective atmospheres to carburize and volatilize the binder, reduce the oxide, and sinter the metal particles into a coherent metallic wire for use as an electrical conductor.

The osmium electric lamp was soon succeeded by tantalum filament lights, which were used widely from 1903 to 1911. The general procedure (Ref 21) was similar to that used for osmium, with the exception that tantalum had to be purified by a vacuum treatment to become ductile. Similar techniques were used for the production of filaments from zirconium, vanadium, and tungsten; with tungsten, especially, extruded wires were bent into hairpin shapes before sintering to shape them for use as filaments. Because lack of ductility was the major shortcoming of these filaments, attempts were made to improve this property by the addition of a few percent of a lower melting, ductile metal. Tungsten powder was mixed with 2 to 3% Ni, pressed into a compact, and sintered in hydrogen at a temperature slightly below the melting point of nickel. The resulting bars could be drawn, and nickel was removed from the final filaments by a vacuum heat treatment at a high temperature (Ref 22). Although this process was not commercially successful, it was an important step toward the industrial development of cemented carbides and composite materials.

Tungsten was soon recognized as the best material for lamp filaments. The problem, however, was to devise an economical procedure for producing these filaments in large quantities. A number of procedures to produce powdered tungsten had been worked out earlier. In 1783, the D'Elhujar brothers (Ref 23) first produced tungsten powder by heating a mixture of tungstic acid and powdered charcoal, cooling the mixture, and removing the small cake, which crumbled to a powder of globular particles. The purification of tungsten powder by boiling, scrubbing, and skimming to remove soluble salts, iron oxide, clay, and compounds of calcium and magnesium was reported by Polte (Ref 24).

Coolidge Process. At the beginning of the 20th century, Coolidge (Ref 25) made the important discovery that tungsten could be worked in a certain temperature range and would retain its ductility at room temperature. Few changes have been made over the years in the Coolidge procedure; it is still the standard method of producing incandescent lamp filaments. In this method, very fine tungsten oxide powder, WO_3, is reduced by hydrogen. The powder is pressed into compacts, which are presintered at 1200 °C (2190 °F) to strengthen them so that they can be clamped into contacts. They receive a final sintering treatment near 3000 °C (5430 °F) by passing a low-voltage, high current density current through the compacts. During sintering, the compacts shrink and reach a density near 90% that of solid tungsten. The sintered compacts can be worked only at temperatures near 2000 °C (3630 °F). When heated to this temperature, they can be swaged into rounds. With increasing amounts of warm work, tungsten becomes more ductile, the swaging temperature can be progressively lowered, and the swaged bars can be drawn into fine wire at relatively low temperatures.

Other Refractory Metals. The procedures developed for the production of tungsten often were adaptable to the manufacture of molybdenum. Lederer (Ref 26) developed a method of making molybdenum using powdered molybdenum sulfide. The sulfide, mixed with amorphous sulfur and kneaded into a paste, was formed into a filament. When exposed to air, the filaments became strong enough to be placed in a furnace. Heating in hydrogen resulted in formation of hydrogen sulfide and sintering of the metal into solid filaments. A similar process was patented by Oberländer (Ref 27), who used molybdenum chloride and other halides as starting materials. When the chloride was treated with a reducing agent such as ether, a paste was obtained.

Tungsten, molybdenum, and tantalum are the three most important refractory metals used today in the lamp, electronics, x-ray, and chemical industries. Other refractory metals of minor significance were developed by the P/M method in the early 1900's, notably niobium, thorium, and titanium. However, at the same time another development, originating in refractory metal processing, took form and rapidly grew to such importance that it far overshadows the parent field. Cemented carbides have become one of the greatest industrial developments of the century.

Cemented Carbides. Ordinary drawing dies were unsatisfactory for drawing tungsten wires and filaments. The need for a harder material to withstand greater wear became urgent. Because it was known that tungsten granules combined readily with carbon at high temperatures to give an extremely hard compound, this material was used as the basis for a very hard, durable tool material known as cemented carbides. The tungsten carbide particles, present in the form of finely divided, hard, strong particles, are bonded into a solid body with the aid of a metallic cementing agent. Early experiments with a number of metals established that this cementing agent had to possess the following properties to permit solidification of the hard metal body:

- Close chemical affinity for the carbide particles
- A relatively low melting point
- Limited ability to alloy with the carbide
- Great ductility (not to be impaired by the cementing operation)

Cobalt satisfied these requirements most closely. The early work was carried out mainly in Germany by Lohmann and Voigtländer (Ref 28) in 1914, by Liebmann and Laise (Ref 29) in 1917, and by Schröter (Ref 30) from 1923 to 1925. Krupp (Ref 31) perfected the process in 1927 and marketed the first product of commercial importance, "Widia." In 1928 this material was introduced to the United States, and the General Electric Company, which held the American patent rights, issued a number of licenses. The process entails carefully controlled powder manufacture, briquetting a mixture of carbide and metallic binder (usually 3 to 13% Co), and sintering in a protective atmosphere at a temperature high enough to allow fusion of the cobalt and partial alloying with the tungsten carbide. The molten matrix of cobalt and partly dissolved tungsten carbide forms a bond, holding the hard particles together and giving the metallic body sufficient toughness, ductility, and strength to permit its effective use as tool material.

Composite Metals. The next development in P/M was the production of composite metals used for heavy-duty contacts, electrodes, counterweights, and radium containers. All of these composite materials contain refractory metal particles, usually tungsten, and a cementing material with a lower melting point, present in various proportions. Copper, copper alloys, and silver are frequently used; cobalt, iron, and nickel are used less frequently. Some combinations also contain graphite. The first attempt to produce such materials was recorded in the patent of Viertel and Egly (Ref 32) issued shortly after 1900. The procedures used either were similar to those developed for the hard metals (Ref 33) or called for introduction of the binder in liquid form by dipping or infiltration. In 1916, Gebauer (Ref 34) developed such a procedure, which was developed further by Baumhauer (Ref 35) and Gillette (Ref 36) in 1924. Pfanstiehl (Ref 37) obtained patent protection in 1919 for a heavy metal, consisting of tungsten and a binder that contained copper and nickel.

Porous Metal Bearings and Filters. In addition to the development of refractory metals and their carbides, another important area of powder metallurgy that gained attention during the early 1900's was that of porous metal bearings. Special types of these porous bearings are referred to as self-lubricating.

The modern types of bearings, usually made of copper, tin, and graphite powders and impregnated with oil, were first developed in processes patented by Loewendahl (Ref 38) and Gilson (Ref 39 and 40). Gilson's material was a bronze structure, in which finely divided graphite inclusions were uniformly distributed. It was produced by mixing powdered copper and tin oxides with graphite, compressing the mixture, and heating it to a temperature at which the oxides were reduced by the graphite and the copper and tin could diffuse sufficiently to give a bronzelike structure. Excess graphite (up to 40 vol%) was uniformly distributed through this structure. The porosity was sufficient to allow for the introduction of at least 2% oil. The process was later improved by Boegehold and Williams (Ref 41), Claus (Ref 42), and many others, primarily by utilization of elemental metal powders rather than oxides.

Metallic filters were the next stage in the development of these porous metals, and patents date back as far as 1923 (Ref 43), when Claus patented a process and machine to mold porous bodies from granular powder.

Modern Developments

Infiltration techniques, porous materials, iron powder cores for radio tuning devices, P/M permanent magnets, and tungsten-copper-nickel heavy metal compositions were developed during the period between 1900, World War I, and the late 1920's. At the beginning of World War II in Europe, iron powder technology began its advance to commercial viability. The most spectacular development of iron parts made by powder metallurgy was during World War II in central Europe, where paraffin-impregnated sintered iron driving bands for military projectiles were extensively used. German powder metallurgists found this technique effective as a substitute for scarce gilding metal, a copper-zinc alloy containing 5 to 10% Zn. Production reached a peak of 3175 metric tons (3500 tons) per month for this application.

The advent of mass production in the automotive industry made possible the use of iron and copper powders in large tonnages and spawned many of the technological advances of the modern P/M industry. The automobile has been the basis for most industrial applications of P/M, even in fields unrelated to the automotive industry. The first commercial application of a P/M product, the self-lubricating bearing, was used in an automobile in 1927. It was made from a combination of copper and tin powders to produce a porous bronze bearing capable of retaining oil within its pores by capillary attraction. At about the same time, self-lubricating bearings were introduced to the home appliance market as a refrigerator compressor component.

Through the 1940's and early 1950's, copper powder and the self-lubricating bearing were the principal products of powder metallurgy. Since then, iron powder and steel P/M mechanical components such as gears, cams, and other structural shapes have become dominant. While copper powder remains an important P/M material, consumed on the order of 27 000 metric tons (30 000 tons) per year, it is overshadowed by iron and iron-based powders with markets of 181 000 metric tons (200 000 tons) per year.

Since the end of World War II, and especially with the advent of aerospace and nuclear technology, developments have been widespread with regard to the powder metallurgy of refractory and reactive metals such as tungsten, molybdenum, niobium, titanium, and tantalum and of nuclear metals such as beryllium, uranium, zirconium, and thorium.

All of the refractory metals are recovered from their ores, processed, and formed using powder metallurgy techniques. With the reactive metals, P/M is often used to achieve higher purity or to combine them with other metals or nonmetallics to achieve special properties. Nuclear power plants use fuel elements often made by dispersing uranium oxide in a metal powder (aluminum, for example) matrix. The control rods and neutron shielding may use boron powder in a matrix of nickel, copper, iron, or aluminum. Tungsten combined with nickel and copper powders is used widely as a shielding component in applications where intricate configuration involving machining is required, such as in cobalt-60 containers. In aerospace, beryllium and titanium are used extensively. Rocket skirts, cones, and heat shields are often formed from niobium. Molybdenum is widely used in missile and rocket engine components. Nozzles for rockets used in orbiting space vehicles often are made from tungsten via the P/M process in order to maintain critical dimensional tolerances.

The 1950's and 1960's witnessed the emergence of P/M wrought products. These are fully dense metal systems that began as powders. Hot isostatically pressed su-

peralloys, P/M forgings, P/M tool steels, roll compacted strip, and dispersion strengthened copper are all examples. Each of these processes and materials is covered in separate articles in this Volume.

The commercialization of powder-based high-performance material emerged as a major breakthrough in metalworking technology in the 1970's by opening up new markets through superior performance, coupled with the cost effectiveness of material conservation and longer operational life.

Recent P/M technology has focused to a great extent on superalloys, primarily because of government-supported research and the desire to produce jet engine components at a lower cost and with higher performance. In the late 1970's, the experimental programs involving P/M wrought products began spilling over into the commercial industrial sector, principally in the form of P/M tool steels and P/M forgings.

P/M forgings marked the beginning: no longer were properties compromised by density. Fully dense components capable of combining the alloying flexibility and the net and near-net design features of P/M were very marketable. The late 1970's and early 1980's witnessed a significant metallurgical breakthrough in the recognition of P/M techniques for eliminating segregation and ensuring a fully homogeneous, fine-grained, pore-free, high-alloy structure. Categorized as P/M wrought metals, they led to the perfection

of extremely high-purity metal powders and improved consolidation techniques such as hot isostatic pressing (HIP). More information on P/M wrought products can be found in the articles "P/M Forging" and "Hot Isostatic Pressing of Metal Powders" in this Volume. The 1980's also saw the commercialization of ultra-rapid solidification and injection molding technology. Both of these developments are also covered in separate articles in this Volume.

Commercial powder metallurgy now spans the density spectrum from highly porous metal filters through self-lubricating bearings and P/M parts with controlled density to fully dense P/M wrought metal systems. A review of major historical developments in powder metallurgy is presented in Table 1.

Powder Metallurgy Literature

A number of literary works are worthy of mention in connection with the background of powder metallurgy. One of the earliest works of significance was *Principles of Powder Metallurgy* by W.D. Jones, published in 1937 in England (Ref 44). It was updated in 1960 and published as *Fundamental Principles of Powder Metallurgy* (Ref 45). The first Russian publication was by Bal'shin (Ref 46) and appeared in 1938; the first comprehensive text in German, *Pulvermetallurgie und Sinterwerkstoffe*, was published by R. Kieffer and W. Hotop in 1943 (Ref 47).

Table 1 Major historical developments in powder metallurgy

Date	Development	Origin
3000 B.C.	"Sponge iron" for making tools	Egypt, Africa, India
1200 A.D.	Cementing platinum grains	South America (Incas)
1781	Fusible platinum-arsenic alloy	France, Germany
1790	Production of platinum-arsenic chemical vessels commercially	France
1822	Platinum powder formed into solid ingot	France
1826	High-temperature sintering of platinum powder compacts on a commercial basis	Russia
1829	Wollaston method of producing compact platinum from platinum sponge (basis of modern P/M technique)	England
1830	Sintering compacts of various metals	Europe
1859	Platinum fusion process	
1870	Patent for bearing materials made from metal powders (forerunner of self-lubricating bearings)	United States
1878-1900	Incandescent lamp filaments	United States
1915-1930	Cemented carbides	Germany
Early 1900's	Composite metals	United States
	Porous metals and metallic filters	United States
1920's	Self-lubricating bearings (used commercially)	United States
1940's	Iron powder technology	Central Europe
1950's and 1960's	P/M wrought and dispersion-strengthened products, including P/M forgings	United States
1970's	Hot isostatic pressing, P/M tool steels, and superplastic superalloys	United States
1980's	Rapid solidification and injection molding technology	United States

In the United States, the first publication was by H.H. Hausner in 1947 (Ref 48), followed closely by P. Schwarzkopf (Ref 49). Two years later, the first of four volumes of a treatise on powder metallurgy, a major work by C.G. Goetzel (Ref 50), was published.

Powder Metallurgy Trade Associations

The advancement of powder metallurgy from a laboratory curiosity to an industrial technology has been influenced greatly by various professional societies and the P/M trade association, whose annual technical conference proceedings chronicle the maturing of the technology. In 1946, an organization called the Metal Powder Association was founded by a group of metal powder producers in the U.S. It was reorganized in 1958 as the Metal Powder Industries Federation, a trade association whose representation embraced the commercial and technological interests of the total metal powder producing and consuming industries. The Federation consists of the following autonomous associations, which together represent the primary elements of the P/M industry:

- *Powder Metallurgy Parts Association* Members are companies that manufacture P/M parts for sale on the open market.
- *Metal Powder Producers Association* Members are producers of metal powders in any form for any use. The Gold Bronze Powder Council is part of this association. Its members manufacture gold bronze flake powders for pigments and printing inks.
- *Powder Metallurgy Equipment Association* Members are manufacturers of P/M processing equipment and supplies, including compacting presses, sintering furnaces, belts, tools, and dies.
- *Refractory Metals Association* Members are manufacturers of powders or products from tungsten, molybdenum, tantalum, or rare earth metals.
- *Powder Metallurgy Industries Association* A conglomerate of companies that utilize the powder metallurgy process in their products, for example, metallic friction materials and high-performance products such as P/M tool steels, electronic cores, metallic filters, and roll compacted products. Within this Association are the P/M Porous Metals Council, whose members manufacture P/M porous products such as filters, and the Iron Powder Core Council, whose members manufacture iron powder cores for electronic use.

- *In-Plant Powder Metallurgy Association* Members manufacture P/M parts for in-plant use, (not for sale on the open market).

The Federation generates industry statistics, process and materials standards, industrial public relations and market development, government programs, research, and various educational programs and materials.

The technology's "professional" society is the American Powder Metallurgy Institute. As distinguished from the Federation, Institute members are individuals, not companies. Members are kept informed of developments in P/M technology through local section activities, conferences, and publications, including the quarterly *International Journal of Powder Metallurgy and Powder Technology* and a monthly *P/M Technology Newsletter*. It is the only professional society organized specifically to serve the powder metallurgist and the P/M industry.

Many of the major professional societies are also active in P/M, usually through committees working on standards, conferences, or publications. This includes the American Society for Metals, the Metallurgical Society, the Society of Automotive Engineers, the American Society for Testing and Materials, and the Society of Manufacturing Engineers.

REFERENCES

1. Carpenter, H.C.H. and Robertson, J.M., The Metallography of Some Ancient Egyptian Implements, *J. Iron and Steel Inst.*, Vol 121, 1930, p 417-448
2. Jones, W.D., *Fundamental Principles of Powder Metallurgy*, London, 1960, p 593
3. Bergsöe, P., The Metallurgy and Technology of Gold and Platinum Among the Pre-Columbian Indians, *Ingeniorsvidenskabelige Skrifter (A)*, Vol 44, Copenhagen, 1937
4. Lewis, W., Experimental Examination of a White Metallic Substance Said to Be Found in the Gold Mines of Spanish West Indies, *Phil. Trans. Roy. Soc.*, Vol 48, 1755, p 638
5. Scheffer, H.T., Det hvita Gullet, eller sjunde Metallen kalladt i Spanien Platina del Pinto...., *Handlingar*, K. Svenska Vetenskaps Academiens, Stockholm, Vol 13, 1752, p 269-275
6. Achard, K.F., Sur l'Arsenic et sur sa Combinaison avec Differents Corps, *Nouveaux Mem. Acad. Roy. Sci.*, Berlin, Vol 12, 1781, p 103-119
7. von Mussin-Puschkin, A., Eine neue Methode die Platina zu schmieden, *Allgem. J. Chem.*, Scherer, Vol 4, 1800, p 411
8. Ridolfi, C., *Quart. J. Sci. Lit. Arts*, Vol 1, 1816, p 259-260 (From Giornale di scienza ed arti, Florence, 1816)
9. Rochon, A., Mémoire qui traite du Platine, de son utilité dans les arts...., *J. Phys. Chem. Arts*, Vol 47, 1798, p 3-15 (Rochon states that this was written in 1786 as part of his voyage to Madagascar)
10. Knight, R., A New and Expeditious Process for Rendering Platina Malleable, *Phil. Mag.*, Vol 6, 1800, p 1-3
11. Tilloch, A., A New Process of Rendering Platina Malleable, *Phil. Mag.*, Vol 21, 1805, p 175
12. Leithner, Letter quoted by A.F. Gehlen, *J. Chem. Phys.*, Schweigger, Vol 7, 1813, p 309 and 514
13. Baruel, M., Process for Procuring Pure Platinum, Palladium, Rhodium, Iridium, and Osmium From the Ores of Platinum, *Quart. J. Sci. Lit. Arts*, Vol 12, 1822, p 246-262
14. Sobolewskoy, P., *Ann. Physik Chem.*, Vol 109, 1834, p 99
15. Marshall, W., An Account of the Russian Method of Rendering Platinum Malleable, *Phil. Mag.*, Vol 11 (No. II), 1832, p 321-323
16. Wollaston, W.H., On a Method of Rendering Platina Malleable, (Bakerian Lecture for 1828), *Phil. Trans. Roy. Soc.*, Vol 119, 1829, p 1-8
17. Streicher, J.S., *Powder Metallurgy*, J. Wulff, Ed., American Society for Metals, 1942, p 16
18. Osann, G., *Ann. Physik Chem.*, Vol 128, 1841, p 406
19. U.S. Patents 101,863; 101,864; 101,866; and 101,867, 1870
20. U.S. Patent 976,526, 1910
21. U.S. Patents 899,875, 1908 and 912,246, 1909
22. Smith, C.R., *Powder Metallurgy*, J. Wulff, Ed., American Society for Metals, 1942, p 4
23. Deller, A.W., *Powder Metallurgy*, J. Wulff, Ed., American Society for Metals, 1942, p 582
24. U.S. Patent 735,293, 1903
25. U.S. Patent 963,872, 1910
26. U.S. Patent 1,079,777, 1913
27. U.S. Patent 1,208,629, 1916
28. German Patents 289,066, 1915; 292,583, 1916; 295,656, 1916; 295,726, 1916. Swiss Patents 91,932 and 93,496, 1919
29. U.S. Patents 1,343,976 and 1,343,977, 1920
30. German Patent 420,689, 1925. U.S.

Patent 1,549,615, 1925

31. British Patents 278,955, 1927, and 279,376, 1928. Swiss Patent 129,647, 1929. U.S. Patent 1,757,846, 1930
32. U.S. Patent 842,730, 1907
33. U.S. Patents 1,418,081, 1922; 1,423,338, 1922; and 1,531,666, 1925
34. U.S. Patent 1,223,322, 1917
35. U.S. Patent 1,512,191, 1924
36. U.S. Patent 1,539,810, 1925
37. U.S. Patent 1,315,859, 1919
38. U.S. Patent 1,051,814, 1913
39. U.S. Patent 1,177,407, 1916
40. Gilson, E.G., *General Electric Review*, Vol 24, 1921, p 949-951
41. U.S. Patents 1,642,347, 1927; 1,642,348, 1927; 1,642,349, 1927; and 1,766,865, 1930
42. U.S. Patent 1,648,722, 1927
43. U.S. Patent 1,607,389, 1926
44. Jones, W.D., *Principles of Powder Metallurgy*, Arnold, London, 1937
45. Jones, W.D., *Fundamental Principles of Powder Metallurgy*, Arnold, London, 1960
46. Bal'shin, M.Y.J., *Metal Ceramics* (in Russian), Gonti, Moscow and Leningrad, 1938
47. Kieffer, R. and Hotop, W., *Pulvermetallurgie und Sinterwerkstoffe*, Springer, Berlin, 1943; Re-issue Springer, Berlin, 1948
48. Hausner, H.H., *Powder Metallurgy*, Chemical Publishing Co., New York, 1947
49. Schwarzkopf, P., *Powder Metallurgy*, Macmillan, New York, 1947
50. Goetzel, C.G., *Treatise on Powder Metallurgy*, Vol 1-4, Interscience, New York, 1949

Production of Metal Powders

Introduction

METAL POWDER PRODUCTION has always been influenced greatly by cost and quality requirements of powder consolidation techniques and applications. This interrelationship has led to the development of many powder-producing processes. Manufacturing methods must be cost effective; powders must possess the required physical and chemical characteristics for use in the various P/M consolidation processes and applications. Generally, powders in current use are cleaner, more uniform in size, and may be produced within narrower specification ranges. Microprocessor control can be used to achieve refined manufacturing methods and improved final product performance. Improved and more complete powder characterization is certain to become more important.

Because of its increasing use in structural parts during World War II, iron powder has become the single most important powder in the P/M industry. Statistics on iron powder shipments have been used to measure the growth of the P/M industry. Figure 1 graphically illustrates iron powder shipments for North America from 1965 to 1981. Table 1 shows iron powder production capacities for 1979 for North America.

Up to 1965, most of the iron powders used in North America were made by reduction of oxides or ore. Due to significant advances in atomization technology in the 1950's and 1960's, most iron powder is currently produced by atomization techniques. Water-atomized iron powders have enjoyed increased usage because they offer reduced production costs, excellent compressibility, and high purity.

High compressibility is of prime importance for the production of structural parts by pressing and sintering. This constitutes the major use of iron powders. Before 1965, electrolytic iron powder, despite its higher cost, was used widely because of its unsurpassed compressibility. Water-atomized low-carbon iron provides a less expensive substitute of almost equal compressibility. With the development of techniques such as powder forging that achieve full density, powder pur-

ity has become equally important for the achievement of maximum dynamic properties.

The development of mechanical alloying in the late 1960's and the demand for finer powders in recent years have led to the use of high-energy milling (attritor and vibratory ball milling). Full-dense sintering of high-alloy steels and injection-molding processes have led to the use of powders with particle sizes in the 5 to 10 μm range, compared to −80 mesh (177 μm) powders used in conventional P/M processes.

Powder producers who supply metal powders to the P/M industry also supply metal powders for manufacturing paints, inks, catalysts, pyrophorics, explosives, food additives, and welding electrodes, and for use in plasma and flame spraying.

Fig. 1 Iron powder shipments and uses for North America
Source: Metal Powder Industries Federation/Metal Powder Producers Association

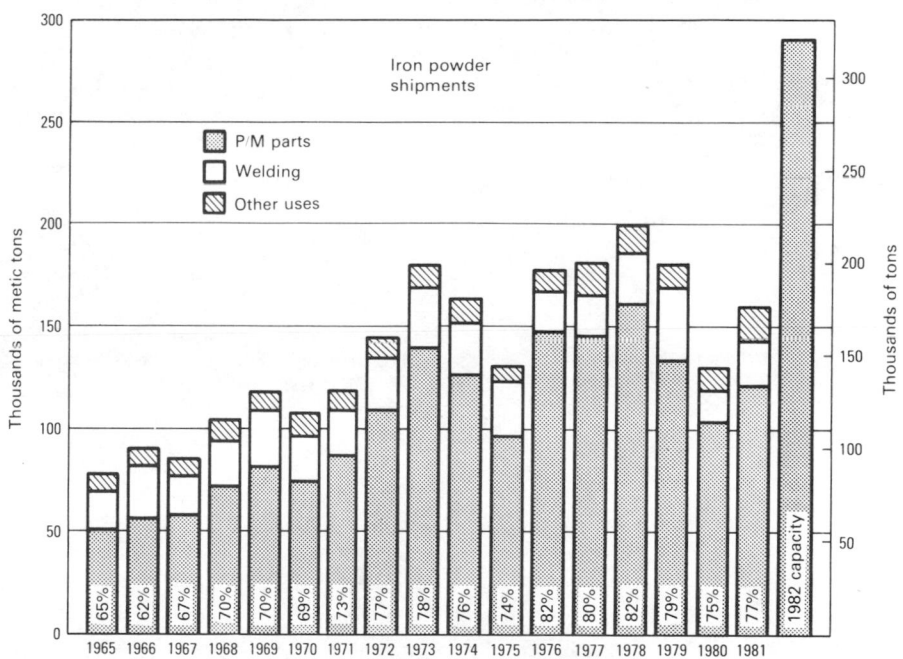

Table 1 1979 production capacities for iron powders in North America

Type of iron powder	Metric tons	Tons
Comminuted sponge iron	63 500	70 000
Other reduced	14 500	16 000
Total reduced	78 000	86 000
Atomized low-carbon steel	91 000	100 000
Ball milled high-carbon shot plus mill scale	62 000	68 000
Air-atomized high-carbon iron	...	...
Total atomized	153 000	168 000
Total all iron powders	231 000	254 000

Source: S.I. Huthén, "Five Decades of Iron Powder Production," *Int. J. Powder Metall. Powder Technol.*, Vol 17 (No. 2), April 1981, p 81-106

Table 2 1982 powder shipments

Metal	Metric tons	Tons
Iron	126 000(a)	139 000(a)
Copper and copper base	15 200(a)	16 765(a)
Atomized aluminum	25 500(b)	28 130(b)
Molybdenum	1 676(b)	1 848(b)
Tungsten (hydrogen reduced)	1 603(b)	1 767(b)
Tungsten carbide	2 133(b)	2 351(b)
Nickel (including flake)	11 008(b)	12 134(b)
Tin	890(b)	981(b)

(a) Data for North America provided by the Metal Powder Industries Federation. (b) Data for United States provided by the U.S. government.

Manufacturing processes for these powders are similar to those used for producing powders for structural parts. Table 2 gives data on metal powder shipments in North America.

In this Section, discussion of metal powder production processes is divided into articles on basic processes and individual powders. The articles on basic processes describe the methods and/or processes of current importance that are applicable to a number of metals. The articles on individual powders give a more detailed description of the various industrial processes used for a particular powder. In some cases, powders produced by different processes can be used interchangeably, provided certain powder properties fall within specified ranges.

Powders used for P/M processing often undergo one or more treatments, such as screening, cleaning, annealing, blending, mixing, and lubricating. These treatments usually are performed by the powder producer. Each of these processing steps is addressed in a separate article in this Section. In addition, articles on the explosivity, pyrophoricity, and toxicity of metal powders are included.

Atomization

ATOMIZATION technology has grown steadily and has become more sophisticated since large-scale production of atomized iron powder first began during World War II. High-quality powders —from aluminum, brass, and iron powders to stainless steel, tool steel, and superalloy powders—in combination with new consolidation techniques, have led to many new applications and to properties that are sometimes superior to wrought counterparts. Currently, many types of powders are produced by atomization. Some of the more commercially important powders and their respective 1981 shipments within North America are shown in Table 1.

Atomization may be defined simply as the breakup of a liquid into fine droplets, typically smaller than about 150 μm. For larger particles, this process is referred to as "shotting." Consequently, any material available as a liquid can be atomized. The breakup of a liquid stream brought about by the impingement of high-pressure jets of water or gas is referred to as "water" or "gas atomization," respectively, or more generally, two-fluid atomization (Fig. 1a and b). The use of centrifugal force to break up a liquid stream is known as "centrifugal atomization" (Fig. 1c). Atomization into a vacuum is known as "vacuum" or "soluble-gas atomization" (Fig. 1d). The use of ultrasonic energy to effect breakup is referred to as "ultrasonic atomization" (Fig. 1e).

The most important properties of an atomized powder include average particle size, particle size distribution or screen analysis, particle shape, chemical composition (including surface composition), and microstructure. These fundamental properties subsequently determine the engineering properties (apparent density, flowability, green strength, compressibility, sintering rate, forgeability, and toughness) and applications of the powders and resultant parts.

Table 1 North American metal powder shipments in 1981

Powder	Total		Atomization	
	Metric tons	Tons	Metric tons	Tons
Iron	157 850	174 000	78 900	87 000
Copper	18 140	20 000	~15 400	~17 000
Copper alloys	3 175(a)	3 500(a)	3 175(a)	3 500(a)
Aluminum	36 300(b)	40 000(b)	36 300(b)	40 000(b)
Stainless steel (303L, 304L, 410L, 430L, 434L)	1 540	1 700	1 540	1 700
Nickel	~4 550(c)	~5 000(c)	0	0
Nickel-based alloys	910	1 000	910	1 000
Cobalt	910	1 000	...	...
Cobalt-based alloys	910	1 000	910	1 000
Tin	910	1 000	910	1 000

(a) Cupronickel for coinage not included. (b) Major uses are as chemical coatings, explosives, pyrotechnics, and solid rocket fuels. (c) Excludes captive use in briquet and strip production

Gas and Water Atomization

By Erhard Klar
Manager of Particle Technology
SCM Metal Products
and
James W. Fesko
Metallurgist
SCM Metal Products

CONVENTIONAL WATER AND GAS ATOMIZATION processes presently account for the bulk of atomized metal powders. Method of production preference is determined to a large extent by the powder properties desired.

Water-atomized powders generally are quite irregular in shape and have relatively high surface oxygen contents. Gas-atomized powders, on the other hand, generally are more spherical or rounded in shape and, if atomized by an inert gas, generally have lower oxygen (oxide) contents. There are, of course, exceptions in each type. For high-volume, low-cost production, water atomization generally is preferred over gas atomization, providing powder character-

istics are compatible with the application. In spite of these important differences, the two processes are similar in many other respects.

A schematic of a typical water atomization production setup is shown in Fig. 2. The major components of a typical installation include a melting facility, an atomizing chamber, and powder drying equipment. Figure 3 provides detailed schematics of modified gas and water atomization processes. A typical gas/metal spray pattern occurring during gas atomization is shown in Fig. 4. Melting of metals follows standard procedures. Air, inert gas and vacuum induction melting, arc melting, and fuel heating are suitable procedures.

Typically, the molten metal is poured, either directly or by means of a runner, into a tundish (Fig. 2 and 3), which is essentially a reservoir that supplies a uniform and controlled flow of molten metal to the tundish nozzle. The nozzle, which is located at the base of the tundish, controls the shape and size of the metal stream and directs it through an atomizing nozzle system in which the metal stream is disintegrated into fine droplets by the high-velocity atomizing medium.

Liquid metal droplets cool and solidify as they settle to the bottom of the atomi-

Fig. 1 Schematic of atomization processes

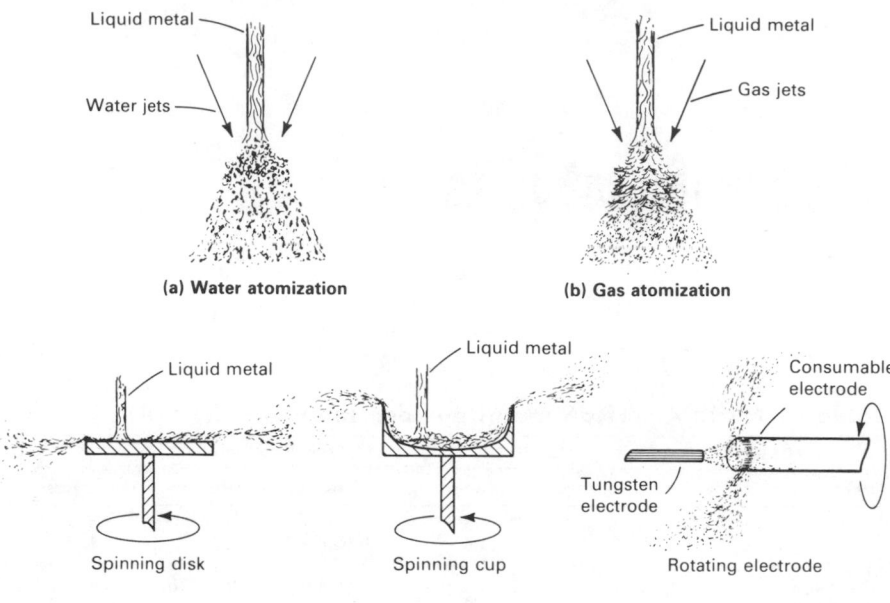

(a) Water atomization

(b) Gas atomization

Spinning disk

Spinning cup

Rotating electrode

(c) Centrifugal atomization

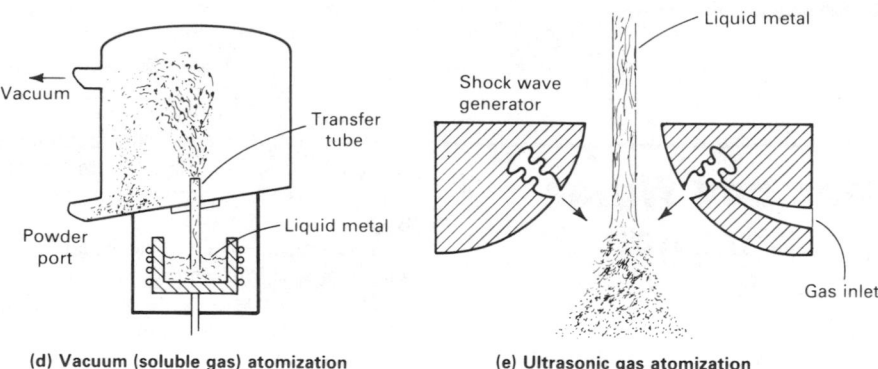

(d) Vacuum (soluble gas) atomization

(e) Ultrasonic gas atomization

Fig. 2 Water atomization setup

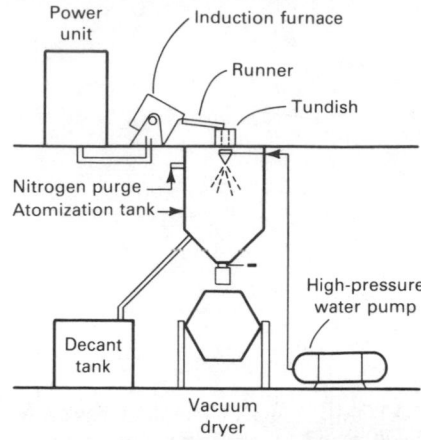

zation tank. This tank may be purged with an inert gas to minimize or prevent oxidation of the powder. In gas atomization, the powder may be collected as dry particles or cooled with water at the bottom of the tank. In so-called dry collection, the atomization tank is tall, usually more than 6 m (20 ft), to ensure solidification of the powder particles before they reach the bottom of the collection chamber. Horizontal gas atomization using long horizontal tanks is also used for the same purpose.

To atomize large amounts of powder or for continuous operation, external cooling of the bottom of the atomizing tank may be supplemented by fluidized bed cooling of the powder to prevent sticking and caking of the powder particles (Ref 1). In some gas atomization processes, especially in the production of low-melting-point metals such as aluminum and tin, molten metal may flow through a horizontal nozzle, or may be aspirated up into a vertical nozzle.

Nozzle Designs

Many different gas and water nozzle designs are in current use, as shown in Fig. 5. Their function is to control the flow and pattern of the atomizing medium to provide for efficient disintegration of the liquid metal and to produce a powder of the desired properties. Many nozzle designs in current use are proprietary. Available performance data are, for the most part, empirical and originate from patent literature.

Nozzles that direct and control the at-

omizing medium generally take the form of either an annulus, which is concentric around the metal stream, or a discrete jet. Nozzles are available in "free-fall" and "confined" designs, as shown schematically in Fig. 6. The confined design is rarely, if ever, used in water atomization.

In free-fall atomization, the liquid metal issues from the tundish, falls by gravity a certain distance (about 5 to 20 cm, or 2 to 8 in.), then interacts with and becomes atomized by the high-pressure medium. In gas and water atomization, a vacuum is created above the impingement point of the atomizing medium, thus causing the liquid metal to form a hollow cone- or umbrella-shaped pattern. This process is referred to as "prefilming," which is shown schematically in Fig. 1(a) and (b). A typical free-fall design incorporating discrete jets is shown in Fig. 5(e). Annular free-fall designs, as shown in Fig. 6(a), are also used. Many water atomization systems use a multiple discrete jet free-fall design. Typical water patterns obtained by such designs are shown in Fig. 7 and 8.

With confined designs, which are used almost exclusively with annular gas jets, atomization occurs at the orifice of the tundish nozzle. In this case, energy transfer from gas to metal is more efficient and more uniform due to prefilming of the molten metal over the end of the tundish nozzle, as well as the shorter distance between gas and metal. Because of the rapid attenuation of energy with increasing distance of high-velocity gas jets, confined, annular designs are frequently used in gas atomization systems (Fig. 5a through d). Free-fall, annular, and multiple discrete free-fall jet designs are also used.

The energy efficiency advantages of confined, annular designs, however, must be weighed against their greater sensitiv-

Fig. 3 Schematic of water and gas atomization processes
Source: Ref 1

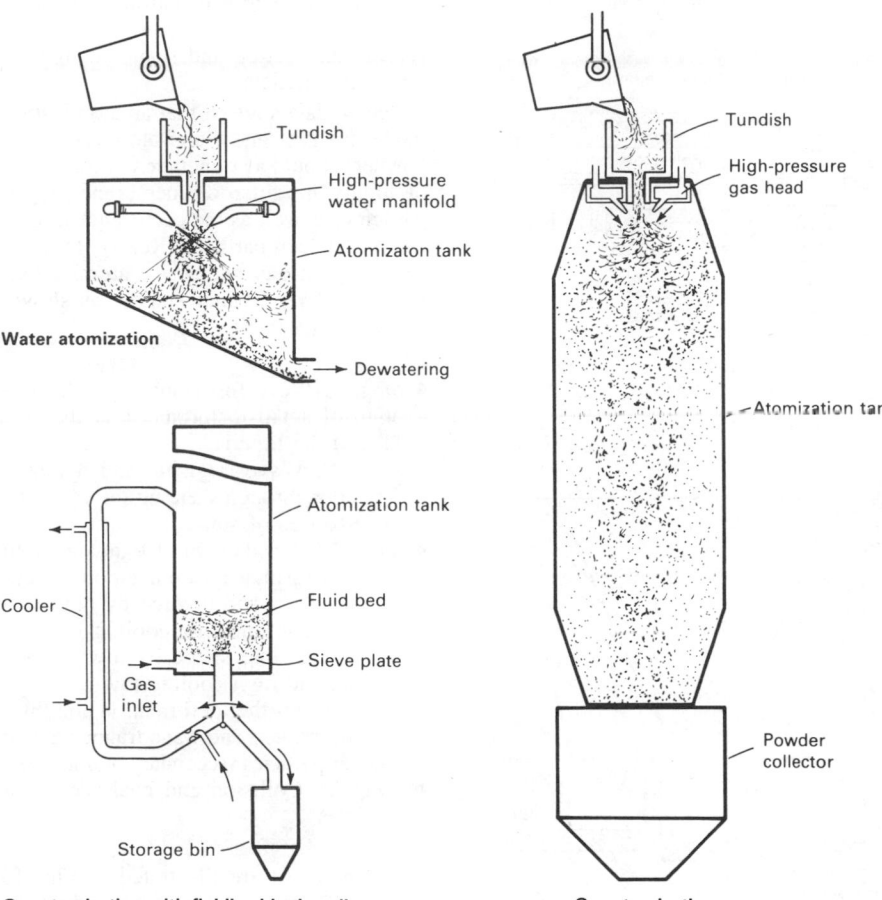

Water atomization

Gas atomization with fluidized bed cooling

Gas atomization

Fig. 4 Melting of alloys and typical gas/metal spray pattern
Courtesy of Nyby Uddeholm AB

lium, the cost of gas consumption is significant, and a means of circulation to facilitate gas reuse is desirable. A schematic of an argon recirculation system to purify the gas for reuse is shown in Fig. 10.

Mechanism of Atomization

In contrast to some of the more recently developed atomization processes discussed later in this article, conventional water and gas atomization are not completely understood in terms of basic principles. Nevertheless, a large body of experimental data exists, including several useful empirical relationships. Some of these data pertain to atomization studies of room-temperature liquids and cover applications such as spray drying, agricultural and insecticidal spraying, paint spraying, fuel injection in internal combustion engines, and oil burner injectors. For more information on atomization, see Ref 12 and 13.

Figure 11 shows a model for the disintegration of a liquid sheet by a high-velocity jet of gas (Ref 14). The initiation of sineous waves, or other disturbances that rapidly increase in amplitude in stage I (see below), is followed by fragmentation, which forms ligaments in stage II. Breakdown of ligaments into droplets occurs in stage III.

The recent use of high-speed and Schli-

ity to metal freeze-up; that is, metal that freezes at the end of the nozzle and blocks further atomization. Careful attention to nozzle design and heat and metal flow balances can prevent freezing. Examples of such metal buildup are shown in Fig. 9.

Most annular gas nozzles have tangential gas inlets that cause the gas to assume a spiral motion, which thus affects the trajectory of the molten metal droplets. The exit configuration of a gas nozzle may be a converging design (Fig. 6a), which limits the gas to subsonic velocities. A properly proportioned dual converging/diverging exit design (Fig. 6b) permits the gas to reach supersonic velocities.

Ultrasonic gas atomization is similar to conventional gas atomization, except that the discrete gas jets (Fig. 1e) impinge on the metal stream in a pulsed fashion, with frequencies of 40 to 100 kHz (40 000 to 100 000 cycles/s) (Ref 10). Improved efficiency and faster quenching compared to conventional gas atomization are claimed.

Atomization Variables

In conventional two-fluid atomization, typical metal flow rates through single orifice nozzles range from about 4.5 to 90 kg/min (10 to 200 lb/min); typical water flow rates range from 110 to 380 L/min (30 to 100 gal/min) at water velocities ranging from 70 to 230 m/s (230 to 750 ft/s) and pressures from 5.5 to 21 MPa (800 to 3000 psi). Typical gas flow rates range from 1 to 14 m³/min (40 to 500 scfm) at gas pressures in the range of 350 to 8400 kPa (50 to 1200 psi). Gas velocities depend on nozzle design and may range from 20 m/s (66 ft/s) to supersonic velocities. The temperature differential between the melting point of the metal and the temperature at which the molten metal is atomized (superheat of the molten metal) is generally about 75 to 150 °C (135 to 270 °F).

In gas atomization with argon or he-

Fig. 5 Atomization nozzle designs

Gas atomization: (a) Mannesmann process design (Ref 2). (b) Thompson design (Ref 3). (c) Naeser design (Ref 4). (d) Probst design (Ref 5). Water atomization: (e) Batten design (Ref 6). (f) Winstrom design (Ref 7)

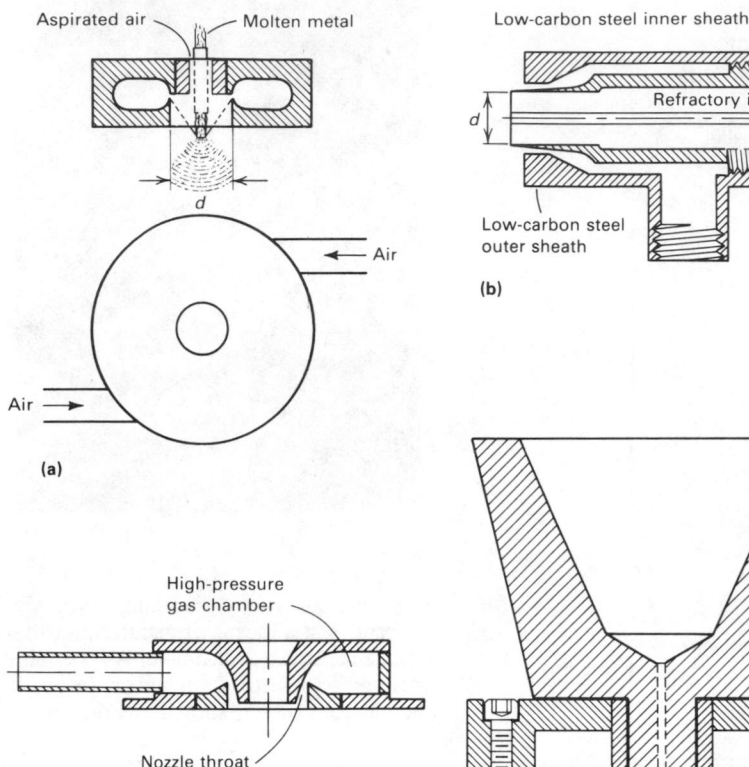

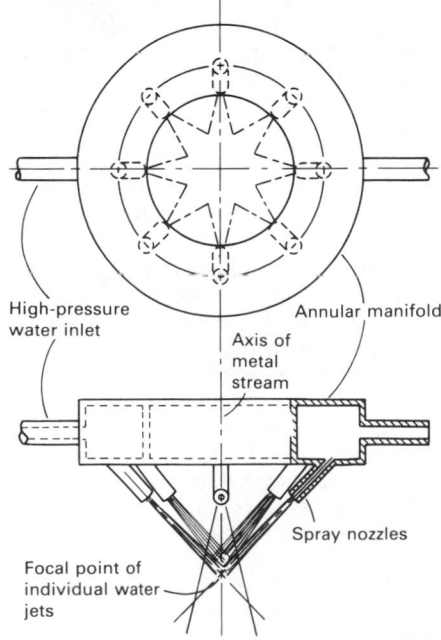

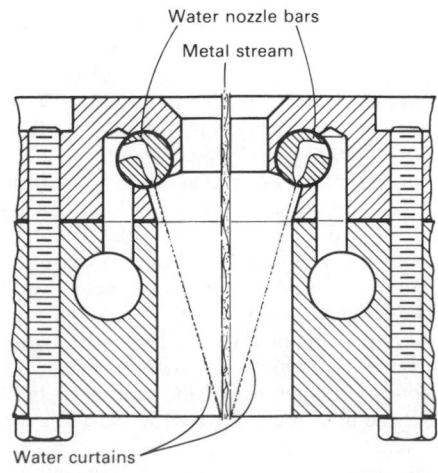

eren photography* has helped to determine velocity distributions in the gas phase, as well as velocities of particles. Resolution is still inadequate, however, for observing the motion and shape changes of individual particles.

Recent data suggest that an atomization model for gas- and water-atomized metal powders should also include the formation of highly irregular or membrane-shaped particles, as well as the occurrence of collisions between particles (Ref 8). Thus, in conventional gas and water atomization, there are five distinct stages, as shown schematically in Fig. 12:

- *Stage I:* Wave formation through initiation of small disturbances at the surface of the liquid
- *Stage II:* Wave fragment and ligament formation through shearing forces on the disturbances of stage I
- *Stage III:* Breakdown of ligaments into droplets (primarily atomization); regular particle shape favored by high surface tension and low cooling rate; irregular particle shape by low surface tension and high cooling rate
- *Stage IV:* Further deformation and thinning of droplets and wave fragments into smaller particles (secondary atomization)
- *Stage V:* Collision and coalescence of particles

These stages are illustrated in Fig. 13 through 17. The stability of a liquid stream —that is, its length prior to breakup— increases with an increase in the diameter of the stream. Decreasing the length-to-diameter ratio of the nozzle from which the liquid issues changes the flow from turbulent to laminar and improves jet stability, as illustrated for a glycerol-water mixture in Fig. 13. Figure 14 shows gas- and water-atomized particles of ligamentary shape (stage II). Figure 15, a water-atomized tin-based powder, illustrates membraneous shape formation (stage III).

Membrane formation has been documented with high-speed photographs of water droplets exposed to a high-velocity stream of air. The membranes were shown to burst into a shower of very fine and rather large droplets, thus forming a bimodal size distribution (Ref 17). For liquid metal, low liquid metal surface tension and high cooling rates favor the

*Schlieren photography is a two-mirror optical system that is sensitive to gas density gradients. It permits determination of the location of shock waves and of regions of compression and expansion.

Fig. 6 Two-fluid atomization designs

Design characteristics: α, angle formed by free-falling molten metal and impinging gas; A, distance between molten metal and gas nozzle; D, diameter of confined molten metal nozzle; P, protrusion length of metal nozzle. Source: Ref 8

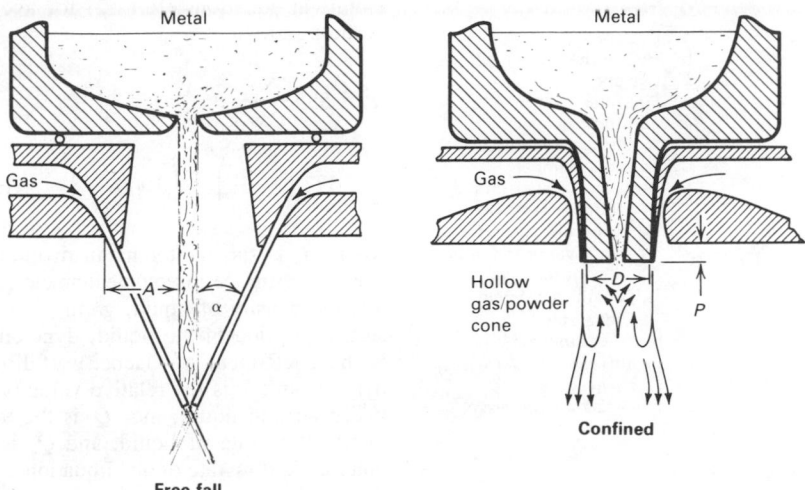

Fig. 7 Water jet configurations

(a) Open flat stream V-jets. (b) Closed V-jets. Source: Ref 9

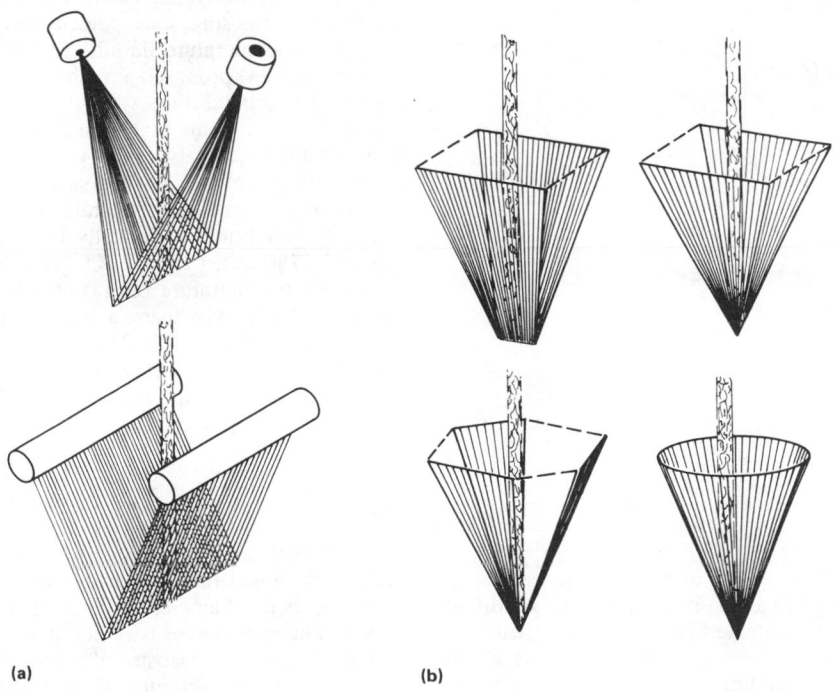

(a) (b)

Fig. 8 Multiple water jet system

Fig. 9 Typical metal buildup obtained with confined two-fluid atomization design

formation of membraneous shapes in the solidified powder, while high surface tension and low cooling rates favor formation of spherically shaped particles. Cooling rate is further increased by the thinness of the membrane.

Further breakdown of droplets occurs when primary particles generated in regions of lower gas velocity (Fig. 16a) enter regions of higher gas velocity. This is usually referred to as "secondary atomization" or "secondary disintegration" (stage IV).

According to Ref 18 and 19, collisions occurring during atomization have a much greater effect on particle size, particle size distribution, and particle shape than previously assumed. Collisions reduce the efficiency of atomization. Satellite-dotted spheres (Fig. 17a), splash-covered particles (Fig. 17b), and partly coalesced aggregates account for a broadening of the particle size distribution and changes in particle shape.

For example, the $-30+60$ mesh water-atomized copper particles that are shown in Fig. 17(e) are composed of thousands of predominantly spherical-shaped primary particles less than 10 μm in diameter. Actual or final particle size and shape are determined by the number of collisions and the degree of coalescence of the aggregate.

The probability of particle collisions increases with increasing metal flow rate. Thus, the maximum of average particle size and geometric standard deviation are achieved in the center of the atomizing stream, where metal concentration generally is highest (Fig. 16b).

Collisions also cause gas entrapment and microinhomogeneity from internal (primary) particle boundaries. Collisions between particles usually overlap with the earlier stages of atomization. These col-

Fig. 10 Argon recirculation and purification system
Source: Ref 11

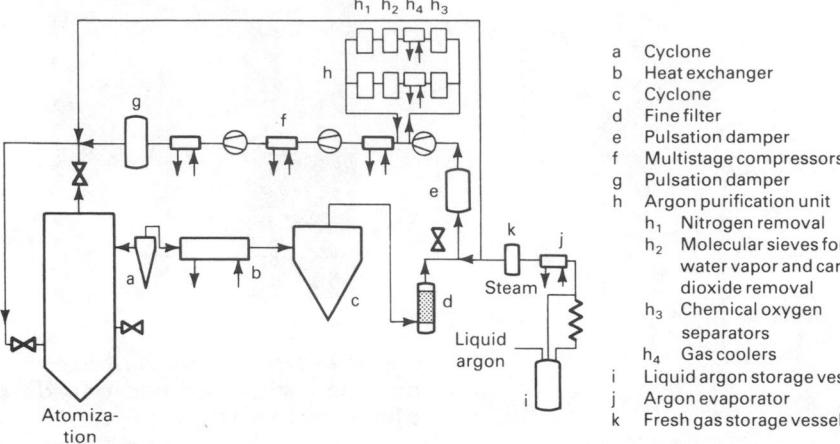

a Cyclone
b Heat exchanger
c Cyclone
d Fine filter
e Pulsation damper
f Multistage compressors
g Pulsation damper
h Argon purification unit
  h_1 Nitrogen removal
  h_2 Molecular sieves for water vapor and carbon dioxide removal
  h_3 Chemical oxygen separators
  h_4 Gas coolers
i Liquid argon storage vessel
j Argon evaporator
k Fresh gas storage vessel

Fig. 11 Model for the disintegration of a liquid sheet by a high-velocity gas jet
Source: Ref 14

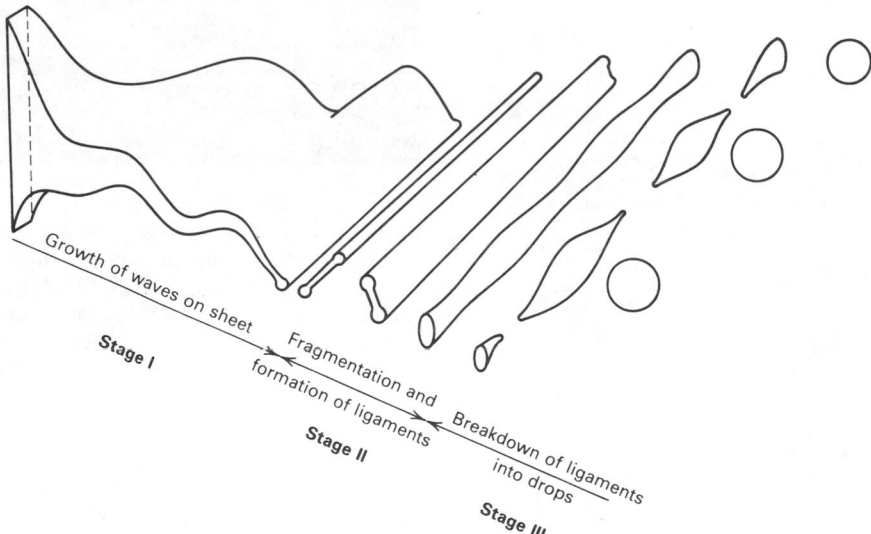

lisions are minimized or practically eliminated in centrifugal modes of atomization, in which traffic density of the particle decreases rapidly with distance.

Droplet formation by impact rather than shear was proposed by Grandzol and Tallmadge for water atomization (Ref 20, 21). This is due to the dispersed nature of the water and an inverse proportionality between particle size and the normal velocity component of the water (with respect to the metal stream axis).

Calculations of atomization efficiency usually are based on a comparison of the surface energy of an atomized powder (surface energy at temperature of atomization times surface area of solidified powder obtained from BET data) with the amount of energy provided by the atomizing medium. For both gas and water atomization, efficiency calculations are low, typically around 1% or less. Maximizing contact area between liquid metal and atomizing medium in the high-velocity space of the atomizing medium increases efficiency. Efficiency improvements could significantly reduce overall cost of gas atomization, where gas consumption is frequently a significant cost factor.

Particle Size and Size Distribution

One of the earliest and most well-known empirical equations (Ref 22) for average particle size is:

$$d_o = 585 \frac{\sqrt{\sigma}}{V\sqrt{\rho}} + 597 \left(\frac{\sqrt{\mu}}{\sqrt{\sigma\rho}}\right)^{0.45} \left(1000 \frac{Q_l}{Q_a}\right)^{1.5} \quad \text{(Eq 1)}$$

where d_o is the Sauter mean droplet diameter, or surface-to-volume diameter, μm; ρ is the density of liquid, g/cm^3; σ is the surface tension of the liquid, dyne/cm; μ is the coefficient of viscosity of liquid, dyne $\cdot$ s/cm^2; V is the relative velocity between air and liquid, m/s; Q_l is the volumetric flow rate of liquid; and Q_a is the volumetric flow rate of air. Equation 1 was derived from a series of atomizing experiments with room-temperature liquids of varying surface tensions and viscosities, using air as the atomizing medium.

In 1955, Bitron (Ref 23) extended the validity of the Nukiyama-Tanasawa equation (Eq 1) to the supersonic velocity range using room-temperature liquids. In 1973, Bradley (Ref 24) proposed a mathematical model for this condition.

Due primarily to the higher surface tensions of liquid metals, energy requirements and, therefore, gas pressures used for atomizing metals are generally higher than for other liquids—typically between 350 and 2750 kPa (50 and 400 psi). By comparing the literature on gas-atomized metal powders and by using a relationship developed by Wigg (Ref 25), Lubanska (Ref 26) proposed the following empirical equation for the average particle size (d_m) of gas-atomized powder:

$$\frac{d_m}{d_{\text{Met Str}}} = K \left[\frac{\nu_m}{\nu_g} \frac{1}{W} \left(1 + \frac{M}{A}\right)\right]^{1/2} \quad \text{(Eq 2)}$$

where W is the Weber number $V^2 d_{\text{Met Str}}/\sigma$; d_m is the mass median particle diameter; $d_{\text{Met Str}}$ is the diameter of metal stream; ν_m is the kinematic viscosity of liquid metal; ν_g is the kinematic viscosity of atomizing medium; V is the velocity of atomizing medium; ρ is the density of liquid metal; σ is the surface tension of liquid metal; M is the mass flow rate of liquid metal; A is the mass flow rate of atomizing medium; and K is a constant.

The individual terms of this equation are dimensionless; the Weber number measures the ratio of inertial to surface tension forces. With the explicit form for the Weber number, Eq 1 and 2 indicate that

Fig. 12 Particle formation stages during atomization

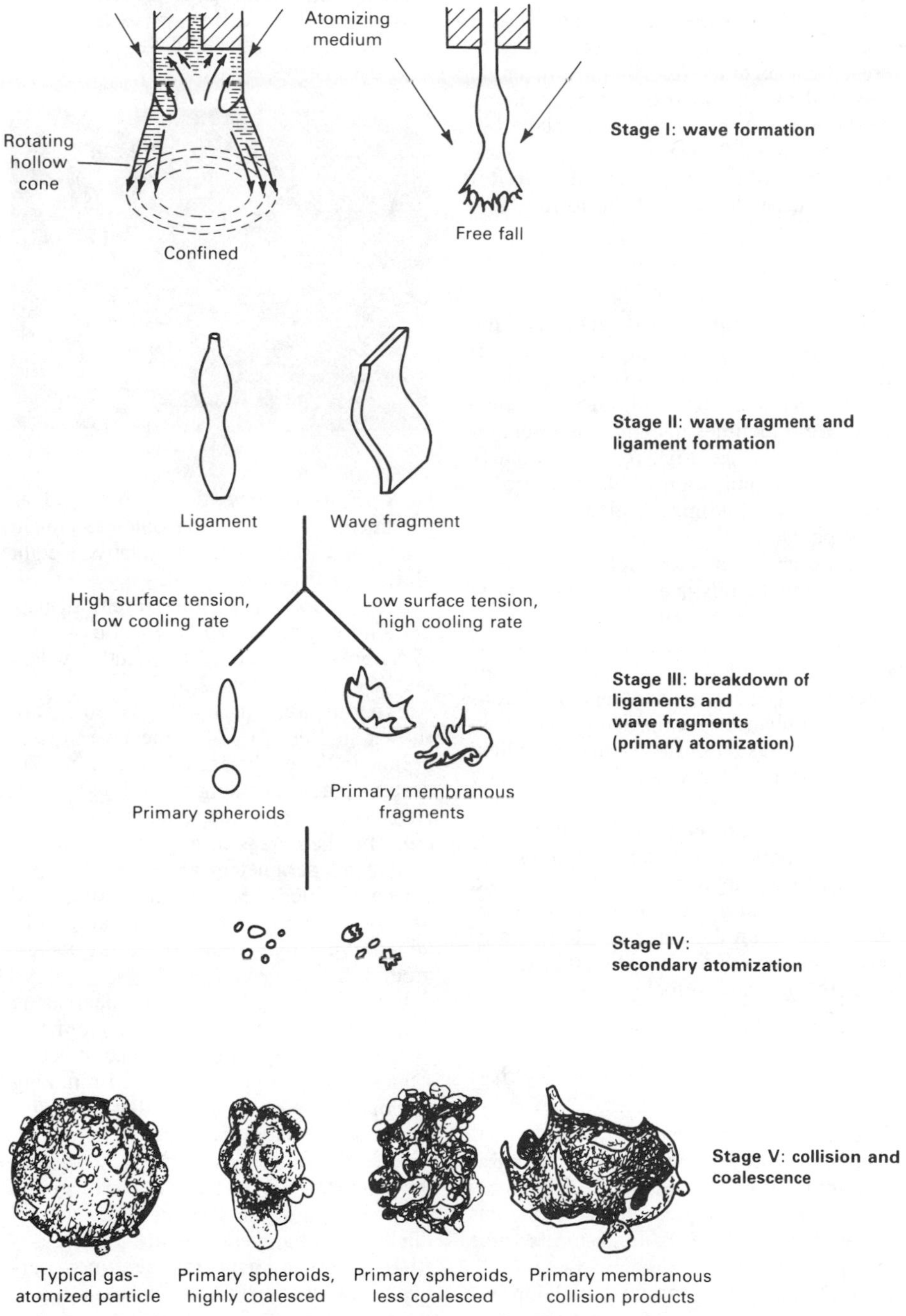

Atomizing medium

Rotating hollow cone

Confined

Free fall

Stage I: wave formation

Ligament

Wave fragment

Stage II: wave fragment and ligament formation

High surface tension, low cooling rate

Low surface tension, high cooling rate

Primary spheroids

Primary membranous fragments

Stage III: breakdown of ligaments and wave fragments (primary atomization)

Stage IV: secondary atomization

Typical gas-atomized particle

Primary spheroids, highly coalesced

Primary spheroids, less coalesced

Primary membranous collision products

Stage V: collision and coalescence

Fig. 13 Effect of nozzle design on the stability of glycerol-water jets
Source: Ref 15

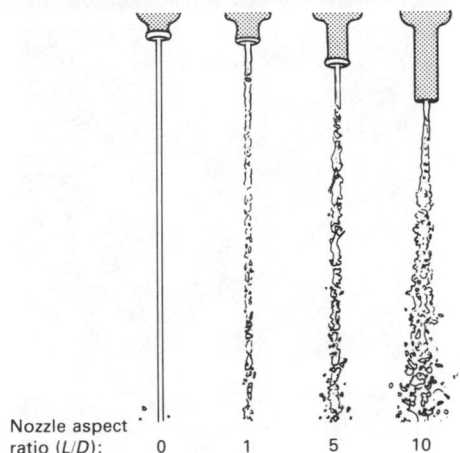

Nozzle aspect ratio (L/D): 0   1   5   10

particle size decreases with decreasing surface tension of the liquid metal and with increasing velocity of the atomizing medium.

Data from Ref 26 for various metals are depicted graphically in Fig. 18. In spite of substantial scatter, Eq 2 is useful for estimating the effect of the major variables from a single atomization experiment, as data from various atomization head designs, including water atomization, pro-

duce linear plots of almost identical slope. Additional empirical relationships between particle size, material, and process parameters may be found in Ref 9, 27-34.

Controlling Particle Size

In practice, for a given gas nozzle design and size, average particle size is controlled by the pressure of the atomizing medium. For converging-type or so-called

sonic nozzles, the velocity of the gas usually reaches sonic velocities of about 300 m/s (1000ft/s) in the narrowest region of the nozzle near its exit (the upstream gas pressure must be at least 1.9 times the external pressure in accordance with the laws of flow of compressible fluids through orifices and nozzles). Consequently, the amount of gas flow (A) depends on gas pressure and temperature. For ideal conditions and zero velocity on the entrance side of the nozzle, gas flow is:

$$A = a \left(\frac{2}{k+1} \right)^{k+1/2(k-1)} \frac{p \sqrt{2g}}{\sqrt{RT}} \qquad \text{(Eq 3)}$$

where a is the cross section of gas nozzle at exit; k equals C_p/C_v, the ratio of the specific heat of the gas at constant pressure and the specific heat of the gas at constant volume; p is the gas pressure in gas reservoir; T is the temperature in gas reservoir; R is the gas constant; and g is the acceleration due to gravity. For nitrogen, with $k = 1.4$ (Ref 35):

$$A \simeq 4 \cdot 10^6 \frac{ap}{\sqrt{T}} \text{ g/s}$$

As a compressible fluid passes through a nozzle, a drop in pressure and a simultaneous increase in velocity result. If the pressure drops sufficiently, a point is reached where, in order to accommodate the increased volume due to expansion, the nozzle unit must diverge. Thus, nozzles for supersonic velocities must converge to a minimum section and diverge again (Ref 23, 36). In water atomization, water velocity and flow rate are proportional to the square root of water pressure.

Fig. 14 Ligament-shaped particles

(a) Water-atomized M2 tool steel. (b) Water-atomized Ni-Cr-Si-B-C alloy. (c) Gas-atomized tin (−60+80 mesh). Source: Ref 16

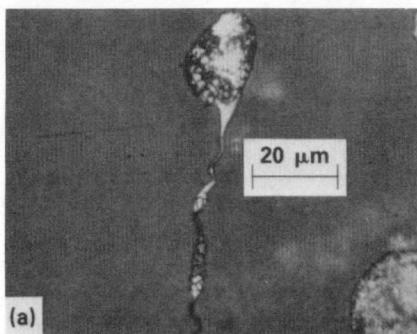

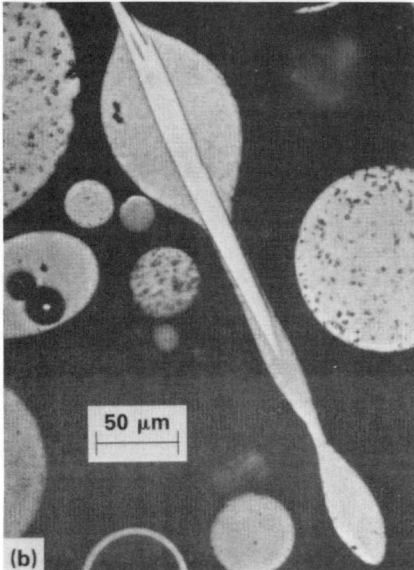

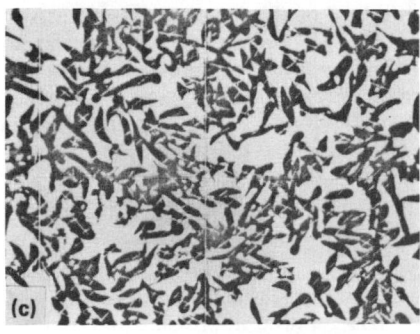

Figure 19 shows experimental data relating water pressure (p) with mass median particle diameter (d_m), according to $d_m \sim p^{-n}$ over a large pressure range (Ref 37).

Some data indicate that straight line semilogarithmic plots of average particle sizes exist versus atomizing pressure for both gas- and water-atomized powders (Ref 38).

In addition to the relationships shown above, particle size decreases with increasing apex angle (α), such as the angle formed between the axes of the atomizing medium jets. Angles greater than about 60° may result in "nozzle freezing" from splash-back of particles. According to Ref 20, this dependence is of the form:

$$d_m \sim \frac{1}{V_{\sin \alpha}}$$

Increasing temperature (superheat) of the liquid metal—similar to its effect on surface tension and viscosity—produces a smaller particle size, although these effects are sometimes insignificant. Smaller jet distances (see Fig. 6a), particularly during gas atomization where pressure attenuation is maximized, also produce a finer powder.

In atomization research on room-temperature liquids, a decrease in gas density (helium versus nitrogen) and an increase in gas viscosity (ethylene versus nitrogen) resulted in coarser particles (Ref 39). The latter phenomenon is contrary to Lubanska's empirical equation in which the gas viscosity term was added to maintain dimensional consistency (Ref 26).

Some of these empirical relationships can be rationalized in terms of physical models for droplet formation. The so-called maximum stability criterion, for instance, gives maximum drop size as a function of surface tension of the liquid metal (σ), gas density (ρ_g), and velocity (V) of the atomizing medium as:

$$d_{crit} \sim \frac{\sigma}{\rho_g V^2} \qquad \text{(Eq 4)}$$

This equation applies to the breakup of ligaments (Ref 16). Secondary disintegration into particles smaller than d_{crit} occurs only if the dynamic pressure due to the gas stream velocity exceeds the restoring force of surface tension.

Gas- and water-atomized metal powders, when atomized at constant pressure, follow log-normal distributions. In other words, screen analysis data or particle size distributions form straight or nearly straight lines when plotted cumulatively on log-probability paper, as shown in Fig. 20. Two numbers—d_m as mean mass diameter and σ_g as geometric standard deviation—define the entire size distribution. The geometric standard deviation represents the slope of the curve. It is a measure of the width of the distribution, or the spread of particle size about the mean

Fig. 15 Scanning electron micrograph of water-atomized Sn-3.3Cu-0.6P alloy powder

Note "hollow bag" membranous fragments typical of those proposed by Lane (Ref 17).

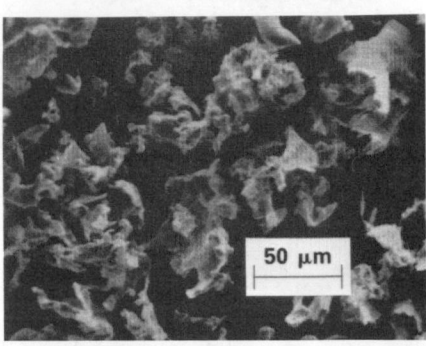

value. Standard deviation is obtained as the ratio of the particle size diameters taken at 84.1 and 50.0% of the cumulative weight plot, respectively (Ref 40).

A σ_g value of 1.0 signifies that all particles are the same size. Typical σ_g values of water- and gas-atomized metal powders may vary from about 2 to 7.

Experimental data relating σ_g to d_m are shown in Fig. 21. In some cases, data points indicate different metals, nozzle designs, and processing conditions, which leads to the scatter range indicated by the cross-hatched areas in Fig. 21. By maintaining all parameters and variables at a constant value, except for atomizing fluid pressure, good reproducibility can be obtained, typically within ≤0.1 for σ_g values around 2.

Usually, σ_g decreases with decreasing diameter of the metal stream. Confined nozzle designs generally produce lower σ_g values than free-fall designs. Minimizing agglomeration or collisions during atomization also reduces σ_g values. The empirical relationships shown in Fig. 21 are of the form $\sigma_g = a d^b_m$. The value a is related to the design of the atomizing nozzle and thickness of the metal stream; a decreases (that is, the distribution becomes narrower) as the interaction between liquid metal and atomizing medium becomes more uniform and more efficient. Such is the case with a thinner metal stream and confined design.

Figures 22 and 23 illustrate the effects of nozzle diameter and pouring temperature for water-atomized copper powders. The large effect of the tank atmosphere—air versus nitrogen (Fig. 22)—suggests that b in the equation above for σ_g is extremely sensitive to chemical reactions taking place during atomization.

The use of discrete values of σ_g, in

Fig. 16(a) Lines of constant gas velocity, metres per second, with superimposed metal stream and indication of stages of particle formation
Source: Ref 16

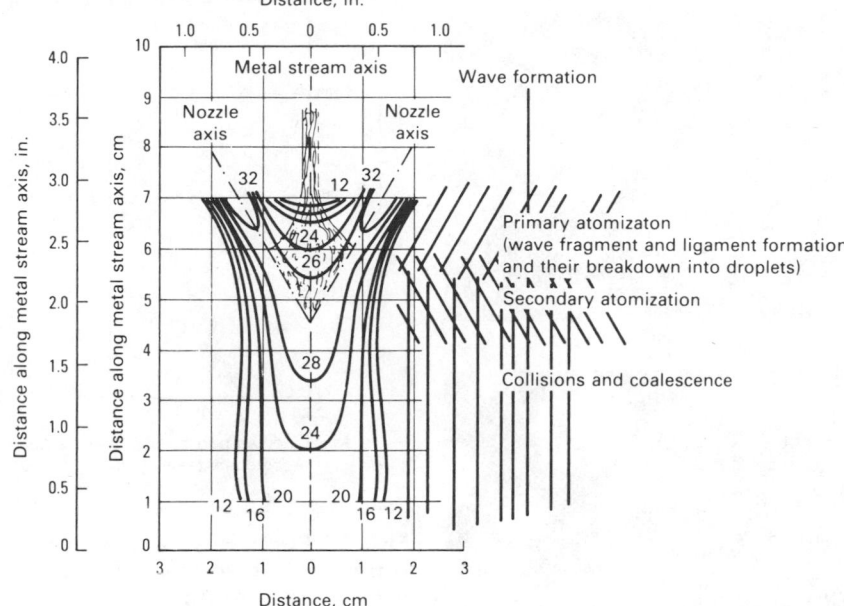

Fig. 16(b) Effect of patternization on mass median particle diameter (d_m) and geometric standard deviation (σ_g)
Distance from gas nozzle: 43 cm (17 in.). Apex angle: 0°. Source: Ref 8

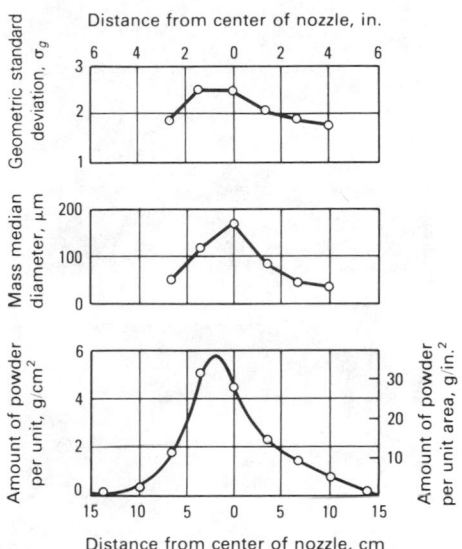

Particle Shape

With the availability of computers, new approaches to quantitative particle shape analysis have become feasible. The state of the art, as well as the theoretical basis of particle formation during atomization, is still being developed. Because of the numerous empirical correlations between engineering properties of powders and their bulk properties (tap density, apparent density, and flow rate), it is practical to use the latter as measures of particle shape. Generally, good consistency is found between the various measures (Ref 41).

An understanding of particle shape in atomized powders requires knowledge of the factors that determine the relative rates of cooling and spheroidizing of liquid droplets. Approximate cooling rates for aluminum alloy particles with diameters of about 100 μm are shown in Fig. 24 and are compared to a large ingot (Ref 42). Usually, cooling rates for water atomization are one to two orders of magnitude larger than for gas (nitrogen or argon) atomization. Cooling rates may be estimated from measurements of secondary dendrite spacings; the smaller the spac-

ings, the greater the cooling rate. Figure 25 shows the effect of particle size on dendrite spacings for three nickel-based superalloy powders.

In atomization, surface tension forces tend to spheroidize an irregular liquid droplet as the forces of the atomizing medium diminish with increasing distance from the atomizing zone proper. Thus, once the droplet leaves the zone of high turbulence, surface tension becomes operative, provided the liquid droplet has not yet cooled to a point where its viscosity has greatly increased, or where it has already solidified. At this point, temperature, cooling rate, size of the liquid droplet, and chemical reactions (affecting surface tension and viscosity) with the atomizing medium are critical.

For example, if the spheroidization time (τ_{sph}) of the liquid droplet is short compared to its solidification time (τ_{sol}), particle shape tends to be spherical; if spheroidization time is long, particles tend to be irregular. Based on this assumption, Nichiporenko and Naida have derived and evaluated expressions for spheroidization and cooling times, τ_{sph} and τ_{sol}, respectively (Ref 44):

$$\tau_{sph} = \frac{3\,\pi^2\mu}{4\,V\sigma}\,(r_1^4 - r_2^4) \qquad \text{(Eq 5)}$$

where r_1 is particle radius after transfor-

mation to spherical shape; r_2 is particle radius before transformation to spherical shape (r_2/r_1 is assumed to be 10); μ is viscosity of liquid metal; σ is surface tension of liquid metal; and V is volume of particle.

The time for cooling a superheated metal sphere to the melting point, assuming heat transfer by convection (Ref 16, 44), including the time required to lose its latent heat of fusion, is:

$$\tau_{sol} = \frac{d\rho_m}{6h_c}\,(C_p)_m \ln \frac{T_i - T_g}{T_m - T_g} + \frac{\Delta H_m}{T_m - T_g}$$
$$\text{(Eq 6)}$$

where d is particle diameter; ρ_m is density of metal; h_c is convective heat transfer coefficient; $(C_p)_m$ is heat capacity of metal; T_i is initial temperature of particle; T_g is temperature of gas; T_m is melting point of metal; and ΔH_m is the specific latent heat of fusion of metal.

In spite of the simplified nature of Eq 5 and 6, they are useful for estimating the effects of variables on particle shape. The wider range of particle shapes observed in water-atomized powders, for instance, may be explained in terms of τ_{sol}/τ_{sph}, assuming values from ≫1 to <1 depending on the density of the atomizing medium, pouring temperature, and other processing parameters (Ref 45). The density of the atomizing medium is determined by relative amounts of water and steam. More spher-

Fig. 17 Particle collisions and particle coalescence

(a) Gas-atomized cobalt-based alloy. Note satellites. (b) Water-atomized 304L stainless steel. Note metal splatted (white arrows) around doublet of primary particles (black arrows). (c) Water-atomized nickel-based alloy. (d) Water-atomized Cu-0.5Li alloy. Note entrapment of particle and multilayered envelopes (see arrows). (e) Water-atomized copper

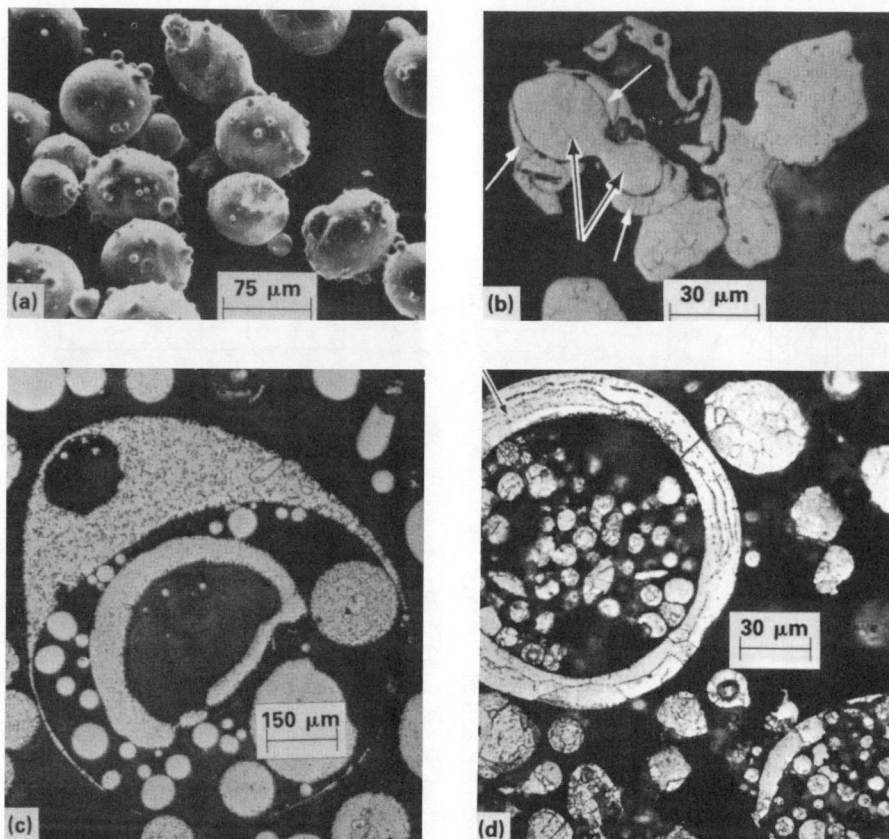

Fig. 18 Correlation of atomization data for liquid metals

See Eq 2 in text for line equation. Source: Ref 26

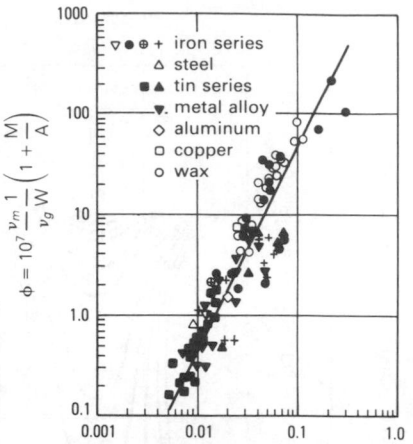

$$\phi = 10^7 \frac{\nu_m}{\nu_g} \frac{1}{W} \left(1 + \frac{M}{A}\right)$$

Mass median particle diameter/diameter of metal stream $d_m/d_{\text{Met. Str.}}$

Fig. 19 Particle size of water-atomized metals as a function of atomizing pressure

Source: Ref 37

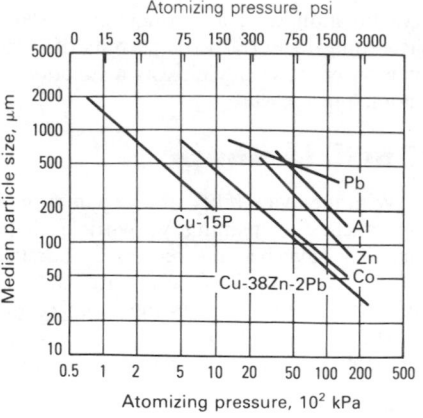

and tin, the ratio is about 0.1. In accordance with these ratios, the particle shape of the iron and copper powders tends to be spherical, while those of lead and tin are irregular. The irregular particle shape of aluminum and zinc, constituting a transition stage, is attributed to process side effects such as oxidation.

Effect of Additions on Particle Shape

The particle shape of a water-atomized copper powder becomes more irregular if small amounts of certain elements (magnesium, titanium, lithium, or calcium), which tend to lower the surface tension of

ically or more regularly shaped high-melting-point metal powders tend to result from the large amount of steam formed at higher temperatures, with attendant decreases in cooling efficiency. Application of this theory to water-atomized metal powders is illustrated in Fig. 26. For iron and copper powders, cooling times are several orders of magnitude higher than spheroidization times ($\tau_{sol}/\tau_{sph} \gg 1$). For aluminum and zinc, cooling times are about 10 to 15 times higher, and for lead

Fig. 20(a) Semi-log plots of cumulative particle size distributions of water-atomized (80Ni-20Cr and type 316L) and gas-atomized (tin) metal powders

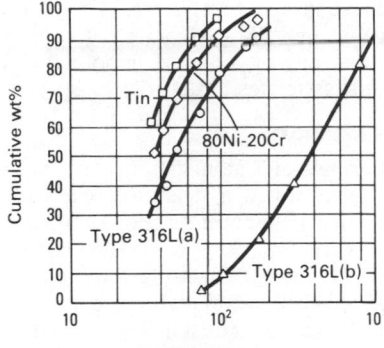

Particle diameter, μm

Mesh No.	Sieve designation, wt% less than	Powder			
	μm	Tin	80Ni-20Cr	316L(a)	316L(b)
20	841	···	···	···	81.9
50	297	···	···	···	40.0
80	177	···	98.3	90.9	21.3
100	149	···	96.6	87.5	···
140	105	97.6	92.3	79.0	9.2
200	74	91.3	81.4	65.0	4.6
270	53	81.3	69.9	52.0	···
325	44	72.5	58.7	40.1	···
400	37	61.9	51.0	34.2	···

Note: (a) and (b) refer to figure.

Fig. 20(b) Log-normal plots of cumulative particle size distributions of water-atomized (80Ni-20Cr and type 316L) and gas-atomized (tin) metal powders

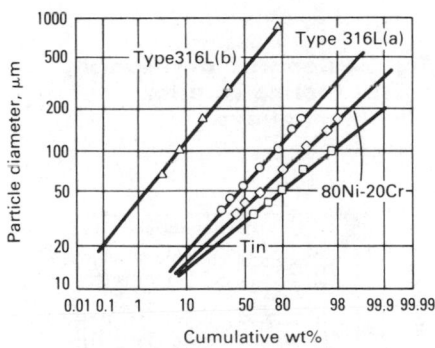

Cumulative wt%

Powder	Mass median particle diameter (d_m), μm	Standard deviation (σ_g)
316L(a)	52.9	2.4
316L(b)	367	2.6
80Ni-20Cr	36.8	2.1
Tin	30.0	1.9

Note: (a) and (b) refer to figure.

liquid copper, are added to the liquid copper prior to atomization, as shown in Fig. 27 (Ref 46 and 47).

According to Eq 5, this lowering of the molten metal surface tension increases the spheroidization time of the particle and thereby results in a more irregular particle shape. The use of low superheat and large water-to-metal ratios for irregular particle shape also can be determined from Eq 5 and 6.

In many atomized powders, particle shape tends to become more spherical for the finer fractions of a powder. This also can be determined from Eq 5 and 6 if one substitutes particle size and gas velocity dependencies of a suitable expression of the heat transfer coefficient (h_c) into Eq 6, although the greater undercooling of finer particles may also play a role (Ref 48):

$$h_c = \frac{k}{d}(2 + 0.6Re^{0.5}Pr^{0.33}) \qquad (Eq\ 7)$$

where k is thermal conductivity of atomizing medium; d is particle diameter; Re is the Reynolds number; and Pr is the Prandtl number. The Reynolds number and Prandtl number are dimensionless expressions used in fluid dynamics to determine flow conditions. The Reynolds number is equal to the density of the fluid, times its velocity, times the particle diameter, divided by the fluid viscosity. The Prandtl number is equal to the absolute (dynamic) viscosity of the fluid, times the specific

heat of the fluid at constant pressure, divided by the thermal conductivity of the fluid.

The coefficient of heat transfer (h_c), through its effect on cooling time (Eq 6), explains the irregular particle shape of water-atomized powders, where h_c is significantly larger than in gas atomization and where higher-melting-point metals tend to form more spherically shaped particles due to the increasing amount of steam production and the lower heat transfer coefficient of the latter (Ref 49).

Effect of Collisions on Particle Shape

The presence of particle collisions during atomization, particularly during water atomization, may have significant effects on particle shape. However, this phenomenon has been overlooked in most research on metal powder atomization. Mathematical models have only dealt with

Fig. 21 Variation of geometric standard deviation with mass median diameter

Source: Ref 38

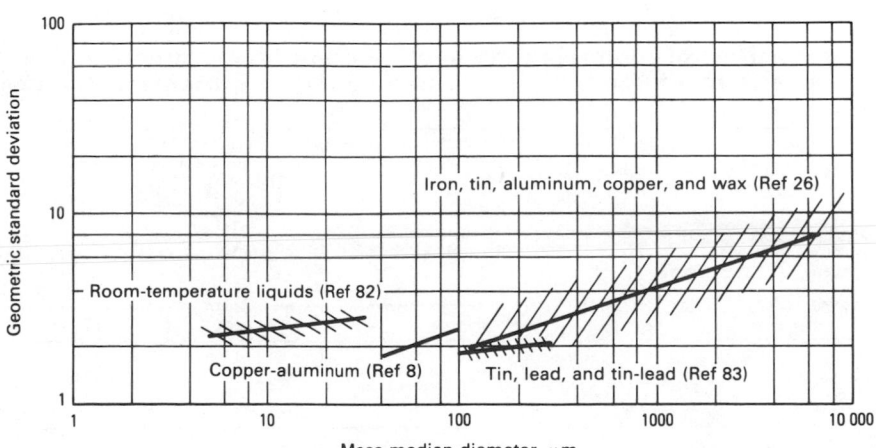

Mass median diameter, μm

Fig. 22 Effect of diameter of metal stream on geometric standard deviation (σ_g) and mass median diameter (d_m) of water-atomized copper powder

Pouring temperature: 1200 °C (2190 °F); figures in parentheses show average metal flow rates in g/s. Source: Ref 39

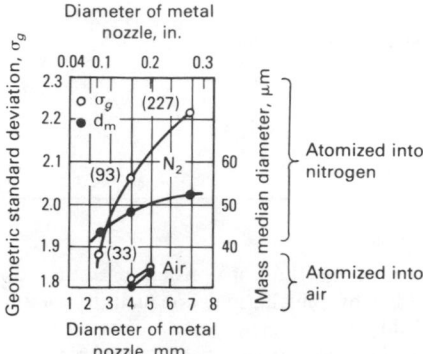

Fig. 23 Effect of pouring temperature on standard deviation (σ_g) and mass median diameter (d_m) of water-atomized copper powder

Metal stream-diameter: 4 mm (0.16 in.). Water flow rate: 320 L/min (84.5 gal/min). Water pressure: 13.2 MPa (1920 psi). Tank atmosphere: nitrogen. Source: Ref 39

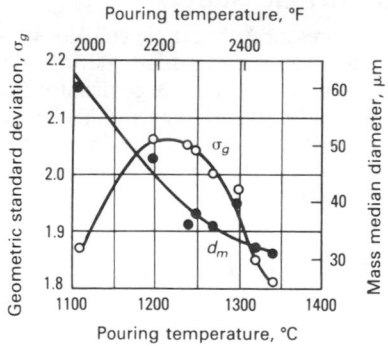

the effects on particle size distribution (Ref 50 and 51). However, based on recent experimental data, these models have apparently underestimated the extent of the phenomenon (Ref 18 and 19).

In contrast to the above model in which particle shape changes cease when the surface of the liquid droplet has become solid, the collision stage extends changes in both size and shape to longer times and distances in the atomization chamber. During water atomization, particularly under conditions of high surface tension and low cooling rate (Fig. 12), particle size appears to be determined by the number of collisions between primary particles of spherical shape, which may amount to several thousand for a final particle of about 125 μm in diameter, while particle shape is determined by the degree of coalescence after collisions (Ref 18). The latter depends on many factors, including particle size, temperature, composition and velocity of atomizing medium, metal composition, degree of oxidation, and solubility of formed oxides.

Chemical Composition and Microstructure of Atomized Powder

The rapid quench rates attainable with atomization, combined with new consolidation techniques for compacting powders to theoretical density, have led to the development of new alloys, with emphasis on reduced segregation, refinement of microstructures, extended solubility, and attendant benefits in hot formability, fatigue, and impact strength, for example. Low-alloy steels, superalloys, and tool steels are three examples where these advances and benefits have reached commercial status during the past two decades.

Although the high cooling rates typical of all atomized powders produce particles

of rather homogeneous and refined microstructures, each atomization process has its own characteristics, which for certain applications can be of critical importance. In conventionally gas- and water-atomized powder, one of the most important characteristics relates to the role of oxygen.

Oxidation

During water atomization, most metals react with water or steam to form oxides, following a chemical reaction of the form: $x\mathrm{Me} + y\mathrm{H_2O} = \mathrm{Me}_x\mathrm{O}_y + y\mathrm{H_2}$. Oxidation generally increases with increasing superheat of the liquid metal and depends on particle size, as shown in Table 2 and Fig. 28. Some materials exhibit a minimum if oxygen content is plotted versus particle size, as shown for water-atomized copper in Fig. 28. Such a minimum may result from two opposing factors. With increasing particle size, cooling time and exposure to oxidation increase, and oxide layers become thicker. Conversely, with decreasing particle size, the specific surface area of a powder increases rapidly.

Fig. 24 Dendrite arm spacing versus cooling rate for aluminum alloys

Source: Ref 42

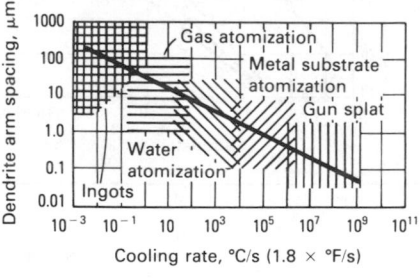

Fig. 25 Variation of secondary dendrite arm spacing with particle size and alloy composition

Source: Ref 43

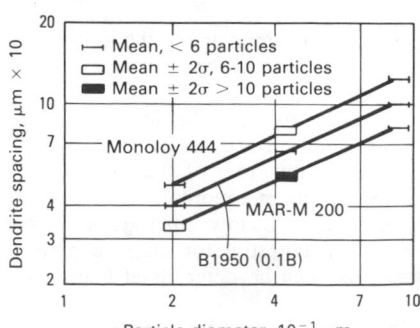

Table 2 Effect of atomizing medium, pouring temperature, and particle size on oxygen, nitrogen, and hydrogen contents of Haynes Stellite No. 31

	Contaminant content, ppm			
	Water atomized at:		Argon atomized at:	
	1500 °C (2700 °F)	1650 °C (3000 °F)	1550 °C (2800 °F)	1650 °C (3000 °F)
Oxygen	4380	7580	210	400
Nitrogen	500	780	50	60
Hydrogen	18	24	3	5
	Oxygen content, ppm			
	−60+150 mesh	−150+400 mesh	−400 mesh	
Inert gas	123	161	255	
Water	3850	4410	5270	

Source: Ref 52

Fig. 26 Comparison of cooling (τ_{sol}) and spheroidization (τ_{sph}) times for various water-atomized metals

(a) Iron and copper. (b) Zinc and aluminum. (c) Lead and tin. Source: Ref 45

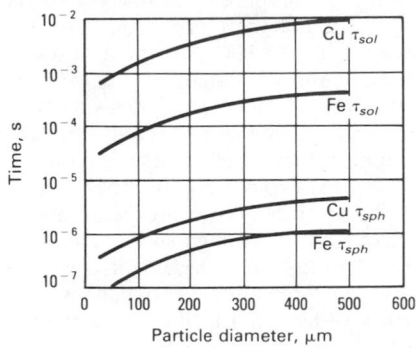

(a)

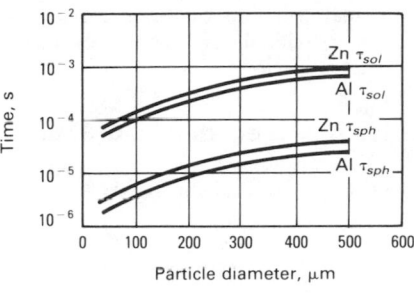

(b)

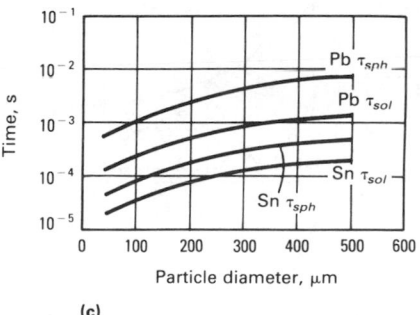

(c)

Thus, even though finer particles cool much faster and produce thinner oxide layers, their large surface area (per unit weight) results in increasing oxygen content with decreasing particle size.

In the Quebec Metal Powder and Mannesmann processes for iron powder production from high-carbon iron melts, oxidation during atomization is intended and controlled to achieve subsequent decarburization and deoxidation in a self-generated atmosphere of carbon monoxide and carbon dioxide during annealing (see the article "Production of Iron Powder" in

this Volume for more information on these processes).

Currently, inert gas atomization systems have been perfected so that the oxygen content of a powder does not exceed that of the liquid metal prior to atomization. To minimize the presence of oxides in the molten metal prior to atomization, particularly in the case of alloys containing readily oxidizable elements, good melting and deoxidation practices are important. Precautions during transfer of the liquid metal from ladle to tundish and avoidance of turbulence inside the tundish are equally important.

In addition to the amount of oxidation, the nature and composition of oxides and their spatial distribution within a particle can have a great effect on the usefulness and subsequent processing requirements of a powder. Surface analytical techniques have shown that the surfaces of water-atomized alloys are typically highly enriched in the oxides of elements that have high negative free energies of oxide formation and high diffusivities at elevated temperature. Oxides penetrate the particle to a depth of several hundred to several thousand angstroms.

This phenomenon is also true for gas-atomized powders in which some oxidation has occurred during atomization. Typical oxygen contents of water-atomized metal powder are:

Metal	Oxygen content, ppm
Silver	100
Gold-nickel	<100
Copper	<800
Copper-tin	<400
Nickel	<500
Iron	1500-3000
Cobalt	2000-3000
Iron-carbon	500
Aluminum	2000
Fe-13Cr	1000-3000
Fe-45Si	2000
Ni-Cr-B-Si	250-1000
Fe-80Mn	30 000

Source: Ref 37

Within the group of alloys that contain elements with intermediate oxygen affinity (chromium, manganese, and silicon), silicon-containing alloys generally produce low-oxygen-content powders, while manganese-containing alloys become heavily oxidized during water atomization. This tendency is attributed to the more protective nature of silicon oxide films compared to manganese oxide films.

Examples of so-called "composition-depth profiles" are shown in Fig. 29 and 30. These profiles show that the surface

composition can vary significantly from the bulk composition. Also, the profiles usually differ from those of wrought samples of the same composition that are surface oxidized at lower temperatures.

Typically, 80 to 90% of the oxygen picked up during atomization is found in the surface layers of powder particles. The remainder is contained in the interior of the particles, both in solution and as submicron-size oxides.

For these reasons, water atomization and gas atomization incorporating powder collection in water is not used for highly reactive metals, such as titanium and superalloys. Water atomization has been used successfully to atomize certain stainless

Fig. 27 Effect of additions of various metals to molten copper on the apparent density of atomized copper powder

Source: Ref 47

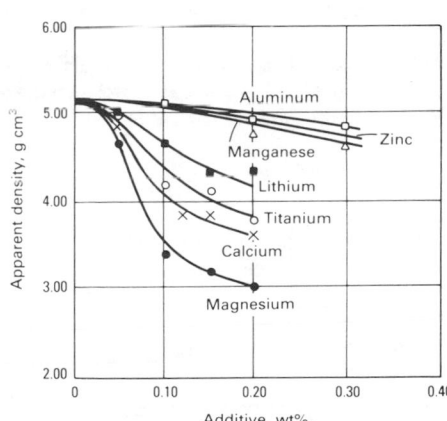

Fig. 28 Effect of particle size on hydrogen loss (oxygen content) of water atomized copper

Melt atomized at 1250 °C (2280 °F) with a water flow rate of about 310 L/min (82 gal/min), and a water pressure of 6.8 MPa (990 psi). Source: Ref 19

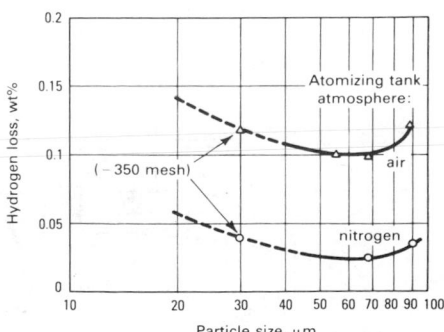

Fig. 29 Auger composition-depth profile of argon-atomized nickel-based superalloy powder

1 min ≃40 Å. Source: Ref 53

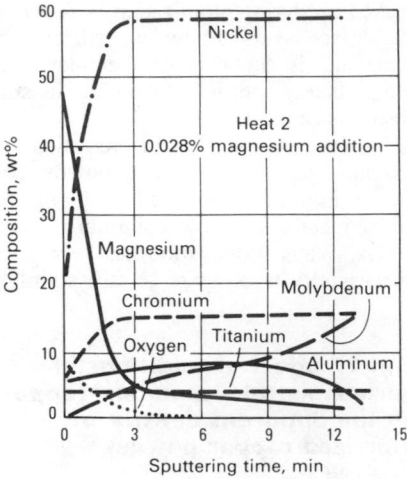

steels and high-speed tool steels, because the oxide films can be reduced during subsequent sintering (see the article "Production of Stainless Steel, Low-Alloy Steel, and Tool Steel Powders" in this Volume).

Microstructural Characteristics

Microstructural characteristics of conventional gas- and water-atomized powders range from amorphous to fine crystalline and dendritic. Dendritic characteristics are frequently found in gas-atomized powders. Glassy or noncrystalline particles are sometimes found in the finer fractions of a powder as a result of the higher cooling rates of fine particles. This effect is increased, because small particles are likely to be undercooled to a lower temperature with subsequent accelerated absorption of the heat of fusion. Glass-forming additives (boron and phos-

phorus) also favor the appearance of amorphous structures.

Collisions between liquid droplets and already solidified particles, as shown in Fig. 31, can produce high quench rates for the liquid splashes. Special techniques for producing amorphous and fine crystalline materials and their properties are discussed in the article "Amorphous Powder Metals" in this Volume.

In alloys that form precipitates on cooling, such precipitates generally are uniform in size and small compared to ingots of the same composition (Ref 55). Figure 32 shows such a comparison of microstructures for T15 high-speed tool steel. The benefits of these differences—improved formability and toughness, for example—have led to the increasing use of P/M tool steels (see the article "P/M Tool Steels" in this Volume).

Gas-atomized metal powders may contain small amounts of the atomizing gas within individual particles, which causes

Fig. 30 Electron spectrographic chemical analyses composition-depth profiles of water-atomized 316L stainless steel powder and sintered part

A: as-received powder; R: reduced at 800 °C (1470 °F) in hydrogen for 30 min; S: powder compacted and sintered at 1120 °C (2050 °F). Source: Ref 54

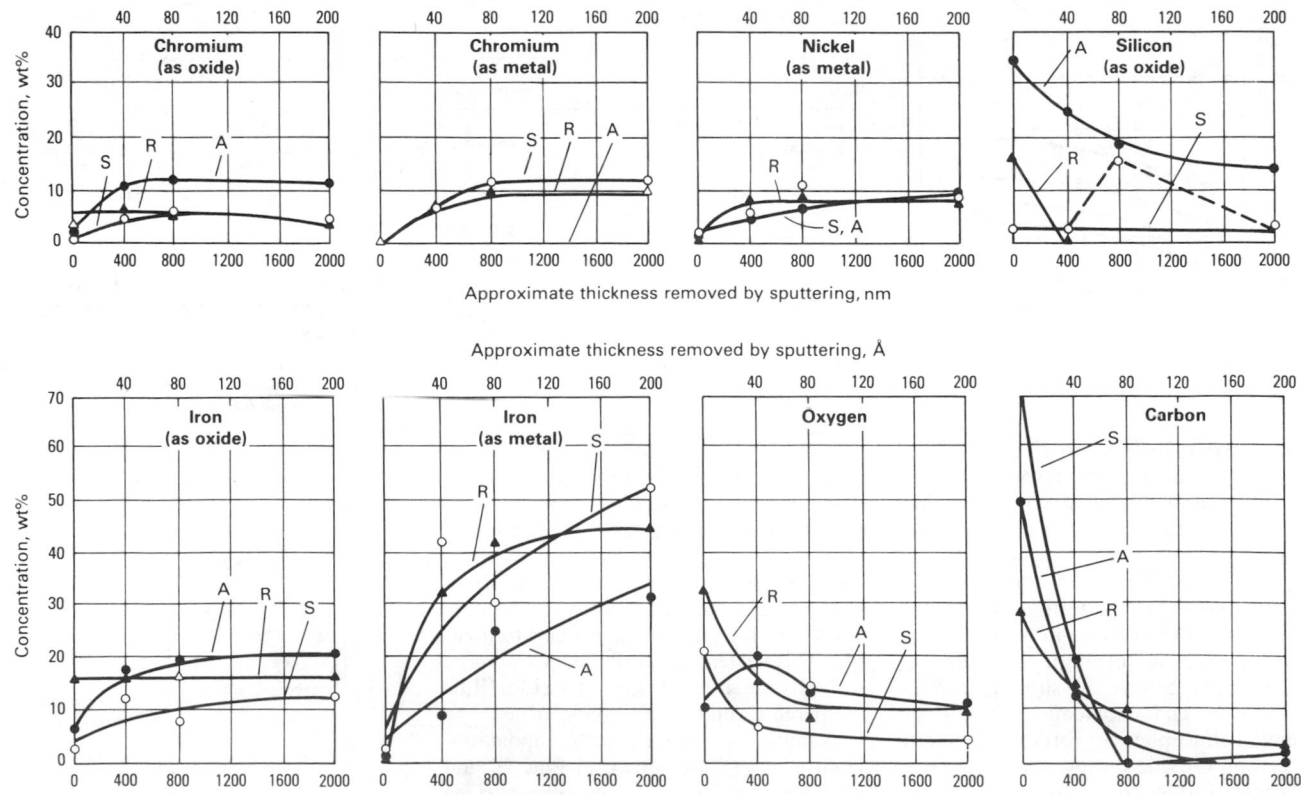

Fig. 31 Water-atomized copper powder

Note evidence of multiple particle collisions of "doublet" from fineness of microstructure. Polished section; ferric chloride etch. Source: Ref 18

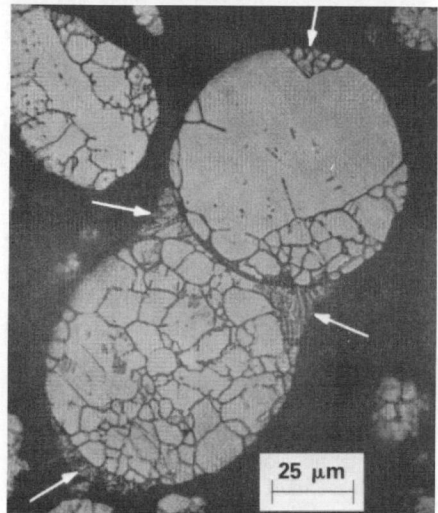

microporosity. This gas originates from mechanical entrapment that occurs when particles collide during atomization (Ref 45). This tendency may be minimized by reducing the number of collisions, or by using an atomization process in which the collision mechanism is not a significant factor, such as centrifugal atomization, vacuum atomization, and the rotating electrode process.

Rotating Electrode Process

By Peter R. Roberts
Engineering Manager
for Specialty Powders
Nuclear Metals, Inc.

THE ROTATING ELECTRODE PROCESS is a method for producing metal powders where the end of a metal bar is melted while it is rotated about its longitudinal axis. Molten metal is centrifugally ejected and forms droplets which solidify to spherical powder particles. The basic process is shown in Figure 33, where the melting method is by electric arc. The rotating electrode process (REP) concept was developed by Nuclear Metals, Inc., and resulted in the granting of patents (Ref 56).

The rotating electrode process has evolved to include equipment that consists of a tank 2440 mm (96 in.) in diameter in which powder is generated. This is mounted with its circular plane section in the vertical position. Figure 34 shows a typical production setup. The consumable rotating electrode is introduced through a seal-and-bearing assembly; the long axis of the electrode is horizontal and centrally located in the tank and is made the anode of a direct current power circuit. The nonmelted permanent cathode may be a simple tungsten-tipped device provided with adequate cooling (REP) or a transferred arc plasma torch (PREP). This equipment projects through the other face of the tank to oppose the rotated electrode. Usually, melting is conducted under inert gas; the preferred medium is helium, which offers improved heat transfer properties and electric arc characteristics.

Equipment

There are two standard types of machines used in the rotating electrode process: short bar and long bar.

Short Bar Apparatus. The short bar machine (Fig. 35) accepts consumable anodes up to 89 mm (3½ in.) in diameter by

Fig. 32 Microstructures of T15 tool steel
(a) Wrought. (b) P/M processed

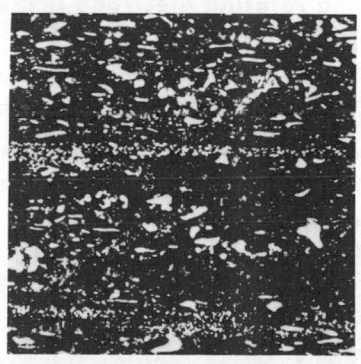

(a)

(b)

Fig. 33 Schematic of rotating electrode process
Courtesy of Nuclear Metals, Inc.

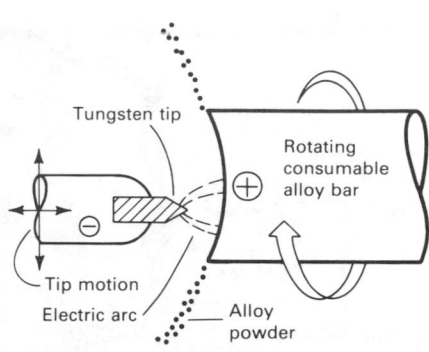

250 mm (10 in.) long. The anode is held in a collet in a precision spindle, the head of which projects into the tank through a rotating seal mechanism.

Usually at least 80% of the length of a short bar is converted to powder. Electrode stub removal and the introduction of new electrodes to the collet is performed manually via a glove port that is located in the front of the machine adjacent to the cathode or plasma torch.

Short bar methodology is appropriate for converting experimental quantities of material and specifically for alloys that are inherently brittle, or materials that have a low specific stiffness where an electrode of long aspect ratio is not practical.

Long Bar Apparatus. Increased productivity and conversion efficiency are realized in the long bar operation (Fig. 36), which is designed to consume 63.5-mm (2½-in.) diam electrodes that are up to 1830 mm (72 in.) long. Typically the assembly includes a precision spindle similar to that used in the short bar machine. Instead of being fixed in position relative to the tank, the spindle is mounted on a table that also carries the drive motor and the electrical transfer brush mechanism. The table moves toward the tank from an extended starting position, feeding the electrode through the special seal-and-bearing assembly.

When the head of the spindle approaches the seal housing, the process is interrupted while the stub of the consumed electrode, now typically 230 to 250 mm (9 to 10 in.) long, is pushed through the seal by another long bar that is mounted in the spindle head when it is retracted to the starting position. Stubs produced in long bar machines may be joined to new bars to obtain effectively 100% conversion to powder. Rotating and sealing equipment is widely used for powder conversion.

The standard rotation rate for 63.5-mm

Fig. 34(a) Rotating electrode/plasma rotating electrode process machine for generation of specialty spherically shaped powders
Courtesy of Nuclear Metals, Inc.

Fig. 34(b) Cleaning of a dedicated plasma rotating electrode process machine in preparation for the generation of titanium powder
Courtesy of Nuclear Metals, Inc.

(1½-in.) diam long bars is 1570 radians/s (15 000 rpm). Smaller diameter short bars have been rotated up to 2620 radians/s (25 000 rpm). Rotation rate is variable and is controlled by the machine operator. Because high-speed rotation rates are employed, electrodes must have precise dimensions; they must also be straight to keep mechanical out-of-balance forces to a minimum.

Particle Size Distribution

Accurately controlled and maintained rotation of the anode is necessary to obtain a desired range of particle size distribution. Molten droplet diameter of a given material is determined by parameters such as surface tension of the liquid metal, by centrifugal ejecting forces (related to rotation speed), and to some extent by the aerodynamics of the droplet trajectory through the inert cover gas. The following equation can be used to obtain an approximate value for median particle diameter, where the centrifugal force on a droplet is almost balanced by the surface tension force, thus causing the drop to adhere to the rim of the electrode face:

$$d = \frac{3.464}{\omega} \cdot \sqrt{\frac{\gamma}{\rho D}} \qquad \text{(Eq 8)}$$

where d is droplet diameter; ω is rotation rate in radians/s; γ is the surface tension; ρ is the density of the alloy being powdered; and D is the electrode diameter (Ref 57).

A useful approximation for particle size distribution may be obtained from a standard screen analysis of a split sample of powder. A cumulative plot of weights passing through progressively finer screens versus screen size is shown in Fig. 37. The effect of rotation speed is illustrated, and the median particle size d_{50} is approximately defined by:

$$d_{50} = \frac{K}{\omega \sqrt{D}} \qquad \text{(Eq 9)}$$

which follows the form of Eq 8, where ω is the rotation rate; D is the diameter of the electrode; and K is a constant for a given alloy for a limited range of arc power. It is possible to preselect powdering conditions to achieve a desired particle size distribution for most materials.

More detailed analyses can be found in Ref 58-60, which describe mechanisms for the formation of drops, ligaments, and sheets as a function of increasing melt rate. Ultimately, these filamentary and laminar forms disintegrate due to liquid instabili-

Fig. 35 Short bar plasma rotating electrode process machine

Courtesy of Nuclear Metals, Inc.

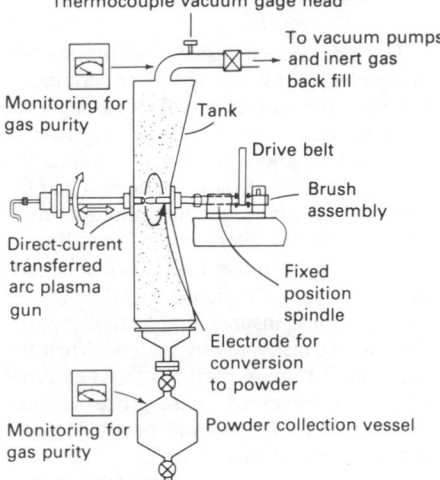

Fig. 36 Long bar plasma rotating electrode process machine

Courtesy of Nuclear Metals, Inc.

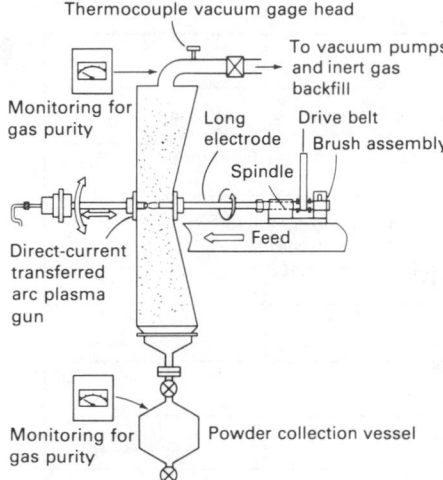

titanium is extremely aggressive and reacts with all container materials; consequently, the rotating electrode process is ideally suited for production of clean titanium alloy powders in commercial quantities. Typical characteristics for Ti-6Al-4V powder made by the rotating electrode process are:

Composition, wt%

Aluminum	5.50-6.75
Vanadium	3.50-4.50
Oxygen(a)	0.13-0.20
Oxygen(b)	0.05-0.13
Iron	0.30 max
Carbon	0.10 max
Nitrogen	0.05 max
Hydrogen	0.0125 max
Tungsten	<10 ppm
Other	0.4 max
Titanium	rem

Screen analysis, μm	% retained
500	0
354	1
250	5.5
177	43
125	38
88	9.4
63	3
44	0.1
<44	0
Median particle size (d_m), μm	175
Particle size range	50-500 μm
Bulk density (60% theoretical), g/cm³	2.65
Tap density (65% theoretical), g/cm³	2.90
Flow rate, s/50 g	24-32
Surface area, m²/g	0.008

(a) Standard grade. (b) ELI grade (extra-low-interstitial, 0.13% oxygen max)
Source: Ref 72

ties and aerodynamic forces to provide predominantly spherical particles with some odd shapes frozen within and an overall tendency toward broader size distribution.

Champagne and Angers (Ref 59, 60) have demonstrated that a bimodal distribution of particle size can occur due to a mechanism in which liquid drops separating from the electrode rim remain connected momentarily by a necked-down column of liquid metal that breaks into smaller drops when the major spheroid is freed. This effect is illustrated in Fig. 38; bimodal particle size distribution is shown in Fig. 39.

Although smaller particles are not present in large quantities by weight, their absolute numbers may be significant. This behavior is superimposed on distributions and is shown as a function of anode melting rate (Fig. 40) and rotation rate (Fig. 41).

The ordinate values of these curves are expressed in terms of the percentage of retained particles on a given sieve size divided by the size range between sieves, which eliminates the effect of this range on the shape of the curve. Additional information can be found in Ref 62 through 68.

Ultra-Clean Powders for Aerospace Applications

Spherical metal powders made by the rotating electrode process or by gas atom-

ization are not well suited for cold pressing into green compacts. Therefore, spherical powders tend to be used in specialized applications where consolidation is achieved by hot isostatic pressing or other high-temperature processing in which interparticle voids are more readily closed. These particles flow well into complex mold shapes and can be tapped to a reproducible density of packing to provide fully dense parts that closely approach the dimensions of the finished component.

Much research has been performed on near-net shape compaction for military airframe parts in titanium alloys (Ref 69). Gas turbine disks are made from hot isostatically pressed nickel-based superalloy powders in a similar fashion. An example of a complex shape that can be made from titanium alloy powders pressed to near-net shape is shown in Fig. 42.

This use of powders offers both economic and metallurgical advantages. Near-net shape technology ensures more efficient material utilization, which is important for high-cost materials such as titanium. Additionally, the metallurgical benefits include improved homogeneity and control of microstructure to achieve enhanced mechanical properties.

Because the rotating electrode process prevents contact of the melted alloy with any container material, it provides a decided advantage over other methods of making contamination-free spherical particles. This process generates powder with cleanliness and composition approximating that of the precursor electrode. Molten

Tensile tests conducted on hot isostatically pressed Ti-6Al-4V compacted from samples of rotating electrode processed powder indicate excellent properties, as shown in Table 3. Subsequent improvements in the process, which included plasma torch melting, have provided plasma rotating electrode processed Ti-6Al-4V P/M compacts with fatigue properties that are comparable or superior to those for cast and wrought materials, as illustrated in Fig. 43.

More recent work has demonstrated that plasma rotating electrode processed Ti-6Al-4V compacts generally have a narrower fatigue distribution than other forms. This allows design engineers to work with higher minimum property values in advanced aerospace structures. Careful control of cleanliness is critical to obtain this type of performance.

Nonmetallic inclusions have been associated with the initiation of fatigue failures in compacted pieces made from these materials, and the overall powder-making

Fig. 37 Cumulative plot of particle size distribution
Weight percent of sample finer than a given screen size versus screen size. Electrode: C-1018 steel, 63.5-mm (2¹/₂-in.) diam

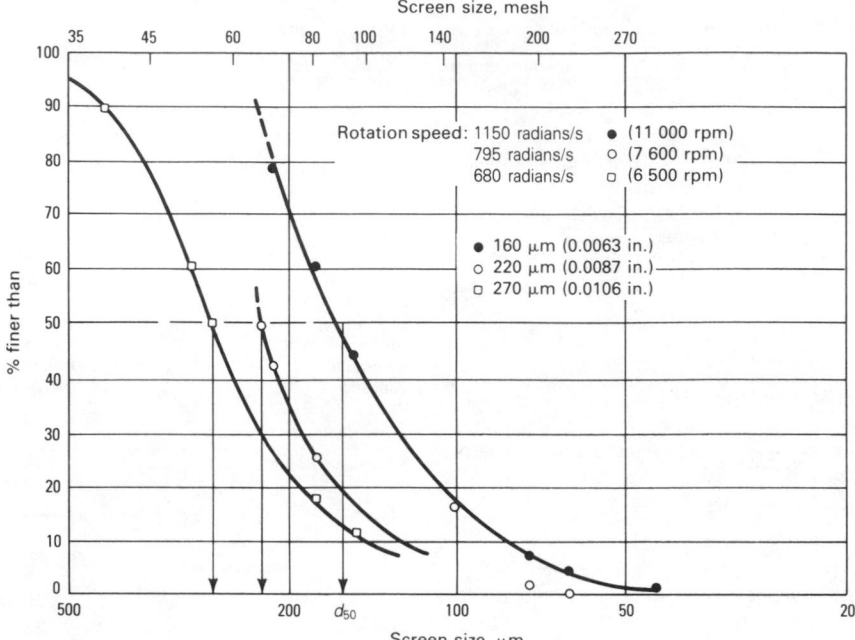

Process Advantages

The rotating electrode and plasma rotating electrode processes offer certain advantages over other powder-making processes. Titanium alloys are optimally produced by these methods, because the corrosive nature of molten titanium and the difficulty involved in containing this molten material are overcome. Generally, all materials to be made into powder benefit in that there is no liquid metal/container contact. This ensures that ceramic particles are not inadvertently added. When the addition of foreign particles from external sources is prevented, these processes provide a method for making powders to exact standards of cleanliness.

Size distribution for powders can be held within tighter ranges than is commonly achieved by gas atomization. In addition, hollow gas-filled particles, which cause thermally induced porosity in compacts, are not generated. Powders made by these techniques have demonstrated a high degree of sphericity along with good surface quality, which is typified by the plasma rotating electrode superalloy powder particles shown in Fig. 45. Particles also pour readily into molds of complex shape. They give consistent packing at approximately 65% full density so that near-net shapes can be obtained by hot isostatic pressing.

Finer powder fractions produced by the rotating electrode process are subjected to cooling rates that range from 10^3 to 10^4 °C/s (1.8×10^3 to 1.8×10^9 °F/s). These particles are considered to be rapidly solidified. The basic concept of rotating electrode and plasma rotating electrode is well suited for further adaptation to rapid solidification technology.

finer size ranges are used for consolidation to near-net shape for various implanted prostheses.

and handling process is designed to ensure that foreign particles are excluded. Raw powder is removed from the plasma rotating electrode machine under inert gas cover in closed collection vessels and is subsequently passed through a processing tower located within a class 100 clean room. This assembly sieves and passes particles whose sizes lie within desired upper and lower limits, and extracts a representative sample from a given lot of powder prior to loading into appropriate containers. The entire powder processing tower may be evacuated and backfilled with inert gas, which is an advantage when working with superalloys. The overall system is shown schematically in Fig. 44.

ical powder acts as a carrier for carbon black used in the printing cycle (see the article "Copier Powders" in this Volume). In this application, the required particle size distribution is narrow enough so that the ability to optimize yields in the desired size range has made the rotating electrode process an attractive method for powder generation.

Coarse particles in narrow size ranges have been made from cobalt-chromium alloys and titanium alloys. These are sintered into the stems of some prosthetic devices used in body implants; the porous structures so generated promote part fixation by bone ingrowth. In addition, the

Other Powder Applications

Although the rotating electrode process and the plasma rotating electrode process are ideally suited to the production of clean powders for subsequent consolidation to fully dense parts, there are other applications for spherical metal particles. Wide application has been found for low-carbon steel powder in photocopier applications, where rotating electrode processed spher-

Table 3 Typical tensile properties of hot isostatically pressed Ti-6Al-4V rotating electrode processed powder

Orientation	Tensile strength MPa	ksi	0.2% offset yield strength MPa	ksi	Elongation (4D), %	Reduction in area, %
L	938.4	136.1	850.8	123.4	20.0	37.0
	936.3	135.8	868.1	125.9	18.0	37.4
T	950.8	137.9	863.3	125.2	18.0	40.2
	936.3	135.8	848.8	123.1	18.0	35.6
S	932.9	135.3	843.3	122.3	23.0	42.2
	941.9	136.6	848.8	123.1	20.0	39.1
AMS 4928-H	896.4(a)	130(a)	827.4(a)	120(a)	10(a)	25(a)

Note: Consolidated material made by hot isostatic pressing at 950 °C (1750 °F) for 10 h at 100 MPa (15 ksi). Vacuum annealed for 10 h at 700 °C (1300 °F). Hydrogen after vacuum annealing equals 0.0057%.
(a) Minimum
Source: Ref 72

Vacuum Atomization

By Charles W. Fox
Technical Director
Homogeneous Metals, Inc.

VACUUM, OR SOLUBLE GAS, ATOMIZATION is a commercial batch process based on the principle that, when a molten metal supersaturated with gas under pressure is suddenly exposed to vacuum, the gas expands, comes out of solution, and causes the liquid metal to be atomized. Alloy powders based on nickel, copper, cobalt, iron, and aluminum can be vacuum atomized with hydrogen. Powders are spherical, clean, and of a high purity compared to powders produced by other powder processing methods. This process was developed and patented by Homogeneous Metals, Inc. (Ref 75).

Equipment

A schematic of the equipment used in vacuum atomization is shown in Fig. 46. Typically, there are two vertical chambers—a lower vacuum induction melting section and an upper powder collection tank. Powder is produced by saturating a molten metal bath with a soluble and/or nonreactive gas. The molten metal stream is atomized by introducing the gas-saturated stream through a ceramic transfer tube into the reduced-pressure collection chamber. The concentration of gases in molten metals increases with the pressure of a gas over the liquid. This phenomenon pro-

Fig. 38 Formation of satellite drops

SAE 1090 steel electrode. (a) $t = 0$. (b) $t = 0.002$ s. $\omega = 314$ radians/s (3000 rpm; $D = 1.91$ cm (0.75 in.); $Q = 0.17 \times 10^{-7}$ m³/s (6×10^{-7} ft³/s). Source: Ref 61

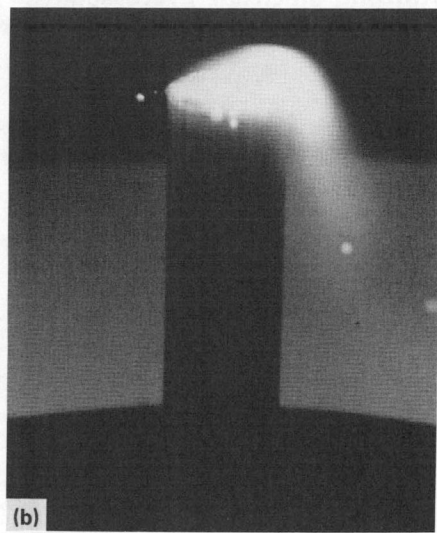

vides the energy to disperse the molten metal stream into droplets. The intimate mixture of gas and metal under pressure in the stream has significant impact on powder yields and is more effective than conventional gas atomization.

The use of a diatomic gas, such as hydrogen, further enhances atomization. As hydrogen evolves from solution in the molten metal, substantial energy is released upon recombination of the hydrogen atoms. This energy disperses the metal stream, as well as releases heat to the atomizing plasma. Heating at the hydrogen-metal interface lowers the droplet surface tension, thus producing more and finer powder particles. An additional advantage of using a soluble gas is a resultant lower

residual content of nonreactive gas when argon and hydrogen, for example, are used in combination.

Melting for atomization is performed in a conventional vacuum induction furnace. Typically, the chamber volume is small for a given melt capacity, as only vertical motion of the furnace is required. Pouring molten metal from a furnace requires room for tilting the furnace assembly; this is unnecessary in vacuum atomization. Because the chamber operates at high positive pressures as well as vacuum, the chamber construction is different from other vacuum induction systems. The collection chamber is designed to maximize yields, minimize contamination sources, and provide for cleaning ease to pre-

Fig. 39 Bimodal particle size distribution attributed to satellite drop formation

Source: Ref 60

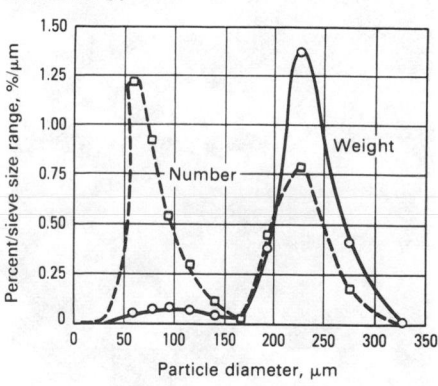

Fig. 40 Particle size distribution as a function of melting rate

Source: Ref 59

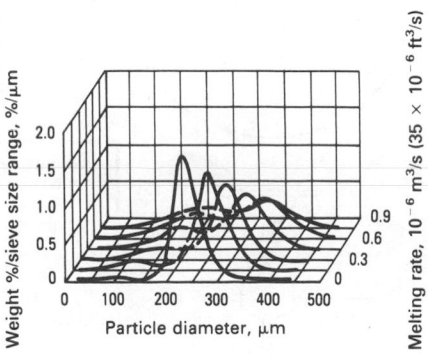

Fig. 41 Particle size distribution as a function of rotation rate

Source: Ref 59

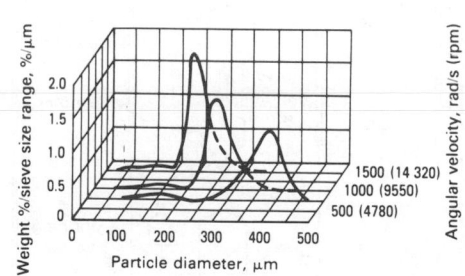

Fig. 42 Gas turbine engine compressor rotor made from hot isostatically pressed plasma rotating electrode processed Ti-6Al-4V powder
Source: Ref 70

Fig. 43 Comparison of fatigue behavior of Ti-6Al-4V compacts with ingot metallurgy material
Source: Ref 73

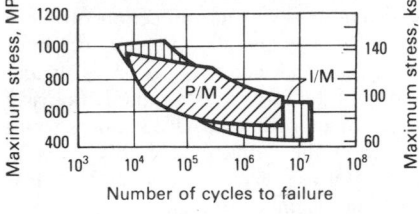

vent cross contamination among powder compositions.

After atomization, the powder is removed through a powder drain-valve assembly and collected in a transfer container. This container may be isolated from the collection chamber by valving, and the container may be removed without exposing the powder to air. Preventing contamination of an inherently clean process is particularly important, as well as excluding air that may oxidize the powder particles and degrade the properties of the consolidated product. All powder-handling operations are performed under inert gas or vacuum atmospheres within clean-room areas. All equipment for powder handling is specially designed and modified for the purpose of preventing powder contamination.

Powder Properties

Powder produced by vacuum atomization is essentially spherical, with few satellite particles (Fig. 47). Typically, the product exhibits a tap density of 65 to 72%, depending on alloy and mesh size. A typical size distribution for −80 mesh superalloy powder is:

Mesh size	Particle size distribution, %
−80+100	3
−100+200	22
−200+325	30
−325+500	25
−500	20

With process modifications, particle size distribution can be varied to the coarser or finer side of the distribution. All particles cannot be produced in a narrow size distribution; however, the distribution can be shifted to achieve desired results. A typical gas analysis for −80 mesh superalloy powder would be 75 ppm oxygen, 25 ppm nitrogen, and 2 ppm argon.

The intimate mixture of gas and molten metal during atomization results in rapid solidification and cooling of the metal particles. The majority of particles produced by vacuum atomization exhibits cellular or microcrystalline structures, which indicates high rates of cooling.

There are several advantages to the vacuum atomization process compared to other atomizing techniques. Pressurizing the melt chamber results in high melt stream velocity, which minimizes exposure to the ceramic transfer tube. The simple configuration eliminates the need for tundish or pouring basins, thus minimizing exposure to refractory materials that can contaminate the product. The furnace crucible and transfer tube are the only remaining refractories.

This system permits consistent production of powder with total ceramic content in the low ppm range. Approximately half of the contamination is inherent in standard vacuum induction melted stock as a carryover from refining and melting operations. Atomizing in an upward direction instead of down doubles the cooling path distance for metal particles. Without

Fig. 44 Ultra-clean powder generation and clean-room processing
Source: Ref 74. Courtesy of Nuclear Metals, Inc.

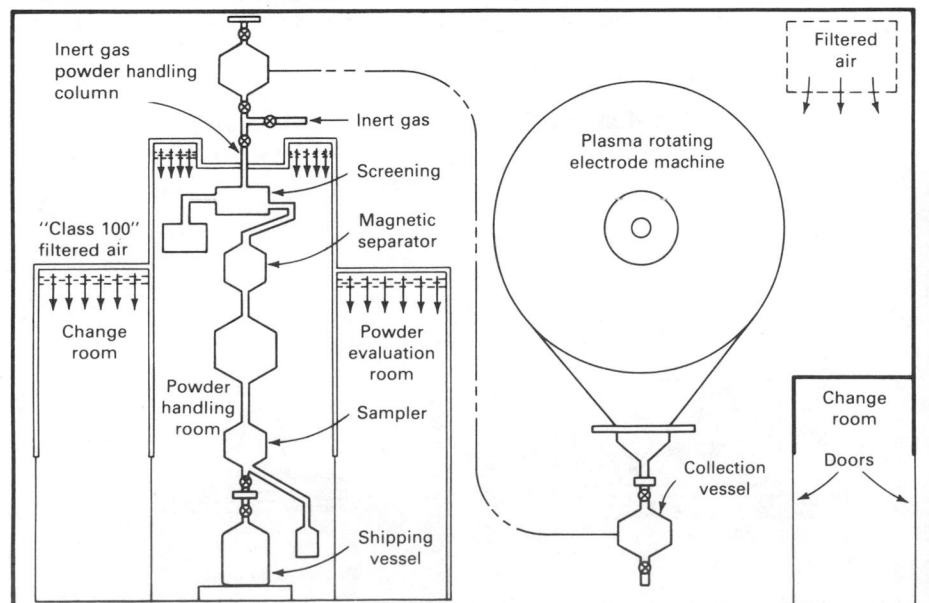

Fig. 45 Scanning electron photomicrograph of plasma rotating electrode processed René 95 powder particles
Magnification: approximately 70×. Courtesy of Nuclear Metals, Inc.

auxiliary cooling, particles are cooled as rapidly as in most other atomization processes.

Powder Applications

Principal use of powder made by vacuum atomization has been for the production of gas turbine disks. Thousands of disks have been manufactured from several million pounds of powder. This process facilitates the use of high-strength alloy compositions that are not hot workable by conventional techniques.

The homogeneous and fine-grained structure imparts hot workability and provides the superplasticity required for hot isothermal forming operations. Other applications for vacuum-atomized powder include its use as a coating for such critical gas turbine parts as blades and vanes and the production of intricate parts by injection molding.

Fig. 46 Schematic of the vacuum atomization process
Source: Ref 75

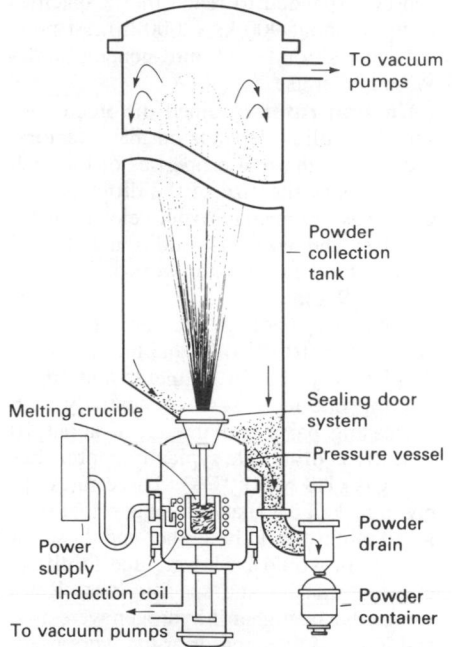

Fig. 47 Scanning electron micrograph of a vacuum-atomized nickel-based powder (−100 mesh)
Magnification: 650×

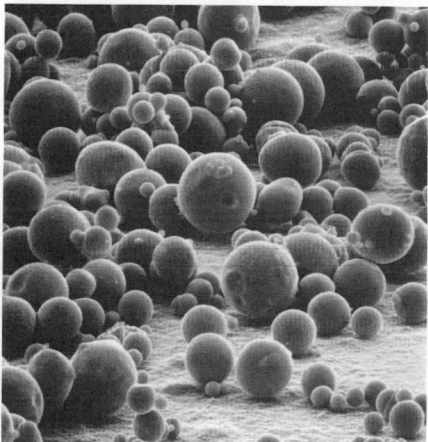

Rotating Disk Atomization

By Robert J. Patterson II
Engineering Specialist
Pratt & Whitney Aircraft Group
Government Products Division

ROTATING DISK ATOMIZATION involves the impinging of a stream of molten metal onto the surface of a rapidly spinning disk. The liquid metal is mechanically atomized and thrown off the edges of the spinning disk. Solidification occurs in flight and can be enhanced by blasting the emerging particles with a stream of helium.

Particles are generally spherical, with the average particle size decreasing with increasing disk speed. Rotating disk atomization is a form of centrifugal atomization in which solidification rates on the order of 10^4 to 10^6 °C/s (1.8×10^4 to 1.8×10^6 °F/s) can be achieved. This phenomenon also places rotating disk atomization under consideration as a form of rapid solidification rate (RSR) powder atomization.

Most rotating disk atomized powders are used in aerospace engine parts. As a result, the powder is handled and processed under vacuum or inert gas. Consolidation to theoretical density is accomplished by hot extrusion.

Rapid Solidification

As the cooling rate of a melt is increased, the time available for solidification is decreased. The degree of chemical

Fig. 48 External view of the first-generation rapid solidification rate powder machine

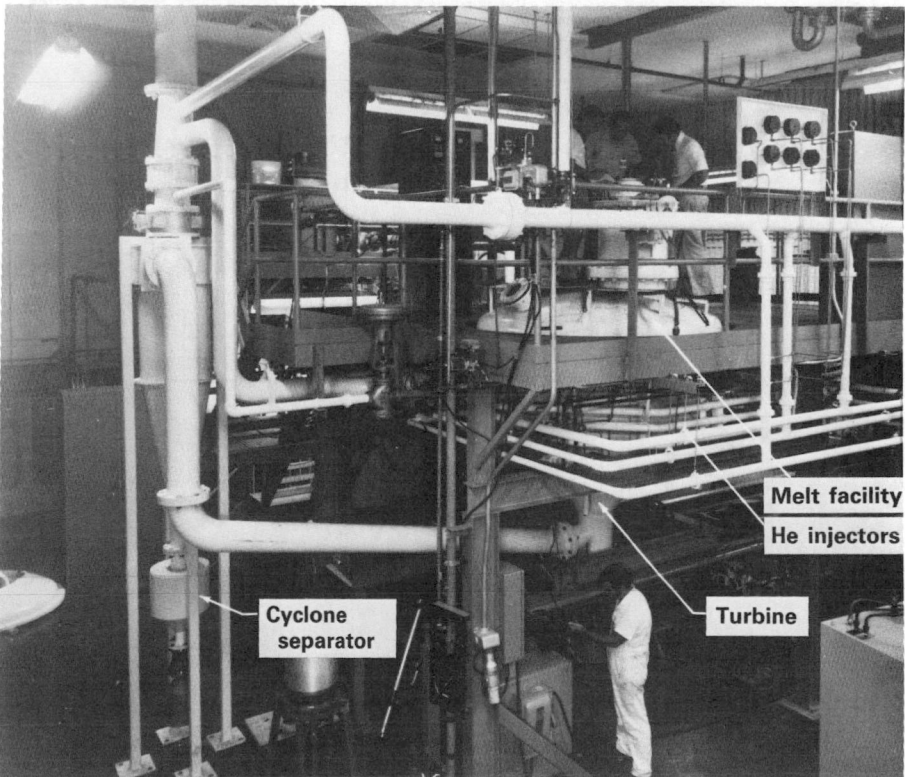

segregation and the tendency to form massive precipitates are reduced, while microstructural features (dendrite spacing and grain size, for example) become finer. The solid solubility of alloying elements also may be increased, and new phases (including glassy or noncrystalline phases) may be formed.

Although no specific cooling or quench rate conveniently separates conventional from rapid solidification techniques, rapid solidification typically ranges from about 10^4 °C/s (1.8×10^4 °F/s) and above (Ref 76). While several powder production processes may thus be considered rapid solidification technology, Duwez *et al.* (Ref 77) are credited with deliberately achieving high cooling rates in order to modify microstructure.

Since that time, many techniques have been developed, most of which produce flake-like materials by quenching small volumes of material on solid substrates that act as heatsinks. Production rates frequently have been as low as a few grams per day, and the irregular shape of the material has imposed difficulties in handling, sizing, and consolidation.

The rapid solidification rate powder process produces materials at quench rates approaching those for solid substrate techniques, with improved production rates and sizing and handling characteristics (Ref 78). Heat transfer studies suggest that spherical droplets with moderate velocity occurring in gases with thermal conductivities similar to those of hydrogen or helium should achieve high cooling rates (Ref 79, 80).

Equipment and Production Sequence

Typically, the metal alloy is melted under vacuum by induction. When the appropriate level of superheat is obtained, the system is backfilled with helium, and the rotary disk atomizer is brought up to speed. Helium quench jets are turned on, and the melt is poured onto the center of the atomizer through a preheated nozzle assembly that regulates the flow.

The centrifugal force imparted on the fluid disintegrates it into droplets that are then quenched to the solid state, while traversing the cross flowing helium. The resulting powder then falls into a collector below the machine, or it is carried by the gas to cyclones that remove the powder and exhaust the gas. An external view and schematic diagram of the first-generation rapid solidification rate powder machine are shown in Fig. 48 and 49. Figure 50 shows an atomizer in operation.

As shown in Fig. 49, the original (first-generation) device had a melt capacity of about 45 kg (100 lb) based on nickel, and the gas was exhausted. Second-generation machines have melt capacities of about 135 kg (300 lb). Generally, third-generation machines are used when a large quantity of material is desired. Based on nickel, third-generation machines have melt capacities of about 450 kg (1000 lb), but they can be expanded to reach melt capacities of up to about 900 kg (2000 lb). At present, only second- and third-generation devices are in use.

Cooling rates obtained in forced convection cooling depends on many factors, such as the thermal properties of the melt and coolant, the size of the droplets, and the relative temperature and velocity of the droplets and coolant. Additional discussion of the heat transfer work is included in Ref 79 and 80.

Typically, cooling rates are above 10^5 °C/s (1.8×10^5 °F/s) for helium-quenched droplets 100 μm in diameter and finer. Cooling rate increases exponentially with decreasing particle diameter; at about 10 μm, the cooling rate typically approaches 10^7 °C/s (1.8×10^7 °F/s). These rates apply to nickel; calculations suggest that rates for titanium would be higher and those for aluminum would be lower, due to differences in density and specific heat. Rates would also be higher if hydrogen were used and lower if nitrogen or argon were used.

Production Rate. Powder production rate has been successfully varied from about 160 to 1000 kg/h (350 to 2200 lb/h), based on nickel, which meets gas turbine requirements. At higher turbine speeds, the

Fig. 49 Schematic of first-generation rapid solidification rate machine

The second-generation machine incorporates closed-loop helium recirculation, higher atomizer speeds, and a three-fold increase in melt capacity. The third-generation machine retains these features, but the melt capacity is increased to 900 kg (2000 lb). Source: Ref 78

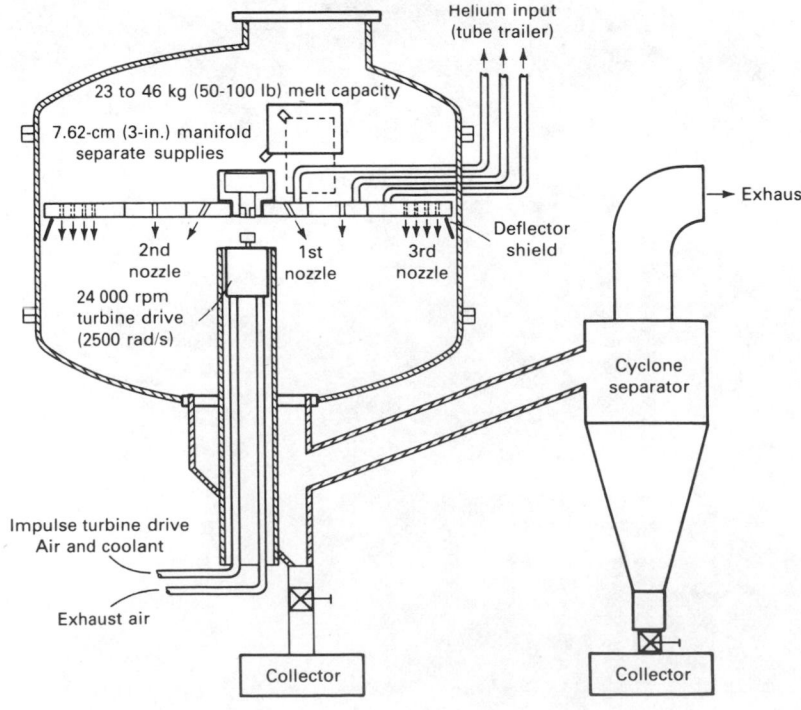

Fig. 50 Internal view of atomizer during operation

Note molten metal stream (bright vertical line) impinging on the rotary atomizer with the resultant disintegration into droplets (bright streaks).

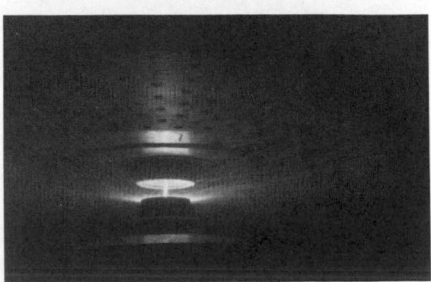

Fig. 51 Cumulative weight distribution

For most iron-, nickel-, and aluminum-based alloys for atomizer speeds of 24 000 rpm and metal flow rates of 0.2 kg/s (0.4 lb/s) (based on nickel). As flow rate is decreased or speed is increased, the curve tends to rotate counterclockwise and shift to the left. Based on −10 mesh = 100%.

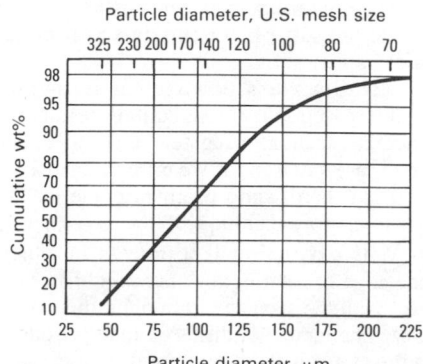

existing size distribution could be maintained at even higher flow rates.

Powder Characteristics

Size Distribution. The cumulative size distribution by weight, obtained with both U.S. standard and precision sieves, for nickel-, iron-, and aluminum-based alloys is shown in Fig. 51 for atomizer speeds of about 24 000 rpm and melt flow rates (based on nickel) of about 0.2 kg/s (0.4 lb/s). Under these conditions, the median particle size by weight is about 88 μm, and except for both ends, the distribution tends to be normal. Particle size tends to become finer, rotating counterclockwise and shifting to the left in Fig. 51, as melt flow rate is reduced or as atomizer speed is increased.

Powder Appearance. As shown in Fig. 52, the powder produced by this process is typically spherical and free of attached satellites. The microstructure depends on both particle size and alloy composition. Structures typical of nickel-based superalloys are shown in the particle cross sections of Fig. 53 and tend to be either dendritic (Fig. 53a) or microcrystalline in nature. The presence of the latter form varies with alloy composition and particle size (Ref 79). In some aluminum alloys, metastable structures have been observed (Ref 80); in some transition metal-metalloid alloys, noncrystalline particles have been produced (Ref 81).

Handling Characteristics. While Hall flow rates have not been determined, powders produced by rotating disk atomization are similar to argon-atomized powders. They exhibit an apparent tap density of approximately 62%.

Process Limitations

Conventional vacuum induction melt practice incorporates oxide ceramic crucibles and is therefore limited to elements and alloys that are nonreactive and that may be melted by induction and poured at temperatures below about 1760 °C (3200 °F). Because particle shape is spherical, the powder produced by this process is not well suited to applications that require the use of cold compacts with usable green strength.

Ultrarapid Solidification Processes

By Ranjan Ray
President
Marko Materials, Inc.

ULTRARAPID SOLIDIFICATION PROCESSING has added a new dimension to alloy design by allowing materials scientists unlimited flexibility in controlling materials properties by varying alloy microstructures and compositions. Ultra-

rapid solidification processes cool liquid metal at a rate of 10^6 °C/s (1.8×10^6 °F/s) or higher. The principal effect of such high cooling rates frequently is suppression of movement or diffusion of atoms in the material being quenched. In many cases, atoms are prevented from occupying the equilibrium (lowest energy) crystal sites that they would normally occupy in a conventional slowly cooled material.

Rapidly solidified alloys generally possess complete, or at least greatly enhanced, chemical homogeneity and metastable phases, such as highly supersaturated solid solutions, metastable microcrystalline phases, and amorphous phases (metallic glasses).

The techniques that are commercially viable and are currently used to prepare ultrarapidly solidified material directly from liquid metal are dominated by either conductive or convective heat-transfer mechanisms. In both mechanisms, solidification rate, which is a function of cooling rate, depends primarily on the heat transfer coefficient of the liquid metal. Typically, the cooling rate of conduction processes ranges between 10^6 and 10^8 °C/s (1.8×10^6 and 10^8 °F/s), while the convection processes may be limited to 10^4 to 10^6 °C/s (1.8×10^4 to 10^6 °F/s).

Conduction Processes

Most conductive rapid quenching techniques are based on the principles of bringing the melt into contact with a cool and highly conducting solid substrate(s) at high relative velocity to promote melt spreading and intimate thermal contact. Rapid cooling rates require a high heat transfer coefficient at the interface of the liquid metal and the solid substrate, and a sufficiently thin metal cross section to facilitate heat being conducted out of the melt

Fig. 52 Scanning electron micrographs of typical rapid solidification rate powder

Note the lack of attached satellite droplets and the generally spherical shape.

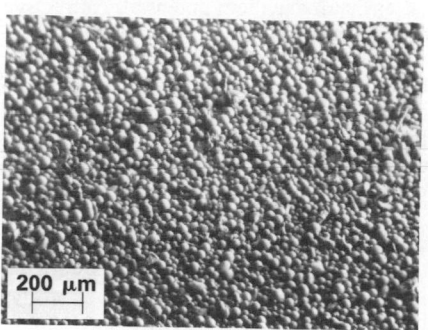

Fig. 53 Representative cross-sectional internal microstructure observed in rapid solidification rate nickel-based superalloy powders

(a) Dendritic. (b) Microcrystalline. The relative occurrence of the two forms has been found to depend on both particle size and alloy compositions. Source: Ref 79

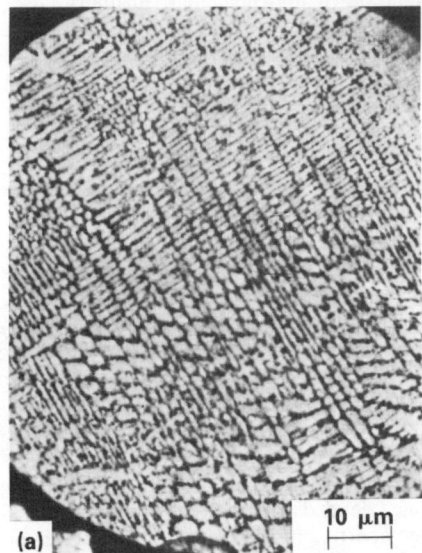
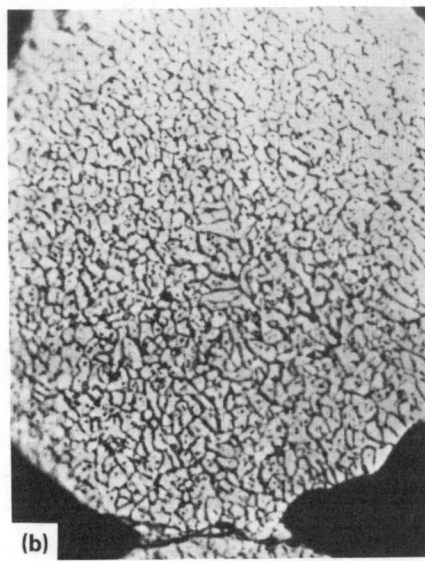

(a) 10 μm
(b)

quickly. Because heat conductivities are generally high for metals, most conductive rapid quenching processes have utilized metal substrates, such as copper.

Various techniques have been explored to produce rapidly solidified materials, generally in the form of continuous or discontinuous thin filaments and ribbons, flakes, particulates, and powders. Some of the processes that appear to be commercially viable are described briefly in the following sections.

Chill-Block Melt-Spinning. In this process, a jet of liquid metal is extruded through a round orifice, and rapid solidification is achieved when the molten jet impinges on the outer rim of a rotating water-cooled solid substrate. Continuous ribbons, several millimeters wide and 25 to 50 μm thick, are formed (Fig. 54). Production of rapidly quenched ribbons with uniform dimensions by melt-spinning requires good control of the liquid metal stream and formation of a stable liquid pool on the moving substrate surface.

The melt-spinning process has become the most widely used technique for producing a wide variety of metallic glass alloys of commercial interest, such as ribbons. Metallic glass alloys characteristically exhibit a unique combination of high strength, ductility, and hardness; excellent soft magnetic properties; and outstanding corrosion resistance. These properties may be further enhanced by selective alloying.

Production of wide metallic glass ribbons is achieved in a process variation of melt-spinning. It involves feeding molten metal through a slotted nozzle that is held in the proximity of a moving chill substrate.

Melt-spinning technology has advanced considerably in the early 1980's, thus demonstrating feasibility of high-speed production (1500 m/min, or 4900 ft/min) of metallic glass ribbon 1, 2.5, 7.5, and 15 cm (0.4, 1.0, 3.0, and 6.0 in.) wide.

Nickel-based metallic glass alloy compositions designated and approved by the American Welding Society (AWS) currently are used as the ductile glassy foils for brazing superalloy turbine components. These foils can be braided, stamped, or cut to fit the grooves of the part to be brazed.

High-strength, large width-to-thickness ratio, and exceptional toughness of metallic glass ribbons are some of the properties that make them potentially attractive as materials for use in epoxy resin matrix composites. Currently, metallic glass ribbons are being considered as viable materials for applications such as soft ferromagnetic laminations for power-distribution transformers to replace the grain-oriented, silicon-containing steel core.

Some iron- and nickel-based metallic glasses with low metalloid elements contents (up to 10 at.%) can be crystallized into high-strength microcrystalline alloys by heat treating at elevated temperatures.

Such glassy alloys are produced as powders by in-line pulverization of ribbons by a hammer mill directly off the rotating chill substrate. Metallic glass powders are consolidated into bulk shapes by P/M techniques such as hot extrusion and hot isostatic pressing. Bulk iron- and nickel-based microcrystalline alloys are available with high hot hardness and wear resistance; these alloys are suitable for cutting and forming tool applications.

Metallic glass powders of iron- and nickel-based alloys that contain refractory metals, such as tungsten and molybdenum, are suitable for wear- and corrosion-resistant hardfacing coatings made by the plasma spray technique.

Melt extraction is a process in which the edge of rotating disk is brought in contact with the surface of a molten melt (Fig. 55). The metal solidifies on the disk edge, adheres for a short time, and then departs the disk. By introducing notches at regular intervals in the extraction disk edge, the continuity of the casting is interrupted, and staple fibers of controlled length are produced. Melt-extracted fibers of steel and stainless steel are produced commercially for use as reinforcing elements in cements or castable refractories.

Convection Processes

The most frequently used processes are based on ultrasonic atomization and centrifugal atomization. Solidification in each of these processes is dominated by convective heat transfer. These atomization processes involve production of metallic particles cooled at rates of 10^4 to 10^6 °C/s (1.8×10^4 to 10^6 °F/s). Average cooling rate is directly proportional to the heat transfer coefficient and inversely proportional to the size of the particle. Consequently, forced convective cooling can be used to increase the cooling rate. Cooling rate may also be increased by decreasing the size of the particles.

Ultrasonic atomization involves production of ultrafine atomized molten droplets that are rapidly cooled by convection in a gaseous medium. A liquid stream is disintegrated, or atomized, by impact with multiple high-velocity gas pulses. The gas is accelerated by a shock wave tube at speeds up to Mach 2 at frequencies ranging from 60 to 120 kHz. The particles of gas traveling at extremely high velocities in pulsed waves strike the metallic stream. On impact, the gas pulses shear the liquid metal stream into fine droplets, usually less than 30 μm. Fine liquid droplets solidify convectively at high cooling rates.

Ultrasonic atomization has been used commercially for production of low-melting alloys such as aluminum alloys. However, only laboratory and pilot-scale operations have been reported for high-melting-temperature alloys (stainless steels and nickel- and cobalt-based alloys). Quench rates experienced by 30- to 70-μm aluminum particles produced by ultrasonic atomization are about 10^5 °C/s (1.8×10^5 °F/s), yielding highly refined structures. Three advantages of this process are (1) high relative velocites between gas and metal droplets, (2) small powder size, and (3) a gas chilling factor due to expansion of the high-pressure gas.

The emergence of new alloys is particularly evident in aluminum, where ultrasonic atomization has enabled use of increasing amounts of lithium in aluminum alloys, thus leading to high strength and low density. New dispersion-strengthened aluminum alloys containing iron, cobalt, molybdenum, and cerium have been developed by rapid solidification processing. These materials characteristically exhibit high strength at elevated temperatures.

Centrifugal atomization involves molten metal that comes in contact with a rotating disk or cup. The molten metal is mechanically atomized and thrown off the edges of the rotating substrate into a cooling gas and then solidified. For detailed examples of centrifugal atomization processes, see the sections on Rotating Electrode Process and Rotating Disk Atomization in this article.

One rapid solidification rate powder process that has been developed for commercial use is based on the principles of rotary atomization and forced convective cooling. In this rapid solidification rate process, molten alloy is transferred through a preheated ceramic tube as a stream on a water-cooled copper rotating cup (24 000 rpm) from which it is ejected as finely atomized molten droplets. The molten particles are quenched convectively by forced streams of helium gas that are traveling at approximately 50 m/s (164 ft/s).

The gas stream has a flow rate of 1 kg/s (2.2 lb/s) and transports the atomized rapidly solidified powder to a cyclone separator, where it is sealed under helium. The collected powder is transferred to a dry box for sizing by screening and stored under vacuum. Powders with particle sizes less than 100 μm undergo a cooling rate of 10^5 °C/s (1.8×10^5 °F) in this process.

The rapid solidification rate centrifugal atomization process has been used to produce powders of nickel-based superalloys, aluminum, and iron-based alloys for ap-

Fig. 54 Schematic of the chill-block melt-spinning process

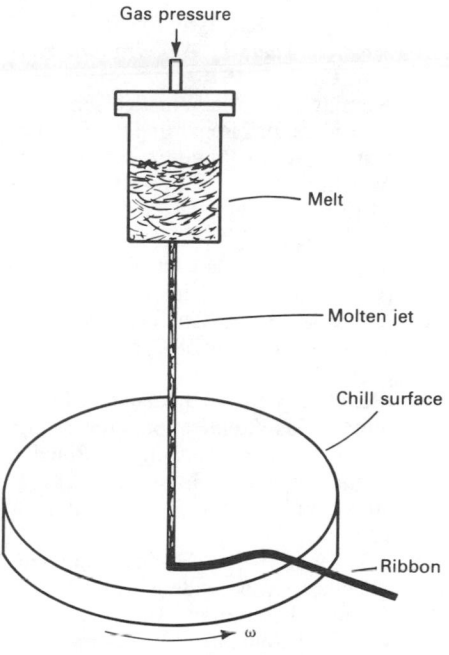

Fig. 55 Schematic of the metal extraction process showing formation of staple fibers

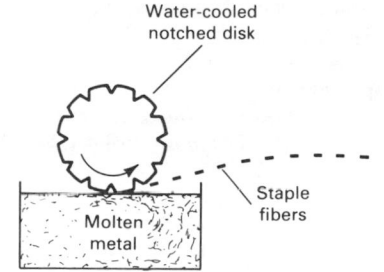

plications such as jet engine hardware components. Some of these alloys appear to be ideal candidates for applications that require improved durability, such as turbine airfoil, tubine and compressor disks, corrosion-resistant bearings, oxidation-resistant lightweight burner liners, and lightweight aluminum fan blades.

A recently developed high-temperature aluminum alloy (Al-8Fe-2Mo) is made by the rapid solidification process. This alloy has improved high-temperature strength up to 260 °C (500 °F). Such an alloy is likely to replace heavier and more costly titanium alloys in jet engine compressor blades.

Rapid solidification rate processed

bearing alloys that have improved rolling contact fatigue life and corrosion resistance compared to conventional M50 bearing steel are being developed.

Turbine blades fabricated from rapid solidification rate processed nickel-based superalloys exhibit significantly improved creep rupture, oxidation, and fatigue properties compared to conventional alloys used for similar applications.

REFERENCES

1. U.K. Patent 1 389 750, 1972
2. Naeser, G., Steffe, H., and Scholz, W., *Stahl Eisen*, Vol 68 (No. 19/20), 1948, p 346-353
3. Thompson, J.S., A Study of Process Variables in the Production of Aluminum Powder by Atomization, *J. Inst. Met.*, Vol 74, 1948, p 101-132
4. German Patent 917 226, 1954
5. Probst, R.L., Production and Use of Spherical Metal Powders, *Met. Prog.*, July 1962, p 107-11; U.S. Patent 4 253 783, "Atomizing Nozzle," 1964
6. U.S. Patent 2 956 304, "Apparatus for Atomizating Molten Metal," 1960
7. U.S. Patent 3 309 733, "Apparatus for Producing Metal Powder," 1967
8. Klar, E. and Shafer, W.M., High Pressure Atomization of Metals, in *Powder Metallurgy for High-Performance Applications*, Burke, J.J. and Weiss, V., Ed., Syracuse University Press, 1972, p 57-68
9. Gummeson, P.U., Modern Atomizing Techniques, *Powder Metall.*, Vol 15 (No. 29), 1972, p 67-94
10. U.S. Patent 2 997 245, "Method and Device for Pulverizing and for Decomposing Solid Materials," 1961
11. Ruthardt, R., Novel Aspects for High Quality Metal Powders Equipment, *Powder Metall. Int.*, Vol 13 (No. 4), 1981, p 175-178
12. Lapple, C.E., Henry, J.P., and Blake, D.E., "Atomization—A Survey and Critique of the Literature," Stanford Research Institute Technical Report No. 6, AD-821 314, April 1967
13. Marshall, W.R., Atomization and Spray Drying, *Chem. Eng. Prog. Mon. Ser.*, Vol 50 (No. 2), 1954, p 112
14. Dombrowski, N. and Johns, W.R., The Aerodynamic Instability and Disintegration of Viscous Liquid Sheets, *Chem. Eng. Sci.*, Vol 18, 1963, p 203-214
15. McCarthy, M.S. and Molloy, N.A., Review of Stability of Liquid Jets and

the Influence of Nozzle Design, *Chem. Eng. J.*, Vol 7, 1974, p 1-20

16. See, J.B. and Johnston, G.H., Interactions Between Nitrogen Jets and Liquid Lead and Tin Streams, *Powder Technol.*, Vol 21, 1978, p 119-133

17. Lane, W.R., Shatter of Drops in Streams of Air, *Ind. Eng. Chem.*, Vol 43 (No. 6), 1951, p 1312-1317

18. Klar, E. and Fesko J., On the Particle Shape of Atomized Metal Powders, *Prog. Powder Metall.*, Vol 37, 1981, p 47-66

19. Takeda, T. and Minagawa K., Effects of Atmosphere during Water Atomization on Properties of Copper Powders, *J. Jpn. Soc. Powder Metall.*, Vol 25 (No. 7), 1978, p 213-218

20. Grandzol, R.J. and Tallmadge, J.A., Effect of Jet Angle on Water Atomization, *Int. J. Powder Metall. Powder Technol.*, Vol 11 (No. 2), 1975, p 103-116

21. Grandzol, R.J. and Tallmadge, J.A., Water Jet Atomization of Molten Steel, *Am. Inst. Chem. Engineers J.*, Vol 19 (No. 6), 1973, p 1149-1158

22. Nukiyama, S. and Tanasawa, Y., Experiments on the Atomization of Liquids, Reports 1-6, translated from *Trans. Soc. Mech. Eng. (Jpn.)*, Vol 4-6, 1938

23. Bitron, M.D., Atomization of Liquids by Supersonic Air Jets, *Ind. Eng. Chem.*, Vol 47 (No. 1), 1955, p 23-28

24. Bradley, D., On the Atomization of Liquids by High-Velocity Gases, *J. Phys. D: Appl. Phys.*, Vol 6, 1973, p 1724-1736, 2267-2272

25. Wigg, L.D., Drop-Size Prediction for Twin-Fluid Atomisers, *J. Inst. Fuel*, Nov 1964, p 500-505

26. Lubanska, H., Correlation of Spray Ring Data for Gas Atomization of Liquid Metals, *J. Met.*, Vol 22 (No. 2), 1970, p 45-49

27. Fraser, R.P., Einsenklam, P., and Dombrowski, N., Liquid Atomization in Chemical Engineering: Part 4—Twin-Fluid Atomisers, *Brit. Chem. Eng.*, Vol 2 (No. 11), 1957, p 610-613

28. Dixon, C.F., Atomizing Molten Metals—A Review, *Can. Metall. Quart.*, Vol 12 (No. 3), 1973, p 309-322

29. Lawley, A., An Overview of Powder Atomization Processes and Fundamentals, *Int. J. Powder Metall. Powder Technol.*, Vol 13 (No. 3), 1977, p 169-188

30. Shinde, S.L. and Tendolkar, G.S., Analyses of Atomization—A Review, *Powder Metall. Int.*, Vol 9 (No. 4), 1977, p 180-184

31. Beddow, J.K., *The Production of Metal Powders by Atomization*, Heiden, London, 1978

32. Schmitt, H., Mathematical-Physical Considerations Regarding the Production of Metal Powders for Powder Metallurgy, *Powder Metall. Int.*, Vol 11 (No. 1), 1979, p 17-21; Vol 11 (No. 2), 1979, p 68-71

33. See, J.B. and Johnston, G.H., Interactions Between Nitrogen Jets and Liquid Lead and Tin Streams, *Powder Technol.*, Vol 21 (No. 2), 1981, p 119-133

34. Mehrotra, S.P., Mathematical Modelling of Gas Atomization Process for Metal Powder Production, *Powder Metall. Int.*, Vol 13 (No. 2), 1981, p 80-84; Vol 13 (No. 3), 1981, p 132-135

35. Schellenberg, R., "Zerstaubung von Metallen, insbesondere von Zink," doctoral dissertation, Technical University of Stuttgart, 1964

36. Marks, L.S., *Standard Handbook for Mechanical Engineers*, 7th ed., Baumeister, T., Ed., McGraw-Hill, New York, 1967, p 11-102

37. Dunkley, J.J., The Production of Metal Powders by Water Atomization, *Powder Metall. Int.*, Vol 10 (No. 1), 1978, p 38-41

38. Small, S. and Bruce, T.J., The Comparison of Water and Inert Gas Atomized Powders, *Int. J. Powder Metall. Powder Technol.*, Vol 4 (No. 3), 1968, p 7-17

39. Lewis, G.C., *et al.*, Atomization of Liquids in High Velocity Gas Streams, *Ind. Eng. Chem.*, Vol 40 (No. 1), 1948, p 67-74

40. Allen, T., *Particle Size Measurement*, 3rd ed., Chapman and Hall, 1981, p 136

41. Rao, P., "Shape and Other Properties of Atomized Metal Powders," Ph.D. thesis, Drexel University, Philadelphia, 1973

42. Grant, N.S., A Review of Various Atomization Processes, in *Rapid Solidification Processing*, Mehrabian, R., *et al.*, Ed., Claitor's Publishing Division, Baton Rouge, 1978, p 230-245

43. Patterson, R.J., II, Cox, A.R., and VanReuth, E.C., Rapid Solidification Rate Processing and Applications to Turbine Engine Materials, *J. Met.*, 1980, p 34-39

44. Nichiporenko, O.S. and Naida, Yu.I., Heat Exchange Between Metal Particles and Gas in the Atomization Process, *Sov. Powder Metall. Met. Ceram.*, Vol 67 (No. 7), 1968, p 509-512

45. Nichiporenko, O.S., Shaping of Powder Particles During the Atomization of a Melt by Water, *Sov. Powder Metall. Metal Ceram.*, Vol 15 (No. 9), 1976, p 665-669

46. U.K. Patent 811 507, "Improvements in or Relating to the Manufacture of Copper Powders," 1959

47. Tamura, K. and Takeda, T., On the Manufacture of Irregular Particle Shaped Copper Powders by Gas Atomization, *J. Jpn. Soc. Powder Powder Met.*, Vol 10 (No. 4), 1965, p 153-159

48. Geldhart, B., *Heat Transfer*, 2nd ed., McGraw-Hill, New York, 1971

49. Nichiporenko, O.S. and Naida, Yu.I., Fashioning the Shape of Sprayed Powder Particles, *Sov. Powder Metall. Metal Ceram.*, Vol 70 (No. 10), 1968, p 753-755 McGraw-Hill, New York, 1971

49. Nichiporenko, O.S. and Naida, Yu.I., Fashioning the Shape of Sprayed Powder Particles, *Sov. Powder Metall. Met. Ceram.*, Vol 70 (No. 10), 1968, p 753-755

50. Dunskii, V.F., On Coagulation During the Atomization of a Liquid, *Sov. Phys. Tech. Phys.*, Vol 1, 1957, p 1232-1239

51. Medvedovskii, A.B., *et al.*, Coagulation of Molten Metal Drops During Atomization, *Sov. Powder Metall. Met. Ceram.*, Vol 154 (No. 10), 1975, p 1-6

52. Small, S. and Bruce, T.J., The Comparison of Water and Inert Gas Atomized Powders, *Int. J. of Powder Metall. Powder Technol.*, Vol 4 (No. 3), 1968, p 7-17

53. U.S. Patent 4 047 933, "Porosity Reduction in Inert-Gas Atomized Powders," 1977

54. Ferriss, D.P., Surface Analysis of Steel Powders by ESCA, *Prog. in Powder Metall.*, Vol 38, 1983

55. Svilar, M., Ambs, H.D., and Klar, E., Powder Metallurgy High Speed Steels, *Carbide Tool J.*, Vol 14 (No. 2), 1982, p 24-32

56. U.S. Patents 3 099 041 and 3 802 816, 1974

57. Munro, W.B., "Rotating Electrode Shot Process," NMI-TJ-30, Internal Report, Nuclear Metals, Inc., Concord, MA, 19 Aug 1960

58. Hodkin, D.J., *et al.*, Centrifugal Shot Casting: A New Atomization Process

for the Preparation of High Purity Alloy Powders, *Powder Metall.*, Vol 16 (No. 32), 1973, p 277-313

59. Champagne, B. and Angers, R., Fabrication of Powders by the Rotating Electrode Process, *Int. J. Powder Metall. Powder Technol.*, Vol 16 (No. 4), 1980, p 359-367

60. Champagne, B. and Angers, R., Size Distribution of Powders Atomized by the Rotating Electrode Process, *Modern Developments in Powder Metallurgy*, Vol 12, Proceedings of the 1980 International Powder Metallurgy Conference, Washington, DC, Hausner, H., Antes, H., and Smith G., Ed., Metal Powder Industries Federation, Princeton, NJ, 1981, p 83-104

61. Champagne, B., Ph.D. thesis, Laval University, 1980

62. Walton, W.H. and Prewett, W.C., The Production of Sprays and Mists of Uniform Drop Size by Means of Spinning Disc Type Sprayers, *Proc. Phys. Soc.*, Sec B, Vol 62, 1949, p 341-350

63. Muraszew, A., Continuous Fuel-Injection Systems with Rotating Fuel Chamber, *Engineering*, Vol 165-166, 1948, p 316-317

64. Hinze, J.O. and Milborn, H., Atomization of Liquids by Means of a Rotating Cup, *J. Appl. Mech.*, Vol 17, 1950, p 145-153

65. Friedman, S.J., Gluckert, F.A., and Marshall, W.R., Jr., Centrifugal Disk Atomization, *Chem. Eng. Progr.*, Vol 48, 1952, p 181-191

66. Dombrowski, N. and Fraser, R.P., A Photographic Investigation into the Disintegration of Liquid Sheets, *Phil. Trans. Roy Soc.*, Vol 247(A), 1954, p 101-130

67. Fraser, R.P. and Eisenklam P., Liquid Atomization and the Drop Size of Sprays, *Inst. Chem. Eng.*, London trans., Vol 34, 1956, p 294-319

68. Fraser, R.P., Dombrowski, N., and Routley, J.H., The Filming of Liquids by Spinning Cups, *Chem. Eng. Sci.*, Vol 18, 1963, p 323-337

69. Fleck, J.N., "Consolidation of Titanium Powder to Near Net Shapes," U.S. Air Force Contract No. F33615-74-C-5114, 1974

70. Petersen, V.C., Chandhok, V.K., and Kelto, C.A., Hot Isostatic Pressing of Large Titanium Shapes, *Powder Metallurgy of Titanium Alloys*, Froes, F.H. and Smugeresky, J.E., Ed., symposium proceedings, 109th Annual Meeting of AIME, Las Vegas, 1980, p 243-265

71. Roberts, P.R. and Loewenstein, P., Titanium Alloy Powders Made by the Rotating Electrode Process, *Powder Metallurgy of Titanium Alloys*, Froes, F.H. and Smugeresky, J.E., Ed., symposium proceedings, 109th Annual Meeting of AIME, Las Vegas, 1980, p 21-35

72. Petersen, V.C. and Chandhok, V.K., "Manufacturing Process for Hot Isostatic Pressing of Large Titanium PM Shapes," Air Force Contract No. F33615-77-C-5005, Interim Report No. AFML-IR-184-7T(1), 1977

73. Eylon, D., Froes, F.H., and Parsons, L.D., "Titanium Powder Metallurgy Components for Advanced Aerospace Applications," 24th Structures, Structural Dynamics and Materials Conference, 2-4 May 1983, Lake Tahoe, Nevada, American Institute of Aeronautics and Astronautics, New York, Publication No. AIAA-83-0982

74. Loewenstein, P., "Superclean Superalloy Powders," Proceedings of the 1980 Powder Metallurgy Superalloys Conference, Nov 1982, Zurich, Switzerland, Vol 1, MPR Publishing Services Ltd., Shrewsbury, England, p 7/1-7/28

75. U.S. Patent 3 510 546

76. Polk, D.E., Giessen, B.C., and Gardner, F.S., *Mat. Sci. Eng.*, Vol 23, 1976, p 309

77. Duwez, P., Willens, R.H., and Klement, W., *J. Appl. Phys.*, Vol 31, 1960, p 1136

78. Holiday, P.R. and Patterson, R.J., II, U.S. Patent 4 078 873, 1978; U.S. Patent 4 343 750, 1982

79. Holiday, P.R., Cox, A.R., and Patterson, R.J., II, in *Rapid Solidification Processing: Principles and Technologies*, Mehrabian, R., Kear, B.H., and Cohen, M., Ed., Claitor's Publishing Division, Baton Rouge, 1978, p 246

80. Patterson, R.J., II, in *Rapidly Solidified Amorphous and Crystalline Alloys*, Kear, B.H., Giessen, B.C., and Cohen, M., Ed., North-Holland, New York, 1982, p 123; Adams, C.M., p 411

81. Patterson, R.J., II, Ledwith, D.L., and Dwyer, J.C., in Proceedings of ACS-AIME Powder Processing Symposium, Lay, K.W. and German, R.M., Ed., TMS-AIME, Warrendale, PA, 1982

82. Gretzinger, J. and Marshall, W.R., Characteristics of Pneumatic Atomization, *Am. Inst. Chem. Engineers*, Vol 7, 1961, p 312-318

83. Rao, K.P. and Mehrotra, J.P., Effort of Process Variables on Atomization of Metals and Alloys, *Modern Developments in Powder Metallurgy*, Vol 12, Hausner, H.H., et al., Ed., Metal Powder Industries Federation, Princeton, NJ, 1981, p 113-130

SELECTED REFERENCES

• Cunningham, R.E., A Novel Technique for the Production of Inorganic Filaments, in *Solidification Technology*, Burke, J.S., et al., Ed., Brook Hill Publishing, 1974, p 241-250

• Lee, D.W. and Spencer, R.C., "Photomicrographic Studies of Fuel Sprays," NACA TR 454, 1933, p 27

• Ozerskii, A.D., Olsson, L., and Fischmeister, J.F., Solidification Structure of Gas Atomized High Speed Steel Powders, 1980 Powder Metallurgy Group Meeting: "High Performance Materials by Powder Metallurgy," The Metals Society, Bristol, England, 27-29 Oct 1980

Chemical Methods of Powder Production

By Erhard Klar
Manager of Particle Technology
SCM Metal Products

CHEMICAL AND PHYSIOCHEMICAL METHODS of metal powder production allow great variations in powder properties. The wide variety of processing variables and production parameters currently available permit close control of particle size and shape. Powders made by reduction of oxides, precipitation from solution or from a gas, thermal decomposition, chemical embrittlement, hydride decomposition, and thermit reactions belong in this classification. The most widely used processes within this category include oxide reduction, precipitation from solution, and thermal decomposition.

Oxide Reduction

The production of iron, copper, tungsten, and molybdenum powders from their respective oxides are well established commercial processes. Detailed process descriptions for these oxide-reduced powders can be found in the articles "Production of Iron Powder," "Production of Copper Powder," and "Production of Refractory Metal and Carbide Powders" in this Volume. On a smaller scale, oxide reduction also is used for production of cobalt and nickel powders.

Oxide-reduced powder grades of iron and copper compete with powder grades made by other processes. Oxide-reduced powders characteristically exhibit the presence of pores within each powder particle and thus are called sponge powders. This sponginess is controlled by the amount and size of the pores and accounts for the good compactibility (high green strength) and sinterability of such powders.

Processing conditions for oxide reduction, which are discussed in detail in Ref 1, are based on the generally known equilibria for the reduction reactions using hydrogen, carbon monoxide, and carbon as the reducing media, as shown in Fig. 1. The ratios of carbon monoxide to carbon dioxide, hydrogen to water, and the partial pressure of oxygen and phosphorus permit determination of the minimum ratios necessary to maintain reducing conditions at a given temperature and at a total gas pressure of 1 atm.

In practice, however, reduction temperatures are usually much higher than indicated by thermodynamic data. Final powder properties such as particle size, particle porosity, and hydrogen loss that critically determine performance properties (apparent density, powder flow, and compacting and sintering properties) primarily depend on purity and size of the starting material and the kinetics of the reduction process. The kinetics of the reduction process depend on composition and flow rate of the reducing gas, reduction temperature, temperature profile in the furnace, and bed depth of the oxide if reduction is performed in a stationary system.

Improved reduction rates are possible with nonstationary reduction in rotary kilns or fluidized bed reactors because of improved access of the reducing gas to the metal oxide particles. For many reducible oxides, however, it is difficult to control sintering and "freezing" of the fluidized bed. The former H-iron process (see the article "Production of Iron Powder" in this Volume), a fluidized bed reduction of iron oxide, used a high hydrogen pressure of approximately 3.5 MPa (500 psi) and a low reduction temperature of approximately 540 °C (1000 °F) to avoid freezing or caking of the bed. The low reduction temperature, however, caused the porous iron powder to be pyrophoric.

Process Variables. The various combinations of processing parameters used by powder manufacturers to produce numerous grades of powder are proprietary. The most important process variable is the reduction temperature. Typically, low reduction temperatures result in powders possessing fine pores, large specific surface areas, and high green strength. High reduction temperatures ($>0.6\,T_m$) produce large intraparticle pores and small specific surface area powders that exhibit high compressibility. Extremely low reduction temperatures ($<0.3\,T_m$) can readily produce pyrophoric powder. High temperatures can cause excessive sintering and agglomeration, which lead to difficulties with the breakup of the sinter cake.

With tungsten and molybdenum, oxide reduction is used partly for economic reasons, because the melting points of these metals are very high. Reduction processes, which use hydrogen as the reducing medium, are similar for both tungsten and molybdenum oxides. For molybdenum trioxide, however, reduction is performed in two stages to control particle size. Due to the high vapor presence of molybdenum trioxide, the first step is carried out at 600 to 700 °C (1110 to 1290 °F). The second step, the reduction of molybdenum dioxide to molybdenum monoxide, is carried out at around 900 to 1100 °C (1650 to 2000 °F).

In contrast to atomized powders where oxides often are enriched on the surface of a particle, oxide-reduced powders, at least when freshly reduced or stabilized against tarnishing, contain most of their residual oxides within the particles.

Fig. 1 Standard free energy of formation of metal oxides
To convert kcal to kJ, multiply kcal by 4.184. Source: Ref 1

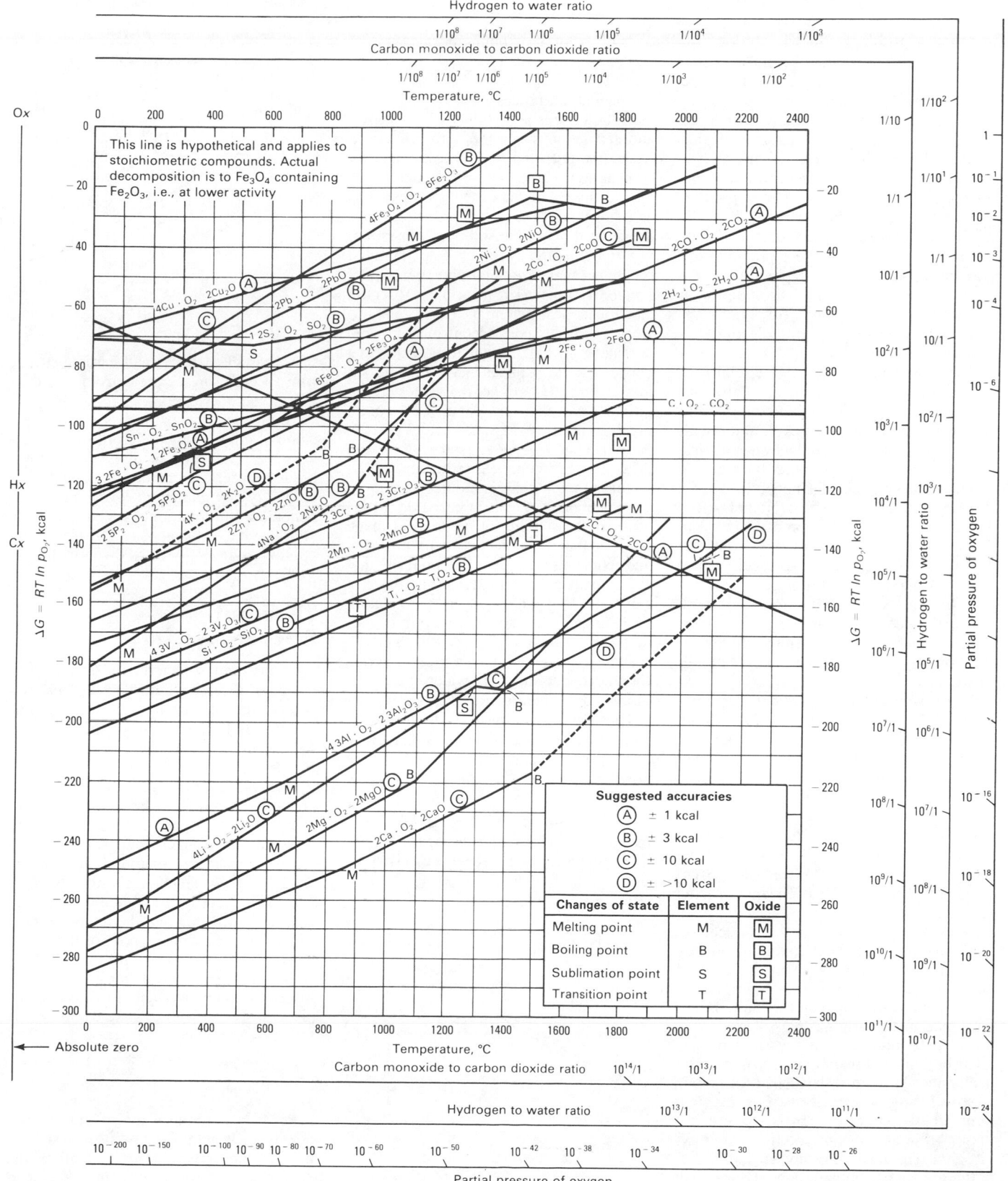

Precipitation From Solution

Production of metal powders by hydrometallurgical processing is based on leaching an ore or ore concentrate, followed by precipitating the metal from the leach solution. Although basic precipitation reactions have been known for more than 100 years, commercial use of this process did not flourish until the 1950's, as a result of increasing interest and work on lower grade ores. Metal precipitation from solution can be accomplished directly by electrolysis, cementation, or chemical reduction. Indirect precipitation may be achieved by first precipitating a compound of the metal (hydroxide, such as carbonate, or oxalate, for example), followed by heating, such as decomposition and reduction.

The most widely used commercial processes based on hydrometallurgy are copper cementation and the separation and precipitation of copper, nickel, and cobalt from salt solutions by reduction with hydrogen (Sheritt Gordon process) (see the articles "Production of Copper Powder," "Production of Nickel and Nickel Alloy Powders," and "Production of Cobalt and Cobalt Alloy Powders" in this Volume). In the 1960's and 1970's, several pilot plants using hydrometallurgical processing (solvent extraction and reduction with sulfur dioxide and hydrogen) were operating temporarily in the United States for the production of high-purity copper powder.

In its simplest form, copper cementation recovers copper from acidic dump leach solutions as an impure powder precipitate. Due to the presence of significant amounts of iron and silicates, low apparent density, and high green strength, such copper powders find use in P/M friction composite components (see the article "P/M Friction Materials" in this Volume). They are not used in conventional structural parts because of insufficient sintering activity.

Nickel powders are now produced in large quantities directly from their ores by precipitation. The powder has a purity of at least 99.8%, with the major impurities being cobalt, iron, and sulfur, due to their presence in the original nickel ore. Powder can be produced in a variety of size distributions and is quite uniform. Similar techniques have been developed for cobalt.

Processing Conditions. For divalent ions, these processes consist of precipitation from an aqueous solution using hydrogen. The basic concept is that a metallic ion such as nickel, copper, or cobalt

in the solution reacts with gas (hydrogen) by the following reaction:

$$M^{++} + H_2 \rightarrow M^o + 2H^+$$

or if the solution is ammoniacal:

$$M^{++} + 2NH_3 + H_2 \rightarrow M^o + 2NH_4^+$$

Generally, processing begins with leaching of ores and includes purification and separation stages prior to reduction. Reduction potential may be estimated by comparing the electrochemical potential of the metal ion and that of the hydrogen ions as a function of the partial pressure of hydrogen and pH of the solution. For reduction to occur, the hydrogen potential must be greater than the metal potential. Metal concentration has a minimal effect on electrochemical potential. These relationships are discussed in detail in Ref 2 and are shown in Fig. 2, which indicates that copper can be reduced in very acidic solutions (low pH). Higher pH values are required for nickel and cobalt. To obtain practical reaction rates, the actual process is carried out at elevated temperatures and pressures. For complete reduction, pH is increased by adding ammonia.

Separation of metals is based on differences in the stability of complexed metal ions (ammines) subjected to hydrolyzing and oxidizing reactions. Potential-pH diagrams with the thermodynamics of oxygen reduction and hydrogen oxidation provide the necessary information to predict separation potential within a given system. Figure 3, a potential-pH diagram for the copper-ammonia-water system, is discussed in detail in Ref 3.

The use of additives and control of nucleation, particle growth, and particle agglomeration allow the production of powders with a wide range of particle

Fig. 2 Potential of 1 and 10⁻³ molar metal solutions and hydrogen potential at varying pH at 25 °C (75 °F)

Source: Ref 2

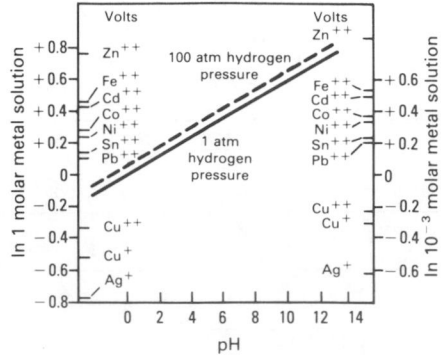

Fig. 3 Potential-pH diagram for the Cu-NH₃-H₂O system

Total NH₃ equals 1 mol/L. Source: Ref 3

sizes, particle density, and particle shape, with specific surface areas from less than 1 m²/g to about 8 m²/g.

Co-precipitation or successive precipitation of different metals from solution allows the production of alloyed and composite powders. Spray drying extends this capability to innumerable combinations. For more information, see the articles "Production of Composite Powders" and "Spray Drying of Metal Powders" in this Volume.

Thermal Decomposition

Of the group of thermally decomposed powders, those produced by thermal decomposition of carbonyls are the most important. Both iron and nickel are produced by decomposition of the respective carbonyls.

Processing Conditions. Carbonyls are obtained by passing carbon monoxide over spongy metal at specific temperatures and pressures. Iron pentacarbonyl, $Fe(CO)_5$, is a liquid at room temperature, boiling at 103 °C (217 °F). Nickel tetracarbonyl, $Ni(CO)_4$, boils at 43 °C (109 °F). When the pressure is reduced to 1 atm and the temperature is raised correspondingly, both of these carbonyls decompose to re-form the metal and carbon monoxide. The latter is recycled to form more carbonyl and to continue the process. These reactions are expressed as follows:

$$Fe + 5CO \rightleftarrows Fe(CO)_5$$

$$Ni + 4CO \rightleftarrows Ni(CO)_4$$

Powder is produced by boiling the carbonyls in heated vessels at atmospheric pressure under conditions that allow the vapors to decompose within the heated space and not on the sides of the container. The powder is collected and sieved and may be milled, followed by an anneal in hydrogen. The chemical purity of the powders can be very high (over 99.5%), with the principal impurities being carbon, nitrogen, and oxygen. Particle size can be controlled very closely. Iron carbonyl powder is usually spherical in shape and very fine (less than 10 μm), while the nickel powder is usually quite irregular in shape, porous, and fine. Detailed descriptions of the iron and nickel carbonyl processes are given in the articles "Production of Nickel and Nickel Alloy Powders" and "Production of Iron Powder" in this Volume.

Other Methods

Precipitation From Salt Solutions. A number of metal powders are made by the precipitation of their soluble salts as insoluble hydroxide, carbonate, or oxalate. On heating, these compounds decompose into the respective metals or metal oxides and gaseous products. Examples include the production of uranium dioxide, platinum, selenium, tellurium, silver nickel, and silver cadmium oxide compounds through co-precipitation. Thorium dioxide dispersion-strengthened nickel, known as TD nickel, used to be made by chemical precipitation of a nickel salt solution containing colloidal thorium dioxide.

Precipitation from a gas is used for certain reactive metals such as titanium and zirconium. In these cases, the gaseous metal chlorides are reduced with liquid sodium or magnesium (Kroll process). The reaction product is leached with dilute hydrogen chloride to remove sodium chloride and/or magnesium chloride. The result is a metallic sponge material that is subsequently pulverized.

Hydride Decomposition. For titanium and zirconium, the so-called hydride-dehydride process also may be used to make powder. The metals are heated in hydrogen in the form of chips or turnings. The brittle hydrides are milled into powder and then dehydrided under vacuum at elevated temperature. The resulting sinter cake is milled into powder.

Chemical embrittlement processes of past commercial importance include the carburization of iron, sulfurizing of 2-81 Permalloy, and the sensitizing (with carbon) of stainless steel (Ref 4). These reactions produce grain-boundary precipitates that make it possible to mill or dissolve the material into powder. None of these processes is of commercial importance today.

Thermit Reactions. Some difficult-to-reduce metals may be produced either by thermit reactions (the reduction of an oxide with a metallic powder that has a very high negative free energy of oxide formation) or by reduction with metal hydrides. Thermit reactions include the reduction of chromium trioxide with magnesium and the reduction of uranium dioxide with calcium or magnesium. Calcium hydride, by means of its nascent hydrogen which forms at elevated temperature by decomposition, can reduce chromium trioxide and other difficult-to-reduce oxides. Presently, neither of these methods is important to the P/M industry.

Chemical Processing of Cermets. Although cermet powders (carbides, nitrides, borides, or silicides) are beyond the scope of this Volume, process descriptions for several refractory carbide powders are included because of their use in the P/M industry (see the article "Production of Refractory Metal and Carbide Powders" in this Volume). These processes are based on reacting the metal or metal oxides with carbon at high temperatures.

REFERENCES

1. Richardson, F.D. and Jeffes, J.H.E., The Thermodynamics of Substances of Interest in Iron and Steelmaking from 0 °C to 2400 °C, *Journal of the Iron and Steel Institute*, Vol 160, 1948, p 261
2. Schaufelberger, F.A., *Journal of Metals*, Vol 8 (No. 5), 1956, p 695-704
3. Halpern, J., *Journal of Metals*, Vol 9 (No. 2), 1957, p 280-289
4. Lenel, F.V., *Powder Metallurgy Principles and Applications*, Metal Powder Industries Federation, Princeton, NJ, 1980, p 33, 44

Milling of Brittle and Ductile Materials

By the ASM Commmittee on Milling*

MILLING OF MATERIALS, whether hard and brittle or soft and ductile, is of prime interest and of economic importance to the P/M industry. Mechanical comminution is the most widely used method of powder production for hard metals and oxide powders. Secondary milling of spongy cakes of oxide-reduced, atomized, or electrolytic powders is the most common milling process; hammer and rod mills are used for this type of milling. Depending on the degree to which the material is sintered, either primary particle size distribution is reestablished during milling or larger agglomerates are produced.

Mechanical comminution is restricted to relatively hard, brittle metals (electrolytic iron or bismuth, for example), some reactive metals such as beryllium and metal hydrides, ductile metals used for producing metal flakes, and chemically embrittled materials such as sensitized stainless steel.

Increasing interest in metal powder with particle sizes that are finer than the particle sizes of powders produced by atomization has reactivated interest in milling, particularly in solid-state alloying or high-energy milling. However, milling of metal powders has received minimal attention to date. Research conducted on ball milling of metals is primarily proprietary and empirical, and thus restricted to specialized P/M applications. Objectives of milling include:

- Particle size reduction (comminution or grinding)
- Particle size growth
- Shape change (flaking)
- Agglomeration
- Solid-state alloying (mechanical alloying)
- Solid-state blending (incomplete alloying)
- Modifying, changing, or altering properties of a material (density, flowability, or work hardening)
- Mixing or blending of two or more materials or mixed phases

In most cases, the objective of milling is particle size reduction.

The milling operation fractures, deforms (cold works), or cold welds the impacted particles. Milling also may produce polymorphic transformations (Ref 1), as is the case with lead oxide and stainless steel (Ref 2). The specific effect that milling has on a powder depends on the physical and chemical properties of the powder, the vacuum, gaseous, or liquid environment in which the operation is conducted, and milling conditions. Selection of the milling process is based on the desired result of the milling operation, the behavior of the powder under milling conditions (if known), the characteristics of the powder, and the physical and mechanical properties of the material.

During milling, four types of forces act on particulate material: impact, attrition, shear, and compression. Impact is the instantaneous striking of one object by another. Both objects may be moving or one may be stationary. Attrition is the production of wear debris or particles created by the rubbing action between two bodies. This type of milling force is preferred when the material is friable and exhibits minimal abrasiveness. Shear consists of cutting or cleaving of particles and usually is combined with other types of force. Shear contributes to fracturing by breaking particles into individual pieces with a minimum of fines. Compression is the slow application of compressive forces to a body (crushing or squeezing of particulate material). This type of milling action usually is associated with jaw crushers and the breaking of large agglomerates of hard, nonductile material.

Design and process improvement has been based primarily on empirical and semi-empirical data. Advancement has been made in understanding the nature of brittle fracture of single particles of relatively homogeneous materials, such as glass. Fundamental research on milling ductile metals is lacking, although much is known phenomenologically about the process. Comminution is not adequately understood as it applies to the detailed mechanism involved when particles are impacted by colliding mediums.

Fracture occurs in hard, brittle materials, with minimal particle deformation and agglomeration by welding. Particle deformation results from attempts to comminute or blend particulate materials. Deformation, cold welding, and fracturing occur in varying degrees with both hard and soft ductile materials. Although powders of <100 mesh (<150 µm) are most frequently milled, particles up to 6 mm (0.2 in.) and larger can be processed.

Principles of Milling

The impact process is shown in Fig. 1. This model represents the moment of collision, at which particles are trapped between two colliding balls within a space occupied by a dense cloud, dispersion, or

*William E. Kuhn, *Chairman,* President, Dymatron, Inc.; Ira L. Friedman, President, Metallurgical Industries, Inc.; Wade Summers, Regional Manager, Sweco; Arno Szegvari, President, Union Process Inc.

Fig. 1 Model of impact event at a time of maximum impacting force showing the formation of a microcompact

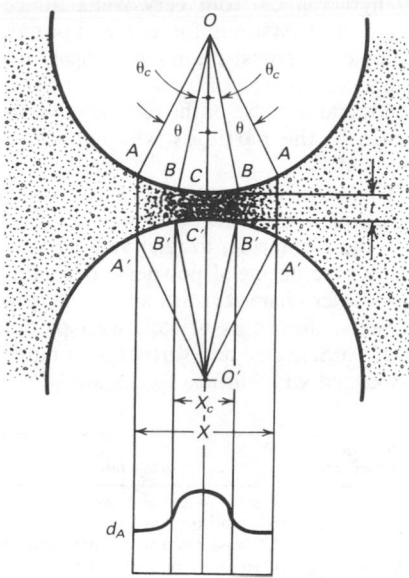

Fig. 2 Process of trapping an incremental volume of powder between two balls in a randomly agitated charge of balls and powder

(a) through (c) Trapping and compaction of particles. (d) Agglomeration. (e) Release of agglomerate by elastic energy

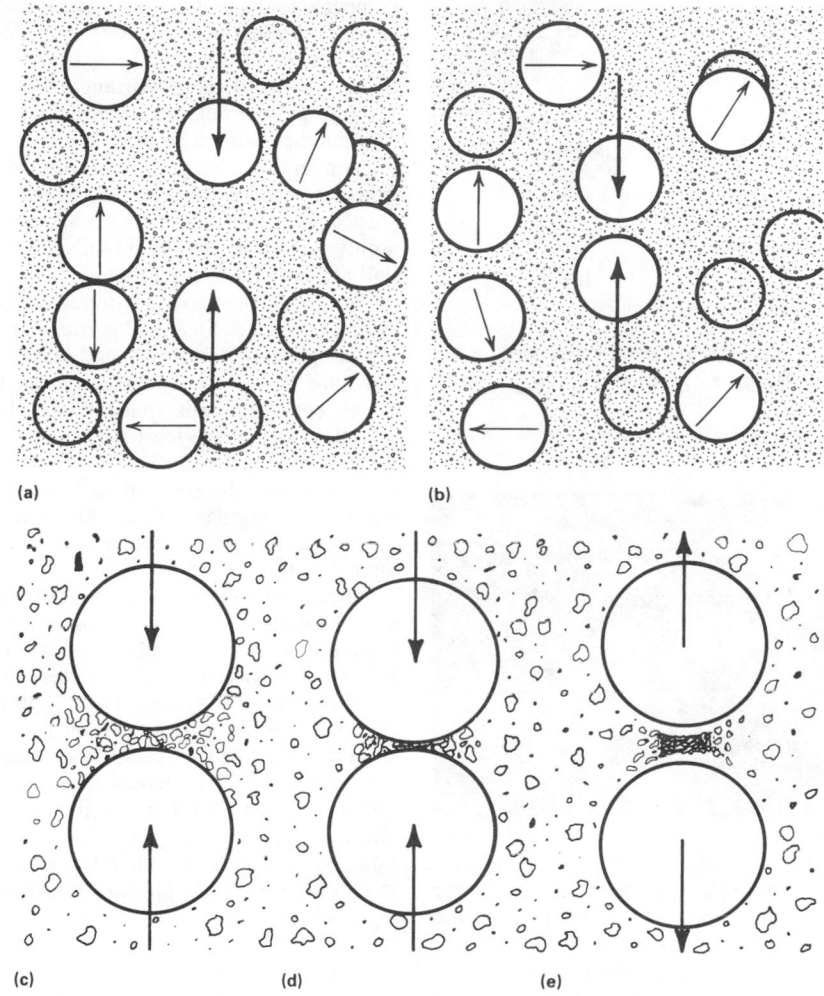

(a)　　　　　　　　　　　　　(b)

(c)　　　　　(d)　　　　　(e)

mass of powder particles. This phenomenon is typical in dry and wet milling operations that use colliding milling mediums such as tumbler, vibratory, and attrition ball mills.

The number of particles within the incremental volume subjected to impact compression stress may range from a single particle to several million, depending on the average particle size and the radius of the ball or curved surfaces contacting the trapped incremental volume. The number of particles trapped between two balls on impact in a loose powder mass should increase as the ratio of the ball diameter (D) to the particle diameter (d) increases. Also, the probability of a particle being trapped and impacted between the two balls increases as powder fill (percentage of space between the balls filled with powder) is increased and apparent density of the agitated or fluidized powder increases.

At the high ratios of D/d encountered in milling of fine particles (<100 mesh, or <150 μm), the trapped volume consists of particles within the microvolume $BB'BB'$ (Fig. 1). Outside this incremental volume, the apparent density of the powder sharply decreases to the density of the agitated loose powder.

The process of trapping an incremental volume between two balls within a randomly agitated charge of balls and powder, assuming no rotation or transverse motion of the curved surfaces, is shown

in Fig. 2. Between the stages shown in Fig. 2(a) and (c), a certain volume of agitated powder is trapped between two impacting balls. The amount of powder trapped and the size of the trapped volume depends on many factors, including particle size, apparent density of the particle cloud, the presence of a slurry and its concentration and viscosity, air or gas pressure within the mill, surface roughness of the balls, and velocity of the balls.

In a gaseous atmosphere prior to impact, the majority of particles is ejected as the balls approach each other, while the remaining fraction becomes trapped and compacted between the decelerating balls at the moment of collision (Fig. 2a-c). If impacting force is sufficient, the trapped incremental volume of powder is compacted to form an agglomerate particle or

pellet (Fig. 2d), which is released (Fig. 2e) when elastic energy impels the balls apart. If bonding by welding or adhesion occurs between contacting surfaces of the particles and bonding forces are sufficient, the agglomerate does not break apart. Similarly, particles may bond to the surfaces of the balls, which may become coated with the powder, as shown in Fig. 3 and 4.

If the compressive impact forces exerted by the balls and particles are sufficient, particles fracture or compact within a critical zone of diameter X_c and thickness (t), within the boundaries B-B' and B-B' (Fig. 1). The initial zone of compaction theoretically does not exceed $AA'AA'$. The incremental volume of compacted powder consequently consists of surviving original particles and fragments of fractured particles. Surviving particles

Fig. 3 Grinding balls coated with powder

(a) Cement. (b) Marble. (c) Silica. (d) Silicon. (e) Corundum

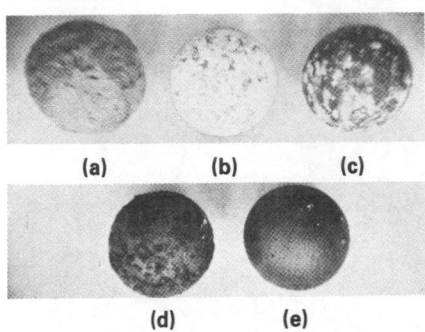

(a)   (b)   (c)

(d)   (e)

Fig. 4 Spikes of magnesium formed on 12.7-mm (¹/₂-in.) diam balls

After prolonged milling under nonreactive milling conditions in a vibratory ball mill

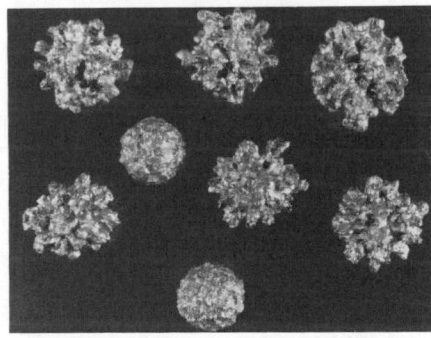

include those particles protected by larger particles that underwent fracture and fine particles that absorbed and dissipated impact force without fracturing. Density of the trapped bed of particles varies from maximum at the point of first contact between the two balls (C-C' in Fig. 1) to the density of the freely agitated powder outside the two A-A' boundaries.

As the balls decelerate during impaction because of the cushioning effect of the trapped powder, a radial displacement of powder particles away from the line of contact O-O' (Fig. 1) occurs in the direction of least resistance to particle flow. The greater the curvature of the balls, the greater the radial wedge effect acting on the trapped particles and the higher the velocity of ejection. Radial displacement of powder is resisted by interparticle friction and friction between particles and the ball surface. Such displacement is dependent on the roughness of the particle and ball surfaces and the degree of adhesive binding or cold welding between agglomerate and ball surfaces.

Comminution is similar to impact compacting of a radially unrestrained micropowder compact. Compaction begins with a powder mass that is characterized by relatively large spaces between particles compared to the particle size. The finished product is a mass of powder with comparatively small pores. The first stage of compaction consists of rearrangement and restacking of particles. Particles slide past one another with a minimum of deformation and fracture, producing some fines—especially from brittle, irregularly shaped particles. During this stage, flowability is reduced, and void spaces are partially filled with fine particles.

Particle shape plays an important role during this stage. Spherical particles, having the greatest flowability and least friction, are almost totally ejected from between colliding balls; pancake or flake particles, having the least flowability and greatest frictional resistance to flow, tend to be retained between ball surfaces. Particles with irregular surfaces also tend to mechanically bond together to form an agglomerate.

The second stage of compaction involves elastic and plastic deformation of particles. During this stage, cold welding may occur between metal particles. Elastic deformation plays a relatively minor role, whereas plastic deformation and cold welding have major effects, depending on the ductility of the materials and the milling environment. For hard, brittle ceramic and metallic materials, plastic deformation and welding have little effect on comminution. However, plastic deformation and cold welding drastically alter the mechanism of milling for ductile metals. Most metals work harden during the second stage of compaction, which increases deformation resistance.

Finally, the third stage of compaction, involving particle fracture, results in further deformation and/or fragmentation of the particles. Densification of brittle ceramic and metallic particulate materials in the microbed occurs when small pores are filled with fragmented particles. Metal powder and soft ceramic particles achieve higher densities when pores are filled by flow processes.

As shown in Fig. 1, the density of the powder bed reaches a maximum at the core of the zone of compaction, C-C'. The strength of the microcompact formed at this point depends on the strength of the bonds between particles, which in turn depends on factors such as conformance of the bonding surfaces, oxide films, surface cleanliness, temperature, and impact force.

The size of the microcompact depends

on the above factors, as well as on particle shape, surface roughness of the balls and particles, and the size of the balls and particles. Thus, particle-to-particle bond strength ranges from very weak for ceramics to very strong for metals. For most ceramic materials, the microcompact disintegrates because of weak van der Waals bonds and because of the release of elastic strain in the particles when the balls rebound.

The microcompact ranges from very small for slightly ductile nonmetals, such as sodium chloride, to relatively large for highly ductile metal powders, depending on powder characteristics and the type of attractive forces associated with particle size. Particle size and surface area ranges associated with surface forces are:

Particle size or surface area	Surface force
50-10 μm	van der Waals and low level chemisorption forces
10-0.1 μm	Chemisorption forces predominate
200-500 m²/g	Chemisorption and valency forces
500-1000 m²/g	Valency forces

If chemical species that tend to chemisorb on the surface of the particles in a dry or gaseous environment can be eliminated or reduced, larger and stronger agglomerates tend to form. If the species in the environment promote adhesion between particle surfaces, agglomerates tend to be larger. In general, chemical reactions between fluid or wet environments and powder inhibit agglomerate formation and assist comminution, especially for metal powders. The less reactive the milling fluid, the greater the tendency for particles to agglomerate by welding. Moisture in the powder promotes packing of large masses of powder in ball mills. Drying the powder at 100 °C (212 °F) or above often alleviates or eliminates this problem, especially with fine aluminum alloy powders.

Brittle Fracture. The Griffith theory assumes that a brittle material has minute cracks dispersed within it and that fracture emanates from these cracks (Ref 3). The stress σ_c at which a crack propagates depends on the size of the crack. The general form of the fracture equation for modes of loading that lead to catastrophic failure is:

$$\sigma_c = A \left(\frac{\gamma E}{c(1 - \mu^2)} \right)^{1/2} \qquad \text{(Eq 1)}$$

where A is a numerical constant that depends on crack geometry and its location, mode of loading, and dimensions of the particle (for example, A is $\sqrt{2/\pi}$ for the slit crack in an infinite sheet); c is the size of the crack; E is the modulus of elasticity; μ is Poisson's ratio; and γ is the surface energy of the milled substance. Thus, measured strength depends on the size of the crack, elastic properties, and surface energy. For brittle materials, γ is between 10^3 and 10^4 erg/cm^2.

The stress σ for fracture of a particle may be represented by:

$$\sigma = \sqrt{\frac{2Er}{L}} \qquad \text{(Eq 2)}$$

where L is the length of the crack, and r is the radius at the crack tip. When stress at the crack tip equals the strength of the cohesion between atoms, all new surface energy is supplied by the stress itself from elastic strain energy and corresponds roughly to the Griffith model.

Three types of flaws (facial, volume, or edge) have been identified by Gilvarry (Ref 4), with respect to location on the surface or interior of single particles. The initial fracture surface in a single particle usually starts in a facial flaw on the surface of the original particle. However, insufficient surface flaws exist to explain the large number of fragments formed in brittle fracture. When an internal flaw is activated, two surfaces in apposition are created, each of which contains flaws intersected by the propagating fracture. New fracture surfaces are produced in turn by fractures propagating from these volume flaws. Similarly, edges formed by intersecting fracture surfaces contain flaws distributed linearly along the edge (edge flaws), which on activation generate new fractures.

According to Gilvarry, the three types of flaws are distributed independently of each other, as a consequence of the randomness of fracture surface orientation and the fracturing stress. Activated flaws are distributed randomly, independent of any stress-producing fracture. More specifically, the position of a flaw within a domain is independent of the position of all other flaws. The probability of a subdomain containing a given number of flaws is independent of the number contained in any other subdomain. Single fractures are initiated by an external stress system that is relieved with the propagation of the first facial flaws. After initial fracture, flaws are activated by stress waves that are generated by the application and withdrawal of the external stress. According to Gilvarry, flaws exhibit a Poisson-type distribution.

During a milling operation, the conditions for single fracture are satisfied only

occasionally. The amount of time and energy required to produce fracture varies with the grinding process and environment. As particle size decreases, flaw concentration and size diminish to a point where edge flaws no longer remain in the particles activated by stress (Ref 4). At this point, fracture stops.

Ultrafine Grinding of Brittle and Hard Materials. The primary function of a mill is to sufficiently stress as many individual particles of the charge as possible, to induce fracture with a minimum expenditure of energy. During milling, a major portion of the energy input is expended on various internal processes that do not contribute to particle failure. As grinding proceeds into the ultrafine region, fracture mode varies, and conditions that were insignificant when particles were relatively large gradually become controlling factors.

Eventually, the milling process reaches an effectiveness limit where the physical and chemical effects associated with surfaces, edges, and corners multiply to inhibit comminution. Edge length per unit volume varies directly with the square of specific surface area, and the number of corners per unit volume varies as the cube of specific surface area. Thus, the probability that an individual particle will be stressed to failure becomes increasingly small.

If the particle deforms plastically, fracture is difficult to induce, and the probability that stress will occur is reduced further. As fragments decrease in size, the tendency to aggregate increases, and fracture resistance increases. Particle fineness approaches a limit as milling continues, and maximum energy is expended. According to Ref 5, the major factors contributing to grind limit are:

- Increasing resistance to fracture
- Increasing cohesion between particles, with decreasing particle size causing agglomeration

- Excessive clearance between impacting surfaces, which is minimized as ball diameter or radius of curvature decreases
- Coating of grinding medium by fine particles that cushion the microbed of particles from impact
- Surface roughness of grinding medium (highly polished, hard mediums that retain a minimum root mean square surface roughness during milling are most effective)
- Bridging of large particles to protect smaller particles in the microbed
- Increasing apparent viscosity as particle size decreases
- Decreasing internal friction of slurry as particle size decreases, thereby reducing production of finer particles by wear mechanism and probability of impacting particles

Generally, as batch comminution proceeds over an extended time, the mean applied stress needed for particle failure increases, while the magnitude of local stresses available to initiate fracture decreases.

Milling of Single Particles. Particles and powders are classified according to particle size, particle size distribution, particle shape, and surface texture. These powder characteristics determine flowability, apparent density, color, sinterability, compactibility, and the properties of sintered products (Fig. 5).

Particle characteristics determine the manner in which particles are deformed or fractured and their resistance to breakage. Thus, the fracture mechanics of single particles must be considered in the selection of milling equipment and milling conditions.

Figure 6 illustrates the crushing of single particles between impacting balls. This condition is encountered in the milling of relatively large particles or a fraction of a powder, as with large granules of a ceramic material, hard metals, or large atomized particles of a ductile metal

Fig. 5 Effect of vibratory milling time on apparent density and Hall flowability of cobalt-based and titanium-based powders

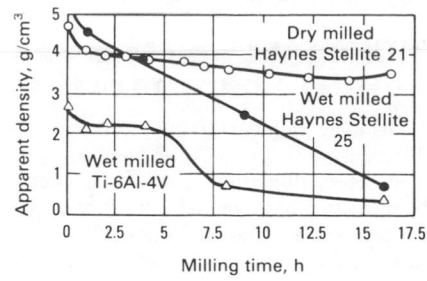

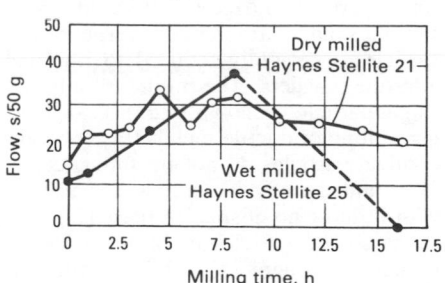

Fig. 6 Effect of impact
(a) Brittle single particle. (b) Ductile single spherical particle

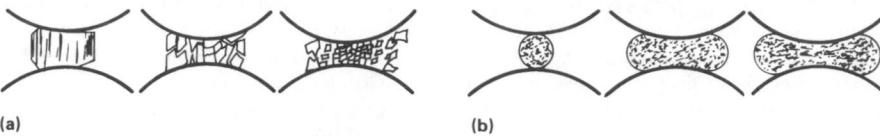

Fig. 7 Effect of milling time on particle shape change of spherical Ti-6Al-4V alloy particles
(a) As-received powder. (b) After 1 h. (c) After 2 h. (d) After 4 h. (e) After 8 h. (f) After 16 h

powder. Under these conditions, many single-particle impact events occur. With ceramics and hard metals, large brittle particles are rapidly reduced to fine and ultrafine powders. The milling of a brittle single particle is shown in Fig. 6(a). During milling of ductile metals (Fig. 6b), individual particles do not fracture, but deform and thereby undergo shape change with little or no change in mass (Fig. 7). Large single particles of brittle materials shatter into fragments, while ductile single particles deform, first into a pancake shape and then into a flake shape.

Milling Parameters and Powder Characteristics

Grinding elements in ball mills travel at different velocities. Consequently, collision force, direction, and kinetic energy between two or more elements vary greatly within the ball charge. Frictional wear or rubbing forces act on the particles, as well as collision energy. These forces are derived from the rotational motion of the balls and movement of particles within the mill and contact zones of colliding balls.

Surface roughness of the balls is a significant factor in milling—the rougher the surface, the greater the frictional energy expended, thus causing increased abrasive action. Frictional forces exert a wearing action between particles in contact with one another and reduce particle size by attrition, producing wear debris. Impact forces effectively reduce particle size in grinding applications involving coarser particle fractions or hard and/or brittle materials.

Impact forces are desirable for deformation processing of metal powders, such as the production of flake powders. Figure 8 shows the change in flake width of iron powder with milling time in a vibratory ball mill. Attrition grinding is most effective in fine grinding both hard and ductile materials and in mechanical alloying. In most ball mills, particles are subjected to both impact and attrition forces. When forces are highly energetic, welding may occur between powder particles, between the powder and the balls, and between powder particles and the walls of the chamber.

Force of impact is directly proportional to the mass of the milling medium. Consequently, processing rate and the forces acting on particles are a direct function of the effective diameter and mass of the milling bodies. Actual collision forces vary widely from theoretical values because of the complex motion of the medium and varying velocity, path lengths, and trajec-

Fig. 8 Relationship between flake size and milling time for electrolytic iron milled in a Megapact vibratory ball mill
Source: Ref 6

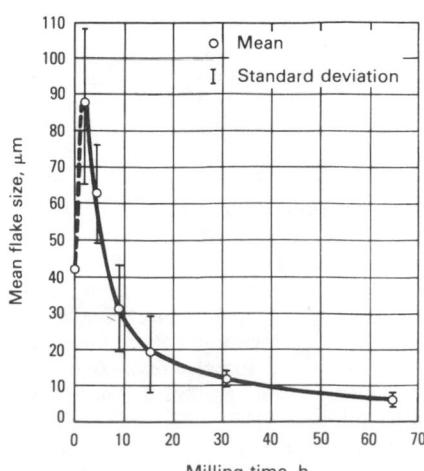

tories of individual milling bodies. Milling fluid viscosity also affects collision rate.

Generally, the size and density of the milling medium selected are determined by the deformation and fracture resistance for metals. For hard, brittle materials, fracture resistance is the only selection criterion. Large, dense grinding mediums are used to mill larger and stronger particulates, whereas smaller diameter mediums are used for finer grinding. For example, ceramic beads having a diameter of 1.6 mm (0.06 in.) and a density of 3.9 g/cm³ may be used for reducing coarse, thin metal flakes to very fine flakes. Milling elements of such a small size cannot be used effectively in conventional tumbler ball mills. For fine grinding of tough, hard materials, tungsten carbide mediums are recommended.

Often, a compromise must be made in the selection of ball size and material. Although tungsten carbide balls are economical for use in small-scale experimental mills or for milling expensive materials, costs may be prohibitive for some large-scale milling operations. Some mediums, such as stainless steel grinding balls, may not be readily available in large quantities in certain sizes; thus, optimum milling conditions may not be satisfied.

The surface area and particle size of nonmetallic materials change continuously. Total surface area of the medium increases as particle size decreases. With metals, however, two additional factors must be considered in addition to comminution. Metal particles may agglomerate by cold welding at impact, thus consuming surface area and causing a shape change because of ductility and the ability to cold weld.

Thus, depending on the dominant process during milling (fracturing, welding, or microforging), a particle may (1) become smaller in size through fracturing, (2) grow in size through agglomeration by welding, or (3) change from an equiaxed shape to a platelet or flake-like particle by microforging. For metals, changes in surface area and particle size measurements do not provide a meaningful criterion for comparing the effect of changes in process parameters, competing milling processes, and equipment.

A more useful criterion for assessing the milling process is one that reflects the spectrum of structural and physical changes occurring in the metal and that is sensitive to differences in processing parameters. Microhardness measurement of individual particles large enough to accommodate the smallest Knoop or diamond pyramid hardness indentation provides a suitable mea-

Fig. 9 Effect of milling time on microhardness of Nickel 123 powder

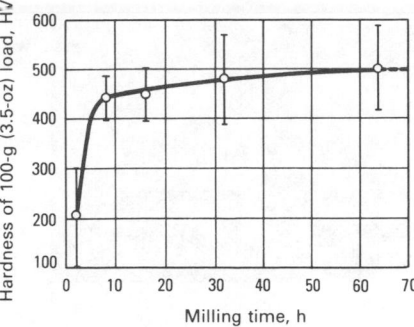

sure of the effect of milling on metal powders, because hardness is a measure of cold work and internal defects produced by milling. As shown in Fig. 9, hardness measurement is most meaningful up to milling times that produce maximum levels of cold work.

X-ray line broadening is sensitive to both the amount of cold work and the refinement in crystallite structure that occurs with continued kneading and working of the metal well after saturation cold work. Changes in the deformation rate of a metal powder produced by a given set of milling parameters are shown by the relationship of x-ray line broadening to milling time (Fig. 10).

Energy Relationship

When a metal is plastically deformed by cold working, most of the mechanical energy of the deformation process is converted into heat. However, a small amount

Fig. 10 Relationship between x-ray line broadening and vibratory milling time for Nickel 123 powder

Numbers in parentheses refer to coordinates of atomic planes in face-centered cubic structure

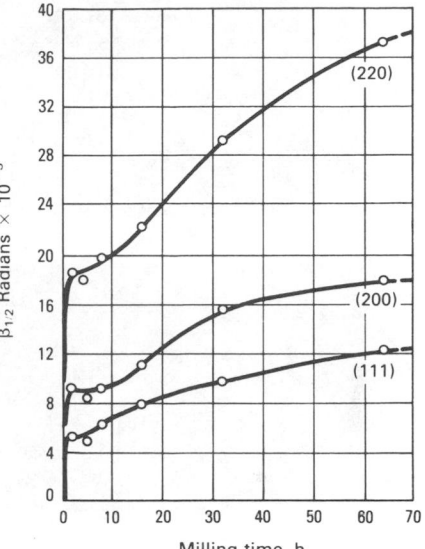

(about 5%) is stored in the metal, thereby raising its internal energy. The energy associated with permanent lattice strain or cold work is minimal for hard, brittle particles, but can be large for ductile materials. The energy expended to overcome the friction between particles is translated to heat and performs no useful work in milling.

If the temperature of the powder rises above a certain point, the cold worked metal particles may undergo recovery and

Fig. 11 Heating curve for 280 cm³ (17 in.³) stainless steel milling chamber during vibration milling of Fe-27Ni-16Cr

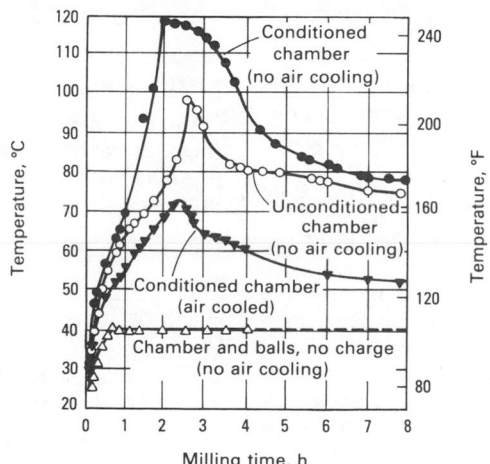

Fig. 12 Scanning electron micrographs of Haynes Stellite 21 cobalt-based alloy powder
Milled in ethyl alcohol with aluminum nitrate grinding aid. (a) As-received powder. (b) After 1 h.
(c) After 2 h. (d) After 4 h. (e) After 8 h. (f) After 16 h. (g) After 32 h. (h) After 64 h

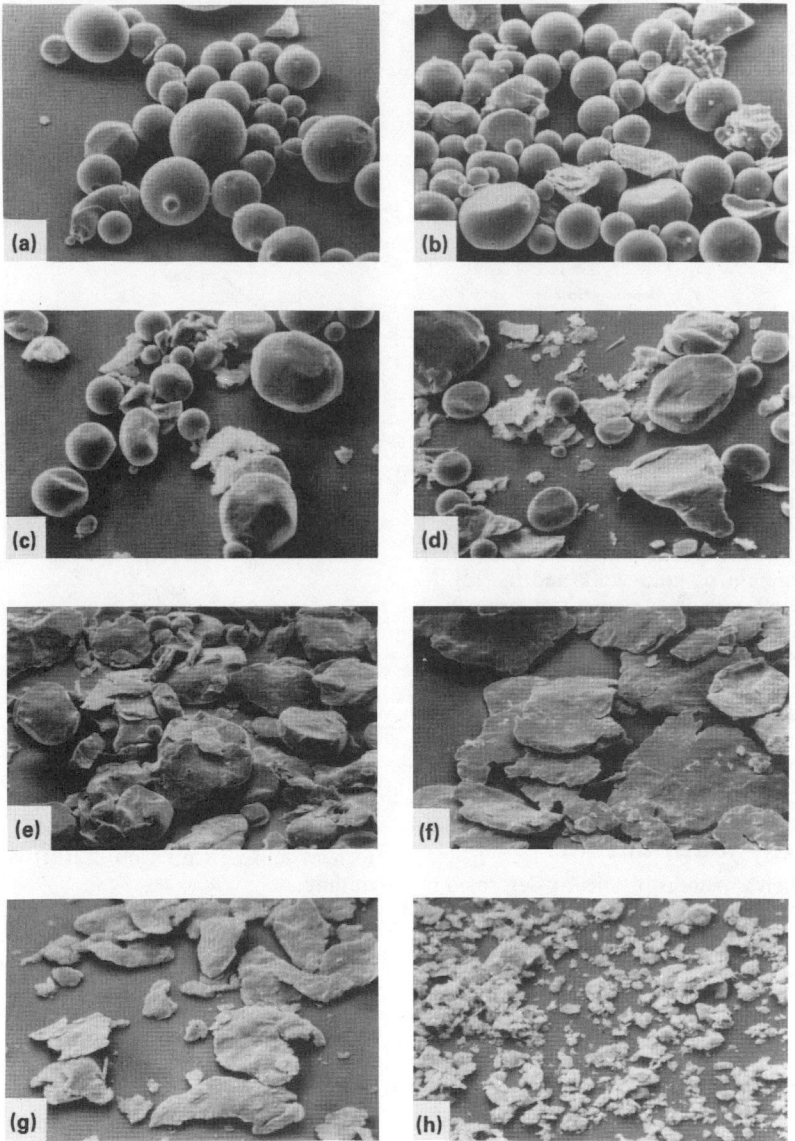

Microforging. The initial and predominant process during milling is compression shape forming of ductile metal particles by impact from the milling medium, as shown in Fig. 6(b). Individual particles, or a cluster of particles, are impacted repeatedly by milling medium so that they deform (flatten) with little or no net change in individual mass.

Fracture. After a period of milling, individual particles deform to the extent that cracks initiate, propagate, and ultimately fracture the particles. Fissures, cracks, defects, and inclusions in particles facilitate fracturing. Particles formed from irregular or spongy particles contain fissures and cracks that facilitate compression, fatigue failure, and fragmentation, compared to the smooth, relatively nonporous spheroidal particles formed by atomization. Some friable metals (antimony, bismuth, silicon, manganese, and chromium, for example) and brittle intermetallic alloys (such as copper-aluminum, aluminum-magnesium, copper-tin, nickel-iron, nickel-zirconium, and phosphorus-copper) can be milled easily into powders.

Agglomeration of particles may occur by welding, mechanical interlocking of spongy or rough surfaces, or autohesion. Autohesion is the molecular interaction of particles among themselves, characterized by van der Waals forces. Deagglomeration is a process that breaks up agglomerates formed by autohesion without further disintegration of the individual powder particles.

Mechanism of Milling

One type of milling mechanism consists of an initial microforging stage in which particles are deformed in the absence of agglomeration by welding and fracture. Eventually, particles become so severely deformed and embrittled by cold work that they enter a secondary stage, during which the particles fracture by a fatigue failure mechanism and/or by the fragmentation of fragile flakes. Fragments generated by this mechanism may continue to reduce in size in the absence of strong agglomerating forces by alternate microforging and fracture. When fracture dominates and the role of microforging is negligible in the absence of welding agglomeration, flakes continue to become smaller until molecular interaction occurs.

As particles (particularly flakes) become finer, coupling forces tend to become greater, and agglomerates become stronger. Eventually, milling forces that deagglomerate the particles reach equilibrium with the coupling forces present, and

recrystallization. Heat is generated by particle deformation, and by elastic deformation of metal grinding balls and grinding chamber walls. Figure 11 illustrates typical temperature versus milling time curves for dry vibratory milled Fe-27Ni-16Cr alloy powder (Ref 7). Generally, the temperature rises during severe cold working, abruptly falls just before cold work attains saturation levels, and then slowly decreases after extended milling times. Water-jacketed milling chambers usually are required for large, high-energy vibratory and attrition mills that reach temperatures above 200 °C (390 °F).

Milling Processes

Changes in powder particle morphology (Fig. 12) that occur during milling of metal powders are produced by the following processes:

Process	Symbolic designation
Microforging	M
Fracture	F
Agglomeration	A
Welding (atomic bonding)	A_w
Mechanical interlocking	A_m
Autohesion (van der Waals forces)	A_a
Deagglomeration	D

an equilibrium agglomerate particle size results.

When cold welding between particles occurs, the initial process is microforging. This is followed by a secondary stage involving a closed loop process of fracture (F), microforging (M), and agglomeration by welding (A_w), as shown below:

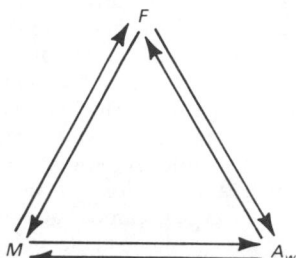

Constant kneading, fracturing, and re-welding produce a composite particle that consists of fragmented original and composite particles. At later stages in the milling process, composite particles are composed entirely of fragmented composite particles. The degree of refinement and blending of the composite structure is a function of milling time, as is shown in Fig. 13 and 14 for wet-milled iron and dry-milled nickel powder, respectively.

After a period of time, cold welding and fracturing attain a steady-state equilibrium. Average particle size obtained at this stage depends on the relative ease with which agglomerates can be formed by welding, fatigue and fracture strength of composite particles, and resistance of particles to deformation. Particle size distribution is narrow, because particles larger than average are reduced in size at the same rate that fragments smaller than average grow through accretion of smaller particles and wear debris produced by attrition.

This process is shown in Fig. 8 and 15. When a large number of fine particles are produced by attrition, the degree of refinement is very high. If particles are spongy, rough, or have surfaces covered with asperities, microforging may be accompanied by agglomeration through mechanical interlocking.

Dispersion of Oxide Films, Hard Particles, and Wear Debris

The strengthening effect produced by ultrafine oxide particles through solid-state or mechanical alloying is an important factor affecting final equilibrium particle size and the time required to reach equilibrium particle size. Oxide particles, derived from oxide films originally present

Fig. 13 Electrolytic iron showing transverse cross sections of particles

Milled in heptane. (a) As-received powder. (b) After 4 h. (c) After 8 h. (d) After 16 h. (e) After 32 h. (f) After 64 h

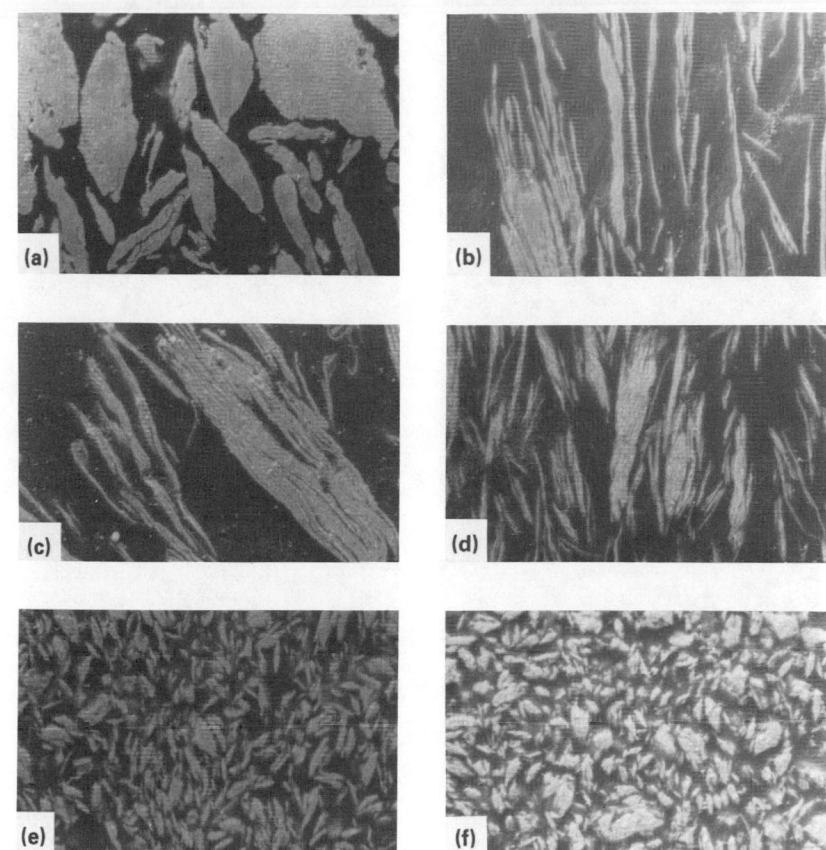

on the particle surface and/or from oxygen in the chamber environment, become entrapped between cold welded surfaces by kneading. This process is illustrated in Fig. 16 using a fluid medium. The strengthening effect of the dispersoid, cold working, and the composite structure greatly increase toughness, strength, and fracture resistance of individual particles. Under these conditions, particles assume a more equiaxed shape. More importantly, the degree of hardening may limit the amount of structural refinement if milling energy is insufficient.

Because the mechanical alloying effect depends on achieving the ultimate degree of refinement, sufficient energy is required for milling. For more information on mechanical alloying as applied to oxide dispersion-strengthened superalloys, see the article "Dispersion-Strengthened Materials" in this Volume.

Milling Environment

Surface-active agents and lubricants generally are used to nullify the forces of welding (or autohesion) and thereby inhibit agglomeration. Surface-active agents adsorbed on a particle surface interfere with welding and lower the surface tension of solid material. Because the energy required for milling is equal to the product of new surface area generated times the surface tension, a reduction in surface tension results in the use of shorter milling times and/or finer powders.

Corrosion reactions between the metal and grinding fluid or environment in reactive grinding also facilitate comminution. The greater the corrosion resistance of a metal, the less effective are corrosive agents. For example, aluminum nitrate, an effective grinding aid for nickel milled in ethyl alcohol, is significantly less effective for either type 316L stainless steel or Haynes Stellite 21.

Milling of iron powder in water, a highly reactive environment for this metal, produces an ultrafine mixture of iron oxide and metal particles. Heptane is one of the least reactive fluid milling environments for metals. However, addition of a grind-

Fig. 14 Nickel 123 showing transverse cross sections of particles
Dry milled. (a) As-received powder. (b) After 4 h. (c) After 8 h. (d) After 16 h. (e) After 32 h.
(f) After 64 h

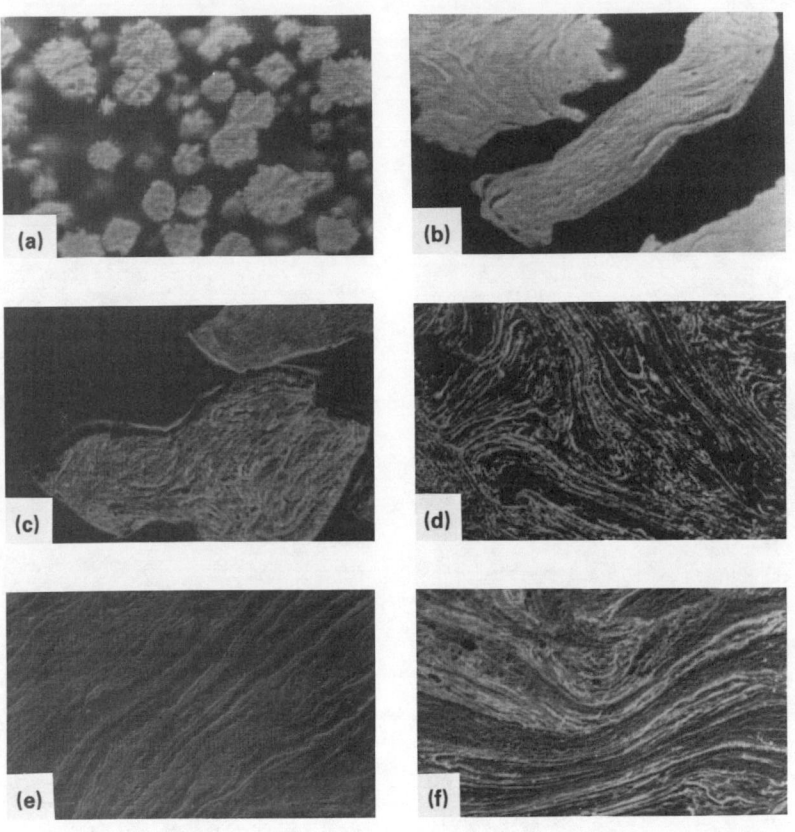

Fig. 15 Narrow particle size distribution caused by tendency of small particles to weld and large particles to fracture

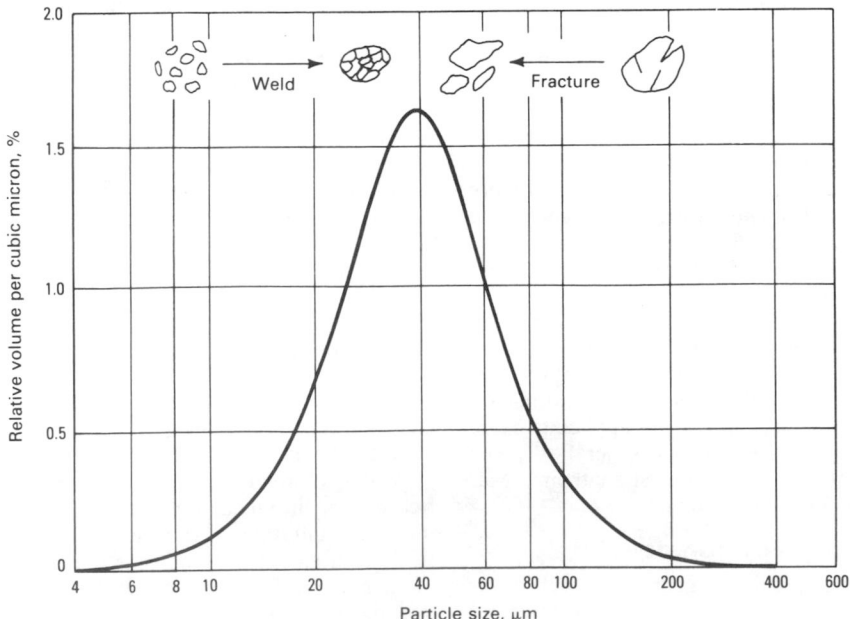

ing aid such as cetyl alcohol may reduce agglomeration by welding. The effectiveness of a given additive depends on its reactivity with the metal being comminuted; the more reactive the metal, the more effective the additive.

Metal powders milled in organic or inorganic fluids retain small amounts of the fluid dispersed throughout each particle (Ref 9). Thus, hydrocarbons containing hydrogen and carbon and carbohydrates containing hydrogen, carbon, and oxygen are likely to introduce carbon and/or oxygen into the particle. The hydrogen subsequently escapes as a gas or is absorbed into the metal lattice on heating or sintering. For example, Fig. 17 shows the amounts of oxygen, carbon, and nitrogen picked up during milling of electrolytic iron in heptane. Hydrated aluminum oxide surface films on aluminum alloy powders produce blisters during subsequent rolling into sheet after solid-state blending, unless removed by heating at 370 to 480 °C (700 to 900 °F) in an inert atmosphere or in a flowing inert gas stream.

Some metals, such as aluminum, nickel, and copper, react with certain alcohols during milling to form complex metalloorganic compounds. For example, aluminum reacts with isopropyl alcohol. Other metals, such as titanium and zirconium, may react explosively with chlorinated fluids such as carbon tetrachloride. Chlorinated fluids should never be used with reactive metals.

According to Arias (Ref 8), two classifications of chemical activity exist between metal powders and the milling environment—reactive and nonreactive milling. In reactive milling, metal powders react extensively with the milling fluid during milling. The metal powder is comminuted to a particle size much smaller than the starting powder. Balls, mill chamber walls, and powder of the same metal do not weld together during milling.

During nonreactive milling, the metal powder either reacts mildly or not at all with the milling fluid during milling. Powder particles weld together to a great extent during ball milling. Additionally, balls, mill chamber walls, and powder of the same metal may weld together during milling. Welding of powders to balls and chamber walls may occur even if they are made of different alloys.

Reactive and nonreactive milling may occur in wet or dry environments. Mechanical alloying consists of dry milling in a vacuum or an inert gas, or in a flowing gas that may be reactive. In mechanical alloying, the welding process may be controlled by monitoring the oxygen concen-

Fig. 16 Nonreactive milling process

(a) Initial milling: nickel flattened into flakes; dispersoid embedded in nickel particles. (b) Intermediate milling: nickel flake formation continues; flakes weld; embedding of dispersoid particles in nickel continues. (c) Final milling: multilayered nickel flakes fold and/or break and reweld; all dispersoid kneaded into milled powder. Source: Ref 8

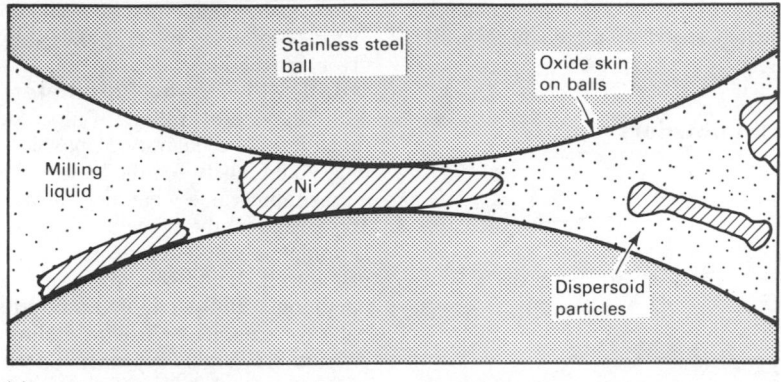

(a)

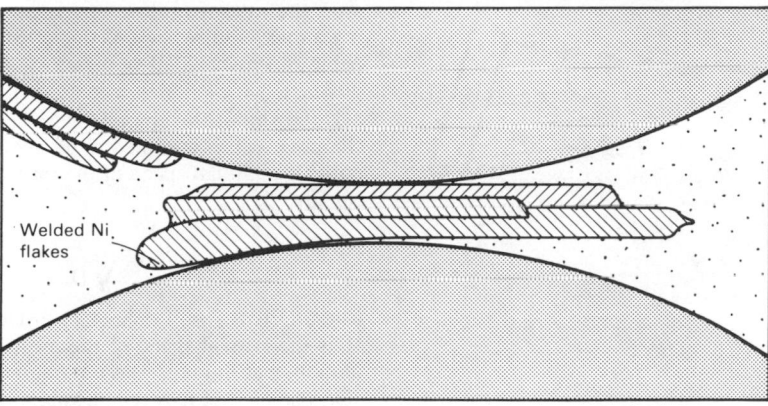

(b)

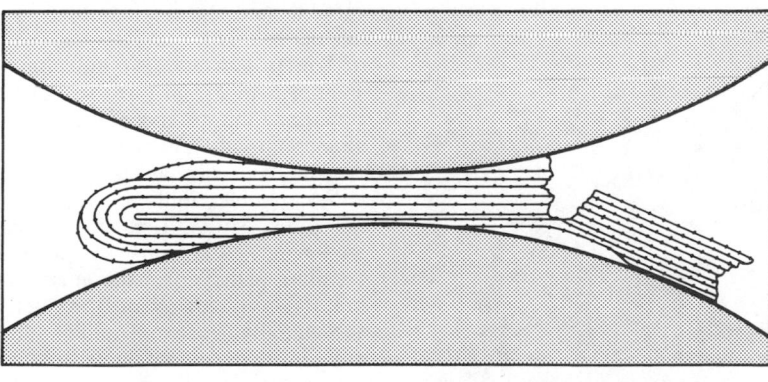

(c)

Fig. 17 Pickup of oxygen, carbon, and nitrogen in electrolytic iron

Milled in heptane in a Megapact vibratory mill. Source: Ref 7

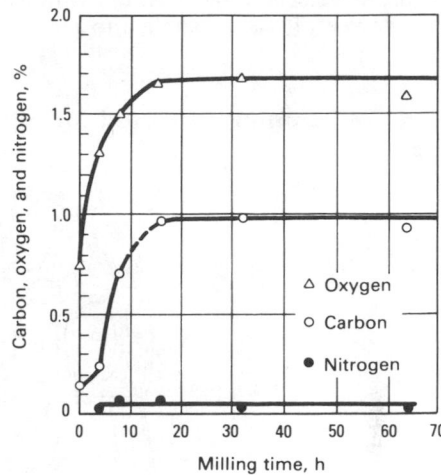

tration in the grinding chamber (Ref 10).

Milling of metal powders produces external shape and textural changes and internal structural changes. The extent of these changes is determined largely by milling parameters, milling environment, and physical and chemical properties of the metal or alloy being milled. These changes in turn affect the physical properties of the metal being processed and the resulting consolidation products. Powder characteristics associated with milling include:

External

- Shape
- Texture
- Reflectivity
- Particle size and particle size distribution

Internal

- Composite structure or composite agglomerates
- Cold work (stored energy)
- High dislocation density
- Improved homogeneity (alloys)
- Finer structure (laminar structure)
- Refined crystallite size
- Chemical composition (incorporation of gases and fluids from the milling environment)

Physical

- Mechanical properties (resistance to deformation and hardness)
- Physical properties (magnetic resistivity)
- Chemical properties (corrosion)
- P/M properties (apparent density, flowability, sinterability, compressibility, and sintered density)

Knowledge of the variables that influence characteristics of a milled product can be used to improve product performance and material properties and to achieve innovative P/M applications not possible by conventional metallurgy. Some of these include silicon carbide whisker-reinforced composites, oxide-dispersion-strengthened alloys, and superconducting compounds such as Nb_3Sn.

Milling Equipment

During comminution, the milling machine stresses the maximum number of individual particles in a powder mass to initiate fracture with a minimum of energy. The motion of milling medium and charge varies between types of mills, with respect to movement and trajectories of individual balls, movement of the mass of balls, and the degree of energy applied to impact, shear, attrition, and compression forces acting on powder particles.

Tumbler Ball Mills

In the tumbler mill shown in Fig. 18, the motion of the balls or rods depends on the rotation speed of the chamber walls.

Balls may roll, or "cascade," down the surface of the charge in a series of parallel layers or they may project, or "cataract," into space, falling on the balls and particles at the toe of the charge below. A detailed mathematical discussion of the trajectories followed by balls and particles is given in Ref 11.

Ball Mill Charge Parameters (Ref 11). The rate of milling of a powder is a function of the quantity of powder in the total volume between the balls and attains a maximum value when the powder fill is 100% (that is, the space between the balls is completely filled with powder). Speed of the mill should be adjustable to ensure operation with the required proportions of balls for cataracting and cascading. This proportion is largely dependent on the speed of rotation and the coefficient of friction of the charge material. With increasing ball size, the rate of milling increases, attaining a maximum when balls reach sizes capable of crushing all particles in the feed powder.

If the slip of the charge against the mill chamber wall is assumed to be negligible, critical rotational speed of the mill may be calculated:

$$N_c = 76.6 \sqrt{\frac{1}{D}}$$

where D is the mill diameter, in feet; and N_c is the critical speed of the mill, in revolutions per minute. These parameters are approximate and may not be valid for metal particles that tend to agglomerate by welding.

Solid-state blending can be achieved in tumbler ball mills using 6- to 25-mm (0.25- to 1-in.) diam balls, if mill diameter is about 1.5 to 1.8 m (5 to 6 ft) and other conditions that promote cold welding agglomeration are satisfied. For large-scale production of mechanically alloyed superalloys, tumbler mills are replacing attrition mills. Milling times in tumbler mills are longer to accomplish the same level of blending achieved in the attrition or vibratory mill, but overall productivity is substantially greater. Tumbler mills usually are used to comminute or flake metals, using a grinding aid or lubricant to prevent cold welding agglomeration and to minimize oxidation (Ref 12 and 13).

Vibratory Ball Mills

Vibratory Tube Mill. In the laboratory vibratory tube mill shown in Fig. 19, oscillatory motion of the balls is complicated. The path of the balls and particles is determined by many factors, including vibrational speed, amplitude, curvature of

Fig. 18 Tumbler mill used for milling metal powders

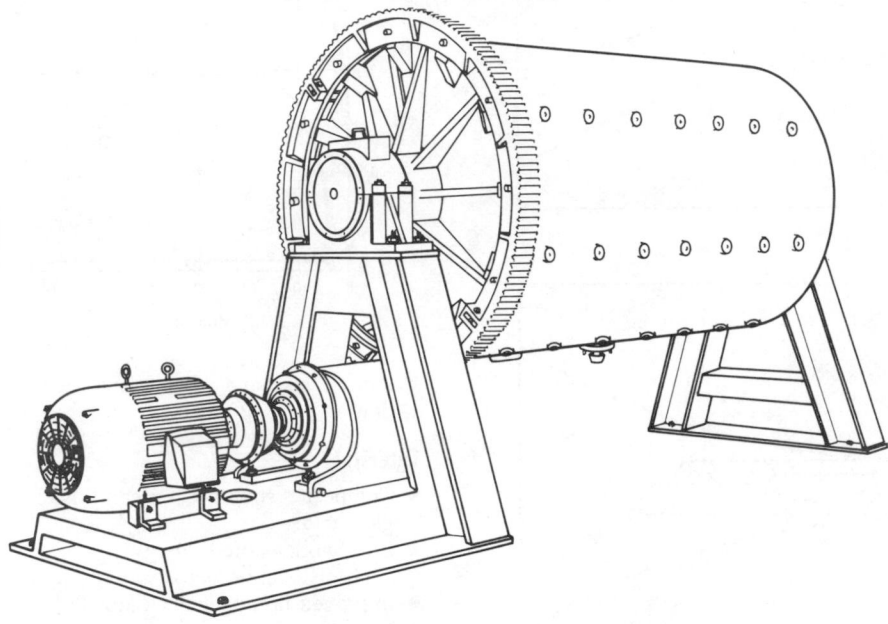

Fig. 19 Megapact vibratory ball mill
Courtesy of Battelle Memorial Institute

the sides of the mill chamber wall, horizontal motion of the mill, and charge contact with the upper surface of the mill chamber. The balls gyrate around the chamber wall, sometimes traveling horizontally in a spiral path. Gyration in cylindrical chambers occasionally produces deformation and/or wear grooves in the inner walls of the chamber. These grooves may extend 360° or less, depending on whether the balls make contact with the upper surface of the chamber wall. Balls also revolve at different rates and directions, causing substantial shearing action, which is desirable in mixing and blending operations.

Impact forces acting on powders in a vibratory mill are a function of the rate of milling, amplitude of vibration, and mass of the milling medium. High-energy milling forces can be obtained by using high vibrational frequencies and small amplitudes of vibration. The mill shown in Fig. 19 operates at 3300 rpm, with a 2-mm (0.08-in.) amplitude, reaching a maximum acceleration rate of 12.2 g, where g is the gravitational acceleration at 9.81 m/s^2 (32.2 ft/s^2). Large production mills operate at relatively low vibrational frequencies and high amplitude (for example, 1000 to 1500 rpm and up to 12 mm or 0.48 in.).

The vibratory ball mill is an excellent means of producing solid-state alloyed and dispersion-strengthened metals in amounts up to 4.5 kg (10 lb) or more, depending on the apparent density of the powder. Figure 20(a) and (b) show the microstructure of an aluminum-iron-cerium alloy and a transmission electron micrograph of oxide dispersion in the same alloy after solid-state alloying in a laboratory vibratory mill of the type shown in Fig. 19.

In large tube mills of the type shown in Fig. 21, vibratory motion of the medium decreases from the chamber walls to the center of the mill tube; consequently, the milling effect is less at the center of the mill than adjacent to the chamber walls. Efficient operation is obtained at a ball fill of 60 to 80% of the volume of the mill chamber for tubes about 500 mm (20 in.) in diameter, with 100% fill of the volume between the balls.

In vibratory mills, the grinding medium receives rapid impulses at a rate proportional to the vibrational frequency of the mill. Impact forces acting on the powder exceed shearing and friction forces. The entire charge slowly revolves counterclockwise to the oscillatory vibrations, so that grinding and intense mixing occur simultaneously. Vibratory mills utilize smaller mediums because of higher impact forces, frequencies, and acceleration;

Fig. 20 Homogenization of Al-Fe-Ce alloy by high-energy milling
(a) Untreated rapidly solidified powder after hot pressing. (b) Hot pressed rapidly solidified powder after high-energy milling in a Megapact mill

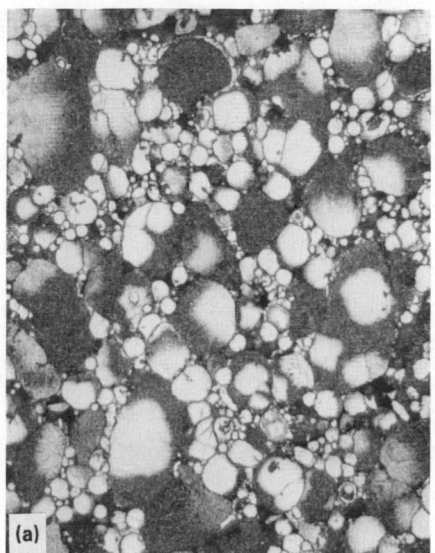

Fig. 21 Pilot- and production-size vibratory mills

thus, a higher specific surface is available for milling. The rate of processing in a vibratory mill is:

- Proportional to the density of balls (diameter constant)
- Proportional to the diameter of the balls (density constant)
- Proportional to the cube of the frequency of vibration
- Negligible for speeds less than 900 to 1000 rpm

- Proportional to the square root of ball diameter/mean particle diameter ratio
- Not significantly affected by chamber diameter
- Increased as amount of powder in mill decreases
- Greater with balls than cylinders or other shapes

Sweco vibratory mills (Fig. 22) are equipped with a grinding chamber in the form of a vertical cylinder with a solid center axis. A double-ended motor with eccentric weights rigidly attached to the bottom of the chamber generates high-frequency, three-dimensional vibrations. The chamber and motor assembly are mounted on a base and are supported by compression springs. Vibration of the grinding medium within the chamber creates the milling action. Vibrations are transmitted from the sides and base of the chamber to the grinding medium. Particles trapped within the medium are broken down by high-frequency impaction.

The medium is packed to provide near-maximum packing density. The packed mass slowly gyrates horizontally, rises near the outer wall of the chamber, and descends as it approaches the inner wall. This motion facilitates distribution of the charge in dry grinding and serves to maintain solids in suspension in wet milling. The Sweco mill is not widely used to mill metal powders. It is particularly ill suited for high-density metals, because high density causes particles to settle to the bottom and become caked.

Attrition Mills

Milling in an attrition mill (Fig. 23) is effected by the stirring action of an agitator that has a vertical rotating shaft with horizontal arms. This motion causes a differential movement between the balls and the material being milled, thus providing a substantially higher degree of surface contact than is achieved in tumbler or vibratory mills.

Milling is accomplished by impact and shear forces. The rotating charge of balls and milling product form a vortex at the upper end of the stirring shaft, into which the milling product and balls are drawn. The milling product is impacted by balls traveling in various trajectories that collide within the dilated charge of medium and powder.

While ball mills use large mediums, normally 12.7 mm (0.5 in.) or larger, and run at low rotational speeds of 10 to 50 rpm, the attrition mill agitator rotates at speeds ranging from 60 rpm for produc-

Fig. 22 Sweco vibratory wet grinding mill

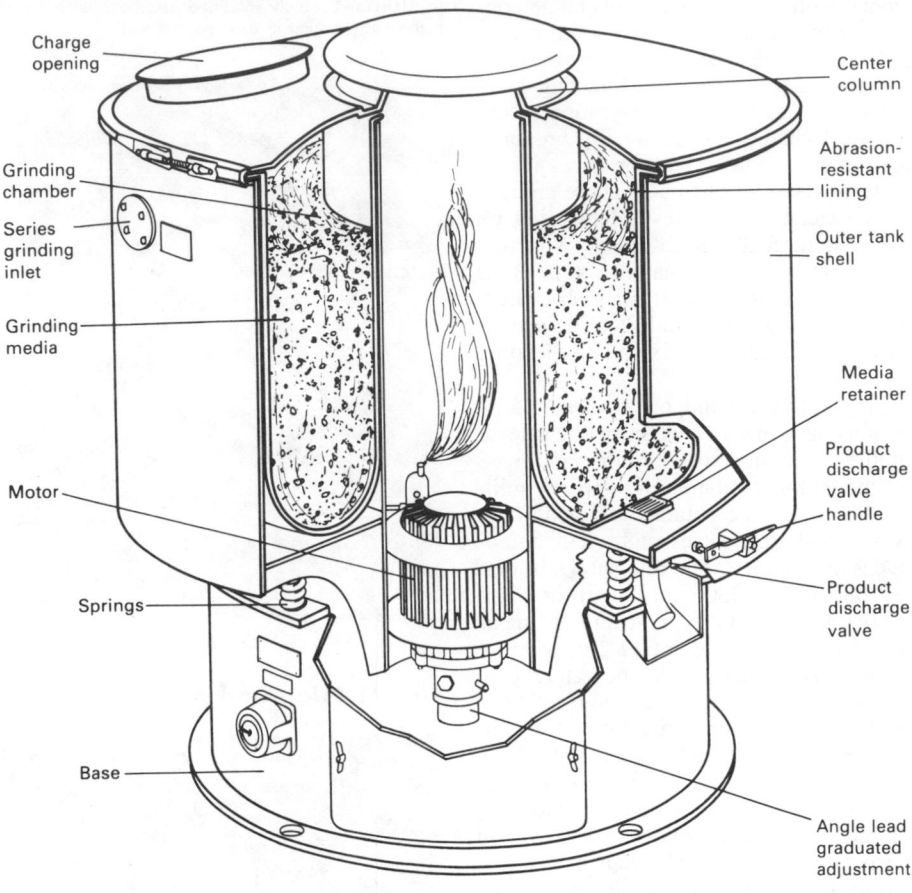

tion units to 300 rpm for laboratory units and uses mediums that range from 3 to 6 mm (0.125 to 0.25 in.). Power input to attrition mills is used to agitate the medium, not to rotate or vibrate the heavy containment vessel.

For hard materials such as ceramics, carbides, and hard metals, attrition mills are more efficient than conventional tumbler and vibratory ball mills. The main advantages of attrition mills for mixing and blending tungsten carbide/cobalt cutting tool powders include short milling time, production of fine particle size (submicron sized), and enhanced smearing of cobalt onto carbide particles. Attrition mills effectively mill metals in inert atmospheres, such as in solid-state or mechanical alloying processes, but are not cost effective. Product output is relatively low with attrition mills, compared to large tumbler and vibratory ball mills. Consequently, tumbler ball mills usually are used for production runs of over 135 to 180 kg (300 to 400 lb) per day.

Mechanism of Attrition Mills. The central rotating shaft of an attrition mill,

Fig. 23 Attrition ball mill

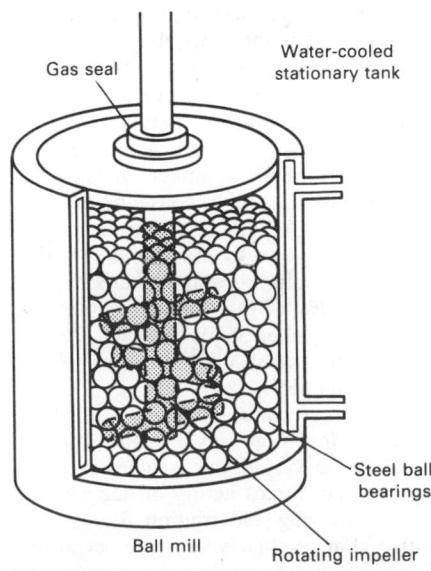

equipped with several horizontal arms, exerts sufficient stirring action to tumble the grinding medium randomly through the entire chamber volume, causing irregular movement by:

- Impact action on the medium
- Rotational force on the medium
- Tumbling force as medium fills the void left by the shaft

For fine grinding, both impact and shearing forces must be present. In attrition mills, impaction is caused by constant impinging of the grinding medium, due to irregular movement. Shearing action is produced by random movement of the balls in different rotational directions, which exerts shearing forces on the adjacent slurry. The strongest medium agitation occurs at a point located two thirds of the way from the center. Grinding does not occur against the chamber walls, which serve as a container rather than a grinding surface. Minimal wear of chamber walls ensures long service life.

In attrition grinding, grinding time is related to medium diameter and agitator speeds, within given limits, as:

$$t = \frac{kd}{\sqrt{n}}$$

where t is grinding time required to reach a certain median particle size; k is a constant that varies with the slurry being processed and the type of medium and mill being used; d is diameter of the medium; and n is shaft movement, in revolutions per minute.

Attrition mills are classified as batch-, continuous-, or circulation-type mills. In the batch mill, material is fed into the jacketed chamber and is ground until the desired dispersion and particle size are achieved. Premixing is accomplished in the grinding chamber. Chamber walls are jacketed so that either hot or cold water can be circulated to control and maintain the temperature of the batch. Batch attrition mills can process high-density material, such as tungsten carbide, as well as viscous materials, and are suitable for dry grinding and for processing dispersion-strengthened metals.

Continuous attrition mills, best suited for large production output, consist of a tall, narrow, jacketed chamber through which a well-premixed slurry is pumped in at the bottom and discharged at the top. Grids located at the bottom and top retain the grinding medium.

The circulation grinding system is comprised of an attrition mill and a large holding tank, generally ten times the volume of the grinding unit. The circulation attrition mill is filled with medium and contains grids that restrain the medium while the slurry passes through. Attrition mills generally have high pumping rates. Usually, the contents of the holding tank are passed through the system at a rate of ten times per hour.

One advantage of the circulation attrition mill is that large quantities of material can be handled with a minimal investment in grinding medium and equipment. The slurry can be monitored continuously, and processing can be stopped when the desired particle size dispersion is achieved.

Coldstream Impact Process

The Coldstream process is a high-velocity process in which material is entrained in a gas stream and projected against a stationary target. After material has struck the target and shattered, it is removed from the impact chamber by suction. Material is then transported to a classifier, which allows oversized products to drop into a storage vessel for subsequent impact against the target. The finished product is separated by the classifier and transported to collection containers.

The process chamber is lined with tungsten carbide or other wear-resistant material. The supersonic nozzle and target generally are made of cemented tungsten carbide, but other wear-resistant materials or the product being comminuted can be used to make these components. Figure 24 shows a high-speed closeup of a raw material stream impacting a target and shattering.

Rapidly expanding gases exiting the nozzle create a strong cooling effect through adiabatic expansion. This effect is greater than the heat generated by pulver-

Fig. 24 Raw material stream impacting a target and shattering in Coldstream impact process

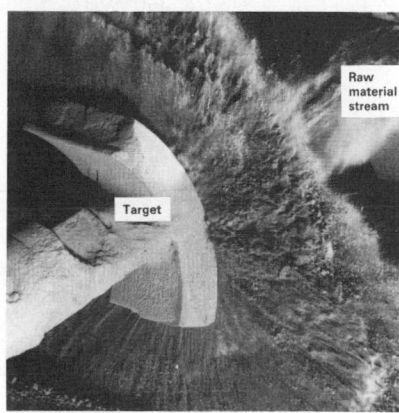

ization. Therefore, the process operates at well below room temperature. A compressor capable of operating at 56 m³/min (2000 ft³/min) at 6.9 MPa (1000 psi) is used. Such a compressor can operate two or three Coldstream impact systems simultaneously.

The Coldstream impact process is used to pulverize hard, abrasive, relatively expensive materials such as tungsten carbide, tungsten alloys, molybdenum, tool steels, beryllium, and other alloys. The process rapidly converts a feedstock of particles 6 mesh or smaller to micron size. Process benefits include simplicity, low operating cost, low operating temperature (which prevents oxidation), and the ability to retain high purity and particle size control over the material being pulverized.

Many standard products are supplied as atomized powders. The higher cost of additional Coldstream processing is justified by production of a finer powder than is commercially available by straight atomization. Most atomized powders have a slight oxide film on the surface of each particle, which is particularly noticeable in coarse powders of nonstainless steel composition. This brittle oxide film is stripped from the metal and broken up into fine dust during Coldstream processing. Lower density and smaller particle size facilitate separation from the metal particles by air classification. Coldstream processing generally produces a relatively oxide-free powder.

High-Energy Milling

Vibratory, attrition, and large-diameter tumbler mills may be classified as high-energy ball mills and, as such, may be used effectively in solid-state or mechanical alloying processes. Figure 25 provides the capabilities of typical size reduction equipment for grinding a material of a given feed size to a desired product size, when the milling objective is comminution of a hard and/or brittle metal or ceramic material. Applications suited to high-energy milling include:

- Flaking metal powders (Ref 12-14)
- Mechanical alloying (Ref 6, 7, 10, 15, 16)
- Dispersion strengthening (Ref 5, 9, 16, 17)
- Composite powders (Ref 18)
- Particle shape control
- Control of powder flowability
- Apparent density control
- Tungsten carbide and tungsten carbide/cobalt alloys
- P/M alloy blending (Ref 18)

Fig. 25 Typical size capabilities of common classes of size reduction equipment

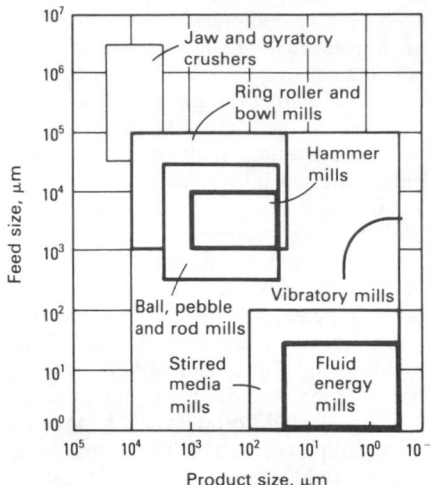

- Alloying of immiscible alloying elements (Ref 10)
- Particle size control
- Ultrafine powders (Ref 5, 19)
- Activated sintering (Ref 7)

Hammer and Rod Mills

Hammer and rod mills are used in the P/M industry to mill large quantities of sinter cakes, which are intermediate products in the production of metal powders by reduction of their oxides. If such powders are to be used in conventional P/M parts processing, it is essential that they possess good green strength and compressibility characteristics. For these reasons, the sinter cakes must be ground into powder, typically −80 mesh, while introducing minimal cold work and densification. Hammer and rod mills with built-in screens or other separation devices are best suited to achieve these properties. Figure 26 shows the "blocky" particle shape of a copper powder produced by hammer milling of a porous sinter cake.

Fig. 26 Particles (+8 mesh) of copper powder hammer milled from oxide-reduced sinter cake

Courtesy of SCM Metal Products

REFERENCES

1. Senna, M. and Kuno, H., Polymorphic Transformation of PbO by Isothermal Wet Ball-Milling, *J. Am. Ceram. Soc.*, Vol 54 (No. 5), May 1971
2. Kuhn, W.E. and Lucky, H., Characterization of Ball-Milled Type 316-L Stainless Steel Powder, *Fine Particles*, W. Kuhn and J. Ehretsmann, Ed., The Electrochemical Society, Inc., 1974, p 95-121
3. Griffith, A.A., *Trans. Roy. Soc. Lond.*, Vol A221, 1921, p 163
4. Gilvarry, J.J. and Borgstrom, B.M., Fracture and Comminution of Brittle Solids, *AIME Trans.*, Vol 220, 1961, p 380
5. Harris, C.C., On the Limit of Comminution, *Trans. Soc. Mining Eng.*, March 1967, p 17-30
6. Kuhn, W.E., High Energy Milling of Electrolytic Iron Powder, *Modern Developments in Powder Metallurgy*, Vol 12, Metal Powder Industries Federation, Princeton, NJ, 1980, p 195-218
7. Patel, A.N. and Kuhn, W.E., Influence of High Energy Vibra-Milling on the Annealed and Sintered Structures and Properties of a P/M Austenitic Stainless Steel, *Modern Developments in Powder Metallurgy*, Vol 13, Metal Powder Industries Federation, Princeton, NJ, 1980, p 27-50
8. Arias, A., "Oxide Dispersion Strengthened Nickel Produced by Nonreactive Milling," NASA TMX-3331, Jan 1976
9. Arias, A., "Chemical Reactions of Metal Powders With Organic and Inorganic Liquids During Ball Milling," NASA TN D-8015, Sept 1975
10. Benjamin, J.S. and Volin, T.E., The Mechanism of Mechanical Alloying, *Met. Trans.*, Vol 5, Aug 1974, p 1929
11. Rose, E. and Sullivan, R.M.E., *A Treatise on the Internal Mechanics of Ball, Tube, and Rod Mills*, Chemical Publishing Co., New York, 1958
12. U.S. Patent 4 172 720, Oct 1979
13. British Patent 1 224 736, Jan 1969
14. U.S. Patent 4 115 107, Sept 1978
15. Gilman, P.S. and Nix, W.D., The Structure and Properties of Aluminum Alloys Produced by Mechanical Alloying: Powder Processing and Resultant Powder Structures, *Met. Trans.*, Vol 12A, May 1981, p 813
16. Babich, B.N., *et al.*, Production of a Dispersion-Strengthened Nickel-Tungsten-Aluminum Alloy By Mechanical Alloying, *Poroshkovaya Metall.*, Vol 235 (No. 7), July 1982, p 44-47
17. Benjamin, J.S. and Bonford, M.V., Dispersion Strengthened Aluminum Made by Mechanical Alloying, *Met. Trans.*, Vol 8A, Aug 1977, p 1301
18. Webster, O., Effect of Lithium on the Mechanical Properties and Microstructure of SiC Whisker Reinforced Aluminum Alloys, *Met. Trans.*, Vol 13A, Aug 1982, p 1516
19. Domazer, H.G., New Milling Process for Fine Grinding of Powder Alloys in a Gas-Tight Attrition Mill, *Powder Metall. Int.*, Vol 10 (No. 3), 1978, p 150

Electrodeposition of Metal Powders

By Pierre W. Taubenblat
Vice President and Associate Director
AMAX Base Metals Research & Development, Inc.

ELECTRODEPOSITION OF MET-ALS from aqueous solutions produces a variety of metal powders. This article discusses the methods used to produce powders electrolytically, and describes the types of metal powders made and their physical and chemical characteristics. Because of the limited number of metals processed electrolytically and the small quantity of material produced by this method, a complete analysis of the electrolytic process is beyond the scope of this article. Detailed information on the processing variables that permit close control of the chemical and physical properties of electrolytic powders can be found in the articles "Production of Iron Powder," "Production of Copper Powder," and "Production of Precious Metal Powders" in this Volume.

There are two practical methods of obtaining powder by electrodeposition: (1) direct deposition of a loosely adhering powdery or spongy deposit that can easily be disintegrated mechanically into fine particles, and (2) deposition of a dense, smooth, brittle layer of refined metal that can be ground into powder. Choice of production method depends mainly on the metal used. For example, copper and silver produce powdery or spongy cathode deposits. Conversely, iron and manganese produce coherent cathode deposits; because these deposits will be crushed and ground into powder, it is highly desirable that they be brittle. Brittleness of the cathode deposit can be achieved by proper control of the electrolytic cell conditions.

Although electrodeposition produces a high-purity powder with excellent properties for conventional P/M processing, current usage of electrolytically produced powders is limited, because the produc-tion involves the control and manipulation of many variables and sometimes is significantly more expensive than other techniques. Currently, only iron, copper, and silver powders are produced commercially to any extent by electrodeposition.

Direct Deposition of Powder or Sponge

Direct deposition of powder or sponge on the cathode is achieved by controlling the composition (concentration of metal and pH), temperature, and rate of circulation of the electrolyte; current density; size and type of anode and cathode and their distance from each other; quantity and type of addition agent; and removal of deposits at the cathode (brush-down interval). These and other variables are discussed elsewhere in this Volume.

The shape of electrolytically produced powder particles depends on the metal deposited and the operating conditions. Electrolytic copper deposited from a sulfuric acid electrolyte forms dendritic (fernlike) particles, as shown in Fig. 1. Silver electrolytic powder deposited from a silver nitrate solution generally is coarse and regularly shaped (Fig. 2). Electrolytic iron, copper, and silver are the most widely used powders made by this technique, but tin, chromium, beryllium, antimony, cadmium, lead, and zinc powders also have been produced by electrodeposition.

Electrolytic copper powder has been produced commercially for at least 70 years; the original process was developed in France. Modern practice, which utilizes a copper sulfate/sulfuric acid electrolyte, closely follows the original procedure.

Electrolytic copper production follows the same electrochemical principles that apply to copper refining. However, the conditions of electrodeposition are modified so that a powdery or spongy material

Fig. 1 Electrolytic copper powder, showing dendritic structure
Magnification: 85×

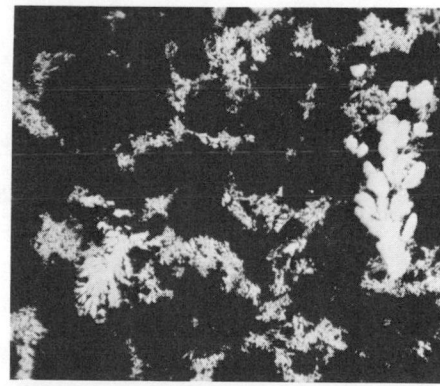

Fig. 2 Electrolytic silver powder
Magnification: 60×

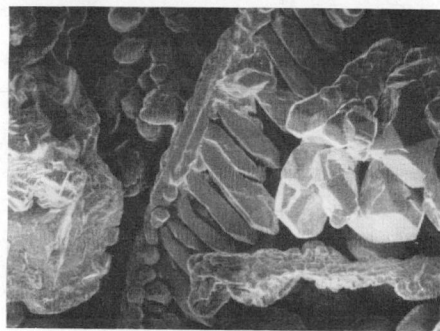

Fig. 3 Scanning electron micrograph of electrodeposited iron sheet, indicating dendritic grain structure
Magnification: 25×

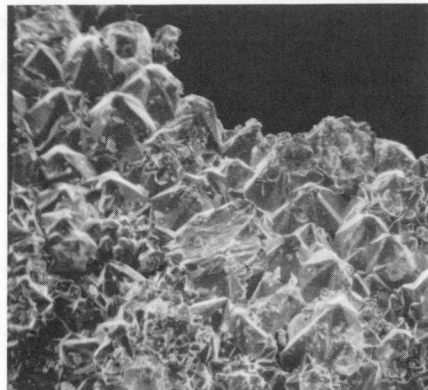

is deposited on the cast lead antimonial cathodes instead of the smooth, adherent deposit desired in electrorefining. Formation of powdery deposits is facilitated by low copper ion concentration and high acid content in the electrolyte. High cathode current density and use of an electrolytically refined copper anode (instead of the impure cast copper anode used in refining operations) also contribute to powder formation.

Powder characteristics can be altered to a considerable extent by control of the operating conditions. Further size control can be achieved by use of addition agents such as colloids, which promote the formation of finer powders. The dendritic structure

of copper particles is shown in Fig. 1. Details of the process variables and properties and applications of electrolytic copper powders are discussed in the article "Production of Copper Powder" in this Volume.

Electrolytic silver powder can be made by using a silver nitrate electrolyte and pure cast silver anodes. The powder particles deposited on the cathode are generally regularly shaped (see Fig. 2) and coarse, with particle sizes ranging from 40 μm to several millimetres, depending on the operating conditions and current densities. Details of the electrolytic and chemical processes are discussed in the article "Production of Precious Metal Powders" in this Volume.

Zinc Powder. The most recent metal powder to be produced electrolytically is zinc powder, which is made by using an electrolyte consisting of sodium hydroxide and sodium zincate. Zinc content is maintained at about 55 g/L to yield a deposit of suitable particle size. Current density is about 1076 A/m^2 (100 A/ft^2). Under these conditions, the product is a loosely adherent sponge that can be easily scraped or brushed off the cathode and processed into powder.

Brittle Cathode Process

Metals characterized by high electrolytic polarization yield coherent deposits that can be made brittle under controlled operating conditions. Metals of this type include iron, manganese, nickel, and cobalt, with only iron and manganese being produced in commercial quantities.

Iron. Although electrolytic iron powders were made experimentally in the late 19th century, the first commercial production occurred in 1904. The electrolyte, composed of ferrous and ammonium sulfates, was maintained in a slightly alkaline condition. As a result, oxides were deposited along with the iron on the cathode, producing a brittle deposit that could be readily powdered. Steel anodes were used. The product was not particularly pure, but the application (electrical loading coils) did not require high purity. A later modification permitted the production of pure iron powder.

In modern practice, a chloride or sulfate electrolyte is used with soluble low-carbon steel or ingot iron anodes and stainless steel cathodes. Cathodes are stripped of deposited iron by flexing and striking them against a table top. The iron deposits are pulverized subsequently in ball or hammer mills, and the powder then is annealed in hydrogen to make it softer. The dendritic structure of the powder (see Fig. 3) is altered by crushing and ball milling to form irregularly shaped, flaky particles with a fairly rough surface. Details of this process can be found in the article "Production of Iron Powder" in this Volume.

Manganese powder is produced by electrodeposition from a purified manganese sulfate electrolyte using insoluble lead anodes and stainless steel cathodes. The brittle cathode deposit is readily removed. Because the powder is pyrophoric, it is ground into powder in ball or hammer mills under a protective atmosphere.

Spray Drying of Metal Powders

By David L. Houck
Advanced Research Engineer
Chemical and Metallurgical Division
GTE Products Corp.

SPRAY DRYING is a powder-producing process in which a slurry of liquids and solids or a solution is atomized into droplets in a chamber through which heated gases, usually air, are passed. The liquids are evaporated from the droplets and the solids are collected continuously from the chamber. References 1 to 15 discuss the fundamentals of the spray drying process in greater detail. Spray drying is used widely in the pharmaceutical, chemical, and food industries. It is used to a lesser extent in the metals-related industries.

Spray Drying Equipment and Process Variables

Figure 1(a) illustrates a spray drying installation that uses a centrifugal (rotating disk) atomizer. Figure 1(b) illustrates the use of a nozzle to create the droplets. When the nozzle is pointed upward, the process is termed "fountain spraying." The slurry or solution can also be introduced through a nozzle (pointed downward) at the top of the spray dryer. Gases (usually air) used in the spray drying chamber can be heated using either direct or indirect heaters. Gas and oil are used with direct heaters. Steam, waste heat, gas, oil, electricity, or a combination of these heat sources is used with indirect heaters.

Atomized droplets dry rapidly because of their high surface-to-volume ratio. Coarse dried solids fall to the bottom of the chamber and are continuously collected, typically through a rotary airlock valve. The finer solids, which are entrained in the gases exiting the chamber, are removed by and collected from a cyclone separator. Baghouse collectors or wet scrubbers can be used to clean ultrafine particles from the air exhausted from the cyclone. Baghouse collectors and electrostatic precipitators also can be substituted for cyclone collectors. An alternate spray dryer design includes the introduction of the air through ductwork above the cyclone exit.

When the directions of the gas and droplets are the same, drying is termed cocurrent. For example, when gas and slurry enter at the top of the drying chamber (Fig. 1a), cocurrent drying occurs. When they are opposite, it is termed countercurrent. Countercurrent fountain spray drying, as shown in Fig. 1(b), occurs when the slurry is introduced from the bottom and gas from the top of the dryer. Droplets dry both in their upward and downward trajectories. Therefore, narrower, shorter dryers can be used. However, if partially dried products cannot be exposed to the high temperatures at the top of the dryer, this mode of spraying may not be applicable.

In cocurrent drying (Fig. 1a), the maximum temperature to which larger dried particles are exposed is approximately 10 °C (18 °F) less than the outlet temperature of the dryer. This is caused by the evaporative cooling effect that occurs when the liquid leaves the droplet while passing through the heated zone. Finer particles, which dry more quickly, experience temperatures approaching those of the inlet gas.

Size of the drying chamber is dictated by (1) product characteristics, (2) product throughput, (3) atomization technique, (4) the properties of the slurry or solution, and (5) the evaporative capacity of the heating unit and associated fans. The evaporative capacity is a function of the allowable inlet and outlet temperatures of the dryer. These temperatures are, in turn, dictated by the heat sensitivity of the product. The interrelationship of these factors is shown in Fig. 2. Generally, a spray drying facility is designed as an integral unit so that all component capacities are compatible.

Some special applications in the cemented carbide industry require the use of organic solvents, such as acetone, ethanol, hexane, and methanol. Because of the flammable nature of these materials, closed-system drying technology (see Ref 17-19) requires that the medium be spray dried using a nonoxidizing gas, such as nitrogen. By recycling the nitrogen that contains organic vapors, organic emissions are eliminated, and the solvent can be recycled.

Closed-cycle systems are gastight and operate slightly above atmospheric pressure. Most of the dried particles are collected from the dryer chamber; a small percentage of fines that pass through the chamber in the gas stream is collected in the cyclone.

The drying gas subsequently passes through a scrubber/condenser to recover the solvent and to recondition the nitrogen. Typically, gas temperature is controlled by a recirculating scrubber and heat exchangers that are cooled by water or brine. For some spray drying applications, a cooling system may be required.

Control of Powder Properties

Spray drying can be used to remove water from a slurry or to create agglomerates of fine particles contained within the slurry. Agglomeration is achieved by using a binder. Discussion of the formulation of slurries is contained in the next section of this article.

Most of the applications of spray drying in the metals industry require the formation of free-flowing agglomerates. Many of these powders are used for producing pressed parts. Therefore, the agglomerate size distribution and bulk density are the

Fig. 1 Basic spray drying installations

(a) Centrifugal (rotating disk) atomizer. (b) Nozzle atomizer. (1) Feed pump, (2) feed pipe, (3) atomizer, (4) inlet fan, (5) air heater, (6) hot air duct, (7) air disperser, (8) drying chamber, (9) outlet duct, (10) duct, (11) cyclone, (12) discharger, (13) duct, (14) exhaust fan, (15) exhaust duct, (16) air hood, (17) instrument panel, and (18) fines recycling (optional). Source: Ref 16

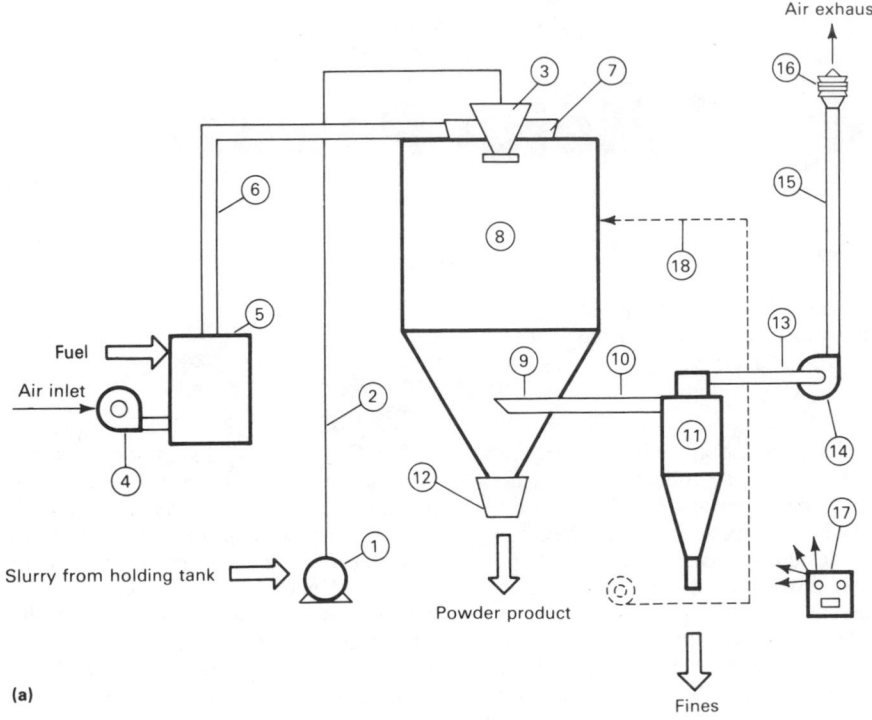

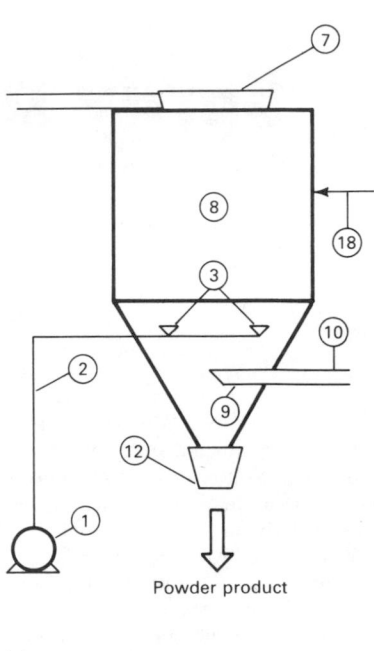

(a)

(b)

two most important properties of spray-dried powders.

Agglomerate size distribution is a function of atomization conditions and the properties of the slurry. Generally, a lower solids content yields a finer average agglomerate size. The maximum attainable solids content varies with material, but usually can be increased by using defloc-culating or suspending agents.

Bulk density is a function of the solids content of the slurry, inlet temperature, and slurry additives. Lower bulk densities generally are achieved from slurries with low solids contents. Also, excessive inlet temperatures can cause lower bulk densities. Rapid evaporation of the liquids causes the partially dried droplets to expand rapidly, thus decreasing density. Introduction of frothing agents may entrap air in the slurry, which also leads to lower bulk densities. Typically, frothing agents are not added in metallurgical applications.

Moisture content of a powder can be controlled by the inlet and outlet temperature of the spray dryer in conjunction with the slurry feed rate. Moisture levels below 0.1% are possible. For a given airflow and inlet temperature, outlet temperature is controlled by the rate of slurry feed and the solids content. With higher percentages of solids, less water must be evaporated, which leads to higher throughput of dry product. If a product can withstand higher inlet temperatures, throughput can be increased (see Fig. 3).

Slurry Formation

Suitable binder materials must be homogeneously dispersable (preferably soluble) in the liquid used to form the slurry. When dry, binders must form a coating and/or adhere to the material being agglomerated. They must impart the required strength and crush resistance to the particle for subsequent handling. In addition to the liquid, solids, and binders used to formulate a slurry, various other additives may be necessary. Typical components of spray drying slurries are:

Organic binders

- Polyvinyl alcohol
- Gum arabic
- Other natural gums
- Carboxy-methyl cellulose salts
- Polyvinyl acetate
- Methyl cellulose
- Ethyl cellulose
- Polyvinyl butyral dispersions
- Protein colloids
- Acrylic resin emulsions
- Ethylene oxide polmyers
- Water-soluble phenolics
- Lignin sulfonates
- Propylene glycol alginates
- Flour
- Starches

Inorganic binders

- Sodium silicate
- Boric acid
- Borax
- Carbonates
- Nitrates
- Oxylates
- Oxychlorides

Plasticizers

- Glycerine
- Ethylene glycol
- Triethylene glycol
- Dibutyl phthalate
- Diglycerol
- Ethanolamines
- Propylene glycol
- Glycerol monochlorhydrin
- Polyoxyethylene aryl ether

Fig. 2 Relationship between chamber volume and evaporation capacity of spray dryers
Data based on retention time of 20 s. Source: Ref 3

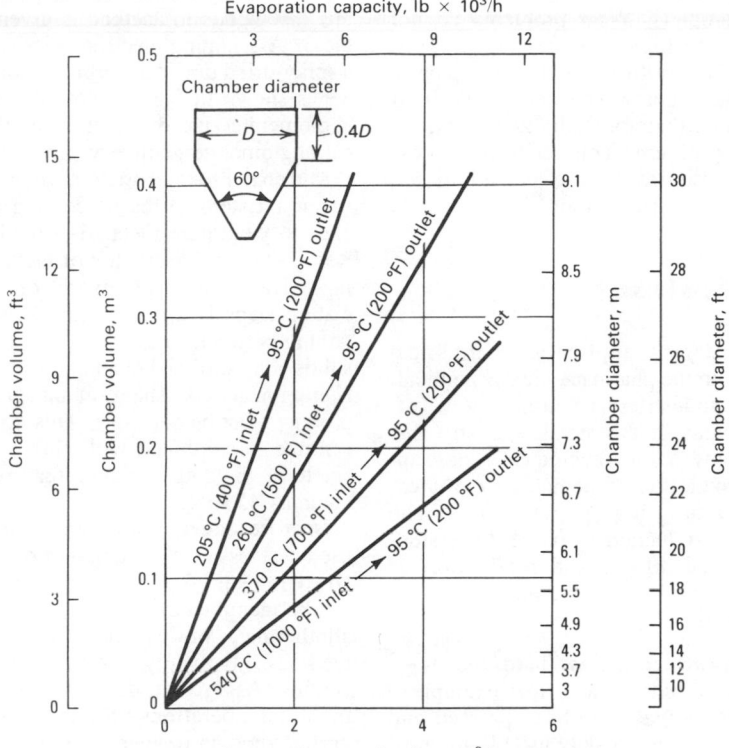

Suspending agents (high molecular weight)

- Sodium carboxymethyl cellulose
- Methyl cellulose
- Ethylene oxide polymers

Fig. 3 Relationship of evaporation capacity of spray dryers to production rate and slurry solids concentration
Source: Ref 3

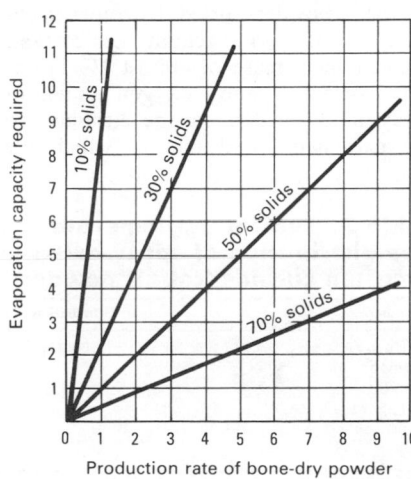

Deflocculating agents

- Sodium hexametaphosphate
- Sodium molybdenate
- Tetrasodium pyrophosphate
- Ammonium citrate
- Ammonium oxalate
- Ammonium tartrate
- Ammonium chloride
- Monoethylamine

Wetting agents

- Synthetic detergents
- Alkylaryl sulfonates
- Alkylaryl sulfates
- Soaps

Plasticizers may be used with binding materials that are hard or brittle and that tend to crack during drying. Suspending agents may be needed to prevent solids from settling within the slurry. Deflocculating agents aid in the formation of slurries by preventing the agglomeration of fine particles. Wetting agents also may be used to maintain solids in suspension. Some slurries have a tendency to foam during mixing. Antifoaming agents or defoamers may be used to control this action. Chemical activators also may be used as additives to aid in subsequent sintering or processing of powders.

Atomization Techniques and Agglomerate Size Distribution

Three standard techniques are used to atomize slurries for spray drying: (1) single-fluid nozzle atomization, (2) centrifugal (rotating disk) atomization, and (3) two-fluid nozzle atomization. Studies of the efficiency, agglomerate size distribution, and trajectory of droplets using these techniques have been conducted (Ref 5-9, 11, 12, 15, 16). Table 1 gives the advantages and disadvantages of these techniques for atomization of slurries. Table 2 gives the relative agglomerate sizes produced by these techniques. For a more detailed description of these techniques, see the article "Atomization" in this Volume.

As shown in Table 2, the largest agglomerate sizes (600 μm) are achieved by the single-fluid nozzle. The centrifugal (rotating disk) atomizer yields agglomerate sizes up to 300 μm, and the two-fluid nozzle produces agglomerates only up to about 200 μm in size. Centrifugal atomization yields the narrowest agglomerate size distribution range, followed by single-fluid atomization and two-fluid atomization.

Centrifugal atomization (Ref 5, 7-9, 16) entails the introduction of the slurry into a horizontally rotating disk that is equipped with vanes or holes through which the slurry exits. The slurry is atomized into fine droplets, the average size of which is a function of the design and peripheral speed of the wheel. Most wheel designs

Table 1 Advantages and disadvantages of various spray drying atomization systems

Atomization system	Advantages	Disadvantages
Centrifugal (rotating disk)	High feed rates; Less downtime; Low-pressure pumps; Never plugs	Larger diameter dryer required; Coarse agglomerate size not obtainable
Single-fluid nozzle	Large agglomerate capability; Smaller dryers can be used	Downtime due to part wear and plugging; High-pressure pumps
Two-fluid nozzle	Less part wear	Broad agglomerate size distribution; Compressed air needed

Table 2   Relative agglomerate sizes produced by various atomizing systems

Atomizing system	Mean agglomerate size, μm
Centrifugal (rotating disk)	
High speed	25-100
Medium speed	50-200
Low speed	100-300
Single-fluid nozzle	
High pressure	25-100
Medium pressure	50-200
Low pressure	100-300
Very low pressure	200-600
Two-fluid nozzle	
High pressure	10-50
Medium pressure	25-100
Low pressure	50-200

Source: Ref 20

incorporate a wear-resistant material for extended life.

Centrifugal atomization has several advantages, as listed in Table 1. Because the wheel openings are sufficiently large, they are unlikely to become plugged, thus ensuring continuity of atomization and minimal downtime. Rotary atomizers are capable of high feed rates; some atomization motors are capable of producing up to 800 hp.

Two major disadvantages of centrifugal atomization are: (1) coarse agglomerate sizes are not obtainable, and (2) a large-diameter dryer usually is required, as most drying must occur in a horizontal plane. Particles therefore must be dried prior to hitting the sidewalls of the chamber. Because of the high momentum imparted to the particles as they exit the wheel, the dryer diameter and the distributor design must be compatible with the required drying time.

Single-fluid atomization is capable of producing the largest diameter agglomerates. However, because of the high pressures required to force the slurry through a single, small orifice in order to produce the desired droplet size, considerable downtime may be required to replace worn parts and to unplug the nozzle. Use of a nozzle in the fountain spray position extends the residence time of the droplets in the chamber. As a result, smaller dryers can be used for single-fluid atomization of materials that are not heat-sensitive.

Two-fluid atomization, which uses a pressurized air blast to break up a slurry

stream into droplets, produces the widest range of agglomerate sizes of all three atomization techniques. However, it is a relatively easy technique for producing atomized droplets. Wear problems do not exist because high pressures are not employed. The orifices used are typically larger than those used for single-fluid atomization; therefore, plugging is not as severe a problem. This technique does, however, require the use and additional expense of pressurized air.

Applications

Spray drying applications are most prevalent in the pharmaceutical, chemical, and food industries. There are, however, several areas in the metals industry that utilize spray drying techniques. These include production of cemented carbides, mineral processing, production of iron powders, production of oxide-dispersion-strengthened alloys, and production of powders used for thermal spraying applications.

Cemented carbides (hard metals)—tungsten carbide/cobalt, for example—are materials that have been pressed and sintered. They are widely used in the production of cutting tools for machining and for tools used in oil drilling and coal mining. An estimated 8 million kg (17.6 million lb) of cemented carbide parts were produced in the United States in 1981.

Historically, cemented carbides have been pressed from powders produced by various agglomeration techniques. With the advent of closed-cycle systems, production of cemented carbides by spray drying has increased considerably. In 1982, an estimated 40% of these powders were produced by spray drying. Closed-cycle spray drying is required for most cemented carbide powders because the binders that are used are soluble only in volatile organic fluids. Because of the proprietary and highly competitive nature of the cemented carbide industry (Ref 21), limited data are available on the spray drying of these materials (Ref 17-19).

The nitrogen drying gas that is used in the spray drying of cemented carbides is heated to 75 to 100 °C (170 to 210 °F), depending on the milling liquid used. The solids content of the slurry varies from 75 to 80%. Viscosity of milled slurries is sometimes modified with stabilizers, such as stearic acid (0.3 to 0.5 wt%). Pressures for single-fluid nozzle atomization range from 590 to 1470 kPa (85 to 213 psi). Pressure is a function of the particular type

of powder, slurry viscosity, and binder content.

Agglomerate size distribution of a spray-dried tungsten carbide/cobalt powder using 78% solids in acetone is given in Table 3. A scanning electron micrograph of a spray-dried tungsten carbide/cobalt powder is shown in Fig. 4. Note the uniform agglomerate size distribution and spherical agglomerate geometry produced.

Mineral Processing. One of the most important steps in the processing of minerals to yield pure metals is smelting (see Ref 22-24 for a discussion of nickel smelting). The product of several ore beneficiation steps is a concentrate slurry containing between 71 and 74% solids. These solids consist of nickel and copper sulfide-bearing materials. The resultant spray-dried product must be bone dry. This dried concentrate is combined with flux and flash smelted, yielding a matte that typically contains 45% Ni.

Iron Powders. Spray drying has been used as a less costly method of drying iron powders (Ref 25). These powders, which can be manufactured by electrolytic, metallothermic, or chemical metallurgical methods, involve hydrometallurgical processing. As a result, dehydration is one of the final operations. Some methods of drying yield an oxygen content as high as 2%. Consequently, this drying step must be followed by a hydrogen anneal to reduce the oxygen level. Conventional methods of dehydration and drying are expensive and difficult to mechanize. As a result, hydrometallurgical processing of metal powders is not recommended. However, if an automated method such as spray drying is used, these methods could be of potential use.

Barak *et al.* (Ref 25) showed that the oxygen content of iron powders spray dried from aqueous solutions did not increase from their initial content of 0.64%. In contrast, powders dried by other techniques, such as vacuum desiccation, showed an increase of almost 1% in oxygen content. The low oxygen content of spray-dried powders can be attributed to the extremely rapid rate at which the

Table 3   Typical agglomerate size distribution of spray-dried tungsten carbide/cobalt powder

Agglomerate size, μm	Distribution, %
>315	Negligible
200-315	16
125-200	60
90-125	13
63-90	8
<63	3

Fig. 4 Tungsten carbide/11% cobalt powder containing 2% wax as a binder

Spray dried with a single-fluid nozzle. Magnification: 50×. Courtesy of GTE Products Corp.

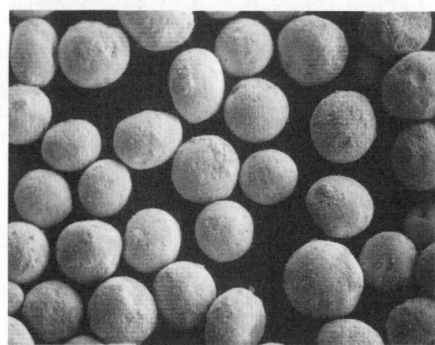

moisture is removed. Processing time within a small spray dryer may be only 5 s.

Oxide-Dispersion-Strengthened Alloys. Cheney and Smith (Ref 26) have described a process for the production of thoria dispersion-strengthened matrices by the selective reduction of spray-dried mixtures. Among the various matrices employed were Ni-15Mo, Ni-20Cr, Co-15Mo, and Co-20Cr-11Ni-8Mo.

Production of these powders involved spray drying a solution that contained all of the alloy components. For the thoria-dispersed nickel-molybdenum, an aqueous ammonium oxylate solution was used. Appropriate amounts of nickel oxylate, molybdenum trioxide, and thorium nitrate were dissolved in this solution. After spray drying, the powder consisted of a crystalline solid containing ammonium oxylate and metal oxylates. On heating, the ammonium oxylate was driven off, and metal oxides formed. After additional closely controlled heating in hydrogen, nickel and molybdenum oxides were reduced to pure metals. The thorium oxide particle size could be varied from 10 to 50 nm (100 to 500 Å) by this treatment.

This method of forming a uniform dispersion of the alloying elements in an oxide-dispersion-strengthened material is successful because of the high level of homogeneity in the initial slurry. Homogeneity is maintained throughout powder production, compaction, and sintering.

Thermal Spray Powders. Reference 27 provides a thorough review of spray drying techniques used to produce powders for thermal spray applications. Spray drying is an ideal process for the manufacture of these materials, because thermal spray processes inherently require a narrow particle and/or agglomerate size distribution to achieve high deposit efficiency. Dittrich (Ref 27) describes techniques that use conventional binders. Cheney et al. (Ref 28, 29) also have described the use of spray drying for the production of plasma spray powders.

Laferty et al. (Ref 30) have described the use of a binder for the production of a molybdenum plasma spray powder. It contains ammonium paramolybdate, a water-soluble salt of molybdenum. This binder is formed by dissolving molybdenum trioxide in a slurry containing water, num. On drying, the dried ammonium paramolybdate serves as a binder for the fine molybdenum particles. On subsequent hydrogen reduction, the molybdate binder decomposes to pure molybdenum, giving off ammonia and water vapor. Consequently, the powder is not contaminated by any residue caused by decomposition of a more conventional carbonaceous binder. The resultant high-density, high-purity material is an excellent plasma spray powder.

Figure 5 is a scanning electron micrograph of molybdenum plasma spray powder that has been centrifugally (rotating disk) atomized using the type of binder described above. These agglomerates have been hydrogen reduced and screened to the desired agglomerate size distribution (−200+325 mesh) for plasma spray applications. Agglomerate size distribution of the as-spray-dried product is typically:

Mesh No.	Agglomerate size distribution, %
+170	18
−170+200	20
−200+270	25
−270+325	19
−325	18

Advantages of Spray Drying

Spray drying offers several advantages over other powder-processing techniques, particularly in applications requiring agglomerates for subsequent pressing and sintering operations.

Spray drying is a continuous rather than a batch process that allows close control of agglomerate size, bulk density, and moisture content. Agglomerate shape is spherical, facilitating excellent flowability. Particles have very short exposure time. Residence time within the dryer may range from 2 to 20 s, depending on dryer size. Therefore, heat-sensitive materials can be

Fig. 5 Molybdenum spray dried with a centrifugal atomizer and partially sintered

Yields free-flowing, spherical agglomerates. Magnification: 250×. Courtesy of GTE Products Corp.

spray dried. Lubricating additives can be added easily for die pressing applications.

Spray drying is one of the most economical ways of drying slurries. Although the cost of equipment is high, the overall cost of spray drying is relatively low compared to other processes (Ref 19, 31-33). Oil and power consumption for various sizes of spray drying chambers is a function of evaporative capacity (Ref 20).

REFERENCES

1. Masters, K., *Spray Drying Handbook*, 3rd ed., John Wiley & Sons, New York, 1979
2. Sherrington, P.J. and Oliver, R., Spray Drying, in *Granulation, A Monograph in Powder Science and Technology*, Heyden, New York, 1981, p 122-139
3. Perry, R.H. and Chilton, C.H., Spray Dryers, in *Chemical Engineers Handbook*, 5th ed., McGraw Hill, New York, 1973, p 58-63
4. Katta, S. and Gauvin, W.H., Some Fundamental Aspects of Spray Drying, *Am. Inst. Chem. Engineers J.*, Vol 21 (No. 1), 1975, p 143-152
5. Gavioli, G. and Palmonari, C., A Theoretical Approach to Particle Trajectory in Mixed-Flow Spray Dryers, *J. Am. Ceram. Soc.*, Vol 59 (No. 7-8), 1976, p 290-294
6. Marshall, W.R., Jr. and Seltzer, E., Principles of Spray Drying, *Chem. Eng. Prog.*, Vol 46 (No. 10), 1950, p 501-508; Vol 46 (No. 11), 1950, p 575-584
7. Adler, C.R. and Marshall, W.R., Jr., Performance of Spinning Disk Atomizers, *Chem. Eng. Prog.*, Vol 47

(No. 10), 1951, p 515-522; Vol 47 (No. 12), 1951, p 601-608

8. Friedman, S.J., Gluckert, F.A., and Marshall, W.R., Jr., Centrifugal Disk Atomization, *Chem. Eng. Prog.*, Vol 48, 1952, p 181-191

9. Herring, W.M., Jr. and Marshall, W.R., Jr., Performance of Vaned-Disk Atomizers, *Am. Inst. Chem. Engineers J.*, Vol 1, 1955, p 200-209

10. Duffie, J.A. and Marshall, W.R., Jr., Factors Influencing the Properties of Spray Dried Materials, *Chem. Eng. Prog.*, Vol 49 (No. 8), 1953, p 417-423, 480-486

11. Gretzinger, J. and Marshall, W.R., Jr., Characteristics of Pneumatic Atomization, *Chem. Eng. Prog.*, Vol 7 (No. 1), 1961, p 312-318

12. Marshall, W.R., Jr., Atomization and Spray Drying, *Chem. Eng. Prog.*, Monograph Series, Vol 50, 1954, p 135

13. Marshall, W.R., Jr., Heat and Mass Transfer in Spray Drying, *Trans. ASME*, Vol 75 (No. 11), 1955, p 1377-1385

14. Crosby, E.J. and Marshall, W.R., Jr., Effects of Drying Conditions on the Properties of Spray Dried Particles, *Chem. Eng. Prog.*, Vol 54 (No. 7), 1958, p 56-63

15. Dickinson, D.R. and Marshall, W.R., Jr., The Rates of Evaporation of Sprays, *Am. Inst. Chem. Engineers J.*, Vol 14 (No. 4), 1968, p 541-552

16. Lee, D.A., Comparison of Centrifugal and Nozzle Atomization in Spray Dryers, *Ceram. Bull.*, Vol 53 (No. 3), 1974, p 232-233

17. Jensen, A.O. and Masters, K., Spray Dryer for Producing Tungsten Carbide Products, Bulletin F-125, Niro Atomizer, Inc., Columbia, MD

18. Closed Circuit Spray Drying Systems, Bulletin 1342, Anhydro, Inc., Attleboro Falls, MA

19. Closed Cycle Systems, Bowen Engineering Bulletin, Bowen Engineering, Inc., Somerville, NJ

20. Spray Drying Processing Plants, Bulletin A-2-ENG, Anhydro, Inc., Attleboro Falls, MA

21. Strube, F.E., Much of the World's Prosperity Depends Upon Carbides, translation of reprint from the Danish Technical Review, *Ingeniorog Byginngsvaesen*, Vol 24 (No. 12-67), March 1972

22. Hoppe, R., Selebi-Phikwe: Concentrating and Smelting Nickel Copper Ore, *Eng. Mining J.*, Vol 181 (No. 5), 1980, p 74-82, 85, 86

23. Kleenman, C., Kambaldas' Nickel Concentrator, *Austral. Miner*, April 1970

24. Greenvale, the First Australian Producer of Nickel from Lateritic Ores, *Mining Mag.*, Vol 129 (No. 6), 1973

25. Barak, B.A., Solov'eva, Z.V., Ku-dryavtsev, E.S., and Sabinin, P.G., Spray Drying of Metal Powders, *Researches in Powder Metallurgy*, Consultants Bureau, New York, 1971, p 124-126

26. Cheney, R.F. and Smith, J.S., Oxide-Strengthened Alloys by the Selective Reduction of Spray-Dried Mixtures, in *Oxide Dispersion Strengthening*, Gordon and Breach, New York, 1968, p 637-674

27. Dittrich, F.J., Flame Spray Powders and Process, U.S. Patent 3 617 358, 1971

28. Cheney, R.F., Lafferty, W.D., and Long, G.J., Tungsten Carbide-Cobalt Flame Spray Powder and Method, U.S. Patent 4 025 334, 1977

29. Cheney, R.F., Port, D.J., and Spencer, J.R., Free Flowing, Sintered Refractory Agglomerates, U.S. Patent 3 881 911, 1975

30. Laferty, J.M., Jr., Ritsko, J.E., and Port, D.J., Free Flowing Powder and Process for Producing It, U.S. Patent 3 973 948, 1976

31. Cook, E.M., Estimating Spray Drying Costs, *Chem. Eng. Prog.*, Vol 62 (No. 6), 1966, p 93-97

32. Quinn, J.J., Jr., The Economics of Spray Drying, *Ind. Eng. Chem.*, Vol 57 (No. 1), 1965, p 35-37

33. Moeller-Hausen, K., Spray Drying Ceramics, *J. Canadian Ceram. Soc.*, Vol 44 (No. 2), 1975

Production of Iron Powder

IRON POWDER represents the largest tonnage of raw materials used in P/M fabricating. Iron powder leads in commercial P/M applications because (1) it is inexpensive to produce compared to nonferrous metals; (2) it possesses superior properties, particularly strength, compared to other metals or nonmetals; (3) it has a favorable strength-to-weight-to-cost ratio; (4) it alloys readily, especially with carbon, and consequently possesses all of the attributes of the iron-carbon system (including heat treatability); and (5) it is available in sufficient quantity. Iron powder is not limited to the manufacture of P/M parts. About one third of all iron powder produced is used for applications such as welding rods, flame cutting, food enrichment, and electronic and magnetic applications.

Uses of iron powder and iron P/M parts are discussed in the "Powder Systems and Applications" Section of this Volume. This article covers the various commercial methods of iron powder production, including reduced-iron powders, atomized iron powders, carbonyl iron powders, electrolytic iron powders, and iron powders produced by fluidized bed reduction. Of these methods, atomizing supplies the largest percentage of iron powders, followed by reduction of oxides, which is the oldest process for production of iron powders.

Production of Iron Powder by the Reduction of Iron Oxide

By Fritz V. Lenel
Professor Emeritus
Department of Materials Engineering
Rensselaer Polytechnic Institute

PRODUCTION OF IRON POWDER by the reduction of iron oxide, and in particular the reduction of iron ore by carbon, is the oldest method of producing iron powder. The Swedish sponge iron process, which was developed in Höganäs, Sweden in the early 1900's, was originally intended to produce metallic iron in sponge form as the raw material for steelmaking. A modified version of the process is still one of the important industrial methods of producing iron powder and is practiced in Sweden and in the United States. A similar adaptation also is used in the Soviet Union.

The Swedish sponge iron process is one of many developed for the direct reduction of iron ore to metallic iron at temperatures below the melting point of iron. However, metallic irons produced by most of these direct reduction processes are not suitable for use as iron powder in the production of P/M iron and steel products, because they contain too high a percentage of impurities. Also, they do not have the required physical and mechanical properties, such as particle size and distribution, particle shape, internal particle porosity, and lack of strain hardening.

Process Conditions

The Höganäs process uses pure magnetite (Fe_3O_4) ore found in northern Sweden, which has an iron content of approximately 71.5% after beneficiation by grinding and magnetic separation. The impurities present in small amounts are not in solid solution in the oxide, but exist as discrete phases. The ore is of consistent quality and is available in sufficient quantities for continued usage.

Coke breeze or another carbon source that provides the reducing agent is required to produce sponge iron powder. Additionally, limestone is used to react with the sulfur contained in the coke and to prevent its inclusion in the iron powder as an impurity. The ore is prepared for reduction and drying in a rotary kiln, followed by magnetic separation (Fig. 1a).

Fig. 1 Flowchart of Höganäs process

(a) Preparation of ore and of coke-limestone mixture and charging into ceramic tubes. (b) Tunnel kiln in which the ground ore is reduced to sponge iron. (c) Annealing furnace for sponge iron. (d) Packaging and mixing of annealed sponge iron powder

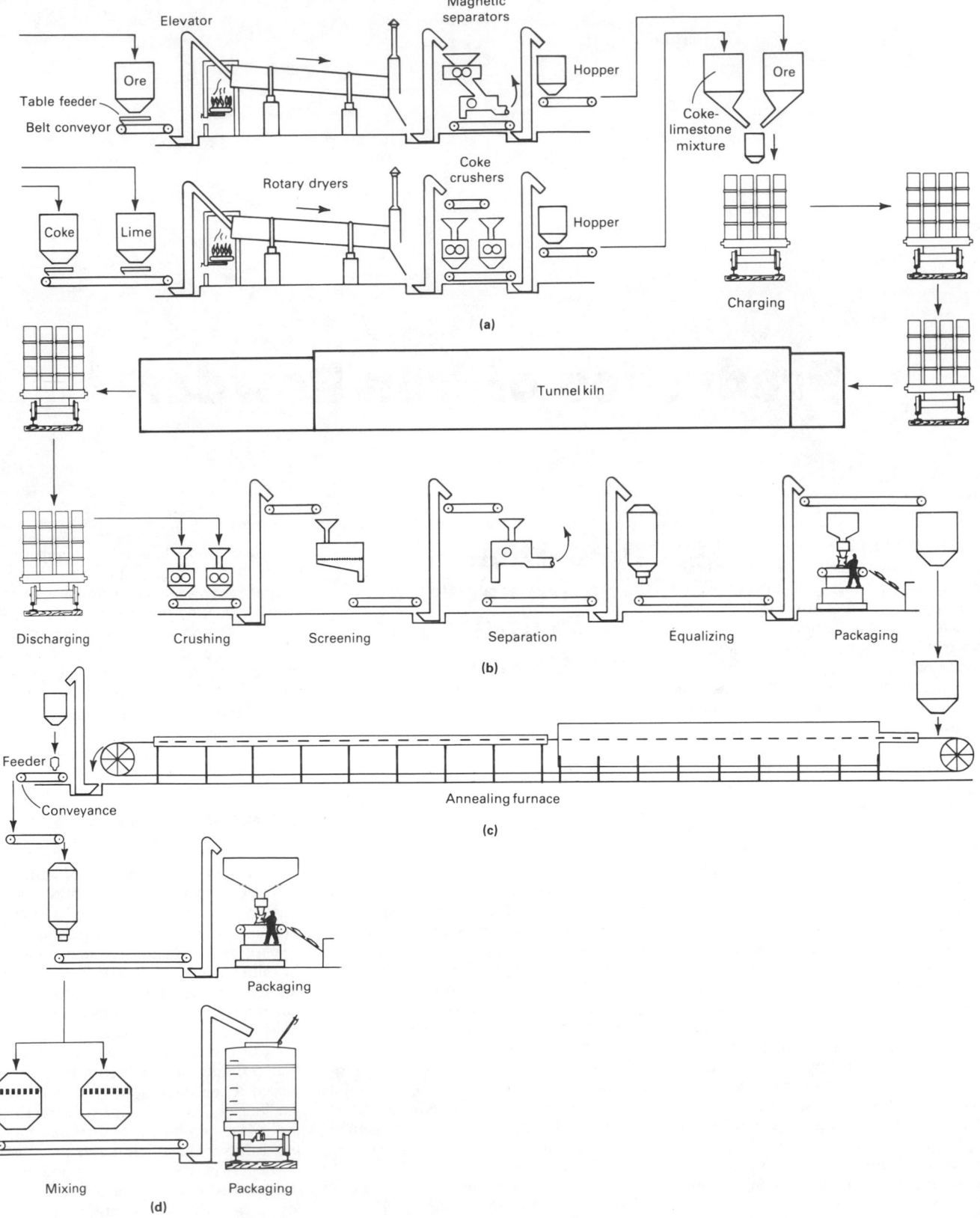

Coke and limestone are mixed in the proportion of 85% coke and 15% limestone. This mixture also is dried in a rotary kiln and then crushed to uniform size (Fig. 1a).

The ore and the coke-limestone mixture are then charged into ceramic tubes, which consist of silicon carbide (Fig. 1a). A pair of concentric steel charging tubes is lowered to the bottom of the ceramic tubes. The ore is fed between the steel tubes. The coke-limestone mixture is fed within the inner of the two concentric charging tubes and between the outer charging tube and the inner wall of the ceramic tube. The charging tubes are then withdrawn from the ceramic tube, leaving the ore and the reduction mixture in contact with one another, but not intermixed.

Charged ceramic tubes are placed on kiln cars that carry them into the kiln (Fig. 1b). Ceramic tubes are arranged in rows of six tubes by six tubes per kiln car. Cars are pushed into a 170-m (560-ft) long tunnel kiln, where the reduction occurs. A total of 60 kiln cars, carrying 36 ceramic tubes each, travel through the kiln, bumper to bumper. Every 53 min, one car is pushed into one end of the kiln, while another car is removed at the other end. The total time that a car is in the kiln is about 68 h. Gas burners heat 150 m (500 ft) of the tunnel to a temperature of 1260 °C (2300 °F); the remaining 19 m (62 ft) are cooled with circulating air. The tops of the kiln cars form the bottom of the heating chamber.

Within the hot zone of the tunnel kiln, several chemical reactions occur. Decomposition of the limestone generates carbon dioxide, which oxidizes the carbon in the coke to form carbon monoxide. Carbon monoxide reacts with magnetite to form additional carbon dioxide and ferrous oxide. The ferrous iron oxide is further reduced by the carbon monoxide to metallic iron. Metallic iron particles are sintered together to form sponge iron cake. Additionally, a reaction occurs between the sulfur contained in the coke and the limestone admixed to the coke.

The thermodynamics and the kinetics of the process by which iron oxide is reduced to metallic iron have been studied repeatedly, but it is difficult to apply the results of these studies to the conditions found in the Höganäs sponge iron process. When the material in the ceramic tubes has been cooled to room temperature at the end of the reduction process, 96% of the oxide has been reduced to iron. The iron contains 0.3% C. The reduced iron particles sinter together to form a hollow cylinder of sponge iron that has pores between and within the iron particles.

Figure 2 shows the cross section of +40 mesh ground magnetite powder. Figure 3 is a micrograph of the cross section of +40 mesh sponge iron powder obtained by reduction of the magnetite at 1250 °C (2280 °F). While the magnetite particles are dense, those of the reduced iron powder are highly porous. Figure 4 illustrates sponge iron particles etched with 2% nital after polishing. Residual unreduced oxide is evident in Fig. 4. Also, etching highlights the very small grain size of the iron produced by reduction of the oxide.

After removal from the kiln, cars are transferred to the discharge section (Fig. 1b). Cylindrical cakes of sponge iron are removed from the ceramic tubes and dropped into a tooth crusher, where they are broken into pieces approximately 25 mm (1 in.) in diameter. The emptied ceramic tubes are cleaned and transferred to the charging station, where they are reloaded for successive reduction runs.

The 25-mm (1-in.) diameter sponge iron lumps undergo a series of grinding, mag-

Fig. 2 Cross section of +40 mesh ground magnetite
Magnification: 30×

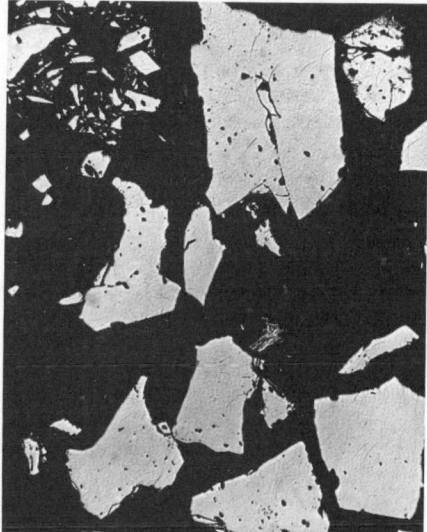

Fig. 3 Cross section of +40 mesh sponge iron obtained by reduction of ground magnetite
Magnification: 125×

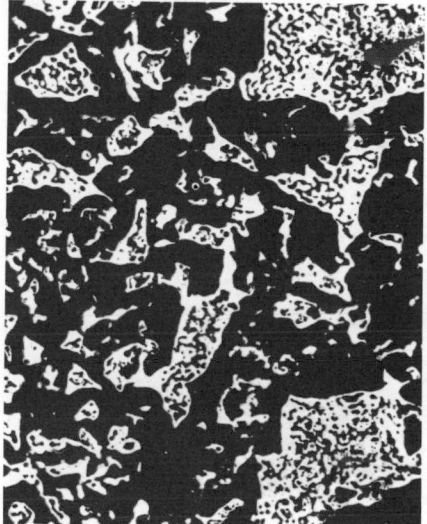

Fig. 4 Cross section of sponge iron particles
Etched with 2% nital. Magnification: 180×

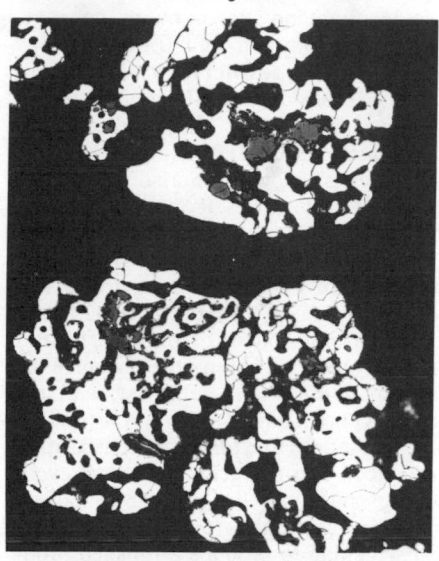

Fig. 5 Scanning electron micrograph of particles of annealed sponge iron powder
Magnification: 180×

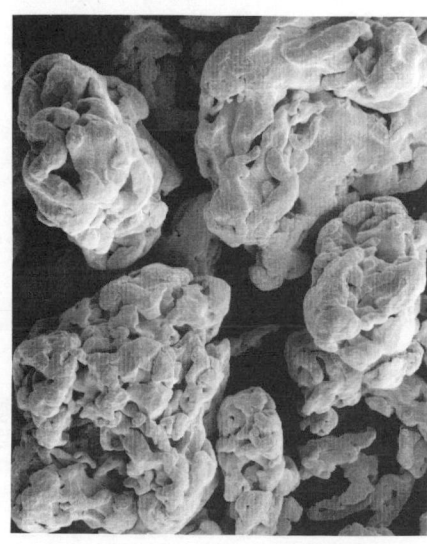

netic separation, and screening steps (Fig. 1b) to produce a powder of the desired particle size distribution and to remove nonmagnetic impurities, while preserving the porous nature of the powder particles that makes them well suited for P/M fabrication. Consequently, grinding is continued only until about 65% of the powder passes through a +100 mesh screen. For P/M purposes, however, a −100 mesh powder is generally required. Oversize powder is separated by screening and is used for coating welding electrodes and other non-P/M applications.

Only minor changes in chemical composition of the powder occur during grinding, magnetic separation, and screening. The resultant powder contains approximately 1% oxygen and 0.3% carbon. However, the powder has been strain hardened considerably. Consequently, it is annealed in a belt furnace that is approximately 55 m (180 ft) long (Fig. 1c).

Powder is loaded onto a solid stainless steel belt that carries it into the furnace through a powder lock arrangement. This mechanism seals the entrance of the furnace and prevents air from entering. The powder is heated to about 870 °C (1600 °F) in an atmosphere of dissociated ammonia. During annealing, the carbon in the powder is essentially removed, and the oxygen content is reduced from 1% to about 0.30%. At the same time, strain hardening introduced during grinding is relieved. During annealing, the powder is loosely sintered, but requires only light grinding and screening to produce a finished prod-

uct ready for packaging and shipping to P/M fabricators (Fig. 1d).

Reduction and annealing steps in the production of sponge iron are essentially continuous, but there are several holding areas between operations. These holding areas assist in producing a powder of uniform quality through an "equalizing" process, which consists of recirculating the powder by withdrawing it from the bottom of the containers in the holding areas and feeding it back at the top.

Figure 5 is a scanning electron micrograph of Hoeganaes Ancor MH-100 iron powder particles. Irregular shape of the particle surfaces and particle porosity are visible. Typical powder properties are:

Apparent density	2.58 g/cm³
Composition, %	
Iron	98.2
Silicon dioxide	0.20
Carbon	0.01
Hydrogen loss	0.26
Sulfur	0.01
Phosphorus	0.01
Green density at compacting pressure of 414 MPa (30 tsi)	6.42 g/cm³
Green strength at compacting pressure of 414 MPa (30 tsi)	16.5 MPa (2400 psi)

Screen analysis (U.S. mesh)

Mesh No.	Percent
+80 mesh	Trace
−80+100	1
−100+150	18
−150+200	26
−200+250	9
−250+325	24
−325	22

Production of Iron Powder by the Pyron Process

By Harold T. Harrison
President
Pyron Corp.

THE PYRON PROCESS provides an alternative for the production of iron powders by the reduction of oxides. Instead of using ground iron ore as the raw material for reduced iron powders, the Pyron process uses mill scale taken from steel mills that produce plain carbon steel products, such as sheet, rod, wire, plate, and pipe. Mill scales containing alloying elements other than manganese are not used. A flowchart of the Pyron process is shown in Fig. 6.

Process Conditions

Mill scales from different mills are mixed by layering in a bed that is large enough to supply the plant with a constant feed for up to 3 months. Scale is scalped as it is fed to the plant to remove large, undesirable objects. It is then magnetically cleaned to remove any sand, dirt, or other nonmagnetic material. The mill scale is then ground in a continuous ball milling operation to −100 mesh. Careful control of this

Fig. 6 Flowchart of Pyron iron powder process

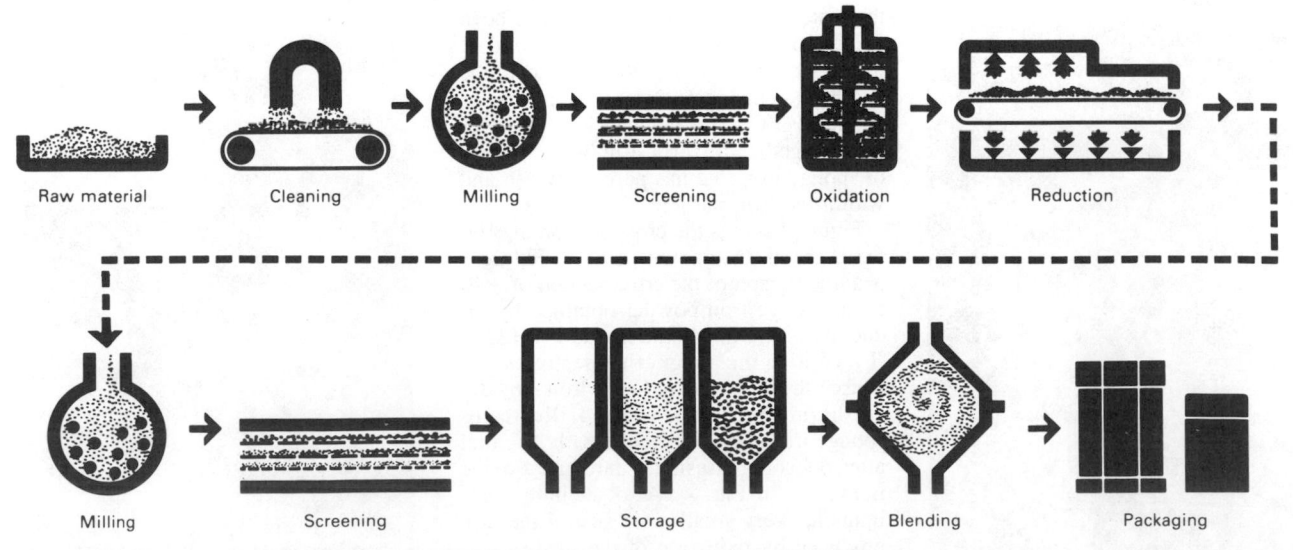

Raw material   Cleaning   Milling   Screening   Oxidation   Reduction

Milling   Screening   Storage   Blending   Packaging

Fig. 7 Scanning electron micrograph of Pyron iron particles
Magnification: 450×

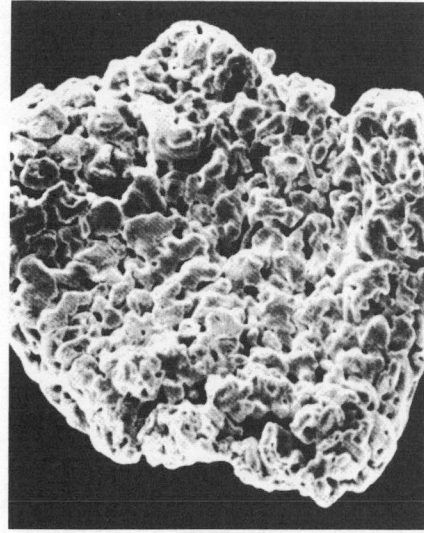

Fig. 8 Optical micrograph of cross section of Pyron iron
Magnification: 750×

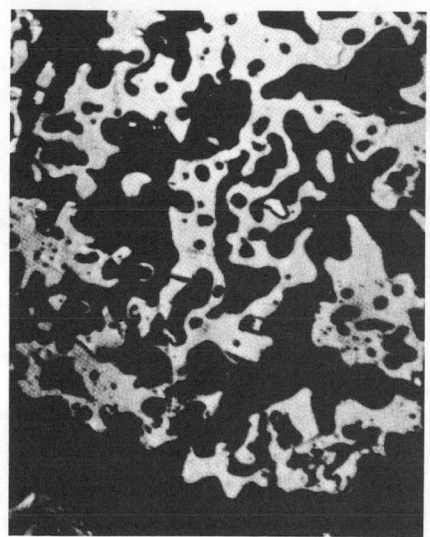

Table 1 Physical properties of Pyron iron powders

Properties	P-100, D-63	LD-80	R-80	R-12	AC-325
Apparent density, g/cm³	2.3-2.5	1.75-2.10	1.0-1.5	1.0-1.5	2.2-2.5
Hall flow rate, s/50 g	27-35	35-poor	Poor	Poor	Poor

Screen analysis, % Tyler mesh No.	Particle size, μm					
+20	850	...	...	...	2 max	...
+35	425	...	...	...	10-20	...
+60	250	...	...	...	20-30	...
+80	180	Trace	2 max	2 max	10-20	...
+100	150	2 max	1-12	1-12	5-15	Trace
+150	106	10-15	15-30	15-30	10-20	Trace
+200	75	15-25	15-30	15-30	5-15	0.2
+325	45	25-40	20-40	20-40	3-10	5 max
−325	<45	28-45	15-35	15-35	12 max	95 min

Table 2 Properties and uses of Pyron hydrogen-reduced iron powders

Property	Uses
High strength-to-weight ratio	Low-density, high-strength iron bearings and structural parts requiring a high percentage of interconnecting porosity for oil retention
High green strength	Low-density fragile parts; additives in blend of other iron powders to increase green strength. Friction material compounds
Low apparent density	Friction material compounds; additive in blends of other iron powders to reduce apparent density
Dimensional stability	Close tolerance, low- and medium-density iron-copper or iron-copper carbon bearings and structural parts requiring good strength, machinability, and hardenability

operation ensures desired particle size distribution.

Oxidation at about 980 °C (1800 °F) converts the iron oxides that are present in the mill scale, FeO and Fe_3O_4, to ferric oxide (Fe_2O_3). This operation is critical to the Pyron process and is necessary to ensure the uniform properties of Pyron iron powders. Gas-fired, multiple-hearth roasters are used for this operation.

Hydrogen reduction is done in electric furnaces 37 m (120 ft) long. The prepared oxide is carried from the roasters through the reduction furnace on 183-cm (72-in.) wide belts. It is converted to iron:

$$Fe_2O_3 + 3H_2 \rightarrow 2Fe + 3H_2O$$

Hydrogen in this reaction is supplied by pipeline from a nearby chemical manufacturer. It circulates through the furnace in a totally enclosed system that removes water and returns the unspent hydrogen to the furnace.

Reduction is done at about 980 °C (1800 °F). Small adjustments in temperature and belt speed are needed to control quality. The resulting sinter cake is friable, and a simple milling operation converts it to iron powder. Particle size distribution is determined by the original grinding of the mill scale. From the milling operation, the powder is conveyed to large storage bins, where it is subsequently scalped and blended.

Pyron iron powder has fine porosity and a sponge-like microstructure (Fig. 7 and 8). The pores in the interior structure of Pyron iron powder particles (Fig. 8) are considerably finer than those in Swedish sponge iron powder particles, because in the latter the long-term reducing treatment at higher temperature coarsens the pores. Because of the fine pore structure, compacts from Pyron iron powder sinter faster than those from other commercial iron powders. Properties of several grades of Pyron iron powder are given in Table 1.

Table 2 lists characteristics and uses of hydrogen-reduced iron powders produced by the Pyron process. Typical chemical compositions of Pyron powders are:

Total iron (all grades)	97.0-98.5%
Carbon (all grades)	0.01-0.05%
Sulfur (all grades)	0.005%
Phosphorus (all grades)	0.012%
Manganese (all grades)	0.40-0.65%
Acid insolubles:	
AC-325	0.20-0.90%
Other grades	0.20-0.45%
Hydrogen loss:	
D-63	0.20-0.50%
AC-325	1.00-1.75%
R-12	2.50% max
Other grades	0.70-1.20%

Production of Iron Powder by Water Atomization of Low-Carbon Iron

By Erhard Klar
Manager of Particle Technology
SCM Metal Products

PRODUCTION OF IRON POWDER by water atomization of low-carbon iron

Fig. 9 Iron powder production by water atomization of low-carbon iron

(1) Selected scrap iron. (2) Arc furnace. (3) Liquid steel. (4) Atomizing: (A) tundish, (B) steel stream, (C) high-pressure water, (D) nozzle, (E) atomized iron powder. (5) Wet magnetic separation. (6) Dewatering. (7) Rotary dryer. (8) Dry magnetic separation. (9) Screening. (10) Equalizing. (11) Annealing and automatic packing. (12) Annealing in belt furnace at approximately 800 to 900 °C (1470 to 1650 °F). (13) Equalizing. (14) Automatic packing. Courtesy of Höganäs Corporation

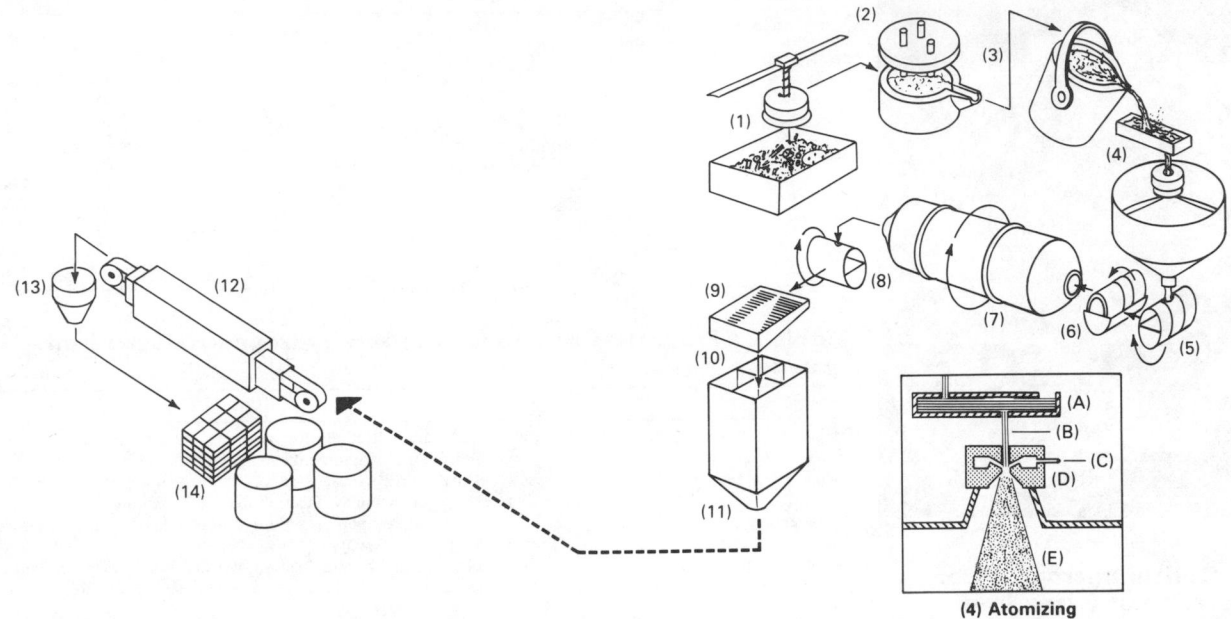

(4) Atomizing

Fig. 10 Water atomization of iron
Source: Ref 1

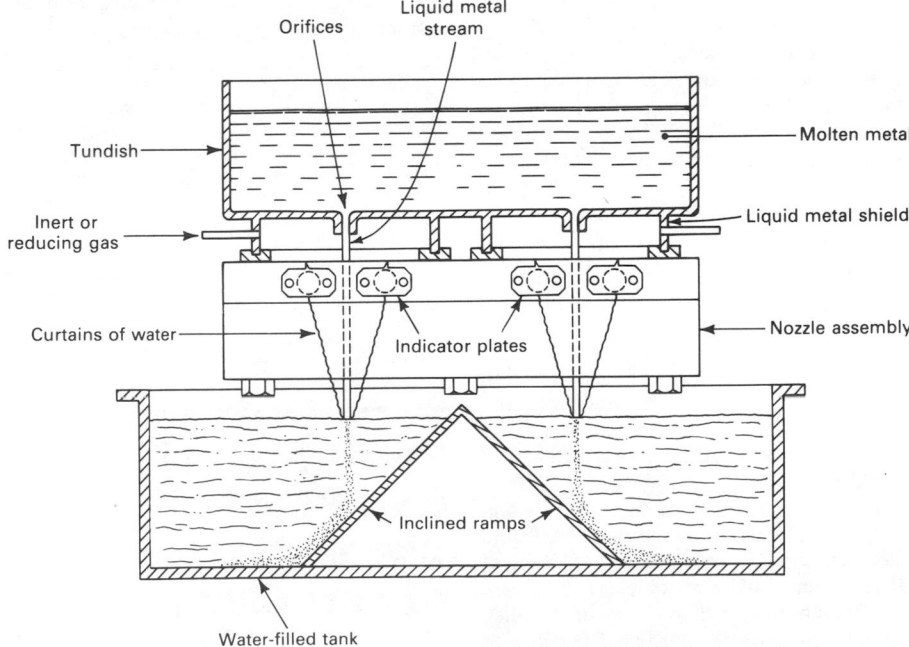

was pioneered by the A.O. Smith Company in the early 1960's. More iron powder currently is produced by this process than by any other process. The high compressibility and high purity of this powder are the primary reasons for its wide usage.

High compressibility, or high pressed and sintered density, is desirable in the many applications that require superior mechanical properties. High purity is essential in powder forging and other full-dense processes, where the presence of even small amounts of oxides is detrimental to dynamic properties.

Process Conditions

Low-carbon steel (typically 0.1% C) is melted in an 18- or 27-metric ton (20- or 30-ton) carbon arc furnace. Through slagging, it is possible to remove or reduce phosphorus, silicon, and other elements to produce a high-purity iron melt. Figure 9 schematically illustrates the process.

Water atomization occurs in ambient air through a multiple metal stream/water jet assembly. Figure 10 shows the original A.O. Smith design, with eight individual free-fall metal streams (four on each side) being atomized by thin curtains of water. Water pressures are approximately 8.3 MPa (1200 psi). The angle formed by the water curtains meeting the metal stream is adjustable for maximum efficiency and control of particle shape. A typical enclosed angle is 40°. Metal flow rate of an individual metal stream is typically 70 kg/min (150 lb/min). Flow rate is adjusted by the size of the nozzle holes in the tundish. To protect against excessive oxidation during atomization, a shield surrounds each row of metal streams, and an inert or reducing gas is introduced (Fig. 10) into the shielded space.

The atomized powder is quenched and collected in a water-filled tank. Water level

Fig. 11 Water-atomized iron powder

(a) Scanning electron micrograph of particles. Magnification: 2000×. (b) Polished cross section of particles. Magnification: 170×

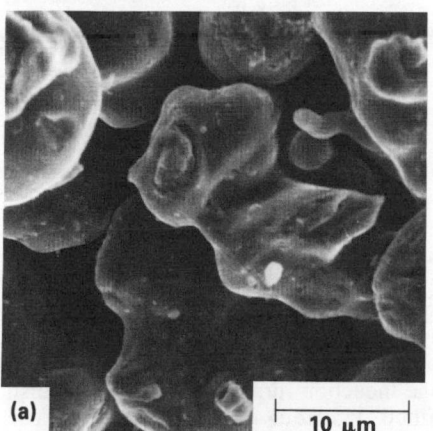 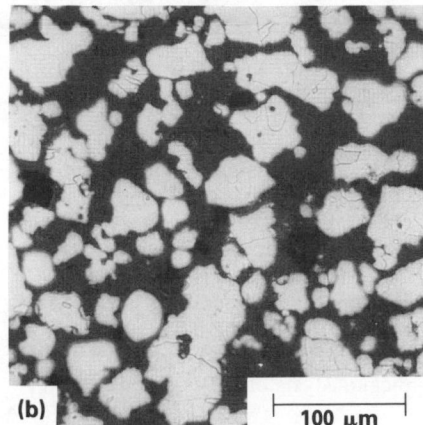

Fig. 12 Effect of internal particle porosity on green density and green strength of solid and porous iron powders

Pressed at 414 MPa (30 tsi) using die wall lubrication. Figures in parentheses signify BET-specific surface areas and average intraparticle pore sizes of powders.

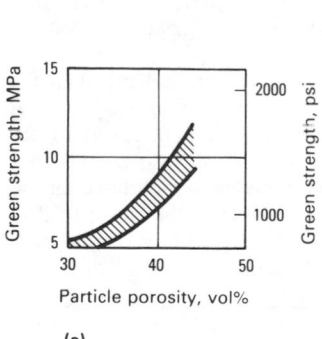 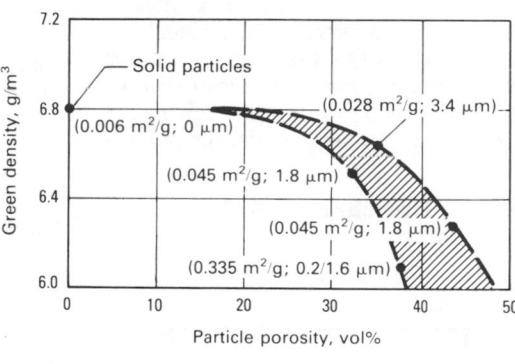

(a)   (b)

of the tank is located in proximity to the zone of atomization to prevent oxidation by air. Two inclined ramps inside the collection tank (1) deflect the powder laterally, (2) prevent the hot powder particles from piling beneath the nozzles, and (3) prevent formation of welded or sintered chunks. The powder is predominantly −60 mesh, with combined oxygen and carbon contents of less than 1%.

After magnetic separation, dewatering, and drying, the powder is heated in a belt furnace to about 800 to 1000 °C (1470 to 1830 °F). Furnace atmosphere is hydrogen or dissociated ammonia. During this process, the oxygen content is lowered to less than 0.2%, and the carbon content is reduced to about 0.01%. At the same time, the powder softens through annealing. Subsequent hammer milling of the sinter cake restores the original as-atomized particle size distribution, with only moderate work hardening of the resultant powder. Sieving and blending follow standard practices.

Powder Properties

Figure 11 is a scanning electron micrograph and polished cross section of iron powder. The solid structure of the particles (Fig. 11b), coupled with their low carbon and oxygen contents, accounts for its outstanding compressibility. This solid particle structure distinguishes this powder from iron powders atomized from high-carbon iron. Powders made from high-carbon iron contain porous or spongy components due either to the addition of iron oxide or to partial oxidation during the atomization, which provides for subsequent decarburization.

Figure 12, which shows the results of a model study, illustrates the effect of par-

ticle structure on green strength and compressibility. Data were obtained from electrolytic-grade iron powders prepared by nitrogen atomization, full oxidation to iron oxide, and reduction under varying conditions to produce powders with varying particle porosities and particle pore size. High compressibility (green density) requires low internal particle porosity; improvement in compressibility is at the expense of green strength (Fig. 12a). However, the irregular particle shape of water-atomized low-carbon iron powders provides adequate green strength at the higher pressed densities for which these powders were developed.

Low residual carbon contents also contribute to high compressibility and green strength, as shown in Fig. 13 for several iron powders produced by decarburization of water-atomized high-carbon iron. This strong hardening effect of residual combined carbon is the reason why many prealloyed powders are made without carbon. Instead, carbon is added to the powder in the form of graphite, which becomes alloyed during sintering to form steel.

Powder properties of typical grades of iron powders made by water atomization of low-carbon iron are given in Table 3. These powders are used primarily in conventionally pressed and sintered parts with

Fig. 13 Effect of residual carbon content on compressibility and green strength of water-atomized high-carbon iron

Pressed at 550 MPa (40 tsi) with 1% zinc stearate admixed. Symbols represent experimental data points.

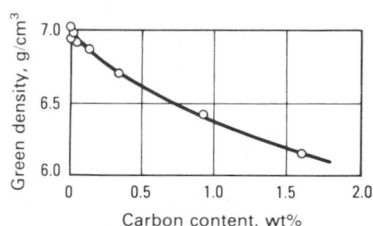

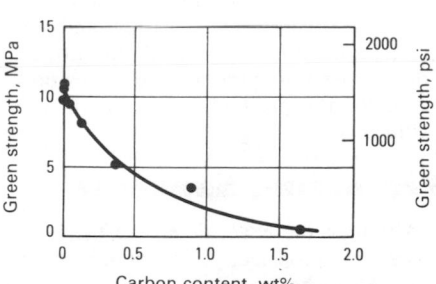

Table 3 Properties of commercial-grade low-carbon water-atomized iron powders

Fe	C	Mn	S	P	Hydrogen loss, %	Hall flow rate, s/50 g	Apparent density, g/cm³	+60	−60+80	−80+100	−325	Compressibility(a) at 414 MPa (30 tsi), g/cm³	MPa	psi
		Composition, %							Sieve analysis (U.S. standard mesh), %				Green strength	
99.0	0.01	0.18	0.02	0.01	0.15	25	2.9	1	5	7	26	6.68	10(b)	1450(b)
99.0	0.01	0.18	0.02	0.01	0.10	25	2.9	1	5	7	26	6.62	12(c)	1750(c)

(a) 0.8% zinc stearate admixed. (b) At 6.8 g/cm³. (c) At 7.0 g/cm³.

densities over 6.6 g/cm³. They also are suitable for hot and cold forming of preforms to near-theoretical density.

The relatively high apparent densities of these powders permit the use of shorter dies and punches. For most applications, these powders are used in mixes containing varying amounts of graphite, copper, and/or nickel, which on sintering and (optional) heat treating form steels with a wide range of mechanical properties (see the article "Mechanical Properties of P/M Materials" in this Volume).

Production of Iron Powder by the QMP Process

By Joseph M. Capus
Technical Director
Quebec Metal Powders Limited

QUEBEC METAL POWDERS LIMITED (QMP) manufactures high-compressibility iron powders by conversion of liquid pig iron from the nearby smelter of its parent company, QIT-Fer et Titane Inc., which is now part of Sohio Chemicals & Industrial Products Company. Shipment of high-purity ferrous powders from its Tracy, Quebec plant began in 1969.

QIT-Fer et Titane operates a large electric smelting plant in which ilmenite ore from a mine on the north shore of the St. Lawrence River is converted to titanium-rich slag and a high-purity low-manganese iron. The titania slag is crushed and sold to the pigment industry, while most of the iron is cast into pigs and sold as Sorel-metal for the production of ductile iron castings.

Processing Sequence

Although the exact details of the QMP process are proprietary, descriptions of the process can be found in Ref 2 to 5. Process steps are illustrated schematically in Fig. 14.

The QMP process uses freshly refined high-purity molten iron as its sole raw material. The hot metal is delivered in 45-metric ton (50-ton) ladle cars after passing through an intermediate injection station, where it is desulfurized and recarburized. Typical analysis of the as-received iron is:

Carbon	3.5%
Manganese	0.01%
Silicon	0.09%
Sulfur	0.008%
Phosphorus	0.025%

Due to the nature of both the ore deposit and the smelting process, the residual element content of QIT iron remains constant on a lot-to-lot as well as a year-to-year basis. The high purity of QIT iron makes it a suitable starting material for production of ferrous powders for a variety of applications.

Granulation and Oxidation of Liquid Iron. On arrival at the iron powder plant, the hot metal is poured into one of two 82-metric ton (90-ton) induction holding furnaces (Fig. 15). After temperature adjustment, the molten iron is poured from the induction furnace into a large tundish fitted with four refractory nozzles. The streams of iron falling from the nozzles are disintegrated with high-pressure horizontal water jets (Fig. 16) directed into a large quench tank. Air drawn through ports in the quench tank causes the resulting coarse granulated iron to be partially oxidized. The wet slurry from the granulation chamber is pumped to a filter, then dried in a rotary kiln dryer that is heated by natural gas.

Carbon content of the hot metal in the tundish ranges from 3.3 to 3.8%. After granulation, carbon content drops to approximately 3%, due to the partial oxidation of the granules, which have an average oxygen content in the range of 3 to 9%. Particle size distribution of the granules falls within the following ranges:

On 30 U.S. mesh	4-16%
−30+100 U.S. mesh	45-55%
−100+325 U.S. mesh	18-38%
−325 U.S. mesh	7-20%

Ball Milling of Oxidized Granules. Coarsely granulated iron is then ball milled into powder. This step determines the differences in particle size among the various iron powder grades. The ball mill circuit has a screening loop that allows grinding to be adjusted to specific particle size ranges. Ball milled iron is then analyzed and stored in large bins according to chemical composition.

Particle size distribution of ball milled, QMP-processed iron falls within the following ranges:

On 100 U.S. mesh	1-7%
−100+200 U.S. mesh	35-45%
−200+325 U.S. mesh	18-28%
−325 U.S. mesh	30-40%

Decarburization/Annealing. Material from the holding bins is weighed out based on chemical analysis and fed to a blender.

Fig. 14 Flowchart of QMP process

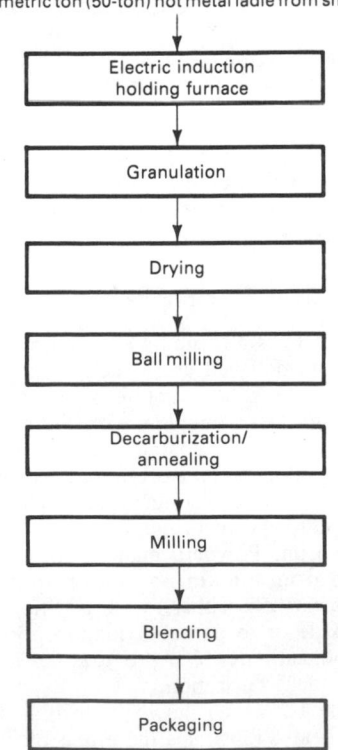

45-metric ton (50-ton) hot metal ladle from smelter

- Electric induction holding furnace
- Granulation
- Drying
- Ball milling
- Decarburization/annealing
- Milling
- Blending
- Packaging

Fig. 15 QMP electric induction furnaces for holding molten iron

Quantities are determined so as to produce a blend of powder with a uniform oxygen-to-carbon ratio close to 2. Ball milled iron is conveyed from the blender to the reduction furnace area, where carbon and oxygen are removed by heat treating. This section houses three continuous belt annealing furnaces, each equipped with a 120-cm (48-in.) wide stainless steel belt that passes through five heating zones and two cooling zones.

Ball milled powder is fed onto the belt to form a bed approximately 25 mm (1 in.) thick and heated to 980 to 1040 °C (1800 to 1900 °F) by gas-fired radiant-tube burners positioned above and below the belt. A protective atmosphere of dissociated ammonia prevents the reduced-iron powder from re-oxidizing.

After cooling, the product emerges from the furnace as a loosely sintered spongy cake of pure iron. Inside the decarburization (reduction) furnaces, carbon and oxygen in the powder react to produce carbon monoxide. The reduced-iron cake emerges from the furnace with carbon and oxygen levels typically below 0.1 and 0.2%, respectively.

Crushing, Milling, and Blending. The iron powder cake falls into a sawtooth crusher. After crushing, it passes through a series of attrition mills for grinding to powder. Screening removes oversize material for recirculation. Dust from the grinding of freshly reduced iron is highly combustible. Consequently, grinding is conducted in an inert atmosphere provided by the combustion of natural gas. Finally, after the powder has passed through a 36 000-kg (80 000-lb) capacity product blender to homogenize individual batches of powder, the finished product is packaged for shipment. For P/M applications, press-ready premixes also are manufactured by blending the raw powder with selected additives, such as lubricants, graphite, and copper powders.

Powder Characteristics

The QMP process, because of its flexibility, allows high-quality iron powders to be produced for all the major applications of ferrous powders, including P/M parts fabrication, welding electrode coatings, flame cutting, and scarfing. Also, special grades are manufactured for magnetic particle inspection, food additives, pharmaceutical applications, and photocopying machines. In several of these applications, the highly porous nature of ATOMET* powders (Fig. 17) confers desirable properties to the finished products.

Powder Metallurgy Grades. Typical chemical and physical properties of ATOMET 25, 28, and 30 (high-compressibility P/M grade powders) are given in Table 4. ATOMET 30 differs from the other grades listed in Table 4, because it is given a separate annealing pass after decarburization. As a result, ATOMET 30 shows the highest purity, in terms of carbon and oxygen levels, and produces parts of a higher density with a single pressing and sintering operation. The porous nature of ATO-

*ATOMET is a registered trademark of Quebec Metal Powders Limited.

Fig. 16 Granulation of molten iron with high-pressure water jets

Fig. 17 Scanning electron micrograph of ATOMET 28 particles

Magnification: 700×

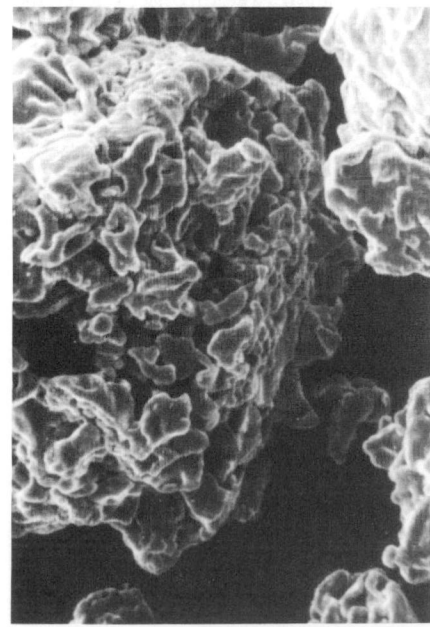

MET powder particles promotes sintering and permits rapid carbon pickup. Table 4 shows the high strength achieved by ATOMET 25 at a pressed density of 6.2 g/cm^3 after only 20 min of sintering at 1120 °C (2050 °F). For higher strength requirements, ATOMET 30 pressed to densities of 7.0 g/cm^3 and above provides transverse-rupture strength in excess of 1030 MPa (150 ksi) after 30 min of sintering.

Welding Electrode Grades. Properties of fine and coarse iron powders manufactured by the QMP process for use as a high-purity iron constituent in the coating of welding electrodes are given in Table 5. Properties obtained through use of these coatings include improved speed and uniformity of welding, resulting in improved productivity and lower production costs. ATOMET powders are suitable for both coated and flux-cored electrodes. See the article "Metal Powders Used for Filler Materials" in this Volume for additional information on welding-grade powders.

Flame Cutting and Scarfing Grades. ATOMET iron powders with closely controlled particle size distribution and free-flowing characteristics also are available for flame cutting, scarfing, and lancing applications. In oxygen or oxyacetylene cutting, heat-enhancing properties resulting from oxidation of the finely divided ferrous powders facilitate cutting of high-melting-point metals and refractories. ATOMET 86 is a −70 mesh iron powder that is used widely in flame cutting; ATOMET 85 is a blend of iron and aluminum powders that is used for cutting and for lancing furnace tap holes. See the article "Metal Powders Used for Flame Cutting" in this Volume for additional information on powder cutting operations.

Photocopying and Miscellaneous Applications. A variety of grades of ATOMET powders are manufactured for use in magnetic particle inspection, chemical processing, food additive and pharmaceutical use, as well as powders for processing into developer for plain-paper copying machines. The latter are included in a group of specialty powders (ATOMET 600 series) that are produced in narrow particle size ranges. Properties of ATOMET 600 series powders are given in Table 6.

Table 4 Typical chemical, physical, and metallurgical properties of ATOMET P/M grade powders

Property	ATOMET 25	ATOMET 28	ATOMET 30
Chemical analysis, %			
Carbon	0.08	0.07	0.01
Oxygen	0.17	0.18	0.10
Iron	99+	99+	99+
Physical properties			
Apparent density, g/cm³	2.55	2.85	2.9
Flow rate, s/50 g	28	26	26
Screen analysis (U.S. mesh), %			
On 100	2	5	5
−100+140	12	28	28
−140+200	27	23	23
−200+325	33	24	24
−325	26	20	20

	P/M properties for an FC 0208 mix composition		
	ATOMET 25	ATOMET 28	ATOMET 30
Density, g/cm³	6.2	6.7	7.0
Compaction pressure, MPa (tsi)	294 (21.5)	476 (34.5)	572 (41.5)
Green strength, MPa (psi)	8.3 (1200)	12 (1800)	11 (1600)
Sintered strength:			
Ultimate tensile strength, MPa (psi)	50 000(a) (345)	60 000(b) (413)	78 000(b) (538)
Transverse rupture strength, MPa (psi)	98 000(a) (675)	124 000(b) (855)	157 000(b) (1082)
Hardness, HRB	71	84	86

(a) Sintered in rich endothermic atmosphere (0.3% carbon dioxide) at 1120 °C (2050 °F) for 20 min. (b) Sintered in rich endothermic atmosphere (0.3% carbon dioxide) at 1120 °C (2050 °F) for 30 min

Table 5 Typical chemical and physical properties of QMP welding-grade powders

Property	ATOMET 67	ATOMET 68
Chemical analysis, %		
Carbon	0.10	0.05
Hydrogen loss	0.4	0.8
Physical properties		
Apparent density, g/cm³	2.8	3.0
Flow rate, s/50 g	26	29
Screen analysis (U.S. mesh), %		
On 30	...	0
−30+100	5 (−70+100)	55
−100+200	51	25
−200+325	24	20(a)
−325	20	...

(a) Represents percentage of powder through 200 mesh

Table 6 Properties of ATOMET 600 series powders used for photocopying applications

Property	ATOMET 602	ATOMET 664	ATOMET 669
Chemical analysis, %			
Carbon	0.1	0.03	0.1
Oxygen	0.1	0.2	0.1
Iron	99+	99+	99+
Physical properties			
Apparent density, g/cm³	3.0	2.7	2.9
Flow rate, s/50 g	26	26	26
Screen analysis (U.S. mesh), %			
On 70	Trace	...	Trace
−70+100	0.3	Trace	8
−100+140	52	1	36
−140+200	45	49	28
−200+325	2.5	48	26
−325	0.2	2	2

Energy Consumption

An analysis of the energy consumption for iron powder production by the QMP process indicates that energy costs are less compared with alternative processes (Ref 3). The comparison was based on actual energy consumption figures for QMP and published estimates for the alternative processes of direct atomization of remelted steel scrap and the reduction of iron oxide (Ref 6). According to these analyses, the QMP process consumes an over-all 16 million kJ/metric ton (14 million Btu/ton) of iron powder produced—20% less than the estimate for atomized steel powder. The difference in consumption rate is mostly due to the energy required for melting scrap in the direct atomization process. From an energy standpoint, the QMP process has a distinct advantage over other methods. The fundamental reason for this benefit is chiefly a matter of location rather than actual process differences, due to the availability of high-purity molten iron on site, thus saving the energy of remelting.

Production of Iron Powder by the Domfer Process

By Louis G. Roy
Manufacturing Manager
Domfer Metal Powders Ltd.

THE DOMFER PROCESS of making iron powder was developed in 1952 at Iberville, Quebec. It consists of water atomizing a high-carbon iron melt to a granular "shot," which is then ball milled to powder size. The ground shot, which contains carbon, is mixed with ground mill scale (iron oxide) that contains oxygen. The carbon and oxygen are chemically combined in a belt furnace at sintering temperatures to form carbon monoxide gas, thus leaving a pure iron cake. The cake is disintegrated to powder size and further refined. A flowchart of the Domfer powder process is given in Fig. 18.

Processing Sequence

Melting and Refining. High-quality selected steel scrap is melted in a 12.7-metric ton (14-ton) coreless induction furnace with an addition of low-sulfur carbon to make a melt of approximately 3.5% C. For low-alloy steel powders, nickel and molybdenum also are added to the furnace.

Approximately 30% of the furnace charge is tapped into a ladle, deslagged, and desulfurized if necessary. Molten metal is poured through a tundish with bottom openings to control the size of the stream. The molten metal is disintegrated by high-pressure jets of water to a granular material approximately 0.8 to 3.2 mm (0.03 to 0.12 in.) in size. This shot is dewatered

Fig. 18 Flowchart of the Domfer powder process

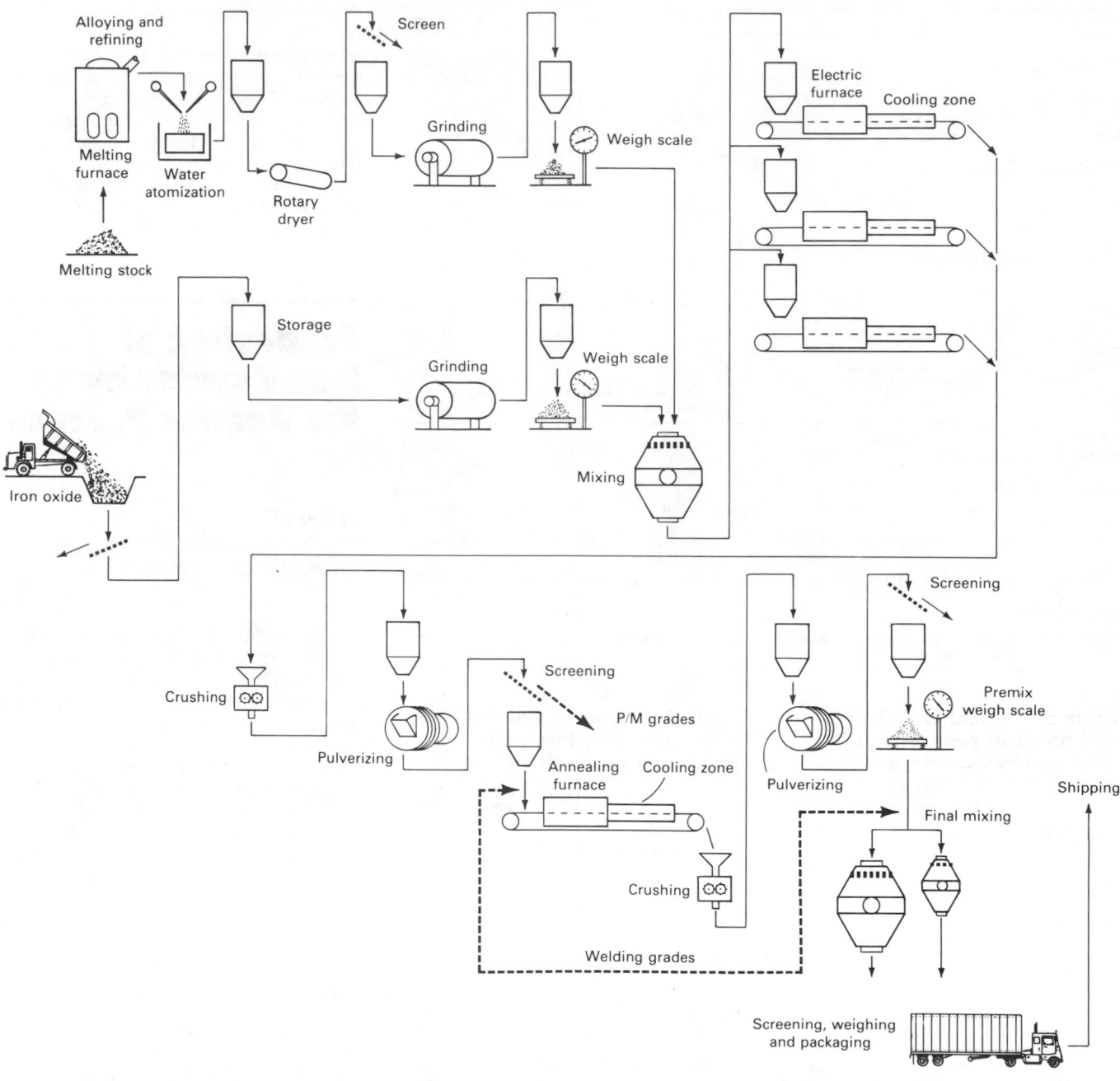

in a cyclone and dried in a rotary gas-fired dryer.

Grinding. Because the high-carbon shot is very brittle, it can be ground to the required powder size in an airswept ball mill. Air flow controls the size of particles that leave the mill, and the powder is recovered from a combination drop-out and cyclone separator. Selected mill scale, which is a by-product of rolling steel, is similarly ground to powder. The ground shot and mill scale are blended to achieve a carbon-to-oxygen ratio that results in de-

carburization and deoxidation, thus leaving essentially pure iron.

Decarburizing. The mixed powder is fed into a steel belt and passed into an electrically heated furnace. The reaction between the high-carbon iron and mill scale occurs at a high temperature. The carbon monoxide gas that is generated provides both a reducing atmosphere and a furnace seal. The resulting pure iron sinters to a cake that is cooled to room temperature before leaving the furnace. The cake is reduced to powder size in an attrition mill

and is screened to the desired size. It is then blended in a double-cone blender to the required specifications.

In the first stage of processing, carbon is reduced to 0.08%, and oxygen is reduced to 0.9%. This powder is suitable for the manufacture of welding rods and for lancing, cutting, and scarfing operations. Table 7 lists several iron powders of welding grade that are produced by the Domfer process.

Annealing. A pure, soft powder is required by the P/M parts industry to reduce

Table 7 Domfer welding-rod grade powders

U.S. mesh No.	MP 61	MP 61LA	MP 62	MP 63	MP 64 (coarse)(a)	MP 64 (fine)(a)	MP 65
Screen analysis, wt%							
+24	...	Nil	...	...	...	...	...
+30	Trace	3	...	...	Nil	...	Nil
+40	...	...	...	...	2	...	...
+50	12	25	...	...	...	...	19
+60	...	...	Nil	...	...	...	...
+80	28	31	4	...	...	...	35
+100	11	10	10	...	21	Nil	9
+150	...	...	...	Trace	27	2	...
+200	28	18	38	8	25	10	22
−200	21	13	...	...	25	...	15
+325	...	...	25	47	...	40	...
−325	...	...	23	45	...	48	...
Physical properties							
Apparent density, g/cm³	2.60	2.86	2.60	2.30	3.35	3.20	3.35
Chemical analysis(b), wt%							
Carbon	0.08	0.08	0.08	0.08	3.2	3.7	0.08
Hydrogen loss	0.85	0.85	0.85	1.00	1.00	1.00	0.85
Total iron	98.0	98.0	98.0	98.0	rem	rem	98.0

(a) MP 64 high-carbon iron powder is used as a source of carbon in hardfacing applications. All other grades in this table are welding electrode powders or electrode coating powders. The manufacturer should be consulted for selection. (b) For all welding-rod grade powders, sulfur is less than or equal to 0.023%; phosphorus is less than or equal to 0.022%; manganese is less than or equal to 0.38%; and silicon is less than or equal to 0.11%.

Table 8 Domfer P/M grade powders

Typical property	MP 32	MP 35HD	MP 36S(a)	MP 39	MP 52	MP 55
Screen analysis, wt%						
+100	4	5	4	4	4	4
+150	20	25	20	20	20	20
+200	26	24	28	28	28	28
+325	32	30	29	30	30	30
−325	18	16	18	18	18	18
Chemical analysis, wt%						
Carbon	0.02	0.02	0.03	0.20	0.02	0.02
Sulfur	0.03	0.02	0.40	0.03	0.03	0.03
Phosphorus	0.03	0.02	0.02	0.03	0.03	0.03
Manganese	0.40	0.30	0.40	0.40	0.40	0.40
Silicon	0.10	0.10	0.13	0.10	0.13	0.13
Molybdenum	...	...	...	...	0.70	0.65
Nickel	...	...	...	...	0.45	1.90
Hydrogen loss	0.25	0.20	0.20(b)	0.20(c)	0.25	0.30
Total iron	rem	rem	rem	rem	rem	rem
Physical properties						
Hall apparent density, g/cm³	2.5	2.9	2.6	2.5	2.6	2.6
Hall flow rate, s/50 g	30	25	28	29	28	28
Powder mix, %						
Powder	97.00	97.00	100.00	97.25	97.25	97.25
Copper	2.00	2.00	...	2.00	2.00	2.00
Graphite	1.00	1.00	...	0.75	0.75	0.75
Lubricant	0.75	0.75	1.00	0.75	0.75	0.75
Mechanical properties						
Green density, g/cm³	6.80	6.80	6.20	6.80	6.80	6.80
Compacting pressure, MPa (tsi)	447	403	314	496	507	606
	(32.4)	(29.2)	(22.8)	(36.0)	(36.8)	(44.0)
Green strength, MPa (psi)	16.7	7.5	11.0	13.7	17.3	15.7
	(2425)	(1090)	(1600)	(1998)	(2516)	(2278)
Modulus of rupture, MPa (psi)	959	893	310	1007	1058	1158
	(139 200)	(129 600)	(45 000)	(146 200)	(153 600)	(168 000)
Hardness	83 HRB	82 HRB	52 HRF	82 HRB	95 HRB	96 HRB
Tool size, in./in.	+0.0006	+0.0029	+0.0009	+0.0009	+0.0007	+0.0006

Note: All grades sintered in endothermic atmospheres for 30 min at 1120 °C (2050 °F) (a) Prealloyed sulfur-containing powder for use in applications where the manganese sulfide assists in machining operations. (b) Hydrogen loss is determined after correction for sulfur loss. (c) Hydrogen loss is determined after correction for carbon loss.

die wear and to lower the pressing tonnage required to attain the desired density. For this purpose, powder from the first stage is blended and fed to a second belt furnace that operates at a lower temperature in a dissociated ammonia atmosphere. The lower temperature produces a very soft cake that can be reduced to powder size without work hardening.

Annealing reduces carbon to less than 0.04% and oxygen to less than 0.40%. It also produces soft powder with good compressibility. Powder premixes containing lubricant, graphite, and copper are blended to customer specification. Table 8 lists several grades of compactible Domfer iron powders.

Quality Control

The Domfer process is a multistep operation that incorporates testing and blending between the stages to permit a wide range of powder qualities. The process sequence allows close control of uni-

Fig. 19 Scanning electron micrograph of MP32 powder
Magnification: 170×

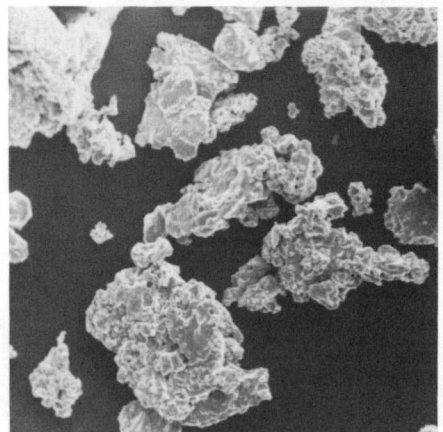

formity from batch to batch. For most users, especially P/M parts manufacturers, uniformity is the most important requirement.

For welding rod applications, essential properties to be controlled include carbon, oxygen, and sulfur content, screen analysis, apparent density, and flow characteristics. For additional P/M parts applications, sample bars are pressed and sintered to measure:

• Pressure to reach a given density
• Green strength at this density
• Dimensional change during sintering
• Strength and hardness of the sintered bar

Figure 19 is a scanning electron micrograph of the MP32 powder particle (see Table 8).

Production of Iron Powder by Carbonyl Vapormetallurgy Processing

By Donald H. Antonsen
Manager, Chemical Industry Sales
The International Nickel Co., Inc.
and
John H. Tundermann
Vice President, Technology
Huntington Alloys, Inc.

IRON POWDERS, like nickel powders, can be produced by carbonyl vapor-metallurgy techniques. The first patents covering the carbonyl process were registered to Ludwig Mond in 1890. A brief review of early metal carbonyl research, with emphasis on his work, is presented in the article "Production of Nickel and Nickel Alloy Powders" in this Volume. In addition, some of the different types of metal carbonyls that can be formed and the formation and decomposition reactions of carbonyls used for the production of primary metal and powder products are described in that article in the discussion of nickel carbonyl powders. This article, however, deals only with the production of iron powders produced by the decomposition of iron pentacarbonyl, Fe(CO)$_5$.

Process Conditions

Iron pentacarbonyl, the raw material from which carbonyl iron powder is produced, is a liquid with a boiling point of 102.8 °C (217 °F). It is formed by passing carbon monoxide over a reduced sponge iron at relatively high pressures and at temperatures ranging from 170 to 200 °C (340 to 390 °F). The physical properties of iron pentacarbonyl are:

Formula	Fe(CO)$_5$
Color and state (room temperature)	Viscous yellow liquid
Molecular weight	195.9
Iron	28.51%
Melting point	−21 °C (−5.8 °F)
Boiling point	102.8 °C (217 °F)
Specific gravity	1.457 (room temperature)
Heat of formation	−964.0 kJ/g·mol (−230.2 kcal/g·mol)

Typically, the feed stocks used are high-surface-area oxidized iron powders or iron turnings that are reduced in hydrogen or another suitable atmosphere prior to carbonylation. The presence of oxygen or oxides on the surface of the iron hinders the reaction, whereas the presence of catalysts improves the rate of formation.

The reaction is exothermic. To achieve commercial production rates, carbonylation is conducted at 130 to 180 atm (1900 to 2600 psi) and temperatures of 170 to 175 °C (340 to 350 °F). As shown in Fig. 20, iron pentacarbonyl formation increases with increasing carbon monoxide pressure. Increasing the temperature increases the rate of reaction. Increased pressure prevents excessive decomposition of the carbonyl. At temperatures above 200 °C (390 °F), carbonyl yield is reduced rapidly by the increased conversion of carbon monoxide to carbon and carbon dioxide by the disproportionation reaction. Iron pentacarbonyl is then condensed and pu-rified by distillation, whereupon it reverts to diiron nonacarbonyl (Fe$_2$(CO)$_9$).

The rate of decomposition for diiron nonacarbonyl is temperature dependent, with maximum rates achieved at about 200 to 250 °C (390 to 480 °F). At higher temperatures, the iron produced oxidizes in the carbon monoxide atmosphere, and high-carbon powder is formed.

Present Commercial Processes

In the United States, the GAF Corporation currently produces high-purity iron powder using carbonyl technology. Iron powder, or sponge, initially is treated under hydrogen to reduce surface oxides to metallic iron. This feed stock is then reacted with carbon monoxide under pressure at elevated temperature to form liquid iron pentacarbonyl. Subsequently, the carbonyl is vaporized and thermally decomposed to form "crude" carbonyl iron powder. The crude powder is refined by chemical processing and mechanically separated to yield various size grades of powder. The carbon content of the powder may reach a maximum of 0.8%, which is reduced to as low as 0.075% to satisfy specific requirements.

Powder Properties

High-purity iron powders produced using carbonyl technology are typically spherical (Fig. 21) and are available in several grades that range from 2 to 9 μm Fisher subsieve size, with apparent densities of 1.2 to 3.2 g/cm^3. The powder can be produced at or near atmospheric pressure. Particles characteristically exhibit an onion-skin structure due to minute carbon deposits in alternate layers. The high pu-

Fig. 20 Effect of system pressure of 100 to 300 atm (1470 to 4400 psi) and temperature on the formation of iron pentacarbonyl
Source: Ref 7

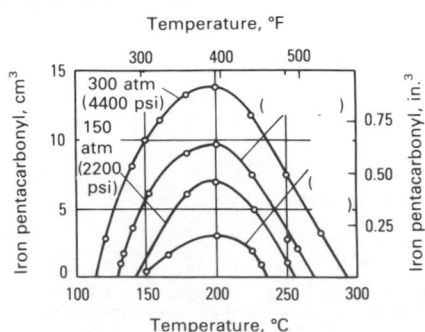

Fig. 21 Scanning electron micrograph of iron powder produced by carbonyl decomposition
Magnification: 3000×

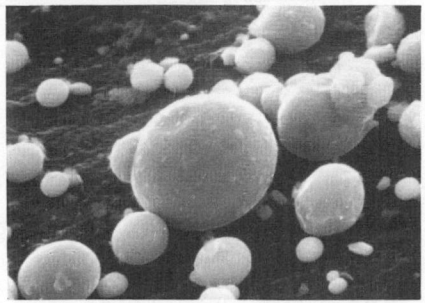

rity of these powders makes them an excellent starting material for the production of magnetic cores and electronic components. This powder oxidizes readily in air and is packaged under an inert gas to facilitate storage.

These powders are essentially free of nonferrous metals and contain 0.3% oxygen, 0.075 to 0.8% carbon (depending on the grade), and 0.05 to 0.9% nitrogen. Alloy powders also are available that nominally contain 92.5% iron (minimum), 0.8% carbon, 0.3% oxygen, and 6.0% nitrogen. Primary applications of these powders include:

- High-frequency cores for radio transmitters, telephones, televisions, direction finders, VHF and UHF circuitry, and radar equipment
- Magnetic-fluid clutch and brake systems
- Carbide and diamond cutting tools
- Chemicals
- Pharmaceuticals
- Catalysts
- P/M materials and alloys
- Cermets
- Visual magnetic recording tapes
- Food enrichment

Production of Iron Powder by Electrolysis

By Prasanna K. Samal
Senior Metallurgist
SCM Metal Products

IRON POWDER made by electrolytic deposition has the highest level of purity among all types of iron powders produced commercially. As a result of its high purity and irregular particle shape, it possesses high compressibility and green strength. Despite these advantages, the current level of usage of electrolytic iron powder is limited, due to its high production cost. The average consumption level of electrolytic iron powder in various applications in the United States is only about 270 metric tons/year (300 tons/year). Before improvements in the atomization of molten metal permitted the production of iron powders with relatively low oxygen contents and with moderate levels of purity, electrolytic iron powder was used widely for the manufacture of conventional P/M parts.

For many years, the telephone industry used electrolytic iron powder for magnetic cores. Some electrolytic iron still is used in magnetic cores and P/M magnets. In recent years, electrolytic iron powder has found new applications because of its high purity. These applications include use as the carrier of the developing medium in dry-toner plain-paper copiers and as iron-enrichment additives in foods (see the articles "Copier Powders" and "Iron Powders for Food Enrichment" in this Volume). Of all the other commercial iron powders, the carbonyl iron powders closely parallel the electrolytic iron powders in purity. However, carbonyl iron powders are somewhat higher in cost and contain higher amounts of carbon and nitrogen.

Process Conditions

Manufacture of metal powder by the electrolytic process can be achieved by one of two methods: (1) a loosely adhering, powdery deposit is formed directly on a stainless steel cathode; or (2) a smooth, dense layer of refined metal is deposited on the cathode, which is subsequently milled to obtain powder. Electrochemical polarization characteristics of the metal being processed largely determine which process can be used. In metals with high electrochemical polarization properties, such as iron, nickel, and cobalt, a large change in the cathode potential is required to cause a relatively small change in current density. Consequently, smooth, coherent deposits are obtained easily with these metals.

In metals where deposition potential varies little with current density, such as in copper, silver, zinc, and cadmium, cathode deposits are obtained readily as spongy or powdery forms, which are periodically scraped off the cathode surface. Because the manufacture of electrolytic iron powder involves crushing and grinding of dense, coherent electrodeposits, brittleness is desired in these deposits. Brittleness of cathode deposits can be enhanced by the proper control of electrolytic cell conditions.

Two types of electrolytes commonly are used for electrodeposition of iron—chloride and sulfate baths. Each type of bath has certain limitations and advantages over the other. Chloride baths have higher concentrations of iron due to the higher solubility of ferrous chloride in water. High concentrations of iron increase the conductivity of the bath, which in turn lowers power consumption. Chloride baths are, however, more corrosive than sulfate baths. The iron powder and chips produced by chloride baths contain small amounts of chloride (about 0.10%), which usually is more objectionable than the residual sulfur found in products of sulfate baths. Ammonium sulfate usually is added to sulfate baths to increase electrical conductivity.

Commercial Processes

Currently, SCM Metal Products is the sole producer of electrolytic iron powder in the United States, with a production capacity of 2700 metric tons/year (3000 tons/year) for electrolytic iron powder and chips. The chips are sold as melting stock for specialty alloys. The production process is based on a sulfate electrolyte with soluble anodes.

The electrolytic cells are made of concrete with fiberglass linings. Cell interiors measure about 3 m (10 ft) long by 0.75 m (2.5 ft) wide by 0.75 m (2.5 ft) deep. The electrolyte consists of mixed ferrous and ammonium sulfates. Each cell contains 17 consumable anodes and 16 cathodes. Current density is 215 A/m² (20 A/ft²). A direct-current power supply is provided by a silicon-controlled rectifier. Plating cycle is typically 96 h, which results in a cathode deposit thickness of approximately 3 mm (0.12 in.). Cathode current efficiency is 95%.

Cells are checked daily for electrical short circuits, which result from fast-growing dendrites, or trees, at the edges of the cathodes. Cell solutions are analyzed weekly and adjusted to specific levels of Fe^{++}, ammonium (NH_4^+), and pH. Anode mud is generated by the impurities present in the anode. It is also caused by a gradual increase in the Fe^{++} ion concentration of the bath, resulting from a slightly higher current efficiency of the anode compared to the cathode. The cathodes and the adhering metal deposits are

Table 9 Chemical and physical properties of compacting-grade iron powder

Property	Electrolytic	Reduced	Atomized
Chemical analysis, %			
Total iron	99.61	98.80	99.15
Insolubles	0.02	0.10	0.17
Carbon	0.02	0.04	0.015
Hydrogen loss	0.29	0.30	0.16
Manganese	0.002	...	0.20
Sulfur	0.01	0.007	0.015
Phosphorus	0.002	0.010	0.01
Physical properties			
Apparent density, g/cm^3	2.31	2.40	3.00
Flow rate, s/50 g	38.2	30.0	24.5
Sieve analysis, %			
+100	0.5	0.1	2.0
−100+150	13.1	7.0	17.0
−150+200	22.6	22.0	28.0
−200+325	29.4	17.0	22.0
−325	34.4	27.7	22.0
Compacting properties(a)			
Green density, g/cm^3	6.72	6.51	6.72
Green strength, MPa (psi)	19.7	19.0	8.4
	(2800)	(2700)	(1200)

(a) At 414 MPa (30 tsi) with 1% zinc stearate

Fig. 23 Mechanical properties of pressed and sintered electrolytic iron powder compacts

Note: SCM A-210 powder was mixed with 0.5% zinc stearate, pressed into standard MPIF test bars, and sintered in dissociated ammonia atmosphere at 1120 °C (2050 °F) for 30 min.

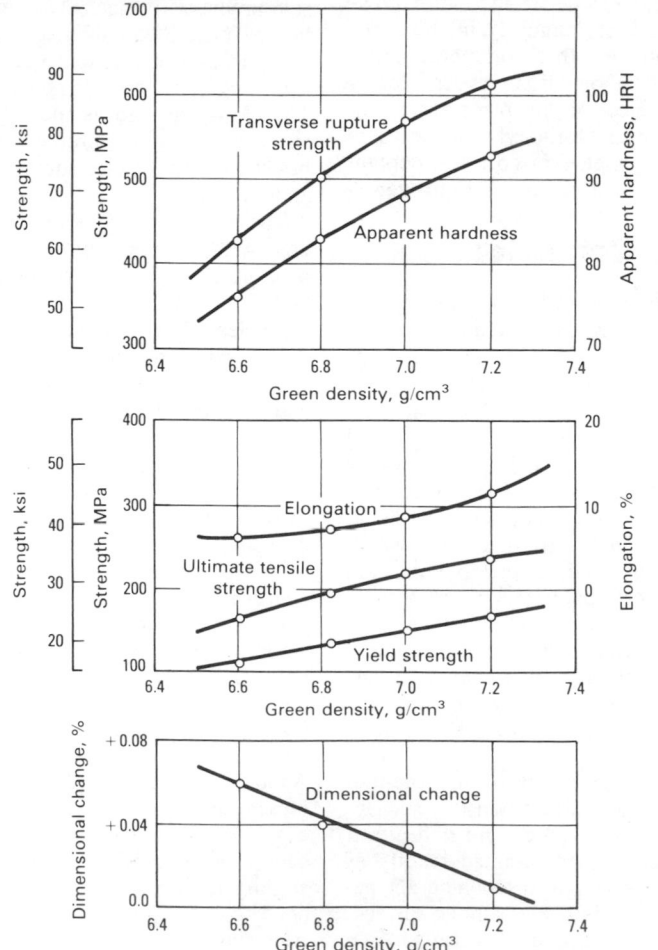

Fig. 22 Compacting properties of electrolytic SCM A-210 iron powder

Powder admixed with 0.5% zinc stearate for lubrication

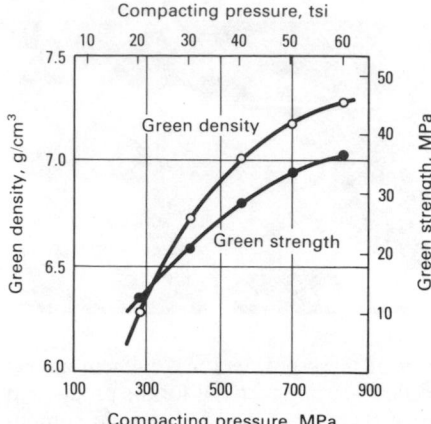

washed thoroughly in water to remove the electrolyte solution. This washing is especially important in achieving low values of total sulfur.

Cathodes are stripped of deposited iron by passing them through an automated stripping machine. The deposits are broken into chips approximately 15 mm ($^5/_8$ in.) in size in a spike roll crusher. These chips are ground in closed-type continuous ball mills using an inert atmosphere such as nitrogen to prevent excessive oxidation and to ensure safety. Typical continuous ball mills contain 1600 kg (3500 lb) of grinding stock and 5000 kg (11 000 lb) of 50-mm (2-in.) diam steel balls. Batch-type mills with 25-mm (1-in.) diam steel balls are used for further grinding to 200 mesh and finer powder. Milled powder is classified using screens and air classifiers. The powder is used in this condition for some chemical and food-enrichment applications.

For most P/M applications, however, powder is annealed in a reducing atmosphere. The annealed powder is bright and soft. Annealing is carried out in hydrogen or dissociated ammonia using belt-type furnaces. Annealing conditions are adjusted to the mesh fraction of powder to prevent formation of excessively hard sinter cakes.

Mixing to ensure uniformity is the final step in processing. This is carried out in 11 400-kg (25 000-lb) capacity mixers. Samples are then taken for various quality control checks, including chemical and physical analyses.

Powder Properties

Table 9 compares the properties of a compacting grade of electrolytic iron

Fig. 24 Scanning electron micrographs of electrolytic iron powders

(a) Electrodeposited iron sheet indicating dendritic grain structure. Magnification: 50×. (b) Electrolytic iron compacting grade powder (SCM A-210) showing particle shape. Magnification: 200×. (c) Electrolytic iron photocopy grade powder (SCM A-277) showing particle shape and narrow size distribution. Magnification: 75×. (d) Surface of electrolytic iron food grade powder (SCM A-131) showing roughness of particle surface. Magnification: 3750×

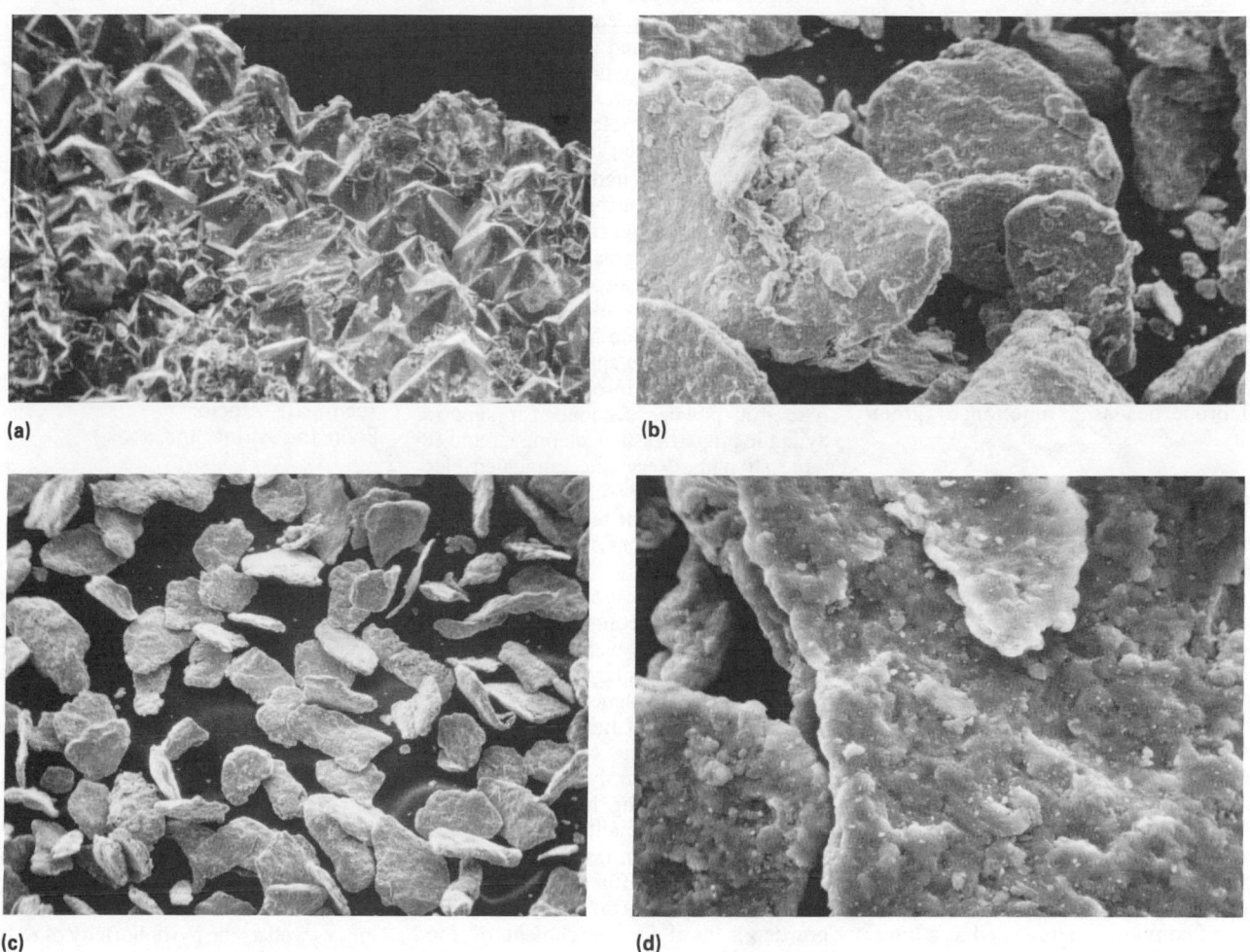

(a)　　　　(b)

(c)　　　　(d)

Table 10 Properties of various grades of electrolytic iron powder

| Property | Compacting powders | | | Ferromagnetic core powders | | | | Photocopier powders | | Food enrichment powder |
	A-210	A-220	A-221	A-233	A-234	A-251	A-270	A-277	A-278	A-131
Chemical analysis, %										
Iron	99.61	99.25	99.50	...	...	...	...	99.50	99.50	98.30
Carbon	0.02	0.02	0.02	0.02	0.02	0.02	0.10	0.02	0.02	Arsenic, <0.0004
Hydrogen loss	0.29	0.65	0.40	0.35	0.65	0.15	0.15	0.40	0.35	Lead, <0.0010
										Mercury, <0.0002
Physical properties										
Apparent density, g/cm^3	2.31	2.30	2.25	2.31	1.75	2.30	2.43	2.50	2.60	2.1
Sieve analysis, %										
−100 mesh	0.5	...	...	...	...	...	...	...	4.0	20-40 μm, 25%
−100+150	13.1	...	...	<50 μm	<40 μm	−150 mesh	−14+100 mesh	4.0	38.0	10-20 μm, 35%
−150+200 mesh	22.6	0.8	0.8	...	...	...	...	41.0	50.0	0-10 μm, 40%
−200+325 mesh	29.4	14.1	14.1	...	...	...	...	51.0	50.0	...
−325 mesh	34.4	85.1	85.1	...	...	...	...	4.0	...	...
Green density(a), g/cm^3	6.72	...	6.6	...	...	...	...	...	...	...
Permeability	...	...	...	Medium-high	Medium-high	High	High	...	...	...
Q-value	...	...	...	Medium-high	Highest	Low	Low	...	...	...

(a) At 414 MPa (60 ksi)

powder (annealed) with properties of typical atomized and reduced-iron powders.* Due to its high purity and irregular particle shape, electrolytic iron powder has excellent compressibility and green strength. Figure 22 graphically depicts the compacting properties of this powder. It is ideal for making intricate and fragile parts that otherwise present molding and handling difficulties. Also, as a result of its intrinsic low hardness and freedom from refractory and oxide contaminants (such as aluminum oxide and silicon dioxide), it does not cause excessive die wear. Electrolytic iron powder also has superior sintered properties (particularly ductility and dimensional accuracy) that are achieved with reasonable compacting pressures. Figure 23 shows the relationship of sintered properties as a function of green density.

Applications

The high green strength, high compressibility, irregular particle shape, and high purity of electrolytic iron powder make it suitable for a number of P/M applications. Due to its extremely low manganese content, the pressed and sintered parts exhibit excellent ductility. This makes it ideal for high-density, close-tolerance applications requiring double press and double sinter processing techniques. Secondary operations can be readily performed on high-density components made from electrolytic iron powder. Because of the irregular particle shape, re-pressed parts made from electrolytic iron powder will have interconnected porosity eliminated at a lower density level than with more regularly shaped atomized iron powders. Hence, there is little chance for entrapment of plating solutions or for infiltration of brazing material.

In applications requiring case hardening heat treatment, electrolytic iron powder is often specified, because it provides a uniform case depth. Because there is no porosity network for the carburizing gas to enter, carbon enrichment can only progress by diffusion from the surface of the part. For optimum control of case hardening, a minimum density of 7.2 g/cm³ is recommended for parts made from electrolytic iron powder.

Electrolytic iron powder is also used for making soft magnetic parts because of its

*The properties and applications discussed in this section are based on iron powder produced by electrodeposition from ferrous sulfate baths. Chloride baths also are used commercially, and iron powders produced from these baths have properties similar to those produced by sulfate baths, except for chloride content.

high compressibility and lack of impurities. For maximum magnetic properties, soft magnetic parts should be pressed to the highest density from a powder mix that contains a minimum amount of lubricant and should be sintered in dissociated ammonia, vacuum, or hydrogen to minimize carbon contamination. For more information, see the article "Electrical and Magnetic Applications" in this Volume.

Most higher strength sintered nickel steels are made by using electrolytic iron powder because of the finer distribution of pores. The superior green strength of electrolytic iron powder also makes it ideal for use as the matrices of abrasive cutting wheels and diamond-bonded saw segments. In fully dense applications, the low oxide content (acid insolubles) of electrolytic iron powder accounts for superior dynamic properties such as impact and fatigue strength. Very fine, annealed electrolytic iron powder (<325 mesh) is used in the manufacture of ferromagnetic cores for radios, televisions, and other audio-frequency devices.

Annealed electrolytic iron powder screened to closely controlled particle size range is used as a developer or carrier in toners for plain-paper copier machines. The particle shape and magnetic quality of high-purity powder result in a magnetic brush of high loading characteristics that allows the toner to be uniformly transferred to the paper, thus providing high-contrast copies. For more information, see the article "Copier Powders" in this Volume.

Unannealed, ultrafine (average particle size of about 10 μm) electrolytic iron powder is used for enrichment of food products. This grade of electrolytic iron powder meets or exceeds all specifications of the *Food Chemical Codex*. In addition to possessing the high purity required for this application, these powders exceed other iron powders, such as carbonyl and reduced-iron powders, in bioavailability because of their large surface area. For more information, see the article "Iron Powders for Food Enrichment" in this Volume.

Microstructure

The morphology of the surface of electrodeposited iron and the particle shape of ball milled electrolytic iron are shown in Fig. 24. As-deposited electrolytic iron consists of an agglomeration of dendritic grains. Figure 24(a) shows the exterior surface of this deposit. Crushing and ball milling of these deposits produce dense, irregularly shaped, and somewhat flaky particles with a fairly rough surface. Fig-

ure 24(d) depicts the surface roughness of ultrafine food-grade electrolytic iron powder. Table 10 lists the properties of various grades of electrolytic iron powder.

Production of Iron Powder by Fluidized Bed Reduction

By Vyacheslav Styskin
Technical Director
Ervin Industries, Inc.

FLUIDIZED BED PROCESSES currently are used extensively in industrial applications. Process variations include fluidized bed combustion, gasification, heat treatment, and catalytic reactions. However, iron powder production by fluidized bed processes is limited to approximately 22 680 metric tons (25 000 tons) per year worldwide. Attractive characteristics of fluidized bed processes include high, uniform heat transfer between gas and solid and high reaction rates due to the large exposed surface area of the fluidized bed solid material. Fluidized bed oxidation of metals and reduction of metal oxides are possible, but defluidization caused by sticking, as well as the pyrophoricity of reduced powders, has restricted industrial usage.

This article describes the process of fluidized bed reduction of iron oxides that was used in the United States from 1958 to 1970. Oxide reduction, atomization, electrolytic deposition, and carbonyl processes have replaced the fluidized bed process for current iron powder production.

Typically, in the fluidized bed process, a granular material is kept in fluid motion by using a fluidizing gas. In many cases, the fluidizing gas reacts chemically with the granular solid. Thus, in fluidized bed oxidation, the fluidizing gas typically is air or oxygen. In fluidized bed reduction, the fluidizing gas contains hydrogen or carbon monoxide, or a mixture of the two.

Principles of Fluidized Reduction Processes

Many of the principles of fluidized bed reduction were developed as a by-product of research aimed at direct ironmaking, that is, bypassing the reduction of iron ores in a blast furnace and replacing this procedure with continuous fluidized bed reduction of iron ore. Iron processed in this manner is referred to as direct-reduced iron. To a great extent, these principles also apply to the reduction of high-purity iron ore or iron oxide, where the reduced iron is used in metallurgical applications without further processing.

Selection of a fluidized bed reduction process should include the following considerations:

- Raw materials must be of granular composition with sufficient green strength.
- Raw materials must be resistant to expansion and sticking during reduction.
- Reduction gas must be prepared.
- Heat may have to be supplied to the reactor.
- Reduced materials should not be pyrophoric.

For direct-reduced iron production, ore feed in the form of pellets is preferred (Ref 8). Fully oxidized pellets, such as hematite (Fe_2O_3) pellets, should be used, because magnetite (Fe_3O_4) forms a relatively compact iron shell during reduction that strongly resists gas diffusion into the pellets, thus increasing reduction time (Ref 9). Pellets made from rich magnetite also disintegrate during reduction because of the high degree of swelling (Ref 10).

Sticking of the particles during reduction is perhaps the most serious problem encountered in fluidized bed reduction of metal oxides, because it results in defluidization of the bed. The susceptibility of iron ore to sticking is a complex relationship between percent reduction, bed temperature, and gas velocity (Ref 11). By reducing low-gangue ore, Agarwal et al. (Ref 11) found that the susceptibility to defluidization is highest between about 620 and 730 °C (1150 and 1350 °F) for products reduced more than 90%. Below about 620 °C (1150 °F), the tendency to defluidize is minimal, but the resulting product is usually pyrophoric. At temperatures above 700 °C (1300 °F), the tendency to defluidize decreases.

Similar findings were observed by Bondarenko et al. (Ref 12), who established that the zone of fluidization expands by using larger pellets and increasing gas velocity and temperature from 550 to 900 °C (1020 to 1650 °F). Evidently, high temperature decreases surface activity of reduced particles, and sticking is decreased.

The extremely high surface area of direct-reduced iron contributes to rapid reoxidation (pyrophoric iron). The reoxi-

dation of direct-reduced iron (the reaction of iron with oxygen and water) is a rapid process in which spontaneous ignition of iron powder may occur.

The tendency of direct-reduced iron to reoxidation depends on reduction temperature. The higher the temperature, the less the degree of reoxidation. Raw material composition also has an effect on the oxidation or degree of reactivity (Ref 13). According to Jensen (Ref 14), the presence of cementite (Fe_3C) has a passivating effect on oxidation at temperatures up to 150 °C (330 °F). At higher temperatures, this material oxidizes faster due to cementite decomposition. As discussed in Ref 14, high-temperature reoxidation may be prevented successfully by:

- Passivation during the manufacture of the product
- Keeping the product dry prior to use
- Avoiding contact with hot objects or other sources of thermal energy above about 200 °C (390 °F)

Advantages of reduction in a fluidized bed include:

- High velocity of reactions due to effective interaction of gases and solids in the bed and high rates of heat and mass transfer between the solid and fluid
- Fluidized bed reduction is an isothermic process, which allows a narrow temperature range to be maintained.
- Gas recirculation is possible.
- Reduction in a fluidized bed can be converted into a fully automatic, continuous process.

Disadvantages include:

- Not all iron ore materials can be reduced in a fluidized bed, because of the tendency of these particles to stick together.
- Particle size is limited from about 6.5 mm (0.25 in.) to a few microns, or about 0.001 in.
- Swelling of iron ore materials during reduction can result in disintegration and excessive loss of particles from the bed by the transport of material of the reaction zone (carry-over).
- Pyrophoricity hinders storage and transport of reduced iron.
- A sharp temperature gradient exists near the bottom of the bed (Ref 15), which causes overheating of the grid.
- Countercurrent rates of fluid throughput are limited to the range over which the bed may be fluidized, the minimum fluidization velocity, or the carry-over velocity.

- The hydrodynamics of fine materials is mismatched to the reduction process, because fluidization occurs at a gas velocity lower than that required for reduction.

Critical Fluidization Velocity. Critical velocity is the reduction gas velocity at which the fluidization of material is achieved without carry-over from the reaction zone. References 16 to 24 discuss empirical expressions for the parameters of the critical fluidization velocity. The derivations consider particle diameter, density of the solid and fluid, viscosity, and voidage.

Application of Fluidized Bed for Reduction Processes

Depending on the degree of reduction* of the product, fluidized bed processes may be divided into three categories: (1) low reduction degree ($r = 11.1\%$), iron ore magnetic roasting; (2) average reduction degree ($r - 70$ to 90%), direct-reduced iron; and (3) high reduction degree ($r = 96$ to 99%), powder metallurgy.

Iron ore magnetic roasting involves the hematite-magnetite transformation. Its most common application is for iron ore beneficiation.

Direct-Reduced Iron Production Processes

Direct-reduced iron produced in a fluidized bed is achieved by three processes—the Nu-iron, high iron briquettes (HIB), and fluid iron ore reduction (FIOR) processes.

Nu-Iron Process. Developed by the United States Steel Company in 1950 (Ref 25), the Nu-iron process uses fine iron ore (1.65 mm, or 0.065 in.) in the fluidized bed reactor at temperatures of 600 to 700 °C (1110 to 1290 °F), which eliminates sticking and pyrophoricity. The reducing gas (73% hydrogen and 16% carbon monoxide) is obtained by catalytic conversion of natural gas with steam. Before entering the reactor, iron ore is preheated to 375 °C (705 °F) by the off gas; then it is preheated to 925 °C (1695 °F) in a special furnace. The sponge iron, reduced to 90 to 95%, is briquetted and used in electric steelmaking furnaces.

---

*Degree of reduction: $r = [(O^0 - O^f)/O^0] \times 100\%$, in which O^0 is oxygen content of raw material and O^f is oxygen content of reduced iron.

High iron briquettes process was developed in the United States and is based on Nu-iron technology (Ref 25). Iron ore (1.65 mm, or 0.065 in.) is reduced at 600 to 700 °C (1110 to 1290 °F) by a gas mixture containing 14% carbon monoxide and 75% hydrogen or by 85% hydrogen gas. The HIB process currently is in commercial operation in Venezuela. Fine iron ore (−10 mesh) is reduced in a two-stage fluidized bed reactor. In the first stage, reduction of ferric oxide (Fe_2O_3) to ferrous oxide (FeO) is accomplished at 870 °C (1600 °F) by the spent gas from the lower stage.

Reduction of ferrous oxide to iron is accomplished by fresh reducing gas in the second stage at 700 °C (1290 °F). The reactor is operated at a pressure of 2 kg/cm^2 (2 atm gauge, or 29.4 psi). The reducing gas is generated by steam reforming of natural gas. The reduced iron, metallized to 75%, is briquetted and then cooled with inert gas in a shaft cooler. The reduced briquettes are used in the blast furnace, thus decreasing coke rate. Capacity of the commercial plant in Venezuela is 591 000 metric tons (651 000 tons) of direct-reduced iron per year.

Fluid iron ore reduction process was developed in the United States by Esso Research and Engineering, in cooperation with Arthur D. Little, Inc., in 1960 (Ref 9). A commercial-size FIOR plant (364 000 metric tons/year, or 400 000 tons/year) is now in operation in Venezuela (Ref 25). The iron ore (−10 mesh) is reduced in four pressurized fluidized bed reactors by a reducing gas that contains a high concentration of hydrogen and carbon monoxide. Operating temperature does not exceed 600 °C (1110 °F), and gas pressure in the reactor equals 10 atm (147 psi). The reduced sponge iron (92 to 93.5% total iron content, 83.5 to 89.6% metallic iron content, 0.03 to 0.1% carbon, 0.16% phosphorus, and 0.01% sulfur) is briquetted and is used in electric steelmaking furnaces.

H-Iron Process

The H-iron process was developed by Hydrocarbon Research Company and Bethlehem Steel Corporation. The process consists of reducing fine iron ore or mill scale (0.04 to 0.8 mm, or 0.0016 to 0.032 in.) by high-pressure hydrogen at 35 atm (515 psi) at a temperature of 540 °C (1000 °F).

Hydrogen is obtained by the water-steam or oxygen reforming of natural gas. High hydrogen pressure is used to increase reduction velocity and decrease reactor size. To increase hydrogen utilization, three- and four-zone reactors are used, with capac-

ities of 45 to 90 metric tons/year (50 to 100 tons/year). The reduced iron powder is annealed at 650 to 870 °C (1200 to 1600 °F) to render the powder nonpyrophoric. Depending on the reduction degree, the resulting reduced iron can be used for P/M structural parts (90 to 95% reduction degree) or in electric steelmaking furnaces (75% reduction degree).

ONIA* Process

Developed in France, this process differs from the H-iron process. A low pressure of 5 atm (73.5 psi) is used. The composition of the reducing gas is 85 to 87% hydrogen and 14 to 16% carbon monoxide. To avoid sticking, the process uses a reduction gas velocity significantly greater than the critical fluidization velocity. The sponge iron is used as a melt stock in electric furnaces and as a molding-grade powder.

Factors Affecting Commercial Viability

As discussed in Ref 26, all of the fluidized bed processes have been limited commercially because of sticking, high degrees of pyrophoricity due to extremely high microporosity, unsatisfactory quality of iron powder for molding applications, or a combination of these. Sticking can be eliminated by using a "semi-fluidized" bed in a reactor with pull-out grids (Ref 27). In this process, iron ore pellets (2 to 6 mm, or 0.08 to 0.24 in.) are reduced by the products of catalytic conversion of natural gas (35% hydrogen, 18% carbon monoxide, 40% nitrogen).

Pellets are reduced in a two-stage fluidized bed process. At first, a reduction degree of $r = 30\%$ is reached. After this stage, the fluidized bed sticks, forming a sponge cake. The second reduction stage ($r = 95$ to 99%) is accomplished in a semi-fluidized pulsating bed. Gas velocity during the impulse fluidization substantially exceeds the critical fluidization velocity, provided that the cake breaks and that there is a uniform gas supply. After cooling by nitrogen, the reduced iron is not pyrophoric because of high reduction temperatures of 800 to 870 °C (1470 to 1600 °F).

Laboratory experiments with a drum-type rotary fluidized bed also have shown methods of eliminating sticking (Ref 28). The advantages of fluidized bed technology can be realized by use of an "improved" process that combines the fluidized bed with a mechanical or hydraulic device for breaking the sticking material.

*Office National Industrial d'Azote

The technical feasibility to expand and develop fluidized bed reduction processes depends on:

- Possibility to increase the temperature level of reduction with a simultaneous elimination of sticking and pyrophoricity
- Possibility to increase gas utilization efficiency
- Elimination of temperature gradient between the top and bottom of the fluidized bed

REFERENCES

1. U.S. Patent 3,309,733, "Apparatus for Producing Metal Powder," B.G. Winstrom, A.O. Smith Corp., 1967
2. Capus, J.M. and Fossen, W.F., The Powder Metallurgy Industry in Canada, *Int. J. Powder Metall. Powder Technol.*, Vol 18 (No. 4), 1982, p 335-346
3. Capus, J.M., Energy Consumption in Iron Powder Production, *Prog. Powder Metall.*, Vol 35, 1978-1979, p 431-439
4. "ATOMET Metal Powders & Quebec Metal Powders Ltd.," Quebec Metal Powders Limited, Sorel, Quebec, 1980
5. Hulthen, S.I., *Int. J. Powder Metall. Powder Technol.*, Vol 17, 1981, p 345-346
6. Kaufman, S.M., Energy Consumption in the Manufacture of Precision Metal Parts from Iron Powder, *Int. J. Powder Metall. Powder Technol.*, Vol 15 (No. 1), 1979, p 9
7. Mond, R.L. and Wallis, A.E., Researches on the Metallic Carbonyls, *J. Chem. Soc.*, Vol 121, 1922, p 29-32
8. Stephenson, J.G., The Influence of Iron Ore Feed Characteristics on Direct Reduction, *16th Ann. Conf. Metall.*, CIM, Vancouver, 1977
9. Brimacoude, J.K. and Venkateswaran, V., Literature Review of the Production and Use of Direct Reduced Iron, *DRI—Technology and Economics of Production and Use*, Iron and Steel Society of AIME, Warrendale, PA, 1980, p 216
10. Bogdandy, V., *et al.*, *Archiv Fur Das Eisen-Huttenwesen*, Vol 34, 1963, p 401-409
11. Agarwal, J.C., *et al.*, Fluidized Bed Technology, *Chem. Eng. Prog. Symp.*, Vol 62 (No. 67), 1966, p 107
12. Bondarenko, E.I., *et al.*, Production of Sponge Iron in Fluidized Bed Reactor, *Visocotemp. Endoter. Proc. Metall.*, Moscow, 1968
13. Pietsch, W., Storage, Shipping and Handling of Direct Reduced Iron,

Society of Mechanical Engineers —American Institute of Mechanical Engineers Fall Meeting, Denver, 1976

14. Jensen, H.B. and Smailer, R.M., The Handling, Storage, and Shipment of Direct Reduced Iron, *DRI— Technology and Economics of Production and Use,* 1980, p 96

15. Fane, A.G. and Wen, C.Y., Fluidized-Bed Reactors, *Handbook of Multiphase Systems,* Hemisphere Publishing, Washington, DC, 1982, p 8-104

16. Rowe, P.N., The Effect of Bubbles on Gas-Solids Contacting in Fluidized Beds, *Chem. Eng. Prog. Symp. Ser.,* Vol 58 (No. 38), 1962, p 42

17. Grace, J.R., Fluidized Bed Hydrodynamics, *Handbook of Multiphase Systems,* Hemisphere Publishing, Washington, DC, 1982

18. Todes, O.M., *Methods and Processes of Chemical Technology,* Moscow, 1956

19. Schenk, H.U.A., *Archiv. Eisen-Hutten.,* Vol 33 (No. 4), 1962, p 211-216

20. Kogevnikov, I.V., *Bekoksovaya Metallurgia Galesa, Metallurgia,* Moscow, 1970

21. Geldart, D. and Abrahamson, A.R., The Effect of Fines on the Behavior of Gas Fluidized Beds of Small Particles, *Fluidization,* Plenum Press, New York, 1980, p 453

22. Bogdandy, L. and Engell, H., Vosstanoulenie Gelesnich Rud, *Metallurgia,* Moscow, 1971, p 197

23. Yerushalmi, J., McIver, A.E., and Squires, A.M., "Preprint of GVC/AI Che-Joint Meeting," E-31, Munich, 1974

24. Yerushalmi, J., Turner, D.H., and Squires, A.M., The Fast Fluidized Bed, *Ind. Eng. Chem. Proc. Design Devel.,* Vol 15 (No. 1), 1976, p 47

25. Direct Reduced Iron, *Technology and Economics of Production and Use,* Iron and Steel Society of AIME, Warrendale, PA, 1980, p 90

26. Hulthen, S.I., Five Decades of Iron Powder Production, *J. Powder Metall. Powder Technol.,* Vol 17 (No. 2), 1981

27. USSR Patent No. 624723, 1978

28. Hagane, T., Countercurrent Reduction of Powder in a Drum-Type Rotary Fluidized Bed, *J. Iron Steel Inst. Jpn.,* Vol 66 (No. 13), 1980, p 1995-2003

SELECTED REFERENCES

● Trout, W.E., The Metal Carbonyls, Parts I & II, *J. Chem. Ed.,* Oct 1937, p 453-459

● Trout, W.E., The Metal Carbonyls, Parts III & IV, *J. Chem. Ed.,* Dec 1937, p 575-581

● Trout, W.E., The Metal Carbonyls, Parts V & VI, *J. Chem. Ed.,* Feb 1938, p 77-83

● Trout, W.E., The Metal Carbonyls, Part VII, *J. Chem. Ed.,* March 1938, p 113-121

● Goetzel, C.G., *Treatise on Powder Metallurgy,* Vol 1, Interscience, New York, 1949, p 49-53

● Actmann, G.O., Carbonyl Iron Powders, *FM-TV Journal #9,* 1949, p 29

● Carbonyl Iron Powders, Chemical Division Brochure, New York, GAF Corp., 1960

● Carlton, H.E. and Goldberger, W.M., Fundamental Considerations of Carbonyl Metallurgy, *J. Metals,* June 1965, p 611-615

● Sale, F.R., Vapour-Phase Transport Reactions in Extractive Metallurgy, *Min. Sci. Eng.,* Vol 3 (No. 4), 1971, p 3-12

● Cambropoulos, M., "The Technology of Metal Powders: Production, Properties and Applications," APMI Powder Short Course, Newport Beach, CA, Feb 1978

● Wagner, F.S., Carbonyls, *Kirk-Othmer: Encyclopedia of Chemical Technology,* Vol 4, 3rd ed., John Wiley & Sons, New York, 1979, p 794-813

Production of Stainless Steel, Low-Alloy Steel, and Tool Steel Powders

By the ASM Committee on Production of Steel Powders*

POWDER METALLURGY stainless steels, low-alloy steels, and tool steels attained commercial importance between 1950 and 1970. Most of these powders currently are produced by water or gas atomization. Information on these processes may be found in the article "Gas and Water Atomization" in this Volume. This article discusses powder production techniques and powder properties specific to these three classes of alloys.

Stainless Steels

Early experimentation concentrated on producing stainless steel powders from elemental components and from alloy powders by sensitization embrittlement and grinding of stainless steel sheet. Water atomization became the established process in the 1950's for producing stainless steel powders for conventional die compaction and sintering. Over the years, slight modifications to the existing wrought compositions led to improved compacting properties. Recently, further modifications in composition, combined with processing precautions, have led to improvements in corrosion resistance.

Melting of virgin raw materials is performed in open air or vacuum induction furnaces. Low manganese concentrations (<0.3%) and deoxidation with ferrosilicon to achieve 0.7 to 1.0% Si in the alloy prevent excessive oxidation of the powder during water atomization in an inert gas (nitrogen) purged atomization chamber. Typical water pressures for producing predominantly −80 mesh powder are about 14 MPa (2000 psi). A considerable amount of powder is produced by proprietary gas atomizing processes.

Typical powder properties of commercial grades of stainless steel powders are given in Table 1. These powders normally are used in the as-atomized condition, although martensitic grades may be annealed to improve green strength and compressibility. Particle shape (Fig. 1) and other powder characteristics are controlled to produce powders with apparent densities ranging from 2.5 to 3.2 g/cm^3, adequate green strength, and good compressibility (Fig. 2a and b). For more detailed

Table 1 Commercial P/M grades of stainless steel powder

Alloy	Cr	Ni	Si	Mo	Cu	Sn	Mn	C	S	P	Fe	Oxygen content, ppm	+100 (>150 μm)	−325 (<44 μm)	Apparent density, g/cm³	Flow rate, s/50 g
Austenitic grades																
303	17-18	12-13	0.6-0.8	...	...	...	0.3(a)	0.03(a)	0.1-0.3	0.03(a)	rem	...	3(a)	40-60	3.0-3.2	24-28
304L	18-19	10-12	0.7-0.9	...	...	...	0.3(a)	0.03(a)	0.03(a)	0.03(a)	rem	1000-2000	1-4	30-45	2.5-2.8	28-32
304LSC	18-20	10-12	0.8-1.0	...	2(b)	1(b)	0.3(a)	0.03(a)	0.03(a)	0.03(a)	rem	...	3(a)	30-45	2.7-2.9	26-30
316L	16.5-17.5	13-14	0.7-0.9	2-2.5	...	...	0.3(a)	0.03(a)	0.03(a)	0.03(a)	rem	1000-2000	1-4	35-45	2.6-3.0	24-32
Martensitic grade																
410L	12-13	...	0.7-0.9	...	...	...	0.1-0.5	0.05(a)	0.03(a)	0.03(a)	rem	1500-2500	3(a)	30-45	2.6-2.9	26-30
Ferritic grades																
430L	16-17	...	0.7-0.9	...	...	...	0.3(a)	0.03(a)	0.03(a)	0.03(a)	rem	...	3(a)	30-45	2.5-2.9	26-32
434L	16-18	...	0.7-0.9	0.5-1.5	...	...	0.3(a)	0.03(a)	0.03(a)	0.03(a)	rem	...	3(a)	30-45	2.5-2.9	26-32

(a) Maximum. (b) Typical

*Harry D. Ambs, Manager, Technical Support, SCM Metal Products; Mark Svilar, Metallurgist, SCM Metal Products; Erhard Klar, Manager of Particle Technology, SCM Metal Products

Fig. 1 Scanning electron micrograph of water-atomized 304L stainless steel

−100 mesh. Magnification: 150×

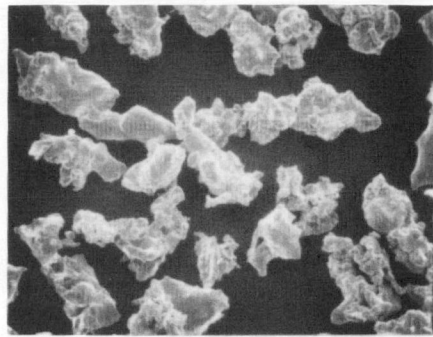

Fig. 2 Typical compressibility and green strength curves of stainless steel powders

Die lubricant: 1% lithium stearate. (a) Austenitic grades. (b) Ferritic and martensitic grades

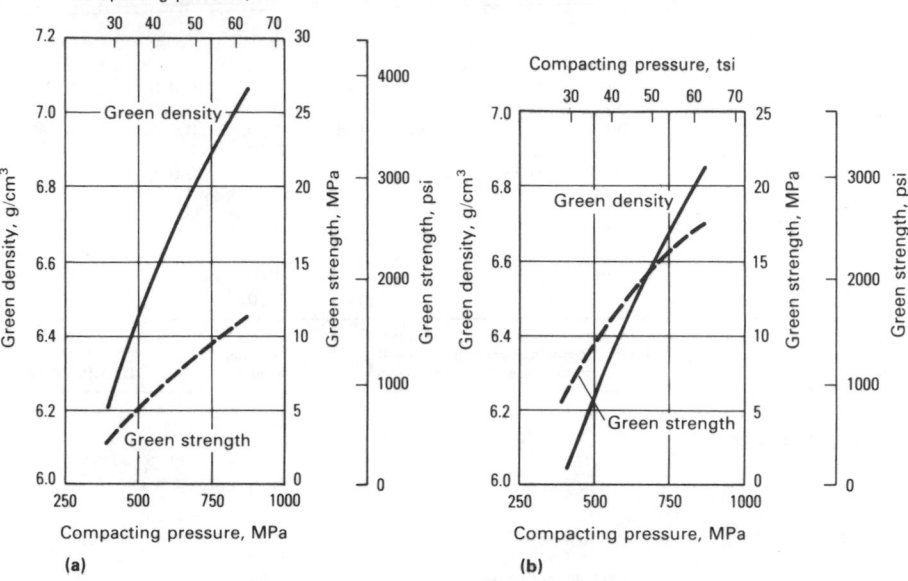

(a)

(b)

information on sintered properties and applications of stainless steel powders, see the article "P/M Stainless Steels" in this Volume.

Gas (nitrogen or argon) atomized stainless steel powders have particles that are spherical in shape (Fig. 3). These powders have high apparent densities of about 5 g/cm³ and excellent flow rates. Except for gravity sintering, they require special methods of consolidation (see the articles "Hot Isostatic Pressing of Metal Powders" and "Cold Isostatic Pressing of Metal Powders" in this Volume). Oxygen contents are less than 200 ppm, and mechanical properties and corrosion resistance of hot isostatically pressed or extruded and fully dense products (bar, billet, and tubes) are equal to or greater than those of conventionally produced wrought products.

Commercially available grades include American Iron and Steel Institute (AISI) grades 410, 440C, 446, 304L, 304, 347, 316L, 316, and 317. Chemical compositions are similar or identical to conventional wrought grades. Special compositional adjustments are unnecessary due to the absence of water during atomization and due to the special consolidation methods.

Low-Alloy Steels

Water-atomized low-alloy steel powders became available in the early 1960's as a direct result of the invention and commercialization of water-atomized iron powders made from low-carbon iron (see the article "Production of Iron Powders" in this Volume). These powders require the same type of melting and atomizing facilities. Until the 1960's, low-alloy P/M parts were produced from mixtures of elemental components, as they still are to a limited extent. Fully alloyed powders, however, require lower sintering temper-

atures and/or shorter sintering times to avoid the risk of component segregation and incomplete alloying. In diffusion-bonded grades (Table 2), powder blends are heated in a reducing atmosphere, which creates a bond between the iron and alloy particles and minimizes segregation due to demixing. At the same time, the superior compressibility typical of elemental powder blends is retained.

Low-alloy steel powders that contain chromium and/or manganese have become available recently and are enjoying increasing usage due to lower production costs and superior hardenability. Water-atomized low-alloy steels containing these elements are produced under conditions that minimize oxidation; residual chromium and manganese oxides, like other nonmetallic inclusions, are very detrimental to dynamic properties of fully dense parts.

Consequently, special precautions, such as minimizing turbulence of the liquid metal within the tundish, are taken to prevent entrapment of slag particles during melting and pouring prior to atomization. An inert atomizing chamber and more rapid quenching of the metal droplets also contribute to lower oxygen content. Melting furnaces with basic magnesia linings are preferred over acid silica linings. Hard water used for atomization may cause the presence of calcium-rich oxides. Silicon concentration is kept low (0.05%) because of the tendency of silicon to oxidize readily during atomization and its difficult reduction during sintering. Higher sintering

Fig. 3 Scanning electron micrograph of nitrogen-atomized 316L stainless steel

Magnification: 65×

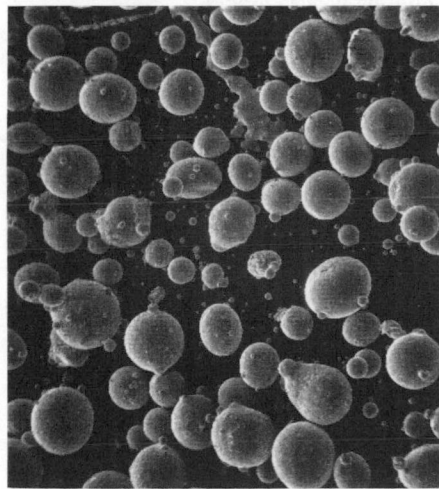

temperatures, combined with a low dew point and carbon as the reducing agent, effectively reduce chromium and manganese oxides that are formed during atomization.

Table 2 lists compositions and properties of typical commerical grades of low-alloy steel powders. Figure 4 is a scanning electron micrograph of a water-atomized low-alloy steel powder containing 0.5% Ni, 0.5% Mo, 0.3% Mn, 0.2% C, and the remainder iron. The particle shape shown is typical of this class of powder.

Table 2 Commercial low-alloy steel powders

Material	C	Mn	Cr	Ni	Mo	S	Composition, % Hydrogen loss	O	Si	P	Cu	Fe
Prealloyed grades												
Alloy A	0.01	0.30		0.45	0.60	0.020	0.20	0.17	0.01	0.01		rem
Alloy B	0.01	0.16		1.80	0.50	0.020	0.20	0.15	0.01	0.01		rem
Alloy C	0.05	0.25	0.08	1.90	0.50	0.015	0.28	0.13	0.01	0.02		rem
Alloy D	0.10	0.40	0.17	0.24	0.30	0.015	0.33	0.22	0.01	0.015		rem
Alloy E	0.10	0.33	0.21	0.23	0.29	0.024	0.32	0.21	0.01	0.014		rem
Diffusion-bonded grades												
Alloy F	0.01			1.75	0.50		0.10				1.50	rem
Alloy G	0.01			4.00	0.50		0.10				1.50	rem

Material	Approximate particle size		Apparent density, g/cm³	Flow rate, s/50 g	Compressibility, g/cm³, at: 414 MPa (30 tsi)	500 MPa (36 tsi)	372 MPa (27 tsi)
Prealloyed grades							
Alloy A	10% >147 μm	23% <43 μm	3.0	22	6.57	...	...
Alloy B	10% >147 μm	25% <43 μm	3.0	22	6.47	...	...
Alloy C	5-10% >147 μm	20% <43 μm	3.0	25	6.45	...	...
Alloy D	5-10% >147 μm	20% <43 μm	3.0	25	6.57	...	...
Alloy E	5-10% >147 μm	20% <43 μm	3.0	25	6.49	...	...
Diffusion-bonded grades							
Alloy F	20-175 μm(a)		3.0	26	6.80	6.98	6.75
Alloy G	20-175 μm(a)		3.0	26	6.70	6.98	6.70

(a) Particle size range

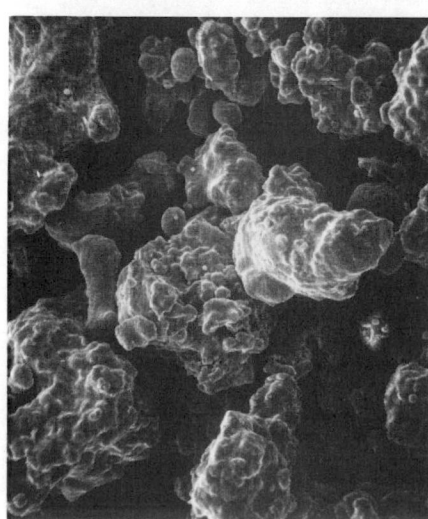

Fig. 4 Scanning electron micrograph of water-atomized low-alloy steel
−80 mesh. Magnification: 325×

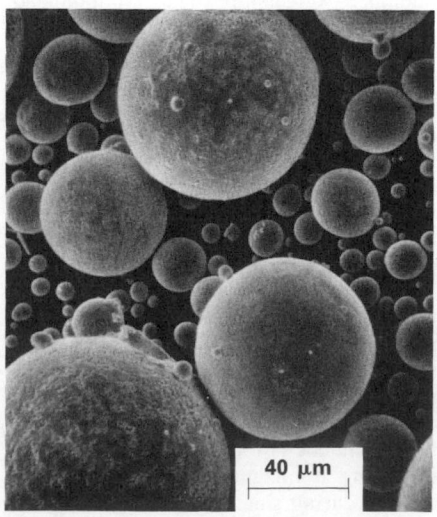

Fig. 5 Scanning electron micrograph of nitrogen-atomized T15 tool steel

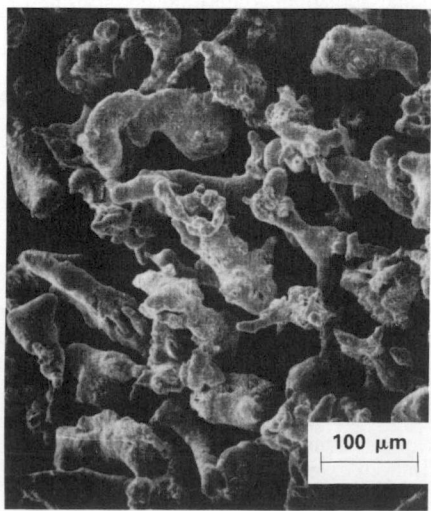

Fig. 6 Scanning electron micrograph of water-atomized high-speed tool steel

Table 3 Nominal compositions of gas-atomized tool steels

Alloy	C	Mn	Si	Cr	W	Mo	V	Co	S	Fe
M2S	1.00	0.30	0.30	4.15	6.40	5.00	1.95	⋯	0.12	rem
M4	1.35	0.30	0.30	4.25	5.75	4.50	4.00	⋯	⋯	rem
M42	1.10	⋯	⋯	3.75	1.50	9.50	1.15	8.00	⋯	rem
T15	1.55	0.30	0.30	4.00	12.25	⋯	5.00	5.00	⋯	rem
CPM 76(a)	1.50	0.30	0.30	3.75	10.00	5.25	3.10	9.00	⋯	rem
CPM 10V(a)	2.45	0.50	0.90	5.25	⋯	1.30	9.75	⋯	0.07	rem
CPM Rex 25(a)	1.80	0.30	0.35	4.00	12.25	6.50	5.00	⋯	0.07	rem
M3 type 2	1.27	0.30	0.30	4.20	6.40	5.00	3.10	⋯	⋯	rem
ASP302(b)	1.27	0.30	0.30	4.20	6.40	5.00	3.10	8.50	⋯	rem
ASP602(b)	2.30	0.30	0.40	4.00	6.50	7.00	6.50	10.50	⋯	rem

(a) Trademark of Crucible Specialty Metals, Division of Colt Industries. (b) Trademark of Uddeholm, Sweden

Table 4 Composition and properties of water-atomized high-speed tool steels

Properties	M2	M3 type 2	M42	T15
Composition, %				
Carbon	0.85	1.20	1.10	1.60
Chromium	4.15	4.10	3.75	4.40
Tungsten	6.30	6.00	1.50	12.50
Molybdenum	5.00	5.00	9.50	⋯
Vanadium	1.85	3.00	1.15	⋯
Cobalt	⋯	⋯	8.00	5.00
Iron	rem	rem	rem	rem
Oxygen content, ppm	<1000	<1000	<1000	<1000
Physical properties				
Apparent density, g/cm³	2.2	2.1	2.3	1.8
Tap density, g/cm³	3.1	3.0	3.3	2.4
Flow rate, s/50 g	45	40	30	50
Sieve analysis (Tyler)				
+100 mesh (<150 μm)	⋯	⋯	⋯	⋯
−100+150	13	13	13	13
−150+200	22	22	22	22
−200+325	30	30	30	30
−325 (<44 μm)	35	35	35	35
Green density(a), g/cm³				
Pressed at 620 MPa (45 tsi)	6.2	6.0	6.0	6.15
Pressed at 830 MPa (60 tsi)	6.6	6.4	6.3	6.55
Green transverse-rupture strength(a), MPa (psi)				
Pressed at 620 MPa (45 tsi)	23 (3300)	24 (3500)	21 (3000)	43 (6200)
Pressed at 830 MPa (60 tsi)	52 (7500)	48 (7000)	41 (6000)	69 (10 000)

(a) Green properties determined using die wall lubrication.

Fig. 7 Microstructures of water-atomized T15 tool steel powder

(a) As-atomized. (b) After annealing

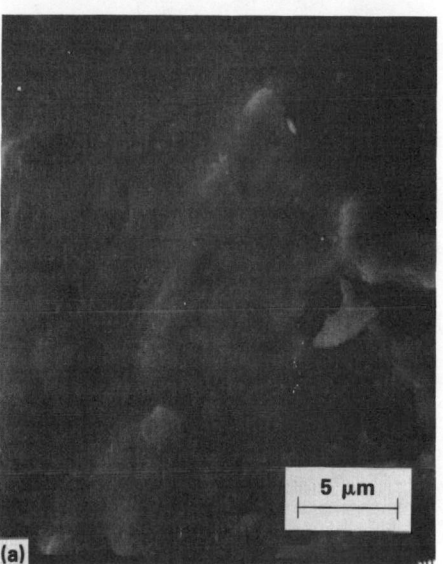

5 μm

(a)

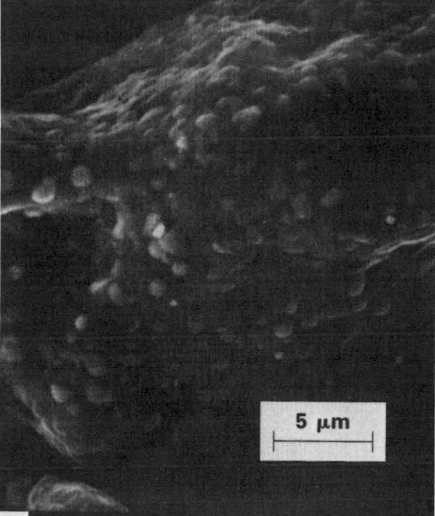

5 μm

(b)

Tool Steels

Powder metallurgy tool steel development in the United States began in the 1970's. Success of P/M tool steels is based mainly on the uniform microstructure obtainable, compared to wrought and conventionally produced products. Superior uniformity of composition leads to excellent toughness and less distortion during heat treatment, which in turn reduces total grinding costs and provides other benefits, such as uniform hardness and increased tool life. For additional information, see the article "P/M Tool Steels" in this Volume.

Tool steel powders are produced by induction melting of virgin raw materials and/or scrap. Gas or water atomization methods are used. Gas (argon or nitrogen) atomized tool steel powders have particles that are spherical in shape (Fig. 5) with high apparent densities (about 6 g/cm³). Oxygen content is less than 200 ppm. These powders are hot isostatically pressed into semifabricated shapes of full density. Complex shapes can be made by cold isostatic pressing, followed by hot isostatic pressing. Standard and proprietary compositions of gas-atomized tool steel powders are given in Table 3.

Water-atomized tool steel powders have irregularly shaped particles (Fig. 6) and are suitable for conventional die compaction and sintering to high or theoretical density. Unlike stainless steel and low-alloy steel powders, water-atomized tool steel powders are vacuum deoxidized and vacuum annealed to improve their compacting properties. Deoxidation occurs at elevated temperature as a solid-state re-

action between the oxygen and carbon within individual particles, in which the oxygen content of the powder is lowered from about 1500 to 3000 ppm to below 1000 ppm.

Annealing lowers hardness from 700 HV to less than 300 HV. Further reduction of oxygen to less than 100 ppm occurs during sintering. Water-atomized tool steel powders may be pulverized for very fine particle size and accelerated sintering characteristics.

Powder cleanliness is important for tool steel powders. Powder producers provide high standards of cleanliness through campaigning of heats, dedication of production equipment, use of powder cleaning

equipment, and meticulous housekeeping.

Typical compositions and powder properties of water-atomized high-speed steel powders are given in Table 4. Small (0.15 to 0.2%) amounts of graphite sometimes are added to the powder to adjust or increase carbon content of the sintered product.

Figure 7 illustrates the microstructures of water-atomized T15 tool steel before and after annealing. Annealing depletes the carbon from the martensitic matrix by forming carbides. For more information on the processing of gas-atomized tool steel powders into net shape parts and related applications, see the article "P/M Tool Steels" in this Volume.

Production of Copper Powder

COMMERCIAL PRODUCTION OF COPPER POWDER originally resulted from the invention and development of the self-lubricating porous bronze bearing in the 1920's. The first two large-scale processes in copper powder production were reduction of copper oxide and electrolysis. Copper powder made by the cementation process was used in copper-based composite friction materials developed in the 1930's. Other hydrometallurgical processes for copper powder production were developed in the 1950's and 1960's, some of which were used commercially for a number of years.

Further growth of copper powder usage occurred with the development, during World War II, of the ferrous structural parts industry, which uses copper additions for alloy strengthening. Later, in the 1950's, atomization of copper and copper alloys became commercially significant.

This article discusses the fundamentals of the four processes used for production of copper powders, as well as the physical and mechanical properties of copper powders produced by these processes, and applications of copper powders. More detailed information can be found in the articles in this Volume that deal with individual powder-producing methods.

Production of Copper Powder by the Reduction of Copper Oxide

By Erhard Klar
Manager of Particle Technology
SCM Metal Products

REDUCTION OF COPPER OXIDE is the oldest and, by tonnage, the most extensively used process for the production of copper powder. In this process, particulate copper oxide is reduced with gaseous reducing agents at elevated temperatures. The product is essentially a sintered porous cake of copper that must be ground to powder. Raw materials include copper scale, cement copper, particulate copper scrap, and atomized copper. The latter two are roasted to form cuprous or cupric oxide, or a mixture of both.

Figure 1 shows various approaches for producing copper powder by oxide reduction. They differ mainly in the degree of oxidation of the starting material prior to reduction. Powders made by atomization and subsequent oxidation and reduction are sometimes designated as atomized or reduced powders. Actually, such powders may be partly spongy and partly solid; that is, they may possess the characteristics of both atomized and oxide-reduced powders. Truly atomized copper powders (those with a negligible degree of oxidation during processing) are described later in this article.

Oxide-reduced copper powders can be tailored to possess the properties needed in all major applications of copper powders. Those applications and significant powder-related property requirements are:

Bronze bearings

- Controlled dimensional change during sintering
- Sintering rate
- K-factor
- PV-factor
- Permeability
- Interconnected porosity
- Green strength

Friction parts

- High green strength
- High thermal conductivity

Carbon brushes

- High green strength
- High electrical conductivity

Structural parts, pure copper

- High electrical conductivity
- Powder flow

Structural parts, bronze

- Green strength
- Compressibility
- Powder flow

Additive to P/M steel

- Green strength
- Compressibility
- Powder flow

Infiltration of porous P/M parts (generally in the form of a copper-based prealloy or multicomponent blend)

- Green strength
- Nonerosive and nonadhesive characteristics
- Efficiency

Conductive paints and decorative fillers

- High surface area (flake copper)

Fig. 1 Processing steps of copper powder manufacture by reduction of copper oxide

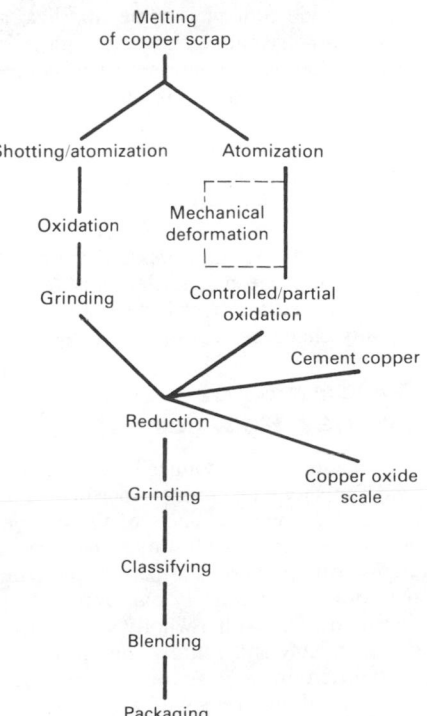

For a discussion of these applications, see the appropriate articles in the Section "Powder Systems and Applications" in this Volume. Integral to the entire oxide reduction process are the stages of copper melting, atomization, and oxidation reduction and post-reduction handling.

Melting of Copper

In recent years, the use of high-quality elemental copper as the starting material has increased at the expense of both cement copper and copper oxide scale. Reasons for this choice include availability of elemental copper and the need for better control of finished product purity. A typical grade of cement copper has rather high levels of iron and acid insolubles that largely limit its use to friction material applications, in which it is valued for its high green-strength characteristics. Chemical analysis of a typical dry (a wet sample has a moisture content of 11%) cement copper is:

Component	Weight, %
Total copper	75
Iron	6
Sulfur	1
Nitric acid insolubles	2
Hydrogen (loss)	16

Copper melting is performed in both fuel-fired and induction-heated furnaces. Strong oxide formers, such as aluminum and silicon, are kept low; thus, the molten metal remains fluid and can be poured without difficulty. Aluminum and silicon oxides also make the powder less compressible and very abrasive. In addition, lead and tin cause problems in pouring the melt due to buildups and clogging in the furnaces and nozzles.

In some applications of copper powder (for example, metal-graphite brushes and friction parts), good electrical or thermal conductivity is important. For this reason, impurity levels must be kept low. The detrimental effect on electrical conductivity caused by some impurities that are present in solid solution is illustrated in Fig. 2. Thermal conductivity is similarly dependent on the impurity content.

Atomization and Shotting of Copper

Large-scale atomization of copper usually is conducted in air as a continuous process. Liquid copper may be atomized directly from a tube in the side of the furnace wall or through a tundish. Both air and water may be used as atomizing mediums, and horizontal atomization with

Fig. 2 Effect of impurities in solid solution on electrical conductivity of oxygen-free copper

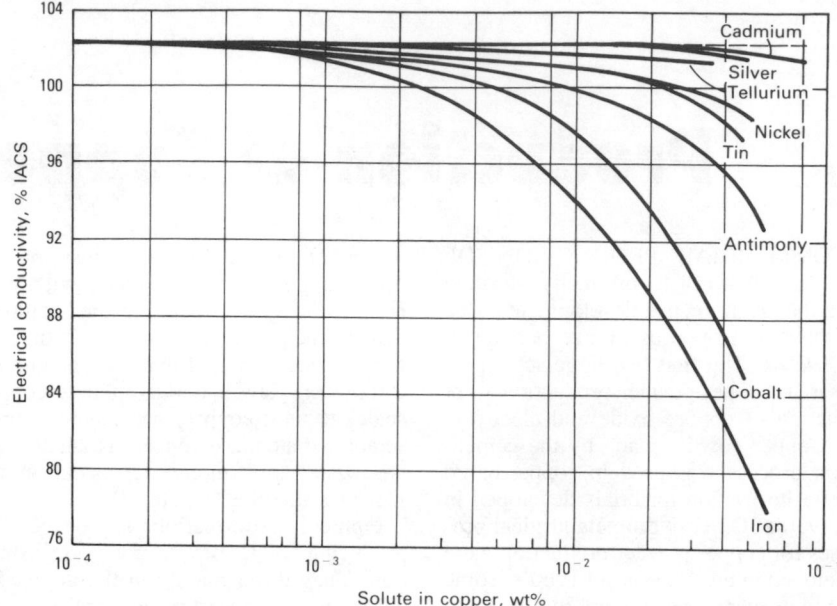

high-pressure air into a rotating drum eliminates the need for drying of the powder. For details of atomization, including nozzle configurations and pressure relationships, see the article "Atomization" in this Volume.

Scanning electron micrographs of air- and water-atomized copper powders are shown in Fig. 3. The water-atomized powder has a more irregular particle shape, and its oxide content is somewhat lower. At this process stage, however, particle shape is not very important. The coarse shot-type powder that is produced by low-pressure air or water atomization (Fig. 4) is the typical starting material for the pure copper oxide process—the approach that relies on complete oxidation of atomized or shotted copper prior to its reduction (see Fig. 1). The fine high-pressure atomized powders may be ball milled and/or partially oxidized to effect further shape and porosity changes.

Oxidation of Copper Powder

Oxidation of air-atomized, water-atomized, or shotted copper is performed to radically change the shape of the powder and thus enhance control over the various engineering properties of parts made from the powder. Completely oxidized and reduced powders with a wholly spongy (porous) structure and entirely solid powders, as typified by a gas-atomized powder, constitute the extremes of available copper powders. Partial oxidation results in intermediate structures.

The oxidation of copper is a well-known process. There are two copper oxides, the red cuprous oxide, Cu_2O, and the black cupric oxide, CuO. At high temperatures and for the so-called thick film range, oxidation follows the parabolic rate law, in which film thickness (y) increases with the square root of time ($y = \sqrt{k_p t + c}$). At low temperatures, linear, logarithmic, and cubic oxidation rates have been observed, depending on the history of the oxide. Free energies, heats of reaction, and rates of copper oxide formation are shown in Table 1.

Figure 5 shows the effect of oxygen pressure and temperature on composition of the outer oxide film. In air of atmospheric pressure, the outer oxide film consists of cupric oxide if oxidation takes place between about 270 and 990 °C (520 and 1815 °F), the dissociation temperature for cupric oxide.

As with other multivalent metals, the oxide in contact with elemental copper is the lower valence cuprous oxide, and the outer oxide is cupric oxide. The relative thicknesses of cuprous oxide and cupric oxide depend on the diffusion rates of the ions through the layers, the gradient of the chemical potential across the layers, and porosity. As shown in Fig. 6, for a plane sheet of copper, the ratio of cuprous oxide to cupric oxide (Cu_2O/CuO) increases with rising temperature of oxidation. For copper powder, deviations from this behavior

Fig. 3 Scanning electron micrographs of copper powder
(a) Air atomized. (b) Water atomized

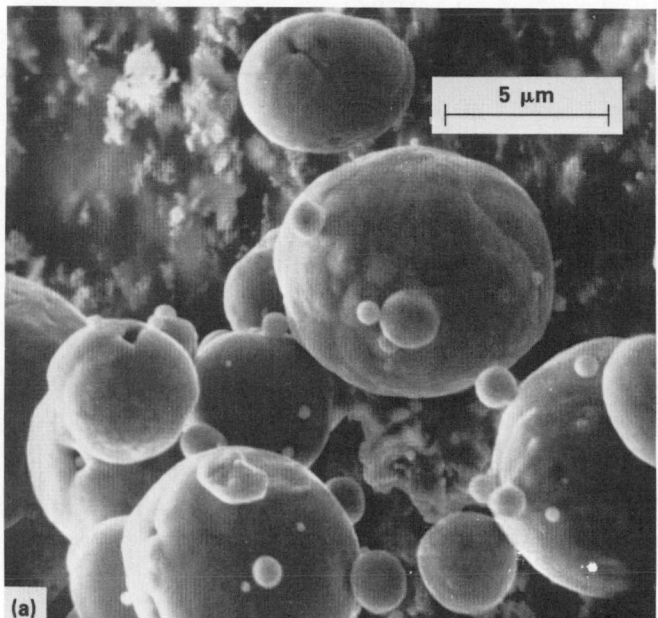

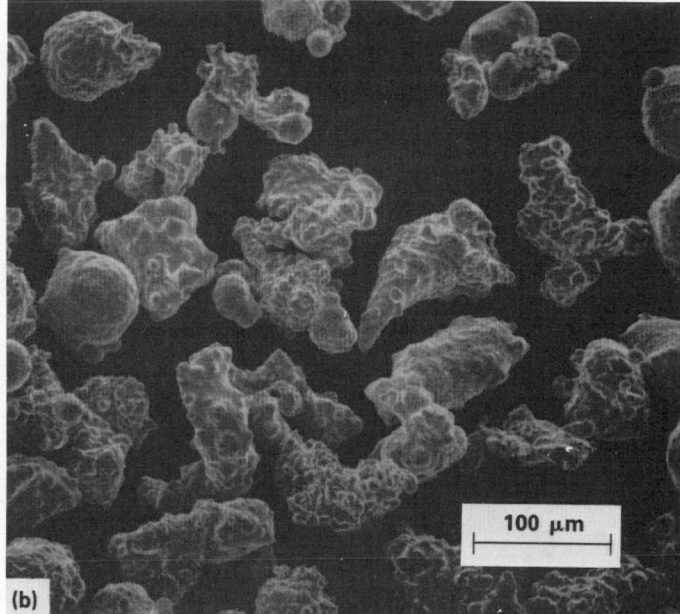

Fig. 4 Air-shotted copper

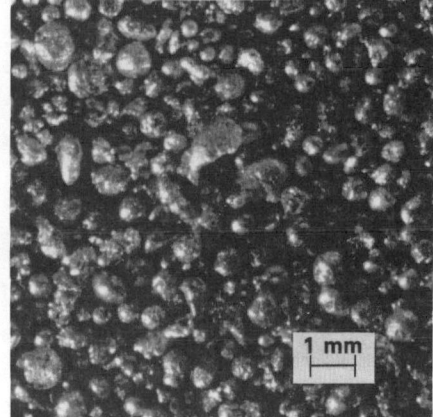

Table 1   Free energies, heats, and rates of copper oxide formation

ΔG is free energy; ΔH is heat; values of ΔG and ΔH are given in calories per gram mole; k is a rate constant mathematically derived; T is the absolute temperature, degrees kelvin; R is the absolute gas constant; ln is the natural logarithm (base e, where e = 2.7182).

$2\langle Cu \rangle + 1/2(O_2) = \langle Cu_2O \rangle$ exothermic
$\Delta G = -41\,166 - 1.27 \times 10^{-3}TlnT + 3.7 \times 10^{-3}T^2 - 1.80 \times 10^{-7}T^3 + 27.881T$
$k = 957\ e^{-37\,700/RT}\ g^2cm^{-4}h^{-1}$

$\langle Cu \rangle + 1/2(O_2) = \langle CuO \rangle$ exothermic
$\Delta G = -37\,353 - 0.16TlnT - 1.69 \times 10^{-3}T^2 - 9 \times 10^{-8}T^3 + 25.082T$
$\Delta H = -38\,170 + 1.30T + 0.99 \times 10^{-3}T^2 + 0.57 \times 10^5T^{-1}$

$\langle Cu_2O \rangle + 1/2(O_2) = 2\langle CuO \rangle$ exothermic
$\Delta G = -33\,550 + 0.95TlnT - 3.75 \times 10^{-3}T^2 + 22.340T$
$\Delta H = -35\,710 + 3.28 \times T - 0.40 \times 10^{-3}T^2 - 0.20 \times 10^5T^{-1}$
$k = 0.0268\ e^{-20\,140/RT}\ g^2cm^{-4}h^{-1}$

occur due to geometric factors and larger stresses within the oxide films that cause detachment of copper scale from the metallic copper. When all copper is consumed, oxidation proceeds with cupric oxide formation.

In commercial practice, oxidation or roasting of copper powder normally is done in air at temperatures above 650 °C (1200 °F). Oxidation in rotary kilns or fluidized beds provides faster oxidation rates by increasing the contact area between powder and oxidizing gas. Due to the strongly exothermic nature of the oxidation reactions, these operations are more difficult to control than roasting in a belt conveyor furnace.

Grinding of Copper Oxide. Both oxides of copper are brittle and easy to grind to −100 mesh powder. The oxide particles themselves are porous. Figure 7 shows oxidized shot before and after grinding.

Reduction of Partially or Fully Oxidized Ground Copper

Reduction of particulate copper oxide is generally accomplished on a stainless steel belt in a continuous belt furnace. The depth of the oxide bed is about 25 mm (1 in.). The typical reduction temperature ranges from 425 to 650 °C (800 to 1200 °F). Reduction occurs gradually from top to bottom of the bed. The reducing atmosphere in the furnace generally flows countercurrent to the motion of the conveyor belt.

Reducing atmospheres may include hydrogen, dissociated ammonia, water-reformed natural gas, or other endothermic or exothermic gas mixtures. Because reduction of copper oxides with hydrogen or carbon monoxide is exothermic, a careful balancing of oxide particle size, reducing gas species, and reduction temperature is necessary to optimize the reduction rate and to control the pore structure. Hydrogen diffuses readily through solid copper and is a more effective reducing agent than carbon monoxide, particularly at low temperatures. At higher temperatures, however, all reduction reactions involving either hydrogen or carbon monoxide proceed almost to completion. Free energies and heats of reaction for copper oxide reduction with hydrogen and carbon monoxide are shown in Table 2.

Fig. 5 Composition of outer oxide film of copper for short periods of oxidation as a function of oxygen pressure and temperature

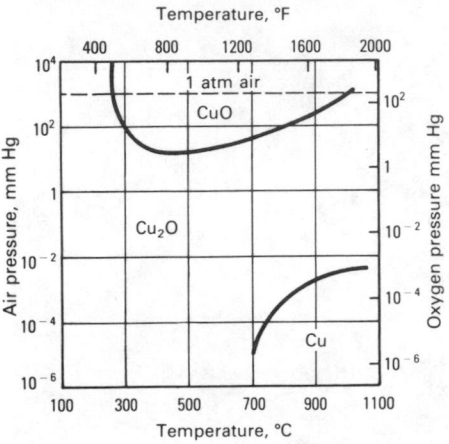

Fig. 6 Effect of oxidation temperature on the weight composition of the copper oxide layer

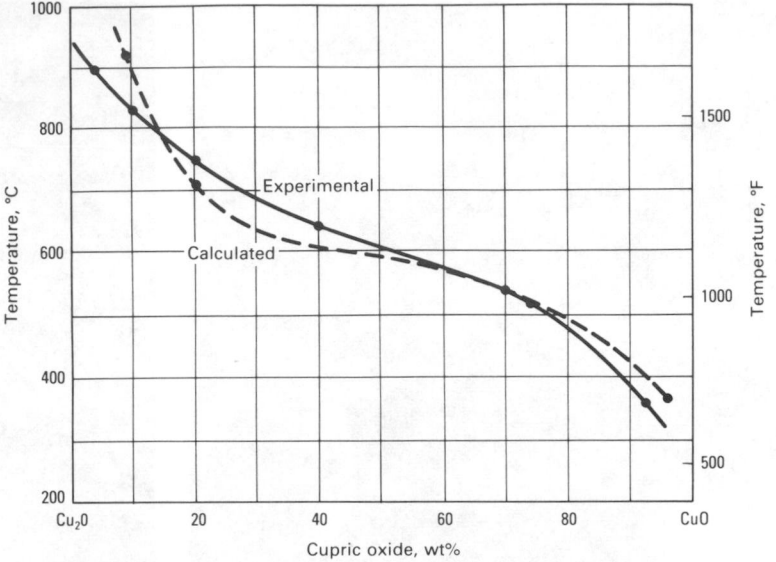

Control of Powder Properties. By manipulating the reduction process, particle porosity, pore size, and particle size distribution of the finished product can be controlled over a wide range. As with other metal oxides, low reduction temperatures generally produce particles having fine internal porosity ranging from 15 to 30%, pore sizes of about 1 μm, and corresponding large specific surface areas of 0.1 m²/g or larger (Fig. 8 and 9). High reduction temperatures of 815 °C (1500 °F) or higher produce particles containing pores as large as several microns and specific surface areas as small as about 0.01 m²/g. High reduction temperatures generally result in more interparticle sintering and more complete reduction.

Figure 10 shows polished cross sections of partially oxidized and reduced particles. These powders are partly spongy and partly solid. Figure 11 illustrates that compacted green strength increases, at the expense of compressibility, with an increasing amount of preoxidation. In other words, green strength of powder improves as its sponginess increases.

Post-Reduction Procedure. The reduced copper oxide emerges from the reduction furnace as a porous cake (Fig. 12). It is broken into smaller pieces in a jaw crusher or similar equipment, followed by fine grinding in hammer mills. Particle size and shape of the ground powder depend on several factors. With coarse copper oxide and low reduction temperatures, interparticle sintering during reduction is moderate; therefore, milling of the copper cake can produce a copper particle size almost that of the original oxide. For fine copper oxide and high reduction temperatures,

Fig. 7 Micrographs of copper oxide
(a) As-oxidized copper shot. (b) Scanning electron micrograph of copper shot after grinding

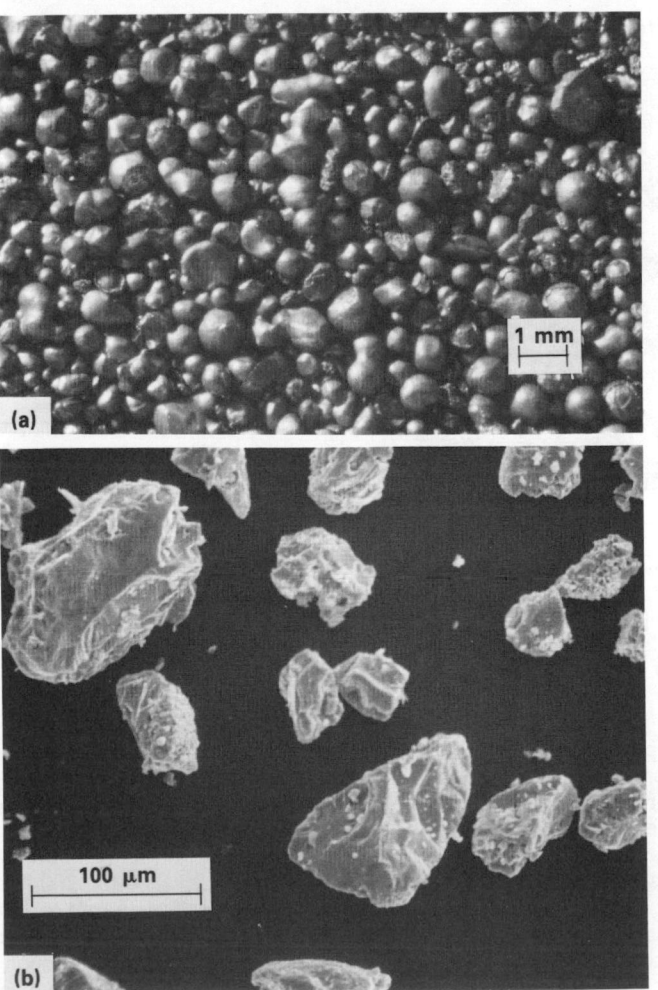

Table 2 Free energies, heats, and partial gas pressures for copper oxide reduction with hydrogen and carbon monoxide

ΔG is free energy; ΔH is heat; values of ΔG and ΔH are given in calories per gram mole; P is pressure; T is the absolute temperature, degrees kelvin; and $\ln$ is the natural logarithm (base e, where e = 2.7182).

$\langle Cu_2O \rangle + (H_2) \rightleftharpoons 2\langle Cu \rangle + (H_2O)$ exothermic
$\Delta G = -16\,260 + 2.21T\ln T + 1.28 \times 10^{-3}T^2 + 3.8 \times 10^{-7}T^3 - 24.768T$
$\Delta H_{298.1\,K} = -17\,023$
Temperature, °C (°F): 450 (840), 900 (1650), 950 (1740), 1000 (1830), 1050 (1920)
P_{H_2}, torr(a):   0.0104,   0.0150,   0.0207,   0.0283

$\langle Cu_2O \rangle + (CO) \rightleftharpoons 2\langle Cu \rangle + (CO_2)$ exothermic
$\Delta G = -27\,380 + 1.47T\ln T - 1.4 \times 10^{-3}T^2 + 0.5 \times 10^{-6}T^3 - 7.01T$
$\Delta H = -27\,380 - 1.47T + 1.4 \times 10^{-3}T^2 - 1.1 \times 10^{-6}T^3$
Temperature, °C (°F): 25 (77), 900 (1650), 1050 (1920), 1083 (1980)
P_{CO}, torr(a):   0.021,   0.068,   0.085

$2\langle CuO \rangle + (H_2) \rightleftharpoons \langle Cu_2O \rangle + (H_2O)$ exothermic
$\Delta G = -24\,000 - 0.01T\ln T + 5.4 \times 10^{-3}T^2 - 3.7 \times 10^{-7}T^3 + 22.896T$
$\Delta H_{298.1\,K} = -23\,543$

$\langle CuO \rangle + (H_2) = (H_2O) + \langle Cu \rangle$ exothermic
$\Delta H_{290\,K} = -31\,766$

$2\langle CuO \rangle + (CO) \rightleftharpoons \langle Cu_2O \rangle + (CO_2)$ exothermic
$\Delta H = -33\,300$

(a) Total pressure is 1 atm.

Fig. 8 Scanning electron micrographs of 180-μm copper oxide particles
Reduced at (a) 315 °C (600 °F), (b) 595 °C (1100 °F), and (c) 980 °C (1800 °F)

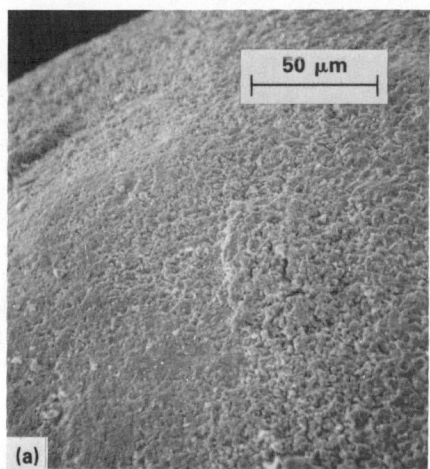

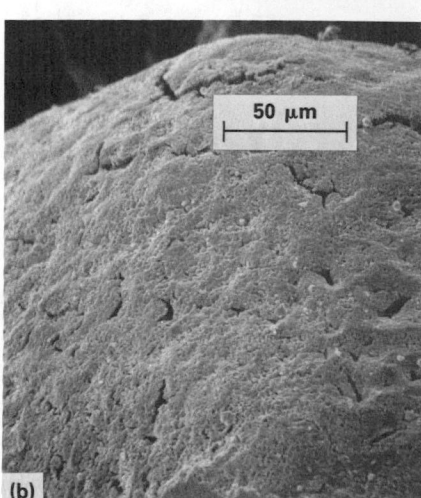

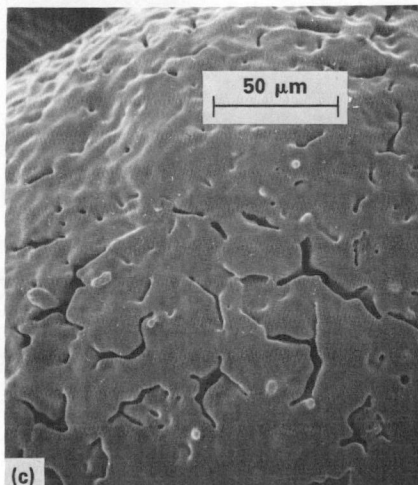

considerable sintering and interparticle bonding take place during reduction. As a result, more intense grinding is necessary and new particle size distributions are generated. In spite of this, the amount of work hardening is still moderate, and the resulting powders have good compressibility and green strength. A scanning electron micrograph of a typical copper powder is shown in Fig. 13.

Various combinations of the controlling parameters during melting, atomization, oxidation, reduction, and grinding permit the manufacture of powders with characteristics tailored to the requirements of the different applications listed at the beginning of this section. For some applications, complete oxidation of copper with subsequent reduction may be substituted by partial oxidation or a combination of mechanical deformation and partial oxidation followed by reduction (see Fig. 1).

The reduced and milled powder is screened and/or classified and, if necessary, blended and lubricated. These processing steps are carefully controlled to avoid losses or uncontrolled changes in characteristics such as apparent density, amount of fines, and powder flow. Some powder grades are treated with proprietary antioxidants to stabilize them against oxidation. Without this treatment, copper powders, particularly when exposed to moist air, generally tarnish, with attendant losses in green strength (Fig. 14) and other side effects. With increasing tarnish, the color of the powder changes from orange to purple to black. Simultaneously, oxygen content increases from a typical 0.1 or 0.2% to several tenths of a percent, and as much as about 1%. Copper powders with

Fig. 9 Effect of reduction temperature on specific surface area of copper
Hydrogen reducing gas

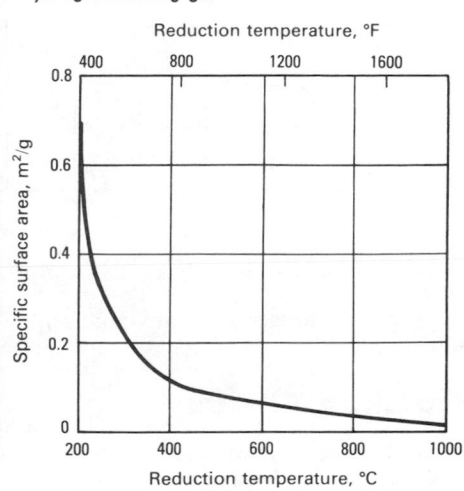

Fig. 10 Copper particles
(a) After partial oxidation. (b) After partial oxidation plus reduction

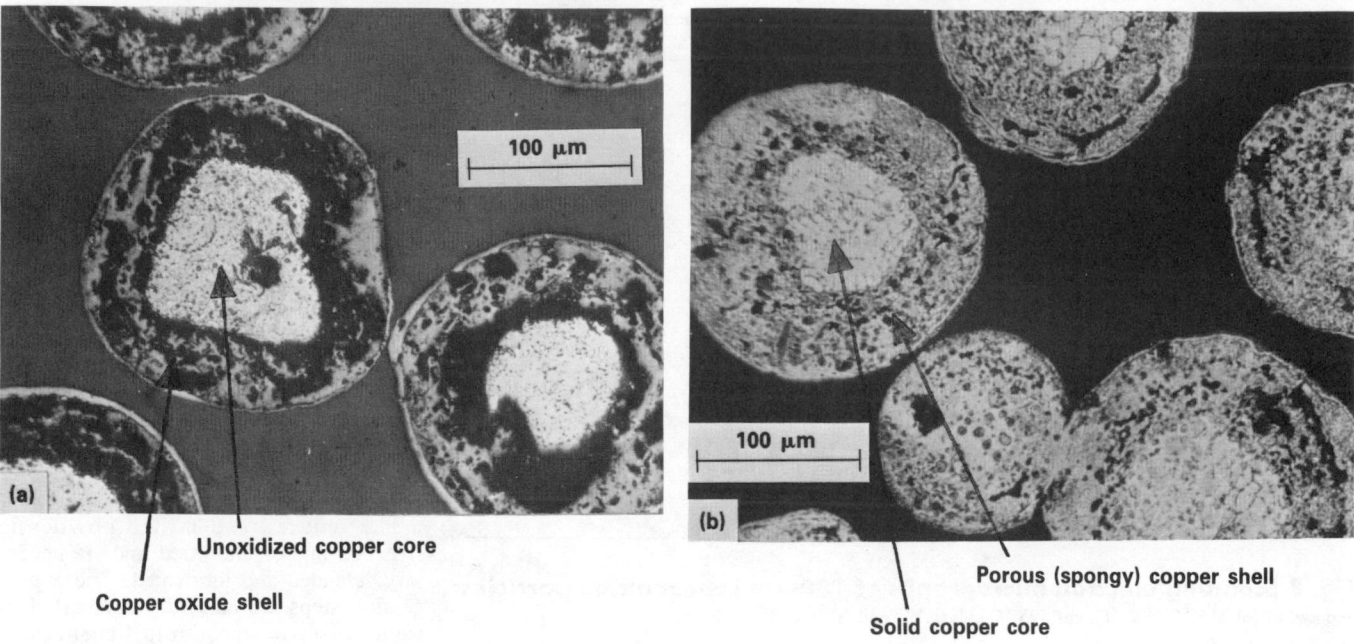

100 μm

100 μm

(a)

(b)

Unoxidized copper core

Copper oxide shell

Porous (spongy) copper shell

Solid copper core

Fig. 11 Effect of the degree of oxidation of copper powder on green density and green strength properties of reduced powders
Material was 100/150 mesh copper oxidized to various levels of oxygen and reduced in hydrogen.

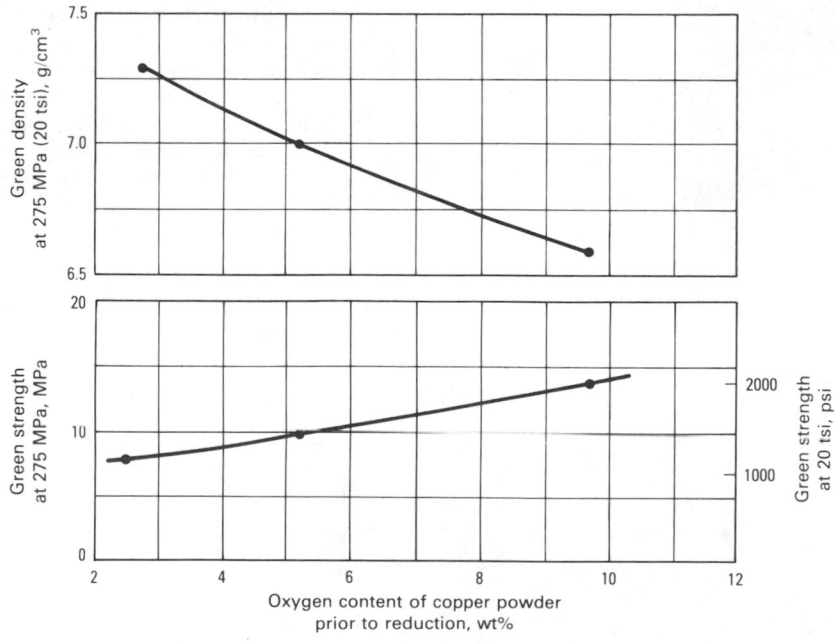

Cu and 10% Sn that have graded dimensional change characteristics during sintering are increasing in use.

Table 3 lists various grades of copper powders produced by the oxide reduction process, including those used in bronze premixes. The product brochures and data sheets of the powder producers generally provide details on the characteristics and performance properties of their powders (including sintered properties), as well as recommendations for specific uses.

Production of Copper Powder by Electrolysis

By Pierre W. Taubenblat
Vice President and Associate Director
AMAX Base Metals Research &
Development, Inc.

large specific surface areas are more sensitive to tarnishing.

Finished Powders

Finished lots of powder are subjected to a series of tests designed to ensure performance in the various applications. The single most important application for copper powder is for self-lubricating bronze bearings. In the United States, these bearings are made from elemental blends of copper and tin. Since about 1960, preblended and lubricated mixtures of 90%

PRODUCTION OF ELECTROLYTIC COPPER POWDER, as shown in Fig. 15, follows the same electrochemical principles that apply to the electrorefining of copper. The conditions of deposition, however, are changed to yield a powdery or spongy deposit, instead of the strongly

Fig. 12 Fragment of reduced copper oxide cake

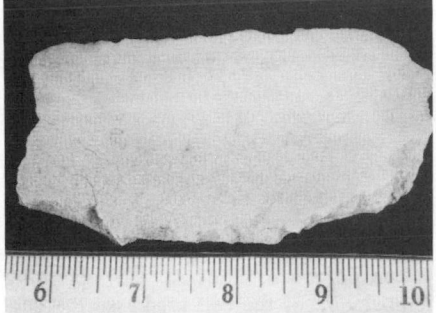

Fig. 13 Scanning electron micrograph of copper oxide reduced copper powder

50 μm

adherent product desired in electrorefining. The formation of powdery deposits is favored by low copper ion concentration and high acid content in the electrolyte, as well as by high cathode current density.

Although these conditions yield spongy deposits, control of additional variables is necessary to produce powders that meet commercial requirements. Other factors are quantity and type of addition agent, temperature and circulation rate of the electrolyte, size and type of anode and cathode, electrode spacing, and brush-down interval. For more detailed information on electrolytic production of metal powders,

see the article "Electrodeposition of Metal Powders" in this Volume.

Effects of Electrolyte Composition

The composition of the electrolyte is a major factor in the production of powder. Copper concentration in the electrolyte must be sufficiently low to prevent adherent deposits. In the desired range, current efficiency rises with increasing copper ion concentration, as shown in Fig. 16, with a maximum of 92% in the range of 23 to 33 g/L (3 to 4.4 oz/gal) copper. Above about 33 g/L (4.4 oz/gal), current efficiency decreases, and a hard deposit is produced instead of powder. Apparent density and particle size also increase with a rise in copper ion concentration.

Acid Concentration. A high acid concentration favors the formation of powder. As shown in Fig. 17, current efficiency

Fig. 14 Effect of tarnishing of copper powder on its green strength

Green strength was determined for copper powder oxidized to various interference colors.

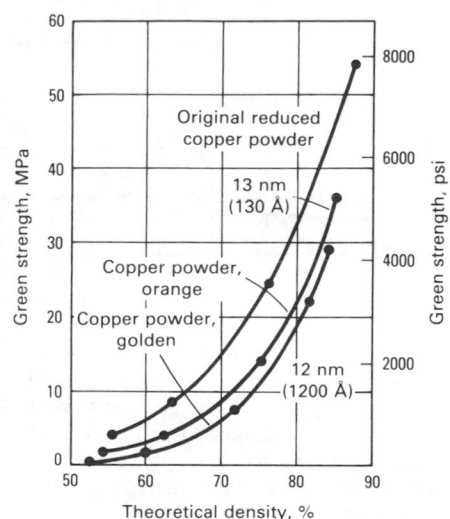

Table 3 Properties of commercial grades of copper powder produced by the copper oxide process

	Chemical properties, %					Physical properties			Tyler sieve analysis, %				Compacted properties		
Copper	Tin	Graphite	Lubricant	Hydrogen loss	Acid insolubles	Apparent density, g/cm³	Hall flow rate, s/50 g	+100	+150	+200	+325	−325	Green density, g/cm³	Green strength, MPa (psi), at: 165 MPa (12 tsi)	6.30 g/cm³
99.53 ..	...	...	...	0.23	0.04	2.99	23	0.3	11.1	26.7	24.1	37.8	6.04	6 (890)	...
99.64 ..	...	...	...	0.24	0.03	2.78	24	...	0.6	8.7	34.1	56.6	5.95	7.8 (1140)(a)	...
99.62 ..	...	...	...	0.26	0.03	2.71	27	...	0.3	5.7	32.2	61.8	5.95	9.3 (1350)(a)	...
99.36 ..	...	...	...	0.39	0.12	1.56	...	0.1	1.0	4.9	12.8	81.2	5.79	21.4 (3100)(a)	...
99.25 ..	...	...	...	0.30	0.02	2.63	30	0.08	7.0	13.3	16.0	63.7	...	...	8.3 (1200)(a)
90	10	...	0.75	...	...	3.23	30.6	0.0	1.4	9.0	32.6	57.0	6.32	...	3.80 (550)
88.5 ...	10	0.5	0.80	...	...	3.25	12(b)	...	...	...	...	...	...	...	3.6 (525)

(a) Measured with die wall lubricant only. (b) Carney flow

Fig. 15 Flowchart for production of electrolytic copper powder

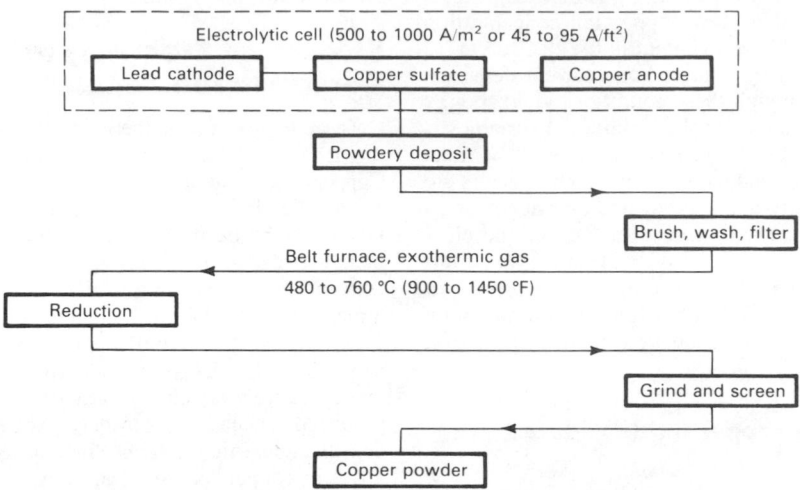

Fig. 16 Effect of copper concentration on current efficiency and apparent density
Source: Ref 1

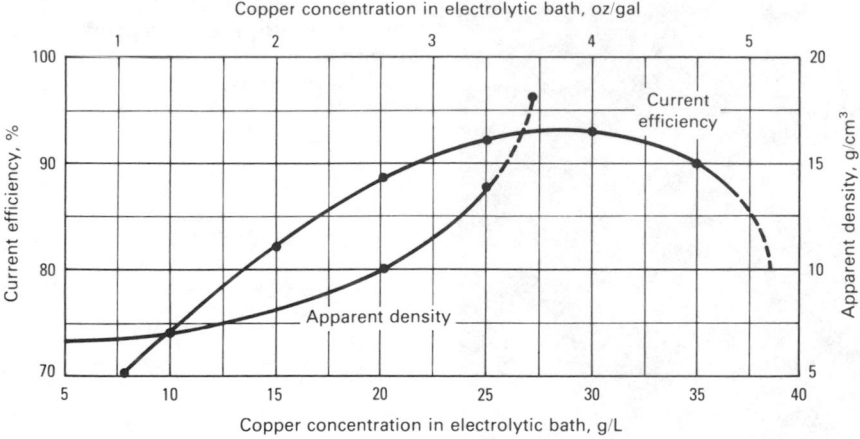

increases to a maximum at a concentration of 120 g/L (16 oz/gal) of sulfuric acid, then gradually falls as the acid concentration increases. Continued rise in acid concentration leads to passivity. Apparent density decreases with increasing acid concentration.

Addition Agents. The copper sulfate/ sulfuric acid electrolyte occasionally is modified to alter powder characteristics. The addition of colloidal materials, such as glue or glucose, leads to the formation of fine powder deposits, possibly because the colloid retards the evolution of hydrogen at the cathode. Table 4 indicates the effects of a number of additions to the electrolyte.

The addition of surfactants is reported to yield a powder with controllable particle size at a current density of 215 A/m² (20 A/ft²), in contrast with the 700 to 1100 A/m² (64 to 100 A/ft²) normally employed, which results in considerable reduction in power cost (Ref 2). Small quantities of copper chloride have been added to the electrolyte to increase the dendritic character of the powder particles and to increase the yield of fine powder due to the polarizing effect of the chloride ions (Ref 3). The addition of sodium sulfate reportedly reduces the cathode current density, and as the sulfate content is increased, the powder becomes finer. (Ref 4). By contrast, replacement of the normal sulfuric acid electrolyte by a sulfamate electrolyte favors the formation of a coarse copper powder (Ref 5).

Effects of Operating Conditions

Alteration of operating conditions can affect process variables such as current efficiency and the formation and size of particles.

Current Density. High current density favors the formation of powder, but has only a minor effect on current efficiency. A marked decrease in particle size occurs with rising current density. For example, in an electrolyte containing 25 g/L (3.3 oz/ gal) copper and 120 g/L (16 oz/gal) free sulfuric acid, increasing the current density from 600 to 1000 A/m² (55 to 90 A/ ft²) raised the quantity of powder passing through a 300 mesh sieve from 20 to 96%.

Temperature. Raising the operating temperature of the cell increases the current efficiency and reduces the cell voltage. However, cell operation is difficult at temperatures higher than 60 °C (140 °F), and powders obtained at high temperatures are coarser than those produced at lower temperatures. Generally, electrolytic cells are operated at temperatures between 25 and 60 °C (77 and 140 °F).

Powder Removal. The method used to remove the powder from the cathode has a significant effect on powder characteristics. Usually the powder is removed mechanically by brushing. The brush-down interval aids in control of the particle size of the deposit and, as shown in Fig. 18, the powder becomes coarser as the interval is increased from 15 to 60 min. Figure 19 indicates that the apparent density increases as the brush-down interval is extended. Frequent brush-down also limits variations in cathode current density. In another approach to powder removal, an organic extract is used in combination with sodium lauryl sulfate to discharge the powder automatically from the cathode without brush-down (Ref 6).

Powder Production

The major producers of electrolytic copper powder generally follow the procedures used in electrolytic refining. However, as has been mentioned, changes in operating conditions are required to obtain a deposit of powder. Typical conditions for the production of copper powder are illustrated in Fig. 15 and can be summarized as:

Condition	Quantity
Copper	5-15 g/L (0.6-2 oz/gal)
Sulfuric acid	150-175 g/L (20-23 oz/gal)
Temperature	25-60 °C (77-140 °F)
Anode current density	430-550 A/m² (39-50 A/ft²)
Cathode current density	700-1100 A/m² (64-100 A/ft²)
Cell potential	1.0-1.5 V

Generally, the anodes are electrolytically refined copper, and the cathodes are

Fig. 17 Effect of acid concentration on current efficiency and apparent density
Source: Ref 1

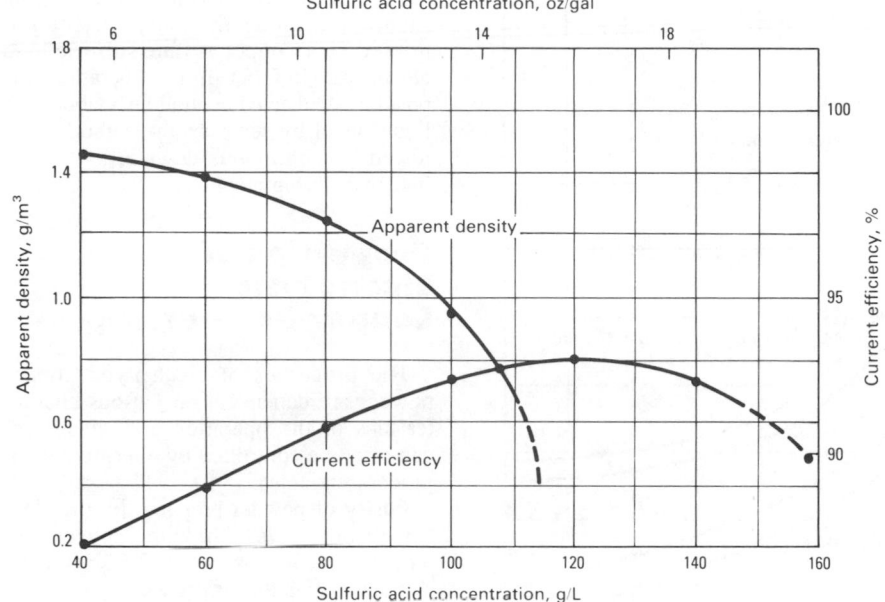

Table 4 Effect of addition agents on current efficiency and particle size

Test No.	Addition agent	Solution strength, %	Voltage, V	Current efficiency, %	Sieve analysis, % −200 mesh	Sieve analysis, % −300 mesh
1	···	···	1.0	95.9	74.6	55.0
2	Boric acid	0.5	1.0	95.2	···	100
3	Glucose	0.5	1.2	85.4	···	100
4	Glycerine	0.5	1.9	94.7	···	100
5	Glue	0.5	1.5	94.5	···	100

Source: Ref 1

lead alloy sheet. In a typical installation, the cathodes are 61 by 86 by 0.95 cm (24 by 34 by 0.37 in.) in dimension (Ref 7). Both anodes and cathodes are short to allow enough space at the bottom of the tank for the collection of the powder. The electrodes are arranged parallel to one another in lead-lined, rubber-lined, or plastic tanks, typically 3.4 m long by 1.1 m wide by 1.2 m high (11 ft long by 3.6 ft wide by 3.9 ft high). Each cell contains 18 cathodes spaced at 16-cm (6.4-in.) intervals and 19 anodes that are hung alternately. To ensure uniform current density and to eliminate short circuits in the cells, the electrodes are inspected frequently for prevention of excessive nodule build-up.

Typically, the electrolyte is pumped to an elevated storage tank, from which it flows by gravity into the tops of the cells and out through the bottoms. Thus, circulation is from top to bottom of the tanks, a type of circulation that yields a more homogeneous powder than that obtained by bottom-to-top circulation. The overflow returns to basement storage tanks to be recirculated.

The copper deposits on the cathode in the form of dendritic particles. To prevent short circuits between anode and cathode and heavy accumulation of powder on the cathode, both of which would decrease the cathode current density, the deposit is removed periodically by brushing.

After operating the cell for several days, the power is turned off. Most of the electrolyte is drained from the cell, leaving enough solution to cover the powder. The anodes and cathodes are washed down and removed, the remaining electrolyte is drained from the cell, and the powder is removed.

Thorough washing of the powder is essential. All traces of the electrolyte must be removed to prevent the powder from becoming oxidized. In addition, any remaining sulfate damages the heating elements if an electric furnace is used for subsequent drying and treatment of the powder. Various methods are employed to wash the powder. Although centrifuging to remove the electrolyte and wash the powder yields a clean product, the particles are compacted, and production of a low-density powder by this procedure is difficult.

In another method, the powder is transferred into a large tank and water is added to produce a slurry that is pumped into a filter. In the filter, the powder is dewatered, washed several times, and again dewatered. Because the wet powder oxidizes readily due to its finely divided state and active surface, addition of a stabilizer is desirable. Treatment with an aqueous solution of gelatin protects the powder from oxidation in the intervals between successive operations (Ref 8). The addition of surface-active agents during washing or subsequent powder treatment also protects the powder from oxidation. Next, the powder must be subjected to a furnace operation to obtain an acceptable grade.

Furnace Operation

After thorough washing and filtering, the wet powder is ready for furnace processing. The furnace operation also alters certain properties, particularly particle size and shape, apparent density, and green strength. In a typical operation, the powder is transferred into the charge box of a mesh belt electric furnace (Ref 7). To prevent the powder from falling through the belt, a continuous sheet of high wet-strength paper is fed to the belt, and then the powder is transferred to the paper. A roller compresses the powder to improve heat transfer. As it enters the furnace, water is driven off and the paper burns—but not before the powder has sintered sufficiently to prevent it from falling through the belt.

The furnace atmosphere is produced in exothermic gas units in which natural gas and air are blended to yield an atmosphere containing 17% H, 12% CO, 4% CO_2, and the balance of nitrogen. The gas is refrigerated to lower the dew point to the range of −22 to −40 °C (−8 to −40 °F). The gas enters the furnace from the discharge end and, because it is refrigerated, aids in cooling the powder cake. The furnace operation dries the powder, alters the particle shape, reduces the oxides, and sinters the fines. The discharge temperature is sufficiently low to prevent reoxidation of the powder cake.

By varying the furnace temperature between 480 and 760 °C (900 and 1400 °F) and altering the time of exposure, considerable change can be made in the content

Fig. 18 Effect of brush-down interval on particle size
Source: Ref 1

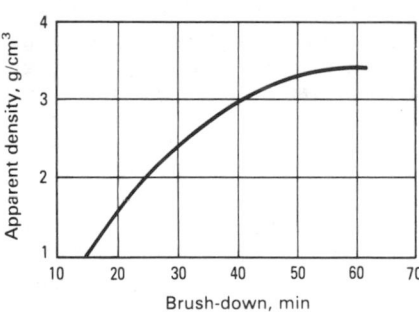

Fig. 18 Effect of brush-down interval on particle size graph showing Cumulative weight retained (%) vs Size of powder particles (μm), with curves labeled Brush-down interval: 60 min, 45 min, 30 min, and 15 min.

Fig. 19 Effect of brush-down interval on apparent density
Source: Ref 1

Fig. 19 Effect of brush-down interval on apparent density graph showing Apparent density (g/cm³) vs Brush-down (min).

of fines, apparent density, and dimensional characteristics. Upon completion of the furnace operation, the cake is broken and is ready for grinding.

Grinding and Finishing

Fine grinding is performed in high-speed, water-cooled hammer mills in which feed rate, mill speed, and screen openings under the mill can be varied to obtain the powder characteristics desired; thus, grinding is another operation in which powder properties can be changed. The powder leaving the hammer mills is fed to screens, where the oversize is separated and returned for additional grinding. The −100 mesh powder is classified in an air classifier and the fines are transferred to the blending operation. Oversize material is returned for regrinding or is used as melting stock.

The products of the grinding and classifying operations range in apparent density from about 1 to 4 g/cm³. They are stored in drums to which a drying agent such as silica gel or camphor is added to prevent further oxidation. To produce high-volume finished powder that meets customer specifications, powder is selected from various batches in the proper proportions and mixed in a blender. The lot is sampled before being removed from the blender, and if required, adjustments of the particle size distribution are made before the powder is packed in shipping drums. Table 5 lists the physical characteristics of several representative blends.

Process Modifications

Although most electrolytic powder is produced as described above, other methods have been used. To obtain a very low density product, one producer employs small-size cells and handles the powder lightly in both the cell and the finishing operations. The resulting powder shape is fern-like and has an apparent density of 0.9 to 1.3 g/cm³.

Another process for the production of copper powder employs a vertical rotating titanium cathode that is partially immersed in a copper sulfate/sulfuric acid electrolyte (Ref 10). In the operation, the powder is collected continuously above the liquid level by scraping the cathode, followed by continuous dewatering, washing, and drying steps.

Properties of Electrolytic Copper Powder

The properties of electrolytic copper powder are dependent on various characteristics of the operation and, therefore, can often be controlled by altering certain process variables.

Purity of powder prepared by the electrolytic process is high, with a copper content that usually exceeds 99.5% (Ref 9). A measure of the oxygen content is obtained by exposing a sample of powder to hydrogen at an elevated temperature as specified in American Society for Testing and Materials standard ASTM E 159 or Metal Powder Industries Federation standard MPIF 02. Generally, the hydrogen loss ranges from 0.1 to 0.5%, depending on the apparent density and particle size distribution of the powder. Nitric acid insolubles are also determined by ASTM or MPIF standard procedures and generally are less than 0.05%.

Particle size distribution for copper powder is selected to meet the requirements of the application and can be varied over a wide range. As noted in Table 5, several combinations are produced, and these are only representative. For example, the −325 mesh fraction can be varied from 5 to 90%.

Apparent densities of powders are available from 1.0 to 4.0 g/cm³. Densities that are somewhat lower and higher can be produced, depending on process and electrolyte conditions.

Flow rate is shown in relationship to apparent density in Fig. 20. Generally, powders with apparent densities of less than about 1.3 g/cm³ do not flow, powders with apparent densities of 1.3 to 2.3 g/cm³ have poor flow rates, and powders with higher apparent densities flow freely. At about 2.2 g/cm³, which is the transition range, flow depends on the content of fine particles of the powder, because relatively fine powders have poor flowability and relatively coarse powders flow freely. The flow rates of the usual blends of electro-

Table 5 Physical properties of typical copper powder blends

Apparent density, g/cm³	Maximum flow, s/50 g	Screen analysis (mesh size), %				
		+100	−100+150	−150+200	−200+325	−325
2.4-2.6	32	0.5 max	5-15	25-35	25-42	22-32
2.5-2.6	32	0.2 max	1-11	13-23	20-37	43-53
2.45-2.55	33	0.2 max	3-13	17-27	23-40	33-43
2.5-2.6	35	0.2 max	1-10	9-19	24-31	55-65
2.7-2.8	32	0.2 max	1-10	7-17	15-32	54-64
2.5-2.6	40	0.2 max	1-6	5-15	11-26	65-75
2.1-2.5	· · ·	0.1 max	0.5 max	4 max	8 max	90 min
High-conductivity powder						
2.5-2.6	32	0.2 max	1-11	13-23	20-37	43-53
Friction-grade powder						
1.7-2.0	· · ·	0.5 max	1-6	5-15	10-26	60-80
High-density powder						
3.25-4.00	24	0.8 max	7-17	17-27	19-35	35-45

Source: Ref 9

Fig. 22 Electrolytic copper powder showing dendritic structure

Magnification: 85×. Source: Ref 9

Fig. 23 Effect of sintering time on dimensional change

Source: Ref 9

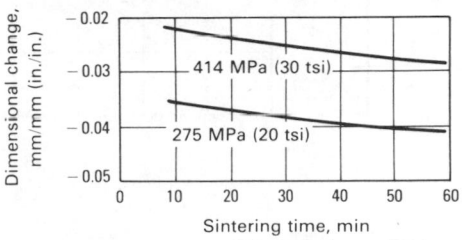

Fig. 20 Relationship between apparent density and flow rate

Source: Ref 11

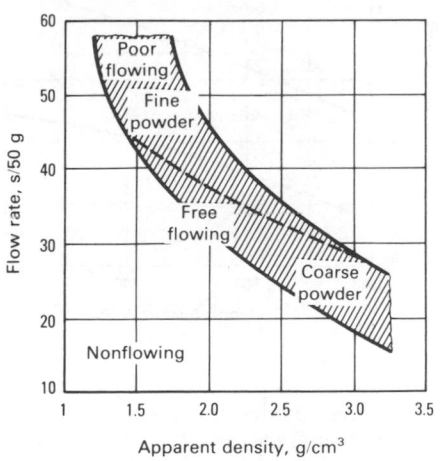

Fig. 21 Effect of compacting pressure on green strength and green density

Source: Ref 9

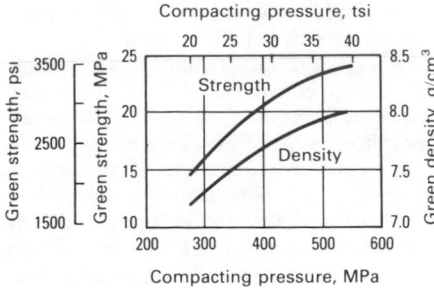

lytic copper powder range from 25 to 40 s/50 g.

Green density is a function of the compacting pressure. As shown in Fig. 21, the green density for the blend used rises from 7.2 to 8.0 g/cm³ as the compacting pressure is increased from 275 to 550 MPa (20 to 40 tsi).

Green strength increases with the compacting pressure, as indicated in Fig. 21. In this example, the green strength rises from less than 15 to 24 MPa (2200 to 3500 psi) as the compacting pressure is increased from 275 to 550 MPa (20 to 40 tsi).

Particle shape of electrolytic copper generally is dendritic when deposited on the cathode (Fig. 22). During subsequent operations, however, the dendrites are rounded somewhat.

Compacting pressure is an important variable to consider, because the pressing and sintering conditions have a marked influence on the properties of sintered compacts. If sound compacts are to be obtained, gases from the atmosphere, from the products of reduction, or from the lubricant must escape during sintering. When the compacting pressure is too high, the flow through interconnected pores may be obstructed, and the gases cannot escape. Compacting pressures no higher than 275 MPa (20 tsi) should be used in the production of large, thick parts from electrolytic copper powder, although higher pressures can be used for thin-walled parts (Ref 12).

Dimensional changes in a typical powder blend lubricated with lithium stearate are shown in relationship to sintering time in Fig. 23. The lubricated powder blend was pressed as indicated and sintered at 1000 °C (1830 °F) in a dissociated ammonia atmosphere. Relatively stable dimensional changes are achieved at normal sintering times.

Tensile strength and elongation of a typical powder blend are shown in Fig. 24 in relationship to compacting pressure and sintering time. The powder was lubricated with lithium stearate, pressed as indicated, and sintered at 1000 °C (1830 °F) in an atmosphere of dissociated ammonia. The graph demonstrates that good tensile properties can be obtained with short sintering times.

Electrical conductivity that is high can be achieved, but a high-purity powder—a characteristic of powder produced by the electrolytic process—must be used. Figure 25 shows the relationship between electrical conductivity and sintered density; high conductivity can be achieved only in high-density compacts. Electrical conductivity can be increased by coining and resintering (Fig. 26).

Fig. 24 Effect of compacting and sintering conditions on tensile properties
Source: Ref 9

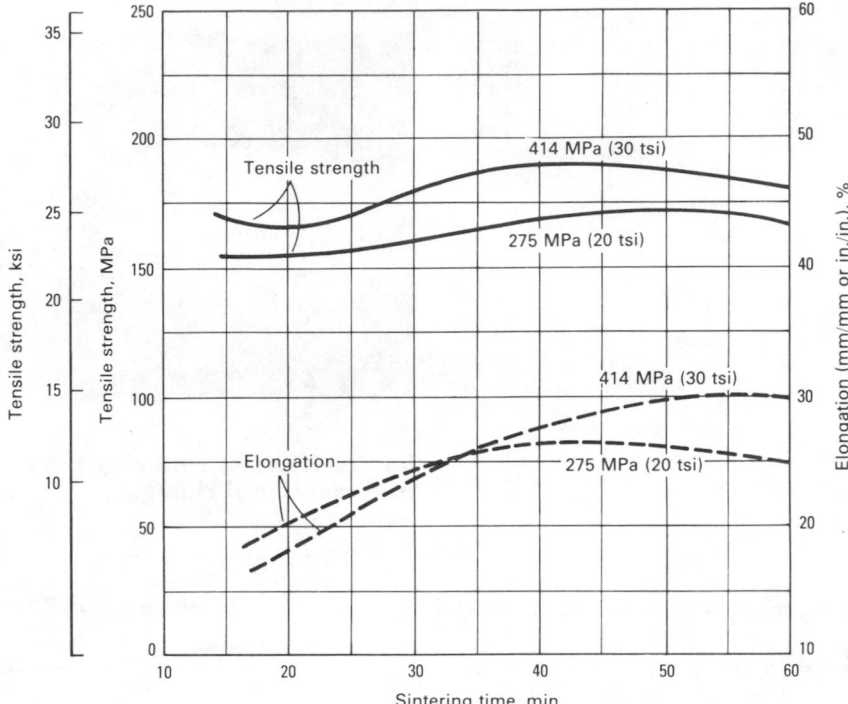

Fig. 25 Effect of sintered density on electrical conductivity
Source: Ref 9

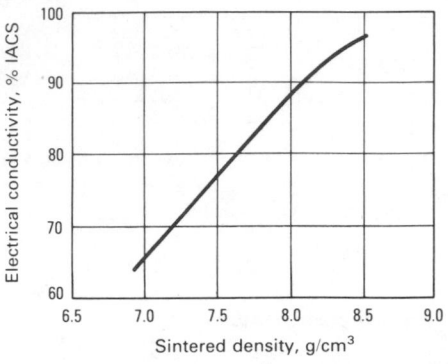

Fig. 26 Effect of coining and resintering on electrical conductivity
Source: Ref 9

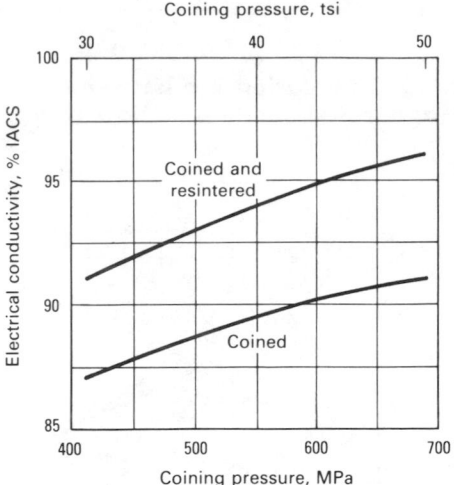

Applications

The major use of electrolytic copper powder is in the production of self-lubricating bearings. For this purpose, copper is combined with tin and graphite, or with tin alone, to obtain parts with interconnected pores that can absorb up to 30% oil and supply a continuous lubricating film. Porous bronze bearings are used widely in automotive service, household appliances, automatic machines, and industrial machines as low-duty shaft bearings or as an alternate for ball bearings.

The high electrical and thermal conductivities that are characteristic of electrolytic copper powder have led to wide use in the electrical and electronic industries. Because of these attributes, the use of suitable fabricating procedures permits the production of parts with conductivities of 90% IACS and higher. Complex parts such as armature bearing blocks, contacts for circuit breakers, shading coils for contactors, heavy-duty contacts for circuit breakers, switch gear components for use in switch boxes with capacities up to 600 A, and components for 150- and 250-A fuse blowouts are in regular production. Heat sinks for diodes used in silicon rectifiers for the alternating current system in automobiles and electrode tools for elec-

trical discharge machining also are produced from electrolytic copper powder.

Electrolytic copper powder is mixed with zinc or with zinc and nickel to produce brass and nickel silver for applications such as gears and cams, industrial parts, hardware, and automotive parts. Additionally, this powder is used with various nonmetallic materials to produce friction parts, such as brake bands or clutch disks. Premixes of iron-copper or iron-copper-carbon are employed in various automotive applications—cams, sprockets, gears, piston rings for small-bore engines—and similar service.

Production of Copper Powder by Atomization

By Erhard Klar
Manager of Particle Technology
SCM Metal Products

ATOMIZATION OF COPPER—that is, the disintegration of a molten stream of

high-quality copper with high-pressure water jets—produces copper powders of compacting-grade quality. The resulting dried powder then may be subjected to an elevated-temperature treatment that further modifies its characteristics and engineering properties.

Nearly spherical powders result from the atomization of liquid copper with inert gas or air. These powders are used in copper flake production and other specialty applications. Spherically shaped copper powders have insufficient green strength for use in conventional P/M. To render them compactible, their particle shape and morphology may be changed by deformation and/or oxidation and reduction, as described in the section "Production of Copper Powder by the Reduction of Copper Oxide" in this article. Particle shapes of gas- and water-atomized copper pow-

Fig. 27 Scanning electron micrographs of gas- and water-atomized copper powders
(a) Nitrogen atomized. (b) Water atomized, apparent density of 3.04 g/cm³. (c) Water atomized, apparent density of 4.60 g/cm³

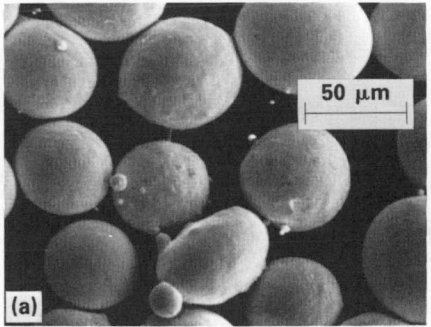

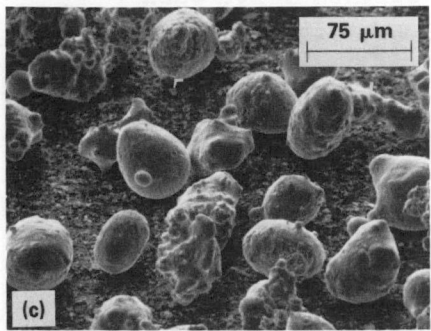

Fig. 28 Effect of partial pressures of hydrogen and steam on oxygen content of liquid copper at 1150 °C (2100 °F)
$Cu + H_2 \rightleftarrows 2Cu + H_2O$ reaction at various water vapor pressures

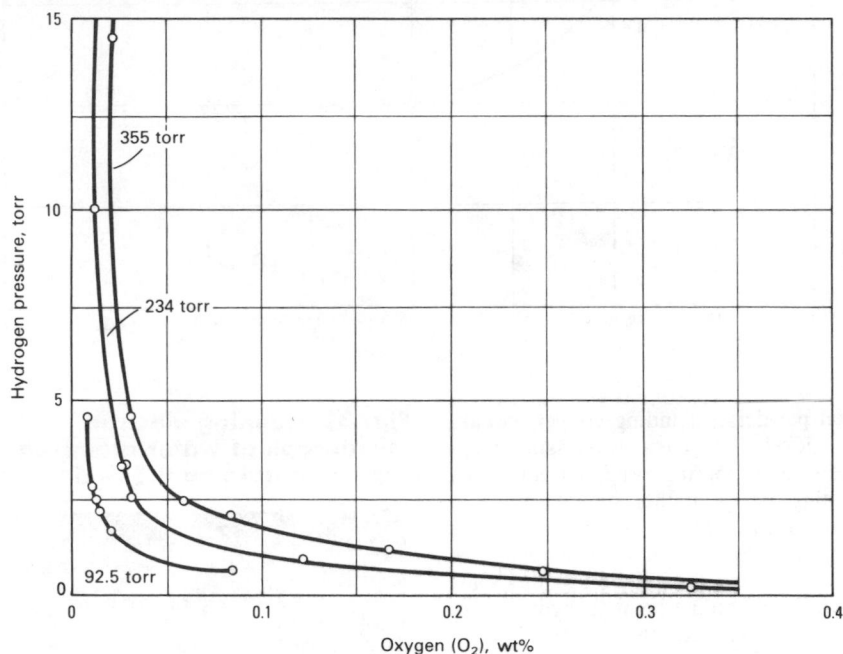

ders are shown in Fig. 27. The following discusses details specific to the water atomization of copper and subsequent processing. For more information, see the article "Atomization" in this Volume.

Commercial Practice

Liquid copper is superheated to about 1150 to 1200 °C (2100 to 2200 °F), utilizing flow rates of 27 kg/min (60 lb/min) or more. Generally, to produce a predominantly −100 mesh powder, water pressures of 10 to 14 MPa (1500 to 2000 psi)

are used. Atomization may be conducted in an air or inert (nitrogen gas) environment.

During melting of the copper, impurity content must be controlled to achieve good fluidity and high conductivity (see Fig. 2). The need for oxygen control at this stage of the process depends on subsequent processing and end use of the powder. In conventional refining of copper, poling and dissociation of steam-generated hydrogen keep copper oxidation under control in accordance with the equilibrium curves shown in Fig. 28. High oxygen content tends to

produce a more irregular powder, and subsequent reduction of the atomized powder further improves compactibility through agglomeration and pore generation. If the powder is used in the as-atomized condition, lower oxygen contents are generally preferable, because of the detrimental effects of oxygen in many applications.

The apparent densities of −100 mesh, gas-atomized copper powders, as a consequence of their spherical particle shape, range between 4 and 5 g/cm³. In comparison, apparent densities of water-atomized copper powders are controllable between about 3 and 4.5 g/cm³. Oxygen picked up during atomization is present partly as surface oxide and partly as copper oxide throughout the bulk of a copper particle. Removal of oxygen requires reduction temperatures of about 700 °C (1290 °F) or higher. At these temperatures, considerable sintering takes place, which in turn requires substantial milling of the sinter cake.

During reduction, hydrogen readily diffuses through solid copper to react with oxygen and form steam. The large steam molecules, unable to diffuse through solid copper, force their way outward through grain boundaries—a phenomenon known as hydrogen embrittlement of copper that manifests itself in the formation of blisters or cracks. Figure 29 illustrates the grain-boundary widening of air-atomized copper particles due to this phenomenon. These defects improve both compactibility and sintering rate during liquid phase sintering of copper mixed with tin.

Alloying Additions

Some applications of copper powders require apparent densities lower than those attainable with water atomization of pure

Fig. 29 Hydrogen-embrittled, air-atomized copper after reduction in hydrogen

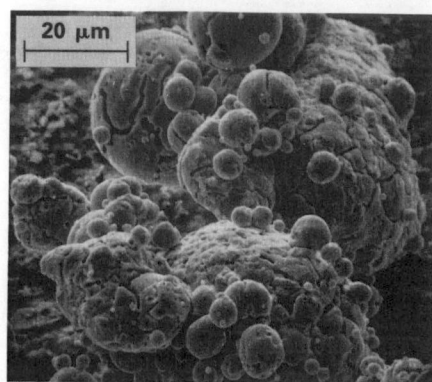

20 µm

Fig. 30 Effect of additions to the molten copper on apparent density of atomized copper powder

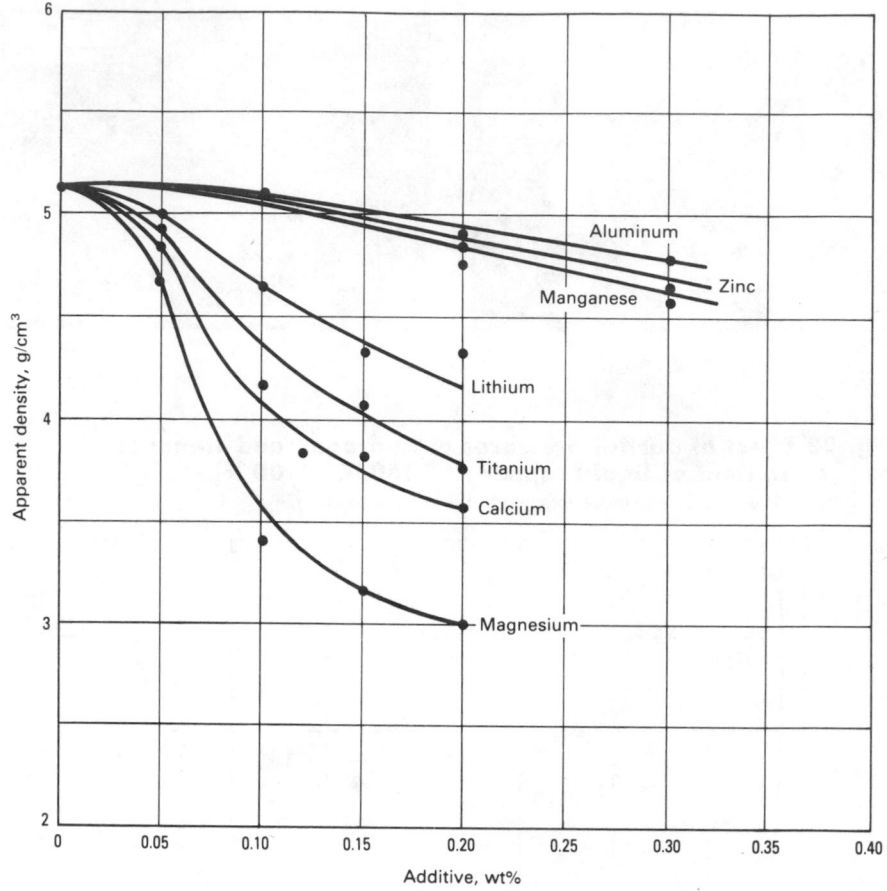

Aluminum
Zinc
Manganese
Lithium
Titanium
Calcium
Magnesium

Apparent density, g/cm³

Additive, wt%

copper. These powders can be produced by addition of small amounts, up to 0.2%, of certain elements (for example, magnesium, calcium, titanium, and lithium) to the liquid copper prior to atomization (Fig. 30 and 31). These metals are believed to decrease the surface tension of copper and/ or to form thin oxide films on the particle surface during atomization. Magnesium additions are used most frequently to produce compacting-grade copper powders for applications such as bronze bearings, filters, structural parts, and additives for iron powders. These powders may have apparent densities as low as 2 g/cm³.

The addition of small amounts (0.1 to 0.3%) of phosphorus to the liquid copper, prior to atomization, allows the production of a powder that is very spherical and very low in oxygen. During atomization, even with air, the phosphorus oxidizes preferentially and forms protective gaseous phosphorus pentoxide (P_2O_5). Such powders have apparent densities up to about 5.5 g/cm³. Table 6 lists powder properties of commercial atomized copper powders. The specific surface areas of these powders are between 0.02 and 0.1 m²/g.

Production of Copper Powder by Hydrometallurgical Processing

By Erhard Klar
Manager of Particle Technology
SCM Metal Products

HYDROMETALLURGICAL METHODS can be used to produce a number of metal powders, including copper, cobalt, and nickel. The basic processing steps consist of preparing pregnant liquor by leaching ore or another suitable raw material, followed by the precipitation of the metal from this solution. For copper, the most important precipitation methods are cementation, reduction with hydrogen or sulfur dioxide, and electrolysis. Use of several leach-precipitation steps or the inclusion of flotation, solvent extraction, or ion exchange improves the purity of the final material.

Despite several process development efforts in the 1950's and 1960's that led to radical improvements in purity and properties of hydrometallurgically produced copper powders, commercialization attempts in the United States have failed. Causes for this failure include the increasing cost of energy, coupled with the high energy requirements of some hydrometallurgical methods and the difficulty in economically producing powders with a wide range of properties suitable for various uses. The only hydrometallurgically produced copper powder in the United States today

Fig. 31 Scanning electron micrograph of water-atomized copper containing 0.5% Li

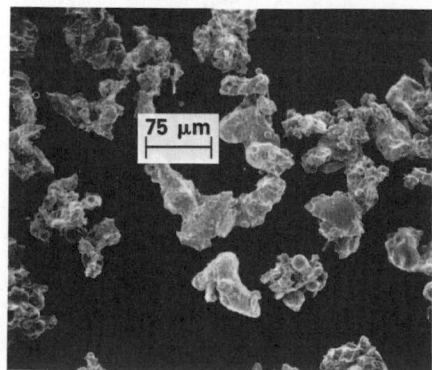

75 µm

is cement copper, an impure copper powder precipitate obtained from copper sulfate solution by the addition of iron.

Cement copper, like other hydrometallurgically produced copper powders, has low apparent density and high specific surface area (~1 m²/g). The particles are spongy, because they are agglomerates of

Table 6 Properties of commercial grades of water- and gas-atomized copper powders

Copper, %	Chemical properties, %		Hall flow rate, s/50 g	Apparent density, g/cm³	Physical properties				
	Hydrogen loss	Acid insolubles			+100	−100+150	−150+200	−200+325	−325
					Tyler sieve analysis, %				
99.65(a)	0.28	. . .	. . .	2.65	Trace	0.31	8.1	28.2	63.4
99.61(a)	0.24	. . .	. . .	2.45	0.2	27.3	48.5	21.6	2.4
99.43(a)	0.31	. . .	. . .	2.70	tr	0.9	3.2	14.2	81.7
>99.1(b)	<0.35	<0.2	~50	2.4	<8	17-22	18-30	22-26	18-38
99.1	0.77	. . .	No flow	4.8	Trace	3	. . .	. . .	. . .
99.2	<0.7	. . .	9-13	4.9-5.5	7-14	←20-30→	←20-30→	15-30	30-50

(a) Water atomized plus reduced. (b) Contains magnesium

Fig. 32 Scanning electron micrograph of cement copper

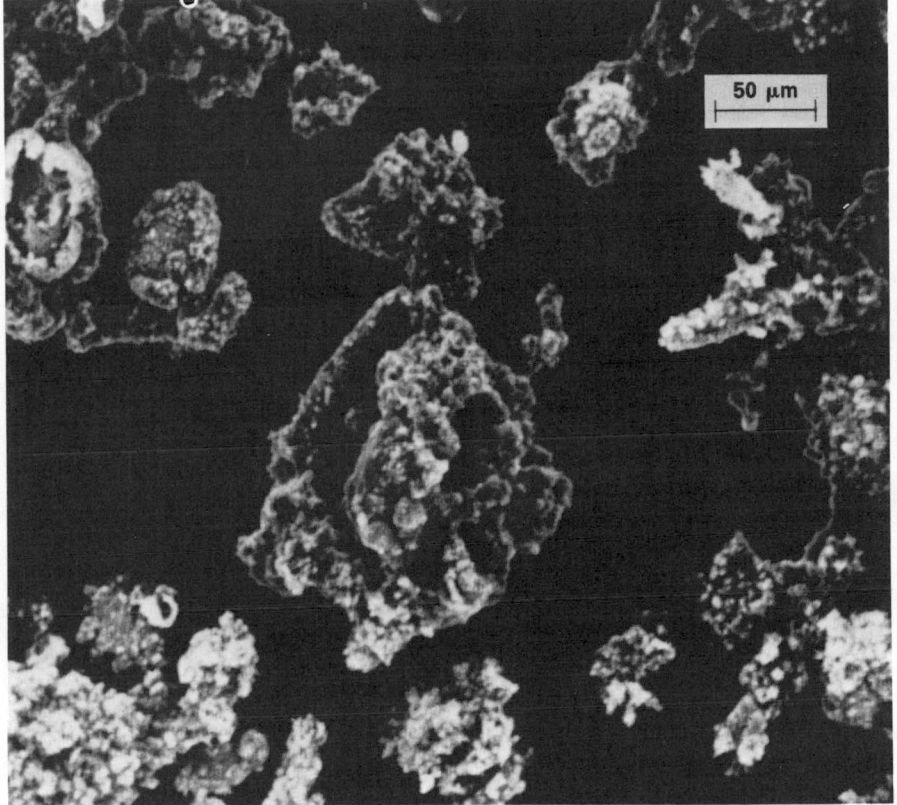

50 µm

Table 7 Chemical analyses of cement copper from various locations (dry basis)

Component	Composition, wt%, for location:			
	A	B	C	D
Total copper	75	83.0	87.4	85.0
Iron	6	2.4	0.7	10.0
Sulfur	1	0.5	. . .	1.1
Nitric acid insolubles . .	2	. . .	0.7	1.9
Hydrogen loss	16	. . .	. . .	. . .
Calcium oxide	. . .	0.08	. . .	. . .
Aluminum oxide (alumina)	. . .	1.2	0.5	. . .
Silicon dioxide	. . .	0.4	. . .	. . .
Lead	. . .	. . .	0.2	. . .
Oxygen	. . .	. . .	9.5	. . .

very small primary particles (Fig. 32). Although cement copper often has higher green strength characteristics than most other copper powders, its sintering activity, when used alone or in 90/10 bronze, is inferior because of the presence of finely divided, unreducible aluminum and silicon oxides. The primary use of cement copper is in composite friction material applications.

Leaching

The leaching of copper oxide and copper sulfide ores entails partial dissolution to cupric sulfate ($CuSO_4$) with sulfuric acid (H_2SO_4) plus iron sulfate ($FeSO_4$). The presence of pyrite (FeS_2) in many ore deposits, and its reaction with water and oxygen to form iron sulfate and sulfuric acid, represents an important source of acid. In dump leaching, the pH of the leach solution is maintained between 1.5 and 3.0, partly to protect bacteria that promote and accelerate the oxidation of pyrite and sulfidic copper minerals and also to avoid hydrolysis of iron salts.

The copper content of the pregnant solutions varies from less than 1 g/L (0.15 oz/gal) to several grams per litre (ounces per gallon); for vat leaching, it can be much higher. Other leaching methods include ammonia leaching, which is used for certain copper oxide ores, and the so-called leach-precipitation-flotation method, which is applied to mixed oxide sulfide ores.

Cementation

The precipitation of a metal from its solution by the addition of another less noble metal is known as cementation. The fundamental equation for copper recovery from copper-bearing pregnant liquors on iron is:

$$Fe + CuSO_4 = Cu + FeSO_4$$

In practice, the copper-bearing solution is passed over scrap iron, such as detinned and shredded cans. Subsequent separation, washing, reduction, and pulverizing produces a copper powder that contains considerable amounts of iron and acid insolubles, such as alumina and silica. Contamination with gangue varies and depends on the nature of the pregnant liquor.

Significant reduction in iron and alumina levels is possible through the use of V-trough or inverted precipitators that allow faster precipitation rates and more efficient iron utilization. Table 7 shows chemical analyses of cement copper from different sources. Copper and iron are present, partly as oxides.

Additional Reduction Processes

Electrowinning may be used to recover copper from leaching solutions containing more than about 25 g/L (3.4 oz/gal) of copper (Harlan process). The electrolytic cells are equipped with insoluble lead-antimony anodes and 99% Ni cathodes. The copper powder does not adhere, but falls to the bottom of the cells. Temperature of the electrolyte is 60 °C (140 °F); cathode current density varies from 1350 to 2700 A/m^2 (125 to 250 A/ft^2). When the copper concentration falls below 15 g/L (2 oz/gal), the electrolyte is drained and used for ore leaching. Particle size of the powder prior to furnace processing is 1 to 25 μm, with excellent purity (>99.9%). The power requirements for electrowinning of copper are about ten times as large as those for electrorefining of copper with soluble anodes.

Copper may be concentrated from low-content leach solutions by solvent extraction, followed by stripping with dilute sulfuric acid into an aqueous solution and electrowinning. Carboxylic acid and hydroxylamine-based compounds have been found to be selective solvents of low water solubility, to have good stability, and to be compatible with cheap diluents. Direct powder precipitation with hydrogen or ammonia is an alternative to stripping the metal from the organic solvent into an aqueous solution.

Metals can be precipitated from their acid or basic solutions by reduction with hydrogen. Sulfuric acid, ammoniacal ammonium carbonate, and ammoniacal ammonium sulfate solutions have been used to produce copper powder by this method. Sulfuric acid leaching of a cement copper and hydrogen reduction of the filtered solution in an autoclave at 120 to 140 °C (250 to 280 °F) and 3 MPa (425 psi) is reported to produce a precipitate with a purity of almost 100% Cu. Drying and furnace processing in a reducing atmosphere at 540 to 790 °C (1000 to 1450 °F) increases particle size, due to agglomeration of the very fine powder.

REFERENCES

1. Kumar, D. and Gaur, A.K., Electrochemical Studies on Production of Electrolytic Copper Powders, *J. Electrochem. Soc. India,* July 1973, p 211-216
2. Harper, S. and Marks, A.A., Electrodeposition of Copper Powder with the Aid of Surfactants, Copper Development Association/American Society for Metals Conference on Copper, Vol 3, 1972
3. Gurevich, L.I. and Pomosov, A.V., The Effect of Chloride on Electrodeposition of Powdered Copper Precipitates, *Soviet P/M Met. Cer.,* Jan 1969, p 10-15
4. Artamonov, V.P. and Pomosov, A.V., Effect of Foreign Electrolytes on the Production of Copper Powder by Contact Deposition, *Isv. V.U.Z. Tsvetn. Metall.,* No. 2, 1976, p 30-34 (in Russian); *Metall. Abstracts,* No. 54-0503, Nov 1976
5. Kotovskaya, S.L., *et al.,* Manufacture of Coarse Copper Powder from Sulfamate Electrolytes, *Soviet P/M Met. Cer.,* Feb 1973, p 93-96
6. Singh, S.K. and Akerkar, D.D., A Continuous Self-Regulating Method of Making Copper Powder by Electrolysis, *NML Tech. J.,* Vol 17, 1975, p 23-26
7. Wills, F. and Clugston, E.J., Production of Electrolytic Copper Powder, *J. Electrochem. Soc.,* Vol 106, 1959, p 362-366
8. Pomosov, A.V., Numberg, M.I., and Krymakove, E.F., Protection of Copper Powder Against Corrosion During Manufacture and Storage, *Soviet P/M Met. Cer.,* March 1976, p 175-177
9. Technical Data—AMAX Metal Powders, AMAX Copper, Inc., 1968
10. Adamson, D.L. and Toddenham, W.M., Production of High Quality Electrolytic Copper Powder, American Institute of Mining, Metallurgical and Petroleum Engineers Annual Meeting, New York, 1-4 March 1971
11. Peissker, E., *Metal Powders,* Norddeutsche Affinerie, Aug 1974
12. Taubenblat, P.W., Smith, W.E., and Evans, C.E., Production of P/M Parts from Copper Powders, *Precision Metals,* April 1972, p 41

Production of Copper Alloy Powders

By Paul E. Matthews
Vice President and
Technical Director
United States Bronze Powders, Inc.

COMMERCIAL COPPER ALLOY POWDERS, including brasses, bronzes, and nickel silvers, are manufactured in a similar manner. Usually, the same integrated manufacturing facilities are used to complete the melting process, atomization, final screening, and blending of a lot or batch.

The powder-producing process is similar to a foundry operation in which high-purity virgin metals are charged to a melting furnace in preweighed batches to be processed at predetermined heating rates and times. High-purity raw materials are required, because melting produces minimal refinement (only a partial transfer of contaminants from the melt to the formed slag occurs). To ensure continuity, homogeneity, and uninterrupted atomization, the molten batch of alloy is transferred to a second furnace with a greater holding capacity than the primary furnace melting rate. Induction heating is preferred for at least one furnace to ensure the constant induced metal movement required for alloy homogeneity and to ensure uniform lead dispersion in lead-bearing alloys.

Atomization is achieved by particulation of a controlled, constant-flowing, molten stream emitted from the secondary furnace by medium-pressure dry air. A typical melting and atomizing sequence is shown in Fig. 1. Subsequent reduction of oxides is not required for standard P/M grades.

Air-cooled, atomized powder is collected and passed over a primary control screen (sieve) to remove oversize particles. Usually, these are remelted concurrently during the processing of each alloy. Adjustment of particle size distribution of each alloy is accomplished by controlled

manipulation of the atomizing air rate and/or molten metal temperature. Finally, the screened alloy powder may be blended with dry organic lubricants, such as lithium and zinc stearate, for use in the manufacture of P/M structural components.

A variety of copper-based alloys can be manufactured by the atomizing process; however, commercial P/M applications normally are confined to a rather narrow

Fig. 1 Flowchart for copper alloy powder atomization

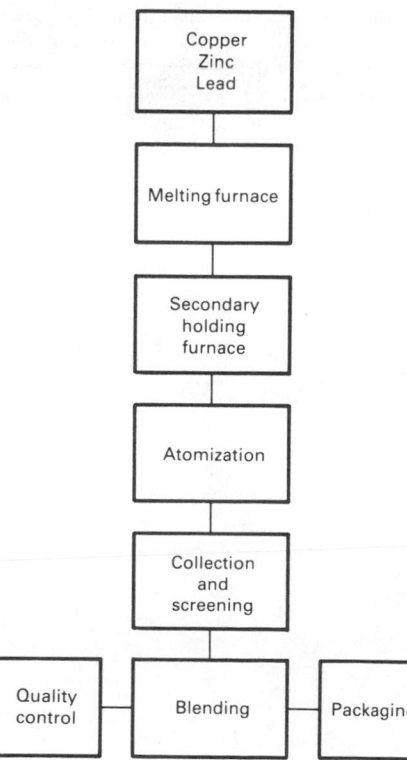

range of specific, single-phase (alpha) compositions.

Brasses

Brasses comprise the major portion of copper-based alloy powders used for parts fabrication, in comparison with prealloyed bronze and nickel silver materials. Typical copper-zinc brass powders contain zinc contents ranging from 10 to 30%. Lead may be added in small amounts (1 to 2%) to improve machinability of the sintered compact. A photomicrograph of an 80%Cu-18%Zn-2%Pb alloy powder is shown in Fig. 2. The melting temperatures of these alloys range from 1045 °C (1910 °F for 90%Cu-10%Zn to 960 °C (1760 °F) for 70%Cu-30%Zn. As zinc content increases, melting temperature is lowered.

Fig. 2 Photomicrograph of prealloyed, air-atomized brass (80%Cu-18%Zn-2%Pb)
Magnification: 165×

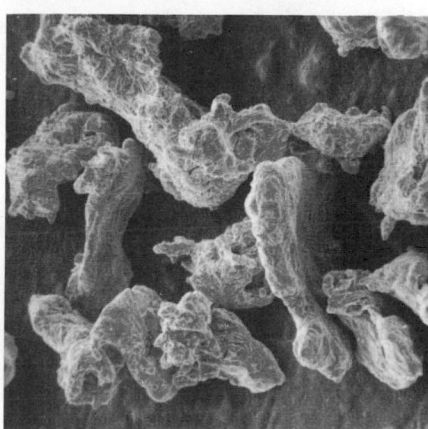

Table 1 Physical properties of typical brass, bronze, and nickel silver alloy compositions

Property	Brass(a)	Bronze(a)	Nickel silver(a)(b)
Sieve analysis, %			
+100 mesh	2.0 max	2.0 max	2.0 max
−100+200	15-35	15-35	15-35
−200+325	15-35	15-35	15-35
−325	60 max	60 max	60 max
Physical properties			
Apparent density	3.0-3.2	3.3-3.5	3.0-3.2
Flow rate, s/50 g	24-26	...	...
Mechanical properties			
Compressibility (c) at 414 MPa (30 tsi), g/cm³	7.6	7.4	7.6
Green strength(c) at 414 MPa (30 tsi), MPa (psi)	10-12 (1500-1700)	10-12 (1500-1700)	9.6-11 (1400-1600)

(a) Nominal mesh sizes: brass, −60 mesh; bronze, −60 mesh; nickel silver, −100 mesh. (b) Contains no lead. (c) Compressibility and green strength data of powders lubricated with 0.5% lithium stearate

Fig. 3 Photomicrograph of prealloyed, air-atomized bronze (89%Cu-9%Sn-2%Zn)
Magnification: 165×

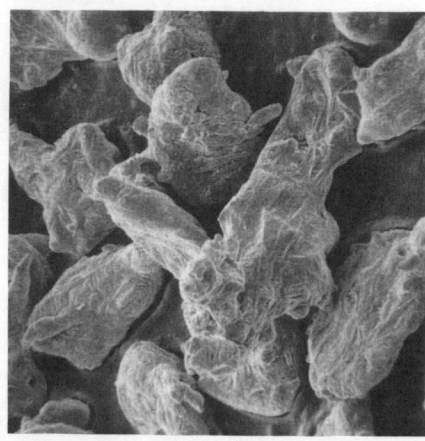

Fig. 4 Photomicrograph of prealloyed, air-atomized nickel silver (63%Cu-18%Ni-17%Zn-2%Pb)
Magnification: 165×

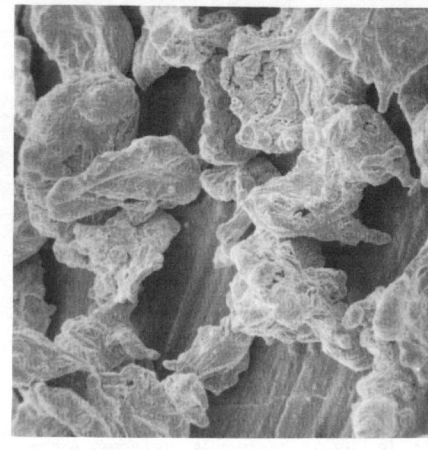

Additional superheat, or the temperature in excess of the alloy melting temperature, depends on heat losses within the manufacturing system and the physical property requirements of the atomized powders. Typical physical properties of brass alloy compositions are given in Table 1.

Bronzes

Prealloyed atomized bronze compositions are not used extensively as base powders for compacted parts fabrication because of their nodular particle form and high apparent density, both of which contribute to poor compacted green strength. Common prealloyed compositions are 90%Cu-10%Sn and 85%Cu-15%Sn, prepared in the same manner as brass powder, except that high-purity elemental copper and tin are used. A photomicrograph of an 89%Cu-9%Sn-2%Zn alloy powder is shown in Fig. 3. Typical physical properties of a bronze alloy composition are given in Table 1.

Nickel Silvers

Only one nickel silver-based alloy composition, 65%Cu-18%Ni-17%Zn, is commonly used in the P/M industry. This alloy can be modified by the addition of lead to improve machinability. Foundry practices are similar to those employed for brasses, except the melting temperature is in excess of 1093 °C (2000 °F). A photomicrograph of a 63%Cu-18%Ni-17%Zn-2%Pb alloy powder is shown in Fig. 4. Typical physical properties of a nickel silver alloy composition are given in Table 1.

Production of Tin Powder

By William J. Ullrich
Chief Metallurgist
Alcan Ingot and Powders

TIN POWDERS are used extensively in the production of porous, self-lubricating bronze bearings and as constituents in soldering and brazing pastes and powders. Low-melting-point tin powders normally are produced by air atomization, although other methods of production such as chemical precipitation and electrodeposition have been used (Ref 1). Tin powders also are used in P/M structural parts, friction disks, clutches, brake linings, metal graphite brushes, diamond abrasive grinding wheels, bronze filters, peen plating, plasma arc spraying, chemical formulations, additives for rubber and plastics, chemical manufacturing, smokeless powder for pyrotechnics, and tin flake.

Melting

High-purity pig tin is melted in a gas-fired or electrically heated crucible (cast iron, clay, graphite, or ceramic). Melt temperature is controlled to maintain the desired degree of superheat above the low melting point of tin (232 °C or 449 °F).

Atomization

Atomization results in a fine powder whose average particle size can be regulated over a wide range. Resultant pow-

Fig. 1 Schematic of apparatus for atomizing tin powder

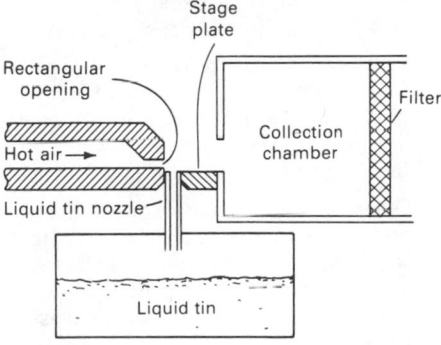

ders are of high purity; excessive oxidation, characteristic of other atomized products, does not occur because of the rapid chilling effect of the expanding gases released through the nozzle. Thus, the oxygen content of atomized tin is normally below 0.2%; the thin film of oxide produced during atomization with steam or air is sufficient to inhibit further oxidation of the particles.

The molten tin can be atomized vertically (up or down) or horizontally, depending on the nozzle direction, to produce a powder (Fig. 1). There are two atomization techniques: annular nozzle and cross-jet atomization. In an annular noz-

zle, the gas stream aspirates liquid tin into the nozzle, where it is disintegrated into tiny droplets by the high-velocity gas stream. In cross-jet atomization, the gas is at right angles to the molten tin stream. This method usually produces coarser particles than annular nozzles.

To atomize uniformly fine tin powders, temperature, stream diameter, and flow rate of the molten tin and temperature, pressure, velocity, and angle of impingement of the atomizing gas must be well controlled. Pressures commonly employed to atomize tin powder range from 345 to 1725 kPa (50 to 250 psi). Generally, finer powders require the higher pressure; however, the exact pressure is related to the nozzle design used.

Compressed air is usually used as an atomizing medium. It is frequently preheated in a gas-fired heat exchanger to prevent solidification of tin within or around the orifice caused by the chilling effect of expanding air as it is released through the nozzle. Nozzles are designed to facilitate the atomization of several different particle size ranges, usually by changing orifice diameter and air pressure.

A blower (fan) at the end of the system pulls the atomized tin powder from the atomization chamber into a cyclone, where it is collected. The finest particles, which have not settled in the cyclone, are then

Table 1 Commercially available tin powders

	Sample A	Sample B	Sample C	Sample D	Sample E	Sample F	Sample G	Sample H
Sieve analysis, %								
+100 mesh	48.0	0.2	0.5 max	0.3 max	0.3 max	...	...	...
+150 mesh	46.5	5.2	5 max	3 max	2 max	...	...	...
+200 mesh	4.0	11.3	4-12	3-8	5 max	0.1	...	...
+325 mesh	1.0	29.0	15-30	12-25	2-8	...	2 max	...
−325 mesh	0.5	53.6	65-75	70-85	90 min	96 min	96 min	98 min
Fisher sub-sieve size								
Average particle size, μm	...	...	12-18	10-15	8-11	8-10	7-9	1-3
Apparent density, g/cm³	3.35	3.90	3.7-4.2	3.7-4.2	3.3-3.8	3.0-3.5	3.0-3.5	1.3-2.0

retained in the cyclone filters. The tin powder collected in the cyclone is sieved to remove the undesirable large particles, most commonly +100, +200, or +325 mesh.

A uniform mixture of tin powder of the desired lot size is produced by tumbling in a blender. A sample of the blended tin is then analyzed for physical and chemical properties. The tin powder is packed in steel or waterproof fiber containers weighing up to 320 kg (700 lb).

Properties of Tin Powder

The high-purity virgin tin melt stock meets the requirements of ASTM grade A:

Element	Composition (max), %
Tin	99.8(a)
Antimony	0.04
Arsenic	0.05
Lead	0.05
Copper	0.04
Bismuth	0.015
Iron	0.015
Sulfur	0.01

(a) Minimum (99.9% typical)

Impurities in the tin melt stock carry over to the powder and usually are measured by atomic absorption spectroscopy. The tin powder purity is directly related to the purity of the ingot melted.

During atomization, a thin oxide film, which inhibits further oxidation of the particles, forms on the particle surfaces. Oxygen pickup commonly is measured according to MPIF 02, "Method for Determination of Hydrogen Loss of Metal Powder," published by the Metal Powder Industries Federation. A typical oxygen content of tin powder determined by this method is 0.05%.

Physical properties of tin powder depend on atomization conditions and sieving procedures. Coarser grades of tin powder are atomized from larger metal streams using lower air pressures. The finer grades of powder require higher pressures and small molten metal streams. Sieving is an independent means of controlling the size of the largest particles in the powder. Generally, 100-, 200-, or 325-mesh sieves are used to remove oversize particles.

Particle Size Distribution. Most grades of tin powder are classified by particle size. Methods of measuring particle size distribution include:

- Sieve analysis, as specified by ASTM B 214 and MPIF 5. Usually 100-, 150-, 200-, and 325-mesh sieves are used.

- Fisher sub-sieve size (FSSS), as specified by ASTM B 330 and MPIF 32, provides a quick measure of the average particle size of the sub-sieve (−325 mesh) fraction of fine tin powder.

- Other particle size analysis methods, such as the use of micromesh sieves, light scattering, sedimentation, and electrical and optical sensing zone instruments, are used to measure sub-sieve size particles.

Table 1 gives sieve analysis data and Fisher sub-sieve size diameters for several commercially available tin powders.

Apparent density varies with particle size and shape. It is measured as specified by ASTM B 212 and B 417 and MPIF 4 and 28. Apparent densities of some com-

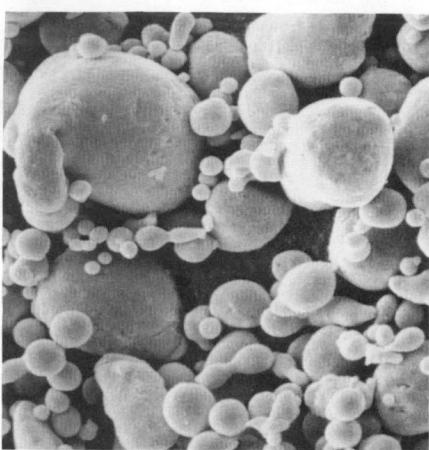

Fig. 2 Scanning electron micrograph of air-atomized tin powder particles
Magnification: 900×

mercially available tin powders are given in Table 1.

Surface area usually is measured by inert gas adsorption, or Brunauer-Emmett-Teller (BET) techniques. The specific surface area of the tin powder varies between approximately 1.0 m²/g for the finest grade to 0.1 m²/g for the coarsest grade.

Particle shape and surface topography can be observed by light or scanning electron microscopy. Air-atomized tin particles are rounded to spherical in shape with relatively smooth surfaces. Figure 2 is a scanning electron micrograph of atomized tin powder particles displaying typical rounded shape.

REFERENCE

1. Goetzel, C.G., *Treatise on Powder Metallurgy*, Vol 1, Interscience, New York, 1949, p 199-200

Production of Aluminum Powder

By James E. Williams, Jr.
Section Head
Fabricating Technology Division
Aluminum Company of America

ALUMINUM POWDERS were first used commercially as flake products. Production of flake aluminum powders began in the United States around 1900, using the Bessemer dry stamping process. Stamp mills were both inefficient and dangerous to operate, because aluminum powder forms an explosive mixture with air over a wide range of metal-to-air ratios. Several fatal fires and explosions marred the early days of the aluminum powder industry.

Historical Background

Two major breakthroughs occurred in aluminum powder production in the late 1920's: the development of safer ball mill production processes that are still employed today for flake aluminum powders, and the introduction of the atomization process, which allowed the manufacture of the forerunners of today's atomized powders.

Initially, atomized aluminum powder was used only as raw material to produce aluminum flake pigments by ball milling. After World War II and the development of aluminized high explosives, a major market for atomized aluminum was created. The application of atomized powders broadened in postwar years from military explosives to such markets as chemicals, metallurgical products, commercial blasting agents, rocket fuels, and pharmaceuticals.

To meet the individual needs of each of these markets, manufacturers now produce many grades of aluminum powders in several different general categories: granules, regular atomized powders and coated atomized powders, spherical powders, high-purity powders, alloy powders, blended powders, and dedusted atomized powders. All are produced by the same basic atomization technique, but starting materials may differ and, as with flake powders, further processing may be required.

General Atomizing Process

Atomized aluminum powder grades are produced by atomizing molten aluminum. A typical process for atomized aluminum production is shown schematically in Fig. 1. The molten aluminum is drawn through an atomizing nozzle; the lower end of the nozzle dips into the molten aluminum, and the upper end terminates in a small orifice. This orifice is surrounded by a small chamber into which a jet of pressurized air or inert gas is introduced. The air jet impinges on the stream of molten aluminum and disintegrates it into small particles. These irregularly shaped (nodular) particles, together with a substantial volume of cooling air, are drawn through a chiller chamber and into a cyclone and/or bag filter collecting system. The atomized powders are then graded by screening. After product sampling and inspection, the powders are ready for packing.

Several techniques can be utilized to provide the energy required to move the molten aluminum through the atomizing nozzle, including systems to force the metal stream through the orifice by metal-head or induced pressure. Metal-head pressure refers to the pressure difference created when the molten metal level in the furnace is higher than that at the atomizing nozzle. However, the process employed by most

major producers of atomized powder achieves the necessary metal flow through air aspiration of the molten aluminum. As the atomizing air or gas jet passes over the orifice of the nozzle, a suction effect is created. If gas pressures and volumes are sufficient, this causes the molten aluminum to be aspirated into the gas stream and disintegrated into droplets, which then rapidly solidify into particulates.

Aluminum powder atomizing can be accomplished by directing the molten metal stream either horizontally or vertically. Systems that utilize metal-head pressure to induce molten metal flow through the nozzle often atomize either vertically downward or horizontally. Aspirating systems, however, usually atomize either horizontally or vertically upward. The system shown in Fig. 1 atomizes vertically upward.

Each of the above systems has its particular advantages and disadvantages. The metal-head, vertically downward system generally produces aluminum powders with more closely controlled particle size distributions, but at relatively low production rates. Horizontal systems allow efficient production of medium- and coarse-sized powders. Vertically upward, aspirating processes permit the production of the widest range of powder grades—fine, medium, and coarse—at high production rates.

Atomizing Nozzle Technology

The basic technology for gas aspirating atomization was developed by E.J. Hall in the late 1920's (Ref 1). In this system, molten metal disintegration is controlled

Fig. 1 Schematic of atomized aluminum powder production
The powder is packed in drums or bins or is loaded for bulk shipment in trucks or railroad cars.

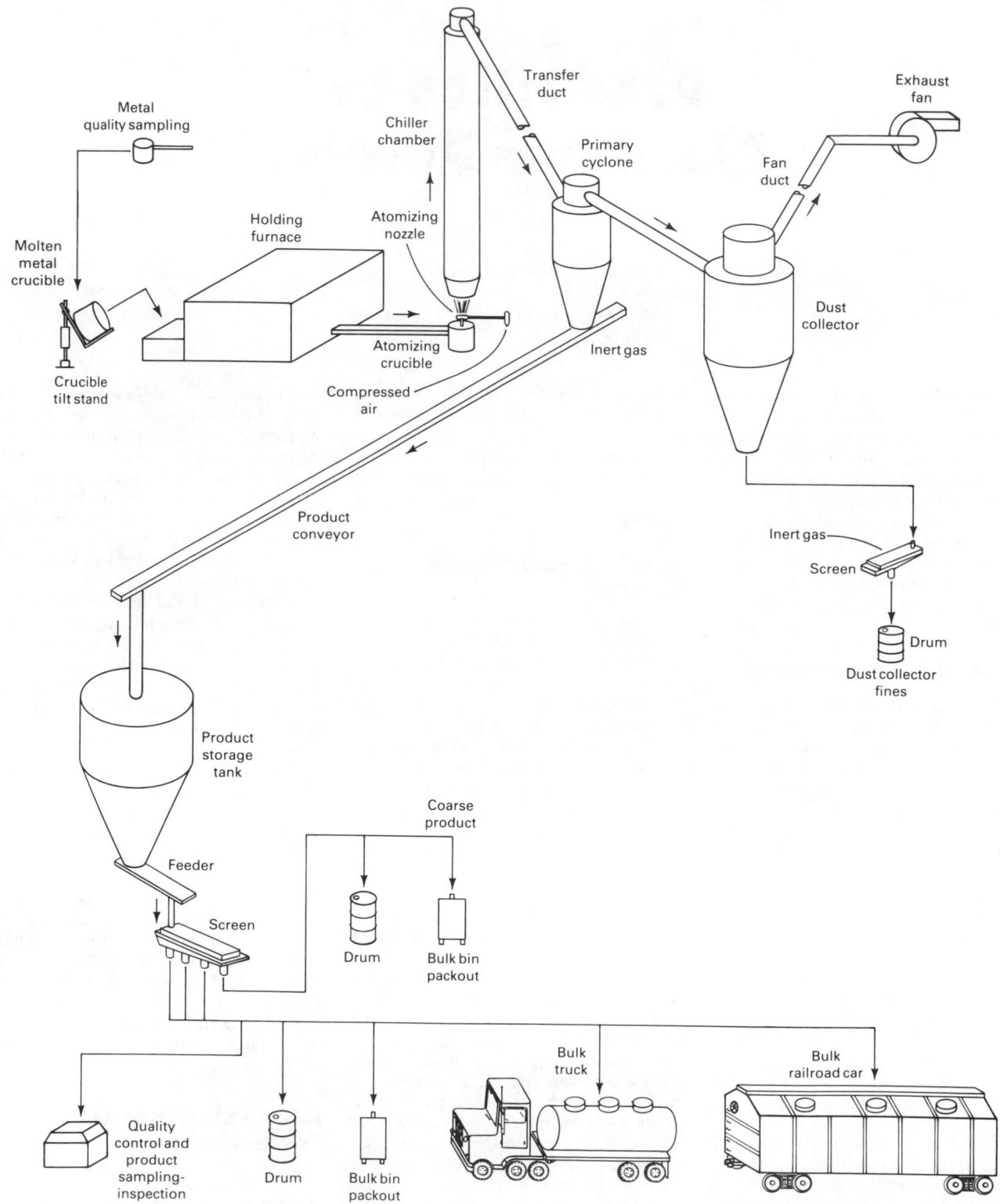

by directing the jet of disintegrating air or gas in converging angles upon the stream of molten aluminum through a rotating annular flow around the nozzle. The rotary motion of the atomizing stream contributes both to aspiration and disintegration of the molten aluminum stream. Rate of metal flow and resultant powder particle size are influenced by the aspirating force, the nozzle metal orifice diameter, and the vertical distance between the nozzle and the molten metal level.

Figure 2(a) is a front view of Hall's nozzle design and jet-producing apparatus. Figure 2(b) is a vertical cross-sectional view, and Fig. 2(c) is a sectional view. Figure 2(d) is an illustration of the jet rotating ring. The aluminum is maintained in molten condition with the nozzle tube immersed in the bath, as shown in Fig. 2(b). The nozzle may have its outer surface tapered toward the end. The amount of taper influences the degree of aspiration; the more acute the angle, the greater the vacuum created.

Molten metal in the trough or crucible under the nozzle is maintained at a temperature sufficient to prevent premature cooling during atomization (~750 °C or ~1380 °F), which leads to nozzle clogging and ineffective atomization. The atomizing air often is preheated to further ensure against heat loss in the nozzle. Atomizing gas is supplied to the annular chamber (Fig. 2c) tangentially to impart an initial rotation to the gas. At the front of this expansion chamber is a vane ring (Fig. 2d) that has inclined blades along the rim. This ring allows the atomizing gas to escape as a hollow, rapidly whirling jet, which is then directed by the distributor on converging lines around the nozzle tip. The condensed gas stream finally escapes through the annular jet orifice (Fig. 2b) surrounding the nozzle tip and orifice.

The flow of atomizing gas striking the stream of molten aluminum simultaneously from all sides and at converging angles breaks it into fine particulates. Although the convergent flow is very effective for atomizing the metal stream, it is less effective than a plain annular jet for producing aspiration. This loss in vacuum, however, is compensated for by the vortex that is created at the end of the nozzle. The whirling motion of the jet not only assists in creating a vacuum at the nozzle tip, but because of the centrifugal force generated, back pressure is prevented, thereby increasing atomization.

The rate of metal flow and powder particle size depend on the amount of vacuum created at the tip of the nozzle. Because this aspirating force depends on the volume and pressure of the atomizing gas, a drop in gas pressure means a decrease in atomizing rate. As discussed earlier, the aspiration force, and therefore atomizing rate and particle size, may also be controlled by regulating the nozzle taper, which affects the angle at which the atomizing air strikes the molten aluminum stream.

Alcoa Process

Because aluminum powder forms an explosive mixture with air over a wide range of metal-to-air ratios, aluminum powder plants have explosion hazards associated with them. During the late 1970's, virtually every major atomized aluminum powder plant in the world had an explosion (Ref 2). As a result of losing two powder plants due to an explosion during that period, one major producer decided to design and engineer a new atomizing facility with significantly improved safety technology. A schematic of the Alcoa process for atomizing aluminum powder is shown in Fig. 3.

This process has several basic differences from the typical system shown in Fig. 1. The bottom of the chiller chamber is closed rather than open as in Fig. 1, and the chiller chamber walls are reinforced. Explosion vents are located in the chiller chamber, along the exhaust duct, and in other critical locations such as the lids of the cyclone abatement chambers. These vents minimize pressure buildup in the event of an explosion. Clean air enters the system through filters, and air flow is created by an eductor so that powder does not have to travel through a rotating fan, which eliminates a significant static electricity ignition source.

Fig. 2 Hall nozzle design and jet-producing apparatus
(a) Front view. (b) Vertical cross-sectional view. (c) Sectional view. (d) Jet rotating ring

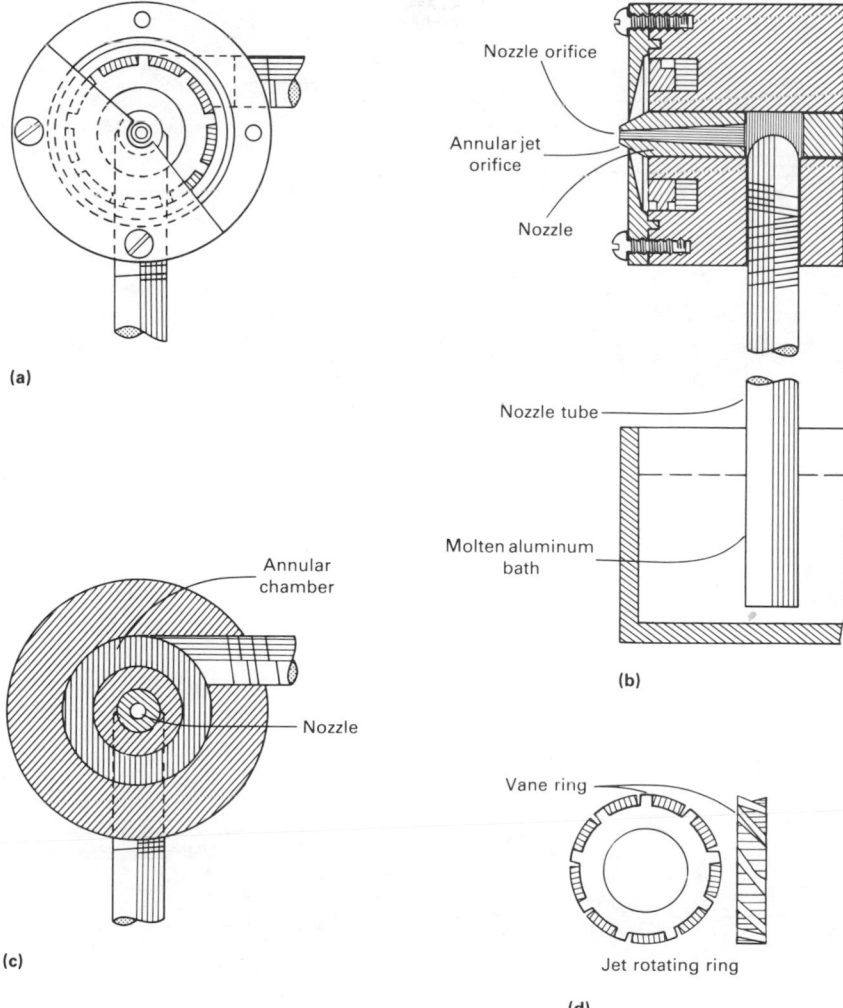

(a)

Nozzle orifice

Annular jet orifice

Nozzle

Nozzle tube

Molten aluminum bath

(b)

Annular chamber

Nozzle

(c)

Vane ring

Jet rotating ring

(d)

Fig. 3 Schematic of the Alcoa process for atomizing aluminum powder

Entire operation is under computer control. Powder is packed in drums or bins or is loaded for bulk shipment in trucks or railroad cars.

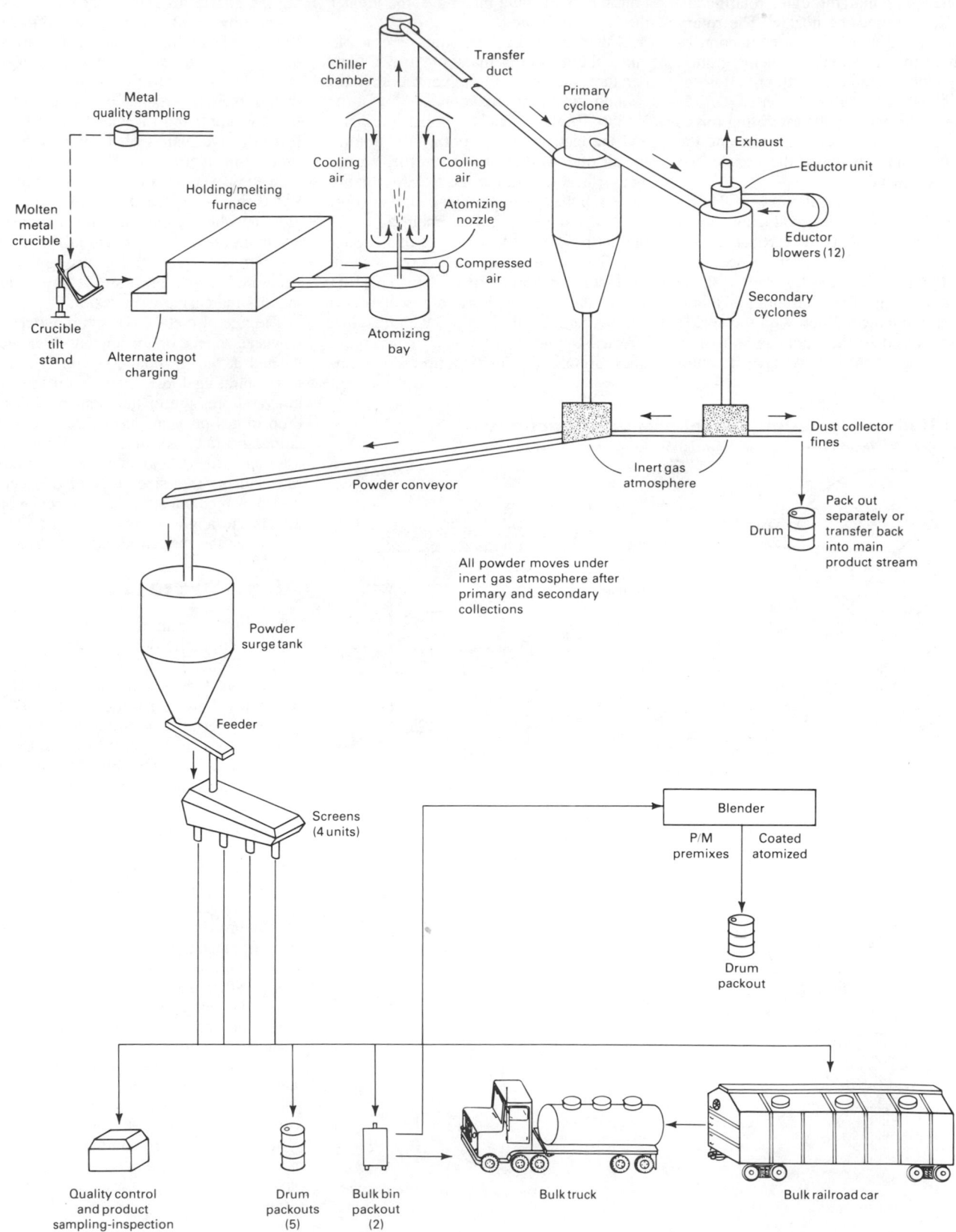

Fig. 4 Electron micrograph of air-atomized aluminum powder with low surface area (0.2 m²/g)
Magnification: 500X

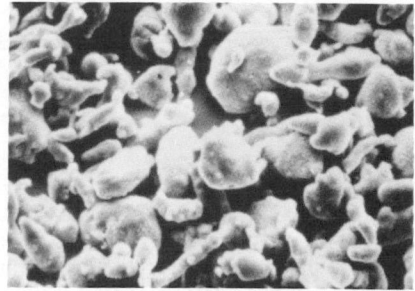

Fig. 5 Particle size distributions for typical atomized powder grades
Determined by Micromerograph analysis

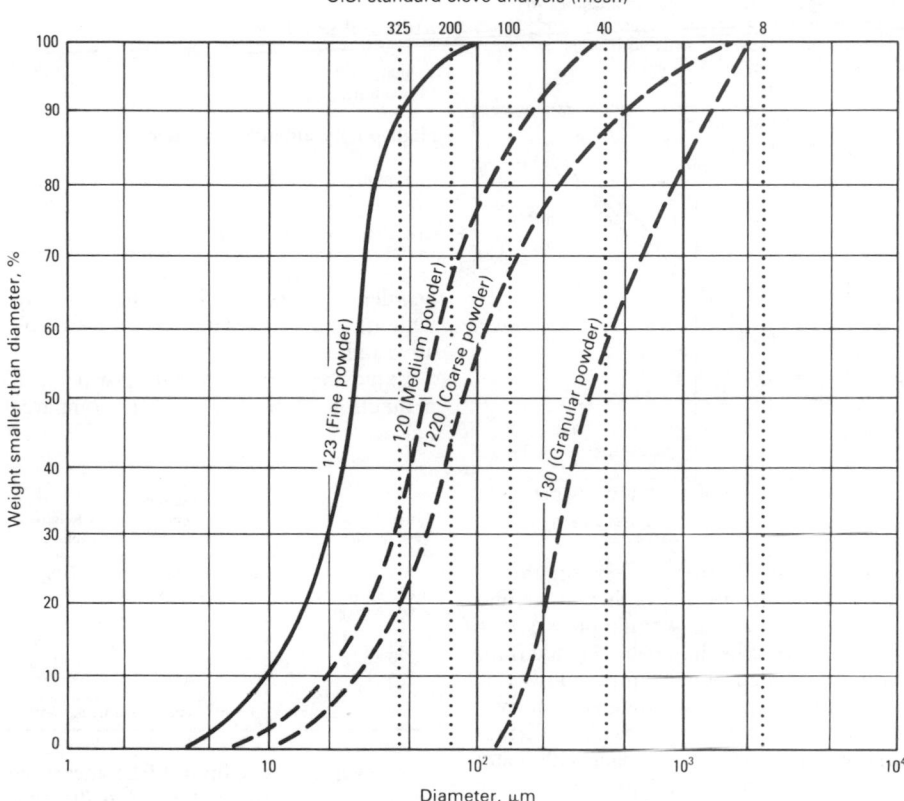

Workers are removed entirely from the areas containing the greatest hazards by substitution of computerized controls, remote TV surveillance, and automated operations. Explosion hazards are reduced between the atomizing nozzle and the cyclones by controlling the airborne concentration of aluminum powder to a level below that which forms an explosive mixture (lower explosive limit).

After the cyclones, the aluminum powder is transported in an atmosphere of inert gas to substantially reduce the explosion hazard. Based on a hazards analysis of serious accidents per manhours worked, the safety factor for the new atomizing operation has been improved 100 times over the previous conventional system.

Powder Size and Shape

The electron micrograph in Fig. 4 shows the irregular, nodular shape of conventional air-atomized aluminum powders. In most aluminum powder manufacturing methods, compressed air is used as the atomizing medium. Inert gases are used only for specialty applications that require a specific particle shape.

By modifying process parameters and screen sizes, the particle size distribution of atomized powders can be adjusted to produce a range of products from aluminum granules (typically +200 mesh) to fine atomized powders (up to 99% −325 mesh). Figure 5 shows the Micromerograph particle size distribution curves for typical granular, coarse, medium, and fine grades (Ref 3). Besides the wide range of grades available from direct atomizing, others are possible from blending screen fractions.

In addition to the Micromerograph, other techniques such as the Coulter counter and the MSA-Whitby centrifuge are used to measure fine aluminum powders (−200 mesh). Each method yields rapid and re-producible particle size distribution curves, but results often vary between methods. Gas-sorption methods are employed for precise measurements of total surface area. The surface areas of various atomized aluminum powders (Ref 4) are:

Atomized aluminum	Gas-sorption area, m²/g
Coarse (40-mesh designation)	0.15
Fine (200-mesh designation)	0.25
Dust collector fines (325-mesh designation)	0.90

Physical Properties

Typical physical properties of atomized aluminum powders (Ref 5) are:

Wrought density	2.7 g/cm³ (metal)
Melting point	660 °C (1220 °F)
Boiling point	2430 °C (4410 °F)
Surface tension at 800 °C (1470 °F)	865 dynes/cm
Apparent density	0.8 to 1.3 g/cm³
Tap density	1.2 to 1.5 g/cm³
Melting point of oxide	2045 °C (3720 °F)
Al₂O₃ content	0.1 to 1.0 wt%

The real density of aluminum powder approaches that of the base metal, but both apparent density and tap density vary as a function of particle size distribution.

Aluminum reacts readily with moisture or free oxygen in the air during manufacture to form an oxide coating on the powder surface. Oxide contents vary as a function of particle size and range from 0.1 to 1.0 wt%. Finer powders, due to their increased surface area per unit of weight, show the highest percentage of oxide by weight. Because the aluminum powder surface is hygroscopic, it also contains moisture in the form of physically absorbed water and hydrated aluminum oxide. The oxide film thickness on aluminum powder is relatively constant, even when atomizing conditions vary. Experiments employing air, argon, and nitrogen as atomizing gases and gas temperatures ranging from ambient to 595 °C (1100 °F) have shown that oxide thickness is 5 ± 0.5 nm (50 ± 5 Å), independent of atomizing gas composition and temperature, as illustrated in Fig. 6.

Chemical Properties

The chemical compositions of unalloyed atomized aluminum powders are

Fig. 6 Oxygen content of powder, as related to surface area

Oxide thickness does not vary significantly with variations in atomizing gas composition and temperature. Source: Alcoa Laboratories, Aluminum Company of America

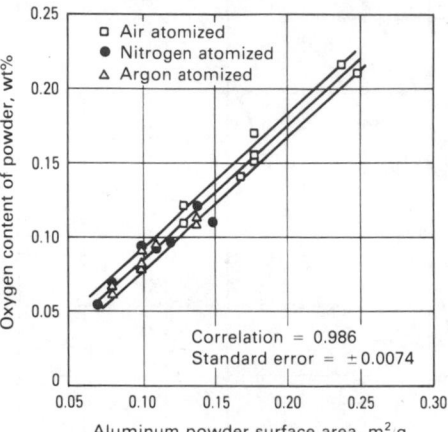

Table 1 Typical chemical analyses of atomized aluminum powders

| Type of powder | Composition, wt% | | | Other metallics | |
	Aluminum	Iron	Silicon	Each	Total
Atomized powders					
Typical	99.7	...	...	...	...
Maximum	...	0.25(a)	0.15(a)	0.05	0.15
High-purity atomized powders					
Minimum	99.97	...	...	...	...
Typical	99.976	0.007	0.008	...	0.009

(a) Iron plus silicon, maximum 0.30 wt%

shown in Table 1. Iron and silicon are the major contaminants for both regular and high-purity powders. Aluminum is stable in air because of its thin, natural oxide film. In finely divided powder form, however, aluminum is more chemically reactive. Aluminum powders can oxidize further after atomization and can react with water to liberate hydrogen and form aluminum hydroxide. Atomized aluminum exposed to moisture-saturated air at room temperature slowly forms beta alumina trihydrate and hydrogen.

Extremely pure aluminum is resistant to attack by acids, but dissolves in a mixture of nitric and hydrochloric acids. Solutions of alkali hydroxides rapidly attack aluminum, with evolution of hydrogen and formation of the corresponding soluble aluminate. Aluminum reacts vigorously with bromine and iodine to form aluminum halides. Dry, oxygen-free chlorine or hydrogen chloride gas also reacts with aluminum to form aluminum chloride. Finely divided aluminum can react violently when exposed to halogenated hydrocarbons such as methylene chloride and carbon tetrachloride, especially if the operation is carried out under pressure.

Explosibility

With certain combinations of oxygen, powder, and an initiating spark, dust clouds of aluminum can ignite and explode. Relative explosibility, as determined by the U.S. Bureau of Mines, is an index that measures the airborne concentration of powder and ignition energy necessary to initiate combustion (Ref 6). Their testing has shown that, although nearly all metal

powders have explosibility, aluminum, titanium, and magnesium have severe relative ratings.

Explosibility of aluminum powders, as a function of particle size, is compared below:

Powder mesh fraction	Lower explosive limit(a) oz/ft³	Relative explosibility
−40+100	No ignition	None
−100+140	No ignition	None
−140+200	No ignition	None
−200+270	0.111	Strong
−270+325	0.090	Strong
−325	0.045	Severe

(a) Values for kilograms per cubic metre are exactly the same.

The lower explosive limit (LEL), measured in ounces per cubic foot of air, is the minimum dust concentration that will permit ignition of an explosion. Coarse particles over 200 mesh are very difficult to ignite regardless of concentration; −200 to +325 mesh particles have strong explosibility in relatively small concentrations, and −325 mesh material has severe explosive characteristics in very small concentrations.

Another way of evaluating the effect of aluminum powder particle size on explosibility is to compare LEL with the percentage of particles that are less than 200 mesh in size (Fig. 7). A summary of the

Fig. 7 Effect of aluminum powder particle size on explosibility

Shown by comparing lower explosion limit with the percentage of particles less than 200 mesh in size

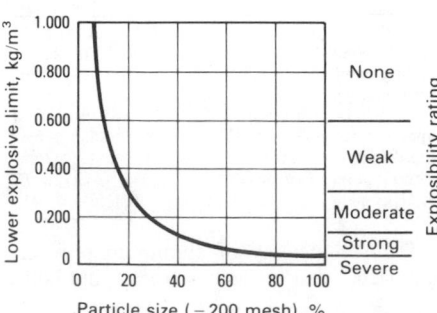

typical explosibility characteristics of aluminum powder as determined by the Bureau of Mines is as follows:

Minimum concentration (lower explosive limit)	0.045 kg/m³ (0.045 oz/ft³)
Minimum igniting energy	50 mJ (0.012 cal)
Ignition temperature of cloud	650 °C (1200 °F)
Ignition temperature of layer	760 °C (1400 °F)
Maximum explosion pressure	503 kPa (gage) (73 psig)
Maximum rate of pressure rise	140 MPa (20 000+ psi/s)

The reactivity and combustibility of atomized aluminum powders is an advantage in many of its applications, but can be a problem in handling. These powders can be shipped, stored, and used with safety, if proper attention is given to good housekeeping and basic rules of safety. The Aluminum Association's booklet, *Recommendations for Storage and Handling of Aluminum Pigments and Powders*, is an excellent reference source on this subject and is available from aluminum powder manufacturers. See the article "Explosivity and Pyrophoricity of Metal Powders" in this Volume for additional information on the explosibility, reactivity, and combustibility of aluminum powders.

REFERENCES

1. Hall, E.J., Process for Disintegrating Metal, U.S. Patent 1,659,291, 14 Feb 1928
2. Thompson, D.B., *Industry Week*, Vol 213, 14 June 1982, p 28-32
3. Aluminum Company of America, *Manufacture of Alcoa Aluminum Powders*, Pamphlet PAP 916 (FA2D-1), June 1982
4. Luyk, K.E., *Aluminum*, Vol 2, K.R. Van Horn, Ed., American Society for Metals, 1967, p 623-642
5. Rolles, R., Aluminum Flake Pigment, *Pigment Handbook*, T.C. Patton, Ed., John Wiley & Sons, New York, 1973
6. Jacobson, M., Cooper, A.R., and Nagy, J., Explosibility of Metal Powders, RI6516, U.S. Dept. of the Interior, Bureau of Mines, 1974

Production of Magnesium Powders

By Robert S. Busk
President
International Magnesium Consultants, Inc.

MAGNESIUM POWDERS gained importance during World War II, when large quantities were used for pyrotechnics and explosives. They currently are used to (1) manufacture Grignard reagents, which are organometallic halides such as ethyl magnesium chloride (C_2H_5MgCl) that are used in organic synthesis to produce pharmaceuticals, perfumes, and other fine chemicals; (2) effect chemical reductions, as in the manufacture of beryllium and uranium; (3) act as a light source in flares and photoflash bombs; (4) modify metallurgical properties of other metals, such as in the production of ductile iron and the removal of sulfur from hot iron products of blast furnaces; and (5) function as additives in electric welding electrode flux. At one time, magnesium powders were extruded into sections for structural applications (Ref 1 and 2); however, they are no longer used to produce such parts.

Almost all powder is produced from pure magnesium, although some magnesium-aluminum alloy powder containing up to 50% aluminum is used for photoflash bombs and for desulfurizing hot metal. Powder is produced by a variety of methods of comminution of solid or liquid metal.

Comminution of Solid Magnesium

Lathe turning of a billet is used to produce relatively coarse (about 10 mesh) particles. These are typically used for producing Grignard reagents. Particle shape is similar to a lathe chip produced with a chip-breaking tool. Chips are produced in air, with no control over the atmosphere. Figure 1 is a scanning electron micrograph of magnesium lathe turnings.

Fig. 1 Scanning electron micrograph of magnesium lathe turnings

Particle size: −4+11 mesh used for Grignard reagents. Magnification: 8×. Courtesy Reade Mfg. Co., a division of Remacor

Scratching. Powder is manufactured by applying a file to a rotating cylinder, resulting in curled particles that vary in size from about 50 to 300 mesh. This powder is suitable for chemical reactions and flares. It is produced in air with no atmospheric control. Although no longer used in the United States, this powder-producing method is still used in Europe. Apparent density of as-filed powder is about 0.3 to 0.5 g/cm^3. A ball milling procedure can be used to round the individual particles and thus increase apparent density to 0.6 to 0.8 g/cm^3.

Chipping, Milling, and Grinding. Solid ingots are fed into a chipping mill and proceed to a hammer mill to produce powders varying from about 10 mesh to less than 325 mesh. Particles vary in shape from highly irregular to almost spherical (Fig. 2). Apparent density varies from less than 0.5 to more than 1 g/cm^3. Powder is produced in air with no atmospheric control.

Magnesium powders produced by chipping and milling are used for military flares, fireworks, and production of uranium. In addition, these powders are used in the chemical and pharmaceutical industries and in the iron- and steel-making industries, where they are used as desulfurizing agents and inoculants for the nodulation of gray iron.

There is a new method to produce powder of less than 325 mesh for flares by grinding. Figure 3 is a scanning electron micrograph of ground magnesium particles.

Fig. 2 Scanning electron micrograph of mechanically comminuted (chipped and milled) magnesium granules

Particle size: −10+50 mesh; apparent density, 1.0 g/cm^3 minimum. Magnification: 13×. Courtesy Reade Mfg. Co., a division of Remacor

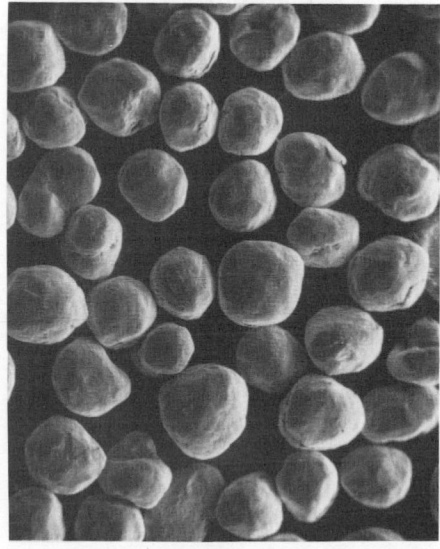

Fig. 3 Scanning electron micrograph of mechanically comminuted (ground) magnesium particles

Particle size: −100 mesh; apparent density, 0.4 to 0.6 g/cm³. Fisher particle size number, 21 μm. Magnification: 65×. Courtesy Reade Mfg. Co., a division of Remacor

Fig. 4 Scanning electron micrograph of atomized magnesium particles

Particle size: −30+50 mesh. Magnification: 18×. Courtesy Hart Metals, Inc.

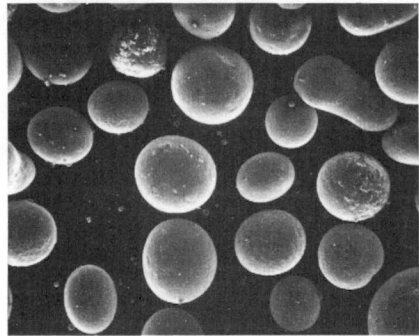

Fig. 5 Scanning electron micrograph of atomized magnesium particles

Particle size: −200+325 mesh. Magnification: 325×. Courtesy Hart Metals, Inc.

Comminution of Liquid Magnesium

Atomizing consists of disintegrating a liquid stream of magnesium into droplets, either by a high-velocity gas jet or through the use of a horizontally rotating disk. Droplets subsequently freeze into solid particles. Particle size varies from about 10 mesh to less than 325 mesh. Two samples of atomized magnesium powders are shown in Fig. 4 and 5.

A nonoxidizing atmosphere is required; helium, argon, and methane are suitable. Nitrogen, which reacts with magnesium to form magnesium nitride (Mg_3N_2), is not used. Particle shape depends on the quality of the atmosphere. If oxygen and water are excluded, particles are spherical. As small quantities of either water or oxygen are added to the atmosphere, the shape becomes irregular, although remaining roughly spherical. Apparent density, regardless of mesh size, is 0.7 to 0.8 g/cm³.

References 3, 4, and 5 describe a Russian technique for co-atomizing magnesium and flux on a spinning cup to produce flux-coated magnesium particles. These powders are used in the ferrous industry to desulfurize iron and steel by injecting the salt-coated magnesium particles into liquid iron and steel. Because the particles are flux-coated, atmospheric control is not required.

Emulsification. References 6 and 7 describe a process for producing salt-coated magnesium particles, which consists of agitating liquid magnesium in a liquid boron-containing flux. The magnesium emulsifies as small spheres in the flux. On freezing, magnesium particles are freed by grinding and screening (Fig. 6). A small amount of flux coats each particle, and the resulting powder is suitable for desulfurizing liquid iron and steel. Particles are spherical and vary in size from about 10 to 50 mesh.

Potential Safety Hazards

The principal safety precautions required for the handling of magnesium powder are related to avoiding and controlling fire and explosion (Ref 8).

Fire. Magnesium burns if heated sufficiently to release significant quantities of magnesium vapor. In effect, magnesium must be melted before ignition occurs, 650 °C (1200 °F) for pure magnesium. Magnesium oxidizes rapidly on the surface; the resulting oxide film varies from a few angstroms at low temperature to a few

Fig. 6 Salt-coated magnesium particles produced from emulsion
Courtesy Dow Chemical Co.

microns at high temperature. The high surface-to-volume ratio of small particles, combined with the high heat of oxidation, results in raising the powder particle temperature above the ambient temperature. Thus, apparent ignition temperature is lower with smaller sized particles. When particles are approximately $0.1 \mu m$ in size, apparent ignition temperature is room temperature, and the powder is considered pyrophoric.

Although magnesium fire is not violent, it is spectacular. Magnesium produces a brilliant white light, and large quantities of white magnesium oxide are given off. Face shields and fire-resistant clothing should be worn whenever there is a potential hazard.

Powder fires can be extinguished by cooling—using cast iron chips, for example. Proprietary extinguishing compounds are also available. Normal melting fluxes are suitable extinguishers. Water should never be used on a powder fire. It scatters the powder, accelerates the burning dra-matically, and produces hydrogen, which is explosive in nature.

Explosion is the greatest hazard associated with magnesium powder. If powder is fine enough so that an air suspension can be obtained, any source of ignition will result in a violent explosion. Air suspensions should be avoided. Powder coarser than 100 mesh is relatively safe, because an air suspension is difficult to maintain. Powder finer than 200 mesh is hazardous, as air suspensions are easily maintained. Sources of ignition should be avoided. Thus, spark-proof motors and switches should be installed, and nonsparking tools should be used. All equipment should be well grounded to prevent sparks from static electricity. Smoking and the carrying of matches should be prohibited.

Equipment presenting an explosion hazard should be isolated from personnel and from other equipment. For more information on fire and explosive hazards associated with metal powders, see the article "Explosivity and Pyrophoricity of Metal Powders" in this Volume.

REFERENCES

1. Busk, R.S. and Leontis, T.E., *Trans. AIME*, Vol 188, 1950, p 297
2. Busk, R.S., The Pellet Metallurgy of Magnesium, *Magnesium Association and Magnesium Industrial Council Transactions*, London, 1960
3. U.S. Patent 3,881,913, 1975
4. Voronova, N.A., External Desulfurization of Hot Metal by Magnesium Injection, *Proceedings of the International Magnesium Association*, May 1977
5. Voronova, N.A., *Desulfurization of Hot Metal by Magnesium*, International Magnesium Association, 1983
6. U.S. Patent 4,186,000, 1980
7. U.S. Patent 4,279,641, 1981
8. Jacobson, M., Cooper, A.R., and Nagy, J., "Explosibility of Metal Powders," U.S. Department of the Interior, Bureau of Mines, ROI 6516, 1964

Production of Nickel and Nickel Alloy Powders

NICKEL POWDERS can be classified into three groups: powders produced by the decomposition of nickel carbonyl, powders produced by the reduction of aqueous solution of a nickel salt with hydrogen under pressure (the Sherritt process), and powders produced by inert gas or water atomization. Of these, decomposition of nickel carbonyl and the Sherritt process, which produce metallic nickel from its ore, are the most important commercially.

The section of this article on the production of nickel powders by carbonyl vapormetallurgy processing reviews early metal carbonyl experimentation, with emphasis on the nickel tetracarbonyl research by Ludwig Mond. The types of metal carbonyls that can be formed and the carbonyl formation and decomposition reactions used for the production of primary metal and powder products, as well as the commercial carbonyl processes used for the production of high-purity nickel, are described.

The section of this article on hydrometallurgically produced nickel powders is presented according to the three processing steps: leaching, solution purification, and metal recovery. The Sherritt process for recovering nickel from sulfide concentrates is described. The final section of this article discusses the production of nickel powders produced by inert gas or water atomization. These powders are used mainly in the production of hardfacing alloys or superalloys.

Production of Nickel Powder by Carbonyl Vapormetallurgy Processing

By Donald H. Antonsen
Manager, Chemical Industry Sales
The International Nickel Co., Inc.
and
John H. Tundermann
Vice President—Technology
Huntington Alloys, Inc.

THE PRODUCTION OF NICKEL POWDERS by the decomposition of nickel carbonyl dates back to a process developed in 1889 by Ludwig Mond and his co-workers Carl Langer and Friedrich Quincke. According to Mond (Ref 1), the firm of Brunner, Mond and Company was endeavoring to prepare chlorine from ammonium chloride, which in turn was obtained as a by-product of the ammonia-soda process, widely known as the Solvay process.

One of the obstacles to the successful production of chlorine was the unusually rapid corrosion of the nickel valves on the brick-lined tanks in which ammonium chloride was vaporized. Laboratory study of this corrosion proved that it resulted from

the small amount of carbon monoxide used to sweep ammonia from the vaporizing tanks. Carbon monoxide was partially converted into carbon dioxide, and a black mixture of nickel and amorphous carbon was deposited. This observation led to a more elaborate study of the reaction of carbon monoxide with finely divided nickel.

Mond and co-workers found that four molecules of carbon monoxide at atmospheric pressure and at temperatures between 40 and 100 °C (105 and 212 °F) react with active nickel to form a colorless gas, nickel tetracarbonyl:

$$Ni + 4CO \overset{50\ ^\circ C}{\rightleftarrows} Ni(CO)_4$$

They further demonstrated that the reaction is readily reversible by heating the nickel tetracarbonyl to temperatures in the 150 to 300 °C (300 to 570 °F) range to yield pure nickel and carbon monoxide:

$$Ni(CO)_4 \overset{230\ ^\circ C}{\rightleftarrows} Ni + 4CO$$

Nickel that has been in contact with air does not react with carbon monoxide, which is why carbonyls were not discovered earlier.

Encouraged by their success in preparing nickel tetracarbonyl, Mond and co-workers immediately attempted to prepare carbonyls from all the metals at their disposal. However, it was Berthelot who first announced the formation of a volatile compound of carbon monoxide and iron, iron pentacarbonyl, in June 1891. Subsequent discoveries were made of many metal carbonyls, including cobalt, iron, molybdenum, and ruthenium carbonyls.

After preliminary laboratory trials, Mond built an experimental plant near Birmingham, England, and subsequently developed the Mond-Langer process for separating nickel from Canadian matte. By 1895, the plant successfully produced $1^1/_2$ tons of nickel weekly from Canadian matte containing 40% Ni. Mond subsequently opened a refinery in Clydach, Wales (Mond Nickel Company), and during the 27 years the refinery operated, the plant processed over 82 000 metric tons (90 000 tons) of nickel in the form of pellets that exhibited greater purity than nickel produced by any other commercial process at that time.

In this process, which in modified form is still used at the refinery in Clydach, nickel oxide produced from roasting nickel sulfide is reduced to nickel sponge by hydrogen, activated by sulfiding, and volatilized as carbonyl in an atmospheric reactor. Nickel carbonyl produced in this manner is decomposed directly in the powder and pellet units as part of a continuous process of refining.

Metal Carbonyls

Many metals form carbonyls; in fact, all of the metals of the first, second, and third transition metal series have been converted into one or more types of carbonyls. Additionally, several of the lanthanides and actinides have formed carbonyls. Nickel forms a single carbonyl, $Ni(CO)_4$, in its zero valent form:

$$OC \rightarrow Ni \leftarrow CO$$

Nickel also forms a hydrocarbonyl, $H_2Ni_2(CO)_6$, where nickel has an oxidation number of -1. The carbon monoxide ligands in nickel carbonyl can be replaced by other ligands such as phosphines, phosphites, and certain unsaturated hydrocarbons where a high degree of electron density is capable of allowing "back bonding π," as well as the conventional sigma bonding typical of donar ligands.

The iron penta compound, iron pentacarbonyl or $Fe(CO)_5$, rapidly condenses into the bimetallic species diiron nonacarbonyl $Fe_2(CO)_9$, which on heating further condenses into the trimetallic triiron do-

decacarbonyl $Fe_3(CO)_{12}$. The nonacarbonyl compound exhibits two types of carbonyl bonding, the donar sigma type and the bonding bridge type (π):

Chromium, molybdenum, and tungsten each form a single hexacarbonyl, a hexacoordinated octahedron. Cobalt forms a binuclear carbonyl, dicobalt octacarbonyl, $Co_2(CO)_8$:

which condenses to the tetranuclear species, containing bridged cobalt atoms:

The condensing continues with the formation of what is known as a cluster carbonyl:

$$Co_8(CO)_{18}C^{-2}$$

Several types of carbonyl clusters exist; one of the largest is the tetracapped pentagonal prism:

$$Rh_{15}(CO)_{28}C_2^-$$

Some metals have avoided becoming "pure" carbonyls to date, but halo derivatives of copper, gold, platinum, and palladium have been synthesized:

$Cu(CO)Cl$	$Pt(CO)_2Cl_2$
$Au(CO)Cl$	$Pd_2(CO)_2Cl_4$

Heterocarbonyls also have been produced, including:

$$(CO)_5MnCo(CO)_4$$
$$[(CO)_5Re]_2Fe(CO)_4$$
$$Re_2Ru(CO)_{12}$$
$$Ru_2Os(CO)_{12}$$

Metal carbonyl carbides also exist, such as $Fe_5(CO)_{15}C$ and the dicarbide clusters $Ru_{10}C_2(CO)_{24}$, $Co_{13}C_2(CO)_{24}$, and $Rh_{15}C_2(CO)_{23}$. Another example of metal carbonyl is the dinitrogen derivative:

A complete list of all the metal carbonyls that can be formed is beyond the scope of this article. The above examples are provided to illustrate the diversity of the chemical forms of metal carbonyls. For a historical perspective and technical overview of metal carbonyl technology, see the list of selected references at the end of this article.

Metal Carbonyl Formation and Decomposition

While the equation for metal carbonyls is generally written as follows, the mechanism is far more complex than the equation implies:

$$xMe + yCO \rightleftarrows Me_x(CO)_y$$

At room temperature, carbonyls typically form liquids, vapors, and colored crystals. Carbonyls may be formed, volatilized, and condensed in a continuous process to separate them from the inert constituents of the feed. Some must be separated with elaborate techniques, while still others are intractable materials. They may be formed as liquids or as solutions in organic solvents and separated from inert solids by filtration. Crude carbonyls are separated from each other and residual impurities by distillation, sublimation, recrystallization, or selective solution of the carbonyl. Purified carbonyls are decomposed by heating into the carbon monoxide and pure metal. When a metal carbonyl can be separated from its impurities, subsequent decomposition generally results in high-purity metals. Exceptions to this include the formation of carbides.

Nickel tetracarbonyl was the first carbonyl discovered 90 years ago. Iron pentacarbonyl and cobalt octacarbonyl followed soon after; these remain the only carbonyls whose singular chemical properties have been applied commercially to extractive metallurgy, although it is speculated that ruthenium separation by carbonyl formation should ensue.

Nickel carbonyl formation is achieved by:

- Oxidation of nickel metal by a "suitable" oxidant to Ni (II)
- Addition of the ligands to the Ni (II) species
- Reduction of the liganded Ni (II) complex with the concurrent expulsion of the suitable oxidant to continue the reaction with additional nickel metal

In the formation of nickel tetracarbonyl, a mixture of freshly prepared nickel metal and nickel sulfide is heated in the presence of carbon monoxide. Nickel sulfide adds the ligands:

$$Ni^{+2}S^{-2} + 4CO \rightarrow Ni^{+2}(CO)_4S^{-2}$$

Additional carbon monoxide causes the reduction of this intermediate with the expulsion of the suitable oxidant, carbonyl sulfide:

$$Ni^{+2}(CO)_4S^{-2} + CO \rightarrow Ni^0(CO)_4 + COS$$

Nickel sulfide is then regenerated:

$$COS + Ni^0 \rightarrow Ni^{+2}S^{-2} + CO$$

Thermal decomposition reactions of commercial interest for cobalt, iron, and nickel carbonyls occur at temperatures about 200 °C (390 °F), where carbonyls can be handled in vapor form. Under these conditions, reaction kinetics permit acceptable powder production rates to be maintained. Carbonyls are heated rapidly to the desired decomposition temperature; at this temperature, nuclei form in the vapor to provide the required sites for metal deposition.

Decomposition products are the pure metal and carbon monoxide. During decomposition in commercial decomposers, the disproportionation reaction of carbon monoxide is catalyzed by the freshly formed metals. Iron is much more active as a catalyst than either nickel or cobalt, and therefore, carbon contents in nickel carbonyl-derived powder are related to the trace iron content:

$$2CO \underset{Fe}{\rightarrow} CO_2 + C$$

Nickel Tetracarbonyl

The versatility of the carbonyl extraction of nickel is characterized not only by selectivity in formation, but also by the relative ease of separation and decomposition to high-purity metal. Under the conditions employed for reaction at atmospheric pressure, carbonyl-forming impurities in crude nickel metal do not enter the gas phase. Thus, extraction of nickel by its volatilization from crude metal as carbonyl vapor is a highly selective process.

For a given condition of surface activity, the rate of gaseous or liquid carbonyl formation is determined by the temperature of the reaction and increases with increasing partial pressure of carbon monoxide. Whereas the rate of carbonyl formation increases with increasing temperature, the equilibrium mole fraction of carbonyl formed decreases sharply with increasing temperature. At a given total pressure, there is an optimum temperature at which metal carbonyl will form, as shown in Fig. 1.

Kinetic data for the nickel tetracarbonyl reaction can be correlated by means of a modifed first-order rate equation. The ultimate degree of metal conversion of "initial surface activity" has a definite value at a given temperature and pressure, which permits the derivation of a functional relationship between surface activity and metal conversion. Substituting this relationship and integrating the first-order rate equation yields a semiempirical model for correlating nickel tetracarbonyl formation rate data at a given temperature (Ref 3):

$$\ln \frac{a^\circ}{a^\circ - x} = K_o \left(P_{coi} - \sqrt[b]{\frac{P_{cai}}{K_e}} \right) t$$

where a° is the initial activity of the surface; b is the number of moles of carbon monoxide per mole of carbonyl; K_o is the rate constant in grams per square centimetre per hour; K_e is the equilibrium constant; P_{cai} is the partial pressure of carbonyl at the interface in torr; P_{coi} is the partial pressure of carbon monoxide at the interface in torr; t is time in seconds; and x is the fraction of metal reacted.

From an engineering design point of view, conversion of nickel to a carbonyl at a fixed temperature depends primarily on the partial pressure of the carbon monoxide present. The effect of increased system pressure on the conversion of nickel to a carbonyl has been shown experimentally and demonstrated commercially. The volume change inherent in the carbonyl reaction—four carbon monoxide molecules combining in one molecule of nickel tetracarbonyl—also points to an increase in carbon monoxide pressure as a means of improving reaction kinetics. In addition to directly increasing the rate of reaction, increased pressure stabilizes the carbonyl, thus permitting reaction at higher temperatures for a further increase in rate. The nickel tetracarbonyl formation reaction is highly exothermic:

$$Ni + 4CO \rightarrow Ni(CO)_4$$
$$+ 622.2 \text{ kJ/g} \cdot \text{mol}$$
$$(148.6 \text{ kcal/g} \cdot \text{mol})$$

Accordingly, high-conversion-rate commercial nickel carbonylation processes require large heat removal systems.

At room temperature, nickel tetracarbonyl is a colorless volatile liquid with a high vapor pressure. It boils at about 43 °C (109.4 °F) and begins to decompose at 60 °C (140 °F) or less, depending on conditions. It is only slightly soluble in water, but is highly miscible with many organic solvents. Some physical properties of nickel tetracarbonyl include:

Chemical formula	$Ni(CO)_4$
Color and state	Colorless liquid
Molecular weight	170.75
Nickel, %	34.37
Melting point	−25 °C (−13 °F)
Boiling point	43 °C (109.4 °F) at 101.1 kPa (751 mm Hg) 42.1 °C (107.8 °F) at atmospheric pressure
Specific gravity	1.36153 at 0 °C (32 °F) 1.27132 at 36 °C (97 °F)
Heat of formation	−622.2 kJ/g · mol (−148.6 kcal/g · mol)

Nickel tetracarbonyl vapor is exceedingly toxic when inhaled; 50 ppb of the compound in the air is the maximum allowable concentration designated by the American Conference of Governmental and Industrial Hygienists. Nickel tetracarbonyl should be handled only by person-

Fig. 1 Influence of system pressure and temperature on reaction rate in the formation of nickel tetracarbonyl
Source: Ref 2

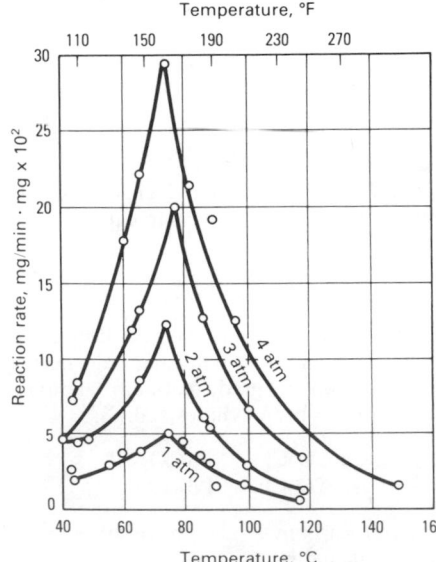

nel who are familiar with its toxic properties.

The stability of metal carbonyls varies considerably. Nickel tetracarbonyl begins to liberate carbon monoxide and nickel at 0 °C (32 °F) in a vacuum, and the decomposition becomes rapid above 60 °C (140 °F) in an inert gas under atmospheric pressure. Thermal decomposition is endothermic and is markedly inhibited by the presence of carbon monoxide. The mechanism of formation of nickel powder by the thermal decomposition of nickel tetracarbonyl is complex. In the formation of powder particles from the gaseous phase, several simultaneously occurring processes exert an influence. These include the formation of complex nuclei, secondary crystallization of nickel on the surfaces of the finest particles, the interaction of particles in the crystallization process, and side reactions such as the decomposition of carbon monoxide. At a given temperature, the homogeneous part of the decomposition of nickel tetracarbonyl follows the rate equation (Ref 3):

$$r = \frac{K_o P_{ca}}{1 + K_g P_{co}}$$

where r is the decomposition rate in grams per cubic centimetre per hour; K_g is the adsorption constant of carbon monoxide on metal in $torr^{-1}$; K_o is the rate constant in grams per square centimetre per hour; P_{ca} is the partial pressure of carbonyl in torr; and P_{co} is the partial pressure of carbon monoxide in torr. The decomposition rate for nickel tetracarbonyl is directly proportional to the partial pressure of the carbonyl and inversely proportional to the partial pressure of the liberated carbon monoxide.

Conditions that influence the formation of the self-nucleating nickel particles during carbonyl decomposition (such as process temperature and the concentration and rate of the nickel tetracarbonyl supply to the decomposer) can vary greatly, which affects the physical and technical properties of the powders produced. Additions to the nickel tetracarbonyl gas stream sent to the decomposer can alter the mechanism by which the powder is formed, as well as its resulting morphology. For example, hydroquinone vaporized with the carbonyl acts as a free radical trap, resulting in the formation of macroparticles of nickel metal. Nickel tetracarbonyl can be photochemically decomposed to form elemental nickel by light of wavelengths shorter than 390 nm (3900 Å).

The most important industrial application of nickel tetracarbonyl is for refining nickel, in which nickel pellets and powder products are formed. Gas-plated products and a wide range of metallic and nonmetallic (nickel-coated graphite) powders can be produced by forming coatings of nickel on various surfaces by decomposing the carbonyl under special conditions. Nickel tetracarbonyl is manufactured on a limited basis to serve the chemical industry. The carbonyl serves as a catalyst for organic synthesis, as well as providing a means for the manufacture of other organonickel compounds. It also is used for vapor plating on smooth surfaces (such as plastic and metal) and to form nickel molds for the glass industry.

Commercial Processes

Currently, Inco Ltd. produces high-purity nickel powders by the thermal decomposition of nickel tetracarbonyl. Gaseous nickel tetracarbonyl is formed by reacting carbon monoxide with nickel concentrates under controlled conditions; subsequent thermal decomposition of the gas permits recovery of the nickel as a fine metallic powder and nickel pellets. The process, which affords a high degree of purity with respect to metallic elements other than nickel, produces nickel powders with extremely uniform particle size and structure.

The original Mond-Langer process has been greatly improved during its continued use. The only two commercial nickel refineries in the Free World employing the carbonyl process to produce nickel powder are located on the site of the original Mond facility in Clydach, Wales and in Copper Cliff, a subdivision of Sudbury, Ontario.

The Clydach refinery began nickel output from Sudbury cupriferous nickel matte in 1902. Current plant feed matte is granular nickel oxide that contains small amounts of copper, cobalt, and iron, as well as siliceous matter. This refinery still uses the basic concept of the Mond-Langer process and produces nickel products using atmospheric pressure carbonyl processing. The latest innovations involve replacement of the many small hearth-type reducers and volatilizers of the original works by large rotary kilns arranged in trains of two kilns with a smaller sulfiding unit. In the initial unit of each train, the nickel feed matte is reduced to metal by a countercurrent flow of preheated hydrogen. The reduced nickel matte is then activated by sulfiding and is volatilized as carbonyl in the final kiln by a countercurrent flow of carbon monoxide. The resulting nickel tetracarbonyl vapor is decomposed directly into pure nickel pellets and powders in reactors or decomposers.

The Copper Cliff Nickel Refinery began operations in 1973 with two newly developed techniques, including the use of top-blown rotary converters that take a 64-metric ton (70-ton) charge of feed materials. The other manufacturing method uses the Inco pressure carbonyl process for the recovery of nickel from a variety of nickel-bearing feed materials, including the removal of nickel from copper, cobalt, and precious metal concentrates. The refinery produces pellets and powder of 99.9+% nickel purity and a co-decomposed iron-nickel powder.

Environmental controls at the plant include a totally enclosed process that constantly recycles all process reagents. The Copper Cliff refinery uses the same basic process chemistries, but forms nickel tetracarbonyl at high pressure, which is necessary to extract nickel in the presence of high copper concentrations.

The plant feed is a mixture of oxides and sulfides of nickel, copper, iron, and cobalt, other crude metallics, and partially processed precious metals and refinery residues that contain nickel. This mixture is charged along with coke into one of two 64-metric ton (70-ton) capacity, top-blown rotary converters, where it is melted and partly blown to slag silicates and some of the iron. The converter-treated hot material is subsequently granulated by pouring it through high-velocity water jets. At this point, the granules contain 65 to 75% Ni, 15 to 20% Cu, 2 to 3% Fe, and 3% S present as sulfides.

The metallic granules are batch-reacted with carbon monoxide at temperatures up to 180 °C (360 °F) and pressures up to 70 atm (1030 psi) in one of three 136-metric ton (150-ton) capacity rotating reactors. Nickel, and some iron, is extracted as a crude carbonyl vapor, while copper, cobalt, precious metals, and impurities are retained in the residue. Removal of iron is controlled to between 20 and 50% of the input iron level. Iron concentration in the granules rarely exceeds 4%. Cobalt in the feed does not form $Co_2(CO)_8$ or $Co_3(CO)_{12}$, because these compounds occur only when carbon monoxide pressures of 150 atm (2205 psi) are achieved.

Formation of nickel tetracarbonyl is quite exothermic; consequently, the reactors are water cooled. Extracted carbonyl vapor is liquefied and stored at atmospheric pressure. The carbon monoxide carrier gas also is recycled. The liquefied product stream, containing mutually soluble nickel and iron carbonyls, subsequently is pumped to distillation columns, where it is separated into nickel tetracarbonyl vapor and an iron-rich liquid carbonyl. Fractionating the crude carbonyl liquid is achieved easily, because the boiling points of the nickel and iron carbonyls are 43 °C (109.4 °F) and 102.8 °C (217 °F), respectively. The system produces nickel tetracarbonyl vapor of 99.998% purity, with the bottom liquor having nickel-to-iron ratios of up to 3 to 7. At the top of the column, nickel tetracarbonyl vapor is siphoned off and fed directly into powder decomposers. Somewhat below the top of the column, liquid nickel tetracarbonyl is taken off and either held in storage or used to feed the nickel pellet decomposers.

The powder decomposers are steel cylinders, the walls of which are heated with high-capacity electrical resistance heaters. Liquid carbonyl vapor is introduced into the top of the decomposer chamber slightly above atmospheric pressure, where it contacts the heated decomposer walls that are preset at a temperature between 250 and 350 °C (480 and 660 °F). The thermal shock decomposes nickel tetracarbonyl into nickel powder, with the release of carbon monoxide. The latter is recycled through filters to the main gas compressor to be recycled to the pressure carbonyl reactors, while the powder is collected, gas purged, stabilized with an oxide coating, and transferred to storage completely free from carbonyl and carbon monoxide.

The dangerous nature of the gaseous and liquid carbonyl compounds formed in the Inco processes is fully countered by established, effective, and practical safety measures. All product discharge systems are purged completely prior to powder and pellet packaging.

Powder Properties and Applications

Currently, there are three types of nickel powders produced by thermally decomposing nickel tetracarbonyl—single spikey particles, filaments, and high-density semismooth particles. These powders exhibit uniform size and structure and high surface areas, and the method of production affords a high degree of purity with respect to other metallic elements.

Spikey particles are a general-purpose type of nickel powder. They are fine and regular in shape, with rough surface projections (Fig. 2). The powder typically contains 700 to 900 mL/m^3 (700 to 900 ppm) oxygen, 3 to 5 mL/m^3 (3 to 5 ppm) iron, 1 mL/m^3 (1 ppm) sulfur or less, and 600 to 700 mL/m^3 (600 to 700 ppm) of graphitic carbon. The Fisher subsieve size is 3 to 7 μm, and apparent density is 1.8

Fig. 2 Scanning electron micrograph of general-purpose nickel powder produced by carbonyl decomposition
Magnification: 3000×

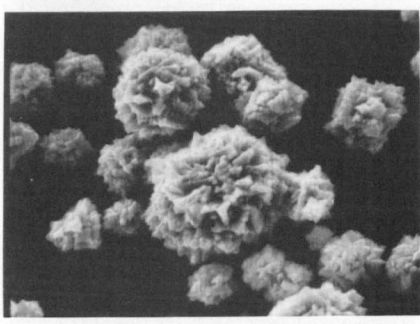

Fig. 3 Scanning electron micrograph of the filamentary-type nickel powder produced by carbonyl decomposition
Magnification: 1000×

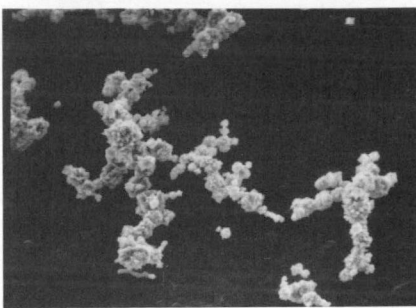

Fig. 4 Scanning electron micrograph of high-density (fine) nickel powder produced by carbonyl decomposition
Magnification: 1000X

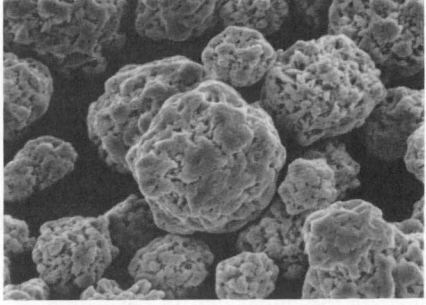

to 2.7 g/cm^3. Surface area is on the order of 0.4 m^2/g (BET).

Filamentary nickel powders are characterized by their unique, spikey, chainlike structure of relatively fine particles, which makes them fluffy in nature, with low apparent density (0.5 to 1.0 g/cm^3) and large specific surface area (0.6 to 0.7 m^2/g). The structure and asymmetrical filament shape of this powder type are illustrated in Fig. 3.

Semi-smooth high-density nickel powders are available in fine and coarse sizes. The fine is 10 to 20 μm in diameter (Fig. 4); the coarse is −16+40 mesh. Powder apparent density ranges from 3.5 to 4.2 g/cm^3.

The properties of these high-purity nickel powders have been tailored to meet the demanding requirements of the chemical, energy, and metal industries. Currently, these products provide the basic nickel source for a wide range of specialized products. Application areas for these powders include:

- Batteries and fuel cell electrodes
- P/M parts
- Ferrites
- Carbide cutting tools
- Catalysts
- Welding products
- Pigments and coatings
- Chemicals
- Electronic alloys
- Getters
- Conductive resins and plastics
- Electromagnetic shielding

Production of Nickel Powder by Hydrometallurgical Processing

By Basil Meddings
Manager, Analytical Services
Sherritt Gordon Mines Ltd.

HYDROMETALLURGICAL PROCESSING of nickel powders incorporates leaching, solution purification, and metal recovery. A typical example is the Sherritt process for recovering nickel from sulfide concentrates, which was developed between 1948 and 1953. The process was first put into commercial use at the Fort Saskatchewan refinery of Sherritt Gordon Mines in 1954. A flowchart of the Sherritt process is shown in Fig. 5.

Leaching. The feed to the process can be either sulfide concentrate or sulfide matte or a mixture of both. The mineralogical and chemical compositions of typical so-

Fig. 5 Sherritt ammonia leach process
Source: Ref 4

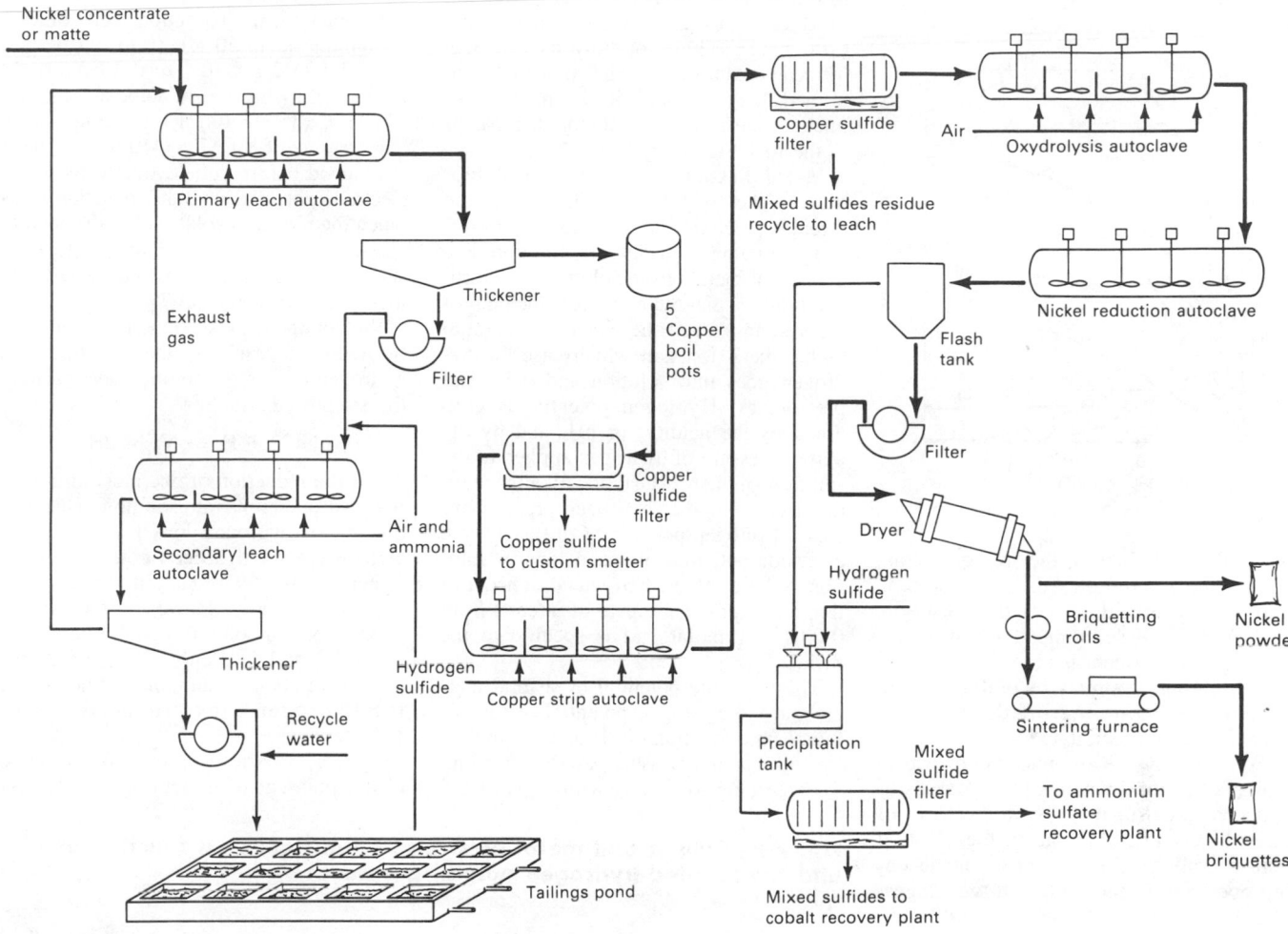

lutions are given in Table 1. Because the four principal metals—iron, cobalt, nickel, and copper—occupy adjacent positions in the periodic table (atomic numbers 26 through 29), they have similar chemical properties that render separation difficult.

The leaching operation takes advantage of the differing stabilities of the ammine complexes under hydrolyzing and oxidizing conditions. The sulfide concentrate (or matte) is leached in autoclaves in an ammoniacal solution of ammonium sulfate at 93 °C (200 °F) under air pressures of 0.8 MPa (116 psig) in a continuous operation. All four metals pass into solution as soluble ammine complexes:

$$NiS + 6\,NH_3 + 2O_2$$

$$\rightarrow Ni(NH_3)_6^{++} + SO_4^{=}$$

The iron complex is the least stable; it is completely hydrolyzed to a hydrated iron oxide (Fe_2O_3), which is re-precipitated and removed by filtration. This hydrated iron oxide can adsorb appreciable quantities of nickel, cobalt, and copper. Adsorption is prevented by using a high concentration of free ammonia in the leach solution. For high-iron copper feeds (such as concentrates), the molar ratio of ammonia to nickel plus cobalt plus copper can be 7 to 1, whereas for low-iron feeds (such as mattes), this molar ratio can be as low as 2 to 1. After the iron residue has been filtered off, the clarified leach solution passes to the solution purification stages.

Solution Purification. During leaching, much of the sulfide is oxidized in several stages to sulfate. At any given time, the leaching solution will also contain sulfur in the following forms:

- Thiosulfate, $S_2O_3^{=}$
- Sulfamate, $SO_3NH_2^{-}$
- Trithionate, $S_3O_6^{=}$
- Thionates, $S_xO_6^{=}$ (x = 2 through 6)

The proportion of these species varies with the duration of the leach, as illustrated in Fig. 6.

Copper Removal. The thiosulfate and trithionate species disproportionate to sul-

Table 1   Mineralogical and chemical compositions of sulfide concentrates

Material	Mineralogical composition, wt%	
	Concentrate	Matte
Pyrrhotite (FeS)	10-11	· · ·
Pentlandite, (FeNi)$_9$S$_8$	75-77	· · ·
Pyrite (FeS$_2$)	2-3	· · ·
Gangue	9-10	· · ·
Hazelwoodite (Ni$_3$S$_2$)	· · ·	96-100

Material	Chemical composition, wt%	
	Concentrate	Matte
Iron	31.0	0.2
Nickel	27.5	70.0
Cobalt	0.4	0.9
Copper	0.1	0.8
Sulfur	31.0	26.0
Other	10.0	2.0

Fig. 6 Batch leaching curves
Source: Ref 5

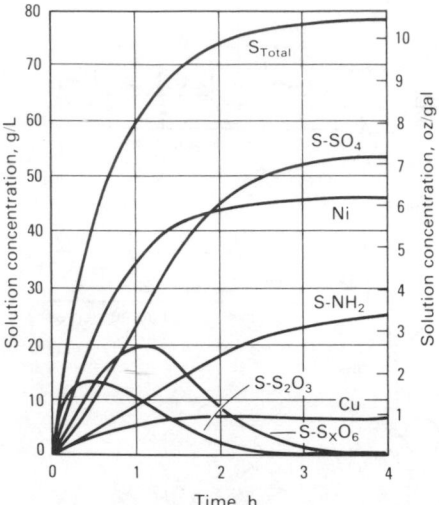

fate and sulfide ions, if the ammonia content of the leaching solution is lowered. Sulfide then precipitates as the least soluble metal sulfide, copper sulfide. Accordingly, the copper is removed from the leaching solution simply by boiling off the ammonia and filtering off the resulting copper sulfide precipitate.

To maximize the amount of copper precipitated, approximately equal amounts of copper and sulfur must be in the form of thiosulfate and trithionate in the solution before boiling. Consequently, the leaching operation is carried out in two stages. The first step, or the adjustment leach, achieves the desired copper-to-sulfur ratio. From this controlled leach, the liquids and solids are separated. The liquid is boiled, while the solids pass on to the final leach, where any residual metals are dissolved completely in ammoniacal ammonium sulfate at 80 °C (175 °F) under 0.9 MPa (128 psig) pressure.

The adjustment leach dissolves about 60% of the nonferrous metals, and the final leach raises the recovery rate to about 96%. The solution from the final leach goes to the adjustment leach as solvent, and the hydrated iron oxide solids are rejected to the tailings pond.

Removal of copper by simple boiling is not complete. Copper concentration in solution usually ranges from 3 to 5 g/L (0.4 to 0.6 oz/gal) and drops to about 0.5 g/L (0.07 oz/gal) in this step; it is finally reduced to about 0.005 g/L (0.0007 oz/gal) by injection of hydrogen sulfide.

Removal of Nonsulfate Sulfur. The solution remaining after copper removal contains appreciable quantities of sulfa-

mate and trace levels of thionates. These are removed by hydrolyzing the sulfamate to sulfate at 245 °C (475 °F) and oxidizing the thionates to sulfate with air at 4.8 MPa (700 psig). These operations are performed continuously in one combined step, termed "oxydrolysis," in a separate continuous autoclave. The solution then passes to the metal recovery stage.

Metal Recovery. In the Sherritt process, nickel is recovered from solution through precipitation as a powder by injecting hydrogen gas into the solution. Reduction of metals from solution by hydrogen requires that the hydrogen potential of the system exceed the electrode potential of the metal ions, in which case the hydrogen goes into solution and the metal precipitates. Hydrogen potential is governed by the acidity, or pH, and by the partial pressure of hydrogen applied to the solution; higher pH (more alkaline conditions) and higher hydrogen pressure increase hydrogen potential. Conversely, the electrode potential of metal ions in solution is little affected by metal concentration in the practical range of interest from 0.001 to 1 molar and is specific to the metal.

The electrode potential of several metals and the hydrogen potential of the solution, as a function of pH and the applied hydrogen partial pressure, are shown in Fig. 7. According to Fig. 7, with a pH of 5.5

or higher, a hydrogen pressure of 1 atm or more will reduce nickel and cobalt from solution.

In practice, the reduction feed solution containing 45 to 50 g/L (6 to 6.7 oz/gal) Ni and 1 to 2 g/L (0.1 to 0.2 oz/gal) Co is reduced by batches in autoclaves at 175 to 205 °C (350 to 400 °F), using hydrogen at a pressure of 3.1 MPa (450 psig). Nickel is reduced before cobalt, due to its lower electrode potential, and the reduction stops when the nickel in solution has dropped to about 1 g/L (0.1 oz/gal). This practice ensures that the cobalt level in the reduced nickel is only about 0.07%.

Reduction of nickel ions from solution by hydrogen results in the dissolution of hydrogen ions in the solution according to the simplified equation:

$$Ni^{++} + H_2 \rightarrow Ni^0 + 2H^+$$

Thus, the reduction of one nickel ion puts two hydrogen ions into solution. This production of hydrogen ions lowers the pH and lowers the hydrogen potential of the system so that the reduction of metal ions eventually ceases. However, the control of the free ammonia to nickel plus cobalt ratio at 2 to 1 was chosen to provide the required amount of ammonia to neutralize the hydrogen ions produced during reduction.

In fact, the ratio of free ammonia to nickel plus cobalt is held at slightly less

Fig. 7 Hydrogen and metal electrode potentials as a function of pH and the applied hydrogen partial pressure
Source: Ref 6

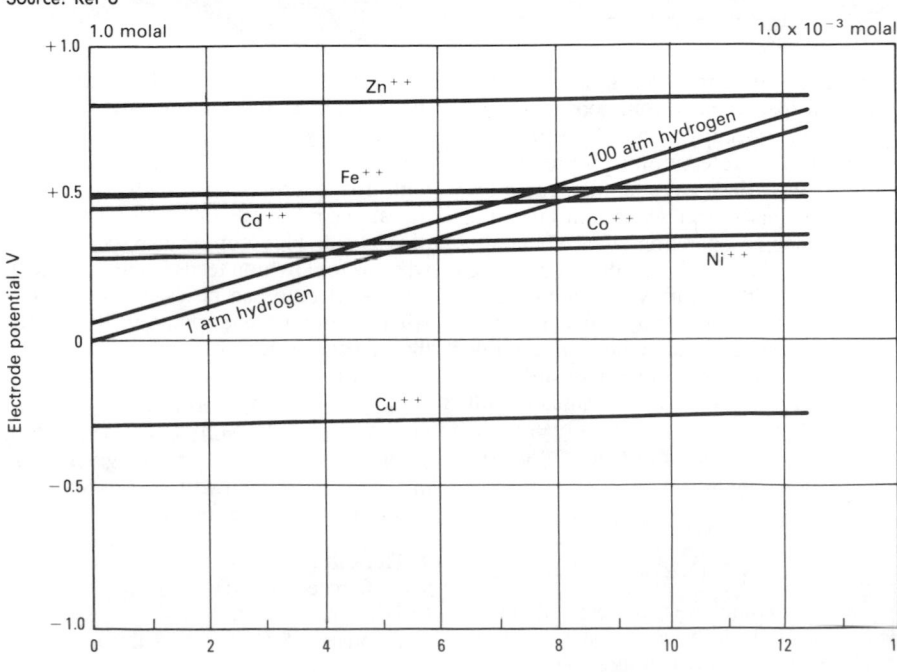

Fig. 8 Nickel powders produced by the Sherritt process
Magnification: 75×

than 2 to 1 to ensure that the reduction reaction stops before all of the nickel is precipitated and before the reduction of cobalt begins. Thus, contamination of nickel by cobalt is minimized.

Reduction of nickel ions to metal occurs rapidly only as a heterogeneous reaction, in which the nickel is precipitated onto an existing surface. For the first reduction, a small amount of ferrous sulfate (1 g/L or 0.1 oz/gal) is added to the solution to form nuclei, on which the fresh nickel precipitates when reduction conditions of temperature and hydrogen pressure are applied.

During the reduction process, the solution in the autoclave is agitated by impellers. When reduction is completed, agitation is stopped, and the fine nickel powder is allowed to settle in the autoclave. The spent solution is discharged,

the autoclave is refilled with fresh nickel solution, the agitators are re-started to suspend the nickel powder in the solution, and the hydrogen pressure is re-applied to reduce more nickel from solution.

Beginning with the second cycle, ferrous sulfate catalyst is not required, as existing nickel powder particles serve as nuclei for the heterogeneous reduction process. This batch-type reduction process is repeated 50 to 60 times to build up consecutive nickel coatings on the powder particles, resulting in a finished product with an average particle size of about 150 μm. Figure 8 is a micrograph of nickel particles produced in this manner.

The time required for each consecutive reduction cycle increases as the powder particle size increases, taking up to about 45 min in the final stages. The overall process is stopped when it becomes impractical to maintain the large powder particles in suspension by means of the mechanical agitators. For the final discharge, the agitators are kept running, and the powder and solution are discharged as a slurry from which the nickel powder is separated, washed, and dried.

The small amount of nickel and cobalt ions remaining in the spent solution after each nickel reduction cycle (about 1 g/L or 0.1 oz/gal of each) is precipitated by hydrogen sulfide and forms the feed material for the cobalt recovery process. The metal-free solution is evaporated to form crystals of ammonium sulfate, which is used as a fertilizer. Typical properties of

nickel powder produced by hydrometallurgy are:

Chemical composition, %

Nickel & cobalt	99.9
Cobalt	0.05-0.10
Copper	0.003
Iron	0.005-0.010
Sulfur	0.03
Carbon	0.006

Screen size, %

+100 mesh	0-10
−100+150 mesh	5-30
−150+200 mesh	20-45
−200+250 mesh	10-25
−250+325 mesh	10-35
−325 mesh	5-25

Physical properties

Apparent density, g/cm³	3.4-4.1
Flow rate, s/50 g	20

By discharging the powder before the maximum number of reduction cycles has been reached, finer powder with lower apparent density can be obtained. Figure 9 shows the relationship of particle size and apparent density to the number of reduction cycles.

Applications of Nickel Powders

Most hydrometallurgically produced nickel powder is compacted into briquettes or rondelles for consumption as an

Fig. 9 Physical characteristics of nickel powder produced in a commercial operation
(a) Particle size distribution of reduced precipitate when discharged from the reactor at the end of nucleation, 30, 45, and 60 reaction cycles. (b) Apparent density. Source: Ref 7

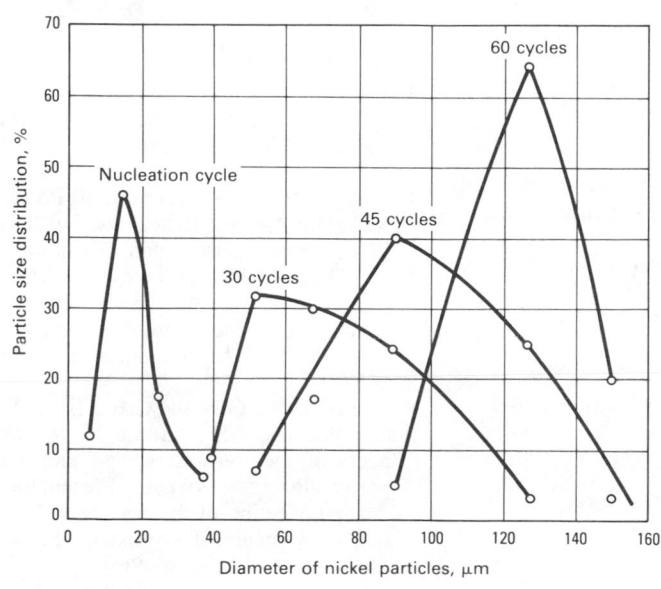

(a)

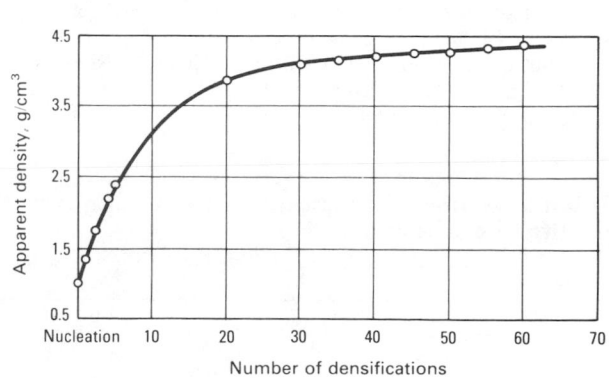

(b)

alloying additive in the steel industry. Smaller amounts of powder are roll compacted into nickel strip or dissolved to produce various nickel salts.

Production of Nickel Alloy Powders by Atomization

By K.M. Kulkarni
Operations Manager
Powders/PM Parts
Cabot Corp.
and
Vidhu Anand
Engineering Associate
Cabot Corp.

ATOMIZED nickel-based alloy powders are used primarily for hardfacing or aerospace components. This article, however, describes materials for hardfacing processes only. The various types of hardfacing processes, equipment, materials, and types of applications are described in greater detail in the article "Metal Powders Used for Hardfacing" in this Volume.

These processes involve deposition of molten or semi-molten materials on a base metal to obtain a coating with high wear resistance. In some coating applications, other properties such as high-temperature abrasion resistance or corrosion resistance are also more important. Hardfacing is a process in which wear resistance is the main objective and includes welding and thermal spraying.

Nickel-based hardfacing powders are produced by gas and water atomization. Unlike other alloy systems, nickel-based hardfacing powders are predominantly spherical in shape, even when produced by water atomization. Oxygen content is somewhat higher than gas-atomized powders, but is still below 1000 mL/m³ (1000 ppm). Generally, melting practice and control during each manufacturing step af-

Fig. 10 Atomized nickel-based hardfacing powder
Magnification: 250×

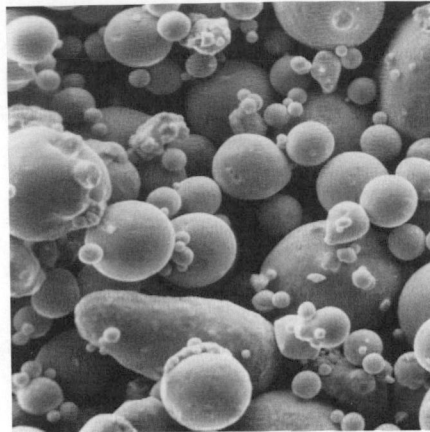

fect product quality more than the atomization process.

Most nickel-based hardfacing powders are of the nickel-chromium-boron-silicon type and are self-fluxing during deposition because of the presence of boron and silicon. These elements also influence the melting temperature range of any specific alloy; melting range is an important factor for hardfacing powders. Table 2 gives typical compositions of several common grades, and Fig. 10 illustrates typical powder particles.

REFERENCES

1. Mond, L., Langer, C., and Quincke, F., *Proc. Chem. Soc.*, Vol 86, 1890, p 112
2. Goldberger, W.M. and Othmer, D.F., The Kinetics of Nickel Carbonyl Formation, *Ind. Eng. Chem. Proc. Des. Devel.*, July 1963, p 202-209
3. Carlton, H.E. and Goldberger, W.M., Fundamental Considerations of Carbonyl Metallurgy, *J. Metals*, June 1965, p 611-615
4. Evans, D.J.I., "The Production of Metals by Gaseous Reduction from Solution, I. Processes and Chemistry," Symposium on Advances in Extractive Metallurgy, Institution of Mining and Metallurgy, London, April 1967
5. Forward, F.A., Ammonia Pressure Leach Process for Recovering Nickel, Copper, and Cobalt from Sherritt Gordon Nickel Sulphide Concentrate, *Trans. Canadian Inst. Mining Metall.*, Vol LVI, 1953, p 363-370
6. Meddings, B. and Mackiw, V.N., The Gaseous Reduction of Metals from Aqueous Solutions, *Applications of Fundamental Thermodynamics to Metallurgical Processes*, Gordon and Breach, New York, 1967
7. Kunda, W. and Evans, D.J.I., "Controlling the Properties of Nickel Powders Produced by the Hydrogen Reduction of Nickel Ammine Sulphate Solutions," Second European Symposium on Powder Metallurgy, Stuttgart, May 1968

SELECTED REFERENCES

- Mond, R.L. and Wallis, A.E., Researches on the Metallic Carbonyls, *J. Chem. Soc.*, Vol 121, 1922, p 29-32
- Trout, W.E., The Metal Carbonyls, *J. Chem. Educ.*, Oct 1937, p 453-459
- Trout, W.E., The Metal Carbonyls, *J. Chem. Educ.*, Dec 1937, p 575-581
- Trout, W.E., The Metal Carbonyls, *J. Chem. Educ.*, Feb 1938, p 77-83
- Trout, W.E., The Metal Carbonyls, *J. Chem. Educ.*, March 1938, p 113-121
- Goetzel, C.G., *Treatise on Powder Metallurgy*, Vol I, Interscience, New York, 1949, p 49-53
- Actmann, G.O., Carbonyl Iron Powders, *FM-TV Journal #9*, 1949, p 29
- Carbonyl Iron Powders, GAF Corp., New York, Chemical Division Brochure, 1960
- Bibliography on Nickel Carbonyl, Part I, Inco Publication, The International Nickel Co., New York, Feb 1962
- Tracey, V.A. and Perks, R.P., The Structure and Properties of Sintered Carbonyl-Nickel Powders, *P/M*, No. 12, 1963
- Tracey, V.A., Production of Porous Nickel for Alkaline-Battery and Fuel Cell Electrodes: Practical and Economical Conditions, *P/M*, Vol 8 (No. 16), 1965, p 241-255
- Boldt, J.R., *The Winning of Nickel*, Queneau, P., Ed., Longmans, Toronto, 1967
- Queneau, P., O'Neill, C.E., Illis, A., and Warner, J.S., "Some Novel Aspects of the Pyrometallurgy and Vapormetallurgy of Nickel," Presentation Annual Meeting of the American Institute of Mechanical Engineers, Parts I (A1) and II (A2), Feb 1969
- Verblovskii, A.M., Aleshina, N.I.,

Table 2 Compositions and melting ranges of typical nickel-based hardfacing alloys

Nominal composition	Melting range °C	Melting range °F	Deposit hardness, HRC
1.5B-2.8Si-Ni	940-1260	1725-2300	19-24
1.7B-0.35C-7.5Cr-3.5Si-Ni	994-1152	1820-2105	35-42
2.4B-0.45C-11Cr-4Si-Ni	976-1063	1790-1945	49-52
3.5B-0.8C-15.5Cr-4.3Si-Ni	964-1003	1770-1840	59-62

Khamarmer, S.I., and Leschinskaya, G.P., The Physical Properties of Nickel Carbonyl and Their Influence on Technical Characteristics, translated from *Poroshkovaya Metallurgiya,* Vol 1 (No. 73), Jan 1969, p 6-12

• Sale, F.R., Vapour-Phase Transport Reactions in Extractive Metallurgy, *Min. Sci. Eng.,* Vol 3 (No. 4), 1971, p 3-12

• Belli Dell-Amico, D. and Calderazzo, F., Convenient Methods for the Preparation of Anhydrous Gold (III) Chloride and Chlorocarbonyl Gold (I), *Gass. Chim, Ital.,* Vol 103, 1973, p 1099

• Technical note, Carbonyl Nickel Powders—Properties and Uses, *Powder Metall. Int.,* Vol 6 (No. 4), 1974, p 203-206

• Jolly, P.W. and Wilke, G., The Organic Chemistry of Nickel, *Organonickel Complexes,* Vol 1, Academic Press, New York, 1974

• "Inco Nickel Powders—Properties and Uses," Publications A-1164, The International Nickel Co., New York, 1975

• Wailes, P.C., Coults, R.D.P., and Weigold, H., *Organometallic Chemistry of Titanium, Zirconium, and Hafnium,* Academic Press, New York, 1974

• Jolly, P.W. and Wilke, G., *The Organic Chemistry of Nickel,* Vol II, Organic Synthesis, Academic Press, New York, 1975

• Shortened Route to Pure Nickel, *Chem. Eng.,* Nov 1975

• "The Technology of Metal Powders: Production, Properties and Applications," GAF paper, American Powder Metal Institute Powder Short Course, Newport Beach, CA, Feb 1978

• Production and Properties of Carbonyl Nickel Powders, *Metal Powder Report,* Vol 33 (No. 4), 1978, p 158-161

• Sikora, D.J., Rausch, M.D., Rodgers, R.D., and Atwood, J.L., Structure and Reactivity of the First Hafnium Carbonyl, $(n^5\text{-}C_5H_5)_2Hf(CO)_2$, *J. Am. Chem. Soc.,* Vol 101 (No. 17), 1979, p 5081

• Wagner, F.S., Carbonyls, in *Kirk-Othmer: Encyclopedia of Chemical Technology,* Vol 4, 3rd ed., John Wiley & Sons, New York, 1979, p 794-813

• Belli Dell-Amico, D. and Calderazzo, F., Synthesis of Carbonyl Halides of Late Transition Elements in Throryl Chloride. A Convenient Synthesis of Dicarbonyldichloroplatinum II, *Gass. Chim, Ital.,* Vol 109 (No. 99), 1979, p 1099

• Pensak, D.A. and McKinney, R.J., Application of MO Theory to Transition Metal Compounds, *Inorgan. Chem.,* Vol 18 (No. 12), 1979, p 3407

• Marco, P., Floriani, C., and Gaetani-Manfredotti, Preparation and Properties of Copper Halocarbonyls, *Inorg. Chem.,* Vol 20 (No. 10), 1980, p 3382

• Antonsen, D.H., Nickel Compounds, in *Kirk-Othmer: Encyclopedia of Chemical Technology,* Vol 15, 3rd ed., John Wiley & Sons, New York, 1981, p 801-819

• Tachikawa, M. and Mutteries, E.L., Metal Carbide Clusters Synthesis Systematics for Hetronuclear Species, *Prog. Inorg. Chem.,* Vol 28, 1981, p 203

• Calderazzo, F. and Belli Dell-Amico, D., Transition-Metal-Catalysed Oxidation of Carbon Monoxide by Dichlorine to Produce Phosgene, *Inorg. Chem.,* Vol 20, 1981, p 1310

• Goldberg, K.I., Hoffman, D.M., and Hoffmann, R., On the Existence of Binuclear Bonded Dinitrogen Complexes, *Inorg. Chem.,* Vol 21 (No. 11), 1982, p 3863

• Hayward, C.T. and Shapley, J.R., Synthesis, Characterization and Crystal Structure of $[Ru_{10}C_2(CO)_{24}]$ Dianion. An Edge-fused Bioctahedral Dicarbide Cluster, *J. Am. Chem. Soc.,* Vol 104, 1982, p 7347

Production of Cobalt and Cobalt Alloy Powders

By Basil Meddings
Manager, Analytical Services
and Quality Control
Sherritt Gordon Mines Ltd.

COBALT POWDER is used primarily in the P/M industry as an alloying element in the production of permanent magnets, superalloys (high-temperature, creep-resistant alloys), wear-resistant hardfacing alloys, tool and die steels, and as a binder in cemented carbides. About 60% of the world's production of cobalt originates from Zaire and Zambia. The rapidly growing dependence of the United States on cobalt imports is illustrated in Fig. 1. Because political upheaval in these countries has disrupted the supply of cobalt, thereby resulting in significant price increases, much research has centered around material substitution of cobalt. For example, high-speed tool steels that do not contain cobalt are being produced by hot isostatic pressing. Such alloys, which cannot be made using conventional wrought processing, are designed to replace cobalt-containing American Iron and Steel Institute (AISI) grades T15 and M42. See the article "P/M Tool Steels" in this Volume for more information on P/M tool steel uses.

Despite the diminished availability of cobalt, it remains an important material to the P/M industry, due to its ability to enhance the high-temperature properties of superalloys, the cutting ability of sintered carbide and high-speed tool steels, and the toughness of high-strength steels. This article discusses the three methods of cobalt powder production: hydrometallurgical processing, reduction of oxides, and atomization. Articles concerning materials and/or components that use cobalt as an alloying element can be found in the Section of this Volume entitled "Powder Systems and Applications."

Hydrometallurgical Processing

All hydrometallurgical processes for the production of a metal entail three processing stages: chemical dissolution (leaching), solution purification, and metal recovery. A typical hydrometallurgical process is the Sherritt Gordon soluble cobaltic pentammine process for the recovery of cobalt from mixed nickel-cobalt sulfides. A flowchart of this process is shown in Fig. 2.

Fig. 2 Flowchart of Sherritt Gordon cobalt refining process
Source: Ref 2

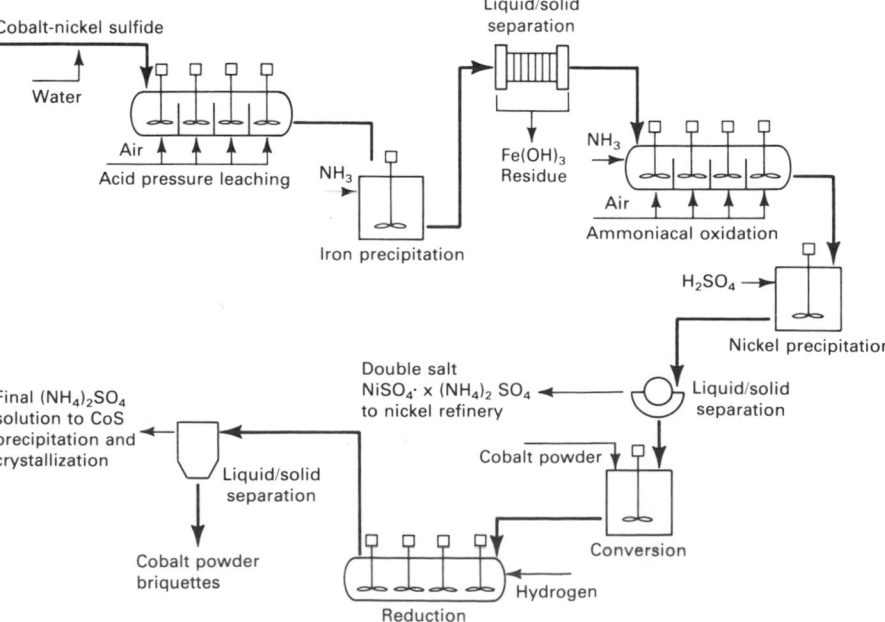

Fig. 1 Projected supply and demand for cobalt in the United States
Source: Ref 1

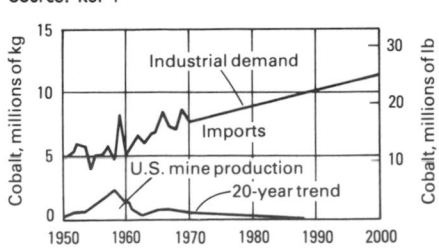

Fig. 3 Micrograph of hydrometallurgically processed cobalt powder

Magnification: 150X

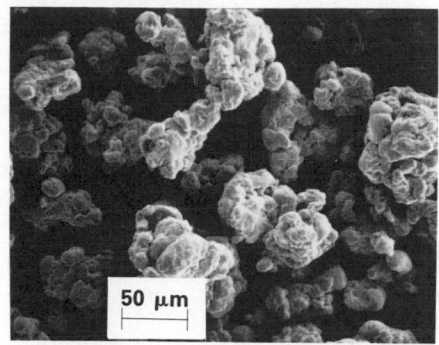

Table 1 Compositions and hardnesses of selected cobalt-based hardfacing alloys

AWS designation or trade name	Nominal composition	Nominal macrohardness DPH	Nominal macrohardness HRC	Matrix, DPH	Hard particles Type	Hard particles DPH
Alloy 21	Co-27Cr-5Mo-2.8Ni-0.2C	255	24-27	250	Eutectic	900
RCoCrA	Co-28Cr-4W-1.1C	424	39-42	370	Eutectic	900(a)
RCoCrB	Co-29Cr-8W-1.35C	471	40-48	420	Eutectic	900(a)
RCoCrC	Co-30Cr-12W-2.5C	577	52-54	510	M_7C_3	900(a)
					M_6C	1540
Alloy 20	Co-32Cr-17W-2.5C	653	53-55	540	M_7C_3	1700
					M_6C	900
Tribaloy T-800	Co-28Mo-17Cr-3Si	653	54-64	800(b)	Laves phase	1100

(a) Matrix and M_7C_3 eutectic. (b) Matrix and Laves phase eutectic

Leaching. The feed material for the Sherritt Gordon process is the mixed nickel-cobalt sulfide produced after the main nickel recovery operation. This stage is described in detail in the article "Production of Nickel and Nickel Alloy Powders" in this Volume. These mixed sulfides are continuously leached in autoclaves at 120 to 135 °C (250 to 275 °F) at an air pressure of 700 kPa (100 psig). This oxidizes all of the sulfides to sulfates, and the metal ions pass into solution.

Solution Purification. Typically, the leach solution has a pH of about 2 and contains small amounts of iron. Iron is removed by increasing the pH to 5.5 by adding ammonia and filtering off the precipitated ferric hydroxide.

The iron-free solution of nickel and cobalt is then mixed with ammonia and passed through a continuous autoclave at 75 to 100 °C (165 to 212 °F) at an air pressure of 700 kPa (100 psig). Cobalt is oxidized to the trivalent state and forms the cobaltic pentammine complex, $[Co(NH_3)_5]^{+++}$. Nickel remains in the divalent state and is largely in the diammine complex, $[Ni(NH_3)_2]^{++}$.

This solution is then acidified to pH 2.5 by adding sulfuric acid. This addition destroys the divalent nickel complex and precipitates the double salt—nickel ammonium sulfate, $Ni(NH_4)_2(SO_4)_2$—which is filtered off while the more stable cobaltic pentammine complex remains in solution. The purified cobaltic solution then passes to the metal recovery stage.

Metal Recovery. Cobalt is recovered by hydrogen reduction from solution. Because this is easier to achieve with divalent cobalt, the initial step consists of reduction of the cobaltic solution to the divalent cobaltous state. This is carried out by adding metallic cobalt powder to the acidified (pH 2.5) cobaltic solution,

which results in reduction to cobaltous cobalt:

$$2 Co^{+++} + Co^\circ \rightarrow 3 Co^{++}$$

Free ammonia concentration of the cobaltous cobalt solution is simultaneously adjusted to a molar ratio of 2 to 1. Hydrogen reduction of cobalt is carried out in the same manner as for nickel (see the article "Production of Nickel and Nickel Alloy Powders" in this Volume). The prime difference between these reduction reactions is that the nucleation catalyst for the initial reduction of cobalt is a mixture of sodium cyanide and ammonium sulfide rather than ferrous sulfate. Another minor difference is that the reduction is forced as far toward completion as possible; consequently, the initial molar ratio of ammonia to cobalt is generally slightly greater than 2 to 1, usually about 2.2 to 1. A micrograph of cobalt particles produced in this manner is shown in Fig. 3.

Typical properties of cobalt powder produced by hydrometallurgical processing are:

Chemical composition, %

Cobalt + nickel	99.9
Nickel	0.10
Copper	0.005
Iron	0.005
Silicon	0.03
Carbon	0.05

Screen size, %

+100 mesh	0-15
−100+150 mesh	5-25
−150+200 mesh	5-15
−200+250 mesh	5-15
−250+325 mesh	20-45
−325 mesh	10-50

Physical properties

Apparent density, g/cm^3	2.5-3.5
Flow rate, s/50 g	23

Applications. Hydrometallurgically produced cobalt powder is used primarily

for the production of various cobalt salts, such as cobalt napthenate, which is used as a paint dryer, and for the production of cobalt-samarium magnets. See the article "Electrical and Magnetic Applications" in this Volume for more information on these applications. Cobalt powders made by this process also are used as alloying elements in superalloys and tool steels. Powder is usually roll compacted into strip before being sold as melting stock.

Thermal Decomposition of Cobalt Oxide

One of the major applications of cobalt powder is its use as a binder in the production of cemented carbides. Most of this powder is produced by the reduction of cobalt oxide with hydrogen at relatively low temperatures, generally below 800 °C (1470 °F) in order to obtain the necessary degree of fineness.

Powder grades used in cemented carbides should pass through 325 mesh. Small particle sizes are required, because the cobalt powder must be intimately mixed with tungsten carbide particles by ball milling (see the article "Cemented Carbides" in this Volume for more information on this application). Typical properties of reduced cobalt powders are:

Chemical analysis, %

Cobalt(a)	99.60
Nickel	0.08
Iron	0.08
Silicon	0.035
Calcium	0.020
Manganese	0.020
Carbon	0.015
Zinc	0.010
Sulfur	0.008
Copper	0.001
Lead	0.003
Hydrogen loss	0.20

Physical properties

Apparent density, g/cm³ 1.8
Tap density, g/cm³ 3

Screen analysis(b), %

+100 mesh 0.01
−100+200 mesh 0.04
−200+300 mesh 0.15
−300+400 mesh 0.20
−400 mesh 99.60

(a) Hydrogen loss not included. (b) Average particle size, 5 μm
Source: Ref 3

Atomization

Atomized powders of cobalt-based alloys find wide usage in high-temperature applications, although this section discusses only hardfacing powders. Cobalt-based alloys for hardfacing applications typically are the cobalt-chromium-tungsten-nickel-carbon types. These powders are gas atomized to minimize oxygen pickup (see the article "Atomization" in this Volume).

Fig. 4 Micrograph of gas-atomized cobalt-based hardfacing powder

Composition: 50Co-19Cr-13Ni-8W-3.0Si-1.7B-0.7C. Magnification: 180×

Some manufacturers prefer vacuum melting and inert-gas atomization over air melting followed by inert-gas (typically nitrogen or argon) atomization. Oxygen content in these atomized powders is significantly lower than 1000 ppm, although the nitrogen level may vary from 600 to 2000 ppm. Typical melting and atomizing practice requires strict control to produce a quality product.

Compositions and hardnesses of several typical alloys are given in Table 1. A typical microstructure of a cobalt hardfacing alloy is shown in Fig. 4. Detailed information on cobalt hardfacing alloys can be found in the article "Metal Powders Used for Hardfacing" in this Volume.

REFERENCES

1. Hurlich, A., Planet Earth's Metal Resources, *Met. Prog.*, Vol 117 (No. 5), Oct 1977, p H6
2. Mackiw, V.N. and Benz, T.W., Application of Pressure Hydrometallurgy to the Production of Metallic Cobalt, in *Extractive Metallurgy of Copper, Nickel and Cobalt*, Queneau, P., Ed., Interscience, New York, 1961, p 503-534
3. Poster, A.R., Ed., *Handbook of Metal Powders*, Reinhold, New York, 1966, p 109

Production of Precious Metal Powders

Production of Silver Powder

By Howard D. Glicksman
Manufacturing Technology Manager
Engelhard Industries
Engelhard Corp.

SILVER POWDERS are used in a variety of electrical, electronic, and industrial applications. In powder metallurgy, silver powders are used in the manufacture of electrodes for primary batteries and storage cells. These powders are also used in the production of "pseudo" alloys, or mixtures of metals used in the preparation of electrical contacts, and other P/M parts containing silver. These mixtures include silver-tungsten, silver-iron, silver/tungsten carbide, silver-molybdenum, silver/cadmium oxide, silver-graphite, and silver-nickel (Ref 1). Electrical applications commonly require a dense, coarse silver powder of high purity with good compressibility. Average particle sizes vary from approximately 10 to 100 µm.

In the electronics industry, silver powders are used in conductive inks, pastes, and adhesives. Conductive silver inks are used in the manufacture of capacitors, potentiometers, printed circuits, and other electronic parts. They also are used in heating applications, such as heated automobile windows and hot trays.

Generally, silver powders used in electronic processes are chemically precipitated, exhibit high purity, and are small in size, with average particle sizes ranging from 1 to 10 µm. Silver powders that are mechanically milled into flat powders or silver flakes are also used. These flakes are well suited for conductive coatings and shielding materials and are mixed with silver powders in some conductive ink systems.

Other industrial applications of silver powders include water purification, catalyst manufacture, dental amalgams, and preparation of silver alloys. In addition, silver powders are used as a protective coating against hostile industrial environments and as a source of high-purity silver (99.99+% pure).

Most of the methods currently used to manufacture metal powders can be applied to the production of silver powders. Chemical, physical (atomization and milling), thermal decomposition, and electrochemical processes are used industrially to make silver powders. These powders generally are characterized by physical properties, such as apparent density, surface area, tap density, particle size and shape, flow, and particle size distribution. A general summary of silver powders and their properties is given in Table 1.

Chemical Processes

Numerous chemical processes are used to produce silver powders. Most silver salts are easily reduced from solution, with limitations imposed only by their solubilities. The most common salts are silver nitrate and silver oxide. Organic reducing agents such as alcohol, sugar, or aldehydes are used to manufacture silver powders from silver oxides, for example:

$$12Ag_2O + C_6H_{12}O_6$$
$$\rightarrow 24Ag + 6CO_2 + 6H_2O$$

These powders can have small particle sizes, with average particles ranging from less than 5 µm, and typical apparent densities ranging from less than 0.5 g/cm^3 to more than 1.5 g/cm^3. Organically precipitated silver powders tend to agglomerate into aggregates of 10 to 50 µm or larger. To avoid agglomeration, the powder should be precipitated in the presence of a surfactant, or high shear stirrers can be used. Typical chemically precipitated silver powders are shown in Fig. 1 and 2.

Inorganic reducing agents such as hydrazine, potassium sulfite, and sodium formate are used to reduce silver nitrate to

Fig. 1 Chemically precipitated silver powder
Magnification: 1200×

Table 1 Physical properties of silver powders

Method of preparation	Apparent density(a), g/cm^3	Tap density(b), g/cm^3	Particle size(c), µm	Surface area, m^2/g
Atomization	3.0-7.0	4.0-7.0	>40	<0.2
Chemical:				
Organically reduced	0.4-1.5	0.8-3.0	0.5-3.0	0.2-4.0
Inorganically reduced	1.0-2.0	1.0-3.0	3-20	<0.3
Galvanic reduction	1.5-4.0	2.0-4.5	>100	<0.2
Electrolytic reduction	1.5-3.0	2.5-3.5	40-1000	<0.2

Note: The properties of thermally decomposed silver powders, which are not given in this table, are dependent on the subsequent mechanical treatment (comminution) of the powder.
(a) Scott volumeter per ASTM B 329. (b) Tap-Pak volumeter per ASTM B 527. (c) Fisher subsieve sizer per ASTM B 330

Fig. 2 Chemically precipitated silver powder
Magnification: 6000×

Fig. 3 Galvanic reduction
Metal + silver nitrate → metal nitrate + silver
$Cu + 2AgNO_3 \rightarrow Cu(NO_3)_2 + 2Ag$

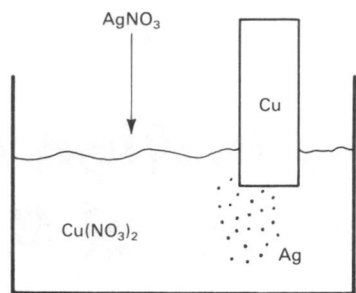

Fig. 4 Irregularly shaped galvanic silver powder
Magnification: 600×

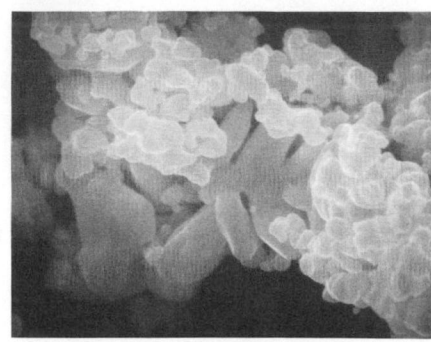

Fig. 5 Electrochemical reduction

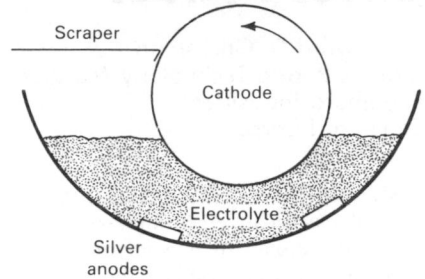

silver powder. Two examples of the reduction process are:

$$2AgNO_3 + 2K_2SO_3$$
$$\rightarrow 2Ag + K_2SO_4 + 2KNO_3 + SO_2$$
$$AgNO_3 + NaCO_3H$$
$$\rightarrow Ag + NaNO_2 + CO_2 + {}^1/_2H_2$$

These powders are coarse in size, resembling those made by electrochemical methods. Free bulk densities range from 1 to 2 g/cm³. Higher densities are also possible. Average particle size varies from 5 to 20 μm. These powders exhibit good flow and good sintering characteristics.

Electrochemical Processes

Several electrochemical processes are used to manufacture silver powders. Galvanic reduction precipitates silver powder from a solution by a more anodic metal such as zinc, copper, or iron (Fig. 3). This process produces an irregularly shaped powder with a free bulk density ranging from 1.5 to 4.0 g/cm³ (Fig. 4). Particle sizes of 100 μm or more are produced by galvanic reduction. Some galvanically reduced powders exhibit good flow characteristics.

In electrolytic reduction an electrical current passes through an electrolyte at room temperature to produce a crystalline silver. Current densities vary from 215 to 2150 A/m² (20 to 200 A/ft²), depending on the desired particle size. Silver anodes, usually as cast silver bars, or granulated silver is dissolved; crystallized silver is subsequently grown on a cathode, as shown in Fig. 5. This crystalline silver powder is regularly shaped and very large, with particle sizes ranging from 40 μm to as large as several millimetres, depending on pro-

cessing conditions and current densities. An example of electrolytic silver powder is shown in Fig. 6.

Physical Processes

The major physical process used in the production of silver powder is the atomization of a molten stream of silver with gas or liquid. The atomized powder generally approaches a spheroidal shape, with a very smooth surface and minimal porosity. Particle sizes generally are greater than 40 μm, and the silver powder has a high apparent density, typically about 6.0 g/cm³.

Mechanical comminution can be used to prepare and alter most types of silver powders with processes such as ball milling, vibratory milling, or attrition milling. These methods are time consuming, and impurities generally are introduced by the equipment and lubricants added during milling. Large silver particles are broken down during the milling process. Continued milling produces flat silver powder or silver flake. In addition, milling can be used to narrow the particle size distribution of a particular silver powder.

Thermal Reduction

Heat is used to reduce silver salts to produce silver powders. Silver oxide and silver carbonate can be reduced to silver powder by thermal reduction in an inert atmosphere at temperatures greater than 250 °C (480 °F) as follows:

$$2Ag_2O + \Delta \rightarrow 4Ag + O_2$$
$$2Ag_2CO_3 + \Delta \rightarrow 4Ag + 2CO_2 + O_2$$

These powders tend to be spongy, agglomerated, and very porous. Generally, they require subsequent mechanical treatment such as ball milling to provide usable silver powder.

Production of Gold, Platinum, and Palladium Powders

By K.C. Yang
Manager—Chemical & Refining
Development
Handy & Harman
and
Bruce D. Rowan
Project Manager—Product
Development
Handy & Harman

GOLD, PLATINUM, AND PALLADIUM and their mixtures and alloys are widely used as conductor materials in electronic applications. Properties of precious metal powders include high conductivity, relative inertness, stability on firing in air,* solderability, and durability. Two

*Firing refers to the high-temperature heating (generally between 600 and 1500 °C, or 1110 and 2730 °F) of a system to remove organics and promote sintering and bonding to the substrate.

Fig. 6 Electrolytic silver powder
Magnification: 60×

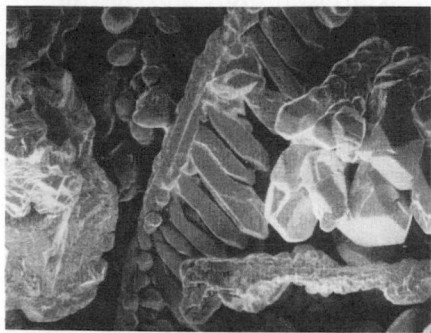

of the major applications that utilize these materials are cermet (ceramic metal) thick films, used in hybrid microcircuits, and multilayer ceramic capacitors.

In both of these applications, which are discussed later in this article, the precious metal powder is mixed in an organic vehicle. Rheology modifiers, wetting agents, glass frit, and other additives may be present. These conductive inks usually are applied by screen printing. In thick-film applications, the most common substrate is alumina (96%). For multilayer ceramic capacitor fabrication, inks are applied to the green (unfired) ceramic sheet.

Manufacturing of Precious Metal Powders

Generally, most precious metal powders used in electronics are chemically precipitated. Mechanical methods are used only to produce plate-like particles (flakes). Other means, such as freeze-drying, vaporization and condensation, and conventional thermal decomposition techniques, are sometimes used to prepare the powders for nonelectronic applications. After precipitation, powders are filtered and washed. Drying normally is performed in a tray oven. Most precious metal powders are screened through 200 mesh. Powders are then ready for use.

Chemical precipitation begins with the preparation of a metal-soluble salt solution by dissolving the metal in an inorganic acid, such as nitric acid or aqua regia. When aqua regia is used, the nitrosyl compound must be removed by continuous boiling by adding hydrochloric acid. In the manufacture of binary or ternary metal powders, such as palladium silver or platinum-palladium-silver, the metal salts are dissolved in a common solvent prior to precipitation.

The solution then is adjusted to the proper concentration, and a pH regulating agent is added to reach the required pH level. With sufficient mixing, the reducing agent usually is added last to the reacting system. To prevent agglomeration of the newly formed precipitated particles, an anti-agglomerating agent may be added either before or after particles are formed, or during their formation. The metal slurry is then filtered and washed, dried at proper temperature, and screened to obtain the final finished product.

According to chemical kinetics, a large class of reactions approximates the simple type of rate law:

$$-r_{PM^+} = -\frac{1}{V}\frac{dN_{PM^+}}{dt} = k\, C_{PM^+}^n$$

where PM^+ is precious metal ions; V is the volume of bulk; N is number of moles; t is time; r_{PM^+} is moles of PM^+; k is the rate constant; C_{PM^+} is the concentration of PM^+; and n varies from 0 to 3.

Reaction rates decrease with time as the concentration of PM^+ decreases. The size of the formed particles is proportional to the reaction rate. Consequently, particle size and particle size distribution, as well as other physical characteristics of powders, are affected mainly by the chemical reduction rate. In addition, the rate of chemical reduction and, hence, particle size and size distribution also are affected by the strength of the reducing agent, nature of the base, temperature of the reaction, initial and subsequent concentration of the metal ions, and the order of addition of the reagents. Theoretically, the equation above demonstrates that better control of the reaction system can be obtained if the precious metal solution is added drop by drop to the mixture of pH regulator and reducing agent. This procedure enables the precious metal solution to react nearly to completion before more reactant is added. In a production environment, this procedure is not practical because it is slow.

Batch reactors commonly are used to precipitate precious metal powders in typical production environments. Plug flow reactors, in which the reaction takes place in a tubular vessel, produce highly uniform powder. Information on reactor design can be found in Ref 2 and 3.

One of the most serious problems encountered in the manufacture of precious metal powders is the tendency of finely divided particles to agglomerate. Gold powder is especially susceptible. Agglomeration can occur due to primary or secondary coagulation (Ref 4-7). To prevent agglomeration, fine powders can be stabi-

lized by organic compounds that are preferentially adsorbed on metal particle surfaces. Other deagglomeration techniques are described in Ref 8-11.

Mechanical Comminution. By applying wet continuous comminution to precious metal powders in the presence of a proper lubricant and solvent, plate-like precious metal flake particles can be successfully produced. Various types of milling equipment can be used. The physical characteristics of the milled particles are primarily controlled by:

- *Charge*: Quantity of metal powder and the ratio of lubricant and solvent to metal powder
- *Lubricating system*: Quantity of the lubricant used rheologically controls the milling conditions, and the type of lubricant affects the organic wettability of the finished product.
- *Milling conditions*: Size and quantity of metal balls used as grinding medium and rotational speed of the mill

For additional information on the principles of milling and the equipment used in such operations, see the article "Milling of Brittle and Ductile Materials" in this Volume.

Gold Powders

For gold powders to be acceptable for thick-film applications, a high bulk density (approximately 4.8 g/cm^3) and small particle size (approximately 1 μm) are required. Large particles usually result in poor performance. A typical gold powder is shown in Fig. 7.

Several processes are well suited to the precipitation of gold, particularly from chloroauric acid ($HAuCl_4$) aqueous solution. Sodium hydroxide or sodium carbonate (or bicarbonate) commonly are used

Fig. 7 Scanning electron micrograph of a typical gold powder used for thick-film applications
Magnification: 2500×

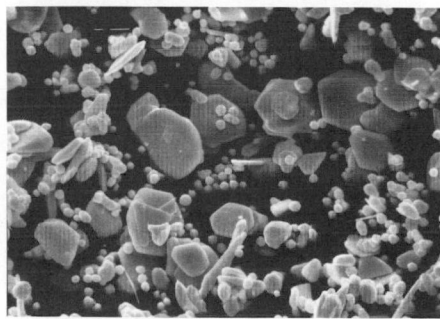

as pH regulators. Sodium carbonate is preferred, because it buffers the solution during the reaction. A variety of reducing agents can be used, including metal powders (zinc, magnesium, lead, aluminum, and iron), sulfur compounds (sulfur dioxide, sodium sulfite, potassium sulfite, and ferrous sulfate), hydrazines (hydrazine hydrate and hydrazine sulfate), aldehydes (formaldehyde), oxalates (oxalic acid and potassium oxalate), hydrogen peroxide, hydroquinone, hydroxylamine, Rochelle salts, and hypophosphorous acid.

The most commonly used reducing agents are sulfur dioxide and ferrous sulfate, which produce a very yellow gold powder. An excess of Rochelle salts effectively precipitates gold powder, but the material agglomerates, causing problems in screening. Both formaldehyde and hypophosphorous acid produce very soft gold powder that smears during screening of the powder. Hydroquinone produces a yellow, rather than a brown, powder. Hydrazine hydrate yields a tan to brown powder with physical properties that vary depending on the precipitation conditions.

Gold powders produced with the reducing agents mentioned above exhibit poor dispersibility in the organic vehicles used to prepare inks and pastes for use in thick-film applications. This problem can be alleviated by adding dispersing agents or by directly precipitating the powder in an organic phase.

Gold powders generally are difficult to screen print, because they tend to agglomerate during precipitation. By doping with base-metal impurities amounting to 200 to 1000 ppm or by adding a suitable surface active agent during the chemical reaction, screenability can be significantly improved.

Platinum Powders

Platinum powders can be prepared easily by conventional techniques. Chloroplatinic acid (H_2PtCl_6) is the most common starting material. It can be prepared by dissolving platinum sponge in aqua regia. After the nitric acid is driven off by continuous boiling with the addition of hydrochloric acid, the solution is then precipitated by a suitable reducing agent at the proper pH level. Workable reducing agents include sodium formate, formic acid, sodium borohydride, sodium hydrosulfite, hypophosphorous acid, and hydrazines. The pH of the solution may be adjusted by the addition of reagents such as sodium, potassium, or ammonium hydroxides, or their carbonates.

Precipitated platinum powders generally are extremely fine and catalytically active (Ref 12). For instance, platinum black, one of the oldest commercially

available platinum powders, is prepared by precipitating platinum from an alkaline solution using formaldehyde. Typical physical characteristics of platinum black are:

Apparent density, g/cm^3	0.6-1.1
Tap density, g/cm^3	1-1.8
Specific surface area, m^2/g	7-15
Average particle diameter, μm	0.2-0.5

Because of its high specific surface area, platinum black contains significant quantities of gases such as diatomic nitrogen. The presence of such gases in powders may cause blistering during the firing process.

A scanning electron micrograph of a platinum powder is shown in Fig. 8. Techniques have been developed to reduce the surface area of the powders. By heat treating at controlled conditions or by using other techniques (Ref 12), typical physical properties of the post-treated powders become:

Apparent density, g/cm^3	1.8-2.4
Tap density, g/cm^3	2-5
Specific surface area, m^2/g	0.5-1.8
Average particle diameter, μm	0.5-2.8

In addition, platinum powders exhibit improved screenability and flowability after thermal post-treatment and are considered to be more suitable for use in the electronics industry.

Palladium Powders

Finely divided palladium powders can be effectively prepared by many methods. The powder products usually are very fine, with an average particle diameter of less than 2 μm. For electronic applications, preferable particle sizes range from 0.1 to 0.5 μm or 0.8 to 1.2 μm. A typical small-size palladium powder particle is shown in Fig. 9.

Generally, as with other precious metal powders, soluble palladium salts such as chloropalladous acid, palladium nitrate, and bromide are chosen as starting materials. Chloropalladous acid is the most frequently used. However, for the production of precious metal alloy powders involving silver by coprecipitation (Ref 13-15), chloride salts should be avoided, because they immediately precipitate the silver ions to form insoluble silver chloride. The nitrate salts of these metals should be used if silver is a coprecipitant. The pH of the aqueous solution of the reactants may be adjusted by the addition of ammonium, sodium, or potassium hydroxide, sodium or potassium carbonate or bicarbonate, or similar compounds.

Reducing agents that are effective for

Fig. 8 Scanning electron micrograph of chemically precipitated platinum powder
Magnification: 2500×

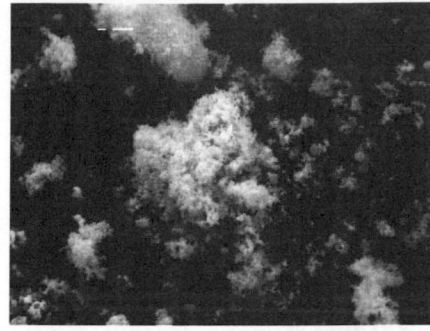

Fig. 9 Scanning electron micrograph of small-sized (≦1.2 μm) palladium powder particle
Magnification: 2500×

platinum or gold precipitation also are applicable for palladium. Hydrazine, formaldehyde, hypophosphorous acid, hydroquinone, sodium borohydride, formic acid, and sodium formate are used. In addition, metal powders such as zinc and copper are useful for preparation of coarse palladium powder from acidic solution. The precipitation of metallic palladium powder from palladium salt solutions by the use of metallic zinc or copper is known as "cementation" or "galvanic precipitation." The reaction is:

$$Pd^{++} + Zn^0 \rightarrow Pd^0 + Zn^{++}$$

However, products prepared by this technique require special treatment to reduce the amount of impurities present. Also, powders prepared through a galvanic process usually have significantly different physical properties than powders made by chemical precipitation. Consequently, they usually are used in different applications.

The starting solution for most palladium powder precipitation is obtained by dissolving palladium sponge in aqua regia.

Fig. 10 Scanning electron micrograph of palladium powder with low specific surface area
Magnification: 2500×

Palladium sponge is usually manufactured by the "red salt/yellow salt" refining process (Ref 16). Ammonium hexachloropalladate (red salt) is first precipitated, followed by the precipitation of diamino palladous chloride (yellow salt). Ignition of this salt produces palladium sponge. The concentration of palladium solution and the pH are adjusted. The reducing agent is carefully metered into the reactor. The palladium powder precipitates and is filtered and washed. Following drying and screening, the powder is ready for use.

Many types of palladium powder are produced by varying reaction conditions. For example, by changing the type of reducing agent used, the surface areas of the finished products may vary from 1 to 80 m²/g. At different addition rates of reducing agent, the bulk density of the powders may change from 0.3 to 3 g/cm³. Also, the physical properties of the powder can be modified by the same techniques used to modify platinum. A typical palladium powder with low specific surface area is shown in Fig. 10.

Palladium oxidizes to palladium oxide between 400 and 800 °C (750 and 1470 °F). This oxidation can present problems due to the volume expansion that accompanies the formation of palladium oxide, especially in the manufacture of multilayer ceramic capacitors.

Thick-Film Applications

Platinum thick films withstand high firing temperatures (750 to 1200 °C, or 1380 to 2190 °F), while gold films exhibit good performance as pads in die (integrated circuit) bonding, but exhibit relatively poor solderability. Palladium thick films provide low-cost substitutes for platinum. Silver films exhibit excellent conductivity and offer the lowest cost. Unfortunately, silver is rapidly dissolved by tin-lead sol-

ders and serious silver migration problems can occur.

For binary systems, platinum-gold thick films exhibit excellent solderability and leach resistance to molten solders. However, the high cost of platinum-gold compositions precludes their use, except in high-performance applications, in which the added cost is not a limiting factor. Therefore, a good cost and performance compromise is an alloy of palladium and silver. This mixture is the most frequently used conductor material for microcircuits. Palladium-silver is more conductive than gold-palladium, with the added benefits of improved adhesion and solderability. This combination of good performance and reasonable cost is especially important for commercial products.

A recent trend in thick-film technology is the use of nonprecious metal conductors to replace higher cost precious metals. In addition to lower costs, nonprecious metal conductors such as copper offer high conductivity, solder-leach resistance, and excellent solder wetting. Also, copper does not migrate like silver. However, the use of a nonprecious metal still requires the inconvenience and expense of firing in a nonoxidizing atmosphere such as nitrogen, nitrogen-hydrogen, hydrogen, or argon.

Multilayer Ceramic Capacitor Applications

Historically, a mixture of platinum-palladium-gold has been used as the metal phase for internal electrodes in the manufacture of multilayer ceramic capacitors. These early devices required very high firing temperatures for the ceramic dielectric to develop the desired electrical properties. Continuing advancements in dielectric development have enabled ceramics to be fired at lower temperatures. As the firing temperature decreased, the nobility of the metals used could be decreased as well. The most widely used powders for the manufacture of these capacitors are mixtures or alloys of palladium and silver. Also, some dielectrics can be fired at about 900 °C (1650 °F), thus eliminating the need for palladium and allowing the use of 100% Ag.

The manufacturing process for multilayer ceramic capacitors requires specific physical properties in the metal powders. These include a fairly uniform particle size, typically about 1 μm. The powder should be free from large particles or nondispersible agglomerates that may damage the ceramic sheet by causing short circuiting after firing the device.

REFERENCES

1. Des Forges, C.D., Sintered Materials for Electrical Contacts, *Powder Metall.*, Vol 22 (No. 3), 1979, p 138-144
2. Levenspiel, O., *Chemical Reaction Engineering*, John Wiley & Sons, New York, 1967
3. Aris, R., *The Optimal Design of Chemical Reactors*, Academic Press, New York, 1961
4. Yang, K.C., *et al.*, Secondary Coagulation, *J. Colloid Interface Sci.*, Vol 56 (No. 3), 1976, p 573
5. Joseph-Petti, A.M., *et al.*, Effect of Particle Size on Stability of Monodisperse Selenium Hydrosols, *J. Colloid Interface Sci.*, Vol 43 (No. 3), 1973, p 649
6. Kruyt, H.R., Ed., *Colloid Science*, Vol 1, Elsevier, New York, 1952
7. Schenkel, J.H., *et al.*, A Test of the Deijaguin Verwey-Overbeck Theory with a Colloidal Suspension, *Trans. Faraday Soc.*, Vol 42, 1966, p 285
8. Adamson, A.W., *Physical Chemistry of Surfaces*, Interscience, New York, 1967
9. Van Olphen, H., *Clay Colloid Chemistry*, Interscience, New York, 1963
10. Yang, K.C., *et al.*, in *Recent Developments in Separation Science*, Vol 4, Li, N., Ed., CRC Press, West Palm Beach, FL, 1978
11. Devereux, O.F., *et al.*, *Interaction of Parallel Plane Double Layers*, MIT Press, Cambridge, MA, 1963
12. U.S. Patent No. 3 667 935, 1972
13. U.K. Patent No. 1 004 652, 1965
14. U.S. Patent No. 3 390 981, 1968
15. U.S. Patent No. 3 385 799, 1975
16. Foo, G., A Critical Analysis of the Processing Parameters in Palladium Refining, *Precious Metals 1982*, El Guindy, M.I., Ed., Pergamon Press, New York, 1982, p 463

SELECTED REFERENCES

● Butts, A. and Coxe, C.D., Ed., *Silver-Economics, Metallurgy, and Use*, Krieger Publishing Co., Huntington, NY, 1975, p 441-445
● Lenel, F.V., *Powder Metallurgy Principles and Applications*, Metal Powder Industries Federation, Princeton, NJ, 1980, p 39-40
● Smith, I.C. and Carson, B.L., Ed., *Trace Metals in the Environment*, Vol 2—Silver, Ann Arbor Science, Ann Arbor, MI, 1977, p 201-202
● Yaverbaum, L.H., Ed., *Technology of Metal Powders—Recent Developments*, Chemical Technology Review No. 153, Noyes Data Corp., Park Ridge, NJ, 1980, p 264-266

Production of Tungsten, Molybdenum, and Carbide Powders

Production of Tungsten and Molybdenum Powders

By Richard F. Cheney
Section Head
Chemical and Metallurgical Division
GTE Products Corp.

TUNGSTEN AND MOLYBDENUM are refractory metals, whose melting points are higher than 2000 °C (3630 °F). Tungsten has the highest melting point of all metals (3410 °C, or 6170 °F) and one of the highest densities (19.3 g/cm^3). Molybdenum has a melting point of 2610 °C (4730 °F) and a density of 10.22 g/cm^3. Of the refractory metals, the refractory powders of primary commercial interest are tungsten, molybdenum, tantalum, and niobium. Tantalum and niobium are discussed in the article "Production of Tantalum and Niobium Powders" in this Volume.

This section discusses the raw materials, processing steps, powder properties, and finishing techniques (blending, handling, inspection, and packaging) involved in the production of tungsten and molybdenum powders. For information on applications of these materials, see the articles "P/M High-Temperature Materials," "Electrical and Magnetic Applications," and "Ordnance Applications" in this Volume.

Production of Tungsten Powder

The most important tungsten-bearing ores are Scheelite and Wolframite (Ref 1, 2).

Primary purification techniques depend on the type of ore being used to produce the powders. Typical purification and reduction reactions to chemically decompose mineral and ore concentrates of tungsten are (Ref 3):

Wolframite (Fe, Mn) (WO$_3$) + 2NaOH
= Na$_2$WO$_4$ + H$_2$O

Na$_2$WO$_4$ + 2HCl = H$_2$WO$_4$ + 2NaCl

Scheelite (CaWO$_4$) + 2HCl → H$_2$WO$_4$ + CaCl$_2$

12 H$_2$WO$_4$ + 10NH$_4$OH = (NH$_4$)$_{10}$H$_{10}$W$_{12}$O$_{46}$
+ 12H$_2$O

H$_2$WO$_4$ + heat = WO$_3$ + H$_2$O

(NH$_4$)$_{10}$H$_{10}$W$_{12}$O$_{46}$ + heat = 10NH$_3$ + 10H$_2$O
+ 12WO$_3$

3H$_2$ + WO$_3$ = W + H$_2$O

A flowchart depicting the chemical processing of tungsten ore to final product form is shown in Fig. 1.

The end products of chemical purification are usually ammonium paratungstate, (NH$_4$)$_{10}$H$_{10}$W$_{12}$O$_{46}$, and tungstic acid, H$_2$WO$_4$. Tungstic acid must be heated in air at 600 to 900 °C (1110 to 1650 °F) to convert it to tungstic oxide (WO$_3$) prior to beginning reduction. Ammonium paratungstate can be similarly treated to form WO$_3$. However, it is usually converted to blue tungsten oxide (WO$_{2.9}$), because the decomposition of ammonium paratungstate results in a slightly reducing atmosphere if oxygen is essentially excluded.

Ammonium paratungstate is available from several commercial sources and is a typical starting material (Ref 5). Very high purity ammonium tungstate can be produced by solvent extraction processes,

Fig. 1 Flowchart of processing tungsten ore to final product form
Source: Ref 4

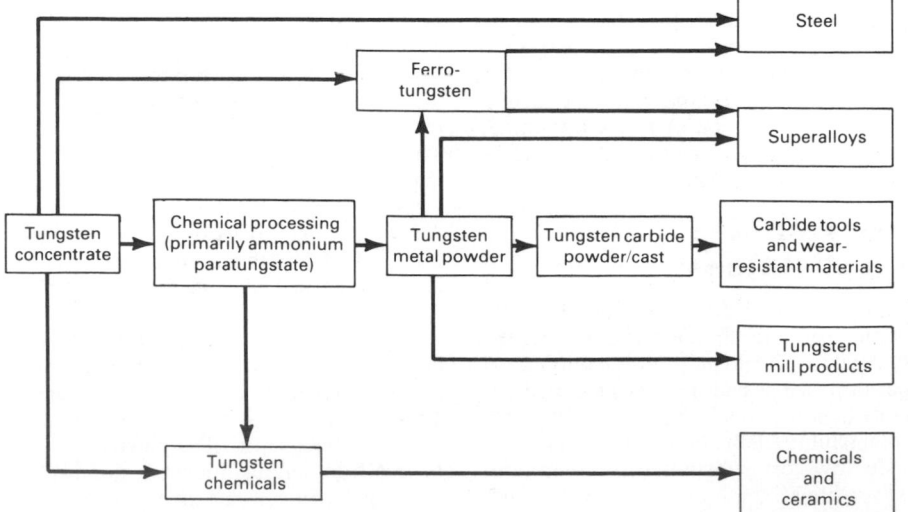

which are highly selective toward the tungstate ion (Ref 6). Therefore, purity of the starting sodium tungstate, which is produced when tungsten concentrate is leached, is not critical. Thus, lower grade tungsten sources can be processed than previously was possible. The solvent extraction process is also more easily automated and is therefore more cost effective.

Doping. It is common practice to add minor amounts of specific elements or compounds to the tungsten oxides prior to reduction (Ref 7-11). These dopants act to control the size and distribution of metal powder particles. The dopants often are substantially vaporized during reduction. The remainder is removed by acid washing of the reduced metal. In some applications, residual dopants are desirable to control the sintering process or to alter the microstructure of the final sintered or wrought P/M product (Ref 12, 13).

One such example is the doping of tungsten oxide with thorium nitrate to render tungsten powder containing 1 to 2% thorium dioxide dispersions (Ref 12). Thorium dioxide is desirable for welding electrodes, because it improves the emissive properties of the electrode (see the article "Electrical and Magnetic Applications" in this Volume for additional information). In wire, it results in a fine-grained, recrystallized microstructure that exhibits good shock resistance (Ref 14). As a result, thoriated tungsten wire is used in appliances and vibration-prone devices.

Another example of doping is the small amounts of aluminum oxide, potassium, and silicon dioxide introduced into the oxide powder used for lamp filaments (Ref 14, 15). The exact mechanism is not fully understood, but the result is a powder that has aluminum, potassium, and silicon incorporated into the tungsten particles. Aluminum and silicon are removed during subsequent acid washing and during sintering. About 50 to 100 ppm of potassium is retained, creating a sag-resistant grain structure in the rolled, drawn, and recrystallized lamp wire (Fig 2). Additional information on tungsten wire filaments can be found in the article "Electrical and Magnetic Applications" in this Volume.

Oxide Reduction. There are several suboxides of tungsten, as shown in Table 1. Reduction typically is carried out by stoking the oxide through tubes with a countercurrent hydrogen flow. The powder is transported through the tubes in containers made from nickel, nickel alloys, or molybdenum sheet.

A rotary furnace consisting of a large cylinder inclined at a slight angle to the horizontal (Ref 16) may be used for the

Table 1 Characteristics of various tungsten oxides

Oxide	Phase	Stable range (x-ray)	Color
WO$_3$	Alpha	WO$_3$ to WO$_{2.95}$	Yellow to blue
WO$_{2.9}$	Beta	WO$_{2.95}$ to WO$_{2.88}$	Blue
	Beta + gamma	WO$_{2.88}$ to WO$_{2.76}$	Red violet
WO$_{2.72}$	Gamma	WO$_{2.76}$ to WO$_{2.65}$	Violet
	Gamma + delta	WO$_{2.65}$ to WO$_{2.03}$	Violet
WO$_2$	Delta	WO$_{2.03}$ to WO$_{2.0}$	Brown
WO$_2$ + W . . . . ⋯		WO$_2$ to W	Gray

Fig. 2 Recrystallized structure of 180-μm (7-mil) diam tungsten wires

(a) Lamp grade, potassium doped. (b) Undoped

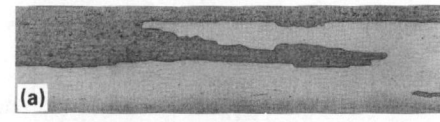

reduction procedure. Furnace rotation and the incline continuously move the powder through the hot zone. This method requires less manpower and heats, as well as reduces the powder more effectively. However, particle size control is more difficult than in tube furnaces. Contamination and lengthy cleanup times are also drawbacks to this method.

Properties of Tungsten Powder

Particle size and purity of tungsten powder greatly affect properties in subsequent operations such as sintering and metalworking. Particle size and purity, as well as other properties of tungsten powder, are discussed in detail below.

Particle Size Control. Control of the particle size of tungsten powder is critical to successful sintering and subsequent metalworking (see the article "Production Sintering Practices for P/M Materials" in this Volume). Particle size determines the grain size of tungsten carbide, which influences the strength and toughness of cemented carbides (Ref 17).

Tungsten powders range in size from less than 0.1 to greater than 15 μm. Typical commercial practice uses particles in the range of 0.5 to 15 μm. Average particle sizes typically are controlled to ±15% of specified size. Typical particle sizes and distributions are shown in Fig. 3. Scan-

ning electron micrographs of several tungsten powders are shown in Fig. 4.

To some extent, particle size is determined by the starting oxide. However, the major controlling variables are reduction temperature and powder depth. Parsons (Ref 18) has derived the following rate equations for reduction time and particle size, respectively:

$$t = K\rho_A T - 5.7 (Rh)^{1.6} \qquad \text{(Eq 1)}$$

$$F = 3.54 + 0.33 \log \rho_A - 1.9 \log T + 0.28 \log h + 0.002T + 0.028D \qquad \text{(Eq 2)}$$

where t is time, s; K is 6.66×10^{20}; ρ_A is apparent density; R is fraction of oxide reduced in time to h; h is powder bed depth, cm; T is temperature, K; F is Fisher subsieve size, μm; and D is tungstic oxide particle size, μm. Reduction time and powder particle size are also greatly affected by impurities and added dopants (Ref 11-19).

The reduction processes and particle growth mechanisms supporting Eq 1 and 2 are:

$$WO_{2(S)} + 2H_{2(G)} \leftrightarrow W_{(S)} + 2H_2O_{(G)}$$

$$\text{(Process 1)}$$

and

$$WO_{2(S)} + 2H_2O_{(G)} \leftrightarrow WO_3 \cdot H_2O_{(G)} + H_2$$

$$\leftrightarrow WO_3 \cdot H_2O_{(G)} \leftrightarrow W_{(S)} + 4H_2O_{(G)}$$

$$\text{(Process 2)}$$

Process 1 is a gas-solid reduction process that results in tungsten particles nucleated within the precursor oxide skeleton. Process 2 is a vaporization reduction deposition process that results in mass transport from the moisture-rich portion to the moisture-lean portion of the powder bed. This can be seen schematically in Fig. 5, which illustrates a cross section of the powder bed in the reduction furnace and the reduction and particle growth mechanisms that occur. Processes 1 and 2 compete with each other. When process 2 dominates, mass transport occurs, resulting in larger particle sizes.

Fig. 3 Typical particle sizes and distributions for tungsten powders used in the manufacture of cemented tungsten powder

Powders shown are M series powder produced by GTE Products Corp. Other manufacturers produce similar grades. (a) Ultrafine. (b) Fine. (c) Medium. (d) Coarse

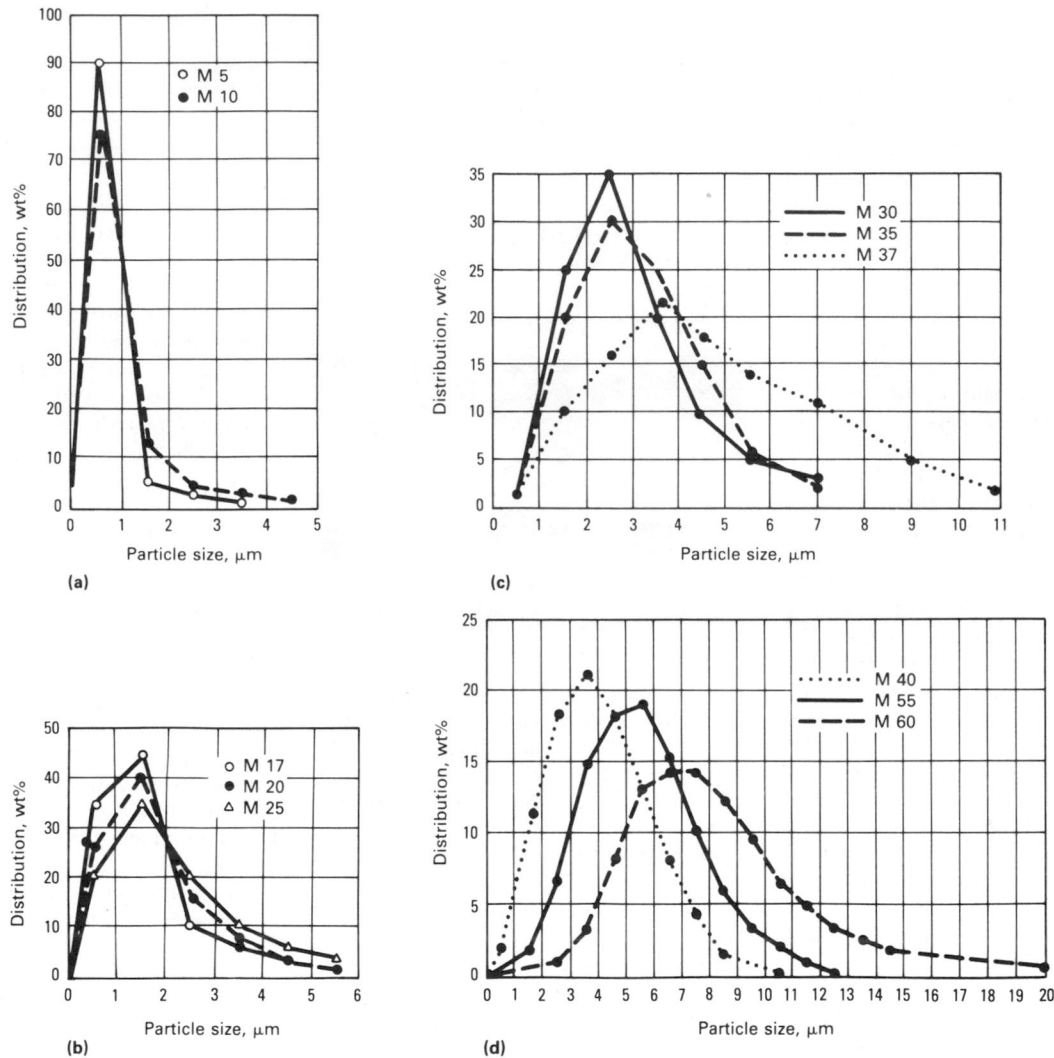

Particle growth in process 2 is controlled by regulating temperature and water vapor content within the powder bed. Thus, bed depth, hydrogen flow over the bed, and the partial pressure of water vapor in the hydrogen affect particle size.

Volatile compounds, present either as dopants or impurities, also influence the kinetics of the reduction and particle growth mechanisms. Inert particles that are present, such as thorium dioxide, may act as nuclei for the deposition of tungsten via the vapor transport mechanism. This also influences particle size.

Purity of tungsten powder affects subsequent operations, such as sintering and metalworking (Ref 20). Typical impurities are listed in Table 2. Powder impurities usually originate in the oxide. However,

contamination may result from contact with construction or container materials, such as iron, nickel, and chromium, in the reduction furnace.

Other Properties. Other properties affecting tungsten powder production include apparent density, surface area, degree of agglomeration, and particle size distribution. These properties are greatly affected by powder size and preparation method.

Finishing of Tungsten Powder

Tungsten powders are available in blended lot sizes of up to 6800 kg (15 000 lb). These powders usually are screened on 200 mesh to ensure removal of contami-

nants. Because the powders are extremely fine (325 mesh), they do not flow well. It is thus impractical to sieve them on screens below 200 mesh.

Typically, tungsten powders are packaged in 45-kg (100-lb) metal containers with polyethylene liners. No special atmosphere is necessary, because the powders do not adsorb atmospheric contaminants to any significant extent. Reblending of powders after transportation is also unnecessary because of their small size and poor flow characteristics.

Production of Molybdenum Powder

Molybdenum disulfide or molybdenite (MoS_2) is the major source of raw mate-

Fig. 4 Scanning electron micrographs of various tungsten powders

(a) 1.93-μm powder. (b) 5.30-μm powder. (c) 8.70-μm powder. (d) 16-μm powder. Magnification: 500×

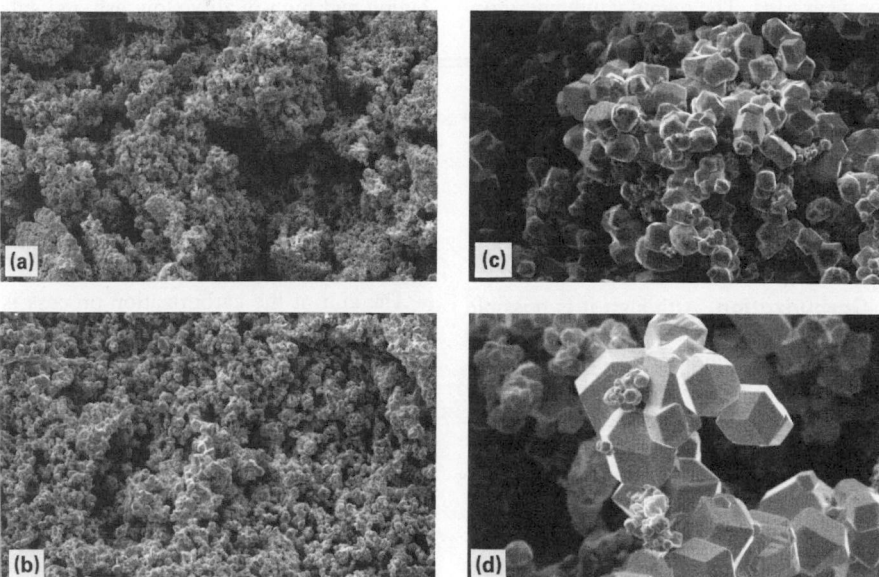

Fig. 5 Schematic cross section of the powder bed in a reduction furnace

h is the depth of powder bed; *x* is the depth of tungsten powder; *dx* is the reaction zone in the section through the bed horizontal plane.

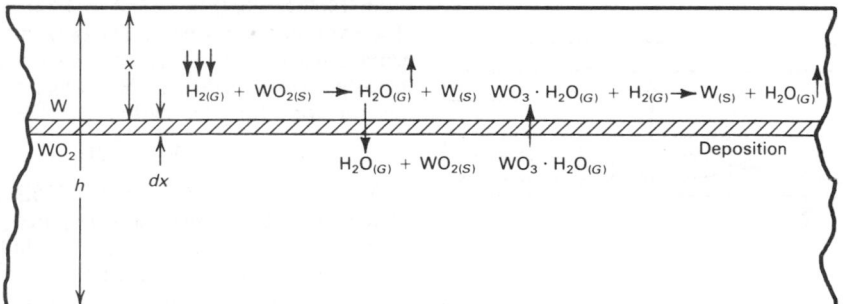

Table 2 Tungsten powder chemical analysis

Element	Maximum, ppm	Typical, ppm
Aluminum	10	<5
Calcium	50	<5
Chromium	25	5
Copper	10	<5
Iron	60	10
Manganese	50	<5
Magnesium	10	<1
Molybdenum	750	250
Nickel	100	15
Potassium	150	<15
Silicon	50	15
Sodium	100	15
Tin	20	<1
LOR(a)	5000	1000

(a) The combination of oxygen plus water is referred to as "loss on reduction" (LOR). In actual application, oxygen is more important to consider, as it ties up tungsten chemically as an oxide and is a factor, for example, in controlling carbon in subsequent carburization. Water is not so attached and is less critical, yet is controlled to very low content.

rial. Molybdenum disulfide is concentrated by flotation and converted to an impure technical oxide (molybdenum trioxide, MoO_3) by roasting, which oxidizes the sulfur and removes it as gaseous sulfur dioxide (SO_2) (Ref 21). Figure 6 is a flowchart showing development from concentrate to final product form.

Further purification commonly is done by taking advantage of the sublimation characteristics of molybdenum trioxide. Above 550 °C (1020 °F), it sublimes readily and can be distilled from its impurities and condensed again as "pure" molybdenum trioxide. Solvent extraction procedures are also available (Ref 22).

Additional purification can also be achieved by dissolving molybdenum trioxide in ammonia to form ammonium molybdate, $(NH_4)_2MoO_4$. The ammonium

molybdate or the pure molybdenum trioxide is reduced in furnaces similar to those used for tungsten reduction. Typical purification and reduction reactions are:

Molybdenite (MoS_2) + $^7/_2O_2$ + heat
= MoO_3 + $2SO_2$

$MoO_3 + 2NH_4OH = (NH_4)_2MoO_4 + H_2O$

$(NH_4)_2MoO_4 + 3H_2 = Mo + 2NH_4OH + 2H_2O$

$(NH_4)_2MoO_4 + H_2 = MoO_3 + NH_4OH$

$MoO_3 + 3H_2 = 3H_2O + Mo$

Molybdenum recovery processes have been reviewed extensively in References 21 to 23.

Reduction is made more difficult by the volatility of molybdenum trioxide. Molybdenum dioxide is more easily controlled and can be formed from molybdenum trioxide below the sublimation temperature. Thus, the reduction sequence should always be $MoO_3 \rightarrow MoO_2 \rightarrow Mo$ (Ref 24).

Properties of Molybdenum Powder

The particle size of molybdenum is affected by the same variables that influence tungsten production, but to a lesser extent. Thus, there are fewer particle sizes available over a smaller range—typically, 1 to 6 μm. Scanning electron micrographs of molybdenum powder are shown in Fig. 7. Typical compositions are:

Element	Content, ppm
Aluminum	5-25
Calcium	3-15
Chromium	5-25
Copper	5-10
Iron	10-100
Lead	5-10
Magnesium	1-10
Manganese	1-5
Nickel	5-50
Potassium	10-30
Silicon	5-250
Sodium	5-10
Tin	15-50
Tungsten	100-300
NVM (summation of Ca, Na, K, Si)	30-400
Carbon	10-50
LOR(a)	500-1000
Molybdenum: 99.9%, exclusive of gases	

(a) The combination of oxygen plus water is referred to as "loss on reduction" (LOR).

Finishing of Molybdenum Powder

Molybdenum powders are available in blended lots of up to 3600 kg (8000 lb).

The powders are screened at either 100, 250, or 325 mesh.

Packaging typically is in 45-kg (100-lb) metal containers with polyethylene liners. The powders are slightly surface active; after several months, oxygen levels may rise from about 0.10 to 0.3%. Reblending of powders after shipment is not necessary.

Production of Metal Carbides

By Peter C. Eloff
Manager—Powder Product
Engineering
General Electric Co.

TUNGSTEN CARBIDE provides the basis for the manufacture of the majority of cemented carbides. Cemented carbides, or hardmetals, are used in metal cutting and forming, in mining and drilling, and for wear parts. The manufacture and properties of cemented carbides are covered in the article "Cemented Carbides" in this Volume. This section discusses the manufacture of tungsten carbide powders and briefly describes the manufacture of titanium and tantalum carbides.

Production of Tungsten Carbide Powder

Carburization. Although it is possible to produce tungsten carbide directly from ore, oxide, or ammonium paratungstate, or by gaseous carburization, the preferred method is to carburize tungsten metal powder by adding controlled amounts of carbon black. This addition helps control particle size and size distribution, which, together with binder content, determine the resultant properties of the sintered carbide.

First, "black mix" tungsten powder of the desired particle size and size distribution and high-quality (low ash and sulfur content) carbon black is prepared. Because the two powders differ significantly in density, great care must be taken to ensure uniform distribution of carbon. Mixing is performed in ball mills, attritors, or specialized blenders. Ball milling times of 24 h or more may be required for adequate mixing. Attritors or properly designed and operated blenders are faster, typically $2^{1}/_{2}$ to 6 h.

The aim of the carburization process is to produce stoichiometric tungsten carbide with 6.13 wt% C or a small excess (0.01 to 0.03 wt%) of free carbon. Carbon deficiency results in the formation of brittle η phase (W_2C) in the final product. The exact amount of the carbon black addition in the black mix is determined in practice and depends on carburization conditions. Finer powders, containing more adsorbed oxygen or water vapor, require more carbon black than coarser powders. Gas flow and size (depth) of the black mix charge in the carburization furnace can also influence carbon composition.

Carburization is performed in the presence of hydrogen at temperatures ranging from 1400 to 2650 °C (2550 to 4800 °F). The hydrogen atmosphere reacts with the carbon black to form gaseous hydrocarbons (primarily methane, CH_4), which then react to form tungsten carbide:

$$W + CH_4 \rightarrow WC + 2H_2$$

As shown in Fig. 8, 1400 °C (2550 °F) is the minimum temperature required to form stoichiometric tungsten carbide. Higher temperatures are used for coarser powders, but temperatures generally are kept as low as possible to prevent grain growth in the newly formed tungsten carbide crystals. A common arrangement for production of large quantities of tungsten carbide is to charge black mix into cov-

Fig. 6 Flowchart of processing molybdenum ore to final product form
Source: Ref 4

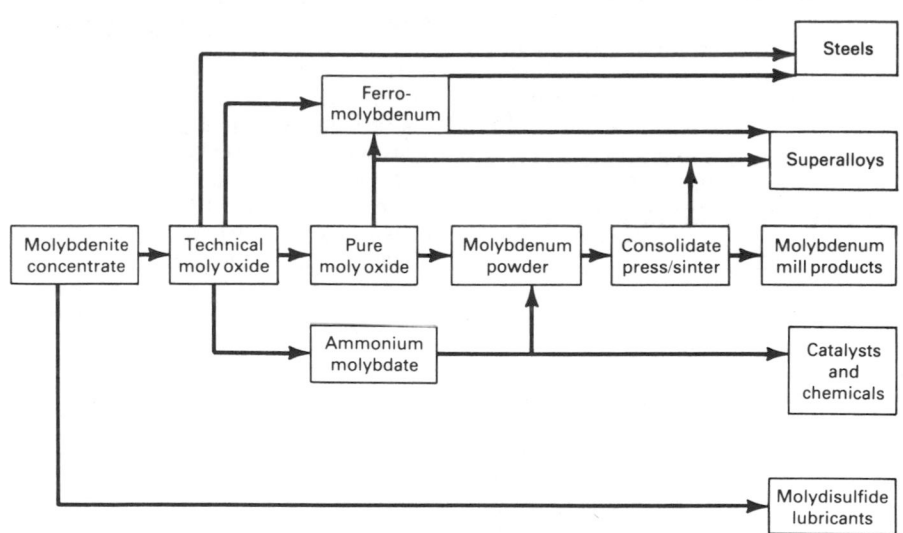

Fig. 7 Scanning electron micrographs of molybdenum powders
(a) 1.8-μm powder. (b) 3.4-μm powder. (c) 5.7-μm powder. Magnification: 1000×

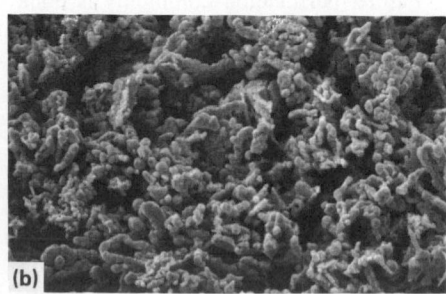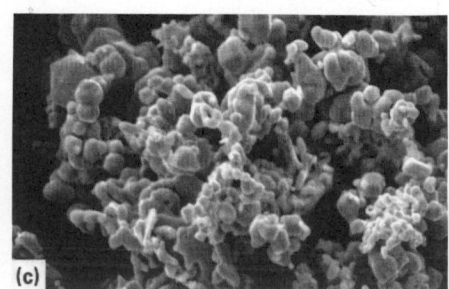

ered graphite boats, which are then conveyed at a controlled stoking rate through a pusher-type tube furnace. Figure 9 shows this arrangement, which incorporates fully automated loading, unloading, and handling of boats. Smaller batches of tungsten carbide can be produced in an induction-heated graphite crucible, as shown in Fig. 10. Graphite containers are used to maintain the proper carbon potential in the powder mass.

Fig. 8 Tungsten carburization
Combined carbon content as a function of temperature

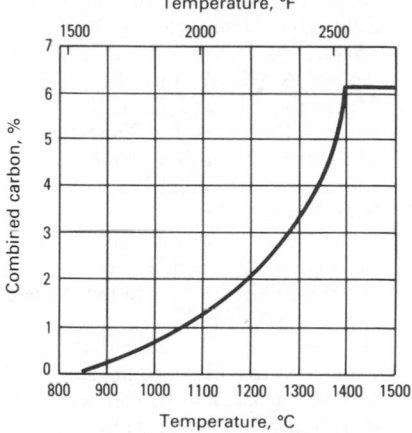

Upon exiting the carburizing furnace, the powders are caked or agglomerated. Therefore, carbides normally are subjected to a milling and sieving operation to produce deagglomerated powders with proper size distributions. Such powders are then ready to be mixed with binder powders (usually cobalt) for the manufacture of cemented carbide parts, as described in the article "Cemented Carbides" in this Volume.

Small amounts of vanadium, tantalum, or chromium carbides are frequently added to inhibit grain growth. These additives may be in the form of oxides or metals and are added to the tungsten/carbon black mix with the appropriate amount of carbon black. They may also be blended with tungsten carbide powders as pure carbides. Typical additions range from 0.5 to 2.0%.

Alternative Production Methods. Although carburization is the most widely used process for tungsten carbide production, two additional processes have also been used commercially on a limited basis. The first, called the Menstruum process, involves the formation of tungsten carbide within a melt of auxiliary metal, such as nickel or cobalt. Tungsten or tungsten ore and carbon are added to the melt and react to form the carbide at temperatures around 2000 °C (3630 °F). With a

ratio of auxiliary metal to tungsten of approximately 2 to 1 and slow cooling of the melt, relatively coarse carbide grains (up to 1 mm, or 0.04 in.) are precipitated. Finer sizes can be obtained by additional dilution and faster cooling. The solidified auxiliary metal is dissolved in hot hydrochloric acid to extract the relatively pure and stoichiometric carbide.

The second alternative process also employs auxiliary metals, in an exothermic thermite reaction. This process involves the reduction of a mixture of tungsten ore concentrates and iron oxide by aluminum metal, and simultaneous carburization through the addition of calcium carbide and carbon. The reactants are added in proportions that sustain an exothermic reaction above 2500 °C (4530 °F).

The reaction is carried out on a large scale (65-metric ton, or 72-ton charge), with the charge subdivided in heavy-gage aluminum bags for progressively charging and maintaining the reaction. Such a charge yields 20 metric tons (22 tons) of macrocrystalline tungsten carbide. The cooled mass is crushed and washed, and the iron-manganese-aluminum matrix is removed by acid leaching. The resulting tungsten carbide is coarse (−40 mesh) but completely carburized, with less than 0.03% free carbon. The carbide can be used as is for some hardfacing applications, or it can be reduced in size for more conventional carbide applications.

Fig. 9 Hydrogen pusher furnace for carburization of tungsten powder
Tungsten/carbon black mix is automatically weighed into carbon boats from large bin in background. Handling, stoking, and emptying of boats are also automated. Courtesy of General Electric Co.

Fig. 10 High-frequency furnace for carburization of tungsten

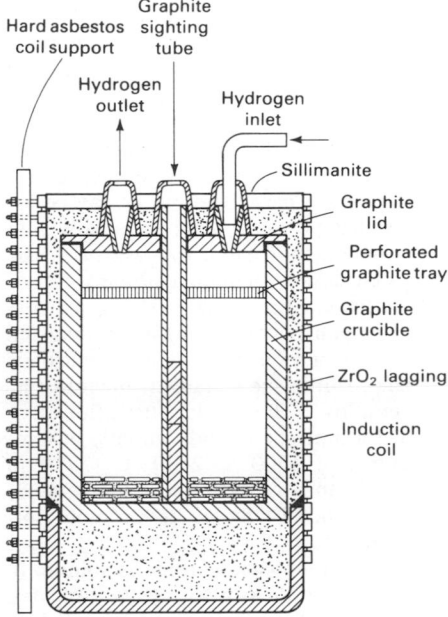

Production of Tungsten/Titanium Carbide Powder

Historically, cemented carbides used for the machining of steels have incorporated titanium carbide, because it increases resistance to cratering, which is adhesive wear caused by a continuous steel chip rubbing against a cutting tool during machining. For this purpose, a tungsten carbide/titanium carbide solid solution or "mixed crystal" is produced by co-carburization, rather than by mixing separately produced carbides and relying on diffusion during sintering. The latter approach tends to produce soft and hard spots, as well as porosity, in a sintered structure. The mixed crystal or double carbide has superior cratering resistance to the diffused mixture. This, along with the difficulty of producing titanium carbide by itself, makes co-carburization the preferred method of manufacture.

Production of mixed crystal carbide is accomplished in much the same way as described for tungsten carbide, with fine titanium dioxide pigment particles and additional carbon black added to the furnace charge. It is necessary to remove almost all of the oxygen to obtain optimum properties in tungsten carbide/titanium carbide; removal of oxygen becomes more difficult as the titanium carbide fraction increases, however. Below about 18% titanium carbide, the residual oxygen can be removed by hydrogen. As the titanium carbide content increases, the hydrogen atmosphere must be drier and faster flowing. Above 18% titanium carbide, vacuum levels of 50 μm Hg or lower must be used to keep oxygen (and nitrogen) at or below acceptable limits. Furnace temperatures between 1500 and 1600 °C (2730 and 2900 °F) are required to form the mixed carbide.

In calculating the amount of carbon black that must be added to titanium dioxide for reduction and carburization, it may be assumed that oxygen is removed equally with carbon monoxide and carbon dioxide. The tungsten carbide/titanium carbide solid solution can exist over a wider range of possible carbon contents than pure tungsten carbide. Consequently, carbon control is enhanced. In some instances, manufacturers employ a double carburization process, forming the carbide initially at temperatures of 1500 to 2100 °C (2730 to 3800 °F), then reheating to a temperature 200 °C (360 °F) higher to promote solid solubility.

Although tungsten carbide is soluble in titanium carbide (up to 82.2% tungsten carbide), titanium carbide is only slightly soluble in tungsten carbide. Therefore, mixed crystal compositions containing less than 17.8% titanium carbide consist of a mixture of double carbide and tungsten carbide. Because of the sensitivity of tungsten carbide/titanium carbide properties to oxygen content, it is preferable to avoid exposure or lengthy storage times and to convert to cobalt-containing "grade" powder, followed by sintering as quickly as possible. The Menstruum process is also effective for the production of mixed carbide, producing very homogeneous crystals of tungsten carbide/titanium carbide.

Production of Titanium and Tantalum Carbide Powders

Pure titanium carbides, as opposed to mixed crystal tungsten carbide/titanium carbides, are used as the basis of a separate class of cemented carbides. Tantalum carbides frequently are added to tungsten carbide for specific applications. The most common method of preparation for both titanium and tantalum carbide is reduction and simultaneous carburizaton of their oxides. While carburization of the metal powders is feasible, the cost and availability of the oxides are much more favorable. The starting material for titanium carbide is pure titanium dioxide, commonly used as a pigment. A mixture of 68.5% titanium dioxide and 31.5% carbon black is prepared by wet or dry milling for extended periods. Carburization can be carried out in very dry, nitrogen-free hydrogen, in its own protective gas (carbon monoxide in a sealed furnace system) or in vacuum. For production of tantalum carbide, a mixture of tantalum pentoxide (Ta_2O_5) and carbon black can be carburized in a carbon-tube furnace under hydrogen or vacuum. For both materials, vacuum processing is preferred because of lower temperature requirements and faster reaction rates.

Alternative methods for producing an inexpensive tantalum carbide also containing niobium carbide are direct carburization of tantalum ores, or melting of ferrotantalum (60 to 70% Ta and Nb) and adding carbon until the melt becomes viscous. In the latter method, the brittle material is crushed after cooling and treated with warm hydrochloric acid, forming tantalum carbide (and niobium carbide). This material is mixed with additional carbon and given a final carburization treatment at 1600 to 1700 °C (2910 to 3100 °F) in hydrogen.

REFERENCES

1. Smithells, C.J., *Tungsten,* Chemical Publishing, New York, 1953
2. Yih, S.W.H. and Wang, C.T., *Tungsten Sources, Metallurgy, Properties and Applications,* Plenum Press, New York, 1979
3. MacInnis, M., Tungsten Chemistry, *Tungsten: 1982,* Proc. 2nd Int. Tungsten Symp., San Francisco, June 1982
4. *What Are Refractory Metals—And How Do They Affect Our Lives?,* Refractory Metals Association of the Metal Powder Industries Association, Princeton, NJ
5. Shamsuddin, M. and Sohn, H.Y., Extractive Metallurgy of Tungsten, in *Extractive Metallurgy of Refractory Metals,* conference proceedings, 110th Annual AIME Meeting, Feb 1981
6. MacInnis, M. and Kim, T., Commercial Processes for Tungsten and Molybdenum, in *Handbook for Solvent Extraction,* John Wiley & Sons, New York, 1983
7. Lenel, F.V., *Powder Metallurgy—Principles and Applications,* Metal Powder Industries Federation, Princeton, NJ, 1980, p 338
8. Heitzinger, F., The Industrial Significance of the Most Important High Melting Metals; Part II, Production and Processing into Pure Metal Alloys, *Powder Metall. Int.,* Vol 10 (No. 3), 1978, p 136-138
9. Eck, R., Kienzl, F., Tiles, B., and Wagner, H., Fabrication, High Temperature and Corrosion Properties of Tungsten Metallizing Wire, in *Modern Developments in Powder Metallurgy,* Vol 11, 1977, p 91-108
10. Wilker, T., Wert, C., Woodhouse, J., and Morcom, W., Reduction of Blue Tungsten Oxide, in *Modern Developments in Powder Metallurgy,* Vol 5, Plenum Press, New York, 1971, p 161-169
11. Mayer, H. and Schreiner, M., *Proc. 10th Plansee Seminar,* Metalwerk Plansee, Reutte, Austria, 1980
12. Lenel, F.V., *Powder Metallurgy—Principles and Applications,* Metal Powder Industries Federation, Princeton, NJ, 1980, p 338, 344
13. Roll, K., Powder Metallurgy, in *Kirk-Othmer Encyclopedia of Chemical Technology,* Vol 19, 3rd ed., John Wiley & Sons, New York, 1982, p 56
14. Lenel, F.V., *Powder Metallurgy—Principles and Applications,* Metal Powder Industries Federation, Princeton, NJ, 1980, p 342-343

15. Moon, D.M. and Koo, R.C., Mechanism and Kinetics of Bubble Formation in Doped Tungsten, *Metall. Trans.*, Vol 2, 1971, p 2115-2122

16. Schwier, G., *Proc. 10th Plansee Seminar*, Metallwerke Plansee, Reutte, Austria, 1980

17. Exner, H.E., Physical and Chemical Nature of Cemented Carbides, *Int. Metals Rev.*, Vol 24 (No. 4), 1979, p 149-173

18. Parsons, D.S., The Reduction of Tungsten Oxides by Hydrogen, *Electrochem. Technol.*, Vol 3 (No. 9-10), Sept-Oct 1965, p 280-283

19. Parsons, D.S., The Reduction of Doped Tungsten Oxides: Particle Size and Distribution Effects, Internal Report TT-405, GTE Products Corp., Towanda, PA, 1964

20. Eck, R., Powder Metallurgy of Refractory Metals and Applications, *Int. J. Powder Metall. Powder Technol.*, Vol 17 (No. 3), 1981

21. Dorfler, R.R. and Laferty, J.M., Review of Molybdenum Recovery Processes, *J. Metals*, Vol 33 (No. 4), 1981, p 48-54

22. MacInnis, M., Kim, T., and Laferty, J., *1st Int. Conf. Chem. Uses of Molybdenum*, AMAX, Greenwich, CT, 1973, p 56-58

23. Molybdenum Ore Treatment—1980 to December 1982, NTIS Published Search, PB83-857250, Dec 1982

24. Kennedy, M.J. and Bevan, S.C., *1st Int. Conf. Chem. Uses of Molybdenum*, AMAX, Greenwich, CT, 1973, p 11-13

SELECTED REFERENCES

- Smithells, C.J., *Tungsten*, 3rd ed., Chapman & Hall, London, 1952

- Schwarzkopf, P. and Kieffer, R., *Refractory Hardmetals*, Macmillan, New York, 1953

- Brookes, K.J.A., *World Directory and Handbook of Hardmetals*, 2nd ed., Engineers' Digest Limited, London, 1979

- Hardmetal Review, *Metal Powder Rep.*, Vol 34 (No. 8), Aug 1979, p 356-383

- Chung, T.E., Coleman, D.S., Dowson, A.G., and Williams, B., Ed., Proceedings: Recent Advances in Hardmetal Production, Loughborough University of Technology, published by *Metal Powder Rep.*, Sept 17-19, 1979

Production of Tantalum and Niobium Powders

By John B. Lambert
Vice President and Technical Director
Fansteel, Inc.
and
Robert E. Droegkamp
Manager of New Products
Metals Division
Fansteel, Inc.

TANTALUM AND NIOBIUM frequently are found in close association in their ores. The most important tantalum-bearing minerals are tantalite and columbite, which are variations of the same compound $(Fe, Mn)(Ta, Nb)_2O_6$. The tantalum mineral is called tantalite when the tantalum pentoxide (Ta_2O_5) content exceeds the niobium pentoxide (Nb_2O_5) content. Columbite is a tantalum ore in which the reverse relationship is true. Cassiterite (SnO_2) deposits found in Thailand, Malaysia, Australia, and other countries also contain significant amounts of tantalum and niobium. By-product slags from tin smelting operations are frequently used as tantalum feedstock.

Tantalum and niobium are separated from the source ore and from one another by digestion of the ore in aqueous hydrofluoric acid, followed by liquid-liquid solvent extraction with methyl isobutyl ketone (MIBK), as shown in Fig. 1. The liquid-liquid-solvent extraction process depends on promoting solute ion exchange between immiscible liquids in intimate contact, typically by varying the concentration of other solution components. In the first contacting, only the fluorides of tantalum and niobium are soluble in the MIBK; they are therefore separated from the contaminant iron, manganese, titanium, and zirconium, which remain in aqueous solution. The solubility of tantalum in MIBK is high over an extensive range of acidity; niobium is soluble only at high acidity. In a series of sub-sequent steps, the MIBK is contacted with water of varying acidity to produce separate streams of tantalum fluoride and niobium fluoride.

Commercially important deposits of niobium also are found in the mineral pyrochlore $[(Na, Ca)_2(Nb, Ta, Ti)_2O_4(OH, F) \cdot H_2O]$, which is mined in Canada and Brazil. These ores are almost tantalum free and are processed to concentrates containing 55 to 60% niobium pentoxide by a series of operations including size reduction, flotation, and leaching. Extraction of niobium by high-temperature chlorination of the ore is also a suitable means of obtaining niobium (Ref 2).

Metal Powder Manufacture

Tantalum

Usually, the starting material for producing tantalum powder is the double salt, potassium tantalum fluoride (K_2TaF_7), which is precipitated by adding potassium fluoride to the aqueous tantalum solution. The metal is freed from the compound by chemically reacting with molten sodium in a stirred retort containing molten sodium chloride diluent under inert atmosphere. After cooling, the resulting salt cake is removed from the retort and crushed. The salt-encased tantalum powder is recovered by thoroughly washing with water, then acid.

Particles, as shown in Fig. 2, are usually spherical in shape, with a tendency to form grape-like clusters during reduction. Individual particles in such clusters range in size from 1 to 10 μm, depending on reduction variables such as temperature, agitation, and salt purity. Electronic capacitor applications normally require fine particle sizes with high surface area, whereas P/M mill products use the easier to handle coarse particle fractions. Additional information on tantalum capacitors can be found in the article "P/M High-Temperature Materials" in this Volume.

In one modification of the process, potassium tantalum fluoride, sodium chloride, and sodium metal are mixed into a paste and are reacted without premelting (Ref 3). Platelet-shaped tantalum particles are produced by this technique. Because platelets have a higher surface-to-volume ratio than spheres, they are well suited for capacitor applications.

Purified tantalum powders also are produced by a process that involves electron beam melting as the means of purification. Sodium-reduced powders are pressed into bars that are subsequently electron beam melted. The resulting ingot is placed in an evacuated and purged furnace into which hydrogen is introduced. The tantalum is fully hydrided on slow cooling from 800 °C (1470 °F) under hydrogen. The brittle hydride is crushed, ground, and classified to yield powder with a mean particle size ranging from 3 to 6 μm. Tantalum powder with angular particles, as

Fig. 1 Tantalum/niobium production flowchart
Source: Ref 1

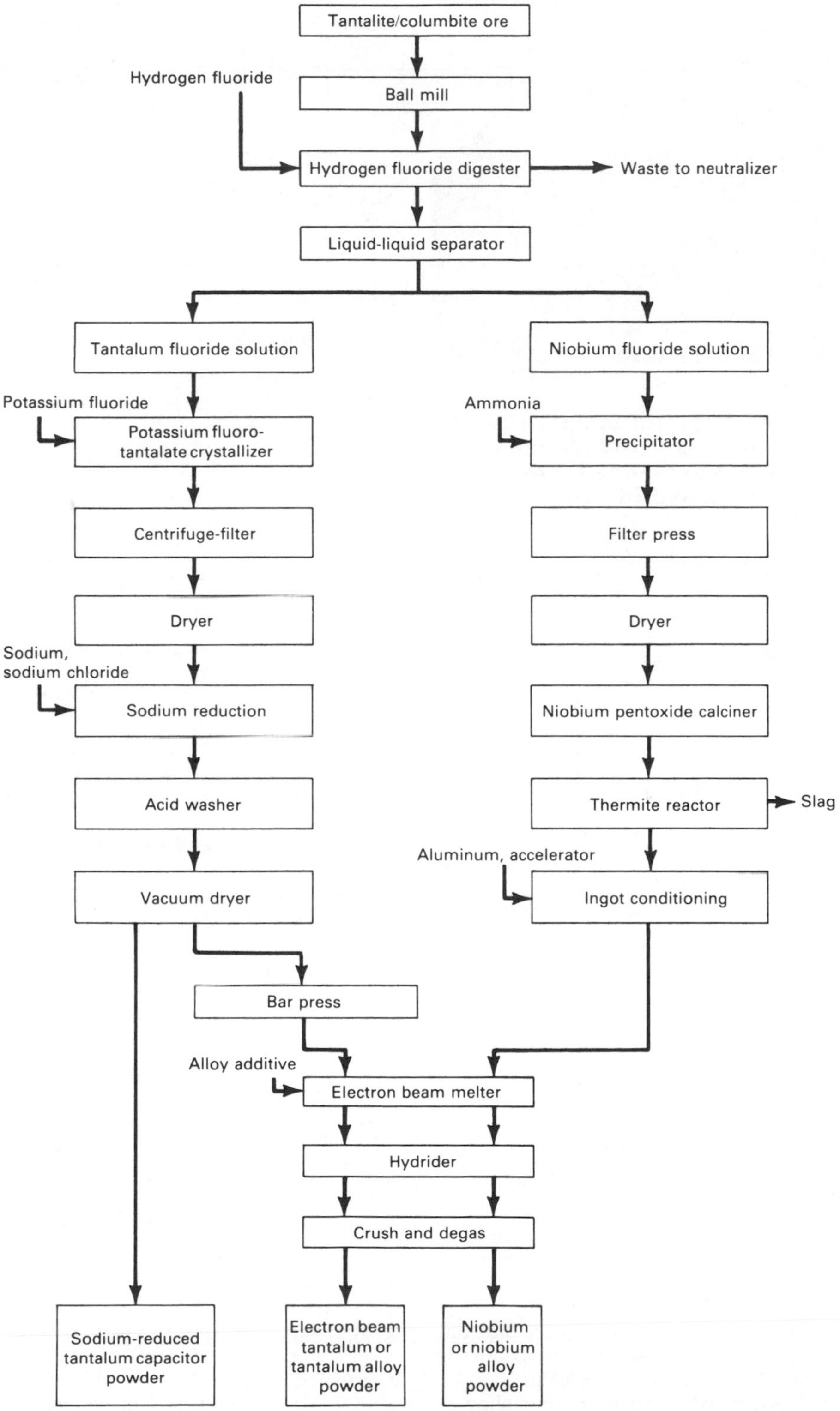

Tantalite/columbite ore
→ Ball mill ← Hydrogen fluoride
→ Hydrogen fluoride digester → Waste to neutralizer
→ Liquid-liquid separator

Tantalum fluoride solution | Niobium fluoride solution

Potassium fluoride → Potassium fluoro-tantalate crystallizer
Ammonia → Precipitator

Centrifuge-filter | Filter press

Dryer | Dryer

Sodium, sodium chloride → Sodium reduction
Niobium pentoxide calciner

Acid washer | Thermite reactor → Slag

Aluminum, accelerator → Ingot conditioning

Vacuum dryer

Bar press

Alloy additive → Electron beam melter

Hydrider

Crush and degas

Sodium-reduced tantalum capacitor powder | Electron beam tantalum or tantalum alloy powder | Niobium or niobium alloy powder

Fig. 2 Particle shape of tantalum powder anode produced by sodium reduction of potassium tantalum fluoride
Magnification: 2600×

strength for handling. Because capacitance is directly proportional to the tantalum surface area accessible to an electrolyte, a porous pellet structure is desirable. In general, sodium-reduced powders can be pressed to lower densities than electron beam melted, degassed-hydride powders. However, electron beam melting purifies the tantalum, thereby allowing capacitor devices to operate at higher voltages. The advantages of the two basic types of tantalum powders when used for capacitor manufacture are:

Sodium-reduced powder

- Higher capacitance per gram of powder
- Better green strength for pressed compacts

Electron beam melted, degassed-hydride powder

- Higher purity
- Better flow characteristics
- Higher voltage capability in capacitor devices

Table 1 shows typical compositions of sodium-reduced and electron beam melted powders.

The "hydride-crush-dehydride" process is also a convenient method for reclaiming tantalum metal scrap and converting it to powder. Platelet particle shapes also can be made by ball or vibratory milling of the pure, ductile particles obtained after degassing.

Although there are few current applications for alloy tantalum powder, a Ta-7.5%W powder has found use as a corrosion-resistant spring material. Such powders usually are produced by the melt-

shown in Fig. 3, is obtained after dehydriding (degassing).

For capacitor manufacture, small tantalum pellets are pressed. It is desirable to press the powders to as low a density as possible, while retaining sufficient green

Fig. 3 Particle shape of tantalum powder made by electron beam melting, hydriding, crushing, and degassing
Magnification: 2900×

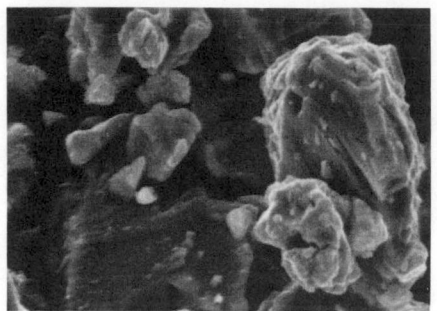

Fig. 4 Particle shape of niobium powder made by electron beam melting, hydriding, crushing, and degassing
Magnification: 250×

Table 1 Typical compositions of sodium-reduced and electron beam melted degassed-hydride tantalum powders

	Powder analysis, ppm	
Element	Sodium-reduced	Electron beam melted degassed-hydride
Aluminum	10	5
Calcium	10	5
Carbon	100	45
Chromium	25	5
Cobalt	10	5
Copper	10	5
Hydrogen	30	30
Iron	50	30
Lead	10	5
Magnesium	10	5
Manganese	10	10
Nickel	50	5
Niobium	50	20
Nitrogen	90	40
Oxygen	2400	1650
Silicon	25	10
Sodium	10	5
Tin	10	5
Titanium	10	5
Tungsten	25	25
Vanadium	10	5
Zirconium	10	5

Source: Fansteel, Inc.

Table 2 Typical composition of niobium and C-103 niobium alloy powder made by the hydride-dehydride process

	Analysis, ppm	
Element	Niobium	C-103
Niobium	99.7+%	87.2+%
Oxygen	1820	1980
Tantalum	800	2800
Hafnium	<20	9.8%
Zirconium	<20	1800
Titanium	20	0.91%
Carbon	500	194
Iron	100	200
Aluminum	<20	<20
Nitrogen	197	62
Silicon	30	<20
Copper	<40	<40
Cobalt	<10	<10
Boron	<1	<10
Hydrogen	150	50
Nickel	<20	<20
Molybdenum	<20	100
Tungsten	<50	1100
Other elements(a)	<20	<20

(a) Other elements include: cadmium, chromium, magnesium, manganese, lead, tin, vanadium, and zinc.
Source: Fansteel, Inc.

Table 3 Physical properties of thermally agglomerated, sodium-reduced and electron beam melted, degassed-hydride tantalum powders

	Sodium-reduced powder	Electon beam melted, degassed-hydride powder
Sieve analysis, %		
−35+80	0-20	15-35
−80+200	10-25	15-25
−200+325	10-20	5-10
−325	45-70	45-60
Physical properties		
Scott bulk density, g/cm^3	1.6-2.0	3.5-4.0
Apparent Fisher subsieve particle size, μm	2.5-3.5	6.5-8.5
Hall flow index(a), s	130	50

(a) ASTM B 213

as titanium, hafnium, or zirconium are frequently added during arc remelting of the electron beam melted ingot prior to hydride-dehydride processing. Typical compositions of niobium and C-103 alloy (Nb-10Hf-1Ti) powder made by this method are given in Table 2.

Physical Properties

Tantalum is unusual among powder materials because its most important use is in capacitors, and optimum capacitor performance requires very low pellet density after sintering. Therefore, many of the conventional metallurgical techniques for improving flowability cannot be used, and the green strength of low-density pellets is a serious concern. Production of a "popcorn ball" structure through vacuum thermal agglomeration of powders results in significant improvements in both characteristics (Ref 4). In fact, unagglomerated capacitor powders are almost obsolete because of inferior flow and green strength. The agglomeration step is carried out by heating the loose powder somewhat below the temperatures used for subsequent pellet sintering. The lightly sintered product is then recrushed and screened to a predetermined sieve analysis. Table 3 compares typical properties of agglomerated sodium-reduced and degassed-hydride tantalum powders. Figure 5 further illustrates the superior green strength of capacitor anodes made from sodium-reduced

ing and hydriding method. The tungsten is added during electron beam melting. The homogeneous ingot is then hydrided, crushed, and dehydrided. Except for the tungsten addition, the chemistry and properties of this alloy powder are typically the same as for unalloyed degassed-hydride (Table 1).

Niobium

Currently, virtually all niobium metal is extracted and purified by aluminothermic reduction of the oxide:

$$3Nb_2O_5 + 10Al \xrightarrow{\text{accelerator}} 6Nb + 5Al_2O_3$$

During the exothermic reaction, oxide impurities slag from the molten niobium. The metal is further purified by electron beam melting. Subsequently, the ingot is converted to powder by hydriding, crushing, and dehydriding. The particle structure, shown in Fig. 4, is completely analogous to degassed-hydride tantalum (Fig. 3). The niobium particle size is coarser than tantalum because its use does not require the generation of high surface area.

Alloy powders are made by adding alloying agents during melting. Low-volatility metals such as tungsten and tantalum may be added during electron beam melting. More volatile alloying agents such

Fig. 5 Crush test results for tantalum capacitor powder anodes made from sodium-reduced and electron beam melted, degassed-hydride powders

Both types were thermally agglomerated.

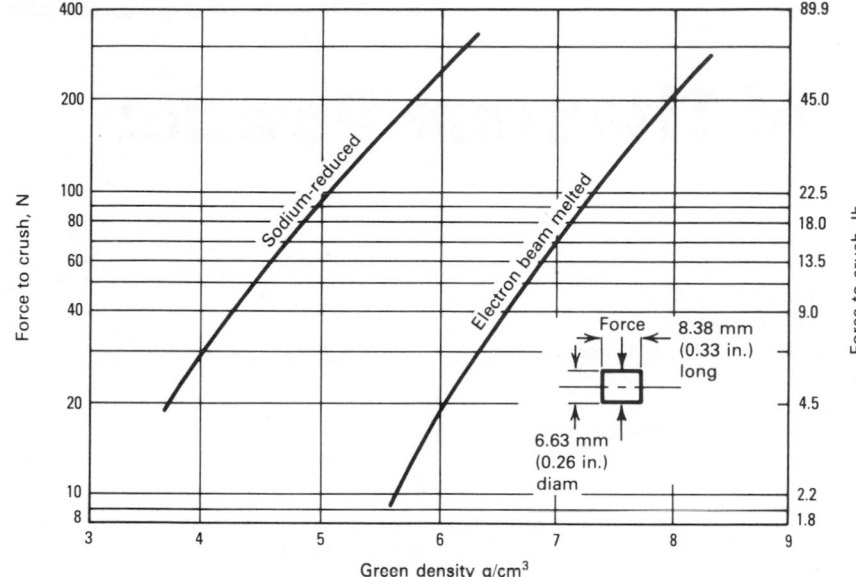

material as measured by means of a standardized crush test.

Because niobium powders produced by the hydride-crush-degas process are not milled as fine as tantalum, they are easier to handle. Normally, powders are crushed to pass an 80-mesh screen, and a mean particle size in the 10- to 15-μm range is typical. Thermal agglomeration, although feasible, is rarely practiced.

REFERENCES

1. Kirk, R.E. and Othmer, D.F., Ed., *Encyclopedia of Chemical Technology,* Vol 22, 3rd ed., John Wiley & Sons, New York, 1983, p 541-564
2. Lerner, B.J., U.S. Patent 3 294 482, 1966, Dominion Gulf Co.
3. Borchers, P., *et al.,* Ed., *Extractive Metallurgy of Refractory Metals,* Conference Proceedings, Metallurgical Society of the American Institute of Mechanical Engineers, Feb 22-26, 1981
4. Pierret, J.A., U.S. Patent 3 418 106, 1968, Fansteel, Inc.

Production of Titanium Powder

By F.H. Froes
Technical Area Manager
U.S. Air Force
AFWAL Materials Laboratory
and
D. Eylon
Metcut-Materials Research Group

TITANIUM is the design choice for many aerospace and nonaerospace applications, because of an attractive combination of low density, good mechanical properties, and general corrosion resistance (Ref 1, 2). However, titanium is expensive to produce, fabricate, and machine. Consequently, much effort has been expended to reduce the cost of titanium parts. The major thrust in cost reduction has been to produce near-net shape parts, which has led to much research and development of P/M techniques. The emphasis of this research has been placed on the commonly used alloys, such as Ti-6Al-4V, rather than on commercially pure titanium. However, the techniques discussed in this article equally apply to commercially pure titanium by elimination of the alloying elements. This article details the two basic methods currently used to produce titanium alloy powders—the blended elemental and prealloyed techniques. Further information can be found in Ref 3 and 4 and in the article "P/M Lightweight Metals" in this Volume.

Blended Elemental Alloy Powder Production

The commercial production of titanium metal involves the chlorination of natural and synthetically produced rutile (TiO_2) in the presence of carbon. The most important reaction is:

$$TiO_2(s) + 2Cl_2(g) + 2C(s)$$
$$\rightarrow TiCl_4(g) + 2CO(g)$$

The resulting titanium tetrachloride is pu-rified by distillation and chemical treatments, and it is then reduced to metallic titanium (termed "sponge" because of its appearance).

Most of the world's production employs magnesium metal as the reducing agent, based on a process developed by Kroll (Ref 5). An alternate process, using sodium metal, was developed by Hunter (Ref 6). These processes are chemically similar, but differ in operating details. The basic reactions are:

$$2Mg(s) + TiCl_4(l) \rightarrow Ti(s) + 2MgCl_2(s)$$

$$4Na(l) + TiCl_4(l) \rightarrow Ti(s) + 4NaCl(s)$$

In some Hunter processes, the initial step involves only a partial reduction, followed by a second step for complete reduction:

$$2Na(l) + TiCl_4(l) \rightarrow TiCl_2 \cdot 2NaCl(s)$$

$$2Na(l) + TiCl_2 \cdot 2NaCl(s)$$

$$\rightarrow Ti(s) + 4NaCl(s)$$

To ensure complete reaction and to provide coarse sponge particles by sintering, the final reactions of both processes are completed at temperatures as high as 1040 °C (1900 °F). In the Kroll process, the bulk of the molten magnesium chloride is then generally tapped off to free additional reactor capacity and to simplify recycling of magnesium chloride. Most of the residual salt is removed by vacuum distillation or, after cooling, by water leaching. In the Hunter process, the sodium chloride is generally removed only by water leaching.

Except for minor differences in purity and particle sizes, these two grades of titanium sponge are interchangeable. Particle sizes are controlled by operating pro-cedures and crushing, and depending on the end use, range from coarse pieces to powders. The -100 mesh powders are generally termed "sponge fines" (Fig. 1) and are the starting stock used in the blended elemental P/M technique. Screen analysis of typical sodium-reduced sponge fines is given in Table 1; chemical analysis is given in Table 2. Alloy additions are made to these fines, normally in the form of a powdered master alloy, to achieve the desired bulk composition.

The blended mixture is then cold compacted, under pressures up to 415 MPa (60 ksi), to a green density of 85 to 90%. Compacting is carried out either iso-statically or with a relatively simple mechanical press and a rigid die. The green compact is typically vacuum sintered at 1260 °C (2300 °F) to increase density to 95 to 99.5% of theoretical density, depending on the compacting practice used, and to homogenize composition.

A further increase in density can be achieved by hot isostatically pressing the sintered part, which generally improves mechanical properties (see the articles "Mechanical Properties of P/M Materials" and "P/M Lightweight Metals" in this Volume for more information). The combined isostatic pressing process is often referred to as cold/hot isostatic pressing. This method produces parts more economically than cast and wrought processes. However, porosity (even after hot isostatic pressing) results in degraded initiation-related properties such as fatigue, which make these products unsuitable for demanding applications.

Pores in a blended elemental Ti-6Al-4V powder compact are shown in Fig. 2. The

Fig. 1 Scanning electron micrograph of commercially pure titanium −100 mesh sponge fines
Fines are porous, with a sponge-like morphology.

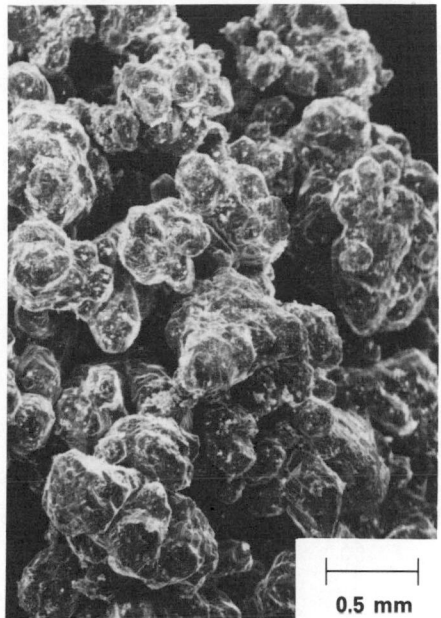

0.5 mm

Fig. 2 Pores in blended elemental Ti-6Al-4V compact after cold pressing and sintering

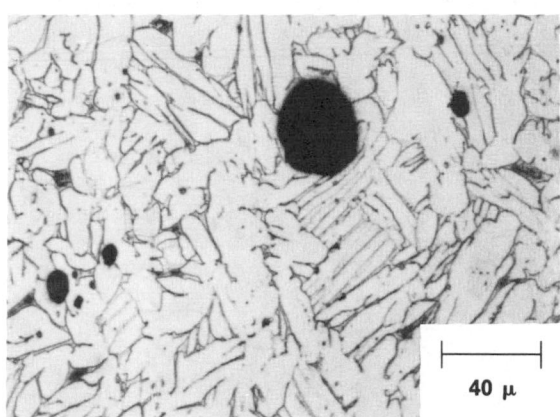

40 μ

Table 1   Screen analysis of sodium-reduced sponge fines

Mesh size (U.S.)	Particle size, μm	Weight percent retained
+80	+177	0
−80+100	−177+149	0.1
−100+140	−149+105	11.2
−140+200	−105+74	32.9
−200+230	−74+64	5.0
−230+325	−64+45	23.3
−325	−45	27.5

Table 2   Typical composition of titanium sponge fines and compacted Ti-6Al-4V

Element	Composition, wt% Powder	Sintered compact
Aluminum	⋯	6.2
Vanadium	⋯	4.1
Oxygen	0.13	0.24
Nitrogen	0.03	0.016
Hydrogen	0.07	0.002
Carbon	0.02	0.02
Iron	0.02	0.18
Sodium	0.10	0.10
Chlorine	0.13	0.12
Titanium	rem	rem

remnant chlorides (magnesium or sodium, depending on whether the Kroll or Hunter process is used to produce the titanium sponge) preclude obtaining 100% density

with this product. When heated during sintering or post-sintering compaction, the chlorides become gaseous; they are not absorbed into the titanium and cannot be fully compressed, resulting in porosity.

Work is presently in progress to reduce chloride levels—by producing titanium sponge by the electrolytic process, for example. Generally, however, information is not available because of proprietary limitations. Chloride level apparently must be below at least 100 ppm to avoid property degradation, but the exact level has not been defined. Further complications arise because of the nonuniform microstructure (aligned and equiaxed alpha), which can also degrade mechanical behavior.

Prealloyed Powder Production

Starting stock for prealloyed titanium powder products initially contains the necessary alloy elements. Two methods that are currently used for prealloyed powder production are comminution of the starting stock and the plasma rotating electrode process (PREP).

Comminution. Commercially pure and common titanium alloys such as Ti-6Al-4V are generally quite ductile and therefore not readily amenable to comminution to fine powder. However, titanium is a hydride former and can be readily converted to a brittle state by the introduction of hydrogen, allowing comminution to be carried out easily. This process is referred to as the hydride-dehydride process.

The hydride-dehydride process uses a variety of feed-stock materials, such as ingot, billet, solid scrap, or machine turnings. However, because of the finite length of time required for hydrogen to diffuse

into the titanium, a light-gage starting stock such as machine turnings (chips) is preferred. Chips are processed carefully to remove the maximum possible amount of extraneous material. This includes x-ray evaluation and removal of high-density contaminants such as tungsten carbide tool bit pieces. Cleaned chips are then hydrogenated and ground in batch lots under an argon atmosphere in a vibratory-type ball mill. Typical hydrogenation conditions for 25 kg (55 lb) of Ti-6Al-4V machine turnings are: 400 °C (750 °F), 0.007 MPa (1 psi), for 4 h.

Powder particles are angular in shape (Fig. 3) and generally show an increase in oxygen content of 700 to 800 ppm over starting stock. Powder may also contain contaminants such as iron (from the grinding equipment), paint chips, rust flakes, soil, dust, and lint, which make them un-

Fig. 3 Scanning electron micrograph of Ti-6Al-4V powder produced by the hydride-dehydride process

0.2 mm

Fig. 4 Schematic representation of the plasma rotating electrode process

(a) Principle of the transferred arc plasma gun. (b) Design of long bar PREP machine. (c) Cleaning and processing equipment for PREP titanium powders

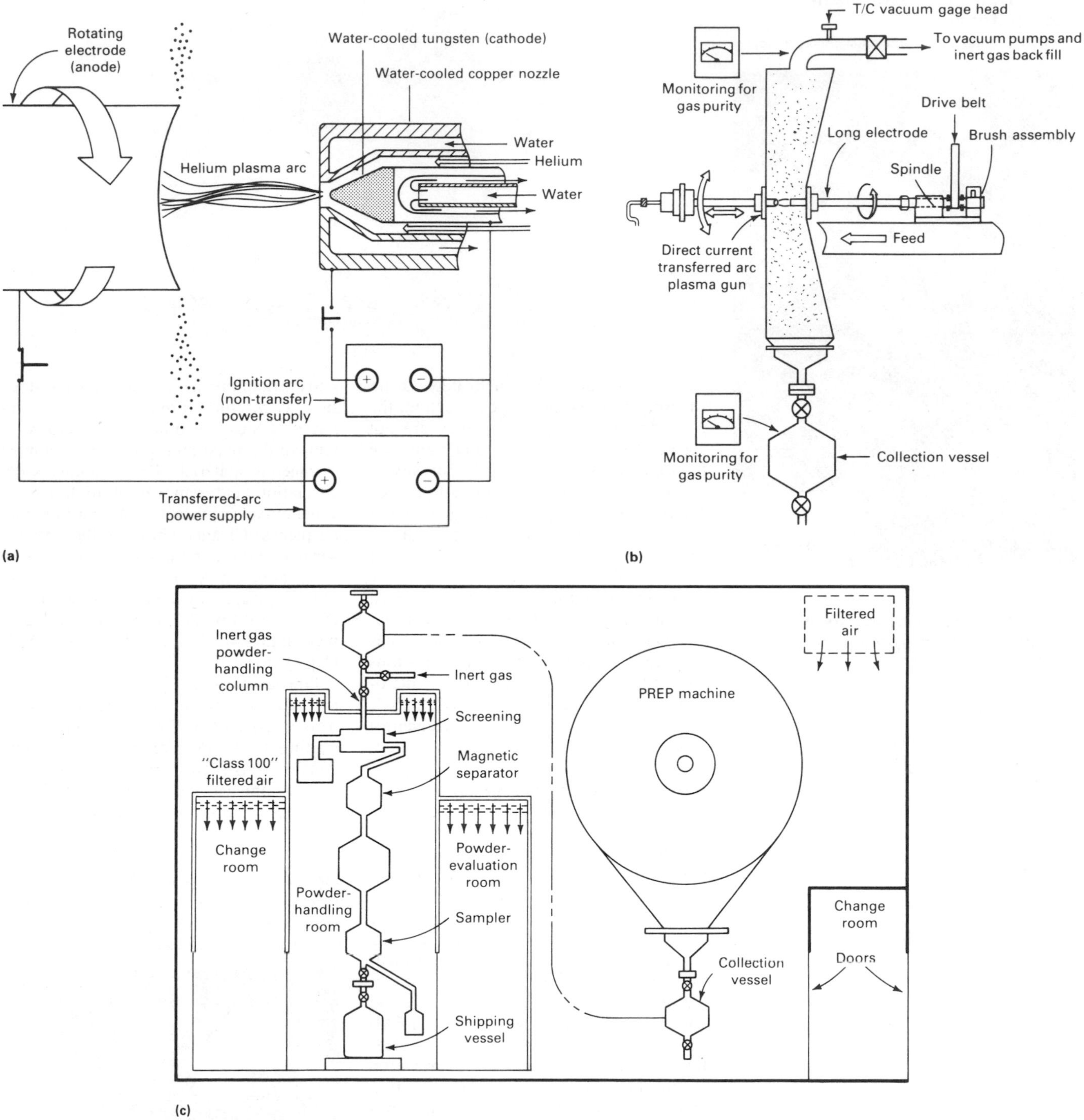

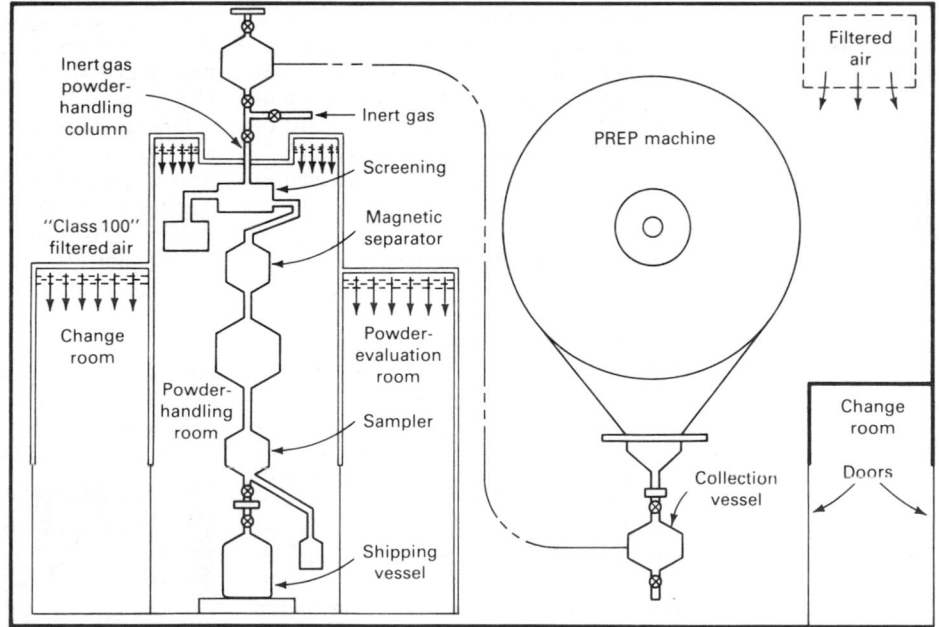

(a)

(b)

(c)

suitable for use in demanding applications such as critical aircraft engine and structural components. The angular shape of the powder particles is a disadvantage for flow and packing. Particle flow is restricted due to bridging of the angular particles, and packing density is lower and somewhat less consistent than with spherical particles. Cold compaction, however, is possible with angular particles. Powder produced by this method is relatively inexpensive.

Traditionally, the hydrided powder is used only after dehydrogenation. Recently, however, it has been demonstrated that starting with powder in the hydrogenated condition provides advantages in processing (lower compaction pressure and/or temperature) and in control of the microstructure for compaction by both hot pressing and hot isostatic pressing. During vacuum hot pressing, the hydrogen is removed during the compaction cycle. In the case of hot isostatic pressing, the hydrogen is retained in the titanium during compaction. After both processes, a final vacuum anneal is required to bring the hydrogen content below specification levels (generally <125 ppm).

Atomization. For production of high-quality, near-net shapes by hot isostatic pressing or other hot pressing processes, spherical prealloyed powder that is free of detrimental foreign particles is required. In contrast to flake or angular particles, spherical powder flows readily, with minimal bridging tendency, and packs to a very consistent density (approximately 65%). These characteristics lead to excellent part-to-part dimensional reproducibility. Because of the high reactivity of titanium in the molten state, powder production from a large mass of liquid metal (gas atomization, for example) has not yet been achieved. Instead, metal is melted locally, ejected radially due to centrifugal forces, and cooled to form spherical particles (centrifugal atomization).

A variety of centrifugal atomization techniques have been evaluated, but currently only PREP, illustrated schematically in Fig. 4, is available commercially. This process has the advantage over the earlier rotating electrode process, in that it produces powder that is not contaminated with tungsten particles from the cathode. A micrograph of Ti-6Al-4V PREP powder is shown in Fig. 5. More information on this process can be found in the article "Atomization" in this Volume.

Other powder-making techniques in the pilot production stage include the

Fig. 5 Scanning electron micrograph of Ti-6Al-4V powder particles produced by PREP

Source: Nuclear Metals, Inc.

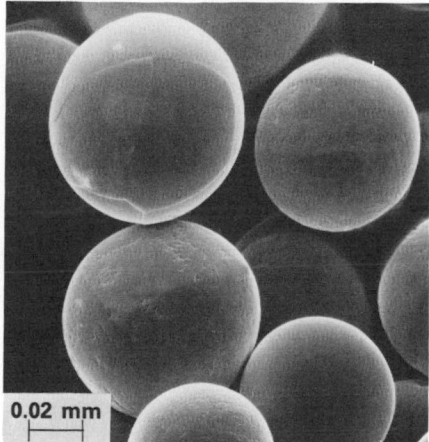

0.02 mm

Leybold-Heraeus electron-beam rotating disc process, the French Atomic Energy Authority powder-under-vacuum process, and the Colt-Crucible process. One concept of the latter process differs somewhat from the other techniques in that the powder is produced by local melting of a hydrogenated billet.

Additionally, melt extraction techniques, such as the pendant drop process (Ref 7), offer a method for potentially producing large quantities of powder. The goal of all these processes is to produce powder that is free of foreign particles, which degrade mechanical properties of the compacted material. Of particular concern are nonmetallics that are present due to poor "housekeeping," either during powder production or subsequent handling.

Developing Processes

Although powder production of titanium alloys by the blended elemental and prealloyed approaches has shown major advances in the past 10 years, a number of new developments are in progress. In the blended elemental approach, the major hurdle to surmount is elimination of the porosity, which degrades initiation-related mechanical properties such as fatigue. Methods of sponge production such as the electrolytic process yield a product that is an order of magnitude lower (to approximately 100 ppm) in chloride content than conventional sponge. However, this level is not sufficiently low to eliminate porosity and enhance fatigue behavior.

Either a process that produces lower chloride sponge or a method for reducing the chloride level of conventional sponge must be developed. Work is proceeding along both lines, but information is proprietary.

In the prealloyed approach, two developmental thrusts are in progress. One is to produce prealloyed powder directly from a beneficiated ore product. A government-funded program (Ref 8) has resulted in production of Ti-6Al-4V powder directly from a mixture of titanium, aluminum, and vanadium chlorides added to a sodium-laden reactor vessel. This vessel is heated to approximately 100 °C (210 °F) and vigorously shaken to initiate the exothermic reaction. This process is known as the Hurd Shaker Process. Work to date has resulted in nearly 100% powder yield with the desired bulk composition. The powder particles show fine beta grains containing a martensitic structure, indicating a fast cooling rate (Fig. 6).

A recent development is the production of unique titanium alloys using rapid

Fig. 6 Photomicrograph of sectioned Ti-6Al-4V powder particle produced directly from mixed chlorides

Note the fine beta grains containing a martensitic structure, indicating a rapid cooling rate.

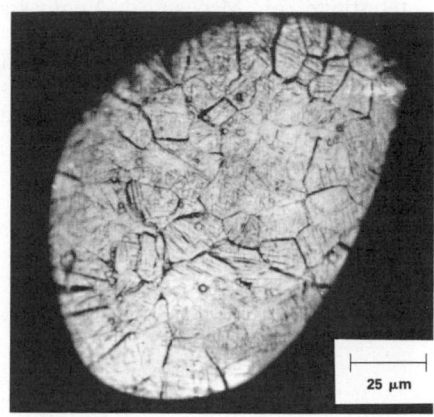

25 μm

solidification rates. Rapid solidification allows extension of solid-solubility limits, extremely fine structures, and production of unique metastable phases.

A government-funded program has demonstrated that a fine dispersion of rare earth dispersoids can be produced in titanium (Ref 9). This is impossible in an ingot metallurgy approach, because the rare earths are virtually insoluble in solid titanium.

To allow this area to be properly exploited, new methods of producing titanium powder must be developed. A large mass of molten metal is required to ensure homogeneity. This must then be finely divided and cooled rapidly either by conduction onto a cooled substrate or by convection (gas atomization directly or in a two-stage centrifugal atomization and forced gas cooling process). Recognizing the potential of rapidly solidified titanium alloys, the government has initiated the development of rapidly solidified titanium powder-making processes of the types discussed above (Ref 10).

REFERENCES

1. Eylon, D., Field, M., Froes, F.H., and Eichelman, G.E., Manufacturing Cost-Affordable High Performance Components for Advanced Air Force Systems, *Society for the Advancement of Materials and Process Equipment Quart.*, Vol 12 (No. 3), 1981, p 19-25

2. Tupper, N.G., Elbaum, J.K., and Burte, H.M., Opportunities for Cost-

Affordable Titanium Aerospace Structures, *Journal of Metals,* Vol 30, 1978, p 7-13

3. Froes, F.H. and Smugeresky, J.E., Ed., *Powder Metallurgy of Titanium Alloys,* Proceedings of the American Institute of Mechanical Engineers Symposium, Las Vegas, Feb 1980

4. Parsons, L., Bruce, J., Lane, J., and Froes, F.H., Titanium Powder Metallurgy Comes of Age, to be published in *Metal Progress,* 1984

5. Kroll, W.J., *Journal of the Electrochemical Society,* Vol 78, 1940, p 35

6. Hunter, M.A., *Journal of the American Chemical Society,* Vol 32, 1910, p 330

7. Maringer, R.E. and Mobley, C.E., Direct Casting of Wire, Filament and Fiber, *Wire Journal,* Jan 1979, p 70-74

8. Dekock, C. and Froes, F.H., work in progress, 1982-1984

9. Sastry, S.M.L., Peng, T.C., Meschter, P.J., and O'Neal, J.E., "Dispersion-Strengthened Powder Metallurgy Titanium Alloys," AFWAL Materials Laboratory Contract F33615-81-C-5011, Dec 1982

10. "Titanium Powder Making and Process Development," AFWAL Materials Laboratory Contract F33615-83-C-5034

Production of Beryllium Powder

By James M. Marder
Supervisor, Beryllium Metal
Research & Development
Brush Wellman Inc.

BERYLLIUM has a high elastic modulus (303.4×10^3 MPa, or 44×10^6 psi) and a low density (1.85 g/cm^3, or 0.067 lb/in.^{3}). It is used extensively in aerospace structures, weight- and inertia-critical structures, precision optics, and nuclear applications. Powder metallurgy is the primary fabrication technology used in the manufacture of beryllium products. This article discusses the techniques used to produce beryllium powder and the resulting material properties.

The production of beryllium powder begins with a vacuum-melted ingot, followed by chipping and mechanical or pneumatic pulverization. The mechanical properties of the final consolidated product depend on grain size and oxygen content, as well as the method of powder production because it affects the crystallographic texture of the billet. Additional information on the properties, consolidation, and applications of beryllium can be found in the articles "Hot Pressing" and "P/M Lightweight Metals" in this Volume.

Powder metallurgy is used primarily to produce beryllium mill products, because the casting process does not produce parts with adequate structural integrity and mechanical properties. In making a beryllium structural casting, the degree of superheat that produces sufficiently low molten metal viscosity required for filling a mold cavity of any complexity also results in a very large as-cast grain size.

Excessive grain size decreases strength and ductility. High superheat conditions may cause an excessive metal-mold reaction; insufficient superheat may result in a lack of mold fill, excessive porosity, or cold shuts. In addition, the reactivity and viscosity of molten beryllium exaggerate problems encountered when using conventional casting materials.

The ingot/wrought billet manufacturing sequence also poses serious manufacturing difficulties. Columnar grain growth, which results from low-temperature ingot pouring, causes exaggerated planes of weakness at the impingement of the sets of columnar grains. Subsequent mechanical breakdown operations, therefore, result in cracking along the weak planes. Ultrasonic vibration (Ref 1) and the use of inoculants (Ref 2) are ineffective grain-refinement techniques. Zero-gravity experiments aboard Spacelab using beryllia and alumina as inoculants have also been ineffective (Ref 3). Powder metallurgy, therefore, is the only viable technique for manufacturing strong, homogeneous billets.

Physical Properties

Beryllium has a close-packed hexagonal crystal structure, with $a = 2.286$ Å and $c = 3.584$ Å and a c/a ratio of 1 to 1.56. This value is low compared to other metals, such as titanium (1 to 1.58). Partly as a result of this c/a ratio, slip occurs almost exclusively on basal planes, with a burger's vector of $b = \frac{1}{3}\langle11\bar{2}0\rangle$. Slip also can occur—with much greater difficulty, however—on first-order prism planes ($b = \frac{1}{3}\langle11\bar{2}0\rangle$), and with even more difficulty on second-order pyramidal planes ($\{11\bar{2}2\}$ with $b = \frac{1}{3}\langle11\bar{2}3\rangle$) (Ref 4).

Basal slip is the predominant deformation mode, however. As with most close-packed hexagonal metals, basal cleavage is the predominant fracture mode. These crystal structure characteristics affect the shape of powder particles and subsequently affect billet properties.

Processing Sequence

Manufacturing a consolidated, fully dense beryllium product involves ingot melting and casting, powder manufacture, and consolidation. The consolidated product usually is a right circular cylinder that is made by vacuum hot pressing. This vacuum hot pressed billet can be readily machined into parts for satellites, aircraft, optical systems, or guidance instruments. Beryllium is particularly well suited to these applications because of its superior strength, modulus, and density. The ductility afforded by P/M processing also facilitates forging, rolling, extruding, and other hot working processes.

Raw Material. Extraction begins with the mining of bertrandite ore, $Be_4Si_2O_7(OH)_2$, and/or beryl ore, $3BeO\text{-}Al_2O_3\text{-}6SiO_2$, from domestic and international sources. The crude beryllium hydroxide product is purified, converted to anhydrous beryllium fluoride, and reduced with magnesium to produce "pebble" or primary beryllium. Residual reduction slag containing these beryllium pebbles is purified by vacuum melting. Vacuum melting uses a magnesium oxide crucible and a graphite mold to produce an ingot weighing 180 to 200 kg (400 to 450 lb). The ingot is "skinned" (machined) to remove the high-carbon material adjacent to the mold wall. Next, a multi-head cutting tool (Fig. 1) is used to reduce the ingot to chips. Standard P/M

Fig. 1 Reduction of beryllium ingot to chips

Fig. 3 Impact grinding system

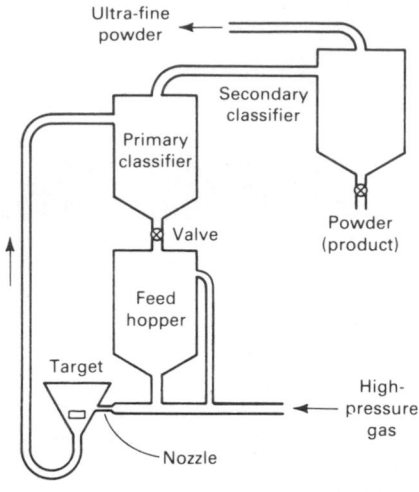

techniques are utilized for powder production at this stage.

Powder Manufacture

Chips produced by the multi-head cutter are relatively coarse and must be ground or comminuted to fine particle sizes before consolidation into a homogeneous, fine-grained billet. Before discussing grinding and comminution, however, it is important to recognize the effects that the close-packed hexagonal structure and chemical reactivity have upon the powder-making processes and the physical and mechanical properties of the final product.

During the late 1960's and early 1970's, mechanical attritioning and ball milling were used almost exclusively to reduce machining chips to powder. Mechanical attritioning is a rotary grinding procedure similar to the grinding of corn in a grist mill. Chips enter an area between a stationary and rotating beryllium plate, where they are ground. The particles then exit the mill at its periphery.

This process results in the occurrence of basal plane cleavage. Therefore, the powder particles produced by attritioning tend to be flakes, with large, flat surfaces corresponding to {0002} basal planes. A scanning electron photomicrograph of attritioned powder is shown in Fig. 2. Ball-milled powders have a similar shape. As a result of their flat particle shape and lack of a truly random crystallographic texture, consolidated billets exhibit anisotropy of both physical and mechanical properties. During vibratory die loading for consolidation, the flat faces of many particles align, forming extended volumes with a common crystallographic feature. The basal plane corresponds to the flat faces.

During consolidation—usually uniaxial vacuum hot pressing—the imposed stress causes these extended areas to rotate so that the c-axis poles align parallel to the stress axis. In billets made with attritioned powder, c-axis poles exist with two to three times the frequency in the longitudinal as compared to the transverse direction. The net result is substantial anisotropy of mechanical and physical properties. For example, ductility is substantially reduced in the longitudinal direction.

Impact grinding is used to minimize anisotropic effects. Figure 3 schematically illustrates the impact grinding process. Chips and powder are introduced into a high-pressure gas stream and are impacted against a beryllium target. The high speed of this pulverization process, coupled with relatively low temperature due to adiabatic gas cooling, activates the secondary and tertiary (nonbasal) cleavage systems.

Resultant particles are blocky, rather than flake-like (Fig. 4). Consequently, there is a substantially reduced correlation between crystallographic basal planes and flat powder particle faces, leading to reduced anistropy and an improvement in longitudinal tensile elongation (Ref 5). These effects are readily seen in Table 1, which illustrates the relationship between the

Fig. 2 Scanning electron micrograph of flake-like attritioned powder
Magnification: 270×

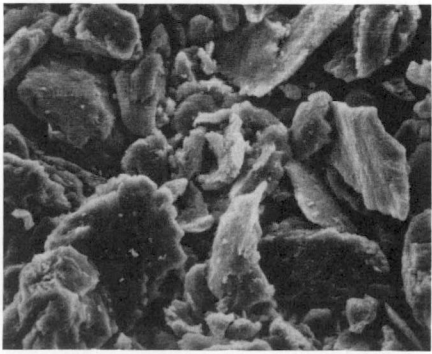

Fig. 4 Scanning electron micrograph of blocky impact ground powder
Magnification: 270×

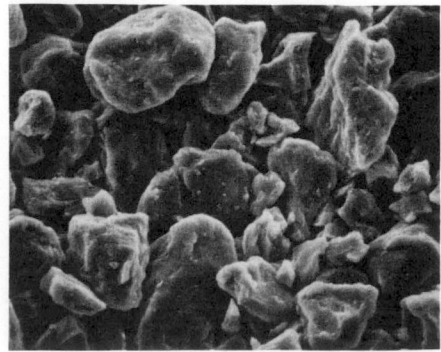

Table 1   Comparison of billets using impact ground and attritioned powder

Ductility and crystallographic orientation effects

Powder lot(a)	Test direction	Tensile elongation, %	Elongation T/L ratio	{0002} pole density T/L ratio
Attritioned powder				
Lot A .... Transverse		4.4	2.93 to 1	2.62 to 1
Longitudinal		1.5	...	...
Lot B .... Transverse		5.0	3.3 to 1	2.5 to 1
Longitudinal		1.6	...	...
Impact ground powder				
Lot A .... Transverse		5.0	1.7 to 1	1.48 to 1
Longitudinal		2.9	...	...
Lot B .... Transverse		5.4	1.5 to 1	1.45 to 1
Longitudinal		3.6	...	...

(a) Powder lots A and B are produced from separate melts.

method of powder comminution, {0002} pole density, and ductility.

The transverse-to-longitudinal (T/L) basal pole density ratio and ductility T/L ratio for attrition milled powder are about 3 to 1. For impact ground powder, these values are approximately 1.6 to 1. The isotropy of the impact-ground powder billet is improved, as is the minimum elongation value in any direction. This demonstrated correlation between crystal texture and ductility is causing attritioning to be gradually replaced by impact grinding.

The relationship of powder particle size to oxide content and grain size also affects billet properties. Beryllium obeys a Hall-Petch relationship, as shown in Fig. 5. As the size of the powder particles decreases, the oxygen content of the powder increases due to the increase in specific surface area. The mean free path between oxide particles also is reduced. Because oxides are a primary grain-boundary pinning agent, fine particle size enhances higher oxide content, finer grain size, and higher strength. The strength imparted by fine grain size and the potential embrittling effect of excessively high oxide content must be balanced against each other.

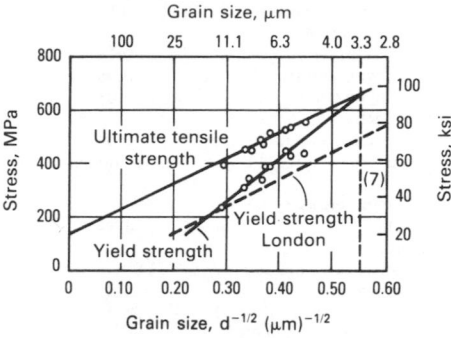

Fig. 5 Hall-Petch diagram of strength versus grain size of vacuum hot pressed beryllium with intermediate purity

----- Yield strength values from Ref 6

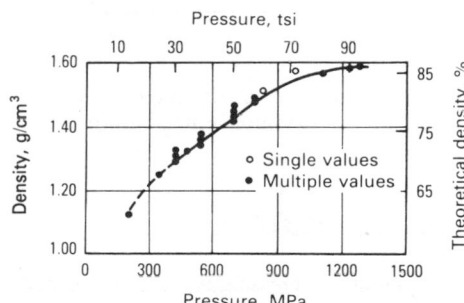

Fig. 6 Green density versus compacting pressure in uniaxially cold pressed beryllium powder

Source: Ref 7

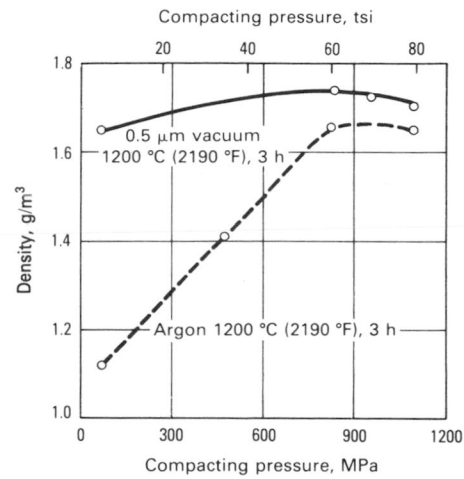

Fig. 7 Effects of compacting pressure on density after sintering in argon and in vacuum

Source: Ref 7

Table 2   Characteristics of powder used for commercial beryllium billets

	Grade			
	SP-65	SP-100E	SP-200E	IP-220
Composition, %				
Beryllium assay, min 99.0		98.5	98.0	98.0
Beryllium oxide, max 1.0		1.2	2.0	2.2
Aluminum, ppm max 600		1400	1600	1000
Carbon 1000		1500	1500	1500
Iron 800		1500	1800	1500
Magnesium 600		800	800	800
Silicon 600		800	800	800
Boron 2		...	...	...
Cadmium 2		...	...	...
Calcium 100		...	...	...
Chromium 100		...	...	...
Cobalt 10		...	...	...
Copper 150		...	...	...
Lead 20		...	...	...
Lithium 3		...	...	...
Manganese 120		...	...	...
Molybdenum 20		...	...	...
Nickel 300		...	...	...
Nitrogen 300		...	...	...
Silver 10		...	...	...
Other metallic impurities(a), ppm max 200		400	400	400
Particle size .... 98% −325 mesh		98% −200 mesh	98% −325 mesh	98% −325 mesh

(a) Each; determined by normal spectrographic methods

Table 3 Properties of beryllium hot pressed block

Property	S-200E	S-200F	S-65A	I-70A	I-220A	I-400
Chemical composition, %						
Beryllium assay (min)	98.0	98.0	98.5	99.0	98.0	94.0
Beryllium oxide (max)	2.0	2.0	1.0	0.7	2.2	4.25 min
Aluminum (max)	0.16	0.16	0.06	0.07	0.10	0.16
Carbon (max)	0.15	0.15	0.12	0.07	0.15	0.25
Iron (max)	0.18	0.18	0.12	0.10	0.15	0.25
Magnesium (max)	0.08	0.08	0.08	0.07	0.08	0.08
Silicon (max)	0.08	0.08	0.06	0.07	0.08	0.08
Other metal impurities (max), %	0.04	0.04	0.04	0.04	0.04	0.10
Powder type	Attritioned	Impact ground	Impact ground	Impact ground	Impact ground	Ball milled
Minimum density, g/cm^3	1.84	1.84	1.84	1.84	1.84	1.84(a)
Theoretical density, %	99	99	99	99.3	99	99
Average grain size (max), μm	25	25	15	20	25	10
Ultimate tensile strength (min), MPa (ksi)	275.8 (40)	275.8 (40)	289.6 (42)	241.3 (35)	310.2 (45)	344.7 (50)
Yield point, 0.2% offset (min), MPa (ksi)	206.8 (30)	206.8 (30)	206.8 (30)	172.3 (25)	241.3 (35)	...
Elongation (min) in 25 mm (1 in.), %	1	2	3	2	2	...

Note: Pressing (billet) sizes can range from 18 to 183 cm (7 to 72 in.) in diameter and 15 to 168 cm (6 to 66 in.) in length, depending on grade and composition.
(a) Special grades available with 9.000 permissible exposure limit (min) or 1.878 g/cm^3 (min)

Characteristics of powder used to produce commercial beryllium billets are given in Table 2.

Consolidation

Vacuum hot pressing is the main method by which most beryllium components are fabricated. Other methods, such as cold pressing and sintering, hot isostatic pressing, and vacuum sintering, have been used successfully to consolidate beryllium powder, but vacuum hot pressing constitutes the highest volume production method.

Vacuum hot pressing typically is carried out at 1000 to 1200 °C (1830 to 3630 °F) and at a compacting pressure of approximately 8.3 MPa (1200 psi) or less. This process yields a product that is essentially 100% dense. Billets up to 183 cm (72 in.) in diameter have been manufactured by this technique. Properties of commercial vacuum hot pressed block are given in Table 3.

A cold pressing/sintering/coining technique has been used successfully for specific applications, such as the production of aircraft brakes, heatsinks, rotors, and stators. Although the relationship of green density to pressure depends strongly on the specific powder used, a typical curve is shown in Fig. 6. After cold pressing to obtain a high green density, sintering is carried out in vacuum. An argon sintering atmosphere severely restricts final density, as shown in Fig. 7.

REFERENCES

1. Polotzki, I.G., Benieva, T.Y., and Khodor, Z.L., Effect of Supersonic Vibrations on the Crystallization Processes, *Akademiha Nauk*, Vol 6, 1953
2. Bibb, A.E. and Bishop, S.M., "Grain Refinement of Cast Beryllium," KAPL-1917, General Electric Co., 1950
3. Gelles, S.H. and Malik, R.K., "Process Development for Producing Fine-Grain Casting in Space," Report NASA-CR-120739, April 1975
4. Aldinger, F., Flow and Fracture of Single Crystals, *Beryllium Science and Technology*, Vol 1, Webster, D. and London, G., Ed., Plenum Press, New York, 1977
5. Paine, R.M. and Stonehouse, A.J., "Investigation into Effects of Microalloying and Thermal Treatment on the Properties of Beryllium," Report BW-TR-549, Brush Wellman, Inc., 1974
6. London, G.J., *et al.*, *Metals Eng. Quart.*, Vol 16 (No. 4), 1976
7. Hausner, H.H. and Pinto, N.P., *Trans. ASM*, Vol 43, 1951, p 1052

Production of Composite Powders

By Basil Meddings
Manager, Analytical Services
Sherritt Gordon Mines Ltd.

A COMPOSITE POWDER, according to the definition provided by the International Standards Organization (ISO 3252), is a powder in which each particle consists of two or more different materials. In order to exclude alloy and oxide-dispersion-strengthened powders from being classified as composite powders, it is necessary to expand the ISO definition by stating that the two or more different materials or components must be present within a particle at sizes that are large enough (typically greater than 0.5 μm) to exhibit the macroscopic properties of the individual components.

The chemistry, or the ratio of the components, from particle to particle in a composite powder may vary considerably, depending on the manufacturing method. Some persons prefer to restrict the term composite powders to powders that possess essentially uniform chemical composition from particle to particle; others prefer to include broader variations.

Some of the reasons for using composite powders rather than mixtures of powders that have been produced by blending include significant improvements in powder flowability, powder homogeneity (lack of segregation of components), and corrosion resistance.

Composite powders with a very high degree of homogeneity are made by spray drying solutions of salts. This process allows good control over particle size and particle shape (spherical). It is used for the manufacture of complex compositions with excellent control of powder flowability (see the article "Spray Drying of Metal Powders" in this Volume).

Thermal agglomeration of powder blends is an important process, yielding so-called diffusion-bonded powders. Examples include iron-copper and iron-nickel-copper-molybdenum powders. The individual composite particles typically are agglomerates of the individual components. Substantial chemistry variability from particle to particle can be tolerated, because the main purpose of this process is to eliminate gross segregation of the components.

Cobalt-covered tungsten carbide is produced by ball milling together tungsten carbide and cobalt powders. With sufficiently long milling, each individual tungsten carbide particle becomes coated with cobalt (see the article "Cemented Carbides" in this Volume).

This article describes the production of nickel and cobalt composite powders by hydrometallurgical processing. In this process, nickel and cobalt are recovered from sulfide concentrates. Leaching under oxidizing conditions in ammoniacal solution yields ammine complexes of these metals, from which nickel and cobalt are recovered as metal powder via hydrogen reduction cycles (see the articles "Production of Nickel and Nickel Alloy Powders" and "Production of Cobalt and Cobalt Alloy Powders" in this Volume).

Because the hydrogen reduction requires nucleation sites, it is possible to use the process to prepare composite powder in which a metallic or nonmetallic inner core is coated with nickel or cobalt. If such powders are consolidated, the continuous matrix of the consolidated part is made up of the coating component, while the core component represents the dispersed phase. This is true even if the coating is very thin in comparison to the core. With such composite materials, many physical and technological properties are entirely different from those obtained by consolidating a mixture of the two components.

Production of Nickel-Coated Composite Powders

Nickel ammonium sulfate solutions can be reduced to metallic nickel powders by processing in hydrogen at 450 psig (3.1 MPa) at 175 to 205 °C (350 to 400 °F). This nickel reduction process is used to produce hydrometallurgical nickel. For additional processing information, see the article "Production of Nickel and Nickel Alloy Powders" in this Volume. Nickel reduction is a heterogeneous reaction that requires a solid surface on which the nickel can precipitate. In the production of nickel, this solid surface is provided in the form of nickel powder inoculant.

During processing, the nickel inoculant is replaced by a foreign, or core, powder. Under hydrogen reduction conditions, for-

Fig. 1 Photomicrograph of a composite powder consisting of 82%Ni/18%Al
Magnification: 150×. Source: Ref 3

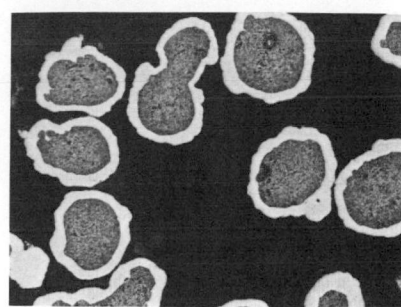

eign powder particles become coated with pure metallic nickel, thereby producing composite powder. Foreign particles must be stable in the nickel ammonium sulfate solution under reducing conditions. Additionally, the size and density of the powder particles must facilitate suspension in the well-agitated solution.

Additionally, to effect precipitation of nickel, the surface of the foreign powder must be active. Several catalysts are available to increase the activity of core powders. The metallurgy of nickel-coated composite powders is described in Ref 1 and 2.

Properties of Composite Powders

Generally, both metallic and nonmetallic core powders with particle sizes ranging from less than 1 μm to greater than 100 μm can be coated. Nickel or cobalt, and sometimes copper, is usually used as a metal coating and ranges in thickness from 2 to 3 μm. This thickness range allows adequate particle suspension within the solution. Consequently, powders may be produced in which the coating weight comprises 20 to 99 wt%, and core weight represents 1 to 80 wt%.

More than 50 types of composite powders consisting of a core and coating have been produced commercially. Examples of typical commercial powders are shown in Fig. 1 and 2. Such composite powders are inherently heterogeneous, in that each particle consists of a core of one material and a coating of another. The powder as a whole, however, is homogeneous, because no segregation of the two components is possible during transportation, storage, or handling, for example.

The nickel or cobalt coating imparts the compacting and sintering characteristics of nickel or cobalt to the composite powder particle, which is advantageous if the core

Fig. 2 Photomicrograph of a composite powder consisting of 85%Ni/15% graphite

Magnification: 180×. Source: Ref 3

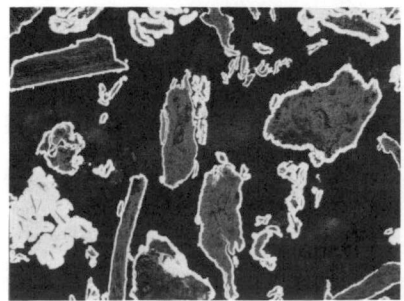

particle is graphite or tungsten carbide. Otherwise, these particles in particular would be difficult to compact. If the core of a composite powder is a reactive material such as aluminum or phosphorus, or a readily oxidizable material such as graphite, the metallic coating protects the core from reaction. This tendency is beneficial in the thermal spraying of nickel-graphite or cobalt/tungsten carbide; without the protective metal coating, these powders would decarburize heavily during thermal spraying (Ref 3).

Selection of core/coating combination also may achieve the joining of two materials uniformly and tightly so that upon heating they react as desired. Examples include nickel-aluminum and nickel-phosphorus, both of which react on heating to form intermetallic compounds that release large heats of formation. A powder heated to 650 °C (1200 °F) rises spontaneously to approximately 3000 °C (5400 °F).

Alloy-Coated Composite Powders. Proprietary diffusion processes can be used to produce alloy-coated composite powders, in which the core particle is coated with an alloy (usually of nickel or cobalt). A standard composite powder is produced that has a monocomponent metal coating. This coating is alloyed by a diffusion process, usually with chromium and/or aluminum, to enhance oxidation resistance for high-temperature applications. Alloying is achieved by pack cementation using a halide carrier.

Nickel-chromium/chromium carbide (Cr_3C_2) and nickel-chromium-aluminum/Bentonite are alloy-coated composite powders. Nickel-chromium/chromium carbide is a thermal spray powder used in wear resistance applications that also require oxidation resistance. The coating-to-core ratio and the nickel-to-chromium ratio of the coating may vary. A typical composition is a 50-to-50 ratio of nickel-chromium coating to chromium carbide core, with a 80-to-20 ratio of nickel to chromium in the alloy coating. Thus, the alloy coating has the basic composition of the well-known nickel-chromium oxidation-resistant alloys that form the binder matrix in a sprayed or sintered composite produced from this powder.

The uniformity of the nickel-chromium coatings is shown in the three electron microprobe scans of a nickel-chromium/diatomite composite powder of Fig. 3. Diatomite is a naturally occurring silicate material, and its presence as the core material is shown in the silicon x-ray image (Fig. 3c). The uniformity of the nickel-chromium alloy coating is depicted in the respective nickel and chromium x-ray im-

ages (Fig. 3a and 3b). Aluminum also can be added by a diffusion process to the nickel-chromium alloy coating. The resulting alloy-coated composite powder provides superior oxidation resistance in high-temperature dynamic environments.

Applications

Composite powders are used extensively in thermal spraying, such as flame,

Fig. 3 Electron microprobe scans of nickel-chromium/diatomite alloy-coated composite powder

(a) Nickel x-ray image. (b) Chromium x-ray image. (c) Silicon x-ray image. Magnification: 180×

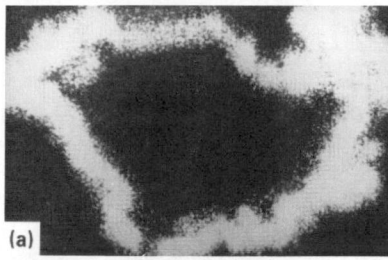

(a)

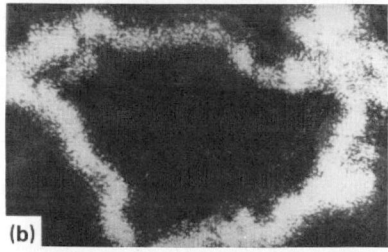

(b)

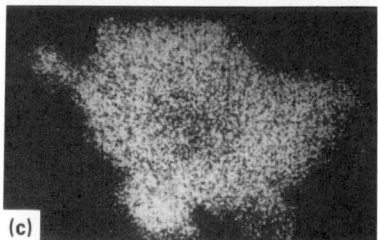

(c)

Fig. 4 Nickel/graphite composite powder

Magnification: 150×

plasma, or detonation gun processes, because they provide good metallic bonding, protected core particles, specific required compositions, and lack of segregation. Because of the wide variety of available spraying processes, deposits can be as large as 2.5 mm (0.100 in.) in thickness, with densities ranging from 70 to more than 95%.

The largest use of composite powders consisting of a core and coating is for the production of nickel/graphite thermal-sprayed abradable seals for clearance control of aircraft turbine engines. This application uses composite powders that are similar to those shown in cross section in Fig. 2 and Fig. 4. The abradable seal maintains a close clearance between rotating and stationary components in the turbine under expansion, creep, or distortion conditions encountered during engine operation. This improved seal results in increased engine efficiency and reduced fuel consumption. A successful abradable seal must exhibit:

- Abradability
- Erosion resistance
- Corrosion resistance
- Compatibility with other engine materials
- Easy repair
- Low cost

Up to 500 °C (930 °F), a thermal-sprayed deposit of nickel/graphite is suitable. For engine temperatures above 500 °C (930 °F), alloy-coated composite powders are required. Typically, more refractory core materials and oxidation-resistant alloy coatings are used, as described previously.

Because of the advantage of joining two diverse materials uniformly by the core/coating combination, composite powders are used in many tribologic applications—from hard wear-resistant materials to low-friction bearing materials. In applications requiring wear resistance, composite powders such as cobalt/tungsten carbide are used to coat cutting edges, knife blades, and rubbing interfaces, for example. For low-friction applications, several unique core/coating combinations have been developed, such as copper/graphite for sliding electrical contacts and nickel oxide/calcium fluoride for a rubbing seal against a ceramic heat exchanger.

The nickel/aluminum powder in Fig. 1 is the product of combining two reactive materials in a composite powder. This powder is used as a thermal spray undercoat, because of the high bond strength that is derived from the exothermic reaction in the formation of nickel aluminide. A review of composite powders in thermal spray applications is given in Ref 3.

REFERENCES

1. Meddings, B., Lund, J.A., and Mackiw, V.N., The Metallurgy of Nickel-Coated Composite Powders, *The Canadian Mining and Metallurgical Bulletin*, July 1963

2. Kunda, W., "New Developments in the Preparation of Composite Powders," 7th Plansee Seminar, Reutte, Austria, June 1971; published in *High Temperatures—High Pressures*, Vol 3, 1971, p 593-612

3. Clegg, M.A., Silins, V., and Evans, D.J.I., "Composite Powders in Thermal Spray Applications," 7th International Metal Spraying Conference, London, Sept 1973

Screening of Metal Powders

By Albert J. Mastrangelo
Manager, Plant Laboratory
SCM Metal Products

MOST METAL POWDER production processes result in powders with relatively broad particle size distributions. Because many powder properties (apparent density, flow, compacting properties, sintering rate, dimensional change, and mechanical strength) depend strongly on the particle size distribution of a powder, most powder applications have size distribution specifications. In the P/M industry, the most widely used method for controlling particle size distribution is screening.

Screening is a mechanical method of separating a mixture of metal powder particles into two or more components by means of a screening surface. The screening surface acts as a multiple "go/no-go" gage, allowing particles smaller than the surface opening to pass through while retaining large particles on the screening surface. The net result is a physical separation based on particle size. Each resulting component is termed a fraction of the original mixture, and is more uniform in size.

The basic process of screening consists of passing the original material or feed over a screen with openings of a definite size. Particles passing through the screening surface are termed undersized, and particles retained on the screening surface are the oversized fraction.

Screen Surfaces

Screens are made of many materials, including woven wire, silk, or plastic cloth; perforated or punched plate; grizzly bars (devices used for screening ores or powders, which consist of a number of parallel bars over which the powders or crushed ores are passed); or wedge wire sectors. The latter three types are normally used to separate very coarse mixtures (≥12.7 mm or 0.5 in.) and have little application in powder metallurgy. The screen openings are usually square, although other shapes such as rectangular, slotted, and parallel rod decks may be used.

Wire cloth generally is specified by "mesh," which is the number of openings per linear inch counting from the center of any wire to a location exactly 1 in. (25 mm) distant from that point, or by an opening specified in inches or millimetres which designates the clear opening or space between the wires. Mesh is generally the preferred term for cloth 2 mesh or finer, while "clear openings" is the term for an opening ≥12.7 mm (0.5 in.) or larger. The screen openings or apertures are defined as the minimum clear space (in inches or millimetres) between edges of the opening in the screening surface. Table 1 lists typical screen (sieve) sizes and related particle sizes. More detailed information on determining the actual size of powders by the screening of metal particles in standardized screens or sieves can be found in the article "Sieve Analysis" in this Volume.

Square mesh cloth is the conventional type of screen cloth, but oblong weave types are available. Oblong weave construction provides a greater open area and

Table 1   Typical screen (sieve) sizes and related particle sizes

Sieve designation Mesh No.	μm	U.S. standard Sieve opening in.	mm	Tyler Standard Mesh No.
80	177	0.0070	0.177	80
100	149	0.0059	0.149	100
120	125	0.0049	0.125	115
140	105	0.0041	0.105	150
170	88	0.0035	0.088	170
200	74	0.0029	0.074	200
230	63	0.0024	0.063	250
270	53	0.0021	0.053	270
325	44	0.0017	0.044	325
400	37	0.0015	0.037	400

capacity and also enables the use of stronger wire for the same size opening. Woven wire screens are available in a variety of metals and alloys, but materials such as high-carbon steel are preferred for coarse openings because of their abrasion resistance. Phosphor bronze, nickel-copper alloys, and stainless steel are also used because of their corrosion resistance or noncontaminating qualities.

In choosing wire cloth screens, several characteristics should be considered, including sharpness of separation, capacity, freedom from block-up or blinding, and life expectancy. Close control of maximum particle size requires reduction in the effective size openings, with a consequent reduction in capacity. Thorough removal of undersized material often requires an aperture larger than the smallest size material acceptable in the oversize. Screens with rectangular openings offer increased capacity, but sacrifice sharpness when handling rounded or cubical materials. Screen cloth selection is also highly dependent on equipment type, the nature of the screened material, and its morphology. Consultation with equipment vendors and experimentation are usually necessary to arrive at an appropriate balance of factors.

Screens tend to block up or blind, a condition that is unavoidable. Interaction between the particle and the screen can be of three types. If the particle is very large, it does not pass through the screen and tends to move off the screen surface as oversize. If the particle is smaller than the opening, it tends to pass through the opening, thus being undersized. However, if the angle θ (Fig. 1) between the particle and the opening is less than $\tan^{-1}\mu$, where μ is the friction coefficient between the particle and the screen material, then the particle tends to be trapped in the opening. The maximum size of this trapped mate-

Fig. 1 Relationship of particle size to screen opening for blinding

If $\theta < \tan^{-1}\mu$, where μ is the coefficient of friction between particle and screen, then the maximum size of the trapped particle = 1.1 D.

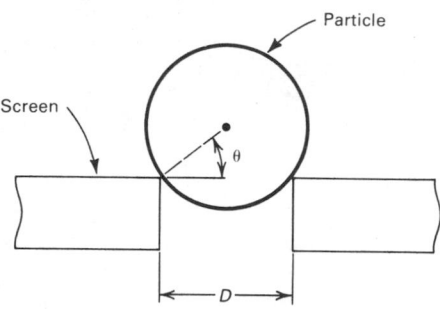

rial is about 1.1 D, where D is the diameter of the screen opening in inches.

An important practical consideration in selecting screening materials is residual or hard blinding, a condition where the screen openings become blocked with particles that are difficult to remove. The degree of hard blinding can be correlated to opening shape and particle shape, both of which affect efficiency. The impact on efficiency is greater with circular or rectangular openings than with square openings. Hard blinding is a progressive condition that results in the need to eventually replace the screen cloth.

Screening Equipment

Selection of equipment is influenced by many factors, including the nature and morphology of the metal powder, process flow sheet, throughput, and efficiency. However, two governing factors are that the width of the screen relates to capacity and the length relates to efficiency.

For powdered metals, screens predominantly are used rather than grizzly bars or trommels, which are normally used for screening materials with larger particles, such as coal or ores. These screens normally range from 100 to 0.05 mm (4 to 0.002 in.). They may be of the vibrating, horizontal, shaking, sifter, centrifugal, or revolving types. Inclined vibrating screens are used in high-capacity operations for separating a wide range of particle sizes. When close sizing is required, a horizontal vibrating screen is applicable. Reciprocating or shaking screens find wide application in size separation and conveying and are useful in sizing larger particles.

There are many types of sifter screens, including circular motion, gyratory motion, and circular vibratory motion types. In general, sifter screens are preferred for the finest size separations, down to 300 mesh. A centrifugal screen consists of a vertical cylinder rotating at a constant speed with a gyratory motion. Gravity moves the oversize particles down the length of the cylinder, as fines are forced through the opening. This type is useful down to 35 mesh. Key variables in screening operations are:

- *Method of feed*: Feed should be evenly spread and designed so as to maximize efficiency and capacity. Materials should approach the screen surface parallel to the longitudinal axis of the screen at a velocity that is as low as practical.
- *Screening surfaces*: Efficient screening results when a single deck is used, as lower decks are not fed uniformly.
- *Angle and slope*: Optimum slope of an inclined vibrating screen is that which will handle the greatest volume of oversize while removing the available undersize.
- *Direction of rotation*: Counterflow rotation yields greater efficiency in circular motion screens, but the screen rotating with the flow will have a greater capacity.
- *Vibration, amplitude, and frequency*: These parameters should be adjusted to convey the material properly and to prevent blinding.

Cleaning of Metal Powders

By L. James Barnard
Materials Development Manager
Kelsey-Hayes Co.

ALLOY PURITY AND POWDER CLEANLINESS are of great importance in the field of fully dense, high-performance powder metallurgy. Typical applications, such as superalloy turbine discs that require very high strength, creep resistance, and low-cycle fatigue resistance at elevated temperatures, have demonstrated the need for low contaminant levels. The most effective means of ensuring the cleanliness and associated high performance of powders and consolidated products is the development and maintenance of powder-making and handling practices that preclude any foreign matter from entering the powder or product. In the absence of a perfect system, however, methods have been developed to clean powders. These powder-cleaning methods improve product properties and reduce the frequency and size of contaminants which serve as fatigue-crack initiation sites.

Sources, types, and the importance of contaminants vary with methods of powder manufacture, handling systems, and product applications. For the nickel-based superalloys used in turbine applications, two general categories of contaminants are important: particulate contaminants (ceramic and organic particles) and gaseous contaminants (adsorbed atomization gases and oxygen, nitrogen, or moisture).

Particulate contaminants can result in reduced fatigue life by creating sites for fatigue-crack initiation either in the contaminant itself or in a zone of reaction between the contaminant and the surrounding matrix. Gaseous contaminants reduce the overall bulk properties of the material by leaving a thin, inherently weak grain-boundary film; by leaving areas of entrapped gas that may be locked into the microstructure under high stress; or in severe cases, by preventing the alloy from

consolidating to full density. The effect of residual gas left on the powder particle and, therefore, inside the consolidated product can be very difficult to detect metallographically.

Particulate Contaminant Removal

Contaminant particles can be categorized as reactive or nonreactive. A reactive contaminant is damaging, particularly as a fatigue-initiation site, because it reacts with the surrounding powder particles or matrix during consolidation and subsequent heat treatment, increasing the effective defect size to many times that of the original contaminating particle.

Many of the particles that concern the P/M turbine component industry are either ceramics introduced during master alloying and subsequent gas atomization or organics introduced during powder handling. Ceramic particles typically form from small amounts of crucible lining, tundishes, and atomization nozzles that erode or react with alloying elements in the melt and remain in the powder either combined with a metal powder particle or isolated as a ceramic particle. Examples are alumina (Al_2O_3), zirconia (ZrO_2), magnesia (MgO), and hafnia (HfO_2). These types of ceramics generally are nonreactive. Organic and some reactive particles usually enter the powder during handling and processing. Sources include bits of O-ring seals, lint from cleaning rags, and plastic particles from storage bottles.

Manufacturers of spherical superalloy powders have made steady progress in reduction of the level of contaminant particles through improved melt practices and powder-handling systems. If further improvements in cleanliness are necessary, the powder can be cleaned by utilizing the

different mass, aerodynamic, and electrostatic properties of the contaminants.

Powder Cleaning. The process developed to clean superalloy powders consists of two stages. First, the superfine powder dust is separated so that the coarse fraction can be fed directly into the second stage, an electrostatic nonmetallic separator (ENS). Figure 1 illustrates a typical powder cleaning flow diagram. The powder is fed at a controlled rate into the superfine particle separator (Fig. 2), which is a vertically rising column of gas or air that separates the coarse fraction of input powder from the very fine constituents. The fine fraction (which also contains poor aerodynamic particles such as splat, flake, and low-density ceramic particles) is carried into the cyclone separator. The discharge from the cyclone separator is screened at 270 mesh (it also may be screened at 325). The +270 particles are rejected, while the −270 particles are retained for reblending later. The coarse particles that fall through this gas elutriator are rerun twice, then fed directly into the ENS.

The ENS consists of a stainless steel rotating drum at ground potential, on which a monolayer of highly charged powder particles falls (Fig. 3). Upon contact with the drum surface, metallic powder particles lose their charge to ground rapidly and fall off the drum into the product bin. The dielectric contaminants, typically ceramics or organics, are slow to lose their charges and remain electrostatically held to the drum, eventually either falling behind the splitter into the middling bin or remaining on the drum surface until they are physically wiped off into the reject bin. Middlings, which are a combination of powder and some contaminants, may be either rerun or rejected. The ENS product is then reblended with the −270 mesh discharge from the cyclone separator to yield

Fig. 1 Powder cleaning flow diagram

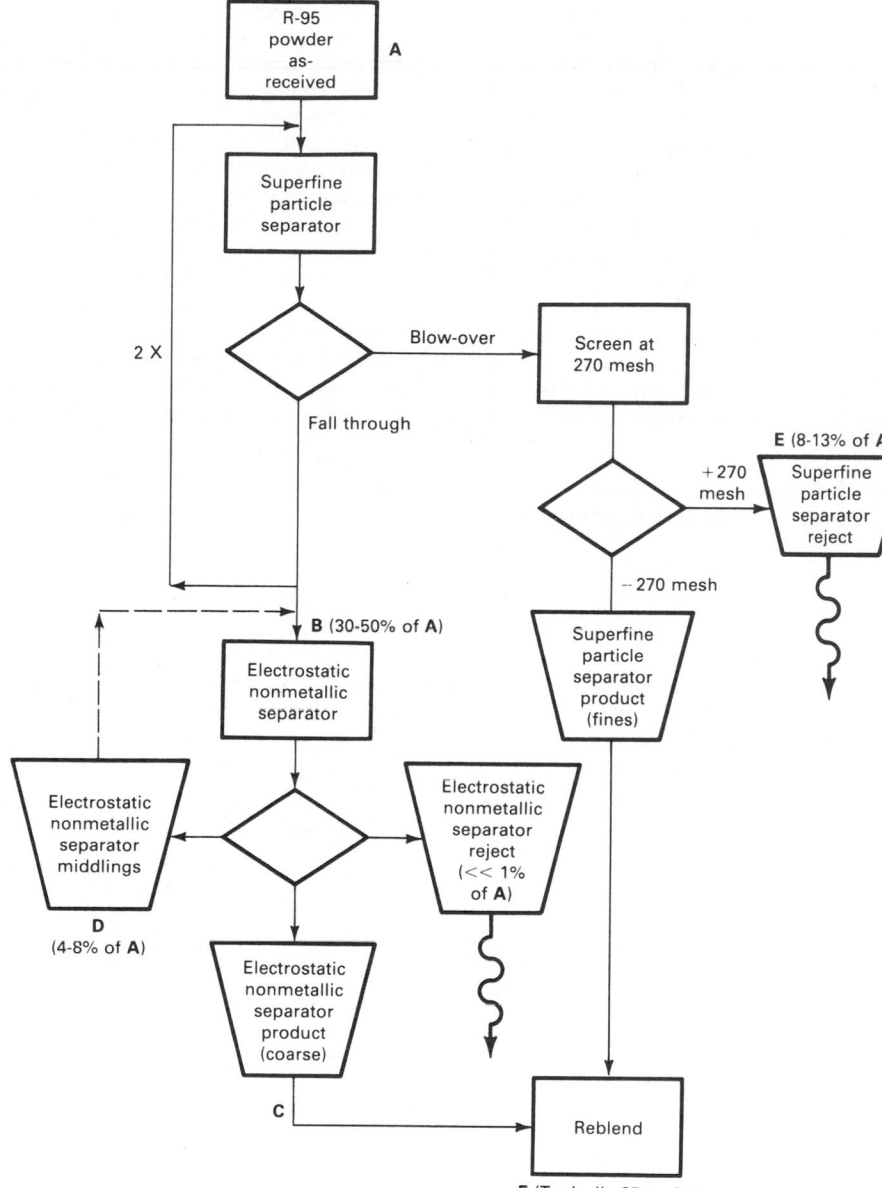

Fig. 2 Superfine particle separator

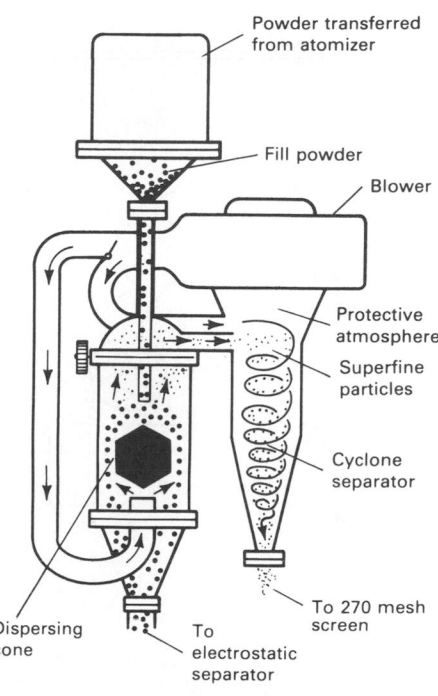

Fig. 3 Electrostatic nonmetallic separator

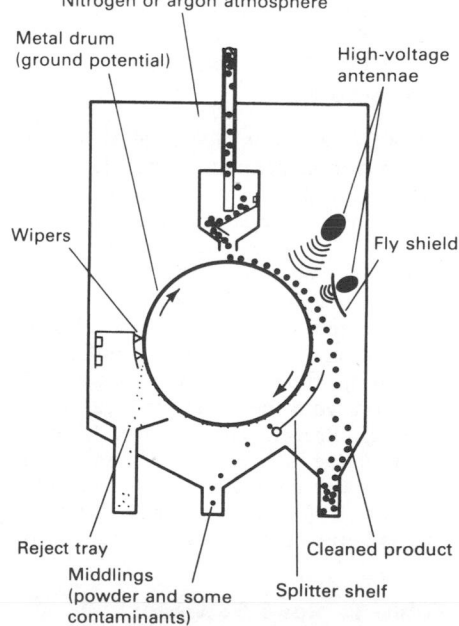

"cleaned" powder that is ready for degassing and placement in containers.

Some types of contaminating particles cannot be removed. For example, a contaminant particle that becomes coated with fine alloy dust or a ceramic particle combined with alloy metal may act much the same as a powder particle and fall into the product bin of the ENS. Also, oxidized powder particles, as well as other contaminants displaying aerodynamic and electrostatic characteristics similar to the base powder, cannot be removed effectively.

Measurement of Particulate Contaminants. The technique used to evaluate powder particulate cleanliness before

and after cleaning is called water elutriation. This method concentrates foreign particles that have a density lower than the powder metal, so that they can be characterized by physical behavior, composition, and relative population.

Figure 4 illustrates the apparatus used for water elutriation. A known amount of powder, typically 0.45 kg (1 lb), is screened into several size fractions: $-100+140$, $-140+170$, $-170+200$, and $-200+230$ mesh. Each size fraction is weighed and placed in water with a wetting agent. The wetted powder then is placed in a vertically rising column of water. The water flow rate of each column

is adjusted according to the particle size, allowing the bulk of the powder to be suspended near the bottom of the column while the lighter particles—such as ceramics, organics, and hollow powder particles—carry to the top overflow and collect at the

Fig. 4 Water elutriation diagram

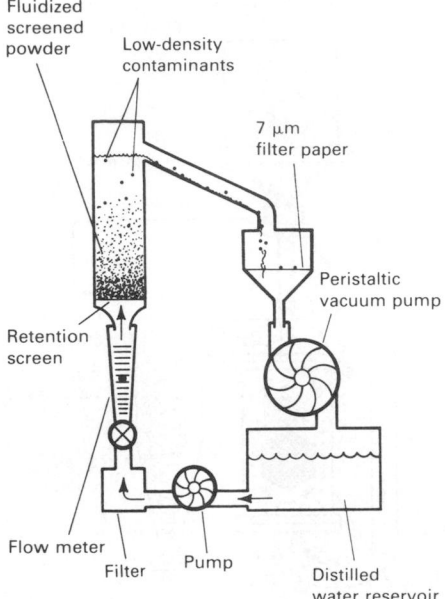

filter. The filters are dried and examined under a low-power microscope for contaminant characteristics such as color, friability or plasticity, and population. Typical water elutriation results for René 95 from three powder suppliers are presented in Table 1.

Gaseous Contaminant Removal

Hot consolidation of superalloy or other high-performance spherical powders, such as titanium alloy powder and high-speed tool steel powders, usually is performed using an evacuated, hermetically sealed container, such as a steel can or glass container, to separate the powder from the pressure-transmitting medium. During atomization and subsequent handling and processing, the powder adsorbs gas, such as argon from inert gas atomization, or oxygen, nitrogen, and moisture from the air. With the high surface area inherent in powders, the gas content can become significant. If allowed to remain in powder that is hermetically sealed during consol-

idation, the gases become part of the consolidated metal product. When heated during consolidation, these entrapped gases can react and form oxide/nitride layers on the particle-boundary surfaces, can remain as porosity, or can be forced into the microstructure under very high pressure.

Any one or all of these effects usually lower the overall performance of the compacted product. Therefore, removal of these adsorbed gases has become a necessary powder preparation practice for consolidation of containerized powder. There are five methods of powder degassing. Figure 5 illustrates conventional degassing systems.

Cold Static Degassing. As shown in Fig. 5(a), this degassing method consists of placing the powder in a suitable vacuum-tight container (such as a consolidation container) and evacuating the powder for a period of time ranging from 1 h for small amounts of powder to 2 or more days for large amounts. The powder usually is evacuated until the vacuum level returns to the minimum system pressure. Degassing is most effective for those powder particles with surfaces that are directly exposed to an open and free path to the evacuation port, such as the top layer of particles in a partially filled container that is evacuated from the top. Gases adsorbed on powder particle surfaces located farther from the top layer of powder have a more difficult escape through the large mass of powder particles. Therefore, powder par-

Table 1 René 95 water elutriation comparisons of as-received versus cleaned powder

Source	Elutriation rate, particles per pound	
	As-received	Cleaned
A	22	9
B	90	2.5
C	55	5

Fig. 5 Degassing methods

(a) Cold static. (b) Cold dynamic. (c) Hot static. (d) Hot dynamic

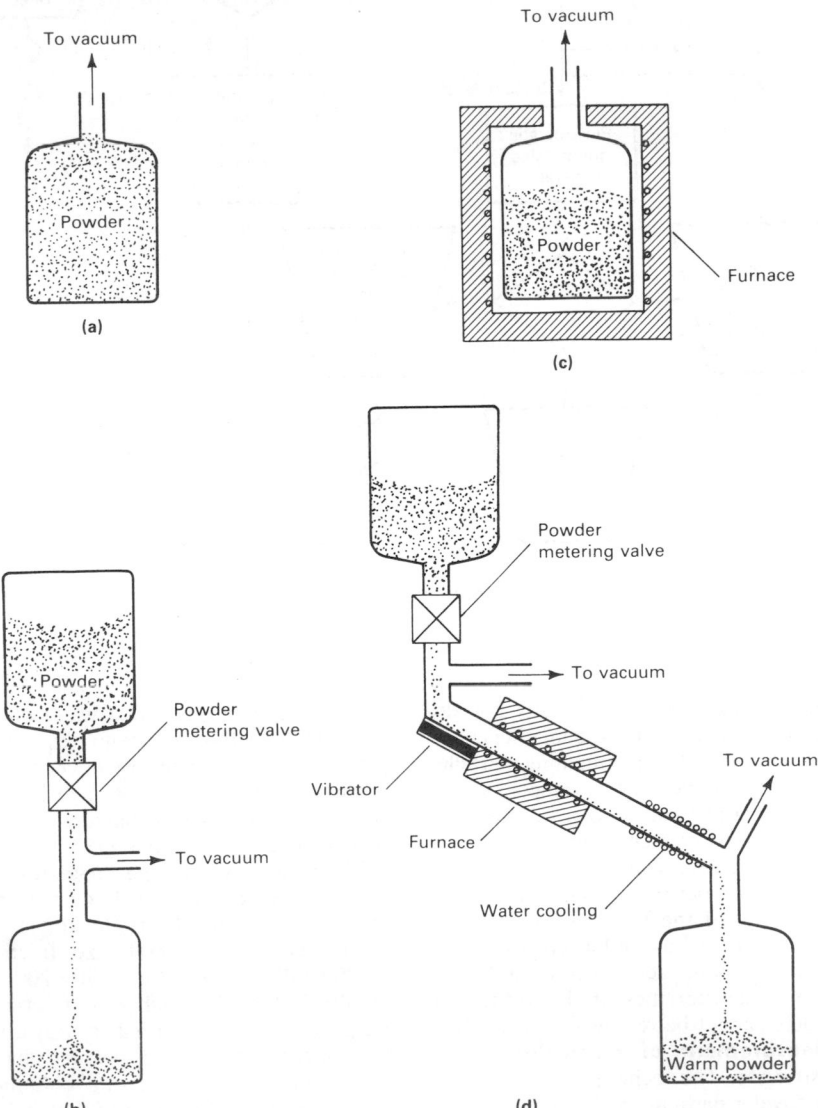

ticles located deep within a large quantity of powder may require long degassing times, possibly several days.

Cold Dynamic Degassing. This method eliminates the mass effect on degassing (described in the previous paragraph) by exposing each powder particle to an open and free path to the vacuum port (Fig. 5b). The path permits gas molecules to escape toward the vacuum pump by transferring the powder from one container to another through an evacuation tee. As the powder flows by the vacuum port, each powder particle surface is exposed equally to the vacuum. This type of degassing is more effective than cold static degassing, but usually does not provide enough time or energy for all the adsorbed gases to escape from the powder particle surfaces.

Hot Static Degassing. Static powder degassing may be enhanced by heating the powder under a vacuum (Fig. 5c). The required temperature and time at temperature depend on many factors, such as method of powder manufacture, powder surface area, powder morphology, and handling or processing history. The degassing temperature and time must be below that at which sintering will take place. Small quantities of typical argon gas atomized, nickel-based superalloy powder that has a particle size distribution of 70% −80+325 mesh and 30% −325 mesh can be satisfactorily degassed at 480 to 540 °C (900 to 1000 °F) for 2 to 4 h. As the powder temperature increases, the vacuum level indicates evolution of adsorbed gases at an increasing rate. Once the powder reaches temperature, and the vacuum level again returns to its minimum level, the degassing, to the extent it can be reasonably achieved, is complete. Then the powder can be cooled under vacuum to room temperature and transferred into a consolidation container.

Hot static degassing has the same basic limitation as cold static degassing: the ability to remove adsorbed gases decreases as the mass and distance from an open and free path to the vacuum port increase. Furthermore, large quantities of powder can take days to degas and days to cool. The process is most commonly used for small quantities of powder or where more elaborate equipment is not available to degas powders that have a residual gas limitation.

Hot Dynamic Degassing. This method combines the application of heat with the exposure of each powder particle to an open and free path to the vacuum port (Fig. 5d).

Powder in an evacuated transfer container is metered into an externally heated inclined plane where the powder flow rate and furnace temperature are controlled so that the actual powder temperature reaches 425 to 540 °C (800 to 1000 °F) during its travel down the incline. The powder is then cooled to less than 93 °C (200 °F), before it is discharged into the receiving transfer container. With this type of unit, the powder is exposed in small quantities to an open vacuum system where any adsorbed gas molecules can be removed from the surface of the powder particles.

If the hot powder is exposed to atmosphere or to various vacuum sealing devices, such as O-rings, the powder particle surfaces can become contaminated. Therefore, before the powder can be transferred into a consolidation container, the powder must cool to nearly room temperature. Cooling times can vary from a few hours to several days, depending on the quantity of powder collected in the receiving container.

Hot dynamic degassing methods are very effective; however, the equipment required is more complex than any other degassing system. The entire system must hold a 10^{-3} torr (or better) vacuum, while metering fine powder into the system, heating the powder, cooling the powder, and finally, discharging it into receiving containers. If any portion of the system does not function properly or does not hold a high vacuum, powder quality can be damaged by oxidation of powder particle surfaces.

Electrodynamic Degassing. A system has been developed that uses electrical energy instead of thermal energy to excite gas molecules to enhance their removal from powder particles. Instead of passing powder through a heated zone, as in hot dynamic degassing, the powder is gravity fed at a controlled rate onto a positively charged gas ionizer (Fig. 6). The powder, radially dispersed by the gas ionizer, continues by gravity toward a bottom transfer container. The adsorbed gas, being in the vicinity of a positively charged ionizer, loses a free electron and becomes a positively charged ion. The gas increases in velocity toward a ground screen, where it can lose its charge and continue in the same direction through the vacuum port.

To further preclude any charged gas ions from following the radially dispersed powder to the bottom transfer container, a magnetic trap is provided around the circumference of the gas ionizer. The powder is ready for immediate vacuum trans-

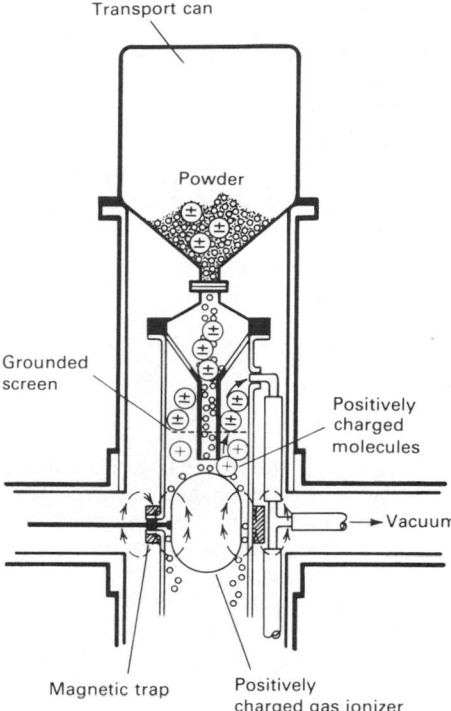

Fig. 6 Electrodynamic degassing

fer into the consolidation container. This system provides ionization of gas at room temperature—equivalent to a thermal excitation of several thousand degrees.

The electrodynamic degassing system is typically used to prepare powders such as nickel-based superalloys that have a low limit on residual gas content. The system can be used virtually anywhere powder has to be transferred under vacuum (for example, from a holding container to the compaction container) and can be used with or without power. Running without power constitutes a cold dynamic degassing system.

Measurement of Gaseous Contaminants. If gaseous contaminants are inert to an alloy, they remain trapped under pressure within the microstructure. A test has been developed to aid in metallographic or pycnometric observation of the gas content within an alloy. A test coupon or part is heated to a temperature greater than the consolidation temperature (typically 93 °C or 200 °F greater) and held for 2 to 4 h. After the part has cooled, sections can be evaluated for a change in density. A decrease of 0.3% or more in density is generally considered unacceptable. The density decrease is referred to as thermally induced porosity.

Annealing of Metal Powders

By William D. Marks
Technical Services Representative
SCM Metal Products

ANNEALING OF METAL POWDERS has become common heat treating practice, because many powder-producing processes yield powders containing residual amounts of carbon, oxygen, and/or nitrogen that reduce green strength and compressibility. Thus, in metal powder production the term "annealing," or the stress relieving and recrystallizing of a cold worked material to increase ductility, includes chemical reactions such as oxygen reduction, decarburization, and denitriding. The term annealing in this article, however, does not apply to the production of metal powder from its oxide.

Annealing also is used to remove thin surface oxide films from powders that tarnished during prolonged storage or exposure to humidity. Use of annealing for this purpose is more prevalent with relatively noble metal powders such as iron and nickel, which are easier to reduce but which frequently oxidize more readily and com-

pletely. Generally, the lower the reduction temperature used to reduce oxide films, the less stable the powder toward reoxidation. Lubrication and/or protective surface treatments improve the stability of powders.

Ferrous Metals

A comparison of the conditions and effects of annealing for various iron powders is shown in Table 1, which provides an overview of the functions of annealing. As shown in Fig. 1, softening of the powder requires the removal of carbon to fairly low levels (0.05% or less). Decarburization generally is accomplished by the reaction of residual oxygen present in the powder with the carbon at elevated temperature. The additional use of a reducing atmosphere such as hydrogen or dissociated ammonia ($N_2 + 3H_2$) supports de-

carburization, lowers the oxygen content further, and prevents reoxidation.

It is sometimes possible to achieve adequate levels of carbon and oxygen in the primary powder process, as is the case with water-atomized high-carbon iron (Table 1). Conversely, a separation of the process into two reduction stages, such as water-atomized high-carbon iron plus mill scale (Table 1), allows the first or main reduction/decarburization reaction to proceed under milder conditions. The economics of this approach are justifiable in that the first-stage material, with relatively high levels of carbon (0.9%), is suitable for the manufacture of welding rods and for cutting and scarfing.

Figures 2 and 3 show polished and etched cross sections of high-carbon (4.2% C) and decarburized (0.5% C) iron particles. The high-carbon material is hard and brittle due to the presence of large amounts of cementite (Fe_3C). It is frequently used in the

Table 1   Function of annealing in various iron powder production processes

Process	Raw material of powder process	Primary process reactions	Annealing			Comments
			Atmosphere temperature	Effect		
Sponge iron (Hoeganaes)	Magnetite; carbon limestone	Reduction at 1260 °C (2300 °F) in self-generated atmosphere of carbon monoxide; crushing of sinter cake	Dissociated ammonia at 870 °C (1600 °F)	Softening, deoxidation, decarburization		Sinter cake is strain hardened from crushing and contains ~1% oxygen and ~0.3% C. Annealing removes all carbon and lowers oxygen to ~0.3%.
Pyron	Mill scale	Reduction with H_2 at 980 °C (1800 °F); crushing of sinter cake	—————No annealing—————			Mill scale(s) initially oxidized from Fe_3O_4 to Fe_2O_3 to generated uniform starting materials. This process does not use annealing, as the sinter cake is very friable and has low carbon (0.01 to 0.05%) and adequate hydrogen loss.
Water-atomized low-carbon iron (A.O. Smith; Hoeganaes)	Low-carbon steel	Water atomization	Hydrogen or dissociated ammonia at 800 to 1000 °C (1470 to 1830 °F)	Softening, deoxidation, decarburization		Combined carbon and oxygen contents of as-atomized powder are less than 1%. After annealing: oxygen is ~0.2%; carbon is ~0.01%.

Table 1 (continued)

Process	Raw material of powder process	Primary process reactions	Annealing		Comments
			Atmosphere temperature	Effect	
Water-atomized high-carbon iron (QMP) . . . . .Pig iron		Water atomization combined with partial oxidation, followed by decarburization in dissociated ammonia at 980 to 1035 °C (1800 to 1900 °F); crushing of sinter cake	——————No annealing——————		Crush atomized/oxidized powder generates own carbon monoxide-reducing atmosphere on heating. Addition of dissociated ammonia lowers carbon from 3 to <0.01%, oxygen from 3 to 9% to <0.2%.
Water-atomized high-carbon iron plus mill scale (Domfer) . . . . .High-carbon (3.5%) steel scrap; mill scale		Water atomization followed by decarburization with mill scale in dissociated ammonia at elevated temperature; crushing of sinter cake	Dissociated ammonia	Softening, deoxidation, decarburization	Crushed atomized powder plus mill scale generate own carbon monoxide-reducing atmosphere on heating, which lowers carbon from ~3.5 to 0.08%, oxygen from >5 to ~0.9%. Annealing produces final carbon content of <0.04%, and oxygen is <0.4%.
Carbonyl iron . . . . .High surface area iron		Vapor phase carbonylation and decomposition	Hydrogen or dissociated ammonia at 800 to 1100 °F	Softening, deoxidation, decarburization	As-decomposed powder typically contains 0.7 to 0.8% carbon and 1 to 2% oxygen. Low-temperature annealing lowers carbon to ~0.02%, oxygen to ~0.15%. Softness and low interstitials are essential for certain magnetic/electronic applications.
Electrolytic iron . . . . .Iron		Electrolysis and milling of chips	Hydrogen or dissociated ammonia	· · ·	As-milled powder typically contains 0.2% C and 0.5 to 1.3% oxygen. Annealed powders typically contain 0.02% C and 0.1 to 0.6% oxygen.

Note: For detailed descriptions of the production of the various iron powders discussed in this table, see the article "Production of Iron Powders" in this Volume.

Fig. 1 Effect of residual carbon content on compressibility and green strength of water-atomized high-carbon iron

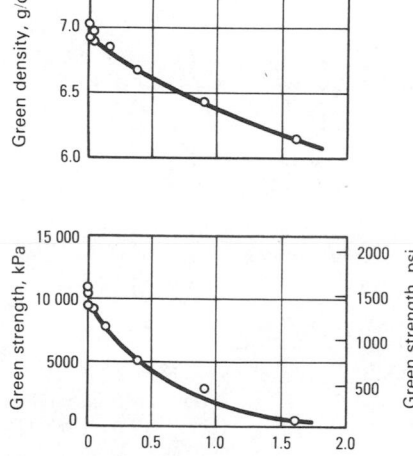

Fig. 2 High-carbon (4.2%) iron (etched)
Note large amount of Fe$_3$C.

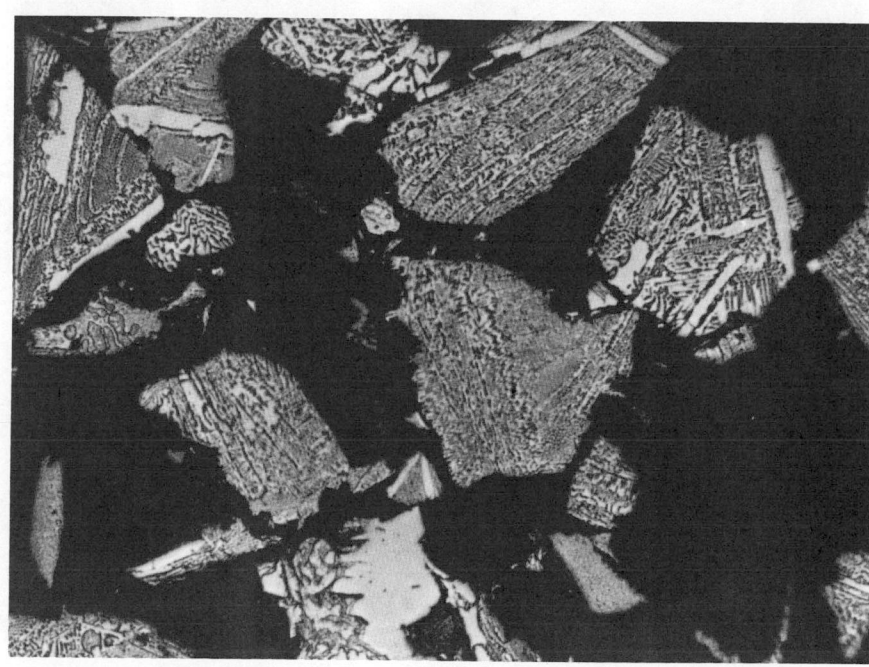

Fig. 3 Annealed (decarburized) iron powder particle showing ferritic grain structure with carbide platelets (etched)

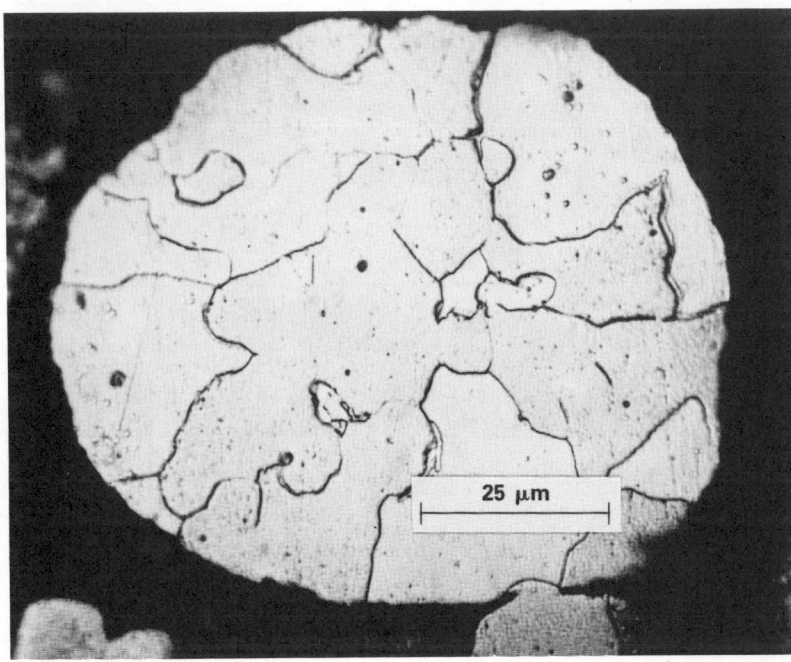

25 µm

Fig. 4 Compactibility of 434-L stainless steel

Annealed versus unannealed condition

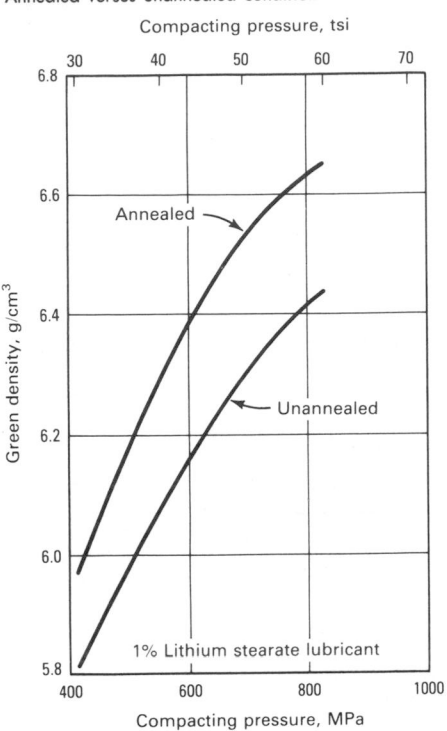

Fig. 5 Scanning electron micrograph of carbide enlargement in T-15 tool steel particles

(a) Before annealing. (b) After annealing

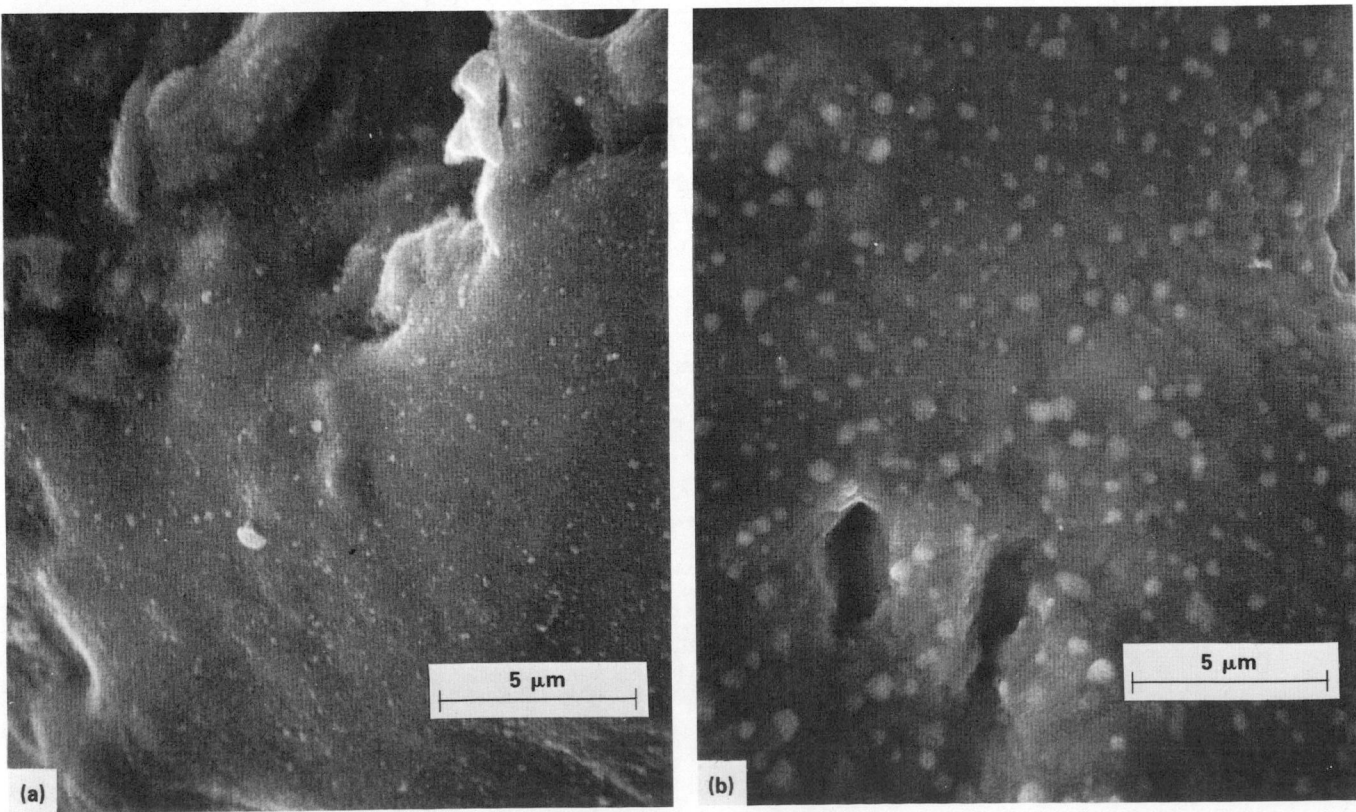

(a)　　　5 µm

(b)　　　5 µm

QMP and Domfer processes (Table 1) because of its lower melting point and, therefore, lower melting temperatures and reduced furnace maintenance costs compared to the low-carbon iron used in the Hoeganaes process (Table 1). Decarburization transforms the hard material into soft and compressible ferrite.

Annealing is used primarily for removing strain hardening introduced during milling in certain grades of electrolytic iron powders (see the article "Milling of Brittle and Ductile Materials" in this Volume).

In all of the powder-producing processes listed in Table 1, annealing conditions are adjusted to obtain the desired decarburization and deoxidation with minimum sintering. Consequently, final milling is mild and produces little work hardening and densification of the powder. More detailed information on the iron powder-producing processes listed in Table 1 can be found in the article "Production of Iron Powders" in this Volume.

Ferrous Alloys

High-speed tool steel powders require an annealing treatment after atomization. Martensitic and ferritic stainless steels sometimes are annealed in either a nitrogen or dry reducing atmosphere such as hydrogen or dissociated ammonia to improve their compacting properties. Annealing temperatures are kept as low as possible to minimize sintering. Austenitic stainless steel powders are generally used in the as-atomized condition. The effect of annealing on the compressibility (green strength) of 434-L ferritic stainless steel is illustrated in Fig. 4.

Water-atomized high-speed tool steel powders are annealed either in vacuum or hydrogen. Martensite and retained austenite are thereby transformed to a mixture of ferrite and spheroidal carbide. This renders the powder softer and more ductile and suitable for compaction.

Generally, the annealing treatment for tool steels combines deoxidation and temper annealing. The particles of as-atomized tool steels are as hard as their fully hardened wrought counterparts, with apparent hardnesses in the range of 63 to 65 HRC, which renders the powder virtually impossible to compact, and oxygen levels of about 2000 to 4000 mL/m^3 (2000 to 4000 ppm). The annealing treatment reduces the hardness to 23 to 25 HRC and oxygen to about 800 to 1000 mL/m^3 (800 to 1000 ppm). A small amount of decarburization is a by-product of annealing.

In the production of T-15 tool steel, annealing improves the compressibility of the powder by drawing dissolved carbon out of the matrix under formation of carbides. Figure 5 shows carbide enlargement on particle surfaces caused by annealing.

Blending and Premixing of Metal Powders

By Russell T. Dotter
Director, Process Laboratories
Patterson-Kelley Co.
Harsco Corp.

BLENDING OF POWDERS is defined as the thorough intermingling of powders of the same nominal composition; premixing is the preparation of a uniform mixture of two or more components. Many properties of powders and sintered parts—such as powder flow, apparent density, ejection stress, delubrication behavior, dimensional change, and mechanical strength—are quite sensitive to even small changes in particle size distribution and to fluctuations in the concentrations of components within a powder mixture.

In many cases, the P/M parts producer is interested in relatively large, uniformly blended powder lots, because costly and time-consuming process adjustments, often of an empirical nature, must be made. Adjustments are often made from lot to lot, even when the various lots are within given specifications.

Lubricated, single-component metal powders comprise the most important two-component systems in powder metallurgy. Important three- and four-component systems include iron-carbon-lubricant, copper-tin-lubricant, iron-copper-carbon-lubricant, and copper-tin-carbon-lubricant. The widespread use of powder mixtures rather than prealloyed powders is attributable to both economic and technical factors: powder mixtures often are less expensive, have better compacting properties, and sometimes permit shorter sintering times.

Blending and Premixing Variables

The degree of blending and premixing should be related to the intended use of the product. As the degree of blending increases toward the theoretically perfect blend, equipment energy use will also increase. Accordingly, equipment operating maintenance, as well as analytical, assay, and other associated costs, will be higher. Thus, the economic importance of defining the degree of uniformity or blending needed for each particular application is obvious.

Most types of mixing equipment achieve their highest degree of particle distribution rather quickly when dealing with dry, free-flowing, noncaking materials. Any extension of the mixing cycle usually results in a plus or minus variation from some acceptable mixing norm. Therefore, it is prudent to determine the minimum processing time needed to achieve the desired result and to terminate processing at that time.

Control over mixing time becomes more important when using higher velocity mixing devices, in which particle fragmentation can introduce severe changes in particle behavior and a tendency toward segregation. Although the determination of minimum or optimum blending is usually done by measuring specific technological powder properties as a function of blending time, a general method based on measuring the uniformity of a blend or mixture follows.

The blending of a powder mixture improves as the number of contacts between the different components of the mixture increases. Figure 1 schematically illustrates various powder dispersions. A truly ordered particle arrangement (Fig. 1a) is a theoretical standard. The agglomerated (Fig. 1b) and the demixed or segregated

Fig. 1 Schematic representation of particle patterns in a powder mixture

(a) Ordered. (b) Agglomerated. (c) Statistical (random) distribution. (d) Demixed or segregated

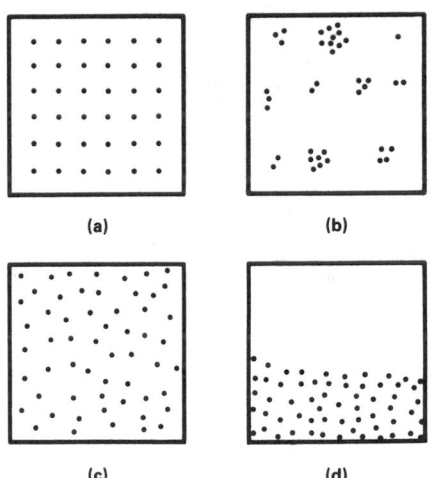

(a)   (b)   (c)   (d)

(Fig. 1d) arrangements represent extreme and undesirable particle dispersions. From a practical point of view, the statistical distribution of the components represents the optimum degree of mixing (Fig. 1c). This does not necessarily mean that a statistical distribution is best or even attainable for a given application, but it provides a reference mark against which the actual distribution can be measured.

Thus, the quality of a mixture can be described quantitatively by comparing the actual distribution of the components with that of a statistical mixture. An example of such a comparison, in terms of the number of contacts between identical par-

Fig. 2 Effect of stabilizer on iron-iron contact formation in binary iron-copper system

Lower curve represents theoretical random mixture.

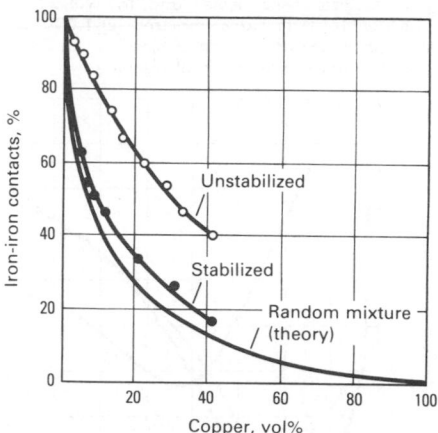

ticles in the binary iron-copper system, is shown in Fig. 2. In this sample the addition of a stabilizer minimizes segregation of components and helps produce a near-random (bottom curve) mixture.

Statistical Analysis of the Blend. Analytical determination of the standard deviation of a mixture is described as a statistical measure of how much the concentration of one constituent in a sample varies from the true concentration that characterizes the entire blend. For ideal conditions, the standard deviation for a random mixture can be calculated if the mechanism of mixing is known. Thus, formulas for the calculation of standard deviation for the random mixture for certain cases of diffusional (Ref 1), convective (Ref 2), and shear mixing (Ref 3) exist. A statistical mixture of a typical two-component P/M premix depends on the component mass ratio, the average component masses, the standard deviations of the component ratios, the standard deviations of the component particle mass distributions, and the sample size (Ref 4). Further details on statistical techniques are found in Ref 5 and 6.

The quality of a mixture may be assessed in terms of its uniformity, either on a macroscale or a microscale. The results of structure-property relationships, often of an empirical nature, usually determine which degree of uniformity must be attained in a given application. When evaluation of the blend or premix is desired, thief probes using spaced compartments are used to obtain multiple samples throughout the cross section of the batch, whether in a mixer, drum, or other storage container.

Sample size should be related to the

method of quality determination. Ideally, the closer the sample size approaches the actual quantity used in analysis, the more accurate and consistent the results will be, because sample segregation will have minimal effect. Subdivisions of samples can yield nonrepresentative results, particularly when the materials are prone to segregate. To eliminate or reduce the effect of segregation related to particle size, reduction of large particle sizes by grinding or milling the sample is recommended. The number of samples taken should be determined by the application of the mix. Other factors such as analysis methods, analysis costs, and how the first few samples analyze to the expected assay ultimately dictate how many samples are required (Ref 7).

If the percentage of one component in a batch of blend or premix is X, and the percentage found in a spot sample is X_i, then the difference is $(X - X_i)$, which can be a plus or minus value. The average of all the X_i readings could approach the value of X if enough samples were checked; the average of all the plus or minus variations of $(X - X_i)$ would approach 0. To obtain a measure of the overall sample—the significance of all the spot samples—the statistical variance and standard deviation are computed. This involves averaging the squares of the individual sample errors, as illustrated in the simple example that follows. The following data represent analytical results of six spot samples (Ref 7) taken from a batch of mixed powder. The percentage of ingredient A in batch:

$$X = 40$$

Percentages measured in the spot samples:

$$X_1 = 41 \qquad X_4 = 42$$
$$X_2 = 45 \qquad X_5 = 36$$
$$X_3 = 38 \qquad X_6 = 38$$

To compute variance $(X - X_1)^2$:

$$(40 - 41)^2 = 1$$
$$(40 - 45)^2 = 25$$
$$(40 - 38)^2 = 4$$
$$(40 - 42)^2 = 4$$
$$(40 - 36)^2 = 16$$
$$(40 - 38)^2 = \underline{4}$$
$$54$$

$$\frac{54}{6} = 9 \qquad \sqrt{9} = 3$$

The number 9 is the biased variance, and 3% is the biased standard deviation. These

values quantitatively characterize the batch of mixed powder. In mathematical form, it can be expressed:

$$S^2 = \frac{\sum_{i=1}^{n}(x - x_i)^2}{n}$$

where S^2 is biased sample variance; S is biased standard deviation; i is spot sample number; X_i is spot sample measurement; X is actual content of ingredient; and n is number of spot samples.

To compensate for the increased uncertainty entailed using very few spot samples, the unbiased variance is used. The difference is that $n - 1$ is placed in the denominator instead of n. The expression then becomes:

$$S^2 = \frac{\sum_{i=1}^{n}(x - x_i)^2}{n - 1}$$

The unbiased sample variance for the foregoing example is calculated to be:

$$S^2 = \frac{54}{6 - 1} = 10.8$$
$$S = \sqrt{10.8} = 3.3$$

Thus, 3.3% is the unbiased standard deviation. It is a statistical measure of the probable variation, plus or minus, from 40% of ingredient A in the blend, as measured by the sampling operation.

For more information on sampling, see the article "Sampling of Metal Powders" in this Volume.

Effect of Powder Characteristics

The present state of the art of blending and mixing does not permit a calculation or prediction of blending results based on available powder component data. Part of the reason for this is that some of the forces and phenomena operating during blending are not measurable and their relationships to blending are not known.

There is a general agreement in the literature (Ref 7-9) that the primary reason for segregation is the different flow rates caused by powder properties, or velocity gradients created by mixing, handling, transporting, or movement of materials. Segregation mechanisms that interfere with random motion of particles operate in combination, and reduction of industrial problems depends on minimizing the more severe ones (Ref 7). Some of the variables affecting flow, and thus segregation, are:

- Particle shape, size, and size distribution
- Surface roughness
- Surface hardness
- Particle density
- Material density
- Batch volume
- Ratio of components
- Electrostatic phenomena
- Mass flow conditions

Examination of the variables causing segregation indicates that segregation is a material, rather than a mechanical, problem. Campbell and Bauer (Ref 8) point out that segregation also occurs in processing equipment. Therefore, the cure lies in control, adjustment, or change of the variables affecting the powder or its mass behavior.

For fine powders, particularly those able to hold static charges, electrostatic charge phenomena often are at the core of the blending problem. In such cases, charge dissipators (such as water) and surfactants should be investigated first. The effects of particle size, shape, density, and surface characteristics (Ref 10) should then be evaluated if the problem persists.

Particle Size. As a general rule, the greater the uniformity of particle size of the material to be blended or components to be premixed, the greater the possibility for consistent and uniform results. As the particle size range and distribution change, a change in blendability should be expected. Larger particles of the same or of different materials have a strong tendency to segregate during blending or premixing. Elimination of the larger particle sizes, if feasible, is beneficial. Slowing down the blender, or the use of vacuum blending (wherein the atmosphere that supports and fluidizes the fines is absent), or mechanical or thermal agglomeration (as in the so-called diffusion-bonded grades of iron-based powders) also are possible solutions to segregation problems. Spray drying (see the article "Spray Drying of Metal Powders" in this Volume) of multicomponent mixtures is another solution.

Particle Shape. The blendability of powder is strongly influenced by particle shape. Figure 3 shows the variability coefficient—that is, the standard deviation divided by the average value of the measured property—of 90%Fe-10%Cu mixtures for different particle sizes and shapes. For spherical iron, the quality of mixing improves quickly (lower coefficient of variability), but results in demixing as mixing is continued. Consequently, blending should be stopped once a random or near-random distribution has been

Fig. 3 Effect of particle size and shape of components of 90%Fe-10%Cu mixtures on degree of blending

Quality of blending improves as variability coefficient decreases. Particle size and shape for components: (a) Cu, 200 to 300 μm; Fe, <63 μm of spherical particle shape. (b) Cu, 200 to 315 μm; Fe, 100 to 200 μm of spherical particle shape. (c) Cu, 200 to 315 μm; Fe, <63 μm of irregular particle shape. (d) Cu, 200 to 315 μm; Fe, 100 to 200 μm of irregular particle shape

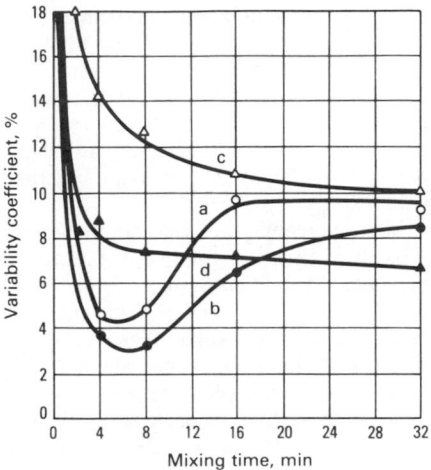

achieved. Demixing or overblending is often caused by the accumulation of electrostatic charges, which often can be dissipated by the addition of a small amount of water. Surfactants can improve the flow of materials. This is illustrated in Fig. 4, which shows the beneficial effect of oleic acid for 90%Fe-10%Al mixtures.

The net effect of both particle size and shape on blendability tends to be the result of a combination of surface conditions (smooth, rough, or jagged) and the basic particle shape.

Density. One of the long-standing problems of blending and premixing has been density differences of the powders being processed. Low-density materials tend to "float" and collect at the top of a mass, while high-density components tend to "sink" to the bottom of the mix. The situation is aggravated if the heavy particles are coarse and the light particles are fines. Size and density differences can balance out, on the other hand, if the fines are denser than the coarse components. This is true of many water-atomized metal powders, in which the coarse fractions have more irregular particle shape, that is, their effective density is lower. For mixtures where small particles tend to coat the larger ones, a good blend usually depends on the uniform size of the small particles and a narrow ratio of the two components (Ref 7).

Fig. 4 Effect of stabilizer on degree of blending of 90%Fe-10%Al mixtures

Particle size for both components: 100 to 200 μm. Stabilizer: benzeneloleic acid. (a) Without stabilizer. (b) With 1.0% stabilizer. (c) 1.0% stabilizer added after 8 min of blending. (d) 3% stabilizer added after 8 min of blending

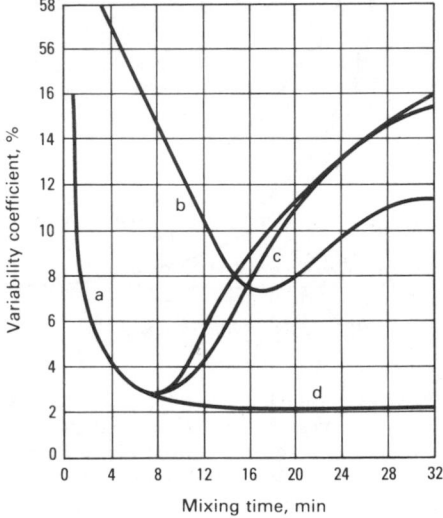

Hausner (Ref 11) has compiled some of the variables affecting loose powder density, as follows:

- Material density
- Particle shape
- Particle density (particle porosity)
- Average particle size
- Particle size distribution
- Average particle shape
- Particle shape distribution
- Specific surface area of the powder
- Oxide films
- Additions, such as lubricants
- Medium surrounding the particles

Hausner has shown the change in apparent and tap densities with differently shaped particles (Tables 1 and 2) and reported on tests made by Kothari (Ref 12) reflecting the apparent and tapped densities as affected by powder particle size (Tables 3 and 4). He has also described the importance of friction conditions in a mass of powders and the effects of particle size and shape on friction, as well as apparent and tap densities.

Table 1 Particle size distribution of the copper powders listed in Table 2

Mesh size	wt%	Mesh size	wt%
+100	1	−200 to +250	10
−100 to +150	5	−250 to +325	30
−150 to +200	35	−325	19

Table 2 Apparent and tap densities of three differently shaped copper powders of similar particle size distribution

Particle shape	Apparent density, $d_{a'}$, g/cm³	Tap density(a), $d_{t'}$, g/cm³	$d_{t'}/d_{a'}$
Spherical	4.5	5.3	1.18
Irregular	2.3	3.1	1.35
Flake	0.4	0.7	1.74

(a) Vibrated in Synchron vibrator, at constant amplitude, for 10 min

Table 3 Apparent and tap densities of tungsten as affected by powder particle size

Particle size, μm	Specific surface, mm²/g	Apparent density, $d_{a'}$, g/cm³	Tap density, $d_{t'}$, g/cm³	$d_{t'}/d_{a'}$
0.5	5480	2.58	4.32	1.68
3-4	1120	3.56	5.64	1.57
14-16	220	3.92	5.10	1.30

Table 4 Apparent and tap densities of three types of iron powders

Particle size, μm	Type of iron powder, wt%		
	Carbonyl	Electrolytic	Reduced
0-10	2.0	6.6	1.8
10-15	59.0	9.2	17.0
15-30	36.0	53.0	41.0
30-44	3.0	31.2	40.2
Apparent density, g/cm³	3.42	2.56	1.89
Tap density, g/cm³	4.09	3.42	2.12
$d_{t'}/d_{a'}$	1.20	1.34	1.12

For more information on particle size and size distribution, particle shape, and density, see the related articles in this Volume.

Equipment for Blending and Premixing

Many different types of machines are available for blending and premixing solids. The high density, abrasive nature, frictional characteristics, and tendency to segregate of metal powders, however, reduce the suitability of some machines. A blender suitable for metal powders should:

- Achieve a maximum blending in minimum time
- Have a gentle mixing action to avoid particle degradation
- Provide repeatability of blending
- Provide for complete batch discharge
- Provide accessibility for ease of cleaning
- Have low power consumption

Fig. 5 Effect of powder fill in mixer on quality of mixing

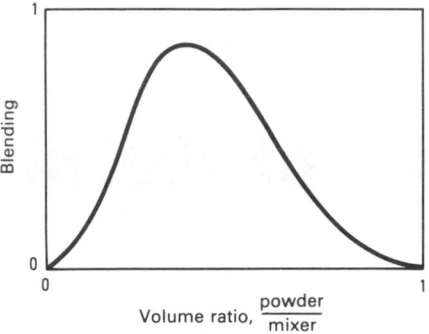

- Require low maintenance
- Provide dust-tight operation

The above list indicates that the use of high-shear paddle- or plough-type devices is inappropriate for metal powders, with few exceptions. Thus, the choice narrows down to tumble-type and low-shear agitated-type machines.

Tumble-type blenders meet the greatest number of the aforementioned criteria, making them the most desirable. Furthermore, scale-up from laboratory size tumblers is very precise. Drum-type, double cone, twin shell, and cross-flow, as well as other tumbler shapes, all produce some quality or degree of blending and also permit straightforward scale-up. Some of these have obvious shortcomings, such as lack of material movement along the axis of rotation, incomplete discharging, and lack of accessibility to the internal surfaces for cleaning. Figures 5 and 6 illustrate the effects of the ratio of total powder volume to mixer volume and the effect of baffle

Fig. 6 Effect of baffles on quality of mixing in cylindrical mixer

Radius of cylindrical mixer: R. Rotational speed: 25 rpm. (a) No baffle plates. (b) Height of baffle plate: 0.66R. (c) Height of baffle plates: 0.5R. (d) Height of baffle plates: 0.33R

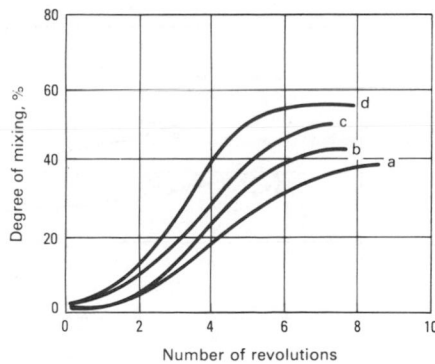

plates in cylindrical mixers on mixing quality.

Low-shear agitated-type blenders, using ribbons, slow-speed paddles, screw-type augers, and other means of moving components in a stationary vessel, are inferior to tumblers. Excessive power consumption resulting from driving the submerged mixing device through the mass of dense metal powders, grinding action caused by the relative motion between the moving and stationary parts, excessive wear of the agitator in an abrasive medium, and lack of precision mixing capability are the main drawbacks. Also, ribbon blender effectiveness does not scale up in direct proportion to laboratory models.

The ideal mixer has not yet been invented. Material variables have such an overwhelming influence on the results obtained from any mixing operation that it is mandatory to sample and analyze to arrive at the best combination of materials and equipment for each application.

REFERENCES

1. Hogg, R., *Chem. Eng. Sci.*, Vol 21 (No. 11), 1966, p 1025
2. Harris, C.C., *Powder Technol.*, Elsevier Sequoia S-A, Vol 12 (No. 85), 1975
3. Brouthman, A., Wollan, G.N., and Feldman, S.M., *Chem. Metall. Eng.*, Vol 52 (No. 4), 1945, p 102
4. *Pulvermetallurgie Sinter und Verbundwerkstoffe*, W. Schatt, Ed., VEB Deutscher Verlag fur Grundstoffindustrie, Leipzig, 1979
5. Volk, W., Industrial Statistics, *Chem. Eng.*, March 1956
6. Weidenbraum, S.S., Mixing of Solids, in *Advances in Chemical Engineering*, Academic Press, New York, 1958
7. Fisher, J.J., Solid-Blending, *Chem. Eng.*, 1960, p 107-128
8. Campbell, H. and Bauer, W.C., Cause and Cure of Demixing in Solids-Solids Mixers, *Chem. Eng.*, 12 Sept 1966, p 179-185
9. Donald, M.B. and Roseman, B., *Chem. Eng.*, Vol 7, 1962, Part I, p 749, Part II, p 823, Part III, p 992
10. Fisher, J.J., Liquid-Solids Blending, *Chem. Eng.*, 5 Feb 1962
11. Hausner, H.H., Friction Conditions in a Mass of Metal Powder, *Int. J. Powder Metall.*, Vol 3 (No. 4), 1967, p 7-13
12. Kothari, Credit by Hausner in Friction Conditions in a Mass of Metal Powders, *Int. J. Powder Metall.*, Vol 3 (No. 4), 1967, p 7-13

Lubrication of Metal Powders

By C.B. Thompson
Copper Products Manager
SCM Metal Products

LUBRICATION is essential to reduce friction between the pressed compact and the rigid tool components when compacting metal powders in steel or carbide tooling. Extremely high part ejection pressures and, ultimately, seizing and excessive tool wear result without proper lubrication. In addition, the part surface quality is detrimentally affected. Although much work has been done with systems where the tooling itself is lubricated, these systems generally have not proven effective in high-volume P/M parts production. Parts produced by isostatic compaction do not require lubrication, due to the absence of high frictional forces during this type of processing.

In the majority of P/M applications, the lubricant is premixed directly with the metal powder. Stearic acid, zinc stearate, lithium stearate, and synthetic waxes, such as Acrawax, are the most popular lubricants. Lubricant selection is based primarily on the ability of the lubricant to adhere to metal particle surfaces. The amount of lubricant added depends on many factors, including composition of the metal powder, type of tooling, compacted density, and complexity of the part. Amounts generally vary from 0.5 to 1.5 wt%.

Ferrous Materials

Although lubricants are a necessary addition to compacting metal powders, they can have some critical, and oftentimes deleterious, effects on the ultimate premix. Hoeganaes Corporation (Ref 1) has made a comprehensive study of lubricants and their influence on the properties of iron premixes. Lubricant properties, such as bulk density and particle size, are very important; for example, high-bulk density and large particle size provide good flow and low stripping pressure, as illustrated in Fig. 1 and 2. To minimize premix segregation, the particle size of the lubricant must be smaller than the size of largest particle of iron powder; for example, a 150- to 200-mesh (104- to 75-μm) lubricant should be used with 100-mesh (150-μm) iron powder.

The melting point of the lubricant must be high enough to prevent melting or softening from heat developed during mixing. If this occurs, flow deteriorates, apparent density decreases, and the mix tends to agglomerate. Higher lubricant content yields poorer flow, lower apparent density, and lower stripping pressure. The effect of increasing amounts of three lubricants on the flow of iron powder is shown in Fig. 3. Stripping pressure also is dependent on the material of which the die is made. Although stearic acid provides a lower stripping pressure in carbide tooling in comparison to zinc stearate, it requires a much higher stripping pressure in a steel die, as shown in Fig. 4 and 5. Thus, different lubricants may be required for steel and carbide tooling materials.

Sintered strength can vary considerably with the type of lubricant used. When

Fig. 1 Relationship of mix flow and bulk density of Ancor MH-100 iron powder with 1.0% zinc stearate lubricant
Source: Ref 1

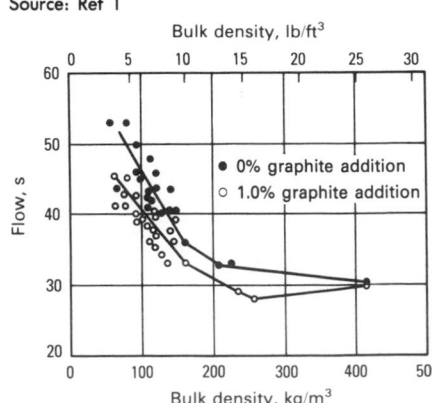

Fig. 2 Relationship of stripping pressure and bulk density of iron powders with 1.0% zinc stearate lubricant
Carbide die with 660 MPa or 48 tsi compacting pressure (○) and with 414 MPa or 30 tsi compacting pressure (●). Source: Ref 1

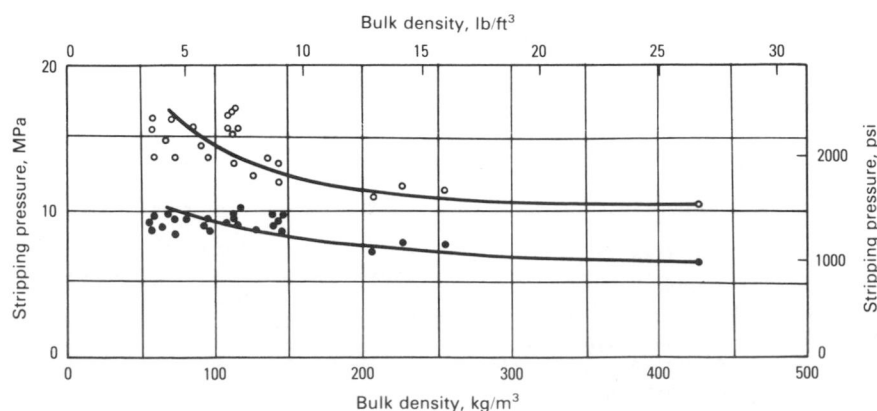

Fig. 3 Effect of increasing amounts of three lubricants on the flow rate of Ancor MH-100 iron powders

Lubricant A-1: 82% stearic acid, 15% palmitic acid, 1.0% oleic acid; lubricant A-2: 49% stearic acid, 50% palmitic acid, 0% oleic acid; lubricant A-3: 41% stearic acid, 51% palmitic acid, 6% oleic acid. Mixing time of 30 min. Source: Ref 1

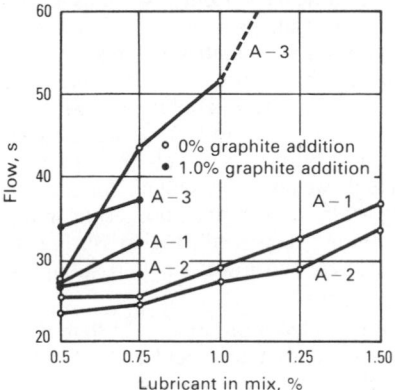

Fig. 4 Effect of lubricant mixing time and compacting pressure on the stripping pressure of Ancor MH-100 iron powder with 1.0% zinc stearate lubricant

Source: Ref 1

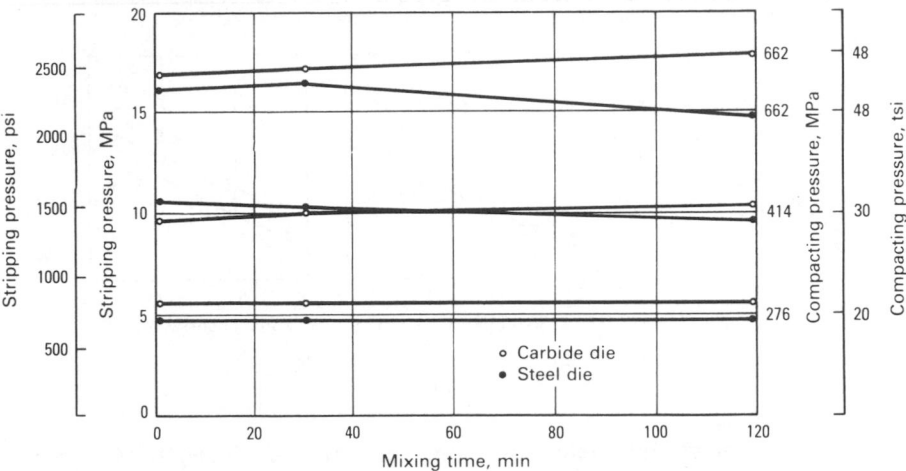

Fig. 5 Effect of lubricant mixing time and compacting pressure on stripping pressure for Ancor MH-100 iron powder with 1.0% stearic acid lubricant

Source: Ref 1

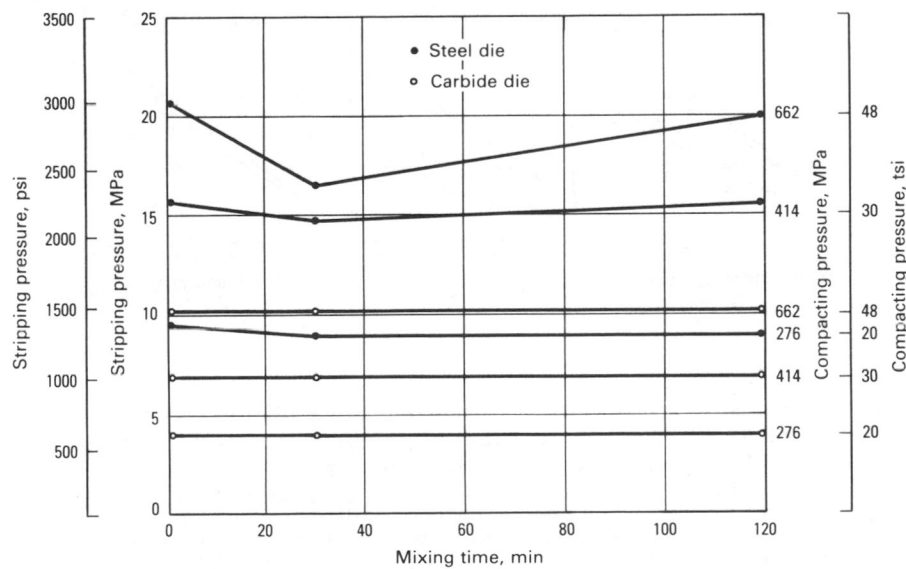

compared to parts pressed without lubricant, zinc stearate and stearic acid decrease sintered strength only slightly, as illustrated by Fig. 6 and 7. However, several lubricants, such as calcium and barium stearates, added to pure iron or to an iron and 1% graphite mixture cause significantly large decreases in sintered strength, as shown in Fig. 8. The appearance of P/M parts after sintering also can be affected by lubricants. Although the effect on iron parts is not as significant as with brass and nickel-silver powders, spotty surfaces and discoloration can be a major concern. Lubricants that vaporize completely during sintering generally leave a clean surface. Dark mottled surfaces occasionally result from zinc and calcium stearate lubricants, due to condensed zinc metal or calcium oxide deposits.

Lubricant removal, prior to the actual sinter operation, is vitally important for several reasons. Generally, this removal is accomplished during the sintering preheat operation and is referred to as "burn-off." Because of highly adverse effects of residual lubricant carbon deposits when sintering stainless steel, resulting in lowered corrosion resistance, Moyer (Ref 2) conducted an extensive study of the burn-off characteristics of common lubricants (lithium stearate, zinc stearate, Acrawax C, and Nopco Wax) in 316L powder compacts. Burn-off temperatures of 370, 425, and 480 °C (700, 800, and 900 °F) were used in both air and dissociated ammonia atmospheres. This study revealed that the waxes leave no residue at 425 °C (800 °F) or higher; the stearates, however, leave approximately 15% residue even when burned

at temperatures up to 540 °C (1000 °F). Maximum lubricant burn-off is achieved at about 425 °C (800 °F). Burn-off is less controllable in dissociated ammonia, and the amount of lubricant removed decreases as the compacting pressure is increased. Carbon burn-off is incomplete when compacts are burned off in a dissociated ammonia atmosphere.

Nonferrous Materials

Many lubricants frequently used in iron-based powder systems have not been

effective in all copper-based prealloyed systems. Without lubricant system development, primarily lithium stearate, it is unlikely that brass and nickel-silver P/M parts would have the strong market position that they hold today (Ref 3). The use of lithium stearate overcame many early sintering problems and provided consistently high properties in brass P/M parts.

Lubricants such as zinc stearate, lithium stearate, stearic acid, and Acrawax have been used successfully with conventional premixed bronze materials (90%Cu-

Fig. 6 Effect of zinc stearate additions on sintered strength of Ancor MH-100 compacted powders sintered 40 min at 1120 °C (2050 °F) in purified exothermic gas

Preheat of 650 °C (1200 °F). Bulk density: (a) 141 kg/m³ (8.8 lb/ft³); (b) 256 kg/m³ (16.0 lb/ft³); (c) 91 kg/m³ (5.7 lb/ft³). Particle size: (a) 1.9 μm; (b) 4.5 μm; (c) 1.0 μm. Graphite addition of 0% (○) or 1.0% (●). ——— without lubricant. – – – with lubricant. Source: Ref 1

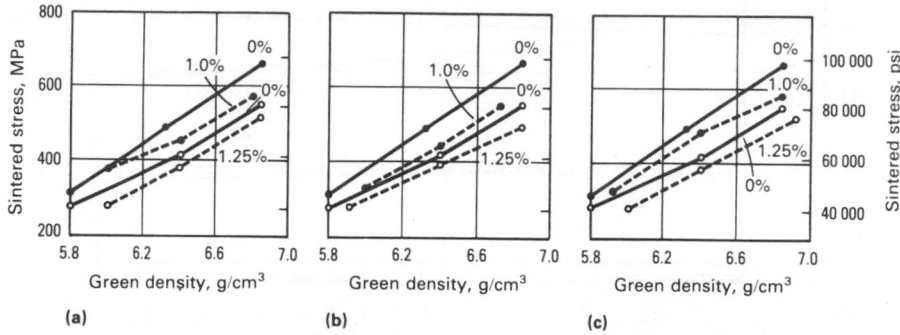

(a)   (b)   (c)

Fig. 7 Effect of stearic acid additions on sintered strength of Ancor MH-100 compacted powders sintered 40 min at 1120 °C (2050 °F) in purified exothermic gas

Preheat of 650 °C (1200 °F). Bulk density: (a) 354 kg/m³ (22.1 lb/ft³); (b) 434 kg/m³ (27.1 lb/ft³); (c) 384 kg/m³ (24.0 lb/ft³). Softening temperature: (a) 64°C (147 °F); (b) 55 °C (131 °F); (c) 54 °C (129 °F). Graphite addition of 0% (○) or 1.0% (●). ——— without lubricant. – – – with lubricant. Source: Ref 1

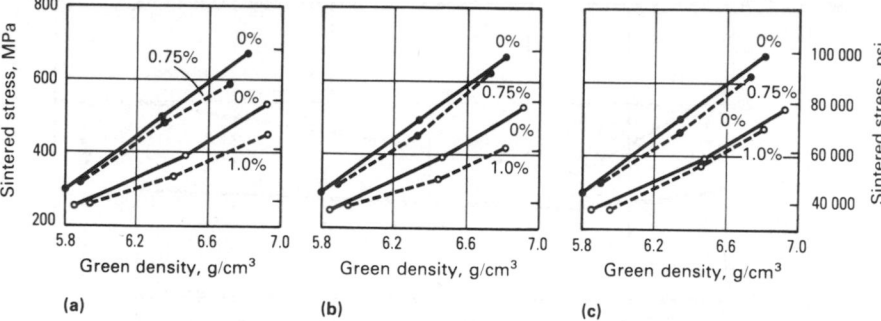

(a)   (b)   (c)

Fig. 8 Effect of calcium and barium stearate additions on sintered strength of Ancor MH-100 compacted powders sintered 40 min at 1120 °C (2050 °F) in purified exothermic gas

Preheat of 650 °C (1200 °F). (a) and (b) Calcium stearate. (c) Barium stearate. Bulk density: (a) 104 kg/m³ (6.5 lb/ft³); (b) 431 kg/m³ (26.9 lb/ft³); (c) 178 kg/m³ (11.1 lb/ft³). Particle size: (a) 11 μm; (b) 18.8 μm; (c) 1.2 μm. Graphite addition of 0% (○) or 1.0% (●). ——— without lubricant. – – – with lubricant. Source: Ref 1

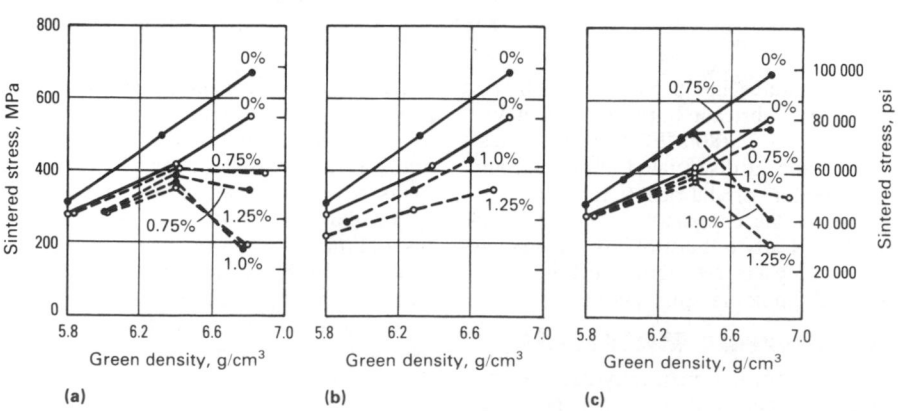

(a)   (b)   (c)

10%Sn). A recent study (Ref 4) comparing the two stearates and Acrawax revealed slightly better compressibility and lower ejection pressures with zinc stearate. Furthermore, green strength increased with decreased amounts of lubricant, and Acrawax provided superior green strength over zinc and lithium stearate.

Bi-lubricant systems, such as lithium-zinc stearate, provide the advantages of high physical properties and relatively clean surface appearance. Lithium stearate and lithium-zinc stearate lubricant additions in the range of 0.5 to 0.75% are used most often in brass and nickel-silver parts. Although waxes are excellent for maintaining green strength, their use results in lower apparent densities, higher briquetting pressures, and lower sintered strengths in brass parts.

Frequently, a bi-lubricant system of zinc stearate and stearic acid at 0.75 wt% is employed for conventional bronze premix blends; these two lubricants complement each other. Stearic acid has excellent lubricating characteristics, but its spherical morphology tends to promote segregation. Therefore, the irregular shape of the zinc stearate particles provides a supportive interlocking effect. Acrawax often is used alone to increase green strength, although it can lower the apparent density and retard the flow of bronze premixes.

Lubricants play a unique role in the production of copper P/M parts for high electrical conductivity applications (Ref 5). Lithium stearate allows the highest electrical conductivity among lubricated compacts, as illustrated in Fig. 9. Compacts pressed at 275 MPa (20 tsi), using a synthetic wax lubricant, have conductivities of about 74% IACS; similar compacts produced with lithium stearate exhibit conductivities of about 86%. The effect of the quantity of lubricant on electrical conductivity of copper P/M parts is shown in Fig. 10. The electrical conductivity reached a maximum with approximately 0.6% lithium stearate in compacts pressed at 275 MPa (20 tsi) and with approximately 0.5% at 415 MPa (30 tsi).

Synthetic organic waxes, which have low moisture and ash content, are employed in the production of aluminum P/M premixes (Ref 6). The addition of 1.5% lubricant (nominally) permits the consistent production of high-density aluminum P/M parts with fine surface quality and eliminates tool problems with seizing and cold welding. High sintered properties are achieved, because these lubricants leave virtually no ash to interfere with particle bonding during sintering.

Fig. 9 Effect of lubricant on electrical conductivity of copper P/M parts

Compacting pressure: (a) 276 MPa (20 tsi); (b) 414 MPa (30 tsi)

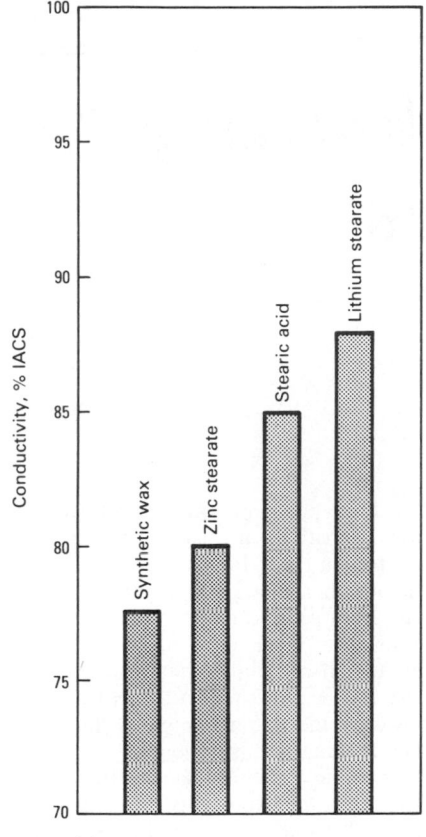

(a)

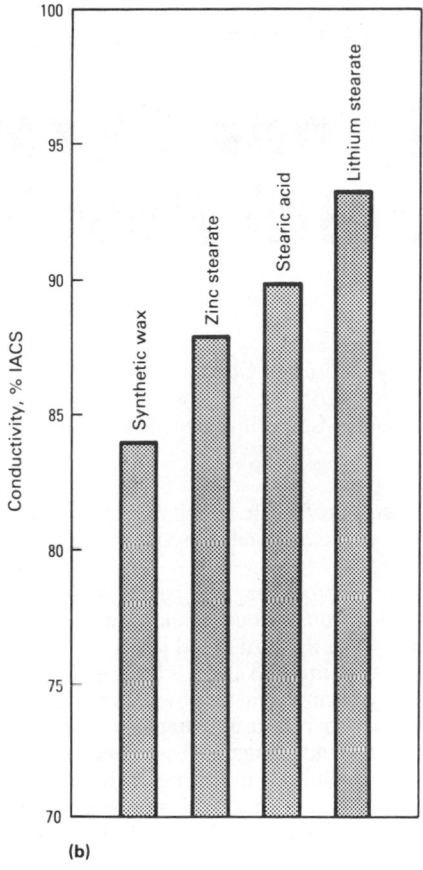

(b)

Fig. 10 Effect of lubricant quantity on the electrical conductivity of copper P/M parts

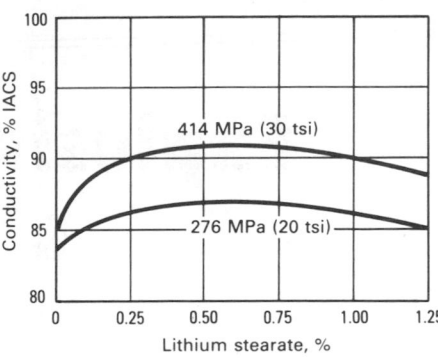

REFERENCES

1. Geijer, E. and Jamison, R.B., ''Lubricants for Powder Metallurgy Parts Manufacturing,'' Hoeganaes Report No. 142 1M, June 1965
2. Moyer, K.H., "The Burn Off Characteristics of Common Lubricants in 316L Powder Compacts," *International Journal of Powder Metallurgy*, Vol 7 (No. 3), 1971, p 33-43
3. Matthews, P.E., "Brass and Nickel Silver Powders," *Copper Base Powder Metallurgy*, Vol 7, Metal Powder Industries Federation, Princeton, NJ, 1980
4. McGraw, J. and Koczak, M.J., "A Laboratory/Production Comparison of Powder Compacting and Ejection Response," *International Journal of Powder Metallurgy*, Vol 16 (No. 1), 1980
5. Taubenblat, P.W., "Production and Properties of High Conductivity Copper P/M Parts," *Copper Base Powder Metallurgy*, Vol 7, Metal Powder Industries Federation, Princeton, NJ, 1980
6. Dudas, J.H. and Dean, W.A., "The Production of Precision Aluminum P/M Parts," *Progress in Powder Metallurgy*, Vol 25, Metal Powder Industries Federation, Princeton, NJ, 1969

Explosivity and Pyrophoricity of Metal Powders

By C. James Dahn
President
Safety Consulting Engineers, Inc.

TO AVOID THE DANGERS of explosions and fires during the processing and handling of metal powders, knowledge of how various powders react in the presence of air is essential. Factors such as chemical composition, particle size and size distribution, particle shape, chemical additives, surface coatings, and variations in gaseous environments have a direct effect on the fire and explosion sensitivity of a material. This article discusses the nature of these hazards, mechanisms of explosive, pyrophoric, and autogenous ignition reactions, classification of metal powders based on their explosive limits and flammability, and safety precautions.

Explosivity of Metal Powders

Normally, a mass of metal powder subjected to localized heating melts or, if sufficient air is present, ignites locally. If the metal powder is extremely reactive in the presence of air, however, the burning rate can accelerate sufficiently to cause suspension of metal particles in the air, creating a dust cloud ahead of the burning surface. As a result of external disturbances, or under proper atmospheric conditions, the dust cloud can ignite and explode.

Mixtures of metals or metals and another material may self-react or ignite from external ignition sources (a metal spark, localized impact energy, or fire). An example of such a reactive mixture is a thermite type of mix, such as aluminum plus iron oxide. An intercomponent reaction can accelerate and cause lofting of incendiary particles into the air at very rapid speeds. Unreacted mixture particles may be dispersed into the air simultaneously, creating a dust cloud and subsequent explosion hazard.

During processing, dust clouds and handling of metal powders, dust clouds usually are generated. Once the dust cloud has formed into the proper mixture and has been ignited by energy from the metal powder or from an external source, a dust explosion occurs. A compound danger exists, because any dust accumulated in the area is lofted into the air by the first explosion and is ignited, causing a secondary explosion or fire which usually has more catastrophic results. Figure 1 summarizes the conditions necessary for a dust explosion and secondary explosion or fire.

In any dust explosion, significant pressure rises occur that last a relatively long time for an explosion (up to 1380 kPa or 200 psi in confined spaces and for many milliseconds). Although values of the explosion pressures (even in confined spaces) are not particularly high, the effects on materials of construction in buildings can be devastating, such as the fracturing and destruction of concrete silos and grain elevators. Similarly, the damage potential of a metal dust cloud explosion is very high.

Factors Influencing Explosivity

Chemical and Physical Factors. The chemical composition of metal powder or metal powder mixture surfaces and high surface areas may cause susceptibility to self-initiation or may compound the danger of propagation from external ignition.

Special chemical treatment or formation of protective oxide can alter the explosivity of a metal powder. For example, very fine flake powders without protective lubricant coatings are particularly susceptible to ignition. Because reactive gases trapped in a metal powder, which originate from processing or a combustion reaction, can reach very high temperatures due to self-heating of the powder and the very low heat-dissipation properties of fine powders, the flammable gases can ignite, thus igniting the dust cloud.

Thermite reactions, as mentioned earlier, affect the explosivity of a metal mixture. Thermite refers to the exothermic chemical reduction of a metal oxide by metal powders of another more reactive metal. Such reactions do not require any external source of oxygen, since it is supplied by the metal oxide. Once the thermite reaction has caused ignition of powder and carried unreacted dust into surrounding air, a dust cloud can form. The lack of homogeneity in the dust-air mixture may reduce the hazard of significant explosion output, but existing dust in the air or equipment can lead to serious complications. For example, the right conditions for a reactive dust cloud may be created by processing equipment, such as pulverizers and hammer mills, which may generate a fairly homogeneous mixture with air because they generate very high turbulence and mixing.

Extreme reactivity in the presence of moisture occurs in some metal powders. Potassium, rubidium, and cesium when contacted by water (acting as the oxidizer) burst explosively into flame (Ref 2). Some metals produce flammable gases or vapors (hydrogen, for example) in the presence of moisture, which compounds the danger of explosion if a dust cloud is present. Metal powders such as magnesium, alu-

Fig. 1 Conditions required for dust explosion
Source: Ref 1

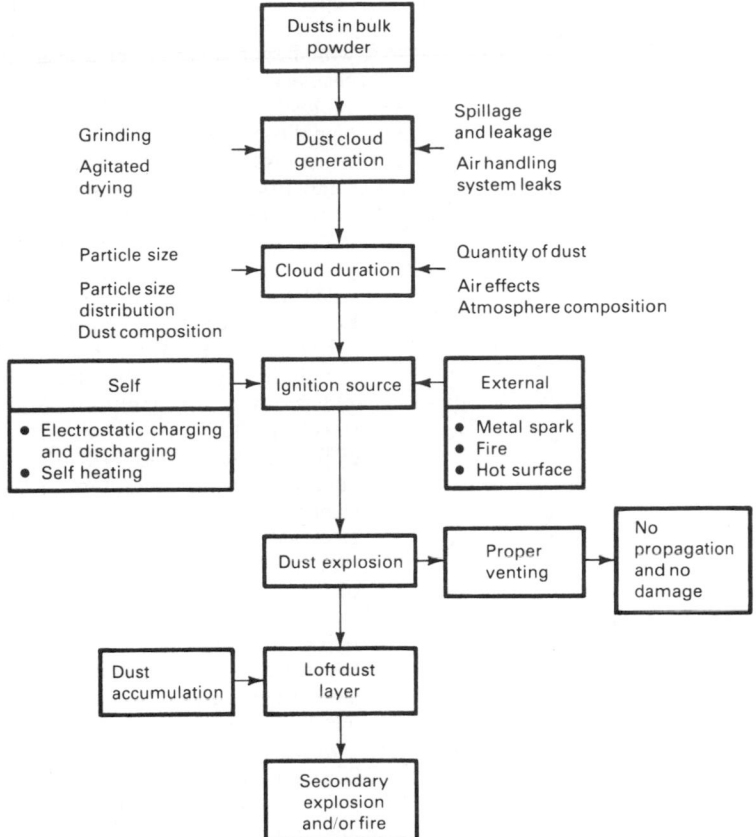

minum, and zinc react with moisture and can cause spontaneous heating, sufficient to cause localized ignitions. In addition, some metal powders such as aluminum may react violently with halogenated hydrocarbons or other chemicals via chemical oxidation/reduction reactions that do not involve oxygen. Pyrophoricity, or spontaneous reactivity in the presence of air, is also a cause of ignition in metal powder explosions and is discussed later in this article.

Many physical characteristics, such as particle size and surface area, determine whether a dust cloud ignites or whether it propagates once ignited. As particle size is reduced, a metal dust cloud becomes easier to ignite and yields higher rates of reaction. Usually, as particle size drops below 149 μm (100 mesh), a metal powder is easier to ignite and has a higher rate of pressure rise on explosion. If the more explosive component in a metal powder mixture is present in smaller particle sizes, the larger and heavier particles may settle from a generated dust cloud, leaving a cloud that is far more explosive than the original powder mixture.

Particle shape also can affect explosivity. For example, flake aluminum and magnesium, which have a greater surface area to volume ratio than other particle shapes, are much more easily ignited than their atomized counterparts. Furthermore, flake aluminum or magnesium breaks into finer particles more easily, producing higher secondary effects. Metal particles (powders) that are more brittle are more likely to generate finer particles for dust clouds and secondary effects.

The concentration of particles in a dust cloud is a major deciding factor in explosivity. For propagation of explosion to occur, a relatively homogeneous mixture of particles with air must be formed, and the particles must stay suspended as a cloud (slow settling). The ability of the particles to stay in homogeneous suspension also is affected by electrostatic forces. Electrostatic forces generated during lofting of the cloud may hinder or help the dispersion of dust in air, depending on material characteristics. The composition of the particle surface determines whether the particle accumulates electrostatic charge and/ or inherently possesses free ions on the

surface. For example, the stearate coating on aluminum particles gives them free ions on the surface, causing them to repel each other in a suspension and to seek oppositely charged ions in the gas space. Atomized aluminum tends to remain more electrostatically inert and settles more quickly.

Turbulence within a cloud increases the rate of propagation of the explosion and its resultant destructiveness. It can be created internally by the initial dust explosion or can be created externally by process equipment, giving particles more energy to remain suspended. Normally, as turbulence increases, the explosion pressure and rate of pressure rise are greater.

The containment conditions (size and shape) of the dust cloud affect explosion output. A cloud contained in a smaller volume has a higher explosion rate of pressure rise due to multiple reflections of the pressure output wave (reflection reinforces the intensity of the pressure wave and accelerates the reaction). Dust cloud size determines the possible extent of propagation, and shape determines the major direction for propagation. In general, chemical and physical parameters affect the susceptibility of powder to ignition, the tendency of powder to loft and form a cloud, and the probability of powder generating secondary explosions.

Atmospheric Conditions. The atmosphere in which the dust is lofted greatly influences dust explosion characteristics. Oxygen content can determine whether metals ignite and propagate an explosion or fire; some metals require very little oxygen content to ignite, while others require a large oxygen content (Ref 3).

Some metal powders that will burn in a dust cloud (lithium, magnesium, and zirconium, for example) react with gases other than air, such as carbon dioxide. The presence of flammable gases and vapors drastically affects the ignition and explosion parameters—usually by reducing the explosive limit significantly.

Depending on the compatibility of the metal with water, the relative humidity of surrounding air may enhance or dampen propagation. Although atmospheric pressure and temperature influence ignitability and output to a certain extent, the effect is not as significant as that of a parameter such as oxygen content.

Ignition Source. The magnitude of energy of an ignition source, as well as the type of energy, determines whether a dust cloud ignites and how strong the explosion output is. Typical ignition sources are heat, mechanical sparks, electric or electrostatic discharge sparks, and flames. A

Table 1 Factors influencing explosivity

Factor	Ignition	Dust lofting	Duration of cloud	Propagation	Peak pressure	Pressure rate rise, dp/dt(a)
Chemical						
Oxide layers	X			X		X
Chemical coating	X			X		X
Gas permeation	X			X	X	X
Chemical additive	X			X	X	X
Physical						
Temperature	X			X	X	X
Particle size	X	X	X			X
Particle shape		X	X			
Brittleness	X			X	X	
Mixtures of materials		X	X	X	X	X
Particle size distribution	X	X	X	X	X	X
Particle density		X	X			
Moisture content	X	X	X	X	X	X
Cloud						
Concentration	X			X	X	X
Turbulence		X	X	X	X	X
Electrostatic force		X	X			
Uniformity of concentration				X	X	X
Containment, size, and shape		X	X	X	X	X
Atmosphere						
Oxygen content	X			X	X	X
Humidity	X			X	X	X
Volatile gas content	X			X	X	X
Inert gas content	X			X	X	X
Gas temperature and pressure	X			X	X	X

(a) Rate of pressure rise for dust explosion in a particular vessel

stronger ignition source (higher energy of ignition) usually causes a higher and faster pressure rise in an explosion of a dust cloud. Table 1 summarizes the factors influencing the explosivity of metal powders.

Estimating Explosion Output

Recently, new methods have been developed to scale up the explosion output of a dust cloud from laboratory tests (Ref 4). The methods are based on a "cubic" law, which takes the form of:

$$K_{st} = (dp/dt)_{max}(V^{1/3})$$

where $(dp/dt)_{max}$ is the maximum rate of pressure rise for dust explosion in a particular vessel (bars per second); V is the volume of the particular vessel (cubic metres); and K_{st} is the Bartknecht constant (Ref 8). The Barknecht constant is used for comparing the explosivity of different materials.

Based on the cubic law, proper full-scale venting area is determined from laboratory scale data by:

$$F_2 = \frac{F_1 V_2^{2/3}}{V_1^{2/3}}$$

where F_1 is the vent area on the test vessel to prevent pressure during combustion from exceeding a given value; F_2 is the vent area of full-scale containment volume to prevent pressure during combustion from exceeding the same given value; V_1 is the volume of the test vessel; and V_2 is the volume of full-scale containment. There are other complicated factors which must be considered for proper containment and venting. Careful study of and strict adherence to available standards is recommended in establishing venting designs.

Classification of Metal Powders

In the past, extensive studies by the U.S. Bureau of Mines have yielded various classifications of metal powders (Ref 5). Tests were conducted using a Hartmann explosion test chamber to determine the minimum ignition energy, explosive limits, relative flammability, limiting atmospheric oxygen content, maximum pressure rise, and rate of pressure rise.

Since that time, researchers have discovered that the Hartmann dust chamber does not yield realistic information for scale-up of the explosion output and venting requirements. Also, the lean limits of flammability (minimum concentration for explosion) were found to be too high, depending on the distribution of the dust cloud and the size of chamber. In recent tests with larger spherical and near-spherical chambers, minimum concentrations for dust explosion were found to be much lower than those previously recorded. Thus, the ranking of explosivity has shifted significantly from the past. Caution must be used, because the actual conditions of processing and handling of the metal powders greatly influence the degree of hazard.

Metal Powders of Significant Dust Explosion Hazard. Included in this classification are metal powders that require very little oxygen content (less than 3%) and that have low ignition temperatures (less than 600 °C or 1110 °F) and very low explosive limits (20 to 50 g/m^3, or 9 to 22 grains/ft^3). In addition, powders that yield significant dust explosion output are in this category. Metals such as zirconium, magnesium, and aluminum, and combinations of these, represent very high dust explosion hazards, because of these characteristics. Zirconium powder is especially hazardous as a dust cloud, because the electrostatic charges developed in dispersing the cloud may ignite the material.

Metal Powders of Moderate Dust Explosion Hazard. This classification includes metal powders that require a minimum oxygen content of 10% or higher to promote dust explosion, that have ignition temperatures of 300 to 800 °C (570 to 1470 °F), and that have lower explosive limits, ranging from 100 to 500 g/m^3 (44 to 218 grains/ft^3). Metal powders such as tin, zinc, iron, and copper fall into this category.

Metal Powders of Low Dust Explosion Hazard. This classification includes powders that have ignition temperatures greater than 700 °C (1290 °F) and that cannot be ignited by electric sparks or strong initiating sources. Also included are metal powders that can only be ignited by strong initiating sources in very high powder concentrations. Examples of metals in this category are molybdenum, cobalt, and lead.

The relative explosivity ranking of some metal powders can be summarized as:

High explosivity

- Zirconium
- Magnesium
- Aluminum
- Lithium
- Sodium

Moderate explosivity

- Tin
- Zinc

- Iron
- Silicon
- Manganese
- Copper

Low explosivity

- Molybdenum
- Cobalt
- Lead

Preventing Metal Powder Explosions

Material Testing. Knowledge of the explosivity characteristics of a metal powder, which are dependent on factors such as particle size and shape, composition, and surface area, is the first step in preventing dust explosions during processing or handling of the powder. Much of the data and test method information available from previous testing are based on use of the Hartmann chamber. Since then, however, spherical-type chambers have been found to yield better information. Spherical bombs ranging from 0.02 to 1 m³ (0.7 to 35 ft³) in volume are available for explosibility testing. Also, a large-scale version of the original Hartmann chamber with a length-to-diameter ratio of 1.0 that yields much better pressure and rate of pressure rise data has been developed (Ref 6).

The ignition source used in the Hartmann chamber was an electric high-voltage arc. Other chambers have used electric matches and squibs. One testing facility uses a fuel vapor/air flame ignition in both the Hartmann chamber and the large-scale 0.0078-m³ (0.28-ft³) Bureau of Mines chamber.

Careful testing procedures can yield good information regarding the lower limit of explosibility (lean limit or minimum concentration for explosion), minimum energy input for explosion, peak pressure, and pressure rise rate. An example of an iron powder dust explosion pressure-time trace is shown in Fig. 2.

Various methods for determining the ignition temperature of a dust cloud are also available. One method uses a Godbert-Greenwald furnace (Ref 7), in which the material is placed in a furnace and heated to a specific temperature. At that point, the dust is lofted into the air and ignition is denoted by the appearance of a flame below the mouth of the furnace.

The dust cloud ignition temperature also can be evaluated by placing single grains of the metal powder into a capsule (with air) and testing the sample in a differential scanning calorimeter or thermal gravimetric analyzer unit. Ignition temperature determined by this method is the onset tem-

Fig. 2 Typical pressure-time trace for iron powder (~40 μm) dust explosion

P_{max} is 600 kPa (90 psig) at 50 ms/cm (125 ms/in.) sweep rate.

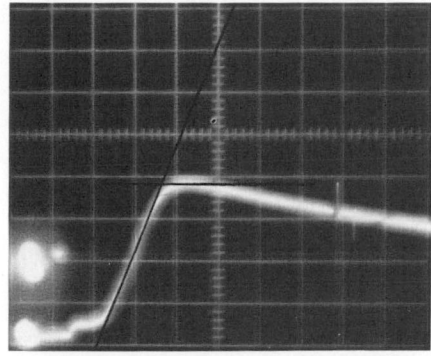

perature for runaway reaction. Caution must be exercised to ensure that the actual kinetics for decomposition of the powder with air can be simulated by this sampling method.

As mentioned previously, data gathered on the explosive output of a material can be scaled up to actual process conditions, according to findings of Bartknecht (Ref 8). Bartknecht also suggests that the testing chamber be larger than 0.02 m³ (0.7 ft³) to provide good data for such extrapolations. Nomographs in Ref 3 allow scale-up of explosion output values.

Again, caution must be exercised that laboratory tests are not immediately interpreted as accurate simulation of the full-scale situation. Handling of volatile material on a large scale may leave room for errors in estimations to provide adequate venting. In addition, K_{st} values for laboratory tests vary greatly according to the strength of the initiating source and the degree of turbulence in the laboratory test apparatus.

Evaluation of Plant Operation Hazards. The second phase of explosion prevention is evaluation of the plant to identify hazards. The evaluation should focus on:

- Ignition sources
- Dust cloud generators
- Dust leaks and buildups
- Localized gaseous atmosphere composition

The methodology for combining the evaluation with laboratory knowledge of the metal powder explosivity is summarized in Fig. 3. Specific types of processing—such as drying, subjection to high temperatures, or mixing with another metal—may warrant further testing in the

laboratory, such as spontaneous heat tests (Ref 9) or tests for thermite reaction.

Hazards created by equipment include dust accumulation near heating, ventilation, and air-conditioning equipment where high temperatures occur; cloud-generating equipment that is close enough to other parts of the process to promote propagation; and dangerous levels of dust concentration in the dust-collecting systems (especially ductwork). These hazards can be corrected by sealing equipment to prevent leakage and installing extra dust-collecting equipment. Cleanup and housekeeping schedules should be adequate to prevent accumulation of dust layers.

No additional hazard should be created by these preventive measures, however. Dust-collecting systems must be designed to handle specific metal powders safely. For instance, in aluminum processing, moisture and water must not be allowed to accumulate in the collected dust, or self-ignition may occur in the dust-collecting bag house. Also, powder accumulations should never be cleared with compressed air nozzles, which only generate additional dust clouds. Only explosion-proof air-moving systems should be employed for cleanup operations.

Ignition sources are a major concern during evaluation of process hazards. Obvious ignition sources, such as lit cigarettes, open flame, and sparking materials, should be removed from potentially dusty areas. Repair work utilizing welding equipment and torches should be performed only after all dust layers and clouds have been thoroughly cleared. Secondary ignition sources consist of rotating equipment (overheated bearings) and mechanical impacting in the process.

The third group of ignition sources includes chemical reactions (self-heat) and electrostatic discharge. Chemical decomposition can promote heating and heat buildup, if a dust layer is thick enough to promote runaway reaction. Metals such as zirconium, magnesium, and aluminum are ignited very easily by electrostatic discharge. In the handling of these metal powders, personnel and equipment should be adequately grounded.

In particularly hazardous metal powder operations, several additional preventive measures can be taken. First, inert gases can be added to the process to bring the oxygen content below the minimum flammable limits. Second, nonflammable and nonexplosive powder compositions such as calcium carbonate or sodium bircarbonate can be introduced into the powder. Caution must be used, however, to prevent the new component from separating

Fig. 3 Materials dust explosion hazard test methodology
Source: Ref 1

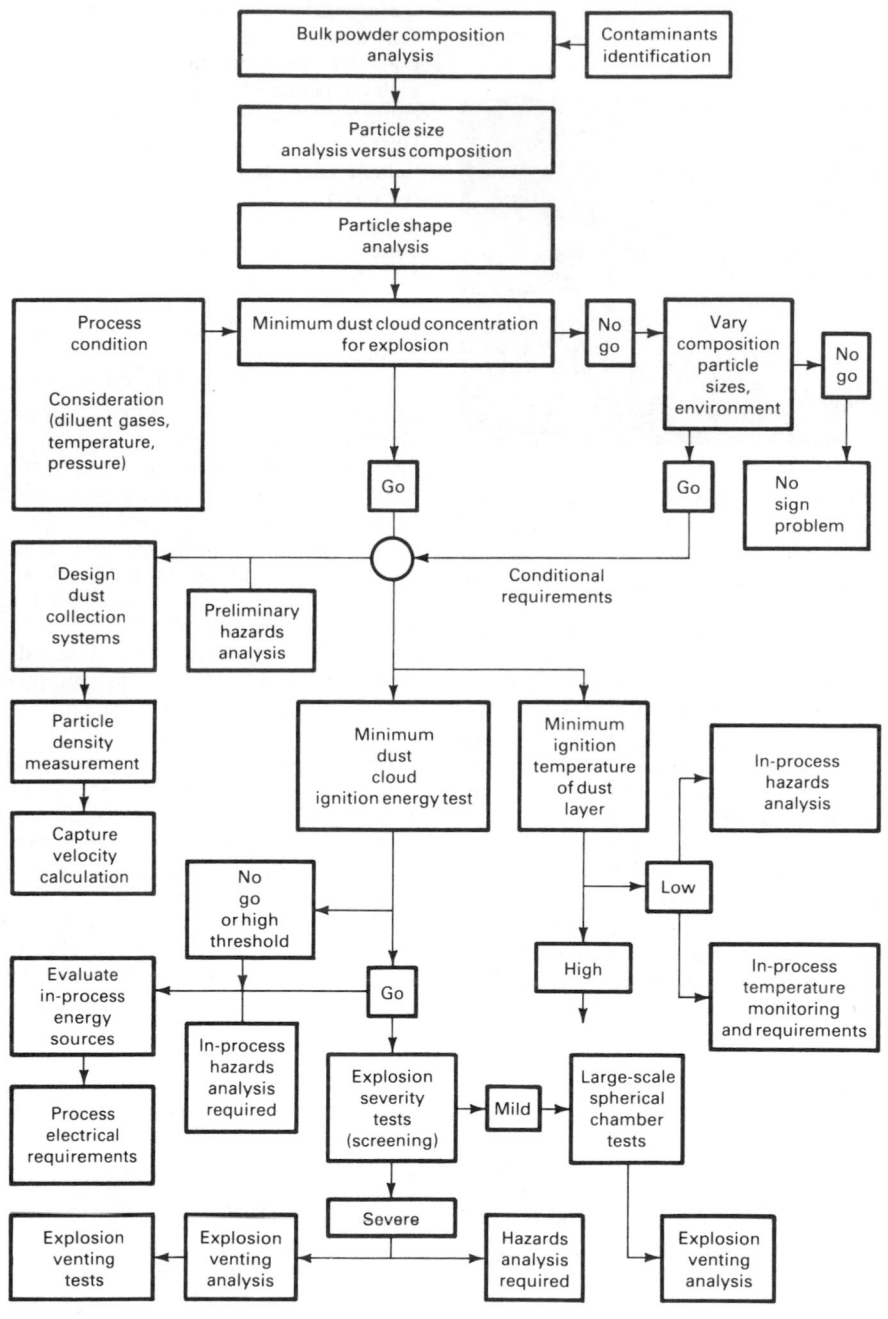

due to turbulence, thereby permitting a dust explosion. The effectiveness of any extinguishing powder composition should be experimentally determined for each specific case.

Explosion Suppression. Explosion suppression systems can be used to prevent propagation of a dust explosion, provided that the initial reaction is sensed in sufficient time. Caution also must be exercised here, because suppression of a localized dust explosion could permit a sec-

ondary dust cloud to form, ignite, and explode. The high rate of reaction can preclude initiation of effective suppression, as is the case with aluminum powders. Success of an explosion suppression system depends on the rate of pressure rise in the initial dust explosion.

In areas where a strong probability of dust explosion exists, venting systems should be incorporated. Vents permit reduction of internal pressures to prevent peak pressures from developing. The vent ports

must open quickly enough to allow dust explosion reaction gases to escape. As mentioned previously, laboratory test data may not scale up perfectly for vent design; therefore, estimates should be conservative. Also, venting of an explosion may be accompanied by lofting of an accumulated dust layer, creating a secondary explosion hazard. Therefore, the size, shape, and direction of vent alignments should be carefully established.

In case of fire in the processing or handling of metal powders, great effort should be made to prevent the use of water for fire extinguishing, because water can react vigorously with the metal powder or can loft metal powder into a dust cloud. Normally, special low-velocity, class D, dry fire-extinguishing powders are recommended to extinguish metal powder fires. Sand, talc, and other inert materials can also be sprayed to suppress a fire if care is taken not to disperse the powder into a cloud.

Because metal powders usually burn at higher intensities than nonmetallic powders, extinguishing such fires is much more difficult. In many cases, no attempt should be made to actively fight such fires due to the risk to personnel and the probability of making the situation worse. Local fire departments should be notified immediately of the nature of the fire and consulted as to the best fire-fighting methods for the powder involved. Prior discussion and planning before the fact is highly recommended. National Fire Protection Association codes for metal powder plant operations, such as production of aluminum, magnesium, and bronze powders, should be consulted on a routine basis to ensure that all precautions to prevent dust explosions and metal powder reactions are being taken (Ref 10, 11).

Pyrophoricity of Metal Powders

Metal powders are considered pyrophoric if they react and ignite in the presence of air at normal ambient temperatures. A distinction, however, should be made between pyrophoricity and autogenous ignition. The latter refers to the lowest temperature at which combustion begins and continues in a metal powder when it is heated in air.

Normally, finely divided powders which do not have a protective oxide film ignite pyrophorically when a critical specific surface area is exceeded. The heat of formation of the oxide layer when exposed to oxygen in the air has a direct control on whether the metal powder is pyrophoric.

The degree of pyrophoricity is directly dependent on particle size, particle size distribution, and surface area. For instance, zirconium powder of about 3-μm size is extremely pyrophoric with air, whereas the powder of 12-μm particle size does not ignite even at red-hot heat.

Uranium powders are pyrophoric in much larger particle sizes than zirconium. Iron, nickel, and copper become pyrophoric at particle sizes 100 times smaller than those of zirconium powders that react pyrophorically. Fresh, very finely ground aluminum is also pyrophoric when particle sizes drop to 0.03 μm. In addition, autogenous ignition may occur in aluminum at moderately warm temperatures of 75 to 95 °C (170 to 200 °F), depending on the fineness of the powder. Powders that are porous may be pyrophoric at much larger particle sizes. Iron, nickel, and copper powders, for example, result in porous or spongy particles when prepared from their oxides at low reduction temperatures. Even if the particle size of such powders is 100 μm or more, specific surface area of the powders, as measured by the gas absorption (BET) method, may approach 1 m²/g; as a result, the powders may be pyrophoric.

Pyrophoricity characteristics are a major concern if a powder is:

- Contacted by air in a process
- Accidentally ground finely in the presence of air
- Subjected to accidental introduction of air into the grinding equipment

If chemical protective layers, such as stearate on aluminum, are suddenly removed from the metal powder, pyrophoric reaction can result.

If the pyrophoric reaction lofts metal powder dusts into the air and creates a dust explosion, a very large, intense fireball, localized intense heating, or violent secondary explosions can result. Pyrophoricity and/or autogenous ignition characteristics of metal powder also contribute to the potential for spontaneous reaction when stored for long durations of time. Surface oxides on the metal tend to mitigate the pyrophoric characteristics. These oxide layers can, however, fracture, dissolve into the melted metal, or react with other chemicals, losing their effectiveness.

The presence of moisture generally tends to increase the potential metal pyrophoricity. In some cases, the hydrogen that is developed during reaction with water results in lowering the ignition temperature of the metal powder, thus permitting it to react pyrophorically. Numerous incidents in which moisture has collected on magnesium and zirconium powders stored in containers have occurred. The metals spontaneously caught on fire and, when self-heating of the interior produced excessive gases, the consequence was a massive eruption of incendiary hot particles, in all directions, from the containers.

Reaction Parameters

Although the mechanisms of pyrophoricity and autogenous ignition in metal powders are not totally understood, it generally is agreed that reduction of the metal powder size increases its likelihood to become pyrophoric. Also, the ignition temperature of the material generally reduces as particle size reduces. Based on present scientific developments, the heat of formation of the oxide layer and the effective rate of heat dissipation control whether a pyrophoric reaction occurs. If the heat of reaction overcomes the heat of oxide formation, an accelerated reaction occurs and promotes the pyrophoric reaction in the air. For example, nonpyrophoric tungsten metal powder, which is prepared by hydrogen reduction of tungsten trioxide at about 750 to 900 °C (1380 to 1650 °F), becomes extremely pyrophoric if prepared by slow reduction, between 600 to 650 °C (1110 to 1200 °F) (Ref 12). The powder prepared at this lower temperature range ignites spontaneously when removed from the hydrogen flame.

The surface brittleness also may play a role in the onset of a pyrophoric effect. The extent of a pyrophoric and/or autogenous ignition reaction is a function of the oxygen content of the gases and, to a certain extent, entrained hydrogen that may be in the material. Although the effects are not well understood to date, the surface layer condition is believed to play an important role in the tendency to pyrophoricity. If chemical deposits and oxide layers can be removed quickly from the metal powder surface by flaking or dissolution, a sudden increase in pyrophoricity may occur.

Generally, pyrophoric reactions follow traditional combustion behavior, in that the material decomposition produces excessive heats of oxidation that cannot be removed rapidly. Consequently, the metal heats to the point where it melts, and possibly vaporizes, on the surface, which starts the combustion reaction similar to liquid vapor combustion (Ref 13). The contribution of hydrogen (from reactions with water) to such mechanisms also can accelerate the reactions. Furthermore, the thermal properties of the bulk metal powder play an important role in terms of removing heat, especially if the metal powder is stored in quantity.

Prevention of Pyrophoric Reactions

Because the mechanisms of pyrophoricity and autogenous ignition are not definitively known for many metal powders, these characteristics should be evaluated relative to the processing, chemical treating, and handling intended for a specific powder. Also, effects of contaminants, chemicals, and variations in gas environment should be studied. Chemical kinetics of reactions that lead to the liberation of heat should be studied prior to start-up of new processes to avoid promoting conditions leading to pyrophoricity or autogenous ignition. The American Society for Testing and Materials (ASTM) spontaneous heating test is helpful in obtaining pyrophoricity characteristics of metal powders (Ref 9).

Precautions employed in the past have recommended using the largest particle size possible and protective coatings. Also, blanket fluids such as hydrocarbons, alcohols, and dry inert gases can be incorporated into processes.

Metals known to be pyrophoric are:

- Aluminum
- Calcium
- Cerium
- Cesium
- Chromium
- Cobalt
- Hafnium
- Iridium
- Iron
- Lead
- Lithium
- Magnesium
- Nickel
- Palladium
- Platinum
- Plutonium
- Potassium
- Rubidium
- Sodium
- Tantalum
- Thorium
- Titanium
- Uranium
- Zirconium

Some metals in their pure form can become pyrophoric in the presence of air and other gases and should be appropriately studied before use in metal powder processes. An example of this is lithium, which reacts in the presence of nitrogen as well as air.

If a fire occurs in a metal powder, the fire should be approached with extreme caution and not disturbed to prevent eruptions of incendiary material or small explosions. Water should not be applied to most metal powder fires. Some burning powders, such as iron and nickel, can be extinguished with a blanket of inert gas if the gaseous blanket can be maintained intact long enough for excess heat to be dissipated. Where insufficient information is available about reaction characteristics of burning metal, the fire-fighting team should stay clear of the fire area and try to contain the fire in a certain zone. Whenever processing materials have pyrophoric or explosive tendencies, individual process operations should be isolated; thus, if fires or small explosions occur, they will not spread.

REFERENCES

1. Dahn, C.J., "Hazards Associated with Solids Handling," Powder and Bulk Solids Conference, May 1980
2. Ellern, H., *Military and Civilian Pyrotechnics*, Chemical Publishing Co., New York, 1968
3. *Fire Protection Handbook*, 14th ed., ch. 8, National Fire Protection Association, 1976
4. *Fire Protection Handbook*, 14th ed., "Guide for Explosion Venting," National Fire Protection Association, 1976
5. Explosibility of Metal Powders, RI 6516, U.S. Bureau of Mines
6. Cashdollar, K., Hertzberg, M., and Opferman, J., Flammability of Coal Dust-Air Mixtures, RI 8360, U.S. Bureau of Mines, 1979
7. Laboratory Equipment and Test Procedures for Evaluating Explosibility of Dusts, RI 624, U.S. Bureau of Mines
8. Bartknecht, W., *Explosions Course Prevention, Protection*, translated from German by Burg, H. and Almond, T., Springer-Verlag, New York, 1981
9. Standard Test Method for Spontaneous Heating Tendency of Materials, ASTM E 771, American Society for Testing and Materials, Philadelphia
10. Aluminum and Magnesium Powders, NFPA 651, National Fire Protection Association, 1980
11. Processing and Finishing Aluminum, NFPA 65, National Fire Protection Association, 1980
12. Kopelman, B. and Compton, V., Spontaneous Combustion of Metal Powders, *Metal Progress*, Vol 63 (No. 2), Feb 1953
13. Smith, R.B., Pyrophoricity—A Technical Mystery Under Attack, *Nucleonics*, Dec 1956

SELECTED REFERENCES

- Palmer, K.N., *Dust Explosions and Fires*, John Wiley and Sons, Inc., New York, 1973
- Pratt, T.H., "Explosions: Prevention and Protection," proceedings of McGraw-Hill Chemical Engineering Seminar, 1982

Toxicity of Metal Powders

By Robert P. Beliles
Associate Professor
Department of Forensic Sciences
George Washington University

TOXICITY, the capacity of a material to produce harmful effects in the body, is determined by the material's biochemical characteristics, the route through which it enters the body, and its dosage. Knowledge of the toxic effects of metal powders is necessary to ensure the health of personnel exposed to them.

The route by which one is exposed to a toxic material has a bearing on the material's site of action (adverse effect), degree of absorption, and concentration in various tissues in the body. For example, the amount of material absorbed into the bloodstream is likely to be less when the route of exposure is through the skin, rather than oral ingestion; at the same time, the likelihood of adverse effects in the form of skin lesions is increased.

When considering metal powders, the inhalation route is the route of greatest concern, because the powders are most likely to enter the body as particulates in the air being breathed in the workplace. Particulates in the atmosphere are classified as dust, fumes, smoke, mist (fog), or smog. Dust is formed during operations that reduce the size of large masses, such as comminution and grinding. Fumes are formed by combustion, sublimation, and evaporation. Smoke is the combustion product of organic materials; it is composed of particles that are less than 0.05 μm and that tend not to settle. Mist or fog is composed of liquid droplets of water condensed into microscopic particles. Smog is similar to fog, but is a function of air pollution. Smog is composed of photooxidant products that form as a result of the interaction of hydrocarbons with other components in the atmosphere, particularly ozone.

Particulate Behavior

Particulates, in addition to their own toxicity, tend to alter the toxicity of other materials that they absorb, or materials into which they are absorbed. For example, sodium chloride particles tend to absorb formaldehyde and thus penetrate more deeply and remain longer in the lungs. Likewise, carbon particles in the presence of sulfur dioxide penetrate more deeply.

All particulates share several physical characteristics that affect their behavior in the atmosphere. They have a tendency, which is directly proportional to particulate concentration, to coagulate, or impact with one another. Sedimentation, or the tendency of particles to settle under the influence of gravity, is the second most important physical characteristic of particulates. Settling is counteracted by air turbulence. Additionally, particles diffuse because of Brownian motion due to kinetic energy of impacting molecules of air or gas or because of electrical charges.

The behavior of particulates also is influenced by their size and density, which is sometimes expressed as aerodynamic diameter or mass median diameter. For example, a 1-μm-diam particle with a density of 1 g/cm³ has a mass median diameter of 1. Density and diameter control the penetration of particulates into the airways, lungs, and deeper portions of the respiratory tract. Larger particles (greater than 12 μm) are impacted largely in the upper respiratory tract, while smaller particles tend to penetrate farther down into the alveolar region of the lungs.

Not all particles inhaled into the lungs remain there. Some are suspended in the air and are exhaled. Particles that impact on the mucous lining covering the ciliated epithelium of the respiratory tract travel through the respiratory tract and are expelled by coughing or are swallowed and either reabsorbed in the gastrointestinal tract or excreted in the feces.

Particles that impact beyond the ciliated area may be transported by the alveolar fluid surface to the ciliated area, or they may dissolve and be transported to the bloodstream, or they may be bound to tissues in the lungs. Such particles also may be sequestered in the lung tissue, or engulfed by phagocytes, which are blood cells that ingest and destroy foreign particles, such as bacteria. Engulfing of the particle may kill the phagocyte, or may cause the particle to travel to the ciliated area, to the lymph nodes, or through the alveolar wall into the interstitial tissue. These reactions cause a multifaceted response to pulmonary clearance. The upper respiratory tract is cleared in a relatively short time. The half-life of the material deposited in this area is 24 h. In contrast, low-solubility particulates deposited deep in the pulmonary system have a slow clearance rate, with half-lives measured in months or years.

The presence of dust in the respiratory tract may cause several types of reactions, depending on the toxicological characteristics of the irritating substance. Particulates may cause secondary infection or inflammation. Their presence may cause (1) allergic responses, (2) fibrogenic responses due to cytotoxicity, (3) immunological responses releasing histamine, or (4) malignant neoplastic lesions. However, the particulate may be absorbed and transported to another organ, or it may be excreted.

Regulatory Agencies and Standards

The Occupational Safety and Health Administration (OSHA) regulates the airborne concentration of contaminants in the workplace. Current OSHA standards (1983) or permissible exposure limits (PEL) are based on the 1968 American Conference of Governmental and Industrial Hygienists (ACGIH) threshold limit values (TLV). Current ACGIH threshold limit values were revised in 1982, and TLV standards for new substances were added. These threshold limit values carry no legal status unless included by reference in federal and/or state statutes; they are recommendations and should be used as guidelines for good practice.

The OSHA permissible exposure limits and ACGIH threshold limit values are the time-weighted average of airborne concentration for an 8-h workday. Recently, ACGIH has, in some cases, adopted short-term exposure limits (STEL) for various materials. A STEL is defined as a 15-min time-weighted average that should not be exceeded at any time during a workday. These limits represent expert opinion as to the airborne concentration to which workers can be exposed continuously for a short time without suffering irritation, chronic or irreversible change, or narcosis of sufficient degree to increase accidental injury, impair self-rescue, or materially reduce work efficiency. Short-term exposure limit values are valid measures to control toxicity only if the daily 8-h TLV also is not exceeded.

Beryllium

Beryl, which is a silicate of beryllium and aluminum ($3BeO \cdot Al_2O_3 \cdot 6SiO_2$), is the chief ore of beryllium. Beryllium is used extensively in the x-ray, nuclear, and aerospace industries and as an alloying element in beryllium copper and beryllium nickel alloys used for electrical and electronic applications.

The acute toxicity of beryllium compounds on the skin and respiratory tract have been well documented and are caused exclusively by water-soluble extraction intermediaries such as beryllium sulfate, beryllium chloride, and beryllium fluoride.

Mere contact with solid beryllium metal, alloys, or ceramic materials does not produce adverse reactions. Only when beryllium particles are deposited in subcutaneous tissues, such as from cuts or puncture wounds caused by broken fluorescent light tubes, granulomatous skin lesions develop, sometimes forming ulcerations which require curettage for successful healing. The use of beryllium oxide for the interior coating of fluorescent tubes, however, was discontinued by the lamp industry in 1950.

Chronic beryllium disease (berylliosis), characterized by a granulomatous interstitial pneumonitis (inflammation and granuloma formation within the supportive lung tissues), results from a hypersensitivity reaction. Unless proper ventilation is provided, berylliosis must be considered a potential hazard whenever the processing of beryllium compounds by melting, grinding, or buffing results in finely dispersed particles. Cases among persons living in the vicinity of beryllium plants have been traced to beryllium oxide and metal dust carried on workers' clothing into homes. Work uniforms should be issued and laundered at the plant site.

Chronic beryllium disease can remain latent for 5 to 30 years after exposure. In contrast to typical pneumoconiosis, such as silicosis or asbestosis, only days or weeks of excessive exposure are required to produce the disease. However, the disease is very selective, affecting only 1 to 2% of those persons exposed. A predisposition to the disease, sensitization, and minimum exposure are prerequisites for berylliosis.

The signs and symptoms of chronic beryllium disease, such as exertional dyspnea, substernal burning pain, and persistent cough, can also be observed with many other types of interstitial lung disease. The same can be stated for the fine nodular opacities that appear on the chest x-ray, the reduction in vital and diffusion capacity, and even the histological changes. An open lung biopsy (which is performed almost routinely in suspected cases) is actually of little differential diagnostic value because of the histological similarities and the fact that higher than background levels of beryllium (0.02 $\mu g/g$ lung tissue) are found in all beryllium workers. High concentrations have been demonstrated in perfectly normal lungs and thus can serve only as evidence of exposure but not of disease.

The lymphoblast transformation test, based on lymphocyte sensitization to beryllium and performed by immunologic tissue laboratories, represents the only differential diagnostic procedure of value. Its use is recommended whenever exposure history and clinical symptoms raise a suspicion of beryllium-related disease and where similar diseases of different origin have to be ruled out. Sarcoidosis, an interstitial lung disease of much greater prevalence than chronic berylliosis, is particularly difficult to distinguish unless a negative lymphoblast transformation test has been obtained.

Chronic beryllium disease, following its initial manifestation, has a tendency to transform portions of the affected lung into scar tissue, with a resulting loss of vital capacity and respiratory reserve. The extent of this disabling development varies significantly from minimal to severe. Treatment with steroids will not effect a cure, but will reduce the respiratory impairment to some degree and make the patient more comfortable.

Beryllium compounds have been found to be carcinogenic in several animal species. Beryl ore, beryllium fluoride and sulfates, and beryllium hydroxide and oxides produce tumors in rats exposed by inhalation or intratracheal injection. Pure beryllium, carbonates, oxides, phosphates, silicates, and zinc silicates have produced osteocarcomata in rabbits after intravenous injection. In monkeys, beryllium oxide and sulfate produce lung tumors after intrabronchial implantation or inhalation (Ref 1).

Epidemiological evidence for beryllium carcinogenicity to man has to date remained unconfirmed. Several investigations, confounded largely by a lack of accurate smoking histories, provide limited evidence that exposure to beryllium may lead to lung cancer in humans. These data for the most part were derived from beryllium production facilities and from the beryllium case registry (Ref 2).

Bony lesions, osteosclerosis in rats and rabbits, rickets in growing rats, and anemia have been produced in experimental investigations, but appear to be without counterparts in man (Ref 3).

Exposure Limits. Despite a long and lengthy hearing in 1975 on a proposal to lower the PEL to 1 $\mu g/m^3$, the current OSHA standard for limiting exposure to workers is a 2-$\mu g/m^3$ time-weighted 8-h average, and a 25-$\mu g/m^3$ peak for 30 min. A 5 $\mu g/m^3$ ceiling was proposed, but not adopted. This standard was adopted by OSHA for indications other than cancer. While the ACGIH threshold limit value is 2 $\mu g/m^3$, beryllium is classified as an industrial substance with suspected carcinogenic potential in man (Ref 4 and 5).

Nickel

Nickel, which generally is not found as a pure metal, is widely distributed in nature and occurs at a concentration of 80 mL/m^3 (80 ppm) in the earth's crust. Nickel is found in sulfide ores, nickel arsenides and nickel antimonides, silicates, and other ores such as morenosite, annabergite, and zaratite. Nickel is used primarily in electroplating applications, electronics, coins, batteries, food processing, alloying elements in alloys and stainless steels, and in nickel alloy and superalloy production. Nickel also is used as a catalyst in the hydrogenation of fats and oils. The most important use of nickel powders is in the manufacture of special parts made of corrosion- and heat-resistant alloys. Small additions of nickel also are used in porous P/M premixes to increase strength and hardness.

Humans normally ingest 300 to 600 μg of nickel per day. The total amount normally present in the body (body burden) is less than 10 mg, but some geographic variations occur. Some nickel is found as a contaminant in food from processing, and relatively large amounts occur in vegetables, legumes, and grains. Nickel may be essential for animal nutrition, although it has not been demonstrated that nickel deprivation produces abnormalities in experimental animals. Most ingested nickel is excreted in the feces, but a small amount is absorbed and can be excreted in the urine, bile, and sweat. As with many essential elements, mechanisms for limiting intestinal absorption may be involved. Additionally, many nickel salts have astringent or irritant properties that limit absorption. Thus, nickel and nickel salts have relatively low toxicity in most animal species when administered orally. However, injection of nickel salts produces indications of toxicity that include hyperglycemia, as well as effects on the gastrointestinal and central nervous systems.

Nickel is contained in the atmosphere in urban areas, as a result of incineration and automobile and industrial exhaust. Little information from animal studies on the effects of inhaled nickel compounds, except for nickel carbonyl, is available. A few nickel-containing substances, including nickel dust, nickel subsulfide, nickel oxide, nickel carbonyl, and nickel biscyclopentadiene, have produced cancer in experimental animals after inhalation or injection. Epidemiologic studies of nickel smelters and refiners have shown an increase in lung cancer and carcinoma of the nasal cavities in workers. Nickel subsulfide and nickel oxide are apparently associated with this increase.

Dermatitis (nickel itch) is the most frequent effect of exposure to nickel. This condition can occur from direct contact with metals containing nickel, including coins and jewelry. An estimated 5% of all eczema

is caused by nickel or nickel compounds. This type of dermatitis is a sensitization reaction; in some cases, the respiratory tract as well as the skin may be affected.

Nickel carbonyl is the most toxic of nickel compounds. It is a volatile liquid that is used as an intermediate in the Mond process of nickel refining. Nickel carbonyl is formed when carbon monoxide is passed over the heated nickel. It also is used for nickel plating in the electronics industry and as a catalyst in the petroleum and rubber industries.

Fatigue, nausea, dizziness, and headaches are the immediate symptoms. These symptoms subside in about half of the subjects, especially when they are exposed to fresh air. Delayed symptoms occur in 12 to 36 h and include difficulty in breathing, pain experienced during inhalation, dry cough, muscular weakness and pain, chills, and chest pain. Interstitial pneumonitis can be detected by x-ray; fever, cyanosis, and increased heart rate may be present. Liver damage may also occur.

Death may occur due to respiratory failure, cerebral edema, or hemorrhage. Frequently, 2 to 3 months may be required for functional recovery, but the effects of nickel poisoning—pulmonary fibrosis, for example—may be experienced for several years.

Chelating agents, such as sodium diethyldithiocarbonate, are currently used to treat nickel carbonyl poisoning. Corticosteroids are used as a supplementary treatment (Ref 6).

Exposure Limits. The current OSHA permissible exposure limit for nickel (metal and soluble compounds determined on the basis of nickel content) is 1 mg/m^3. The PEL for nickel carbonyl is 0.007 mg/m^3. The ACGIH has adopted a TLV of 0.35 mg/m^3 for nickel. They designate a separate TLV for pure nickel and soluble salts of 1 and 0.1 mg/m^3, respectively. Furthermore, ACGIH suggests a STEL (15 min) of 0.3 mg/m^3 for soluble nickel salts. Nickel sulfide, roasting, fumes, and dust are rated separately by ACGIH. They indicate that these substances are associated with industrial processes and are recognized to have carcinogenic or cocarcinogenic potential; a TLV of 1 mg/m^3 for nickel is suggested by ACGIH (Ref 4).

Iron

Iron is an abundant metal in the earth's crust. The most common iron ore is hematite (Fe_2O_3), from which iron is obtained by reduction with carbon. Other forms of commercial iron ore are mag-

netite (Fe_3O_4) and taconite, an iron ore consisting of a ferruginous chert derived from ferrous silicate. Pure iron is seldom used commercially. Iron usually is alloyed with other metals and minerals.

Iron is basic to the production of steel. With carbon, it is used for casting major parts of machines, and it can be machined, cast, formed, and welded. Powdered iron is the most important metal in the P/M industry for the production of structural parts. Elemental iron and iron compounds are used for food enrichment. Iron compounds also are used to precipitate other metals and undesirable minerals from industrial waste streams.

Every tissue of the body contains iron. In normal concentrations, the body burden is about 4 g, which is highly concentrated in red blood cells as hemoglobin. Sixty-seven percent of the total iron is contained in hemoglobin, and 27% is stored (mainly in the liver) as ferritin or hemosiderin in case of excess intake.

Oral absorption of iron is physiologically controlled. The principal site for limiting absorption of iron is the intestinal mucosa. In this homostatic mechanism, the trivalent form is absorbed into the gastrointestinal mucosa and converted to the divalent form and attached to the ferritin. The ferritin passes into the bloodstream and is then converted and transferred to the trivalent form or is transported to the liver or spleen for storage as ferritin or hemosiderin. The absorption of iron from the gastrointestinal tract may depend on hepatic and pancreatic secretions. The adequacy of iron stores in the body, however, seems to be the major controlling factor in the absorption of iron by the gastrointestinal tract. Vomiting may be the first sign of gastrointestinal tract irritation.

Long-term inhalation exposure to iron, particularly iron oxide, results in mottling of the lungs, or siderosis. This condition is considered a benign pneumoconiosis and does not ordinarily cause significant physiological impairment. However, hematite miners in certain areas have been reported to show up to 70% higher death rates due to lung cancer. It has been suggested that at least part of this increase may be due to radioactivity in the mine fields. Injectable and organic irons are much more toxic than the inorganics, because they bypass the physiological mechanisms controlling absorption (Ref 6).

Exposure Limits. The current OSHA permissible exposure limit for iron oxide fumes is 10 mg/m^3. However, ACGIH has adopted a TLV for iron oxide of 5 mg/m^3

and has set a TLV of 1 mg/m^3 for soluble iron salts (Ref 4).

Magnesium

The primary ores of magnesium are magnesite (MgCO$_3$) and dolomite CaMg(CO$_3$)$_2$. Magnesium also may be obtained from brine wells and salt deposits. Its concentration in the earth's crust is about 21 000 mL/m^3 (21 000 ppm). Magnesium alloys are used for applications requiring lightweight materials. Additionally, wire and ribbon used in radios contain magnesium. In the P/M industry, it is used mainly in flares and other incendiary devices and as an alloying element in aluminum alloys.

Magnesium is an essential nutrient. Magnesium citrate, oxide sulfate, hydroxide, and carbonate are used widely as antacids, or at higher doses as cathartics. Normal intake is high (500 mg), mostly from foods. Nuts, cereals, seafoods, and meats contain substantial amounts of magnesium. The average body burden is 20 000 mg. Excretion via the kidneys is the principal physiological mechanism for regulating the magnesium content of the body (Ref 7).

Other than causing diarrhea, high doses of magnesium salts rarely cause poisoning, unless kidney damage exists. Indications of magnesium intoxication include a sharp drop in blood pressure and respiratory paralysis caused by depression of the central nervous system.

The risk of ill effects from occupational exposure to magnesium, its alloys, and salts is relatively low. Magnesium oxide fumes have caused metal fume fever, although exposure to zinc oxide fumes, generated while welding magnesium-zinc alloys, is the most frequent cause of this condition. Initial exposure to magnesium oxide fumes may cause chills; repeated exposure causes no reaction. In the most severe cases, recovery from this allergic-like reaction usually is complete in 24 to 48 h. Some investigators have reported that there is a high incidence of digestive disorders in magnesium plants. Slivers of magnesium penetrating the skin may cause damage from the evolution of hydrogen. The greatest danger in the handling of magnesium occurs in grinding operations, where small fragments of the metal are ignited and burned at high temperatures (Ref 6).

Exposure Limits. The OSHA permissible exposure limit is 15 mg/m^3 for magnesium oxide fumes. The ACGIH threshold limit value is 10 mg/m^3. Additionally, ACGIH has established a TLV of 10 mg/m^3 for magnesite (magnesium carbonate). A STEL of 20 mg/m^3 for magnesite also has been suggested (Ref 4).

Cobalt

Cobalt is produced chiefly as a by-product of copper and nickel. Its concentration in the earth's crust is 25 mL/m^3 (25 ppm). Cobalt is extracted primarily from smaltite or cobalt speiss, (FeCoNi)As$_2$; cobaltite, CoAsS; and cobalt bloom or erythrite, Co$_3$(AsO$_4$)$_2 \cdot$8H$_2$O. It is used in high-temperature alloys and in permanent magnets. Important applications of cobalt in the P/M industry include powders for plasma spraying and binders in cemented carbides. Cobalt salts are used in paints, in pigments, and as catalysts.

Cobalt salts generally are well absorbed after oral ingestion, probably from the small intestine. Fish, cocoa, and mollusks contain relatively high concentrations of cobalt. The average daily intake is 0.3 mg, but body burden remains relatively constant at 1 mg because of rapid excretion, mainly in the urine (80%) and secondarily in the feces by an enterohepatic pathway.

Polycythemia, an increase in red blood cells, is the characteristic response of most mammals, including man, to ingestion of excessive amounts of cobalt. In addition, epidemiological studies have suggested that an increased incidence of goiter occurs in areas that have higher levels of cobalt in the water and soil. Cardiomyopathy has been caused by excessive intake of cobalt, particularly in beer to which cobalt salts have been added to enhance foaming qualities (Ref 6).

Occupational exposure to cobalt in the metallurgical industry increases the risk of asthma and chronic bronchitis, and decreases ventilating capacity. Airborne concentrations of 0.1 mg/m^3 cobalt sulfate increase the risk of asthma about five-fold in exposed workers, with increased phlegm and wheezing (Ref 8).

Exposure Limits. The current OSHA permissible exposure limit and ACGIH threshold limit value is 0.1 mg/m^3. However, ACGIH has set 0.1 mg/m^3 as the TLV for cobalt hydrocarbonyl and carbonyl. For metal dust and fumes, they propose 0.05 mg/m^3 for the 8-h time-weighted average and 0.1 mg/m^3 for a short-term exposure limit (Ref 4). These changes in policy reflect increased respiratory indications at levels of 1 to 2 mg/m^3 in the cemented carbide industry.

Molybdenum

Molybdenite (MoS$_2$) is the most important ore of molybdenum. Industrial uses include the manufacture of catalysts, lubricants, dyes, and high-temperature-resistant steel alloys used in pressure-vessel components, turbines, and jet aircraft engines. In the P/M industry, molybdenite is widely used in powder premixes as an alloying element because during sintering it is readily kept in its reduced state.

Molybdenum is distributed widely in nature, about 1 mL/m^3 (1 ppm) in the earth's crust, and is an essential element. Total body burden averages about 9 mg, and the daily intake is about 0.35 mg. Relatively high concentrations are found in shellfish and foods of plant origin. The role of various valence forms of molybdenum is not well understood. After oral ingestion, the hexavalent forms are well absorbed, and the liver concentration of molybdenum is increased. Excretion of molybdenum in the urine is relatively rapid.

Ingestion of forage containing high levels (20 to 100 mL/m^3, or 20 to 100 ppm) of molybdenum by cattle and sheep may result in anemia, poor growth, and diarrhea. Continual high levels of ingestion may cause joint deformities. This can be prevented by including large amounts of copper in the diet.

Exposure to soluble molybdenum salts at air concentrations of about 9.5 mg/m^3, as an 8-h time-weighted average, has been reported to increase serum ceruloplasmin and serum uric acid. These changes are probably due to the interaction of molybdenum with copper and the role that molybdenum has with xanthine oxide, the enzyme which metabolizes uric acid. High serum uric acid levels normally are associated with gout. At apparently high levels of intake (particularly in mining operations), joint conditions similar to gout may occur (Ref 9).

Exposure Limits. The current OSHA permissible exposure limit for molybdenum takes into consideration the potential hazard for compounds of different solubility. The permissible level for soluble molybdenum compounds is 5 mg/m^3, while that for insoluble compounds is 15 mg/m^3. The ACGIH recommendation is also 15 mg/m^3; however, they recommend a STEL of 10 or 20 mg/m^3 for soluble and insoluble compounds, respectively.

Copper

Copper, as well as silver, gold, and mercury, occurs in nature as a free metal (not combined with other materials). In addition to its free metal state, copper occurs in several oxides and as carbonate and sulfide ores. As an oxide, it occurs in small amounts as cuprite (Cu_2O) and as melaconite (CuO). In carbonate and hydroxide forms, copper is found in malachite and azurite. The most important sources, however, are sulfide ores, such as copper pyrite ($CuFeS_2$) and erubescite (Cu_3FeS_3).

The average concentration of copper in the earth's crust is 45 mL/m³ (45 ppm). It is used widely in industry because of its superior conductivity and malleability. Various copper salts have been used as pesticides and coloring agents. Copper provides the base of many alloys. For example, beryllium-copper alloy (1 to 2% Be) is used where toughness and high tensile properties are required. Other copper alloys include various types of bronzes, brass, gunmetal, and Nevsilber (German silver). In the P/M industry, pure copper powder is used as a constituent in bronze bearings, friction materials, carbon brushes, and ferrous structural parts. Copper alloy powders include bronzes, brasses, and nickel silvers.

Copper is an essential element that is a constituent of many enzymes. Dietary deficiency is associated with many disorders, including anemia, depressed growth, neonatal ataxia, and impaired reproductive performance. Ceruloplasmin, which acts as an oxidase, is the major copper compound in the body and is also essential to plant growth. Thus, most foods contain some copper. Shellfish have particularly high concentrations. The normal body content is about 100 mg for a 70-kg (150-lb) man.

Copper apparently prevents toxicity induced by high oral intake of molybdenum and zinc (Ref 10). High oral doses of copper salts induce systemic toxicity; the liver is the main target organ. Oral intake of most copper salts produces only gastroenteritis, with accompanying nausea and intestinal irritation.

Copper poisoning is not cumulative or systemic, because it is readily excreted. However, acute copper sulfate poisoning with severe gastrointestinal symptoms may occur under some circumstances. Death caused by acute poisoning and subsequent liver necrosis has been reported (Ref 11).

Industrial exposure may cause "brass chills," another form of metal fume fever. However, this condition may be attributed to other factors, such as exposure to zinc oxide. The increased incidence of lung cancer in coppersmiths generally is not accepted as significant evidence of the carcinogenic potential of copper. Increased risk of lung cancer among workers at copper smelters probably is attributable to the arsine trioxide released from the ore (Ref 1 and 6).

Exposure Limits. The current ACGIH threshold limit value is 0.2 mg/m³ and 1 mg/m³ for dust and mist. A short-term exposure limit of 2 mg/m³ has been adopted. The OSHA permissible exposure level is 0.1 mg/m³ for copper fumes and 1 mg/m³ for dust and mist. The ACGIH increased the TLV for copper fumes from 0.1 mg/m³ in 1973, because of data suggesting fumes were not as hazardous as previously believed (Ref 4). However, the OSHA permissible exposure limit is legally enforceable.

Silver

Silver occurs in small amounts in many areas; average concentration in the earth's crust is 1 mL/m³ (1 ppm). Its primary ore is argentite, or silver glance (Ag_2S). Silver also is obtained as a by-product of copper, lead, and other metal-smelting operations. It is used in electrical applications because of its conductivity. The largest commercial application of silver is photography, as silver halides. It is also used to produce tableware and jewelry. In powder form, silver is used to make composite electrical contacts.

Large oral doses of silver nitrate cause some gastrointestinal irritation due to its caustic action. Chronic occupational exposure to silver may cause argyria. Local argyria is characterized by the formation of gray-blue patches on the skin, or manifests itself in the eye. The generalized form of this disease is characterized by widespread blue pigmentation on the face, which eventually spreads from the face to most uncovered parts of the body. Interaction of silver and light cause this change in skin coloration. In some cases, the skin may become blackish, with almost a metallic luster. The eyes may be affected to such an extent that the lenses and vision are disturbed. X-rays of the chest may reveal radiopaque changes, but these are benign. There is no known method of removing silver from the tissues once deposited, and the effect is cumulative (Ref 6).

Exposure Limits. The OSHA permissible exposure limt is 0.01 mg/m³. The ACGIH threshold limit value for soluble silver compounds is the same, but the TLV for pure silver was raised to 0.1 mg/m³ (Ref 5).

Gold

Gold is distributed in small quantities, but the major economically usable deposits occur as a free metal in quartz veins as alluvial gravel. It is recovered from the alluvial deposits by some form of water concentration, followed by amalgamation. Gold is used in industrial applications only where its superior electrical and heat conductivity, malleability, and ductility outweigh its cost. Gold also is used in jewelry, dental alloys, and solders.

Toxicity of gold and its salts seems to be largely associated with its therapeutic or medical rather than industrial use. For example, organic gold salts used in the treatment of arthritis are highly toxic. Dermatitis is the most frequently reported toxic reaction. Occasionally, renal lesions occur (Ref 6).

South African gold miners have been reported to develop a respiratory disability with bronchitis. However, silicosis has been well documented, and the roles excess temperature, humidity, and altitude may play are unknown (Ref 7).

Exposure Limits. No regulations or acceptable levels for gold exist, possibly because the cost of control is outweighed by the recovery of the material.

Aluminum

Aluminum is a silvery white metal that is very abundant in the earth's crust (8.1%), but is never found free in nature. Its principal ore is bauxite, a hydrated aluminum oxide ($Al_2O_3 \cdot 2H_2O$). Alumina is extracted from bauxite that is dissolved in molten cryolite. Aluminum is produced by electrolysis of this melt.

Aluminum is light, malleable, ductile, possesses high thermal and electrical conductivity, and is nonmagnetic. It can be machined or cast. Although very reactive, aluminum forms a protective oxide that, under many conditions, prevents corrosion.

Aluminum is one of the principal basic metals used in the coil-coating industry. In addition, aluminum cast products are widely used in the automotive industry. Wrought alloys, such as the 2xxx and 7xxx series alloys, are widely used in the aerospace industry. Wrought alloys also are used in the building-construction industry. In powdered form, the most important use of aluminum is in the pyrotechnics industry, e.g., in rocket fuels. Aluminum and aluminum alloy powders are used to make P/M structural parts.

Aluminum is commonly used for cooking utensils, but there have been no reported adverse effects from low concentrations in the diet or drinking water, normally in the range of 10 to 100 mg and leading to the 50- to 150-mg body burden. Massive oral doses are reported to be toxic, interfering with phosphate absorption and producing gastrointestinal irritation (Ref 1).

Shaver's disease is the only suggested human manifestation of aluminum-induced industrial disease. It is caused by bauxite fumes and by the use of abrasive wheels containing aluminum. Exposure to aluminum fumes produces weakness, fatigue, and respiratory distress. Silica may play an important role in this syndrome. However, in experimental animals, similar changes can be reproduced by intratracheal injection. Repeated skin contact with the soluble salts results in irritation from the acid of hydrolysis. In addition, an anesthetic condition of the fingers can occur from long contact with aluminum sulfate. Fumes from aluminum soldering may produce a delayed type of asthma (Ref 12).

Exposure Limits. No specific exposure limit for aluminum or its compounds has been established by OSHA; thus, for regulatory purposes, these materials are considered inert dust (15 mg/m^3 total or 5 mg/m^3 respirable). The ACGIH currently has four threshold limit values: 10 mg/m^3 for the metal and its oxides, 5 mg/m^3 for pyro powders, 5 mg/m^3 for welding fumes, and 2 mg/m^3 for soluble salts. In addition, ACGIH has proposed a TLV of 2 mg/m^3 for aluminum alkyls (Ref 4).

Tungsten

The concentration of tungsten in the earth's crust is 70 mL/m^3 (70 ppm). Wolframite, $(Fe,Mn)WO_4$, and scheelite, $CaWO_4$, are the most abundant ores. Tungsten powders are used in cemented carbides, in electrical contacts, and as alloying elements in high-speed tool steels. Additional applications include uses in welding rods, x-ray tubes, light bulbs, pigments, and waterproof textiles. Although not a normal complement of animal tissues, tungsten is absorbed to some extent from the gastrointestinal tract and retained largely in bone and to a lesser extent in the spleen, liver, and kidneys. Toxicity of tungsten compounds after oral ingestion is a function of the particular salt. While nervous prostration, coma, and death due to respiratory paralysis have been caused in experimental animals following oral administration, this route of exposure is apparently not of significance in man. In addition, it has been noted experimentally that tungsten interacts with two other metals, selenium and molybdenum, reducing the hepatic deposition of molybdenum in rats and reducing the hepatic toxicity of selenium.

The metal dust and tungsten carbide, both highly insoluble in body fluids, deposit in the connective tissue and lymph nodes after intratracheal injection. However, tissue response is minimal, and no fibrogenic response is present. Industrial experience has suggested that tungsten inhalation may be a component of "hard metal disease." This syndrome is characterized by a moderate incidence of cough, difficult breathing, and wheezing, an asthma-like response, and minor radiographic changes (Ref 6).

Exposure Limits. The ACGIH adopted a TLV of 1 mg/m^3 for soluble compounds and 5 mg/m^3 for insoluble compounds in 1969. More recently, they have suggested short-term exposure limits of 10 and 3 mg/m^3 for the insoluble and soluble compounds, respectively (Ref 4). Permissible exposure limits for tungsten or any of its compounds have not been developed by OSHA.

Tin

The average concentration of tin in the earth's crust is about 3 mL/m^3 (3 ppm). Tin is obtained almost entirely from the mineral cassiterite (SnO_2). The major use of tin is for the manufacture of tinplate. Other uses are in food packaging, dental alloys, solder, bronze, and brass. Stannous and stannic chlorides are used in dyeing textiles. Various organic tin compounds have been used for pesticides as well as stabilizers for plastics. Stannous fluoride is used as an anticavity agent in toothpaste. This action is unrelated to the tin ion. In the P/M industry, the most important use of tin is in the manufacture of self-lubricating bronze bearings.

The daily intake of 3.5 to nearly 40 mg of tin is largely from dietary sources, mainly due to increased levels in processed canned food as a result of the packaging. This results in a body burden of 30 mg, mostly in the gastrointestinal tract, with only a minor part due to inhalation of tin compounds in urban air. Only a very small portion of ingested inorganic tin salts is absorbed. While organic tin compounds tend to absorb more readily when ingested and can result in death due to effects on the central nervous system, 4 mg/kg of inorganic tin salts taken by volunteers produced only nausea and diarrhea, probably as a result of local gastrointestinal irritation.

Inhalation of tin oxides as dust or fumes leads to a benign pneumoconiosis. Many case reports of accompanying minor radiograph changes in the lungs are in the medical literature. No necrosis or foreign body response is reported.

Exposure Limits. The TLV for tin and its inorganic salts, excluding tin oxide, is 2 mg/m^3, which is the same as the OSHA permissible exposure limit. A recommended STEL is 4 mg/m^3. The TLV for organic tin is 0.1 mg/m^3, which illustrates the more hazardous nature of these materials (Ref 4).

Titanium

Titanium is considered the fourth most abundant metal in the earth's crust. Rutile (TiO_2) and ilmenite ($FeTiO_3$) are the primary ores. Industrial uses include deoxidizers in corrosion-resistant alloys, welding rods, aerospace components, electrodes, and lamp filaments. Titanium dioxide salve has been used in the treatment of burns. Also, about half of the titanium produced as titanium dioxide is used in pigments for paint, lacquers, and varnishes. The paper industry uses large quantities of titanium because of its ability to lend brightness and clearness to paper. In the P/M industry, titanium powder is used to make structural parts; both titanium-based and titanium-containing powders are used for high-temperature applications.

Although titanium has been detected in some foods (butter, corn oil, shrimp, and lettuce), most of the average body burden (15 mg) is in the lungs because of inhalation of urban air. About 90% of the titanium emission is from the burning of coal. Oral toxicity of titanium is minimal; only about 3% is absorbed (Ref 6).

A slight fibrosis of lung tissue has been reported following inhalation exposure to titanium dioxide, the principal industrial compound. Titanic acid, another important industrial compound, is an irritant.

Exposure Limits. The ACGIH threshold limit value for titanium oxide dust was 15 mg/m^3 until 1970; the TLV for "inert" particulates was reduced to 10 mg/m^3 at that time. The Occupational Safety and Health Administration has not acted in a similar fashion; they have set 15 mg/m^3 as an acceptable PEL. In summary, except for certain organo titaniums, titanium and its compounds are almost physiologically inert (Ref 7).

Tantalum

The concentration of tantalum in the earth's crust is less than 1 mL/m^3 (1 ppm). Oral absorption of tantalum and its salts is low. It is frequently found with niobium in tantalite, $(Fe,Mn)(Nb,Ta)_2O_6$, or columbite, $(Fe,Mn)Ta_2O_6$. The most important application for tantalum is in electrolytic capacitors. Tantalum also is used in the form of strip, sheet, tubing, wire, and fabricated shapes in the chemical industry because of its corrosion resistance, in the electronics industry as emitters in electronic tubes, and as high-temperature heating elements and radiation shields. Medically, tantalum is used as supporting gauze in the repair of hernias, in dressings for burns, and in prosthetic appliances, because it is biologically nonirritating.

Exposure Limits. Several animal experiments have suggested slight pulmonary effects of a benign and nonfibrotic nature. Adverse effects resulting from industrial exposure are minimal. The current OSHA permissible exposure limit is 5 mg/m^3, the same as the ACGIH threshold limit value. In addition, ACGIH has suggested a 10 mg/m^3 STEL (Ref 4).

Zirconium

Zircon (ZrSiO$_4$) is the primary zirconium-containing ore. Hafnium and zirconium commonly occur together because of their chemical similarities. In nuclear applications, hafnium must be removed from zirconium because of the high neutron absorption characteristics of hafnium. This is not the case with most other commercial applications, which include dyes, pigments on ceramics, and abrasives. The daily oral intake is about 1 to 6 mg. The average body burden is estimated at 250 mg for a 70-kg (155-lb) man. Meats, grains, and dairy products are the highest source. Plants take up zirconium from the soil.

Inhalation of zirconium for 60 days produced slight decreases in hemoglobin and red blood count in dogs and increased mortality rates in rats and guinea pigs. No evidence of industrial diseases related to zirconium exposure has been documented. Oral toxicity of zirconium is low. Granulomas of allergic epithelial origin have been described, principally from use of sodium zirconium lactate in deodorants. Removal of the sensitizing zirconium compounds results in healing of the lesion (Ref 6).

Exposure Limits. The ACGIH threshold limit value and the OSHA permissible exposure limit for zircon compounds are 5 mg/m^3.

Platinum Group Metals

Six elements (platinum, palladium, rhodium, ruthenium, iridium, and osmium) are called the platinum group metals. They are found in very low concentrations in the earth's crust. Palladium is the most abundant, but its concentration is only 0.01 mL/m^3 (0.01 ppm).

Platinum occurs naturally in alluvial deposits and is recovered as a by-product from nickel-copper ores. Palladium occurs naturally in crude platinum and in small quantities in cupriferous pyrites, particularly those containing nickel and iron sulfide.

Rhodium is mainly produced as a by-product in the extraction of platinum from the copper-nickel refining residues. Iridium is obtained from the extraction of platinum from electrolytic copper and nickel refinery residues. Osmium occurs naturally in crude platinum and is alloyed with iridium in osmiridium. It usually is extracted from this latter source.

Because of their resistance to oxidation and other properties, these metals and compounds are extremely useful as catalysts in the chemical and petroleum industries, as conductors in the electrical industry, and in dental alloys, medical prostheses, and jewelry. Generally, platinum and palladium are frequently used for catalysts, alloying, and electrical and high-temperature applications. Rhodium also is used in high-temperature applications and where corrosion resistance is required. Organic platinum complexes have been found to have some antitumor activity. Osmium tetraoxide has been used as a histologic stain.

The platinum metals are largely nontoxic. Presently, significant occupational exposure is limited to employees in platinum refineries and plants that synthesize noble metal catalysts. Asthmatic or dermal allergies (platinosis) disappear when exposure is discontinued. Ionic water-soluble platinum salts cause these reactions (Ref 13).

Exposure Limits. The ACGIH has set a TLV of 1 and 0.002 mg/m^3 for platinum and soluble platinum salts, respectively. Rhodium metal has a TLV of 1 mg/m^3. A TLV of 0.001 mg/m^3 and a STEL of 0.003 mg/m^3 for the soluble rhodium salts have been suggested. Osmium tetraoxide has a TLV of 0.02 mg/m^3 and a 0.006 mg/m^3 STEL (Ref 4). A PEL of 0.002 mg/m^3 for osmium tetraoxide has been set by OSHA. Rhodium metal fumes and dust have a PEL of 0.01 mg/m^3; the PEL for soluble rhodium salts is 0.001 mg/m^3. Soluble platinum salts have a PEL of 0.002 mg/m^3.

Lead

Lead ores containing from 3 to 10% Pb are prevalent, mostly in the form of sulfides. Galena (PbS) is the most common ore. Overall concentration in the earth's crust is 15 mL/m^3 (15 ppm). Lead smelters produce about two thirds of the lead used in the United States. Lead is also produced by reclaiming scrap lead, primarily from automobile batteries; lead is used mainly for storage batteries. Use of lead as an anti-knocking gasoline additive is declining. It also is used to manufacture solder, electrical cables and pipes, ammunition, vibration damping, radiation shielding, and paint pigments. In powder form lead is used in composite friction materials and in solders.

The toxicity of lead has long been established. The Greeks recognized the abdominal pain of lead colic and noted the relationship between lead exposure and pallor, constipation, and paralysis. The decline of the Roman Empire has been suggested to have been due, at least in part, to the effects of chronic lead poisoning caused by food containers and wine. Epidemics in the Middle Ages were caused by the ingestion of acidic beverages from lead-glazed pottery or from lead used to improve the taste of bad wine.

The normal human body burden of lead is 120 mg, resulting from daily intake of 0.3 mg. About 40% of inhaled lead is absorbed, while 90% of ingested lead is excreted. The absorbed lead in the blood comprises about 4% of the body burden and has a half-life of about 30 days. The portion of the body burden in the soft tissues is about 2% and has a half-life of approximately 35 days. The majority of the lead body burden is in the bone, with a half-life of 10 000 days.

Lead is a cumulative poison. Thus, excessive levels of lead in air, food, or water can cause a buildup in the body, leading to lead poisoning. Lead poisoning results from interference with enzymatic processes, which causes impairment of blood formation, kidney function, and reproductive function, and injury to the nervous system. Exposure to lead resulting in blood concentrations of more than 60 μg/100mL leads to lead poisoning or plumbism. Signs and symptoms of the disease include neurological disorders, chronic kidney disease, anemia, and various gastrointestinal problems.

Prolonged blood levels in the range of 30 to 60 μg/100 mL have been associated with the so-called subclinical signs of lead poisoning, which include changes in blood and urine enzymes, a slowing of nerve conduction, and slight behavioral or neurological alterations. It has been suggested that prolonged elevated blood lead concentrations can result in premature aging. Clearly, adverse effects on children and reproductive performance can occur in the 30 to 60 μg/100 mL blood lead range.

The major source of environmental pollution has been automotive exhaust; with greater use of unleaded gasoline, this is decreasing. The major industrial sources of lead pollution include:

- Dust from mines
- Drainage and runoff from mines
- Lead particulates emitted from smelting operations (particularly the sinter machines and blast furnaces)
- Wastewater from smelting operations

Near some smelting operations, the environmental burden has risen to such an extent that farm animals grazing nearby are afflicted with the typical signs of lead poisoning.

Exposure Limits. The Occupational Safety and Health Administration adopted complete health standards for inorganic lead and its compounds in 1978, based on the acute and chronic toxicity of lead. Two lead compounds, lead phosphate and lead acetate, have been shown to produce cancer in experimental animals. Furthermore, lead arsenate is regulated by OSHA standards. Currently, OSHA limits on lead are being reviewed, and changes may be forthcoming.

Fifty μg of lead per cubic metre of air has been established as the PEL value. Other provisions of the OSHA standard—requiring personal protective equipment, engineering and work practice controls, and medical surveillance with provisions for medical removal—are triggered by an action level of 30 μg/m³ measured as an 8-h time-weighted average. The OSHA standard requires biological monitoring for elevated blood lead levels. Workers whose blood lead exceeds 50 μg/100 mL must be removed to low lead areas (below 30 μg/m³) until their blood lead level drops to below 40 μg. Areas that have lead levels exceeding the PEL must be regulated, with signs posted. Smoking and eating are prohibited in these areas. When the PEL is exceeded, change rooms and showers must be provided to limit the amount of lead workers can accidentally carry on their clothes and skin.

Organic leads, specifically tetramethyl and tetraethyl lead, have a PEL of 75 μg/m³ to limit workers' exposure. Organic lead compounds can cause classical lead poisoning, but generally they produce acute poisoning due to more rapid and selective penetration of the central nervous system. These organic lead compounds are also absorbed through the skin (Ref 1, 2, 7, 14).

REFERENCES

1. IARC Monographs on the Evaluation of the Carcinogenic Risk of Chemicals to Humans, Supplement 4, International Agency for Research on Cancer, Lyons, France, 1982
2. IARC Monographs on the Evaluation of the Carcinogenic Risk of Chemicals to Humans, Vol 23, International Agency for Research on Cancer, Lyons, France, 1979, p 143-204
3. Hammond, P.B. and Beliles, R.P., Metals, in *Casarett and Doull's Toxicology*, 2nd ed., Doull, J., Klaassen, C.D., and Amdur, M.O., Ed., Macmillan, New York, 1980
4. Threshold Limit Values of Airborne Contaminants, American Conference of Governmental and Industrial Hygienists, Cincinnati, OH, 1982
5. Documentation of the Threshold Limit Values, American Conference of Governmental and Industrial Hygienists, Cincinnati, OH, 1980
6. Beliles, R.P., The Lesser Metals, in *Toxicity of Heavy Metals in the Environment*, Oehme, F.W., Ed., Marcel Dekker, New York, 1979
7. Stokinger, H.E., The Metals, in *Patty's Industrial Hygiene and Toxicology*, Vol IIA, 3rd ed., Clayton, G.D. and Clayton, F.E., Ed., John Wiley & Sons, New York, 1981
8. Roto, P., Asthma Symptoms of Chronic Bronchitis and Ventilating Capacity Among Cobalt and Zinc Production Workers, *Scandinavian Journal of Work and Environmental Health*, Supplement 6, 1980
9. Walrarens, P.A., Eraso, R.M., Solomons, C.C., Chappell, W.R., and Bently, G., Biochemical Abnormalities in Workers Exposed to Molybdenum Dust, *Archives of Environmental Health*, Vol 27, 1979, p 302-308
10. Underwood, E.J., Copper, in *Trace Elements in Human and Animal Nutrition*, 3rd ed., Academic Press, New York, 1971
11. Chuttani, H.K., Gupti, P.S., and Gultati, F., Acute Copper Sulfate Poisoning, *American Journal of Medicine*, Vol 39, 1965, p 849-854
12. Norseth, T., Aluminum, in *Handbook on the Toxicology of Metals*, Ch. 13, Elsevier/North Holland Biomedical Press, Amsterdam, 1979
13. Platinum-Group Metals, National Research Council, National Academy of Sciences, Washington, DC, 1972
14. Drill, S., Konz, J., Mahar, H., and Moore, M., The Environmental Lead Problem: An Assessment of the Lead in Drinking Water From a Multi-Media Perspective, MTR-7802, The MITRE Corporation, Metrek Division, McLean, VA, 1978

Characterization and Testing of Metal Powders

Introduction

THE SUCCESS of a P/M processing technique depends to a great extent on understanding and evaluating the physical and chemical properties of metal powders, both as individual particles and in bulk form. Relationships between these powder properties and their behavior during processing (for example, compressibility, sinterability, dimensional change during sintering) in most cases are qualitative in nature. Nevertheless, a good description of powder properties is important for several reasons:

- To improve understanding of powder behavior
- To determine tolerances in property specifications
- To ensure reproducibility of powder behavior during processing

The processing of a powder into a compact and the resultant properties of P/M parts are affected by such physical characteristics as particle size, particle size distribution, particle shape and structure, and surface condition. Other characteristics, such as apparent density and flowability, depend on these basic physical properties.

Equally important are the chemical properties of a powder; contaminants have a deleterious effect on both compressibility and sinterability. The nature and extent of these effects depend primarily on the type of contaminant, which may be classified as solid or gaseous.

This Section of the Handbook examines the physical and chemical properties of metal powders by dividing the topic of characterization and testing of powders into four subject areas. In order for a lot of powder to be tested, a sample has to be taken. Methods for taking a representative sample are described in the first part of this Section.

The second topical area discusses powder properties and concentrates on tests performed on powder particles. These test methodologies begin with measurement of particle size and particle size distribution. Many of these tests are not specific to metal powders, but have been developed for other materials, such as pigments and cements, in powder form. However, emphasis is placed on the application of these methods to metals. Optical and electrical sensing techniques, as well as microscopic methods, are described. Microscopic and image analysis techniques for determining particle shape are discussed in addition to chemical and microstructural analyses and methods to determine surface area, particle density, and porosity of metal powders.

Bulk properties of metal powders is the third subject area examined in this Section. Tests for apparent density and flow rate are included. Many of these tests are described in industry testing standards and are used extensively to characterize metal powders for commercial use.

The final part of this Section describes test methods for determining green strength and compressibility, which measure the processing behavior characteristics of metal powders. These tests are also industry standards and are important to the commercial use of metal powders in the manufacture of structural parts and bearings.

Sampling of Metal Powders

By Albert J. Neupaver
Manager of Metal Powder Technology
Pfizer Inc.
and
John C. Witsberger
Production Supervisor
Pfizer Inc.

ACCURACY OF MEASUREMENT of a property that is statistical in nature is dependent on the sampling procedures used. The properties of metal powders are statistical in nature, thus their measurement requires proper sampling techniques if the test results are to be representative of the properties of the powder as a whole.

Great care must be taken to secure accurate samples. First, the material to be tested should be properly blended or mixed. An ideal mix is one in which all particles and constituents are distributed uniformly throughout the mixture. This condition is accomplished only through proper application of the variables involved in the mixing process, such as type of blender, volume of metal powder, rotational speed of the blender, and blending time (see the article "Blending and Premixing of Metal Powders" in this Volume).

Obtaining a representative sample presents different problems for the metal powder producer and the powder consumer. The major task of the producer is to ensure that an adequate blend is accomplished and that the powder meets predetermined specifications for quality. The powder consumer must be able to obtain representative samples from a shipment often consisting of several drums. Mixture segregation caused by dumping, shaking during transit, and other vibrations complicate the consumer's task. Improper sampling techniques are the major cause of inconsistent test results.

Sampling Procedures

The methods and procedures for sampling of powders for powder metallurgical purposes are covered in MPIF 01, published by the Metal Powder Industries Federation; ASTM B215, published by the American Society for Testing and Materials; and ISO 3954, published by the International Standards Organization. These standards describe similar procedures for obtaining representative samples either from powder packaged in containers or from a powder discharged in a continuous stream from a blender. They also list equipment used for various test methods.

The specific sampling procedure should be selected after adequately reviewing the method of taking samples, sample location, size, and number. The sample should not disturb the total blend if possible. If taken from a blended lot, samples should be obtained at various times while the mixture is being discharged from the blender. If the material to be tested is in drums or containers, a sampling "thief" or a special device should be used to obtain

Fig. 1 Sample splitter

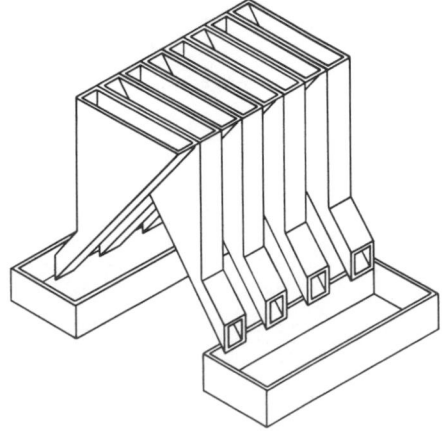

samples. In all cases, a sample splitter should be used to reduce the sample to the desired size.

The size of the sample need not be large. Generally, not more than 5% of the blend need be removed for the sample. More importantly, the sample should be truly representative of the mixture.

Several simple statistical methods can be used to analyze data obtained from sampling. To compare average property results to expected values, the t-test (test

Table 1 Recommended sampling procedures

No. of containers in lot	No. of containers to be sampled
ISO 3954(a)	
1 to 5	All
6 to 11	5
12 to 20	6
21 to 35	7
36 to 60	8
61 to 99	9
100 to 149	10
150 to 199	11
200 to 299	12
300 to 399	13
MPIF 01	
1 to 5	All
6 to 15	5
16 to 35	7
36 to 60	8
61 to 99	9
100 to 149	10
150 to 199	11
200 to 299	12
300 to 399	13
More than 400	13 + 1 per 100 additional containers

(a) For every additional 100 containers or portion thereof in the lot, one additional container shall be sampled.

Fig. 2 Sampling thieves

(a) For loosely packed powder. (b) For hard-packed powder

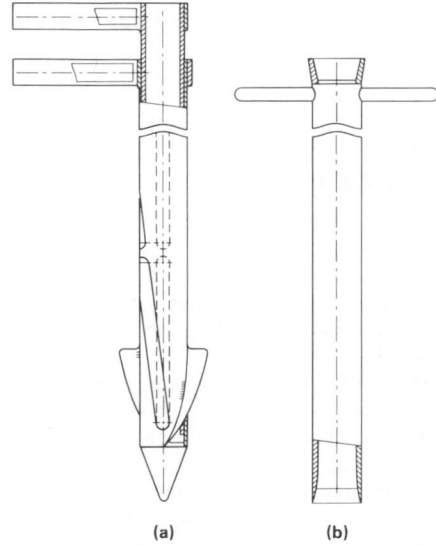

(a)   (b)

Fig. 3 Sampling from packaged containers

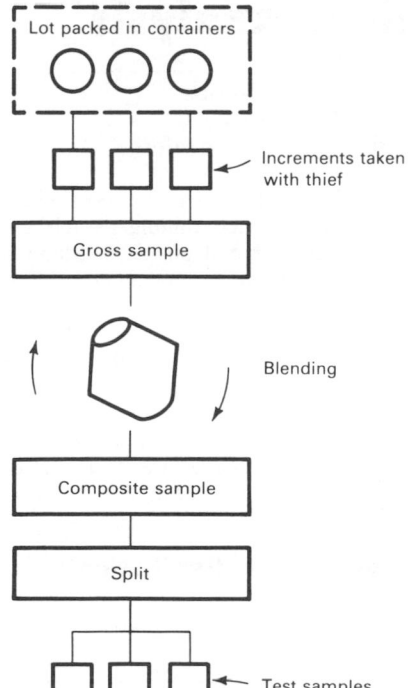

of the significance of means) is useful, because the number of samples does not have to be large to obtain meaningful results (Ref 1-4). A simple test to measure the uniformity of a lot of powder is the F-test (significance test for variances) (Ref 1-4). As with any statistical sampling, when

Fig. 4 Sampling during discharge in continuous stream

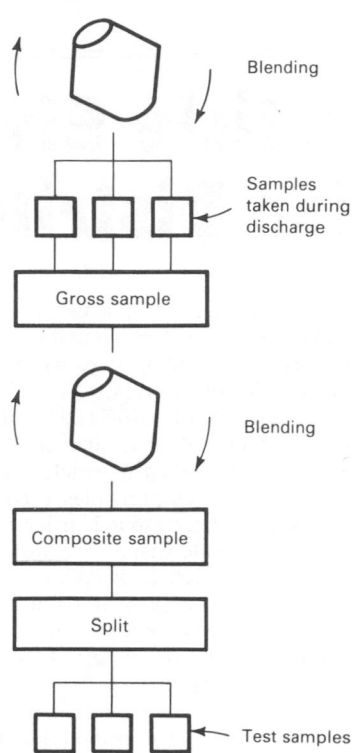

more test samples are taken, more precise information can be obtained, thus improving statistical accuracy, but at an increased cost.

Sampling During Discharge in a Continuous Stream. Samples from a blender discharge should be taken during the filling of the first shipping container when it is half full. Another sample should be obtained when the blending apparatus is half discharged. Finally, an additional sample should be taken when the last shipping container is half filled. Powder from the entire cross section of the blender discharge stream should be collected for each sample. This is accomplished by passing a rectangular receptacle through the entire discharging stream at a constant speed. Portions of these samples are blended; then a sample of the proper size for the desired test is obtained using a sample splitter. Figure 1 schematically illustrates a typical sample splitter.

Sampling From Packaged Containers. For powders packaged in containers, samples from a number of random containers should be obtained with the use of a sampling thief. The number of samples required depends on the number of containers (see Table 1). The flow characteristics of the powder dictate the type of sampling thief required. A sampling thief

constructed with an inner and outer tube with a closed end and longitudinal windows is recommended for loosely packed powders (Fig. 2a). Figure 2(b) illustrates a sampling thief with a single tube and open end, which is suitable for hard-packed powders.

The sampling thief allows powder to be taken from various depths in a container without emptying the container. The thief is inserted vertically to the container bottom at a location 70% along the radius from the center of the container. The inner tube of the thief (Fig. 2a) is rotated to open the windows until the tube is filled with powder, then rotated to close the window. The thief is removed and its contents emptied through the open top into a storage bin or blender. The single-tube thief (Fig. 2b), when withdrawn from the container, retains the powder that has entered the tube. The withdrawn sample is emptied with the aid of a ramrod. Samples should be blended and split to the desired test sample size. Figures 3 and 4 show flowcharts for sampling from packaged containers and during discharge in continuous stream, respectively.

REFERENCES

1. Dixon, W.J. and Massey, F.J., *Introduction to Statistical Analysis*, 3rd ed., McGraw-Hill, New York, 1969
2. Johnson, N.L. and Leone, F.C., *Statistics and Experimental Design*, Vol I, John Wiley and Sons, New York, 1964
3. Duncan, A.J., *Quality Control and Industrial Statistics*, 3rd ed., Richard D. Irwin, Inc. Homewood, IL, 1965
4. Natrella, M.G., "Experimental Statistics," National Bureau of Standards Handbook 91, Government Printing Office, Washington, DC, 1963

SELECTED REFERENCES

- ASTM standard B215, "Sampling Finished Lots of Metal Powders," American Society for Testing and Materials, Philadelphia
- ISO standard 3954, "Powders for Metallurgical Purposes—Sampling," International Standards Organization, New York
- MPIF standard 01, "Sampling Finished Lots of Metal Powders," Metal Powder Industries Federation, Princeton, NJ
- Lenel, F.V., *Powder Metallurgy Principles and Applications*, Metal Powder Industries Federation, Princeton, NJ, 1980
- Bradbury, S., *Source Book on Powder Metallurgy*, American Society for Metals, 1979

Particle Size and Size Distribution

PARTICLE SIZE AND SIZE DISTRIBUTION have a significant effect on the behavior of metal powders during processing; thus to a considerable extent, they govern the properties of the final products made from powder. Consequently, characterization of such properties is essential.

In the P/M industry, the traditional and most widely used method of particle size measurement is by sieving. Sieves or screens are used not only for particle size measurement, but also for separation of powders into different sieve fractions. This twofold use of sieves, in addition to the fact that most P/M powders are −80 mesh (smaller than about 177 μm in diameter, with only minor amounts smaller than 10 μm) has been well suited to industrial applications. For powders with large percentages of −400 mesh (37 μm) particles, sieve distribution data are often complemented with Fisher sub-sieve size analysis, microsieve data, or specific surface area data.

Several electronic methods of particle size analysis have been developed during the past two decades. These generally provide higher measurement speed, resolution, and small-size sensitivity than sieves, saving time and labor while yielding more precise data. However, there has also been some confusion; particle size data for the same powder can vary greatly depending on the instrument and method used.

The basic reason for such discrepancies is the phenomenon of particle shape. A definition of particle size is simple for spherical particles in which particle diameter generally is used to characterize size. Most metal powders, however, are of irregular shape. Determination of particle size and size distribution of such powders usually is based on the measurement of certain properties, or a combination of powder particle properties—length, volume, mass, settling rate, total or projected surface area, light scattering, and electric field perturbation, for example.

From such measurements, so-called equivalent spherical particle diameters can be calculated. Equivalent spherical diameters are the diameters of spheres that have the same, or equivalent, lengths, volumes, masses, or settling rates as the irregular particles themselves.

Various methods of particle size analysis provide identical or similar analysis results for spherically shaped particles only. Typically, the various principles used to measure particle size respond differently to changes in particle shape. The following sections of this article discuss the different methods of analysis and illustrate that very different results can be achieved for the same particle size, if particle shape deviates from spherical.

One method does not provide more accurate results than another method, but one method may indeed be more appropriate or discriminatory in a given application. If the application of the powder is such that its projected surface area is important, as in paint pigments, then a particle size method based on the principle of light obscuration would be more appropriate than one based on particle volume or other particle characteristics. If the application is one in which the volume, surface area, linear dimensions, permeability, or settling rate of a powder is indicative of a significant relationship with the technological properties of interest, then an instrument that directly measures this property should be used.

With appropriate precautions, useful empirical correlations can be obtained between particle size and size-sensitive technological properties. If particle size and size distribution data do not adequately predict the behavior of a powder, additional powder characteristics should be used. See the article "Particle Shape Analysis" in this Volume.

Common methods of determining particle size and size distribution and their limits of applicability are summarized in Table 1.

Sieve Analysis

By William J. Ullrich
Chief Metallurgist
Alcan Ingot and Powders

SIEVE ANALYSIS is the most widely used method of determining particle size distribution of metal powders. Standard sieves have been in use in the United States since 1910.

Particle size distribution is controlled and certified by the powder producer and is frequently checked by the end user. Typically, a series of sieves is selected that spans the full range of particle sizes present in a powder. Sieves are stacked in order, with the largest mesh size at the top

Table 1 Typical methods of particle size and size distribution measurement

Measuring principle	Method	Approximate useful size range, μm
Mechanical or ultrasonic agitation	Sieving	5 ~ 800
Microscopy	Optical	0.5 ~ 100
	Electron	0.001 ~ 50
Electrical resistivity	Coulter counter	0.5 ~ 800
	Electrozone	0.1 ~ 2000
Sedimentation	Sedigraph	0.1 ~ 100
	Roller Air Analyzer	5 ~ 40
	Micromerograph	2 ~ 300
Light scattering	Microtrac	2 ~ 100
Light obscuration	HIAC	1 ~ 9000
Permeability	Fisher sub-sieve sizer	0.2 ~ 50
Surface area	Gas adsorbtion (BET)	0.01 ~ 20

Fig. 1 Schematic of sieve series stacked in order of size

and a pan at the bottom (Fig. 1). An appropriate sample weight of metal powder is spread on the top sieve and covered. The stack of sieves is agitated in a prescribed manner (shaking, rotating, or tapping) for a specified period of time. The powder fractions remaining on each sieve and contained in the bottom pan are weighed separately and reported as percentages retained or passed by each sieve.

Sieving Equipment

Sieves

Wire cloth sieves are woven in a square mesh pattern, usually from phosphor bronze or stainless steel wires. The mesh count (number of openings per unit length) and the wire diameter determine the actual nominal sieve opening.

In 1970, the American Society for Testing and Materials (ASTM), the American National Standards Institute (ANSI), and the International Standards Organization (ISO) agreed upon a standard sieve series (see Table 2). Wire diameters and manufacturing tolerances are specified in ASTM standard E 11 and ISO standard 565. Sieves can be purchased with the following types of certification:

● Statement by the sieve manufacturer that the wire cloth has been made in compliance with ASTM E 11
● Certification from the manufacturer listing the optical measurements at several random spots on the sieve
● Certification from the National Bureau of Standards that the individual sieve has been optically examined and meets the qualification of ASTM E 11

The National Bureau of Standards supplies calibrated glass spheres for evaluating the effective openings of testing sieves. Sieve manufacturers supply "matched sieves," where extra precision is desired for comparing the results of several locations. Frequently, a laboratory performing many sieve analyses may purchase a matched pair of sieves. One sieve is used as the routine working sieve, and the other is kept as a "master" to periodically check the accuracy of the working sieve.

Sieve Agitators

Several methods of advancing the powder particles through the sieve openings have been devised that rely on one or more of these mechanisms:

● Hand shaking
● Gyratory motion
● Gyratory motion with tapping
● Mechanical shaking
● Mechanical vibration

Table 2 Standard U.S. sieve series

Sieve designation, mesh	Sieve openings	
	μm	in.
30	600	0.0232
40	425	0.0164
50	300	0.0116
60	250	0.0097
80	180	0.0069
100	150	0.0058
140	106	0.0041
200	75	0.0029
230	63	0.0024
325	45	0.0017

Source: ASTM E 11

● Electromagnetic vibration
● Oscillating air column
● Vacuum

Precision and accuracy must be assessed for each test method. Microscopic checks of the end sieve fractions and reproducibility must be evaluated. The length of sieving time must be established for each method as well. Usually, the sieving period ends when the quantity passing through the sieve retaining the largest fraction of the test portion in 1 min is less than 0.1% of the test portion.

Standard Test Method

Testing standards have been established for the sieve analysis of metal powders. These include ASTM standard B 214, published by the American Society for Testing and Materials; MPIF standard 05, formulated by the Metal Powder Industries Federation; and ISO standard 4497, developed by the International Standards Organization. All of the standards are very similar and specify the following requirements:

● Standard wire cloth sieves conforming to ASTM E 11 or ISO 565
● Mechanical sieve shakers combining a rotary motion and tapping action (Fig. 2)
● Test specimen size of 100 g (3.53 oz) for metal powders with an apparent density greater than 1.50 g/cm^3 and a 50-g (1.76-oz) test specimen for lower apparent density powders
● Mechanical sieve shaking for 15 min or another specified time
● Weighing the powder retained on each sieve and in the bottom pan to the nearest 0.1 g (0.0035 oz) and totaling all weights

Fig. 2 Mechanical sieve shaker combining rotary motion and tapping action

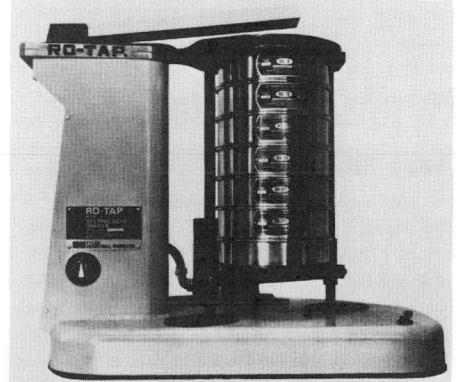

- Calculating and reporting the percentages of each sieve fraction to the nearest 0.1%

Sieving Problems

Overloaded Sieves. In this condition, the sieve surface becomes crowded with oversized and near-mesh particles after material finer than one half the size of the mesh opening has passed through. On an overloaded sieve, the weight of the oversized material tends to wedge the near-mesh particles into the openings, thus blocking passage of additional particles.

Blinded sieves (blanked sieves) result when near-mesh size particles become entrapped in the openings. When brushing fails to remove the particles, an ultrasonic wash with a wetting agent must be used to dislodge them.

Damaged sieves may be torn or have stretched and distorted mesh openings. A microscope should be used to inspect sieves regularly. Epoxy can be used to repair small defects.

Irregularly shaped particles may pass through sieve openings in a specific orientation only. For these particles to pass successfully through the openings, vibrating or shaking should be sufficient to change the orientation of all particles. Some methods of sieve agitation are more efficient than others in achieving this goal, so they must be evaluated individually.

Agglomeration of particles prevents fine particles from passing through the mesh openings. Electrostatic charges may cause fine particles to be attracted to one another, frequently forming small balls. Fine particles also can attach themselves to larger particles. The use of sieving aids, such as powdered magnesium carbonate, tricalcium phosphate, hydrophobic fumed silica, activated carbon, or commercial antistatic agents, may alleviate agglomeration. Sieving aids are usually very fine powders that pass through the sieve and collect in the pan. Thus, the amount of sieving aid used can be subtracted from the −325 mesh fraction collected.

Types of Sieving

Wet Sieving

Some difficult-to-screen materials can be handled successfully by wet sieving techniques. Apparatus is available to handle a single sieve or a stack of sieves. Ultrasonic wet sieving devices are also available. Wet sieve analyses are cumbersome to perform and require the sieve fractions to be dried before weighing. Difficult-to-screen materials are usually very fine powders, such as silicon or aluminum, that agglomerate poorly and will not separate using dry sieving techniques.

Micromesh Sieving

Micromesh sieves are produced by photoetching and electroforming techniques. Electroformed nickel produces precise square openings with planar surfaces. This thin mesh is supported by a coarser square-etched grid of nickel-plated cupronickel. These sieves are mounted in 75- or 200-mm (3- or 8-in.) stainless steel rings.

Sieve openings may be as small as 5 μm at frequent intervals. The American Society for Testing and Materials specification ASTM E 161 permits a maximum standard deviation of only 1.0 μm for sieves ranging from 125 to 5 μm.

Sieving techniques include the same mechanical shakers and vibrators used for wire cloth sieving. However, for sieves finer than 325 mesh (45 μm), most dry sieving techniques are inadequate. For these smaller size sieves, a vacuum-type siever should be used (Fig. 3). This type of apparatus uses suction coupled with a rotating blowback nozzle. Fine particles are suctioned through the mesh, while the blowback nozzle redistributes the powder sample on the sieve, thus breaking up agglomerates and purging the mesh openings.

Fig. 3 Micromesh vacuum siever
(a) Apparatus. (b) Sectional view

(a)

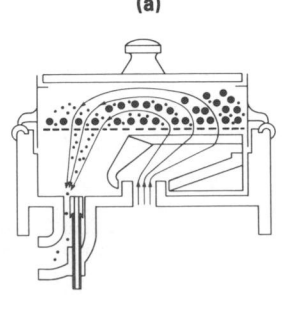

(b)

The sieving operation is rapid and allows sieving down to 10 μm. Due to the smaller sieve diameter (75 mm, or 3 in.) and the smaller amount of open area of the finer micromesh sieves, samples of only 1 or 2 g (0.035 or 0.070 oz) are used to prevent overloading. It is also possible to perform wet sieve analyses using micromesh sieves. For more information on sieve analysis, see the ASTM manual (STP 447) entitled "Test Sieving Methods."

Light Scattering

By Harold N. Frock
Applications Engineer
Leeds & Northrup Instruments

LIGHT SCATTERING provides a rapid, reproducible, and convenient method of determining particle size and size distribution. Powder particles used in P/M applications vary in size from 0.1 to 300 μm. This range of sizes conveniently corresponds to the capabilities of light-scattering techniques.

Previous optical instruments based on light scattering enjoyed only limited use for particulate measurements. Transmissometer-type instruments have been calibrated to measure particulate mass, but such instruments can maintain calibration only as long as the particle size distribution remains unchanged from the calibration distribution. Other instruments are available that rely on scattering theory to provide mass measurement over a wider particle size range. These instruments historically have required extensive computational capability and thus have not been suitable for continuous measurements. However, an instrument is now available that employs an optical transform and masking technique that performs the necessary computation directly and continuously.

Principle of Light Scattering

The diffraction phenomenon on which these analyzers are based is well known. For particles that are large compared to the wavelength of light used to make the measurement, Fraunhofer diffraction theory applies. Light incident on small particles is scattered or diffracted through angles that are inversely proportional to particle size. Scanning of the angular distribution of scattered light flux provides information on particle size and size distribution.

This section mathematically describes the optical system. Using the light scattered from a single illuminating beam, data regarding particle dimensions can be calculated from the angular distribution of scattered light flux. Information relating to the cross-sectional area of the particle is contained in a measure of the total scattered flux. Volume information, which is directly related to the mass of the particle, can be calculated from the scattered flux signal by forming the Fraunhofer diffraction pattern in the focal plane of a positive lens. An optical filter is placed in the plane of the Fraunhofer diffraction pattern, which is shaped to pass a fraction of the collected flux proportional to the cube of the diameter of the particle which produces the scattering.

A simple demonstration of the scattering principle is shown in Fig. 4. A series of randomly located pinhole apertures produces Fraunhofer patterns. These diffraction patterns are essentially identical to those produced by a series of similarly located opaque spots or particles. In Fig. 1, a point source of light is collimated by lens L_1. At some point between the collimator lens and a collecting lens (L_2), an opaque screen is located in which circular apertures or pinholes of equal diameter (d) are randomly located.

Light diffracted by these openings and collected by lens L_2 is displayed in a plane behind the lens as the Fraunhofer diffraction pattern. The precise location of this plane is defined by the plane of the image of the source. For collimated light, the plane is located a distance (X_f) behind the lens L_2, where X_f is the focal length of the lens.

Theory of Operation

Successful operation of the instrument is based on the assumptions that particle concentration is low enough for multiple scattering to be negligible and that parti-cles are spherical. The wide range of particle sizes discussed in this article requires the use of low-angle, forward light scattering, or Fraunhofer diffraction for particles ranging from 2 to 300 μm. Mie scattering theory is applied for sizes ranging from 0.3 to 2 μm. A combination of 90° scatter and light polarization of several different wavelengths extends the size capability down to about 0.1 μm. The following discussion pertains only to Fraunhofer diffraction.

For a particle whose radius a is sufficiently large compared with the wavelength, the intensity distribution of the Fraunhofer diffraction pattern is given by the Airy formula (Ref 1):

$$I(w) = Ek^2a^4\left[\frac{J_1(kaw)}{kaw}\right]^2 \qquad (Eq\ 1)$$

where E is the flux per unit area of the incident beam; $k = 2\pi/\lambda$, with λ the wavelength of light; $w = \sin\theta$, with θ the angle relative to the direction of the incident beam; and J_1 is the first order Bessel function of the first kind.

If the total diffracted flux is collected and measured, a signal will be generated that is proportional to the square of the particle diameter, thus permitting a direct measurement of the cross-sectional or surface area of the sphere.

According to Eq 1, the total flux diffracted by the particle is:

$$F_2(a) = \int_0^1 I(w)2\pi w\, dw$$

$$= 2\pi Ek^2a^4\int_0^1\left[\frac{J_1(kaw)}{kaw}\right]^2 w\, dw$$

$$= 2\pi Ea^2\int_0^{ka}\frac{J_1{}^2(x)}{x}\, dx$$

and, because for values of a for which Eq 1 is valid, the integrand is negligibly small beyond the upper limit:

$$F_2(a) = 2\pi Ea^2\int_0^{\infty}\frac{J_1{}^2(x)}{x}\, dx = \pi Ea^2 \qquad (Eq\ 2)$$

Many industrial processes, however, require measurement of the total mass or volume of particulates. Selective attenuation and collection of the diffracted flux can produce a signal proportional to the cube of the particle diameter and hence a measure of the volume, or mass, if the specific gravity is known. A Fraunhofer plane mask that passes flux in proportion to the diameter cubed is desirable.

Attenuation of this kind can be accomplished by a fixed spatial filter whose properties to a first approximation do not depend on the particle radius. As shown by Eq 1, the radial variation of diffracted flux depends only on the product (aw) so that attenuating the flux is proportional to $1/a$. Thus, after passage through a filter having transmission $T(w) = C/w$, where C is constant, the integrated flux, using Eq 1, is:

$$F_3(a) = \int_0^1 I(w)T(w)2\pi w\, dw$$

$$= 2\pi Ek^2Ca^4\int_0^1\left[\frac{J_1(kaw)}{kaw}\right]^2 dw$$

$$= 2\pi EkCa^3\int_0^{ka}\left[\frac{J_1(x)}{x}\right]^2 dx$$

and, assuming as before that a is large enough for the upper limit of the integral to be taken as infinity:

$$F_3(a) = \frac{8}{3}EkCa^3 \qquad (Eq\ 3)$$

Actual measurements deviate from theory at the large particle end, because the angle of the diffracted flux is not sufficient to be distinguishable from the unscattered beam for very large particles. Deviation at the small particle end occurs as the particle size approaches the wavelength of the light source. These limitations define a working range of particle size from about 2 to 300 μm for a standard midrange instrument design.

Instrumentation

The light-scattering measurement system is comprised of three major discrete sections—optics, electronics, and sample handling. Figure 5 shows an exploded view of the optical components. Typically, a helium-neon laser illuminates a sample zone, which consists of particles suspended in a liquid carrier. The particles cause light to be scattered, collected by a

Fig. 4 Schematic of diffraction setup producing the Fraunhofer diffraction pattern

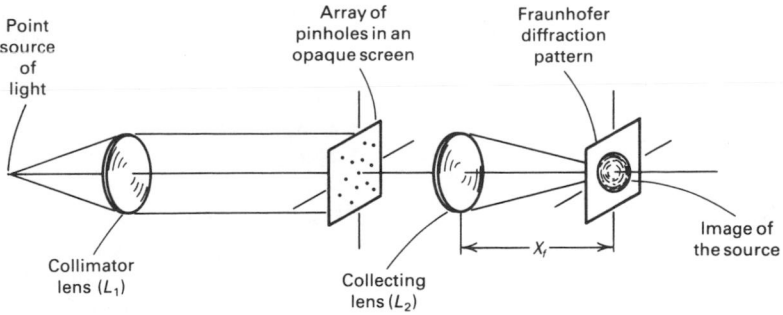

Point source of light

Array of pinholes in an opaque screen

Fraunhofer diffraction pattern

Collimator lens (L_1)

Collecting lens (L_2)

X_f

Image of the source

Fig. 5 Optical components of particle analyzer based on the light-scattering principle

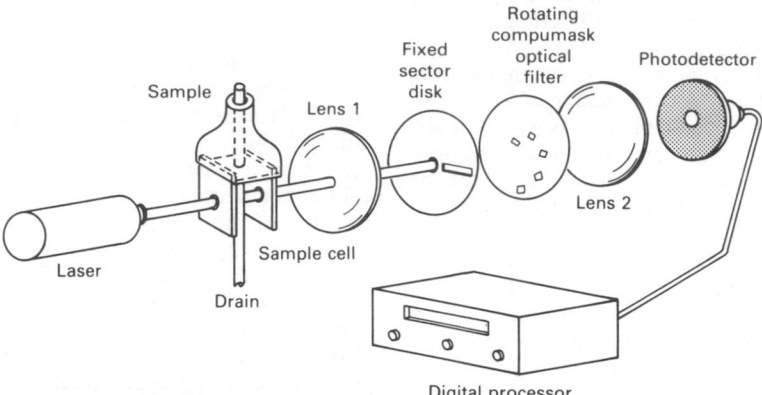

Fig. 7 Particle analyzer based on the light-scattering principle

lens, and directed onto the Fraunhofer plane, where the optical filter is located. The transmitted scattered light is focused onto the detector, which produces electrical currents that are converted to digital signals and used to compute particle size distribution.

The electronic section (Fig. 6) performs the analog-to-digital conversions, after which the raw data are stored in random access memory (RAM). The program steps reside in the programmable read-only memory (PROM) and direct the Z-80 central processor unit (CPU) to collect data to produce the complete particle size distribution, which is printed on paper tape.

Sample handling is accomplished by means of an aqueous system that consists of a 4-L (1.06-gal) mixing chamber, a sample pump to circulate the liquid, and a sample cell (Fig. 6). Alternate sampling arrangements can be provided to analyze

materials in organic solvents, or to introduce dry powders into the sensing zone. Analysis time is typically one to several minutes for most materials. A laboratory analyzer based on the light-scattering principle is shown in Fig. 7.

Sedimentation

SEDIMENTATION is a mechanism for classifying metal powders according to their rate of settling in a fluid. This measurement technique is based on Stoke's law of fluid dynamics, which states that, at low velocities, the frictional force on a spherical body moving through a fluid at constant velocity is proportional to the product of the velocity, the fluid viscosity, and

the radius of the sphere. Sedimentation is routinely used to determine particle size and size distribution.

Among the number of sedimentation methods available, only a few are commonly used for metal powders—the micromerograph and light and x-ray turbidimetry. Another analyzer using the sedimentation method is the Roller Air Analyzer; however, it is no longer manufactured, and its standards are being withdrawn.

To obtain valid results, convection currents must be avoided in the suspending fluid, and the relative rate of motion between the fluid and the powder particles must be slow enough to ensure laminar flow. This generally restricts the particle sizes that can be determined to the subsieve range.

Particles should be large compared with inhomogeneities in the fluid, which for sedimentation in air restricts the available methods to particles larger than 5 μm. For sedimentation in liquids, particle sizes down to 0.1 μm can be determined.

Particles in the suspension must be perfectly dispersed, and the suspension must be dilute enough to guarantee independent motion, which translates to a maximum concentration of about 1 vol% of particles in the suspending medium. Finally, wall effects should be minimized; consequently, the inside diameter of the sedimentation chamber should be sufficiently large to overcome this phenomenon.

Methods and Equipment

Micromerographs

This equipment is a sedimentation balance occasionally used for determining the particle size distribution of subsieve metal

Fig. 6 Electronic flow diagram of particle analyzer based on the light-scattering principle

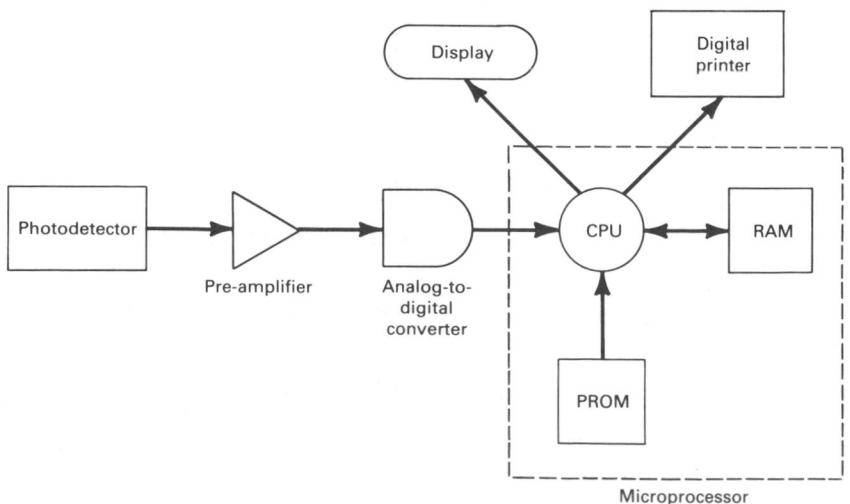

powders. The powder is suspended in air by projecting the sample with a burst of nitrogen through a deagglomerating device consisting of a conical annulus into the settling chamber.

The chamber consists of a thermally insulated vertical aluminum tube with a 10-cm (3.94-in.) inside diameter that is 2.5 m (8.20 ft) high. The pan of an automatic balance that weighs the amount of powder settling on it is located at the bottom of the chamber. A recorder calculates the cumulative weight of powder settled as a function of time. Particle size distribution is calculated from this value on the basis of Stokes' law. Particle size distribution ranging from 2 to 100 μm may be determined with this instrument. One limitation of the method is the tendency of powder particles of various sizes to cling to the wall of the column.

Light and X-Ray Turbidimetry

Turbidimetry methods are widely used to determine the particle size distribution of refractory metal powders, such as tungsten and molybdenum, and of refractory metal compound powders, such as tungsten carbide. The turbidimeter is standardized in ASTM B 430. Because turbidimetry is used for refractory metals and compounds, several additional factors must be considered to obtain reproducible size and size distribution data. The particles of fine refractory metal powders are often agglomerated by sinter bonds. In ASTM B 430, determination of particle size distribution of the powder is described in the as-supplied condition and after deagglomeration by rod milling (laboratory milling). To obtain consistent, reproducible results by turbidimetry, not only the turbidimetric procedure, but also the deagglomeration procedure, must be standardized. Dispersion of the powder in the liquid before turbidimetric analysis must also be standardized.

Figure 8 shows a typical turbidimeter. The procedure for determining particle size distribution of refractory metal powders with this instrument is included in ASTM B 430. A sample of powder dispersed in a liquid is poured into a glass cell and allowed to settle. A collimated beam of light is passed through the cell at a level having a known vertical distance (h) from the liquid level. The intensity of the light beam is determined by the current generated in a photocell. The current is passed through a potentiometer, where the voltage drop across it is measured by a recording millivoltmeter.

The reading of the millivolt recorder for the intensity of the light beam shining

Fig. 8 Schematic of turbidimeter

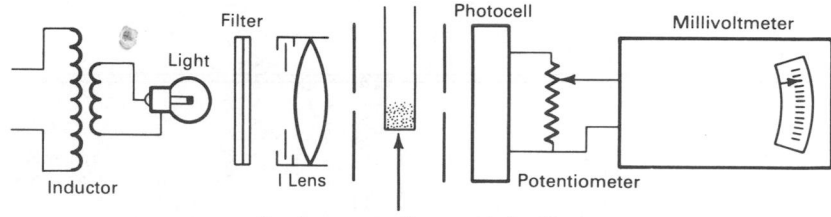

Powder sample dispersed in liquid

through the clear solution is adjusted to 100%, while the concentration of the suspension is adjusted so that the reading for the intensity of the light beam through the suspension before any settling has occurred ranges from 20 to 40% of that through the clear solution. As the suspension settles, the projected area of the particles in the suspension decreases, and the intensity of the light beam increases.

At the beginning of settling, all particle sizes are uniformly distributed through the volume of the sedimentation cell. As settling proceeds, large particles settle faster than small ones. After a given time (t_x), all particles larger in diameter than x have settled below the level of the light beam. The concentration of particles at the light beam level is now equal to the original concentration of particles, minus all particles with diameters equal to or larger than x. The projected surface of the particles at time (t_x) is therefore smaller than that of the particles in the original suspension, and the intensity of the transmitted light is greater.

Determination of Particle Size Distribution. To obtain information on particle size distribution from the plot of light intensity versus time, use is made of the relationship between the total weight of n particles of size x, which is proportional to nx^3, and the projected surface area of the n particles, which is proportional to nx^2. Therefore, the cumulative weight of particles up to a given particle size x_{lim}, which is:

$$\int_0^{x_{lim}} dW$$

is proportional to

$$\int_0^{x_{lim}} x\,dS$$

which is the integral of the product of particle size and projected surface area integrated from 0 to size x_{lim}.

Turbidimetric measurements using white light are relatively inexpensive and have proven quite reproducible and useful. They are used in research and in routine anal-

yses for comparing different lots of refractory metal powders. These methods generally are used for comparative work on the same type of material.

When x-rays instead of white light are used to determine particle size distribution of a subsieve particle suspension, the attenuation of the x-ray beam intensity is proportional to the mass of the powder particles rather than their projected area.

Roller Air Analyzers

This apparatus determined the particle size distribution of metal powders by suspension in a stream of air. With this apparatus (Fig. 9), a powder was classified into particle size fractions that ranged from 5 to 40 μm. A stream of high-velocity air flowing through a nozzle of suitable size impinged on the powder sample contained in a U-tube so that the powder became dispersed within the stream. The suspension of powder in air rose through a cylindrical settling chamber. The velocity (v) of the air stream through the chamber balanced the settling velocity of particles with diameter and density determined by a Stoke's law calculation:

$$v = 29.9 \times 10^{-4} \rho x^2$$

where v is the velocity of the air stream, cm/s; ρ is the particle density, g/cm³; and x is the particle diameter, μm.

Given this velocity, particles with a size smaller than x were carried through the settling chamber into the collecting system, which consisted of an extraction thimble. Larger particles fell back into the U-tube. By using a series of vertical settling chambers with diameters in the ratio 1-to-2 to 4-to-8 and a constant volumetric rate of flow, the powder was classified into particle size fractions with the maximum sizes in a ratio of 1-to-2 to 4-to-8 (for example, 5, 10, 20, and 40 μm).

The apparatus was equipped with manometers to regulate the volume rate of air flow. Means to vibrate the U-tube to ensure proper suspension of the powder in the air stream and to vibrate the settling chamber to dislodge particles adhering to the side of the chamber were also provided.

Fig. 9 Schematic of Roller Air Analyzer

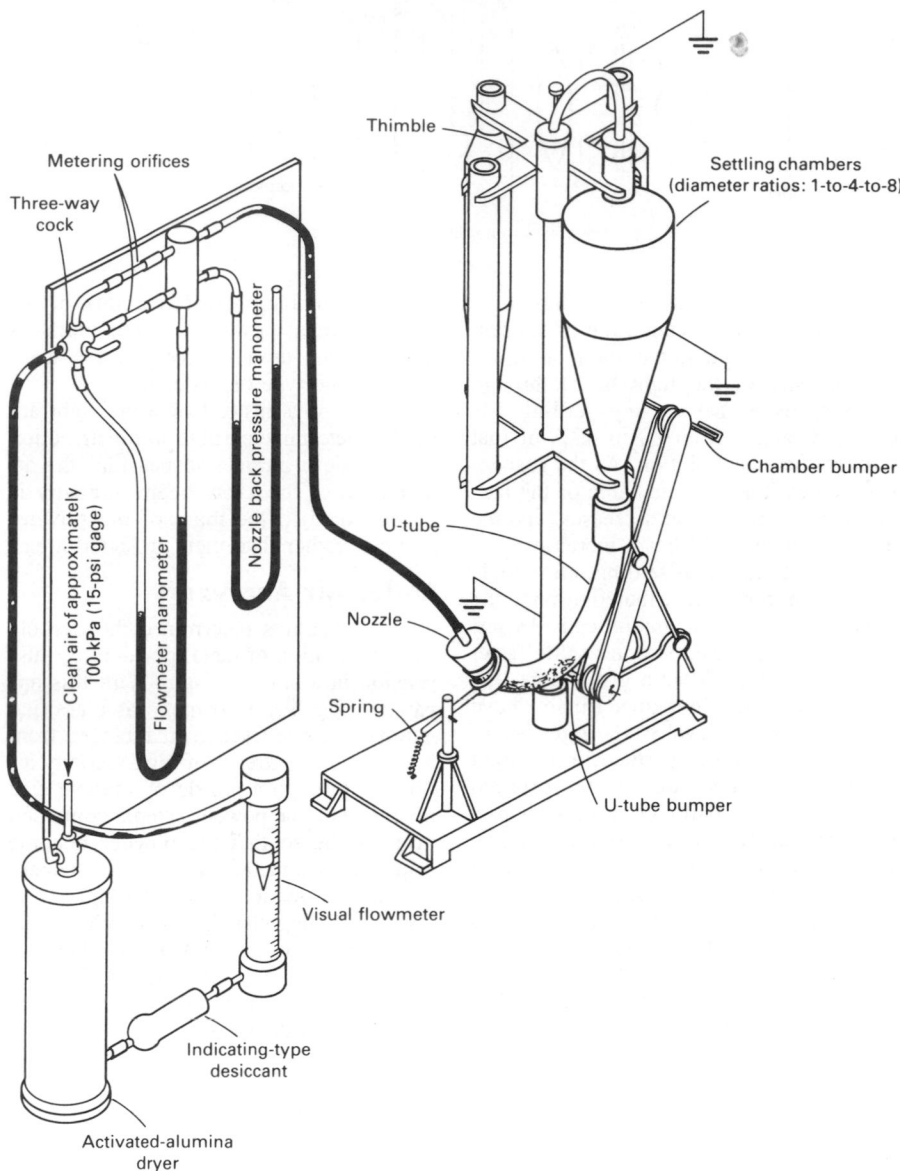

velope" volume. Because this voltage pulse is independent of particle shape, diameter is expressed as that of a sphere of equal volume, or equivalent spherical diameter.

The size distribution measured by a given testing setup depends on orifice diameter, orifice current, amplifier gain, and logarithmic distribution. The latter provides a distribution selection from 2 to 3% up to 50 to 75% of the orifice diameter, which is full range for a given orifice. Overall, the electrozone method is capable of measuring particle diameters ranging from 0.1 to 2000 μm, depending on particle density. For wide distributions, the computer program can blend data sets, which is facilitated by logarithmic scales.

Collection of size distribution data normally requires several minutes and consists of transfer of the digitized values for many thousands of particle-pulses to their proper locations in computer memory. The initial data format is thus a frequency histogram, usually of 128 channels.

Data processing includes smoothing, scale normalizing or converting (both size and quantity axes), marking, editing, extrapolating and blending, subtracting, and ratioing. Throughout data acquisition and processing, a videoscope display provides interactive, operative monitoring of data and associated items.

Sample Preparation

Sample preparation requires a particle-free, filtered electrolyte. Many electrolytes have the required characteristics to adequately test various materials. Powdered metals are best analyzed by aqueous solutions of common salts, such as sodium or potassium phosphates (basic) or chlorides, sulfates, or nitrates (neutral). Electrolyte conductivity (strength) may be approximately 2000 mmhos/orifice diameter (1% sodium chloride conductivity is about 20 mmhos/orifice diameter).

Anionic or nonionic dispersants are also used, sometimes with ultrasonic dispersants as needed. Solvating organic electrolytes are available for oily samples. To avoid excessive spurious pulses from coincidence in the sensing zone, sample concentration is limited to 1 to 50 ppm by volume, depending on distribution and shape of the size distribution.

Metal Density

Metal powders have relatively high density, which limits the maximum particle size for uniform electrozone sample suspension. Even if the suspension has ef-

Electrozone Size Analysis

By Robert H. Berg
President
Particle Data, Inc.

ELECTROZONE (electric sensing zone) measurement of particles detects the volumes of individual particles in a liquid suspension, at rates ranging from tens to thousands of particles per second. Elec-

trical pulse amplitudes, representing individual particle volumes, are processed via microcomputer to yield particle size distribution data in whatever form may be required.

Figures 10 and 11 show the main components of a computerized electrozone system. The powder to be evaluated is suspended in a conducting liquid. The suspension is circulated through a sampling chamber, which contains a sensing orifice with immersed electrodes on either side. Within the sensing orifice, particles displace the electrolyte to cause a resistance change (and consequently a voltage pulse) proportional to its "electrical en-

Fig. 10 Schematic of the instrumentation of the computerized electrozone system

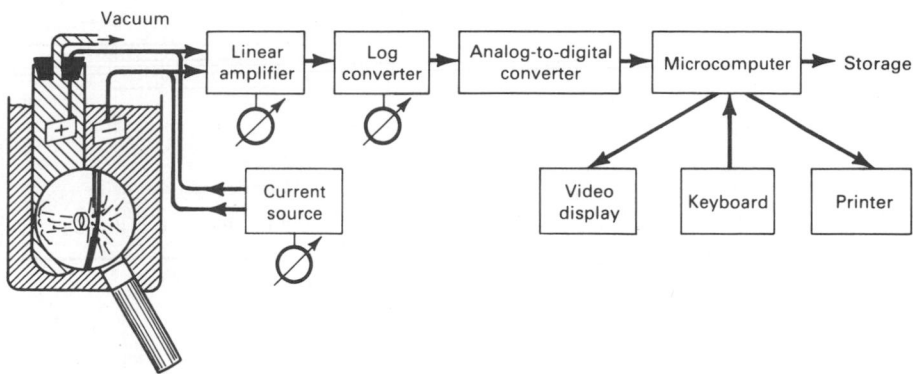

Fig. 11 Computerized electrozone system

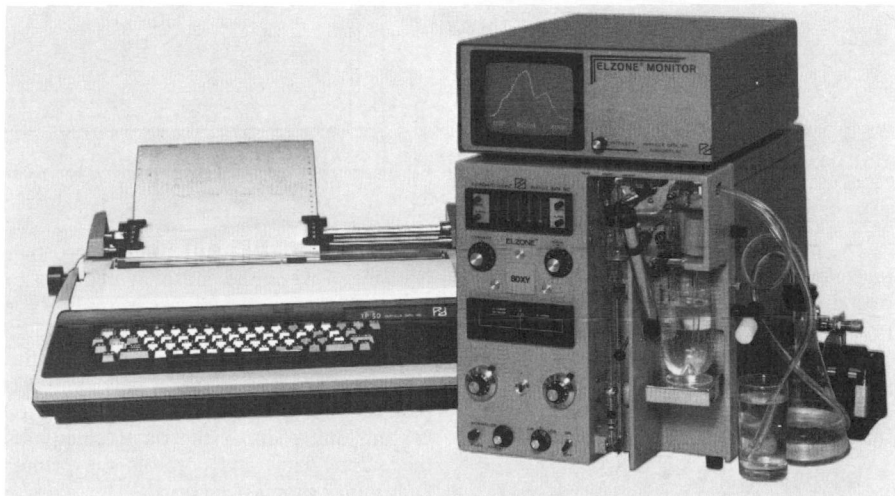

ficient agitation and high viscosity (50% glycerol, for example), electrozone data generally will not be acceptable if the particle size-density function exceeds 500, expressed by:

$$(d_m)|\rho - \rho_0|^{1.5} > 500$$

or

$$(d_m)_{max} \leqq 500|\rho - \rho_0|^{-1.5}$$

where d_m is diameter, μm; and ρ and ρ_0 are densities of particle and liquid, respectively. Typical values are:

Density, g/cm³	2	5	10	17
d_m (max), μm	500	62	19	8

Below a value of 100 for this function, sieving becomes increasingly effective. Thus, some samples may require both sieve and electrozone analyses, with blending of the two data sets in the overlap region of 100 to 500 for this size-density function.

Metal Conductivity/ Magnetism

The oxide surface layer present on most metal particles causes them to appear to be nonconductive, in effect creating a barrier impedance. Electrozone currents high enough to produce 1 to 3 V across the surface layer overcome this barrier impedance, thus causing conduction for only part of the pulse cycle (prior to the occurrence of surface polarization and ion depletion). Consequently, the particle pulse will erroneously appear to be smaller.

This effect can be detected by test observation of the total relative volume of particles per unit volume of suspension with increased current value. Samples may be treated with coating reagents or surfactants to increase surface layer impedance. Orifice current settings then must be kept below the level where this effect appears. Similarly, porous particles appear to be too

small, due to the conductance of the pores aligned with the electric field, unless they are plugged with nonconductive material.

Magnetic powders may be dispersed for a sufficient time to permit electrozone measurement if they are degaussed for 20 min at temperatures as high as 750 °C (1380 °F) (but below the sintering point, however) and/or spatulated with viscous honey to inhibit reagglomeration. A microscope slide covered with the honey suspension is then dipped into the electrolyte in the stirred sample breaker. In this manner, data acquisition is initiated instantly, while the suspension is being formed from the dissolving honey.

Advantages

The electrozone method rapidly measures particle size distribution with excellent low-end sensitivity and with good resolution (volume response) and precision (minimal side effects).

By contrast, photozone (photicsensing zone) methods, which include light-beam scattering, blockage, and diffraction, have much lower resolution and precision. This is due to their area response and several significant side effects, including photic properties of particles, high coincidence levels, and beam/sensor instabilities. Sieving and sedimentation methods have similar limitations as to response and side effects, and are much slower in producing data.

Optical Sensing Zone

By Mary A. Pao
Metallurgist II
SCM Metal Products
and
Erhard Klar
Manager of Particle Technology
SCM Metal Products

THE LIGHT BLOCKAGE PRINCIPLE (obscuration) was used by Carver in 1958 to develop a method for counting and sizing particulate contamination in hydraulic fluids (Ref 2, 3). To date, this principle has found application in other particle-fluid systems and has been used for the measurement of metal powder particles.

In apparatus using the light-obscuration principle, particles are suspended in a fluid with a refractive index that differs from the refractive index of the particles. The suspension is then passed through a restricting orifice—the sensing zone—across which a collimated white light beam passes. Figure 12 illustrates the sensing zone. The beam of light falls on a photodetector, which measures the intensity of the light passing through the zone.

Large particles block the light beam, thus lowering the intensity with which the beam strikes the sensor. Small particle sizes (<3 μm) do not block the light rays; the rays bend around the particles. The sensor can be designed so that a slight bend that is equal to an index of refraction change of 0.05 causes the light to miss the detector, thereby detecting particles.

Instrumentation

The particle size analyzer based on the obscuration principle is shown schematically in Fig. 13. The suspension is prepared in the reservoir cup, which is equipped with a stirrer. A vacuum pump pulls the suspension up through the sampling tube and through the sensing zone located in the back of the sampler stand. Once analyzed, the suspension is collected in a waste chamber.

The sensing zone is designed to prevent laminar flow, which causes each particle to tumble through the zone. The sensor output is a voltage signal, which is proportional to the amount of light blocked by a particle at a given instant, superimposed on a constant baseline voltage. The height of the pulse generated above the baseline as a particle traverses the sensing zone is proportional to the particle's largest cross section.

Fig. 12 Sensing zone for (1- to 45-μm) particle sensor based on light obscuration

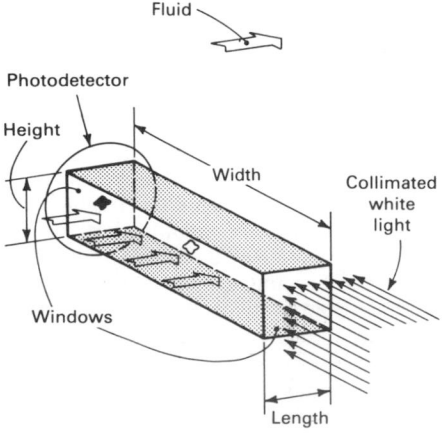

Fig. 13 Schematic of particle size analysis apparatus based on light obscuration

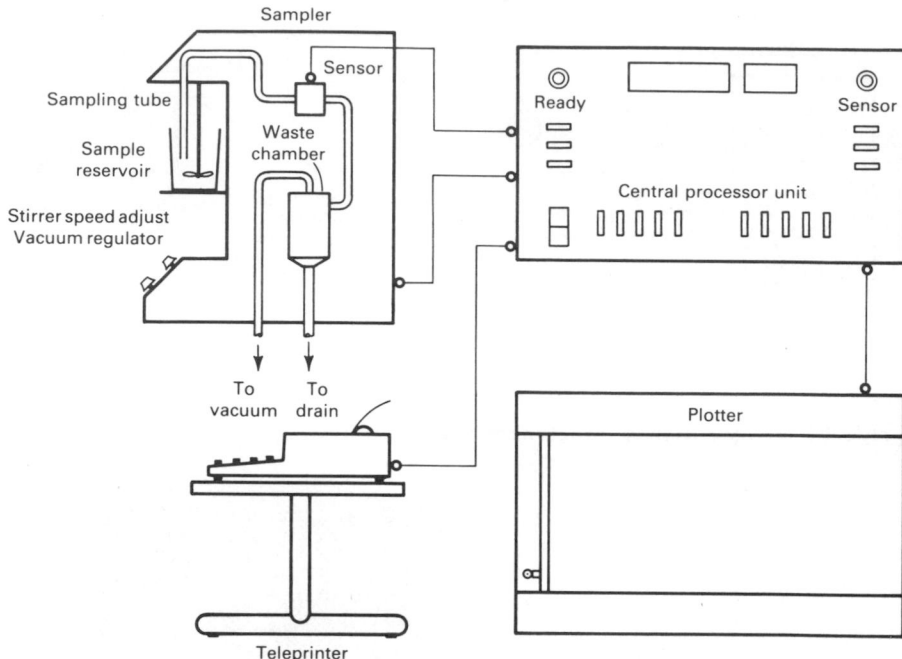

The relationship between the size of the particle and amplitude of the pulse produced can be determined theoretically. In practice, however, calibrated spheres are used to determine this relationship and to calibrate the sensor (Ref 4, 5). Sensors are available in the following particle size ranges:

- 1 to 45 μm
- 1.5 to 67.5 μm
- 2 to 90 μm
- 2.5 to 112.5 μm
- 5 to 225 μm
- 10 to 450 μm
- 16.7 to 750 μm
- 41.7 to 1875 μm
- 150 to 6750 μm

The first five ranges are the most useful for studying metal powders used in the P/M industry. The actual sensor dimension (height in Fig. 12) is one third larger than the largest particle size for which the sensor is rated; consequently, sensor blockage is minimized (Ref 4, 5).

The microprocessor scales the sensor output pulses and sorts them into 23 size ranges, or channels, in a geometric progression of 1 to 1.18 by using 24 analog comparators. Within the limits of the physical size of the sensing zone, the thresholds can be adjusted to change the widths of the channels to provide more or less detail of certain areas of the particle size distribution, or to extend the range slightly. The threshold adjustment circuits are housed on a printed circuit board that is changed when the sensor is changed to monitor a new particle size range, for example.

The system memory accumulates the number of pulses generated between each pair of thresholds, and the central processing unit reduces the data, calculates the output data, and controls the printer and plotter output functions.

Accuracy and Interpretation

Instrument-related factors must be considered, even with perfectly dispersed, spherical particles (Ref 6). The oversized particle indicator has another purpose in addition to detecting large particles that exceed the range of the sensor—it indicates an overconcentrated solution. If several particles overlap while passing through the sensor, the oversize indicator flashes, and the oversize counter is incremented. If this tendency occurs frequently (more than once or twice per analysis), the solution may need to be diluted, or the sensor may be blocked. If dilution does not eliminate the oversize indication, but does reduce its frequency by an amount proportional to the dilution, the large particles are most likely part of the actual distribution.

Even if dilution eliminates the oversize indication, small particles may overlap and skew the distribution toward the coarse side. To eliminate this occurrence, a maximum recommended number of particles per cubic centimetre of liquid is given for each sensor. High-concentration sensors that allow twice the concentration to be analyzed by restricting the volume of fluid, and therefore the occurrence of two particles in the sensing zone at the same time, are available.

To increase detail in some parts of the distribution and to extend the analysis range, channel thresholds can be adjusted. Adjustable thresholds, however, can slip out of adjustment. The appearance of a sharp peak and an adjacent dip in the same place on several different derivative distributions may indicate the need for threshold adjustment. When this condition occurs, counts are being placed in the channel above or below the proper one. For the cumulative percent under or over distributions, the effect is cancelled out when both of these channels are passed. Threshold adjustment can be performed with a printed circuit extender board and a digital voltmeter. The overall detection level of the instrument should be checked periodically using calibrated spheres.

Application to Metal Powders

Other factors to be considered when interpreting particle size distribution results arise because the particles being measured are metal powders. Most metal powders have a higher specific gravity than the particulates for which the obscuration-principle apparatus, as well as other particle size analyzers, was originally designed. As a result, a greater tendency exists for particles to settle in the sample reservoir before being sampled, thus causing the distribution to be skewed toward the fine side. This phenomenon is greater for coarser and more regularly shaped particles. Good reproducibility of a distribution does not guarantee the absence of particle settling, as settling of coarse particles also occurs and is reproducible.

Settling of particles in the sample reservoir can be reduced by using a more viscous fluid such as glycerine. Pure glycerine, however, may be too viscous for the vacuum to draw and also may prevent the dispersion of particles that tend to agglomerate. Glycerine can be thinned with water, but antisettling benefits are diminished.

A more effective way to eliminate settling is to use reservoir shapes, such as round-bottom beakers, and stirrer configurations that maximize turbulence and prevent "dead" spots in the stirring cur-

rent, without the generation of foam or bubbles that appear as particles to the sensor. Even if reservoir suspension is ideal, particle settling can still occur in the sampler tube (the tube connection between the reservoir and the sensor).

The higher the linear velocity of the fluid, the larger the size of the particles that can be entrained and carried along. The linear velocity of the fluid in the sampler tube is determined by the rate of flow through the restricted sensing zone and by the diameter of the sampler tube. The small sensor has a smaller sensing zone; consequently, the flow rate through the sampler tube is lower, causing the larger particles to settle before reaching the sensor. This tendency can be controlled by fitting the sensor with a smaller sampler tube. This increases the linear velocity of the fluid and keeps the larger particles entrained.

Small particles may have a tendency to loosely agglomerate due to moisture, or van der Waal's forces, for example. Stirring the water suspension of such a powder may be only partially successful in separating these particles. Addition of a surfactant or type IC dispersing aid to the powder before water is added helps the water wet the particles and allows them to separate with the stirring action. Ultrasonically vibrating the solution for several minutes also helps separate particles.

Many metal powders for P/M use are prelubricated by the powder manufacturer. Typical lubricant additions range from 0.5 to 1.0 wt%. While the presence of lubricants or other chemical additives on particle surfaces may inhibit wetting and cause agglomeration, lubricants have another significant effect. Lubricants or other chemical additives are much less dense than metal particles. Consequently, a small amount of lubricant by weight may contribute greatly to the particle count (population) distribution.

When the volume distribution is calculated from the population distribution, the small amount of lubricant (by weight) is weighted the same (per particle) as the metal powder. In this case, the volume distribution is not the true weight distribution. The same occurs during the analysis of a powder mix in which the constituents have different specific gravities. Single-component powders provide the best test results, unless only the population distribution is desired. Powders should be unlubricated, or the lubricant should be removed prior to analysis.

In interpreting particle size data accumulated by the obscuration-principle particle analyzer, or by any other testing method, the interaction of particle shape

and the particular technique must be considered. For example, rod-shaped particles appear smaller than spheres of the same volume to testing sieves, because the rod-shaped particles can pass lengthwise through sieve openings that are too small for spheres to pass through.

In this comparison, the obscuration-principle particle analyzer tends to characterize rod-shaped particles as larger than they actually are; it calculates the volume of a spherical particle with the same cross-sectional area as the elongated view of the rod-shaped particle, which is larger than the cross-sectional area of the sphere. This has led to confusion as to which method is correct. Actually, it is not the method that is correct or incorrect, but that one method may be more appropriate for a given application than another one.

In the following applications, some of the effects of particle density, shape, and instrument resolution are described. Comparison with various methods is made, including sieving. Synthetic bimodal powder distributions were chosen to highlight some of these factors. For normalization purposes, all cumulative undersized distributions were plotted, and the relative differential volume curves were then constructed using a uniform spacing of 12.7 mm (1/2 in.) on standard analyzer output paper.

Figure 14 shows the cumulative and relative volume curves of 9.95-μm-diam (measured by microscopy) polystyrene microspheres measured with HIAC (light-obscuration principle) and Microtrac (light-scattering principle) analyzers. Good agreement of the volume mean diameters with the microscopically determined value of 9.95 μm indicates adequate instrument calibration. The narrower distribution of the HIAC curve indicates superior resolution in comparison to the Microtrac.

Figure 15 shows a scanning electron micrograph of a bimodal gas-atomized copper powder and the cumulative and relative volume curves obtained with HIAC and electrozone particle size analyzers. Bimodal distribution was obtained by sieving the gas-atomized powder (using Tyler sieves), which itself followed a log-normal distribution, and blending the two sieve fractions, $-100+150$ and $-270+325$ mesh, using a weight ratio of 3 to 7. The particle shape of this powder is nearly spherical. Both curves show good resolution of the two peaks as a result of the large number of channels and the distinct sizing of each particle.

Electrozone instruments have improved resolution through hydrodynamic focusing and pulse editing (that is, particles trav-

Fig. 14 HIAC versus Microtrac analyses of 9.95-μm standard polystyrene microspheres

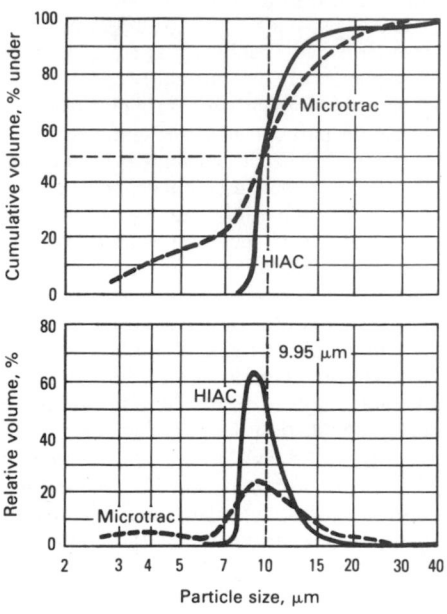

peaks are displaced to the right of the sieves, reflecting the tendency of HIAC to oversize irregular or nonequiaxed particles. The quantification of the coarse particle size range has improved, as the more irregularly shaped particles tend to settle less than the spherical particles in Fig. 15.

Fig. 15 HIAC versus electrozone analyses of a bimodal distribution of gas-atomized copper powder

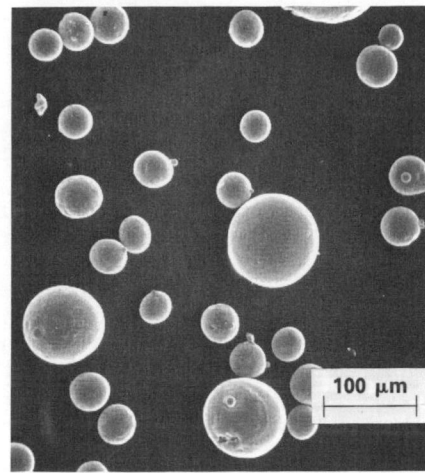

The Microtrac curve, while producing peaks roughly congruent with sieve modes, gives a less distinct resolution of the peaks. This is partly due to the use of fewer channels, but is due mainly to the inherently lower resolution that is possible with low-angle forward-scattering of light. The weighting of the two particle size ranges appears to be significantly influenced by resolution characteristics. In broader dis-

Fig. 16 HIAC versus Microtrac analyses of a bimodal distribution of water-atomized copper powder

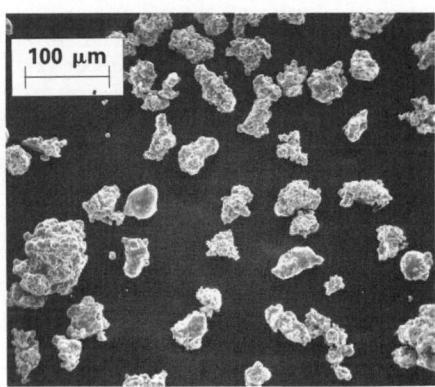

eling through the sensor near the sensor walls are excluded in the analysis). Although particle size analysis instruments usually show excellent agreement for spherically shaped particles—in fact, they generally are calibrated with spherical powders—the HIAC distribution, in this example, is displaced to the right of the sieves.

Both curves show significantly less coarse powder—9 and 16%, respectively—compared to the 30% present in the sieve distribution. Such discrepancies may be caused by several factors. First, preferred settling of the larger particles in the sample reservoir or in the sampling tube connecting the reservoir and sensor results in an under-counting of coarse particles. Secondly, calibration errors of the various channels, as well as particle shape effects, under- or over-represent a particle size range, depending on whether the recorded size is smaller or larger than that of the referenced method. Thirdly, nonuniform resolution (the quality of separating peaks and the degree of peak broadening) can affect the relative weighting of a particle size range. In Fig. 15, the major reasons for under-representation of the coarse particles was found to be preferred settling of the coarse particles. Population distributions can provide valuable information on settling problems.

Figure 16 compares the HIAC and Microtrac analyses of a bimodal water-atomized copper powder. Both HIAC

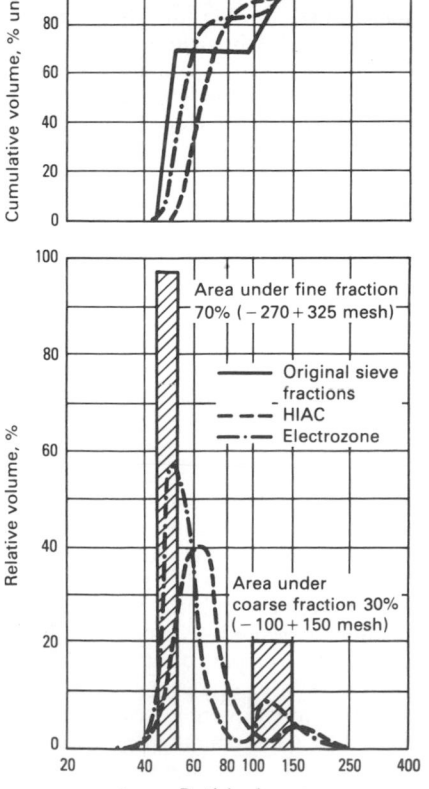

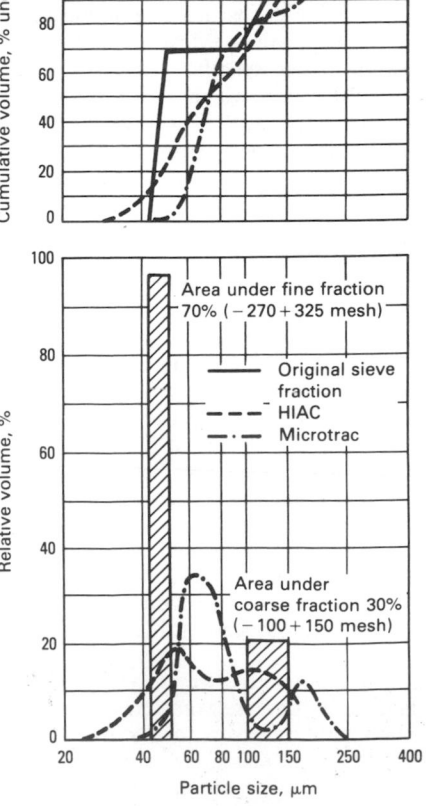

tributions, distortions caused by low resolution are less severe due to averaging effects.

Figure 17 compares HIAC, Microtrac, and electrozone analyses for −400 mesh brass flakes. Compared with HIAC, electrozone analysis is expected to show a smaller size, as it measures the change in conductivity of the suspending liquid due to the presence of a particle. As a first approximation, this change is proportional to the volume of the particle, provided the particles are equiaxed. Because of the turbulent flow through the sensing zone, HIAC measures the largest projected area of a particle.

The two methods, however, give almost identical results in this case. This is most likely caused by the tendency of particles to orient as they pass through the sensor. For nonequiaxed particles, the effect of particle orientation on the conductivity of its surrounding medium can be significant. If the particles offer the highest resistance to flow in the sensing cross section, conductivity analysis approximates an electrical analog of the light-obscuration method. Microtrac analysis shows a less distinctive peak due to a lower resolution.

Figure 18 compares HIAC and Microtrac analyses for a −400 mesh dendritic copper powder. Both analyses show particles with diameters greater than 38 μm (400 mesh). This is partly due to the tendency of long particles to pass lengthwise through the 38-μm opening of the sieve used to prepare the powder (see scanning electron micrograph in Fig. 18).

The HIAC instrument shows a higher average particle size than the Microtrac, which indicates a number of small peaks. This may be due to a tendency for the small dendrite arms on a large particle to scatter light in a manner that is typical of small particles.

Figure 19 shows the relative volume curves of 2.02-μm-diam polystyrene microspheres measured with HIAC and Microtrac analyzers. The HIAC instrument, for particle diameters of 3 μm or less, under-counts particles so that the number of particles approaches zero in the first channel.

Fig. 19 HIAC versus Microtrac analyses of 2.02-μm standard polystyrene microspheres

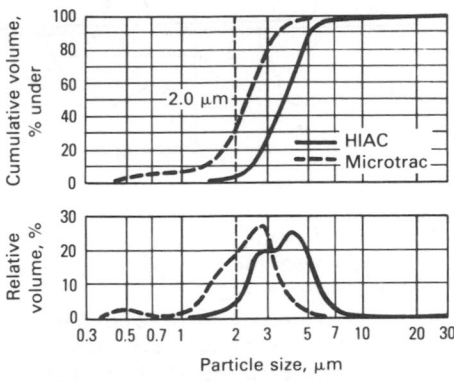

Microscopy and Image Analysis

By James Lee Hubbard
Senior Research Scientist
Georgia Institute of Technology

MICROSCOPY is the most definitive method of particle size analysis, because individual particles are observed and measured. Some of the techniques used are more art than science, and some measurement techniques are subjective. However, if the basic principles of sampling, preparation, and counting are followed, a precise count can be made with a thorough understanding of the nature of the particles being studied. ASTM standard E 20 details the use of microscopy for particle sizing.

Particle Size

Various techniques are used to measure the size of irregularly shaped particles when viewed through a microscope. This has resulted in different measurements, which are used to classify these two-dimensional particle images in terms of an equivalent spherical particle. Accepted measurements (Fig. 20) include:

- *Feret's diameter* (*F*): The maximum length of the particle measured in a fixed direction
- *Martin's diameter* (*M*): The length of a line that bisects the area of the particle image; all particles measured in the same direction
- *Projected area diameter* (d_a): The diameter of a circle with the same area as the two-dimensional image of the particle

Fig. 17 HIAC, Microtrac, and electrozone analyses of −400 mesh brass flakes

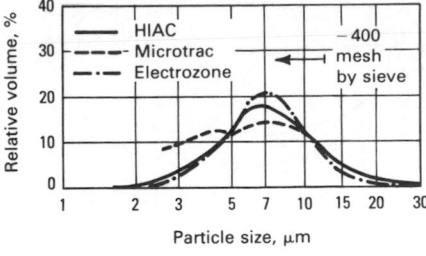

Fig. 18 HIAC versus Microtrac analyses of −400 mesh dendritic copper powder

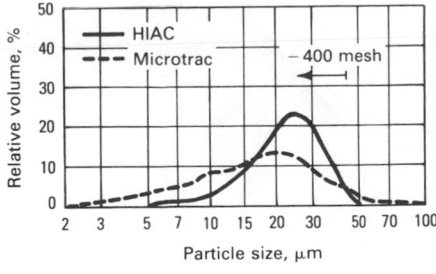

Fig. 20 Techniques for measuring sizes of irregularly shaped particles
See text for identification of variables.

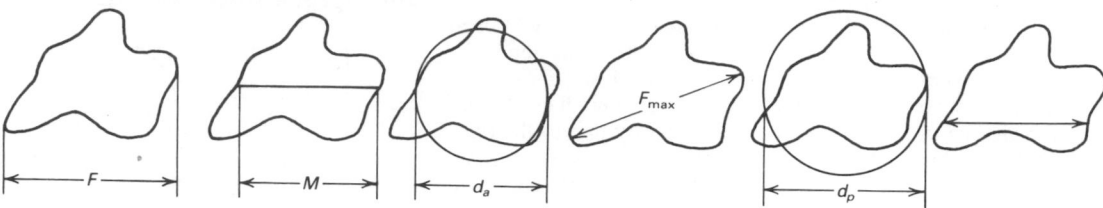

- *Longest dimension*: The maximum Feret's diameter for each particle; no set direction
- *Perimeter diameter* (d_p): The diameter of a circle having the same circumference as the perimeter of the particle
- *Maximum horizontal intercept*: The length of the longest line that can be drawn through the particle in a fixed direction

Feret's diameter is the easiest to measure manually. The average Feret's diameter is related to the perimeter of the convex hull of the particle, as shown in Fig. 21, by the relationship:

$$F_{av} = \frac{P_{CH}}{\pi}$$

where P_{CH} is the perimeter of the convex hull. Martin's diameter is related to the specific surface S_v of the particle by:

$$M = \frac{4}{S_v}$$

The projected area diameter gives the best estimate of the true cross-sectional area of the particle.

Generally, the three diameters above are related to one another by the expression, $M < d_a < F$. Their ratios remain fairly constant for a given material. The expression F/M has been used as a shape function, which is equal to 1 for spherical particles and increases in magnitude as particle shape becomes more acicular. Perimeter diameter and maximum horizontal intercept (or longest chord) are measured easily by some of the automatic particle sizing analyzers.

Sampling Techniques

The most important step in any particle size analysis, but especially by microscopic techniques, is sampling from the bulk. Because an extremely small quantity of material is used to determine the particulate size, an accurate analysis cannot be obtained if the bulk material is not properly sampled. Particles tend to segregate according to size. If handling has

Fig. 21 Feret's diameter

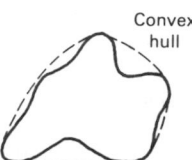

Convex hull

caused vibration of the sample, coarse material tends to collect near the surface. When free-flowing material is poured into a pile, the coarse portion collects near the outside of the pile, and the fines concentrate in the center. The following sampling techniques are recommended for obtaining a valid representation of particle sizes in a laboratory sample. See the article "Sampling of Metal Powders" in this Volume for more information.

Chute rifflers (Fig. 22) consist of V-shaped hoppers that feed a series of chutes, which alternately feed into trays on opposite sides of the hopper. A sample that is properly poured into the hopper is halved, and the process is repeated until a suitable sample size is obtained. Sampling efficiency depends on an even feed of the sample into the hopper and on the width of the chutes. Numerous narrow chutes provide more accurate sample statistics than a few wide chutes, provided the powder can flow easily through them.

Fig. 22 Chute riffler

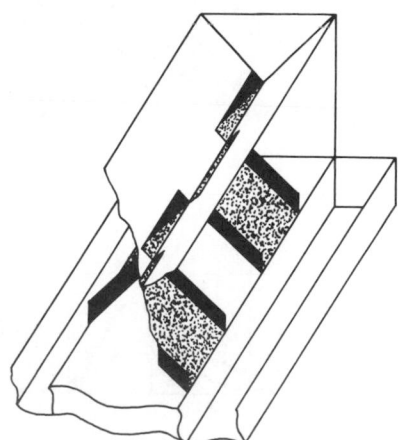

Spinning rifflers (Fig. 23) divide a sample by feeding it into a number of containers that rotate on a table under the feed. This method is efficient, provided the table rotates at least 100 times during the flow of the sample.

Oscillating Sampler. The feed hopper oscillates rapidly between two sampling cups, thereby halving the sample. Feed times must be long compared to oscillation times for optimum efficiency.

Cone and Cup Sampler. In this device, the sample is fed onto a cone that disperses the powder in all directions along the cone surface. A sample cup is rotated about the base of the cone, gathering portions of the sample at every position. The width of the cup and length of the cone periphery determine the sample reduction.

Sampling tables (Fig. 24) are used to reduce larger quantities of sample. Sampling tables consist of an inclined plane with a series of holes that halve the sample in each of four stages, leaving $1/16$ of the original volume as a final sample.

Cone and quartering involves pouring the sample into a cone-shaped pile on a glass plate, flattening the top, and dividing it into quarters with a thin metal

Fig. 23 Spinning riffler

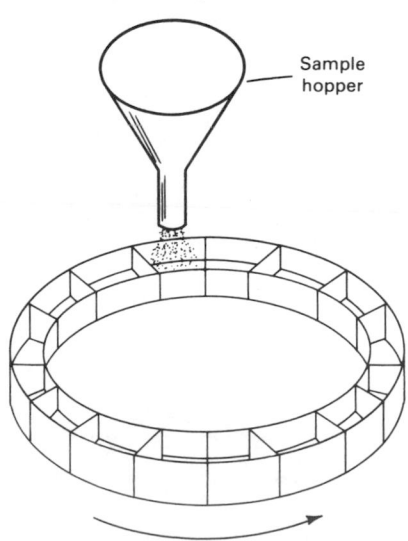

Sample hopper

Fig. 24 Sampling table diagram

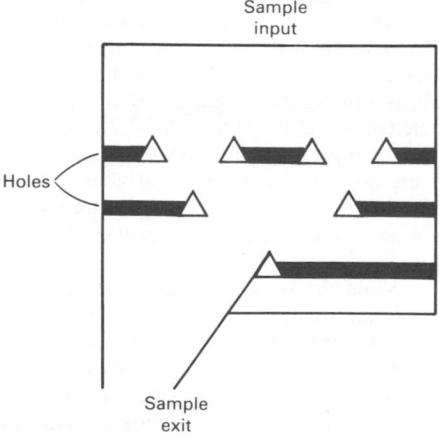

blade. One of the quarters can then be mixed, repoured, and requartered until a suitably small sample is obtained. This method requires some technique to ensure symmetry of particle sizes in the pile and accurate quartering.

Scoop sampling is the least precise, but perhaps the most frequently used, sampling technique. The sample container is shaken, and a small portion is extracted with a scoop or spatula. However, segregation can be imparted to the sample by the shaking method. The scoop sampling technique that provides the least variance is shown in Fig. 25 and consists of shaking the container back and forth while rotating. Because particle sampling is the initial and most important step to particle size analysis, miniature sampling devices should be used whenever possible.

Optical Microscopy

The optical microscope is used to count and size particles ranging from 100 to 0.5 μm diam. The resolution of an optical microscope is optimally about 0.25 μm, but diffraction effects at the edges of small particles result in significant errors in measurements within this size range. Res-

Fig. 25 Scoop sampling apparatus showing direction of shaking

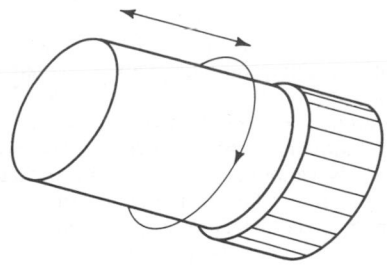

olution down to 0.1 μm can be improved by using ultraviolet light and quartz optics.

A severe limitation of the optical microscope is its small depth of field, which is about 10 μm at 100× and 0.5 μm at 1000×. This requires that the specimen powders be mounted carefully in one plane and eliminates the use of automatic counting devices on samples with wide particle size ranges. Another limitation is that due to the small number of particles usually counted, only the fields of view can be characterized, not the entire lot from which the specimen was obtained.

Preparation of slides for optical microscopy is an art, and the quality of a dispersion depends more on the skill of the preparer than on the particular technique used. Care must be taken to prevent segregation of particle sizes and to prevent the formation of agglomerates. Preparation techniques vary with the type of powder and depend on particle size, particle size range, and particle composition. Typical dispersion techniques are described below.

Dry Mounting. When a permanent slide is not required, and the powder exhibits good flow characteristics, slides may be prepared by "puffing" dry powder onto the slide or by applying it with a brush.

Dispersing Fluids. In applications that require shearing action to disperse the powder, a dispersing fluid such as cedar oil or glycerol may be used. The simplest technique involves placing a drop of dispersing fluid on a glass slide and placing a small amount of powder onto the drop. A cover slip is then placed carefully over the mounting fluid and pressed evenly with an eraser, while gently rotating the cover slip back and forth.

Another method involves placing some of the powder on a glass slide, adding a few drops of dispersing fluid, and working the mixture with a flexible spatula or soft brush. More fluid is added until the proper concentration is obtained. A drop of this mixture is then transferred to a new slide and covered with a cover slip.

Permanent Slides. A small amount of powder is placed in a beaker to which 2 or 3 mL (0.07 or 0.10 fluid oz) of a solution of 1 to 2% collodion in butyl acetate is added. The mixture is stirred vigorously, and a drop of the suspension is placed on the surface of distilled water. The drop spreads and forms a thin film as the solvent evaporates. The film is then picked up on a microscope slide and dried. Other films can be produced by dissolving parlodion in amyl acetate, Canada balsam in xylene, and polystyrene in xylene. Films are formed by casting a 1% solution on water or a ½% solution di-

rectly on a glass slide. After the films dry, the powder is applied by spraying or by letting a drop of aqueous suspension dry on the surface. The particles sink into the medium, forming a permanent slide. Melted Canada balsam or glycerol jelly also can be used to form permanent samples.

Other Methods. Certain applications may require samples prepared by specialized techniques, such as heating magnetic particles above their Curie point for dispersion. These examples generally serve as guidelines to technique development.

Particle sizing usually is performed directly in the optical microscope, because the small depth of field frequently requires refocusing on individual particles. If the size range is narrow and the preparation is carefully made in a single plane, photographs or automatic particle counting devices can be used. Projection microscopes are also available that reduce operator fatigue for analyses that require numerous counts.

Transmission Electron Microscopy (TEM)

The transmission electron microscope is used for counting particles that range from 0.001 to 5 μm in diameter. This instrument has a large depth of field; consequently, all particles in the field of view are in focus regardless of size. Particles are usually not counted directly from the viewing screen of the transmission electron microscope. Photographs are normally taken, and counts are made from prints or projected images using the negatives.

The electron beam is easily absorbed, and films greater than 100 to 200 nm (1000 to 2000 Å) are completely opaque. It is therefore necessary to produce very thin support films on which powders can be dispersed for counting. These films usually are made of carbon. One of the best techniques for producing strong, flat carbon support films involves cleaving high-quality mica and placing it immediately in a vacuum evaporation unit. After a hard vacuum has been obtained, carbon is evaporated onto the mica surface to a thickness of about 10 nm (100 Å).

Film thickness can be calculated by including a white porcelain slide containing a drop of vacuum oil in the vacuum chamber. When the porcelain that is not covered by the oil appears as a light chocolate shade, the carbon film is about 10 nm (100 Å) thick. The mica is then removed from the vacuum and slowly lowered at an angle into a beaker of distilled water. The carbon film floats off the mica onto the water surface, where it can be picked up on an electron microscope grid.

Another technique involves casting a thin film of parlodion onto the surface of distilled water by depositing two or three drops of 1% parlodion dissolved in amyl acetate onto a clean glass slide. The slide is allowed to drain and dry, and the film is floated from the slide onto distilled water. The plastic film is then picked up on TEM grids, placed in a vacuum evaporator, and coated with carbon.

The parlodion can be carefully dissolved in amyl acetate by placing the grids on a fine stainless steel mesh bridge suspended in the center of a petri dish. Amyl acetate is added to the petri dish until the level of the liquid reaches the bottom of the mesh. The dish is then covered and left for a few hours, after which the grids are carefully removed.

If the parlodion film is thin enough, it can be left on the grids. However, this requires fresh, water-free solutions and care in handling, or holes will form in the thin plastic film as the solvent evaporates. Prepared carbon substrates on grids are available.

Powders are dispersed on the surface of support films by puffing from an aspirator or by allowing a drop of aqueous suspension to dry. Carbon films are hydrophobic; use of an aqueous suspension requires that they be rendered hydrophilic by either treatment with a thin solution of albumen or exposure to reactive oxygen in a low-temperature oxygen asher. The latter must be done at a very low setting (4 W) for several seconds, or the carbon film will be destroyed.

Powders that require additional dispersion can be prepared by placing a small amount of powder on a clean glass slide and adding several drops of 1% parlodion in amyl acetate. This mixture is sandwiched between a second glass slide. As the amyl acetate evaporates, the mixture becomes more viscous. Immediately before the parlodion dries, the slides are sheared apart. This preparation can be viewed in the optical microscope to find the area of best dispersion. This area is then scribed into 3.2-mm ($^1/_8$-in.) squares and floated from the glass slide onto distilled water. The squares are picked up on electron microscope grids and coated with carbon to add stability to the electron beam.

For some samples that are difficult to disperse, ethylene glycol may be a suitable dispersing agent. A small amount of powder is added to a few drops of this viscous liquid on a glass slide and mulled until a good dispersion is obtained. A carbon substrate grid is then touched to the thin liquid film, which sticks to the grid. The grid is placed in a vacuum chamber and pumped for a few minutes, during which time the ethylene glycol sublimates, leaving the particles on the carbon substrate.

Particle thicknesses can be measured in the transmission electron microscope by "shadowing" particle dispersions in which the particles sit on a substrate and are not embedded in a plastic film. This is done by evaporating a small amount of metal placed at an angle to the substrate surface in a vacuum chamber. The metal coats the surface and particles in a line-of-sight fashion, leaving a "shadow" cast behind the particles. Commercially available precision-sized latex spheres can be included with the powder sample so that the shadow length to particle height ratio can be calculated.

Of the pure metals, platinum is best suited for shadowing because of its fine structure in an evaporated film. About 12 mm (0.5 in.) of 250-μm (10-mil) platinum wire, evaporated from a point source at an angle of 30° and distance of 100 mm (4 in.), casts a good shadow.

Many other preparation techniques are used to observe powders in transmission electron microscopy, and most analysis methods are modified to suit the application and the operator's ability.

Scanning Electron Microscopy (SEM)

The scanning electron microscope has a resolution of about 10 nm (100 Å) and is capable of very low magnification (about 10×) up to about 50 000×. It therefore can be used to count particles ranging in size from 1 mm to 0.1 μm. Particles smaller than 0.1 μm usually have too low a contrast with the background to be counted efficiently. The scanning electron microscope has about 300 times the depth of field of an optical microscope.

The image in the scanning electron microscope usually is obtained by using the secondary electron output of the sample as it is scanned by a very narrow electron beam. The contrast of the image depends more on the topography of the sample than on differences in atomic number. Therefore, prepared powders must not be embedded in films, but dispersed on a smooth substrate. Any smooth surface can be used as a substrate. However, if energy dispersive x-ray analysis (EDXA) is to be performed for particle identification, a carbon or polystyrene surface is preferred.

An excellent substrate can be made by placing a polystyrene pellet on a glass slide and heating it on a hot plate until it softens. A second glass slide is then placed over this slide and pressed until the pellet forms a thin disk. The slides are removed from the hot plate and pressed together until the polystyrene sets. The disk thus formed is as smooth as the glass and contains no elements that may hinder EDXA analysis. For sample preparations using aqueous suspensions, polystyrene surfaces can be rendered hydrophilic by a brief treatment in an oxygen asher at low power (5 to 10 W for 5 s).

While the substrates for scanning electron microscopy do not have to be as thin as those used for transmission electron microscopy, they must be conductive. Consequently, if glass or plastic surfaces are to be used, they must be coated with an evaporated metal (or carbon, for EDXA) film. This coating is usually applied after the particles have been dispersed on the surface.

Many of the dispersing techniques used for transmission electron microscopy can be applied to scanning electron microscopy. Particle dusting, drying from liquid suspensions, and mulling in liquids that can be sublimed in a vacuum are suitable dispersing methods, depending on the powder. If the technique of mulling in parlodion and amyl acetate is used, parlodion can be removed in an oxygen asher, thus leaving the particles on the substrate. Suitable substrates include glass or metal, because they are not affected by the ashing.

The prepared sample should always be placed in the scanning electron microscope with the surfaces normal to the electron beam so that the magnification, which changes with working distance, will be the same on all areas of the viewing screen. Particle counting can be done directly from the viewing screen, from photographs, or by using an automatic image analyzer.

Measurement Techniques

Direct Measurements. Particle diameters can be measured directly in an optical microscope with the use of a filar micrometer eyepiece. This eyepiece contains a scale and a movable cross hair that is operated by a calibrated knob on the side of the eyepiece. A particle is moved so that one side touches one of the fixed scale markings, and the cross hair is moved to touch the other side of the particle. The difference between the two readings is the Feret's diameter of the particle. The eyepiece is calibrated with an optical stage micrometer. This technique is time consuming, and because absolute measurements are not required for accurate size analysis, eyepiece graticules are usually used for direct measurements.

Typical examples of eyepiece graticules are shown in Fig. 26. These patterns are etched on glass disks that are positioned in the back focal plane of the microscope ocular and are therefore in the same focal plane as the particle images. Using one of these gauges, the size range into which the actual diameter of a particle falls can be measured easily. Feret's diameter, projected area diameter, or perimeter diameter are measured conveniently with graticules. Where measurements are made from photographs, projected images, or cathode ray tube screens, simple scales or plastic overlay graticules can be used to measure any of the accepted diameters.

Image shearing eyepieces are used directly within the optical microscope. This

Fig. 26 Eyepiece graticules

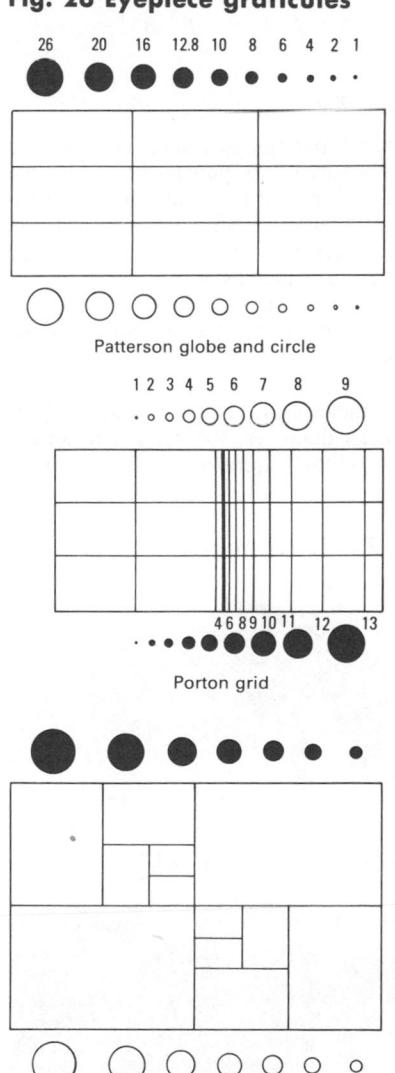

Patterson globe and circle

Porton grid

British standard graticule

device divides the optical beam into two parts by using mirrors and/or prisms. The distance between the two formed images is adjusted by shearing of the prisms, which is controlled by a micrometer dial on the side of the eyepiece. The maximum horizontal intercept is measured when the two images barely touch. With this device, a particle does not have to be moved to a particular position within the field of view to be measured.

The micrometer can be set to a given diameter, and the number of particles larger or smaller than this measurement can be counted, as determined by whether or not the images are touching. Red and green filters facilitate this by coloring the two images.

Semiautomatic Techniques. Many devices are available to shorten the time required to complete a particle size analysis. Most of these instruments record the number of particles in a size range, as judged by the analyst.

Adjustable light spot analyzers use a circle of light projected onto a photograph. The diameter of the circle can be adjusted by a diaphragm; a foot switch causes one of a bank of registers to record the diameter according to the diaphragm setting. Each particle on the photograph is passed under the light, which is adjusted to measure the projected area diameter or the perimeter diameter.

A transparent electronic graticule is a plastic sheet on which various size circles have been drawn. Under each circle, an electrical contact is connected to one of a bank of registers. When a contact is touched by the analyst with a pencil contact, the register advances one number. This device can be used with photograhs, back-projected images and cathode ray tube screens. The analyst decides which circle best represents the diameter of a particle and touches the corresponding contact.

Recording calipers work in the same manner, with a bank of registers connected to the caliper spread adjuster. Any particles of the prescribed diameter can be recorded with this device.

Recording micrometer eyepieces use a similar recording mechanism, but work directly within the optical microscope. Two movable cross hairs that move toward or away from each other with the turn of a micrometer screw are used to bracket a particle; a button is subsequently pushed to record the diameter. This technique is particularly useful when the depth of field prevents photography of the particles, because each particle can be sharply focused before measuring.

Sensitive surfaces connected to computers facilitate counting from photographs. The computer can be programmed to record diameters when the opposite sides of particles are touched with a pointer, or specific areas of the particle are touched. The computer consequently compiles data to suit user requirements.

Automatic techniques are used for counting large numbers of particles directly within the microscope. The main limitations of these techniques are the inability to distinguish particles that are touching and the fastidious sample preparation that is required.

Spot scanning devices use the moving spot from a cathode ray tube, which is projected through the microscope onto the specimen. When the spot passes over a particle, the light beam is interrupted, and a photocell records the particle. Particles are sized by scanning with different spot sizes. This technique also has been used on photographs, so electron micrographs can be used.

The slit scanning method projects the microscope image onto a slit. The microscope slide is mechanically scanned, and the signal produced as the particle images pass the slit is recorded. The width of the slit can be varied to eliminate coincidence and overlap.

Quantitative image analyzers are computer-controlled devices that use television cameras for direct analysis within an optical microscope. Analysis of photographs also can be accomplished. A quantitative image analyzer can be connected directly to an scanning electron microscope to control the scanning system. Particles can be counted according to their maximum horizontal chords, vertical chords, perimeters, or areas.

Data Presentation. The most precise form of presenting particle size data is in tabular form, as shown in Table 3. However, graphical representations are more concise and visually show the mean and deviation from the mean. Some graphical presentations can yield specific values for descriptive constants such as the mean, median, and the standard deviation.

A histogram is a bar graph that illustrates the frequency of occurrence as a function of the size range. Figure 27 shows a typical histogram for a log-normal size distribution. The smooth curve drawn through the histogram is a valid size-frequency curve if sufficient particles are counted and the number of size intervals is at least 10.

In cumulative plots, such as Fig. 28, the median particle size can be noted easily. These plots are made by plotting the per-

Table 3 Particle size data

Particle size range, μm	No. of particles	Particles, %	Cumulative percent less than size
1-2 0		0.0	0
2-3 3		0.6	0.6
3-4 8		1.6	2.2
4-5 15		3.0	5.2
5-6 79		15.9	21.1
6-7 163		32.9	54.0
7-8 121		24.4	78.4
8-9 64		12.9	91.3
9-10 28		5.7	97.0
10-20 13		2.6	99.6
20-30 2		0.4	100

Fig. 27 Histogram and size frequency curve for log-normal size distribution

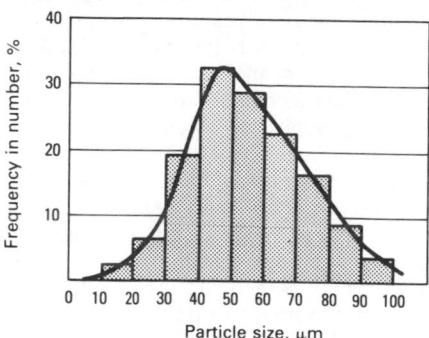

Fig. 28 Cumulative plot used in determining median particle size

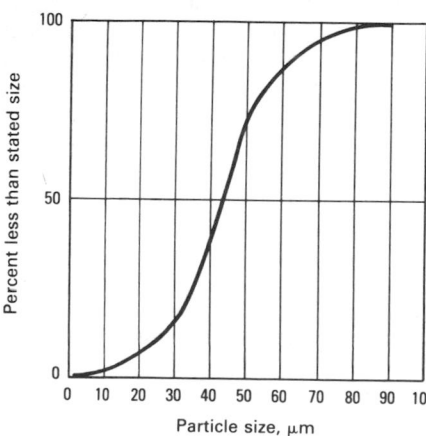

centage of particles less than (or greater than) a given particle size against the particle size.

A special grid for plotting size-frequency data was developed so that the resulting curve is a straight line. This grid consists of a coordinate based on probability. Cumulative percentages are plotted on the probability scale against particle size on either an arithmetic or a log scale.

Symmetrical size distributions, characterized by a bell-shaped size-frequency curve symmetrical to a vertical line through the mode (apex) point, plot as a straight line on arithmetic-probability paper. Most natural particle distributions do not follow this normal probability, but follow a skewed or log-normal distribution. These distributions provide a symmetrical size-frequency plot if the particle size scale is logarithmic and a straight line on log-probability paper, as shown in Fig. 29. The median particle size is found at the 50% point on these probability plots, and the standard deviation for arithmetic-probability plots can be calculated as:

$$\sigma = 84.13\% \text{ size} - 50\% \text{ size}$$
$$= 50\% \text{ size} - 15.87\% \text{ size}$$

and for log-probability plots can be calculated as:

$$\sigma = \frac{84.13\% \text{ size}}{50\% \text{ size}} = \frac{50\% \text{ size}}{15.87\% \text{ size}}$$

Fisher Sub-sieve Sizer

By Donald S. Parsons
Head, Physical Testing Laboratory
Chemical and Metallurgical Division
GTE Products Corp.

THE FISHER SUB-SIEVE SIZER is a simple, inexpensive permeameter, which measures the flow of air through a packed powder bed. From this measurement, average particle size and porosity can be calculated from a chart that is an integral part of the instrument. The Fisher sub-sieve sizer is used extensively throughout industry, especially for process control purposes where only relative values are required.

Fig. 29 Log-probability plot of symmetrical size distributions

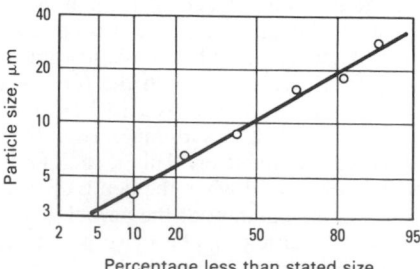

Percentage less than stated size

Frequently, however, the results are often misinterpreted to designate an exact particle size. Consequently, analysis results refer to a Fisher "number" rather than a size. Fisher sub-sieve numbers should not be compared with size results from other measuring techniques, as frequently no correlation exists. Figure 30 illustrates a typical Fisher sub-sieve sizer, and Fig. 31 provides a schematic of the method.

Permeameters accurately and precisely measure permeability. They do not measure surface area or particle size. The flow through a porous compact depends on the structure of the pores. Flow, therefore, depends on the porosity, particle shape, particle size, particle size distribution, and method of compaction, all of which affect pore structure. Conversion of the measured permeability to a surface area is accomplished by the Kozeny-Carman equation:

$$K = E^3/(1.0 - E^2)BS_w^2\rho^2$$

and

$$E = (V_{bed} - V_{powder})/V_{bed}$$

where K is permeability of the powder bed; E is porosity of the powder bed; V is volume, cm^3; ρ is true, or absolute, density of the powder material, g/cm^3; S_w is surface area per unit weight of powder, cm^2/g; and B is constant.

From this equation, it can be seen that permeability is related to three macroscopic quantities: bed porosity, surface area, and particle density. Average particle diameter (d_m), also known as the Sauter mean diameter, the surface mean diameter, or the volume surface mean particle

Fig. 30 Fisher sub-sieve sizer

Fig. 31 Schematic of Fisher sub-sieve sizer

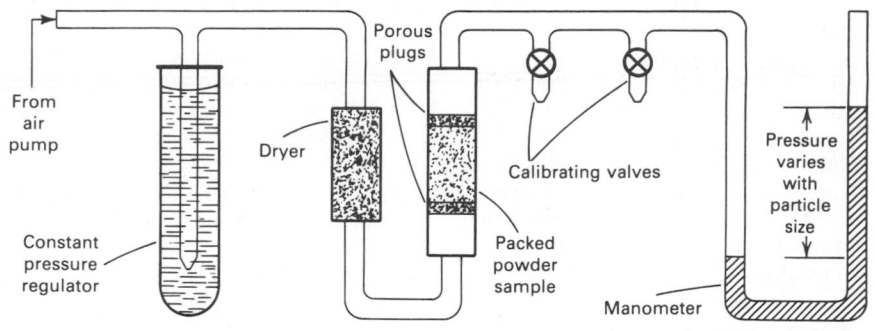

size, is calculated from the surface area by:

$$d_m = 6/\rho S_w$$

For powders of the same composition and of similar size distributions, the Kozeny-Carman equation is probably satisfactory. However, if particles can be packed differently and exhibit the same porosity, but with different pore structures, this relationship may not hold. For instance, two powders of the same chemical composition with different particle size distributions have given the same Fisher sub-sieve numbers. Comparison of results from the Fisher sub-sieve sizer with those from other surface area and size distribution techniques may show no agreement and should not be undertaken.

When particle shape, size, and distribution are essentially the same and do not vary significantly from one production lot to another, Fisher sub-sieve analysis provides adequate comparisons for quality control purposes. The major limitations include constant sample packing and instrument calibration.

Operation

The American Society for Testing and Materials standard ASTM B 300, "Average Particle Size of Powders of Refractory Metals and Compounds by the Fisher Sub-sieve Sizer," provides calibration and operating instructions for Fisher sub-sieve analysis. Standard ASTM C 721, "Average Particle Size of Alumina and Silica Powders by Air Permeability," also provides guidelines. These specifications are based on the Fisher sub-sieve sizer instruction manual (Ref 7).

The output of an air pump is regulated to a pressure head of 50 cm (19.7 in.) of water by the water level in the standpipe. At this constant pressure, the flow through the packed powder bed is measured by the water manometer. Calibration is accom-

plished by varying the leakage rate through the calibrating valves.

The powder sample is weighed to 0.1 g (0.003 oz) and placed in the sample tube. Using a rack and pinion, it is compacted with a 222-N (50-lb) force applied with a torque wrench. This technique minimizes sample-to-sample packing variation. Porosity is determined from the compacted sample height and the instrument calculator chart.

The sample tube is then placed in the air system, and after equilibrium is reached, the manometer water height is measured. The instrument calculator chart is used to obtain the average size. If the sample height is such (high or low) that the calculator chart cannot be used, the actual height (in centimetres) above the base line is measured for both sample height and manometer water height. These values are used in the equation below to provide calculated values. The calculator chart uses a modification of the Kozeny-Carman equation:

$$\text{Porosity} = [LA - (M/\rho)]/LA$$

and

Fisher number =

$$CL(M/\rho)\sqrt{2H/(P - 2H)[AL - (M/\rho)]^3}$$

where L is sample height after compaction, cm; A is cross-sectional area of sample tube, cm^3 (1.267); M is weight of sample, g; ρ is true, or absolute, density of powder material, g/cm^3; H is height of water column above the base line, cm; P is overall air pressure, cm (50 cm, or 19.7 in., of water); and C is instrument constant (3.80).

If the sample weight is made equal to the true density, the equation is simplified:

$$\text{Porosity} = (LA - 1.0)/LA$$

and

Fisher number =

$$CL\sqrt{2H/(P - 2H)(AL - 1.0)^3}$$

Standardization

The Fisher sub-sieve sizer must be standardized daily (or more frequently, if necessary) according to the procedures given in the instrument manual (Ref 7). Monitoring of the sample post height, which sets the chart base line as the instrument zero, is frequently overlooked. This can be verified easily by placing two filter paper sheets between the two sample tube porous plugs located on top of the sample post. The rack is brought down in contact with the upper porous plug, and the pointer is checked to see that it is on the base line. If it is not, post height is adjusted until the measurements coincide. This is an infrequent, but critical, adjustment.

A method of standardization uses a Fisher ruby orifice mounted in a sample tube, which is calibrated by the manufacturer against the primary standard, the National Bureau of Standards Cement No. 114. Operating the Fisher sub-sieve sizer with the calibrating tube in place and the chart positioned to the proper porosity produces a water column of a predetermined height. This indicates a size on the chart that is equal to that engraved on the side of the tube. If it does not, the internal calibration valves are adjusted until the correct value is obtained. Other secondary standards may be used and include any uniform powder that has an accepted Fisher sub-sieve number.

Results

Because analysis results are obtained by interpolation of the pointer position between lines on the calculator chart, they should be reported with varying intervals and places after the decimal point, as indicated in ASTM C 721, Section 9.1. Reading and chart line intervals for given particle ranges are:

Particle range, μm	Chart line interval, μm	Reading interval, μm
0.2 to 4	0.1	±0.02(a)
4 to 8	0.2	±0.1
8 to 15	0.5	±0.1(b)
15 to 20	1.0	±0.2
20 to 25	5.0	±1.0

(a) For porosities less than 0.55, the reading interval for the 0.5- to 4-μm range is +0.05 μm. (b) For porosities greater than 0.6, the reading interval for the 11- to 15-μm range is ±0.2 μm.

In one laboratory, a 5-year weekly evaluation of the ASTM B 459 tungsten powder gave an average of 1.23 μm and a standard deviation of 0.012 μm.* Fisher sub-sieve numbers are generally agreed to be precise within ±5%. Because of the care taken in making the measurements, many laboratories report ±3% precision.

*GTE Products Corp., Towanda, PA

REFERENCES

1. Born, M. and Wolf, W., *Principles of Optics*, Pergamon Press, New York, 1969, p 395.
2. West, G.C., "The Application of the Light Obscuration Principle to the Particle Size Analysis of Powdered Materials," Third Particle Size Analysis Conference, University of Salford, England, Sept 1977
3. Kupfer, H.A., Automatic Particle Size Distribution Analysis of Powdered Material by the Light Blockage Principle, *Powder Metall.*, Vol 10 (No. 2), 1979, p 96-97
4. Allen, T., *Particle Size Measurement*, 3rd ed., Chapman and Hall, London, 1981

5. "Operation and Maintenance Manual for HIAC Model PA-720," HIAC/Royco Instruments, Menlo Park, CA
6. Frock, H.N. and Plantz, P.E., "Correlation Among Particle Sizing Methods," *Proceedings of the Technical Program*, International Powder and Bulk Solids Handling and Processing, Rosemont, IL, May 1982, p 77-88
7. Fisher Sub-sieve Sizer Instruction Manual No. 14579, 7th ed., Fisher Scientific Co., Pittsburgh, Feb 1975

SELECTED REFERENCES

- Beddow, V.K. and Melay, T.P., *Testing and Characterization of Powders and Fine Particles*, Heyden, London, 1980

- Dallavalle, J.M., *Micromeritics*, Pitman, Chicago, 1943

- Irani, R.R. and Callis, C.F., *Particle Size: Measurement, Interpretation and Application*, John Wiley & Sons, New York, 1963

- Jelinck, Z.V., *Particle Size Analysis*, John Wiley & Sons, New York, 1970

- Kaye, B.H., *Direct Characterization of Fineparticles*, John Wiley & Sons, New York, 1981

- Stockham, J.D. and Fochtman, E.G., *Particle Size Analysis*, Ann Arbor Science, Ann Arbor, MI, 1977

Particle Shape Analysis

PARTICLE SHAPE, like particle size, is a primary powder characteristic that must be considered when determining the optimum utilization of a P/M material. Behavioral characteristics of a metal powder, such as flow rate, apparent density, compressibility, and sinterability, are all influenced by particle shape and size.

To date, the most common approach to describe and differentiate particle shapes has been the use of qualitative concepts. Two fundamental concepts have been used: (1) the dimensionality of the particle, and (2) the surface contour of the particle. Using these concepts, a model system of shape characterization is presented in Fig. 1. Photomicrographs of several types of loose powders described in the International Standards Organization standard ISO 3252 are shown in Fig. 2.

One-dimensional particles are essentially rod-like in nature (Fig. 2a and 2d). The length of such particles, which are often termed fibrous, is of primary significance and of much greater magnitude than the cross-sectional area. The aspect ratio (the ratio of the length to the diameter) is also sometimes used to describe metal and ceramic fibers for use in fibrous composites. Two different types of one-dimensional particles can be considered on the basis of their surface contour. One is smooth and the other rather roughened, with an irregular type of surface.

Two-dimensional particles are very flat in nature. Lateral dimensions are considerably greater than the thickness. The surface contour of such particles is usually rather irregular. The dendritic type (Fig. 2c) is characterized by a tree-like shape and is often associated with electrolytic powders. However, secondary mechanical treatments often destroy such a shape. Flake particles (Fig. 2e) are also considered two-dimensional particles. The length and width are the most important parameters of flake particles, both of which are significantly greater than the thickness.

Most powders are three dimensional in nature. These powders may be equiaxed and nodular. The simplest type of particle in this category is the spherical type (Fig. 2j). By departing from this perfect shape and contour, irregular particles (Fig. 2g) and nodular types (Fig. 2h) are obtained.

It should be noted that porous particles are different than irregular particles because of the presence of porosity, which may also be very irregular in both size and shape. Internal porosity may be isolated or interconnected. Large amounts of porosity make shape characterization difficult. It is best observed by examination of cross sections of mounted powder.

However, the ideal conditions shown in Fig. 1 and 2 do not always exist. As a result, difficulties arise when analyzing particle shape, particularly shape parameters of three-dimensional particles, by using normal light or electron microscopy. Additionally, uniformity of shape does not completely exist within any one type of powder. For example, in atomized nonferrous metals and ferrous alloys, such as aluminum, tin, and stainless steel, smaller particles usually exhibit a greater degree of sphericity than larger particles. Some believe that atomization always produces rounded particles; however, this is not true. The method of atomization used has an important effect on particle shape. Gas or air atomization tends to produce more rounded particles, while water atomization yields more irregularly shaped particles.

Frequently, shape and size analyses are difficult because of the bonding of several small particles into a large agglomerate (Fig. 2f). Experimental observation of such powders requires adequate dispersion of the particles.

Due to the limitations of qualitative analysis when determining three-dimensional shape parameters, many attempts have been made to place shape characterization on a quantitative basis by the use of various mathematical descriptions. The goal of quantitative shape analysis is to define mathematically particle shape, which in turn can be used to predict behavioral characteristics of powder.

This article reviews recent developments in qualitative and quantitative particle shape analyses. Advancements in qualitative analysis are based on the use of the scanning electron microscope (SEM) for determination of particle shape. In addition to its high magnification capability, this device produces an image with a three-dimensional appearance. Because of its advantages, scanning electron microscopy has replaced optical microscopy to a large extent in the area of shape and topographical analyses of metal powders.

Quantitative shape analysis has posed numerous difficulties. However, with the advent of advanced computers and graphics systems, interest in quantitative analysis has increased. There are a number of alternate approaches, which include, in order of their historical development, conventional shape factors, stereological characterization of shape, morphological analysis, and use of fractals as descriptors of P/M systems. Because a detailed description of the mathematical and statistical concepts inherent in these quantitative methods is beyond the scope of this article, the reader should consult the appropriate bibliographic references for additional information.

Fig. 1 Various shapes of powder particles and their methods of manufacture

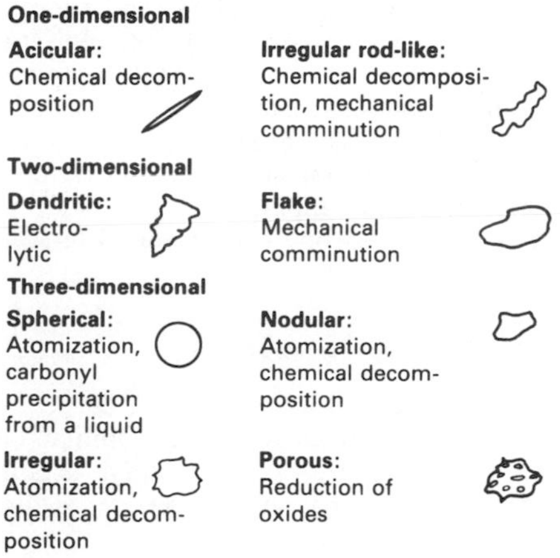

One-dimensional

Acicular: Chemical decomposition

Irregular rod-like: Chemical decomposition, mechanical comminution

Two-dimensional

Dendritic: Electrolytic

Flake: Mechanical comminution

Three-dimensional

Spherical: Atomization, carbonyl precipitation from a liquid

Nodular: Atomization, chemical decomposition

Irregular: Atomization, chemical decomposition

Porous: Reduction of oxides

Fig. 2 Common particle shapes as depicted in ISO 3252
(a) Acicular powder particles. (b) Angular powder particles. (c) Dendritic powder particles. (d) Fibrous powder particles. (e) Flaky powder particles.
(f) Granular powder particles. (g) Irregular powder particles. (h) Nodular powder particles. (j) Spheroidal powder particles

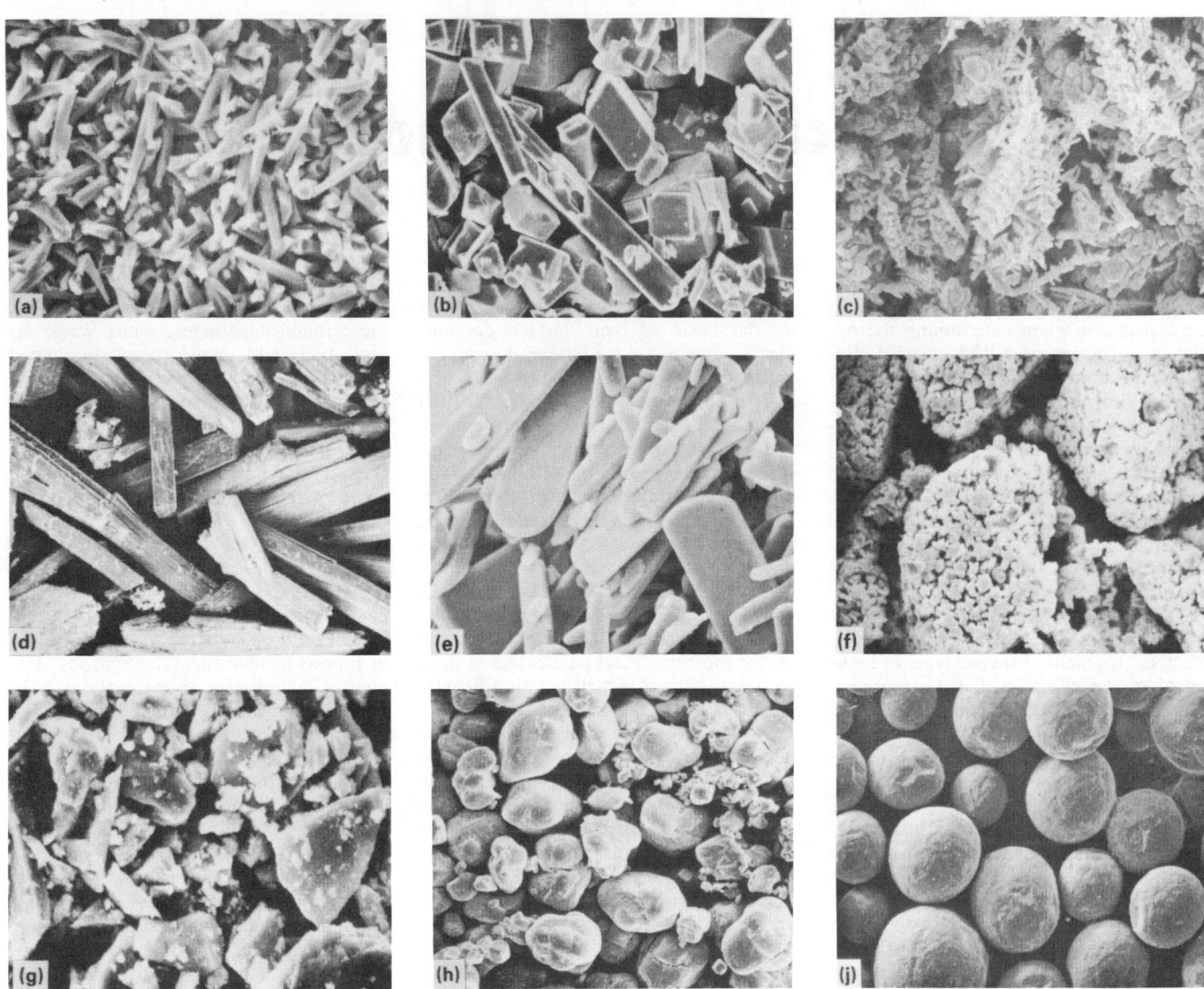

Qualitative SEM Examination of Particles

By Peter J. Heinzer
Coordinator—Technical Services
Imperial Clevite, Inc.

THE FIRST SCANNING ELECTRON MICROSCOPE was developed by von Ardenne in 1938 by adding scanning coils to a transmission electron microscope (Ref 1, 2). The electron beam was focused to a fine spot, which scanned across the surface of a very thin specimen. Electrons transmitted through the sample formed a corresponding image on the photographic plate below it. Resolution was between 50 and 100 nm (500 and 1000 Å). Modifications of this basic concept over the years have resulted in sophisticated modern instruments with resolutions greater than 10 nm (100 Å).

The scanning electron microscope (shown schematically in Fig. 3a) is composed of an electron gun, an electron column with a series of apertures and magnetic lens assemblies, a sample chamber, various detector systems (primarily for secondary and back-scattered electron detection), and an instrument console that provides beam control and image display. Figure 3(b) illustrates a typical installation.

Additional features of scanning electron microscopes include the attachment of computerized image analysis and energy dispersive x-ray spectrometry systems to provide elemental analysis and mapping of the sample surface as it is being scanned. Detailed information on the physical fundamentals, instrument details, and various techniques and applications of scanning electron microscopy can be found in Ref 3 through 8.

P/M Applications

The most significant feature of the scanning electron microscope, in addition to its high magnification capability (useful magnifications beyond 10 000×), is its

Fig. 3 Typical SEM installations

(a) Schematic representation of electron gun, electron column, and sample chamber. (b) Photograph of typical installation. Courtesy of Case Western Reserve University

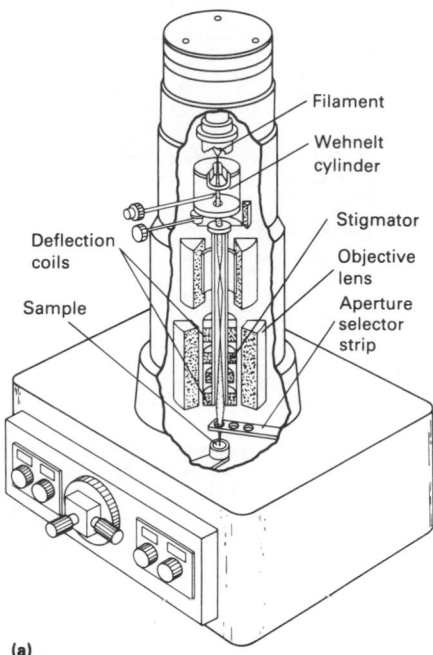

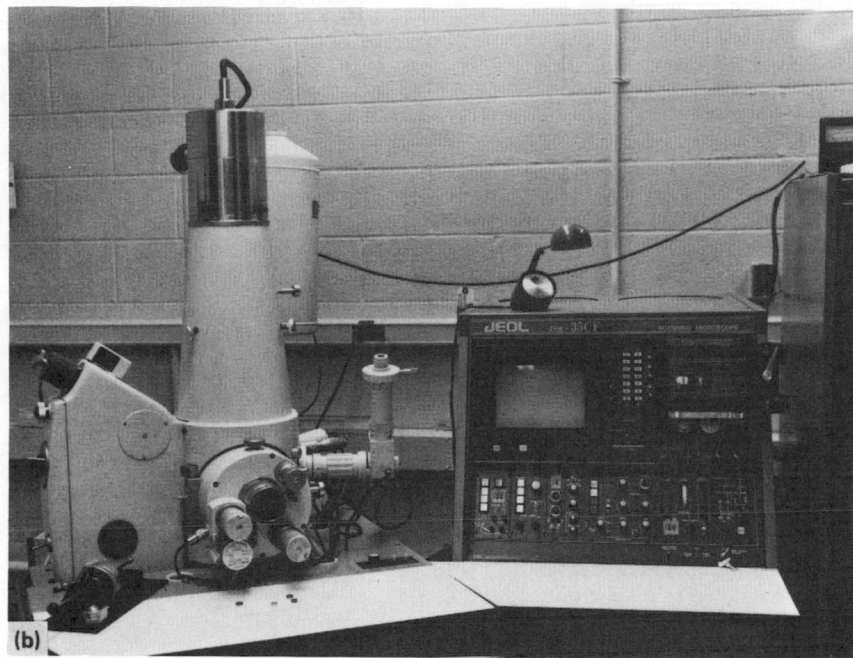

ability to produce images with a three-dimensional appearance (Ref 2). This ability derives from the fact that the depth of field for the scanning electron microscope is over 100 times that of the optical microscope. This increased depth of field (ranging from 1 μm at 10 000× to 2 mm at 10×) also accounts for the extensive particulate analysis performed on the scanning electron microscope at magnifications that are within the capabilities of optical microscopy.

The imaging capabilities of the scanning electron microscope make it a useful tool for P/M applications concerned with all phases of powder production and processing. Scanning electron microscopy has been used to study all aspects of particle morphology, including size, shape, surface topography, surface structure (crystal, grain, and dendrite), coating or thin film characteristics (oxides), inclusion, void, and agglomeration characteristics, and satellite formation. The scanning electron microscope has also been used to study surface topography, effect of oxides or other coatings, porosity, inclusions, and other contaminants on P/M materials.

One of the more useful applications of the scanning electron microscope in powder metallurgy is qualitative particle characterization in terms of visual appearance. Despite the development of automated instruments for quantitative particle characterization, no substitute has been found for the interpretive capability of man. The use of the scanning electron microscope extends this capability into the microscopic domain. The following examples provide a representative sample of the usefulness of scanning electron microscopy in the field of P/M technology.

Inspection of powder particles to be used in P/M processing is essential, because individual and agglomerate characteristics can have significant effects on final material properties. Figure 4 shows representative alloy powder particles used in the direct rolling of aluminum P/M strip (Ref 9). The irregular shape promotes interlocking of particles, which provided suf-

Fig. 4 Scanning electron micrographs of 7091 aluminum alloy particles used in direct powder rolling of strip

(a) Magnification: 70×. (b) Magnification: 700×. Source: Ref 9

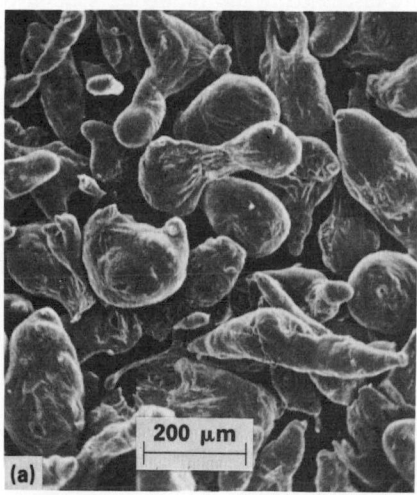

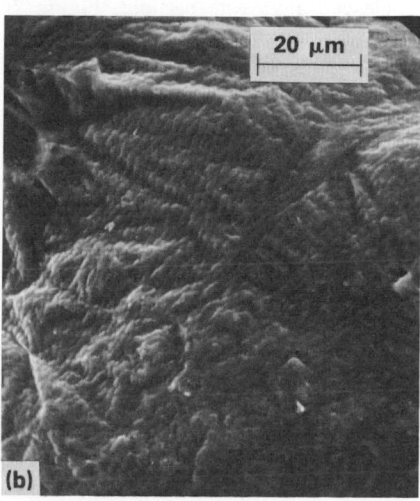

ficient green strength to permit handling of the strip during subsequent processing.

Scanning electron microscopes are widely used to investigate green strength differences between pressed powder compacts of porous and solid copper powder (Ref 10). The scanning electron microscope was used to obtain fractographs of the green specimens. The shape of the compacted individual particles revealed that the porous powder had greater contact area than the solid powder (Fig. 5). Higher magnification micrographs show a surface interpenetration effect, which further increases contact area and resultant green strength.

In another study, use of the scanning electron microscope provided evidence to support a proposed mechanism of the formation of particles during water atomization. As shown schematically in Fig. 6, four stages comprise the mechanism: (1) disturbance of the liquid metal stream by the atomizing medium, (2) fragment detachment and ligament formation, (3) further breakdown into spheroidized droplets and membranous fragments, and (4) collision and coalescence. See the article "Atomization" in this Volume for detailed information concerning the mechanisms of particle formation during atomization.

Particle shapes observed by SEM examination of various atomized materials are shown in Fig. 7. Figure 7(a) shows gas-atomized spheroidal particles with spheroidal satellites attached. In this example, the time required for spheroid formation is less than that required for solidification, and the collisions between large and small particles are occurring near the melting point. Figure 7(b) illustrates the irregular shapes of typical water-atomized particles.

In Fig. 7(c), the smooth features of the small satellite particle are contrasted against the surface grain structure of the highly coalesced aggregate particle. Figure 7(d) illustrates large aggregate particles with little coalescence. Hollow particles and membraneous fragments (collision products) are shown in Fig. 7(e), (f), and (g).

Conventional Shape Factors

By Chi-Ren Chang
Software Engineer
Industrial Vision Systems, Inc.

PARTICLE SHAPE is a fundamental characteristic of powder particles and thus

Fig. 5 Scanning electron micrographs of pressed copper powder

Porous copper particles at lower (a) and higher (b) magnifications. Solid copper particles at lower (c) and higher (d) magnifications. Lower magnifications reveal greater contact area of the individual porous particles. Higher magnifications show surface penetration effect, which further increases the contact area, resulting in enhanced green strength of the porous powder.

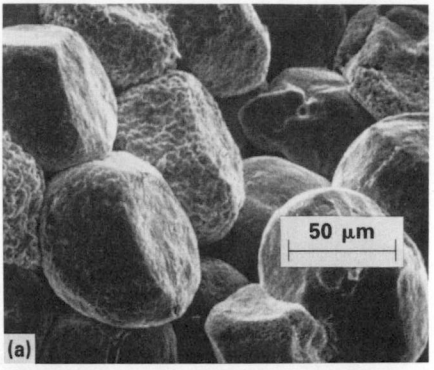
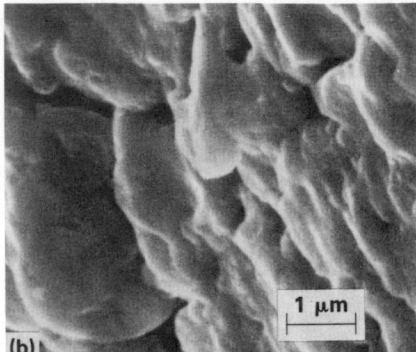
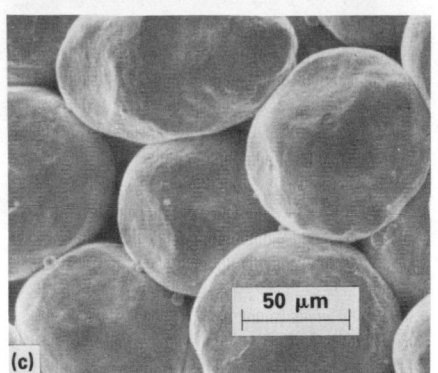
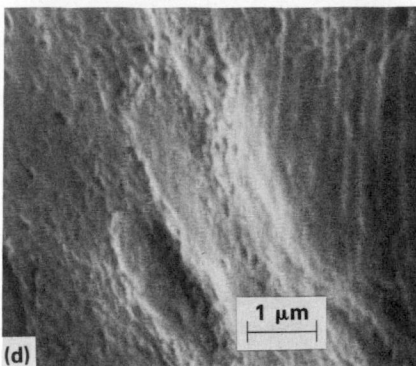

influences the properties of particulate systems. Various shape terms have been proposed to quantitatively represent particle shape. Early systems tended to measure one specific feature of a particle.

Table 1 lists some of the most frequently used shape terms. The applicability and/or limitations of various shape factors, also referred to as shape parameters, are discussed in the following section of this article on stereological characterization of shape. For discussion purposes, the following method developed by Hausner (Ref 16) will be used to demonstrate the applicability of shape factors in quantitative analysis of particle shape.

A rectangle of minimum area is drawn around the cross section of a particle (particle projection) as it is observed under the microscope (Fig. 8). The ratio of the rectangle side lengths permits calculation of particle elongation:

$$\text{Elongation factor, } x = \frac{a}{b}$$

The ratio of the area (A) of the projected particle to the area of the enveloping rectangle of minimum area ($a \times b$) indicates

the bulkiness of the particle:

$$\text{Bulkiness factor, } y = \frac{A}{a \times b}$$

In this way, the cross-sectional area of the particle is correlated with some of its linear dimensions.

To characterize the surface configuration (or surface area), which is an essential factor of shape, the surface of the respective particle should be compared with the surface of a sphere of identical volume, or a cross section should be compared with that of the particle:

$$\text{Surface factor, } z = \frac{c^2}{12.6A}$$

where c is the perimeter of the projected profile of the particles. For a spherical particle, $z = 1$; and for particles of any other shape, $z > 1$.

Although the three ratios (x, y, and z) do not permit exact characterization of particle shape, they are nevertheless descriptive. It is possible, therefore, to correlate particle behavior with the three ratios—elongation factor, bulkiness factor, and surface factor. As indicated, however, there have been other shape factors proposed to characterize particle shape (see Table 1 and the following section of this article on stereological characterization of shape).

Stereological Characterization of Shape

By H.E. Exner
Max-Planck-Institut für
Metallforschung
Institut für Werkstoffwissenshaften

STEREOLOGY refers to the study of the three-dimensional structure of materials from two-dimensional sections or projections. This discipline is primarily concerned with the geometrical evaluation of microstructural features, based primarily on geometrical probabilities. However, statistics, topology, and projection geometry also contribute to the overall quantitative characterization (Ref 19). Quantitative characterization of shape of microstructural features (pore space, precipitates, phase regions, and grains, for example) and of particles on a substrate or embedded in a resin by means of data derived from planar images (sections or projections) is the prime consideration of stereology.

Stereology usually does not apply to the geometric nature of single particles, but rather to the geometric characteristics of a large number of particles or to the geometric characteristics of the "average particle" in a powder mass. Detailed information on the fundamentals, instrumentation, and applications of quantitative stereology can be found in Ref 20 through 23.

Numerous shape parameters have been proposed, but only a few are practical. A useful stereological shape parameter (factor) must fulfill several requirements:

- *Shape sensitivity:* The value of a particular parameter must vary systematically with changes in shape—it should be sensitive to specific aspects such as elongation, bulkiness, and symmetry.
- *Independence of other geometric properties:* Size, size distribution, volume fraction, or other non-shape-related geometric characteristics should not influence the value for a given shape parameter.

Fig. 6 Schematic of particle formation stages occurring during atomization

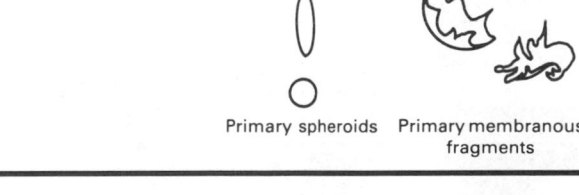

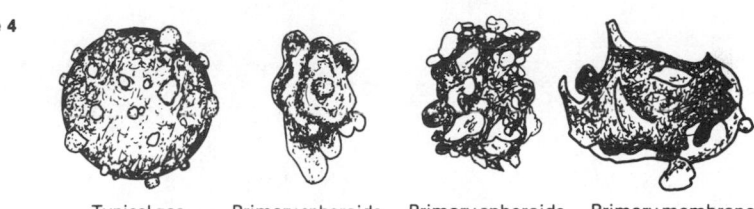

- *Accessibility:* Quantities from which a stereological shape parameter is calculated must be available by simple measurements performed on planar images.

Other less significant requirements are that shape parameters must be dimensionless and independent of rotation or translation of the objects. Furthermore, it should be easy to visualize the significance of a shape parameter in terms of interpretable microstructural shape changes. The range for typical microstructural changes should be wide compared to statistical fluctuations and measuring errors.

From the parameters listed in standard textbooks (Ref 20, 21) and from those proposed in literature (Ref 23-26), none conforms to these requirements com-

pletely. In the following sections, some of the more useful stereological shape parameters, their limitations, and their practical alternatives (fingerprinting by means of simple two-dimensional parameters) are discussed.

Image Data for Shape Characterization

With manual or semiautomatic data acquisition (in which counting and measurements are performed by the operator), only average quantities for the measured fields of view (field quantities), such as area fraction, total and mean intercept length, total and mean perimeter length, mean curvature, and number of features per unit

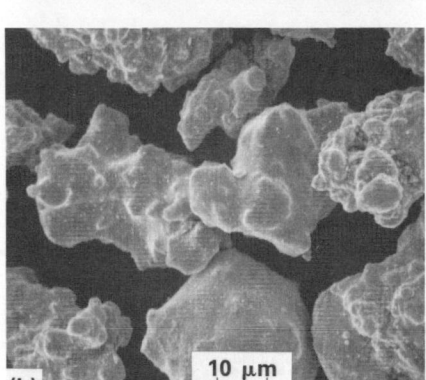

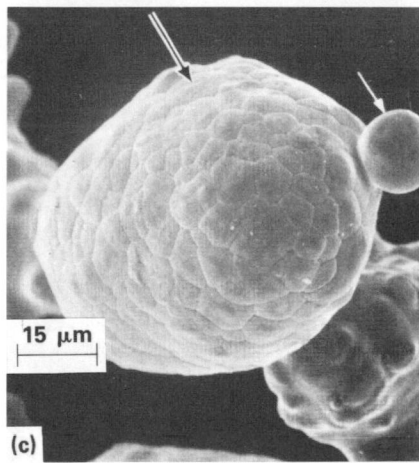

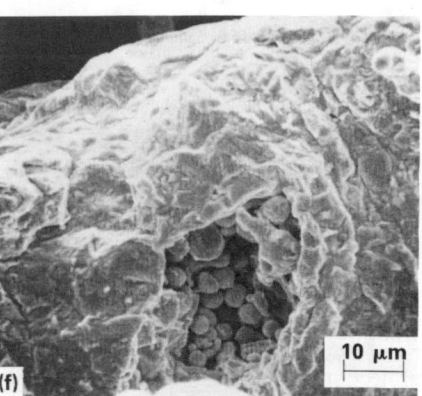

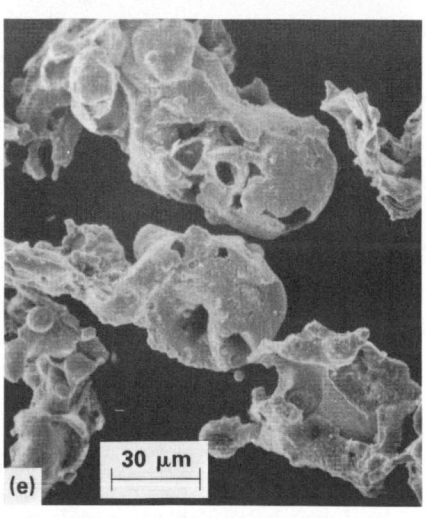

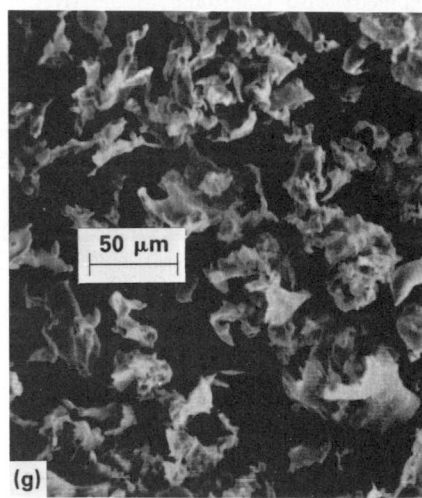

Fig. 7 Particle shapes observed by SEM examination of atomized materials

(a) Gas-atomized cobalt-based alloy particles. Note attached satellites. (b) Water-atomized low-carbon steel. Note coalescence. (c) Large copper particle with evident grain structure. Note attached satellite with comparatively smooth surface. (d) Water-atomized copper aggregate particles exhibiting little coalescence. (e) Hollow particles of a water-atomized Cu-9.3P alloy. (f) Large hollow particle with small entrapped particles from water-atomized bronze powder. (g) Membraneous fragments (collision products) from a water-atomized Sn-3.3Cu-0.6P alloy powder

area or unit length, can be obtained in a timely manner. Automatic instruments facilitate measurement of parameters describing the geometric properties of each individual planar feature, as seen in the fields of view at a high speed. From those parameters that can be used for shape description, the following have been shown to be of practical use: individual intersect areas, intercept length, tangent (Feret) diameters, as a function of direction and their extreme (maximum and minimum) for each closed planar feature; individual curvature; and moments of inertia.

Concise characterization of three-dimensional shape is possible only if the spatial coordinates of the particle surfaces or three-dimensional feature parameters (spatial Feret diameters, spatial moments of inertia, surface areas, and volumes of the particles) are available. Direct acquisition of three-dimensional data is possible by (1) serial sectioning, (2) stereometric measurements using SEM stereo-pair photographs and instrumented stereometers, and (3) additional specialized techniques, including x-ray topography, shadowgraphy, or densitometry

(for semi-transparent materials). However, these techniques are time consuming and not yet well established.

Stereology of Shape

It is generally agreed that three-dimensional shape cannot be quantitatively assessed from planar images without severe limitations. For example, topological parameters, such as the number of separate parts (the number of pores in a porous particle, for example) or the degree of connectivity (the number of channels between

Table 1   Shape terms and their definitions

Volume specific surface (S_v)

$$S_v = \frac{S}{V}$$

V: Particle volume
S: Particle surface area

Heywood ratios (Ref 11)

Elongation ratio (n)
$$n = L_h/B_h$$

B_h: Breadth—the minimum distance between two parallel planes that are perpendicular to planes defining T_h

Flakiness ratio (m)
$$m = B_h/T_h$$

T_h: Thickness—minimum distance between two parallel planes tangential to the particle, one of which is the maximum stability plane

L_h: Length—distance between two parallel planes that are perpendicular to both T_h and B_h planes

Heywood shape factor (ϕ_h) (Ref 12)

$$\phi_h = f/k$$

where $f = \dfrac{S}{d_a^3}$

$$k = \frac{V}{d_a^3}$$

d_a: Diameter of a circle having the same projected area
S: Particle surface area
V: Particle volume

For spheres: $n = m = 1$, $\phi_h = 6$

For spheres: $n = m = 1$, $\phi_h = 6$

Wadell

Sphericity (ψ_s) (Ref 13)

$$\psi_s = (d_v/d_s)^2$$
$$\approx 4.84\,(k/f)^{2/3}$$

d_v: Diameter of the sphere that has the same volume of the particle
f, k: Diameter of the sphere having the same surface areas as the particle
f, k: Defined in Heywood ratios

Roundness (ψ_r) (Ref 14)

$$\psi_r = \frac{\sum\limits_{i=1}^{N} r_i}{NR}$$

r_i: ith radius of curvature along the particle profile which is smaller than the radius of the largest inscribed circle
N: Total number of radii of curvature smaller than the inscribed circle
R: Radius of the largest inscribed circle

Krumbein (Ref 15)

Sphericity (ψ_k)

$$\psi_k = \sqrt[3]{\left(\frac{T_k}{B_k}\right)\left(\frac{B_k}{L_k}\right)^2}$$

L_k: Longest dimension of the particle
B_k: Breadth—measured perpendicular to L
T_k: Particle thickness

Hausner (Ref 16)

Elongation ratio (x)
$$x = a/b$$
Bulkiness factor (y)
$$y = A/ab$$
Surface factor (z)
$$z = c^2/12.6A$$

a: Length of the enveloping rectangle that has the minimum area
b: Width of the rectangle
A: Projected area of the particle
c: Perimeter of the projected profile

Church shape factor (ϕ_c) (Ref 17)

$$\phi_c = E(d_M)/E(d_F)$$

d_M: Martin's diameter—length of the chord which divides the profile into two equal areas with respect to a fixed direction
$E(d_M)$: Expectation of Martin's diameter
d_F: Feret's diameter, the distance between a pair of parallel tangents of the particle profile with respect to a fixed direction
$E(d_F)$: Expectation of Feret's diameter

Centroid aspect ratio (CAR) (Ref 18)

$$CAR = \frac{d_m}{d_p}$$

d_m: Longest chord passing through the centroid
d_p: Length of the chord passing through the centroid and perpendicular to the one defining d_m

Fig. 8 Determination of particle size characteristics

(A) Surface area of the projected particle.
(a) length of the enveloping rectangle, which has a minumum area. (b) Width of the rectangle.
(c) Circumference of the projected particle

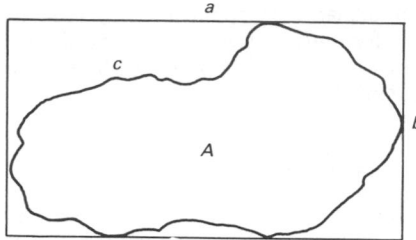

the pores), cannot be obtained from a single cross section. From projections, especially from stereo-pair micrographs, a qualitative shape description can be easily derived. For stereological quantification, however, the data obtained from projections are generally less suitable than those obtained from the cross sections of particles embedded in a resin, a metal, or in a glass prepared by normal metallographic techniques.

Thorough analysis of stereological shape parameters obtained from a single cross section through a multitude of particles, as proposed in the literature (Ref 27), intended to show that parameters that depend exclusively on the shape of the three-dimensional objects, and not on size or other non-shape-related features, can be traced to one simple combination of field data, the number of intersecting points between a measuring line and the perimeters of the particle cross section (P_L), which, for convex particle features, is twice the number of particles per unit length of the measuring traverse—N_L ($P_L = 2\,N_L$)—the volume fraction (V_V) measured as area fraction (lineal fraction or point fraction) and the number of objects per unit area (N_A). Fischmeister (Ref 27) derives the universal shape parameter:

$$F_F = \frac{1}{6\pi}\cdot\frac{N_L^2}{V_V\cdot N_A} \qquad \text{(Eq 1)}$$

The numerical factor aims to normalize the value to 1 for spheres. However, it can be easily demonstrated that this shape parameter is not independent of the size distribution, and deviations from a value of 1 for spheres of varying size are possible. A shape parameter that is actually independent of other geometric properties was proposed by De Hoff (Ref 28) in 1964:

$$F_D = \frac{1}{6\pi}\cdot\frac{N_L^2}{V_V\cdot N_A}\cdot\frac{M_1\cdot M_3}{M_2^2} \qquad \text{(Eq 2)}$$

where M_1, M_2, and M_3 are the first three

moments of the spatial size distribution, which generally are not accessible from planar data without *a priori* information on shape (the shape of the particles must be simple and known in order to calculate three-dimensional size distributions from planar measurements). F_D reduces to F_F for uniform size, while logarithmic-normal size distributions with known standard deviation $ln\ \sigma$ yields:

$$F_D = \frac{1}{6\pi} \cdot \frac{N_L^2}{V_V \cdot N_A} \exp (ln\ \sigma)^2 \qquad \text{(Eq 3)}$$

The only other parameter that appears to meet the three basic requirements outlined above has been derived by Hilliard (Ref 29) from a geometric theorem relating the total surface and the total volume of three-dimensional objects to the first and the fourth moment of the intercept length distribution—$M_1(1)$ and $M_4(1)$ (see also Ref 23). After normalizing to 1 for spheres again, this yields the shape parameter:

$$F_H = \frac{2}{3} \frac{N_L^2}{V_V} \sqrt{\frac{2}{3} M_4(1)} \qquad \text{(Eq 4)}$$

N_L and V_V are defined above. The main limitation of F_H is that it fulfills Eq 3 in principle only; in practice, $M_4(1)$ is subject to large experimental errors due to the fact that the longest intercepts, which usually are present with a low numerical frequency, contribute most to $M_4(1)$.

Numerous other shape parameters have been proposed by Underwood (Ref 24, 25), Ministr (Ref 30), and others. However, most equations are in pronounced contradiction to one or more of the basic requirements for true shape factors. Due to the problems outlined in this section, two-dimensional shape characteristics provide practical alternatives and as such are used almost exclusively in practical work.

Quantitative Description of Planar Shape

Two-dimensional shape parameters provide a means to monitor shape changes occurring in microstructures of P/M products due to variations in processing parameters and a means to compare the particle shape of powders obtained with different materials and production techniques. For this method, Fischmeister (Ref 27) uses the term "fingerprinting," because if applied properly, it yields indirect but accurate information on three-dimensional shape. The basic requirement outlined previously, however, must also be observed for two-dimensional shape factors.

Hausner's shape parameters, described earlier in this article in the section on conventional shape parameters, are an example of "fingerprinting," as are other methods of planar shape description such as Fourier coefficients and fractals of the sectioned or projected particle circumferences. In these latter two methods, parameters are described that are sensitive to particular shape aspects and are easily determined with semiautomatic or fully automatic image analyzers. In the Hausner method, the fitting of the minimum area rectangle requires repeated fitting, which in some cases leads to ambiguous results.

By far the most frequently used planar shape parameter (implemented in most image analyzing devices) combines the area and the perimeter of a planar feature into a dimensionless number normalized to 1 for circles of the same size:

$$f_{PL} = 4\pi a/b^2 \qquad \text{(Eq 5)}$$

where a and b are the area and the perimeter of the planar features (intersects or projections), respectively. Averaging can be completed easily if a and b are measured individually for each feature:

$$F_L = \sum_{i=1}^{n} f_{PL}/n$$
$$= 4\pi \sum_{i=1}^{n} (a/b^2)/n \qquad \text{(Eq 6)}$$

where n is the total number of planar features. However, if a and b are averaged, or if total area and perimeter length are used (as provided by field analysis), the average shape parameters obtained depend on size distribution, with values deviating from unity for a system of circles of varying diameter.

If Eq 6 is used, deviations from unity are interpreted as deviations from circularity. Elongation (elliptical deformation), as well as concave deformations of the perimeter ("rugged" outlines), yields smaller values approaching zero for highly elongated features or highly rugged perimeters.

Thus, F_1 combines different aspects of shape. To differentiate between these aspects, simple combinations of moments of inertia, Feret diameters, or curvature have been derived by Schwarz (Ref 31). Exner and Hougardy (Ref 23) propose the following parameters for individual features, which also can be averaged for any number of features in the same way as f_{PL} (see Eq 6):

Elongation, $f_{EL} = \sqrt{i_1/i_2} \qquad \text{(Eq 7)}$

Compactness, f_{COM}

$$= \frac{1}{2\pi} \cdot \frac{a^2}{\sqrt{i_1^2 + i_2^2}} \qquad \text{(Eq 8)}$$

Ruggedness (waviness), $f_{RU} = b_c/b$

$$\text{(Eq 9)}$$

where i_1 and i_2 are the two principal moments of inertia; b_c is the convex perimeter; and b is the actual perimeter.

These parameters are sensitive to the respective shape aspects, are easily measured by computer-aided image analysis, and are independent of other geometric properties; their interpretation through visualization is easily accomplished.

Consideration should be given to the method of averaging, however. Averaging on a volume basis, rather than on a number basis as in Eq 6, may be appropriate in some practical applications. If three-dimensional measurements are performed, similar shape parameters can be derived for the various aspects of spatial shape.

Multidimensional Shape Characterization

In order to analyze differences in shape precisely, more than one shape aspect must be monitored. Depending on the number of aspects considered using mutually independent shape parameters, two-, three-, or multidimensional representation of

Fig. 9 Characterization of the shape of twelve typical partical projections (or sections) in a two-dimensional shape space
f_{RU} is ruggedness; f_{EL} is elongation.

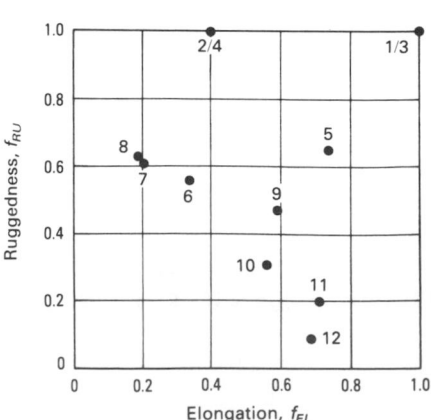

shape can be useful for comparison, and cluster analysis may be appropriate to quantify the significance of differences. Figure 9 shows a two-dimensional shape characterization of idealized particles in a ruggedness/elongation diagram.

These procedures require a computer with a large capacity, but which is modest compared to Fourier analysis, for example. The assessment of the individual parameters involves straightforward arithmetic calculation, and cluster analysis is performed on a few numbers for each feature measured.

Simple shape characterization, normally accomplished using Eq 6, prevails in practice, because the multidimensional procedures referred to previously are not yet implemented in commercial image analyzers. At present, the best approach to the difficult problem of quantitatively characterizing the shape of a system of particles in a powder mass is to use one or several shape parameters carefully adjusted to the problem under investigation.

Morphological Analysis

By J.K. Beddow
Professor
Fine Particle Research Group
University of Iowa

MORPHOLOGICAL ANALYSIS is primarily the characterization of particle shape. Particle size may also be obtained in the analysis (Ref 18). Typically, a Shape Analyzer is used to conduct the analysis. This instrument, which consists of a high-quality imaging (graphics) system that is augmented with computer equipment, must provide a clear image of the object for effective and accurate morphological analysis.

The image is digitized to obtain the (x, y) data set of the profile. This data set is then converted to the polar form. The polar data set (R, θ) constitutes a line, and the Fourier equation that represents this line can be constructed and the set of coefficients extracted from this equation. In turn, the coefficients are converted to invarient forms called "morphic features." Statistical properties of these features are also determined.

The rigorous definition of various morphological descriptors are given in Eq 10

through 14, and statistical properties are given in Eq 15 through 17. The size term is defined as the equivalent radius, R_0, of the particle profile, such that R_0^2 is the area of the circle of equivalent area to that of the particle profile.

$$R_0 = \sqrt{a_0^2 + \frac{1}{2}\sum_{}^{\infty}(a_n^2 + b_n^2)} \quad \text{(Eq 10)}$$

The mean radius a_0 and R_0 are related as follows:

$$L_0 = \frac{a_0}{R_0} \quad \text{(Eq 11)}$$

$$L_1(n) = 0 \text{ for all } n \quad \text{(Eq 12)}$$

The next set of terms, called the $L_2(n)$ terms, where n is the order of the Fourier coefficient, is useful in describing morphic features such as aspect ratio, triangularity, squareness, roughness, and other shape aspects.

$$L_2(n) = \frac{1}{2R_0^2}(a_n^2 + b_n^2),$$

$(n = 1, 2, 3, \ldots N$ is the order

of the coefficient) (Eq 13)

$$L_3(m,n) = \frac{3}{4R_0^3}(a_m a_n a_{n+m} - b_m b_n a_{n+m} +$$

$$a_m b_n b_{n+m} + b_m a_n b_{n+m}) \quad \text{(Eq 14)}$$

The mean, standard deviation, and skewness of the radial distribution of the particle are defined below. They are standard statistical properties and each has a distinct physical meaning. For example, the standard deviation is an indicator of the not-roundness of the particle profile.

$$\mu_0 = L_0 R_0 \text{ (mean radius)} \quad \text{(Eq 15)}$$

$$\mu_2 = R_0^2 \sum^{\infty} L_2(n)$$

(second moment about the mean)

(Eq 16)

$$\mu_3 = R_0^3 \sum_{m=1}^{\infty=1}\sum_{n=1}^{\infty} L_3(m,n)$$

(third moment about the mean)

(Eq 17)

Morphological Descriptors and Statistical Terms

Morphological descriptors are unique in that they can be used to regenerate the form of the original particle profile. A set of de-

scriptors therefore, contains all of the information in the original profile. Because morphological analysis has been developed to solve real problems, it is important that the morphological features can be identified with a corresponding physical meaning wherever possible.

The equivalent radius R_0 is defined so that πR_0^2 is the area of the particle. Therefore, R_0 can be interpreted as a size term; a_0 is the mean radius of the particle.

The series of terms $L_2(n)$ are defined in terms of the Fourier coefficients a_n and b_n, and they are normalized by dividing by R_0^2 to make them independent of size. Therefore, $L_2(n)$ terms are the equivalent of the Fourier coefficients A_n. In terms of their physical meaning, an indication of the relative strength of the $L_2(3)$ term can be obtained by dividing the value of $L_2(3)$ by the sum of $L_2(n)$. This value provides an indication of the angularity (triangularity) of the particle.

There is a relationship between the $L_3(m,n)$ terms and the radial skewness as shown in Eq 17. Therefore, if all of the $L_3(m,n)$ terms are summed and multiplied by R_0^3, the skewness of the radial distribution can be obtained. A pear-shaped drop or a teardrop shape, for example, would provide a high value of skewness. However, the physical meaning of the individual $L_3(m,n)$ terms is not understood at present.

The statistical properties of the morphological features defined in Eq 15 and 17 have been discussed above. The second moment of the radial distribution (the radance), which is defined in Eq 16, is interpreted as indicating the not-roundness (out-of-roundness) of the particle profile.

The equivalent radius squared minus the mean radius squared is equal to the sum of all of the $L_2(n)$ terms multiplied by R_0^2. Equation 10, therefore, relates intrinsic and extrinsic properties of the particle profile quantitatively. In addition, the radance (the sum of the $L_2(n)$ terms) can be equated to a difference between the square of the size minus the square of the mean radius.

This discussion indicates that there is a basis for the physical interpretation of the morphological features of a particle profile. The exception to this is the physical meaning of the $L_3(m,n)$ terms. This problem requires further study

Figure 10 provides a set of morphic templates that may be used to relate a physical particle characteristic to the specific values of select morphic features and statistical properties. Features used include the equivalent radius, aspect ratio, triangularity, radance (not-roundness), roughness, and skewness.

Fig. 10 Morphic templates used to obtain a physical indication of the specific values of selected morphic features and statistical properties

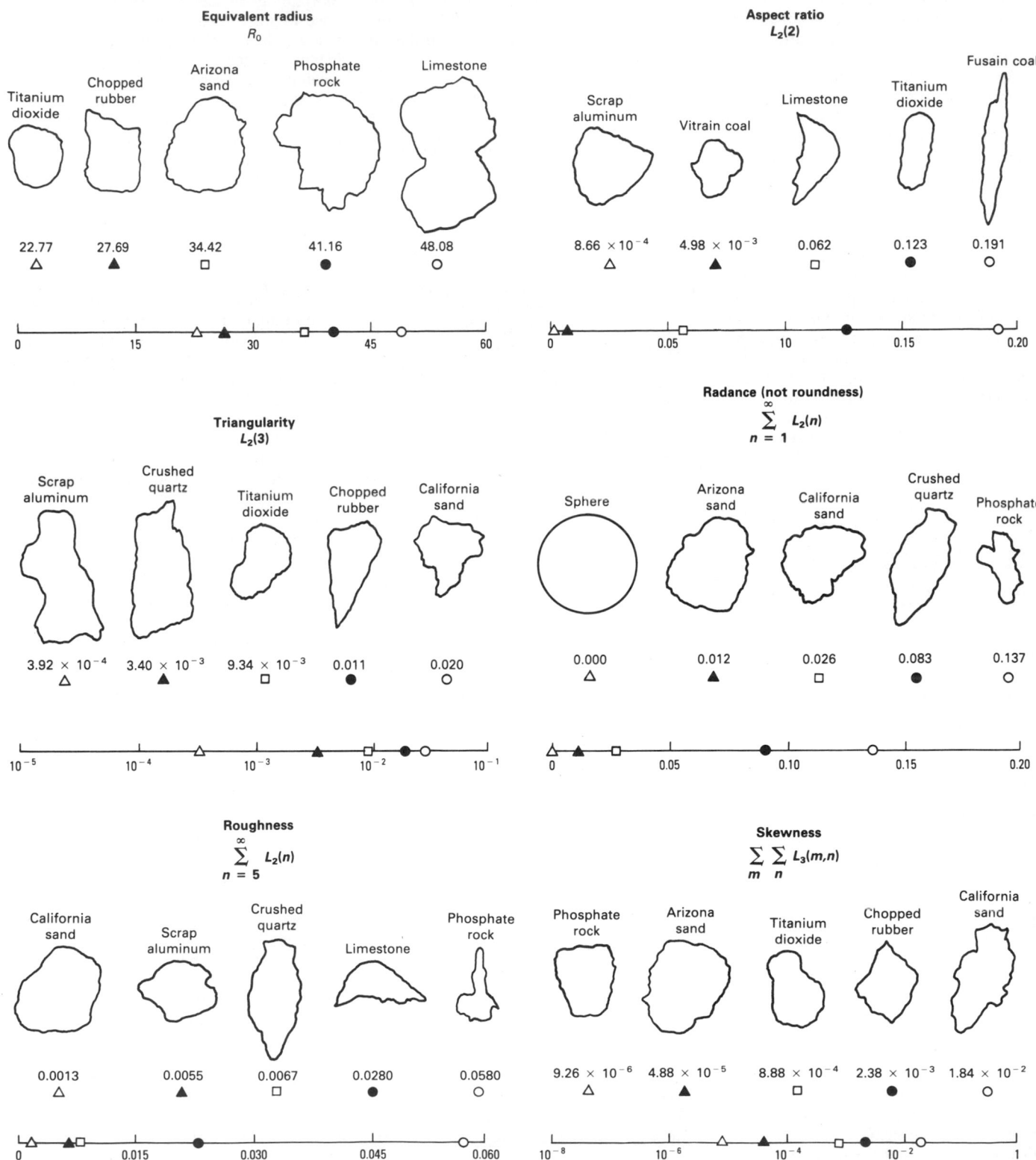

Fractals as Descriptors of P/M Systems

By Brian H. Kaye
Director
Institute for Fineparticles Research
Professor of Physics
Laurentian University

THE USE OF FRACTALS was promulgated in 1977 by Mandelbrot, who suggested that it was mathematically useful to extend dimensional descriptions of systems by using numbers that were not integers to describe rugged systems (Ref 32). The outline of a profile similar to an island could be described in terms of a dimension between 1 and 2, in which the fractional part of the dimension is an estimate of the curve's ability to fill space.

These new descriptors were termed "fractal dimensions" or "fractals." A normal Euclidean boundary represented by a circle or an elipse has a dimension of 1, whereas an extremely rugged boundary has a dimension approaching 2. Thus, the mathematical approach of Mandelbrot provided a useful description of rugged boundaries. In a discussion of the structure of metal particles, the term "rugged boundary" generally refers to the overall surface structure of, for example, a porous sponge iron particle, or to its projected boundary as viewed under a microscope. A scanning electron micrograph of a sponge iron particle is shown in Fig. 11. As shown, viewing a particle at increasing levels of magnification appears to reveal an increasing (infinite) surface area.

It has been shown that fractals are useful in describing various P/M systems. Thus, Kaye and co-workers have used fractals to describe the grains of metal powder encountered in sponge iron powders and aluminum shot pigments (Ref 33). Pfeifer and co-workers have used fractal dimensions to describe surfaces (Ref 34-36). Chermant and co-workers have also used fractals to describe boundaries of metal specimens (Ref 37, 38).

Recently, it has been shown that fractals provide an alternative method to describing the structure of some types of holes in a metal sponge (Ref 39). The roughness of a polished metal surface can also be described using fractal dimensions (Ref 40).

On first use of fractal dimensions to describe rugged systems, one resists the idea

Fig. 11 Scanning electron micrograph of a sponge iron particle

Viewing a particle at increasing levels of magnification appears to reveal an increasing (infinite) surface area. (a) 115×. (b) 575×. (c) 2850×

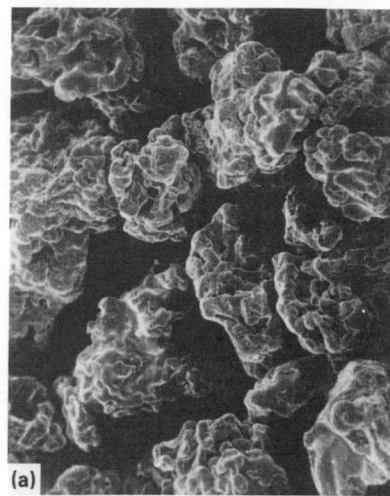

(a)

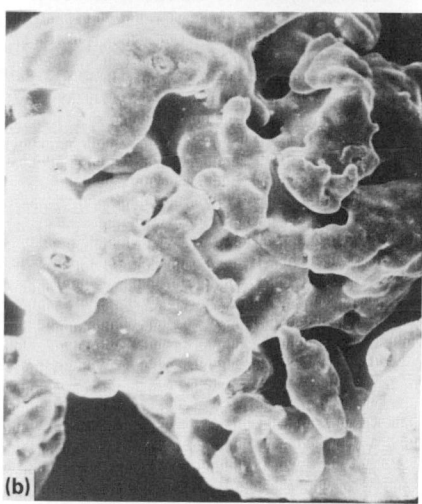

(b)

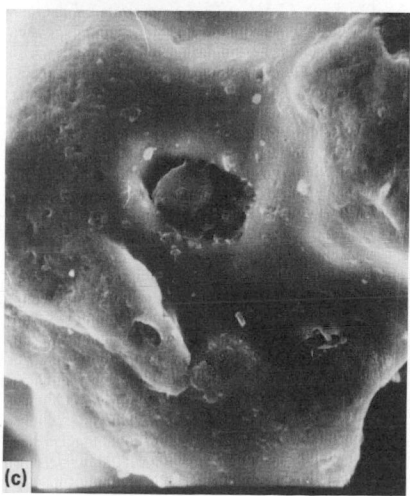

(c)

that the surface area of all rough metal powder fine particles tend to infinity (approach infinity). However, it should be realized that the use of fractal dimensions shifts attention from the extrapolation trend of data to a study of how fast a border or a surface examined with increasing resolution appears to tend to infinity.

According to Mandelbrot, the traditional allocation of one, two, or three dimensions to a system is itself an artifact conditioned by convenience and perspective. Thus, the allocation of a dimension of two or three to a sheet of metal foil depends on one's perspective. When a sheet of aluminum foil is examined from a reflective point of view, it is considered a two-dimensional body; however, if it is used as an insulator on the outside of a house, for example, its thickness must be taken into account as well as its surface, and it must be considered a three-dimensional body. In the same manner, the description of a boundary or a surface by a fractal dimension is useful within the constraints of the measurement technique used to estimate the fractal dimension.

It can be shown that different types of powders can be usefully described by different fractal dimensions when examined at different scales. For example, in viewing a piece of aluminum shot, a low-resolution examination of the rugged boundary describes its overall configuration, whereas very high-resolution examination of the contours of the profile shows that the system can be regarded as a Euclidean boundary. This is because surface tension forces operating during the cooling stage ensure that the ultimate surface is smooth, even though the overall configuration of the fine particle when viewed at low resolution is rugged.

One of the simplest methods for evaluating the fractal dimensions of a rugged profile of a grain of metal powder is illustrated in Fig. 12. The structured-walk technique is used for evaluating the boundary. Computer-based image analyzers have been programmed to give fractal boundaries using erosion dilation techniques. A discussion of the logic of such automated systems can be found in Ref 41 to 43. Manual techniques are discussed in this article to illustrate the type of data generated and the basic physical principles of the procedure for evaluating a fractal dimension.

The procedure for evaluating a profile, such as that shown in Fig. 12, by the structured-walk technique is as follows. The points of a pair of compasses are set at the distance λ. The compasses are then

Fig. 12 Typical Richardson plot for the evaluation of the fractal dimension of the rugged boundary of the profile of a sponge iron fine particle

F_D is the maximum Feret diameter, P is the normalized perimeter estimate, λ is the normalized stride length, and δ is the structure of the boundary over the range of 0.28 to 0.07 normalized stride units. Source: Ref 34

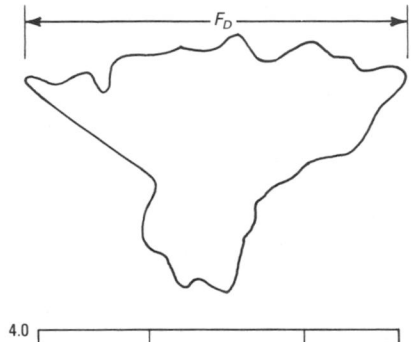

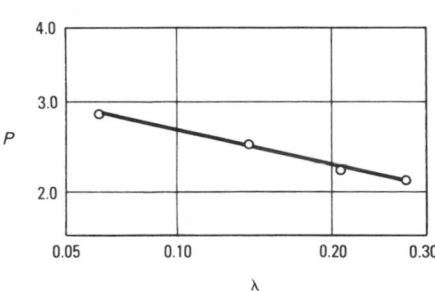

"walked" around the profile to create a polygon constructed on the profile. It is often necessary to have a fractional step to complete the walk, and in such a situation, the estimate of the perimeter of the polygon is $(N + \alpha)\,\lambda$, where α is the fractional step, and N is the number of complete strides to form the polygon. The perimeter of the polygon now becomes a perimeter estimate of the irregular boundary.

It is common practice to normalize the perimeter estimate by using the maximum projected length of the profile as a normalizing factor. The stride length λ is also normalized using the same factor. A graph of the logarithm of the perimeter estimate against the logarithm of λ is then plotted. If the boundary can be described by a fractal dimension, the data points exhibit a linear region on such a graph. This type of graph is known as a Richardson plot. Richardson was the first to draw attention to the problem of evaluating rugged boundaries using finite measuring techniques.

Mandelbrot has shown that the fractal dimension of a rugged boundary is related to the slope of a linear region of a Richardson plot by the relationship $1 - \delta =$

m, where m is the slope of the line. The dimension 1.20 can be used to describe the structure of the boundary (δ) over the range of 0.28 to 0.07 normalized stride units for the Richardson plot of the sponge iron fine particles in Fig. 12.

It is possible that examination of the profile at even greater resolutions using a scanning electron microscope would indicate that a different fractal dimension would be appropriate at the higher resolution. The rapid development of image analysis systems for evaluating fractals should facilitate the use of this technique for describing rugged boundaries and surfaces.

REFERENCES

1. Van Essex, C.G., Scanning Electron Microscopy, in *Electron Microscopy and Microanalysis of Crystalline Materials*, Blek, J.A., Ed., Applied Science Publishers, London, 1979, p 99
2. Goldstein, J.I., Yakowitz, H., and Newbury, D.E., Evolution of the Scanning Electron Microscope, in *Practical Scanning Electron Microscopy*, Goldstein, J.I. and Yakowitz, H., Ed., Plenum Press, New York, 1975, p 3-8
3. Thornton, P.R., *Scanning Electron Microscopy: Applications to Materials and Device Science*, Chapman and Hall, London, 1968
4. Heywood, V.H., Ed., *Scanning Electron Microscopy—Systematic and Evolutionary Applications*, Academic Press, London, 1971
5. Cross, P.M., Hearle, J.W.S., and Sparrow, J.T., *The Use of Scanning Electron Microscopy*, Pergamon Press, Oxford, 1972
6. Oatley, C.W., *The Scanning Electron Microscope*, Cambridge University Press, 1972
7. Wells, O.C., *Scanning Electron Microscopy*, McGraw-Hill, New York, 1974
8. Goldstein, J.I. and Yakowitz, H., Ed., *Practical Scanning Electron Microscopy*, Plenum Press, New York, 1977
9. Ro, D.H. and Toaz, M.W., Structure-Property Relationships in Direct-Rolled Powder Alloys, in *Progress In Powder Metallurgy 1982*, 1982 National Powder Metallurgy Conference Proceedings, Bewley, J.G. and McGee, S.W., Ed., 1982
10. Klar, E. and Shafer, W.M., On the Nature of Green Strength, *Int. J. Powder Metall.*, Vol 5 (No. 4), 1969
11. Heywood, H., Symposium on Particle Size Analysis, *Trans. Inst. Chem. Eng.*, Vol 25, 1947, p 14-24
12. Heywood, H., *J. Imperial Coll. Eng. Soc.*, Vol 8, 1954
13. Wadel, H., *J. Geol.*, Vol 40, 1932, p 443-357
14. Wadel, H., *J. Geol.*, Vol 43, 1935, p 250-280
15. Krumbein, W.C., *J. Sedimen. Petrol.*, Vol 11 (No. 2), 1941, p 64-72
16. Hausner, H.H., *Planseeber*, Vol 14 (No. 2), 1966, p 75-84
17. Church, T., *Powder Technol.*, Vol 2, 1968-69, p 27-31
18. Beddow, J.K., *Particulate Science and Technology*, Chemical Publishers, New York, 1980
19. Underwood, E.E., Stereological Analysis of Particle Characteristics, in *Testing and Characterization of Powders and Fine Particles*, Beddow, J.K. and Meloy, T.P., Ed., 1980, p 77
20. Underwood, E.E., *Quantitative Stereology*, Addison-Wesley, Boston, 1971
21. Weibel, E.R., *Stereological Methods*, Vol 2, Theoretical Foundations, Academic Press, London, 1980
22. Serra, J., *Image Analysis and Mathematical Morphology*, Academic Press, London, 1982
23. Exner, H.E. and Hougardy, H.P., *Quantitative Image Analysis of Microstructures*, A Practical Guide to Techniques, Instrumentation and Assessment of Materials, Deutsche Gesellschaft für Metallkunde, Oberursel, 1984
24. Underwood, E.E., Quantitative Shape Indices by Stereological Methods, in *Quantitative Analysis of Microstructures in Medicine, Biology and Materials Development*, Riederer-Verlag, 1975, p 223-241
25. Underwood, E.E., "Three-Dimensional Shape Parameters from Planar Sections," *Proc. Fourth Int. Cong. Stereology*, NBS Special Publication 431, National Bureau of Standards, Gaithersburg, MD, 1976, p 91-92
26. Underwood, E.E., Stereological Analysis of Particle Characteristics, in *Testing and Characterization of Powders and Fine Particles*, Heyden and Sons, London, 1980, p 77-96
27. Fischmeister, H.F., Shape Factors in Quantitative Microscopy, *Z. Metallkunde*, Vol 65, 1974, p 558-562
28. DeHoff, R.T., The Determination of the Geometric Properties of Aggregates of Constant Size Particles from Counting Measurements Made on

Random Plane Sections, *Trans. AIME,* Vol 230, 1964, p 764-769

29. Hilliard, J.E., private communication, 1975

30. Ministr, Z., The Determination of Parameters for Two and Three Dimensional Microstructures with Aid of Specific Perimeter or Specific Surface Area of the Microstructural Features, *Practical Metallogr.,* Vol 8, 1971, p 333-359, p 407-423 and Vol 12, 1975, p 244-258

31. Schwarz, H., Two Dimensional Feature-Shape Indices, *Proc. Fifth Int. Cong. Stereology,* Mikroskopie, Vol 37 (Suppl.), 1980, p 64-67

32. Mandelbrot, B.B., *Fractals: Form, Chance and Dimension,* W.H. Freeman, San Francisco, CA, 1977

33. Kaye, B.H., Fractal Description of Fine Particle Systems, *Modern Methods in Fine Particle Characterization,* Beddow, J.K., Ed., CRC Press, Boca Raton, FL, 1983

34. Pfeifer, P. and Avner, D., Fractal Dimension in Chemistry: An Intensive Characteristic of Surface Irregularity, *Nou. J. Chemie,* Vol 7 (No. 2), 1983, p 71

35. Pfeifer, P. and Avner, D., Chemistry in Non-integer Dimensions Between 2 and 3, Part 1: Fractal Theory of Heterogenous Surfaces, Part 2: Fractal Surfaces of Absorbants, *J. Chem. Phys.,* Vol 79, 1983

36. Pfeifer, P., Avner, D, and Farin, D., New Developments in the Application of Fractal Theory to Surface Geometric Irregularity," Symposium on Surface Science, Technical University of Wein, Brown, P., Betz, G., Husinsky, W., Sollner, E., Stori, A., and Varga, P., Ed., 1983

37. Chermant, J.L. and Coster, M., "Fractal Object in Image Analysis," Proceedings of the International Symposium on Quantitative Metallography, Florence, 1978, p 125-137

38. Coster, M. and Deschanvres, A., Fracture, Object Fractal et Morphologie Mathematique, *Sonderbands Praktischen Metallorgraphie,* Vol 8, 1978, p 61-73

39. Kaye, B.H., "Swiss Cheese, Deposited Droplets and other Holey Fractals," editorial in *Particulate Technology,* in preparation

40. Kaye, B.H., Multifractal Description of Rugged Fineparticle Profiles, *Particle Characterization,* Vol 1 (No. 1), 1984

41. Flook, A.G., The Use of Dilation Logic on the Quantimet to Achieve Fractal Dimension Characterization of Textured and Structure Profiles, *Powder Technol.,* Vol 21, 1978, p 295-298

42. Reid, A., Institute of Earth Resources, Port Melbourne, Victoria, Australia

43. Schwarz, H. and Exner, H.E., The Implementation of the Concept of Fractal Dimensions on a Semiautomatic Image Analyzer, *Powder Technol.,* Vol 27, 1980, p 207-213

Bulk Chemical Analysis of Metal Powders

CHEMICAL COMPOSITION of a powder is a significant factor in determining powder reactions to conditions of initial compaction—specifically, the ability of the powder to densify. During sintering, many chemical reactions are possible within the compact between the compact and the sintering atmosphere.

The exact composition of the material determines the type and extent of these reactions. Composition dictates the temperature and type of sintering atmosphere required. Subsequently, the properties of the final sintered material greatly depend on the composition of the material.

Usually, chemical analyses are given in terms of metallic and nonmetallic impurity content. Such impurities may be present in the elemental form. For example, silicon in iron powder may be in the form of dissolved silicon atoms in solid solution. It is also possible for impurities to be in the powder in the form of a compound. In iron powder that contains silicon, for example, many powders produced from ores or mill scale may contain silica (silicon dioxide) as fine dispersed particles rather than silicon in solid solution. This type of phenomenon is important for two reasons. The actual chemical analysis used to detect silicon may not be suitable or accurate if the silicon (or any such element) is present in the form of a compound.

Of greater consequence is the difference in powder behavior depending on the physical form of the impurity. If the element is in the form of a compound, which may be a hard ceramic, difficulties may be encountered in compaction because of the presence of an abrasive phase. Also, such a compound may remain unchanged during subsequent processing and may seriously affect the final properties of the sintered material. If the element is present in solid solution, then its effect on compaction may vary significantly, and chemical reactions may occur during sintering. Actual powder analyses do not always in-dicate the form in which an element is present.

Although many analytical methods are available to determine the chemical composition of powders, most of these procedures are geared toward the analytical chemist. However, special analytical procedures have been developed by the P/M industry for specific P/M applications. This article concentrates on methods of sampling and analysis germane to the P/M industry. A proper understanding of these methods is required if chemical analyses are to be interpreted properly.

Sampling

Before a lot of metal powder can be analyzed, a representative sample must be taken from the lot. In most chemical analyses, only a small sample is required. For metal powders, it is critical that the sample represent the full range of particle sizes within the powder lot. For example, in a sulfurized powder, the sulfur content is concentrated at the surface. Because smaller particles have a larger surface area, the sulfur concentration is greater in the smaller particles. This also holds true for oxygen content, which is largely surface related. The physical characteristics of a powder also vary with particle size. Consequently, it is essential to obtain a representative sample of the particle size distribution.

The preferred method for obtaining a representative sample is stream sampling. A container is passed completely through a stream of flowing powder. Several passes provide a representative sample. If the material is prepackaged, a representative sample can be taken with a sample thief.

The sample should then be divided into a test quantity by using a sample splitter. These procedures are described in the article "Sampling of Metal Powders" in this Volume and can also be found in American Society for Testing and Materials standard ASTM B 215 and Metal Powder Industries Federation standard MPIF 01.

Hydrogen Loss Testing

Hydrogen loss value provides an approximation of the oxygen content of the powder. Oxygen content of powders is an important variable in that many metals are strongly hardened and embrittled by relatively small amounts of dissolved oxygen or by compounds containing oxygen. Thus, the ability of a powder to densify during compaction and to attain a minimum required green strength may be degraded. For most metals, a reducing atmosphere is preferred during sintering to prevent oxidation so that diffusion can take place.

Hydrogen loss testing usually is based on either MPIF 02 or ASTM E 159. These specifications are similar in approach. A sample of powder, usually 5 g, is placed in a combustion boat and exposed for a specified time and temperature in an atmosphere of dry (dew point less than −40 °C, or −40 °F), purified hydrogen. Testing conditions and methods are given in Table 1.

Testing procedures include placement of the sample in the combustion boat, temperature measurement and control, and purging of the combustion tube with nitrogen both before and after exposure to the hydrogen. Other factors to consider include type of combustion tube, method of weighing the residue, and purity and control of the gases. The weight loss of the powder specimen due to the hydrogen exposure is measured according to the following:

$$\text{Hydrogen loss, wt\%} = [(A - B)/(A - C)] \times 100$$

where A is the original weight of the boat and specimen, in grams; B is the weight of the boat and specimen after reduction, in grams; and C is the original weight of

Table 1 Standard conditions for hydrogen loss testing

Metal powder	Temperature of reduction(a) °C	°F	Reduction time, min	Combustion boat	Standard
Cobalt	1050	1920	60	Alundum	MPIF
Copper	875	1605	30	Quartz	MPIF, ASTM
Copper-tin	775	1425	30	Zircon	ASTM
				Fused alumina	ASTM
				Quartz	MPIF
Lead	550	1020	30	Alundum	MPIF
Lead-tin	550	1020	30	Alundum	MPIF
Iron	1150	2100	60	Alundum	MPIF
				Fused alumina	ASTM
				Zircon	ASTM
Nickel	1050	1920	60	Alundum	MPIF
Tin	550	1020	30	Alundum	MPIF
Tungsten	875	1605	30	Fused alumina	ASTM
				Quartz	ASTM
				Zircon	ASTM

(a) Temperatures have a permissible range of ±15 °C (±27 °F).

the specimen, in grams. The result is reported to the nearest 0.01%.

Testing Limitations. There are several potential sources of error in hydrogen loss testing. It is possible to obtain a weight loss that is significantly lower than the actual value of oxygen content, if oxygen-bearing compounds are present in the material that are not chemically reduced by the hydrogen during testing. These compounds include silicon dioxide, aluminum oxide, magnesium oxide, calcium oxide, beryllium oxide, titanium dioxide, chromium oxide, and manganese oxide. A low oxygen content value also may result from incomplete reduction of potentially reducible compounds. This could result from insufficient time at temperature, or if the temperature used is too low for some compounds.

Standardization of testing ensures accurate results. A lower than actual oxygen content reading may also be related to the particle size in the sample or the presence of particle porosity. Larger particles require greater paths for material transport for the hydrogen to come into contact with the oxygen-bearing compound or dissolved oxygen. Particle porosity promotes intimate contact of the hydrogen with the oxygen or oxide and leads to shorter times.

Too high of a weight loss value may be obtained if elements contained in the powder either are in solid solution or are present in a compound that reacts with the hydrogen and contributes to weight loss of the specimen. Two common such impurities are carbon and sulfur.

Analytical testing to determine the content of such elements can be used to correct the hydrogen loss value. If the metal or alloy is volatile at the designated temperature, then weight loss is caused by another factor. Cadmium, zinc, and lead, as

pure metals or alloys, are examples of such metals.

It also is possible to detect the presence of oxides in powder particles by metallographic examination of mounted, polished, and etched particles. Figure 1 shows gray, unreduced iron oxide (Fe_3O_4) in the interior of an iron powder particle produced by hydrogen reduction of iron mill scale. In reduced powders, small oxide inclusions near the center of a particle are typical, resulting from the incomplete reduction of the original oxide.

Acid Insoluble Test

Another widely used test method for bulk analysis is the acid insoluble content test of iron and copper powder, which is described in ASTM E 194 and MPIF 06. This

Fig. 1 Microstructure of iron powder particle produced by hydrogen reduction of iron mill scale

Interior of particle shows some gray, unreduced iron oxide (Fe_3O_4) in the unetched condition. Magnification: 475×

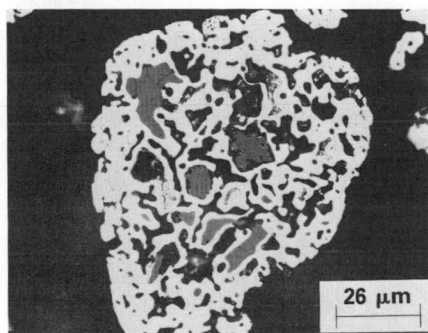

26 μm

test indicates the amount of nonmetallic substances that do not dissolve in the mineral acid used to dissolve the metal. In copper powder, such insolubles include silica, insoluble silicates, alumina, clays, and other refractory materials. Lead sulfate also may be present in the insoluble materials. In iron powder, the insoluble content may include carbides in addition to the substances listed previously for copper.

Iron powder is dissolved with hydrochloric acid, and copper powder is dissolved with nitric acid. A 5-g sample is reacted with 100 mL (3.4 fluid oz) of suitable acid, then boiled, diluted with water, and boiled again. The solution is filtered, and the residue is then washed. Finally, the residue and filter paper are ignited at 980 °C (1795 °F) for 1 h. What remains after firing is cooled and weighed. The difference in weight before and after firing is used to calculate the amount of insoluble matter. The weight percent of insoluble matter in the powder is obtained by dividing the weight of the final residue by the weight of the original sample. Any tin in the copper interferes with the test, and special procedures to eliminate this phenomenon must be followed.

Testing Limitations. The value of the data generated by this test is limited. While the test determines the acid insoluble content present in the material, the precise nature of such substances is not determined. There is, however, a specialized procedure to obtain the amount of silica present in the powder, but testing errors are possible. There may be ceramic impurities contained in the powder, which are either soluble in the acid used or which are eliminated during ignition of the residue. This test is generally limited to amounts of acid insolubles less than 1%. However, with improvements in powder production methods that result in purer, cleaner powders, the acid insoluble test is becoming less widely used.

Iron and Iron Oxide Content of Iron Powders

Although used infrequently, there is a test to determine the iron content of iron powder. The objective of this method, which is described in MPIF 07, is to separate the total iron content into two contributions: actual metallic iron and ferric and ferrous oxides. The testing procedure is based on the use of wet chemical techniques involving dissolution in acid, followed by several titration analyses.

Table 2 Thickness of oxide film on 99.6% pure aluminum on exposure to atmosphere

Time of exposure, days	Thickness of oxide film, μm
1	0.0025
2	0.005
4	0.0065
8	0.008
16	0.0088

Effect of Adsorbed Gases and Water Vapor

One practical problem concerning the significance of chemical analyses should be recognized. Many materials, both metallic and nonmetallic, adsorb significant quantities of gases and water vapor from the atmosphere. This may not be indicated by chemical analyses that are performed on freshly produced powders. Such adsorption can lead to the formation of surface oxides on metals, which may interfere with compaction and sintering and sometimes remain in the sintered material.

Some metals, such as aluminum and copper, are particularly susceptible to this type of surface oxidation at room temperature. The amount of such surface contamination increases with decreasing particle size, because of the increased surface-to-volume ratio, and with increasing chemical activity of the surface. Data for the formation of oxide film on aluminum powder are given in Table 2.

Additional Methods of Analysis

As stated earlier in this article, no attempt has been made to describe all of the methods of analysis used in the metal powder industries. These procedures are fully described in ASTM and International Standards Organization (ISO) standards. If a pressed or sintered part is to be chemically analyzed, milling, drilling, or crushing may be used to obtain a representative sample. If the part contains oil, it should be removed by using a Soxhlet extractor with a solvent as detailed in ASTM B 328. Oil may also be removed by heating in a protective atmosphere at 705 to 815 °C (1300 to 1500 °F). However, this method cannot be used for materials such as sintered aluminum, because they melt. A list of available ASTM standards on sampling and chemical analysis of metals includes the following:

Aluminum and aluminum alloys

- Methods for Chemical Analysis of Aluminum and Aluminum Alloys (ASTM E 34)
- Test Method for Aluminum in Iron Ores by Complexometric Titration (ASTM E 738)
- Test Method for Aluminum Oxide in Iron Ores (ASTM E 464)

Cadmium

- Methods for Chemical Analysis of Cadmium (ASTM E 396)

Copper and copper alloys

- Methods for Chemical Analysis of Special Brasses, and Bronzes (ASTM E 54)
- Methods for Chemical Analysis of Copper Alloys (ASTM E 478)
- Methods for Chemical Analysis of Copper (Electrolytic Determination of Copper) (ASTM E 53)
- Photometric Methods for Chemical Analysis of Copper and Copper Alloys (ASTM E 62)
- Methods for Chemical Analysis of Copper-Beryllium Alloys (ASTM E 106)
- Methods for Chemical Analysis of Copper-Chromium Alloys (ASTM E 118)
- Methods for Chemical Analysis of Copper-Nickel and Copper-Nickel-Zinc Alloys (ASTM E 75)
- Methods for Chemical Analysis of Copper-Tellurium Alloys (ASTM E 121)
- Methods for Chemical Analysis of Manganese-Copper Alloys (ASTM E 581)

Niobium

- Methods for Chemical Analysis of Reactor and Commercial Niobium (ASTM E 195)

Ferroalloys

- Methods for Chemical Analysis of Ferroalloys (ASTM E 31)
- Methods for Chemical Analysis of Ferroboron (ASTM E 371)
- Methods for Chemical Analysis of Ferrochrome-Silicon (ASTM E 364)
- Methods for Chemical Analysis of Chromium and Ferrochromium (ASTM E 363)
- Methods for Chemical Analysis of Ferrocolumbium (ASTM E 367)
- Methods for Chemical Analysis of Ferromanganese and Spiegeleisen (ASTM E 361)
- Methods for Chemical Analysis of Ferromolybdenum (ASTM E 368)
- Methods for Chemical Analysis of Silicon and Ferrosilicon (ASTM E 360)
- Methods for Chemical Analysis of

Ferrovanadium and Vanadium Alloying Additives (ASTM E 365)
- Methods for Chemical Analysis of Magnesium Ferrosilicon (ASTM E 568, redesignated as E 372)
- Methods for Chemical Analysis of Silicomanganese and Ferrosilicon Manganese (ASTM E 362)

Iron, nickel, and cobalt alloys

- Methods for Chemical Analysis of High-Temperature, Electrical, Magnetic, and Other Similar Iron, Nickel, and Cobalt Alloys (Including Tentative Revision) (ASTM E 354)

Lead, tin, antimony, and their alloys

- Methods for Chemical Analysis of Antimony (ASTM E 86)
- Methods for Chemical Analysis of Lead- and Tin-Base Solder (ASTM E 46)
- Photometric Methods for Chemical Analysis of Lead, Tin, Antimony, and Their Alloys, (ASTM E 87)
- Methods for Chemical Analysis of Pig Lead (ASTM E 37)
- Methods for Chemical Analysis of White Metal Bearing Alloys (ASTM E 57)

Magnesium and magnesium alloys

- Methods for Chemical Analysis of Magnesium and Magnesium Alloys (ASTM E 35)
- Methods for Spectrochemical Analysis of Magnesium Alloys for Calcium by Flame Photometry (ASTM E 226)

Metal powders

- Test Method for Acid-Insoluble Content of Copper and Iron Powders (ASTM E 194)
- Test Method for Hydrogen Loss of Copper, Tungsten, and Iron Powders (ASTM E 159)

Molybdenum

- Methods for Chemical Analysis of Molybdenum (ASTM E 315)

Nickel and nickel-copper alloys

- Methods for Chemical Analysis of Nickel (ASTM E 39)
- Methods for Chemical Analysis of Nickel-Copper Alloys (ASTM E 76)
- Methods for Chemical Analysis of Electronic Nickel (ASTM E 107)

Beryllium

- Methods for Chemical Analysis of Beryllium (ASTM E 439)

Nickel-chromium-iron alloys

- Methods for Chemical Analysis of Nickel-Chromium and Nickel-Chro-

mium-Iron Alloys (ASTM E 38)
- Methods for Chemical Analysis of Stainless, Heat-Resisting, Maraging, and Other Similar Chromium-Nickel-Iron Alloys (Including Tentative Revision) (ASTM E 353)

Silver solders and other brazing alloys

- Methods for Chemical Analysis of Silver Brazing Alloys (ASTM E 56)
- Test Method for Photometric Method for Determination of Phosphorus in High-Phosphorus Brazing Alloys, Photometric Method for Determination of (ASTM E 156)

Steel, cast iron, open-hearth iron, and wrought iron

- Methods for Chemical Analysis of Carbon Steel, Low-Alloy Steel, Silicon Electrical Steel, Ingot Iron, and Wrought Iron (ASTM E 350)
- Methods for Chemical Analysis of Cast Iron—All Types (ASTM E 351)
- Methods for Chemical Analysis of Steel, Cast Iron, Open-Hearth Iron, and Wrought Iron (ASTM E 30)
- Methods for Chemical Analysis of Steel, Tool, and Other Similar Medium- and High-Alloy Steels (ASTM E 352)

Titanium and titanium alloys

- Methods for Chemical Analysis of Titanium and Titanium Alloys (ASTM E 120)

Tungsten

- Methods for Chemical Analysis of Tungsten (ASTM E 397)

Zinc and zinc alloys

- Methods for Chemical Analysis of Slab Zinc (Spelter) (ASTM E 40)
- Methods for Chemical Analysis of Zinc and Zinc Alloys (ASTM E 536)
- Methods for Chemical Analysis of Zinc Die-Casting Alloys (ASTM E 47)

Zirconium and zirconium alloys

- Methods for Chemical Analysis of Zirconium and Zirconium Alloys (ASTM E 146)

General practices and specifications

- Recommended Practices for Apparatus, Reagents, and Safety Precautions for Chemical Analysis of Metals (ASTM E 50)
- Recommended Practices for Conducting Interlaboratory Studies of Methods for Chemical Analysis of Metals (ASTM E 173)
- Recommended Practices for Flame Atomic Absorption Analysis (ASTM E 663)
- Recommended Practices for Indicating Which Places of Figures Are To Be Considered Significant in Specified Limiting Values (ASTM E 29)
- Recommended Practices for Photometric Methods for Chemical Analysis of Metals (ASTM E 60)
- Recommended Practices for Sampling Copper and Copper Alloys for Determination of Chemical Composition (ASTM E 255)
- Recommended Practices for Acceptance of Evidence Based on the Results of Probability Sampling (ASTM E 141)
- Recommended Practices for Choice of Sample Size to Estimate the Average Quality of a Lot or Process (ASTM E 122)
- Recommended Practices for Probability Sampling of Materials (ASTM E 105)
- Specification for Gravity-Convection and Forced Ventilation Ovens (ASTM E 145)

Sampling methods

- Methods of Sampling Finished Lots of Metal Powders (ASTM B 215)
- Methods of Sampling Ferroalloys for Determination of Chemical Composition (ASTM E 32)
- Methods of Sampling Nonferrous Metals and Alloys in Cast Form for Determination of Chemical Composition (ASTM E 88)
- Methods of Sampling Wrought Nonferrous Metals and Alloys for Determination of Chemical Composition (ASTM E 55)
- Methods of Sampling Steel and Iron for Determination of Chemical Composition (ASTM E 59)

Surface Chemical Analysis of Metal Powders

By the ASM Committee on Surface Chemical Analysis*

IMPROVED SURFACE ANALYSIS TECHNIQUES for the elemental analysis of the first few atomic layers of a solid surface are available. Detection sensitivities of more than one tenth of a monolayer are possible with Auger electron spectroscopy, x-ray photoelectron spectroscopy (also known as electron spectroscopy for chemical analysis), and ion-scattering spectroscopy (Ref 1). Higher sensitivities are achievable by using secondary ion mass spectrometry.

These techniques have wide chemical and metallurgical usage in solving problems related to lubrication, catalysis, adhesion, corrosion, wear, and fracture applications in which resultant properties are determined by the composition at the surface of the solid. However, little has been reported concerning the use of such techniques in powder metallurgy.

With the application of these techniques, there is increasing evidence that a powder particle may be homogeneous within its bulk but very inhomogeneous near its surface. Both type and degree of inhomogeneity appear to be sensitive to powder processing methods. Application of surface analysis techniques to P/M products is essential for an improved understanding of the surface chemistry of the powder and the consolidated product and its effect on engineering properties.

The many problems associated with elemental identification on a surface preclude the use of any one method. The investigator instead must determine the specific information required from the surface and subsequently which testing techniques are most useful. Each surface analysis technique has its own capabilities

and limitations. Frequently, the use of combined techniques is necessary to provide comprehensive and complementary information. For instance, Auger electron spectroscopy combined with x-ray photoelectron spectroscopy permits a comprehensive analysis in which both elemental composition and chemical bonding information can be obtained from the same surface.

Some specific features and characteristics of the various surface analysis techniques are summarized in Table 1. Advanced computerized systems for each of the techniques are being perfected to aid in quantitative interpretation of information. This article reviews the fundamentals and provides an aid to selecting techniques for surface analysis of P/M products. The reader is referred to the selected references at the end of this article for additional information.

Auger Electron Spectroscopy

Historical Development. In 1925, Pierre Auger observed the phenomenon of Auger electrons emitted from argon atoms being irradiated with x-rays (Ref 2). In 1953, Lander described a technique for using Auger electrons to study the surface chemistry of a solid. It was not until 1967 that electronic methods were sufficiently developed to monitor and measure the Auger electron spectra. Further development has led to the widespread use of Auger electron spectroscopy as a surface analytical tool.

Auger electron emission may be stimulated by bombardment of a substance with

electrons, x-ray photons, or ions. Electron inducement is more suitable in metallurgical applications because electrons can be focused to a very small beam, allowing localized analysis and rastering (scanning Auger microscopy) (Ref 2). The application of electron-induced Auger electron spectroscopy to powders is particularly appropriate because of the large specific surface area of powders. Also, many fundamental and engineering properties of powders and their products greatly depend on surface properties.

Theory. In Auger electron spectroscopy, the sample to be analyzed is bombarded with electrons. If an incident electron (Fig. 1a) strikes a k-shell (core-level) electron with sufficient energy to free it (Fig. 1b), the atom is left in a singly ionized state with a core-level electron vacancy. If the atom is near the surface, both the incident electron and the core-level electron are emitted from the sample as backscattered electrons, with energies below approximately 25 eV.

The singly ionized atom undergoes electron rearrangement to achieve a more stable energy configuration, where an outer shell electron fills the core level vacancy (Fig. 1c). The energy thus released is either emitted as a photon (x-ray fluorescence) or given to another outer shell electron, which is emitted if the energy is great enough (called the Auger transition).

If the atom is within approximately 5 nm (50 Å) of the surface, the electron is likely to escape from the sample and be available for measurement. The energy of the ejected Auger electron is characteristic of the particular element, its chemical state, and the transition between energy levels

*David H. Ro, *Chairman,* Principal Scientist, TRW Inc.; Warren J. Haws, Senior Metallurgist, Brush Wellman, Inc.; Erhard Klar, Manager of Particle Technology, SCM Metal Products; Mary A. Pao, Metallurgist I, SCM Metal Products

Table 1 Surface analytical techniques

Testing parameters	Anger electron spectroscopy/ scanning Auger microscopy	X-ray photoelectron spectroscopy or electron spectroscopy for chemical analysis	Secondary ion mass spectroscopy	Ion-scattering spectroscopy
Principle	Emission of Auger electrons by core-hole recombination	Emission of photoelectrons by core-level excitation	Sputtering of surface atoms by ion beam	Elastic ion backscattering
Probe	1 ~ 30 keV electrons	X-ray, photons	1 ~ 3 keV ions	0.5 ~ 3 keV ions
Signal	Derivative of electron emission energy	Electron current vs electron binding energy	Ion current vs mass	Ion current vs energy
Elements detected	All except H and He	All except H	All	All except H and He
Sensitivity, general	Variable	Variable	Variable	High
Sensitivity, low atomic number	High	High	High	Low
Spatial resolution	High	Low	High	Low
Spectral shift due to chemical bonding	Yes	Yes	No	Generally no
Molecular information	Moderately effective	Yes	Usually no	Not effective
Quantitative analysis	Yes	Yes	Probably no	Yes
Beam-induced damage	Yes	No	Slight	Slight
Materials to be studied	Excellent for metals, damaging to polymers	Good for metals and polymers	Good for metals and polymers	Good for metals and polymers
Analysis, depth	<2 nm (20 Å)	1 ~ 3 nm (10 ~ 30 Å)	Monolayer	Monolayer
Depth-profiling analysis versus sputter	Simultaneous or sequential	Usually sequential	Usually sequential	Usually sequential

Fig. 1 Representative electron configuration and Auger transition

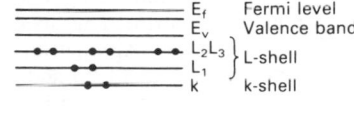

(a)

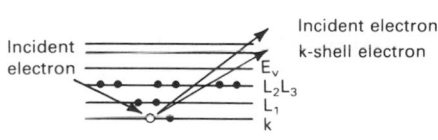

(b)

(c)

involved (KLL, LMM, or MNN). Thus, the energy and shape of these Auger features can be used to identify the composition of the surface of a material.

Auger Apparatus. Typical Auger equipment is illustrated in Fig. 2 (Ref 3). An ultrahigh vacuum ($<10^{-8}$ torr) is used to prevent contamination of the material and to reduce collisions of emitted Auger electrons with gas ions. The sputter ion gun is used to "sputter away" successive layers of sample between analyses so that

Fig. 2 Schematic of experimental arrangement used in obtaining standard Auger spectra
Source: Ref 3

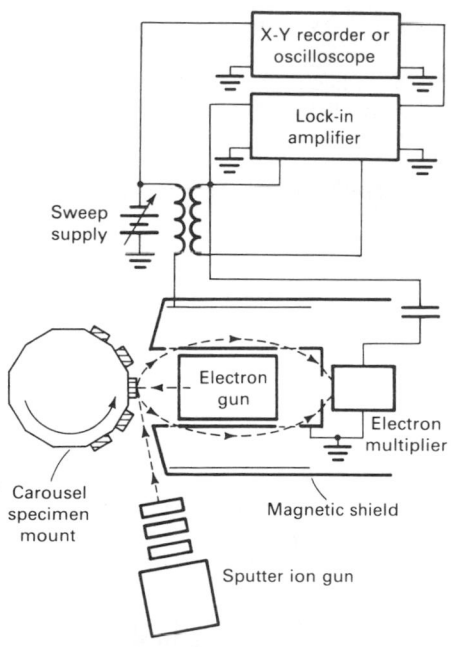

a concentration-depth profile may be obtained.

The cylindrical mirror analyzer sweeps the relevant voltage range. At any particular mirror voltage, only electrons with a specific energy will be deflected sufficiently to pass through the opening in the inner cylinder and into the multiplier. The electron multiplier converts the number of

Fig. 3 Wide-beam Auger electron spectroscopy spectrum of fracture surface of sintered titanium-based P/M alloy
Source: Ref 4

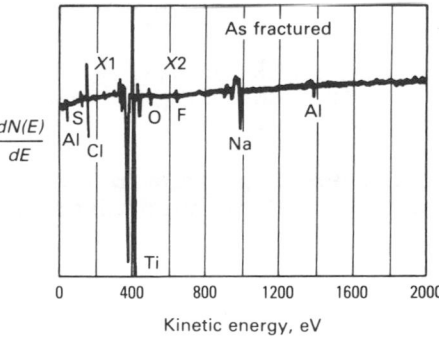

electrons, $N(E)$, into an electrical signal.

Because there is a large background signal of scattered electrons, the occurrence of a large number of Auger electrons appears as a small peak on a plot of frequency versus energy. For this reason, the derivation of the curve, $N(E)/dE$, is often calculated by a lock-in amplifier and plotted on an X-Y recorder. Figure 3 is an example of the Auger spectrum obtained (Ref 4).

Instead of scanning the entire range of mirror voltages, the mirror may be tuned to one voltage and the beam rastered across the sample surface. The intensity output of the electron multiplier can then be applied to a cathode ray tube. This produces a sample image with bright spots that correspond to the element whose Auger electron energy is the same as the chosen en-

Fig. 4 Auger composition-depth profiles of green and sintered type 316L stainless steel

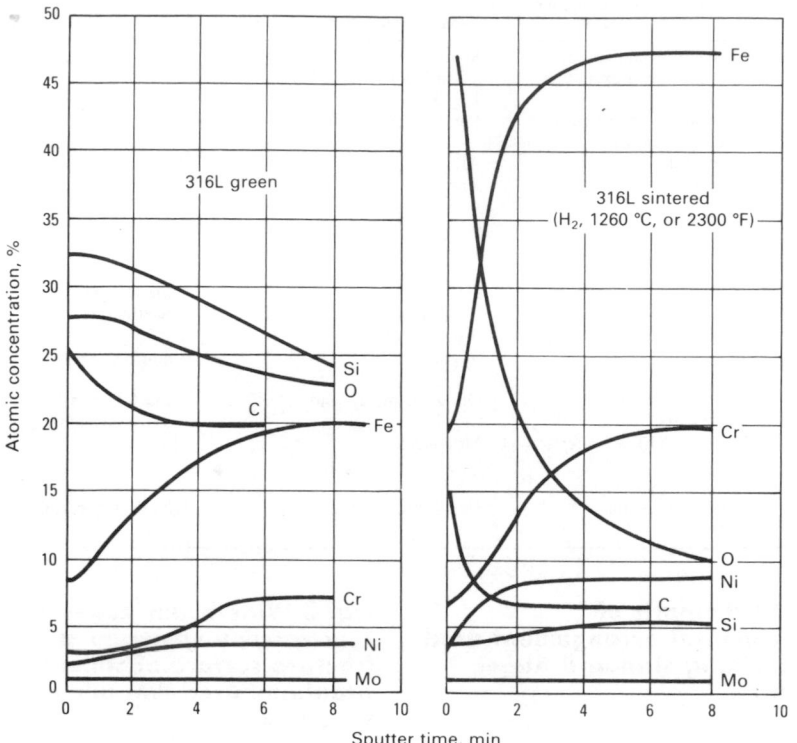

ergy window. The technique is know as scanning Auger microscopy.

P/M Applications. Figure 4 shows the composition-depth profiles of green and sintered compacts of a water-atomized type 316L stainless steel (Ref 5). The surface of water-atomized particles is highly enriched in silicon (as silicon dioxide determined by electron spectroscopy chemical analysis spectra). Prior to this analysis, the surface oxides of such powders were believed to consist mainly of chromium oxides. The presence of silica explains the unique role of small silicon additions (0.8 to 1%) during water atomization of many alloys.

Driven by its high affinity for oxygen and facilitated by its high mobility in the liquid state, silicon diffuses during atomization to the surface of a particle to form a thin, protective silicon oxide layer. For a type 316L stainless steel powder, this layer limits total surface oxidation during water atomization to about 1500 to 2500 ppm oxygen. Without silicon, the surface oxidation contains 5000 to 10 000 ppm oxygen. (The high carbon content in Fig. 4 comes from the die wall lubricant used in compaction of the green part.)

Figures 5 and 6 show Auger spectra and composition-depth profiles of the surfaces of compacted water-atomized type 316L stainless steel sintered in dissociated ammonia (Ref 5). Figure 5 refers to a standard type 316L powder, and Fig. 6 to a type 316L powder containing 1.5% Sn in fully prealloyed form. The spectra and profiles suggest an explanation for the beneficial effect of small tin additions on corrosion resistance, particularly if sintering is done in dissociated ammonia. As is evident from Fig. 6, tin is highly enriched on the free surfaces of the part.

Fig. 5 Auger spectrum (a) and composition-depth profile (b) of standard P/M type 316L stainless steel sintered in dissociated ammonia
Note presence of nitrogen on surface.

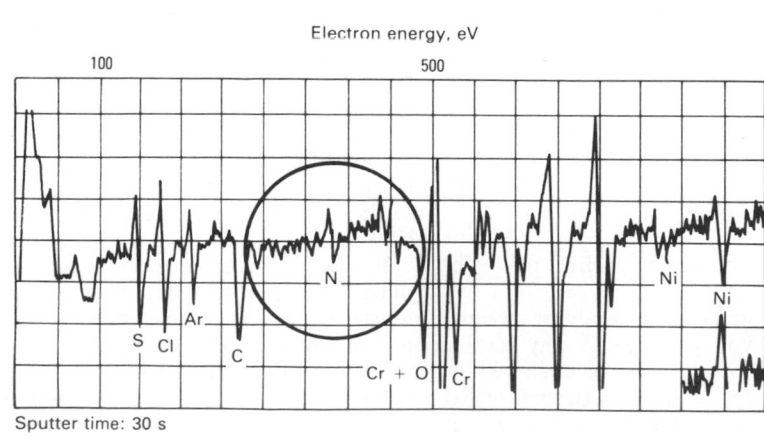

(a)

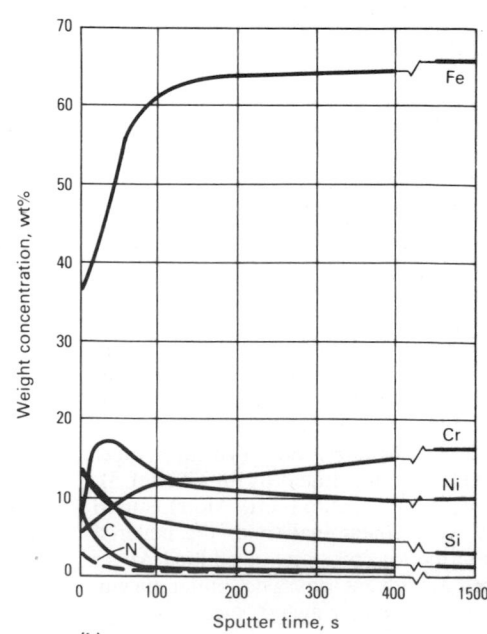

(b)

Fig. 6 Auger spectrum (a) and composition-depth profile (b) of tin-modified P/M type 316L stainless steel sintered in dissociated ammonia
Note presence of tin and absence of nitrogen on surface.

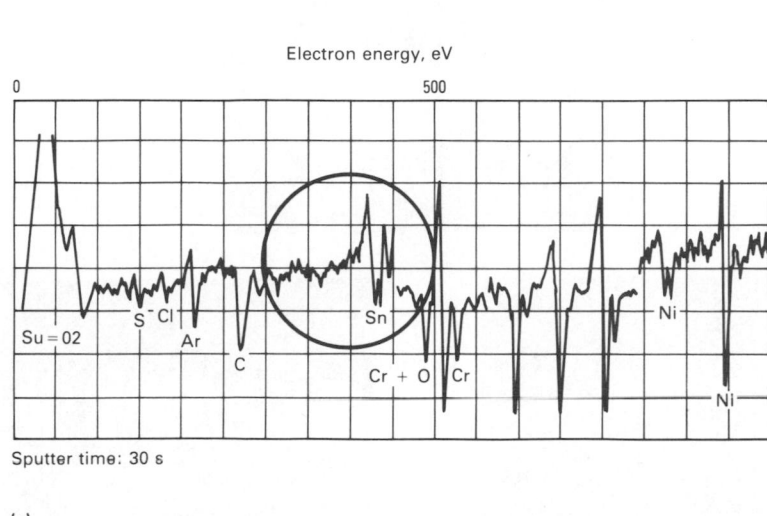

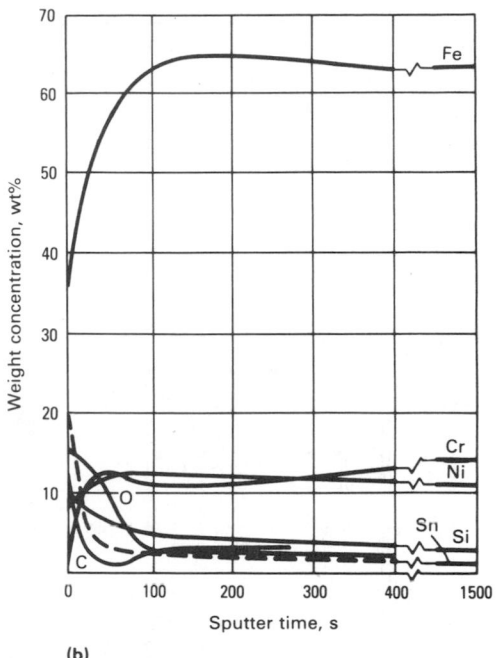

(a)

(b)

Fig. 7 Fractured surface of standard P/M type 316L stainless steel sintered in dissociated ammonia
(a) Auger absorbed current micrograph. Magnification: 350×. (b) Auger nitrogen image. Magnification: 350×. (c) Auger chromium image. Magnification: 350×. (d) Nitrogen spectrum. Magnification: 350×. (e) Chromium spectrum. Magnification: 350×

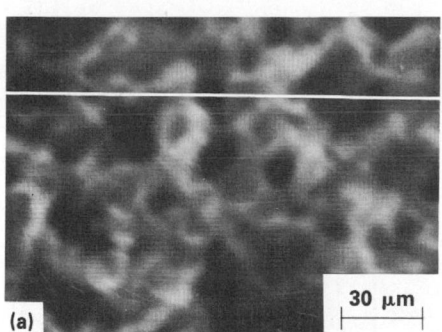

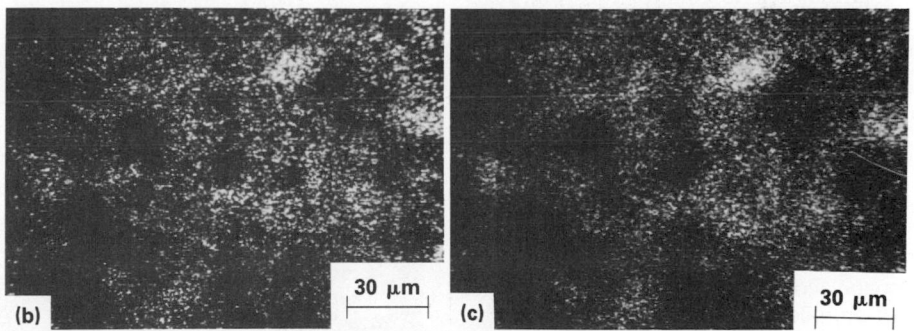

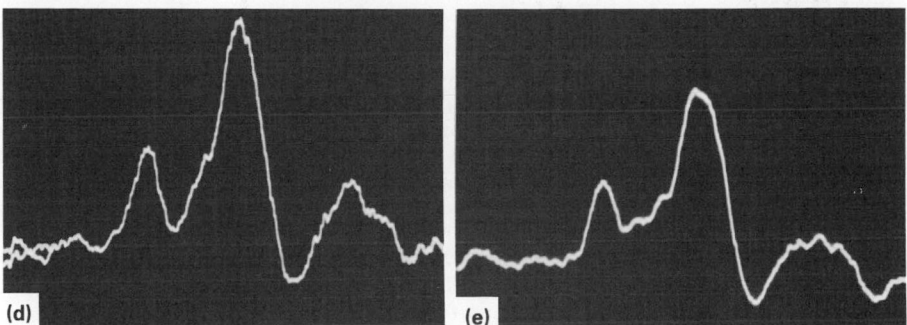

Fig. 8 Auger composition-depth profile of argon-atomized nickel-based superalloy powder
Source: Ref 6

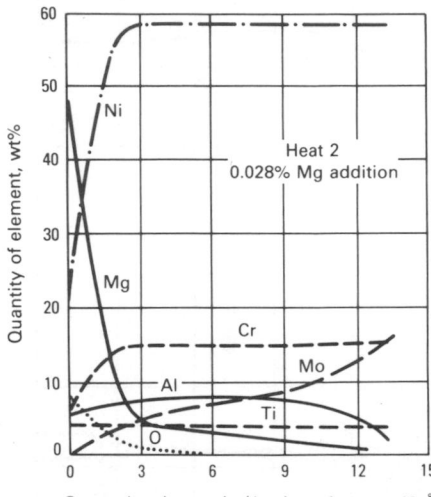

Fig. 9 Scanning Auger microscopy map of fracture surface of sintered P/M titanium alloy
(a) Fracture area. (b) Chlorine map. (c) Sodium map. Source: Ref 4

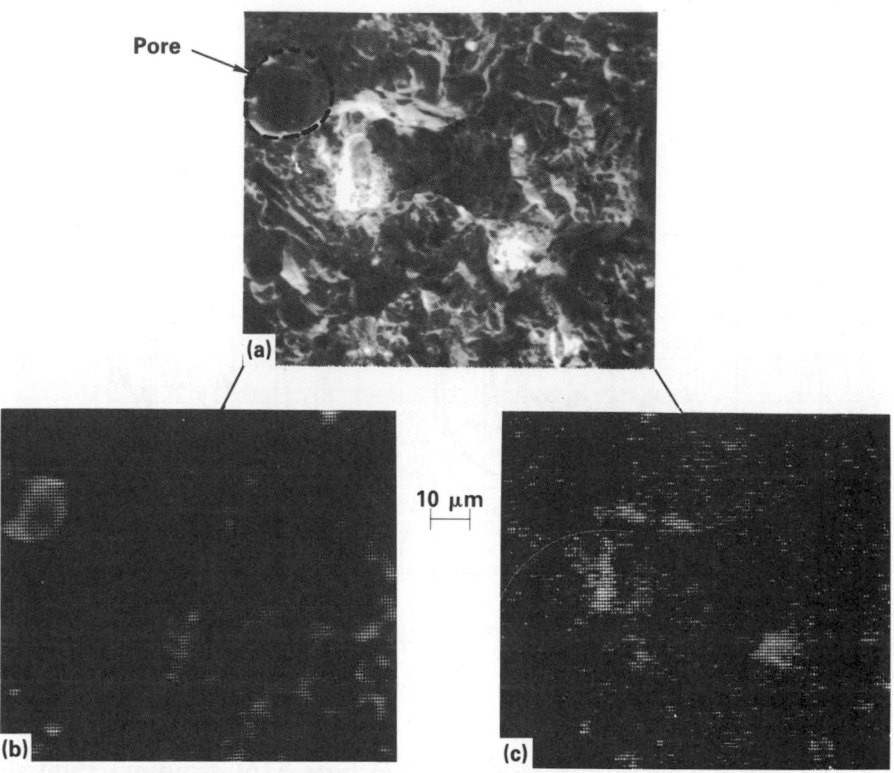

Fig. 10 Focused-beam Auger electron spectroscopy spectrum of particle found in a pore of sintered P/M titanium alloy
(a) Secondary electron image. (b) Auger spectra from particle. Source: Ref 4

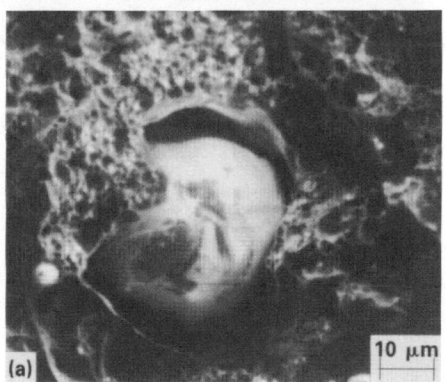
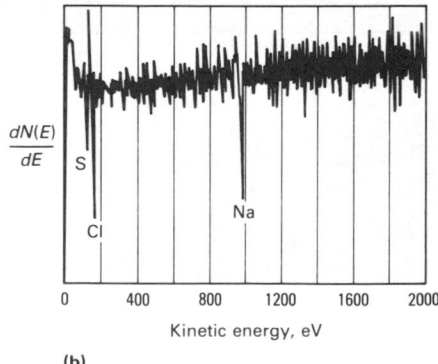

The improved corrosion resistance of tin-containing P/M stainless steels may be explained in terms of more stable passive films. The presence of tin also prevents or minimizes nitrogen absorption during cooling, as demonstrated by the absence of nitrogen peaks at 379 eV. Standard type 316L stainless steel powder (Fig. 5) readily absorbs nitrogen during cooling, resulting in chromium nitride formation and attendant deterioration of corrosion resistance. Chromium nitride presence is supported by the Auger elemental images and line scans of the fractured surface shown in Fig. 7.

A composition-depth profile of an argon-atomized superalloy is shown in Fig. 8. Occluded argon can cause undesirable, thermally induced porosity in parts consolidated from powder containing argon. Addition of a small amount of magnesium to the alloy markedly reduces the presence of occluded argon in the powder particles. The strong enrichment of magnesium in the surface of the powder particles indicates that the formation of magnesium-oxide-enriched surface films reduces the occurrence of particle collisions during atomization, resulting in fewer particles containing mechanically entrapped gas.

Auger electron spectroscopy and scanning Auger microscopy may be used to study the presence of a weakening or embrittling species at an interface. To avoid contamination, the sintered part is broken under vacuum and immediately analyzed. This technique provides information explaining low elevated-temperature ductil-

ity observed in titanium-based blended elemental P/M alloys (Ref 4).

A wide-beam Auger electron spectroscopy spectrum (Fig. 3) clearly indicates the presence of sodium and chlorine on the as-fractured surface. Scanning Auger microscopy indicates that the chlorine was associated with some pores in the fracture surface (Fig. 9). Finally, a focused-beam Auger electron spectroscopy spectrum shows that a particle found within a pore

was sodium chloride (Fig. 10). The low ductility is attributed to stress corrosion cracking, caused by the presence of salt, at elevated temperatures. Bulk chemical analysis indicates that the salt comes from the titanium sponge used in blended elemental powders.

Figure 11 shows Auger profiles of argon-atomized and rotating electrode process atomized powders. Figure 12 shows rapid solidification rate process atomized

Fig. 11 Auger sputtering profiles of Astroloy powders
(a) Argon atomized. (b) Rotating electrode process atomized. Source: Ref 7

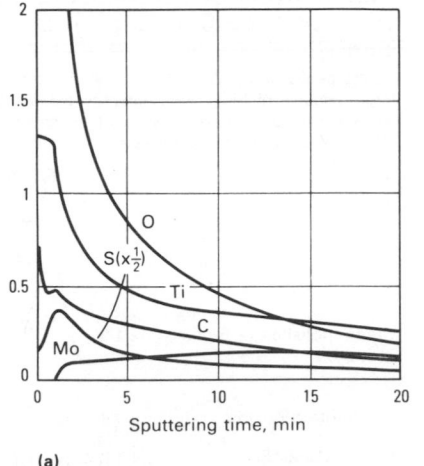

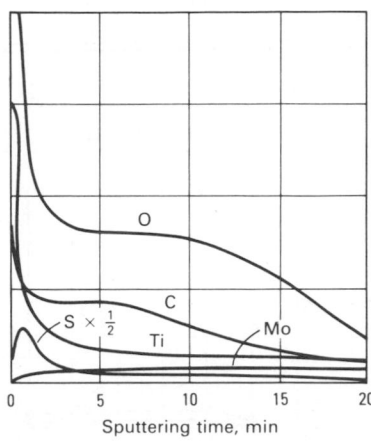

(a)   **(b)**

Fig. 12 Auger depth profile of rapid solidification rate process atomized nickel-based alloy B-1950
Source: Ref 8

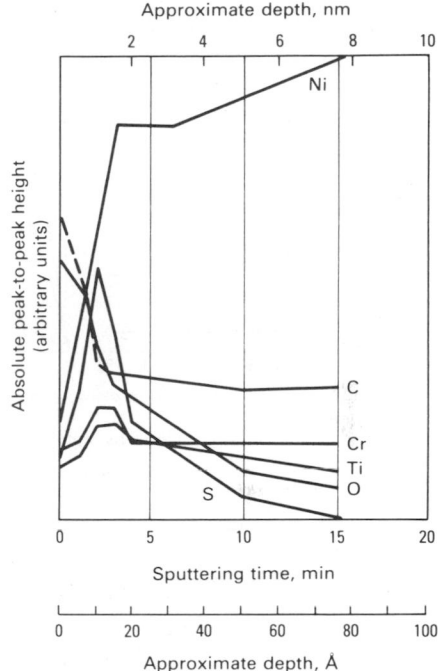

Fig. 13 Photoelectron emission process

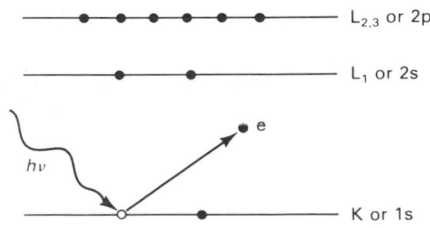

powders. In all cases, including the inert and rapid solidification conditions of rapid solidification rate technology, segregation of certain elements is present and persists to large depth profiles. Differences in surface composition were successfully correlated to differences in the mechanical properties of parts consolidated from these powders (Ref 7, 8).

Interpretation of Auger spectra, particularly spectra of irregular surfaces such as powders or porous structures, is difficult due to the number of variables present, including surface contamination and roughness, dissociation effects of chemical species due to the incident electron beam itself (beam damage), and differences in sputtering rates of different chemical species. Thus, conversion of peak heights into atomic concentrations presumes an accuracy that may not exist. However, even as a qualitative or semiquantitative tool, Auger electron spectroscopy/scanning Auger microscopy has already proved invaluable in the solution of many problems.

X-Ray Photoelectron Spectroscopy

Historical Development. Photoemission studies began in 1914 when electrons were observed to be released from a solid under x-ray irradiation (Ref 9). The practicality of x-ray photoelectron spectroscopy as a quantitative analysis technique was demonstrated much later (Ref 10). In 1958, x-ray photoelectron spectroscopy was used to identify not only elemental but also chemical species (Ref 11) by means of the low-energy photoelectrons found to escape from the outer surface layers of solids (Ref 12).

In 1967, the relative heights or areas of spectral peaks were used with suitable calibrations to determine relative amounts of elemental components in alloys and compounds (Ref 13). These successive discoveries led to development of commercial x-ray photoelectron spectroscopy systems in the late 1960's and early 1970's, which were designed to measure surface chemical characteristics of solids (Ref 14).

Theory. When a photon is absorbed in a solid, a photoelectron is emitted near the surface, such that:

$$hv = E_k + E_b + \psi$$

where hv is the photon energy, E_k is the electron kinetic energy, E_b is the electron binding energy, and ψ is the work function of the spectrometer. Mechanisms for emission of photoelectrons under x-ray irradiation are illustrated in Fig. 13. The binding energy (E_b) is defined as the energy needed to remove an electron from a core level to the Fermi level (the 1s level in Fig. 13).

Knowing the photon energy and measuring the kinetic energy with the electron spectrometer permit direct determination of the binding energy of electrons, which is unique and can be used to identify specific elements. Because the mean free path of the electrons is very small, only electrons that originate in the top atomic layers are detected.

Apparatus. Because x-ray photoelectron spectroscopy is a surface-sensitive technique, an inert environment such as an ultrahigh vacuum system—for example, 10^{-10} torr—is required for measurements. A typical x-ray photoelectron spectroscopy analytical chamber consists of an x-ray source, an electron energy

analyzer, an electron multiplier, a pulse-counting detector, and a computer system. An ion gun is employed for sputtering and compositional depth profiling. For good resolution during profiling, the ion beam—for example, argon—must be defocused over a surface area of approximately 1 cm^2 (0.15 in.^{2}), because photoelectrons are collected from a surface area of several square millimetres during analysis.

Many x-ray photoelectron spectroscopy systems incorporate computer systems for data reduction purposes (Ref 14). It is desirable for computer software to include routines such as smoothing, differentiation, deconvolution, and curve fitting, which contribute significantly to precise measurements.

Fig. 14 Chemical analyses by electron spectroscopy of atomized steel powder versus depth below original surface

A, powder as received; R, powder reduced in hydrogen for 30 min at 800 °C (1470 °F); S, powder compacted and sintered at 1120 °C (2050 °F). Source: Ref 15

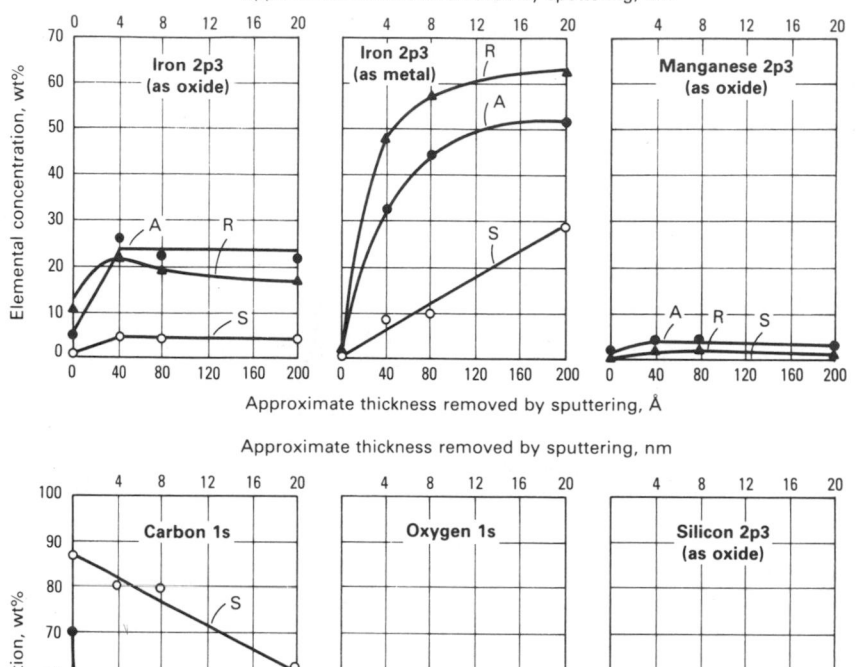

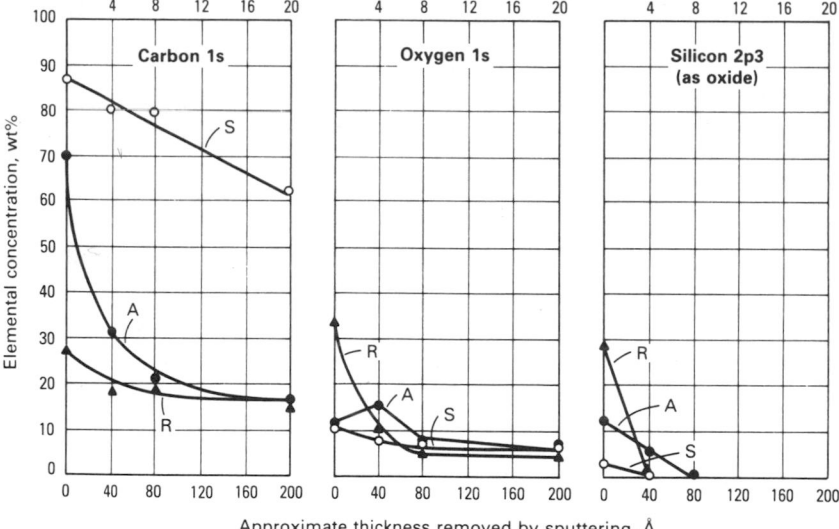

Fig. 15 Si-2p₃/₂ spectra obtained by x-ray photoelectron spectroscopy of as-atomized and sintered type 316L stainless steel

Strong peaks at 103.2 and 103.8 eV indicate large amounts of SiO_2. Peak intensity decreases and shifts toward elemental state of silicon at 99.15 eV as sputtering progresses.

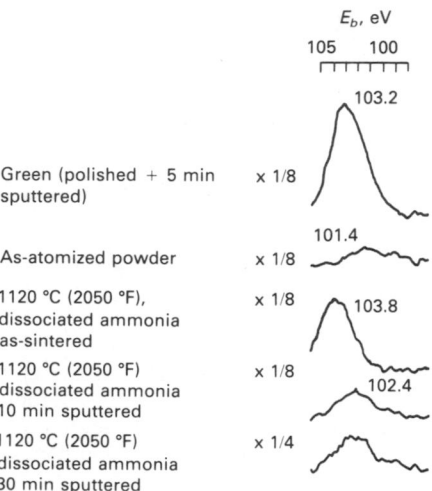

P/M Applications. Characteristics of x-ray photoelectron spectroscopy include detecting of all elements except hydrogen, response to surface materials only, and detection of chemical compounds. This sensitivity to chemical compounds is beneficial to P/M applications, because it enables determination of surface chemical bonds from precise measurements of the photoelectron peak energies (the identification of oxidation states). This is particularly useful because surfaces of atomized powder particles are often enriched with complex chemical compounds that include oxygen, nitrogen, or carbon. Despite these advantages, use of x-ray photoelectron spectroscopy in P/M materials has been minimal.

Figure 14 shows how x-ray photoelectron spectroscopy is used to determine the composition-depth profiles of water-atomized steel powder (Hoeganaes 1000B) under three different processing conditions (Ref 15). High concentrations of carbon, oxygen, and silicon in the form of oxides exist at or near the powder particle surfaces. The very high level of carbon on the surface of the sintered compact was presumed to be a result of the residual die wall lubricant.

Considering the profiles of iron, oxygen, and silicon, it is apparent that there is a thin iron oxide layer and a thin (<10 nm or 100 Å) silicon oxide layer on the surface of the as-received powder. The presence of even higher levels of oxides after reduction in hydrogen at 800 °C (1470 °F) with a −40 °C (−40 °F) dew point suggests that the freshly reduced surface atoms were active in air and immediately reoxidized. Manganese as an oxide is also present near particle surfaces. Metallic iron is found just under the surface, increasing gradually through the subsurface thickness.

As was discussed, the surfaces of atomized steel powders are largely comprised of carbon and silicon oxide, not metallic iron. As surface atoms are removed by ion sputtering, the bulk composition is approached slowly. However, at about 20 nm (200 Å), the subsurface composition varies from that of the interior of the particle. Reduction of such oxide layers in the powder through advanced atomization techniques or thermomechanical processing is expected to improve the properties of consolidated products.

X-ray photoelectron spectroscopy can be used to correlate changes in surface composition with corrosion resistance of water-atomized and sintered stainless steels (Ref 5). With the binding energy sensitivity of x-ray photoelectron spectroscopy, the large silicon enrichment in the surface of water-atomized type 316L stainless steel (see Fig. 4) was identified as consisting mainly of silicon dioxide, as shown in Fig. 15. The beneficial effect of silicon during water atomization—preventing excessive oxidation by forming silicon dioxide—was found to be detrimental to corrosion resistance.

Sintering in dissociated ammonia at low dew points (−35 °C or −30 °F and below), however, leads to substantial reduc-

Fig. 16 N-1s$_{1/2}$ spectra obtained by x-ray photoelectron spectroscopy of type 316L stainless steel

Sintered at 1120 °C (2050 °F) in dissociated ammonia with dew point of −29 °C (−20 °F)

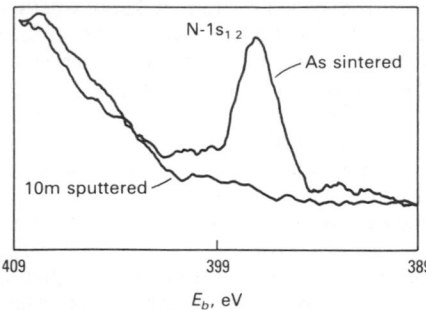

Fig. 17 Schematic of apparatus for secondary ion mass spectroscopy
Source: Ref 16

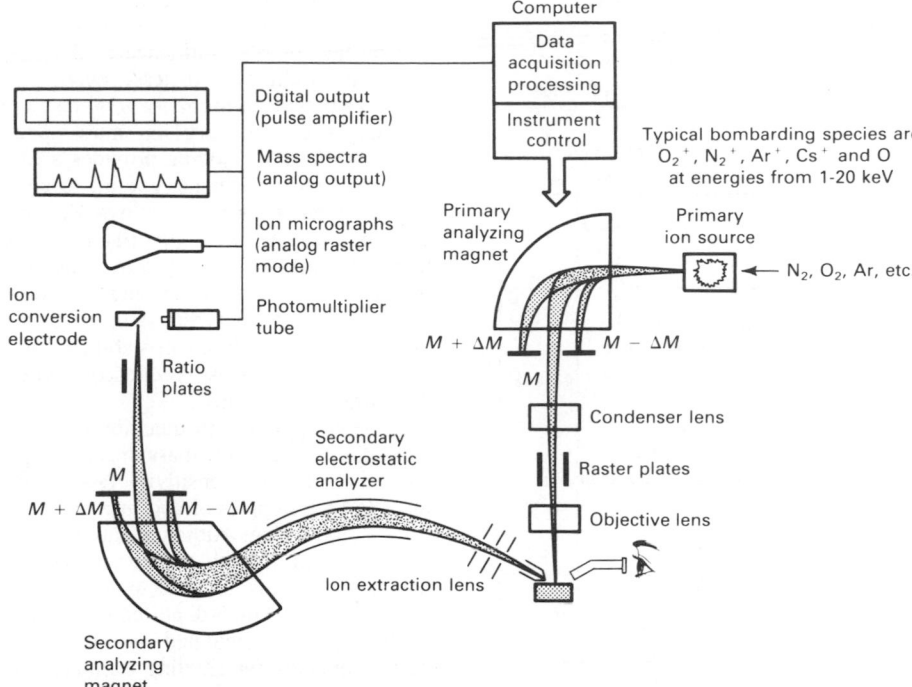

tion of silicon dioxide and enrichment of chromium and nitrogen in the surface (see Fig. 16). The nitrogen, which is mainly absorbed from the sintering atmosphere during cooling of the sintered part, is present as chromium nitride.

The resulting solute chromium depletion of the stainless steel is very detrimental to corrosion resistance. This reaction can be minimized by employing rapid cooling rates, or, perhaps more practically, by using the tin-modified type 316L stainless steel mentioned earlier in this article. Use of tin-modified type 316L prevents nitrogen absorption, because the enriched concentration of surface tin allows only low solubility of nitrogen.

X-ray photoelectron spectroscopy yields useful information about chemical bonding of detected elements on surfaces of P/M products from peak shift measurements. Although x-ray photoelectron spectroscopy penetrates only 3 nm (30 Å) deep or less into the top surface, its use for microanalysis is limited due to poor spatial resolution. The minimum spot size is about 2 mm (0.08 in.) in diameter. X-ray photoelectron spectroscopy is the least destructive of the four surface analysis techniques.

Secondary Ion Mass Spectroscopy

Historical Development. The technique of secondary ion mass spectroscopy was introduced in the early 1960's. At that time, point analysis of solids was done using an electron microprobe. The low yield of x-ray photons from this technique made it impossible to image an extended area of a sample. The higher sensitivity of the secondary ions to the exciting beam made is possible to analyze simultaneously the entire area being excited. Using this tech-

nique, elemental maps were obtained directly by an imaging technique with the need for a scanning excitation beam.

Description. Secondary ion mass spectroscopy consists of probing the sample surface with an ion beam, which excites and ejects atoms or clusters of atoms from the surface. The ejected atoms leave the surface as positive ions, negative ions, or neutrals. These positive or negative secondary ions are then mass analyzed in the spectrometer, depending on which mode the detector is set. The two modes are complementary due to differing sensitivities to different elements. Secondary ion mass spectroscopy is a true surface technique because the secondary ions must come from the surface monolayer.

When compared to other surface analysis techniques, secondary ion mass spectroscopy offers several advantages. It is noted for its high degree of sensitivity and inherently greater sensitivity to some elements than other surface analysis techniques. Elements with a high electron affinity or low ionization potential are particularly sensitive. Trace quantities as low as 1 ppb have been measured for the more detectable elements, with 1 ppm being the typical sensitivity limit.

The technique can also directly measure all elements, from hydrogen to uranium. In addition, isotopes, molecular fragments, and molecular clusters can be mea-

sured, yielding information regarding the structural/chemical state of the surface. Modes of operation include point analysis, line scan, and rastering to obtain elemental maps. Secondary ion mass spectroscopy gives good spatial resolution (0.5 to 1.0 μm or 20 to 40 μin.). Auger electron spectroscopy is the only surface analysis technique having a finer resolution.

Unlike other surface analysis techniques that analyze nondestructive photon or electron energy emissions from an excited surface, secondary ion mass spectroscopy analysis destroys the test surface. Thus, depth profiling is intrinsic to this technique. The ion milling rate can be reduced and the beam defocused at the expense of spatial resolution to analyze a monolayer surface.

The main limitation of secondary ion mass spectroscopy is inaccuracy in performing quantitative analysis. Large variations in ionization probability due to matrix effects limit quantitative interpretation of profiles when large differences in concentration are present.

Recent developments in instrumentation and experimental techniques provide greater evaluation accuracy, and linear relations between measured counts and composition can be obtained over a wide composition range. In addition, computer programs that compensate for nonlinear behavior have been developed.

258/Characterization and Testing

Fig. 18 Comparison of data obtained using high and low mass resolution detector

Source: Ref 22

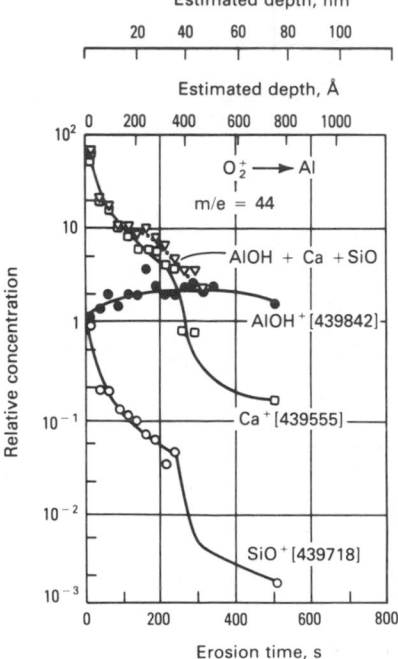

Theory and Apparatus. Basic equipment for secondary ion mass spectroscopy consists of a high-vacuum system (10^{-8} torr) with an ion source and a mass spectrometer. A schematic of a typical apparatus is shown in Fig. 17 (Ref 16). Pressure rises during analysis due to primary and secondary ions in the chamber.

Theoretically, any ion can be used as the ion source, but in practice the commonly used ions are O_2^+, N_2^+, Ar^+, Cs^+, and O^-. Energies range from 1 to 20 keV, with 6 to 10 keV being typical. Resolution of the instrument strongly depends on the ability to focus the ion beam. The beam can be focused to a spot diameter of 1 μm or more for microprobe analysis or hardened to a diameter of several millimetres for macroprobe analysis. Behavioral differences of the various exciting ions are described in Ref 17 to 21.

The secondary ions that are emitted are mass analyzed in the mass spectrometer. Quadrupole mass spectrometers or magnetic sector mass spectrometers are used for improved mass resolution. Figure 18 illustrates the type of data confusion that can occur if the detector has insufficient mass resolution. The improvement in detection resulting from the high resolution is shown in Fig. 18. The line labeled AlOH + Ca + SiO is a composite line produced at low resolution. The lines labeled $AlOH^+$, Ca^+, and SiO^+ were produced at high resolution.

Quantitative analysis of secondary ion mass spectroscopy spectra is made difficult by the complex influence of the matrix on ion yield. Nonetheless, several approaches produce reasonably good results. For negative secondary ion mass spectroscopy, cesium flooding provides a linear dependence of intensity ratios to concentration ratios for binary alloys. Several standards may be used to verify the relation in an untested alloy system. Linearity is only necessary over the range of composition of the analyzed sample. Similarly, oxygen flooding can stabilize positive secondary ion mass spectroscopy when the surface is saturated.

Another approach to quantitative analysis of secondary ion mass spectroscopy spectra establishes sensitivity factors using reference samples of approximately the same composition. Although this method is tedious and depends on the accuracy of the standards, it is fairly accurate. Ion implantation, for which depth concentration profiles can be calculated, is one of the best techniques for creating standards. It is necessary to suppress poly ions for elemental analysis (Ref 21).

The local thermal equilibrium model for calculation of ion yields as a function of relevant physical parameters permits use of semiquantitative and, in some cases, quantitative analysis (Ref 22). Despite successful uses of all these methods, the applicability to new types of samples must be studied and proven.

A desirable feature of secondary ion mass spectrometry is multielement detection. This capability is particularly important when high ion milling rates are used to probe a complex surface, because the surface is continually being sputtered. A mass range of up to 300 to 500 atomic mass units is typical for a given detector, but the range can be extended if necessary. This can be important when examining complex fragments.

P/M Applications. Secondary ion mass spectroscopy has been applied to P/M materials on a limited basis. Wider usage of this technique in the P/M field will occur as more confidence is generated in its ability to provide accurate quantitative analysis.

Surface analysis of a material in powder form presents a unique problem. The surface roughness of a powder specimen generates problems with uniform sputtering for depth profiling. Roughness influences the area of sample surface that is excited and the quantity of the secondary ions that can reach the detector. This problem has been addressed for thickness measurements by x-ray photoelectron spectroscopy using ion sputtering (Ref 23).

Secondary ion mass spectroscopy is able to measure elemental surface chemistry and extract structural and chemical information. It may be important to know whether an element that has migrated to the surface exists as a precipitate or as a compound. Such information may suggest an approach for surface modification to achieve desired properties.

One application of secondary ion mass spectroscopy in powder metallurgy is the measurement of adsorbed or chemisorbed species on a powder surface. In this case, the technique can distinguish whether a molecular species is adsorbed intact or has dissociated at the surface. Secondary ion mass spectroscopy was used to determine whether carbon monoxide on the surface of nickel powder was present as molecular or dissociated carbon monoxide (Ref 24). The presence of the former leads to $NiCO^+$ and Ni_2CO^+ species, while the latter leads to NiC^+ and NiO^+.

Powder processing in general gives rise to the possibility of contamination by foreign particles. Identification of powder contamination is difficult because of the dilution that occurs as a result of the thermal processing used to consolidate the powder. Thus, a contaminating particle only 44 μm in diameter in the powder may, through interaction with the matrix, diffuse to a zone of 760-μm diameter.

Secondary ion mass spectroscopy was used to measure the concentration of a copper-enriched area in a consolidated beryllium part (Ref 25). The analysis was done using a step-scan across the enriched copper region exposed by sectioning the sample. An area of about 250 μm^2 was sputtered, but only ions from a central 30-μm diameter area were collected. This eliminated the matrix effect caused by enhancement of beryllium detection due to surface oxide (see Fig. 19). This analysis showed that copper was the only impurity in the affected zone and that the maximum concentration was 7000 ppm (Fig. 20). This analysis was useful in identifying and eliminating the contamination source.

Another application of secondary ion mass spectroscopy to powder metallurgy was the evaluation of an atomized copper-aluminum alloy (Ref 26). Depth profiling of the powder surface showed aluminum segregation on the surface of the as-atomized powder. Larger concentration gradients were observed for inert gas-atomized powder than for water-atomized powder. The qualitative data obtained by secondary ion mass spectros-

Fig. 19 Surface oxide enhancement of Be⁺ ion yield using Cs⁺ bombarding ion
(a) Raw data. (b) Data normalized to beryllium concentration. Source: Ref 25
The number 9 refers to beryllium isotope; 63 and 65 refer to isotopes of copper.

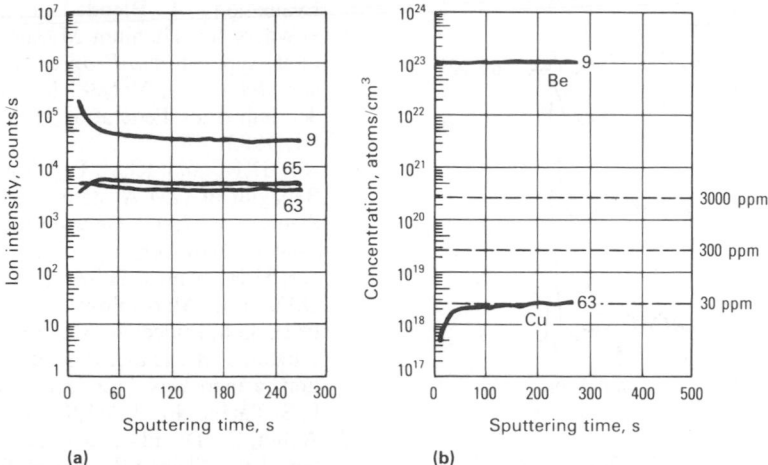

(a)   (b)

Fig. 20 Line scan across high-density region of beryllium sample
(a) Raw data. (b) Data normalized to beryllium concentration. Source: Ref 25
The number 9 refers to beryllium isotope; 63 and 65 refer to isotopes of copper.

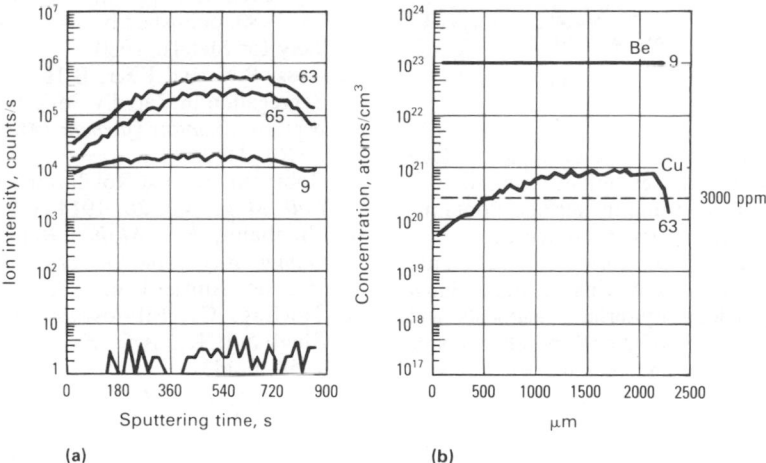

(a)   (b)

copy confirmed quantitative data measured by Auger electron spectroscopy.

Atomic, elemental, or molecular clusters are important because they provide information regarding the chemical state of the surface atoms. A study of corrosion films on chromium at room temperature illustrates what can be gleaned from secondary ion mass spectroscopy spectra (Ref 27). The first layer of oxidation was found to be very thin and characterized by the emission of CrO_2^- ions. Cr_2^+ and Cr_3^+ ions were also found, indicating that adjacent chromium atoms still existed on the surface. Presence of these ions declined as oxidation proceeded. The second layer of oxidation was thicker and characterized by CrO^+ ions. The initial growth of the CrO^+ oxide phase was characterized by a decline in the CrO_2^- ions.

Other applications of secondary ion mass spectroscopy to powder metallurgy include measuring the thickness of an intentionally applied coating, such as an electroplated or vapor-deposited layer. The effectiveness of an organic inhibitor in controlling surface oxidation in powders exposed to hostile environments prior to consolidation also can be measured.

Ion Scattering Spectroscopy

Historical Development. Use of mega electron volt ions, as an indirect means of identifying surface atoms, was first demonstrated in 1959 through analysis of peaks in the energy distribution of elastically scattered primary ions (Ref 28). Despite its potential, high-energy scattering was not used for quantitative surface analysis. This was partly due to the relative insensitivity of the scattering signals to the surface composition and partly to the unavailability of the required high-energy accelerators (Ref 29).

The first analytical applications of ion-scattering spectroscopy to surface analysis used low-energy (0.5 to 3 keV) ions under well-defined experimental conditions (Ref 30-32). For noble gas ions in the range of a few hundred electron volts, it was observed that the kinematics of the collision process can be described as a simple elastic binary collision, and a substantial number of backscattered ions are not neutralized by the surface. Thus, the masses of the surface atoms can be determined from the recoil momentum of the backscattered ions. This unique property of low-energy noble gas ions to produce simple backscattered ion energy spectra permits their utilization as a practical method of surface analysis.

Theory and Apparatus. In ion-scattering spectroscopy, the surface to be analyzed is bombarded by a monoenergetic beam of positive ions such as $^3He^+$, $^4He^+$, or $^{20}Ne^+$ that undergo elastic collision with the surface atoms. The number of primary ions scattered from the surface at a given angle is measured as the ratio of final energy to incident energy. The energy of the recoil ion is a function of the mass of the atom it encountered; hence, there is a measurable peak at a unique energy.

The distribution of scattered ion energy at a scattering angle of 90°—that is, an electrostatic sector located at 90° from the incoming primary ion beam—appears as the energy ratio E_1/E_o according to:

$$E_1/E_o = (M_2 - M_1)/M_2 + M_1$$

(provided $M_2 > M_1$)

where E_o is kinetic energy of the primary ion; E_1 is the kinetic energy of the primary ion after scattering; M_1 is the mass of the primary ion; and M_2 is the mass of the surface atom. Thus, M_2 is determined by selection of the mass of the primary ion M_1 and by measuring the energy ratio E_1/E_o. The height of this measurable peak of unique energy for each surface element is related to the concentration of that element in the outer monolayer.

Modern instrumentation (Ref 33) permits incorporating a cylindrical mirror analyzer for sensitivities 500 times greater than those of earlier ion-scattering spectroscopy methods. This analyzer also allows detection of scattered ions throughout a complete 360° solid angle. High-performance ion-scattering spectroscopy

Fig. 21 Adsorption isotherms and ion-scattering spectra of two catalysts for nitric oxide
Source: Ref 36

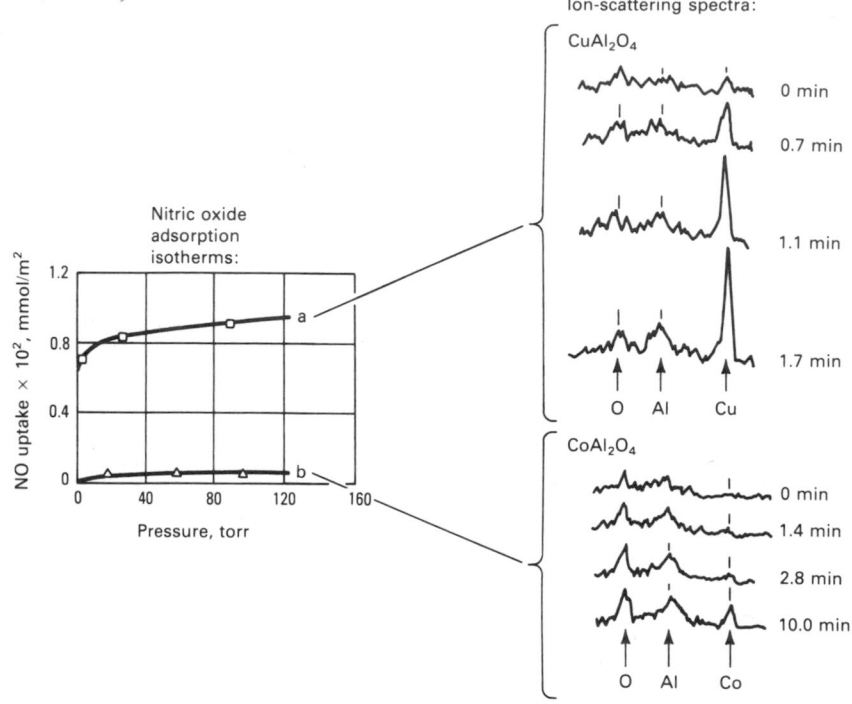

provides detection sensitivities greater than 10 000 counts per second per nanoampere of beam current and surface sensitivities approaching 100 ppm of a species in the top monolayer, especially for high atomic number elements (Ref 32).

P/M Applications. The high sensitivity of low-energy noble gas ion scattering to the first monolayer of surface atoms allows accurate identification of surface species. Ion-scattering spectroscopy appears to require fewer corrections than other surface analytical methods to obtain quantitative information (Ref 29, 32). However, at present, this technique is considered semiquantitative.

Although ion-scattering spectroscopy has been used for surface analysis in both metals and nonmetals such as glass and polymers, information directly related to P/M materials is scarce. Ion-scattering spectroscopy offers minimal information for chemical bonding studies. Like Auger electron spectroscopy, it is used for elemental identification. Because the primary ion beam serves the dual role of surface analysis and sputtering surface atoms, a composition-depth profile at the surface can be obtained with ion-scattering spectroscopy.

There is growing evidence that the combination of ion-scattering spectroscopy and secondary ion mass spectroscopy offers

several advantages. For example, surface analysis by this combined technique is useful in determining surface contaminants, verifying the presence of surface inhomogeneity, comparing surface-to-bulk composition, identifying compositional changes induced by surface treatments, and determining the relationship between surface chemistry and bonding mechanisms (Ref 34, 35). In the absence of a P/M specific example, Fig. 21 illustrates how ion-scattering spectroscopy was used to explain the behavior of two potential catalysts (Ref 36). The cobalt-bearing spinel had no cobalt in the outermost layer and low nitric oxide uptake in the adsorption isotherms. The high level of nitric oxide uptake in the copper-bearing spinel was attributed to the presence of copper in the outermost layer.

REFERENCES

1. McIntryre, N.S., Ed., *Quantitative Surface Analysis of Materials,* ASTM STP 643, American Society for Testing and Materials, Philadelphia, 1978, p 1

2. Alford, N.A., *et al.,* Auger Electron Spectroscopy (AES): An Appraisal, *Surface and Interface Analysis,* Vol 1 (No. 1), 1979

3. Davis, L.E., *et al., Handbook of Au-ger Electron Spectroscopy,* 2nd ed., Physical Electronics Industries, Inc., Eden Prairie, MN, 1976

4. Bonini, J.J. and Sankaran, K.K., Evaluation of Blended Elemental Powders for Titanium Missile-Structural Applications, *Progress in Powder Metallurgy,* Vol 38, Metal Powder Industries Federation, Princeton, NJ, 1982

5. Ro, D.H. and Klar, E., Corrosion Behavior of P/M Austenitic Stainless Steels, *Modern Developments in Powder Metallurgy,* Vol 13, Hausner, H.H., Antes, H.W., and Smith, G.D., Ed., Metal Powder Industries Federation, Princeton, NJ, 1981, p 147

6. Larson, J.M., *et al.,* Porosity Reduction in Inert-Gas Atomized Powders, U.S. Patent No. 4,047,933, 1977

7. Aubin, C., Davidson, J.H., and Trottier, J.P., "The Influence of Powder Particle Surface on the Properties of a Nickel-Based Superalloy Produced by Hot Isostatic Pressing," Superalloys 1980, Champion, PA, Sept 21-25, 1980; published by American Society for Metals, 1980

8. Ross, P.N. and Kear, B.H., Surface Segregation in Rapidly Solidified Superalloys, *Rapidly Quenched Met.,* Vol 3 (No. 1), 1978

9. Robinson, H. and Rowlinson, W.F., *Phil. Mag.,* Vol 28, 1914, p 277

10. Siegbahn, K., *Alpha, Beta, and Gamma-Ray Spectroscopy,* North-Holland, Amsterdam, 1955

11. Norling, C., Silolowski, E., and Siegbahn, K., *Ark. Fys.,* Vol 13, 1958, p 483

12. Larsson, K., *et al., Acta Chem. Scand.,* Vol 20, 1966, p 2880

13. Siegbahn, K., *et al., ESCA, Atomic Molecular and Solid State Structure Studied by Means of Electron Spectroscopy,* Almquist and Wiksells, Uppsala, 1967

14. Davis, L.E., Technique Fundamentals of AES and XPS, *Modern Surface Analysis: Metallurgical Applications of AES and XPS,* TMS-AIME, 1980, p 1

15. Ferriss, D.P., Surface Analysis of Steel Powders by ESCA, *Int. J. Powder Metall. Powder Tech.,* Vol 19 (No. 1), 1983, p 11

16. White, G.W. and Christie, W.H., The Use of RBS and SIMS to Measure Dopant Profile Changes in Silicon Caused by Pulsed Laser Annealing, *Sol. State Tech.,* Sept 1980, p 109-116

17. Brundle, C.R., Chemical Characterization of Surfaces and Interfaces of

Industrial Materials by X-Ray Photoelectron Spectroscopy, Auger Electron Spectroscopy and Secondary Ion Mass Spectroscopy, *Thin Solid Films,* Vol 72, 1980, p 3-8

18. Hammer, G.E., Matrix Effects in SIMS: Ion Yield Enhancement in Copper Sulfide Films on Copper and Copper Alloys, *J. Vac. Sci. Technol.,* Vol 20 (No. 3), March 1982, p 403-405
19. Castaing, R. and Slodzian, G., Analytical Microscopy by Secondary Ion Imaging Techniques, *J. Phys. E: Sci. Instrum.,* Vol 14, 1981, p 1119-1127
20. Yu, M.L. and Reuter, W., Matrix Effect in SIMS Analysis Using an O_2^+ Primary Beam, *J. Vac. Sci. Technol.,* Vol 17 (No. 1), Jan/Feb 1980, p 36-39
21. Morgan, A.E. and Werner, H.W., Quantitative Analysis of Low Alloy Steels by Secondary Ion Mass Spectroscopy, *Anal. Chem.,* Vol 48 (No. 4), 1976, p 699-708
22. Werner, H.W., New Developments in Secondary Ion Mass Spectroscopy, *Applied Surface Analysis,* ASTM STP 699, Barr, T.L. and Davis, L.F., Ed., American Society for Testing and Materials, Philadelphia, 1980, p 81-110
23. Cross, Y.M. and Dewing, J., Thickness Measurements on Layered Materials in Powder Form by Means of XPS and Ion Sputtering, *Surf. Interface Anal.,* Vol 1 (No. 1), 1979, p 26-32
24. Hopster, H. and Brundle, C.R., Use of SIMS for Studies of Adsorption on Well-Defined Metal Surfaces (1) Combined XPS-LEED-SIMS Studies of O_2, CO, H_2O, and H_2 on Ni (100), *J. Vac. Sci. Technol.,* Vol 16, 1979, p 548-551

25. Haws, W.J., unpublished research, Brush Wellman, Cleveland, 1981
26. Haws, W.J., unpublished research, Glidden Metals, Cleveland, 1978
27. Benninghoven, A., Surface Investigation of Solids by the Statistical Method of Secondary Ion Mass Spectroscopy, *Surf. Sci.,* Vol 35, 1973, p 427-457
28. Rubin, S., *Nucl. Instr. Meth.,* Vol 5, 1959, p 177
29. Smith, D.P., Analysis of Surface Composition with Low-Energy Backscattered Ions, *Surf. Sci.,* Vol 25, 1971, p 171
30. Datz, S. and Snoek, C., *Phys. Rev.,* Vol 134, 1964, p A347
31. Smith, D.P., *Bull. Am. Phys. Soc.,* Vol 11, 1966, p 770
32. Smith, D.P., *J. Appl. Phys.,* Vol 38, 1967, p 340
33. McKinney, J.T. and Ruscj, T.W., High-Performance ISS Surface Analysis with a Cylindrical Mirror Analyzer, presented at the Pittsburgh Conference on Analytical Chemistry and Applied Spectroscopy, Cleveland, 1976
34. Sparrow, G.R. and Mishmash, H.E., *Surface Analysis of Polymer and Glass, Quantitative Surface Analysis of Materials,* ASTM STP 643, McIntyre, N.S., Ed., American Society for Testing and Materials, Philadelphia, 1978, p 164
35. Madura, A.R., *Ind. Res. Dev.,* Aug 1982, p 98
36. Shelef, M., et al., *Surf. Sci.,* Vol 47, 1975, p 697

SELECTED REFERENCES

- Carlson, T.A., *Photoelectron and Auger Spectroscopy,* Plenum Press, New York, 1975

- Kane, P.F. and Larrabee, C.B., Ed., *Characterization of Solid Surfaces,* Plenum Press, New York, 1974
- Czanderna, A.W., Ed., *Methods and Phenomena: Their Applications in Science and Technology,* Vol 1, Elsevier, New York, 1975
- Carbonara, R.S. and Cuthill, J.R., Ed., *Surface Analysis Techniques for Metallurgical Applications,* ASTM STP 596, American Society for Testing and Materials, Philadelphia, 1976
- McIntyre, N.S., Ed., *Quantitative Surface Analysis of Materials,* ASTM STP 643, American Society for Testing and Materials, Philadelphia, 1978
- Davis, L.E., et al., *Handbook of Auger Electron Spectroscopy,* 2nd ed., Physical Electronics Industries, Inc., Eden Prairie, MN, 1976
- Brundle, C.R. and Baker, A.D., Ed., *Electron Spectroscopy: Theory, Technique, and Applications,* Vol 1, 2, and 3, Academic Press, New York, 1977, 1978, and 1979
- Dobryzynski, L., Ed., *Handbook of Surfaces and Interfaces,* Garland Publishing, 1978
- Briggs, D., Ed., *Handbook of X-Ray and Ultraviolet Photoelectron Spectroscopy,* Heyden and Son, 1977
- Wagner, C.D., et al., *Handbook of X-Ray Photoelectron Spectroscopy,* Perkin-Elmer Corp., Physical Electronics Division, Eden Prairie, MN, 1978
- Heinrich, K.F.J. and Newbury, D.E., Ed., *Secondary Ion Mass Spectrometry,* National Bureau of Standards Special Publ. 427, Washington, DC, 1975
- Smith, D.P., *J. Appl. Phys.,* Vol 38, 1967, p 340
- Smith, D.P., *Surf. Sci.,* Vol 25, 1971, p 171

Surface Area, Density, and Porosity of Metal Powders

IN DETERMINING the behavior of a metal powder during processing and the mechanical properties of the resulting P/M material, the surface area (m²/kg, or cm²/g), density (g/cm³), and porosity (% porosity = 100 − % of theoretical density) can be useful indicators. For example, surface area is helpful in understanding sintering behavior; density permits the distinction between different materials and/or the determination of inaccessible porosity of a powder or P/M part; and porosity can greatly influence the hardness of P/M materials as measured by macroscopic methods.

The values for these three characteristics can be determined by a variety of methods; in reporting data, the method should also be indicated, because the values will often differ depending on the test method used.

The methods used to determine surface area, density, and porosity are not specific to metal powders; many have been developed for testing other materials in powder form. The major techniques used to determine these powder characteristics are summarized below. Expanded discussions of several test methods can be found in subsequent sections of this article, including testing parameters, specifications, and instrumentation.

- *Gas adsorption method:* Determines the surface area of a powder sample by measuring the amount of gas adsorbed by a monomolecular coverage of the powder. Nitrogen is the most common gas used; krypton has also been used for very small surface areas. This method can also be used to determine pore volume and size distribution for pore diameters ranging from 60 to 1.4 nm (600 to 14 Å).
- *Permeametry:* Measures the resistance to fluid flow through a compacted powder bed. This information is used to determine related properties of a powder, such as specific surface area and average particle size. This method does not

give information on surface area of internal pores. Gas permeametry is the most commonly used testing procedure.
- *Pycnometry:* Determines density by measuring the difference between the specific and bulk volumes of a sample. This method is based on the displacement principle, using the powder as the solid body and helium or mercury as the displaced medium. Pycnometry can be used to determine total pore volume. However, it does not provide quantification of pore size or distribution.
- *Mercury porosimetry:* Measures the volume of mercury intruded into the pores of a powder sample as a function of the pressure applied to the mercury. This method gives pore size and distribution over a wide range—5 × 10⁵ nm to 3 nm (5 × 10⁶ Å to 30 Å)—depending on the capability of the apparatus used.

Although surface area, density, and porosity of powders are interrelated, a given powder sample may require the use of several testing methods to provide a complete analysis of these characteristics. A comprehensive treatment of each of the methods outlined above follows.

Gas Adsorption*

THE BRUNAUER-EMMET-TELLER (BET) method of measuring specific surface area is based on the determination of the amount of gas that is adsorbed on the surface of a powder in a monomolecular layer. The BET method is widely used to determine the specific surface of catalysts. Its use for metal powders is limited primarily to fine powders. The specific surface (m²/kg or cm²/g) determined by this method includes the surface of interior

*Adapted from Ref 1.

surface-connected pores. Therefore, BET measurements can sometimes be helpful in characterizing the porosity of a metal powder.

Theory

To determine specific surface by the BET method, the amount of gas adsorbed in a monomolecular layer on the surface of a powder and the area occupied by a molecule of the adsorbed gas must be known. The latter value is generally taken as 16.2 × 10⁻²⁰ m² for nitrogen, the most widely used adsorbate.

The amount of gas adsorbed in a monomolecular layer (V_m), measured in m², is calculated from an adsorption isotherm—a series of measurements of the volume (V) of gas adsorbed as a function of the pressure (p). From these measurements, the volume (V_m) may be calculated using a relationship derived by Brunauer, Emmett, and Teller.

$$\frac{p}{V(p_o - p)} = \frac{1}{V_m C} + \frac{(c - 1)p}{V_m C p_o}$$

where p_o is the vapor pressure of the adsorbed gas at the adsorption temperature, and C is a constant. A plot of $p[V(p_o - p)]$ versus p/p_o generally gives a straight line.

From the slope and intercept of this line, the amount of gas adsorbed in a monomolecular layer (V_m) is determined. Frequently, C is large enough so that the intercept of the straight line can be taken as 0.

Apparatus

Figure 1 shows the adsorption apparatus originally developed by Emmett, which is still widely used. Before measurements can be made, any gas adsorbed on the metal powder sample must be desorbed by heating the sample in a vacuum generated by a diffusion pump. After desorption, a low-temperature bath, generally liquid nitro-

Fig. 1 Schematic of BET apparatus for determining specific surface area

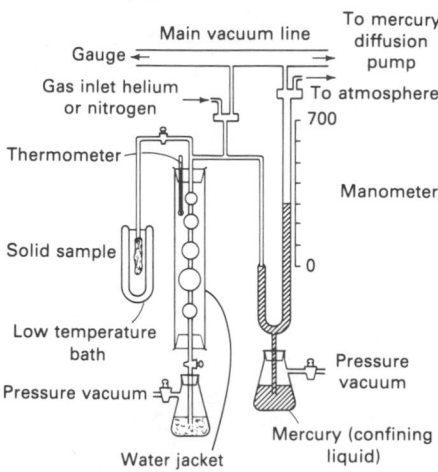

gen in a Dewar's flask, is raised to surround the sample.

Procedure

Adsorbate gas, usually nitrogen, is taken into the burette, and its pressure is measured on the manometer. The stopcock between the sample and the burette is then opened, and after allowing time for equilibrium to be established, the new pressure is read on the manometer.

The volume of the gas admitted to the sample chamber is proportional to the difference in the pressures before and after opening the stopcock. The latter pressure is also the equilibrium adsorption pressure. The volume adsorbed is equal to the volume admitted, less the volume of gas required to fill the dead space in the sample chamber and burette connections.

To obtain more adsorption points, the mercury level is raised to the next volume mark, and a new pressure is established. Helium is used to calibrate the dead-space. In all adsorption calculations, a correction for the nonideal behavior of nitrogen at liquid nitrogen temperature is included. When low specific surfaces of less than 1 m^2/g are to be determined, krypton may be used as an adsorbate instead of nitrogen.

Process Modifications

Recently, a modified BET method of measuring specific surface area has been developed, which uses a continuous flow of gases. This method is based on the gas chromatographic technique, which is a separation technique involving passage of a moving gaseous phase through a column containing a fixed adsorbent phase.

Typically, a helium-nitrogen stream of known composition is flowed continuously over a sample cooled by liquid nitrogen; the sample adsorbs a quantity of nitrogen proportional to its surface area. When the coolant is removed and the sample warms, the nitrogen adsorbed on its surface is desorbed and transferred to the flowing gas stream. Its concentration is measured with a thermal conductivity cell and displayed as a peak on a strip-chart recorder; peak size is a direct measure of the quantity of nitrogen liberated.

Repeating this procedure at two additional helium-nitrogen ratios gives the data required for a standard BET plot to determine the surface area of the sample. If accuracy is not required, one measurement often produces satisfactory results.

Permeameter

By Peter J. Heinzer
Coordinator—Technical Services
Imperial Clevite Technology Center

PERMEAMETRY is the measurement of resistance to fluid flow through a compact powder bed. Its main purpose in powder metallurgy is not so much to quantify resistance as it is to measure the related properties of a particle population—namely, specific surface area and average particle size. Fluids may be in liquid or gaseous form. Liquid permeametry was prevalent in the early stages of development and is the simplest method if the minimum diameter of any appreciable size fraction is 5 μm (Ref 2). Settling and segregation, aggregation, and the difficulty of removing bubbles make liquid permeametry unsatisfactory for smaller sized particles. Gas permeametry is now the preferred method, because it extends the size measurement capability down to 0.1 to 0.5 μm.

Commercial permeametry is applicable in the following ranges (Ref 3, 4):

Specific surface area	70 to 20 000 cm^2/g
	(10.8 to 3100 in.^{2}/g)
Particle size	0.5 to 50 μm

In production practice, these ranges are subdivided into smaller segments to improve accuracy, because the behavior of fluid flow through powder beds with diverse characteristics changes considerably.

A typical sample occupies a volume of about 5 cm^3 (0.30 in.^{3}) and, depending on density, weighs from 5 to 20 g (0.18 to

0.70 oz). Reproducibility in the surface area range cited for commercial instruments is ±1%. Accuracy varies with the type of sample, as discussed later in this section.

Inherent problems in the use of permeametry for particle size measurement involve its use of semiempirical relationships with parameters that can only be evaluated indirectly. Additionally, permeametric methods provide only an average particle size—not the more useful size distribution information. Despite these disadvantages, permeametry is popular due to its simplicity of operation, reproducibility, speed of analysis (2 to 15 min compared to 1 h or more with other methods), and low cost (one fifth to one tenth of the equipment cost of competing methods).

History and Theory

The original study of fluid flow through compacted particulate matter is attributed to D'Arcy (Ref 5), who examined water flow rates from the public fountains of Dijon, France through sand beds of varying thicknesses in 1856. He formulated the basic principle behind permeability, showing that the average flow rate is proportional to the pressure gradient and inversely proportional to the thickness of the bed.

In 1927, Kozeny (Ref 6) published the first derivation showing correlation among porosity, permeability, and particle surface area. Following the lead of Blake (Ref 7), Kozeny treated the flow of fluid through a particulate bed as being equivalent to the flow of fluid through a comparable volume of parallel pipe channels of circular cross section. This simplification resulted in determining an equivalent diameter of pipe channel to characterize flow rate through the powder bed.

In 1938, Carman (Ref 8) and Dallavalle (Ref 9) independently proposed the determination of specific surface area for powders using permeability methods. Carman published related experimental work in 1941 (Ref 10). He developed a liquid flow technique to determine surface area for coarse materials by taking into consideration (1) the dependence of permeability on the number of permeable pores of the particle bed; (2) the pore or void volume fraction contribution to total bed volume; (3) the friction of the gas or liquid flowing through the bed; and (4) the adsorption of immobile liquid layers that effectively reduce the capillary diameter, thus causing less permeability and therefore a greater apparent surface area.

Fig. 2 Lea and Nurse permeability apparatus with manometer and flowmeter

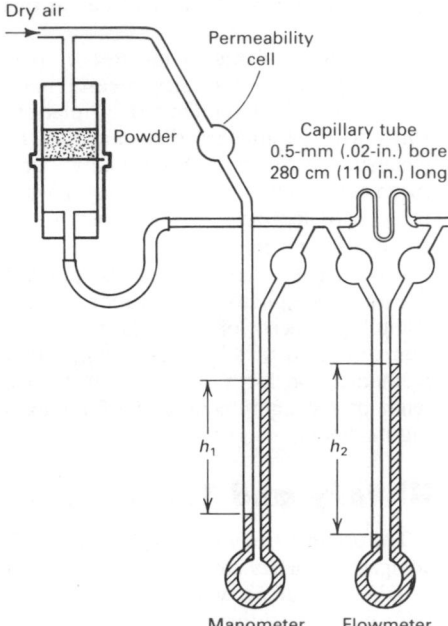

Fig. 3 Schematic of Fisher subsieve sizer operation

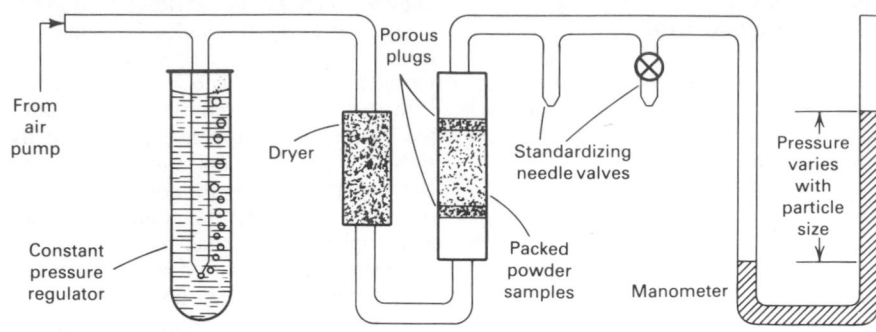

Fig. 4 Blaine air permeability apparatus
Source: Ref 13

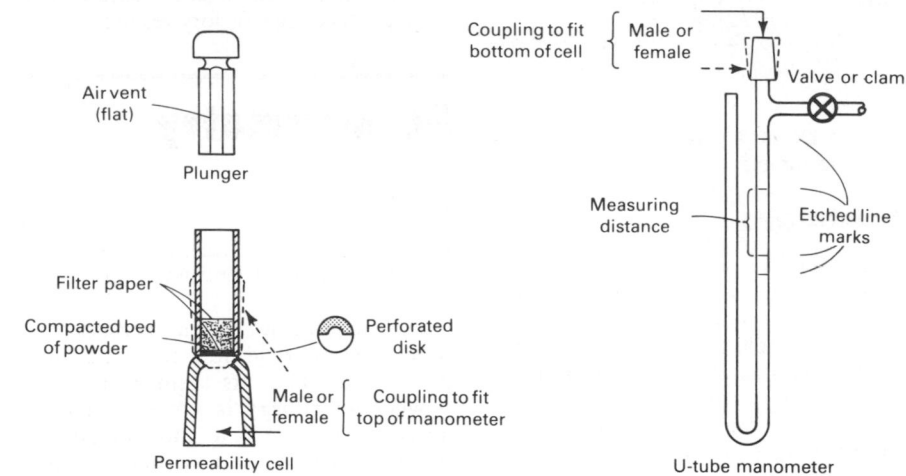

The Kozeny-Carman equation, given below, has served as the most widely used basis for all permeability variations:

$$S^2 = \frac{L^2}{2\eta \upsilon L_e^2 \rho^2} \frac{\Delta P}{L} \frac{f_v^2}{(1 - f_v)^3}$$

where S is surface area per unit weight of the powder; ΔP is the pressure drop across the powder bed; f_v is void fraction of packed sample; υ is velocity of fluid flow; ρ is density of the powder material; η is viscosity of the fluid; L is length of the powder bed; and L_e is average path length through the powder bed.

Apparatus

Lea and Nurse (Ref 11) developed the apparatus shown in Fig. 2 to provide permeability measurements. The powder was compacted in the sample cell to a predetermined porosity. Air was permitted to flow through the bed, and the pressure drop (h_1) was measured on the first manometer; the air then passed through a capillary flowmeter, across which another pressure drop was measured as h_2 on a second manometer.

The capillary permitted the system to operate under a constant pressure. The volume rate of flow through the flowmeter, the pressure drop across the bed as measured by the manometer, and the constants associated with the apparatus permitted determination of the specific surface area (surface area per unit volume).

Gooden and Smith (Ref 12) added a self-calculating chart to a modified Lea and Nurse apparatus to enable direct readout of the specific surface. The commercial version of their modification is known as the Fisher subsieve sizer (Fig. 3).

A simplified version of the air permeameter, known as the Blaine permeameter (Fig. 4), relied on a variable pressure technique (Ref 14). A vacuum was used to displace the oil in a U-tube connected in series with the powder cell. The resultant pressure caused air to flow through the powder bed, and the time required for the displaced oil to fall back to its equilibrium position was measured. This method resulted in a measured specific surface area, which decreased with porosity. Usui (Ref 15) showed that log t and the void fraction exhibited a linear relationship and that a plot of these parameters gave a value for surface area.

Limitations

For very fine powders, the basic Kozeny-Carman equation is not accurate.

This is due to the fact that the laminar flow assumption on which it is based is no longer valid. Compressed fine particles result in a powder bed with very small channel widths. If these widths are comparable to the mean free path length of the gas molecules, laminar flow conditions are not maintained.

Such a situation, involving molecular flow or diffusion conditions, is known as Knudsen flow (Ref 16) and can occur with very fine powders, or with coarser particles at low pressures. Figure 5 shows a typical apparatus used to measure fine particles under molecular flow conditions (Ref 17, 18). In some powders, both laminar and molecular flow may be significant. This is known as the transitional region.

Evaluating surface areas with steady-state flow conditions historically excluded noninterconnected blind pores. A method for including blind pores by utilizing transient-state flow measurements is the principle behind the apparatus shown in Fig. 6 (Ref 19, 20). A typical flow rate curve, showing extrapolation of the

Fig. 5 Modified Pechukas and Gage apparatus for fine powders
Source: Ref 17

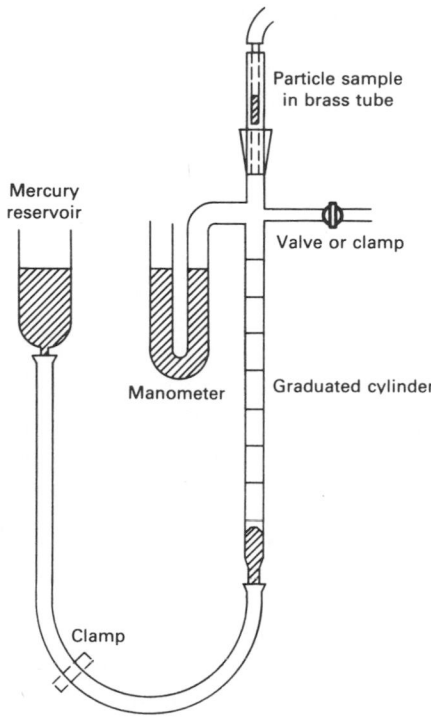

Fig. 6(a) Transient flow apparatus
Source: Ref 2

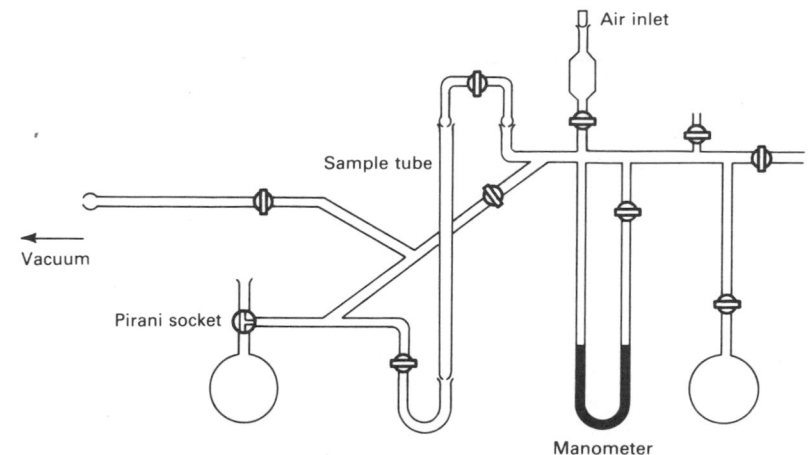

steady-state portion to determine the time lag, is given in Fig. 6(b).

Pycnometry

By Peter J. Heinzer
Coordinator—Technical Services
Imperial Clevite Technology Center

PYCNOMETRY is used to determine the true density of P/M materials. Based on the displacement principle, pycnometry is actually a method of determining the volume occupied by a solid of complex shape, such as a powder sample. For commercial pycnometers (Fig. 7), typical sample sizes range from 5 to 135 cm^3 (0.30 to 8.24 in.^{3}). A properly prepared specimen can be analyzed in 15 to 20 min.

The pycnometric determination of density can be quite useful in P/M applications. In addition to its primary use in measuring the true density of a P/M part or product, it can be used to distinguish among different crystalline phases or grades of material, different alloys, compositions, or prior treatments.

Information on the porosity of a material can be obtained from pycnometry if the sample has a uniform geometry, or if the bulk volume is known. Pore volume is the difference between the bulk volume (1/bulk density) and the specific volume (1/true density). Finally, pycnometry can be useful in determining properties that relate to density. Often, P/M materials have no solid counterpart to use for measuring true density, making percentage of theoretical density measurements questionable. Pycnometric measurements of the true density of the powder has provided a good point of reference.

Density is one of the most important properties of P/M materials. Critical processing parameters, such as applied force and pressure, and properties of the resulting P/M product, such as strength and hardness, usually depend on the density of the materials being processed. Standard industry practice compares the achieved density of a P/M product with the full, or theoretical, density.

Theory and Apparatus

Archimedes devised the first method for determining true density by using the displacement principle. Modern pycnometry represents a refinement of the displacement principle and uses either a liquid or gaseous substance as the displaced medium.

Absolute densities of solids can be measured by the displacement principle using either liquid or gas pycnometry. In liquid pycnometry, volume displacement is measured directly, as liquids are incompressible. Inability of the liquid to penetrate

Fig. 6(b) Flow rate curve for the transient flow apparatus
Source: Ref 2

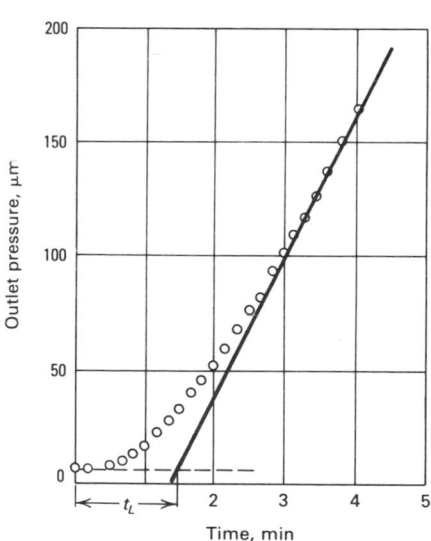

Fig. 7 Gas pycnometer

pores and crevices, chemical reaction or adsorption onto the sample surface, wetting or interfacial tension problems, and evaporation contribute to errors in density measurement. Therefore, gas pycnometry is usually preferred for P/M applications.

In gas pycnometry, volume displacement is not measured directly, but determined from the pressure-volume relationship of a gas under controlled conditions. Gas pycnometry requires the use of high-purity, dry, inert, nonadsorbing gases such as argon, neon, dry nitrogen, dry air, or helium. Of these, helium is recommended, because it:

- Does not adsorb on most materials
- Can penetrate pores as small as 0.1 nm (1 Å)
- Behaves as an ideal gas

In commercial pycnometers, the sample is first conditioned or outgassed to remove contaminants that fill or occlude pores and crevices, thus changing surface characteristics. This is accomplished by evacuating the system and heating to elevated temperatures, followed by purging with an inert gas such as helium.

The helium-filled sample system (Fig. 8) is "zeroed" by allowing it to reach ambient pressure and temperature. At this point, the sample cell and reference volume are isolated from each other and from the balance of the system by valves.

The state of the system can then be defined by:

$$PV = nRT \qquad (Eq\ 1)$$

for the sample cell and:

$$PV_R = n_R RT \qquad (Eq\ 2)$$

for the calibrated reference cell. In these equations, P is ambient pressure, Pa; V is volume of the sealed empty sample cell, cm^3; V_R is volume of a carefully calibrated reference cell, cm^3; n is moles of gas in the sample cell volume at P; n_R is moles of gas in the reference cell volume at P; R is the gas constant; and T is ambient temperature, K.

A solid sample of volume (V_s) is then placed in the sample cell:

$$P(V - V_s) = n_1 RT \qquad (Eq\ 3)$$

where n_1 is moles of gas occupying the remaining volume in sample cell at P. The system is then pressurized to P_2, about 100 kPa (15 psi) above ambient:

$$P_2(V - V_s) = n_2 RT \qquad (Eq\ 4)$$

where n_2 is moles of gas occupying the remaining volume in the sample cell at P_2. The valve is then opened to connect the sample cell with the calibrated reference

volume, and the pressure drops to a system equilibrium level P_3:

$$P_3(V - V_s) + P_3 V_R = n_2 RT + n_R RT \qquad (Eq\ 5)$$

Substituting PV_R from Eq 2 for $n_R RT$ in Eq 5 and substituting $P_2(V - V_s)$ from Eq 4 for $n_2 RT$ results in:

$$P_3(V - V_s) + P_3 V_R = P_2(V - V_s) + PV_R \qquad (Eq\ 6)$$

Simplifying:

$$(P_3 - P_2)(V - V_s) = (P - P_3)V_R \qquad (Eq\ 7)$$

$$V - V_s = \frac{(P - P_3)V_R}{(P_3 - P_2)} \qquad (Eq\ 8)$$

$$V_s = V + V_R/[1 - (P_2 - P)/(P_3 - P)] \qquad (Eq\ 9)$$

Because P is "zeroed" at ambient before pressurizing, the working equation becomes:

$$V_s = V + V_R/[1 - (P_2/P_3)] \qquad (Eq\ 10)$$

Mercury Porosimetry

By A. Ashurst
Senior Metallurgist
SCM Metal Products
and
Erhard Klar
Manager of Particle Technology
SCM Metal Products

PENETRATION OF POROUS SOLIDS by mercury under pressure is used to characterize the pore volume and pore size distribution of interconnected pores ranging from 500 to 0.003 μm in diameter. The porous material to be analyzed is placed into a mercury porosimeter, in which the volume of mercury absorbed by the pores is measured as a function of the pressure applied to the mercury.

The technique has been applied to a variety of porous materials, such as concrete, bricks and tile, activated charcoal, textile fabrics, rubber, leather, paper, wood, minerals, membranes, catalysts, and porous metals and powders. Mercury porosimetry was developed from the need to measure the pore size distribution in porous construction materials with a wide range of pore sizes.

Theory

The basic equation used in mercury porosimetry was derived by Washburn (Ref 21) in 1921 from Young and Laplace's (Ref 22) equation for the rise or depression of liquids in capillaries:

$$P = \frac{-4\gamma \cos \theta}{D} \qquad (Eq\ 11)$$

where P is the pressure (Pa) required to force the mercury through a pore of diameter D (m); θ is the contact angle between the mercury and the porous material; and γ is the surface tension (N/m) of the mercury.

This equation indicates that the pressure required to force mercury into a pore depends on the surface tension of the mercury, the contact angle, and pore size. Mercury is used as the intrusion liquid, because it does not wet most materials (contact angle $180° > \theta > 90°$), a fundamental requirement of the method.

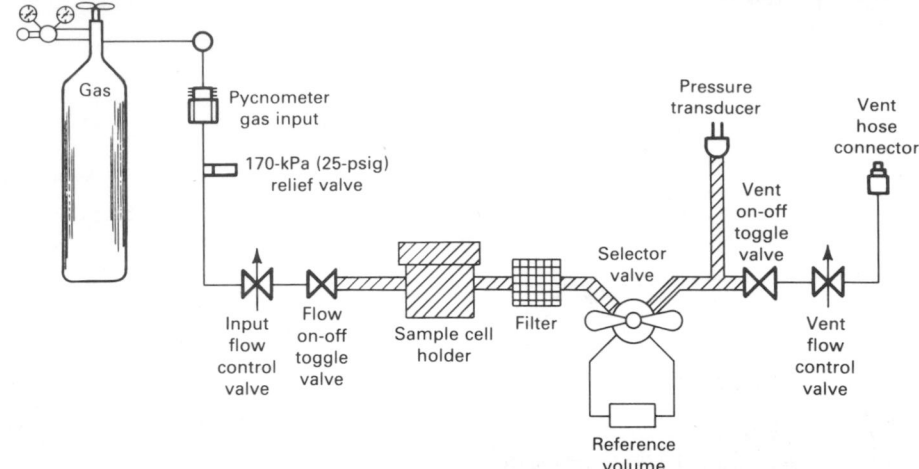

Fig. 8 Flowchart for typical pycnometer

Fig. 9(a) Direct-reading mercury porosimeter
Source: Ref 23

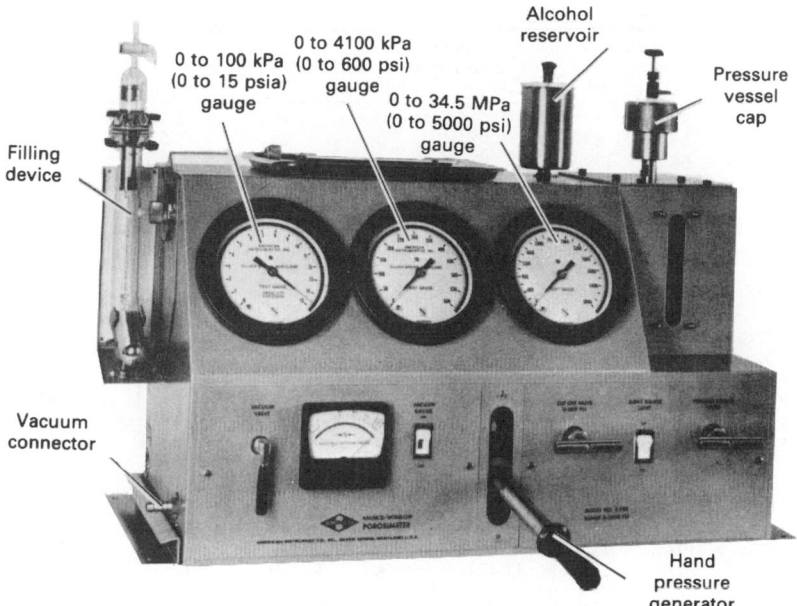

Fig. 9(b) Filling device for typical porosimeter
Source: Ref 23

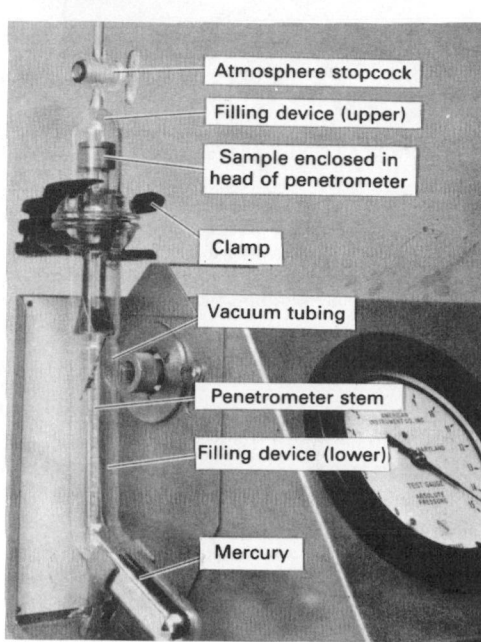

Fig. 10 Flowchart of pressure system for typical porosimeter
Source: Ref 23

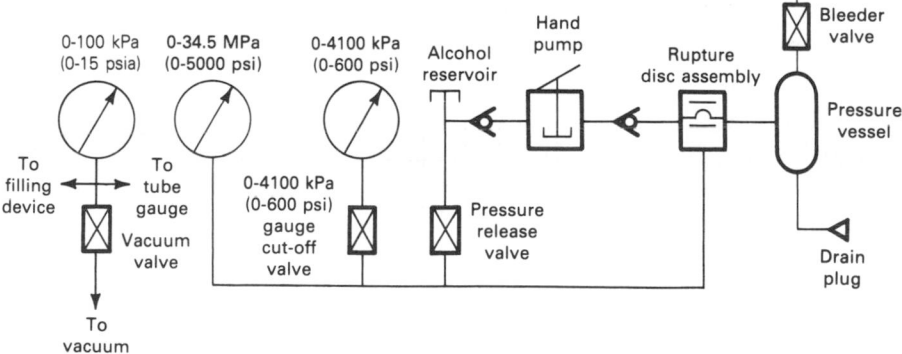

Principles of Operation

To develop a pore size distribution, a weighed sample of powder or porous solid material is placed in the sample space of the penetrometer (a glass tube with graduated capillary stem). The penetrometer assembly is then placed into the mercury filling device. After evacuation, the penetrometer is filled with mercury so that the sample is completely covered. The system is now ready for the analysis. Typical direct-reading mercury porosimeters are shown in Fig. 9 and 10. Sample cells, or penetrometers, for solid and powder samples are shown in Fig. 11.

The volume of the coarse pores (100 to 17 μm) is measured by volume changes in the mercury in the capillary stem as the pressure on the mercury is increased from the evacuated state to atmospheric pressure. The penetrometer is then transferred to the hypobaric pressure chamber. The volume of smaller pores is determined by the change in volume in the penetrometer stem as the pressure on mercury in the sample increases from atmospheric to the pressure limit of the instrument.

Typical pressure ranges for commercial porosimeters are 35, 205, and 410 MPa (5, 30, and 60 ksi). A pressure of 410 MPa (60 ksi) corresponds to a pore size of 0.003 μm. Low-pressure porosimeters for analysis of materials with large pores increase pore size analysis capability to about 500 μm for coarse particles.

Typical pore size distribution data from mercury porosimeter measurements are given in Table 1 for an atomized-iron powder compacted to a density of 6.6 g/cm³. The volume of mercury forced into the porous iron compact is monitored continuously as the absolute pressure is increased from 17 to 34 500 kPa (2.5 to 5000 psi).

The total pore volume measured (0.132 cm³, or 0.008 in.³) is indicated by the last reading at 34 500 kPa (5000 psi). Because pore sizes usually range over several orders of magnitude, data are plotted on semilog paper, as shown in Fig. 12.

Measuring Displacement Volumes

Mercury volume displacements may be measured by direct visual observation of the mercury level in a glass penetrometer stem with graduated markings. Remote methods are also used, including:

● *Precision capacitive bridges:* Measure changes in the capacitance between the column of mercury in a dilatometer stem

Fig. 11 Penetrometer assembly

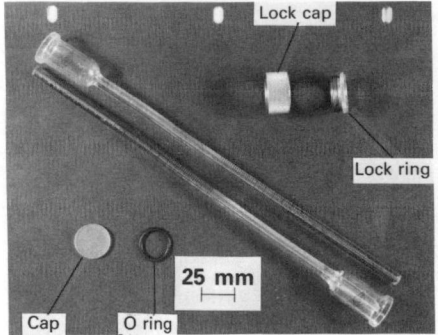

and coaxial sheath surrounding the column
- *Mechanical transducers:* Indicate the change in height of the mercury column in the stem
- *Submerged wires:* Measure change in resistivity corresponding to the change in length of the mercury column

Reliability

Pore size distribution data generated by mercury porosimetry are primarily useful in comparative studies of similar materials. This is true because the accuracy of the data depends on various assumptions and experimental factors, some of which cancel out for relative comparisons of similar materials. However, it has been shown that, with certain precautions and corrections, mercury porosimetry data have sometimes shown satisfactory agreement with data obtained from other methods. Some of the major precautions and corrections are discussed below.

Contact Angle. The contact angle in Eq 11 between mercury and the sample being tested is frequently assumed to be 130° for porosimetry calculations. This assumption is probably the largest source of error, as contact angles of different materials may differ significantly, as shown in Table 2.

Because pore size according to Eq 11 is directly proportional to the cosine of the contact angle, the respective pore size errors for iron ($\theta = 115°$) and glass ($\theta = 153°$), using the values from Table 2 versus a constant value of 130° for γ, would be:

$$52\% = \frac{\cos 130 - \cos 115}{\cos 115} \times 100$$

and

$$28\% = \frac{\cos 153 - \cos 130}{\cos 153} \times 100$$

Table 1 Mercury porosimetry determinations on Ancorsteel 1000 powder

Compacted to a density of 6.60 g/cm^3

Mercury head pressure kPa	psi	Pressure data Gauge reading (0 to 100 kPa, or 0 to 15 psi) kPa	psi	Pressure gauge reading kPa	psi	Total absolute pressure(a) kPa	psi	Penetrometer stem readings, cm^3
31.37	4.55	48.3	7	0	0 ...... 17.2		2.5	0.004
30.34	4.40	55.2	8	0	0 ...... 24.8		3.6	0.010
29.65	4.30	68.9	10	0	0 ...... 39.3		5.7	0.016
29.65	4.30	82.7	12	0	0 ...... 53.1		7.7	0.019
28.96	4.20	101.4	14.7	0	0 ...... 72.4		10.5	0.021
28.96	4.20	103	15	68.9	10 ...... 144.8		21	0.024
28.96	4.20	103	15	206.8	30 ...... 282.7		41	0.026
28.96	4.20	103	15	413.7	60 ...... 489.5		71	0.028
28.61	4.15	103	15	689.4	100 ...... 765.3		111	0.032
27.58	4.0	103	15	900	130 .... 972.2		141	0.037
27.58	4.0	103	15	1 030	150 .....1 110.0		161	0.040
27.58	4.0	103	15	1 380	200 .....1 454.8		211	0.057
27.58	4.0	103	15	2 410	350 .....2 490.0		361	0.093
27.58	4.0	103	15	3 100	450 ....3 178		461	0.100
27.58	4.0	103	15	4 136	600 ....4 213		611	0.106
27.58	4.0	103	15	6 890	1000 ....6 971		1011	0.112
27.58	4.0	103	15	8 270	1200 ....8 350		1211	0.118
27.58	4.0	103	15	9 650	1400 ....9 728		1411	0.120
27.58	4.0	103	15	12 410	1800 ...12 486		1811	0.123
27.58	4.0	103	15	16 550	2400 ...16 623		2411	0.126
27.58	4.0	103	15	22 060	3200 ...22 140		3211	0.128
27.58	4.0	103	15	34 470	5000 ...34 550		5011	0.132

(a) Absolute pressure is the sum of the following pressure components: reading on 0 to 100 kPa (0 to 15 psi) upper gauge; mercury head pressure (subtract); reading on the pressure gauge (applicable only while penetrometer is inserted in the pressure chamber).

Fig. 12 Mercury porosimetry curve for Ancorsteel 1000 powder

Compacted to a density of 6.60 g/cm^3

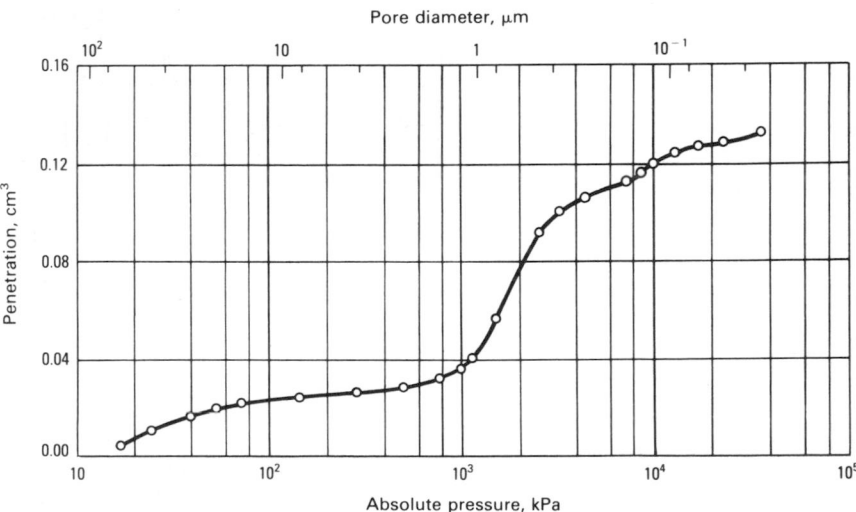

Compressibility of Mercury. Due to the slight compressibility of mercury, the measured pore volume of a porous material appears larger than its actual volume. Accurate corrections must be made to compensate for the change in the compressibility of mercury with pressure (Fig. 13) and the total amount of mercury in the penetrometer. Consequently, the larger the sample and pore volume of the sample in comparison to the amount of mercury in the penetrometer, the smaller the error from this source.

Table 3 lists percentage errors due to mercury compression for several penetrometers, for the maximum pressure of the instrument 34 500 kPa (5000 psi) and without a sample in the bulb of the penetrometer. A pressure of 413 700 kPa (60 000 psi) increases the error to over 40% for the 6.1-mL (0.2-fluid oz) penetrometer.

Table 2 Contact angle between mercury and select P/M materials determined by indirect measuring

Powder	Contact angle (θ), degrees
Aluminum	140
Copper	116
Glass	153
Iron	115
Tungsten	135
Tungsten carbide	121
Zinc	133

Source: Ref 24

Fig. 13 Isothermal compressibility of mercury at 20 °C (68 °F)
Source: Ref 25

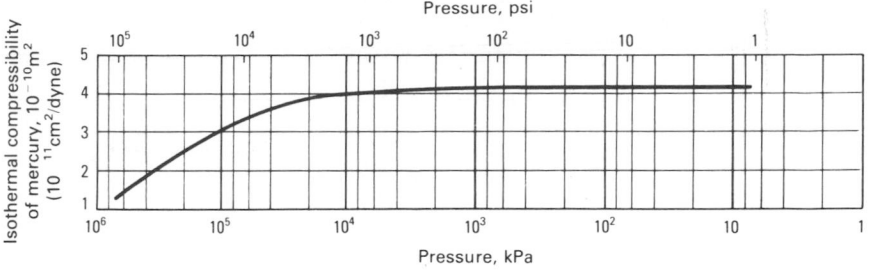

Constrictive Pores. Another source of inaccuracy involves the intrusion of mercury into a large pore through a much smaller constriction (neck), or what is sometimes termed an "ink-bottle" pore. In this case, the instrument measures the volume of the large pore as a correspondingly greater number of pores the size of the constrictive pore. Therefore, the pore size distribution curve is biased toward small pore sizes. Many P/M materials, whether tested as loose powder or as pressed and sintered parts, contain constrictive pores; therefore, pore size distribution curves may be skewed.

Another example of this phenomenon may be seen in certain atomized spherical hollow particles. These may contain a few small pores in the surface, which lead to cavernous volumes inside the particle.

Mercury extrusion from pores upon release of pressure provides an indication as to the amount of constrictive pores contained in the powder sample and the pore volume present behind these constrictions. For more information on this technique, see Ref 26.

Surface Tension of Mercury. Values for the surface tension of mercury can vary with atmosphere, temperature, and purity of the mercury used. The effect of temperature is minimal, because the temperature coefficient of the surface tension of mercury is only 2.1×10^{-4} N/m $\cdot$ °C (0.21 dynes/cm $\cdot$ °C).

The purity of mercury has a significant effect on surface tension. Reported values vary by about 0.1 N/m (100 dynes/cm). The lower values are reportedly due to contamination of the mercury.

Entrapped Air. Residual air in the penetrometer, in particle pores, and adsorbed air on particle surfaces may lead to minor errors in reported values (Table 3). These errors can be reduced by heating the powder during evacuation of the penetrometer bulb. Depending on the volume of the bulb holding the test sample, errors

Table 3 Selection of glass penetrometers

Penetrometer selection	Type sample	Bulb volume mL	fluid oz	Porosity range(a), %	Mercury compression(b), %	Entrapped air(c), % of full scale
1	Solid	1.68	0.06	12-100	1.3	1.1
2	Powder	1.68	0.06	12-100	1.3	1.1
3	Solid	1.68	0.06	1.2-100	11	9.4
4	Powder	1.68	0.06	1.2-100	11	9.4
5	Solid	6.1	0.21	3.3-100	4.0	3.5
6	Powder	6.1	0.21	3.3-100	5.0	3.5

Note: Stem graduations (100 total divisions) range from 0 to 0.2 mL (0 to 0.007 fluid oz). (a) Full scale, depends on sample size. For example, both a 0.5-mL (0.017-fluid oz) sample of 40% porosity and a 1.7-mL (0.06-fluid oz) sample of 12% porosity will yield full-scale readings (0.2 mL or 0.007 fluid oz) in penetrometer No. 4. (b) Full scale at 34 500 kPa (5000 psi). Assumes no sample in bulb. (c) Assumes no sample in bulb. Values are based on the residual air that occupies the entire volume of the penetrometer when it is evacuated to 100 μm, and the air, upon filling the penetrometer with mercury at 12.1 kPa (1.75 psi), is subsequently compressed by a factor of 90 500 times. If large pores are of no interest, penetrometer can be filled at atmospheric pressure (70.3 kPa, or 10.2 psi sample pressure).

in pore volume distribution due to entrapped air are generally less than 10%.

Other Factors. The so-called kinetic hysteresis effect is related to the time that is required for the mercury to flow into pores. Volume penetration readings taken before equilibrium has been reached may result in a shifting of the distribution toward smaller pore sizes.

Fragile porous materials may be subject to a breakdown of pores during pressurization. Knowledge of the compressibility or fracture strength of the material to be analyzed is desirable to properly estimate whether mercury intrusion will occur before deformation or fracture of the porous material occurs.

Equation 11 applies to cylindrically shaped pores. Most porous materials, however, have irregularly shaped pores. For corrections based on a toroidal void model, see Ref 27.

Applications

Mercury porosimetry measures only interconnected porosity. This may be a large or small percentage of the total porosity of the part, depending on powder characteristics, fabrication methods, and so forth. An alternative determination of density should be used with the mercury intrusion method to distinguish interconnected porosity from total void volume.

Fig. 14 Effect of compacting pressure on porosities of compacts made from porous copper powders
Void, intraparticle porosities, and average intraparticle pore diameters (values in graph refer to average pore diameters of uncompacted powder) were determined by mercury porosimetry. ○, 0.18 μm; ●, 10 μm; □, 1.2 μm

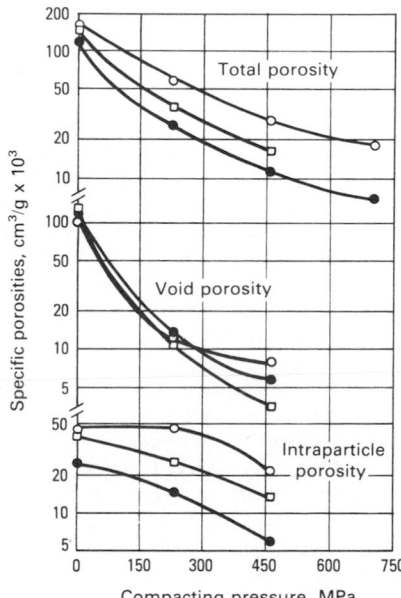

Fig. 15 Mercury porosimetry curves of sintered steel parts of varying densities

Green skeletons were sintered at 1093 °C (2000 °F) for 20 min. Total porosity (ϕ) is determined from sample weight and dimensions.

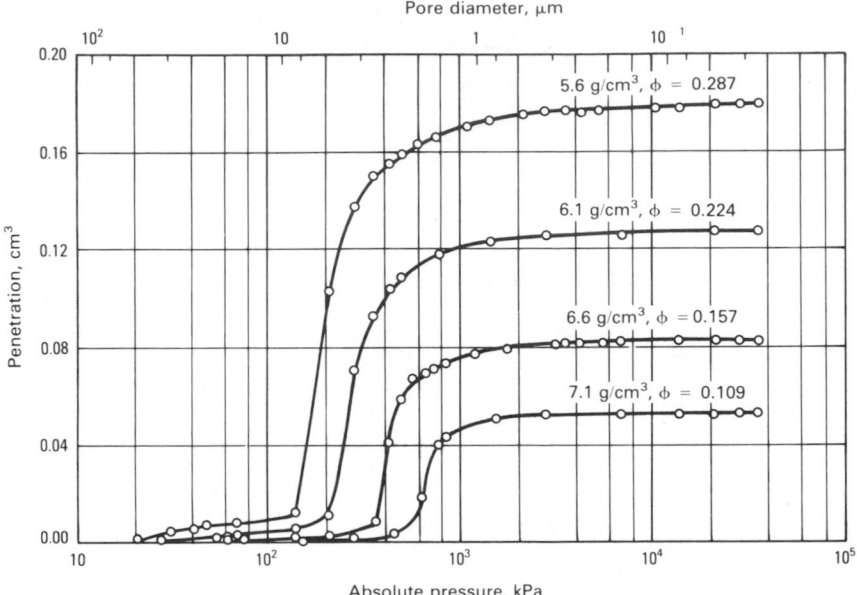

The mercury porosimetry method has been used for the evaluation of particle size, surface area, and contact angle (Ref 24). Volume, pore size, and pore size distribution may have a significant effect on the performance of P/M products such as structural parts, bearings, filters, fuel cells, energy absorbers, sound attenuators, catalysts, and prosthetic devices. A narrow pore size distribution in filters ensures reproducible permeability and high filter efficiency.

A fine pore size distribution in structural parts provides greater strength compared to parts with the same pore volume composed of coarser pores. Permanent, natural fixation of bone implants is more successful if the surface of the prosthetic device contains coarse pores that are greater than 100 μm in diameter (Ref 28, 29).

Iron and copper powders manufactured by the oxide-reduction method yield porous or spongy powders. Such powders are utilized for their high green strength characteristics. In many applications, however, compressibility is also desirable. Figure 14 shows the effect of internal particle pore size of spongy copper powders on compressibility (Ref 30). In this case, specific porosity is inversely proportional to compressibility. Mercury porosimetry has proved to be an effective tool for separating interparticle voids and internal particle porosity. It also quantifies the detrimental effect of small pores on compressibility.

Figure 15 shows mercury porosimetry curves of sintered steel parts of varying densities. The entry pore sizes decrease with increasing part density. This characteristic, in combination with other pore characteristics, results in high penetrating capillary pressure, thus explaining the greater ease with which dense parts can be infiltrated with liquid copper (Ref 31).

An obvious limitation of this technique to P/M applications is that powder materials that alloy with mercury should not be used. The user should also remember that the closed pores in the sintered materials do not contribute to the measurement of the pore size distribution.

REFERENCES

1. Lenel, F.V., *Powder Metallurgy—Principles and Applications*, chapt 3, Metal Powder Industries Federation, Princeton, NJ, 1980
2. Allen, T., Permeametry, in *Particle Size Measurement*, Chapman and Hall, London, 1968
3. Callis, C.F. and Irani, R.R., Miscellaneous Techniques, in *Particle Size: Measurements, Interpretation, and Application*, John Wiley & Sons, New York, 1963
4. "Permaran Specific Surface Area Meter," Outokumpu Oy, Instrument Division, Tapiola, Finland, 1973
5. D'Arcy, H.P.G., *Les Fontaines Publiques de la Ville de Dijon*, Victor Dalmont, Paris, 1856
6. Kozeny, J., *Akad. Wiss. Wein. Math. Naturwiss. K.I.*, Sitzungsbor., Abstr. IIA, Vol 136, 1927, p 271-306
7. Blake, F.C., *Trans. Am. Inst. Chem. Eng.*, Vol 14, 1922, p 415
8. Carman, P.C., *J. Soc. Chem. Ind. Lond.*, Vol 57, 1938, p 225-234; *Trans. Inst. Chem. Eng.*, Vol 15, 1937, p 150-166
9. Dallavalle, J.M., *Chem. Met. Eng.*, Vol 45, 1938, p 688
10. Carman, P.C., ASTM Symposium on New Methods for Particle Size Determination in the Sub-sieve Range, American Society for Testing and Materials, Philadelphia, 1941, p 24
11. Lea, F.M. and Nurse, R.W., *J. Soc. Chem. Ind.*, Vol 58, 1939, p 277-283; Symposium on Particle Size Analysis, *Trans. Inst. Chem. Eng.* (suppl.), Vol 25, 1947, p 47-56
12. Gooden, E.L. and Smith, C.M., *Ind. Eng. Chem. Anal. Ed.*, Vol 12, 1940, p 479-482
13. Niesel, K., External Surface of Powders From Permeability Measurements—A Review, in *Silicates Industrials*, 1969, p 69-76
14. Blaine, R.L., ASTM Bull. No. 123, 1943, p 51-55; also see ASTM Bull. No. 108, 1941, p 17-20
15. Usui, K., *J. Soc. Mat. Sci. Jpn.*, Vol 13, 1964, p 828
16. Knudsen, M., *Ann. Physik*, Vol 28, 1909, p 75-130
17. Pechukas, A. and Gage, F.W., *Ind. Chem. Eng. Anal. Ed.*, Vol 18, 1946, p 37
18. Carman, P.C. and Malherbe, P.R., *J. Soc. Chem. Ind.*, Vol 69, 1950, p 134
19. Barrer, R.M. and Grove, D.M., *Trans. Faraday Soc.*, Vol 47, 1951, p 826, 837
20. Kraus, G., Ross, R.W., and Girifalco, L.A., *J. Phys. Chem.*, Vol 57, 1953, p 330
21. Washburn, E.W., Note on a Method of Determining the Distribution of Pore Sizes in a Porous Material, *Proc. Nat. Acad. Sci.*, Vol 7, 1921, p 115-116
22. Adamson, A.W., *Physical Chemistry of Surfaces*, 2nd ed., Interscience, New York, 1967, p 546-549
23. Aminco Porosimeter Instruction Manual No. 877-A, American Instrument Co., Silver Springs, MD, July 1969
24. Rootare, H.M., A Review of Mercury Porosimetry, *Advanced Techniques in Powder Metallurgy, Perspectives in Powder Metallurgy*, Vol 5, Plenum Press, New York, 1970
25. Weast, R.C., Ed., *Handbook of*

Physics and Chemistry, 47th ed., CRC Press, Boca Raton, FL, 1967

26. Lowell, S., *Powder Technology,* Vol 25, 1980, p 37-43
27. Frevel, L.K. and Kressley, L.J., Modifications in Mercury Porosimetry, *Anal. Chem.,* Vol 35, 1963, p 1492-1501
28. Pilliar, R.M., MacGregor, D.C.,

Macnob, I., and Cameron, H.U., *Modern Developments in Powder Metallurgy,* Vol 11, Metal Powder Industries Federation, Princeton, NJ, 1976, p 263-278

29. German, R.M., Porous Materials, *Advances in Powder Technology,* Chin, G.Y., Ed., American Society for Metals, 1982

30. Klar, E. and Petrosh, A., Pore Structure and Compressibility of Iron Powders, *J. Mat.,* Vol 7 (No. 1), 1972, p 57-59
31. Chen, P.C. and Klar, E., New Copper Infiltrating Powders, *Modern Developments in Powder Metallurgy,* Vol 13, Metal Powder Industries Federation, Princeton, NJ, 1981, p 371-384

Apparent Density of Metal Powders

By the ASM Committee on Apparent Density*

APPARENT DENSITY of a metal powder, or the weight of a unit volume of loose powder expressed in grams per cubic centimetre, is one of the fundamental properties of a powder. This characteristic defines the actual volume occupied by a mass of loose powder, which directly affects processing parameters such as the design of compaction tooling and the magnitude of the press motions required to compact and densify loose powder.

In most compacting operations, dies are filled by volume measure, and presses operate either to a fixed position or a fixed pressure. If the press operates to a fixed position, pressure can be maintained at a constant level only if the apparent density of the powder does not change. If, however, the press operates to a fixed pressure, consistency in apparent density is necessary to ensure compacts of equal height. Small fluctuations in apparent density can be compensated for by adjustments of pressure or stroke of the

presses, but large-scale compacting requires that the apparent density of the powder be controlled within close limits.

Factors Affecting Apparent Density

Apparent density of a metal powder depends on the density of the solid material, particle size, particle size distribution, particle shape, surface area and roughness of individual particles, and particle arrangement. Apparent density is strongly affected by particle size. It generally (1) decreases with decreasing particle size, (2) decreases as the particle shape becomes less spherical and more irregular, (3) decreases with increasing surface roughness,

and (4) is frequently controlled by mixing various sizes of particles.

Particle Size. Decreasing particle size generally decreases apparent density. The smaller the particles, the greater the specific surface of the powder. This phenomenon increases the friction between particles and subsequently decreases the apparent density. Powder particles that exhibit very low friction because of their rounded shape, such as gas-atomized (spherical) stainless steel powder, do not demonstrate this characteristic. The effect of decreased particle size on density is particularly significant for particle sizes of less than 20 μm. Table 1 shows the effect of particle size on apparent density for several metal powders.

Particle Shape. As particle shape becomes less spherical, apparent density de-

Fig. 1 Effect of particle shape on apparent density of a metal powder

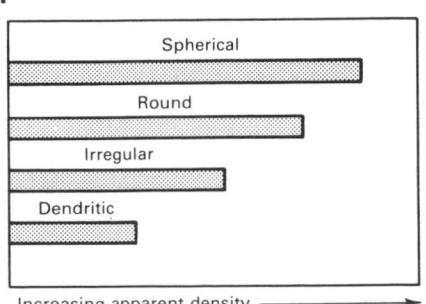

Increasing apparent density ⟶

Fig. 2 Effect of three different shapes of −325 mesh powder addition to a +325 mesh distribution on apparent density of 316 stainless steel powder

Source: Hirschhorn, J.S., *Introduction to Powder Metallurgy*, American Powder Metallurgy Institute, Princeton, NJ, 1969

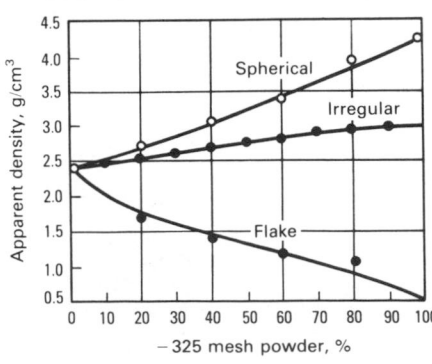

Fig. 3 Hall flowmeter

*Leo A. Adams, Technical Director, Metallurgical Industries, Inc.; Norbert A. Arnold, Manager, Powdered Metal Engineering, Keystone Carbon Co.; C.B. Thompson, Copper Products Manager, SCM Metal Products

Table 1 Effect of particle size on apparent density for several metal powders

Material	Average particle diameter(a), μm	Apparent density, g/cm³
Aluminum		
Atomized	5.8	0.62
	6.8	0.75
	15.5	0.98
	17.0	1.04
	18.0	1.09
	60% above 44 (+325 mesh)	1.22
	75% above 44 (+325 mesh)	1.25
Copper		
Electrolysis	90% min, −325 mesh	1.5-1.75
Hydrometallurgical	81.9%, −325 mesh	1.69
Oxide reduced	95% min, −325 mesh	2.10-2.50
Hydrometallurgical	49.1%, −325 mesh	2.42
Oxide reduced	50-65%, −325 mesh	2.65-2.85
Electrolysis	60-75%, +100 mesh	4.0-5.0
Atomized	70% min, −325 mesh	4.5-5.1
Atomized	50-60%, −325 mesh	4.9-5.5
Nickel		
Carbonyl	3.2	0.61
Precipitation	3.5	1.81
Carbonyl	3.8	1.87
Carbonyl	4.1	2.10
Precipitation	4.4	2.09
Precipitation	8.0	2.60
Precipitation	−40+325 mesh	3.60
Tungsten		
Oxide reduced	1.20	2.16
	2.47	2.52
	3.88	3.67
	6.85	4.40
	26.00	10.20
Stainless steel		
Atomized, spherical	−325 mesh	4.3
	−270+325 mesh	4.5
	−200+270 mesh	4.4
	−150+200 mesh	4.5
	−100+150 mesh	4.5
Iron		
Reduced	6	0.97
Carbonyl	7	3.40
Reduced	51	2.19
Electrolytic	53	2.05
Electrolytic	63	2.56
Reduced	68	3.03
Electrolytic	78	3.32

(a) From Fisher subseive sizer for single values and screens for size fractions

Fig. 4 Dimensions for apparent density equipment

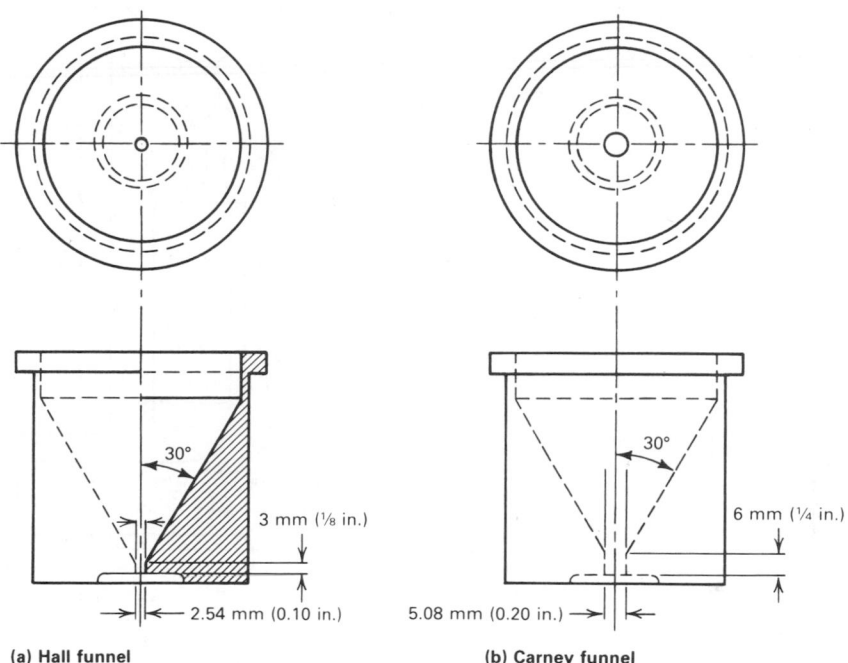

(a) Hall funnel   (b) Carney funnel

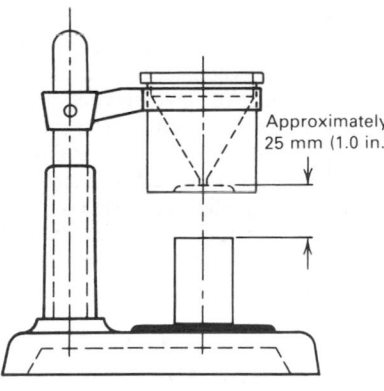

(c) Equipment assembly   (d) Density cup

creases, due to both the increase in frictional surface area and less uniformity of powder particles during packing. Spherical powders, which are normally produced by atomizing, frequently have high apparent densities, about 50% of the density of the wrought metal. Spheres are most likely to pack without bridging or arching to create empty spaces; they tend to move easily past each other because of their smooth surfaces. At the other extreme in particle shape are flake powders, which often have apparent densities less than 10% of the wrought density. These powders are useful primarily as pigments, because their low apparent density aids in obtaining mixtures in paint.

Most powders used for compacting have irregular, somewhat equiaxed particle shapes with apparent densities that fall in the range between those of spherical and flake powders. Apparent densities of these particles range from 25 to 35% of the wrought density of the metal. Figure 1 illustrates the effect of particle shape on apparent density.

Surface Roughness. Decreasing surface area-to-volume ratios and decreasing surface roughness tends to reduce frictional forces between settling particles. This tendency thus increases apparent density by allowing the particles to move more ef-

Fig. 5 Scott volumeter

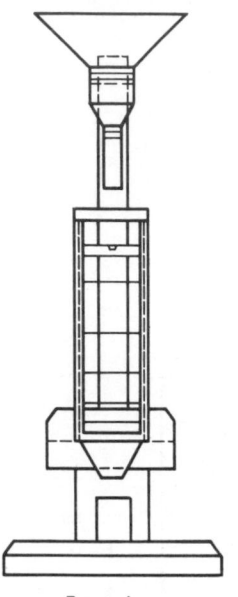

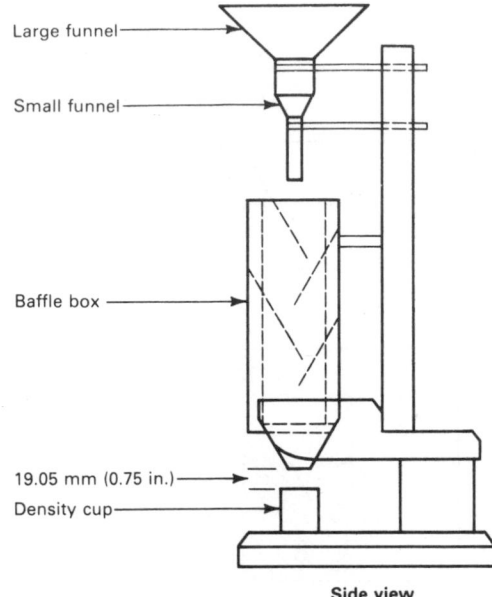

Large funnel

Small funnel

Baffle box

19.05 mm (0.75 in.)

Density cup

Front view

Side view

Fig. 6 Arnold meter

Table 2   Effect of mixture of spherical coarse and fine stainless steel particles on apparent density

Particle size (mesh)	Particles, %					
−100+150	100	80	60	40	20	⋯
−325	⋯	20	40	60	80	100
Apparent density, g/cm³	4.5	4.9	5.2	4.8	4.6	4.3

Table 3   Apparent densities and flow rates of electrolytic iron powders of three particle size distributions

Particle size (mesh)	Particles, %		
	Powder A	Powder B	Powder C
+100	4	3	15
−100+150	11	26	10
−150+200	18	18	30
−200+250	16	6	25
−250+325	18	16	5
−325	33	31	15
Apparent density, g/cm³	2.6-2.8	3.2-3.4	3.8-3.9
Flow rate, s/50 g	29	24	20

fectively to fill the free spaces between previously settled particles.

Particle Size Distribution. An effective way to increase the apparent density of a powder is to fill the spaces between particles with smaller particles. Figure 2 shows the effects of adding differently shaped −325 mesh powder to a standard +325 mesh blend of stainless steel pow-

der. Table 2 shows this effect for mixtures of fine and coarse spherically shaped stainless steel powders, where a mixture of about 60% coarse and 40% fine particles is optimal.

The addition of fine spherical powder effectively increases apparent density, while the opposite is true of flake powders. Distribution of a variety of particle sizes greatly affects apparent density. The relative amount of coarsest and finest particles and the percentage of particles between the two extremes determine apparent density. An example of this is shown in Table 3 for three particle size distributions.

Hall Flowmeter and Carney Funnel

The most common method for determining apparent density of metal powders uses the Hall flowmeter (Fig. 3). Both American Society for Testing and Materials standard ASTM B 212 and Metal Powder Industries Federation standard MPIF 04 describe this determination.

Critical equipment dimensions are illustrated in Fig. 4(a), (c), and (d). Apparent density determinations are made by pouring powder into the funnel and allowing it to flow into the 25-cm³ (1.5-in.³) density cup. After the cup is filled, the funnel is moved away and the excess powder is carefully leveled off using a spatula or straight edge. Care must be exercised to prevent physical densification of the powder in the cup when leveling. The apparent density in grams per cubic centimetre

is then determined by weighing the powder in the cup in grams and dividing by 25 cm³ (1.5 in.³) (cup volume).

For powders that do not flow freely, a second method, described in ASTM B 417 and MPIF 28, has been devised. This is similar to the Hall flowmeter procedure, except that a different funnel, the Carney funnel, which has an orifice diameter twice that of the Hall funnel, is used (see Fig. 4a and b). This larger opening permits a greater variety of powders to flow. Powders that do not flow readily can be freed by poking a wire up and down in the hole. The wire must not enter the density cup at any time. This second method is fast and correlates well with the Hall flowmeter evaluation of free-flowing metal powders.

A wire is not used with the Hall flowmeter funnel, because it may scratch the orifice and ruin the calibration of the funnel for flow tests. The Carney funnel is often used when measuring apparent density of lubricated powders, because lubricant adhering to the smaller orifice of the Hall funnel temporarily affects the calibration of the Hall flowmeter.

Scott Volumeter

Another instrument frequently used for determining apparent density is the Scott volumeter, described in ASTM B 329, which was originally developed by Scott, Schaeffer, and White for the determination of the density of dry pigment for paint. As shown in Fig. 5, the device consists of:

● A *large brass funnel* with a metal screen and a smaller funnel with a straight stem for directing the powder into the baffle box
● A *baffle box* with glass sides and two wooden sides containing a series of glass baffle plates and a funnel at the bottom to collect the powder and direct it into the density cup
● A *square density cup* with a capacity of 16.4 ± 0.032 cm³ (1 ± 0.002 in.³) or a

Fig. 7 Arnold meter steel block dimensions

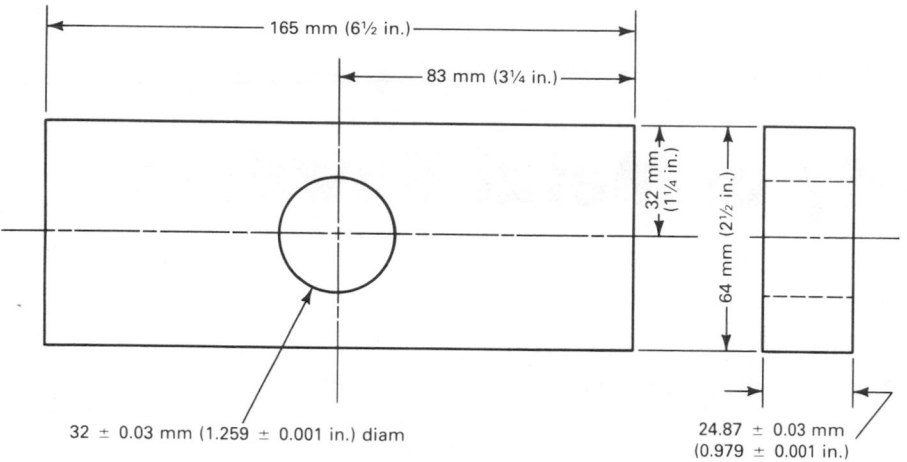

32 ± 0.03 mm (1.259 ± 0.001 in.) diam

24.87 ± 0.03 mm
(0.979 ± 0.001 in.)

cylindrical cup with a capacity of $25.00 \pm 0.05\,cm^3$ $(1.5 \pm 0.003\,in.^3)$ with an inside diameter of 30.00 ± 2.00 mm $(1.2 \pm 0.08$ in.)

- A *stand* to support the funnels and baffle box concentric with the density cup, so that the bottom of the baffle box funnel is 19 mm (0.75 in.) above the top of the density cup when the apparatus is assembled
- A *level, vibration-free base* to support the funnels and baffle box
- *An analytical balance* having a capacity of at least 100 g (3.5 oz) and a sensitivity of 0.1 g (0.0035 oz)

Operating Procedure. The test specimen is carefully poured into the funnel. Ultrafine powders may require light brushing with a nylon brush to initiate powder flow through the screen in the funnel. Powder is allowed to run into the density cup until it completely fills and overflows the periphery of the cup. The funnel and baffle box should then be rotated approximately 90° in a horizontal plane to clear the cup.

Excess material should be removed from the cup by passing a spatula blade in flat contact with the top of the cup. The spatula is moved smoothly back and forth along the top of the cup until all excess powder has been removed. When insufficient powder is left for the first reverse pass to smooth the surface completely, powder on the spatula should be gently replaced on top of the cup. The spatula must be kept level at all times to prevent packing or pulling out of the powder.

After the leveling operation, the side of the density cup should be tapped lightly to settle the powder to avoid spilling. The powder is transferred to a balance and weighed to the nearest 0.1 g (0.0035 oz). The density of the powder in the density cup is given in grams per cubic centimetre if a metric cup is used, or grams per cubic inch if a nonmetric cup is used.

Arnold Meter

Another device developed to determine the apparent density of metal powders is the Arnold meter, described in ASTM B 703 and MPIF 48, which is designed to duplicate the action of a feed shoe filling a die cavity on a P/M press. A hardened, fully demagnetized steel block with a round hole having a volume of 20 cm^3 (1.2 in.^{3}) is placed on a sheet of glazed paper. A bronze bushing filled with powder is slid across the hole. The powder collected in the hole is then removed and weighed. Apparent density is calculated by dividing the weight by the volume of the hole.

Apparent densities obtained with this procedure, which takes less than 5 min to perform, are quite close to those values measured on powder that has filled a die cavity from a filling shoe in an automatic compacting press. The Arnold meter is shown in Fig. 6, and the dimensions of the steel block are given in Fig. 7.

Tap Density of Metal Powders

By Willard E. Soper
Metallurgical Quality Control Manager
Adamas Carbide Corporation

TAP DENSITY is defined as the density of a powder when the volume receptacle is tapped or vibrated under specified conditions. Tapping or vibrating a loose powder induces movement and separation and lowers the friction between the powder particles. This short-term lowering in friction results in powder packing and in a higher calculated density of the powder mass. Tap density is always higher than the free-flow apparent density.

Tap density is a function of particle shape, particle porosity, and particle size distribution. It is commonly included as a control specification for metal powder, but is used in other industrial applications as a practical measure of the degree of powder packing that occurs in containers.

The amount of increase from apparent to tap density depends to a great extent on particle shape. Table 1 compares the density increases for three types of copper powders. Usually, the lower the apparent density, the higher the percentage increase in density on tapping.

Equipment and Test Procedures

Three pieces of equipment are needed to determine tap density:

- A balance with the capacity of weighing up to 100 g with an accuracy of 0.1 g
- A graduated glass cylinder with a capacity of 100 mL and an accuracy of 0.2 mL. A smaller graduated cylinder may be used for high-density powders.
- A mechanical apparatus, such as the Tap-Pak volumeter, capable of tapping the

graduated cylinder at a rate of 100 to 250 impacts per minute or, alternatively, a hard rubber slab approximately 100 × 100 × 5 mm (4 × 4 × $^1/_4$ in.)

To determine tap density, a standard weight (usually 50 g) of powder is weighed to ±0.01 g. The powder is poured into a clean, dry graduated cylinder, taking care that a level surface of powder is obtained. For refractory metal powders that have high apparent densities (above 4 g/cm³), it is preferable to use a reduced-volume graduated cylinder (25 mL) to improve the accuracy of the results.

The powder is settled in the cylinder by mechanical or hand tapping. If mechanical tapping (Fig. 1 and 2) is used, the filled cylinder is placed in the mechanical apparatus, which is operated until no further decrease in the volume of the powder is observed. If hand tapping is used, the base of the filled cylinder is tapped squarely upon a hard rubber slab until no further decrease in volume is observed. Care must be exercised to avoid loosening the surface layers of the sample during this procedure.

Table 1 Effect of particle shape of copper powders

Particle size distribution is the same for apparent and tap density values.

Particle shape	Apparent density, g/cm³	Tap density, g/cm³	Increase, %
Spherical	4.5	5.3	18
Irregular	2.3	3.14	35
Flake	0.4	0.7	75

Source: Poster, A. R., Ed., *Handbook of Metal Powders*, Reinhold, New York, 1966, p 17

Fig. 1 Diagram of tapping apparatus

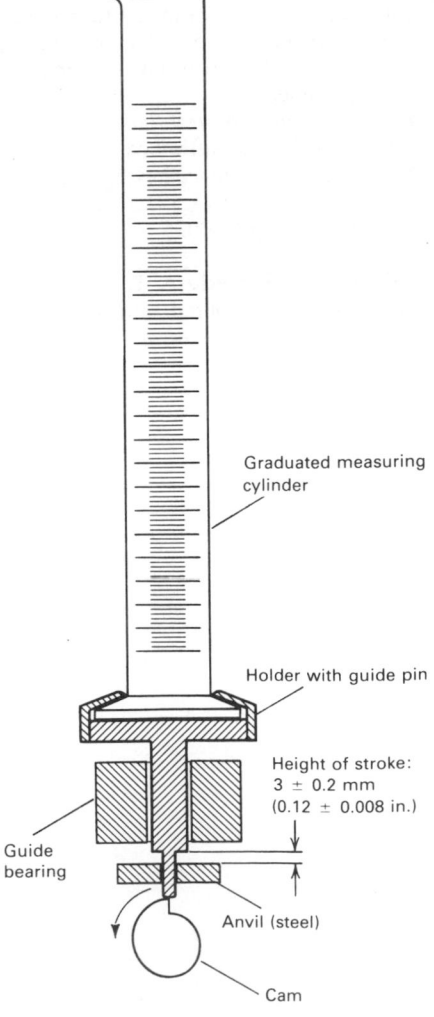

Graduated measuring cylinder

Holder with guide pin

Height of stroke: 3 ± 0.2 mm (0.12 ± 0.008 in.)

Guide bearing

Anvil (steel)

Cam

The volume of the fully densified powder sample in the graduated cylinder is read and used in the following calculation of tap density:

$$\text{Tap density } \rho_t = \frac{m}{v}$$

where m is the mass of powder in grams and v is the volume of tapped powder in cubic centimetres. Results should be reported to the nearest 0.1 g/cm^3.

Representative tap densities of several metal powders and compounds are listed in Table 2, with their respective Fisher subsieve sizes. Mechanical tapping is performed in accordance with American Society for Testing and Materials standard ASTM B 527, Metal Powder Industries Federation standard MPIF 46, or International Standards Organization standard ISO 3953. Manual tapping is performed in accordance with either the MPIF or ISO standard. Interlaboratory reproducibility of tap density values can be expected to fall within a standard deviation of about 3.5%.

Table 2   Typical tap densities of metal and metal carbide powders

Powder	Fisher subsieve size, μm	Tap density, g/cm³
Aluminum	5.05	1.30
Chromium	3.20	3.10
Chromium carbide (Cr$_3$C$_2$)	3.70	3.50
Cobalt	1.50	1.60
Hafnium carbide (HfC)	3.50	5.95
Iron	5.40	3.55
Manganese	3.40	3.05
Molybdenum	4.30	3.75
Molybdenum carbide (Mo$_2$C)	4.50	3.45
Nickel	3.00	1.90
Tantalum carbide (TaC)	2.65	8.00
Tin	2.45	3.15
Titanium carbide (TiC)	3.20	3.65
Tungsten	...	
Fine	1.15	4.45
Coarse	6.00	6.10
Tungsten carbide (WC)	...	
Fine	1.45	4.20
Coarse	6.50	6.50
Tungsten–titanium carbide	...	
(WC-TiC)	3.90	4.30
Vanadium carbide (VC)	4.50	2.70
Zirconium	3.70	2.50

Fig. 2 Tap-Pak volumeter
Courtesy of Shandon Southern Instruments Inc.

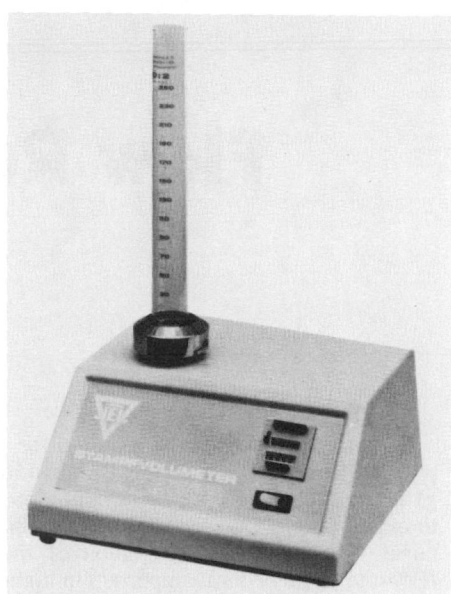

Flow Rate of Metal Powders

By Krishnakant B. Patel
Metallurgist
Alcan Ingot and Powders

FLOW RATE is the time required for a powder sample of a standard weight (50 g) to flow under atmospheric conditions through a funnel into the cavity of a container or mold. A determination of the flow rate of a powder is important in high-volume manufacturing, which depends on rapid, uniform, consistent filling of the die cavity. Poor flow characteristics cause slow and nonuniform press feeding and difficulty in ensuring even fills of the die cavity.

Before a powder is used in production, its flow characteristics must be known, because some compacting tools require a free-flowing powder, while others can be used with a relatively poor-flowing powder. The term free-flowing refers to those physical properties of a powder, such as composition, particle fineness, and particle shape, that permit the powder to flow readily into the die cavity.

Fig. 1 Hall flowmeter

If a compacting tool is designed to handle a free-flowing powder, the use of a poor-flowing powder will necessitate modification. Compacting press manufacturers provide modified hopper designs and feeding shoe arrangements to accommodate finer, poor-flowing powders such as tungsten, molybdenum, or lighter aluminum powders.

The flow of powder from the feeding shoe into the die cavity can be increased by tapping or by changing the design of the filling device. These factors, however, are not taken into consideration in conventional flow rate test procedures, which consist of determining the time required for a given weight or volume of powder to flow through a standardized funnel-shaped cup with a small orifice at the bottom.

Testing of Flow Rate

Flow of metal powders is determined by standard methods developed by the American Society for Testing and Materials (ASTM) and the Metal Powder Industries Federation (MPIF).

Hall Flowmeter. The device most commonly used for measuring flow rate is the Hall flowmeter (Fig. 1 and 2), taken from ASTM B 213 and MPIF 3 (equivalent standards include ISO 4490, Japanese standard JIS 7-2502-1966, and German standard 82-69). The test equipment consists of a funnel with a calibrated hole 2.5 mm (0.1 in.) in diameter. The funnel, which is made of aluminum alloy 6061-T6, is supplied with a smooth finish to minimize wall friction.

With the help of a stopwatch and weighing balance, the flow rate of metal powders can be easily determined. A dry 50-g weight sample is transferred to the funnel, the orifice of which is covered with the operator's fingertip. The stopwatch is started when the fingertip is removed and is stopped when the last quantity of the powder leaves the funnel. The flow rate (s/50 g) of the sample is reported as the elapsed time in seconds for 50 g of powder to flow through the orifice. A powder that does not flow through a 2.5-mm (0.1-in.) orifice Hall funnel, with or without an external impulse, is said to be a nonfree-flowing powder (as per ASTM B 213 and MPIF 3 method).

The Hall funnel is calibrated using a standardized powder (150-mesh Turkish emery grit), a sample of which is supplied with the equipment. The desired hole size is obtained by precisely honing a drilled hole until a satisfactory flow rate of emery powder is obtained.

A change in surface finish and the radius of the orifice (at the junction of the wall), buildup of material on the sidewalls of the orifice, or enlargement of the hole size due to continuous use can alter the standardization of the funnel. Verification

Fig. 2 Cross-sectional view of Hall flowmeter
Brass density cup, 25-cm³ (1.5-in.³) capacity

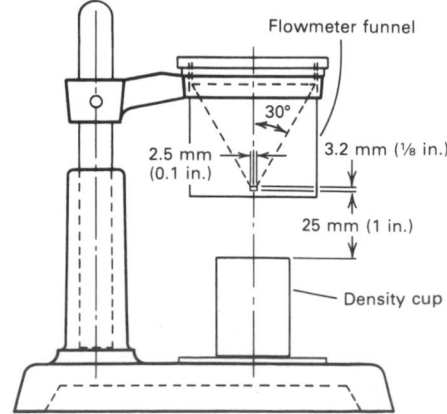

Table 1 Flow rate of metal powders through Hall and Carney funnels

Material	Metal powder Grade	Lubricant Type	Additions, wt%	Apparent density, g/cm³	Flow rate of 50 g (2 oz) powder, s Hall funnel	Carney funnel	Flow rate of 25 cm³ (1.5 in.³) powder through Hall funnel, s	Weight of 25 cm³ (1.5 in.³) powder used in volumetric flow rate study g	oz	Calculated flow rate for 50 g (2 oz) powder based on volumetric flow rates, s
Iron	MP-35HD	Zinc stearate	None	2.81	25.77	4.62	37.74	70.27	2.46	26.86
		Zinc stearate	0.25	3.12	23.37	4.16	...	...	...	...
		Zinc stearate	0.50	3.05	25.93	4.30	...	...	...	...
		Zinc stearate	0.75	3.02	26.80	4.41	...	...	...	...
		Zinc stearate	1.00	3.00	27.57	4.59	40.55	74.52	2.61	27.21
Iron	MH-100	Zinc stearate	None	2.48	30.14	5.26	38.61	62.06	2.17	31.11
		Zinc stearate	0.25	2.97	23.23	4.14	...	...	...	...
		Zinc stearate	0.50	2.93	26.39	4.47	...	...	...	...
		Zinc stearate	0.75	2.86	28.97	4.80	...	...	...	...
		Zinc stearate	1.00	2.87	30.42	5.12	42.20	71.31	2.50	29.59
Iron	A-Met 1000	Zinc stearate	None	2.94	26.24	4.34	39.16	73.91	2.59	26.49
		Zinc stearate	0.25	3.27	23.89	4.04	...	...	...	...
		Zinc stearate	0.50	2.98	28-30	4.55	...	...	...	...
		Zinc stearate	0.75	3.18	25.46	4.41	...	...	...	...
		Zinc stearate	1.00	3.18	25-58	4.45	41.70	79.57	2.78	26.20
Stainless steel	304-L	Lithium stearate	None	2.61	30.62	4.92	39.65	65.45	2.29	30.29
		Lithium stearate	0.50	3.08	29.43	4.80	...	...	...	...
		Lithium stearate	0.75	3.01	33.20	5.42	...	...	...	...
		Lithium stearate	1.00	3.02	37.51	6.13	59.89	75.54	2.64	39.64
Premix bronze (90% Cu-10% Sn)	5099	Stearic acid	None	2.96	21.68	3.99	32.69	74.92	2.62	21.82
		Stearic acid	0.25	3.54	24.01	5.17	...	...	...	...
		Stearic acid	0.50	3.54	24.66	5.38	...	...	...	...
		Stearic acid	0.75	3.42	27.91	5.24	49.84	85.68	3.00	29.08
		Stearic acid	1.00	3.38	34.75	7.20	...	...	...	...
Brass	B-126	Lithium stearate	None	2.89	33.26	5.51	48.56	72.67	2.54	33.41
		Lithium stearate	0.25	3.06	33.52	5.77	...	...	...	...
		Lithium stearate	0.50	3.14	38.70	6.38	64.01	78.25	2.74	40.90
Aluminum	...	None	...	1.19	66.43	...	39.23	29.73	1.04	65.97

of the calibration should be performed periodically by using the standardized emery powder. Calibration of the Hall funnel with Ballotini solid glass spheres, having particle size ranges of 0.090 to 0.102 mm (0.0036 to 0.004 in.) and 0.065 to 0.090 mm (0.0026 to 0.0036 in.) diameters, has yielded flow rates of 35.6 and 33.4 s/50 g, respectively.

The feed of powder to the die is handled on a volume basis. Thus, differences in apparent densities of powders can lead to considerable variations in the weight of material filling a given volume. A test for volumetric flow rate determination is under investigation by ASTM. Table 1 lists the flow rates of metal powders for both weight and volume basis. The volumetric flow rate tests were carried out using the Hall flowmeter. The data in Table 1 indicate that the readings for volumetric flow rate lie within a narrower range, as compared to the Hall flow rate.

Many factors can affect the accuracy of the results obtained with the Hall flowmeter, such as the moist finger of an operator (use of gloves can eliminate this effect), vibration of the surface on which the flowmeter is placed, humidity and temperature, condition of the sample, uniformity of the powder mix, and alternate use of the device for unlubricated and lubricated powders. Residual lubricant film left by a lubricated powder on the flowmeter wall or orifice can affect the subsequent flow test results of an unlubricated powder.

Carney Funnel. Certain characteristics of some metal powders, such as particle shape and size distribution and lower specific gravity, may affect the powders to such an extent that they will not flow through the Hall funnel. In this case, the Carney funnel (Fig. 3), which has the same dimensions as that of the Hall, except for a larger orifice diameter of 5 mm (0.2 in.), can be used to obtain a relative measure of the flowability of nonfree-flowing metal powders. The use of the Carney funnel,

Fig. 3 Carney funnel

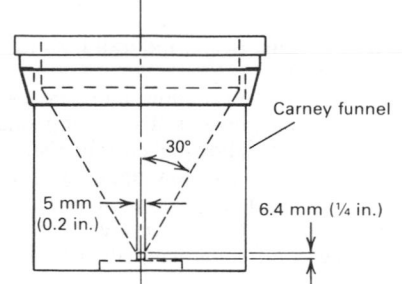

which is further described in ASTM B-417 and MPIF 28, is not a standardized test method in that there is no standard calibration procedure. However, it is used in industry to compare flow rates through a 5-mm (0.2-in.) orifice for a variety of materials. Because there is no correlating factor to relate the data obtained using the Carney funnel with that of the Hall, the user must establish an empirical relationship between the two methods.

Other Testing Methods. A number of other devices or methods for measuring the flow rates of metal powders have been developed. Efforts have been made to design test methods for powders that do not flow through either the Hall or Carney funnel and to reflect the shop floor conditions of these powders.

One method currently under consideration by the International Standards Organization determines the filling characteristics of metal powder into cavities of increasing sizes. A powder-filled shoe is slid back and forth once on the surface of a 40-mm (1.6-in.) thick plate (placed on a paper) into which bores with diameters of 5, 7, 10, 15, 20, 25, and 30 mm (0.2, 0.3, 0.4, 0.6, 0.8, 1.0, and 1.2 in.) have been drilled.

The plate is then lifted, and the powder that has fallen onto the paper from each of the seven die cavities is weighed. When the mass of the powder is divided by the volume of the respective cavity, the filling density for each cavity is obtained. The test provides the critical diameter, that is, the dividing line between cavity diameters which will be filled at a constant apparent density and those through which the powder may not flow at all or which may result in incomplete cavity fill.

In 1956, a second test method for determining the flow rate of metal powders that do not flow through the Hall funnel was developed by Chrysler Corporation as an internal standard. The test equipment consists of a powder shoe of 102 cm³ (6.2 in.³) in volume and a pivoted lever attached to the shoe at one end and to a cam follower at the other. A four-lobed cam is connected to a gearbox and is driven by a motor. A circular opening of 12.7 mm (0.5 in.) diam (with a 3.0 mm (0.12 in.) wide bar at center) drilled on a 2.4-mm (0.1-in.) thick metal plate acts as a die opening.

To conduct the test, the powder-filled shoe is moved back and forth over the opening four times. This is referred to as a one-cycle operation. The quantity of powder passing through the opening is collected on a balance pan and weighed. The test equipment simulates the action of a production press, and flow is measured in terms of the quantity of powder that passes through the opening in one filling cycle of a shoe.

In one production example, the normal range for a bronze powder mix was 80 to 95 g (2.8 to 3.3 oz) per cycle, whereas the data of all the lots ranged from 30 to 125 g (1 to 4 oz) per cycle of operation. The test method was used to check the flow of a production mix under conditions comparable to those encountered in the production presses. The test indicated the flowability of a powder mix accurately.

The die-filling operation of a production press can be simulated by a third test method in which actual bearing die cavities are used. For the production of bearings with varying wall thicknesses and complicated shapes, the tool design should accommodate the variations in filling properties of the powder mixes. The following test procedure was found to be especially useful for complex powder mixes that may or may not flow through the Hall flowmeter.

The equipment consists of a powder shoe, sliding over a metal plate with a recessed cavity in which a series of cups (with or without core rods) can be fitted. The sizes of the resulting bearing-shaped cavities vary between 3.3 to 8.2 cm³ (0.2 to 0.5 in.³) in volume with wall thicknesses ranging from 1 to 3 mm (0.04 to 0.12 in.) and corresponding core rod diameters from 19 to 15 mm (0.8 to 0.6 in.) with a constant height of 45 mm (1.8 in.). A reference cavity of the same height is coreless, with a volume of 16 cm³ (1.0 in.³).

With the selected size of the bearing die in position, the powder shoe is mechanically traversed through the die (or cup) opening and back to the original position. The cup is then removed and weighed. The quantity of powder is divided by the total volume of the cup, and the fill density of the powder is calculated. The process is repeated for the remaining die sizes by changing the core rod inserts. The filling density of the reference cavity is always the highest and is designated 100 for a given sample. This value is decreased with a reduction in wall thickness.

Tests with iron, nickel, copper, tin, and Turkish emery grit indicate that the results are consistent. The powders tested either flowed into the cup or did not enter at all. By evaluating filling densities in different die sizes, the test provides meaningful information in designing tools for a production shop. This test method was used to study blending variables and yielded close agreement with the performance of the production presses. Unfortunately, the test is not simple and requires the use of precise equipment.

Variables Affecting Flow Rate

Flow characteristics are dependent on several variables, including (1) interparticle friction, (2) particle shape and size, (3) type of material, (4) environmental factors, and (5) weight of the bulk.

Characteristics of powder surfaces, such as surface oxide films and lubricant films, also affect flow characteristics. The presence of oxide films on powder particle surfaces alters the friction between particles and increases flow rate. Powders with lower surface oxide contents flow more slowly than powders with higher oxide levels. Minute additions of lubricants may increase the flow rates of metal powder, but further additions will reduce flow rates. For practical purposes, the higher the lubricant level, the slower the Hall flow rate (see Table 1).

In general, reduced flow rates are encountered with powders that exhibit one or more of the following characteristics: low specific gravity, low apparent density, high friction coefficient of fine particles, high specific surface area, a complex blend of different materials, and high moisture content.

Interparticle Friction. The resistance to flow depends primarily on the regions in which one particle hampers the free movement of other particles, either by direct contact or indirectly. This is mainly determined by the coefficient of interparticle friction. Particles may be prevented from moving separately by temporary adherence or interlocking. In this manner, clusters are formed that may occupy considerable volume. The phenomenon of cluster formation depends on the movement and type of powder, the flow varying markedly with the size and structure of the particles. If the particles were all truly spherical, they would generally roll readily into a die cavity. This can rarely be achieved in commercial powders; differences in size and shape are unavoidable.

Particle Size and Shape. Subsieve powders, those with particle sizes less than 44 μm, generally have poor flow rates. For this and other reasons, very fine powders are not used for compacts that are pressed on automatic presses. The particles of most powders used for compacting have irregular equiaxed shape, with flow rates between those of spherical (high flow rates) and flaky (low flow rates) powders.

Type of Material. Flow is influenced by the type of material, whether it be copper, aluminum, or iron. The major influence is the theoretical density. Other characteristics, such as adhesive and cohesive surface properties and magnetic or electrostatic interactions, are also factors.

Environmental Factors. Powders exposed to air containing high relative humidity absorb moisture on particle surfaces, resulting in reduced flow rate. Seasonal changes in temperature do not affect the flow rate of metal powders considerably. Very low temperatures, however, can cause condensation of moisture; very high temperatures may partially melt the lubricants in the powder mix. Such extreme temperature conditions may cause interruptions in the flow of material through conveying systems.

Weight of the Bulk. A metal powder with a lower specific gravity, such as aluminum, generally exhibits slower flow rates compared to high specific gravity powders such as iron. At the same time, the higher the apparent density of a given material, the faster the flow. The ratio of apparent density to specific gravity can be used to correlate the bulk properties of various metal powders.

Angle of repose is defined as the basal angle of a pile formed by a powder when freely poured under specified conditions onto a horizontal surface. This was once considered an indication of the flow characteristics of powders; however, it has been confirmed that there is no direct correlation between the flow rate and the angle of repose of metal powders.

SELECTED REFERENCES

- Haertlein, J. and Sachse, J.F., The Flow Rate of Metal Powders, *Handbook of Metal Powders*, Reinhold Publishing Corp., New York, 1966
- Adler, A., Flow Properties of Metal Powders, *International Journal of Powder Metallurgy*, Vol 5 (No. 1), 1969, p 7-20
- Carson, J., Applying Bulk Solid Flow Principles to Metal Powders, *Progress in Powder Metallurgy*, Vol 31, Metal Powder Industries Federation, Princeton, NJ, 1975, p 101-110
- Matei, G., Claussen, N., and Hausner, H.H., Influence of Relative Humidity on Flow of Metal and Ceramic Powders, *Modern Developments in Powder Metallurgy*, Vol 8, 1974, p 5-11
- Zborovski, A.V., Correlation Between the Apparent Density of Iron Powder and Its Flow Rate and Pressing Behavior, *Poroshkovaya Metallurgia*, Vol 84 (No. 12), Dec 1969, p 5-9
- Grey, R.O. and Beddow, J.K., On the Hausner Ratio and Its Relationship to Some Properties of Metal Powders, *Powder Technology*, Vol 2, 1968-69, p 323-326
- Mahoney, M.J., Method for Determination of Flow Rates of Metal Powders Blends, Amplex Div., Chrysler Corp., Laboratory Test No. 301-302, 1956
- Determination of Filling Characteristics, International Standards Organization, Report TC-119/SC-2N 234, 1981
- Matthews, P.E., Volumetric Flow Rate—Experimental, American Society for Testing of Materials—No. ASTM B-09:02, 1980
- Oakley, J., A Method of Assessing the Die-Filling Characteristics of Powders, *Journal of the Institute of Metals*, Vol 87, 1958-59, p 26-28
- Hausner, H.H., Environmental Effects on the Behavior of Metal Powders, *International Journal of Powder Metallurgy and Powder Technology*, Vol 14 (No. 4), 1978
- Yarnton, D., Effect of Lubricant on the Flow and Packing Density of Cu Powders, *Engineering Materials & Design*, Sept 1970, p 1103-1107
- "Flow Rate of Metal Powders," ASTM Standard No. B 213, American Society for Testing and Materials, Philadelphia, 1977
- "Determination of Flow Rate of Metal Powders Using the Hall Apparatus," MPIF Standard No. 03, Metal Powder Industries Federation, Princeton, NJ, 1972

Angle of Repose

By Louise H. Hua
Research Assistant
Fine Particle Research Group
University of Iowa
and
J.K. Beddow
Professor
Fine Particle Research Group
University of Iowa

THE ANGLE OF REPOSE of an aggregate, or the angle of the surface of an unconstrained pile of solids with the horizontal, is of practical interest rather than of theoretical concern. Of the various physical properties of bulk powders, the angle of repose is the easiest to obtain. It is closely related to interparticle friction and the flowability of cohesionless material. This property is frequently used to characterize powdered materials.

The angle of repose of a powdered material does not always give satisfactory reproducibility and is often masked by other factors that are not inherent to the material (most frequently, the presence of a liquid). Thus, before a standard measuring method that provides reasonable reproducibility is developed, its usability as a measure of powdered material property must be established.

However, although the angle of repose is frequently used as a convenient characterization of powdered material, its ultimate reliability depends on the measuring method used. Therefore, the measuring method should be carefully selected, especially when the measured angle of repose is to be used to determine another property, so that it best reflects the property to be quantified.

Methods of Measurement

Several methods have been used to measure the angle of repose of materials. Train (Ref 1) studied the angle of repose of a number of grades of glass spheres, lead shot, and silver sand using four different methods, as illustrated in Fig. 1. These methods are:

- *Fixed height cone:* The powder is carefully poured through a funnel at a fixed height until the apex of the heap formed by the powder reaches the tip of the funnel. The tangent of the angle of repose is the ratio of the height to the mean radius of the base of the powder heap.

- *Fixed base cone:* The powder is allowed to flow through a funnel, which is raised vertically until the heap covers a circular base of fixed size. The tangent of the angle of repose is calculated in the same manner as the fixed height cone method.

- *Tilting table:* A rectangular box filled with powder is tilted until the contents begin to slide. The angle that the upper surface of the box makes with the horizontal is equal to the angle of repose.

- *Rotating cylinder:* A sealed hollow cylinder half full of powder is rotated until the surface of powder exhibits its maximum angle with the horizontal. This maximum angle is the angle of repose.

Brown (Ref 3) reported three methods to measure the drained angle of repose (Fig. 2) and a rotating drum method to measure the dynamic angle of repose, as follows:

- *Ledge:* Material is first charged into a rectangular Perspex box that is 30 cm (12 in.) in height with a 10 by 10 cm (4 by 4 in.) base. A slot at the base of one vertical wall can be closed by a board. The closure board is then removed to allow the material to flow slowly through the narrow slot. The angle with the horizontal of the surface of the material remaining when the flow stops is measured as the angle of repose.

- *Crater:* A circular Perspex tube with a 14.5-cm (5.7-in.) diam is placed vertically on a flat, horizontal base plate having a 1.5-cm (0.59-in.) diam orifice in the center. The powder is discharged through the orifice. The height of the remaining material resting against the wall of the tube is measured at eight equidistant points around the circumference to determine the angle of repose.

- *Circular heap:* A circular platform 7.6 cm (3 in.) in diameter is supported horizontally over a circular hole in a flat base plate and surrounded by a cylindrical tube having a 17.8-cm (7-in.) diam and sufficient height (35 cm or 13.8 in.) to ensure that when it is filled with

Fig. 1 Four methods used to measure the angle of repose
Source: Ref 2

Fixed height cone

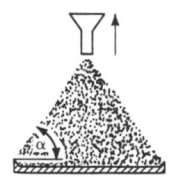
Fixed base cone

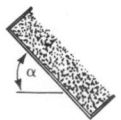
Tilting table

Rotating cylinder

Fig. 2 Apparatus used to measure the drained angle of repose
Source: Ref 2

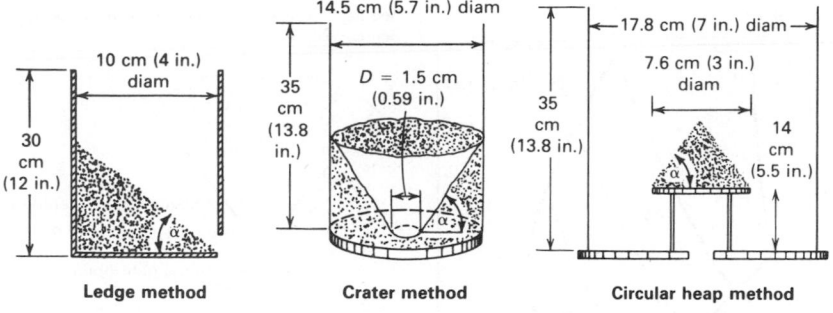

Ledge method   Crater method   Circular heap method

Fig. 3 Dynamic angle of repose and shear angle of material in a rotating cylinder

B is shear plane. Angle ϕ_R is upper angle of repose (dynamic angle of repose). Angle ϕ_S is lower angle of repose (shear angle).

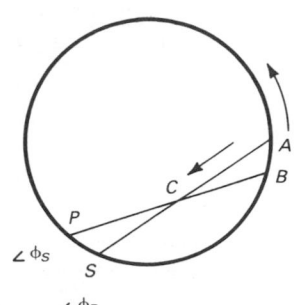

Table 1 Drained angles of repose (degrees to the horizontal)

Descriptive class	Material	Circular heap ($\pm 1/2°$), α	On a ledge ($\pm 1 1/2°$), α	Crater ($\pm 2°$), α	Dynamic(a), α
Smooth, spherical	Beads	$17^1/_2$	25	27	$25^1/_2$
	Beads	20	23	$21^1/_2$	24
Rough, nearly spherical	Sand	$32^1/_2$(b)	34	35-$35^1/_2$	$34^1/_2$
	Tapioca	30	34	$37^1/_2$	32
	Rice	35	37	42	...
Angular	Sand	37	37	39	$36^1/_2$
	Sand	$35^1/_2$	37	$38^1/_2$	$38^1/_2$
	Sand	$35^1/_2$	36	$37^1/_2$	$38^1/_2$
	Durite	37	40	41	...
	Charcoal	...	38	...	...
	Charcoal	...	$38^1/_2$	...	...
	Charcoal	...	$42^1/_2$	...	...
	Charcoal	...	$42^1/_2$	...	...
	Coal	$37^1/_2$	37	41	34
	Coal	$35^1/_2$	$37^1/_2$	...	...
	Coal	36	38	...	...
	Coal	$36^1/_2$	$38^1/_2$	...	...
	Coal	$38^1/_2$	$38^1/_2$	...	...
Containing fine particles	Coal	...	52	...	...
	Coal	54	59-61	...	...
	Coal	...	$47^1/_2$	...	...
	Fine coal	...	67	...	...
	Limestone	...	64	...	...

(a) In a drum rotating at 6 rpm. (b) On a 5.1-cm (2-in.) diam platform in a 12.7-cm (5-in.) diam cylinder
Source: Ref 2

powder the platform and any heap that may form on it is completely immersed in the powder. The powder in the cylindrical tube is then allowed to flow slowly out of the circular hole in the base plate. The height of the resulting heap on the circular platform is then obtained. The tangent of the angle of repose is calculated as in the fixed height methods.

• *Dynamic angle of repose:* A drum 15 or 30 cm (6 or 12 in.) in diameter and 10.2 cm (4 in.) long with Perspex end faces and roughened internal surfaces is half filled with powder and slowly rotated counterclockwise, with its axis horizontal. Within a certain range of rotation speeds (usually 2.5 to 6 rpm), the surface of the powder in the drum becomes substantially steady. The angle of inclination of the surface to the horizontal is measured at various speeds to determine the angle of repose.

Table 1 lists the angles of repose measured using these methods. Henein and co-workers (Ref 4, 5) used a method similar

Table 2 Upper and lower angles of repose of several materials

Material	Average size mm	Average size in.	Particle shape	Particle density, kg/m³	Loose bulk density, kg/m³	Dense bulk density, kg/m³	Static angle of repose, degrees	Dynamic angle of repose, degrees	Cylinder diameter m	Cylinder diameter ft	Shear angle, degrees
Gravel	3.0	0.12	Angular	2870	1560	1690	40.7	37.5	0.40	1.3	34.7
								37.0	1.06	3.5	34.4
Iron oxide	11.6	0.46	Spherical	...	...	...	31.5	35.2	0.40	1.3	33.3
Limestone B	4.3	0.17	Irregular	2700	1450	1610	40.3	39.6	0.40	1.3	37.7
								36.5	1.06	3.5	34.5
Limestone C	1.5	0.06	Irregular	2690	1520	1600	37.8	36.0	0.40	1.3	33.6
									1.06	3.5	32.5
Limestone D	0.58	0.02	Irregular	2680	1490	1570	35.6	34.9	0.40	1.3	33.5
Limestone F	8.1	0.32	Angular	2690	...	...	42.8	41.5	1.06	3.5	38.5
Nickel oxide	4.9	0.19	Spherical	...	870	900	32.5	30.2	0.40	1.3	29.9
Sand B	0.50	0.02	Nodular	2660	1640	1740	33.4	33.6	0.40	1.3	32.2

Note: 1000 kg/m³ = 1 g/cm³
Source: Ref 5

Fig. 4 Typical plot of mixture composition versus angle of repose

Curve extrapolated to determine angle of pure material. Source: Ref 6

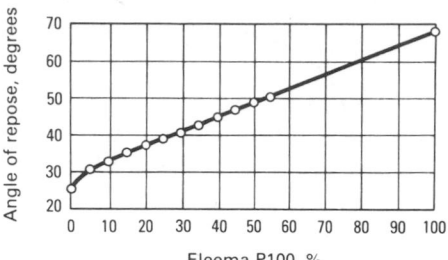

Powder	Tilting box	Fixed bed
Calcium carbonate	83.0	78.0
Eleema P100 (microfine cellulose)	64.5	68.0
Avieel PH101 (microcrystalline cellulose)	57.5	51.5
Methylcellulose 20 BPC	50.0	52.0
Methylcellulose 450 BP	63.0	55.0
Hydroxypropyl methylcellulose 4500	64.5	. . .
Hydroxypropyl methylcellulose 5000	59.5	58.5
Hydroxyethyl methylcellulose 3500	61.0	61.0
Magnesium stearate	69.0	66.0
Stearic acid	. . .	63.5

Source: Ref 6

to that used by Brown to determine the dynamic angle of repose and lower angle of repose (shear angle) of several materials. Rotating cylinders of 40 and 106 cm (16 and 42 in.) diam were lined with 24-3 grit type E silicon carbide abrasive paper. One of the two end plates of each cylinder were made of plexiglas, to observe and photograph the bed. The maximum angle of bed inclination just before slump occurred was measured with a long-arm protractor, and this angle was designated the upper angle of repose, or the dynamic angle of repose. The angle relative to the horizontal of the shear plane that separated the slumping solids of the bed surface from the material moving with the cylinder wall was considered the lower angle of repose, or shear angle (see Fig. 3). Table 2 summarizes the upper and lower angles of repose of several materials.

Riley and co-workers (Ref 6) reported methods to measure the angles of repose of cohesive powders. They used both tilting box and fixed bed methods to measure the angle of repose of mixtures of glass ballotini and cohesive powders of different compositions and then extrapolated the results to obtain the angles of repose of pure cohesive material. Figure 4 shows a

Fig. 5 Effect of moisture content on angle of repose of loose and compacted coal

Source: Ref 2

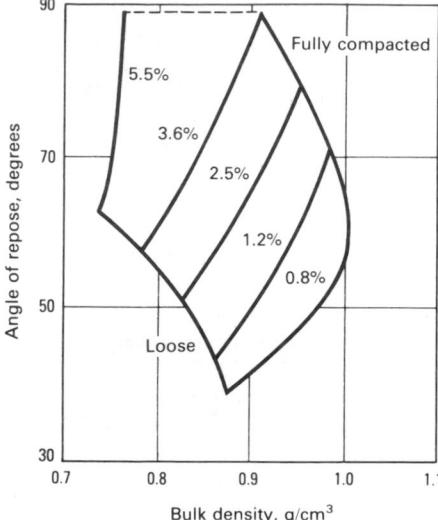

sample plot and gives the angle of repose of several cohesive powders.

Factors Affecting Angle of Repose

Internal factors are those inherent to powders, or characteristic of the nature of powders. These include particle size, particle shape, and cohesiveness. In general, larger particles have higher angles of repose. However, very small particles may exhibit cohesiveness due to the electrostatic effect, which increases the angle of repose (Ref 3-5). Because spherical particles have a greater tendency to roll, they typically have smaller angles of repose than irregularly shaped particles.

External factors are introduced by the environment and may include the method of measurement or the presence of other materials (either solid or liquid). The effect of measuring method is clearly evident as discussed above. In general, the more momentum introduced by the measuring method, the smaller the angle of repose. As a consequence, the drained angle of repose is higher than that obtained from the heap formation method.

The angle of repose of loosely packed dry powder may be increased by compacting, as well as by introducing liquid to the powder. Figure 5 demonstrates the significant effect these factors have on the angle of repose (Ref 2, 7).

Frydman (Ref 8) reported that the angle of repose of potash pellets may be increased by more than 100% by applying a light spray of moisture to the material sur-

Fig. 6 Vertical straight-walled channel hopper for determining flow rates of powders

See Eq 1 in text for explanation of symbols. Source: Ref 9

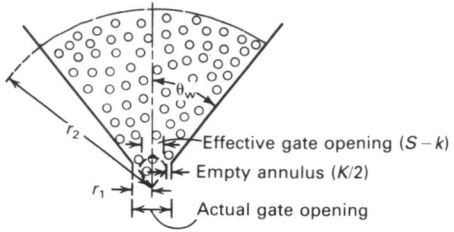

face. The effect of the presence of another solid is shown in Fig. 4.

Relationship of Angle of Repose to Other Powder Properties

Flow Rate. Harpavat and Frantz (Ref 9) derived the following equation to predict the rate of a powdered material flowing through a hopper:

$$\text{Flow rate } (Q_R) = \rho \left[\frac{g(S-k)^3}{2 \sin^3 \theta_\omega} \left(\frac{K+1}{K-2} \right) \cdot \frac{\left[1 - \left(\frac{r_2}{r_1} \right)^{(2-k)} \right]}{\left[1 - \left(\frac{r_2}{r_1} \right)^{(-k-1)} \right]} \right]^{1/2} \cdot \int_0^{\theta_\omega} \sqrt{\cos} \, d\theta$$

(Eq 1)

where Q_R is the flow rate per unit length of gate opening; ρ is the bulk density of the granular material in the hopper; g is the acceleration due to gravity; θ_ω is the hopper angle as shown in Fig. 6; K is $(1 + \sin \delta)/(1 - \sin \delta)$; δ is the angle of repose for the granular material; S is the actual gate opening; k is the width of the empty annulus; r_1 is the lower radius of the hopper; and r_2 is the upper radius of the hopper.

Mobility. Neumann (Ref 10) studied the mobility of several powders, mostly cohesive, by mixing the powder with clean −60+100 mesh sand to enable the mixture to flow steadily through an 3.2-mm (0.13-in.) orifice. The amount of sand required increased with the angle of repose of the powders. Table 3 lists mobility

Table 3   Mobility and angle of repose of various powders

Material	Particle shape	Mobility(a)	Angle of repose, degrees
Relatively free-flowing hard particles			
Silica	Amorphous microspheres	0	27.7
Calcium fluoride	Cubes and octahedra	0	31.0
Sodium carbonate	Anhydrous elongated crystals	0.23	42.6
Cohesive powders:			
Titanium oxide pigment	. . .	0.26	38.9
Refined Kaolin	Soft, porous aggregates of 0.5 μm crystallites	12.0	45.2
Natural fuller's earth	Aggregates of 0.01-0.1 μm crystallites	0.25	42.3
Activated fuller's earth	More spongy than natural fuller's earth	0.60	43.9
Portland cement	Irregular particles	1.00	44.8
Self-rising flour	Disk-shaped	0.30	41.0

(a) Ratio of weight of sand to weight of powder in a mixture that will flow steadily through a 3.2-mm (0.13-in.) diam orifice in the base of a wide container.
Source: Ref 2

Fig. 7 Angles of repose and flow rates versus Hausner ratio for three copper powder samples

α is the angle of repose measured by the fixed height cone method (see Fig. 1). α' is the angle of repose, defined as the maximum angle that the surface plane of the powder made upon rotation in a cylinder (rotating cylinder method, as shown in Fig. 1). β is the angle between the surface of the powder mass and the horizontal, after the powder mass had collapsed as a result of further rotation. Source: Ref 11

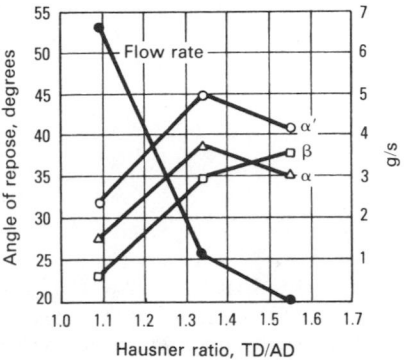

and angle of repose values for several hard and cohesive powders.

Hausner Ratio. The ratio between the tap density (*TD*) and the apparent density (*AD*) of a powder mass is termed the Hausner ratio. This ratio is used to measure the relative magnitude of the friction between the particles in that mass. Figure 7 shows the relationship between the *TD/AD* ratio and the angles of repose and also between the *TD/AD* ratio and the Hall flowmeter rates for three copper powders tested in the dry condition. In general, the Hausner ratio increases as the value of angle of repose increases (Ref 11).

REFERENCES

1. Train, D., Some Aspects of the Property of Angle of Repose of Powders, *J. Pharm. Pharmac.*, Vol 10 (Suppl), 127T-135T, Dec 1958
2. Brown, R.L. and Richards, J.C., *Principles of Powder Mechanics*, Pergamon Press, London, 1970
3. Brown, R.L., Flow Properties, Powders in Industry, *Soc. Chem. Ind.*, Monograph No. 14, 1960, p 150-166
4. Henein, H., Brimacombe, J.K., and Watkinson, A.P., Experimental Study of Transverse Bed Motion in Rotary Kilns, *Metall. Trans. B*, Vol 14B, 1983, p 191-205
5. Henein, H., Brimacombe, J.K., and Watkinson, A.P., The Modeling of Transverse Solids Motion in Rotary Kilns, *Metall. Trans. B*, Vol 14B, 1983, p 207-220
6. Riley, G.S., Mann, S., and Jesse, R.O., Angles of Repose of Cohesive Powders, *J. Powder Bulk Solids Technol.*, Vol 2, 1978, p 15-18
7. Wolf, E.F. and Hohenleiten, H.L., Experimental Study of the Flow of Coal in Chutes, *Trans. Am. Soc. Mech. Eng.*, Vol 67, 1945, p 585-599
8. Frydman, S., The Angle of Repose of Potash Pellets, *Powder Technol.*, Vol 10, 1974, p 9-12
9. Harpavat, G. and Frantz, C.L., An Experimental and Theoretical Study of the Flow of Xerographic Developer Powder from Small Hoppers, *IEEE Trans. Ind. Appl.*, Vol 1A-15 (No. 6), Nov/Dec 1979, p 688-694
10. Neumann, B., Powder, in *Flow Properties of Disperse Systems*, Herman, J.J., Ed., North-Holland Publishing Co., Amsterdam, 1953, p 382-422
11. Grey, R.O. and Beddow, J.K., On the Hausner Ratio and Its Relationship to Some Properties of Metal Powders, *Powder Technol.*, Vol 2, 1968/1969, p 323-326

Compressibility of Metal Powders

By John E. Davidson
Manager of Quality and Technical Service
Quebec Metal Powders Ltd.

COMPRESSIBILITY AND COMPACTIBILITY are terms used to describe the extent to which a mass of powder can be densified by the application of pressure. Compressibility of a powder is a major factor in the design of pressing tools, the part density attainable, and the size of press needed to press to that density. A related term, compression ratio, is the ratio of the final pressed density to the apparent density of the powder. This ratio determines how deep the part die must be in order to hold all the powder needed to meet the specified part density. Powders of high apparent density are preferred, because tooling can be made shorter and thus stronger.

The compressibility of a powder is influenced by several factors:

- *Inherent hardness of the metal or alloy.* Also, some metals tend to work harden to a greater extent than others. Since considerable deformation of a powder takes place during pressing, its work-hardening characteristics have a strong influence on compressibility.
- *Particle shape.* In general, the more irregular a powder, the lower its compressibility.
- *Internal porosity.* Fine internal porosity in a powder tends to become noninterconnected during pressing, thus trapping air within particles. Although air is highly compressible, it takes up volume and contributes little to weight. Nonporous powders have the highest compressibility.
- *Particle size distribution.* Powders of a uniform size exhibit relatively poor compressibility. Mixtures of particle sizes designed to occupy the most space have better compressibility.

- *Presence of nonmetallics.* Nonmetallics such as unreduced oxides reduce compressibility because of their hardness and low specific gravity.
- *Use of solid lubricants.* Solid lubricants added to aid the pressing of a metal powder affect compressibility because of their light weight and the volume they occupy (5 to 7%). Some lubricants perform better than others at pressing pressures.
- *Addition of alloying elements.* Alloying additions such as graphite and sulfur generally degrade compressibility.

More detailed information on factors that affect compressibility of metal powders can be found in the articles "Mechanical Fundamentals of Consolidation" and "Shape Fundamentals: Rigid Tool Compaction" in this Volume.

Standard test methods for compressibility have been issued as ASTM B331, "Compressibility of Metal Powder in Uniaxial Compaction," by the American Society for Testing and Materials; as MPIF 45, "Determination of Compressibility of Metal Powders," by the Metal Powder Industries Federation; and as ISO 3927, "Metallic Powders Excluding Powders for Hardmetals—Determination of Compactibility (Compressibility) in Uniaxial Compression," by the International Standards Organization.

Typically, a cylindrical or rectangular test piece is made by pressing powder in a die, with pressure applied simultaneously from top and bottom. The pressure required to achieve a specified density is a measure of compressibility. Compressibility can also be specified as the density achievable at a given pressure. By plotting the density obtained by a se-

ries of increasing levels of pressure against these pressures, a compressibility curve is developed.

Two shapes of test pieces have been standardized. One is a cylinder having a diameter of 25 mm (1 in.) and a height of 12.7 to 25 mm (0.5 to 1 in.). The other is a rectangular bar 12.7 mm (0.5 in.) wide by 31.8 mm (1.25 in.) long and 5 to 7 mm (0.2 to 0.3 in.) high. Sufficient powder to form the test bar is loaded into the appropriate die, and a pressing load is applied to the opposing punches. When the specified pressure is reached, the part is ejected from the die and its height and peripheral dimensions measured to determine volume. The weight of the part divided by its volume is a measure of the density achieved at the specified pressure. Alternatively, a given weight of powder is pressed to a predetermined thickness, thus

Fig. 1 Compacting tools and rectangular test specimen

Fig. 2 Compressibility curves for various metal powders

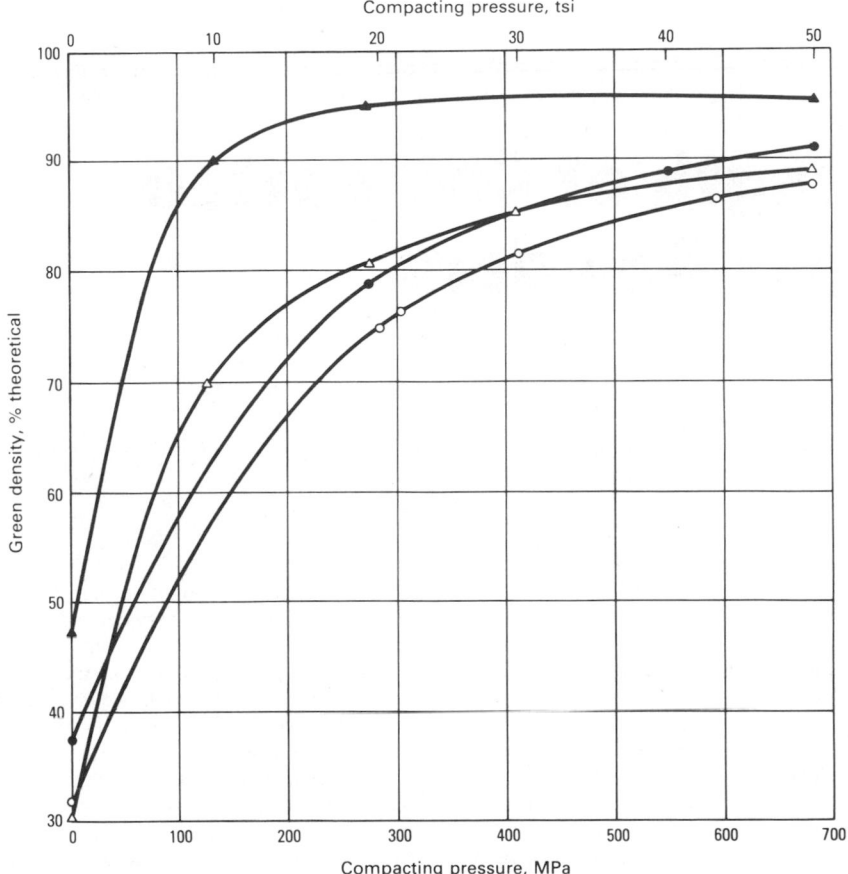

Symbol	Metal powder	Theoretical density, g/cm³
▲	Atomized aluminum powder	2.69
△	Electrolytic copper powder	8.86
●	Atomized iron powder	7.86
○	Sponge iron powder	7.86

to a specific density. The pressure needed to reach this density is the measure of compressibility.

Tooling used to make these specimens consists of a die, preferably of cemented carbide, or alternatively of tool steel, and two steel punches for producing either cylindrical or rectangular compacts. A typical set of tools for compacting the rectangular test specimen is shown in Fig. 1. The upper punch (left) and compacted rectangular test piece (right) are shown in the foreground. Exact tool dimensions required for making the cylindrical test piece and the rectangular bar are described in MPIF 45 and ISO 3927.

Lubrication is used to assist ejection of the compacted specimen from the die. There are two lubrication techniques. One method is to blend a dry lubricant with the powdered metal. The other method, commonly referred to as die wall lubrication, is to lubricate the die walls and punches of the compacting tooling prior to introducing the powder metal into the die cavity. Failure to use a lubricant may result in tool seizure and excessive die wear, particularly at high compacting pressures. Common solid lubricants include zinc stearate, amide wax, and stearic acid. With the powder lubrication method, the level of lubricant addition may range from 0.5 to 1.5%. With die wall lubrication, the solid lubricant (for example, zinc stearate, 100 g) is mixed with a volatile organic liquid (for example, methylchloroform, 1 L) and is either painted or sprayed on the tooling. The organic liquid evaporates, leaving a thin film of dry lubricant on the working surfaces of the die cavity and punches.

Density of the ejected test specimen varies with the method of lubrication, as well as with the type and amount of lubricant used. These parameters should be specified when reporting density values that are to be used to determine the compactibility of a metal powder. For more information, see the article "Lubrication of Metal Powders" in this Volume.

Compressibility is reported as the density in grams per cubic centimetre, rounded to the nearest 0.01 g/cm³, at a specified compacting pressure, or as the pressure needed to reach a specified density. Density is calculated by dividing the weight of the test specimen by its volume, as determined by micrometer measurement. This is commonly referred to as green density.

Typical compressibility data for various kinds of metal powders are given in Fig. 2 as a percentage of theoretical density to demonstrate the relative compressibility of various powders.

Green Strength of Compacted Metal Powders

By C. Evans
Technical Sales Engineer
Pyron Corporation

GREEN STRENGTH is the mechanical strength of a green—that is, unsintered—powder compact. This property is very important, as it determines the ability of a green compact to maintain size and shape during handling prior to sintering. Strength of green compacts results mainly from mechanical interlocking of irregularities on the particle surfaces, which is promoted by plastic deformation during pressing. Much information on green strength is available, especially regarding iron and copper, which are the most commonly used powders in the P/M industry.

Testing for Green Strength

Although metal powder compacts are seldom used in the green condition, green compacts must be strong enough to resist abrasion and breakage while being transferred from a compacting press to a sintering furnace. This is particularly important for thin parts, thin sections of large parts, low-density parts, and part edges.

The Rattler test, adapted from a method used to test paving bricks, was used in the early P/M industry to determine green strength. Five compacts of predetermined weight and size were placed in a bronze-screened cylinder and rotated for a certain number of revolutions for a specified time. The compacts were then removed from the cylinder, weighed, and the weight loss determined. The Rattler test provided information on the abrasion resistance of a compact and its ability to retain its shape (edge stability).

Current test methods to determine green strength of an unsintered compact are specified in ASTM B 312 and MPIF 15.

The standard green strength test is a transverse bend test of a 12.7 by 31.7 mm (0.50 by 1.25 in.) rectangular specimen 6.35 mm (0.25 in.) thick. A simple rectangular die with an upper and lower punch is needed to produce this transverse-rupture specimen. Prior to pressing the compact, the die wall is lubricated with a mixture of zinc stearate in a solvent. The die is then drained, dried, and filled with the test powder. Alternatively, the powder and lubricant may be mixed prior to compaction. Comparisons between the two methods are not recommended, because green strength, green density, compressibility, and rupture characteristics vary depending on whether the die or the powder particles are lubricated. ASTM B 312 and MPIF 15 provide all necessary information on test procedures, die dimensions, transverse-rupture test fixture dimensions, transverse-rupture bar dimensions, and required formulations. The term "green strength" in these standards defines the stress, calculated from the flexure formula, required to break the specimen, and relates to the resistance of a pressed compact to abrasion and breakage.

Variables Affecting Green Strength

Many theories have been proposed to explain the precise mechanics of green strength. Because the strength of green compacts results mainly from mechanical interlocking of particle surface irregularities, particle shape is the most important factor contributing to green strength. Powders with irregularly shaped particles produce compacts of higher green strength than powders with spherical particle shape.

Spherical particles provide the lowest degree of mechanical strength because of low initial surface contact between adjacent particles and a low surface-to-volume ratio.

Generally, green strength is increased by increasing the powder surface area. This can be achieved by increasing the particle surface roughness and/or reducing the average particle size, thus providing more sites for mechanical interlocking. These characteristics also result in decreased apparent density. Figure 1 illustrates the dependence of green strength on apparent density for various iron powders.

Green strength also is increased when oxidation and contamination of particle surfaces are reduced. Optimum mechani-

Fig. 1 Dependence of green strength on apparent density for various iron powders

Source: Hirschhorn, J.S., *Introduction to Powder Metallurgy*, American Powder Metallurgy Institute, Princeton, NJ, 1969, p 140

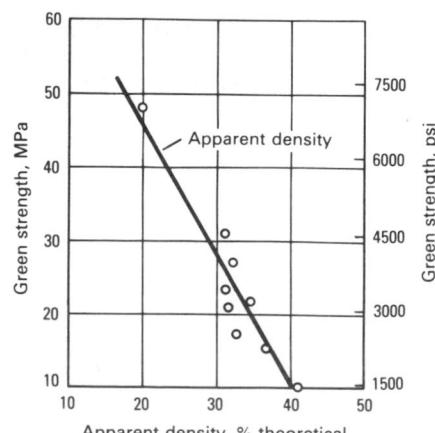

Table 1 Green density and green strength for various types of iron powders

Powder	Apparent density, g/cm³	Compaction pressure		Green density, g/cm³	Green strength	
		MPa	tsi		MPa	psi
Sponge(a) 2.4		410	30	6.2	14	2100
		550	40	6.6	22	3200
		690	50	6.8	28	4100
Atomized sponge(b) 2.5		410	30	6.55	13	1900
		550	40	6.8	19	2700
		690	50	7.0	...	...
Reduced(a) 2.5		410	30	6.5	16	2300
		550	40	6.7	21	3000
		690	50	6.9	24	3500
Sponge(a) 2.6		410	30	6.6	19	2700
		550	40	6.8	25	3600
		690	50	7.0	27	3900
Electrolytic(c) 2.6		410	30	6.3	32	4600
		550	40	6.7	43	6200
		690	50	6.95	54	7800

(a) Powders contained 1% zinc stearate blended in. (b) Powder contained 0.75% zinc stearate blended in. (c) Isostatically pressed.
Source: Hirschhorn, J.S., *Introduction to Powder Metallurgy*, American Powder Metallurgy Institute, Princeton, NJ, 1969, p 142

Table 2 Effect of lubricant on density and strength of iron and copper compacts pressed at 550 MPa (40 tsi)

Material	Metal stearate lubricant, %	Green density, g/cm³	Green strength	
			MPa	psi
Reduced iron .... None		6.47	32	4600
	1	6.57	23	3300
Electrolytic copper .... None		7.97	67	9700
	1	8.11	35	5100

Source: Hausner, H.H. and Mal, M.K., *Handbook of Powder Metallurgy*, 2nd ed., Chemical Publishing, New York, p 160

Fig. 2 Effect of compaction pressure on green strength of various iron powders

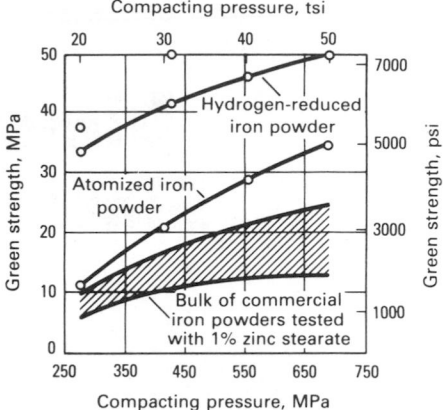

cal interlocking is impeded when particles are covered with heavy oxide films and adsorbed gases.

Green strength also is affected by other variables such as green density and compressibility. Increasing green density or compacting pressure promotes increased particle movement and deformation, which are the bases for mechanical interlocking.

Table 1 shows the relationship among green strength, apparent density, compacting pressure, and green density for several types of iron powders. Figure 2 shows the effect of compaction pressure on green strength of various iron powders.

Increasing the amount of additives also decreases green strength. The addition of some alloying elements, such as adding graphite to iron, or the addition of lubricants can reduce the green strength of a compact. Compacts from mixtures of base metal powders and a lubricant generally have considerably lower green strength than those from the powder alone, since the lubricant, which has been mixed in to facilitate ejection from the die wall, does not act as a binder. Instead, the lubricant interferes with particle-to-particle binding. Therefore, green strength tests should be performed on the powder-lubricant mixture as well as on the powder alone. Table 2 illustrates the effect of a lubricant on green strength and green density of iron and copper compacts. Using a different lubricant, such as an amide wax rather than a metal stearate, can also improve green strength values.

For more information on green strength, see the article "Mechanical Fundamentals of Consolidation" in this Volume.

Dimensional Change of Sintered Metal Compacts

By Cynthia L. Freeby
Technical Service Representative
The New Jersey Zinc Co.

CHANGES IN DIMENSIONS of metal powder compacts can occur during the production process. The ability to control or forecast the changes is an important facet of the fabrication process, because the attractive economics achieved through near-net shape manufacturing is the principal drive for the process. The metal powder blend employed, as well as various processing parameters, determines whether shrinkage, growth, or virtually no change in dimension will be exhibited by the finished metal powder compact. Generally, the size change is defined as the linear difference between die size and the size of the sintered part after cooling to room temperature, expressed as a percentage. However, the size change is occasionally specified as the difference between the green part size and the size of the sintered, cooled P/M part.

This article describes the importance of dimensional tolerances, factors that influence dimensional change, and test procedures that must be followed to determine dimensional change. Production examples of dimensional change data for specific powders are included. For additional information on the fundamentals of sintering and changes in dimensions during sintering, see the article "Physical Fundamentals of Consolidation" in this Volume.

Standard Test Method

Specifications that describe the standard test methodology for determination of dimensional change of metal powder specimens as a result of sintering include American Society for Testing and Materials standard ASTM B 610, Metal Powder Industries Federation standard MPIF 44, and International Standards Organization standard ISO 4492. These standard methods employ a rectangular test bar that is pressed to a uniform specified density from the powder being tested. The test specimen is sintered simultaneously with bars from an approved standard powder lot of the same composition, preferably under actual production conditions. Dimensional changes in the bar made from the powder being tested are compared with those of the standard powder test bars to determine whether the test powder is satisfactory.

The recommended test specimen is a transverse-rupture test specimen (Fig. 1), having dimensions 12.7 mm (0.5 in.) wide by 31.75 mm (1.25 in.) long by 6.35 mm (0.25 in.) thick. This is the same test bar described in the article "Green Strength of Compacted Metal Powders" in this Volume.

Equipment Requirements. Determination of dimensional change in sintered P/M parts requires precise control of numerous variables. A suitable test facility should include:

- Balance suitable for weighing accurately to 0.01 g
- Mixer capable of thorough and controlled mixing of metal powders and lubricants
- Compression testing machine or compacting press capable of applying the required pressure to produce the test specimen
- Punches and die to produce the test bar (typical tools for producing the preferred specimen are shown in Fig. 2)
- Micrometers capable of measuring the necessary dimensions to 0.0025 mm (0.0001 in.)
- Sintering furnace that can reach the temperature needed to sinter the metal powder specimens in a controlled atmosphere and that can cool them in a controlled atmosphere, usually a water-jacketed cooling zone

Test Procedure. The length of the die cavity is measured to the nearest 0.0025 mm (0.0001 in.). This dimension (L_D) is recorded for future reference. A representative sample of powder is obtained according to the procedures set out in MPIF 01 or ASTM B 215, "Sampling Finished Lots of Metal Powders." The procedure is summarized in the article "Sampling of Metal Powders" in this Volume.

Three specimens each from the test powder and from the control powder are compacted to a uniform specified density. The procedure for pressing the bars should be agreed on by the supplier and the user, or should be in accordance with MPIF 15

Fig. 1 Standard transverse-rupture test specimen

(a) Width. (b) Length. (c) Thickness. Source: ASTM B 312

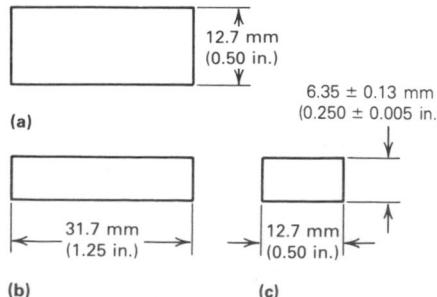

12.7 mm (0.50 in.)

(a)

6.35 ± 0.13 mm (0.250 ± 0.005 in.)

31.7 mm (1.25 in.)

(b)

12.7 mm (0.50 in.)

(c)

Fig. 2 Tooling to produce standard transverse-rupture test specimen

Mating parts should be fitted and lapped to 5 root mean square or better. Source: ASTM B 312

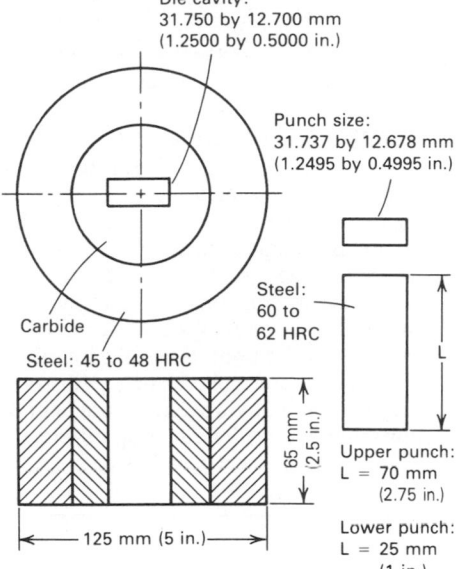

Die cavity:
31.750 by 12.700 mm
(1.2500 by 0.5000 in.)

Punch size:
31.737 by 12.678 mm
(1.2495 by 0.4995 in.)

Steel:
60 to
62 HRC

Carbide

Steel: 45 to 48 HRC

Upper punch:
L = 70 mm
(2.75 in.)

Lower punch:
L = 25 mm
(1 in.)

125 mm (5 in.)

65 mm (2.5 in.)

L

or ASTM B 312, "The Determination of Green Strength of Compacted Metal Powder Specimens."

The test and control bars are sintered simultaneously, using processing conditions agreed to by the customer. Processing conditions include type of furnace, atmosphere, lubricant burn-off time and temperature, sintering time and temperature, and cooling rate.

The length of the sintered bars is measured to the nearest 0.0025 mm (0.0001 in.). This dimension (L_S) is recorded for use in subsequent calculations.

Calculations. Dimensional change for both the test powder and standard powder is calculated as follows:

$$\text{Dimensional change, \%} = \frac{L_S - L_D}{L_D} \times 100$$

where L_D is the die cavity dimension and L_S is the sintered specimen dimension. It should be noted that the above equation expresses percent dimensional change. Deletion of the factor 100 gives the dimensional change in mm/mm (in./in.).

Dimensional changes for the test and control bars are reported to the second decimal place as a percentage of dimensional change from die size. Growth is indicated by a plus sign (+), and shrinkage is indicated by a minus sign (−). The reported value is the average of at least three test bars. The dimensional changes of the

control bars should be within the agreed upon limits to ensure test validity.

Other Reporting Requirements. In addition to the percent dimensional change, the following supplementary information must also be reported for clarification:

- Brand, grade, and lot number of test powder
- Chemical composition of powder mix, if other than elemental powders are being tested
- Type, brand, and percent of lubricant
- Green specimen density
- Preheat (lubricant burn-off) time and temperature
- Sintering time and temperature
- Furnace atmosphere
- Use of a boat, belt, or carrier slab in the furnace for specimen transport

Factors Influencing Dimensional Change. There are numerous variables that can affect the dimensional change of a metal powder component (Ref 1):

Factors associated with pressing

- Powder mix
  Apparent density
  Compressibility
  Ability to work harden
  Chemical composition
  Flowability
  Particle size distribution
  Adsorbed gases
  Segregation of constituents
  Nature of lubricant (particle size and shape)
  Amount of lubricant
  Method of mixing lubricant
  Completeness of mixing
  Temperature rise in mixing
  Changes in storage after mixing and before pressing
- Press tools   • Presses
  Materials   Alignment
  Stiffness   Pressure variations
  Design   Speed
  Clearances   Ejection stresses
  Finish   Vibration
  Wear   Wear
  Tolerances   Design
                 Method of filling
                 Temperature rise

Factors associated with sintering

- Activity of powder
- Chemical composition
- Gas content of compacts
- Amount and nature of lubricant
- Thermal expansion of metals involved
- Density of compact
- Time and temperature
- Phase transformations
- Rate of heating
- Rate of cooling

- Type of atmosphere
- Impurities in atmosphere
- Flow rate of atmosphere

Factors associated with sizing

- Physical properties of compacts
- Density of compacts
- Surface condition of compacts
- Tool tolerances (wear, finish, design, and materials)
- Press alignment, speed of operation, and design
- Amount of reduction in dimensions
- Temperature rise
- Method of lubrication
- Amount and nature of lubrication

Factors associated with heat treatment

- Nature of heat treatment practice
- Atmosphere
- Rates of heating and cooling
- Types of quenching mediums
- Change in composition

Factors associated with dimensional inspection methods

- Changes in personnel
- Temperature
- Gage tolerances
- Condition of inspection tools
- Design of gages
- Lubrication

Dimensional Tolerances

Virtually every powdered metal part produced is manufactured according to dimensional tolerances specified on the part drawing. Depending on the part and its subsequent application, these tolerances may vary in degree of precision. Usually, the more narrow or "tighter" the tolerance, the higher the cost of production, which can raise the cost of the component significantly. Therefore, dimensional precision is important in the price of a P/M part. Generally, tolerances should be specified as wide as possible to minimize production costs by reducing the part rejection rate.

Materials can be prepared so as to exhibit a predictable dimensional change during sintering. This allows for the joining of two P/M parts. Green parts are generally placed in direct contact with each other. During sintering, one portion of the assembly may shrink while the other expands, or one part may simply shrink to a much greater degree than the other to produce a joining effect.

For example, Tabata and Masaki (Ref 2) successfully joined a cylindrical P/M electrolytic copper part to the inside of a ring-shaped compact composed of the same material at a lower density. The procedure was accomplished by pressure fitting the two parts and sintering in a vacuum furnace. The joining strength was enhanced by the difference of shrinkages between the cylinder and ring and the relative sliding under a high contact pressure.

Complex or long P/M parts can be produced easily by applying such a joining method. This joining method has been successfully used with several materials, including brass, nickel-silver, stainless steel, and copper-steel alloys. High-temperature vacuum sintering is often employed for joining stainless steel (Ref 3). Brass, nickel-silver, and iron can be processed in conventional furnaces under protective atmosphere (Ref. 4).

Production Examples

Brass Powder. Different alloy compositions of brass exhibit different dimensional characteristics. In general, as the zinc content of the brass powder increases, shrinkage increases. Typically, brass powder is lubricated with 0.5 to 1% lubricant and pressed at 414 MPa (30 tsi). Sintering is performed in conventional belt-type or pusher furnaces under protective atmospheres, such as dissociated ammonia, endothermic, and nitrogen based. Temperature ranges for sintering brass powders are:

70Cu-30Zn	860 to 880 °C (1580 to 1620 °F)
70Cu-30Zn (leaded)	860 to 880 °C (1580 to 1620 °F)
80Cu-20Zn (leaded)	860 to 890 °C (1580 to 1635 °F)
90Cu-10Zn	870 to 890 °C (1600 to 1635 °F)
90Cu-10Zn (leaded)	870 to 890 °C (1600 to 1635 °F)

Higher lubricant contents and longer sintering times also result in increased shrinkage. Typical data for the above materials (0.5% lubricant) compacted at 414 MPa (30 tsi) are:

Leaded brass	Green density, g/cm³	Dimensional change, %
70Cu-30Zn	7.30	−2.30 to −2.60
80Cu-20Zn	7.60	−1.40 to −1.60
90Cu-10Zn	7.80	−1.10 to −1.40

Nickel-Silver Powder. The two different alloy compositions of nickel-silver powders exhibit different dimensional characteristics. The unleaded grade, which contains a higher percentage of zinc, tends to shrink more than the leaded grade. Nickel-silver powder, like brass powder, is commonly lubricated with 0.5 to 1% lubricant and pressed at 414 MPa (30 tsi). Sintering is performed in belt-type or pusher furnaces under protective atmo-spheres, such as dissociated ammonia, endothermic, and nitrogen based. The temperature range for sintering both materials is 910 to 914 °C (1670 to 1678 °F). Higher lubricant contents and longer sintering times also result in increased shrinkage. Typical data for these powders compacted at 414 MPa (30 tsi) are:

Nickel-silver	Green density, g/cm³	Dimensional change, %
64Cu-18Zn-18Ni	7.4	−1.25 to −1.45
64Cu-16.5Zn-18Ni-1.5Pb	7.5	−1.15 to −1.25

Iron powder sintering is generally performed in conventional belt-type or pusher furnaces under protective atmospheres such as endothermic and, more recently, nitrogen based. A typical sintering temperature for iron powder is 1120 °C (2050 °F). Zinc stearate and synthetic waxes are commonly used lubricants that are added to iron powder in percentages varying from 0.75 to 1%. Carbon additions affect the dimensional change of iron powders in that growth is greater at lower carbon levels. Copper, which is added for strength purposes, increases growth at increasing levels. Typical data for different types of iron powder (0.75% lubricant) are:

Grade	Compacting pressure MPa	tsi	Green density, g/cm³	Dimensional change, %
A	300	21.5	6.20	−0.38
B	455	33	6.00	−0.02
C	450	32.5	6.70	−0.34
D	430	31.2	6.70	−0.02
E	525	38	7.00	−0.09

Note: Sintered in rich endothermic atmosphere at 1120 °C (2050 °F) for 30 min

Bronze Powder. Typically, bronze powders for bearing applications are lubricated with 0.75% lubricant and pressed to a density in the range of 6.0 to 6.4 g/cm³. Sintering is performed in conventional belt-type furnaces in which the preheat zone is at least two thirds of or equal to the high heat zone. Exothermic atmosphere is most commonly used. The recommended temperature range for sintering bronze powder is 815 to 840 °C (1500 to 1540 °F). Generally, slow heating rates result in less growth. For each 5 °C (10 °F) increase in temperature between 815 and 830 °C (1500 and 1525 °F), a 0.2 to 0.3% increase in growth occurs. At temperatures higher than 830 °C (1525 °F), shrinkage occurs. Growth increases approximately 0.1% per 0.1 g/cm³ increase in green density. Maximum growth usually takes place during the first 10 min of sintering; after that, shrinkage is evident. Typical data for a series of premixed bronze powders are:

Grade	Green density, g/cm³	Sintered density, g/cm³	Dimensional change, %
A	6.3	6.27	−0.11
B	6.3	6.12	+0.42
C	6.3	6.02	+1.08
D	6.3	5.90	+1.57
E	6.3	5.85	+1.90

Note: Sintered in exothermic gas atmosphere at 815 °C (1500 °F) for 10 min

Stainless Steel Powder. Different alloy compositions of stainless steel powder exhibit varying dimensional changes. Stainless steel powder can be lubricated with lithium stearate, stearic acid, or synthetic waxes such as Acrawax. Sintering is performed in conventional belt-type or pusher furnaces or in vacuum furnaces. Sintering must be done in a noncarburizing, protective atmosphere such as dissociated ammonia, hydrogen, or vacuum. Temperature ranges for sintering different types of stainless steel powders are:

Types 303L, 304L, and 305L	1120 to 1316 °C (2050 to 2400 °F)
Types 410L and 434L	1120 to 1232 °C (2050 to 2250 °F)

Increased shrinkage occurs when sintering in hydrogen or in vacuum rather than in dissociated ammonia. Typical dimensional change data for stainless steel grades (lubricated with 1% lithium stearate) are:

Grade	Compacting pressure MPa	tsi	Green density, g/cm³	Dimensional change, %
303L	496	36	6.4	−0.80 to −1.60
304L	455	33	6.4	−0.40 to −1.80
316L	469	34	6.4	−0.60 to −1.85
410L	469	34	6.2	−0.65 to −1.95
434L	510	37	6.2	−0.65 to −2.40

Note: Sintered in dissociated ammonia atmosphere at temperature for 45 min

REFERENCES

1. Jones, W.D., *Fundamental Principles of Powder Metallurgy*, Edward Arnold Ltd., London, 1960, p 887-889
2. Tabata, T. and Masaki, S., A New Way of Joining P/M Compacts, *Int. J. Powder Metall. Powder Technol.*, Vol 15 (No. 3), 1979, p 239-244
3. Johnson, P.K., Award Winning Parts Show P/M's Precision and Cost Savings, *Int. J. Powder Metall. Powder Technol.*, Vol 19 (No. 4), 1983, p 312, 313
4. Johnson, P.K., Award Winning Parts Demonstrate P/M Developments, *Int. J. Powder Metall. Powder Technol.*, Vol 15 (No. 4), 1979, 325, 326

Consolidation of Metal Powders

Introduction

CONSOLIDATION in powder metallurgy refers to the production of coherent metal structures using metal powders as the major raw material. Although net shape machine parts such as gears, bearings, filters, and magnetic cores are usually thought of as the typical structures produced, the principles also apply to billets, rod, strip, wire, tubing, and most mill products. The same principles are also involved to some degree in the flame and plasma surface coating processes, where powder metallurgy merges with welding technology.

Once suitable powders have been made, consolidation becomes the second vital step in the development of controlled microstructure materials vital to modern engineering and technology. Such materials range from grain-stabilized tungsten through the enhanced stress/rupture performance superalloys and carbide-dispersed tool steels to bimetal strip and self-lubricating bearings.

The technology of metal powder consolidation includes two basic bodies of theory. These may be drawn upon to solve problems in existing systems. The critical concepts, moreover, are vital to extend technology and thus create new systems.

The first body of theory concerns the mechanics of metal powder aggregates and is dealt with in the article "Mechanical Fundamentals of Consolidation." This deals with the stage of consolidation in which metal powders are converted from more or less free-flowing granular particles into the "green" or unsintered state, having some of the characteristics of a rigid solid.

The second body of consolidation theory is concerned with the further conversion of metal powder aggregates from the weak "green" condition into coherent structural materials through the process of sintering. This is dealt with in the article "Physical Fundamentals of Consolidation," which explains the mechanisms of sintering and the effect of sintering on dimensional changes and mechanical and physical properties of powder compacts.

For industrial, scientific, and economic reasons, the attainment of the desired final shape accompanied by adequate material properties is of great importance in the consolidation process. As is true in other industrial evolution, shape attainment in powder metallurgy is the result of historical interaction among scientific reasoning, engineering principles, and craft knowledge. This is the subject of the article "Shape Fundamentals: Rigid Tool Compaction." When studying this section, one should be aware that rigid tool compaction is the most widely used method for shape attainment in powder metallurgy. This is because no other shaping method has yet equaled it in speed, dimensional accuracy, and economy for similar products.

"Production Presses and Tooling" extends the shape principles of rigid tool compaction to the press requirements to provide the complex force-motion cycles necessary to densify powder metal parts into the green state. The article provides guidance for selection among several systems of mounting tools in a press and discusses capabilities as well as limitations for tool members operating in the compaction cycle. Discussion covers several major types of presses that have been developed and the areas of best application for each.

The next three articles—"Production Sintering Atmospheres," "Production Sintering Equipment," and "Production Sintering Practices for P/M Materials"— extend the discussion of physical fundamentals into contemporary industry. The practical requirements for sintering and development of full sintered properties begin with the technology of gas-metal thermochemistry in providing suitable protective atmospheres in which sintering can proceed.

Once a protective atmosphere requirement is recognized and understood, the time-temperature cycle must be provided precisely by the furnace design. Specific time-temperature-atmosphere practice for a number of major metal groups is discussed in the sintering practices article.

"Roll Compacting of Metal Powders" covers the growing importance of P/M techniques for the large-scale production of strip and sheet. The reader previously acquainted only with conventional cast ingot fed rolling should find the emerging P/M mill practice enlightening.

"P/M Forging" covers an important net shape P/M manufacturing process that has emerged during the 1970's, from technological roots going back to the 1940's. This section will provide both users and producers of P/M forgings with an enhanced data source for effective decisions.

The articles on hot and cold isostatic pressing present an extensive dissertation on these newly emerged consolidation processes. They are historically contemporary with P/M forging. In many instances, the isostatic process is justified on the basis of unique material properties alone. Additionally, the net shape and size capabilities definitely equal or exceed those of rigid tool compaction.

"Secondary Operations Performed on P/M Parts and Products" summarizes the vital guidelines for routing P/M parts and products through conventional heat treating, machining, finishing, and joining operations in manufacturing. Mistakes and waste can be avoided by careful attention to this article.

"Mechanical Properties of P/M Materials" provides a basis for assessment of design capabilities using P/M products. This data section is important for either planning new designs or upgrading older designed equipment, using P/M materials.

"Inspection and Quality Control for P/M Materials" is of basic importance to both manufacturing and design engineers. For designers, this article complements the articles on tooling and presses. For manufacturing engineers, the inspection guidelines should be considered in choice of equipment, tests, and variance ranges.

Following this Section of the Handbook is a Section devoted to special and developing consolidation processes. This is a unique compilation of consolidation processes that have been or are currently close to large-scale emergence. In addition to enhanced material properties, the use of these processes, such as injection molding, results in the fabrication of complex shapes that are difficult or impossible to produce by applying conventional P/M techniques. A review of these processes will be of great value to those involved in further creative development.

Mechanical Fundamentals of Consolidation

By Fritz V. Lenel
Professor Emeritus
Department of Materials Engineering
Rensselaer Polytechnic Institute

CONSOLIDATION OF A METAL POWDER into useful product forms generally begins by transferal of the powder into a container; use of slip casting and subsequent sintering, however, suspends the powder in the slip casting liquid. Density of the powder transferred into the container depends on the properties of the powder and on the method of transfer, which may include tapping and vibrating. Powder density is important when the consolidation step is loose powder sintering, as well as when the powder is subsequently subjected to pressure. If the powder will be subjected to pressure, it can be transferred into a flexible envelope prior to isostatic pressing, or it may be poured into a rigid die prior to unidirectional compacting.

Packing of Metal Powders

Considerable research has been done on the density of packed powders (Ref 1 and 2). Spherical particles of uniform size may be packed into regular arrays. Types of arrays and densities that can be obtained, (expressed as a percentage of the solid density of the powder material) and the coordination number (defined as the number of other particles closest to a given particle for each of the arrays) are:

Array	Density, %	Coordination No.
Simple cubic	52	6
Body-centered cubic	68	8
Face-centered cubic	74	12
Hexagonal closed packed	74	12

When powders of spherical particle shape and uniform size are poured into a container, the density is lower than if the particles were arranged in a regular face-centered cubic or hexagonal close-packed array; that is, the density is lower than 74%. Densities can be increased by vibration or by adding the powder in small quantities with intermediate tapping. The effect of packing method on density and mean coordination number of lead shot with a particle size of 3.78 mm (0.15 in.) is:

Method of packing	Density, %	Mean coordination No.
Poured into glass beaker	55.3	6.92
Poured and vibrated	56.0	7.34
Vibrated to maximum density	57.4	8.06
Added in small quantities and tapped intermittently	62.8	9.51

Source: Ref 3

The number of contacts with other particles for each of approximately 1000 particles in each experiment was determined as a function of density. Mean coordination numbers were calculated from these contacts by averaging. Vibration increases not only the density, but also the mean coordination number. However, even with the most careful vibrating, the highest density is only 62.8%, and the mean coordination number is well below the theoretical maximum of 12. The reason why higher densities are not obtained in packing spherical particles of uniform size is explained in Ref 1.

To increase the packed density of particles of a single size that are poured into a container, particles of a smaller size that fit into the interstices between the larger particles are added. This principle is illustrated in Fig. 1, in which the effect of adding fine (−325 mesh) spherical stainless steel powder to coarse (−100+150 mesh) spherical stainless steel powder on apparent density is shown. When increasing amounts of finer particles are added to the coarser powder, the apparent density first increases and then decreases.

Most powders used in powder metallurgy have irregular particle shapes. Therefore, the density of the powders in the packed state are determined by empir-

Fig. 1 Effect of adding fine (−325 mesh) spherical powder to coarse (−100+150 mesh) stainless steel powder on apparent density
Source: Ref 4

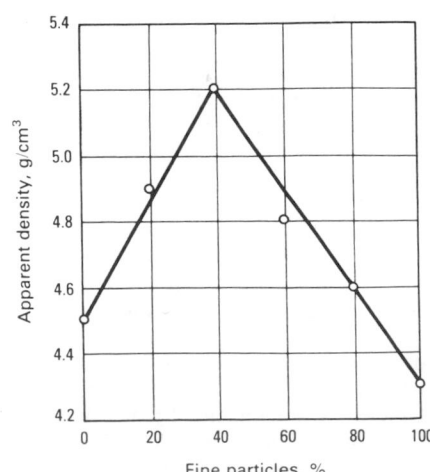

ical methods. Two types of packed densities are apparent and tap densities of powders. Another important property of metal powders determined by empirical methods is flow rate, or the rate at which powders flow from storage containers into dies. Methods for determining apparent density, tap density, and flow rate of metal powders are described in the Section "Characterization and Testing of Powders" in this Volume.

Apparent density depends on the method of producing a powder and its particle size distribution. Table 1 provides apparent densities of some widely used metal powders. Apparent density usually decreases with decreasing particle size, because an increase in surface area causes interparticle friction. Apparent density also decreases as particle roughness increases and the particles become less spherical. Apparent density is one factor that determines compression ratio, or the ratio of the height of loose powder to that of the

green compact, when the powder is compacted at a given pressure. A high apparent density yields a low compression ratio, which may be desirable. However, other factors, such as powder flow rate and green strength, also must be considered.

Flow rate depends on parameters similar to those for apparent density. Fine powders with a particle size smaller than 44 μm (−325 mesh) generally do not flow. For coarser powders, flow rate, as with apparent density, is highest for spherical powders and decreases with increasing roughness of the particle surface.

A comparison of apparent and tap densities of lots of copper, iron, and aluminum powders is shown in Table 2. Tap density is higher than apparent density in each case. The difference depends markedly on the shape of the powder particles. The increase is small for spherical copper and electrolytic iron powders and large for flake copper powder. In Fig. 2, the effect of repeated tapping on the density of an atomized aluminum powder and an atomized and subsequently reduced iron powder is shown. Density is plotted against the number of taps used to increase tap

density. The atomized aluminum powder, with a spherical particle shape, reaches its maximum density with fewer taps than the more irregular iron powder.

Tapping or vibrating is used to obtain high packed densities in ceramic powders that are used in the as-packed condition, uranium oxide and a combination of uranium and plutonium oxide powders packed in cans for use as nuclear fuels (Ref 8). Elaborate studies of ways to obtain high densities of packed powders were made for this application. Oxide powders were produced by crushing billets formed by elevated-temperature, high-energy pneumatic impaction. Three sizes of powder, with particles ranging from −6+10 mesh, −20+65 mesh, and −200 mesh, were used. Quantities of each size range were chosen so that the finest particles would fill the interstices between the medium-size particles, and the medium-size particles would fill the interstices between the coarse particles. Loading of the particles into the cans and the characteristics of applied vibration—frequency, acceleration, and wave form—were adjusted to obtain packing densities of 86 to 88% of theoretical.

Tapping and vibration also are used to fill flexible envelopes prior to isostatic pressing. Vibration before pressing powders in rigid dies increases the density of the loose powder in the die and also makes the powders flow more readily, allowing controlled volumetric fill of the die, even for powders with poor flowability. However, vibrating the die prior to compaction makes the process of filling the die more complicated.

Behavior of Metal Powders Under Pressure

The major processes by which powders are compacted by the application of an exterior force are:

- Pressing of powders in rigid dies
- Isostatic pressing of powders in flexible envelopes
- Roll compacting of powders into sheet or strip
- Extrusion of metal powder products

Only the first two of these processes are discussed in this article. Fundamentals and applications of powder rolling can be found in the article "Roll Compacting of Metal Powders" in this Volume. Hot extrusion of metal powders is covered in the Appendix to the article "Hot Pressing" in this Volume.

Compaction of loose powder by either pressing in rigid dies or isostatically has

Table 1 Apparent densities of various sizes of several different powders

Material	Average particle diameter, μm	Apparent density, g/cm³
Aluminum		
Atomized	5.8	0.62
	6.8	0.75
	15.5	0.98
	17.0	1.04
	18.0	1.09
	(a)	1.22
	(b)	1.25
Nickel		
Carbonyl	3.2	0.61
Precipitation	3.5	1.80
Carbonyl	3.8	1.87
Carbonyl	4.1	2.10
Precipitation	4.4	2.09
Precipitation	8.0	2.60
Precipitation	(c)	3.60
Tungsten	1.20	2.16
	2.47	2.52
	3.88	3.67
	6.85	4.40
	26.00	10.20
Stainless steel		
Atomized, spherical	(d)	4.3
	(e)	4.5
	(f)	4.4
	(g)	4.5
	(h)	4.5
Iron		
Reduced	6	0.97
Carbonyl	7	3.40
Reduced	51	2.19
Electrolytic	53	2.05
Electrolytic	63	2.56
Reduced	68	3.03
Electrolytic	78	3.32

(a) 60% above 44 (+325 mesh). (b) 75% above 44 (+325 mesh). (c) −40+325 mesh. (d) −325 mesh. (e) −270+325 mesh. (f) −200+270 mesh. (g) −150+200 mesh. (h) −100+150 mesh
Source: Ref 5

Table 2 Apparent and tap densities of various powders

Material	Apparent density, g/cm³	Tap density, g/cm³	Increase, %
Copper			
Spherical	4.5	5.3	18
Irregular	2.3	3.14	35
Flake	0.4	0.7	75
Iron (−100+200 mesh)			
Electrolytic	3.31	3.75	13
Atomized	2.66	3.26	23
Sponge	2.29	2.73	19
Aluminum (−200 mesh)			
Atomized	0.98	1.46	49

Source: Ref 6

Fig. 2 Density of loose powder as a function of the number of taps for atomized aluminum and atomized and reduced iron powder
Source: Ref 7

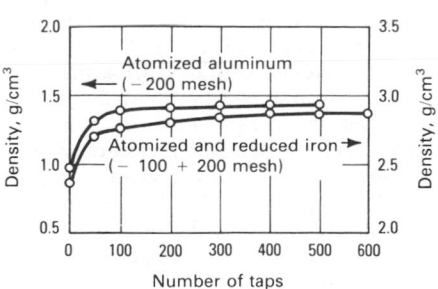

been described qualitatively by Seelig and Wulff (Ref 9), who postulated three stages. In the first stage, powder particles are restacked or rearranged so that bridging, which always occurs in a randomly arranged stack of loose powder particles, is partially eliminated.

The second stage involves elastic and plastic deformation of the particles. The amount of plastic deformation depends on the ductility of the powder material. Deformation is more extensive in the compaction of a ductile metal powder than in that of a brittle ceramic powder. Even in powders composed of the same ductile metal, compressibility, or the density obtainable in a green compact with a given pressure, may vary considerably. Lower compressibility results when powder particles are strain hardened during disintegration. Powder particles that have internal porosity (powders reduced from oxide), rather than solid (atomized and electrolytic) powders also exhibit characteristically low compressiblity.

In the third stage, the powder particles—which are inherently brittle, as in ceramic powders, or have been embrittled through work hardening—fracture under the applied load and form smaller fragments. The three stages usually overlap.

Fischmeister, Arzt, and Olsson (Ref 10) re-examined the rearrangement or sliding stage and the plastic deformation stage during compacting using quantitative metallography. They worked primarily with spherical powders of ductile metals, and they found that deformation and sliding occur concurrently from the beginning. Sliding, or nonradial particle motion, stopped for spherical particle compacts at a relative density of about 80%; for particles with complex contours, sliding may continue to higher densities. The plastic flow stage of densification for spherical powders may be divided into two substages: deformation in the region surrounding the particle contacts, and filling of remaining cavities by local extrusion.

Pressure-Density Relationship in Powder Compacts

With increasing applied pressure, the density of the powder mass increases, or porosity decreases. The relationship between applied pressure and density or porosity of the powder compact has been studied by many researchers, who have attempted to establish mathematical relationships between pressure and relative density. In much of the earlier work on this relationship, powders were compacted in rigid dies. This method of compaction is of greatest interest to the practical powder metallurgist who wishes to produce parts to closely controlled dimensions.

However, both the distribution of applied pressure through the compact and the density distribution within the compact after compacting are uneven. Mathematical relationships have been established between the nominal pressure applied to the upper punch and the average density of the compact produced. This relationship has much practical importance, but because only nominal or average values of pressure and density are used in the equations, it is difficult or impossible to develop fundamental bases. Therefore, in more recent investigations in which pressure-density relationships in powder compacts have been determined, the loose powder generally is compacted isostatically in flexible envelopes, so that a uniform density throughout the compact may be assumed.

Mathematical Relationships. Past literature on the pressure-density relationship has been reviewed by Bockstiegel and Hewing (Ref 11). The following relationship, which has been widely used, was developed by Balshin (Ref 12):

$$\ln P = -AV + B$$

in which $\ln P$ is the natural logarithm of the applied pressure; V is the relative powder volume (the ratio of the volume of the powder compact and the volume of solid metal of the same mass); and A and B are constants. This equation cannot be valid at high pressures, because it would predict a relative powder volume smaller than one at these pressures. Also, for zero pressure, the equation would predict an infinitely large relative powder volume.

The results of Balshin's experiments (Ref 13) show that the linear relationship between $\ln P$ and V is valid only for certain powders and, even for these powders, only within narrow pressure ranges. The equation cannot be explained on the basis of fundamental principles, but is the result of curve fitting.

A second relationship between pressure and density was first developed empirically by Athy (Ref 14), Konopicky (Ref 15), and Shapiro (Ref 16). Later, Torre (Ref 17) attempted to develop a theoretical basis for the equation by using a formula for the densification of a hollow sphere under hydrostatic pressure, assuming that the pores in the green compact could be represented by the single large pore in the center of the hollow sphere. This relationship was developed by Heckel (Ref 18) as:

$$P = \frac{1}{K}\left(\ln\frac{1}{1-D} + B\right)$$

where D is the relative density of the compact (ratio of green density to solid density); P is the applied pressure; and K and B are constants. Figure 3 presents these experimental results for a number of powders, which show that for pressures between 150 and 700 MPa (20 and 100 ksi), the relationship applies reasonably well. The powders were compacted in a rigid die.

A comparison of the pressure-density relationship for compacts made from MH-100 sponge iron powder pressed in a rigid die and sponge iron powder pressed isostatically is shown in Fig. 4. It is evident that higher densities can be obtained by

Fig. 3 Relationship between pressure and green density for an iron, a nickel, a copper, and a tungsten powder

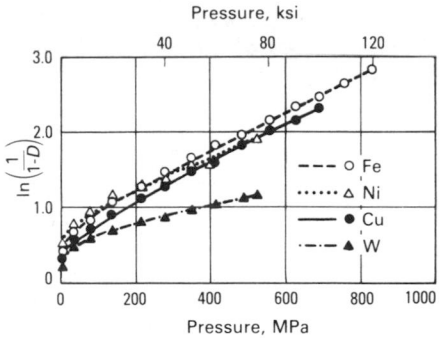

Fig. 4 Relationship between pressure and green density for Hoeganaes MH-100 iron powder pressed isostatically and in rigid dies

Source: Ref 19

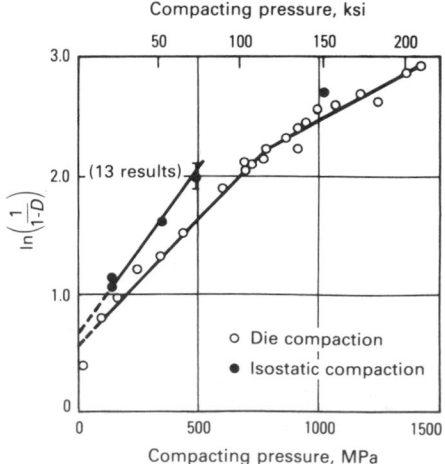

isostatic pressing than by compaction in rigid dies. Also, the relationship between ln $(1/1-D)$ and pressure is best represented by two straight lines—one for pressures up to 700 MPa (100 ksi) and the other for higher pressures.

James (Ref 20) studied the pressure-density relationship for isostatically pressed compacts of several powders. Results are shown in Fig. 5. Two linear relationships of ln $(1/1-D)$ versus P occur for low- and for high-pressure ranges, respectively, for all powders. The break between these two relationships, corresponding to two specific slope factors (K_1 and K_2), occurs at different pressures for different powders. James related the slope factor K_1 for low pressures to the yield strength of the metals with an approximately linear relationship between K_1 and the reciprocal yield strength $1/\sigma$. Such a relationship would be expected from Torre's (Ref 17) theoretical model for the isostatic compaction of a hollow sphere, as indicated by Heckel (Ref 18).

The slope factor K_2 at higher pressure is related by James to the work hardening of the metals. He showed an approximately linear relationship between the slope factor K_2 and the Meyer work hardening index η. For relatively soft powders of aluminum, copper, zinc, and even iron, a terminal density appears to be reached at very high pressures; further increases in pressure do not increase density.

When a powder mass is subjected to increasing pressure, the total amount of porosity in the mass decreases; however, the size and size distribution of the pores will change as well. These changes in pore size

and pore size distribution of green compacts as a function of pressure have been studied by Bockstiegel (Ref 21-23) and by Klar (Ref 24, 25). Bockstiegel studied pore size distribution in green compacts of electrolytic iron powder, whose particles are solid and do not contain interior pores (Ref 21). He found by quantitative metallography that, with increasing pressure, pores are eliminated by size class; the largest pores disappear at the lowest pressure and the smallest at the highest. This does not mean that when the pressure with which a green compact is pressed is increased, only pores larger than a given critical size decrease in volume. Pores of all sizes shrink simultaneously so that the fraction of pores that, due to shrinkage, is pushed from one pore size class to the next smaller one is replaced by the pores that have shrunk from the next larger pore size class.

Bockstiegel (Ref 22, 23) and Klar (Ref 24, 25) have investigated changes in pore size distribution when powders are compacted in which the individual particles are porous. This has great practical importance, because the particles of powders produced by the reduction of oxide generally are porous. Green compacts of such reduced powders contain two types of porosity that is either due to internal pores

within the particles (I-pores), or due to voids between particles (V-pores).

In most cases, I-pores are much smaller than V-pores. The difference often amounts to an order of magnitude. Upon compaction at increasing pressures, both the size of the V-pores and the porosity caused by them decrease much faster than the size of the I-pores and the porosity they cause.

This is illustrated in Fig. 6, in which the logarithm of the internal particle porosity (I-pores), that of the interparticle porosity (V-pores), and that of the total porosity, all expressed as a volume percent, are plotted against compacting pressure for two powders. Powder A consists of highly porous iron powder particles produced by reduction from magnetite, and powder B is a nonporous electrolytic iron powder. The interparticle porosity of powder A decreases more rapidly with increasing pressure than its internal porosity. For nonporous powder B, total porosity is identical with interparticle porosity. The curve for the decrease in total porosity for this powder is very close to that for the decrease in interparticle porosity of powder A.

Klar made similar investigations on porous iron (Ref 24) and porous copper (Ref 25) powders. He found that the relative shrinkage of the small I-pores increased with increasing compacting pressure, while the relative shrinkage of the large V-pores decreased with increasing pressure. According to Klar, maximum compressibility is found in powders without internal pores.

Internal porosity of powders produced by oxide reduction may be eliminated by precompacting the powder to a high density, followed by grinding and annealing. However, maximum compressibility achieved by eliminating internal porosity must be balanced against adequate green strength, which decreases with decreasing internal porosity.

Among the relationships between pressure and green density of powder compacts, only Torre's equation (Ref 17) was based on an actual physical model (the isostatic compaction of a hollow sphere). Whether the derivation of this equation is valid has been disputed by Bockstiegel (Ref 11, 21) and Hewitt, Wallace, and de Malherbe (Ref 26). Heckel's equation (Ref 18), although nominally based on Torre's model, must be considered empirical, because it uses two adjustable parameters that are not related to the properties of the powder. The adjustable parameters are selected empirically based on best fit results with the compaction data.

Recently, Fischmeister and Arzt (Ref 27) developed a new model for the densifi-

Fig. 5 Compacting pressure versus ln $(1/1-D)$ for various powders (isostatic compaction)

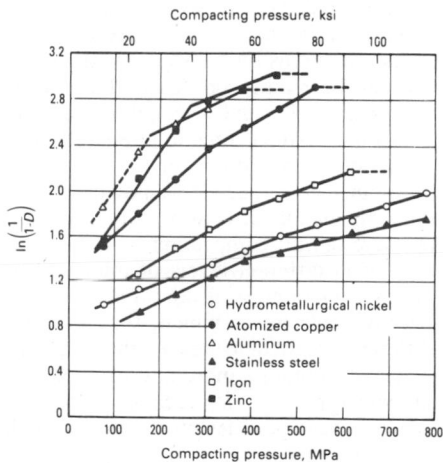

Fig. 6 Change of total porosity, porosity inside particles, and porosity between particles with increasing compacting pressure

Powder A is iron powder obtained from pure magnetite reduced at 500 °C (930 °F) in hydrogen, particle size −24+100 mesh, with a high degree of internal particle porosity. Powder B is electrolytic iron powder, particle size −40+150 mesh, with minimal internal particle porosity.

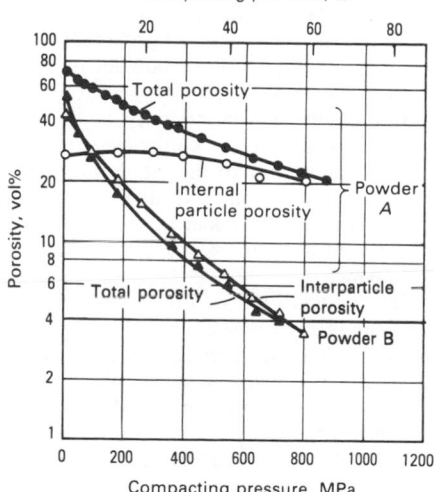

cation of compacts made from ductile metal powders. The model is directly applicable only to spherical particles of uniform size. It applies to the stage of deformation of the surroundings of the particle contacts and the hydrostatic extrusion stage. In the first of these stages, the applied pressure acts to increase the area of contact between particles through the flow of material at right angles to lines between particle centers. At the same time, the average number of contacts (coordination number) per particle, which is 7.3 in the loose powder arrangement, increases and reaches values of more than 12 at relative densities of 92%.

The change in the total area of all contacts can be analytically related to the change in relative density of the compact. The force required to increase the contact area also increases with increasing densification. This increase can be related to the total contact area and to the flow stress of the material. The material work hardens, and the flow stress subsequently increases. Its value can be calculated from the yield stress of the annealed material and the degree of deformation as the contact area increases.

At relative densities of over 92%, individual contact areas between particles begin to overlap. Further densification occurs through a process of hydrostatic extrusion of material through openings with sizes equal to the size of the pores between particles. Flow stress increases rapidly, and it can be calculated from the shape of the model at this stage.

The equation for the two stages of densification—increase in contact area between particles and extrusion of material into the pores—can be combined additively. Pressure is shown on the left side of the equation. On the right side, the product of the yield strength of the material (increasing through work hardening) and a factor that depends exclusively on the shape of the model (which changes through increases in density during the contact deformation and the extrusion stages of densification) are shown.

The equation does not contain parameters that can be arbitrarily adjusted. The agreement between the equation and the data for relative density as a function of pressure during the isostatic compaction of spherical bronze powder particles is quite satisfactory in the range between 63% relative density for zero pressure and 98% relative density for a pressure of 1000 MPa (145 ksi).

The model of Fischmeister and Arzt and the equation derived from it cannot be applied to powders of nonspherical particles of nonuniform size. The equation may become the basis for adjusted equations for other powders if the parameters necessary for the adjustment can be related to the actual properties of the powders.

Stress and Density Distribution in Compacts Pressed From Powder

Of the two methods of compacting powders discussed in this section—pressing in rigid dies and isostatic pressing—isostatic pressing benefits from reduced friction between die wall and powder, which is characteristic of rigid die compaction. Isostatic pressing—in which the powder is sealed in a flexible envelope and the assembly of envelope and powder is immersed in a fluid, which is subsequently pressurized—is shown schematically in Fig. 7. Cold isostatic pressing, generally in elastomeric flexible envelopes or bags immersed in liquids, and hot isostatic pressing, in thin, flexible, metallic envelopes with a gas as the medium of pressure application, are discussed in separate articles in this Volume.

The transmittal of stress through the mass of the powder may not be completely uniform, even in isostatic pressing. Uniformity depends on the design details of the flexible envelope. The uniformity of stress transmitted through the powder mass and the density distribution in the green compacts produced in rigid dies are much less uniform than in isostatic compacting.

When a liquid is subjected to hydrostatic pressure inside a confined, rigid die, the stress transmitted by the liquid upon the interior surfaces of the die is uniform, regardless of whether the liquid must flow around corners. This is not so when a powder is pressed; it flows only in the direction of the applied pressure and not around corners. This phenomenon is illustrated in Fig. 8, which schematically shows the compaction of powder in a die with a sidearm.

If pressure is applied only to the top punch, the powder is compacted only in the vertical section of the die, but remains loose in the horizontal section. To obtain compaction in the sidearm, pressure must be applied to both the top punch and the side punch.

In producing parts from metal powder, pressure is applied to the powder only with punches that move in the vertical direction. This is because in stripping a pressed compact that has been pressed by punches moving horizontally and vertically, the die must be disassembled after each compression stroke. This procedure would not be

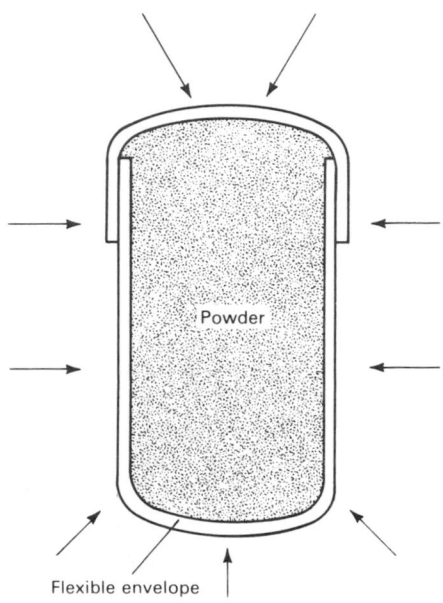

Fig. 7 Isostatic pressing with powder sealed into a flexible envelope and the assembly subjected to uniform pressure

Powder

Flexible envelope

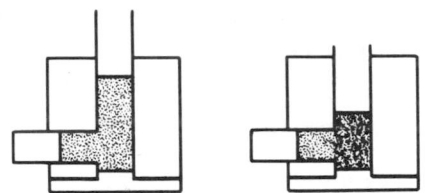

Fig. 8 Density distribution of compacts pressed in dies with a sidearm

practical in large-scale, high-rate parts production. It also would be difficult to prevent a plane of weakness in the compact where the horizontal and the vertical sections meet. Compacting metal powder parts in rigid dies, in which the punches move only in the vertical direction, imposes limitations on the shape of the parts that can be readily produced. Parts with re-entrant angles and with holes at an angle to the vertical direction generally are not produced by compacting.

Another consequence of powder flowing only in the direction of the applied pressure during compaction is that, when parts with different levels of thickness in the direction of pressing are compacted with only a single lower punch, they develop different green densities in the different levels. To obtain a more uniform density, individual punches for each level are necessary. These punches must travel in such a way that the ratio of the height of the

Fig. 9 Density distribution in a two-level compact

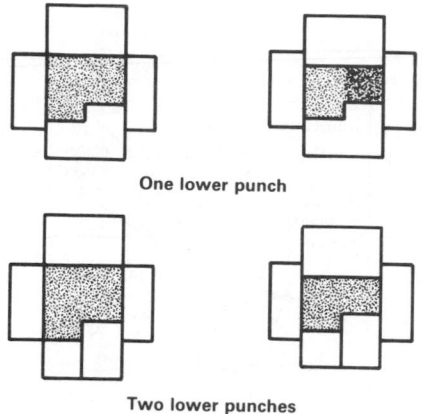

One lower punch

Two lower punches

Fig. 10 Density distribution in compacts

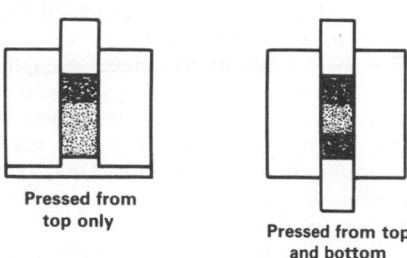

Pressed from top only

Pressed from top and bottom

Fig. 11 Density distribution in a cylindrical nickel powder compact

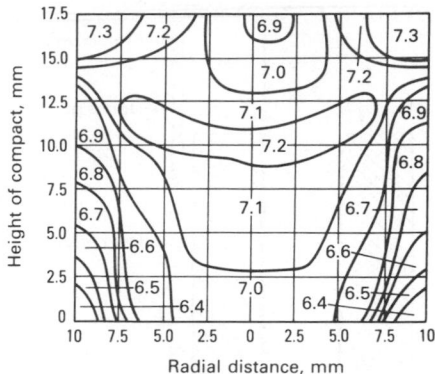

loose powder to that in the green compact (the compression ratio) is the same for all levels. This type of punch arrangement is illustrated in Fig. 9.

Even compacts with only one level in the direction of pressing show variation in density in the pressing direction due to the friction between the die wall and the powder. This friction is reduced as much as possible by lubricants (see the article "Lubrication of Metal Powders" in this Volume). Usually, lubricants are added to the metal powder in powder form. Lubricants reduce friction between the powder and the side wall, as well as between individual powder particles. They also may be applied as a thin coating to the walls of rigid dies. In large-scale production, this die wall lubrication is not commonly used.

Variation in density in single-level compacts depends on the efficiency of the lubrication and whether pressure is applied only from the top or from both the top and bottom. Improved density distribution with both top and bottom pressure application is illustrated schematically in Fig. 10. The ratio of the area of the compact to which pressure is applied to the area of contact between the powder and the side walls of the die also affects density. The smaller the ratio, the lower the density of the compact pressed at a given pressure (Ref 28).

A straightforward method of determining density distribution in a compact is to slice the compact into small sections both radially and axially and to determine the density of each slice. This method was first suggested by Unckel (Ref 29). When such an experiment is made on a single-level cylindrical compact, density gradients are found in the vertical and horizontal direction.

An alternative to Unckel's time-con-

suming method of determining density distribution was developed by Kamm, Steinberg, and Wulff (Ref 30). They inserted a thin lead grid with square or round holes into the powder prior to compacting, with the plane of the lead grid parallel to the direction of pressing. The grid must not touch either the punches or the side walls of the die. During compaction, the lead grid maintains its position on a plane parallel to the direction of pressing, but is deformed as the holes decrease in area proportionally to the increase in density from loose powder to green compact. By measuring the areas of the holes by x-ray radiography of the compact, as they depend on their vertical and horizontal positions in the compact, density distribution can be determined.

A third method of measuring density distribution is to slice the pressed compact vertically and measure the hardness distribution over the sliced surface of the compact. Hardness values can be translated into density values by a calibration curve, which is obtained by measuring the hardness and density of thin compacts of the same material pressed at a series of pressures.

As an example, the density distribution of a nickel powder compact, pressed only from the top, was determined by this method by Kuczynski and Zaplatinsky (Ref 31). These results are shown in Fig. 11, which depicts a density distribution that is typical for compacts pressed from only one direction. The highest density is found at the top of the outer circumference, where wall friction causes maximum relative motion between particles. The density at the circumference decreases rapidly from the top downward, with the lowest compact density near the bottom. Density distribution at the centerline of the compact, where the influence of friction between die wall and powder is at a minimum, is more uniform. Density is at a maximum about halfway between the top and bottom faces of the compact.

The results of these density distribution measurements in metal powder compacts

clearly show that uniform normal compression stress applied to the upper punch when a single-level cylindrical compact is pressed from one direction is not transmitted uniformly throughout the compact. Instead, both the direction of stress (normal and shear components) and the magnitude of the stress is quite complex. Earlier attempts to treat the distribution of stress in powder compacts theoretically have been discussed by Bockstiegel and Hewing (Ref 11). Recent reviews include that of Broese van Groenou (Ref 32).

A number of experimental methods and apparatus for measuring stress distribution have been developed. The simplest one was developed by Unckel (Ref 29), which is illustrated schematically in Fig. 12. This method measures vertical stress distribution. The apparatus consists of a die with a fixed lower punch and a movable upper punch. The die barrel rests on three hardened steel balls, which in turn rest on a soft iron ring. The Brinell-type impression on the ring measures how much of the stress exerted on the upper punch is transmitted to the bottom of the compact and how much is lost due to friction between the compact and die barrel. The size of the Brinell impression can be translated into units of stress by suitable calibration.

In Unckel's experiment, the average axial stress across the cross section of the compact is measured. Duwez and Zwell (Ref 33) showed how the radial distribution of this axial stress can be determined. An experimental method of measuring die wall friction is described by Broese van Groenou (Ref 32).

The problem of measuring stress distribution in compacts of more complex shape, that is, those with more than one level of thickness in the direction of pressing, was addressed by Morimoto, Hayashi, and

Fig. 12 Apparatus to measure die wall friction during compacting

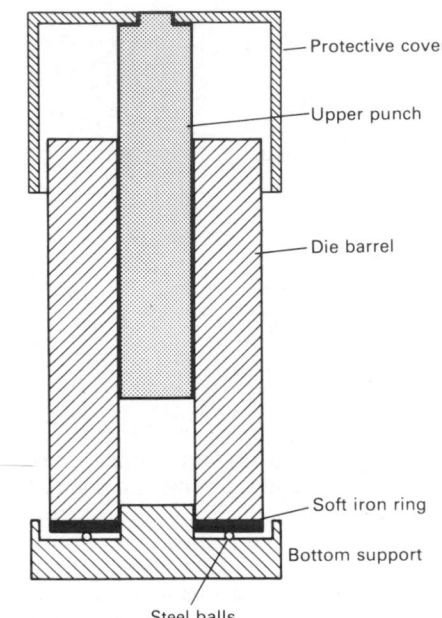

- Protective cover
- Upper punch
- Die barrel
- Soft iron ring
- Bottom support
- Steel balls

Takei (Ref 34). They determined stress distribution with moire patterns.

Lubrication affects density and stress distribution when metal powders are pressed in rigid dies. More uniform distribution is achieved with better lubrication. A method of evaluating the effectiveness of various lubricants by measuring the coefficient of friction between metal powder and die wall was developed by Tabata, Masaki, and Kanata (Ref 35).

Green Strength of Powder Compacts

Green strength refers to the mechanical strength of a cold pressed powder compact. Adequate mechanical strength of a green part permits cold pressing of the powder, ejection of the green part from the die, and transfer of the green part to the sintering furnace without breakage. Standard tests for green strength measurement include the American Society for Testing and Materials standard ASTM B 312 and Metal Powder Industries Federation standard MPIF 15, which are based on pressing the powder to be tested into a rectangular bar of specified dimensions and measuring the transverse-rupture strength of the bar. For more detailed information, see the article "Green Strength of Compacted Metal Powders" in this Volume.

The Rattler test for green strength measurement (Metal Powder Association standard 15-517) is still widely used in Japan.

It is based on tumbling green parts under specified conditions and is particularly well suited to characterize the resistance of the edge of a part to crumbling.

For most applications, green strength values of about 5.5 MPa (800 psi) or more permit pressing and safe handling of green parts. Lower green strength values may produce insufficient strength and horizontal lamination in the part. These laminations appear when the part is ejected from the die as a result of a small expansion of the part, due to the release of elastic stresses. This type of defect can be minimized or avoided by providing the die with a slight taper (draft angle) or a small radius at the upper rim. Parts with sharp contours or thin sections, or parts made from a blend of several powders in which some of the components detract from the overall strength of the part, require powders with higher green strength characteristics.

Theories on green strength include cold or friction welding and interlocking of particles as the mechanisms responsible for the strength of a green part. While these theories do not permit prediction of green strength from basic powder characteristics, they provide principles for the development of powders that require specific green strength characteristics. The large body of empirical green strength data that is available may be conveniently separated into intrinsic, surface-related, and geometric powder properties (Ref 36).

Intrinsic Powder Properties. Figure 13 shows green strength/compacting pressure curves of several metal powders. These curves are typically linear or concave upward (with respect to the pressure axis) and illustrate that green strength increases rapidly with increasing compacting pressure. These curves may be converted into green strength/green density curves by means of the corresponding compressibility (green density/compacting pressure) curves of the powders.

At high compacting pressures or with high-speed compaction, green strength may be lower than expected due to air entrapment during compaction. Such parts may have lamination defects with an appearance similar to those produced by differential expansion during ejection from the die.

In general, green strength increases with the increasing intrinsic softness, or plasticity, of a powder. This relationship is, however, often masked by the influence of geometric powder characteristics. Annealing of a work-hardened powder raises its green strength/compacting pressure curve; it may lower its green strength/green

Fig. 13 Green strength versus compacting pressure curves for various metal powders

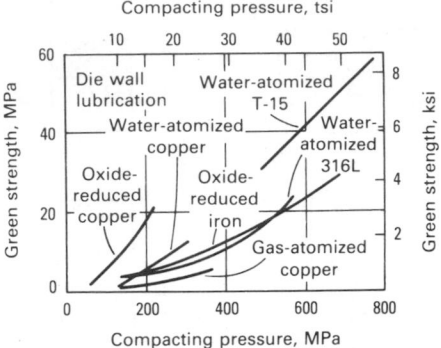

density curve, because the annealed powder requires a lower compacting pressure to achieve a given density. If compaction exceeds the work-hardening capacity of a powder, particle fracture may occur and may cause green strength irregularities.

Surface-Related Factors and Powder Mixtures. The most important commercial examples in this category include the use of mixtures of powders (such as copper and tin or the addition of copper and nickel to iron), the addition of graphite and lubricants, and the presence of oxide films due to tarnishing. The green strength of mixtures of similar metal powders may be estimated from the green strength of its components by applying the rule of mixtures.

The addition of graphite and most lubricants reduces green strength to a much greater extent than indicated by the rule of mixtures, however. The effect of a 1% addition of zinc stearate on the green strength of a water-atomized low-alloy steel powder is shown in Fig. 14. The detrimental effect of the lubricant is more pronounced at high compacting pressures, because it causes the shape of the curve to change from concave to convex.

Because lubricants affect many other powder properties, most notably compressibility and powder flow, selection of amount and type of lubricant addition is made on the basis of the best compromise for a given application (see the article "Lubrication of Metal Powders" in this Volume). Typically, lubricants that are less detrimental to green strength reduce both powder flow and compressibility.

Most metal powders possess thin oxide films—either directly from manufacturing, as in the case of water-atomized powders, or from subsequent tarnishing in air. The effect of these oxide films on green strength appears to depend on whether these

Fig. 14 Effect of admixed lubricant on green strength of water-atomized 4600 low-alloy steel powder

Source: Product Bulletin 182, Hoeganaes Corp, Riverton, NJ

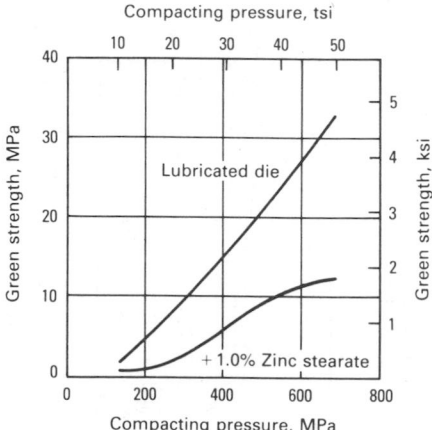

Fig. 15 Effect of surface oxide films on green strength of copper and type 316L stainless steel powders

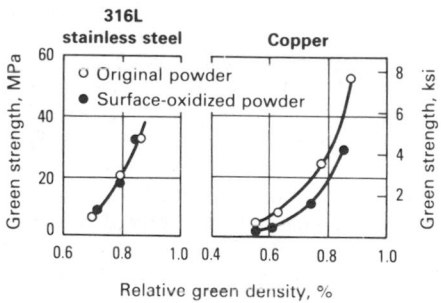

films break down during compaction. For both iron and copper powders, tarnishing or surface oxidation produces significant losses in green strength. Figure 15 illustrates the extent of such losses for copper powder and also shows that oxide films on water-atomized stainless steel powders do not impair green strength. These oxide films were produced by low-temperature oxidation with air in a fluid bed.

Geometric powder properties include particle shape, particle size, and particle porosity. Spherically shaped solid particles (typical of many gas-atomized powders) have such low green strength that they cannot be used in conventional die compaction. Many water-atomized powders have an irregular particle shape with a solid bulk structure. Such powders typically provide intermediate green strength. Oxide-reduced powders are of irregular

Fig. 16 Fractographs of copper parts made from spherical copper powders

(a) Gas-atomized solid and spherical powder (−140+200 mesh). (b) Oxide-reduced porous and spherical copper powder (−140+200 mesh)

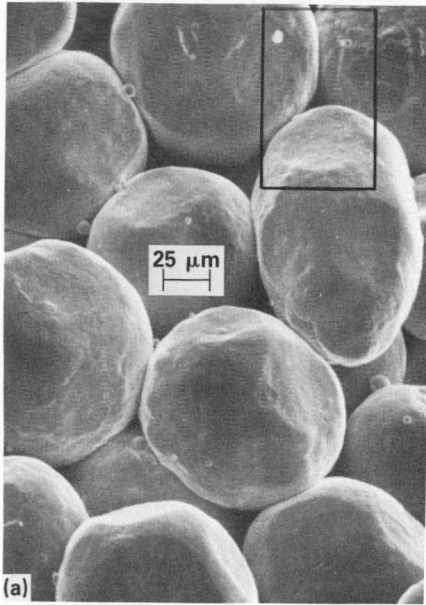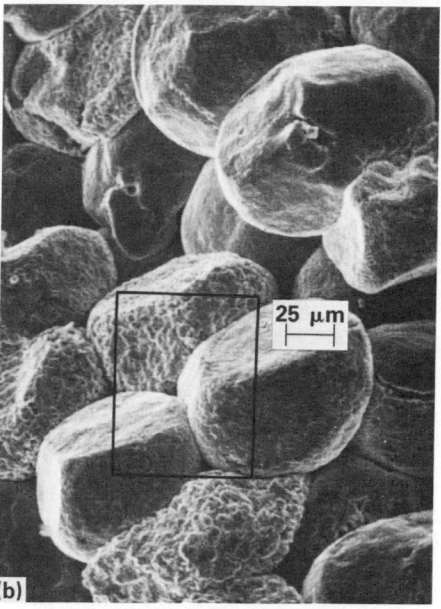

particle shape and possess internal particle porosity. Such powders give maximum control over green strength. Green strength increases with increasing particle porosity and with decreasing particle pore size. The fractographs of green parts shown in Fig. 16 illustrate the larger nominal contact areas and greater particle interpenetration that can be obtained with such powders.

Other geometric factors that are beneficial to green strength include decreasing particle size and increasing specific surface area of a powder. Figure 17 shows the effect of particle size for isostatically compacted iron powders. The true effect of particle size is sometimes masked by other factors. Figure 18, for example, shows a maximum for the green strength of a water-atomized steel powder of intermediate particle size. In this instance, particle shape becomes more regular with decreasing particle size; oxygen content also varies with particle size. This is typical of many water-atomized powders. In such cases, the apparent density of a powder is frequently a useful measure of particle shape and green strength. All geometric factors that are beneficial to green strength are detrimental to compressibility.

Green Strength Theories. As mentioned previously, the two mechanisms used to explain the origin of green strength are cold welding and mechanical interlocking of particles. Cold welding refers to the formation of metallic contacts be-

Fig. 17 Effect of particle size on green strength of isostatically pressed electrolytic iron powder

Fine: 100% −325 mesh, 90% 10 to 44 μm. Medium: 22% −325 mesh, 78% −65+325 mesh. Coarse: 100% −42+100 mesh. Source: Ref 37

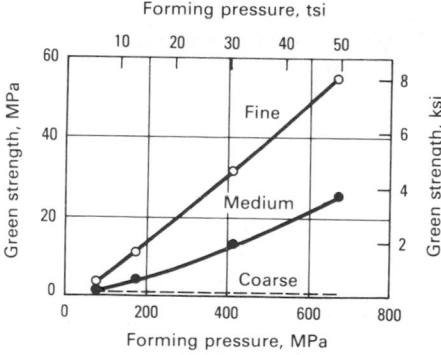

tween neighboring particles during pressing; mechanical interlocking refers to their entangling in the manner of a three-dimensional jigsaw puzzle.

Because both of these mechanisms respond similarly to many powder properties, it is difficult to provide conclusive experimental evidence for either theory. For instance, the important influence of a low apparent density of a powder and the accompanying increase of shearing and heat development during compaction is beneficial to the formation of metallic bonds,

Fig. 18 Green strength, green density, and apparent density of water-atomized steel powder
Source: Technical Bulletin on Green and Sintered Properties of 316L and Other High Alloys, Hoeganaes Corp, Riverton, NJ, 1969

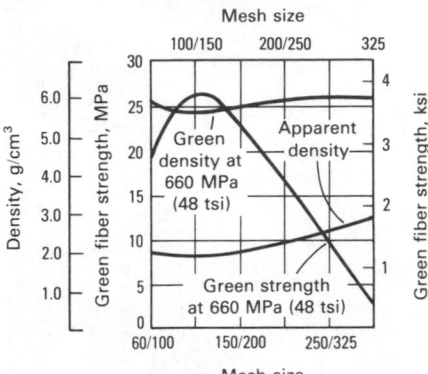

Fig. 19 Effect of thin film of oleic acid on green strength and electrical conductivity of copper parts

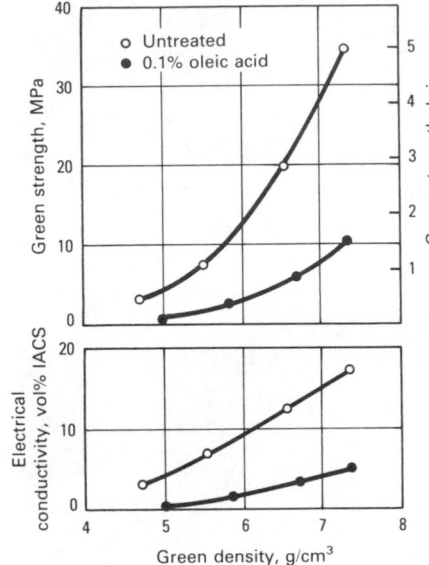

Fig. 20 Experimental triaxial compaction chamber
Source: Ref 44

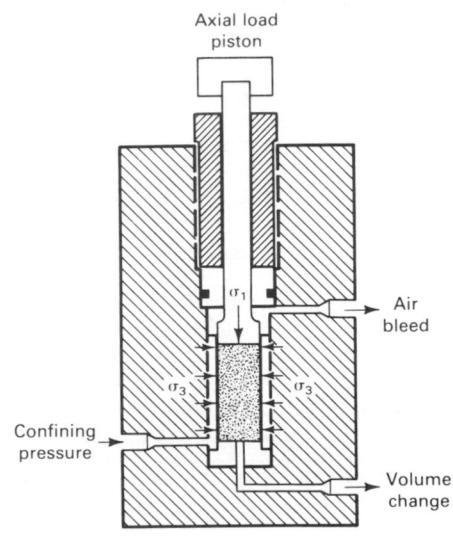

as well as to improved keying and interlocking of particles. Nevertheless, some of the features and relationships believed to be specific to each of the two mechanisms are summarized below.

Proponents of the cold welding theory argue that the well-known fact that clean metal surfaces form strong attractive junctions when joined (Ref 38) also applies to the pressing of parts from metal powders. In fact, experimental evidence of such interatomic or metallic bonding forces between powder particles has been presented (Ref 39).

Large and similar reductions in both green strength and electrical conductivity of copper compacts pressed from pure and oleic acid-coated copper powders (Fig. 19) were interpreted to indicate that green strength, like electrical conductivity, depends on the number and size of metallic microcontacts formed between particles. Thin oxide films on copper powder particles gave similar results (Ref 40).

High green strength powders pressed and sintered to the same density as low green strength powders of the same chemical composition produce significantly stronger parts. Because the strength of sintered parts is explainable in terms of metallic bonding and microstructural features, such as grain size and pore structure, and because proportionality exists between green and sintered strength, it is reasonable to assume that bonding forces responsible for strength are identical for both green and sintered parts.

Proponents of the interlocking theory of green strength argue that scanning electron micrographs of fractured surfaces of green parts do not show any evidence of

metallic contacts and that parts pressed from spherical powders have no measurable green strength, in spite of substantial deformation and nominal contact areas between particles. Furthermore, roll-compacted green strip has good mechanical strength, yet is quite flexible (Ref 41). Some evidence shows that interlocking may be prevalent at low densities and that cold welding may be dominant at high densities (Ref 40, 41).

Methods for Compacting Powders to Higher Density

Many experiments have been reported that attempt to press compacts to higher densities or to produce a more uniform density and stress distribution in pressed compacts. Although none of the techniques developed has led to large-scale industrial use, they are discussed below.

Die Barrel Rotation. Rotating the die barrel while the powder is being pressed, using a fine atomized aluminum powder, was reported by Hammond and Schwartz (Ref 42). They lubricated the die barrel with a suspension of lithium stearate in didecyl alcohol. Annular compacts, 12.7 mm ($\frac{1}{2}$ in.) high, were pressed, with outer diameters of 38 mm ($1\frac{1}{2}$ in.) and inner diameters of 25 mm (1 in.). The core rod was stationary, but the die barrel could be rotated during compression.

Compaction with a stationary and a rotating die barrel were compared. While in static compression, 20% of the applied stress was consumed in die wall friction, while only 2% of the stress was consumed when the die barrel was rotated. In addition, the pressure necessary to eject the compact from the die was reduced to approximately one half of the pressure for compacts pressed with a stationary die barrel. Similar experiments on the effect of die barrel rotation on the density of iron powder compacts were reported by Rutkowski *et al.* (Ref 43).

Triaxial compression by simultaneous isostatic and uniaxial compression is obtained by applying pressure to the circumference of a cylindrical specimen confined in a flexible envelope, while an axial load is superimposed by a vertical piston, as shown in Fig. 20. With this method, the level of pressure necessary to obtain a given density is less than with isostatic or uniaxial compression alone. For example, to compact an atomized iron powder (Ancor 1000) to a relative density of 85%, uniaxial compression at 540 MPa (78 ksi) or isostatic compression at 415 MPa (60 ksi) is necessary. The same density can be obtained by combining a confining (isostatic) pressure of 83 MPa (12 ksi) with a uniaxial pressure of 470 MPa (68 ksi). The principles involved in this method of compaction have been reviewed by Broese van Groenou (Ref 32).

High-Energy-Rate Compacting. The rate at which pressure is applied in compacting in a hydraulic press is slow. Compacting in certain mechanical presses is somewhat faster. The effects of the rate of pressure application in compacting was studied by Davies and Elwakil (Ref 45). They found that somewhat higher densities can be obtained for a given pressure and for a given energy input when compacts from iron powder are pressed in high-speed presses (Petro-forge presses). They also determined the effects of multiple blows during pressing.

In the fabrication of sheet metal products, techniques were developed that formed sheet metal at rates considerably higher than those obtained in fast-acting mechanical presses. This is the high-energy-rate forming technique, which generally uses explosives.

The success of high-energy-rate forming in fabricating sheet metal led to extensive experimental work on high-energy compacting of metal powders. The most common means to achieve high velocity is by the use of explosives. Three experimental set-ups for high-energy-rate compacting are shown schematically in Fig. 21 to 23.

In the apparatus of Fig. 21 (Ref 46), compaction takes place in a rigid die. Pressure is applied by a projectile propelled by an explosive charge that moves through a barrel and impacts the powder at velocities of 200 to 350 m/s (655 to 1150 ft/s). The experiments showed that the following high relative densities (percent of solid density) can be obtained:

Powder	Density, %
Aluminum	99+
Aluminum alloy 7075	99+
Dendritic copper	99
Carbonyl iron (10-μm diam)	98
Nickel (2.5-μm diam)	97
Molybdenum	93
304 stainless steel	97

In experiments similar to those of Hagemeyer and Regalbuto (Ref 46), but using compressed gas to actuate the projectile, Rusnak (Ref 47) showed that the density of copper powder is not so much a function of projectile velocity, but depends primarily on the kinetic energy of the projectile, which can be varied by proper selection of gas pressure and projectile mass. For copper compacts with a volume of 0.86 cm³ (0.052 in.³), relative densities of 95% and higher were obtained with energies of 150 J (110 ft · lb).

Figure 22 shows an apparatus for explosive isostatic compacting. The energy of the moving piston is transferred in the form of pressure applied to the liquid, which in turn transfers the pressure to the powder in the flexible envelope. In con-

Fig. 21 High-velocity compacting in a rigid die

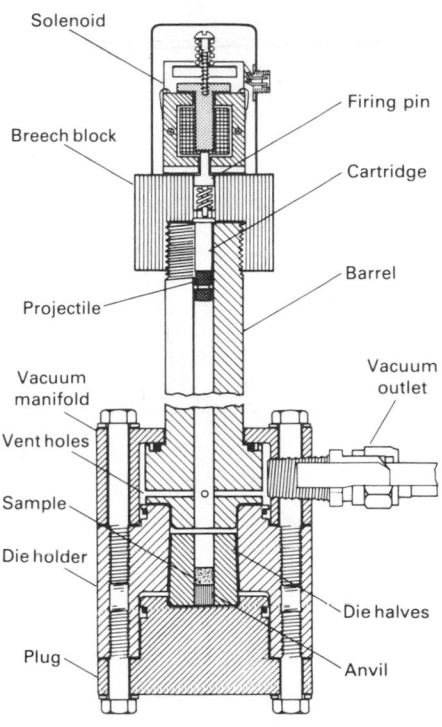

Fig. 22 Explosive isostatic compacting
Source: Ref 48

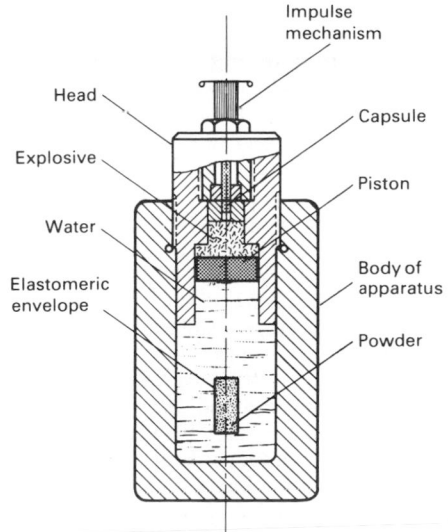

trast to conventional isostatic compacting, the piston need not be carefully fitted into the head of the apparatus. This method of explosive isostatic compaction has been used experimentally for producing parts from iron and tungsten powder.

Fig. 23 Explosive compacting with powder contained in a steel tube

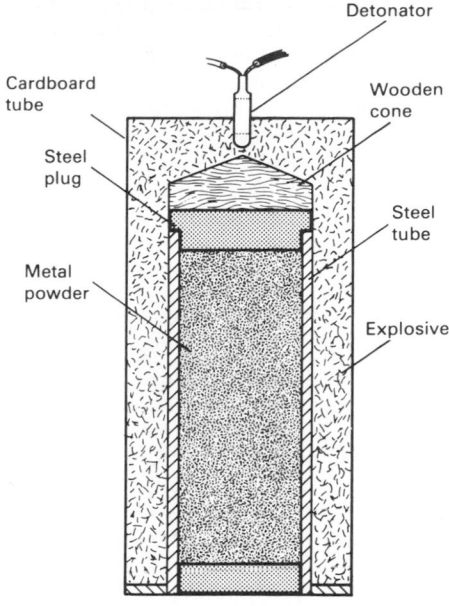

The most widely used method of explosive compacting is shown schematically in Fig. 23. The powder to be compacted is placed in a steel tube, which is closed at each end by steel plugs. The steel tube is surrounded by an explosive that is set off by a detonator located so that on detonation the tube collapses uniformly inward. Experiments by Lennon et al. (Ref 49) showed that density is a function of energy. They developed the equation:

$$D_c = D_T - \Delta D^{-(\beta E^\gamma)}$$

where D_c is the compacted density; D_T is the full density of the material; ΔD is the difference between full and compacted density; β and γ are constants; and E is the net energy absorbed by the powder. The highest relative densities obtained for iron, nickel, copper, and aluminum powders and the corresponding net energies absorbed per unit volume of compact in their experiments were:

Powder	Density(a), %	Net energy J/cm³	J/in.³
Iron	98.1	261	4280
Nickel	98.1	556	9110
Copper	98.5	285	4670
Aluminum	99.0	182	2980

(a) Percent of theoretical density

This method of explosive compacting is not necessarily confined to cylindrical compacts. Cones and hollow cylinders have also been explosively compacted by this technique (Ref 50).

Two other methods for producing shock waves have been applied to high-energy-rate compacting of metal powders. One method uses a pulsed magnetic field (Ref 51). The principle involved is the transformation of electrical energy in a condenser into mechanical work. As in the method of explosive compacting shown in Fig. 23, powder is confined in a steel tube that is surrounded by a coil. The magnetic field produced by the sudden discharge of a condenser through the coil causes the tube to collapse and to compact the powder. This method has been recommended for long rod-like or tube-like compacts.

In the other method, the energy produced by the discharge of a condenser through two electrodes located in a chamber filled with water is transferred to a piston, which in turn applies pressure to the powder held in a rigid die (Ref 48). As in other methods of dynamic pressing, the coefficient of friction between powder and die is lower than in conventional static pressing.

Vibratory Compacting of Powders. In the section of this article on packing of metal powders, vibration was shown to be very effective in obtaining higher packed densities in powders. Vibrating and simultaneous compacting of powders to obtain higher densities has been studied primarily by Soviet scientists (Ref 52).

The relative densities of powders vibrated under carefully controlled conditions are much higher than those obtained by simply pouring the powder into the container. Therefore, much lower compaction pressures are required to reach a given density for a vibrated than for a poured powder. This is illustrated in Fig. 24 for a carbonyl iron powder. The density of 5.53 g/cm³ (71% relative density), reached by compacting under a pressure of 245 MPa (35 ksi), is due to the plastic deformation of the iron powder particles, while the 5.37 g/cm³ (69% relative density) obtained by vibrating at 167 oscillations per second is due mainly to vibratory packing. Plastic deformation during the simultaneous compacting at 2.4 MPa (0.36 ksi) is minimal. The method of consolidating powders by vibrating and simultaneous compacting is, therefore, primarily applicable to hard powders, such as refractory metal and cemented carbide powders, which can be densified relatively little by pressure application alone.

Experiments in which iron powder was compacted while subjected to supersonic vibrations has been described by Siwkiewicz *et al.* (Ref 53). The effects of supersonic vibrations on average density of the compacts, their structural homogeneity, and the reduction of die wall friction were investigated.

Fig. 24 Comparison of densities of carbonyl iron compacts

(1) Obtained in static pressing. (2) Vibratory compacting at a frequency of 233 oscillations per second. (3) Vibratory compacting at a frequency of 167 oscillations per second

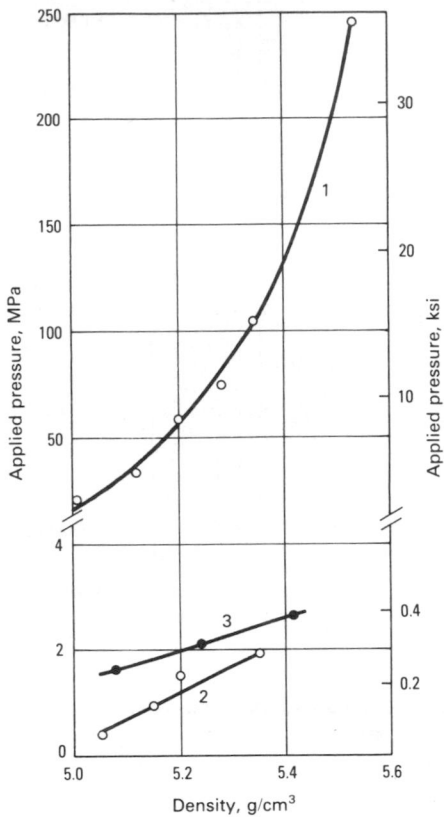

Compacting in Rigid Dies at High Pressures. Pressures on the order of 3000 MPa (435 ksi), which are approximately four times those customarily used for compacting in rigid dies (750 MPa, or 109 ksi), have been suggested. Gutmanas (Ref 54), developer of this technique, terms it "cold sintering." Useful products can be produced either without subsequent sintering or by sintering at much lower temperatures than those ordinarily used. According to Gutmanas, through plastic deformation at high pressures, clean surfaces are freshly formed that provide extensive atomic contact between powder particles at room temperature.

Gutmanas has "cold sintered" iron, stainless steel, and aluminum powders, as well as vanadium, niobium, and tantalum powders. For stainless steel powders, Gutmanas showed that no defects, typical for compacts pressed to very high density, are found when stainless steel powder is pressed at 4 GPa (580 ksi) to a density of over 95% of solid and that good bonding occurs (Ref 55). However, no data on reduced die life, one of the principal drawbacks of compacting at very high densities, have been published.

REFERENCES

1. McGeary, R.K., Mechanical Packing of Spherical Particles, *J. Am. Ceramic Soc.*, Vol 44, 1961, p 513
2. Evans, P.E. and Millman, R.S., The Vibratory Packing of Powders, in *Perspectives in Powder Metallurgy*, Vol 2, Hausner, H.H., Roll, K.H., and Johnson, P.K., Ed., Plenum Press, New York, 1967, p 237-251
3. Smith, W.O., Foote, P.D., and Busang, P.F., Packing of Homogeneous Spheres, *Phys. Rev.*, Vol 34, 1929, p 237-251
4. Hirschhorn, J.S., *Introduction to Powder Metallurgy*, American Powder Metallurgy Institute, Princeton, NJ, 1969, p 81
5. Hirschhorn, J.S., *Introduction to Powder Metallurgy*, American Powder Metallurgy Institute, Princeton, NJ, 1969, p 79
6. Hirschhorn, J.S., *Introduction to Powder Metallurgy*, American Powder Metallurgy Institute, Princeton, NJ, 1969, p 80
7. Hirschhorn, J.S., *Introduction to Powder Metallurgy*, American Powder Metallurgy Institute, Princeton, NJ, 1969, p 82
8. Hauth, J.J., Vibrational Compacting of Nuclear Fuels, in *Perspectives in Powder Metallurgy*, Vol 2, Hausner, H.H., Roll, K.H., and Johnson, P.K., Ed., Plenum Press, New York, 1967, p 253-276
9. Seelig, P.R. and Wulff, J., The Pressing Operation in the Fabrication of Articles by Powder Metallurgy, *Trans. AIME*, Vol 166, 1946, p 492-500
10. Fischmeister, H.F., Arzt, E., and Olsson, L.R., Particle Deformation and Sliding During Compaction of Spherical Powders: A Study by Quantitative Metallography, *Powder Metall.*, Vol 21, 1978, p 178-187
11. Bockstiegel, G. and Hewing, J., Critical Review of the Literature on the Densification of Powders in Rigid Dies, *Archiv. Eisenhüttenwesen*, Vol 36, 1965, p 751-767
12. Balshin, M. Yu., Theory of Compacting, *Vestnik Metalloprom.*, Vol 18, 1938, p 137-147
13. Wretblad, P.E. and Wulff, J., Sintering, in *Powder Metallurgy*, Wulff, J., Ed., American Society for Metals, 1942, p 36
14. Athy, L.F., Density, Porosity and Compaction of Sedimentary Rocks, *Bull. Am. Assoc. Petroleum Geologists*, Vol 14 (No. 1), 1930, p 1-34
15. Konopicky, K., *Sintereissen und Sinterstahl*, Springer, Vienna, 1948, p 126

16. Shapiro, I. and Kolthoff, M., *J. Phys. Colloid Chem.*, Vol 51, 1947, p 483-493

17. Torre, C., Theory and Behavior of Pressed Powders, Berg-und Hüttenmännische Monatshefte, *Montan. Hochschule Leoben*, Vol 93, 1948, p 62-67

18. Heckel, R.W., An Analysis of Powder Compaction Phenomena, *Trans. AIME*, Vol 221, 1961, p 1001-1008

19. Morgan, V.T. and Sands. R.L., Isostatic Compacting of Metal Powders, Metallurgical Reviews No. 134, *J. Met.*, May 1969, p 87

20. James, P.J., Particle Deformation During Cold Isostatic Pressing of Metal Powders, *Powder Metall.*, Vol 20, 1977, p 199-204

21. Bockstiegel, G., The Porosity-Pressure Curve and its Relation to the Pore Size Distribution in Iron Powder Compacts, *Modern Developments in Powder Metallurgy*, Vol 1, Hausner, H.H., Ed., Plenum Press, New York, 1966, p 155-187

22. Bockstiegel, G., Relation Between Pore Structure and Densification Mechanism in the Compaction of Iron Powder, Compaction Properties in Relation to the Pore Structure Inside and in Between Powder Particles, *Int. J. Powder Metall.*, Vol 2 (No. 4), 1966, p 13-26

23. Bockstiegel, G., Relation Between Pore Structure and Densification Mechanism in the Compaction of Iron Powder, Theoretical Considerations About the Change of Pore Size Distribution in Compacting, *Int. J. Powder Metall.*, Vol 3 (No. 1), 1967, p 29-37

24. Klar, E. and Petrosh, A., Pore Structure and Compressibility of Iron Powders, *J. Met.*, Vol 7, 1972, p 57-59

25. Klar, E., Relationship Between Pore Characterization and Compacting Properties of Copper Powders, *J. Met.*, Vol 7, 1972, p 418-424

26. Hewitt, R.L., Wallace, W., and de Malherbe, M.C., The Effects of Strain Hardening in Powder Compaction, *Powder Metall.*, Vol 16 (No. 31), 1973, p 88-106

27. Fischmeister, H.F. and Arzt, E., "A Model for the Compaction of Ductile, Spherical Powders," Proceedings of 7th International Powder Metallurgy Symposium, Dresden, Vol 1, Sept 1981, p 105-117

28. Squire, A., Density Relationship of Iron Powder Compacts, *Trans. AIME*, Vol 171, 1947, p 487-503

29. Unckel, H., Mechanical Properties of Sintered Iron for Porous Bearings, *Archiv. Eisenhüttenwesen*, Vol 18, 1944, p 125

30. Kamm, R., Steinberg, W., and Wulff, J., Plastic Deformation in Metal Powder Compacts, *Trans. AIME*, Vol 171, 1947, p 439-453; and Kamm, R., Steinberg, W., and Wulff, J., Lead-Grid Studies of Metal Powder Compaction, *Trans. AIME*, Vol 180, 1949, p 694-706

31. Kuczynski, G.C. and Zaplatinsky, I., Density Distribution in Metal Powder Compacts, *Trans. AIME*, Vol 206, 1956, p 215

32. Broese van Groenou, A., Pressing of Ceramic Powders, A Review of Recent Work, *Powder Metall. Int.*, Vol 10, 1978, p 206-211

33. Duwez, P. and Zwell, I., Pressure Distribution in Compacting Metal Powders, *Trans. AIME*, Vol 185, 1949, p 137-144

34. Morimoto, Y., Hayashi, T., and Takei, T., Mechanical Behavior of Powders During Compaction in a Mold with Various Cross Sections, *Int. J. Powder Metall. Powder Technol.*, Vol 12, 1982, p 129-145

35. Tabata, T., Masaki, S., and Kanata, K., Coefficient of Friction Between Metal Powder and Wall During Compaction, *Powder Metall. Int.*, Vol 13, p 179

36. Klar, E. and Shafer, W.M., Green Strength and Compressibility in Metal Powder Compaction, *Modern Developments in Powder Metallurgy*, Vol 9, Metal Powder Industries Federation, Princeton, NJ, 1976, p 91-113

37. van Buren, C.E. and Hirsch, H.H., Hydrostatic Pressing of Powders, *Powder Metallurgy*, Leszynski, W., Ed., Interscience, New York, 1961, p 403-441

38. Bowden, F.P. and Tabor, D., *Friction and Lubrication of Solids*, Clarendon Press, Oxford, 1954

39. Easterling, K.E. and Thölén, A.R., The Role of Surface Energy and Powder Geometry in Powder Compaction, *Powder Metall.*, Vol 16 (No. 31), 1973, p 112-118

40. Klar, E. and Shafer, W.M., On the Nature of Green Strength, *Int. J. Powder Metall.*, Vol 5 (No. 2), 1969, p 5-10 and Vol 5 (No. 4), 1969, p 5-16

41. Lund, J.A., Origin of Green Strength in Iron Compacts, *Int. J. Powder Metall. Powder Technol.*, Vol 18 (No. 2), 1982, p 117-127

42. Hammond, L.F. and Schwartz, E.G., The Effect of Die Rotation on the Compaction of Metal Powders, *Int. J. Powder Metall.*, Vol 6 (No. 1), 1970, p 25-36

43. Rutkowski, W., Bialo, D., and Dusz-czyk, J., Problems of Increase of Homogeneity and Density of Powder Metal Compacts Produced Under Conditions of Pressing Process Assisted by Additional Movements of the Die, *Planseeberichte für Pulvermetallurgie*, Vol 28, 1980, p 194-203

44. Koerner, R.M., Triaxial Stress State Compaction of Powder, *Powder Metallurgy Processing*, Kuhn, H.A. and Lawley, A., Ed., Academic Press, New York, 1978, p 33-50

45. Davies, R.I. and Elwakil, S., "Comparison of Slow Speed, High Speed and Multiple Compaction in Ferrous Powders," Proceedings of 17th Machine Tool Design and Research Conference, sponsored by University of Birmingham, Vol 3, 1976, p 483-488

46. Hagemeyer, J.W. and Regalbuto, J.A., Dynamic Compaction of Metal Powders with a High Velocity Impact Device, *Int. J. Powder Metall.*, Vol 4 (No. 3), 1968, p 19-25

47. Rusnak, R.M., Energy Relationship in High Velocity Compacting of Copper Powder, *Int. J. Powder Metall.*, Vol 12 (No. 2), 1976, p 91-99

48. Borginsky, L., Directions of New Development in the Impulse Processing of Powders, *Planseeberichte für Pulvermetallurgie*, Vol 17, 1969, p 225-236

49. Lennon, C.R.A., Bhala, A.K., and Williams, J.D., Explosive Compacting of Metal Powders, *Powder Metall.*, Vol 21, 1978, p 29-34

50. Porembka, S.W., Explosive Compacting, *Ceramic Age*, Dec 1963

51. Mironov, V., Densification of Powder with Magnetic Impulses, *Planseeberichte für Pulvermetallurgie*, Vol 24, 1976, p 175-190

52. Shatalova, G., Gorbunov, N.S., and Likhman, V.I., Physicochemical Principles of Vibratory Compacting, in *Perspectives in Powder Metallurgy*, Vol 2, Hausner, H.H., Johnson, P.K., and Roll, K.H., Ed., Plenum Press, New York, 1967, p 1-206

53. Siwkiewicz, Z., Filipowicz, W., and Stolarz, S., Investigation on the Influence of Ultrasonic Sound Waves in Pressing of Iron Powder, *Planseeberichte für Pulvermetallurgie*, Vol 25, 1977, p 255-262

54. Gutmanas, E.Y., Cold Sintering Under High Pressure—Mechanism and Application, *Powder Metall. Int.*, Vol 15, 1983, p 129-132 (contains references to earlier work)

55. Gutmanas, E.Y., High Pressure Compaction and Cold Sintering of Stainless Steel Powders, *Powder Metall. Int.*, Vol 12 (No. 4), 1980, p 178-182

Physical Fundamentals of Consolidation

By the ASM Committee on
Physical Fundamentals of Consolidation*

CONSOLIDATION OF METAL POW-DERS generally entails compacting, followed by sintering. However, metal powders sometimes are sintered without compacting, which is called "loose powder sintering." This process is suitable for certain specialized P/M applications such as bronze filters and porous electrodes, as well as for model experiments to study the mechanism of sintering. During sintering, the powder particles, as compacts or as loose powder aggregates, bond into a coherent body at elevated sintering temperatures, in the presence of a suitably controlled atmosphere. For more information, see the article "Production Sintering Atmospheres" in this Volume.

During the simplest type of sintering, compacts of a single metal powder, or compacts of homogeneous solid-solution alloy powders (alpha brass powder or an austenitic stainless steel powder), are heated to sintering temperature. Sintering temperature is below the melting point of the single metal powder or the solidus temperature of the homogeneous alloy powder.

Use of prealloyed powders, in which each powder particle has the same composition as the final part produced by compacting and sintering, is one method of P/M alloy production. Such prealloyed powders typically are produced by atomization of a molten alloy by impingement of a high-pressure gas or liquid stream. The fluid disperses the liquid alloy into small droplets to solidify into powder particles.

Powder metallurgy alloys also can be produced by blending two or more constituent powders in proportions required to achieve the desired overall alloy composition. Constituent powders of the mixture may be composed of pure elements (elemental powders) or alloys containing two or more elements (master alloy powders). If a homogeneous alloy is to be produced from a mixture of powders, sintering must be controlled to achieve the required degree of homogenization.

Homogenization occurs through interdiffusion of the chemical elements among the constituent powders of different composition. It typically is a solid-state diffusion process (Ref 1). Complete homogenization during sintering is achieved during the sintering of mixtures of iron and graphite powder. For example, graphitic carbon goes into solid solution in the gamma modification of iron that is formed at the sintering temperature; austenite, a solid-solution of carbon in gamma iron, is formed.

Homogenization circumvents some of the disadvantages of prealloyed powder processing. Hardness and strength of prealloyed powder particles result in powder compacts with low densities and low green strengths when pressed at room temperature. Mixtures of powder with the major component of the alloy present as an elemental powder (usually soft and ductile) typically exhibit high densities and high green strengths in the as-compacted condition. Furthermore, adjustments to alloy composition can be made readily by varying the proportions of constituent powders in the mixture, thus eliminating the need for additional powders of different composition. Powder mixtures composed of elemental powders normally can be obtained with high purity. Prealloyed powders, on the other hand, may contain elements other than the main components of the alloy because of pickup of impurities and trace elements during melting prior to atomization.

The disadvantages of homogenization processing center around the need to remove compositional heterogeneities by thermal processing (often in conjunction with mechanical working) subsequent to compaction. Homogenization often involves more extensive elevated-temperature exposure than normal sintering; this added processing expense must be minimized by limiting processing conditions to levels that provide the required homogeneity (Ref 1).

Sometimes, compacts pressed from a mixture of a high-melting-point metal with a small addition of a lower melting point metal are sintered, such as compacts of tungsten powder to which small amounts of nickel powder have been added. During sintering at temperatures below the melting points of both metals, the compacts densify at much lower temperatures than those of the high-melting-point metal. This is one type of activated sintering (Ref 2), which is discussed later in this article.

Compacts of mixtures of metal powders also may be sintered at temperatures that form a certain amount of liquid phase. The liquid phase must be small enough to be held by capillary force within the skeleton of the remaining solid phase so that the compacts retain their shape. In this process the liquid phase may be present through the duration of the sintering cycle. This is the case in the sintering of compacts from a mixture of tungsten, nickel,

*Fritz V. Lenel, *Chairman*, Professor Emeritus, Materials Engineering Department, Rensselaer Polytechnic Institute; Randall M. German, Professor, Materials Engineering Department, Rensselaer Polytechnic Institute; Richard W. Heckel, Professor, Department of Metallurgical Engineering, Michigan Technological University; Prakash K. Mirchandani, Graduate Student, Department of Metallurgical Engineering, Michigan Technological University

Fig. 1 Shrinkage of copper powder compacts from −74+43 μm sieve fraction

Compacted at 138 MPa (20 000 psi) and sintered at various temperatures as a function of sintering time. Source: Ref 4

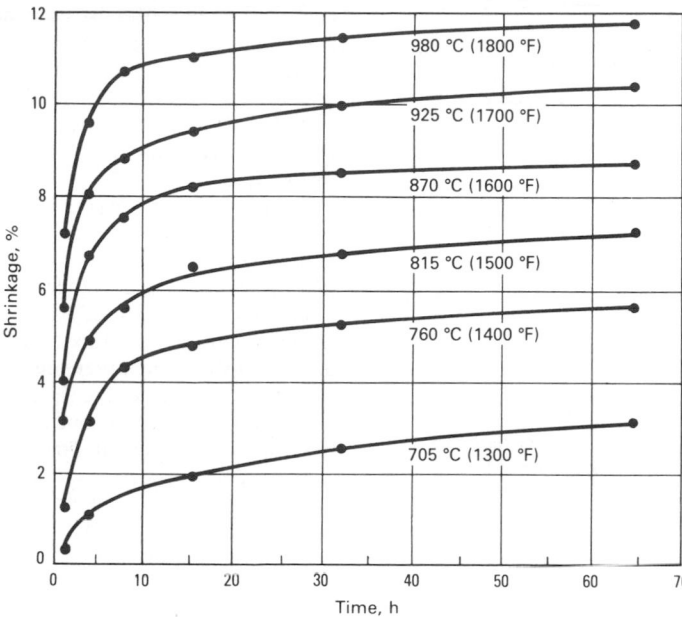

and iron powders (the so-called heavy alloys), and in the sintering of cemented carbides (combinations of tungsten and other refractory metal carbides with cobalt). Because the solution of small particles of the solid phase in the liquid and its reprecipitation on the large particles during cooling plays an important role, this method often is referred to as the solution and reprecipitation process of liquid-phase sintering.

In transient liquid-phase sintering, however, a liquid is formed when the compact is heated to the sintering temperature. The liquid is transformed into a solid by interdiffusion while the compact is at the sintering temperature, such as in the sintering of compacts from mixtures of copper and tin powders, in which an alpha bronze solid solution is formed.

The simplest form of pressure-assisted sintering is that which occurs by uniaxial hot pressing. A refractory mold (die) is inserted into a furnace, and both the powder and the die are heated while the punches are driven by an external pressure source. Hot isostatic pressing is the most widely used process (Ref 3). This process is also called gas pressure bonding, because a high-pressure gas assists densification. In hot isostatic pressing, the powder is enclosed in a thin-walled container that is flexible but pressure-tight at sintering temperatures. A pressurized gas is uniformly applied to the container in a high-pressure chamber. The chamber contains

a furnace to heat both the material and the gas.

The hot consolidation of metal powders, where sintering and densification are combined by applying pressure to the powder at elevated temperature, is not discussed in detail in this article. However, hot consolidation processes are discussed in the articles "Hot Pressing," "Hot Isostatic Pressing of Metal Powders," and "P/M Forging" in this Volume.

This article discusses the important changes that take place when metal powder compacts in the as-pressed condition (green compacts) are sintered. These changes are described phenomenologically and from the point of view of the driving forces and the material transport mechanisms which produce these changes. In addition, the mechanisms of homogenization during sintering and of activated and liquid phase sintering are examined.

Sintering of Compacts From Single Metal Powders

Changes in the dimensions and density of compacts are important considerations in sintering. In studies of the fundamentals of sintering, dimensional changes are determined as the difference in dimensions between green and sintered compacts, parallel and perpendicular to the direction of pressing, and are expressed in percent of the green dimensions. From a

technological standpoint, the dimensional change during sintering of greatest importance is the difference between the dimensions of the die and the sintered compact in the radial direction, that is, the direction perpendicular to that of pressing. Expansion of compacts on ejection from the die, or springback, should be subtracted from any shrinkage or added to any growth measured as the difference between green and sintered dimensions. Standards for determining the dimensional change of metal powder specimens during sintering are discussed in the article "Dimensional Change of Sintered Metal Compacts" in this Volume.

Green and sintered densities are the ratios of the masses of the compacts to their volumes in the green and sintered conditions, given in grams per cubic centimetre. Changes in density during sintering are primarily due to changes in volume (dimensions), but changes in mass, which may be caused by the loss of lubricant added during compacting, volatilization of a component such as zinc from brass, and the reduction of oxide skins on the powder particles during sintering also should be considered.

When compacts from a single-phase metal powder are sintered, they usually shrink and density increases. Shrinkage of copper powder compacts from a −74+43 μm sieve fraction, pressed at 138 MPa (20 000 psi), as a function of sintering time and temperature is shown in Fig. 1. These compacts were heated rapidly to the isothermal temperature and, after being held at this temperature, cooled rapidly to room temperature. Data represent the "isothermal" shrinkage of the compacts, even though shrinkage during heating and cooling is included. As shown in Fig. 1, shrinkage of the copper powder compacts is primarily a function of sintering temperature. The higher the temperature, the greater the shrinkage. Shrinkage also increases with increasing sintering time. The rate of shrinkage is initially quite high, but then decreases with increasing sintering time. The higher the sintering temperature, the more rapid the decrease in shrinkage rate. Accordingly, high sintered densities can be obtained more readily by increasing sintering temperature than by increasing sintering time.

Another factor affecting shrinkage is the particle size of the powder. As shown in Fig. 2, an increase in density during sintering of copper powder compacts pressed at 276 MPa (40 000 psi) and sintered at 865 °C (1590 °F) is a function of sintering time. The rate of densification of compacts from −44 μm powder is consider-

Fig. 2 Density of compacts from electrolytic copper powder

Two particle size fractions: −75+100 μm and −44 μm sintered at 865 °C (1590 °F) as a function of sintering time. Source: Ref 5

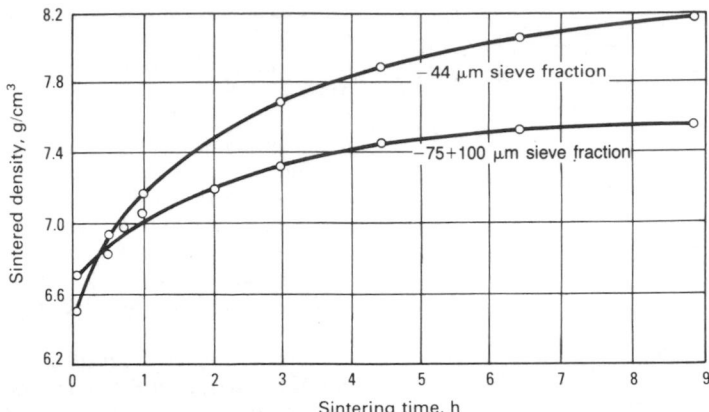

Fig. 3 Density of carbonyl nickel powder compacts

Sintered at a series of temperatures for 2 h as a function of compacting pressure. Source: Ref 6

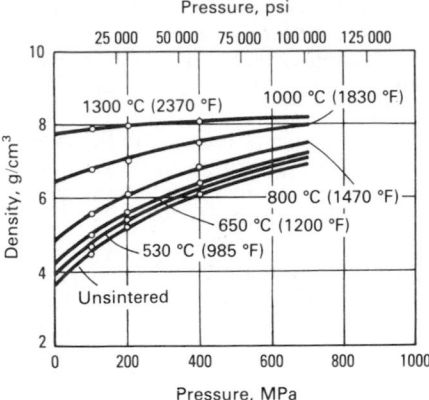

Fig. 4 Expansion of compacts from −74+43 μm copper powder

Pressed at 138 MPa (20 000 psi) and heated at a rate of 3.9 °C/min (7 °F/min) to 925 °C (1700 °F) and then cooled at the same rate

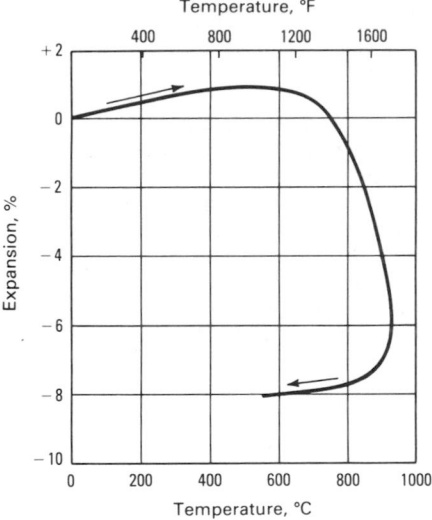

ably faster than that of compacts from a −75+100 μm sieve fraction.

A final factor affecting the densification of compacts from a single-phase metal powder is the green density of the compact or the pressure at which the compact is pressed, which determines green density. As shown in Fig. 3, density of compacts from carbonyl nickel powder is plotted as a function of the compacting pressure for green compacts and for compacts sintered at five different temperatures after pressing. The higher the compacting pressure, the smaller the sintering shrinkage, or the change from green to sintered density.

Another approach to studying shrinkage and densification during sintering is to determine the dimensional changes with a dilatometer (Ref 4). A compact is heated at a constant rate of heating to a given sintering temperature and then cooled at this rate from the maximum temperature to room temperature. A dilatometer curve is

shown in Fig. 4, which represents the dimensional changes in the axial direction of the same type of copper compact as Fig. 1. Compacts were pressed at a pressure of 138 MPa (20 000 psi) and were heated and cooled at a uniform rate of 3.9 °C/min (7 °F/min).

The first portion of the curve up to 315 °C (600 °F) is identical to that of a specimen of solid copper and represents the thermal expansion of the metal with increasing temperature. When sintering begins, shrinkage counteracts thermal expansion, and the curve deviates from that for solid copper. As the temperature increases, the rate of shrinkage increases and the compact contracts. Immediately after reaching the maximum sintering temper-

ature of 925 °C (1700 °F), the compact is cooled at the same rate at which it was heated. During cooling, contraction initially is faster than normal thermal contraction, because the compact is still sintering. Eventually, however, the dilatometer curve slope follows normal contraction. This type of experiment is particularly well suited for studies of the rate of densification as a function of temperature.

During sintering, most compacts pressed from a single-phase metal powder shrink rather than expand. The higher the sintering temperature, the longer the sintering time, the finer the particle size of the powder from which the compacts are pressed, and the lower the green density of the compacts, the greater the shrinkage. Compacts from powders of body-centered cubic metals (molybdenum and tungsten) of a given particle size pressed to a given density and sintered a given length of time at a given homologous temperature (ratio of sintering temperature to melting temperature of the metals in degrees kelvin) show higher shrinkage than compacts from face-centered cubic metals (copper and nickel) sintered under corresponding conditions.

Compacts made of iron powder generally are sintered in the temperature range where iron has the gamma (face-centered cubic) structure. Consequently, compacts made of iron powder with conventional commercial particle size distribution, pressed to green densities in the range from 6.6 to 7.1 g/cm³ and sintered at temperatures ranging from 1100 to 1200 °C (2010 to 2190 °F) for ¹/₂ to 1 h, experience minimal shrinkage, often less than 1%. Despite this small amount of shrinkage, iron powder compacts are sintered adequately with regard to their mechanical and physical properties. This is important in the fabrication of structural parts from iron and steel because of the need for control of dimensions. Large amounts of shrinkage during sintering may be undesirable, thus rendering control of final dimensions of the sintered part difficult (Ref 7).

In other P/M applications, large amounts of shrinkage are desirable to obtain sintered densities near those of the solid metal. For instance, compacts from very fine tungsten powder (body-centered cubic) sintered at temperatures near the tungsten melting point increase in density from 11 g/cm³ in the green condition to 18 g/cm³ in the sintered condition; that is, they shrink 15% linearly (Ref 8). Because the sintered compacts are subsequently swaged and drawn into wire, close control of sintered dimensions is not required.

Fig. 5 Microstructure of green compact

Pressed from copper powder at 180 MPa (26 000 psi). Unetched. Magnification: 450×

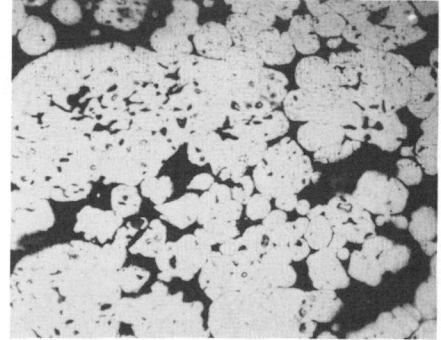

Fig. 6 Microstructure of copper powder compact

Pressed at 180 MPa (26 000 psi) and sintered 15 min at 705 °C (1300 °F). Unetched. Magnification: 450×

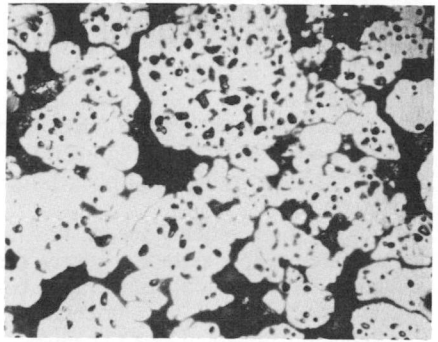

Fig. 7 Microstructure of copper powder compact

Pressed at 180 MPa (26 000 psi) and sintered 30 min at 980 °C (1800 °F). Potassium dichromate etch. Magnification: 500×

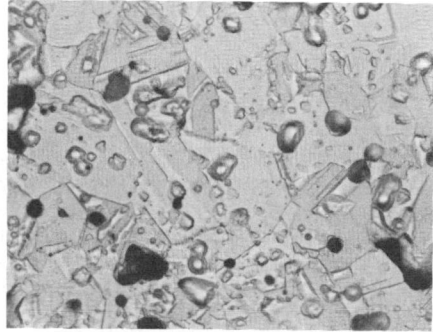

Expansion during sintering is common in compacts made of soft metal powders, such as copper, that are pressed at high pressures (Ref 9). This expansion is due to gases entrapped in closed-off pores of the compact. Gases may be entrapped during compacting or may be formed by chemical reactions, such as the formation of water vapor by reaction of oxide skins on the surface of the particles with the hydrogen of the sintering atmosphere which diffuses into the closed pores. Such expansion or growth generally is not observed in compacts that are made of harder metal powders.

Microstructural Changes. The microstructure of green metal powder compacts can be observed, if the compacts are carefully sectioned, polished, and etched. Examination of this microstructure reveals the outlines of the original powder particles and the pores. Depending on the particle size distribution of the powder from which the compact is pressed, powder particles and pores that are visible in the microstructure may have narrow or wide size distributions. If the powder particles are polycrystalline, grain boundaries may be observed within the particles. Particles may be flattened and distorted, depending on the nature of the powder and the compacting pressure.

The microstructure of a green compact pressed from copper powder at a pressure of 180 MPa (26 000 psi) in the unetched condition is shown in Fig. 5. This microstructure indicates that the copper powder had a fairly wide particle size distribution and that some of the particles had interior porosity. Most of the boundaries between particles can be readily discerned. When this compact was sintered 15 min at 705 °C (1300 °F), a temperature too low for adequate sintering, the microstructure shown in Fig. 6 was obtained. The principal difference between the microstructures of the green compact and the one sintered at 705 °C (1300 °F) is in the shape of the pores within the powder particles, which have become rounded during sintering. The sintering temperature is too low to produce adequate bonding between powder particles.

Figure 7 shows the microstructure of the copper powder compact sintered 30 min at 980 °C (1800 °F). The sample was etched to reveal the grain boundaries in the microstructure. Boundaries between particles can no longer be discerned; instead, a network of grain boundaries similar to those in wrought and annealed copper is illustrated. Pores of different sizes filled with the medium used to mount the specimen for microstructural examination are seen. These pores are no longer cusp-shaped and irregular, but have become rounded. In compacts made of fine metal powders, complete spheroidization of the pores often is observed.

The transition in microstructure from that of a green compact to that of a well-sintered compact is a function of sintering temperature and time. During annealing of cold worked wrought material, a recrystallization temperature is reached, where nuclei of strain-free grains are formed that grow into the recrystallized structure. Subsequently, grain growth of the recrystallized structure occurs. This temperature for recrystallization and grain growth is a function of the time during which the cold worked material is annealed and the amount of prior cold work. The temperature range in which the typical microstructure of a well-sintered powder compact is developed is higher than the recrystallization temperature of the wrought material. Compacting pressure frequently has a relatively minor effect on the temperature range at which the sintered structure is developed.

Development of the typical well-sintered structure of a metal powder compact requires that grain growth occur across prior particle boundaries. This grain growth is restricted until material transport that occurs during sintering has progressed to a point where a substantial increase in the contact area between particles has taken place. This increase in contact area is impeded by the network of pores. It is not responsive to the effects of strain hardening during compacting, which are relieved at temperatures below those at which the development of a sintered microstructure is observed. The principal effect of higher compacting pressures on grain growth during sintering is to facilitate extensive contact between particles.

Changes in mechanical properties of compacts that occur during sintering may be illustrated by comparing the properties of green compacts with those of sintered compacts. The strength of green compacts usually is determined by the transverse-rupture strength of rectangular compacts according to the American Society for Testing and Materials standard ASTM B 312 and the Metal Powder Industries Federation standard MPIF 15, as described in the article "Green Strength of Compacted Metal Powders" in this Volume. The level of strength determined by this test is quite low, generally below 7 MPa (1000 psi). Strength must be adequate enough to allow handling of green compacts during transfer from the compacting press to the sintering furnace.

The transverse-rupture strength of sintered metal powder specimens may be determined by a test very similar to the green strength test. It is described in ASTM B 528 and MPIF 35. Strength values obtained are much higher than those for green specimens. They vary, depending on the type of material (lower for weak metals and alloys and higher for strong metals and alloys) and specimen density (ratio of sintered compact density to that of the solid material). The highest values will be achieved for the densest sintered compacts. The sintered strength values are always at least 20 times the transverse-rupture strength of green compacts. For many sintered materials, transverse-rupture strength is about 1.8 to 2.2 times the ultimate tensile strength.

Although transverse-rupture strength is not a design consideration, it is often used to measure the strength of specimens from a lot of metal powder and to compare its strength with a standard strength value. The transverse-rupture strength test of sintered metal powder specimens is applicable only to a relatively brittle material whose permanent deflection, as a result of testing, is less than 0.25 mm (0.010 in.). However, low-density, high-carbon P/M steels generally are relatively brittle. The transverse-rupture strength test is therefore more applicable to such P/M steels than to wrought materials.

Yield and ultimate tensile strengths also can be determined on appropriate specimens of sintered materials. Strength properties depend on relative density; materials with near full density have strength properties comparable to those of wrought materials of the same composition. Strength decreases with decreasing density (that is, with increasing porosity), but the rate of decrease is not nearly as rapid as the decrease in ductility (elongation and reduction in area) and particularly impact energy. These properties are generally much lower than properties of corresponding wrought materials. Only sintered mate-

rials with densities that closely approximate wrought materials (on the order of 99% or more of theoretical density) have ductility and impact energy properties that approach properties of corresponding wrought materials. However, the mechanical properties of many cast products are equaled or exceeded.

The large difference in mechanical properties between green and sintered specimens does not apply to conventional indentation hardness testing, particularly Brinell hardness. Green compacts pressed at high pressures often exhibit high Brinell hardness values, which decrease when the compacts are sintered at increasing temperatures, because the strain hardening introduced during compacting is relieved.

At still higher sintering temperatures, compacts shrink and hardness increases due to their higher density. Conversely, the hardness of compacts pressed at low pressures increases with increasing sintering temperature, because the effect of increase in compact density predominates. This is illustrated in Fig. 8, in which the hardness of compacts made of carbonyl iron powder pressed at three different pressures is plotted as a function of sintering temperature for a sintering time of 2 h. More detailed property comparisons for P/M materials, in particular the effect of density on the properties, is presented in the article "Mechanical Properties of P/M Materials" in this Volume.

Driving Forces and Material Transport in Sintering of Compacts From Single Metal Powders

As discussed above, important changes in the dimensions of compacts occur when green compacts are heated to sintering temperature. In many sintering operations, green compacts shrink and may even reach the density of corresponding wrought materials. Even when compacts shrink only minimally or not at all, changes in interior geometry occur. In pressed compacts, the approach of the surfaces of the individual powder particles toward each other to distances on the order of metal lattice dimensions is quite limited.

One of the important occurrences in the early stages of sintering is the increase in the contact area between particles. This may occur without significant changes in the overall density of the compact. Another change within the green compact that does not necessarily involve large density change is the rounding and spheroidiza-

tion of the pore structure, which is best observed by metallography. Changes in mechanical properties during sintering are, of course, a direct consequence of these geometrical changes.

During compacting, powder particles are subjected to strain hardening that increases with higher pressure. Sintering causes annealing, during which strain hardening is relieved, primarily through recrystallization. A small portion of the geometrical changes that occur in compacts during sintering results from recrystallization. However, the fact that a well-developed sintered microstructure is formed only at temperatures above the recrystallization temperature range indicates that the major driving force for sintering must be sought elsewhere.

The principal driving forces are capillary forces due to the surface and interfacial tensions, also called specific surface and specific interfacial free energies, of the free surfaces and the interfaces between grains (Ref 10). These forces cause stresses that are related to the curvature of the surfaces and interfaces. Stresses due to surface and interfacial tension forces tend to decrease the surface and the interfacial areas and are the principal driving forces in sintering.

In single-phase systems, the existing interfaces are the free surfaces between the solid and gas phase and the interfaces between grains with different orientations. The simplest case would be a single-phase, monocrystalline system, for which the relationship between surface tension (specific surface free energy) and stress is given by the LaPlace equation:

$$\sigma = \gamma \left(\frac{1}{r_1} + \frac{1}{r_2} \right)$$

where σ is the stress; γ is the surface tension; and r_1 and r_2 are the radii of curvature of the surfaces. Under convex surfaces for which the curvature is negative, the stress is tensile ($\sigma < 0$). For concave surfaces with positive curvature, the stress is compressive. Experimental methods for readily determining the surface tension of metals and alloys at sintering temperatures are available (Ref 11). The stress defined by the LaPlace equation causes a gradient in chemical potential between surfaces with different radii of curvature:

$$\Delta\mu = \sigma\Omega$$

where Ω is the atomic volume.

As discussed above, single-phase metal powders shrink rather than expand during sintering, except for soft metal powders

Fig. 8 Brinell hardness of compacts from carbonyl iron powder

Pressed at 76, 138, and 386 MPa (11 000, 20 000, and 56 000 psi) and sintered for 2 h as a function of sintering temperature. Source: Ref 6

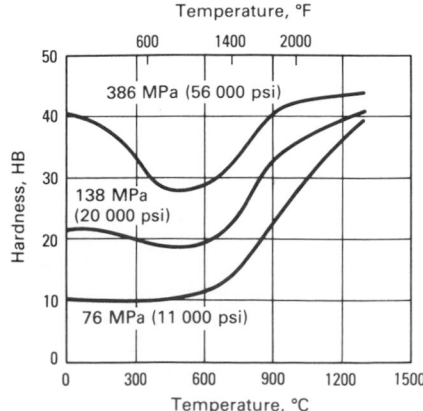

compacted at high pressures. Gases entrapped in these compacts may expand when heated during sintering. Stress caused by gas pressure, which increases with increasing temperature, counteracts surface tension stresses that cause shrinkage.

In this discussion of the driving forces that cause sintering—or, more specifically, material transport during sintering—compacts are assumed to be sintered without the application of external pressure. If external pressure is applied, the capillary forces are of secondary importance compared to external forces, particularly in metallic systems.

Several possible mechanisms have been identified that cause material transport under the influence of a chemical potential gradient. In single-phase systems, the chemical potential gradient is due to the stress gradient between surfaces and interfaces.

One of the early mechanisms postulated is evaporation and condensation. Atoms evaporate from flat surfaces that have a higher vapor pressure and are transported through the gas phase to highly curved convex surfaces that have a lower vapor pressure, where they condense. Because most metals do not have high vapor pressures at temperatures near their melting points, evaporation and condensation contribute nominally to material transport in metallic systems.

Diffusional flow is the most important mechanism of material transport (Ref 12). It is based on the concept that a certain concentration of vacancies exists in the crystal lattice of a metal. This concentration is a function of temperature and may reach values on the order of one vacancy for every 10 000 occupied lattice positions at temperatures near the melting point of the metal. Vacancy concentration also is a function of the chemical potential or stress to which the surface of the metal is subjected. Consequently, a gradient of vacancies exists between a highly curved convex surface, which has a higher vacancy concentration, and an adjacent flat surface, which has a lower vacancy concentration.

Studies of diffusional flow have concentrated on the first or initial stage of sintering, where the small necks between particles become larger. This causes rounding of the pores and some decrease in total pore volume. However, the original particles are distinguishable. The studies were based on theoretical analysis and on experiments on models, such as those in which a sphere is sintered to a flat surface (Ref 13) or in which two spheres are sintered together (Ref 14).

Fig. 9(a) Schematic cross section through two spheres sintering together

Radius, ρ, is at the end of the neck.

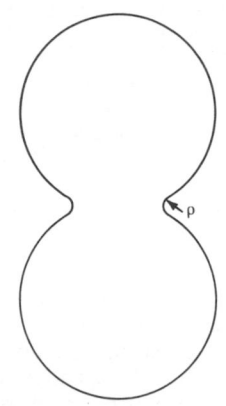

Fig. 9(b) Types of material transport when two single crystal spheres with a grain boundary at the interface sinter together

Path *a*, volume diffusion from the flat surface to the neck. Path *b*, surface diffusion from the flat surface to the neck. Path *c*, grain-boundary diffusion from the grain boundary to the neck. Path *d*, volume diffusion from the grain boundary to the neck

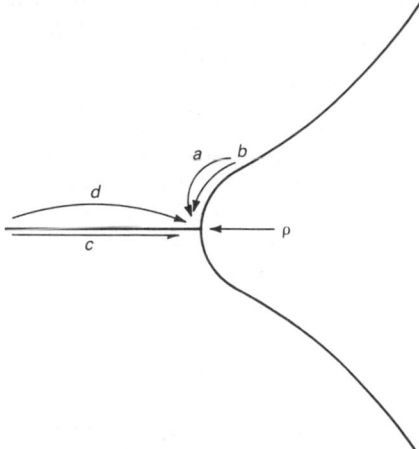

Figure 9(a) schematically illustrates a cross section of a two-sphere model. The radius, ρ, is formed at the end of the neck between the two particles in Fig. 9(a). Figure 9(b) shows the geometry near this radius in more detail. Assuming that the two particles are single crystals, with different orientations, a grain boundary is formed at the neck. The difference in curvature at the neck with ρ and the adjacent flat surface causes a difference in stress and chemical potential between the two points, which in turn produces a gradient in the concentration of vacancies between

the highly curved neck surface, which has a high vacancy concentration, and the adjacent flat surface, which has a lower concentration.

The difference in vacancy concentration under surfaces with different radii of curvature can be expressed mathematically (Ref 12). This difference causes a flux of vacancies away from the highly curved surface to the flat surface, which is equivalent to a diffusional flow of atoms in the opposite direction.

There are two types of diffusional flow possible under the influence of vacancy concentration differences between flat and highly curved surfaces. They are volume diffusion, indicated schematically by path *a* in Fig. 9(b), and surface diffusion, indicated by path *b*. As shown in Fig. 9(b), these two types of diffusional flow cause rounding and spheroidization of irregular pores in a compact. However, they do not explain shrinkage, which would be represented by a decrease in the distance between the centers of the two spherical particles.

To explain shrinkage, flow of atoms from the grain boundary to the neck must be considered. This is indicated by path *c*, which represents grain-boundary diffusion, and path *d*, which represents volume diffusion from the interior of the grain boundary to the neck. When atoms flow from the grain boundary to the neck, vacancies flow in the opposite direction and are thus eliminated at the grain boundaries. This causes the centers of the two particles to approach each other, and shrinkage occurs. The rates of the diffusional mechanisms (volume diffusion, surface diffusion, and grain-boundary diffusion) vary significantly, and the geometries that govern transport also vary and are quite complex even for simple model systems, such as sintering a sphere to a sphere. Nevertheless, the calculated rates of pore rounding and shrinkage in model systems, based on known values of specific surface and interface free energy, and the various types of diffusivities closely approximate those observed experimentally (Ref 15).

Rates of material transport during the initial stage of sintering can be calculated for model systems such as the two-sphere model discussed above. However, these calculations presently cannot be extended to more complex systems consisting of large numbers of particles, where the presence of neighboring particles has constraining effects on neck growth and shrinkage (Ref 16).

Mathematical models have been constructed not only for the initial stage of

Fig. 10 Microstructure of a compacted blend of copper particles (−100+140 mesh)

Dispersed in a matrix of nickel (−325+400 mesh) particles. Mean composition: 0.19 atom fraction copper

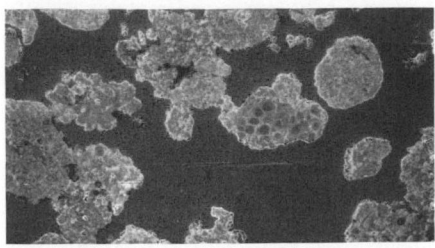

Fig. 11 Variation of sintered density during homogenization in nickel-copper compacts

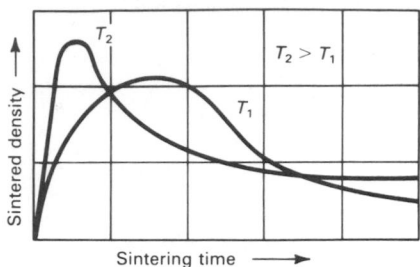

sintering, but also for the intermediate and final stages. During the intermediate stage of sintering, the original particles are no longer distinguishable, because the pore channels between particles are pinched off and become closed. Grain boundaries between the original particles migrate due to grain growth. The pores form a more or less connected continuous network throughout the compact. They become isolated in the third or final stages of sintering. Models for the intermediate and final stages of sintering are complex (Ref 17). Results of the calculations and their verification by experimental observation are not generally accepted and therefore are not discussed in this article.

Homogenization During Sintering

Alloys may be produced by P/M techniques either by sintering compacts pressed from homogeneous prealloyed powder particles or by first mixing different powders for compacting and sintering. Typically, as-compacted blends of powder exist as solute-rich particles (either elemental or master alloy) surrounded by a matrix of solvent particles (normally elemental particles of the base element of the alloy). The density to which the powder is pressed determines the degree of residual interparticle porosity.

Figure 10 shows the microstructure of a green compact made from a mixture of nickel and copper powders. Dispersion of the solute-rich particles, copper in this case, facilitates rapid homogenization and can be optimized through use of proper mixing procedures and determination of the best particle size ratio (the finer the matrix particle size, the greater the dispersion of solute-rich particles). Overall alloy composition and composition of the solute-rich particles also can affect the dispersion, but sometimes are specified and cannot be

considered independent variables. Density differences between two different types of particles often cause mixing difficulties. Sometimes, such problems are minimized by varying the size of the particles used.

Sintering causes solid-state diffusion (lattice diffusion) of the solute atoms into the solvent matrix and diffusion of solvent atoms into the solute-rich particles. Homogenization, as a function of time at temperature, may be described in terms of concentration/distance profiles that exhibit discontinuities in the as-compacted condition and that approach the average compact composition as the concentration gradients are eliminated viã interdiffusion. Generally, the compact behaves as a collection of many small diffusion couples that are allowed to undergo interdiffusion to the point where atomic fluxes become negligible and homogenization is achieved.

Variation in sintered density with sintering temperature and time during homogenization is frequently quite different compared to the variation that exists if compacts are composed of particles of identical composition. This is due to two phenomena, which have opposing effects on the densification of the compact, that occur simultaneously. The matrix densifies due to pore shrinkage. Additionally,

a net flow of vacancies from one particle to the other occurs if there is a difference in the fluxes of the atomic species to and away from a specific type of particle. These vacancies can accumulate in the form of fine pores, which is called Kirkendall porosity (Ref 18). The formation of Kirkendall porosity can result in expansion of the compact, with a subsequent reduction in density of the compact.

The variation of sintered density with time for compacts made from a blend of nickel and copper powders is shown schematically in Fig. 11. Density of the compact first increases and then decreases until a minimum is reached, at which point the density appears to stabilize. In the initial stages of sintering (short sintering times), the densification mechanism dominates, and density increases sharply. However, a point is reached where density begins to decrease, indicating that the formation of Kirkendall porosity (expansion) has begun to dominate.

With increased sintering time, density reaches a minimum, after which it remains almost stable (Ref 18). The time at which the density reaches a minimum indicates the point at which the two opposing mechanisms have balanced each other and that an advanced level of homogenization has been reached. The formation of Kirkendall porosity and its eventual stabilization are shown in Fig. 12(a) and (b), respectively. At higher sintering temperatures, the observed changes in density occur earlier.

The formation of austenite during sintering of compacts from a mixture of iron and graphite powders is an example of complete homogenization. This is due to the high diffusivity of carbon in gamma iron at the sintering temperature. In this case, compacts generally exhibit only slight shrinkage during sintering. At the end of the sintering treatment at temperatures near

Fig. 12 Formation and coarsening of Kirkendall porosity in compacts made from a blend of nickel (−325+400 mesh) and copper (−100+140 mesh) powders

Mean composition: 0.19 atom fraction copper. Magnification: 100×. (a) T = 850 °C (1560 °F); t = 1.0 h. (b) T = 850 °C (1560 °F); t = 24.0 h

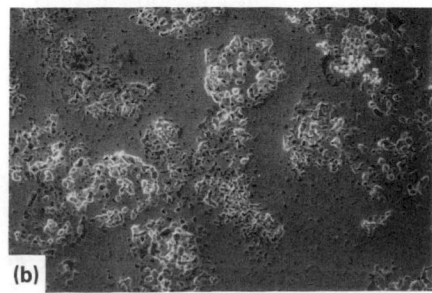

Fig. 13 Microstructure of compact from mixture of iron and graphite powder

Pressed to a density of 6.1 g/cm³, sintered 30 min at 1175 °C (2150 °F). 4% picral etch, plus 0.5% nitric acid. Magnification: 400×

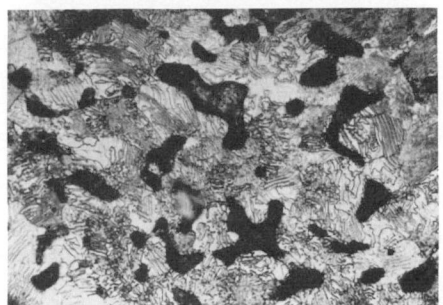

1100 °C (2010 °F), compacts consist of homogeneous austenite. When cooled to room temperature at a moderate rate, the austenite transforms. The microstructure of a sintered steel of eutectoid composition (0.8% C content), which consists of the typical lamellae of ferrite and cementite (pearlite), is shown in Fig. 13. It is similar to that of an annealed wrought eutectoid steel, except for the presence of pores in the sintered compact.

The following additional processing parameters can affect the rate and extent of densification or expansion during homogenization:

- Extent of the difference in the diffusivities of the particles in the compact
- Size of the particles in the compact
- Mean composition of the compact
- Compacting pressure
- Oxygen content of the starting powders
- Entrapped gases within powder particles

Thus, the form of the density-time relationship for a specific homogenization treatment may vary and is frequently more complicated for multi-phase systems.

Powder, Alloy System, and Homogenization Variables

Characteristics of the powder particles that compose the mixture exert a significant effect on homogenization kinetics (Ref 1, 19). The sizes of the particles in the mixture establish the distance over which interdiffusion must occur to achieve homogeneity. In alloy mixtures where the elemental solvent particles form a continuous matrix around the dispersed solute-rich particles, particle size of the dispersed particles becomes the important parameter in establishing the interdiffusion distance (assuming that the proportion of the powders in the mixture is already established by the alloy composition desired). Very fine, intimate mixtures of powders, such as those produced by reduction of submicron mixtures of oxides or alternatively by mechanical alloying, result in diffusion distances that are small enough to preclude the consideration of homogenization.

The composition of the dispersed particles also can influence homogenization kinetics; for example, solvent alloyed into the solute-rich particles reduces the extent of inhomogeneity in the mixture and can shorten the interdiffusion distance. If the dispersed particles in Fig. 10 contained only 50% Cu, the amount of nickel needed in the mixture to achieve the same overall composition would be less, thus shortening interdiffusion distance.

Alloy systems with more than two components that contain more than a single phase are more difficult to analyze in terms of homogenization kinetics. Interdiffusion in single-phase ternary systems must be described by four diffusion coefficients in order to consider the diffusion of both solutes. Each is influenced by its own concentration gradient and that of the other solute.

Multi-phase binary systems must be described by the solute solubilities and interdiffusion coefficients of each of the phases present (Ref 1, 19, 20). In such complexities, the parameters required to analyze the homogenization process are not available, and reasonable simplifications (assumption of binary and/or single-phase behavior) are made. Similarly, the presence of a transient liquid phase at the outset of the homogenization process also is difficult to analyze, and reasonable approximations are necessary.

The duration of sintering required to achieve a desired degree of homogeneity is critically dependent on temperature, because interdiffusion coefficients ($\tilde{D}$) (units of length squared/time) are exponentially dependent on temperature:

$$\tilde{D} = D_o e^{-Q/RT} \qquad \text{(Eq 1)}$$

where D_o is the pre-exponential factor; Q is the activation energy; R is the universal gas constant; and T is the absolute temperature.

For example, the interdiffusion coefficient for most solid-solution elements in nickel doubles for every 50 K increase in temperature near the liquidus. The total duration (t) of elevated-temperature treatment to achieve essentially complete homogenization is inversely proportional to $\tilde{D}$:

$$t = k\ell^2/\tilde{D} \qquad \text{(Eq 2)}$$

where ℓ is the average diameter of the dispersed particles and k is a constant, depending on alloy system parameters, powder particle compositions, and overall alloy composition. For example, k is approximately 0.2 for a 4-to-1 mixture of elemental powders that exhibits complete solid solubility (Ref 19). Thus, it can be seen that both the homogenization temperature and the size of the dispersed particles are the most significant variables in controlling homogenization kinetics.

Blending and mechanical working during powder fabrication of an alloy also can influence homogenization kinetics (Ref 19, 21). Ideal mixing is assumed in Eq 2 to predict homogenization time. Inadequate blending technique or "incompatible" powder particles (significantly different sizes and/or densities) result in poor mixing and greatly enlarged interdiffusion distances, which therefore dictate longer homogenization treatment. The serious problems of poor mixing stem from homogenization time being proportional to the square of the interdiffusion distance.

Effective interdiffusion distance may be reduced and homogenization kinetics increased by mechanical working (Ref 19). Such working has the same effect as a reduction in particle size of the dispersed particles. However, the effect of a given mechanical working reduction is less than that indicated by the same reduction in ℓ in Eq 2, because the spherical diffusion flux geometry in a compacted mixture of powders is changed to the less efficient unidirectional flux (by rolling) or cylindrical flux (by extrusion). Any enhancement of homogenization kinetics will result only if mechanical working reduces the thicknesses of the solute-rich regions. If mechanical working merely causes the matrix to flow past the solute-rich regions, little, if any, increase in homogenization kinetics is exhibited during subsequent thermal processing.

Determination of the Degree of Homogeneity

Direct experimental techniques for the determination of the progress of homogenization or the evaluation of the state of homogeneity include microscopy (qualitative and quantitative), compositional analysis of the microstructure via electron microprobe analysis or scanning electron microscope instrumentation, and x-ray compositional line broadening techniques (Ref 1, 19, 20, 22).

X-ray compositional line broadening is a technique for the rapid determination of

Fig. 14 X-ray diffraction peaks (311) for compacts of blended nickel and copper powders

Mean composition: 0.28 atom fraction copper. Nickel: −400 mesh. Copper: −200+270 mesh. (a) Initial condition. (b) and (c) Various partially homogenized conditions

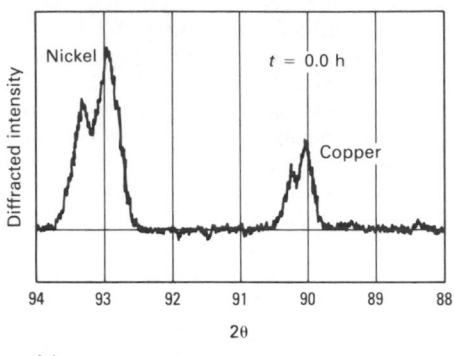

(a)

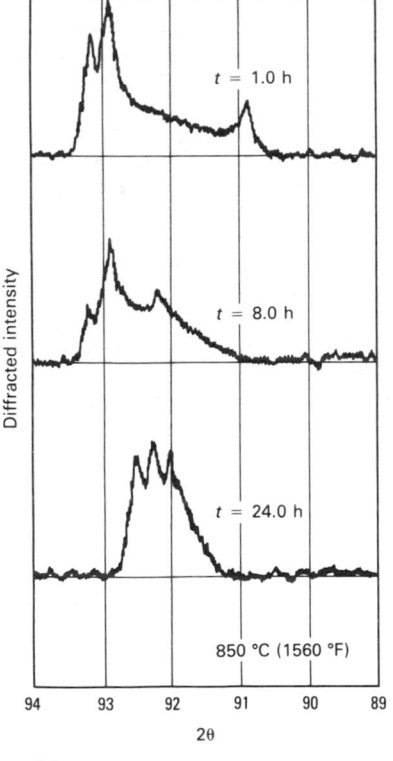

(b)

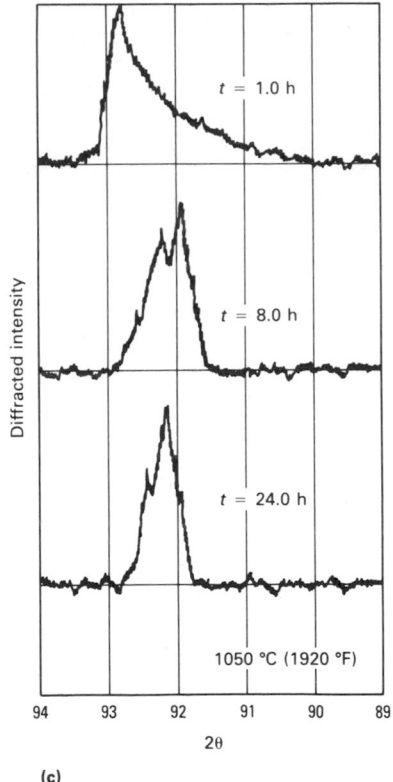

(c)

the degree of homogeneity. It displays the spectrum of compositions existing in a given inhomogeneous phase from the shape of a diffraction peak broadened by a range of lattice parameters in the phase. This technique may be used for the analysis of single-phase or multi-phase binary alloys. The experimental procedure involves obtaining x-ray diffractometer patterns of the specimen whose degree of homogeneity is to be determined.

The progress of homogenization is seen clearly, although qualitatively, in Fig. 14. The Ni (311) peak occurs at a diffraction angle $2\theta \simeq 93°$ and the Cu (311) peak at $2\theta \simeq 90°$. Figure 14(b) indicates that the initial stages of homogenization (low temperatures and short times) proceed by forming the complete range of compositions from nickel to copper. This is the manifestation of the initial interdiffusion across the nickel-copper interparticle boundaries.

As shown in Fig. 14(c), later stages of homogenization (high temperatures and long times) result in the formation of a single peak at a diffraction angle between those of pure nickel and pure copper. This corresponds to the approach to a single homogeneous phase having the mean composition of the system. The diffraction angle of the single peak is determined by this mean composition. Quantitative analysis of peak shapes may be used to obtain the spectrum of phase compositions within a specimen.

Microscopy is also useful to determine the degree of homogeneity, particularly for multi-phase systems that are difficult to analyze using x-ray compositional line broadening techniques. Figure 15 shows the development of microstructures and the progress of homogenization in compacts made from a blend of nickel and tungsten powders. In a partially homogenized compact (Fig. 15b), three distinct regions are apparent:

- Dispersed light gray particles (the undissolved tungsten-rich phase that originated as tungsten particles)
- Dark gray unetched regions surrounding the tungsten-rich particles (the nickel-rich phase that has tungsten concentrations ranging from about 0.05 atom fraction of tungsten at the interface between the etched and unetched regions up to 0.17 atom fraction of tungsten at the interface between the unetched region and the tungsten-rich particles)
- Etched matrix (the nickel-rich phase that has tungsten concentrations less than 0.05 atom fraction)

Quantitative microscopy can be used to determine the extent of homogenization in multi-phase systems. This technique is compatible with x-ray compositional line broadening (XCLB) in providing data on the presence of second phases and may be performed on the same test specimens used for XCLB measurements.

Activated and Liquid-Phase Sintering

In activated sintering, the rate of densification is enhanced compared to that observed in compacts of a single metal powder or a homogeneous alloy powder. The processes to which the term activated sintering is applied are the solid-state sintering processes. In contrast, liquid-phase sintering refers to processes in which a liquid phase is formed. Of the two types of liquid-phase sintering discussed in this article, the solution and reprecipitation process frequently is used to provide enhanced liquid-phase sintering. Transient liquid-phase sintering, however, may be used for applications where rapid densification is undesirable, or where alloy formation is desired.

Fig. 15 Scanning electron micrographs showing the progress of homogenization in compacts made from a blend of nickel (−400 mesh) and tungsten (−325+400 mesh) powders
Mean composition: 0.05 atom fraction tungsten. Magnification: 75×. (a) $t = 0.0$ h. (b) $T = 1150$ °C (2100 °F); $t = 10.0$ h. (c) $T = 1150$ °C (2100 °F); $t = 48.0$ h

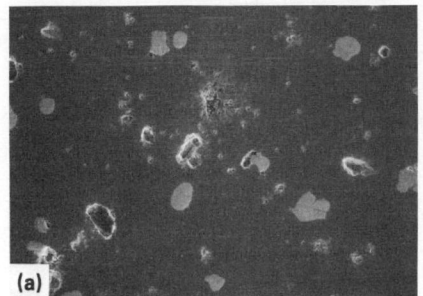
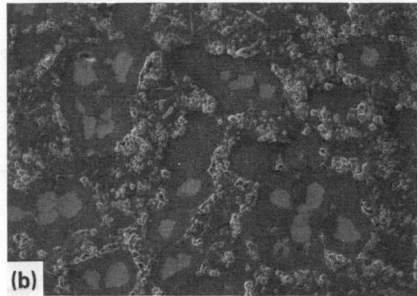

Enhanced sintering, whether achieved by activated or liquid-phase sintering, is commonly used with refractory metals (Ref 2). This is due to the difficulties associated with the extremely high sintering temperatures of these metals. The problems in understanding conventional sintering are compounded in enhanced sintering by the presence of a second phase or supplemental treatment. Most enhanced sintering studies have been conducted with little or no theory to predict beneficial treatments. The eventual success of this approach cannot be denied. Fortunately, a theory to explain such behavior is beginning to evolve and should aid future applications (Ref 2, 23, 24).

Sintering enhancement generally results from an increased driving force, through physical or chemical treatments. Many such processes are well known. While most attention has focused on tungsten, several other materials, including molybdenum, rhenium, iron, tantalum, uranium, tin, copper, aluminum, titanium, and several ceramic materials, have been investigated.

Enhancement of the sintering process generally is attributed to one or more changes in the fundamental material properties resulting from a special treatment (Ref 25). The strongest effects are those associated with changes in the interfacial properties (higher surface energy or lower grain-boundary energy). Alternatively, a less common means is to induce the operation of a normally dormant mass transport mechanism.

To the ceramist, many of these phenomena are commonplace. Impurities and stoichiometry departures can provide enhanced sintering of many ceramic materials. Likewise, the sintering atmosphere can have a profound influence on sintering rate, as well as the sintering mechanism. Any change in a material that induces an enhanced defect concentration or higher atomic mobility or that promotes the operation of new mass transport processes is considered enhanced sintering.

Typically, enhanced mass flow during the sintering cycle is beneficial. However, in the fabrication of filters, porous bearings, and flow restrictors, enhanced sintering can be detrimental. For most P/M materials fabricated for structural, magnetic, radiation, thermal, or electrical applications, improved service properties are associated with greater mass flow during sintering. Hence, any technique that delivers a greater degree of sintering is beneficial to these applications.

Usually, superior sintered properties are ensured by a higher sintered density. Pressing to high densities (above 90% of theoretical) before sintering is difficult. Consequently, specific techniques have evolved to enhance densification of powder compacts during sintering. Although the mathematics describing densification are somewhat formidable, they provide a concise description of the effects of the various process parameters. In a qualitative sense, there are some useful concepts worthy of review. For example, a smaller grain size (or smaller particle size) aids sintering densification and final properties. Higher sintering temperatures have a significant effect (because of an exponential term), thus increasing the rate of densification.

In a similar manner, a lower process activation energy has the same effect as an increase in sintering temperature. By contrast, a large pore size inhibits densification. Sintering time has a nominal effect on densification; generally, prolonged sintering is not advantageous, because grain size is increased. The grain size increase reduces the amount of grain-boundary area, thus reducing the beneficial effects of the enhanced diffusion rates at grain boundaries. Furthermore, time at temperature is expensive and usually is avoided for economic reasons.

Phase Changes

As previously discussed, compacts made from powders of body-centered cubic metals exhibit more rapid sintering than compacts made from face-centered cubic metals. This is directly related to the higher diffusivities in metals with a body-centered cubic lattice structure compared to metals with a face-centered cubic lattice structure.

In sintering compacts of iron-based compositions, it may be desirable to achieve maximum shrinkage rather than dimensional control. The body-centered cubic phase of iron (ferrite) is unstable at the usual sintering temperatures of 1000 to 1300 °C (1830 to 2370 °F). However, this phase may be stabilized in iron alloys with silicon or molybdenum. This is achieved by sintering compacts of mixtures of iron powder and ferrosilicon or molybdenum powders. Densification increases with the amount of ferrite stabilized at the sintering temperature. Increased densification is most likely due to the higher diffusivity and the fact that the phase boundary is a good vacancy sink. Additionally, the mixed phase microstructure resists grain growth during sintering.

Activated Sintering

Activated sintering refers to any of several techniques that lower the activation energy for sintering (Ref 23). Several techniques have been developed to achieve this goal, including chemical additions to the powder and the use of special atmospheres. In this respect, the sintering treatments described previously in this article that achieve stabilization of the body-cen-

Fig. 16 Microstructure of 2-μm tungsten powder compacts

Sintered 1 h at 1400 °C (2550 °F). (a) Without any special treatment. Magnification: 275×. (b) Treated with 0.4% Pd. Magnification: 770×

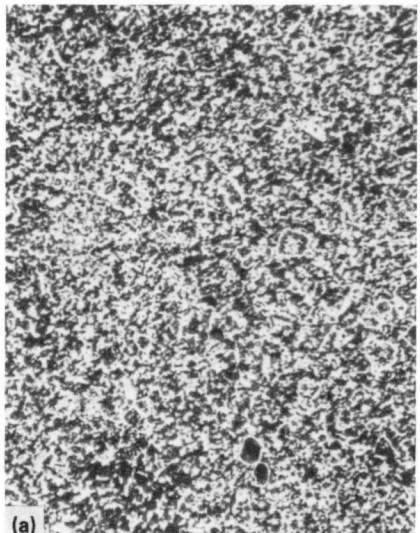
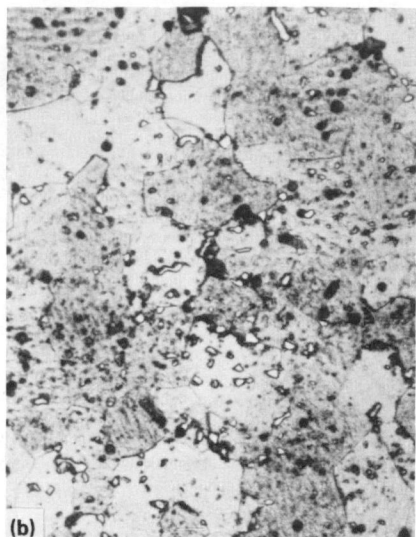

Fig. 17 Compressive strength of compacts from 0.8-μm tungsten powder

Treated with varying amounts of palladium as a function of sintering temperature

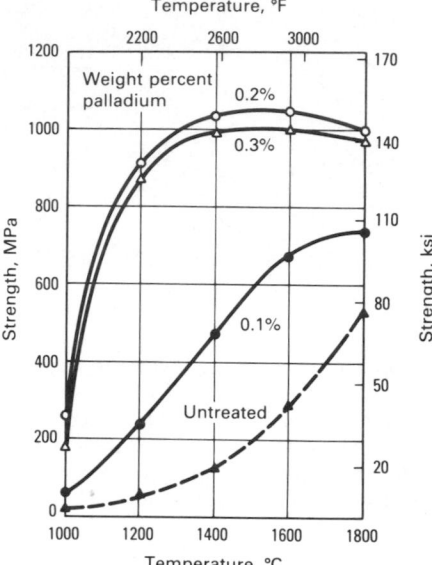

tered cubic crystal structure in iron are considered a form of activated sintering.

Many of the detailed investigations of activated sintering have been conducted on tungsten. Figure 16 shows the contrast between the microstructures of tungsten obtained by conventional sintering and by activated sintering. Tungsten without special treatment exhibits a high level of porosity and is quite weak. Treatment with 0.4 wt% Pd promotes activated sintering. Diminished porosity produced with enhanced sintering increases strength significantly.

Fine tungsten powder coated with a uniform layer of certain transition metals undergoes rapid densification at unusually low temperatures (Ref 2). The amount of additive required to promote low-temperature sintering is equivalent to one atomic layer on the powder surface. Additive contents above this amount produce minimal further enhancement in sintering and tend to lessen the degree of activation. Densification occurs in two stages. The second, slower stage begins with the onset of grain growth. Thus, the importance of grain boundaries is well demonstrated in activated sintering studies.

The effect of palladium as an activator in the sintering of tungsten on the compressive strength of tungsten compacts is shown in Fig. 17. The strength of the compacts is plotted as a function of palladium content and sintering temperature. In addition to the amount of activator and sintering temperatures, particle size of the powder is an important parameter. In Fig. 18, the density of sintered compacts of two

types of tungsten powder with 0.5- and 5-μm particle size doped with nickel are plotted as a function of nickel content.

The type of additive that proves successful as an activator must meet several criteria. First, it must form a phase that has a lower melting temperature than the base metal being sintered. Second, the activator must have a high solubility for the base metal, while the base metal should have a low solubility for the activator. The function of the activator is to remain segregated to the interparticle interfaces during sintering. Such a segregated layer provides a high diffusivity path for rapid sintering. A lower melting temperature ensures a lower activation energy for diffusion, while the solubility ensures that the activator is not dissolved into the base metal during sintering. Typically, an activator that decreases the liquidus and solidus of the base metal remains segregated to the interface between particles. Figure 19 shows an ideal phase diagram for activated as well as liquid-phase sintering systems. Note that at temperatures slightly above the activated sintering range, a liquid forms. The formation of a liquid phase is another means of enhancing sintering.

The kinetics of activated sintering are dependent on the rate of diffusion of the base metal through the thin activator layer. It is necessary to form sufficiently thick layers to provide significant diffusion fluxes by sintering activation. Concentrations above this level frequently do not prove beneficial. The measured activation energies for shrinkage in activated systems closely approximate those for self-diffu-

Fig. 18 Density of compacts of two tungsten powders treated with nickel with 0.5- and 5-μm particle size

Sintered 1 h at 1400 °C (2550 °F) as a function of the amount of nickel

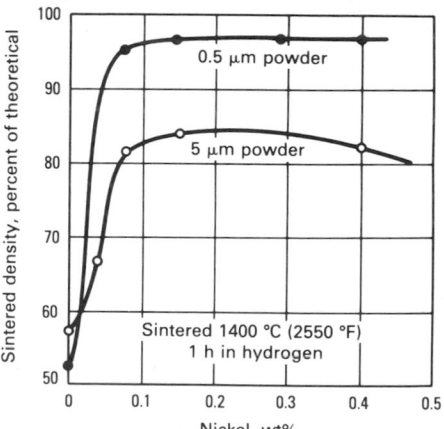

sion in the activator. Because the process has a low activation energy for diffusion, temperature is the most sensitive process control. The mechanism resembles grain-boundary diffusion-controlled sintering, thus shrinkage initially depends on the cube root of time. During the later stages of sintering, the rate of grain growth appears accelerated because of the low porosity and the high grain-boundary motion. Consequently, rapid grain growth during long-

Fig. 19 Idealized phase diagram showing favorable conditions for enhanced sintering by activated and liquid-phase sintering

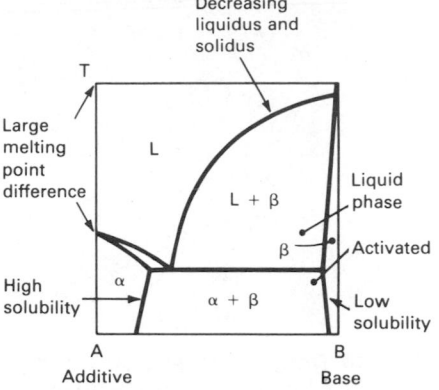

Table 1 Effect of hydrogen chloride on iron sintered in hydrogen

Temperature °C	°F	Time, min	Atmosphere, % hydrogen chloride	Density, g/cm³	Strength MPa	ksi	Elongation, %
950	1740	30	0	6.20	131	19	6
		30	1	6.30	159	23	10
		120	0	6.30	138	20	6
		120	1	6.30	159	23	10
1375	2505	30	0	7.00	193	28	11
		30	1	7.20	234	34	20
		120	0	7.50	234	34	17
		120	1	7.80	283	41	25

Fig. 20 Microstructure of compact from a mixture of 90 wt% Cu, 10 wt% Sn powders

Sintered by transient liquid-phase sintering below the solidus temperature of a 90%Cu-10%Sn bronze at 830 °C (1525 °F). Ammonia-hydrogen peroxide etch. Magnification: 140×

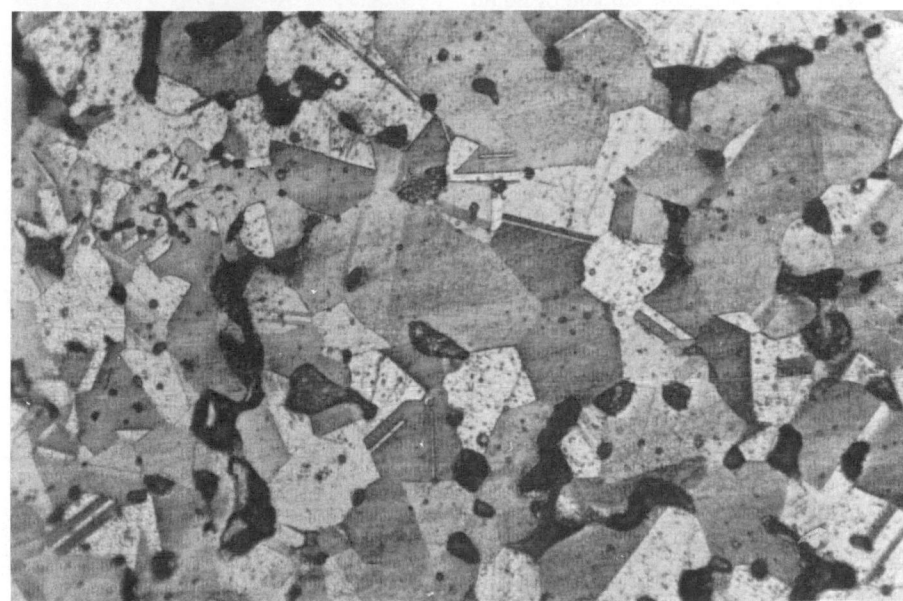

term sintering actually degrades the sintering rate and sintered properties.

Activated sintering also refers to sintering processes in which activation is produced through control of the sintering atmosphere. For example, the addition of a halide to the sintering atmosphere (Ref 26) aids transport during sintering by the formation of high-vapor-pressure molecules. Consequently, major changes in pore shape are possible. In such cases, the sintered product has greater strength and significantly higher ductility.

Table 1 provides a comparison of the strengths and the ductilities for iron compacts sintered in hydrogen. The addition of 1% hydrogen chloride to a hydrogen atmosphere results in improved properties. Metallographic examination indicates that the main effect of the hydrogen chloride is in promoting more rapid vapor phase transport by iron chloride molecules. Improvement in mechanical properties has been demonstrated in other systems. It is thus possible to achieve unique benefits from the sintering atmosphere in addition to control of oxide or carbon content.

Chemical additions are the most successful means of effecting activated sintering (Ref 2, 23, 24, 27, 28). Other processes such as radiation treatments have been successful in promoting rapid sintering. However, there has been little interest in this process. Sintering activation treatments can be classified as means of altering either the kinetics or the driving force of sintering. In radiation bombardment, sintering kinetics are altered by the creation of a vacancy excess. This eliminates the vacancy formation energy from the activation energy for diffusion. Alternatively, treatments such as cyclic heating of a material such as iron through the poly-

morphic phase transformation represent a change in the driving force. In the latter case, cyclic heating generates an internal stress, which effectively raises the driving force. Such a treatment is analogous in some respects to external stresses with hot pressing (Ref 25, 29).

Liquid-Phase Sintering

In systems involving mixed powders, liquid formation is possible because of the differing melting ranges of the components. Common systems involving liquid-phase formation during sintering include copper-cobalt, tungsten-copper, tungsten-nickel-iron, tungsten-silver, copper-tin, iron-copper, tungsten carbide/cobalt, and copper-phosphorus (Ref 30).

Heating of a mixed-phase system begins with a mixture of elemental powders. When the liquid forms, it flows and wets the solid particles and begins acting on the solid phase. Melt penetration between the solid particles can cause swelling of the

compact, especially if the green density is high and the particles are coarse (Ref 31).

However, along with melt penetration, there is possible rearrangement of the particles. The combination of wetting, liquid flow, and particle rearrangement contributes to a rapid change in the volume of the compact (Ref 31, 32). With continued heating in the presence of a liquid phase, the solid phase begins to dissolve. Depending on the solubility limit, amount of liquid, phase diagram, and composition of the alloy, one of two actions occurs. If a solid has a high solubility in the liquid, the liquid composition may recross a solidus boundary of the phase diagram and solidify. This is termed transient liquid-phase sintering.

A commerical application of transient liquid-phase sintering is the production of self-lubricating bronze bearings by sintering compacts from a mixture of copper and tin powders. In these bearings, porosity must be controlled. The goal of this process is not to achieve rapid densification, but to control carefully the changes in density during sintering. This is achieved by proper selection of the initial particle size distribution of the powders, compacting pressure, sintering temperature, and rate of heating the compacts. The microstructure of a bronze bearing produced by transient liquid-phase sintering, which consists of large grains of homogeneous copper-tin alpha bronze solid solution interspersed with pores, is shown in Fig. 20.

An alternative type of liquid-phase sintering is the solution and reprecipitation method, in which a limited solubility of the liquid phase in the solid is evident. In this type of sintering, the amount of liquid grows until saturation for the base phase is achieved. With continued time at temperature, the liquid phase becomes a carrier for the solid-phase atoms. Solution and reprecipitation does not change the amount of liquid and solid; however, it may lead to coarsening of the solid phase. Accordingly, the solid phase achieves a higher packing density through particle shape accommodation, with a flattening of faces.

An example of phase accommodation in liquid phase sintering is given in Fig. 21, which depicts a tungsten-nickel-iron alloy containing 95% tungsten. This material has achieved a high tungsten packing density (and full densification) by shape accommodation of the tungsten grains. The grains are several times larger than the tungsten particles in the original powder mixture.

Another example of liquid-phase sintering with solution and reprecipitation is shown in Fig. 22, which represents a cemented tungsten carbide with 10% cobalt. In this example, the solid-phase tungsten carbide grains are triangular and rectangular, rather than rounded. Their size is only slightly larger than that of the tungsten carbide powder particles in the powder mixture, indicating a slow rate of coarsening.

The degree of dimensional change in the solution and reprecipitation type of liquid-phase sintering can be quite large. A schematic plot of density versus sintering time is given in Fig. 23. Initially, the liquid forms and contributes to rapid densification through rearrangement. Subsequently, the rate of sintering slows as the solution and reprecipitation stage begins. With continued sintering, the solid phase typically forms an interconnected network, which further reduces the densification rate. The use of fine particles and large volume contents of liquid phase aids the achievement of high sintered densities in short times. With prolonged sintering, coarsening of the microstructure results,

Fig. 21 Compact of 95 wt% W, 3.5 wt% Ni, 1.5 wt% Fe

Sintered 1 h at 1470 °C (2680 °F). Murakami's etch. Magnification: 350×

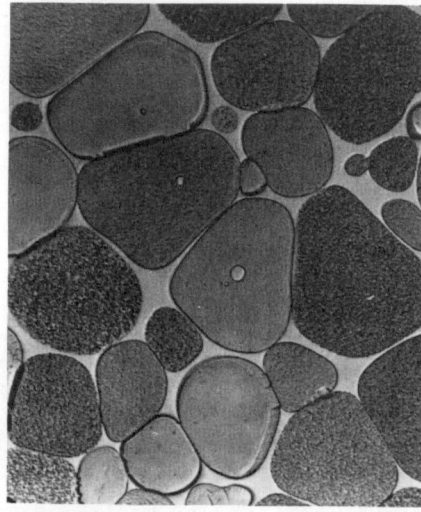

and shape accommodation is evident. Generally, coarsening should be avoided, because it degrades the mechanical properties.

Liquid-phase sintering kinetics generally are more sensitive to temperature than to sintering time. For solution and reprecipitation shrinkage, the transport path is through the liquid phase. In the final stage, solid-state sintering is dominant, and the rate of densification is quite low. Hence, the major focus is on the solution and reprecipitation stage, where larger amounts of liquid and higher solubilities aid densification. Increasing the liquid content up to approximately 35 vol% aids initial densification. Beyond this amount, little benefit is gained, as full density has been achieved already. Such large amounts of liquid phase are not used, because the compacts lack sufficient rigidity during sintering to maintain dimensional or shape control.

Green density and initial particle size can have a profound effect on densification. Coarse particle size and high green density offset the favorable effects of the liquid. In many systems, swelling rather than densification results. The melt penetrates along the interparticle junctions and causes particle separation. Expansion occurs if there is not rapid dissolution of the solid phase.

Throughout liquid-phase sintering, material is transferred through the liquid phase. When the interfacial tension between the solid and the liquid phase is isotropic (liquid solidifies to same crystal as solid), a microstructure of smooth, rounded grains

Fig. 22 Microstructure of tungsten carbide/cobalt alloy with 10 wt% Co

Murakami's etch. Magnification: 1500×

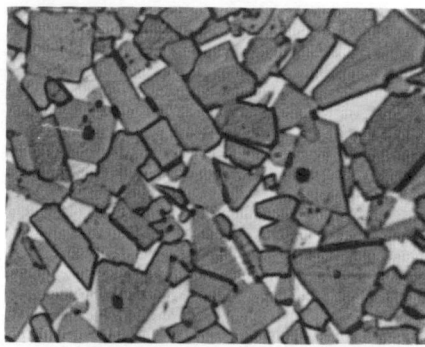

Fig. 23 Density versus sintering time for liquid-phase sintering

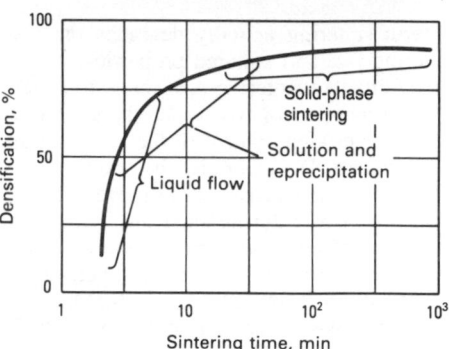

of the solid phase is produced in the matrix, as shown in Fig. 21 for a tungsten-nickel-iron alloy. When the interfacial tension between the solid and the liquid phase is highly anisotropic (the liquid precipitates at least one crystal type different from the solid), as in the tungsten carbide/cobalt system, a microstructure of rectangular or triangular particles is produced in the matrix, as shown in Fig. 22. The initial green compacts exhibit considerable porosity and irregular particle shape. After liquid-phase sintering, the microstructure shows essentially full density.

In many applications of the solution and reprecipitation type of liquid-phase sintering, the solid phase coarsens considerably with increasing sintering time, as shown in Fig. 21. In other cases, such as the tungsten carbide/cobalt system, change in the size of the solid-phase particles is minimal once characteristic shape has been established. In general, higher sintering temperatures cause increased microstructural coarsening.

The occurrence of swelling in liquid-phase sintering has been the subject of extensive study. In the iron-copper system, the onset of swelling is best explained by melt penetration along the grain boundaries within the iron particles. As a consequence, iron particles disintegrate and separate. If the amount of liquid is large, green density is high, and the particles are coarse, then swelling is more likely. Swelling can be controlled by selecting a finer particle size, lower compaction pressure, and slower heating rate. In the iron-copper system, the addition of carbon also offsets swelling. Carbon inhibits melt (copper) penetration of the iron grain boundaries.

The solution and reprecipitation types of liquid-phase sintering and activated sintering are similar in many respects. Both involve the use of a second phase at the sintering temperature to provide rapid mass transport by an effective short circuit path. As shown in Fig. 19, the phase diagram characteristics applicable to both enhanced sintering techniques are similar. Furthermore, the use of fine initial particle sizes and short sintering times produces the best sintered particles. Grain coarsening via rapid diffusional processes is experienced with both forms of enhanced sintering. Closing of large pores is slow by either technique.

A major difference between liquid-phase and activated sintering is the amount of second phase present at the sintering temperature. Liquid-phase sintering systems typically have several times more second phase than activated sintering systems. In many respects, this is beneficial, because the second phase is the continuous phase in the sintered product and is better able to sustain strain and deformation, leading to improved mechanical properties.

REFERENCES

1. Heckel, R.W., Diffusional Homogenization of Compacted Blends of Powders, *Powder Metallurgy Processing*, Kuhn, H.A. and Lawley, A., Ed., Academic Press, New York, 1978, p 51-97
2. German, R.M. and Munir, Z.A., Activated Sintering of Refractory Metals by Transition Metal Additions, *Prog. Powder Met. Phys. Ceram.*, Vol 2, 1982, p 9-43
3. Fischmeister, H., Isostatic Hot Compaction—A Review, *Powder Metall. Int.*, Vol 10, 1978, p 119-123
4. Duwez, P. and Martens, H., A Dilatometric Study of the Sintering of Metal Powder Compacts, *Trans. AIME*, Vol 185, 1949, p 571-577
5. Lenel, F.V., *Powder Metallurgy*, Metal Powder Industries Federation, Princeton, NJ, 1980, p 219
6. Grube, G. and Schlecht, H., Sintering of Metal Powders and Properties of Metal Compacts, *Z. Electrochemie*, Vol 44, 1938, p 367-374
7. Lenel, F.V., *Powder Metallurgy*, Metal Powder Industries Federation, Princeton, NJ, 1980, p 224
8. Lenel, F.V., *Powder Metallurgy*, Metal Powder Industries Federation, Princeton, NJ, 1980, p 340
9. Lenel, F.V., *Powder Metallurgy*, Metal Powder Industries Federation, Princeton, NJ, 1980, p 221
10. Exner, H.E., Principles of Single Phase Sintering, *Rev. Powder Metall. Phys. Ceram.*, Vol 1 (No. 1-4), 1979
11. Udin, H., Shaler, A.J., and Wulff, J., The Surface Tension of Solid Copper, *Trans. AIME*, Vol 185, 1949, p 186-190
12. Lenel, F.V., *Powder Metallurgy*, Metal Powder Industries Federation, Princeton, NJ, 1980, p 248
13. Kuczynski, G.C., Self-Diffusion in Sintering of Metallic Particles, *Trans. AIME*, Vol 185, 1949, p 169-178
14. Gessinger, G.H., Lenel, F.V., and Ansell, G.S., Continuous Observation of the Sintering of Silver Particles in the Electron Microscopy, *Trans. ASM*, Vol 61, 1968, p 598-604
15. Johnson, D.L., New Method of Obtaining Volume, Grain Boundary and Surface Diffusion Coefficients from Sintering Data, *J. Appl. Phys.*, Vol 40, 1969, p 192-200
16. Eloff, P.C. and Lenel, F.V., The Effect of Mechanical Constraints Upon the Early Stages of Sintering, *Modern Developments in Powder Metallurgy*, Vol 4, Hausner, H.H., Ed., Plenum Press, New York, 1971, p 291-302
17. Kuczynski, G.C., Statistical Approach to the Theory of Sintering, in *Sintering and Catalysis*, Kuczynski, G.C., Ed., New York, 1975, p 325-337; Kuczynski, G.C., Statistical Theory of Sintering, *Z. Metallkunde*, Vol 67, 1976, p 606-610
18. Fisher, B. and Rudman, P.S., Kirkendall Effect Expansion During Sintering in Cu-Ni Powder Compacts, *Acta Metallur.*, Vol 10, 1962, p 37-43
19. Heckel, R.W. and Balasubramaniam, M., The Effects of Heat Treatment and Deformation on the Homogenization of Compacts of Blended Powders, *Met. Trans.*, Vol 2, 1971, p 379-391
20. Tanzilli, R.A., Lanam, R.D., and Heckel, R.W., Techniques for the Study of Homogenization in Compacts of Blended Powders, *Advanced Experimental Techniques in Powder Metallurgy (Perspectives in Powder Metallurgy)*, Vol 5, Hirschhorn, J.S. and Roll, K.H., Ed., Plenum Press, New York, 1970, p 139-188
21. Masteller, M.S., Heckel, R.W., and Sekerka, R.F., A Mathematical Model Study of the Influence of Degree of Mixing and Particle Size Variation on the Homogenization Kinetics of Compacted Blends of Powders, *Met. Trans.*, Vol 6A, 1975, p 869-876
22. Rudman, P.S., An X-ray Diffraction Method for the Determination of Composition Distribution in Inhomogeneous Binary Solid Solutions, *Acta Cryst.*, Vol 13, 1960, p 905-909
23. Reshamwala, A.S. and Tendolkar, G.S., Activated Sintering, *Powder Metall. Int.*, Vol 2, 1970, p 15-19, 58-61
24. Munir, Z.A. and German, R.M., A Generalized Model for the Prediction of Periodic Trends in the Activation of Sintering of Refractory Metals, *High Temp.*, Vol 9, 1970, p 275-283
25. Coble, R.L., Diffusion Models for Hot Pressing with Surface Energy and Pressure Effects as Driving Forces, *J. Appl. Phys.*, Vol 41, 1970, p 4798-4807
26. McIntyre, R.D., The Effect of HCl-H Sintering Atmospheres on the Properties of Compacted Iron Powder, *ASM Trans. Quart.*, Vol 57, 1964, p 351-354
27. Hwang, K.S. and German, R.M., High Density Ferrous Components by Activated Sintering, *Processing of Metal and Ceramic Powders*, German, R.M. and Lay, K.W., Ed., The Metallurgical Society, Warrendale, PA, 1982, p 295-310
28. Jandeska, W.F., Activated Low Temperature Sintering of Iron Powder Structures, *Prog. Powder Met.*, Vol 37, 1981, p 233-253
29. Notis, M.R., Smoak, R.H., and Krishnamachari, V., Interpretation of Hot Pressing Kinetics by Densification Mapping Techniques, *Sintering and Catalysis*, Kuczynski, G.C., Ed., Plenum Press, New York, 1975, p 493-507
30. Eremenko, V.N., Naidich, Y.V., and Iavrinenko, I.A., *Liquid Phase Sintering*, Consultants Bureau, New York, 1970
31. Huppmann, W.J. and Petzow, G., The Elementary Mechanisms of Liquid Phase Sintering, *Sintering Processes*, Kuczynski, G.C., Ed., Plenum Press, New York, 1979, p 189-201
32. Kingery, W.D., Densification During Sintering in the Presence of a Liquid Phase, *J. Appl. Phys.*, Vol 30, 1959, p 301-306

Shape Fundamentals: Rigid Tool Compaction

By Leonard L. Confer
General Manager
National Carbide Die
and
Jerome P. Straub
Chief Design Engineer
Keystone Carbon Co.

RIGID TOOL SYSTEMS compact unconsolidated metal powders that are granular and exhibit some of the properties of natural sands and soils, such as distinct angle of repose when freely heaped and a limited ability to flow under the influence of gravity. Unlike massive solids, unconsolidated powders have no tensile ability, and they do not transmit force hydrostatically like liquids. When external mechanical forces are applied, metal powders generally lose the ability to flow granularly and begin to develop some measurable tensile properties as applied force is increased.

The article "Mechanical Fundamentals of Consolidation" in this Volume describes the mechanism by which an unconsolidated, free-flowing metal powder changes to the green (compacted) state when subjected to external compressive forces of sufficient magnitude (generally exceeding the plastic flow stress of at least one species of particle contained in the powder mix). The green strength of metal powders facilitates production of tooling systems that produce parts with precise dimensions and complex shapes in one rapid compaction cycle. Because no waste material is generated, this method is more cost effective than forming methods that involve cutting, grinding, chemical etching, and other types of metal removal to obtain the finished (net) shape.

Some metallic powders, such as the carbide and tool steel types, and some gas and centrifugally atomized specialty powders, such as spray-dried tungsten carbide, do not develop significant green strength, because their individual particles are predominantly spherical or they lack plasticity. To compact such powders in rigid tool systems, wax or wax-stearate binders are added, which may occupy up to 20 vol% of the green compacted shape (see the article "Lubrication of Metal Powders" in this Volume for additional information). The development of full metallic properties during sintering requires a volume shrinkage (see the article

Fig. 1 Basic geometries of MPIF class I (simple) and MPIF class IV (complex) parts

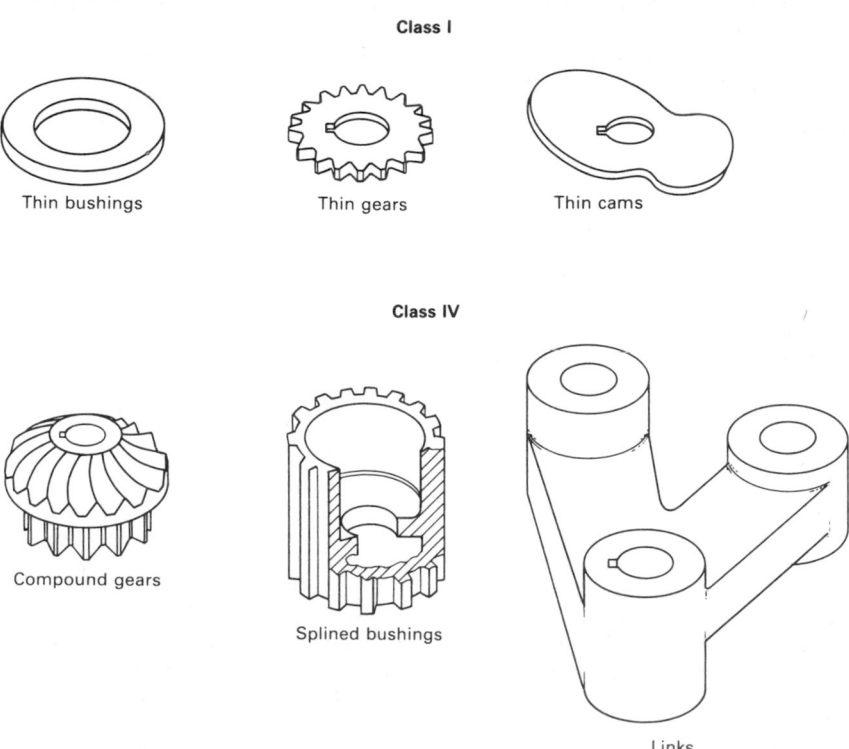

Class I

Thin bushings   Thin gears   Thin cams

Class IV

Compound gears

Splined bushings

Links

Fig. 2 Methods of fill control

(a) Spring-mounted lower outer punch. (b) Air-mounted lower outer punch. (c) Independent moving press platens

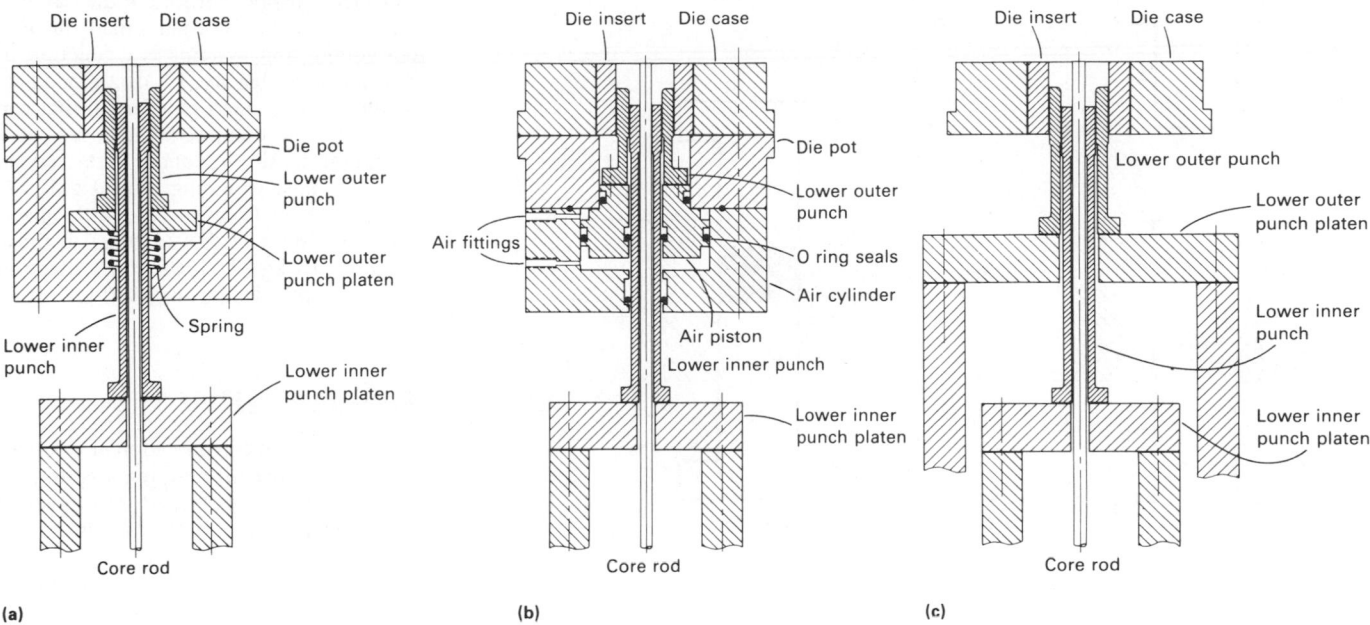

(a) (b) (c)

"Physical Fundamentals of Consolidation" in this Volume). Therefore, the rigid tool set must be constructed oversize, with exact linear dimensions, to compensate for the final volume change. Although theoretical computations are useful, most successful rigid tool sets are based on shrink-

age allowances developed from existing tooling and the dimensional histograms developed for particular powders.

Rigid tool compaction differs from roll compaction, isostatic compaction, hot isostatic pressing, and injection molding in that a quantity of powder (fill) is confined

in a rigid die cavity at ambient temperature. The die cavity is entered by one or more punches, which apply compaction pressure to the fill powder. As a result of the compaction pressure, the fill powder densifies, develops green strength, and assumes the exact shape of the die cavity and

Fig. 3 Methods of achieving fixed fill levels

(a) Fixed fill on an upper level using a step die. (b) Fixed fill using a splash pocket to permit a projection feature on an upper punch. (c) Stepped core rod forming an internal shoulder

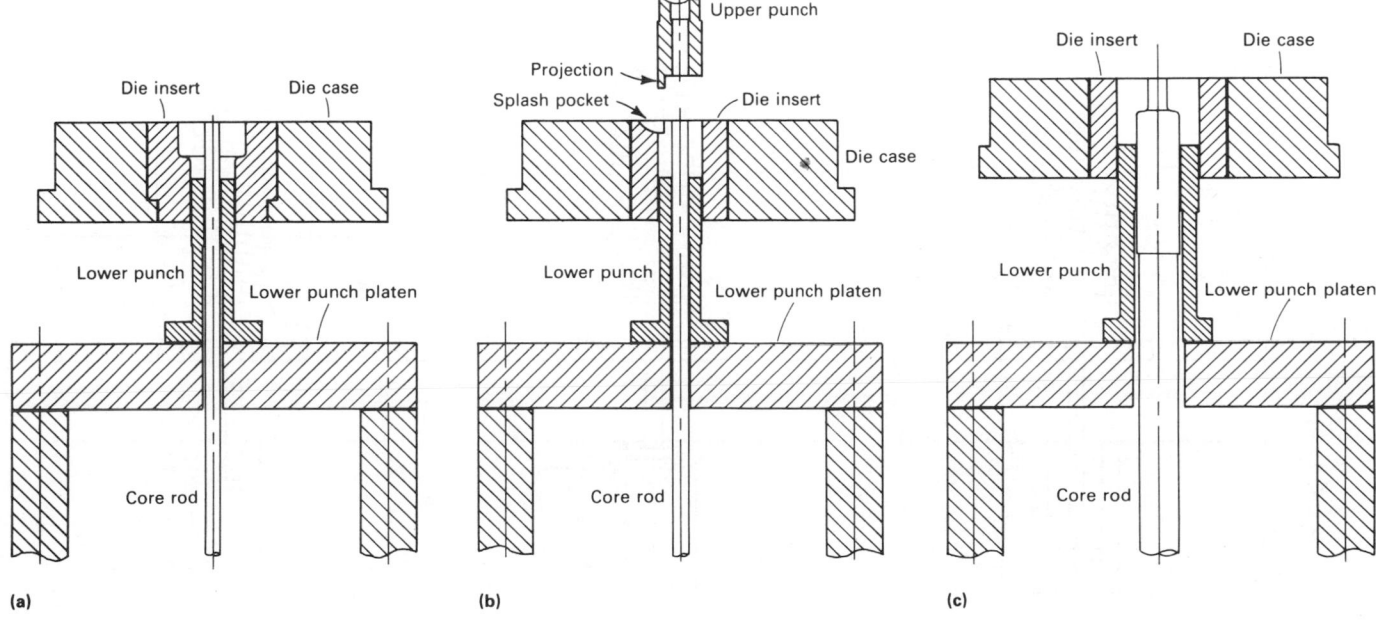

(a) (b) (c)

Fig. 4 Fill depth and length of ejection stroke for pressing compacts

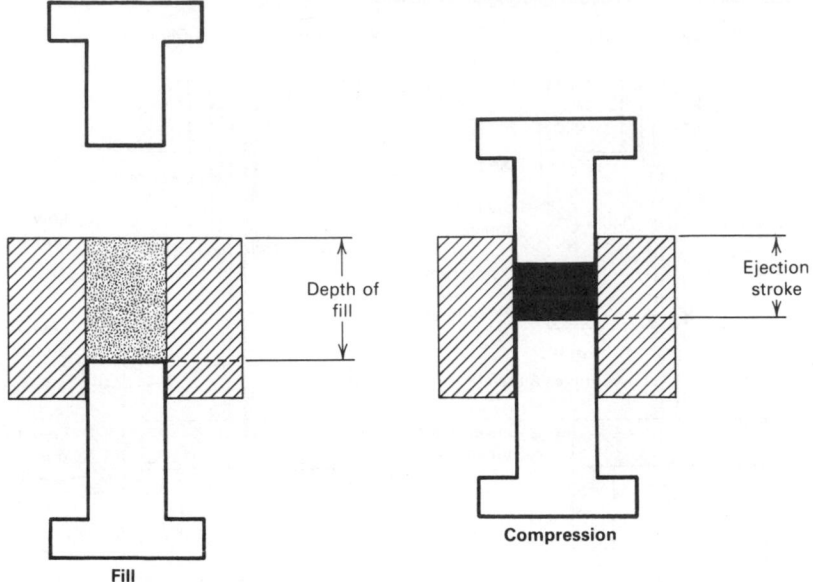

Depth of fill

Ejection stroke

Fill

Compression

punch faces. Following the pressure cycle, the shaped powder fill, now a piece part, is ejected (stripped) from the die cavity.

The physical size of parts made in rigid tool compaction systems is a function of press tonnage capacity, fill depth, and also the length of a green powder fill that can be effectively compacted in terms of a maximum density variance. Parts vary in size from those weighing about 1 g (0.035 oz) that are made in presses with capacities as small as 35 kN (4 tons) to those weighing 10 kg (22 lb) that are made in presses with capacities of 8900 kN (1000 tons).

Powder Fill

The important consideration in P/M part production is the fill ratio required to produce parts to a density that is compatible with its end use requirements. The fill ratios must remain constant for a given part to maintain dimensional reproducibility.

Fig. 5 Helical gear compaction using anvil die closure

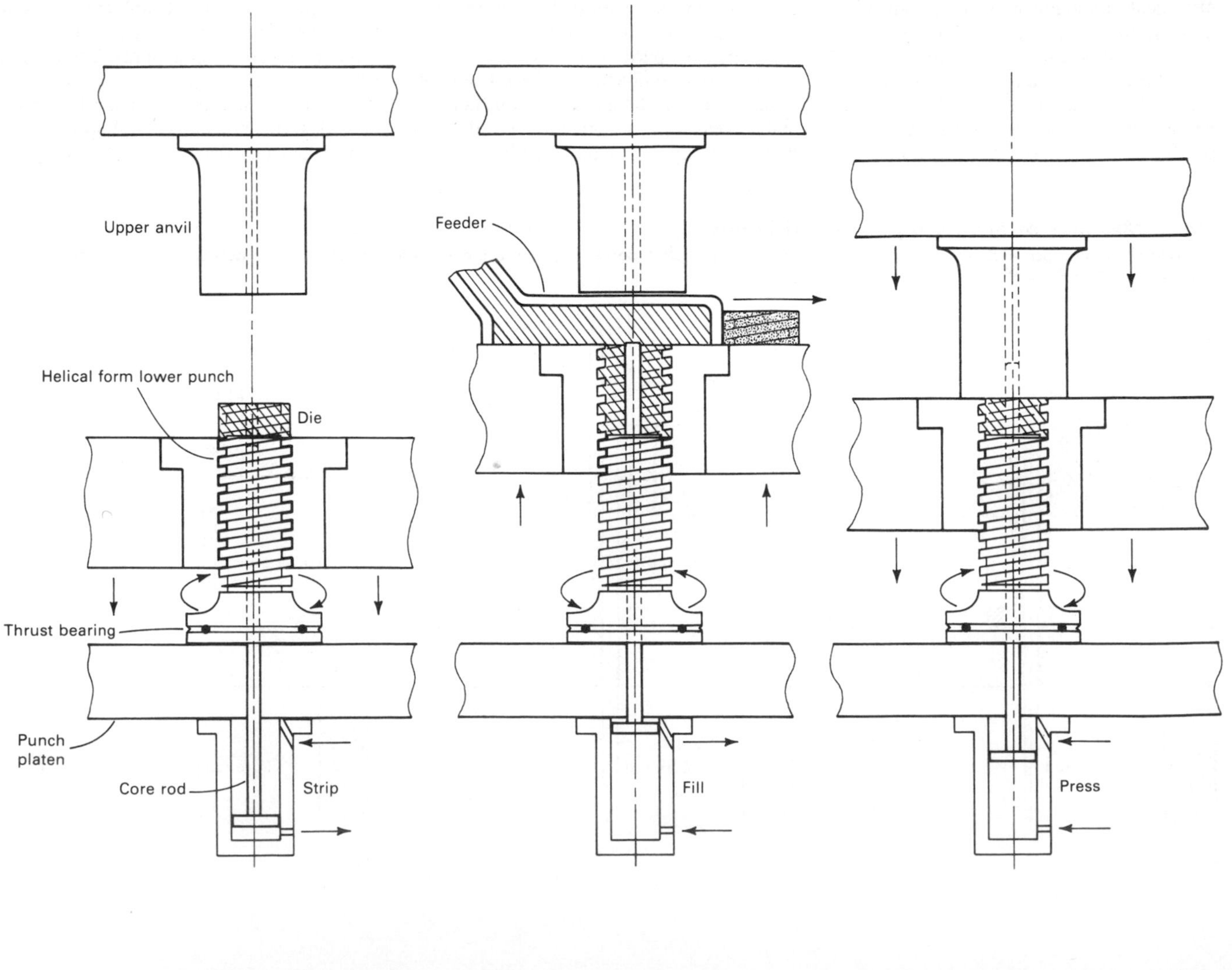

Upper anvil

Feeder

Helical form lower punch

Die

Thrust bearing

Punch platen

Core rod

Strip

Fill

Press

Parts may be of single-level or multilevel design.

Single-level parts, designated as class I by the Metal Powder Industries Federation (MPIF), present the least difficulty to the tool designer, regardless of size or part configuration. The main consideration is designing a die that is long enough to provide guidance for the lower punch (usually 25 mm or 1 in.) and providing adequate fill depth for compacting the powder to the required density. This, coupled with the primary mechanical consideration of locating the center of mass in the press center, provides the best potential for producing a uniform quality part. Figure 1 shows basic geometries of MPIF class I parts (simple) and MPIF class IV parts (complex). See the article "Production Presses and Tooling" in this Volume for more information on the MPIF classification system.

Multilevel parts, with industry classifications II through IV, present two additional complications to the tool designer: powder fill and part ejection. Because metal powders tend to compact in vertical columns and generate little hydraulic flow, the tool designer must create fill levels in the tools that compensate for the thickness variations present in the final part configuration. Uniform density, neutral axis of compaction, and part ejection must be considered to determine the need to vary fill levels and the manner in which these variations are achieved. Excessive density variations contribute to green cracks and sintered distortion (see the article "Mechanical Fundamentals of Consolidation" for additional information on density distribution in metal powder compacts).

A common method of varying fill levels is by using multiple lower punches, which are timed to react to one another either through the use of springs or air, or by mounting on separate press platens (Fig. 2). Other methods are less effective, because punches are not adjustable and are fixed on one of the tool members, such as the die or core rod.

Fixed fill levels are commonly referred to as die chokes, core rod steps, or splash pockets (Fig. 3). Fixed fills are sensitive to the apparent density of the material being compacted. In operations that control compacting pressure, such as in hydraulic pressing, fixed fills cause dimensional variations in part thickness. Because mechanical presses are set to operate to a fixed position relative to the die, the variation created by the apparent density of the powder causes overdensification or underdensification, resulting in a corresponding oversize or undersize peripheral area on

the part. Green expansion occurs as a part is stripped from the die. Ideally, the part returns to die size through shrinkage during sintering.

Press Tonnage and Stroke Capacity

The capacity (kilonewtons or tons) a press must have to produce compacts in rigid dies at a given pressure (megapascals or tons per square inch) depends on the size of the part to be pressed and is equal to the required pressure multiplied by the projected area of the part in square metres or square inches, respectively. Compacting pressure depends on the desired green density of the part, which in turn is determined by requirements for mechanical and physical properties of the sintered part. Compacting pressures may be as low as 70 to 140 MPa (5 to 10 tsi) for tungsten powder compacts or as high as 550 to 830 MPa (40 to 60 tsi) for high-density steel parts.

When a part is pressed from the top and bottom simultaneously, the press should

apply the required load to the upper and lower ram of the press. To eject the pressed compact, an ejection capacity must be available that is sometimes divided into the load for the breakaway stroke, that is, the first 1 to 12 mm ($1/32$ to $1/2$ in.) of the ejection stroke, and the load for a sustained stroke, which is generally one fourth to one half of the breakaway stroke.

The stroke capacity of a press, or the maximum ram travel, determines the length of a part that can be pressed and ejected. In presses used for automatic compacting, the stroke capacity is related to the length available for die fill and ejection stroke. As shown in Fig. 4, the depth of fill is the distance from the top of the surface of the die to the top surface of the lower punch in the filling position. It is equal to the length of the part multiplied by the compression ratio of the powder.

The ejection stroke is the distance from the top of the lower punch to the top of the die in the compression position, as seen in the right portion of Fig. 4. The press must be able to eject the pressed part by moving the lower punch up until it is flush with the die or by moving the die down

Fig. 6 Split die compacted parts

until its upper surface is flush with the lower punch.

Tooling

Dies. In P/M tooling, the die normally controls the outer peripheral shape and size of the piece part. Typically, it is constructed from materials such as tungsten carbide or high-alloy tool steels, such as T15, D2, or CPM-10V, with high hardness and good wear resistance (see the article "P/M Tool Steels" in this Volume for more information on compositions, properties, and applications of P/M tool steel materials). Dies are usually constructed in one or more sections and compressed into a retaining ring made of a low-alloy steel, such as AISI 4340 or 6150.

Considerations in die design and material selection include initial tool cost, shear strength of the die material, and die shape. A large die may require tungsten carbide, which costs ten times as much as tool steel materials. Tungsten carbide may be the best material for a set of gear tools with a relatively steep helical angle. Sectional die construction may be required for specific shapes such as sharp corners or projections into the die cavity.

Core Rods. Basically, the core rod is an extension of the die that controls the inner peripheral shape and size of the piece part. Tungsten carbide and M2 or M4 high-speed steels are the most common materials used for core rods. Primary factors in materials selection include wear resistance and hardness, which enable the core rod to resist the high compressive force exerted during compaction and the abrasive action sustained during part ejection. Core rods larger than 25 mm (1 in.) in diameter or area are held to a base by mechanical means such as a screw, while smaller core rods are held by means of silver solder or braze.

Punches may perform the function of a die or a core rod and carry the full load of the compressive force required to compact the P/M part. Wear resistance and toughness are the most important factors in materials selection. The most commonly used materials are A2, D2, S7, and H13 tool steels. Dimensional control, especially in areas such as concentricity and hole-to-hole location, depends on the amount of clearance that can be maintained between the punches, die, and core rods. Clearance should be calculated for each specific range and size of part. It is important to note that thermal size changes occur during operation, primarily because of the friction created by stripping the

compacted part and the speed of the pressing cycle.

Compaction

Pressing of P/M parts at pressures exceeding 690 MPa (50 tsi) presents unique considerations as to size and tolerance in multilevel parts. A variety of tool members must be utilized to establish proper fill ratios, and deflection and springback may occur. Deflection occurs because of the column loading effect on the compacting tools during the briquetting cycle. For column load consideration, the bottom section of the lower punch is considered fixed, while the top section or working end of the lower punch can be considered free to rotate. The amount of deflection on the tool member will be determined by the column slenderness ratio of the punch and its adapter. When the column load is released after the press goes through the bottom dead center compaction point, the deflected punches will return to their original lengths, if their elas-

Fig. 7 Split die compaction sequence

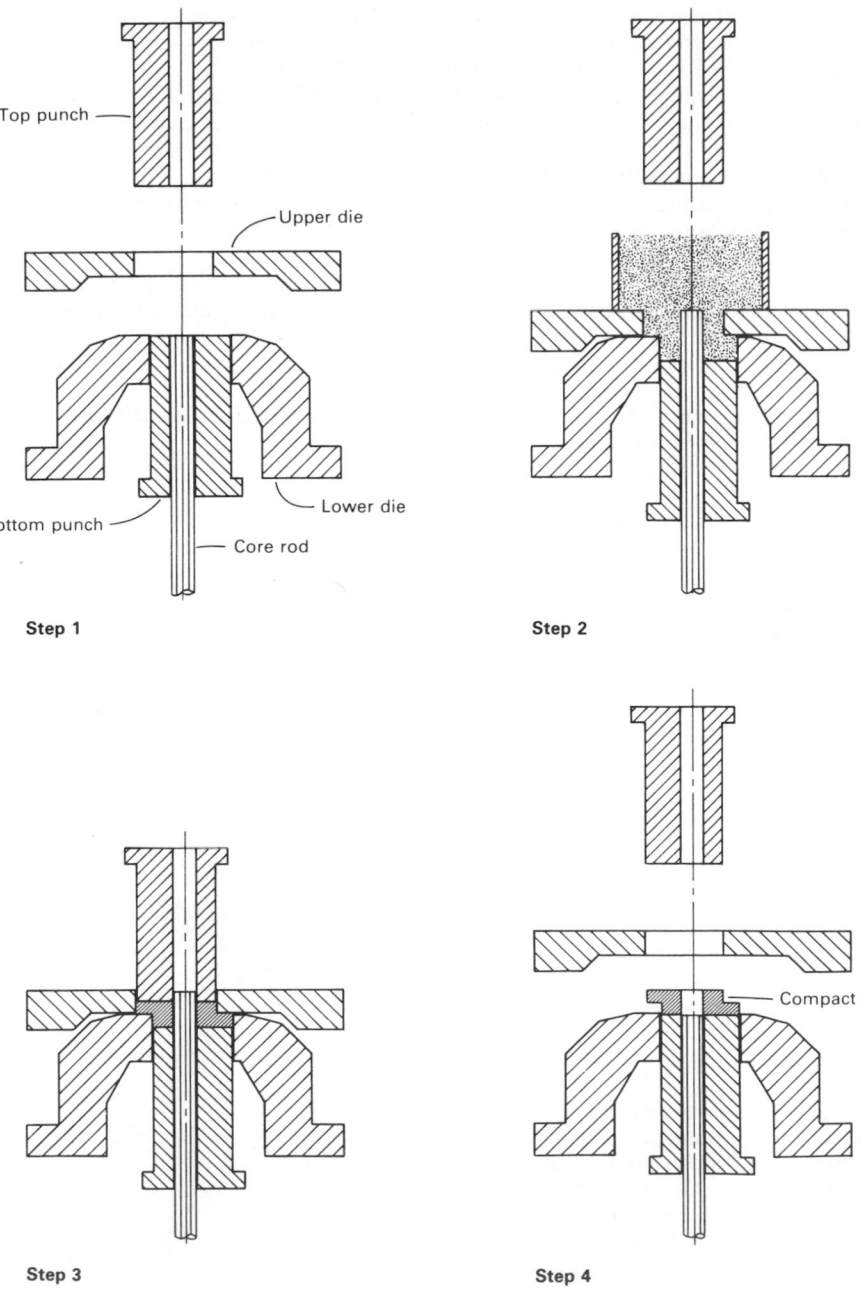

Step 1

Step 2

Step 3

Step 4

tic material property limits have not been exceeded. This return movement is generally called springback and can be deleterious to the green part, depending on the fragileness of the green part section geometry involved.

Deflection may be minimized by strengthening the various tool members through changes in physical size or shape and/or by changes in material selection. The most common method of minimizing deflection effects is to equalize deflection using tool members and adapters that are designed to match the deflection characteristics of the most critical member. The ability of the tool designer to find the proper balance is paramount for production of crack-free parts.

Advanced Tool Motions

A common limitation of some rigid tooling systems is that part features not perpendicular to the direction of pressing cannot be compacted and stripped. Frequently, it is cost effective to form features such as cross holes and threads by machining (see the article "Secondary Operations Performed on P/M Parts and Products" in this Volume). Other nonperpendicular features, notably helix shapes and hidden flanges, can be formed using complex tool motions. Another type of advanced tooling system permits production of complex shapes with magnetic orientation of the microstructure.

Helical shapes, typically helical spur gears, are produced in rigid compaction tool sets with punch rotation capability. A simple system is shown in Fig. 5, where a helical form lower punch is engaged in a die with a matching gear form. In this system, the lower punch remains engaged in the die at all times, as is common practice for all rigid tool systems, so that indexing rotation of the punch to the die is avoided. The die acts a guide. Rotation is carried out on a thrust bearing which rests on the punch platen that supports the lower punch. An upper punch is not required, because the top of the die cavity is closed by an upper anvil, which does not enter the die cavity. Central core rods, with or without additional features such as splines and key forms, are commonly operated in this helical tool system.

Helical gears made in this manner are limited to helix angles of about 25° and a thickness of 32 mm (1¼ in.) due to fill limitations along the helix tooth form. More complex helical gear tooling systems have

been proposed and may eventually be developed for routine production. These systems involve helical upper punches, driven by follower cams for indexed die entry, with inner and outer lower helical punches for stepped helical gears. For more information on complex helical gear tooling systems, see Ref. 1.

Split Die Systems. Another rigid tooling system that avoids some through-cavity limitations is known as the split die, or "double die," system. It enables the compaction of parts with completely asymmetric upper and lower sections in the pressing direction. Typical parts are illustrated in Fig. 6. Typical tool motions

in split die compaction are shown in Fig. 7. This system requires two die-holding platens to carry the upper die and lower die. Each platen is controlled and moved independently.

Wet magnetic compaction, as shown in Fig. 8, has enjoyed wide usage in the production of magnetically oriented ferrite shapes. In this production process, a feed shoe is not required. Instead, the die cavity is injected with an aqueous slip (slurry) that has a high concentration of ferrite powder, with the addition of green binders as required. Typically, the die filling pressure is 35 MPa (5000 psi). By using an aqueous slip, many of the gravity die fill

Fig. 8 Wet magnetic compaction

(a) Force-time diagram for magnet presses. (b) Schematic of press tool for chamber-filling method designed for withdrawal operation

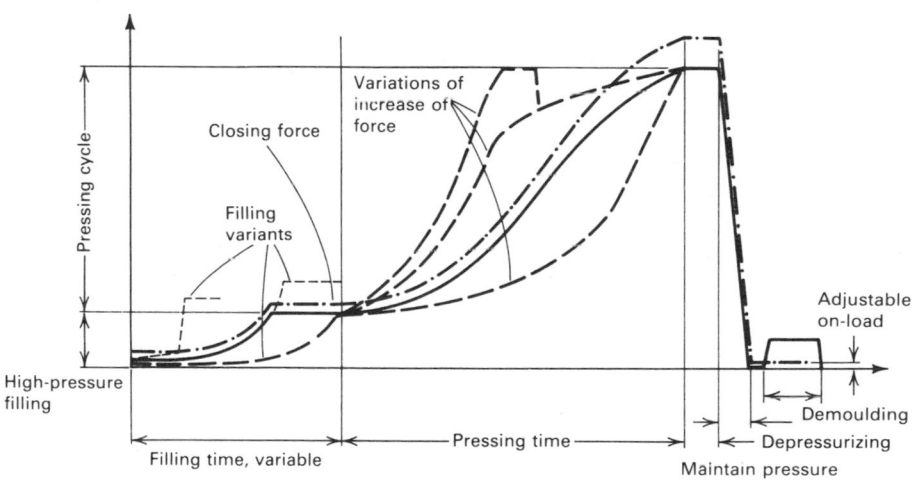

problems, such as attainment of uniform powder density and filling the areas that are difficult for the powder to reach, are avoided.

Following die fill injection, an orienting magnetic field is applied to the slip, resulting in magnetic polarization of the individual ferrite particles, which remain mobile at this point. The optimum orientation of the ferrite particles directly determines the quality of the finished permanent magnet. After magnetic orientation, the main pressing load is applied, densifying the ferrite mass and causing the suspending aqueous carrier to be expelled through drainage ports. The compact is imparted with the precision shape and dimensions of both the upper and lower dies, plus any core rods that may be inserted. The cycle is completed by separation of the press platens and ejection of the compacted ferrite shape.

REFERENCE

1. MPIF Newsletter, Vol 12 (No. 2), Feb-March 1983

Production Presses and Tooling

By Robert Burns
Manager, Engineering
P/M Products
Cincinnati Incorporated

POWDER METAL COMPACTING PRESSES, equipped with appropriate tooling, frequently are used for producing P/M components. Although commonly called P/M presses, use is not limited to the pressing of metal powders. Almost any alloy or mixture of materials produced in powder form can be compacted into suitable end products. The majority of components fabricated by P/M presses, in number of pieces and pounds of product produced, consists of compacted metals. Ferrous-based metals constitute the largest usage. Powder metallurgy compacting presses usually are mechanically or hydraulically driven, but can incorporate a combination of mechanically, hydraulically, and pneumatically driven systems.

Compacting Press Requirements

Although P/M presses resemble stamping and forming presses, several significant differences exist. Press frames generally have straight sides. Gap-type or "C" frame presses are not suitable because the frame deflects in an arc under load, resulting in a slight out-of-alignment condition between the bed and side of the press. This arrangement produces a compacted part that is slightly out of parallel, top to bottom. Because P/M tooling clearances are generally about 0.025 mm/25 mm (0.001 in./1 in.) total, bending deflection may cause broken tooling or excessive tool wear. Press frames and drive mechanisms for machines used in other press-working applications are described in Volume 4 of the 8th edition of *Metals Handbook*.

Powder metallurgy presses apply sufficient pressure from one or both pressing directions (top and bottom) to achieve uniform density throughout the compact. Design should include provision for ejecting the part from the tooling. Pressing and ejection take place during each cycle of the press and must be accurately synchronized.

Presses must have sufficient connected horsepower to compact and eject the part. In most press-working applications, the working stroke is a small portion of the total stroke of the press. In P/M presses, the working stroke during the compaction portion of the cycle is usually greater than the length of the part being produced, and the ejection portion of the cycle has a working stroke equal to or greater than the length of the part by a factor of about 2 or 3. In some cases, the power required during the ejection cycle is greater than that required during compaction.

Presses should provide for adjustable die filling (the amount of loose powder in the tooling cavity). Automatic powder feeding systems that are synchronized with the compaction and ejection portion of the press cycle are desirable. Finally, P/M presses must meet federal, state, and local design and construction safety laws. Metal Powder Industries Federation (MPIF) standard 47 details safety standards for P/M presses.

Mechanical presses are available in top-drive and bottom-drive arrangements. In top-drive presses, the motor, flywheel, and gearing system are located in the crown or upper structure of the press. Presses with pressing capacities of approximately 1780 kN (200 tons) are floor mounted, requiring little or no pit. Top-drive presses with pressing capacities above 1780 kN (200 tons) usually require a pit to maintain a convenient working height for the operator.

In bottom-drive presses, the drive mechanism, motor, and flywheel are located in the bed of the press. These presses usually are "pulled down"; that is, the top ram of the press is pulled downward by draw bars or tie rods. Bottom-drive presses with pressing capacities above 445 kN (50 tons) usually require pits. Top-drive and bottom-drive presses are comparable in terms of partmaking capability, reliability, and equipment cost.

Mechanical Presses

In most mechanical P/M compacting presses, electric motor-driven flywheels supply the main source of energy used for compacting and ejecting the part. The flywheel normally is mounted on a high-speed shaft and rotates continuously. A clutch and a brake mounted on the flywheel shaft initiate and stop the press stroke. To initiate a press stroke, the brake is disengaged and the clutch is engaged, causing the energy stored in the rotating flywheel to transmit torque through the press gearing to the final drive or press ram.

Clutch and brake systems should be of the partial revolution type that can be engaged and disengaged at any point in the pressing cycle. The clutch usually is pneumatically engaged with a spring release, and the brake is pneumatically released with a spring set, thereby providing full stopping ability in the event of loss of air pressure. An adjustable speed device normally is supplied with the electric drive motor, providing production rate adjustment as indicated by pressing and ejection conditions.

On presses that have main motor capacities up to approximately 19 kW (25 hp), the adjustable speed drive is usually of the variable-pitch pulley or traction-drive type. Above 19 kW (25 hp), direct current or eddy-current control devices are preferred. The motor and drive must be totally enclosed to prevent contamination by metal powder dust.

Gearing systems usually are either single-reduction (Fig. 1) or double-reduction

Fig. 1 Single-reduction gearing systems for P/M compacting press

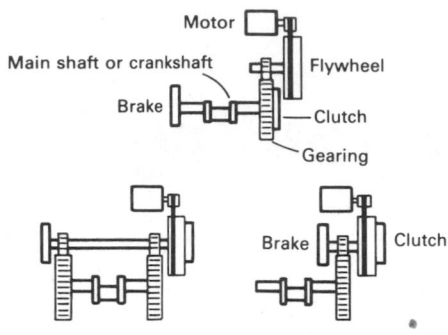

Fig. 2 Double-reduction gearing systems for P/M compacting press

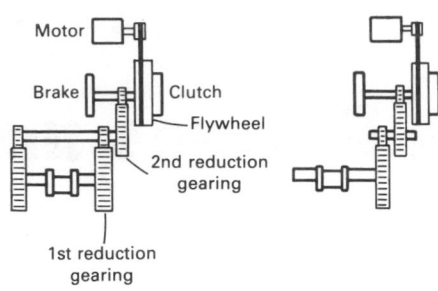

Fig. 3 Schematic of cam-driven compacting press

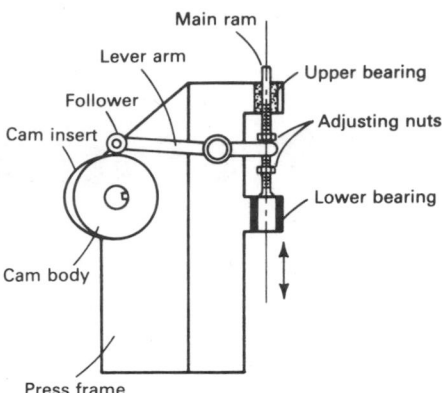

(Fig. 2) arrangements. Single-reduction gearing frequently is used in lower tonnage presses (up to 445 kN, or 50 tons) that have stroking rates of about 50 strokes/min. Higher tonnage presses use double-reduction gearing and commonly have maximum stroking rates of 30 strokes/min.

The low-speed shaft of the press, normally called the main shaft, is linked to the press ram, causing motion of the tooling for the compacting and ejection cycles. Ram driving mechanisms may be either cam- or eccentric-driven arrangements.

Cam-driven presses generally are limited to pressing capacities up to 890 kN (100 tons). The main shaft of the press has two cams—one operates the upper ram and the other operates the lower ram for compacting the part. The cam that operates the lower ram also controls the powder feed into the die and ejects the part from the die after compacting. Cams normally operate linkages that convert the main shaft rotary motion into the linear motion of the tooling.

A cam-driven press is shown schematically in Fig. 3. The cams in this type of press can be adjusted or arranged with removable sections, thus allowing cam motion to be varied to produce special motions to compact the part. Pressure can be applied either simultaneously or sequentially to the top and bottom of the compact. Anvil and rotary presses are types of cam-driven machines. These presses are described in more detail later in this article.

Eccentric-driven Presses. Presses that have a final drive mechanism consisting of an eccentric or crank on the main shaft are the most widely used type of mechanical press. A connecting rod is used to convert the rotary motion of the main shaft into the reciprocating motion of the press ram. Generally, an adjustment mechanism is built into the connecting rod or press ram assembly, thus permitting the height position of the press ram to be changed

with respect to the main shaft or press frame, thereby controlling the final pressing position of the ram. This adjustment mechanism may be used to control the length of the compacted part. Standard eccentric-driven presses have pressing capacities ranging from 6.7 to 7340 kN (0.75 to 825 tons).

Figure 4 shows a typical main shaft, connecting rod, and press ram assembly. The main shaft is fitted with cams that operate the ejection system. These cams normally are not adjustable, because the timing of the ejection portion of the cycle must be synchronized with the pressing portion. Presses of this type can produce a wide variety of parts, from a single-level flat washer to a part that has as many as five individual pressing levels or surfaces.

Hydraulic Presses

Hydraulically driven compacting presses are available with pressing capacities ranging from 445 to 11 100 kN (50 to 1250 tons) as standard production machines, although special machines with capacities up to 44 500 kN (5000 tons) have been used in production. Hydraulic presses normally can produce longer parts in the direction of pressing than mechanical presses; longer stroke hydraulic machines are less expensive compared to an equivalent stroke produced in a mechanical press. The maximum depth of powder fill in mechanical presses is approximately 180 mm (7 in.), while 380 mm (15 in.) of powder fill is common in hydraulic presses.

The maximum production rate for hydraulic presses producing a single part per stroke is approximately 650 pieces per hour. The slower speed of a hydraulic press when pressing long parts is preferable, because the longer time during pressing permits trapped air within the powder to escape through the tooling clearances.

Most hydraulic presses are considered top-drive machines, because the main operating cylinder is centrally located in the

top of the press. This main cylinder provides the force for compacting the part. Hydraulic presses have three distinct downward speeds:

- *Rapid advance:* Produces minimal pressing force, used for rapid punching of the die cavity
- *Medium speed:* Pressing capacities about 50% of full-rated capacity, used during initial compaction when lower pressing force is required
- *Slow speed:* Maximum capacity available for final compaction

Two types of hydraulic pumping systems are commonly found in P/M presses: the high-low system and the filling circuit system. The high-low system has a double-acting main cylinder. A regenerative circuit is used for rapid approach. Initially, the piston of the cylinder is activated by a high-volume, low-pressure pump, the fluid from the bottom of the cylinder being directed into the top of the cylinder in addition to the low-pressure pump volume. At medium speed, the regenerative circuit is deactivated, while the piston remains activated by the low-pressure pump. In a full-tonnage press, the low-pressure pump is deactivated, and a high-pressure pump activates the piston.

The filling circuit hydraulic pumping system has a single-acting main cylinder, and ram motion is controlled by small double-acting cylinders. The ram control cylinders are smaller than the main cylinder, so only a low flow rate of fluid is needed to cause rapid movement of the ram. During approach and return cycles, however, the fluid flow rate into and out of the main cylinder is high. The main cylinder is fitted with a large two-way valve that allows fluid to flow at low pressures (usually gravity feed). During pressing, the two-way valve is closed, and high pres-

Fig. 4 Typical eccentric-driven assembly for a P/M compacting press

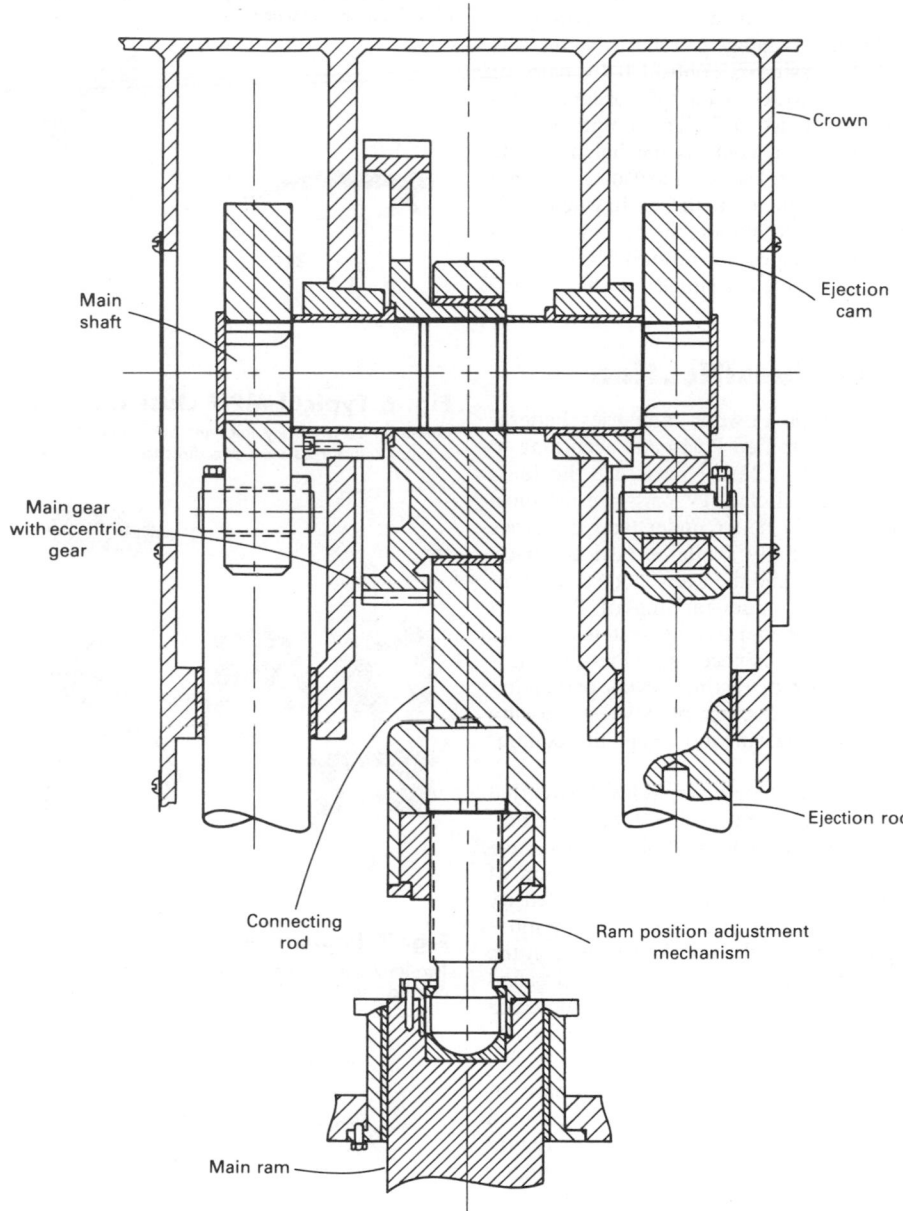

Main shaft

Main gear with eccentric gear

Connecting rod

Main ram

Crown

Ejection cam

Ejection rod

Ram position adjustment mechanism

draulic press must supply energy directly during the pressing and ejection portion of the cycle.

Comparison of Mechanical and Hydraulic Presses

In terms of partmaking capability, no distinct advantage is gained by using either a mechanical press or a hydraulic press. Any part can be produced to the same quality on either type of machine. However, the following parameters influence press drive selection.

Production Rate. A mechanical press produces parts at a rate one and one half to five times that of a hydraulic press as a result of inherent design of the energy transfer systems and stroke length.

Operating cost of a hydraulic press is higher, because the total connected horsepower of a hydraulic press is one and one half to two times that of an equivalent mechanical machine. Theoretically, the required energy to compact and eject a part is the same for a hydraulic or a mechanical press, except that the overall efficiency of a mechanical press is slightly higher than that of a hydraulic one. Also, the kilowatt usage of the larger motor on a hydraulic press is greater than that of a mechanical press during the idle portion of the machine cycle.

Machine overload protection is an inherent feature of a hydraulic press. If the hydraulic system is operating properly, the machine cannot create a force greater than its rated capacity. Consequently, overload of the machine frame is not possible, even if a double hit or operator error occurs in adjusting the machine. Misadjustment or double hits may cause a mechanical press to overload and can damage the machine or may cause tooling overload and failure if the tooling cannot withstand full machine capacity.

Equipment cost of a hydraulic press generally is one half to three quarters that of an equivalent mechanical press. Facility, foundation, installation, and floor space costs generally are comparable.

Die Sets. The mounting into which the tooling is installed is known as the die set. Generally, the die set must be well guided because of the close tooling clearances used. Guide bearings must be protected with boots or wipers to prevent powder particles from entering guiding surfaces. Tooling support members should have high stiffness to minimize deflection.

The die set must be free of residual magnetism. Approximately 2 gauss is the maximum acceptable level. To ensure press

sure from the pump is applied to the main cylinder piston.

Ejection of the part usually is accomplished by a cylinder that is centrally located in the bed of the press. The cylinder either upwardly ejects the part or pulls the die downward from the part, depending on the type of tooling used.

When pressing parts to a given thickness, positive mechanical stops are used on hydraulic presses to control downward ram movement. When pressing parts to a desired density, downward ram movement is controlled by adjustment of the pressure to the cylinder. When the part is pressed

to the desired unit pressure, the press ram stops and returns to its retracted position. Some types of P/M materials, such as P/M friction materials, are always pressed to density rather than size, because uniform density provides uniform friction and wear properties.

The drive-motor horsepower on a hydraulic press is considerably larger than on an equivalent mechanical press. A mechanical press has a flywheel from which energy is taken during the pressing and ejection of the part. Energy is restored to the flywheel during the die feeding portion of the cycle. The motor on a hy-

operator safety, die sets should be adequately guarded. In a complex tooling arrangement, as many as seven independent tooling members and supports are moving relative to one another during the pressing and ejection cycles.

Die sets can be classified as removable or nonremovable. Both types are used in mechanical and hydraulic presses. Nonremovable die sets are used throughout the entire tonnage requirements of available presses. Removable die sets are used primarily in presses with pressing capacities up to approximately 2670 kN (300 tons). Above this press size, the die set assembly becomes too large and heavy to be moved in and out of the press easily, although removable die set presses with capacities of up to 17 800 kN (2000 tons) are available.

The major advantage offered by nonremovable die sets is flexibility in setup and operation. Presses equipped with nonremovable die sets usually have all adjustments required for setup and operation built into the press and die set, including:

- *Part length adjustment:* Any dimensions of the part in the direction of pressing can be quickly changed during production.
- *Part weight:* Material weight in any level of the part can be changed easily during production.
- *Tooling length adjustment:* Adjustments are provided to accommodate shortening of punch length due to sharpening or refacing.

Another advantage of nonremovable die sets is the greater space available for tooling, compared to the removable type. This space provides more freedom in tooling design. However, presses incorporating nonremovable die sets must be shut down during tooling changes or maintenance. Tooling change and setup time generally is from 1 to 4 h—but sometimes substantially longer, depending on the complexity of tooling.

Nonremovable die sets are well suited for developing new P/M parts, because press and tooling adjustments can be made quickly to achieve the desired weight, density, and part dimension. Adjustment features of nonremovable die sets make them desirable on long production runs, where changes in powder quality among lots require frequent tooling adjustment to maintain part quality.

Users of removable die sets normally have two or more die sets per press. Tooling can be set up in a spare die outside the press. Removable die sets normally can be changed in less than 30 min, so loss of production time is minimal. On small presses where the die set is also small, the die set is restricted to a given set of tools and is considered semidurable tooling.

One disadvantage of removable die sets is that pressing is controlled by pairs of pressing blocks made of hardened tool steel, such as D-2. The height of the pressing block controls the height of the part. If the part length dimension is changed due to design, or if the tooling length is changed due to repair, the pressing blocks must be changed accordingly. Removable die sets are ideally suited for shorter production runs.

Part Classification

The Metal Powder Industries Federation has classified P/M parts according to complexity. Class I parts are the least complex, and class IV parts are the most complex. To better understand the types of commercially available P/M compacting presses, and their advantages and limitations, an understanding of P/M part classification and tooling systems used to produce parts is necessary. Part thickness and number of distinct levels perpendicular to the direction of powder pressing determine classification—not the contour of the part.

Class I parts are single-level parts that are pressed from one direction, top or bottom, and that have a slight density variation within the part in the direction of pressing (Fig. 5). The highest part density is at the surface in contact with the moving punch, and the lowest density is at the opposite surface. Parts with a finished thickness of approximately 7.5 mm (0.3 in.) can be produced by this method without significant density variation.

Class II parts are single-level parts of any thickness pressed from both top and bottom (Fig. 6). The lowest density region of these parts is near the center, with higher density at the top and bottom surfaces.

Class III parts have two levels, are of any thickness, and are pressed from both top and bottom (Fig. 7). Individual punches are required for each of the levels to control powder fill and density.

Class IV parts are multilevel parts of any thickness, pressed from both top and bottom (Fig. 8). Individual punches are required for each level to control powder fill and density.

Tooling Systems

High-production P/M compacting presses are available as standard production machines in a wide range of pressing capacities and production rate capabilities.

Fig. 5 Typical MPIF class I parts
Thin, one-level parts of any contour pressed with a force from one direction

Fig. 6 Typical MPIF class II parts
One-level parts of any thickness and contour, pressed with forces from two directions

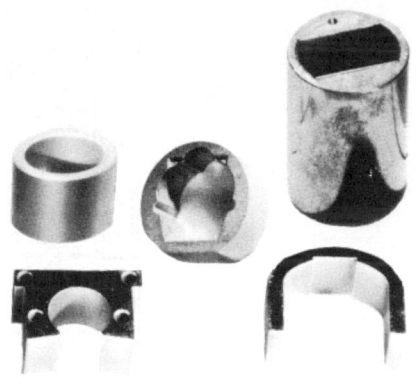

Fig. 7 Typical MPIF class III parts
Two-level parts of any thickness and contour pressed with forces from two directions

Presses are designed to produce parts of a specific classification, as discussed above.

Single-action tooling systems generally are limited to production of class I parts. During the compacting cycle, the die, core rod, and one of the punches (usually the lower punch) remain station-

Fig. 8 Typical MPIF class IV parts
Multilevel parts of any thickness and contour
pressed with forces from two directions

ary. Compacting is performed by the moving punch, which is driven by the action of the press. One or more core rods may form any through holes in the part.

During ejection, the upper punch moves away from the formed part, and the part is ejected from the die by the lower punch. The core rod (Fig. 9) is stationary, and the part is ejected from the die and core rod simultaneously. On some presses, the core rod is arranged so that it is free to move upward (float) with the part as it is ejected. The compacted part experiences slight elastic expansion on ejection from the die, which causes the part to free itself from the core rod. The core rod is then free to move downward to the fill position. This floating core rod arrangement reduces ejection forces and core rod wear.

Double-action tooling systems primarily are used to produce class I and II parts. Force is applied to the top and bottom of the part simultaneously, because the punches have the same travel rate. The die and core rod are stationary. Densification takes place from the top and bottom, with the lowest density region near the center of the part. Although the core rod is fixed in this system, it can be arranged in a floating position. Figure 10 shows the compacting sequence of a double-action tooling system.

Floating die tooling systems are similar to double-action arrangements. As shown in Fig. 11, the die is mounted on a yielding mechanism (springs). However, pneumatic or hydraulic cylinders usually are used, because they offer an easily adjustable resisting force. As the

Fig. 9 Compacting sequence utilizing single-action tooling
Dashed line indicates motion of lower punch.

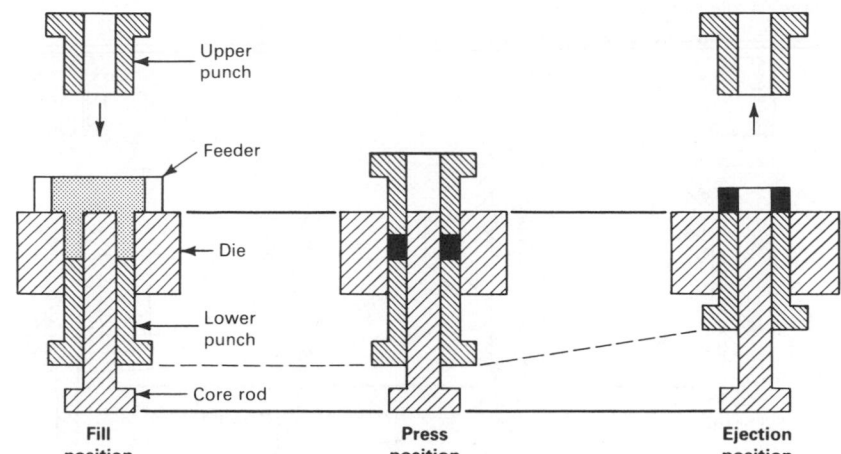

Fig. 10 Compacting sequence utilizing double-action tooling
Dashed line indicates motion of component parts.

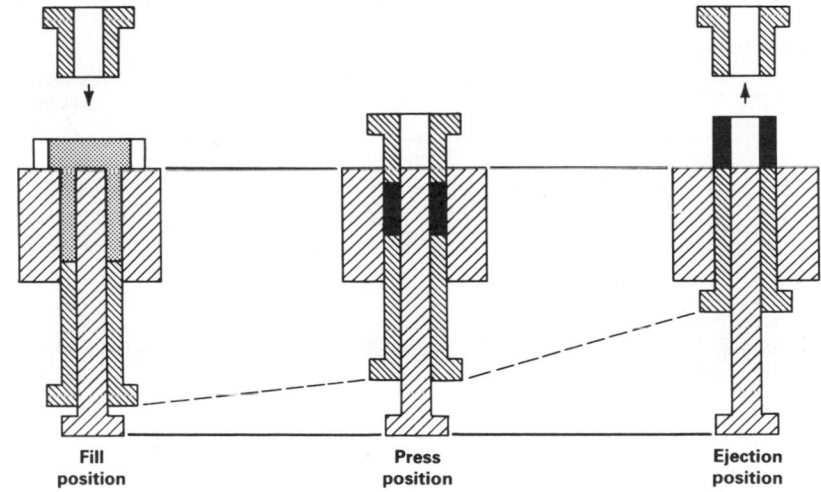

Fig. 11 Compacting sequence utilizing floating die tooling
Dashed lines indicate motion of component parts.

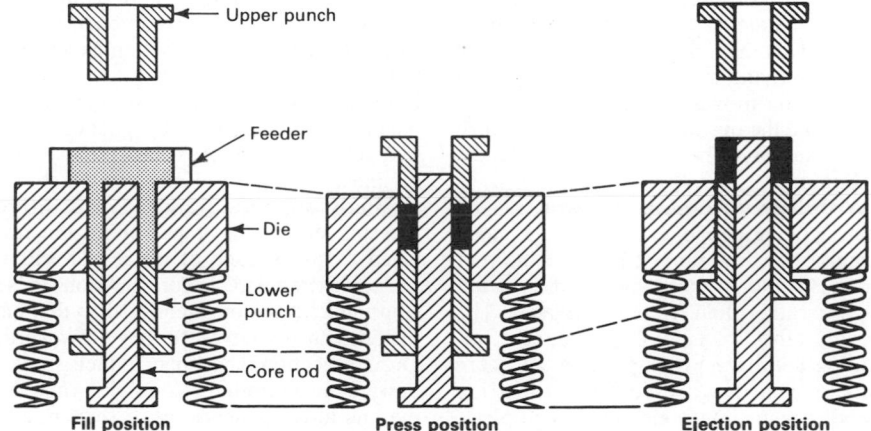

Fig. 12 Compacting sequence utilizing floating die withdrawal double-action tooling

Dashed lines indicate motion of component parts.

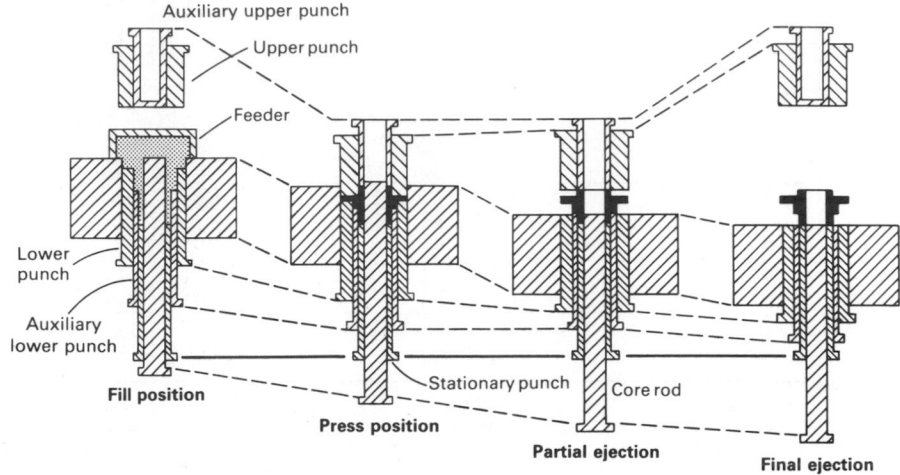

Auxiliary upper punch

Upper punch

Feeder

Lower punch

Auxiliary lower punch

Fill position

Stationary punch Core rod

Press position

Partial ejection

Final ejection

Fig. 13 Compacting sequence utilizing sliding anvil single-action tooling

Dashed line indicates motion of component parts.

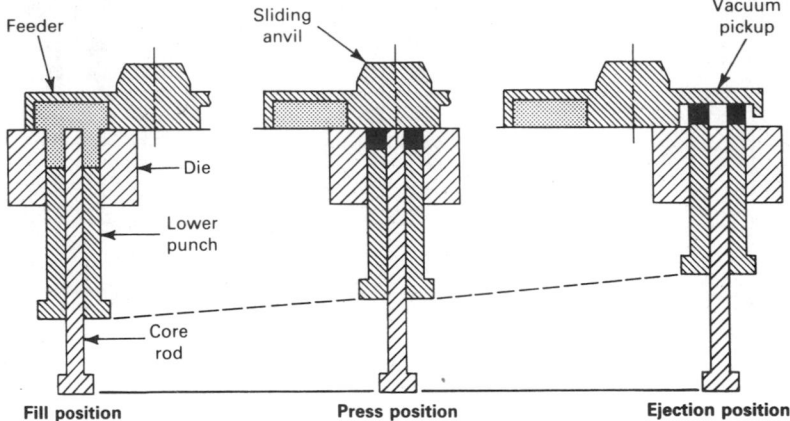

Feeder

Sliding anvil

Vacuum pickup

Die

Lower punch

Core rod

Fill position

Press position

Ejection position

upper punch enters the die and starts to compact the powder, friction between the powder and die wall causes the die to move down. This has the same effect as an upward-moving lower punch. After pressing, the die moves upward to its fill position, and the upward-moving lower punch ejects the part. The core rod can be fixed or floating.

Withdrawal tooling systems use the floating die principle, except that the punch forming the bottommost level of the part remains stationary and that the die motion is press rather than friction activated. The die and other lower tooling members, including auxiliary lower punches and core rods, move downward from the time pressing begins until ejection is complete.

Figure 12 shows the compacting sequence in a multiple-motion withdrawal tooling system. During compaction, all elements of the tooling system except the stationary punch move downward. The die is mounted on the top press member of platen and is supported by pneumatic or hydraulic cylinders. Auxiliary punches are mounted on additional platens, which are similarly supported and have positive pressing stops. The stops control the finished length of each of the levels within the compacted part. Before ejection, these stops are released or disengaged so that the platens can be moved further downward. During ejection, the upper punch moves upward, away from the compact, while the die and lower punches move sequentially

downward until all tool members are level with the top of the stationary punch. The compact is fully supported by the tooling members during ejection, resting on the stationary punch as the die and lower punches are lowered to release it.

The core rod can be provided with pressing position stops to allow a part to be produced with blind or counterbored holes. The core rod is held stationary until the part is free of all other tooling members before moving downward to its ejection position.

At this point in the machine cycle, the feeder moves across the die, pushing the compacted part from the die area and covering the die cavity. The die and auxiliary lower punch move upward to their respective fill positions. The core rod then moves upward, displacing the excess powder into the partially empty feed shoe. The feeder retracts, wipes the top fill level, and readies the press for the next cycle.

Types of Presses

Anvil presses generally are limited to compaction of class I parts in a single direction. Anvil presses do not have an upper punch; a moveable, solid, flat block seals the top of the die. Compacting is done by the lower punch, which, after the anvil is released and moved, moves farther to eject the compact from the die.

Anvil presses are available with pressing capacities ranging from 6.7 to 310 kN (0.75 to 35 tons), with maximum depth of fill ranging from 1 to 75 mm (0.040 to 3 in.). Multiple-cavity pressing frequently is used in anvil presses, with possible production rates of over 100 000 pieces per hour. Some anvil presses can be converted to double-action, using an upper punch entry system. Anvil presses usually are mechanically driven. Anvil press operation is shown schematically in Fig. 13.

Rotary presses generally are limited to compaction of single-level class II parts, although some class III parts such as flanged bushings are produced. Rotary machines are available with pressing capacities ranging from 36 to 590 kN (4 to 66 tons), with a depth of fill up to 75 mm (3 in.). Production rates of over 60 000 pieces per hour are possible, depending on machine size and the number of tooling stations. Rotary presses are mechanically driven. Figure 14 schematically illustrates rotary press action.

Single-Punch Opposing Ram Presses. Like rotary presses, these machines are limited to production of class II and some class III P/M parts. These presses are available in top- and bottom-drive models,

Fig. 14 Rotary press compacting sequence

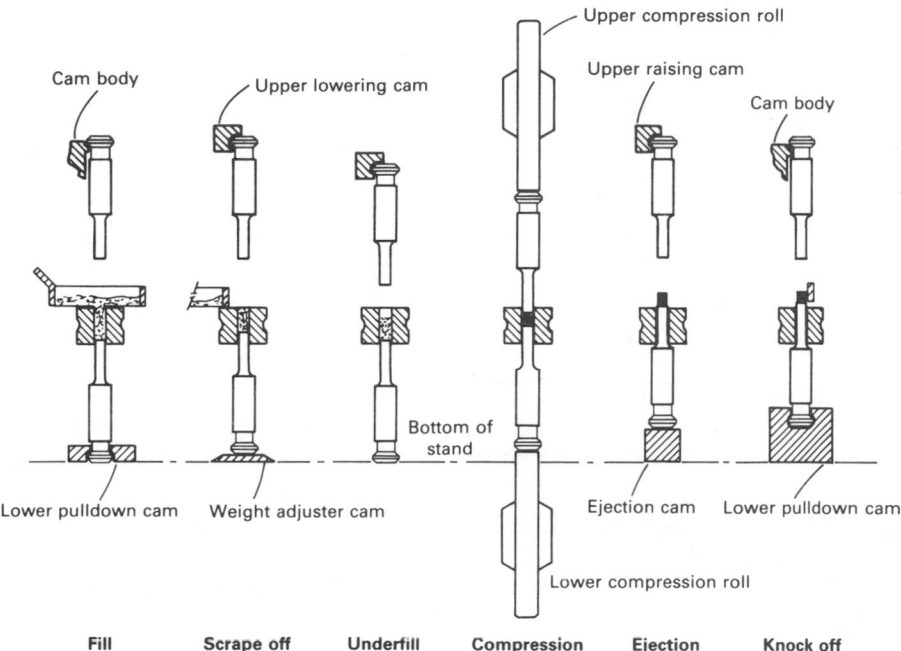

Fill   Scrape off   Underfill   Compression   Ejection   Knock off

with pressing capacities ranging from 36 to 980 kN (4 to 110 tons) and with a maximum depth of fill up to 100 mm (4 in.).

Production rates of up to 3000 parts per hour are possible using mechanical presses with single-cavity tooling, although production rates of 900 to 1800 pieces per hour are more common. Hydraulic presses produce about 900 pieces per hour. Ejection of the part is accomplished by the lower punch moving upward. Mechanical and hydraulic presses are available.

Single-punch withdrawal presses have essentially the same partmaking capabilities as the single-punch opposing ram system in terms of pressing capacity, depth of fill, and production rate. The major difference is that floating dies are used to achieve top and bottom pressing. The die is moved downward to eject the part.

Multiple-motion die set presses can be designed to produce the most complex P/M parts. These presses use floating die and withdrawal tooling methods. Machines are available with either bottom- or top-drive arrangements. Pressing capacities range from 27 to 4900 kN (3 to 550 tons), with a maximum depth of fill of 180 mm (7 in.). Production rates vary from over 6000 pieces per hour on smaller machines to 1800 pieces per hour for 1960-kN (220-ton) presses.

In addition to producing complex parts, the removable die set (tool holder) minimizes press downtime for part changeover if the die set for the next part to be pro-

duced is set up outside the press and is ready for installation. Pressing position for each level being produced by a separate tooling member is controlled by fixed-height tooling blocks (stop blocks), which usually are ground to the proper height to produce a given dimension on the part. A change in part dimension requires the respective tooling blocks to be changed accordingly.

Multiple-motion adjustable stop presses have the same partmaking capability as multiple-motion die set presses and use the same tooling methods. Pressing capacities range from 980 to 7340 kN (110 to 825 tons), with a maximum depth of fill of 150 mm (6 in.). These presses do not incorporate removable die sets; however, press stop positions are adjustable, and a change in any dimension of the part in the direction of pressing is easily accomplished.

Tooling Design

Traditionally, P/M tooling was designed on the basis of production experience. In simple parts, such as single-level class I and II parts, these determinations proved successful. As state-of-the-art materials and presses advanced to the production of complex, multilevel parts, the "cut-and-try" method of tool design became obsolete. The high cost of complex tooling and adapters, plus downtime to redesign and rebuild tooling, requires the

partmaking system, including the press, to be carefully analyzed in terms of load, stress, and deflection.

Tooling layout is required to design a suitable set of tools and to determine the physical dimensions (length and thickness) of tooling members. A preliminary layout helps to determine fill, pressing, and ejection positions and to eliminate interference at these positions.

The die space drawing supplied with every compacting press, which usually starts with the ejection position, is the basis of the tooling assembly layout. Generally, tooling members are never closer than in the ejection position, which constitutes the minimum space available to contain all components and their adapters.

Die Design. Dies are commonly constructed by using inserts that are held in the die case by shrink fitting. The amount of interference between the insert and the die case depends on the inside and outside diameter of each member and on the compacting pressure used. The powder can be considered a fluid in a closed container that transmits the compacting pressure in all directions; therefore, the die must be designed as though it were a pressure vessel with internal pressure.

In actual practice, radial pressure on the die walls due to compacting rarely exceeds 50% of compacting pressure. The interference fit of the die case and die insert should be such that the stress on the insert always remains in compression for round dies. However, for shaped dies such as gears, cams, and levers, the use of finite element analysis is the best method for accurately determining stress and deflection.

Punch Component Stress. Compacting powder causes compressive stress in the punch. This stress must be below the yield strength of the punch material. Calculation of buckling stability should be made for long, thin-walled punches. Fig. 15 shows a schematic of a stationary punch with mountings.

The effect of axial compressive force on a tubular punch is shown in Fig. 16. A tubular punch is subjected to internal pressure during compacting of multilevel parts. In this case, the resulting circumferential tensile stress in the punch wall should be calculated. If the stress and accompanying deflection is excessive, tooling clearances should be designed so that when the outer punch wall expands, it is supported by the die wall before the stress reaches the yield limit (Fig. 17).

During ejection, the punch is subjected to compressive stresses by resisting the stripping action of the die and to tensile

Fig. 15 Stationary punch and mountings

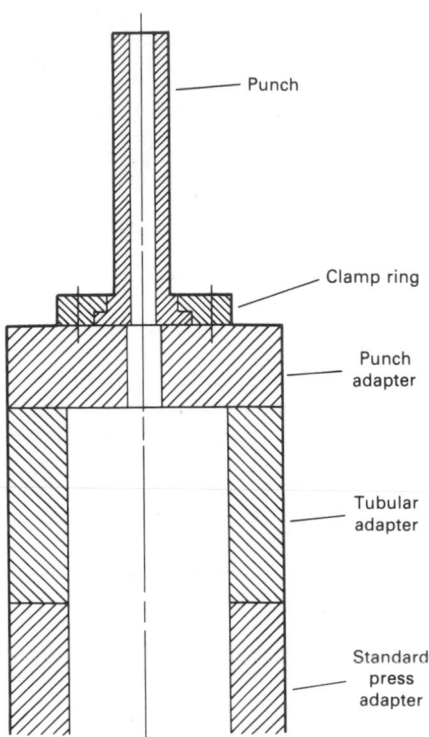

Fig. 16 Effect of compressive stress on tubular punch

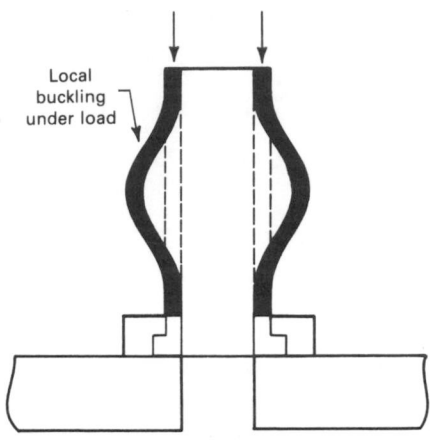

Fig. 18 Core rod mounting

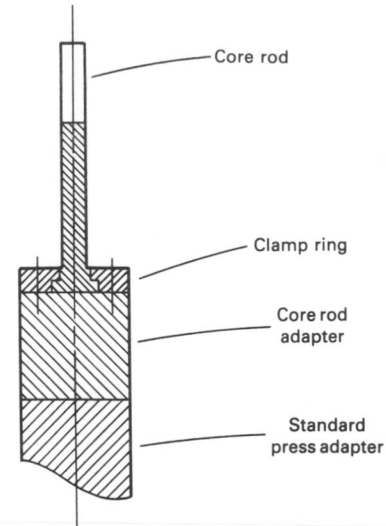

Fig. 17 Tensile stresses in a tubular punch during compacting

Large arrows indicate action of powder on walls of punch.

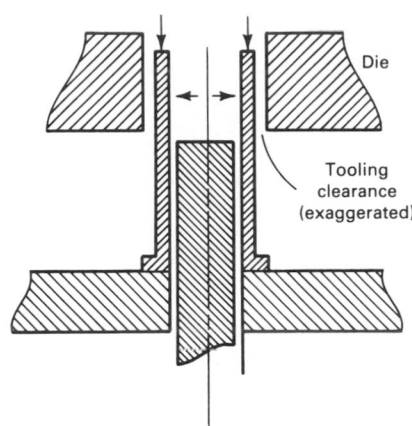

Fig. 19 Deflection of core rod from nonuniform pressure of powder during compacting

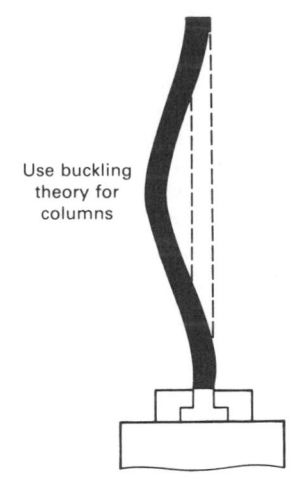

stresses from the stripping action of the punch. These stresses normally are lower than compacting stresses. Components of the punch subjected to stress include the punch clamp ring and bolts (Fig. 15), which should resist the ejection of the punch without permanent deformation. Punch adapters are subjected to bending loads that create a tensile stress around the center hole during compacting. This stress should not exceed the fatigue limit of the adapter material.

The tubular adapter (Fig. 15) must have sufficient cross-sectional area to withstand the pressing load without permanent deformation. A stepped core rod, or a core rod forming a blind hole, must not buckle during compacting. The base of the core rod must resist, without permanent deformation, whatever ejection loads are imposed on the core rod. A typical core rod mounting is shown in Fig. 18. Figure 19 illustrates the effect of compressive stresses on the core rod.

The core rod clamp ring (Fig. 18) and its retaining bolts should be sized to withstand the ejection force on the core rod without permanent deformation. The core rod adapter (Fig. 18) generally is strong enough to resist both pressing and ejection loads, due to the size of the adapter when space is provided for clamp ring fasteners.

The upper punch (Fig. 20) is stressed by compression during compacting of the powder. The upper punch clamp ring (Fig. 20) should retain and immobilize the punch between strokes. It should withstand the required clamping force without permanent deformation. The upper punch adapter (Fig. 20) must be of sufficient hardness to prevent indentation of the punch base into the adapter.

Deflection Analysis. When designing tools for production of parts other than single-level class I or II parts, deflection analysis of the tooling, tooling adapters, and press is desirable. These members are essentially stiff springs, each with a different spring rate or modulus. When the compacting load is applied, the parts deflect. When the load is released, they return to their original length. If the press contains two or more separate lower punches, the total deflection of each punch and its supporting members must be the same. Otherwise, the compacted part will move with the punch that has the greatest total deflection, leaving a portion of the part unsupported. This condition is likely to cause cracking during part ejection.

A punch under load normally is in pure compression and therefore will follow Hooke's law. If the punch has varying cross-sectional areas, each length having the same cross-sectional area is calculated individually. The total punch compression is the sum of these calculations. For a long, thin-walled punch, local buckling of the punch wall under load should be investigated. Compression of punches and their

Fig. 20 Upper punch mounting

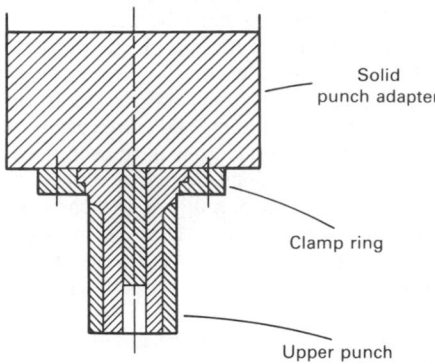

Solid punch adapter

Clamp ring

Upper punch

Fig. 21 Punch compression

P is total punch load; *L* is length; *Y* is deflection; *A* is area of punch; and *E* is Young's modulus.

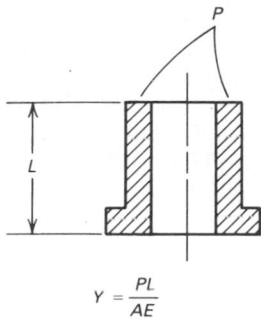

$$Y = \frac{PL}{AE}$$

supporting members may be calculated using the equation given in Fig. 21.

Adapter Bending. The adapter, on which the punch is mounted, usually is a flat plate with the punch and its load positioned at its center, around a hole through which either another punch or a core rod passes. This plate, if supported at the outer edge, is subjected to the pressing load around the center hole. Two forms of deflection—bending and shearing—occur in this area. Adapter deflection is linearly proportional to force. Calculated adapter stress should be compared with the allowable adapter material stress to evaluate design suitability.

Press Deflection. Like the tooling and its support, the press is subject to deflection. This tendency is considerably less than that of punch compression or adapter bending, but must be considered in total tool design. Data regarding press deflection should be obtained from the press manufacturer. Deflection is linearly proportional to the amount of force exerted:

$$Y = C \times W$$

where *C* is the equipment constant; *W* is pressing force; and *Y* is deflection.

Tooling Materials

Die Inserts. The type of material selected for die inserts is dictated by the desired die life and part configuration. Wear frequently is the limiting factor in die insert life. Prototype and low-production-requirement die inserts normally are made from an air- or oil-hardening grade of high-carbon, high-chromium tool steel, such as American Iron and Steel Institute (AISI) D-2 or D-3. Inserts normally are heat treated to 60 to 64 HRC. Inserts for medium production requirements of up to 500 000 pieces are made from high-speed tool steel, such as AISI M-2 or M-4, heat treated to 62 to 66 HRC.

Long production runs usually require the use of tungsten carbide die inserts. The type of carbide used depends mainly on die shape. Simple round shapes use 6% Co binder for maximum wear resistance, with cobalt increasing up to 30% for maximum toughness in irregularly shaped dies such as gears, cams, and levers. For additional information on the properties of tungsten carbide/cobalt materials, see the article "Cemented Carbides" in this Volume.

Core rod materials, like die materials, are selected for wear resistance. However, core rods are smaller than dies, are not as well supported, and are more prone to failure due to tooling misalignment or incomplete powder fill, which causes side loadings. Materials for core rods, which are similar to those used for die inserts, generally are selected for toughness, with some sacrifice in wear resistance.

Punches for simple-shaped class I and II parts normally are made from AISI D-2 or D-3 tool steel. However, if the punches have projections to form chamfers on the P/M part, a tougher material such as AISI L-6 tool steel is required. Unlike dies and core rods, punches can be refurbished by grinding punch faces to restore their original shape. Wear on the outside or inside diameter of a punch used for producing simple parts generally is minimal.

Punches that produce complex parts also must serve as a die. Forming occurs inside the punch (Fig. 17). Ideally, these punches should be made of materials similar to those used for die inserts; however, punches do not have simple round shapes in most cases. Often they have gear teeth, splines, or other shapes on the inside and outside diameters. The extremely hard, abrasion-resistant materials used for die inserts are too notch sensitive to be used in these applications. Wear resistance is thus sacrificed for toughness. Shock-resistant tool steels, such as AISI S-4 and S-5 heat treated to 52 to 56 HRC, commonly are used for

producing complex parts. In especially difficult applications, 18% Ni high-strength maraging steels have produced acceptable components in applications where conventional tool steels have not provided a satisfactory wear life.

Tooling support adapters normally are made from medium-carbon alloy steel, such as AISI 4140 or 6150, heat treated to a hardness sufficient to resist brinelling of the punches into the adapter surface without failing due to fatigue. Adapters should be heat treated to a minimum hardness of 28 to 32 HRC to reduce damage to critical mounting surfaces when handling. Die cases and adapters usually are made from similar materials and are similarly heat treated.

Punch clamp rings normally are not highly stressed members, but they should be made from a heat treatable alloy to prevent damage when handling. Heat treating of the clamp ring is optional.

Tooling Clearance and Finish. As in many other manufacturing operations, process variables—such as the type of material being processed, density of the part being produced, amount and type of powder or die lubrication, and the production rate—dictate operating conditions. Density and production rate greatly affect tool clearances during a continuous production run in which tooling temperature increases as compacted density and/or production rate increases. Temperature variations and corresponding dimensional changes within the various tooling members must be considered.

Standard tooling clearance is 0.016 mm/25 mm (0.0006 in./1 in.) on the diameter total. Minimum clearance should be used initially, because material can always be removed from the punch or die to provide additional clearance as needed. The surface finish of all tooling members should be 0.10 μm (4 μin.). Final finishing of tooling surfaces should consist of lapping in the direction of pressing.

Secondary Pressing Operations

Sizing, re-pressing, and coining involve plastic deformation of a part and normally take place at room temperature. Sizing, re-pressing, and coining frequently are used synonymously; however, each operation varies in procedure and objectives. These secondary pressing operations use a press to restrike a part after it has been sintered.

Sizing is used to improve the dimensional accuracy of a part. During sintering, a part may experience a slight change in size, or it may warp if not properly supported in the sintering furnace. Deformation can be corrected by sizing. Only a small amount of metal deformation is required; thus, the force required for sizing normally does not exceed compacting force. Because minimal deformation occurs, only slight improvement in density and material properties results. Material type and part shape and size make a general estimation of part tolerance after sizing difficult.

Re-pressing is done primarily to increase density, which increases mechanical and physical properties and hardness. Improvements in part dimensions also can be achieved by re-pressing. The amount of material deformation achieved with re-pressing is greater than in sizing, because the forces used are greater than sizing forces. Reduction in height of a ferrous material sinter generally ranges from 3 to 5%. Up to 7% reduction in height can be achieved for nonferrous materials. As with sizing, part tolerance after re-pressing depends on material type and part size.

Coining is used to change the top and/or bottom surface configuration of P/M parts such as medallions and coins. Loads are difficult to predict in coining. Although the overall measured force may appear to be low, some high localized pressure may occur in areas of high deformation. Wear on the punch face is high in coining operations.

Presses used for secondary pressing operations may be either hydraulic or mechanical. The advantages and limitations discussed previously for compacting presses also apply to secondary presses. Some parts producers use compacting presses as secondary presses, but this is not the most efficient use of equipment for high-volume production.

With an automatic feeding mechanism, a secondary operation restrike or coining press operates faster than a P/M compacting press, because the workpiece is shorter and thus the working stroke is shorter. Press feeders may be of either the shuttle or rotary dial type. Tooling designs and constructions are similar to P/M compacting tooling. Punch and core rod lengths are shorter than compacting tools due to shorter part length and working stroke.

Lubrication. Unlike compacting presses where die lubrication is provided by admixing dry lubricants into the powder, external die lubricant must be used in secondary pressing operations. Liquid or dry lubricants may be used.

Suitable liquid lubricants range from cutting fluids to automobile crankcase oil. Extreme pressure additions have been used for severe applications. Application method varies from hand dipping of the parts to spraying of the tooling. Dipping of parts is unsuitable, because parts may absorb or be coated with excessive fluid. Excessive lubricant can cause a hydraulic effect within the tooling, causing nonrepeatability of the pressing results and excessive tool loading. Lubricants absorbed by the part can be detrimental to further processing, such as plating. Spraying the tooling directly with a controlled amount of lubricant is preferred.

Solid lubricants, such as stearic and palmetic acid, do not cause excessive lubrication or lubrication absorption when applied directly to the part. However, solid lubricants are difficult to remove for subsequent secondary operations, such as brazing or plating.

Production Sintering Atmospheres

By Harbhajan S. Nayar
Manager, Particle Technologies Group
Airco Industrial Gases
Division of Airco, Inc.
A Member of the BOC Group

DURING SINTERING, compacted metal powders are bonded or sintered by heating in a furnace to a temperature that is usually below the melting point of the major constituent. Sintering occurs in a series of overlapping, but balanced phases, all of which depend on temperature, time, and atmospheric composition, flow, circulation, and direction.

Furnaces used for sintering typically are similar to protective atmosphere furnaces used in brazing and heat treating of steel. However, the significant differences between heat treating and sintering require careful selection of furnace atmospheres for sintering. Primarily, P/M compacts are porous; therefore, a much greater surface area is exposed to the furnace atmosphere than with solid parts. Sintering temperatures are considerably higher than heat treating temperatures (1120 °C, or 2050 °F, for iron and steel compared to 900 °C, or 1650 °F, for carburizing and neutral hardening of steel).

This combination of greater exposed surface area and higher temperatures enhances chemical reactivity between the surface and atmosphere during sintering. Subsequent operations that the sintered parts will undergo must also be considered. For example, an atmosphere must be selected that will produce a surface composition that can be machined or electroplated.

Sintering Atmosphere Requirements (Ref 1-3)

Atmospheres used for sintering must perform several functions. During sintering, they must:

- Prevent air from entering the furnace
- Delube or dewax the parts
- Reduce surface oxides on the powder particles
- Control carbon on the surface and in the core of steel parts
- Remove carbon in special applications
- Provide controlled oxidation during cooling in special applications
- Convey or remove heat efficiently and uniformly

Because furnace atmospheres affect the sintering process and the material being treated, sintering is never performed in air or in an oxygen-rich atmosphere. A basic function of a sintering atmosphere is to protect metal parts from the effects of contact with air. Atmospheres must be maintained at a sufficient pressure and flow rate to prevent infiltration of air through furnace openings. In addition to protecting the part, atmospheres provide sufficient conduction/convection for uniform heat transfer to ensure even heating or cooling within the various furnace zones.

Another important function of the atmospheres is to burn and remove hydrocarbon lubricants used to compact the powdered metal. Removal occurs early in the sintering process, in the low-temperature delube zone of the furnace (see Fig. 1). Small amounts of oxidizing gases in the atmosphere cause effective lubricant burning without oxidizing parts or decarburizing steel parts. The most effective oxidants are controlled amounts of water and carbon dioxide.

Without such oxidizing gases, lubricants tend to decompose thermally into

Fig. 1 Typical furnace schematic for sintering steel

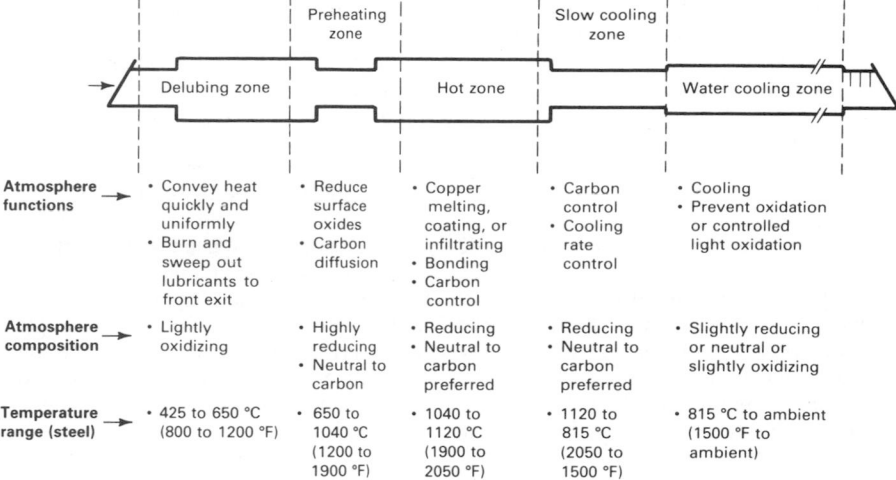

	Deluking zone	Preheating zone	Hot zone	Slow cooling zone	Water cooling zone
Atmosphere functions	• Convey heat quickly and uniformly • Burn and sweep out lubricants to front exit	• Reduce surface oxides • Carbon diffusion	• Copper melting, coating, or infiltrating • Bonding • Carbon control	• Carbon control • Cooling rate control	• Cooling • Prevent oxidation or controlled light oxidation
Atmosphere composition	• Lightly oxidizing	• Highly reducing • Neutral to carbon	• Reducing • Neutral to carbon preferred	• Reducing • Neutral to carbon preferred	• Slightly reducing or neutral or slightly oxidizing
Temperature range (steel)	• 425 to 650 °C (800 to 1200 °F)	• 650 to 1040 °C (1200 to 1900 °F)	• 1040 to 1120 °C (1900 to 2050 °F)	• 1120 to 815 °C (2050 to 1500 °F)	• 815 °C to ambient (1500 °F to ambient)

hydrogen and carbon above 540 °C (1000 °F). Hydrogen is given off as a gas, but carbon remains as a sooty deposit. If a suitable oxidant is present, however, a chemical reaction (oxidation) rather than thermal decomposition occurs. The carbon in the lubricant combines with the oxidant and is removed as a gaseous mixture of carbon monoxide and carbon dioxide. For example, assuming C_xH_y is a wax-like lubricant and water is the oxidizing agent, the reaction is:

$$C_xH_y + H_2O \xrightarrow{\text{approx 550 °C (1020 °F)}} CO + CO_2 + H_2$$

Delubing in a sintering atmosphere is performed most efficiently when a wet (high dew point) atmosphere is used. Atmosphere circulation should also be provided; atmosphere flow toward the furnace entrance is desirable. Ideally, the maximum temperature of the P/M compacts during delubing should range from 425 to 650 °C (800 to 1200 °F).

As the sintering process progresses, the furnace performs another function in addition to delubrication. At about 650 °C (1200 °F), generally near the end of the delube zone and through the hot zone (about 1120 °C, or 2050 °F, for iron and steels), the atmosphere begins to strip surface oxides on the metal particles that comprise the part.

In the preheat zone (Fig. 1), the atmosphere must reduce surface oxides on the metal particles to ensure a clean, metallic surface on the particles comprising the P/M compact. This permits admixed graphite to diffuse into the iron particles, thus creating a pearlitic microstructure during cooling. Graphite diffusion into the particles is essentially completed when parts reach about 1040 °C (1900 °F).

In sintering processes using either admixed or infiltrated copper, oxide-free particle surfaces allow molten copper to wet and coat the iron particles uniformly. This effect, which occurs at about 1080 °C (1980 °F) in the hot zone, ensures a sound weld or braze between the particles and consequently greater strength within the part.

Clean particle surfaces further improve bonding of the particles above 1100 °C (2010 °F). Thus, the greater the oxide reducing effect of the atmosphere, the stronger the sintered bond. Furthermore, porosity between the particles becomes rounded, thus improving the structural integrity and toughness of the part.

Under most operating conditions, metal oxides (MO) are reduced to metal (M) pri-

marily by the hydrogen component of the atmosphere:

$$MO + H_2 \rightarrow M + H_2O$$

The effectiveness of an oxide-reducing atmosphere can be determined by the ratio of hydrogen to water, as shown for iron in Fig. 2(a) and 2(b). In the P/M industry, water content is generally determined by measuring the dew point of the atmosphere. Water content as a function of sintering atmosphere dew point is shown in Fig. 3. As the ratio of hydrogen to water increases, the reducing reaction is more effective. The ratio of hydrogen to water, not the absolute levels of these constituents, determines the reducing effectiveness of the atmosphere. Furthermore, for a given dew point, the hydrogen content of the atmosphere becomes more reducing to oxides as temperature increases.

During sintering, atmospheres control the surface and core carbon contents in steel. The reactions relating to this function occur primarily in the hot zone and the slow cool zone of the furnace (Fig. 1). If the atmosphere is very active—an endothermic atmosphere with 60% combustibles (hydrogen and carbon monoxide) at about 5 °C (40 °F) dew point, for example—a steel surface with about 0.8% combined carbon is decarburized to about 0.2% combined carbon (Fig. 4) in the hot zone (typically 1120 °C, or 2050 °F) to a depth of 0.5 to 2.5 mm (0.02 to 0.1 in.).

The degree and depth of decarburization depend on the dew point of the active endothermic atmosphere, the processing time and temperature in the hot zone, and the porosity of the steel part. Higher values for these parameters lead to a greater degree of decarburization.

During slow cooling from about 1120 to 815 °C (2050 to 1500 °F) under an active endothermic atmosphere, carbon is more or less restored to the surface of the part. Decarburization in the hot zone followed by restoration of carbon in the slow cooling zone also occurs in other active atmospheres, such as nitrogen-diluted endothermic or nitrogen-diluted dissociated methanol, with a combustible content of 20% or more hydrogen and carbon monoxide. The slow cooling section of the furnace is designed to aid carbon restoration.

If the atmosphere is neutral to carbon (a low dew point synthetic nitrogen-based atmosphere with less than 10% combustibles such as hydrogen, for example), decarburization in the hot zone and subsequent carbon restoration in the slow cooling zone do not occur at all or occur to a much lesser extent than they do under active endothermic atmospheres. Carbon

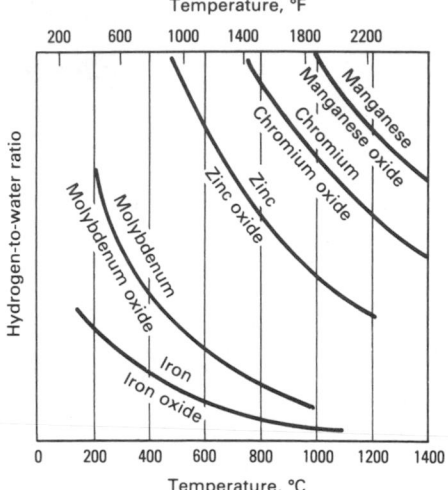

Fig. 2(a) Metal-to-metal oxide equilibria for common elements used in powder metallurgy
Copper, lead, cobalt, nickel, and tin oxides are easier to reduce than iron oxide.

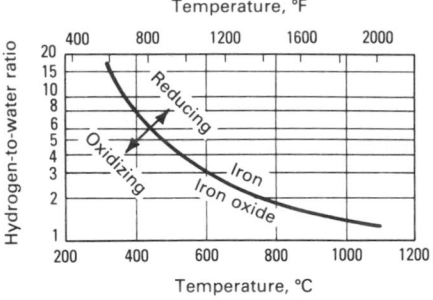

Fig. 2(b) Iron-to-iron oxide equilibria

potential for neutral atmospheres is unrelated to temperature. Consequently, neutral atmospheres, unlike endothermic atmospheres, are suitable for high-temperature sintering of steel P/M parts.

Furthermore, parts with different carbon levels can be sintered in the same tray or in random tray sequence under neutral atmosphere without the need for furnace conditioning. Carbon content in the steel part depends on the amount of admixed graphite in the powder and the dew point of the neutral atmosphere. With a lower dew point, less carbon is lost.

In some specialty sintering applications, such as those used for magnetic and stainless steels and high-ductility iron, carbon must be reduced to the lowest possible level during sintering. In such applications, atmospheres with carbon-containing gases such as carbon monoxide should be avoided. A relatively high dew point is required to decarburize the part without ad-

Fig. 3 Water content as a function of dew point of the atmosphere

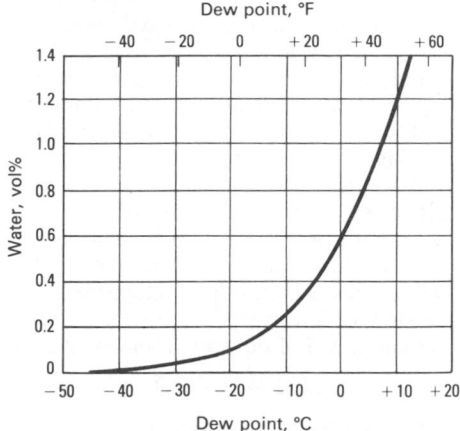

Dew point		Water,
°C	°F	vol%
−80	−112	0.00005
−60	−76	0.00106
−50	−58	0.00388
−40	−40	0.01270
−30	−22	0.03760
−20	−4	0.1020
−10	14	0.2570
0	32	0.6020
10	50	1.2120
20	68	2.3080

versely affecting the sintered quality of the part.

In some applications, an atmosphere that is slightly oxidizing to sintered parts during cooling below 540 °C (1000 °F) is desired to provide controlled oxidation for improved corrosion resistance and increased service life.

Atmospheres also assist cooling by transferring heat efficiently from the part to the furnace walls, which are usually water cooled. The three modes of heat

Fig. 4 Carbon equilibrium in endo gas at different dew points

Endo gas composition: 40% hydrogen, 20% carbon monoxide, and 40% nitrogen

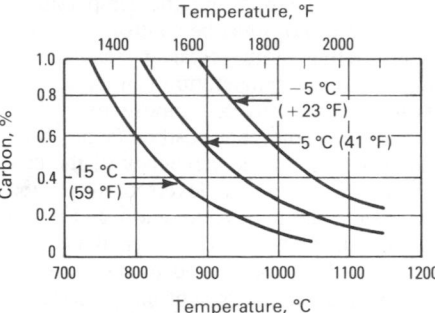

transfer are radiation, conduction, and convection. Radiation heat transfer is least effective at temperatures below about 700 °C (1300 °F). Success of conduction heat transfer depends on thermal conductivity of the atmosphere or gases that separate the heat-emitting parts and heat-receiving furnace walls. Hydrogen-rich atmospheres are more conductive than nitrogen-rich atmospheres. Convection heat transfer is based on the circulation or motion of the atmosphere between the hot and cold bodies. With greater atmosphere circulation, the heat transfer rate increases. Atmosphere circulation and high-thermal-conductivity gases enhance cooling. Circulation of atmosphere in the cooling zone can offset the lower thermal conductivity of a nitrogen-rich atmosphere.

In addition to these process functions, atmospheres have other effects. They protect the furnace and its components from the effects of thermal and chemical reactions that occur during sintering. Atmospheres must also provide a safe working area for the furnace operator. They must be readily available in consistent compositions at an economical price.

Types of Atmospheres

Conventional atmospheres used in sintering are endothermic gas, exothermic gas, dissociated ammonia, hydrogen, and vacuum. According to the Metal Powder Industries Federation, the most common atmosphere, endothermic gas, was used by approximately 83% of the industry in 1976 for sintering carbon steels (Ref 4). Endothermic gas is generated directly from partial combustion of natural gas. Exothermic gas, also a product of incomplete combustion of natural gas, was used by 5% of the industry for sintering nonferrous materials, such as copper-based metal components (Ref 4).

Dissociated ammonia and hydrogen are indirectly derived from natural gas. These atmospheres are used for sintering stainless steels, magnetic alloys, and other specialty materials.

Among the more recently adopted innovations, synthetic nitrogen-based atmospheres were used by about 30% of the industry in 1983. Recent data show that nitrogen-based atmospheres, which are more versatile and relatively independent of natural gas, are becoming more economical and thus more widely used.

Tables 1 and 2 list typical physical properties of various gases and liquids used to produce sintering atmospheres. Table 3 summarizes the composition of conventional and synthetic nitrogen-based atmospheres. Various chemical requirements —burning of lubricant vapors, reduction of surface oxides, and carburization and decarburization, for example—are controlled by the ratios of various atmospheric constituents, as shown in Table 3.

Endothermic Atmospheres (Ref 5-7)

Endothermic, or endo gas is generated by reacting relatively rich mixtures of hy-

Table 1 Typical physical properties of various gaseous feedstocks at 20 °C (68 °F) and 1 atm

Property	Natural gas (CH₄)	Ammonia vapor(a)	Air	Propane gas	Methanol vapor(a)	Nitrogen gas	Hydrogen gas
Specific gravity(b)	0.59	0.60	1.00	1.55	1.11	0.97	0.07
Gas density, kg/m³(lb/ft³)	0.72	0.74	1.20	1.91	1.33	1.17	0.08
	(0.045)	(0.046)	(0.075)	(0.119)	(0.083)	(0.073)	(0.005)
Specific volume, m³/kg (ft³/lb)	1.39	1.35	0.83	0.52	0.75	0.85	12.50
	(22.2)	(21.7)	(13.3)	(8.43)	(12.05)	(13.8)	(192.0)
Molecular weight	16.08	17.03	28.95	44.00	32.04	28.01	2.02
Specific heat vapor, J/kg·°C × 10⁻³ (Btu/lb·°F)	2.48	2.20	1.01	1.69	1.38	1.04	14.3
	(0.593)	(0.525)	(0.241)	(0.404)	(0.33)	(0.249)	(3.42)
Caloric value, J/m³ × 10⁷ (Btu/ft³)	3.837	1.814	⋯	9.501	3.029	⋯	1.211
	(1030)	(487)	⋯	(2550)	(813)	⋯	(325)
Caloric value, J/kg × 10⁷ (Btu/lb)	5.343	2.549	⋯	5.001	2.279	⋯	14.210
	(22 970)	(10 958)	⋯	(21 500)	(9800)	⋯	(61 095)
Air/fuel ratio at 100% combustion	10.3	3.6	⋯	24.0	7.14	⋯	2.4

(a) 1 kg (1 lb) of liquid ammonia yields 1.4 m³ (22.5 ft³) of ammonia vapor. Ammonia vapor doubles in volume on dissociation; 1.4 m³ (22.5 ft³) of ammonia vapor becomes 2.8 m³ (45 ft³) of nitrogen and hydrogen mixture (1-to-3 ratio). 1 kg (1 lb) of liquid methanol yields 0.68 m³ (11 ft³) of methanol vapor. Methanol vapor triples in volume on dissociation. 0.68 m³ (11 ft³) of methanol vapor becomes 2.0 m³ (33 ft³) of carbon monoxide and hydrogen mixture (in 1-to-2 ratio). (b) Air equals 1.0.

drocarbon gas (usually natural gas, predominantly methane) and air over a clean catalyst such as nickel oxide in an externally heated chamber (Fig. 5). The volumetric reaction between natural gas and air is:

$$2CH_4 + 4.8 \text{ air } (O_2 + 3.8N_2)$$
$$\rightarrow 2CO + 4H_2 + 3.8N_2$$

The theoretical ratio of air to natural gas is 2.4 to 1; however, generator ratios are generally between 2.5 and 3 to 1. This reaction is endothermic. Consequently, the generator must be heated by an external source. Operating temperature in the catalyst must range from 980 to 1000 °C (1800 to 1830 °F); typically, the generator operates at about 1060 to 1100 °C (1960 to 2010 °F).

Air-to-gas ratios and the temperatures in the generator must be controlled to prevent sooting of the catalyst. Insufficient air or too low temperature can cause carbon monoxide to decompose to carbon dioxide and amorphous carbon (soot). Soot deposits destroy the efficiency of the catalyst and cause high levels of unreacted methane, carbon dioxide, and water in the output atmosphere. When a methane-rich atmosphere is introduced to the furnace, the methane deposits soot on the parts.

A sooted catalyst also causes loss of control over the ability of the generator to maintain a definite carbon potential in the atmosphere. When higher concentrations of carbon monoxide are required, endo gas can be generated from propane instead of natural gas. Commercial propane containing more than 5% unsaturated hydrocarbons does not react properly and causes sooting.

After passing over the catalyst, the reaction is "frozen" by rapid chilling of the gases to below 315 °C (600 °F) in a water-jacketed zone surrounding the top of the retort. This prevents the reaction from reversing and forming carbon (soot) and carbon dioxide from carbon monoxide:

$$2CO \rightarrow C + CO_2$$

The reaction in the direction indicated above predominates at 700 to 480 °C (1300 to 900 °F). The soot-free reacted gas can be further cooled for metering and distribution.

Changes in the air to natural gas ratio also change the composition of the output gas and hence its dew point and carbon potential. Sulfur (as H_2S) reacts with the catalyst to deactivate it. More than 0.65 g (10 grains) of sulfur per 2.83 m^3 (100 ft^3) of gas renders the catalyst less effective; consequently, carbon monoxide and hydrogen contents of the endo gas are lowered, and water, methane, and carbon dioxide contents are increased.

A higher concentration of hydrogen is found in endo gas generated by the hydrocarbon steam process. In this generator, natural gas and water are reacted to produce carbon monoxide and hydrogen:

$$CH_4 + H_2O \rightarrow CO + 3H_2$$

$$(-12 \text{ °C, or } 10 \text{ °F, dew point})$$

The result is a higher proportion of hydrogen and an absence of nitrogen.

Regardless of the process or feedstock used to produce the endo gas, it is mon-

Table 2   Typical physical properties of various liquid feedstocks at 20 °C (68 °F) and 1 atm

Property	Propane	Methanol	Ammonia	Nitrogen	Hydrogen
Liquid density:					
kg/m³ × 10²	5.09	7.94	6.09	8.09	0.71
lb/gal	4.25	6.63	5.08(a)	6.75	0.5906
lb/ft³	31.8	49.5	38.0	50.5	0.071
Specific gravity(b)	0.509	0.794	0.609	0.809	4.42
Molecular weight	44.0	32.04	17.03	28.01	2.02
Freezing point, °C (°F)	−123.3	−97.8	−77.8	−209.9	−252.1
	(−190)	(−144)	(−108)	(−345)	(−434.5)
Boiling point, °C (°F)	−46.1	64.5	−33.3	−195.6	−252.7
	(−51)	(148)	(−28)	(−320)	(−423)
Vapor pressure:					
At 20 °C (68 °F),					
kPa (mm Hg)	855.2	13.3	889.5	Very high	Very high
	(6413)	(100)	(6670)		
At 50 °C (120 °F),					
kPa (mm Hg)	1692.4	53.5	2033.9	⋯	⋯
	(12 691)	(400)	(15 252)	⋯	⋯
Specific heat (liquid), J/kg·°C × 10⁻³					
(Btu/lb·°F)	2.46	2.55	4.73	2.00	7.32-9.76
	(0.588)	(0.61)	(1.13)	(0.475)	(1.75-2.33)
Heat of vaporization at boiling point, J/m³ × 10⁸ (Btu/gal)	2.19	8.75	8.36	1.62	0.318
	(786)	(3136)	(2997)	(580)	(114)
Heat of dissociation, J/m³ × 10⁸ (Btu/gal)	12.05	31.71	16.59	⋯	⋯
	(4318)	(11 360)	(5944)	⋯	⋯
Heat of combustion, J/m³ × 10⁸ (Btu/gal)	255.03	179.49	155.37	⋯	100.71
	(91 375)	(64 310)	(55 670)	⋯	(36 083)

(a) At 889 kPa (129 psi). (b) Water equals 1.

Table 3   Compositions of principal furnace atmosphere constituents

Atmosphere	AGA class	Air-to-natural gas ratio	Dew point °C	Dew point °F	Nominal composition, vol% Nitrogen	Hydrogen	Water	Carbon monoxide	Carbon dioxide	Methane	Key ratios(b) Hydrogen to water	Carbon monoxide to carbon dioxide	Hydrogen to carbon monoxide
Lean exothermic	101	9.0	20(a)	68(a)	84.7	1.2	2.5	1.4	10.2	⋯	0.5	0.1	0.9
Rich exothermic	102	6.0	20(a)	68(a)	69.8	12.2	2.5	10.2	4.9	0.4	4.9	2.1	1.2
Endothermic	302	2.5	5	40	38.2	40.4	0.8	19.8	0.3	0.5	51	66	2.0
Dissociated methanol	⋯	⋯	15.5	60	⋯	65.6	1.7	32.4	0.4	⋯	39	81	2.0
Dissociated ammonia	601	⋯	−50	−60	25.0	75.0	0.004	⋯	⋯	⋯	19 000	⋯	⋯
Hydrogen	⋯	⋯	−60	−80	⋯	100	0.001	⋯	⋯	⋯	100 000	⋯	⋯
Nitrogen	⋯	⋯	−60	−80	100	⋯	0.001	⋯	⋯	⋯	⋯	⋯	⋯

Note: For nitrogen-based atmospheres, composition, dew point, and key ratios can be synthesized over a broad range, depending on which of the above generated atmospheres is blended with the high-purity nitrogen gas.
(a) Dew point is about 6 °C (10 °F) above temperature of cooling water; dew point may be reduced to 5 °C (40 °F) by refrigeration, or to −45 °C (−50 °F) by absorbent tower dehydration. Hydrogen-to-water changes accordingly. (b) Control carburization or decarburization, oxidation or reduction, or combination of these

Fig. 5 Schematic of endothermic gas generator

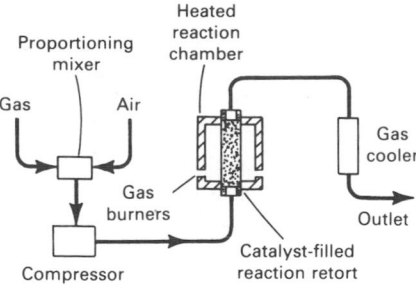

itored by measuring dew point and carbon dioxide content upon exiting the generator. Oxygen probes are also used to monitor the composition of the endo gas.

Variations in composition and peak shaving of the natural gas feedstock make control difficult. Consequently, input ratios of air and methane or propane are adjusted to maintain the dew point and carbon dioxide at a constant level. Frequent monitoring and adjustment are required to maintain a consistent endo gas composition. Automatic ratio adjusters are available to ensure consistent composition of the endo gas (Ref 8). However, these controls require careful calibration and maintenance.

Endo gas is typically 40% nitrogen, 40% hydrogen, and 20% carbon monoxide, with small amounts of water, carbon dioxide, and methane (see Table 3). Typically, 60% of its gaseous constituents are combustibles (carbon monoxide, hydrogen, and methane), which are oxide-reducing agents. Carbon monoxide and methane are carburizing; carbon dioxide and water decarburize, oxidize, and delube. The carbon dioxide and water content of endo gas generally burns off the lubricant vapors by chemical reaction in the delubing section of the furnace. Other atmospheres containing higher oxidant levels also are capable of achieving optimum results.

Typically, hydrogen-to-water ratio determines the reduction-oxidation function of an atmosphere (Fig. 2). Endo gas is reducing to iron above 260 °C (500 °F), because its typical 40% hydrogen and 0.8% water (dew point: 5 °C, or 40 °F) contents provide a hydrogen-to-water ratio of 50 to 1. Although this ratio achieves acceptable oxide reduction, other atmospheres—notably those based on hydrogen or dissociated ammonia or synthetic nitrogen-based atmospheres—can provide higher ratios.

The ability of typical endo gas to transfer carbon depends on the temperature and water or carbon dioxide content (Ref 9). Figure 4 shows carbon potential at various temperatures for three different dew points. Typically, a high-carbon part loses carbon at and near its surface as it approaches the peak hot zone temperature (1120 °C, or 2050 °F) in the heat-up cycle. At this peak temperature, the carbon potential corresponds to approximately 0.2% C, depending on the dew point of endo gas in the furnace. As the part slowly cools, the carbon potential increases, as shown in Fig. 4. The carbon thus tends to be restored.

The extent of decarburization and recarburization determines the finished carbon content at or near the surface of the P/M part. Because the sintering temperature profile is determined by the sintering process, carbon potential must be controlled by the dew point or the carbon dioxide content in the endo gas. Most sintering under endothermic atmosphere is done at dew points ranging from −5 to 15 °C (23 to 59 °F).

High-carbon steel parts are sintered at low dew points, and low-carbon parts are processed at high dew points. Endo generators tend to operate soot free at higher dew points; consequently, natural gas is sometimes added to the high dew point endo gas atmosphere to secure a lower dew point in the furnace for sintering high-carbon parts.

Exothermic Atmospheres (Ref 5-7)

Exothermic, or exo, atmospheres typically contain 67 to 87% nitrogen. The remainder is carbon monoxide, hydrogen, carbon dioxide, and water. Exothermic atmospheres are produced by the partial combustion of hydrocarbon fuels (generally natural gas) with air. A 6-to-1 ratio of air to natural gas produces rich exo gas, American Gas Association (AGA) class 102, which is high in carbon monoxide and hydrogen.

In contrast, lean exo gas (AGA class 101), produced from a 9-to-1 ratio of air to gas, contains low percentages of carbon monoxide and hydrogen, but correspondingly higher percentages of carbon dioxide and nitrogen (Table 3). Partial combustion in a gas generator equipped to control the ratio of air to fuel gas is capable of producing any desired composition between rich and lean exothermic gases.

The exothermic gas generator is a refractory-lined, water-jacketed combustion chamber that operates at 1100 to 1425 °C (2010 to 2600 °F). The internal temperature of the chamber depends on (1) the ratio of air to fuel gas, (2) the volume of gas being burned, and (3) the effectiveness of the heat transfer from the inside of the chamber to the water-jacketed exterior. A typical exo gas generator is shown in Fig. 6.

The gas mixture enters the two-zoned converter, which is the main portion of the generator. In the first zone, where combustion occurs, most of the fuel and air react to produce hydrogen, carbon monoxide, and nitrogen. In the second zone, which is heated by the earlier combustion, the unburned hydrocarbon and combustion products are cracked over the catalyst into hydrogen and hydrogen by-products. The turn-down volume (expressed as a percentage of the rated capacity) is limited by the minimum exo gas output that will maintain the chamber at 1030 °C (1890 °F) or above. Lean ratios permit lower turn-down volumes than rich ratios.

The hot gases are quenched in the hot gas cooler to separate water as condensate from the gas stream for removal through a condensate trap. The amount of moisture in the cooled gas varies according to the temperature of the incoming cooling water; the temperature of the exit gas should

Fig. 6 Schematic of exothermic gas generator

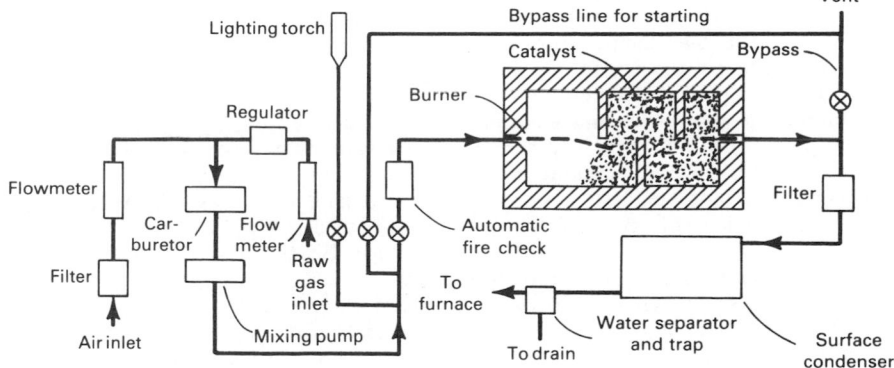

be within about 5 °C (10 °F) of the cooling water.

More moisture can be removed by refrigerating to a dew point of about 4 °C (40 °F) or by adsorbent tower dehydration to a dew point of −45 °C (−50 °F) when relatively high oxide reducing capability (hydrogen-to-water ratio) is desired.

Most exo gas (lean or rich ratio) has a typical dew point of 21 °C (70 °F), corresponding to 2.5% water. It contains 1.2 or 12.2% hydrogen, depending on whether it is a lean or rich exo gas. The hydrogen-to-water ratios for lean and rich exo gases are 0.5 and 4.9, respectively (Table 3). These atmospheres are significantly less oxide reducing than endo gas or synthetic nitrogen-based atmospheres.

Exo gas is highly decarburizing to steel above about 700 °C (1300 °F). Because of its high water and carbon dioxide content, it is a very effective delubing agent.

Exo gas generally is used to sinter bronze and some other copper-based parts. Copper and most copper-based alloys do not require high oxide-reducing power (Fig. 2a). Exo gas requires much less natural gas than endo gas; consequently, it is one of the most economical atmospheres.

Dissociated Ammonia (Ref 5-7)

Heating metallurgical-grade anhydrous ammonia in the presence of a catalyst causes the ammonia to dissociate into a mixture of hydrogen and nitrogen molecular gases in a 3-to-1 ratio (75% hydrogen and 25% nitrogen):

$$2NH_3 \xrightarrow[\text{catalyst}]{\text{heat}} 3H_2 + N_2$$

Dissociated ammonia is pure, consistent, and dry; its dew point is typically lower than −51 °C (−60 °F). Residual undissociated ammonia is generally well below 250 ppm.

Although ammonia molecules begin dissociating at 315 °C (600 °F), 980 °C (1800 °F) provides a practical operating compromise between dissociation rate and equipment life. Higher temperatures must be used to ensure that virtually no undissociated ammonia molecules remain. Generally 1 kg (1 lb) of liquid ammonia yields approximately 1.4 m³ (22.5 standard ft³) of undissociated ammonia vapor or about 2.8 m³ (45 standard ft³) of dissociated ammonia (hydrogen and nitrogen in a 3-to-1 ratio).

A typical ammonia dissociator produces 2 to 140 m³/h (70 to 4950 ft³/h) of dissociated ammonia at an outlet pressure of 35 to 100 kPa (5 to 15 psig). Liquid am-

monia is fed from a storage tank into a vaporizer, which transforms it into ammonia vapor. A heater in the storage tank maintains gas operating pressure if the temperature of the liquid ammonia falls below the recommended level.

The vapor then passes through a pressure regulator, is preheated, and enters an externally heated retort (typically made from Inconel) that contains a nickel- and iron-bearing catalyst. The ammonia molecules dissociate into molecular hydrogen and nitrogen at 925 to 1040 °C (1700 to 1900 °F). The resulting mixture leaves the retort and passes through a heat exchanger; heat is subsequently transferred to the incoming ammonia vapor. If necessary, the dissociated ammonia may be pressure-regulated and piped to the furnace.

Output can be easily controlled between 0 and 100% of capacity by regulating flow at any point along the processing line. This control feature is unavailable on exo and endo gas generators, which typically operate between 70 and 100% of their rated capacity. Some recent designs of endo generators allow greater output reduction.

The most important component of the dissociator is the catalyst chamber or retort, which is generally fabricated from Inconel. This chamber is heated either electrically or by natural gas. If electrically heated, a dissociator requires 0.5 to 1.0 kW (depending on size of dissociator) of electricity per cubic metre (14 to 28 W/ft³) of dissociated ammonia produced. A typical ammonia dissociator is shown schematically in Fig. 7. Ammonia dissociators are relatively maintenance free compared to endo and exo gas generators.

Because dissociated ammonia is very dry (low dew point), it is highly reducing to surface oxides, as the high hydrogen-to-water ratio given in Table 3 indicates. Consequently, it is frequently used where the high reducing capability of stable oxides is desirable, such as in sintering of stainless steel. Thermodynamically, the high hydrogen content in dissociated ammonia is decarburizing to carbon steel; kinetically, however, the decarburizing reaction is very slow when the dew point in the furnace remains low (below −40 °C, or −40 °F).

The decarburizing reaction can be further slowed by diluting dissociated ammonia with a lower dew point inert gas such as nitrogen when sintering carbon steel that cannot have any decarburization at its surface. Because dissociated ammonia is very dry, it does not provide optimum burning of the lubricant during delubing in the preheating zone. However, because dissociated ammonia is 75% hydrogen, it

Fig. 7 Schematic of ammonia dissociator

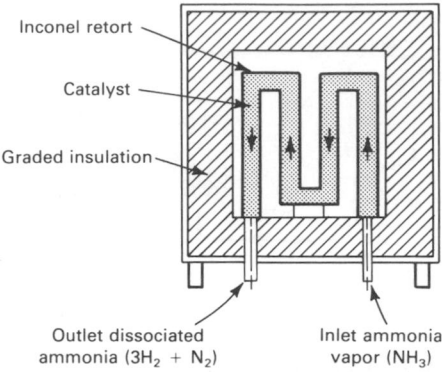

Inconel retort

Catalyst

Graded insulation

Outlet dissociated ammonia (3H₂ + N₂)

Inlet ammonia vapor (NH₃)

has a thermal conductivity higher than nitrogen-rich atmospheres.

Hydrogen (Ref 10)

Hydrogen, the lightest of all the gases listed in Table 1, is produced for the most part by steam reforming natural gas and light hydrocarbon liquids over a catalyst. Metallurgical hydrogen is refined from excess hydrogen-gas feedstocks from petrochemical steam-reforming plants. The gas supplier removes hydrocarbons and moisture from the raw gas stream and liquefies the partially refined gas at −253 °C (−423 °F). Other impurities such as nitrogen and argon are removed during the liquefaction process.

Hydrogen is delivered either as compressed gas or liquid to volume and purity requirements. Commercial grades of compressed hydrogen gas, with typical pressure of 17 MPa (2500 psig), are delivered by trailer in cylinders with capacities between 1000 and 4000 m³ (35 000 and 140 000 standard ft³). Smaller quantities of compressed hydrogen usually are available in cylinders containing 7 m³ (250 ft³).

Most liquid hydrogen is delivered by trailer to liquid receivers or storage vessels with capacities between 5000 and 75 000 L (1500 and 20 000 gal). Liquid hydrogen usually is vaporized as required. One litre (one gallon) of liquid hydrogen yields 0.85 m³ (113.4 ft³) of hydrogen gas. Smaller quantities of liquid hydrogen are available in 150-L (40-gal) cylinders.

Hydrogen specifications vary, depending on the grade. Typical hydrogen is 99.999% pure. Commercial-grade hydrogen meets the following specifications:

Hydrogen .... 99.995% min
Water .... 8 ppm max at −68 °C (−90 °F)
Oxygen .... 1 ppm max

Because hydrogen is highly flammable, care must be taken to ground major stationary and mobile equipment to remove static electricity and the potential for spark ignition of the hydrogen. Electrical equipment must meet applicable national electric codes and fire prevention standards.

As indicated in Table 3, hydrogen is the most effective reducing atmosphere available, because it has the largest hydrogen-to-water ratio. It is essentially nondecarburizing, like dissociated ammonia, to carbon steels as long as the dew point remains below about -40 °C (-40 °F). It is similarly inefficient at burning lubricant vapors because of its low dew point. Of all atmospheres commonly used in the P/M industry, hydrogen is most heat conductive.

Vacuum (Ref 11-12)

Vacuum is essentially a lack of atmosphere. It is used mainly for sintering stainless steels, tool steels, carbides, magnetic alloys, and metals such as titanium, zirconium, uranium, tantalum, and other refractory metals and compounds that react with hydrogen-, nitrogen-, and carbon-monoxide-bearing atmospheres.

Vacuums are being used increasingly for high-temperature sintering of conventional ferrous P/M parts. Under vacuum, care must be taken not to lower the pressure in the furnace below the vapor pressure of the constituents of the alloy to be sintered so that depletion does not occur.

Vacuum cannot be used in the conveyor furnaces conventionally used by the sintering industry. Currently, almost all vacuum sintering furnaces are batch types. Some continuous (compartmentalized) conveyor vacuum furnaces are being introduced to the P/M industry.

Green compacts of carbides, refractory metals, and other materials are sometimes heated under a protective or reducing atmosphere in a separate furnace to drive off volatile lubricants before entering the vacuum furnace for sintering. This minimizes the contamination of heater elements and heater element supports that is frequently associated with vacuum furnaces.

Nitrogen-Based Atmospheres (Ref 13-22)

The main constituent of the recently developed nitrogen-based systems is molecular nitrogen. Molecular nitrogen is obtained from air, which consists of approximately 78% nitrogen, 21% oxygen, 0.93% argon, 0.03% carbon dioxide, and small amounts of such rare gases as neon and helium.

Nitrogen is produced commercially by cryogenic and noncryogenic methods. Air is first filtered to remove particulates and is then compressed and passed through a heat exchanger for cooling to remove water and carbon dioxide. The air is then rapidly depressurized to an ultracold liquid that is distilled to separate out the high-purity molecular nitrogen gas. The gas is either piped directly to the sintering plant or liquefied for long-distance shipping. One cubic metre of liquid nitrogen equals 697 m³ of nitrogen gas.

Nitrogen is also produced in noncryogenic generators located at the sintering plant. In noncryogenic production, a stream of compressed air is passed through a bed of special adsorbent, such as a carbon molecular sieve maintained at ambient temperature. Oxygen, water vapor, and carbon dioxide are preferentially retained on the adsorbent, but nitrogen (with a small amount of oxygen) flows through it. As the adsorbent bed becomes nearly saturated with oxygen, water, and carbon dioxide, the system automatically switches to a second bed without interrupting nitrogen delivery.

The system regenerates the first bed by reducing the bed pressure and releasing the adsorbed gases. The regenerated bed is then ready to repeat the cycle. All adsorption and desorption activities occur at ambient temperatures rather than at the low temperatures used in cyrogenic plants. To reduce the oxygen content of the nitrogen, the product stream with added hydrogen is passed through a catalytic deoxidizer and a dryer to produce high-purity nitrogen. Powder metallurgy fabricators with a nitrogen consumption greater than about 100 m³/h (3500 ft³/h) could use an on-site noncryogenic plant for economic reasons. These plants have been introduced into the market only recently.

Whether produced cryogenically or noncryogenically, nitrogen is readily available and economical. Additionally, it provides a consistent source of high technical quality, high-purity product—maximum oxygen content is 10 ppm (typically 2 ppm) and dew point is -65 °C (-85 °F). Because nitrogen is molecular rather than atomic, it is essentially inert to steel and other compositions that are usually sintered in P/M plants.

Inert nitrogen has a density close to that of air (Table 1); consequently, it effectively prevents air from entering the furnace, thus protecting porous P/M parts from harmful contact with air during sintering. Assuming an adequately air-tight furnace, an atmosphere of 100% nitrogen is suitable for sintering aluminum and some less

critical ferrous and nonferrous metal parts. Nitrogen alone, however, will not reduce particle surface oxides or effectively control surface carbon of steel parts.

With small, controlled additions of active gases, nitrogen atmospheres can perform all functions required of sintering atmospheres. Furthermore, the amount and type of active ingredients can be varied to change the level of reactivity of the atmosphere. The most important active ingredient required for sintering of commonly produced P/M parts is hydrogen.

Systems that use low dew point hydrogen sources and that maintain other oxidants at low levels require only small amounts of reducing gas (hydrogen) to create an effective and efficient nitrogen-based atmosphere. Such atmospheres are potentially highly reducing to metal oxides. Typically, nitrogen-based systems consist of essentially inert nitrogen and small amounts of one or more active gaseous ingredients.

Conventional atmosphere generators are not required in such systems; instead, feedstock gases (nitrogen and active gases) are piped through a flow panel into the furnace. The gases entering the panel regulation system exit as the atmosphere—the metering panel becomes, in effect, an atmosphere generator.

These systems exhibit great flexibility by enabling the ratio and types of active ingredients to be varied easily to suit the various sintering requirements within the plant. Nitrogen systems can also provide different atmospheres in different sections of a furnace, as shown in Fig. 8. The volume output reduction for nitrogen is 100%, considerably more than for endo and exo atmospheres. Nitrogen-based systems contain three basic types of active ingredients, as discussed below.

Oxide-Reducing Agents. Hydrogen is the most desirable ingredient for reduction of particle surface oxides, which thus improves bonding and provides maximum integrity to the sintered part. It is supplied by compressed hydrogen tube trailers or liquid hydrogen storage tanks, dissociated ammonia, endo gas, or dissociated alcohols such as methanol. Each supply source has advantages and disadvantages. Technically, hydrogen sources can be divided into two systems, as discussed below.

System 1. Hydrogen is supplied from endo gas or dissociated methanol. The total amount of combustibles (primarily hydrogen and carbon monoxide) in system 1 (nitrogen-endo or nitrogen/dissociated methanol) atmospheres varies between 15 and 30%, typically 24%.

Hydrogen is accompanied by carbon

Fig. 8 Schematic of nitrogen-based zoned atmosphere system

Atmosphere and temperature profiles corresponding to the nitrogen-based atmosphere system are shown below.

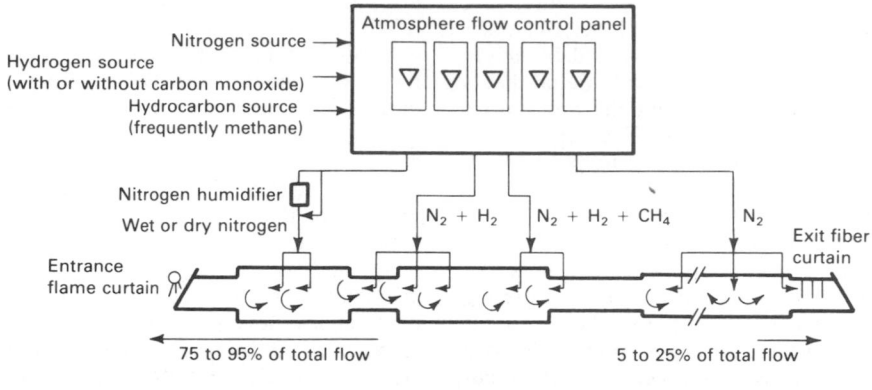

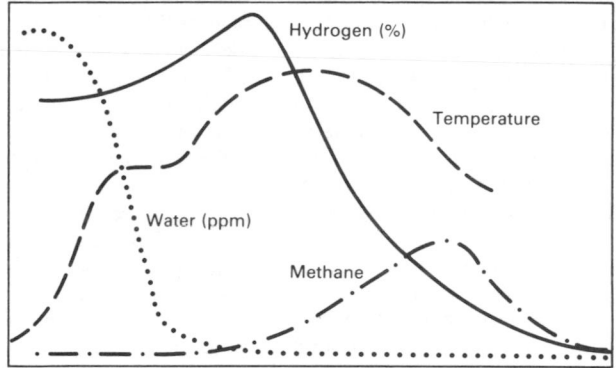

monoxide in a 2-to-1 ratio. The dew point is relatively high. Consequently, the hydrogen-to-water ratio is relatively low (40 to 50, as shown in Table 3). Although this ratio achieves acceptable oxide reduction in steel parts, other atmospheres—notably system 2, which uses dissociated ammonia in a nitrogen base—provide better ratios and consequently greater oxide reduction.

Methanol, as a source of hydrogen, is mixed with nitrogen before introduction into the furnace. Methanol molecules dissociate in the hot zone of the furnace as:

$$CH_3OH \xrightarrow{heat} 2H_2 + CO$$

$$+ \text{ small amounts of } H_2O, CO_2,$$

$$\text{and } CH_4$$

For this reaction to occur, furnace temperatures must be above 800 °C (1470 °F), preferably above 900 °C (1650 °F). The reaction is endothermic and requires about 5 kW of heat energy per gallon of methanol dissociated.

One litre of methanol produces approximately 1.7 m³ of hydrogen and carbon monoxide in a 2-to-1 ratio (1 gal produces 225 ft³). Typical water, carbon dioxide,

and methane contents in dissociated methanol (without nitrogen additions) are 1.7, 0.4, and 0.2%, respectively. The carbon potential of dissociated methanol (without nitrogen additions) is 0.2 to 0.3% C at 925 °C (1700 °F). Composition control is generally improved when hydrogen is derived from methanol rather than from endo gas.

Because hydrogen produced from endo gas or dissociated methanol is always accompanied by carbon monoxide, its use is limited to sintering P/M compositions that are not adversely affected by its presence. Endo gas or methanol is not recommended as a hydrogen source in P/M plants that use only one basic nitrogen system and have capabilities to sinter not only carbon steels but also bronze (at relatively low temperatures), magnetic steel, stainless steel, high-ductility iron parts, and high-carbon high-density parts that blister, or "ripple," during sintering.

System 2. Hydrogen is supplied from tube trailers, liquid storage tanks, or from dissociated ammonia. This atmosphere does not contain carbon monoxide and is pure and dry (very low dew point). The total amount of combustibles (primarily hydrogen) in system 2 (nitrogen-hydrogen or nitrogen/dissociated ammonia) atmospheres

varies between 3 and 10%, typically 5%.

Hydrogen-to-water ratios are much higher than in endo gas or dissociated methanol atmospheres (Table 3). Tube trailers and storage tanks are the most flexible hydrogen supply sources, but are approximately two to three times as expensive as dissociated ammonia per unit of furnace atmosphere.

A system 2 using dissociated ammonia provides the optimum combination of oxide-reducing power, long-range economics, technical flexibility, operator safety, furnace life, and ease of operation. Because these atmospheres do not contain carbon monoxide, a broader range of P/M compositions can be sintered compared to system 1.

Carbon-Control Agents. Systems using hydrogen or dissociated ammonia (system 2) as the only active ingredient in nitrogen-based atmospheres do not experience surface decarburization of P/M steel parts if the dew point in the furnace is kept below about −30 °C (−20 °F). However, if the dew point in nitrogen-hydrogen (dissociated ammonia) atmosphere is slightly higher, small amounts of noticeable surface decarburization occur, and wear resistance of the finished part is lowered. In such situations, small amounts of hydrocarbon, preferably from natural gas, can be used with an oxide-reducing agent (hydrogen or dissociated ammonia) in system 2 to prevent decarburization or to add small amounts of carbon to the surface of the P/M part.

Generally, the amount of carbon monoxide present in system 1 is adequate to prevent decarburization of P/M steel parts. Small amounts of hydrocarbon derived from natural gas can be added to nitrogen-endo gas or nitrogen/dissociated methanol (system 1) atmospheres if additional carbon is desired in the part surface.

Oxidants. Oxidizing agents enhance burning of lubricants in P/M parts. Water and/or carbon dioxide can be added or generated within the furnace in controlled amounts for use in the delube section of the furnace to accomplish this purpose.

Oxygen from air is a relatively strong oxidant compared to water and carbon dioxide. Generally, adding oxygen or air to the atmosphere entering the delube section is not recommended for burning of lubricants.

Small amounts of oxidizing agents such as water can be added to the atmosphere entering the hot zone to achieve limited surface decarburization of parts requiring subsequent pressing and sizing. Small amounts of oxygen (air) or water can be added to nitrogen-based atmospheres with some circulation in the cooling section of the furnace to provide controlled but limited surface oxidation of the sintered parts. This procedure improves the appearance of the parts by imparting a straw, blue, or gray finish, depending on the amount of air or water added and the temperature of the part in contact with the oxidant-containing atmosphere. This also improves corrosion resistance and service life of sintered parts.

Furnace Zones Concept (Ref 23-28)

Typical sintering furnaces can be thought of as having three or more interconnected zones (depending on the material being sintered), each with a separate function. The sintering process consists of several sequential phases, each requiring a unique combination of temperature, time, and atmosphere composition, flow, direction, and circulation. Each phase of the sintering process occurs in a specific zone of the furnace. Separating these zones and phases conceptually enhances design flexibility. A close match between the temperature and atmosphere of each zone and the function of each phase results in an optimum overall sintering process. In a single system, the base nitrogen can be modified with other gases or active ingredients to produce an appropriate and optimum atmosphere composition for each sintering phase before introduction into the proper furnace zone.

In order to understand the optimum combination of temperature, time, and atmosphere for each sintering phase, the processing sequence of a P/M part through the sintering furnace must be visualized. If the first sintering phase is completed effectively and efficiently in the first furnace zone, then the second phase in the second furnace zone, and each successive phase, is likely to be successful. The net result is improved productivity (production rate, quality, consistency, safety, versatility, and cost). Figure 1 shows a schematic of the processing sequence for sintering steel.

For example, if a P/M part composed of iron, 1% graphite, 2% copper, and 1% lubricant was processed in this furnace, at least five phases would be required to complete the sintering process. These would occur in five different zones of the furnace. The requirements for each sintering phase and furnace zone are outlined in Fig. 1 and are summarized in Table 4.

Prime objectives in sintering this green part include the absence of sooting (internal or external). Admixed graphite should be effectively and uniformly diffused into iron particles as combined carbon. This results in an optimum pearlitic microstructure during cooling, with minimum car-bon loss. Admixed copper, after melting, should be uniformly distributed over all particle surfaces in the P/M part.

High structural integrity (maximum possible number of bonds or necks between particles, high average bond or neck area, and relatively rounded pores) is desirable. Surface carbon should be approximately equal to the carbon content in the core or interior. Emerging parts should be clean and bright, or should have a controlled thin surface oxide to improve corrosion resistance. To meet these technical objectives economically, the total flow of atmosphere should be held to a minimum, with the least amount of added active gases.

For the purposes of the example, the atmosphere introduced into the furnace is nitrogen-based system 2 (carbon monoxide free), with 5% hydrogen and 0.5% methane. If the furnace is a conventional one, all of the atmosphere is introduced at one inlet—generally in the slow-cooling section between the hot zone and the water-cooled section. As a consequence, atmosphere composition (water, hydrogen, and methane content) is generally uniform throughout the different zones or sections as the atmosphere flows from the inlet toward both ends. This means that each active ingredient is at the same level whether or not the atmosphere is in a zone where a particular ingredient is needed. For the furnace to function at all, it is necessary

Table 4   Furnace zones, sintering phases, and atmosphere requirements to sinter an iron, 2% copper, 1% graphite, and 1% lubricant part

Furnace zone	Temperature °C	°F	Time, min	Sintering phase	Sintering subphases	Atmosphere requirement	Atmosphere composition(a), % Hydrogen(b)	Methane	Water	Flow	Circulation
1 delube	425-650	800-1200	10-30	Delubing	Lubricant vaporizing; vapor burning; flushing vapors to entrance	Fast, uniform heat transfer; slightly oxidizing	2-7	0-0.1	0.5-1.5	High toward entrance	Highly desirable
2 preheat	650-1070	1200-1960	5-15	Elimination of particle surface oxides	Oxide reducing; graphite diffusing; copper melting; and coating particle surfaces	Highly reducing to surface oxides; neutral to carbon fast, uniform heat transfer	5-15	0.1-0.3	0.01-0.03	High toward entrance	Desirable
3 high temperature	1070-1150	1960-2100	10-30	Particle bonding	Copper diffusing; neck (bond) growing; pore rounding	Neutral to carbon reducing	3-8	0.2-0.5	0.01-0.02	Medium toward entrance	Desirable
4 slow cool	1150-815	2100-1500	5-15	Carbon restoring	Carbon transferring, homogenizing	Uniform slow cooling; slightly carburizing	2-7	0.3-1.0	0.01-0.02	Medium toward entrance	Highly desirable
5 cool down	815-50	1500-120	30-90	Cooling down; optionally oxidizing	Pearlite forming; part cooling down; preventing oxidation or controlling light oxidation	Slightly reducing or neutral (optionally oxidizing) to iron; uniform fast cooling	0-2	0-0.1	0.01-0.02	Low to medium partially toward exit	Highly desirable

(a) Balance is essentially nitrogen with or without small amounts of carbon monoxide and carbon dioxide. (b) Hydrogen can be derived from hydrogen storage vessel, dissociated ammonia, endothermic gas, or dissociated methanol. (c) If slightly oxidizing atmosphere is required, water is relatively higher or a small amount of air is introduced along with nitrogen in water-cooled section at a selected location.

Table 5 Potential hazards and functions of heat treating atmosphere constituent gases

	Potential hazard			
Gas	Flammable	Toxic	Simple asphyxiant	Atmosphere function
Nitrogen	...	...	Yes	Inert
Hydrogen	Yes	...	Yes	Strongly reducing
Carbon monoxide	Yes	Yes	...	Carburizing and mildly reducing
Carbon dioxide	...	Yes	Yes	Oxidizing and decarburizing
Natural gas	Yes	...	Yes	Strongly carburizing and deoxidizing
Ammonia	Yes	Yes	...	Strongly nitriding
Methanol	Yes	Yes	...	Carbon monoxide and hydrogen generating

to compromise between the atmosphere needed in one part of the furnace and that needed in another.

In the atmosphere system used in a zoned sintering furnace, however, different atmospheres (composition, flow, direction, and circulation) are introduced to different zones of the furnace (Fig. 8) to obtain optimum results. Wet nitrogen is introduced near the loading end of the furnace, hydrogen in the center, and dry nitrogen near the exit end. Custom-designed injectors provide proper flow, circulation, and direction to these atmospheres. Injectors are also designed to act as gas barriers and reduce overall atmosphere flow. As a result of this zoning, the constituents peak where they are needed (Fig. 8). In this example, water (dew point) is at its highest level near the entrance in the preheat zone, where delubing takes place. Hydrogen is introduced at its peak level at the mouth of the hot zone and in the hot zone, where it efficiently reduces surface oxides. Methane is added in sufficient amounts to reach its maximum levels at the end of the hot zone and at the beginning of the slow cooling zone, where it is needed for carbon control. Only nitrogen is added in the cooling zone, constituting nearly 100% of the atmosphere near the exit end of the furnace; furthermore, there is no need for a flame at the exit end. Therefore, exiting parts are cooler, and operators are safer and more comfortable.

In contrast to the conventional system, no compromise is needed to balance atmosphere makeup and furnace function; the zoned sintering atmosphere facilitates optimum sintering regardless of makeup or other part specifications. In zoned systems, then, atmosphere is introduced as needed with respect to composition, quantity, and location. Atmosphere zoning, combined with suitable temperature and time, allows each phase of the sintering process to be completed effectively, efficiently, and economically at the lowest possible atmosphere flow rates.

Nitrogen-based zoned atmosphere systems are compatible with existing furnaces, and the constituents can be varied to permit sintering of a wide variety of materials. A well-designed nitrogen-based system can meet the needs of a plant that sinters a wide range of ferrous and nonferrous parts. Technically, zoned atmospheres can achieve the same results as endo, dissociated ammonia, and exo atmospheres, while being more energy efficient, easier to operate, safer, and more flexible.

Nitrogen-based zoned atmosphere systems are capable of meeting the varying requirements imposed by furnace components. Atmospheres must be mild to furnace components. Conveyor belt alloys, for example, oxidize easily at sintering temperatures. However, under very dry and reducing atmospheres, the protective oxide layers are removed, and belt links tend to weld together. The belt "crackles" and gradually loses strength. Furthermore, belts, heating elements, and muffles become brittle and have short service lives under carburizing atmospheres. In addition, insulating brick oxides are reduced if the atmosphere is too reducing, which shortens service life.

Safety Precautions (Ref 29-35)

Sintering atmospheres (endo, exo, dissociated ammonia, dissociated alcohols, and nitrogen based) normally consist of a mixture of gases, which may be flammable, toxic, asphyxiant, or a combination of these. Explosion, fire, and poisoning are potential hazards. National Fire Protection Association standard 86C covers various safety considerations pertaining to continuous conveyor belt furnaces used in the P/M industry.

Four intrinsic dangers are associated with producing and using the common atmosphere gases. A mixture of atmosphere gas and air can accumulate in a confined area

and explode. Relatively small quantities of atmosphere gas can unexpectedly burn or flash out of control. Personnel may be poisoned by carbon monoxide, ammonia, or methanol. "Simple" asphyxiation is possible when asphyxiants are present in high concentrations.

Table 5 describes the characteristics of the common sintering atmosphere constituent gases. Most of the major constituents are flammable; four are toxic; and four are simple asphyxiants. Carbon monoxide, ammonia, and methanol are both flammable and toxic. The percentage by volume of the flammable ingredients ranges from 75% hydrogen in dissociated ammonia atmosphere to only a few percent in purified exothermic and nitrogen-based atmospheres.

Dangers are present even when heat treating with atmospheres that contain relatively small percentages of hazardous ingredients, because these gases may accumulate or concentrate. Under typical operating conditions, however, the active ingredients in nitrogen systems are diluted below the level that is flammable, even if all the nitrogen is replaced by air. Explosion, burning, poisoning, and asphyxiation hazards are thus reduced, but not eliminated.

The explosive ranges of typical atmosphere constituents are:

Atmosphere constituents	Concentration in air, %
Hydrogen	4.0-74
Carbon monoxide	12.5-74
Methane	5.3-14
Ammonia	15.0-28
Methanol	6.7-36

Any mixture between the high and low limits of flammability will burn when ignited and, under certain conditions, detonate or explode. The destructive power of the pressure wave from an ignited flammable mixture depends on the amount of gas and the heat of combustion of the fuel gas, the combustion mode (deflagration or detonation), and the configuration of the confinement space. The energy released is either absorbed by the surroundings or destroys them.

The explosive potential of the large volumes of flammable gases used in sintering furnaces poses an important safety consideration. To ensure the safety of sintering and heat treating, the equipment and systems that handle gases must be designed, operated, and maintained to prevent accumulation of explosive mixtures. Dangerous accumulations that cannot be readily detected by personnel can occur. Properly designed safety systems, maintained and operated by well-trained and competent personnel, substantially reduce explosion hazards.

Generally, sufficient volumes of atmosphere gas flowing through the furnace leave the charge and discharge doors, mix with air, and burn uniformly and completely. However, several conditions may exist that prevent the atmosphere gas from burning off properly. Sometimes, a combustible mixture of air and gas may form in the furnace throats or vestibules. It may burn rapidly and forcefully exhaust, or flash, flames and hot gases through the furnace door. Unprotected personnel in the vicinity may be burned. Eyes are particularly sensitive to damage.

Although flashes of this type are more common than explosions, they are potentially less destructive. Nonetheless, personnel who approach these areas without safety glasses, protective face shields, gloves, and flameproof clothing risk serious flash burns. If safe methods are established and used for clearing jam-ups, inspecting the furnace interior, and removing products from the furnace, this hazard is substantially reduced.

Protective guards and shields used for routine operation are no substitute for the required protection of personnel working in the vicinity of the furnace doors. However, the possibility of flash-back under nonflammable, nitrogen-rich atmospheres is remote.

Ammonia, carbon monoxide, and methanol, which are highly toxic, are used routinely in the sintering and heat treating industries. Liquid ammonia produces dissociated ammonia or ammonia vapor for nitriding, and carbon monoxide is a constituent of exothermic, endothermic, and dissociated methanol gases. Methanol is used to produce dissociated methanol, which consists of hydrogen and carbon monoxide. Concentrations of less than $1/2$% ammonia or carbon monoxide in air are considered fatal in less than $1/2$ h of exposure. The physiological effects of various concentrations of these gases are given in Tables 6 and 7.

Threshold limit values, published by the American Conference of Governmental and Industrial Hygienists, lists carbon monoxide, carbon dioxide, ammonia, and methanol as commonly used toxic chemicals. Concentrations as small as 35 ppm of ammonia, 400 ppm of carbon monoxide, and 250 ppm of methanol are harmful; therefore, only short exposures are allowed.

Protection from poisoning by raw ammonia or methanol vapor or from contact with liquid ammonia or methanol is pro-

Table 6 Physiological effects of ammonia

Concentration, ppm	Physiological effects
20	First perceptible odor
40	Slight eye irritation in a few individuals
100	Noticeable irritation of eyes and nasal passages after a few minutes of exposure
400	Severe irritation of the throat, nasal passages, and upper respiratory tract
700	Severe eye irritation; no permanent effect if the exposure is limited to less than $1/2$ h
1700	Serious coughing, bronchial spasms; less than $1/2$ hr of exposure may be fatal
5000	Serious edema, strangulation, asphyxia; almost immediately fatal

Table 7 Physiological effects of carbon monoxide

Concentration, ppm	Physiological effects
100	Allowable for an exposure of several hours
400	Can be inhaled for 1 h without appreciable effect
600	Causes a barely appreciable effect after 1 h of exposure
1000	Causes unpleasant symptoms, but not dangerous after 1 h
1500	Dangerous for exposure of 1 h
4000	Fatal for exposure of less than 1 h

vided partly by the design of the storage and delivery system. Suppliers are excellent sources of detailed safety information on ammonia and methanol systems and ammonia dissociators.

Carbon monoxide is not quite as toxic as ammonia; however, because it is odorless, it poses a greater safety hazard. Lethal concentrations can collect in isolated areas and remain undiscovered until personnel are overcome by the fumes. According to Ref 30, "... in concentrations far short of the asphyxiation level, carbon monoxide is still dangerous, especially if exposure is prolonged. Carbon monoxide (poisoning) can be a contributing factor to death that ultimately results from other causes, such as fatal accidents, and is a health as well as safety problem." Carbon monoxide interferes with the ability to breathe and, as a result, the body's ability to utilize oxygen. It is over 200 times more attractive than oxygen to the blood hemoglobin that delivers oxygen throughout the body. A relatively small amount of carbon monoxide, therefore, depletes the body of a large amount of oxygen.

To ensure personnel safety, piping between the generator and the furnace must be leaktight, and all gas that enters the furnace must be either burned off or properly vented. The highly toxic nature of carbon monoxide requires use under carefully controlled conditions. The safest and most convenient disposition of carbon monoxide is to ensure complete mixing of the furnace atmosphere effluent with air to combustible proportions and subsequent ignition of the mixture. Products of combustion are far less toxic, but should still be properly vented.

In some sintering applications, carbon monoxide cannot be burned directly because it has been diluted in inert gas to a level that is nonflammable when mixed with air. Special precautions must be taken to vent and dilute it to below toxic limits. To ensure the continued effectiveness of the venting, carbon monoxide levels near the furnace setup must be monitored continuously. Consequently, apparatus using or emitting gas containing carbon monoxide that will not burn when mixed with air should not be operated unless the user is willing to accept the added liability. Additionally, special gas monitoring and disbursing equipment should be installed.

Personnel should also be familiar with the early symptoms of carbon monoxide poisoning and trained in appropriate first aid. Early symptoms include slight dizziness, weakness, or headache. In the later stages of poisoning, the victim's lips and skin turn a characteristic cherry red.

Finally, the danger of asphyxiation is not intrinsic to the gases used in the P/M industry, but is a concern. Asphyxiation can be caused by a harmless gas (for example, one that has no significant physiological effect) if it is present in high concentrations in the air. Normal oxygen content in air is about 21 vol%. Minimum oxygen content should be 18 vol% under normal atmospheric pressure.

REFERENCES

1. German, R.M., *Powder Metal Science*, Metal Powder Industries Federation, Princeton, NJ, 1984
2. Lenel, F.V., *Powder Metallurgy*, Metal Powder Industries Federation, Princeton, NJ, 1980
3. Heckel, R., Effects of Gas Composition on Gas-Solid Equilibrium During Sintering of Ferrous Powders, *Prog. Powder Metall.*, Vol 35, Princeton, NJ, 1979
4. Nayar, H.S., Status Report on Energy Consumption and Distribution in Powder Metal Industry, *Prog. Powder Metall.*, Vol 34, 1978
5. *Furnace Atmospheres and Carbon*

Control, American Society for Metals, 1964

6. *Metals Handbook,* Vol 4, 9th ed., American Society for Metals, 1981
7. Nayar, H.S. and Drew, J., Furnace Atmospheres: Feedstock, Make-Up, Properties, *Heat Treating,* July-Sept 1980
8. Stevenson, R.W., Alabrese, C., and Fredette, P.E. Generator Control System Gives Endo an Edge, *Heat Treating,* March 1982
9. Kaufman, S., The Three-Zone Sintering Furnace, *Prog. Powder Metall.,* Vol 33, 1977
10. *Matheson Unabridged Gas Data Book, Hydrogen,* Matheson Gas Products, 1974
11. Doak, D.W., State of the Art in Vacuum Sintering Furnace Technology, *Carbide Tool J.,* March 1980
12. Doekler, W.J., Sintering Powder Metallurgy Compacts in Vacuum Furnaces, *Prog. Powder Metall.,* Vol 35, 1979
13. U.S. Patent 4 106 921, 1978
14. Nayar, H.S., Nitrogen-Based Sintering Atmospheres, *Modern Developments in Powder Metallurgy,* Vol 9, Metal Powder Industries Federation, Princeton, NJ, 1976
15. Nayar, H.S., Sintering and Infiltration of Steel P/M Parts in Nitrogen-Based Atmospheres, *Prog. Powder Metall.,* Vol 33, 1977
16. Nayar, H.S., Endo Vs. Nitrogen, Some Technical Considerations, *Heat Treating,* March-April 1980

17. Becker, J.S., Comparison of Endothermic and Nitrogen-Based Sintering Atmospheres, *Prog. Powder Metall.,* Vol 35, 1979
18. Muzik, J., Smith, R., and Marsden, J., Comparative Results of Nitrogen-Based and Conventional Sintering, *Modern Developments in Powder Metallurgy,* Vol 12, Metal Powder Industries Federation, Princeton, NJ, 1980
19. Sibley, A.T., Marsden, J.G., and Buck, D.M., High Temperature Sintering with a Nitrogen-Methanol Atmosphere, *Prog. Powder Metall.,* Vol 38, 1982
20. Kostelitz, M., Heat Treatment Processes in Nitrogen-Based Atmospheres, *Powder Metall. Int.,* Vol 13 (No. 3), 1981, p 140-144
21. Gaines, G.W. and Kohler, R.H., Powder Metallurgy Sintering with a Nitrogen Methanol Atmosphere, SAE Paper No. 810245, Society of Automotive Engineers, Warrendale, PA, 1981
22. Adams, S.M. and Nayar, H.S., Advantages of Nitrogen as a Brazing Atmosphere, *Heat Treating,* April 1981
23. Nayar, H.S. and Schaeffer, D., How Furnace Zoning Can Optimize Atmosphere Efficiency, *Heat Treating,* March 1981
24. U.S. Patent No. 4 294 395, 1981
25. Nayar, H.S., The Concept of Furnace Zoning: Its Use in Developing Highly Effective Sintering Atmospheres, International Powder Metallurgy Conference, Florence, Italy, Associazione Italiana di Metallurgica, June 1982
26. Nayar, H.S., Atmosphere Zoning to Optimize Productivity During Sintering of Steel—Part 1, *Prog. Powder Metall.,* Vol 39, 1983
27. Frey, R., Effects of Furnace Atmosphere Compositions on Sintering Ferrous P/M Parts, *Prog. Powder Metall.,* Vol 38, 1982
28. Nowotarski, M., Reduction of Atmosphere Requirements in Continuous Sintering Furnaces, *Prog. Powder Metall.,* Vol 38, 1982
29. McKinley, W. and Nayar, H.S., Safety Considerations in Sintering Atmospheres, *Prog. Powder Metall.,* Vol 35, 1979
30. Holtzberg, J.T., Requirements for Monitoring Carbon Monoxide, *Ind. Heat.,* March 1980
31. Industrial Furnaces Using a Special Processing Atmosphere, ANSI/NFPA 86C, National Fire Protection Association, Quincy, MA, Dec 1983
32. "Threshold Limit Values of Substances in Workroom Air," American Conference of Governmental and Industrial Hygienists, Cincinnati, 1979
33. Safety and Health Standard 29CFR 1910, Occupational Safety and Health Administration, June 1981
34. Bodurtha, F.T., *Industrial Explosion, Prevention and Protection,* McGraw-Hill, New York, 1980
35. *Handbook of Industrial Loss Prevention,* Factory Mutual Engineering Co., McGraw-Hill, New York, 1967

Production Sintering Equipment

By A.P. Crease, Jr.
Vice President—Sales
Drever Co.
(deceased)

CONTINUOUS PRODUCTION FURNACES for sintering of P/M compacts normally are composed of two heating chambers: a burn-off, preheat, or delubrication zone for removal of compacting lubricants and a sintering zone in which the parts are sintered. The parts then enter a slow cooling zone and a final cooling zone, where they are cooled to below their oxidizing temperature before being discharged from the protective atmosphere of the furnace. Figure 1 shows the various heating and cooling zones of a muffle-type continuous furnace.

Four distinct types of continuous furnaces that are used for sintering are the mesh-belt conveyor furnace, the pusher furnace, the roller-hearth furnace, and the walking-beam furnace (Fig. 2). The humpback furnace, which is an adaptation of the mesh-belt conveyor furnace, is not shown in Fig. 2, but will be discussed later in this article. In addition, bell and elevator furnaces, which are batch-type furnaces, and vacuum furnaces are used for special applications. This article discusses the general characteristics of production sintering equipment and describes these various types of continuous and batch furnaces.

Furnace Characteristics

Burn-Off Zone. The first heating section of a continuous sintering furnace is the burn-off zone, sometimes referred to as the preheat or delubrication zone. This section of the furnace heats the green P/M compact to drive off the compacting lubricant. The effective heating length of the burn-off zone generally ranges from 66 to 100% of the effective length of the sintering chamber; 100% is the most desirable length.

Some of the more commonly used lubricants are zinc stearate, lithium stearate, paraffin, and Acrawax, which melt in the general range of 40 to 220 °C (100 to 430 °F) (see Table 1). The low melting point of the lubricants suggests that the maximum continuous operating temperature of the burn-off zone could be limited to about 420 °C (800 °F); however, this is not the case.

The density of a P/M compact has a direct bearing on the ease with which the lubricant exits the compact. As density increases, it becomes increasingly more difficult to drive out the lubricant. Because of this density factor, the continuous maximum operating temperature of the burn-off zone ranges from a low of 650 °C (1200 °F) to a high of 980 °C (1800 °F). Because it is also important that the heating rate of a densely compacted part be controlled to avoid thermal stress rupturing, the burn-off zone generally has a minimum of three controlled heating zones.

The burn-off zone usually contains a muffle to help channel lubricant residue to the entrance door of the furnace. The muffle also allows for gas firing to keep flue products out of the furnace atmosphere, or for electric heating to protect elements from attack by the vaporized lubricant.

As an alternative to the conventional burn-off zone, the rapid burn-off zone has been gaining acceptance (Fig. 3). This unit normally is only about 25% of the effective length of the sintering zone and consists of a muffle that has a panel of gas-fired radiant cup-type burners mounted on top. The protective atmosphere in the rapid burn-off zone is a combination of the furnace atmosphere and the combustion products from the burners. Heating in this unit is very rapid, with parts reaching 815 °C (1500 °F) in approximately 7 min.

Sintering Zone. The sintering zone is either gas fired, with a heat-resistant alloy muffle to isolate the products of combus-

Fig. 1 Longitudinal section of a muffle-type continuous production furnace

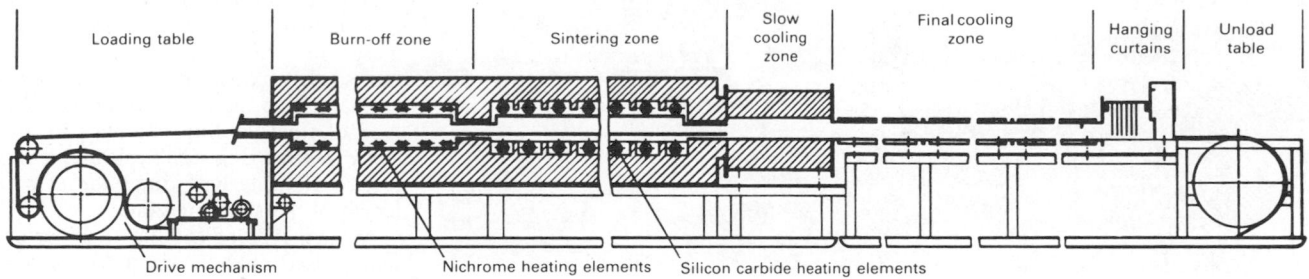

Loading table | Burn-off zone | Sintering zone | Slow cooling zone | Final cooling zone | Hanging curtains | Unload table

Drive mechanism   Nichrome heating elements   Silicon carbide heating elements

Fig. 2 Schematic of the four types of continuous furnaces used for sintering

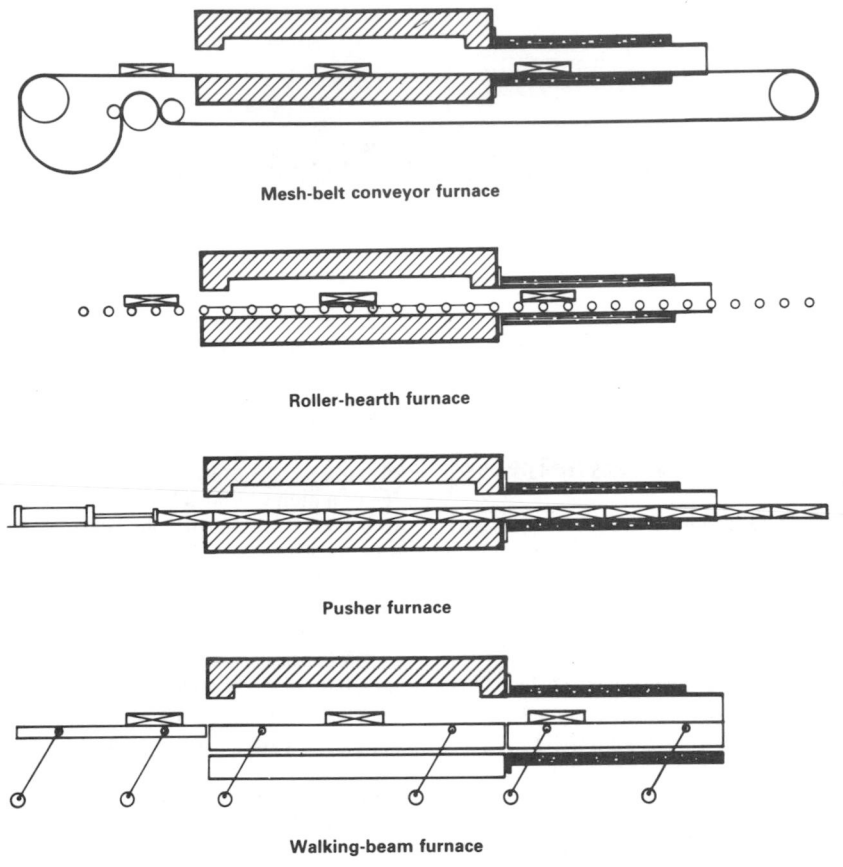

Mesh-belt conveyor furnace

Roller-hearth furnace

Pusher furnace

Walking-beam furnace

Fig. 3 Mesh-belt conveyor furnace equipped with rapid burn-off system and high convection cooling system

Table 1 Commonly used compacting lubricants

Lubricant	Chemical formula	Melting point, °C (°F)
Zinc stearate	$Zn(C_{18}H_{35}O_2)_2$	130 (266)
Lithium stearate	$LiC_{18}H_{35}O_2$	221 (430)
Paraffin	$C_{22}H_{46}$ to $C_{27}H_{56}$	40-60 (104-140)
Acrawax	...	140-143 (284-290)

tion from the furnace atmosphere, or is electrically heated and constructed with or without a muffle. When constructed without a muffle, the furnace atmosphere is contained by the gastight steel casing of the furnace and is referred to as an open brickwork chamber furnace (Fig. 4). In this type of furnace, the atmosphere is free to react with the brickwork in the furnace, which normally gives less precise control of carbon potential than is achieved in a muffle furnace.

When the 1150 °C (2100 °F) sintering zone is electrically heated, silicon carbide rod-type elements, sometimes called globars, usually are used. When higher temperatures are required (1290 to 1650 °C, or 2350 to 3000 °F, for example), heating elements normally are made of molybdenum rod. When lower temperatures (less than 1150 °C, or 2100 °F) are required for the sintering of nonferrous or aluminum compacts, electric heating elements can be made of 80%Ni-20%Cr alloy ribbon.

Because sintering is a function of time at temperature, an adequate sintering zone provides a temperature profile in which the parts reach the sintering temperature in the shortest possible time. This is a function of belt loading and heating chamber length under appropriately powdered heating zones (Fig. 5).

Slow Cooling Zone. Typically, a short slow cooling zone follows the sintering zone. This permits some cooling to occur before the work enters the final cooling zone, thereby reducing thermal shock. Like the sintering zone, the slow cooling zone can be constructed with a muffle or open brickwork.

Final Cooling Zone. The final cooling zone cools the work in a protective atmosphere to a sufficiently low temperature so that oxidation does not occur on discharge into the air. The most common type of final cooling chamber in use is water jacketed (Fig. 6). Typically, it is two to two and one half times the effective length of the sintering zone.

Another type of cooling chamber commonly used operates on the principle of

Fig. 4 Open brickwork and muffle-type furnaces

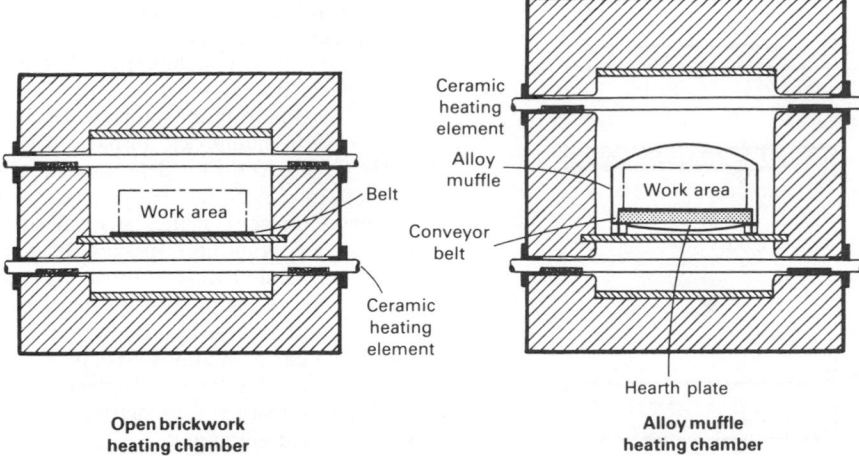

Open brickwork heating chamber

Alloy muffle heating chamber

Fig. 5 Production nomograph for continuous production furnaces

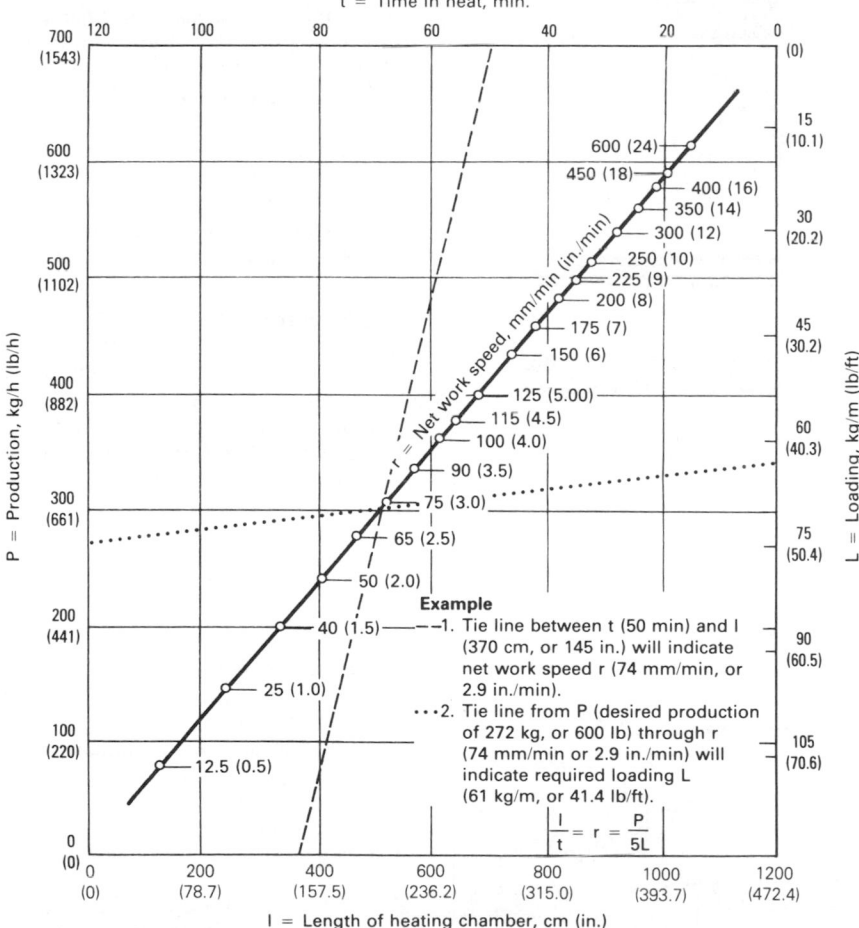

high convection, using the furnace atmosphere as the convection medium. In this unit, cooled atmosphere is directed through nozzles onto the work from above and below (Fig. 7). This produces more rapid cooling than a water-jacketed cooler; as a

result, the convection cooler length is lowered by about 50%.

The convection cooler produces a fine-grained pearlite structure in sintered iron-carbon materials. When air-hardening alloys, including iron-copper-carbon ma-

terials, are cooled by convection, greater hardness is achieved than when they are cooled by a water-jacketed cooler. The ability to control cooling rate in a convection cooler adds a new metallurgical dimension to the sintering process.

Continuous Furnaces

Mesh-belt conveyor furnaces are the most commonly used equipment for the sintering of P/M compacts (Fig. 8). They provide a continuous, reproducible time-temperature-atmosphere (thermal) profile, which is very important in powder metallurgy. When the belt speed of a furnace has been made constant, when all temperature control zones have been set at the proper temperature, and when the atmosphere flow has been established, all compacts processed through that furnace will be subject to the same set of process conditions. As a result, a stable thermal profile will have been established.

Mesh-belt conveyor furnaces consist of a charge and belt-driven table, a burn-off zone, a sintering zone, a slow cooling zone, a final cooling zone, and a discharge table (Fig. 9). Parts are loaded onto a continuously driven mesh belt on the charge table, carried through the zones, and then removed from the belt and placed on the discharge table. The speed of the belt can be adjusted to obtain different cycle times in the sintering furnace. Because of the limitations of alloy strength at elevated temperatures, the belt normally limits the maximum temperature in the furnace to 1125 to 1150 °C (2050 to 2100 °F).

The mesh-belt conveyor furnace has intermediate production capabilities, with normal belt loading limited to about 49 kg/m² (10 lb/ft²). With special roller-assisted belt drives (Fig. 10), belt loading can be increased to about 98 kg/m² (20 lb/ft²).

Humpback furnaces (Fig. 11 and 12), an adaptation of the mesh-belt conveyor furnace, are used when high atmosphere purity and low atmosphere consumption are desired, usually in conjunction with a low-density atmosphere such as dissociated ammonia or hydrogen. A long, inclined entry section, which is gas tight, carries the belt and work from the charge table up to the sintering zone. The incline generally is at an 11 to 13° angle. After passing through the sintering and slow cooling zones horizontally, the final cooling zone descends to the proper level to balance the atmosphere in the furnace. The belt and work usually are assisted up the entry incline by an escalator drive.

Low-density atmospheres such as dissociated ammonia or hydrogen tend to rise

Fig. 6 Schematic of waterjacketed cooling sections

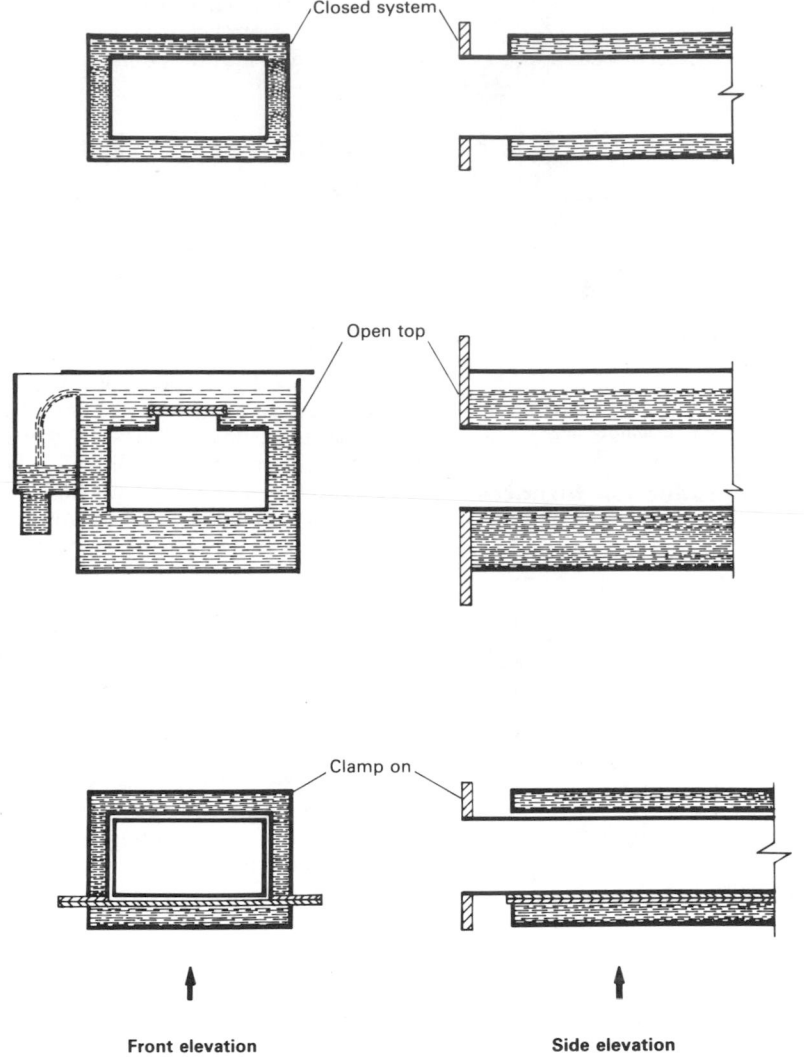

Front elevation   Side elevation

Fig. 7 Schematic of high-convection cooling chamber

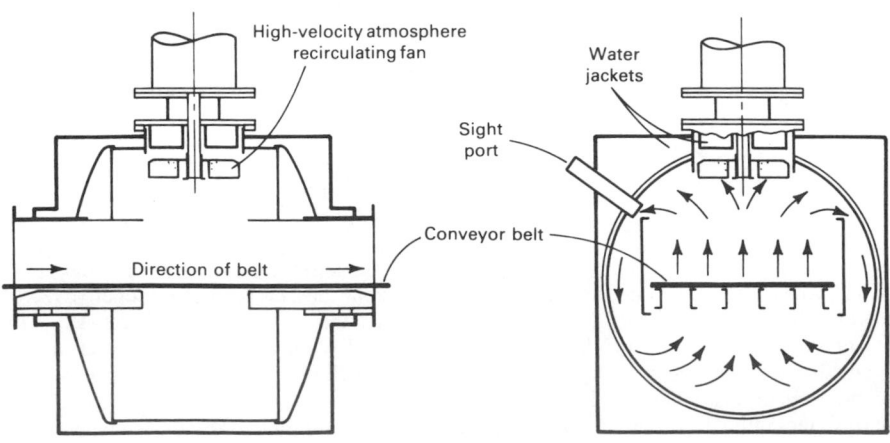

in the hump, allowing for substantially reduced atmosphere flow compared to a flat-hearth, mesh-belt conveyor furnace. Humpback furnaces permit higher furnace door openings without significantly higher atmosphere flows.

The maximum temperature that can be employed in humpback furnaces (up to 1150 °C, or 2100 °F) is determined by the alloy from which the belt is made. The humpback furnace is particularly well suited for low dew point atmospheres. In many cases, a separate burn-off furnace is used, which may be coupled to the humpback furnace with a conveyor, thus ensuring maximum purity of atmosphere in the sintering zone. This type of furnace is particularly well suited for sintering of stainless steel and aluminum.

Pusher Furnaces. In a pusher furnace (Fig. 13 and 14), parts to be processed are loaded on trays or ceramic plates that are pushed through the stationary hearth furnace. Generally, this type of furnace has a lower output than a belt furnace. Furnaces may have a manual, hydraulic, or mechanical pusher system. The pusher mechanism may operate intermittently or continuously. When it is continuous, a reproducible thermal profile of the work can be achieved. Pusher furnaces operate at temperatures up to about 1650 °C (3000 °F).

Roller-Hearth Furnaces. In a roller-hearth furnace (Fig. 15 and 16), parts are carried in trays through the furnace on driven rolls. Roller-hearth furnaces can be made to greater lengths and widths than most conveyor-belt furnaces, so they have large output potential. Generally, these furnaces are capable of heavier loading (pounds per square foot) on the hearth than mesh-belt furnaces. The roller-hearth design allows for fast entry and exit of the work, which minimizes open door time and reduces atmosphere consumption. This furnace also allows a continuous, reproducible thermal profile. The maximum achievable temperature is about 1150 °C (2100 °F), because of the limitations imposed by available alloys for rolls.

Atmosphere quality in a roller-hearth furnace is poor. The large cross-sectional area of this furnace gives the lowest forward velocity of atmosphere, which may allow lubricant burn-off residue to diffuse backward into the sintering furnace.

Walking-beam furnaces (Fig. 17 and 18) are particularly well suited for applications in which sintering temperatures are above the limitations of the mesh-belt conveyor and roller-hearth furnaces. These furnaces operate up to 1650 °C (3000 °F).

Fig. 8 Mesh-belt conveyor furnace

Fig. 11 Humpback furnace for sintering stainless steel P/M parts
Burn-off is accomplished in a separate furnace.

Fig. 9 Longitudinal section of a mesh-belt sintering furnace

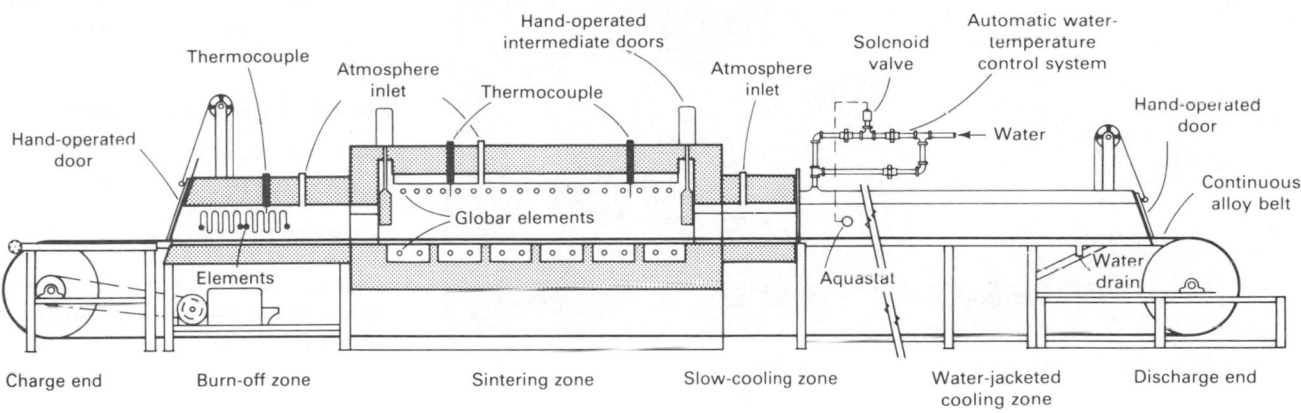

Fig. 10 Schematic of belt drives for mesh-belt conveyor furnaces
(a) Normal belt drive. (b) Roller-assisted belt drive, which allows increased belt loading capacity

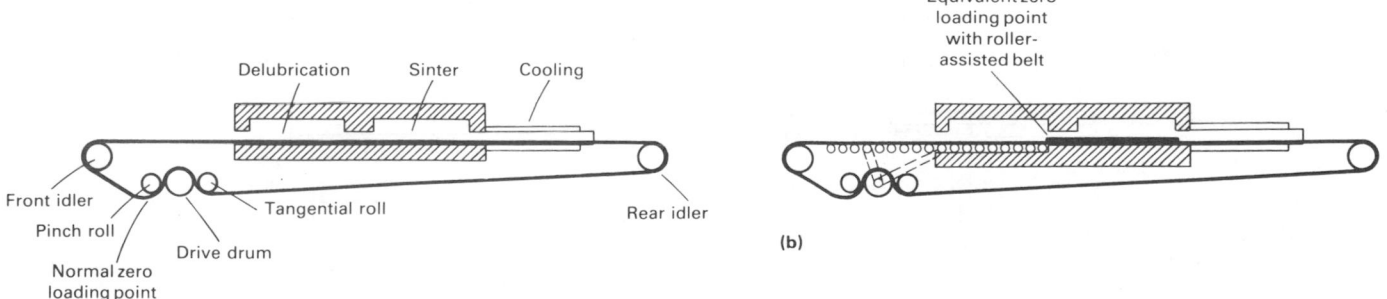

Fig. 12 Longitudinal section of humpback sintering furnace

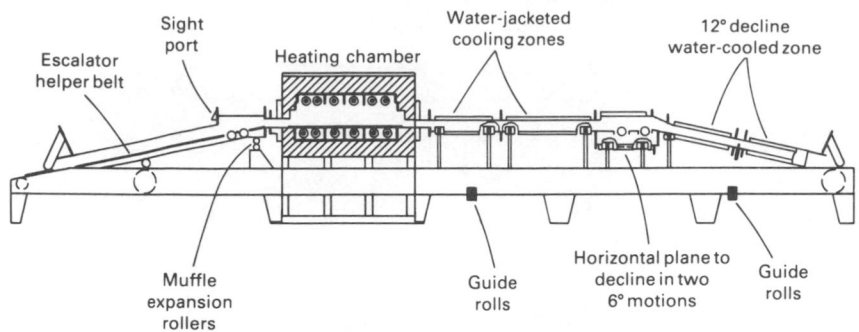

Fig. 13 Molybdenum-heated high-temperature pusher-type furnace for sintering metal carbides

The work to be conveyed through the furnace is placed on ceramic carrier plates. In a walking-beam furnace, the middle section of the hearth is a movable beam system that extends the entire length of the furnace. This beam can be raised above the hearth level, lowered below the hearth level, and moved longitudinally. To convey the work through the furnace, the beam is raised, lifting the ceramic carrier plates off the stationary hearth section.

The beam then moves forward a fixed distance, taking the carriers with it, and drops below the stationary hearth level, where it retracts to the initial position. With this rectilinear movement, the beam intermittently moves the carriers containing the work through the furnace and cooling zones.

Design of the walking-beam furnace permits all-ceramic construction of the furnace interior, the walking beam, and the carrier plates, thereby facilitating operation at high temperatures. In addition, walking-beam furnaces are capable of carrying extremely heavy loads at temperatures up to 1650 °C (3000 °F), which is unattainable in mesh-belt conveyor, roller-hearth, and pusher furnaces.

Batch-Type Furnaces

Bell and elevator furnaces (Fig. 19) are batch-type rather than continuous furnaces. These furnaces consist of a base and work support, a retort, and a bell-shaped or cylindrical furnace.

The base of the bell furnace is permanently installed on the floor. Work to be processed is loaded onto the base and then covered with a heat-resistant alloy retort to contain the protective atmosphere. The

Fig. 14 Longitudinal section of a mechanical pusher sintering furnace

Fig. 15 Electrically heated roller-hearth sintering furnace

Fig. 17 Walking-beam sintering furnace

furnace bell then is lifted and placed over the retort and base.

The elevator furnace has the furnace bell above the mill floor on a fixed structure. The base and work load are covered with a retort and rolled on tracks under the furnace. An elevator system then raises the base, work, and retort into the furnace.

Bell and elevator furnaces commonly are used for sintering friction materials. They can be equipped with work-pressing devices to apply heavy pressure to loads such as clutch and brake facings. They also are useful for sintering heavy loads or large pieces and for applications that require long programmed heating and cooling cycles. These furnaces also can be used when extremely pure atmosphere conditions are required.

Vacuum Furnaces

Vacuum sintering can be conducted in either batch-type or continuous furnaces. Vacuum sintering does not differ from conventional atmosphere sintering, except for one modification: a vacuum furnace operates in the absence of an internal atmosphere.

Batch-type vacuum sintering furnaces (Fig. 20) were first developed for sintering compacts from refractory and reactive metals such as tantalum, niobium, titanium, zirconium, and their alloys (which react with any but the noble gases, such as argon or helium), for sintering cemented carbides containing titanium carbide and tantalum carbide, and for sintering Alnico magnets. A schematic drawing of a high-frequency batch-type furnace for sintering cemented carbides is shown in Fig. 21.

Fig. 16 Longitudinal section of a roller-hearth sintering furnace

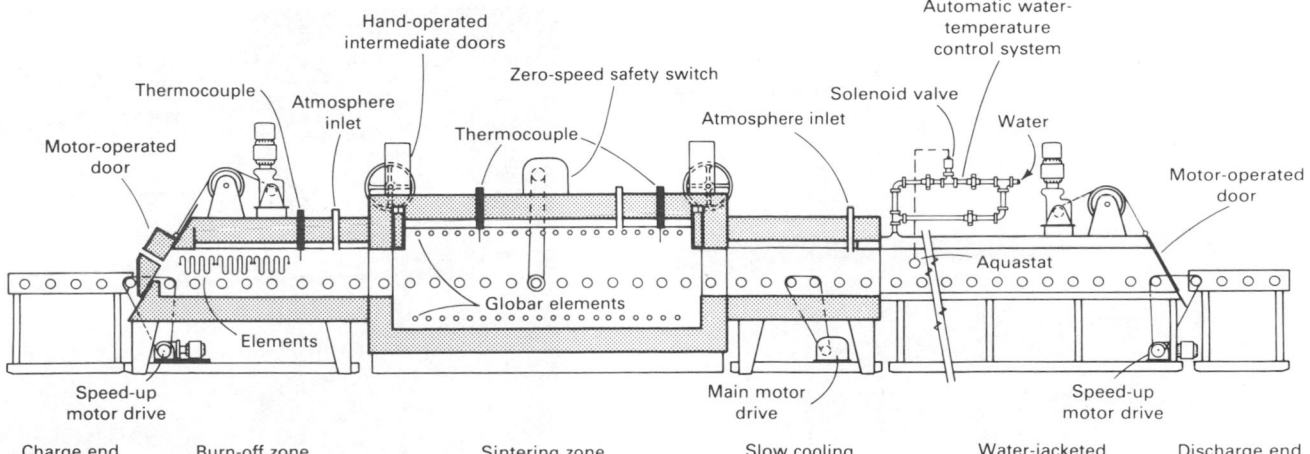

Fig. 18 Longitudinal section of a walking-beam sintering furnace

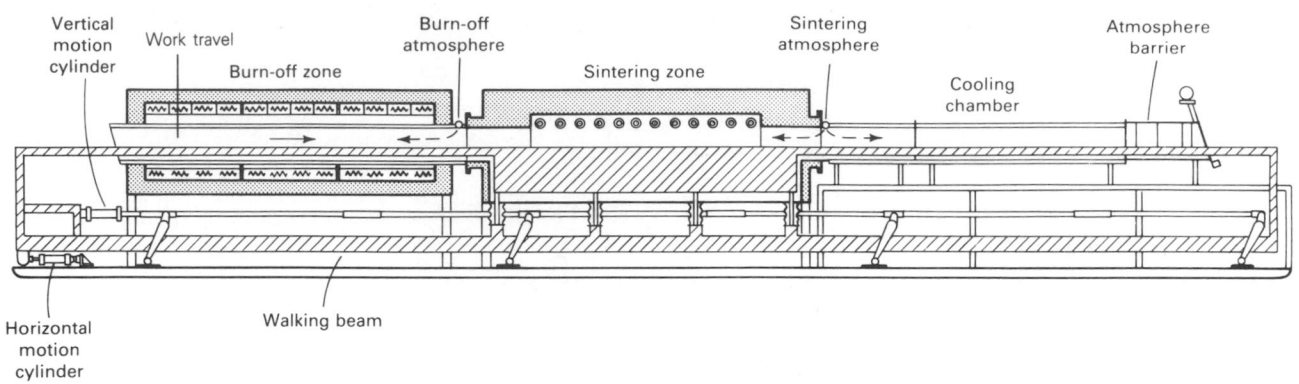

Fig. 19 Batch-type sintering furnaces
(a) Bell furnace. (b) Elevator furnace

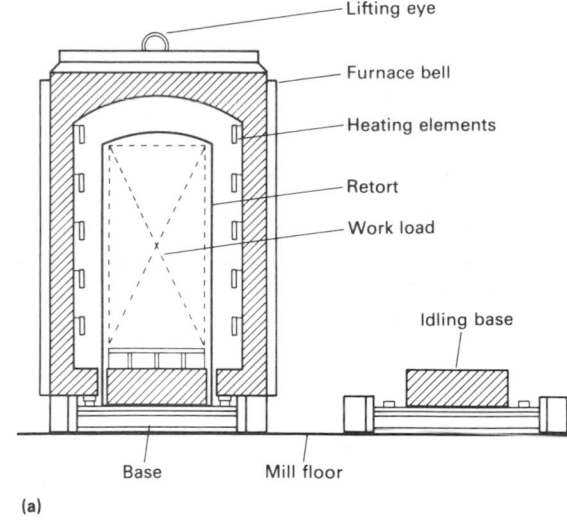

(a)

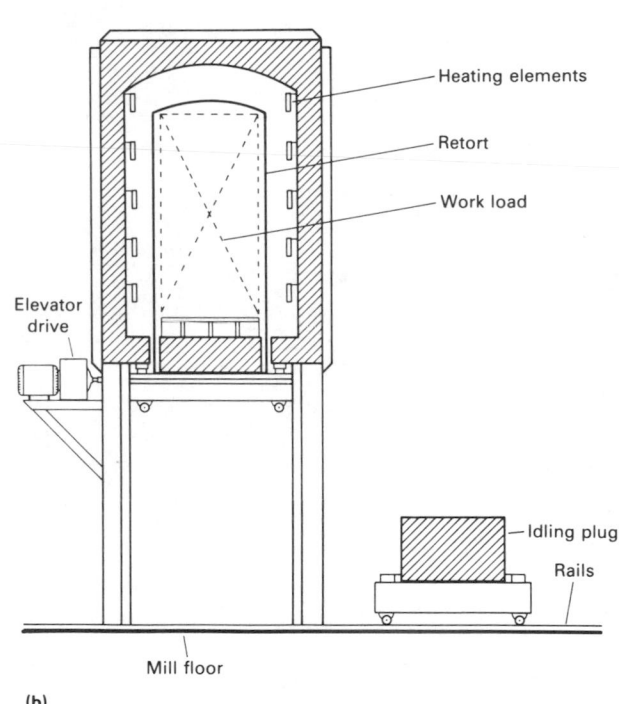

(b)

Fig. 20 Batch-type cold wall vacuum furnace for operation at 1315 °C (2400 °F)
Front door and control module are shown.

Fig. 22 Continuous vacuum sintering furnace

Generally, a batch-type vacuum furnace consists of an outer vacuum-tight cylindrical casing that contains a furnace with radiation shields or other types of insulation, work support, and heating elements. This casing is fitted with roughing and diffusion pumps to achieve the desired vacuum levels.

Vacuum furnaces are designed for use at a variety of temperatures. Units capable of reaching 2760 °C (5000 °F) are available. Because all heat transfer in a vacuum is accomplished by radiation, cooling cycles can be quite long. Therefore, in many cases, the furnace is designed for back filling with nitrogen or argon, which when combined with natural convection shortens cooling time.

Continuous vacuum sintering of iron and steel compacts and silicon-iron mag-netic material is accomplished in furnaces like those shown in Fig. 22 and 23. The furnace is comprised of an external loading table, followed by an atmosphere delubrication chamber, a transfer station, and an "atmosphere to vacuum" vestibule section, followed by a heating chamber, vacuum cooling chamber, a combination fan cooling and "vacuum to atmosphere" vestibule, and an unloading table. The operation of the furnace is completely automatic, under the supervision of an operator.

The advantages of a continuous vacuum sintering furnace are a more flexible temperature range and a choice of slow cooling or rapid cooling in a pressure quench. Most importantly, there is absolute carbon control and no need to continuously monitor an atmosphere for dew point or carbon concentration.

Fig. 21 Schematic of batch-type vacuum furnace for sintering cemented carbides

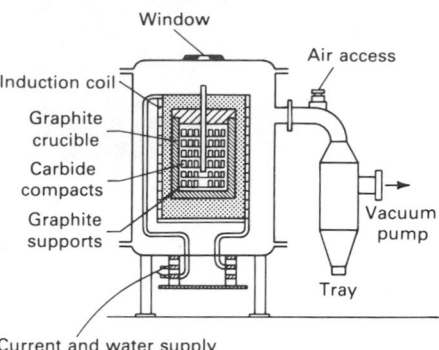

Fig. 23 Longitudinal section of a continuous vacuum sintering furnace

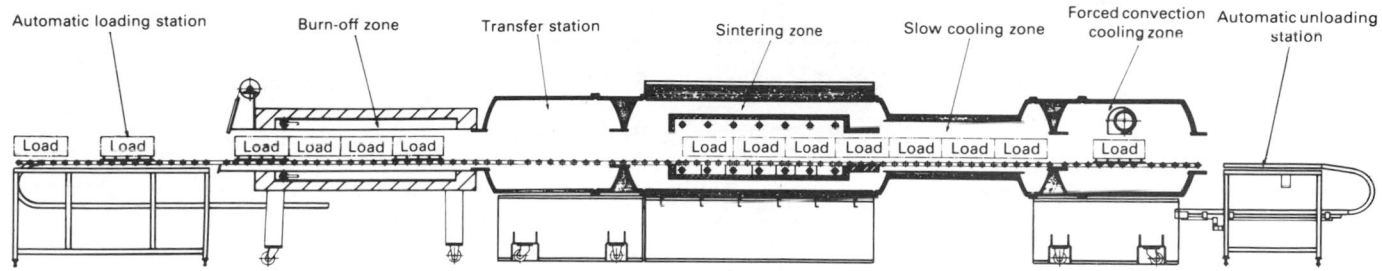

Production Sintering Practices for P/M Materials

SINTERING is a process in which an assembly of loose or compacted particles metallurgically bond into a coherent body at elevated temperatures. During the processing of a P/M component, the sintering step subjects the fragile green compact to heating in a protective atmosphere to establish the desired mechanical properties by causing the powder particles to form coherent bonds and to alloy any admixed elements. Temperatures below the melting point of the major constituent are usually used.

Although sintering has been the subject of extensive experimental and theoretical work, there is no universally applicable theory of sintering. Much is not understood about the process, and no single applicable explanation or understanding of the process has emerged. Sintering generally consists of six distinct, but not necessarily sequential, stages:

- Initial particle bonding
- Neck growth
- Pore channel closure
- Pore rounding
- Densification or pore shrinkage
- Pore coarsening

Material transport mechanisms involved include surface diffusion, volume diffusion, evaporation and condensation, and grain boundary diffusion. Several factors affect sintering. However, the most significant are time and temperature, with temperature being the most important single variable. Particle size, compact porosity, and prealloying also affect the sintering process.

This article examines production sintering practices for a variety of ferrous, non-ferrous, and metal carbide materials. No attempt has been made to explain the aforementioned sintering stages or the material transport mechanisms involved in the process. This information can be found in the article "Physical Fundamentals of Consolidation" in this Volume. In addition, detailed information on sintering fur-

naces and atmospheres can be found in the articles "Production Sintering Equipment" and "Production Sintering Atmospheres" in this Volume. Methods for producing all of the materials discussed in this article can be found in the Section of this Handbook entitled "Production of Metal Powders."

Sintering of Ferrous Materials

PRESSED AND SINTERED IRON-BASED MATERIALS represent the largest segment of the P/M parts industry. Ferrous P/M parts are made from iron powders, alloy steel powders, or mixtures of (1) iron and graphite powders, (2) iron (or steel) and copper powders, or (3) iron, copper, and graphite powders (with or without powders of other metals).

During sintering of ferrous P/M parts, the following reactions occur:

- Metallurgical bonds are developed from the mechanical interlocks between powder metal particles in the compact.
- Metal oxides in the powder compact are reduced by reaction with the carbon from the blended graphite powders.
- Desired final carbon level of the P/M compact is obtained by diffusion of carbon from graphite powder.
- Densification of P/M compacts can be achieved during sintering. The degree of densification depends on sintering parameters and the alloy(s) involved. Generally, higher sintering temperatures and longer sintering times promote densification of ferrous sintered parts.

For conventional sintering, which generally is conducted at 1120 °C (2050 °F), mesh belt conveyor furnaces are widely used. Green compacts are carried on a

conveyor mesh belt made from nickel-chromium alloy wire. Mesh belt furnaces can be used at temperatures up to 1150 °C (2100 °F) only, due to the temperature limitation of the belt material.

For high-temperature sintering applications, which are discussed later in this section, walking beam furnaces have been successfully used for continuous operation at temperatures up to 1315 °C (2400 °F). Vacuum furnaces also can be used for high-temperature batch-type sintering of ferrous P/M parts. However, any lubricant blended into P/M compacts must be burned off in a separate controlled-atmosphere furnace before the compacts are sintered in a vacuum furnace.

This section reviews sintering practices for several ferrous materials, with an emphasis on the types of atmospheres used and the resulting as-sintered microstructures and properties.

Sintering Atmospheres

Sintering atmospheres primarily control chemical reactions between the materials being processed and the furnace surroundings. Additionally, atmospheres may be the source of one or more chemical elements that alloy with the material being sintered. An example of this is the addition of carbon to iron in a carburizing furnace atmosphere. Sintering atmospheres may also prevent the loss of alloying elements existing in the material being processed. Atmospheres are routinely used to flush the decomposition products of the admixed lubricants from the furnace to prevent deposit on furnace interiors, where they rapidly degrade the performance of the equipment.

In common operating practice, the most important functions of select atmospheres are to aid in the reduction of oxides on the surfaces of the metal particles in the compact and to control carburization and decarburization in iron and steel. Oxidation and decarburization of iron and iron-based compacts are caused by oxygen, water vapor, and carbon dioxide when present in improper proportions with respect to the hydrogen and carbon monoxide contents of the sintering atmosphere. Iron oxides are reduced by hydrogen, carbon monoxide, and carbon. Carburization is caused by carbon monoxide and by hydrocarbons such as methane.

The most frequently used atmospheres in commercial sintering of P/M iron and steel materials are endothermic, exothermic, dissociated ammonia, pure hydrogen, and nitrogen based. Of these, endothermic gas is the most widely used, followed by dissociated ammonia.

Endothermic and Exothermic Atmospheres

Although other hydrocarbon gases may be used, methane and propane are the most commonly used bases for the production of endothermic gas and exothermic gas atmospheres. These furnace atmospheres consist of mixtures of nitrogen, water vapor, carbon monoxide, hydrogen, carbon dioxide, and methane. The degree of combustion that occurs is controlled by the amount of air admitted to the process. This air-to-gas ratio determines the properties of the resultant gas. With methane, endothermic gas is produced with air-to-gas ratios of about 2.4, while exothermic gas ratios are usually above the 9.0+ value required for complete combustion (10-to-1 is the air-to-gas ratio for complete combustion).

For an endothermic-type gas produced from methane, the reaction can be approximated by:

$$2CH_4 + O_2 + 3.8N_2$$
$$\rightarrow 2CO + 4H_2 + 3.8N_2$$

This gas is strongly reducing to iron oxide. Control of carbon level, however, is difficult because the carbon potential of endothermic gas varies with temperature and dew point, as shown in Fig. 1. The carbon potential of the endothermic gas with -1 °C (30 °F) dew point varies from about 0.2% at 1120 °C (2050 °F) to 1.0% at about 850 °C (1560 °F). Dew point is a measure of the dryness of the atmosphere gas. The lower the dew point, the lower the water vapor content of the atmosphere. Generally, low dew point gases increase the reduction potential of the atmosphere. The relationship between dew point and water vapor content is shown in Fig. 2.

When generating exothermic-type atmospheres, the reaction is self-supporting and generates heat, which in the presence

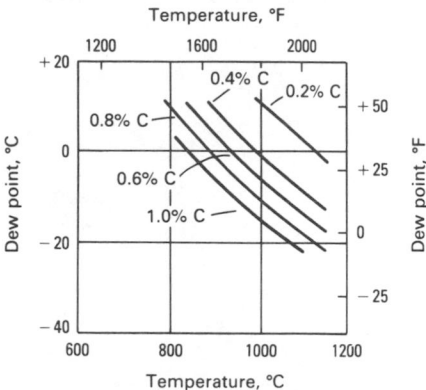

Fig. 1 Carbon potential of endothermic gas as a function of dew point and temperature

Fig. 2 Relationship between dew point and water vapor content

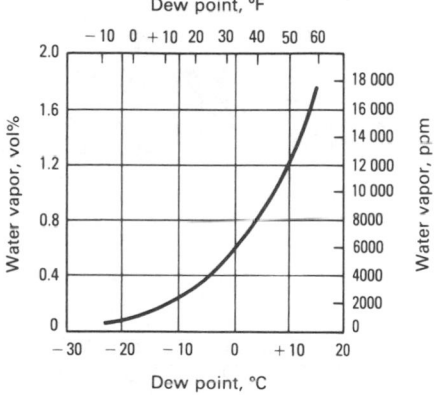

of a nickel catalyst promotes cracking of any of the hydrocarbons remaining unburned into hydrogen and carbon monoxide. The two reactions that occur are:

$$CH_4 + 2O_2 \rightarrow 2H_2O + CO_2 + heat$$
$$2CH_4 + O_2 + heat \rightarrow 4H_2 + 2CO$$

Exothermic atmospheres are not strongly reducing to iron and are decarburizing at normal sintering temperatures. Their use in sintering iron-based materials is limited to applications that do not require a residual carbon content. Removal of water and carbon dioxide can improve the properties of this protective atmosphere.

Nitrogen-Based Atmospheres

Protective atmospheres based on nitrogen obtained from the fractional distillation of liquid air are enjoying increased usage, not only in sintering, but in heat treating as well. This is primarily due to the increasing cost of hydrocarbon fuels.

Basically, nitrogen can be taken from a bulk storage facility or directly from the distillation tower and can be used as a replacement for all, or nearly all, of the carbon monoxide and/or hydrogen-containing gases previously used. Typical usage consists of utilizing nitrogen to extend either endothermic gas or dissociated ammonia through dilution. Hydrogen content may be reduced to a few percent of the total, versus a normal 40% presence in conventional endothermic gas.

Because there is less hydrogen available to reduce the oxide present on the surface of powder particles, greater demand is placed on the graphite present to take part in the reduction process. As a result, greater amounts of admixed graphite are required if the sintered carbon content is to remain the same.

Dissociated Ammonia Atmospheres

Dissociated ammonia, consisting of 75 vol% hydrogen and 25 vol% nitrogen, is produced from gasified ammonia by the reaction:

$$2NH_3 = N_2 + 3H_2$$

Dissociated ammonia directly from the dissociator is very dry, generally yielding dew points of -40 °C (-40 °F) or lower.

It is common practice to generate dissociated ammonia with higher nitrogen contents by reacting ammonia with air either catalytically, or through partial burning. The water formed in such reactions is removed from the gas by absorption.

Hydrogen Atmospheres

Typical methods used to produce metallurgical-grade hydrogen include the electrolysis of aqueous solutions and the reaction of hydrocarbon gases with steam or oxygen to form mixtures of hydrogen and carbon monoxide. Carbon monoxide is subsequently converted to carbon dioxide and scrubbed from the mixture. Hydrogen is used primarily in the sintering of iron-based magnetic materials. Alnico permanent magnets, for example, are frequently sintered in an atmosphere of very dry hydrogen at 1300 °C (2370 °F). See the article "Electrical and Magnetic Applications" in this Volume for additional information on sintering of iron-based soft and permanent magnets.

Sintering of Iron Powder

Sintering of plain iron powder sequentially involves the establishment and growth

Fig. 3 Effects of sintering temperature on porosity of iron compacts sintered in dissociated ammonia for 30 min

See text for description of microstructural changes. Etchant: 2% nital. Magnification: 300×. (a) As-pressed condition. (b) Sintered at 1065 °C (1950 °F). (c) Sintered at 1120 °C (2050 °F). (d) Sintered at 1175 °C (2150 °F)

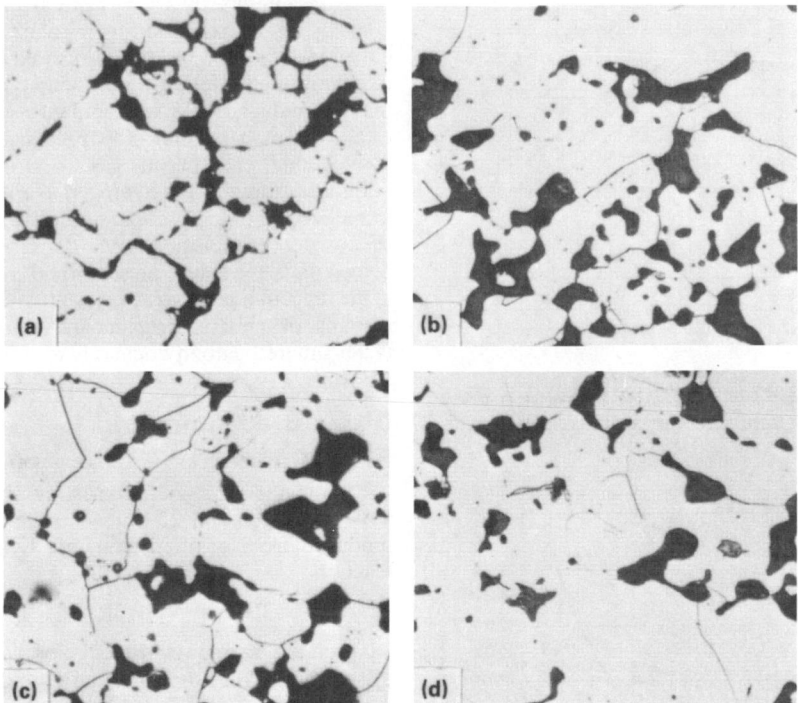

Fig. 4 Effect of combined carbon content on the tensile strength of wrought (rolled) steel

Source: Ref 1

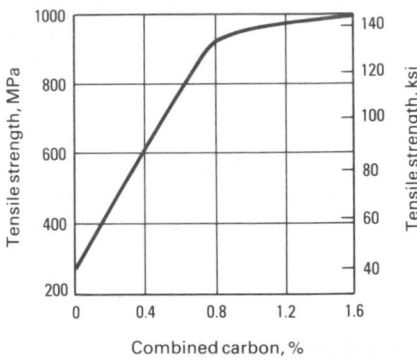

Fig. 5 Effect of combined carbon content on the transverse rupture strength of sintered steel

Test bars were pressed to a density of 6.3 g/cm^3, then sintered for 30 min at 1120 °C (2050 °F) in dissociated ammonia. Source: Ref 1

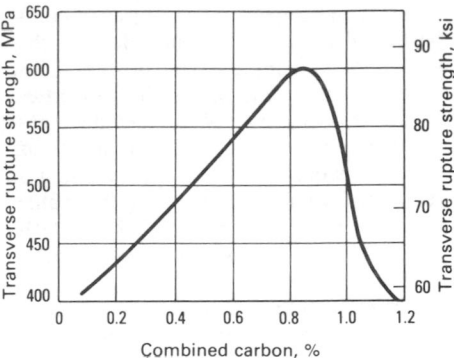

of bonds between the particles of powder at their areas of contact, grain growth and migration of the grain boundaries formed at the bonds, spheroidization of the pores between the particles, and the elimination of small pores (and, possibly, the growth of large pores). The formation of bonds is opposed by residual material from the lubricant, by impurities and surface oxides, and by poor contact.

The density and size of the green compact change during sintering. As the sintering temperature increases, porosity decreases and shrinkage increases. Changes in the pore structure of an iron powder compact due to sintering temperature are illustrated in Fig. 3. Figure 3(a) shows an iron powder mixed with 0.75% lubricant and pressed to a density of 6.5 g/cm^3. The multigranular iron particles are completely separated by porosity.

After sintering the same iron compact in dissociated ammonia for 30 min at 1065 °C (1950 °F), many bonds form between the particles during sintering, but numerous particle boundaries remain (Fig. 3b). Sintering at 1120 °C (2050 °F) further increases the number of particle bonds, as shown in Fig. 3(c). In addition, some grain growth and spheroidization of pores oc-

cur. Finally, as depicted in Fig. 3(d), only a few particle boundaries remain after sintering at 1175 °C (2150 °F), and considerable grain growth, grain boundary migration, and spheroidization have occurred.

Sintering of Iron-Graphite Powder

Mixtures of iron and graphite powders are sintered by (1) establishment and growth of iron-to-iron bonds, (2) diffusion of carbon into and combination of carbon with iron, and (3) spheroidization of pores. The oxide content of iron is an important factor in determining the reactivity of iron with graphite.

During sintering, the major portion of the oxide must be reduced by the graphite and the sintering atmosphere before the iron and graphite can combine. To ensure high reactivity with the iron for rapid reduction of oxide and fast carburization, fine graphite powder—free of silicon carbide and having low ash content—usually is used.

Although iron powder of low oxide content is not required, determination of the oxide content (hydrogen loss value) is essential, because the graphite addition

must be large enough to allow for reaction with the oxide and hydrogen loss to the atmosphere and still provide a sufficient quantity to produce the desired iron-carbon alloy.

Effect of Combined Carbon. Aside from bonding, the main factor affecting the properties of a sintered steel part is the amount of combined carbon formed in the steel. Overall expansion during sintering is directly proportional to the amount of combined carbon. As with wrought steel, the strength of sintered steel increases rapidly with increasing combined carbon content. For wrought steel, strength increases rapidly up to 1.00% combined carbon, then gradually reaches a maximum at the highest carbon content of 1.6%, as shown in Fig. 4.

A similar plot for sintered steel is shown in Fig. 5. Combined carbon was varied by adding increasing amounts of graphite. Test bars were pressed to a density of 6.3 g/

Fig. 6 Effect of combined carbon content on microstructure and transverse rupture strength of sintered iron-carbon alloys

Iron powder mixed with increasing amounts of graphite to produce various contents of combined carbon. Etchant: 4% picral plus 0.5% nitric acid. Magnification: 600×. (a) Combined carbon: trace. Strength: 407 MPa (59 ksi). (b) Combined carbon: 0.4%. Strength: 483 MPa (70 ksi). (c) Combined carbon: 0.6%. Strength: 538 MPa (78 ksi). (d) Combined carbon: 0.8%. Strength: 600 MPa (87 ksi). (e) Combined carbon: 1.0%. Strength: 503 MPa (73 ksi). (f) Combined carbon: 1.2%. Strength: 400 MPa (58 ksi)

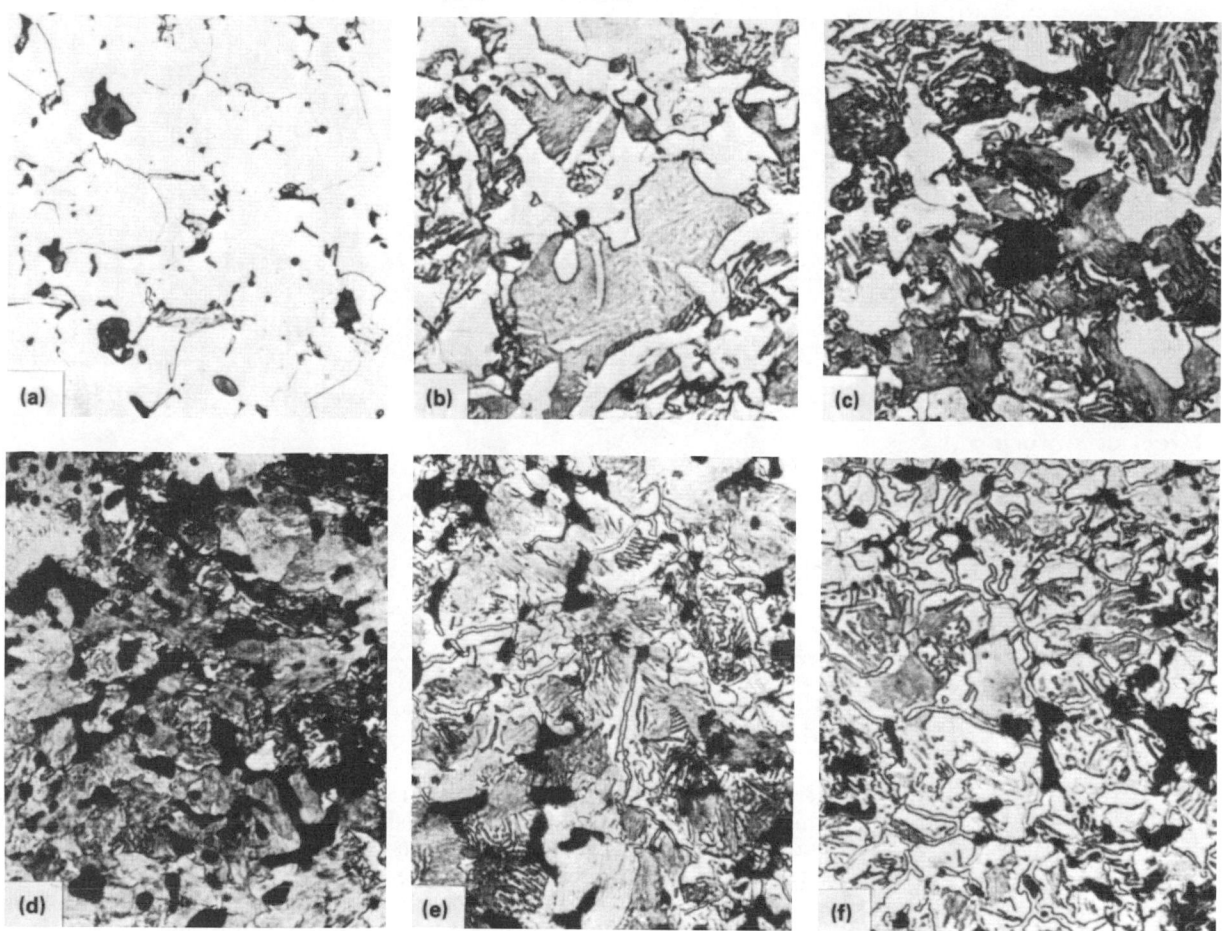

cm³, then sintered 30 min at 1120 °C (2050 °F) in dissociated ammonia. Figure 5 shows transverse rupture strength as a function of combined carbon. As with solid steel, there is a rapid initial increase in strength with increasing combined carbon content, but a maximum is reached at about the eutectoid composition of 0.85% C. Between 0.90 and 1.00%, strength drops markedly. This sharp decrease is illustrated in Fig. 6. As the amount of pearlite that contains the combined carbon increases, strength and hardness increase correspondingly until a combined carbon content of 0.8% is reached, at which point the structure consists entirely of fine pearlite (Fig. 6d). At 1.0% combined carbon, free cementite forms at the grain boundaries (Fig. 6e), which reduces the transverse rupture strength. At 1.2% combined carbon, the network of free cementite becomes con-

tinuous (Fig. 6f), further reducing the transverse rupture strength.

Thus, most sintered steel has a combined carbon content of 0.8 to 0.9%, and a graphite addition greater than 0.9%. Carbon loss is controlled by maintaining the carbon potential of the sintering atmosphere at 0.7 to 0.9%.

As the sintering temperature or time is increased, spheroidization of the pores causes the strength of a compact to continue to increase after carburization of the iron to an all-pearlite structure is complete. The resulting densification of the compact causes a reduction in overall growth of the part.

Although the influence of combined carbon on the strength of sintered steel has been illustrated, there are several factors that further determine the amount of combined carbon in addition to the type and

amount of graphite added. The furnace atmosphere must be controlled at a high enough carbon potential to prevent excessive loss of carbon to the atmosphere, because decarburization results in correspondingly poor properties.

Effect of Atmosphere Composition. The tendency of parts to oxidize or reduce and to carburize or decarburize, as well as the rates at which these reactions occur during sintering, depends on the sintering temperature and the proportion of various gases contained in the sintering atmosphere. The sintering atmosphere has a strong effect on the amount of combined carbon formed during sintering. If the carbon potential is not controlled, excessive carbon may be lost to the atmosphere, or excessive carburization may occur.

The carbon potential is determined by the ratio of water vapor to hydrogen, car-

Fig. 7 Equilibrium ratios at various temperatures for oxidation-reduction and decarburization-carburization reactions

Equilibrium reactions: R, reducing; C, carburizing; O, oxidizing; D, decarburizing. Source: Ref 1

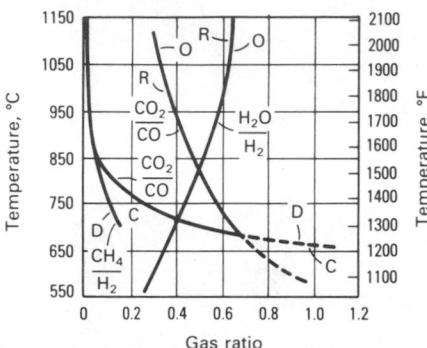

Fig. 8 Effect of sintering temperature on transverse rupture strength of iron plus 1.25% graphite test bars

Compacted to a density of 6.1 g/cm³ and sintered for 30 min at temperature. Source: Ref 1

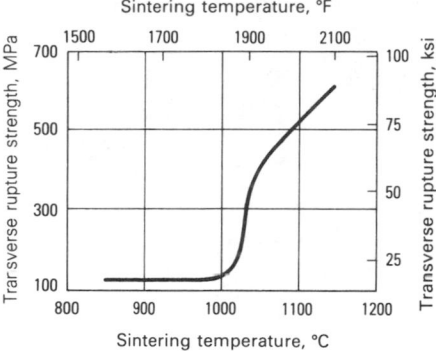

Fig. 9 Effect of sintering temperature on the microstructure and transverse rupture strength of iron plus 1.25% graphite test bars

Etchant: 4% picral plus 0.5% nitic acid. Magnification: 800×. (a) Sintering temperature: 1010 °C (1850 °F). Combined carbon: 0.1%. Strength: 138 MPa (20 ksi). (b) Sintering temperature: 1040 °C (1900 °F). Combined carbon: 0.75%. Strength: 352 MPa (51 ksi). (c) Sintering temperature: 1120 °C (2050 °F). Combined carbon: 0.75%. Strength: 552 MPa (80 ksi). (d) Sintering temperature: 1175 °C (2150 °F). Combined carbon: 0.75%. Strength: 655 MPa (95 ksi)

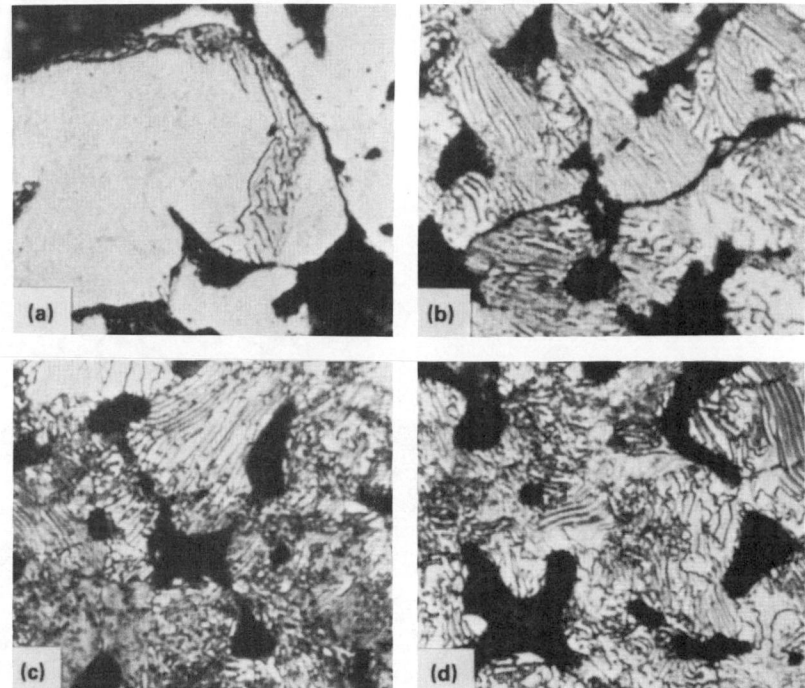

bon dioxide to carbon monoxide, and methane to hydrogen. The amount of graphite in the parts also contributes to the carbon potential in the furnace atmosphere. Figure 7 illustrates equilibrium ratios at various temperatures for oxidation-reduction and decarburization-carburization reactions. Typically, the carbon to carbon monoxide ratio can be fairly high at all temperatures without causing oxidation. However, it is necessary to keep the carbon dioxide content low to prevent decarburization.

Carbon dioxide content is kept low by maintaining a low enough dew point (water vapor content) to balance the otherwise continuously reversible "water-gas" reaction:

$$CO_2 + H_2 \rightarrow H_2O + CO$$

A dew point of −4 to −1 °C (25 to 30 °F)

at the generator is usually low enough to maintain a 0.70 to 0.90% carbon potential during continuous sintering of iron-graphite mixes. If lower combined carbon is desired, then a smaller graphite addition may be made or a higher dew point may be used to raise the carbon dioxide content.

The ratio of methane to water vapor also determines the tendency of parts to gain or lose carbon. At normal sintering temperatures, even small amounts of methane in the furnace atmosphere increase carburizing tendency. Increased amounts of methane further increase this tendency.

Atmospheres having better than equilibrium ratios are required unless long sintering times or high sintering temperatures are permitted. Also, nitrogen content should be kept as low as possible. For example, purified exothermic gas (high nitrogen content) can be used, but frequently larger graphite additions are required to obtain the desired combined carbon. Endothermic gas (low nitrogen content), however, promotes rapid carbon pickup, and less graphite is required.

Effect of Temperature. Although sintering temperature has a profound effect on the amount of combined carbon formed

for a given sintering time, graphite combines readily under normal sintering temperatures and times. Figure 8 illustrates the effect of sintering temperature on the transverse rupture strength of bars made of iron plus 1.25% graphite, pressed to a density of 6.1 g/cm³ and sintered for 30 min at temperature. No strengthening occurs as the sintering temperature is raised from 900 to 1010 °C (1650 to 1850 °F). Above this temperature, however, a substantial increase is noted.

Figure 9 illustrates the changes in microstructure. At low sintering temperatures, no carbon is combined, as evidenced by the complete absence of pearlite. At 1010 °C (1850 °F) some carbide has begun to form (Fig. 9a), while at 1040 °C (1900 °F) the maximum combined carbon content is achieved (Fig. 9b). Additional strengthening at higher temperatures is caused by increased sintering, as evidenced by the elimination of grain boundaries and spheroidization of pores—shown in Fig. 9(c) at 1120 °C (2050 °F) and Fig. 9(d) at 1175 °C (2150 °F).

Effect of Sintering Time. Time of sintering also affects the amount of combined carbon formed. At the sintering times nor-

Fig. 10 Effect of sintering time on properties of iron-graphite test bars pressed to a density of 6.1 g/cm³

(a) Transverse rupture strength. (b) Dimensional change. Source: Ref 1

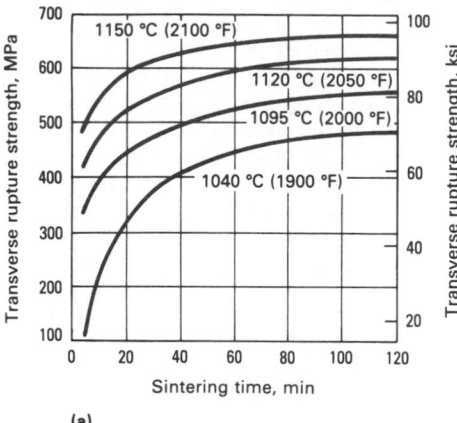

(a)

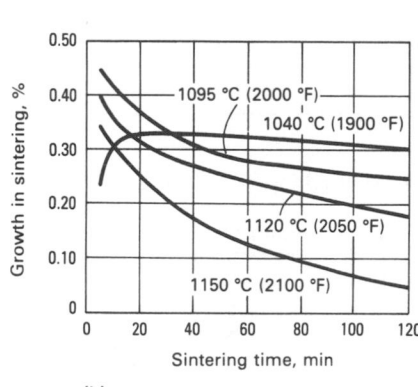

(b)

Fig. 11 Effect of sintering time on the microstructure and transverse rupture strength of iron-graphite test bars pressed to a density of 6.1 g/cm³

(a) As-pressed condition. Combined carbon: 0%. Strength: about 124 MPa (18 ksi). (b) Sintering time: 5 min at 1120 °C (2050 °F). Combined carbon: 0.7%. Strength: about 421 MPa (61 ksi). (c) Sintering time: 30 min at 1120 °C (2050 °F). Combined carbon: 0.7%. Strength: about 552 MPa (80 ksi). (d) Sintering time: 120 min at 1120 °C (2050 °F). Combined carbon: 0.7%. Strength: about 627 MPa (91 ksi)

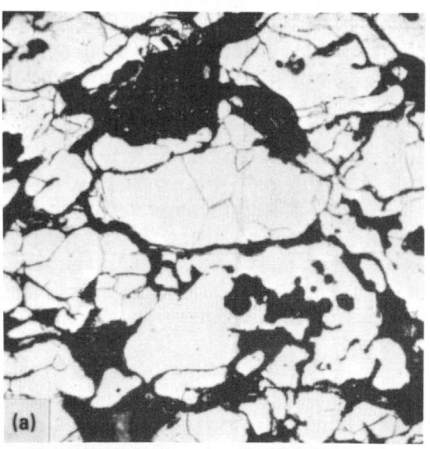

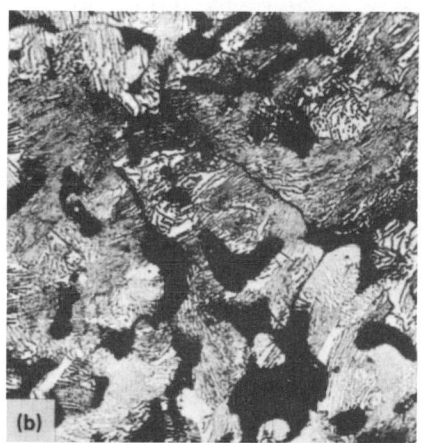

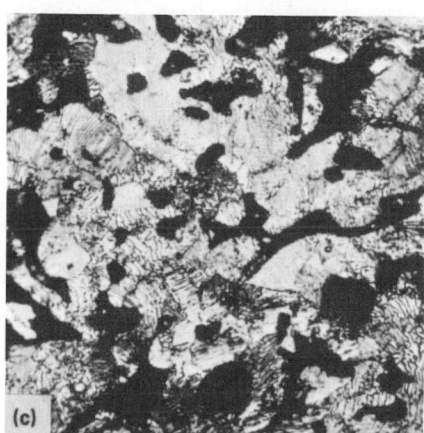

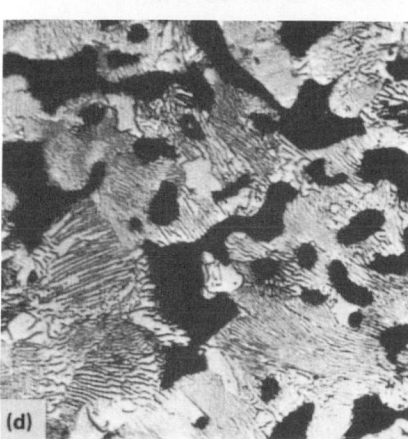

mally used, however, the maximum amount of combined carbon is usually formed. A noticeable effect on strength and dimensional change during sintering was apparent for a series of test bars sintered at several temperatures for times varying from 5 to 120 min. Figure 10(a) is a plot of sintering time versus transverse rupture strength. A rapid increase in strength occurs up to 30 min, followed by continuing increases at longer times.

Figure 10(b) is a plot of sintering time versus dimensional change for several temperatures. These are typical size change curves showing a maximum growth at relatively short times, followed by a gradual reduction at longer times. Figure 11 illustrates how microstructure is influenced by sintering time at 1120 °C (2050 °F). Figure 11(a), a microstructure of an unsintered bar, shows the individual particles, grains within particles, and graphite located primarily at the pores formed between particles. Figure 11(b) shows almost complete formation of pearlite in 5 min; numerous grain boundaries are visible, and the porosity is quite angular. Figure 11(c) shows some disappearance of grain boundaries and slight spheroidization of pores after 30 min, while Fig. 11(d) shows an almost complete absence of grain boundaries and substantial spheroidization of pores after 120 min. Sintering for 10 min at 1200 °C (2200 °F) typically results in strength equal to that attained after 120 min at 1120 °C (2050 °F).

Effect of Cooling Rate. The influence of cooling rate on strength in wrought steel is well known. Iron-graphite samples similar to those used in the study of effect of time and temperature were sintered for 30 min at 1120 °C (2050 °F), followed by cooling at different rates ranging from 730 to 535 °C (1350 to 1000 °F). The three rates used and the resulting mechanical properties are shown in Table 1. Typically, faster cooling rates cause increased strength and hardness.

Sintering of Iron-Copper Powders

The sintering of iron-copper mixtures involves (1) solid bonding of iron to iron, (2) solid bonding of copper to iron, (3) melting of copper, (4) solution and diffusion of copper in solid iron, and (5) solution and precipitation of iron in liquid copper.

At typical sintering temperatures of 1095 to 1120 °C (2000 to 2050 °F), 7.5 to 9.0% Cu is soluble in iron. However, with this copper content and typical sintering conditions, some of the molten copper re-

Table 1   Effect of cooling rate on iron-graphite samples
Pressed to a density of 6.1 g/cm³ and sintered 30 min at 1120 °C (2050 °F)

Sample No.	Cooling rate		Transverse rupture strength		Hardness, HRB	Pearlite spacing
	°C/min	°F/min	MPa	ksi		
1 . . . . .1.9		3.5	462	67	37	Very coarse
2 64		115	565	82	50	Medium
3 125		225	600	87	57	Very fine

Source: Ref 2

Fig. 12 Transverse rupture strength of iron, copper, and graphite powder compacts

Sintered to a density of 6.8 g/cm³ in endothermic gas. Lines represent compositions having the same transverse rupture strength, given in MPa with ksi equivalent values in parentheses; combined carbon in alloys is about 80% of amount of graphite included in mixture.

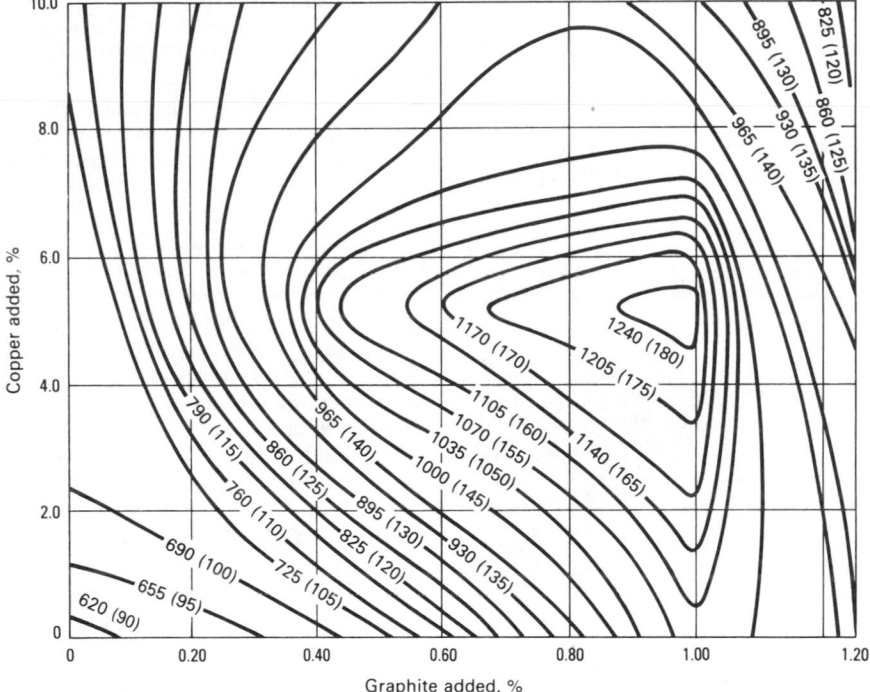

The combined effects of graphite and copper powder additions on iron compacts sintered in an endothermic atmosphere to a density of 6.8 g/cm³ are illustrated in Fig. 12. Transverse rupture strength is plotted as a function of the amounts of graphite and copper powder added to the iron powder. The amount of combined carbon in these compacts is about 80% of the graphite added. The highest value of transverse rupture strength of 1240 MPa (180 ksi) is obtained with additions of 0.9% graphite and 5% copper.

A steel of this composition is quite brittle and is used in applications that do not require superior toughness. The high strength of iron-copper-carbon compacts in the as-sintered condition is essentially due to the combined effect of copper causing precipitation strengthening and increasing the hardenability of the steel.

Sintering of Alloy Steels

Sintered alloy steels are made from three types of powders: admixed, semialloyed, and fully alloyed. Sintering of a common admixed iron-copper-nickel-graphite alloy involves (1) diffusion of carbon; (2) melting, solution, and diffusion of copper; and (3) solution of nickel in liquid copper, and solution and diffusion of nickel in solid iron.

Diffusion of nickel is comparatively rapid along the surface and along grain boundaries, but sluggish within grains. Nickel frequently diffuses incompletely, leaving nickel-rich areas of austenite. The periphery of these nickel-rich areas sometimes dissolves sufficient carbon to form martensite during normal cooling from sintering.

Semialloyed sintered steel usually contains nickel, molybdenum, and copper powders that have been partly alloyed during co-reduction of oxides. The diffusion of copper and molybdenum is relatively rapid, whereas the diffusion of nickel is slow. This leaves numerous nickel-rich areas of austenite, sometimes surrounded by high-carbon martensitic areas.

Fully alloyed sintered steel usually is made of atomized alloy steel powder. This powder commonly contains nickel, molybdenum, and manganese. Copper powder is sometimes mixed with the steel powder to limit shrinkage of the compact.

High-Temperature Sintering

Interest in high-temperature sintering of ferrous components continues to grow in

mains undissolved, or free, and can dissolve about 3% Fe.

The solubility of copper in iron decreases with decreasing temperature and is less than 0.1% at room temperature. Therefore, copper dissolved in iron at the sintering temperature must precipitate as the compact is cooled, thus hardening it. Faster cooling rates lower the temperature at which the precipitate forms, which makes the precipitate finer and increases its hardening effect.

The solution of copper in iron causes growth of the compact; the solution of iron in the free copper causes shrinkage. These processes go on simultaneously, with solution of copper predominating only in the early stages of sintering. For example, the addition of 7.5 to 10% Cu causes significant amounts of growth, but a 20% addition increases the amount of free copper to the extent that growth is no greater than for a 7.5% copper addition.

Sintering of Iron-Copper-Graphite Powder

The sintering of iron-copper-graphite mixtures involves the mechanisms associated with both the iron-graphite and iron-copper mixtures mentioned above. Diffusion of carbon usually is complete before the melting point of copper is reached. Carbon has little effect on the solubility of copper in solid iron, but it decreases the rate of solution. Thus, there is usually more free copper remaining when carbon is present, which acts to reduce the amount of growth.

Fig. 13 Temperature profile of sintering cycles

Profile A shows the high-temperature cycle used in study. Profile B shows the conventional sintering cycle used as a standard. Source: Ref 2

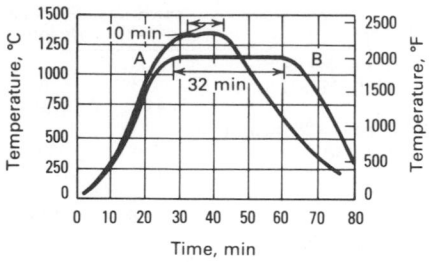

Table 2 Effect of sintering temperature on transverse rupture strength of test bars pressed to 6.8 g/cm³

Alloy	Transverse rupture strength at 1120 °C (2050 °F)		Transverse rupture strength at 1290 °C (2350 °F)	
	MPa	ksi	MPa	ksi
Fe-0.9C	693	100.5	728	105.6
Fe-4Ni-0.7C	910	132.0	924	134.0
Fe-2Cu-0.9C	975	141.5	964	139.8

Note: Transverse rupture strength of bars sintered for 32 min at 1120 °C (2050 °F) and average transverse rupture strength for all atmosphere tests of bars sintered for 10 min at 1290 °C (2350 °F)
Source: Ref 2

Table 3 Effect of sintering temperature on dimensional change of test bars pressed to 6.8 g/cm³

Alloy	Change at 1120 °C (2050 °F), %	Change at 1290 °C (2350 °F), %
Fe-0.9C	+0.24	+0.04
Fe-4Ni-0.7C	0.0	−0.2
Fe-2Cu-0.9C	+0.4	+0.4

Note: Dimensional change of bars sintered for 32 min at 1120 °C (2050 °F) and average for all atmosphere tests of bars sintered for 10 min at 1290 °C (2350 °F)
Source: Ref 2

Table 4 Effect of sintering temperature on the apparent hardness of test bars pressed to a density of 6.8 g/cm³

Alloy	Apparent hardness, HRB, at: 1120 °C (2050 °F)	1290 °C (2350 °F)
Fe-0.9C	57.9	63.2
Fe-4Ni-0.7C	73.1	75.9
Fe-2Cu-0.9C	75.6	81.3

Note: Apparent hardness of bars sintered for 32 min at 1120 °C (2050 °F) and 10 min at 1290 °C (2350 °F) in comparable atmospheres
Source: Ref 2

the P/M industry. Improvements in both production rates and properties are possible as sintering temperatures increase above 1120 °C (2050 °F). Rounding of the porosity results in increased strengths, especially impact strength. Higher diffusion rates also increase the strength and hardenability of admixed powder compositions. The greater oxide reduction that occurs at higher temperatures proves valuable in powder forging and other applications.

To demonstrate the feasibility of high-temperature sintering of ferrous materials on a production basis, a study was conducted on three iron-based compositions sintered in a nitrogen-methanol atmosphere (Ref 2). Alloys tested had the following nominal compositions: Fe-0.9C, Fe-4Ni-0.7C, and Fe-2Cu-0.9C.

Transverse rupture bars were pressed to a density of 6.8 g/cm³. Methanol was blended with nitrogen to form the following compositions: 1% carbon monoxide and 2% hydrogen, remainder nitrogen; 3% carbon monoxide and 6% hydrogen, remainder nitrogen; and 5% carbon monoxide, 10% hydrogen, remainder nitrogen. Methane additions of 0, 0.25, and 0.5% were made to each of the atmospheres. Atmosphere composition in the hot zone of the furnace was monitored by infrared analysis, gas chromatography, and dew point analysis.

Lubricant was burned off in a mesh belt furnace at 760 °C (1400 °F) for 35 min. An atmosphere of 90% nitrogen and 10% hydrogen with a dew point of −12 °C (10 °F) was used for lubricant burn-off. Sintering was performed in a pusher furnace with a ceramic muffle and wound molybdenum heating elements.

A temperature cycle was selected (Fig. 13) that held the parts above 1290 °C (2350 °F) for 7 min and above 1280 °C (2340 °F) for 10 min. The maximum temperature reached was 1301 °C (2374 °F). For comparison with conventional sintering, a set of test bars was sintered at 1123 °C

(2053 °F) for 32 min (Fig. 13). A single nitrogen-methanol atmosphere forming 5% carbon monoxide, 10% water vapor, 0.25% methane, and the remainder nitrogen was chosen for conventional temperature sintering.

The sintered bars were tested for transverse rupture strength, apparent hardness, and dimensional change. Surface and core carbon contents were determined metallographically and by combustion and thermal conductivity analysis.

Effect of Sintering Temperature. The 10-min sinter at 1290 °C (2350 °F) exhibited equivalent or slightly better properties than the longer 30-min sinter at 1120 °C (2050 °F) in comparable atmospheres. Transverse rupture strengths were very similar, with only a 1 to 1.5% variation in the Fe-Ni-C and Fe-Cu-C alloys and 5% for the Fe-C steel (Table 2).

Metallographic analysis revealed a slightly more spheroidized pore structure and increased sintering in the 1290 °C (2350 °F) specimen. Dimensional change (Table 3) was more negative by approximately 0.2% for the Fe-C and Fe-Ni-C samples sintered at 1290 °C (2350 °F). Consequently, increased particle diffusion may have occurred at the higher temperatures even though a shorter sintering time was used. No difference in dimensional change was evident for the Fe-Cu-C samples.

Apparent hardness increased slightly for all three alloys at the higher sintering temperature (Table 4). Generally, the 10-min sinter at 1290 °C (2350 °F) was found to

be an adequate substitute for a 30-min sinter at 1120 °C (2050 °F). This type of cycle may be used if an increased production rate that yields equivalent mechanical properties is desired.

Effect of Atmosphere Composition. Variations in methanol (forming carbon monoxide and hydrogen) and methane additions to the nitrogen carrier gas did not result in any significant change in transverse rupture strength for any of the alloys. This is attributed to the consistent carbon content (0.6 to 0.7%) at the core of the samples for all atmospheres tested.

Dimensional change of the alloys was unaffected by the atmosphere composition. The Fe-Cu-C and Fe-Ni-C alloys were consistent for all atmospheres tested. Some variations were evident in the Fe-C compacts, although no trend was evident.

Atmosphere composition was found to significantly affect surface carbon content and apparent hardness. Methanol enrichment without methane additions resulted in the most decarburization. Increases in methanol alone did not significantly increase the surface carbon. The surface of an Fe-0.9C sample was decarburized to a combined carbon content of 0.45 to 0.55% when sintered for 10 min at 1290 °C (2350 °F) in 15% dissociated methanol. Methane additions were required to increase the surface carbon.

Although some decarburization was still evident, acceptable results were obtained with an atmosphere containing 15% dissociated methanol (5% carbon monoxide and 10% hydrogen) with 0.5% methane additions. The dissociation of methanol and the resulting carbon dioxide and water vapor levels were found to be affected by localized temperature, residence time, and availability of local catalytic surfaces. When introduced properly, the methanol dissociation was demonstrated to be very efficient at 1290 °C (2350 °F).

Concentrations of carbon dioxide and water vapor were low, and the formation of methane was too low to be detected by gas chromatography. The atmosphere formed with nitrogen-methanol compared favorably to an equivalent atmosphere formed by blending nitrogen and endothermic gas.

These results indicate that efficient dissociation of methanol was achieved and that methane additions should be used to further reduce carbon dioxide and water vapor levels. As a result, nitrogen-methanol-methane atmospheres can be considered as viable alternatives to nitrogen-hydrogen-carbon monoxide and nitrogen-hydrogen-methane atmospheres for high-temperature sintering of ferrous alloys.

Sintering of Stainless Steel

SINTERING is the most critical step in processing stainless steel parts. During this treatment, the lubricant must be removed, and the particles must bond together. Average temperatures for sintering stainless steels range from 1120 to 1150 °C (205 to 2100 °F). When improved mechanical properties and corrosion resistance are required, however, temperatures up to 1315 °C (2400 °F) and higher are used.

Continuous mesh belt sintering furnaces are suitable to about 1150 °C (2100 °F). At higher temperatures, manual or automatic pusher, walking beam, or vacuum furnaces are used.

This section reviews sintering practices for the production of stainless steel parts in dissociated ammonia, hydrogen, and vacuum. The influence of sintering temperature, time, and cooling rate on the properties and dimensional change are emphasized.

Presintering

Green compacts usually are presintered in air or nitrogen at 425 to 540 °C (800 to 1000 °F) to volatilize and burn off the

Fig. 14 Oxidation-reduction curve of chromium oxide with hydrogen

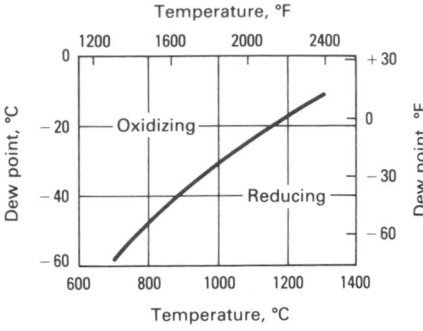

pressing lubricant. If higher temperatures are used, a protective atmosphere is required. Lubricant must be removed, because it may cause carburization of the alloy during sintering, which thus reduces machinability and corrosion resistance of the austenitic alloys. Those martensitic alloys that contain carbon for hardenability must be presintered in nitrogen.

Sintering Atmospheres

The most widely used commercial atmosphere for sintering of stainless steel is dissociated ammonia, although due to price increases, nitrogen-based atmospheres that contain as little as 3% hydrogen are being used more frequently. The principal alternative to dissociated ammonia and nitrogen-based systems is vacuum. Hydrogen atmospheres, although viable, are not used commercially to a large extent because of higher cost.

Dissociated Ammonia. When sintering stainless steels in dissociated ammonia, a dew point of −45 to −50 °C (−50 to −60 °F) is required to prevent oxidation. The oxidation-reduction curve for chromium oxide (Fig. 14) shows that lower dew points are required at lower temperatures if a reducing environment is to be maintained. To ensure that no discoloration occurs during cooling and to allow some latitude in the sintering process, dissociated ammonia is sometimes dried to a dew point of −62 °C (−80 °F) or less before being introduced into the furnace. Mechanical properties of type 316L stainless steels sintered in dissociated ammonia and hydrogen atmospheres are compared in Table 5.

Nitrogen-based atmospheres are enjoying increased usage due to the high price of ammonia. It has been demonstrated that many parts can be sintered successfully in production with as little as 3% hydrogen

Table 5 Mechanical properties of type 316L
Pressed to 6.85/cm³ and sintered for 30 min at 1120 °C (2050 °F)

Property	Sintered in dissociated ammonia	Sintered in hydrogen
Yield strength (0.2% offset), MPa (ksi)	274 (39)	183 (26)
Ultimate tensile strength, MPa (ksi)	365 (53)	288 (41.8)
Elongation in 25 mm (1 in.), %	7.0	10.9
Apparent hardness, HRB	67	47

in the atmosphere, in contrast to the 75% in dissociated ammonia.

Nitrogen content is proportional to the square root of the positive pressure of nitrogen. Thus, sintering in an atmosphere of 90% or more nitrogen results in almost twice the amount of nitrogen as that obtained in dissociated ammonia. Slow cooling results in additional nitrogen pickup, partly because of increasing nitrogen solubility down to a temperature of about 1095 °C (2000 °F) and partly because of reduced nitrogen solubility coupled with chromium nitride precipitation below that temperature.

Vacuum. The principal alternative to dissociated ammonia or nitrogen-based atmospheres is vacuum. The acceptance and wide usage of vacuum sintering is increasing largely because of energy conservation and ecological considerations. Although conventional cold walled vacuum furnaces are used, the operation is more properly described as partial pressure sintering. The vapor pressure of some elements (such as chromium) at the sintering temperature is near the pressure that can be achieved in commercial vacuum furnaces equipped only with mechanical pumping systems.

Chromium evaporates if the furnace pressure falls below its vapor pressure, and corrosion resistance is seriously reduced. Chromium content in a stainless steel can be virtually depleted in a typical vacuum sintering cycle if the vacuum level is not properly controlled.

The required control is achieved by backfilling the vacuum vessel with a suitable gas to a partial pressure above the vapor pressure of any of the elements in the alloy. A gas pressure of 27 to 67 Pa (200 to 500 µm Hg) is typical for sintering of stainless steel at 1315 °C (2400 °F). If argon is used as backfill gas, mechanical properties are similar to those obtained in hydrogen. When nitrogen is used for backfilling, sintered properties are comparable to those achieved in dissociated ammonia (see the article "P/M Stainless Steels" in this Volume for properties of vacuum sintered stainless steels using argon and nitrogen backfilling). A circulating gas quench generally is used to provide the desired high cooling rate. Gas composition should be the same as that used for maintaining a partial pressure during heating and soaking.

Hydrogen is the most strongly reducing of all commercially available sintering atmospheres. The principal disadvantage of using hydrogen is higher cost. Mechanical properties of hydrogen-sintered type 316L stainless steel are given in Table 5. Typically, the ratio of hydrogen to water vapor determines the extent of surface chromium oxide reduction during sintering and cooling.

The dew point controlling the reduction-oxidation reaction depends on temperature. As the sintering temperature increases, the dew point required for reduction is less critical, as shown in Fig. 14. Theoretically, at 1120 °C (2050 °F), a dew point of less than −22 °C (−8 °F) at the surface of the powder is required for reduction. At 1315 °C (2400 °F), a dew point of −10.6 °C (13 °F) is satisfactory.

To maintain reducing conditions in actual practice, a dew point of at least −34 to −40 °C (−30 to −40 °F) is required in the furnace. Because water vapor is formed during reduction, a sufficient flow of gas is required to continually remove this water and the water formed by the reaction between the hydrogen and the air introduced with the parts and through furnace openings.

Sintering Cycles

Tensile strength increases with sintering temperature and sintering time, while yield strength decreases with increasing temperature and increases with increasing time. After surface oxides have been reduced in the initial stage of sintering, the particles bond together by solid-state diffusion. This is followed by a gradual increase in the amount of bonding and an increase in grain size. The two phenomena occur simultaneously, but their rate of growth is influenced differently by sintering time and temperature.

The growth of the bond area depends on both time and temperature, while grain size depends primarily on temperature. As the bond areas increase, yield and tensile strengths increase because of an increase in actual cross sectional area of the bonds.

Fig. 15 Effect of sintering temperature on tensile and yield strengths and apparent hardness of type 316L stainless steel
Pressed to 6.85 g/cm³ and sintered for 30 min in various atmospheres

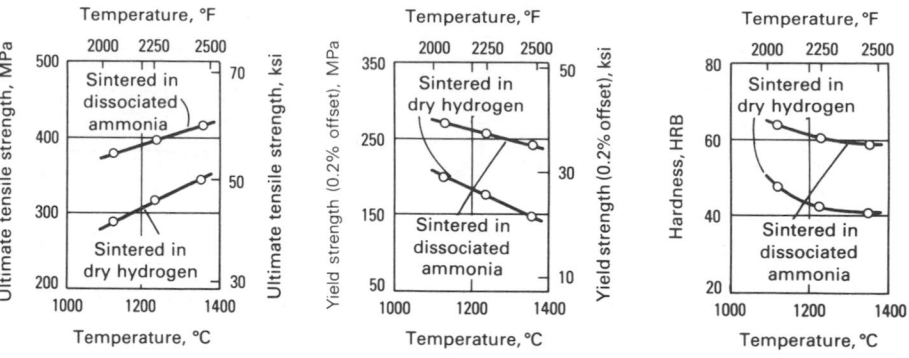

Fig. 16 Effect of sintering temperature on elongation and dimensional change during sintering of type 316L stainless steel
Pressed to 6.85 g/cm³ and sintered for 30 min in various atmospheres

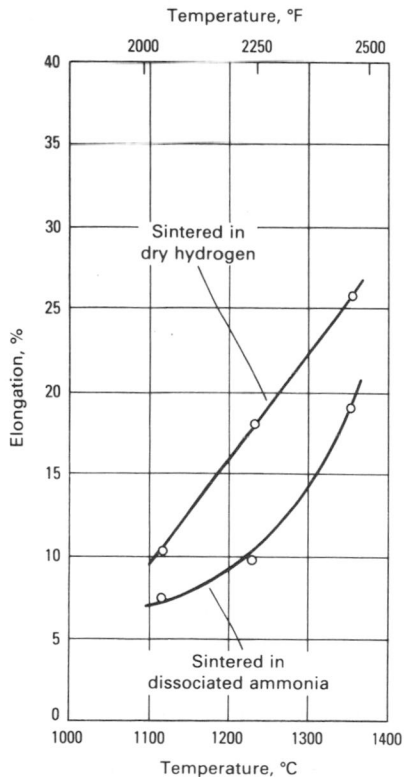
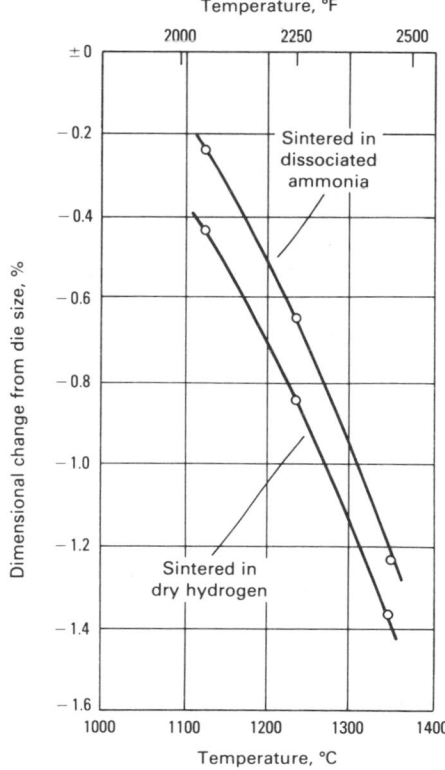

However, for a given material, yield strength decreases as grain size increases. These two effects combine to increase tensile strength of sintered material with increased sintering time and temperature and to increase yield strength with time, but to decrease it with increased temperature.

Figures 15 and 16 show the effect of sintering temperature and atmosphere on the mechanical properties and the dimensional change (calculated from die size) of type 316L stainless steel. The decrease in yield strength and increase in ductility are readily apparent. A high degree of shrinkage is apparent when sintering at higher temperatures, especially in hydrogen. Lengthening the sintering time increases all tensile properties and the amount of

Fig. 17 Effect of sintering time on tensile and yield strengths of type 316L stainless steel

Pressed to 6.85 g/cm³ and sintered at various temperatures in dissociated ammonia atmosphere

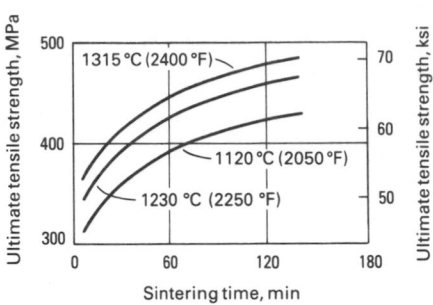 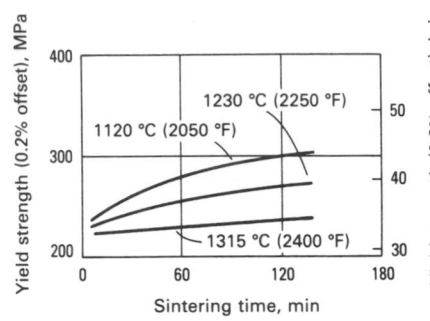

Fig. 18 Effect of sintering time on elongation and dimensional change during sintering of type 316L stainless steel

Pressed to a density of 6.85 g/cm³ and sintered at various temperatures in dissociated ammonia atmosphere

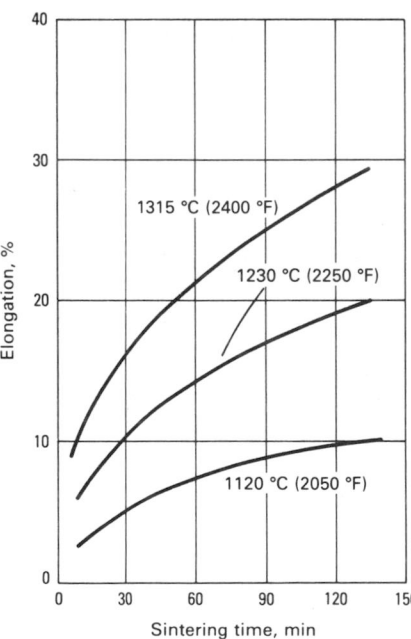 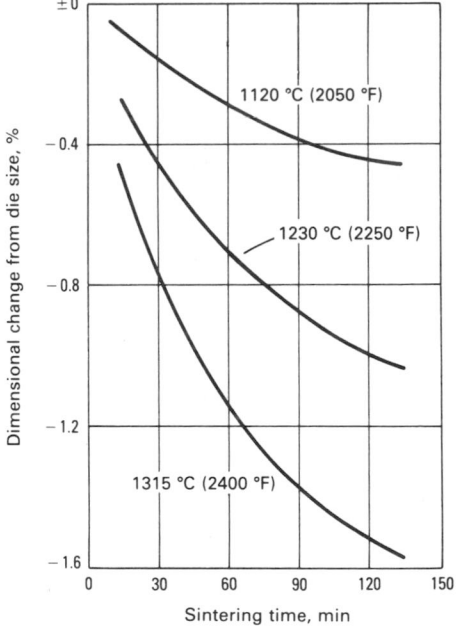

shrinkage. Figures 17 and 18 illustrate the influence of sintering time on type 316L stainless steel in dissociated ammonia. Sintering in hydrogen or vacuum produces similar curves.

Cooling Rates

Regardless of the sintering atmosphere, the rate of cooling from the sintering temperature has a significant effect on the properties of sintered austenitic stainless steels. The rates of cooling normally used provide the best combination of strength and ductility. All stainless steels contain some carbon, and when dissociated ammonia is used, nitrogen is also present. The amount of these elements normally present is beyond the limits of room-temperature solubility. On cooling, carbides and nitrides tend to precipitate in the form of fine particles. With extremely slow cooling, such as furnace cooling, precipitation occurs preferentially at grain boundaries. As a result, there is a noticeable increase in strength and loss of ductility, even though the amount of the precipitate is minimal.

However, carbide precipitation is detrimental to machinability. With the normal rates of cooling (10 to 30 min) used in typical P/M production, less precipitation occurs at grain boundaries. Consequently, strength is lower, but ductility is higher. Rapid cooling, such as obtained by water quenching, suppresses the precipitation of carbides and nitrides, which produces maximum ductility.

The cooling rate effect in vacuum sintering is largely controlled by the atmosphere in which the parts are cooled. If a nitrogen backfill is used, properties are similar to those obtained by sintering in dissociated ammonia. If argon is used, properties resemble those obtained when sintering in hydrogen.

Control of Carbon Content

Austenitic grades of stainless steel powders are made with less than 0.03% C. To retain this low carbon content after sintering, which ensures maximum corrosion resistance, weldability, and machinability, carbon pickup must be avoided.

The solid lubricant used for pressing must be completely removed by presintering, because stearates and waxes are a source of carbon. Frequently, small loads can be sintered directly by using excessive flow of atmosphere to flush away the volatilized lubricant. Generally, two-step sintering is the most efficient practice.

Other common sources of carbon include soot, residual lubricant in the furnace, and traces of previous furnace atmosphere. Dissociated ammonia or hydrogen atmospheres develop a carburizing potential when small amounts of residual carbonaceous materials are present in the furnace. Care must be taken to ensure that the furnace is clean and that the atmosphere is pure.

Sintering of High-Speed Steels and Tool Steels

By Mark Svilar
Metallurgist
SCM Metal Products

POWDER METALLURGY processing offers several advantages to costly and highly alloyed tool steel materials. These advantages include uniform and finer microstructure, improved grindability (Ref 3), improved cutting performance (Ref 4), and capabilities of high-speed steels and tool steel alloys that cannot be made by conventional ingot metallurgy.

The use of conventional press-and-sinter technology offers the additional advantage of net shape or near-net shape capability. Cutting tools, bearings, and wear parts are being produced commercially by fully dense sintering. Wear parts also are produced by conventional P/M techniques to densities of 80 to 90%.

Current commercial fully dense sintering uses high green strength and compressible water-atomized tool steel powders that are compacted in rigid dies using uniaxial pressing or in flexible rubber molds using cold isostatic pressing to make green tools and parts. These parts are then sintered in a microprocessor-controlled vacuum furnace near the solidus temperature of the alloy to virtually full density (at least 98% and frequently 99+% of theoretical). The flexibility of this process is illustrated in the pressed and sintered parts shown in Fig. 19. Additional information on pressed and sintered tool steel powders can be found in the article "P/M Tool Steels" in this Volume.

Generally, parts are pressed from 70 to 85% of theoretical density before sintering to full density. Pressing to lower green densities tends to require longer times at temperature to obtain full density and results in coarser microstructures. Figure 20 shows microstructures of fully dense sintered parts that were pressed at 207 MPa (15 tsi) and 827 MPa (60 tsi), respectively.

Pressing at higher pressures, which is required for increased green densities, results in increased tool wear and breakage. Parts with high green density also may require extra care in sintering. The surface can sinter rapidly to high density and entrap gases from the center of the part.

Sintering Mechanisms

Important contributions to sintering arise from diffusion and viscous flow. Diffusion rates increase with increasing temperature because of the increased number of vacancies that promote diffusion of substantial alloying elements. Sintering temperature must be held very close to the solidus temperature to attain full density in a reasonable time. A sintering temperature 5.5 to 11 °C (10 to 20 °F) above the solidus reportedly forms a small amount of liquid, which allows high diffusion rates for enhanced sintering. These results are based on relatively rapid densification at high temperatures. However, metallo-

Fig. 19 Typical parts pressed from high-speed steel powders

Fig. 20 Sintered microstructures of T15 specimens
Pressed and sintered to a density of 8.2 g/cm³. (a) Pressed at 207 MPa (15 tsi). (b) Pressed at 827 MPa (60 tsi). Note: The microstructure of (a) is much coarser than that of (b) due to the higher temperatures and/or longer sintering times needed to achieve full density. Magnification: 275×

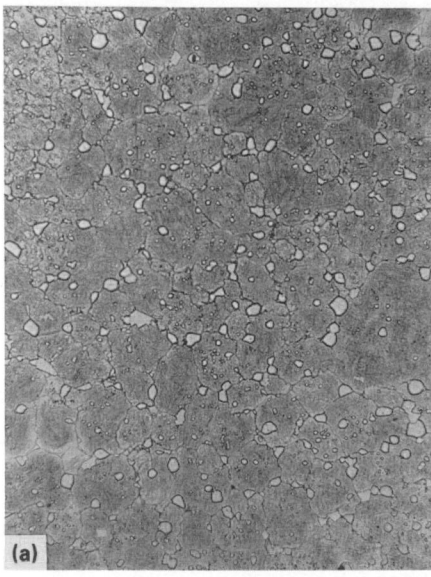

graphically, there is no resemblance between a successfully sintered high-speed steel and a typical liquid-phase sintered material, such as tungsten carbide or tungsten heavy metal (Ref 5). Additional research is required to fully understand high-temperature sintering.

Factors Affecting Sintering

Sintering is a series of complex processes, of which densification is only one phase. As green parts are heated to the sintering temperature, gases can be ad-

Fig. 21 Relationship among sintered density, sintering time, and sintering temperature

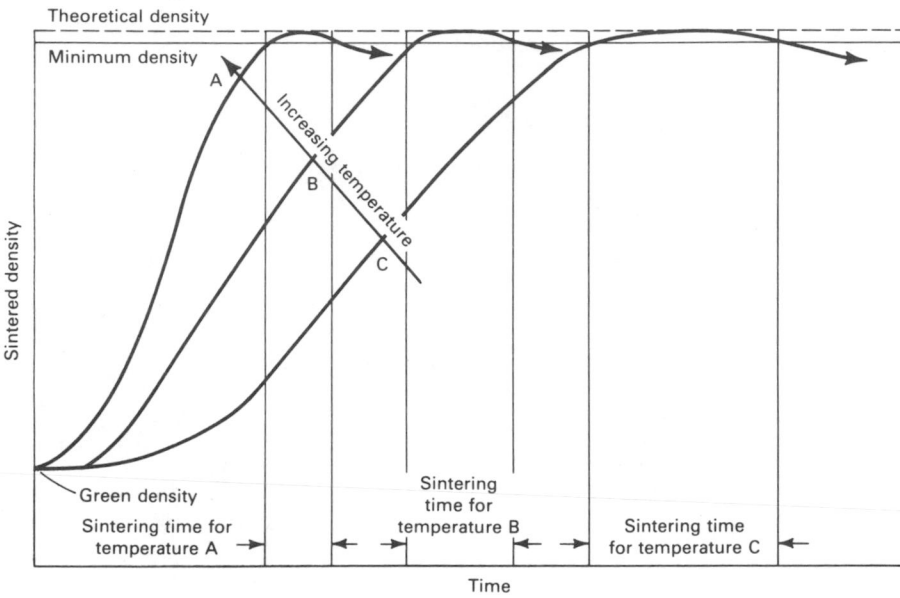

Fig. 22 Sintering curves for compacted M2 high-speed steel
Sintering time for all materials is 1 h.

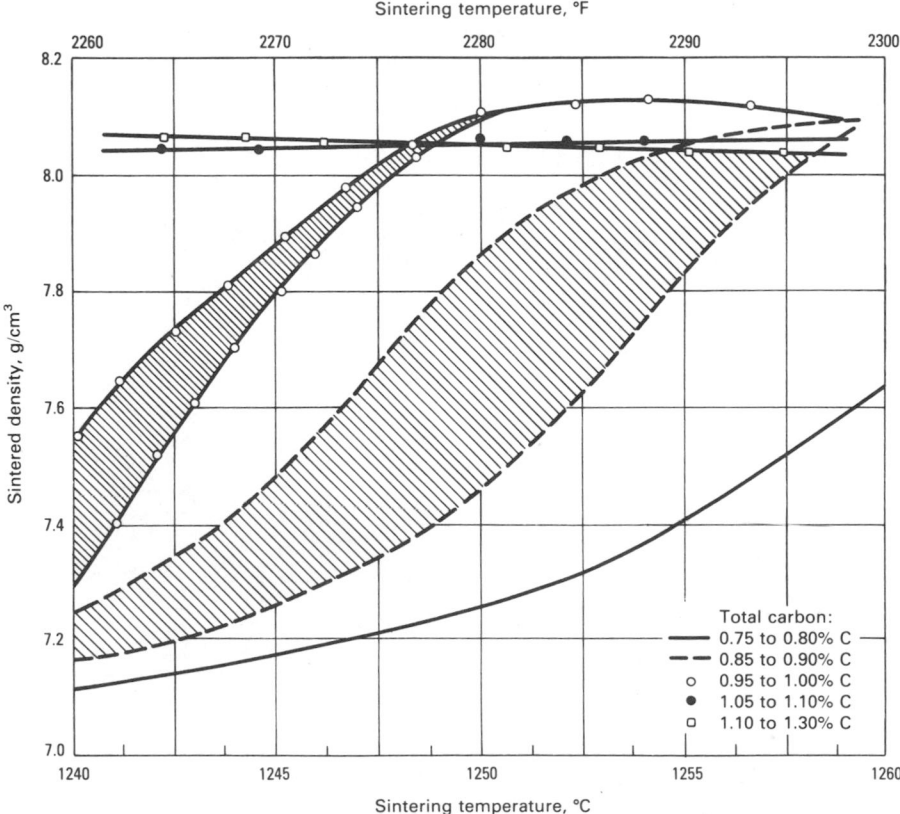

Total carbon:
— 0.75 to 0.80% C
-- 0.85 to 0.90% C
○ 0.95 to 1.00% C
● 1.05 to 1.10% C
□ 1.10 to 1.30% C

sorbed (nitrogen, hydrogen, or oxygen) or evolved (nitrogen, hydrogen, or carbon monoxide). Admixed carbon dissolves and homogenizes, while carbides dissolve and

grow. Grain growth also occurs as pores shrink and virtually disappear.

Sintering Time and Temperature. Increasing the sintering temperature de-

creases the amount of time required to achieve full density. Higher temperatures also reduce the time between reaching full density and oversintering. These relationships are shown schematically in Fig. 21. A sintering curve for M2 high-speed steel is shown in Fig. 22. Table 6 provides typical properties of M2 tool steel for various sintering temperatures.

Effect of Carbon Content. Carbon content has a significant effect on sintering temperature. Increasing carbon content reduces the sintering temperature by lowering the solidus temperature.

Carbon can be controlled closely by blending graphite or lampblack into the powder before pressing. Carbon dissolves rapidly, and its high diffusivity enables rapid homogenization above 980 °C (1800 °F). However, blending of more than 0.15 to 0.2% C may cause several detrimental effects, such as nonuniform distribution of carbon, variable response to sintering, variable part-to-part density, and part distortion.

More than 0.15% C may be admixed with 0.05 to 0.30% blending oil, such as mineral oil. However, blending oils may create powder flow difficulties and may lead to excessive outgassing during vacuum sintering.

The effect of carbon content on microstructure is shown in Fig. 23 for M2 high-speed tool steel. The amount of mixed carbon critically affects the microstructure. For a given sintering cycle, insufficient carbon produces incomplete sintering, and excessive carbon produces oversintering. The effect of carbon additions for five heats of M2 tool steel and three sintering temperatures are given in Table 6.

Effect of Oxygen Content. Oxygen affects the sintering of tool steels by reducing carbon content during sintering. Oxygen from the annealed powder reacts with carbon during sintering to form carbon monoxide. About 0.01% C is lost in the atmosphere for every 0.01% of oxygen present in the annealed powder. This reaction reduces the oxygen to less than 200 ppm during vacuum sintering.

Effect of Silicon Content. Silicon aids sintering by suppressing the melting point of tool steels. A carbon equivalent (CE) for estimating sintering temperature can be calculated as:

$$CE = \%C \text{ (powder)} + \%C \text{ (graphite)} + \frac{\%Si}{30}$$

Admixed silicon does not homogenize rapidly because of its low diffusivity; con-

Table 6   Sintering data for M2 tool steel
Testing was conducted on five lots. All compacts were pressed at 827 MPa (60 tsi) and sintered for 60 min.

Total carbon(a), %	Added graphite, %	At 1240 °C (2260 °F) Density, g/cm³	Characteristics	At 1250 °C (2280 °F) Density, g/cm³	Characteristics	At 1260 °C (2300 °F) Density, g/cm³	Characteristics
Lot 1							
0.77	0.0	7.19	MLP, FG	7.29	FG, SP	7.47	MSP, VFG
0.87	0.1	7.25	VFG, LP	8.03	FP, LP	8.04	FP, SP
0.97	0.2	7.48	VFG, LP	8.11	FP	8.09	FP
1.07	0.3	8.01	FG	8.09	FP, SP	8.07	E, LP
Lot 2							
0.79	0.0	7.00	VFG, MP	7.25	FG, MP	7.77	MP
0.89	0.1	7.18	VFG, MP	7.76	MP	8.08	FP
0.99	0.2	7.58	MP, VFG	8.08	FP	8.06	FP, LG
1.09	0.3	8.07	VFP	8.00	LG, SP	8.05	E, LP, FP
Lot 3							
0.86	0.0	7.17	VFG, MP	7.47	MP	8.09	FP
0.96	0.1	7.29	VFG, MP	8.08	MLP	8.08	FP, LP
1.06	0.2	7.84	VFG, MP	8.09	SP	8.06	E, LP, LG
1.16	0.3	8.07	FG, VFP	8.07	SP, E	8.07	E, LG
Lot 4							
0.87	0.0	7.22	VFG, MP	7.87	MP	8.09	FP, LP
0.97	0.1	7.47	FG, MP	8.08	FP, SP	8.07	SP
1.07	0.2	8.07	FP	8.06	FP, SP	8.06	E, MSP
1.17	0.3	8.08	FP	8.06	E, SP	8.05	E, MSP
Lot 5							
1.00	0.0	7.82	VFG	8.09	FP, SP	8.06	E, FP, LG
1.10	0.1	8.08	FG	8.07	FP, SP	8.05	E, FP, LG
1.20	0.2	8.07	VFP	8.06	LG, E	8.05	E, FP, LG
1.30	0.3	8.06	LG	8.06	LG, E	8.04	E, FP, LP

Note: M, many; V, very; FG, fine grain; LG, large grain; SP, small pores; LP, large pores; FP, few pores; MP, many pores; E, eutectic
(a) wt% C in powder plus wt% admixed graphite; does not account for carbon loss during sintering due to deoxidation

sequently, it usually causes incipient melting.

Effect of Boron Content. Boron forms a low-melting-point eutectic with steels. Only small amounts can be added without changing carbide morphology, as shown in Fig. 24. Boron can be added to powder as an alcoholic solution of boron oxide, followed by drying at about 95 °C (200 °F). This results in an even coating of the particle with boron oxide (Ref 6). The treated powder thus sinters uniformly.

Effect of Carbide Formers. The amount of carbon required for sintering and heat treating increases with the concentration of tungsten, molybdenum, and vanadium. Tungsten, molybdenum, and vanadium combine with carbon to form carbides that deplete the matrix of carbon. Vanadium is especially potent, because the vanadium-rich MC carbide is very stable and resists dissolution at high sintering temperatures.

The stability of the vanadium carbide effectively retards grain growth. Generally, increasing vanadium content (when properly balanced with carbon) facilitates sintering. High-speed steels containing 3% or more vanadium (M3 type 2 and T15) can be sintered to full density with microstructures finer than wrought.

In high-speed steels, chromium content does not affect sintering of alloys due to the diminished stability of chromium carbide compared to molybdenum, tungsten, and/or vanadium carbides. Chromium carbide is formed only in annealed material. Chromium content affects the sintering of alloys that contain only minor amounts of vanadium, molybdenum, or tungsten, such as the high-chromium die steels.

Vacuum Sintering to Full Density

The prime object of vacuum sintering to full density is to uniformly expose all parts to sufficient temperature and time. This is typically done to produce the minimum required density and to prevent overheating and/or excessive carbide growth.

Sintering Cycles. Typical sintering cycles are shown in Fig. 25. These sintering cycles consist of a lower temperature hold to compacts, followed by one or more high-temperature holds to densify compacts.

Deoxidation. Heating rates to the deoxidation temperatures are not critical. Parts should be heated at the maximum practical heating rate to minimize furnace time, provided the temperature of the load does not greatly surpass the deoxidation soak temperature. Batch-type vacuum furnaces can be backfilled with inert gas or hydrogen to above 46 kPa (350 torr) to improve the heating rate and temperature uniformity of the load.

The primary function of the deoxidation step is to react oxygen with carbon to form carbon monoxide, which is removed by the vacuum pumps. Oxygen, nitrogen, and other gases that may be present in the powder compact must be removed before sintering closes the interconnected porosity to the surface. Once the porosity is no longer interconnected, gas evolution results in blistering.

The deoxidation soak or hold also provides an opportunity for load temperature to become uniform. Temperature in any part of the load should not vary more than ±9 °C (±16 °F) before heating to the sintering soak. The load may be soaked at 1040 °C (1900 °F) for several hours to improve temperature uniformity without seriously affecting the microstructure.

Heating to sintering temperature is slow and closely controlled to ensure temperature uniformity. Typical ramp rates are 0.5 to 5.5 °C/min (1 to 10 °F/min).

Sintering Soak. Sintering treatments may consist of an isothermal soak or a very slow ramp (typically 3 to 33 °C/h, or 5 to 60 °F/h). Temperature uniformity is essential for successful sintering. Temperature range should not vary by more than 8 °C (15 °F) at any location within the load. Typical sintering temperatures for two-soak sintering cycles are given in Table 7 for several alloys.

Cooling. Generally, cooling from the sintering temperature is performed as rapidly as possible by backfilling the furnace with inert gas and using forced convection cooling (also known as fan cooling or gas quenching) to minimize furnace time. Typically, the hardnesses of gas-quenched fully dense tool and high-speed steels are 50 to 60 HRC. These hardnesses are substantially lower than the austenitized and quenched hardnesses of 55 to 65 HRC because of excessive dissolution of carbides, which results in excessive retained austenite.

Fig. 23 Effect of increasing carbon content on the sintered microstructure of M2

All samples were sintered in the same run. Carbon was increased by blending in graphite. Note the microstructure changes from being undersintered (round pores) for 0.0 and 0.1% C to being oversintered (sharp angular pores, eutectic) for 0.2 and 0.3% C. Magnification: 250×

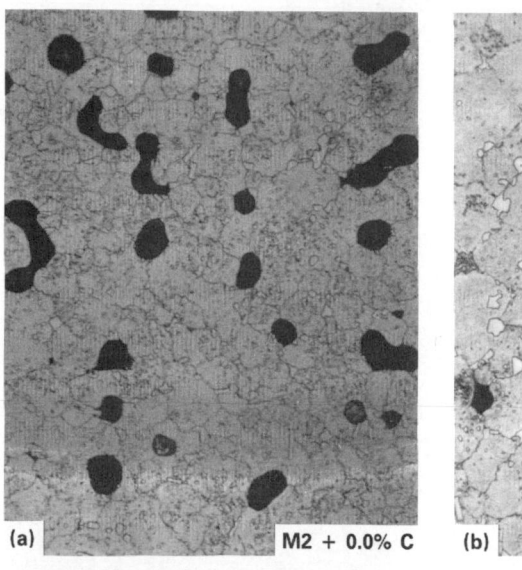
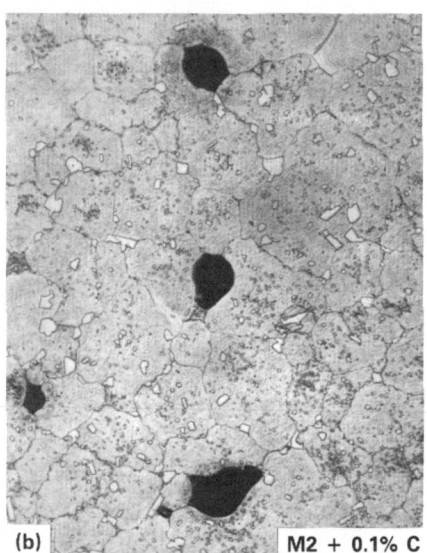

(a)   M2 + 0.0% C   (b)   M2 + 0.1% C

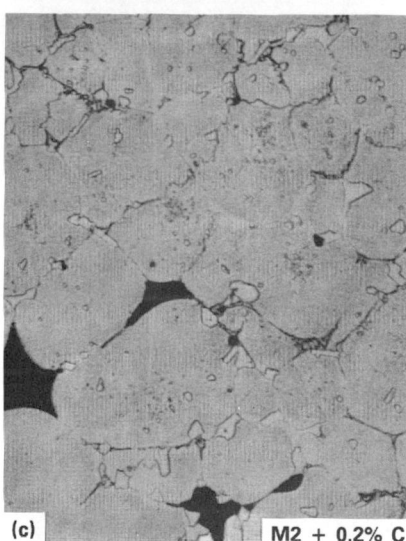
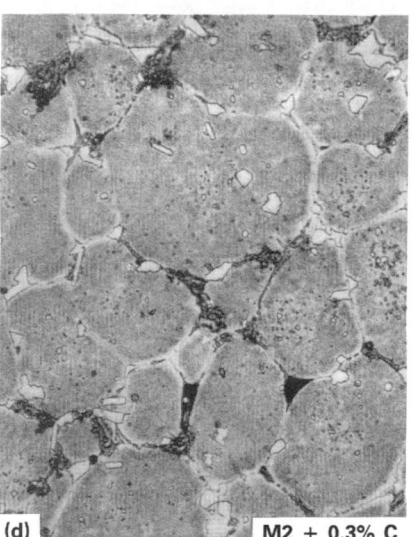

(c)   M2 + 0.2% C   (d)   M2 + 0.3% C

Fig. 24 Effect of 0.05% boron on carbide morphology of sintered M2 steel

Magnification: 250×

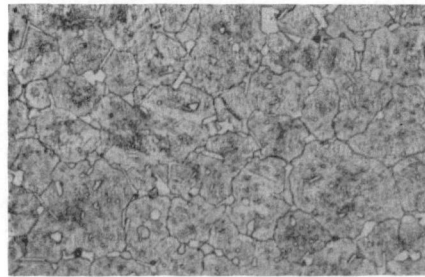

the quenching gas during forced convection cooling does not result in any significant nitriding.

Heat Treatment

Heat treatment of sintered high-speed steels is similar to the heat treatment of wrought counterparts. In both cases, heat treatment consists of hardening an annealed structure, followed by tempering to achieve desired properties. The finer microstructure of P/M high-speed steels may require slightly lower temperatures than wrought components to optimize performance.

Annealing. Sintered parts should be annealed before austenitizing. This treatment provides grain refinement and transforms large amounts of retained austenite. Sintering temperatures and times are higher and longer than optimum austenitizing heat treatment times. Consequently, parts quenched from the sintering temperature contain large amounts of retained austenite, which lower material properties. A suitable annealing cycle for sintered parts is heating to 900 °C (1650 °F) for 4 h and cooling at a rate of 50 °C/h (90 °F/h) to 500 °C (930 °F), followed by rapid cooling to ambient temperature (Ref 7).

Austenitizing. The austenitizing temperature of high-speed steels is influenced by the exact composition of the alloy (particularly carbon), as well as the carbide size. Fine carbides dissolve more rapidly than coarse carbides; consequently, parts with fine carbides should be heat treated at a lower austenitizing temperature. Table 8 gives hardness values, hardening and tempering temperatures, and the average Snyder-Graaf intercept grain size for several high-speed steels.

The Snyder-Graaf method for determining intercept grain size is based on an actual count of the grains. In the as-quenched condition, the grain boundaries of high-alloy tool steels are clearly revealed by deep etching in nital. The test method is conducted on a metallograph, with the structure shown on a ground-glass screen at a magnification of 1000×. A 127-mm (5-in.) line drawn on the ground glass represents a length of 127 μm (0.005 in.) on the sample. The number of grains crossed or touched by this 127-mm (5-in.) line is counted; the average of ten readings at random points on the sample gives the intercept grain size.

Tempering temperatures of P/M parts are similar to those used for wrought materials. Depending on property requirements, temperatures range from 540 to 595 °C (1000 to 1100 °F). Tempering curves

Nitriding. Tool steels and high-speed steels can be alloyed with nitrogen by maintaining a nitrogen partial pressure after deoxidation, but before sintering. Significant alloying does occur with partial pressure above 133 Pa (1 torr). Increasing the partial pressure increases the nitrogen content. Depending on the alloy, nitrogen contents from 4000 ppm (M2) to over 8000 ppm (T15) can be produced by backfilling to atmospheric pressure.

Nitrided cases can be produced by introducing a suitable nitrogen partial pressure after sintering has closed off the interconnected porosity. Use of nitrogen as

Fig. 25 Sintering cycles for fully dense sintering of high-speed steels
(a) British Patent 1 562 788. (b) U.S. Patent 4 063 940

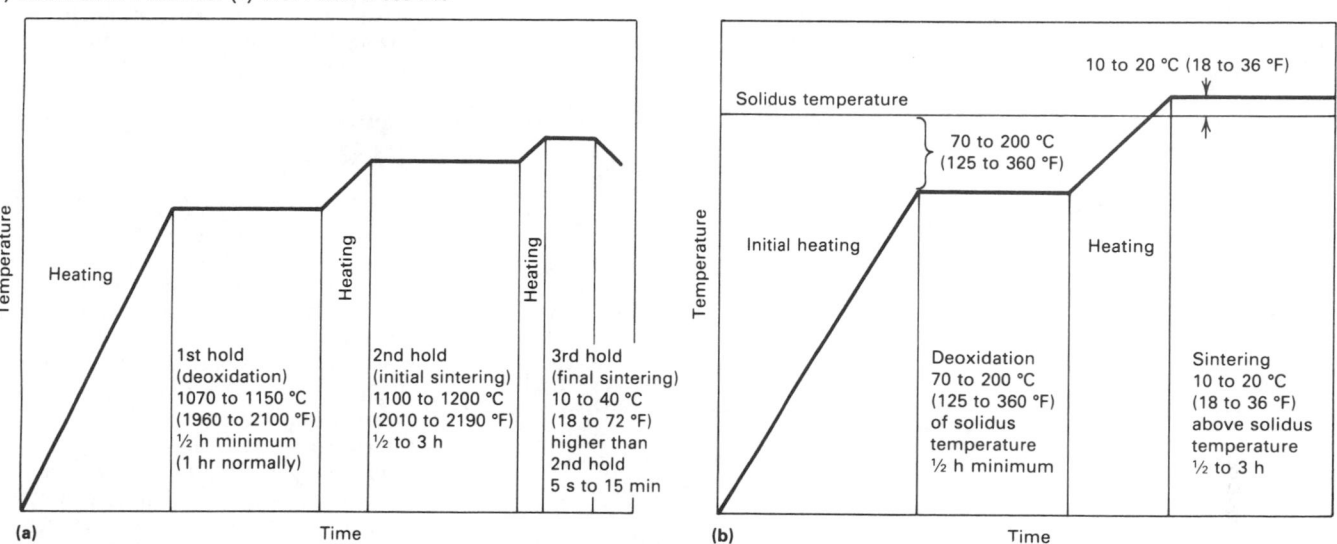

Table 7 Typical sintering temperatures and compositions for several high-speed steels

Alloy	Sintering temperature(a) °C	°F	C	Cr	Composition, % Mo	W	V	Co	Relative sinterability
M2	1245	2270	0.85	4.2	5.0	6.3	1.9	...	3
M2 (high carbon)	1240	2260	1.00	4.2	5.0	6.3	1.9	...	3
M3 type 2	1255	2290	1.20	4.1	5.0	6.0	3.0	...	2
M4	1260	2300	1.32	4.5	4.5	5.5	4.0	...	3
M35	1225	2240	1.15	4.2	5.1	6.4	2.0	5.0	4
M42	1220	2230	1.10	3.8	9.5	1.5	1.2	8.0	5
T15	1265	2310	1.60	4.4	...	12.5	4.8	5.0	1

(a) Approximate sintering temperature for powders annealed at atmospheric pressure and pressed at 830 MPa (60 tsi). (b) 1 represents the easiest, ranging up to 5, which is the most difficult

Table 8 Recommended conditions for salt bath hardening of sintered high-speed steels
Based on 20-g (0.7-oz) specimen size

Alloy	Required hardness HRC	DPH	Hardening temperature(a) °C	°F	Tempering temperature(b) °C	°F	Intercept grain size
M2	63-64	790-815	1170	2138	570	1058	...
	64-65	815-840	1180	2156	560	1040	...
	65-66	840-870	1200	2192	550	1022	...
M3 type 2	64-65	815-840	1170	2138	570	1058	15
	65-66	840-870	1180	2156	570	1058	15
	66-67	870-905	1200	2192	560	1040	15
	67-68	905-940	1220	2228	550	1022	14
M4	63-64	790-815	1180	2156	570	1058	10.5
	64-65	815-840	1200	2192	560	1040	8
	65-66	840-870	1200	2192	540	1004	8
M15	63-64	790-815	1180	2156	560	1040	17
	64-65	815-840	1200	2192	580	1076	17
	65-66	840-870	1200	2192	570	1058	17
	66-67	870-905	1210	2210	560	1040	15
M35	63-64	790-815	1180	2156	580	1076	12
	64-65	811-840	1180	2156	560	1040	12
	65-66	840-870	1200	2192	560	1040	12
T15	65-66	840-870	1170	2138	570	1058	17
	66-67	870-905	1180	2156	565	1049	17
	67-68	905-940	1200	2192	550	1022	16
	68-69	940-980	1220	2228	520	968	13.5
T42	66-67	870-905	1160	2120	570	1058	15
	67-68	905-940	1180	2156	570	1058	15
	68-69	940-980	1210	2210	560	1040	14

(a) All samples were hardened for 150-s soak time. (b) All samples were triple tempered × 3 × 1 h.

for M2 and T15 are given in Fig. 26. Heat treated properties of sintered M2, M35, and T15 are given in Table 9.

Sintered Microstructures

Sintered Versus Wrought. Fully dense sintered tool steel and high-speed steel metallurgy and microstructures are similar in most respects to wrought counterparts. Lower significant differences in carbide size and uniformity of carbide distribution may have significant consequences. A schematic of the microstructure of an as-sintered tool steel is shown in Fig. 27. Typical microstructures for as-sintered M2 and T15 are compared to wrought alloys in Fig. 28. Sintered T15 is capable of finer carbides and grain size than the wrought alloy. Sintered M2 exhibits carbide and grain sizes that generally are not as fine as sintered T15 and are frequently coarser than wrought M2. T15 contains a large amount of vanadium-rich carbides (MC) that are very stable and do not coarsen readily. These carbides also inhibit grain growth during sintering.

Photomicrographs of as-sintered and wrought D2 steel are shown in Fig. 29. The sintered D2 microstructure exhibits somewhat finer carbides than wrought D2. Although not apparent in Fig. 29, sintered D2 has a very uniform dispersion of carbide.

Wrought D2, like wrought T15, frequently has a segregated or "banded" structure in which large carbides group together. Improved properties of P/M tool steels are attributed to their finer and more uniform microstructure.

Fig. 26 Tempering curves for high-speed steels at varying hardening temperatures

(a) Mean hardening response of sintered M2. (b) Mean hardening response of sintered T15. Source: Ref 7

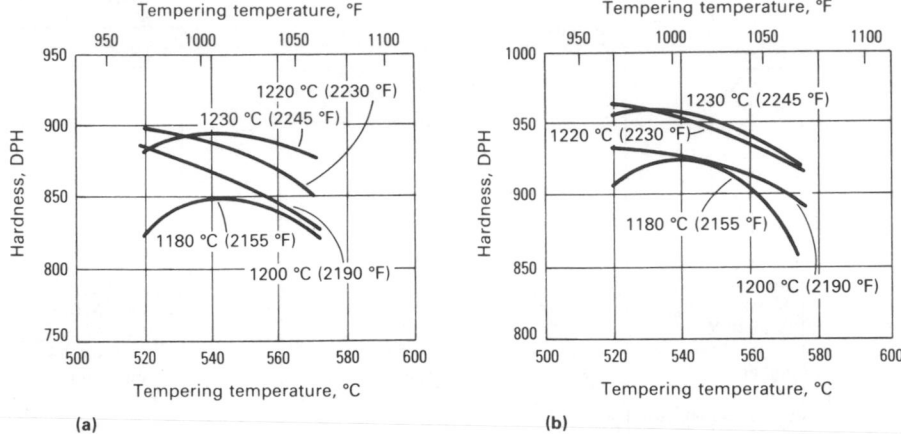

(a)   (b)

Table 9   Typical mechanical properties of commercially sintered M2, M35, and T15 high-speed steels

Property	Grade M2	Grade M35	Grade T15
Density g/cm³	8.05-8.2	8.05-8.2	8.15-8.3
Ultimate tensile strength(a), MPa (ksi)	750-800 (108-116)	770-820 (112-119)	770-830 (112-120)
Elongation(a), %	12-14	6-9	3-6
Ultimate tensile strength(b), MPa (ksi)	750-2000 (108-290)	770-2000 (112-290)	770-2000 (112-290)
Hardness, HRC	62-65	63-66	64-67

(a) Fully annealed. (b) Depending on heat treatment
Source: Ref 4

As-sintered microstructures consist of untempered martensite within prior austenite grain boundaries and large amounts of retained austenite. Etching in nital reveals the presence of prior austenitic grain boundaries during sintering. Etching also may reveal gray or white areas in the matrix, as shown in Fig. 30. Gray areas are reported to be caused by the very fine carbide dispersions that result from marginal or insufficient vacuum quenching. White areas indicate locations in which precipitation has not occurred. Reference 9 discusses this phenomenon for M2, M3 type 2, and M7 heat treated in vacuum.

Undersintered, correctly sintered, and oversintered microstructures are shown in Fig. 31. Undersintered microstructures (Fig. 31a) have fine or very fine carbides and grains in addition to porosity. Porosity is rounded (without sharp ends) and can be irregular in shape when sintered density is low. Correctly sintered microstructures have little or no porosity, no eutectic structure or evidence of melting, uniformly dispersed carbides, and uniform grain size, as shown in Fig. 31(b).

The oversintered microstructure has large carbides and grains. Porosity with sharp ends, usually at grain triple points, is evident. A eutectic structure from localized melting and a continuous carbide network around grains may also be present in oversintered microstructures (Fig. 31c). Oversintered parts have slightly lower sintered densities (about 0.2 to 0.5 g/cm³) than correctly sintered parts because of the formation of porosity at the grain triple points.

Contamination. Metallic contamination can be readily detected in sintered tool steel microstructures. Contamination appears as carbide-free areas with porosity and/or carbide networks, as shown in Fig. 32. Such defects occur when lower melting alloy particles or alloy particles that react with carbon from the tool steel powder to form low-melting alloys are present in higher melting tool steel alloys. Iron, stainless steel, and low-alloy steel powders form such defects in M2, T15, and other high-speed steels. These defects also occur when particles of T15 are present in M2 high-speed steel.

Tool steel powder producers have significantly reduced typical levels of contamination by using powder cleaning, compaction, and sintering equipment only for tool steel powders and by keeping powder-producing environments meticulously clean. Tool steel powder users must also clean powder hoppers and blenders to maintain high-quality P/M part production.

Atmospheric Pressure Sintering

Although vacuum is the preferred mode of sintering, wear parts made of tool steels and high-speed steels can be sintered to conventional P/M densities or to full density at atmospheric pressure and in atmospheres with dew points below −40 °C (−40 °F). Atmosphere-sintered parts have higher oxygen and nitrogen levels than vacuum-sintered parts.

Nitrogen-based, dissociated ammonia, and hydrogen atmospheres are viable alternatives to vacuum. Atmosphere composition has little effect, however, on sintered density. Parts sintered in pure nitrogen or nitrogen-based atmospheres are nitrided. Increasing the nitrogen content decreases the sintered transverse rupture strength. Consequently, as-sintered transverse rupture strength is lowest for parts sintered in nitrogen and highest for parts sintered in hydrogen. Sintered high-speed steels and tool steels should be double tempered to maximize transverse rupture strength.

Sintering of Copper-Based Materials

By Paul E. Matthews
Vice President and
Technical Director
United States Bronze Powders, Inc.

COPPER-BASED P/M MATERIALS rank second only to iron-based parts in terms of commercial applicability. As with other P/M materials, the final properties and related performance of copper-based parts depend on successful sintering techniques. This section reviews sintering practices for copper-tin, copper-zinc-lead, and copper-nickel-zinc alloys. For information on the manufacture of copper and copper alloy powders, see the articles "Production of Copper Powder" and "Production of Copper Alloy Powders" in this Volume. Numerous applications of copper and copper-based P/M parts can be

Fig. 27 Schematic of microstructure of as-sintered P/M tool steels
Source: Ref 8

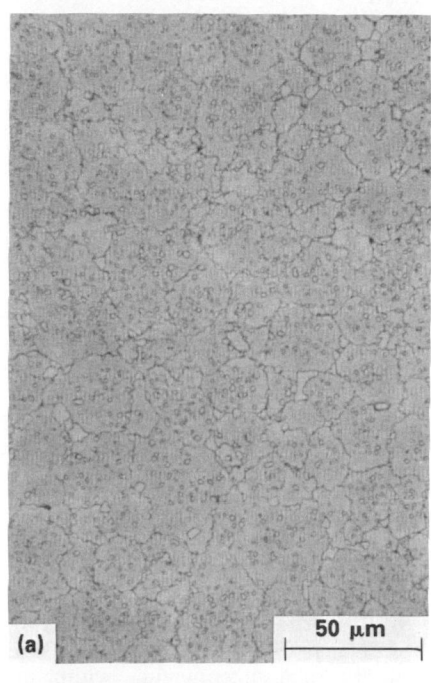

Fig. 28 Photomicrographs of as-sintered and wrought high-speed steels comparing primary carbide size and distribution
(a) As-sintered M2. (b) Annealed wrought M2. (c) As-sintered T15. (d) Annealed wrought T15. Magnification: 350×

Carbides within grains

• Carbides of tungsten, vanadium, and molybdenum
• Retard martensitic transformation
• Go into solution or precipitate slowly
• Resist grain growth

Grain boundary carbides

Continuous carbide network around grain boundary and large grains cause brittleness

Large carbides
• Cause poor grindability
• Wear resistance may improve

Grain boundary

Most likely location of porosity

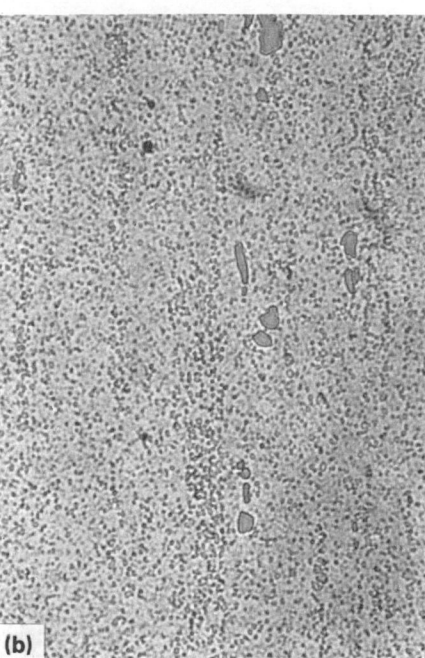

found in the Section of this Volume entitled "Powder Systems and Applications."

Sintering of Bronze

Powder metallurgy bronze typically originates as premixes consisting of elemental copper and tin powders plus 0.5 to 0.75% dry organic lubricants such as stearic acid or zinc stearate. The nominal composition of 90Cu-10Sn may be complemented with other constituents such as graphite, lead, and iron, depending on the specified grade. Bronzes are used extensively in the manufacture of porous, self-lubricating bushings and bearings and for more complex structures requiring superior bearing and mechanical strength.

Self-lubricating bushings and bearings are produced at nominal densities (oil impregnated) of 6.0 to 6.7 g/cm³, with oil contents ranging from 27 to 19 vol%, respectively. Corresponding radial crush values (K strength constant) is approximately 100 MPa (15 ksi) of the lower density, increasing to 183 MPa (26.5 ksi) at the highest nominal density.

The basic manufacturing procedure consists of compacting the premixed powder shapes to the appropriate green density, sintering to achieve a homogeneous metallurgical alpha bronze structure, followed by oil impregnation. A sizing operation completes the process to ensure dimensional precision and general surface integrity.

Dimensional Change. Effective sintering is essential, because the homogeneity of the sintered structure affects the resultant secondary forming and operational characteristics of the finished part. A va-

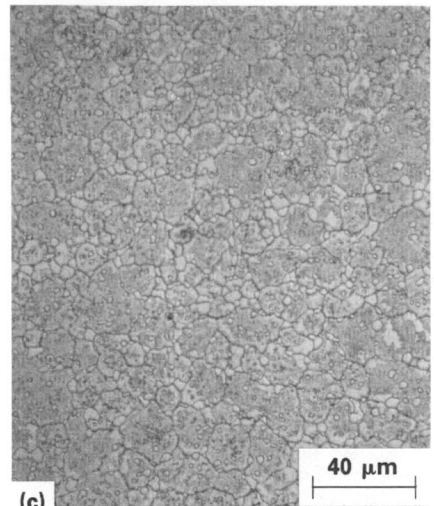

(a) 50 μm

(b)

(c) 40 μm

(d) 50 μm

riety of premixed powders are available with specific sintered dimensional patterns to satisfy customer design and tooling needs. Despite the dimensional magnitude of the particular premix being sintered, compositions of this type exhibit a common sintered dimensional pattern.

To include all dimensional patterns with related absolute dimensional change values for each commercially available premix system is prohibitive, however. For discussion purposes, a typical sintering/dimensional change pattern is shown in Fig.

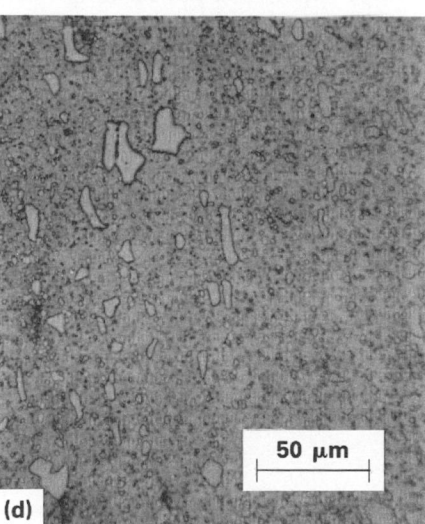

33, which illustrates the relationship of a "medium growth" copper-tin system as a function of total time in the furnace hot zone.

Absolute sintered dimensional characteristics typically are unique to a specific source of copper and tin powders. For example, sintered dimensional consistency may be obtained by blending two or more base copper powders that exhibit different growth characteristics and/or by use of tin powders that also exhibit different growth characteristics.

Fig. 29 Photomicrographs of as-sintered (a) and wrought (b) D2 tool steel

Magnification: 325×

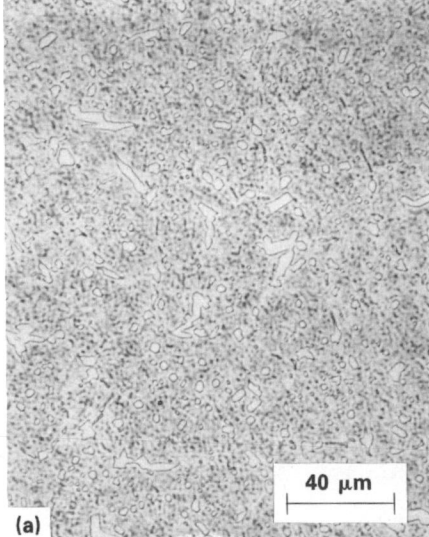

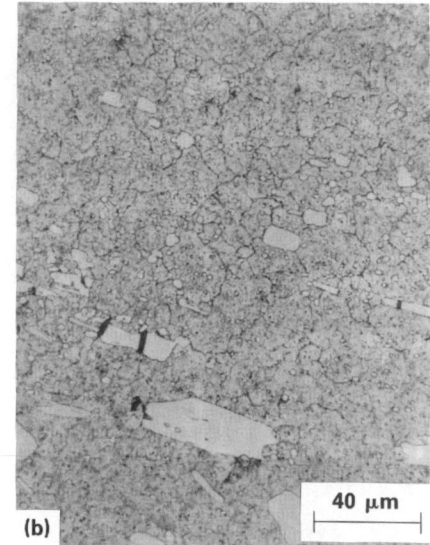

Fig. 30 Microstructure of as-sintered M2 quenched in vacuum furnace

Gray areas are caused by fine carbide dispersion due to marginal or insufficient quenching. White areas are where carbide precipitation has not taken place. Etchant: nital. Magnification: 300×

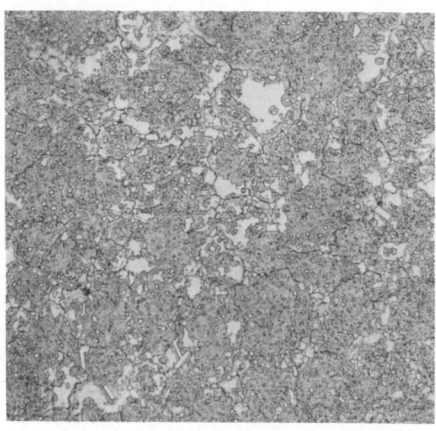

Generally, copper-tin blends composed of relatively coarse powder sinter to higher growth values than a blend composed of finer powders. After powder blends have been tested and adjusted to provide an approximation of target dimensions, final adjustments are made during production sintering to obtain dimensional precision.

Factors affecting the ultimate, or peak, dimensional values include physical characteristics of the constituents and compacted density. Control of sintered dimensions in premix systems is achieved by manipulating sintering time and/or temperature.

Sintering Time and Temperature. Typical sintering furnace temperatures for bronze range from 815 to 860 °C (1500 to 1580 °F); total sintering time within the hot zone may range from 15 to 30 min, depending on the furnace temperature selected, required dimensional change, and most importantly, the presence of an optimum alpha grain structure (Ref 10).

Sintering atmospheres should be protective and reducing to facilitate sintering. Reduction of the copper oxides that may surround each copper powder particle and reduction of tin oxide formation allow for increased diffusion rates. Consequently, faster sintering rates and more homogeneous structures can be obtained.

Sintering of Brass and Nickel Silvers

Powders of brasses and nickel silvers are prealloyed, single-phase (alpha) powders that, on sintering, yield moderate mechanical strength, excellent ductility, and good corrosion resistance. Parts may be subsequently burnished to improve surface finish. The various alloy compositions produced also provide suitable color, or shade, selection for applications that require a high degree of surface finish and appearance.

Fig. 31 Photomicrographs of as-sintered T15 high-speed steel

(a) Undersintered structure. Magnification: 190×. (b) Correctly sintered structure. Magnification: 190×. (c) Oversintered structure. Magnification: 190×

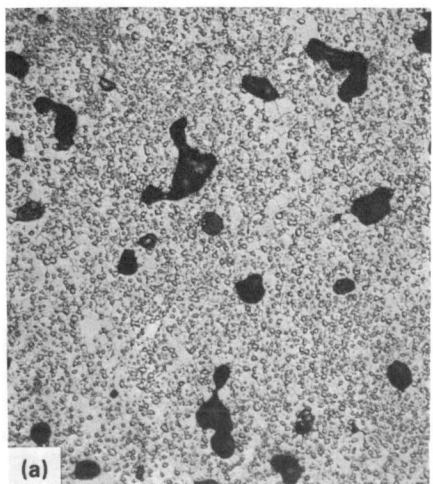

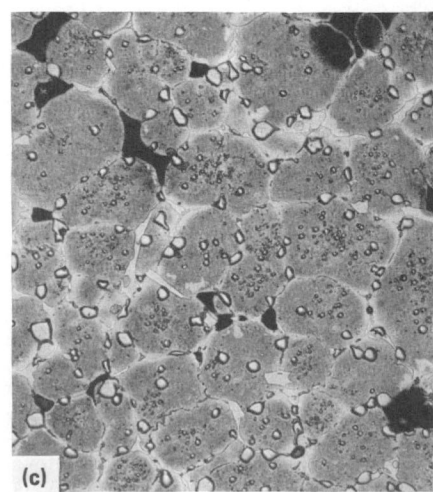

Fig. 32 Metallic contamination in as-sintered T15 high-speed steel
(a) Microstructure at 70× showing three particle contaminants. (b) Microstructure at 275× showing voids, carbides at grain boundaries, and lack of carbides in contaminant grains

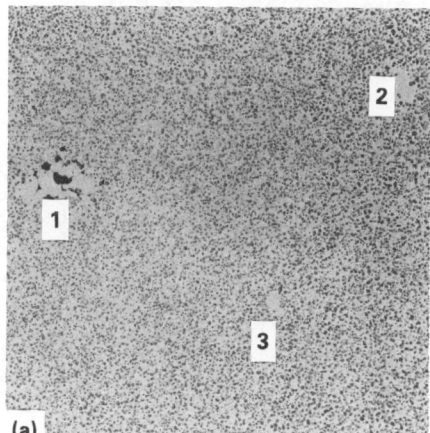

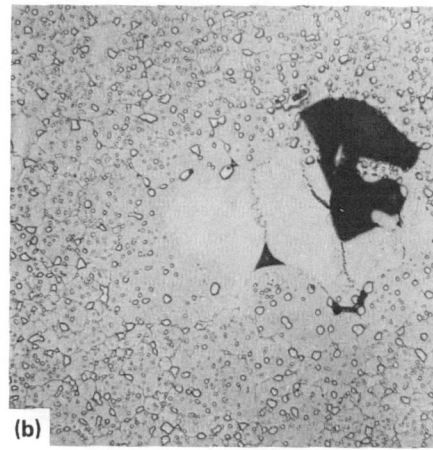

Fig. 33 Dimensional change in an elemental bronze blend (90Cu-10Sn) as a function of time and temperature
Sintered in hot zone at 845 °C (1550 °F) in dissociated ammonia atmosphere

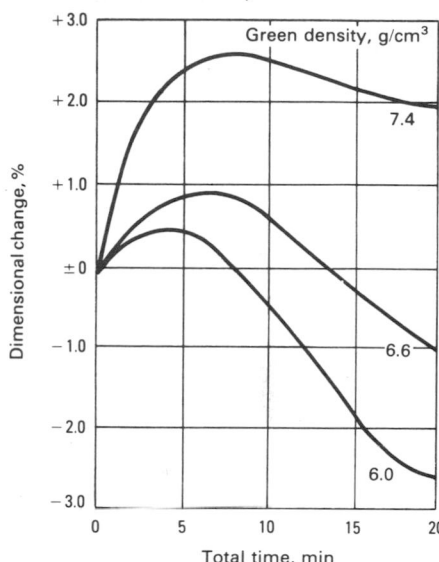

During alloy preparation, lead may be added to improve machinability of the sintered forms. Typical machining operations include drilling, tapping, turning, threading, and grinding. Excellent sintered ductility also facilitates secondary operations, such as sizing, cold densification, swaging, and staking. By using multiple pressing and sintering operations, the yield strength and hardness of the P/M structure may approach those of its wrought alloy counterpart (Ref 11).

Standard prealloyed brass and nickel silver powder compositions are controlled to conform to existing materials standards. This conformance precludes the additions of foreign metallic constituents, such as tin and iron, that affect sintered mechanical and dimensional characteristics. In spite of the various compositions comprising these standard alloy powders, most exhibit similar characteristics. After blending with lubricant, powders densify approximately 10% more than their as-atomized apparent density. For example, an as-atomized powder with an apparent density of 3.0 g/cm³ usually blends to a density of 3.3 g/cm³ minimum with the addition of dry lubricant. Compressibilities are excellent, as lubricated powders compact to 85% of wrought counterpart densities at 414 MPa (60 ksi). Compression ratios of lubricated powders range from 2.0-to-1 to 2.2-to-1.

Effect of Lubricant. Blending of powders with dry organic lubricants is normally accomplished in a double cone-type blending unit. To minimize the inclusion of large lubricant agglomerates and other undesirable particles, the material is passed through a 40-mesh sieve, or screening is recommended prior to blending. The pri-

mary lubricant employed with brasses and nickel silvers is lithium stearate.

Lithium stearate provides an apparent scavenging or cleansing effect that enhances the sinterability of these powders. It may also result in spotty or speckled superficial stains on sintered surfaces. These phenomena affect the appearance of the components, but are not detrimental to mechanical properties. To minimize staining, lithium stearate additions of less than 0.5 wt% are recommended, and zinc stearate may be added to provide the additional required lubricity.

Different types of lubricant have a marked effect on the physical and mechanical properties of nonferrous prealloyed powder (Ref 12). To illustrate the effects on mechanical properties, sintered data are shown in Fig. 34 for three frequently used lubricants in the P/M industry—lithium stearate, zinc stearate, and stearic acid. The beneficial effect of 1 wt% lithium stearate and the deleterious effect of 1 wt% stearic acid on mechanical properties are shown.

Compacting of lubricated powders is performed with standard types of compacting presses employing steel or carbide dies and punches. Excellent compressibility and good green strength permit compacting to 75% of theoretical density at pressures as low as 207 MPa (30 ksi).

Although powders are normally free of gangue and other inclusive, abrasive material that may cause tool wear, their relative softness does cause tooling difficulties. If powders are not adequately lubricated, fines within tool clearances gall the die wall and adjacent punch areas, thereby requiring tool removal and cleaning. The amount of lubricant added to the

powder should be proportional to the total surface area of the die assembly requiring lubrication during forming and ejection. Low-profile, minimum die wall contact parts may only require 0.5 wt% added lubricant, whereas a high die wall contact part having core rods for holes or internal cavities may require 1.0 wt% lubricant.

Generally, there are no restrictions on the compacted configuration of these powders. In tool design, particularly with regard to die fill, consideration must be given to the relatively higher apparent densities of lubricated brass and nickel silver powder. Typical apparent densities for lubricated powders range from 3.3 to 3.6 g/cm³.

Sintering of brasses and nickel silvers typically is not difficult; however, basic sintering practices do differ from those employed with other common alloy systems, such as elemental copper-tin blends and iron powder blends. These differences include sintering temperature, time at sintering temperature, and atmosphere protection.

Sintering temperatures for standard brasses range from 760 to 925 °C (1400 to 1700 °F). Temperature selection depends on the brass alloy being sintered and the mechanical properties desired after sintering. Lower brasses with higher zinc contents and lower melting points are sintered at the lower temperature. Generally, a starting temperature of 100 °C (180 °F) below the solidus temperature (as deter-

Fig. 34 Effect of lubricants and sintering time at temperature on tensile properties, sintered density, and dimensional change of brass compacts

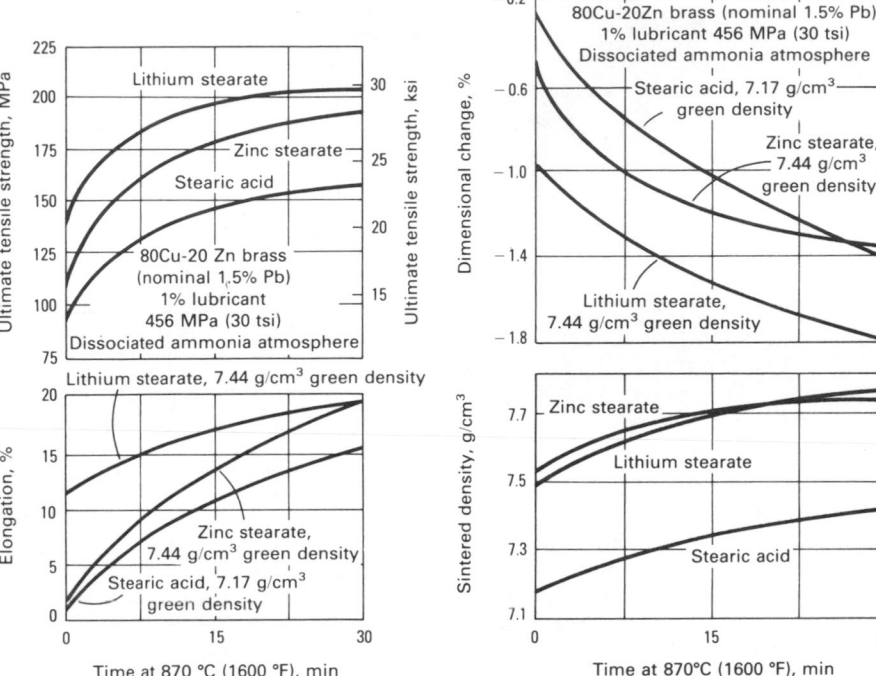

Fig. 35 Effect of varying sintering time on properties of prealloyed 80Cu-20Zn leaded brass (nominal 1.5% Pb)

Lubricant: 0.375% lithium stearate and 0.375% zinc stearate. Compaction pressure: 414 MPa (30 tsi). Green density: 7.6 g/cm³. Sintering temperature and atmosphere: 870 °C (1600 °F) in dissociated ammonia.

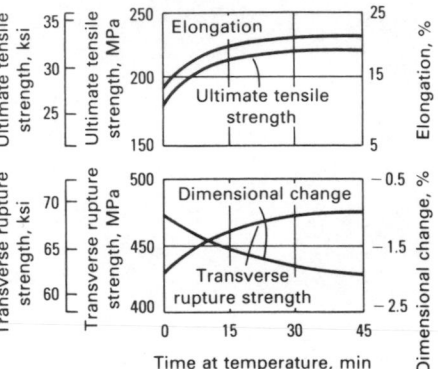

Fig. 36 Effect of varying sintering time on properties of prealloyed 70Cu-30Zn leaded brass (nominal 1.5% Pb)

Lubricant: 0.375% lithium stearate and 0.375% zinc stearate. Compaction pressure: 414 MPa (30 tsi). Green density: 7.3 g/cm³. Sintering temperature and atmosphere: 870 °C (1600 °F) in dissociated ammonia.

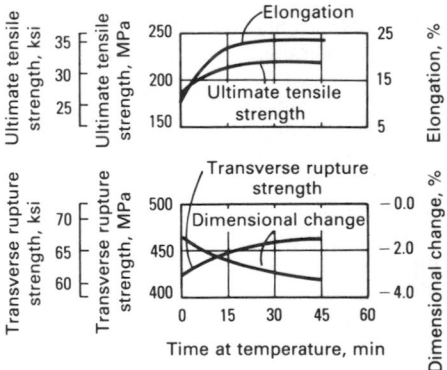

mined from any copper-zinc binary alloy constitutional diagram) is suitable.

Nickel silver may be sintered at 870 to 980 °C (1600 to 1800 °F). Currently, only one base alloy is used for the manufacture of P/M structural parts; it has nominal composition of 64Cu-18Ni-18Zn. The leaded alloy composition contains 1.5% Pb. Sintering characteristics are similar to those of the brasses; therefore, responses to sintering parameters that affect dimensional and mechanical properties of brass are equally applicable to nickel silver.

Sintered Properties. Dimensional and mechanical properties of brasses and nickel silvers are primarily affected by compact density and the amount of time at temperature, as well as the sintering temperature itself. As mentioned above, other elements that affect dimensional and mechanical properties usually are not added to powders. However, sintered properties, especially dimensional change, may be effectively controlled by manipulation of sintering time at the appropriate temperature. Each alloy exhibits unique dimensional characteristics—a 90Cu-10Zn brass compacted at 414 MPa (30 tsi) and sintered for 30 min at 870 °C (1600 °F) may shrink 0.5%, while a 70Cu-30Zn brass similarly treated may shrink 2.5%.

Figures 35 and 36 show typical property relationships that can be controlled through manipulation of time at temperature. The leaded brasses shown in Fig. 35 and 36 (80Cu-20Zn and 70Cu-30Zn) are commonly used for structural parts fabrication. Densities shown are "average" for compacting lubricated prealloyed powders containing 0.375% lithium stearate and 0.375% zinc stearate at 414 MPa (30 tsi). As shown, close dimensional control may be obtained with a minimum reduction in mechanical properties after 15 min at temperature. Ductility is increased for subsequent forming operations, such as sizing, cold re-pressing for densification, or coining, by increasing sintering time.

Atmosphere protection is required for sintering brasses and nickel silvers to prevent oxidation and to ensure effective sintering. Use of lithium stearate as the base lubricant allows the use of most common sintering atmospheres over a wide range of dew points. Although dry hydrogen or dissociated ammonia provides the best sintering atmosphere, comparable properties can be obtained with nitrogen-based or partially combusted hydrocarbon gas atmospheres.

When sintering, compacts should be protected from direct impingement of furnace flame curtains and atmosphere gases by partially or fully covering loaded trays to minimize zinc loss. Because it has a high vapor pressure at standard sintering tem-

perature (boiling point of pure zinc is 906 °C, or 1663 °F), zinc may be lost to the atmosphere as it diffuses through to the particle surfaces. Loss of excessive surface zinc results in a change in surface composition. In the case of brasses, pink copper or zinc-depleted areas are apparent. Although superficial zinc losses do not adversely affect sintered properties, surface finish is diminished; finished parts may be rejected because of color differences.

Furnace Design. A variety of protective sintering tray arrangements can be used, including graphite trays with full covers and graphite base plates with steel covers, for example. Sintering of brasses on open belts is not common practice for several reasons; for example, overheating may melt brass parts that may fuse to the mesh belt. Also, excessive heat and direct impingement of flame curtain gas through the mesh may impair effective sintering due to oxidation and premature partial lubricant removal. Direct exposure to the sintering atmosphere may also cause objectionable surface dezincing.

Conversely, extremely gastight protective sintering setups should be avoided, because lubricant decomposition products may not be completely volatilized. Excessive residual lubricant within the compact inhibits sintering; consequently, mechanical properties are adversely affected. Typically, compacts with excessive internal lubricant residuals may not be salvaged by additional passes through a full sintering cycle. To avoid or minimize the deleterious effect of residual lubricants, covered tray setups should facilitate free venting of lubricant gases to the flowing furnace atmosphere.

Sintering of Aluminum and Aluminum Alloys*

ALUMINUM P/M PARTS are enjoying increased usage because of their combination of high strength, light weight, corrosion resistance, and high thermal and electrical conductivities. Successful sintering of aluminum P/M alloys requires proper selection of sintering conditions such as atmosphere, dew point, and temperature. This section discusses sintering practices for the production of aluminum P/M parts in nitrogen, dissociated ammonia, and vacuum atmospheres. The influence of sintering atmosphere, dew point, and temperature on the mechanical properties and dimensional changes of aluminum P/M alloys 601AB and 201AB are emphasized. For information concerning production of aluminum powder and properties and applications of various aluminum P/M alloys, see the articles "Production of Aluminum Powder" and "P/M Lightweight Metals" in this Volume.

*This section is adapted from Dudas, J.H. and Thompson, C.B., Improved Sintering Procedures for Aluminum P/M Parts, in *Modern Developments in Powder Metallurgy*, Vol 5, Plenum Press, New York, 1971, p 19-36. Studies referred to in this section originally appeared in this publication.

Table 10 Nominal compositions of aluminum P/M alloys

Element, %	601AB	201AB
Copper	0.25	4.4
Silicon	0.6	0.8
Magnesium	1.0	0.5
Lubricant	1.5	1.5
Aluminum	rem	rem

Table 11 Thermal treatments for sintered aluminum P/M alloys

Temper	Description
T1, as-sintered	Cooled from sintering temperature to room temperature at an uncontrolled rate
T4	Heat treated 30 min either at 520 °C (970 °F) for 601AB or 505 °C (940 °F) for 201AB in air, cold water quenched and aged 4 days minimum at room temperature
T6	Heat treated 30 min either at 520 °C (970 °F) for 601AB or 505 °C (940 °F) for 201AB in air, cold water quenched and aged 18 h at 160 °C (320 °F)

Materials and Test Procedures

Sintering tests were conducted employing commercial aluminum P/M alloys 601AB and 201AB (see Table 10 for nominal compositions). Both alloys are premixed powders containing aluminum with alloying powder additions of copper, silicon, and magnesium, plus 1.5 wt% lubricant. Sintering atmospheres tested included nitrogen, dissociated ammonia, and vacuum.

Sintered tensile properties were determined using Metal Powder Industries Federation flat powder metal tension test bars (MPIF 10). A description of the thermal treatments used is given in Table 11. Dimensional changes were measured on 75-mm (3-in.) long by 100-mm (4-in.) wide by 16-mm (5/8-in.) thick compacted bars.

Sintering Furnaces

Batch and conveyor atmosphere furnaces are suitable for sintering aluminum if provisions exist for lubricant burnoff, sintering, and cooling. While batch furnaces have lower investment costs and moderate atmosphere consumption, production rates are low to medium. Continuous furnaces provide high production rates, but have higher equipment costs and require increased atmosphere flow. Both furnace types provide strong, ductile alu-

minum P/M parts for a variety of applications, if atmosphere dew point and sintering temperature are controlled.

Batch sintering furnaces provide an economical means for prototype and low-to medium-production sintering of aluminum parts. Capital equipment costs for such furnaces are low, and atmosphere flow requirements usually are less than for conveyor furnaces, because door openings can be sealed easily.

For testing, the furnace used in this study was fitted with a 305-mm (12-in.) by 305-mm (12-in.) by 457-mm (18-in.) Inconel* muffle. An internal positive pressure of 3.5 to 7 kPa (0.5 to 1 psi) achieved with a nitrogen flow of 1.7 to 2.8 m³/h (60 to 100 ft³/h) maintains a −40 to −50 °C (−40 to −60 °F) furnace atmosphere dew point, with the entering gas at −57 to −62 °C (−70 to −80 °F) dew point.

A typical heating cycle for a batch-type furnace is shown in Fig. 37(a). Lubricant is expelled from the parts at 345 to 425 °C (650 to 800 °F), after which the temperature is raised to 620 °C (1150 °F) for sintering. Heating rate is not critical. After sintering, parts are normally cooled to 425 °C (800 °F) by an air blast directed around the outside rear of the furnace muffle.

The parts are then removed and air cooled to room temperature. Cycle time in the furnace is approximately 90 min. If heat treatment of the parts is required, they are cooled in the furnace to the solution heat treating temperature (520 °C, or 970 °F for 601AB). They are then cold water quenched and either naturally or artificially aged to the T4 or T6 temper.

Many types and sizes of commercial batch furnaces are available for sintering aluminum P/M parts, with capacities up to 45 kg (100 lb) of aluminum per hour. Designs include bell, elevator, pit, and box-type furnaces.

Continuous conveyor sintering furnaces are widely used in the P/M industry for high-volume parts production. The continuous humpback conveyor furnace used in this investigation was designed specially for sintering aluminum in nitrogen or dissociated ammonia atmosphere. In operation, parts were conveyed in anodized aluminum trays through the 150-mm (6-in.) wide by 100-mm (4-in.) high muffle opening at a preselected belt speed ranging from 50 to 200 mm/min (2 to 8 in./min). Nitrogen curtains at each end purged air from the parts and assisted in maintaining a low furnace dew point.

A typical sintering cycle is illustrated in Fig. 37(b). After the lubricant was vola-

*Trade name of the International Nickel Co., Inc.

Fig. 37 Heating cycles for sintered aluminum P/M parts
(a) Batch sintered parts. (b) Continuous sintered parts. (c) Vacuum sintered parts

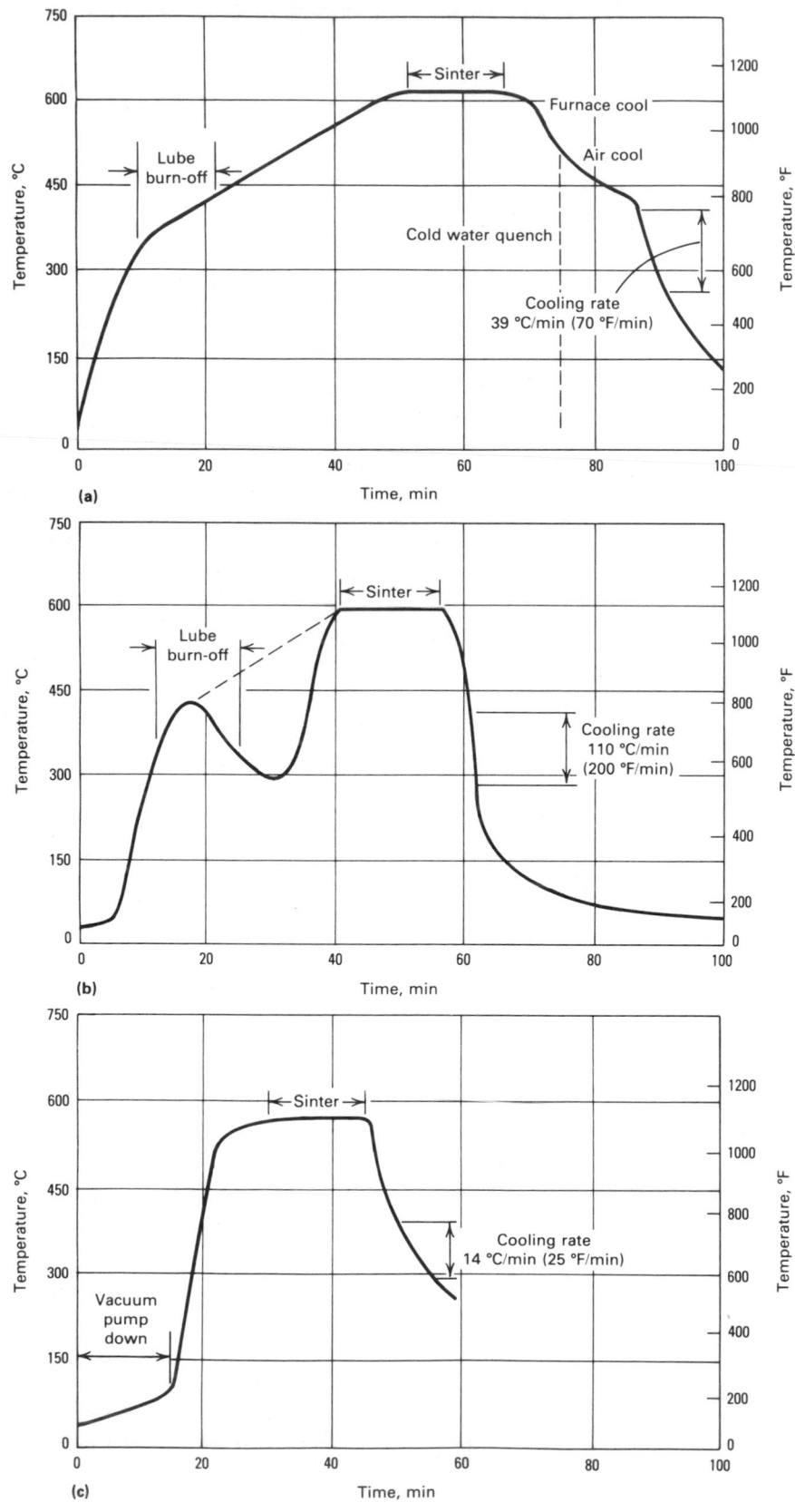

tilized at 345 to 425 °C (650 to 800 °F), parts were cooled to approximately 300 °C (570 °F) in the section separating the preheat and high heat chambers. To eliminate this inefficiency, the furnace was modified to allow continued heating of the parts to the sintering temperature (dashed line in Fig. 37b). The rate of part heating was not found to influence sintered properties.

Parts traveled into the 2.3-m (90-in.) long sintering chamber, which was heated by nickel-chromium ribbon elements mounted above and below the Inconel muffle. Temperature uniformity within ±2.8 °C (±5 °F) across the belt was achieved through adjustment of 12 saturable core reactors that proportionally controlled 305-mm (12-in.) individual heating sections along the sintering chamber. Sintered parts were cooled quickly to 150 °C (300 °F) in the 610-mm (24-in.) long water-jacketed section, then slowly cooled to approximately 65 °C (150 °F) in the nonwater-cooled inclined tunnel.

Humpback furnace designs help to achieve atmosphere "savings" when using atmospheres such as dissociated ammonia, which contains a high percentage of hydrogen. With nitrogen atmosphere, which has about the same density as air, the humpback feature is neither required nor desirable. An atmosphere flow of 14 to 20 m³/h (500 to 700 ft³/h) is required with nitrogen, but only 11 to 14 m³/h (400 to 500 ft³/h) is required with dissociated ammonia.

The production rate of sintered aluminum P/M parts in conveyor furnaces depends on belt width and speed, as well as on part shape and size. Belt loading is not a limitation because of the light weight of aluminum and lower sintering temperatures used.

Sintering rates of 18 kg/h (40 lb/h) have been achieved with a 150-mm (6-in.) wide furnace. On a per piece basis, this is equivalent to 54 kg/h (120 lb/h) of iron parts in the same furnace. Continuous aluminum furnaces with conveyor widths up to 915-mm (36 in.) capable of production rates up to 225 kg/h (500 lb/h) of aluminum P/M parts are available.

Vacuum Furnaces. Aluminum P/M parts also may be sintered in conventional cold walled or hot walled vacuum furnaces. Inert atmosphere or gas drying equipment is not required. Roughing pump vacuums of about 27 Pa (200 μm Hg) are suitable for higher density aluminum parts. Consequently, vacuum cold traps and diffusion pumps are not required.

A typical sintering cycle for the 205-mm (8-in.) diam hot walled vacuum furnace used in this study is shown in Fig. 37(c).

The parts, which were presintered separately to remove lubricant, were heated directly to the sintering temperature once pumpdown was complete. After sintering, parts were furnace cooled to approximately 260 °C (500 °F) before the vacuum was released.

Vacuum sintering, which is gaining increased acceptance in the production of stainless steel P/M parts, should not be overlooked as a method for sintering aluminum. Various sizes of batch and semi-continuous vacuum furnaces are available.

Sintering Atmospheres

Sintering atmosphere studies were conducted in nitrogen (the most frequently used atmosphere for aluminum), in dissociated ammonia, and in vacuum. The type of atmosphere and sintering conditions directly affect tensile properties, dimensional change, and microstructural change.

Nitrogen at high purity and low dew point is particularly well suited for sintering aluminum P/M parts because of its ready availability and moderate cost. Special handling is not required; nor are a generator and adsorbent dryer required to convert nitrogen to a dry, gaseous form. The highest sintered strength in both 601AB and 201AB parts is achieved in nitrogen.

Production rates for continuously sintered aluminum parts in nitrogen varied from 4.5 to 18 kg/h (10 to 40 lb/h). Atmosphere flow of 14 to 20 m³/h (500 to 700 ft³/h) produced consistently well-sintered parts in simulated production runs that lasted 4 to 6 h. Sintering time varied with part section thickness. Parts up to 6.4 mm (¹/₄ in.) thick sintered in 10 to 15 min, whereas parts 25 to 50 mm (1 to 2 in.) thick required 30 to 40 min. Extended sintering times caused some surface nitriding of parts, but had no significant effect on tensile properties. Cooling rate from the sintering temperature varied with part size and belt speed, causing some variation in as-sintered (T1) strengths. However, heat treating to the T4 or T6 temper raised properties to an acceptable level.

Table 12 lists the tensile properties of alloys 601AB and 201AB sintered in nitrogen. Tensile strengths of 601AB parts were 140 to 240 MPa (20 to 35 ksi), depending on density and thermal treatment. As-sintered (T1) parts had moderate strength and high ductility for easy repressing.

The heat treated and naturally aged T4 temper provided the best combination of strength and ductility, whereas highest strength was obtained in the fully heat treated T6 temper. Alloy 201AB extended the strengths to approximately 340 MPa (49 ksi) in the 95% density parts that were heat treated to the T6 temper. This composition is particularly well suited for P/M applications requiring the highest aluminum strengths.

Dissociated ammonia is used in many P/M applications for sintering brass and bronze parts. It may be used for aluminum as well. The atmosphere used in this investigation was generated from premium-grade anhydrous ammonia with a minimum purity of 99.99%. It was dissociated in a 14-m³/h (500 ft³/h) dissociator at 954 °C (1750 °F). The dissociated product —75% hydrogen and 25% nitrogen— had a dew point of −50 °C (−60 °F) and contained 75 to 90 ppm of undissociated ammonia. A 14-m³/h (500 ft³/h) dual-tower adsorbent dryer further dried the gas to a dew point of −62 to −73 °C (−80 to −100 °F) and reduced the undissociated ammonia to less than 1 ppm.

Dissociated ammonia contains high concentrations of flammable hydrogen; consequently, care must be taken in handling, particularly for aluminum processing during which sintering temperatures are not high enough to ensure self-ignition upon contact with air. One precautionary method involves purging the furnace with an inert gas such as nitrogen prior to introducing dissociated ammonia.

Electric ignitors are installed in the burn-off stand pipe to provide positive ignition of the exiting hydrogen. Although some hydrogen escapes at the furnace doors, it is quickly diluted to a noncombustible concentration with nitrogen from the end curtains and air. Burning of hydrogen at the furnace doors is not recommended because the resultant moisture may back-diffuse into the furnace, thus raising the dew point.

Conditions and production rates for aluminum P/M parts sintered in dissociated ammonia were similar to those sintered in nitrogen; however, lower atmosphere flow rates were possible because of the humpback furnace design. A dissociated ammonia flow of 11 to 14 m³/h (400 to 500 ft³/h) was satisfactory to maintain a −40 to −50 °C (−40 to −60 °F) furnace dew point during simulated production tests.

Tensile properties of 601AB and 201AB alloys sintered in dissociated ammonia are given in Table 13. Both tensile strength and ductility were lower than for nitrogen sintered parts, but were satisfactory for most aluminum P/M applications. Strength levels up to 207 MPa (30 ksi) were obtained in 601AB alloy and up to 290 MPa 42 ksi for 201AB alloy. For 601AB parts, relative tensile strengths were 65 to 85% of those produced in a nitrogen atmosphere; 201AB parts were 85 to 95% as strong as those sintered in nitrogen.

The lower properties of parts sintered in dissociated ammonia appear related to the presence of hydrogen and/or undissociated ammonia in the sintering atmosphere. The tendency of hydrogen to cause gassing of aluminum has been well documented (Ref 13). Also, ammonia reacts with aluminum to liberate hydrogen. When 100% ammonia vapor was used as a heat treating atmosphere for aluminum, the tensile strength of 2024 alloy sheet was reduced 29%, and the elongation was reduced 82% (Ref 14).

Gas analysis performed during this investigation indicated the level of undissociated ammonia in the sintering atmosphere was less than 1 ppm at the dryer, but it re-formed to 8 to 15 ppm in the furnace. Both hydrogen and undissociated ammonia apparently can be sources for hydrogen absorption by the liquid phase at the sintering temperature. This can lead to increased porosity in the compact during solidification, thus reducing the tensile properties of P/M parts.

Table 12 Properties of P/M aluminum alloys sintered in nitrogen

Alloy	Green density %	Green density g/cm³	Temper	Tensile strength MPa	Tensile strength psi	Yield strength MPa	Yield strength psi	Elongation in 25 mm (1 in.), %
601AB(a) 90	90	2.42	T1	138	20 100	88	12 700	5.0
			T4	172	24 900	114	16 600	5.0
			T6	232	33 600	224	32 500	2.0
	95	2.55	T1	145	21 000	94	13 700	6.0
			T4	176	25 600	117	17 000	6.0
			T6	238	34 500	230	33 400	2.0
201AB(b) 90	90	2.50	T1	167	24 200	147	21 300	3.0
			T4	201	29 200	167	24 300	3.0
			T6	265	38 500	262	38 000	1.5
	95	2.64	T1	209	30 300	177	25 700	3.0
			T4	253	36 700	198	28 700	2.5
			T6	336	48 800	322	46 700	2.0

(a) Sinter 10 to 30 min at 620 °C (1150 °F) at a dew point of −40 to −50 °C (−40 to −60 °F). (b) Sinter 10 to 30 min at 595 °C (1100 °F) at a dew point of −40 to −50 °C (−40 to −60 °F)

Table 13 Properties of P/M aluminum alloys sintered in dissociated ammonia

Alloy	Green density %	Green density g/cm³	Temper	Tensile strength MPa	Tensile strength psi	Yield strength MPa	Yield strength psi	Elongation in 25 mm (1 in.), %
601AB(a)	90	2.42	T1	93	13 500	76	11 000	2.5
			T4	108	15 700	88	12 700	3.5
			T6	159	23 100	. . .	. . .	1.0
	95	2.55	T1	121	17 600	87	12 600	3.5
			T4	146	21 200	99	14 300	5.0
			T6	207	30 100	205	29 700	1.5
201AB(b)	90	2.50	T1	161	23 300	141	20 500	2.0
			T4	198	28 800	163	23 700	2.5
			T6	247	35 800	. . .	. . .	0.5
	95	2.64	T1	174	25 200	152	22 000	2.0
			T4	221	32 000	180	26 100	3.0
			T6	288	41 800	287	41 600	1.0

(a) Sinter 10 to 30 min at 620 °C (1150 °F) at a dew point of −40 to −50 °C (−40 to −60 °F). (b) Sinter 10 to 30 min at 595 °C (1100 °F) at a dew point of −40 to −50 °C (−40 to −60 °F).

Table 14 Effect of vacuum level on tensile strengths of 601AB and 201AB specimens

Alloy	Density, %	Tensile strength(a) MPa	Tensile strength(a) psi	Tensile strength(b) MPa	Tensile strength(b) psi
601AB	85	176	25 500	208	30 200
	90	223	32 400	225	32 600
	95	227	32 900	245	35 500
201AB	85	186	27 000	263	38 200
	90	303	43 900	304	44 100

Note: All specimens heat treated to T6 temper after sintering
(a) 27 Pa (200 μm Hg). (b) 1.3 mPa (0.01 μm Hg)

Table 15 Properties of P/M aluminum alloys sintered in vacuum

Alloy	Green density %	Green density g/cm³	Temper	Tensile strength MPa	Tensile strength psi	Yield strength MPa	Yield strength psi	Elongation in 25 mm (1 in.), %
601AB(a)	90	2.42	T1	112	16 300	68	9 900	4.5
			T4	140	20 300	91	13 200	4.0
			T6	223	32 300	211	30 600	2.0
	95	2.55	T1	131	19 000	80	11 600	5.0
			T4	161	23 400	99	14 400	7.0
			T6	230	33 300	219	31 800	2.0
201AB(b)	90	2.50	T1	185	26 800	143	20 700	4.0
			T4	241	35 000	187	27 100	5.5
			T6	296	43 000	287	41 600	2.0
	95	2.64	T1	184	26 700	146	21 200	4.0
			T4	250	36 300	185	26 900	6.5
			T6	312	45 300	290	42 000	2.0

(a) Sinter 10 to 30 min at 605 °C (1125 °F) at 0.0013 to 27 Pa (0.01 to 200 μm Hg) vacuum level. (b) Sinter 10 to 30 min at 580 °C (1075 °F) at 0.0013 to 27 Pa (0.01 to 200 μm Hg) vacuum level

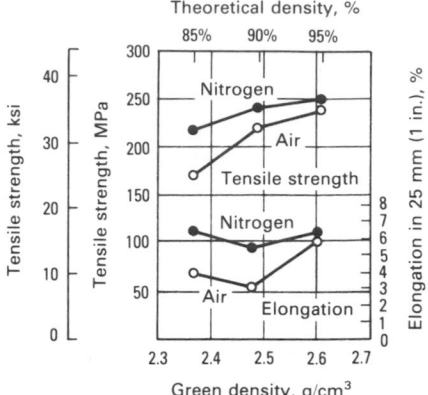

Fig. 38 Effect of presintering atmosphere on properties of 201AB (T4)

Sintered in vacuum 20 min at 580 °C (1075 °F)

Vacuum studies attempted to establish optimum sintering conditions for alloys 601AB and 201AB. Sintering times for 6.4- to 12.7-mm (¼- to ½-in.) thick specimens were comparable to those for parts sintered in either nitrogen or dissociated ammonia. Sintering temperatures in vacuum had to be 14 °C (25 °F) lower to prevent excessive distortion and melting.

The effect of presintering in air or nitrogen to remove lubricant prior to vacuum sintering was examined using a burnoff time of 15 min at 370 to 425 °C (700 to 800 °F). Figure 38 shows that 85% density 201AB parts presintered in nitrogen developed significantly higher properties than parts presintered in air. At 95% density, strength and ductility were comparable for both atmospheres. Although air presintering appears practical for higher density parts, the time should be kept as short as possible to prevent excessive oxidation.

The influence of vacuum level on sintered properties of 601AB and 201AB is shown in Table 14. A low pressure of 1.3 mPa (0.01 μm Hg) produced substantially stronger parts at 85% density, but only slightly stronger parts at 90 and 95% density. Most mechanical roughing pumps can

easily achieve a vacuum level of 7 to 27 Pa (50 to 200 μm Hg), eliminating the need for diffusion pumps and minimizing pumpdown time.

Average mechanical properties of vacuum-sintered 601AB and 201AB alloys developed over a range of vacuum levels, sintering times, and presintering conditions are given in Table 15. In all cases, strength and ductility were higher than for parts sintered in dissociated ammonia.

Compared to property values in nitrogen atmospheres, vacuum-sintered 601 AB parts exhibited properties of 80 to 95%. Alloy 201AB parts, however, had higher sintered strengths in vacuum at 90% density and were only slightly lower than nitrogen values at 95% density. Elongations for vacuum-sintered 201AB parts were the highest obtained.

The slightly lower properties for most vacuum-sintered parts most likely result from magnesium loss by vaporization at elevated temperatures and a higher porosity than nitrogen-sintered parts. Chemical analyses of several specimens revealed that the magnesium content of the part surface was reduced by 25 to 50% during vacuum sintering.

Dimensional Change During Sintering

Dimensions of sintered aluminum P/M parts are affected by compact density, sintering atmosphere, temperature, and dew point. The effects of green density and atmosphere on sintered dimensions of 601AB and 201AB alloys are illustrated in Fig. 39(a) and (b). Dimensions increased with increased green density in all atmospheres. Shrinkage or a lack of growth was exhibited by 85% green density parts,

Fig. 39 Effect of green density and atmosphere on sintered dimensions
(a) 601AB. (b) 201AB

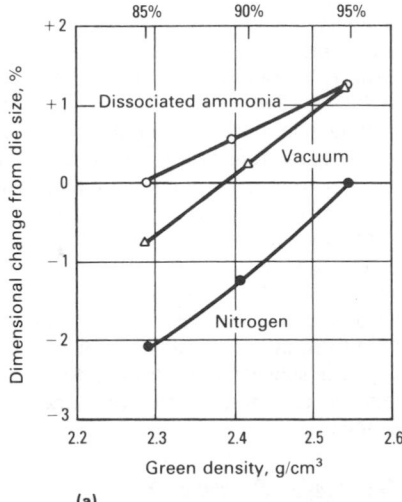

(a)

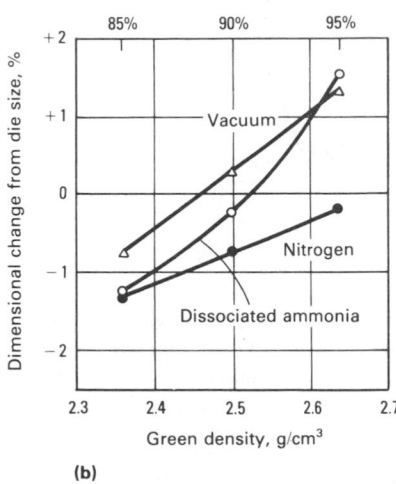

(b)

whereas high-density compacts exhibited growth when sintered in dissociated ammonia and vacuum. Nitrogen-sintered parts experienced shrinkage over the full range of densities—except for 95% density 601AB, in which no change was noted.

These dimensional changes were consistent as long as the sintering temperature was constant and the dew point in the furnace was at least −40 °C (−40 °F). Higher than normal temperatures caused excessive shrinkage and distortion or even melting in extreme cases. Lower than normal temperatures produced parts with increased dimensions and reduced properties.

Thus, for example, lowering the sintering temperature of 201AB alloy from 595 to 570 °C (1100 to 1060 °F) caused 95% density specimens sintered in dissociated ammonia to change from 0.25% shrinkage to 1.0% growth. At 580 °C (1080 °F), 0.25% growth was observed. Mechanical properties were less affected, although a reduction in tensile strength of 2 to 10% was observed, depending on thermal treatment. High dew points in the furnace resulted in excessive part expansion and a significant reduction in properties.

Microstructure

Representative samples of 201AB and 601AB were examined metallographically after sintering in nitrogen, dissociated ammonia, and vacuum. No significant differences that could be related to atmosphere were observed in the general microstructure. However, higher porosity was noted in specimens sintered in dissociated ammonia and vacuum, due to the

fact that these specimens exhibited growth after sintering.

Sintering of Cemented Carbides

By Thomas E. Hale
Manager—Materials Research
Carboloy Systems Department
General Electric Co.

CEMENTED CARBIDES consist of various mixtures of carbides and iron-group metals, such as tungsten carbide/cobalt, tungsten carbide/titanium carbide/tantalum carbide/niobium carbide/cobalt, or titanium carbide/molybdenum carbide/nickel. These materials typically are sintered to essentially 100% of theoretical density using a liquid-phase sintering process. The successful application of cemented carbides to high-stress operations such as metal cutting, oil well drilling, or metal forming dies depends on achieving low levels of residual porosity during sintering. Additionally, sintering must be controlled carefully to obtain desired microstructure and chemical composition.

Cemented carbides are used in many applications in the as-sintered condition. Frequently, the as-sintered surface acts as the critical wear/stress-bearing surface. In

most metal-cutting applications, the tool is considered worn out when the wear scar depth exceeds 0.2 to 0.4 mm (0.008 to 0.016 in.). Consequently, control of surface properties is essential.

Two basic methods are used to sinter cemented carbides. Hydrogen sintering uses a hydrogen-based atmosphere at atmospheric pressure to dynamically control part composition. Vacuum sintering uses a vacuum or reduced-pressure environment to control composition through slowing of the reaction kinetics. The vacuum method enjoys wide commercial usage.

Sintering of cemented carbides consists of removal of pressing lubricant, densification, and microstructure development. Each of these phases is discussed below for hydrogen and vacuum sintering modes.

Hydrogen Sintering

During hydrogen sintering, an atmosphere of hydrogen provides a reducing environment which, when subsequently modified by reactions with the furnace walls and carrier devices, provides the correct carburizing and oxidizing potentials to maintain thermodynamic equilibrium with the cemented carbide. In conventional hydrogen sintering, carbon content of the carbide powder is adjusted to the nominal theoretical value, and the hydrogen atmosphere subsequently maintains that value throughout the sintering cycle. For example, a 94WC-6Co composition may enter the furnace with a carbon content of about 5.70 to 5.80 wt% and exit the furnace with 5.76 ± 0.04%.

The atmosphere control capability of the hydrogen process is adequate for straight tungsten carbide/cobalt grades, but the oxidizing potential usually is too high to provide high-quality compositions containing titanium carbide, tantalum carbide, and/or niobium carbide additions suitable for steel machining. Such compositions usually are sintered in vacuum to achieve lower oxygen contents. Hydrogen sintering frequently is accomplished by mechanically stoking the parts through the furnace; a presintering furnace is required to remove pressing lubricant to prevent contamination from evolved lubricant vapors during the high-temperature stage of sintering. Presintering can also be used to partially sinter parts to green strengths that permit green forming operations, such as lathe turning or drilling. Peak presintering temperatures range from 500 to 800 °C (930 to 1475 °F), depending on the level of green strength desired.

A typical time-temperature curve for a mechanically stoked hydrogen sintering run

386/Consolidation of Metal Powders

Fig. 40 Typical time-temperature cycle for hydrogen stoking furnace

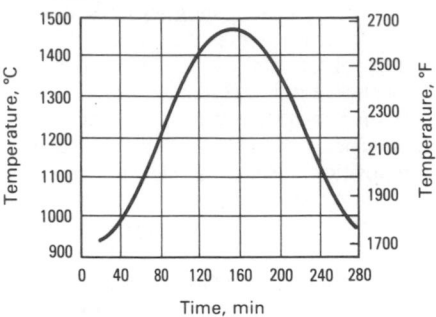

Fig. 41 Cemented carbide vacuum sinter time-temperature cycle

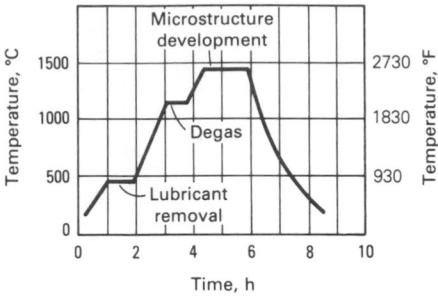

is shown in Fig. 40. The combined hot zone profile and stoking rate resulted in a heating rate of about 6 °C/min (11 °F/min) up to 1350 °C (2460 °F), holding above 1350 °C (2460 °F) for 75 min, and a peak temperature of about 1460 °C (2660 °F).

Vacuum Sintering

Vacuum sintering is the predominant industrial process for sintering cemented carbides. Compared to hydrogen atmosphere sintering, vacuum sintering has several major advantages. This process allows superior control of product composition. At pressures of 1.3 to 133 Pa (10^{-2} torr to 1 torr), the rate of carbon and oxygen exchange between the atmosphere and the cemented carbides is very low. The main factor controlling composition is the oxygen content of the carbide powder, not the rate of reaction with the atmosphere.

Because of the low oxidizing potential of the vacuum sintering atmosphere, the vacuum process is superior to the hydrogen process for the sintering of titanium carbide, tantalum carbide, and niobium carbide compositions that are sensitive to oxidizing potential. While capable of attaining low oxidizing potentials, most commercial hydrogen sintering furnaces have relatively high oxidizing potentials due to air leakage and reactions with ceramic furnace parts and fixturing.

Secondly, vacuum sintering offers greater flexibility and control of the thermal cycle, particularly the heating cycle. This is due to the fact that vacuum sintering is a batch process, while most hydrogen sintering is a continuous process. Slow heating rates are essential, usually followed by an intermediate temperature hold to obtain optimum product quality when titanium carbide, tantalum carbide, and niobium carbide additives are used in the carbide composition.

This allows sufficient time for the carbon-oxygen reactions to occur; a suffi-

cient open-pore network exists to allow escape of the carbon monoxide gas that forms. If the temperature is raised too rapidly, gas becomes entrapped in the structure, thus forming pores. The heating cycle can be adjusted easily for vacuum furnaces; for mechanically stoked hydrogen furnaces, however, it is fixed by the hot zone design and the peak temperature time required to obtain the desired microstucture.

Conventional vacuum sintering offers low operating cost. Previous vacuum equipment used relatively small induction-heated furnaces that were characterized by high power consumption, long cooling times, and the need for separate lubricant removal prior to sintering. Commercial vacuum sintering furnaces currently cost less to operate than hydrogen furnaces due to:

- Large batch size
- Resistance heating
- Forced gas cooling
- Built-in lubricant removal

A typical vacuum process time-temperature cycle is shown schematically in Fig. 41. The most significant features are three temperature holds—the first at about 500 °C (930 °F) for lubricant removal, the second at about 1200 °C (2190 °F) for degassing, and a third at 1400 to 1500 °C (2550 to 2730 °F) for microstructure development.

Lubricant Removal

Pressing lubricants used for cemented carbides range from low levels (1 to 2 wt%) of simple paraffins for pressed parts up to high levels of various synthetic polymers for extruded and slurry cast parts. The heating rate used for lubricant removal in a hydrogen atmosphere depends on the type and amount of lubricant used and the size of the pressed part. Heating rates as high as 15 to 20 °C/min (25 to 35 °F/min) to 500 °C (930 °F) can be used for pressed

cutting inserts containing 1 to 2% paraffin. Much slower rates (0.5 to 4 °C/min, or 0.9 to 7 °F/min) are required for larger parts or for parts with higher lubricant contents.

Commonly used lubricants, such as paraffin, evaporate readily in vacuum at 100 to 250 °C (210 to 480 °F). Consequently, heating rate should be controlled in this temperature range to prevent excessive pressure within the compact due to evolving lubricant vapors. A condenser is used in the vacuum line to remove the portion of the lubricant vapor that condenses at about 10 to 90 °C (50 to 195 °F). The interior walls of the furnace are sometimes used to condense paraffin vapor to liquid, which is subsequently drained to the bottom of the furnace to enter the main condenser.

Carbon Control

Carbon content of cemented carbides must be controlled to within very narrow limits, typically ±0.04 wt%, to prevent formation of brittle lower carbides, such as the eta phase in tungsten carbide alloy systems on the low carbon side and carbon precipitates on the high carbon side.

Because carbon content can be changed substantially during sintering by reactions with oxygen-containing phases in the powder and by carbon exchange reactions with the furnace atmosphere, control of the furnace atmosphere carbon potential and the oxygen content of the powder is essential to produce high-quality P/M sintered parts.

During hydrogen presintering, reactions between the hydrogen atmosphere and cemented carbides are relatively sluggish at the lower end of the peak temperature range (400 to 600 °C, or 750 to 1110 °F), but reaction rates are sufficiently high at temperatures above 600 °C (1110 °F) to cause significant changes in the carbon content of the part if the atmosphere is not properly adjusted.

This phenomenon is illustrated in Fig. 42, which shows the effect of varying the carburizing potential of the hydrogen atmosphere by methane additions on the carbon content of a 94WC-6Co composition held 1 h at the peak presinter temperature. The carburizing potential can be described as:

$$\frac{1}{k} \cdot \frac{CH_4}{H_2^2}$$

where k is the equilibrium constant for the reaction $2H_2 + C \leftrightarrows CH_4$, and CH_4/H_2^2 is a partial pressure ratio.

Fig. 42 Effect of carburizing potential of hydrogen-methane mixtures on carbon content of WC-6Co

1-h hold; 1 atm pressure. A_c is the thermodynamic carbon activity. Total carbon content is the sum of the compounded carbon in the carbide phase plus any graphite present. It does not include carbon present in the lubricant.

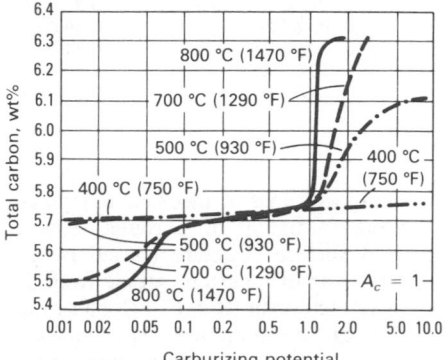

At temperatures up to 500 °C (930 °F), reactions causing carbon loss are too slow to noticeably change compositions. At 700 and 800 °C (1290 and 1470 °F), however, large carbon losses occur when the carburizing potential (relative to pure carbon) is less than about 0.05.

Thus, even pure hydrogen can cause large carbon losses above about 600 °C (1110 °F). When the carburizing potential exceeds 1.0 (saturation value), large carbon increases occur at 500 °C (930 °F) and above, thus allowing buildup of the carbon-containing volatilized lubricant in the peak temperature portion of the furnace.

Above about 900 °C (1650 °F), hydrogen-based atmospheres can change the carbon content of cemented carbides substantially by:

$$3WC + 3Co + 2H_2O$$
$$\rightarrow W_3Co_3C + 2CO + 2H_2$$
$$WC + Co + CH_4$$
$$\rightarrow WC + Co + C + 2H_2$$

To prevent the formation of undesirable carbon reaction phases, the carburizing potential (carbon activity) of the atmosphere must be carefully controlled. Table 16 shows experimentally determined activities required to maintain the desired carbon contents in tungsten carbide/cobalt cemented carbides at 1000 to 1450 °C (1830 to 2640 °F), with examples of typical gas compositions required to provide such carbon activities (Ref 15). The three major constituents of cemented carbide sintering atmospheres are hydrogen, carbon monoxide, and methane.

Typical commercial practice for hydrogen sintering carbon control is based on partial reaction of the incoming gas (hydrogen plus impurity levels of water vapor and oxygen), which has zero carbon activity. Graphite fixturing is used to carry the carbide parts through the furnace to increase the atmosphere carbon activity. Fixturing may thus contain openings that are empirically sized to react the incoming gas to the proper degree with the graphite to obtain the desired carbon activity.

Alternately, the graphite carrier may completely enclose the parts, thus causing carbon activity to increase to near unity. Modifiers such as aluminum oxide sand are included within the carrier to react with the atmosphere to thus lower the carbon activity to the desired level. A common practice uses a packing mixture of aluminum oxide sand and a small amount of carbon that surrounds the cemented carbide parts inside graphite carriers. By adjusting the carbon addition, carbon activity can be adjusted to the desired level.

Because the rates of carbon and oxygen exchange between the atmosphere and cemented carbides are low during vacuum sintering, the final carbon content of the cemented carbide when vacuum sintered is determined primarily by the initial carbon content of the powder and the amount of carbon lost during heating due to reactions between carbon and chemisorbed oxygen in the powder. Much of the oxygen present as oxides of tungsten and cobalt reacts with carbon in the powder during the vacuum heating cycle and evolves as carbon monoxide and carbon dioxide gases, whereas in hydrogen sintering, those oxides are reduced by hydrogen and evolve as water vapor. When titanium carbides and tantalum carbides are present, much of the oxygen contained in these compounds is also removed by carbon reduction. To compensate for this carbon loss, vacuum-processed powder usually contains about 0.1 to 0.3 wt% added carbon than is used for hydrogen-processed powder.

Densification

During hydrogen sintering, densification of pressed cemented carbide parts usually is accomplished by mechanically stoking through a muffle-type electrically heated furnace held at a peak temperature of about 1400 to 1500 °C (2550 to 2730 °F). A densification curve for WC-10Co composition (Fig. 43) shows that significant densification begins at about 1000 °C (1830 °F) and that greater than 90% theoretical density can be obtained at 1300 °C

(2370 °F). Liquefaction of the cobalt phase occurs between 1280 and 1350 °C (2335 and 2460 °F) as a result of the formation of a tungsten carbide/cobalt eutectic or eutectic-like phase (approximately 35 wt% tungsten carbide that forms during heating, as discussed in Ref 16). Essentially, the part becomes fully dense soon after the liquid phase forms.

During densification, the pressed part undergoes a volumetric shrinkage of about 40 to 45% (16 to 18% linear shrinkage). Complete densification is readily accomplished for typical cemented carbide compositions that contain about 3 to 25 wt% Co (5 to 40 vol%). Complete densification is more difficult to achieve for cobalt contents of less than 3%. For cobalt contents greater than about 25%, part shape is difficult to maintain due to the presence of high liquid phase fraction, which allows the part to sag.

Densification proceeds during vacuum sintering at about the same rate as it does for hydrogen sintering, reaching nearly 100% of theoretical density soon after the cobalt or nickel phase liquefies at 1280 to 1350 °C (2385 to 2460 °F); the exact temperature of liquefaction depends on the carbon content. When titanium carbide, tantalum carbide, and/or niobium carbide are present in the cemented carbide, the carbon-oxygen degassing reactions must be completed before densification closes the pores. If this does not occur, the evolving gases become entrapped in the interior of the compact, thus causing high residual porosity. Holding at about 1100 to 1250 °C (2010 to 2280 °F) for about 30 to 90 min ensures proper degassing of titanium carbide, tantalum carbide, and niobium carbide compositions.

Microstructure

At the beginning of peak temperature holding, the microstructure is nearly fully dense, and the carbide particles are nearly unchanged in size and shape from the original milled powder state. They are relatively small, irregularly shaped, and poorly dispersed, tending to agglomerate. The cobalt phase is also poorly dispersed, with many pools or lakes present (see Fig. 44a). A noticeable amount of residual porosity also exists.

The main purpose of the final stage of the sintering operation is to develop the microstructure by holding at a temperature above the cobalt melting point for a time sufficient to develop a more uniform carbide structure with good cobalt phase dispersion and minimal residual porosity (Fig. 44b). This is usually accomplished by

Table 16 Atmosphere compositions required to maintain neutral carburizing potentials for tungsten carbide/cobalt

As a function of temperature and total oxygen content

Temperature °C	°F	Carbon activity	Hydrogen, %	Carbon monoxide, %	Water, ppm	Carbon dioxide, ppm	Methane, %
1450	2640	0.6-0.25	99.944-99.976	None	None	None	0.056-0.023
1450	2640	0.6-0.25	99.45-99.47	0.5	3.4-8	4.7×10^{-3}-1.1×10^{-2}	0.055-0.023
1450	2640	0.6-0.25	89.96-89.98	10	60.0-144.0	1.7-4.0	0.045-0.019
1200	2190	0.6-0.25	99.84-99.93	None	None	None	0.24-0.10
1200	2190	0.6-0.25	99.27-99.40	0.5	12.2-29.2	3×10^{-2}-7.2×10^{-2}	0.24-0.10
1200	2190	0.6-0.25	89.79-89.87	10	220.0-570.0	12.2-29.2	0.19-0.08
1000	1830	0.5-0.2	99.56-99.82	None	None	None	0.44-0.18
1000	1830	0.5-0.2	99.05-99.29	0.5	114.0-290.0	0.34-0.85	0.44-0.18
1000	1830	0.5-0.2	89.43-89.34	10	2080.0-5200.0	135.0-338.0	0.36-0.14

Note: At pressure of 1 atm

Fig. 43 Density-temperature relationship for WC-10Co

30-min hold at each point

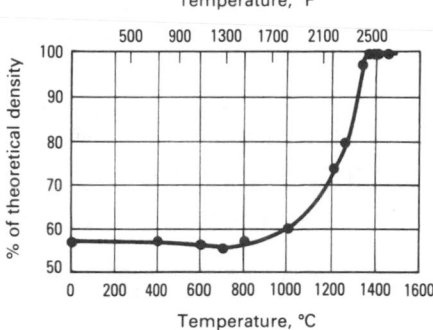

holding for 30 to 90 min above 1350 °C (2460 °F), reaching a peak temperature at about 1425 and 1500 °C (2595 and 2730 °F). During this period, the cobalt phase, driven by capillary forces, disperses more evenly. This process also improves carbide particle distribution.

Carbide distribution is further improved during holding by the dissolution of small particles into the liquid phase, with sub-sequent precipitation onto the larger particles during cooling. This results in a gradual increase in average particle size (Fig. 45a). Increasing the sintering temperature has a similar influence on grain growth (Fig. 45b).

One undesirable consequence of grain growth by the solution and precipitation process is the tendency of large tungsten carbide grains to grow at a disproportionately high rate. This discontinuous grain growth (Fig. 44c) occurs more readily at lower cobalt contents (3 to 6 wt%) with finer average particle size mixes. It is most pronounced when tungsten carbide is the only carbide phase present. Small additions (0.1 to 0.5%) of group VB carbides (vanadium carbide, niobium carbide, and tantalum carbide) undergo significant grain growth in fine-grained tungsten carbide/cobalt compositions. Titanium carbide is also a strong grain growth inhibitor (Ref 18).

During vacuum sintering, cobalt losses by evaporation should be controlled. If uncontrolled, as much as 10 to 20% of the cobalt content of the part may be lost, thus resulting in a loss of mechanical strength and the formation of a rough, coarse-grained surface structure caused by the precipitation of tungsten carbide from the evaporating cobalt (Ref 19). Cobalt evaporation can be minimized by completely enclosing the pressed parts in graphite fixturing, the walls of which are maintained at the sintering temperature.

The enclosure causes a buildup of cobalt vapor pressure around the cemented carbide parts, which reduces the evaporation rate. Cobalt losses can also be controlled by operating at pressures ranging from 13 to 133 Pa (0.1 to 1 torr), rather than in the range of 1.3 to 13 Pa (0.01 to 0.1 torr) that mechanical vacuum pumps are capable of maintaining. Higher pressures can be maintained by placing a throttling valve between the furnace and the pump or by injecting an inert gas into the furnace.

Cooling

After holding at sintering temperature, loads containing small parts can be cooled

Fig. 44 Microstructure of WC-6Co at various stages of sintering

(a) Undersintered. (b) Normal structure. (c) Discontinuous tungsten carbide grain growth. Magnification: 1000×

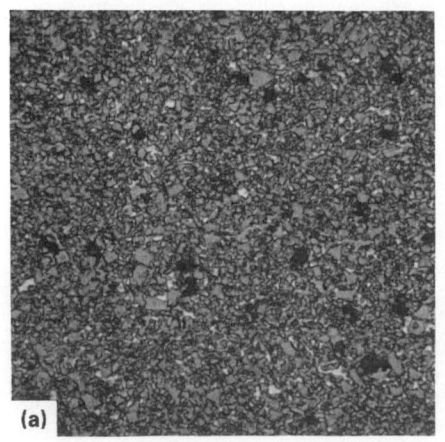
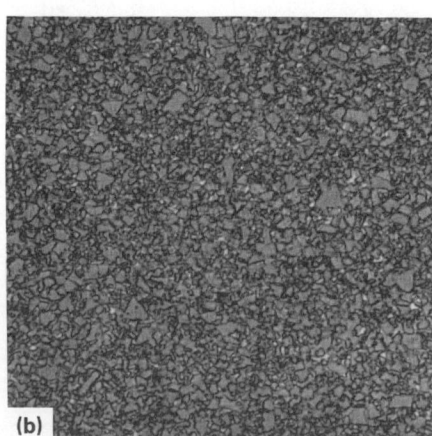
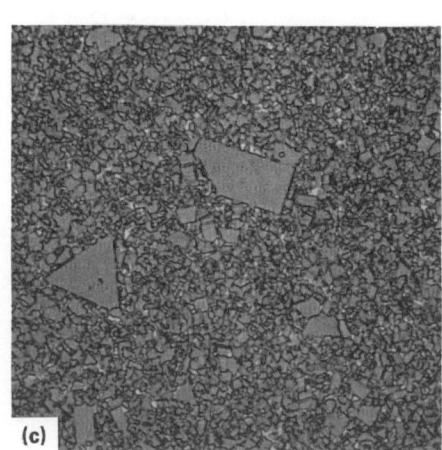

(a)   (b)   (c)

Fig. 45 Effect of sintering on grain growth of WC-25Co compacts
(a) Effect of sintering time at 1400 °C (2550 °F). (b) Effect of sintering temperature (1-h hold).
Source: Ref 17

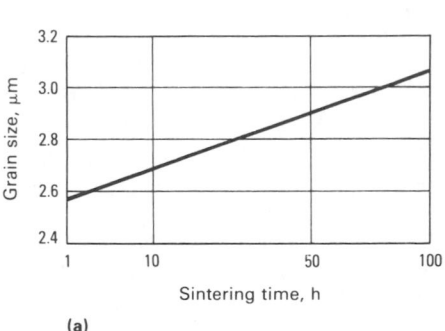

(a)

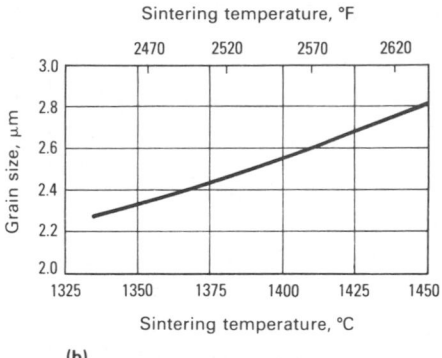

(b)

to room temperature in about 2 to 4 h by circulating inert gas (argon or helium) through the load at or near atmospheric pressure. Cooling of large parts is prolonged to minimize thermal stresses. When all significant material and process parameters, such as powder preparation, chemical composition, heating rate, holding times and temperatures, and cooling procedures, are properly adjusted, the as-sintered surfaces of all pieces in the load achieve the same composition and structure as the interiors of the pieces. This eliminates the need for costly surface grinding operations to remove the "sinter skin" formed by compositional imbalances and reactions with the furnace atmosphere.

Sintering of Refractory Metals

By Richard F. Cheney
Section Head
Chemical and Metallurgical Division
GTE Products Corp.

TUNGSTEN AND MOLYBDENUM sintering practices are similar; consequently, they will be described together in this section. The sintering of tungsten heavy alloys (tungsten-iron-nickel and tungsten-nickel-copper) and the sintering of tantalum are described separately. Of the other commercially important refractory metals, rhenium is sintered like tungsten, and niobium is sintered in a manner similar to tantalum.

Sintering of Tungsten and Molybdenum

The achievement of high density is a primary goal when sintering tungsten and molybdenum, because high density is frequently required for performance. It is also required if metalworking operations are necessary, in which case at least 90% of theoretical density is essential. High density values indicate that compact shrinkage has been significant, because pressed densities are seldom above 75% of theoretical. In addition to ensuring high density, purification occurs during sintering, and the resultant microstructure is established.

Sintering Variables

Several factors influence the sintering of tungsten and molybdenum, including time, temperature, particle size, atmosphere, purity of powders and compacts, compacted density, thermal gradients, heating rate, cooling rate, particle size distribution, compact weight, gravity, and friction, and additives such as dispersed oxides and sintering activators (Ref 20). These variables and their interrelationships are discussed in the following sections.

Time and temperature are the fundamental determinants of the sintering kinetics for tungsten and molybdenum—in addition to particle size, which is discussed below. Mass transport by diffusion is very slow for refractory metals at most conventional sintering temperatures that are economically or technically feasible. Therefore, the effects of time, temperature, and particle size are more significant for these metals than for metals where temperatures up to and exceeding their melting points are readily attainable.

Figure 46 shows sintering rates, in terms of relative density, achieved for tungsten

and molybdenum at temperatures from 1400 to 1800 °C (2550 to 3270 °F) for particle sizes of 4 μm. The temperature of 1800 °C (3270 °F) is a practical limit for furnaces that use molybdenum or tungsten heating elements and aluminum oxide refractory materials. Note the long sintering time (50 h at 1800 °C, or 3270 °F) required to attain 92% density for tungsten. At the same temperature, molybdenum sinters to a similar density in 1 to 2 h.

Powder particle size is extremely important to the sintering of tungsten, but less so for molybdenum. Figure 47 shows the effect of particle size on the density of tungsten sintered in hydrogen at 1800 °C (3270 °F). Figure 48 compares the effect of several particle sizes on the sinterability of tungsten, as expressed by the relative shrinkages of compacts pressed to the same starting density and sintered at 1800 °C (3270 °F).

Induction and tungsten mesh heating can be used to attain temperatures up to 2400 °C (4350 °F). At these temperatures, sintering may proceed to high density in 1 to 2 h if other factors are not rate limiting. These other factors include heating and cooling rates and billet outgassing.

Self-resistance heating permits attainment of temperatures greater than 3000 °C (5430 °F), and complete sintering cycles require only 20 to 30 min, depending on compact size. Compact size is limited by the available electrical power; the largest practical compact has a maximum cross section of 645 mm² (1 in²) and ranges from 610 to 915 mm (24 to 36 in.) in length.

Atmosphere. Tungsten and molybdenum are always sintered in a hydrogen atmosphere. Vacuum sintering generally is unsuccessful. The only exception to this rule is initial exposure of compacts to hydrogen up to about 1200 °C (2190 °F) before furnace chamber evacuation. Tungsten and molybdenum are highly reactive with carbon and oxygen at almost all sintering temperatures, and the presence of either is usually catastrophic to sintering and billet properties.

Dissociated ammonia generally can be used as a source of hydrogen with no detrimental effects. Nitride formation appears nonexistent for either pure molybdenum or tungsten. However, in some applications, entrapped or dissolved nitrogen may cause a reduction in properties in worked rods and wire.

Powder and Compact Purity. Impurities greatly affect the sintering of tungsten and molybdenum. Because tungsten and molybdenum are sintered at such high temperatures, most impurities are molten, and many exhibit high vapor pressures.

Fig. 46 Relative density versus the logarithm of sintering time for 4-μm tungsten and molybdenum powders

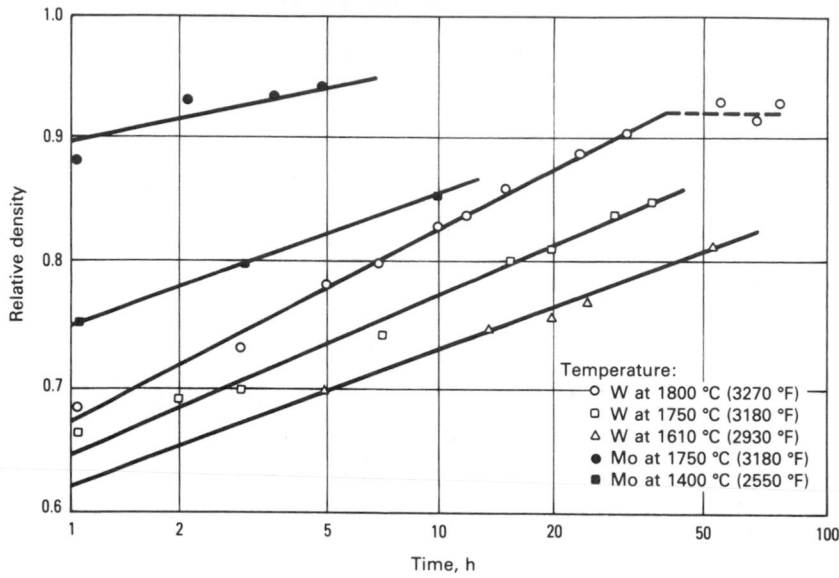

Fig. 48 Relative shrinkage versus tungsten powder particle size

Relative shrinkage ($\Delta L/L_o$) is the change of length over the original length.

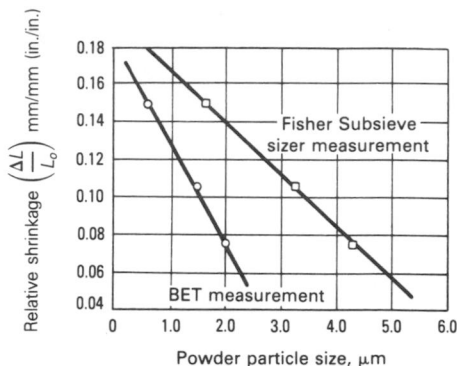

Fig. 47 Relative density versus the logarithm of sintering time for several tungsten powder sizes

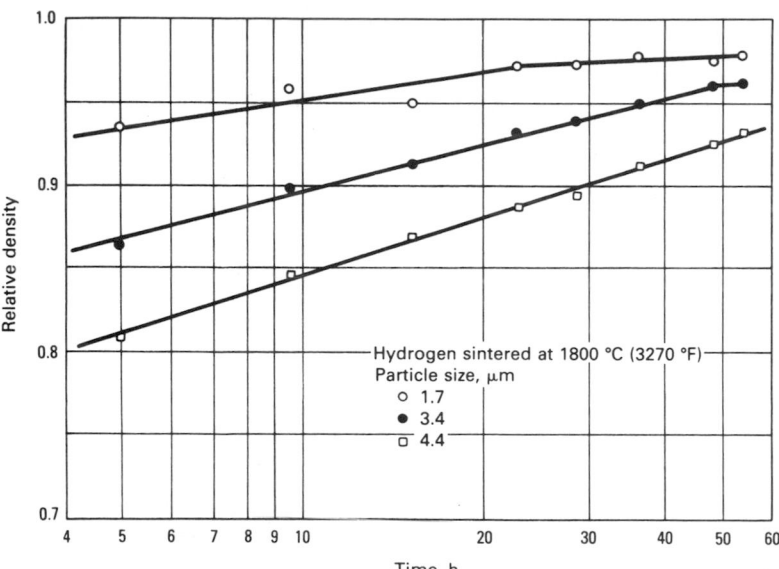

Other impurities, however, may react with the sintering atmosphere to form new compounds, while some may alloy with the base metal—all leading to undesirable influences on properties.

Tungsten and molybdenum powders are fine, typically with particle diameters of 1 to 10 μm and have 0.1 to 0.2% of adsorbed gases (oxygen, nitrogen, and water vapor) on particle surfaces. Entrapment of gases prevents achievement of high density. Additionally, densification does not proceed satisfactorily until surfaces are purged of these gases. Heating and hold-

ing periods of sintering schedules should be designed to allow sufficient time for entrapped gases to be released, particularly for large billets. An adequate hold at about 1200 °C (2190 °F) ensures uniformly low gas concentrations and thus uniform density and microstructure. Large compacts and loads require significantly longer hold times.

Control of Carbon Content. Typically, carbon should be avoided. Carburization begins at 800 °C (1470 °F) for molybdenum and at 1100 °C (2010 °F) for tungsten. Small amounts of carburization

greatly inhibit densification and may greatly affect the ductility and/or workability of the part. Common carbon sources include vacuum pump oils and heating elements.

Metallic elements are used as sintering activators, or they may form alloys or oxides. As liquids, metallic elements such as nickel, cobalt, and iron activate localized grain growth and may leave large voids where the contaminant was originally located. As alloys or oxides, they usually retard local grain growth and frequently inhibit sintering. Generally, contaminants make uniform densification impossible and further working difficult.

Compacted Density. Generally, higher compacted densities are desirable, as long as entrapped gases do not result. Figure 49 shows the effect of cold isostatic compaction pressure on the compacted density of three tungsten powders of different sizes. Figure 50 shows the effect of higher compacted density on the sintered density for a 4.4-μm tungsten powder after various sintering times in hydrogen at 1800 °C (3270 °F). Higher compacted density significantly shortens the sintering times needed to produce a given density. From a practical viewpoint, pressures greater than 517 MPa (75 ksi) are economically difficult to obtain, and compacted billet densities therefore usually range from 65 to 75% of theoretical.

Thermal gradients and heating rates are particularly important when sintering large furnace loads, which is frequently the case with tungsten and molybdenum. Because sintering rates are so sensitive to temperature, thermal gradients greatly affect the density attained in a given time. This in turn affects property and dimensional control. From empirical rate equations, it can be calculated that sintering at

Fig. 49 Relative compact density versus the logarithm of compaction pressure for tungsten powders of various sizes

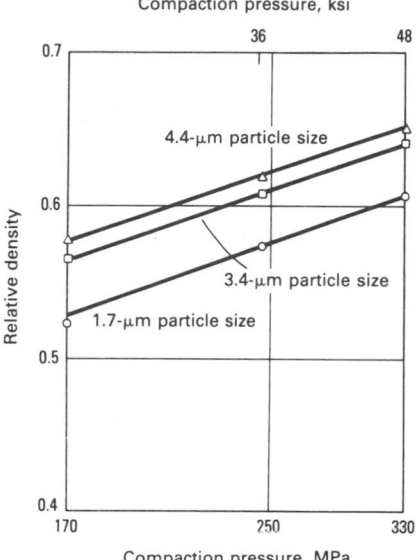

Fig. 50 Relative density versus the logarithm of sintering time for tungsten powders compacted at different pressures

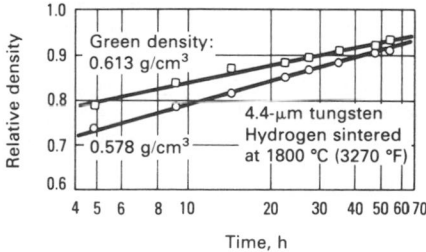

1775 °C (3225 °F) rather than at 1825 °C (3315 °F) will require 110 h (4.6 days) rather than 40 h (1.66 days) to reach 92% density, a difference of 275%.

Temperature gradients are particularly detrimental when sintering large compacts such as 130-mm (5-in.) diam by 250-mm (10-in.) long tungsten billets for forging. At 1800 °C (3270 °F), radiation is the predominant mode of heating, and billet surfaces reach sintering temperatures rapidly compared to billet interiors. Sintering thus begins sooner at the billet perimeter; if uncontrolled, a thin high-density shell forms, which effectively restricts further densification of the interior of the billet.

Furnace heating rate is the primary method of controlling exterior shell formation. Proper choice of powder particle

size is also important—coarser sizes decrease the effect of temperature gradients at the expense of lowering the initial rate of sintering.

The rate of outgassing of impurities is also related to temperature gradients in large compacts. Surface outgassing proceeds more rapidly than outgassing from interior volumes. The inhibiting effect of these gases allows surface volumes to sinter more rapidly, thus creating a high-density shell. Proper control of furnace heating prevents shell formation.

Cooling rate is particularly important with very large parts (305-mm, or 12-in. diam or more). Excessively rapid cooling causes cracking due to thermal stresses.

Particle size distribution is of secondary importance to the sintering variables mentioned above. It is also difficult to quantify these effects to provide general guidelines. Consequently, they will not be discussed in this section.

Compact weight, gravity, and friction are important considerations when large or heavy parts are being sintered. Tungsten cylinders weighing 23 kg (50 lb) or more shrink less at their bases than at their tops when sintered on end. This tendency is mainly due to restricting friction forces. A 12.7-mm (0.5-in.) thick pedestal of compacted tungsten placed under the cylinder alleviates this problem.

Gravity is a significant factor in only a few sintering applications. Rounds may become slightly elliptical when sintered on their sides. Rods suspended from their top ends shrink more across top sections and less along lengths because of their suspended weight. These shrinkage disparities diminish as the bottom of the rod is approached.

Furnace Loading. The use of radiative heating necessitates loading of furnaces to minimize nonuniform heating. Heating rates are important. Parts partially shielded by other parts heat slower and with less uniformity. Parts partially buried in a refractory sand also experience nonuniform heating.

Oxide-Dispersed Additives. Typically, these additives, or dopants, may be thorium dioxide (Ref 21) or the so-called KAS dopant, which is a combination of potassium oxide, aluminum oxide, and silicon dioxide (Ref 22). Their presence inhibits sintering, which can be useful in promoting outgassing of other impurities, in minimizing the effects of thermal gradients, and in controlling microstructure. The silicon dioxide and KAS compounds usually vaporize when sintering occurs at 2000 °C (3630 °F) or above.

Fig. 51 Logarithm of grain diameter versus relative density for tungsten powders of various sizes

Data cover a range of sintering temperatures (1600 to 1800 °C, or 2910 to 3270 °F).

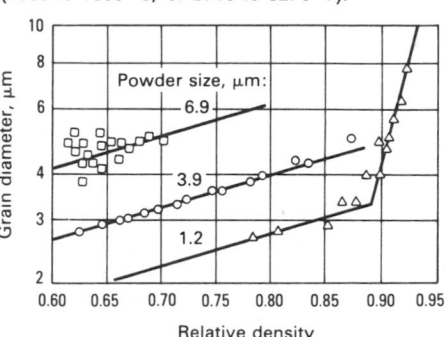

Sintering Activators. The sintering of tungsten and molybdenum can be activated to a considerable degree, but at the expense of significant losses of ductility and workability (Ref 23). Nickel additions to tungsten ranging from 0.2 to 0.5% result in densities of more than 90% of theoretical in only several hours at temperatures as low as 1100 °C (2010 °F), compared to 50 h for unactivated tungsten at 1800 °C (3270 °F). Molybdenum is also activated, but to a lesser extent. The activating elements for both tungsten and molybdenum include palladium, nickel, cobalt, iron, rhenium, ruthenium, iridium, and osmium.

Effects on Microstructure. Sintered grain size is primarily a function of sintered density and starting powder size, as shown in Fig. 51 for tungsten. Grain growth typically proceeds at a more rapid rate above about 95% density, as shown by the lower line. This is due to the decreased effectiveness of grain boundary pinning by the pores. Once the grain boundaries break away from the pores, sintering rates decrease significantly.

The workability of compacts of tungsten and molybdenum is highly dependent on sintered density and, to a lesser extent, on grain size. Densities should be a minimum of 90% of theoretical, and values ranging from 92 to 95% are more desirable. Typical grain sizes are 10 to 30 μm.

Sintering Equipment

Self-resistance heating has been used extensively for tungsten because of the extremely high temperatures required for practical sintering. Typically, rods with cross sections less than 645 mm² (1 in.²) are heated to 3000 °C (5430 °F) or higher. Densification to greater than 90% of theoretical can occur in less than $1/2$ h.

Sintering is done under a hydrogen atmosphere in a water-cooled, copper-jacketed enclosure. Tungsten contacts are required, and one contact must be able to move as the bar shrinks. Bar cross sections must be uniform or sintering temperature will vary along the bar length. Temperatures and densities are always low within several inches of the clamped ends because of the heatsink effect of the contacts. Sintering of molybdenum by this method has been replaced by resistance and induction heated furnaces.

Resistance element heating furnaces are frequently used for hydrogen sintering of molybdenum and tungsten at 1800 °C (3270 °F). Furnaces having uniformly hot zones up to 1.2 m (4 ft) long and 0.14 m^2 (1.5 ft^2) are available. Loads of up to several thousand pounds can be sintered.

A typical furnace consists of water-cooled entrance and exit zones and a central zone of alumina refractory brick that supports molybdenum heating element rods. Furnace loads are supported on molybdenum slabs that are conveyed through the entrance and exit zones by powered rolls. The load is moved through the hot zone by a walking beam that oscillates above and below the hearth and that carries the load forward during the portion of the cycle that is above the hearth. Varying the frequency of oscillation varies the rate of forward movement of the parts in the furnace. Smaller furnaces with lighter loads can be operated by mechanically stoking a series of boats through the furnace.

Induction radiant heating is used when temperatures above 1800 °C (3270 °F) are required and also when very large (greater than 405-mm, or 16-in. diam) parts are involved. Molybdenum alloy billets weighing up to 4500 kg (10 000 lb) have been sintered with this type of heating mode.

A typical furnace consists of a large-diameter water-cooled chamber that contains a cylindrical susceptor ring of tungsten or molybdenum. The susceptor is inductively heated, and the compact is heated by radiation from the susceptor. Depending on the construction materials and loads, sintering temperatures up to 2400 °C (4350 °F) may be obtained.

Sintering of Tungsten Heavy Alloys

Tungsten heavy alloys are tungsten-nickel-iron or tungsten-nickel-copper alloys containing 90 to 98% tungsten. Typical nickel-to-iron ratios range from 1-to-1 to 4-to-1, while nickel-to-copper ratios generally range from 3-to-2 to 4-to-1. Minor amounts of other metals may also be added. Because of the liquid-phase sintering, these alloys achieve essentially full theoretical density. They exhibit good machinability and can exhibit substantial ductility. For example, a 90W-7Ni-3Fe alloy may exhibit elongation as high as 40%. Copper-containing alloys are less ductile and are more difficult to sinter to full density, especially in larger cross sections. One advantage of copper-containing alloys is that they are nonmagnetic. Primary uses for tungsten heavy alloys are as counterweights, radiation shielding, and kinetic energy penetrators.

Tungsten heavy alloys are made from blends of elemental powders. Homogeneity is of prime concern, so fine tungsten powders in the 2- to 8-μm range are used. Standard production procedures for tungsten powder yield fine particle sizes. Carbonyl powders are used in iron and nickel blends, and fine electrolytic powder is used in copper blends. As a precaution against impurities and agglomerates, powders are screened to −200 mesh prior to blending. Any standard powder blending procedure can be used.

Likewise, almost any P/M procedure for compacting powders can be used. If die pressing is used, a binder must be added to the powder. Like pure tungsten powder, high pressed densities cannot be achieved in tungsten heavy alloys. Consequently, allowance must be made for considerable shrinkage during sintering. A die-pressed piece typically is about 55% dense, while an isostatically pressed piece is about 65% dense.

Sintering Furnaces and Atmospheres. Sintering can be carried out in any furnace capable of sustaining the temperature and atmosphere required. Generally, the pieces are supported on aluminum oxide sand or aluminum oxide fixtures. Good control of temperature and heating rates is essential to produce consistent, high-quality tungsten heavy alloys.

Tungsten heavy alloys usually are sintered in pure hydrogen, but hydrogen and nitrogen mixtures, dissociated ammonia, or vacuum also can be used. Argon or mixtures of argon and hydrogen are not used, because the presence of argon during sintering results in a lower density product. With some tungsten powders, blistering of the heavy alloy parts occurs during sintering. The use of wet hydrogen through the sintering schedule usually corrects this problem.

Regardless of the atmosphere used during sintering, a cleanup step in hydrogen at about 1000 °C (1830 °F) is advised. For small pieces, only a few minutes at temperature is required; large pieces may require an hour or more.

During vacuum sintering, hydrogen cleanup is essential. For die-pressed pieces, hydrogen cleanup must be used to remove any residual carbon remaining from lubrication removal. If substantial carbon is still present when the liquid phase forms, serious density and mechanical property problems result.

Solid-State Sintering. At 1200 to 1430 °C (2190 to 2605 °F), it is possible, and sometimes desirable, to solid-state sinter tungsten-nickel-iron alloys to near-theoretical density. Solid-state sintering is not possible with tungsten-nickel-copper alloys because of the low melting point of copper. Once near-theoretical density is achieved, only minimal holding at liquid-phase temperatures is required to complete sintering.

Liquid Phase Sintering. During liquid-phase sintering, alloys densify as tungsten particles grow by dissolution and subsequent precipitation and coalescence of the tungsten grains. Tungsten particles grow from 3 to 8 μm to rounded grains 50 to 150 μm in diameter. As the tungsten content of the alloy increases, grain size tends to increase.

In tungsten-nickel-copper alloys, the low melting point of the copper and the wide liquidus temperature range of the matrix make it difficult to achieve full density in large parts. Normally, these alloys are sintered at 1380 to 1450 °C (2515 to 2640 °F). Very low heating rates that begin near the melting point of copper are required to properly sinter these alloys. Sintering times vary from less than an hour for small parts to several hours for large billets. One advantage of sintering these alloys is that they do not have the strong tendency to slump (distort) as do tungsten-nickel-iron alloys.

Liquid-phase sintering of tungsten-nickel-iron alloys is carried out at 1450 to 1600 °C (2640 to 2910 °F). Generally, sintering temperature increases with tungsten content. Heating rates are not as critical as with tungsten-nickel-copper alloys. With large parts, however, heat should proceed slowly to the temperature where the liquid phase first forms (1450 to 1560 °C, or 2640 to 2840 °F). Large temperature gradients within the part resulting from too rapid heating may cause compositional variations and porosity.

Sintering times are relatively short; 30 min at temperature is sufficient in most cases. Some large parts and alloys with tungsten contents above 97% may require sintering times as long as 2 h. One limitation experienced in sintering of tungsten-nickel-iron alloys is slumping of the parts. As tungsten content decreases, slumping increases. For example, a 97% tungsten alloy experiences minimal slumping, while a 90% tungsten alloy may exhibit severe slumping. Slumping can be controlled by using the lowest sintering temperature possible and by supporting the parts during sintering.

Cooling. On cooling from liquid-phase temperatures, some problems may be experienced. If a long bar is directionally cooled, the composition of both the solute and the resulting solid are likely to change during solidification, thus causing end-to-end variations in composition. Gases released during solidification also can collect near the tail end of the bar, which is the last section to solidify.

If small pieces are cooled rapidly from the liquid phase, a solid shell forms around the part, thus causing the liquid matrix to be drawn from the center of the part in an effort to compensate for solidification shrinkage. Porosity is subsequently produced at the centers of thicker sections.

Heat Treatment. For maximum tensile elongation, tungsten-nickel-iron alloys must be heat treated after sintering. Typical heat treatment involves heating in nitrogen, vacuum, or inert atmosphere at 900 to 1300 °C (1650 to 2370 °F). Time varies with temperature and cross section. One hour at 1200 °C (2190 °F) is sufficient for a 12.7-mm ($1/2$-in.) diam bar, but a 50-mm (2-in.) diam bar requires about 24 h.

Mechanical properties of tungsten alloys vary greatly, depending on processing. For example, a MIL-T-21014B specification calls for 2% minimum elongation for a 95W-Ni-Fe alloy. However, if this alloy is properly processed, it exhibits a typical elongation of 28 to 32%.

As-sintered hardness and yield strength are affected only slightly by tungsten content. Typical hardness is 28 to 31 HRC, and yield strength ranges from 550 to 620 MPa (80 to 90 ksi). Elongation ranges from a maximum of 40% for a 90% tungsten alloy to 18% maximum for a 97% tungsten alloy.

Sintering of Tantalum

Large tantalum products are usually made by vacuum arc casting. Smaller products may be produced economically by vacuum sintering of powders produced by the methods described in the article "Production of Tantalum and Niobium Powders" in this Volume. Powder is compacted and self-resistance sintered at 2500 to 2700 °C (4530 to 4890 °F) in a high vacuum. Alternatively, compacts can be

sintered in resistance-element furnaces at up to 2400 °C (4350 °F) in a vacuum of 1.3 mPa (10^{-5} torr). Additional information on sintering of tantalum capacitors can be found in the article "P/M High-Temperature Materials" in this Volume.

Sintering of Titanium

By Peter C. Eloff
Manager—Powder Product Engineering
General Electric Co.

SINTERING practice (as opposed to hot pressing techniques) generally is applied only to fine titanium powders, which are the by-products of the production of titanium sponge. Such powders, referred to as "sponge fines," are produced by the reduction of titanium tetrachloride with sodium or magnesium. Particles are soft and irregularly shaped, thus well suited to cold compaction. Although sodium-reduced sponge fines are preferred for titanium P/M sintering in the United States, similar powders made by molten salt electrolytic reduction are reported in Russian literature.

Another class of titanium powder, prealloyed powder, is made from alloy ingot by ultraclean processes such as the plasma rotating electrode process, which is described in the article "Atomization" in this Volume. Such powders are spherical and are not compactible at room temperature by ordinary means. These powders can, however, be sintered in loose form to produce a porous structure or coating (Ref 24). This section concentrates primarily on the sintering practices and resultant microstructures and property relationships for parts manufactured from sponge fine types of powders.

Titanium Powder and Compacts

Titanium is very reactive, forming very stable oxides and nitrides. Unlike many metals, titanium is capable of dissolving its own surface oxides at sintering temperatures (Ref 25-27). Consequently, these oxides do not have to be reduced through the use of a reactive atmosphere. The formation of additional oxides should be avoided; consequently, the use of high vacuum or high-purity inert gas atmosphere is common practice.

Table 17 Screen analysis of titanium sponge fines

U.S. mesh	Mesh size μm	Weight percent retained
+80	+177	0
−80+100	−177+149	0.1
−100+140	−149+105	11.2
−140+200	−105+73.7	32.9
−200+230	−73.7+63.5	5.0
−230+325	−63.5+44.5	23.3
−325	−44.5	27.5

Titanium sponge fines used for P/M parts typically range from −80 to −100 mesh. Table 17 shows a typical screen analysis for −100 mesh powder. Sodium-reduced sponge fines contain about 0.23% residual sodium chloride entrained in the particles. Although it is impossible to remove this residual salt by leaching, some salt usually is liberated at high temperatures and condenses in cooler parts of the sintering system.

Sponge fines can be compacted at room temperature in rigid dies or isostatically pressed in elastomeric tooling. Due to the reactive nature of titanium, admixed lubricants are not used, but die-wall lubrication is required with rigid dies. The irregular particle shape of sponge fines provides compacts with excellent green strength, and green densities can range from 60 to 85%, the latter at die pressures of 552 MPa (40 tsi) or isostatic pressures of 414 MPa (60 ksi).

When additional sintered strength is required, alloying powders may be elemental or master alloys; in the case of Ti-6Al-4V, for example, a master alloy powder of 60Al-40V is frequently used. Sponge fine powders exhibit strengthening from sintering at temperatures as low as 1000 °C (1830 °F), but higher temperature and sufficient time must be used when alloying powders are added to allow for complete interdiffusion and to obtain the strengthening benefit of the alloy additions. Temperatures in the range of 1200 to 1260 °C (2200 to 2300 °F), with holding times of 1 to 4 h, are sufficient to homogenize admixed alloy powders.

Vacuum Sintering

The use of cold walled vacuum furnaces generally is preferred for sintering of titanium powders. The furnace should be equipped with a diffusion pump that has a sufficient capacity to accommodate outgassing during heatup and maintenance of high vacuum (133 mPa, or 1×10^{-3} mm Hg) at peak temperatures. Heating elements of molybdenum or graphite cloth with a temperature limit of at least 1245 °C (2275 °F) can be used.

The use of cold traps is necessary to condense vaporized sodium chloride and other impurities that may impair diffusion pump performance. Because significant outgassing occurs during heating, a control system that cuts power to the furnace when a preset pressure level (400 mPa, or 3×10^{-3} mm Hg) is reached should be used. Heating may be restored when higher vacuum is achieved again.

This type of system automatically compensates for variations in furnace load, lubricants, or compact density, for example. The use of an inert gas recirculating system for cooling is optional, except for alloys requiring rapid cooling through the β-α temperature range. While a cooling system can improve furnace throughput, the added difficulties and costs of additional valving (sources of leaks) and maintenance of a water-cooled heat exchanger should be considered carefully before making the installation.

Furnace Loading

Due to its reactivity, titanium adheres to most support materials after sintering. The preferred support or racking material is molybdenum, which has a very low solubility in titanium and is able to maintain its strength at high temperatures. Ti-6Al-4V plate can also be used by placing sheets of felted ceramic between the work and support. At lower temperatures (1000 to 1080 °C, or 1830 to 1975 °F), high-density graphite, coated with a wash of yttrium oxide, can be used. It must be dried separately and outgassed prior to sintering. To help ensure against contamination of the parts being sintered, some loose titanium powder may be included with each load to act as a "getter" for gases. Molybdenum foil placed on top of the load can be used to contain the powder.

Leak Testing. Prior to heating, the furnace should be leak tested for each load. This can be accomplished by pumping the furnace down at room temperature to diffusion pump levels, then isolating the furnace chamber from the pumping system. The leakage, or drop in vacuum, is read on the tank vacuum gauge. The leakage rate constant should not exceed a value of 40 calculated by multiplying the observed leakage rate (μm Hg/h) by the furnace chamber volume (ft³). Thus, for a furnace 3 ft in diameter by 3 ft long, with a volume of 21.2 ft³, the maximum allowable leakage rate would be 1.89 μm Hg/h. This procedure helps to ensure against contamination of the work load, as a high-

Fig. 52 Microstructure of sintered Ti-6Al-4V alloy
(a) 95% dense showing Widmanstatten pattern and porosity (dark areas). Magnification 95×.
(b) 99+% dense showing finer equiaxial structure and finer porosity. Magnification: 95×

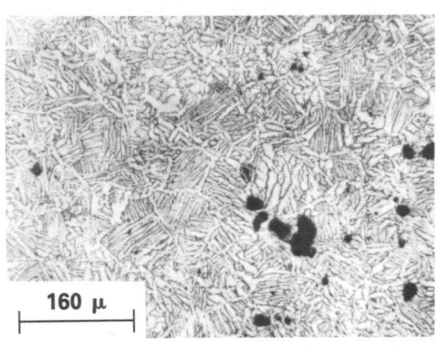

(a)

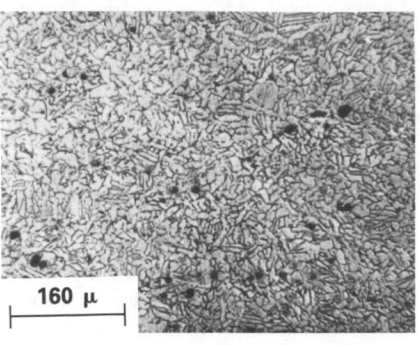

(b)

capacity pumping system can maintain a good vacuum even with a leak. However, continuous introduction of a stream of air over the work rapidly saturates the getter and begins to contaminate the work.

Sintering Cycles. Care must be taken during heating to avoid excessive outgassing of the work, furnace walls, and elements, unless the system pumping capacity is sufficiently large compared to the furnace volume. Heat transfer in vacuum furnaces is primarily by radiation; therefore, a "shadowing" effect occurs with workpieces close to the heating elements—they heat before pieces in the center of the load are heated. Because a heating cycle can be relatively long (about 8 h), peak temperature should be maintained for 2 to 3 h to ensure equalization of the load temperature before cooling.

The use of load-monitoring thermocouples at several locations in the load is recommended, particularly with large loads. Sintering temperatures of 1000 to 1300 °C (1830 to 2370 °F) are commonly used. Lower temperatures are usually used for porous and commercially pure products, and higher temperatures are used to produce alloy products (Ref 28-36).

Alternative Practices

While vacuum sintering is the most widely used sintering practice for titanium, some alternative methods have been used. References 31 and 37 discuss the use of purified argon for sintering powder-rolled titanium strip product. Argon is purified by passing it over titanium chips heated to 800 to 900 °C (1470 to 1650 °F) (Ref 37). Densification in argon is reported to be slightly less than in vacuum (Ref 31).

The use of temperature cycling through the α-β transformation—880 °C (1620 °F) for unalloyed titanium—can promote den-

sification (Ref 38); the effect is the most significant when heating from α to β, which lasts for about 2 min.

Because large thermal lags exist in most vacuum furnaces, this technique is not practical for large production loads. In addition to normal sintering, a supplementary containerless hot isostatic pressing cycle can be used to improve density and mechanical properties (Ref 35, 39). Because the as-sintered density of P/M Ti-6Al-4V is 94 to 96% of theoretical, the porosity is not interconnecting, and densification can be achieved through the application of gas pressure and heat. This procedure is discussed in the article "Hot Isostatic Pressing of Metal Powders" in this Volume.

Microstructure

Because diffusion rates are higher in the beta-phase (bcc) region, sintering is generally done above 880 °C (1620 °F). The microstructure following slow cooling is a transformed beta microstructure, which is considered an inferior structure for conventional wrought material. In the case of commercially pure titanium parts, this limitation is not significant, because static strength and corrosion resistance are of prime importance. Only fatigue and fracture toughness-related properties are affected by this microstructure.

Figure 52 shows microstructures of Ti-6Al-4V at sintered densities of 95% (Fig. 52a) and 99+% (Fig. 52b). In Fig. 52(a), the slowly cooled Widmanstatten, or basketweave, microstructure is apparent, as is residual porosity. The material in Fig. 52(b) was produced by an activation technique (Ref 40) and shows a finer, more nearly equiaxed microstructure. The small, rounded pores in this structure are due to the residual sodium chloride in the sponge fine powder.

Table 18 Comparative properties of P/M and wrought titanium

Property	P/M	Commercially pure Wrought grade II	Ti-6Al-4V Cold pressed/sintered blend		Wrought
Density, %	95.5	100	94	99	100
Modulus of elasticity, GPa (10^6 psi)	103 (15)	102.7 (14.9)	103 (15)	110 (16)	110 (16)
Reduction in area, %	14	35	3	23	43
Ultimate tensile strength, MPa (ksi)	414 (60)	345 (50)	754.3 (109.4)	960.5 (139.3)	948 (137)
Yield strength, MPa (ksi)	324 (47)	345 (50)	643.3 (93.3)	884.6 (128.3)	869 (126)
Elongation, %	15	5	3	12	19

Source: Ref 34, 42, 43

Properties

Static properties of pressed-and-sintered P/M titanium are comparable to those of wrought material, as shown in Table 18. Fatigue resistance is not quite as good as for wrought alloys, primarily due to the presence of salt-related residual porosity and to a lesser extent the transformed beta microstructure. For a further discussion of properties, see Ref 34 and 40 to 42 and the articles "Mechanical Properties of P/M Materials" and "P/M Lightweight Metals" in this Volume.

Sintering of Nickel and Nickel Alloys

By the ASM Committee on Sintering of Nickel*

SINTERED NICKEL-BASED MATE-RIALS enjoy wide commercial usage due to their unique properties, such as corrosion resistance, wear resistance, mechanical strength at low and elevated temperatures, thermal expansion, electrical conductivity, and magnetic permeability. Commercial products represent a wide range of shapes, sizes, and microstructures. Sintered products with up to 90% porosity are technologically as important as fully dense products. Sintering practices aimed at achieving such diverse properties represent a variety of manufacturing technologies.

Depending on the end product, sintering operations may be performed on a loose powder bed, a thin layer applied to a sub-strate, a lightly compacted coating, a roll-compacted strip, or die-compacted preforms. Sintering temperatures range from those at which the material is completely solid to those at which up to 60% of the material is in the liquid state.

Conventional sintering practices generally are useful, although more recent techniques such as hot isostatic pressing are required in specialized applications. Sintering atmospheres are generally reducing; in some applications, however, nonreactive environments such as vacuum or an inert gas are required. Physical and chemical characteristics of the starting powders, such as moisture content, degree of oxidation, segregation, morphology, and purity, are equally as important as the green compact microstructure prior to sintering.

Sintered Porous Nickel Poducts

Sintered porous nickel products of commercial interest include electrodes of alkaline nickel-cadmium, nickel-zinc, and nickel-iron rechargeable batteries, electrodes of alkaline fuel cells and metal-air (zinc-air or iron-air) cells, electrolyzers, and filter elements. Table 19 lists the functions and structural requirements these components must meet for efficient operation.

Highly porous electrodes for alkaline rechargeable batteries are made from nickel powders that have low bulk densities (<1 g/cm³). These powders are produced by carbonyl processing, which is described in the article "Production of Nickel and Nickel Alloy Powder" in this Volume. Slurries made from these powders are coated onto a support strip and then sintered, as shown in Fig. 53. Alternatively, loosely packed beds of powder can be sintered. For all other porous structures, a carbonyl nickel powder with higher bulk density is suitable. Finer nickel powders can also be used by lightly compacting in a die or by roll compacting to produce the desired high green density.

Characteristics of three carbonyl nickel powders are listed in Table 20. The microstructures of these alloys are shown in Fig. 54. The dependence of the porosity of a loosely packed nickel powder bed on powder bulk density is shown in Table 21.

Sintering temperatures between 850 and 1050 °C (1560 and 1920 °F) are commonly used to produce porous nickel products. The reducing conditions required for sintering are obtained easily with nitrogen-hydrogen gas mixtures or a burnt natural gas with low combustibles.

During sintering, loosely packed beds or compacted preforms of carbonyl nickel powders do not densify below 600 °C (1110 °F). However, considerable strengthening of compacted materials occurs below 600 °C (1110 °F) due to surface diffusion. As the temperature increases above 600 °C (1110 °F), volume diffusion increases. Around 950 °C (1740 °F), filaments in the porous structures become smooth and broadened, while in consolidated structures the pores become rounded. To attain adequate strength without excessive loss of porosity, sintering temperatures between 950 and 1000 °C (1740 and 1830 °F) are optimum.

Typical porosity, strength, and electrical resistivity values for loose sintered and pressed-and-sintered materials are given in Table 22. Reduction in surface area as a result of sintering is shown in Table 23. Pore size distributions of sintered porous nickel structures are shown in Fig. 55.

Sintered Dense Nickel Products

Sintered high-density nickel products generally are used as preforms for further working operations. Additional densification occurs as the preform is hot and/or cold worked into a useful commercial shape. Preforms may be round or square billets or strips. Conventional compacting and sintering of individual small components is also used commercially.

A typical example of a high-density billet for further hot working involves extremely high-purity (99.97% Ni), inclusion-free nickel for electronic applications. Commercial melting and casting practices

*By Arun D. Jatkar, Chairman, Metallurgist, Inco Alloy Products Co.; Victor A. Tracey, Manager, Powder Metallurgy, Inco Europe Ltd.; David O. Gothard, Senior Metallurgist, Huntington Alloys, Inc.; J.J. Fischer, Principal Metallurgist, Huntington Alloys, Inc.

Table 19 Functional, service, and structural requirements for sintered porous nickel products

Product	Function	Service conditions of service requirements	Structural requirements
Electrodes of alkaline nickel-cadmium, nickel-zinc, and nickel-iron rechargeable batteries	Hold active mass (nickel hydroxide or cadmium hydroxide) of the battery	Pores in the electrodes are impregnated with the active mass before battery assembly. Electrodes are in contact with a highly caustic solution of potassium hydroxide. Contraction and expansion of the active mass during discharge and charge cycles leads to ~25 vol% change	75-85% porosity for holding the active mass, mechanical strength to withstand handling during assembly and volume changes during battery operation
Electrodes of alkaline fuel cells and metal-air cells	Provide a barrier between the electrolyte and gases	Must provide the largest electrolyte/gas interface. Must avoid complete wetting of pores or bubbling of gases through pores	Small pore sizes; very narrow pore size distribution; total pore volume of about 48%
Filter elements	Filter highly caustic solutions	Depends on specific applications	Depends on specific applications. Both pore volume and pore size distribution important

Fig. 53 Schematic of continuous production of porous nickel for alkaline batteries by slurry sintering process

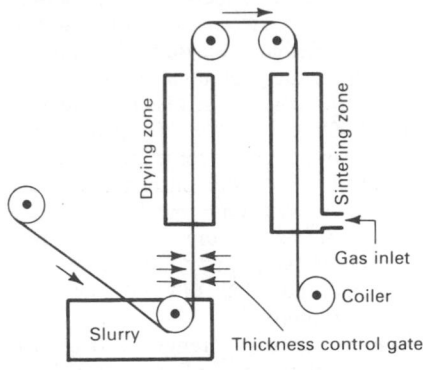

Table 20 Properties of carbonyl nickel powders

Powder type	Bulk density, g/cm³	Average particle size(a), μm	Typical surface area, m²/g
Nickel 255	0.5-0.65	2.2-2.8	0.6(b) 0.7(c)
Nickel 287	0.75-0.95	2.6-3.3	0.5(b) 0.6(c)
Nickel 123	1.80-2.70	3.0-7.0	0.3(b) 0.4(c)

(a) Fisher subsieve analysis. (b) BET method using krypton. (c) BET method using nitrogen

previously used could not guarantee the high purity and freedom from inclusions required for this application. Powder metallurgy processing, however, provides extremely high-purity nickel powders produced by techniques other than atomization. As a result, various mill shapes such as plates, rod, bar, and tube were manufactured from sintered billet stock.

The optimum starting powder for this application is a high-purity nickel powder with a carbon content below 0.1%, oxygen content below 0.15%, and sulfur content below 7 ppm. A high degree of compressibility is desirable. Powder is typically compacted by cold isostatic pressing to a density of approximately 6 g/cm³. Sintering is carried out in a sulfur-free, reducing atmosphere such as hydrogen, dissociated ammonia, or a mixture of 90% nitrogen and 10% hydrogen. The prime objectives of the sintering operation are densification, development of strength (for subsequent hot working operations), and reduc-

Fig. 54 Microstructures of carbonyl nickel powders
(a) Nickel 287. (b) Nickel 255. (c) Nickel 123. See Table 20 for properties of these powders. Magnification: 4000×

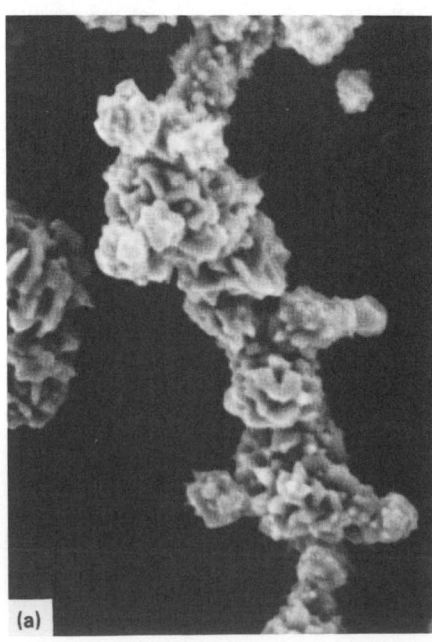
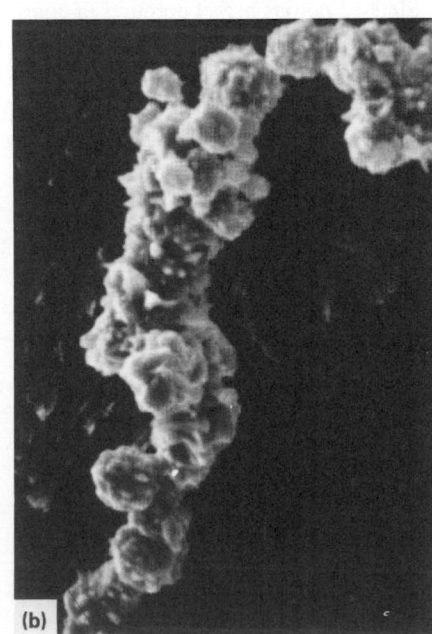
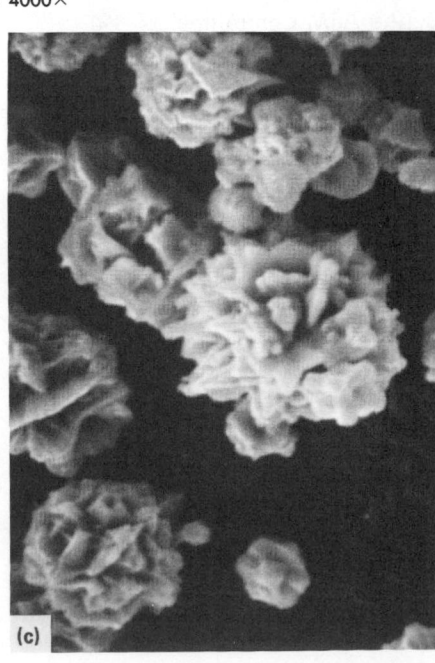

Table 21   Dependence of loosely packed nickel powder bed density on bulk density of powder

Bulk density of nickel powder, g/cm³	Approximate porosity of a loosely packed bed, %
0.5	94.4
1.0	88.8
1.5	83.1
2.0	77.5
2.5	71.9
3.0	66.3
4.5	49.4

Table 22   Typical physical properties of sintered porous nickel structures

Material	Condition	Porosity, %	Electrical resistivity, $\mu\Omega \cdot cm$	Bend strength MPa	psi
Nickel 287	Loose sintered	83	150-200	5.8	840
Nickel 255	Slurry sintered	82	100-150	12.7	1 840
Roll compacted strip	Sintered (49% porosity)	38	30-40	137	19 865
Nickel 123	Loose sintered	60	60-70	24.5	3 550

Note: All materials were sintered at 950 °C (1740 °F) for 5 min.

Table 23   Surface area of sintered porous nickel structures

Powder type	Loose powder	Surface area, m²/g As-sintered at 850 °C (1560 °F) for 5 min	As-sintered at 950 °C (1740 °F) for 5 min
Nickel 255	0.6	0.18	0.16
Nickel 123	0.3	0.11	0.10

Note: Surface area determined by BET method using krypton

Fig. 55 Pore size distributions in sintered porous nickel structures

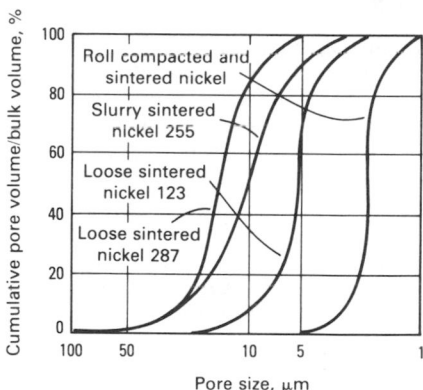

Fig. 56 Effect of sintered density and sintering temperature on transverse rupture strength of water-atomized nickel

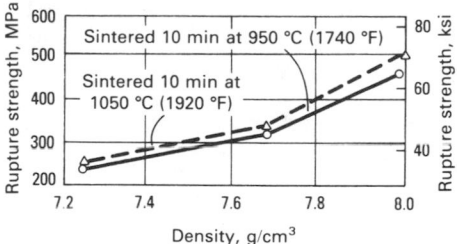

tion of carbon and oxygen levels to below 0.01% and 0.005%, respectively.

During heatup of the green billet, gases adsorbed on powder surfaces are liberated. Carbon and oxygen in the powder then begin to react and produce gaseous products such as carbon dioxide and carbon monoxide. Sufficient time must be allowed at low and intermediate temperatures to permit the escape of these gases and the resulting decarburization before densification begins. For billets to reach a density above 96% of theoretical, sintering temperature is raised to 1100 to 1200 °C (2010 to 2190 °F) following decarburization. Depending on sintering temperature, sintering time may vary from 4 to 12 h.

Coinage strip represents another application in which conventional P/M operations are combined with conventional metalworking operations to yield a unique product. While coin blanks can be compacted from powders using high-speed presses, considerations such as economy and productivity may favor punching of the coin blanks from powder-rolled strip. Green strip is produced by roll compacting nickel powder that is sintered in a continuous sintering furnace to a reasonably high density. The density must be high enough (generally above 80% of theoretical) to facilitate handling and coining operations. Following sintering, the strip is hot and/or cold worked to full density and to the desired thickness of the final product. See the article "Roll Compacting of Metal Powders" in this Volume for additional information.

Nickel powders with consistent high flow rates and compressibility are required so that high volume production units with automatic powder-feeding mechanisms can be used. Water-atomized nickel powders usually are used for this purpose, although powders produced by other techniques are also used commercially. Water-atomized nickel powder typically contains approximately 0.1 to 0.2% oxygen. A major por-

tion of this oxygen is present on or near the surfaces of the powder particles.

Conventional reducing atmospheres (hydrogen, dissociated ammonia, or burnt natural gas) can be used for sintering. The purpose of sintering is to increase the strength of the roll-compacted strip. The effect of sintering temperature on the strength of a sintered strip is shown in Fig. 56. Sintering time-temperature cycles are designed to obtain the desired strength and density, while matching the speed of the roll compacting operation.

The same considerations that apply to high-purity electronic-grade nickel (high purity and freedom from inclusions) are responsible for commercial interest in the production of magnetic and low-expansion P/M alloys. Soft magnetic alloys such as Ni-50Fe, Ni-17Fe-4Mo, and Ni-17Fe-5Cu-4Mo are commercially produced from high-purity elemental metal powders. Car-

bonyl nickel and iron powders and high-purity copper and/or molybdenum powders are mechanically blended, isostatically compacted, and sintered in hydrogen at 1200 to 1400 °C (2190 to 2550 °F). During sintering, the density increases from about 70% in the as-pressed condition to 90 to 95% in the sintered condition. Bulk liquid hydrogen gas with an incoming dew point of about −73 °C (−99 °F) is used for sintering. Sintered billet is subsequently converted to a wrought product by hot and cold working.

Selection of Sintering Atmosphere

Composition and dew point of a sintering atmosphere determine whether reducing or oxidizing conditions prevail during sintering of a given material. The Richardson free energy chart should be consulted for proper conditions. For example, in sintering of nickel at 1000 °C (1830 °F), according to the Richardson free energy chart, the hydrogen-to-water and carbon monoxide-to-carbon dioxide ratios must be greater than 0.01 to prevent oxidation of nickel. A reducing gas containing approximately 5% hydrogen, 5% carbon monoxide, 8% carbon dioxide, and the remainder nitrogen easily meets these requirements.

However, if nickel is alloyed with reactive elements such as silicon, manganese, chromium, vanadium, titanium, aluminum, zirconium, magnesium, and calcium, much higher ratios must be maintained to prevent oxidation of the reactive elements. Thus, for instance, the hydrogen-to-water and carbon monoxide-to-carbon dioxide ratios must be greater than 8×10^3 to prevent oxidation of chromium. Very pure hydrogen with a dew point of -45 °C (-49 °F) or lower is required for sintering nickel-chromium alloys. Also, time and temperature during heatup to sintering temperature must be carefully controlled so that the material being sintered is surrounded by an atmosphere with the required ratios. Additionally, alloys containing nitride-forming elements (chromium, titanium, and zirconium, for example) must be sintered in nitrogen-free atmospheres. In some cases, vacuum sintering may be necessary.

Nonconventional Sintering Practices

Many modifications to conventional pressing and sintering to produce dense bodies from metal powders are commercially important, as discussed below.

Hardfacing and brazing rods of nickel-based alloys classified by the American Welding Society as AWS BNi-1 (13 to 15% Cr, 2.75 to 3.50% B, 4.0 to 5.0% Si, 4.0 to 5.0% Fe, 0.6 to 0.9% C, rem Ni), AWS BNi-3 (2.75 to 3.50% B, 4.0 to 5.0% Si, 0.5% Fe, rem Ni), AWS BNi-4 (1.5 to 2.2% B, 3.0 to 4.0% Si, 1.5% Fe, rem Ni), and low-carbon variations of AWS BNi-1 are customarily cast products.

These alloys can be produced more economically from water-atomized powders of these compositions. The process consists of packing graphite molds with powders to a packed bed density of 50 to 60% and heating them in a reducing atmosphere to a temperature between the solidus and liquidus temperatures—977 to 1030 °C (1790 to 1886 °F) for BNi-1, 982 to 1038 °C (1800 to 1900 °F) for BNi-3, and 982 to 1066 °C (1800 to 1950 °F) for BNi-4. Typically, 40 to 60% mushy liquid is produced at the sintering temperature. Sintering cycles are adjusted so that cooling begins as soon as the mushy liquid is produced. Sintered product is typically 85 to 95% dense. Sintered rods up to 1350 mm (53 in.) long are routinely produced. In some applications, further mechanical working, such as extrusion and wire drawing, is required.

Production of high-temperature corrosion- and creep-resistant P/M nickel-based alloys is also an established practice. The presence of large amounts of highly reactive elements in these alloys makes it difficult to produce fully dense shapes by the conventional press-and-sinter method. Powders of these alloys are produced by inert gas (argon) atomizing a vacuum-induction-melted alloy, so as to exclude any oxygen and nitrogen. In addition, they are sized and encapsulated in an inert atmosphere such as argon.

Consolidation by hot isostatic pressing to produce fully dense, complex near-net shapes components, such as gas turbine engine disks, has become commercially viable. Other consolidation practices for nickel-based superalloy powders include hot isostatic pressing plus forging, hot extrusion, and extrusion plus rolling. Techniques of producing thin, hollow shapes of reactive metals by plasma spraying onto a disposable mandrel are being applied to production of large integral gas turbine engine components. Alloy powder, usually -400 mesh, is sprayed onto acid-leachable steel preforms by low-pressure plasma spraying.

REFERENCES

1. Gummeson, P. and Stosuy, A., "Iron-Carbon Behavior During Sintering," Technical Bulletin D 164, Hoeganaes Corp., Riverton, NJ, 1972
2. Buck, D.M., Marsden, J.G., and Sibley, A.T., High Temperature Sintering With a Nitrogen-Methanol Atmosphere, *Progress in Powder Metallurgy 1982*, Vol 38, Metal Powder Industries Federation, Princeton, NJ, 1983, p 53-64
3. Carlson, E.A., Hansen, J.E., and Lynn, J.C., Characteristics of Full-Density P/M Tool Steel and Stainless Steel Parts, *Modern Developments in Powder Metallurgy*, Vol 13, Hausner, H.H., et al., Ed., Metal Powder Industries Federation, Princeton, NJ, 1981, p 121-136
4. Huppman, W.J. and Beiss, P., "Sintering of P/M Tool Steels to Full Density," Powder Metallurgy Short Course: Fully Dense P/M Material for High Performance Applications, Metal Powder Industries Federation, New Orleans, 1982
5. Podob, M.T. and Woods, L.K., The Mechanism of Sintering High Speed Steel to Full Density, *Modern Developments in Powder Metallurgy*, Vol 13, Hausner, H.H., et al., Ed., Metal Powder Industries Federation, Princeton, NJ, 1981, p 71-92
6. U.S. Patent 3 7094 508
7. Brewin, P.R., Reed, B.W., and Maurer, H.J., "The Influence of Chemical Composition and Production Conditions on the Metallurgical Properties of Sintered High Speed Steels," International Powder Metallurgy Conference (Proceedings), Plansee, Austria, June 1981
8. Ashurst, A., Kulkarni, K.M., and Svilar, M., Role of Additives in Full Dense Sintering of Tool Steels, *Modern Developments in Powder Metallurgy*, Vol 13, Hausner, H.H., et al., Ed., Metal Powder Industries Federation, Princeton, NJ, 1981, p 97
9. Leckie-Ewing, P. and Jacobsen, W.A., The High Temperature Transformation of High Speed Steel, *Met. Trans.*, Vol 1, May 1970, p 1427-1435
10. Farmer, J.I., The Metallography of Bronze P/M Materials, *Copper Base Powder Metallurgy*, Vol 7, Metal Powder Industries Federation, Princeton, NJ, 1980
11. Matthews, P.E., "The Mechanical Properties of Brass and Developmental Non-Ferrous P/M Materials," Technical Bulletin, United States Bronze Powders, Inc., Flemington, NJ
12. Matthews, P.E., Brass and Nickel Silver Powders, *Copper Base Powder Metallurgy*, Vol 7, Metal Powder Industries Federation, Princeton, NJ, 1980
13. Van Horn, K.R., *Aluminum*, Vol 1, American Society for Metals, 1967, p 26-28
14. Van Horn, K.R., *Aluminum*, Vol 3, American Society for Metals, 1967, p 313-314
15. Hale, T.E., private communication
16. Jaffrey, D., Lee, J.W., and Browne, J.D., Co-WC Pseudobinary Eutectic Reaction, *Powder Metall.*, No. 3, 1980, p 140-144
17. Lee, J.W. and Browne, J.D., Influence of Process Variables on Sintering of WC-Wt% Co, *Powder Metall.*, Vol 2, 1980, p 50
18. Tulhoff, H., On the Grain Growth of WC in Cemented Carbides, *Modern Developments in Powder Metallurgy*, Vol 14, Hausner, H.H., et al., Ed., Metal Powder Industries Federation, Princeton, NJ, 1980, p 269-277
19. Rudiger, O. and Rottger, H., "On the Problem of Evaporation of the Auxiliary Metal During Vacuum Sintering of Hard Metals," International Symposium on Powder Metallurgy, Paris, June 1964
20. Cheney, R.F., "Sintering of Tungsten and Molybdenum," GTE Technical Report TT-702, GTE Products Corp., Towanda, PA, 1970
21. Lenel, F.V., *Powder Metallurgy—Principles and Applications*, Metal

Powder Industries Federation, Princeton, NJ, 1980, p 344

22. Lenel, F.V., *Powder Metallurgy—Principles and Applications*, Metal Powder Industries Federation, Princeton, NJ, 1980, p 338

23. Zovas, P.E., *et al.*, Activated and Liquid Phase Sintering—Progress and Problems, *J. Metals*, 1983, p 28-33

24. Shimizu, H. and Gibbon, W.M., Green and Sintered Compacts of Ti-6Al-4V Powders, *Modern Developments in Powder Metallurgy*, Vol 13, Hausner, H.H., *et al.*, Ed., Metal Powder Industries Federation, Princeton, NJ, 1980, p 513-521

25. Hansen, M., Constitution of Binary Alloys, McGraw-Hill, New York, 1958, p 1068-1072

26. Elliot, R.P., *Constitution of Binary Alloys, First Supplement*, McGraw-Hill, New York, 1965, p 697-699

27. Munir, Z.A., Analytical Treatment of the Role of Surface Oxide Layers in the Sintering of Metals, *J. Mat. Sci.*, Vol 14 (No. 11), 1979, p 2733-2740

28. Fukube, Y. and Kashu, S., "Vacuum Sintered Titanium and Its Properties," Proceedings of the Fourth International Conference on Vacuum Metallurgy, Japan Institute of Metals, 1974, p 272-275

29. Pavlov, V.A., *et al.*, Processes Suitable for the Manufacture of Parts from Ti Powders, *Sov. Powder Metall. Met. Ceram.*, Vol 13 (No. 5), 1974, p 380-383

30. Isserow, S. and Roderick, D.J., Large Ti Tube Fabrication by Cold Isostatic Pressing and Sintering of Powders, *Powder Metallurgy in Defense Technology*, Vol 3, Metal Powder Industries Federation, Princeton, NJ, 1977, p 151-155

31. Maramatsu, Y. and Tamura, K., Production of Titanium Strip by Powder Rolling, *Trans. Nat. Res. Inst. Met. (Japan)*, Vol 20 (No. 3), 1978, p 172-177

32. Bezruchko, V.P., *et al.*, Oxidation as a Means of Increasing the Fatigue Strength of Sintered Titanium, *Sov. Powder Metall. Met. Ceram.*, Vol 17 (No. 3), 1978, p 246-249

33. Antsiferov, V.N., *et al.*, Long-Time Strength of Sintered Titanium Alloys, *Sov. Powder Metall. Met. Ceram.*, Vol 18 (No. 3), March 1979, p 196-199

34. Gessinger, G.H., Titanium Powder Metallurgy and Composites, *Titanium '80, Science and Technology*, Vol 1, TMS-AIME, Warrendale, PA, 1980, p 243-251

35. Abkowitz, S., Isostatic Pressing of Complex Shapes from Titanium and Titanium Alloys, *Titanium '80, Science and Technology*, Vol 3, TMS-AIME, Warrendale, PA, 1980, p 2321-2330

36. Pavlov, V.A. and Lyashenko, A.P., Forging and Extrusion of Products from Powder Titanium Billets, *Titanium and Titanium Alloys, Scientific and Technological Aspects*, Vol 1, Plenum Press, New York, 1982, p 355-358

37. Katrus, O.A. and Asleshina, A.V., Sintering of Rolled Strip from Titanium Powders, *Sov. Powder Metall. Met. Ceram.*, Vol 10 (No. 12), 1971, p 965-966

38. Akechi, K. and Hara, A., Influence of Transformation on Sintering of Titanium Powder, *Titanium '80, Science and Technology*, Vol 3, TMS-AIME, Warrendale, PA, 1980, p 2255-2265

39. Abkowitz, S., Isostatic Pressing of Complex Shapes from Titanium and Titanium Alloys, *Powder Metallurgy of Titanium Alloys*, TMS-AIME, Warrendale, PA, 1980, p 26-28

40. Andersen, P.J. and Eloff, P.C., Development of Higher Performance Blended Elemental Powder Metallurgy Titanium Alloys, *Powder Metallurgy of Titanium Alloys*, TMS-AIME, Warrendale, PA, 1980, p 175-187

41. Mahajan, Y., *et al.*, Microstructure Property Correlation in Cold Pressed and Sintered Elemental Ti-6Al-4V Powder Compacts, *Powder Metallurgy of Titanium Alloys*, TMS-AIME, Warrendale, PA, 1980, p 189-202

42. Boyer, R.R., *et al.*, Characterization of Pressed and Sintered Ti-6Al-4V Powders, *Powder Metallurgy of Titanium Alloys*, TMS-AIME, Warrendale, PA, 1980, p 203-216

43. Garriott, R.E. and Thellmann, E.L., Titanium Powder Metallurgy—A Commercial Reality, *Modern Developments in Powder Metallurgy*, Vol 11, Metal Powder Industries Federation, Princeton, NJ, 1976, p 63-78

SELECTED REFERENCES

● Bitzer, J.D., "Recent Developments in the Production of Strip, Rod, and Wire by Powder Metallurgy," Specialty Metals Division, Magnetics, Inc., Butler, PA, Aug 1969

● Blore, M.H.D., *et al.*, Pure Nickel Strip by Powder Rolling, *Metals Eng. Quart.*, Vol 6, 1966, p 54-60

● "British Achievements with Magnetic Tape Rolled from Powdered Metals," Henry Wiggin and Company Ltd., England, 1957

● Davis, G.L., P/M Nickel Alloys for Thermionic Emitters, *Powder Metall.*, Vol 25 (No. 4), 1982, p 216-220

● Durdaller, C., Furnace Atmospheres, in *Source Book on Powder Metallurgy*, American Society for Metals, 1979, p 179

● Exner, H.E., Physical and Chemical Nature of Cemented Carbides, *Int. Metals Rev.*, No. 4, 1979

● Froschauer, L. and Fulrath, R.M., Direct Observation of Liquid Phase Sintering in the System Tungsten Carbide-Cobalt, *J. Mat. Sci.*, Vol 11, 1976

● "Inco Nickel Powders: Properties and Applications," International Nickel Company, Inc., Dec 1983

● Johansson, T. and Uhrenius, B., Phase Equilibria, Isothermal Reactions, and a Thermodynamic Study in the WC-Co System at 1150 °C, *Metal Science*, Feb 1978

● Meridith, B. and Milner, D.R., Densification Machining in the Tungsten Carbide-Cobalt System, *Powder Metall.*, No. 1, 1976

● Murrell, D.L. and Enoch, R.D., The Sintering Process of a High Permeability Ni-Fe-Cu-Mo Alloy Made by Powder Metallurgy, *J. Mat. Sci.*, Vol 5 (No. 6), 1970, p 478-486

● Nayar, H.S., Hardfacing and Welding Rods by P/M, in *Modern Developments in Powder Metallurgy*, Vol 11, Hausner, H.H. and Taubenblat, P.W., Ed., Metal Powder Industries Federation, Princeton, NJ, 1977, p 109-141

● Rautala, P. and Norton, J.T., Tungsten-Cobalt-Carbon System, *Trans. AIME*, Oct 1952

● Snowball, R.F. and Milner, D.R., Densification Processes in the Tungsten Carbide-Cobalt System, *Powder Metall.*, Vol 11 (No. 21), 1968

● Tracey, V.A., The Production of Porous Nickel for Alkaline Battery and Fuel Cell Electrodes: Practical and Economic Considerations, *Powder Metall.*, Vol 8 (No. 16), 1965, p 241-255

● Tracey, V.A., Sintering of Porous Nickel—Theoretical and Practical Considerations, in *Modern Developments in Powder Metallurgy*, Vol 12, Hausner, H.H., *et al.*, Ed., Metal Powder Industries Federation, Princeton, NJ, 1981, p 423-438

● Tracey, V.A., Factors Affecting the Properties of Sintered Structures for Alkaline Batteries, I. and E.C. Product Research and Development, No. 21, 1982, p 626-629

- Tracey, V.A. and Cutler, C.P., High Temperature Alloys from Powders, *Powder Metall.*, Vol 24 (No. 1), 1981, p 32-40
- Tracey, V.A. and Perks, R.P., The Properties of Sintered Carbonyl Nickel Powders, *Powder Metall.*, No. 12, 1963, p 54-71
- Walker, E.V., Worn, D.K., and Walters, R.E.S., "Application of Powder Metallurgy to the Production of High Permeability Magnetic Alloy Strip," Iron and Steel Institute Special Report No. 58, 1956, p 204-208
- Warren, R. and Waldron, M.B., Microstructural Development During the Liquid-Phase Sintering of Cemented Carbides, *Powder Metall.*, Vol 5 (No. 30), 1972
- Williams, N.J. and Tracey, V.A., Porous Nickel for Battery and Fuel Cell Electrodes: Production by Roll Compaction, *Int. J. Powder Metall.*, Vol 4 (No. 2), 1968, p 47-62
- Worn, D.K., The Continuous Production of Strip by the Direct Rolling Process, *Powder Metall.*, No. 1/2, 1958, p 85-93
- Worn, D.K. and Perks, R.P., Production of Pure Nickel Strip by the Direct Rolling Process, *Powder Metall.*, No. 3, 1959, p 45-71

Roll Compacting of Metal Powders

By the ASM Committee on Roll Compacting*

ROLL COMPACTING, or powder rolling, refers to the continuous compaction of metal powders by a rolling mill. In this process, metal powders are fed from a hopper to a set of compacting rolls that produce a continuous green (unsintered) strip or sheet. These materials undergo further processing by sintering (heating) and re-rolling to produce an end product with the desired material properties. Material can be produced fully dense or with a desired porosity and pore size.

Ideal powder characteristics for roll compaction are comparable to those desired in powders for die compaction: flowability is important to ensure consistency in the final product; softness and irregularity of powder particle shape promote green strength; and particle size and size distribution influence green density. Metal powders suitable for roll compacting include elemental powders, blends of elemental powders or elemental and alloy powders that are homogenized during sintering, blends that contain additives that are nonreactive and that cannot be produced by standard melting and casting processes, and prealloyed atomized powders.

Atomization has the capability to produce powder particles that are individually uniform in composition and unique in internal structure. Strip produced by roll compacting of powders subsequently produces materials that are clean and completely uniform throughout the coil, with a fine grain size.

Production Procedures

Roll Position. The type and position of the rolls vary with the type of system chosen to produce the desired finished strip. Rolls can be positioned vertically (as in conventional rolling mills), horizontally, or at an inclined angle, as shown in Fig. 1.

Roll position selection is determined by several factors: the characteristics of the material, the amount of strip to be produced, the desired end product, and special structural requirements. Horizontal roll position (Fig. 1a) using gravity feed offers the most flexibility in controlling powder feed and is a convenient way of introducing several powders for multiple-layer strip. A conventional vertical mill can also be used to roll compact powder. However, vertical roll position (Fig. 1c) requires a very tall sintering furnace, or the green strip must be turned horizontally to enter a conventional furnace. This action could disrupt green bonds and inhibit sintering. The inclined angle position (Fig. 1d) offers a compromise between vertical and horizontal roll positions. Using this approach, a conventional mill can usually be used and powder can be gravity fed.

Powder Feeding. The initial step in the roll compacting process is the feeding of powder to the rolls. The green strip should exhibit uniform density from edge to edge; otherwise, problems may be encountered during processing, and nonuniform conditions cannot be rectified. If strip has a higher density on one side, it may exit the re-rolling mill at an angle, exhibiting camber or a long edge. However, if the center of the strip has a lower density, it may crack on re-rolling due to densification and elongation of the edges. If the edges of the strip have a lower density, the higher density center area may densify and elongate during re-rolling, causing edge cracks.

Two methods of feeding powder to the rolls are illustrated in Fig. 2. The powder is fed to the hopper by a belt or a vibrating feeder from the storage bin. During saturated feeding (Fig. 2a), the powder falls directly from the hopper into the roll gap. The head of powder in the hopper should be held constant, because it applies pressure on the powder in the roll gap. A variation in pressure at the gap affects the feed rate of the powder, which results in varying density along the strip length. In the unsaturated, or "starved," feeding system (Fig. 2b), the amount of powder entering the roll gap is controlled by adjustable gates. Variations in compacted density can be made by adjusting the gates. Adjustable gates also can be used for saturated feeding.

Cover plates, formed to fit the curvature of the roll diameter, are available in sections so that powder feed can be varied at different areas of the roll surface as required. Figure 3 illustrates a typical powder feeder. This design incorporates a lateral deflector plate that distributes the powder evenly across the powder hopper. At present, strips of any roll-compacted powder can be produced by both saturated and unsaturated feeding in widths up to 53 cm (21 in.) and thicknesses of 2.5 to 3.2 mm (0.100 to 0.125 in.). Lateral deflector plates in the powder feeder are not necessary to achieve these results in all cases.

Edge Control. The green strip rolled from powder should have uniform thickness and density across the width of the strip, and its edges should be well formed and as dense as the center of the strip. Therefore, a method of containing the powder being fed to the compacting rolls is essential to process control. Without edge

*Walter V. Knopp, *Chairman,* President, P/M Engineering & Consulting Co., Inc.; William R. Duncan, Superintendent, Fabricated Metal Products, Sherritt Gordon Mines Ltd.; Alan J. Moses, Plant Manager, Pfizer Inc.; Milton W. Toaz, Program Manager, Engine Parts Division, Imperial Clevite Inc.

Fig. 1 Typical roll positions

(a) and (b) Horizontal positioning. (c) Vertical positioning. (d) Inclined angle positioning

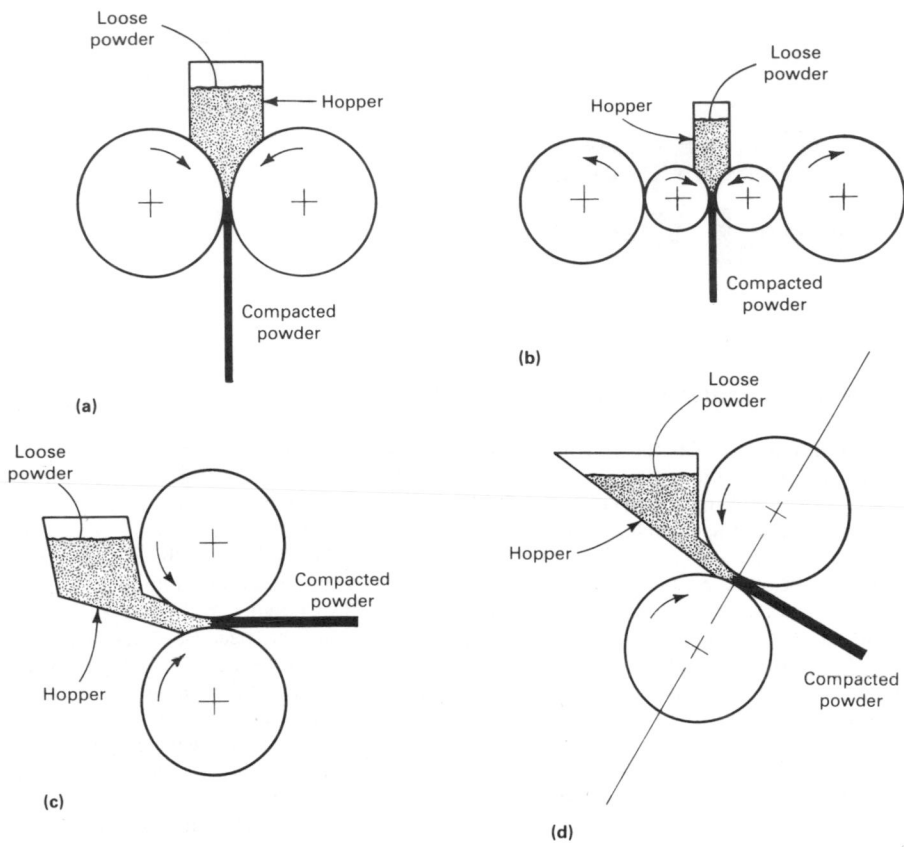

Fig. 3 Typical powder compaction mill feeder arrangement

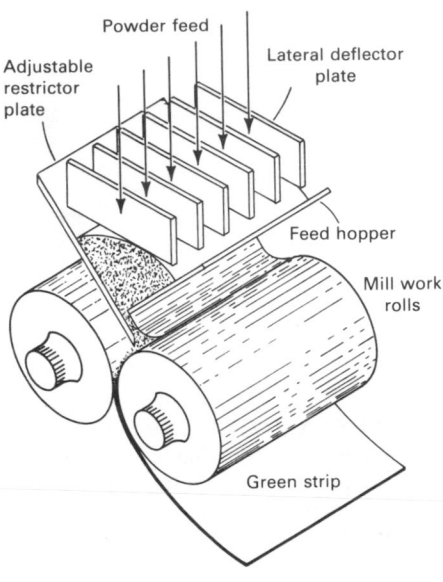

Fig. 2 Methods of feeding powder to the rolls

(a) Schematic of roll compaction with saturated feed, rolling mill rolls arranged horizontally. (b) Roll compaction with unsaturated feed controlled by adjustable gates

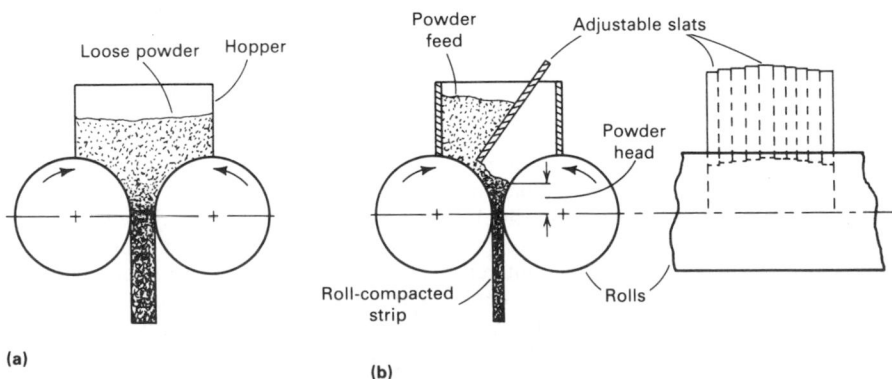

control, powder escapes from the roll gap. Consequently, a strip with low-density edges is produced, which necessitates edge trimming after the strip is re-rolled.

Figures 4 and 5 illustrate typical methods of retaining powder in the roll gap. The method shown in Fig. 4(a) uses floating flanges that are attached to one roll and that overlap the other roll. Pressure is applied to the flanges as they approach the roll gap, thus preventing powder loss from the gap. A continuous belt that covers the gap at the edge of the rolls also is effective in preventing powder loss (Fig. 4b and 5). Other systems are available that produce strip with a tight edge that requires minimal edge trimming.

Roll Diameter. In powder rolling, strip thickness is mainly determined by the roll diameter used. Roll diameter varies from 12.7 mm (0.5 in.) to 920 mm (36 in.). Figure 6 illustrates the effect of roll diameter. The nip, or gripping, angle for conventional metal rolling is the same for powder rolling (around 7 to 8°). Note that on the larger diameter roll, a larger arc is included by the nip angle on the roll surface. Therefore, a larger amount of powder is pulled into the roll gap than in the small-diameter rolls. For a given powder, a thicker strip can be produced with the larger diameter rolls.

Figure 7 shows the strip thickness that can be obtained with different roll diameters with two types of nickel powder. The atomized nickel is soft, while the hydrometallurgical nickel powder is hard. The ratios of roll diameter to optimum green strip thickness vary widely for different powders. They depend on the flow and apparent density of the powder, the coefficient of friction between the powder and roll, roll temperature, and whether the powders produced are hard or soft.

If the coefficient of friction between the powder and the rolls is increased by roughening the roll surface, a larger amount of powder will be pulled into the roll gap. For the same roll setting, higher compacted densities will be obtained. An increase in roll temperature, achieved through preheating or increased pressure, will also increase the coefficient of friction with the same results.

The roll gap is normally set to provide a green density about 80 to 90% of the theoretical density at a thickness that will

Fig. 4 Methods of controlling powder feed to compacting rolls
(a) Flange edge control. (b) Belt edge control

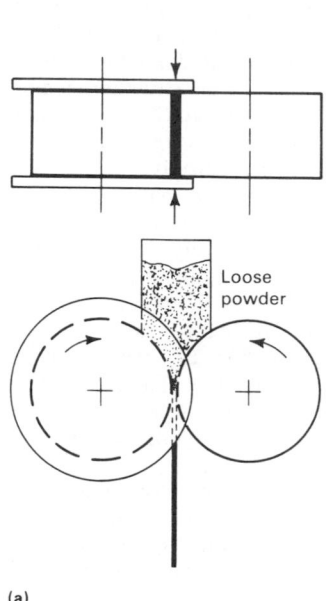

(a)

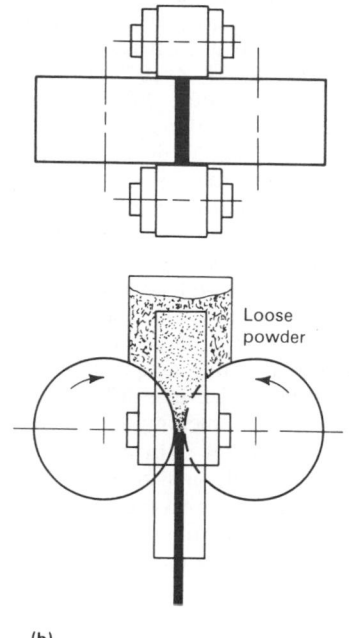

(b)

Fig. 5 Edge restriction device
V-belts and pulleys are used to contain powder at the edge of roll face and in the roll gap of a powder compaction mill.

Fig. 6 Effect of roll diameter

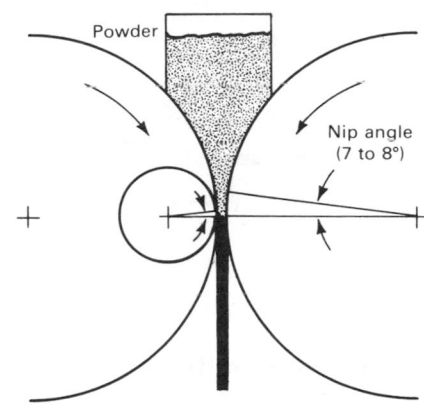

Fig. 7 Effect of work roll diameter on green strip thickness for two types of nickel powder
Type of powder and mill geometry affect strip thickness.

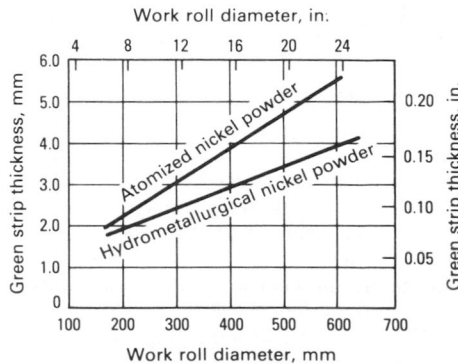

provide full density when processed to final gage. Hot and cold workability of the material will determine the compacted gage.

Finishing. Uniform green strip produced by the compacting mill must be sintered before further densification. Green density usually is between 75 and 90% of theoretical density. A higher density renders the green strip hard and brittle, while a lower density produces a weak strip that is difficult to handle. Finishing of strip can be accomplished by several methods, depending on the type of material being produced, quantity required, sintering temperature and time, atmosphere requirements, and hot or cold working characteristics.

Figure 8 shows two methods of handling strip when densification is achieved by cold rolling. In Fig. 8(a), strip is coiled green and then sintered in a bell-type furnace. In Fig. 8(b), strip is passed through a sintering furnace, cooled, and then coiled. If necessary, additional sintering and rerolling steps are completed in the same manner.

In Fig. 9, strip from the compacting mill is passed through the sintering furnace. While still hot, strip is hot rolled to full density. Strip is then cooled before leaving the furnace and subsequently coiled.

Commercial Production

Powder rolling was described in a number of early patents (Ref 1), but the first in-depth analysis of the process was undertaken by Naeser and Zirm (Ref 2), who documented their work on rolling RZ iron powder into strip. Powder rolling of copper powder is described in Ref 3 and 4.

From 1958 to 1968, powder rolling processes received considerable attention in Great Britain, the United States, Canada, the Soviet Union, and Japan. Principal interest centered around combining low-cost processes for producing metal powders, such as copper, nickel, cobalt, iron, and aluminum, with the powder rolling process to develop methods of fabricating thin sheet or strip more economically.

Nickel Powder Strip. The Sherritt Gordon process of rolling nickel powders into strip and sheet was one result of these efforts (Ref 5). This strip was used to make coinage blanks for the Canadian Mint. Typical mechanical properties of roll compacted nickel strip at various stages of manufacturing are given in Table 1.

After nickel powder is compacted into a green strip, the strip is sintered between 1000 and 1200 °C (1830 and 2190 °F) in a muffled furnace. Hydrogen, dissociated ammonia, or even less reducing gases provide a suitable atmosphere. Strip is conveyed through the furnace on a mesh belt or rollers.

Densification of the roll-compacted nickel strip can be achieved by hot or cold rolling. Hot rolling of nickel strip is carried out above 800 °C (1470 °F), with a

Fig. 8 Powder rolling processes

(a) Cold rolled green strip is coiled and then sintered in a bell-type furnace. (b) Cold rolled strip is reeled into individual coils after first sintering treatment.

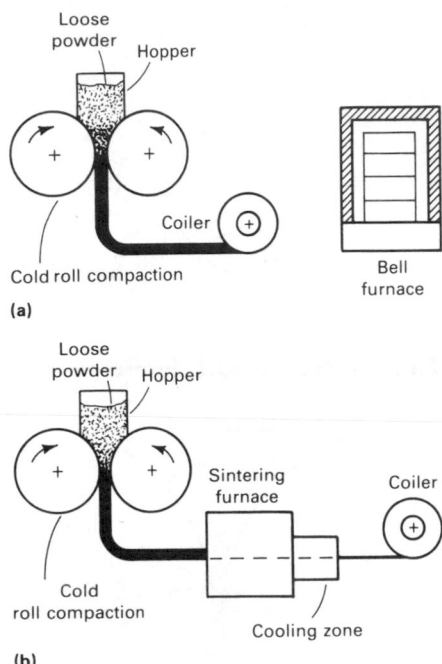

Fig. 9 Powder rolling process

Shown are horizontal arrangement of rolls, sintering, hot rolling, cooling, and coiling of strip

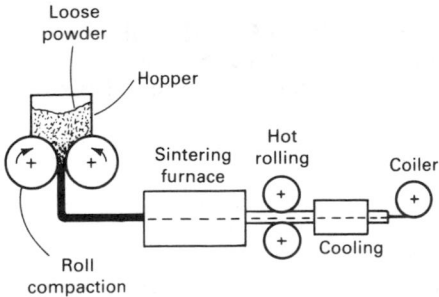

reduction in thickness of approximately 50%. Because of its porous nature, the sintered strip should be protected from oxidation when heating for hot rolling. This can be accomplished by using an inert or exothermic atmosphere.

Cold rolling and annealing cycles also may be used to densify the sintered strip. To be successful, sintering must be carried out at 1100 °C (2010 °F) or higher. Limited reduction is achieved on the first cold mill pass. The strip must then be fed directly to a furnace without coiling, where it is annealed and cold rolled to full den-

sity. Reductions in excess of 35% are necessary to achieve full density.

Several advantages are gained from producing high-purity nickel strip by roll compacting. Lower electrical resistivity is possible (73 to 79 $\times$ 10^{-6} $\Omega \cdot$ m, or 44 to 48 $\Omega \cdot$ cir mil/ft). Stability can be maintained at $\pm 2\%$ throughout the coil and from heat to heat. Wrought nickel offers an erratic $\pm 6\%$ tolerance. Work hardening rates for roll-compacted nickel strip are 25% less than for wrought nickel strip.

The lower softening (annealing) temperature coupled with high purity makes this nickel useful in clad metal combinations. In these applications, low and closely controlled annealing temperatures are required to minimize interdiffusion and to prevent incipient melting reactions.

Finished nickel strip produced from powder is virtually indistinguishable from strip produced from an ingot. Differences in physical properties are the result of compositional variances, rather than the method of fabrication. Typical physical properties for strip made from nickel powder are:

Density . . . . 8.90 g/cm³
Coefficient of thermal expansion
  at 20 to 100 °C . . . . 14 μm/m · °C
Coefficient of thermal expansion
  at 20 to 500 °C . . . . 15 μm/m · °C
Thermal conductivity . . . . 86.23 W/m · K
  (0.206 cal/cm · s · °C)
Cold working capacity . . . . Good
Hot forming capacity . . . . Good
Hot work temperature . . . . 800 to 900 °C
  (1470 to 1650 °F)
Annealing temperature . . . . 700 to 900 °C
  (1300 to 1650 °F)

Magnetic properties

Curie temperature . . . . 353 °C (667 °F)
Initial permeability 130
Maximum permeability 1240
Saturation induction . . . . 6.05 T
Remanance . . . . 3.25 T
Coercivity . . . . 23.87 A/m
Magnetostriction (soft) 1590 A/m 0.000032
  mm/mm

Cobalt Powder Strip. Other metal powders can be roll compacted using methods similar to that of nickel powder. Powder characteristics and material properties alter processing parameters and conditions, however. For example, cobalt has a close-packed hexagonal crystal structure, which changes the surface morphology of the cobalt powder. Roll-compacted cobalt powder produces a green strip that is stronger and denser than the comparable green nickel strip. Typical mechanical

properties of powder rolled cobalt strip at each stage of the production process are given in Table 2.

After sintering at 1100 to 1150 °C (2010 to 2100 °F), cobalt may be densified by hot or cold rolling. Cold rolling of pure cobalt is inhibited by the close-packed hexagonal crystal structure, which characteristically produces rapid work hardening and restricts cold reduction to about 25% between anneals.

The use of high-purity cobalt powders for the feed to a powder rolling mill results in a relatively ductile strip. Use of controlled conditions during annealing results in increased amounts of cubic phase (usually associated with cobalt above 445 °C, or 835 °F), which is retained at room temperature. The retained phase is stable and shows little tendency to transform to the hexagonal structure at room temperature.

The rapid work hardening characteristics of cobalt make it an ideal candidate for powder rolling. The initial strip can be rolled to a thickness very close to the required final gage. Because initial strip thickness is thin (1.5 to 2.0 mm, or 0.06 to 0.08 in.), cold rolling is minimized. Typical physical properties of powder rolled cobalt strip are:

Density . . . . 8.85 g/cm³
Coefficient of thermal expansion
  at 20 to 100 °C . . . . 13 μm/m · °C
Coefficient of thermal expansion
  at 20 to 500 °C . . . . 14 μm/m · °C
Thermal conductivity
  at 70 °C (160 °F) . . . . 115.9 W/m · K
  (0.277 cal/cm · s · °C)
Cold working capacity . . . . Poor
Hot forming capacity . . . . Good
Hot work temperature . . . . 600 to 800 °C
  (1110 to 1470 °F)
Annealing temperature . . . . 800 to 1000 °C
  (1470 to 1830 °F)

Magnetic properties
(Unworked high-purity powder rolled cobalt strip)

Curie temperature . . . . 1121 °C (2049 °F)
Initial permeability 11.6
Maximum permeability 29.2
Saturation induction . . . . 1.9 T
Saturation field strength . . . . 310 000 A/m
Remanance . . . . 0.3 T
Coercivity . . . . 3 600 A/m
Coefficient of friction at 70 °C
  (160 °F) cobalt/cobalt 0.3

Production of heterogeneous alloys is possible with roll compacting processes. In cobalt-iron materials, for example, the presence of the iron promotes a suppression of the cubic to hexagonal phase transformation in cobalt. The ductility of the cobalt-iron mixture increases with in-

Table 1 Typical mechanical properties of hydrometallurgical powder rolled nickel strip

Strip	Strip thickness		Strip density, %	Ultimate tensile strength		Yield strength		Elongation, %
	mm	in.		MPa	ksi	MPa	ksi	
Green strip	4.0	0.158	79	4	0.6	4	0.6	0
Sintered strip	4.1	0.161	79	138	20	136	19	0
Hot rolled strip	2.1	0.084	100	358	52	165	24	38
Cold rolled strip	1.3	0.052	100	579	84	572	83	5
Annealed strip	1.3	0.052	100	362	53	83	12	48

Note: Compacting roll diameter, 560 mm (22 in.). Roll speed, 2.2 rpm. Roll gap (green strip), 3.5 mm (0.140 in.).

Fig. 10 Properties of cobalt-iron alloys
Manufactured by powder rolling of blended cobalt and iron powders

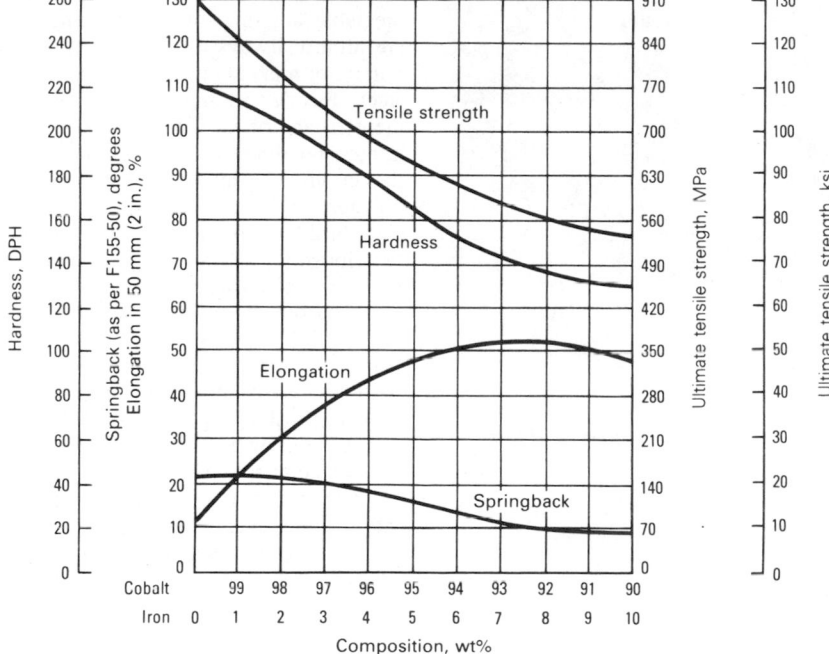

creased iron content. The effects of adding iron powder to cobalt powder on the properties of cobalt strip are shown in Fig. 10.

Specialty Applications

Roll compacting facilitates the production of small coils in which minor compositional changes can be made during blending. Because roll compacting produces materials with uniform and reliable properties, many specialty materials are produced by these methods.

Specialty P/M strip produced by roll compacting include various compositions of nickel-iron strip produced for controlled expansion properties and special copper-nickel-tin alloys. The latter must meet stringent property requirements, such as strength, formability, stress relaxation, electrical conductivity, solderability, and corrosion resistance. Casting of this material produces elemental segregation during solidification, which can be avoided by using P/M fabrication techniques. Figures 11 through 14 illustrate typical applications of roll-compacted strip materials.

Figure 15 illustrates a method for producing porous strip of pure nickel for alkaline battery and fuel cell electrodes. In this procedure, an electroformed nickel screen or mesh is fed into the center of the powder hopper and incorporated into the rolled strip to produce highly porous electrodes with adequate green strength. The same system can be used to feed wires into a green strip for extra strength.

Composite Bearings

Roll compacting also can be used for producing composite, or "sandwich," materials. An example of such a roll-compacted composite material is bimetallic strip used in producing main and connecting rod bearings (Ref 6). A change in automotive emission standards led to the use of these new materials to replace traditional copper-lead sleeve-bearing components. The rolled strip consists of a layer of Al-8.5Pb-4.0Si-1.5Sn-1.0Cu prealloyed powder "sandwiched" to a pure aluminum layer.

The production setup for roll compacting such a composite structure is shown in Fig. 16. Three powder hoppers are required, as well as a powder flow blade that controls the flow of the powders into the roll gap. The coil of the composite strip is then sintered, and eventually the pure aluminum layer is roll bonded to a steel backing material—an operation that is common in the bearing industry.

Powder Melting. An induction melting furnace is charged with elemental aluminum, copper, and silicon (Ref 6). At an intermediate elevated temperature, lead and tin are added to the melt. Furnace temperature is raised to the single-phase temperature (925 °C, or 1700 °F), and 38 °C (100 °F) of superheat is added to provide a safety margin. The induction current of the furnace creates a stirring action that ensures complete dissolution of the lead and tin to form a true single-phase solution.

From the induction furnace, the molten alloy is poured through a launder into a gas-fired tundish furnace. An extension on the bottom of the tundish crucible holds a ceramic nozzle that meters the molten alloy at a controlled rate.

Atomization. The thin metal stream falls vertically into the atomizing chamber, where it is disintegrated into discrete particles and rapidly solidified. Powder particles fall into a 6 m (20 ft) high by 1 m (4 ft) diam collector. They then pass through a cyclone separator to remove the fines and a powder screen to separate oversized particles.

Small, finely dispersed lead-tin alloy particles are contained in a hypoeutectic aluminum-silicon-copper matrix. Because of solubility considerations, the lead-tin bearing constituent remains intact

Fig. 11 Resistor cap and band terminations made from nickel-iron strip

Fig. 12 High-purity nickel strip parts for heart pacemaker battery

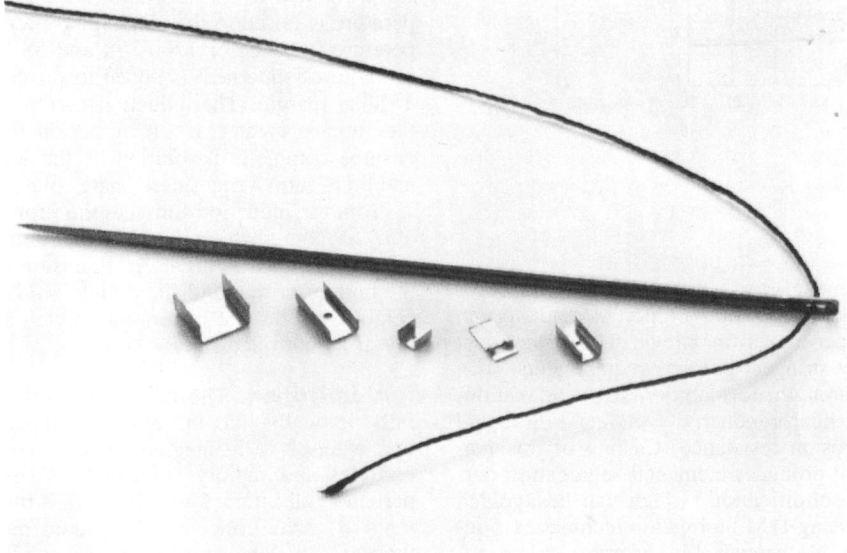

minum-based bearing materials has proven them unsuitable for direct bonding to a steel liner. Ultimate fatigue resistance of the bearing material depends on both the intrinsic strength of the alloy and the integrity and strength of the bond between alloy layer and steel backing.

When aluminum alloys with soft phases such as lead or tin are bonded directly to steel, the bond interface necessarily contains microscopic discontinuities that occur where lead or tin precipitates are in direct contact with the steel. These imperfections create brittleness in the interface and act as sites for accelerated fatigue crack propagation and bond separation under service conditions. Consequently, wrought aluminum alloys containing appreciable amounts of soft bearing phase are almost always used with a pure aluminum or nickel bonding layer next to the steel. Powder rolling provides a convenient means of incorporating a pure aluminum bonding layer during the consolidation of the aluminum-lead prealloyed powder.

A third layer on the opposite side of the prealloyed layer from the bonding layer is necessary to ensure that the strip exits cleanly from the rolls of the powder rolling mill. This process layer, which is made from a blend of aluminum and lead-tin particles, is machined off in subsequent bearing finishing operations.

The three distinct compositions are fed into the compacting mill, as shown in Fig. 16. The work rolls compact these powders into a three-layer strip, which is close to 100% of theoretical density as it leaves the roll nip. A flowchart of the fabrication of powder rolled sleeve bearings is shown in Fig. 17. The production mill used to fabricate these bearings, which resembles a conventional four-high rolling mill (Fig. 1b), is shown in Fig. 18. A cross-sectional view of the unsintered structure of the aluminum-lead alloy strip is shown in Fig. 19.

Sintering. The green strip is then sintered, which improves the morphology of one of the phases (silicon) and provides the necessary diffusion across particle boundaries to achieve excellent strength and ductility. The finely distributed lead phase is unaffected. Further processing consists of cladding the aluminum strip to a steel backing, followed by blanking, forming, and machining of the bearing half-shells.

Bearing materials must possess a unique combination of properties, including fa-

throughout subsequent processing. Controlled distribution of the lead-tin ensures the development of desired mechanical properties.

Consolidation. A consolidation process is selected to produce a strip of prealloyed aluminum-lead alloy that can be ultimately roll bonded to a low-carbon steel backing strip.

Previous experience with wrought alu-

Fig. 13 Typical roll-compacted nickel strip applications
(a) Pure nickel semiconductor caps. (b) Nickel-iron lead frame. (c) Nickel-iron Dual Inline Package integrated circuit. (d) Nickel-iron resistor end caps

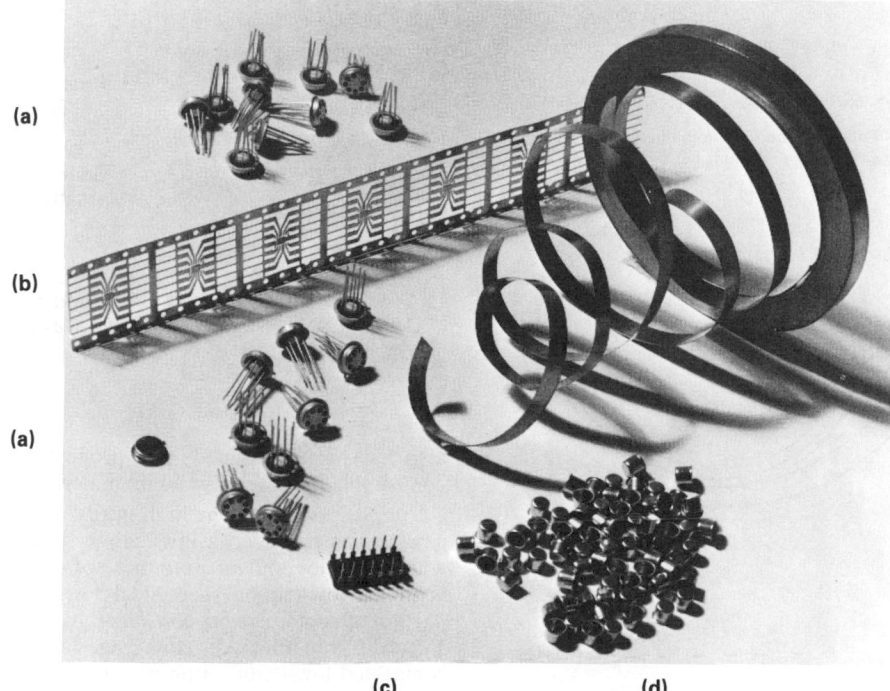

(a)
(b)
(a)
(c)   (d)

Fig. 14 Typical roll-compacted nickel strip applications
(a) Ni-Fe-Co F-15 electronic part. (b) Nickel semiconductor cans. (c) Nickel battery can. (d) Nickel-iron semiconductor lead frame. (e) Ni-Fe-Co F-15 electronic cans for glass-sealed application

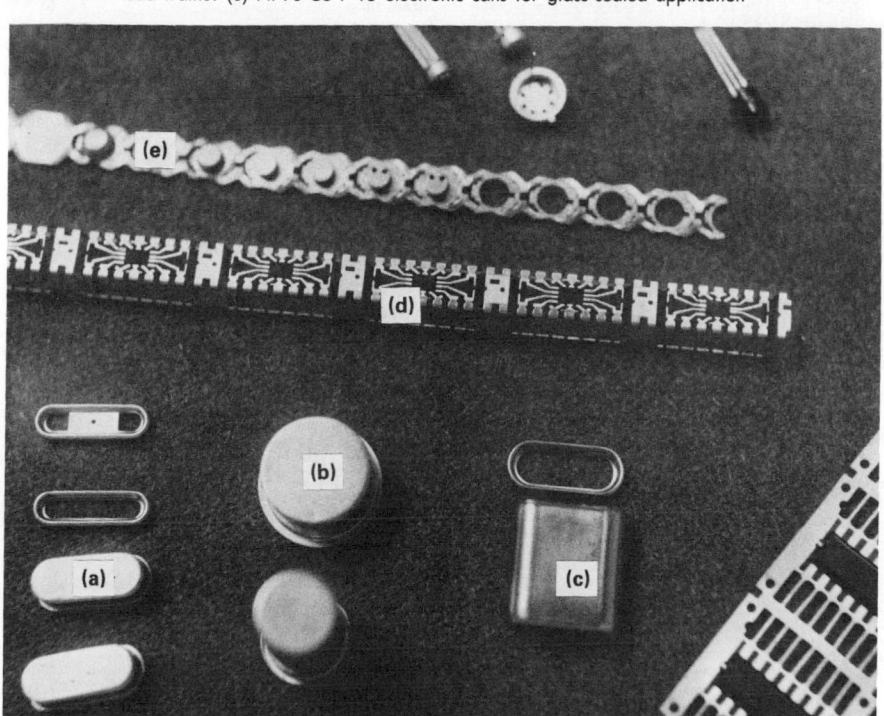

Fig. 15 Rolling of strip from nickel powder
Designed to incorporate mesh into strip

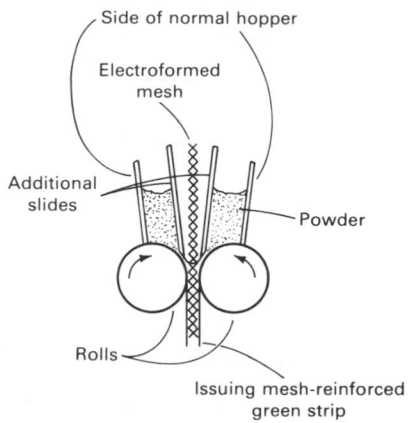

Fig. 16 Rolling of strip for bearings
Using bonding powder, prealloyed bearing powder, and process powder

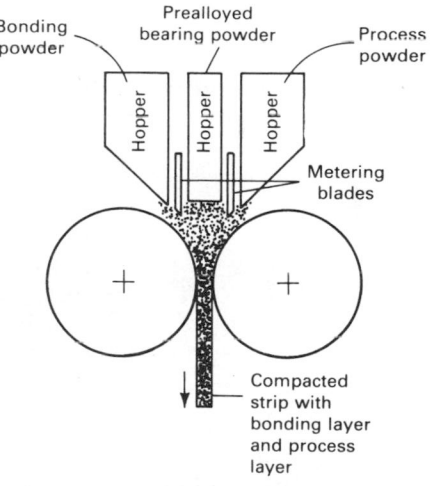

tigue strength, seizure resistance, wear resistance, and corrosion resistance. The performance characteristics of several bearing alloys are given in Table 3.

Recent Developments

Roll compacting of metal powders into strip is a relatively new process. Research and development currently is being conducted on many different materials to produce a superior material as well as to achieve properties not possible with conventional ingot metallurgy (Ref 7). One of these materials is alloy 7091, a high-strength aluminum sheet, with properties comparable to those of wrought materials (Ref 8). Titanium alloy sheet (Ti-6Al-4V)

Fig. 17 Fabrication of powder-rolled sleeve bearing

(a) Flowchart. (b) Sleeve-bearing component

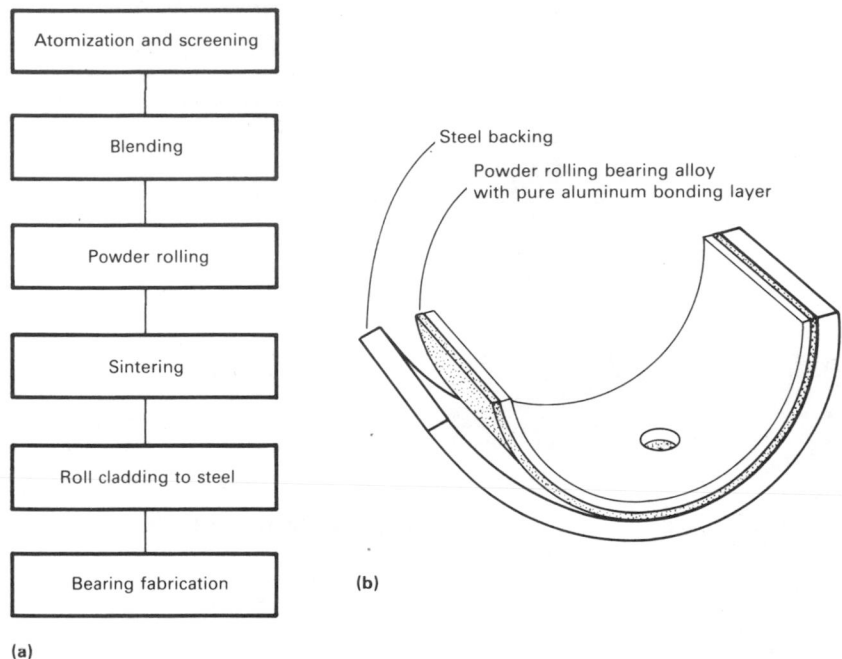

(a)

(b)

Fig. 18 Powder rolling mill

Used to fabricate composite bearing materials. Courtesy of Imperial Clevite Inc.

Fig. 19 Cross section of unsintered powder rolled aluminum-lead strip

Original powder particle boundaries are visible. Magnification: 25×

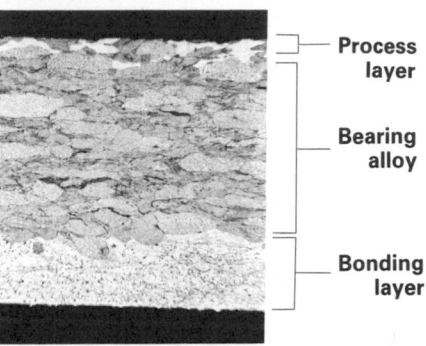

also has been produced with properties comparable to those of wrought materials.

Iron strip made from high-purity iron powder has produced a low-carbon sheet with properties equivalent to those of conventional low-carbon sheet steel for production of containers for consumer goods. Typical properties of thin-gage roll-compacted low-carbon iron powder are:

Gage .... 0.30 mm (0.012 in.)
Yield strength .... 290 MPa (42 ksi)
Tensile strength .... 330 MPa (48 ksi)
Elongation 34%
ASTM grain size .... 11 to 12
Limiting draw ratio 2.10
Strain ratio .... 1.3 to 1.6
Earing 3%
Work hardening exponent 0.220

Source: Ref 9

Further information on ferrous roll compaction can be found in Ref 10 and 11.

REFERENCES

1. Jones, W.D., *Fundamental Principles of Powder Metallurgy*, London, 1960, p 924
2. Naeser, G. and Zirm, F., *Stahl und Eisen*, Vol 70, 1950, p 995-1003
3. Franssen, H., *Z. Metallkunde*, Vol 45, 1954, p 238
4. Evans, P.E. and Smith, G.C., "The Continuous Compacting of Metal Powders," Symposium on Powder Metallurgy, The Iron and Steel Institute, London, Special Report No. 58, 1956, p 131-136
5. Blore, M.H.D., *et al.*, Pure Nickel Strip by Powder Rolling, *Metals Eng. Quart.*, Vol 6 (No. 2), 1966, p 54-60

Table 2 Typical mechanical properties of hydrometallurgical powder rolled cobalt strip

Strip	Strip thickness mm	in.	Strip density, %	Ultimate tensile strength MPa	ksi	Yield strength MPa	ksi	Elongation, %
Green strip	2.1	0.084	86	22	3	22	3	0
Sintered strip	2.1	0.084	86	201	29	195	28	5
Hot rolled strip	1.2	0.048	100	758	110	413	60	15
Cold rolled strip	0.9	0.036	100	1103	160	1100	159	1
Annealed strip	0.9	0.036	100	793	115	345	50	20

Note: Compacting roll diameter, 254 mm (10 in.). Roll speed, 6.0 rpm. Roll gap (green strip), 1.5 mm (0.06 in.)

Table 3 Typical properties of common bimetal bearing alloys

Alloy composition	Fatigue (Underwood life)	Coefficient of friction	Wear scar width mm	in.
		┌─Surface action at 150 °C (300 °F)─┐		
Al-8Pb-5Si-1.5Sn-2Cu				
(Clevite 66) .... 200 h at 49 MPa (7 ksi)		0.21	4.52	0.178
Cu-10Pb-10Sn (SAE 792) .... 200 h at 69 MPa (10 ksi)		0.22	5.15	0.203
Cu-23Pb-3Sn-2Zn (SAE 794) .... 200 h at 49 MPa (7 ksi)		0.17	5.08	0.200

Note: The Underwood test simulates the cyclic loading of a rod bearing resulting from the power stroke in an internal combustion engine. Pressure is based on unit load. Wear scar and coefficient of friction are obtained from the standard LFW-1 test machine, Faville LeVally Corp.

6. Mackay, M.L., Innovation in P/M: An Engine Bearing Material, *Met. Prog.*, Vol 111 (No. 6), 1977, p 32-35

7. Ro, D.H., Toaz, M.W., and Moxson, V.J., The Direct Powder Rolling Process for Producing Thin Metal Strip, *J. Metals*, Vol 35 (No. 1), Jan 1983, p 34-39

8. Ro, D.H. and Toaz, M.W., Direct Powder Rolling of High Strength Aluminum Alloy Strip, in *Processing of Metal and Ceramic Powders*, R.M. German and K.W. Lay, Ed., The Metallurgical Society of AIME, 1981

9. Klein, A.J., *et al.*, A Modern Mini-Mill Process for Powder Sheet Rolling, *Prog. Powder Metall.*, Vol 31, 1981, p 329-345

10. Tundermann, J.H., The Roll Compaction of Iron Powder, Ph.D. dissertation, University College, Swansea, 1967

11. Ayers, M.D., "New Technology for Steel Strip Production," OTA Seminar on Technology and Steel Industry Competitiveness, May 2, 1979, p 213

P/M Forging

By Peter W. Lee
Senior Research Specialist
The Timken Co.
and
Howard A. Kuhn
Professor of Metallurgical Engineering
University of Pittsburgh

POWDER FORGING involves fabrication of a preform by conventional press-and-sinter processing, followed by forging of the preform, which is quite porous, into a final shape with substantial densification. Forging is generally performed in one blow in confined dies to eliminate flash formation and achieve net shapes. Thus, parts containing through holes and complex configurations can be manufactured with little or no material loss due to trimming, hole punching, machining, or grinding. Powder forging is used to obtain higher density in P/M parts for high-stress applications, where voids must be minimized or completely eliminated.

Plastic deformation of sintered powder materials is similar to that of conventional fully dense materials, but is influenced by the volume fraction of voids (interstices between powder particles) in the preform material. Voids are eliminated during deformation so that a sound metallurgical structure is obtained. However, voids are sites of weakness at which ductile fractures may initiate during deformation. Also, a preform with voids has less workability than one free of voids. Another consideration is that the existence of voids leads to volume change during plastic deformation, so that die design practice cannot be approached through conventional guidelines, and classical plasticity theory cannot be used for plasticity analysis.

Deformation, Densification, and Fracture

Plastic deformation of sintered powder material involves collapse of voids and distortion of the solid material. Although low porosity is desirable to achieve good

mechanical properties, other factors are also important. The collapse of voids should be accompanied by shear deformation, so that opposite sides of a pore slide over each other. This breaks up any contaminant film on the internal surfaces and results in greater structural integrity. Dynamic properties, such as impact resistance, are strongly influenced by this effect.

Figure 1 shows the result of Izod impact testing on hot forged disks of 4620 steel powder that underwent various amounts of deformation. Full density was achieved in each forged disk, but increasing levels of deformation and, consequently, shearing led to enhanced dynamic properties, such as impact strength (Ref 1). Similar results have been obtained in fatigue testing on material forged to full density, but with varying amounts of deformation (Ref 2).

Fracture during plastic deformation of material can occur if voids develop around inclusions or other inhomogeneities. The voids coalesce as deformation continues and form a crack. In sintered powder material, large voids already exist, so crack formation during forging of powder preforms is a strong possibility.

In deformation processing of materials, tensile ductility is not always a true indicator of the ability of a material to be deformed. During processing, the material is generally subjected to combined stresses; therefore, an understanding of fracture under such stress states is required.

Lee (Ref 3) has developed a useful technique for evaluation of deformation to fracture (workability) under conditions of deformation processing. Known as the upset test, it has been successfully applied to P/M parts. During axial compression of a cylinder, friction at the die contact surfaces retards radial outward flow of material at these surfaces and leads to barreling of the cylindrical free surface, as shown in Fig. 2. Curvature of the bulge surface increases with increasing friction

Fig. 2 Nature of stresses on the bulge surface of upset cylinders

Fig. 1 Increase in impact resistance with increasing forging deformation of sintered nickel steel (4600 series) powder

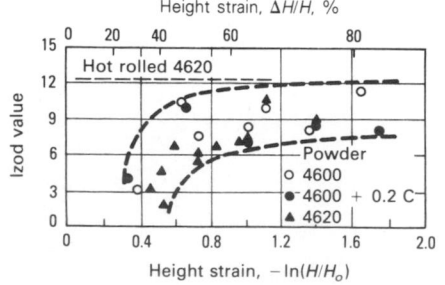

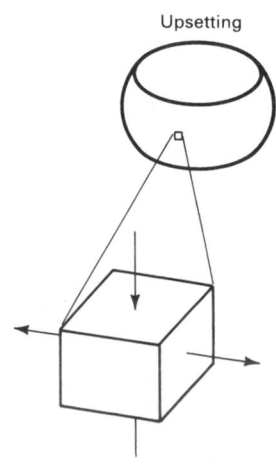

Upsetting

and decreasing aspect ratio (H/D, height-to-diameter ratio). The nonuniform deformation under this condition results in a secondary tensile stress in the circumferential direction accompanying the axial compressive stress. Increasing barreling curvature due to increasing friction or decreasing H/D increases the circumferential tensile stress. Because fractures occur as a result of this tensile stress, the upset test may be utilized to study fracture during deformation processing of sintered powder materials.

In the Lee-Kuhn (Ref 4, 5) workability test, measurements are made of the axial and circumferential strains on small grid marks at the equator of the bulge surface of the upset cylinders, shown in Fig. 2. The ratio of these strains during compression is altered by modifying the friction condition and aspect ratio of the cylinder. At fracture, the surface strains fit a straight line having a slope of one half on tensile strain/compressive strain axes. Figure 3 gives the results of such tests for 601AB aluminum alloy powder and for 4620 low-alloy steel powder.

Fracture strain loci, shown in Fig. 3, can be considered as a fracture criterion for evaluation of the deformation to fracture in more complex deformation processes. The progressing deformation strain paths in potential fracture regions of the process under consideration are first determined through plasticity analysis or measurements on a model material. Any strain path crossing the fracture locus before the deformation process is complete indicates that fracture is likely. Alterations of the process can be utilized to change the strain paths so that they do not cross the fracture locus. The overall effects of friction and

Fig. 3 Tensile strain vs. compressive strain at fracture of 4620 steel powder cylinders and 601AB aluminum powder cylinders

Solid line shows 601AB at room temperature; data points are for 4620 steel at 980 °C (1800 °F); dashed line represents homogeneous deformation.

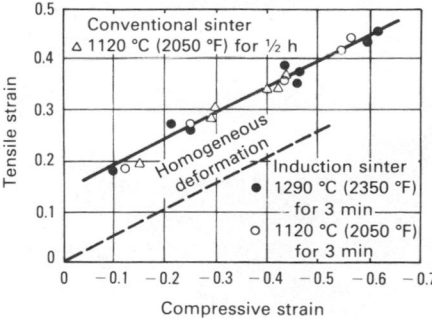

Fig. 4 Height strain at fracture as a function of cylinder height to diameter ratio for unlubricated and lubricated 601AB aluminum powder

Test run at 370 °C (700 °F).

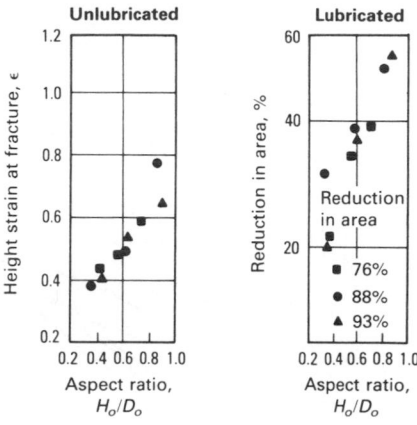

aspect ratio on fracture are shown in Fig. 4 for 601AB aluminum powder.

Preform Design

A major key in successful powder forging is proper preform design. Preform design has a significant effect on metal flow and the distribution of stresses in the material. These factors, in turn, affect densification and avoid fracture of the material. In general, sufficient metal flow must take place to achieve full density and good bonding; however, increasing the amount of metal flow also increases the possibility of fracture. Thus, preform shape must fall between the fracture limits (Fig. 3) and the threshold for the best properties (Fig. 1). This concept is illustrated for a simple disk forging in the following example.

For a given disk diameter and height, it is required to specify the cylindrical preform geometry (height and diameter) and the forging process conditions that will lead to sufficient deformation for maximum (or required) properties without the occurrence of fracture during forging. Fracture in upsetting (see Fig. 4) can be avoided if the expanding free surface of the cylinder reaches the die sidewalls before the reduction is reached at which fracture would otherwise occur. This prevents further development of diametral tensile stresses.

As an example, consider a material requiring 50% reduction in height to reach maximum properties. Equating the mass of the preform to the mass of the forged disk, the required disk aspect ratio (height-to-diameter ratio) is:

$$H_f/D_f = (H_o/D_o)\rho_o^{-1/2}(H_f/H_o)^{3/2}$$

where H_o, D_o, and ρ_o are the height, diameter, and relative density of the preform, and H_f and D_f are the height and diameter of the forged disk.

Similarly, if the material exhibits a relation between height strain at fracture and preform aspect ratio H_o/D_o as in Fig. 4, then the forged disk aspect ratio at fracture is:

$$H_f/D_f = (H_o/D_o)\rho_o^{-1/2}\exp[-3/2\epsilon_H]$$

where height strain ϵ_H is determined, for a particular preform aspect ratio H_o/D_o, from Fig. 4.

These equations are plotted in Fig. 5 on axes of preform aspect ratio versus forged disk aspect ratio. For a given preform aspect ratio, the forged disk aspect ratio must be to the left of the solid line to achieve the 50% reduction required for maximum properties ($H_f/H_o = 0.5$ in.). In addition, the forged disk aspect ratio must be to the right of the dashed line in order to avoid fracture.

In other words, starting at the reference line for a given H_o/D_o, move horizontally to the left as deformation proceeds. For maximum properties, the solid line must be crossed; to prevent fracture, the dashed line must not be crossed. It is clear that, for the unlubricated case, forged disk H_f/D_f must be greater than 0.3, and the corresponding preform H_o/D_o must be greater than 0.78. For the lubricated case, any forged disk aspect ratio can be formed, but there is a narrow range of allowable preform aspect ratios for successful forg-

Fig. 5 Design curves for disk forging for optimum properties without fracture

H_o/D_o is ratio of original height to original diameter; H_f/D_f is ratio of final height to final diameter.

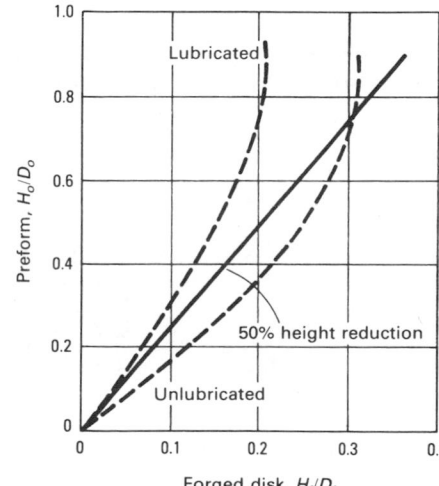

ing. These limits will be different, of course, for materials having characteristics other than those in Fig. 4.

Preform design for each complex part must be handled on an individual basis. Nevertheless, some guidelines can be obtained from the forging of simple, generic shapes. One example is a flanged hub shape formed by partial extrusion.

As illustrated in Fig. 6, extrusion forging of a hub involves compression of the flange section, with radial flow inward and up into the hub sections. The top surface is a free surface that undergoes bulging and tensile strains due to friction along the sides of the die. These strains are nominally equal, and fracture occurs on the top surface when the strains reach the fracture limit.

Typically, for small draft angles, the hubs reach a height roughly equal to the hub base diameter before fracture occurs. Use of large draft angles (>40°) drastically reduces the hub height at fracture. Increasing friction or decreasing the die corner radius increases the tensile strains at the top surface and thus decreases the hub height at fracture.

An alternative approach involves use of a preform that partially fills the hub section of the die. The metal flow occurs upward by extrusion and at the same time radially outward by upsetting (lateral flow). The nature of the strains on the top surface, however, depends on the die angle and friction, as illustrated in Fig. 7. For a friction coefficient of 0.1, draft angles of 40° and 50° lead to tensile strains at the hub top surface, and fracture occurs as the strain paths cross the fracture line. Hub heights at fracture are very small in this case.

For die angles of 10°, 20°, and 30°, the strains on the top free surface are compressive initially, then reverse and move toward the origin. Decreasing die angle

Fig. 6 Strains at the top free surface in partial hub extrusion

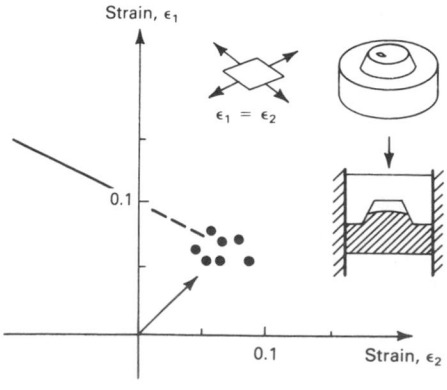

increases the compressive strain to reversal. Reversal occurs because friction along the conical die surfaces overcomes the compressive radial strains resulting from the decrease in diameter of the top surface. Fracture occurs when the tensile strain after the strain path reversal is approximately equal to the tensile strain to fracture in the first quadrant. Fracture did not occur in the 10° die, because the hub section reached the top of the die before the strains became large enough for fracture. For the small die angles, hub heights reach approximately twice the hub base diameter before fracture occurs.

For any given friction coefficient, there is a specific draft angle above which the hub top surface strains are tensile and below which the strains are initially compressive. These angles have been predicted through slipline field analysis and confirmed experimentally by Suh and Kuhn (Ref 6).

Internal defects, often called central burst, may also occur in hub forgings. Unlike the other types of fracture, internal fractures can remain undetected, because they rarely penetrate to the surface. Although internal fractures may be closed up in subsequent deformation of the part, there is no certainty that the bond will be sound. Service failures due to internal defects are particularly catastrophic, because they cannot be detected in their early stages of

growth by surface inspection. These aspects emphasize the need for investigation of internal fracture, even though its occurrence is much less frequent than other types of fracture.

Internal fracture takes the form of a large crack at the center of a forged part, as shown in Fig. 8. Axial tension at the central point, caused by opposed axial flow from the center toward each hub, leads to such defects. The magnitude of tension at

Fig. 7 Strains at the top free surface in compression between tapered dies

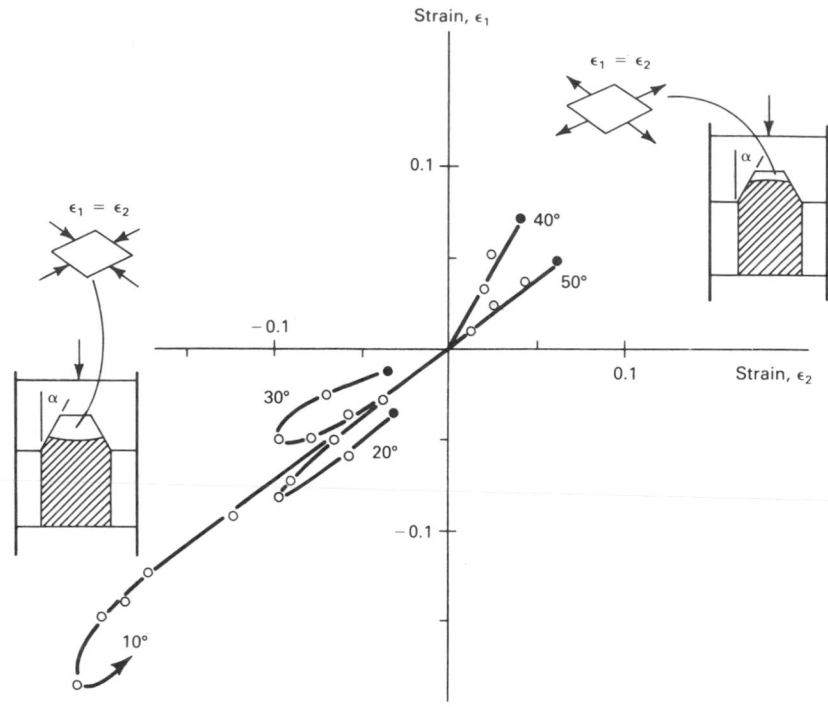

Fig. 8 Central fracture in a double hub forging

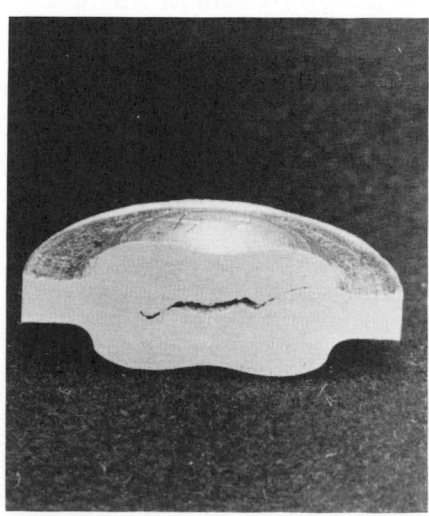

Fig. 9 Deformation limits at fracture for a double hub forging

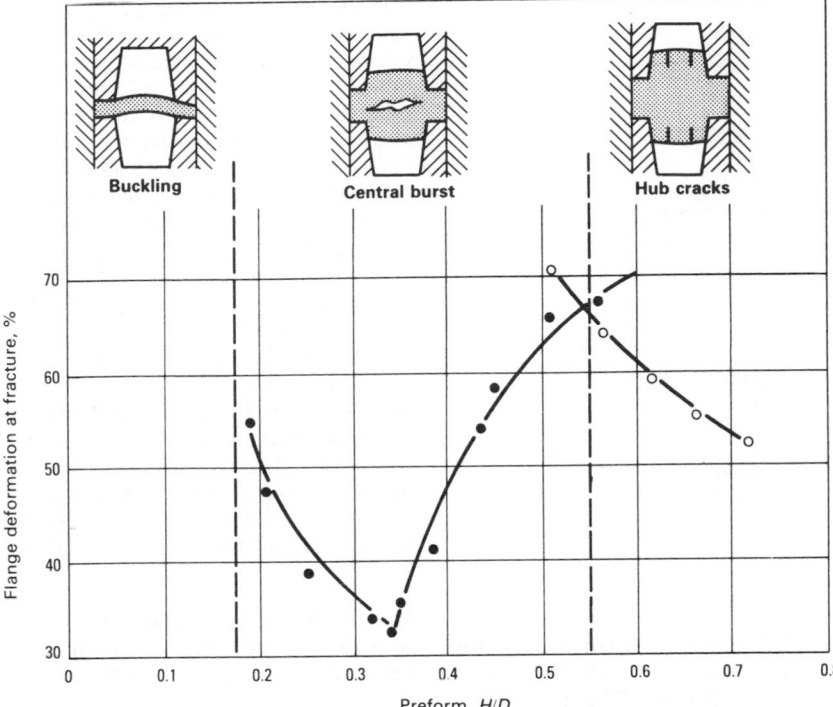

Buckling   Central burst   Hub cracks

Fig. 11 Preform options for forging the part shown in Fig. 10

See text for discussion of (a) through (d).

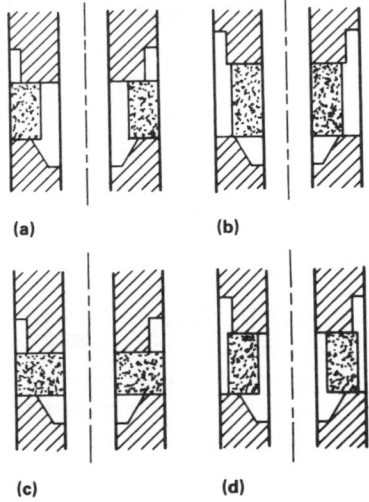

(a)   (b)

(c)   (d)

the center depends on the ratio of preform aspect ratio H_o/D_o. Friction has very little effect on this type of fracture.

Experimental studies on 601AB sintered aluminum powder indicates the amount of deformation possible before internal crack formation (Fig. 9). A minimum amount of deformation to fracture occurs for $H/D = 0.35$, below which a double bulge forms on the top hub surface and above which a single bulge forms. No theoretical explanation of this result has yet been developed. Also shown in Fig. 9 are the forming limits due to surface cracks at the hub top surface, and due to buckling that occurs in thin preforms.

An example of a complex part that illustrates the principles of preform design is a hub-flange-rim combination (Fig. 10) examined by Downey and Kuhn (Ref 7). The re-pressing alternative is rejected because of its limited development of mechanical properties. Flat ring preforms provide large amounts of metal flow in reaching the final shape.

Even though the general ring shape is specified, its dimensions must be determined. As shown in Fig. 11, the preform may have:

● Clearance at the bore diameter but no clearance at the outside diameter (Fig. 11a)
● Clearance at the outside diameter but no clearance at the bore diameter (Fig. 11b)

Fig. 10 Prototype part for illustration of preform design

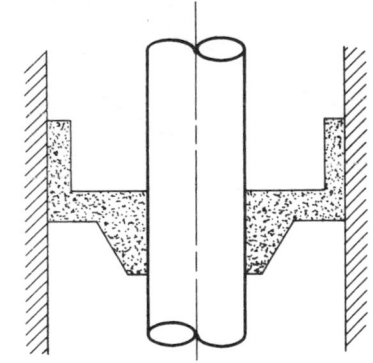

● No clearance at either the outside diameter or the bore diameter (Fig. 11c)
● Clearance at both the outside diameter and bore diameter (Fig. 11d)

Both preforms in Fig. 11(b) and (d) can be rejected, because cracks will occur at the outer rim as metal flows around the upper punch radius. This occurs because the metal is expanding in diameter as it flows around the corner, although there is axial compression to help compensate for the circumferential tension. This type of cracking can be avoided by using a preform that fills the die—that is, a preform that has no clearance at the outside diameter, as in Fig. 11(a) and (c).

The preform in Fig. 11(c) can be rejected because it is similar to hub extrusion, and may lead to cracking at the hub top surface. Use of clearance between the bore diameter of the preform and the mandrel eliminates this type of crack. Therefore, the preform shown in Fig. 11(a) is selected for the flange-hub-rim part. Use of this preform has led to successful, defect-free parts, while the expected cracking occurred with the use of the other preforms (Ref 7).

Forging Modes

There are two modes of densification during powder forging—re-pressing and upsetting (Fig. 12). In the re-pressing mode, an extensive flow of material in the lateral direction is prevented by die wall constraints. The outside diameter of the preform for the re-pressing is close to the outside diameter of the forged part.

The stress state during re-pressing consists of a small difference between vertical and horizontal stresses and results in very little movement in the horizontal direction, thus resulting in small lateral flow. As full density is approached, the state of stress approaches a pure hydrostatic condition. A typical pore simply flattens, and the opposite sides of the pore are brought together under pressure. Thus, a spherical pore becomes ellipsoidal on collapse, with the long axis close to that of the original pore diameter. Because of the insignificant amount of the lateral flow, re-pressed parts do not exhibit marked anisotropic properties.

Fig. 12 Forging modes and stress conditions on pores for (a) re-pressing and (b) upset forging

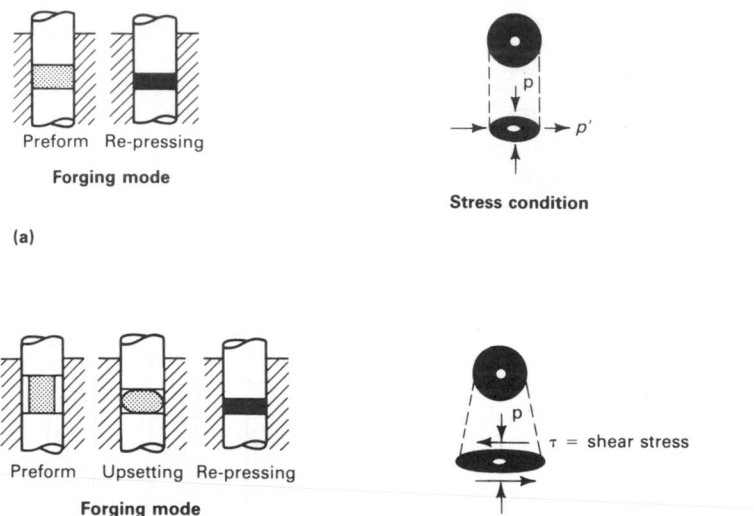

In the upset forging mode, an extensive unconstrained lateral flow of material is allowed during forging. The stress state around the pore is the combination of normal stress and shear stress. A spherical pore becomes flattened and elongated in the direction of lateral flow.

During upset forging, the oxide films that encapsulate the powder particles are broken up and expose the clean material. The virgin material and sliding due to shear stress enhance strong metallurgical bondings across collapsed pore interfaces and consequently improve dynamic properties such as impact strength and fatigue strength.

Because of the substantial amount of lateral flow during upset forging, non-metallic inclusions will be elongated along the lateral direction and result in anisotropic in mechanical properties. However, the amount of anisotropy in upset forged P/M parts is generally much smaller than that in wrought material.

Poisson's ratio is a measure of the lateral flow behavior and is expressed as the diameter strain (ϵ_d) divided by the height strain ($-\epsilon_z$). Typical results of increasing density and diameter strain with increasing height strain are given in Fig. 13. Densification and lateral flow (diameter strain) occur simultaneously rather than in sequence.

Because density increases during compression, the Poisson's ratio also increases and is determined from the slope of the curve ϵ_d versus $-\epsilon_z$,

$$\nu = d\epsilon_d/d\epsilon_z$$

Fig. 13 Effect of height strain on relative density and diameter strain of sintered 601AB aluminum compacts

○ is 82% of theoretical density; ● is 93% of theoretical density.

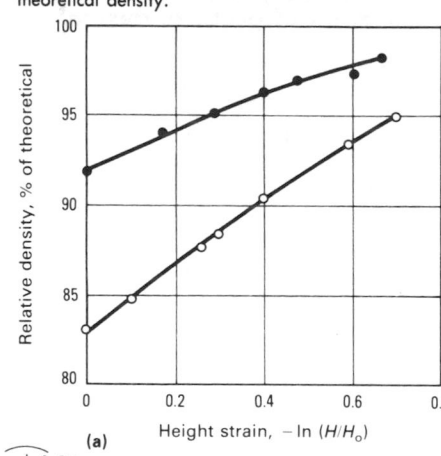

(a)

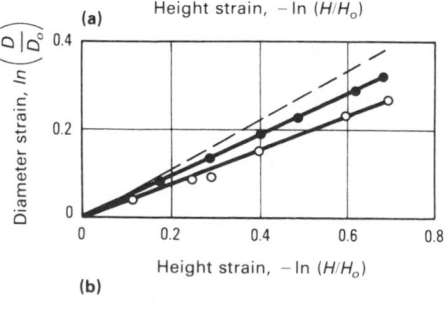

(b)

The dashed line in Fig. 13 represents the relationship between diameter strain and height strain for a fully dense material (slope of one half). Note that the curves for porous material are below the dashed

line, but gradually become parallel to it as full density is approached.

From data such as that in Fig. 13, the variation of the Poisson's ratio with density can be plotted. Figure 14 gives the results for room-temperature deformation of sintered iron powder. As full density is reached, the Poisson's ratio extrapolates to one half. Note that the data follow the same curve, regardless of the initial density. Similar results have been determined for room-temperature deformation of copper and aluminum powders. In general, the Poisson's ratio relationship to density is given by $\nu = 0.5\,\rho^2$.

Full densification of the porous P/M preform can be achieved by either re-pressing or upset forging. The re-pressing mode, however, requires a higher forging pressure to obtain a given density level than the upset forging mode, as shown in Fig. 15. Mechanical properties, such as strength and ductility, of the fully dense P/M parts

Fig. 14 Poisson's ratio versus theoretical density for iron powder at room temperature

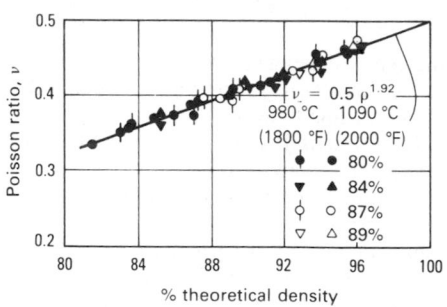

Fig. 15 Effect of forging mode on densification of hot forged P/M parts

Solid line represents upset forging (plane strain) at 870 °C (1600 °F). Dashed line represents re-pressing at 870 °C (1600 °F). Source: Ref 8

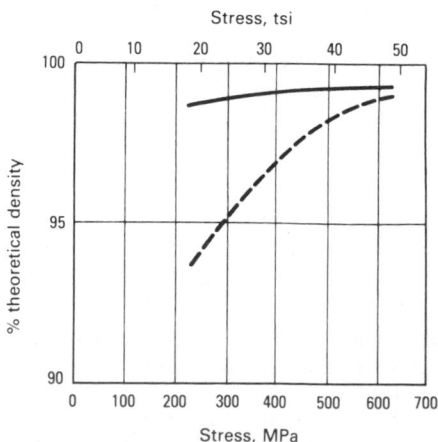

are equal to or greater than those of the wrought material.

Effects of Process Parameters on Mechanical Properties

Mechanical properties of P/M forged parts depend on the number of processing parameters for compaction, sintering, and hot forging operations. Particle size distribution has effects on green compacts as well as on forged parts. For example, it has been shown that preforms compacted from large-size particle distributions have greater green strengths than those from smaller size particle distributions (Ref 9). This increase in green strength occurs because the "popcorn" shape of the large particles can achieve better interlocking among the individual particles than the relatively smooth shape prevalent in small particles. For the smaller particles, sizes down to 80 μm substantially increase elongation and reduction in area in forged parts. Particle sizes smaller than 80 μm only slightly increase ductility.

Density of preforms does not have a significant effect on density of forged parts. However, lower density preforms with more interconnecting porosity may obtain better reduction of oxides during sintering than higher density preforms. On the other hand, lower density preforms are more prone to internal oxidation and decarburization on exposure to the atmosphere prior to forming or densifying. The control of preform weight is a critical parameter, because hot forging of P/M preforms is a closed die process without flash. An excessive preform weight may lead to tooling breakage or stalling of the press. Conversely, an underweight preform will not achieve full densification by forging in a fired strobe press.

Sintering of the preform is a critical step in obtaining the mechanical properties of the final forged part. The oxygen level of the forged part is determined primarily by the oxygen level of the sintered preform. The oxygen level of the sintered preform is dependent on the sintering temperature, type of sintering atmosphere, dew point of sintering atmosphere, and the furnace type and its conditions. Oxygen level of the sintered preform decreases with higher sintering temperature and lower dew point of the furnace atmosphere.

It has been shown that mechanical properties in general, and dynamic properties in particular (such as impact and fatigue strength), are strongly influenced by the oxygen level of the forged part (Ref 10, 11). The lower the oxygen level of the forged part, the higher the dynamic properties of the forged part, because the metallurgical bonding between particles during forging is enhanced by the cleaner particle boundaries (see Fig. 16).

The mechanical properties of the forged parts are also influenced by forging parameters, such as forging temperature, forging pressure, preform chilling by tooling, speed of the press, amount of lateral flow during forging, and chemical composition. Generally, higher forging temperature and less preform chilling enhances the densification of the forged part and consequently increases mechanical properties. Crowson et al. (Ref 12) evaluated the effect of forging temperatures of 980, 1090, and 1200 °C (1800, 2000, and 2200 °F) and forging pressures of 276 and 552 MPa (20 and 40 tsi) on the densification of the forged part. They demonstrated that by increasing the forging temperature from 425 to 1200 °C (800 to 2200 °F), the forging pressure required to obtain the same level of densification was reduced by 50% to 276 MPa from 552 MPa (20 from 40 tsi). Higher forging speed reduces tool contact time and minimizes preform chilling.

The amount of lateral flow is particularly important for dynamic properties such as fatigue and impact strength (Ref 13), as shown in Fig. 17 and 18. The data in these figures are obtained from fully dense P/M forged 4620 steels. Low height strain and high height strain represent re-pressing with no lateral flow and upset forging with a substantial amount of lateral flow. Dynamic properties are believed to be improved with increasing lateral flow because of the extensive particle shearing caused by lateral flow.

Commercial Ferrous P/M Forging

Because commercially produced hot forged near-net shape P/M parts are, for the most part, produced from ferrous alloy powders, discussion in this section will be limited to the ferrous P/M forging process. Aluminum powders produced by rapid solidification and titanium powders produced by the rotating electrode process are processed into fully dense P/M billets by either hot extrusion or hot isostatic pressing. These billets are then hot forged into final parts, primarily for aerospace applications. They are not produced by the near-net shape process used for ferrous P/M forgings.

Preform Compaction

After water-atomized powder is blended with graphite and lubricant powders, the powder is compacted into preforms within a die. The exact compacting pressure used depends on the compressibility of the powder and desired preform density. Although cold isostatic pressing is sometimes used for compacting preforms, rigid die compaction is the most widely used process because of its speed and ability to produce fairly complex parts. A lubricant decreases the required ejection force, and graphite provides carbon for reduction of oxides during sintering and the final car-

Fig. 16 Variation of reduction of area and impact strength as a function of oxygen content
Source: Ref 11

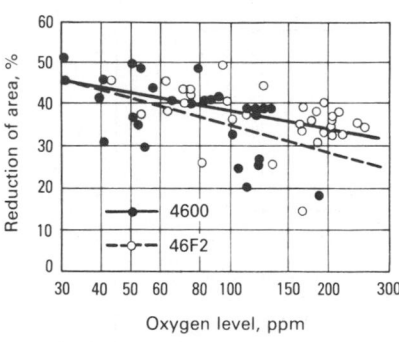
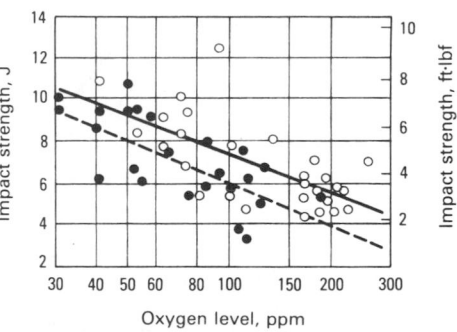

Fig. 17 Charpy impact response for fully dense as-forged 1000 iron powder preforms as a function of extent of height strain
Source: Ref 2

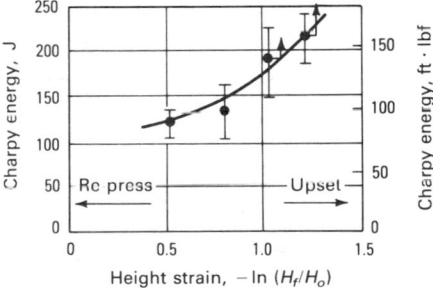

Fig. 18 Axial fatigue of P/M forged 4620 steel for various levels of forging deformation

Fatigue limit increases as deformation level (height strain) increases. Source: Ref 13

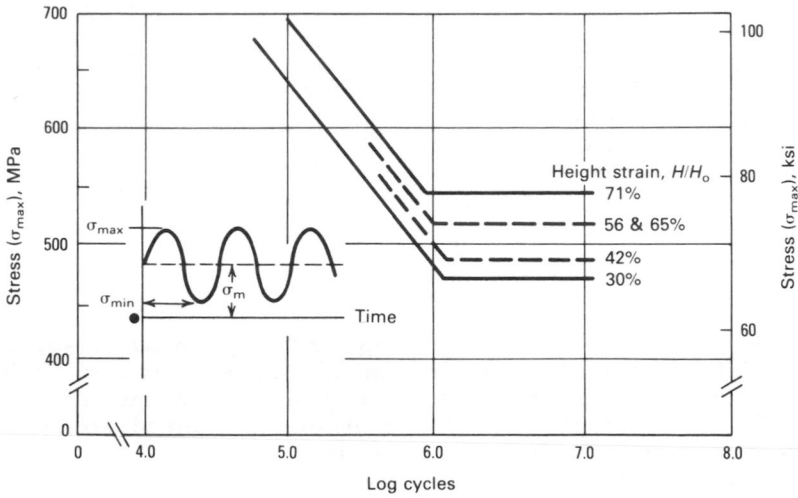

bon level of the P/M forged parts. Admixed solid lubricant is easy to use; however, it tends to slightly diminish the green strength. Die wall lubricants are also available.

During die compaction, friction develops from the relative motion between die sidewalls and powder particles, between powder particles within the powder mass, and between powder particles and punch faces. Because of this friction, the local pressure within the powder mass caused by the upper punch load is not uniform. Irregularly shaped powder particles will be mechanically interlocked during compaction and provide sufficient green strength for handling. The life of the compaction tooling is influenced by the required tolerances of the part (Ref 14). Close tolerances tend to decrease tool life by decreasing the amount of tool wear that can be tolerated. Therefore, tolerances of P/M parts should be as liberal as possible. Table 1 lists some tolerances of powder forged parts.

The density of preform should be selected on an individual basis. A lower density part requires a smaller press and provides better sintering reaction; however, it will have lower green strength and a greater chance of internal oxidation of preforms prior to hot forging. Normally, the density of preforms for hot forging ranges from 70 to 85% of the theoretical density (5.5 to 6.7 g/cm^3).

Preform Sintering

Sintering of preforms is a critical step in the P/M forging process; the properties of the sintered preform influence the final product. During sintering, the following

Table 1 Tolerances of powder forged parts

	Nominal dimension		Tolerance(a)	
	mm	in.	mm	in.
Outside diameter	50.8	2.0	0.127	0.005
	50.8	2.0	0.254	0.010
	50.8	2.0	0.127	0.005
	50.8	2.0	0.102	0.004
	76.2	3.0	0.127	0.005
	76.2	3.0	0.381	0.015
	95.3	3.76	0.254	0.010
	203.2	8.00	0.508	0.020
Inside diameter	38.1	1.50	0.204	0.008
	63.5	2.50	0.254	0.010
Thickness	15.9	0.625	0.254	0.010
	25.4	1.0	0.381	0.015
	25.4	1.0	0.254	0.010
Concentricity on outside diameter	95.3	3.75	0.102	0.004
Out of round on outside diameter	95.3	3.75	0.102	0.004

(a) Different tolerances for a given nominal dimension were provided by different producers of forged P/M parts. Source: Ref 15

reactions take place. First, metallurgical bonds between powder particles are developed from the simple mechanical interlocks presented in the preform compacts. The mechanical properties of a forged part generally increase with improved metallurgical bonding during sintering. Second, the oxides in the preform are reduced and the final carbon level of the preform is obtained. The mechanical properties of forged parts, particularly dynamic properties such as fatigue and impact strengths, improve with lower oxygen levels. A typical sintering cycle for ferrous P/M parts consists of preheating or delubrication at about 535 to 760 °C (1000 to 1400 °F) and sintering at 1120 °C (2050

°F) for 30 min in an endothermic atmosphere, followed by cooling to room temperature under the atmosphere (Ref 14).

A high-temperature sintering process in which preforms are sintered at temperatures of 1230 °C (2250 °F) or higher is presently being developed. High-temperature sintering will improve performance of forged P/M parts by lowering oxygen content. High-temperature sintering also will enable part fabricators to use powders with low-cost alloying elements such as manganese or chromium to achieve the same hardenability requirements.

Forging Preforms

There are two basic routes for production forging processes: induction heating and furnace heating. The steps involved for both routes are shown in Fig. 19. The primary difference between these routes is the method of heating preforms prior to forging.

In the induction route, preform compacts are sintered and cooled to room temperature. The sintered preforms are then heated to about 150 °C (300 °F) and sprayed with a protective coating, which can be either graphite based or nongraphite based. This coating provides protection against oxidation as the part is transferred from

Fig. 19 Flow diagram of P/M hot forging process routes

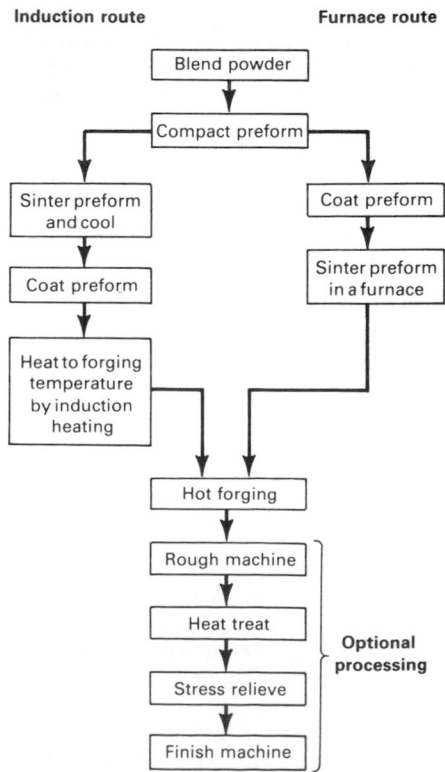

the furnace to the press and also provides lubrication during the forming operation. The coated preforms are heated from room temperature by an induction heating system under a protective atmosphere to a forging temperature of 980 to 1090 °C (1800 to 2000 °F). Tooling is heated to 260 to 315 °C (500 to 600 °F) in order to minimize preform chilling. The preforms are forged by a mechanical press, which provides a short contact time and high production rate. The induction heating route is used for symmetrically shaped parts, such as bushings and stator clutch races for automatic transmissions, because of its ability to fast-feed automatically.

In the furnace heating route, the preform compacts are sprayed with a protective coating prior to sintering. After the preforms are sintered and cooled to the forging temperature, they are forged directly out of the sintering furnace. Tooling is sprayed with water-based lubricant to cool and lubricate the tooling.

The forging process using P/M preforms was initially done by the induction heating method. However, the furnace heating method provides better temperature control, better protection of the parts in a controlled atmosphere, and more versatility, with its ability to produce nonsymmetrically shaped parts. The life of hot forging tooling depends on forging parameters such as forging pressure, ejection pressure and temperature, tool clearance, and level of densification. Forging pressure and ejection pressure increase exponentially when density of the forged part approaches full density, as shown in Fig. 20. Lower forging and ejection pressures reduce the required press size and prolong tool life. Therefore, it is important to determine whether full density is necessary for a given application.

Fig. 20 Forging pressure and ejecting force as a function of density

Iron + 0.4%C; preform temperature 1100 °C (2010 °F). Source: Ref 16

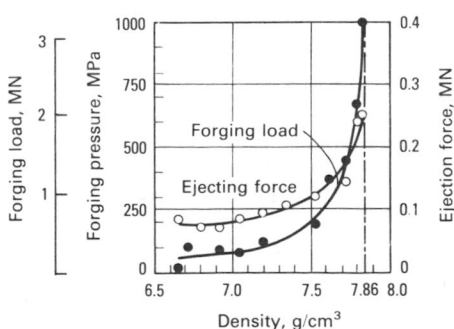

Advantages of P/M Forging

Forging of powder metallurgy preforms produces near-net shape or net shape parts and provides several advantages:

- P/M forging increases materials utilization by eliminating completely or substantially reducing machining operations. Actual material savings depends on the complexity of the part and its manufacturing process.
- P/M forging is an energy-efficient operation, eliminating energy-intensive operations such as rolling for bar stocks and rolling and piercing for tubings.
- P/M forging uses lower temperatures (870 to 1090 °C, or 1600 to 2000 °F) than conventional forging (1090 to 1200 °C, or 2000 to 2200 °F). Lower forging temperatures mean longer tooling life and lower energy consumption.

Higher materials utilization, substantially reduced machining costs, and efficient energy usage result in lower overall manufacturing costs. For example, an economic study of a machine gun accelerator (Ref 17) shows that producing the part by near-net shape P/M forging would result in a cost savings of more than 50%. Currently, the part is produced by hot forging of 4340 steel bar stock to a rough shape in six operations, followed by 27 machining operations. The P/M forging process consists of compacting a preform with water-atomized modified 4600 steel powder, sintering the preform, and hot forging. Only seven machining steps are required.

In addition to reducing manufacturing costs, P/M forging has several other advantages:

- P/M forging provides flexibility in alloy selection; as with conventional P/M parts, the material can be blended to minimize material cost and maximize properties.
- Forged P/M parts exhibit less anisotropic than conventional ingot cast materials due to a substantially smaller amount of metal flow.
- Forged P/M parts have fine grain size, uniformly distributed nonmetallic inclusions and carbides, and chemical homogeneity.
- The P/M forging process can be readily automated.
- P/M forging is sometimes the only viable manufacturing technique for materials with poor workability, such as tungsten, beryllium, carbides, and ceramic materials.

Although P/M forging is generally considered an economically attractive process, analysis of cost as well as technical feasibility is recommended for each individual part. The economics of P/M forging strongly depend on current manufacturing methods for a particular part.

Applications

Forged P/M parts for mass-produced structural parts were developed because of performance requirements that exceeded the strength levels of conventionally pressed and sintered P/M parts. Applications of P/M forged parts have been increasing due to the economic advantages of the P/M process and the availability of prealloyed powders. A number of different parts were attempted for P/M forging; however, not all of these parts were successful because of either economic or technical reasons. As early as 1941, an ammunition feed pawl for antiaircraft guns was hot forged to a density of 7.80 g/cm³ from sponge iron powder.

Currently, P/M forged parts are produced mainly for automotive applications—primarily for transmissions, but also for engines and differentials. Some of the P/M forged parts in current commercial production and use include:

- Connecting rods for automotive engines
- Stator cams for automatic transmissions
- Races for tapered roller bearings
- Valve seats for automotive engines
- Various gears for automotive applications
- Torque converter clutch hubs
- Chainsaw sprockets
- Hand wrenches

REFERENCES

1. Kuhn, H.A. and Downey, C.L., How Flow and Fracture Affect Design of Preforms for Powder Forging, *Int. J. Powder Metall. Powder Technol.*, Vol 10, 1974, p 59
2. Ferguson, B.L., Suh, S.K., and Lawley, A., Impact Behavior of P/M Steel Forgings, *Int. J. Powder Metall. Powder Technol.*, Vol 11, 1975, p 263
3. Lee, P.W., "Fracture in Cold Forming of Metals—A Criterion and Model," Ph.D. thesis, Drexel University, Philadelphia, 1972
4. Lee, P.W. and Kuhn, H.A., Fracture in Cold Upset Forging—A Criterion and Model, *Met. Trans.*, Vol 4, 1973, p 969
5. Lee, P.W. and Kuhn, H.A., Cold Upset Test, in *Workability Testing*

Techniques, G.E. Dieter, Ed., American Society for Metals, 1984

6. Suh, S.K. and Kuhn, H.A., Three Fracture Modes and Their Prevention in Forming P/M Preforms, in *Modern Developments in Powder Metallurgy*, P.W. Taubenblat and H.H. Hausner, Ed., Vol 9, Metal Powder Industries Federation, Princeton, NJ, 1977, p 407

7. Downey, C.L. and Kuhn, H.A., Application of Forming Limit Concept to the Design of Powder Preforms for Forging, *Trans. ASME*, Vol 97, 1975, p 121

8. Hanejko, F. and Muzik, J., "Successful Applications and Processing Considerations for Powder Forming," Powder Metallurgy Technical Conference, Hoeganaes Corp., Philadelphia, 1978

9. Crowson, A. and Anderson, F.E., Properties of P/M Steel Forgings, *Powder Metall. Defense Technol.*, Vol 3, 1977

10. Hanejko, F., Mechanical Properties of Powder Forged 4100 and 1500 Type Alloy Steels, *Modern Developments in Powder Metallurgy*, Vol 12, Metal Powder Industries Federation, Princeton, NJ, 1981

11. Dower, R.J., and Campbell, W.E., "The Toughness of Powder Metallurgy Forgings as a Function of Processing Route," National Engineer Laboratory Report No. 632, Glasgow, Scotland, March 1977

12. Crowson, A., Grandzol, R.J., and Anderson, F.E., "Properties of P/M Steel Forgings," Technical Report No. R-TR-76-039, Rock Island Arsenal, May 1976

13. Ferguson, B.L., Kuhn, H.A., and Lawley, A., Fatigue of Iron-Base P/M Forgings, *Modern Developments in Powder Metallurgy*, Vol 9, Metal Powder Industries Federation, Princeton, NJ, 1977

14. Lee, P.W., Ferrous Powder Metallurgy, in *Powder Metallurgy—Applications, Advantages and Limitations*, E. Klar, Ed., American Society for Metals, 1983

15. Pease, L.F., III, "An Assessment of Powder Metallurgy Today and Its Future Potentials," SAE Paper No. 831042, 1983

16. Bockstiegel, G. and Stromgren, M., "Höganäs Automatic P/M Forging System, Concept, and Application," SAE Paper No. 790191, 1979

17. Lally, F.T., Toth, I.J., and DiBenedetto, J., "Forged Metal Powder Products," Final Technical Report No. SWERR-TR-72-51 on Army Contract No. DAAF01-70-C-0654, Nov 1971

SELECTED REFERENCES

● Adams, J.S. and Glover, D., SAE Paper No. 750856, 1975
● Chmura, W., U.S. Patent No. 4 059 879, Nov 1977
● Eloff, P.C. and Wilcox, L.E., in *Modern Developments in Powder Metallurgy*, Vol 7, Metal Powder Industries Federation, Princeton, NJ, 1974
● Ferguson, H.A., U.S. Patent No. 3 874 049, April 1975
● Gleixner, J.R., *Met. Prog.*, Dec 1983
● Halter, R.F., in *Modern Developments in Powder Metallurgy*, Vol 7, Metal Powder Industries Federation, Princeton, NJ, 1974
● Imahashi, K., Suzuki, T., and Tsumiki, C., SAE Paper No. 750409, 1975
● "Imperial Clevite Canada, Inc.," Metal Powder Report, March 1983, p 129
● Korbrin, C.L., *Iron Age Metalworking Int.*, Vol 6, Nov 1967
● Kuhn, H.A., *Powder Metallurgy Processing—New Techniques and Analyses*, H.A. Kuhn and A. Lawley, Ed., Academic Press, New York, 1978
● Lusa, G., in *Modern Developments in Powder Metallurgy*, Vol 7, Metal Powder Industries Federation, Princeton, NJ, 1974
● McGee, S.W. and Waller, G.M., *Met. Prog.*, May 1977
● McGee, S.W., private communications
● Tsumuki, C., *et al.*, in *Modern Developments in Powder Metallurgy*, Vol 7, Metal Powder Industries Federation, Princeton, NJ, 1974
● Vaccari, J.A., *Am. Machinist*, May 1983
● Wick, C., *Manuf. Eng.*, May 1978
● Williams, V.A., "Production," Production Publicity Co., Bloomfield Hills, MI, Oct 1981
● Wisker, J.W. and Jones, P.K., in *Modern Developments in Powder Metallurgy*, Vol 7, Metal Powder Industries Federation, Princeton, NJ, 1974

Hot Isostatic Pressing of Metal Powders

By Peter E. Price
Director of Engineering
Industrial Materials Technology, Inc.
and
Steven P. Kohler
Graduate Student
Massachusetts Institute of Technology

HOT ISOSTATIC PRESSING is a materials processing technique in which high isostatic pressure is applied to a powder part or compact at elevated temperatures to produce particle bonding. This process usually results in the manufacture of a fully dense body, although partially dense bodies also can be intentionally produced. During processing, the compact is subjected to equal pressure from every side.

Elevated temperature, in reference to hot isostatic pressing, ranges from approximately 480 °C (895 °F) for aluminum alloy powder processing to approximately 1700 °C (3090 °F) for tungsten powder processing. High-density argon gas is the most common medium used in the process, and pressures range from approximately 20 to 300 MPa (3 to 45 ksi), with 100 MPa (15 ksi) as the average pressure.

The hot isostatic pressing process was invented at Battelle Memorial Institute in 1955 by Saller et al. (Ref 1). Early designs utilized the "hot wall" configuration; the furnace surrounded the pressure vessel. Material limitations precluded scale-up, and the development of a cold-wall vessel design, now used throughout industry, took place.

Hot isostatic pressing was initially used for diffusion bonding of clad nuclear fuel elements. Consolidation of beryllium metal powder "to-shape" was first carried out in 1964. High-volume hot isostatic compaction of high-speed tool steel was achieved in the United States and Sweden by 1972. The U.S. Air Force Materials Laboratories expanded hot isostatic pressing technology to include forging of preforms and net shapes of nickel-based superalloy and titanium alloy powders from 1970 to 1980. A comprehensive review of early hot isostatic pressing applications is given in Ref 2.

Current applications of hot isostatic pressing technology in P/M processing include net shapes in nickel-based superalloys for aircraft engine turbine disks and shafts (such shapes are "squared-off" cross sections suitable for sonic inspection), nickel-based P/M forging and rolling preforms, and nickel-based P/M integral pump and turbine impeller wheels; titanium alloy P/M billets, forging preforms, and shapes; tool steel billets (for mill processing), large die blocks, and composite structures; net shapes in P/M beryllium, niobium alloys, and other refractory metals; and dispersion- and fiber-strengthened P/M aluminum alloys. Small parts processed by a combination of cold compaction of metal powder, sintering, and hot isostatic pressing include tool steel shapes, rare earth magnets, and tools, dies, rolls, wear parts, and seals manufactured in tungsten carbide/cobalt and other carbide compositions. Nickel-based P/M aircraft engine applications represent the highest technology level of the method and tool steels the highest production tonnage. Hot isostatic pressing of P/M tungsten carbide/cobalt parts is employed worldwide.

Process Equipment

Typical hot isostatic pressing units consist of a pressure vessel, gas storage and handling system, furnace, tooling, power supply, controls, and instrumentation. Figure 1 presents a schematic of a complete hot isostatic pressing system. A typical large-sized unit is shown in Fig. 2.

Pressure Vessel. Proper design of the pressure vessel is essential. The explosive energy stored inside an operating unit can be equivalent to several hundred pounds of high explosives in destructive capability, depending on unit size. Pressure vessels must be designed to meet appropriate American Society of Mechanical Engineers (ASME) codes and American Society for Testing and Materials (ASTM) standards for materials.

Consideration must be given to the maximum allowable stress that the vessel can sustain for a designed service life. Typically, the general primary membrane stress intensity in the vessel wall cannot exceed one third of the ultimate tensile strength. Due to the cyclic loading of the hot isostatic pressing unit, fatigue life of the vessel is an important design criterion. The possibility of plastic failure of the vessel caused by the applied test and operating pressures must be considered in the vessel design.

Provisions must be made for suitable ports into the pressure vessel to accommodate instrumentation, electrical power, and pressurization equipment. These penetrations are usually located at the bottom of the vessel in present designs.

The first hot isostatic pressing units were designed with the furnace surrounding the vessel. These hot-wall vessels were constrained to operate within a limited range of pressures and temperatures, up to about

Fig. 1 Simplified schematic of hot isostatic pressure system
Courtesy of Industrial Materials Technology, Inc.

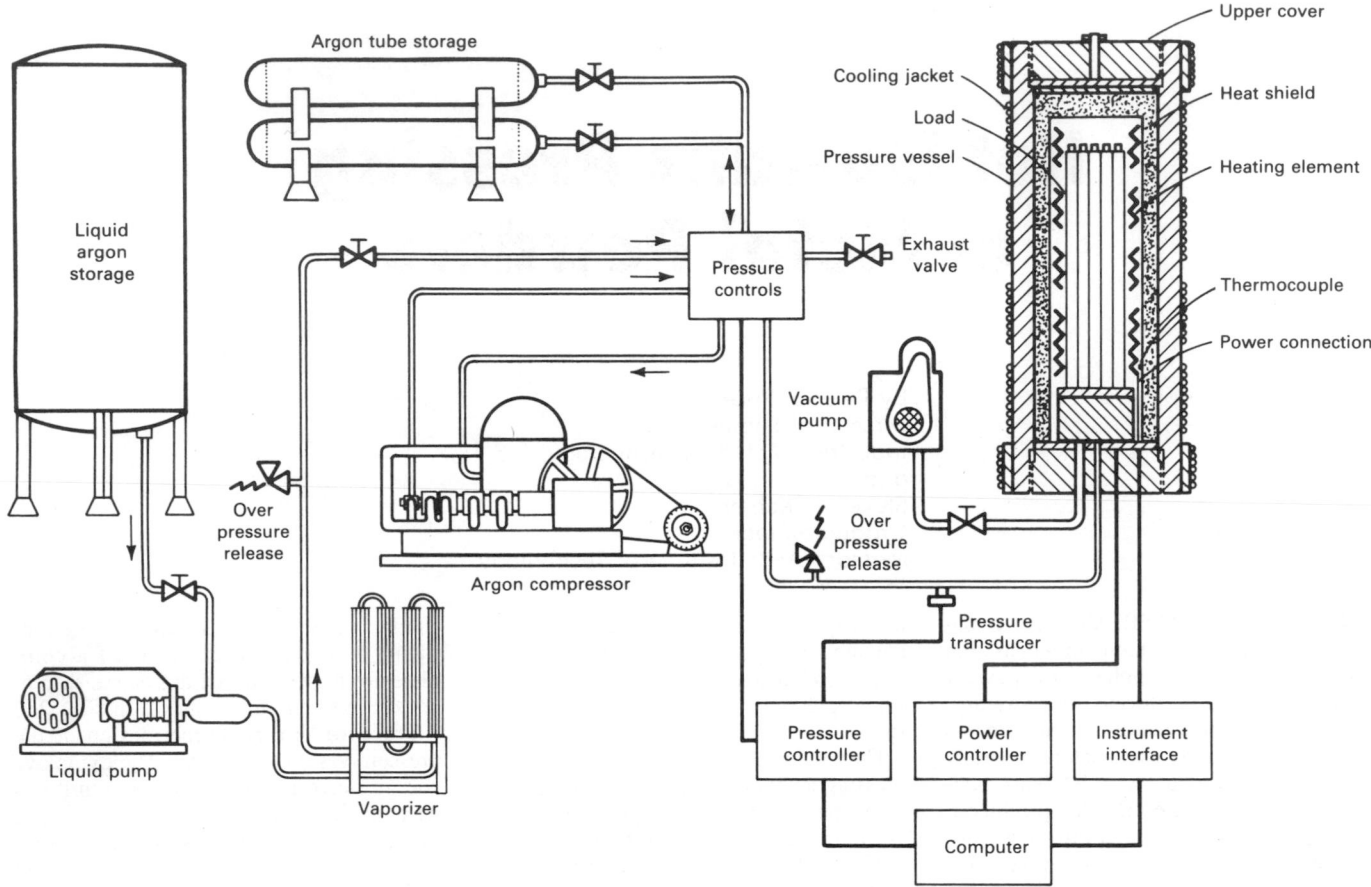

207 MPa (30 ksi) and 790 °C (1460 °F). Currently, all hot isostatic pressing units are cold-wall designs (furnace inside the vessel), with pressure vessel components operating up to about 80 °C (150 °F). A heat shield is placed between the furnace and the vessel to limit heat losses and control vessel body temperature. Additionally, the wall of the pressure vessel is cooled by water or other suitable heat transfer fluids.

The cylindrical portion of the vessel may be cooled using several methods. Cooling fluid may be contained by a jacket and come into direct contact with the vessel wall, plate coils with heat transfer mastic may be clamped around the vessel, or an interior-cooled bore liner may be used. In all cases, careful control of the cooling fluid composition must be maintained to prevent corrosion or fouling of the cooling channels.

Hot isostatic pressure vessels are constructed by several methods: multi-wall forged, monolithic forged, wire wound, or multiple ring-liner. An example of a monolithic forged pressure vessel is shown

in Fig. 3. Pressure is sealed within the vessel by Bridgman seals, metal-to-metal seals, single or double O-rings, or a combination of these devices.

Closure of the pressure vessel is accomplished by either threaded closure, pin closure, or nonthreaded closure supported by a yoke frame. In pressure vessels using threaded or pin-type closures, additional consideration must be given to the end loads carried by the vessel and closures. Threaded closures must be designed carefully to distribute the load evenly over all of the threads so that stress is not concentrated on the first two threads. Use of resilient threads or a specially designed thread form and undercut controls stress concentrations. For rapid opening of the vessel, interrupted threads are available that open the vessel in less than one rotation of the threaded closure.

In pin-type closures, a transverse pin is inserted through the pressure vessel and the end cap to close the vessel. This is accomplished quickly and easily; however, high unit stresses occurring at the pinholes in this type of design limit its application.

Designs that reduce stress concentrations by using two pins have been proposed. Currently, this design is used primarily with smaller cold isostatic presses.

As an alternative to threaded end closures, a yoke frame may be used (Fig. 4) to restrain the end loads. Opening and closing a pressure vessel with a yoke frame requires two actions. First the yoke frame (or the cylindrical body) is moved horizontally; then the end closure is extracted axially and moved to the side to allow access to the vessel. Large production units of both types (threaded closure and yoke frame) are currently in operation. User selection of design (yoke frame or threaded end closure) depends on such factors as system size, design pressure, system application, cost, inspectability, and preference.

Gas System. The pressure medium used in hot isostatic pressing is gas—usually argon, although nitrogen and other mixtures may be employed. The gas supply system must be able to quickly deliver the gas at the desired high pressure. Additionally, argon must be delivered at a very high pu-

Fig. 2 Large-sized hot isostatic pressing unit

96.5-cm (38-in.) diam by 244-cm (96-in.) work zone. Operating pressure: 100 MPa (15 ksi). Courtesy of Industrial Materials Technology, Inc.

Fig. 3 Hot isostatic pressure vessel

Design conditions: 100 MPa (15 ksi) at 200 °C (400 °F). Material: ASTM A 723, grade 3, class 3. Courtesy of Industrial Materials Technology, Inc.

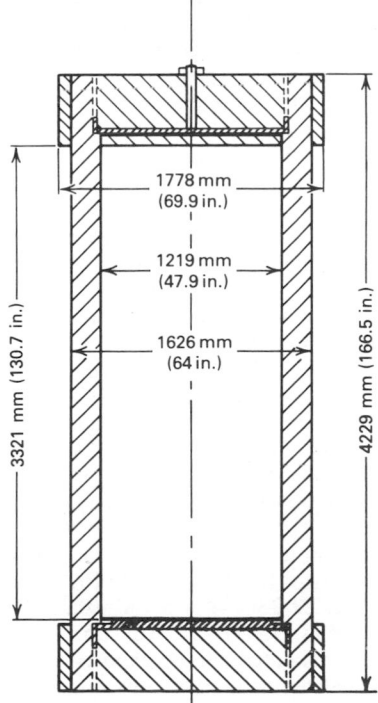

rity level to protect the work load and process equipment from detrimental chemical reactions with contaminants. Gas can be delivered from storage tanks and/or a liquid argon storage vessel.

Supplying argon from liquid storage (Fig. 5) ensures that the gas is of very high purity. Argon is delivered from liquid storage to a cryogenic pump by another smaller transfer pump or by gravity. Pressurized liquid argon moves from the cryogenic pump to a vaporizer, which delivers pressurized gas to the vessel.

Argon can be introduced into the system at a high rate. Unfortunately, there is no small-scale economic method of reliquefying argon and returning it to liquid storage after the hot isostatic pressing cycle.

The used and possibly contaminated gas can either be stored in a tube bank (Fig. 6) or vented to the atmosphere.

Argon in gas tube banks must be pumped by a compressor for storage. Multiple-stage, oil-lubricated piston pumps can be used for this purpose. The use of multiple stages provides high-pressure gas at a high-volume rate. However, hydrocarbon contamination of the argon is possible with this type of compressor. Process techniques can minimize this carryover problem.

Diaphragm compressors are used in hot isostatic pressing gas systems. They provide a very clean operation. Contamination of the gas can occur only after gross diaphragm failure. The use of multiple

Fig. 4 Hot isostatic pressing unit with wire-wrapped pressure vessel and yoke frame closure support

Courtesy of ASEA Pressure Systems, Inc.

Fig. 5 Liquid argon storage system

Courtesy of Industrial Materials Technology, Inc.

Fig. 6 Tube bank storage for argon gas

Courtesy of Industrial Materials Technology, Inc.

Fig. 7 High-pressure gas compressor

Suction pressure: 14 MPa (2000 psig). Working pressure: 200 MPa (30 000 psig). Courtesy of Hydro-Pac, Inc.

diaphragms and failure sensors and routine replacement of the diaphragm prevent this occurrence. Diaphragm compressors have a low volumetric rate capacity and low gain. High inlet pressure maximizes the mass pumping rate and outlet pressure.

Another type of compressor that has been introduced in the hot isostatic pressing industry is of relatively recent design. A double-acting hydraulic cylinder with a dry-lubricated, high-pressure gas cylinder on each end has been used to compress argon to very high pressures (70 to 400 MPa or 10 to 60 ksi) at a fairly high rate, without contamination problems. Figure 7 shows such a typical high-pressure gas compressor.

The purity of the gas entering or leaving the hot isostatic pressing unit must be monitored to ensure that the work load and internal process equipment are not contaminated during a cycle. Possible contaminants include nitrogen, oxygen, water vapor, carbon monoxide, carbon dioxide, methane, and hydrocarbons. Analysis for these gases can be performed with an on-line gas chromatograph.

Different hot isostatic pressing applications require gases of varying purity levels. For example, when hot isostatically pressing powder in a can, protection of process equipment is the only factor that must be considered in determining acceptable gas purity.

Furnace. During hot isostatic pressing, a controllable hot work zone is created and maintained by the furnace. The size of this zone is governed by vessel size and by furnace and insulation system design. Furnaces utilize primarily convection and radiation mechanisms to transfer heat. Convection dominates at lower temperatures, whereas radiation may be the dominant form of heat transfer at elevated temperatures. Presence of high-density gas as a heat transfer medium allows operation at a higher power density than is possible at atmospheric pressure or in a vacuum for comparable furnace elements.

Pressure vessel body temperatures are kept within design conditions by use of a heat shield between the furnace element and the vessel wall. This heat shield must meet several requirements. In addition to minimizing heat transfer to the vessel wall to reduce power consumption, it must shield the vessel wall as well. During cooldown, the shield should allow heat loss to reduce cooldown time. Density gradients in the argon (caused by thermal gradients between the vessel wall, furnace, and work load) can produce convection loops within the unit. These loops must be controlled by the heat shield to maintain uniform operating temperature within the hot zone. The heat shield also must withstand the thermal gradient from the work zone to the vessel wall. These requirements must be satisfied with as little mass as possible to minimize heatup and cooldown times and with the smallest volume to provide maximum space within the vessel for the work load. Multiple concentric metal shells, some of which are fabricated of molybdenum or molybdenum alloys, are frequently used for the heat shield elements.

Typically, the furnace consists of resistance heaters arranged in multiple, independently controlled zones below and beside the work load for optimum temperature control A hot zone operating at 1750 °C (3200 °F) has been achieved with a single-level element beneath the work load. Natural convention in the high-density gas medium with the pressure vessel is the functional heat transfer mechanism in this design. Forced convection within the pressure vessel has been incorporated in this base heater design, producing an increased work load length as well as improved heating and cooling rates. A heat shield, radiation shields, and baffles are also utilized in this design.

Installation and loading of the furnace, mantle, and work load into the pressure vessel can be accomplished by several methods. The furnace and mantle can have the capability to automatically connect the power supply and load instrumentation within the vessel. The work load can be introduced into the vessel before, after, or along with the furnace. Figure 8 shows introduction of the work load with the furnace already in place. Some hot isostatic pressing units permit loading and unloading of the work load while it is still hot, thus requiring a shorter cycle time.

The type of material used for furnace elements varies, depending on the range of operating temperatures. For applications up to 1230 °C (2250 °F), Kanthal A-1 (Fe–22-23Cr–4.5-5.7Al–0.5-2.0Co), or Hoskins 875 (Fe–22.5Cr–5.5Al–0.5Si) is used. Kanthal can be exposed to air at elevated temperatures; consequently, the work load can be hot loaded into a furnace

Fig. 8 Loading of the vessel with P/M parts
Courtesy of Industrial Materials Technology, Inc.

of this type. However, this material is prone to oxidation and may embrittle. Broken elements can be repaired by conventional welding. Because Kanthal is a relatively poor thermal conductor, low power densities must be used or overheating can occur, causing sagging and melt out. To compensate for this tendency, Kanthal frequently is manufactured and used in ribbon form.

Molybdenum furnace elements can operate in the temperature range of 500 to 1600 °C (930 to 2900 °F). They cannot be exposed to air at elevated temperatures because rapid oxidation occurs. Molybdenum usually is used as stranded wire or a band. During extended use, molybdenum recrystallizes, thus causing an embrittlement condition at ambient temperature. However, stranded molybdenum that becomes damaged can be repaired by lashing, a technique of splicing across the break using molybdenum wire wrapping. In the frequent presence of carbon monoxide in a hot isostatic pressure system, molybdenum case carburizes. The electrical resistivity of the wire becomes greater, increasing the heating capability of the wire.

Graphite is also used as both structural and resistance heating elements for hot isostatic pressing furnaces operating in the range of 400 to 2200 °C (750 to 3990 °F). Graphite is advantageous because of its excellent strength and dimensional stability at very high temperatures, its low cost compared to refractory metals, and its ready

availability in various grades and stock sizes.

Two forms of graphite are used: bulk and fiber reinforced. Bulk graphite is machined into the required furnace element shapes. Fiber-reinforced graphite is used because it can be fabricated into thinner (and higher electical resistance) sections, while retaining good resistance to mechanical damage and breakage.

The principal shortcoming of graphite furnaces is that unless the system is dried by a combination of heating, purge, and evacuation, carbon monoxide can be generated by reaction of water vapor and graphite. Carburization of work-load pieces can occur unless proper care is taken in the evacuation cycle. Contact between graphite and most metals must be avoided to prevent melting reactions. Currently, graphite furnaces up to 75 cm (30 in.) diam are used in routine production at temperatures up to 1750 °C (3180 °F).

Tooling. The type of tooling used in a hot isostatic pressing unit depends on the parts being processed. Tooling has a direct effect on the economics of a hot isostatic pressing cycle, because it determines the number of workpieces that can be packed into the uniform hot zone. Generally, tooling must provide adequate support for the workpieces to minimize distortion during the processing cycle. Structural thickness of the tooling must be kept to a practical minimum so as not to increase the thermal load and to maximize available space.

Tooling is constructed from materials that do not melt or interact with the workpiece at the temperatures reached during hot isostatic pressing. Carbon steel frequently is used, because it offers versatility, is readily available, and is cost-effective. It does, however, undergo a volume change due to phase transformation that contributes to eventual tooling distortion. Other tooling materials include molybdenum, Inconel 600, graphite, or oxide ceramics.

Tooling configuration varies, depending on whether P/M parts are encapsulated. If the part is in a container, a simple modular tray set (Fig. 9) may be used. If the P/M parts are not containerized, they may require separate supports to prevent them from diffusion bonding together. Often, a tray set and parting material, such as aluminum oxide or zirconium oxide, are used to support and separate the parts. All parts must be properly supported to prevent bending from their own body weight.

Controls and Instrumentation. The control system of the hot isostatic pressing unit must provide for the safe and reliable

Fig. 9 Modular tray set for multiple small parts
Courtesy of Industrial Materials Technology, Inc.

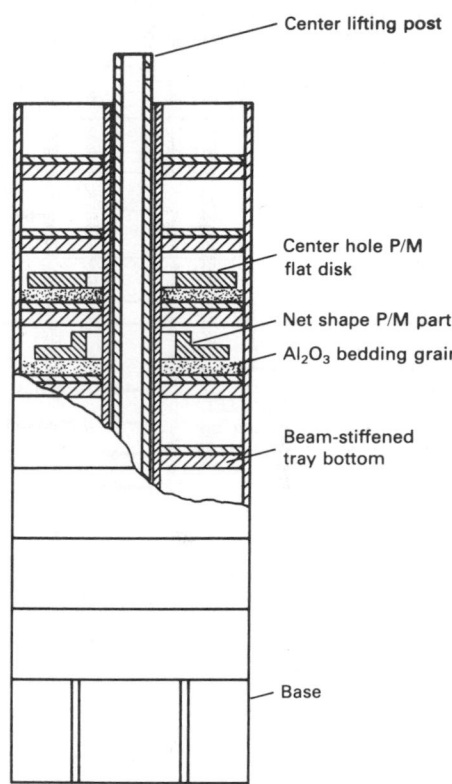

operation of the entire system, as well as accurate measurement and control of pressures and temperature parameters throughout the entire processing cycle. Pressure inside the vessel is measured by a Bourdon tube gauge or a strain-gauge pressure transducer. Temperature is measured with thermocouples that are distributed throughout the hot zone. With a multizoned furnace, each furnace element or zone must be monitored and controlled to adequately control the entire hot zone. Measurements are taken at several points on large workpieces, as well as throughout the working volume, to guarantee that process conditions are met. Permanent pressure and temperature records should be maintained.

The power supply to the furnace is usually a silicon-controlled rectifier system. Phase angle firing is used to control power input for molybdenum element furnaces. Zone current, zone voltage, and total power are monitored to control furnace performance. Power consumption of a well-designed 25-cm (10-in.) diam hot isostatic pressing unit is about 15 kW; a 38-cm (15-in.) diam unit consumes about 40 kW, and a 102-cm (40-in.) diam unit uses about 200 kW. Ground fault protection may be pro-

Fig. 10 Schematic of minicomputer control of hot isostatic pressing facility
Courtesy of Industrial Materials Technology, Inc.

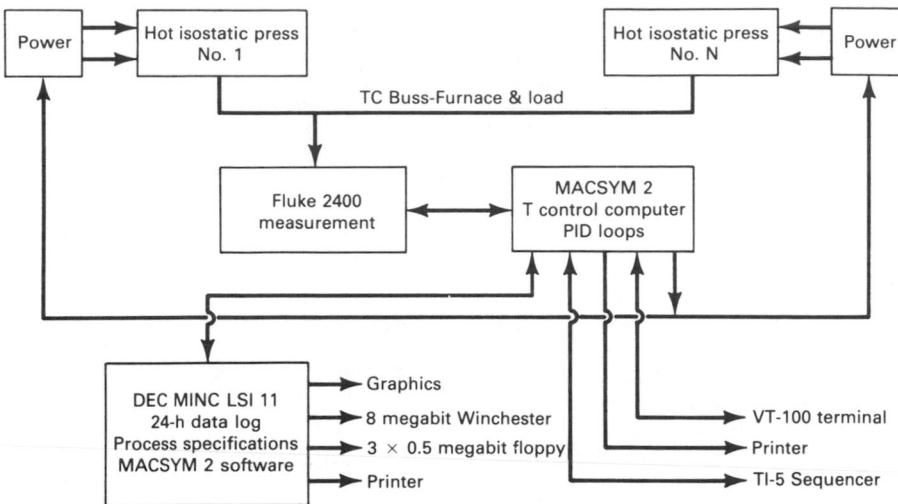

Fig. 11 Sample computer output showing process parameters
Courtesy of Industrial Materials Technology, Inc.

```
06/30/83 15:46:04      ASEA      G.E. HIP#8989

   ELAPSED TIME    0.00

   WINDING TC  1 = 1185. C     LOAD TC  3 = 1184. C
   WINDING TC  2 = 1185. C     LOAD TC  4 = 1183. C
     LOAD TC  3 = 1184. C     LOAD TC  5 = 1184. C

   PRESSURE =      0. PSI


06/30/83 15:47:03      ASEA      G.E. HIP#8989

   ELAPSED TIME    0.00

   WINDING TC  1 = 1185. C     LOAD TC  3 = 1184. C
   WINDING TC  2 = 1185. C     LOAD TC  4 = 1183. C
     LOAD TC  3 = 1184. C     LOAD TC  5 = 1184. C

   PRESSURE =      0. PSI
```

vided to prevent extensive furnace damage in the event of electrical insulation failure.

Other system parameters that should be monitored include gas purity, high-pressure leaks, vessel temperatures, and water system conditions. If a system failure is detected, the control system should provide a warning, take corrective action, and/or initiate system shutdown.

Control equipment consists of automatic controllers, analog data presentation devices, and remotely operated valves and switches. Minicomputers are capable of automatically monitoring and controlling all system parameters. Programs written to administer specified heating and cooling rates are carried out on systems such as that shown in Fig. 10. This system allows the operator to specify a variety of process cycles with accurate reproducibility. Process verification is possible, as shown in Fig. 11. Minicomputers also can provide complete and reliable capability to detect process errors or system malfunction and take corrective action.

Processing Sequence

Figure 12 presents the basic process flowchart for production of various sized encapsulated and unencapsulated P/M parts. This chart applies to both production and development routes, wherein interactive modification of process variables is required to obtain design objectives. Included is a provision for a subroutine where parts can be low-temperature hot isostatically pressed to a closed porosity condition, decanned, and re-hot isostatically

pressed, usually at higher temperatures. This option can be employed when the powder/container interaction (melting, alloying, contamination, etc.) is unacceptable at the preferred higher hot isostatic pressing temperature. This technique has been used, for example, for niobium alloys that are initially hot isostatically pressed at 1205 °C (2200 °F) in low-carbon steel containers, decanned, and re-hot isostatically pressed at 1595 °C (2900 °F) to circumvent an iron-niobium eutectic reaction at 1360 °C (2480 °F). The key aspects of the various elements of the process flowchart (part size and shape, powder properties, encapsulation, and preliminary powder processing, and containerless hot isostatic pressing) and their interrelationships are discussed in the following sections of this article.

Part Size

The upper size limitation for densification of encapsulated parts is governed primarily by the processing unit uniform temperature working zone diameter and length. Tool steel billets approximately 60 cm (24 in.) in diameter by 300 cm (120 in.) long and larger have been fully densified by hot isostatic pressing. Nickel-based superalloy P/M turbine disks greater than 1 m (3.3 ft) in diameter have been successfully densified. For sheet metal encapsulation of P/M parts weighing more than approximately 20 kg (44 lb), attachment of handling lugs is recommended. For large-diameter parts (greater than 0.5 m or 1.6 ft in diameter) and weights greater than 100 kg (220 lb), sheet metal bending

stresses due to enclosed powder weight must be considered. Careful consideration must be given to the support of large parts in hot isostatic press tooling to prevent bending during heating prior to complete densification. For small net shape parts (1 to 1000 g, or 0.03 to 35 oz), particularly with thin sections, tooling that permits separate setting of each part is required.

Production of small (less than 10-kg, or 22-lb) net shape parts by hot isostatic pressing using encapsulation techniques is not generally economical because of the cost of container fabrication and leak checking. This applies particularly to P/M tool steels. Exceptions include experimental parts and manufacture of specialty parts in P/M refractory metals, composites, and precious metals, where metal cost is a controlling factor. Small net shape parts (less than 0.5 kg, or 1 lb) are best manufactured by containerless hot isostatic pressing, particularly for tool materials, provided satsifactory process procedures can be developed.

Part Shape

Cylindrical billets (Fig. 13) are the most cost-effective product form for further mill processing to bar or rounds, because maximum use can be made of the working volume of the hot isostatic pressing unit. Flat rectangular sheet bar billets for straight rolling, or combined with cross rolling to control width and texture, also can be produced by hot isostatic pressing. Sheet bar can components are shown in Fig. 14.

Production of P/M parts to near-net shape by hot isostatic pressing with dimensions suitable for nondestructive examination, typically ultrasonic inspection, depends entirely on shape and dimension transfer. In the encapsulation method, part shape is generated by the powder container. In containerless hot isostatic pressing, shape and dimension are derived from the powder pressing operation, usually press com-

Fig. 12 Flowchart for P/M manufacture of metal parts by hot isostatic pressing
Courtesy of Industrial Materials Technology, Inc.

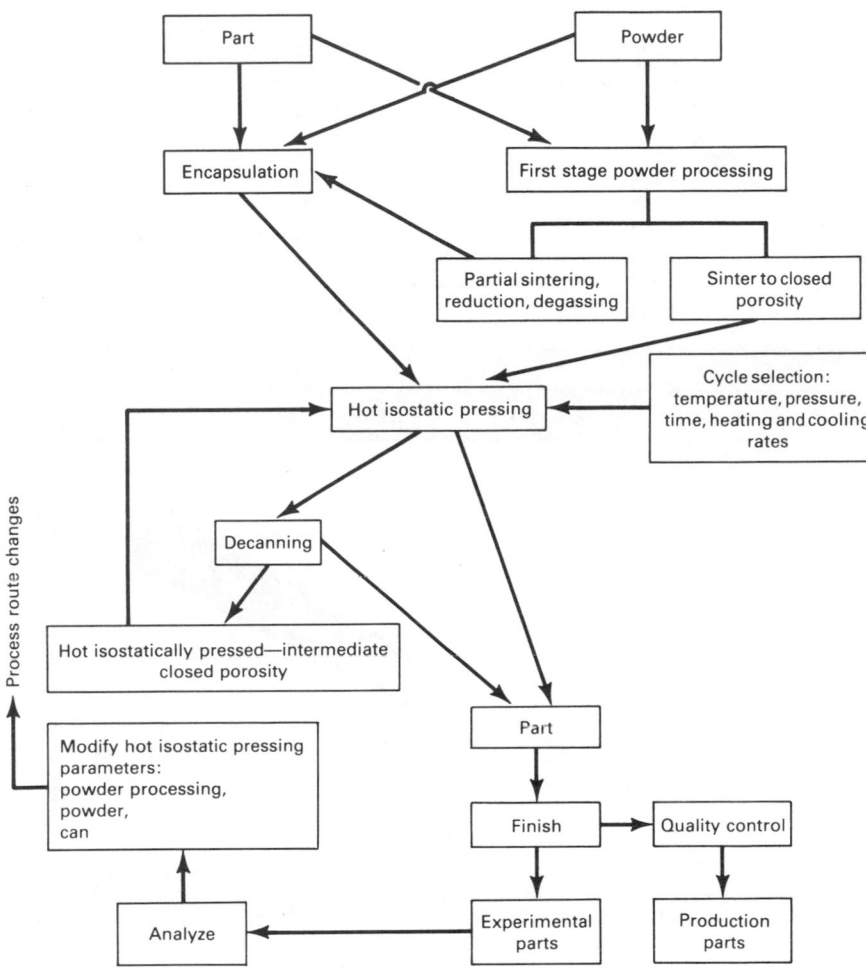

Fig. 13 Large-sized cylindrical high-speed steel billet
Courtesy of Crucible Materials Research Center

Fig. 14 Sheet bar can
(a) Components. (b) Assembled. Courtesy of Industrial Materials Technology, Inc.

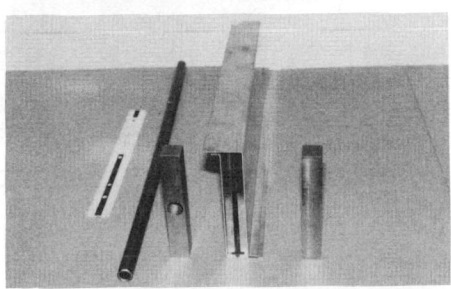

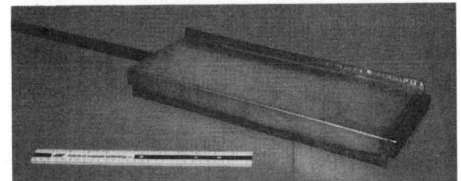

paction in tool steel dies or cold isostatic pressing. In all net shape hot isostatic pressing methods, an underlying requirement exists for "single-point" machining of the basic reference tooling. Sheet metal powder cans require machining of either dies for pressing or hydroforming of can components, or laminated wood or steel tooling for metal spinning.

In the ceramic mold technique, shape is transferred from wax patterns that are produced by either single-point machining (usually experimental quantities) from wax billets or wax injection molding in precision-machined molds. Unless desired powder container shapes can be built from standard stock sheet metal and/or seamless metal tubing elements, special tooling will be a cost element of the technique.

Solid net shapes with a complex external geometry are easily produced by a ceramic mold technique (Fig. 15) that uses injection molded wax patterns, ceramic investment, dewaxing, powder filling, outside canning with pressure transmittal medium, evacuation, sealing, and hot isostatic pressing. Large solid shapes, such as the turbine disk in Fig. 16, usually are produced using sheet metal containment. For complex shapes with internal cavities, sheet metal encapsulation with internal filler pieces that can be removed by machining or chemical etching is a possible alternative (Fig. 17 and 18).

Initial mold dimensions to produce a finished part from encapsulated uncompacted powder must include machining allowance, powder/can interaction, bowing (particularly in thin sections), end effects, integral sample material allowance, and densification shrinkage. This shrinkage, the largest single factor affecting dimensional accuracy, varies with powder characteristics and the methods of settling loose powders in containers. Figure 19 illustrates typical shrinkage in a billet during

hot isostatic pressing. Containerless processing requires normal allowances for sintering shrinkage and cold densification shrinkage during cold isostatic pressing.

Powder Properties

Powder properties that affect the results of hot isostatic pressing include particle shape, mean particle size, particle size distribution, composition, and impurity content. Blended, prealloyed, or elemental powders for encapsulation hot isostatic pressing are prepared without lubricants (waxes or binders). Partial separation of

Fig. 15 Ceramic mold hot isostatic pressing process
Courtesy of Crucible Materials Research Center

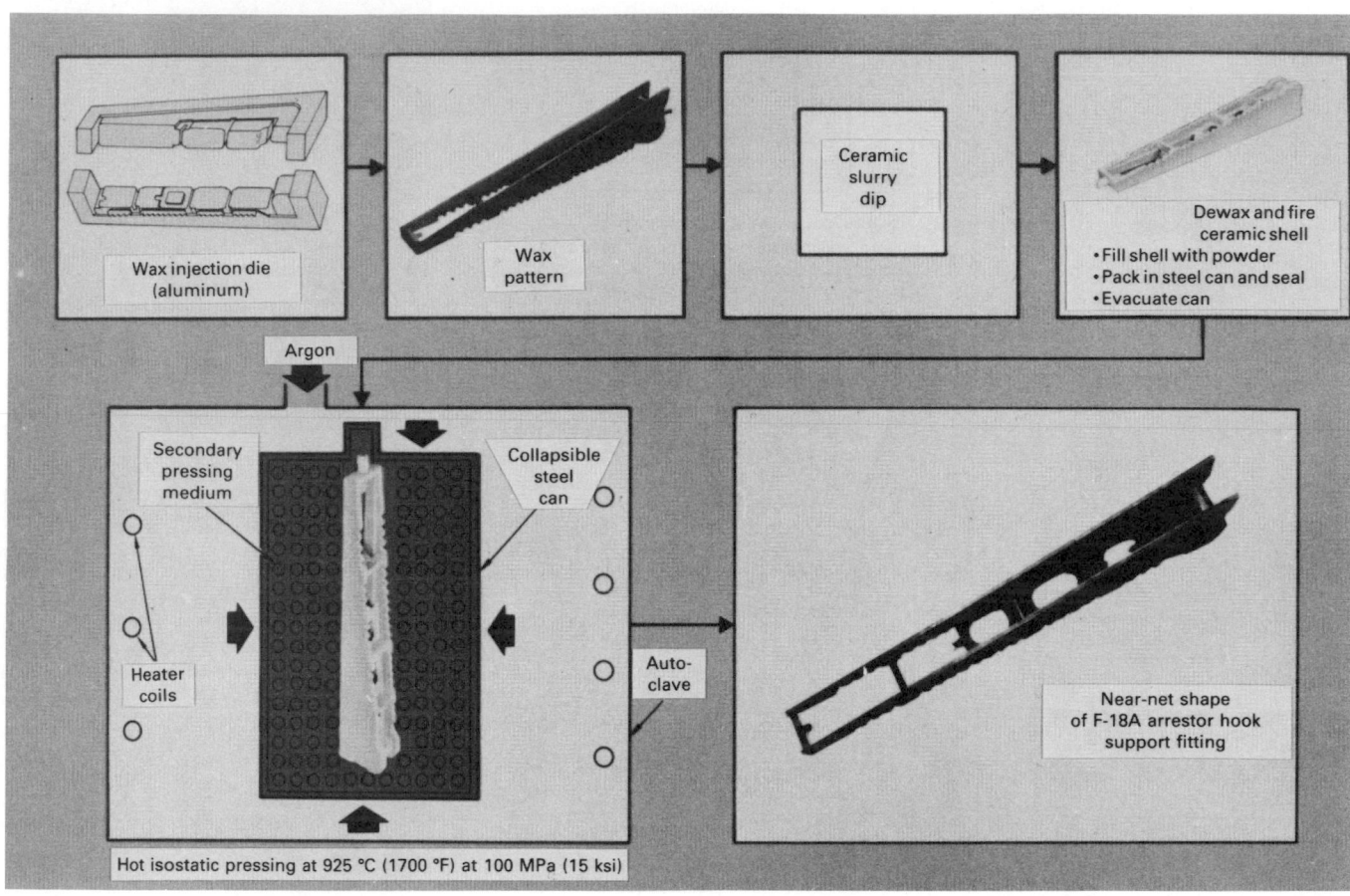

powder constituents due to size differential and specific gravity effects during flow and settling in a hot isostatic pressing container is possible.

Specification of packing density of metal powder after "tapping" or vibration (impact) settling is required for design of containers used to produce parts to dimension. Powder packing density is defined as the ratio of apparent density of settled (tapped) powder to 100% dense material. The reference value is taken for the cast or wrought form with allowance for variations due to volume percentages of various phases (austenite, ferrite, gamma, gamma prime, etc.) depending on thermal history. Thus:

$$D = \frac{\rho'}{\rho}$$

where D is the packing density (fraction of theoretical); ρ' is the apparent density of settled powder, g/cm^3; and ρ is the density of 100% dense material, g/cm^3.

In hot isostatic pressing of encapsulated powders, the "empty space" between powder particles is "squeezed out." The fractional volume shrinkage (percent initial volume, 1-g basis) during hot isostatic pressing is given by:

$$\frac{\Delta V}{V_i} = \frac{V_i - V_f}{V_i} = \frac{\dfrac{1}{\rho'} - \dfrac{1}{\rho}}{\dfrac{1}{\rho'}}$$

$$= 1 - \frac{\rho'}{\rho} = 1 - D$$

where V_i and V_f are specific volumes of initially encapsulated and fully dense metal powder, respectively.

Linear dimensional shrinkage can be estimated from the relationship for a unit volume using the defined packing density:

$$V = l^3$$

$$\ln V = 3 \ln l$$

$$\frac{\Delta V}{V_i} = \frac{3\Delta l}{l_i}$$

$$\frac{\Delta l}{l_i} = \frac{1}{3}\frac{\Delta V}{V_i} = \frac{1}{3}(1 - D) = \frac{1 - D}{3}$$

With metal powder of a given packing

Fig. 16 MERL 76 disk produced by hot isostatic pressing

(a) Top view, (b) bottom view, and (c) side view of compacted disk with unremoved can. Courtesy of Pratt & Whitney Aircraft Group, Division of United Technologies Corp. and NASA

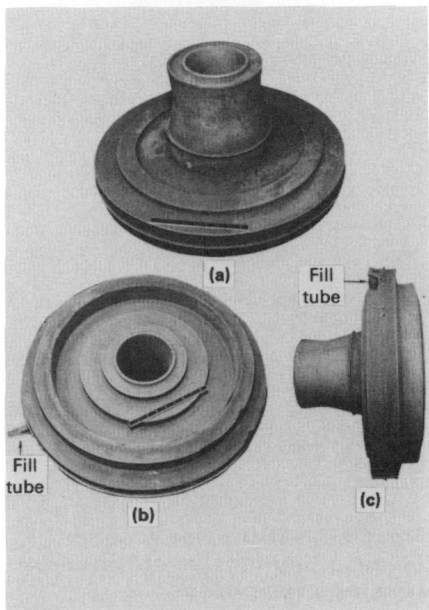

Fig. 17 Beryllium substrate with internal cavities

(a) Cell structure of mirror blank. (b) Arrangement of beryllium and tooling components. Courtesy of Battelle-Columbus Laboratories

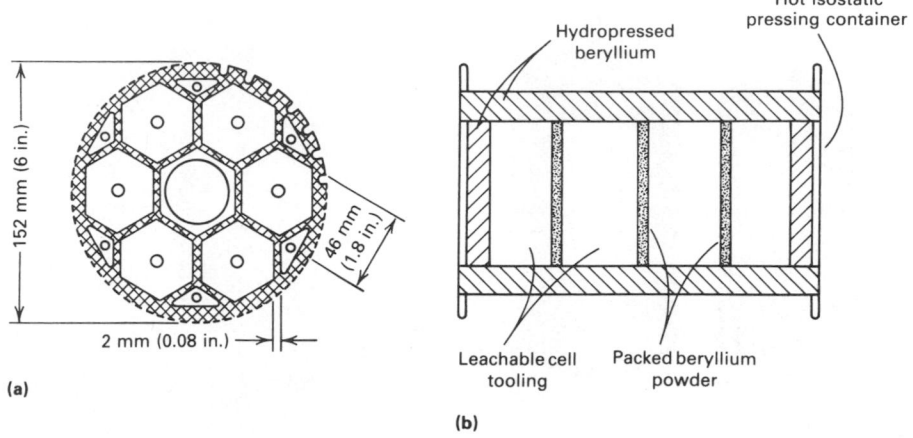

Fig. 18 Net shape disk with internal cavities

Courtesy of Industrial Materials Technology, Inc.

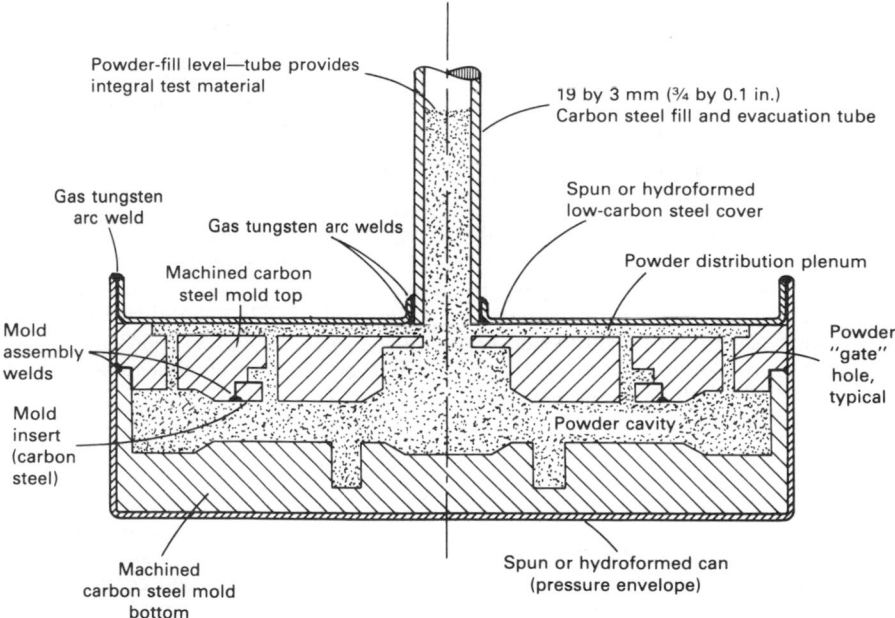

density (D), the starting dimension (l_i) to produce a specific fully compacted powder dimension (l_f) is estimated as:

$$l_i = \frac{3}{(2 + D)} \, l_f$$

This equation is valid in the range $0.55 < D \leq 1$. Additional allowances of stock for post-hot isostatic pressing operations may be required.

Metal powders with a mean particle size on the order of microns and packing densities of $D < 0.55$ require special techniques for encapsulation hot isostatic pressing. Without precompaction before canning, essentially more empty space than powder within the container will, in effect, be hot isostatically pressed. Elastic/plastic collapse of thin-walled cans usually occurs when pressure is applied (either hot or cold), resulting in severely distorted shapes and loss of dimensions. Simple shapes can be hot isostatically pressed in thin-walled containers by using die compacted or cold isostatically pressed "slugs" or shapes ($0.7 < D \leq 0.8$) close packed in the can interior. Alternately, heavy-walled containers that resist buckling under external pressure can be used, but the space efficiency of powder densification is lower.

Four categories of powder impurities are encountered in P/M hot isostatic process-

ing: nonmetallic, metallic (other P/M metals inadvertently present from process operations), interstitials (carbon, oxygen, and nitrogen), and argon (from inert gas atomization). Because of extended time at high temperature during hot isostatic pressing densification of powders, diffusion fields surrounding impurities can be extensive and can adversely affect certain properties. Figure 20(a) shows a hot isostatically pressed densified superalloy powder containing a nonreactive oxide inclusion; Fig. 20(b) shows a reactive steel inclusion with its surrounding diffusion zone.

For net shape hot isostatically pressed parts with no post-pressing hot work, interstitial elements originating from atomization may concentrate at powder particle boundaries, forming nonmetallic compound aggregates such as carbides and preferred crack paths, as shown in Fig. 21. Argon trapped in hollow powder particles during gas atomization (Fig. 22) is compressed during hot isostatic pressing densification, but it can re-expand during post-hot isostatic pressing heat treatment operations, producing re-formed porosity. Porosity is particularly detrimental in tool

Fig. 19 Cylindrical billet can

(a) After hot isostatic pressing. (b) Before hot isostatic pressing. Courtesy of Industrial Materials Technology, Inc.

(a)   **(b)**

Fig. 20 Impurities in hot isostatically pressed Astroloy powder

(a) Nonreactive oxide inclusion. (b) Reactive steel inclusion and surrounding diffusion zone. Courtesy of Motoren-und Turbinen—Union Munchen GmbH

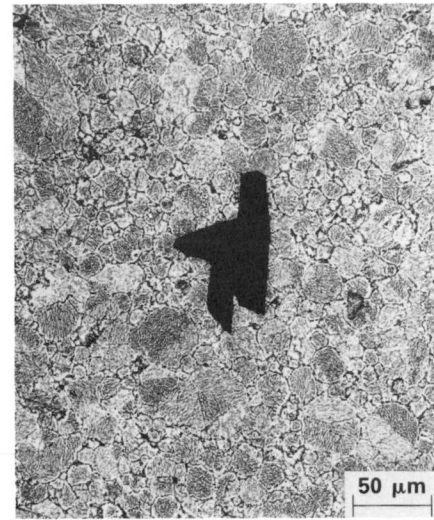
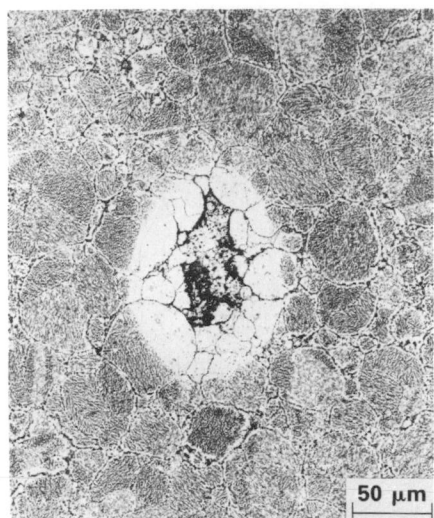

steels. Argon in metal powder can be detected by a vacuum fusion/mass spectrometer analysis method.

Encapsulation

For metal powder hot isostatically pressed using encapsulation techniques, the functional requirements of the powder container include: (1) shape and dimension control prior to and during processing; (2) maintenance of leaktightness against low and high pressure (typically up to 100 MPa, or 15 ksi) during evacuation, sealing, and densification, where substantial elastic and plastic container deformation occurs during shrinkage; (3) noncontamination of powder due to abradable nonmetallic materials; and (4) minimal interaction with powder by diffusion processes during the hot isostatic pressing cycle. Additionally, the container must be removable by machining, grinding, or chemical etching.

Encapsulation Methods. Sheet metal, ceramic investment molds, and glass molds are used for simple or net shape production. Currently, glass molds are not used extensively in production of P/M hot isostatically pressed components. Glass mold technology is described in Ref 3. The ceramic investment mold method, an adaption of investment casting techniques (metal powder is poured into the mold instead of liquid metal), is described in Ref 4 and 5. This method is depicted schematically in Fig. 15. It is particularly advantageous for production of complex precision net shapes with compound curved surfaces and re-entrant configurations (Fig. 23). Because the ceramic mold containing metal powder is surrounded by a large volume of pressure-transmitting medium

(usually granular oxides), outgassing and heating cycles (slow heat transmission through a packed granular bed) are long.

Currently, metal encapsulation is used widely for experimental and production-scale hot isostatic pressing of P/M parts. Compound curved surfaces are costly to produce from sheet metal, because multistage forming die sets are required as well as intermediate anneals for deep drawn parts. Metal encapsulation is used for spun cylindrical, simple cylindrical, or rectangular shapes. However, additional shapes can be produced by combinations of metal forming and welding.

Encapsulation Materials. Sheet materials commonly used for metal powder encapsulation include 1010, 1018, and 1020 low-carbon steel, extra-low-carbon killed steel, type 304 austenitic stainless steel, commercially pure titanium, and Ti-6Al-4V alloy. Nickel and copper can be used, but thermodynamic reduction of sheet metal grain boundaries and weld metal oxides may occur in high-pressure argon during processing. Reduction of "continuous path" oxides leads to subsequent leakage.

Powder/can metal combinations must be evaluated or tested for detrimental interactions at processing temperatures. Binary phase diagrams can be consulted for potential melting reactions. Commercially pure titanium encapsulation of refractory metals (molybdenum, niobium, tantalum, and tungsten) permits processing of these alloy systems between 1400 and 1650 °C (2550 and 3000 °F) without adverse melting reactions. Solid-state reactions be-

Fig. 21 IN-100 prior powder particle boundaries outlined by carbide precipitates

Courtesy of Industrial Materials Technology, Inc.

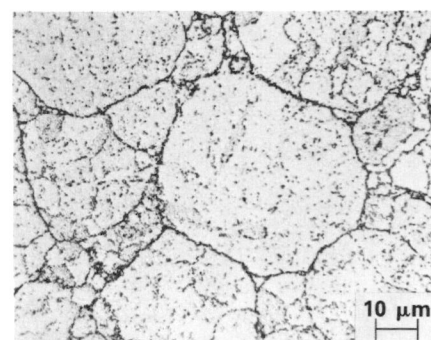

Fig. 22 Pores in −80 mesh Astroloy powder particles produced by argon gas atomization

Courtesy of Industrial Materials Technology, Inc.

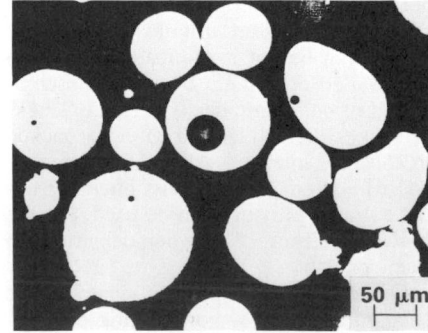

Fig. 23 Stages in ceramic mold shapemaking of pivot shaft

(a) Wax pattern. (b) Shell mold. (c) As-hot isostatically pressed. (d) Machined. Courtesy of Crucible Materials Research Center

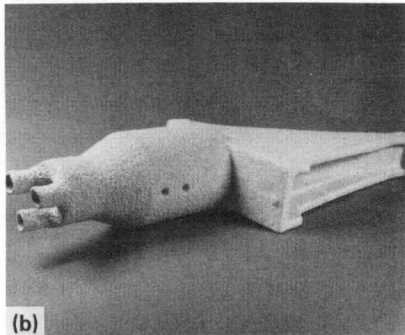
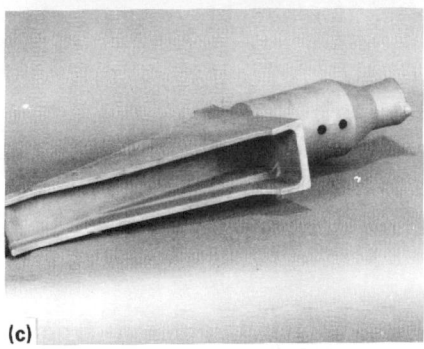
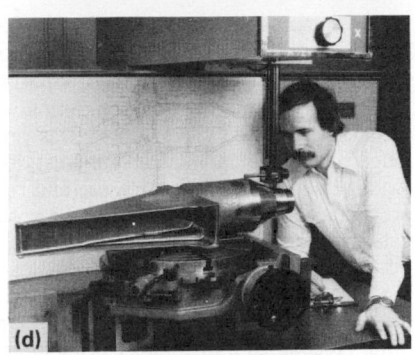

Fig. 24 Titanium can/molybdenum powder interdiffusion zone

Courtesy of Industrial Materials Technology, Inc.

200 μm

Fig. 25 Interaction between can and powder causing carbide formation

Courtesy of Industrial Materials Technology, Inc.

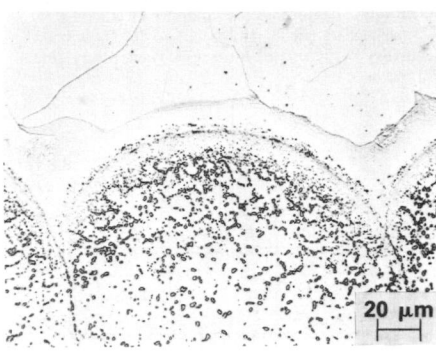

20 μm

tween can and powder include interdiffusion alloying (Fig. 24) and carbide formation (Fig. 25).

In some boron- and carbon-containing materials (cemented carbides, for example), thermodynamic activity of these elements may approach unity, leading to can/powder system melting points approximated by binary phase diagram eutectics. The iron-boron eutectic melts at 1149 °C (2100 °F) and the iron-carbon eutectic melts at 1153 °C (2107 °F), for example. The latter case precludes steel canning of tungsten carbide/cobalt powder and hot isostatic pressing above approximately 1100 °C (2010 °F), because tungsten carbide/cobalt contains approximately 6 wt% C.

Form of materials that can be used include flat sheet, seamless tubing, and pipe. Thicknesses less than 16 gauge (1.5 mm, or 0.06 in.) are impractical because of limited tensile ductility, whereas appreciable plasticity without through fracture may be required in certain container configurations that stretch or bend during hot isostatic pressing compaction. Use of container envelope components machined from heavy mill products (plate or bar) is not advised, because pinhole piping originating in steel ingot solidification occurs frequently and persists through mill operations.

Low-carbon steel parts made from plate or bar (free-machining grades are not recommended) may be used for can interior shapes or cavity-forming components. As compact weight increases, container sheet gauge should be increased to provide structural support during pre-hot isostatic pressing operations. Usually 14-gauge (1.9-mm, or 0.07-in.) sheet is adequate for parts weighing approximately 200 kg (440 lb), and 12-gauge (2.7-mm, or 0.1-in.) sheet is suitable for parts weighing approximately 1000 kg (2200 lb).

Container Design. Container dimensions are determined by design requirements of the P/M part being fabricated, with an added allowance for anticipated powder shrinkage. A typical can design for a simple cylinder configuration is shown in Fig. 26. End covers are manufactured by spinning, hydroforming, or drawing.

The cylinder also may be made from seamless tubing or pipe, or it may be rolled from sheet and longitudinally seam welded. Minimum wall thickness available from stock in tubing increases with diameter, making rolled sheet metal preferred for some large sizes.

The "matched-lip" weld design successfully accommodates severe deforma-

Fig. 26 Spun sheet metal powder container

75 mm (3 in.) ≤ Diameter (*D*) ≤ 1000+ mm (40+ in.). Height (*H*) is limited by sheet composition and thickness, diameter, and cost and feasibility of intermediate anneals in spinning. Corner radius (*R*) = 1*t* for *t* < 6 mm (¼ in.). Courtesy of Industrial Materials Technology, Inc.

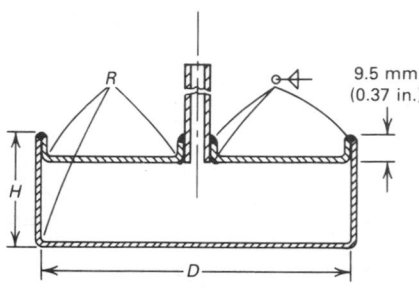

Fig. 27 Typical can corner designs

(a) Recommended design. (b) Poor design. Courtesy of Industrial Materials Technology, Inc.

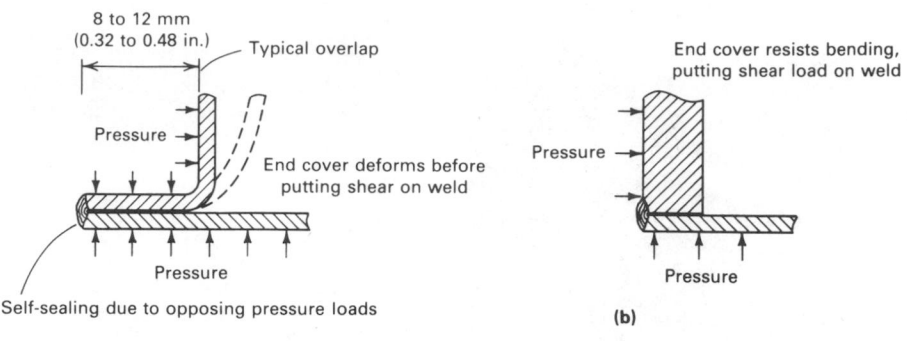

(a)

(b)

Fig. 28 Cleanable corner design

No crevices or corners for dirt entrapment. Good only where inner side is accessible for welding. Courtesy of Industrial Materials Technology, Inc.

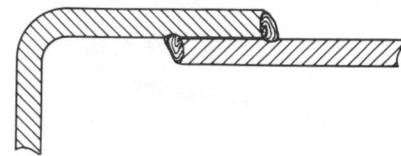

Fig. 29 Rectangular cans from sheet metal

(a) End view: self-sealing design matching can halves. (b) End view: alternate rectangular can design. (c) End view: detail section. (d) End cover before edge bending and cover welding. (e) Perspective view of can edge welded, but not cover welded. Courtesy of Industrial Materials Technology, Inc.

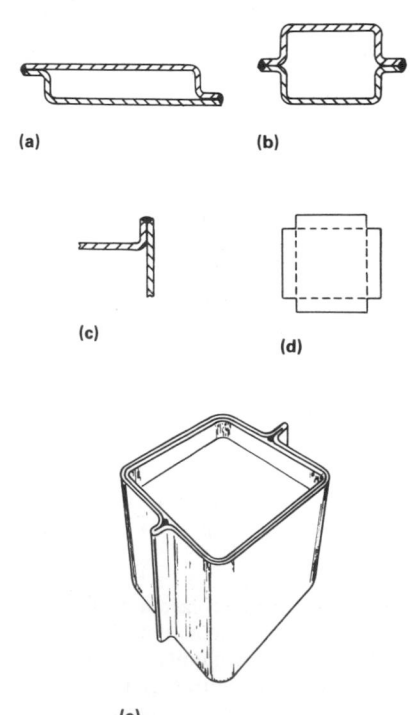

(a)

(b)

(c)

(d)

(e)

tion during consolidation, without leakage. Applied process pressure creates a self-sealing action and eliminates shear loads on the weld. The small volume of accessible weld metal also permits repair by grinding and rewelding. Recommended and poor can corner designs are shown in Fig. 27. An alternate corner detail that facilitates cleaning operations is shown in Fig. 28. The design of self-sealing rectangular cross section cans is shown in Fig. 29. Proven can designs are shown in Fig. 30.

The fill, evacuation, and seal tube attachment to the powder envelope is designed to prevent can buckling and tearing in the transition area (Fig. 31). Thick-walled fill tubes are preferred, typically 19-mm outside diameter by 3 mm (¾ in. outside diameter by 11 gauge) for compacts weighing 5 to 200 kg (11 to 440 lb) and 9.5-mm outside diameter by 2.3 mm (⅜ in. outside diameter by 13 gauge) for smaller compacts. A thick tube wall prevents buckling under applied pressure and facilitates metal flow during the forge welding pinch-off step. Loss of powder during evacuation and degassing (after can filling) is prevented by inserting a stainless steel wool plug, a metal plug, and partial crimping (Fig. 32).

Special shapes, with one or more critical dimensions carefully controlled, can be produced using sheet metal fabrication in combination with insert pieces. Typical configurations are illustrated in Fig. 33.

Container Fabrication. Fabrication of thin-walled or thick-walled containers from sheet metal can be achieved by using the metalworking methods described in Volume 4 of the 8th edition of *Metals Handbook*. Utilization of standard stock seamless tubing is recommended for cylindrical configurations wherever possible. Selec-

tion of spinning versus die forming for can covers is dictated by size, configuration complexity, and number of parts required.

Cleaning Procedures. Contamination of encapsulated powders, as well as prolonged outgassing, is inevitable, unless dirt, oxides, metalworking lubricants, and rust preventatives used on sheet and tube products are removed. See Volume 5 of the 9th edition of *Metals Handbook* for cleaning procedures applicable to various metals. Proper cleaning, storage, and handling procedures immediately prior to any welding operation are necessary to prevent dirt entrapment in can corners.

Powder metallurgy alloys that are particularly sensitive to contamination (titanium, nickel-based alloys, and refractory metals) require controlled humidity and stringent cleanliness for final can preparation, assembly, welding, and filling. Electropolishing of stainless steel can components and nonchlorinated solvent cleaning (usually acetone, methyl ethyl ketone, or methanol) of titanium can components represent typical cleaning processes for specialized applications. Additionally, hydrogen (vacuum) annealing is recommended for low-carbon steel container components. A typical processing sequence includes annealing at 1000 °C (1830 °F) for 1 h, followed by vacuum or inert gas cooling to 200 °C (390 °F) before unloading.

Carbon steel sheet metal parts must be supported carefully or partially assembled to prevent distortion during oxide reduction. Gloved handling, dry temporary storage, plus immediate assembly, welding, and filling are recommended to prevent rusting of carbon steel containers in normal plant atmospheres. Similar precautions are recommended for the outer sheet metal container used in a ceramic mold process.

Container Assembly. Containers for loose powder are assembled, welded, leak

Fig. 30 Sheet metal powder cans

(a) Rectangular. (b) Cylindrical. Courtesy of Industrial Materials Technology, Inc.

Fig. 31 Fill tube attachment detail

(a) Sheet metal cover punched and turned up. Interior and exterior weld. Thick-walled fill and evacuation tube. (b) Machined adapter to strengthen large-diameter tube connections to sheet metal. Courtesy of Industrial Materials Technology, Inc.

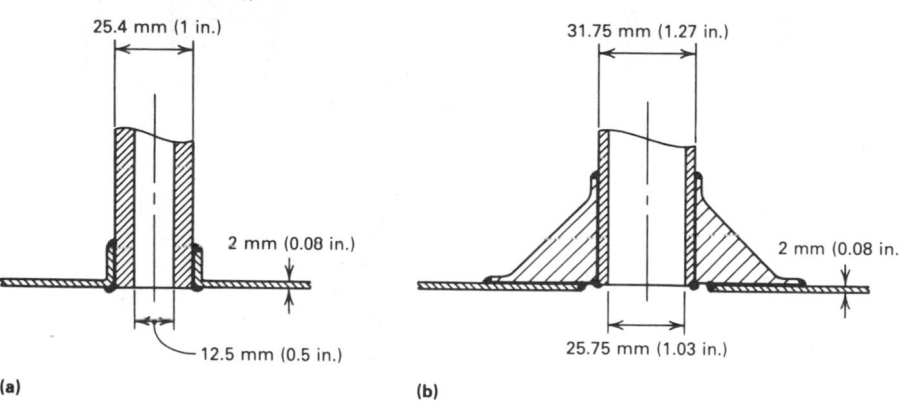

25.4 mm (1 in.)

2 mm (0.08 in.)

12.5 mm (0.5 in.)

(a)

31.75 mm (1.27 in.)

2 mm (0.08 in.)

25.75 mm (1.03 in.)

(b)

Fig. 32 Insertion of plugs to prevent loss of powder during evacuation

Courtesy of Industrial Materials Technology, Inc.

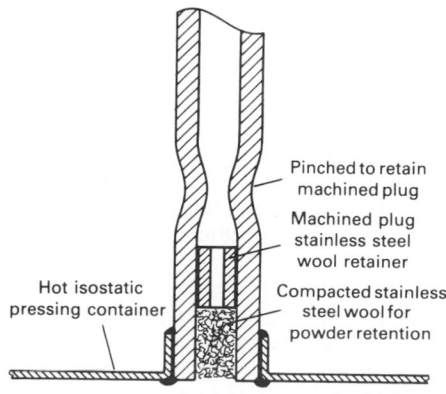

Pinched to retain machined plug

Machined plug stainless steel wool retainer

Hot isostatic pressing container

Compacted stainless steel wool for powder retention

tested, and filled in sequence. Containers with interior spacers (mandrels), powder/solid composites, or precompacted and sintered P/M compacts are filled with at least one cover removed. This procedure results in an extensive assembly weld area that cannot be leak tested in the vacuum mode because of the slow response time

of helium through the interior of the filled can. Consequently, careful removal of loose powder from the weld area is necessary, and use of precision and reproducible (preferably machine) weld techniques is required to prevent leakage in cans. Leaktightness of hot isostatic pressing containers is a major process consideration; leak testing methods are discussed later in this article.

Container Welding. Electron beam and gas tungsten arc welding are used for container assembly. See Volume 6 of the 9th edition of *Metals Handbook* for preferred welding practices. Argon dry box and electron beam welding are used for titanium alloy cans, because nonoxidizing conditions are required. Gas tungsten arc welding with and without filler is used for carbon steel and stainless steel cans. Carbon steel cans may require a final reducing anneal after weld assembly. Because weld metal is essentially a solidified casting, shrinkage and gas porosity are the fundamental causes of leakage at welds.

Matched weld lip can configuration is designed to promote directional plane front solidification with good liquid metal feed-

ing in the solidifying weld metal. Certain oxides (iron, nickel, and copper) can be reduced at high temperature in a high-pressure argon environment. This process may produce leaky cans during hot isostatic pressing if oxides extend through weld metal or can wall materials. Use of stainless steel filler metal for carbon steel can repairs is recommended, because chromium oxides essentially are stable under processing conditions up to 1200 °C (2200 °F) in argon. Gas tungsten arc welding of nickel cans with stainless steel filler metal also is advised.

Loose powder filling can be accomplished on a small scale by pouring powder through a funnel into the fill tube, as shown in Fig. 34. Precautions to control metal dust and prevent inhalation are required. Simultaneous tapping, shaking, or "g" loading is used to settle powder. Organic materials such as rubber tubing are not recommended in the powder flow path, as they are an obvious contamination source.

Particular attention should be paid to fill completion. Figure 35 illustrates recommended and poor container filling practice. An incompletely filled container results in loss of shape control and may result in collapse and tearing of the can under external process pressure. Compacted powder in the fill tube provides integral contiguous test material.

Advanced filling systems have been developed to ensure clean, dry handling of powder for critical aerospace applications. Magnetic particle separation, screening, outgassing, and settling have been incorporated in these systems. Figure 36 illustrates a commercial degassing and capsule filling station.

Filling of a container with predensified compacts is shown in Fig. 37. This method can be employed when powders have very low tap densities. Compacts must be shaped to permit close packing (cubes or hexagonal cylinders) in the container. Open spaces situated around the rim can be filled with loose powder. This method also requires cover assembly and welding after filling.

Leak Testing. Containerized hot isostatic pressing of metal powders can be achieved successfully only with leak-free containers. Location of leaks by use of valid leak testing procedures and subsequent repair are fundamental requirements of this hot isostatic pressing technology. Leak detection is based on characteristics of helium and argon flow through small capillaries when compared at 1 atm (0.1 MPa) and 1000 atm (100 MPa) total pressure.

Fig. 33 Shape and dimensional control by can design
Courtesy of Industrial Materials Technology, Inc.

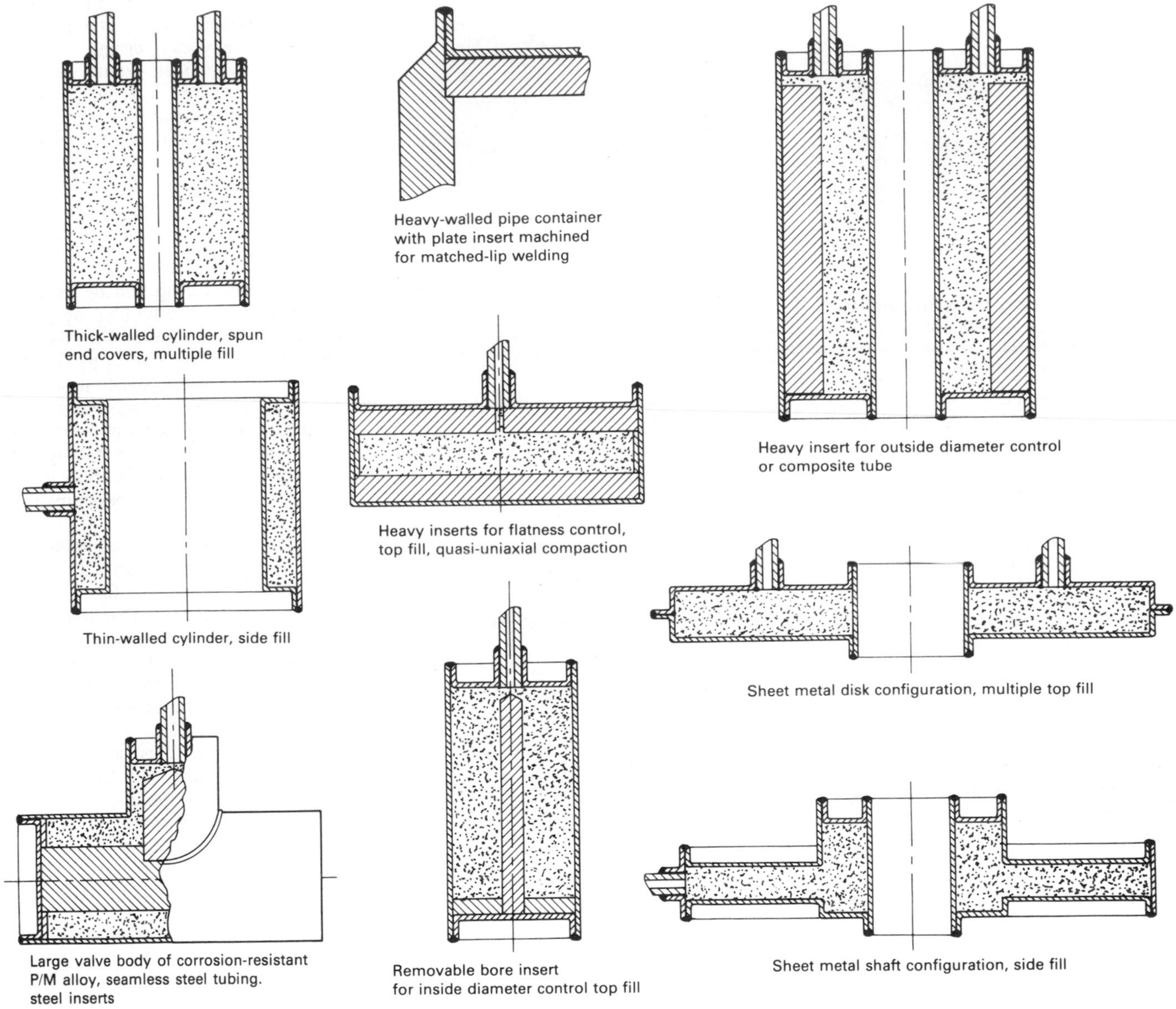

Thick-walled cylinder, spun
end covers, multiple fill

Heavy-walled pipe container
with plate insert machined
for matched-lip welding

Heavy insert for outside diameter control
or composite tube

Thin-walled cylinder, side fill

Heavy inserts for flatness control,
top fill, quasi-uniaxial compaction

Sheet metal disk configuration, multiple top fill

Large valve body of corrosion-resistant
P/M alloy, seamless steel tubing.
steel inserts

Removable bore insert
for inside diameter control top fill

Sheet metal shaft configuration, side fill

Flow characteristics of a cylindrical capillary have been described by Guthrie and Wakerling (Ref 6):

$$Q = \frac{1}{L} [C_1 P^2 + C_2 P + C_3 \ln (1 + C_4 P)]_{P_2}^{P_1}$$

(Eq 1)

where Q is the flow rate, cgs units; P_1 and P_2 are the exterior and interior pressure, cgs units; C_1, C_2, and C_3 are constants; and L is the capillary length, cm.

For $P_2 = 0$ (evacuated container interior) and P_1 large:

$$Q = \frac{C_1 P_1^2}{L} = \frac{\pi D^4}{256 \, \eta} \cdot \frac{P_1^2}{L}$$

(Eq 2)

where η is the gas viscosity and D is capillary diameter, both in cgs units. This applies strictly in the viscous flow region, when Reynolds number (Re) < 1200:

$$Re = \frac{DV\rho}{\eta} < 1200$$

(Eq 3)

where V is the gas velocity, cgs units; and ρ is the gas density, cgs units.

Equation 2 indicates the relationship of container design, manufacturing, and leak testing. Leakage flow is proportional to the exterior pressure squared. Whereas leak testing is conducted with pressure differences across the can wall of one to several atmospheres, hot isostatic pressing typi-

cally uses 1000 atm (100 MPa). Thus, a leak occurring just below the detectability limit of a selected method permits leakage flow rates 10^6 times greater during hot isostatic pressing. Consequently, sensitivity of the leak detection method is of utmost importance.

Capillary length (L) can be identified with can wall thickness, and with all other variables being constant, a capillary leaks ten times faster through a 0.25-mm (0.01-in.) wall than a 2.5-mm (0.1-in.) wall. The self-sealing, matched weld lip design is advantageous, because capillary path length through the weld increases rapidly as weld flanges deform and solid-state bonding occurs. The fourth power dependence of

Fig. 34 Container filling with powder and settling

Courtesy of Industrial Materials Technology, Inc.

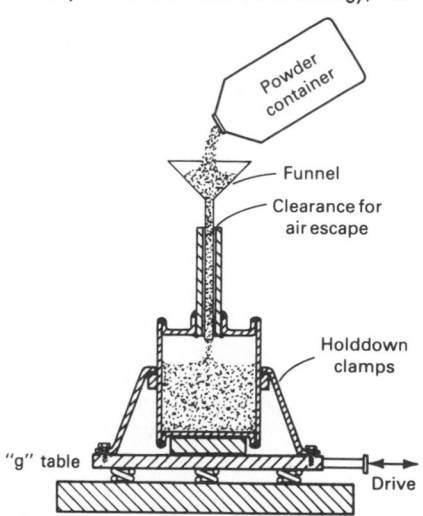

Fig. 35 Container filling practices

(a) Poor practice. (b) Recommended practice. Courtesy of Industrial Materials Technology, Inc.

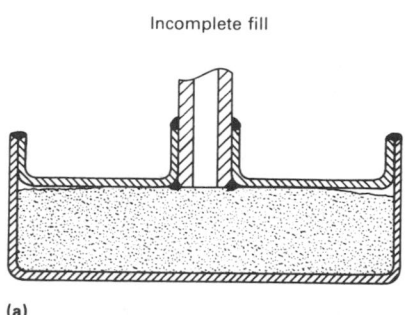

(a)

(b)

Incomplete fill

Tube partially filled ensures geometry control and provides integral sample material

Fig. 36 Degassing and capsule filling station

Courtesy of Leybold-Heraeus Powderline

Fig. 37 Cans filled with predensified compacts

Courtesy of Industrial Materials Technology, Inc.

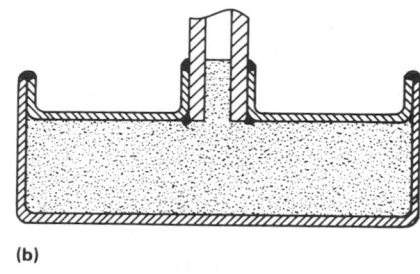

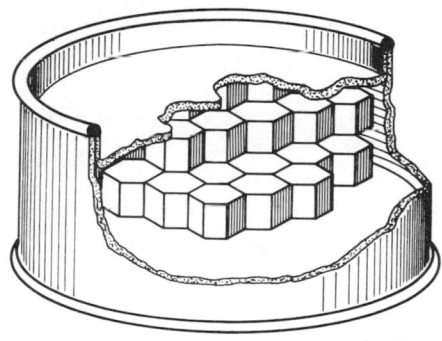

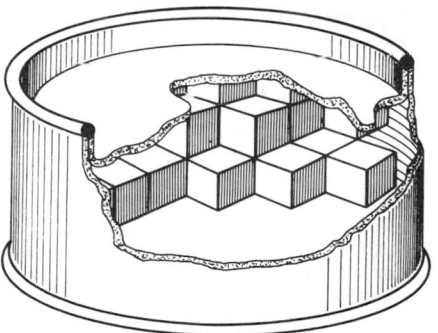

leakage flow rate on capillary diameter indicates the necessity for procedures that eliminate weld porosity.

Leakage flow rates for helium at 1 and 10 atm and for argon at 1000 atm (100 MPa) for a set of capillary leak sizes and D^4/L parameters that could occur in practice are given in Table 1. This illustrates the major problem in leak testing of hot isostatic pressing containers: the leakage flow rate of argon through a capillary hole at 1000 atm (100 MPa) process pressure is approximately 10^5 times greater than the flow rate during a 1-atm (0.1-MPa) leak testing procedure such as use of the helium mass spectrometer in the vacuum mode—i.e., evacuated can and/or atmospheric helium surrounding can exterior. This flow rate difference defines a requirement for maximum sensitivity of the leak testing method that is satisfied only by use of the helium mass spectrometer method in the vacuum mode.

Figure 38 illustrates a typical leak testing setup for hot isotatic pressing containers. Figure 39 illustrates a weld-assembled hot isostatic pressing container that is connected to a commercial helium mass spectrometer for leak testing in the vacuum mode.

The effect of argon in powder compacts is estimated here to emphasize the importance of using only leak-free cans for powder encapsulation. Total leakage (in argon at standard temperature and pressure, assuming constant leak conditions and no appreciable pressure rise inside the container) for 1000 and 10 000 s flow times is given in Table 2. Distribution of this "leaked" argon within compacts of var-

Table 1   Leakage rate as determined by capillary (hole) diameter, gas type, and pressure

Capillary length L = 0.1 cm

Capillary diameter, D, μm	Leakage rate, cm³/s, for:		
	Helium at 0.1 MPa (1 atm)	Helium at 1 MPa (10 atm)	Argon at 100 MPa (1000 atm)
0.001	3.8×10^{-16}	3.8×10^{-14}	3.3×10^{-11}
0.01	3.8×10^{-12}	3.8×10^{-10}	3.3×10^{-7}
0.1	3.8×10^{-8}	3.8×10^{-6}	3.3×10^{-3}
1.0	3.8×10^{-4}	3.8×10^{-2}	3.3×10^{-1}

Note: Leakage rate is inversely proportional to the capillary length; that is, if the capillary length is twice as long, the leakage rate will be half as much. For example: if D = 0.1 μm, helium pressure = 1 MPa, and L = 0.3 cm, then leakage rate = $(3.8 \times 10^{-6}/3)$ cm³/s = 1.3×10^{-6} cm³/s.
Courtesy of Industrial Materials Technology, Inc.

Fig. 38 Leak testing setup

Acceptance criterion: Q (flow rate) $< 10^{-9}$ standard cm^3/s. (a) Test piece evacuated and hooded with helium atmosphere to determine overall leakage rate. (b) Test piece evacuated; helium jet probe used to locate leak. Courtesy of Industrial Materials Technology, Inc.

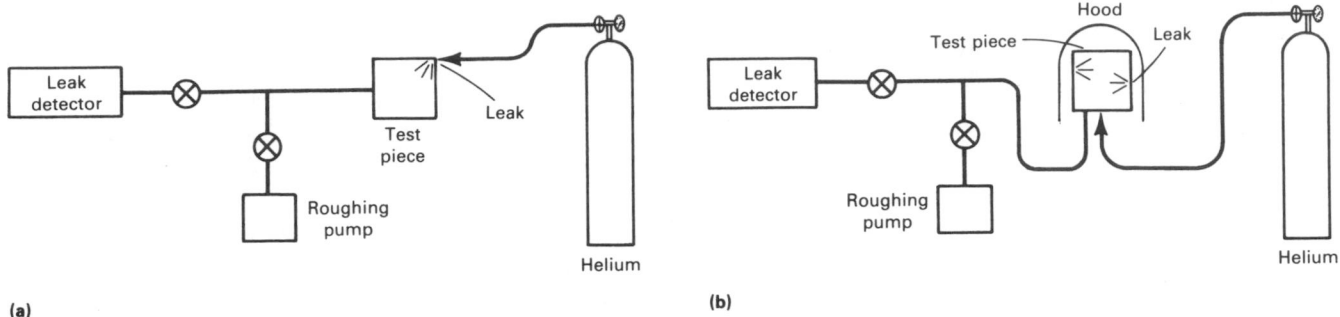

(a)   (b)

Fig. 39 Assembled hot isostatic pressing container connected to helium mass spectrometer

Courtesy of Industrial Materials Technology, Inc.

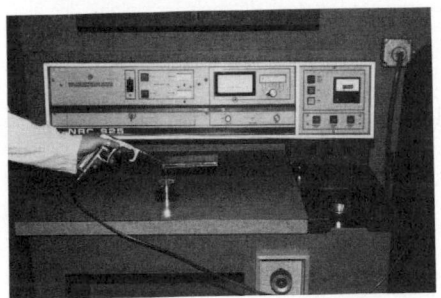

Table 2   Total argon leakage flow at standard temperature and pressure

Capillary (hole) length $L = 0.1$ cm

Capillary diameter, D, μm	Leakage flow, cm^3, at a leakage time of:	
	1000 s	10 000 s
0.001	3.8×10^{-8}	3.8×10^{-7}
0.01	3.8×10^{-4}	3.8×10^{-3}
0.1	3.8×10^{0}	3.8×10^{1}

Note: Leakage flow is inversely proportional to capillary length; that is, if the capillary length is twice as long, the leakage flow will be half as much. For example, if $D = 0.1$ μm, argon pressure $= 100$ MPa (1000 atm), leakage time $= 1000$ s, and $L = 0.3$ cm, then leakage flow $= 3.8 \times 10^0/3$ $cm^3 = 1.3 \times 10^0$ cm^3.
Courtesy of Industrial Materials Technology, Inc.

ious sizes permits estimates of argon contamination in parts per million (ppm) by weight, as shown in Fig. 40.

Contained argon, although compressed during early powder densification stages in the hot isostatic pressing cycle, can limit end point densification by "pressure balance" within small remaining pores. Regrowth of pores in subsequent heat treating operations, with related adverse effects on properties, may occur at levels as low as 0.1 mL/m^3 (0.1 ppm) for tool steels and 1 to 5 mL/m^3 (1 to 5 ppm) for superalloys. Leaks representing argon contamination at

Fig. 40 Argon contamination level versus total leakage flow for various compact sizes

Courtesy of Industrial Materials Technology, Inc.

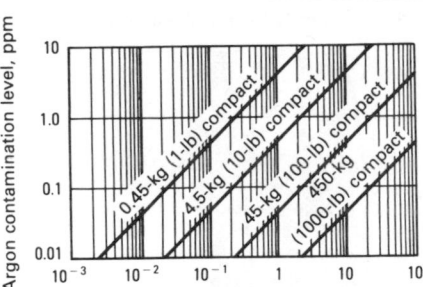

the 10 to 100 mL/m^3 (10 to 100 ppm) level generally prevent full densification, and larger leaks usually result in partial or no hot isostatic pressing densification.

Outgassing. The functional requirement of encapsulated powder vacuum outgassing is to remove air and water vapor (free and adsorbed) from the packed powder bed to prevent formation of particle surface oxide and nitride films, which reduce workability and/or mechanical properties. Outgassing has been integrated into large production-scale container filling equipment. Behavior of powder in a heating/vacuum cycle (T is approximately 400 °C, or 750 °F maximum), for the purpose of defining process specifications, can be determined by thermogravimetry, combined with limited-range mass spectrometry techniques. Vacuum outgassing does not remove gas entrapped in hollow powder particles originating from inert gas atmosphere atomization operations.

Evacuation time for a packed powder bed can be estimated using viscous and molecular flow concepts. Elevated temperature is used to raise gas pressure within a bed and to promote desorption of water vapor. Packed metal powder beds are poor thermal conductors; therefore, an exces-

sively high heating rate and temperature gradient in the compact during outgassing can result in redistribution of gas by chemisorption and reaction in the outer zone before all the gas is pumped out of the bed. This can occur because the center of the bed evolves gas at "low" temperature, which diffuses and reacts with the outer "high" temperature portion of the bed before it leaves the compact.

The required practical end point for degassing a powder-filled hot isostatic pressing container can be estimated from the residual bed pressure (assuming air composition), which contributes oxygen and nitrogen levels ten times less than the base level of the powder. Based on powder packing density, temperature, bed pressure, and ideal gas laws, parts per million by weight is given by:

$$ppm = 1.32 \, Pf \frac{(1 - \rho'/\rho)}{\rho' RT} M$$

where P is bed pressure, μm Hg; ρ is full density of metal, g/cm^3; ρ' is apparent (tap) density of powder, g/cm^3; f is fractional composition of gas, oxygen $= 0.21$ (air); M is molecular weight, g/mole; R is gas constant, 82.06 cm^3-atm/K; T is absolute temperature, K.

For $T = 600$ K, oxygen in air ($f = 0.2$) $\rho = 8.0$ g/cm^3, and $\rho' = 5.2$ g/cm^3 (65% packing density):

$$ppm_O = 1.8 \times 10^{-4P}$$

and for $ppm_O = 1$:

$$P \cong 5.5 \times 10^3 \text{ μm Hg} = 5.5 \text{ mm Hg}$$

The normal oxygen level of commercial superalloy powders ranges from 10 to 50 mL/m^3 (10 to 50 ppm) by weight. Thus, relatively high finishing evacuation pressures are acceptable in some cases. Other alloys, particularly refractory metals, may be more sensitive to residual gas.

A vacuum degassing setup for powder containers weighing up to approximately

Fig. 41 Small-sized container heating and evacuation

Courtesy of Industrial Materials Technology, Inc.

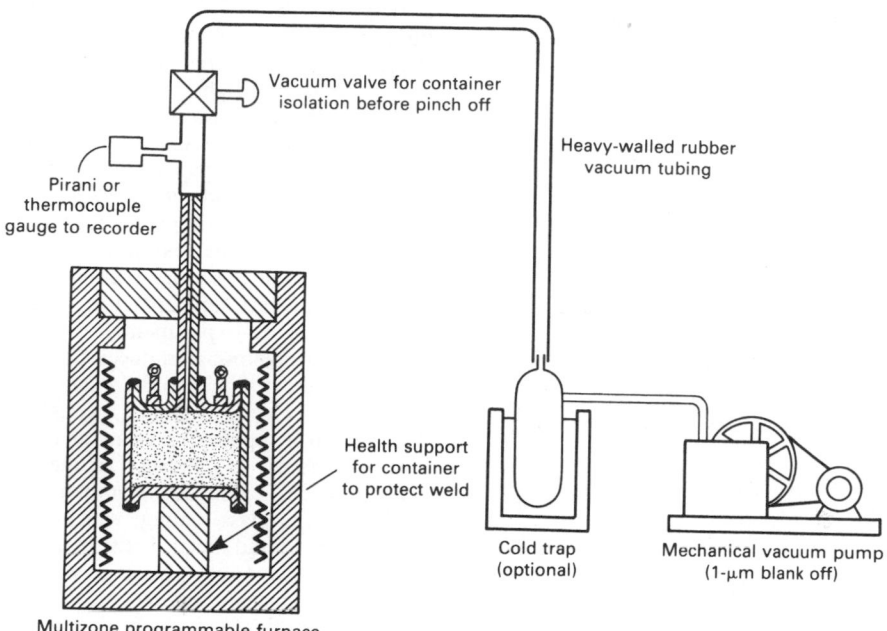

Fig. 42 Vacuum degassing cycle

Courtesy of Industrial Materials Technology, Inc.

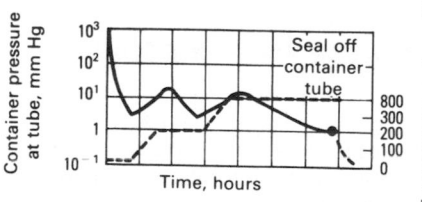

Fig. 43 Combination of cold isostatic pressing, sintering, and hot isostatic pressing to fully densify titanium alloy P/M product

(a) Before hot isostatic pressing. (b) After hot isostatic pressing. Courtesy of Industrial Materials Technology, Inc.

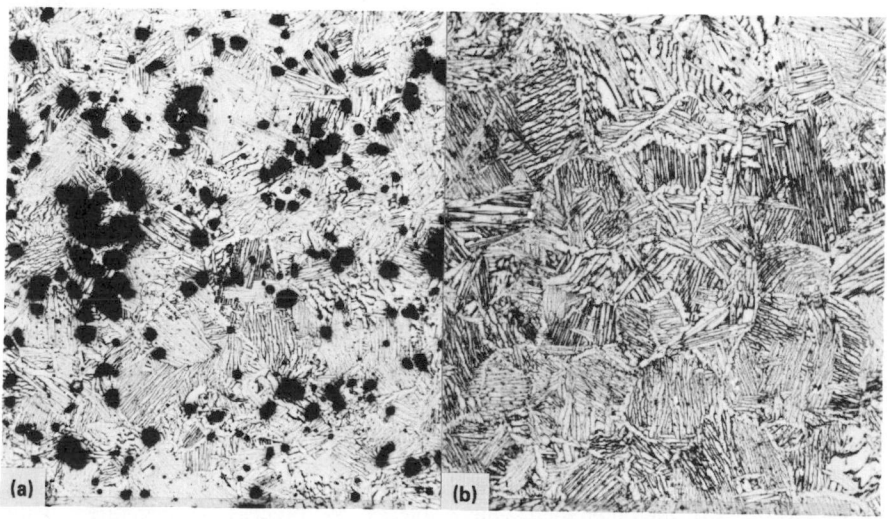

powder degassing. Stainless steel containers offer greater oxidation resistance during powder degassing and do not require descaling.

Loss of part dimensional control in large compacts (greater than about 25 kg, or 55 lb) also can occur during degassing, because the sheet metal container heats faster than the contained packed powder. New empty space is created inside the can at the bottom, into which powder flows from top areas. An oversize diameter at the bottom and uneven top geometry result from this type of powder movement during can heating without applied pressure. This particular problem also can occur in hot loading hot isostatic pressing operations.

Preliminary Powder Processing

For encapsulation, hot isostatically pressed metal powders may be blended, screened, compacted, and sintered. For loose powder containerization, lubricants, binders, or waxes are not used in blending operations because of the extreme difficulty encountered in subsequent removal. Partial separation of constituents due to particle size and/or specific gravity may occur during pouring operations after blending, and process changes may be required to control these effects.

Preliminary powder processing may include pressing and sintering to a closed porosity condition that is suitable for final densification by hot isostatic pressing. The concept of containerless hot isostatic pressing was established in 1971; see Ref 7 for a comprehensive description of this method. A closed porosity condition is required because of the dynamics of argon leakage through capillaries. Rapid internal/external pressure equalization within a P/M part by capillary flow prevents any net densification during hot isostatic pressing. Volume percent residual compact porosity that can be removed by hot isostatic pressing ranges from 0.1 to approximately 8%, depending on the type of pores. Powder compaction and sintering process combinations that produce isolated (non-surface-connected) pores enable greater amounts of porosity to be removed by hot isostatic pressing.

Figure 43 illustrates a Ti-6Al-4V powder that was cold isostatically pressed and sintered to 95% density, then hot isostatically pressed to full density. Figure 44 illustrates a high-density P/M part that was produced with apparently isolated residual pores. Hot isostatic pressing of this part did not produce density increases. Evidently, the pores were interconnected

50 kg (110 lb) is shown in Fig. 41. Figure 42 (with a variable time scale, time dependent on powder weight) illustrates a practical vacuum degassing cycle for titanium and superalloy powders. Carbon steel containers can be through oxidized in an air bake-out furnace with prolonged exposure. Can wall thickness, therefore, should be increased with increasing container size and weight. For steel cans, grit-blast descaling is recommended after

Fig. 44 Isolated residual pores in die compacted, sintered, and hot isostatically pressed stainless steel compact
(a) Before hot isostatic pressing. (b) After hot isostatic pressing. Courtesy of Industrial Materials Technology, Inc.

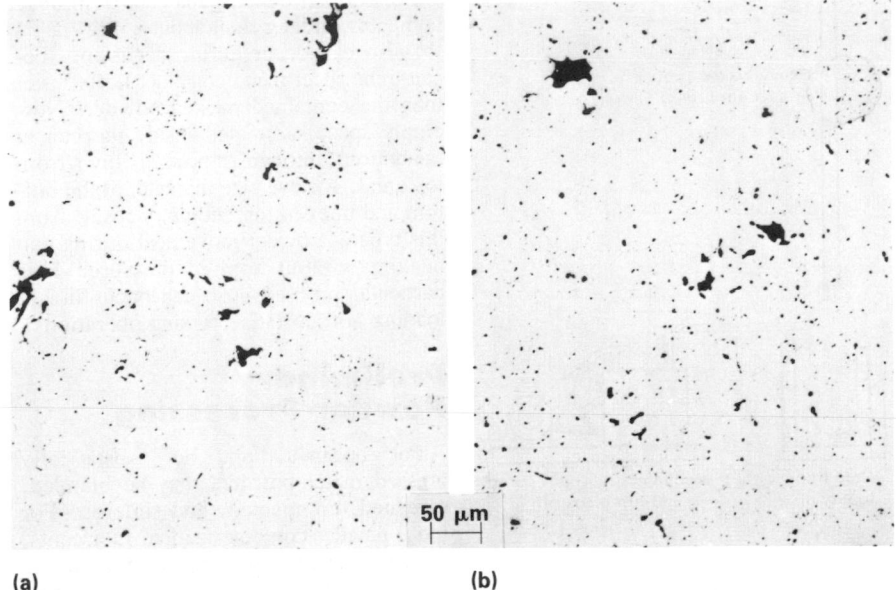

(a)   (b)

Fig. 45 Leak testing of P/M material for containerless hot isostatic pressing
(a) Criterion: Q (flow rate) $<10^{-9}$ standard cm³/s helium. (b) Sealed sample test configuration. Courtesy of Industrial Materials Technology, Inc.

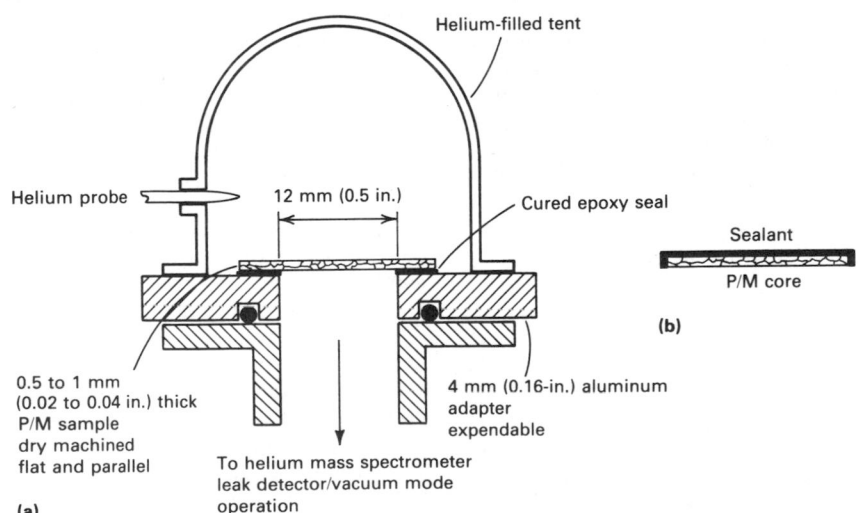

(a)

microscopically in capillary paths to the surface, and internal pressure equalization occurred.

Containerless Hot Isostatic Pressing

Suitability of P/M parts for containerless hot isostatic pressing can be established by correlation of pores in unetched micrographs with results of helium mass spectrometer leak testing of sections machined from compacted and sintered P/M parts. A typical test setup is shown in Fig. 45. Dry machining of specimens and/or vacuum baking at approximately 200 °C (400 °F) to avoid invalidation of the test by fluid impregnation of pore structures is required.

Sample material can be taken from one external surface of the part to determine leakage properties of the outer shell. One to two small, low-density surface areas contiguous with a low-density core will not comply with the 10^{-9} standard cm³/s helium leak criterion for containerless hot isostatic pressing, even though all other surface areas are "tight."

Use of nonagglomerating, fine powders, compacted to high green density (more than 80% theoretical) and sintered under vacuum to avoid entrapped gas, improves the chances of producing parts suitable for containerless hot isostatic pressing. Currently, surface-sealing methods for small parts are difficult, expensive, and largely experimental. For example, electroplating introduces possible electrolyte absorption into the P/M part.

Likewise, chemical vapor deposition does not cover support points and may not bridge all pores. A mismatch of thermal expansion coefficients between metal and glass occurs in glass sealing, which leads to cracking. Liquid metal, braze metal, or transient liquid metal phase in vacuum sealing of surfaces offers technically feasible possibilities, provided compatible metal systems can be identified. Any sealing method must be capable of passing the leak criterion of Q (flow rate) $< 10^{-9}$ standard cm³/s helium using the method shown in Fig. 45.

Process Parameters

Hot isostatic pressing of P/M material produces fully dense material in the shortest process time, using pressures within the rating of equipment large enough to accommodate the parts being processed. Strict grain-size control also may be required for P/M parts subjected to subsequent isothermal forging or other hot working operations.

Process Cycle. Several types of hot isostatic pressing cycles can be used. These vary primarily in the magnitude and sequence of pressure and temperature application, as shown in Fig. 46. Additional variations are possible with staged heating and cooling operations.

A type I cycle is carried out by cold pressurization to approximately one fourth to one third of final process pressure (T equals approximately 1100 to 1200 °C, or 2000 to 2200 °F), isolation of the pressure vessel, and heating to process temperature. Pressure rises, due to increased gas temperature under constant volume constraint. This process constitutes a typical cold loading cycle and is particularly useful for geometry control of large net shapes in sheet metal encapsulation.

Type II cycles involve heating workpieces to process temperature, while

Fig. 46 Process cycles
Courtesy of Industrial Materials Technology, Inc.

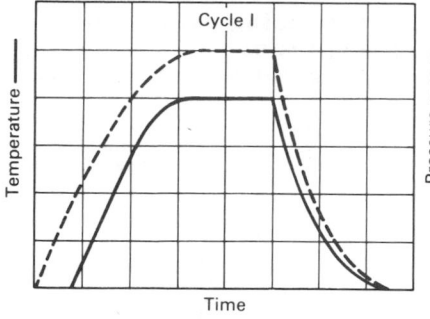

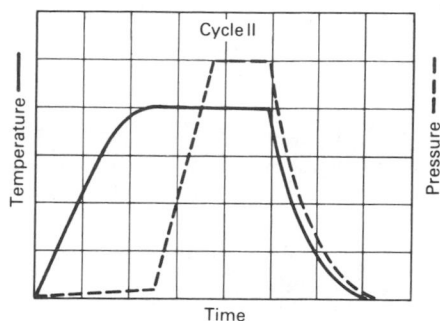

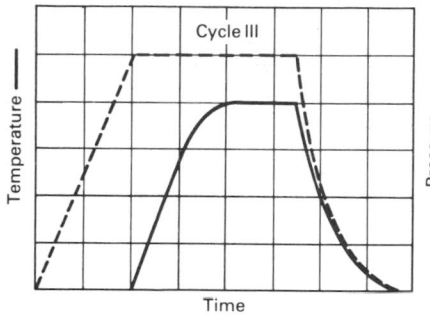

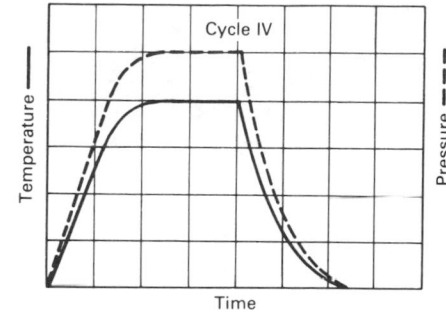

maintaining a low vessel pressure (0.7 to 7 MPa, or 100 to 1000 psi). Pressurization is accomplished by equalization from storage cylinders and pumping. This type of processing is equivalent to a hot loading cycle and is used for glass-encapsulated powder to prevent container breakage.

Type III cycles involve full pressurization, followed by heating, while process gas is bled back to storage to maintain pressure within the equipment rating. This cycle tends to enhance recrystallization of powder particles through plastic deformation at the lowest possible temperature.

A comparison of microstructures for type II and type III cycles is shown in Fig. 47. For subsequent isothermal forging operations, a fine recrystallized powder structure is preferred to reduce upset load levels. The preheat cycle (type II) tends to maintain or decrease the grain size variation in as-hot isostatically pressed parts.

A type IV process cycle, with the accelerated cooling capabilities of hot isostatic pressing equipment, is the most cost-effective processing mode in terms of reduced process time for a cold loading cycle. Pumping and heating occur concurrently to a level where sufficient pressure exists within the system to attain the required final pressure when the process temperature is achieved.

Selection of process parameters, such as temperature, time, and heating and cooling rates, to achieve material property and dimensional requirements usually is verified by processing through a matrix and testing. Guidelines to process temperatures for hot isostatic pressing of P/M parts may be obtained from published literature. Rapid diffusion and power law creep, in which deformation rate is proportional to applied stress raised to a power (n), occur in most cases near these temperatures. General temperature guidelines indicate that metal powders can be densified at 100 MPa (15 ksi) pressure in 2 to 4 h, as indicated:

- *Nickel-based alloys:* 1100 to 1200 °C (2000 to 2200 °F)
- *Iron-based tool steels:* 1000 to 1100 °C (1800 to 2000 °F)
- *Titanium-based alloys:* 850 to 950 °C (1550 to 1750 °F)
- *Aluminum powders:* 480 to 530 °C (900 to 1000 °F)
- *Copper-based alloys:* 750 to 850 °C (1400 to 1550 °F)
- *Molybdenum and niobium fine powder:* 1200 to 1650 °C (2200 to 3000 °F)

Generally, the maximum pressure of the equipment (usually 100 MPa, or 15 ksi) should be used, unless intentionally porous parts are desired. This approach minimizes process time, but scale-up requirements must be considered to avoid requiring pressures not available in larger hot isostatic pressure vessels.

Process times for most powders listed above are 2 to 4 h at 100 MPa (15 ksi). This processing time is based on temperature after soak-through. Because of capacity and transfer lags, particularly for large parts weighing more than 25 kg (55 lb), workpiece temperature must be measured by closely coupled thermocouples (Ref 8). Figure 48 shows a recommended thermocouple installation within a hot isostatic pressing unit to eliminate overriding heat transfer effects from the hot high pressure gas. Final-stage densification depends primarily on diffusion, because pores shrink to radii where the applied pressure driving force is insignificant compared to surface tension effects (Ref 9).

Evaluation of the time parameter effect on densification must be done on the basis of a structure-sensitive mechanical property, such as notched stress rupture. High magnification microscopy of grain boundaries in polished sections, if grain-boundary structure is critical for the particular application, will also disclose residual pores. For an alternate approach to process parameter selection, see Ref 10 for mechanism maps of hot isostatic pressing.

Special constraints on process parameter selection may relate to controlled grain-size requirements, avoidance of powder/container interactions, and solid-state phase transformations. Micron-sized (0.1 to 1.0 μm) powders (usually precompacted and partially sintered) can be densified at approximately 100 to 200 °C (180 to 360 °F) below coarser (100-μm mean particle size) powders at equivalent pressure and time. This effect is derived from greater contribution of grain-boundary sliding during creep deformation. It is useful in processing of refractory metal powders.

Applications

The ability of hot isostatic pressing to produce near-net shapes has been a primary impetus behind the development of hot isostatically pressed P/M parts in nickel-based and titanium alloys. Conventional manufacturing methods for these alloys typically utilize only 10 to 30% of the material purchased in the final product; the remainder becomes scrap during machining. Hot isostatic pressing to near-net shape improves materials utilization significantly. A hot isostatically pressed near-net shape part may lose only 10 to 20% during final machining.

The inability to provide nondestructive inspection of near-net shape parts for certification has somewhat inhibited application of this technology, particularly for

Fig. 47 Comparison of microstructures produced by type II and type III hot isostatic pressing cycles
Courtesy of Industrial Materials Technology, Inc.

Type II   Type III

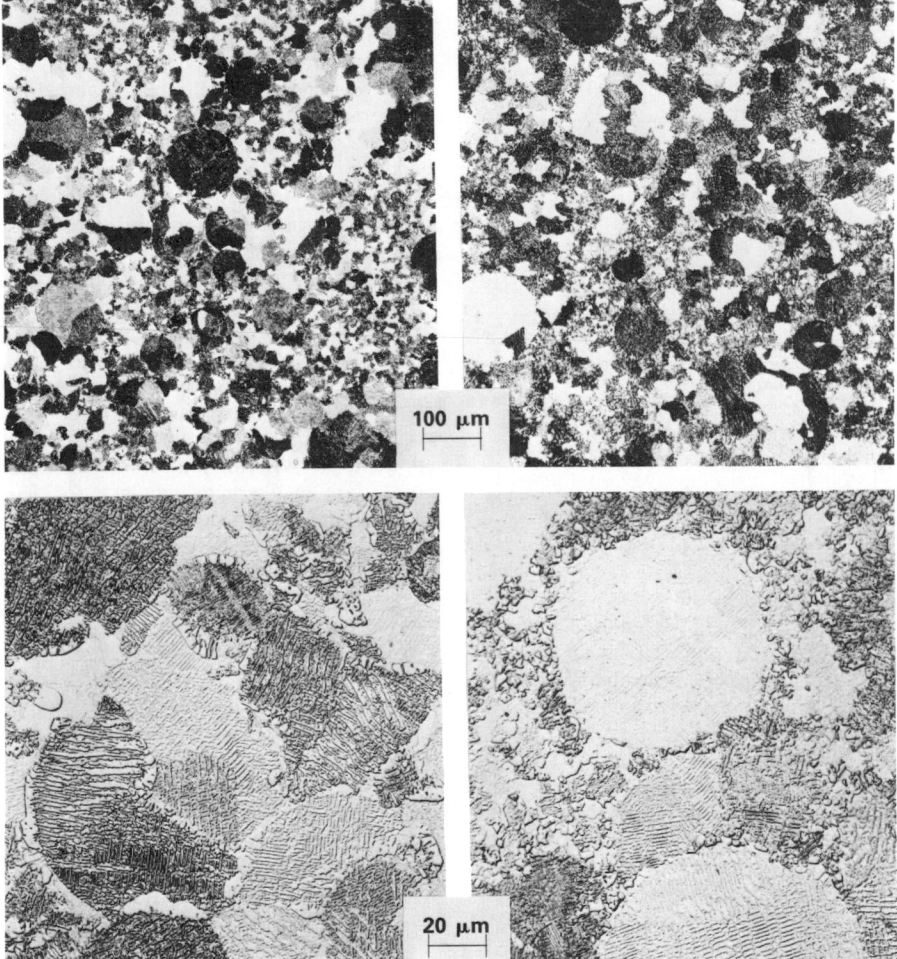

Fig. 48 Workpiece thermocouple installation designed to eliminate transfer and lag errors in timing hot isostatic pressing cycles
Courtesy of Industrial Materials Technology, Inc.

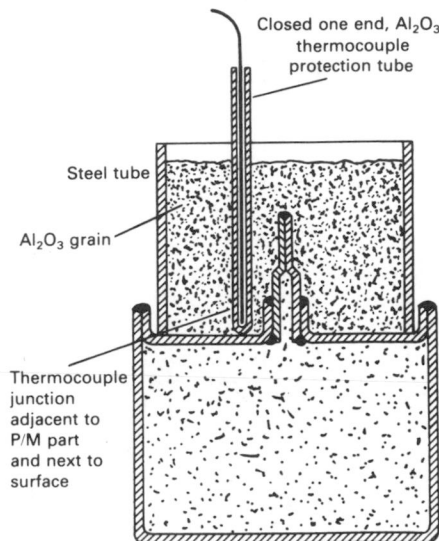

Fig. 49 Relation between the sonic, near-net, and final shapes for a typical P/M turbine disk
Source: Ref 11

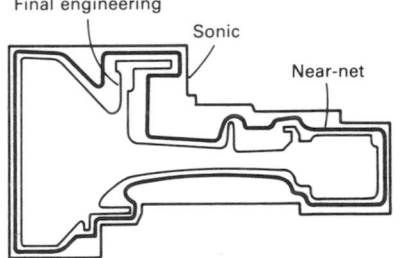

engine applications. Sonic, near-net, and final shapes are compared in Fig. 49 for a P/M turbine disk. The sonic configuration is currently of most interest for aircraft turbine parts.

Titanium-Based Alloys

Powder production techniques for titanium alloys are still under development. The potential of using inexpensive blended elemental powder rather than prealloyed powders is also under investigation. Blended cold isostatically pressed and sintered hot isostatically pressed blended elemental powder compacts typically have inferior fatigue properties compared to prealloyed counterparts. Tensile properties of blended elemental powder com-

pacts compare favorably with prealloyed compacts, however.

Material properties of hot isostatically pressed titanium alloys are comparable to properties of wrought counterparts, except for inferior fatigue behavior. Tensile properties of P/M Ti-6Al-4V alloy are given in Table 3. Fatigue properties of wrought, prealloyed, blended elemental, and cast Ti-6Al-4V are compared in Fig. 50. The inconsistent fatigue life of hot isostatically pressed P/M titanium is often caused by the presence of inclusions in the compact. These inclusions can be introduced during powder manufacture and/or handling. For example, tungsten inclusions in early generation rotating electrode processed titanium powder were caused by deterioration

of the tungsten electrode. The plasma rotating electrode process (PREP) has eliminated this problem. Clean handling techniques have been defined that are expected to expand the applications of hot isostatically pressed P/M titanium.

Due to its high reactivity and the economics of producing near-net shape parts, titanium alloy powder usually is hot isostatically pressed to near-net shape by the ceramic mold process or by cold isostatic pressing and sintering, followed by hot isostatic pressing. Because of superior material properties, titanium P/M alloys are usually consolidated below the beta transus. Typical processing conditions for Ti-6Al-4V include pressing at 950 °C (1750 °F) at 100 MPa (15 ksi) for 4 h.

Table 3 Typical room-temperature tensile properties of Ti-6Al-4V

Process	Ultimate tensile strength MPa	ksi	Yield strength MPa	ksi	Elongation, %	Reduction in area, %
Cast	930	135	841	122	7	15
Isothermally forged	965	140	875	127	15	40
Diffusion bonded(a)	924	134	855	124	11	23
Prealloyed P/M(b)	979	142	882	128	14	26
Blended elemental P/M(c)	917	133	834	121	11	18

Composition: (a) 0.18 wt% oxygen, 0.009 wt% nitrogen for all materials; bond line oxygen and nitrogen are higher; failure typically is remote from bond joint. (b) Hot isostatically pressed. (c) Mechanically pressed and sintered
Source: Ref 12

Fig. 50 Comparative fatigue bands for Ti-6Al-4V product processed by various fabrication techniques
Source: Ref 12

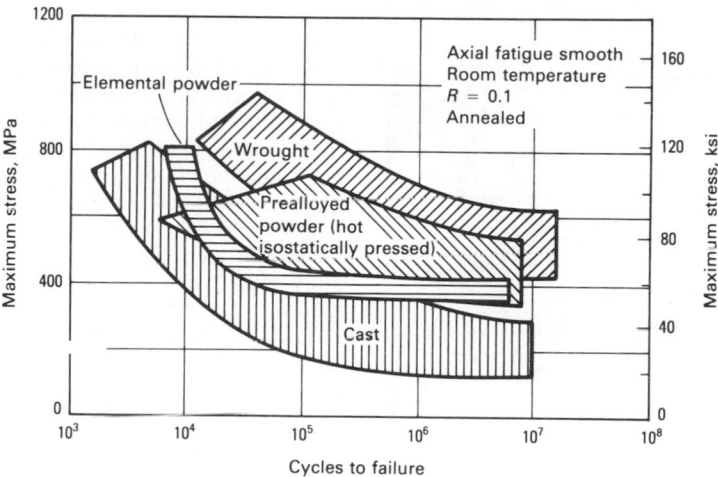

Fig. 51 Comparison of Ti-6Al-4V alloy keel splice preforms made by forging and hot isostatic pressing

(a) Forging, 2.12 kg (4.67 lb). (b) Hot isostatically pressed near-net shape, 0.45 kg (1.0 lb). (c) Final part, 0.18 kg (0.4 lb). Courtesy of Crucible Materials Research Center

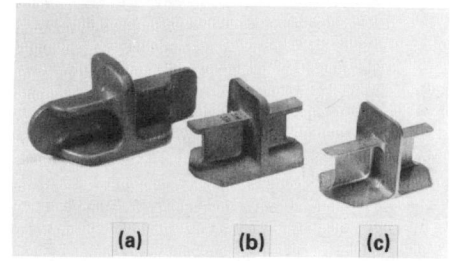

static pressing of forging preforms represents a significant portion of current production.

The use of P/M hot isostatic pressing consolidation for superalloys is economically attractive because of its near-net shape capabilities. High alloy content superalloys can be produced with attractive properties. Superalloys strengthened by a large volume fraction of second phase gamma prime undergo severe segregation during ingot formation. Such ingots would be virtually unworkable by conventional hot working techniques for large size parts. The division of the melt into small powder particles during atomization eliminates macrosegregation, and the dimensional scale of microsegregation is reduced because of high cooling rates during particle solidification. Hot isostatic pressing of these powders produces a homogeneous microstructure that improves mechanical properties and workability.

Superalloy powders typically are made by inert gas atomization processing or rotating electrode processing. The presence of stable nonmetallic compounds on the surface of the powder particles is ultimately detrimental to the properties of consolidated superalloy powders. This prior particle boundary contamination (Fig. 21) weakens interparticle boundaries. The presence of carbides, oxides, and nitrides on the powder particle surfaces has a direct influence on the extent of prior particle boundary decoration. Low carbon content in P/M alloys reduces carbide buildup on the powder particle surfaces.

Tensile strength, creep resistance, and stress rupture of consolidated superalloys are comparable to or exceed the conventionally processed ingot material, as shown in Table 5 and Fig. 53. Fatigue properties of P/M superalloys are comparable to conventional material, although the presence

Because limited improvement in P/M material properties over wrought properties has been gained by hot isostatic pressing, developments in this area have been spurred primarily by the economics of near-net shape processing. A keel splice former (Fig. 51) weighing 0.18 kg (0.4 lb) is conventionally manufactured from a forging weighing 2.12 kg (4.67 lb). By hot isostatic pressing to near-net shape, it can be made from rotating electrode processed powders to near-net shape (0.45 kg, or 1 lb) using the ceramic mold process.

Parts were hot isostatically pressed at 950 °C (1750 °F). Prior to machining, parts were heat treated by annealing at 700 °C (1300 °F) for 2 h. Static load testing indicated that yield and failure loads were slightly above forged counterparts. Notched fatigue life of hot isostatically pressed material superior to comparable forged material has been observed. This is ascribed to the fine grain size of the hot isostatically pressed product. Spectrum fatigue testing of finished parts indicates a fatigue life equivalent to wrought counterparts.

Economic benefits from hot isostatic pressing to near-net shape indicate a cost savings of 16% compared to forged and machined counterparts. The keel splice former shown in Fig. 51 has material properties that are equivalent to forged counterparts, while being cost-effective.

The potential for hot isostatic pressing large titanium components to near-net shape has been demonstrated. A large (860 by 250 by 100 mm, or 34 by 10 by 4 in.) horizontal stabilizer pivot shaft was made by hot isostatic pressing to near-net shape using the ceramic mold process. The first near-net shape attempt made had a 3.2-mm (0.125-in.) envelope over the target shape. Typical parts produced by hot isostatic pressing to near-net shape are shown in Fig. 52. Table 4 illustrates the material savings realized by this near-net shape production method.

Nickel-Based Superalloys

Consolidation of P/M nickel-based superalloy parts by hot isostatic pressing enjoys wide usage. In 1979, over 450 metric tons (500 tons) of P/M superalloy components were processed. Hot isostatically pressed parts are used in many military and commercial aircraft applications. Hot iso-

Fig. 52 First hot isostatic pressing trial of airframe shapes
(a) Horizontal stabilizer pivot shaft (General Dynamics). (b) Arrestor hook support fitting (Northrop). (c) Walking beam support fitting (Boeing). (d) Drop-out link (MCAIR). Source: Ref 12

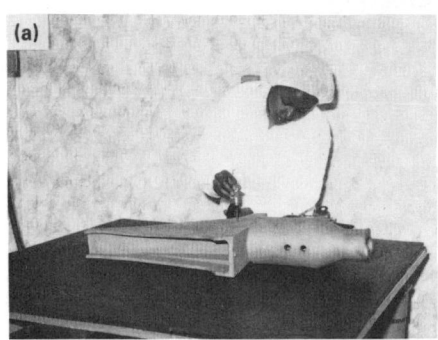

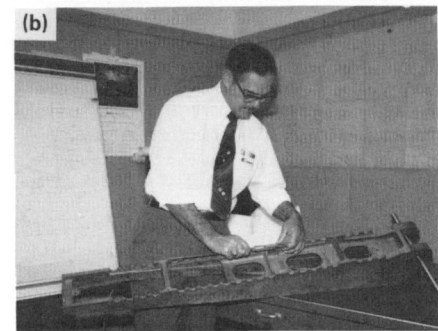

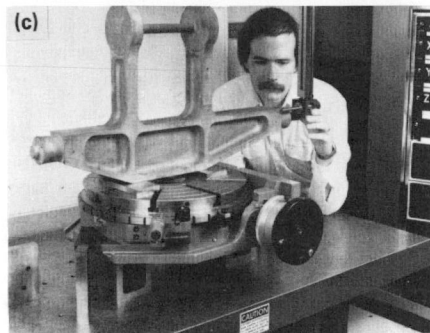

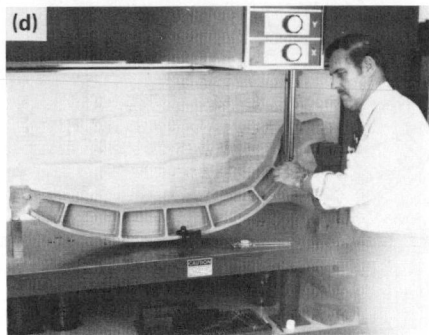

Table 4 Part selection and weight comparison

	Forging billet		Hot isostatic pressing		Final part	
Part	kg	lb	kg	lb	kg	lb
Walking beam	25	55	14	30	9.5	21
Pivot shaft	67	148	24	53	15	32
Compressor spool	67	147	30	65	6.8	15
Drop-out link	52	115	25	56	6.3	14
Keel splice	2.7	6	0.4	0.8	0.2	0.4
Arrestor hook	82	181	25	55	13	28
Fan disk	54	120	30	65	13	27
Compressor rotor	14	32	2.8	6.2	1.6	3.6

Source: Ref 6

of inclusions in P/M parts produces variations in these properties.

Thermoplastic processing of the powder prior to hot isostatic pressing can improve material properties of superalloys. The powder is passed through a feed rolling mill, during which strain energy is imparted to the powder. This causes grain refinement by recrystallization during hot isostatic pressing. Prior particle boundaries are also broken up. Fine grain size increases workability through reduced elevated temperature flow stresses. Consequently, thermoplastically processed powder also can be hot isostatically pressed at somewhat lower temperatures than usual.

Heat treatment after hot isostatic pressing can have significant effects on material properties, as shown in Table 6. Material response to post-hot isostatic pressing treatment depends on the processing conditions. Near-net shape parts also may be subject to distortion during post-hot isostatic pressing heat treatment.

Superalloys typically are hot isostatically pressed in stainless steel cans, particularly for rotating aircraft turbine parts. If complex shapes are required, the ceramic mold process is suitable, particularly for static parts. If a stainless steel can is used for powder consolidation, a 0.5-mm (0.02-in.) diffusion zone may sur-

round the part. This does not cause a problem in the final part, because hot isostatic pressing envelopes usually exceed this dimension. Hot isostatic pressing conditions are alloy dependent. Processing temperatures may be keyed to the gamma prime solvus temperature for purposes of grain-size control in nickel-based alloys.

Oxide-dispersion-strengthened superalloys also can be consolidated by hot isostatic pressing. Prior to processing, alloy powders, additives, and oxide dispersoids are put in a high-attrition ball mill and mechanically alloyed, which ensures fine grain size and uniform oxide distribution throughout the powder. Hot isostatic pressing produces fully dense material with these microstructural features maintained.

Additional forming operations may be used to increase properties of hot isostatically pressed superalloys. Forging following hot isostatic pressing increases the low-cycle fatigue of Nimonic alloys.

Figure 16 illustrates an alloy composition (MERL 76) designed specifically for hot isostatic pressing consolidation. This alloy was designed to have mechanical properties equivalent to isothermally forged IN-100. To accomplish this goal, the composition of IN-100 was modified in three ways. Carbon content was reduced to minimize prior particle boundary effects, hafnium and niobium were added to increase ductility and carbide structure control, and vanadium was eliminated to improve hot corrosion resistance.

This alloy was hot isostatically pressed to near-net shape at 1180 °C (2160 °F) at 100 MPa (15 ksi) for 3 h. Tensile, stress-rupture, and low-cycle fatigue properties, determined from tests conducted after heat treatment, met process goals. The potential benefits of using hot isostatic processing to improve material utilization are illustrated in Fig. 54. A 15% cost savings resulted from using hot isostatic pressing P/M techniques rather than conventional processing.

Alloy AF115 is used for disks in advanced gas turbine engines. Desired material properties include high tensile strength in the bore area and high creep strength at 760 °C (1400 °F) in the disk rim. AF115 is highly alloyed and specifically designed for high-temperature creep applications. It is not producible by conventional cast and wrought methodologies.

The hot isostatically pressed creep strength of AF115 at 760 °C (1400 °F) was superior, but tensile strength was not adequate at high temperatures. Tensile strength could be increased, but only with a reduction in creep properties. By subjecting the bore area to post-hot isostatic

Table 5 Comparison of mechanical properties of several P/M superalloys

Condition	Test temperature °C	Test temperature °F	0.2% offset yield strength MPa	0.2% offset yield strength ksi	Ultimate tensile strength MPa	Ultimate tensile strength ksi	Reduction in area, %	Total elongation, %
René 95								
Hot isostatically pressed(a)	23	74	1214	176	1636	237	15	16
Hot isostatically pressed and forged	23	74	1179	171	1629	236	23	18
Cast and wrought	23	74	1144	166	1434	208	12	10
Minimum hot isostatically pressed	650	1202	1120	162	1514	220	17	16
Hot isostatically pressed and forged	650	1202	1122	163	1480	215	14	13
Cast and wrought	650	1202	1055	153	1282	186	10	8
Astroloy								
Hot isostatically pressed	23	74	936	136	1379	200	31	27
Hot isostatically pressed and forged	23	74	1055	153	1517	220	23	27
Hot isostatically pressed	650	1202	881	128	1234	179	36	31
Hot isostatically pressed and forged	650	1202	975	141	1261	183	25	38
IN-100								
Hot isostatically pressed	650	1202	1286	187	942	137	...	21
Hot isostatically pressed and forged	650	1202	1200	174	1000	145	...	8
Hot isostatically pressed and extruded	650	1202	1350	196	1000	145	...	18

(a) 1120 °C (2050 °F) at 103 MPa (15 ksi) for 3 h, solution treated at 1150 °C (2100 °F) 1 h, hot salt quench to 535 °C (1000 °F), aged at 870 °C (1600 °F) 1 h, then 650 °C (1200 °F) 24 h; air cooled
Source: Ref 14

pressing forging and rapidly cooling the forging, an increase in tensile properties and reduction of creep properties was achieved without affecting creep strength of the rim area.

High-Speed Tool Steels

The development of gas-atomized prealloyed steel powders in the 1960's led to hot isostatic pressing of tool steels. This represented the first production application of hot isostatic pressing of a relatively low-cost material. Hot isostatic pressing improves the microstructure of tool steels by preserving the fine grain size and carbide distribution present in the atomized powder during consolidation. Increased homogeneity of the fine carbides throughout the material is an added benefit.

Superior tool properties result from the improved microstructure. Shape stability during subsequent heat treatment is superior in hot isostatically pressed material. Grindability, wear resistance, and uniformity of hardness also are improved.

Additionally, cutting performance of high-speed tool steels is improved by this processing treatment, due to the increased toughness related to fine austenite grain size.

New high-alloy content steels with enhanced material properties can be produced. High-speed tool steels generally are consolidated in billet form in a container. Typical processing conditions are consolidation at 1100 °C (2000 °F) at 100 MPa (15 ksi) for 1 h. A hot isostatically pressed high-speed steel billet is shown in Fig. 13.

Cemented Carbides

Tungsten carbide/cobalt tools are the premier example of containerless hot isostatic pressing to achieve full density by removing residual porosity. Superior transverse-rupture strength results from hot isostatic pressing. The wear performance of cutting tools at high speeds is not significantly improved, however, because this behavior is governed by the hardness of the material rather than by its fracture properties. Low cobalt content (3%) alloys that are very hard can be produced by hot isostatic pressing to give enough toughness to be used in drawing dies.

Fully dense cemented carbide can be finished to give a perfectly smooth surface, which is required for high-quality rolls, dies, mandrels, and extrusion tools. Generally, tungsten carbide/cobalt tool materials are manufactured by cold isostatic pressing and sintering of blended powders, followed by hot isostatic pressing. Typical conditions for hot isostatic pressing are 1290 °C (2350 °F) at 100 MPa (15 ksi) for 1 h. Cemented carbide parts produced using hot isostatic pressing are shown in Fig. 55.

Refractory Metals

Consolidation of refractory metals by hot isostatic pressing has recently received increased attention. Processing these materials to net and near-net shape promotes conservation of these critical resources. Niobium alloy C-103 (Nb-10Hf-1Ti-5Zr) has been successfully hot isostatically pressed using a duplex cycle. Hydride/dehydride and plasma rotating electrode processed powders were consolidated in a plain carbon steel container filled with powder at 1260 °C (2300 °F) at 100 MPa (15 ksi) for 3 h. The can was then removed in a nitric acid solution and further chemically milled in a nitric-hydrofluoric acid solution to remove the alloy/container interaction layer. The material was then finished by a hot isostatic pressing step at 1590 °C (2900 °F) at 100 MPa (15 ksi) for 3 h to a final density in excess of 99% of theoretical.

Room-temperature and high-temperature (1650 °C, or 3000 °F) tensile and ductility properties compared favorably with wrought alloy properties. The ductile/brittle transition temperature was higher (−18 °C versus 160 °C, or 0 °F versus 320 °F, for standard products) in the hot isostatically pressed material due to increased oxygen content. Gas content of the hydride/dehydride material gave poor weldability, as shown by the weld area. Hydrogen embrittlement also occurred in the hydride/dehydride alloy C-103. Vacuum baking at 870 °C (1600 °F) for 2 h eliminated embrittlement, and the alloy failed in a ductile manner in tensile and Charpy tests.

Near-net shape forward bowls manufactured by consolidation of C-103 in the duplex hot isostatic pressing cycle are shown in Fig. 56. The diameter of the bowls was within 0.13 mm (0.005 in.) of final dimensions. The P/M net shape weighed 0.8 kg (1.8 lb). This, compared with a rough

Table 6 Mechanical properties of hot isostatically pressed plus conventionally forged Nimonic alloy AP1

Processing temperature		Size of sample disk			Tensile properties(a)						Notched tensile strength		Stress rupture(b)			
					Yield point, 0.2% offset		Ultimate tensile strength		Elonga-tion, %	Reduction in area, %			Plain life, h	Elonga-tion, %	Notch life, h	Low-cycle fatigue(c), cycles
°C	°F	mm	in.	Solution treatment	MPa	ksi	MPa	ksi			MPa	ksi				
1150	2100	150	6	4 h at 1110 °C (2030 °F), air cool	971	141	1307	190	30.4	31.6	1869	271	42	30.1	195	>276 000
1150	2100	150	6	4 h at 1080 °C (1980 °F), oil quench	1120	162	1513	219	23.2	24.2	1992	289	64	15.3	159	>307 000
1150	2100	150	6	4 h at 1110 °C (2030 °F), quenched and aged(d)	1037	150	1381	200	30.4	46.7	1776	258	88	20.4	163	>214 000
1220	2230	150	6	4 h at 1110 °C (2030 °F), air cool	999	145	1328	193	28.6	32.7	1868	270	45	20.5	188	>155 000
1220	2230	150	6	4 h at 1080 °C (1980 F), oil quench	1085	157	1463	212	23.2	23.4	1941	281	66	17.2	247	>228 000
1220	2230	150	6	4 h at 1110 °C (2030 °F), quenched and aged(d)	1052	153	1383	201	25.0	25.8	1844	267	74	16.9	315	>242 000
1150	2100	475	19	4 h at 1110 °C (2030 °F), air cool	952	138	1320	191	29.5	31.4	1521	221	85	22.9	>500	>35 000
1150	2100	475	19	4 h at 1080 °C (1980 °F), oil quench	993	144	1356	197	26.1	28.0	1785	259	113	20.3	>450	>100 000

(a) At 650 °C (1200 °F). (b) 760 MPa (110 ksi) at 705 °C (1300 °F). (c) 1080 MPa (157 ksi) at 600 °C (1110 °F). (d) 50% water-soluble polymeric compound, 50% water. All material aged 24 h; 650 °C (1200 °F); air cooled, 8 h; 760 °C (1400 °F); air cooled
Source: Ref 15

Fig. 53 Comparison of tensile and creep properties of hot isostatically pressed P/M and cast and wrought René 95
Source: Ref 11

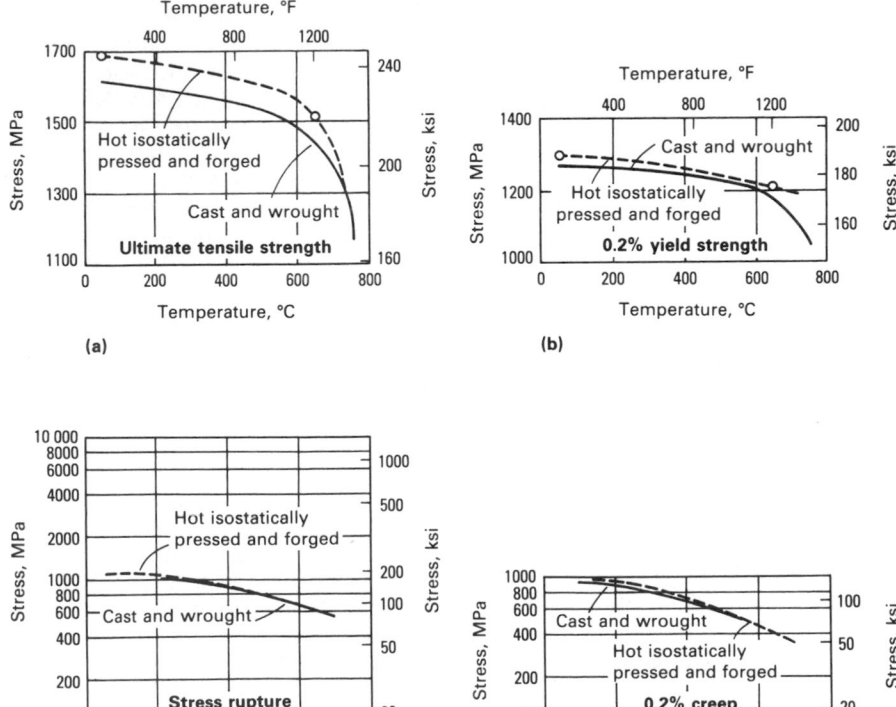

$$P = (T + 460)(\log T + 25) = 10^{-3}$$

(c)

$$P = (T + 460)(\log T + 25) = 10^{-3}$$

(d)

forging weighing 1.7 kg (3.8 lb) and a final part weighing 0.6 kg (1.4 lb), illustrates the material savings achieved by hot isostatic pressing to near-net shape.

Composites

Composite materials constitute another application of hot isostatic pressing. Parts requiring a combination of mechanical properties that cannot be attained by a single material may be fabricated as a composite.

Composites made from powder metals can be manufactured in two ways. A shaped container is filled with the necessary quantities of powder, each in its own location, and then hot isostatically pressed. The major problem encountered with this method is ensuring that each powder material is filled and remains in its desired location. This tendency can be prevented by preconsolidating a portion of the shape, which will subsequently be bonded to the remainder of the part during the hot isostatic pressing cycle. The solid component is loaded into the finished part can, and the remainder of the container is filled with powder of desired composition.

During the hot isostatic pressing cycle, loose powder is consolidated and bonded to the prefabricated component. Composites can be susceptible to cracking because of the different rates of thermal expansion of the materials and/or because of volume changes during phase transformations on cooling.

Figure 57 illustrates a high-strength composite extrusion die. MAR M509 powder was hot isostatically pressed in the center of a hollow cylindrical tool steel backer ring. A carbon steel insert was used to provide the taper geometry in the bore of the die. The composite was machined to finish.

Useful discussions of the application of hot isostatic pressing technology are available in Ref 11 to 16.

Fig. 54 Improved material utilization through hot isostatic pressing of superalloys
Courtesy of AVCO, Lycoming Division

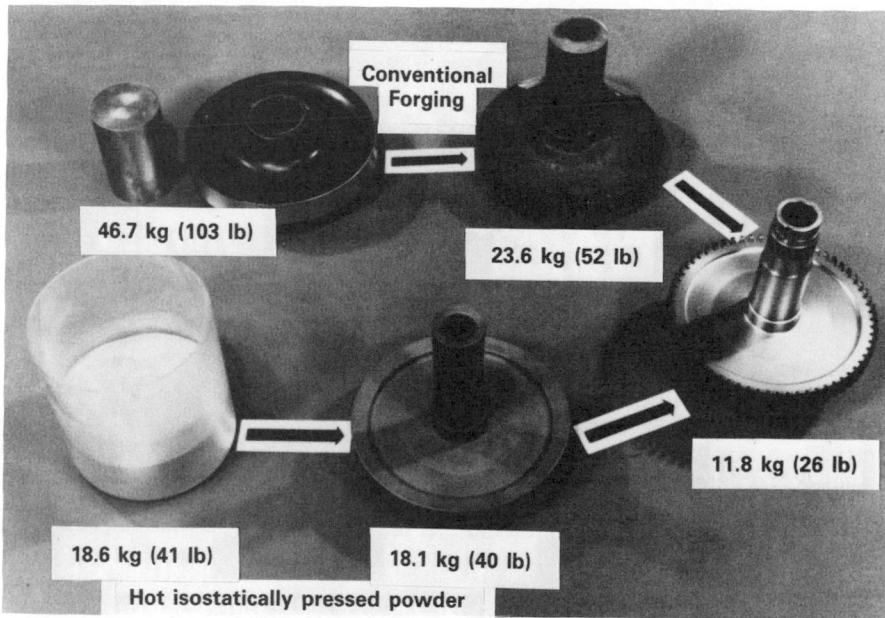

Fig. 57 Composite extrusion die
Courtesy of Industrial Materials Technology, Inc.

Fig. 55 Tungsten carbide/cobalt parts produced by hot isostatic pressing
Courtesy of Industrial Materials Technology, Inc.

Fig. 56 Niobium forward bowls hot isostatically pressed to shape
Courtesy of Industrial Materials Technology, Inc.

REFERENCES

1. U.S. Patent 3 687 842, pending
2. Claver, A.H., Meiners, K.E., and Boyer, C.B., "Hot Isostatic Pressing," Report No. MCIC-82-46, Metals and Ceramics Information Center, Battelle Memorial Institute, Columbus, OH
3. U.S. Patent No. 3,622,313, Nov 1971
4. U.S. Patent No. 3,700,435, Oct 1972
5. U.S. Patent No. 3,804,575, April 1974
6. Guthrie, A. and Wakerling, R.K., *Vacuum Equipment and Techniques*, McGraw-Hill, New York, 1949, p 191
7. U.S. Patent No. 3,562,371, Feb 1971
8. Schumann, R., Jr., *Metallurgical Engineering Principles*, Vol I, Addison Wesley, Boston, 1952, p 277-282
9. Coble, R.L., Diffusion Model for Hot Pressing with Surface Energy and Pressure Effects as Driving Forces, *J. Appl. Phys.*, Vol 41 (No. 12), 1970, p 4798
10. Arzt, E., Ashby, M.F., and Easterling, K.E., Practical Applications of Hot Isostatic Pressing Diagrams: Four Case Studies, *Met. Trans. A*, Vol 13A, Feb 1983, p 211-221
11. "Superalloys From Powder: Production and Properties," National Materials Advisory Board Commission on Sociotechnical Systems, National Research Council, Report No. NMAB-369, 1981
12. Kelto, C., *et al.*, Powder Metallurgy of Titanium Alloys—A Perspctive, in *Powder Metallurgy of Titanium Alloys*, Froes, F. and Smugersky, J., Ed., American Institute of Mechanical Engineers, Warrendale, PA, 1980, p 5
13. Peterson, V., Chandhok, V., and Kelto, C., Hot Isostatic Pressing of Large Titanium Shapes, in *Powder Metallurgy of Titanium Alloys*, Froes, F. and Smugersky, J., Ed., American Institute of Mechanical Engineers, Warrendale, PA, 1980, p 251
14. Evans, R., "Review of European Powder Metallurgy of Superalloys," Report No. EOARD-TR-80-10, prepared for the U.S. Air Force by the University of Swansea, 1979
15. Symonds, C., *et al.*, Properties and Structures of Hot Isostatic Pressed and Hot Isostatic Pressed plus Forged Superalloys, *Powder Metall. Int.*, Vol 15 (No. 1), 1983, p 35
16. Van Tyne, C.J. and Avitzur, B., *Production to Near Net Shape*, American Society for Metals, Metals Park, OH, 1983

Cold Isostatic Pressing of Metal Powders

By Peter E. Price,
Director of Engineering
Industrial Materials Technology, Inc.
and
Steven P. Kohler
Graduate Student
Massachusetts Institute of Technology

COLD ISOSTATIC PRESSING is a materials processing technique in which high fluid pressure is applied to a powder part at ambient temperature to compact it into a predetermined shape. The powder part is consolidated into a partially dense compact. Water or oil is usually used as the pressure medium. Compacting pressure ranges from 210 to 410 MPa (30 to 60 ksi), although pressures as high as 760 MPa (110 ksi) have been used. Compacting is performed at ambient temperature. Powder parts typically are densified to between 60 and 80% of their theoretical density.

There are two major types of cold isostatic pressing techniques, both of which utilize a preshaped elastomeric mold to hold powder that is compacted in the press. If the elastomeric mold is fixed to the pressure vessel, the technique is termed "drybag" isostatic pressing. If the mold can be removed from the vessel, the process is called "wetbag" isostatic pressing. These two methods are shown schematically in Fig. 1.

Parts in the as-pressed condition are said to be in the "green," or unsintered, state. After cold isostatic pressing, the parts are sintered to increase part density and improve mechanical properties. If further deformation is required, the part can be fully densified by forging, extruding, or rolling. If the part is already made to net shape, it can be fully densified to improve material properties by hot isostatic pressing. An advantage of this method is that a cold isostatically pressed and sintered part fre- quently can be hot isostatically pressed without a container.

Isostatic pressing was developed by H.D. Madden of the Westinghouse Lamp Co. in 1913 (U.S. patent 1 081 618) to process tungsten and molybdenum billets. A semi-automated process for the production of ceramic parts using the drybag technique was patented by B. Jeffrey of the Champion Spark Plug Co. in 1942 (U.S. patent 2 290 910).

During World War II, interest centered on the pressing of explosives, special metals, and ceramics, which moved isostatic pressing from the laboratory into production. With the development of larger pressure vessels, the ability to compact powders at high pressures expanded cold isostatic pressing applications into the P/M field.

Although conventional methods of metal forming are well developed, traditional ceramic process techniques are limited. Consequently, early isostatic pressing production applications dealt primarily with ceramics. Recently, cold isostatic pressing applications have expanded into P/M fabrication because:

- Powder can be pressed to near-net shape economically.
- New alloys can be processed that are not possible with ingot metallurgy techiques.
- Difficult-to-process materials, such as tungsten carbide, can be easily formed.

Currently, cold isostatic pressing is performed mainly on metals that are difficult to fabricate using conventional metalworking techniques. Typical cold isostatic pressing applications include tungsten, tungsten carbide, tantalum, high-speed steels, titanium, and aluminum. Cold isostatically pressed parts always undergo further processing to improve material properties. Some applications, such as metallic filters, require a specific level of porosity, but in most applications, it is desirable for the part to approach theoretical density.

Advantages

Because pressure is applied isostatically over the entire surface of the part, more uniform density is achieved throughout the entire part than by uniaxial pressing. Consequently, shrinkage is fairly uniform and predictable during compaction and sintering, thus allowing attainment of close tolerances for relatively complex shapes.

There are virtually no residual stresses in the compacted material, because there is no die wall friction. Adequate material strengths for handling can be obtained in the green state. In some materials, machining of the part in the green or partially sintered state is possible or easier than in the fully densified condition.

The addition of binders or lubricants to the powder is not required. Powder metallurgy compressed and sintered material has a homogeneous microstructure. Additionally, tooling costs are lower than the die sets used for mechanical forming. Elimination of intermediate forming op-

Fig. 1 Schematic of wetbag and drybag processes
Source: Ref 1

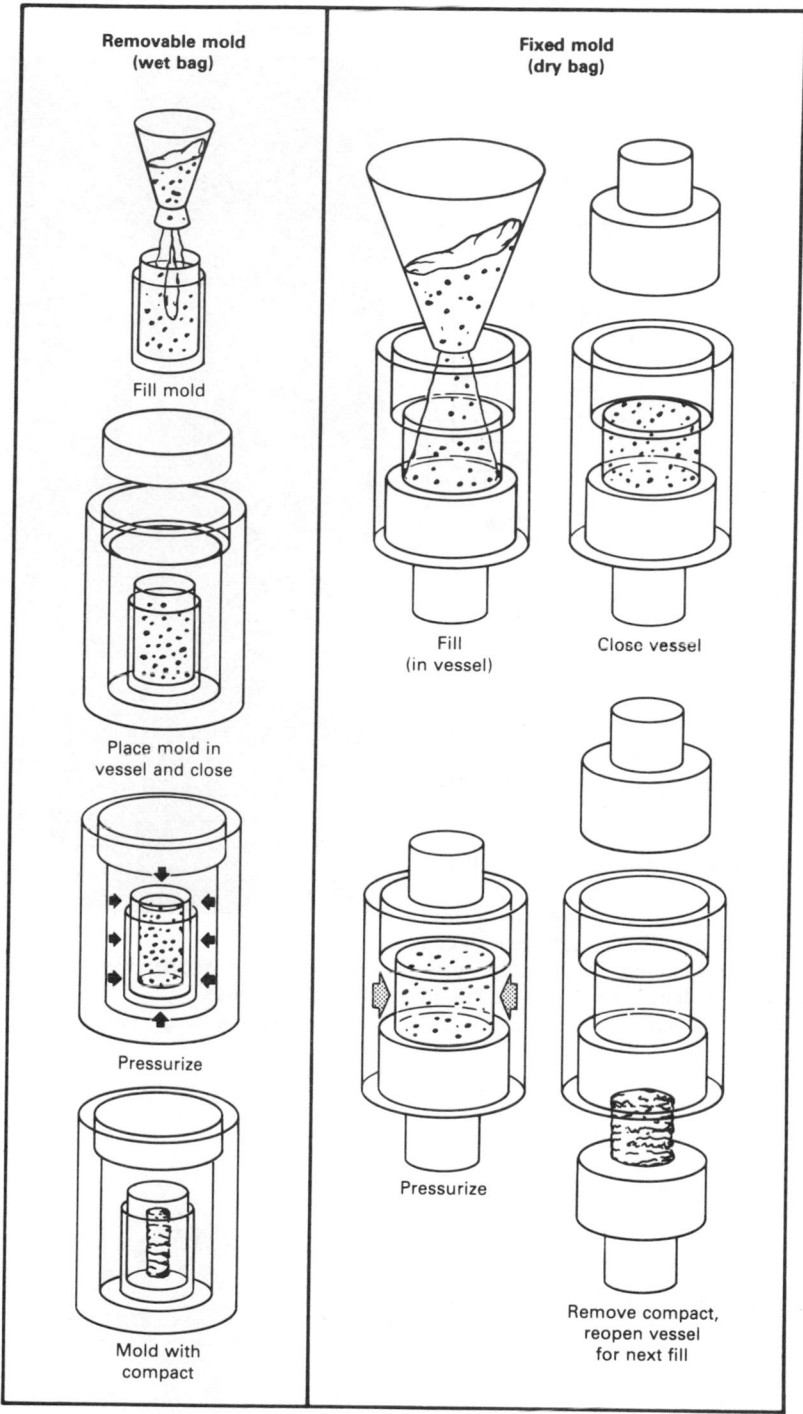

erations by cold isostatic pressing to near-net shape reduces the total number of dies required to produce a part. Cold isostatic pressing units require lower capital investment than a mechanical forming process of comparable capacity. Also, cold isostatic pressing to near-net shape or to a preform reduces the amount of material required to fabricate a part.

Process Equipment

Equipment used for both drybag and wetbag cold isostatic pressing processes includes the pressure vessel, pressure generator, control system, depressurization equipment, and tooling. A typical cold isostatic pressing unit is illustrated schematically in Fig. 2. Figures 3 and 4 provide examples of actual units.

Pressure Vessels. The design and construction of pressure vessels used in cold isostatic pressing are discussed in the article "Hot Isostatic Pressing of Metal Powders" in this Volume. Hot isostatic pressing and cold isostatic pressing pressure vessels are substantially equivalent in design and construction.

Pressure Generators. Pressure is generated in the pressure medium through the use of air-driven and hydraulically driven pumps and pressure intensifiers. Air-driven units typically are used with small presses and for applications that can accommodate a slow rate of compaction. Air-driven units are inexpensive compared to hydraulic units. Hydraulic pumps are used in large presses and are capable of rapid compaction. They are widely used with automated production presses. The pressure medium typically is oil for drybag processes; water containing additives (water-soluble oil or rust inhibitors) is used for wetbag processes. A filtering system should be included with all systems to protect the pressure-generating equipment from particulate contamination.

Control Systems. The control system of a cold isostatic processing unit is fairly simple. A pressure gage can be mounted close to the pressure vessel, and the pressure can be controlled by manually operated valves. Alternatively, remote pressure-sensing devices can easily be coupled with an automatic controller that is capable of monitoring and adjusting the system to maintain desired process parameters. As with all high-pressure equipment, adequate safety precautions should be taken to protect the system from overpressure.

Depressurization Systems. Depressurization can be accomplished with a single metering valve. If this is set at one position, depressurization rate varies greatly with system pressure, and no change of rate occurs at critical pressures, such as the pressure at which entrapped air expands from pores in the compact. An automated system that retards depressurization rate when the vessel approaches a critical process pressure and rapidly releases the pressure at low system pressure is recommended.

Tooling. Cold isostatic pressing tooling is composed of two parts—the elastomeric mold, or bag, and a mandrel. Elastomeric molds are made of a variety of materials; some are flexible, while others are fairly

Fig. 2 Schematic of cold isostatic pressing process
Source: Ref 2

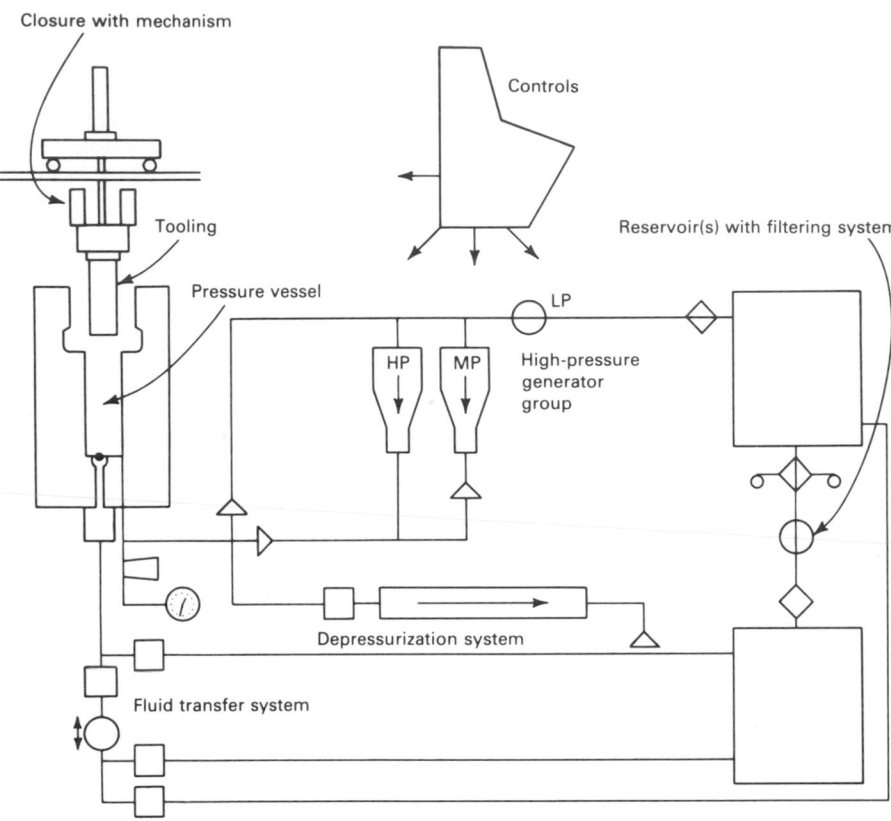

Fig. 3 Small laboratory cold isostatic press
Courtesy of National Forge, Europe, Pressure Systems, Systems Division

Fig. 4 Large-production cold isostatic press
Courtesy of National Forge, Europe, Pressure Systems, Systems Division

rigid. An outside fixture to hold the loaded form is required if the mold is exceptionally flexible.

A number of factors must be considered in the selection of a mold material. Primarily, the mold material must not interact with either the powder or the pressure medium. Materials have a range of durabilities. Depending on the size of the production run and the abrasiveness of the powder, a material with the appropriate wear resistance should be selected. Mold materials used include natural rubber, neoprene, urethane, polyvinyl chloride, butyl, nitrile, and silicone. Properties of various bag materials are given in Table 1.

Molds are produced by a variety of methods, including dipping or spraying, mechanical pressing, form casting, and injection molding. The steps in producing cold isostatic pressing tooling are shown in Fig. 5. Depending on the degree of accuracy required of the compact dimensions, several mold iterations may be required to form the exact shape. Bag life varies with mold material, wall thickness, and powder characteristics. For example, a neoprene mold with a Shore hardness ranging from 40 to 50 (26 to 36 HRC) and

with a compaction ratio of 2 to 1 will last approximately 1000 cycles.

If internal shapes or cavities are desired in the compacted part, a rigid mandrel can be used in the tooling. The mandrel usually is made of hardenable steel. It should have a very smooth surface finish to encourage part release. Complex shapes with reasonable tolerances can be achieved with the use of a mandrel.

Tooling also is classified in terms of movement of the bag. If the pressure compresses a bag inward or toward a mandrel, it is referred to as the "collapsing bag technique." If pressure inflates the bag inside the hollow member, it is referred to as the "dilating bag technique." The collapsing bag technique is the most common.

Drybag Isostatic Pressing

In drybag isostatic pressing, the mold is fixed to the inside of the pressure vessel. After the elastomeric mold is filled with powder, pressure is applied by introducing pressurized oil between the fixed mold and the vessel wall. Contamination of the pressure medium is not a likely problem,

Table 1 Properties of potential bag materials

Property	Natural rubber (molded)	Natural rubber (latex)	Butyl	Polyvinyl chloride	Polyurethane	Neoprene	Nitrile	Silicone
Tensile strength, MPa (psi)	21 (3000)	21 (3000)	14 (2000)	14-21 (2000-3000)	28 (4000)	14 (2000)	10 (1500)	7 (1000)
Shore hardness	30-90	40	40-75	...	20-98	40-95	40-95	40-85
Tear resistance	Very good	Very good	Good	Fair to good	Excellent	Good	Fair	Poor
Abrasion resistance	Excellent	Good	Good	Fair	Excellent	Good	Good	Poor
Resilience	Excellent	Excellent	Bad	Bad	Good	Good	Fair	Excellent
Compression set	Good	Fair	Fair	Poor	Poor	Good	Good	Fair
Solvent resistance								
Aliphatic hydrocarbon	Poor	Poor	Poor	Fair	Excellent	Good	Excellent	Poor
Aromatic hydrocarbon	Poor	Poor	Poor	Fair	Fair to good	Fair	Good	Poor
Oxygenated hydrocarbon	Good	Good	Good	Fair	Poor	Poor	Poor	Fair
Resistance to								
Lubricating oil	Poor	Poor	Poor	Poor to good	Excellent	Good	Very good	Fair
Petrol	Poor	Poor	Poor	Poor	Excellent	Fair	Excellent	Fair
Animal and vegetable oils	Good	Good	Excellent	Poor	Excellent	Good	Excellent	Fair
Heat	Good	...	Excellent	Good	Poor	Excellent	Excellent	Outstanding

Source: Ref 3

Fig. 5 Production of cold isostatic pressing tooling
Courtesy of Trexler Rubber Company

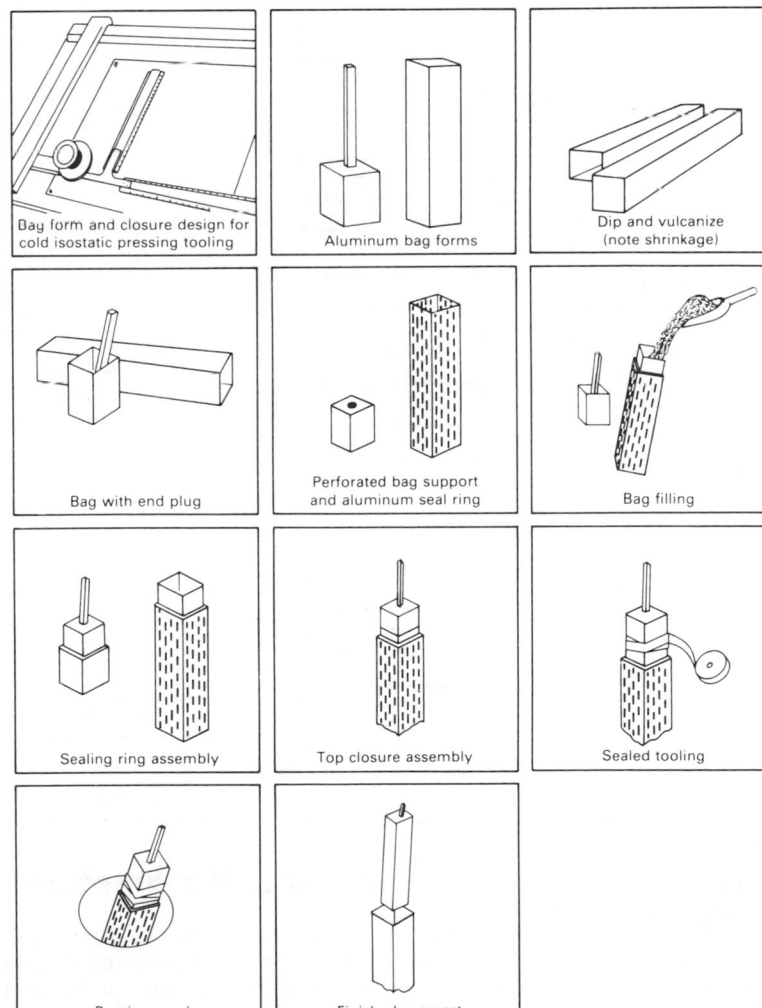

unless catastrophic failure of the elastomeric mold occurs.

The drybag process is characterized by fast cycle time and high production rates. Fairly simple shapes usually are compacted with this method to facilitate removal of the compact from the mold after pressing. Presently, the drybag process typically is used to compact ceramic powders in high-output applications, such as spark plugs. As the isostatic pressing of powder metals enjoys wider usage, development of highly automated drybag presses capable of rapid, economical production rates is likely.

Wetbag Isostatic Pressing

In wetbag isostatic pressing, the mold is filled and sealed outside the pressure vessel. After the mold is introduced into the pressure vessel, it is completely immersed in the pressure medium, usually water containing lubricating and corrosion-preventive additives, such as Mobil Met C250. Pressure is applied isostatically by pressurizing the water around the mold. Even if the mold is washed under high pressure prior to introduction into the vessel, contamination of the pressure medium may still result. A filtering system must be included in the pressure-generating system.

The wetbag process typically is used for items that are large, have complex shapes, and are produced in small quantities. This process is very flexible, accepting various sizes and shapes of parts, and is limited only by the size of the pressure vessel. Parts of different sizes and shapes that require the same process parameters can be pressed in the same cycle. The wetbag process currently is used in most P/M applications.

Part Size or Shape

The only limitation on the size of the P/M part is the volume capacity of the press. Parts are compacted in three shapes—billets, preform, and near-net shape.

Billets are simple shapes, usually cylindrical, that are easily manufactured. Billets undergo further processing to final form.

Forging preforms have a more complicated shape, but dimensional constraints are not severe, because they frequently can be compensated for in the forging step. Forging preforms are economical, because they reduce the amount of material needed to produce a part and eliminate the need for intermediate forging operations and costly dies. Satisfactory material properties can be realized in the finished product, because the final forging operation fully densifies the part. Other metal forming operations that can fully densify a cold isostatically pressed and sintered part include rolling and extrusion.

Near-Net Shape. Cold isostatic pressing to near-net shape is economical; it minimizes the mass of material and the amount of machining required to manufacture a part. Because the exact location and degree of shrinkage during processing is difficult to predict quantitatively, several mold iterations may be necessary before the exact shape is attained. Tolerances on the order of 1% are attained. Shapes are sintered after cold isostatic pressing to between 90 and 97% of theoretical density. If properties of a fully dense part are required of this shape, it can be hot isostatically pressed to 99 to 100% of theoretical density.

Powder Properties

The powder material greatly influences the parameters of the cold isostatic pressing cycle. Powder particle size, shape, density, and mechanical properties influence the ability of the powder to flow into a mold (flow rate), as well as the packing density (tap density) in the mold. When cold isostatic pressing to shape, the mold must be filled uniformly and repeatably for dimensional tolerances to be met. Hardness and abrasiveness, as well as the shape and size of the particle, influence mold wear.

Compacts pressed from blended elemental powders usually are less expensive than compacts made of prealloyed powders. Current research and development is aimed at producing parts from blended elemental powders that will have equivalent properties (fatigue life), compared to parts compacted from prealloyed powder.

Fig. 6 Cold isostatic pressing flowchart
Courtesy of Industrial Materials Technology, Inc.

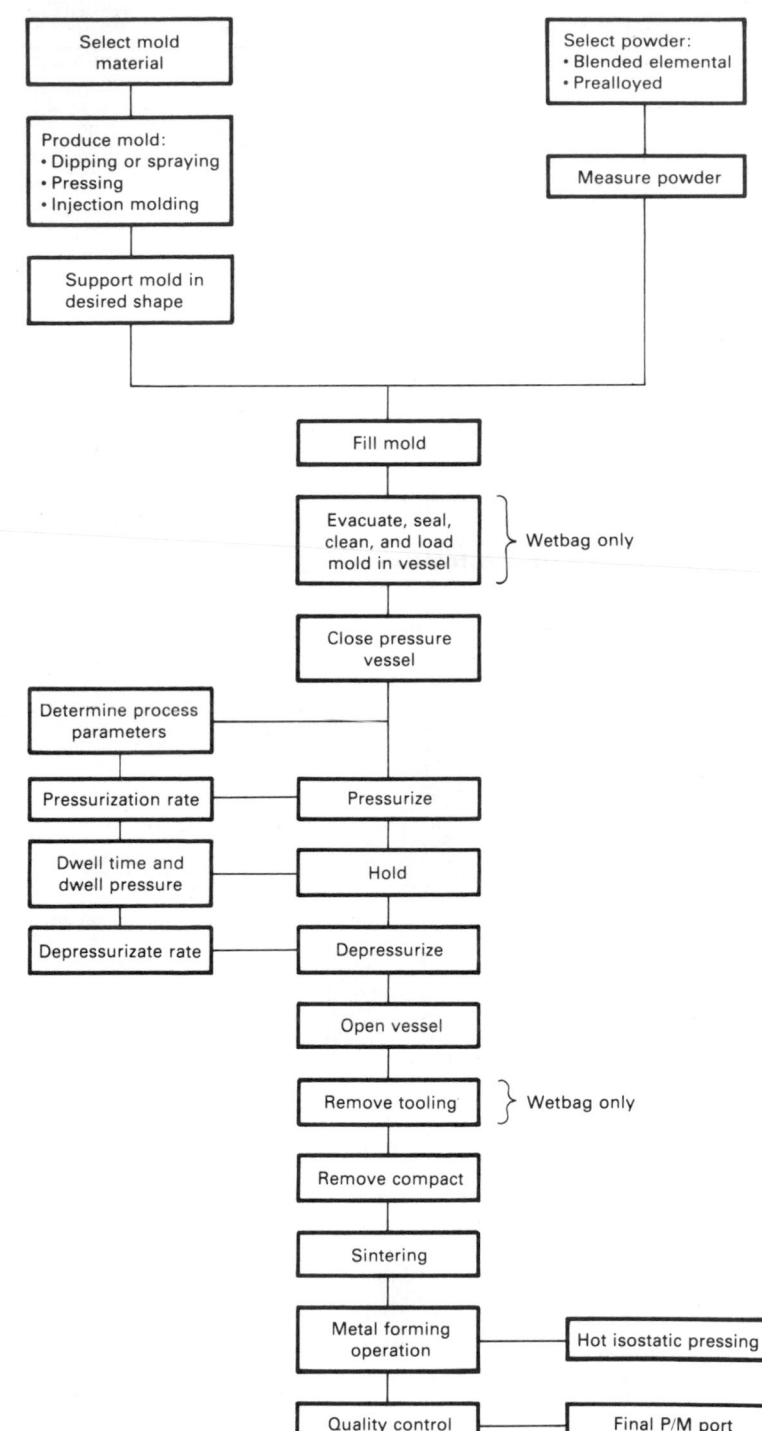

Process Parameters

A typical process flowchart is shown in Fig. 6. After the design specification of a part is known, tooling can be produced, the powder can be prepared, and process

Fig. 7 Density as a function of pressure

For cold isostatically pressed and conventionally pressed powder metals

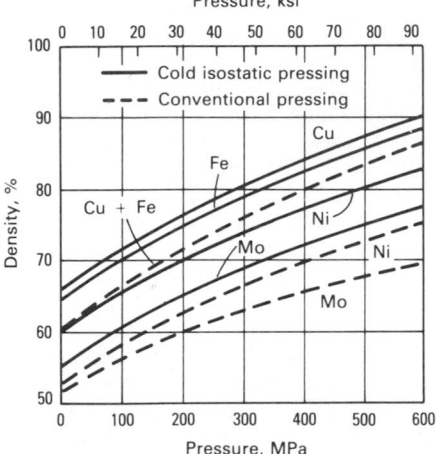

Fig. 8 Porous applications of cold isostatic pressing

Courtesy of Dynamet Technology, Inc.

parameters can be determined. After the mold is filled, gas entrapped in the mold may need to be evacuated. This gas may be between the metal particles or adsorbed on the particle surface. Also, it may be mechanically entrapped during pressing.

Entrapped gas can have two deleterious effects on the pressed compact. During depressurization, the expansion of entrapped gas may crack the compact as it expands. Gas entrapped in the compact may appear in the final part as porosity and can adversely affect material properties.

Dwell Pressure. For isostatic compacting, a lower pressure is needed than for conventional powder pressing methods. The density stays the same even when the parts are large and complex. Figure 7 shows density as a function of pressure for isostatically and conventionally pressed powder metals. Pressure conditions vary with powder material characteristics. Typically, pressures in excess of 200 MPa (30 ksi) are required to consolidate metal powders. Typical cold isostatic pressing

dwell pressures for a variety of materials are:

Powder	Pressure range, MPa (ksi)
Aluminum	55-140 (8-20)
Iron	310-415 (45-60)
Stainless steel	310-415 (45-60)
Copper	140-275 (20-40)
Tungsten carbide	170-205 (25-30)
Titanium	310-415 (45-60)
High-speed tool steel	240-345 (35-50)
Tungsten	240-415 (35-60)

Dwell time at pressure varies with powder compaction characteristics. Dwell time is not as important a criterion in determining part properties as is dwell pressure. Typical dwell time is only a few minutes, because negligible time-dependent deformation takes place for most metals at the usual room process temperature.

Depressurization Rate. During depressurization, critical process pressures that release the elastic energy stored in the mold and the gas entrapped in the compact are reached. When released, this elastic energy generates a shock wave that may be forceful enough to cause cracks in the compact. The amount of elastic energy stored in the mold depends on mold wall thickness and compression ratio. The potentially detrimental effects of the mold and the gas can be alleviated by controlling the depressurization rate.

Material Properties

The effectiveness of any cold isostatic pressing operation is evaluated by measuring the density and green strength achieved. In most cases, green density of the cold pressed compact is higher than the density obtained by die pressing under similar conditions. Furthermore, density is more uniform throughout the entire part.

Green strength of the isostatically pressed compact usually is higher than can be obtained by die pressing. Cold isostatically pressed parts usually can be handled and sometimes machined in the green state. After sintering, a cold isostatically pressed part may reach around 95% of theoretical density, resulting in improved mechanical properties, such as tensile strength and ductility, compared to conventionally formed P/M parts of similar composition. Further working of the materials, or a combination of cold and hot isostatic

Fig. 9 Ti-6Al-4V alloy housing finished component and preform

Courtesy of Dynamet Technology, Inc.

Fig. 10 Titanium forging preform and finished component

Courtesy of Dynamet Technology, Inc.

pressing, usually produces a part with satisfactory material properties.

Applications

Parts generally are not used in the as-cold isostatically pressed condition. Sintering following normal P/M practices increases density and strength to final values. If closed porosity is achieved in sintering, a hot isostatic pressing may be used for final densification. Steel-bonded titanium carbide and tungsten carbide/cobalt are two P/M product types where both process steps are employed.

Porous Applications. Figure 8 illustrates porous applications of cold isostatic pressing. Metallic filter cartridges in the shape of long tubes with very thin and constant wall thicknesses and specific porosity throughout the material can be made by cold isostatic pressing. Cartridges of this type are pressed from bronze, brass, stainless steel, Inconel, Hastelloy, Monel, and titanium.

Titanium. Because titanium and its alloys are expensive and somewhat difficult to machine, substantial economic benefits

Fig. 11 Titanium perforated disk

Contains over 1000 holes. Courtesy of Dynamet Technology, Inc.

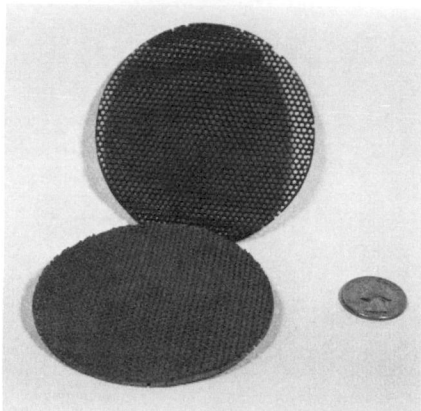

Fig. 12 Cold isostatically pressed near-net shape titanium parts

Courtesy of Dynamet Technology, Inc.

can be derived by pressing titanium to near-net shape or to a forging preform. Cold isostatic pressing also is used to produce billet preforms that can be extruded to a precise shape. Titanium parts of specific porosity are finding increased usage as surgical implants. Otherwise, cold isostatically pressed titanium parts are used either in the sintered state (94% dense) or are fully densified (99 to 100%) by subsequent operations.

Figure 9 illustrates a part that is currently in production (300 parts per month). The titanium alloy housing had been ma-

Fig. 13 Tool steel parts made by cold isostatic pressing and sintering

Courtesy of SCM Metal Products

chined from a billet blank weighing 3.6 kg (8 lb) or from a tubular blank weighing 2.3 kg (5 lb). A cold isostatically pressed near-net shape part can be produced that weighs 0.6 kg (1¼ lb). This weight reduction also decreased materials cost by 40%, compared to the billet. By using cold isostatic pressing, energy consumption required to produce this part was reduced by 75%. Additional savings were realized in the final machining steps, because the cold isostatically pressed and sintered part was easier to machine than its wrought counterpart.

The titanium preform shown in Fig. 10 eliminated six forging operations required to shape the part. With advances in tooling, a perforated disk (Fig. 11) can be cold isostatically pressed with over 1000 holes. Examples of other cold isostatically pressed and sintered titanium preforms and near-net shape parts are shown in Fig. 12.

High-Speed Tool Steel. Another difficult-to-fabricate material that can be cold isostatically pressed effectively is high-speed tool steel. Billets that are 60 to 65% dense can be pressed from annealed powders (less than 500 μm particle size) at pressures between 200 and 300 MPa (30 and 40 ksi). Tool tests (continuous turning, interrupted turning, drilling, and milling) comparing conventional and extruded cold isostatically pressed tools show equal or superior performance for the cold

isostatically pressed material. Typical high-speed steel parts cold isostatically pressed to near-net shape are shown in Fig. 13.

Tungsten carbide/cobalt is another difficult-to-process material that can be cold isostatically pressed effectively. Large laminating rolls have been cold isostatically pressed from tungsten carbide that contains high cobalt content to increase shock resistance. Due to the size of the rolls (35 by 10 cm, or 14 by 4 in.), an oversized hydraulic press with expensive tooling would have been needed. Cold isostatic pressing of the part also allowed for machining of the roll in the presintered condition.

Pure tungsten is cold isostatically pressed into the shape of long rods that are extruded to manufacture tungsten wire for lamp filaments. These rods must have constant density and a uniform cross section for subsequent sintering and extruding.

Cold isostatic pressing can be used to produce parts with complicated shapes. Threaded steel bushings that do not require further machining can be manufactured by this process. Trapezoidal threads also have been pressed using iron-based P/M material.

REFERENCES

1. Raes, H.D.B. and Papen, E.L.J., "Isostatic Dry Bag Production of Powder Metallurgy Parts," paper read in Scarborough, Nov 1981

2. Papen, E.L.J., Isostatic Pressing, in *High Pressure Technology*, Spain, I.L. and Paauwe, J., Ed., Marcel Dekker, New York, 1977, p 361

3. Popper, P., *Isostatic Pressing*, Heyden & Sons, New York, 1976, p 42

4. Papen, E.L.J., Isostatic Pressing, in *High Pressure Technology*, Spain, I.L. and Paauwe, J., Ed., Marcel Dekker, New York, 1977, p 356

5. Abbrowitz, S., Isostatic Pressing of Complex Shapes from Titanium and Titanium Alloys, *Production to Near Net Shape Source Book*, American Society for Metals, 1983, p 290-299

Secondary Operations Performed on P/M Parts and Products

By Howard E. Boyer
Consultant

SECONDARY OPERATIONS—including heat treating, cleaning, coating, machining, and joining—are performed on P/M parts in much the same manner as on conventional cast or forged parts. However, certain characteristics of P/M parts, principally inherent porosity, frequently impose sharp limitations on secondary operations.

For many applications, such as bearings, the porosity developed in P/M parts is highly desirable; in fact, this porosity is the prime reason for utilizing powder metallurgy instead of another method. Under these production conditions, it is not only essential to preserve porosity but also to preserve exposure of the porosity to the surfaces, thus facilitating absorption of the intended lubricant for use in service. Any finishing operation that tends to close surface porosity may defeat the principal reason for using P/M processes.

Entrapment of liquids, such as molten salt, cutting fluids, cleaners, or plating solutions, in the pores during performance of secondary operations is also important. Directly related to porosity, entrapment of liquids usually results in corrosion, often to the extent of ruining the parts. Therefore, a prime consideration in planning secondary finishing operations for P/M parts is to avoid the use of liquids. If the use of liquids is essential, procedures must be used to ensure that they are totally removed from the pores or are totally inert if allowed to remain.

This article is principally oriented to the processing of iron-based P/M parts. However, much of the information can be used as guidelines for processing of nonferrous metal parts.

Heat Treating

Heat treating procedures used for P/M parts are theoretically similar to those used for wrought or cast counterparts. There are, however, practical limitations that must be considered when processing iron-based alloy (steel) P/M parts. If a powder compact has the exact composition as its wrought or cast counterpart, response to heat treatment should be the same. In both cases, two principles apply: the maximum hardness that can be obtained depends on the amount of carbon in solution at the austenitizing temperature (Fig. 1); and the depth of hardening (hardenability) depends largely on the alloy content and density, for example, nickel, manganese, chromium, or molybdenum (Fig. 2).

The first principle must be thoroughly understood before heat treating procedures are planned for P/M parts, because response to hardening depends entirely on the amount of combined carbon, not necessarily on total carbon. Iron-carbon compacts are typically produced by mixing iron powder and graphite. Graphite forms cementite on sintering and then dissolves in the austenite. On cooling, the composition of the compact should be that of a true steel. Certain variables must be controlled to ensure successful results. The amount of carbon that dissolves in the austenite depends primarily on the amount added to the green compact. Sintering time and temperature also affect the amount. Consequently, iron-graphite compacts require control of the same variable as in the heat treating of cast irons; that is, it is not always known what form the carbon takes, or how much is combined in the form of cementite and available to provide the potential hardening response. If too little combined carbon (excessive graphite) is known to be present before heat treatment, this condition can be compensated for by using higher austenitizing temperatures to dissolve most or all of the graphite.

Such a procedure has drawbacks, however, because too much cementite may result in the parts being excessively brittle after heat treating. Consequently, the need for close processing control cannot be overemphasized. It is just as important to know the amount of combined carbon in a sintered compact as it is to know the carbon content of a forging undergoing heat treatment.

Hardenability of P/M parts, as for wrought or cast parts, is controlled by the amounts of alloying elements. This phenomenon is illustrated in Fig. 2, which compares the hardenability of carbon steel 1038H with an alloy steel (4137H), both of which have approximately the same carbon content. The increased hardenability is due to about 1% Cr and 0.2% Mo in the 4137H alloy. Hardenabilities of P/M parts closely parallel those of wrought counterparts with similar compositions. It

Fig. 1 Relation of carbon content and percentage martensite to Rockwell C hardness

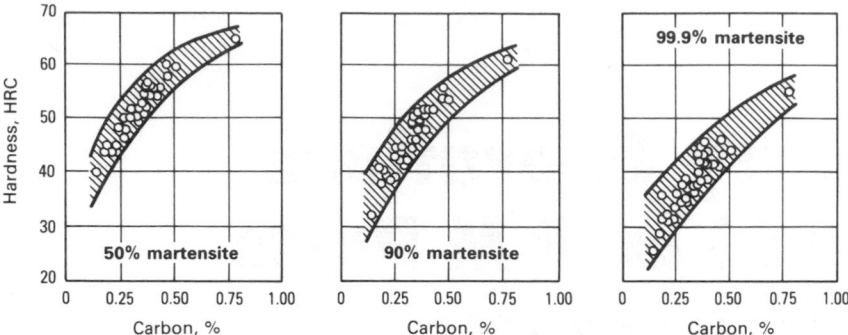

resintered register minimal variation, which decreases as part density increases.

For cast irons, charts have been developed that plot observed indentation hardness versus actual hardness as measured by a microhardness tester (Vickers or Knoop readings). These charts also include a third variable—carbon equivalent. This system provides reasonable accuracy for estimating true, or actual, hardness. See Volume 4 of the 9th edition of *Metals Handbook* for additional information on hardness measurement of cast irons.

For P/M parts, determination of actual hardness is more complex because of the wide variations in density even though the composition is established. Fluctuations in density may vary widely from one part to another, from location to location within the same part, or from batch to batch of the same parts. Charts that provide accurate hardness conversions for P/M parts are difficult to develop, although such charts can be developed on an empirical basis. Such charts should be developed empirically by taking many conventional and microhardness readings on parts that are identical in design, composition, and density.

The preferred procedure, in view of the variables associated with P/M parts, is to

is essential to determine precise alloy content, because this affects heat treating procedures, particularly austenitizing and quenching.

Hardness Evaluation

Sintered compacts are nonhomogeneous materials, like cast iron. In cast iron, the graphite phase represents a dispersion of voids, because graphite exhibits no hardness when measured by indentation methods. Therefore, indentation readings indicate lower hardness values than are actually present. The extent of this depends on the amount of carbon present in the form of graphite. Frequently, iron-carbon P/M compacts also contain some free graphite, which causes erroneous hardness readings. However, in P/M parts, the presence of actual voids causes most of the measurement error.

The variation between observed hardness (actual reading taken from the test instrument) and the actual hardness (measured on individual constituents) depends almost entirely on density. Powder metallurgy parts that have been re-pressed and

Fig. 2 Hardenability of a carbon steel (1038H) with an alloy grade (4137H) of similar carbon content
(a) 1038H end-quench hardenability. Normalized at 870 °C (1600 °F); austenitized at 845 °C (1550 °F). (b) 4137H end-quench hardenability

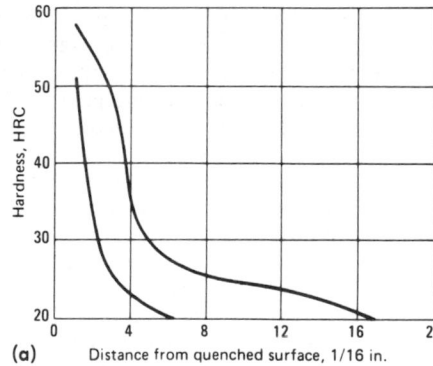

Distance from quenched surface		Hardness, HRC		Distance from quenched surface		Hardness, HRC	
1/16 in.	mm	max	min	1/16 in.	mm	max	min
1	1.58	58	51	6	9.48	28	21
1.5	2.37	56	42	6.5	10.27	27	20
2	3.16	55	34	7	11.06	27	···
2.5	3.95	53	29	7.5	11.85	26	···
3	4.74	49	26	8	12.64	26	···
3.5	5.58	43	24	9	14.22	25	···
4	6.32	37	23	10	15.80	25	···
4.5	7.11	33	22	12	18.96	24	···
5	7.90	30	22	14	22.12	23	···
5.5	8.69	29	21	16	25.28	21	···

Distance from quenched surface		Hardness, HRC		Distance from quenched surface		Hardness, HRC	
1/16 in.	mm	max	min	1/16 in.	mm	max	min
1	1.58	59	52	13	20.54	51	35
2	3.16	59	51	14	22.12	50	34
3	4.74	58	50	15	23.70	49	33
4	6.37	58	49	16	25.28	48	33
5	7.90	57	49	18	28.44	46	32
6	9.48	57	48	20	31.60	45	31
7	11.06	56	45	22	34.76	44	30
8	12.64	55	43	24	37.92	43	30
9	14.22	55	40	26	41.08	42	30
10	15.80	54	39	28	44.24	42	29
11	17.38	53	37	30	47.40	41	29
12	18.96	52	36	32	50.56	41	29

evaluate actual hardness of heat treated samples by microhardness testing. These samples should also be tested by a suitable indentation method such as Rockwell A, after which an acceptable range of observed hardness can be established. Conventional indentation hardness testing may then be used as a quality control tool, but it is advisable to continually monitor the total process by use of microhardness testing.

Austenitizing

Practices used for direct hardening of P/M compacts do not differ significantly from those used for wrought counterparts. Generally, P/M compacts are less homogeneous compared with forged or rolled parts; therefore, somewhat higher temperatures, or longer times at temperature, are required to achieve complete austenitization. For carbon steel compacts, the following austenitizing temperatures are recommended:

Combined carbon, %	Temperature °C	°F
0.25	915	1675
0.35	900	1650
0.45	885	1625
0.55	870	1600
0.65	860	1575

These austenitizing temperatures are slightly higher than those generally recommended for wrought parts. Time at temperature depends on the type of furnace and maximum section thickness of the parts, but the use of time periods about 50% longer than those used for wrought parts under the same conditions is generally recommended. When alloying elements are added, time and temperature should be increased slightly.

Furnaces and Furnace Atmospheres. Generally, selection of a furnace for austenitizing P/M parts is governed by the same conditions that affect selection of furnaces for other types of parts—that is, furnace availability, size of parts, total production quantities, and production flow. A wide variety of batch-type and continuous furnaces can be used successfully. See Volume 4 of the 9th edition of *Metals Handbook* for more information on furnaces and furnace atmospheres.

A prime limitation is the type of heating medium used. Powder metallurgy parts to be through hardened normally should be heated in a gaseous atmosphere such as that provided by an endothermic generator. Any of the various nitrogen-based atmospheres may be used, as long as the carbon po-

tential is controlled to match the carbon content of the workpiece.

Vacuum atmospheres have also proved to be satisfactory. Heating of P/M parts in molten salt is not recommended, even for parts that have maximum density. These parts are extremely difficult to wash free of salt, and if any salt remains in the voids, the parts will corrode immediately.

Quenching

For quenching of plain carbon steel P/M parts with sections greater than about 5 mm (0.2 in.), an aqueous (water and brine, water and caustic, plain water, or a synthetic) type of quenching medium is usually required to achieve full hardness. However, this type of quenching may result in cracking, because P/M parts exhibit a greater tendency toward cracking than wrought counterparts. An aqueous medium may also cause corrosion. Therefore, oil quenching is preferred for P/M parts because it is a much less drastic quench.

Thicker sections should contain sufficient amounts of alloying elements, such as nickel manganese and/or chromium, to impart the required hardenability. The use of a "fast-quenching" oil enhances the maximum hardness by oil quenching. "Fast oils" are generally so identified and are readily available.

Tempering

All austenitized and quenched P/M parts, like their wrought counterparts, should be tempered immediately after quenching. Forced air convection-type furnaces are preferred for tempering. Neither molten salt nor molten metal is recommended as a tempering medium. Tempering temperature should be at least 150 °C (300 °F); higher temperatures can be used if some decrease in hardness can be tolerated. Tempering at 150 °C (300 °F) converts white martensite to tempered martensite and has little or no effect on as-quenched hardness.

Steam Treating

Steam treating of steel parts is not a new procedure; it has been used as a protective atmosphere for tempering at temperatures up to about 540 °C (1000 °F). In addition to serving as a protective atmosphere, the steam induces a generally bluish oxide coating, which serves as a finish, increasing wear and corrosion resistance as well as improving appearance; it is often an acceptable final finish.

Because steam treatment frequently has been used in conjunction with tempering, it is often referred to as "steam temper-

ing." Steam treating is not confined to tempering; in fact, it is frequently used to produce a blue-black finish on various steel parts—heat treated or not heat treated.

The steam process can be applied to P/M parts in the same manner as it is used for wrought or cast parts. However, in P/M parts, ferrous-ferric oxide forms in the pores and tends to close them, thus eliminating or minimizing entry of contaminants. This phenomenon is also effective in resisting hydraulic pressures. Additionally, the adjunct equipment required to convert a forced air tempering furnace to a steam treating furnace is minimal.

Processing Sequence. Although not as precise as other processing methods, the following procedure has proved successful. Degreased parts are loaded in a cold (or nearly cold) furnace in baskets so that parts are separated. The furnace is then closed and set at 370 °C (700 °F). When the temperature reaches 345 °C (650 °F), steam is introduced into the furnace at full pressure. At 370 °C (700 °F), the furnace chamber is purged with steam for 15 min. Furnace pressure is then adjusted and maintained at a slight positive pressure for another 15 min.

The load is then heated to 540 °C (1000 °F) and held for $1^1/_2$ h with steam on. The charge is then cooled to 290 °C (550 °F). After the steam is turned off, the parts are removed from the furnace. Frequently, parts are quenched in mineral oil from 290 °C (550 °F) to improve appearance as well as corrosion resistance.

Case Hardening

Powder metallurgy parts can be case hardened by several processes, although various available processes are not equally suited to every application. A clear case/core relationship can be obtained only with parts of at least 7.2 g/cm^3.

Carburizing. Powder metallurgy parts with relatively low combined carbon contents of 0.10 to 0.20% (that are free of any graphite) can be carburized by conventional pack or gas methods. Liquid carburizing is not recommended because of the difficulty of washing the parts free of salt.

Pack carburizing can be used; however, it is uneconomical for small parts produced in large quantities. This method is seldom used for case hardening of P/M parts. See Volume 4 of the 9th edition of *Metals Handbook* for a detailed description of this process.

Gas carburizing is more practical for P/M parts than pack carburizing. For this process to be successful, however, density as well as the precise composition of the parts should be known. Low-density parts should not be subjected to gas carburizing, because the carburizing gases penetrate the voids. Consequently, a distinct case is not achieved compared with the case developed on wrought parts under the same conditions. Instead, the carbon penetration in P/M parts is generally deeper and relatively nonuniform. The extent of this condition varies with density. In parts that have been re-pressed and resintered, this condition may be tolerable or even negligible, but for parts of lesser density, the depth of carbon penetration may be so great that thin sections of quenched parts will be brittle.

Another reason that conventional gas carburizing enjoys only limited usage is because it does not increase hardenability; thus, plain carbon grades usually must be quenched in an aqueous medium. This may result in cracking, especially if carbon penetration is excessive.

As with wrought parts, the depth of carburized case on P/M parts depends on time and temperature. Typical case depths for wrought parts for three carburizing temperatures at various lengths of time are shown in Table 1. Results such as those shown in Table 1 can be expected for the highest density P/M parts.

Carbonitriding is a modified form of gas carburizing, rather than a form of nitriding. The principal process modification consists of introducing ammonia into the gas carburizing atmosphere, which results in the addition of nitrogen to the carburized case as it is being produced. Nascent nitrogen forms at the workpiece surface by dissociation of the ammonia in the furnace atmosphere. Nitrogen diffuses into the steel surfaces simultaneously with the carbon. Typically, carbonitriding is carried out at a lower temperature (at least 55 °C, or 100 °F, lower) and for a shorter time (half an hour or longer) than gas carburizing.

Carbonitriding is widely used for case hardening of P/M parts made from ferrous powders. Densities of the sintered compacts vary from approximately 6.5 g/cm³ to values that approach those of wrought steel, up to approximately 7.9 g/cm³. Parts may be infiltrated with copper prior to carbonitriding. Carbonitriding is extremely effective for case hardening high-density (7.2 g/cm³) parts made from sintered iron compacts. Additionally, it is reasonably effective for case hardening parts of lower density.

To a great extent, carbonitriding overcomes the disadvantages of carburizing. With carburizing, high transformation temperatures coupled with inherent porosity produce high rates of diffusion. Carbonitriding at 790 to 845 °C (1450 to 1550 °F) eliminates this problem. The lower rates of diffusion that occur at these temperatures permit better control of case depth and allow adequate buildup of carbon in the case.

Nitrogen also has a marked effect on retarding the rate of transformation from the austenite by lowering the critical cooling rate. This provides a marked increase in hardenability, which enhances development of full hardness by oil quenching, whereas a more drastic quench, such as brine or water, is required for carburized cases.

File-hard cases with microhardnesses equal to 60 HRC and with predominantly martensitic structures are consistently obtained by carbonitriding, although the allowable case depth range must be increased over that prescribed for wrought steels. Typical ranges of case depth are 0.08 to 0.30 mm (0.003 to 0.012 in.).

The high rate of carbon and nitrogen penetration that occurs as a result of porosity is demonstrated in Fig. 3 for parts made from American Society for Testing and Materials (ASTM) B 310 class A iron powder. Case depth (penetration) decreases with increasing density. Case depths for higher densities (7.2 to 7.3 g/cm³) are deeper than for wrought steel at a density of 7.87 g/cm³. Although most commercial iron powders exhibit this type of response to carbonitriding, copper-infiltrated parts are more resistant to penetration of carbon and nitrogen, thus approaching wrought parts.

Equipment and Techniques. Procedures for carbonitriding of P/M parts are essentially the same as those used for similar wrought parts. Control of temperature and time is generally more critical than for wrought parts because of the porosity. In one proprietary process, P/M parts are carbonitrided at 790 °C (1450 °F) in a batch-type furnace for all copper-free iron parts. Lower temperatures are avoided to minimize the potential danger of explosion, and higher temperatures are avoided because case depth control is more difficult.

The processing cycle, including composition of the atmosphere, is critical. Ammonia content (usually 10%) increases hardenability and affects dimensional stability. Because dimensional changes in heat treating are often crucial to the economic justification of producing parts by powder metallurgy, gas composition, temperature, and quenching medium must be closely controlled.

In addition to batch-type furnaces, shaker-hearth and belt-type continuous furnaces are also well suited to carbonitriding of P/M parts. When processing to a new specification, furnace adjustments are usually made on small production quantities, thereby requiring sacrifice of only a few parts to arrive at optimum conditions. Such adjustments are then recorded so they can be used when processing the next lot of similar parts.

Tempering of carbonitrided parts requires special consideration, largely because the quenching oil they may contain may pose a fire hazard. Although tempering is capable of removing oil held by the pores of the workpiece, air tempering of P/M parts generally is limited to 205 °C (400 °F) because of the danger of fire. Generally, P/M parts are tempered at temperatures higher than those used for similar wrought parts. If tempering temperatures above 205 °C (400 °F) are required, parts should be cleaned by ultrasonic degreasing, for example, prior to placing in the tempering furnace.

In addition to removing the quenching oil in the pores of P/M parts, handling during tempering may eliminate the need for tumbling and deburring. Additionally, although there is little danger that oil-

Table 1   Values of case depth calculated by the Harris equation

| | Case depth(a), after carburizing at: | | | | | |
| | 870 °C (1600 °F) | | 900 °C (1650 °F) | | 925 °C (1700 °F) | |
Time (t), h	mm	in.	mm	in.	mm	in.
2	0.64	0.025	0.76	0.030	0.89	0.035
4	0.89	0.035	1.07	0.042	1.27	0.050
8	1.27	0.050	1.52	0.060	1.80	0.071
12	1.55	0.061	1.85	0.073	2.21	0.087
16	1.80	0.071	2.13	0.084	2.54	0.100
20	2.01	0.079	2.39	0.094	2.84	0.112
24	2.18	0.086	2.62	0.103	3.10	0.122
30	2.46	0.097	2.95	0.116	3.48	0.137
36	2.74	0.108	3.20	0.126	3.81	0.150

(a) Case depth: mm = 0.635 $\sqrt{t}$ (case depth: in. = 0.025 $\sqrt{t}$) for 925 °C (1700 °F); 0.533 $\sqrt{t}$ (0.021 $\sqrt{t}$) for 900 °C (1650 °F); 0.457 $\sqrt{t}$ (0.018 $\sqrt{t}$) for 870 °C (1600 °F). For normal carburizing (saturated austenite at the steel surface while at temperature)

quenched P/M parts will crack, toughness can be significantly increased by tempering.

Hardness Testing of Carbonitrided P/M Parts. Carbonitrided parts cannot be accurately tested by the common indentation methods because of their very thin cases. File testing is the prime method used for routine evaluation. For development work and for monitoring of the process, microhardness testing is used for determining true hardness, however.

Nitrocarburizing

There are two general classes of nitrocarburizing processes—salt bath and gaseous. Discussion in this article is limited to gaseous nitrocarburizing, because the salt bath process is not well suited to processing of P/M parts, primarily for the same reasons discussed for other heat treating processes.

Gaseous ferritic nitrocarburizing is a diffusion heat treatment that involves the addition of nitrogen and carbon to the surfaces of steel parts. To this extent, the process resembles carbonitriding. It differs from carbonitriding mainly in that the temperatures used are completely within the ferritic phase field—typically about 570 °C (1060 °F)—a temperature just below the austenite range for the iron-nitrogen system. However, carbonitriding is done in the austenitic range.

Gaseous nitrocarburizing can be used for producing thin, hard, wear-resistant cases on iron-based P/M parts. This process typically utilizes sealed quench batch furnaces of the same design as is used for gas carburizing and carbonitriding (Fig. 4). Furnace operating temperatures are low enough to maintain the workpieces in the ferritic condition.

Atmosphere compositions may vary somewhat, depending on the specific process. In one proprietary process, the atmosphere is formed from equal amounts of ammonia and endothermic gas—American Gas Association (AGA) type 302. In another process, a typical atmosphere consists of 35% ammonia and 65% refined exothermic gas (AGA type 301, nominally 97% nitrogen), which may be enriched with a hydrocarbon gas such as methane or propane.

Time cycles for gaseous nitrocarburizing generally range from 1 to 5 h. However, time cycles of no more than 1 h usually are recommended for P/M parts, because deep penetration may cause excessive embrittlement, as is the case for gas carburizing.

The prime objective of the gaseous nitrocarburizing is to produce a thin layer of iron carbonitride and nitrides (the "white layer" or compound zone), with an underlying diffusion zone containing dissolved nitrogen and iron (or alloy) nitrides. The white layer provides resistance to galling, corrosion, and wear. The diffusion zone serves to significantly increase fatigue properties in carbon and low-alloy P/M parts. The white layer is composed primarily of the epsilon nitride phase. The precise composition of the material and the furnace atmosphere determine ultimate case composition.

Optimum results in gaseous nitrocarburizing are obtained by oil quenching from the processing temperature, thus emphasizing the suitability of an internal quench furnace, as shown in Fig. 4. Because the cases formed by nitrocarburizing are extremely thin, the process should be applied to finished parts only, because no stock is removed after nitrocarburizing.

The white layer (extreme interior) of nitrocarburized parts is relatively brittle, but has less tendency to spall compared to the white layer formed on parts that have been subjected to ammonia gas nitriding. This tendency is attributed to an essentially single-phase white layer of the less brittle epsilon phase. Additionally, nitrocarburized layers generally are thinner.

Gas nitriding is another case hardening process in which the processing temperature is maintained within a range in which all steels remain in the ferritic phase, usually 510 to 565 °C (950 to 1050 °F). For a detailed description of the gas nitriding process, see Volume 4 of the 9th edition of *Metals Handbook*.

The application of gas nitriding to P/M parts has been extremely limited. Not unlike gas carburizing, the nitriding gas penetrates too deeply into parts that contain voids, partly because of the long cycles used. Such a condition is difficult to control and often results in excessive embrittlement. For optimum results, parts should be first austenitized, quenched, and tempered, because successful gas nitriding is achieved when the microstructure consists of tempered martensite. When this preliminary operation is performed, parts must be finished and/or thoroughly cleaned to produce the type of surface condition required for nitriding.

Another reason for the limited use of gas nitriding for P/M parts relates to composition requirements. The combined carbon content should be maintained within 0.30 to 0.50%. In addition, certain alloying elements that are capable of forming hard, stable nitrides must be present in the P/M compact. Chromium promotes the formation of nitrides, as does aluminum. Volume 4 of the 9th edition of *Metals Handbook* lists steel compositions that are suitable for nitriding.

Fig. 4 Typical furnace for gaseous nitrocarburizing

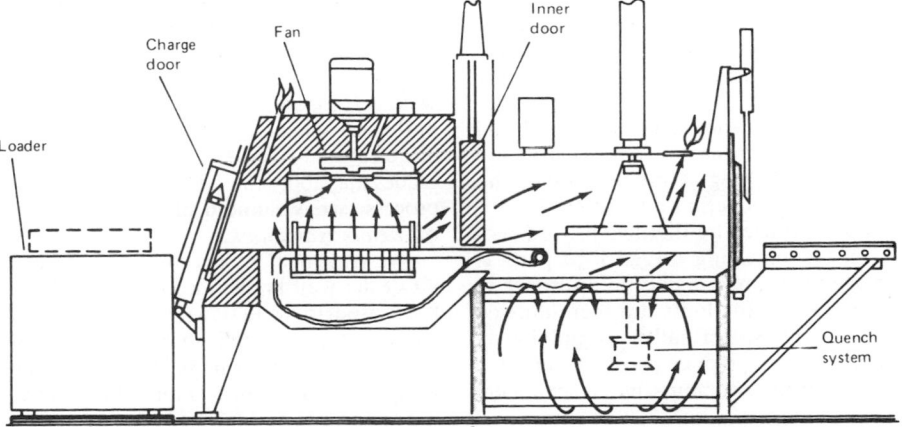

Fig. 3 Case depth versus time for carbonitriding P/M parts

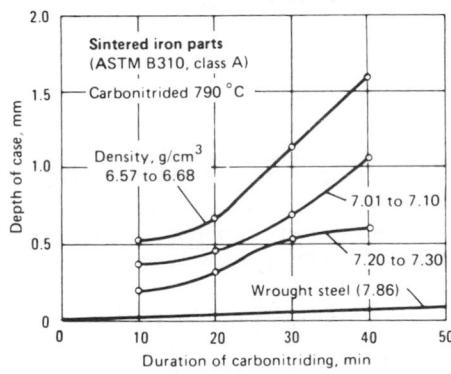

Annealing

Generally, annealing practices for P/M parts are of limited practical significance; annealing is infrequently required. Notable exceptions, however, are P/M forgings, which frequently require annealing for improved machinability. For plain carbon-iron compacts, optimum austenitizing temperatures for annealing depend on the combined carbon content:

	Temperature	
Combined carbon, %	°C	°F
0.20	900	1650
0.30	885	1625
0.40	870	1600
0.50	860	1575

Forged compacts are soaked thoroughly at the appropriate temperature for the combined carbon content. Cooling rates from the austenitizing temperature depend on composition and section thickness, as is true for the annealing of forgings.

For parts that contain no more than approximately 0.20% combined carbon, an annealed structure that is predominantly pearlitic (usually preferred for machining) usually can be obtained by air cooling from the austenitizing temperature. However, as combined carbon content and/or alloy content is increased, air cooling results in the formation of considerable quantities of bainite, which is undesirable for optimum machinability.

Generally, the preferred procedure for annealing P/M forgings is to austenitize at a temperature that is appropriate for the specific combined carbon content (above), followed by cooling as rapidly as possible to approximately 705 °C (1300 °F) and transforming isothermally at this temperature. Depending on the exact composition, holding time at 705 °C (1300 °F) generally ranges from $1/2$ to 2 h. To select optimum holding time, the time-temperature transformation (TTT) curve for the specific composition to be annealed should be reviewed. Additionally, because of the variables involved with P/M forgings, several samples should be processed and examined for hardness and microstructure before establishing a precise annealing practice for production quantities.

Another procedure that has proved successful for annealing forged P/M parts is to place them immediately in a furnace at 705 °C (1300 °F). Parts are then allowed to transform isothermally, as previously described. This practice produces a coarse ferrite-pearlite structure that is excellent for machining.

Joining

One of the main objectives in utilizing P/M techniques to produce parts is to eliminate the need for joining operations. However, there are several notable exceptions. Designers of P/M parts frequently must produce shapes that are not amenable to standard compaction techniques. Therefore, a means of joining two or more parts together must be considered. Mechanical means are frequently used, such as bolts, screws, rivets, and sometimes staking.

With some restrictions due primarily to the inherent porosity of P/M parts, most joining methods that are suitable for cast and wrought parts may be used for similar P/M parts—namely, welding, brazing, and adhesive bonding. In addition, for P/M parts, sinter-bonding offers a means of joining.

Welding

Basically, a P/M part consists of many welds; consequently, they exhibit superior weldability. Because of the voids (porosity) in P/M parts, the commercially available welding processes are not equally applicable to all P/M parts. When a part is porous or "full of cracks," these voids tend to spread during welding.

Often, certain welding processes are impractical and should not be considered. However, there are many variations among P/M parts, specifically in density. Therefore, evaluation of a particular welding process for P/M applications cannot be assessed without full knowledge of the part density and composition. For example, a part with less than about 70% theoretical density may be totally impractical for welding, whereas one that has been repressed and resintered may be satisfactory for welding.

Compositions of P/M parts must also be considered, regardless of the welding process being considered. Similar to wrought materials, as the hardenability of the compact increases, the need for special precautions such as preheating and postheating increases. Generally, selection of a welding process for P/M parts must minimize heat spread, or the width of the heat-affected zone. For equipment and techniques that apply to parts for each specific process, see Volume 6 of the 9th edition of *Metals Handbook*.

Arc Welding. At one time or another, all of the well-known arc processes have been used to weld P/M parts. Figure 5 shows a typical application of gas metal arc welding on P/M parts. In view of the need for restricted heat spread, the following arc processes are listed in order of de-

creasing suitability for welding of P/M parts:

- Plasma arc welding
- Gas tungsten arc welding
- Gas metal arc welding
- Shielded metal arc welding

The most common problem encountered in arc welding P/M parts is cracking at or near the weld interface. This tendency is predominant in low-density parts that have undergone any arc process. Generally, particle-to-particle bonding in the part is insufficient to withstand the stresses imposed by the heating and cooling in the weld zone.

Every effort should be made to keep the heat input as low as possible, thus minimizing heat spread. The importance of minimizing heat spread increases as part density decreases. Consequently, some development work should precede the adaptation of any arc process to a production application.

Electron beam welding has proved successful as a means of joining P/M parts and is generally superior to arc welding. Figure 6 shows an electron beam welded P/M part. The heat-affected zone associated with electron beam welding is substantially less than with any arc process, even plasma arc welding.

Friction Welding. Under conditions permitted by part design, friction welding has proved to be one of the best methods for joining P/M parts. The heat developed by spinning one part against another is sufficient to obtain a highly localized melting and subsequent bonding. In addition, welding occurs across the entire section, which minimizes cracking. As with other welding methods, the higher the density of the P/M part, the better the weldability, in terms of producing sound welds.

Resistance welding has not proved totally successful for joining P/M parts, although projection welding has, in many cases, provided acceptable results. Projection welding allows improved control of heat spread, as illustrated in the following example.

As shown in Fig. 7, three P/M parts were welded to a wrought steel rim with four projection welds. Truncated-cone projections on the sintered iron backing (see section A-A in Fig. 7) were designed so that their conical surfaces would contact the holes in the rim to produce an annular nugget with only shallow penetration into the sintered metal.

This design prevented the formation of porous, weak welds that could have resulted from excessive melting of the sintered metal. For welding, the rim was clamped in an indexed position on the fixture, as shown at the right of Fig. 7. An air-operated ram lowered the upper electrode to make contact with the lining segment. The upper electrode, made from Resistance Welder Manufacturers Association (RWMA) class 2 material, was radiused to fit the lining segment. The lower electrode consisted of a base made of class 2 material with four 4.8-mm ($^3/_{16}$-in.) diam inserts made from class 11 material. Additional processing details are given in Fig. 7.

Brazing

Basically, P/M steel (powdered iron compacts) are readily brazeable if inherent porosity is considered. However, not unlike some other processes involving P/M parts, there are practical limitations. Powder metallurgy parts have been brazed by all of the well-known methods—that is, furnace, torch, induction, resistance, and salt bath. However, salt bath brazing is not recommended, primarily because of the difficulties encountered in postbraze cleaning.

Furthermore, resistance brazing is not recommended because of the difficulty encountered in controlling nugget size (heat-affected zone). Torch brazing and induction brazing of P/M parts are performed infrequently because of the difficulties encountered in precise control and postbraze cleaning.

Furnace brazing is the preferred method of brazing P/M parts. Conducted in a controlled atmosphere, furnace brazing is adaptable to several types of brazing techniques for mass production. However, certain limitations must be recognized.

A primary consideration is the fact that a P/M compact readily absorbs filler metal. Frequently, brazing and infiltration occur simultaneously. For example, low-density parts may absorb several times the amount of filler metal required to braze the joints. Because many filler metals contain substantial amounts of precious metals, this practice is expensive. There are two means of preventing, or at least minimizing, this waste of filler metal: (1) brazing only P/M parts that have been infiltrated with copper, and (2) brazing only P/M parts that have been re-pressed and resintered. Some filler metal will still penetrate these parts, but waste will be minimized.

Another method of reducing production costs involves the use of a filler metal that has an initial melting temperature of about 1050 °C (1922 °F), but which rises gradually as the molten alloy reacts with the iron. Consequently, the flowability of the filler metal decreases, and penetration into the pores is reduced. Flowability of the filler metal is also restricted, as it can penetrate the clearances between mating parts.

As with welding, development work should precede production brazing applications to ensure that a reproducible process with acceptable quality is selected.

Copper brazing in a furnace poses the most practical method for brazing P/M parts to produce assemblies. Copper is relatively inexpensive, so losses resulting from excessive penetration into the part are minimal. Also, copper infiltration is frequently desirable for strengthening and/or sealing the part.

Copper brazing in a prepared atmosphere does not require flux. Therefore, the simultaneous joining and infiltration with copper in an atmosphere furnace presents an excellent means of producing assemblies. For applications in which porosity must be retained, brazing is unsuitable.

Brazing Atmospheres. A lean exothermic atmosphere is the most widely used atmosphere for furnace brazing, because it is the least expensive. This type of atmosphere provides bright, clean surfaces on iron-based alloy parts and is equally well suited for processing nonferrous alloys.

A lean exothermic atmosphere should not be used for iron-carbon compacts with combined carbon contents higher than 0.20%, because decarburization occurs. For higher carbon compacts, either a rich exothermic, an endothermic, or a vacuum atmosphere should be used. Vacuum atmospheres are well suited to most furnace brazing applications. For a complete discussion of furnace atmospheres, see Volume 4 of the 9th edition of *Metals Handbook*.

Brazing Procedures. Typical alloys, temperatures, and techniques used for brazing P/M parts are generally the same as those used for brazing wrought or other high-density parts. Detailed information on these facets of joining are presented in Volume 6 of the 9th edition of *Metals Handbook*.

Adhesive Bonding

The use of adhesive bonding for producing assemblies of P/M parts has been enhanced by the development of anaerobic adhesives. In one application, a gear and sprocket assembly was made by brazing two P/M parts; a change to adhesive bonding allowed a significant reduction in manufacturing cost. Both the gear hub and the sprocket were made from 0.6 to 1.0% iron-graphite powder compacted to a density of 6.1 to 6.5 g/cm³. The green compacts were sintered in an endothermic atmosphere at 1130 °C (2065 °F), followed by steam treating to attain an oxide coat and to close the pores. Parts were assembled by means of an arbor press. The parts were fitted with a clearance of 0.03 to 0.08 mm (0.001 to 0.003 in.); the press was used primarily to ensure proper fit.

The assembly was then placed on a parallelism gage, while the anaerobic liquid adhesive was applied to the joint with an applicator. Assemblies were cured sufficiently in 30 min to permit handling and were cured completely in 24 h.

Sinter Bonding. In some cases, P/M parts can be joined simultaneously with sintering. Two or more green parts can be inserted within one another, provided tolerances enable them to fit tightly together. This is not usually a problem, because green size of the compact is easily controlled. Pressure must be applied to the interfaces during sintering, which is usually accomplished by using a copper-alloy powder that expands on sintering for the inner part. With this procedure, two or more parts that comprise the assembly "grow together" during sintering.

Fig. 5 P/M sprocket reduction gears welded by gas metal arc welding

Fig. 6 P/M magnetic brake assembly joined by electron beam welding

Fig. 7 Setup for joining three brake-lining segments

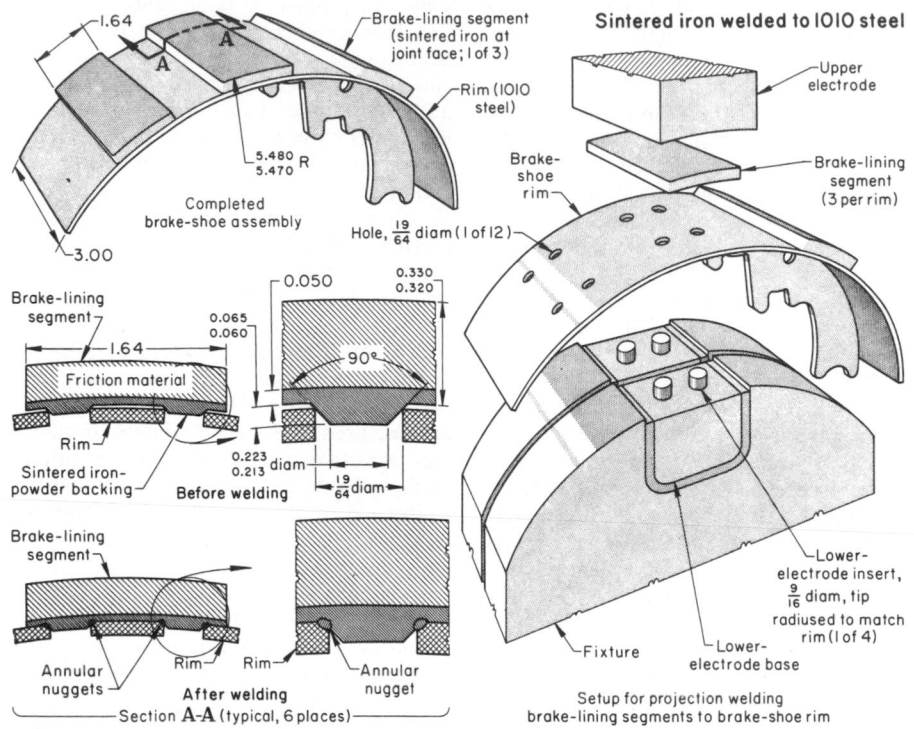

Setup for projection welding
brake-lining segments to brake-shoe rim

Equipment details

Welding machine .... Manual press type, with
16-tap transformer
Rating at 50% duty cycle .... 175 kV·A
Electrodes .... RWMA classes 2 and 11(a)
Electrode force, max .... 20 kN (4500 lb)
Heat control .... Electronic tube type

Welding conditions

Welding current .... 35 000 A
Heat-control setting 90%
Electrode force .... 4.5 kN (1000 lb)
Squeeze time .... 50 cycles
Weld time .... 22 cycles
Hold time .... 35 cycles

(a) Upper electrode was made of RWMA class 2 material and was radiused to fit the surface of the brake-lining segment; lower electrode consisted of four $^9/_{16}$-in.-diam RWMA class 11 inserts radiused to conform to the inside surface of the rim and silver brazed to a class 2 base.

Deburring

Although cleaning and deburring generally are considered different operations, they are often accomplished simultaneously. Therefore, much of the discussion on deburring is applicable to the subsequent section of this article on cleaning.

The inherent porosity in P/M parts demands special considerations in all secondary operations. This is also true for cleaning and deburring; the relatively small size and complex shape of the parts also require special procedures and/or precautions that are not required for wrought or cast parts. The P/M parts shown in Fig. 8 are typical of small, intricate parts that frequently present deburring problems.

Deburring Methods. Due to the nature of the P/M process, burrs typically form on the edges and surfaces of P/M parts. In many simple shapes, deburring is almost automatic—that is, burrs are broken off during handling operations. If parts are carbonitrided or steam treated, subsequent deburring may be unnecessary. However, for intricate parts such as those shown in Fig. 8, separate deburring operations generally are required.

The use of liquid deburring methods is not usually suitable, especially if such liquids are corrosive. Thus, acid pickling is not recommended, because acid may be entrapped in the pores, resulting in severe corrosion. Tumbling in a wet medium is used frequently as a deburring method, but removal of the liquid from the pores requires extra drying time. Preferred methods of deburring include:

- Rotary tumbling (self or with abrasive)
- Vibratory
- Abrasive blasting
- Centrifugal or high-energy methods

These methods, which are used for deburring and sometimes for cleaning, are described in detail in Volume 5 of the 9th edition of *Metals Handbook*. The discussions that follow are unique to P/M parts.

Rotary Tumbling. Self-tumbling, tumbling with dry abrasive, and tumbling with abrasive in a liquid medium are suitable for deburring of P/M parts. Wet tumbling is not suitable for deburring P/M parts because of the difficulty of removing the tumbling liquid from the pores of the parts.

During self-tumbling, the workpieces are tumbled in a revolving barrel. This method provides an economical and efficient means of deburring, but is effective only on relatively simple parts. For parts such as those shown in Fig. 8, the internal surfaces and recesses are not completely deburred by this method. For more complete deburring, an abrasive is added. Size of the abrasive is important. At least a portion of the added abrasive should have a mesh size that is smaller than the smallest hole or recess in the workpiece; otherwise, not all surfaces will be reached.

Over-tumbling of P/M parts must be avoided, because it peens the surfaces and may partially close pores (not necessarily desirable). Over-tumbling also may damage gear teeth or other protrusions by removing too much metal or by excessive peening. Tumbling cycles should be based on the minimum time that will provide acceptable deburring.

Vibratory processing is similar to rotary tumbling in principle. However, the shaking involved in the vibratory method is faster and provides more uniform results compared to rotary tumbling. As in rotary tumbling, care must be taken to prevent over-tumbling.

Abrasive blasting, in which various materials are propelled by air or centrifugal force, offers another method of deburring. For practical reasons, it is used less frequently than tumbling or vibrating. The abrasive must be selected carefully. Coarse shot or grit tends to peen the surfaces and close the pores. Also, abrasive blasting can "hammer" bits of abrasive into the workpiece, thus "charging" it.

Another disadvantage of conventional abrasive blasting is that, especially for large volumes of small workpieces, results are likely to be nonuniform. One type of blasting machine, which tumbles and blasts simultaneously, has been used successfully for deburring of P/M parts. Silica sand or a milder abrasive is suitable and is less likely to damage intricate workpieces. As with other deburring methods, over-processing must be avoided.

Fig. 8 Typical shapes of P/M parts that present deburring or cleaning difficulties

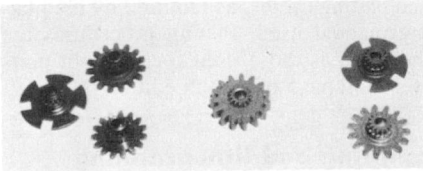

Centrifugal or High-Energy Methods. Centrifugal finishing combines rotating action with high centrifugal force, which results in a more severe abrading action than can be obtained by conventional rotary tumbling. This action is obtained by revolving several rotating barrels around the periphery of a large carrier disk.

As a result, the action within one barrel consists of a combination of rotating motion and high centrifugal forces, which provides pressures up to 25 times the weight of the abrasive medium (if used) and the workpiece. As the disk rotates in one direction, the barrels rotate at a faster speed in the opposite direction. This counter rotation causes a high-pressure, sliding movement within the entire mass, thus accomplishing the desired results in a shorter time compared to other abrading processes. An advantage of this process is that it drives the abrasive into relatively

inaccessible areas where burr removal may present problems.

Cleaning

Some of the deburring methods discussed above also may be considered as methods of cleaning. Frequently, however, methods such as tumbling and blasting are considered as preliminary cleaning operations to be followed by a more thorough cleaning, especially if the parts are to be coated.

Cleaning Methods. The inherent porosity in P/M parts imposes restrictions on selection of cleaning method. The use of a cleaning solution that is corrosive to the metal being cleaned is not recommended, because even the most thorough washing is not likely to remove all of the fluid, which presents a corrosion problem. Acid pickling is therefore not recommended.

Because of porosity, thorough cleaning of P/M parts is more difficult than their wrought counterparts; P/M parts require more attention than is provided in many conventional cleaning systems. Preferred methods are hot caustic washing, ultrasonic degreasing, and electrolytic alkaline cleaning.

Ultrasonic Degreasing. Oils, greases, and other shop soil may be removed by vapor degreasing techniques such as vapor phase, vapor-spray-vapor, warm liquid-vapor, or boiling liquid-warm liquid-vapor techniques. These procedures are described in detail in Volume 5 of the 9th edition of *Metals Handbook*. For most P/M parts, especially if the degree of soiling is severe and/or part density is low, the boiling liquid-warm liquid-vapor process is preferred. This technique should be used in conjunction with an ultrasonic transducer, which literally shakes all entrapped contaminants out of the pores, resulting in a thorough and safe method of cleaning.

Electrolytic Alkaline Cleaning. Ferrous P/M parts can be cleaned, deoxidized, and stripped of nonmetallic coatings by subjecting them to electrolysis in a strongly alkaline aqueous solution. In this method the base metal is not attacked, and the possibility of rusting is minimal. Typical alkaline solutions are comprised of a:

- Source of caustic to aid cleaning
- Chelating agent to detach scale or rust
- Complexing agent to hold relatively large amounts of iron in solution

Solution concentration is generally 0.18 to 0.24 kg (1.5 to 2 lb) of caustic and 0.06 to 0.12 kg (0.5 to 1 lb) of cyanide per litre (gallon) of water. Processing time may vary

from 5 to 30 min, depending on type and degree of soil. Temperature should be maintained at 21 to 49 °C (70 to 120 °F) at a current density of 270 to 1075 A/m^2 (25 to 100 A/ft^2) of part surface. Best results are obtained by periodically reversing the current during the processing cycle.

Electrolytic alkaline cleaning is described in Volume 5 of the 9th edition of *Metals Handbook*. This cleaning process is well suited to cleaning of P/M parts, because the electrolytic action provides additional energy required to dislodge contaminants from pores or from relatively inaccessible areas.

Coating

In addition to the surfaces provided by steam treatment, P/M parts are frequently coated by mechanical means, painting, or electroplating. The blue-black oxide-covered surface produced by exposure to steam is often the final finish for a variety of hardware items. It may be also used as a preliminary coating for a final finishing process, such as painting.

Regardless of the method used for coating, major emphasis must be placed on initial cleaning. If liquid contaminants are allowed to remain in the pores of parts, bleeding occurs, and defective coatings result. Steam treatment provides an excellent paint base.

Mechanical coating uses kinetic energy to deposit metallic coatings on parts. This process is also known as mechanical plating, or peen plating, when the coating is less than 25 μm (1 mil) thick. Coating is accomplished by placing the workpiece, glass beads, water, and the metal plating powder in a tumbling barrel.

Zinc is most commonly used as a plating material, although a wide range of metals and mixtures of metals can be mechanically plated on ferrous metal parts. For example, a mixture of 75% Zn and 25% Sn is commonly used. Metal powders are added to the mixtures to be tumbled. Complete details on the mechanical plating process may be found in Volume 5 of the 9th edition of *Metals Handbook*.

Powder metallurgy parts with densities not less than 83% can be mechanically plated without special considerations for porosity. When density drops below 83%, tests should be conducted to determine whether moisture is entrapped, which is detrimental to the finished parts. Generally, when density is below 83%, parts must be impregnated with wax or resin.

Painting. Usually, P/M parts are ideal candidates for coating by painting; the porosity enhances paint adhesion. Furthermore, P/M parts can be painted by spraying, dipping, or the contact transfer method. Air-drying types are suitable only for indoor protective coatings. Baking produces finishes of higher quality that are well suited for outdoor exposure.

Spray painting has several advantages over dipping, including improved control of dimensions and coating quality, and the ability to coat localized areas. However, higher labor costs and more paint loss from overspray are associated with spraying.

For spraying, baking types of alkyds are reduced with solvents to a spraying viscosity of 35 s through a No. 4 Ford Cup. Parts are sprayed and then air dried for 10 min, after which they may be baked for 30 min at a temperature compatible with the type of paint being used. This practice results in a dry film coating 38 to 46 μm (1.5 to 1.8 mils) thick.

In paint dipping, the parts to be coated are placed in baskets or on racks, immersed in the paint, and then allowed to drain. Dipping saves labor and paint, compared to spraying, but general quality of dipped parts is lower, notably because of edge buildup.

Roll painting and lithographing (transfer coating) is a process in which paint is applied to external surfaces of cylindrical P/M parts, followed by the application of lithographing ink. This process is done with special machinery (see Fig. 9). Typically parts are roll coated and oven baked. The initial coating is usually a background color. Numbers or characters, as required, are then roll coated over the background coating, followed by baking dry.

Advantages of this painting procedure include:

● Precision painting can be achieved without applying paint to areas that do not require a coating.
● With proper design of parts, areas can be painted with sharply defined edges.
● Coating thickness can be closely controlled by varying the number of revolutions the part is permitted to make.
● An unlimited number of character forms can be applied at relatively low cost.

Principal limitations of the process are:

● Special handling is required.
● Internal surfaces are not coated.

Types of paint used and general procedures for painting of P/M parts are similar to those used for wrought counterparts, all of which are covered in detail in Volume 5 of the 9th edition of *Metals Handbook*.

Electroplating. Powder metallurgy parts can be electroplated with various metals like their wrought counterparts. Methods used for plating of cast parts generally can be used for very dense parts (95% or more of theoretical). For parts of lower densities, special preparation procedures are required.

During plating of P/M parts, the pores act as thermal pumps. Plating solutions are released from or absorbed by the pores, depending on the temperature differential between the workpiece and the solution. Interconnecting pores entrap solutions, which are then released slowly. Part density should be known before the sequence of cleaning and plating operations begins, and suitable precautions should be taken to prevent solution entrapment.

Entrapped solutions not only cause spotty plating and staining, which may develop within days, but also can cause contamination and depletion of all solutions used in the production process. It is therefore necessary that, if part density is below about 95%, pores must be closed before coating by electroplating.

Methods of closing pores that have proved successful include burnishing, buffing, rolling, heat treating, steam treating, and impregnation. All of these methods, except impregnation, provide varying degrees of closure. Mechanical methods are often excluded because of dimensional tolerances. Consequently, impregnation is the most suitable approach to closing pores.

Impregnation of iron compacts with metals such as copper is common practice and completely solves the porosity problem for subsequent electroplating. However, the cost of metal impregnation (frequently done to achieve pressure tightness) usually cannot be justified only to ensure satisfactory electroplating.

Impregnation with plastic seals P/M parts for further processing, such as electroplating. Pressure tightness and frequently an improvement in machining characteristics are added benefits derived from plastic impregnation. The process is not unlike the plastic impregnation process used to attain pressure tightness in porous castings. Optimum results are obtainable with various types of plastic sealants, although the most commonly used are polyester types 19V5M and 95-145 and anaerobic type PMS-10E.

A typical processing cycle consists of:

● Cleaning thoroughly
● Baking at 120 to 150 °C (250 to 300 °F) to drive off all moisture or solvent
● Applying sealant under vacuum, such as in an autoclave
● Removing excess sealant by means of an emulsion cleaner
● Curing at 120 to 150 °C (200 to 250 °F)
● Tumbling, polishing, or abrasive blasting to remove excess cured sealant

Parts are now ready for routine cleaning and plating cycles, as required by the plating method used. Plating procedures are the same as those used for wrought parts (see Volume 5 of the 9th edition of *Metals*

Fig. 9 Special machine used for roll painting and lithographing

Handbook for details of the various plating procedures).

Machining

Parts made by P/M techniques generally require machining methods that differ from those used for wrought or cast parts of similar composition because of the inherent porosity of P/M parts. See Volume 3 of the 8th edition of *Metals Handbook* for an explanation of the machining methods discussed in this section.

The structure of P/M parts causes machining problems; pores may be closed by smearing of the metal surface. Cutting fluids also may cause difficulties by entering the pores, and parts may become charged with abrasives when ground, honed, or lapped.

Preserving Porosity. Many parts, such as bearing components, are made by P/M techniques, because the desired degree of porosity can be obtained and controlled. Because such parts either have been or will be impregnated with a lubricant, minimum machining is preferred because of the danger of closing the pores. Even if pores are only partly closed, it is difficult to obtain adequate impregnation after machining. Furthermore, if parts are impregnated before machining, and if pores are closed to a significant extent, lubricant will not be released as required for service.

Under conditions in which porosity must be maintained, the following procedure is recommended:

● Review the manufacturing sequence; the principal objective is to minimize the number of machining operations.
● Use the sharpest possible cutting tools.
● Ascertain that all cutting tools have a good surface finish and are free from the "saw-tooth" effect.
● Use light cuts—never more than 0.38 mm (0.015 in.).

Cutting fluids are required less frequently in the machining of P/M parts compared to similar wrought parts, because their porosity automatically renders them free-machining. If a specific machining operation indicates the need for a cutting fluid, a stream of low-pressure air usually suffices to cool the tools and remove the chips.

Liquid cutting fluids should not be used for machining of P/M parts that have been or will be impregnated with lubricant. If lubricant is present before machining, it will be contaminated by the cutting fluid. If cutting liquid is used on nonimpregnated parts, the pores will absorb the cutting fluid, and unless thoroughly removed, impregnation will be impaired.

When porosity is incidental or minimal and no impregnated lubricant is required for the end use, a cutting liquid can be used if required. However, subsequent processing must be considered. When machining is followed by any type of coating process, the pores must be free from any contamination.

If a cutting fluid is used, it must be compatible with the metal being machined to ensure against corrosion. Under some conditions, a light mineral oil is used, partly as an aid to machining and partly to prevent corrosion during subsequent operations.

Turning or boring frequently is used for finishing P/M parts to make projections, re-entrant angles, and tapers and to obtain accuracy that is not possible by pressing. For turning or boring of parts in which surface porosity must be maintained, carbide or diamond tools should be used—most commonly carbide. After grinding, the tool faces should be honed or polished to obtain the sharpest possible tool edges. Saw-tooth edges can be avoided on tools by advancing the grinder perpendicular to the tool face, not at an angle. All sharpened edges of turning or boring tools should be examined at a magnification of 25 to 40 diameters to ensure that smooth, sharp edges have been obtained.

Some development work is usually required to determine optimum tool shape and feed. The design shown in Fig. 10 is representative of carbide tools used for turning or boring. The design shown in Fig. 11 has proved satisfactory for boring porous bronze bearings in high production.

Optimum feeds are influenced by the shape of the tool. For instance, when a nose radius of 1.6 mm ($1/16$ in.) is used, a speed of 0.25 to 0.40 mm/rev (0.010 to 0.015 in./rev) is suitable. If the nose radius is decreased to 0.25 mm (0.010 in.), feed should be between 0.075 and 0.130 mm/rev (0.003 and 0.005 in./rev) to obtain clean-cut surfaces. Nose radii between 0.25 and 1.6 mm (0.010 and $1/16$ in.) require feeds between the above ranges. Feeds at 0.13 to 0.38 mm/rev (0.005 to 0.015 in./rev) are usually selected when preservation of porosity is critical. For roughing, or when a porous surface is not required, feeds of up to 0.75 mm/rev (0.030 in./rev) can be used.

Speeds of 55 to 90 m/min (175 to 300 sfm) are frequently used for turning or boring with sharp carbide tools. If maintenance of porosity is not critical, speeds up to 150 m/min (500 sfm) can be used.

Planing and Shaping. A carbide or high-speed tool steel, (usually carbide), ground as shown in Fig. 10, illustrates the

configuration for planing or shaping of P/M parts at the maximum speeds available on most planers and shapers. If automatic lifting of the tool on the return stroke is not available, the tool must be lifted manually to prevent marring of the finish of the workpiece.

Planer and shaper feeds typically are too coarse to provide good cutting finishes on P/M parts. However, it may be possible to improve the surface finish by changing the end cutting edge angle of the tool (Fig. 10) from 5 to 2° so that the heel barely clears the work and feed marks are removed.

Drilling. Speeds of 20 m/min (70 sfm) for high-speed steel drills and 60 m/min (200 sfm) for carbide drills produce satisfactory holes in P/M parts. A low right-hand helix angle, or sometimes a left-handed helix angle, prevent the drills from "digging in." Cutting edges should be "dubbed," or ground, to reduce axial rake sufficiently to prevent digging into the work.

Mechanical feed should be used wherever possible and should be selected to produce the desired finish, if the drill is used for the final sizing operation. This is particularly true for larger drills. Satisfactory feeds for various drill sizes are:

Drill diameter		Feed	
mm	in.	mm/rev	in./rev
3.2-6.4	$1/8$-$1/4$	0.5	0.002
8.0-12.8	$5/16$-$1/2$	0.10	0.004
14.3-19.0	$9/16$-$3/4$	0.15	0.006
20.6-25.4	$13/16$-1	0.25	0.010

Reaming. To control bore accuracy in P/M parts, reaming is sometimes used instead of pin sizing, ball sizing, or burnishing. Standard reamers generally are satisfactory, although left-handed reamers also have proved successful. Cutting edges should have the best possible finish to minimize edge buildup, which results in oversized holes. The drilled hole should have a reaming allowance that is based on hole size:

Hole diameter		Allowance	
mm	in.	mm	in.
6.4 or less	$1/4$ or less	0.05	0.002
6.4-12.8	$1/4$-$1/2$	0.05-0.10	0.002-0.004
12.8-25.4	$1/2$-1	0.10-0.15	0.004-0.006

When possible, reamers should be operated in floating holders and run at 7.6 to 12.2 m/min (25 to 40 sfm). Recommended feeds are:

Hole diameter		Feed rate	
mm	in.	mm/rev	in./rev
Up to 6.4	Up to $1/4$	0.13	0.005
8.0-12.8	$5/16$-$1/2$	0.18	0.007
14.3-19.0	$9/16$-$3/4$	0.25	0.010

Fig. 10 Tool for turning, boring, shaping, or planing of P/M parts

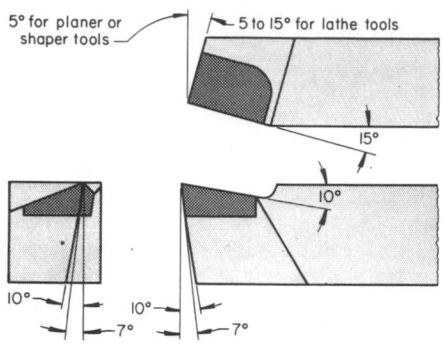

Fig. 11 Tool for boring porous bronze bearings in high production

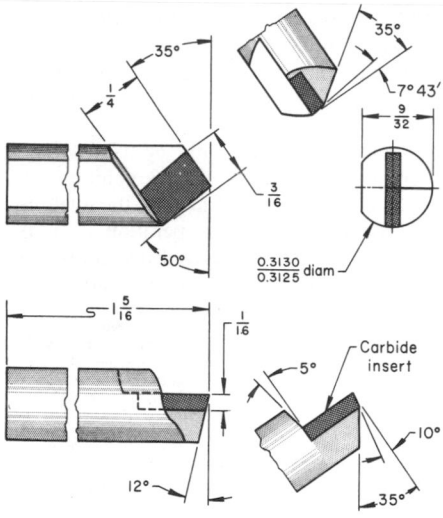

Fig. 12 Ball broach for burnishing bores in P/M parts

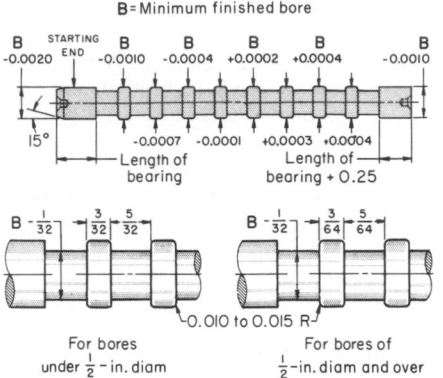

Fig. 13 Tool for high-speed burnishing of holes in P/M parts

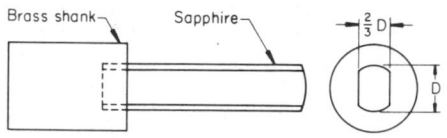

Burnishing. When the clearance between a shaft and a P/M bearing is ±0.13 mm (±0.0005 in.) or less, burnishing of the bearing bores after installation in the housing is the preferred method of sizing. No more than 0.05 mm (0.002 in.) per 25 mm (1 in.) of diameter should be displaced. The smallest possible amount of displacement that will produce the true diameter is desirable. Details of the type of burnishing tool recommended for this purpose are presented in Fig. 12.

A method for burnishing 12.7-mm (¹/₂-in.) diam holes or smaller in P/M parts is shown in Fig. 13. This method utilizes a sapphire tool that can be operated in a high-speed lathe or drill press. A drop of lubricating oil serves as a fluid, and the

workpiece is manually pushed onto the rotating tool.

Tapping. For routine tapping of P/M parts, conventional tap drill charts should be followed to maintain a 65 to 75% depth of thread. Two-flute taps are recommended for bore diameters up to 8.0 mm (⁵/₁₆ in.). Three-flute taps are preferred for tapping of bore diameters within 8.0 to 12.7 mm (⁵/₁₆ to ¹/₂ in.). Spiral-point taps are desirable, because they throw the chips out of, instead of driving them into, the pores of the workpiece. Frequently, trial and error is required in tapping of P/M parts to determine which type is best suited for a specific composition and/or density.

Milling is done infrequently on P/M parts, although in some applications it is the preferred method of providing slots or undercuts that would be difficult or impossible to form by pressing. For best results, carbide-tipped and helical-tooth cutters should be used. Cutters must be kept sharp and ground with minimum corner radii, or preferably with beveled corners.

For general-purpose milling with carbide cutters, speeds of 75 to 90 m/min (250 to 300 sfm) are recommended. When high-speed steel cutters are used, speeds should not exceed 20 m/min (70 sfm). Feeds of

0.25 to 0.38 mm/tooth (0.010 to 0.015 in./tooth) for rough milling and 0.05 to 0.13 mm/tooth (0.002 to 0.005 in./tooth) for finish milling are recommended.

If wear lands as much as 0.4 mm (¹/₆₄ in.) wide appear on the back of the cutter teeth, or if wear becomes visible on the cutting edges, cutters should be replaced or sharpened. Dull cutters smear workpiece surfaces.

Grinding operations are commonly required to maintain precise dimensions of P/M parts. Grinding may cause parts to lose surface porosity if parts become charged with abrasive particles. This problem increases as density of the parts decreases.

Heat treated high-density iron parts can be ground successfully with aluminum oxide wheels that have a medium grit size, porous structure, and a vitrified bond. Operating speeds of 535 to 580 m/min (1750 to 1900 sfm) with about 0.05 mm (0.002 in.) per pass are satisfactory. Grinding fluids used for similar wrought parts are generally satisfactory for grinding of high-density P/M parts.

Parts that have been copper-infiltrated can be ground successfully, but the copper or copper alloy causes the grinding wheels to load. Wheel loading can be minimized by using coarse-grit, open-structure types of wheels.

In special applications that require grinding of porous parts, thorough cleaning is essential; ultrasonic methods are recommended. Surfaces of P/M parts that have become severely smeared can be opened by acid etching. However, if the acid is not thoroughly removed, serious corrosion results. Therefore, acid etching is seldom advisable.

Honing and Lapping. Holes in P/M parts that require extreme accuracy can be honed or lapped by normal techniques, if retention of porosity is not critical. However, size control in P/M parts usually can be obtained more economically by reaming or burnishing.

High-density ferrous metal parts, especially those that have been hardened, may be honed and/or lapped successfully with conventional procedures.

Honing of parts infiltrated with nonferrous metals is seldom practical, because the honing stones become loaded. Neither honing or lapping is recommended for porous parts, because these processes cause pores to become charged with abrasive particles.

Mechanical Properties of P/M Materials

By Leander F. Pease III
President
Powder-Tech Associates, Inc.
and
Vernon C. Potter
Chief Metallurgist
Merriman Inc.
Division of Litton Industrial Products

MECHANICAL PROPERTIES of P/M structural materials depend on the composition, density, and heat treatability of the material, as well as processing and design considerations. Final density, which is expressed in g/cm^3 as well as percent of theoretical density—the ratio of the density of a P/M material to that of its wrought counterpart—has the greatest effect on properties of P/M materials. Powder metallurgy parts with theoretical densities less than 75% are considered to be low density, those above 90% are high density, and those in between are classified as medium density. Generally, structural parts have densities ranging from 80% to above 95%. Oil-impregnated bearings have densities of approximately 75%; densities of filter parts can be as low as 50%.

This article reviews mechanical properties of a wide range of ferrous and nonferrous P/M materials. The data reflect the influence of composition, processing method, and heat treatment on properties of P/M structural materials. For additional information on mechanical properties, see the "Powder Systems and Applications" Section in this Volume.

Medium- to High-Density Ferrous Materials

Ferrous materials of medium to high density are produced by blending iron powder with copper, carbon in the form of graphite, and nickel. These materials are pressed and sintered at about 1120 °C

(2050 °F) for 20 to 30 min. Graphite dissolves rapidly to form steel. Copper melts and partially diffuses into the iron particles. Nickel diffuses to a lesser extent, producing nickel-rich islands. To achieve densities above approximately 7.1 g/cm^3, materials frequently are pressed, presintered at 845 °C (1550 °F), re-pressed to a high density of 7.2 to 7.6 g/cm^3, and final sintered.

Many medium- to high-density ferrous materials have been standardized by the American Society for Testing and Materials (ASTM), the Society of Automotive Engineers (SAE), and the Metal Powder Industries Federation (MPIF). As shown in Table 1, the ASTM and SAE designations use the term "class" to indicate the carbon content of the finished ferrous P/M part, and the term "grade" to differentiate between related alloys bearing the same designation. The term "type" is used to indicate density of the finished part, as shown in Table 2.

The MPIF designations for ferrous P/M materials, described in detail in Ref 1, include a prefix of one or more letters (the first of which is "F" to indicate an iron-based material), four numerals, and a single-letter suffix. The second letter in the prefix identifies the principal alloying element (if one is specified); the percentage of that element is indicated by the first two digits. The third and fourth digits indicate the amount of carbon in the compacted and sintered part; the code designation "00" indicates less than 0.3%, "05" indicates 0.3 to 0.6%, and "08" indicates 0.6 to

1.0%. The suffix is used to indicate the density of the compacted and sintered part, as described in Table 2.

Table 3 provides typical mechanical properties for ferrous P/M materials. Values given are not minimums. However, MPIF standard 35 is being revised to include minimum yield strengths for as-sintered material and minimum ultimate tensile strengths for heat treated material.

As shown in Table 3, mechanical properties generally increase with increases in density. Figure 1 illustrates the effects of density on the mechanical properties of as-sintered nickel steels. The data of Fig. 1 apply only when electrolytically produced iron-based powder is used and the processing includes presintering and re-pressing of all materials above 6.6 g/cm^3. Other iron powders and processing methods produce different results. Under the revised MPIF system based on specified minimum tensile properties, suppliers may select suitable powders, density, and processing to meet minimum properties.

Young's Modulus, Poisson's ratio, and linear coefficients of thermal expansion are determined primarily from the density of the low-alloy ferrous parts, under average sintering conditions. Table 4 illustrates the effects of density on elastic modulus, Poisson's ratio, and coefficient of thermal expansion.

Compressive strength and apparent hardness of sintered ferrous materials can be increased by steam treating at 480 to 595 °C (900 to 1100 °F) for 1 to 4 h, as

Table 1 Typical compositions of ferrous P/M structural materials

Material	Designation(a) MPIF	Designation(a) ASTM	Designation(a) SAE	MPIF composition limits and ranges(b), % C	Ni	Cu	Fe
P/M iron	F-0000	B 310, class A	853, class 1	0.3 max	...	...	97.7-100
P/M steel	F-0005	B 310, class B	853, class 2	0.3-0.6	...	...	97.4-99.7
	F-0008	B 310, class C	853, class 3	0.6-1.0	...	...	97.0-99.1
P/M copper iron	FC-0200	...	...	0.3 max	...	1.5-3.9	93.8-98.5
P/M copper steel	FC-0205	...	...	0.3-0.6	...	1.5-3.9	93.5-98.2
	FC-0208	B 426, grade 1	864, grade 1, class 3	0.6-1.0	...	1.5-3.9	93.1-97.9
	FC-0505	...	...	0.3-0.6	...	4.0-6.0	91.4-95.7
	FC-0508	B 426, grade 2	864, grade 2, class 3	0.6-1.0	...	4.0-6.0	91.0-95.4
	FC-0808	B 426, grade 3	864, grade 3, class 3	0.6-1.0	...	6.0-11.0	86.0-93.4
	...	B 426, grade 4	864, grade 4, class 3	0.6-0.9	...	18.0-22.0	75.1 min
P/M iron-copper	FC-1000	B 222; B 439, grade 3	862	0.3 max	...	9.5-10.5	87.2-90.5
P/M iron-nickel	FN-0200	B 484, grade 1, class A	...	0.3 max	1.0-3.0	2.5 max	92.2-99.0
P/M nickel steel	FN-0205	B 484, grade 1, class B	...	0.3-0.6	1.0-3.0	2.5 max	91.9-98.7
	FN-0208	B 484, grade 1, class C	...	0.6-0.9	1.0-3.0	2.5 max	91.6-98.4
P/M iron-nickel	FN-0400	B 484, grade 2, class A	...	0.3 max	3.0-5.5	2.0 max	90.2-97.0
P/M nickel steel	FN-0405	B 484, grade 2, class B	...	0.3-0.6	3.0-5.5	2.0 max	89.9-96.7
	FN-0408	B 484, grade 2, class C	...	0.6-0.9	3.0-5.5	2.0 max	89.6-96.4
P/M iron-nickel	FN-0700	B 484, grade 3, class A	...	0.3 max	6.0-8.0	2.0 max	87.7-94.0
P/M nickel steel	FN-0705	B 484, grade 3, class B	...	0.3-0.6	6.0-8.0	2.0 max	87.4-93.7
	FN-0708	B 484, grade 3, class C	...	0.6-0.9	6.0-8.0	2.0 max	87.1-93.4
P/M infiltrated steel	FX-1005	...	...	0.3-0.6	...	8.0-14.9	80.5-91.7
	FX-1008	...	...	0.6-1.0	...	8.0-14.9	80.1-91.4
	FX-2000	B 303, class A	870	0.3 max	...	15.0-25.0	70.7-85.0
	FX-2005	B 303, class B	...	0.3-0.6	...	15.0-25.0	70.4-84.7
	FX-2008	B 303, class C	872	0.6-1.0	...	15.0-25.0	70.0-84.4

(a) Designations listed are nearest comparable designations; ranges and limits may vary slightly between comparable designations. (b) MPIF standards require that the total amount of all other elements be less than 2.0%, except that the total amount of other elements must be less than 4.0% in infiltrated steels.

Table 2 Typical density designations and ranges of ferrous P/M materials

Density of pure iron is 7.87 g/cm^3.

Designation MPIF density suffix	ASTM type(a)	SAE type	Density, g/cm^3
N	I	1(b)	.... Less than 6.0
P	II	2	.... 6.0 to 6.4
R	III	3	.... 6.4 to 6.8
S	IV	4	.... 6.8 to 7.2
T	V(c)	5(c)	.... 7.2 to 7.6
U	...	...	.... 7.6 to 8.0

(a) ASTM B 426 only; different density ranges used in ASTM B 310 and B 484. (b) Density range of 5.6 to 6.0 g/cm^3 is specified. (c) Minimum density of 7.2 g/cm^3 is specified.

shown in Table 5. During steam treating, surface and some subsurface pores become filled with black iron oxide, thus making the parts pressure tight to 690 kPa (100 psi) with increased corrosion resistance. Iron oxide, however, may lower impact resistance, although abrasive wear resistance is increased.

Medium-Density Stainless Steels

Medium-density stainless steels usually are produced by pressing at 414 to 827 MPa (30 to 60 tsi) and sintering at 1120 to 1150 °C (2050 to 2100 °F) in dissociated ammonia, hydrogen, or vacuum. Typical mechanical properties of medium-density stainless steels, selected from MPIF standard 35, are given in Table 6. Superior elongation, impact resistance, and corrosion resistance are obtained by sintering at 1260 to 1315 °C (2300 to 2400 °F). The Metal Powder Industries Federation currently is standardizing materials based on high-temperature sintering.

Low- to Medium-Density Copper-Based Materials

Bronzes generally are made by pressing a mixture of copper and tin powders at 138 to 207 MPa (10 to 15 tsi) and sintering at 815 to 845 °C (1500 to 1550 °F) for 10 to 20 min. The resultant microstructure is predominantly alpha bronze, with little or no copper-tin intermetallic compounds. The important properties to consider are radial crushing strength and pore volume (percent), for holding oil in bearing applications. For additional information, see the article "P/M Bearings" in this Volume. Brasses and nickel silvers are pressed and sintered prealloyed powders. Table 7 lists typical mechanical properties of selected copper-based P/M materials.

P/M Forged Low-Alloy Steels

Hot forming and P/M forging processes are suitable for many types of powders. Powders are preformed into medium-density shapes. The compacts are sintered, reheated, and restruck while hot to achieve near-theoretical density. Hot forming consists of re-pressing the preform without substantial lateral flow. If a 10 to 50% increase in area results through lateral flow, the process is termed "hot forging."

Final mechanical properties of forged parts do not depend solely on density and heat treatment; dynamic properties (impact and fatigue resistance, for example) depend on powder cleanliness and low final oxygen content. Composition of the starting powder, alloying elements, and method of sintering significantly affect the final oxygen content. High-temperature sintering at 1205 to 1315 °C (2200 to 2400 °F) in a reducing atmosphere can produce low oxygen contents (100 to 300 ppm) and acceptable properties.

Lateral flow may break up powder oxide films and may partially substitute for highly reductive sintering. Table 8 illustrates the effects of processing variables on final properties of selected low-alloy steels. Fatigue ratio (endurance limit/ultimate tensile strength) typically is 0.3 to 0.4 for as-forged material and 0.4 to 0.5 for heat treated materials.

Nearly Dense and Fully Dense Stainless Steels

High-density stainless steels can be produced by P/M forging, but high-temper-

Table 3 Typical mechanical properties of ferrous P/M materials

Designation	MPIF density suf-fix(a)	Condi-tion(b)	Tensile strength MPa	ksi	Yield strength MPa	ksi	Elongation in 25 mm (1 in.), %	Fatigue strength MPa	ksi	Impact energy(c) J	ft·lb	Apparent hardness	Elastic modulus GPa	10⁶ psi
F-0000	N	AS	110	16	75	11	2.0	40	6(d)	4.1	3.0	10 HRH	70	10.5
	P	AS	130	19	95	14	2.5	50	7(d)	6.1	4.5	70 HRH	90	13
	R	AS	165	24	110	16	5	60	9(d)	13	9.5	80 HRH	110	16
	S	AS	205	30	150	22	9	80	11(d)	20	15	15 HRB	130	19
	T	AS	275	40	180	26	15	105	15(d)	34	25	30 HRB	160	23
F-0005	N	AS	125	18	105	15	1.0	45	7(d)	3.4	2.5	5 HRB	70	10.5
	P	AS	170	25	140	20	1.5	65	10(d)	4.7	3.5	20 HRB	90	13
	R	AS	220	32	160	23	2.5	85	12(d)	6.8	5.0	45 HRB	110	16
		HT	415	60	395	57	0.5	155	23(d)	···	····	100 HRB	110	16
	S	AS	295	43	195	28	3.5	110	16(d)	12	9.0	60 HRB	130	19
		HT	550	80	515	75	0.5	210	30(d)	···	···	25 HRC	130	19
F-0008	N	AS	200	29	170	25	0.5	75	11(d)	2.7	2.0	35 HRB	70	10.5
		HT	290	42	···	···	<0.5	110	16(d)	···	···	90 HRB	70	10.5
	P	AS	240	35	205	30	1.0	90	13(d)	4.1	3.0	50 HRB	90	13
		HT	400	58	···	···	<0.5	150	22(d)	···	···	100 HRB	90	13
	R	AS	290	42	250	36	1.5	110	14(d)	4.7	3.5	65 HRB	110	16
		HT	510	74	···	···	<0.5	195	28(d)	···	···	25 HRC	110	16
	S	AS	395	57	275	40	2.5	150	22(d)	9.5	7.0	75 HRB	130	19
		HT	650	94	625	91	<0.5	245	36(d)	···	···	30 HRC	130	19
FC-0200	P	AS	160	23	115	17	2.5	60	9(d)	7.5	5.5	80 HRH	90	13
	R	AS	205	30	145	21	4	80	11(d)	9.5	7.0	15 HRB	110	16
	S	AS	255	37	160	23	7	95	14(d)	23	17	30 HRB	130	19
FC-0205	P	AS	275	40	235	34	1.0	105	15(d)	4.7	3.5	35 HRB	90	13
	R	AS	345	50	260	38	1.5	130	19(d)	7.5	5.5	60 HRB	110	16
		HT	585	85	560	81	<0.5	220	31(d)	···	···	30 HRC	110	16
	S	AS	425	62	310	45	3.0	160	24(d)	13	9.5	75 HRB	130	19
		HT	690	100	655	95	<0.5	260	38(d)	···	···	35 HRC	130	19
FC-0208	N	AS	225	33	205	30	<0.5	85	13(d)	3.4	2.5	45 HRB	70	10.5
		HT	295	43	···	···	<0.5	110	16(d)	···	···	95 HRB	70	10.5
	P	AS	310	45	280	41	<0.5	115	17(d)	4.1	3.0	50 HRB	90	13
		HT	380	55	···	···	<0.5	145	21(d)	···	···	25 HRC	90	13
	R	AS	415	60	330	48	1.0	155	23(d)	6.8	5.0	70 HRB	110	16
		IIT	550	80	···	···	<0.5	210	30(d)	···	···	35 HRC	110	16
	S	AS	550	80	395	57	1.5	210	30(d)	11	8.0	80 HRB	130	19
		HT	690	100	655	95	<0.5	260	38(d)	···	···	40 HRC	130	19
FC-0505	N	AS	240	35	205	30	0.5	90	13(d)	4.1	3.0	50 HRB	70	10.5
	P	AS	345	50	290	42	1.0	130	19(d)	6.1	4.5	60 HRB	90	13
	R	AS	455	66	380	55	1.5	170	25(d)	6.8	5.0	75 HRB	116	16
FC-0508	N	AS	330	48	295	43	<0.5	125	18(d)	4.1	3.0	60 HRB	70	10.5
	P	AS	425	62	395	57	1.0	160	24(d)	4.7	3.5	65 HRB	90	13
		HT	480	70	480	70	<0.5	185	27(d)	···	···	30 HRC	90	13
	R	AS	515	75	480	70	1.0	195	29(d)	6.1	4.5	85 HRB	116	16
FC-0808	N	AS	250	36	···	···	<0.5	···	···	···	···	55 HRB	···	···
FC-1000	N	AS	205	30	···	···	0.5	···	···	···	···	70 HRF	···	···
FN-0200	R	AS	195	28	125	18	4	75	11	19	14	38 HRB	115	17
	S	AS	260	38	170	25	7	105	15	43	32	42 HRB	145	21
	T	AS	310	45	205	30	11	125	18	68	50	51 HRB	160	23
FN-0205	R	AS	255	37	160	23	3.0	105	15	14	10	50 HRB	115	17
		HT	565	82	450	65	0.5	225	33	8.1	6	32 HRC	115	17
	S	SS	345	50	215	31	3.5	140	20	24	18	70 HRB	145	21
		HT	760	110	605	88	1.0	305	44	22	16	42 HRC	145	21
	T	SS	420	61	255	37	4.5	165	24	43	32	85 HRB	160	23
		HT	925	134	725	105	2.0	370	54	38	28	46 HRC	160	23
FN-0208	R	AS	330	48	205	30	2.0	130	19	11	8	62 HRB	115	17
		HT	690	100	650	94	0.5	275	40	8.1	6	34 HRC	115	17
	S	AS	450	65	280	41	3.0	180	26	19	14	79 HRB	145	21
		HT	930	135	880	128	0.5	370	54	16	12	45 HRC	145	21
	T	AS	545	79	345	50	3.5	220	32	30	22	87 HRB	160	23
		HT	1105	160	1070	155	0.5	415	60	24	18	47 HRC	160	23
FN-0400	R	AS	250	36	150	22	5	95	14	22	16	40 HRB	115	17
	S	AS	340	49	205	30	6	140	20	47	35	60 HRB	145	21
	T	AS	400	58	250	36	6.5	160	23	68	50	67 HRB	160	23
FN-0405	R	AS	310	45	180	26	3.0	125	18	14	10	63 HRB	115	17
		HT	770	112	650	94	0.5	310	45	8.1	6	27 HRC	115	17
	S	AS	425	62	240	35	4.5	165	24	20	15	72 HRB	145	21
		HT	1060	154	880	128	1.0	415	60	14	10	39 HRC	145	21
	T	AS	510	74	295	43	6.0	205	30	41	30	80 HRB	160	23
		HT	1240	180	1060	154	1.5	450	65	19	14	44 HRC	160	23

(continued)

(a) For density range, see Table 2. (b) AS, as sintered; SS, sintered and sized; HT, heat treated, typically austenitized at 870 °C (1600 °F), oil quenched and tempered 1 h at 200 °C (400 °F). (c) Unnotched Charpy test. (d) Estimated as 38% of tensile strength. (e) X indicates infiltrated steel; see Table 1.
Source: Ref 1.

Table 3 (continued)

Designation	MPIF density suffix(a)	Condition(b)	Tensile strength MPa	ksi	Yield strength MPa	ksi	Elongation in 25 mm (1 in.), %	Fatigue strength MPa	ksi	Impact energy(c) J	ft·lb	Apparent hardness	Elastic modulus GPa	10⁶ psi
FN-0408	R	AS	395	57	290	42	1.5	160	23	8.1	6	72 HRB	115	17
	S	AS	530	77	390	57	3.0	215	31	14	10	88 HRB	145	21
	T	AS	640	93	470	68	4.5	255	37	22	16	95 HRB	160	23
FN-0700	R	AS	560	52	205	30	2.5	145	21	16	12	60 HRB	115	17
	S	AS	490	71	275	40	4	195	28	28	21	72 HRB	145	21
	T	AS	585	85	330	48	6	240	34	35	26	83 HRB	160	23
FN-0705	R	AS	370	54	240	35	2.0	150	22	12	9	69 HRB	115	17
		HT	705	102	550	80	0.5	280	41	11	8	24 HRC	115	17
	S	AS	525	76	330	48	3.5	205	30	23	17	83 HRB	145	21
		HT	965	140	760	110	1.0	385	56	20	15	38 HRC	145	21
	T	AS	620	90	390	57	5.0	250	36	33	24	90 HRB	160	23
		HT	1160	168	895	130	1.5	500	65	27	20	40 HRC	160	23
FN-0708(e)	R	AS	395	57	280	41	1.5	160	23	8	6	75 HRB	115	17
	S	AS	550	80	380	55	2.5	220	32	16	12	88 HRB	145	21
	T	AS	655	95	455	66	3.0	260	38	22	16	96 HRB	160	23
FX-1005(e)	T	AS	570	83	440	64	4.0	...	...	19	14	75 HRB	135	20
		HT	830	120	740	107	1.0	...	...	9.5	7.0	35 HRC	135	20
FX-1008(e)	T	AS	620	90	515	75	2.5	...	...	16	12	80 HRB	135	20
		HT	895	130	725	105	60.5	...	...	9.5	7.0	40 HRC	135	20
FX-2000(e)	T	AS	450	65	...	...	1.0	...	...	20	15	60 HRB	...	...
FX-2005(e)	T	AS	515	75	345	50	1.5	...	...	12.9	9.5	75 HRB	125	18
		HT	790	115	655	95	<0.5	...	...	8.1	6.0	30 HRC	125	18
FX-2008(e)	T	AS	585	85	515	75	1.0	...	...	14	10	80 HRB	125	18
		HT	860	125	740	107	<0.5	...	...	6.8	5.0	42 HRC	125	18

(a) For density range, see Table 2. (b) AS, as sintered; SS, sintered and sized; HT, heat treated, typically austenitized at 870 °C (1600 °F), oil quenched and tempered 1 h at 200 °C (400 °F). (c) Unnotched Charpy test. (d) Estimated as 38% of tensile strength. (e) X indicates infiltrated steel; see Table 1.
Source: Ref 1.

Table 4 Effects of density on elastic modulus, Poisson's ratio, and coefficient of thermal expansion of P/M steels

MPIF density suffix(a)	Density, g/cm³	Elastic modulus GPa	10⁶ psi	Poisson's ratio	Coefficient of thermal expansion, 10⁻⁶/K
N	5.6-6.0	72	10.5	0.18	8.1
P	6.0-6.4	90	13	0.20	8.7
R	6.4-6.8	110	16	0.21	9.2
S	6.8-7.2	130	19	0.23	9.8
T	7.2-7.6	160	23	0.26	10.4
Theoretical	7.86	205	30	0.28	11-12

(a) See Table 2 for density designations.

Table 5 Effects of steam treating on density and apparent hardness of ferrous P/M materials

MPIF designation	MPIF density suffix(a)	Density, g/cm³ Sintered	Steam treated	Apparent hardness Sintered	Steam treated
F-0000	N	5.8	6.2	7 HRF	75 HRB
	P	6.2	6.4	32 HRF	61 HRB
	R	6.5	6.6	45 HRF	51 HRB
F-0008	M	5.8	6.1	44 HRB	100 HRB
	P	6.2	6.4	58 HRB	98 HRB
	R	6.5	6.6	60 HRB	97 HRB
FC-0700	N	5.7	6.0	14 HRB	73 HRB
	P	6.35	6.5	49 HRB	78 HRB
	R	6.6	6.6	58 HRB	77 HRB
FC-0708	N	5.7	6.0	52 HRB	97 HRB
	P	6.3	6.4	72 HRB	94 HRB
	R	6.6	6.6	79 HRB	93 HRB

(a) For density range, see Table 2.

ature sintering is the preferred fabrication method. At high temperatures (1315 to 1425 °C, or 2400 to 2600 °F), austenitic and ferritic stainless steels densify during sintering. With minor variations in composition, near-theoretical density can be achieved in pressed and sintered parts. Table 9 provides typical mechanical property data for nearly dense (>98% of theoretical density) stainless steels.

An alternative process for forming stainless steel into mill shapes, such as tubing, is to begin with clean, gas-atomized powder. Powder is packed in cans, sealed, cold isostatically pressed to reduce bulk, and extruded into tubing. The cold isostatically pressed cans may be hot isostatically pressed to full density, thus eliminating extrusion. Table 10 gives mechanical property data for extruded and hot isostatically pressed materials compared to wrought (ingot metallurgy) materials.

Injection Molded Materials

Finely divided (1- to 10-μm) powders can be mixed with organic binder and injection molded like plastics. After binder removal, metal preforms can be sintered to near-full density. Mass transport mechanisms during sintering are enhanced by the high surface area of the powder and the elevated temperatures of 1150 to 1315 °C (2100 to 2400 °F). Densification is accompanied by more than 10% linear shrinkage. High final densities, often 95 to 99%, result in improved dynamic properties compared to conventionally pressed and sintered materials. Injection molding allows greater flexibility in part design. Currently, parts less than 75 mm (3 in.) in diameter are being produced by injec-

Fig. 1 Effects of density on mechanical properties of as-sintered 4% nickel steel (FN-04XX)

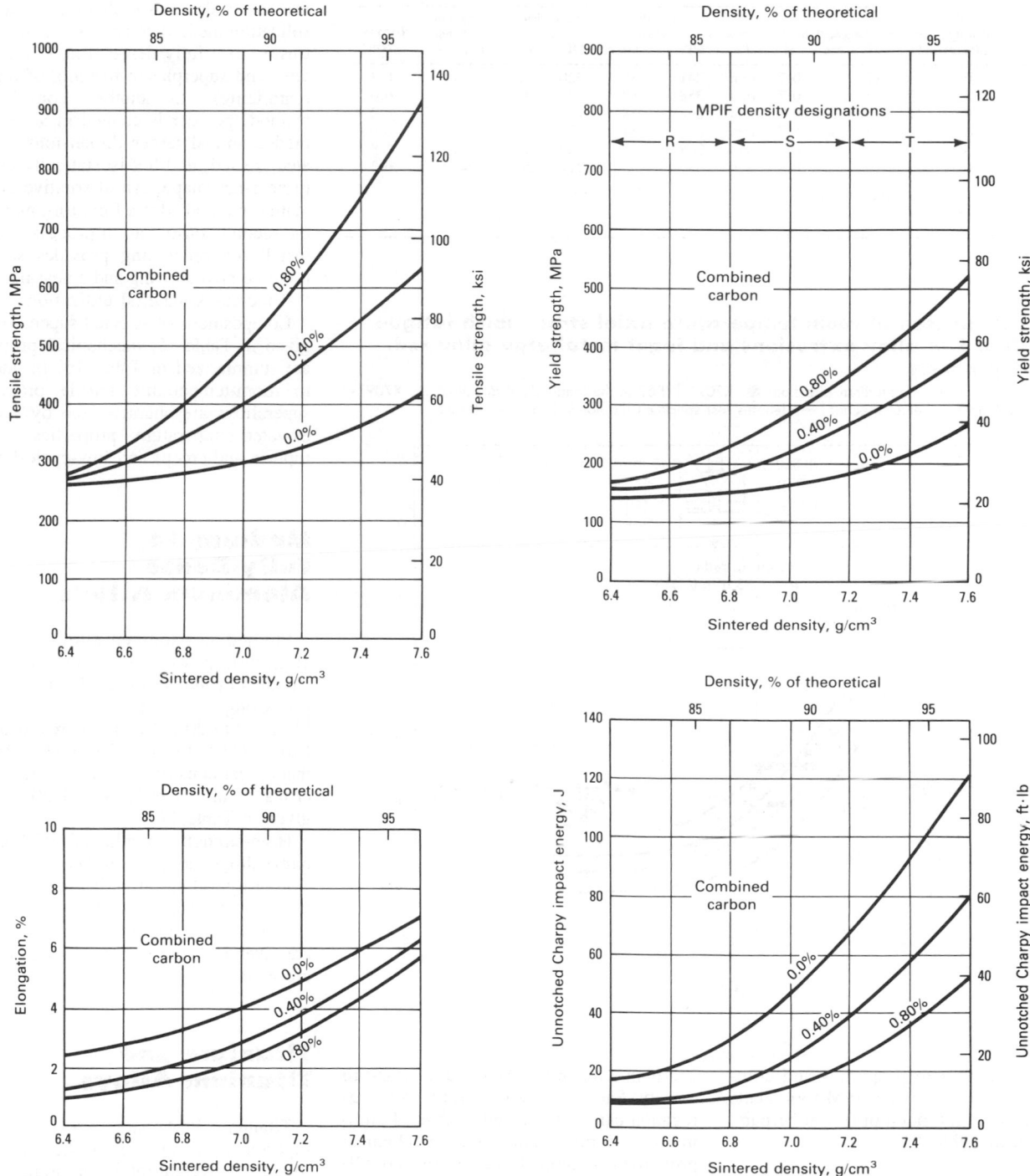

tion molding techniques. Table 11 summarizes properties for four different alloys.

Tool Steels

Fully dense P/M tool steels are made either from hot isostatically pressed gas-atomized tool steel powder or from cold compacted and vacuum sintered water-atomized tool steel powder. In both cases, the resulting billets have a fine, uniform carbide distribution with no segregation. This results in a material with excellent grindability that is, a high ratio of metal volume removed to volume of wheel worn. These materials also have improved hot workability and exhibit less distortion during heat treating than ingot metallurgy counterparts. Special grades can easily be produced, with higher volume fractions of carbide than are available by conventional processing.

Table 12 lists mechanical properties of tool steels made from water-atomized

Table 6 Typical mechanical properties of medium-density P/M stainless steels

MPIF designation	MPIF density suffix(a)	Composition, %					Tensile strength		0.2% yield strength		Elongation in 25 mm (1 in.), %	Density, g/cm³
		Cr	Ni	Mo	Si	Fe	MPa	ksi	MPa	ksi		
SS-303 ....P		17	12	···	0.7	rem	241	35	220	32	1	6.2
	R	17	12	···	0.7	rem	358	52	324	47	2	6.6
SS-316 ....P		16	13	2	0.7	rem	262	38	220	32	2	6.2
	R	16	13	2	0.7	rem	372	54	275	40	4	6.6
SS-410 ....N		12	···	···	0.8	rem	289	42	283	41	<1	5.8
	P	12	···	···	0.8	rem	379	55	372	54	<1	6.2

Note: All materials sintered in dissociated ammonia.
(a) For density range, see Table 2.
Source: Ref 1

Fig. 2 Comparison of room-temperature axial stress notch fatigue strength of P/M alloy extrusions and ingot metallurgy alloy rod, bar and products

○, X7090-T7E71 in the longitudinal direction; ●, X7091-T7E69 in the longitudinal direction; △, X7091-T7E69 in the long transverse direction; → denotes test specimen did not fail. Source: Ref 25

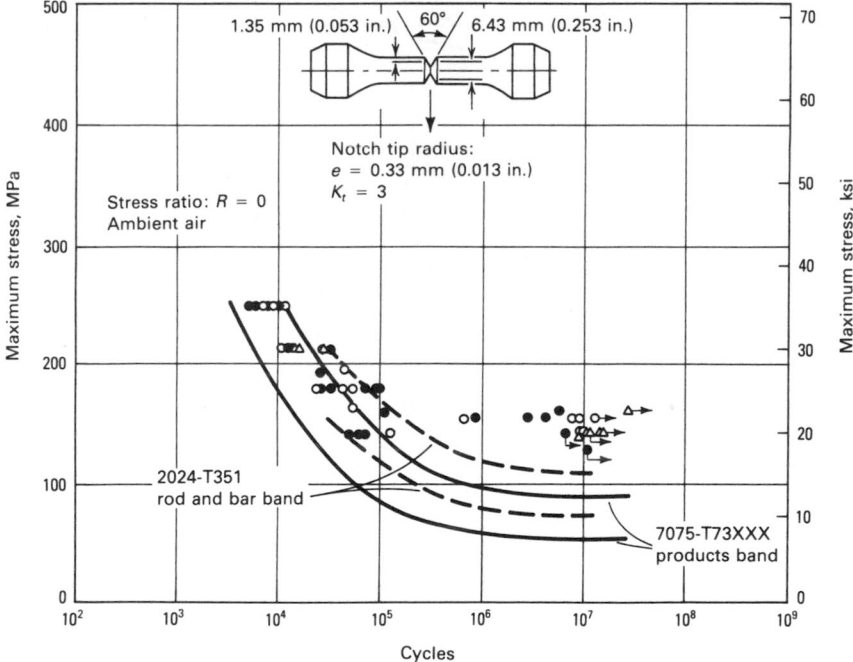

powder. Mechanical properties and compositions of gas-atomized P/M tool steels compared to ingot metallurgy counterparts are given in Table 13.

Nickel- and Cobalt-Based Alloys

Most small- to medium-size parts made from these wear- and corrosion-resistant alloys traditionally have been castings. Prealloyed powders are now used as the starting material. Powders are cold pressed and high-temperature sintered to near-full density. Parts that are free from alloy segregation can be produced. Table 14 summarizes compositions and mechanical properties of typical nickel- and cobalt-based P/M alloys.

Superalloys

These highly alloyed materials are produced from powders to prevent alloy segregation and to provide better forgeability (freedom from cracking). Several consolidation methods may be used, but all require the use of very clean powder that is free of oxides and inclusions. One consolidation method consists of canning, extrusion to a fully dense, fine grain structure, and superplastic forming of the part (gatorizing). In another consolidation method, powder is canned in a vessel of predetermined larger dimensions, evacuated, sealed, and hot isostatically pressed to near-net shape. An alternative consolidation method adds a final forging step to the second consolidation process. Powder metallurgy processing provides segregation-free high alloying and a near-net shape that increases material utilization.

Compositions of selected superalloys are given in Table 15; mechanical properties are summarized in Table 16. In addition to room-temperature tensile properties, superalloys are characterized by superior elevated-temperature properties (stress-rupture and creep) and low-cycle fatigue.

Medium- to Fully-Dense Aluminum Alloys

Pressed and sintered alloys consist of aluminum powder admixed with small amounts of magnesium, silicon, and copper. Compacts are pressed at 138 to 276 MPa (10 to 20 tsi) and sintered at 595 to 620 °C (1100 to 1150 °F) to 90 to 95% of theoretical density. The mechanical properties of alloys 601AB and 201AB are given in Table 17.

High-strength, prealloyed P/M aluminum alloys are consolidated by extrusion or by rolling into mill shapes. Room-temperature properties are given in Table 17, and elevated-temperature properties are given in Table 18. Typical fatigue data are given in Fig. 2 to 4.

Titanium and Titanium Alloys

Titanium P/M parts are favored for superior corrosion resistance and high strength-to-weight ratio (for aircraft applications). Selection of the P/M process depends on the method of powder production (blended elemental or prealloyed) and desired final density. Medium- to high-density parts with close tolerances are made by pressing and sintering. Starting powder is usually sponge fines with master alloy additions of vanadium and aluminum to make Ti-6Al-4V. These are referred to

Fig. 3 Comparison of axial stress notch fatigue strength of P/M alloy X7091-T7E69 die forgings and ingot metallurgy alloy 7175-T736 die forgings

○, longitudinal direction, one lot; ●, short transverse direction, two lots; → denotes test specimen did not fail in number of cycles indicated. Stress ratio: R = 0.1. Source: Ref 25

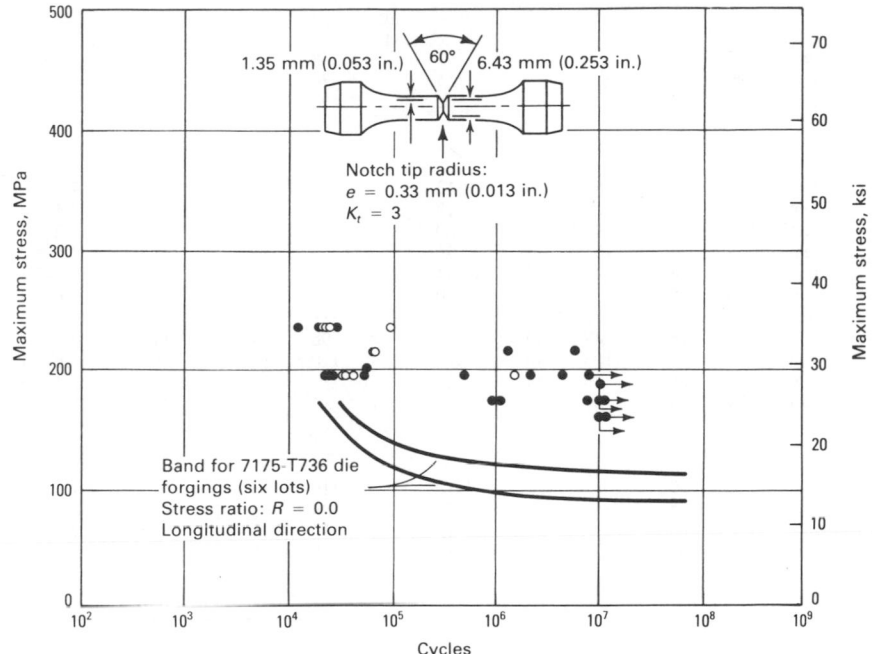

Fig. 4 Rotating beam fatigue strength for die forgings of P/M alloy X7091-T7E76 and ingot metallurgy alloys 7075-T7352 and 2014-T61

For P/M X7091-T7E76: ○, smooth, transverse direction; ●, notched, transverse direction; → denotes test specimen did not fail in number of cycles indicated. Source: Ref 25

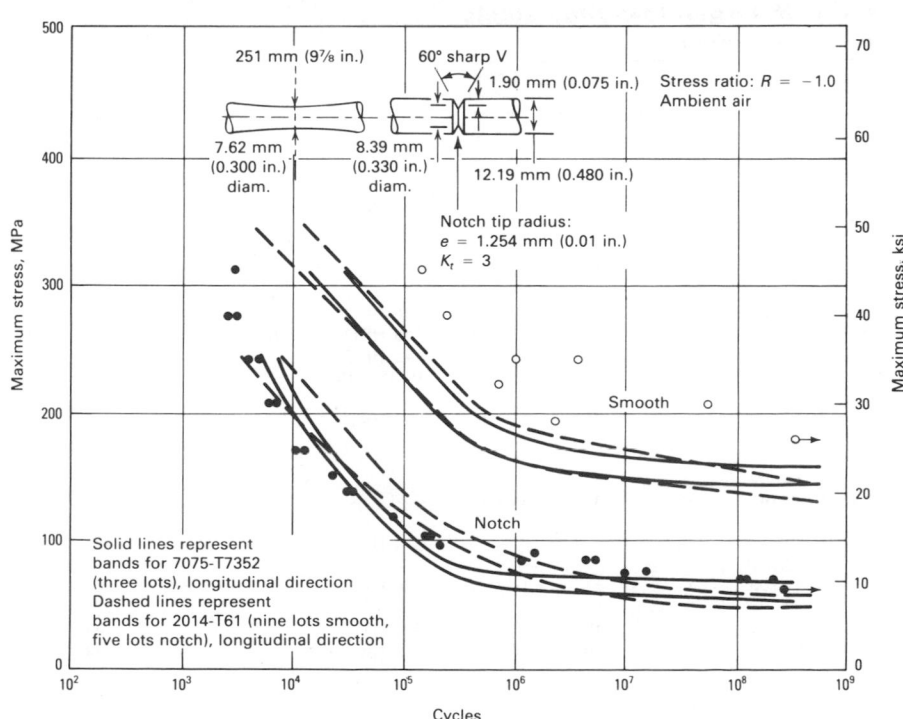

as blended elemental powders. Sintering causes the vanadium and aluminum to diffuse into the titanium. Table 19 compares properties of wrought and P/M titanium materials. For aircraft applications, which require fully dense parts, the starting powder should be a high-purity, prealloyed material. Such powder is often made by the plasma rotating electrode process. These powders are canned in oversized containers and hot isostatically pressed to near-net shape, resulting in full density with reduced material utilization and machining.

Cemented Carbides

Mixtures of tungsten carbide, cobalt binder, and paraffin are cold pressed to produce cemented carbides. Sintering above the melting point of the cobalt results in liquid-phase sintering and shrinkage to near-full density. Other carbides, such as titanium carbide and tantalum carbide, frequently are admixed to achieve special properties. Table 20 gives hardness, transverse-rupture strength, compressive yield, and fracture toughness values for various chemical compositions of cemented carbides.

Tungsten Alloys

Tungsten heavy metals are made from elemental blends of tungsten, nickel, iron, copper, and molybdenum. These mixtures contain less than 10% alloy additions and are sintered in a liquid phase formed by the additions. Parts shrink to near-theoretical density, resulting in very dense materials (17 to 18 g/cm³). High density makes these materials suitable for use in gyroscopes, in armament penetrators, and as counter weights. Typical mechanical properties are given in Table 21.

Refractory Metals and Alloys

Refractory metals and alloys of molybdenum, rhenium, and tantalum generally are made from metal powders. These powders are cold isostatically pressed into bar forms that are subsequently heated to the sintering temperature (2200 to 2700 °C, or 3990 to 4890 °F) by direct current passage at high current densities. Sintering takes place in hydrogen or vacuum. Following sintering, bars are swaged, drawn, or rolled into mill shapes. Powder metallurgy techniques are used for these high-melting point materials to prevent problems encountered in conventional casting techniques. Tables 22 through 24 give typical mechanical property data for selected refractory metals and alloys.

Table 7 Typical mechanical properties of copper-based P/M materials

MPIF designation	MPIF density suffix(a)	Composition, %					Tensile strength		0.2% yield strength		Elongation in 25 mm (1 in.), %	0.1% compressive yield strength		Apparent hardness, HRH	Density, g/cm³
		Sn	Zn	Ni	Pb	Cu	MPa	ksi	MPa	ksi		MPa	ksi		
Bronzes															
CT-0010	N	10	...	...	...	rem	55	8	...	...	1	48	7	...	5.6-6.0
	R	10	...	...	...	rem	96	14	...	...	1	76	11	...	6.4-6.8
	S	10	...	...	...	rem	124	18	...	...	2.5	121	17.5	...	6.8-7.2
Brasses															
CZ-0010	T	...	10	...	...	rem	138	20	62	9	13	...	...	57	7.2-7.6
	U	...	10	...	...	rem	186	27	69	10	18	...	...	70	7.6-8.0
CZ-0030	T	...	30	...	...	rem	214	31	89	13	20	...	...	76	7.2-7.6
	U	...	30	...	...	rem	255	37	103	15	26	...	...	85	7.6-8.0
CZP-0210	T	...	10	...	2	rem	124	18	48	7	14	...	...	46	7.2-7.6
	U	...	10	...	2	rem	176	25.5	55	8	20	...	...	60	7.6-8.0
CZP-0220	T	...	20	...	2	rem	165	24	76	11	13	...	...	55	7.2-7.6
	U	...	20	...	2	rem	193	28	89	13	19	96	14	68	7.6-8.0
	W	...	20	...	2	rem	221	32	103	15	23	110	16	75	8.0-8.4
CZP-0230	T	...	30	...	2	rem	193	28	76	11	22	...	...	65	7.2-7.6
	U	...	30	...	2	rem	234	34	89	13	27	...	...	76	7.6-8.0
Nickel silvers															
CZN-1818	U	...	18	18	...	rem	206	30	...	...	10	110	16	75	7.6-8.0
	W	...	18	18	...	rem	255	37	...	...	12	124	18	85	8.0-8.4
CZNP-1818	U	...	18	18	1.5	rem	206	30	...	...	10	110	16	75	7.6-8.0
	W	...	18	18	1.5	rem	241	35	...	...	12	117	17	85	8.0-8.4

(a) For density range, see Table 2.
Source: Ref 1

Table 8 Typical mechanical properties of P/M forged low-alloy steels

All materials are in the hardened and tempered condition unless otherwise indicated

Material	Processing	Ultimate tensile strength		0.2% yield strength		Elongation in 25 mm (1 in.), %	Reduction in area, %	Charpy V-notch impact energy		Hardness	Fracture toughness (K_{Ic})		Density, % of theoretical	Ref
		MPa	ksi	MPa	ksi			J	ft·lb		MPa $\sqrt{m}$	ksi $\sqrt{in.}$		
Fe-2MCM-0.67C(a)(b)	...	960	139.3	590	86	...	12	...	...	98 HRB	...	...	...	2
Fe-2MCM-0.67C(a)	...	1900	275.6	1500	218	...	4.5	...	...	49 HRC	...	...	...	...
4120	Sintered at 1315 °C (2400 °F), re-pressed	701	101.7	616	89.4	14	46	38	28	20-25 HRC	...	...	100	3
1520	Sintered at 1315 °C (2400 °F), re-pressed	936	135.7	...	...	9	13	39	29	20-25 HRC	...	...	100	...
4130	Gas atomized, −65 mesh	1586	230	1303	189	5	3	10	7.5	46 HRC	49	45	100	4
4640	Gas atomized, −65 mesh	...	...	...	...	...	...	7	5	55 HRC	36	33	100	4
	Water atomized	...	...	...	...	...	...	7	5	42 HRC	37	34	100	4
	Sintered at 1200 °C (2190 °F)	1040	150.8	1000	145	20	40	36	26	310-350 HV	...	...	99	6
Fe-2Ni-0.35C	Mixed elemental powders	938	136	600	87	13	44	...	13	31 HRC	...	...	99	5
Fe-0.55Ni-0.32Mo-0.47Mn-0.23Cr-0.30C	Sintered at 1200 °C (2190 °F)	1020	147.9	970	141	17	37	46	34	...	...	...	...	6
Fe-3Cu-0.5C-0.3S	...	873	127	...	...	6.5	...	...	...	274 HV	...	...	99	7
Fe-9Cu-0.34Mn-0.43Ni-0.65Mo-0.31C	...	1675	245	1410	205	13	31	19	14	49 HRC	...	...	99	8
Fe-0.35Mn-0.57Mo-1.95Ni-0.5C	...	1200	174	1120	162	10	19	30	22	475 HV	...	...	99	9
4630 modified	Sintered at 1205 °C (2200 °F)	148	215	1331	193	6	10	8	6	42 HRC	...	...	98	10

(a) MCM is a master alloy containing 20% Mn, 20% Cr, 20% Mo, and 7% C. (b) As-sintered condition

Table 9 Typical mechanical properties of nearly dense P/M stainless steel

Based on high-temperature sintering

Alloy	Condition	Ultimate tensile strength MPa	ksi	0.2% yield strength MPa	ksi	Elongation in 25 mm (1 in.), %	Hardness	Impact strength J	ft·lb	Density, g/cm³	Theoretical density, g/cm³
Ultimet 04, 304	Sintered	593	86	248	36	36	80 HRB	10.8(a)	8(a)	7.8	7.9
Ultimet 16, 316	Sintered	687	99.6	308	44.7	26	94 HRB	8.1(a)	6(a)	7.7	7.8
	Solution treated and quenched	684	99.3	329	47.7	45	90 HRB	5.4(a)	40(a)	7.7	7.8
Ultimet 40C, 440C	Sintered	...	...	...	...	...	20-30 HRC	2.7(b)	2(b)	7.6	7.7
	Hardened and tempered	...	...	...	...	...	50-60 HRC	2.7(b)	2(b)	7.6	7.7

(a) Charpy V-notch. (b) Unnotched. Source: Ref 11

Table 10 Typical mechanical properties of fully dense stainless steel

Property	P/M material	Wrought material
Extruded 0.3- by 15.5-mm (0.1- by 0.61-in.) 317LM tube(a)		
Ultimate tensile strength, MPa (ksi)	693 (100)	693 (100)
0.2% yield strength, MPa (ksi)	324 (47)	353 (51)
Reduction in area, %	71	73
Elongation in 25 mm (1 in.), %	47	50
Hot isostatically pressed type 316		
Ultimate tensile strength, MPa (ksi)	579 (84)	...
0.2% yield strength, MPa (ksi)	288 (42)	...
Elongation in 25 mm (1 in.), %	58	...

(a) Gas-atomized powder, canned, cold isostatically pressed, and extruded. Source: Ref 12

Table 11 Typical as-sintered mechanical properties of injection molded P/M materials

Material	Density, g/cm³	Tensile strength MPa	ksi	0.2% yield strength MPa	ksi	Elongation in 25 mm (1 in.), %
Fe-2Ni	7.7	380	55	241	35	25
316L	...	517	75	345	50	18
17-4 PH	7.63	1028	149	966	140	12
IN-100	...	1083	157	904	131	12

Source: Ref 13

Table 12 Mechanical properties of tool steels made from cold compacted and vacuum sintered water-atomized powders

Grade	Hardness, HRC	Ultimate tensile strength(a) MPa	ksi	Elongation(b), %	Impact strength(c) J	ft·lb	Ref
M2	62-65	750-2000	109-290	12-14	9-12	7-9	14, 15
M35	63-66	770-2000	112-290	6-9	8-11	6-8	14, 15
T15	64-67	770-2000	112-290	3-6	8-11	6-8	14, 15

(a) Values depend on heat treatment. Lowest values are for fully annealed condition. (b) Fully annealed. (c) Triple temper. Izod unnotched impact bar (ASTM E 23)

Table 13 Comparison of mechanical properties and compositions of gas atomized and hot isostatically pressed P/M tool steels and ingot metallurgy tool steels

Alloy	Hardness, HRC	Bend fracture stress MPa	ksi	Bend deflection at fracture mm	in.	Charpy V-notch impact strength(a) J	ft·lb	Fracture toughness MPa√m	ksi√in.	0.2% compressive yield strength MPa	ksi	Ref
Ingot metallurgy alloys												
M2	64-65	3819	554	...	...	23	17	...	...	...	...	16
M4	64-65	3585	520	...	...	16	12	...	...	...	...	16
M42	67-68	2565	372	...	...	7	5	...	...	...	...	16
T15	66-67	2151	312	...	...	5	4	...	...	...	...	16
D2	62	2068	300	...	...	23	17	...	...	...	...	16
P/M alloys												
M2 CPM	64-65	4991	724	...	...	41	30	...	...	...	...	16
M4 CPM	64-65	5377	780	...	...	43	32	...	...	...	...	16
M42 CPM	67-68	4005	581	...	...	16	12	...	...	...	...	16
T15 CPM	66-67	4674	678	...	...	...	14	...	...	...	...	16
CPM Rex 20	67-68	4005	581	...	...	19	12.5	...	...	...	...	16
CPM Rex 25 (M61)	66-67	4323	627	...	...	15	11	...	...	...	...	16
CPM Rex 76 (M48)	69	4088	593	...	...	14	10	...	...	...	...	16
CPM 10V	63	4240	615	...	...	23	17	...	...	...	...	16
ASP 23	66	4800	696	...	...	...	...	13	12	3500	508	17
ASP 23	62	...	...	...	...	...	...	19	17	2800	406	17
ASP 30	66	5100	740	2.1	0.083	...	...	...	...	3600	522	18
ASP 60	67	4600	667	...	...	...	...	...	...	...	...	19

(continued)

Table 13 (continued)

AISI designation	Composition, wt%							
	C	Mn	Si	Cr	V	W	Mo	Co
M2	0.85	0.30	0.30	4.0	2.0	6.0	5.0	...
M4	1.30	0.30	0.30	4.0	4.0	5.5	4.5	...
M42	1.10	0.30	0.30	3.75	1.15	1.5	9.5	8.0
CPM Rex 20	1.30	0.30	0.30	3.75	2.0	6.25	10.5	...
T15	1.55	0.30	0.30	4.0	5.0	12.25	...	5.0
CPM Rex 25	1.80	0.30	0.35	4.0	5.0	12.5	6.5	...
CPM Rex 76	1.50	0.30	0.30	3.75	3.1	10.0	5.25	9.0
CPM 10V	2.45	0.50	0.90	5.25	9.75	...	1.30	...
D2	1.55	0.35	0.45	11.5	0.9	...	0.8	...
ASP 23	1.3	...	...	4.2	3.1	6.4	5.0	...
ASP 30	1.3	...	...	4.2	3.1	6.4	5.0	8.5
ASP 60	2.3	...	...	4.0	6.5	6.5	7.0	10.5

(a) 12.7-mm (1/2-in.) radius notch
Source: Ref 16, 17

Table 14 Typical mechanical properties and compositions of nickel- and cobalt-based P/M alloys

Alloy designation	Room temperature			540 °C (1000 °F)			650 °C (1200 °F)			760 °C (1400 °F)		
	Ultimate tensile strength, MPa (ksi)	Elongation, %	Hardness, HRC	Ultimate tensile strength, MPa (ksi)	Elongation, %	Hardness, HRC	Ultimate tensile strength, MPa (ksi)	Elongation, %	Hardness, HRC	Ultimate tensile strength, MPa (ksi)	Elongation, %	Hardness
Stellite 3	863 (125)	<1	54	725 (105)	<1	40	690 (100)	<1	39	621 (90)	1	28 HRC
Stellite 6	897 (130)	<1	40	828 (120)	1	37	766 (111)	1	30	518 (75)	10	15 HRC
Stellite 19	1035 (150)	<1	49	... (...)	...	...	... (...)	...	...	... (...)	...	...
Stellite 31	828 (120)	4	...	676 (98)	14	...	614 (89)	13	...	... (...)	...	...
Stellite 190	621 (90)	<1	58	518 (75)	<1	54	518 (75)	<1	46	518 (75)	<1	34 HRC
Star J Metal	523 (76)	0.1	56	539 (78)	0.1	52	569 (82)	0.1	43	573 (83)	0.1	31 HRC
Stellite 98 M2	794 (115)	0.3	58	725 (105)	0.3	...	690 (100)	0.5	...	656 (95)	0.5	...
Haynes 208	690 (100)	<1	44	552 (80)	<1	41	552 (80)	<1	37	483 (70)	1	25 HRC
Haynes N-6	656 (95)	2	30	545 (79)	3	25	545 (79)	4	20	428 (62)	7	82 HRB
Haynes 711	559 (81)	<1	50	490 (71)	<1	43	490 (71)	<1	43	504 (73)	<1	27 HRC

	Composition, %											
	Ni	Si	Fe	Mn	Cr	Mo	W	C	V	B	Co	Other (total)
Stellite 3	3(a)	1(a)	3(a)	1(a)	31	...	12.5	2.4	...	1(a)	rem	1(a)
Stellite 6	3(a)	1.5(a)	3(a)	1(a)	29	1.5(a)	4.5	1.2	...	1(a)	rem	2(a)
Stellite 19	3(a)	1(a)	3(a)	1(a)	31	...	10.5	1.9	...	1(a)	rem	2(a)
Stellite 31	10.5	1(a)	2(a)	1(a)	25.5	...	7.5	0.5	...	...	rem	2(a)
Stellite 190	3(a)	1(a)	5(a)	1(a)	26	1(a)	1.4	3.1	...	1(a)	rem	2(a)
Star J Metal	3(a)	1(a)	3(a)	1(a)	32.5	...	17.5	2.5	...	1(a)	rem	2(a)
Stellite 98 M2	3.5	1(a)	5(a)	1(a)	30	0.8(a)	18.5	2	4.2	1	rem	2(a)
Haynes 208	rem	1(a)	12.5	0.75(a)	26	10	10	2.6	...	1(a)	10	2(a)
Haynes N-6	rem	1.5(a)	3(a)	1(a)	29	5.5	2	1.1	...	0.6	3	...
Haynes 711	rem	...	2.3	...	27	7	3	2.7	...	1(a)	12	2

(a) Maximum
Source: Ref 20

Table 15   Nominal compositions of P/M superalloys

Alloy	C	Cr	Mo	Fe	Co	Al	Ti	B	Nb	V	Hf	W	Zr	Ni
IN-100	0.1	10.0	3.5	1.0	14.0	4.5	5.5	0.01	···	1.0	···	···	0.05	rem
René 95	0.1	14.0	3.5	···	8.0	3.5	2.5	0.01	3.5	···	···	3.6	0.05	rem
Astroloy	0.05	15.0	5.0	···	18.0	4.0	3.5	0.03	···	···	···	···	···	rem
MERL 76	0.025	12.5	3.0	···	18.5	5.0	4.3	0.02	1.4	···	0.4	···	0.06	rem
AF-115	0.05	10.5	2.8	···	15.0	3.8	3.9	0.02	1.8	···	0.8	5.9	0.05	rem
Udimet 100	0.03	14.72	4.90	0.62	17.72	3.86	3.53	0.026	···	···	···	0.04	0.03	rem

Table 16   Typical mechanical properties of P/M superalloys

Property	René 95(a) (Ref 21)	Low-carbon Astroloy(b) (Ref 22)	Low-carbon Astroloy(c) (Ref 22)	Low-carbon Astroloy(d) (Ref 22)	IN-1000(e) (Ref 23)	MERL 76(f) (Ref 23)	Udimet 700(g) (Ref 24)	Udimet 700(a) (Ref 24)
0.2% yield strength at 210 °C (410 °F), MPa (ksi)	1257 (182)	973 (141)	928 (135)	994 (143)	1095 (159)	1188 (172)	860 (125)	1115 (162)
Ultimate tensile strength at 210 °C (410 °F), MPa (ksi)	1671 (242)	1376 (200)	1338 (194)	1359 (197)	1594 (231)	1674 (243)	1355 (197)	1515 (219)
Elongation, %	20	22	26	28	26	21	25	18.5
Reduction in area, %	20.3	23	28	32	27	22	27	18.5
Creep at 595 °C (1110 °F) at 1034 MPa (150 ksi), 100 h/% strain	0.15	···	···	···	···	···	···	···
Stress rupture at 650 °C (1200 °F) at 1034 MPa (150 ksi), service life (hours)/% elongation	29.5/5.4	···	···	···	···	···	···	···
	28.4/4.7	···	···	···	···	···	···	···
Strain-controlled low-cycle fatigue at 535 °C (1000 °F), strain/cycles to failure	0.78/26 948	···	···	···	···	···	···	···
	0.66/94 447	···	···	···	···	···	···	···
Stress rupture at 621 MPa (90 ksi) at 730 °C (1350 °F) at 151 h, % elongation/% reduction in area	···	14/17	17/22	16/21	···	···	···	···
0.2% yield strength at 705 °C (1300 °F), MPa (ksi)	···	···	···	···	1044 (151)	···	···	···
Ultimate tensile strength at 705 °C (1300 °F), MPa (ksi)	···	···	···	···	1265 (184)	···	···	···
% elongation/% reduction in area at 705 °C (1300 °F)	···	···	···	···	19/23	···	···	···
Stress rupture at 730 °C (1350 °F) at 655 MPa (95 ksi), hours to failure/% elongation	···	···	···	···	35.6/16.6	···	···	···
	···	···	···	···	25.5/11.0	···	···	···
	···	···	···	···	37.0/14.5	···	···	···
Creep at 705 °C (1300 °F) at 551 MPa (80 ksi), time for 0.1%/time for 0.2%	···	···	···	···	140.5/193.5	···	···	···
	···	···	···	···	100.0/142.0	···	···	···
	···	···	···	···	91.0/125.0	···	···	···
0.2% yield strength at 620 °C (1150 °F), MPa (ksi)	···	···	···	···	···	1136 (165)	···	···
Ultimate tensile strength at 620 °C (1150 °F), MPa (ksi)	···	···	···	···	···	1492 (216)	···	···
% elongation/% reduction in area at 620 °C (1150 °F)	···	···	···	···	···	18.5/17	···	···

(a) Hot isostatically pressed and hardened and tempered. (b) Produced by rapid omnidirectional compaction. Consolidated at 811 MPa (58.8 tsi); 0.5-s dwell in composite of copper-nickel fluid dies. Preheated to 1075 °C (1970 °F); held at temperature 1 h. Powder was electrodynamically degassed prior to vacuum filling. Post-consolidation solution treated at 1165 °C (2125 °F) for 4 h, fan air cooled. (c) Hot isostatically pressed at 1150 °C (2100 °F) at 104 MPa (15 ksi) for 4 h. Hot loaded at 925 °C (1700 °F). Hot unloaded at 980 °C (1800 °F). Post-consolidation solution treated at 1120 °C (2050 °F) for 2 h, fan air cooled. (d) Hot isostatically pressed and forged. Forging conditions: open die side upset at 1095 °C (2000 °F). Average reduction is 52%. Aging heat treatment for all processes: 650 °C (1200 °F) for 24 h, air cooled plus holding at 760 °C (1400 °F) for 8 h, air cooled. All tensile specimens tested normal to the forging direction. (e) Hot isostatically pressed and gatorized billet. (f) Hot isostatically pressed and gatorized. (g) As hot isostatically pressed

Table 17 Room-temperature mechanical properties of aluminum alloys

Alloy designation	Nominal composition	Oxygen content, %	Temper designation	Test direction	Density, % of theoretical	Tensile strength MPa	ksi	Yield strength MPa	ksi	Elongation, %	Fracture toughness (K_{Ic}) MPa $\sqrt{m}$	ksi $\sqrt{in.}$	Ref
Extrusions													
X7090	0.12 Si, 0.15 Fe, 0.6-1.3 Cu, 2-3 Mg, 7.3-8.7 Zn, 1.0-1.9 Co	0.2-0.5	T6E192	Longitudinal	100	676	98	641	93	10(a)	···	···	25
				Long transverse	100	648	94	600	87	10(a)	···	···	25
	0.12 Si, 0.15 Fe, 0.6-1.3 Cu, 2-3 Mg, 7.3-8.7 Zn, 1.0-1.9 Co	0.2-0.5	T7E71	Longitudinal	100	621	90	579	84	9(a)	30.8	28	25
				Short transverse	100	558	81	496	72	8(a)	19.8	18	25
X7091	0.12 Si, 0.15 Fe, 1.1-1.8 Cu, 2-3 Mg, 5.8-7.1 Zn, 0.2-0.6 Co	0.2-0.5	T6E192	Longitudinal	100	614	89	558	81	11(a)	···	···	25
				Long transverse	100	586	85	538	78	13(a)	···	···	25
	0.12 Si, 0.15 Fe, 1.1-1.8 Cu, 2-3 Mg, 5.8-7.1 Zn, 0.2-0.6 Co	0.2-0.5	T7E69	Longitudinal	100	593	86	545	79	11(a)	46.2	42	25
				Long transverse	100	545	79	496	72	9(a)	33.0	30	25
				Short transverse	100	524	76	455	66	9(a)	26.4	24	25
	0.12 Si, 0.15 Fe, 1.1-1.8 Cu, 2-3 Mg, 5.8-7.1 Zn, 0.2-0.6 Co	0.2-0.5	T7E70	Longitudinal	100	538	78	483	70	11	···	···	25
				Long transverse	100	510	74	462	67	10	47.3	43	25
Pressed bars													
601AB	1.0 Mg, 0.6 Si, 0.25 Cu, rem Al	···	T1	Longitudinal	91.1	110	16	48	7	6(b)	···	···	26
			T4	Longitudinal	91.1	141	20.5	97	14	5(b)	···	···	26
			T6	Longitudinal	91.1	183	26.5	176	25.5	1(b)	···	···	26
			T61	Longitudinal	···	241	35.0	237	34.4	2(b)	···	···	26
			T1	Longitudinal	93.7	121	17.5	55	8	7(b)	···	···	26
			T4	Longitudinal	93.7	148	21.5	100	14.5	5(b)	···	···	26
			T6	Longitudinal	93.7	224	32.5	214	31	2(b)	···	···	26
			T61	Longitudinal	···	252	36.5	247	35.8	2(b)	···	···	26
			T1	Longitudinal	96.0	124	18.0	59	8.5	8(b)	···	···	26
			T4	Longitudinal	96.0	152	22.0	103	15.0	5(b)	···	···	26
			T6	Longitudinal	96.0	252	36.5	241	35.0	2(b)	···	···	26
			T61	Longitudinal	···	255	37.0	248	36.2	2(b)	···	···	26
201AB	0.5 Mg, 0.8 Si, 4.4 Cu, rem Al	···	T1	Longitudinal	91.0	169	24.5	145	21.0	2(b)	···	···	26
			T4	Longitudinal	91.0	210	30.5	179	26.00	3(b)	···	···	26
			T6	Longitudinal	91.0	248	36.0	···	···	···	···	···	26
			T61	Longitudinal	···	343	49.7	339	49.2	0.5(b)	···	···	26
			T1	Longitudinal	92.9	201	29.2	170	24.6	3(b)	···	···	26
			T4	Longitudinal	92.9	245	35.6	205	29.8	3.5(b)	···	···	26
			T6	Longitudinal	92.9	322	46.8	···	···	···	···	···	26
			T61	Longitudinal	···	349	50.6	342	49.6	0.5(b)	···	···	26
			T1	Longitudinal	97.0	209	30.3	181	26.2	3.0(b)	···	···	26
			T4	Longitudinal	97.0	262	38.0	214	31.0	5.0(b)	···	···	26
			T6	Longitudinal	97.0	332	48.1	327	47.5	2.0(b)	···	···	26
			T61	Longitudinal	···	356	51.7	354	51.3	2.0(b)	···	···	26

Note: T1, as sintered; T4, solution heat treated for 30 min at 520 °C (970 °F) for 601AB or 505 °C (940 °F) for 201AB, quenched, aged 4 weeks at room temperature; T6, solutionized and quenched as T4, aged 18 h at 160 °C (320 °F); T61, re-press bars at 345 MPa (25 tsi), temper as T6
(a) Elongation in 50 mm (2 in.). (b) Elongation in 25 mm (1 in.)

Table 18 Typical mechanical properties of fully dense aluminum alloy extrusions

Alloy	Nominal composition	Oxygen content, %	Temper	Temperature °C	°F	Time at temperature, h	Tensile strength(a) MPa	ksi	Yield strength MPa	ksi	Elongation in 50 mm (2 in.), %	Ref
X7091	0.12 Si, 0.15 Fe, 1.1-1.8 Cu, 2-3 Mg, 5.8-7.1 zn, 0.2-0.6 Co	0.2-0.5	T7E69	24	75	0	579	84	545	79	11	25
				149	300	10	427	62	414	60	19	25
				149	300	100	352	51	345	50	26	25
				177	350	10	290	42	283	41	33	25
				177	350	100	200	29	186	27	37	25
				204	400	10	200	29	193	28	35	25
			T7E70	24	75	···	524	76	469	68	13	25
				149	300	10	372	54	359	52	22	25
				149	300	100	324	47	324	47	25	25
				177	350	10	283	41	276	40	31	25
				177	350	100	207	30	193	28	37	25

(continued)

Table 18 (continued)

Alloy	Nominal composition	Oxygen content, %	Temper	Temperature °C	°F	Time at temperature, h	Tensile strength(a) MPa	ksi	Yield strength MPa	ksi	Elongation in 50 mm (2 in.), %	Ref
XAP001 ......6.0 oxide, 0.2 Fe, 0.40 C, 0.08 Si		...	...	24	75	(b)	255	37	186	27	13	27
				204	400	(b)	159	23	131	19	13	27
				316	600	(b)	110	16	97	14	11	27
				427	800	(b)	76	11	69	10	4	27
				538	1000	(b)	41	6	34	5	6	27
XAP002 ......8.0 oxide, 0.28 Fe, 0.19 C, 0.10 Si		...	...	24	75	(b)	248	36	145	21	12	27
				204	400	(b)	152	22	117	17	12	27
				316	600	(b)	97	14	76	11	8	27
				427	800	(b)	62	9	55	8	3	27
				528	1000	(b)	41	6	34	5	4	27
XAP004 ......14.0 oxide, 0.29 Fe, 0.39 C, 0.10 Si		...	...	24	75	(b)	372	54	262	38	5	27
				204	400	(b)	234	34	200	29	2	27
				316	600	(b)	159	23	138	20	4	27
				427	800	(b)	103	15	97	14	1	27
				538	1000	(b)	55	8	48	7	2	27

(a) Longitudinal direction. (b) Properties are independent of time at temperature.

Table 19 Mechanical properties of P/M and wrought titanium and alloys

Alloy	Processing	Density, %	Ultimate tensile strength, MPa (ksi)	Yield strength MPa (ksi)	Elongation, %	Reduction in area, %	Elastic modulus, GPa (10^6 psi)	Fatigue limit, notched MPa (ksi)	Fracture toughness, MPa $\sqrt{m}$ (ksi $\sqrt{in.}$)	Ref
Wrought commercial purity titanium grade II	...	100	345 (50)	344 (50)	5	35	103 (14.9)	...	...	28
	...	95.5	414 (60)	324 (47)	15	14	103 (15)	...	...	28
Sponge commercial-purity P/M titanium(a)	...	94	427 (62)	338 (49)	15	23	...	...	...	29
	Forged	100	455 (66)	365 (53)	23	30	...	...	...	29
Wrought Ti-6Al-4V (AMS 4298)	...	100	896 (130)	827 (120)	10	20	114 (16.5)	427 (62)	55(e) (50)(e)	28
P/M Ti-6Al-4V	Blended elemental alloy, cold pressed	95.5	876 (127)	786 (114)	8	14	117 (17)	193 (28)	45(e) (40)(e)	28, 30
	...	98+	919 (133)	839 (121.6)	10.9	19.0	...	262 (38)	56(e) (51)(e)	30, 31
	Blended elemental alloy, forged preforms or vacuum hot pressed	99 min	937 (136)	862 (125)	12-18	15-40	116 (16.8)	414 (60)	61(e) (56)(e)	28, 30
	Hot isostatically pressed prealloy	100	947 (137.4)	868 (125.9)	18.8	43.2	117 (17)	414 (60)	...	28
	Solution treated and aged	99	1103 (160)	1013 (147)	4.9	7.6	...	...	...	32
	Rapid omnidirectional compacted(c)	100	1014 (147)	944 (137)	18.4	40.9	...	...	...	22
Plasma rotating electrode processed Ti-6Al-6V-2Sn ...	Hot isostatically pressed	100	1053 (152.7)	1008 (146.3)	18	36.5	110 (16)	448 (65)	...	33
Plasma rotating electrode processed Ti-6Al-4V	Hot isostatically pressed	100	951 (138)	910 (132)	15	39	...	414(d) (60)(d)	83(f) (76)(f)	34
P/M Ti-6Al-4V(a)	...	94	827 (120)	738 (107)	5	8	...	...	...	29
	Forged	100	920 (133.5)	841 (122)	11.5	25	...	...	...	29
P/M Ti-6Al-4V(b) ..	Hot isostatically pressed	100	917 (133)	827 (120)	13	26	...	...	...	29
P/M Ti-6Al-6V-2Sn	...	99	1067 (155)	977 (142)	10	14	...	...	...	32

(a) 0.12% oxygen. (b) 0.2% oxygen. (c) Consolidated at 811 MPa (58.8 tsi), 0.5-s dwell in low-carbon steel fluid dies. Preheat temperature was 940 °C (1725 °F), held at temperature $^3/_4$ h. Powder was vacuum filled into fluid dies following cold static outgassing for 24 h. (d) K_t = 3. (e) K_c. (f) K_{Ic}.

Table 20   Typical mechanical properties of cemented carbides

Tungsten carbide	Cobalt	Composition, % Titanium carbide	Tantalum carbide	Hardness, HRA	Transverse rupture strength(a) MPa	ksi	Ultimate compressive strength(b) MPa	ksi	Fracture toughness(c) MPa $\sqrt{m}$	ksi $\sqrt{in.}$
97	3	0	0	92.9	1827	265	5860	850	···	···
94	6	0	0	92.8	2103	305	5929	860	···	···
94	6	0	0	92.0	2378	345	5446	790	10	9.2
94	6	0	0	91.0	2516	365	5170	750	12.5	11.5
91	9	0	0	89.5	2654	385	4550	660	14.17	13.0
90	10	0	0	89.0	2895	420	4343	630	···	···
87	13	0	0	88.2	3102	450	4136	600	15.8	14.5
84	16	0	0	86.8	3102	450	3860	560	17.2	15.8
75	25	0	0	84.0	2758	400	3102	450	22.9	21.0
64	6	25.5	4.5	93.0	1262	183	4895	710	···	···
71	4.5	12.5	12	92.4	1620	235	5756	835	···	···
72	8.5	8.0	11.5	91.2	1965	285	5136	745	···	···

(a) Three-point bending, 15.87-mm ($^5/_8$-in.) span, 6.35 by 6.35 mm (0.25 by 0.25 in.) cross section. (b) 9.5 mm ($^3/_8$ in.) diam by 25 mm (1 in.) high, 0.025 mm (0.001 in.) shims between sample ends and loading plates to minimize end effects. (c) K_{Ic} values using double cantilever beam test
Source: Ref 35

Table 21   Typical mechanical properties of tungsten alloys

Nominal composition	Density, g/cm³	Ultimate tensile strength, MPa (ksi)	0.2% yield strength, MPa (ksi)	Elongation in in 25 mm (1 in.), %	Hardness, HRC	Young's Modulus, GPa (10⁶ psi)	0.2% compressive yield strength, MPa (ksi)	Transverse rupture strength, MPa (ksi)
90W-6Ni-4Cu	17	758 (110)	482 (70)	5	25	276 (40)	683 (99)	1520 (220)
95W-3.5Ni-1.5Fe	18	827 (120)	634 (92)	12	30	365 (53)	621 (90)	2000 (290)
89.5W-3Ni-3Cu-1.5Fe-3Mo	16.95	965 (140)	827 (120)	2	32	317 (46)	827 (120)	1790 (260)
90W-4Ni-2Fe-4Mo	17.25	1030 (150)	965 (140)	2	36	331 (48)	827 (120)	1930 (280)
90W-4Ni-2Fe-4Mo	17.25	965 (140)	862 (125)	3	34	338 (49)	···	1520 (220)
90W-10Cu	17.2	758 (110)	···	···	27	···	···	···

Nominal composition	Fatigue limit, rotating beam, MPa (ksi)	Charpy V-notched impact strength, J (ft·lb)	Charpy unnotched impact strength, J (ft·lb)	Short time tensile strength, MPa (ksi), at elevated temperature at: 650 °C (1200 °F)	815 °C (1500 °F)	980 °C (1795 °F)	1095 °C (2005 °F)	Ref
90W-6Ni-4Cu	276 (40)	6.1 (4.5)	14.9 (11)	241 (35)	138 (20)	90 (13)	72 (10.4)	36, 37
95W-3.5Ni-1.5Fe	465 (67.5)	4.7 (3.5)	54.2 (40)	496 (72)	324 (47)	207 (30)	138 (20)	36, 37
89.5W-3Ni-3Cu-1.5Fe-3Mo	379 (55)	2.0 (1.5)	10.8 (8)	517 (75)	379 (55)	241 (35)	138 (20)	37
90W-4Ni-2Fe-4Mo	448 (65)	2.7 (2)	13.6 (10)	869 (126)	662 (96)	379 (55)	239 (34.6)	37
90W-4Ni-2Fe-4Mo	···	2.7 (2)	23 (17)	724 (105)	517 (75)	···	193 (28)	37
90W-10Cu	···	2.7 (2)	5.4 (4)	···	···	···	···	37

Table 22   Typical mechanical properties of molybdenum and molybdenum alloys

Alloy	Density, g/cm³	Density, %	Ultimate tensile strength MPa	ksi	0.2% yield strength MPa	ksi	Elongation parallel to rolling, %	Young's modulus GPa	10⁶ psi	Hardness(a), HV10
99.95% molybdenum(b)	10.2	100	685	99	590	86	15	317	46	260-360
Molybdenum (doped with potassium plus silicate)	···	100	640	93	570	83	10	317	46	260-360
TZM (Mo-0.5Ti-0.07Zr-0.05C)	···	100	1080	157	960	140	7	317	46	300-450

(a) For sheet reduced 70% or more. (b) 0.45-mm sheet
Source: Ref 38

Table 23   Room- and elevated-temperature mechanical properties of fully dense rhenium and rhenium-containing alloys

Property	Test temperature °C	°F	Rhenium		Molybdenum—50% rhenium		Tungsten—25% rhenium	
			Wrought	Recrystallized	Wrought 95%	Recrystallized	Wrought	As-Sintered
Modulus of elasticity in tension, GPa (10^6 psi)	−65	−85	...	...	350 (50.8)	384 (55.7)	...	...
	20	70	...	469 (68)	360 (52.3)	367 (53.3)	431 (62.5)	370 (53.6)
	200	290	...	445 (64.5)	...	...	...	...
	400	750	...	424 (61.5)	...	...	...	...
	600	1110	...	403 (58.5)	...	...	...	...
	800	1470	...	383 (55.5)	...	...	...	...
			Wrought 15%	Recrystallized	Wrought	Recrystallized	Wrought	Recrystallized
Ultimate tensile strength, MPa (ksi)	20	70	1930 (280)	1069 (155)	1655 (240)	1034 (150)	2137 (310)	1310 (190)
	800	1470	1000 (145)	620 (90)	827 (120)	482 (70)	1310 (190)	1034 (150)
	1200	2190	551 (80)	414 (60)	345 (50)	241 (35)	758 (110)	724 (105)
	1600	2910	207 (30)	207 (30)	138 (20)	103 (15)	310 (45)	228 (33)
	2000	3630	124 (18)	124 (18)	...	...	...	...
			Wrought 15%	Recrystallized	Wrought 50%	Recrystallized	Wrought	Stress-Relieved
0.2% yield strength, MPa (ksi)	20	70	1758 (255)	290 (42)	1447 (210)	800(a) (116)(a)	...	1717(a) (249)(a)
	800	1470	...	...	...	...	...	903(a) (131)(a)
	1200	2190	...	...	...	...	...	407(a) (59)(a)
	1600	2910	...	...	...	...	...	1041(a) (15)(a)
	2000	3630	...	...	...	...	...	41(a) (6)(a)
	2500	4530	...	...	...	...	...	10(a) (1.5)(a)
Elongation in 75 mm (3 in.), %	20	70	2	15-20	4	19	...	15-20
	800	1470	1	5	2	18	...	...
	1200	2190	1	2	4	18	...	...
	1600	2910	1	2	8	17	...	...
	2000	3630	1	2	...	17	...	...
			Wrought 10%	Recrystallized	Wrought 95%	Recrystallized	Wrought 50%	Stress Relieved
Room-temperature hardness, HV	...	...	450	250	600	350	590	450
			Wrought 20%				Wrought 90%	
	...	...	530	...	...	...	740	...
			Wrought 30%					
	...	...	580	...	...	...	...	...

Note: Percentages following wrought designation denote reduction in area during final cold working. (a) minimum values
Source: Ref 39

Table 24 Properties of P/M tantalum wire

Property	Diameter 0.250 mm	0.010 in.	Diameter 0.380 mm	0.015 in.	Diameter 0.510 mm	0.020 in.	Diameter 0.635 mm	0.025 in.
Ultimate tensile strength, MPa (ksi)								
Annealed	469	68	414	60	400	58	379	55
Stress relieved	517	75	496	72	482	70	448	65
Unannealed	965	140	813	118	758	110	689	100
0.2% yield strength, MPa (ksi)								
Annealed	372	54	324	47	296	43	275	40
Stress relieved	482	70	448	65	414	60	393	57
Unannealed	965	140	772	112	724	105	648	94
Elongation, %								
Annealed	18	18	22	22	25	25	29	29
Stress relieved	10	10	15	15	18	18	21	21
Unannealed	···	<2	···	<2	···	<2	···	<2

Source: Ref 40

REFERENCES

1. Metal Powder Industries Federation standard 35, "P/M Materials Standards and Specifications," Metal Powder Industries Federation, Princeton, NJ

2. Huppmann, W.J. and Albano-Muller, L., Production of Powder Forged Parts of Complex Geometry, in *Modern Developments in Powder Metallurgy*, Vol 12, Metal Powder Industries Federation, Princeton, NJ, 1981, p 631

3. Hanejko, F., Mechanical Properties of Powder Forged 4100 and 1500 Type Alloy Steels, in *Modern Developments in Powder Metallurgy*, Vol 12, Metal Powder Industries Federation, Princeton, NJ, 1981, p 689

4. Pilliar, R.M. *et al.*, Fracture Toughness Evaluation of Powder Forged Parts, in *Modern Developments in Powder Metallurgy*, Vol 7, Metal Powder Industries Federation, Princeton, NJ, 1974, p 51

5. Badia, F., Heck, F., and Tundermann, J., Effect of Composition and Processing Variations on Properties of Hot Formed Mixed Elemental P/M Nickel Steels, in *Modern Developments in Powder Metallurgy*, Vol 7, Metal Powder Industries Federation, Princeton, NJ, 1974, p 255

6. Lindskog, P., Reduction of Oxide Inclusions in Powder Preforms Prior to Hot Forming, in *Modern Developments in Powder Metallurgy*, Vol 7, Metal Powder Industries Federation, Princeton, NJ, 1974, p 285

7. Tsumuki, C. *et al.*, Connecting Rods by P/M Hot Forging, in *Modern Developments in Powder Metallurgy*, Vol 7, Metal Powder Industries Federation, Princeton, NJ, 1974, p 385

8. Mocarski, S. and Hall, D.W., Properties of Hot Formed Mo-Ni-Mn P/M Steels with Admixed Copper, in *Modern Developments in Powder Metallurgy*, Vol 9, Metal Powder Industries Federation, Princeton, NJ, 1977, p 467

9. Saritas, S., James, W.B., and Davies, T.J., Influence of Preforging Treatments on the Mechanical Properties of Two Low Alloy Powder Forged Steels, *Powder Metall.*, Vol 3, 1981, p 131

10. Pietrocini, T.W. and Gustafson, D.A., Fatigue and Toughness of Hot Formed Cr-Ni-Mo and Ni-Mo Prealloyed Steel Powders, in *Modern Developments in Powder Metallurgy*, Vol 4, Plenum Press, New York, 1971, p 431

11. High Technology Materials, Amstead Research Laboratories, Bensenville, IL, private communication, 1983

12. Aslund, C., "Fully Dense Stainless Steel Products Compete Successfully With Forged Products," Metal Powder Industries Federation National Powder Metallurgy Conference, New Orleans, 1983

13. Billiet, R., "Plastic Metals from Fiction to Reality With Injection Molded P/M Materials," Metal Powder Industries Federation National Powder Metallurgy Conference, Montreal, June 1982

14. Consolidated Metallurgical Industries Inc., Farmington Hills, MI, private communication, 1983

15. Beiss, P. and Huppman, W.S., "Sintering of P/M Tool Steels to Full Density," presented at Powder Metallurgy Short Course "Fully Dense P/M Materials for High Performance Applications," sponsored by Metal Powder Industries Federation, New Orleans, Feb 1983

16. Stasko, W., Crucible Research Center, Pittsburgh, private communication, 1983

17. Hellman, P., "Potential of High Strength P/M High Speed Steel," in *Processing and Properties of High Speed Tool Steels*, Ferrous Metallurgy Committee of the Metallurgical Society of American Institute of Mining, Metallurgical and Petroleum Engineers 109th Annual Meeting, Las Vegas, 26 Feb 1980

18. Uddeholm High Speed Steel, UHB ASP 30, Uddeholm Steel Corp., Totowa, NJ, private communication, 1983

19. Powder Metallurgy High Speed Steel, ASP 60, Uddeholm Corp., Totowa, NJ, private communication, 1983

20. Stellite Powder Metallurgy Products, Cabot Wear Technology Division, Cabot Corp., Kokomo, IN, private communication, 1983

21. Coyne, J.E. *et al.*, "Superalloy Powder Engine Components: Controls Employed to Assure High Quality Hardware," in *Powder Metallurgy Superalloys*, Metal Powder Report Conference, Zurich, Nov 1980

22. Kelto, C.A., Kelsey-Hayes Powder Technology Center, Traverse City, MI, private communication

23. Coyne, J.E. *et al.*, "Superalloy Turbine Components, Which Is the Superior Manufacturing Process, As HIP'd, HIP'd plus Isoforged, or Gatorizing™ of Extrusion Consolidated Billets," in *Powder Metallurgy Superalloys*, Metal Powder Report Conference, Zurich, Nov 1980

24. Moser, G. *et al.*, Hot Isostatic Pressing of Superalloys, presented at Powder Metallurgy Superalloys, Aerospace Materials for the 1980's, Vol 2, Metal Powder Report Conference, Zurich, Nov 1980

25. Hart, R.M., "Wrought P/M Aluminum Alloys X7090 and X7091," Aluminum Company of America, Alcoa Technical Center, Alcoa Center, PA, Aug 1981, unpublished

26. "Aluminum Powder for Powder Metallurgy Parts," Aluminum Company of America, General Information, Pittsburgh, 1971

27. Properties of Powders and Powder Metallurgy Products, in *Aluminum*,

Vol 1, *Properties and Physical Metallurgy*, Kent R. Van Horn, Ed., American Society for Metals, 1967

28. Garriott, R.E. and Thellmann, E.L., Titanium Powder Metallurgy—The Commercial Reality, 1976 International Powder Metallurgy Conference, Metal Powder Industries Federation, Princeton, NJ, 1976

29. Abkowitz, S., Isostatic Pressing of Complex Shapes from Titanium and Titanium Alloys, in *Titanium Alloys Science and Technology*, Proceedings of the 4th International Conference on Titanium, Kyoto, Japan, 19 May 1980

30. Anderson, T.J. *et al.*, "Fracture Behavior of Blended Elemental P/M Titanium Alloys," United State Air Force contracts F 33615-76-5227, F 33615-79-C5151, F33615-77-C-5008, 1979

31. Toaz, M.W., Blended Titanium P/M Alloys—Current Developments, *Metal Powder Rep.*, Vol 37 (No. 2), Feb 1982, p 85

32. Eloff, P.C., "Development of Low Cost Titanium Structures Using Blended Elemental Powder Metallurgy," Winter Annual Meeting of American Society of Mechanical Engineers, Washington, DC, 15 Nov 1981

33. Witt, R.H. and Bruce, J.S., Progress on Hot Isostatic Pressing of Titanium, 28th National Society for Advancement of Material and Process Engineering Symposium, 12 April 1983

34. Bruce, J.S. *et al.*, Evaluation of Titanium P/M Technology for Naval Aircraft Components, 28th National Society for Advancement of Material and Process Engineering Symposium, 12 April 1983

35. Hale, T.E., General Electric Co., Carboloy Systems Div., Detroit, private communication, 1983

36. Kurtz, R.A., Kulite Tungsten Corp., Ridgefield, NJ, private communication, 1983

37. CMW Inc., Indianapolis, private communication, 1983

38. Anderson, J., Schwarzkopf Development Corp., Holliston, MA, private communication, 1983

39. Kish, J.A., Rhenium Alloys Inc., Elyria, OH, private communication, 1983

40. Fansteel Metals, Chicago, private communication, 1983

SELECTED REFERENCES

- Hoffmann, G. and Dalal, K., Correlation Between Individual Mechanical Properties and Fracture Analysis of Hot Formed P/M Steel, in *Modern Developments in Powder Metallurgy*, Vol 10, Metal Powder Industries Federation, Princeton, NJ, 1977, p 171

- "New Wrought P/M Alloys for Lighter High Performance Parts," Aluminum Company of America, Pittsburgh, 1983

Inspection and Quality Control for P/M Materials

By Leander F. Pease III
President
Powder-Tech Associates, Inc.

DIMENSIONAL ACCURACY of P/M sintered parts is determined with the same measurement techniques that are used for wrought materials. Other testing methods for P/M materials are specialized, however, such as determination of surface finish, for example. For sintered parts, a chisel-pointed stylus is used to de-emphasize the effects of porosity. A conical stylus tends to engage porosity, thus giving an exaggerated measurement of roughness. Pores do not interfere mechanically with mating parts (Ref 1).

During the manufacture of sintered parts, dimensional change must be accommodated for during each processing step. Causes of these changes are described in the articles "Physical Fundamentals of Consolidation" and "Shape Fundamentals: Rigid Tool Compaction" in this Volume. Dimensional change occurs during P/M processing as:

- Elastic springback during ejection from tooling used for cold pressing
- Growth or shrinkage during delubrication, presintering, and sintering
- Elastic springback from tooling during cold re-pressing or sizing
- Thermal contraction from the tools used in hot forging or hot re-pressing
- Tool wear in cold or hot compacting
- Machining tolerances at secondary machining and associated tool wear
- Distortion during annealing
- Growth or shrinkage during carburizing, nitriding, or neutral hardening
- Shrinkage during tempering
- Growth during steam blackening

Parts manufacturers must be familiar with the amount of dimensional change to expect for the materials and equipment in use so that tooling can be produced that accommodates these changes and produces accurate parts. Parts ejected from appropriately made tooling may experience the changes discussed above, but can be completed to specification. Understanding and controlling these factors is essential to commercial P/M parts manufacturing.

Springback at Molding

Metal powders have varying yield points, and green compacts have varying elastic moduli. Thus, even if a tool set is perfectly rigid, the size of the ejected compact is larger than the tool cavity. This amount also varies, depending on molding pressure and powder characteristics. Furthermore, expansion is not uniform or isotropic, except for right circular cylinders. Dies exhibit some compliance, and split dies distort slightly in use, resulting in further changes in green dimensions. Thus, green dimensions depend on die design and construction.

Sintering Dimensional Change

Compacted elemental powders generally shrink during sintering. The compacts begin as unsintered objects that are larger than die size and eventually shrink below die size. This phenomenon is illustrated in Fig. 1 for pure iron (F-0000), based on the 89.61-mm (3.528-in.) dimension of a Metal Powder Industries Federation (MPIF) standard 10 tensile bar. Prealloyed powders generally shrink during sintering (see the curve for AISI 4680 steel in Fig. 2). Graphite additions to iron do not inhibit shrinkage from green dimensions, but do cause additional elastic springback from the mold dimension. Such materials thus

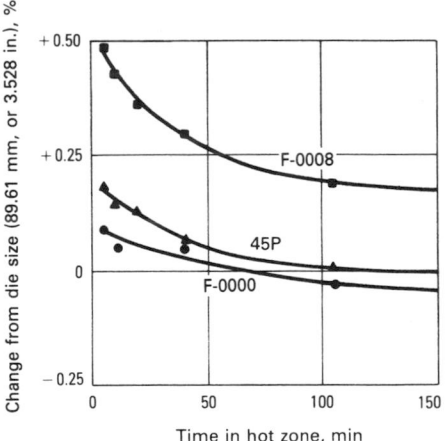

Fig. 1 Dimensional change on sintering of F-0000, F-0008, and 45P irons

Sintered at 1120 °C (2050 °F) in dissociated ammonia. Green density: 6.8 g/cm³. 45P: atomized iron plus 0.45% phosphorus added as ferrophosphorus master alloy. F-0000: pure atomized iron. F-0008: atomized iron plus 0.9% graphite

begin sintering about 0.5% larger than die size (see Fig. 2).

Elements such as carbonyl nickel accelerate the shrinkage of iron mixes (see FN-0208 in Fig. 2). In iron mixes, copper tends to cause growth. The higher the density of an iron particle (low surface area), the greater its dimensional growth with copper additions. Carbon dissolved in iron prior to copper diffusion inhibits growth (see Fig. 3 and 4). Growth of copper-tin premixes is shown in Fig. 5.

Growth of these bronze premixes is strongly density dependent, with densities near 7.0 g/cm³ providing greater growth and densities near 6.0 g/cm³ providing

Fig. 2 Dimensional change on sintering of FC-0208, 4680, and FN-0208 irons

Sintered at 1120 °C (2050 °F) in dissociated ammonia. Green density: 6.8 g/cm³. FC-0208: sponge iron plus 2% Cu plus 0.97% graphite. 4680: prealloyed AISI type 4600 steel plus 0.9% graphite. FN-0208: atomized iron plus 2% Ni plus 0.9% graphite

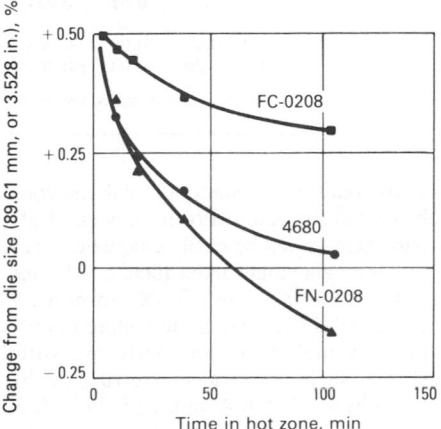

Fig. 3 Dimensional change on sintering of MH100 iron

Values inside the graph indicate the percentage change from die size. MH100 iron sintered 30 min at 1120 °C (2050 °F) in dissociated ammonia. Green density: 6.4 g/cm³. Source: Ref 2

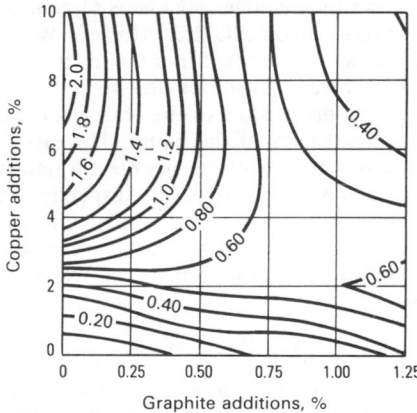

significantly less. Dimensional change data during sintering are complicated by the fact that dimensional change parallel to the pressing direction is not the same as dimensional change occurring perpendicular to the pressing direction.

Annealing Dimensional Change

Annealing dimensional change is similar to change occurring during extra sintering or stress relaxation, as compacts continue to shrink. Lower density com-

pacts experience the most shrinkage. With pure copper that has been pressed, sintered, and re-pressed to high density, annealing in a hydrogen atmosphere that contains gases can cause growth or blistering. The hydrogen diffuses into the pores that are isolated from the surface. If hydrogen encounters residual oxygen, water vapor is formed, which causes expansion and growth. This phenomenon also can be caused by residual sulfur or the presence of lubricant. Consequently, the initial sintering of pure copper should remove as many impurities as possible while the pores are still open.

Dimensional Change During Heat Treating

Carbon and nitrogen absorbed during carburizing and nitriding of steel cause growth in areas in which they dissolve. During quenching, regions that form martensite experience growth. Thus, if a small part is through hardened to all martensite, all regions (inside and outside diameter) experience outward expansion. Many sintered steels have modest hardenability, and only the outer 3.2 mm (0.125 in.) adjacent to surfaces forms martensite and expands.

On quenching a medium- or large-sized part (with a section greater than 9.5 mm, or 0.4 in.), the interior remains as ferrite and fine pearlite, experiencing neither shrinkage nor growth. The outer surfaces expand outward, and the inner surfaces shrink inward. This phenomenon also is evident in case-hardened wrought parts. Thus, prediction of exact size change during heat treatment is difficult.

Steam Blackening

Sintered parts frequently are treated in steam at 540 to 595 °C (1000 to 1100 °F) for 1 to 4 h to fill the pores and coat the surface with a hard coating of black iron oxide. The coating causes a uniform growth of 0.0025 to 0.0050 mm (0.0001 to 0.0002 in.) similar to electroplating. The amount of blackening should be controlled, as measured by hardness and destructive break tests; excessive oxide coating thickness may lower impact properties. Thickness of the oxide layers also can be measured metallographically with a polishing procedure described in the section of this article on density measurement.

Evaluation of Dimensional Change in Incoming Powder

New lots of blended or raw powder are checked against internal standard lots to

Fig. 4 Dimensional change on sintering of ATOMET 28 iron

ATOMET 28 powder plus 0.9% graphite plus 0.75% zinc stearate plus copper as shown. Green density ranges from 6.2 to 7.0 g/cm³. Sintered 30 min in dissociated ammonia at 1120 °C (2050 °F). Source: Ref 3

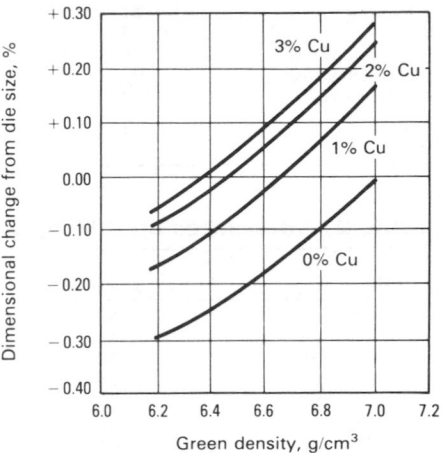

Fig. 5 Dimensional change on sintering of 90-10 bronze

PMB 18 premixed bronze sintered 15 min in hydrogen at various temperatures. Green density: 6.3 g/cm³. Source: Ref 4

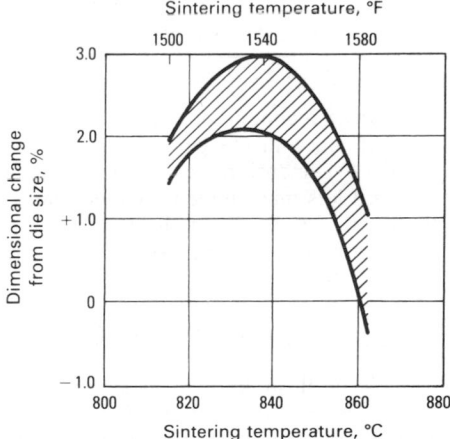

ensure consistent sintered dimensional change. Transverse-rupture bars 31.8 by 12.7 by 6.4 mm (1.25 by 0.50 by 0.25 in.) are molded at a fixed density or pressure from both the standard and test lot of powder. The two sets of bars are sintered simultaneously in a laboratory or production furnace. Dimensional change in the 31.8-mm (1.25-in.) length are checked against the requirements of American Society for Testing and Materials (ASTM) standard B 610.

Although dimensional change from sintering a bar made from the standard powder may differ from previous tests, comparable dimensional changes in the test bar

482/Consolidation of Metal Powders

Table 1 Typical P/M tolerances (other than length)
Up to 12.7 mm (0.500 in.)

Material	As-sintered mm	in.	As-sized mm	in.	As-heat treated mm	in.
Brass	±0.089	±0.0035	±0.013	±0.0005	...	...
Bronze	±0.089	±0.0035	±0.013	±0.0005	...	...
Aluminum	±0.051	±0.002	±0.013	±0.0005	±0.013	±0.0005
Iron	±0.025	±0.001	±0.013	±0.0005	...	...
Copper alloy steel	±0.038	±0.0015	±0.025	±0.001	±0.038	±0.0015
Nickel alloy steel	±0.038	±0.0015	±0.025	±0.001	±0.038	±0.0015
Stainless steel	±0.025	±0.001	±0.013	±0.0005	...	...

Note: Length tolerance, ±0.102 mm (±0.004 in.), unless machined or ground
Source: Ref 1

Table 3 Dimensional tolerances of parts in the as-high-temperature sintered condition

Material	Nominal dimension mm	in.	Tolerance mm	in.
Composite	25.4	1.00	0.05	0.002(a)
3Si-Fe	19.0	0.75	±0.08	±0.003
4600	76.2	3.00	0.38	0.015
M-2	70.3	2.77	0.61	0.024
Low-alloy steel	22.2	0.88	0.08	0.003
Stellite	25.4	1.00	0.03	±0.001(b)

(a) Roundness. (b) Inside diameter sintered against a mandrel
Source: Ref 5

Table 2 Tolerances on P/M forged parts

Parameter	Nominal dimension mm	in.	Tolerance mm	in.
Outside diameter	50.8	2.00	0.13	0.005
Outside diameter	50.8	2.00	0.25	0.010
Inside diameter	38.1	1.50	0.20	0.008
Thickness	25.4	1.00	0.38	0.015
Spline	25.4	1.00	0.23	0.009
Outside diameter	95.25	3.75	0.25	0.010
Inside diameter	63.5	2.50	0.25	0.010
Concentricity	95.25	3.75	0.10	0.004
Roundness	95.25	3.75	0.10	0.004
Thickness	15.8	0.625	0.25	0.010
Outside diameter	50.8-76.2	2.00-3.00	0.13	0.005
Outside diameter	25.4-50.8	1.00-2.00	0.10	0.004
Outside diameter	76.2	3.00	0.38	0.015
Outside diameter	50.8	2.00	0.13	0.005
Outside diameter	203	8.00	0.51	0.020
Thickness	25.4	1.00	0.25-0.634	0.010-0.025

Source: Ref 5

made from incoming powder demonstrate the difference in the performance of the powders. Dimensional change in test and standard lots must agree to within a specified range (±0.1% of the bar length). These bars also can be used to evaluate sintered strength and hardness.

Dimensional Control

Table 1 illustrates typical dimensional tolerances of P/M materials. Separate tolerances apply to as-sintered, as-sized, and as-heat treated conditions. For concentricity between an inside diameter and an outside diameter, a total indicator reading of up to 0.075 mm (0.003 in.) is permitted. The distance between holes may be as great as 0.075 mm + 0.013 mm/mm (0.003 in. + 0.0005 in./in.). Gears can be molded to American Gear Manufacturers Association (AGMA) class 7, which is limited primarily by the concentricity of the bore to pitch line. If gears are held on the pitch line and bored more concentrically, AGMA class 10 or 11 is achieved.

Other processes such as P/M hot forging, injection molding, and high-temperature sintering produce wider tolerances than presented in Table 1. Powder metallurgy forged dimensional tolerances are given in Table 2. High-temperature sintering tolerances are given in Table 3. Injection-molded tolerances range from 0.075 to 0.10 mm/mm (0.003 to 0.004 in./in.), even though parts have experienced 12 to 15% linear shrinkage (Ref 5).

Measurement of Density

Density is the ratio of mass to volume. For a given material, degree of sintering, and heat treatment, density determines mechanical and physical properties. For example, higher density in sintered steels results in higher tensile strength, elongation, and impact resistance values. As-pressed, or green, density also influences growth or shrinkage that occurs during sintering. With nonuniform green density, parts grow or shrink nonuniformly, as in a thin-walled bronze bearing with a low-density region equidistant from the ends. This results in a significantly smaller diameter at midlength than at the ends and necessitates re-pressing or sizing for close dimensional control.

If cubes or right cylinders could be extracted from actual parts, linear dimensions could be measured and volume could be calculated easily. From the weight of a part, density can be easily calculated. This yields a value that, under ideal conditions, differs by 0.04 g/cm^3 (0.5%) from a reference (Ref 6). Unless the sintered part is directly molded to an easily measured shape, such as a transverse-rupture bar (31.8 by 12.7 by 6.4 mm, or 1.25 by 0.50 by 0.25 in.), this method of measuring linear dimensions is used infrequently.

Typical methods of measuring density depend on Archimedes' principle, in which hydrostatic forces in liquids exert buoyant forces proportional to the part volume. This measurement is standardized in ASTM B 328 (Ref 7), MPIF test method 42 (Ref 8), and International Standards Organization test method ISO 2738 (Ref 9). When an object is immersed in a liquid, the liquid exerts an upward buoyant force that is equal to the product of the object volume and the density of the liquid. The difference in weight between an object weighed in air and its weight when suspended in water is equal to the object volume in cubic centimetres times the density of water. Approximating the density of water as unity:

$$V = W_{air} - W_{water}$$

where V is the volume, cm^3; W_{air} is the weight in air, g; and W_{water} is the weight of object suspended in water less the weight of the suspending wire in water (tare), g. Density in g/cm^3 is then:

$$Density = \frac{W_{air}}{W_{air} - W_{water}}$$

For unsintered materials molded with 0.75% lubricant, pores are well sealed, and water cannot penetrate. For such parts, the above calculation is suitable. It is also suitable for materials with pores that are sealed off from the surface (materials close to theoretical density). For most sintered materials that are 70 to 95% dense, water tends to infiltrate the pores during weighing in water. This minimizes the buoyancy effect of the water (that is, the liquid is acting on a smaller volume) and results in an erroneous calculation of low volume.

This low volume then causes an erroneously high density value. Infiltration of water into pores usually is accompanied by air bubbles escaping from the part. If the part is blotted to remove surface water and reweighed in air after weighing in water, any weight gain indicates that water has entered the pores. Although not a standard procedure, volume can be approximated as the weight in air after removing the part from the water, minus the weight in water.

To prevent infiltration of water, all three standard test methods require that the pores of the part be filled with oil. Oil impregnation is done after the part is weighed in air; this is carried out under vacuum or by immersion in hot oil. Oil prevents the water from entering the pores. The volume of the part is then determined as the part weight in air with oil in the pores, minus the weight of the oiled part suspended in water. Care should be taken to select an oil that is not soluble in water or not soluble in water plus wetting agent. Such oils also must exhibit superior demulsibility.

The precision of the ISO method is ±0.25%, regardless of sample density, and assumes a water density of 0.997 g/cm³. Moyer (Ref 6) has reviewed the literature on precision methods of density determination (Ref 10-16) and has devised a method that provides accuracy to two or three decimal places, depending on sample porosity. The basic measuring apparatus is shown in Fig. 6. Requirements of precision density measurement include:

- Balance capable of measuring to the nearest 0.0001 g
- Vibration- and draft-free atmosphere
- Measurement of the density of the immersing liquid (water) by checking the density of a substance of accurately known density (four decimal places)
- Conversion of all densities back to 20 °C (68 °F) by compensating for thermal expansion of the sintered part
- Maintenance of liquid level at a constant height on the suspending wire
- Careful brushing of all bubbles from the test object

Using the above procedures, Moyer reports standard deviations of 0.0130 to 0.0005 g/cm³ on 17-g parts with densities ranging from 5.12 to 7.85 g/cm³, respectively.

To determine density variation from one point to another in a complex part, the

Fig. 6 Density measurement apparatus

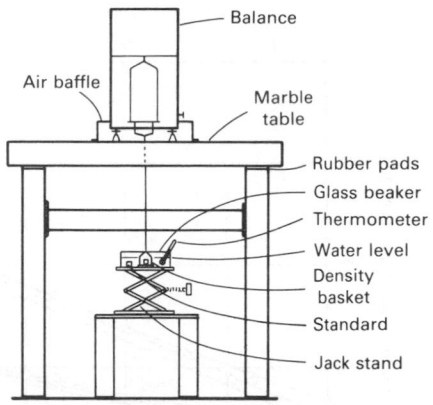

available samples must be considerably smaller than 17 g. According to ASTM B 328, a minimum sample of 2 g is recommended, because a relatively high error rate results from measuring small samples. Table 4 shows density errors that can occur because of weighing errors, if all weighing errors are assumed to occur so as to maximize density errors. A 1-g sample on a balance accurate to 0.01 g (per ASTM B 328) could result in a density range of from ±1.05 g/cm³ to a mean of approximately 6.5 g/cm³.

Metallographic estimates can be made of the area fraction of porosity, which is numerically equal to the volume porosity, and thus the density of sintered materials. The method is not standardized, and accuracy of results depends on the skill of the metallographer to define the correct area fraction of porosity. Frequently, the amount of porosity is exaggerated or minimized because of overpolishing or underpolishing (Ref 17). A suitable method for sintered iron-based materials is to:

- Remove oil and cutting fluids from the pores by Soxhlet extraction or heating in air or atmosphere to 315 °C (600 °F).
- Impregnate the pores with epoxy resin; generally does not fill all the pores, and polishing is required.
- Mount the sample in epoxy, bakelite, or other suitable medium.
- Wet grind through 600-mesh silicon carbide-coated paper.
- Open smeared pores by immersion etching in 2% nital for 1 min.
- For samples up to 12.7 by 12.7 mm (¹⁄₂ by ¹⁄₂ in.), polish with 1-μm alumina on Struers felt cloth or an equivalent long-napped cloth for 3 min on a 250-rpm, 20-cm (8-in.) diam wheel, using moderate to heavy hand pressure. This opens pores and exaggerates their presence.

Table 4 Effect of sample size and weighing errors on density measurement

Balance accuracy (weighing error), g	Density range, g/cm³, for a sample weight (at 6.50 g/cm³) of:		
	0.1 g	1.0 g	10.0 g
0.01 0		±1.0459	±0.0922
0.001 .... ±1.0459		±0.0922	±0.0091
0.0001 .... ±0.0922		±0.0091	±0.0009

- Restore pores to true area fraction by polishing for 3 min by hand, using a 1-μm diamond on Struers MOL cloth or equivalent, and moderate to heavy pressure. Use a 20-mm (0.8-in.) strip of diamond paste per specimen.
- Final polish for 30 to 60 s, using 0.05-μm alumina with Buehler Microcloth or equivalent long-napped cloth on a 20-cm (8-in.) diam polishing wheel at 125 rpm with light pressure; an automatic polisher is suitable.
- Wash, dry, and examine unetched at 200×.

A simple manual method to estimate the area fraction of porosity uses a 100- by 120-mm (4- by 5-in.) photograph of the sample taken at a magnification of about 200×. A grid of 13 by 17 thin black lines 6.4 mm (0.25 in.) apart should be constructed on clear vinyl to fit on the photograph. The grid contains 221 intersections. By placing the grid over the photograph, then counting the number of intersections that lie completely over the pores and adding one half of the number of intersections that are on the edge of the pores to obtain the effective number of intersections, the area fraction of porosity is calculated:

$$Ap = \frac{221 - Np}{221}$$

where Ap is the area fraction of porosity; and Np is the effective number of grid intersections overlaying pores. For common iron-based materials, theoretical density is 7.87 g/cm³; density (ρ) can be estimated as:

$$\rho = (1 - Ap) \times 7.87 \text{ g/cm}^3$$

This method has a precision of ±0.1 g/cm³ in the author's laboratory. The advent of metallography with a television monitor and quantitative metallographic functions allows rapid measurement of area fractions of pores. This method is highly dependent on proper sample preparation.

Crack Detection

Proper press setup for molding P/M parts is critical to prevent cracking. In a flanged

Fig. 7 Cracks and unbonded particles at the junction (radius) of a horizontal flange and vertical hub resulting from shear during compaction
Unetched

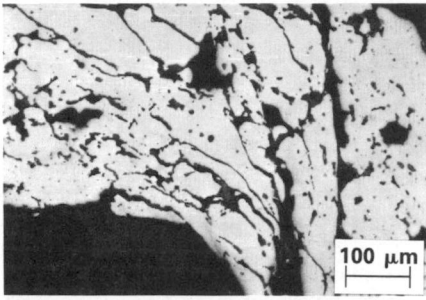

100 μm

Fig. 8 Relationship of ultrasonic velocity and strength for Ancorsteel 1000B

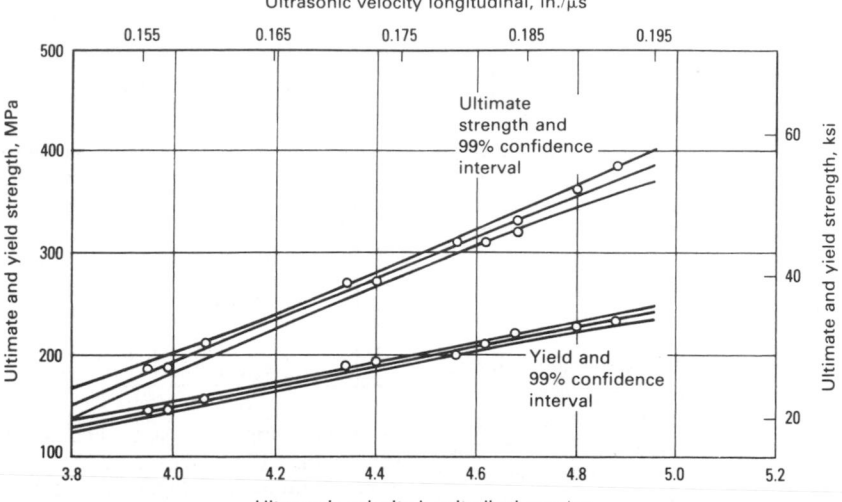

part that experiences a change in diameter, density in the hub and flange should be nearly equal. Unequal density leads to powder displacement from one part level to the next and to the formation of shear cracks. Such cracks often occur at 45° to the pressing direction and at surfaces at the junction (radius) between the hub and flange. At press setup, equal density should be obtained in the hub and flange. A high green strength powder and a press that maintains a small counter pressure on the top of the part during ejection from the tools (top punch hold-down) should be used.

Eventually, ultrasonic or eddy current methods may be developed to check green or sintered parts for cracks. Currently, crack detection is accomplished by mechanical proof testing, metallography, and filtered particle or magnetic particle inspection.

Mechanical Proof Testing. A sampling of sintered parts can be broken to confirm the presence of a suspected cracking problem—for example, pushing flanges off hubs in such a manner that in addition to shear effect there is some stretch effect at the crack. The presence of a few unexplained low readings indicates that an initiating crack is present.

Metallography. Low-powered binocular microscopes can be used to detect cracks at changes in diameter. Metallography is a more time-consuming method. A sampling of parts are sectioned parallel to the pressing direction. When mounted and carefully polished to expose open pores and cracks, the presence of minute cracks is readily apparent (Fig. 7).

Liquid-Penetrant Crack Detection. During liquid-penetrant testing, the surface of the part is covered with a colored or fluorescent penetrating fluid that fills all cracks and some pores. Surface liquid is then washed off, leaving the penetrant in the cracks. A layer of developer powder, which acts as a blotter, is applied to the surface. The penetrant leaves the cracks and is absorbed by the developer layer near the crack. Crack indications are visible by color contrast or by fluorescence under ultraviolet light. Most sintered parts have porous surfaces that absorb and then release sufficient penetrant in all regions so that it is impossible to distinguish the crack from the porosity background.

Filtered Particle Crack Detection. One proprietary process of filtered particle crack detection (Partek) involves brief immersion of the test piece in a suspension of fluorescent particles (Ref 18). Particles are filtered and collect near the surface of cracks as the fluid enters. Cracks are clearly visible under black light. This one-step method is used to detect cracks in presintered porous tungsten carbide blanks (Ref 19).

To the extent that an unsintered part has open porosity, this method also can be used on green parts. Density cannot be too high, however, and excessive lubricant tends to clog the pores. Successful use of this method on presintered porous tungsten carbide blanks indicates that it may be suitable for sintered P/M parts with open pores into which fluid can enter. Small cracks fluoresce brightly, while large cracks are darker than the surrounding fluorescing surface.

Magnetic Particle Crack Detection. A magnetized part demonstrates abrupt changes in magnetic field when the field crosses a crack. When dry or wet magnetic particles are passed over the part, any magnetic leakage field at the part surface attracts and holds iron particles. These assemblies of particles over cracks are visible with the unaided eye or with black light on fluorescent particles.

This method also detects some near-surface cracks. Unsintered parts, however, are not adequately bonded to support a magnetic flux, and the method is consequently unsatisfactory. Magnetic particle detection methods (Ref 18) have been successfully used for many years for inspecting medium-density sintered automotive parts, both at the P/M parts producer and at the automotive manufacturer.

Relationship of Ultrasonic Velocity to Mechanical Properties

Recent work has demonstrated a linear relationship between ultrasonic velocity and ultimate tensile strength or tensile yield strength in test bars (Ref 20). Figure 8 illustrates this relationship for as-sintered 0.65% carbon steel. Table 5 gives sintering times and densities that resulted in various tensile properties and velocities. With similar data for a specific part, ultrasonic velocity can be easily measured, thereby determining the state of sintering or mechanical properties. Future testing involves additional alloys, heat treated parts, geometries, and automating the procedure. It remains to be seen if this process is capable of detecting cracks and other flaws.

Hardness Testing and Electromagnetic Sorting

Low-frequency comparator-bridge instruments frequently are used to sort steel

Table 5 Characteristics of as-sintered FN-0208 alloy specimens prepared with Anchorsteel 1000B powder, 0.75% carbon, and 0.75% Acrawax C

Sintering time, min	Final density, g/cm³	Ultrasonic velocity		Yield strength		Tensile strength	
		mm/μs	in./μs	MPa	ksi	MPa	ksi
5	6.26	3.95	0.1557	143	20.8	182	26.4
15	6.27	3.98	0.1568	142	20.6	182	26.4
30	6.27	4.06	0.1600	154	22.4	210	30.4
30	6.29	...	...	...	...	...	...
30	6.45(a)	4.42	0.1741		...	273	39.6
45	6.48	4.35	0.1711	183	26.6	266	38.6
5	6.76	4.56	0.1795	200	29.0	307	44.6
15	6.74	4.63	0.1821	210	30.4	307	44.6
30	6.74	4.69	0.1847	220	32.0	330	47.8
30	6.77	...	...	...	...	...	...
...	6.81(a)	4.88	0.1922	234	34.0	379	55.0
30	6.75	4.69	0.1846	218	31.6	317	46.0
45	6.89	4.83	0.1902	232	33.6	363	52.6

(a) Restruck

Fig. 9 Luminous points on oscilloscope screen for hardness investigation

Magnetic bridge system applied to sorting of parts by hardness

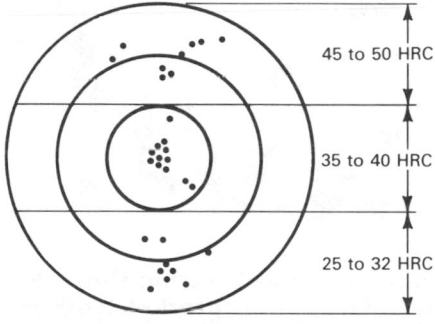

parts based on hardness (Ref 21). Reference and test coils are balanced initially with sample parts of known hardness. In subsequent tests, parts of unknown hardness are substituted for one of the reference parts. The degree of unbalance that results is then correlated with an increase or decrease in hardness. Unbalance may be displayed on a properly connected oscilloscope.

Luminous point patterns, representing three ranges of hardness, are shown as they appeared on the screen of an oscilloscope in Fig. 9. These patterns were developed in connection with the sorting of mixed lots of tapered studs, using a comparator-bridge instrument. The required hardness range established for the steel studs, following hardening and tempering, was 34 to 40 HRC. However, some of the studs exceeded this hardness range because they were not tempered, whereas others fell short of it because of surface decarburization. The mixed lots were successfully sorted into three hardness ranges based

on the luminous point patterns shown in Fig. 9.

Sintered parts can be sorted by the same technique using low-frequency (roughly 60 cycles) comparator-bridges. In one case, parts with hardnesses ranging from 75 to 85 HRB were to be sorted from parts with substantially higher or lower hardnesses. The feeding mechanism to the test coil presented 100 pieces per minute. These were graded into groups with hardnesses below 75 HRB, between 75 and 85 HRB, and above 85 HRB. At the coil exit, gates deflected the parts into the appropriate category.

Measurement of Electrical Conductivity

Electrical conductivity of nonmagnetic materials can be measured with instruments (Ref 18) that capitalize on the electromagnetic phenomena associated with induced eddy currents. This technique is suitable for copper, aluminum, brass, molybdenum, nickel, tin, titanium, and zinc.

The instruments are suitable for conductivities ranging from 8 to 100% IACS. Typical accuracy is ±3% of instrument scale reading. The resistivity of the material being tested may be obtained by:

Specific resistivity

$$= \frac{172.41}{\%\text{IACS}} \; \mu\Omega \cdot cm$$

$$= \frac{1037.1}{\%\text{IACS}} \; \Omega/\text{circular-mil ft}$$

The probe, energized by 60 kC alternating current, induces circulating currents in the part to be measured. These currents, which increase with conductivity, react back to the probe. The dial-operated main capacitor, probe, calibrating controls, and var-

ious fixed impedances form the impedance arms of the bridge circuit. The impedance of the probe varies, depending on the conductivity of the test sample, which unbalances the bridge. The bridge is restored to balance by adjusting the main capacitor, and the balance condition is shown as "0" on a meter. The instrument is calibrated before use on a high-conductivity copper sample and a low-conductivity brass sample.

Sintered copper alloys are subject to the formation of a well-sintered, high-conductivity case on a lower conductivity core. Because the penetration of the eddy currents is only 0.75 to 1.25 mm (0.030 to 0.050 in.), the instrument can provide an erroneous impression of the internal conductivity.

With proper sintering, conductivities of 90 to 95% IACS are possible. One procedure to increase conductivity uses a high-purity, low-oxygen powder and 0.50 to 0.75% lithium stearate lubricant molded at 6.4 to 6.8 g/cm³, presintered or declubricated for 30 min at 760 to 815 °C (1400 to 1500 °F), sintered for 30 min at 980 to 1040 °C (1800 to 1900 °F) with 1 to 3% shrinkage, and re-pressed at 550 to 690 MPa (40 to 50 tsi). This procedure should provide over 90% conductivity. Final annealing at 815 to 980 °C (1500 to 1800 °F) for 30 min further increases conductivity, but atmospheres containing hydrogen may cause blistering in the presence of residual oxygen.

Metallography and Microstructure

Microstructural analysis is essentially a diagnostic tool to be used when parts do not function properly, rather than for routine quality control. Metallographic preparation of a porous material is significantly different from preparation required for wrought materials; several precautions are warranted.

A section parallel to the original pressing direction is examined. For some purposes, such as measuring case depth on gear teeth, a section perpendicular to the pressing direction also is used. Usually the cut-out section contains oil or cut-off fluid, which should be removed before mounting. For as-sintered parts containing cut-off fluid, heating on a hot plate until they turn tan or light blue drives off the liquid. Oil can be removed in a Soxhlet apparatus (ASTM B 328) or by ultrasonically cleaning in heated 1-1-1 trichloroethane for 30 min under a hood.

Pores are best revealed by impregnating with epoxy resin. The most thorough method involves suspending the parts above the resin, with a magnet on the outside of a bell jar. After evacuating the chamber to less than 25.4 torr (1 in. Hg), the sample is dropped into the resin by removing the magnet. The bell jar is brought back to atmosphere, and the pressure is applied for 15 min to force the resin into the pores. Because many epoxy resins boil at these low pressures, the jar must not be evacuated to that point.

Once impregnated and cured, the sample can be metallographically mounted. A simple one-step method consists of placing the specimen and resin in a silicone-coated two-piece cup and placing the entire assembly in the bell jar. Common epoxy resins can be evacuated to 50 torr (2 in. Hg), which permits filling of most pores. The sample is allowed to outgas for 10 min, before pressure is reapplied to drive in the resin. This method fills most, but not all pores.

The mounted sample is then wet ground on 400- and 600-mesh silicon carbide-coated paper, with a 90° sample rotation between papers. This grinding closes or smears many of the pores. The following procedure opens pores and polishes the surface:

- Etch 1 min by immersion in 2% nital (for ferrous materials).
- Rough polish 3 to 4 min by hand, using moderate pressure with 1-μm aluminum oxide on a long-napped cloth, such as Buehler Microcloth, on a 20-cm (8-in.) diam wheel at 250 rpm. This procedure opens pores and exaggerates pore size.
- Fine polish by hand with moderate pressure for 3 min with a 1-μm diamond on a short-napped cloth, such as Struers MOL, at 250 rpm. Use approximately a 25-mm (1-in.) strip of diamond paste per mount, which removes scratches and restores the true area fraction of porosity by eliminating the bell-mouthed effect where the pore is exposed at the polished surface.
- Final polish (optional) for 30 s on 0.05-μm aluminum oxide or Microcloth by hand with light pressure, or on an automatic polisher, using a 100- to 200-g load at 125 rpm.

This method prevents the most common problem encountered in metallographic etching of porous materials—smearing the pores, causing the structure to appear denser than it actually is (Ref 17). With all the pores open during microhardness testing, fewer tests will penetrate undisclosed pores. Also, pore shape and the degree of

sintering can be determined only if pores are open.

Porous materials usually are examined in the unetched condition, generally at 100 to 200×. Higher magnification is used for clarification of significant findings. In the unetched condition, visual examination includes review of:

- Original particle boundaries in the undersintered condition (see Fig. 10); under conditions of average sintering, very few thin, gray boundaries should exist between contacting particles (see Fig. 11).
- Pore rounding (see Fig. 11b)
- Microlaminations and cracks (Fig. 12)
- Density variations between points
- Oxide particles or thickness and penetration of steam-blackened oxide layers
- Type of base metal powder used
- Unmelted and undissolved additives (Fig. 13)
- Interconnected or isolated pores (Fig. 14)
- Surface pore condition (whether open or shut) (Fig. 15)
- Presence of sulfides or manganese sulfide added to steel to increase machinability (Fig. 16)

Fig. 10 Undersintered atomized and diffusion alloyed iron powder

Sintered for 5 min in the hot zone at 1120 °C (2050 °F) in dissociated ammonia. Arrows indicate original particle boundaries. (a) Unetched (b) Unetched

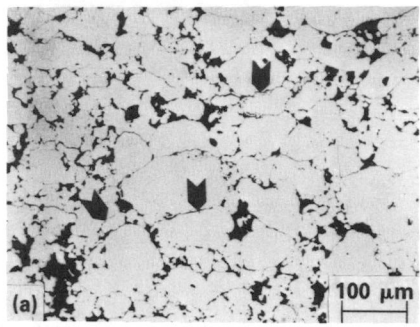
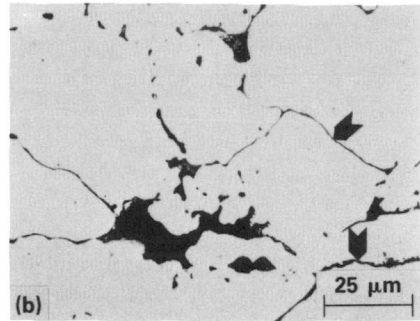

Fig. 11 Atomized Fe-2Ni-0.5C alloy under conditions of average sintering

Sintered for 45 min in the hot zone at 1120 °C (2050 °F) in dissociated ammonia. Note that very few particle boundaries exist between contacting particles. (a) Unetched (b) Rounded pores are shown. Unetched

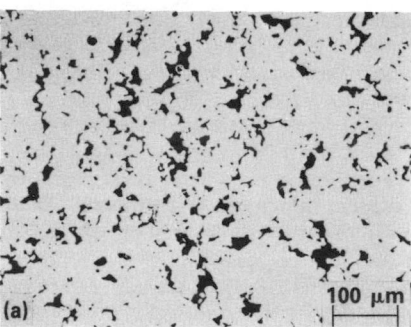
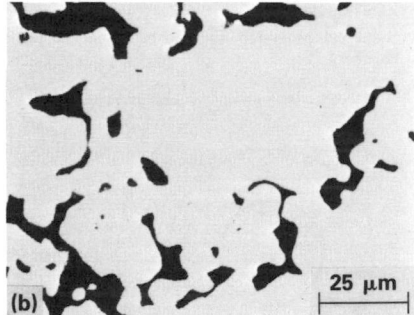

Fig. 12 Vertical section through a sintered iron sprocket tooth at tip of tooth

Microlaminations and particle boundaries resulting from overpressing are shown. Unetched

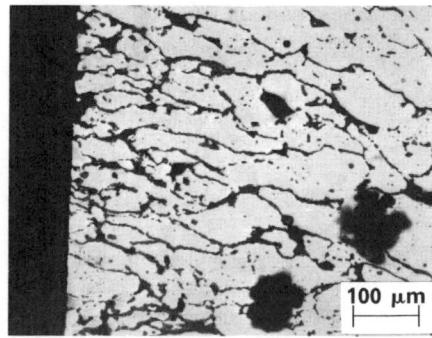

Etching makes grain size visible. In carbon steels, the area fraction of pearlite multiplied by 0.8 equals the approximate combined carbon content for hypoeutectoid steels. Steels are etched in 2% nital (low carbon) or 4% picral (high carbon). Stainless steel etches well in Glyceregia, and copper-based materials etch satisfac-

Fig. 13 Undissolved graphite in sintered atomlzed and diffusion alloyed iron

Pores containing gray flakes of graphite are indicated by arrows. Sintered for 5 min in hot zone at 1120 °C (2050 °F) in dissociated ammonia. 0.5% graphite added. Unetched

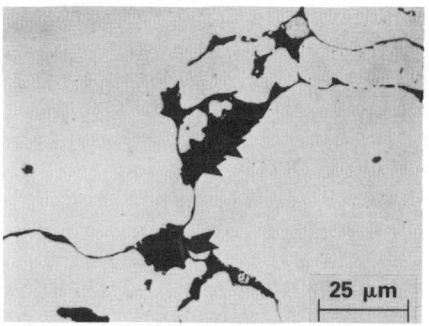

Fig. 14 Pore distribution in sintered type 316L stainless steel

Rounded pores are reduced in size and isolated as a result of high temperature sintering. Unetched

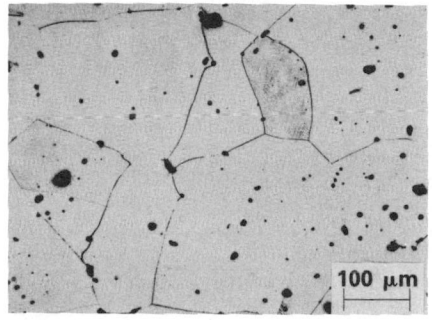

Fig. 15 Pore distribution in sintered and sized bronze bearing

Cross section showing pores not connected to the inside diameter surface at left. Unetched

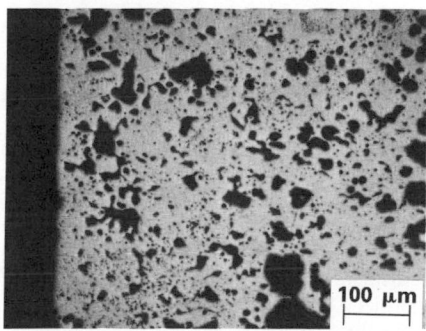

Fig. 16 Inclusions in atomized diffusion alloyed iron

Gray areas are manganese sulfide added for machinability. Sintered for 45 min in hot zone at 1120 °C (2050 °F) in dissociated ammonia. Unetched

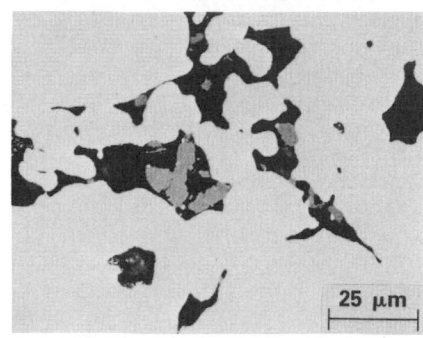

Fig. 17 Atomized iron wlth 0.8% graphite addition

Average sinter produced all-pearlite structure. 4% picral etch

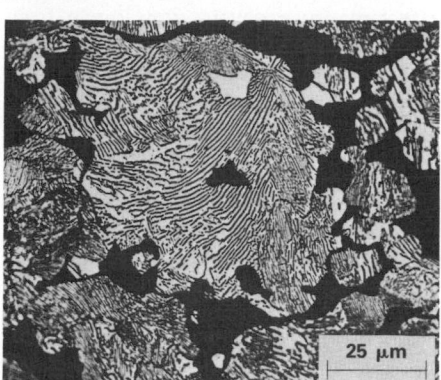

torily in a mixture based on ammonium hydroxide.

Plain Carbon Steels. Carbon (graphite) goes into solution rapidly in iron (5 min at 1040 °C, or 1900 °F); therefore, free graphite is rarely visible. Figure 17 shows all pearlite in an atomized iron material.

Copper Steels. Copper is added in amounts of 2 to 20% to iron or to iron and carbon. In practice, up to 3% Cu dissolves under conventional sintering conditions. Copper should melt at 1083 °C (1981 °F) and run into the small capillaries between or inside iron particles. From there, it gradually diffuses into the iron particles. If copper is present as irregular, unmelted lumps after sintering, the part is clearly undersintered. Furthermore, 2 to 3% Cu should be dissolved, not present as melted islands of copper. Alloys with 5 to 20% Cu exhibit melted islands of copper and sometimes surround the iron particles.

Copper generally causes growth during sintering, even though it forms a liquid phase. Melted copper migrates to the interparticle capillaries, including those at the points of contact between the iron particles. From there, it diffuses volumetrically into particles and along grain boundaries, resulting in lattice expansion or swelling of the iron at points of contact. Growth of the compact is rapid at this point and can occur in 3 to 10 min above 1083 °C (1981 °F). This effect is most pronounced in low-surface-area, regular particles (atomized iron). In the sponge irons, which have high surface area and significant internal porosity, some copper enters internal pores and less migrates to the interparticle contacts. This results in significantly less growth. The higher surface area also causes the shrinkage reaction. The addition of carbon (graphite) interferes with copper growth and frequently is used to control parts to near die size.

Although 2 to 3% Cu can disappear into solid solution, copper concentration in iron is nonuniform. Examination of 4% picral etched iron-copper and iron-copper-carbon materials exhibits a coring effect, with a tan staining of ferrite particle surfaces due to copper solution and precipitation.

Infiltrated Steels. During infiltration, an iron-graphite matrix at 6.4 to 6.8 g/cm^3 is partially filled with copper alloy. This is usually accomplished with one furnace step, in which a wafer of copper alloy is placed on or under the green matrix. The assembly passes through the furnace, where the copper alloy melts and infiltrates the pores by capillary action. The smallest pores are filled first (in zones of highest density). The infiltrant wafer may leave a residue, which usually does not adhere to the part. However, under conditions of low dew point or a high degree of infiltration, the residue may adhere. Most common infiltrants are alloyed to saturation so that the iron compact does not dissolve or erode in the liquid copper as it enters the compact. If the infiltrant is not properly alloyed, it dissolves the iron and leaves an eroded network of shallow cracks on the part surface.

Final densities usually range from 7.2 to 7.6 g/cm^3, which indicates that parts are 89 to 94% dense and exhibit 11 to 6% porosity. Distribution of the porosity, when examined metallographically, should be uniform. If a part has a low-density zone before infiltration, that zone may remain the least dense region after infiltration. Parts should not vary more than 0.3 g/cm^3 between points. This can be determined metallographically if the sample is polished properly.

Infiltrants commonly contain alloying agents such as manganese, iron, copper, and zinc, some of which increase hardenability. Consequently, small parts and thin sections of larger parts may air harden during normal furnace cooling. This may cause difficulties in subsequent machining. Alloying to achieve this hardenability effect may use matrix irons made from prealloys, partially diffused alloys, or elemental additions of nickel, molybdenum, and manganese. Infiltrants with cobalt may produce a gray iron-cobalt intermetallic that precipitates on the iron particles in the region where the infiltrant enters.

Prealloyed Steels. Type 4600 alloys containing 1.8% Ni, 0.5% Mo, and up to 0.8% C added as graphite enjoy wide usage, as well as the 2000 types containing 0.45% Ni, 0.30% Mn, and 0.6% Mo. Molybdenum sometimes is combined with 1.5% elemental carbonyl nickel for improved hardenability and toughness. Prealloyed steels provide good hardenability, uniform surface hardness, and uniform dimensional change during heat treating. They frequently are forged to full density, but also are used in densities ranging from 6.6 to 7.4 g/cm³. Heat treated structures usually are completely martensitic in the outer 5 mm (0.2 in.) of the part, with mixtures of fine pearlite and martensite in the interior. Sintered products, which contain eutectoid products, are examined infrequently. In the as-sintered condition, the carbide platelets tend to be distributed uniformly throughout the ferrite, with no obvious proeutectoid ferrite. Consequently, it is difficult to determine carbon content, compared to the method used for wrought plain carbon steels, based on the area fraction of pearlite.

Prealloyed powders are subject to contamination by unalloyed iron powders. During sintering, carbon preferentially alloys with the prealloyed material, leaving behind low-carbon ferrite grains. The latter grains do not harden during subsequent heat treatment. Up to 3% of such ferrite impurities is not detrimental. More is acceptable, especially in less than fully dense materials.

Nickel steels commonly are made from elemental additions of 2 to 4% carbonyl nickel. Carbon contents ranging from 0.4 to 0.6% are best suited for heat treated parts, while 0.6 to 0.8% produces the strongest sintered parts. Up to 2% Cu may be added for size control during sintering, because nickel mixtures tend to shrink during sintering. Copper does more than control size, however; it helps the nickel go into solution and increases the hardenability of the iron alloy by its own dissolution.

Nickel steels exhibit strengths of up to 1400 MPa (200 ksi), because they are inhomogeneous mixtures of hard and soft phases. A fully martensitic medium-density material is not the strongest alloy form, because tensile strength is ductility limited. The material is notch sensitive; the pores are the notches, and the material fails prematurely. Typically, martensite structures with 15 to 30% fine pearlite increase tensile strength. Fine pearlite is softer than martensite and interferes with crack propagation. Nickel particles do not fully alloy and create austenitic islands with nickel contents of 12%. These austenitic islands also decrease the notch sensitivity of the steels and raise the tensile strength.

The microstructure of an as-sintered 0.7 to 0.8% carbon steel with 2% admixed nickel exhibits essentially all pearlite and 15 to 25% by area of nickel-rich regions (light-colored islands). Edges of the nickel-rich areas are an air-hardening alloy, often showing needles of martensite with bainite (Fig. 18). These regions may have hardness values of 50 to 60 HRC in the as-sintered condition. Copper additions also promote the formation of hard island surfaces. Pearlite colonies usually are surrounded by a carbide-free white diffusion layer. This layer must contain sufficient nickel to prevent eutectoid decomposition. This layer is not proeutectoid ferrite, such as exhibited by 0.2 to 0.6% C (Fig. 18).

Iron-Phosphorus and Iron-Silicon Soft Magnetic Alloys. Iron-phosphorus materials contain 0.45 or 0.8% P blended in as ferrophosphorus master alloy. During sintering at 1120 °C (2050 °F), the compound melts and the phosphorus diffuses into the iron, improving the soft magnetic properties. This temporary liquid phase results in rounding of the porosity to the greatest degree of any of the common alloys. Phosphorus dissolved in the iron increases hardenability and toughness, as well as machinability.

Silicon is admixed as iron-silicon intermetallic compounds containing 17 to 31% Si to create final alloys with silicon contents of 1 to 3.5%. Sintering must be done above 1230 °C (2250 °F) to melt the intermetallic compounds and allow the silicon to dissolve. This requires a strongly reducing atmosphere or a vacuum to prevent the formation of silicon dioxide. Properly sintered materials exhibit no original particle boundaries or undissolved iron-silicon compound, and pores are well rounded.

Stainless Steels. The common wrought AISI stainless steel grades 303, 304, 316, 410, and 434 are available in approximately the same composition in powder form. All require sintering in hydrogen or low-dew-point dissociated ammonia, or in a vacuum of 0.150 to 0.500 torr to prevent

Fig. 18 Microstructure of sintered Fe-2Ni-0.5C alloy
Sintered for 30 min at 1120 °C (2050 °F). Arrows marked Ni outline nickel-rich particle. Arrow M: martensite or bainite at nickel-rich boundary. Arrows marked L: diffusion layer between nickel and pearlite. This is not unalloyed ferrite. Arrow F: ferrite. 4% nital etched

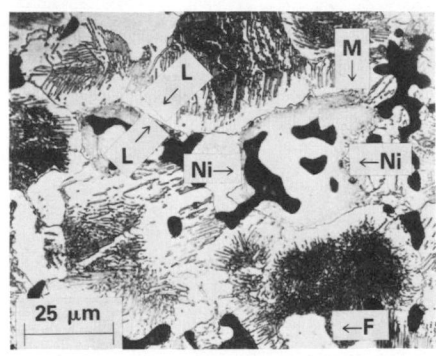

oxidation of chromium and silicon. Sintering temperatures of 1120 to 1150 °C (2050 to 2100 °F) result in a stainless steel with only average corrosion resistance, because of the high surface area after sintering. Improved corrosion resistance and much higher shrinkage during sintering result from sintering at 1260 to 1370 °C (2300 to 2500 °F). The higher temperature results in more rounded pores, with a tendency to close off pores (Fig. 14). Grain growth also results. With either sintering temperature, few or no original particle boundaries or chrome carbides remain. Type 410 stainless steel frequently has 0.15% C in solution, which allows the formation of martensite during cooling in the furnace. Martensite should be tempered for 1 h at 175 °C (350 °F) to improve strength.

Bronzes are made by admixing copper and tin and sintering for 10 to 20 min at 815 to 845 °C (1500 to 1550 °F). As it sinters into the copper, tin forms successively higher-melting-point compounds with the copper and ultimately goes into solution to form single-phase alpha bronze. Simultaneously, the cold worked copper grains recrystallize and grow as alpha bronze. When sintering is complete, the blue-gray copper-tin compounds are dissolved, and the small grain clusters of copper grow into fewer but larger alpha bronze grains. Re-pressing of bronze bearings results in alpha bronze with numerous slip lines and evidence of cold working.

Brasses and Nickel Silvers. Brasses contain 10 to 30% Zn prealloyed at atomization, and nickel silver contains 18% Zn and 18% Ni, also prealloyed. During sintering, these materials shrink rapidly in the

Table 6   Common hardness scales

Material	Sintered hardness scale	Heat treated hardness scale
Iron	HRH, HRB	HRB, HRC
Iron-carbon	HRB	HRB, HRC
Iron-nickel-carbon	HRB	HRC
Prealloyed steel	HRB	HRC
Bronze	HRH	...
Brass	HRH	...

presence of a liquid phase, resulting in 1 to 3% shrinkage. The resultant microstructure shows the absence of original particle boundaries and well-rounded pores. These alloys may exhibit 10 to 30% elongation when properly sintered.

Apparent Hardness and Microhardness

Porous materials exhibit wider variation in hardness testing than wrought counterparts. The entrance of the indenter into pores or groups of pores generally causes this effect. The seller and user of P/M materials should agree on which area or areas of a part are to be hardness tested. At least five consistent readings should be taken, in addition to any obviously high or low readings, which are discarded. The remaining five readings should be averaged. Because most published data show typical hardness values, the buyer and seller must agree on specified or minimum values. The average of five or more consistent readings must meet the standard hardness, not any single reading.

The article "Mechanical Properties of P/M Materials" in this Volume contains tables with hardness values for sintered materials; use of these scales is recommended for taking accurate measurements. These scales are summarized in Table 6.

Unlike fully dense materials, high hardness does not mean high tensile strength. As discussed previously, 15 to 30% fine pearlite in the iron-carbon sintered materials improves tensile strength; consequently, it is evident that high hardness values are not an indicator of high strength. Because P/M steels exhibit low hardenability, hardness specifications should be based on parts that have demonstrated successful service.

Although not yet widely used, the Rockwell B scale may be combined with a carbide ball for testing and controlling hardened parts. Scattering of data is minimized with the Rockwell B 1.58-mm ($^1/_{16}$-in.) diam ball, and it is useful up to 120 HRB (Ref 22, 23).

Fig. 19 Hardened and tempered Fe-0.8C alloy

Light areas are martensite (60 HRC); dark areas are fine pearlite (40 HRC). 4% picral etch.

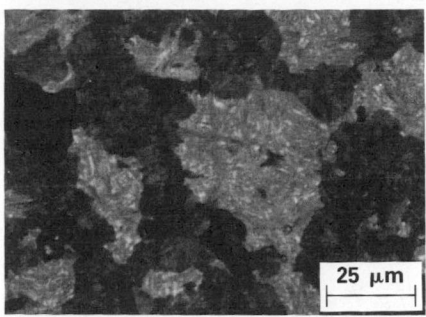

Microhardness of porous materials can best be measured with Knoop or diamond pyramid hardness indenters at loads of 100 g or greater. In atomized irons, particles exhibit minimal porosity; consequently, the Knoop indenter is suitable. It makes a very shallow indentation and is only infrequently disturbed by entering undisclosed pores. Care should be exercised in preparing the sample surface (see the guidelines given in the metallography section of this article). Use of the diamond pyramid indenter is particularly well suited to irons, which contain numerous fine internal pores. Because of its greater depth of penetration, the diamond pyramid indenter frequently encounters hidden pores. Microhardness testing and measurement of case depth are covered by Metal Powder Industries Federation standard MPIF 37.

For examining tempered steels, a light etching in 2% nital for 6 s is recommended. Fine pearlite appears as dark areas, unresolved at 1000×. Martensite appears light colored, with little or no evidence of platelets or needles (Fig. 19). This degree of etching enhances clarity and contrast for measuring the diamond indentation. Subsequent darker etching clearly shows the martensite needles and any retained austenite or carbides. Picral (4%) is recommended for subsequent etching (15 s) by immersion.

For microhardness testing of sintered steels, care must be exercised in the definition of a case and the choice of microconstituent to be tested. For example, in a carburized and quenched nickel steel with 0.3% C in the core, martensite, fine pearlite, ferrite, and nickel-rich areas are visible. Nickel-rich areas should be ignored. The surface is nearly all martensite with some fine pearlite. If 50 HRC is defined as the case depth cut-off, the existence of a fine pearlite colony, which is below 50 HRC, would seem to indicate that the

reading was taken beneath the case. Case may be defined as the point where there is less than 50% martensite by area, or where the hardness falls below the 50 HRC equivalent. By comparison, in testing a wrought material without porosity, a heavier load that makes a large indentation should be used, which averages the microstructures while pinpointing the 50 HRC equivalent. In a P/M steel, this would result in the pores being averaged in with the other phase.

The use of light loads (under 100 g) causes errors when converting to the Rockwell C scale if the 500-g Rockwell conversion table is used (Ref 24). An alternative is to provide individualized calibrations by removing pieces from several Rockwell standard blocks, with hardnesses ranging from 20 to 60 HRC, and metallographically mounting the working faces. These standards are carefully polished, allowing a direct calibration to be made between the Rockwell C scale and the length of the indentation. When this calibration is graphed, an unknown sample can be taken, the length (in filar units) of the hardness indentation made at 100 g can be measured, and the filar units can be converted to Rockwell C values.

Mechanical Testing/ Tensile Testing

The MPIF standard molded "dogbone" test bar is shown in Fig. 20. Recent work (Ref 25) has shown that a 3.6-mm (0.140-in.) thick bar exhibited improved gripping characteristics and a greater tendency to break within the gauge region. For testing to be meaningful, it is important to verify that such bars are free of microlaminations (Ref 25), which requires careful metallographic evaluation. Highest quality bars are molded in well-bolstered die sets (890-kN, or 100-ton rating), with generous exit taper, high green strength powder, and top punch hold-down, if possible. The formation of microlaminations is accentuated by high molding pressures of 415 to 1380 MPa (30 to 100 tsi).

With all test bars, it is necessary to determine that substantial carburization or decarburization resulting from sintering or hardening does not exist. If bars are heat treated, the microstructure at the surface and in the interior should be described in the test report, because many P/M steels have low hardenability. Heat treated bars with 15 to 30% fine pearlite in the martensite have been found to have the highest tensile strength (Ref 25). Ideal sintering conditions lower the tolerable fine pearlite content.

Fig. 20 Flat unmachined MPIF standard 10 tensile test bar

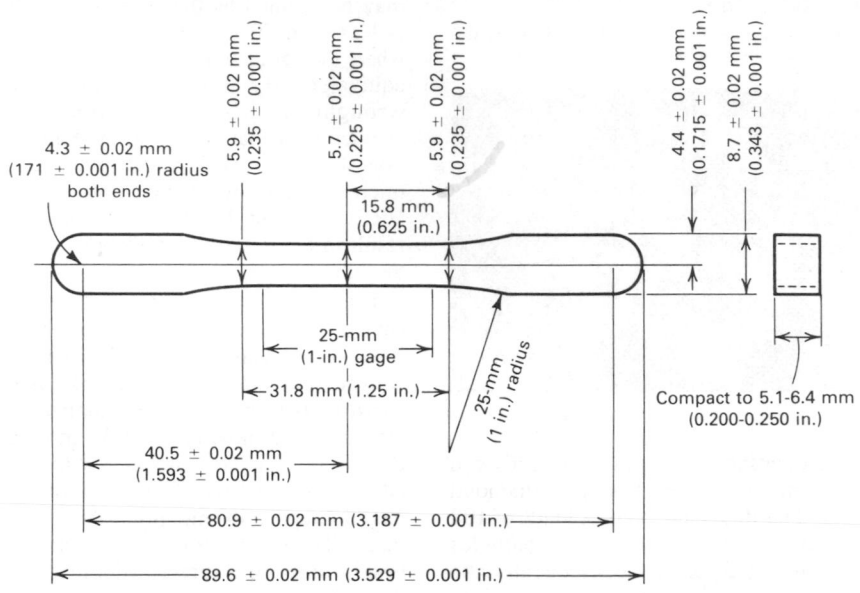

Fig. 21 Standard MPIF machined tensile test bar

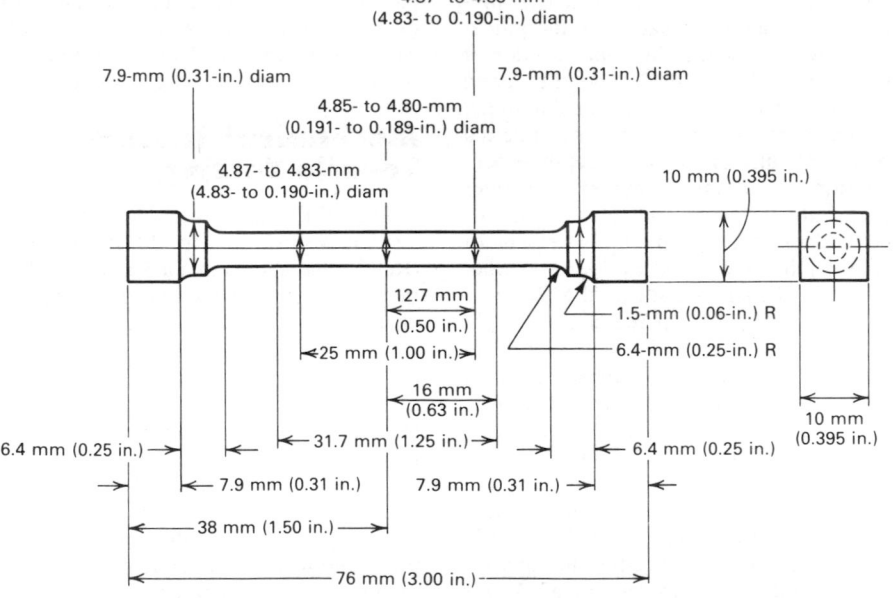

Fig. 22 Sorting grid categories arbitrarily assigned to measuring screen

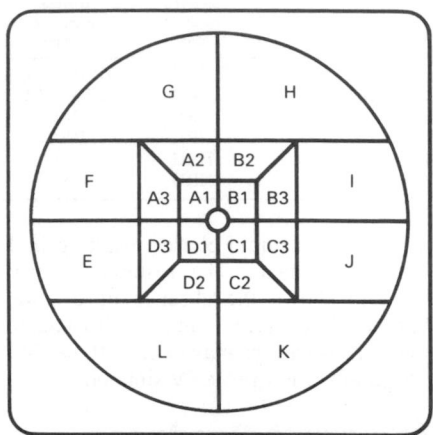

For heat treated materials, MPIF standard bars tend to slip in the grips, and a machined buttonhead bar (Fig. 21) provides an increase in accurate data. The machined bar shows an increase in apparent tensile strength of up to 50% compared to molded bars. Even with machined bars, some heat treated materials exhibit such low elongation that failure occurs prior to reaching 0.2% permanent deformation.

Transverse Rupture Strength. The transverse-rupture test breaks a 31.8- by 12.7- by 6.4-mm (1.25- by 0.5- by 0.25-in.) test bar as a simple beam. The test is theoretically valid only for perfectly brittle materials and measures the stress in the outer fiber at fracture. For many sintered steels, transverse-rupture stress is considered to be equal to twice the ultimate tensile strength. This is only an approximation, and a revised MPIF standard 35 (1984 edition) will present transverse-rupture stress values that correspond with ultimate tensile stress values for common P/M materials.

The transverse-rupture test is useful for comparing and evaluating materials for strength, even if the bar bends before fracture. These bars are preferred, because they can be molded and sintered conveniently. Testing is faster than when using a tensile bar. When heat treated, the test bar does not experience distortion. This testing procedure is used mainly as a quality control tool to ensure the maintenance of minimum mechanical properties. During P/M part production, this method is used to evaluate properties, such as compressibility, sintered strength, and dimensional change, of incoming powder.

Unnotched Charpy Impact Strength. A 10- by 10- by 76-mm (0.35- by 0.35- by 3.00-in.) molded bar is used for impact testing. The unnotched bar provides a more sensitive test, suitable for use on materials with an impact strength below 14 J (10 ft · lb). The bar is conventionally struck on the surface that contacted the die at molding. This concept originated with the similar impact loading direction of the gear teeth on P/M parts (Ref 26). Minimal fracture toughness data currently exist on common sintered materials.

Proof Testing. The most common method of demonstrating the strength of sintered parts is through mechanical tests that stress parts to failure. Qualification samples or first production lots are used to establish desired strength values; these data are incorporated in the part design specification.

For testing gears, several teeth are removed. The remaining teeth are loaded in a predetermined arrangement on a fixture. The load to fracture is recorded. To be meaningful, the destructive test must imitate service loading on the part.

Impact or drop weight tests also are used to evaluate materials. A drop weight test does not only use an acceptance or rejection evaluation. This testing procedure in-

Fig. 23 Frequency distribution for components from a forging batch relative to a reference sample chosen at random

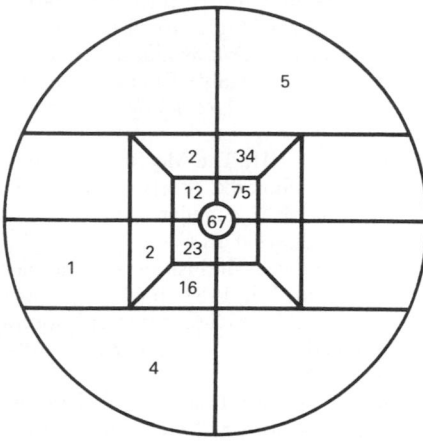

Fig. 24 Effect of changing the reference sample on the frequency distribution shown

(a) A sample originally in zone H (see Fig. 22) has been used as a reference. (b) A sample originally in zone D3 (see Fig. 22) has been used as a reference.

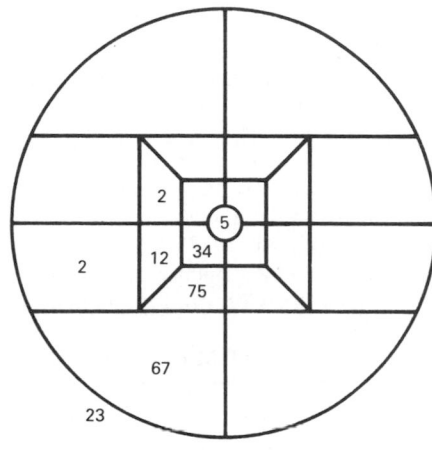

(a)

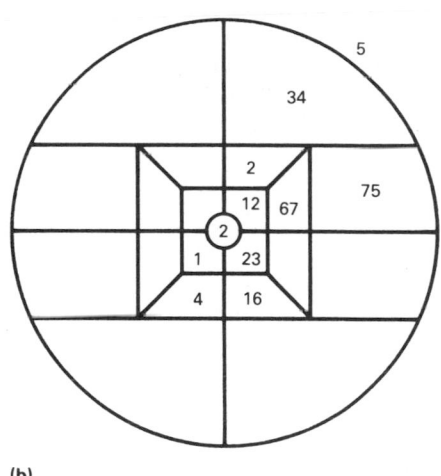

(b)

vestigates the impacts above the acceptance level and below the rejection value. When a part does not break, each succeeding load is increased until breakage occurs. Thus, if load P does not break a part, $1.05 \, P$ is used on the succeeding impact.

With static testing, the forthcoming update of MPIF standard 35 (Ref 27), which includes minimum tensile yield point and ultimate tensile data, may be used to calculate yield and fracture forces on a part. Testing to these loads demonstrates the presence of materials with specified minimum strength values.

Eddy Current/ Nondestructive Testing of Powder Forged Components*

Samples generally are taken on a statistical basis during powder production and consolidation, as part of process and dispatch control procedures. Dimensional checks and surface crack detection procedures similar to those used for wrought and cast components may be applied to P/M forged parts. Additional verification of the microstructural integrity of P/M forged components can be obtained by magnetic bridge sorting.

Nondestructive Testing Examination. Magnetic bridge sorting, in conjunction with an oscilloscope, can be used to compare the eddy currents developed within a forging as it is placed in a coil that carries an alternating current with currents produced in a reference sample un-

*Adapted from Ref 28

der similar conditions. Differences are indicated by the displacement of a light spot from its balanced position in the center of the measuring screen of the system.

If the part being tested is similar to the reference sample, the light spot returns to the center of the screen. The screen can be arbitrarily divided into a number of zones, as illustrated in Fig. 22. A typical frequency distribution of components within a forging batch, relative to the reference sample, is shown schematically in Fig. 23. Reference samples are selected randomly. Figure 24 illustrates changes in the frequency distribution when the reference sample is changed.

Once the frequency distribution has been established for a limited number of components within a forging batch, selected

components that are representative of several zones on the screen are subjected to metallographic examination. Limited metallographic testing thus can be used to check the metallurgical integrity of parts from various zones.

Once acceptable zones have been defined, the entire forging batch can be assessed via the magnetic bridge; components in unacceptable categories are automatically rejected. Experience with the technique minimizes the number of parts requiring sectioning for metallographic examination. Core hardness, surface decarburization, surface oxide penetration, and porosity also can be evaluated by this technique.

Magnetic bridge sorting, an adaptation of the technique used to test drop forged parts, enables potentially defective components to be eliminated from a batch of forgings. It also can be used to provide 100% inspection of the metallurgical integrity of a forging batch.

REFERENCES

1. *P/M Design Guidebook,* Metal Powder Industries Federation, Princeton, NJ, 1983, p 15
2. "Anchor MH100 Standard Molding Powder," Hoeganaes Corp., Riverton, NJ
3. "Atomet 28, Sintered Properties of P/M Copper Steels," Quebec Metal Powders Ltd., Sorel, Quebec, Canada
4. "Controlled Dimensional Change," SCM Metal Products, Cleveland
5. Pease, L., III, "An Assessment of Powder Metallurgy Today, and Its Future Potential," Paper No. 831042, Passenger Car Meeting, Society of Automotive Engineers, Warrendale, PA, 1983
6. Moyer, K., Measuring Density of P/M Materials with Improved Precision, *Int. J. Powder Metall. Powder Technol.,* Vol 15 (No. 1), 1979, p 33
7. "Standard Test Method for Density and Interconnected Porosity of Sintered Metal Powder Structural Parts and Oil Impregnated Bearings," ASTM B 328, American Society for Testing and Materials, Philadelphia, 1983
8. "Determination of Density of Compacted or Sintered Metal Powder Products," MPIF standard 42, Metal Powder Industries Federation, Princeton, NJ, 1977
9. "Permeable Sintered Metal Materials—Determination of Density and Open Porosity," International Standards Organization Standard No. ISO

2738, available from American National Standards Institute, New York, 1982

10. Githens, N. and Kane, J., "Study of Density Changes in Age-Hardening An Aluminum—5% Copper Alloy at 190 °C (374 °F)," senior thesis, Drexel University, Philadelphia, 1970

11. Tilton, L. and Taylor, J., Accurate Representation of the Refractivity and Density of Distilled Water as a Function of Temperature, Research Paper RP971, *J. Res. Nat. Bureau Stand.,* Vol 18, Feb 1937, p 205-214

12. Hemins, I., Precision Density Measurements of Silicon, *J. Res. Nat. Bureau Stand.—Anal. Phys. Chem.,* Vol 68A (No. 5), 1964, p 529-533

13. Bowman, H., Gallagher, W., and Schoonover, M., "The Development of a Working Density Standard," Instruments Society of America Preprint 14.8-4-65, 20th Annual ISA Conference and Exhibit, Los Angeles, Oct 1965

14. Bowman, H. and Schoonover, R., Procedure for High Precision Density Determinations by Hydrostatic Weighing, *J. Res. Nat. Bureau Stand. —Chem. Eng. Instrum.,* Vol 71C

(No. 3), 1967, p 179-198

15. Bowman, H., Schoonover, R., and Carroll, C., A Density Scale Based on Solid Objects, *J. Res. Nat. Bureau Stand.—Anal. Phys. Chem.,* Vol 78A (No. 1), 1974, p 13-40

16. Bowman, H., Schoonover, R., and Carroll, C., The Utilization of Solid Objects as Reference Standards in Density Measurements, *Metrologica,* Vol 10, 1974, p 117-121

17. Pease, L., III, Metallography and Properties of Sintered Steels, in *Progress in Powder Metallurgy,* Vol 33, Metal Powder Industries Federation, Princeton, NJ, 1977

18. "Model FM Conductivity Meter," Magnaflux Corporation, Chicago, 1971

19. Crack Test for P/M Compacts, *Prec. Metal Mold.,* March 1967, p 66

20. "Nondestructive Quality Control on Ferrous P/M Parts," Proposal E-83-217, Southern Research Institute, Birmingham, AL, Aug 1983

21. *Metals Handbook,* Nondestructive Inspection and Quality Control, Vol 11, 8th ed., American Society for Metals, 1976, p 104

22. Judge, W.Q., private communication, Engineered Sinterings and Plastics, Watertown, CT

23. Wood, G.E., *et al.,* Apparent Hardness Testing for Ferrous P/M Parts, *Prec. Metal Mag.,* Nov 1967, p 86

24. *Metals Handbook,* Nondestructive Inspection and Quality Control, Vol 11, 8th ed., American Society for Metals, 1976, p 17

25. Pease, L., III, The Mechanical Properties of Sintered Steels and Derivation for MPIF Standard 35, in *Progress in Powder Metallurgy,* Vol 37, Metal Powder Industries Federation, Princeton, NJ, 1981, p 113

26. Kravic, A., private communication, International Nickel Company, Suffern, NY

27. Metal Powder Industries Federation Standard No. 35, "Mechanical Properties," 1984 (to be published)

28. James. W., Quality Assurance Procedures for Powder Forged Materials, in *Powder Metal Parts, Design and Implementation for Economics and Reliability,* International Congress and Exposition, Detroit, March 1983, Society of Automotive Engineers, Warrendale, PA

Special and Developing Consolidation Processes

Injection Molding

By Arnold R. Erickson
Director, Materials Research
Brunswick Corp.
Technetics/Witec
and
Raymond E. Wiech, Jr.
Physicist
Geneva Materials Research, Ltd.

INJECTION MOLDING provides a unique method of consolidating metal or ceramic and cermet powders to produce a new range of metal parts that extends the scope of P/M techniques. Injection molding differs substantially from conventional P/M techniques. Much finer powders are used; particle size is 10 μm, rather than 100 μm as in conventional P/M processes. Large quantities of thermoplastic binders are used to permit injection molding into dies rather than gravity filling of dry powders into dies. Low uniform pressures are applied isostatically in injection molding dies; no other compaction pressures are used unless secondary coining or sizing operations are required.

Intricate shapes can be made with injection molding. However, a more complex and costlier process is required to add and remove binders. Dimensional shrinkage is large and isotropic, and usually occurs during sintering rather than compaction. Parts are denser and have higher mechanical properties for a given set of sintering conditions. Properties are essentially isotropic.

Because injection molding is more costly than conventional P/M techniques, injection molded parts are not likely to displace conventional P/M parts that fulfill their design function. This technology permits production of stronger, more uniform, and more complex P/M parts as required. Some typical injection molded parts are shown in Fig. 1 and 2.

Processing

Individual process steps in injection molding are selection and production of

Fig. 1 Hardened nickel steel gun parts produced by injection molding
Courtesy of Multimaterial Molding Corp.

Fig. 2 Type 316L stainless steel parts produced by injection molding
(Top) Spring guide for check valve; (center) orthodontic parts; (bottom) lock parts

metal powders, mixing, molding, debinding, and sintering. These are discussed in detail below.

Metal Powder Selection and Production. Because the rate of diffusion is inversely proportional to the square of the particle diameter, shrinkage and densification of a porous powder part proceeds much more rapidly by minimizing particle size, with remnant pores in the sintered part being smaller. Typical powder sizes used for injection molding are in the range of 0.5 to 20 μm, in contrast to the relatively coarse free-flowing powders having particle sizes greater than 40 μm used in conventional P/M. In this article, the finer powders used for injection molding are referred to as "dusts," and conventional free-

flowing materials are referred to as "powders."

The high free surface energy per gram and chemical reactivity of metal dusts have made their industrial use unpopular. Metal dusts are pyrophoric and relatively easy to disperse in air; consequently, they can be explosive. Airborne metal dusts are irritants to workers and may be carcinogenic. For more information, see the articles "Explosivity and Pyrophoricity of Metal Powders" and "Toxicity of Metal Powders" in this Volume.

The conventional compact-and-sinter approach has not been particularly effective with metal dusts due to difficulties in mechanically compacting them. Because of their high sintering potential, the final sintered density of metal dusts is substantially independent of the initial density of the green compact. For example, the sintering of a low-alloy steel with an initial particle size of 3 μm produces a product with a final sintered density of about 95% of the theoretical density for green parts that had an initial porosity of 35 to 55%; linear shrinkage of the green part during sintering thus is about 20%. The part will severely warp unless the initial density of the green part is very uniform.

Green parts that have been formed in a compacting die in a press generally vary in density at different sections of the compacted green part. This effect primarily is due to wall friction in the die and is compounded by the internal pressure distribution in the compacting dust. For example, green density of a compacted cylinder is greater at the ends, producing a sintered part with an hourglass shape.

The initial density of metal dusts is low and nonuniform within a poured mass, so large volume compaction ratios result. The entrapped air frequently causes cracks in the green part, because the small pore size restricts rapid dissipation of internal pressure after compaction. Fires and explosions also may occur in conventional binder-free compacts. Vacuum presses are available, but the long cycle time required makes production low.

The high realizable densities of sintered metal dusts make their use very attractive, as many other properties—for example, elongation, yield strength, and tensile strength—increase as density increases. However, the difficulties encountered in handling have prevented metal dusts from being effectively utilized by the compact-and-sinter approach of conventional P/M parts.

In injection molding, metal dust is mixed with a thermoplastic binder so that the pore volume of the molded part is occupied by a thermoplastic material rather than atmospheric air. By coating the dust particles with a thermoplastic, the particles are effectively removed from atmospheric contact. This greatly inhibits the chemical reactivity of the metal dust and virtually eliminates the pyrophoric hazard of metal dusts. The thermoplastic effectively binds the particles into a uniform mass, eliminating dust explosions and airborne particle toxicity problems. Pore volume is occupied by an incompressible material, so green density of a part is determined by controlled, measurable ratios of thermoplastic and metal dust.

Green part density is established by a controlled mixing process instead of by an uncontrolled compaction process. With the correct conditions, the mixture of thermoplastic and metal dust will form a thermoplastisol feedstock, which can be formed into complex geometries by conventional plastic forming techniques. A complex geometry green part can be injection molded or vacuum formed at relatively low pressures and temperatures while maintaining extremely uniform densities throughout.

In most cases, production of metal dusts is more difficult than production of metal powders. As particle size decreases, the free surface energy per unit weight of the mass increases. Conceptually, there are two ways to make dusts: by breaking down a large mass into dust by some comminution process, or by generating dust particles from condensation of vapors.

The most common way to produce dusts from massive forms is to spray molten metal, usually by gas atomization techniques, to form a population of solid particulates. This technique permits the production of prealloyed powders and dusts. In injection molding, spherical particles are desirable, rather than the irregular particles used in compact-and-sinter manufacture. The larger particles are removed from the atomized particulate population, leaving the desired dust. Because of the broad size distribution of the atomized particulate population, the amount of dust produced is small, making it substantially higher in cost than metal powders on a weight basis.

The technique most commonly used to manufacture metal dusts by condensation is the carbonyl process. A metal carbonyl gas such as iron pentacarbonyl is introduced into a reactor chamber, where it is thermally decomposed to iron and carbon monoxide. The iron is deposited under carefully controlled conditions to build up a particulate population of spherical particles several micrometres in diameter. The dusts produced by the carbonyl process are also considerably more expensive than metal powders on a weight basis.

A third technique of metal dust manufacture, extensively used for producing refractory metal dusts, reduces a metal oxide dust in a hydrogen reduction furnace. For example, molybdenum dust is produced by hydrogen reduction of high-purity molybdic oxide.

Metal dusts as manufactured (virgin dusts) are pure, very reactive, and can be quite hazardous. In some cases, the dust is slowly exposed to an oxidizing atmosphere to passivate it by slowly creating an oxide film on the surface of the particles. The dusts are normally stored in steel drums weighing several hundred pounds. Frequently, the dusts are stored under a dry inert gas, such as argon, during shipment or if they are to be stored for long periods. They usually are classified as combustible material when being shipped.

Mixing of the various materials to make the molding feedstock is critical, because it is the base for the other process steps. All of the parameters of the rest of the process, as well as the characteristics of the final sintered material, are established here.

The feedstock material designer must weigh many, sometimes conflicting, requirements to produce an effective feedstock material. The feedstock must perform like a stable thermoplastic during the molding process. During the debinding, it must readily lose the plastic. The thermoplastic binder system used must be stable and repeatable. Considerable experience is involved in designing the binder, and the various binders employed are considered to be proprietary to the feedstock manufacturers.

An example in Ref 1 describes the production of a molded alumina part in which the pores of the 0.6 μm alumina powder are vacuum impregnated with binder. The volume percent particulate will vary from 40 to 90%, depending on the density and pore volume of the powder. The binder cited in the example contained 50% caranuba wax and 50% low-density polyethylene.

Because the green part undergoes a substantial linear shrinkage of about 20% during processing and sintering, the feedstock manufacturer must hold the volume ratio of binder to dust within a tight tolerance, as it is this ratio that establishes shrinkage and, therefore, final part dimensions. Shrinkage, under standard processing conditions in a standard test mold, is a part of the feedstock specifications. The chemistry of the final part is also established in the mixing step. The metal dusts can be elemental metal or mixtures, or can be prealloyed. Because the particles are so small, complete diffusion of powder mixtures is achieved, so the original particles cannot be identified in the sintered parts.

The rheological properties of the feedstock are established in the mixing. Proper material dispersion is essential. The feedstock producer breaks up all dust agglomerates and uniformly coats each particle with binder. A primary objective of the mixing of feedstock materials is to ensure that the mix is absolutely uniform so that the density of the dust on a microscopic scale is completely uniform through the molded part when the feedstock is injected into the mold cavity. This makes part shrinkage uniform and isotropic.

In general mixing procedure, the dusts selected for the mix are weighed carefully. The dusts are placed in a blender, usually under inert gas to prevent fires or explosions that may be ignited by frictional or triboelectric effects, and blended to a uniform dispersion. The binder is placed in a temperature-controlled mixer and heated to its melting temperature. The blended dust is added to the molten binder and mixed for a prescribed period to achieve a uniform blend with the consistency of toothpaste. The mixture is then cooled and removed from the mixer. The feedstock mass can be extruded and cut into pellets for ease in feeding into the molding machine.

Molding. Essentially conventional plastic molding equipment is used to mold the feedstock material into the desired shape. The thermal conductivity of the metal dust feedstock is many times greater than that of thermoplastics, so mold temperatures are much warmer than are normally found in thermoplastics. Even though the heating capacity per unit volume for these mixtures is higher than for thermoplastics because of the high density of the metal, the thermal diffusivity (thermal conductivity divided by heat capacity and density) is still very much higher for these feedstocks. Mold temperatures of 60 °C (140 °F) and higher are common, compared to about 20 °C (70 °F) for plastics. Molding temperature of the feedstock is correspondingly low, about 160 °C (320 °F). Injection pressures are generally lower than for plastics, the rule being that the minimum pressure required to fill the mold cavity is desired. Injection velocity is much less than for plastics.

Mold dimensions are calculated by applying a "shrinkage factor," which is supplied by the feedstock manufacturer as a feedstock property, to the part drawing to determine cavity dimensions. Only linear dimensions are scaled, with angles being unchanged. Thus, when using a binder having a shrinkage factor of 1.181, all linear dimensions on the part would be multiplied by 1.181 to determine the mold cavity dimensions.

As with plastic molding, shrinkage varies somewhat from the planned shrinkage factor depending on the direction of the dimension relative to the feedstock flow from the mold gate. Because various parts tend to be different, mold makers usually modify their molds after the initial trial part manufacture. Conventional plastic molds constructed by conventional methods are used, except that the ejection pins, runners, and gates are larger than those used for plastics.

Due to the modest temperatures and pressures involved in molding, it is possible to use soft tooling for short runs and experimental parts. Low-carbon steel and aluminum are used successfully. However, it is preferable to use a hardened cavity to avoid scratches, which frequently cause the parts to stick in the mold. Silicone mold releases should be avoided, because they can leave an unacceptable residue that contaminates the sintered part.

The feedstock material can be recycled through the molding machine many times (many more times than for plastic) without affecting its properties or the properties of the final sintered part. Normal molding practice recycles the sprues, runners, and gates, so that essentially 100% of the feedstock is utilized in P/M parts. The process is repeatable and reliable to the point where essentially 100% yield is realized. Green parts of questionable quality should be recycled.

Debinding. Before sintering can take place, the thermoplastic binder must be removed from the pore volume. The debinding process opens the green body by converting the binder to a fluid (liquid, gas, vapor, or a combination) in stages so that the binder can flow from the green body to an external region that is maintained at a lower physical-chemical potential than the interior of the body. This can be accomplished by a variety of methods. The particle-to-particle tensile forces should not be exceeded by any of the binder removal forces or cracking, deformation, or bloating of the body will occur. This requirement places narrow limits on the time-temperature relationships for the debinding operation, which results in slower and much closer process control requirements when compared to conventional compact-and-sinter processing.

Two methods of debinding are in use. The first (Ref 1) is a two-stage method in which the part is first processed in a solvent extraction unit, which removes a portion of the binder and opens the green body. Examples given in Ref 1 discuss the use of trichloroethylene in the liquid phase for the solvent. The extracted part is then placed in an oven and the remainder of the binder burned out. The second method (Ref 2) of debinding is an evaporative method in which the binder is designed to provide the characteristics of reasonable rheology and stability in molding, yet can be rapidly and thoroughly removed from the green body under debinding conditions.

Sintering is accomplished in a programmable atmosphere-controlled furnace. Because some binder is left in the parts to provide green strength so they may be handled, the sintering process must finish the debinding and then provide the necessary physical-chemical environment to sinter the parts.

Typically, the parts are sintered on racks in batch furnaces. If parts are racked at the injection molding machine, the racks then may be transferred on for debinding and sintering. Each part may have to be supported differently to maintain uniform part exposure to the debinding and sintering atmospheres.

The sintering furnace (and debinding units) typically are controlled by microprocessors to maintain product uniformity. Parts with thick walls require more processing time than thin-walled parts. The sintering cycle removes residual binders and oxygen prior to raising the temperature to sintering temperature. The process is designed to be decarburizing, utilizing appropriate time, temperature, and atmosphere. As with conventional P/M parts, carbon content can be controlled by adding graphite to the mix and controlling sintering time, temperature, and atmosphere. Alternatively, the low-carbon parts from sintering can be put through secondary heat treating operations, such as conventional carburizing or carbonitriding.

Miscellaneous Considerations. Applications for injection molding with the greatest cost effectiveness are for those parts that are difficult to form or have a high material value. Generally, the materials have a melting point in excess of 1300 °C (2370 °F) and represent those metals that are difficult to melt and cast. The fact that metals with high engineering properties can be processed on the same equipment with the same type of molds as plastics means that the same exotic geometries that are found in plastics are also now achievable in metals. Prior to plasticized metal dust technology, cost-effective shape production of metals was only practical in die casting and, to a lesser degree, in investment casting. Designers of metal parts and components must realize that all of the complex plastic parts found in a toy store, such as model airplanes, also can be manufactured in a cost-effective way in alloy steels, tool steels, stainless steels, Monel alloys, and so forth.

Shape, as a cost factor in design, has become much less limiting. Because essentially 100% of the feedstock is utilized, scrap also has become much less of a consideration in parts design.

The parts that have been most applicable to injection molding have been those that are small (under 2 cm, or 0.8 in.) and highly configured. The tolerance that is held is normally ±0.3% of the dimension. With care, a tolerance of ±0.1% is practical. One production job running at the time of this writing was holding a tolerance of 25 μm over a dimension length of 33 mm (1.3 in.) in a steel that had been heat treated to 54 HRC in a separate heat treatment operation after sintering. Sections as thin as 1.25 mm (0.05 in.) are readily fabricated, and holes as small as 0.5 mm (0.02 in.) are readily molded into parts.

Co-molding of different materials is possible when the materials have either compatible or overlapping process parameters. For example, parts have been made that are double shot in the molding machine so that a portion of the part is a nickel steel while another portion is type 316 stainless steel. After processing, the part is nonmagnetic and non-heat treatable in the type 316 stainless steel portion and magnetic and heat treatable in the other portion. Many such combinations are possible. The design of double shot mold gating is learned through experience.

Secondary operations are possible at all production stages. Parts may be welded in the green condition if they have the same processing parameters. This permits weldments that do not have large heat-affected zones. This is done by placing the desired faying surfaces together on the fixturing tray prior to debinding. The cross-flow of binders and dusts at the joint during debinding will create the weld. Care must be taken to avoid welding during normal processing by elimination of contact between adjacent parts.

Although parts may be machined in the green or debound state, they are quite fragile. They may be welded, plated, or otherwise treated as any forged metal part after sintering. Because the sintered parts have small, discontinuous pores, they may be hot isostatically pressed to a very high density without containerization.

Process Variations

Several variations of the injection molding process are currently being developed and used by industry. The Witec process consists of mixing powders and thermoplastic binders, injection molding to desired shape, debinding, sintering, and optional secondary operations. Battelle is developing an injection molding process, scheduled for completion in mid-1984. Details are proprietary, but the process is known to be very similar to the Witec process, using the same basic process steps.

Cabot Corporation has developed a unique modification of the injection molding process, which is described in a U.S. patent (Ref 3). Parts are formed from metal powders or dusts by mixing with a plastic medium composed of an organic binder

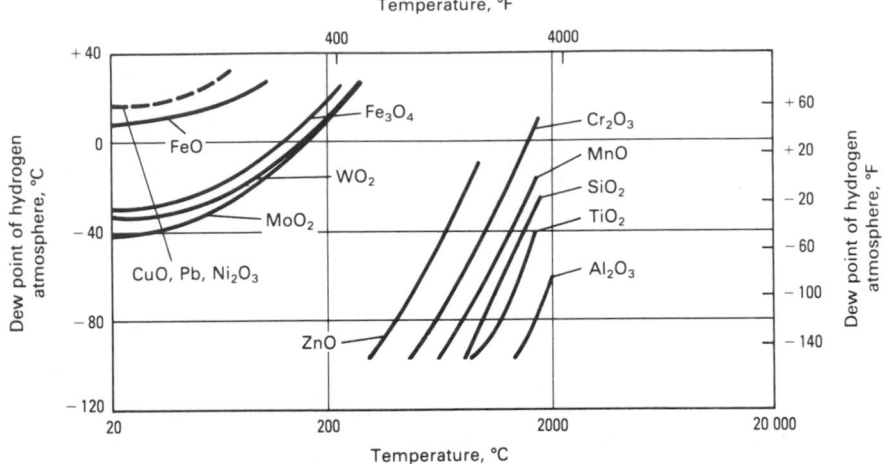

Fig. 3 Metal/metal oxide equilibria in hydrogen atmospheres

(methyl cellulose is preferred) dissolved in a solvent (water is preferred) in which the binder is soluble at room temperature. The binder is less soluble at a higher temperature and becomes very viscous. This binder system behaves like a thermosetting resin rather than a thermoplastic.

The plastic mixture is injected under pressure into a closed die preheated to a higher temperature, whereby the rejection of solvent and increase in viscosity of the plastic medium produce a compact sufficiently self-supporting to hold its molded shape and be ejected from the die. The compact is then dried to evaporate the remaining solvent, thus leaving interconnecting pores in the compact for the escape of gases that form during subsequent burning out of the binder during the sintering operation. Modification of conventional injection molding machines is necessary to accomodate the Cabot process, in which some solvent is removed during molding.

A binder composition provided as an example in Ref 3 for −325 mesh metal powder is:

	wt%
Methyl cellulose	2.0
Glycerin	1.0
Boric acid	0.5
Water	4.5

Materials and Properties

Materials. Plasticized metal dust feedstocks are classified into three groups indicative of their relative difficulty in processing. In general, the most difficult processing parameter is maintaining a sufficiently reducing sintering atmosphere to

completely eliminate all oxides from the dust surfaces prior to any significant or incipient sintering that occurs after oxides are removed and temperatures are high enough so that diffusion bond necks start to develop at contacting particle surfaces. The metal/metal oxide equilibria for several metals in hydrogen as a function of dew point are shown in Fig. 3.

The most commonly used atmosphere in sintering metal dusts is a mixture of a carrier inert gas of argon and a reducing gas of hydrogen. The reduction of the oxides on the particulate surfaces results in the formation of water, which raises the dew point of the system. Thus, the dew point must be maintained below the equilibrium value for the alloy being processed.

Because the sintering potential of metal dusts is very high, the temperature at which sintering begins is lower than for the equivalent metal powders. This means that the temperature of a reducing atmosphere is lower for metal dusts than for powders. The referenced metal/metal oxide equilibrium curves show that the dew point must be lowered to achieve a net reducing atmosphere as the atmosphere temperature is lowered.

Typical ferrous metal dusts begin to sinter at about 980 °C (1800 °F). The metal must be thoroughly reduced prior to sintering to avoid trapping oxides in the metal matrix, which causes embrittlement and reduction in diffusivity.

Feedstock materials are divided into groups according to processing difficulty. The first group can be processed at dew point levels above about −25 °C (−15 °F). Carbon steels, nickel steels, and Monel alloys are in this class. Alloys that are not commonly produced by the metals indus-

Table 1 Tensile strengths of injection molded tool steels in the as-sintered condition

MPIF designation	Chemical composition, wt%			Tensile strength		Yield strength		Elongation in 25 mm (1 in.), %
	Cu	Ni	C	MPa	ksi	MPa	ksi	
FC-0200	2	...	...	384	56	334	48	23
FC-0400	4	...	...	334	48	290	42	26
FC-0600	6	...	...	316	46	265	38	16
FC-0800	8	...	...	354	51	295	43	23
FN-0200	...	2	...	310	45	207	30	28
FN-0500	...	5	...	388	56	254	37	33
FN-0600	...	6	...	403	58	294	43	31
FN-0800	...	8	...	445	65	318	46	30
FN-0405	...	4	0.5	347	50	224	32	33
FN-0605	...	6	0.5	383	56	303	44	26

Note: Typical densities are 95 to 97% of theoretical. All materials are in dead soft condition, resulting from very slow furnace/cooling after sinter. Large increases in strength and hardness are obtainable with appropriate heat treatment.

Fig. 6 Sintered type 414 stainless steel

93% dense. Etched with Kallings reagent. Magnification: 100×

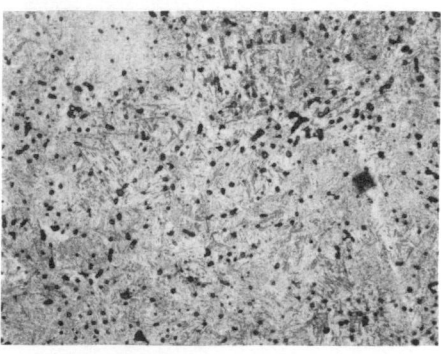

try due to specific chemistry or segregation problems can be made. For example, materials such as manganese that are added as deoxidizers are not necessary, because the sintering atmosphere provides that function.

Other materials such as molybdenum that are not normally added due to their high cost are introduced into the alloy very economically. Special alloys in quantities as small as 10 kg (22 lb) are mixed economically, so many mill-run problems are avoided and special, higher cost alloys can be specified by the designer.

The second group of feedstock materials requires a dew point of about −50 °C (−60 °F) to process. Chromium is the primary metal of concern, so those alloys containing significant quantities of it are in this group. Stainless steels, many tool steels, and Nichromes are in this class. Processing requirements for these materials are more restricted due to the lower dew point. Therefore, a part made from a material in this group requires a longer processing time than the same type of part made from a non-chromium-bearing feedstock because of the longer time required to achieve equilibrium. This adds to the production cost.

The third group of materials, such as titanium and aluminum, requires extremely low dew points and is primarily composed of laboratory materials at this time. The partial pressure of oxidizing gases such as water, carbon dioxide, and oxygen must be less than 10^{-5} torr.

Properties and Microstructure. As with conventional powder metals, the mechanical properties of injection molded parts increase with density, so the typical range of 93 to 97% density obtained with injection molded parts will provide strengths and elongations approaching wrought properties. Typical strengths and elongations for selected copper- and nickel-

Fig. 4 Sintered low-carbon iron plus 2% nickel steel

Iron + 2% nickel steel, 97% density, unetched. Tensile strength, 310 MPa (45 ksi). Magnification: 100×

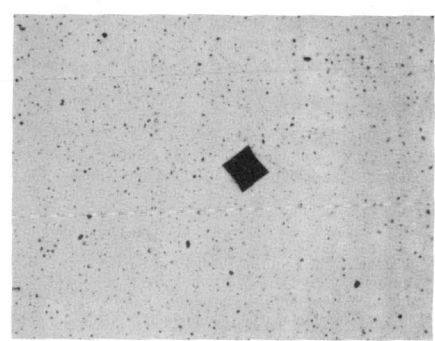

Fig. 5 Sintered low-carbon iron plus 2% nickel steel

Same as Fig. 4, etched with 10% ammonium persulfate. Shows ferrite with grain boundaries. Magnification: 100×

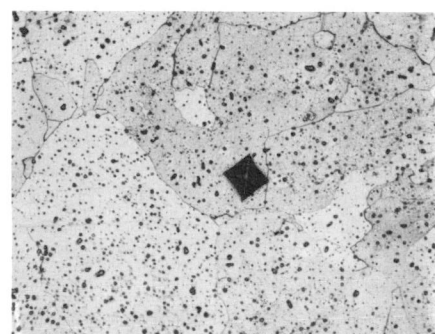

bearing steels in the as-sintered condition are given in Table 1. Tensile bars had necked down sections in the centers with gauge lengths of 18 mm (0.7 in.), overall lengths of 43 mm (1.7 in.), and thicknesses of 1.6 mm (0.064 in.). Widths of

Fig. 7 Sintered Hastelloy B-2

93% dense. Etched with Flat reagent. Magnification: 100×

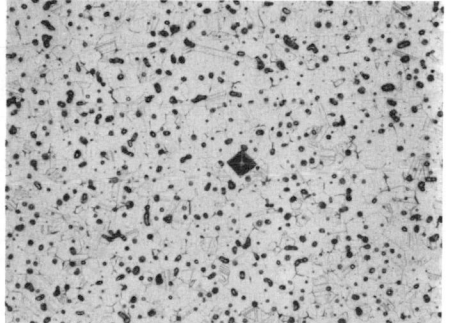

the grip and gauge sections were 10.0 mm (0.394 in.) and 2.4 mm (0.096 in.), respectively. Increased strengths and hardnesses are obtained by conventional carburizing or carbonitriding, followed by heat treating. The strength is uniform in all directions because of the uniform isostatic pressure applied during injection molding.

Figure 4, a photomicrograph of an unetched specimen of low-carbon iron plus 2% nickel steel illustrates the typical pore structure of injection molded parts. With the very small particles used, the pores remain small and isolated, with no interconnected porosity. Figure 5 shows the same field in the etched condition, displaying the typical enlargement of pores obtained by etching. There is no evidence of the presence of the original metal particles in the ferritic structure.

The etched structure in the photomicrograph in Fig. 6 shows an as-sintered specimen of type 414 stainless steel in which the retained austenite has not been completely converted to martensite. The properties of this material are:

Density	93%
Tensile strength	896 MPa (130 ksi)
Yield strength	606 MPa (88 ksi)
Elongation	1.2%
Hardness	48 HRC

Figure 7 shows an etched sample of Hastelloy B-2 with its austenitic grain structure and no evidence of the original metal particles. Typical properties of this material are:

Molybdenum	28.0 wt%
Iron	2.0 wt%
Nickel	70.0 wt%
Density	93%
Tensile strength	724 MPa (105 ksi)
Elongation	20%
Hardness	20 HRC

REFERENCES

1. Wiech, R.E., Jr., U.S. Patent 4 197 118, 1980
2. Wiech, R.E., Jr., U.S. Patent 4 305 756, 1981
3. Rivers, R.D., "Method of Injection Molding Powder Metal Parts," U.S. Patent 4 113 480, 1978

Hot Pressing

By Claus G. Goetzel
Consultant, Lecturer
Department of Materials Science
and Engineering
Stanford University

FULLY DENSE METAL POWDER COMPACTS with controlled microstructures can be produced by hot consolidation, in which pressure and heat are applied simultaneously rather than sequentially, as in conventional P/M processing. Pressure is applied statically or dynamically to the heated powder in one or two opposing directions along a single axis or from all sides. A controlled atmosphere is required to protect the hot powders or prepressed compacts from oxidation or nitridation by air.

Hot consolidation encompasses many diverse operations, including uniaxial hot pressing and pressure sintering, hot isostatic pressing, hot extrusion, and hot forging. Hot extrusion and hot forging are not suitable for heated loose powders, but are applicable for sintered preforms. For more information on these processes, see the articles "Extruded P/M Products" and "P/M Forging" in this Volume.

Hot pressing of metal powder in closed dies is one of the oldest methods of P/M hot consolidation. Reference 1 describes hot pressing of cemented carbides. Cylindrical drawing dies have been produced since the late 1920's in Europe and the United States. Originally, ball-milled powder mixes were compacted in electrically heated graphite molds at the eutectic temperature of the tungsten carbide/cobalt system. Air entrapped in the powder reacted with the graphite to form carbon monoxide, which protected particle surfaces against oxidation, thus permitting uninhibited consolidation by liquid-phase sintering. Concurrently, alloy steel dies were used for hot pressing of copper and iron powders (Ref 2) and copper and gold compacts (Ref 3). Additional information on the hot pressing of bronze powders (Ref 4), brass powders (Ref 5, 6), and iron powders with carbon additions (Ref 7, 8) is available in the literature.

Although hot pressing produces high-quality products, it has not proved economically feasible for steel, brass, and other common metals. Disadvantages include:

- Low production rates
- Hot die loading
- Welding of the compact to metallic die walls
- Tool wear
- Protective atmosphere

However, uniaxial hot pressing is used to fabricate cylindrical cemented carbide dies, rolls, and wear parts. This method is also used to fabricate beryllium (Ref 9, 10) and silicon nitride (Ref 11, 12), but hot isostatic pressing has enjoyed increased usage for these materials as well as for P/M superalloys, tool steels, and the like. Use of hot isostatic pressing techniques eliminates some of the problems caused by interaction of the powder compact and die. Pressure can be applied more uniformly, thus producing a product with more homogeneous microstructure and properties. For more information on this process, see the article "Hot Isostatic Pressing of Metal Powders" in this Volume.

Process Principles

Although the terms "pressure sintering" and "hot pressing" are used interchangeably, distinct differences exist between the two processes. In pressure sintering, the emphasis is on thermal processing; in hot pressing, applied pressure is the main process variable.

During pressure sintering, a contained powder or compact is passed through a heating cycle and subjected to moderate static pressure. Furnace chambers must be relatively large and compacts must have small cross sections to achieve an effective unit pressure. Like conventional sintering, pressure sintering is readily adaptable to mass production. Resistance sintering under pressure, electrically activated pressure sintering, and spark sintering are variations of hot pressing. These processes produce one compact at a time by molding under heat.

During hot pressing, a mold provides the compact with its desired shape. Because pressures are higher than those produced in a sintering furnace, molds must be designed accordingly. Press tools and machinery are similar to those used in small-scale or special P/M operations. Facilities for applying the required heat usually can be adapted to the design features of the press and die.

A variety of techniques may be employed for hot pressing. They differ primarily in the methods used to bring a powder or prepressed compact to a desired temperature and in the material used for press tools. Some methods provide an uncontrolled atmosphere around the powder, such as that created by reaction of mold material with air, while other methods provide carefully controlled atmospheres or a vacuum. Selection of processing method is determined primarily by the physical properties of the material to be processed and by the composition of the hot pressing system, particularly the interaction of the workpiece and tools.

All hot pressing techniques have several common features. Apparatus and operations to consolidate powders and remove end products do not differ substantially from conventional cold pressing and sintering operations. Processing cycle varies greatly with the time required for plastic deformation and creep under the particular stress and temperature conditions employed. Generally, hot pressing requires:

- Machines that can provide the required pressure for powder consolidation. Presses should be capable of slow action in one or two directions and should provide ejection ram pressure and movement. Die and punch constructions should possess sufficient strength and shape retention capabilities under maximum powder consolidation stresses at hot pressing temperatures.
- Loading facilities for loose powder or previously fabricated porous preforms are required for hot pressing operations. These components may be separate or an integral part of the die.
- Knock-out provisions and cooling receptacles or quench tanks are needed to hold the hot pressed product.

Additionally, mechanisms for monitoring and accurately controlling the temperature of the compact throughout the pressing cycle must be provided. Heat must be supplied either directly to the preform or compact or indirectly to the die or punches for transmission to the powder or compact. Type of power source depends on the mode of heating selected. A controlled atmosphere should be available throughout the entire hot pressing cycle.

Process Equipment

Presses. Full densification of a powder compact is achieved by plastic deformation, creep, and diffusion; consequently, time is an important processing factor. Whether consolidation pressure is applied during heating or after the selected hot pressing temperature has been reached, time must be allowed for heat transfer and equilibrium to occur. Because production rates are low, hydraulic or pneumatic presses are more suitable than mechanical presses. High-speed tableting machinery cannot be readily adapted to hot pressing.

Selection of press type depends on the design of the hot pressing tooling; both single- and double-action ram movements can be used for consolidation, as well as ejection of the solid compact. A side ram facilitates opening of split dies for extraction of multilateral products. Platen size and daylight, the clearance between the platens when the press is fully open, must accommodate press tools and pressure blocks, as well as the heat source (resistance wire loops or induction coils and electrical connections). The press stroke must be large enough to permit downward ejection of the compact, and ram speed must complement the hot pressing cycle.

Presses range in size from a simple hydraulic jack built inside a heavy steel frame to heavy double-action hydraulic presses that are capable of providing hydraulic or pneumatic side-ram action for lateral compaction or ejection.

Dies and Punches. Selecting a suitable material for these press tools is one of the most critical aspects of hot pressing. Maintaining the physical integrity of dies and punches during the high-temperature interval of the hot pressing cycle is essential. Punches must have adequate compressive strength to prevent plastic deformation and welding to the die cavity wall.

If the compact is completely solid, only a fraction of this pressure is transmitted to the die wall, even when full densification is achieved. If the material to be processed develops a liquid phase at the hot pressing temperature, isostatic pressure acts on the die cavity walls. Plastic deformation and creep of the die may occur, resulting in ejection difficulties and component distortion. Consequently, dies must be designed for sufficient wall thickness to withstand the high hoop stresses reached under peak temperature and pressure loading.

Generally, metal dies and punches tend to plastically deform unless processing temperatures are kept low by water cooling. Tools made of ceramics or graphite are subject to brittleness, and frequently catastrophic failure occurs if overloaded.

Chemical reactivity of tool parts subjected to high temperatures is another important consideration. In graphite tools, oxidation occurs when a controlled atmosphere is not provided. Burning of the graphite punches at the point of entry into the die may reduce the load-bearing cross section of the punches so that mechanical failure subsequently results. Frequent replacement of the punches in successive production runs becomes necessary.

Reaction of tools with the air inside the die cavity is less detrimental. Gaseous oxides of carbon protect the material to be hot pressed from oxidation, but may cause carburization of the compact surfaces. Carburization does not present a problem in cemented carbides or beryllium products with surfaces to be finished, but may prove undesirable for materials such as iron or titanium where deep finishing cuts would be uneconomical.

If metallic tools are allowed to heat to the maximum temperature of the cycle, interactions with the powder may occur. The high compressive stress may cause the compact to diffusion bond to the punch faces or to weld to the die cavity wall. The presence of a liquid phase in the compact aggravates this effect. Sometimes, punches

and compacts become inseparable and must be removed from the hot die in one piece. Welded or bonded spots may only be broken with great force, if at all, and may cause irreparable damage. Under less severe conditions, erosion and abrasion of the die walls intensify with repeated pressings, leading to a loss of dimensional control of the compact as well as to progressively higher ejection force.

Die wall lubricants, such as colloidal graphite suspensions in oil or water, that are applied to the metallic die cavity surface reduce wear. Colloidal ceramic or mica suspensions in volatile hydrocarbons that are spray coated on punch faces and parts of the die cavity walls inhibit diffusion bonding and welding. Lining of tool faces with thin sheets of defoliated pyrolytic graphite (Grafoil) effectively reduces wear in compacts that develop a liquid phase during hot pressing. The need for replacing the liners for each compacting operation is offset by easing or eliminating ejection difficulties.

Metals suitable for hot pressing tools include molybdenum, TZM, tungsten, Stellite alloys, and heat-resistant alloy steels. Dies and punches should be made of the same alloy to eliminate thermal expansion differences between stationary and moving parts. The die cavity and outside dimensions must prevent thermal contraction that shrinks the die around the punches during cooling from the hot pressing temperature. Loose fits between the moving and stationary die parts may have the adverse effect of extruding part of the compact into the clearance gap, which may cause welding of the work to the tools, causing ejection problems.

Charging. Feeding of loose powder is possible only if the die is cold. Once the die becomes hot, powder particles heat up rapidly on entry into the die cavity, causing two adverse effects. Increased atomic mobility in the particle surface area tends to weld the particles together, impeding powder flow by agglomeration and sticking to the hot tool surfaces. Also, the possibility of oxidation becomes greater because of the exposed large free surface area of the powder. Consequently, the practice of filling the cold die with the loose powder at a separate station has been widely adopted. When radioactive or toxic powders are used, filling and closing of the mold is carried out in glove boxes.

To facilitate charging the die cavity, the powder is preformed into a porous green compact of simple shape that can be handled safely. This permits maintaining the mold at the hot pressing temperature, thus expediting successive pressings. The preform method eliminates flow and minimizes oxidation problems; consequently, advantages of this approach outweigh the additional step of low-pressure preform pressing.

Ejection of the completed hot pressed component is a delicate operation. Conventional knock-out procedures are generally easier for a simple configuration of the product, such as a right circular cylinder or rectangular block. Stripping the compact from adjacent punches may not be feasible within the die assembly and may require delay until cooling to room temperature.

Ejection of the compact from the die should be performed at the hot pressing temperature. Most die assemblies are designed to provide a receptacle for cooling and storage of the compact while hot pressing of the next compact is in progress. A quench tank provides an alternative cooling receptacle, but care must be taken to prevent vapors of the quenching medium from interfering with the hot pressing operation.

Ceramic or water-cooled metal dies that split into two or more parts are also used. The compact can then be removed without difficulty, either while still hot or after having cooled. Cooling a one-piece die with the compact inside is difficult, especially if the die is made of an alloy with a high coefficient of thermal expansion. In this case, shrinkage of the die around the compact may prevent easy extraction and may lead to destruction of the product to save the die. Even if graphite dies are used, this procedure is too time consuming to be economical, except for expensive products.

Heating. Bringing the powder compact to the specified temperature before or during pressing is the primary function of the hot pressing process. This operation differs from conventional P/M heating procedures. To achieve acceptable results, the hot pressing system must (1) provide a means of directing or transferring heat to the material to be pressed, (2) allow thermal equilibrium conditions, and (3) monitor and control the temperature of the compact. To achieve these requirements, the powder compact can be heated directly, or the tools may be heated first, thus transferring heat to the compact.

Direct heating of the material in the dies can be accomplished by induction or resistance heating. In direct induction heating, three requirements must be fulfilled to ensure a workable process: (1) the powder must be electrically conducting and prepressed to a highly dense compact; (2) frequencies in the megacycle range must be used for the induction; and (3) ceramic

dies and punches must surround the compact. Correct tuning and coupling of the magnetic field is essential. Under these conditions, induction heating can be rapid and eliminates waiting for temperature equilibrium between work and adjacent tools.

Temperature control may prove difficult; direct optical pyrometer sighting of the compact is possible only with transparent ceramic dies. Contacting the hot junction of a thermocouple with the compact is mechanically difficult and may give erroneous readings due to self-induction. Quenched spark gap high-frequency converters can provide the required high frequencies more cost effectively than motor-generator-type power sources.

Resistance heating of the powder is accomplished by passing a low-voltage, high-amperage current through the powder while simultaneously applying pressure (Ref 13). The powder or compact must have relatively low resistance that is uniformly distributed throughout the mass so that isothermal conditions prevail. The current is passed through water-cooled conducting plungers and thin wafers of higher thermal and electrical resistivity and heat resistance. The compact is confined in a ceramic die or die liner backed with a stronger steel die that may be water cooled.

Because extremely short processing times (seconds) are required for this type of hot pressing, conventional methods of measuring temperature and temperature distribution in the compacts, such as pyrometers or thermocouples, cannot be used. Instead, a trial-and-error approach, followed by product examination for density, microstructure, and properties, must be performed for each material. Spot welding machines, with high-conductivity, high-strength copper alloy electrodes that serve as plungers, have been used for hot pressing a variety of materials (Ref 13).

A variation of this hot pressing method uses electric spark discharges between the powder particles to activate the surrounding surface gases and thereby promote interparticle bonding (Ref 14). A high-frequency alternating current is passed quickly through the loose or precompressed powder mass before contact pressure is applied, and the direct current for resistance heating and consolidation is passed through the plungers and compact.

Indirect heating of the powder or compact by convection or heat transfer from the tools is widely used. If the plungers or punches are made of graphite or a refractory metal of relatively high electrical resistance, heating is accomplished with a high-amperage current. Heat is then trans-

ferred to the powder, while applied pressure provides contact and compaction.

Holding time is required to obtain temperature equilibrium between the compact and the adjacent parts of the punches and die cavity wall. Die designs that incorporate insulating backups behind liners or sleeves may reduce the time required to reach heat balance. Thermocouples can be used to control temperature. Thermocouple leads can run between tool members or through a blind hole into the die body. High-capacity direct current power sources are required for hot pressing products that exceed laboratory-size specimens.

Heating the die is the most common and convenient method of heating the powder or compact. The entire die assembly may be heated in a furnace chamber and then transferred to the press. The die may also be torch or gas heated in the tool assembly built into the press, or it may be surrounded with a resistance wire-heated muffle furnace or with a high-frequency induction susceptor made of graphite or resistor-type refractory metal.

Enclosing the die with a resistance wire-heated muffle furnace or a high-frequency induction susceptor is commonly used because of its simple design and versatility. Time required to reach temperature equilibrium depends on the mass of the die. Temperature controls are similar to those of other systems and may include optical pyrometry, provided adequate calibration measures are taken. Motor-generator power sources provide suitable induction currents for most applications.

Atmospheric Environment. Because most metal and metal alloy powders react strongly with oxygen (some react equally with nitrogen), a controlled environment must surround the material at elevated temperature. Depending on the powder to be consolidated and the hot pressing setup, the atmosphere may be a gas contained in the die cavity, entrance throat or vestibule, and cooling receptacle. A gas that not only fills the inner part of the die, but also surrounds the entire tool assembly inside a suitable chamber, may be suitable, or a vacuum that is connected to a gas purging system may accommodate processing requirements.

Graphite dies and punches create a semireducing atmosphere inside the heated portion of the cavity, which provides adequate protection for cemented carbides and high-carbon iron-based alloys such as powdered cast iron. In the hot pressing of more reactive metal and alloy powders, a controlled atmosphere is required. Originally, hydrogen or a hydrogen-nitrogen gas mixture including dissociated ammonia was used. Use of these gases has given way to less hazardous and easier-to-handle gases, such as highly desiccated argon or argon mixed with small amounts of hydrogen.

When the powder to be hot pressed is a metal or compound that is subject to contamination by interstitial elements, use of a vacuum offers the best protection. Vacuum hot pressing currently is an accepted procedure for such diverse materials as titanium, silicon nitride, cermets, and superalloys. Vacuum technology can be used effectively, without sacrificing other principal features of the hot pressing process.

Process Variables

The quality of the hot pressed compact is determined by the interrelationship of three process variables: pressure, temperature, and time. These variables can have a major effect on the microstructure, physical properties, dimensional accuracy, and surface conditions of the product. However, optimum results require an understanding of the compensatory effects of these variables, which contribute to the consolidation of the compact by particle rearrangement and plastic flow.

For powders heated to low temperatures, pressure has the same effect as in cold pressing. Particles are brought closer together, and they may be rotated and pushed into openings, deformed, sheared, or fractured. At elevated temperatures, plastic deformation becomes the dominant mechanism. If a liquid phase is formed at the hot pressing temperature, consolidation is further enhanced by the isostatic action of the compressive stress on the compact inside the dies. Additionally, diffusion rates are increased by the liquid phase, and densification is enhanced by good wetting between liquid and solid components of an alloy system.

The magnitude of the pressure used in hot pressing should be based on properties of the powder substance, such as hardness at ambient and elevated temperatures, particle size and surface conditions, possible oxide films, capacity for plastic deformation at various temperatures, and the potential for liquid phase formation. Generally, pressure ranges from about three fourths to one tenth of that required for cold pressing. When a liquid phase appears, pressure must be kept low to avoid squeezing out the liquid, which results in changes in the composition and produces porosity in the hot pressed body.

Because most solid metals and metalloidal compounds soften and become more plastic with increased temperature, a corresponding decrease in specific pressure is

Table 1 Effect of hot pressing temperature and pressure on density of copper, brass, and iron powders

Material	Pressure MPa	ksi	Temperature °C	°F	Density, g/cm³	Density, % of theoretical
Copper	1380	200	250	480	8.87	99.3
	1380	200	300	570	8.87	99.3
	1380	200	400	750	8.90	99.6
	690	100	300	570	8.77	98.3
	690	100	400	750	8.90	99.6
	690	100	500	930	8.91	99.7
	345	50	400	750	8.63	96.6
	345	50	500	930	8.74	97.9
Brass (55Cu-45Zn)	690	100	300	570	8.19	98.7
	690	100	500	930	8.30	100.0
	345	50	500	930	8.24	99.3
Brass (65Cu-35Zn)	690	100	300	570	8.12	96.0
	690	100	500	930	8.44	99.6
	345	50	500	930	8.35	98.6
Brass (60Cu-40Zn)	70	10	700	1290	8.38	100.0
	70	10	800	1470	8.38	100.0
Iron	415	60	500	930	7.47	95.0
	415	60	600	1110	7.87	100.0
	275	40	700	1290	7.87	100.0
	250	36	600	1110	7.50	95.4
	150	22	700	1290	7.44	94.5
	140	20	780	1435	7.58	96.5
	140	20	800	1470	7.85	99.7
	70	10	800	1470	7.51	95.5

Source: Ref 3, 5-8, 15

required for consolidation at higher temperatures. For practical reasons, hot pressing temperatures should be sufficiently high to achieve complete densification with moderate pressures in a reasonably short time. Under these conditions, metals that do not differ appreciably in microstructure or mechanical properties from fine-grained precision castings or forgings can be produced.

If the hot pressing temperature is kept below the temperature range for recrystallization and grain growth, pressures for complete densification must be much higher. The microstructures of compacts produced in this manner are characterized by an outline of the original powder particles that are closely packed and may be fractured, distorted, tilted, or rearranged. The particles may exhibit twinning and deformation. Such a structure yields mechanical properties that are inferior to material that is hot pressed at higher temperatures.

Plastic behavior of the material to be processed is affected by processing time. Plastic flow of solids is a function of time, as well as of temperature and force. For practical reasons, the time during which the hot compact is maintained under pressure should be kept at a minimum. Densification can be more reliably achieved and physical properties greatly improved by the prolonged application of pressure and heat. This principle also applies to the hot pressing of many metals in the solid state, as well as to liquid-phase systems such as cemented carbides.

However, short processing times also may produce fully dense compacts. For example, subjecting a hot compact to an impact blow in a closed die hammer or simultaneously pressing and resistance heating a compact in a spot welder can produce full densification. The effects of pressure, temperature, and time on densification of hot pressed copper, brass, and iron are shown in Tables 1, 2, and 3.

Hot Press Setups

When cemented carbides were first hot pressed more than 50 years ago, the process consisted of simply filling graphite molds with powder, closing the mold cavities with graphite plungers, followed by heating, while applying constant pressure, to sintering temperature. Since that time, many hot pressing setups have been developed and used experimentally, as well as in production. Various types of apparatus and die assemblies incorporating different die materials, modes of pressing and heating, and atmospheric environments are illustrated in the following examples.

Indirect Furnace Heating

Figure 1 shows a cross section of a hot pressing arrangement used to consolidate nonferrous powders. Principal features include a metallic die and plunger, with provision in the die for temperature control

Table 2 Pressure requirements for hot pressing iron powders to specified hot pressed densities

	Swedish sponge, −100 mesh			Electrolytic, −100 mesh			Hydrogen-reduced, −325 mesh		
	90%	95%	99+%	90%	95%	99+%	90%	95%	99+%
At 20 °C (68 °F)									
	1030 (150)	1720 (250)	2760(a) (400)	1380 (200)	2070(a) (300)	2930(a) (425)	1720 (250)	2760(a) (400)	3450(a) (500)
At 500 °C (930 °F)									
	290 (42)	370 (54)	1030(a) (150)	260 (38)	470 (68)	830(a) (120)	330 (48)	480 (70)	960(a) (140)
At 600 °C (1110 °F)									
	170 (24)	250 (36)	550(a) (80)	125 (18)	230 (34)	390 (56)	140 (20)	260 (38)	420 (60)
At 700 °C (1290 °F)									
	110 (16)	150 (22)	280 (40)	80 (12)	125 (18)	260 (38)	95 (14)	140 (20)	280 (40)
At 800 °C (1470 °F)									
	55 (8)	80 (12)	170(a) (24)	40 (6)	55 (8)	125 (18)	40 (6)	70 (10)	140 (20)
At 900 °C (1650 °F)									
	35 (5)	55(a) (8)	110(a) (16)	20 (3)	35 (5)	55(a) (8)	20 (3)	40 (6)	60(a) (9)

(Required pressure, MPa (ksi))

(a) Pressure is estimated for English and metric values; may exceed die safety limit
Source: Ref 7

Table 3 Effect of dwell time on density of electrolytic iron powder compacts hot pressed at 140 MPa (20 ksi)

Temperature °C	°F	Dwell time at temperature and pressure, s	Density, g/cm³	Density, % of theoretical
500	930	50	6.31	80.2
		150	6.38	81.1
		450	6.71	85.3
600	1110	50	6.70	85.2
		150	6.89	87.5
		450	7.05	89.6
700	1290	50	7.32	93.0
		150	7.52	95.6
		450	7.58	96.4
780	1435	50	7.59	96.5
		150	7.71	98.0
		450	7.76	98.6

Source: Ref 8

Fig. 1 Cross section of electric resistance wire furnace-heated alloy steel die assembly built into hydraulic press

Arrangement provides single-action hot pressing under protective atmosphere for copper and copper alloy powder compacts. Source: Ref 16

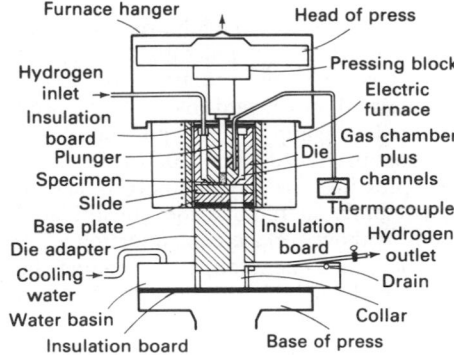

and containment of an atmosphere that can be directed to the compact; a suspended electric resistance wire-wound furnace surrounding the die; hydraulically applied pressure acting in a single, downward direction; and a load-bearing, slidable base with an offset opening that permits ejection of the compact into a quench bath.

The hardened tool steel used for the die and plunger begins to soften above 600 °C (1110 °F); consequently, this temperature constitutes the limit for safe operation when high unit pressures are used. The die is kept at hot pressing temperature for continuous pressing of porous preforms that are dropped into the cavity from the top, requiring a brief removal of the plunger. Because of the small size of the compact (16 mm, or ⅝ in. in diameter and height) in relation to the die mass, 5 min brings the compact to temperature.

A similar setup for high-temperature hot pressing of iron cylinders of about the same size is illustrated in Fig. 2. The main differences in hot pressing arrangements include the materials used for the press tools and the modes of pressing and heating. This die is made of high-speed tool steel, and the plungers are made of cemented carbide for compacting at moderately elevated temperatures requiring high unit pressures for densification of iron powder.

For hot pressing temperatures ranging from 900 to 1100 °C (1650 to 2010 °F), the die material is graphite, whereas plunger composition varies with temperature and pressure. At 900 °C (1650 °F), cemented carbide plungers can be used repeatedly up to 20 MPa (3000 psi), and molybdenum can be used up to 35 MPa (5000 psi) and higher.

Pressing is possible from two directions, and the die assembly containing the compact is heated by induction and maintained at temperature for repeated pressings. The hot pressed compact is ejected into cooling and quenching receptacles. The use of a reducing or inert atmosphere shields the iron compact from oxidation and prevents surface carburization of the iron compact by the graphite die.

Indirect or Direct Resistance Heating

Figure 3 illustrates a different approach to hot pressing at extremely high temperatures. This method is well suited to making bar-shaped steel compacts at sufficient unit pressures to yield fully dense bodies. The machinery is a modified resistance welding machine that accommodates the die containing the powder compact. The die body is split into L-shaped halves that are clamped in a vise. The split die facil-

itates rapid insertion of a preform and rapid removal of the pressed bar by lifting it from the die after raising the punch and press ram or hammer.

Channels for inert gas entry into the die cavity are positioned in the split of the die halves. The die halves are made of graphite, low-binder-containing cemented carbide, or molybdenum and are backed with current-conducting spacer blocks. The vise is formed from water-cooled copper electrodes through which a low-voltage, high-amperage current (6 to 8 V, 15 kA) is passed to heat the compact to 1125 to 1250 °C (2055 to 2280 °F). Punches are made of the same material as the die halves, and the ram or hammer is made of a molybdenum alloy.

Die cavity walls and punch faces in contact with the powder preform are spray coated with a refractory ceramic, such as

Fig. 2 Cross section of induction-heated alloy steel or graphite die assembly built into hydraulic press

a, compact; b, carbide punches; c, graphite cover; d, spacer punch; e, high-speed tool steel die. Arrangement provides double-action hot pressing under protective atmosphere for iron powder compacts. Source: Ref 16

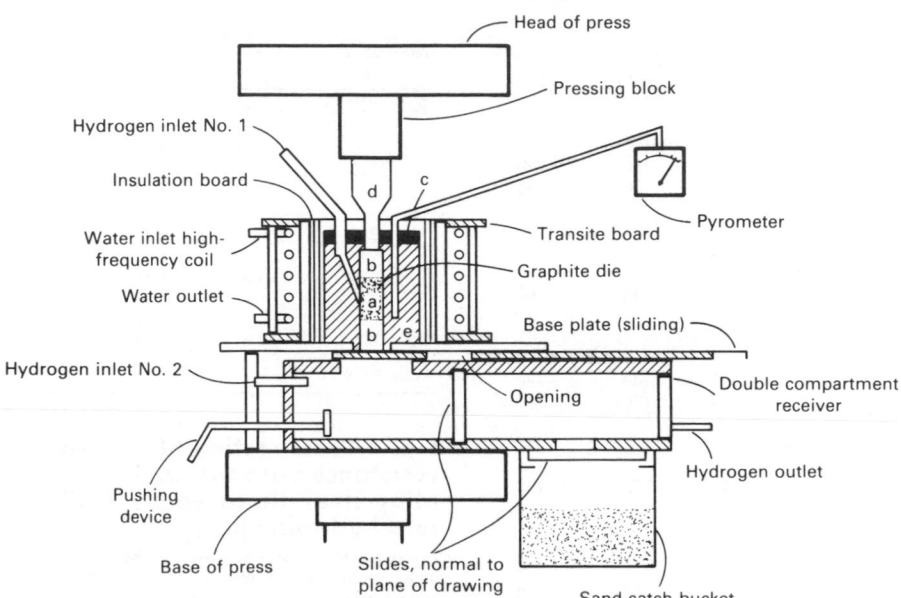

Fig. 3 Modified spot welding machine and split die assembly heated by electrical resistance

Arrangement provides single-action hot pressing under protective atmosphere for iron and steel powder compacts. (a) Elevated view and section through vise and working tools. (b) Top view. (c) Cross section through compact and die assembly. Source: Ref 16, 17

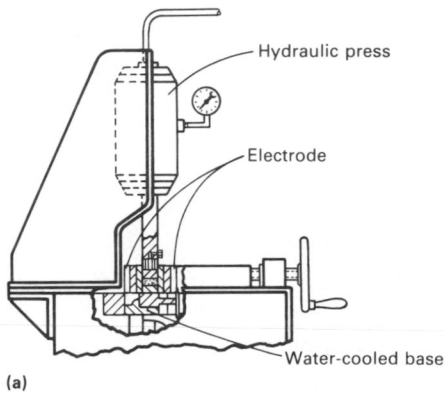

(a)

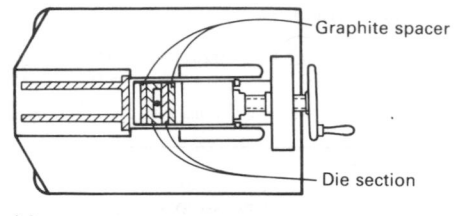

(b)

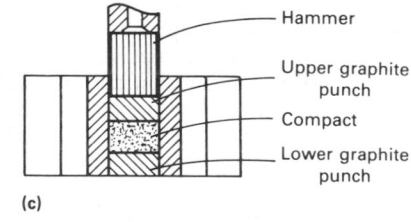

(c)

Fig. 4 Cross section of ceramic-lined die assembly resistance heated by welding machine

Arrangement provides double-action hot pressing to consolidate various metal powders. Source: Ref 13

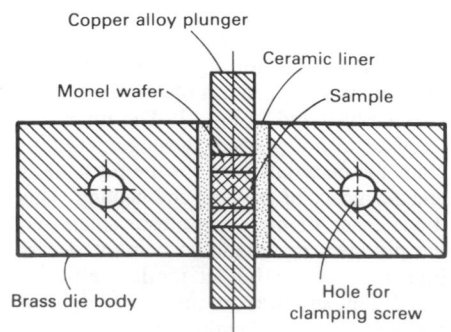

magnesium oxide, to inhibit reactions and welding. Heat and pressure are applied simultaneously for 1 to 10 s; the ram is accelerated downward by a high-speed hydraulic pump. Unit pressures are approximately 20 MPa (3000 psi). After hot pressing, the die assembly is cooled sufficiently to remove the consolidated body without surface contamination.

The entire operation, from loading the preform through hot pressing, cooling, and product withdrawal, requires less than 5 min. The dimensions of the bar permit microtensile testing. A typical bar is 60 mm (2.4 in.) long and 10 mm ($^3/_8$ in.) per side square.

An alternative hot pressing arrangement is shown in Fig. 4. In this pulse resistance heating setup, a spot welding machine is used. Heating of the compact is accomplished by its electrical resistance to the low-voltage (10 to 40 V), high-amperage (300 kV·A) current, which is passed in the form of one to three pulses, each of several cycles, from the electrodes of the spot welder through high-strength copper alloy plungers and wafers.

These wafers are made of a high-melting-point alloy steel or refractory metal and have higher thermal and electrical resistivities than the plungers that aid equalization of the heat within the compact. The die wall consists of a refractory ceramic liner, such as steatite ($MgSiO_3$), which is backed with a brass die body. This is split

to facilitate opening and insertion of new liners in the event of failure by thermal shock.

The system accommodates loose powder or preforms of many types and hot presses them into cylindrical compacts 12.7 mm ($^1/_2$ in.) in diameter and height. Pressure applied from top and bottom by suitable tool arrangement (Fig. 4) ranges from 17 to 70 MPa (2.5 to 10.5 ksi), while time at temperature is about 1 s.

A more sophisticated hot pressing system using direct current electric resistance heating provides a preliminary powder activation phase through high-frequency alternating current spark discharges (Ref 14). Several machines of different capacities have been developed for laboratory and production use and are called "spark sintering" units (Ref 18).

A typical laboratory or small-lot production machine has a maximum power output of 35 kV·A, a maximum alternating current/direct current amperage of 3 kA, and a hydraulic press capacity of 90 kN (10 tons). Machines to produce large billets, up to 250 mm (9.6 in.) in diameter and 110 mm (4.4 in.) in thickness, have power outputs up to 1050 kV·A, maximum alternating current/direct current of 70 kA, and press capacities up to 900 kN (100 tons). Parts 100 mm (4 in.) in diameter and height of such metals as beryllium, titanium, or superalloys can be hot pressed routinely in a machine having

Fig. 5 Cross section of resistance heated graphite die assembly built into hydraulic press

Arrangement provides for telescoping controlled atmosphere enclosure and double-action hot pressing for steel, nickel alloy, and titanium alloy powders. Source: Ref 19

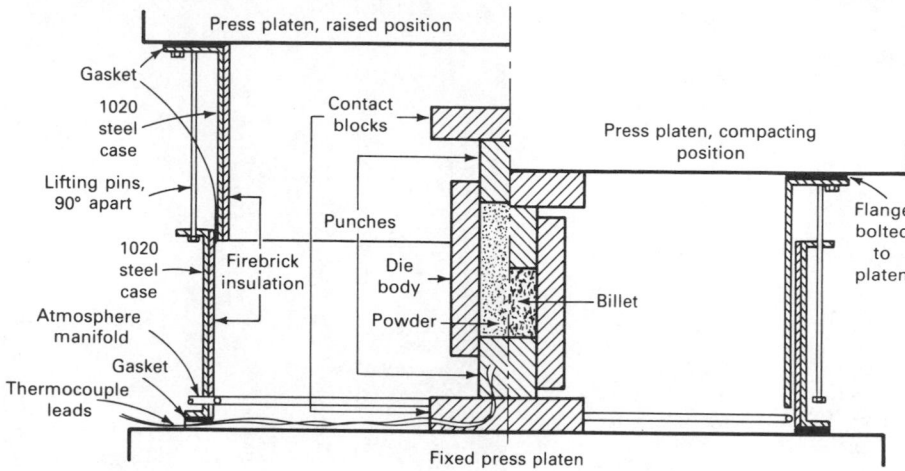

Fig. 6 Cross section of hydraulically activated graphite die assembly surrounded by gas-fired furnace inside vacuum chamber

Arrangement provides for single-action hot pressing of titanium powder compacts. Source: Ref 20

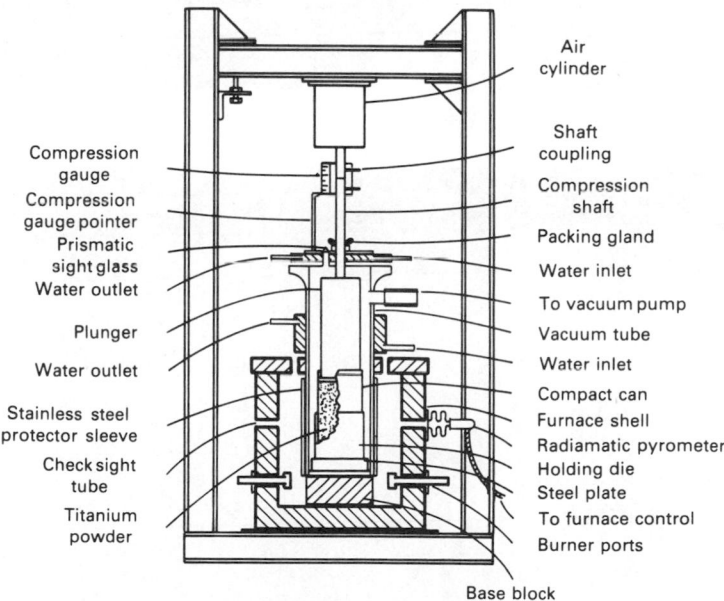

a cooling vessel, can be accomplished in less than 15 min. Heating of the preforms occurs through heat transfer from the plungers and die liner, which in turn are heated through their own resistance. Suitable materials for the plungers and die liner are degassed nuclear reactor graphite or TZM molybdenum alloy.

The thermoelectric efficiency of the system and the strength of the die are enhanced by using a composite die construction in which the relatively thin liner is slightly tapered and shrunk-fit into an insulating ceramic backup mass, which in turn is structurally supported by a steel pressure shell. Solid contact between liner and ceramic is established by the pressure on the taper during hot pressing.

Temperature is controlled by insertion of a thermocouple in the lower plunger near the face contacting the compact. A controlled atmosphere, such as argon, that penetrates the die cavity and surrounds the press tools is contained in a telescoping sheet metal enclosure, as shown in Fig. 5.

Vacuum Hot Pressing

Hot pressing of powders that are particularly sensitive to contamination by oxygen or nitrogen requires strict control of the atmospheric environment to ensure high-quality products. The use of a technical vacuum generally is preferred for production hot pressing of blocks and shaped components of beryllium or titanium.

Figure 6 shows the cross section of a vacuum hot pressing installation for small-lot production of titanium parts (Ref 20). The setup consists essentially of graphite dies filled with a weighed amount of the powder that are closed with a graphite plunger. The assembly is installed in a gas-fired sintering furnace that is placed inside a shell for evacuation to 1 to 2 μm pressure. The die and tools are heated by convection until the compact reaches 850 to 900 °C (1560 to 1650 °F). Hydraulic force then is transferred through a rod and packing gland in the top of the furnace and maintained until full consolidation of the compact is achieved. Stripping of the die takes place after cooling the furnace with inert gas.

Vacuum hot pressing usually is used to produce fine-grained blocks from −200 mesh beryllium powder (Ref 21). Rectangular or cylindrical billets up to 350 kg (770 lb) are produced in a vacuum greater than 133 Pa (1 mm Hg) at 1050 to 1100 °C (1920 to 2010 °F) and at pressures from as low as 0.5 MPa (75 psi) for small blocks weighing up to about 12 kg (26 lb) to as high as 4 MPa (600 psi) for large blocks

225 kV·A power output and 15 kA combined alternating current/direct current.

The power supply consists of a motor-generator and a phase-stabilizing capacitor bank for the alternating current and a direct current rectifier, in addition to the hydraulic pump system for the press. A typical cycle for the 100-mm (4-in.) diam part would have a 15-s alternating current power input at a starting voltage of 15 V, followed by 4 min of direct current input of a maximum current on the order of 235 A/cm² (1500 A/in.²), under steadily diminishing voltage to about 5 V. A unit pressure of 14 MPa (2000 psi) applied throughout the 4-min direct current power input yields fully consolidated products.

The entire operation, including loading of the die with a prepackaged powder charge and ejection of the finished part into

Fig. 7 Vacuum hot pressing of fine-grained beryllium blocks

(a) Die filling station with vibrating table. (b) Filled die and plunger. (c) Small vertical hot pressing unit. (d) Large horizontal hot pressing unit. Source: Ref 21

weighing 250 kg (550 lb) or more. Pressure is applied either mechanically by a heavy weight or hydraulically in a press frame built into a vacuum furnace.

Powder is brought to temperature through heat transfer from the graphite or steel die and plungers, which in turn are heated through their own resistance or by induction. Time for a complete cycle, including die loading, hot consolidation, and product discharge, depends on the mass of the billet and usually ranges from 4 to 24 h.

The vacuum hot pressed block operations shown in Fig. 7 illustrate (a) loading and packing of the powder in a steel die, (b) the filled die and a heavy plunger for applying the load in a vacuum furnace, (c) a small unit with the vacuum furnace built into a hydraulic press frame, and (d) a large vacuum hot pressing unit for producing heavy blocks.

The conventional hot pressed block technique can be adapted to complex shapes. Envelope hot pressing (Ref 22) is such a variation. Powder is filled and pressed into a shaped thin-walled steel tube that is welded shut and evacuated to 50 μm Hg. The tube is then inserted into a pressure chamber, and the space is filled with granular silicon carbide. The vessel has a high-pressure gas line. The assembly is placed in a gas-fired furnace and heated to 1000 to 1100 °C (1830 to 2010 °F), while argon at 1.4 MPa (200 psi) pressure is maintained in the chamber and applies pressure on the sheet metal tube. A cross section of this installation is shown in Fig. 8(a). Figure 8(b) illustrates a star-shaped sheet steel envelope for the hot pressing of finned products.

Vacuum hot pressing is ideally suited to the consolidation of extremely high-purity beryllium powder, such as attritioned electrolytic flake (Ref 10). Figure 9 illustrates a vacuum chamber through which an upper and lower ram of a hydraulic press can move in opposite directions. The double-walled chamber has dished heads on top and bottom, with vacuum seals for the rams and ports for electrical, thermocouple, and vacuum connections. Cooling water flows between the chamber shells and inside the rams. Heating occurs by inductive coupling derived from a motor-generator power source. The die, punches, and pressure-distributing spacer blocks are made of high-purity graphite, and refractory ceramic bricks are built between the die perimeter and the induction coil.

The system is capable of sustaining a vacuum of 500 μm Hg or more throughout the hot pressing cycle. During heating of the powder column, a moderate pressure of about 1.4 MPa (200 psi) is applied

Fig. 8 Sheet metal envelope hot pressing of beryllium

(a) Cross section through powder containment inside gas-fired furnace. (b) Outline of steel sheet envelope used for producing finned product. Source: Ref 22

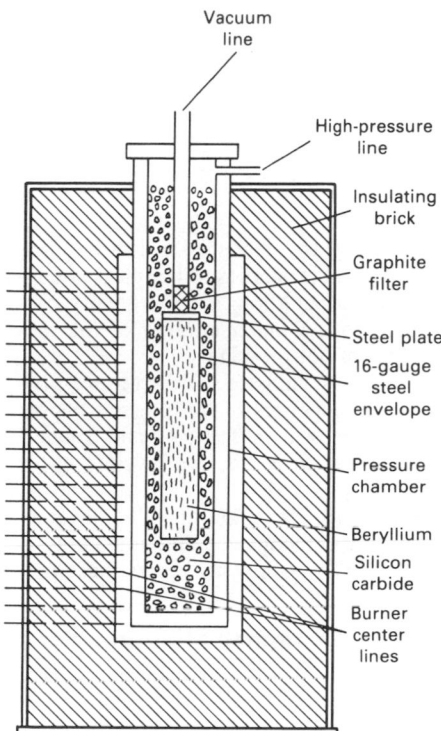

(a)

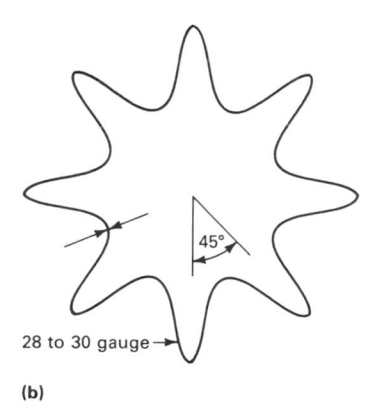

(b)

to both rams; the pressure is increased to 7 to 14 MPa (1000 to 2000 psi) at the hot pressing temperature of 1050 °C (1920 °F) and maintained for several hours until full densification of the billet is achieved. Stripping of the billet from the graphite die can be safely accomplished after furnace cooling of the assembly below 600 °C (1110 °F). The hot pressed products made in this arrangement usually are right circular cylinders, but other shapes such as rectangular blocks or truncated cones also can be fabricated.

Fig. 9 Cross section of hydraulically activated, induction-heated graphite die assembly inside vacuum chamber

Arrangement permits double-action hot pressing of beryllium powder into blocks. Source: Ref 10

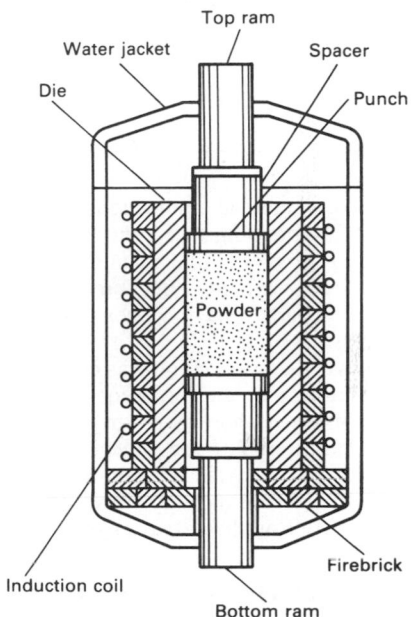

Products

Aluminum Alloys. Prealloyed powder can be hot pressed into fully dense bodies by applying relatively low pressures (42 to 70 MPa, or 6000 to 10 000 psi) for short periods of time (less than 2 min). Because aluminum alloys have the capacity to plastically deform at temperatures above 400 °C (750 °F), hot die steel tools can be used. To avoid powder sticking and flow problems, porous preforms are essential. If an inert atmosphere such as nitrogen is passed through the die cavity, the hot pressed products do not differ in purity from the original powder. This method is limited to simple geometric shapes, such as extrusion billets or forging blanks; more complex shapes require hot isostatic pressing.

Another method, which is especially suitable for producing large aluminum alloy powder billets, involves cold compacting and canning. Vacuum degassing follows at a selected temperature prior to mechanically hot pressing at specified pressure, temperature, and time. After decanning, processing to final mill products is done by conventional techniques (Ref 23).

Copper and Copper Alloys. Hot pressed copper or copper alloys produced from elemental or prealloyed powders have

Table 4 Mechanical properties of unalloyed hot pressed copper powder compacts

Temperature °C	°F	Pressure(a) MPa	ksi	Yield strength MPa	ksi	Tensile strength MPa	ksi	Elongation, %	Compressive strength MPa	ksi	Compression, %	Increase in area, %
500	930	350	50	...	...	...	...	...	2220	322	55	245
500	930	700	100	...	...	...	...	...	2600	377	70	270
610	1130	330	48	...	...	230	33.5	...	...	...	...	...
715	1320	330	48	...	...	210	30.5	...	...	...	...	...
800	1470	70	10	186	27	186	27	4	...	...	...	...
810	1490	330	48	...	...	203	29.5	...	...	...	...	...
940	1740	70	10	76	11	207	30	60	...	...	...	...

(a) Dwell time: 1 min
Source: Ref 2, 15, 25

found no commercial applications. Efforts to adapt hot pressing to a continuous operation for cost-effective manufacture of commutator segments and other small electrical parts have been largely unsuccessful.

Even when consolidation was combined with preforming and subsequent cold coining using a short-cycle drop-forging type of hot pressing, Jones (Ref 24) was unable to gain general acceptance of this process by the P/M industry. However, his work, as well as that of other leaders in this field, has established a data base for the physical properties of hot pressed copper and copper alloys and has greatly contributed to an understanding of the mechanics of hot pressing.

Hardness, tensile, and compressive strength properties of hot pressed copper are summarized in Table 4, and Table 5 gives these properties for several copper alloys.

Iron and Steel. During P/M development, it was recognized that iron and low-carbon or low-alloy steel parts mass pro-

duced by P/M techniques, despite inherent material and labor savings, could compete with stampings or forgings only if their mechanical properties were comparable. This realization gave impetus to hot pressing as a means of producing fully dense material. The advantage of making net shapes by the conventional P/M approach of controlling powder characteristics, press stroke, and die configuration, however, was overshadowed by uneconomical low production rates. As a result, hot pressing of ferrous metal powders historically has been confined to establishing a data base for mechanical properties.

The potential of using hot pressed, high-density billets as a starting material for further hot working, especially forging and coining, also has been explored (Ref 7, 19, 26). This has been superseded, however, by the more economical preform production by cold pressing and sintering in P/M forging technology. For more information on this subject, see the article "P/M Forging" in this Volume.

Hot isostatic pressing appears to be a viable alternative to uniaxial hot pressing of

Table 5 Mechanical properties of hot pressed copper alloy powder compacts

Material	Temperature °C	°F	Pressure(a) MPa	ksi	Yield strength MPa	ksi	Tensile strength MPa	ksi	Elongation, %	Compressive strength MPa	ksi	Compression, %	Increase in area, %	Hardness, HB
Elemental powder mixture														
90Cu-10Zn	900	1650	60	9	117	17	210	31	22	...	...	...	...	...
85Cu-15Zn	500	930	700	100	...	...	...	...	...	1450	210	52	128	103
80Cu-20Zn	900	1650	60	9	124	18	255	37	34	...	...	...	...	...
75Cu-25Zn	500	930	700	100	...	...	...	...	...	2150	312	46	112	105
70Cu-30Zn	800	1470	60	9	152	22	262	38	16	...	...	...	...	...
65Cu-35Zn	500	930	700	100	...	...	...	...	...	2300	333	43	108	111
55Cu-45Zn	500	930	700	100	...	...	...	...	...	1460	212	33	100	160
50Cu-50Zn	775	1425	60	9	145	21	145	21	0	...	...	...	...	...
95Cu-5Sn	500	930	700	100	...	...	...	...	...	2200	319	55	103	114
95Cu-5Sn	700	1290	60	9	179	26	240	35	9	...	...	...	...	110
95Cu-5Sn	800	1470	60	9	165	24	310	45	47	...	...	...	...	114
93Cu-7Sn	800	1470	60	9	165	24	325	47	75	...	...	...	...	114
91Cu-9Sn	800	1470	60	9	207	30	290	42	17	...	...	...	...	...
90Cu-10Sn	500	930	700	100	...	...	...	...	...	2300	333	53	98	130
85Cu-15Sn	500	930	700	100	...	...	...	...	...	1360	197	33	37	165
80Cu-20Sn	500	930	700	100	...	...	...	...	...	880	128	17	23	211
86Cu-10.5Zn-3.5Sn	900	1650	60	9	124	18	262	38	53	...	...	...	...	...
83Cu-10.5Zn-2.5Sn-4Ni	900	1650	60	9	124	18	270	39	32	...	...	...	...	...
89Cu-5.5Sn-4.5Ni-1Si	900	1650	60	9	220	32	310	45	13	...	...	...	...	...
88.5Cu-5.5Sn-5Ni-1Si	900	1650	60	9	303	44	358	52	5	...	...	...	...	...
Prealloyed powder														
85Cu-15Zn	500	930	700	100	...	...	...	...	...	2635	382	66	193	115
75Cu-25Zn	500	930	700	100	...	...	...	...	...	3000	435	69	192	129
70Cu-30Zn	900	1650	60	9	117	17	206	30	21	...	...	...	...	...
65Cu-35Zn	500	930	700	100	...	...	...	...	...	3075	446	67	200	122
55Cu-45Zn	500	930	700	100	...	...	...	...	...	2425	352	54	173	125
95Cu-5Sn	500	930	700	100	...	...	...	...	...	2950	428	66	210	111
90Cu-10Sn	500	930	700	100	...	...	...	...	...	3215	466	59	197	133
85Cu-15Sn	500	930	700	100	...	...	...	...	...	2500	362	44	66	153
80Cu-20Sn	500	930	700	100	...	...	...	...	...	1035	150	25	35	217
83Cu-12Zn-4Sn-1Fe	870	1600	60	9	172	25	296	43	46	...	...	...	...	...

(a) Dwell time: 5 min for powder mixtures, 1 min for prealloyed powder
Source: Ref 4, 5, 24

Fig. 10 Density and hardness of hot pressed iron powder compacts
(a) Swedish sponge iron. (b) Electrolytic iron. (c) Hydrogen-reduced iron. Source: Ref 7

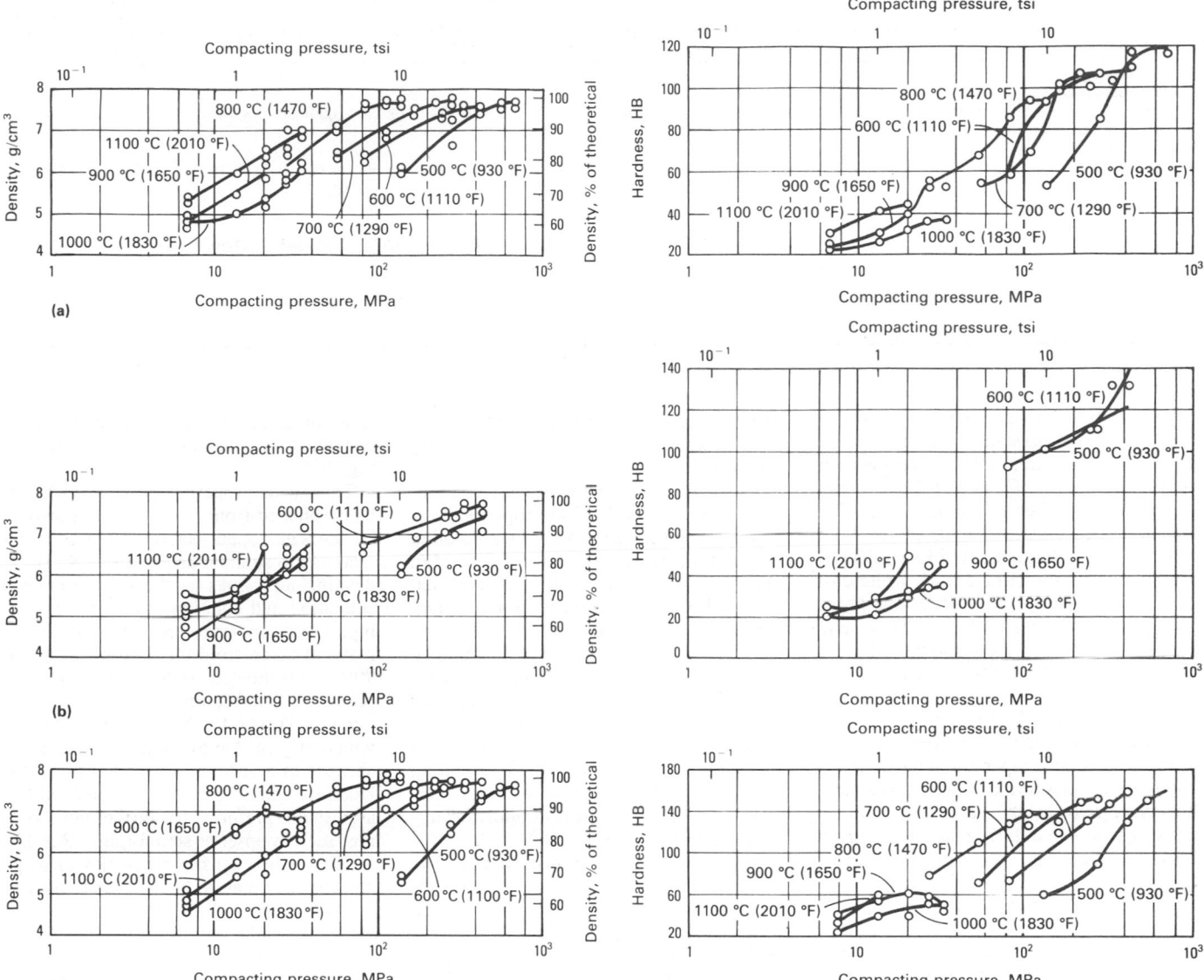

net or near-net shape parts for expensive materials, such as high-alloy steels, heat-resistant steels, and tool steels. For more information, see the articles "Hot Isostatic Pressing of Metal Powders" and "Cold Isostatic Pressing of Metal Powders" in this Volume.

An increase in density and hardness with hot pressing temperatures and pressure is shown graphically in Fig. 10 for several iron powders. Table 6 gives density, hardness, and tensile properties for various hot pressing conditions for electrolytic iron powder. Tensile strength values of 350 MPa (50 ksi) for a hot pressing temperature of approximately 800 °C (1470 °F) are comparable to earlier data (Ref 2). A low-

Table 6   Mechanical properties of electrolytic iron powder compacts hot pressed at 140 MPa (20 ksi)

Temperature		Dwell time at temperature and pressure, s	Tensile strength		Elongation in 25 mm (1 in.), %	Hardness, HB
°C	°F		MPa	ksi		
500	930	50	180	26.2	0	50
		150	176	25.5	0	51
		450	274	39.8	1	63
600	1110	50	254	36.9	0.5	62
		150	281	40.8	1	77
		450	336	48.8	2	80
700	1290	50	330	47.8	1	90
		150	395	57.3	12	95
		450	397	57.5	27	100
780	1435	50	373	54.1	22	101
		150	361	52.4	32	93
		450	365	52.9	37	96

Source: Ref 8

Table 7 Properties of electric spark-activated hot pressed elemental titanium powder compacts

Material	Billet analysis (center), %				Density, g/cm³	Density, % of theoretical	Yield strength		Tensile strength		Elongation in 12.7 mm (0.5 in.), %	Reduction in area, %
	Oxygen	Hydrogen	Magnesium	Sodium			MPa	ksi	MPa	ksi		
Titanium sponge, magnesium-reduced, graded	0.187	0.052	0.212	...	4.47	99.2	470-520	68-75	600-680	87-99	4-6	4-6
Titanium sponge, sodium-reduced, high-purity	0.120	...	0.008	0.156	4.47	99.2	310-390	45-56	430-520	63-75	14-17	27-29
Titanium sponge, sodium-reduced, extra-low-interstitials	0.070	...	...	0.097	4.47	99.2	205-235	30-34	290-345	42-50	41-47	55-66
Titanium powder, electrolytic, acicular, porous	0.196	0.053	...	0.066	4.44	98.5	335-365	49-53	455-510	66-74	10-14	20-25

Source: Ref 30

carbon steel hot pressed at 1000 °C (1830 °F) in the presence of a liquid phase derived from a mixture of iron powder containing 0.3% C and 5% of iron-phosphorus powder (1.3% C and 10.2% P) can reach a tensile strength of 730 MPa (106 ksi) and a hardness of 230 to 270 HB after slow cooling (Ref 27). A white cast iron powder with high silicon and phosphorus contents (3.16% C, 1.13% Si, and 1.154% P), similarly consolidated by hot pressing at 975 °C (1785 °F), attains a tensile strength of 500 MPa (72 ksi) and a hardness of 270 HB (Ref 27).

Superalloys exhibit superior creep resistance at high temperatures under loads encountered in service. Creep resistance, however, reduces the capacity of superalloy powders to deform plastically during hot consolidation. Accordingly, higher temperatures, as well as higher pressures or longer times under load, are required for hot pressing nickel or cobalt superalloys to full density than for iron and low-alloy steels. Lenel (Ref 13) found that a simple binary 80Ni-20Cr alloy powder mixture required the hot pressing temperature to be raised above the nickel-chromium eutectic temperature of 1345 °C (2455 °F) to obtain fully dense compacts. A ceramic die liner was used in conjunction with the short cycle resistance heating technique to minimize a reaction between the liquid phase and the tools.

With more reactive superalloys, such as those containing boron or large additions of aluminum and titanium, and with extended hot pressing cycles, interaction with the dies becomes more severe and more difficult to control by ceramic mold washes or spray coatings. Consequently, uniaxial hot pressing of superalloy powders in inert atmospheres in dies made of TZM molybdenum alloy (Ref 28) or graphite (Ref 19,

29) has given way to hot isostatic pressing, where interaction is more easily controlled.

However, uniaxial hot pressing can produce adequate mechanical properties. For example, an Inconel 718 hot pressed billet can attain a tensile strength of 790 MPa (115 ksi), a shear strength of at least 520 MPa (75 ksi), and elongation and reduction in area values of 10 to 20% (Ref 19, 29). The electric discharge-type activated hot pressing technique used in this case has been applied successfully to composite structures, such as a powder compact bonded to a sheet or other bulk metal of the same composition (bosses protruding from engine cases or root sections integrated with airfoil sheet stampings for turbine blades) (Ref 29).

Titanium and Titanium Alloys. The vacuum hot pressing technique for titanium powder described above has been used successfully to produce not only a variety of small parts of simple shape, but also large ingots weighing 55 kg (125 lb) or more (Ref 20). This material is of high quality, with critical oxygen and nitrogen contents falling below 1000 and 150 ppm, respectively. Representative tensile properties are 415 MPa (60 ksi) ultimate strength, 275 MPa (40 ksi) yield strength, 35 to 45% elongation, and 50 to 60% reduction in area.

Table 7 summarizes the tensile properties of 160-cm³ (10 in.³) billets produced from several elemental titanium powders by electric spark-activated resistance hot pressing (Ref 30). These data constitute optimum values, representing relatively uncontaminated material taken from the cores of the billets. Top, bottom, and peripheral surface regions have lower ductilities and higher strength properties due to penetration of interstitial elements, in

spite of a protective argon atmosphere contained in the enclosure shown in Fig. 5.

The tensile properties of Ti-6Al-4V billets produced from two different powders by the same spark-activated hot pressing method are summarized in Table 8. These data include values for the alloy after consolidation and solution treatment and aging. These properties are representative of material taken from the core of the 98 to 99% consolidated billets. All strength properties are comparable to those of wrought Ti-6Al-4V, but ductility is lower, which may be due to contamination in the absence of a vacuum during hot pressing. Ductility can be improved if alloy billets are used as preforms for further hot working, such as upset or step forging (Ref 32).

Table 9 lists tensile properties of forgings from hot pressed billets from two types of alloy powders. Elongations of 10% and higher and reduction in area values above 10% are obtained with annealed forgings of either kind; the upset forged alloy retains these ductilities after solution treatment and aging.

Vacuum hot pressed Ti-6Al-4V products made from the same two types of alloy powder have more uniform properties than the spark-sintered material. Battelle Columbus Laboratories has made billets and complex-shaped fittings. Values for elongation, reduction in area, and fracture toughness are equal to or better than the minimum specification values for wrought Ti-6Al-4V (Ref 33). Recent investigations also have found that compaction rate, and hence process economics, in vacuum hot pressing of Ti-6Al-4V powders can be enhanced by hydrogenation of the powders (Ref 34). To ensure removal of hydrogen during the latter phases of vacuum hot pressing, the hydrogen level must be be-

Table 8 Properties of electric spark-activated hot pressed Ti-6Al-4V alloy powder compacts

Powder	Billet condition	Billet analysis range, % Oxygen	Hydrogen	Nitrogen	Yield strength MPa	ksi	Tensile strength MPa	ksi	Elongation in 12.7 mm (0.5 in.), %	Reduction in area, %
Hydrided and dehydrided, angular, porous	As hot pressed	0.121-0.128	0.0056-0.0091	...	855-882	124-128	980-1000	142-145	6-7	7-12
	Annealed(a)	...	...	...	896-910	130-132	966-980	140-142	11-13	18-23
	Solution treated and aged(b)	...	...	...	952-986	138-143	1124-1186	163-172	4-5	5-7
Rotating electrode, spherical, solid	As hot pressed	0.120-0.150	0.0035-0.0075	0.004-0.081	828-882	120-128	958-1006	139-146	9-11	11-21
	Annealed(a)	...	...	...	868-802	126-128	930-945	135-137	12-14	20-25
	Solution treated and aged(b)	...	...	...	1055-1068	153-155	1172-1192	170-173	5-7	8-10

(a) Heated in air at 705 °C (1300 °F) for 2 h, air cooled. (b) Heated in air at 955 °C (1750 °F) for 1 h, water quenched; aged at 540 °C (1000 °F) for 4 h, air cooled
Source: Ref 31

Table 9 Tensile properties of forgings made from electric spark-activated hot pressed Ti-6Al-4V alloy powder preforms

Billet material	Forging	Condition	Yield strength MPa	ksi	Tensile strength MPa	ksi	Elongation in 25 mm (1 in.), %	Reduction in area, %
Hydrided and dehydrided powder, 97.6-98.0% dense	Upset	Annealed(a)	1006-1020	146-148	1028-1042	149-151	11-16	24-46
		Solution treated and aged(b)	1192-1206	173-175	1254-1268	182-184	10-12	27-42
	Step, 55% reduced	Annealed(a)	944-958	137-139	980-986	142-143	14-16	35-40
		Solution treated and aged(b)	1130-1138	164-165	1220-1228	177-178	8-10	20-15
	Step, 95% reduced	Annealed(a)	930-938	135-136	972-980	141-142	15-17	29-35
Rotating electrode powder, 97.6-98.5% dense	Upset	Annealed(a)	952-986	138-143	980-1028	142-149	12-15	43-47
		Solution treated and aged(b)	1124-1138	163-165	1186-1200	172-174	12-14	39-46
	Step, 55% reduced	Annealed(a)	924-930	134-135	958-972	139-141	15-17	38-42
		Solution treated and aged(b)	1152-1158	167-168	1214-1220	176-177	7-9	11-15
	Step, 95% reduced	Annealed(a)	966-972	140-141	1006-1014	146-147	16-18	48-52
Rotating electrode powder, 99.5-99.7% dense	Upset	Annealed(a)	952-992	138-144	1000-1028	145-149	10-14	37-40
		Solution treated and aged(b)	1110-1130	161-164	1166-1200	169-174	10-12	32-38
	Step, 55% reduced	Annealed(a)	882-890	128-129	938-958	136-137	9-11	15-18
		Solution treated and aged(b)	1192-1200	173-174	1254-1262	182-183	5-7	11-14
	Step, 95% reduced	Annealed(a)	992-1000	144-145	1028-1042	149-151	17-19	40-44

(a) Heated in air at 705 °C (1300 °F) for 2 h, air cooled. (b) Heated in air at 955 °C (1750 °F) for 1 h, water quenched; aged at 540 °C (1000 °F) for 8 h, air cooled
Source: Ref 32

low 7000 ppm at 705 °C (1300 °F).

Beryllium. Hot pressing is the ideal and, therefore, most commonly used method of consolidating beryllium powder. Hot pressing prevents the grain coarsening and brittleness observed in castings, or if conventional P/M techniques were used, the high compacting pressures and metal loss due to evaporation during sintering. Attrition-milled powder (see the article "Production of Beryllium Powder" in this Volume) is hot pressed at around 1050 °C (1920 °F) in graphite dies under atmospheric control or a vacuum into fully dense products (Ref 35, 36). It also may be processed at much lower temperatures in steel dies under a vacuum (Ref 37). In this case, pressures of 345 MPa (50 ksi) at 600 °C (1100 °F) are required for complete densification.

Beryllium is used primarily as moderator material in nuclear reactors and in missile and aerospace applications. Beryllium pieces of many sizes and shapes are produced by hot pressing, but most are blocks of standard dimensions, such as 1370 mm (54 in.) in length, 510 mm (20 in.) in width, and 100 mm (4 in.) in thickness (Ref 38). Single pieces may weigh between 10 and 1100 kg (25 and 2500 lb). Time required to produce these components varies according to size from 8 to 48 h.

Other pieces made by tamping powder into steel containers or graphite molds and then hot pressing under vacuum include disks up to 1650 mm (65 in.) in diameter and 460 mm (18 in.) in thickness and right circular cyclinders up to 750 mm (30 in.) in diameter and 920 mm (36 in. in length.

Truncated solid cones also can be produced by vacuum hot pressing; they can be machined by trepanning into thin-walled conical rings (Ref 39). Such pieces also have been produced by spark-discharge hot pressing without controlled atmosphere for nose cone prototypes (Ref 18). Table 10 lists typical mechanical properties of hot pressed beryllium (Ref 36).

Cemented Carbides. Hot pressing of cemented carbides produces products with high and uniform density that are harder than cold pressed and sintered material of comparable composition. The increase in hardness may be 1 to 3 points on the Rockwell A scale (Ref 40). The cemented carbide powder is hot pressed in graphite molds that are heated by resistance or by induction to about 1400 °C (2550 °F). An alternate heating method provides for di-

Table 10 Mechanical properties at ambient and elevated temperatures of vacuum hot pressed and vacuum hot pressed and extruded beryllium

	Vacuum hot pressed powder	Powder vacuum hot pressed and extruded at 425 °C (800 °F), 2.25 to 1 reduction		Powder vacuum hot pressed and extruded at 1050 °C (1920 °F), 12 to 1 reduction	
		Longitudinal(a)	Transverse(b)	Longitudinal(a)	Transverse(b)
Tensile strength, MPa (ksi)					
At 25 °C (75 °F)	225-350 (33-51)	440 (64)	260-315 (38-46)	565-620 (82-90)	345-435 (50-63)
At 300 °C (570 °F)	160-240 (23-35)	330 (48)	235-250 (34-36)	340 (49)	295-310 (43-45)
At 500 °C (930 °F)	150-170 (22-25)	250 (36)	205-240 (30-35)	240 (35)	240 (35)
At 700 °C (1290 °F)	95 (14)	115 (17)	90-110 (13-16)	90 (13)	90 (13)
Tensile yield strength, MPa (ksi)					
At 25 °C (75 °F)	220 (32)	...	...	310 (45)	...
Modulus of elasticity in tension, GPa (10^6 psi)					
At 25 °C (75 °F)	305 (44)	...	...	285 (41)	...
Elongation in 50 mm (2 in.), %					
At 25 °C (75 °F)	1-3	4	1	11-17	1
At 300 °C (570 °F)	12-30	13	2-4	23	2
At 500 °C (930 °F)	23-40	14	3-11	15	3-8
At 700 °C (1290 °F).....	10-14	15	11-13	7	4-6
Contraction, %					
At 25 °C (75 °F)	1-4	1	1-4	17	1
At 300 °C (570 °F)......	15-35	5	2-25	28	1-2
At 500 °C (930 °F).....	40-53	33	14-29	24	3-12
At 700 °C (1290 °F).....	10-13	10	14-17	5	3
Compression yield strength, MPa (ksi)					
At 25 °C (75 °F)	170 (25)	...	...	260 (38)	...
Unnotched Charpy impact strength, J (ft·lb)					
At 25 °C (75 °F)	1.1 (0.8)	...	...	5.6 (4.1)	...
Tensile impact strength, J (ft·lb)					
At 25 °C (75 °F)	1.9 (1.4)	...	...	6.1 (4.5)	...

(a) Properties in direction longitudinal to extrusion axis. (b) Properties in direction transverse to extrusion axis
Source: Ref 36

rect passage of a high-amperage current through the slightly precompacted powder between graphite electrodes.

Pressures are a fraction of those needed for cold pressing and range from 3.5 to 17 MPa (500 to 2500 psi). Pressure either is gradually increased during heating until a maximum is reached at sintering temperature, or it is applied in full force after the powder in the mold has reached peak temperature. Selection of the correct pressure is critical. If pressure is too high, the mold may burst, the plungers may crack, or the liquid binder metal phase may squeeze out and penetrate the space between the moving and stationary tools. This causes jamming, ejection difficulty, and surface porosity. If pressure is too low, however, the product remains nonuniformly porous throughout, and uneven shrinkage may result in distortions.

Hot pressing is the preferred production method for heavy-duty drawing and stamping dies and wear-resistant parts of simple geometric configuration. The alloy consists of 9% tungsten carbide and 13% cobalt binder. In the production of very large dies for the deep drawing of steel cylinders, the hot pressing machine has an integrated facility for precompacting the powder and presintering the preform (Ref 40). Die nibs with an internal diameter of 340 mm (13.5 in.) and a weight of more than 45 kg (100 lb) can be processed in such a unit. Deep drawing dies made in this manner are suitable for working many metals in addition to plain carbon steels, such as low-alloy steels, copper, brasses and bronzes, aluminum alloys, and various low-melting-point metals (Ref 41). The properties of several hot pressed carbides cemented with different binder metals are listed in Table 11 (Ref 42).

Intermetallics and Ceramics. Currently, many intermetallic and ceramic materials are produced by hot pressing. Vacuum hot pressing is preferred when contamination of the powder by interstitial elements is possible. In addition to metallic or ceramic carbides, borides, nitrides, and mixtures of these compounds with one another and with metal binders can be consolidated to near-theoretical density for aerospace, nuclear, electronics, and metallurgical applications. Intermetallic compounds such as a 80 at.% Si and 20 at.% Ge composition can be vacuum hot pressed into thermoelectric generator components with densities greater than 3 g/cm^3 (Ref 43, 44).

Vacuum hot pressing is particularly well suited for consolidation of boride-based cermets and composites. These systems include ZrB_2-Ta, ZrB_2-Mo, and ZrB_2-W, in which densification is affected by the content of the metallic phase acting as a lubricant for rearrangement of the boride particles (Ref 45).

Hot pressing is widely used for nitride ceramics, such as boron nitride (Ref 46) and silicon nitride (Ref 46, 47) or Sialons, which are combinations of Si_3N_4 and Al_2O_3. Hot pressing produces virtually fully dense silicon nitride, of essentially the β-modification, if certain sintering additives such as magnesium oxide, aluminum oxide, and yttrium oxide are added (Ref 48). This material has sufficient strength, as well as resistance to oxidation, corrosion, and thermal shock up to 1300 °C (2370 °F) to provide improved performance in diesel engine parts such as cylinder liners, pistons, and valves (Ref 48). With expected further development, the potential of hot pressing components for a pure ceramic gas turbine for vehicular applications may be realized in the near future.

Table 12 lists typical mechanical properties of a commercial-grade silicon nitride (Ref 46) and of several oxide additive-containing compositions (Ref 48). These properties are affected by powder characteristics. Silicon nitride powder, in which the α-modification predominates and is transformed to β-silicon nitride during hot pressing, has approximately twice the flexural strength (656 MPa, or 95 ksi) as material hot pressed under identical conditions from predominantly β-silicon nitride powder (375 MPa, or 55 ksi). Similar relations prevail for critical stress intensity factors and fracture energy (Ref 47).

Vacuum hot pressing also can be used to attain the highest degree of densification in oxide ceramics, such as alumina (Ref 49) or magnesia (Ref 50). The latter has been consolidated successfully in a continuous vacuum hot pressing facility.

Representative examples of current facilities for the vacuum hot pressing of metal, intermetallic, and ceramic powders are shown in Fig. 11. Figure 12 provides an interior view of the vacuum chamber, with press punches, graphite die and plungers, and induction coil in place. Adaptation of the equipment for diffusion bonding is shown in Fig. 13.

Appendix: Hot Extrusion*

By Fritz V. Lenel
Professor Emeritus
Department of Materials Engineering
Rensselaer Polytechnic Institute

HOT EXTRUSION combines hot compacting and hot mechanical working, yielding a fully dense product (Ref 51). The type of deformation occurring in hot extrusion is compared with that in hot isostatic compacting and in hot powder rolling in Fig. 14. In hydrostatic pressing, forces are equal in all directions, and little deformation takes place beyond that necessary for consolidation. In hot powder rolling, compressive forces are applied locally, resulting in densification as well as plastic elongation. In extrusion, large hydrostatic compressive forces occur, and a unidirectional force component makes the compact flow through the die. Frictional forces produce a shear component, which results in a characteristic shear pattern in the extruded metal. The energy expended in shear represents almost one half of the

* Reprinted from *Powder Metallurgy: Principles and Applications,* Metal Powder Industries Federation, Princeton, NJ, 1980, p 324-329. With permission

Table 11 Properties of hot pressed carbides with 10 wt% metal binder

Material	Density, g/cm³	Transverse rupture strength MPa	ksi	Hardness, HRA	Color of fracture
Fourth group of periodic system					
90TiC-10Co 4.9		786	114	92	Mouse gray
90ZrC-10Fe 6.8		786	114	90-91	Light gray
Fifth group of periodic system					
90VC-10Co 5.4		690	100	89	Silvery
90NbC-10Co 7.7		979	142	88	Brownish violet
90TaC-10Co...........13.0		737	107	85	Golden yellow
Sixth group of periodic system					
90Cr₃C₂-10Ni 5.7		490	71	84-85	Silvery
90Mo₂C-10Co 8.6		586	85	87	Light silvery
90WC-10Co 14.4		1790	260	91+	Bluish gray

Source: Ref 40, 42

Fig. 11 Cross section of all-purpose vacuum hot pressing system mounted in a press frame
Courtesy of Vacuum Industries Division, GCA Corp.

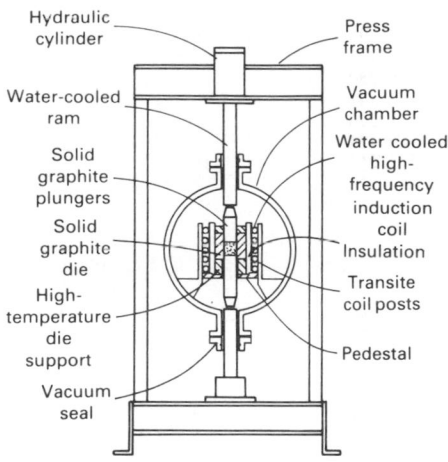

Fig. 12 Interior view of vacuum hot pressing and sintering furnace
Courtesy of Vacuum Industries Division, GCA Corp.

total energy needed for extrusion. The total amount of deformation in extrusion is much larger than in any other single metalworking step.

The three basic methods of hot extrusion of powder are shown in Fig. 15. In the first method, loose powder is placed without preheating into the heated extrusion container and extruded directly through the die. This method has been developed for the extrusion of certain magnesium alloy powders (Ref 52, 53)—or magnesium alloy pellets, as they have been termed, because of their relatively large particle size in the range of 70 to 450 μm. No atmospheric protection is provided, and the heat of the container is used to raise the tem-

perature of the powder sufficiently to allow extrusion. The temperature rise occurs during the 15 to 30 s needed to advance the ram prior to extrusion.

In the second method, which is used for hot extrusion of aluminum alloy powder billets, the powder is cold compacted and then hot pressed (Ref 54). The hot pressed compact is extruded by techniques used for extruding cast aluminum alloy billets. Cold isostatically pressed compacts of molybdenum powder, preheated to the extrusion temperature, can be extruded without canning the compacts, according to the results of recent research (Ref 55).

In most applications of hot extrusion of metal powders, the third method is used.

Table 12 Mechanical properties at ambient and elevated temperatures of hot pressed commercial silicon nitride and experimental silicon nitrides with various oxide additives

	Silicon nitride(a) Parallel(b)	Silicon nitride(a) Perpendicular(c)	SiMgON	α-SiAlON	β-SiAlON	SiYON(d)	SiYON(e)
Flexural strength, MPa (ksi)							
At 25 °C (75 °F)	510 (74)	710 (103)	940 (136)	580 (84)	460 (67)	1015 (147)	1400 (203)
At 800 °C (1470 °F)	490 (71)	690 (100)	895 (130)	630 (91)	515 (75)	985 (143)	1290 (187)
At 1000 °C (1830 °F)	475 (69)	675 (98)	815 (118)	600 (87)	500 (73)	925 (134)	1150 (167)
At 1200 °C (2190 °F)	460 (67)	610 (88)	600 (87)	470 (68)	480 (70)	835 (121)	1020 (148)
At 1300 °C (2370 °F)	445 (65)	535 (76)	320 (46)	300 (44)	435 (63)	720 (104)	950 (138)
At 1400 °C (2550 °F)	300 (44)	350 (51)	...	...	400 (58)	500 (73)	...
Tensile strength, MPa (ksi)							
At 25 °C (75 °F)	360 (52)	415 (60)	...	...	...	...	...
At 800 °C (1470 °F)	320 (46)	355 (51)	...	...	...	...	...
At 1000 °C (1830 °F)	290 (43)	340 (49)	...	...	...	...	...
At 1200 °C (2190 °F)	260 (38)	300 (44)	...	...	...	...	...
At 1300 °C (2370 °F)	225 (32)	230 (33)	...	...	...	...	...
Modulus of elasticity, GPa (10^6 psi)							
At 25 °C (75 °F)	300 (44)	290 (43)	...	...	...	...	...
At 800 °C (1470 °F)	275 (40)	280 (41)	...	...	...	...	...
At 1000 °C (1830 °F)	260 (38)	270 (39)	...	...	...	...	...
At 1200 °C (2190 °F)	240 (35)	255 (37)	...	...	...	...	...
At 1300 °C (2370 °F)	215 (31)	...	...	...	...	...	...
Compressive strength, MPa (ksi)							
At 25 °C (75 °F)	2760 (400)	690 (100)	...	...	...	...	...
Impact strength, J (in.·lb)							
At 25 °C (75 °F)	0.23 (2)	0.40 (3.5)	...	...	...	...	...

(a) Grade HS-130, contains impurity levels of 0.8 wt% Mg, 0.9 wt% Fe, 0.04 wt% Ca, 0.1 wt% Al, 0.5 wt% C, and less than 1 wt% W. (b) Maximum stress parallel to pressing direction. (c) Maximum stress perpendicular to pressing direction. (d) Amorphous grain boundary phase. (e) Crystalline grain boundary phase
Source: Ref 46-48

The powders are placed into a metallic capsule or "can," heated, and extruded with the can. The "canning" step is generally also part of hot consolidation by hot isostatic pressing and hot forging of powders. It permits handling of powders that are toxic, radioactive, pyrophoric, or easily contaminated by the atmosphere.

In hot extrusion of powders, a green metal powder compact may be canned or the powder may be cold pressed into the metal can under moderate pressure, as shown in Fig. 16. The can is prevented from bulging by use of a packing die. For spherical powders that can be vibrated to a high green density, precompacting may not be necessary. An end plate with an evacuation tube is placed over the powder and welded to the can. The can is outgassed by evacuation at room or elevated temperature and sealed off before can and powder are heated for extrusion.

To prevent turbulent flow during extrusion, the end of the can is conical and fits into an extrusion die with a conical opening (Fig. 17). To prevent folding when the powder in the can is not packed very densely, a penetrator ram may be used (Fig. 18). It enters the inside diameter of the can and densifies the powder before actual extrusion takes place.

The material of the can must have as close as possible the same stiffness at the extrusion temperature as the powder to be extruded; it should not react with the powder and should be removable by etching or mechanical stripping. Copper and low-carbon steel are most frequently used as material for cans.

When metal powders are canned and then hot extruded, the resulting extrusions will usually have simple circular, elliptical, or

Fig. 13 180-kN (20-ton) vacuum hot pressing installation adapted for diffusion bonding
Courtesy of Vacuum Industries Division, GCA Corp.

Fig. 14 Forces and deformation of powders in hydrostatic pressing, rolling, and extrusion

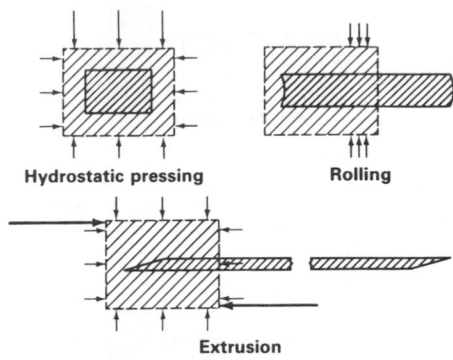

Hydrostatic pressing   Rolling

Extrusion

rectangular cross sections. However, a method of producing extruded structural shapes from superalloy powder, the "filled billet" technique (Ref 56), was developed by Gorecki and Friedman. An enlarged replica of the desired shape is produced within a round extrusion billet (Fig. 19). Figure 19 also shows the dimensions of the shape after extrusion. The ratio of the dimensions in the filled billet and in the final extrusion is determined by the extrusion ratio. The filler is produced from low-carbon steel by milling or electric discharge machining. The cavity is filled with a powder of the superalloy composition to be produced. Most superalloy powders have spherical shape and can, therefore, be vibrated to densities above 60% of theoretical. The machined filler is placed in a carbon steel can. Superalloy powder is poured into the cavity, and the billet is sealed, evacuated, and heated for extrusion. The billet is extruded through a round orifice conical approach die. The filler is then dissolved, leaving the desired shape 100% dense.

Hot extrusion of powders encapsulated in cans was first developed as a method for hot consolidating powders of beryllium and powders of dispersions of fissile material in a matrix of zirconium and stainless steel. The method has been almost universally used for copper- and nickel-based dispersion-strengthened alloys. This includes copper dispersion strengthened by aluminum oxide (Ref 57),

Fig. 15 Hot extrusion methods

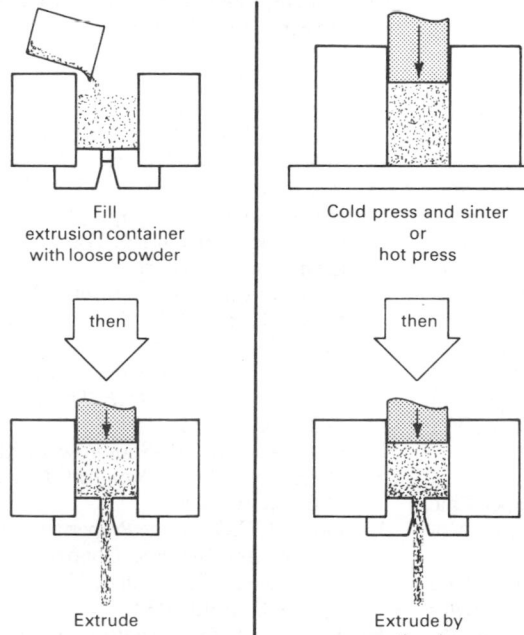

Fill extrusion container with loose powder

then

Extrude

Cold press and sinter or hot press

then

Extrude by conventional means

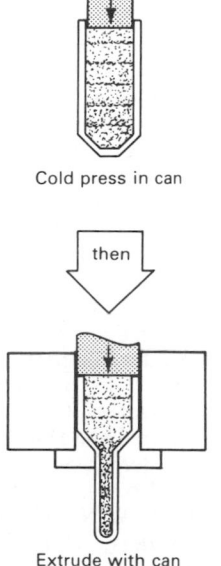

Cold press in can

then

Extrude with can

Fig. 16 Packing of powder into a metal can

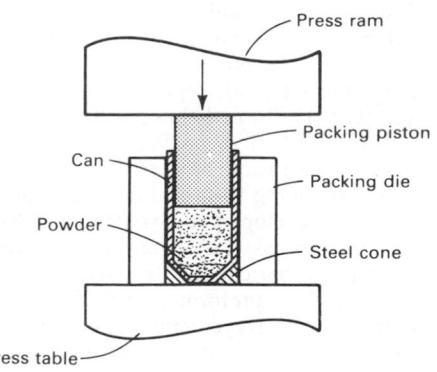

Press ram

Packing piston

Can

Packing die

Powder

Steel cone

Press table

Fig. 17 Extrusion with flat billet and die and with conical billet and die

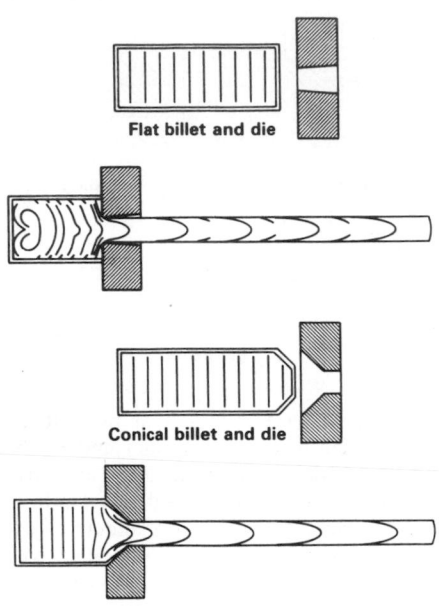

Flat billet and die

Conical billet and die

Fig. 18 Penetrator technique in powder extrusion to prevent can folding

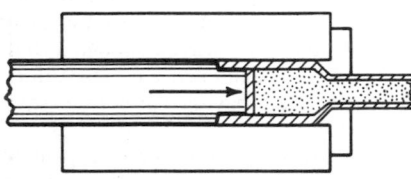

Fig. 19 Outline of filled billet process for extrusion

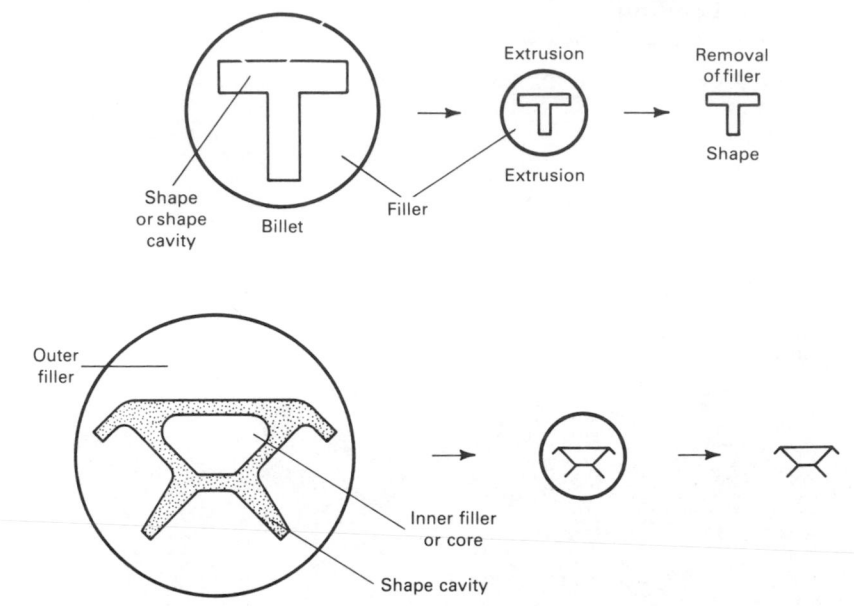

in which an internally oxidized copper-aluminum alloy powder is extruded in a copper can at 930 °C (1700 °F). The alloy TD nickel, a dispersion of thorium oxide in nickel, was consolidated from powder by hot extrusion (Ref 58). Alloy powders of nickel with 15% Mo and varying amounts of thorium oxide were canned in low-carbon steel and extruded into flat bars at 870 °C (1600 °F) at extrusion ratios ranging from 7.1 to 10.4 (Ref 59). For the first of the mechanically alloyed compositions of nickel-based superalloys dispersion strengthened with yttrium oxide, Benjamin (Ref 60) reported extrusion at 100 °C (212 °F) or below, with extrusion ratios of 12 to 1 or greater.

A hot extrusion process for producing seamless tubing from stainless steel powder was developed in Sweden (Ref 61). In this process, an argon- or nitrogen-atomized spherical stainless steel powder is filled into preformed molds of low-carbon steel. By careful control of filling, densities of 70% of theoretical can be obtained. The molds, weighing 35 to 120 kg (75 to 265 lb), are cold isostatically pressed in groups of 20 to 40 molds at pressures of 400 to 500 MPa (58 to 72 ksi) to a density of 85 to 90% of theoretical and are then hot extruded at temperatures near 1200 °C (2190 °F), using a glass lubricant as in the Sejournet process. With extrusion ratios of 4 to 1 or higher, completely dense seamless tubing is produced. The process has been applied to austenitic stainless steels and particularly to a ferritic steel with low interstitial content.

REFERENCES

1. Hoyt, S.L., Hard Metal Carbides and Cemented Tungsten Carbides, *AIME Trans.*, Vol 89, 1930, p 9-58
2. Sauerwald, F. and Hunczek, J., Synthetic Metal Bodies, *Zeitschrift Metallkunde*, Vol 21 (No. 1), 1929, p 22-23
3. Trzebiatowski, W., Hot-Press Experiments with Highly Dispersed Metallic Powders, *Zeitschrift Physikalische Chemie*, Vol A169, 1934, p 91-102
4. Goetzel, C.G., Some Properties of Sintered and Hot-Pressed Copper-Tin Powder Compacts, *AIME Trans.*, Vol 161, 1945, p 580-595
5. Goetzel, C.G., Some Properties of Sintered and Hot-Pressed Copper-Zinc Powder Compacts, *ASM Trans.*, Vol 30 (No. 1), 1942, p 86-123
6. Pratt, W.N., Hot Pressing of Metal Powders, *Symposium on Powder Metallurgy*, American Society for Testing and Materials, Philadelphia, 1943, p 49-55
7. Schwarzkopf, P. and Goetzel, C.G., Hot Pressing of Iron Powders, *Iron Age*, Vol 148 (No. 10), 1941, p 37-44
8. Henry, O.H. and Cordiano, J.J., Hot Pressing of Iron Powder, *AIME Trans.*, Vol 166, 1946, p 520-532
9. Beaver, W.W. and Larson, H.F., The Powder Metallurgy of Beryllium, *Powder Metallurgy*, Leszynski, W., Ed., Interscience, New York, 1961, p 747-773
10. Lidman, W.G. and Moyer, K.H., Hot Pressing of Electrolytic Grade CR Beryllium, *Modern Developments in Powder Metallurgy*, Vol 3, Plenum Press, New York, 1966, p 60-68
11. Lindop, T.W., Engineer's Guide to Silicon Nitride, *Materials Engineering*, Vol 75 (No. 1), 1972, p 28-31
12. Notis, M.R. and Spriggs, R.M., Emerging Areas in Hot Pressing, *Sintering—New Developments, Materials Science Monographs*, Vol 4, M.M. Ristic, Ed., Elsevier Scientific, New York, 1979, p 295-306
13. Lenel, F.V., Resistance Sintering Under Pressure, *AIME Trans.*, Vol 203, 1955, p 158-167
14. de Groat, G., One-Shot Powder Metal Parts, *Am. Mach.*, Vol 109 (No. 21), 1965, p 107-109
15. Goetzel, C.G., Some Properties of Hot-Pressed and Sintered Copper Powder Compacts, ASM Trans., Vol 28 (No. 4), 1940, p 909-932
16. Goetzel, C.G., *Treatise on Powder

Metallurgy, Vol 1, Interscience, New York, 1949, p 475-481

17. Willey, F.H., U.S. patent 2 143 495, "Method of Making a Hard and Compact Metal for Use in the Formation of Tools, Dies" (applied 1937, issued 1938)

18. Boesel, R.W., Jacobson, M.I. and Yoshioka, I.S., Spark Sintering Tames Exotic P/M Materials, *Mat. Eng.*, Vol 70 (No. 4), 1969, p 32-35

19. Courtis, W.F., "Spark Sintering Missile Component Feasibility," U.S. Naval Ordnance Contract NOO197-73-C-0081, Report LMSC-D-326418, 23 Aug 1973

20. Dodds, H.W., A Fabricator Views Titanium Powder Metallurgy, *Proceedings of the Eleventh Annual Meeting*, Vol 1, Metal Powder Association, New York, 1955, p 108-113

21. Beaver, W.W., Fabrication of Beryllium by Powder Metallurgy, *The Metal Beryllium*, D.C. White, Jr. and J.E. Burke, Ed., American Society for Metals, 1955, p 152-201

22. Hoffman, C.G. and Beaver, W.W., "Progress Report on Powder Metallurgy, U.S. Atomic Energy Commission Report BBC-47, 1949; *Nucl. Sci. Abst.*, Vol 11, No. 11692, 1957

23. Tietz, T.E. and Palmer, I.G., Advanced P/M Aluminum Alloys, *Advances in Powder Technology*, G.Y. Chin, Ed., American Society for Metals, 1982, p 194

24. Jones, W.D., Powder Metallurgy, *Metal Industry (London)*, Vol 56 (No. 3), 1940, p 69-71

25. Jones, W.D., Hot-Pressing of Metal Powders, *Metal Industry (London)*, Vol 56 (No. 10), 1940, p 225-228

26. Goetzel, C.G., Sintered, Forged and Rolled Iron Powders, *Iron Age*, Vol 150 (No. 14), 1942, p 82-92

27. Jones, W.D., Powder Metallurgy of Iron with Particular Reference to Pacteron, *Foundry Trade J.*, Vol 59 (No. 1163), 1938, p 401-402

28. Lenel, F.V., *Powder Metallurgy: Principles and Applications*, Metal Powder Industries Federation, Princeton, NJ, 1980, p 323

29. Goetzel, C.G., Electric Discharge-Type Activated Pressure Sintering and Bonding of a Superalloy Composite Structure, *High Temperatures—High Pressures*, Vol 3 (No. 4), 1971, p 425-438

30. Goetzel, C.G. and de Marchi, V.S., Electrically Activated Pressure Sintering (Spark Sintering) of Titanium Powders, *Powder Metall. Int.*, Vol 3 (No. 2, 3), 1971, p 80-87, 134-136

31. Goetzel, C.G. and de Marchi, V.S., Electrically Activated Pressure Sintering (Spark Sintering) of Titanium-Aluminum-Vanadium Alloy Powders, *Modern Developments in Powder Metallurgy*, Vol 4, Plenum Press, New York, 1971, p 127-150

32. Goetzel, C.G., Tensile Properties of Titanium Alloy Forgings Made from Spark-Sintered Preforms, *Metals Eng. Quart.*, Vol 11 (No. 2), May 1971, p 53-61

33. Malik, R.J., Vacuum Hot Pressing of Titanium Alloy Powders, *Progress in Powder Metallurgy*, Vol 31, Metal Powder Industries Federation, Princeton, NJ, 1975, p 277-288

34. Kao, W.H., Eylon, D., Yolton, C.F., and Froes, F.H., Effect of Temporary Alloying by Hydrogen (HYDROVAC) on the Vacuum Hot Pressing and Microstructure of Titanium Alloy Powder Compacts, *Progress in Powder Metallurgy*, Vol 37, Metal Powder Industries Federation, Princeton, NJ, 1982, p 289-301

35. Beaver, W.W., Fabrication of Beryllium by Powder Metallurgy, *Met. Prog.*, Vol 65 (No. 4), 1954, p 92-94, 97, 168, 170-173

36. Beaver, W.W. and Wikle, K.G., Mechanical Properties of Beryllium Fabricated by Powder Metallurgy, *AIME Trans.*, Vol 200, 1954, p 550-573

37. Pinto, N.P., The Warm Pressing of Beryllium Powder, *AIME Trans.*, Vol 200, 1954, p 629-633

38. Beaver, W.W. and Lillie, D.W., Beryllium Metal Alloys and Compounds, *Reactor Handbook*, Vol 1, 2nd ed., C.R. Compton, Jr., Ed., Interscience, New York, 1960, p 903

39. Darwin, G.E. and Buddery, J.H., *Beryllium*, Academic Press, New York, 1960, p 116

40. Goetzel, C.G., *Treatise on Powder Metallurgy*, Vol 2, Interscience, New York, 1950, p 106-107, 164

41. Glen, A.E., Carbide Die Applications, Present and Future, *Iron Age*, Vol 157 (No. 17), 1946, p 51-54

42. Kieffer, R. and Kölbl, F., Tungsten Carbide-Free Hard Metals, *Powder Metall. Bull.*, Vol 4 (No. 1), 1949, p 4-17

43. Lefever, R.A., McVay, G.L., and Baughman, R.J., Preparation of Hot-Pressed Silicon-Germanium Ingots. Part 3, Vacuum Hot Pressing, *Mat. Res. Bull.*, Vol 9 (No. 7), 1974, p 863-872

44. Rowe, D.M. and Bunce, R.W., Vacuum Hot Pressing of Germanium and Silicon-Germanium Alloys, *Sintering—New Developments, Materials Science Monographs*, Vol 4, M.M. Ristic, Ed., Elsevier Scientific, New York, 1979, p 307-318

45. Hirabayashi, M. and Nakagawa, H., Vacuum Hot Pressing and Some Properties of the System Zirconium Diboride—Refractory Metal Composites, *Proceedings of the Fourth International Conference on Vacuum Metallurgy*, Section 6, Japan Institute of Metals, 1973, p 267-271

46. "Engineering Property Data on Selected Ceramics," Vol 1, "Nitrides," Report MCIC-HB-07, Metals and Ceramics Information Center, Battelle Columbus Laboratories, Columbus OH, March 1976

47. Lange, F.F., Mechanical Behavior of Hot-Pressed Si_3N_4 and SiC, *Review of Ceramic Technology*, No. 29, Metals and Ceramics Information Center, Battelle Columbus Laboratories, Columbus, OH, May 1974, p 1-12

48. Lorenz, J., Weiss, J., and Petzow, G., Dense Si_3N_4 Alloys: Phase Relations and Consolidation, Microstructure and Properties, *Advances in Powder Technology*, G.Y. Chin, Ed., American Society for Metals, 1982, p 289-308

49. Rossi, R.C. and Fulrath, R.M., Final Stage Densification in Vacuum Hot Pressing of Alumina, *J. Am. Cer. Soc.*, Vol 48 (No. 11), 1965, p 558-564

50. Coates, D.J.G., Pointon, A.J., and Newman, H.E., Continuous Vacuum Hot Pressing, *Vacuum*, Vol 27 (No. 9), 1977, p 531-535

51. Loewenstein, P., Aronin, L.R., and Geary, A.L., Hot Extrusion of Metal Powders, in *Powder Metallurgy*, W. Leszynski, Ed., New York, 1961, p 563-583

52. Busk, R.S. and Leontis, T.E., The Extrusion of Powdered Magnesium Alloys, *Trans. AIME.*, Vol 188, 1950, p 297-306

53. Busk, R.S., The Pellet Metallurgy of Magnesium, *Light Metals*, Vol 23, 1960, p 197-200

54. Lyle, J.P., Jr. and Cebulak, W.C., Fabrication of High Strength Aluminum Products from Powder, in *Powder Metallurgy for High-Performance Applications*, J.J. Burke and V. Weiss, Ed., Syracuse, 1972, p 231-254

55. Tuominen, S.M. and Dahl, J.M., "Properties of Unsintered Molybdenum Powder Extrusions," presented at 108th annual meeting of AIME, New Orleans, Feb 1979

56. Bufferd, A.S., Complex Superalloy Shapes, in *Powder Metallurgy for High-Performance Applications*, J.J. Burke and V. Weiss, Ed., Syracuse, 1972, p 303-316

57. Nadkarni, A.V., Klar, E., and Shafer, W.M., A New Dispersion Strengthened Copper, *Metals Eng. Quart.*, Vol 16 (No. 3), 1976, p 10-15

58. Anders, F.J., Alexander, G.B., and Wartel, W.S., A Dispersion Strengthened Nickel Alloy, *Met. Prog.*, Vol 82 (No. 6), 1962, p 88-91, 118, 122

59. Cheney, R.F. and Scheithauer, W., Jr., The High-Temperature Strength of Oxide Strengthened Nickel Alloys, *Modern Developments in Powder Metallurgy*, Vol 5, H.H. Hausner, Ed., Plenum Press, New York, 1971, p 137-148

60. Benjamin, J.S., Dispersion Strengthened Superalloy by Mechanical Alloying, *Met. Trans.*, Vol 1, 1970, p 2943-2951

61. Aslund, C., A New Method for Producing Stainless Steel Seamless Tubes From Powder, *Preprints*, Vol 1, 5th European Symposium on Powder Metallurgy, Stockholm, 1978, p 278-283

SELECTED REFERENCES ON HOT EXTRUSION

● Ashlund, C., Gemmel, G., and Anderson, T., Extruded Tubes Based on Powder Metallurgy, *Bander Bleche Rohre*, Vol 22 (No. 9), Sept 1981, p 223-226

● Avitzur, B. and Blum, P., Forging and Extrusion of P/M Preforms, *Modern Developments in Powder Metallurgy*, Vol 7, Metal Powder Industries Federation, Princeton, NJ, 1974, p 73-90

● Breneiser, D., Continuous Extrusion of Powder Into Structural Shapes, *Progress in Powder Metallurgy*, Vol 33, Metal Powder Industries Federation, Princeton, NJ, 1977, p 107-121

● Causton, R.J. and Dunkley, J.J., Extrusion of Powder Metallurgy High-Speed Steel, proceedings of Hot Working and Forming Processes Conf., Sheffield, England, 1980, p 244-253

● Chare, P.J.M. and Sheppard, T., Densification and Properties of Extruded Al-Zn-Mg Atomized Powder, *Int. J. Powder Metall.*, Vol 10 (No. 3), July 1974, p 203-215

● Chare, P.J.M. and Sheppard, T., Powder Extrusion as a Primary Fabricating Process for Al-Fe Alloys, *Powder Metall.*, Vol 16 (No. 32), 1973, p 437-458

● Chellman, D.J. and Wald, G.G., Age Hardening Behavior of Al-Li-(Cu)-(Mg)-Zr P/M Alloys, *Progress in Powder Metallurgy 1982*, Vol 38, Metal Powder Industries Federation, Princeton, NJ, 1983, p 361-381

● Clark, J.B. and Wright, R.N., Laboratory Extrusion of Nb_3Sn, *Metall. Trans. A*, Vol 14A (No. 11), Nov 1983, p 2295-2299

● Dower, R.J. and Miles, G.I., Cold Extrusion of Mild Steel Billets Produced by Powder Metallurgy Techniques, *Modern Developments in Powder Metallurgy*, Vol 7, Metal Powder Industries Federation, Princeton, NJ, 1974, p 175-201

● Dunkley, J.J. and Causton, R.J., P/M Extrusion of Tool Steel Bar, *Int. J. Powder Metall.*, Vol 13 (No 1), Jan 1977, p 13-15, 18-20

● Eck, R., Powder Metallurgy of Refractory Metals and Applications, *Int. J. Powder Metall. Powder Technol.*, Vol 17 (No. 3), July 1981

● Fleming, R.P.H., Hot Impact Extrusion and Subsequent Processing of Some High-Temperature Nickel-Base Alloys, proceedings of Hot Working and Forming Processes Conf., Sheffield, England, 1980, p 254-258

● Foner, S., Development of High-Performance Superconductors by Powder Metallurgy, in *Progress in Powder Metallurgy 1982*, Vol 38, Metal Powder Industries Federation, Princeton, NJ, 1983, p 107-114

● Friedman, G., Properties of Extruded Metal Powders, *Int. J. Powder Metall. Powder Technol.*, Vol 16 (No. 1), Jan 1980, p 29-35

● Halter, R.F. and Rajan, S.S., Die Wall Lubrication for P/M Parts and Preforms, in proceedings of 1971 Fall Powder Metallurgy Conf., Metal Powder Industries Federation, New York, 1972, p 83-95

● Hassler, H. and Schreiner, H., Backward Cold Extrusion of P/M Composite Materials, *Int. J. Powder Metall. Powder Technol.*, Vol 16 (No. 1), Jan 1980, p 21-27

● Hirose, S. and Fine, M.E., Fatigue Crack Initiation and Microcrack Propagation in X7091 Type Aluminum P/M Alloys, in proceedings of High-Strength Powder Metallurgy Aluminum Alloys, Dallas, Feb 1982, The Metallurgical Society/AIME, Warrendale, PA, 1982, p 19-39

● Hunderi, O. and Save, T., Optimization of Processing Conditions for Some Rapidly Quenched Aluminum-Based Alloys, *Scand. J. Metall.*, Vol 10 (No. 5), 1981, p 231-237

● Kandeil, A.Y., et al., On the Hydrostatic Extrusion of Nickel-Base Superalloys, *Can. Metall. Quart.*, Vol 19 (No. 2), April-June 1980, p 245-249

● Kumar, A., Jain, P.C., and Mehta, M.L., Modification Formulation for Extrusion of Porous Powder Preforms Using Finite Element Method, in *Numerical Methods in Industrial Forming Processes*, Pineridge Press, Swansea, England, 1982

● Kumar, A., Jain, P.C., and Mehta, M.L., Aluminum Powder Extrusion Through Wedge-Shaped Dies as an Effective Use of Energy, *J. Mech. Work. Technol.*, Vol 6 (No. 2-3), March 1982, p 235-251

● Kumar, A., et al., Hot Extrusion of Aluminum Powder at Low Reduction Ratios, *Int. J. Powder Metall. Powder Technol.*, Vol 17 (No. 3), July 1981

● Kursetz, E., Extrusion of Tubes From Ferrous Powders, *Bander Bleche Rohre*, Vol 15 (No. 11), Nov 1974

● Langenbeck, S.L., Investigation of the Fatigue and Crack Propagation Properties of X7091-T7E69 Extrusion, in proceedings of High-Strength Powder Metallurgy Aluminum Alloys Conf., Dallas, Feb 1982, The Metallurgical Society/AIME, Warrendale, PA, 1982

● Manegin, Y.V., et al., Hot Extrusion and Properties of Rods From Sintered Mo and W Blanks, *Researches in Powder Metallurgy*, Vol 2, Consultants Bureau, New York, 1972, p 87-90

● Marsh, H.G. and Pierret, J.A., Method for Producing an Embrittlement-Resistant Tantalum Wire, U.S. Patent 4 235 629, 1973

● Martin, P.L., Lipsitt, H.A., and Williams, J.C., The Structure of As-Extruded RSR Ni-Al-Mo and Ni-Al-Mo-X Alloys, in *Rapid Solidification Processing, Principles and Technologies*, Vol 2, Claitor's, Baton Rouge, 1980

● Merrick, H.F., Effect of Heat Treatment on the Structure and Properties of Extruded P/M Alloy 718, *Metall. Trans. A.*, Vol 7A (No. 4), April 1976, p 505-514

● Nakagawa, T., Amano, T., and Nagase, M., Cold Forging and Extrusion of Green Metal Powder, in *Modern Developments in Powder Metallurgy*, Vol 7, Metal Powder Industries Federation, Princeton, NJ, 1974, p 411-422

● Okhrimenko, Y.M., et al., New High-Rate Extrusion Process for Alloys of Low Ductility, *Light Met. Age*, Vol 31 (No. 3-4), April 1973

● Pardoe, J.A., "Conform": Continuous

Extrusion of Metal Powders into Products for Electrical Industry: Development Experience, *Powder Metall.*, Vol 22 (No. 1), 1979, p 22-28
- Park, S.C. and Park, W.K., Properties of Extruded Aluminum-Iron P/M Materials, *Int. J. Powder Metall. Powder Technol.*, Vol 14 (No. 4), Oct 1978
- Schultz, L. and Bormann, R., Superconductivity in Filamentary CuNb Composites Produced by Powder Metallurgy, *J. Appl. Phys.*, Vol 50 (No. 1), Jan 1979, p 418-424
- Sheppard, T., *et al.*, The Extrusion of Atomized Aluminum Alloy Compacts and Composites, *J. Mech. Work. Technol.*, Vol 8 (No. 1), June 1983, p 43-70
- Sheppard, T. and Chare, P.J.M., Extrusion of Atomized Al Powders, *Powder Metall.*, Vol 15, 1972, p 17-41
- Shima, S. and Oyane, M., Cold Back-Pressure Extrusion of Al Powder, Bull. *Jpn. Soc. Mech. Eng.*, Vol 14, Oct 1971, p 1126-1132
- Skogsberg, J.W., Harth, G.H., and Fels, M., Stainless Steel Tubing by P/M, *Precis. Met.*, Vol 31 (No. 2), 1973, p 32-33

Forging and Rolling of P/M Billets

By B. Lynn Ferguson
Partner
Deformation Control Technology

POWDER METALLURGY TECH-NIQUES can be combined with cast and wrought metalworking technology to produce mill products, billets, or near-net shapes. This article describes some of the manufacturing processes that may be used to produce these products. Such production techniques capitalize on the advantages of powder metallurgy for wrought products.

Powder is first consolidated to near-theoretical density to form a workpiece. Suitable primary consolidation techniques include:

- Cold pressing and sintering to high density (cold isostatic pressing and sintering is typical)
- Hot pressing to high density (vacuum hot pressing is typical)
- Hot isostatically pressing containerized powder to high density
- Direct sintering of containerized powder to high density
- Injection molding of powder mass and sintering to high density
- Extrusion of containerized powder to high density

The workpiece is then subjected to a secondary deformation process to produce a mill shape, billet, or near-net shape of theoretical density. Suitable bulk deformation processes include:

- Rolling of mill shapes
- Extrusion of mill shapes
- Forging of billets to shapes
- Isothermal forging of near-net shapes
- Isothermal rolling of near-net shapes

Deformation to a final shape is the equivalent of plastic working of a conventional material into an intermediate or final product shape, because the initial workpiece is of near-theoretical density. Selection of the proper thermomechanical treatment depends on the material being processed and the desired final shape. Regardless of the particular process, thermomechanical treatment ensures the desired final shape, complete densification, and massive deformation, which disrupts any prior particle boundaries and contaminants.

Powder metals are processed into wrought P/M products, including high-strength and high-temperature aluminum alloys, high-speed tool steels, superalloys, refractory metals, and many dispersion-strengthened alloys. Alloy design and processing techniques for these materials are discussed in other articles in this Volume. Improvement of performance by producing a wrought product from a powder metallurgy billet rather than from an ingot metallurgy billet is the prime reason for acceptance of these processes. Alloys that are subject to heavy segregation, poor castability, poor workability, or that cannot be produced by conventional cast and wrought practice are often candidates for wrought P/M processing.

Superalloys

As the operating temperatures of superalloys in turbine engines have increased, the volume fraction of dispersoids and/or precipitates used for strengthening at elevated temperatures also has increased. Superalloys capable of withstanding higher operating temperatures have limited workability due to the higher volume fraction of the second phase present. Also, segregation becomes more significant as alloy complexity increases. As a result, conventional ingot metallurgy practices, such as casting followed by forging, are not practical for many current superalloys if optimum properties are required. Primarily, this affects the production of disks and similar turbine engine components that traditionally have been produced by conventional techniques.

Powder metallurgy offers a method of producing a preform shape that may be worked into a desired final shape, thus preventing segregation and workability problems encountered in casting an ingot and working it into a preform shape. For superalloys, conventional press forging or new processes such as isothermal forging can be used for the final working operation. Examples of manufacturing practices for disk production are described below.

Hot Isostatic Pressing Plus Conventional Forging. One approach for producing a workpiece from powder is to hot isostatically press prealloyed powder that has been containerized into a preform shape (Ref 1). These powders are usually spherical, with low oxygen contents; typically, they are produced by inert gas atomization, by soluble-gas atomization, or by the rotating electrode process. For more information, see the article "Atomization" in this Volume.

Powder is hermetically sealed in a suitable container (containerization is described in the article "Hot Isostatic Pressing of Metal Powders" in this Volume) and then subjected to an appropriate temperature/pressure/time cycle in an autoclave to produce densification and particle bonding. Microstructure also can be controlled by proper selection of these variables.

Temperatures are selected that promote the formation of intergranular rather than interparticle carbides ($M_{23}C_6$ carbides instead of MC carbides). In practice, densities above 95% and normally above 98% of theoretical are achieved in this preforming step. At this point, the container is removed. The hot isostatically pressed billet may be an individual preform, or it may be sectioned into many preforms for subsequent working into final shapes.

Conventional forging is used next to produce a forged blank that is then machined to final dimensions. Preforms are heated to the forging temperature and forged in one step into a blank. A hydraulic press usually is used because of the size of the part and the reduced flow stress. Forging temperatures and deformation levels are selected to optimize the mechanical properties of the disk; low-cycle fatigue resistance is of prime importance. Optimum microstructure is a partially warm worked structure, with carbides distributed in a necklace pattern, as shown in Fig. 1. Full density is ensured by forging, as is disruption of any particle-boundary films. A schematic flowchart for production of nickel superalloy turbine disks is shown in Fig. 2. Superalloy parts of alloy compositions listed in Table 1 are produced by this process.

Hot Isostatic Pressing Plus Isothermal and Hot Die Forging. These processes are similar to hot isostatic pressing

Fig. 1 Necklace microstructure in superalloy for optimum performance

Magnification: 250×. Source: Ref 2

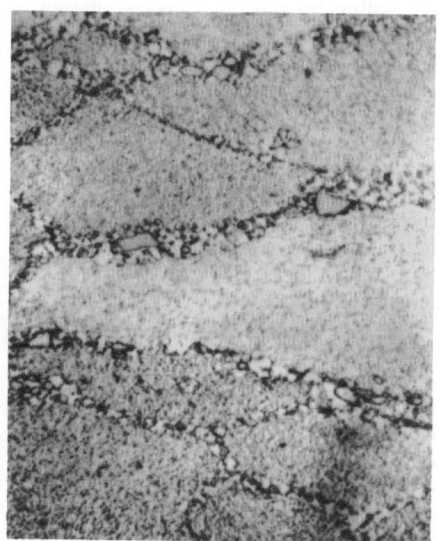

Fig. 2 Flowchart of hot isostatic pressing followed by forging for producing turbine disks

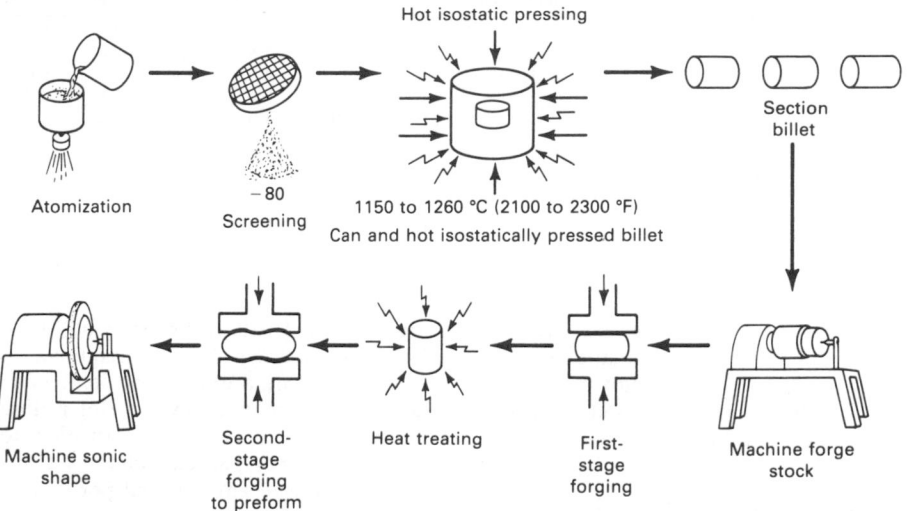

plus conventional forging, except that near-net shape practices of isothermal forging or hot die forging are substituted for conventional press forging. Isothermal forging involves heating the workpiece and tooling to the same temperature and forging at low strain rates. In hot die forging, the workpiece is heated to temperatures above that of the tooling, and the tooling is heated to temperatures well above those of typical forging tooling. Low strain rates are necessary. Sometimes, these processes are referred to as creep deformation processes because of the low strain rates used.

Hot die and isothermal forging processes, which are described in the article "P/M Forging" in this Volume, have greater shapemaking capabilities than conventional press forging for these alloys, due to the use of low strain rates and hot tooling, which eliminates die chill problems. Advantages of these processes over conventional press forging due to low strain rates include:

- Reduced flow stresses in the workpiece
- Enhanced workability
- Greater control over final microstructure
- Greater dimensional precision

The low strain rates capitalize on the fine grain size of P/M preforms for improved workability. These advantages allow production of near-net shapes, which results in significant cost savings due to reduced materials usage and reduced machining. For example, a weight savings of 223 kg (490 lb) of material was realized in the F-100 turbine engine. Conventional press forging of disks required a total of 450 kg (990 lb) of disk material, while isother-

mally forged disks were produced from only 227 kg (500 lb) of starting stock (Ref 3).

Disadvantages of hot isostatic pressing plus isothermal and hot die forging include:

- Long forging cycles (typically 15 min or longer dwell times)
- High cost of forging equipment (atmosphere chambers and integral heating stations)
- Cost of forging dies (expensive alloys such as TZM molybdenum alloy, rather than lower cost tool steels)
- Need for an inert atmosphere or vacuum to protect tooling

Despite these costs, this process combination is economical for producing parts from expensive alloys, where near-net shape production minimizes expensive machining and material losses.

Powder metallurgy alloys processed by these methods are listed in Table 1. IN-100 and alloys such as René 95 and MERL 76 comprise the bulk of isothermally forged P/M tonnage.

Hot Extrusion Plus Forging. Superalloy powders also can be consolidated into billet shapes by extrusion. Powder, usually containerized, is hot extruded at a reduction ratio of at least 9 to 1 to achieve a fully consolidated billet. The extruded length is then sectioned into workpiece billets for subsequent hot working by conventional press forging, isothermal forging, or hot die forging. For press forging, two steps are required. First, the billet is forged into a preform shape. Then a forged blank is produced from the preform. Preform shaping operations also may be added to the isothermal and hot die forging steps

Table 1 Typical compositions of selected P/M superalloys

Alloy designation	Carbon	Chromium	Molybdenum	Tungsten	Tantalum	Titanium	Niobium	Cobalt	Aluminum	Hafnium	Zirconium	Boron	Nickel	Iron	Vanadium	Y₂O₃
								Composition, %								
IN-100	0.07	12.5	3.2	. . .	. . .	4.3	. . .	18.5	5.0	. . .	0.04	0.02	rem	. . .	0.75	. . .
René 95	0.07	13.0	3.5	3.5	. . .	2.5	3.5	8.0	3.5	. . .	0.05	0.01	rem	. . .	. . .	. . .
MERL 76	0.02	12.4	3.2	. . .	. . .	4.3	1.4	18.5	5.0	0.4	0.06	0.02	rem	. . .	. . .	. . .
AF 115	0.05	10.5	2.8	6.0	. . .	3.9	1.7	15.0	3.8	2.0	. . .	. . .	rem	. . .	. . .	. . .
PA101	0.1	12.5	. . .	4.0	4.0	4.0	. . .	9.0	3.5	1.0	. . .	. . .	rem	. . .	. . .	. . .
Low-carbon Astroloy	0.04	15.0	5.0	. . .	. . .	3.5	. . .	17.0	4.0	. . .	0.04	0.025	rem	. . .	. . .	. . .
MA754	0.05	20.0	. . .	. . .	. . .	0.5	. . .	. . .	0.3	. . .	. . .	. . .	rem	. . .	. . .	0.6
MA956	. . .	20.0	. . .	. . .	. . .	0.5	. . .	. . .	4.5	. . .	. . .	. . .	. . .	rem	. . .	0.5
MA6000	0.05	15.0	2.0	4.0	2.0	2.5	. . .	. . .	4.5	. . .	0.15	0.1	rem	. . .	. . .	1.1
Stellite 31	0.5	25.5	. . .	7.5	. . .	. . .	. . .	rem	. . .	. . .	. . .	. . .	10.5	2.0	. . .	. . .

to reduce cycle time in the low-strain rate deformation step. A schematic flowchart of the production of superalloy turbine disks by extrusion followed by isothermal forging is shown in Fig. 3.

High-Speed Steels and Tool Steels

Segregation and alloy limitations in ingot metallurgy processing of high-speed steels and tool steels have provided impetus for P/M processing of these materials. For more detailed information on P/M tool steel production, see the articles "Consolidation by Atmospheric Pressure" and "P/M Tool Steels" in this Volume.

Hot Isostatic Pressing Plus Hot Rolling to Mill Shape. Powder metallurgy tool steel and high-speed steel mill products are produced from gas-atomized powder by hot isostatic pressing of containerized powder to near-theoretical density and then hot rolling the billet to a final mill shape using multipass rolling (Ref 4-6). Containerization, as described in the article "Hot Isostatic Pressing of Metal Powders" in this Volume, is required.

In this application, sheet steel containers are used exclusively. Hot rolling provides sufficient shear deformation to disrupt any prior particle-boundary films and to eliminate any residual porosity in the material. Finished parts may be produced from these mill shapes by subsequent deformation processing or machining operations.

This process combination is used by suppliers of tool steel and high-speed steel products to produce wrought P/M mill shapes. Although any conventional ingot metallurgy composition can be produced, the most commonly produced P/M grades are M2, M4, M42, T15, and special alloys such as CPM-10V, CPM Rex 20, and CPM Rex M42. Chemical compositions of common P/M tool steels are given in Table 2.

As the alloy content increases, wrought P/M products become more economical to produce because of the alloy price premium and the benefits of P/M processing. Highly alloyed P/M tool steel grades best capitalize on the economic/performance benefits of wrought P/M products. As alloy content increases, the cost of powder increases at a lower rate than cast and wrought bar stock. Generally, at combined levels of tungsten, molybdenum, and cobalt above 20%, powder metallurgy product cost is less than ingot metallurgy product cost.

Hot isostatic pressing plus hot rolling to mill shape produces a premium grade of tool steel. Absence of segregation and the achievement of a fine, uniform microstructure result in optimum performance of these products. Property improvements include enhanced grindability, improved wear resistance, improved combinations of hardness and toughness, improved response to heat treatment, and less distortion during heat treatment. For more information, see the performance comparisons in the article "P/M Tool Steels" in this Volume.

Hot extrusion can be used to consolidate high-speed and tool steel powders into fully dense billets with a functional shape. Water- and gas-atomized powders may be extruded; different process steps are used, depending on powder type, as shown in Fig. 4. Spherical powders are containerized, outgassed, and extruded. Water-atomized powders, with their irregular shapes, can be cold pressed into billets prior to hot extrusion. Sintering of water-atomized powder to reduce oxygen content is optional, as extrusion produces a uniform dispersion of fine oxides that do not degrade toughness (Ref 6). Sintering can be used to significantly densify the cold pressed billet prior to extrusion (Ref 7).

Typical extrusion temperatures range from 1100 to 1200 °C (2010 to 2200 °F). Extrusion ratio should be a minimum of 9 to 1 to produce fully consolidated product. Reduction ratios of less than 9 to 1 result in cracking of the extrusion as it exits the die, due to low density and density gradients. Typical ratios for commercial operations are approximately 13 to 1, with

Fig. 3 Flowchart of extrusion followed by isothermal forging for production of turbine disks

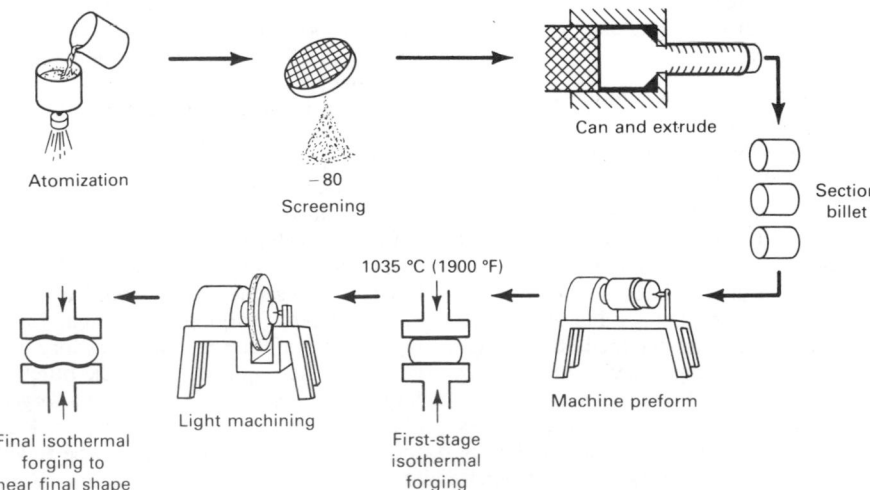

Atomization — ~80 Screening — Can and extrude — Section billet — Machine preform — First-stage isothermal forging — 1035 °C (1900 °F) — Light machining — Final isothermal forging to near final shape

Table 2 Chemical compositions of common P/M tool steels

Alloy	Carbon	Manganese	Silicon	Chromium	Molybdenum	Tungsten	Vanadium	Cobalt	Iron
AISI O1	0.90	1.2	0.3	0.5	...	0.5	0.2	...	rem
AISI S7	0.50	0.7	0.3	3.3	1.4	...	...	...	rem
AISI A2	1.00	0.6	0.3	5.3	1.1	0.2	0.2	...	rem
AISI D2	1.50	0.3	0.3	12.0	0.8	...	0.8	...	rem
AISI D6	2.05	0.8	0.3	13.0	...	1.3	...	...	rem
AISI M2	0.85	0.3	0.3	4.0	5.0	6.4	1.9	...	rem
AISI M4	1.30	0.3	0.3	4.0	4.5	5.5	4.0	...	rem
AISI M42	1.07	0.3	0.3	3.8	9.5	1.5	1.2	8.0	rem
AISI T15	1.50	0.3	0.3	4.0	...	12.0	5.0	5.0	rem
ASP23	1.27	0.3	0.3	4.2	5.0	6.4	3.1	...	rem
ASP30	1.27	0.3	0.3	4.2	5.0	6.4	3.1	8.5	rem
ASP60	2.30	0.3	0.4	4.0	7.0	6.5	6.5	10.5	rem
CPM Rex 20	1.30	...	...	3.75	10.5	6.25	2.00	...	rem
CPM Rex M42....	1.10	...	...	3.75	9.5	1.5	1.10	8.0	rem
CPM Rex 76	1.50	...	...	3.75	5.25	10.0	3.0	9.0	rem
CPM-10V	2.40	...	...	5.3	1.3	0.3	9.8	...	rem

Fig. 4 Extrusion processes for consolidation of high-speed steels and tool steel powders

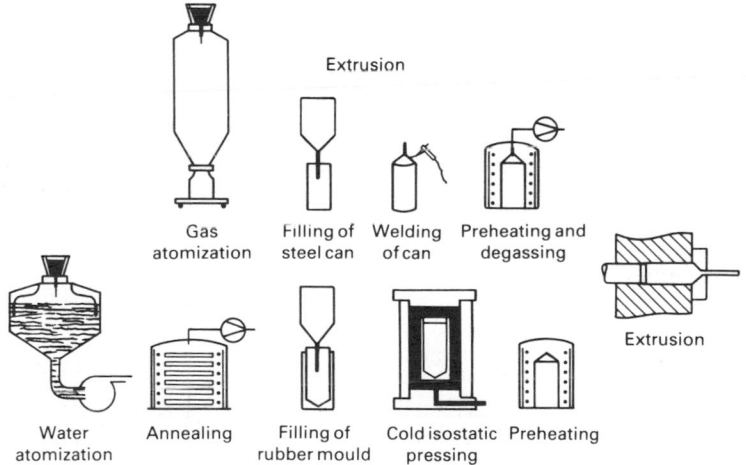

Gas atomization

Water atomization

Annealing

Filling of steel can

Welding of can

Preheating and degassing

Filling of rubber mould

Cold isostatic pressing

Preheating

Extrusion

Extrusion

extrusion pressures being on the order of 620 to 690 MPa (45 to 50 tsi) (Ref 7).

After extrusion, the billet must be pickled to clean the surface and to remove the remains of the container. Residual extrusion lubricant also must be removed. The billet is treated as a wrought material and subjected to hot rolling to produce a mill shape.

Sintering to High Density Followed by Hot Working. Powder can be containerized, sintered to high density, and then hot worked to final shape. The consolidation by atmospheric pressure (CAP) process is an example of such processing (see the article "Consolidation by Atmospheric Pressure" in this Volume). As applied to specialty steels, inert-gas-atomized powder is charged into a glass container and sintered at a temperature near the solidus. Rapid densification occurs. Final density depends on sintering tem-

perature and time. Typically, at least 95% of theoretical density is achieved.

Billets are then processed by hot rolling into a mill shape. As detailed in the article "Consolidation by Atmospheric Pressure," press forging, isothermal hot die forging, or containerless hot isostatic pressing can be used to produce near-net shapes. Although high densities can be achieved by sintering alone, for materials that are subjected to high stress in service, the incorporation of the CAP process (and other sintering processes) with a subsequent deformation process is required to ensure full density and high performance (Ref 8).

Typical thermal consolidation methods based on sintering of cold pressed compacts include the Fuldens process, the HTM process (Ref 9), and the Powdrex process (Ref 10). Each is used commercially for tool steels and stainless steels. Water-

atomized powder is cold pressed by die compaction or by cold isostatic pressing into a shape that can be handled. This is then vacuum sintered to high density.

These processes can be used to produce net or near-net shapes without secondary deformation. Alternatively, a final deformation step, such as forging or containerless hot isostatic pressing, may be added to ensure complete consolidation and optimum part performance. Containerless hot isostatic pressing is possible because sintered densities are well above the level required to ensure the absence of interconnected porosity (92% of theoretical).

Of future potential for P/M tool steels is the use of injection molding plus sintering to produce preform shapes for subsequent hot working into mill or part shapes. See the article "Injection Molding" in this Volume for more information on this process.

Processing of Wrought P/M Aluminum Alloys

Wrought P/M aluminum alloys generally fall into the 2xxx and 7xxx series of alloys. These are high-strength alloys that have the additional benefit of maintaining strength at elevated temperature. Standard compositions are processed to mill shapes from ingots by conventional techniques. Property improvements are possible by using powder starting stock rather than ingot stock (Ref 11). Particularly, fatigue resistance can be improved by starting with powder stock. When testing small, multicomponent joint fatigue specimens with load transfer and fretting, the P/M material may be inferior to cast or wrought material. In addition, the crack growth rate characteristics of P/M aluminum alloys are generally inferior to those of ingot metallurgy materials. As a result, the subject of wrought powder metallurgy has become increasingly important for production of mill shapes from high-strength alloys.

Alloy development has extended wrought P/M technology further than substitution of powders for ingot metallurgy alloys of similar composition. New alloys (see Table 3) have been developed that rely on powder processing and, in particular, on rapid solidification technology for strengthening. Increased solubility of alloy components, such as iron or zinc, controlled precipitation and dispersion strengthening, and refined microstructures are metallurgical benefits gained by producing wrought aluminum alloys from powder stock. Advantages of these prod-

Table 3 Compositions of wrought P/M aluminum alloys

| Alloy | Composition, % | | | | | | | | | |
	Copper	Magnesium	Chromium	Zinc	Cobalt	Iron	Cerium	Carbon	Oxygen	Aluminum
7090(a)	1.0	2.5	...	8.0	1.5	...	...	...	0.35	rem
7091(a)	1.5	2.5	...	6.5	0.4	...	...	...	0.35	rem
IN-9021(a)	4.0	1.5	...	...	...	...	...	1.2	0.75	rem
IN-9051(a)	...	4.0	...	...	...	...	...	0.7	1.4	rem
IN-9052(a)	...	4.0	...	...	...	...	...	1.0	0.75	rem
Al-C(b)	...	...	...	...	...	...	...	3.0	...	rem
Al-Cu-C(b)	1.0	...	...	...	...	...	...	1.5	...	rem
Al-Mg-C(b)	...	2.0	...	...	...	...	...	1.5	...	rem
Al-Fe-Co(b)	...	...	...	...	2-7	3-8	...	...	0.35	rem
Al-Fe-Ce(b)	...	...	...	...	...	8.0	3.5	...	0.35	rem
Al-Fe-Cr(b)	...	...	1.5	...	...	8.5	...	...	0.35	rem
Al-Li(b,c)	2-3.5	1-2	...	...	...	...	...	...	...	rem

(a) Commercially available. (b) Under development. (c) 2.5 to 3.2% Li, 0.6% Mn

ucts over conventional ingot metallurgy grades include:

- High strength
- Improved resistance to stress corrosion cracking and other forms of corrosion
- Excellent fatigue strength, particularly notched fatigue strength, for simple specimens

The following sections describe the manufacturing practices used to produce mill shapes from these alloys. Although powder may be consolidated directly, such practices are not discussed in this article. Because many of these alloys are still under development, standard processes have not been established. These alloys must be processed within small temperature and deformation limits to provide the proper level of working to the alloy and to avoid growth of dispersoids and precipitates, the formation of embrittling films along particle boundaries, and weakened denuded zones along grain boundaries. Additional information on the properties and applications of wrought P/M aluminum alloys can be found in the article "P/M Lightweight Metals" in this Volume.

Degassing of High-Performance Aluminum Powder. For these high-performance alloys, degassing of the powder prior to complete consolidation is critical to the final performance of the alloy. Water vapor and adsorbed gases on particle surfaces must be removed from the powder mass to produce a sound billet. Generally, powder is degassed by evacuating the container holding loose powder to a pressure below 10^{-5} torr while the powder mass is heated. Ideally, powder should be degassed at a temperature above either the forming or service temperature to avoid blistering caused by gas evolution. Although, high temperatures tend to degrade strength, a temperature of 450 °C (850 °F) has been shown to be adequate to promote water vapor removal from the powder. Degassing of the powder is of

concern because of the effect of gas and water vapor on properties and economics and because of safety hazards due to possible hydrogen gas evolution at elevated temperatures.

Various degassing methods have been investigated. Soviet workers experimented with degassing aluminum powders at high temperatures in a dry argon atmosphere (Ref 11). On exposure to air, rehydration did not occur for 5 days. This may allow lots of loose powder to be degassed and then compacted after air exposure. Such handling is simpler than containerization practices. Properties of material degassed in this manner are not as high as properties of material subjected to vacuum degassing.

Other degassing work has attempted to determine minimum temperatures and times for degassing. Roberts (Ref 12) has determined that degassing at a temperature of 310 °C (590 °F) until a vacuum greater than 4×10^{-4} torr is achieved produces blister-free material.

A unique containerization practice, known as "Sinter-Seal," has been patented, which is used in conjunction with degassing practice. After sufficient degassing, the surface temperature of a cold compacted billet is rapidly raised by induction heating to a level that promotes sintering. For aluminum alloys, sintering requires the presence of a transient liquid phase that is composed of an alloy phase with a low melting temperature. The bulk of the powder in the billet is not overheated, and overall strength properties are not degraded. Densification of the surface layer effectively seals the billet to prohibit rehydration of the billet interior for a period of a few days.

The billet may be handled in air during subsequent processing without the need for special containerization. This process may be performed only for alloys that support the formation of a liquid phase during surface heating. Conversely, a special sur-

face alloy treatment may be possible to allow use of this type of approach to containerization. An effective skin may be formed in this manner. More detailed information on degassing techniques can be found in the article "Cleaning of Metal Powders" in this Volume.

Consolidation of Powders and Hot Working. Aluminum alloy powders may be consolidated by either hot or cold compaction methods. Because of nonspherical particle shape and the malleability of the material, cold compaction by cold isostatic pressing, pressing in rigid die cavities, roll compaction, and extrusion are suitable powder consolidation techniques. Hot consolidation methods include hot pressing, hot isostatic pressing, hot extrusion, and hot rolling.

Following consolidation, the material may be porous; this is particularly true for cold compacted material. As a result, degassing is desirable after this compaction and prior to subsequent hot working. Conversely, if the material is fully consolidated, the powder should be degassed prior to consolidation. Powder metallurgy processing is flexible due to this ability to either hot or cold compact the powder and to degas either before or after initial consolidation, depending on porosity. A flowchart of several processing techniques is shown in Fig. 5.

Commercial practices used for these alloys include:

- Cold isostatic pressing, degassing, hot pressing to full density, and extruding or rolling to mill shape
- Cold isostatic pressing, degassing, and hot extruding to mill shape
- Degassing containerized powder, cold pressing or cold isostatic pressing, and extruding to mill shape
- Degassing containerized powder, hot pressing to billet shape, extruding or rolling to mill shape
- Direct rolling of powder, followed by degassing and continued hot rolling

Cold die compaction and cold isostatic pressing usually produce densities of 80 to 90% of theoretical. Hot consolidation results in 100% of theoretical density with appropriate selection of temperature and pressure. For dispersion-strengthened alloys, a pressure of 620 MPa (90 ksi) and temperatures between 510 and 540 °C (950 and 1000 °F) are suitable for hot pressing.

Extrusion is performed at ratios of at least 10 to 1. Extrusion ratios of 20 to 1 and as high as 50 to 1 are commonplace. For dispersion-strengthened alloys, temperatures of 340 to 480 °C (650 to 900 °F) and ratios up to 50 to 1 are used for producing mill

Fig. 5 Processing steps for producing wrought P/M aluminum alloys

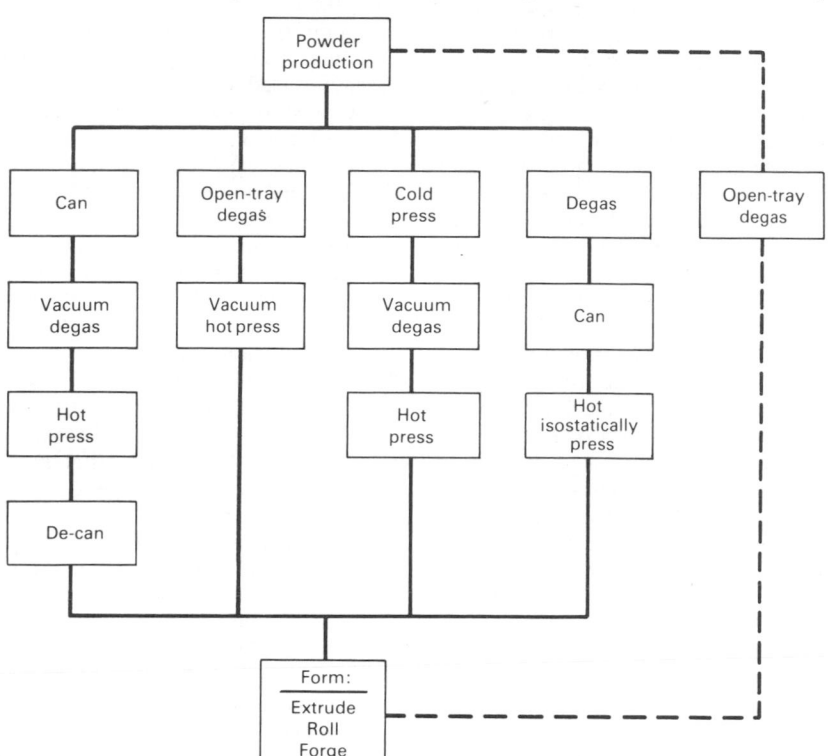

shapes. At lower temperatures, press stalling occurs; at higher temperatures, blistering due to gas evolution and strength loss due to dispersoid agglomeration occur. Strain rates during extrusion are typically 10 to 30 cm/cm/s (10 to 30 in./in./s). Similar conditions are used for other alloy systems, with emphasis on the low-temperature working region below 450 °C (840 °F). Impact extrusion equipment can be used, but the effects of strain rate on these alloys are not fully understood at this time.

Some laboratory success has been reported for consolidation of these alloys using high-energy rate forming (Ref 11). This process involves high strain rates (typically higher than 100 cm/cm/s, or 100 in./in./s). The equipment used for high-energy rate forming is specially

constructed to generate these high deformation rates. Gas guns and controlled explosions can be used to generate high strain rates. Apparently, high-energy rate forming causes localized melting at particle surfaces to generate metallurgical bonds between particles without overheating the bulk of the powder mass. Once consolidated to full density, conventional presses, extrusion equipment, or rolling equipment is employed for mill shape production.

Processing of Wrought P/M Dispersion-Strengthened Alloys

Dispersion-strengthened P/M alloys of commercial importance include iron- and

nickel-based superalloys, aluminum-based alloys, and copper-based alloys. The production and applications of these materials are covered in the article "Dispersion-Strengthened Materials" in this Volume. This section deals with production of wrought forms of these materials.

Wrought oxide-dispersion-strengthened P/M superalloys are produced from mechanically alloyed powders. Mechanical alloying is a dry, high-energy ball milling process in which the attritor charge of balls and powder is held in a stationary, vertical, water-cooled tank and agitated by impellers attached to a rotating central shaft (Ref 13). The key features of this process are the high-energy milling (ball mill operated at 130 to 140 rpm) and the omission of any surface active agent other than the air sealed into the attritor tank with the charge.

This practice promotes particle welding, in contrast to conventional ball milling in which welding is inhibited by the use of liquids and surfactants. Recurrent cold welding of constituents to the ball surfaces occurs, while the composite aggregates flake until all of the constituents are finely divided and uniformly distributed through the interior of each powder particle. Typical nickel-based oxide-dispersion-strengthened compositions are given in Table 4.

Mechanical alloying is also used to produce dispersion-strengthened iron-based alloys. Basically, the combined use of dispersion strengthening and P/M consolidation technologies permits a wrought product with a uniform dispersion of fine oxide particles to be produced on a commercial basis.

A flowchart for manufacturing of mill shapes of these superalloys is shown in Fig. 6. After mechanical alloying, powder is containcrized in low-carbon steel tubes and extruded to full density. Degassing is not critical for these alloys, because the oxide level of the powder is already very high. In fact, removal of residual air from the container is not necessary in practice for mechanically alloyed powder. However, good can-filling and powder-handling techniques should be used to prevent contamination of the powder mass with foreign particles. Extrusion temperature and reduction are critical because final microstructure of the product is affected, and the ability to control this microstructure is important. For nickel-based alloys, reduction ratios of 13 to 1 and temperatures of 1065 to 1120 °C (1950 to 2050 °F) are typical.

The extruded stock, now fully consolidated, is then hot rolled into mill shapes. As with extrusion, the thermomechanical

Table 4   Compositions of experimental dispersion-strengthened nickel-based superalloys

Alloy	Carbon	Aluminum	Titanium	Chromium	Molybdenum	Niobium	Zirconium	Boron	ThO$_2$	Y$_2$O$_3$	Al$_2$O$_3$
A	0.061	0.92	2.46	20.4	...	...	0.029	0.005	...	1.22	0.37
B	0.049	0.96	2.77	18.7	...	...	0.09	0.003	...	1.33	0.99
C	0.056	0.90	2.33	20.6	...	...	0.065	0.005	...	1.22	0.83
D	0.055	0.74	2.10	19.0	...	...	0.025	0.002	2.71	...	1.30
E	0.069	4.19	0.82	10.4	3.0	1.6	0.03	0.007	3.00	...	1.38

Source: Ref 13

Fig. 6 Flowchart of production of oxide-dispersion-strengthened superalloy products by mechanical alloying

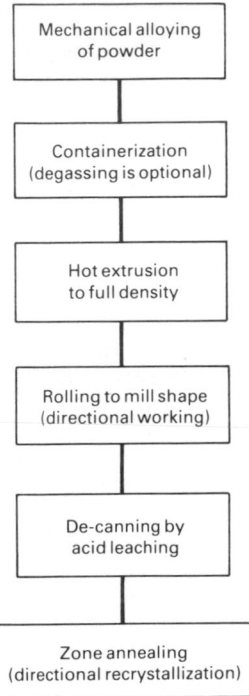

treatment followed during rolling is critical to final microstructure and properties of the product. For iron- and nickel-based superalloys, rolling temperatures are normally between 950 and 1065 °C (1750 and 1950 °F), with rolling resulting in highly directional working of the product. For these superalloys, rolling is performed with the low-carbon steel can remaining on the consolidated workpiece, due to sensitivity of the products to thermal shock. Directional working is required to promote development of the proper microstructure in the final part. For elevated-temperature service, which is the main application of these materials, an elongated, coarse grain structure is required to maximize creep resistance and related properties.

Can removal after rolling is accomplished easily by acid leaching of the low-carbon steel can from the oxide-dispersion-strengthened product. Machining is also possible, but the thermal shock sensitivity of these products in the as-rolled microstructural condition may cause cracking.

The sequence of processing steps after can removal depends on part shape and application. In most cases, the oxide-disperion-strengthened bar can be heat treated prior to final shaping of the part.

Heat treatment consists of zone annealing the rolled product to develop a directionally recrystallized microstructure with the large, elongated grains necessary to achieve optimum elevated-temperature properties.

Zone annealing consists of establishing a hot zone by an induction coil, or other appropriate localized heating method, and passing the workpiece through the hot zone at a controlled rate. The direction of workpiece travel coincides with the rolling direction. With proper levels of stored energy in the workpiece (resulting from the thermomechanical processing during extrusion and rolling), directional recrystallization and grain growth occur.

Parts are produced by machining of bar stock to final shape. Electrochemical machining and electric discharge machining are the preferred methods for final machining due to cracking susceptibility during grinding or abrasive cutting. Deformation processing to near-net shapes is possible, provided that proper thermal and strain rate considerations are followed.

For forging, susceptibility to cracking due to die chill must be countered either by isolating the workpiece from low temperatures or by heating the tooling to the same temperature at which the workpiece is deformed. The former is possible by simply forging with the can remaining on the workpiece, or by sandwiching the workpiece in heated shims. The latter involves hot die or isothermal deformation techniques. Isothermal rolling procedures and equipment are being developed to roll sheet oxide-dispersion-strengthened products. The key to the success of any deformation process for these materials is the effects of temperature and of mode, level, and rate of deformation on heat treat response.

Processing of Dispersion-Strengthened Aluminum Alloys.

Dispersion-strengthened aluminum alloys may contain oxide and/or carbide particles. Methods of producing these alloys range from mechanical alloying to reactive sintering. Both methods effectively strengthen the alloy, enabling higher temperature applications for aluminum-based alloys. Processing these alloys into wrought P/M products is discussed in the section of this article on wrought P/M aluminum alloys.

Processing of Dispersion-Strengthened Copper Alloys.

Dispersion-strengthened copper-based alloys are produced by internally oxidizing containerized powder (Ref 14). Additions of an oxidizable element such as aluminum are made to the copper alloy prior to atomization. Alloy powder is blended with an oxidant such as cupric oxide and is charged into a container. Typically, for copper alloys, the container is a copper tube with a wall thickness of approximately 1.6 mm ($^{1}/_{16}$ in.). After filling, the container is partially sealed, with a leak site present to allow pressure release during subsequent heating for internal oxidation.

The container is heated to above 760 °C (1400 °F) and normally about 955 °C (1750 °F) for internal oxidation. During this treatment, the oxidant is reduced to its base elements. The liberated oxygen combines with the stable oxide-forming element to form a dispersion of oxide particles. After internal oxidation, the billet is extruded at 870 to 925 °C (1600 to 1700 °F), with extrusion ratios ranging from 20 to 1 to 50 to 1. Extrusion may follow the oxidation step immediately, or the billet may be cooled, stored for a desired period of time, and then hot extruded.

After extrusion, the material may be forged into part shapes, rolled into mill or flat rolled shapes, or cold worked into wire shapes by swaging, or other wire-making methods. Additional information on dispersion-strengthened copper-based alloys can be found in the articles "Dispersion-Strengthened Materials," "Electrical and Magnetic Applications," and "P/M Copper Alloys" in this Volume.

REFERENCES

1. Symonds, C.H. and Thompson, F.A., "Nickel Superalloy Production and Fabrication of Turbine Disks," Advanced Fabrication Techniques in Powder Metallurgy and Their Economic Implications, AGARD-CP-200, 1976, p 3.1-3.14
2. Bartos, J.L., P/M in Defense Technology, Vol 5, Metal Powder Industries Federation, Princeton, NJ, 1980, p 81-113
3. Allen, M.M., "Iso-Forging of Powder Metallurgy Superalloys for Advanced Turbine Engine Applications," Advanced Fabrication Techniques in Powder Metallurgy and Their Economic Implications, AGARD-CP-200, 1976, p 5.1-5.15
4. Wick, C., Better Tools From P/M High-Speed Steels, Manuf. Eng., Sept 1980, p 52-54
5. Kasak, A., Steven, G., and Neumeyer, T.A., High-Speed Tool Steels by Particle Metallurgy, Society of Automotive Engineers Paper 720182, 1972
6. Beiss, P., P/M Methods for the Production of High Speed Steels, Metal Powder Rep., Vol 38 (No. 4), 1983, p 185-194

7. Kawai, N. and Takigawa, H., Methods for Producing P/M High-Speed Steels, *Metal Powder Rep.*, Vol 37 (No. 5), 1982, p 237-240

8. Dax, F.R., P/M High-Speed Steels by the CAP Process, *Metal Powder Rep.*, Vol 38 (No. 4), 1983, p 200-202

9. Zickefoose, E.J., Full Density P/M Parts by the HTM Process, Society of Mechanical Engineers Paper MF82-342, 1982

10. Anonymous, Cost Efficient Cutting Tools and Wear Parts from P/M High-Speed Steels, *Metal Powder Rep.*, Vol 38 (No. 4), 1983, p 217-218

11. Pickens, J.R., Aluminum Powder Metallurgy Technology for High-Strength Applications, *J. Mat. Sci.*, Vol 16, 1981, p 1437-1457

12. Roberts, S.G., U.S. Patent 4,104,061

13. Benjamin, J.S. Dispersion Strengthened Superalloys by Mechanical Alloying, *Met. Trans.*, Vol 1, Oct 1970

14. Nadkarni, A.V. and Klar, E., "Dispersion Strengthing of Metals By In-Can Processing," U.S. Patent 3,884,676, 1975

Spray Deposition of Metal Powders

By B. Lynn Ferguson
Partner
Deformation Control Technology

MELT-SPRAY DEPOSITION of powders encompasses a wide variety of materials and product forms. One of the largest applications for melt-spray deposition is welding, including hardfacing and plasma spraying techniques. These techniques are discussed in the article "Metal Powders Used for Hardfacing" in this Volume. This article describes processes for producing preforms by a buildup of sprayed metal powder. These preforms and billets subsequently can be consolidated into various mill shapes. Of these processes, the Osprey process, developed in Wales by Osprey Metals Ltd. (Ref 1-3), and the controlled spray deposition process (Ref 4) are in commercial use. Several other methods are being developed; plasma spray buildup has high commercial potential. Laser techniques such as laser glazing also have commercial potential, especially when combined with rapid solidification technologies (Ref 5, 6).

Osprey Process

Facilities for production of preforms made by the Osprey process consist of induction melting equipment and a preform production unit, including:

- An atomizer
- A chamber for spraying under inert conditions
- A cyclone collector for recycling overspray
- A hydraulic unit for mold positioning and manipulation during atomization
- Molds for preform formation
- Instrumentation

In the Osprey process, which is shown schematically in Fig. 1, an alloy is melted and subjected to gas atomization under in-

Fig. 1 Osprey process for production of parts from spray-formed preforms
Source: Ref 2

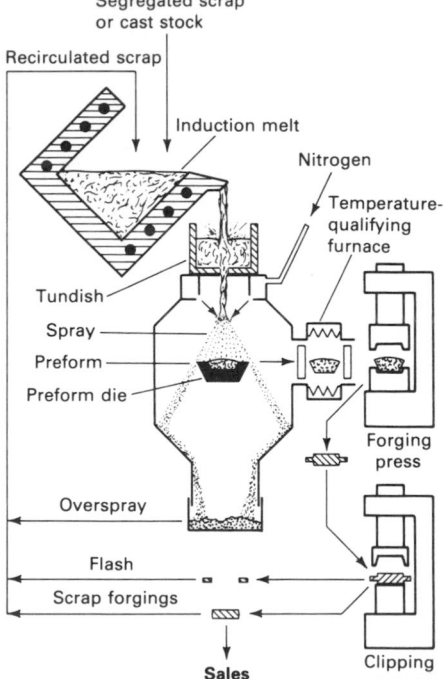

ert conditions (usually nitrogen or argon is used). The atomized droplets are collected in a mold or group of molds, in which final solidification occurs. Molds are normally copper cooled by water. High-temperature ceramics offer other material options for molds. During solidification, welding of particles causes buildup of alloy in the mold. A preform having a density above 96%, and normally above 99%, of theoretical is generated by this buildup

of alloy. The preform then can be consolidated to full density and formed into a mill or near-net shape part.

Material Capabilities. Alloys that have been processed by the Osprey process include stainless steels, high-speed steels, and nickel-based superalloys, although many materials appear to be compatible with the process (Ref 7). Alloy development has centered on high-alloy ferrous metals, Stellite alloys, superalloys, and composite materials. Because an inert atmosphere is maintained during spraying, oxygen levels similar to conventional ingot metallurgy products are attained, typically in the range of 20 to 40 ppm for superalloys. The high preform density ensures that no interconnected porosity is present in the preform, preventing internal oxidation during transfer of the material to subsequent consolidation and forming operations.

Reference 8 gives details on some physical and mechanical properties of two commonly used aerospace alloys processed using the Osprey method. Within the limited property range explored, Osprey processed materials have been shown to have similar properties to wrought materials. The Osprey processed results are also isotropic. A comparison of transverse properties shows the Osprey forgings to have superior elongation and reduction in area values.

Size and Shape Capabilities. The Osprey process is used to produce a wide variety of preform shapes and sizes. Typical preform shapes are tubes, rings, cylinders, disks, or simple billets. Size is dictated by economics, with the melt facility, atomizer, and inert chamber sized for a particular product line. The largest pre-

Fig. 2 Osprey-atomized nickel-based superalloy MAR M002
(a) Magnification: 400×. (b) Magnification: 4000×

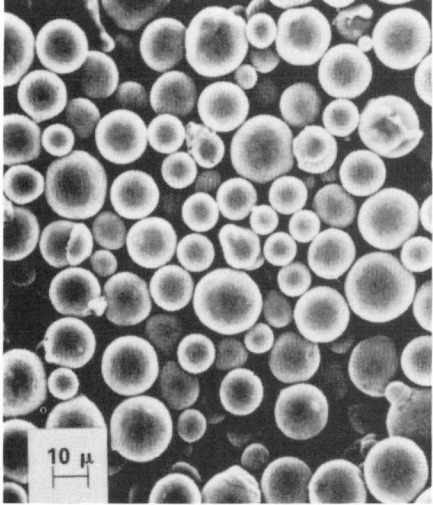

(a)

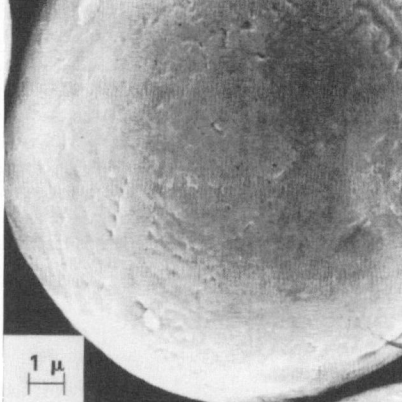

(b)

Fig. 3 Controlled spray deposited M15-type high-speed steel
Magnification: 500×. Source: Ref 4

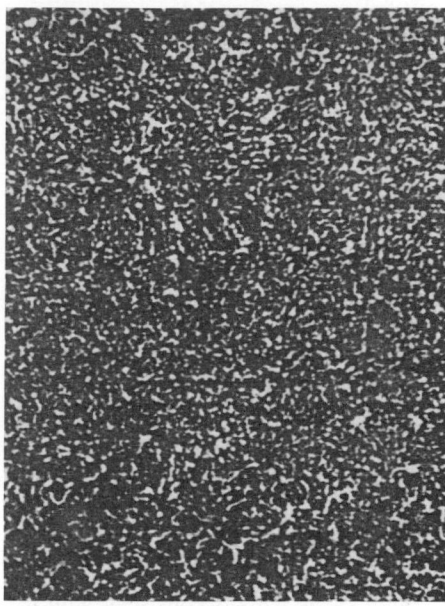

form size produced weighed 540 kg (1200 lb) (Ref 7). Typical deposition rates range from 10 to 90 kg/min (20 to 200 lb/min).

Deformation Processing of Preforms. Final consolidation and shaping processes may include forging, extrusion, or rolling. Because of high microstructural uniformity, fine grain size, absence of prior particle boundaries, low oxygen levels, and lack of macrosegregation, workability of these preforms is improved over that of conventionally cast material. This allows extensive deformation processing of difficult-to-work alloys, such as high-speed steels and Stellite alloys.

Process Variations. Although Osprey process atomizing devices were originally developed for the production of spray-deposited preforms, they also can be used to produce high-purity powders in a wide controlled range of particle size distributions (Ref 9). Powders produced using Osprey atomizers include silver and gold brazing alloy powders, tin-lead soldering powders, aluminum-magnesium powders for explosives, copper-tin-lead bearing alloys, and nickel- and cobalt-based hardfacing alloys. In addition, high-melting-point alloys such as nickel superalloys (Fig. 2), Alnico permanent magnet alloys, and stainless steel powders have been produced using Osprey atomizing equipment.

Controlled Spray Deposition Process

The controlled spray deposition process is similar to the Osprey process in principle, but uses different machinery. Controlled spray deposition uses centrifugal atomization, while the Osprey process involves gas atomization. This process is used for production of mill shapes from high-alloy steel, which utilizes the enhanced workability of P/M workpieces. Highly alloyed metals may suffer from macrosegregation, which reduces the material workability. Elimination of segregation on a macroscale, coupled with a uniform distribution of fine carbides (2- to 3-μm range for M-2 high-speed steel), results in improved workability for P/M workpieces produced by spray deposition. These billets are processed subsequently into mill shapes and sheet products.

Metal Spraying. By atomizing liquid metal into droplets 0.5 to 1.5 mm (0.02 to 0.06 in.) in diameter, solidification rates three or more orders of magnitude higher than those of conventional ingot solidification are achieved. Impacting liquid droplets of metal onto a cooled substrate increases the solidification rate, resulting in solidification rates of 10 000 to 1 000 000 °C/s (18 000 to 1 800 000 °F/s). Controlled spray deposition relies on this type of splat solidification to build up a solidified deposit that becomes a workpiece for subsequent deformation processing. As the thickness and temperature of the built-up material increases, the solidification rate decreases, but it remains much higher than that of conventional ingot solidification. Heating prior to hot working can remove any microstructural variations that may exist throughout the thickness of the built-up deposit.

Along with the metallurgical benefits of controlled spray deposition, economic advantages of direct spraying of powder into preform shapes are attained by eliminating sieving, blending, and other powder preparation steps. Also, primary compaction of powder into a green shape is eliminated. Controlled spray deposition proponents claim that these reductions in equipment needs and processing steps allow more efficient utilization of energy, compared to conventional pressing and sintering P/M technology.

Microstructure. The high cooling rates achieved during controlled spray deposition processing result in segregation-free products. Figure 3 is a micrograph of a controlled spray deposited material. In as-solidified material, no prior particle boundaries or "splat" boundaries exist. Densities of the as-solidified material are around 96 to 98% of theoretical density; residual porosity is isolated and is not interconnected. Grain structures of controlled spray deposition material are significantly different from those of ingot metallurgy material. At highest cooling rates, an equiaxed microcrystalline grain structure is achieved. At slightly lower cooling rates, dendritic solidification occurs, with secondary dendrite arm spacings of less than 0.5 μm.

Carbide morphology and size are dependent on cooling rate as well, with highest rates causing formation of micro-

Table 1 Properties of end mills produced by controlled spray deposition and conventional processing

Tool type(a)	Hardness, HV30	Average wear band		Average diameter loss	
		μm	μin.	μm	μin.
Controlled spray deposition M-15	880	99	3900	150	5 900
Conventional M-15	900	129	5075	521	20 500
Conventional M-42	900	117	4600	373	14 700

(a) Averaged results for five tools

cellular carbides. Even at lower rates of solidification, controlled spray deposition workpieces have higher MC-type carbide contents than conventional ingot metallurgy products of the same composition. Oxygen levels of controlled spray deposition material are typically in the range of 200 to 250 ppm, and nitrogen levels are under 0.04%, typical of air-melted ingot products.

Hot Working. Reductions of 30 to 40% in cross-sectional area by hot working are sufficient to eliminate residual porosity in controlled spray deposition material (Ref 4). High reductions result in fracture resistance improvements to levels above those of ingot stock. This is consistent with results of P/M forging studies (Ref 10, 11), which showed increases in toughness and fatigue properties as the level of deformation increased.

Because mill shapes and difficult-to-form products, such as sheet products of high-speed steel, are typical controlled spray deposition products, hot working normally means hot rolling. The improved workability of the controlled spray deposition workpiece in comparison to a cast ingot permits extensive deformation, if desired. However, a benefit of the process is the attainment of full density with low deformation levels, producing fully consolidated products with uniform, fine microstructures without extensive hot work-

ing. Also, individual preforms can be formed or sectioned from larger workpieces to allow near-net shape parts to be forged from controlled spray deposition stock.

Testing of Controlled Spray Deposition Materials

Product testing has been concentrated on high-speed steel applications using controlled spray deposition bar, sheet, and forgings (Ref 4). End user results for machining type 316 stainless steel with controlled spray deposition M15 and conventional M15 and M42 16-mm (0.63-in.) diam end mills are given in Table 1. Operating conditions for producing these materials were:

```
Speed ......................... 612 rpm
Feed ....................... 170 mm/min
                              (6.7 in./min)
Section of cut ... 7.93 mm axial by 5.3 mm radial
                  (0.31 in. axial by 0.2 in. radial)
Material ............. Type 316 stainless steel
Total cutting time .......... 39 min (18 passes)
```

REFERENCES

1. U.K. Patent No. 1 379 261, 1972
2. U.S. Patent No. 3 826 301, 1972
3. Brooks, R.G., et al., "Osprey Technology for Spray-Deposited Preforms and Powders in Superalloys," Powder Metallurgy Superalloys Conf., Zurich, Nov 1980
4. Rickinson, B.A., Kirk, F.A., and Davies, D.R.G., CSD: A Novel Process for Particle Metallurgy Products, Powder Metall., Vol 24 (No. 1), 1981, p 1-6
5. Breinan, E.M., et al., "New Developments in Laser Glazing Using Continuous Prealloy Powder Feed," Reston Conference, 1980
6. Breinan, E.M. and Kear, B.H., Rapid Solidification Laser Processing of Materials for Control of Microstructure and Properties, in Rapid Solidification Processing—Principles and Technologies, R. Mehrabian, Ed., Claitor's, Baton Rouge, 1978, p 87
7. Williams, B., Preforming Process for Superalloys and High Alloy Steels, Met. Powder Rep., Vol 38 (No. 1), 1983, p 15-17
8. Brooks, R.G., et al., "The Structures and Properties of Some Spray Forged Aerospace Alloys, International Powder Metallurgy Conference, Chicago, 1976
9. Dunstan, G.R., et al., The Osprey Gas-Atomizing Powder Production Process, Progress in Powder Metallurgy, Metal Powder Industries Federation, Princeton, NJ, 1981, p 23-38
10. Ferguson, B.L., Suh, S.K., and Lawley, A., Impact Behavior of P/M Steel Forgings, Int. J. Powder Metall. Powder Technol., Vol 11 (No. 4), 1975, p 263-275
11. Ferguson, B.L., Kuhn, H.A., and Lawley, A., Fatigue of Iron Base P/M Forgings, in Modern Developments in Powder Metallurgy, Vol 9, H.H. Hausner and P.V. Taubenblat, Ed., Metal Powder Industries Federation, Princeton, NJ, 1977, p 51-76

Consolidation by Atmospheric Pressure

By B. Lynn Ferguson
Partner
Deformation Control Technology

CONSOLIDATION OF ATOMIZED POWDERS, without use of expensive hot isostatic pressing equipment, can now be accomplished with a newly emerged, commercially successful process. Developed as an alternative method for consolidating P/M high-speed tool steels, consolidation by atmospheric pressure, or the CAP process, expands the use of conventional P/M technology. Billets for subsequent hot working into mill shapes, billets for forging of blanks or near-net shapes, preforms for isothermal forging, or complex-shaped parts are candidates for consolidation by atmospheric pressure. This article describes process details, current commercial status, material advantages, and properties of materials that have been consolidated by the CAP process.

Process Sequence

High-quality argon or nitrogen gas atomized powder is the starting stock for the CAP process. Powder is spherical and has a low oxygen content. Powder is screened prior to consolidation by atmospheric pressure; −100 mesh is typical powder size. Finer mesh sizes may be used to capitalize on their higher cooling rates, which produce more efficient use of alloying elements and microstructure refinement. The elements of the CAP process are illustrated in Fig. 1. After screening, the powder is loaded into a vacuum blender-dryer unit, where it is blended with a solution containing a sintering activation agent such as boric acid in methanol, as shown in Fig. 1(a). The solution coats each particle during blending. Methanol, for example, is evaporated by application of vacuum and

heat during drying, leaving a thin film of boric acid on all particle surfaces.

The coated powder is then poured through fill tubes into inexpensive borosilicate glass molds, such as Kimax or Pyrex glass. Molds are vibrated during filling to produce a fill density of 65% of theoretical. Mold shape may vary, depending on subsequent treatment of the consolidated part. Examples of mold shapes (Fig. 1b) are simple cylinders, round-corner squares, tubes, or various complex shapes that provide near-net shape consolidated parts.

Powder is then degassed by attaching an evacuation pump to the mold through the fill tube, as shown in Fig. 1(c). During this operation residual gases are removed from the powder mass. The glass mold is then sealed, as shown in Fig. 1(c), by heating the fill tube to the softening point of the glass. The vacuum inside the mold causes the soft glass to collapse and seal, leaving the powder under vacuum.

The sealed container is then placed in a standard air atmosphere furnace used for heat treating, as shown in Fig. 1(d). Sintering occurs during this heating operation; the activation agent accelerates sintering. For example, boric acid chemically combines with metallic oxides present on powder particle surfaces to form borates. This fluxing action produces clean particle surfaces (oxide-free surfaces), which accelerate particle bonding and densification during sintering. The chemical presence of the activation agent is undetectable by standard chemical analysis techniques; consequently, material performance is not affected.

The degree of densification is affected by sintering temperature. Temperatures

near the solidus of an alloy promote more nearly complete densification. Consolidation by atmospheric pressure typically is performed at temperatures close to the solidus to capitalize on this. During sintering, the glass mold softens and contracts, as densification of the part occurs. The mold must be supported during sintering to maintain part shape; simple placement of the mold in a sand medium provides sufficient support, as shown in Fig. 1(d).

When sintering is completed, molds are removed from the furnace and air cooled, as shown in Fig. 1(e). The glass molds are self-stripping and spall from the consolidated parts at around 315 °C (600 °F), thus protecting the parts from excessive oxidation during cooling. Minor residual glass can be removed easily by sandblasting. Density of the consolidated part ranges from 95 to 99% of theoretical. Consolidation products may serve as billet stock for hot working to full density by hot forging, hot rolling (Fig. 1f), extrusion, or isothermal processing. Alternatively, parts may be hot isostatically pressed to full density without containerization.

Process Advantages

Lower cost is the prime advantage of the CAP process over P/M consolidation techniques such as hot isostatic pressing and extrusion. The simplicity of the process, which requires standard air atmosphere sintering furnaces, is largely responsible for the improved economics. Complicated equipment to generate extremely high pressures, such as hydraulics, high-pressure compressors, and protective gas atmospheres, as well as heavy

Fig. 1 CAP process

(a) Blending of alloy powder and activating agent. (b) Borosilicate glass molds and mold loading station. (c) Degassing and sealing of filled glass molds. (d) Sintering in air atmosphere furnace, with mold support. (e) Glass spalls from consolidated P/M billets during cooling in air. (f) Full density and final shape are achieved during subsequent hot working. Source: Ref 1

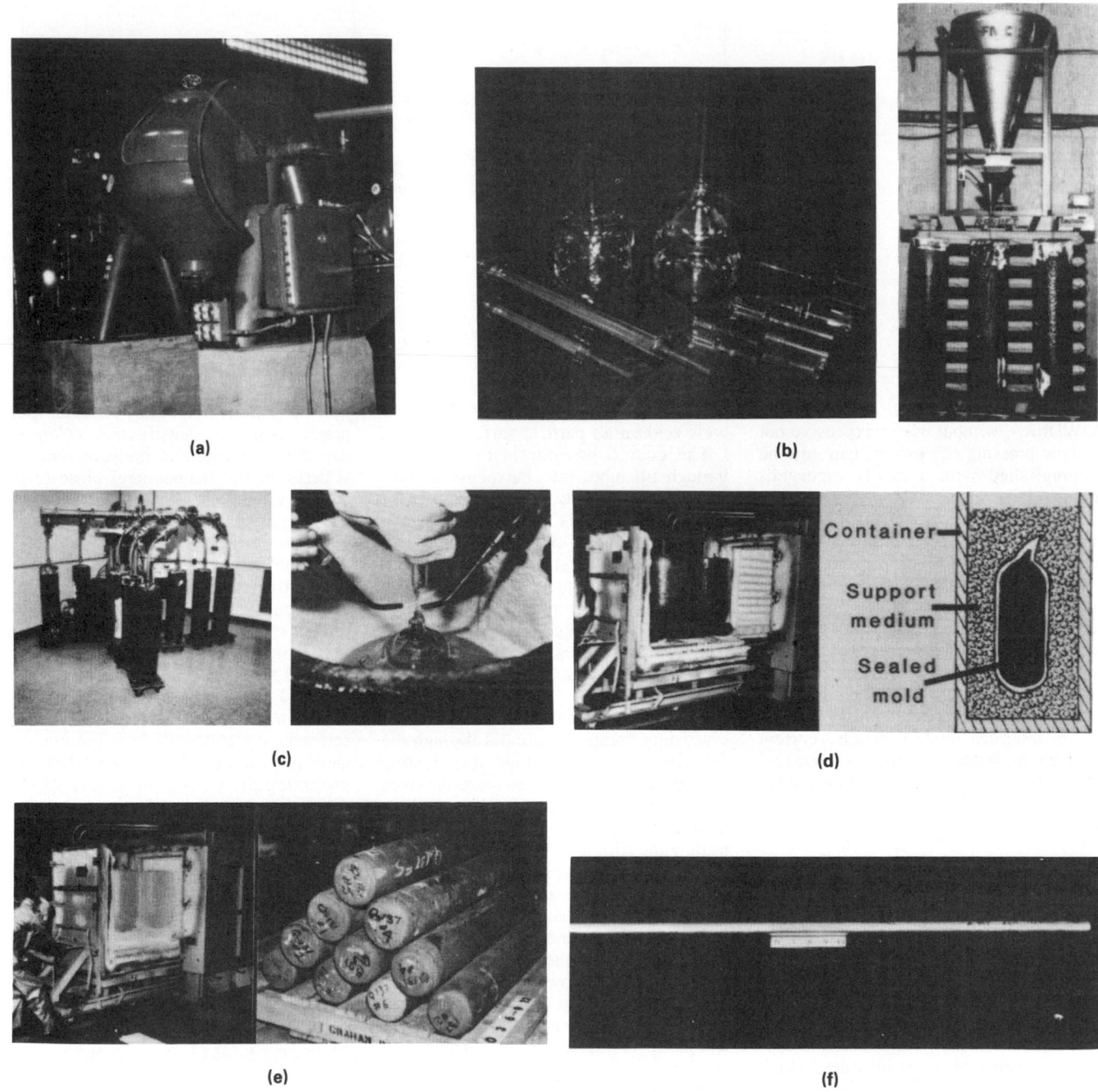

equipment that can withstand the high pressures and stresses generated, are required for the other consolidation processes. This in turn requires greater capital investment and higher operating and maintenance costs, whereas the costs of standard sintering furnaces for the CAP process are significantly lower.

Consolidation by atmospheric pressure also offers an advantage in the area of molds. Hot isostatic pressing and other processes depend on metal cans that must be fabricated, welded, and leak checked both statically and under pressure. Consolidation by atmospheric pressure uses glass molds that are less expensive than

their metal counterparts. Additionally, they are monolithic, thus involving no welds or potential areas for leakage. Vacuum degassing of glass molds can be accomplished more quickly and thoroughly than metal molds, because inside surfaces are extremely clean and free of surface pores. Because of their self-stripping nature, glass molds are less expensive to use than metal containers that require removal by machining, grinding, or chemical methods.

Consolidated Materials

Most materials that have been consolidated by atmospheric pressure are either tool steels or superalloys. These materials are produced in powder form primarily by argon or nitrogen gas atomization, although powders from other atomization techniques, such as centrifugal atomization or the rotating electrode process, may be used. Use of a high-quality, low-oxygen-content powder stock is essential for successful processing. For more information, see the article "Atomization" in this Volume.

P/M tool steel products have many advantages over ingot metallurgy products, as discussed in the article "P/M Tool Steels" in this Volume. Advantages are metallurgical in nature and include the absence of macrosegregation, microstructural uniformity, fine grain size, and fine, uniform carbide size in the P/M products.

Consolidation by atmospheric pressure is used to produce tool steel billets that subsequently are hot worked into mill shapes. Prealloyed powder is produced, sieved, and containerized in borosilicate glass. Consolidation by atmospheric pressure produces billets having densities ranging from 95 to 99% of theoretical. Material utilization to reach the billet stage is approximately 98% for consolidation by atmospheric pressure, versus approximately 60% for ingot practice, as shown in Fig. 2.

Due to the microstructural advantages of P/M processing, the CAP billet has im-proved hot workability over ingot metallurgy billets of identical chemical composition. Less cracking during deformation processing occurs; consequently, less scrap is generated when working CAP-processed material. This tendency further enhances the material utilization factor of CAP processing compared to conventional ingot metallurgy practice. Hot working (rolling) to mill shape produces a fully dense CAP product.

Mill shapes are fabricated into components. Grindability and tool performance of CAP material is improved in relation to ingot metallurgy material. Improved grindability results from finer carbide sizes and their more uniform distribution in the P/M product than in ingot-produced tool steel. Improvements in performance are realized by improved response to heat treatment, higher fracture toughness, and higher resistance to wear in P/M products, due primarily to microstructural advantages.

Typical microstructures of ingot metallurgy T-15 and CAP T-15 products are shown in Fig. 3. Uniformity of carbide distribution and the fine carbide size of the CAP product is evident. The fine microstructure of CAP T-15 versus ingot metallurgy T-15 products results in increased hardness for equivalent austenitizing temperatures, as shown in Fig. 4(a). Another advantage of this type of microstructure is that high hardness levels can be achieved in P/M processed materials in shorter aus-

Fig. 3 Microstructural comparison of T-15 products

Carbides (white) are finer, more uniform, and more uniformly distributed in the CAP product (a) than in the ingot metallurgy product. (b) Magnification: 500×

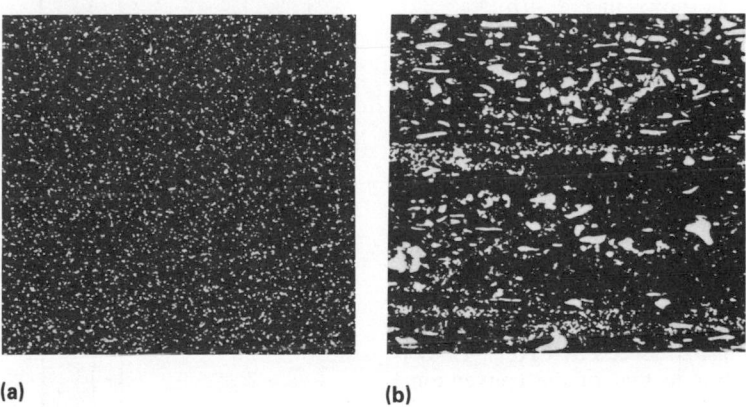

(a)　　　　　　　　　　　　　　　(b)

Fig. 4 Effect of austenitizing temperature on tempered hardness of ingot metallurgy and CAP-produced T-15 tool steel

(a) 25-mm (1-in.) diam bar, austenitized 2 min at indicated temperature and tempered three times for 2 h at 550 °C (1025 °F). Source: Ref 1. (b) 100-mm (4-in.) diam bar austenitized at 1220 °C (2225 °F) for time plotted, oil quenched, and tempered three times for 2 h at 565 °C (1050 °F). Source: Ref 2

Fig. 2 Comparison of ingot metallurgy and CAP production of tool steel mill products

Source: Ref 1

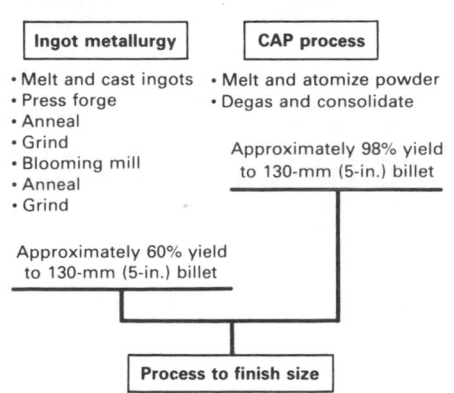

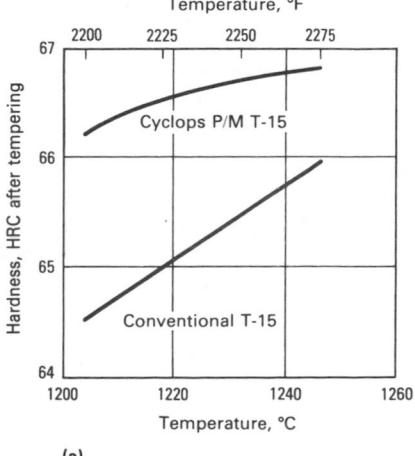

(a)

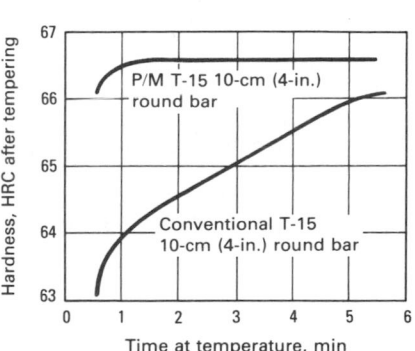

(b)

Fig. 5 Comparison of cutting performance of type 303 stainless steel

Feed rate: 0.05 mm/rev (0.002 in./rev). Speed: 41 m/min (134 sfm). Source: Ref 2

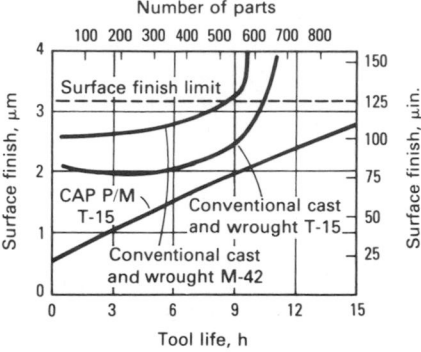

Fig. 6 Comparison of mechanical properties of Inconel 718

Source: Ref 3

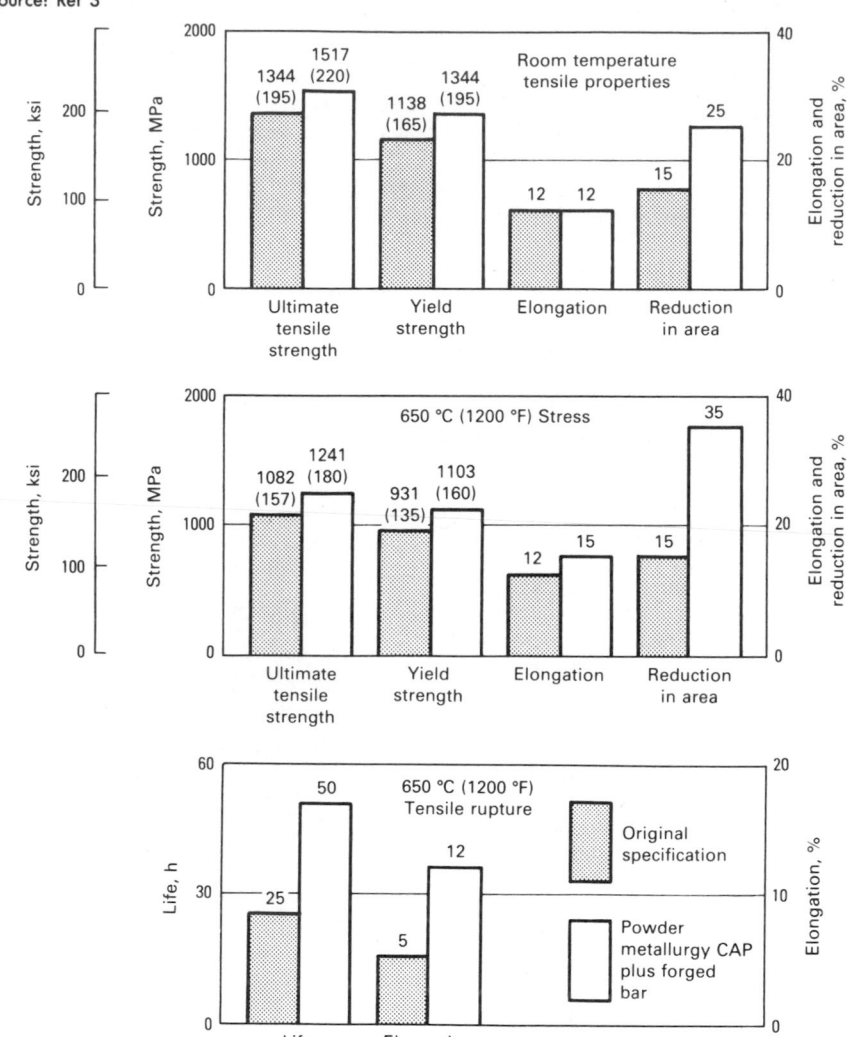

tenitizing time than for ingot metallurgy material, as shown in Fig. 4(b).

The overall cutting performance of CAP-processed T-15 tool steel is superior to conventional cast and wrought material. Testing conducted on an automatic screw machine based on American Society for Testing and Materials (ASTM) standard test procedures showed that CAP T-15 tools used for cutting 25-mm (1-in.) diam type 303 stainless steel bar stock had a life of over 15 h versus approximately 10 h for ingot metallurgy T-15 tools and less than 9 h for ingot metallurgy M-42 tools. The surface finish produced by CAP T-15 tools was superior to that of the conventional tools for all test times. Figure 5 shows the cutting performance comparison for these materials at a speed of 41 m/min (134 sfm) and a feed rate of 0.05 mm/rev (0.002 in./rev).

Superalloys. Jet engine components are candidates for parts processed from hot worked CAP stock. Consolidation by atmospheric pressure is a viable process for consolidation of superalloy powders such as René 95 and Inconel 718. Great potential is envisioned for consolidation of rapidly solidified powders. Isothermal forging of CAP billets to near-net shapes offers performance and cost advantages.

Figure 6 compares mechanical properties of Inconel 718 produced by CAP and subsequent hot forging to original property specifications. Both room-temperature and elevated-temperature tensile property requirements and stress-rupture requirements are surpassed by CAP plus forged P/M alloys. For direct aged material, there is no sacrifice in ductility at higher strength levels, which indicates metallurgically sound particle-to-particle bonding in the CAP plus forged material. For this material, the small amount of boron helps prevent grain-boundary carbide films by promoting the formation of $M_{23}C_6$-type carbides. Also, boron inhibits the formation of cellular orthorhombic delta Ni_3Nb, which tends to form on extended elevated-temperature exposure, thus degrading strength.

REFERENCES

1. Dax, F.R., P/M High Speed Steels by the CAP Process, *Metal Powder Report*, Vol 38 (No. 4), 1983, p 200-202
2. Kent, W.B., An Alternative Method of Producing High Speed Steel Powder, *Processing and Properties of High-Speed Tool Steels*, The Metal Society/American Institute of Mining, Metallurgical, and Petroleum Engineers, 1980, p 159-166
3. Lasday, S.B., Improved Properties of 718 Alloy Fashioned by P/M Process and Heat Treatment, *Industrial Heating*, June 1982, p 22-26

Ceracon Process

By B. Lynn Ferguson
Partner
Deformation Control Technology
and
O.D. Smith
President
Ceracon Corp.

CERACON PROCESSING involves consolidation of a porous metal preform to theoretical density under pseudoisostatic conditions. The term "Ceracon" is derived from ceramic granular consolidation. During processing, a granular ceramic medium transfers pressure from an advancing ram of a press to a P/M preform. This process differs from true isostatic pressing, in which a fluid medium is used to transfer pressure to the powder. This article describes process details, viable materials and parts for Ceracon processing, typical properties, and process economics.

Fig. 1 Ceracon process

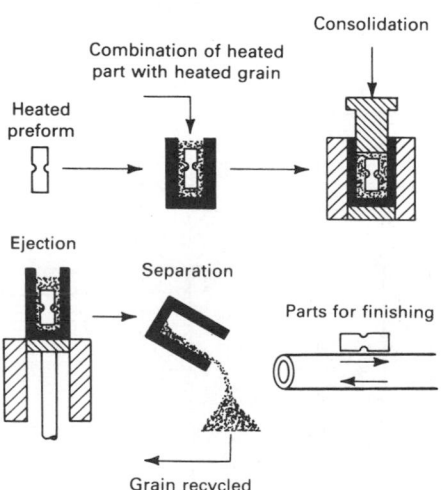

Processing Sequence

To begin processing, a porous preform is prepared by any suitable P/M technique, such as cold pressing, injection molding, pressureless sintering, or slip casting. Once a preform has been prepared, it is heated in a controlled atmosphere in an induction furnace to consolidation temperature.

Simultaneously, a granular ceramic medium is heated to the same or higher temperature. As shown in Fig. 1, heated ceramic grain is charged directly into the die, and the heated preform is inserted into the hot ceramic. Individual preforms or groups of preforms may be loaded into the die for consolidation. Alternatively, preforms can be loaded into the ceramic grain while cold and heated in place. The preform must be totally surrounded by the ceramic grain, which acts as a pressure transfer medium. Preforms are not containerized, although they may be coated to optimize the as-consolidated surface finish and/or to prevent intrusion of the ceramic into the part surface.

The heated assembly is transferred to the consolidation press, and an axial load is applied to the ceramic grain by a moving ram of either a hydraulic or mechanical press. As the axial load is applied to the ceramic grain, the preform experiences both axial and lateral force. Densification and deformation of the preform result, with 100% of theoretical density being achieved with selection of proper processing variables.

Cycle time under pressure is a matter of seconds, with the ram being retracted once the desired load has been achieved. Alternatively, a dwell time under load can be accommodated. After consolidation, the ceramic granular medium and the now fully dense parts are separated by ejecting the content of the die. A shaker table serves to separate the ceramic grain and parts. The near-net part or group of parts can then be heat treated and finished as necessary.

A flowchart of the Ceracon system is shown in Fig. 2. Note that ceramic grain and heat are conserved and reused when possible. Also indicated in the flowchart

Fig. 2 Flowchart of the Ceracon process

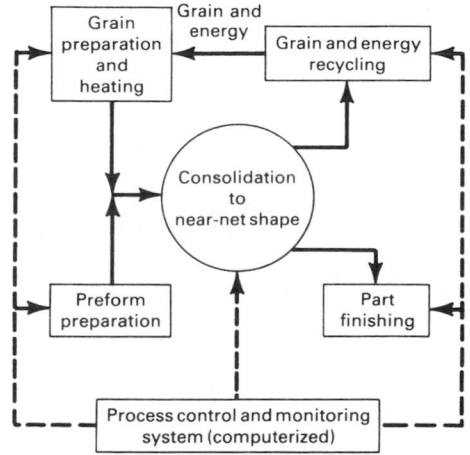

is the system control, with sensors providing data feedback to a central computer for both process control and data storage. The system is designed to be fully automatic, with robotized part and ceramic grain handling, and computer control of material throughput and energy input for both ceramic grain and preform heating. Alternatively, the system can be operated manually, with hand loading of preforms into the ceramic grain and manual operation of the press.

Process Fundamentals

With any emerging process, it is necessary to develop an understanding of process fundamentals. Results of a study* on the consolidation of 4650 steel P/M preforms using an alumina grain as the pressure transfer medium are discussed below to illustrate the fundamental features of the Ceracon process. The process variables requiring consideration to tailor the Ceracon process to specific manufacturing applications include:

- Characterization of the ceramic grain to be used for consolidation
- Proper design of preforms
- Accurate control of temperature and pressure

Cylindrical and rectangular preforms were compacted and sintered to approximately 80% of theoretical density for this study. The preforms were consolidated in a laboratory environment, using conditions representative of actual production.

*Work conducted by Metal Alloys, Inc., Signal Hill, CA

Fig. 3 Pressure-density curves for consolidation of 4650 steel P/M preforms at various temperatures in an alumina granular medium

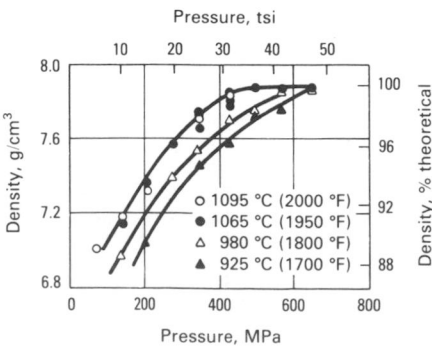

Pressure, tsi

o 1095 °C (2000 °F)
● 1065 °C (1950 °F)
△ 980 °C (1800 °F)
▲ 925 °C (1700 °F)

Fig. 4 Deformation of 4650 steel P/M preforms during consolidation

Full density is indicated by solid data points. Strain path for grain C showing slope change at full density is indicated by open data points. Grain A: angular particle shape. Grain B: spherical (unlubricated) particle shape. Grain C: spherical (lubricated) particle shape

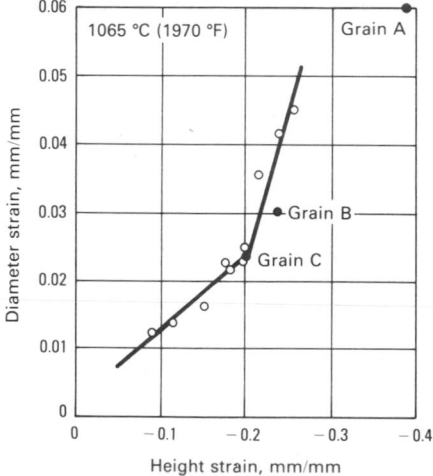

Pressure-Density Relationship. A typical set of pressure-density curves are shown in Fig. 3 for the consolidation of 4650 steel P/M preforms. These curves indicate that an increase in consolidation temperature results in increasing density for a constant consolidation pressure. Although not shown, many combinations of consolidation temperatures and pressures were used with a large number of ceramic materials. For the production parameters and economics desired with the alumina grain, consolidation variables of 441 MPa (32 tsi) and 1065 °C (1950 °F) were chosen. Obviously, other combinations that would result in full density could have been chosen.

The physical size of the ceramic particles was evaluated for each material. The materials were screened using Tyler standard sieves, and the percentage of each size volume was determined. The apparent optimum sizes (and the ones used in this evaluation) were as follows:

Mesh size	Size range, μm	Weight percentage
+80	>177	0
−80+100	177-149	16.72
−100+120	149-125	73.08
−120+140	125-105	9.73
−140+200	105-74	0.46
−200+325	74-44	0

Deformation During Consolidation. Figure 4 shows the shear deformation be-

havior of the 4650 preforms during consolidation. With the alumina grains A, B, and C used in these trials, preform axial contraction and lateral expansion accompanied densification. The relationship between height strain and diametral strain is approximately linear, with full density achieved at a diametral strain/height strain combination of 0.025 to 0.20 for grain C, which was lubricated. Other ceramic grains require different deformation levels to achieve 100% theoretical density, as indicated by the values for grain A (angular particle shape) and grain B (spherical particle shape).

Once full density is achieved, the lateral-to-axial strain ratio changes to the plastic Poisson ratio of −0.5. Deformation during consolidation is affected by the characteristics of the ceramic grain used to transfer pressure to the preform. Such deformation is not a result of isostatic pressure, as lateral and axial strains are not equal. Using an instrumented die holder (Fig. 5), lateral pressures within the preform were measured, as a function of height from the bottom of the die cavity, to assess the pressure on the preform during consolidation. Results for the three ceramic grains tested are shown in Fig. 6 for an axial pressure of 414 MPa (30 tsi). The top location (pin 1) lies above the top punch position in the die cavity and is not measuring lateral pressure transmitted by the ceramic grain. Position 2 is approximately level with the top punch face in the die cavity. These results show that pressure is not isostatic, but that lateral pressure tends to drop as distance from the top punch face increases. Different ceramic grains promote different pressure states during consolidation.

Fig. 5 Location of pins for measurement of lateral pressure during Ceracon process

Pins located 12.7 mm (0.5 in.) apart

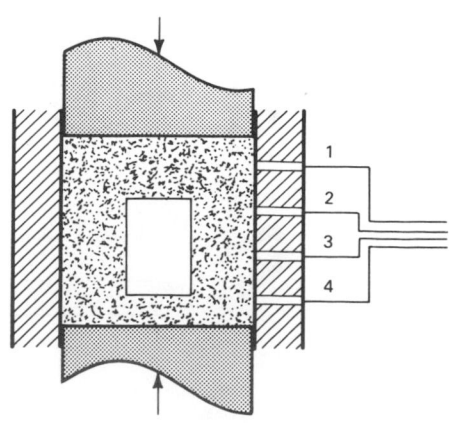

Fig. 6 Lateral pressures in the die cavity during consolidation at 414 MPa (30 tsi) at 1065 °C (1950 °F)

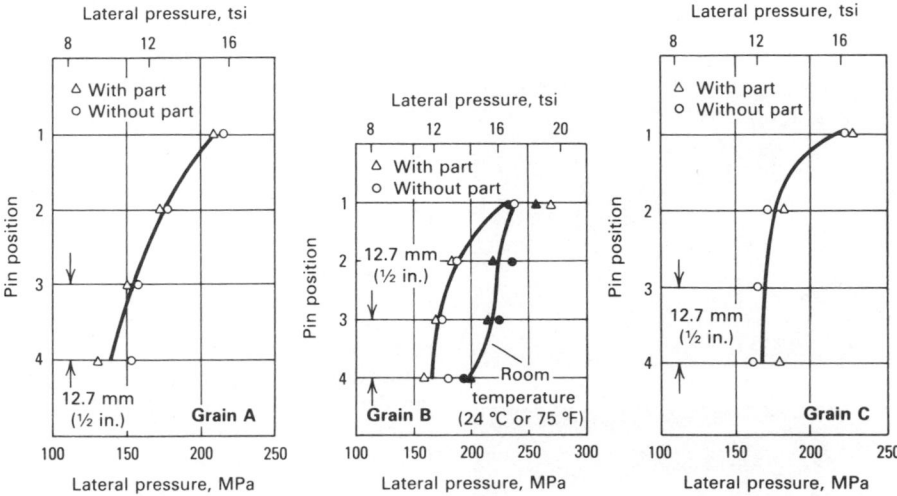

Ceramic Grain Characteristics

A wide variety of ceramic and nonceramic materials have been used as the pressure transfer media. While the perfect material (which would have no internal friction over the range of temperatures and pressures used) has not been found, many single-component systems and mixtures of materials have been used with near-isostatic results. The material selected for each application is proprietary in formation and is the same for all materials tested; however, the "ideal" materials would have the following properties over the range of consolidation temperatures and pressures:

- Very low internal friction, behaving as a gas
- Chemical stability with preform material
- No chemical reactions with the preform
- Density such that heating could be performed in a fluidized bed furnace
- Maintenance of hardness
- No sintering or fracturing

Particle shape affects grain particle movement. Grains must be able to move past one another as the preform deforms during consolidation. The grain also interacts with the surface of the preform. Concepts pertinent to powder flow during densification for conventional P/M materials are applicable to ceramic grain materials, with added emphasis on physical movement under pressure. Surface finish of the part is directly related to size and shape of ceramic granules, as consolidation pressure causes the particles to press

the part surface as a function of the particle size and shape. Embedding of granules can result. Proper selection of ceramic grain and consolidation conditions minimizes such tendencies.

Treatment of preforms with a surface coating also optimizes the surface finish of the part, particularly in applications where indentations of a microscopic size on the part surface is undesirable from a fatigue standpoint. The ceramic grain geometry that provides optimum consolidation is spherical and has a controlled size variation. Without a protective coating, these particles may yield microscopic dimples in the part surface. A wide range of coatings may be used, including plasma-coated magnesium oxide, alumina, titania, zirconia, and magnesium zirconate. The optimum coating must be thin—on the order of 0.127 mm (0.005 in.).

The coating on the surface of the part does not require special removal processing after consolidation, because the part is cleaned prior to finishing. Most coatings literally fall away upon removal from the consolidation media. The part surface finish obtained is typically 0.81 μm (32 μm in.) or more. Refinements directed at controlling thickness to less than 0.127 mm (0.005 in.) have been made and surface finish has been reduced to approximately 0.41 μm (16 μin.).

Chemical compatibility of the ceramic grain and the P/M preform material to be consolidated is necessary for successful parts production. For example, alumina with a high silica content may cause problems with a low-alloy steel. Silica, iron, and carbon in a reducing atmosphere

such as that used in the Ceracon process may form a silicate on part surfaces, requiring additional finishing operations.

Hardness of the ceramic granules at temperature affects the internal pressure distribution in the die cavity. Hardness of the ceramic grain is approximately 4000 to 5000 HV at room temperature. Plastic flow of the ceramic introduces granule interactions that create nonuniform distributions similar to those that can occur during die compaction of metal powders if bridging or particle size segregation occurs. Plastic flow of ceramic grain disturbs internal flow, which may cause preform deformation and distortion and/or incomplete densification.

Resistance to sintering is essential to facilitate separation of the grain and parts after consolidation. If the ceramic grain sinters together, the part is trapped in the grain and must be removed by mechanically breaking the ceramic form from the part. This reduces the economic advantages of Ceracon processing and promotes part distortion during removal. Proper selection of ceramic grain and processing conditions (temperature, pressure, and time) allows separation of the part from the grain by simple dumping, as shown in Fig. 7. Light abrasive blasting sufficiently cleans the part surface for subsequent finishing operations.

Resistance to fracture of the ceramic granules under pressure helps maintain pressure uniformity during consolidation and optimize process economics. If granules fracture, particle characteristics change, as do the resulting pressure distribution and deformation/densification relationships. Nonuniformity of pressure can become severe, causing distortion and incomplete consolidation of parts. Also, fractured granules must be removed from the grain mass before reuse, which increases the cost of removal and the cost of ceramic grain additions.

Preform Design

Although the Ceracon process is a near-net shape process, preform design is critical, because the final dimensions of consolidated parts must fall within certain limits for economical finishing. Preform dimensions can be determined from densification/deformation relationships developed for the particular preform material, preform density, and ceramic grain medium used. Using finished part dimensions, preform dimensions can be determined from these relationships. Slight modification of these calculated dimen-

Fig. 7 Demonstration of Ceracon process sequence

(a) The container holding the forged parts and the hot grain after removal from the press. (b) Dumping the hot grain and the part out of the container. (c) The part (a connecting rod) lying in the loose grain during the cooling cycle. (d) The connecting rod cooling from the forging cycle shown lying in the bed of cooling grain. In actual practice, the tray is not used, and the part is loaded directly into the consolidation die, pressed, and dumped out the bottom.

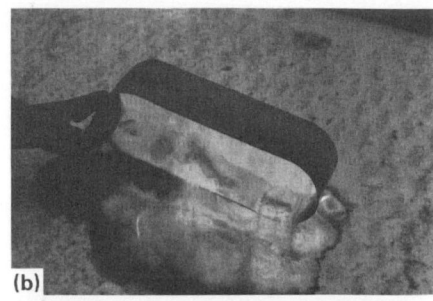
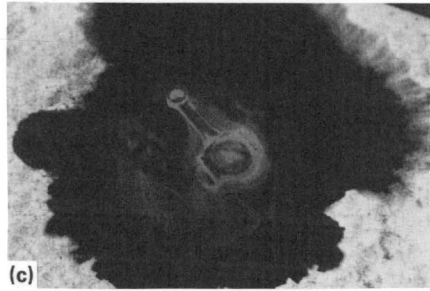
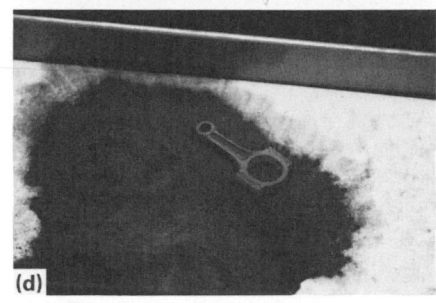

sions is required to accommodate internal flow of the ceramic grain during deformation of the preform.

In contrast to design of preforms for P/M forging, in which cracking during preform deformation may be a major problem, preform design for Ceracon processing is not concerned with cracking. Because it is a soft tooling process, metal flow around die corners is not restricted. However, despite the predictive capabilities of preform design guidelines, dimensional variations do occur as a result of the process. Without rigid tooling to accurately define final part dimensions, the initial preform density must be tightly controlled. For example, if the incoming density variation for a batch of preforms is 0.1 g/cm³, the resulting parts will have a variation of 1.25% in length. Similarly, variations in consolidation temperature and pressure cause variations in part dimensions. The position and orientation of the part in the grain are also important. Computer control of these processes reduces these variations.

Fig. 8 Comparison of preform shape and consolidated part for an open-end wrench

(a) Preform on top and the forged wrench on bottom. (b) Preform on left and forged wrench on right. Material is 4650 steel.

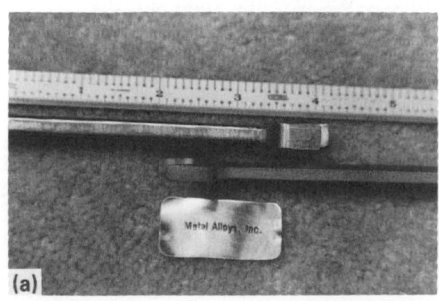
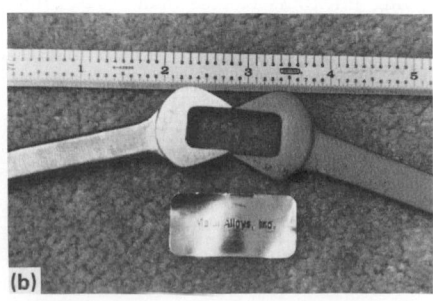

Ceracon Properties and Applications

A variety of part shapes and materials have been consolidated by the Ceracon process. Examples of powder metals that have been consolidated include low-alloy steels, stainless steels, nickel-based superalloys (such as IN100), copper and brass alloys, and commercial-purity and alloyed titanium. The flexibility of the Ceracon process is such that material limitations exist only where preform/ceramic grain chemical interactions cannot be prevented or when the consolidation temperature is so high that no suitable ceramic grain can be found.

Mechanical Properties. As with other powder processes, the mechanical properties of materials consolidated by the Ceracon process are dependent upon the quality of the as-received powder. The presence of oxygen, nonmetallics, or other impurities affects final mechanical properties. For the 4650 steel material in this study, the following average properties were obtained:

Tensile strength	1327 MPa (192.4 ksi)
Yield strength	1242 MPa (180.1 ksi)
Elongation (25.4 mm, or 1 in.)	10%
Reduction in area	26.7%

Heat treatment involved quenching in oil, followed by tempering at 400 °C (750 °F) for 1 h. Chemical analysis indicated an oxygen content of 329 ppm. Finished part composition was 0.48% C, 0.22% Mn, 0.013% P, 0.15% S, 0.02% Si, 0.04% Cr, 0.23% Mo, 1.84% Ni, and 0.08% Cu.

Part Geometries. Many types of part geometries can be consolidated. A preform and consolidated open-end wrench are shown in Fig. 8. Thickness reduction and width expansion are evident by comparing the preform and consolidated part. This part illustrates some of the capabilities and limitations of the process. Through holes and cross slots may be present in the preform to be consolidated without detrimental effects. However, dimensional precision of these features may be no closer than ±0.25 cm (±0.010 in). Use of inserts can improve precision.

Sharp steps and inside corners of preforms do not cause cracking during consolidation. Long, thin parts may bow during consolidation and require post-consolidation straightening, depending on the geometry of the part, its position in the ceramic grain, and distance of the preform from the die wall during consolidation. Blind holes, slots, and undercuts may be present in parts without affecting successful consolidation. Insert technology can be

Fig. 9 End view of connecting rods made by the Ceracon process

Top three are forged; the fourth (bottom) connecting rod is a centered, unforged preform. The material is 4650 steel.

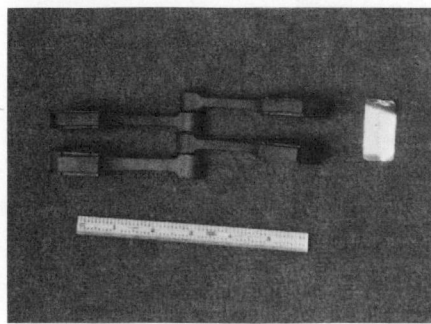

applied to tighten dimensional control. Geometry capability of the process allows a wide range of part shapes to be processed. In addition to the adjustable wrench handle, consolidated parts have included a variety of straight spur and bevel gears with either rack or involute teeth, connecting rods for internal combustion engine applications (Fig. 9), hardware for door locks, and parts with bores, internal splines, and recessed slots.

Currently, markets are developing to capitalize on the use of high-cost powder metals and the economic advantages of soft tooling that does not require expensive forge tooling. High-volume parts with low dimensional requirements, such as hand tools, fall into the latter category. Typically, high-cost powder metals have low volume requirements. Material savings result from net shape processing. Aerospace and medical products are included in this group.

SELECTED REFERENCES

- Ferguson, B.L., *et al.*, Hot Consolidation of Porous Preforms Using "Soft" Tooling, *Int. J. Powder Metall. Powder Technol.*, to be published
- Hailey, R.W., U.S. Patent No. 3 356 496, "Method of Producing High Density Metallic Products," Dec 1967
- Hailey, R.W., U.S. Patent No. 3 689 259, "Method of Consolidating Metallic Bodies," Sept 1972
- Kuhn, H.A., Ferguson, B.L., and Smith, O.D., Pseudo-HIP Using Conventional Presses, *Metal Powder Rep.*, Vol 38 (No. 6), 1983, p 315-323

Rapid Omnidirectional Compaction

By Clifford A. Kelto
Product/Market Development Manager
Kelsey-Hayes Powder Technology Ctr.

RAPID OMNIDIRECTIONAL COMPACTION is a low-cost proprietary process for consolidating high-performance prealloyed powders into fully dense parts. Its ability to produce intricate or simple shapes reproducibly makes rapid omnidirectional compaction suitable for producing near-net shape components or preforms. The relatively low thermal exposure given the powder during rapid omnidirectional compaction processing results in retention of a very fine microstructure and excellent mechanical properties.

The term "rapid omnidirectional compaction" is derived from key features of the process: it is very fast compared to other methods such as hot isostatic pressing and sintering of parts, and the pressure that effects powder consolidation acts in an omnidirectional fashion (as does isostat-ic pressing), despite the fact that this pressure results from a uniaxial ram force applied in a typical compaction (forging) press and tooling.

Rapid omnidirectional compaction utilizes a unique type of thick-walled powder container, a fluid die (Fig. 1). It is comprised of a powder-filled cavity (shown in Fig. 1 in the shape of a gas turbine disk) surrounded by a mass of fluid die material (for example, AISI 1020 steel) that is capable of plastic flow at the temperature and pressure of consolidation (for AISI 1020, 1095 °C at 827 MPa, or 2000 °F at 120 ksi). When pressure is applied to the fluid die exterior, the thick walls act as incompressible fluid elements and transmit that pressure to the powder. These thick fluid elements permit consolidation of complex shapes, including holes, re-entrant angles, and irregular contours.

In rapid omnidirectional compaction, consolidation pressure is applied to the fluid die in a forging press (Fig. 2). The filled and sealed fluid die is preheated to the consolidation temperature, placed in a pot die, and subjected to uniaxial ram force by a close-fitting punch. The ram force is reacted (through the plastic incompressible fluid die elements) by the pot die inner surface, thus applying pressure to the powder from all directions. Complete consolidation and interparticle bonding are accomplished, without pressure dwell, in a single ram stroke producing pressures of 345 to 895 MPa (50 to 130 ksi).

For some powders, preheat temperatures are about the same as hot isostatic pressing temperatures, but some materials

Fig. 2 Fluid die in forging press

(a) Loading of die before consolidation. (b) Die during consolidation

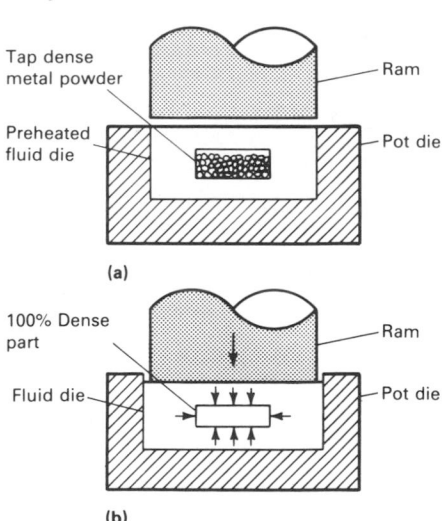

can be rapidly omnidirectionally compacted at temperatures that are several hundred degrees Fahrenheit cooler than hot isostatic pressing temperatures. Time at temperature is limited to the length of the rapid omnidirectional compaction preheat cycle. For heating loosely packed powder, a heating time of 90 min to 3 h is used, depending on temperature, size, etc.

Thermally sensitive materials such as rapidly solidified powders can retain their metastable microstructures and resultant property benefits throughout a rapid omnidirectional compaction consolidation cycle because of the combination of short

Fig. 1 Cross section of filled fluid die

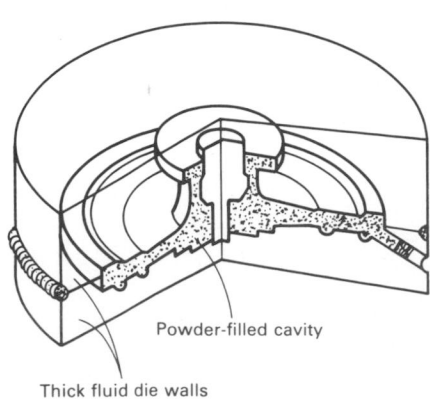

Powder-filled cavity

Thick fluid die walls

Fig. 3 Comparison of rapid solidification technology microstructures (Devitrium 3065)

(a) Produced by hot isostatic pressing. Temperature: 1050 °C (1920 °F). Pressure: 104 MPa (15 ksi) with a 2-h hold. Magnification: 1000×. (b) Produced by rapid omnidirectional compaction. Temperature: 980 °C (1800 °F). Pressure: 810 MPa (58.8 tsi) with a 0.5-s dwell. Magnification: 1000×

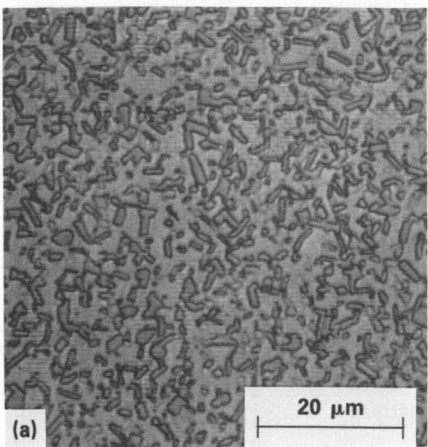

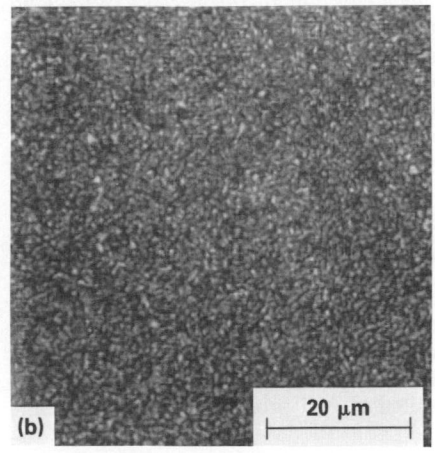

time and low temperature possible for die pressing (see Fig. 3). Only very high rate dynamic compaction, which cannot form large or complex shapes, is capable of producing full theoretical density from powder with lower thermal exposure.

Fluid Die Systems

The development of several types of fluid die materials and fabrication methods to meet technical and economic requirements has brought about improvements in the flexibility, applicability, and cost effectiveness of rapid omnidirectional compaction.

Low-Carbon Steel Fluid Dies. The first dies for rapid omnidirectional compaction were made from low-carbon steel by forging, casting, or machining. This approach is still used for some types of powder. Once impressions of parts are formed in die halves, they are welded together to provide a cavity to be filled with powder. The welded fluid die assembly is vacuum filled with powder, processed by rapid omnidirectional compaction, and removed from the consolidated part by machining and leaching (Fig. 4).

Recyclable Fluid Die. This system utilizes a Cu-10Ni fluid die material that is removed rapidly by melting in a copper-nickel bath. The fluid die material thus is reusable by recasting fluid die segments or blanks for forging. This approach retains the optimum near-net shape capability of steel fluid dies, while reducing the cost of fluid die material and removal.

Can/Cast Fluid Dies. The need for simpler shapes such as forging preforms has led to development of the can/cast fluid die (Fig. 5). A sheet metal can of the same type as that used in hot isostatic pressing is filled with powder, placed in a permanent mold, and embedded in recyclable copper-nickel material by casting. Following rapid omnidirectional compaction and fluid die melt-off, the part is decanned by turning or leaching (as are hot isostatically pressed or extruded components). The can/cast method recycles the fluid die material and reduces the cost of forming the powder cavity, but limitations of can-forming technology restrict the complexity of shapes that can be produced.

Composite fluid dies are the most common types of containers used in rapid omnidirectional compaction. Composites of glass cullet and refractory core mix in various ratios are lightweight, inexpensive, very fluid, and highly insulating. This type of fluid die is fabricated at room temperature, cured, processed by rapid omnidirectional compaction, and mechanically removed (Fig. 6). The composite fluid die process has a broad temperature range of application (315 to above 1650 °C, or 600 to 3000 °F). At least one component of the composite material must be capable of plastic flow at impact temperature to resist shattering of the die.

Composite fluid dies typically are one fourth the weight of metal fluid dies and are lower in cost. They are recyclable up to a 50% ratio with a blended materials mix that has not been cycled through the rapid omnidirectional compaction process.

Cold Compaction and Rapid Omnidirectional Compaction

For some materials, rapid omnidirectional compaction can be used in conjunction with cold pressing to capitalize on the shape-making capability of the latter process. Following cold mechanical or isostatic compaction, the green compact is set at room temperature into a composite fluid die, which is mechanically stripped from the part after rapid omnidirectional compaction. The result is a fully dense near-net shape of very fine microstructure, superior in performance to a sintered piece, which contains very large grains, coarsened microconstituents, and residual porosity. Figure 7 depicts the use of this process for making a high-speed tool steel

Fig. 4 Processing steps using carbon steel fluid die

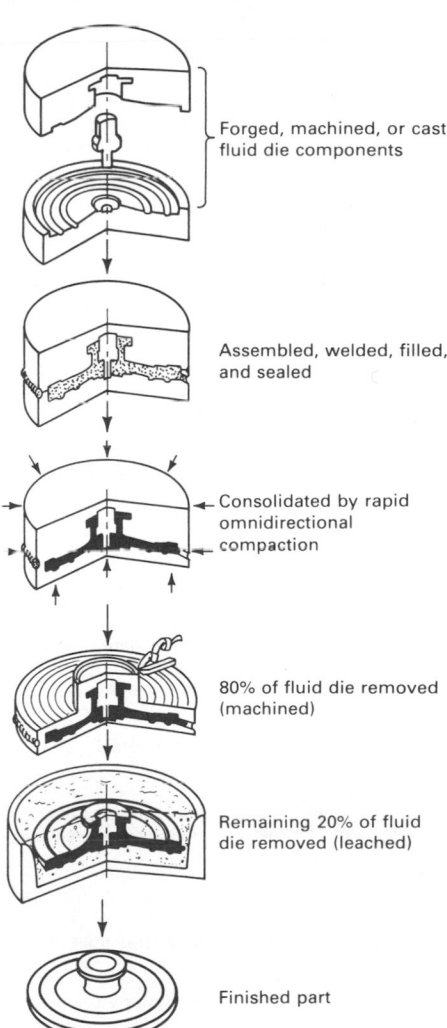

Forged, machined, or cast fluid die components

Assembled, welded, filled, and sealed

Consolidated by rapid omnidirectional compaction

80% of fluid die removed (machined)

Remaining 20% of fluid die removed (leached)

Finished part

milling cutter. Figure 8 compares the microstructure of a rapid omnidirectionally compacted part with that of a pressed and sintered part.

Post-Compaction Processing

Rapid omnidirectional compaction is useful in the manufacture of fully dense P/M forging preforms because of the fine-grained structure of the consolidated preform that is produced. In materials that are difficult to forge or for processes such as isothermal forging that may require superplastic behavior in the workpiece, fine grain size is essential for producing crack-free shapes.

Among other preform consolidation methods, only extrusion produces a fine-grained structure comparable to that produced by rapid omnidirectional compaction, and extrusion cannot make complex preform shapes. Structures produced by hot isostatic pressing, consolidation by atmospheric pressure, and sintering all have much larger grain sizes because of the long thermal cycles and higher consolidation temperatures of these processes. Figure 9 compares inert atomized superalloy microstructures consolidated by rapid omnidirectional compaction and hot isostatic pressing.

Processing of Bimetallic Parts

Fabricating components of two or more elemental powders or of a single alloy is not new or unique to powder metallurgy. However, many bimetallic parts are well suited to P/M processing, and rapid om-

nidirectional compaction offers the particular advantage of dimensional precision. For example, an insert of wrought material can be placed exactly within a metal fluid die and indexed by recesses, notches, or locating pins. The wrought piece remains precisely located throughout subsequent powder filling around the insert, sealing, preheating, and processing by rapid omnidirectional compaction, which is difficult to accomplish in hot isostatic pressing using sheet metal containers.

Bimetallic parts for rapid omnidirectional compaction exhibit complete bonding and good joint properties for most materials with no discernible reaction or diffusion zone at the interface. Most experience to date has been with dual alloy/dual property titanium parts and tool steel components. The impeller shown schematically in Fig. 10 has been successfully

Fig. 5 Processing steps using can/cast fluid die

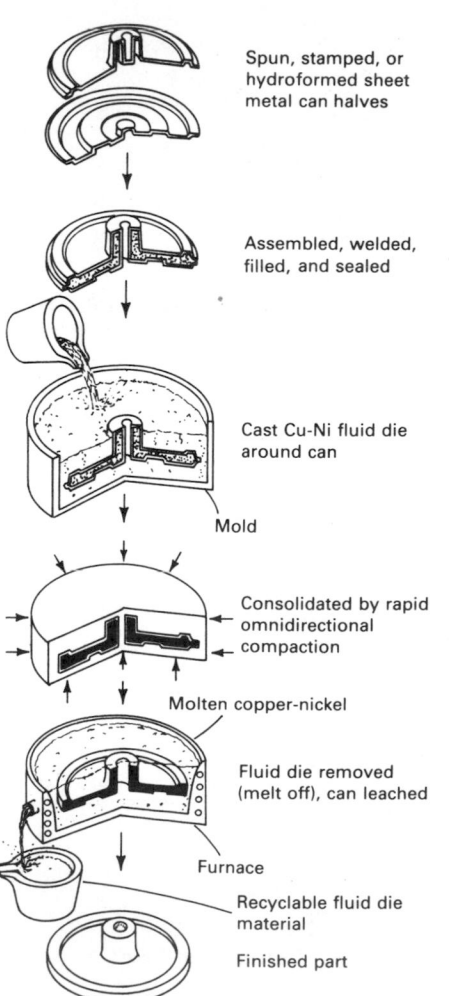

Spun, stamped, or hydroformed sheet metal can halves

Assembled, welded, filled, and sealed

Cast Cu-Ni fluid die around can

Mold

Consolidated by rapid omnidirectional compaction

Molten copper-nickel

Fluid die removed (melt off), can leached

Furnace

Recyclable fluid die material

Finished part

Fig. 6 Processing steps using composite fluid die

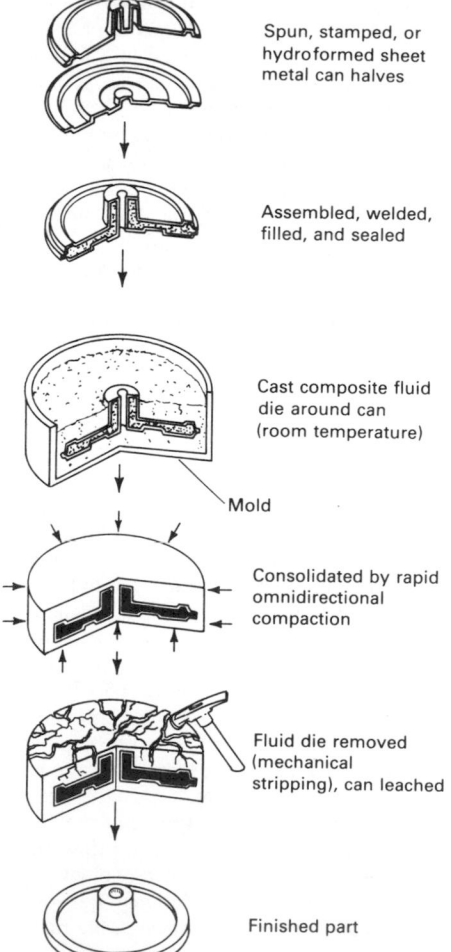

Spun, stamped, or hydroformed sheet metal can halves

Assembled, welded, filled, and sealed

Cast composite fluid die around can (room temperature)

Mold

Consolidated by rapid omnidirectional compaction

Fluid die removed (mechanical stripping), can leached

Finished part

Fig. 7 Processing steps using composite fluid die in which powder is preconsolidated into green compacts

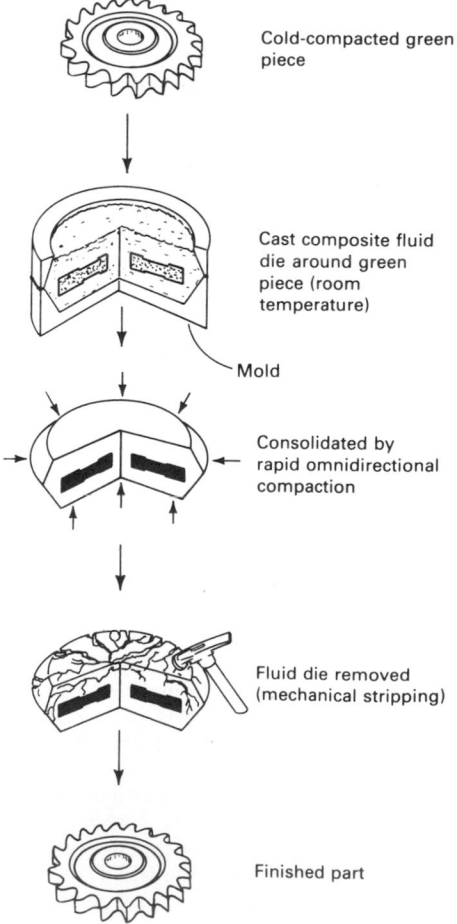

Cold-compacted green piece

Cast composite fluid die around green piece (room temperature)

Mold

Consolidated by rapid omnidirectional compaction

Fluid die removed (mechanical stripping)

Finished part

Fig. 8 Comparison of T15 tool steel powder microstructures

(a) Cold pressing and sintering. (b) Cold pressing and rapid omnidirectional compaction. Magnification: 1000×

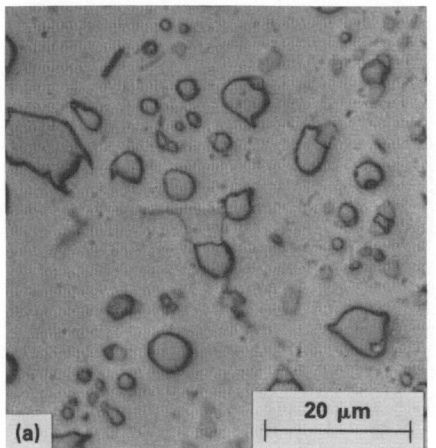

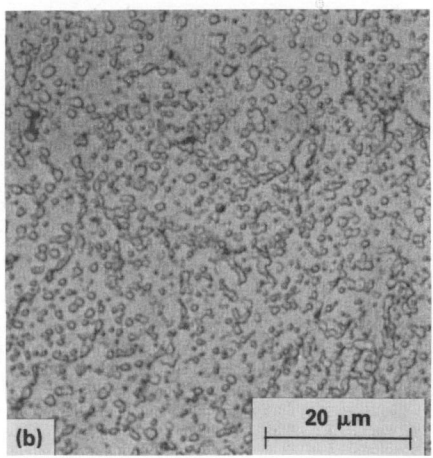

fabricated by two different approaches: all Ti-6Al-4V powder, and Ti-6Al-4V powder formed around and bonded to a wrought Ti-6Al-4V hub. This is an example of a dual property component, with the fatigue-resistant wrought material located in the highly stressed region of the part. This approach leads logically to dual alloy parts as well, for example, Ti-6Al-2Sn-4Zr-2Mo-Si airfoils formed from powder on a Ti-6Al-2Sn-4Zr-6Mo hub.

Process Advantages

In addition to high-pressure, low-temperature consolidation, rapid omnidirectional compaction offers the following production advantages. It provides excellent dimensional control, making it an ideal near-net shape production method, with mechanical properties at least equal to those produced by hot isostatic pressing. For more information, see the article "Mechanical Properties of P/M Materials" in this Volume. Rapid omnidirectional compaction utilizes conventional forging equipment (hydraulic, mechanical, hammer, or screw presses) and normal forge press cycles. Throughput rates are comparable to a standard forging operation; for example, on a single large press, approximately 300 parts per day (weighing several hundred pounds each) can be produced.

Capital costs are very low if an in-place forging press is available. Production costs, which are highly component specific, are typically 50% of those of hot isostatic pressing.

Process Limitations

Because rapid omnidirectional compaction is applicable to relatively high-rate production forging equipment, press setup charges and tool amortization can be prohibitive for low-volume parts. This applies also to tooling for fabrication of cans or fluid dies to make parts with complex shapes. For example, the tooling required to form a can to make a net shape medical hip implant (Fig. 11) can involve a substantial investment. If the demand for that part is low, perhaps 100 parts per year, then the tool amortization charge per piece becomes a major portion of the part cost.

For low-cost alloys, improved materials utilization may not be cost-effective if input material costs are low and machining is not difficult. For materials that are expensive to convert to powder, if powder costs are prohibitively high, the added material costs may offset the advantages gained by rapid omnidirectional compaction. Standard metal forming operations may be more economical in these cases.

Low thermal exposure, an advantage for rapidly solidified and other prealloyed powders, is a drawback for processing low-cost elemental mixtures by rapid omnidirectional compaction. These powders require diffusion to homogenize their compositions into alloys. Unless it is followed by high-temperature diffusion treatment, rapid omnidirectional compaction produces a fully dense mixture of base metal and master alloy particles rather than a uniform alloy of the intended composition.

Table 1 Press requirements for producing cylindrical preform shapes by rapid omnidirectional compaction

| Part diameter | | Minimum press rating(a): | | | |
| | | At 345 MPa (25 tsi) | | At 690 MPa (50 tsi) | |
cm	In.	kN	Tons	kN	Tons
5	2	2 200	250	4 500	500
15	6	11 000	1 250	22 000	2 500
25	10	25 000	2 800	50 000	5 600
50	20	84 500	9 500	169 000	19 000
75	30	178 000	20 000	356 000	40 000

Note: Complex near-net shapes require thicker fluid die walls and thus larger presses than those listed.
(a) Press ratings, a measure of force, are given in kN and tons (force). Forging pressure is given in MPa and tsi.

Fig. 9 Comparison of microstructures of low-carbon Astroloy

(a) Produced by hot isostatic pressing. Temperature: 1150 °C (2100 °F). Pressure: 104 MPa (15 ksi) with a 4-h hold. Grain size: ASTM 7. Magnification: 200×. (b) Produced by rapid omnidirectional compaction. Temperature: 1095 °C (2000 °F). Pressure: 828 MPa (60 tsi) with a 0.5-s dwell. Grain size indeterminate. Magnification: 200×

Fig. 10 Formation of dual property titanium impeller

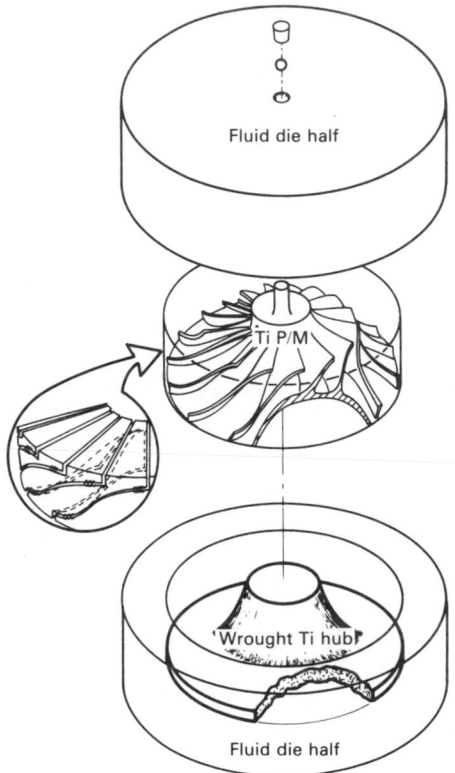

Fig. 11 Prosthetic hip joint can halves and assembly of 1010 steel

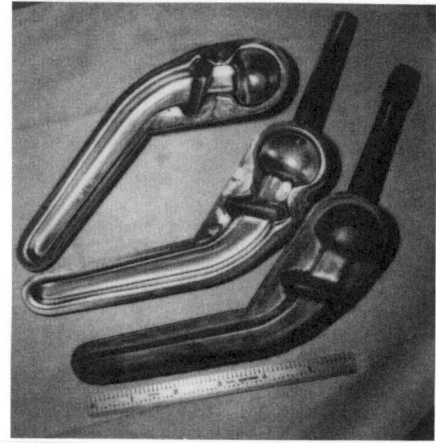

Fig. 12 Titanium alloy gas turbine impeller

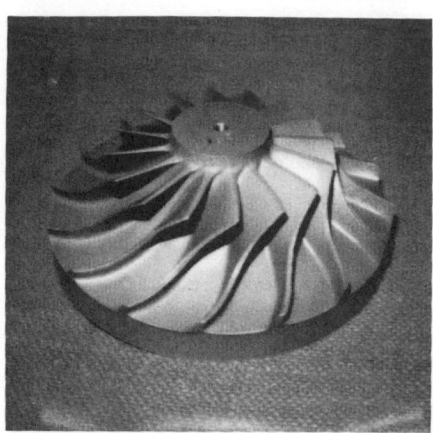

Probably the most significant limitation of rapid omnidirectional compaction is press size. Consolidation pressure is applied over the plan area of the fluid die, including wall thickness surrounding the part. Table 1 gives part diameter and press capacity requirements for consolidating cylindrical preforms at 345 and 690 MPa (25 and 50 tsi). For powders consolidated at 690 MPa (50 tsi), final part diameters greater than about 25 cm (10 in.) require presses rated in excess of 49 000 kN (5500 tons).

The largest hydraulic presses in the United States can produce parts up to about 81 cm (32 in.) in diameter and 0.6 m (2 ft) in length at 690 MPa (50 tsi). Reduction of the compaction pressure to 345 MPa (25 tsi) could increase that part size to 112 cm (44 in.) in diameter by 0.6 m (2 ft) long. This compares with billets of 101+ cm (40+ in.) in diameter and 2.4+ m (8+ ft) in length made in the largest autoclaves used for hot isostatic pressing.

Plan area, tool size, and complexity also affect the cost of specialized shapes made from powder. Parts with varying section size and lengths greater than about 46 cm (18 in.) and high aspect ratios can be made by rapid omnidirectional compaction, but pot/punch tooling may be difficult and expensive. These components may be made more economically by hot isostatic pressing.

The economics of rapid omnidirectional compaction depend on the size of the part, the powder alloy, the complexity of the shape to be made, and allowable dimensional tolerances. The impeller shown in Fig. 12 is a near-net shape Ti-6Al-4V component. To make this part with optimum dimensional control, a steel fluid die is used; Cu-10Ni material would form a eutectic with the titanium, and a composite fluid die would require a precise impeller-shaped can to hold the powder. The fins are formed contoured. The mold can be machined by electron discharge machining or cast or machined by conventional techniques. The mold cost is obviously high.

The cost of fabricating and subsequently removing the steel fluid die from each part contributes significantly to the overall cost of the process and is high compared to the composite fluid die rapid omnidirectional compaction method. However, this method is still very cost effective when compared to the conventional wrought/machine approach, which cuts about 90% of the forging input material into chips. The steel fluid die rapid omnidirectional compaction approach may not be cost competitive with cast/wrought manufacturing of other parts with higher material utilization factors.

STAMP Process

By B. Lynn Ferguson
Partner
Deformation Control Technology

THE STAMP PROCESS, developed by ASEA in cooperation with Surahammars Bruk, both Swedish companies, is used to produce billets and semifinished workpieces that are subsequently hot worked to final shape. The process consists of gas atomization of powder in a horizontal arrangement, followed by preheating and rapid consolidation at high pressure.

Because the method produces defect-free billets, thereby facilitating a high material yield, it is suitable for products made from highly alloyed heat-resistant steels, stainless steels, and high-speed tool steels. This article describes the processing procedure, applications, and economics of the STAMP process.

Processing Sequence

The STAMP process is an integrated procedure that produces fully dense billets from molten metal. It can be added to an existing melt facility, or an independent STAMP facility may be established. The melt facility may be either a vacuum-induction melting furnace or an air-melting furnace, depending on the materials to be processed. A ladle treatment for control of slag inclusions is optional. The STAMP process is shown schematically in Fig. 1.

Molten metal is poured into the tundish of a horizontal atomization unit, which uses nitrogen or inert gas to atomize a descending liquid stream exiting from the tundish, as shown in Fig. 1(a). Instead of allowing the atomized liquid droplets to solidify while falling freely in a vertical atomization tower, the gas blows the droplets horizontally. Solidified powder is collected at a horizontal displacement from the nozzle. By substituting a horizontal displacement for a vertical displacement for powder particle solidification and cooling, a

low-profile atomization unit can be constructed.

Screening stations are integrated into the horizontal chamber so that screened powder is fed into storage vessels without exposure to air. Powder particles are spherical in shape and have oxygen contents representative of the melting method (typically <100 ppm).

Screened powder then is fed into cylindrical containers of steel sheet metal, which are sealed hermetically by welding (Fig. 1b and c). Container size is based on desired billet size. Billet weights range from about 50 to 2500 kg (100 to 5000 lb), depending on the press capacity available for STAMP consolidation.

The containerized powder is heated to approximately 1100 °C (2010 °F) and is then transferred to a hydraulic press for consolidation (Fig. 1d and e). Figure 2 shows the relationship between consolidation pressure, temperature, time under pressure, and final density for three different consolidation methods. If minimal hot working is to be performed after initial compaction, combined pressure, temper-

ature, and time must produce 100% dense material. In Fig. 2, the curve to the right (P_3) represents sintering at air pressure; the center curve (P_2) represents increasing pressure, such as that obtained in hot isostatic pressing; and the curve to the left (P_1) represents high pressure, such as that obtained in the STAMP process at 1100 °C (2010 °F). The required compaction

Fig. 1 STAMP process

(a) Horizontal gas atomization. (b) Filling of steel container. (c) Welding of container. (d) Preheating and degassing. (e) Consolidation under high pressure. (f) Hot working

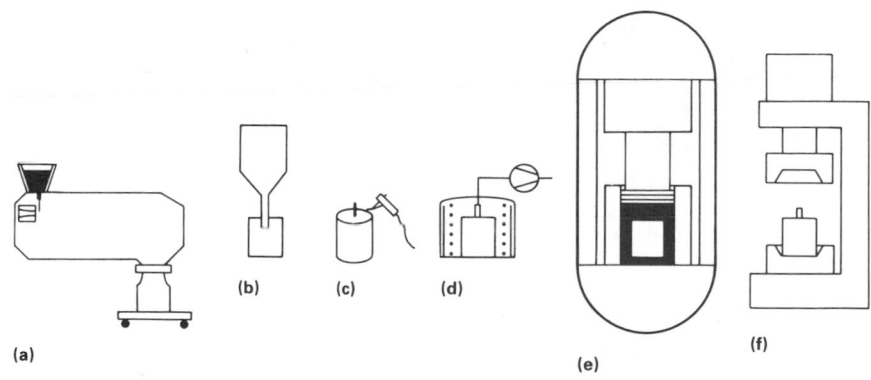

Fig. 2 Relationship between pressure (P), temperature (T), and time to achieve 100% dense steel powder

P_1 represents the STAMP process; P_2 represents hot isostatic pressing; and P_3 represents sintering at air pressure. Source: Ref 1

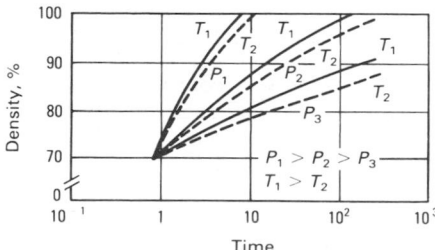

cycle is completed in less than 10 min. Processing cycle times are shorter than for hot isostatic pressing, due to the low pressures associated with hot isostatic pressing.

Because metal flow is limited by the fit of the powder mass in the die cavity during consolidation, the resulting billet is metallurgically similar to a hot isostatically pressed billet. For billets that are to be hot worked after STAMP consolidation (Fig. 1f), densities above 95% of theoretical are acceptable in the as-consolidated product. Microstructures of STAMP-consolidated material are homogeneous and isotropic.

After a billet has been consolidated, it can then be hot worked to final shape. Hot workability of the product is improved by the fine uniform microstructure associated with a P/M product. Figure 3 compares the microstructure of a ferritic 23% Cr steel made conventionally and by the STAMP process.

Materials To Be Processed

ASEA has produced low-alloy steels, including chromium-manganese grades such as 4150, tool steels, and ferritic and austenitic stainless steels by the STAMP process.

Low-Alloy Steels. To determine the structure and properties of low-alloy steels, railroad wheel rings were fabricated from an iron-chromium-molybdenum 4150 steel and tested. The chemical composition was:

Carbon	0.54%
Silicon	0.41%
Manganese	0.77%
Phosphorus	0.021%
Sulfur	0.031%
Chromium	0.89%
Molybdenum	0.32%
Iron	rem

The powder was consolidated by the STAMP process and subsequently forged and rolled into wheel rings. The rings were hardened, tempered, and partly turned prior to testing.

No defects were detected by ultrasonic testing. The microstructure was very fine and uniform and showed no signs of segregation and porosity. The inclusion level was similar to vacuum-degassed steel. Table 1 compares properties of a STAMP-processed wheel ring and a conventionally processed wheel ring of vacuum-degassed low-alloy chromium-molybdenum steel.

Tool Steels. A molybdenum-cobalt high-speed steel, with a nominal composition of Fe-1.3C-4Cr-5Mo-6.5W-8.5Co-3.0V,

Fig. 3 Austenitic ferritic 23% Cr steel
(a) Made by conventional process. (b) Made by STAMP process. Source: Ref 1. Magnification: 100×

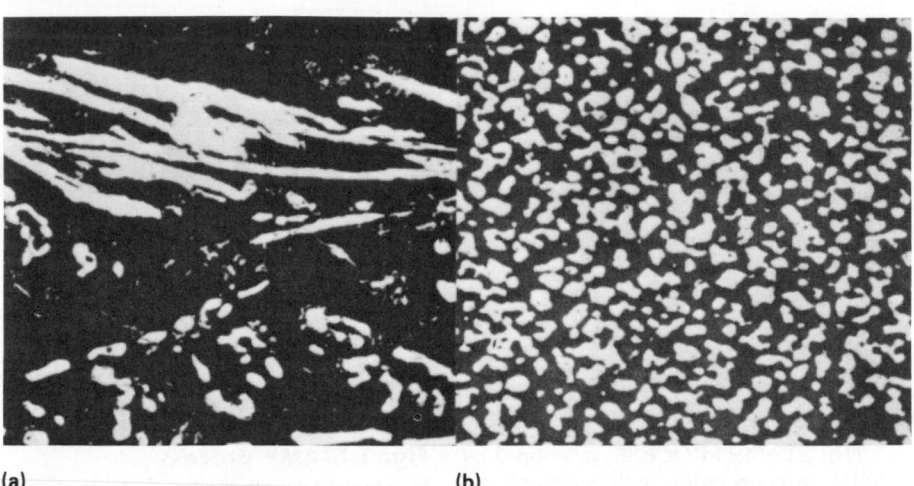

(a)   (b)

Table 1 Tensile properties of low-alloy chromium-molybdenum steel railroad wheel rings made by P/M and conventional processing

Process type	Tensile strength MPa	ksi	Yield strength (0.2% offset) MPa	ksi	Elongation in 50 mm (2 in.), %	Reduction in area, %
STAMP	1060	154	890	129	15	50
Vacuum degassing	1050	152	850	123	15	52

Source: Ref 2

was processed into round bar stock by a STAMP plus forging and hot rolling process (Ref 2). This high-speed steel ingot metallurgy product has poor workability and is susceptible to cracking during conventional processing. The P/M product produced by the STAMP process did not crack during hot working, producing material equivalent in both microstructure and mechanical properties to P/M high-speed steel processed by hot isostatic pressing plus hot working.

Stainless Steels. Steam turbine disks and rings were processed from AISI 422 ferritic stainless steel powder by a STAMP plus hot working process. The chemical composition of the steel was:

Carbon	0.22%
Silicon	0.32%
Manganese	0.61%
Phosphorus	0.017%
Sulfur	0.006%
Chromium	12.3%
Nickel	0.71%
Molybdenum	1.11%
Vanadium	0.38%
Niobium	0.11%
Nitrogen	0.05%
Iron	rem

Oxygen levels below 30 ppm were achieved in this product, which are below normal levels for vacuum-degassed or electroslag-remelted 422 stainless steel. Mechanical properties for STAMP-processed 422 and electroslag-remelted 422 steel are shown in Table 2.

The properties obtained in the axial and tangential test directions of the STAMP-processed material are comparable to those of the electroslag-remelted steel. The impact toughness in the axial test direction of the STAMP-processed steel is nearly equivalent to the tangential direction, indicating that STAMP-processed steels are more isotropic than electroslag-remelted steel.

Creep-rupture data for these products are shown in Fig. 4. The P/M product produced by the STAMP process has creep-rupture resistance equivalent to the electroslag-remelted product, and less scatter at lower stress levels (longer lifetimes) is exhibited.

A ferritic-austenitic stainless steel, similar to AISI 329, was fabricated into rings and bowl bodies for separators, using STAMP-consolidated billets. The chemical composition of the steel was:

Carbon	0.035%
Silicon	0.70%
Manganese	0.50%
Phosphorus	0.018%
Sulfur	0.012%
Chromium	23.3%
Nickel	5.1%
Molybdenum	2.9%
Nitrogen	0.16%
Iron	rem

Ultrasonic inspection showed sound, metallurgically clean material. Mechanical properties are compared in Table 3 for forgings of this alloy produced by the STAMP process, by hot isostatic pressing plus forging, and by conventional electroslag remelting. Mechanical properties of the STAMP product are comparable to those of the electroslag-remelted product. Toughness and ductility are more isotropic in both P/M products. The STAMP product had corrosion resistance similar to the electroslag-remelted product. Absence of segregation in the P/M STAMP product resulted in high resistance to intergranular attack and less sensitivity to temper embrittlement than in conventionally produced products.

New P/M Alloys. The STAMP process can be used to produce new alloys that are only producible by P/M techniques. An example is a replacement for the AISI 329 stainless steel discussed above. The unique P/M composition is based on high alloy additions of chromium, molybdenum, and nitrogen. Due to susceptibility to ingot cracking and poor workability in the cast state, this new steel cannot be produced by ingot metallurgy techniques. However, as a powder metallurgy product, the alloy is workable, due to the homogeneous, fine microstructure associated with similar products.

Fatigue properties of the unique P/M alloy produced by the STAMP process are compared to properties of conventional AISI 329 stainless steel in Fig. 5. Static strength and fatigue strength for rolled plates approximately 20 mm (0.75 in.) thick are higher for the P/M product than for the conventional product. Moreover, the P/M product has improved corrosion resistance, especially against intergranular attack and pitting.

Economic Implications

The STAMP process was designed to be economically justifiable as either an add-on to an existing melt facility or as a stand-alone system. By eliminating the need for an atomization tower, costs are minimized. The horizontal atomizer can be located in existing facilities without extensive building modifications.

Table 2 Property comparisons of 422 stainless steel made by the STAMP process and electroslag remelting

Used to make gas turbine rings

Process type	Tensile strength MPa	ksi	Yield strength (0.2% offset) MPa	ksi	Elongation in 50 mm (2 in.), %	Reduction in area, %	Impact energy(a) J	ft·lb
STAMP(b)	830-840	120-122	650-680	94-99	19-21	62-69	70-80	52-59
Electroslag remelting(b)	840	122	670	97	19	58	75	55

(a) Values shown for tangential test direction. Axial test direction impact energy value for STAMP processing is 65 to 70 J (48 to 52 ft·lb). For electroslag remelting, axial test direction impact energy value is 50 J (24 ft·lb). (b) Axial and tangential test direction Source: Ref 2

Fig. 4 Creep-rupture relationship for AISI 422 stainless steel

At 550 °C (1020 °F). Source: Ref 1

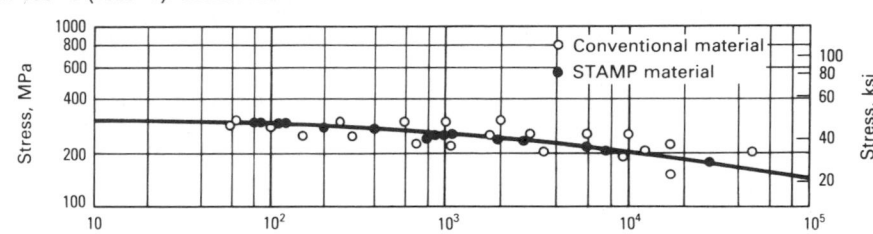

Table 3 Property comparisons of ferritic-austenitic stainless steel forgings made by P/M and conventional processing

Process type and test direction	Tensile strength MPa	ksi	Yield strength (0.2% offset) MPa	ksi	Elongation in 50 mm (2 in.), %	Reduction in area, %	Impact energy J	ft·lb
STAMP:								
Tangential	680	99	510	74	30	62	37	27
Radial	670	97	500	73	28	58	32	23
Hot isostatic pressing and forging:								
Tangential	670	97	490	71	29	59	36	26
Radial	660	96	490	71	28	58	34	25
Electroslag remelting:								
Tangential	630	91	480	70	26	54	32	23
Radial	620	90	470	68	22	46	28	20

Source: Ref 2

Fig. 5 Fatigue test results and mechanical properties of STAMP-processed 600 stainless steel and electroslag-remelted AISI 329 stainless steel

Fatigue ratio ($\sigma\ 10^7/Rm$) for 600 steel: 0.57. Fatigue ratio for electroslag-remelted 329 steel: 0.51. Source: Ref 2

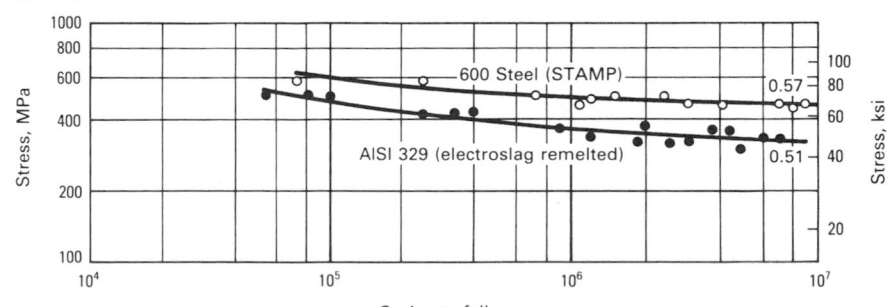

Steel	Tensile strength, MPa (ksi)	Yield strength (0.2% offset), MPa (ksi)	Elongation in 50 mm (2 in.), %	Reduction in area, %	Impact energy, J (ft·lb)	Fatigue strength, MPa (ksi)
STAMP 600	760 (110)	600 (87)	26	54	25 (18)	430 (62)
Electroslag-remelted 329	630 (91)	500 (73)	29	65	35 (25)	320 (46)

By replacing a low-throughput hot isostatic pressing consolidation process with rapid consolidation using a hydraulic press, high production rates can be achieved with the STAMP process. Also, the cost of a hot isostatic pressing facility is greater than the cost of the hydraulic press used for STAMP consolidation.

ASEA estimates that the energy used to produce a billet by the STAMP process is approximately 50% of that used to produce a billet by conventional ingot metallurgy techniques. Additional economic benefits are gained from more efficient material utilization and improved hot workability and homogeneity.

REFERENCES

1. Göransson, M. and Hede, A., Method of Special Steel Production via the STAMP Process, *Met. Powder Rep.,* April 1983, p 206-208
2. Göransson, M. and Hede, A., Breakthrough in Special Steel Production by P/M, *Modern Developments in Powder Metallurgy,* Vol 13, Metal Powder Industries Federation, Princeton, NJ, 1980, p 307-324

SELECTED REFERENCES

- Grinder, O., Swedish Powder Metallurgy Industry—An International Comparison, *J. Scand. Metall.,* Vol 11 (No. 5), 1982, p 216-222
- Karmholt, H.L. and Friburg, S., New High-Production Powder Technology, *Met. Powder Rep.,* May 1981, p 228-230
- Steel Billets Produced by New P/M Process Using Horizontal Gas Atomization, *Ind. Heating,* May 1982, p 33-36

Infiltration

By Claus G. Goetzel
Consultant, Lecturer
Department of Materials Science
and Engineering
Stanford University

DURING INFILTRATION, a liquid metal mass or front moves through or penetrates the pore system of a solid-phase powder compact. This process is similar to liquid-phase sintering, in which one of the constituent powders of a compact melts and disperses throughout the compact *in situ*. Voids are filled by shrinkage and particle rearrangement. During infiltration, liquid externally contacts the porous solid, and capillary forces draw it inward. Reduction of total surface free energy of the system is a prerequisite during infiltration and determines whether a particular solid-liquid phase system is suitable for infiltration.

A variation of capillary-action infiltration at ambient pressure consists of penetration of pores by liquid assisted by external force. Impregnation treatment of this type is widely used to fill pores of sintered products with hydrocarbons, organics, and nonmetallic fillers. For a solid-liquid phase system with suitable surface free energy, infiltration can be used to attain:

- Full density without high compacting or re-pressing pressures or subsequent forging, rolling or extrusion, or hot pressing
- Precise and complex shapes, including large sizes, using conventional P/M operations, tooling, and equipment
- Layered P/M products of different composition, or of one P/M section and a cast or forged element
- Surface characteristics that permit standard joining and coating methods

- Improved machinability due to uninterrupted chip forming, facilitated by minimization of pores or due to favorable chip-breaking caused by a duplex microstructure
- Reasonable control of phase distribution that may result in a uniform or a purposely graded microstructure
- Good mechanical properties due to a minimum of angular stress-raising pores
- Increased strength through subsequent effective heat treatment

Mechanism of Infiltration

The principles of infiltration mechanisms are discussed in Ref 1 to 6. Infiltration rate is related to Poisseuille's law defining the rate of flow or rise of a liquid in a glass capillary (Ref 2, 3). The parabolic rate law was confirmed for the early penetration stages of infiltration for a number of systems. However, bent and twisted channels in powder compacts made comparison with the straight capillaries of the model almost impossible. Furthermore, in many systems involving metals, penetration is followed by a second infiltration stage when the pore filling liquid reacts with the solid.

Successful infiltration of a porous, skeletal solid with a liquid phase requires that the total surface free energy of the system after infiltration be lower than the total surface free energy before infiltration. The total includes the surface free energies of the solid and the liquid phases, as well as the interfacial energy between the solid and liquid. This relationship is expressed by:

$$\gamma_{s/l} = \gamma_s - \gamma_l \cos \Theta$$

where $\gamma_{s/l}$ is the specific surface free energy of the solid-liquid interface; γ_s and γ_l are the respective surface free energies of the solid and liquid; and Θ is the liquid-solid contact or wetting angle. A low contact angle is essential for infiltration. The rate and depth or height of infiltrant penetration of the compact pore system can be enhanced by lowering the contact angle with the use of detergents (Ref 7). Solid, liquid, or vapor-forming surfactants mixed with the powder or passed through the compact prior to infiltration are suitable additives.

With a sufficiently low contact angle, surface tension forces in many systems cause the liquid metal to spread on the outside faces of the skeleton body. Because of diminished resistance, complete envelopes may form before capillary forces draw the liquid into the interior.

A number of conditions other than a low liquid-solid contact angle must be fulfilled to ensure successful infiltration and useful end products. These conditions have been discussed by many investigators (Ref 8-10) and are reviewed below.

Skeleton. The compacted powder matrix, or skeleton, should consist of a network of solid particles or grains throughout the infiltration process. It should provide a system of interconnected pores and channels of a size range that permits unimpeded capillary force action. Closed-off pores must be avoided; they cannot be penetrated by the liquid infiltrant under normal circumstances. An attempt should be made to remove oxide or nitride films prior to infiltration.

Infiltrant must have a melting temperature below that of the skeleton. High fluidity in the liquid state is desirable, because it aids the driving force of the surface tension. Thermal expansion characteristics of the infiltrant affect ultimate strength of the infiltrated body. If the infiltrant skin surrounding the infiltrated skeleton expands during rapid solidification, the strength of the finished part may be decreased; however, the opposite reaction occurs if the skin contracts on cooling from the liquid state. This phenomenon was found for iron infiltrated with copper and bismuth, respectively (Ref 11).

System Compatibility. The ideal liquid infiltrant should have a contact angle with a solid that is near zero, and reaction between solid and liquid should be kept to a minimum. Some refractory metal-conductor metal systems fulfill this condition. If the reaction products formed during infiltration (intermetallic compounds, eutectics, or solid solutions) have a specific volume equal to or greater than the combined initial specific volumes of the skeleton and infiltrant, penetration of the liquid is halted before completion, or it is blocked completely. A similar effect is obtained if the reaction products are dissolved in the liquid phase and the infiltrant becomes less fluid. In either case, external infiltrant residue and unfilled pore space remain.

Solubility. When solid solution formation between components is minimal at low temperatures and when equilibrium conditions at infiltration temperature produce minimal dissolution of the skeleton in the liquid infiltrant, the effects of solubility generally are beneficial. Maximum operating temperature may be lower and kept at only slightly above the liquidus of the infiltrant alloy. Strong bonds at the interface between the two phases are formed, and nearly full density of the product is obtainable.

If the infiltrant dissolves more than traces of the skeleton material, the infiltration rate may be lowered, requiring longer times for complete penetration. Also, at the point of contact with the liquid infiltrant mass, the exterior surface of the skeleton body may erode. Presaturation of the infiltrant with the skeleton metal or alloying with a solubility-inhibiting element eliminates this problem. Prealloying of copper with iron for infiltration into iron skeletons is a typical example.

Fig. 1 Capillary infiltration methods

(a) Dip of a small part of a skeleton body into a melt. (b) Complete submersion of a skeleton body into a melt. (c) Positioning of solid infiltrant on top of a skeleton, followed by pore penetration on melting. Source: Ref 8

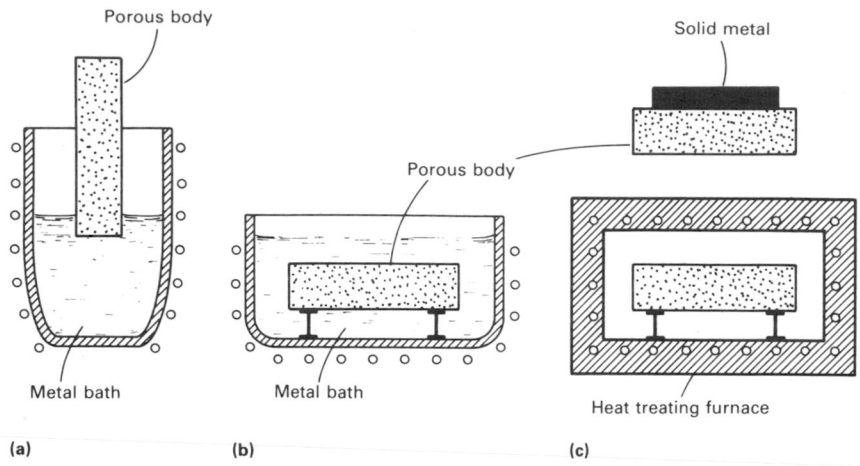

The stability of the compact becomes impaired as the solubility of the solid in the liquid increases. Consequently, wide pore channels and short infiltration times should be utilized. Diffusion barriers deposited on the pore walls of the skeleton may cause sufficient delay in dissolving the solid surfaces to permit penetration of the liquid phase through the entire pore system.

In applying copper films on the inner surfaces of iron skeletons by precipitation from a copper sulfate solution, carbon diffusion through the resulting copper-based solid solution may be slowed sufficiently to prevent solidification of a cast iron infiltrant before the entire compact is penetrated. Unless these precautions are taken, excessive solubility of the solid in the liquid rapidly leads to the collapse of the skeleton and to a meltdown of the partly or fully penetrated compact.

Microstructure. Structural integrity of the infiltrated body is closely associated with the microstructure. If the dihedral angle at the intersection of the boundary between two solid grains or particles and a liquid phase approaches zero, the liquid metal tends to penetrate along the grain boundary. An example of this is the tungsten-nickel-copper heavy alloy system. During the penetration stage of infiltration of the nickel-copper alloy, closed-off pores may become filled. This mechanism may lead to a loosening of the contacts between the solid tungsten grains and a loss of coherence of the skeleton. Consequently, volume expansion may occur during infiltration, while some of the original pores remain unfilled.

New pores also may be generated by particle rearrangement (Ref 12), or as a result of liquid-solid phase reactions.

Control of dimensional changes of the compact during infiltration is difficult in systems where the dihedral angle is zero and may even require the confinement of the skeleton in a mold. Also, the ultimate properties of the infiltrated body may be affected by the cohesive strength level of the skeleton that remains after boundary penetration by the liquid phase.

Vacuum. Capillary infiltration may be enhanced by subjecting the system to a vacuum. Volatile impurities are more readily removed from the infiltrant melt and the free surface of the solid under vacuum. For alloy infiltrants, the solidus temperature may be raised, and premature freezing in the skeleton channels may occur, if the composition is changed as a result of a high vapor pressure of one of the elements. In that event, vacuum operations should be avoided, and pressure should be applied in a sealed vessel during infiltrant melting and melt penetration of the skeleton.

Pressure Gradient. If a vacuum is applied to a confined skeleton on the side opposite the face contacting the infiltrant, penetration of the liquid is enhanced. By increasing the pressure gradient further with the aid of an external force acting on the liquid, compacts with a less controlled pore structure or wider pore spectrum may be penetrated. Infiltration becomes progressively less dependent on capillary action and good wetting as the pressure gradient becomes larger.

Systems in which the liquid forms high

Fig. 2 Cermet turbine blade infiltration mold assembly
Source: Ref 13

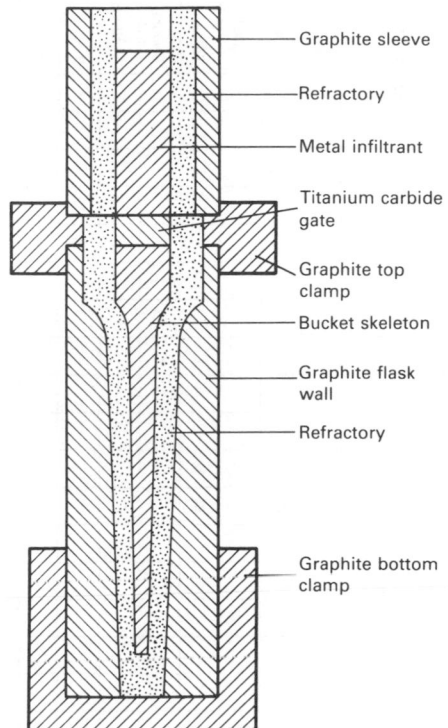

- Graphite sleeve
- Refractory
- Metal infiltrant
- Titanium carbide gate
- Graphite top clamp
- Bucket skeleton
- Graphite flask wall
- Refractory
- Graphite bottom clamp

contact angles with the solid can be infiltrated if the pressure differential is sufficient, but the liquid may not be retained by the skeleton and may be exuded partially or completely. In that case, the skeleton functions as a filter.

Infiltration Techniques

Throughout the development of metal infiltration technology, numerous methods have been employed. All acceptable methods for production work utilize, at least in part, procedures, tooling, and equipment common to the P/M industry. However, depending on the system involved and the quality specifications of the product, specific tools, devices, fixtures, supports, or molds generally are required. For example, the shape complexity of a component or its potential for reaction with either the skeleton or liquid metal may determine whether a carbon base or ceramic container or mold is used during infiltration.

The type of furnace used must permit control of time and temperature for both stages of infiltration—penetration of the liquid through the pore system and reaction of the liquid with the skeleton after filling of the pores. The particle size distribution of the powder must be selected or custom blended, and the compacting pressure must be adjusted to produce a large number of interlocking pores, as well as free paths between them, required to receive and pass the liquid metal.

Capillary-Dip Infiltration. The skeleton body is partly immersed into a molten metal bath contained in a crucible, where it acts as a wick (Fig. 1a). Liquid is drawn in by capillary forces and rises, expelling the gas volume contained in the pores. The bath may be metered to fill exactly the available pore volume of the solid.

Full-Dip Infiltration. The skeleton is completely submerged in the melt (Fig. 1b). Penetration occurs from all sides toward the core; therefore, the gas volume can be displaced only by diffusion through the liquid. To avoid gas entrapment, immersion of the skeleton must proceed slowly or in stages. The melt may be poured onto the porous body from one side with a tilting bath container. Application of a vacuum facilitates degassing. Excess molten metal is unavoidable, but may be used as a cast metal extension of the infiltrated compact.

Contact Infiltration. The infiltrant is initially placed in juxtaposition with the free-standing skeleton. After melting, a liquid film is formed that penetrates the pores. If the infiltrant is placed on top of the skeleton (Fig. 1c), surface tension may spread the film over all faces of the skeleton before penetration into the core occurs. Gas displacement from the pores is aided by placing the infiltrant beneath the skeleton or by applying a vacuum. Exact allotment of the metal required to fill the pore volume is possible, thus eliminating the need for liquid metal containers. This method is an adaptation of conventional sintering and elevates infiltration to a production process.

Gravity-Feed Infiltration. In this method, capillary forces are augmented by external pressure exerted by the high head of an infiltrant melt reservoir above the skeleton, which in turn is contained in an investment casting inert mold. Auxiliary force is a function of the height of the liquid metal head; if the mass of the head is large enough, more than one skeleton can be fed, and cluster arrangements similar to those of the investment casting process are possible.

Separation of the infiltrated product from excess infiltrant of the head or sprues is aided by suitable gating, made of coarse powder compacts or fiber felts to facilitate passage of the molten metal. If gate and skeleton materials are the same composition, the liquid may dissolve some of the gate and become partially saturated with the alloying element, thus preventing surface erosion of the infiltrated product. This process is suitable for producing precision-shaped and graded cermet turbine blades. For more information on this process, see the article "Cermets" in this Volume. The infiltration mold assembly used in this technique is shown in Fig. 2.

Fig. 3 Mold arrangement for pressure impregnation of low-carbon steel compacts with a high-carbon steel melt (experimental process)
Compact dimensions: 6.3-mm (0.25-in.) diam, 80 mm (3 in.) in length. Source: Ref 14

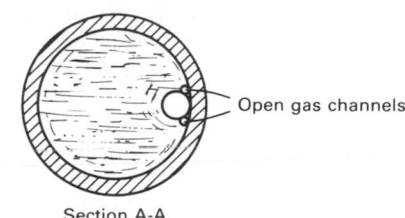

- Open gas channels

Section A-A

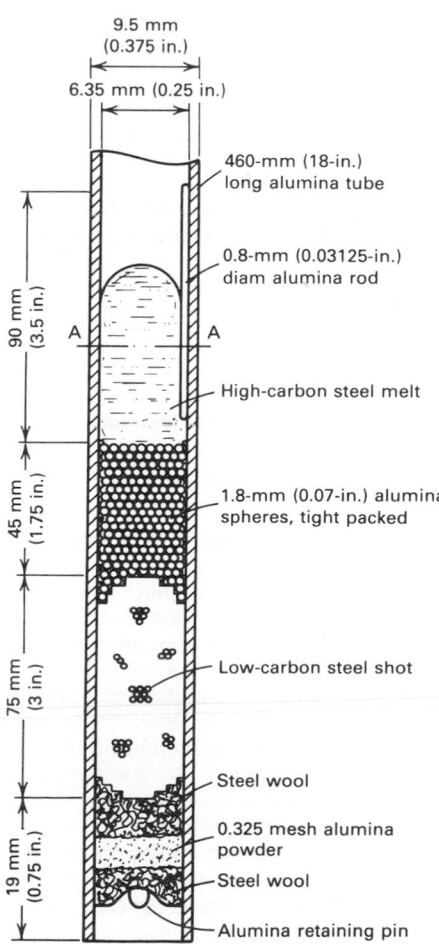

- 9.5 mm (0.375 in.)
- 6.35 mm (0.25 in.)
- 460-mm (18-in.) long alumina tube
- 0.8-mm (0.03125-in.) diam alumina rod
- 90 mm (3.5 in.)
- High-carbon steel melt
- 45 mm (1.75 in.)
- 1.8-mm (0.07-in.) alumina spheres, tight packed
- 75 mm (3 in.)
- Low-carbon steel shot
- Steel wool
- 0.325 mesh alumina powder
- 19 mm (0.75 in.)
- Steel wool
- Alumina retaining pin

External Pressure Impregnation. When capillary forces are ineffective because of poor wetting, unsuitable pore size and distribution, or high viscosity of the liquid, the solid body may be impregnated by the molten metal only with the help of a substantial external force. This force may be supplied by a pressurized gas or liquid, a dead load, or a piston inside a cylinder, and must exert pressure on the melt. The infiltrant mass may be limited to the exact amount required for filling the pores and providing a continuous film surrounding the end product, thereby eliminating the need for cropping excess material.

Molten infiltrant may make contact with the skeleton at any desired face, although the top or bottom usually is the most practical contact surface. The liquid metal must be contained in a suitable pressure vessel before and during impregnation. This method is applicable to systems in which the molten infiltrant tends to solidify before penetration is completed, due to diffusion-induced changes in composition.

An example of such diffusion-solidification can be found in iron/cast iron systems. High-speed steel has been produced by impregnating molten high-carbon iron into porous low-carbon steel compacts made from shotted granules or wire clippings with 0.5 MPa (75 psi) pressure (Ref 14). The closed mold arrangement used for this process is shown in Fig. 3.

Centrifugal pressure impregnation is an adaptation of external pressure-aided impregnation of liquid metal into pores of a solid. This process uses the centrifugal investment casting method, but requires special features such as more refractory, chemically inert, and tight-fitting ceramic molds and atmospheric control. The centrifugal force on the molten infiltrant may cause segregation due to exudation of a heavy liquid from the skeleton. This phenomenon may require special mold adapters or inserts.

Vacuum Infiltration. Two methods can be employed to create a vacuum in metal infiltration. One method creates a pressure gradient by applying suction to the liquid phase through the interconnected pore system of the skeleton, allowing atmospheric pressure on the molten infiltrant to be the driving force. A closed mold system that contains the molten infiltrant and the skeleton is required. The vacuum then is applied to the end of the skeleton, away from contact with the melt.

The second method simply provides for placement of the entire infiltration arrangement in a vacuum furnace. This method is useful for systems with strong outgassing components, but may cause difficulties in controlling the infiltration cycle of monitoring skeleton and infiltrant temperatures.

Infiltration Systems

Even if essential infiltration conditions are strictly met, the possible binary systems in which a high-melting single metal or compound may be infiltrated by a lower melting single metal are many. Table 1 lists systems that are partially or fully immiscible in the liquid state (Ref 8), but also includes workable systems that form terminal or even complete solid solutions (for example, iron-copper and nickel-copper). Systems for which infiltration had gained acceptance for past production tasks or is presently in use by industry are shown by closed circles; open circles indicate systems capable of infiltration in the laboratory or possessing infiltration potential on the basis of established criteria.

The material combinations listed in Table 1 are incomplete, because the possibilities for infiltration combinations are expanded if binary or more complex alloys are used for the skeleton and/or infiltrant. It is possible to produce nonequilibrium alloys of a final composition that cannot be made by melting and casting; if

Table 1 Binary metal infiltration systems

Skeleton	Aluminum	Antimony	Bismuth	Cadmium	Calcium	Cobalt	Copper	Gold	Iron	Lead	Magnesium	Manganese	Mercury	Nickel	Silver	Sodium	Thallium	Tin	Zinc
Aluminum			○	○						●						○	○		
Beryllium	○								○										
Chromium	○		○	○		○	●			○				○	○			○	○
Cobalt			○				○			○				○	○			○	○
Copper		○	●							●				○			○	●	○
Iridium								○							○				
Iron	○	○	○	○	○		●			●	○			○	○	○	○	○	○
Lead															○		○		
Magnesium															○				○
Manganese			○							○					○		○		
Molybdenum	○		○	○			●	○		○		○	○	○	●			○	○
Nickel			○				○			○	○		●		●		○	○	
Niobium				○						○					○				○
Platinum								○							○				
Rhodium								○							○				
Silicon		○	○	○						○							○	○	
Silver			●											○					
Tantalum				○				○		○				○	○				○
Titanium	○						○											○	
Titanium carbide						●	○		●					●					
Tungsten	○	○	○	○	○	○	●		○	●		○		○	●	●			○
Tungsten carbide						●	○	○							●	●			
Vanadium							○								○				
Zinc			○							○						○	○	○	
Zirconium							○			○					○				

(a) Open circle (○) indicates experimental and potential combinations; closed circle (●) indicates industrially significant combinations
Source: Ref 8

the respective phase diagram indicates some solubility relationships in the liquid or solid state, diffusion heat treatments just below or above the melting temperature of the infiltrant may result in truly equilibrated alloys, usually distinguished in their microstructure by rounded grains embedded in a matrix (Ref 8).

In principle, homogeneous alloys may be produced by infiltration if solubility of the skeleton in the infiltrant (or vice versa) is unlimited in the liquid or solid state. The nickel-copper system is an example; Monel compositions were successfully produced with 63 to 67% Ni by complete penetration of nickel skeletons with copper and binary alloys of copper with 10% Al or 10% Si.

Another example is the capillary infiltration of iron or low-carbon steel powder compacts of predetermined porosity with molten cast iron under conditions of close time, temperature, and carbon-diffusion control. When a homogeneous alloy is produced, process variables such as pore and capillary channel, size of the powder compact, time for melting the infiltrant and its complete penetration of the skeleton, and type of diffusion inhibitor must be carefully controlled to ensure homogeneity.

Once the penetration stage has been successfully completed, sintering with the liquid phase in place can proceed at the infiltration temperature, and full homogenization of the alloy can be accomplished by subsequent heat treatment in the solid state. The principal advantage of this procedure lies in the potential for obtaining alloys of nearly full density without mechanical working. Unfortunately, process control problems have made this approach impractical.

Refractory metal-based composite structures of tungsten-copper, tungsten-silver, molybdenum-copper, and molybdenum-silver systems are among the oldest P/M products produced by infiltration. The solubility of refractory metals in molten silver is minimal, and solubility in molten copper is practically nil. Along with excellent wetting characteristics, these properties constitute ideal conditions for infiltration.

The ratio of the two phases that can be combined by this method can vary considerably (from about 35 to 40 vol% up to 85 to 90 vol% of the refractory metal). The upper limit is set by the requirement of an interconnected pore structure in the compacted and sintered refractory metal, while the lower limit can be controlled by selection and treatment of the powder, which is piled loosely in a suitable container (graphite, for example).

Generally, refractory metal powder is of the reduced oxide type with particle sizes ranging from about 1 to 50 μm. However, coarser powders ranging from 50 to about 400 μm with more sharp-edged crystalline particle shape, such as that obtained by mechanical comminution of sintered or swaged scrap metal, have produced sintered skeleton structures with improved erosion and wear resistance (Ref 15). The desired porosity in the loose pile is obtained by tapping, vibration, jolting, or ramming, whereas that of compacted powders is controlled by the static pressure of the press tools.

The refractory metal skeleton usually is sintered prior to infiltration to improve strength and cohesion. Sintering temperature must be selected in conformity with powder particle size and shape, compact-

ing pressure, and specific final composition; it is usually about 100 °C (180 °F) above the melting point of the infiltrant metal. Higher temperatures tend to produce severe grain growth and shrinkage and may even destroy the interconnected pore system needed for successful infiltration, especially in very fine starting powders. Figure 4 schematically presents the growth of tungsten grains with rising sintering temperature and the change in the distribution of a subsequently infiltrated copper phase, especially above 1600 °C (2900 °F).

Infiltrated bodies that have a continuous skeleton phase above approximately 65 vol% can be shaped only by machining, while those having a smaller proportion of refractory metal dispersed as loose grains in the ductile metal matrix are plastically

Fig. 4 Schematic drawing of microstructure of tungsten sintered at different temperatures before being immersed into molten copper
(a) 1100 °C (2010 °F). (b) 1350 °C (2460 °F). (c) 1600 °C (2910 °F). (d) 2800 °C (5070 °F). Magnification: 500×. Source: Ref 16

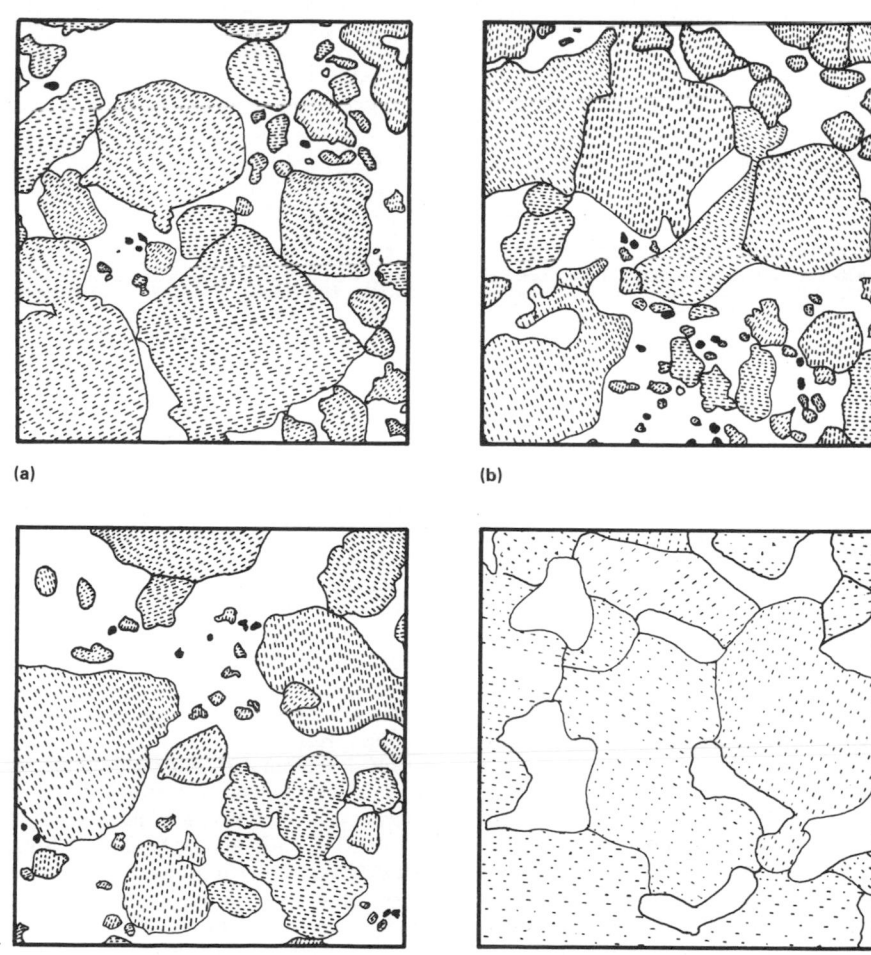

(a)

(b)

(c)

(d)

Table 2 Carbide infiltration test matrix and evaluation

Specimen No.	Skeleton(a)			Infiltrant composition, %	Infiltration(b)				Time, min
	Composition	Density, g/cm³	Pore volume, %		Type	Temperature			
						°C	°F		
1a	WC (6.1%C)	11.05-13.35	29.3-14.6	100Co	Contact, opposing sides(d)	1500	2730		15
1b				95Co-5WC		1460	2660		5
1c				75Co-25WC		1390	2530		5
1d				60Co-40WC		1350	2460		5
2a	80WC-20TiC(e)	6.65-8.34	37.1-21.3	100Co	Contact, opposing sides(d)	1500	2730		15
2b				95Co-5WC		1460	2660		15
3a	TiC(18.8%C)	3.01-3.63	33.2-19.3	80Co-20Cr	Contact, opposing sides(d)	1500	2730		15
3b				66Co-28Cr-6Mo		1450	2640		15
3c				72.7Co-17.3Cr-10TiC		1400	2550		15
3d				80Ni-20Cr		1450	2640		15
4a	97TiC-3Mo₂C(e)	3.38-4.03	25.7-11.4	80Co-20Cr	Contact, one side(d)	1500	2730		5
4b				66Co-28Cr-6Mo		1450	2640		5
4c				72.7Co-17.3Cr-10TiC		1400	2550		5
4d				80Ni-20Cr	Capillary dip in molten infiltrant	1550	2820		3
5a	95TiC-5Mo₂C(e)	3.46-4.05	24.8-11.6	80Co-20Cr	Contact, one side(d)	1500	2730		5
5b				66Co-28Cr-6Mo		1450	2640		5
5c				72.7Co-17.3Cr-10TiC		1400	2550		5
5d				72.7Ni-17.3Cr-10TiC	Capillary dip in molten infiltrant	1550	2820		3
6	90TiC-10Mo₂C(e)	3.54-4.14	26.0-14.6	80Ni-20Cr	Contact, opposing sides(d)	1400	2550		15
7	70TiC-30Mo₂C(e)	4.09-4.75	22.9-9.9	80Ni-20Cr	Contact, opposing sides(d)	1400	2550		15
8	50TiC-50Mo₂C(e)	4.69-5.58	21.3-8.3	80Ni-20Cr	Contact, opposing sides(d)	1400	2550		15

(a) All skeletons were presintered at 950 °C (1740 °F) and high sintered at 1500 °C (2730 °F) for 2 h in a carbon tube resistor furnace under hydrogen, except No. 1 to 3, which were high sintered in vacuo in a carbon susceptor induction furnace. (b) In vacuum induction furnace. (c) A qualitative assessment of resistance against fragmentation by hammer blows. (d) Infiltrant mass was 40 to 45% of mass of infiltrated product. (e) Solid solution
Source: Ref 19

deformable at elevated temperatures (Ref 8). Soldering, brazing, or plating of the infiltrated product is aided by generally smooth surface films or high contents of infiltrant metal.

The preceding binary systems illustrate combinations of two metals of widely differing melting temperatures that can be advantageously produced to near-net shape and full density by infiltration. Nickel also may be infiltrated into tungsten, but equilibrium at liquid-phase temperature causes severe attack of the refractory metal and requires careful process control to prevent incomplete penetration due to diffusion-solidification. Coarse tungsten powder helps to produce compacts with larger capillary channels for better penetration. High heating rates, especially after about 90% of the absolute melting temperature of nickel is reached, improve infiltration conditions.

If nickel is alloyed with copper, the solubility of tungsten in the liquid phase decreases as the copper percentage is increased. Infiltration is more practical (Ref 8) for heavy alloy systems, because tungsten powder and process control requirements are less stringent. If nickel is alloyed with iron, however, the solid-liquid phase interaction is similar to the binary tungsten-nickel system, and infiltration again is more difficult.

The same principle applies to the more complex refractory metal systems with nickel-chromium and cobalt-chromium alloys. Nevertheless, skeleton bodies of tungsten and molybdenum, as well as of binary 85W-15Cr and 75W-25Cr alloys, can be successfully infiltrated with superalloys of the Nichrome-V, Hastelloy-C, Stellite, and Vitallium compositions into shapes simulating mechanical and engine test specimens (Ref 17).

Several other refractory metal-based composite structures can be readily produced by infiltration. These include the high-density tungsten-lead system to produce materials suitable for shielding against radiation and the chromium-copper system to produce compositions for welding electrodes (Ref 8). To retain the low liquid-solid contact angles in these systems, a strong reducing atmosphere is necessary to prevent oxide films on the molten lead or on the solid chromium. A free metallic surface requires sintering of the skeleton above 1250 °C (2280 °F) to reduce any oxide film on the solid chromium.

Carbide-Based Systems. Liquid-phase sintering of tungsten carbide/cobalt or titanium carbide/nickel systems capitalizes on the eutectics of the two phases. A limited solubility of the carbide in the matrix metal facilitates bonding; carbon and metal

Table 2 (continued)

Analysis, %	Hardness, HRA	Infiltrated product — Surface condition	Microstructure	Bench test(c)	Specimen No.
	86-87	Contact face erosion, slight residue	Porous in core, graphite precipitates, fairly uniform grain size	Very tough	1a
					1b
	85-86	No erosion, heavy residue		Tough	1c
					1d
	86.5-87	Contact face erosion	Uniform phase distribution, some porosity	Tough	2a
					2b
	87.5-88	Slight erosion		Very tough	3a
	89.5-90	Slight porosity		Fairly tough	3b
	88	No erosion, smooth		Very tough	3c
24.6Ni, 6.1Cr, rem TiC	83.5-85	Slight erosion	Dense, uniform	Very tough	3d
	88+	Heavy contact face erosion and residue	Higher matrix concentration near contact face, porosity increasing toward far end	Very tough	4a
	89.5			Tough where dense, brittle where porous	4b
	88+	Less contact face residue than in 4a			4c
22.5Ni, 5.7Cr, 2.1Mo$_2$C, rem TiC	84.5-85	Alloy skin becoming heavier toward bottom, forming excess on bottom end	Uniform phase distribution, generally dense	Tough	4d
	88+	Contact face erosion, some residue	Similar to 4a-c	Tough	5a
	90				5b
	88.5	Slight contact face residue			5c
	85	Similar to 4d	Similar to 4d	Tough	5d
22.9Ni, 5.5Cr, 7.1Mo$_2$C, rem TiC	85-86	Slight contact face erosion, small residue, slightly porous	Porous in core, less uniform phase distribution than in 4d	Less tough than 4 and 5	6
22.6Ni, 5.6Cr, 21.4Mo$_2$C, rem TiC	86-87	Similar to 6, but more porous	More porous in core, less uniform phase distribution than in 6	More brittle than 6	7
22.3Ni, 5.7Cr, 35.8Mo$_2$C, rem TiC	86-87	Similar to 7, but more porous	Very porous, nonuniform phase distribution	More brittle than 7	8

diffusion through the liquid are less important in densification than reactions at the carbide-metal phase boundaries. It is unknown whether the carbide particles in these systems form a rigid skeleton, but the interfacial tension between crystals of the carbide and the liquid metal appears to be anisotropic (Ref 18).

During cooling from the sintering temperature, some or most of the carbon and metal dissolved in the liquid precipitates on grains that had remained solid during the process. This mechanism can be altered somewhat if the rigid skeleton is formed first and the liquid phase subsequently infiltrates into the pore system. By first saturating the matrix metal with carbon and skeleton metal, the liquid phase dissolves less skeleton materials, and shape distortion is diminished. Carbide coalescence and grain growth also are decreased.

The earliest attempts to produce cemented carbides were made by infiltrating carbide skeletons with unalloyed binder metals (Ref 19). Later, binder metal prealloyed with elements of the skeleton to inhibit contact face erosion was used in a broad investigation of infiltrating single and double carbides with many cobalt and nickel alloys (Ref 20). Some of the alloys used are listed in Table 2. The large number of feasible combinations includes several noteworthy successes, especially for titanium carbide-based systems with Nichrome and Vitallium infiltrants. A feasibility study (Ref 17) produced similar results for the same infiltration systems and laid the foundation for an extensive development program to utilize these materials for heat-resistant applications.

The structure of infiltrated carbides reflects infiltration mechanics on a macroscale. In zones penetrated by the infiltrant, fully dense regions and slight expansion of the skeleton due to carbide grain separation are observed. Substantial porosity and some shrinkage occur in areas inadequately penetrated by the liquid alloy, such as the side opposite the contact face in unidirectional infiltration, or in the center for infiltration from opposite sides. Subsequent heat treatment is ineffective in eliminating porosity. Erosion at the contact faces is greatly diminished by infiltrating skeletons composed of tungsten-based multicarbides, so that saturating the infiltrant with skeleton elements by prealloys frequently is not required.

The microstructure reflects the crystallographic characteristics of the carbide, and fully infiltrated regions do not differ in grain size and morphology from material whose liquid phase was sintered in situ. Rectangular and triangular grains are retained in infiltrated tungsten carbide, whereas for infiltrated titanium carbide, the cubic lattice is reflected by distinctly rounded grains. Grains of solid-solution carbides of tungsten and titanium or titanium and molybdenum are slightly rounded at the corners. Graphite precipitates accompany porosity in poorly infiltrated regions for tungsten carbide and titanium carbide skeletons, especially if starting powders contain more than a trace of free carbon.

Fig. 5 Infiltration-brazed butted iron-copper bars

Green bars were clamped together end-on-end and one free end surface contacted with molten infiltrant. Rupture of composite bar occurred away from the joint, evidence of the high strength of the brazed bond. (a) Butt joined bars after infiltration. (b) Machined tensile bar. (c) Tested tensile bar

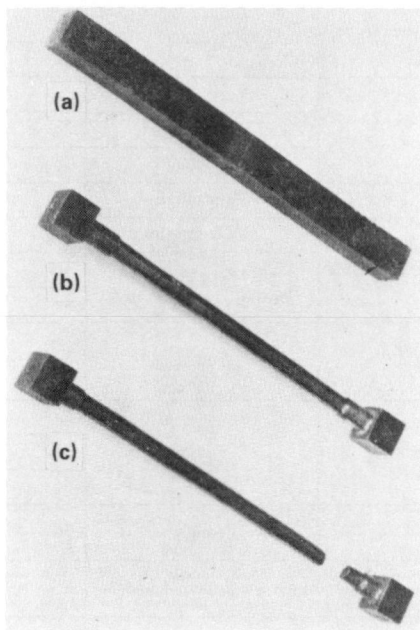

Ferrous-Based Systems. The thermodynamic affinity between solid iron and liquid copper offers the potential for virtually pore-free P/M products by infiltration. Moreover, the excellent wetting characteristics that exist in the brazing of steel also can be utilized for joining disparate bodies during infiltration. A powder compact and a casting or forging, or halves of complex or offset configurations, can be joined. This self-brazing capacity without strength degradation is illustrated in Fig. 5 (Ref 10). Finally, the generally smooth cupric film surrounding the infiltrated body serves as a base for surface coating or plating.

The ability of the infiltration process to combine major proportions of normally unalloyable industrial premier metals (iron and copper) was recognized as early as World War I (Ref 21, 22), but it was only in the late 1940's, through refinements in technique, that sound products could be made (Ref 10 and 23-25). These products, in turn, have culminated in the present advanced state of the art. While copper content must be higher than for most commercially sintered iron-copper alloys because of the need to maintain an inter-

connected pore system for complete infiltration, good mechanical properties can be realized.

This is apparent from the tensile strength-elongation data given in Fig. 6 for commercial iron powder with and without graphite additions (Ref 24). Strength is enhanced because the infiltrated structure, with a minimal amount of isolated pores, is virtually free of internal notches. The iron-copper system permits a precipitation-strengthening mechanism. If carbon is diffused into the iron to produce a hypoeutectoid structure of the skeleton, hardening by martensite transformation is possible. Where other metals are alloyed with the copper, solid solution strengthening of the matrix can be achieved.

In the binary iron-copper system, about 3.8 to 4% Fe is dissolved by the liquid copper under equilibrium conditions at an infiltration temperature of 25 to 50 °C (45 to 90 °F) above the peritectic temperature of 1090 °C (2000 °F) at the copper side of the phase diagram, while the γ-iron dissolves about 8 to 8.5% of the copper. At 900 °C (1650 °F), the solubility of α-iron in copper is about 1.5%; it diminishes to less than 0.04% at room temperature, at which point the solubility of copper in iron is equally low.

These thermodynamic relations form the basis for precipitation hardening. However, macrodispersion of the two phases in the infiltrated alloys causes concentration of precipitates in thin zones at the phase boundaries, as shown in Fig. 7 (Ref 26). Consequently, conventional hardness tests

are not precise enough to show a noticeable increase in macrohardness after a precipitation treatment, such as quenching from 900 °C (1650 °F) followed by prolonged tempering at 600 °C (1110 °F). The precipitation mechanism produces increases in strength, elongation, and impact resistance (Ref 8) and also can be traced through changes in the electrical conductivity (Ref 25).

If carbon is diffused into the iron skeleton and if copper is infiltrated afterward, two processes compete with one another during cooling. During quenching, normal martensitic transformation occurs inside the skeleton structure. At the same time, however, precipitation of the dissolved iron in the copper and copper in iron, respectively, is suppressed in the phase boundary zones. During reheating, martensite decomposition causes a decrease in hardness of the steel skeleton structure, with simultaneous increases in hardness and strength of the boundary zones.

Table 3 shows the effect of such a heat treatment on the mechanical properties of a copper-infiltrated 0.3% carbon steel. Water quenching produces an appreciable increase in hardness and brittleness. With increasing reheating temperatures, however, the material becomes softer and tougher, without loss in strength (Ref 8).

When a copper alloy is used as an infiltrant, other benefits may accrue. When precipitates form in the matrix, such as copper alloys with beryllium, chromium, or silicon, the resulting strength increase during heat treating of the infiltrated body

Fig. 6 Mechanical property ranges of copper-infiltrated iron and hypoeutectoid steel compacts before and after heat treatment

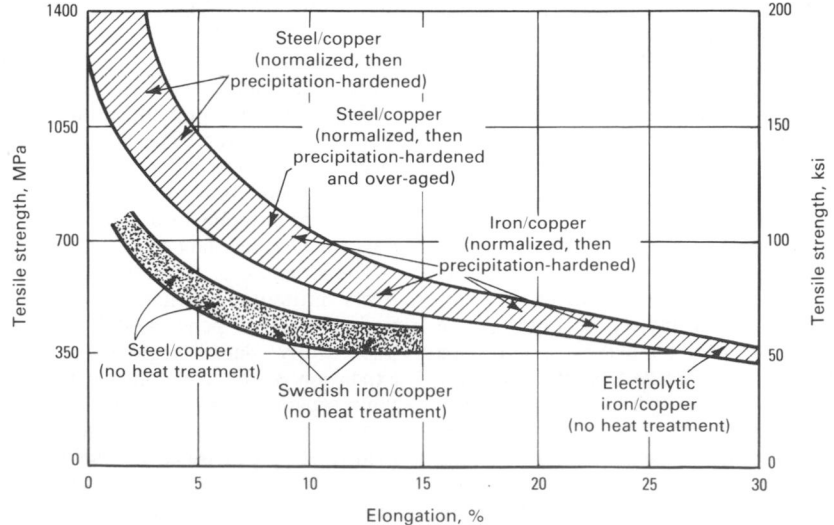

Fig. 7 Microstructure of 25Cu-Fe compact

Sintered at 1100 °C (2010 °F) for 30 min. Large, rounded, dark areas are α-iron, separated by a diffusion layer from the light copper phase containing the fine, dark, iron-rich precipitate. Magnification: 1000×. Source: Ref 26

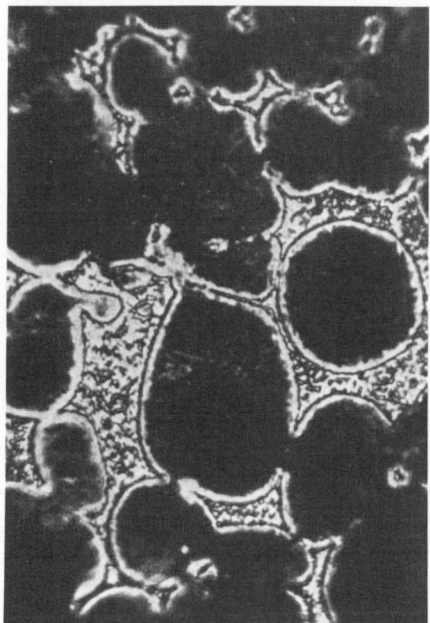

Table 3 Effect of heat treatment on the mechanical properties of 0.3% carbon steel

Infiltrated with 11 vol% Cu

Treatment(a)	Hardness, HV	Ultimate tensile strength		Elongation, %	Impact resistance(b)	
		MPa	ksi		J	ft·lb
Copper-infiltrated and furnace cooled	262	704	102.1	7.3	22.16	16.35
Reheated to 900 °C (1650 °F) and water quenched	437	790	114.6	5.2	11.76	8.68
Reheated for 2 h at 400 °C (750 °F)	360	...	...	...	14.71	10.85
Reheated for 2 h at 500 °C (930 °F)	302	771	111.8	7.6	23.14	17.07
Reheated for 2 h at 600 °C (1110 °F)	255	750	108.8	13.5	62.75	46.30

(a) Hametag iron powder was pressed to 88% of theoretical density, sintered at 1220 °C (2230 °F) for 1 h in hydrogen, and infiltrated with electrolytic copper at 1100 °C (2010 °F) for 30 min. (b) Unnotched test bar of 1-cm² cross section
Source: Ref 8

reinforces iron-copper precipitation zones at the boundaries. The strengthening effect of precipitates is independent of heat treatment and augments strength increases obtained with unalloyed copper infiltrant in solid-solution alloys. However, some solid-solution alloy infiltrants minimize erosion at the point of initial contact with the iron skeleton, because these alloys melt slowly over a range of temperatures. A brass containing 20% Zn exhibits this phenomenon.

A copper alloy with manganese in amounts up to 5%, especially if it also contains sufficient iron to inhibit a severe attack of the skeleton contact faces, has a different beneficial effect. The manganese

oxidizes preferentially in the commercial atmospheres in which infiltration takes place. A nonadhering porous crust results that can be removed during finishing more easily than the tenacious residue formed by a binary copper-iron infiltrant (Ref 27, 28).

Many ferrous metal-based infiltration systems have been explored experimentally. Kieffer and Benesovsky (Ref 8) have investigated the iron-gold, iron-bismuth, iron-cadmium, iron-lead, iron-antimony, and iron-tin systems for bearings, and the iron-cobalt-silicon, iron-copper-silicon, and iron-manganese-silicon systems for magnetic or structural parts. Alloys of the iron-zinc system have also been produced by infiltration, but treatment in a pressure vessel is required to overcome the high vapor pressure of the zinc (Ref 29).

Austenitic stainless steel skeletons infiltrated with silver possess excellent corrosion resistance, thus making them suitable for food processing applications. Ferritic stainless steel and high-manganese steel compacts with varying carbon contents also display improved corrosion resistance when infiltrated with cupric alloys. These alloys also exhibit extraordinary hardness and wear resistance, coupled with a considerable toughness. Mechanical and technological properties of several ferrous-based infiltrated materials are listed in Table 4.

Nonferrous-Based Systems. The major nonferrous metals with higher melting points are thermodynamically compatible with many low-melting metals in the liquid state. Consequently, skeleton bodies of cobalt and nickel can be readily infiltrated with gold, as well as with many of the low-melting heavy metals, such as bismuth, lead, or antimony (Ref 8). Copper can also be infiltrated into cobalt and nickel skeletons; however, because of the formation of solid solution in all proportions, infiltration into nickel powder compacts requires a narrow particle size range, wide capillaries within a pore volume not exceeding about 35%, short infiltration time, and a vacuum to assist the capillary forces. Mercury wets nickel well without forming an amalgam and is easily impregnated into skeleton bodies, provided the pore structure prevents exudation of the heavy liquid metal (Ref 30).

Copper is another skeleton metal whose pores can be readily filled with liquid low-melting-point metals, such as lead (Ref 31, 32) or bismuth (Ref 33). Vacuum impregnation is suitable for incorporating lead-based alloys, such as those containing 15% Sb and 5 to 10% Sn, into spongy structures of nickel-copper or nickel-iron supported by steel backing (Ref 34).

Combinations of aluminum or aluminum alloys with such low-melting metals as bismuth, lead, thallium, or thallium-lead

Table 4 Properties of some ferrous metal-based infiltrated alloys

Skeleton composition, %	Infiltrant composition, %	Infiltrant, vol%	Density, g/cm³		Hardness, HV	Ultimate tensile strength		Elongation, %	Impact resistance(a)		Comment
			Calculated	Determined		MPa	ksi		J	ft·lb	
100Fe	100Pb	10	8.15	7.95	93.5	251	36.4	14	...	...	Free machining, extrudable
99Fe-1Cu	100Ag	11	8.09	8.00	178	378.5	54.9	11	...	...	...
100Fe	80Cu-20Ni	17	8.04	7.76	213	419.5	60.9	8	21.57	15.91	...
100Fe	65Cu-35Mn	13	7.92	7.63	256	446	64.7	10	31.37	23.14	...
93.2Fe-6Mn-0.8C	100Cu	13	7.90	7.87	740	...	...	...	...	...	Naturally hard, wear resistant
87.2Fe-12Mn-0.8C	100Cu	9	7.89	7.69	310	562	81.5	6	...	...	Wear resistant, work hardening
93.5Fe-3Cr-3Mn-0.5C	100Cu	14	7.96	7.93	502	957	138.8	4	...	...	

(a) Unnotched test bar of 1-cm² cross section
Source: Ref 8

Fig. 8 Effect of composition on physical properties of tungsten-copper contact material

(a) Density. (b) Hardness. (c) Electrical conductivity. Heavy framed areas are composition ranges for liquid copper infiltration or sintering of compacts from powder mixtures. Source: Ref 37

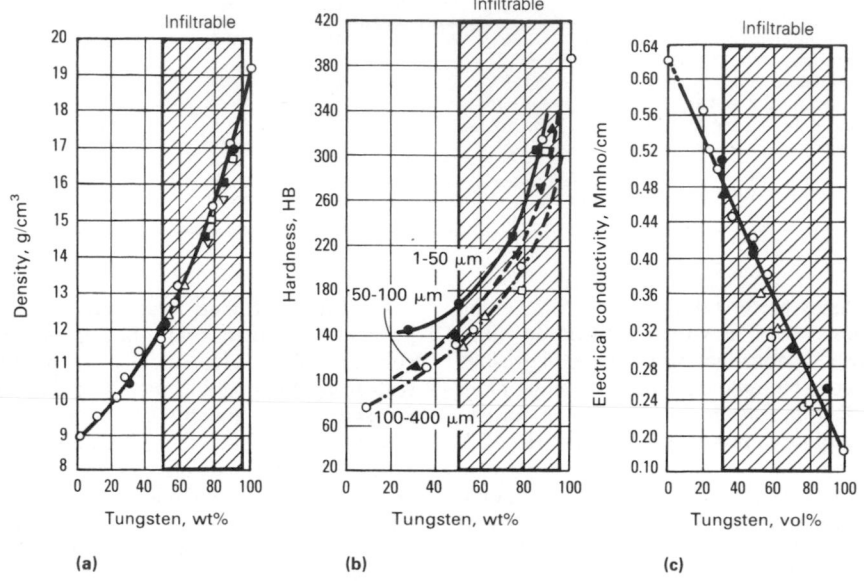

(a)   (b)   (c)

Production method	Particle size of tungsten powder, μm	Symbol
Mixing of the component elemental powders, briquetting, sintering, subsequent working	Coarse, 100-400	○
	Medium, 50-100	▲
	Very fine, 1-50	●
Impregnation of the loosely filled tungsten powder in a mold, extrusion into shape	Coarse, 100-400	△
Dip-impregnation of a pressed (and presintered) tungsten compact, machining to size	Coarse, 100-400	□
	Medium, 50-100	▼
	Very fine, 1-50	■

by means of infiltration in hydrogen or a vacuum have been proposed (Ref 33, 35). However, strict control of powder characteristics, especially particle shape and surface condition, to maximize wetting appears to put infiltration at a disadvantage over *in situ* liquid-phase sintering. To overcome this problem, zinc or cadmium may be added to the aluminum of the skeleton, followed by cleaning and activating the free surface of the pores by evaporation of the lower boiling metal before proceeding with the impregnation of a metal such as lead (Ref 36).

Infiltration Products

Electrical Contacts and Electrodes. Composite materials for heavy-duty service contacts consist of (1) a refractory metal component to supply the necessary hardness, wear resistance, burning and material transfer through arcing, and high-temperature strength, and (2) a matrix of

an electrically and thermally conducting metal. Tungsten and molybdenum are the principal refractory metals, but tungsten carbide, tungsten-niobium double carbide, and titanium carbide also have been used for this purpose (Ref 8). Conductor metals are copper and silver.

Because the two metal types do not alloy with one another, the rule of mixture can be applied to determine the density of a specific composition. Figure 8 shows the effect of increasing tungsten content on density, hardness, and electrical conductivity for copper-tungsten contact material (Ref 37). The straight-line relationship between volume ratio and conductivity is typical for this system. The change in thermal expansion coefficient with tungsten content is shown in Fig. 9; principal mechanical properties versus tungsten content are plotted in Fig. 10 (Ref 38). The boldface portions of the curves represent composition ranges of material that can be readily produced by infiltration with a low-

melting metal content of approximately 10 to 40 vol% (5 to 25 wt%).

Table 5 lists mechanical and electrical property data for tungsten-silver and molybdenum-silver contact materials (Ref 39). Also included are data for nickel-silver contact materials that can be produced by infiltration.

In infiltrated materials, the soft conductor metal matrix is the minor constituent, and extrusion for profile generation is not practical except for nickel-silver composites. All the materials are readily machinable into a variety of shapes, especially caps, platelets, and rings. Usually, the heavy-duty contact metal, as a facing, is joined by clamping or brazing onto a support structure. The facing also may be penetrated by an infiltrant whose major portion, on cooling, solidifies like a casting; an example is copper that is strengthened by small amounts of beryllium, chromium, or nickel and is cast into the form of a backing, arm, or other support structure.

Primary applications for infiltrated tungsten-copper composites are resistance welding electrodes and make-and-break contact facings in oil or air circuit breakers or transformer taps. Tungsten-silver contacts are used in switch gear and low-voltage regulators (Ref 37). Low contact resistance and nonsticking properties of nickel-silver contacts make them suitable for high-voltage disconnect switches (Ref 39). Porous, cup-shaped nickel compacts have been impregnated with mercury as part of a short circuiting safety switch contact in proximity fuses for radio-controlled detonation (Ref 40, 41).

Fig. 9 Effect of composition on thermal expansion of tungsten-copper contact material

Boldface portion of curves designates liquid copper-infiltrated tungsten or sintered compacts from powder mixtures. Light portion of curves designates powder mixtures; compositions are not infiltrable. Source: Ref 38

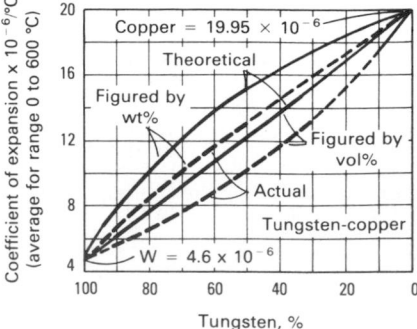

Fig. 10. Effect of composition on mechanical properties of tungsten-copper contact material

Boldface portion of curves designates liquid copper-infiltrated tungsten or sintered compacts from powder mixtures. Light portion of curves designates powder mixtures; compositions are not infiltrable. Rockwell B hardness. Source: Ref 39

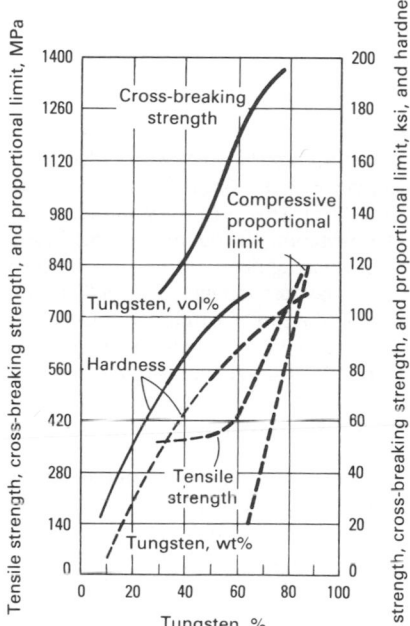

Fig. 11. Silver-infiltrated tungsten billet for rocket nozzle throat liner
Source: Ref 42

Fig. 12 Microstructure of silver-infiltrated 80W-20Ag billet

Angular microconstituents are silver. Magnification: 500X. Source: Ref 42

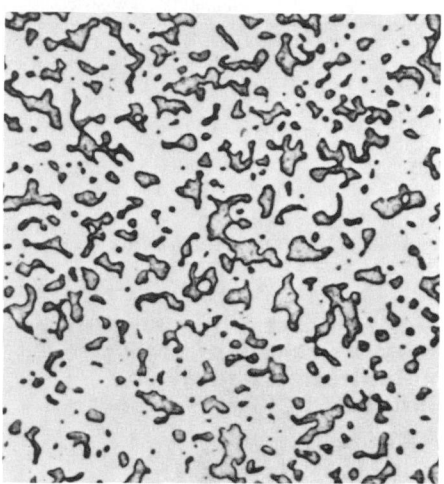

Rocket Nozzles. The manufacture of silver-infiltrated tungsten contacts formed the basis for the development and production of large billets, from which rocket nozzle throat liners have been fabricated. This material has proved successful in actual service in rockets of underwater-launched ballistic missiles (Ref 42) because of:

- Extreme stability of the structure (virtually zero change in critical dimensions due to the high-temperature, high-pressure gaseous environment in the nozzle throat)
- High resistance to thermal shock during heating and cooling at the extremely high rates experienced by the throat surface in contact with rocket propulsion gas. This in turn requires high tensile strength and resistance to high hoop stresses at the outer perimeter of the throat liner, as well as good thermal conductivity to provide adequate heat transfer to the backup structure.
- Fabricability, especially machinability, on an economical production scale

- Optimum reliability and reproducibility of the nozzle throats in spite of their relatively large size, compared to similarly produced P/M parts in electrical switch gear

An additional advantage of using silver as the matrix metal is its relatively high vapor pressure. Silver evaporation aids infiltration of large capillaries by forming deposits on the pore walls before the liquid metal enters individual pores. Also, the exposed surface region of the throat liner is kept at a temperature considerably below that of the propulsion gas due to evaporative transpiration cooling (Ref 43).

Figure 11 shows a typical silver-infiltrated tungsten ring machined from a large billet that weighed nearly 90 kg (200 lb). It contained about 80 vol% W; the conductor metal was uniformly dispersed throughout the entire cross section, filling

Table 5 Properties of contact materials containing silver

Ag	W	Mo	Ni	Density, g/cm³	Brinell Hardness	Transverse rupture strength MPa	ksi	Electrical conductivity Mmho/cm	%IACS	Estimated contact resistance(a)
10	90	...	...	17.5	230-250	1240-1400	180-200	0.20-0.23	35-40	1000
15	85	...	...	17.0	210-230	1100-1240	160-180	0.24-0.25	42-44	700
20	80	...	...	16.3	200-220	965-1100	140-160	0.26-0.28	45-49	500
25	75	...	...	15.8	160-180	860-965	125-140	0.28-0.30	49-53	350
30	70	...	...	15.2	140-160	790-825	115-120	0.30-0.33	53-57	250
20	...	80	...	10.25	190-210	1100-1240	160-180	0.28-0.30	49-53	500
25	...	75	...	10.27	180-190	965-1100	140-160	0.30-0.32	53-56	400
30	...	70	...	10.28	160-170	860-965	125-140	0.32-0.34	56-60	325
35	...	65	...	10.29	140-150	525-860	110-125	0.34-0.36	60-63	275
40	...	60	...	10.30	120-130	690-525	100-110	0.36-0.38	63-67	250
40	...	...	60	9.30	(b)	450	65	0.245	43	...
60	...	...	40	9.60	(c)	350	50	0.33	57	...

(a) As compared to bulk silver = 100. (b) 30T68 Hardness, Rockwell Superficial (c) 30T46 Hardness, Rockwell Superficial
Source: Ref 42

Table 6 Physical and mechanical properties of silver-infiltrated tungsten

Determined at strain rate of 50 μm/mm/min (0.05 in./in./min) for 20 vol% Ag/80 vol% W

Test temperature °C	°F	Thermal expansion, μm/m	Thermal conductivity W/m·K	Btu·in/ ft²·h·°F	Modulus of elasticity GPa	ksi	0.2% offset yield strength MPa	ksi	Ultimate tensile strength MPa	ksi	Elongation in 31.75 mm (1.25 in.), %	Reduction in area, %	Ultimate shear strength MPa	ksi
25	75	...	...	...	262	38 100	...	...	510	74.0	0	0	294	42.7
260	500	1.1	206	1430	250	36 400	...	...	463	67.2	0	0	247	35.8
540	1000	2.7	138	960	193	28 000	308	44.7	331	48.0	9.3	11.2	140	20.3
815	1500	3.8	108	750	206	30 000	...	...	315	45.7	2.9	3.1	155	22.5
1095	2000	4.7	92	640	197	28 500	232	33.6	243	35.2	4.0	8.0	123	17.9
1370	2500	6.0	80	560	97.2	14 100	105	15.2	135	19.6	7.0	17.5	80.6	11.7
1650	3000	7.6	72	500	74.5	10 800	71.7	10.4	80.6	11.7	7.3	18.5	46.9	6.8
1925	3500	9.1	62	430	57.2	8 300	53.8	7.8	57.9	8.4	4.7	14.7	28.3	4.1
2215	4000	10.3	59	410	38.6	5 600	40.0	5.8	42.7	6.2	5.9	19.7	15.9	2.3

Source: Kattus, J.R., Southern Research Institute, Birmingham, AL, Report 5525-1428 II, 19 Sept 1962; 6523-1526-VI, 9 Dec 1963; and 6606-1498-X, 31 Jan 1964

Fig. 13 Effect of temperature on tensile strength of porous uninfiltrated and silver-infiltrated (20 vol%) tungsten

Source: Ref 44

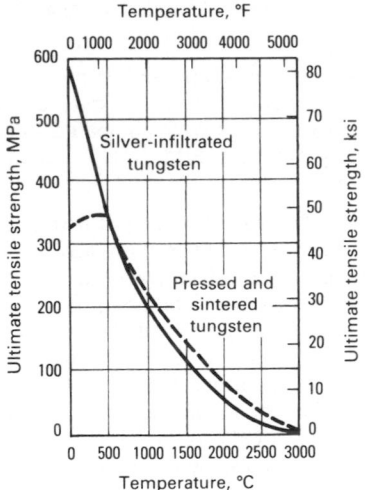

Fig. 14 Graded cermet turbine blade

(a) After superalloy infiltration of titanium carbide skeleton. (b) After machining of root

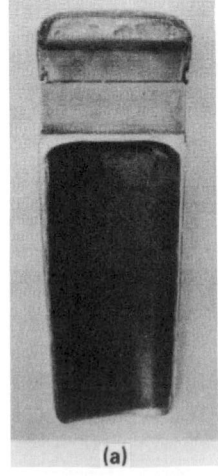

(a) (b)

all interconnected pores, as shown in Fig. 12. Processing parameters, production details, and starting material characteristics are described in Ref 44 and 45.

Toensing and Zalsman also have reported on the effect of these parameters on mechanical properties at ambient and elevated temperatures. The change in tensile strength with temperature for a tungsten skeleton with 20% pore volume before and after silver infiltration is shown in Fig. 13 (Ref 44). Data represent average values, but strengths as high as 620 to 700 MPa (90 to 100 ksi) have been reported. As the melting temperature of the silver is approached, the strength of the composite converges with that of the tungsten skeleton. Table 6 summarizes mechanical and physical properties of 20 vol% silver-infiltrated material for different test temperatures within the operating temperature range experienced by the throat liner during rocket propulsion (Ref 42).

Jet Engine Components. The quest for materials to withstand the high temperatures and stresses imposed by combustion gases in the gas turbines of jet engines led to the development of cermets in the late 1940's and early 1950's. Cermets offered increased operating temperatures and engine efficiency when used in rotating blades. Their inherent brittleness prevented the use of cermets for this purpose, however, and their use for stationary gas-

conducting nozzle vanes also failed to win acceptance. For more information, see the article "Cermets" in this Volume.

To overcome brittleness, the infiltration process was applied to the production of gas turbine components. Sintered and preformed titanium carbide skeletons were infiltrated with a nickel- or cobalt-based superalloy in a vacuum. Graded products that had the following structural characteristics were produced:

● Ductile and tough roots and airfoil tips containing nearly 100% superalloy that withstood combined tensile and bend stresses at moderately high temperatures
● Ductile and tough superalloy-rich leading and trailing edges that resisted impact from small solid particles at temperatures ranging from ambient to operational
● Strong, creep-resistant airfoil portions containing 60 to 80 vol% titanium carbide that withstood the centrifugal force-induced tensile stresses at the highest temperature zones near the radial centroid of the foil, about midway between the root and tip
● Oxidation-protective, ductile, superalloy-rich airfoil encasements integrally joined with the metallic matrix

A graded turbine bucket for a J-47 jet engine after infiltration is shown in Fig. 14(a)

Table 7 Stress-rupture properties of infiltrated graded cermet bucket

Bucket section	Material	Average service temperature °C	°F	Average service stress MPa	ksi	Stress-rupture life (100 h) Temperature °C	°F	Stress MPa	ksi
Root	Superalloy-rich region	650-760	1200-1400	140-170	20-25	705	1300	455	66
Airfoil tip	Superalloy-rich region	870-930	1600-1700	55	8	870	1600	138	20
Airfoil body	Superalloy-infiltrated TiC	1000	1800	70-100	10-15	980	1800	83	12

Source: Ref 44

Fig. 15 Microstructure of graded turbine blade across airfoil

Angular and rounded microconstituents are titanium carbide. Magnification: 150×. Source: Ref 43

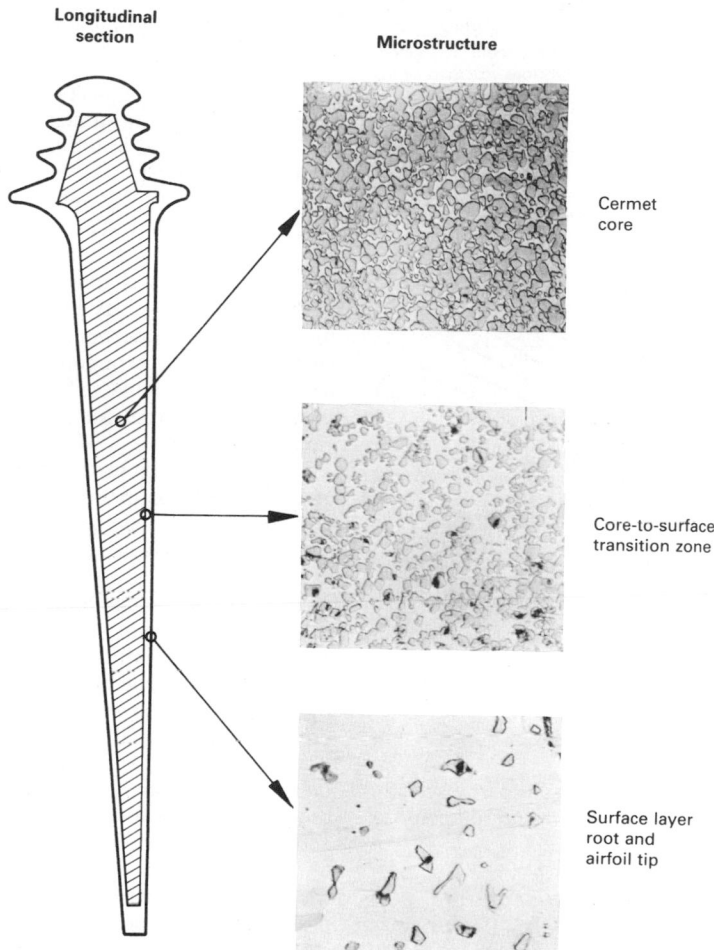

Longitudinal
section

Microstructure

Cermet
core

Core-to-surface
transition zone

Surface layer
root and
airfoil tip

to 1500 MPa (220 ksi). Cutting speeds were twice those of molybdenum high-speed steel, one and one half times those of Stellite, but only one fourth those of commercial steel-cutting grades of cemented carbides.

Because the solubility of titanium carbide in liquid steel is high, the rate of infiltration is slowed so that penetration of the liquid tends to be confined to the regions near the original contact face. The remaining porosity cannot be filled because of diffusion solidification, regardless of the time allowed for penetration of the liquid. As a result, severe size limitations are required for the infiltrated product to be sound and uniform in structure. This disadvantage has caused the substitution of *in situ* liquid-phase sintering of titanium carbide and steel powder mixtures in the production of such tool materials. For more information, see the article "Cermets" in this Volume.

Fig. 16 Infiltrated titanium carbide cermet jet engine turbine components

(a) J-35 nozzle vane. (b) Experimental hollow vane. (c) J-57 turbine bucket. (d) J-47 turbine bucket

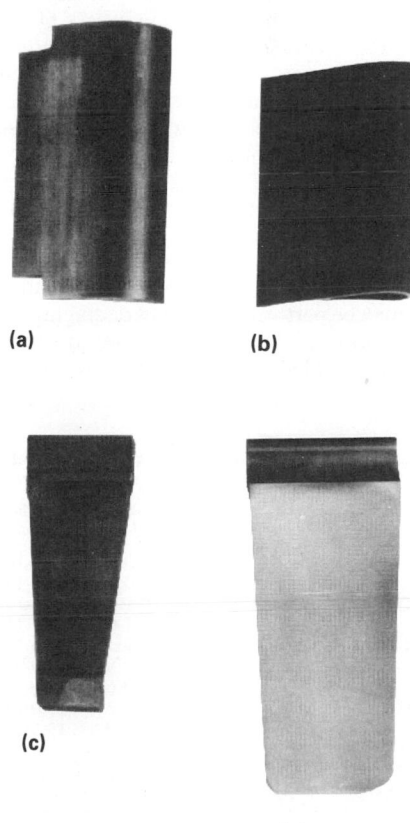

(a)

(b)

(c)

(d)

and after machining of the root configuration in Fig. 14(b). The change in microstructure across the airfoil is shown in Fig. 15 (Ref 43). Table 7 gives stress-rupture properties for different zones of the graded bucket (Ref 43, 46). Other infiltrated titanium carbide turbine components are shown in Fig. 16.

While these infiltrated carbide turbine components, like their sintered cermet counterparts, did not reach commercial production, blades for several stages of the compressor for a J-33 jet engine were mass produced from copper alloy infiltrated steel compacts (Ref 47). These blades were heat treated to a yield strength of 620 MPa (90 ksi), with an elongation of 5%, and withstood three to four times as much vibration at an operating temperature of 370 °C (700 °F) as blades for these stages made from martensitic type 403 stainless steel precision forgings.

Tools. In parallel development with turbine blades, titanium carbide skeletons were infiltrated with a variety of liquid steel alloys to manufacture tools and wear-resistant parts. The infiltrant varied from simple low-carbon steel to alloy and high-speed steels (Ref 48). Angular or rounded titanium carbide grains resulted, depending on carbide content and matrix composition. Infiltrated materials could be heat treated to produce specific properties. For instance, hardness ranged from 90.1 to 90.6 HRA after water quenching and from 86.0 to 90.3 HRA after tempering. At temperatures up to 750 °C (1380 °F), hot hardness for the titanium carbide that was infiltrated with type T6 tungsten high-speed steel was equivalent to commercial grades of cemented carbides. This material had better oxidation resistance than cemented carbides up to 870 °C (1600 °F) and room-temperature transverse-rupture strengths up

Mechanical Parts. Infiltration is widely used in the production of ferrous structural parts requiring densities in excess of 7.4 g/cm^3 and mechanical properties superior to those obtained by compacting, sintering, and coining. Depending on the application, porous skeletons of iron or steel may be fully or partially infiltrated with copper alloy.

There are several advantages to infiltration of iron-based structural parts with copper alloys (Ref 49):

- *Increased mechanical properties.* Higher tensile strengths and hardnesses and greater impact energies and fatigue strengths are obtained through infiltration. The effect of infiltration on strength is shown in Fig. 17.
- *Uniform density.* Parts that contain non-uniform and/or heavy sections can be infiltrated to obtain more uniform density; infiltration tends to even out density variations.
- *Higher density.* Infiltration is a useful method to increase sintered part weight without increasing the size of the part. Given the press size limitations and restrictions in pressing technique and powder compressibility, it is often easier to obtain high density through infiltration. Certainly, when considering normal P/M operations, it would be difficult to obtain densities in excess of 7.2 g/cm^3 without resorting to additional pressing and sintering operations. Infiltration makes densities in excess of 7.2 g/cm^3 possible in a single pressing and sintering operation.
- *Removal of porosity for secondary operations.* Infiltration may be used in place of impregnation as a method to seal surface porosity so that secondary operations such as pickling and plating may be performed without damaging the interior of the part and creating subsequent "bleeding" problems. It is also

Fig. 17 Effect of infiltration on transverse-rupture strength of iron-carbon alloys sintered to a density of 6.4 g/cm^3

Combined carbon in alloys was about 80% of graphite added to iron powder; amount of copper infiltrant was adjusted to fill various fractions of void space. Source: Ref 50

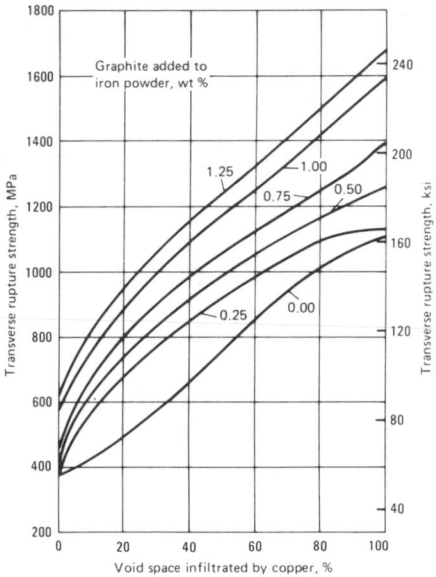

a method of sealing a part used for applications in which no porosity is desired.

- *Selective property variation.* It is possible by infiltrating only selected areas of a part to obtain, within limits, a controlled variation of properties in the part—for example, variations in density, strength, and hardness. This is known as localized infiltration. Infiltration to considerably less than the full density in the part (for example, 7.1 g/cm^3) is known as starve infiltration.

- *Assembly of multiple parts.* Different sections of the final part, pressed separately, can be assembled by sintering the individual pieces together and bonding the pieces into one part through common infiltration.

Table 8 gives composition and typical properties of P/M infiltrated steels.

The usual method of infiltrating iron and steel skeletons is to place a compact pressed from the powder of the infiltrant material next to the skeleton. The compact of infiltrant powder may be positioned on top or underneath the skeleton compact, or two infiltrant compacts may be used—one on top, the other underneath the skeleton compact. The exact amount of infiltrant needed may be compacted in the same die in which the powder for the porous skeleton is pressed.

After the skeleton has been sintered, the green compact or compacts of infiltrant powder are positioned next to the skeleton and the assembly heated to the infiltration temperature. Sintering of the skeleton and infiltration may be combined into one operation, in which the green compact or compacts of the infiltrant are positioned next to the green compact of the skeleton. By controlling the rate of heating, the skeleton compact will be adequately sintered by the time the melting point of the infiltrant is reached. This operation has been called "sintrating."

If only part of the porous skeleton is to be infiltrated—for example, the teeth of an infiltrated gear—the skeleton may be positioned in a graphite container in which space is provided adjacent to the gear teeth to be preferentially infiltrated. This space is then filled with the appropriate amount of infiltrant in powder form. Because iron has some solubility in copper, the liquid infiltrant will attack the surface of the skeleton when it first comes in contact with it, and severe erosion may take place at

Table 8 Composition and properties of P/M infiltrated steels

MPIF designation	Condition(a)	MPIF composition limits(b), % Carbon	MPIF composition limits(b), % Copper	MPIF composition limits(b), % Iron	Tensile strength MPa	Tensile strength ksi	Yield strength MPa	Yield strength ksi	Elongation in 25 mm (1 in.), %	Impact energy(c) J	Impact energy(c) ft · lb	Apparent hardness	Elastic modulus GPa	Elastic modulus 10^6 psi
FX-1005	AS	0.3-0.6	8.0-14.9	80.5-91.7	570	83	440	64	4.0	19	14	75 HRB	135	20
	HT	0.3-0.6	8.0-14.9	80.5-91.7	830	120	740	107	1.0	9.5	7.0	35 HRC	135	20
FX-1008	AS	0.6-1.0	8.0-14.9	80.1-91.4	620	90	515	75	2.5	16	12	80 HRB	135	20
	HT	0.6-1.0	8.0-14.9	80.1-91.4	895	130	725	105	60.5	9.5	7.0	40 HRC	135	20
FX-2000(d)	AS	0.3 max	15.0-25.0	70.7-85.0	450	65	...	...	1.0	20	15	60 HRB	...	...
FX-2005(e)	AS	0.3-0.6	15.0-25.0	70.4-84.7	515	75	345	50	1.5	12.9	9.5	75 HRB	125	18
	HT	0.3-0.6	15.0-25.0	70.4-84.7	790	115	655	95	<0.5	8.1	6.0	30 HRC	125	18
FX-2008(f)	AS	0.6-1.0	15.0-25.0	70.0-84.4	585	85	515	75	1.0	14	10	80 HRB	125	18
	HT	0.6-1.0	15.0-25.0	70.0-84.4	860	125	740	107	<0.5	6.8	5.0	42 HRC	125	18

Note: All materials have a density range of 7.2 to 7.6 g/cm^3.
(a) AS, as sintered; HT, heat treated (typically austenitized at 870 °C (1600 °F), oil quenched and tempered 1 h at 200 °C (390 °F). (b) MPIF Standards require that the total amount of all other elements be less than 2.0%. (c) Unnotched Charpy. (d) ASTM B 303, Class A; SAE 870. (e) ASTM B 303, Class B. (f) ASTM B 303, Class C; SAE 872.

this point. One method to minimize erosion is to use a copper alloy (an 80%Cu-20%Zn brass, for example) as infiltrant. Because the brass does not melt at one temperature, but over a range of temperatures, less erosion takes place.

Another method is to use an alloy of copper with iron as an infiltrant. However, it is difficult to specify the exact amount of iron; too little iron may cause erosion, and too much iron an undesirable adherent deposit on the infiltrated part. A third method is to use an alloy of copper, iron, and a third alloying constituent, such as manganese, which oxidizes in the atmosphere in which infiltration takes place. In this case, a crust containing an oxide of the oxidizing alloying ingredient of the infiltrant is formed, which does not adhere to the skeleton, but which can be more or less readily removed.

Figure 18 shows a small foot holder that required strength, machinability for tapping a thread, and plateability. These characteristics were achieved by infiltrating a sintered iron-rich skeleton (7% Cu) with a copper-rich alloy containing 5% Mn and 5% Fe. The compacts were infiltrated under furnace conditions similar to those used for sintering, using infiltrating slugs previously pressed. The infiltrated parts were electroplated with chromium to a satin finish. Processing details for the two-step procedure are given in the table that accompanies Fig. 18.

The control of dimensions during infiltration may cause problems. It depends on the composition of the steel; carbon-free iron shows the greatest growth during infiltration. It also depends on the exact temperature of infiltration and the time during which the infiltrated compact is above the liquidus temperature of the infiltrant.

Bearings. The best example for the industrial use of infiltration in the bearing field is steel strip-backed, precision-type main and connecting rod bearings for automobiles. In this application, the antifriction babbitt alloy is used to infiltrate the pores of a skeletal cupronickel strip that is bonded to the steel backing and produces an extra layer on the surface facing the crankpin (Ref 7, 34).

Infiltration techniques also may be applied to high-temperature bearing materials, such as bearing retainers in gas turbines for supersonic aircraft. To combat the high wear of these cages, an ideal structure consists of a hard, load-bearing phase intertwined with a liquid metallic lubricant. Structures that closely approach

Fig. 18 Small compact infiltrated with copper alloy to provide machinability, strength, and plateability

Skeleton		Infiltrating slug	
Reduced iron	92.0%	Copper	89.0%
Copper	7.0	Manganese	5.0
Zinc stearate	1.0	Iron	5.0
		Lubricant	1.0

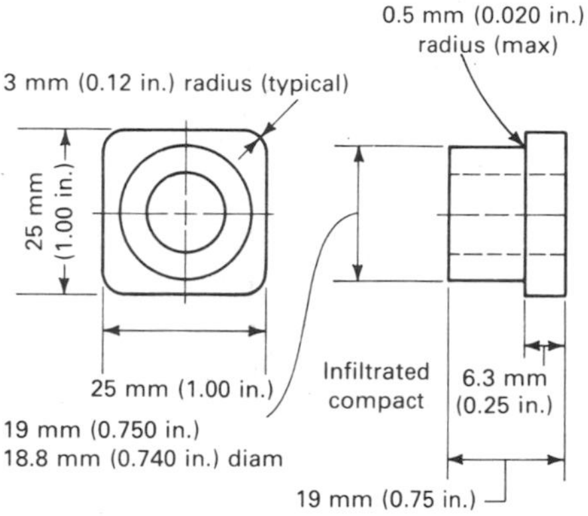

0.5 mm (0.020 in.) radius (max)

3 mm (0.12 in.) radius (typical)

25 mm (1.00 in.)

25 mm (1.00 in.)

Infiltrated compact

6.3 mm (0.25 in.)

19 mm (0.750 in.)
18.8 mm (0.740 in.) diam

19 mm (0.75 in.)

Press and tool details

Type of press . . . . Mechanical(a)
Press capacity . . . . 45.4 metric tons (50 tons)
Die material . . . . Tungsten carbide (6% Co)
Punch material . . . . D2 tool steel
Core-rod material . . . . . Tungsten carbide (6% Co)

Processing details

Compacting pressure:
    Skeleton . . . . 415 MPa (60 ksi)
    Infiltrating slugs . . . . 276 MPa (40 ksi)
Preheating treatment . . . . 15 min at 1100 °C (2040 °F)
Sintering treatment . . 15 min at 1100 °C (2040 °F)
Infiltrating treatment . . . . 15 min at 1100 °C (2040 °F)
Atmosphere for sintering
    and infiltrating . . . . Endothermic

(a) Two movements above and three below

the ideal concept and operate up to 370 °C (700 °F) have been developed (Ref 50).

Materials exhibiting improved antifriction properties against steel under simulated service conditions involving very high rotational speeds are efficiently fabricated by P/M techniques that include an infiltration step. These materials are characterized by a duplex structure (a soft, metallic phase of about 15 vol%) that is uniformly dispersed throughout the hard matrix. The matrix may be either Monel or a 48Ni-48Cr-4Si alloy, with 10 wt% molybdenum disilicide added in some cases to reduce wear. The soft phase consists of silver that is infiltrated after sintering the hard alloy powder compact.

REFERENCES

1. Schwarzkopf, P., The Mechanism of Infiltration, *Symposium on Powder Metallurgy 1954,* Special Report No. 58, The Iron and Steel Institute, London, 1956, p 55-58
2. Semlak, K.A., Spencer, C.W., and Rhines, F.N., Rate of Capillary Rise of Liquid Metal in a High Melting Metal Powder Compact, *Trans. AIME,* Vol 209, 1957, p 63-64
3. Semlak, K.A. and Rhines, F.N., The Rate of Infiltration in Metals, *Trans. AIME,* Vol 212, 1958, p 325-331
4. Jones, W.D., Fundamental Principles of Powder Metallurgy, Edward Arnold, London, 1960, p 505-512
5. Goetzel, C.G. and Shaler, A.J., Mechanism of Infiltration of Porous Powder Metallurgy Parts, *J. Met.,* Vol 16 (No. 11), 1964, p 901-905
6. Shaler, A.J., Theoretical Aspects of the Infiltration of Powder Metallurgy Products, *Int. J. Powder Metall.,* Vol 1 (No. 1), 1965, p 3-14
7. Kimura, T., Kosco, J.C., and Shaler, A.J., Detergency During Infiltration in Powder Metallurgy, *Proceedings of 15th Annual Meeting,* Metal Powder Industries Federation, New York, 1959, p 56-66
8. Kieffer, R. and Benesovsky, F., The Production and Properties of Novel Sintered Alloys (Infiltrated Alloys), *Berg- und Hüttenmännische Monatshefte,* Vol 94 (No. 8/9), 1949, p 284-294
9. Goetzel, C.G., Infiltration Metallurgy, *Research,* Vol 4 (No. 12), 1951, p 555-561
10. Lenel, F.V., *Powder Metallurgy,* Metal Powder Industries Federation, Princeton, NJ, 1980, p 313-319
11. Matsumura, G., Stress Infiltration in Two-Phase Alloys, *Planseeberichte f. Pulvermetallurgie,* Vol 8 (No. 3), 1960, p 110-118
12. Kaysser, W.A., Takajo, S., and Petzow, G., Skeleton Dissolution and Skeleton Formation During Liquid Phase Sintering of Fe-Cu, *Modern Developments in Powder Metallurgy,* Vol 12, Metal Powder Industries Federation, Princeton, NJ, 1981, p 473-482

13. Lavendel, H.W. and Goetzel, C.G., Recent Advances in Infiltrated Titanium Carbides, in *High Temperature Materials*, Hehemann, R.F. and Ault, G.M., Ed., John Wiley & Sons, New York, 1959, p 140-154
14. Langford, G., High Speed Steel Made by Liquid Infiltration, *Mat. Sci. Eng.*, Vol 28, 1977, p 275-284
15. Goetzel, C.G., *Treatise on Powder Metallurgy*, Vol 2, Interscience, New York, 1950, p 196
16. Schröter, K., Border Regions of Metallography, *Zeitschrift Metallkunde*, Vol 23 (No. 7), 1931, p 197-201
17. Krol, J.M. and Goetzel, C.G., "Refractory Metal Reinforced Super Alloys," USAF Technical Report No. 5892, ATI No. 57154, May 1949
18. Lenel, F.V., *Powder Metallurgy*, Metal Powder Industries Federation, Princeton, NJ, 1980, p 383-400
19. Kieffer, R. and Kölbl, F., Production of Hard Metals by Infiltration, *Berg- und Hüttenmännische Monatshefte*, Vol 95 (No. 3), 1950, p 49-58
20. Baumhauer, H., "Hard Tools and Process for Making Them," U.S. Patent 1 512 191 (applied 1922, issued 1924); German Patent 443 911 (applied 1922, issued 1927)
21. Reimann, L., "Production of Bodies from Metallic Compounds," German Patent 300 669 (applied 1914, issued 1917); Reimann, L. and Leiser, H., "Metallic Alloy," British Patent 148 533 (applied 1917, issued 1921)
22. Gebauer, C.L., "Process of Producing Metal Bodies," U.S. Patent 1 342 801 (applied 1917, issued 1920); Gebauer, C.L., "Production of a Composite Metallic Article," U.S. Patent 1 395 269 (applied 1918, issued 1921)
23. Peters, F.P., Cemented Steels—A New High-Strength Powder Metallurgy Product, *Materials and Methods*, Vol 23 (No. 4), 1946, p 987-991
24. Kopecki, E.S., Cemented Steels, *Iron Age*, Vol 157 (No. 18), 1946, p 50-54
25. Goetzel, C.G., Cemented Steels —Infiltration Studies with Pure Iron and Copper Powders, *Powder Metall. Bull.*, Vol 1 (No. 3), 1946, p 37-43
26. Northcott, L. and Leadbeater, C.J., Sintered Iron-Copper Compacts, *Symposium on Powder Metallurgy*, Special Report No. 38, The Iron and Steel Institute, London, 1947, p 142-150
27. Schwarzkopf, P., Infiltration of Powder Metal Compacts with Liquid Metal, *Met. Prog.*, Vol 57 (No. 1), 1950, p 64-68
28. Stern, G., The Effect of Infiltration on Physical Properties of Sinterings, *Prec. Met. Mold.*, Vol 11 (No. 6), 1953, p 92-102
29. Schramm, J. and Mohrnheim, A., Precipitation Hardening of Iron-Zinc and Cobalt-Zinc Alloys, *Zeitschrift Metallkunde*, Vol 39, 1948, p 71-78
30. Hensel, F.R., "Treatment of Bearing," U.S. Patent 2 364 713 (applied 1943, issued 1944)
31. Fetz, E., Bearings from Metal Powder—A New Art, *Metals and Alloys*, Vol 8 (No. 9), 1937, p 257-260
32. Fetz, E., "Manufacture of Composite Bearings," U.S. Patent 2 234 371 (applied 1938, issued 1941)
33. Hensel, F.R., "Impregnation of Metallic Composition with Bismuth," British Patent 590 412 (applied 1944, issued 1947)
34. Boegehold, A.L., Copper-Nickel-Lead Bearings, in *Powder Metallurgy*, J. Wulff, Ed., American Society for Metals, 1942, p 520-529
35. Hensel, F.R. and Larson, E.I., "Sintered Porous Aluminum-Base Bearings," U.S. Patent 2 418 881 (applied 1944, issued 1947)
36. Hensel, F.R., "Method of Making Porous Bearing Surfaces," U.S. Patent 2 447 980 (applied 1945, issued 1948)
37. Kieffer, R. and Hotop, W., *Powder Metallurgy and Sintered Materials*, Springer-Verlag, Berlin, 1943, p 324-326, 329
38. Hensel, F.R., Larsen, E.I., and Swazy, E.F., Physical Properties of Metal Compositions with a Refractory Metal Base, in *Powder Metallurgy*, J. Wulff, Ed., American Society for Metals, 1942, p 483-492
39. Goetzel, C.G., *Treatise on Powder Metallurgy*, Vol 2, Interscience, New York, 1950, p 207-209, 216-217
40. Schwarzkopf, P., *Powder Metallurgy*, Macmillan, New York, 1947, p 167-168
41. Goetzel, C.G., *Treatise on Powder Metallurgy*, Vol 2, Interscience, New York, 1950, p 539-540, 634
42. Goetzel, C.G. and Rittenhouse, J.B., The Influence of Processing Conditions on the Properties of Silver-Infiltrated Tungsten, *Symposium sur la Métallurgie des Poudres*, Éditions Métaux, Paris, 1964, p 279-288
43. Goetzel, C.G. and Lavendel, H.W., Infiltrated Powder Components for Power Plant and Propulsion Systems, *Metals for the Space Age, Plansee Proceedings 1964*, F. Benesovsky, Ed., Springer-Verlag, New York, 1965, p 149-162
44. Toensing, C.H. and Zalsman, S., "Silver-Infiltrated Tungsten Characterization Study," Firth Sterling, Inc., McKeesport, PA, 31 Dec 1962
45. Warga, J.J. and Matt, R.E., Infiltration of Tungsten Using Silver, *Materials Science and Technology for Advanced Applications*, Vol 2, American Society for Metals Golden Gate Metals Conference, San Francisco, 12-15 Feb 1964, p 552-568
46. Lavendel, H.W. and Goetzel, C.G., "A Study of Graded Cermet Components for High Temperature Turbine Applications," Report WADC-TR 57-135, May 1957
47. Stern, G. and Gerzina, J.A., Making Jet Engine Compressor Blades by Powder Metallurgy, *Iron Age*, Vol 165 (No. 8), 1950, p 74-77
48. Goetzel, C.G. and Skolnick, L.P., Some Properties of a Recently Developed Hard Metal Produced by Infiltration, *Sintered High-Temperature and Corrosion-Resistant Materials, Plansee Proceedings 1955*, F. Benesovsky, Ed., Pergamon Press, London, 1956, p 92-98
49. Durdaller, C., "Copper Infiltration of Iron-Based P/M Parts," Hoeganaes Corp., Riverton, NJ, 1969
50. Burwell, J.T., Wear Behavior of High Temperature Bearing Materials, *Prec. Met. Mold.*, Vol 14 (No. 10), 1956, p 40, 41, 87, 88, 90, 91

Powder Systems and Applications

Introduction

By Kempton H. Roll
Executive Director
Metal Powder Industries Federation
American Powder Metallurgy Institute

THE METAL POWDER PRODUC-ING and consuming industries historically have followed the general trends of the economy and the automotive industry in particular. For many years, powder metallurgy enjoyed wide usage because of the ever-increasing production of larger automobiles with more gadgetry, coupled with the emergence of the American automobile as a "necessity" for almost everyone, rather than a luxury for the wealthy few.

Modern powder metallurgy, in many respects, owes its existence to the automotive industry and the recognition of the merits of this process by early automotive engineers, who began with the self-lubricating bearing and graduated to the structural component. Porous self-lubricating bearings could only be made by P/M processing. Structural parts, on the other hand, gained acceptance because they offered cost effectiveness—a competitive advantage over conventional metal-shaping techniques. Acceptance of P/M techniques by the automotive industry led to utilization of P/M processes in other sectors of manufacturing.

Some P/M applications and markets have come and gone, succumbing to technological obsolescence. Mechanical calculators and cash registers, which once represented important applications for P/M parts, have given way to electronics. Nevertheless, with the high volume of information being processed and the rate at which business is conducted, business machines remain an important market for P/M components. These applications require moving parts that operate at high speeds with extreme dependability. These requirements, combined with the need for cost savings, productivity, and quality assurance, have resulted in increased markets for P/M components. For additional information on this application, see the article "P/M Parts for Business Machines" in this Section.

This transition from traditional metalworking and planned obsolescence has had an impact on all industry. Although some items are designed to be disposable, others are designed to last. An overriding factor, however, is the need to produce parts rapidly and accurately that will perform reliably at the lowest possible cost. This has been a direct benefit to the P/M industry.

Significant commercial acceptance and applicability of modern powder metallurgy came at a time when society was concerned about preserving the ecology and protecting the environment. The P/M process does not generate fumes or chemical pollutants. Most nations are becoming keenly aware of the need to conserve materials—especially metals, because they are a nonreproducing resource that is becoming increasingly more expensive and difficult to recover. Metal powders often are regenerated from scrap metals, and the P/M process results in minimal waste. Parts produced by net shape or near-net shape technology require little or no machining, resulting in minimal scrap loss.

Another concern was conservation of energy resources. Heat, whether from coal, oil, gas, or electricity, is costly. Powder metallurgy is one of the most energy-efficient methods of forming metal shapes, because of the solid-state fusion phenomenon known as sintering. The final P/M part is made by compacting and sintering. The metal to produce the part is not melted.

It appeared that P/M applications had reached their limit in the late 1970's. For example, there were no additional parts in a washing machine that could be converted to P/M components. Additionally, doubts were associated with a process that begins with powder and purportedly produces a solid, strong metal structure. Some early P/M proponents were unfamiliar with production methodologies and sometimes promoted P/M components for applications that were bound to lead to failure.

Plastics became available that were capable of operating at higher temperatures than earlier materials. Other more traditional metalworking methods began to become competitive. Development of thin-walled die castings salvaged that segment of metalworking technology. Foundries continued to lag in production, but automation offered assistance. Fine blanking, progressive die stamping, computer-assisted machining, investment casting, and closed die forging all competed with P/M technology.

Concerned by an apparent plateau in growth rate, the metal powder producing and consuming industries began examining various P/M markets. These market studies determined that better standards were needed and that more education and a clearer understanding of the capabilities of the powder metallurgist were required. The studies also revealed untapped potential, not only in existing markets, but in new areas as well. Extensive advertising campaigns, trade show exhibits, sponsorship of design clinics, production of films detailing P/M technology, and publication and distribution of design guidebooks were an integral part of an effort to revitalize the industry. New standards with guaranteed minimum strength values were developed by the Metal Powder Industries Federation (MPIF) and the American Society for Testing and Materials (ASTM). Property data for computer-assisted design were generated. Existing property values were reaffirmed, and new P/M materials were introduced.

P/M Forgings

There has been a rediscovery of P/M forging as a viable method of producing fully dense P/M products that are cost effective and possess the unique strength characteristics of forgings, coupled with the precision and productivity of powder metallurgy. P/M forgings were conceived in the 1960's, explored in the early 1970's, and seemingly rejected in the late 1970's.

The concept of taking a P/M part, heating it, and then forging it in a close die originally appeared simple and logical. There were problems, however. The powder had to have a higher purity. There was less tolerance for oxide content. The mechanics of heating and forming with appropriate lubrication to facilitate ejection were difficult. Die wear had a great impact on productivity and cost effectiveness. Problems were compounded by early design engineers, who insisted on using difficult forging applications to test new P/M techniques. If components were difficult to produce by conventional forging, in most instances, they seemed virtually impossible to produce by P/M techniques. That attitude, too, has changed.

Powder metallurgy forgings currently are produced with enough efficiency and cost effectiveness to replace conventional forgings. Ultimately, P/M forgings will evolve into a substantial market for metal powders and will provide an effective means of meeting the increasing demand for high-strength, high-performance, and inexpensive metal components. For additional information, see the article "P/M Forging" in this Volume.

Unconventional Metallurgy

Since the mid-1970's, a new world of metallurgy has been evolving—metal systems. Some of these systems cannot be formed by melting and casting; they are fully homogeneous, fully dense, fine-grained, segregation-free, high-strength materials. In most instances, these systems do not compete with existing materials.

Compare P/M tool steel and conventionally processed tool steels, for example. Both may possess the same chemical composition, but their microstructures differ greatly. The P/M tool steel grain size is small and uniform, a direct function of the original powder particle size. Each particle is in itself a tiny ingot of the alloy.

Because the P/M tool steel does not enter into the liquid state, there is no segregation. By using hot isostatic pressing techniques, these particles can be consolidated to form a fully dense, pore-free structure. The first true test of this new concept was the ball bearing cup and race, which provided a service life five times longer than its conventionally produced predecessor. For additional information, see the article "P/M Tool Steels" in this Section.

The newest advance in P/M technology toward the creation of space-age metals and materials is the development of rapid-solidification technology. Glass has an amorphous structure. Metal, on the other hand, has a crystalline structure. Rapid-solidification technology creates an amorphous metal that combines the best features of both materials:

• Forming capabilities at moderate temperature and force
• Strength and ductility of metals

After forming, heat treatment converts the amorphous metal into ultra-fine crystalline metal that possesses the tensile properties and resistance to deformation normally associated with metals. For additional information, see the article "Atomization" in this Volume.

Fig. 1 Iron powder shipments
Courtesy of Metal Powder Industries Federation/Metal Powder Producers Association

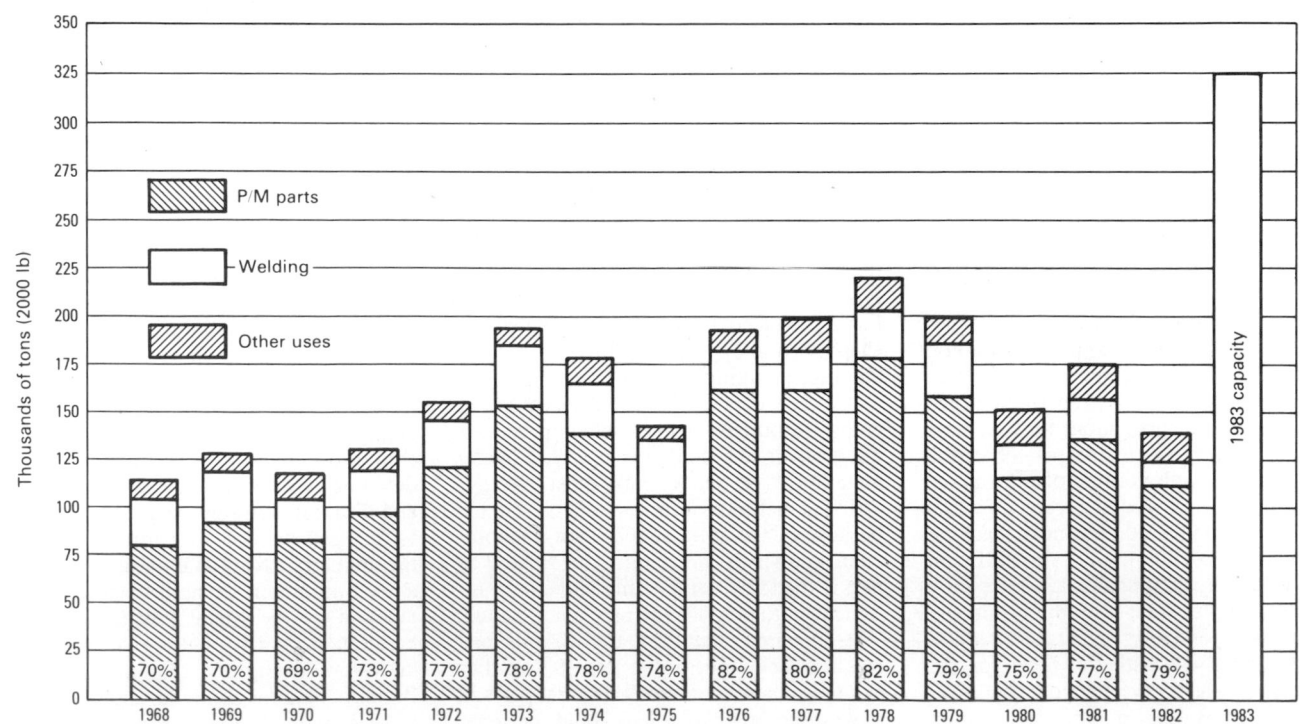

Fig. 2 Shipments of copper and copper-based powders
Courtesy of Metal Powder Industries Federation/Metal Powder Producers Association

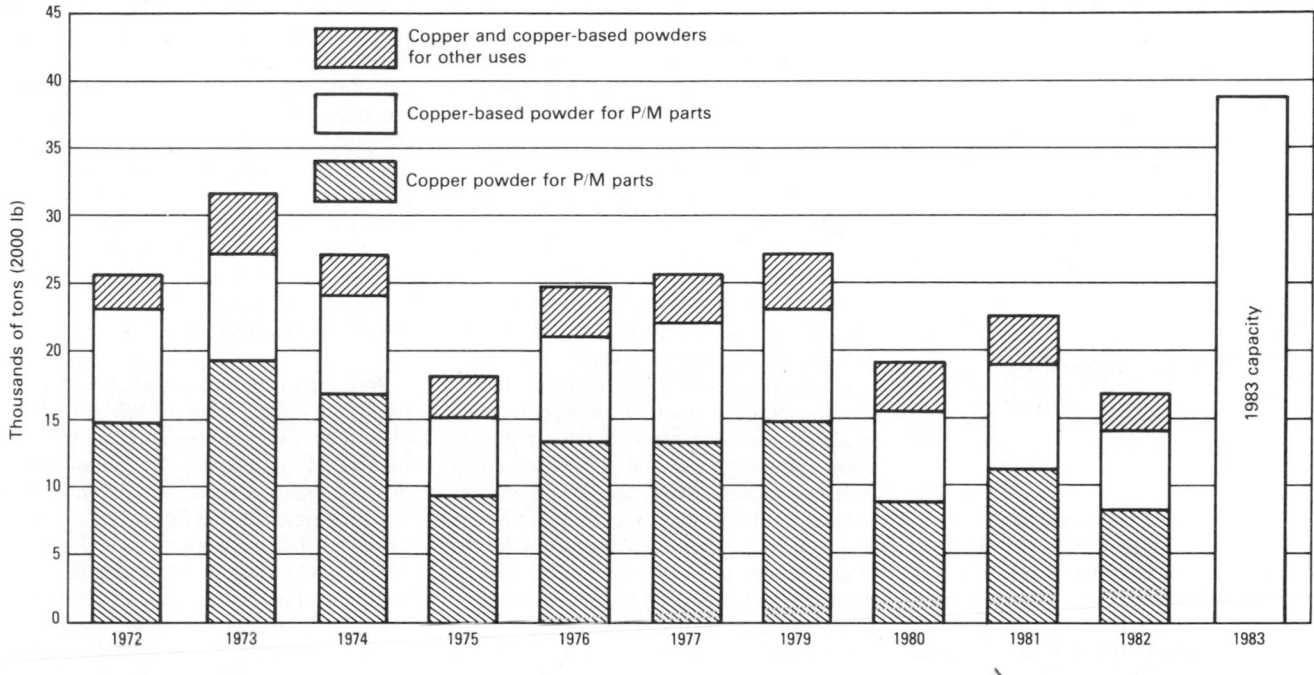

Metal Powder Characteristics and Applications

Metal powders are defined as discrete particles of elemental metals or alloys, normally within the size range of 0.1 to 1000 μm. Particles may be tiny spheres, such as those used in porous metal filters, or they may be tiny flakes, such as those used in paint pigment. Particles may be of irregular shape, such as those used in compacting and sintering P/M parts. They may be dendrites or fibers.

Metal powders represent a small, almost insignificant, portion of the total metals industry. Iron powder, essential to the P/M industry because it represents the largest tonnage product, reached its peak at only about 205 000 metric tons (225 000 tons) in 1978 (Fig. 1). The capacity to produce iron powder is presently about 295 000 metric tons (325 000 tons) per year, well in excess of consumption. On a different order of magnitude, the same may be said for copper and copper-based powders. The production peak for these powders was slightly in excess of 27 000 metric tons (30 000 tons) in 1973, with a capacity of 36 000 metric tons (40 000 tons) per year (Fig. 2).

The primary market for iron powder is the manufacture of P/M parts by P/M techniques. This market represents about 80% of the total. Of the balance, about 10% has been allocated to the welding

Fig. 3 Iron powder markets
1982 shipments, tons (2000 lb). Courtesy of Metal Powder Producers Association

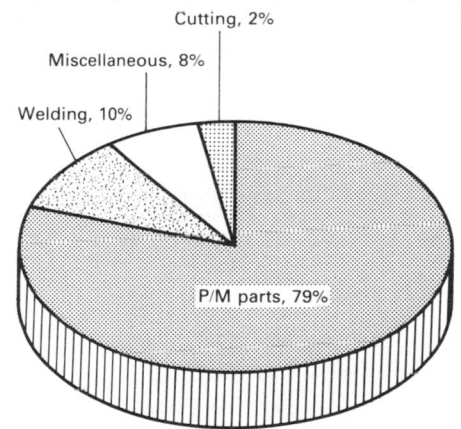

Fig. 4 1982 P/M parts production
Includes bearings and friction materials. Based on metal powder shipments for custom and in-plant P/M parts manufacture. Stainless steel estimated. Courtesy of Metal Powder Industries Federation/Metal Powder Producers Association

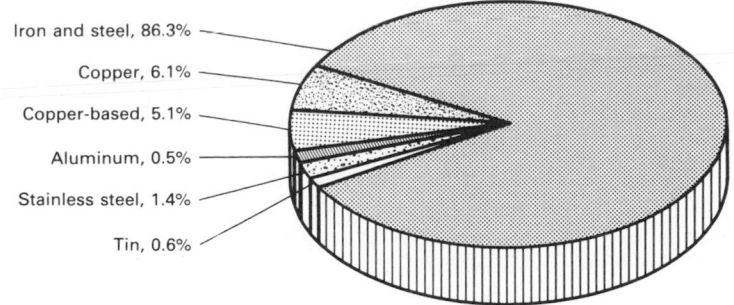

Fig. 5 Copper and copper-based powder markets

1982 shipments, tons (2000 lb). Courtesy of Metal Powder Producers Association

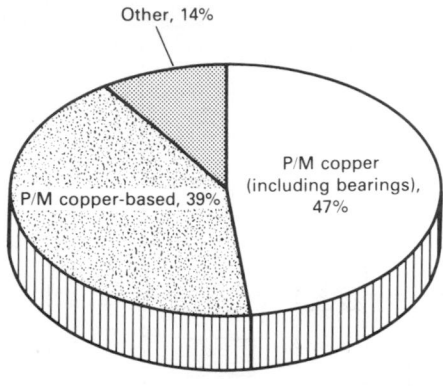

the article "Metal Powders Used for Flame Cutting" in this Section).

The remainder (8%) is used in various applications, including pharmaceuticals and food enrichment (see the article "Iron Powders for Food Enrichment" in this Section). Americans consume about 900 metric tons (1000 tons) of iron powder per year in iron-enriched bread and cereal. Emerging as a major application of P/M iron powder is its use as copier powders—fine iron particles that carry copier toners used in photocopier processes (see the article "Copier Powders" in this Section). Figure 3 provides a breakdown of the major iron powder market in the United States.

In the P/M parts manufacturing industry, the principal metal powders used are made of iron and steel and their alloys, copper and copper-based alloys (including brasses, nickel silver, and bronze), aluminum, stainless steel, nickel, and tin. Tin is used as an elemental addition in the manufacture of self-lubricating bronze bearings; nickel is used in low-alloy nickel steel and stainless steel parts. Figure 4 provides a breakdown of powder usage by alloy.

Copper and copper-based powders follow iron in volume of powder used in P/M fabrication. About 45% of all copper powder is used in the manufacture of self-lubricating bearings. It also is used as elemental additions in ferrous P/M parts. Another 40% is used for making P/M parts for builders' hardware, such as locksets, door latches, and other components that require a decorative finish as well as corrosion resistance. The remainder is used for pigment for marine paint, for peen plating, and for metallic flake paints (Fig. 5) (see the article "Metallic Flake Pigments" in this Section).

Although iron and copper powders represent the largest share of P/M powder applications, they do not represent the entire range of P/M usage. The major applications of metal powders are given in Table 1. The articles in this Section are intended to reveal the diversification and potential of the P/M industry. More detailed information on processing, characterization and testing of powders, and sintering and consolidation practices can be found in previous Sections of this Volume.

market for use in coated electrodes (see the article "Metal Powders Used for Filler Materials" in this Section). About 2% has been allocated for cutting, scarfing, and lancing operations, where the iron powder is injected into oxyacetylene flames to raise the temperature of the cutting zone (see

Table 1   Major applications of metal powders

Application	Powder used	Application	Powder used
Abrasive finishing		**Aerospace (continued)**	
Shot cleaning	Iron, steel, stainless steel	Rocket fuels	Aluminum
		Repairing of components	Nickel chromium, nickel aluminum, stainless steel
Abrasive wheels	Iron, steel, copper, tin, nickel		
		Fretting/wear spraying	Molybdenum, copper nickel alloys, cobalt alloys, chrome carbide
Alloy production			
Steels, electroslag steel, free-machining steel, cast iron, ductile iron	Nickel, lead, ferro-tungsten, ferro-silicon, ferro-molybdenum, ferro-manganese, iron, aluminum, tungsten, molybdenum, selenium, tellurium	**Automotive(a)**	
		Air conditioners	Iron, steel
		Alternator regulator, contacts, pole pieces	Platinum alloy, iron
		Body solder	Steel, aluminum, lead, lead alloys
Agriculture		Bushings, bearings	Copper, lead, tin, iron, aluminum, graphite, bronze
Animal feed	Iron		
Animal medication	Cobalt	Contacts	Tungsten, silver
Chelate fertilizers	Iron	Filters	Stainless steel, bronze
Farm machinery	Iron, steel, copper, bronze	Fuel pump parts	Iron, copper, tool steel
Food enrichment	Iron, copper, manganese	Polychrome body finishes	Aluminum
Fungicides	Copper	Shock absorbers	Iron, aluminum
Lawn and garden equipment	Iron, steel, iron alloy, copper, bronze	Spark plug (body)	Iron
		Spark plug (corrosion protection)	Zinc
Seed cleaning	Iron	Transmission parts	Iron, copper, steel
Seed coating	Aluminum	Truck signal flares	Aluminum
Soil conditioning	Iron, copper	Brake bands, linings	Copper, iron, lead, tin, graphite, brass, zinc
Aerospace		Tire studs	Tungsten carbide
Brake linings	Copper, lead, tin, high-nickel alloys, graphite, iron	Valve inserts	Tool steels, stainless steel
Cores (electronic components)	Iron, MolyPermalloy	**Building and construction**	
Counterweights	Tungsten, copper, nickel, iron	Aerated concrete	Aluminum, iron
Filters for fuel, hydraulic fluids, air	Bronze, stainless steel, nickel	Asphalt roof coating	Aluminum
		Caulking compound	Aluminum
Hardware	Aluminum, beryllium, titanium, iron	Conductive and non-spark flooring	Copper
Heat shields	Beryllium, tungsten	Decorative plastics and linoleum for floors, walls, and counter tops	Iron, brass, copper, aluminum, stainless steel
Heat shield coating	Aluminum		
Jet engine components	Superalloys		
Magnetic shields	Aluminum, nickel, cobalt		
Measuring aircraft drift	Aluminum		

(continued)

Table 1 (continued)

Application	Powder used	Application	Powder used
Building and construction (continued)		**Electrical and electronic (continued)**	
Lancing	Iron, aluminum	Solenoids	Iron
Protective coatings for canvas awnings and decks	Aluminum, zinc	Telephone components	Iron, brass, bronze, palladium, stainless steel
Pipe joint compounds	Zinc, lead, copper	Terminals	Nickel
		Thin-film resistors	Palladium, silver
Carbides	Tungsten, titanium, tantalum, niobium	**Electrical discharge machining**	
		Electrodes, preforms	Tungsten, copper, silver
Catalysts		**Electrochemical machining**	
Gasoline synthesis	Platinum, nickel	Electrodes	Copper, silver, tungsten
Hydrogenation	Platinum, platinum alloys, tungsten, molybdenum, rhenium	**Hardware**	
Promote combustion in space heaters	Platinum	Lock components	Brass, bronze, iron, stainless steel
Alkyls	Aluminum	**Heat treatment**	
Purifying hydrogen	Ruthenium, palladium	Furnace heating elements	Platinum alloys, tungsten, molybdenum
Reacting hydrogen and oxygen	Palladium	Furnace shielding	Aluminum, tungsten
Raney catalyst	Nickel, aluminum	Infiltration	Copper, brass, silver, manganese
Rocket fuels	Iron	Protection tubes	Platinum alloys
General chemical catalysts	Iron, copper, copper oxides	Thermocouples	Platinum alloys, tungsten, molybdenum, rhenium
Cemented carbides	Cobalt, molybdenum, titanium, tungsten, nickel, tantalum, niobium	**Industrial (general)**	
		Air tool exhaust muffler	Stainless steel, bronze
Chemical	Aluminum, lead, tin, copper, iron, nickel, manganese	Bearings and bushings	Copper, tin, lead, bronze
		Bonded asbestos brake linings	Zinc, graphite, brass
Coatings		Cutting tools	Tungsten, cobalt
Anti-fouling paints	Copper, copper oxide	Filters (liquids, gases)	Bronze, nickel, stainless steel, superalloys, titanium
Conductive paints and plastics	Silver, copper, nickel	Flame cutting and scarfing	Iron, aluminum
Corrosion-resisting paints	Stainless steel, aluminum, zinc, lead	Fluids for magnetic clutches	Stainless steel (magnetic)
Decorative paints	Aluminum, brass, bronze, zinc, stainless steel, lead, copper	Friction material	Copper, tin, zinc, iron, lead, graphite
Fabric coatings	Aluminum	Gas engine exhaust purification	Platinum
Heat-reflecting paint	Aluminum	Grinding wheels	Brass, bronze
Hard facing	Cobalt alloys, nickel alloys, tungsten, molybdenum, stainless steel, tungsten carbide	Sound deadening plastic	Lead, graphite
Lacquers	Silver, brass, bronze, aluminum	**Industrial explosives**	
Mechanical (peen) plating	Zinc, lead, tin, brass, copper, cadmium	Mining	Aluminum
Slurry coating	Nickel	**Joining**	
Roof coatings	Aluminum	Brazing	Copper, nickel, silver, cobalt, brass, gold, aluminum, nickel-cobalt-iron-chromium alloy, nickel-silicon-boron alloys
Spray coating	Iron, aluminum, zinc, tin, nickel alloys, copper, bronze, stainless steel, tungsten carbide, titanium carbide, silver, cadmium	Coated and tubular electrodes for arc welding	Iron, nickel, chromium
		Resistance welding electrodes	Tungsten, copper, silver, molybdenum, manganese
Vacuum metallizing	Aluminum, copper, zinc	Soldering	Lead, tin, solder alloys
Coinage	Nickel, copper-nickel, titanium	Thermit welding	Aluminum
Diffusion coating	Aluminum-titanium alloys, chromium, zinc, hafnium	**Lubricants**	
		Anti-galling pipe joint compound	Copper, lead, zinc, graphite
		Copper lubricants	Copper
Electrical and electronic		Greases	Lead, graphite
Acoustical elements to improve microphone response	Stainless steel	High-temperature lubricants	Aluminum, graphite
Batteries	Nickel, zinc, silver, iron, lead, graphite	Plastic-filled metals	Iron, copper
Breathers	Stainless steel	**Magnetic**	
Brushes	Copper, silver, graphite	Clutches	Stainless steel (magnetic), iron
Cathode-ray tubes	Graphite, nickel	Inks for paper and tape	Iron
Contacts	Copper, silver, tungsten, molybdenum, iron, tin, platinum alloys, graphite, carbon	Permanent magnets	Iron, nickel, cobalt, aluminum, barium, ferrite, iron-aluminum alloys, molybdenum
Crystal supports	Silver, copper, tungsten	Pole pieces, relay parts	Iron, iron-cobalt, silicon-iron
Delay lines, computers	Iron, MolyPermalloy	Soft magnetic parts	Iron, nickel-iron, silicon-iron
Fuel cells	Nickel, zinc	Traveling wave tube magnets	Platinum, cobalt
Motor pole pieces	Iron, silicon-iron	**Medical and dental**	
Printed circuits	Copper, silver, palladium, gold, platinum	Dental amalgam	Silver, gold, alloys
Relays	Iron, nickel, molybdenum	Prosthetics	Superalloys
Semiconductors	Lead		

(continued)

Table 1 (continued)

Application	Powder used	Application	Powder used
Medical and dental (continued)		**Petroleum and petrochemical**	
Insulin production	Zinc	Hydrocracking	Platinum alloys
Operating room air filters	Stainless steel, Monel, cobalt alloys	Hydrogenation of olefines	Platinum
Pharmaceutical final filters	Stainless steel, Monel	Petroleum fuels reforming	Platinum
Porcelain overlay dental restorations	Palladium	Purifying hydrogen	Ruthenium
Prevention of infection of open wounds	Aluminum	Oil drilling	Tungsten carbide, tungsten alloys
Surgical gauzes and pins	Tantalum, zirconium	**Plastic reinforcing**	
Surgical implants	Nickel and cobalt alloys	Body solders	Steel, aluminum lead
Orthopedic external fixation system	Titanium	Cements for repairing castings and metal parts	Iron, stainless steel, aluminum, silver
Radiation shielding	Tungsten	Hull-smoothing cements	Iron, aluminum
		Tools and dies	Iron, aluminum
Metal recovery		**Printing inks**	
Copper cementation	Iron	Metallic inks for offset, letter-press, rotogravure, etc.	Copper, brass, aluminum
Gold cementation	Zinc		
Metals from solution	Aluminum	**Process industries**	
		Aluminum alkyls	Aluminum
Nondestructive testing		Catalyst for promoting synthesis of water	Palladium
Magnetic particle inspection	Iron	Coatings for papers	Aluminum, copper, tin
		Control of beer fermentation	Stainless steel
Nuclear engineering		Decolorizing alum solutions	Aluminum
Control rods and reflectors	Zirconium, beryllium, hafnium, uranium	Filters in chemical reactors	Stainless steel
Filters for liquid sodium-potassium, gas, heavy water	Stainless steel, nickel alloys	Hydrogenation, etc.	Platinum metals, aluminum, nickel
Fuel elements	Iron, stainless steel	Optical glass processing	Platinum alloy
High-density concrete	Iron	Rayon production	Platinum alloy
High-density rubber	Lead, brass	Synthesis of caffeine	Iron
Shielding	Aluminum	Synthesis of hydrocarbons	Iron
Shielding (gamma ray)	Tungsten, nickel, copper, lead		
Shielding (neutron)	Boron, nickel, iron, lead	**Pyrotechnics**	Aluminum, graphite, iron, magnesium
Computers	Iron		
		Radio and television	
Office equipment		Cores	Iron, MolyPermalloy
Copying machine parts	Iron, stainless steel, bronze, aluminum	Getters (TV tubes)	Barium, nickel
Facsimile equipment	Aluminum, steel, bronze, stainless steel	Permanent magnets	Iron, nickel, cobalt, aluminum, barium, ferrites, rare earths
Recording tapes	Iron	Printed circuits	Copper, silver
Toner for electrostatic copying machine	Iron	Receiving tubes	Tungsten, molybdenum, nickel
Business machines	Steel, iron, stainless steel, brass, aluminum	**Recreation**	
		Golf clubs	Tungsten, iron, brass
		Sporting darts	Tungsten
Ordnance (Military)		Hunting knives	Brass, stainless steel, iron, nickel silver
Ammunition	Graphite	Shotguns	Iron, stainless steel, steel
Armor-piercing cores	Tungsten, copper, nickel, cobalt	Fishing rod reels	Iron, brass, stainless steel
Anti-personnel bombs	Iron	Outboard motors	Brass, steel
First fire mixes and fuzes	Aluminum, graphite	Sailboat hardware	Stainless steel, bronze
Frangible bullets	Iron, lead		
Fuze parts	Brass, stainless steel, iron, steel	**Self-lubricating parts**	
Incendiary bombs, flares, tracers, torpedoes, etc.	Iron, aluminum, copper, cerium, magnesium, zinc, silicon	Oil-filled	Copper, tin, lead, iron, aluminum, graphite, bronze
Missile filters	Nickel-base alloy	Plastic-filled	Iron, copper, silver
Projectile rotating bands	Copper, iron, brass		
Proximity fuze cup	Nickel	**Sheet and strip**	
Rocket launcher parts	Stainless steel, aluminum	Alloy	Cobalt, iron, nickel, aluminum, tungsten, molybdenum
Solid missile fuel	Aluminum, magnesium	Glass sealing	Iron, cobalt, nickel alloy
		Nickel, cobalt	Nickel, cobalt
Personal products		Steel	Iron
Cigarette lighter flint	Cerium alloy		
Cordless electric toothbrush and razor	Copper, nickel, brass, iron, bronze	**Ships**	
Cosmetics	Zinc, aluminum	Anti-fouling paints	Copper, copper oxide
Enriched foodstuffs and cereals	Iron	Hull-smoothing cements	Iron
Fingernail lacquer	Aluminum, copper	Ship-bottom paints	Aluminum, zinc
Floating soap	Aluminum		
Mirror silvering	Iron	**Soundproofing**	
Pen points	Platinum, ruthenium, tungsten, stainless steel	Acoustical plastics	Lead, graphite
Photography flash bulbs	Zirconium, cerium	**Waterproofing**	
Poker chips	Brass, bronze, copper-nickel	Concrete	Iron, aluminum
Vitamins	Iron	Roof coatings	Iron, aluminum

(a) About 50% of all P/M parts are used in automobiles, many in engine and transmission applications.

Magnetic Particle Inspection

By J. Thomas Schmidt
Manager, Chemical Services
Magnaflux Corporation

MAGNETIC PARTICLE INSPEC-TION is a nondestructive method of detecting flaws or discontinuities in ferromagnetic parts that may cause failure in service. This inspection method reveals discontinuities that are located on or near the surface of the metal, even if the crack or defect is filled with dirt, paint, or other foreign matter. Magnetic particle inspection provides expedient, reliable, and economical production testing of applicable parts. However, some operator skill is required for the procedure to be performed successfully. Parts to be inspected are not damaged in any way and may be used, processed, or reinspected after use, usually without additional treatment. Magnetic particle inspection has contributed greatly to product reliability and safety with minimal cost.

When a material or part is magnetized, magnetic discontinuities that lie in a direction generally transverse to the direction of the magnetic field cause a leakage field to be formed at and above the surface of the part. The presence of this leakage field, and therefore the presence of the discontinuity, is detected by the use of finely divided ferromagnetic particles applied over the surface. Some of the particles adhere to the leakage field. This magnetically held collection of particles forms an outline of the discontinuity and generally indicates its location, size, shape, and extent. Magnetic particles are applied over a surface as dry particles, or as wet particles in a liquid carrier such as water or oil. Additional information on this nondestructive testing method can be found in the article "Magnetic-Particle Inspection" in Volume 11 of the 8th edition of *Metals Handbook*.

Magnetization

The part being inspected must be magnetized so that the magnetic lines of force

that are present in the specimen are perpendicular to the discontinuity. Parts may be magnetized by placing the poles of a magnet on a part at a 90° angle to the expected defect. Figure 1 shows one type of portable magnetic particle inspection equipment. A permanent magnetization source is often used for portable inspections, because additional power sources or heavy equipment is not required. However, permanent magnets can magnetize only small areas at a time. Also, a powerful magnet is needed, which may be hard to remove from the part. Repeated blows or vibration, which occur during normal

use and transportation of the equipment, eventually may demagnetize the magnet, thus reducing magnetizing power. With a permanent magnet, field strength cannot be varied easily, as the magnet has a fixed strength. Therefore, it is not possible to match the magnetization level to the part thickness, and optimum magnetization is seldom obtained.

Parts also may be magnetized by passing electrical current through them. Figure 2 shows typical equipment for this purpose, including a head and tailstock to which current is supplied. Current flows through the part, producing a magnetic field

Fig. 1 Permanent magnet yoke assembly being used to magnetize a test part

Fig. 2 Wet horizontal magnetizing equipment of the head and tailstock type
This equipment produces circular magnetization electrically.

Fig. 3 Part undergoing longitudinal magnetization in coil-type electrical magnetization equipment

at right angles to the current flow. Magnetization of this type can detect flaws that are approximately parallel to the current flow, normally along the length of the part. This equipment must be capable of producing high-amperage, low-voltage current. With this type of magnetization, contact usually is made by either clamping the test piece in a unit equipped with a head and tailstock, or placing hollow parts on a rod that is clamped in a unit with head and tailstock, or by clamps or prods that are attached to the part.

Electrical current can also be passed through a coil encircling the part to effect magnetization. Figure 3 shows a test specimen being magnetized by a coil that is attached to a magnetizing unit. This method detects discontinuities across the width of the part. Magnetizing equipment must be capable of producing high-amperage, low-voltage current, which may be costly. With this type of magnetization, electrical contact usually is not made directly with the part.

Magnetizing current may be either alternating or direct. Figure 4 shows a typical bench-type unit that is capable of producing alternating or direct current, or both types of current interchangeably. Although alternating current is easy to obtain, only surface discontinuities can be revealed by this inspection method. Consequently, alternating current should not be used for inspection of subsurface discontinuities. However, it is very useful for portable inspection when used in the form of an alternating current yoke.

Direct current, however, is capable of revealing surface and subsurface discontinuities and is usually obtained by rectifying alternating current. Therefore, direct current may be either half wave or full wave. Half wave direct current consists of a series of pulses, which may increase particle mobility. Magnetizing equipment required to produce this type of direct current is less expensive and less complex than equipment required to produce full wave rectified direct current. If pulsed current is not desirable, full wave rectified current should be used. Full wave rectified current produced from three-phase alternating current is nearly ripple-free.

The type of magnetizing effect obtained depends on the amperage of the current and the number of coil turns. Amperage must be maintained at a high level to obtain the maximum magnetization, while voltage must be kept as low as possible. This balance minimizes power usage and the amount of heat produced. Safety hazards are reduced, because low-voltage current

Fig. 4 Wet horizontal-type magnetic particle inspection unit
This type of unit is capable of magnetizing parts with either alternating or direct current.

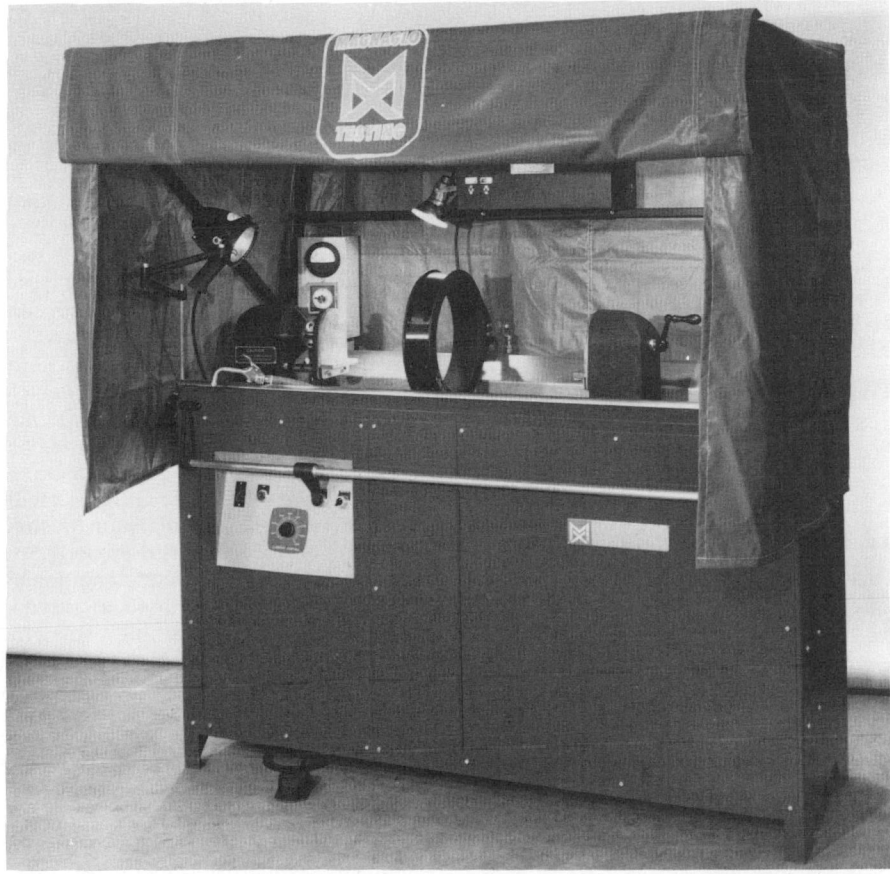

Fig. 6 Buildup of magnetic particles over a defect to indicate its presence

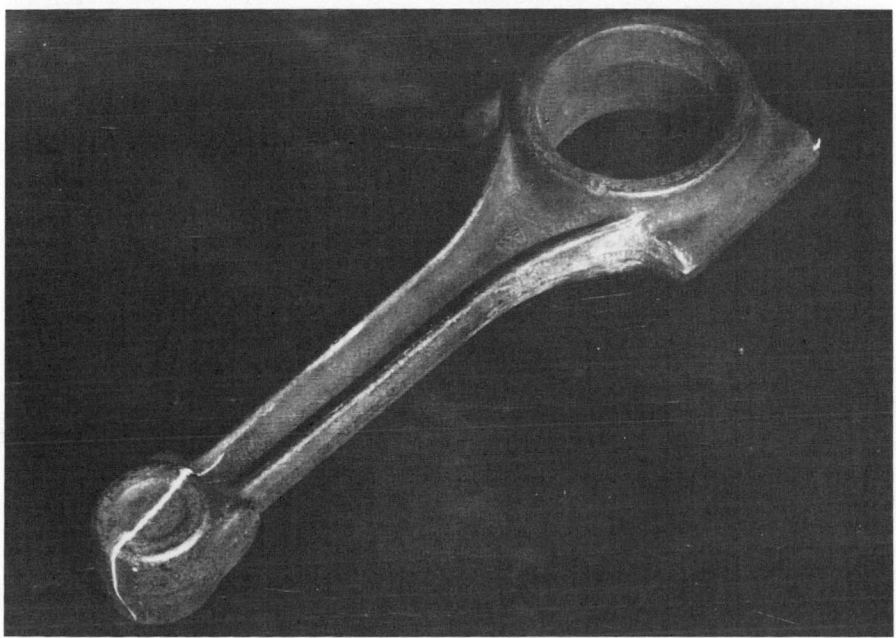

Fig. 5 Bridging of the air gap at a crack
(a) Leakage field. (b) Magnetic particle indication

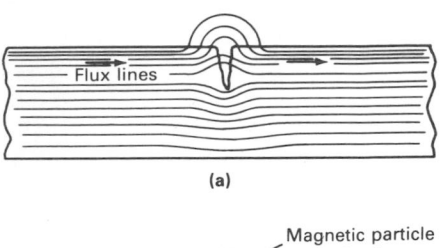

(a)

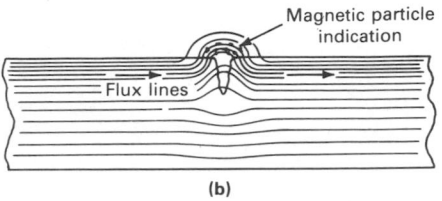

(b)

does not penetrate the skin and cause electrical shock.

Particle Application

The next step in magnetic particle inspection is the application of magnetic powder over the surface of the part to reveal the presence of a discontinuity. Figure 5 shows a diagram of a typical magnetic particle inspection application. The magnetic lines of force flow through the part in a direction induced by the magnetization process. If a part is continuous, the lines of force can flow through the part easily without interruption. If a part is not continuous—due to a defect—some lines of force are forced into the air over the defect. The external magnetic field is thus stronger over the defect than elsewhere on the part. The level of magnetization must be strong enough to hold particles over the defect, but not over other areas of the specimen surface. When particles are applied to the surface of the part, they build up over the defect, indicating its presence. Figure 6 shows a typical defect indication.

Magnetic Particles

Magnetic particles are classified according to the vehicle by which they are carried to the part—air (dry-particle method) or liquid (wet-particle method). Iron powder produced by filing originally was used for magnetic particle inspection. Frequently, iron oxide is used for current industrial applications. Particles must have high magnetic permeability so that they can be readily magnetized by the low-level leakage fields occurring around discontinuities and can be drawn by these fields to the discontinuities to form indications.

Fig. 7 Typical mechanical powder blower for application of dry-powder inspection material

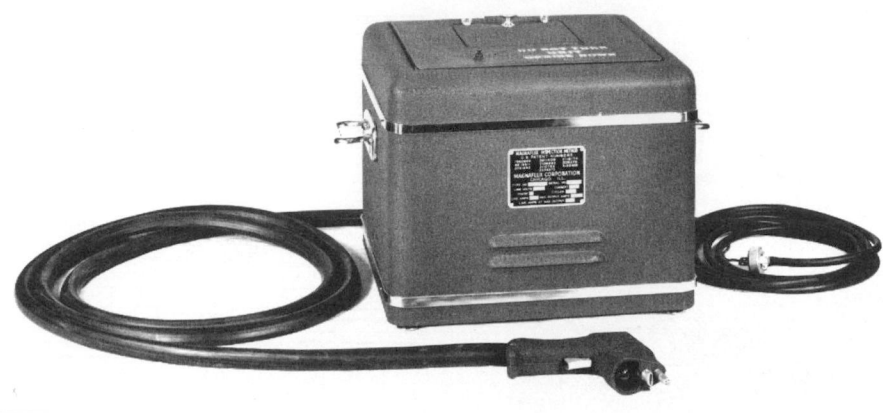

Fig. 8 Defect in engine block indicated by dry-powder magnetic particles

Dry-method particles usually are fine iron powders colored with pigment to aid detection. Typical powder colors include gray, yellow, red, or black. Powders may be applied by shaking from a shaker or blowing from a dustbulb or powder-blowing machine. Figure 7 shows a typical powder-blowing machine.

Excess powder may be shaken or blown from the part with a low-velocity air stream. Examination usually is conducted under room light. Dry powder particles are relatively coarse (20 to 175 μm) and are suitable for detecting large discontinuities rather than small ones. Dry powder particles are usable at all temperatures below 150 °C (300 °F). Certain types are stable up to 350 °C (660 °F) due to more thermally stable components used in their manufacture. Dry powders are not messy. This is particularly advantageous when large areas must be inspected, making recovery of wet-particle slurry difficult. Dry powders are also well suited to inspection for subsurface defects. Figure 8 shows a defect in an engine block detected by dry powders.

Wet-method particles usually are based on magnetic iron oxide, but some types are based on very fine commercially pure iron powder. Typical powder colors are red, black, or green-yellow fluores-cent. Typical particle sizes range from 1 to 40 μm. Powders are supplied in the form of a concentrate that is suspended in a liquid. Wet particles are best suited for detection of fine discontinuities, such as fatigue cracks. In addition, the ability to detect discontinuities in less accessible locations is enhanced by using wet methods. Wet particles commonly are used in stationary equipment where the bath can be circulated and remain in use until contaminated. They also are used in field operations with portable equipment. In such cases, the bath is expended on a single use. Care must be taken to agitate the bath constantly when the wet method is used.

Suspending Liquids. A light kerosene-type oil is a common liquid vehicle for particle suspension. This type of medium can be used on finished parts, because it is not corrosive and often leaves a protective residual film. Oil-type mediums are reusable and maintain desired concentration if recirculated properly.

Water containing a water-conditioning agent also is frequently used to suspend particles for inspection testing. This type of medium often is used on unfinished parts, where fluid loss is anticipated. A water conditioner is required to disperse the particles properly, to enhance wettability of smooth surfaces, and to provide complete, even coverage of the surface. Water-conditioning additives also prevent corrosion of tested specimens and testing equipment. Water baths tend to evaporate rapidly, thus causing fluctuations in bath concentration, and must be closely monitored.

Wet baths usually consist of very fine particles and are best suited for detection of fine discontinuities. Bath temperatures range from 0 to 100 °C (32 to 212 °F), because suspending liquids tend to boil or emit excessive fumes at higher temperatures. Such mediums usually freeze or become too viscous for adequate performance at lower temperatures. Wet baths must be agitated or the particles tend to settle out. Test specimens should be cleaned thoroughly to remove oil and grease; such contaminants reduce the effectiveness of the bath.

Fluorescent Particle Inspection

Many types of inspection particles are coated with fluorescent material that glows when illuminated by black (near ultraviolet) light. Ultraviolet light is not visible, but the induced fluorescence is. Figure 9 depicts several typical fluorescent indications. With fluorescent particles, inspec-

Fig. 9 Typical fluorescent magnetic particle indications

tion is conducted under ultraviolet light in the absence of visible light. Under these conditions, indications appear as light sources on a dark field. This inspection method produces high contrast, and indications are very visible.

Particles are made of pure iron or iron oxide-based powders. Because iron and iron oxide do not fluoresce (absorb light of one wavelength and re-emit it at another), particles must be coated with fluorescent materials. Fluorescent particles may be used in wet or dry inspection techniques. Most wet-particle inspection applications use fluorescent particles, because their high visibility enhances detection of small defects. These inspections usually are done in a permanent setup where electricity to power the black light is available. Few dry-method inspections are done with fluorescent particles, because dry powder indications are usually large and visible. Additionally, the larger particles used produce more background, which reduces contrast. Also, dry-method inspection is frequently done outdoors, where a darkened area is not readily available.

Some inspections are done under combined visible and ultraviolet light. In these cases, highly fluorescent particles are required, and the visible light should be subdued, if possible. Indications are more visible than those produced by nonfluorescent methods, but are less visible than those produced by fluorescent inspection. This procedure is used when complete darkening or lighting of the inspection area is not feasible. It is also used when parts will be handled in an inspection area that requires lighting for further processing operations.

Copier Powders

By Arthur S. Diamond
President
Diamond Research Corp.
and
Lewis O. Jones
Vice President—Engineering
D.M. Steward Manufacturing Co.

COPIER POWDERS are metal powders that play an important role in office copying and related reprographic applications, such as high-speed duplicating and computer output printing. These powders are combined with a finer black or colored thermoplastic powder called "toner" to form the developer mix used in many electrostatic imaging systems.

Because metal powders in a two-component electrostatic developer mix perform a "carrier" function—electrifying and transporting the dry ink (toner) particles that ultimately form the developed image—they are called "carrier" or "developer" powders. Ferrite granules or beads, which also act as electrostatic carriers, directly compete with metal powders for reprographic applications. A coated, spherical developer bead is shown in Fig. 1.

The term "copier powders" describes applications of metallic and nonmetallic developer powders in reprographic equipment. Although this broad term could be applied to each of the components in a developer mix—toner, carrier, and any additives—the term in this article is limited to metal powder carriers used in electrostatic image development.

Although two-component developer mixes are used in most plain paper office copier and computer printers, a one-component system also is widely used. In this system, carrier particles are not used; the toner contains sufficient magnetic pigment to be guided through the development process by a magnetic roller.

The history of electrostatic copier powders is shown in Table 1. The origins of electrostatic copying can be traced to 1777, when the German scientist Lichtenberg

produced unusual designs on glass by exposing the insulating surface to an electrical discharge and developing the invisible charge pattern by dusting with a powdered substance.

Other researchers pursued experiments with these curious "Lichtenberg figures," culminating with the work of Carlson and the invention of xerography in 1938 (Ref 1). This process is generically termed "electrophotography," because it depends on the action of light to create a latent electrostatically charged image.

The first commercial xerographic equipment became available in the early 1950's, when a manually operated Model D machine was introduced by Xerox for offset platemaking. This system used a two-component electrostatic developer. While Xerox and other copier manufacturers pursued the development of machines that

Fig. 1 Cross-sectional view of a coated spherical developer bead

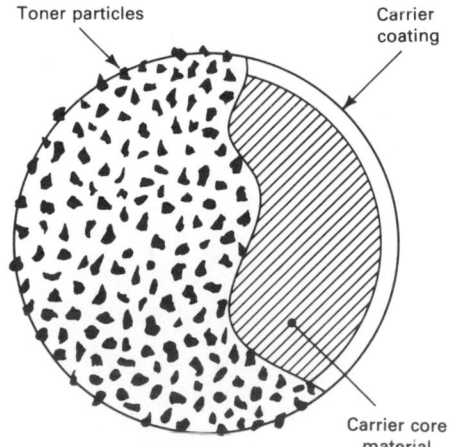

Toner particles

Carrier coating

Carrier core material

transferred the developed image to plain paper, early competitors (American Photocopy Equipment Co., The Charles Bruning Co., Inc., and SCM, Inc.) marketed copiers using direct electrophotography. In this method, the photoconductive layer is embodied in a coating contained on each sheet of copy paper that is passed through the process. The copier field was subsequently divided into plain and coated paper copiers.

In the ensuing years, as electrostatic copiers have become faster and more automated, a duplicating technology advanced to incorporate a means of outputting text and graphics from computers. The most significant advances in electrostatic copying have been in the area of image quality. Original equipment manufacturers, such as Kodak, Xerox, IBM, and a number of Japanese firms, have developed the state of the art to the point where image quality is competitive with offset lithography. Heavy, solid areas in original materials can be reproduced as uniform, rich, black images. Halftones and continuous-tone photographs also are reproducible with near-photographic quality.

Image resolution, another dimension in print quality, also has been improved since the early days of electrophotography. Laser beam xerographic printers are now capable of outputting copy with an image resolution of over 15 line pairs per millimetre (380 line pairs per inch).

Evolution of Copier Powders

In the early stages of electrophotography, off-the-shelf metal powders were

Table 1   History of electrostatic copier powders

Year	Event	Copier powder	Year	Event	Copier powder
1777	Lichtenberg demonstrates electrostatic image development	Lycopodium powder	1972	Xerox markets a plain paper copier based upon the finest spherical steel carrier ever used for xerography—the Xerox 3100, a 12-copy/min machine	Lacquered, 100-μm spherical iron carrier and toner
1920	Selenyi develops a facsimile recording process using an electrified stylus, writing on an insulating surface, and powder development	Lycopodium powder	1972	3M Company revolutionizes electrostatic image development with the VQC series of copiers—the first to use a single-component magnetic toner	Magnetic toner, 20 μm
1938	Chester Carlson invents "xerography," an electrostatic imaging process for plain paper copying	Uncoated iron carrier and toner	1973	Xerox introduces the first nonimpact computer printer based upon xerography—the Xerox 1200	Lacquered, 450-μm spherical steel carrier and toner
1950	Haloid introduces manually operated xerographic equipment for offset platemaking—the Model D	Lacquered, 600-μm sand carrier and toner	1973	Xerox rolls out the first xerographic color copier—the Xerox 6500	Lacquered, 100-μm nodular nickel carrier and toner
1954	RCA researchers Greig, Giaimo, and Young invent Electrofax, an electrostatic imaging process for coated paper copying using a magentic brush	Uncoated, irregular iron carrier and toner	1975	Xerox demonstrates the fastest xerographic copier-duplicator operating at 120 copies/min—the Xerox 9200	Lacquered, 80-μm spherical ferrite carrier and toner
1959	Haloid-Xerox introduces the first automated plain paper copier to use Carlson's xerographic technology—the Xerox 914, a 7-copy/min machine	Lacquered, 600-μm spherical glass carrier and toner	1975	Eastman Kodak enters the office copying field with equipment using nonspherical iron carrier—the Ektaprint 100/150	Lacquered, 170-μm irregular iron carrier and toner
1961	American Photocopy Equipment Co. markets the first automated plain paper copier to use RCA's Electrofax technology—the "Dri-Stat"	Uncoated, irregular iron carrier and toner	1976	IBM commercializes the first laser beam xerographic computer output printer—the IBM 3800 operates at over 150 pages/min	Teflon-coated, 170-μm spherical iron carrier and toner
1964	Xerox announces the first copier-duplicator to operate at 40 copies/min—the Xerox 2400	Lacquered, 600-μm sand carrier and toner	1979	Minolta markets a copier using "micro-toning" technology, the first electrostatic developer to use a fine particle carrier—the EP-310	Thermoplastic, 40-μm magnetic carrier and toner
1969	Xerox demonstrates a major advance in xerographic copier quality using a conductive steel bead carrier and cascade development—the Xerox 3600-III, a 60-copy/min machine	Lacquered, 450-μm spherical steel carrier and toner	1979	Canon introduces Ion Projection Development in the first practical plain paper copier to use one-component toner—the Canon NP-200	Magnetic toner, 10-15 μm
1970	Xerox unveils a compact high-speed plain paper copier—the Xerox 4000, a 43-copy/min machine	Lacquered, 250-μm spherical steel carrier and toner	1980	Xerox offers the first copier in its line to use irregular iron carrier—the Xerox 3300	Lacquered, 100-μm irregular iron carrier and toner
1971	IBM enters the office copying field with a 10-copy/min machine—the Copier I	Teflon-coated, 170-μm spherical carrier and toner	1983	Xerox debuts its first office copier to use an organic photoreceptor—the Xerox 1075	Lacquered, 130-μm irregular iron carrier and toner

adopted for use in electrostatic copying equipment. Some of these powders were abrasive steels originally intended for shot blasting or scarfing operations, such as the cleaning of metallic or nonmetallic parts. Additionally, metal powders designed for compaction into pressed parts were used in the developer apparatus of some electrostatic copying machines. As a rule, original equipment manufacturers purchased tailings or by-products from metal powder producers rather than design a custom copier powder.

As the industry grew, specifications evolved that placed tolerances on particle size and shape, contamination, and other properties. Eventually, some metal powder manufacturers produced photocopier powders or carrier core materials. Currently, three types of powders are manufactured for the copying industry: spherical iron powders produced by the rotating electrode process, spherical ferrite powders produced by a spray drying and firing operation, and plasma spheroidized mag-

netite powders. Other spherical and irregularly shaped carrier beads usually consist of select grades of existing metal powders.

Efforts continue to fabricate the ultimate copier powder—one with minimum apparent density, maximum surface area, good flow characteristics, high conductivity, and optimum magnetic properties. The industry is searching for a lightweight particle (low bulk density) that can carry a maximum amount of toner. However, for any given diameter, spherical particles, which have the best mixing and flow properties, also have the smallest specific surface.

As shown in Table 1, technology is moving toward finer particle size. Available surface area increases by the square of the inverse ratio of carrier diameters. Another trend is the construction of more sophisticated developer mechanisms that are capable of handling finer spherical and nonspherical copier powders.

Market Size and Growth. Almost 13 000 metric tons (14 200 tons) of copier

powder were estimated to be consumed domestically in 1983 in all electrostatic imaging applications (Table 2), as compared to 10 830 metric tons (11 900 tons) in 1981 (Ref 2, 3). Approximately 72% of that volume is shared by spherical and irregular iron and steel copier powders; approximately 24% is attributable to ferrite and magnetite powders; and the balance of the copier powder is sand or flint glass shot. The majority of these copier powders are coated with a thin resin layer to improve copy quality and useful life.

Metal powders probably will continue to dominate xerographic applications in the near future. Indications are that middle- and low-volume copiers will utilize irregularly shaped iron and steel powders in the period from 1983 to 1986. A 15% annual growth rate was assumed in calculating the 1986 data for irregularly shaped iron and steel powders shown in Table 2. The demand for spherical iron and steel beads is expected to increase at an annual growth rate of about 10% during this same interval.

Table 2   Estimated U.S. consumption of copier powders

Powder shape	1981		1983		1986	
	Metric tons	Tons	Metric tons	Tons	Metric tons	Tons
Spherical powders						
Iron	4 080	4 488	4 800	5 280	6 400	7 040
Ferrite	2 165	2 381	2 700	2 970	4 140	4 550
Magnetite	300	330	360	396	450	495
Nickel	8	9	10	11	8	9
Subtotal	6 553	7 208	7 920	8 657	10 998	12 094
Irregularly shaped powders						
Iron and steel	3 742	4 116	4 500	4 950	6 840	7 524
Sand and glass	535	588	540	594	500	550
Subtotal	4 277	4 704	5 040	5 544	7 340	8 276
Total	10 830	11 912	12 960	14 201	18 338	20 370

Copier Powder Recycling. Exhausted ferrite powders can be reclaimed by solvent washing or by refiring to burn off the lacquer coating and any impacted toner. These bead-cleaning operations are followed by a relacquering step. Reclamation of spherical ferrite powders in 1981 probably accounted for less than 15% of total production. By 1986, at least one third of the demand for these ferrite powders is expected to be filled by recycled carrier core material.

Spherical iron and steel powders also may be recovered from spent developer mixes returned by field service personnel. Solvent washing is used to prepare these beads for recoating. This process is not as cost-effective as the refiring of ferrite powders. It is more expensive to operate, and the cost of the metal powder is less than ferrite spheres.

Only about 10% of the spherical iron and steel powders used in 1981 were recycled materials. Recycling may not increase above 15% for these copier powders within the next 3 years.

Development Processes

There are two commercially important electrostatic copying methods—the direct (Electrofax) or coated paper process and the transfer (xerography) or plain paper process. Both processes utilize a photoconductive layer, on which an invisible pattern of electrostatic charges is formed. In the direct process, a zinc oxide-coated paper is given a blanket electrostatic charge (300 to 500 V) in the dark. The dark-adapted zinc oxide is a suitable insulator and supports the charge. Next, the sheet is exposed to a light image. The surface charge is instantaneously dissipated in light-struck areas where the zinc oxide layer be-

comes conductive. An invisible pattern of charges remains—a latent electrostatic image corresponding to the dark regions of the original document.

In the development step, this latent image is made visible as the charged areas attract toner particles. Dry, powdered toners are subsequently fixed in place on the zinc oxide sheet by the application of heat and/or pressure. This completes the direct process, in which the photoconductor is disposable.

In the transfer process, a reusable photoconductive drum or endless belt is employed. The developed image is transported intact from this "photoreceptor" to paper, film, or other substrate on which it is subsequently fixed.

Four methods are used for dry toner development: cascade, magnetic brush, powder cloud, and touchdown. Cascade and magnetic brush are the most widely used and provide dry toner development for most plain paper electrostatic copying, duplicating, and printing equipment. These processes are the only ones that employ copier powders. Powder cloud is limited to xeroradiography, and touchdown is used in experimental or noncommercial applications. More information on image development can be found in Ref 4 and 5.

Cascade development uses spherical or near-spherical beads to carry finely divided toner powder to the imaging surface. These carrier particles strike the charged layer, rolling and bouncing across the surface. Both effects are important. With bouncing contact, impact shakes some toner loose, causing it to become airborne. Airborne particles are readily available for development, being attracted to areas of even low surface potential.

The rest of the carrier beads maintain their coating of toner as they roll or slip across the imaging surface. They only give up toner to areas of relatively high charge

that are capable of stripping particles from the bead. Thus, the relative ratio of airborne particles to stripping particles controls image contrast, that is, the shape of the density-charge curve.

In cascade development, the shape of the carrier bead is more critical than that of the toner particle, and a spherical or a near-spherical carrier generally is preferred. However, all commercially important toners for cascade development have nonspherical or fragment-shaped particles.

One limitation of cascade development, particularly when an insulating carrier such as sand or spherical glass shot is used, is the inability of the system to reproduce solid black areas. Cascade development does not provide a conductive path for electrical discharge or developer biasing, a field-enhancement technique that produces solid area fill-in. Improvements have been made with the introduction of conductive metal copier powders and the use of a closely spaced development electrode. However, the greatest improvement in solid area fill-in has been realized by a switch to magnetic brush development.

Magnetic brush equipment was invented by Giaimo and Young (Ref 6-8). Based on copier powder consumption, two-component magnetic brush development has become the most important dry toner imaging process. This method uses a mixture of magnetic carrier beads with about 1 to 4 wt% toner particles. The size of the magnetic bead ranges from about 40 to 450 μm; toners range from 10 to 15 μm in diameter. A comprehensive review of magnetic brush development is given in Ref 9.

Figure 2 illustrates magnetic brush development. Developer mix is attracted to

Fig. 2 Magnetic brush development

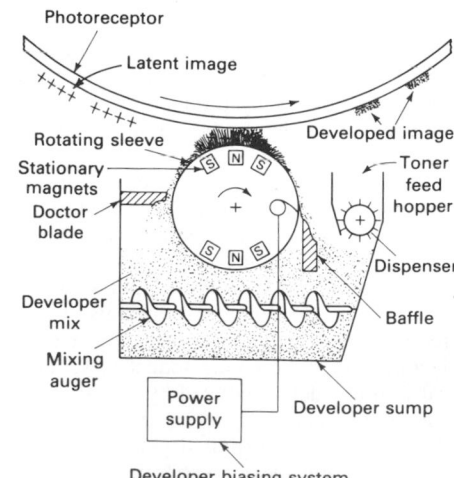

the surface of the rotating roller, which contains stationary magnets positioned inside. The thickness of this surface layer of developer is controlled by a doctor blade. The carrier material aligns itself in a brush-like configuration under the influence of the applied magnetic field. Bristle formations emanate from the surface of the rotating member and contact the photoconductive layer. Toner is stripped from the carrier beads when the electrostatic force of the image exceeds the triboelectric bond by which toner clings to the copier powder. As the sleeve rotates past the upper magnets, centrifugal and gravitational forces cause the developer mix to return to the sump. A baffle sometimes is positioned at the point where the field is relaxed to guide the spent mix toward the incoming stream of replenishment toner.

Fresh toner is added automatically to the developer sump by a hopper and dispensing apparatus to replenish the dry ink consumed in the imaging process. This toner is distributed uniformly throughout the developer mix in a powder blending operation performed by mixing screws, augers, or ribbon blending devices. The measurement, or sensing, of toner concentration in the developer mix has proven difficult for equipment designers. Electrical resistivity, magnetic properties, and optical reflectance are methods that have been used to determine toner concentration. When a lean mix is detected, the control device signals the feed hopper to add toner to the developer housing.

Brush structure is important in two-component magnetic brushes, as this determines the coverage or degree of intimate contact with which toner is presented to the charged surface. The fibrous mat, or pile, that is formed may be either "soft" or "hard," depending on the magnetic properties and particle shape of the carrier powder. A soft brush structure is generated by lower magnetic moment, low-bulk density materials and yields good copy quality, but may produce poor solid area fill-in. A harder brush structure generated by higher magnetic moment materials may provide better solid fill-in, but also tends to generate "raking" lines in the image and to develop copy with ragged rather than sharp edges.

Higher peripheral speed increases the degree of image development, but brush speed is limited by the discharge of toner into the air by centrifugal force. Carrier shape is another key variable. Possible shapes vary from perfect spheres to rod-like shapes, flakes, and twisted interlocking ribbons. Spherical carriers are free flowing, thereby enabling replenishment

toner to be rapidly mixed and uniformly blended into the developer. Also, the absence of sharp-edged particles prolongs developer life by minimizing attrition and extends photoreceptor life by reducing abrasion.

Powder cloud development uses air as the carrier or toner transport medium; solid carrier particles are not used. Air suspends extremely fine toner particles (about 1 to 5 μm) and transports them to the latent electrostatic image for development. Air impact is also the scavenging mechanism used to dislodge background toner deposits that are not electrostatically locked in place. Powder cloud development currently is used in medical xeroradiography to develop mammograms for breast cancer diagnosis.

Touchdown development is essentially a carrierless or one-component process in which a cylinder applicator roll or sheet surface that is bearing a layer of toner powder contacts the imaging member. Excellent solid area fill-in is possible due to the effectiveness of biasing in such a configuration. However, touchdown development remains a laboratory development process. It has not found application in commercial copying or printing equipment.

Other methods of dry toner development include three types that do not involve the use of metallic copier powders: one-component toners, the Toner Projection Development system developed by Canon, and the Microtoning Process produced by Minolta. Monocomponent toners, which do not use carriers, are magnetic, thermoplastic particles that function both as an imaging ink and as a carrier material. They are applied by means of a magnetic brush formed on a magnetized roller, similar to two-component mixtures. Particle size of the early monocomponent toners was somewhat larger than conventional dry, electrostatic ink (about 18 to 25 μm). Monocomponent toners offer the advantages of two-component toners, but eliminate some of the limitations. They can be applied with a relatively simple developer apparatus in which mixing is not required during the replenishment, and measurement and control of toner concentration are not necessary.

The so-called "jumping developer" (Toner Projection Development) introduced by Canon is a monocomponent system. In this process, toner particles "jump" across an air gap, thus reducing background dusting (Ref 10). A key feature of the Canon system is the use of a toner particle with high resistivity, which enables the developed image to be transferred from the photoreceptive drum to plain paper.

Prior to this technology, single-component toners were limited to coated paper copier systems. Furthermore, the toner used in the Canon Toner Projection Development Process has a particle size of 10 to 15 μm. This finer particle size affords image resolution that is competitive with conventional two-component developer systems.

The Microtoning Process was introduced in the Minolta EP-310 copier in 1979. It offers an improvement in image resolution and gray scale rendition, because the carrier particles are much smaller than conventional two-component carriers (approximately 20 to 30 μm in diameter). This enables use of a finer toner particle (5 to 10 μm) than existing systems. Additionally, the nature of the copier powder is different. Because it is a thermoplastic material capable of heat fixing, particles or carriers can be fused into the same image areas, if they are deposited with the toner. This prevents the carrier particles from being carried out on each sheet as a loose powder.

Charge Polarity

The magnitude and polarity of the static charges generated between toner and carrier are determined by the relative positions of the two materials in a triboelectric series.* The following triboelectric series lists components that range from electron donors to electron acceptors (Ref 3):

- Air
- Glass
- Nylon, wool, and silk
- Aluminum
- Cellulose
- Cellulose acetate
- Polymethyl methacrylate
- Iron
- Polyester
- Polyurethane
- Polystyrene
- Polyethylene
- Polypropylene
- Polytetrafluoroethylene

Because triboelectric charging is a surface phenomenon, it is affected by impurities, additives, and adsorbed species. This dependency is believed to account for the differences among triboelectric series published by various authors (Ref 11).

Due to charge polarity, toners com-

---

*A triboelectric series is a list of materials that produce an electrostatic charge when rubbed together, arranged in such an order that a material has a positive charge when rubbed with a material below it in the list, and has a negative charge when rubbed with a material above it in the list.

pounded with polystyrene resin are more electronegative than uncoated iron beads or iron containing a polymethyl methacrylate coating and tend to develop a negative charge when dry blended with such a carrier particle. Selenium and selenium alloy photoreceptors are normally given a blanket positive charge prior to image exposure. Therefore, development must be accomplished with negatively charged toner. Organic photoconductors, in most cases, are given a blanket negative charge; hence, they require a positively charged toner particle for development.

Charge polarity and the charge-to-mass ratio can be changed by the use of certain charge control agents, usually dyes, metallic soaps, surfactants, or other electroactive materials. Thus, the relative positions of toner and carrier in a triboelectric series are not the sole determinant of charge polarity.

Positive and Negative Development

Although "positive" and "negative" are used to describe the polarity of a toner particle, these terms also are used frequently, sometimes with confusion, to describe the mode of development. In positive development, light and dark areas of the original document are reproduced in a similar manner. In negative (or reversal) development, dark areas of the original are reproduced as light areas on the copy and vice versa. Charged toner particles are deposited in uncharged areas of the latent image. Usually, this is accomplished with a toner that has the same polarity as the electrostatic image. Thus, in reversal development, particles are repelled out of the charged image areas and into the uncharged (exposed) areas of the photoreceptor.

Reversal development is often used in laser beam xerographic systems, in which the scanning laser is used to "write" alphanumeric or graphic information. If a positive development system is employed, the laser beam must remove all charges on the photoreceptor surface except those in image areas. This requires that the beam be "on" for a longer period of time, because the image area usually occupies only a small percentage of each page.

Plain and Coated Paper Copiers

Generally, the simpler the equipment, the more complex and costly are the supplies. Thus, plain paper electrostatic copiers, which rely on image transfer to re-

locate the developed image from the photoreceptor surface to a sheet of bond paper, are more complex and more expensive than coated paper devices. However, as machine cost is amortized over a larger number of impressions, the plain paper process becomes the most economically attractive; it yields the lowest cost per copy in both office copying and computer printing.

For many years, coated paper copiers were popular among low-end users. Machine costs were typically below $2000, a price that was affordable for small offices. Because plain paper copiers were two to three times as expensive, economic considerations outweighed consumer objections to the texture and weight of coated paper. The popularity of these machines has declined considerably, however, since the peak in the early 1970's. Currently, plain paper copiers have dropped sharply in price and are capturing an increasingly larger share of the low-end market.

Copier Powders

Copier powders perform three essential functions in electrostatic image development: (1) they charge the finer toner particles to the proper level and appropriate polarity so that they can be attracted to the latent electrostatic image; (2) they physically transport toner from the developer reservoir, or sump, to the photoreceptor surface, where they are presented for development and, ultimately, are deposited in charged image areas; and (3) they act as a cleansing medium to remove toner particles that have been spuriously deposited in nonimage areas. In the cleansing function, copier powders also serve to scavenge background toner deposits, depending on machine design configuration. To accomplish these functions, electrostatic carrier powders must have the proper chemical, electrical, physical, and magnetic properties.

Carrier Core Composition

In powder development, the carrier is a particle formed from sand, glass, iron, steel, nickel, aluminum, or ferrite materials. Ranging from about 40 to 600 μm in diameter, these granules can be irregular or spherical (often called "shot") in shape. As illustrated in Fig. 1, these developer beads carry much smaller toner particles (from 5 to 30 μm in diameter) on their surfaces. They bring the toner powder close enough to the charged, latent image for development to occur. Electrostatic attraction causes the fine toner particles to be stripped from the carrier bead

and to be deposited in image areas on the photoreceptor drum or copy sheet.

Iron and Steel. Irregularly shaped iron powders were first used in the early Electrofax machines of the 1960's. These magnetic brush devices required a conductive, ferromagnetic granule to carry toner. Cascade developers that used spherical steel powders became prominent in 1969, when the Xerox 3600-III copier-duplicator was introduced. Currently, magnetic brush devices use a variety of metal powder shapes, including sponge, fragment, and flake iron powders (Ref 12, 13).

Ferrite. The introduction of spherical ferrite copier powders was a major advancement in the field of carrier bead technology. References 14 to 18 list several important patents and the zinc-nickel ferrites used in the processes they describe. Spherical ferrites and spheroidized magnetite powders offer certain advantages over iron carrier core materials. Due to low remanence, these powders have improved flow properties, which enable rapid release from the magnetic brush roller when the field is relaxed. They are more resistant to changes in resistivity, even when uncoated, because the material is already in its fully oxidized state. An increase in resistivity causes a decrease in the solid fill. Furthermore, spherical ferrite copier powders are easier to reclaim from an exhausted developer mix, either by solvent washing or by high-temperature firing. As metal oxides, however, they have higher resistivities than metallic copier powders.

Sand. Uniformly graded sand was the first carrier core material used in office copiers based on cascade development. In these machines, the raw sand, or "flint shot," is coated with a dyed resin lacquer composition. In the United States, many of these older copiers and duplicators have been refitted with magnetic brush-type developer assemblies. Because the newer models use iron beads, the requirement for flint shot is relatively small. However, sand cascade machines are still in operation in many overseas markets.

Glass. Glass shot is a substitute for sand in cascade developer systems. This shot consists of extremely fine spherical beads. Although improved shape offers certain advantages in mixing and flow characteristics, higher cost has limited its use.

Aluminum. Another copier powder used in cascade development is aluminum shot. Initially, its low bulk density had strong appeal for equipment designers, because it required minimal torque for mixing and blending. Also attractive were its low cost and conductive properties that yielded improved print quality over insulating-type cascade carrier core materials. Aluminum is not widely used as a core material, however, because it is nonmagnetic and has been superseded by the more advanced magnetic brush-type xerographic systems.

Nickel. Spherical or nodular-shaped nickel powder enjoys limited application as a carrier core material (Ref 19). Although relatively expensive, this material is not prone to oxidation and tends to maintain a uniform resistivity throughout the life of the developer mix.

Electrical Properties

Triboelectric charge is measured in terms of a charge-to-mass ratio and, normally, is expressed in units of microcoulombs per gram. A variety of charge-to-mass ratio tests exist, of which the Faraday cage, or "blow-off" technique (Ref 20), is perhaps the most common. In this test, carrier particles are mixed with a small percentage of toner, approximating the toner concentration of an operating developer mix. Next, this mixture is placed in a metallic cylinder fitted with a fine-mesh screen at each end. The cylinder usually is constructed of brass or stainless steel and is designed to be opened easily. The cylinder and its contents are accurately weighed, before placement on a ground-insulated fixture, where it is held firmly.

Next, a sensitive electrometer is connected between the Faraday cage and ground, and dry air or nitrogen gas is used to blow the toner off the carrier beads. This can be accomplished manually by using a spray nozzle directed at the contents of the cylinder that discharges against one of the screened openings. After all traces of toner have been blown from the beads, the Faraday cage and its contents are reweighed. The difference in weight, divided by the recorded charge level, yields the charge-to-mass ratio. The same readings also measure charge polarity and toner concentration. Typical values for the charge-to-mass ratio range from 12 to 20 μC/g.

Conductivity. The conductive nature of the metallic developer bead produced the first improvement in solid area development of xerographic imaging systems. The metal roller that transports the developer mix serves as a counter electrode. Either an applied bias voltage or ground connection enhances the electric field of the charged latent image, producing rich, black solids. A grounded electrode is often positioned near the developer; this electrode enhances the electric field due to its close proximity. Copier powder resistivity

is affected by core material composition and by the nature of coatings applied to prolong developer life. Developer bead design involves a trade-off in achieving good toner release characteristics while maximizing conductivity to maintain a conductive path for field enhancement.

Physical Properties

Particle size and shape determine the amount of active surface area available for (1) charging the toner by contact electrification, (2) carrying toner to image areas of the photoreceptor, and (3) scavenging spurious toner deposits from background areas. They also control magnetic brush bristle structure, image resolution, and continuous-tone reproduction.

Spherical particles are preferred. Their shape enhances powder flow and reduces the torque needed to agitate the mix reservoir. Although spheres may accelerate the dry blending of toner added to replenish the mix, irregularly shaped particles have other advantages. In magnetic brush development, they tend to form longer, more flexible bristles, providing good tone rendition and solid area fill-in.

More importantly, irregularly shaped particles have larger specific surface areas than spherical particles. Thus, they are capable of carrying a higher percentage of toner in a developer mix. Consequently, the mix is less prone to being "starved," a condition that results from printing multiple impressions of an image with heavy, solid areas.

Bulk Density. Table 3 compares the true density, bulk (apparent) density, and calculated surface area values for the major copier powders. Low bulk density is desirable, because it requires less torque to agitate the mix. Therefore, less power is consumed by the device and less heat is dissipated in the mixing chamber. At elevated temperatures, a greater tendency toward toner filming exists. Furthermore, a low-bulk density material (2.5 to 3.0 g/cm³) enables the use of high magnetic filler

velocities, which are necessary for high-volume printing applications, without subjecting the developer bead to excessive centrifugal forces.

Specific Surface. Calculated values for specific surface area are more representative of toner carrying capability than those obtained by nitrogen absorption (BET) methods. This is especially true of porous copier powders that have "internal" surfaces that are unavailable as toner carrying or contact charging sites.

Magnetic Properties

In a two-component magnetic brush system, the carrier particle must be magnetizable. This enables the magnetic brush roller, which contains permanent magnets mounted inside the rotating sleeve, to gather the toner-encrusted developer particles from the developer reservoir and bring them to the photoreceptor for image development (Fig. 2).

The magnetic properties of the carrier, particularly remanence and specific moment (or saturation magnetization), are important factors in the performance of the developer system. Ideally, a low-remanent material is desired. Carrier particles should not have residual magnetism when the brush collapses as the sleeve rotates out of the influence of the magnetic field. Any tendency for clumping or "chaining" of carrier beads due to attractive forces inhibits flow and blending of the developer mix by the mixing augers. Magnetic properties for iron, steel, ferrite, and magnetite copier powders are given in Table 4.

Coatings and Surface Treatments

Carrier coatings serve as a release layer to prevent toner from impacting (filming) on the bead surface. Additionally, they provide a smooth, uniform carrier surface for toner attraction. Carrier coatings inhibit oxidation of steel powders and re-

Table 3 Typical density and surface area for electrostatic copier powders
Based on 100-μm particles

Copier powder	True density, g/cm³	Bulk density(a), g/cm³	Calculated surface area, cm²/g
Spherical iron, rotating electrode process	7.9	4.4-4.5	100
Spherical iron, water atomization process	7.9	3.0-3.5	100-150
Sponge iron	7.9	2.0-2.8	150-200
Electrolytic iron	7.9	2.4-2.7	150-200
Spherical ferrite	4.9-5.1	2.2-2.7	175-200
Spheroidized magnetite	5.1-5.2	2.6-2.8	(b)

(a) ASTM B 213 (Hall flowmeter method). (b) Not available; nitrogen absorption (BET) area is 375.

Table 4 Magnetic properties of electrostatic copier powders

Copier powder	Specific moment, emu/g	Remanence, emu/g
Pure iron	210	1-3
Carbon steel	170-190	5-10
Spheroidized magnetite	83-90	1.3
Ferrite	40-60	<1.0

The electromagnetic system of units is a centimetre-gram-second system of electric and magnetic units in which the unit of current is defined as the current which, if maintained in two straight parallel wires having infinite length and being 1 cm apart in vacuum, would produce between these conductors a force of 2 dynes/cm of length; other units are derived from this definition by assigning unit coefficients in equations relating electric and magnetic quantities.

duce the moisture sensitivity of the developer mix.

Carrier coatings prolong the life of the developer mix and protect the photoreceptor from abrasive carrier core materials. They control charge polarity through the use of certain dyes or other charge control agents that may be incorporated in the coating layer. Also, carrier coatings enhance the triboelectric effect.

Controlled electrical discharge is the basis of the xerographic process. Proper carrier coating is an effective way of exercising this control. Coatings vary from 0.2 to 4.0% of the core material weight, depending on the porosity of the carrier bead. The more porous carriers, such as reduced (sponge) iron, require a heavier coat weight.

The use of external additives that are dry blended with the developer mix is another method of controlling electrical discharge. These additives usually are incorporated in small amounts, from 0.25 to 2.0 wt%. They improve powder flow, modify triboelectric charge generation, increase carrier life, and improve photoreceptor performance. Furthermore, the proper additive can prevent toner accumulation on the carrier surface, the photoreceptor drum, or both. Amorphous silica is the most frequently used external additive, because of its excellent performance as a promoter of powder flow, or as a flow lubricant. Zinc stearate is used frequently in the toner industry as an anticaking and drum cleaning agent.

Carrier coatings and treatments include a broad variety of methods and materials, ranging from the use of raw or solvent-washed developer beads to the use of a controlled oxidation process to treat the surface of iron and steel powders. "Blued" iron and steel powder is frequently used as a carrier in the electrostatic copier industry.

Surface resistivity is lowest with uncoated or oxidized carriers. As the surface coating becomes thicker, the resistivity increases, thus making the particle less capable of effecting complete discharge of the developed image through the grounded magnetic brush. Lacquer-coated carriers provide the best compromise between print quality (low resistivity) and developer life (good release properties). A variety of polymeric materials have been used effectively for carrier core coatings. Styrene-acrylic copolymers, vinyl chloride-acetate copolymers, cellulose acetate butyrate, and nitrocellulose compositions are perhaps the most popular. A unique carrier coating resin that consists of an organosilicon terpolymer is described in Ref 21 and 22. Additional information on powder coating processes is included in the last section of this article.

Yields

Copier yield, in terms of copies per kilogram of toner or developer, varies considerably, depending on the imaging application and the type of carrier used. For example, most toners deliver approximately 20 000 to 25 000 copies/kg (9 000 to 14 000 copies/lb) in copying and duplicating applications, but as many as 35 000 copies/kg (16 000 copies/lb) on nonimpact computer printers. On the computer printer, lower image density is acceptable, and the image area on a page is typically less for alphanumeric printouts than it is in copier applications, in which more graphics and heavy solids may be present.

Carrier yields vary even more widely, from approximately 10 000 to 60 000 copies/kg (4 500 to 27 000 copies/lb). The lower figure represents uncoated core material; the upper number applies to fluoropolymer (polytetrafluoroethylene) coated beads. Generally, a yield of about 26 000 copies/kg (12 000 copies/lb) for dry toner and for carrier core materials can be expected—approximately 1 kg of developer to every kilogram of toner that is consumed.

Typical developer yields for magnetic brush development are illustrated in Table 5, which shows the advantages of coating metal core materials. The nature of the de-

velopment process also determines the useful life of a carrier powder. The same materials used in a cascade development system, for example, tend to fail sooner because of toner filming.

Powder Manufacturing Processes

Three major processes are used to produce spherical powders, and three major processes are used to manufacture irregularly shaped powder for the copying industry. Spherical powder manufacture utilizes the rotating electrode, water atomization, and spray drying (used for the production of spherical ferrite powders) processes. Additionally, electrolytic production of spherical (nodular) nickel powders serves a narrow segment of the market for copier powders. Irregularly shaped copier powders are manufactured by water atomization, reduced iron, and electrolytic deposition processes.

Spherical Powders

Rotating Electrode Process. Nuclear Metals, Inc. developed the rotating electrode process, a proprietary method for producing high-purity spherical powders. Figure 3 shows the procedure, in which a solid bar of the metal to be made into powder is rotated around its axis at very high speeds, typically 15 000 rpm.

The end of the rotating bar is melted by a powerful electric arc. Liquid metal is thrown from the face of the rotating bar by centrifugal force, forming small droplets that solidify rapidly into perfect spheres in sizes ranging from 35 to 325 mesh. Their purity is equivalent to that of the starting material. The solid, spherical powder is removed at the end of a run from the bottom of the rotating electrode processing machine through the powder collection port.

The rotating bar electrode may be from 25 to 75 mm (1 to 3 in.) in diameter and up to 1.8 m (6 ft) long. It enters the machine chamber, which is filled with an inert gas, through a special rotating gastight seal. As the arc is struck and the face of the bar is melted, the bar is fed continuously into the chamber until all but the very

Table 5 Typical developer yields for magnetic brush systems

Copier powder	Developer yield Copies/kg	Copies/lb
Raw iron	5 000-7 000	2 300-3 200
Blued iron and steel	8 000-10 000	3 600-4 500
Lacquered iron and steel	15 000-25 000	6 800-11 000
Blued and lacquered iron and steel	30 000-50 000	13 600-23 000
Lacquered ferrite	40 000-90 000	18 000-41 000
Teflon-coated steel	50 000-100 000	23 000-45 000

Fig. 3 Rotating electrode process
Courtesy of Nuclear Metals, Inc.

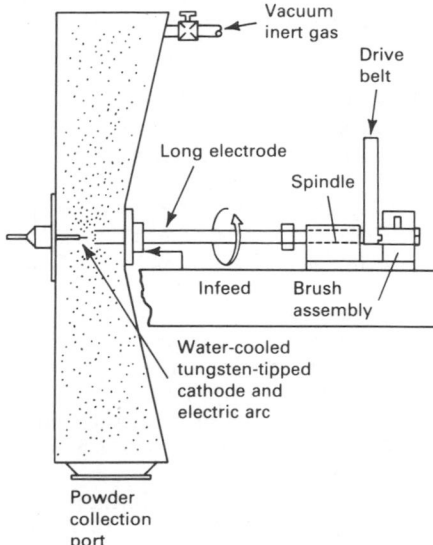

end is consumed. The chamber has a diameter sufficiently large to permit the liquid droplets to solidify before striking the wall. For more information, see the article "Rotating Electrode Process" in this Volume.

Water Atomization. In the water atomization process, a molten stream of metal is interrupted with a water jet, flowing almost coaxially, that disperses the liquid into a spray of droplets. Surface tension acts on these molten particles as they are propelled, giving them their roughly spherical shape. This is the minimum free energy state of a particle in motion.

The impacting jet of water does not cool the stream measurably, serving primarily as a dispersing force. The spherical droplets subsequently are quenched in a water bath. Nonspherical particles are later separated by one of several methods, such as by the difference in their ability to be retained on an upward moving, smooth inclined belt. This difference is one of friction based on contact area. Vibrating and rotating tables also are used for this separation, frequently coupled with a high-velocity air stream. See the article "Gas and Water Atomization" in this Volume for additional information on this process.

Hydrometallurgical Nickel. Nickel powders are produced by hydrogen reduction of nickel particles from an aqueous nickel ammonium sulfate solution. The solution containing nickel compounds is pumped into an autoclave, where small amounts of nickel metal particles are added to act as seed crystals. The mixture is agitated and hydrogen is introduced, reducing the nickel sulfate to nickel metal on the seed crystal surfaces. This process is continued, with new solution added until the nickel particles reach the desired size, which in this case is approximately 100 μm in diameter. For detailed information on the production of hydrometallurgically produced nickel powder, see the article "Production of Nickel and Nickel Alloy Powders" in this Volume.

Spray Drying. Spherical ferrites are produced by dispersing a mixture of iron and other metal oxide pigments in a water solution that contains an adhesive binder. This "grog" is fed to a spray dryer in which spherical particles are formed and collected as a dry powder. The resulting green powder is fired in a tunnel kiln at 1200 to 1315 °C (2200 to 2400 °F) and then classified. The end product is a spherical ceramic particle with controlled magnetic properties. For additional information, see the article "Spray Drying of Metal Powders" in this Volume.

Plasma Spheroidized Magnetite. Spherical carrier particles are produced in a plasma furnace by subjecting mixed magnetite ore to temperatures of approximately 5600 °C (10 000 °F) (Ref 23). The material has a sphericity of 70 to 85% (by number), depending on particle size. The powder is finished by magnetic separation and degaussing steps.

Nonspherical Powders

Water Atomization. The same process used to produce spherical metal powders can be modified to generate irregularly shaped particles. This is accomplished by adjusting water pressure and flow rate, or the ratio of water to metal.

Reduced Iron Process. In one form of this process, sponge iron powders are prepared from iron ore (magnetite) by a carbon monoxide reduction that is conducted in a tunnel kiln. The primary reaction occurs between iron oxide and coke. Because it takes place below the melting point of iron, a sponge-like material is produced. When these spongy clumps are disintegrated into a powder, their porous structure is retained. In another process variation, a hydrogen reduction furnace is used to convert magnetite into a porous iron powder. See the article "Production of Iron Powder" in this Volume for more information on this process.

Electrolytic Iron Process. Powders manufactured by this process have the highest purity, because they are produced by electrodeposition from a solution of iron salts. The metal is harvested as plates by periodically scraping the cathode surface. These chips are crushed, usually in an attrition mill, to form a free-flowing powder. For additional information, see the article "Production of Iron Powder" in this Volume.

Mechanical Attrition. Several crushing and grinding operations have been used with coarse, brittle, high-carbon steels to produce copier powders. In one process, the coarse fragments are ball milled before annealing and decarburizing in a combined operation. In the past, twisted, ribbon-like particles were obtained as a by-product of pin manufacturing (pin dust) and other grinding and sharpening operations. These unusual shapes offer certain advantages in magnetic brush systems, but are often contaminated with oil or other substances. Another limitation of this process is the uncertainty of a continuing, reliable supply of material, as with any by-product material.

Rapidly Solidified Powder. Amorphous, or "glassy," metals are made by cooling molten mixtures of such materials as iron, nickel, molybdenum, boron, and silicon at rates that exceed 1 000 000 °C/s (1 800 000 °F/s). These quick-chilled alloys are superior to conventional metals in strength, magnetic properties, and corrosion resistance. Techniques for manufacturing these rapidly solidified powders include spraying molten metal droplets onto chilled rollers that spin at approximately 80 km/h (50 miles/h) and atomizing a thin film of molten metal by centrifugal ejection from the surface of a rapidly spinning, water-cooled disk. See the article "Atomization" in this Volume for additional information.

Powder Coating Processes

Four processes are used frequently for copier powder coating: surface oxidation, lacquer coating, fluorinated polymer coating, and powder coating. Of these, the first three currently are the most popular. Some original equipment manufacturers use a combination of oxidation and lacquer coating techniques.

Surface Oxidation. Iron and steel copier powders can be "blued" by a heat treatment to produce a surface layer of iron oxide. The process is carried out in an oven at about 285 to 290 °C (550 °F) for 15 to 20 min. Control of both air flow and moisture content is critical for ensuring color uniformity.

Equally important is powder circulation, which provides uniform exposure to the oxidizing environment for the entire surface of each bead. This can be accomplished by treating the metal powder in shallow, open trays with periodic agitation. Vibrating trays can be used to keep the steel powder in constant motion. Alternately, tube furnaces have been employed in which the carrier is contained inside a slowly rotating cylinder.

Lacquer Coating. Resin solutions, usually in the range of 10 to 20% solids, are applied to raw or oxidized carrier core materials in one of four equipment types: vibrating tub, cement mixer, vacuum tumble dryer, or fluidized bed coater. In each process, the lacquer is prepared separately by dissolving the appropriate resin, charge-orienting dyes, and other additives in a volatile solvent, typically toluene or methyl ethyl ketone, at elevated temperature.

The raw copier powder is heated separately to about 80 to 95 °C (175 to 200 °F) prior to introducing the hot lacquer solution. The lacquer is added rapidly to the agitated powder in a vibrating tub or cement mixer. Good ventilation is essential for safe operation. As the lacquer dries, the solid-liquid blending operation passes through a critical phase in which a tendency for agglomeration exists. The dry powder must be screened to remove doublet, triplet, and other multi-particle clumps.

Particle agglomeration can be controlled by using either a vacuum tumble dryer or a fluidized bed coater. Because they are closed systems, these devices afford a means for recovering the lacquer solvent. They are more economical to operate than either the vibrating tub or the cement mixer, both of which discharge to the atmosphere. The lacquer solution is normally injected slowly, and the highly agitated powder charge provides a better mechanism for separating individual granules during the coating and drying cycles.

Fluorinated Polymer Coating. The fluidized bed coater is used almost exclusively to apply polytetrafluoroethylene coatings to carrier core materials. Polytetrafluoroethylene resins require a high-temperature curing cycle, usually ranging from 285 to 345 °C (550 to 650 °F). These temperatures are beyond the capability of most tumble dryers.

Powder coating is a recent innovation in copier powder manufacture. This electrostatic coating process is the same as the powder painting process used to coat appliances, lighting fixtures, and various metal parts. It is a solventless coating method in which the coating resin, in the form of a finely divided powder, is sprayed onto or blended with the agitated carrier beads at room temperature. An applied electrical field or triboelectric force causes the fine resin particles to cling to the surface of each bead. During subsequent heating, the resin particles melt and coalesce to produce a uniform coating.

REFERENCES

1. Schaffert, R.M., Electrophotography Yesterday, Today and Tomorrow, *Photo. Sci. Eng.*, Vol 22, May/June 1978, p 149-153
2. Diamond, A.S., *et al.*, Reprographic Chemicals—USA, *Specialty Chemicals—Strategies for Success*, SRI International, Menlo Park, CA, Oct 1982
3. Diamond, A.S., The Market for Dry Toners, *Am. Ink Maker*, Vol 59 (No. 10), Oct 1981, p 32-34, 38, 165-166
4. Dessauer, J.H. and Clark, H.E., *Xerography and Related Processes*, Focal Press, New York, 1965
5. Schaffert, R.M., *Electrophotography*, 2nd ed., Focal Press, London, 1975
6. Young, C.J., "Electrophotographic Developing Apparatus," U.S. Patent 2 786 439, 26 March 1957
7. Giaimo, E., "Electrophotographic Developing Apparatus," U.S. Patent 2 786 440, 26 March 1957
8. Young, C.J., "Apparatus for Applying Electrostatic Developer Powder by Means of a Magnetic Brush," U.S. Patent 2 786 441, 26 March 1957
9. Van Engeland, J., Special Characteristics of Magnetic Brush Development: A Review, *Photo. Sci. Eng.*, Vol 23, March/April 1979, p 86-92
10. Takahashi, T., *et al.*, Mechanism of Canon Toner Projection Development, *Photo. Sci. Eng.*, Vol 26, Sept/Oct 1982, p 254-261
11. Winkelmann, D., Electrostatic Aspects of Electrophotography, *J. Electrostatics*, Elsevier, Amsterdam, Vol 4, 1977/1978, p 193-213
12. Hagenbach, R., "Highly Shape-Classified Oxidized Steel Carrier Particles," U.S. Patent 3 849 182, 19 Nov 1974
13. Hagenbach, R. and Forgensi, R., "Xerographic Carriers by Two Wire Spheroidization," U.S. Patent 4 018 601, 19 April 1977
14. Jones, L.O., "Stoichiometric Ferrite Carriers," U.S. Patent 4 042 518, 16 Aug 1977
15. Jones, L.O., "High Surface Area Carrier," U.S. Patent 4 040 969, 9 Aug 1977
16. Jones, L.O., *et al.*, "Humidity Insensitive Ferrite Developer Materials," U.S. Patent 3 996 392, 7 Dec 1976
17. Jones, L.O., "Stoichiometric Ferrite Carriers," U.S. Patent 3 929 657, 30 Dec 1975
18. Berg, A.C., *et al.*, "Production of Ferrite Electrostatographic Carrier Materials Having Improved Properties," U.S. Patent 4 075 391, 21 Feb 1978
19. Hagenbach, R., "Nickel Carrier Beads," U.S. Patent 4 018 601, 19 April 1977
20. Dessauer, J.H. and Clark, H.E., *Xerography and Related Processes*, Focal Press, New York, 1965, p 289
21. Jacknow, B.B. and Moriconi, J.H., "Organosilicon Terpolymers," U.S. Patent 3 467 634, 16 Sept 1969
22. Jacknow, B.B. and Moriconi, J.H., "Coated Carrier Particles," U.S. Patent 3 526 533, 1 Sept 1970
23. Hagenbach, R. and Forgensi, R., "Apparatus for Forming Magnetite Electrostatographic Carriers," U.S. Patent 4 019 842, 26 April 1977

Magnetic Separation of Seeds*

By Ruth Bellows
Sales Representative
Engineered Product Sales Corp.

MAGNETIC SEPARATION OF SEEDS is accomplished by pretreating a seed lot and subjecting it to a magnetic field. Pretreatment of the seed mixture consists of dripping water into the lot, adding a measured small amount of fine iron powder, and mixing well. The moist weed seeds and other rough-textured or sticky components pick up the iron powder, but smooth crop seeds do not.

Two major types of magnetic separating systems are used in seed conditioning: one type has magnetic drums over which the seed mixture passes; the second type contains conveyor belts that transport the mixture through magnetic fields. The magnetic drum system (Fig. 1) is used most often. When the pretreated lot passes over the magnetic drum, the iron-coated seeds are attracted to the drum and withdrawn from the seed mixture. The drums have either permanent magnets that provide magnetic fields of fixed strength or electromagnets of variable field strength, regulated by the magnet-energizing electric current.

Regardless of magnet type, most separators have two or more energized drums in series so that the cleaned seeds from the first drum pass over the second, where any missed contaminants may be removed. This arrangement allows capacity to be increased without lowering purity. Depending on machine size and the lot being cleaned, capacities of commercial machines range from 91 to 907 kg (200 to 2 000 lb) of cleaned seed per hour.

Purity of Seed Lots

The Federal Seed Act has established purity guidelines for seed sold in domestic and foreign trade. For example, this act requires that alfalfa seed be a minimum of 99.5% pure before being packaged for trade. The amount of material removed by magnetic conditioning varies, but many operations show figures of 1 to 3% contamination for otherwise clean seed. The amount of material rejected tends to increase as the amount of broken, cracked, or scarified seed in the lot increases. Seed loss also may increase if inert material such as stems, chaff, or dirt is present, because more moisture and powder then are needed for magnetic cleaning.

The greater the amount of moisture and/or powder in the seed lot, the greater the chance that some crop seeds will pick up enough powder to be attracted to the drum and removed with the contaminants. For best results, the seed lot first should be well cleaned with basic separators, such as sieves, shakers, and velvet rolls. The magnetic separator then can best perform its primary function as a finishing machine. At optimum operation, seed crop loss usually is less than 0.5%.

Pretreatment Before Magnetic Separation.
Proper pretreatment depends on the amount of moisture and iron powder used and on the mixing operation. Moisture and powder requirements are determined by the amount of contaminant to be removed, the roughness and water-absorbing characteristics of the contaminant, and the surface texture of the desirable crop seeds. Thus, no quantitative iron powder and moisture addition data are available.

Too little moisture or powder prevents proper coating of the contaminant, while too much moisture causes clustering of seeds, discoloration, and excessive crop loss. Use of more powder than needed is wasteful, if not reclaimed, and can cause crop loss. Crop seeds with hard, smooth coats like alfalfa tend to take less powder

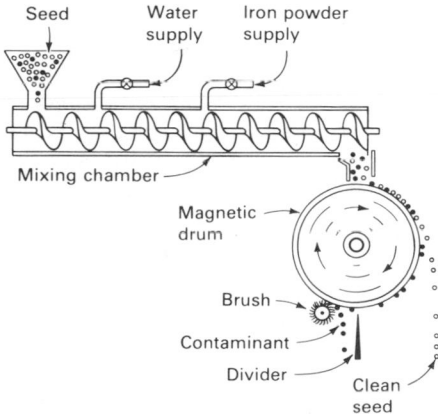

Fig. 1 Schematic diagram of a magnetic separator

than seeds with rougher coats, such as sweet clover. Screw conveyors and batch mixers are used to mix moisture and powder with the seed. Length of mixing time is important, because too much time may cause evaporation of the water and loosening of the adhered powder. Ideal pretreatment requires that all of the contaminants and none of the crop seeds be coated with powder.

The most common crop seeds cleaned magnetically are alfalfa, clover, and onion. Contaminants that can be removed include dodder, buckhorn plantain, sorrel, skinned dock, wild geranium, knapweed, mallow, watergrass, whitetop, morning glory, hulled johnson grass, broken seeds, dirt clods, trash, and other undesirable components. The most common separations performed in the United States are dodder and buckhorn plantain from alfalfa and clovers. Dodder seeds are rough textured, and buckhorn plantain seeds are sticky, especially when moistened.

---

*Portions of this article have been excerpted from Ref 1 with permission.

Powder Size and Shape

In a research study of seed separation conducted by USDA-ARS in cooperation with the Agricultural Experiment Station, Oregon State University (Ref 1), 10 powder lots identified by a number keyed to the percentage of powder passed by a 325-mesh sieve were used.

Powder No.	Amount passing through 325-mesh sieve(a), wt%
1	55.9
2	89.9
3	20.7
4	74.4
5	33.7
6	97.3
7	30.0
8	16.2
9	14.5
10	99.8

(a) Determined by standard sieving techniques specified in ASTM B 214

Fig. 2 Particle size distribution curves for eight iron powder lots

All sieves are square hole.

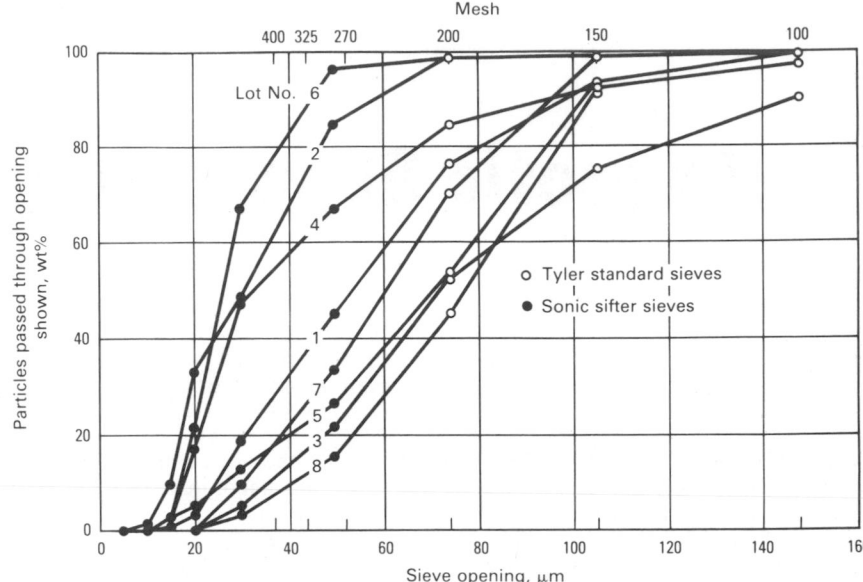

Each powder lot contained a range of particle sizes and varying amounts of each size. Initial particle size distributions were determined for each powder lot by use of conventional small-mesh screens and standard sieving methods. Supplemental air-assisted sieving was then carried out with small micron sieves, which are manufactured by photoengraving and electroplating techniques and are recommended as a primary standard for testing size distributions because of their precise openings.

Sieve analysis data from both procedures were used to develop distribution curves (Fig. 2), which present a complete comparison of different powders with particle sizes above 5 μm. These curves indicate clearly that powders 6, 2, and 4 contained relatively high portions of fine particles. In contrast, powders 3 and 8 contained low portions of fine particles and therefore were classed as coarse in this comparison. The plotted data are useful in ranking powders according to fineness, based on amounts that pass through a given sieve.

Particle size curves demonstrate a relationship between fineness of powder and its performance in removing contaminants. This correlation was explored further in the study (Ref 1), using 50-g (2-oz) lots to permit an actual count of contaminants and to eliminate sample errors. Powders were ranked according to fineness and performance (Table 1). The resulting correlation shows a marked relationship between fineness and performance of the powders, verifying that finer powders do the best job of removing dodder.

This contaminant is best removed from alfalfa and clover with powder that has a 70% passage rate through a 325-mesh sieve. Seed lots contaminated with dodder should not be milled excessively before magnetic separation, because this process smooths the seeds and reduces powder retention. Yields of cleaned crop seed approach 99.8% when fine powder is used at optimum moisture levels.

The relationship between powder fineness and removal of buckhorn plantain was examined by similar ranking methods, but no good correlation was found, suggesting that powder fineness is relatively unimportant in its removal. The surface coating of buckhorn plantain seeds, when moistened, becomes very sticky. Coarse iron particles probably adhere to the surface as readily as fine particles.

Particle Shape. Shapes of powder particles vary considerably, so tests were conducted to learn whether particle shape was related to performance in the removal of contaminants (Ref 1). Results indicate that flake particles perform better than blocky ones. Better results are produced with flake particles because the flat, sliver shapes enter and wedge into pits and crevices of rough-textured seeds more securely than do blocky shapes, and flake particles characteristically exhibit greater

Table 1 Correlation of fineness and performance rank of iron powders used in magnetic separation

Powder fineness rank(a)	Powder performance rank(b)
4 (finest)	4 (best)
6	6
2	2
5	5
1	1
7	7
3	3
8 (coarsest)	8 (worst)

(a) Based on percentages that passed through a 20-μm sieve. (b) Based on dodder seed removal at 0.45 mL (0.02 oz) water and 50 g (2 oz) seed. All water additions were made on a weight rather than volume basis, a standard practice in seed conditioning.

Fig. 3 Powder particle shapes

(a) Flake. (b) Blocky

Fig. 4 Contaminant removal and yield of cleaned red clover seed, as influenced by water addition
Iron powder: 1 g (0.04 oz) of powder 1 per 50 g (2 oz) of seed

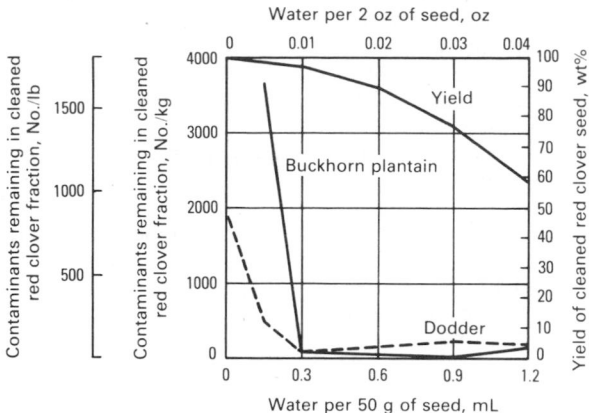

Moisture Tests

Moisture and powder were mixed in duplicate batches with seed lots in a batch tumbler for 8 min per test (Ref 1). About 150 batches were tested, and seed yields and remaining contaminants in final cleaned fractions were determined, recorded, and analyzed. These values generally show similar patterns for all tests. Moisture level has a large influence on contaminant removal, seed yield, and powder consumed in the magnetic separation process. These measures, when plotted for all trials, show the trends illustrated in Fig. 4, 5, and 6.

Up to a certain amount, increased moisture improves the removal of contaminants, but additional moisture tends to reduce the efficiency of dodder removal, with relatively little effect on the removal of buckhorn plantain. The importance of moisture level is shown by the nearly vertical slope of the contaminant curves at low moisture levels, where even a small change in moisture content causes a large variation in contaminant removal. The reason for these curve shapes is not known, but high moisture levels caused undesirable clustering of seeds and powder, thus preventing removal of individual seeds. The minimum moisture level at which essentially complete removal of both contaminants was accomplished in these trials was 0.30 mL (0.01 oz) per 50 g (2 oz) of seed, or about 4.2 L/m^3 (5 fluid oz) per bushel of seed.

Seed yields decrease steadily as moisture levels increase (Fig. 4). This result is reasonable, because increased moisture levels cause a greater quantity of seed in a test lot to pick up powder and be re-

Fig. 5 Dodder removal and yield of cleaned alfalfa seed, as influenced by water addition
Iron powder: 1 g (0.04 oz) of powder 5 per 50 g (2 oz) of seed

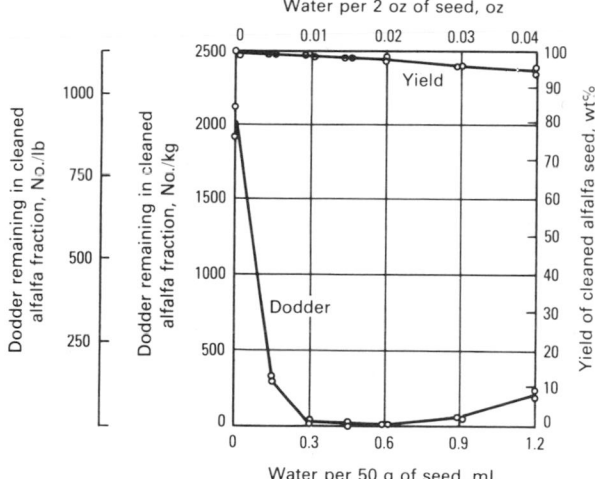

surface contact than blocky particles and have stronger adhesion tendencies. Removal of both buckhorn plantain and dodder is better with flake particles than with blocky ones. Examples of flake and blocky particles are shown in Fig. 3.

Reuse of Powder. Some processing plants salvage and reuse the portion of iron powder that does not adhere to seeds in the magnetic conditioning process, while other processors use a powder addition only once and then discard it. Because fine particles of powder are most effective in removing dodder, reuse of powder depletes available fines and lowers the performance of the powder. The disappearance of fines is the result of fine particles entering readily into pits and crevices and being removed from the powder. In addition, fine particles may consolidate with

reuse into coarser units through the effects of moisture, rust, or magnetic attraction.

Reuse was examined in a controlled study (Ref 1). Progressive changes in powder performance and particle size took place as powder was reused. The amount of fine particles (the percentage that passed through a 40-μm sieve) decreased steadily with reuse. The first use of powder effectively removed dodder, but the amount of dodder remaining in the cleaned fraction increased with each successive use. Particle shape tended to change as powder was reused. Many original particles had a dendritic or tree-like shape that is characteristic of some iron powders. After several uses, particles appeared more rounded, without branches or arms, and these particles could not adhere to the seed as well as the original jagged shapes.

Fig. 6 Removal of buckhorn plantain, as influenced by water addition
Iron powder: 1 g (0.04 oz) of powder 5 per 50 g (2 oz) of seed

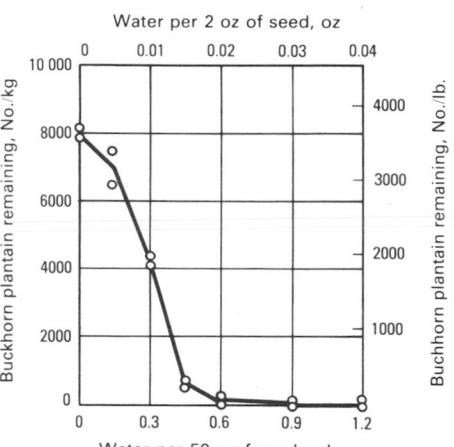

moved by the magnetic drum. More powder also is picked up at higher moisture levels than at lower moisture levels. Figure 5, for powder 5 (medium-grade powder), illustrates that increased water levels, up to a certain amount, are beneficial in removing dodder. A water content of more than 0.60 mL (0.02 oz), however, decreases efficiency of dodder removal. Other disadvantages associated with excessive water content are seed discoloration, mold formation, and germination loss.

The performance of powder 5 in removing buckhorn plantain from alfalfa is shown in Fig. 6. The removal of buckhorn plantain, like the removal of dodder, is improved with added moisture. However, higher moisture levels do not hinder removal of buckhorn plantain and swing the curve upward as in the removal of dodder. The optimum water level for buckhorn plantain removal is 0.90 mL (0.03 oz) per 50 g (2 oz) of seed, and the yield at this level is 97%. If both dodder and buckhorn plantain are present in a seed mixture, the best water level to use is 0.60 mL (0.02 oz) in an attempt to remove all of the dodder, which is more objectionable than buckhorn plantain.

Moisture levels from 4.2 to 9.3 L/m^3 (5 to 11 fluid oz) per bushel of crop seed produce essentially complete removal of dodder and buckhorn plantain with low crop loss (1 to 4%). More moisture may be required on extremely hot or dry days in warm climates because of evaporation. The type of moisture added is relatively unimportant; water produces essentially the same results as water containing additives such as soluble oil or wetting agents.

REFERENCE

1. Brandenburg, N.R., Magnetic Separation of Seeds, Technical Bulletin 137, Agricultural Experiment Station, Oregon State University, Corvallis, OR, May 1977

Metallic Flake Pigments

By Barry Van Hoose
Research & Development Project Manager
Alcan Ingot and Powders

METALLIC FLAKE PIGMENT production originated from the methods used to produce gold leaf. In this process, gold was manually hammered into extremely thin, flat particles and rubbed through a fine mesh screen to produce gold powder. Metallic pigments currently are produced mechanically and may simulate gold or silver, but are less expensive. Flake pigments can be mixed directly into the coating vehicle, without additional grinding, which is required with other inorganic pigments.

"Gold bronze" flake powder, which is a brass (copper-zinc), was mechanically produced in the late 19th century by Henry Bessemer. Since that time, significant improvements have been made in mechanical production methods.

Originally, all metallic flakes were referred to as "bronze" pigments. The term "bronze" described the metallic luster produced by the pigment. "Gold bronze" described brass powder that simulated gold. All other metallics were also referred to as "bronzes"—for example, aluminum bronze. Currently, only gold bronze retains its original name; all other pigments have dropped the term "bronze" from usage.

This article discusses mechanically manufactured metallic pigments. Aluminum flake, copper, and gold bronze pigments enjoy wide usage. Other commercially available metallic flake pigments include stainless steel, nickel, zinc, and silver.

Flake particles are leaf-like, with flat surfaces and lengths and widths that are larger than their corresponding thicknesses (Fig. 1). All flake pigments discussed in this article utilize a lubricant surface coating, which causes leafing. Flakes rise to the surface of the coating, where they remain even though they ex-

Fig. 1 Electron micrograph of flake aluminum powder

Surface area: 5 m²/g. Magnification: 450×

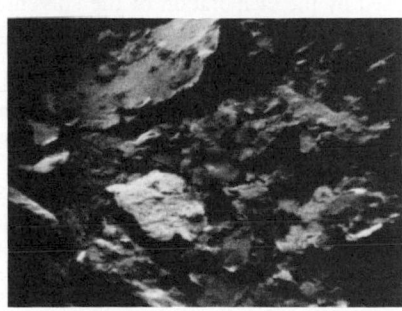

hibit greater specific gravity than the coating vehicle.

Pigment Production

In the late 19th century, a stamping process was developed to produce metallic flake powders. Stamp mills consist of hammers that are raised by cams. Gravity causes the hammers to fall onto steel anvils while powder is passed across the anvils. The hammers turn as they fall, thus flattening and smearing the particles.

Splash metal or foil scrap was transported to large stamp mills that completed the initial milling. This product output was sent through lighter stamps for further processing. At the end of the stamping process, the flake powder was classified by screen or other means and polished.

Polishing was originally done in a brush polisher, which consisted of a cylindrical drum with a ribbed steel interior. Brushes rubbed lightly against the interior of the drum to spread lubricant on the flake powder. This method is infrequently used, because it is inefficient compared to modern ball polishing.

Stamp mills have been largely replaced by ball mills, which are cylindrical containers that use steel balls as the grinding medium. The mill rotates fast enough to create a cascading action of the balls onto the powder. The cylinders have internal metallic ribs that aid in lifting the balls.

Ball milling can be accomplished in a dry inert atmosphere or in air. It can also be accomplished in a wet atmosphere by using a lubricant carrier such as mineral spirits. In either process modification, lubricants such as stearic acid, oleic acid, or lard are used to help the powder particles form flakes under impact and to prevent welding of the flakes. The type and amount of lubricant used are determined by the properties desired in the finished flake.

Most dry ball milling is a continuous-feed process in which the amount of powder in the mill (ballast) is kept nearly constant. When powder is removed from the drum, it is replaced with feedstock and lubricant. Powder is removed from the mill by circulating air through the mill trunnions. The powder is then classified, and the fines are removed. Oversized particles are returned to the mill for further size reduction.

Ball mills vary in size, mill speed, ball size, and the powder capacity of the mill. Most fine grades of flake produced by dry ball milling require a multi-step process. Coarse initial flake typically is manufactured in a large mill from an atomized powder feed, while finer grades require the coarse flake as feed in smaller mills. This prevents abrasive grinding, which forms small granular particles with dark, low-luster finishes.

Most aluminum pigments are produced by wet ball milling as opposed to dry milling because of the potential explosion hazard associated with dry aluminum dust-air mixtures. Other methods of milling include continuous wet milling, attrition milling, and vibratory milling; these are used only to a limited extent. Wet milling uses a lubricant (stearic acid, for example) in solution with a solvent such as mineral spirits. This is added to the mill with the feed metal and milled for a predetermined time before removal from the mill. This process is referred to as "batch milling."

After a batch-milled product is removed from the mill, it is classified, typically by wet screening, and is then placed in a filtering unit that removes most of the solvent. The semi-dry product may be marketed as a paste or placed in a vacuum dryer for removal of residual solvent and marketed as a dry flake powder. Usually, the dry powder is polished and screened, or simply screened to remove oversize particles.

Polishing is done mainly with ball polishers, which are miniature ball mills that contain small balls that rotate at a slow speed. Powder and lubricant are fed slowly into the polisher by gravity, allowing the lubricant to spread thinly over the surface of the flakes. Flakes are further flattened during this operation. This type of polishing is a continuous process that proceeds at a much higher rate than brush polishing.

Aluminum Flake Pigments

Aluminum pigments for coatings generally are classified into four types—leafing paste, leafing powder, nonleafing paste, and nonleafing powder. These types are further classified by fineness. The American Society for Testing and Materials (ASTM) has issued specifications for aluminum flake pigments (ASTM D 962, Part 28). Testing methods are described in ASTM D 480. Specifications are given for screen analysis of the coarse portion of the product, percentage of nonvolatiles, and leafing values; however, specific surface area of the flake and optical properties of each type are not covered. Aluminum flake powders used in noncoating applications, such as pyrotechnics and commercial explosives, also are not covered by specifications.

Finely divided aluminum is extremely reactive with acids, bases, chlorinated solvents, and water. One of the by-products of the reaction is hydrogen gas, which can produce bulging and possible rupture of closed production containers. As an airborne dust cloud, finely divided aluminum is potentially explosive. A dust cloud of aluminum powder will flash if an ignition source is present, and an explosion will occur if the concentration is within explosive limits. For more information, see the article "Explosivity and Pyrophoricity of Metal Powders" in this Volume.

Aluminum flake pigments should not be exposed to high-shear, high-speed agitation during mixing, because this tears the delicate flakes, strips the protective lubricant coating from the particle surfaces, and creates reaction sites. This type of agitation reduces the amount of leafing and reduces the brilliance of the coating of leafing aluminum pigments. High temperatures during mixing affect the thin stearic acid layer on leafing pigments, resulting in loss of leafing properties and luster.

Leafing, as previously described, is the ability of the flake to rise to the surface of a coating and remain there in spite of a higher specific gravity than the coating vehicle. Flakes have a thin layer of a lubricant, such as stearic acid, and when they rise to the coating surface through currents caused by the evaporation of the solvent, they are held there by the surface tension of the vehicle. A minimum of 25 dynes/cm surface tension is required for successful leafing. Higher surface tension solvents such as high aromatic solvents are suitable for leafing.

Lead and zinc dryers deleaf the flakes and have been used to change leafing grades to nonleafing paste. Good leafing vehicles typically are simple compositions, such as a varnish. Nitrocellulose is complex and renders a leafing grade of aluminum nonleafing.

Roof coatings comprise the largest market for leafing aluminum paste. Aluminum roof coatings are either fibrated or nonfibrated asphalt. In nonfibrated systems, a coarse grade of aluminum paste (ASTM D 962 type 2 class C) is used. Coarse leafing grades have a rapid rate of leafing, resulting in a continuous metallic film on the surface of nonfibrated asphalt. The use of fine grades of flaked powder tends to produce a spotty appearance. Typical aluminum paste pigmentation quantities for nonfibrated asphalt are 240 to 360 g/L (2 to 3 lb/gal).

In fibrated systems that contain approximately 5% fiber, such as asbestos, a fine grade of aluminum paste is required for a bright metallic appearance. Fibrated coatings are applied in thick layers so that speed of leafing is not as important as the surface area of the pigment flakes. However, different asphaltic cutbacks may be used in roof coatings. In fibrated coatings, typical pigmentation amounts are in the range of 120 to 360 g/L (1 to 3 lb/gal), with the majority around 180 g/L (1.5 lb/gal).

Aluminum is well suited for roofing applications, because it provides a moisture barrier and has high reflectivity. Black asphalt roof coatings absorb and transmit heat into the building, resulting in increased air conditioning costs in the summer. Aluminum reflects much of the heat, resulting in lower interior temperatures, as well as extending the life of the roof.

Aluminum paints exhibit special properties, such as moisture resistance, good durability, and high reflectivity, which accounts for their wide industrial usage. Moisture resistance enhances protection of structural steel, while high reflectivity increases the usefulness of aluminum paint coatings on highway bridges and guard rails. Usually, aluminum paint is used as a topcoat, because it does not exhibit the corrosion resistance of pigments such as zinc dust.

However, urethane primers containing nonleafing aluminum are gaining acceptance. Urethane paints increase coating life and chemical resistance. Structural steel coatings usually contain a standard lining paste (ASTM D 962 type 2 class B). Aluminum pigmentation rates vary between 180 and 240 g/L (1.5 and 2.0 lb/gal).

Aluminum pastes are used in heat-resistant coatings on surfaces such as smoke stacks. The use of silicone resins has resulted in coatings with good durability at temperatures up to 400 °C (750 °F). Aluminum pastes should be a standard leafing lining paste (ASTM D 962 type 2 class B) or a fine lining paste (ASTM D 962 type 2 class A). For these paste coatings, aluminum pigmentation is typically between 120 and 180 g/L (1 and 1.5 lb/gal).

Water-dispersible aluminum flake pastes are available for use in water-reducible paints. Flake grades have added surfactants that allow the mineral spirits in the paste to mix with water.

Automotive coatings comprise the largest market for nonleafing pastes. Properties required for this application include acid stain resistance, sparkle, seeding, hiding power, and gloss. Standard grades of nonleafing aluminum are produced from commercially pure aluminum (a minimum of 99% Al). When aluminum particles in a coating are exposed to the atmosphere, as in the one-coat finishes normally used in automotive coatings, they may react and stain the coating. High-purity aluminum (99.98% Al) is less reactive than commercially pure grades and offers some protection against staining. Grades of nonleafing aluminum pigment using high-purity aluminum are referred to as "acid stain resistant."

When flake particles align parallel to the surface of the paint film, the appearance is typical of a standard leafing pigment.

When the flake orients at various angles to the surface, light reflects in a scattering pattern, which is described as "sparkle." The sparkle effect causes extremely fine flakes to appear much larger, giving a more metallic appearance without seeding problems. Seeding problems occur when flakes project above the surface of the paint film.

High gloss in automotive coatings is achieved by using an aluminum flake pigment with a short flake diameter. To obtain the combination of a highly metallic appearance and high gloss, more than one aluminum pigment frequently must be used in a coating.

Printing inks use aluminum pigment to simulate the richness of silver at a much lower cost. Aluminum pigmented inks may be used in offset, letterpress, flexogravure, and rotogravure printing. Rotogravure inks pigmented with 10 to 20% Al are printed at extremely fast speeds and use fast-drying solvents. The brilliance of the aluminum depends on the size of the flake. The larger the flake size, the higher the brilliance of the coating. However, maximum usable flake dimensions depend on the depth and width of the etching on the printing cylinder surface.

Ink is picked up in the etching and transferred to the paper. Coarse particles overfill the etching and are removed by a doctor blade before reaching the paper. Leafing is not important in this application, because the ink is applied and dried quickly. However, most tinted metallic inks used in rotogravure printing require leafing grades of aluminum flake powder.

Flexogravure inks are similar to those used in rotogravure. A standard lining grade is recommended. If maximum hiding effect is required, fine lining grades should be used.

Letterpress inks pigmented with 20 to 30% Al require a fine lining grade of leafing paste or flake. There is more transferal of the ink in this application, and optimum transfer properties are obtained with pigments that have short flake diameters. Letterpress inks dry more slowly than rotogravure and flexogravure inks, so leafing properties are important. All pastes contain mineral spirits or a combination of mineral spirits and a high aromatic solvent. Offset printing requires fine lining grades and is similar to letterpress in that a short flake diameter and high leafing are important.

Plastics. Aluminum is used in plastics to provide a silver color and opacity. Aluminum flake powders are normally blended to a paste consistency with a compatible plasticizer and sold to a compounder. This product is mixed with a plastic and ex-truded, usually in pellet form. The pelletized colorant is then mixed at low concentrations with a higher melting plastic, followed by heating and extruding into final form. The fineness of powder used depends on the effect desired; coarse grades provide high brilliance, and fine grades provide higher opacity at the sacrifice of higher brilliance.

Pyrotechnics and Explosives. Other noncoating applications for aluminum flake powder include pyrotechnics (fireworks) and commercial explosives. Pyrotechnic powder grades are much coarser than pigment grades and contain a lower level of lubricant. These grades are desired because of chemical reactivity. In commercial explosives, standard lining grades of flake powder (ASTM D 962) are used at relatively low concentrations (2 to 5%) to increase the ignition sensitivity of watergel type explosives. Dedusted flake powders are gaining popularity in this application because they substantially reduce aluminum dust hazard, but do not adversely affect the sensitivity of the explosive. For more information, see the article "Metal Powders for Fuel Propellants, Pyrotechnics, and Explosives" in this Volume.

Copper and Gold Bronze Pigments

Gold bronze pigments are brass alloy flake powders ranging in composition from 70Cu-30Zn to 90Cu-10Zn. Copper and gold bronze flake pigments are made by either stamping or dry ball milling. The recent trend has been toward ball milling.

The color shades of gold bronze are determined by the alloy composition used and by subsequent heat treating. Heating brass flakes results in partial oxidation of the flake surface, imparting a deepening of color. The higher the level of zinc, the greener the shade; the lower the zinc level, the redder the shade. The red shade is generally termed "pale gold," and the green shade is called "rich gold." Partially oxidized or heat treated powders are referred to as "deep golds" or "colored bronze."

Unlike aluminum, gold bronze powders are not color stable and may tarnish on exposure to the surrounding environment. Tarnish-resistant grades of gold bronze are produced, but heat treating usually results in lower brilliance and opacity.

Decorative Applications. Gold bronze is used primarily for decorative purposes, such as cigarette packages, cigar boxes, wrapping paper, greeting cards, and picture frames. Coarse grades (having up to 60% +325 mesh) are used in lithograph dusting, in which an adhesive is applied to the surface to be printed, and dry powder is dropped onto the wet adhesive. Excess powder is then lightly brushed off, and the powder is retained only where the adhesive was applied.

Paints. Ready mixed paints pigmented at approximately 360 g/L (3 lb/gal) use standard fineness pigments, which yield a smooth, brilliant surface. Leafing is required for this application, and the standard fineness grades (having 1 to 5% +325 mesh) exhibit good leafing properties. Aluminum ready mix paints require a nonreactive coating vehicle, such as varnish.

Aerosol-propelled paints demand a fine lining powder (having less than 0.1% +325 mesh) with good leafing properties. Coarser grades are to be avoided because they may contain large particles that would clog the spray nozzle. The sprayed paint film should have a continuous film of metal flake on its surface (leafing). Brilliance is determined by the size of the flake, leafing value, and amount of polishing performed on the flake during manufacturing.

Gold bronze and copper powders react with vinyl resins, which causes gelling. Many types of stabilizers have been used, such as organotin, organozinc, barium, cadmium, and zinc complexes. Stabilizer should be chosen according to application.

Gold bronze and copper react with nitrocellulose lacquers, causing greening and gelling of the lacquer. Citric acid has been useful in improving stability, but must be evaluated for each application. Most gold bronze nitrocellulose lacquer mixes are used immediately after production because of inherent product stability problems.

Plastics utilize gold bronze and copper pigments for aesthetic reasons. Unlike aluminum, these pigments usually undergo one heating cycle. Gold bronze and copper powders do not pose the same explosion hazard as aluminum powders and can be mixed dry with the plastic, rather than molded into a pellet. Also, gold bronze oxidizes with heat, so heating must be kept to a minimum or flake discoloration will occur. The fineness of the powder used depends on the desired effect. Coarse grades produce brilliance, and fine grades produce high opacity.

Printing ink pigmentation is similar to that of aluminum inks, but the pigmentation rate is much higher. Typical rotogravure and flexogravure inks contain between 30 and 40% gold bronze, due to the difference in density between brass and aluminum. Letterpress and offset ink pigmentation is approximately 50%. Almost all of the gold bronze powders used in printing inks are 99.9% −325 mesh.

One method of evaluating flake pigments is by a water coverage technique, in which the surface area of flake coatings may be approximated. A fixed weight of flake is spread on water and mixed until the flake layer is approximately one flake thick. The area of the flake is then measured and calculated into square centimetres per gram. Powders can also be evaluated by measuring the particle size by optical or mechanical methods.

Both particle size and water coverage are important controlling factors. In rotogravure and flexogravure printing, a large size particle results in high brilliance, while a high water coverage powder results in high opacity. In letterpress and offset inks, small flakes have the best transfer properties, producing higher opacity and a more brilliant finish.

Other Metallic Pigments

Aluminum, copper, and gold bronze constitute the vast majority of metallic flake products sold in the United States. However, other metallic flake pigments are produced.

Stainless steel flake pigments have been available since the late 1940's. Originally, stainless steel flake was available in a variety of alloys, but currently only the 316L alloy is available. Stainless steel flake is used in high-durability coatings that are exposed to highly corrosive atmospheres. Weathering improves the appearance of stainless steel by actually polishing the coating, thus making it smoother and brighter.

Zinc flake has been used in corrosion-resistant paints because of the chemical properties of zinc; however, zinc dust is inexpensive and can perform the same function.

Nickel and silver are also available in flake form. Because of their superior conductivities, they are employed mainly in electrical applications.

SELECTED REFERENCES

- ASTM standard D 480, "Sampling and Testing Aluminum Powder and Paste," *The 1977 Annual Book of ASTM Standards*, Part 28, American Society for Testing and Materials, Philadelphia, 1977

- ASTM standard D 962, "Aluminum Pigments, Powder and Paste for Paints," *The 1977 Annual Book of ASTM Standards*, Part 28, American Society for Testing and Materials, Philadelphia, 1977

- Hay, G.N., Stainless Steel Flake Pigment, *Pigment Handbook*, Vol 1, Patton, T.C., Ed., John Wiley & Sons, New York, 1973

- Noel, D.O., Bronze Powders, *Protective and Decorative Coatings*, Matiello, J.J., Ed., John Wiley & Sons, New York, 1942

- Payne, H.F., Metal Pigments and Metallic Stearates, *Organic Coating Technology*, Vol 2, John Wiley & Sons, New York, 1961

- Rogers, P.E., Greenwald, F., and Butters, W.L., Copper and Copper Alloy Flake Powders, *Pigment Handbook*, Vol 1, Patton, T.C., Ed., John Wiley & Sons, New York, 1973

- Rolles, R., Aluminum Flake Pigment, *Pigment Handbook*, Vol 1, Patton, T.C., Ed., John Wiley & Sons, New York, 1973

- Wray, R.I., *Aluminum Powder and Paint*, Matiello, J.J., Ed., John Wiley & Sons, New York, 1942

Metal Powders for Fuel Propellants, Pyrotechnics, and Explosives

By Mae I. Fauth
Research Chemist
Naval Surface Weapons Center

METAL POWDERS are used as fuels in solid propellants, pyrotechnic devices, explosives, and other applications in which the properties of high density, high flame temperature, and high energy output per unit of expended fuel are desirable. The most widely used type of metal powder —aluminum powder—is applied in propellants, pyrotechnics, and explosives. Other metals used, particularly in pyrotechnics, include magnesium, zirconium, titanium, tungsten, manganese, beryllium, and cerium. Table 1 lists properties of aluminum, magnesium, zirconium, and titanium.

In fuel applications, the metals undergo highly exothermic (heat-producing) chemical reactions to yield stable oxides. Table 2 gives data for the heats of formation of some oxides. Comparison of the values for the metals given with those for carbon shows that all four produce more heat per unit than carbon. Carbon is the basis of many common fuels because it is easily ignited, safe, reasonably stable, abundant, inexpensive, and its products of combustion are gaseous.

Grosse and Conway (Ref 2), who investigated combustion mechanisms of

metals in oxygen and the history of metals combustion, found that metals with low boiling points, such as magnesium, volatilize and that the vapor undergoes combustion above the metal surface. For the high-boiling metals, such as aluminum, the combustion zone is close to the surface.

Because the melting point of aluminum is below the combustion reaction zone temperature of the metal (see Table 1), a pool of molten metal accumulates and floats on the molten aluminum oxide and burns. The temperature of the aluminum at the surface is near the boiling point of the metal, 2467 °C (4472 °F), while the temperature of the flame above the surface of the boiling metal is about 3500 °C (6330 °F). Flame temperature increases as total pressure is increased. High pressure increases the boiling point of the oxide, so that higher flame temperatures are obtained even when the oxide decomposes on vaporization. For example, the combustion of aluminum at 1000 kPa (10 atm) results in a flame temperature of 4127 °C (7460 °F), compared to 3527 °C (6380 °F) at 100 kPa (1 atm).

For many metal-oxygen systems, the combustion temperature is limited to the

boiling point of the oxide. Because aluminum and magnesium form oxides with high heats of formation and very low dissociation pressures, they are extremely stable. Dissociation pressure for a reaction such as

Al_2O_3 (solid)

$$\rightarrow 2Al \text{ (solid)} + 3/2 \text{ } O_2 \text{ (gas)} \quad \text{(Eq 1)}$$

is the partial pressure of oxygen in equilibrium with the metal and oxide phases at a given temperature. This equilibrium oxygen pressure is referred to as the dissociation pressure of the oxide and is an indication of its thermal stability.

Applications of Metals as Fuels

Reference 3 provides a summary of the applications of metal powders as fuels. The oxides produced by combustion of metals are almost always liquids or solids. Finely divided metals such as magnesium have long been used in pyrotechnics to produce light or smoke. Magnesium, when burned alone, produces a highly visible white smoke. When burned with metal salts or

Table 1 Properties of reactive materials

Material	Density, g/cm³	Melting point (approx) °C	°F	Boiling point °C	°F	Heat of fusion(a) kJ/mol	cal/g mol	Heat capacity J/g·°C	cal/g·°C	Combustion temperature(b) °C	°F
Aluminum	2.70	660	1220	2467	4472	0.790	2550	0.90	0.215	3500	6300
Magnesium	1.74	649	1200	1090	1994	8.954	2160	1.02	0.244	3100	5600
Titanium	4.54	1660	3020	3287	5948	6.90	5000	0.27	0.064	3000	5400
Zirconium	6.51	1852	3365	4377	7910	15.450	5500	0.52	0.124	4500(c)	8100(c)

(a) Source: Ref 1, p B-231 to B-235. (b) The adiabatic (no gain or loss of heat) combustion temperature in oxygen at 100 kPa (1 atm) pressure. (c) Metals such as zirconium produce temperatures that approach the maximum possible in a chemical reaction.

Table 2 Heats of formation of inorganic oxides

Reaction	Heat of formation, ΔH_o (a)		
	g cal/mol	kJ/mol	g cal/g
2 Al(solid) + 3/2 O_2(gas) = Al_2O_3	−404 080	−1691.5	−3962
Mg(solid) + 1/2 O_2(gas) = MgO	−144 090	−603.2	−3602
Ti (α) + O_2(gas) = TiO_2	−228 360	−955.9	−2858
Zr (α) + O_2(gas) = ZrO_2	−262 980	−1100.8	−2134
C (graphite) + O_2 = CO_2	−93 690	−392.2	−2129

(a) Negative values mean heat is given off.
Source: Ref 1

dyes, magnesium can be used for colored smokes and flares. In addition, titanium is used for generating showers of sparks.

Thin, shredded zirconium foil is used often for photographic flash bulbs. Although aluminum and magnesium are less expensive, they provide less light intensity. Powdered fuels or slurries of solid particles in liquid suspension can be used for rockets. Solid rockets, especially, can benefit from metal additives, which are used mainly to increase the temperature of the combustion gases, which in turn increases the specific impulse (energy) of the propellant.* Powdered aluminum, and to a lesser extent magnesium, titanium, and zirconium, are employed in high-energy propellants, in the range of 5 to 20 wt%.

Explosive performance depends on total energy released, rate of release, and volume of gas produced. Use of metals in explosives increases the energy density (energy per unit volume) and, hence, the total energy (energy produced by the entire volume of material) available from a given charge. The upper limit of performance improvement is dictated by the fact that, above a certain concentration, metals become more difficult to ignite and do not burn as rapidly or completely as at lower concentrations. The maximum combustion temperature is increased by metal addition, which in turn increases the volume of gas given off by the other components.

Because the products of metal combustion are liquid or solid, the metal yields no product gases; therefore, the metal addition is beneficial only in small amounts. For example, aluminum gives maximum efficiency in improving the performance of 2,4,6-trinitrotoluene (TNT) at about 15%. At this level, the energy content and hot gas volume are increased by 20 and 30%, respectively. Because the reaction rates of metals used are slower in comparison to the main charge materials, the effect on reaction rate is negligible. However, the final effect—duration of the detonation pressure—is longer.

*Specific impulse (I_{sp}), also known as specific thrust, is a performance parameter of a rocket propellant. See Eq 2 and 3.

Safety Precautions. The metal powders used as fuels are often water-reactive and easily oxidized in air, and some are pyrophoric (self-igniting) in air (see the article "Explosivity and Pyrophoricity of Metal Powders" in this Volume). Because the particles are usually very small and chemical reaction may be extremely rapid, fire and explosion hazards exist. Therefore, proper storage, handling, and usage of the metal powders are absolutely essential. Manufacturers and suppliers of explosive materials usually provide data sheets or other literature that outlines handling precautions.

The use of any finely divided metal powder as a fuel should not be undertaken without obtaining detailed information on the recommended procedures for the operations planned. Furthermore, metal powders should not be incorporated into energetic materials, such as solid propellants, explosives, and pyrotechnic devices, until persons experienced in this field have been consulted and supervisors responsible for the program can approve details of the operations. Safety information on specific materials can be found in Ref 4 and 5.

Solid Propellants

Solid propellants used in rocket propulsion systems are composite structures that consist of a solid oxidizer, solid fuel, and polymeric fuel binder, mixed and formed into a monolithic grain or unit. These grains, which range from several millimetres to several metres in length, are usually cylindrical in shape and have various configurations to facilitate the type and rate of burning desired. Typical solid propellant grain shapes and configurations are illustrated in Fig. 1 and 2.

It is possible to limit the exposed burning surface of a grain by the use of inhibitors, which are chemicals that are essentially inert or that burn very slowly (Fig. 2). They are applied by dipping the grain, by bonding sheets of inhibitor material to the grain surfaces that are to be kept from burning, or by wrapping the grain with a special tape. When an inhibitor is applied

to the inner surface of a chamber, it reduces heat transfer to the wall and is referred to as a liner. Because inhibitors restrict the burning of grains, solid propellant units with inhibitors are termed restricted burning rockets; those without inhibitors are called unrestricted burning rockets.

Ammonium perchlorate is the major oxidizer used in solid propellants, but ammonium nitrate also has limited special applications. Examples of binders are polyesters, polyethers, and polybutadienes terminated with chemically reactive groups such as carboxyl or hydroxyl.

Fig. 1 Solid propellant grain shapes

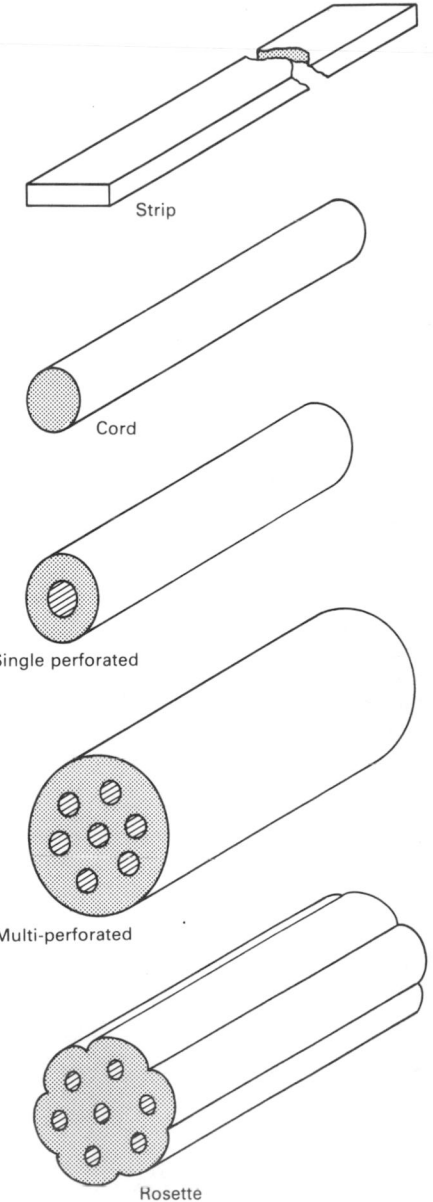

Strip

Cord

Single perforated

Multi-perforated

Rosette

Fig. 2 Typical solid propellant grain configurations

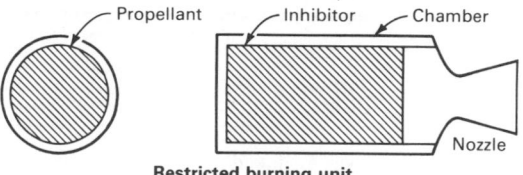

Restricted burning unit

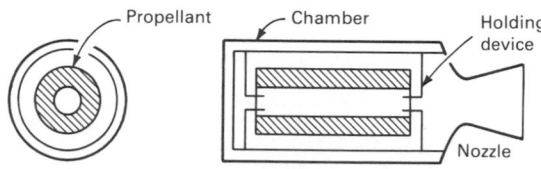

Unrestricted burning unit, single hollow charge

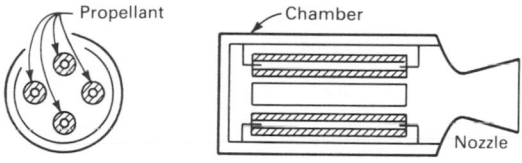

Unrestricted burning unit, quadruple hollow charge

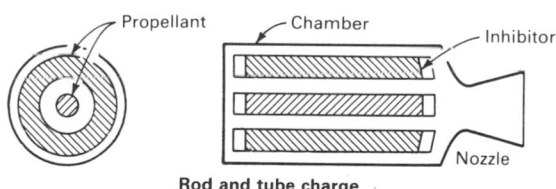

Rod and tube charge

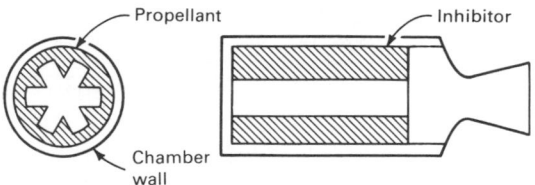

Typical internal burning star configuration

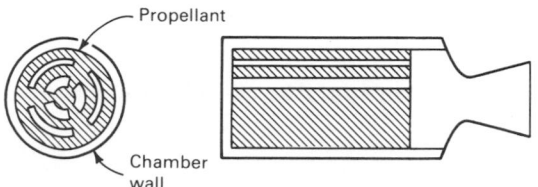

Typical internal burning multi-perforated configuration

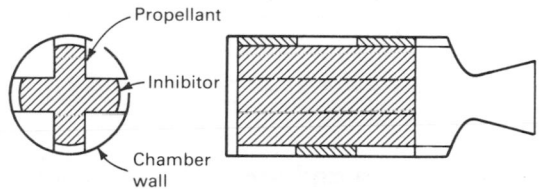

Typical partially restricted external burning cruciform configuration

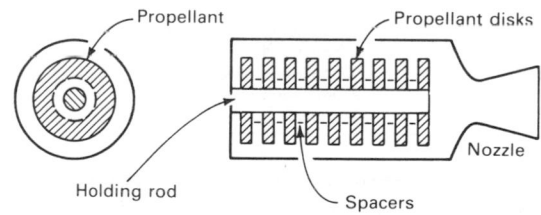

Multiple disk charge

Metal Powder Applications. To obtain higher energy in solid propellants, metal powders such as aluminum or beryllium may be added. Beryllium, however, is not currently used, because of toxicity problems. Two advantages of using aluminum as a fuel in solid propellants are its high combustion energy and its ability to reduce water and carbon dioxide to lower molecular weight (MW) gases during burning. Low molecular weight gases are desirable because the energy output of a propellant is inversely proportional to the square root of the average molecular weight of product gases. Thus, a mixture of diatomic hydrogen (MW = 2), carbon monoxide (MW = 28), and diatomic nitrogen (MW = 28) is more energetic than a mixture of water (MW = 18), carbon dioxide (MW = 44), and diatomic nitrogen (MW = 28). The disadvantages are vaporization and ignition difficulties, slow burning rate, formation of reaction products of higher density than the metal, and, often, lower

combustion efficiencies, which can be defined as:

Combustion efficiency

$$= \frac{I_{sp} \text{ (delivered)} \times 100}{I_{sp} \text{ (theoretical)}} \quad \text{(Eq 2)}$$

where I_{sp} is the specific impulse, defined in Eq 3.

A potential problem is the formation of large, long-burning agglomerates that consist of 10 000 or more particles. When aluminum particles (1 to 50 μm) reach the burning surface, some burn while others agglomerate. In rocket systems, aluminum may cause performance loss of about 1% in the specific impulse for every 10% of unburned aluminum. Impulse is the product of force and duration and is expressed as Newton-seconds (pound-seconds). Specific impulse is Newton-second per kilogram (pound-second per pound) of propellant, or:

$$I_{sp} = \frac{\text{N} \cdot \text{s}}{\text{kg}} = \frac{\text{lb} \cdot \text{s}}{\text{lb}} \quad \text{(Eq 3)}$$

The effects of metal additives on the specific impulse of some ammonium perchlorate compositions are given in Table 3.

Magnesium has been used to a small extent as a fuel in solid propellants. Compared with aluminum, it has a lower specific impulse, and there are hazards in handling the fine powder. Substitution of particles of a magnesium-aluminum blend reduce the tendency to agglomerate, but the specific impulse decreases with increasing magnesium content of the particles.

Specifications for Aluminum. Atomized aluminum is the most widely used metal powder in the propellant, explosive, pyrotechnic, and incendiary fields. The most important factors affecting properties of aluminum are method of manufacture, chemical purity or grade, particle size or class distribution, surface area, and ap-

Table 3 Effect of metal additives on specific impulse of ammonium perchlorate propellants

Binder	Binder composition, wt%	Additive	Specific impulse, I_{sp} N·s/kg	lb·s/lb	Chamber temperature(a) °C	°F	Density, g/cm³
Hydrocarbon-based 15		None	2417	246.5	2677	4850	1.66
Double-based, nitrocellulose and nitroglycerin 40		None	2459	250.7	2778	5032	1.73
Hydrocarbon-based 15		19 wt% Al	2590	264.1	2906	5263	1.74
Double-based 45		20 wt% Al	2584	263.5	3487	6308	1.80
Double-based 40		16.8 wt% Be	2739	279.3	3798	6868	1.72

(a) Chamber temperature is that of the case surrounding the propellant, as shown in Fig. 3.
Source: Ref 6

Table 4 Typical properties of aluminum flake powder

Grade	Typical surface area, m²/g	Typical stearic acid coating, %	Typical particle size range prior to de-dusting (−325 mesh), %	Typical water covering prior to de-dusting, cm²/g Apparent (10³)	Available (10³)
1651 3-4		2.0	94.0	6-7	10-11
1663 5-6		2.5	99.5	13-16	16-20

Note: Nonvolatile content of all grades is 99.6% minimum. These grades are fabricated into standard Alcoa Aluminite Aluminum products. Grades 1651 and 1663 have been treated under license granted by the Harshaw Chemical Company and by an Alcoa modified process. They will contain a nominal 0.1% teflon.

Table 5 Preparation and uses of aluminum powder

Aluminum powder	Manufacturing method	Intended use
Type I, grade A . . . . .Atomization and milling		Primer composition
Type I, grade B . . . . .Atomization and milling		Pyrotechnics
Type II, grade C . . . . .Atomization or grinding		Pyrotechnics
Type II, grade D . . . . .Atomization or grinding		Plain incendiary thermite
Type II, grade E . . . . .Atomization or grinding		High-explosive incendiary projectiles
Type III, grade F . . . . .Atomization		Heavy explosives

Table 6 Classification and apparent density of aluminum powder used for explosives

Classification	Mesh size, nominal	Apparent density, g/cm³ min	max
Type I, flaked			
Grade A(a), class 1 . . . .−325		···	0.30
Grade B(b), class 2 . . . .−100		(c)	(c)
Grade B(b), class 3 . . . .−20		···	0.50
Type II, grained or atomized			
Grade C(d), class 4 . . . .−50		0.90	1.10
Grade D(e), class 5 . . . .−12		0.90	1.10
Grade E(f), class 6 . . . .−100		0.95	1.20
Type III, atomized, grade F(g)			
Class 6 . . . .−100		0.95	1.20
Class 7 . . . .−40		0.95	···
Class 8 . . . .−12		0.95	···
Class 9(h) . . . .−325		0.95	···

(a) 85% Al. (b) 93.0% Al. (c) No determination. (d) 91.5% Al. (e) 92.5% Al. (f) 96.0% Al. (g) 98.75% Al. (h) For hexachloro-ethane white smoke mixtures

parent density. Both spheroidal and flake powders are used. Aluminum flake powders are produced by milling regular atomized powders under controlled conditions. Some flake powders are coated with stearic acid and show good stability against gassing in an aqueous environment.

De-dusted grades, which are described in Table 4, have been specially treated to reduce the dust hazards associated with flake powders. The treatment consists of adding a nominal 0.1% teflon, which, after processing, weakly binds the aluminum flake powder and prevents dust formation. Further processing in slurries breaks the teflon bonds, restoring the powder to its effective state. This treatment does not disturb the integrity of the stearic acid coating on the flakes, and the powders maintain good sensitizing properties.

Tables 5, 6, and 7 provide data on aluminum powder. Additional information can be obtained from Ref 7, 8, and 9. Table 8 provides American Society for Testing and Materials (ASTM) standards that are applicable to aluminum powder. For detailed information on the processing of these powders, see the article "Production of Aluminum Powder" in this Volume.

Explosives

An explosive is an energetic substance that undergoes a rapid chemical change whereupon it produces a large quantity of energy accompanied by the evolution of hot gases. The propagation rate through the material may be subsonic (deflagration) or supersonic (detonation). Aluminum powder may be added to explosives to increase their heat output. Because of the high heat of formation of aluminum oxide (Table 1), there is an increase in the heat of explosion, which is the heat produced when an explosive is detonated in an inert atmosphere, and the products of reaction have a higher temperature. Examples of aluminized explosives are given in Table 9.

Properties of Aluminized Explosives. Table 10 compares properties of explosives with and without added aluminum powder. The aluminized explosives have a higher density and heat of explosion, but the volume of gases is decreased. The loss in gas volume is more than compensated for by the greater production of heat. None of the possible replacements for aluminum in explosives

has proven as effective. Aluminum lowers the detonation velocity of TNT and of 60% RDX-40% TNT. Some aluminum can be added advantageously to almost every explosive, provided that it does not react with any of the compounds present.

Pyrotechnics

The term pyrotechnics refers to substances or mixtures of materials that on ignition undergo chemical reaction at a controlled rate. These materials can produce specified time delays or desired amounts of heat, smoke, noise, light, or infrared radiation. For most applications they must burn and not detonate. A formulation consists of one or more fuels and oxidizers and may contain inert ingredients or binders. A simple example is 10% amorphous boron and 90% barium chromate. Pyrotechnic devices that use metals as fuels are:

- Flares and signals
- Smokes
- Tracers
- Incendiaries
- Delay cartridges
- Photoflash compositions
- Igniters

Specification Requirements for Metals

Specifications for aluminum were described earlier in this article. In addition, specifications have been selected for a number of other metals that have pyrotechnic applications. Often, more than one

Table 7 Chemical requirements of aluminum powder

Chemical requirement, %	Aluminum classification					
	Grade A	Grade B	Grade C	Grade D	Grade E	Grade F
Aluminum, min	85.0	93.0	91.5	92.5	96.0	98.75
Copper, max	...	...	...	...	...	0.5
Iron, max	...	0.8	...	...	(a)	0.5
Silicon, max	...	...	...	...	(a)	0.5
Magnesium, max	...	...	...	...	...	0.1
Zinc, max	...	...	...	...	0.5	0.25
Total other elements, max	...	...	...	...	...	0.5
Total impurities, max	15.0	7.0	8.5	7.5	4.0	1.25
Alkalinity as $Mg(OH)_2$, max	...	...	...	...	...	0.07
Nonvolatile matter, min	99.0	99.5	99.80	99.80	99.80	99.95
Easily extracted fatty and oily matter, max	3.0	1.5	...	...	0.2	0.2

(a) Total of iron plus silicon must not exceed 1.5%.

Table 8 American Society for Testing and Materials (ASTM) standards for aluminum powder

Standard No.	
B 214	Method of Test for Sieve Analysis of Granular Metal Powders
B 329	Method of Test for Apparent Density of Refractory Metals and Compounds by the Scott Volumeter
B 417	Apparent Density of Non-Free-Flowing Metal Powder
D 480	Methods of Sampling and Testing Aluminum Powder and Paste
D 1192	Reagent Water
E 34	Methods for Chemical Analysis of Aluminum and Aluminum-Base Alloys
E 101	Spectrochemical Analysis of Aluminum and Aluminum Base Alloys by the Point to Plane Spark Technique

Table 9 Explosives containing aluminum powder

Explosive	Constituents	%
DBX(a)	RDX(b)	21
	Ammonium nitrate	21
	TNT(c)	40
	Aluminum powder	18
Minol(a)	TNT(c)	40
	Ammonium nitrate	40
	Aluminum powder	20
MOX-1(a)	Ammonium perchlorate	35
	Aluminum (fine grain)	26.2
	Magnesium (fine grain)	26.2
	Tetryl(d)	9.7
	Calcium stearate	1.9
	Graphite	1.0
Torpex(a)	RDX(b)	41
	TNT(c)	41
	Aluminum powder	10
	Wax	1
Japanese Type 1(e)	Ammonium picrate	81
	Aluminum powder	16
	Wood pulp	2
	Petroleum	1
Tritonal(e)	TNT(c)	80
	Aluminum flakes	20

(a) Source: Ref 8. (b) Cyclo-1,3,5-trimethylene-2,4,6-trinitramine. (c) 2,4,6-Trinitrotoluene. (d) 2,4,6-Trinitrophenylmethylnitramine. (e) Source: Ref 9

specification or purchase description exists. Prospective suppliers should contact the user to determine whether special requirements exist.

Magnesium (Ref 11). Important properties include form, chemical composition, granulation, and apparent density. Information on the forms and chemical requirements is given in Tables 11 and 12, respectively. For information concerning the processing of these powders, see the article "Production of Magnesium Powder" in this Volume.

Manganese (Ref 12). Manganese powder is employed in delay cartridges (grade I) and pyrotechnic mixtures (grades II and III). Delay cartridges, as shown in Fig. 3, consist of a primer (an explosive charge that can be set off by percussion), an ignition charge (the delay column or delay train) that burns at a predetermined rate, a transfer charge that transfers the energy of the delay column to the main charge, and the main charge itself. Requirements of manganese powders are given in Table 13.

Tungsten (Ref 13). The major use of tungsten powder is as fuel in long-burning delay compositions. These compositions, which are discussed later in this article, have burning times greater than 8 s/cm (20 s/in.). The particle size distribution is shown in Table 14.

Zirconium (Ref 14). This powder is used in pyrotechnic and special ignition compositions. Its combustion behavior differs from that of aluminum in several respects. In the combustion of zirconium, the oxide—ZrO_2—a stable combustion product, dissolves in molten zirconium metal, which may react with nitrogen to form unstable intermediates. Zirconium is more easily ignited than aluminum, possibly because the oxide is a less protective coating than Al_2O_3. While aluminum melts before it ignites, zirconium particles ignite before they melt. Zirconium is highly reactive with a variety of oxidizing agents. For example, the reaction

$$Zr + halogen \rightarrow ZrX_4 \qquad (Eq\ 4)$$

is self-sustaining and exothermic and can be initiated at relatively low temperatures.

The chemical requirements for zirconium are given in Table 15. Particle size and particle size distribution for the two types, which are extremely important, are shown in Table 16. The average particle size requirements of type II powders are 2.5 ± 1.0 μm for class 1, 2.0 ± 0.3 μm for class 2, and 3.0 ± 1.0 μm for class 3. The ignition gain percentage, which is the percent gain in weight when a sample of zirconium is ignited in a crucible over the full heat of a Meker burner for 15 min, is another critical parameter of zirconium powders. The ignition gain requirements for type II zirconium powders are 30.2 to 33.0% for classes 1, 2, and 3 and 30.0 to 33.0% for class 4.

Because of the safety problems encountered in storing and handling zirconium, special labeling and marking of containers are essential. These special labeling and marking requirements for zirconium are:

- Zirconium metal powder (minimum 25% water)
- Danger! Flammable solid. May explode if water content is 10% or below
- Keep wet in storage—dry powder may be ignited by friction, static electricity, or heat
- Do not attempt to loosen or remove material from container with any tool
- Keep away from heat, sparks, and open flame
- Keep from freezing

Table 10 Comparison of properties of explosives with and without aluminum

Reduction to carbon monoxide and hydrogen (H_2) in aluminum compositions

Explosive	Maximum loading density, g/cm³	Volume of gas		Heat of explosion	
		L/kg	ft³/lb	kJ/kg	kJ/lb
TNT(a)	1.62	684.0	10.9	3970	1800
81.8% TNT(a) and 18.2% Al	1.78	559.0	8.95	6153	2790
Blasting gelatin(b)	1.61	771.0	12.3	6740	3057
68.7% blasting gelatin and 31.3% Al	1.92	488.7	7.825	9564	4338

(a) TNT is 2,4,6-trinitrotoluene. (b) 91.3% nitroglycerin and 8.7% nitrocellulose of 12.24% N
Source: Ref 10

Table 11 Types and forms of magnesium powder

Type	Manganese(a), %	Form
Type I, grade A	99.80 min	Shavings, turnings, flakes, or combinations of these
Type I, grade B	(b)	Same as above
Type II	99.80 min	Oblong chip-like particles with rounded edges
Type III	99.80 min	Granular or spheroidal particles
Type IV	95.50 min	Shavings, turnings, flakes, or combinations of these

(a) Present in manufacturing stock. (b) To meet specification

- In case of spillage, keep wet and remove carefully
- In case of fire, smother with foam-type extinguisher or sand
- Do not use carbon tetrachloride, carbon dioxide extinguishers, or water
- Wear goggles or face shield and fire-retardant clothing when handling

Pyrotechnic Applications

Flares and Signals. Flares provide intense illumination; signals require that the light source be clearly distinguished from the background. Signals are smaller and faster burning than illuminating flares. Some applications require emission in the infrared as well as in the visible range. Infrared or near-infrared emission is more important in flares than in signals, because the ground cover is more reflective in the infrared than in the visible part of the spectrum.

Light intensity is a function of temperature, which usually should exceed 2800 °C (5100 °F). Although magnesium and aluminum produce the most satisfactory products, in terms of heat production, cost, and transparency to visible radiation, zirconium, titanium, and hafnium also are used. Examples of pyrotechnic compositions (Ref 15) are:

White flare material

- Magnesium, 17%
- Aluminum, 15%
- Barium nitrate, 55%
- Strontium nitrate, 5%
- Linseed oil, 3%
- Asphaltum, 5%

White hexachloroethane smoke material

- Hexachloroethane, 45 to 47%
- Zinc oxide, 47 to 48%
- Aluminum, 5 to 8%
- Zinc carbonate, 0 to 1%
- Zinc borate, 0 to 1%

Table 12 Chemical requirements for magnesium powders

Chemical requirement	Type I Grade A	Type I Grade B	Type II	Type III	Type IV
Volatile at 105 °C (221 °F), max	0.10	0.10	0.10	0.10	0.10
Oil and grease, max	0.02	0.02	0.02	0.02	0.02
Carbides, max	0.004	0.004	0.004	0.004	0.004
Material insoluble in sulfuric acid, max	0.15	...	0.15	0.15	0.15
Metallic iron, max	0.05	...	0.05	0.05	0.05
Ferric oxide, max	0.25	...	0.25	0.25	0.25
Aluminum(a), max	...	10.0	...	...	2.50
Free metallic magnesium, min	96.0	...	96.0	98.0	95.50
Aluminum plus free metallic magnesium, min	...	96.0	...	...	...

(a) Calculated as percentage of pure aluminum, not as aluminum oxide (Al_2O_3) or other compound

Table 13 Requirements for manganese powder

Chemical or size requirement	Composition, min % Grade I	Grade II	Grade III
Total manganese	99.5	98.0	95.0
Free metallic manganese	99.0	...	...
Granulation			
−200 sieve	...	98.0	99.0
−230 sieve	100	...	...
−325 sieve	98	...	...
Particles			
<10-μm diam	50 (max)	...	...

Table 14 Particle size distribution for tungsten powder

Particle size distribution, μm	Weight, % As supplied min	max	Lab milled min	max
Type I, agglomerated				
0-5	6	18	...	...
5-10	22	38	...	...
10-15	20	35	...	...
15-20	12	25	...	...
20-25	0	25	...	...
Type II, de-agglomerated				
0-1	0	2	0	5
1-2	1	5	1	11
2-3	5	14	10	20
3-4	10	20	15	27
4-5	11	23	15	27
5-6	11	21	10	20
6-7	8	17	4	12
7-25	rem	rem	rem	rem

Fig. 3 Delay cartridge

Manganese powder is employed in the delay column.

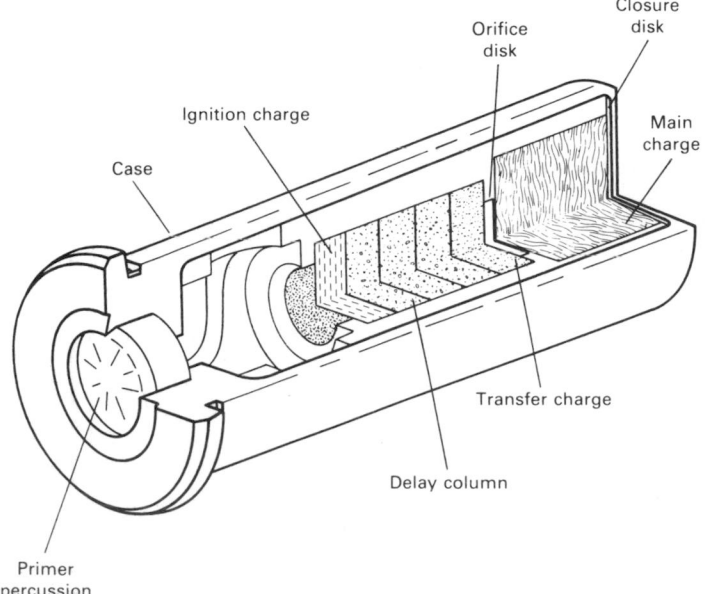

Table 15 Chemical requirements for zirconium powder

Chemical requirement	Type I	Type II Class 1	Class 2	Class 3	Class 4
Total zirconium, min	96.0	94.0	95.0	95.0	94.0
Calcium, max	0.05	0.05	0.05	0.05	0.10
Iron, max	0.30	0.20	0.03	0.05	0.20
Aluminum, max	0.10	0.30	0.15	0.10	0.30
Hydrogen, max	0.20	0.20	0.20	...	0.17
Chloride as Cl, max	0.03	...	...	...	...
Silicon, max	0.10	...	...	...	...
Tin, max	0.75	...	...	...	...

Table 16 Particle size requirements for zirconium powder

Particle size, µm	Sieve No.	Composition, wt% Class 1	Class 2	Class 3
Type I				
840 ...	−20	99 min	...	...
250 ...	−60	5 max	...	...
177 ...	−80	...	100	...
149 ..	−100	...	98 min	...
88 ...	−170	...	...	100
74 ...	−200	...	50 max	98 min
44 ...	−325	...	25 max	...
<20	(a)	...	...	...
<10	(a)	...	2.0 max	10.0 max
Type II				
125 ..	−120	100	100	99.5 min
74 ...	−200	99 min	99 min	...
20	−26(b)	96 min	96 min	...
<9	(a)	85 min	85 min	...
3	(a)	70-90 min	70-90 min	...
0.75	(a)	12-30	12-30	...

(a) Subsieve particle size. (b) Buckbee Mears sieve

Red tracer material

- Magnesium-aluminum alloy, 27%
- Strontium nitrate, 56%
- Polyvinyl chloride, 7%

Smokes. This type of pyrotechnic is used for screening and signaling. An important type of smoke generator employs metals such as aluminum or magnesium with hexachloroethane and a zinc salt, as in the compositions listed above. Zinc chloride is formed upon combustion. Colored smokes are produced by incorporating a volatile organic dye into the pyrotechnic mixture. A typical formula for a colored smoke is:

- Dye, 30 to 50%
- Confectioner's sugar, 20 to 35%
- Potassium chlorate, 22 to 33%
- Sodium bicarbonate, 3 to 10%
- Kerosene or paraffin oil, 2 to 4%
- Infusorial earth, 0 to 4%

Some of the dyes that can be used are auramine (yellow), rhodamine B (red), methylene blue (blue), and 1,4-di-p-toluidine-anthraquinone (green).

In the above example, the heat-producing mixture is the potassium chlorate and confectioner's sugar. The sodium bicarbonate is a coolant. These ingredients form a cool-burning, gas-forming mixture that melts and vaporizes the dye with minimum loss of dye. The dye, which is vaporized by the heat source, then recondenses to yield a brilliant cloud or trait of dye molecules. Some dye is destroyed by the heat, but in a proper formulation enough will be left to produce the desired effect.

Tracers. The compositions of tracers function as incendiaries; they are used for visual control of trajectory for small-arms applications. The color of tracers is based on visual observation. Smoke tracers that produce a colored smoke trail are used by aircraft gunners and create a trace length of about 600 m (1970 ft). Formulation is similar to that listed under pyrotechnic compositions for flares and signals.

Incendiaries. Certain materials have the property of undergoing spontaneous ignition in air. These pyrophoric substances may be used to ignite materials or used as weapons. Many metals in a finely divided form have pyrophoric properties and several emit sparks when abraded (rubbed). Cerium and its mixture with rare earth metals, known as Mischmetal, are used commercially for lighter flints. Zirconium is the pyrophoric metal most often used for military purposes. It may be alloyed with titanium or other metals.

When metal powders are used in flammable compositions intended as military weapons, the essential criteria are easily ignitable particles, high temperature and heat output, burning time not more than 10 s, and dense distribution of burning particles over the target areas. The reaction products should be solids or liquids. The fuels are usually metallic and include elements such as aluminum, iron, magnesium, titanium, zirconium, and antimony. Suitable oxidizers are certain metal oxides, nitrates, chromates, perchlorates, and permanganates and highly fluorinated organic compounds, such as polytetrafluoroethylene. The materials are formed into

pellets and ignited by a suitable heat source. Examples of two fast-burning compositions (Ref 16) are:

Material A

- Potassium perchlorate, 43 to 65%
- Aluminum powder, 22 to 33%
- Iron powder, 2 to 28%
- Nitrocellulose, 1.5 to 2.5%
- Glass powder, 0 to 9%

Material B

- Polytetrafluoroethylene, 45 to 56%
- Magnesium powder, 22 to 28%
- Iron powder, 12 to 26%
- Nitrocellulose, 1.5 to 2.5%
- Glass powder, 0 to 8%

Delay Compositions. Generally, these are mixtures of powdered metals and inorganic oxidizers that produce nongaseous products upon combustion. Their function is to ignite the main charge of a delay cartridge at a definite predetermined rate and within a specified precision (see Fig. 3). Delay compositions must burn for a definite time to produce a delay and also must provide the heat required at the end of that time for ignition of the powder train that it contacts.

The practical range of burning time for chemical delay trains is from 0.004 to 16 s per linear centimetre, and limits on the lengths of the delay train restrict their range to about 5 ms to 5 min. Applications for various time ranges of chemical delay compositions (Ref 17) are:

- *1 to 24 ms*: bombs that explode on impact
- *25 to 1000 ms*: commercial blasting in quarries and mines
- *1 to 6 s*: chemical munitions, such as grenades
- *7 to 60 s*: delay switches

Burning times of delay formulations are given in Table 17. Conventional delay elements produce gases that present problems due to the effect of pressure on burning rate. This problem is reduced in the so-called gasless delays, because the quantity of gas is very small and the burning rate is less affected by pressure. This property of gasless delays is valuable for longer delay times, particularly where the delay required is 1 s or greater. See Table 18 for compositions of gasless delay elements.

Problems that occur in delay trains include limited shelf life and possible carcinogenic properties of the widely used

Table 17 Burning times of delay formulations

Fuel	Fuel composition, %	Oxidant	Inert ingredient	Burning time s/cm	Burning time s/in.
Molybdenum	80-89	Potassium perchlorate, 11-20%	...	0.004-016	0.01-0.04
Boron (amorphous)	10	Barium chromate, 90%	...	0.2	0.5
50Zr-50Ni	54	Barium chromate, 31% Potassium perchlorate, 15%	...	1.0	2.5
Manganese	39	Lead chromate, 47% Barium chromate, 14%	...	2.3	5.8
Tungsten	34	Barium chromate, 52% Potassium perchlorate, 9%	Superfloss(a), 5%	7.1	18
Tungsten	27	Barium chromate, 58% Potassium perchlorate, 10%	Superfloss(a), 5%	16	40

(a) A type of diatomaceous earth
Source: Ref 18

Table 18 Compositions of gasless delay elements

Fuel	Fuel composition, %	Oxidant	Inert ingredient
Boron	10	Barium chromate, 90%	...
Boron	30	Lead dioxide, 70%	...
Zirconium	33	Ferric oxide, 50%	...
Titanium	17	Ferric oxide, 50%	...
Zirconium	65	Ferric oxide, 25%	Diatomaceous earth, 10%

Source: Ref 19

chromate oxidizers. Small laboratory amounts of pyrotechnic compositions can be handled without severe risk; accidents due to the highly reactive, explosive nature of these materials are more likely in large-scale operations. Storage is a concern, because the metal fuel powders corrode in the presence of moisture. Factors affecting burning times are the chemical composition used and the relative amounts of various ingredients, particle size (particularly of the fuels), ambient temperature and pressure, degree of compaction, and diameter of the column. The burning times for a tungsten delay composition in three particles sizes are:

- *4 μm*: average burning time of 3.5 s/cm (8.9 s/in.)
- *8 μm*: average burning time of 6.8 s/cm (17.2 s/in.)
- *15 μm*: average burning time of 11.1 s/cm (28.1 s/in.)

Photoflash Compositions. Materials for these applications require small particle size and must burn violently for periods of less than 0.1 s. The compositions consist of mixtures of fuels (such as magnesium, aluminum, or zirconium) and oxidizers (such as nitrates and perchlorates). Aluminum and magnesium perform best at sea level, but zirconium is better at high altitudes. Because of the explosive nature of the combustion and the large amount of material used in a single device, photoflash compositions are considered the most hazardous class of pyrotechnic mixture. A sample photoflash powder composition is:

- Atomized aluminum, 40%
- Barium nitrate, 30%
- Potassium perchlorate, 30%

Igniters. Devices that are described as igniters can be used to initiate the chemical reactions in explosives, propellants, or pyrotechnic compositions. Because some compositions are difficult to ignite, a more effective mixture is used to generate higher temperatures. Although many ignition devices do not involve metal powders, some use metals such as aluminum, magnesium, titanium, and zirconium. Certain igniters utilize nonmetals such as boron and silicon. Some compositions used as igniters also are applied in other areas, such as delay trains. Three igniter compositions are:

Ignition pellet

- Aluminum, 35%
- Potassium perchlorate, 64%
- Vegetable oil, 1%

Mixture A1A

- Zirconium, 65%
- Red iron oxide, 25%
- Calcined diatomaceous earth, 10%

Boron type

- Amorphous boron, 10%
- Barium chromate, 90%

Miscellaneous Pyrotechnics. Metals that react with water may be used to generate motive power. An example has been described in which granules of aluminum, magnesium, or zinc are injected as a spray by high-pressure steam into a heated reaction chamber (Ref 20). High-temperature steam (>205 °C or 400 °F) is then injected into the metal spray to cause an exothermic reaction between the metal, which is now molten, and water. The reaction is completed by injection of additional water that is vaporized by the heat generated, producing high-temperature, high-pressure steam for motive power.

REFERENCES

1. *Handbook of Chemistry and Physics*, 60th ed., CRC Press, Boca Raton, FL, 1979, p D45-50
2. Grosse, A.V. and Conway, J.B., Combustion of Metals in Oxygen, *Industrial and Engineering Chemistry*, Vol 50, 1958, p 663-672
3. Bardon, M.F. and Lambert, J.J.R.G., Applications of Metal Powders as Fuels, Combustion Institute, Canadian Section, Spring Technical Meeting, Queens University, Kingston, Ontario, 3-4 May, 1979, p 9-1 to 9-3
4. Sax, I.N., *Dangerous Properties of Industrial Materials*, 5th ed., Van Nostrand Reinhold, New York, 1979
5. *Riegel's Handbook of Industrial Chemistry*, 7th ed., Van Nostrand Reinhold, New York, 1974, p 570-596
6. *Encyclopedia of Explosives and Related Items*, Vol 8, U.S. Army Armament Research and Development Command, Dover, NJ, 1978, p P414
7. Aluminum Powder, Flaked, Grained, and Atomized, MIL-A-512B (proposed)
8. Meyer, R., *Explosives*, Verlag Chemie, New York, 1977, p 13, 58, 182, 184, 295
9. Military Explosives, TM9-1300-214, Dept. of the Army, Washington, DC, 28 Nov 1967
10. *Encyclopedia of Explosives and Related Items*, Picatinny Arsenal, Dover, NJ, 1965, p A145-152
11. Magnesium Powder for Use in Ammunition, MIL-M-382C(AR), 10 Aug 1978
12. Manganese, Powdered, for Use in Ammunition, JAN-M-476A, 13 Aug 1948
13. Tungsten Powder (For Pyrotechnics), MIL-T-48140(PA), 14 May 1974
14. Zirconium (Granular and Powdered), MIL-Z-399D, 5 April 1973

15. Ellern, H., *Modern Pyrotechnics,* Chemical Publishing, New York, 1961, p 273, 276, 277
16. Beckert, W.F. and Dengel, O.H., Fast Burning Compositions of Fluorinated Polymers and Metal Powders, U.S. Patent 4000022, 28 Dec 1976
17. Ellern, H., *Military and Civilian Pyrotechnics,* Chemical Publishing, New York, 1968, p 198-200
18. *Encyclopedia of Explosives and Related Items,* Vol 4, Picatinny Arsenal, Dover, NJ, 1969, p D873
19. *Encyclopedia of Explosives and Related Items,* Vol 4, Picatinny Arsenal, Dover, NJ, 1969, p D867
20. Kaiho, Y., Generation of Motive Power with Metal and Water as Fuel, U.S. Patent 3771313, 13 Nov 1973

SELECTED REFERENCES

- Davis, T.L., *The Chemistry of Powder and Explosives,* John Wiley & Sons, New York, 1956
- Fordham, S., *Explosives and Propellants,* Pergamon Press, New York, 1966
- Holzmann, R.T., *Chemical Rockets,* Marcel Dekker, New York, 1969
- James, R.W. *Propellants and Explosives,* Noyes Data, Park Ridge, NJ, 1974
- Kit, B. and Douglas, S.E., *Rocket Propellant Handbook,* Macmillan, New York, 1960
- McLain, J.H., *Pyrotechnics: From the Viewpoint of Solid State Chemistry,* Franklin Institute Press, Philadelphia, 1980
- Sarner, S.F., *Propellant Chemistry,* Reinhold, New York, 1966
- Shidlovsky, A.A., *Fundamentals of Pyrotechnics,* 3rd ed., TM1615, Picatinny Arsenal, Dover, NJ, May 1965; translated from Osnovy Pirotekhniki, 1964
- Shidlovsky, A.A., *Principles of Pyrotechnics,* 4th ed., FSTC-HT-23-0380-75, Army Foreign Science and Technology Center, Charlottesville, VA, 1974; translation of Mashinostroyeniye Press, Moscow
- Warren, F.A., *Rocket Propellants,* Reinhold, New York, 1958

Metal Powders Used in Polymers

By Ihab L. Kamel
Associate Professor
Department of Materials Engineering
Drexel University

METAL POWDERS are added to polymers primarily to change the appearance and/or the physical properties of the polymer. Increased demand in the automotive industry and related fields for high-performance parts that are lighter in weight than conventional metals has led to the use of polymer-based composites. A large number of these composites contain metal powders that match the metallic appearance of the components they replace.

In other applications where the physical properties of the polymers are modified for a specific end use, metal powders are used to change the insulative nature of the covalently bonded polymer or to increase electron density. Physical properties and functions that are affected in specialty polymers include:

- Electrical conductivity
- Thermal conductivity
- Electromagnetic shielding
- Anti-static surfaces

Despite their relative importance, metal-filled polymer composites have received only limited research attention. References 1 to 6 review the influence of particle shape, size, and distribution on the performance of composites. However, the importance of polymer-metal compatibility or adhesion and the differences that exist between the physical properties of these materials are not completely explained by theoretical analysis in relation to the macroscopic properties of the composite.

Compounding Techniques

Powder metal-filled plastics are manufactured in various shapes and sizes, depending on end use. Sheet, films, rods, fibers, molded parts, and liquids are available. Metal forming operations such as compression molding, extrusion, injection molding, cold drawing, deep drawing, thermoforming, and stamping are used to produce polymers. Where a special network of metal powder is desired within the polymer, compression molding is preferred, because it does not disturb the relative position of the particles. Typical polymers used in these composites include polyvinyl chloride, polymethyl methacrylate, polyethylene, polycarbonate, polyurethane, polyvinyl acetate, teflon, nylon, epoxy, and bakelite (Ref 7-10). Filler powder metal varies depending on the application, but may be made of pure metals such as copper, nickel, silver, aluminum, and iron, or alloys such as stainless steel.

Molded articles are usually made by mixing the metal and polymer powders thoroughly (in a V-mixer, for example) for sufficient time to allow the build-up of static charge on the surface of the polymer powder, which aids in attracting the smaller metal particles. To minimize shear, compression molding of the homogeneous mixture follows by applying sufficient heat and pressure to fuse the polymer particles together in the desired shape. The polymer temperature is controlled to be above the glass transition temperature, but not

high enough above the melting point to cause degradation of the polymer. With ultrahigh molecular weight polyethylene (average molecular weight of 2 to 8 million), the molding temperature must be at least 50 to 70 °C (120 to 160 °F) above its melting point to guarantee maximum particle-particle fusion and minimum porosity. Generally, high molecular weights result in high polymer viscosities that require high molding temperatures. Typical molding conditions for polyvinyl chloride are 150 °C (300 °F) and 70 to 100 MPa (10 to 15 ksi) (Ref 2). Control of pressure, temperature, and time is important to minimize porosity in the molded part. This process is particularly suited for molding thermoset polymers such as bakelite where the final curing occurs during the heating cycle.

Extrusion and injection molding of the metal and polymer mixture also are widely used because these processes provide continuous production and superior economics compared to compression molding. The polymer, however, experiences high shear rates in these processes, which cause metal particle segregation and orientation along the flow lines. Shear usually results in the appearance of crack-like flaw lines in the filled injection molded parts when two separate flow patterns of molten polymer merge. These flaw lines are areas with minimum density and are almost devoid of filler particles. In injection molded parts, a skin of unfilled polymer exists where the liquid polymer adheres to the cold mold surface during solidification. This surface layer may have to be removed if a conductive outer surface is desired.

Metal powders also may be dispersed in liquid polymers or solutions, such as in metallic paints containing aluminum flakes. Additionally, cast resins such as epoxy, polyesters, polyurethanes, and silicone rubber may contain metal particles in the liquid phase. Dispersion is accomplished by stirring before the viscous slurry is poured into the mold. Some segregation is expected, although viscosity of the resin can be adjusted to minimize segregation. The resin is then cured by the addition of a suitable catalyst at either ambient or elevated temperature. An added refinement of this technique calls for the orientation of nickel particles in the liquid phase by an external magnetic field, which has been reported to enhance the electrical conductivity of the composite (Ref 11).

Powder Polymer Processing

Specialty polymers, such as ultrahigh molecular weight polyethylene with very high viscosity or polymers with relatively high melting points, are difficult to process in the liquid state. These polymers are best suited for processing by P/M techniques, including cold powder compaction, sintering, and forging (Ref 12, 13). Of these processing operations, sintering is the most time consuming and hence the cost-determining step. Polymers used as insulators are resistant to heat flow. If the part being processed is over 25 mm (1 in.) thick, sintering time may exceed 30 min to obtain a uniform temperature throughout the part. The cost of this process can be greatly reduced by admixing powder metals.

The presence of metal particles in the polymer powder also enhances the stability of the cold compacts, because the ductile polymer must flow around the rigid metal particles. Plastic strain is, therefore, enhanced during compaction, which helps particle-particle interlocking and results in a strong preform.

An alternative to sintering polymer compacts in a furnace by slow heating from the outside is to use inductive heating of the dispersed metal phase within the matrix. This process is still experimental, but is quite attractive because of its speed and the uniformity of heat generation within the part. In induction heating, thermal stresses are eliminated, and thick sections may be processed economically without the need for annealing. In addition, if the metal particles form a segregated network in which the polymer particles are coated,

then induction heating of the metal phase generates heat at the particle-particle interface. This energy can be used to fuse these surfaces locally, minimize the heat transfer distance, and reduce the processing time.

Electrically Conductive Polymers

Synthetic polymers are good electrical insulators because of their covalently bonded chain structure in which the valence electrons are tightly bound to specific orbits. This characteristic is in contrast to the high mobility of the valence electron cloud in the electrically conductive metal structure. There are, however, many applications where a lightweight, flexible conductive material is required. Conductive adhesives to join electronic parts, anti-static material to dissipate dangerous surface charges, polymeric panels with controlled resistivity to generate heat, and semi-conducting walls and floors for hospital operating rooms are typical industrial applications that require conductive polymers. Incorporation of a conductive metal powder as a filler in polymers improves the electrical conductivity of the plastic matrix.

Generally, two types of possible arrangements of the metal particles within a polymer matrix exist: random and segregated. A randomly distributed filler is easy to produce by thorough mixing within the matrix. Choice of filler particles is not crucial to the end product. In the segregated network, however, the filler is not randomly distributed, but forms a three-dimensional network within the polymer matrix (Fig. 1). This arrangement of the metal particles is achieved when these relatively small particles are located at the boundary between the relatively large polymer particles (Ref 1, 14). When this condition exists, the ratio of particle size of polymer and metal is an important factor in the development of composites.

Figure 2 illustrates that the metal-filled plastic composites in random or segregated networks show a sudden drop in resistivity when a continuous conductive path is formed at a critical volume fraction of the metal filler. For a given particle size ratio, this critical composition is only 6 to 8% metal if the distribution is segregated, but increases to 35 to 40% metal if the distribution is random.

Because the segregated network has improved properties and lowered metal consumption, it has been investigated by sev-

Fig. 1 Compaction of polymer and surrounding metal powder filler

(a) Schematic showing the predicted filler segregated network. (b) Photomicrograph of 7 vol% nickel (4 to 7 μm) in polyvinyl chloride. Magnification: 400×. Source: Ref 1

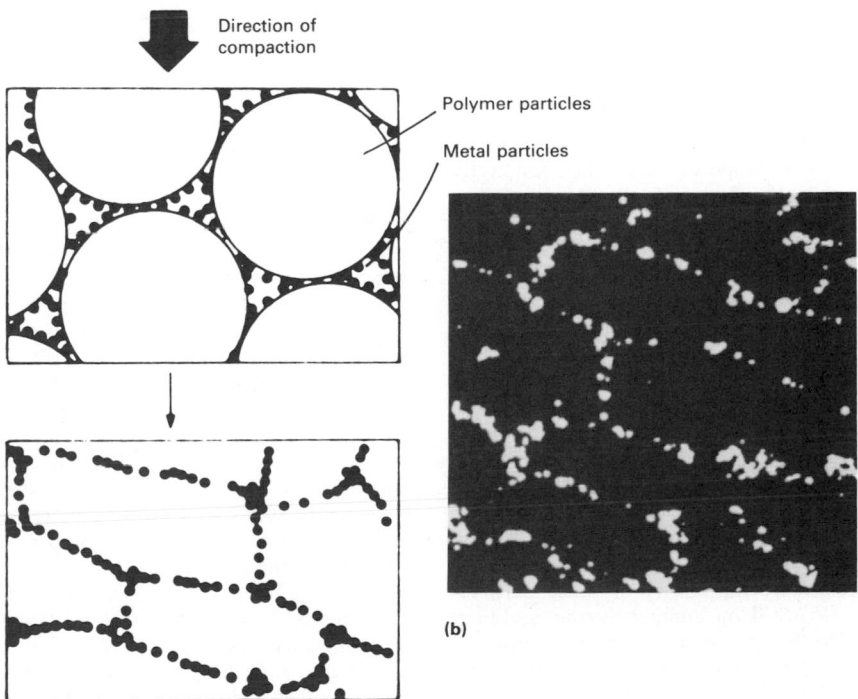

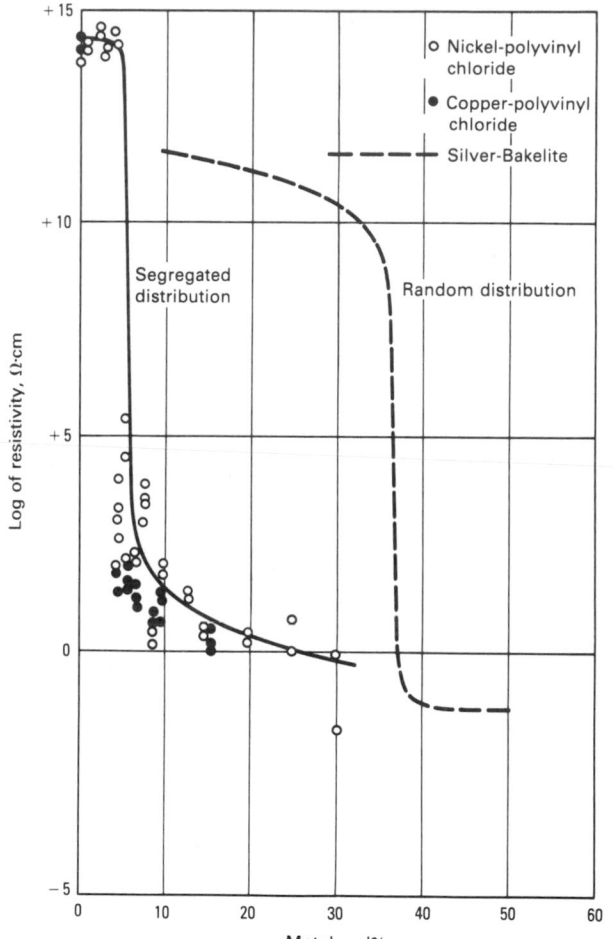

Heating or cooling the composite may result in a change in conductivity. Generally, polymers have a higher coefficient of thermal expansion compared to metals. Therefore, an increase in temperature may cause a net matrix expansion and a concomitant reduction in the metal particle contact area. The net result is a reduction in composite conductivity with temperature.

Thermally Conductive Polymers

The incorporation of metal particles in polymers enhances thermal conductivity. However, development of this property for the replacement of conventional materials is lagging behind polymers with improved electrical conductivity. One reason for the slow development is the modest improvement in thermal conductivity of a basically insulating material. Additionally, heat causes the development of internal stresses due to the large differences in the coefficient of thermal expansion between polymers and metals. This physical mismatch in component properties has led to the development of various theoretical models to predict the composite behavior (Ref 19).

These models, however, are not compatible with all systems. One cause for this deviation from theory is the lack of adhesion between the polymer and the metal phases. Another is the influence of thermally induced stresses. Lack of adhesion between the matrix polymer and the metal filler causes incomplete wetting of the particulate surfaces and promotes the formation of voids at the interfaces. When the composite is heated, the matrix partially expands into these voids, causing lower experimental thermal expansion values than those predicted from theory.

The influence of metal particles on the thermal conductivity of the polymer has been the subject of many proposed models (Ref 20-24). Experimental data on the thermal conductivity of metal-filled polymers are limited. Reference 25 discusses an epoxy with a filler of gold, silver, aluminum, copper, lead, tin, bronze, or stainless steel. Generally, thermal conductivity is influenced by filler size, particle shape and distribution, adhesion between the phases, and level of metal loading.

Figure 6 shows that the thermal conductivity of copper-filled epoxy increases with temperature and filler content. In addition, filler size, which influences the interfacial area, may be important at lower temperatures, as shown in Fig. 7 (Ref 26). Furthermore, a qualitative assessment of the rate of thermal expansion between

eral researchers, and theoretical models have been proposed to predict resistivity and critical volume of filler as a function of the polymer and metal particle size ratio (Ref 2, 15, 16). These models are depicted graphically in Fig. 3 and 4. Experimental agreement with theory appears consistent. Lower critical compositions are achieved in these systems as particle size ratio of the polymer and metal increases. In an extreme case, chemical deposition of ultrafine copper particles onto polyvinyl chloride powder resulted in a low critical composition value of only 0.5 vol% copper (Ref 17).

The formation of a stable segregated network of metallic filler particles is influenced by the ability of these particles to cold weld on contact during molding. Dendritic-type particle structure provides better results due to mechanical interlocking of the particles, as compared to smooth spherical powders. Also, surface hardness

of the metal particles may interfere with their ability to interlock.

Both strength and conductivity of composites with a segregated filler network improve with compacting pressure. Improvement in strength can be directly attributed to a reduction in porosity or an increase in density. In the polyvinyl chloride-copper system, a relative density of 70% of theoretical was found at a molding pressure of approximately 10 MPa (1.5 ksi). At approximately 60 MPa (9 ksi), the relative density was 97% (Ref 18). Improvement of conductivity with pressure is shown in Fig. 5, in terms of increased resistivity with reciprocal compacting pressure for the polyvinyl chloride-copper system. The increase in pressure does not affect the relative position of the network of metal particles to any significant extent, but it may help improve metal-to-metal contact and hence the conductivity of the composite.

Fig. 3 Experimental and calculated effect of polymer and filler particle size ratio (r_1/r_2) on the resistivity of polyethylene-nickel composites of the segregated network type
Source: Ref 1

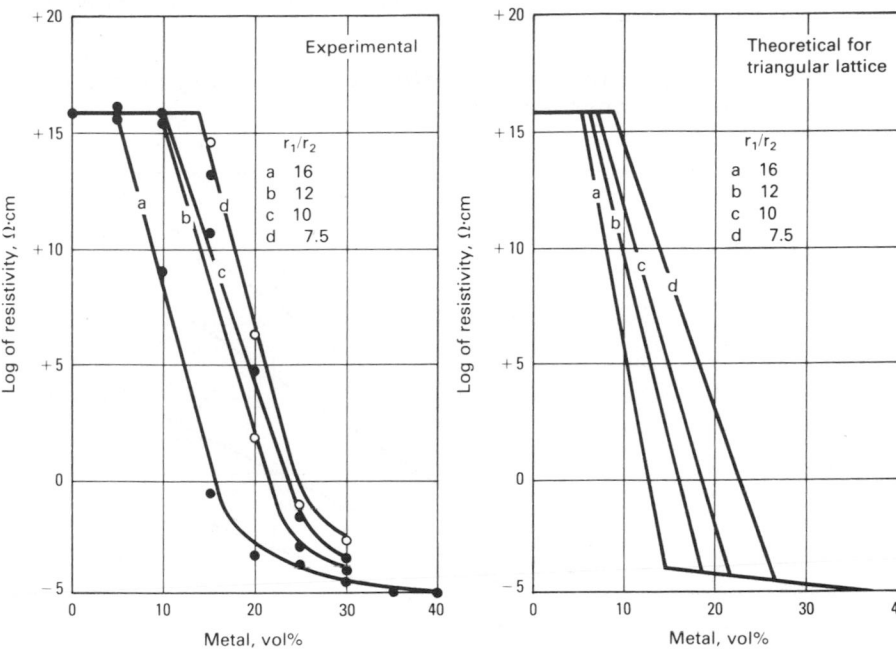

localized stresses at the interface between the two phases.

Electromagnetic Shielding

The ever-increasing use of computers and other electronic devices that generate timing signals and pulses has caused a dramatic growth in radiated electromagnetic and radiofrequency energy. A typical industrial city of 100 000 population is estimated to have over half a million sources of electromagnetic interference. Generally, radiated signals are contained at or near their source. This, however, is quite difficult to do with the extensive use of plastics and their structural foams in computer and electronic housings and cabinets. Plastics are transparent to radiofrequency energies, while metals are the most effective electromagnetic interference reflectors.

Electromagnetic interference waves may be absorbed, reflected, or transmitted, depending on the electrical conductivity of the material. Incorporation of metal powders in plastics improves conductivity and provides effective electromagnetic interference shields. The effectiveness of a shield is measured in decibels, as the ratio of field intensities before and after shielding. A 10-dB attenuation means that the

polymers, metals, and other materials is given in Fig. 8 to illustrate the problem encountered in these composites on heating (Ref 27). The large coefficient of thermal expansion of the polymer matrix is significantly reduced when metal particles are introduced into that matrix. This change in the average composite property causes

Fig. 4 Effect of polymer/filler particle size ratio on the resistivity of polyvinyl chloride-copper composites of a segregated network
Source: Ref 16

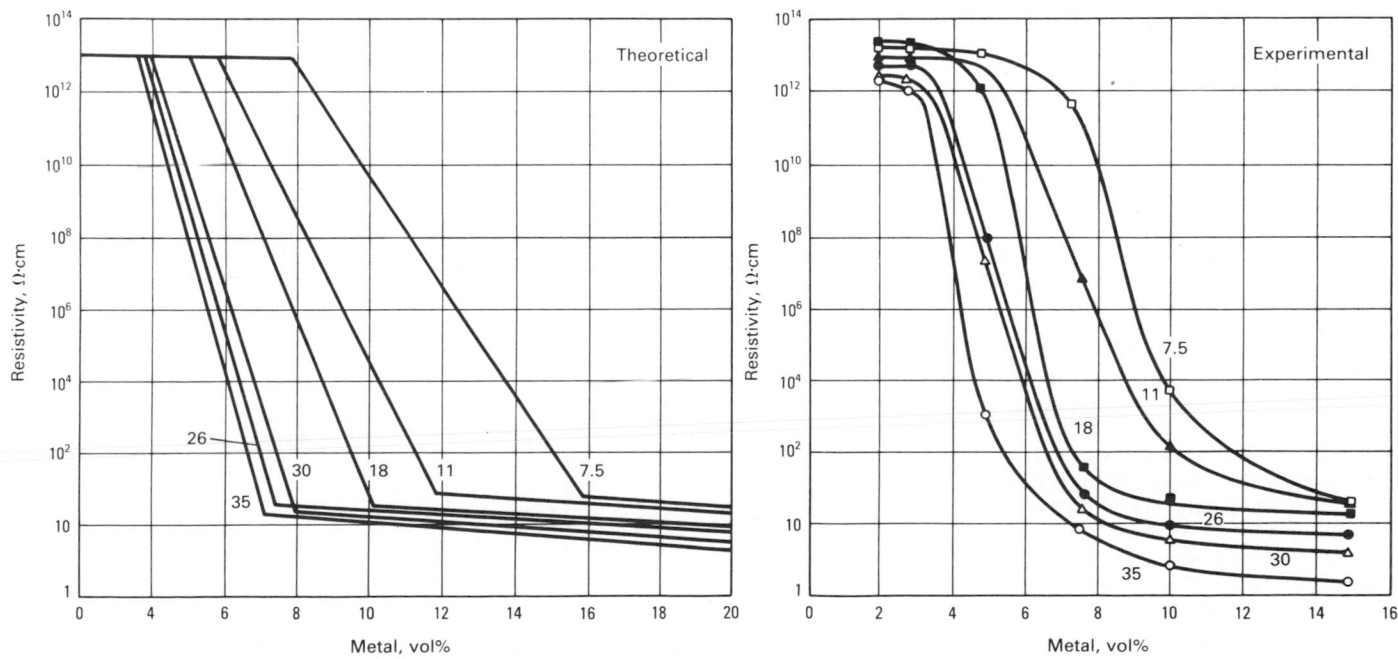

signal is 10% of source energy; 20 dB is 1% of source, 40 dB is 0.01%. According to Ref 28, the range of shield quality is:

Attenuation, dB	Shield quality
0-10	Very little
10-30	Minimal
30-60	Average
60-90	Above average
90-120	Maximum

It is estimated that an attenuation of 30 to 40 dB meets approximately 95% of the commercial needs.

Equipment manufacturers prefer plastic housings for electronic devices because of the strength-to-weight ratio, cost, corrosion resistance, and ease of fabrication. These plastics, however, must be converted to effective electromagnetic interference shields by, among other techniques, the incorporation of metal powders by applying a conductive coating (paint) or by making the plastic part itself conductive, which is a simple one-step operation that reduces the cost of the part. Metal powders or flakes are used in the conductive plastic approach; sometimes metallized glass fibers are preferred because of their mechanical reinforcement. Care must be taken to ensure the uniformity of filler dispersion and to avoid resin-rich areas that may permit inconsistent shielding.

When a conductive coating is applied, a large quantity of conductive metallic particles is incorporated in the paint to provide the necessary surface conductivity. Silver is the most effective metal for this application, but also the most expensive. Copper and nickel also are effective, but are prone to oxidation and adhesion to the substrate. Nickel presents a health problem, as it has been linked to cancer in some animal studies. These conductive paints usually are sprayed with conventional equipment directly onto the molded parts. A comparison of the various approaches to electromagnetic interference shielding is given in Table 1.

In evaluating the most suitable approach, a manufacturer must consider shielding reliability, surface resistivity, overall cost, stability in the environment, coating thickness, and expected government changes or trends. With increased use of microcomputers, government regulations must be expected to tighten as the electromagnetic interference problem intensifies. This will require the production of more efficient and reliable conductive plastics at a lower cost.

Anti-Static Surfaces

Another potential safety hazard that exists when polymers are used as insulators is static charge buildup due to friction. Moving of conveyor belts, filling of plastic bags, or even walking on carpets may result in a rapid accumulation of electrical charge in excess of 30 000 V. This phenomenon is hazardous in hospitals, grain

Fig. 5 Resistivity as a function of reciprocal compacting pressure for polyvinyl chloride-copper composites
Source: Ref 18

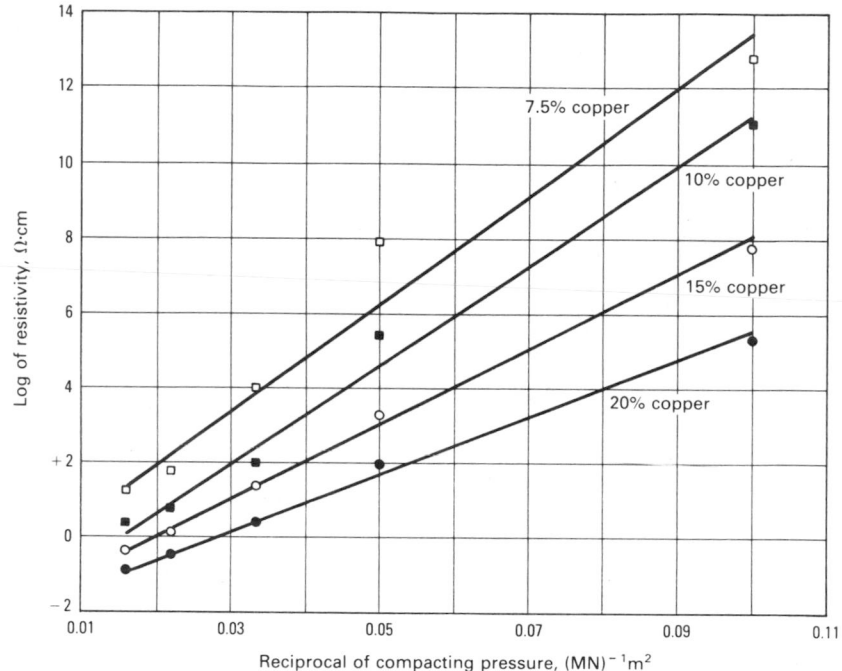

Fig. 6 Thermal conductivity as a function of temperature for epoxy-copper composites at various volume percentages of spherical 100-μm copper

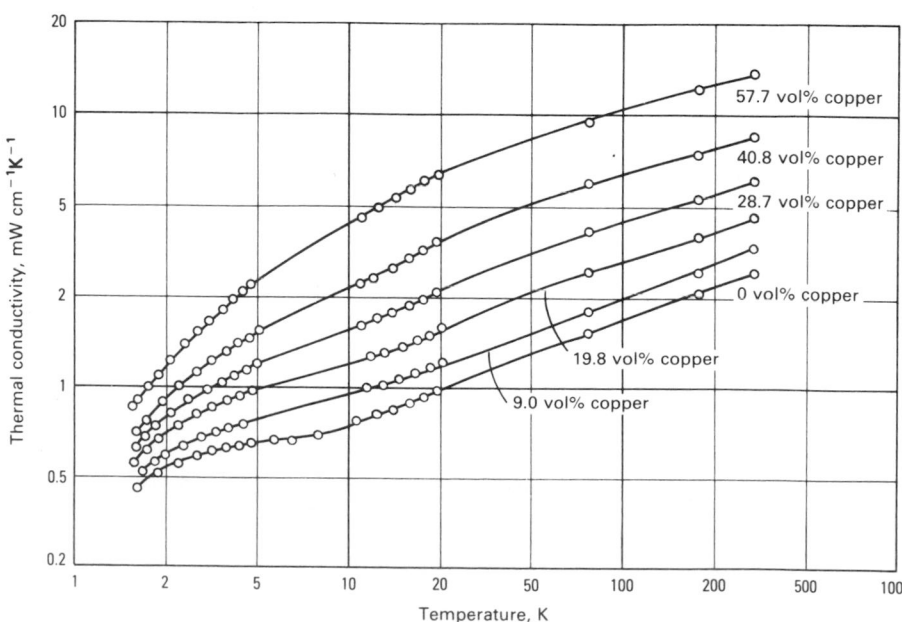

Fig. 7 Thermal conductivity of polymers as a function of the volume percentage of spherical copper at selected temperatures and copper particle sizes. Solid lines are calculated values after Ref 26

Solid lines indicate theoretical values. ○, 11 μm; ●, 100 μm; □, 46 μm; ■, red. 100 μm; ▲, 25 μm. Source: Ref 25

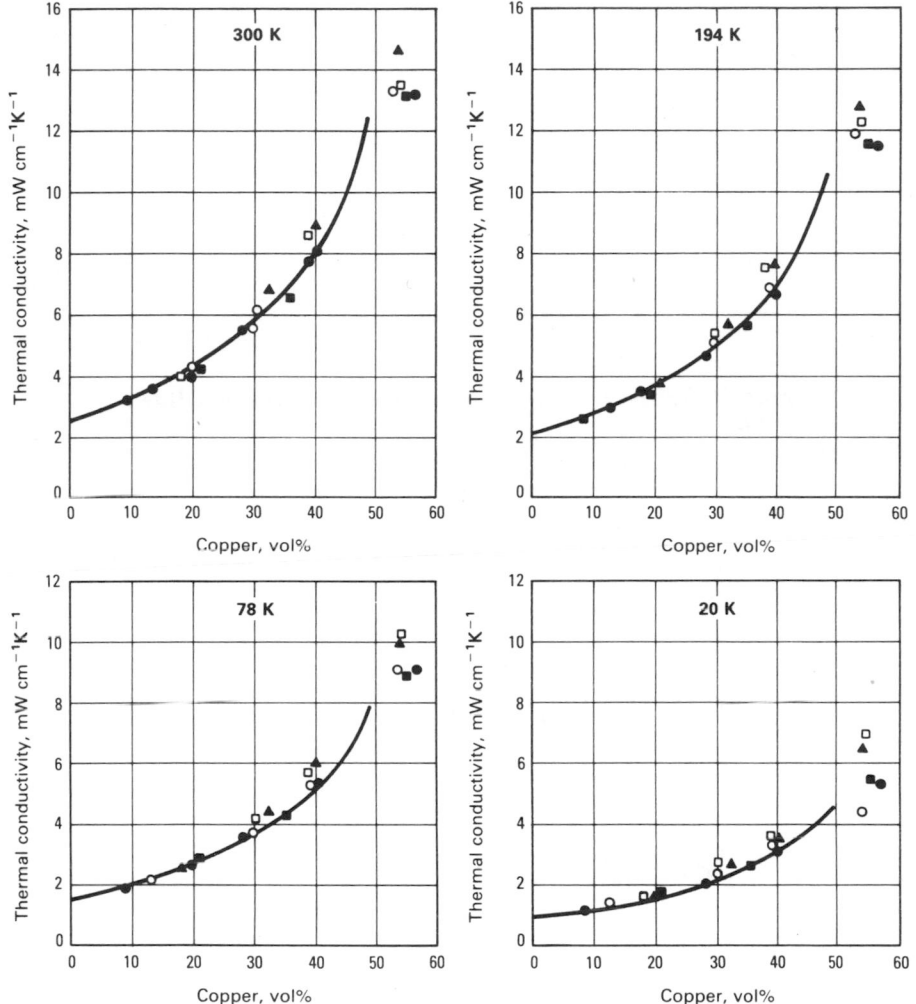

Fig. 8 Thermal expansion coefficients for various classes of materials

Source: Ref 27

elevators, and explosives manufacturing facilities. Electronic integrated circuits also must be protected from static charge.

The ability to dissipate a charge depends on inducing bulk or surface conductivity in the plastic material by providing a continuous path of conductive filler particles to bring the charge potential to the ground. The technique of using conductive fillers (metals or carbon black) enjoys the reliability of material performance; the induced conductivity is permanent and unaffected by changes in humidity, for example. Conductive flakes and fibers have a definite advantage over spherical particles, because they provide higher surface conductivity at lower concentrations. Aluminum flakes appear to be the most popular metallic fillers in commercial products for this application.

The efficiency of charge dissipation can be determined by the time it takes to ground an induced charge of ±5000 V, according to the Federal Test Method standard 101B, method 4046. Table 2 shows experimental static decay data for selected examples of commercially available conductive plastics. The results indicate an instantaneous decay rate for all filled plastics, while pure polycarbonate, which is an insulator, exhibits a decay rate in excess of 100 s. Selection of a particular conductive system, therefore, is governed by the aesthetic appearance and the cost of the filler.

Effect of Filler on Polymer Properties

Introduction of hard, nondeforming particles into a soft polymeric matrix alters physical and mechanical properties of the matrix. If the filler and polymer are chemically compatible, adhesion due to secondary bonds of the van der Waals type can be effective, particularly with submicron filler particles. An important modification of the polymer behavior is an increase in the glass transition temperature. This is usually explained in terms of segmental immobilization of the polymer chains and an increase in entanglement of these chain segments with filler surface. The rigidity of the matrix should be expected to increase with the incorporation of this hard particle. For example, polyvinyl chloride was found to show a 5 °C (9 °F) increase in its glass transition temperature by the addition of 10 vol% copper powder and a similar effect with tungsten powder.

Thermally stable filler particles also improve matrix thermal stability. The onset of matrix thermal degradation is raised to higher temperature levels by the addition

Table 1 Shielding alternatives

Process	Conductor	Surface resistivity(a), max	Equipment cost	Secondary processing steps
Vacuum metallizing	Aluminum	1.0	High	Three
	Silver	1.0		
	Copper	2.0		
Cathode sputtering	Aluminum	1.0	Very high	One
	Silver	1.0		
	Copper	2.0		
Metallic paint	Silver	1.0	Low	Three
	Copper	2.0		
	Nickel	5.0		
Conductive fillers	Metallized fibers and flakes	20-5000	None	None

(a) The surface resistivity of a material, as described in ASTM D 257, is the ratio of the potential gradient parallel to the current along its surface to the current per unit width of the surface. Surface resistivity of a material is equal to the surface resistance between two electrodes forming opposite sides of a square. The size of the square is immaterial.
Source: Ref 29

Table 2 Static decay rate of selected filled plastics

Material	Applied voltage, %	Decay rate, s	
		At +5 kV	At −5 kV
Polycarbonate	50	>100	>100
	10	>100	>100
	0	>100	>100
Nickel painted polycarbonate	50	0.02	0.02
	10	0.03	0.04
	0	0.05	0.05
Polycarbonate/ 30% metal	50	0.02	0.02
	10	0.04	0.03
	0	0.06	0.05
Polycarbonate/ 40% carbon	50	0.02	0.02
	10	0.04	0.03
	0	0.05	0.05

Source: Ref 30

of copper to both polyvinyl chloride and polymethyl methacrylate. This increase in thermal stability is based on an increase in the activation energy for the degradation reaction due to interfacial adhesion between the polymer and the filler.

One of the major effects of filler on polymer matrices is the increase in their stiffness or elastic moduli. Figure 9 shows the result of various theoretical models predicting this increase. Experimental data lie in the middle region and are greatly affected by filler dispersion and agglomeration. Smaller particles appear to be more pronounced in their effect, due to a more efficient load transfer with the increase in the interfacial surface area.

Unlike elastic moduli, tensile strength is affected detrimentally by particulate filler addition. Figure 10 shows the relative reduction in matrix strength with poor and good adhesion. Under stress, particles act as local stress concentration sites, causing a substantial decrease in the load-bearing capability of the composite. The use of a good coupling agent and the presence of a strong interfacial adhesion allows some load transfer from matrix to filler. Because adhesion depends on particle surface area, decreasing particle size coupled with good matrix and filler adhesion increases relative composite strength.

Addition of particulate filler also is detrimental to ultimate elongation of the polymer matrix, because filler particles interfere with matrix deformation and may act as crack initiation sites. As a result, elongation at break decreases in a pseudologarithmic manner with increasing composite filler content. Also, because decreased elongation is due to constraints on the matrix by the filler, theoretical models predict that good matrix and filler adhesion increases constraint and greatly reduces elongation at break compared to poor

Fig. 9 Composite/matrix moduli ratio as a function of filler concentration

Calculated values from various theoretical models reviewed in Ref 31

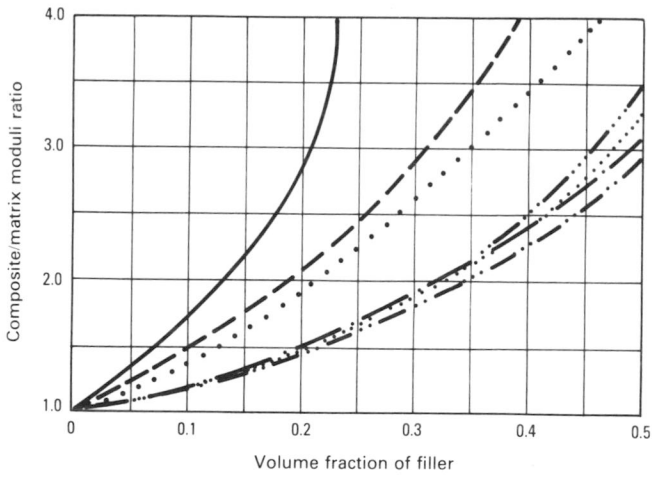

Fig. 10 Composite/matrix tensile strength ratio as a function of filler concentration

Calculated values from various theoretical models reviewed in Ref 31. Lower three plots represent poor adhesion.

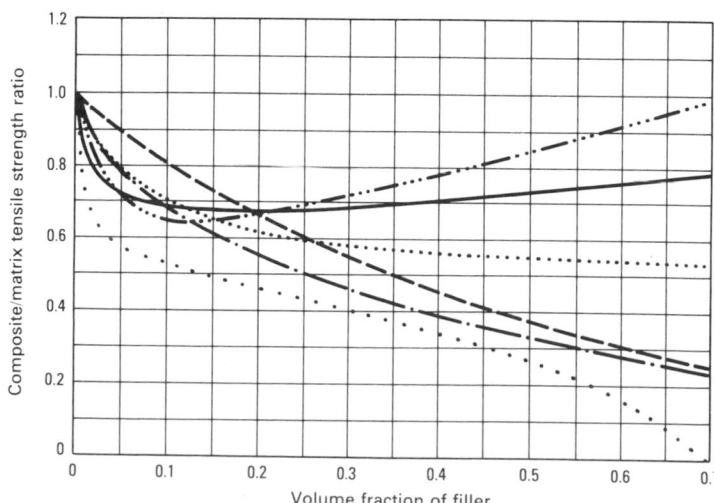

Fig. 11 Composite/matrix ultimate elongation ratio as a function of filler concentration

Calculated values from various experimental data reviewed in Ref 31. Top curve is based on no adhesion at the interface.

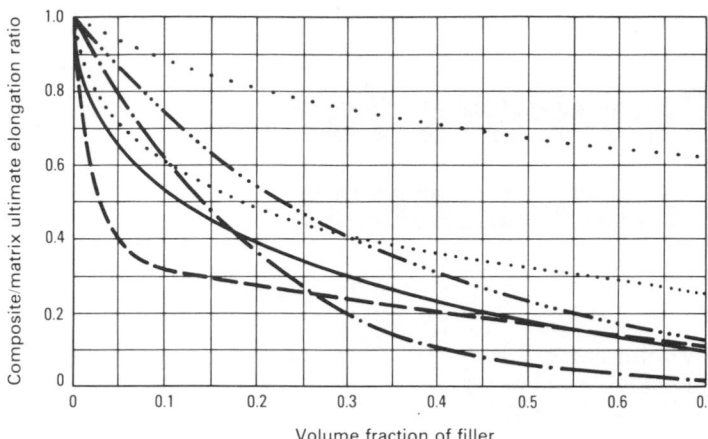

filler adhesion, as shown in Fig. 11. Decreasing filler particle size decreases elongation at break, with good adhesion.

REFERENCES

1. Malliaris, A. and Turner, D.T., *J. Appl. Phys.*, Vol 42, 1971, p 614
2. Kusy, R.P., Ph.D. dissertation, Department of Metallurgical Engineering, Drexel University, Philadelphia, 1973
3. Nielsen, L.E., *Mechanical Properties of Polymers and Composites*, Vol 1 and 2, Marcel Dekker, New York, 1974
4. Bhattacharya, S.K. and Chaklader, A.C.D., *Polymer Plas. Technol. Eng.*, Vol 19 (No. 1), 1982, p 21
5. Bueche, F., *J. Appl. Phys.*, Vol 43, 1972, p 4837
6. Kusy, R.P. and Turner, D.T., *Soc. Plast. Eng. J.*, Vol 29 (No. 7), 1973, p 56
7. Gurland, J., *Trans. Met. Soc. AIME*, Vol 236, 1966, p 642
8. Aharoni, S.M., *J. Appl. Phys.*, Vol 43, 1972, p 2463
9. Fleming, R.J. and Ramicar, J.H., *J. Macromolec. Sci. Chem.*, Vol A14 (No. 5), 1970, p 1223
10. Delmonte, J., *Metal Filled Plastics*, Reinhold, New York, 1961
11. Gul, V.E. and Golumbera, M.G., *Kolloidn. Zh.*, Vol 29 (No. 1), 1967, p 62
12. Halldin, G.W. and Kamel, I.L., *Polymer Eng. Sci.*, Vol 17 (No. 1), 1977, p 21
13. Halldin, G.W. and Kamel, I.L., 39th Proceeding of the Society for Plastics Engineers Annual Technical Conference, Montreal, 1977, p 298
14. Coler, M., U.S. Patent 2,761,849 and 2,761,854, 1956
15. Kusy, R.P., *J. Appl. Phys.*, Vol 48, 1977, p 5301
16. Bhattacharya, S.K., Basu, S., and De, S.K., *Composites*, Vol 9 (No. 3), 1978, p 117
17. Das, D., Basu, S., and Paul, A., *J. Mater. Sci.*, Vol 15, 1980, p 1719
18. Bhattacharya, S.K., *Polymer*, Vol 20, 1979, p 1166
19. Holliday, L. and Robinson, J.D., in *Polymer Engineering Composites*, Richardson, M.O.W., Ed., Applied Science Publishers, London, 1977
20. Gorring, R.L. and Churchill, S.W., *Chem. Eng. Prog.*, Vol 57 (No. 7), 1961, p 53
21. Godbec, H.W. and Ziegler, W.T., *J. Appl. Phys.*, Vol 37, 1966, p 56
22. Cheng, S.C. and Vachon, R.I., *Int. Heat Mass Transfer*, Vol 13, 1970, p 537
23. Nielsen, L.E., *J. Appl. Polymer Sci.*, Vol 17, 1973, p 3819
24. Bigg, D.M., *Composites*, Vol 10, 1979, p 95
25. de Araujo, F.F.T., Garrett, K.W., and Rosenberg, H.M., paper presented at the International Conference on Composite Materials, Geneva, April 1975
26. Meredith, R.E. and Tobias, C.W., *J. Appl. Phys.*, Vol 31, 1960, p 1270
27. BiBenedetto, A.T. and Nicolais, L., in *Advances in Composite Materials*, Piatti, G., Ed., Applied Science Publishers, London, 1978
28. Regan, J., *Polymer Plas. Technol. Eng.*, Vol 18 (No. 1), 1982, p 47
29. Ross, G.S., *Mod. Plast.*, Vol 56, 1979, p 67
30. Litman, A.M. and Fowler, N.E., *Polymer Plast. Technol. Eng.*, Vol 18 (No. 1), 1982, p 93
31. Gordon, R.J., M.S. thesis, Department of Materials Engineering, Drexel University, Philadelphia, 1982

Iron Powders for Food Enrichment

By John Patrick, Jr.
Manager, Purchasing
SCM Metal Products

DIETARY IMPORTANCE OF IRON—recognized long ago—remains a major concern of nutritionists. Widespread prevalence of iron-deficiency anemia exists in the world population today, particularly afflicting infants, young children, and menstruating women. Although cereal-grain products (wheat, corn, and rice) provide a significant portion of the daily caloric intake, modern milling and manufacturing techniques remove about two thirds of the natural iron content. Furthermore, in the home, iron pots and skillets that were once a source of dietary iron have been replaced by aluminum, copper-clad stainless steel, and special glass cookware.

Iron-deficiency anemia results when the blood loses its ability to carry oxygen and carbon dioxide. Capillaries carry oxygen to the cells of the body and pick up carbon dioxide waste, which the blood exchanges for oxygen in the lungs. This carrying capacity depends on an iron protein, called hemoglobin, that constitutes about one third of the mass of red blood cells. Anemia is confirmed by a low red blood cell count.

Early attempts to promote the use of vitamin and mineral supplements and whole-grain products were unsuccessful. In 1941, the fortification of flour with iron and other nutrients began as a public health measure in the United States. Since then, iron enrichment has been extended to a wide range of food products, including ready-to-eat and cooked cereals, noodles, cakes, cookies, crackers, puddings, candies, and diet and liquid foods.

Iron Additives

Iron added to food products is dissolved in the stomach (about 0.1 normal solution of hydrochloric acid or 3.65 g of hydrochloric acid per litre) and adsorbed in the small intestine. Some forms of additive iron, such as ferric orthophosphate, are not readily soluble during the residence time in the stomach. Most of these have been identified through human, animal, and *in vitro* studies and generally are not used for enrichment. The usefulness of an iron source to combat or prevent anemia is expressed in terms of relative bioavailability (RBV), in which the adsorption of the iron source is compared to that of ferrous sulfate, which has an RBV arbitrarily assigned at a value of 100. A higher RBV number indicates a better iron source.

Recently, the usage of iron for food enrichment has been estimated at approximately 40% ferrous sulfate (dried), 50% elemental iron powder, and 10% other iron salts. Although water-soluble ferrous sulfate, which is abundant and inexpensive, may seem to be the favored iron source, ferrous sulfate and other salts are too reactive, leading to food enrichment problems such as off flavors, odors, color changes, and chemical change or reaction during the baking or cooking process. Some of these difficulties reportedly are overcome by encapsulation of the ferrous sulfate with food triglycerides such as hydrogenated vegetable oils, but the practice is not widely used, possibly due to the added cost. Normally, the use of ferrous sulfate is limited to food products that have a short expected shelf-life.

By contrast, elemental iron powder has enjoyed a 50% usage in food enrichment, mostly because of its stability in all food environments. By far, the greatest usage is in enriched flour and bread, in which the amount of iron added is about 37 and 29 mg/kg (17 and 13 mg/lb), respectively. Although the quantity seems small, these and other food product applications amount to a 450 metric tons/year (500 tons/year) elemental iron powder requirement in the United States alone. Negative properties of elemental iron powder, although not considered serious, include its dark gray color, which has a slight graying effect on the color of flour and bread, and its higher density compared to the milled cereal grain, which may lead to segregation if incorporation is not performed carefully. Finally, the reported RBV of the different forms of commercial elemental iron powder varies anywhere from 10 to 90% that of ferrous sulfate.

Elemental Iron Powders

Much of the early work on evaluating elemental iron as a dietary iron source provided confusing RBV data, because investigators did not fully describe or characterize the iron used. The data simply indicated ferrum reductum, or reduced iron. Recently, the studies have been repeated and much has been learned about the food iron source that is now called elemental iron. Today, a family of elemental iron powders is recognized for use in food enrichment, including reduced iron, electrolytic iron, and carbonyl iron.

Reduced iron powder is made by reduction of ground iron oxide with hydrogen or carbon monoxide at an elevated temperature. Purity of the product is dictated by the purity of the iron oxide. Reduced irons are produced from either iron ore or mill scale. These products, however, have the lowest purity of the food-grade iron powders when compared to electrolytic or carbonyl powders. Basic impurities include carbon, magnesium, aluminum, silicon, phosphorus, sulfur, chromium, manganese, nickel, and copper, many of which are present as oxides and appear in the acid insolubles analysis. Because of the nature of the process, iron mill scales reduced in carbon monoxide tend to have higher carbon (in the form of residual or free carbon or graphite) and sulfur. By far, the most common impurity in iron powder produced by any process is oxygen, most of which occurs as a thin film of surface oxide. In the case of reduced iron, however, the probability also exists that some of the original iron oxide remains as a core in the center of the iron particles.

Reduced iron powder is suitable for comminution by ball, hammer, or attrition milling. The particle shape is irregular and porous, and it consists of a number of small, equiaxed grains. Impurity inclusions found along the boundaries of these grains, together with iron oxide inclusions, provide discontinuities that contribute to the friability of reduced iron powder particles.

Electrolytic iron powder is produced by electrolytic deposition of a hard, brittle metal that is mechanically comminuted. Iron is produced domestically, using chemically pure iron anodes, a ferrous sulfate bath (electrolyte), and thin stainless steel cathode sheets, onto which the iron is deposited. These sheets are removed from the bath after a standard plating cycle, washed to remove soluble salts, dried, then flexed to remove the brittle deposit. The resulting fragments then are ground mechanically to a finely divided powder. Insoluble contaminants originating from the anode fall to the bottom of the bath as a sludge, and electrochemical conditions are established to favor the migration, primarily of iron ions, to the cathode. Impurities that remain are usually at levels of hundreds of parts per million or less, with surface oxide as the major impurity.

The particle shape of electrolytic iron powder is irregular, dendritic, or fernlike, from which it receives its high surface factor. Unlike reduced iron, the grains in electrolytic particles are less symmetrical. Iron obtained by the electrolytic method generally is somewhat harder than that produced by reduction; therefore, powders can be ground with greater yield of subsieve size.

Carbonyl iron powder is the third elemental iron powder used for food enrichment. Manufacture of these powders, which

are much finer than other iron powders, entails the treatment of reduced iron with carbon monoxide under heat and pressure. The resulting iron pentacarbonyl, $Fe(CO)_5$, later is decomposed under controlled conditions, yielding an iron powder and carbon monoxide gas. At this point, the major impurity is carbon (about 1%), and further reduction in wet hydrogen is necessary to remove most of it. The powder has particles ranging in size from 0.5 to 10 μm in diameter and is of high purity.

The carbonyl iron particle is close to spherical in shape and often appears as a cluster of several spheres bonded together. The structure of the particle is characterized by concentric shells arranged in onion-skin fashion. The particle is very dense, smooth, and hard-skinned, and not readily prone to surface oxidation (oxygen content is about one fifth that of reduced iron or electrolytic iron). Impurities found in low levels in carbonyl powders include oxygen, carbon, nitrogen, silicon, chromium, manganese, and nickel. The carbonyl process is the most costly of those discussed in this article.

Oxygen is the major impurity in iron powder, even though milling of granular iron usually is done under an inert atmosphere, because feed stock to the grinding equipment contains trapped air that leads to particle surface oxidation. With freshly milled and subsequently package-protected iron powder, the surface oxide is present as ferrosoferric ($FeO \cdot Fe_2O_3$). The ferrous oxide portion is very soluble (high RBV) in stomach acid and conveniently provides attack sites for particle dissolution; ferric oxide has little or no bioavailability.

Regulation of Iron Powders

Because each of the three members of the elemental iron powder family is produced differently, each exhibits its own special set of properties. These variations caused the confusion in early bioavailability studies that used ferrum reductum. In addition, little attention was paid to particle size, the only requirement being a powdered form of iron. Although a powder generally is considered to consist of discrete particles of dry material with a maximum dimension of 1 mm, iron particles that large may never dissolve in the stomach during normal residence time.

Some usage of coarse elemental iron powder in the food industry is known to exist, but most usage now has been restricted to −100 mesh (149-μm or smaller) powder, as specified in the Food Chemi-

Table 1 Influence of particle size on bioavailability of iron powders used for food enrichment

Production method and test laboratory	Particle size, μm	Relative biological value
Electrolytic (AOAC)	7-10	63.5
Electrolytic (FDA)	0-10	76
	10-20	75
	20-40	48
	>40	45
Hydrogen reduction (FDA)	10-20	54
	325 mesh	34
	100 mesh	18
Carbon monoxide reduction (AOAC)	7-10	36
	14-19	21
	27-40	13
Carbonyl iron (FDA)	<4	69
	3-5	69
	4-8	64

cals Codex, third section, 1981. Monographs are included for the three basic forms of elemental iron powder: carbonyl iron, electrolytic iron, and reduced iron. The first two forms are required to be 95% minimum −325 mesh; reduced iron is identified only as −100 mesh. The monographs also list the chemical specifications (total iron, acid insolubles, arsenic, lead, and mercury) and the prescribed methods of analysis. Data on the relative biological values for iron powders used for food enrichment of known particle sizes as determined by the Federal Drug Administration (FDA) and the Association of Official Analytical Chemists (AOAC) are given in Table 1.

Federal regulations or guidelines are lacking to limit the use of elemental iron to only those forms specified in the Food Chemicals Codex or to identify which properties are essential to providing good bioavailability. Industry groups and governmental agencies continue to work toward the development of standards. Meanwhile, greater consumer nutrition awareness and competition in the food enrichment industry has led to a continual upgrading of food products. Generally, it is agreed that high purity, fine particle size, and high unit surface area are necessary for elemental iron to perform as an acceptable dietary iron source.

SELECTED REFERENCES
● Fortmann, K.L., Joiner, R.R., and Vidal, F.D., Uniformity of Enrichment in Baker's Flour Applied at the Mill, Paper presented at 58th Annual Meeting of the American Association of Cereal Chemists, St. Louis, Nov 1973
● Darby, W.J., The Case for the Pro-

posed Increase in Iron Enrichment of Flour and Wheat Products, *Nutrition Reviews*, Vol 30, 1972, p 98-102

- Proposed Fortification Policy for Cereal-Grain Products, Food and Nutrition Board, National Academy of Sciences—National Research Council, Washington, DC, 1974
- Recommended Dietary Allowances, 8th ed., Publication No. 2216, Food and Nutrition Board, National Academy of Sciences—National Research Council, Washington, DC, 1974
- Amine, E.K. and Hegsted, D.M., Biological Assessment of Available Iron in Food Products, *Journal of Agricultural and Food Chemistry*, Vol 22, 1974, p 470-476

- Sapers, G.M., Panasiuk, O., Jones, S.B., Kalan, E.B., and Talley, F.B., Iron Fortification of Dehydrated Mashed Potatoes, *Journal of Food Science*, Vol 39, 1974, p 552-554
- Report of the *ad hoc* Committee on Iron Enrichment of Wheat Flour and Baked Foods, American Bakers Association, Washington, DC, 1972
- Bass, E.J., Borenstein, B., Brooke, C.L., Gortner, W.A., O'Neill, R.D., Rusoff, I., Titus, D., and Bandle, M.R., Report of the *ad hoc* Committee on Iron Enrichment of Wheat Flour and Baked Foods, American Bakers Association, Washington, DC, and Millers' National Federation, Chicago, 1972
- Wilder, R.M., A Brief History of the

Enrichment of Flour and Bread, *Journal of the American Medical Association*, Vol 162, 1956, p 1539
- Olsson, K.S., Heedman, P.A., and Staugard, F., Preclinical Hemochromatosis in a Population on a High-Iron-Fortified Diet, *Journal of the American Medical Association*, in press
- Cook, J.D., Minnich, V., Moore, C., Rasmussen, A., Bradley, W.B., and Finch, C.A., Absorption of Fortification Iron in Bread, *American Journal of Clinical Nutrition*, Vol 26, 1973, p 861
- Waddell, J., The Bioavailability of Iron Sources and their Utilization in Food Enrichment, Life Sciences Research Office, Bethesda, MD, FASEB, 1973, p 63

Automotive Applications

AUTOMOTIVE APPLICATIONS of powder metallurgy products began in the mid-1920's with the development of self-lubricating bearings, which were porous, oil-retaining bushings (see the article "P/M Bearings" in this Volume). Since that time, the automotive industry has been the main consumer of well over 50% of the iron powder used in the United States. Similar proportions are used in the automotive industries in Europe and Japan. Even after World War II, when powder producers broadened their markets to appliances, business machines, and farm machinery, use of P/M parts in the automotive industry continued to dominate the market with the introduction of new applications.

Figure 1 identifies a wide range of P/M applications in modern automobiles. Because parts usage varies among manufacturers, it is difficult to estimate the number and weight of P/M parts in the average automobile. Powder metallurgy parts in different engines produced by a single manufacturer may vary by a factor of three or four to one. Some manual transmissions contain virtually no P/M materials in terms of weight, while others may utilize over 1 kg (2.2 lb) of iron powder. This is dictated by the economics of installed manufacturing facilities, perceptions of durability, and cost.

The powder metallurgy parts content in automatic transmissions made by one manufacturer varies by a ratio of 2 to 1, depending on transmission size and the total number of parts used. Vehicles without power steering, power brakes, and automatic transmissions, as well as those containing certain engines, may contain minimal amounts of P/M materials. Thus, an estimate of P/M weight in automobiles may range from 2 kg (4.4 lb) up to 7 to 8 kg (15.4 to 17.6 lb). With the acceptance of P/M connecting rods and camshafts, P/M weight in American automobiles should increase to about 16 kg (35 lb) by 1987 (Ref 1).

Uniform application of P/M parts among the various types of transmissions and engines represents the largest potential growth of conventional P/M materials in automobiles. Other perceived growth applications include parts that are hot formed (hot forged, sinter forged, or powder forged), parts that are sintered at high temperatures (resulting in properties that lie between those obtained by hot forming and those obtained by conventional pressing and sintering), and soft magnetic parts made of low-carbon iron re-pressed to high density.

There are many P/M applications in the automotive industry that parallel nonautomotive applications. These applications are covered in the articles "P/M Parts for Farm, Lawn, and Garden Equipment" and "P/M Bearings" in this Volume. In addition, material on processing and applications of P/M parts for military vehicles can be found in the article "Ordnance Applications." This article deals with current P/M parts applications that are unique to the automotive industry.

Compacted and Sintered Parts

Ferrous Applications. Powder metallurgy usage in automobiles is dominated by iron-based powder alloys. Primary usage, in descending order of importance, has been in automatic transmissions, engines, steering gears, and chassis components. Some ferrous P/M parts for automotive applications are heat treated; those used in the as-sintered condition are coined after sintering. See the articles "Secondary Operations Performed on P/M Parts and Products" and "Production Presses and Tooling" in this Volume for additional information on heat treating and coining.

Alloy composition, density, sintering practice, and secondary operations vary greatly among the manufacturers. References 2 and 3 describe these parts applications in detail. Most conventional P/M parts are pressed at about 414 to 483 MPa (30 to 35 tsi) and sintered in a belt furnace at about 1120 °C (2050 °F).

Automatic transmission parts, such as drive gears, oil pump gears, turbine hubs, and clutch reaction pressure plates (Fig. 2), have been produced by P/M techniques for many years; the longevity of these parts is proof of their reproducibility and reliability.

Automotive engines have almost exclusively used P/M oil pump gears (Fig. 3), but other applications vary among manufacturers and engines. Crankshaft and camshaft sprocket gears (Fig. 4) frequently are made from ferrous metal powder. Powder metallurgy rocker arm pivot balls (Fig. 5) are widely used, as are valve guides, particularly in aluminum cylinder heads that are produced from iron powder. Valve seat inserts from special P/M iron alloys are made by some foreign manufacturers. Steering gear oil pumps are made by most manufacturers by P/M processing, the pump pressure plate being an excellent example of a complex P/M part (Fig. 6). Chassis parts such as shock absorber pistons, piston rod guides, and the control arm ball joint ball shown in Fig. 7 have been made from P/M materials since the late 1950's and have enjoyed excellent durability. Nearly all automotive manufacturers use these parts.

Manual transmissions contain relatively few P/M parts, the most common being synchronizer gears and keys made of copper-infiltrated iron powder (Fig. 8). A number of new, relatively narrow applications for P/M parts are typified by a shifter yoke lug array found in one American manual transmission.

Nonferrous Applications. A relatively large number of nonferrous P/M parts are used in automobiles, but they represent nominal weight and generally are similar to parts used in nonautomotive applications. These include bushings for radio controls, windshield wiper motor bushings and gears, distributor bushings, and other annular parts that are similar to bushings in other applications. Figure 9 is a copper alloy P/M clutch lever bearing, which is considered a large P/M part. A considerable volume of copper metal powder has been used in the past as a journal bearing matrix material.

Aluminum P/M parts have raised considerable interest in the automobile indus-

Fig. 1 Typical P/M parts used in automobiles
Courtesy of Cincinnati Inc.

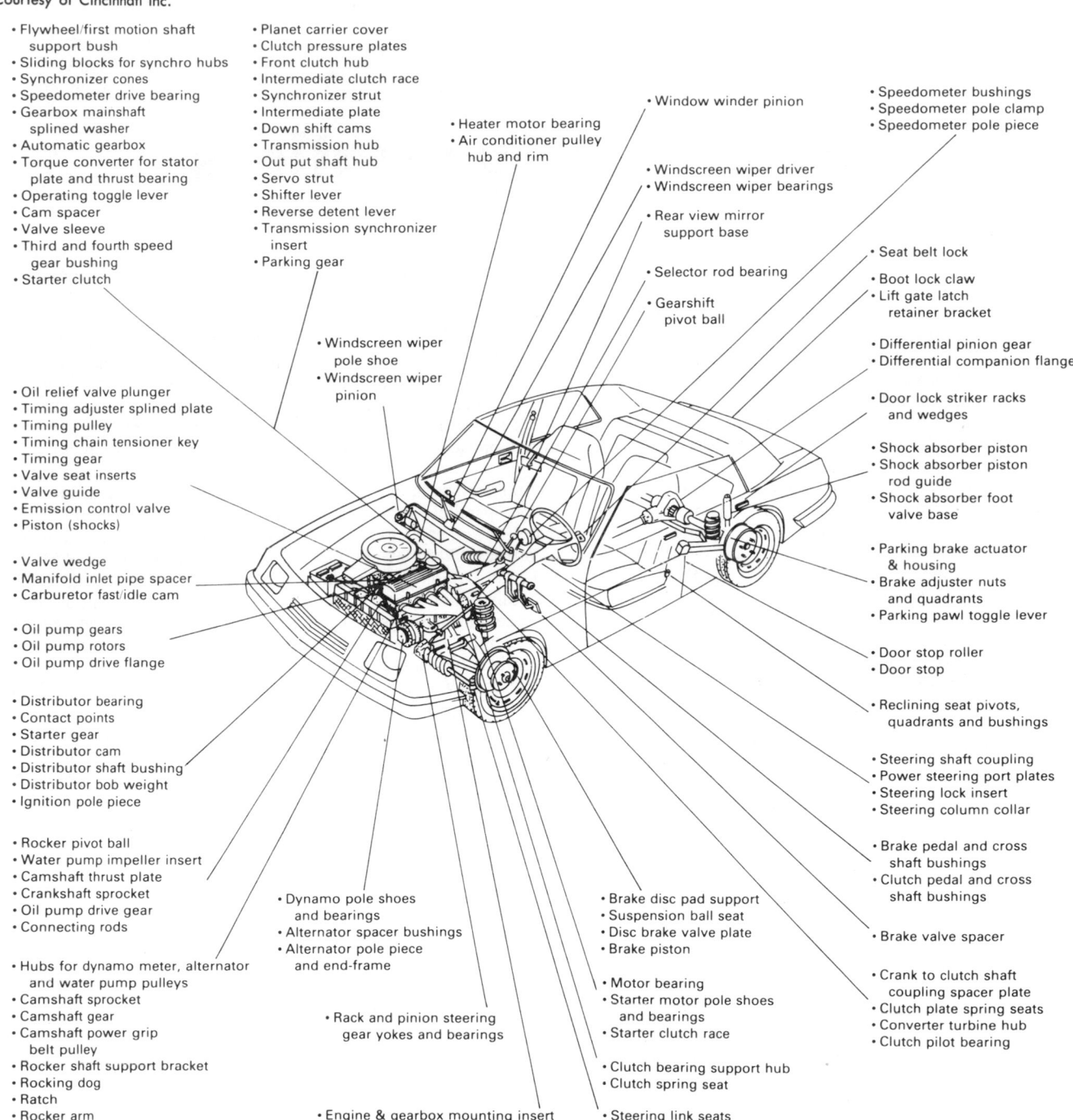

- Flywheel/first motion shaft support bush
- Sliding blocks for synchro hubs
- Synchronizer cones
- Speedometer drive bearing
- Gearbox mainshaft splined washer
- Automatic gearbox
- Torque converter for stator plate and thrust bearing
- Operating toggle lever
- Cam spacer
- Valve sleeve
- Third and fourth speed gear bushing
- Starter clutch

- Planet carrier cover
- Clutch pressure plates
- Front clutch hub
- Intermediate clutch race
- Synchronizer strut
- Intermediate plate
- Down shift cams
- Transmission hub
- Out put shaft hub
- Servo strut
- Shifter lever
- Reverse detent lever
- Transmission synchronizer insert
- Parking gear

- Heater motor bearing
- Air conditioner pulley hub and rim

- Window winder pinion
- Windscreen wiper driver
- Windscreen wiper bearings
- Rear view mirror support base
- Selector rod bearing
- Gearshift pivot ball

- Speedometer bushings
- Speedometer pole clamp
- Speedometer pole piece
- Seat belt lock
- Boot lock claw
- Lift gate latch retainer bracket
- Differential pinion gear
- Differential companion flange
- Door lock striker racks and wedges
- Shock absorber piston
- Shock absorber piston rod guide
- Shock absorber foot valve base
- Parking brake actuator & housing
- Brake adjuster nuts and quadrants
- Parking pawl toggle lever
- Door stop roller
- Door stop
- Reclining seat pivots, quadrants and bushings
- Steering shaft coupling
- Power steering port plates
- Steering lock insert
- Steering column collar
- Brake pedal and cross shaft bushings
- Clutch pedal and cross shaft bushings
- Brake valve spacer
- Crank to clutch shaft coupling spacer plate
- Clutch plate spring seats
- Converter turbine hub
- Clutch pilot bearing

- Windscreen wiper pole shoe
- Windscreen wiper pinion

- Oil relief valve plunger
- Timing adjuster splined plate
- Timing pulley
- Timing chain tensioner key
- Timing gear
- Valve seat inserts
- Valve guide
- Emission control valve
- Piston (shocks)

- Valve wedge
- Manifold inlet pipe spacer
- Carburetor fast/idle cam

- Oil pump gears
- Oil pump rotors
- Oil pump drive flange

- Distributor bearing
- Contact points
- Starter gear
- Distributor cam
- Distributor shaft bushing
- Distributor bob weight
- Ignition pole piece

- Rocker pivot ball
- Water pump impeller insert
- Camshaft thrust plate
- Crankshaft sprocket
- Oil pump drive gear
- Connecting rods

- Hubs for dynamo meter, alternator and water pump pulleys
- Camshaft sprocket
- Camshaft gear
- Camshaft power grip belt pulley
- Rocker shaft support bracket
- Rocking dog
- Ratch
- Rocker arm

- Dynamo pole shoes and bearings
- Alternator spacer bushings
- Alternator pole piece and end-frame

- Rack and pinion steering gear yokes and bearings

- Engine & gearbox mounting insert

- Brake disc pad support
- Suspension ball seat
- Disc brake valve plate
- Brake piston

- Motor bearing
- Starter motor pole shoes and bearings
- Starter clutch race

- Clutch bearing support hub
- Clutch spring seat

- Steering link seats

try, but limited application has been implemented at this time. As aluminum P/M technology matures, the significant advantages of light weight and ductility should lead to a variety of applications (Ref 4).

Hot Formed Parts

The advent of water-atomized high-purity powders in the early 1970's made

it possible to produce differential side pinions and automatic transmission internal ring gears with durability equal to or better than that of carburized steel (Ref 5). These applications, however, did not prove economical, and the parts are no longer produced by powder metallurgy.

A full-density automatic transmission one-way roller clutch race (Fig.10) is one example of a hot formed P/M part that is

still in production. Such parts are manufactured by compacting and sintering of plain ring preforms, followed by lubricating the preforms, reheating in an induction furnace, and hot forming in a closed die. Only residual flash is generated. The parts are subsequently carburized and quenched. Grinding of the thrust faces is the only machining operation required. These parts, which are subjected to high

Fig. 2 Automatic transmission clutch reaction plate

Composed of iron-carbon-copper alloy. Weight: 500 g (1.1 lb). Secondary operation: face grinding

Fig. 5 Engine rocker arm pivot ball

Composed of iron-carbon alloy. Weight: 10 g (0.02 lb)

Fig. 8 Manual transmission synchronizer gear and keys

Composed of iron-carbon-copper alloy. Weight: 700 g (1.54 lb). Weight of key: 10 g (0.2 lb). Secondary operation: face grinding of gear

Fig. 3 Engine oil pump driven gear

Composed of iron-carbon alloy. Weight: 200 g (0.44 lb). No secondary operations were required.

Fig. 6 Power steering pressure plate

Composed of iron-carbon alloy. Weight: 300 g (0.65 lb). Secondary operation: Steam treated

Fig. 9 Clutch lever bearing

Composed of 90Cu-10Sn bronze. Weight: 20 g (0.05 lb). Secondary operation: oil impregnation

Fig. 4 Engine camshaft sprocket gear

Composed of iron-carbon-nickel alloy. Weight: 800 g (1.75 lb). Secondary operations: carbonitriding and vibratory finishing

Fig. 7 Control arm ball joint ball

Composed of iron-carbon-copper alloy. Weight: 60 g (0.13 lb). Secondary operations: carburizing (carbonitriding) and vibratory finishing

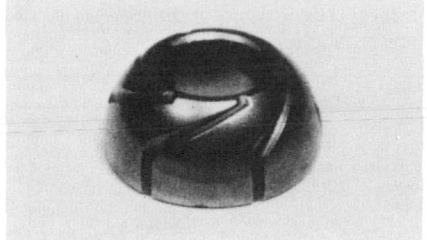

Fig. 10 Hot formed automatic transmission roller

Composed of iron-carbon-nickel-molybdenum alloy. Weight: 500 g (1.1 lb). Secondary operation: parallel grinding

Fig. 11 Typical Weibull plot of a P/M roller bearing cup

Shown is a 10% life (L_{10}) of 563 h compared to ingot steel material L_{10} life of 192 h. Source: Ref 6

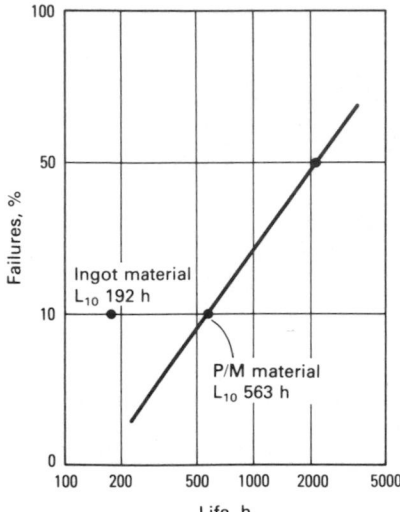

Fig. 12 Connecting rod hot formed from a P/M preform

Weight: 643 g (1.42 lb)

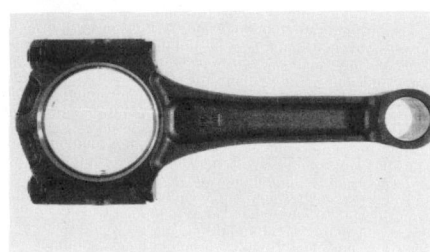

Fig. 13 Production process of sintered composite camshaft

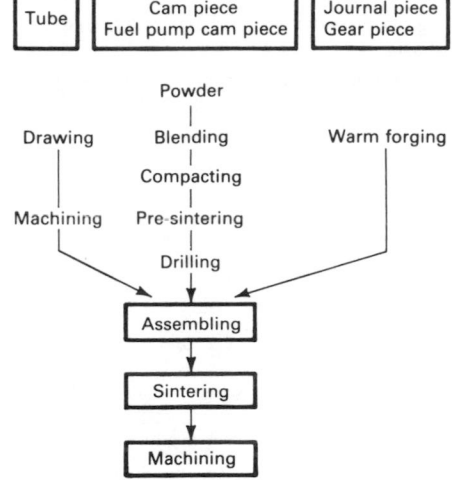

Fig. 14 Sintered composite camshaft

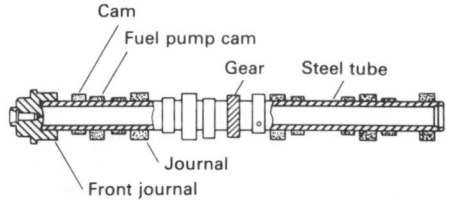

Fig. 15 Starter motor frame and pole shoes

Composed of iron powder. Weight: 900 g (2 lb). Secondary operations: sized, radial holes punched

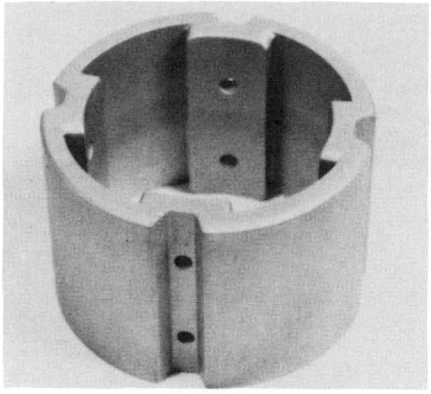

Hertz stresses, are highly durable, with excellent rolling contact fatigue resistance.

Intensive development of powder manufacturing techniques and new alloy compositions resulted in further improvements in the purity of the powder and in freedom from large nonmetallic inclusions. As a result, additional one-way clutch races have been developed, some using the economical sintercarburizing process, which combines sintering with carburizing. In addition, roller bearing outer races have been developed. Fatigue life characteristics of water-atomized steel roller bearing cups similar to AISI 4625 are compared to ingot steel bearings in Fig. 11. Using a modified logarithmic scale, the Weibull plot shows the percentage of bearings that are expected to fail at any particular time.

In several foreign automobiles, engine connecting rods are produced by hot forming P/M preforms (Fig. 12). Improved fatigue life and machining savings are reported (Ref 7, 8).

Hot forming technology is well developed, producing highly durable parts. Currently, the cost of P/M hot formed parts is not competitive with cold or warm formed steel parts made to near-net shape. However, hot forming is frequently competitive with conventional forgings, depending on the amount of flash generated during forging, the amount of subsequent machining necessary, and the production volume of the parts. In some cases, the introduction of P/M hot formed parts cannot be economically justified due to the high cost of in-plant capital equipment.

Nonconventional Applications

One application of powder metallurgy is the use of a composite camshaft assembly consisting of a medium-carbon tubular shaft, P/M pressed and sintered cams and fuel pump eccentric cams, and warm forged steel journal bearings (Ref 9). The P/M cams are made from a high-chromium iron-based alloy that contains an evenly distributed, hard, wear-resisting phase. Chemical composition of the powder used in the production of the cams is:

Element	wt%
Chromium	5
Molybdenum	1
Phosphorus	0.5
Copper	2
Carbon	2.5
Iron	rem
Others	<2

This alloy was selected because of its excellent wear resistance (scuffing resis-

tance) compared with that of conventional chilled cast iron.

The alloy used for the cams has a sintered density of 7.6 g/cm^3 and hardness of 550 HV. The structure after sintering at 1110 °C (2030 °F) in dissociated ammonia consists of fine chromium-molybdenum carbides evenly distributed in a bainite-martensite matrix.

For final assembly, cams and the fuel pump eccentric cam are positioned, located, and temporarily secured to the tube before sintering. Warm forged 4140 steel journals are brazed to the tube while the P/M parts are being sintered. A weight reduction of 26% resulted from the new design. The processing sequence of the sintered composite camshaft is shown in Fig. 13. The camshaft and its component parts are shown in Fig. 14.

Reference 10 describes the use of iron powder in a starter motor case and integral pole piece (Fig. 15). The required electrical properties were achieved by creating a high-density material in a single compacting operation by means of a proprietary die design. Powders specifically designed for high density and maximum electrical permeability promise continued growth in electrical and soft magnetic applications.

Other unusual applications of powder metallurgy in automobiles include cold formed spark plug shells produced to near-theoretical density in a proprietary system and antiwear applications of tool steels made from either isostatically compacted tool steel powder or a tool steel produced by highly controlled sintering (Ref 11). The latter application involves components for an automotive diesel fuel injection pump. A highly stressed ring for the pump is made by compacting and sintering to a minimum density of 98% of theoretical using M2 steel powder. Minor grinding operations are performed before hardening and installation of the part. The potential for growth in diesel engine production and the success of these parts highlight the versatility of powder metallurgy.

REFERENCES

1. Johnson, P.K. and Roll, K.H., State of the P/M Industry and a Marketing Plan for the '80s, *Int. J. Powder Metall. Powder Technol.*, Vol 19 (No. 4), 1983

2. Koehring, R., The Role of Powder Metallurgy in the Automotive Industry—Present and Future, *Prog. Powder Metall.*, Vol 19, 1962, p 7

3. Winquist, L.A., Automotive Applications of Standard P/M Parts, *Prog. Powder Metall.*, Vol 29, 1974, p 87

4. Daver, E., Production and Application of Aluminum P/M Products, *Metal Powder Reports,* Vol 31 (No. 4), 1976, p 115

5. Eloff, P.C. and Wilcox, L.E., Fatigue Behavior of Hot Formed Differential Pinion Gears, *Mod. Devel. Powder Metall.*, Vol 7, 1973, p 213-234

6. Adams, J.S. and Glover, D., Improved Bearings at Lower Cost via Powder Metallurgy, *Metal Prog.*, Vol 112 (No. 3), Aug 1977, p 42

7. Brown, G.T., Powder Metallurgy, Promises and Problems, Fourth European Symposium on Powder Metallurgy, Proceedings of Soc. Francais de Metallurgie, Paris, 1975, p 96

8. Tsumuki, C., Connecting Rods by P/M Hot Forging, *Mod. Devel. Powder Metall.*, Vol 7, 1973, p 385-391

9. Thumaki, C., *et al.*, Development of Sintered Integral Camshaft, SAE paper 830254, Society of Automotive Engineers, Dearborn, MI, 1983

10. Ashton, C.R., Delco Remy 5MT Cranking Motor, SAE paper 790195, Society of Automotive Engineers, Dearborn, MI, 1979

11. Rassenfoss, J.A., Production of Full Density M2 Automotive Diesel Parts Via the HTM TM Powder Metal Process, SAE paper 800309, Society of Automotive Engineers, Dearborn, MI, 1980

P/M Parts for Appliances

By George Otto
Supervisor, Process Engineering
The Maytag Company

POWDER METALLURGY PARTS enjoy wide usage in domestic appliances, particularly washers and dryers, to reduce costs and increase service life of various components. These parts, which have solved strength, wear, and corrosion problems, fall into three general categories: stainless steel structural parts, iron-copper-graphite structural parts, and copper-tin bronzes used for bearings. This article describes the processing steps, applications, and the reasons for conversion to P/M technology for such parts.

Stainless Steel Parts

Functional stainless steel parts frequently are used as a seal surface in wear and corrosive conditions. The seal surface must be a high-density material to minimize leakage and wear against the mating seal surface. Type 304L or 316L stainless steel usually is used, but type 430 or 434 also is satisfactory. Sintering in vacuum or hydrogen is preferred practice to prevent nitrogen hardening, particularly if sizing operations are required, and to retain maximum corrosion resistance.

Figure 1 is the hub for a water pump impeller made with −100 mesh type 304L stainless steel powder to which a lubricant (1% Acrawax C) is added. This part was compacted in a shelf-type die, which requires a relatively high apparent density powder to obtain a high initial density in the flange formed by the die shelf. Acrawax C is used as a pressing lubricant rather than a metallic stearate, because it leaves minimal residue in the part to contaminate the vacuum furnace. The unsintered (green) compact is heated to 540 °C (1000 °F) in an inert or slightly reducing furnace atmosphere, which is independent of the vacuum sintering furnace. This operation is carried out so that the volatile pressing lubricant is burned off prior to the sintering operation.

The parts are vacuum sintered in a batch furnace for 50 min at 1175 °C (2150 °F) and at a vacuum of 50 μm Hg using a nitrogen backfill. Total sintering cycle is about $2^1/_2$ h. Rapid quenching is necessary to prevent formation of chromium nitrides in the microstructure. To enhance quenching, the parts are stacked on a coarse wire mesh screen to promote rapid cooling by the circulating nitrogen quench gas.

The hub is sized to improve the length and inside diameter dimensions, density, and wear resistance of the flange. A water-soluble wax is used as a sizing lubricant. The flange surface is machined to improve surface finish, which provides a seal surface during service. A rubber impeller is molded to the outside of the hub and a hardened stainless steel shaft is pressed into the bore with a 0.050-mm (0.002-in.) interference fit.

This part was originally produced as a brass insert machined from bar stock. It was redesigned using P/M techniques to produce a part with improved wear and corrosion resistance. Flange design allows for improved material utilization.

Fig. 1 Stainless steel (304L) insert for pump impeller
(a) Before insertion. (b) After insertion

(a)   (b)

Fig. 2 Eccentric gear for a washing machine made of iron alloy powder containing 2% copper and 0.5% graphite

Center hole inside diameter: approximately 19 mm (0.76 in.); eccentric outside diameter: approximately 57 mm (2.28 in.); total thickness: approximately 28.50 mm (1.14 in.); eccentric thickness: approximately 15.90 mm (0.63 in.); pitch diameter: approximately 115 mm (4.59 in.)

Iron Functional Parts

Most iron powder metal parts used in appliances are made from iron-graphite or iron-copper-graphite compositions. Choice of material and density depends in part on wear, physical, and impact properties. Other factors, such as machining, pressing operations, and cost, also must be considered. Metal Powder Industries Federation standard MPIF 35 is useful in materials selection.

Parts are compacted, sintered, and sometimes sized. If a component is to be subjected to high stress, the powder materials may be compacted, presintered, sized, and fully sintered to produce a high-density, high-strength part. Low-alloy powder parts that are heat treated achieve high strength.

Some appliance parts are subjected to high-level alternating stresses. Porosity, which decreases as density increases in a metal component, acts as sites of fatigue initiation (or notches); therefore, parts made to a high-density specification are less likely to fail due to fatigue than are lower density parts. Infiltration of the part with copper fills the pores, effectively eliminating porosity and thereby minimizing fatigue failure problems.

Figure 2 shows an eccentric gear for a washing machine that is made of a blend of 2% copper, 0.5% graphite, and the remainder iron, which is compacted, sintered, sized, and machined. An intermediate level of graphite was chosen to produce an as-sintered part that could be sized readily to ensure an accurate gear-tooth profile. Lower carbon content also improves machinability. Density of this part is 6.5 g/cm^3.

This gear is made from atomized powder to obtain optimum compressibility. Because this gear is a two-level part, green strength is critical to eliminate cracking at the hub juncture. Press motion and tool design also are critical in preventing hub cracking.

Sintering is done in an endothermic atmosphere for 30 min at 1105 °C (2025 °F). Dew point is monitored to prevent carburization of the surface layers so as not to affect sizing or machining. After sintering, the part is deburred in a steel medium to facilitate sizing. Prior to sizing, the part is dipped in soluble sizing wax, which also acts as a long-term storage and rust protection medium.

Sizing enhances the tooth form and pitch diameter tolerances. After sizing, the center hole is bored for concentricity, and the outside of the hub is machined to correct any ovality. This part replaced a class 30 gray iron casting that was machined from a rough blank.

Bearings

Powder metal bearing material for appliances may be any one of several compositions, typically 90Cu-10Sn bronze. This composition may be diluted up to 60% with iron, or the material may be a copper-iron blend or straight iron. Selection depends on a combination of economic, loading, wear, and corrosion considerations. Porosity ranges from 18 to 22%; these parts should be impregnated with a high grade of circulating oil containing oxidation stabilizers to ensure long service life.

The washing machine tub spin bearing shown in Fig. 3 requires a high-purity copper powder with high green strength due to the part's thin wall (1.91 mm or 0.075 in.), good flow, and low dimensional growth characteristics during sintering. The powder is blended into a mix-

Fig. 3 Washing machine tub spin bearing made of 89% copper, 10% tin, and 1% graphite

Length: 51 mm ± 0.20 mm (2.03 in. ± 0.0076 in.); outside diameter: approximately 48 mm (1.9 in.); inside diameter: approximately 44 mm (1.75 in.)

ture containing 89% copper, 10% tin, and 1% coarse graphite, plus zinc stearate as a pressing lubricant. Typical tin addition is comprised of half coarse and half fine tin, but this may be adjusted to control the dimensional growth during sintering as needed.

The part is sintered at a relatively high temperature of 845 °C (1550 °F) to obtain a suitable microstructure for machining and to obtain good bearing properties. An endothermic or low-dew-point purified exothermic gas is used during the sintering operation. Due to the high sintering temperature, distortion is severe, and the ends must be "rounded out" in a presizing operation to fully size the part to final dimensions. The impregnating lubricant is used as a sizing lubricant for these operations.

The part is then pressed into a rubber-mounted retainer sleeve and is bored to final dimensions. Care must be taken to retain open surface pore structure; for instance, a tool that has become dull will smear the porosity shut. This design replaced a ball bearing assembly.

Electrical and Magnetic Applications

Resistance Welding Electrodes

By Anil V. Nadkarni
Manager, Non-Ferrous Full Dense
Technology
SCM Metal Products
and
James R. Woodruff
Senior Metallurgical Engineer
CMW Inc.

IN RESISTANCE SPOT WELDING, metal workpieces are joined by the heat generated by resistance to the flow of electric current. Workpieces are held together under force by a pair of electrodes. A low-voltage, high-amperage current pulse passes through the electrodes to the workpieces. The contacting surfaces between the workpieces, in the region of current concentration, are heated by this current to form a fused weld nugget.

When the flow of current ceases, the electrode force is maintained, while the weld metal cools rapidly and solidifies. Electrodes are retracted after each weld, which is usually completed in a fraction of a second. The electrodes or workpieces are then positioned for the next weld. Figure 1 shows a typical resistance spot welding setup. The electrodes used in resistance spot welding:

- Conduct the welding current to the workpieces
- Transmit the required force to the workpieces to produce a satisfactory weld
- Dissipate heat rapidly from the weld zone

During welding, the electrode face encounters large compressive stresses at high temperatures, which cause deformation and enlargement (mushrooming) of the electrode tip. This phenomenon decreases current density in the weld zone and leads to inferior welds. To maintain consistent weld quality, the electrode tip generally is

Fig. 1 Resistance spot welding setup

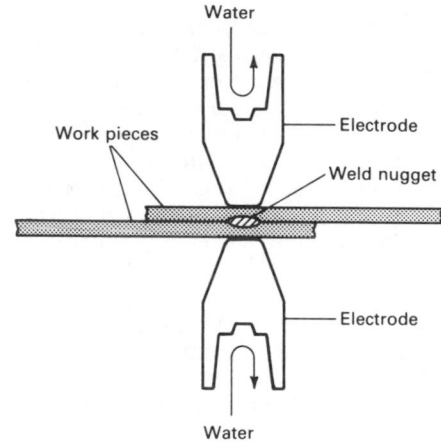

Water

Work pieces

Electrode

Weld nugget

Electrode

Water

dressed back to its original face diameter. Dressing and changing of electrodes cause interruptions in production, which can be minimized by selecting electrodes that do not mushroom as rapidly.

Alloying between the electrode tip and the workpiece metal can greatly increase the rate of mushrooming. This occurs rapidly when copper alloy electrodes are used for welding work metal coated with metals that alloy readily with copper, such as zinc, tin, or aluminum. Sticking of electrodes to the workpiece is a serious problem when welding such materials and causes interruptions. Such interruptions are very costly in high-speed assembly-line manufacturing operations, such as automotive and appliance applications, and can be minimized by selecting electrodes that do not stick to the workpiece.

Materials for spot welding electrodes should have high thermal and electrical conductivities, low contact resistance, and high strength to resist deformation at operating temperatures and stresses. The Resistance Welder Manufacturers Association (RWMA) has classified electrode materials into two composition groups:

copper-based alloys (group A) and refractory metal compositions (group B). These classifications cover a wide range of materials that are suitable for most resistance welding applications.

The copper-based alloys are divided into classes 1, 2, and 3 and account for most conventional high-volume resistance welding applications. Among these, class 2 electrodes are the most versatile and the most commonly used. Copper chromium (C18200), a precipitation-hardened alloy, has been an industry standard in automotive and appliance applications for many years. A special alloy, dispersion-strengthened copper, has gained wide acceptance in these industries at the expense of copper chromium because of superior performance on galvanized and bare steels. This alloy is manufactured by powder metallurgy and is the subject of the next section of this article.

The refractory metal compositions are divided into classes 10 through 14. Classes 10, 11, and 12 are copper-tungsten composite materials and cover a range of hardnesses, compressive strengths, and electrical conductivities. Classes 13 and 14 consist of unalloyed tungsten and molybdenum, respectively. The refractory metal compositions generally are used in specialty applications, in which the high heat, long weld time, inadequate cooling, and high pressure involved may cause rapid deterioration of copper-based alloys. They are also used in conjunction with copper-based alloy electrodes when welding dissimilar metals of different electrical conductivities. These compositions are also manufactured by powder metallurgy and are discussed in the section on refractory metal compositions in this article.

Dispersion-Strengthened Copper Electrodes

Dispersion-strengthened copper is a major resistance welding electrode material. It is widely used in RWMA class 2 welding applications and in some class 3 applications. It offers a unique combination of high strength and high electrical and thermal conductivities, and more importantly, it retains a large portion of these properties when exposed to temperatures approaching the melting point of the copper matrix.

The properties of dispersion-strengthened copper arise from a fine and uniform dispersion of aluminum oxide particles in a pure copper matrix. These particles range in size from 3 to 12 nm (30 to 120 Å) and usually are spaced 50 nm (500 Å) apart. To obtain optimum properties, the dispersoid particles must be small and dispersed uniformly in close proximity to one another.

The aluminum oxide particles in dispersion-strengthened copper are thermally stable at high temperatures, and unlike the chromium precipitate particles in copper chromium, they retain their original particle size and interparticle spacing at temperatures approaching the melting point of copper. This thermal stability of aluminum oxide particles is responsible for the superior high-temperature properties of dispersion-strengthened copper. The quality of the dispersion depends greatly on the method of manufacture.

Manufacture

While there are many techniques available to make dispersion-strengthened copper, internal oxidation produces the best product in terms of dispersoid size and distribution. The manufacturing process involves melting a dilute solid-solution alloy of aluminum in copper and atomizing the melt into powder under a high-pressure nitrogen gas. The powder is blended with a proprietary oxidant, comprised mainly of cuprous oxide, sufficient to oxidize all of the aluminum in the alloy. The blend is heated to 870 °C (1600 °F) for about 1 h. The cuprous oxide in the oxidant dissociates, and the oxygen diffuses into the solid-solution alloy powder to preferentially oxidize the aluminum without oxidizing the copper.

Internal oxidation is a diffusion-controlled reaction and follows a parabolic rate law (the reaction time is proportional to the square of the distance through which the oxygen must diffuse). To maintain reaction times within practical limits and to obtain a fine and uniform dispersion, the size of the part being internally oxidized must be as small as possible. Use of powder offers a convenient way to accomplish both of these objectives. Any excess oxygen remaining in the powder after com-

Table 1   Room-temperature properties of dispersion-strengthened C15760 compared to RWMA class 2 standard

Material	Hardness, HRB	Ultimate tensile strength MPa	ksi	Elongation, %	Electrical conductivity, %IACS
C15760	83	572	83	16	78
RWMA class 2 standard (min)	72	448	65	13	78

Fig. 2 Effect of annealing temperature on hardness of dispersion-strengthened copper and copper chromium

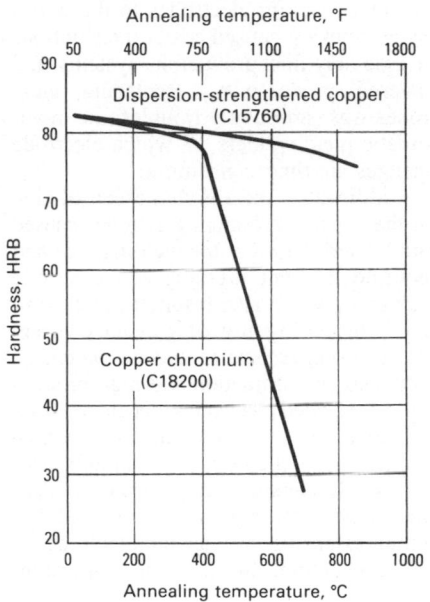

plete oxidation of all of the aluminum is reduced by heating the powder in a dissociated ammonia atmosphere.

The powder is canned in a copper container and hot extruded into a fully dense rod. This rod is then cold drawn to the desired size. Hot extrusion of powder produces a highly oriented fibrous structure in the rod, with the fibers running parallel to the rod axis. Superior electrode performance is obtained when this fibrous grain structure is converted into an equiaxed grain structure. This is accomplished by a patented process, which involves cold upsetting of an appropriate size slug, cut from the drawn rod, in a closed die to expand the cross-sectional area by 50% or more. The upset slug is then converted into a finished electrode by cold forming of the nose and the waterhole.

Properties

Although dispersion-strengthened copper is available in two grades, C15760 is the most widely used grade in resistance welding electrodes. The chemical composition of this grade is 1.1 wt% (2.7 vol%)

aluminum oxide and the remainder copper. Typical room-temperature properties of drawn C15760 rod are given in Table 1, along with the minimum values required to qualify as a RWMA class 2 electrode material.

C15760 grade dispersion-strengthened copper meets all the requirements of a RWMA class 2 electrode material. During welding, the electrode face encounters high temperatures and high compressive stresses. A good measure of the ability of an electrode to resist deformation or mushrooming under these conditions is hardness retention after exposure to elevated temperatures. Figure 2 shows the hardness response of C15760 alloy to annealing temperatures up to 870 °C (1600 °F). RWMA class 2 copper-chromium alloy (C18200) also is shown for comparison.

C15760 alloy exhibits superior hardness retention capability compared to the copper-chromium alloy. The copper-chromium alloy shows a drastic drop in hardness when exposed to temperatures above 450 °C (840 °F), which is close to the initial aging temperature for this alloy. When heated to these temperatures, the second-phase particles of chromium begin to grow and lose their effectiveness. C15760 alloy retains its hardness even after exposure to temperatures of 870 °C (1600 °F) and higher. This is because the aluminum oxide particles do not grow and maintain their effectiveness.

Figure 3 shows the results of a laboratory welding test performed on dispersion-strengthened copper C15760 and copper-chromium C18200 electrodes. Welding test conditions were:

Sheet materials . . . . SAE 1010
   Top . . . . Bare
   Bottom . . . . Galvanized
Sheet thickness, mm (in.)
   Top 1.9 (0.075)
   Bottom 2.0 (0.080)
Welding current, A 13 000
Welding force, N (lb) 5785 (1300)
Welding time, cycles 16
Squeeze time, cycles 84
Hold time, cycles 100
Weld rate, No./min 12
Cooling water, L/min per electrode
   (gal/min per electrode) 3.8 (1)

The electrode length and diameter were measured at the beginning of the test and

Fig. 3 Comparative electrode wear of dispersion-strengthened copper and copper chromium

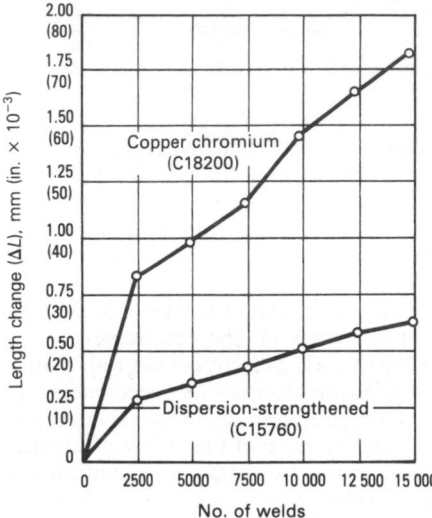

after each 2500 welds. Changes in length and diameter from the initial dimensions were determined and were very similar for each material. The change in length (ΔL), which is a measure of wear, for the upper electrode is plotted in Fig. 3 as a function of the number of welds. The lower electrodes followed a similar wear pattern.

C15760 electrodes had a much lower wear rate than the copper-chromium elec-trodes. In fact, they had less wear after 15 000 welds than the copper-chromium electrodes had after 2500 welds. Similar behavior was later confirmed in several production welding situations.

Applications

Dispersion-strengthened copper C15760 electrodes have received widespread ac-ceptance in automotive and appliance ap-plications. Welding of galvanized steel has been a persistent problem in the automo-tive industry because of sticking of the copper-chromium electrodes to the work-piece. Sticking caused costly interruptions in assembly-line production systems and lowered productivity. Replacing elec-trodes was particularly troublesome in au-tomatic press welders, in which electrode changes are time consuming.

Additionally, the alloying effect of zinc on the copper-chromium electrodes caused rapid deterioration of the welding face and required frequent dressing and electrode changing, which also resulted in interrup-tions. Increasing use of galvanized steel for corrosion resistance and the increased emphasis on automation and high produc-tivity demanded electrodes that did not stick to galvanized steel and did not deterio-rate as fast as the copper-chromium elec-trodes. Dispersion-strengthened copper fulfills both these requirements and has found extensive use as an electrode in this application. Because the dispersion-strengthened copper electrodes do not mushroom as rapidly as the copper-chromium electrodes, they do not re-quire current increases as frequently. This results in energy savings of up to 15%.

Table 2 lists some typical dispersion-strengthened copper C15760 electrode ap-plications. These uses cover cold rolled (bare) steel, galvanized steel, zincrometal, and mixed steel components. Typical life multiples obtained over copper-chromium electrodes are also shown.

As discussed previously, the main ben-efit realized with the use of dispersion-strengthened copper electrodes is a reduc-tion in the cost of welding through re-duced downtime associated with sticking, electrode dressing, and electrode chang-ing. This also leads to increased produc-tivity. Due to slower mushrooming rates, these electrodes reduce the frequency of current increases, which in turn leads to substantial energy savings.

Refractory Metal Compositions

Refractory metals such as tungsten, mo-lybdenum, and tungsten carbide combine excellent strength at high temperatures with adequate electrical conductivity. These characteristics make them desirable for applying high compressive forces, as well as directing electric current to workpieces

Table 2 Typical applications of dispersion-strengthened copper electrodes

Application	Thickness welded, mm (in.)	Material	Force N	Force lb	Current, A	Weld time, cycles	Welding rate, welds/h	Life comparison with copper-chromium electrodes
Rocker panel	0.79-0.79 (0.031-0.031)	Galvanized steel	2670	600	15 000	15	3024	10 to 1
	1.37-1.73 (0.054-0.068)	Galvanized steel	4805	1080	15 500	15	2600	5 to 1
Pillar assembly	1.09-1.37 (0.043-0.054)	Galvanized steel	3115	700	14 000	13	228	6 to 1
Light truck box	1.09-1.37 (0.043-0.054)	Galvanized steel-cold rolled steel	3560	800	16 000	18	160	65 to 1
Cowl side to rocker	1.35-1.27 (0.53-0.050)	Galvanized steel	4360	980	15 500	13	1750	5 to 1
Rocker panel to body	1.73-1.37 (0.068-0.054)	Galvanized steel-cold rolled steel	4360	980	15 000	13	1720	6 to 1
Floor panel extension	0.79-0.79 (0.031-0.031)	Galvanized steel-cold rolled steel	2450	550	13 500	12	645	4.5 to 1
Windshield header	0.91-1.73 (0.036-0.68)	Cold rolled steel	3785	850	14 000	13	645	6 to 1
Dash extensions	0.79-0.79 (0.031-0.031)	Cold rolled steel	2450	550	13 500	12	1144	5.5 to 1
Station wagon assembly	1.37-0.79 (0.054-0.031)	Cold rolled steel	2940	660	12 000	9	3120	4 to 1
Van assembly	1.09-1.42-0.91 (0.043-0.056-0.036)	Cold rolled steel	3450	775	13 500	18	180	6 to 1
Wheelhouse assembly	0.79-0.84 (0.031-0.033)	Zincrometal	3425	770	12 500	11	3300	4 to 1
	0.79-0.79 (0.031-0.031)	Zincrometal-cold rolled steel	2805	630	12 265	10	546	8 to 1
Dash panel	0.79-1.73 (0.031-0.068)	Zincrometal-galvanized steel	2450	550	13 500	12	1806	6 to 1

Fig. 4 Direct heating refractory metal sintering furnace design

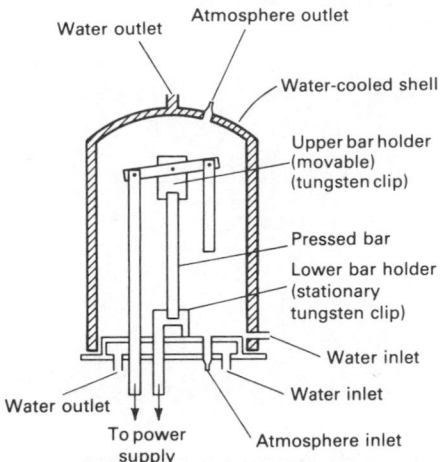

Fig. 5 Wrought tungsten bar
Magnification: 300×

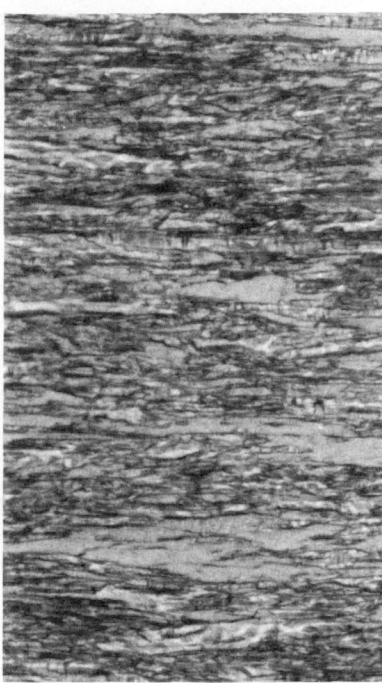

Longitudinal

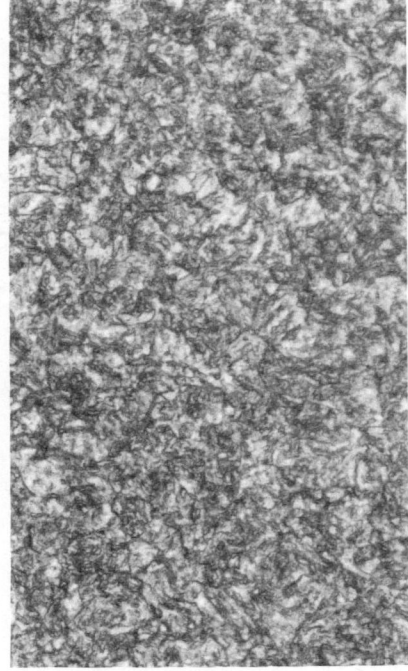

Transverse

being resistance or flash welded. By producing composites of such refractory metals with high electrical conductor metals, such as copper and silver, compromises are achieved between power transfer and high-temperature strength. For economic reasons, copper generally is used for resistance welding of composite materials rather than silver. In resistance welding applications, these characteristics allow the welding engineer to achieve a compromise among welding machine capacity, forging pressure, electrode life, and overall economics.

Manufacture

Resistance welding P/M refractory metal materials generally are produced in compositions ranging from pure refractory metals (consolidated, sintered, and hot/cold worked to 99.9%+ dense) to composites containing up to 65 vol% Cu. Such materials are produced by P/M processing by necessity and tradition rather than by adaptation. One of the earliest applications of modern P/M technology was the production of ductile refractory metal filament wire. This product was introduced by Coolidge at the beginning of the 20th century (Ref 1).

The only commercially viable refining techniques for high-melting refractory metals include hydrorefining and low-temperature chemical reactions, which produce metal powder. Powder metallurgy consolidating techniques and sintering processes were first developed in the production of these materials.

Wrought tungsten bar typically is made by compacting blends of various sized tungsten powders (0.6 to 9 μm) to bars at pressures of 170 to 310 MPa (25 to 45 ksi). Sintering is performed in two stages in a

high-purity reducing atmosphere. In the first sintering stage, bars are placed on molybdenum trays and sintered at 1000 to 1200 °C (1830 to 2190 °F) to improve bar strength for handling. Final sintering temperatures reach 1800 to 3100 °C (3270 to 5610 °F) by direct (Fig. 4) or indirect heating. Sintering time depends on the temperature and equipment selected.

Sintered bars are preheated for hot rolling in a reducing atmosphere to about 1650 °C (3000 °F) initially. Reheat temperatures are lowered with decreasing thickness and successive reductions. Rod is rolled on contour rolls with preheats ranging from 1700 to 1450 °C (3090 to 2640 °F). Rod less than 25 mm (1 in.) in diameter is finished by hot swaging. Swaging temperatures decrease with the section size used, from 1600 to 1200 °C (2910 to 2190 °F). The typical microstructure of heavily worked wrought tungsten rod is shown in Fig. 5.

Wrought molybdenum bar is produced in a similar manner as tungsten. Final sintering occurs around 2100 °C (3810 °F). Rolling and swaging typically require temperatures 200 to 300 °C (360 to 540 °F) less than tungsten processing.

For composite refractory metals, the sintering temperature for the refractory metal powder is reduced and controlled to retain a specified amount of intercon-

nected porosity. Copper or other infiltrant is allowed to fill these pores by melting in a high-purity reducing atmosphere. Continuous pusher furnaces heated by molybdenum elements frequently are used for these processes (Fig. 6).

Frequently, solid copper pieces are placed against presintered bars and allowed to infiltrate by capillary action. The microstructure of a tungsten-copper material produced in this manner is shown in Fig. 7. An alternative method entails *in situ* infiltration in which a blend of refractory metal and copper powders are consolidated prior to sintering. This practice results in chemically equivalent materials with lower densities and inferior properties.

Controlled sintering usually occurs at 1150 to 1400 °C (2100 to 2550 °F) in reducing atmospheres; time and temperature used are the primary variables in obtaining the desired refractory metal porosity levels for the copper to occupy. Precipitation-hardened copper alloys may be used for infiltrating some grades of electrode materials. This increases strength and load capabilities of the materials up to about 500 °C (930 °F), with some sacrifice in electrical conductivity (Fig. 8a and d).

For resistance welding applications, these materials are supplied as bar, rod, presized inserts, or silver-brazed attached

Fig. 6 Refractory metal composite sintering and infiltration furnace design for continuous production

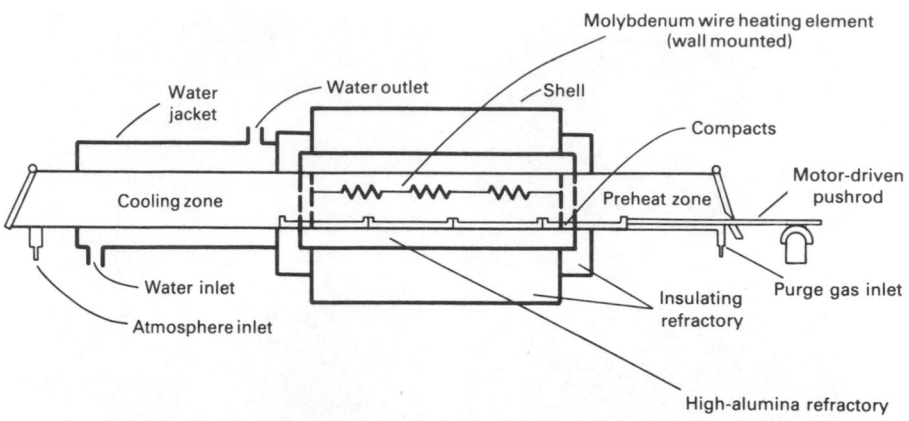

Fig. 7 Photomicrograph of a tungsten-copper composite electrode material

Magnification: 700×

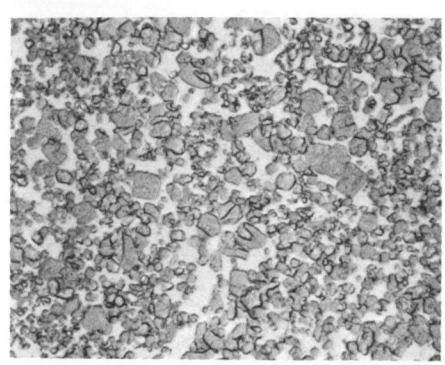

Fig. 8(a) Effect of infiltration with a precipitation-hardened copper alloy on hardness at elevated temperatures

Converted from Rockwell A measurements.

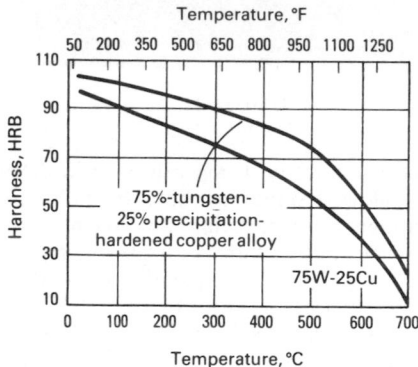

Fig. 8(c) Hardness at elevated temperatures for select tungsten carbide/copper refractory metal composite materials

Converted from Rockwell A measurements. Produced by sintering and separate infiltration

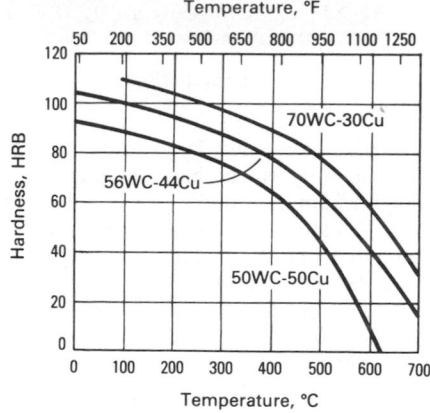

Fig. 8(d) Room-temperature properties as a function of composition

Modulus of rupture in bending is calculated using approximately 7 by 7 mm (0.28 by 0.28 in.) bars bent on 38-mm (1.5-in.) centers. ○, Tungsten-copper; ●, tungsten carbide/copper; △, tungsten precipitation-hardened copper alloy

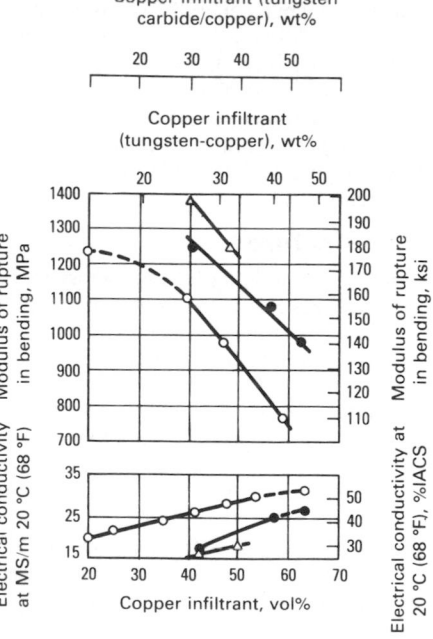

Fig. 8(b) Hardness at elevated temperatures for select RWMA group B refractory metal electrode materials

Converted from Rockwell A measurements. Class 11 and 12 materials produced by sintering and separate infiltration

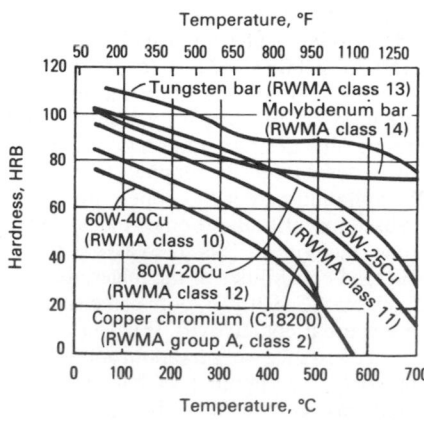

facings on standard copper alloy electrode shapes (Fig. 9). Inserts of tungsten and molybdenum are cut from wrought bar. Cutting and grinding require the use of relatively soft open abrasives (grades K through M) and relatively slow contact speeds (6 to 12 m/s, or 1180 to 2360 sfm).

Lubricant is used, and shallow cuts of 0.03 mm (0.001 in.) are made during grinding to prevent large thermal gradients and surface fissures. Composite materials of tungsten copper machine readily and can tolerate heavier feeds. Carbide tools are required for turning without excessive tool wear.

Composite material inserts frequently are pressed so as to be near-net size and shape upon sintering. Composite material bars are pressed oversize for finishing with minimal machining. Bar size is restricted

primarily by pressing capacity. Transverse dimensions up to 5 cm (2 in.) and lengths up to 20 cm (8 in.) are frequently produced.

For economic reasons, these materials are used only for the working faces of the resistance welder tooling in thicknesses from about 3.8 to 15 mm (0.15 to 0.6 in.). They are readily attached to copper-based tooling by silver brazing, which ensures maintenance of electrical continuity in use. In some applications, inserts may be attached mechanically, but caution must be

Fig. 9 Typical refractory metal-faced welding electrodes

Fig. 10 Application of P/M refractory metal materials in resistance welding

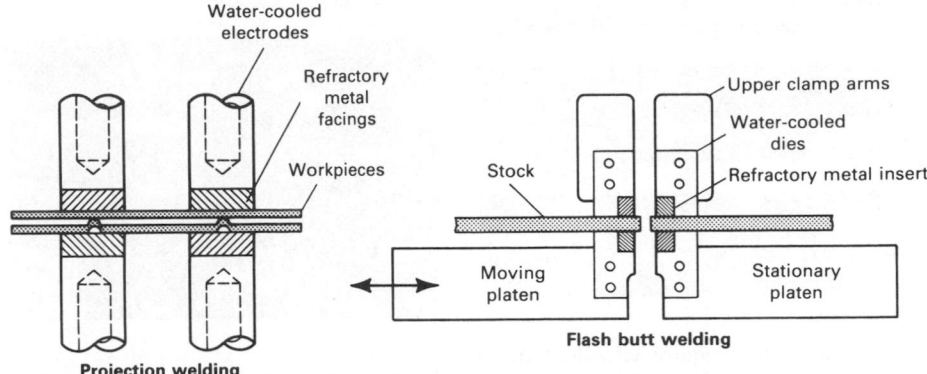

Projection welding

Flash butt welding

Table 3 Properties of RWMA group B refractory metal rod, bar, and inserts

Refractory metal	Hardness	Conductivity, %IACS	Ultimate compression strength MPa	ksi
Class 10	72 HRB	35	930	135
Class 11	94 HRB	28	1100	160
Class 12	98 HRB	27	1170	170
Class 13	69 HRA	30	1380	200
Class 14	85 HRB	30	...	...

Source: Ref 2

used to ensure an adequate electric current path at such a joint.

Properties and Applications

Properties for five RWMA standard classes of refractory metal group B electrode materials are listed in Table 3. Classes 10, 11, and 12 represent composites of tungsten and copper. Class 13 is tungsten rod or bar, and class 14 is molybdenum rod or bar.

These group B materials are used advantageously for projection welding, electrical upsetting, electroflash or butt welding, electroforging, cross-wire welding, and resistance welding and brazing of high-conductivity metals such as copper and brasses. Class 10 is also used for select stainless steel spot welding applications. For more information on these processes, see Volume 6 of the 9th edition of *Metals Handbook*. Tooling for most of these processes is highly specific for the workpiece and setup. Typical use of the materials for projection welding and flash butt welding is shown in Fig. 10.

As pressure requirements on the electrode faces are increased, the refractory metal content of the electrode tip is increased to prevent failure. For maximum wear resistance in electroforging and upsetting dies, tungsten carbide/copper composite materials are chosen. Hardnesses at elevated temperatures of various refractory metal composites are given in Fig. 8. Select room-temperature properties are also shown.

The wide range of electrical conductivities available with group B materials permits adjustment of the weld nugget location or heat-affected zone by careful mismatching of electrode types. Other conditions being equal, heating always concentrates toward the electrode of lower conductivity. This principle is used to correct for electrically dissimilar workpiece metals, or for workpieces of unmatched thicknesses. It is also applied in electroupsetting applications to achieve desired asymmetrical metal flow. For example, spot welding 0.46-mm (0.018-in.) C27000 yellow brass with a conductivity of 28% IACS to 0.46-mm (0.018-in.) C77000 nickel silver with a conductivity of 5.5% IACS is accomplished satisfactorily when a class 2 C18200 copper-chromium electrode is used against the nickel silver, and a class 11 tungsten-copper electrode is used against the yellow brass.

Class 13 (tungsten) and class 14 (molybdenum) electrodes are primarily used for welding and electrobrazing nonferrous metals with high electrical conductivities. These high-melting electrode materials resist welding to the workpieces. In effect, the refractory metal faces become heating elements and transfer heat to the workpieces in this application.

Except for very low power applications of less than 10 kV · A (jewelry welds or small electric leads), electrode designs should always provide water cooling behind the refractory metal faces. This prevents heat buildup in the electrodes and loss of the facing material due to remelting of the attachment brazes.

Resistance Welder Manufacturers Association Bulletins 16 and 34 list suppliers and trade names for all classes of resistance welding electrode materials. Most suppliers offer a wider selection of refractory metal materials than the 11 RWMA classes listed.

Tungsten Filaments

By Sam Leber
Manager, Material Characterization
General Electric Co.
and
R.J. Arena
Senior Product Engineer
General Electric Co.
and
Donald L. Bly
Manager, Wire Engineering
General Electric Co.

TUNGSTEN WIRE exhibits unique properties that have been critical to the origin and growth of the lamp industry and the electronic industry (the electron tube). These properties are developed and controlled by additions of small amounts of second phases such as potassium and thoria. Powder metallurgy is the only method by which such alloys can be made, as other consolidation methods eliminate the added constituents or render them ineffective. Similar advantages can be obtained by using P/M methods to retain volatile phases in molybdenum.

Properties

The properties of tungsten wire can be modified to suit a specific application. The largest commerical use of tungsten wire is in incandescent lamps and other filament

Fig. 11 Recrystallized structure of 180-μm (7-mil) diam tungsten wires

(a) Lamp grade, potassium doped. (b) Undoped

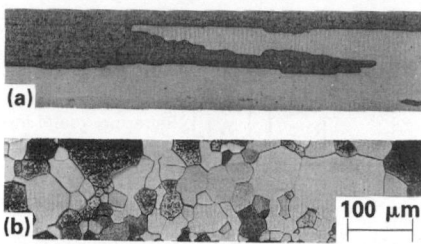

Fig. 12 Transmission electron micrographs of 800-μm (31-mil) diam doped tungsten wires

(a) As drawn. (b) Annealed

Fig. 13 Finger-type grain growth obtained during the recrystallization of 180-μm (7-mil) diam doped tungsten wire

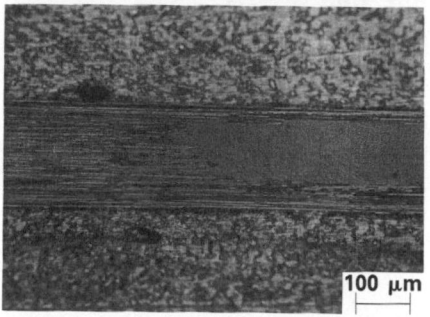

applications that require excellent resistance to creep at elevated temperatures. High-temperature creep resistance is controlled by the recrystallized structure. The large, interlocking, sag-resistant grain structure of doped lamp-grade wire is compared with a normal equiaxed structure of undoped wire in Fig. 11. Rows of very fine bubbles that contain potassium inhibit lateral movement of grain boundaries during annealing (Fig. 12). The resultant finger-type growth is shown in Fig. 13. The efficient long life of a lamp depends on the mechanical stability and thus retention of the filament configuration during operation at high temperatures (Fig. 14).

Another application for tungsten wire requires a fine-grained, equiaxed structure to dissipate shock and vibration energy during rough service usage. Wire containing a dispersion of thoria particles is commonly used for such applications. Thoriated filaments have a recrystallized structure similar to the undoped wire in Fig. 11(b). Thoria particles maintain the fine-grained structure during operation by inhibiting grain growth.

Thoria additions also lower the work function and improve the electron emission of tungsten. Thoriated tungsten wire and rod are therefore used in electronic applications and for electrodes in arc lamps and in gas tungsten arc welding.

For high-temperature shock-resistant and vibration-resistant applications, rhenium additions are made to non-sag type wire. Rhenium also improves the ductility of partially recrystallized wire and increases electrical resistance, thus allowing the use of heavier, more mechanically stable wire.

Process

Structures and properties of tungsten filaments depend on careful process control. The incorporation and retention of volatile metallic potassium in doped tungsten is a specialized operation. To be ef-

fective, dopants must be added prior to the formation of the metal powder. Oxide powders doped with solutions of aluminum, potassium, and silicon are shown in Fig. 15.

The effects of the dopants are also visible after reduction of the oxide to metal powder (Fig. 16). Dopants retain their particulate form, as evidenced by the pock-marked surface of the metal powder. The dopants are effective in the control of the structure and are enclosed within the tungsten grains (Fig. 17).

The doped powder is pressed into ingot form, partially sintered in hydrogen to establish sufficient strength for handling, and then final sintered by direct electrical resistance heating at temperatures approaching the melting point of tungsten. Sufficient density and strength are obtained to permit subsequent fabrication by rolling, swaging, and wire drawing. In addition, the high-temperature sinter decomposes the dopant particles and removes excess aluminum, silicon, and oxygen by diffusion and volatilization. Potassium is insoluble in tungsten and is retained as a fine dispersion of metallic particles.

Mechanical working elongates the potassium particles, as shown in Fig. 18. With working, potassium becomes strung out into very fine tubes. Their visibility decreases, and they become lost in the complex matrix dislocation structure (Fig. 12a). On annealing, the potassium tubes break up into strings of partially filled bubbles (Fig. 12b). Establishment of an optimum recrystallized structure depends on the proper dispersion of these bubbles and on the work history of the wire.

Very high total strain (cold work) is necessary to adjust the relative nucleation and growth rates to produce the required large, high aspect ratio recrystallized grains (Fig. 11a). The ingot is therefore warm worked at progressively lower temperatures to maintain the material below the recrystallization temperature. Extensive

working below the recrystallization temperature also forms a fibrous structure and progressively lowers the ductile brittle transition temperature. Room-temperature ductility permits final fabrication by coiling or other forming operations.

Thoriated products are processed in a similar manner. Thoria can be added directly to the metal powder, or it can be added as a nitrate solution to the powder precursors. A subsequent calcination is used to decompose the nitrate to the oxide. Rhenium is added as a metal powder prior to pressing. The amount, particle size, and sintering conditions must be controlled to prevent the formation of brittle tungsten-rhenium intermetallic phases.

Electrical Contacts

By Norman S. Hoyer
Manager, Contact Operations
Westinghouse Electric Corp.

ELECTRICAL CONTACTS are metal devices that make, carry, and break elec-

Fig. 14 Comparison of the effect of a large, interlocking grain structure from doped tungsten and equiaxed grain structure on the sag of a lamp filament
(a) Doped. (b) Undoped

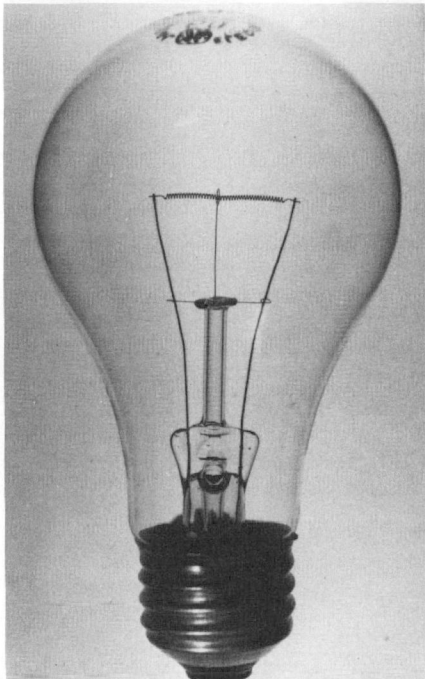
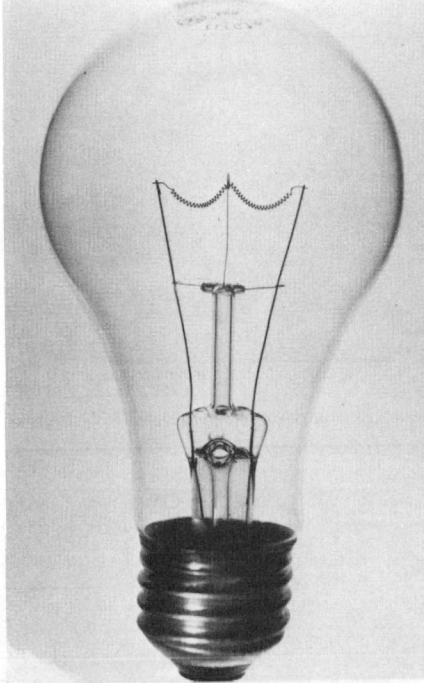
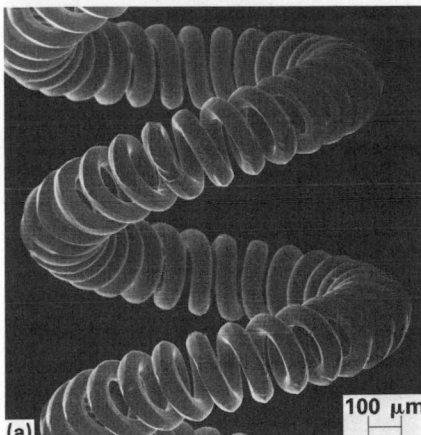
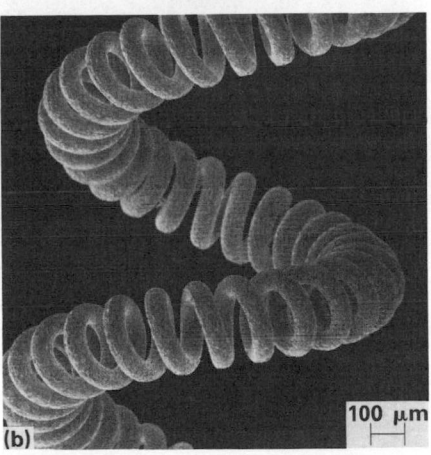

- Semiconductors
- Switches

Silver nickel

- Circuit breakers (air and oil)
- Contactors
- Motor starters
- Relays (aircraft, light and heavy duty)
- Switches
- Tap changers

Silver graphite

- Circuit breakers (air)
- Contactors
- Relays (light duty)
- Rheostats
- Sliding devices
- Switches

Silver-nickel-graphite

- Circuit breakers (air)
- Rheostats
- Sliding devices
- Switches

Copper graphite

- Contactors
- Motor starters
- Rheostats
- Sliding devices
- Switches

Silver/cadmium oxide

- Contactors
- Motor starters
- Relays (aircraft, light and heavy duty)
- Switches

Copper-tungsten

- Circuit breakers (oil)
- Semiconductors
- Tap changers

Silver/tungsten carbide

- Circuit breakers
- Relays (aircraft and light duty)
- Switches

trical circuits. Contacts are made of either elemental metals, composites, or alloys that are made by the melt-cast method or manufactured by P/M processes. The ideal metal or metal combination that can function as the perfect contact material under all conditions does not exist. A thorough evaluation and understanding of the operating conditions of an electrical contact device, as well as consideration of economy, is necessary before selecting the most suitable contact material. Table 4 gives properties of typical P/M contact materials. Applications include:

Silver tungsten

- Circuit breakers (air)
- Contactors
- Motor starters
- Relays (aircraft, light and heavy duty)
- Semiconductors
- Switches

Silver molybdenum

- Circuit breakers (air)
- Contactors
- Motor starters
- Relays (light and heavy duty)

Historically, contact materials have consisted almost entirely of silver, silver alloys, and P/M sintered combinations. Exceptions include some beryllium copper, phosphor bronze, and nickel materials that are also used as contacts. Silver-type contacts, including the pure metal, alloys, and metal powder combinations, comprise the majority of contact applications in the electrical industry. Other types of contacts used include the platinum group metals, tungsten, molybdenum, copper, copper alloys, and mercury. For more in-

Fig. 15 Tungsten oxides
(a) Undoped. (b) Coated with dopants

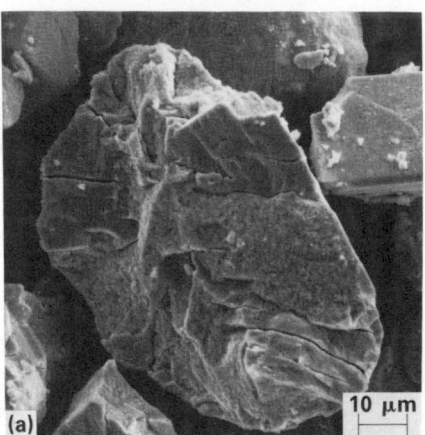 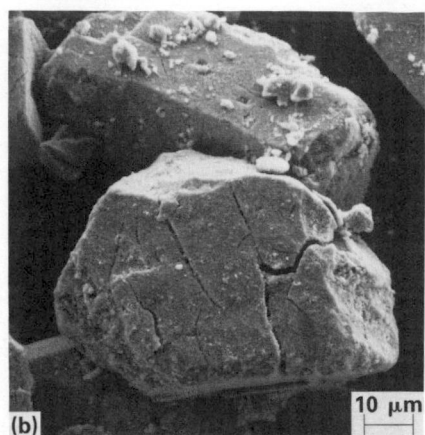

Fig. 16 Tungsten powders
(a) As reduced from undoped oxide. (b) Pock-marked powders made from reduction of doped oxides

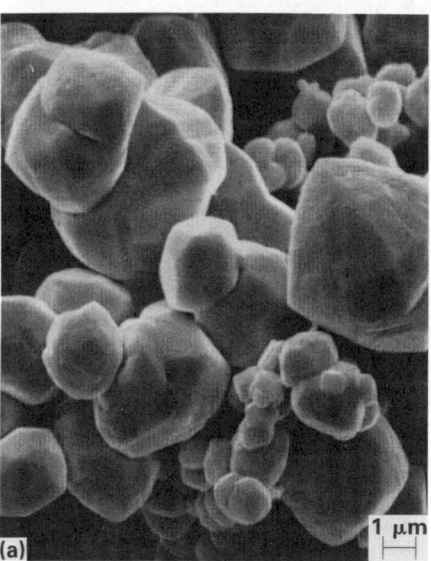 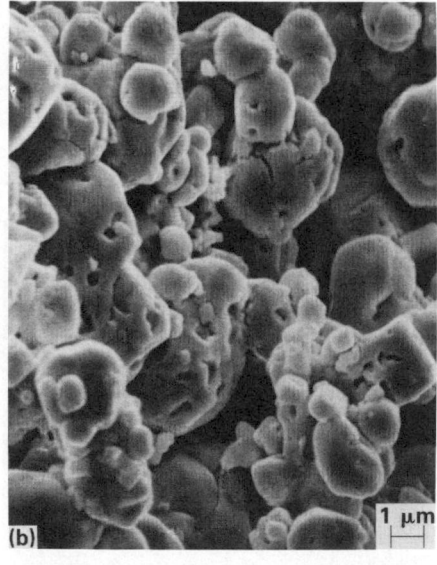

Fig. 17 Cross section of tungsten P/M particle showing internal retention of dopants

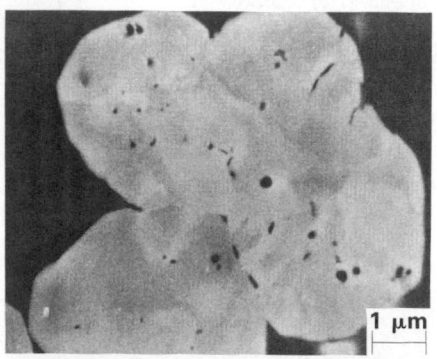

Fig. 18 Longitudinal fracture surface of 2.0-mm (0.08-in.) diam tungsten rod showing elongated, potassium-filled dopant tubes

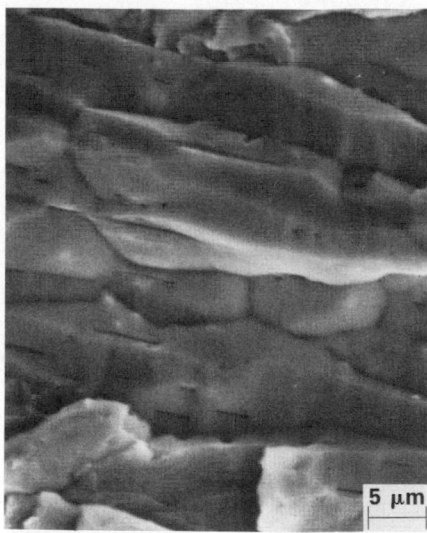

formation on electrical contact materials, see the article "Electric-Contact Materials" in Volume 3 of the 9th edition of *Metals Handbook*.

Powder metallurgy facilitates combination of silver and copper with other metals. These diverse combinations ordinarily cannot be achieved by alloying. When silver is combined with other metals with which it does not conventionally alloy, the resulting P/M materials combine the characteristics of silver with the other metals in a manner in which true alloys cannot duplicate. Furthermore, the chemical characteristics of silver remain unchanged in P/M combinations. The conductivity of silver in P/M combinations is unchanged, so that the resulting conductivity may be only moderately less than that of pure silver.

Silver Nickel

One of the elements typically combined with silver by P/M processes is nickel. Nickel is more effective as a hardening agent than copper; consequently, silver nickel is considerably harder than coin silver. At the same time, nickel does not increase contact resistance appreciably, particularly in combinations that include 15 wt% Ni or less. Silver nickel is combined in proportions ranging to about 40 wt%.

The combinations most widely used are 60Ag-40Ni and 85Ag-15Ni. These materials are very ductile and can be formed in all of the shapes in which silver contacts are used, including very thin sheets for facing large contact areas. This material is ideal for use under heavy sliding pressures. It does not gall like fine silver and coin silver, but instead takes on a smooth polish. It is therefore suitable for sliding contact purposes, as well as for make-and-break contacts. Silver nickel can handle much higher currents than fine silver before it begins to weld. It has a tendency to weld when operated against itself. Therefore, it is frequently used against silver graphite.

The silver nickel material 60Ag-40Ni is the hardest material in the silver nickel series. It is the most suitable for sliding contact in which pressure is high. This alloy

Table 4 P/M contact properties

Nominal composition(a)	Density, g/cm³	Hardness(b)	Condition	Conductivity, % IACS
50Ag-50W	13.06-13.33	50-65 HRB	...	57-70
35Ag-65W	14.32-14.62	75-85 HRB	...	50-58
25Ag-75W	15.32-15.64	85-95 HRB	...	45-52
50Ag-50Mo	9.94-10.14	70-80 HRB	...	45-54
35Ag-65Mo	9.90-10.10	82-92 HRB	...	38-45
85Ag-15Ni	9.60-9.90	48 HR15-T	Annealed	79-84
		78 HR15-T	Cold worked	
60Ag-40Ni	9.21-9.51	63 HR15-T	Annealed	53-65
		85 HR15-T	Cold worked	
98Ag-2C	9.19-9.49	23 HR15-T	Annealed	82-87
		67 HR15-T	Cold worked	
95Ag-5C	8.30-8.50	20 HR15-T	Annealed	54-65
		63 HR15-T	Cold worked	
90Ag-10C	7.23-7.46	10 HR15-T	Annealed	40-47
		32 HR15-T	Cold worked	
88Ag-10Ni-2C	9.04-9.33	22 HR15-T	Annealed	58-68
		60 HR15-T	Cold worked	
77Ag-20Ni-3C	8.61-8.88	47 HR15-T	Annealed	50-57
		70 HR15-T	Cold worked	
95Cu-5C	7.33-7.57	38 HR15-T	Annealed	40-46
		64 HR15-T	Cold worked	
90Ag-10CdO	9.60-10.1	45 HR15-T	Annealed	70-86
		76 HR15-T	Cold worked	
85Ag-15CdO	9.46-9.96	60 HR15-T	Annealed	65-75
		81 HR15-T	Cold worked	
75Cu-25W	9.72-10.03	20 HRB	Annealed	55-65
		60 HRB	Cold worked	
50Cu-50W	11.75-12.00	65-75 HRB	...	50-58
30Cu-70W	13.76-14.04	85-95 HRB	...	40-48
20Cu-80W	15.04-15.36	95-105 HRB	...	30-40
65Ag-35WC	11.15-11.50	45 HRB	Annealed	50-60
		60 HRB	Cold worked	
50Ag-50WC	12.05-12.30	77-87 HRB	...	43-50
35Ag-65WC	12.80-13.06	95-105 HRB	...	32-37

(a) wt%. (b) Superficial Rockwell tester: indentor (T) is a 1.6-mm (1/16-in.) diam steel ball. Major load: 15 kg

also has the lowest rate of wear under sliding action. It is less ductile than silver nickel materials containing less nickel, but it is still sufficiently ductile for all conventional manufacturing processes.

The silver nickel material 85Ag-15Ni is the most widely used material in this series. When selecting contact materials, the advantages of a particular material are weighed against its disadvantages for a particular set of conditions. However, 85Ag-15Ni is comparable to fine silver and coin silver, while silver nickel offers the most advantages. It carries higher currents before welding begins. The contact resistance of silver nickel is about the same as fine silver, and in some tests, it has been found to be lower.

The contact resistance of clean contacts that have not operated under load tend to be slightly lower for fine silver. However, in make-and-break circuits, silver tends to gradually increase contact resistance. This increase is not necessarily permanent, as contact resistance varies with the effects of arcing on the contacts. Generally, average resistance is higher than the initial resistance before the contacts operate. The contact resistance of 85Ag-15Ni is similar, except that it usually varies within a narrower range. Exhibiting nearly constant contact resistance is more important than possessing low contact resistance.

85Ag-15Ni exhibits a lower contact resistance and is also harder than coin silver. Another advantage of 85Ag-15Ni is its low flammability; that is, it makes a smaller arc than other materials. In testing of more than 40 contact materials, 85Ag-15Ni exhibited the lowest arc energy. Low arc energy is important in that the ability to break a circuit with as little flame as possible is desirable. This characteristic was primarily responsible for the adoption of 85Ag-15Ni for relays in aircraft electrical systems.

Because of its ideal mechanical properties, 85Ag-15Ni is an ideal material for motor-starting contactors and is superior in this type of application to fine silver, coin silver, and copper. It is also suitable as a general-purpose contact for various types of relays and switches.

Silver Graphite

Graphite is also combined with silver by P/M techniques. The most frequently used composition is 95Ag-5C, although graphite compositions ranging from 0.25 to 90% with the remainder silver have been used. This material was developed as a circuit breaker contact material. The addition of graphite prevents welding. Frequently, 95Ag-5C is used in combination with silver nickel or silver tungsten contacts. It is also used in combination with pure nickel contacts and with fine silver contacts. Silver graphite is soft compared to other types of contact materials, and electrical and mechanical erosion is more rapid.

95Ag-5C has been widely used as a material for contacts in molded-case circuit breakers, sliding contacts, and contact brushes. The graphite serves as a lubricant and ensures a smooth, low-friction contact surface. This material is only moderately ductile and can be rolled into sheets and punched into contacts of various shapes. However, it cannot be headed to make solid rivets or bent to any great extent without cracking. It can be coined to a moderate extent. 95Ag-5C contacts can be individually molded. Depending on size, shape, and quantity, contacts of this material are either punched from rolled slabs, extruded, or individually molded from powders. Copper is combined with graphite as a substitute for silver in certain applications.

A modified form of silver graphite is silver-nickel-graphite. Typical compositions are 88Ag-10Ni-2C and 77Ag-20Ni-3C. These materials are substantially harder than 95Ag-5C and exhibit superior wear resistance, but offer less protection against welding. Like 95Ag-5C, they can be manufactured from slabs or by molding individually.

Silver/Cadmium Oxide

The silver/cadmium oxide group of P/M materials is the most widely used of all the silver semi-refractory contact materials. The addition of 5 to 15% cadmium oxide to silver imparts excellent nonsticking and arc quenching qualities. Silver/cadmium oxide contact materials are well suited for contactors and motor starters, but are also used in circuit breakers, relays, and switches. This material is made by the press-and-sinter method, with subsequent cold or hot working to improve densification.

Silver/Refractory Metals and Copper/ Refractory Metals

The development of duplex contact materials involving silver or copper with tungsten or molybdenum or their carbides has resulted in materials that can withstand higher currents and more arcing than other contact materials, without experiencing sticking or rapid erosion. Development of these materials has been somewhat difficult. However, there has been a trend in recent years toward using these materials, particularly in circuit breakers that use only one set of contacts, in contrast to other types of circuit breakers that use arcing tips or horns that are separate from the current-carrying contacts. To meet the requirements of this type of apparatus, contact materials are required that can interrupt short-circuit currents without failure, while possessing contact resistance that is low enough to carry continuous heavy currents without overheating. Development has been directed largely toward the production of contact materials of this type.

The refractory metal content may vary from 10 to 90%, although 40 to 80% usually is used in air- and oil-immersed circuit breaker devices. Refractory metals offer good mechanical wear resistance and resistance to arcing. Silver and copper provide good electrical and thermal conductivities.

Depending on the composition, refractory metals containing silver or copper contact materials are made either by pressing and sintering or by the press-sinter-infiltrate method. When infiltration is used, either all refractory metal powder is compacted to shape, or a small amount of silver or copper powder is blended with the refractory metal, compacted, and sintered in a reducing atmosphere. The sintered compact is then returned to the furnace; silver or copper is added to act as the infiltrant.

At temperatures above the melting point of the infiltrant, the liquid metal penetrates and fills the interconnecting voids of the pressed and sintered compact. Densities of 96 to 99% of theoretical can be achieved by this process. This group of materials finds use as current-carrying contacts in air- and oil-immersed circuit breakers, heavy-duty relays, automotive starters, and switches. Lower properties can be obtained by pressing and sintering.

Production of P/M Contact Materials

Silver-nickel contacts are produced from a mixture of electrolytic or precipitated silver powder and carbonyl nickel powder. A compound silver-nickel powder may also be produced by precipitating an intimate mixture of the metals as carbonates with sodium carbonate from an aqueous solution of silver and nickel nitrate. The carbonate mixture is reduced in

hydrogen at 400 °C (750 °F), and the compound powder is washed free of alkali. The compound powder is pressed at relatively low pressure (140 MPa, or 20 000 psi). This avoids trapping of gases during the subsequent sintering at about 800 °C (1470 °F) and re-pressing. This production procedure results in a very fine distribution of nickel in the silver matrix.

Silver-graphite contacts are produced by mixing electrolytic or precipitated silver powder with graphite powder. The mixture is pressed at approximately 275 MPa (40 000 psi), sintered between 700 and 900 °C (1300 and 1650 °F) in an inert or reducing atmosphere, and re-pressed at 600 to 900 MPa (87 to 130 ksi). Nearly complete densification can be achieved in re-pressing, particularly in materials with low graphite contents.

Copper-Tungsten Contacts. Production methods for these contact materials depend on the composition ratio. For compounds with 60% or less tungsten, the classical method of mixing the powders, pressing, sintering (generally below the copper melting point), and re-pressing may be used. Materials with 60 to 80% W are generally produced by infiltration, either of loose tungsten powder or of a pressed and sintered tungsten compact.

Other Contacts. Silver/cadmium oxide contacts may be produced in a similar manner as silver-graphite materials, by mixing silver and cadmium oxide powders, compacting, sintering in an oxidizing atmosphere, and re-pressing. Fusion metallurgy may also be used.

The fabrication of tungsten-silver, molybdenum-silver, and copper/tungsten carbide compound materials is similar to the production of copper-tungsten contact materials. Pure tungsten contacts may be cut from swaged round rods up to 15 mm (0.6 in.) in diam and 0.5 to 3 mm (0.02 to 0.12 in.) thick.

Metal-Graphite Brushes

By Charles P. Farago
Director, Development and
Engineering
Kirkwood Carbon Co.

ELECTRIC MOTOR BRUSHES, the components that transfer electrical current between stationary and rotating elements in electric motors and generators, have been used for over 100 years. Originally, they

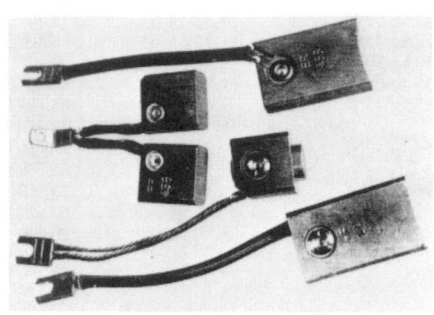

Fig. 19 Industrial metal-graphite brushes
Courtesy of General Electric Co., Carbon Products Operation

consisted of bundles of copper wires that resembled paint brushes. In 1885, the first patent on carbon brushes was issued to George Forbes (Ref 3). It soon became apparent that carbon brushes, due to their self lubrication and lower electrical conductivity, lasted longer and outperformed copper counterparts.

As electrical power became more available, electric motors began finding increased usage, and motors were used in low-voltage (<36 V) direct current applications as well. Carbon and graphite materials became inadequate, because these motors required higher currents to operate satisfactorily. Thus, metal was again introduced into brushes, allowing brush current densities from 90 000 to 250 000 A/m² (60 to 150 A/in.²).

Originally, several pieces of wire gauze were molded into carbon blocks, thus increasing electrical conductivity. This method proved to be cumbersome and uneconomical, because it could only be used in the manufacturing of machined brushes. Furthermore, it was difficult to produce the wide variety of grades needed for different applications. This led to the blending of metal powders with a carbon lubricant (graphite). Typical metal-graphite brushes are shown in Fig. 19.

Types of Materials

Table 5 lists several typical metal-graphite brush grades. Metal contents range from 20 to 99 wt% metal. The most common metals used are copper, silver, lead, tin, antimony, zinc, and iron; copper and silver are the most predominant. Other metals usually are used to modify the properties of copper.

Metal powders are used in most of the available forms. Copper powder may be used in the form of reduced oxide powder, atomized spheriods, electrolytic (dendritic) powder, or flake powder. Silver usu-

Table 5 Typical graphite brush-grade characteristics

Grade No.	Nominal composition	Density, g/cm³	Specific resistance Ω·m	Specific resistance Ω·in.	Maximum current density A/m²	Maximum current density A/in.²	Typical voltage, V	Scleroscope hardness
261C	21Cu-79C	2.2	0.024	0.0006	125 000	80	<72	28
261D	35Cu-65C	2.5	0.016	0.0004	125 000	80	<72	28
FQ	50Cu-50C	2.75	0.006	0.00015	130 000	85	<36	28
179P	65Cu-35C	3.5	0.0016	0.00004	190 000	125	<18	20
179V	75Cu-25C	4.0	0.0008	0.00002	235 000	150	<15	18
GHB	94metal-6C	6.0	0.0003	0.0000065	235 000	150	<6	6
GD	97metal-3C	6.5	0.0001	0.0000025	235 000	150	<6	5
22A-S	40Ag-60C	2.7	0.008	0.0002	150 000	100	<36	30
246	65Ag-35C	3.8	0.001	0.000025	190 000	125	<18	20
2-S	80Ag-20C	4.6	0.0008	0.00002	235 000	150	<9	23
1-S	93Ag-7C	7.0	0.0001	0.0000025	270 000	175	<6	10

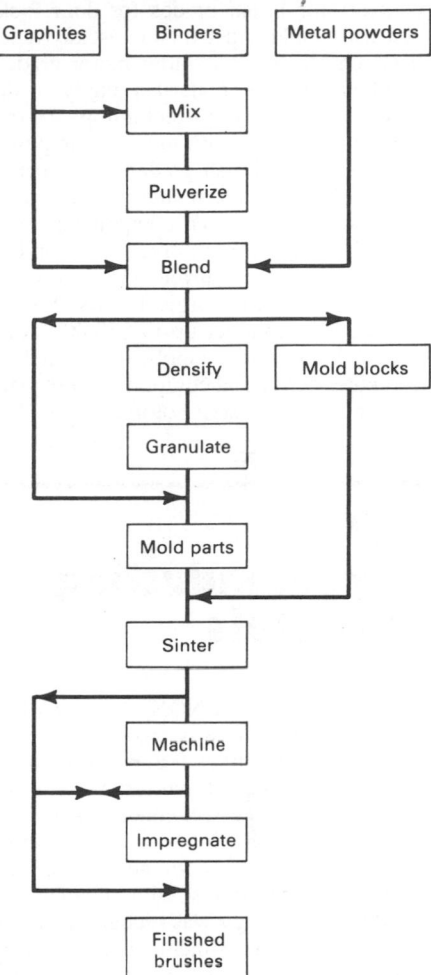

Fig. 20 Flowchart of metal-graphite brush production

ally is used either as electrolytic or flake powder. Particle sizes may vary from 150 μm to less than 1 μm.

Brush grades containing 80% or less metal usually are not blended with pure graphite. Instead, they are blended with a synthetic, coal, or petroleum by-product that is bonded to graphite, which usually is formulated by the brush manufacturer. The mechanical integrity of the brush material comes from these organic binders. Conversely, alloy grades containing more than 80% metal (copper, lead, antimony, zinc, tin, and iron) that is blended with pure graphite derive their strength from the sintered metal matrix.

Manufacturing

Figure 20 is a flowchart of the processes used to manufacture metal-graphite brushes. Bonded or plain graphite is blended with the appropriate metal powders until a uniform mix is attained. Material may then be:

● Molded into brushes on a mechanical or hydraulic press
● Densified on a mechanical or hydraulic press, granulated, and then molded into brushes
● Molded into large blocks

These blocks usually are hot or cold molded on hydraulic or isostatic presses. Brushes usually are cold molded. Typically, molding pressures range from 100 to 200 MPa (15 to 30 ksi), providing green densities of 2.0 to 8.0 g/cm³ depending on the metal content.

After compaction, all brush materials are sintered at 500 to 1000 °C (950 to 1850 °F). Sintering cycles vary significantly, as large parts with low metal contents containing excessive volatiles must be heated at a much slower rate than small parts with high metal contents. Frequently, parts with low metal contents are sintered in an atmospherically protected batch furnace, while small parts or those containing a high

amount of metal are fired in a continuous furnace if the production quantity is large enough to make this economically feasible. Although parts have been protected by packing in coke or sand, they are currently sintered in nitrogen- and hydrogen-containing protective atmospheres. Typically, these may be bottled gases, dissociated ammonia, or exothermic and endothermic atmospheres.

Many of the materials are produced for subsequent molding to size to reduce machining after sintering. Machining is required on some higher metal content grades that have undergone hot molding or that have been hot or cold re-pressed after sintering. However, molding to size usually is desired.

These brush materials also may be impregnated with various organic liquids, such as oil, to modify their frictional properties. Typically, a standard pressure and vacuum system is used.

Applications

Powder metallurgy brush materials are selected for characteristics different from those typically exhibited by most P/M parts—resistivity (conductivity) and coefficient of friction. Tensile and compressive strengths are only important in that parts must be able to withstand operating forces.

Brushes operate either as part of a rotating switch (commutator) or as part of a rotating contact (slip ring). Commutator brushes not only conduct external power to the armature, but they also must dissipate energy stored in the armature windings while neighboring commutator segments are being short circuited. Slip ring brushes need carry only the external current. However, this current may be as high as 300 000 A/m² (200 A/in.²), compared to the 200 000 A/m² (125 A/in.²) typically the carried by commutator brushes.

All metal-graphite brush applications must operate at lower voltages than non-

metallic brush materials. Typical voltages range from near 0 to 36 V and may be a direct-current power source emanating from a battery, the output of a rectifier, or an analog signal from a sensing device, such as a thermocouple.

Widespread application is found in portable battery-operated power tools, which require high outputs in small, lightweight packages. Typically, input voltage influences the metal content required. High voltages require a low metal content, whereas low voltages require a high metal content. Below 9 V, for example, metal contents greater than 80% are required. Above 18 V, grades with more than 50% metal are seldom used.

Automotive applications almost exclusively use metal-graphite brush grades. Starter motors require high-metal alloy grades to enable them to handle extremely high current densities for short periods of time. Blower motor brushes, however, contain 65% or less metal to extend service life to several thousand hours. Metal contents for brush grades for door locks and windshield wiper motors usually range between blower and starter motor grades.

Silver-graphite materials enjoy widespread usage in high-technology instrument applications with low-voltage power sources. These brush grades must operate in a uniform, stable manner for extended periods of time. Some components, such as tachometer generators, must conduct small amounts of current (150 A/m^2, or 0.1 A/in.^{2}), while surgical saws, for example, may conduct up to 500 000 A/m^2 (325 A/in.^{2}). Other applications include computer peripheral equipment, aerospace equipment, and servomotors.

P/M Superconducting Materials

THE PHENOMENON OF SUPER-CONDUCTIVITY was first documented in 1911, when Onnes found that mercury abruptly lost all electrical resistance at 4.1 K and became a perfect conductor (Ref 4). Although the possibility of using superconducting materials for generating very large magnetic fields with a small expenditure of energy was recognized at the time, the first experiments with superconducting solenoids were unsuccessful, and superconductivity research did not progress for the next 50 years.

In 1954, Matthias and co-workers (Ref 5) reported that the intermetallic compound Nb$_3$Sn becomes superconducting at 18 K, a new reported high for the transition temperature (T_c) from the normal to the superconducting state. In 1960, Bell Telephone Laboratories announced that Kunzler et al. (Ref 6) had found that Nb$_3$Sn could sustain a resistanceless current density in excess of 100 000 A/cm^2 (15 500 A/in.^{2}) in magnetic fields up to 8.8 T (88 kG), which is the highest field available. Currently, the critical magnetic field of Nb$_3$Sn is known to be more than 20 T (200 kG). (The earth's magnetic field is about 200 000 times smaller.) Figure 21 compares the behavior of a normal metal and a superconducting metal.

Superconductivity permits electric power generators and transmission lines to have capacities many times greater than currently possible It also allows the development of levitated transit systems ca-

pable of jet speeds and provides an economically feasible way of producing the large magnetic fields required for the confinement of ionized gases in controlled thermonuclear fusion.

Although known superconducting compounds have higher transition temperatures than Nb$_3$Sn (T_c = 18 K), such as Nb$_3$(Al, Ge) (T_c = 21 K) and Nb$_3$Ge (T_c = 23 K), practical methods for producing wire or tape have been developed only for Nb$_3$Sn. Select properties of niobium-based superconductors are shown in Table 6.

The most widely used superconducting material is a niobium and titanium alloy. Although this alloy does not have the high field properties of Nb$_3$Sn, it is extremely ductile and can be fabricated by conventional metalworking processes. Intermetallic compounds with superior superconducting properties are extremely brittle, however. Therefore, special procedures must be used to produce them in the required forms. This section discusses some of the methods used to produce Nb$_3$Sn wire and tape.

Conventional Production Methods

Wind and React Method. The first practical method for producing a high field superconductor was developed by Kunzler et al. (Ref 6), who used P/M techniques. This procedure was refined by Olsen et al. (Ref 7). Elemental powders of niobium and tin were mixed and packed into a niobium tube. An outer jacket of Monel was used to cover the niobium tube to facilitate wire drawing.

This composite could be mechanically reduced to fine wire with a 0.2-mm (0.008-in.) diam. The wire contained a ductile core of the unreacted powders, which could be wound easily to form solenoids. These were then heated for several hours at 950 to 1000 °C (1740 to 1830 °F). During heat treating, the niobium and tin reacted to form a core of the brittle compound Nb$_3$Sn.

Many superconducting magnets have been made successfully with this wire. The coils required winding before the diffusion reaction could be completed, however. This limitation eventually led to replacement by other methods.

General Electric Tape Process. To eliminate heat treating of the entire solenoid, the General Electric Research and Development Center developed a superconductor that could be wound after the reaction to form Nb$_3$Sn was completed. This requirement implied a configuration in which the superconducting compound was present as a thin layer. In 1966, Benz

Fig. 21 Comparison of the behavior of a normal metal and a superconductor

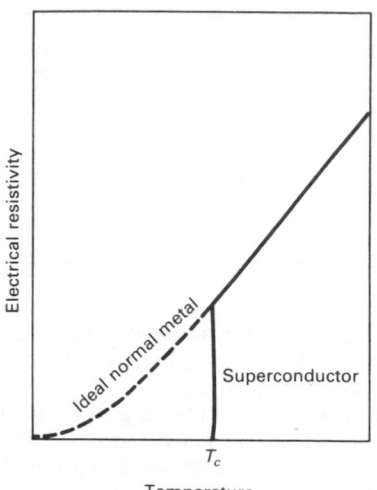

Table 6 High field niobium-based superconductive compounds

Compound	T_c, K	H_{C2} at 4.2 K	
		T	kG
Nb$_3$Ge 23		38	380
Nb$_3$(Al, Ge) 21		40	400
Nb$_3$Ga 20		20	200
Nb$_3$Al 19		30	300
Nb$_3$Sn 18		23	230
NbTi 10		12	120

(Ref 8) presented a procedure for fabricating a flexible tape based on previous work (Ref 9), which is still used on a commercial scale with modifications.

Niobium tape is passed through a molten tin bath and then heated at 950 to 1000 °C (1740 to 1830 °F) to produce a surface layer approximately 2 μm thick of the Nb$_3$Sn compound. This tape is then built up as a laminated composite to provide reliable performance by adding a layer of copper to each side. An additional layer of stainless steel may be applied if high strength is required. One benefit derived from this laminated configuration is the fact that the brittle Nb$_3$Sn layer is close to the neutral axis of bending. Consequently, flexibility is not diminished by the application of the additional layers. Laminated conductors of this type have been used to construct a superconducting magnet with a 25-mm (1-in.) bore capable of generating a magnetic field of 10 T (100 kG).

Bronze Process. To produce materials with stable superconducting properties, a number of conditions must be fulfilled, one of which is to have a maximum size of 10 μm for the superconducting component of the assembly. This is possible in the so-called "bronze process," which does not involve powder metallurgy. This process derives its name from the use of a dilute copper-based alloy matrix at some point during processing. The solute is either gallium, tin, or silicon.

Originally, an ingot of copper and 10 wt% tin was cast in a cylindrical shape. A number of axial holes were drilled, into which niobium rods were inserted. Because of the tendency of the bronze to work harden rapidly, frequent intermediate anneals were used. The composite was thus mechanically reduced to a convenient size.

The resulting product was then cut into short lengths, which were bundled together. The new composite was reduced in the same manner. This sequence was repeated as many times as was required to reduce the niobium cores to filamentary size.

After mechanical reduction, the multifilamentary wire was heated at 650 to 750 °C (1200 to 1380 °F) for up to 120 h. During this heat treatment, tin diffused from the bronze to the surface of the niobium filaments, where a layer of Nb_3Sn formed between the niobium cores and the bronze matrix. The thickness of the Nb_3Sn layers was typically 2 μm.

In one process modification (Ref 10), pure copper is used instead of the bronze matrix to capitalize on the ductility of copper and to eliminate the necessity for time-consuming intermediate anneals. This procedure is otherwise identical to the original, except that the niobium-copper composite wire is passed through a molten tin bath to apply a coating of tin. During heat treating, the tin diffuses first into the copper to form a bronze and from the bronze to the niobium, as in the original process.

P/M Production of Multifilamentary Tapes

Although the GE tape process is more widely used commercially, another process for fabricating niobium-tin tape using P/M methods is enjoying increasing usage. The sequence of operations used in P/M processing of tape is shown schematically in Fig. 22.

Niobium powder produced by the hydride-dehydride process is gravity fed to a pair of 50-mm (2-in.) diam compacting rolls. The emerging tape, with a porosity of 30%, can support approximately 2030 cm (800 in.) of its own weight. After vacuum sintering for 3 min at 2250 °C (4080 °F), the tape is quite ductile and can

Fig. 22 Infiltration process for multifilamentary tape

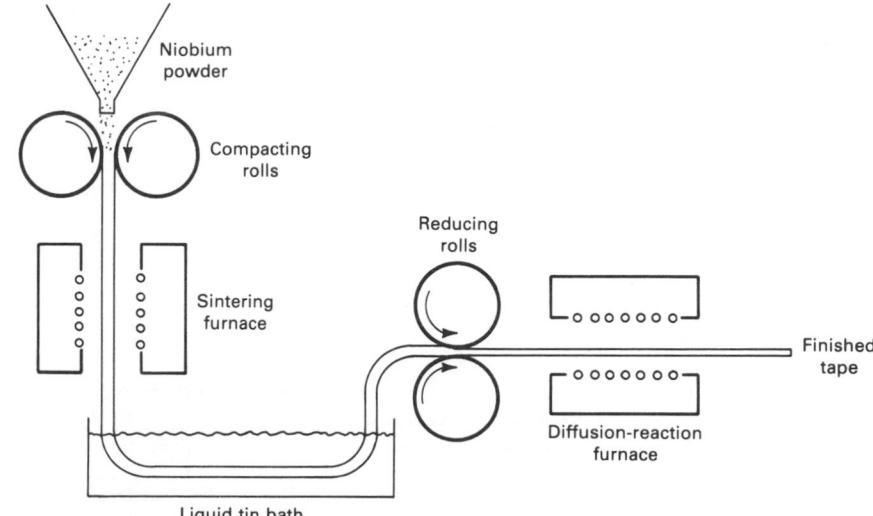

sustain a bend of 2.5-mm (0.1-in.) radius. Porosity is moderately reduced to a level of about 25%.

The sintered tape is immersed for 1 min in a molten tin bath that is maintained at a temperature of 850 °C (1560 °F) to infiltrate the interconnected porous network with tin. While infiltration can be carried out at temperatures as low as 600 to 650 °C (1100 to 1200 °F), difficulties were encountered in subsequent rolling. As a result of the low flow stress of pure tin, some tin is extruded during mechanical deformation. This loss of tin reduced the volume fraction of the superconducting compound (Nb_3Sn) in the final product.

The kinetics of the reaction between niobium and tin are such that, after infiltrating at 650 °C (1200 °F), the only phases present are niobium and pure tin. Raising the infiltration temperature to 850 °C (1560 °F) resulted in the formation of a significant amount of intermediate phases, predominantly as Nb_6Sn_5. The presence of these phases prevented any tin loss during deformation.

A reaction treatment above 930 °C (1700 °F) immediately after infiltration produces Nb_3Sn in massive form, and the tape is extremely brittle. By adding a cold rolling operation between the infiltration and the diffusion reaction stages, the tin is elongated and thinned to a filamentary morphology. The greatly increased interfacial area between the niobium and tin reduced the time required for complete reaction of the tin from more than 20 h to a few minutes.

The filamentary structure also provides the flexibility required for coil winding. A suitable structure was obtained by cold

rolling the infiltrated tape to a reduction in thickness of 75 to 85%. The final diffusion reaction is carried out at 930 to 1350 °C (1700 to 2500 °F). Current-carrying capacity decreases rapidly when reaction temperatures exceed 1000 °C (1850 °F) as a result of the thermally induced disorder in the niobium chains.

The reaction should be carried out at as low a temperature as possible, thus balancing the superconducting properties against the required time at temperature. The tapes usually are heated for 3 min at 950 to 975 °C (1750 to 1790 °F). Completed tapes exhibit high superconducting characteristics, with a critical temperature of 18.1 K and an overall current-carrying capacity at 4.2 K of approximately 100 000 A/cm^2 (1550 $A/in.^2$) in an applied field of 10 T (100 kG).

However, less than optimum filament morphology exists in the tapes. The filaments have a typical cross section 30 μm wide and 5 μm thick. During powder rolling, minimal control over the filament width is possible, which is directly related to the original pore size.

P/M Production of Multifilamentary Wire

One P/M approach to producing multifilamentary Nb_3Sn wire was developed at the Lawrence Berkeley Laboratory of the University of California. The process is shown schematically in Fig. 23. While the "bronze process" can be used only if Nb_3Sn or V_3Sn is the superconducting compound, the Lawrence Laboratory process can also be used with Nb_3Al,

Fig. 23 Infiltration process for producing multifilamentary superconducting wire

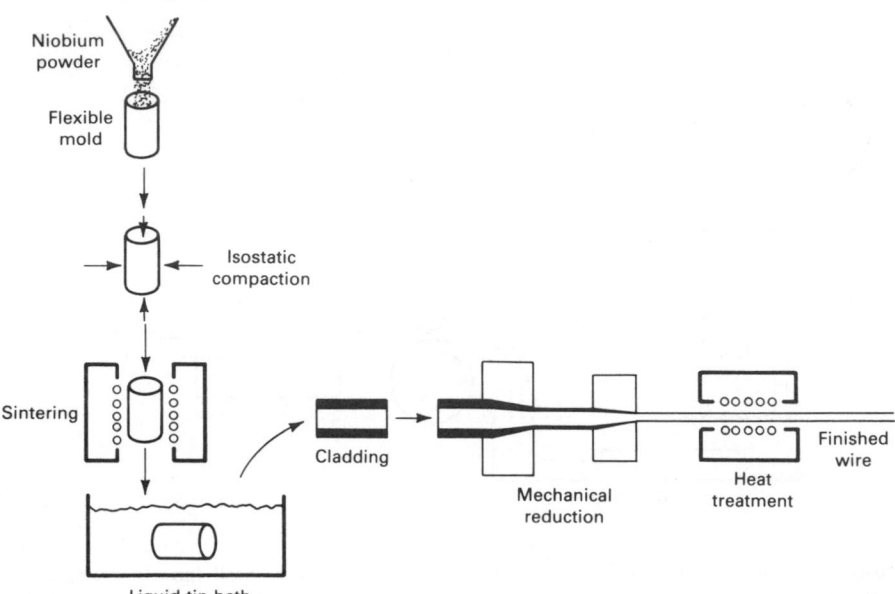

Nb₃(Al, Si), and Nb₃(Al, Ga) as the superconducting compounds.

In a modification of the tape procedure, infiltration of a porous niobium form with tin was retained, because the results were satisfactory. The powder rolling step, however, was replaced by a more appropriate method of compaction. In preliminary experiments, two variations of pressureless sintering (Ref 11) were tried. First, niobium powder with a particle size range of −325+400 mesh was packed into niobium tubes and sintered for 10 min at 2250 °C (4080 °F).

Several difficulties were encountered, however. Both the size and the volume fraction of the pores were so large that it was impossible to obtain a suitably fine filamentary structure. Also, shrinkage was nonuniform, and circumferential cracks were observed adjacent to the niobium inner wall.

A two-stage pressureless sintering operation was then used with a wider range of particle sizes that included all that passed through a 400-mesh screen. Preliminary sintering of the loose powder in a mullite tube at approximately 1600 °C (2900 °F) resulted in a self-supporting niobium rod. After removal of the mullite tube, a final sintering for 10 min at 2250 °C (4080 °F) yielded an acceptable product. However, two sintering operations were deemed impractical, and pressureless sintering was replaced with isostatic compaction.

Niobium powder with a sieve size ranging from −250 to 400 mesh is cold isostatically compacted at a pressure of 207 MPa (30 ksi) to produce a rod form. After vacuum sintering for 12 min at 2250 °C (4080 °F), the rods are infiltrated with tin by immersion in a tin bath at 750 °C (1380 °F) for 30 s. Niobium is difficult to wire draw, because it tends to gall. This is compounded by the fact that tin tends to be squeezed out. Consequently, a double cladding is employed. The inner jacket is niobium, which serves to retain the tin and prevent undesired diffusion reactions between the infiltrated core and the outer jacket, which is usually made of Monel or copper tubing. These precautions greatly facilitate wire drawing.

Composite rods are reduced to wire with diameters ranging from 250 to 375 μm by form rolling followed by wire drawing. Heat treating for 2 min at 950 °C (1750 °F) is sufficient to cause all of the tin to react with the niobium matrix to form filaments of the superconducting niobium-tin compound. These filaments are well within the desired size range, having transverse dimensions of about 1 to 2 μm.

Superconducting Properties of Nb₃Sn P/M Wire and Bronze Process Wire

Both the P/M process and the bronze process for fabricating Nb₃Sn wire are in current commercial use. However, the method of fabrication has a significant effect on the properties of the wire produced. This section compares the superconducting properties of wires made by the two processes, as tested under identical pulsed field conditions.

Overall critical current density, as a function of transverse magnetic field, is shown in Fig. 24 for representative wires. The P/M wire, including cladding, has a diameter of 0.33 mm (0.013 in.). It is a single-core wire with a core diameter of 0.23 mm (0.009 in.). The size of the Nb₃Sn filaments ranges from 1 to 2 μm. The wire, which has a critical temperature of 18.1 K, was reacted for 1 min at 930 °C (1700 °F).

The bronze process wire, which has a diameter of 0.30 mm (0.012 in.) and contains more than 1000 Nb-1Zr cores, was made by repetitive wire bundling and mechanical deformation. A reaction time of 70 h at 700 °C (1300 °F) was required. The thickness of the Nb₃Sn layer around each core is 2 μm.

Bend tests performed on both wires indicated that the P/M wire suffered less degradation of current-carrying capacity for a given diameter bend. The bronze process wire showed a 20% degradation after a 25-mm (1-in.) bend. To produce the same degradation, the P/M wire had to be bent to a diameter of 15 mm (0.6 in.).

Magnetic Applications*

By Fritz V. Lenel
Professor Emeritus
Department of Materials Engineering
Rensselaer Polytechnic Institute

POWDER METALLURGY is used in the production of soft magnetic materials for direct current applications and for permanent magnets. In many of these applications, parts are produced by powder metallurgy, because this method permits production to final shape with a minimum of subsequent machining and grinding, while at the same time achieving desirable magnetic properties. Before individual applications are discussed, magnetic properties that are important in ferromagnetic materials will be reviewed.

Soft ferromagnetic materials for direct-current applications and permanent mag-

*Reprinted from *Powder Metallurgy: Principles and Applications,* Metal Powder Industries Federation, Princeton, NJ, 1980, p 531-548. With permission

Fig. 24 Overall critical current density versus applied field for infiltration process and bronze process Nb₃Sn wire

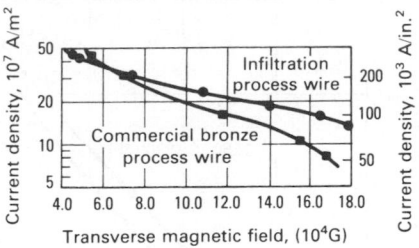

Fig. 25 Schematic of magnetic hysteresis curve for a soft magnetic material

B: Magnetic induction, G; H: magnetizing force, Oe; B_r: residual magnetic induction, G; H_c: coercive magnetizing force, Oe

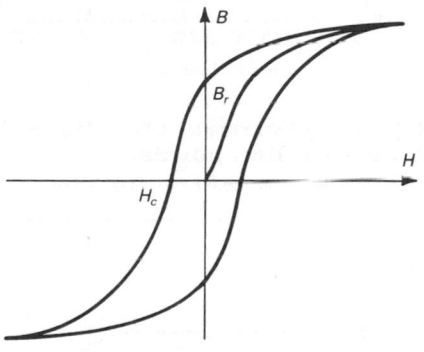

nets are characterized by their hysteresis curves, in which the magnetizing force H in oersteds (79.5 A/m = 1 oersted) is plotted versus the magnetic induction B in gauss (10^{-4} Tesla = 1 gauss).

A typical hysteresis curve for a soft magnetic material is shown in Fig. 25. It consists of the normal induction curve, which shows how induction varies as the magnetizing force is increased starting at zero, that is, the demagnetized state. The slope of this curve (the ratio of B/H) is the permeability of the material.

With increasing magnetizing force, permeability increases, reaches a maximum, and then decreases. Maximum permeability is one of the characteristics of soft magnetic materials. The magnetic saturation, or the value of induction or flux density when the magnetizing force has reached a specified value (for example, 1200 A/m or 15 Oe), is another important characteristic of the material. When the magnetizing force is decreased from this value, the induction curve does not follow the original normal induction curve, but lags behind, as shown in Fig. 25. The value of magnetic induction in gauss, when the

Fig. 26 Schematic of demagnetization curve for a permanent magnet alloy

B: Magnetic induction, G; H: magnetizing force, Oe; B_r: residual magnetic induction, G; BH: energy product, GOe; H_c: coercive magnetizing force, Oe; B_D: demagnetizing coercive remanence; H_D: demagnetizing coercive force

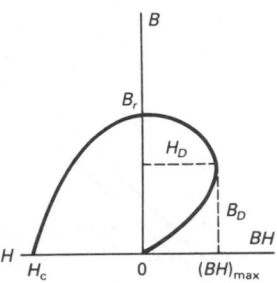

magnetizing force has been reduced to zero, is called the residual induction or remanence B_r. When a magnetizing force in the opposite direction is applied, the magnetic induction decreases further and reaches a value of 0 with a given magnetizing force in oersteds, called the coercive force H_c. Remanence and coercive forces are additional characteristics of soft magnetic materials.

Figure 26 shows the demagnetizing portion of the hysteresis curve from the residual induction B_r to the coercive force H_c for a permanent magnet material. In addition to the values B_r and H_c, the product of magnetizing force and induction at any point on the curve, the so-called energy product in gauss-oersted, is important for permanent magnets. In Fig. 26, the value of this energy product is plotted as a function of induction; it reaches a maximum at a specified value of induction.

In the first part of this section, soft magnetic applications, such as pole pieces, armatures, relay cores, and computer printers made from iron, iron-silicon, iron-phosphorus, and iron-nickel compositions, compacted from powder to shape and sintered, are described. They have adequately high permeability and saturation and low residual induction and coercive force. As described in the second part of this section, powder metallurgy is also used to a limited extent to produce wrought soft magnetic nickel-iron-molybdenum and nickel-iron alloys, in the form of strip, rod, and wire. These materials exhibit superior magnetic properties, in particular higher maximum permeability, which can be obtained by starting with very pure powders instead of cast ingots.

One type of permanent magnet material, the so-called Alnico magnets, generally is produced by casting. Small Al-

nico magnets, however, are also produced from powder. Their production and properties are discussed in the section of this article on Alnico magnets. Another type of recently developed permanent magnets is the cobalt/rare earth magnets, which are always produced from powder. During fabrication of these magnets, particles of the alloy powder must be magnetically oriented before they are pressed and sintered. Production and properties of these magnets are described below.

A third type of permanent magnet material—elongated single-domain permanent magnets—is also produced from powder. Fabrication of these magnets involves electrodeposition of powder particles on a mercury cathode, magnetic alignment, compacting, embedding in a matrix, grinding into a coarse powder, realigning, and compacting into the final magnet shape. Sintering is not required, however.

Because sintering is the distinguishing characteristic in the fabrication of P/M products, elongated single-domain permanent magnets cannot be strictly classified as P/M components. This also applies to another group of magnetic materials produced from alloy and metal powders—core materials for radiofrequency alternating current applications produced from carbonyl iron powder and audiofrequency core materials made from Permalloy powder. In these core materials, powder particles are covered by an insulating film to prevent metallic contact between the particles. Even when cores are treated at elevated temperatures, the insulating film prevents sintering.

Soft Magnetic Materials for Parts Pressed to Shape

The four types of compositions used for soft magnetic parts pressed to shape are pure iron, iron-silicon alloys containing 3% Si, iron-phosphorus alloys containing 0.45 to 0.75% P, and iron-nickel alloys containing 50% Fe and 50% Ni. Many of these alloys are difficult to machine in wrought form because of their softness. Production by investment casting or powder metallurgy has definite advantages if machining is not required.

Pure iron powder is widely used to produce soft magnetic parts. Examples of such parts are shown in Fig. 27. Atomized iron powder that has low oxygen, carbon, and nitrogen contents is used. Before pure atomized iron powder was available, electrolytic iron powder was used. The slight increase in magnetic properties of parts

Fig. 27 Soft magnetic components from iron powder

Courtesy of Remington Arms Co., Inc.

Fig. 28(a) Magnetic induction

At a magnetizing force of 1200 A/m (15 Oe) for cores from iron powder sintered 30 min in hydrogen at 1120 °C (2050 °F) and sintered 30 min at 1260 °C (2300 °F), cooled at a rate of 5.5 °C (10 °F) per minute as a function of sintered density

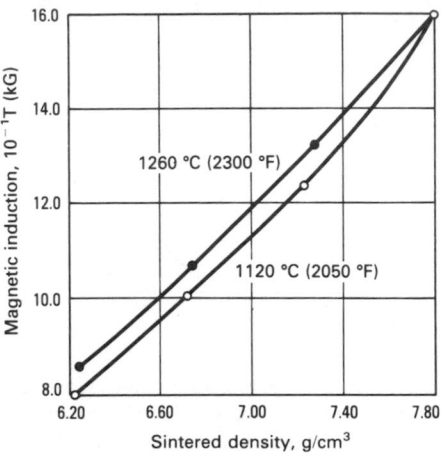

Fig. 28(b) Remanent magnetization of cores from iron powder

Sintered under the same conditions as Fig. 28(a)

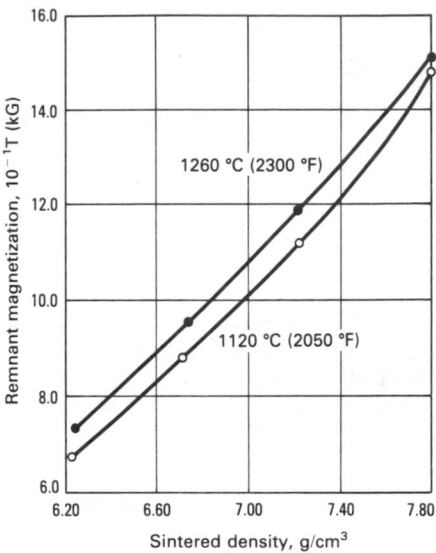

Fig. 28(c) Maximum permeability of cores from iron powder

Sintered under the same conditions as Fig. 28(a)

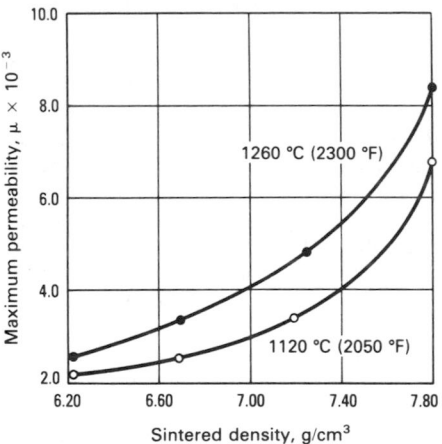

Fig. 28(d) Coercive force (H_c) of cores from iron powder

Sintered under same conditions as Fig. 28(a)

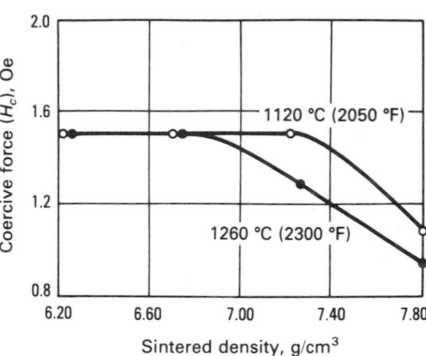

made from electrolytic powders compared to those made from atomized-iron powders generally do not warrant the use of the more expensive electrolytic powder.

Parts are produced by compacting, sintering, and, if necessary, re-pressing and annealing. Magnetic properties are a function of sintered density, sintering temperature and atmosphere, and particle size of the iron powder (Ref 12). Properties generally are determined by the ballistic testing of ring-shaped specimens (Ref 13).

The effect of sintered density and sintering temperature on saturation flux density at a magnetizing force of 1200 A/m (15 Oe) residual induction, maximum permeability, and coercive force of specimens sintered 30 min in hydrogen is shown in Fig. 28(a) through (d) (Ref 14). Improvement in these properties (higher saturation, residual induction, and maximum permeability and lower coercive force) with increasing density and increasing sintering temperature is evident.

Sintering temperatures as high as 1370 °C (2500 °F) are common in the fabrication of magnetic parts. Properties of fully dense material (density of 7.83 g/cm^3) are shown in Fig. 28. Parts were produced by hot re-pressing—a technique that is not widely used in commercial practice.

Maximum permeability and coercive force for this fully dense powder material are higher than those found in wrought annealed ingot iron (μ = 4000 and H_c = 100 A/m, or 1.30 Oe) and wrought annealed 1010 steel (μ = 3800 and H_c = 140 A/m, or 1.80 Oe). Pure iron magnetic parts generally are sintered in dissociated ammonia or hydrogen, which lowers the residual carbon and oxygen content in porous compacts. Parts sintered in endothermic and exothermic atmospheres have inferior magnetic properties.

Pure iron magnetic parts produced from coarse particle size fractions have pores with a larger mean pore area and a larger mean free path between pores than those

made from fine powder. Moyer (Ref 15) has shown that pore characteristics are reflected in the values of maximum permeability and coercive force. For example, specimens from a −60+100 mesh fraction of atomized-iron powder fabricated to a sintered density of 7.4 g/cm^3 and sintered at 1260 °C (2300 °F) have a mean pore area of 102 μm^2 and a free path between pores of 143 μm. Their maximum permeability is 4900, and their coercive force is 71.6 A/m (0.9 Oe). Specimens from the −400 mesh fraction of the powder fabricated in the same manner have a mean pore area of 43 μm^2 and a mean free path between pores of 62 μm. Maximum permeability is 3200, and coercive force is 135 A/m (1.7 Oe).

Iron-silicon alloys for magnetic applications (Ref 16) are produced from powder by mixing the same pure atomized-iron powder used for pure iron parts with ferrosilicon powder that may contain from 17 to 33% Si. The ferrosilicon powder is hard and brittle, but because it constitutes only a minor constituent of the powder mixture for a 3% Si alloy, the mixture can be compacted without major difficulties.

In sintering compacts from this mixture, control of sintering atmosphere, which is often dissociated ammonia, is critical. A low dew point is necessary to keep the atmosphere reducing to the ferrosilicon particles. Otherwise, the oxide coatings on these particles remain during sintering and prevent interdiffusion between the iron and ferrosilicon and prevent the formation of a homogeneous iron-silicon solid-solution

Fig. 29 Actuator base (right) pressed and sintered from 3% silicon steel (top) and armature from pure iron powder (bottom) with assembly (left) for use in computer printers
Courtesy of Remington Arms Co., Inc.

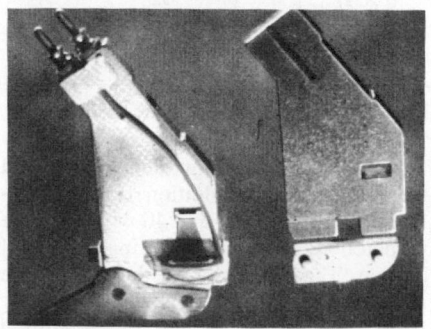

Fig. 30 Pole piece from 50Fe-50Ni prealloyed powder used in aerospace guidance control unit
Courtesy of Supermet, Division of Stanadyne Corp.

alloy. Under suitable reducing conditions, such a homogeneous alloy is formed when compacts are sintered 1 h at 1260 °C (2300 °F). Because of the shift in the equilibrium constant for the oxidation-reduction reaction with increasing temperature, higher dew points are permissible at higher sintering temperatures. At sintering temperatures of 1260 to 1315 °C (2300 to 2400 °F), iron-silicon compacts shrink linearly 2 to 3%, which makes dimensional control during sintering difficult. Sintering in vacuum is a suitable alternative to sintering in dry dissociated ammonia.

Iron-phosphorus alloys are produced by mixing ferrophosphorus powder with pure atomized-iron powder. Alloys with phosphorus contents of more than 0.6% exhibit considerable shrinkage during sintering.

The magnetic properties of iron-silicon and iron-phosphorus alloys differ from those of pure iron (Ref 16, 17). Increasing silicon and phosphorus contents lowers saturation induction slightly. However, maximum permeability is increased, and coercive force is decreased. Typical values of maximum permeability and coercive force for specimens with a density of 7.3 g/cm^3 are μ = 3000 and H_c = 110 A/m (1.4 Oe) for pure iron; μ = 5000, and H_c = 71.6 A/m (0.9 Oe) for a 3% Si alloy; and μ = 6000 and H_c = 103 A/m (1.3 Oe) for a 0.75% P alloy. A beneficial increase in electrical resistivity is also experienced with increasing silicon and phosphorus contents. Typically, increases of from 10 to 20 $\mu\Omega \cdot$ cm for pure iron to 41 to 52 $\mu\Omega \cdot$ cm for a 3% Si-alloy, and 27 to 28 $\mu\Omega \cdot$ cm for a 0.75% P alloy, are common.

These alloys are used when a rapid decay of the magnetic field is required, for example, when the current producing the

magnetizing force is switched off. Rapid decay is due to the higher reluctance and is related to the higher electrical resistivity and lower coercive force of the materials compared to pure iron. A typical application is the actuator base in high-speed printers for computers produced from a 3% Si alloy (Fig. 29).

The 50Fe-50Ni P/M alloys for magnetic applications (Ref 18) are produced from a prealloyed atomized powder. These powders are used for applications requiring high permeability. To obtain the desired high density of 7.5 g/cm^3, powder is compacted at a pressure of 690 MPa (100 ksi) and sintered above 1200 °C (2200 °F) in vacuum. With this processing, a maximum permeability of 40 000 and a coercive force of 16 A/m (0.2 Oe) may be achieved, which are comparable to those of wrought nickel-based magnetic alloys. Figure 30 illustrates a pole piece made from this material.

Wrought Permalloy From Powder

Permalloy materials are a group of soft magnetic nickel-based alloys with very high permeabilities. They generally are produced from cast ingots, but powder metallurgy has been used occasionally for producing a 77Ni-14Fe-4Mo-5Cu alloy from pressed and sintered powder compacts (Ref 19, 20).

Raw materials used to produce these compacts include carbonyl nickel powder, carbonyl iron powders, molybdenum powder produced by hydrogen reduction of pure oxide, and copper powder produced by atomizing oxygen-free high-conductivity copper. These powders are thoroughly mixed and pressed without the addition of a lubricant at pressures between 310 and 520 MPa (45 and 75 ksi) into compacts weighing about 7 kg (15 lb), presintered several hours at 600 °C (1100 °F), and finally sintered at 1200 °C (2200 °F) for several hours.

The sintered compacts are forged and

then rolled into strip, rod, or wire. With P/M techniques, control of compositions of the alloys, which greatly affect magnetic properties, is easily achieved. Additions of small amounts of silicon and manganese, used in the cast alloys for deoxidation purposes, are not necessary in P/M alloys. For this reason, P/M alloys may be annealed in wet hydrogen without impairing initial permeability. Also, P/M alloys do not develop a layer of fine grains on the surface of the bars, which is also detrimental to initial permeability. These tendencies in alloys made from cast ingots are due to internal oxidation of residual silicon and manganese in the alloys.

In addition to Permalloy, a 50Fe-50Ni alloy strip treated to give a cube texture, and therefore a rectangular hysteresis loop, has been produced from sintered compacts from mixtures of carbonyl iron and carbonyl nickel powders (Ref 21). The cube structure is produced by cold rolling 97 to 99% and then annealing. Before the cube structure is produced, the grain size of the coarse-grained sintered compact must be refined by light reductions, with intermediate annealing at 700 to 800 °C (1300 to 1480 °F). In strip produced from powders, development of the cube texture does not depend on the temperature of the final anneal.

Alnico Permanent Magnets

The excellent permanent magnet properties of alloys of iron, nickel, and aluminum—often with additions of cobalt, copper, and titanium—were discovered by Mishima in 1932 (Ref 22). In contrast to the older permanent magnet alloys, which were steels that were magnetically hardened by quenching, the Alnico alloys are precipitation-hardened alloys. These alloys derive their properties through cooling of the alloy, which is isotropic at high temperatures, at a controlled rate with or without subsequent aging. The high coercive force of these alloys is connected with the spinodal precipitation of elongated single-domain particles of a strongly magnetic iron-rich phase (α) in a weakly magnetic or nonmagnetic (aluminum-rich) matrix (α' phase) (Ref 23). The properties of cobalt-rich Alnico alloys can be further improved by cooling alloys through the Curie point in a strong magnetic field.

Alnico magnets are produced by casting, generally in sand molds. If necessary, they are shaped by grinding, as the alloy is too brittle to shape by plastic deformation or by machining. The cast alloy is also mechanically weak. These conditions

Table 7 Compositions and properties of sintered Alnico magnets

Designation	Nominal composition	Condition	Residual induction (B_r) T	G	Coercive force (H_c) kA/m	Oe	Maximum energy product kT·A/m	MG·Oe	Induction at maximum energy product T	G	Demagnetizing force at maximum energy product (H_d) kA/m	Oe	B_D/H_D μT·m A	G Oe
Alnico II	Fe-10Al-19Ni-13Co-3Cu-0.5Ti	Unoriented	0.7100	7 100	43.8	550	11.9	1.5	0.4500	4500	26.7	335	17.0	13.5
Alnico 5	Fe-8Al-14Ni-24Co-3Cu	Oriented	1.0900	10 900	49.3	620	27.8	3.5	0.8000	800	35.0	440	22.6	18
Alnico 6	Fe-8Al-16Ni-24Co-3Cu-1Ti	Oriented	0.9400	9 400	62.9	790	23.9	3.0	0.6000	6000	39.8	500	15.1	12
Alnico 8B	Fe-7Al-13Ni-39Co-3Cu-6Ti	Oriented	0.8300	8 300	131.3	1650	39.8	5.0	0.5000	5000	79.6	1000	6.3	5
Alnico 8H	Fe-7Al-14Ni-37Co-3Cu-8Ti	Oriented	0.7500	7 500	151.2	1900	43.8	5.5	0.4500	4500	97.1	1220	4.6	3.7

limit the manufacture of small, intricately shaped magnets by casting. For this reason, production of small Alnico magnets by powder metallurgy was developed in the 1930's (Ref 24).

Sintered Alnico magnets are produced (Ref 25) from a mixture of powders, all of which pass through a 200-mesh (75-μm) sieve. The mixture consists of about 50% soft elemental powders, iron, nickel, cobalt, and copper. The remaining material is a master alloy powder that contains cobalt, nickel, titanium, and iron in combination with aluminum, produced by grinding of a cast alloy. The master alloy has a melting point below 1200 °C (2200 °F), because a 50 wt% Fe, 50 wt% Al alloy has a liquidus temperature of 1165 °C (2129 °F). The master alloy provides a liquid phase during sintering, but this liquid phase is transient. When the final alloy composition, which is homogeneous at the sintering temperature, has been produced by interdiffusion at the end of the sintering cycle, the compacts are completely solid.

The powder mixture is pressed at a pressure of 1100 MPa (160 ksi) to produce compacts with sufficient green strength. Because the powder mixture does not flow well, attention must be paid to proper filling of the die cavity. Generally, Alnico compacts should have a maximum diameter-to-length ratio of 1 to 1. Most pressed and sintered Alnico magnets have relatively simple shapes.

Alnico magnets are sintered at 1300 °C (2370 °F) in an atmosphere of very dry hydrogen, gettered by a titanium alloy. Sintering Alnico magnets in vacuum has been reported in Germany (Ref 26). Compacts shrink during sintering between 5 and 7% linearly and have a final density of 7.0 to 7.2 g/cm³, depending on configuration.

As in cast Alnico magnets, some grades are produced oriented by cooling the material through the Curie temperature in a magnetic field. Table 7 gives the compositions and magnetic properties of P/M Alnico alloys (Ref 25). Sintered Alnico magnets have certain advantages over cast magnets. Because of their fine grain size, they are stronger than coarse-grained cast

magnets. Also, they do not exhibit the cracking, cold shuts, and segregation of impurities at grain boundaries encountered in cast magnets, and therefore have a more uniform flux distribution.

Sintered magnets can be produced to closer dimensional tolerances than cast magnets. However, because of their slight porosity, the magnetic characteristics of sintered magnets are generally not comparable to those of cast magnets. This does not apply to Alnico 8 alloys, which easily form a nonmagnetic gamma phase and require quenching from a normalizing temperature. The appearance of a non magnetic phase can be more readily avoided in P/M fabrication than in casting. Sintered Alnico 8 grades therefore have magnetic properties equal to those of cast grades. Because of their excellent properties, Alnico 8 grade magnets have become the most widely used sintered Alnico magnets.

Cobalt/Rare Earth Permanent Magnets

The development of cobalt/rare earth permanent magnet materials began in the early 1960's, when Nesbitt and Wernicke (Ref 27-29) established the structure, magnetic moment, and Curie temperature of cobalt/rare earth intermetallic compounds. These compounds have hexagonal crystal structures. Their usefulness for producing permanent magnets is based on their extremely high magnetic crystalline anisotropy in addition to high saturation magnetization and Curie temperature. Among the cobalt/rare earth compounds, the one containing samarium ($SmCo_5$) has the highest anisotropy constant, equal to 17.5×10^7 erg/cm³ at 300 K (Ref 30-33).

The first cobalt/rare earth magnets were produced by bonding crushed powders of the compounds with plastic (Ref 34). Beginning in 1969, a method of producing magnets was developed in which compacts of the intermetallic compound powders were sintered after they had been magnetically aligned and pressed (Ref 35, 36). This method of production is shown schematically in Fig. 31.

Alloys of two compositions, one corresponding to the compound $SmCo_5$ and the other a sinter addition richer in samarium (60 wt% Sm and 40 wt% Co), are prepared by induction melting and chill casting. These alloys are reduced to a fine powder by crushing, pulverizing, and either ball milling under toluene or jet milling with dry nitrogen. During milling, oxidation of the compound should be avoided. The average particle size of the powders should be comparable to the critical diameter for a single domain particle (6 to 8 μm). Additionally, the particle size distribution should be narrow.

The powders with the two compositions are blended by tumbling to give an average composition of 62.6% Co, slightly richer in samarium than the stoichiometric $SmCo_5$ composition. The milled powder is enclosed in a plastic envelope and magnetically aligned in a magnetic field on the order of 4760 A/m (60 kOe) so that the c-axis of the single-crystal particles is parallel to the direction of the applied field.

The powder is then hydrostatically compacted at a pressure of approximately 1400 MPa (200 ksi) to a density of 6.9 g/cm³, which is 80% of the solid density of the compound. The compacts are sintered in purified argon at 1090 to 1150 °C (1995 to 2100 °F). They densify to approximately 7.7 g/cm³, 90% of solid density. Porosity in the sintered compacts is noninterconnected.

During sintering, a liquid phase is formed from the two cobalt-samarium alloys. Densification of the compacts is not due to the heavy alloy liquid-phase sintering mechanism, as the appearance of the liquid phase is only transient. Instead, densification is primarily attributed to sintering within the composition range of the solid $SmCo_5$ phase, but at an off-stoichiometric composition (Ref 37, 38). It is accelerated by the presence of vacant cobalt lattice sites in a composition containing more samarium than stoichiometric $SmCo_5$ phase, together with pores and inclusions of samarium oxide.

Following sintering, a thermal treatment at 900 °C (1650 °F) is applied to the

Fig. 31 Flowchart of the production of cobalt-samarium permanent magnets

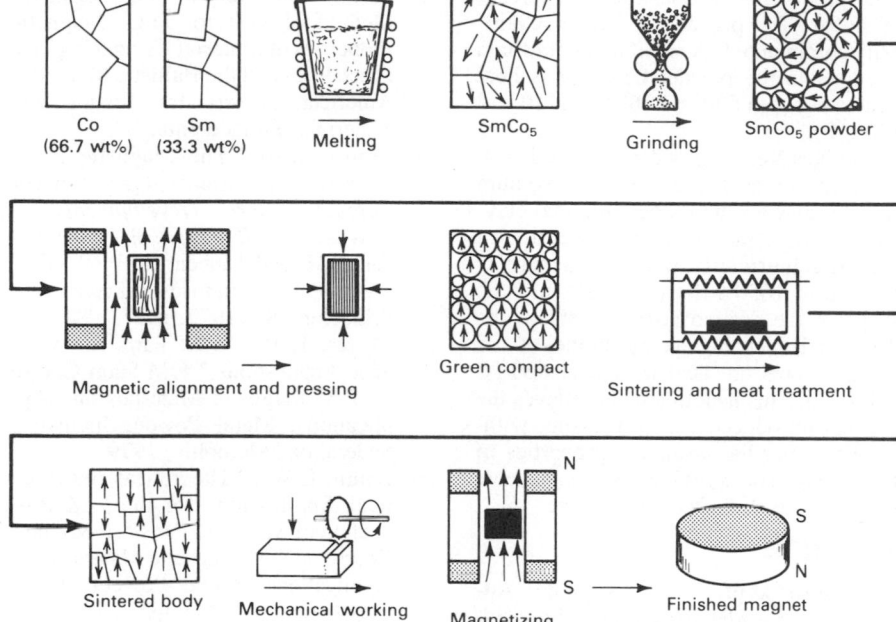

Co (66.7 wt%) Sm (33.3 wt%) Melting SmCo₅ Grinding SmCo₅ powder

Magnetic alignment and pressing Green compact Sintering and heat treatment

Sintered body Mechanical working Magnetizing Finished magnet

Table 8   Magnetic properties of RE-Co₅ permanent magnets

	$SmCo_5$	$Sm_{0.4}Pr_{0.6}Co_5$	$Sm_{0.2}MM_{0.8}Co_5$
Remanence (B_r), T (kG)	1.0 (10.0)	1.63 (16.3)	0.8 (8.0)
Coercive force (H_c), kA/m (kOe)	3024 (38)	1321 (16.6)	1273 (16)
Maximum energy product, kT·A/m (MG·Oe)	605.2 (24.6)	206.9 (26.0)	127.3 (16)

magnets to improve coercive force before they are magnetized in a field of 4780 kA/m (60 kOe). Because of their high remanence and coercivity, magnetizing the magnets and measuring their magnetic properties is difficult. Methods of making the measurements using a superconducting solenoid are described in Ref 39.

Coercivity, remanence, and maximum energy product are extremely high. Table 8 (Ref 40, 41) gives values for commercial permanent magnets of the compositions SmCo₅, samarium-praseodymium-cobalt compound ($Sm_{0.4}Pr_{0.6}Co_5$), and a compound of samarium, Mischmetal, and cobalt ($Sm_{0.2}MM_{0.8}Co_5$). The main reason for replacing some of the samarium with other rare earth metals is to lower cost. Mischmetal is a less expensive cerium-rich rare earth alloy with 50 wt% Ce, 27 wt% La, 16 wt% Nd, and 5 wt% Pr.

Recently, significant research and development effort has been expended on the cobalt/rare earth permanent magnet materials. Besides the RE-Co₅ compositions, those based on the intermetallic compound RE_2Co_7 have been investigated. In spite of the outstanding properties of permanent magnet materials, commercial application has been limited because of the high cost of raw materials.

Elongated Single-Domain Permanent Magnets and Magnetic Core Materials for High Frequencies

Alnico alloys exhibit good permanent magnet properties because their structures consist of magnetically aligned, elongated single-domain precipitated particles in a matrix that is weakly magnetic or nonmagnetic. Production of unsintered permanent magnets from single-domain particles of a magnetic material aligned with, but magnetically insulated from, each other grew out of the advances in the domain theory of magnetization in the early 1940's. The experimental work of Neel (Ref 42) was based on the low-temperature reduction of ferrous formate to which small

amounts of calcium or magnesium formate were added. The resulting very fine powder was then compacted.

These magnets were difficult to produce, were unstable, and did not have adequate magnetic properties for commercial production. Considerable improvement in magnetic properties was achieved with elongated powder particles that had shape anisotropy. These so-called (elongated single-domain) permanent magnets were developed in the late 1950's (Ref 43). The basic steps in the production of these magnets, according to Ref 43, are:

- Highly elongated dendritic particles 10 to 20 nm (100 to 200 Å) in diameter of iron or an iron-cobalt alloy are electrolytically deposited from a sulfate electrolyte into a mercury cathode.
- Dendritic branches of the particles are removed by thermal treatment of the particles in the mercury suspension, and the desired rod-like shape is obtained.
- An alloy layer is formed on the surface of the particles by adding a metal such as tin as a mercury-tin amalgam to the dispersion. This may act as the magnetically insulating layer between particles upon which the quality of the compacted magnet depends.
- Particles are magnetically aligned in a gradually increasing magnetic field, with a maximum value of 318 kA/m (4 kOe), and are compacted to a packing density of approximately 50%.
- All traces of remaining mercury are removed by vacuum distillation.
- The porous bar may be impregnated with an organic binder, which acts as an alternate matrix material in place of a metallic matrix. The binder is at least partially polymerized.
- The mixture of aligned particles and matrix material is ground to a coarse powder.
- Magnets are pressed to final shape after the powder particles have been magnetically realigned.

The magnetic properties of commercial elongated single-domain iron and iron-cobalt permanent magnets are shown in Table 9. Improved properties can be obtained in the laboratory.

While the principal requirement for direct-current applications of soft ferromagnetic materials is a low hysteresis loss, eddy current losses are of primary importance for alternate-current applications. Hysteresis losses increase with the first power of the frequency of the alternating current; eddy current losses increase with the square of the frequency. One method of decreasing eddy current losses is to in-

Table 9 Magnetic properties of permanent magnets from elongated single-domain particles

	Iron powder	Iron-cobalt powder
Remanence (B_r), T (kG)	0.79 (7.9)	0.9 (9)
Coercive force (H_c), kA/m (kOe)	44.6 (0.56)	67.6 (0.85)
aximum energy product, kT · A/m (MG · Oe)	17.5 (2.2)	28.6 (3.6)

crease the electrical resistance of the material. For power frequency applications, high resistance is obtained by dividing the iron-silicon cores of alternate current motors, generators, or transformers into laminations, each separated from the next by a thin layer of electrically insulating material.

There is, however, a practical limit to the reduction of losses by using laminations. For cores in applications with frequencies higher than power frequencies, cores from compressed insulated ferromagnetic powders are widely used. These core materials are described in Ref 44. In many current applications of high-frequency cores, insulated ferromagnetic oxide powders with high intrinsic electrical resistivities have replaced metal powder cores.

Several metal powder cores are still used for some applications, including cores for radiofrequency applications in which the so-called "E" type carbonyl iron powder is used. Particles of this powder are spherical, with diameters ranging from 3 to 9 μm and with onion-like structures. They contain 1% oxygen, 0.6% C, and 0.5% nitrogen and are extremely hard. Particles are coated with a thermosetting phenolic resin; small amounts of an insulating material are added, and the mixture is pressed into the desired shape of the core.

Because of the high intrinsic mechanical and magnetic hardness of the powder, strains introduced by the pressing operation, which are not relieved by subsequent low-temperature polymerizing heat treatment of the core, do not have much effect on the magnetic properties of the cores. Permeability is on the order of 10, which is low compared to materials for direct-current magnetic applications, although this is higher than the permeability of air or vacuum, which is 1.

A second application using insulated alloy powder particles as the core material is the Permalloy powder cores used in loading coils and inductors at audiofre-

quencies (Ref 45, 46). The composition of the alloy is 81% Ni, 17% Fe, and 2% Mo. Powder is produced by embrittling by the addition of a few thousandths of a percent of sulfur. The powder is then mechanically comminuted. A small amount of talc is added to the powder before it is heat treated between 630 and 760 °C (1170 and 1400 °F).

The heat treated powder is coated with an inorganic insulating mixture of sodium silicate, magnesium oxide, colloidal clay, and kaolin. Zinc stearate is added as a pressing lubricant, and the mixture is pressed at 1650 MPa (240 ksi) into the shape of the desired core. To obtain the desired relatively high permeabilities of 190 to 220, cores are heat treated at 630 °C (1170 °F). This heat treatment relieves the stresses introduced during pressing without impairing the insulating properties of the coating. The cores have acceptable core losses in the audiofrequency range.

REFERENCES

1. Schwarzkopf, P.S., Keynote Address, *Powder Metallurgy,* Proceedings of International Conference, June 1960, Interscience, New York, 1961, p 1
2. "Resistance Welding Equipment Standards," Bulletin 16, Resistance Welder Manufacturers Association, Philadelphia, 1976, p 5
3. "Carbon Brushes for Electrical Machines," National Electrical Manufacturers Association, New York, 1954, p 3
4. Onnes, H.K., *Comm. Phys. Lab. Univ. Leyden,* Vol 119 (No. 120), 1911, p 122
5. Matthias, B.T., Geballe, T.H., Geller, S., and Corenzwit, E., *Phys. Rev.,* Vol 95, 1954, p 1435
6. Kunzler, J.E., Buehler, E., Hsu, F.S.L., and Wernick, J.H., *Phys. Rev. Letters,* Vol 6, 1961, p 89
7. Olsen, K.M., Fuchs, E.O., and Jack, R.F., *J. Metals,* Vol 6, 1961, p 724
8. Benz, M.G., *IEEE Trans. Magnetics,* Vol 2 (No. 4), 1966, p 760
9. Martin, D.L., Benz, M.G. Bruch, C.A., and Rossner, C.H., *Cryogenics,* Vol 3, 1963, p 161
10. Suenaga, M. and Sampson, W.B., *Appl. Phys. Letters,* Vol 20 (No. 11), 1972, p 443
11. Hemachalam, K. and Pickus, M.R., *J. Less-Common Metals,* Vol 46 (No. 2), 1976, p 297
12. Bolze, G.A. and Capus, J.M., Factors Affecting Magnetic Properties of Sintered Iron, *Modern Developments in Powder Metallurgy,* Vol 11, Haus-

ner, H.H. and Taubenblat, P.W., Ed., Metal Powder Industries Federation, Princeton, NJ, 1977, p 355-370
13. ASTM standard A 596, "Standard Method of Test for D.C. Magnetic Properties of Material Using Ring Test Procedures and the Ballistic Method," American Society for Testing and Materials, Philadelphia, PA
14. Moyer, K.M., The Magnetic Properties (DC) of Atomized Iron Powder Cores, *Modern Developments in Powder Metallurgy,* Vol 11, Hausner, H.H. and Taubenblat, P.W., Ed., Metal Powder Industries Federation, Princeton, NJ, 1977, p 371-384
15. Moyer, K.H., "P/M Parts for Magnetic Applications," P/M Short Course on P/M Magnetic Materials and Applications, Metal Powder Industries Federation, Memphis, 1979
16. Baum, L.W., "Theoretical and Practical Applications for the P/M Production of Magnetic Parts," P/M Short Course on P/M Magnetic Materials and Applications, Metal Powder Industries Federation, Memphis, 1979
17. Reen, O.W. and Waltenbaugh, J.R., "Properties and Applications of Ferrous Magnetic P/M Materials," P/M Short Course on P/M Magnetic Materials and Applications, Metal Powder Industries Federation, Memphis, 1979
18. Sanderow, H.L., "Soft Magnetic Materials—P/M Applications," P/M Short Course on P/M Magnetic Materials and Applications, Metal Powder Industries Federation, Memphis, 1979
19. Walker, E.V., Worn, D.K., and Walters, R.E.S., "Application of Powder Metallurgy to the Production of High Permeability Magnetic Alloy Strip," Symposium on Powder Metallurgy, Special Report No. 58, Iron and Steel Institute, London, 1956, p 204-208
20. "Wiggin Nickel Alloys by Powder Metallurgy," Henry Wiggin and Co., Birmingham, England
21. Walker, E.V. and Walters, R.E.S., The Production of Grain-Oriented 50-50 Nickel-Iron Magnetic Strip by Cold Rolling From Sintered Compacts, *Powder Metall.,* No. 4, 1959, p 23-31
22. Mishima, T., *Ohm,* Vol 19, 1932, p 353; U.S. Patents 2 027 996 and 2 027 994, 1936
23. Tebble, R. and Craik, D.J., *Magnetic Materials,* John Wiley & Sons, New York, 1969, p 430
24. Howe, C.H., Sintered Alnico, in *Powder Metallurgy,* Wulff, J., Ed.,

American Society for Metals, 1942, p 530-536

25. Hendron, R., personal communication, Arnold Engineering Co., Marengo, IL

26. Oerstit Permanent Magnets, Deutsche Edelstahlwerke A.-G. Magnetfabrik Dortmund

27. Nesbitt, E.A., Williams, H.J., Wernicke, J.H., and Sherwood, R.C., *J. Appl. Phys.*, Vol 32S, 1961, p 342-343

28. Nesbitt, E.A., Williams, H.J., Wernicke, J.H., and Sherwood R.C., *J. Appl. Phys.*, Vol 33, 1962, p 1674-1678

29. Wernicke, J.H. and Geller, S., *Acta Cryst.*, Vol 12, 1959, p 662-665

30. Tatsumoto, E., Okamoto, F., Fujii, H., and Inore, C., *J. Phys.* Vol 32, 1971, p 550-551

31. Kleis, H.P., Menth, A., and Perkins, R.S., *Physica*, Vol 80B, 1975, p 153-163

32. Ermolenko, A.S., *IEEE Trans. Mag.*, Vol MAG12, 1976, p 992-996

33. Sankar, *et al.*, *Phys. Rev.*, Vol 11, 1975, p 435-439

34. Buschow, K.H.J., Luiten, W., Naastepad, P.A., and Westendorp, F.F., *Philips Tech. Rev.*, Vol 29, 1968, p 336

35. Das, D.K., *IEEE Trans. Mag.*, Vol MAG5, 1969, p 214-216

36. Benz, M.G. and Martin, D.L., Cobalt-Samarium Permanent Magnets Prepared by Liquid Phase Sintering, *Appl. Phys. Letters*, Vol 17, 1970, p 176-177

37. Benz, M.G. and Martin, D.L., Mechanism of Sintering in Cobalt-Rare Earth Permanent Magnet Alloys, *J. Appl. Phys.*, Vol 43, 1972, p 3165-3170

38. Gessinger, G.H. and DeLamotte, E., The Sintering Mechanism of Samarium-Cobalt Alloys, *Z. Metallkunde*, Vol 64, 1973, p 771-775

39. Martin, D.L. and Benz, M.G., Measurement of Magnetic Properties of Cobalt-Rare Earth Permanent Magnets, *IEEE Trans. Mag.*, Vol MAG7, 1971, p 285-291

40. Martin, D.L., Geertsen, J.T., Laforce, R.P., and Rockwood, A.C., *Proceedings of the 11th Rare Earth Res. Conf.*, Michigan, 1974, p 342-352

41. Ratman, D.V. and Wells, M.G.H., *AIP Conf. Proc.*, Vol 18, 1974, p 1154-1158

42. Neel, L., *Cahiers Phys.*, Vol 25, 1944, p 1; U.S. Patent No. 2 651 105, 1953

43. Luborsky, F.E., Paine, T.O., and Mendelson, L.T., Permanent Magnets From Elongated Single Domain Particles, *Powder Metall.*, No. 4, 1959, p 57-59

44. Jones, W.D., *Fundamental Principles of Powder Metallurgy*, London, 1960, p 680-710

45. Schumacher, E., Magnetic Powders and Production of Cores for Induction Coils, *Powder Metallurgy*, Wulff, J., Ed., American Society for Metals, 1942, p 166-172

46. Pringel, V.J., Heat Treatment of 2-81 Molybdenum Permalloy Powder Cores, *Western Electric Eng.*, Vol 14 (No. 2), 1970, p 2-10

SELECTED REFERENCES

- Agte, C. and Vacek, J., *Tungsten and Molybdenum*, NASA, Washington, DC, 1963

- *Mallory Resistance Welding Data Book*, P.R. Mallory Co., Indianapolis, 1951

- Del Vecchio, E.J., Ed., *Resistance Welding Manual*, Vol 1 and 2, 3rd ed., Resistance Welder Manufacturers Association, Philadelphia, 1961

- Stanley, W., *Resistance Welding*, McGraw-Hill, New York, 1950

- *Welding Handbook*, Vol 3, 7th ed., American Welding Society, New York, 1980

- Yih, S.W.H. and Wang, C.T., *Tungsten*, Plenum Press, New York, 1979

- Gavaler, J.R., *Appl. Phys. Letters*, Vol 23, 1973, p 480

- Gavaler, J.R., Janocko, M.A., and Jones, C.K., *J. Appl. Phys.*, Vol 45, 1974, p 3009

- Testardi, L.R., Wernick, J.H., and Royer, W.A., *Solid State Comm.*, Vol 15, 1974, p 1

- Wilson, M.N., Walters, C.R., Lewin, J.D., and Smith, P.F., *J. Phys.*, Vol 3, 1970, p 1517

- Larbalestier, D.C., *et al.*, *IEEE Trans.*, Vol 2, 1975, p 247

- Tachikawa, K., *Proc. 3rd Int. Cryogenic Eng. Conf.*, 1970, p 339

- Kaufmann, A.R. and Pickett, J.J., *J. Appl. Phys.*, Vol 42 (No. 1), 1971, p 58; *Bull. Am. Phys. Soc.*, Vol 15, 1970, p 8318

- Suenaga, M. and Sampson, W.B., *Appl. Phys. Letter*, Vol 18 (No. 12), 1971, p 443

- Pickus, M.R., Zackay, V.F., Parker, E.R., and Holthuis, J.T., *Int. J. Powder Metall.*, Vol 9 (No. 1), 1973, p 3

- Pickus, M.R., Hemachalam, K., and Babu, B.N.P., *Mat. Sci. Eng.*, Vol 14, 1974 p 265

- Pickus, M.R., Parker, E.R., and Zackay, V.F., U.S. Patent 3 815 224, 1974

Aerospace Applications

By B. Lynn Ferguson
Partner
Deformation Control Technology

AEROSPACE APPLICATIONS of high-strength P/M materials have grown rapidly over the last 40 years. Powder metallurgy technology produces alloys with improved mechanical properties that can be fabricated cost effectively into near-net shape parts. Near-net shape part production capitalizes on the improved workability inherent in powder products and employs such processing techniques as hot isostatic pressing and isothermal forging. For more information on these processes, see the articles "Hot Isostatic Pressing of Metal Powders" and "P/M Forging" in this Volume. Utilization of near-net shape processes has lowered costs by reducing the input weight of critical raw materials and the amount of secondary machining operations. Powder metallurgy processing also allows closer control of microstructure within a part and permits compositional control of alloys, which is difficult or impossible in cast and wrought products.

Although introduction of P/M parts into aerospace systems has been limited because of costly and lengthy qualification requirements, significant research and development effort is underway. As powder quality, availability, and manufacturing techniques improve, implementation is expected to increase significantly. For jet engine disk applications (Fig. 1), powder superalloy usage is projected to increase to meet the higher mechanical property requirements of future engines.

Powder metallurgy materials used for aerospace applications have followed traditional alloy development. Powder metallurgy superalloys that are currently in volume production for use in jet engine components exhibit high strengths at an acceptable cost. Wrought P/M aluminum is being explored for skin, structural, and engine (compressor section) applications, although actual implementation to date has

been limited to high-strength wrought P/M aluminum grade 7090 landing gear assemblies.

Powder metallurgy titanium applications are also being explored. However, P/M titanium applications are limited to the pressed and sintered and hot isostatically pressed parts described later in this article. Ferrous alloys have been investigated for engine applications, with iron-aluminum alloys receiving most of the attention because of potential savings in weight and material utilization.

Jet Engine Applications of P/M Superalloys

The demands for improved jet engine efficiency and performance (thrust-to-weight ratio) have resulted in continued development of higher strength superalloys. Compositions of typical P/M superalloys are given in Table 1. Concurrent with the increase in strength exhibited by these superalloys is a significant decrease in conventional hot workability. Reduction

Fig. 1 Advancement in disk materials

Note growing role for P/M materials as higher strengths at elevated temperatures are required. Source: Ref 1

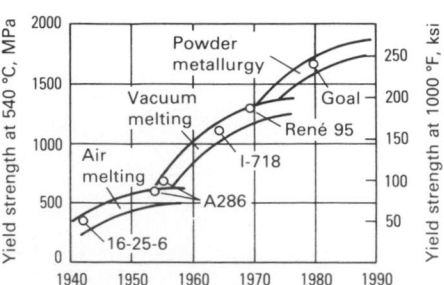

in hot workability adversely affects overall component cost. Powder metallurgy techniques offer a means of producing these highly alloyed compositions with a uniform and homogeneous microstructure that is superplastic and, therefore, suitable for hot working. This has resulted in the replacement of cast and wrought materials by P/M counterparts in many high-performance applications. The following examples describe typical jet engine applications of P/M superalloys.

René 95 Turbine Disks

Several applications for P/M René 95 have been implemented into production engines (Ref 1-3). René 95 was initially developed under Air Force Contract in 1967 as an advanced high-strength cast and wrought alloy. The first components produced were compressor and turbine rotor components for the F-101 engine to be used for fighter/bomber aircraft. This work was followed closely by René 95 turbine disks and cooling plates for T-700 engines used on military helicopters.

The rotor components (compressor disks, seals, high- and low-pressure turbine disks and blade retainers) for the F-101 applications were initially produced by conventional cast and wrought processing, which included forging, ring rolling, and plate cross rolling for thin components. Direct forged ingot and extruded billet were used for these parts. Forge cracking, alloy segregation, carbide clustering, and the inability to completely eliminate the cast ingot structure in the final product were continuing problems, which combined to create cost and production difficulties.

Development work for T-700 engine components was also initiated at this time, using cast and wrought processing with similar results. Extruded small-diameter billet produced from cast ingot exhibited

Table 1 Compositions of several P/M superalloys

Alloy	C	Cr	Mo	W	Ta	Ti	Nb	Co	Al	Hf	Zr	B	Ni	Fe	V	Y_2O_3
IN-100	0.07	12.5	3.2	...	...	4.3	...	18.5	5.0	...	0.04	0.02	rem	...	0.75	...
René 95	0.07	13.0	3.5	3.5	...	2.5	3.5	8.0	3.5	...	0.05	0.01	rem	...	...	...
MERL 76	0.02	12.4	3.2	...	...	4.3	1.4	18.5	5.0	0.4	0.06	0.02	rem	...	...	...
AF 115	0.05	10.5	2.8	6.0	...	3.9	1.7	15.0	3.8	2.0	...	...	rem	...	...	...
PA101	0.1	12.5	...	4.0	4.0	4.0	...	9.0	3.5	1.0	...	...	rem	...	...	...
Low-Carbon Astroloy	0.04	15.0	5.0	...	...	3.5	...	17.0	4.0	...	0.4	0.025	rem	...	...	...
MA 754	0.05	20.0	...	...	...	0.5	...	...	0.3	...	...	...	rem	...	...	0.6
MA 956	...	20.0	...	...	...	0.5	...	...	4.5	...	...	...	...	rem	...	0.5
MA 6000	0.05	15.0	2.0	4.0	2.0	2.5	...	...	4.5	...	0.15	0.1	rem	...	...	1.1
Stellite 31	0.5	25.5	...	7.5	...	...	...	rem	...	...	...	...	10.5	2.0	...	...

excessive carbide stringers that resulted in poor mechanical properties in forgings transverse to the forging flow. This tendency was alleviated by reducing the carbon content. However, producibility difficulties and poor yields due to cracking were never overcome.

Based on the inadequate results with cast and wrought processing of René 95, efforts were initiated to explore P/M processing as a more cost-effective approach to the production of these components. Potential processes included hot isostatically pressed preforms that were hot die forged to final shape and direct hot isostatically pressed disks. Figure 2 compares various processing sequences for fabrication of aircraft engine compressor disks.

Powder metallurgy processing proved to be a viable solution for F-101 and T-700 engine parts. The first P/M process developed for production of the T-700 engine turbine disks utilized a hot isostatically pressed preform for the initial forging stock. The process consisted of producing prealloyed argon-atomized René 95 powders that were sieved to −60 mesh, blended, loaded into containers, and hot isostatically pressed to 100% of theoretical density. Two hot isostatic pressing temperatures were tried (1200 and 1120 °C, or 2200 and 2050 °F) at a hot isostatic pressing pressure of about 100 MPa (15 ksi). The product of the lower hot isostatic pressing temperature was heat treated at 1175 °C (2150 °F) after hot isostatic pressing. A grain size of American Society for Testing and Materials (ASTM) 5 was produced by these conditions. Standard methods for estimating the average grain size of metals can be found in ASTM E 112. Consolidated powder billets were then heated to 1120 °C (2050 °F) and upset forged to a height reduction of 40%. The die temperature was approximately 315 °C (600 °F).

Mechanical properties of cast and wrought products and hot isostatically pressed plus forged P/M products are compared in Fig. 3. Tensile yield and ultimate tensile strength levels of the P/M materials are superior to those of the ingot metallurgy material. Creep properties are similar for the two materials over the temperature range shown. In addition to producing material with improved microstructural homogeneity and improved mechanical properties, hot isostatic pressing plus forging proved to be more economical than conventional cast or wrought practices. Material utilization also was higher, and fewer forging steps were required.

Although the hot isostatic pressing plus forging process yielded suitable mechanical properties, increased material costs prompted development of a less expensive, reliable as-hot isostatically pressed P/M production process for T-700 engine applications. The as-hot isostatically pressed production process was established after 3 years of extensive development and testing. This method was considerably less expensive than the original cast and wrought process, as it used 40% less input material. It was also less expensive than the hot isostatic pressing plus forging process, because fewer processing steps were involved. Currently, the successful production of as-hot isostatically pressed T-700 engine components continues. High-pressure turbine blade retainers

Fig. 2 Processing sequences for the production of F-101 compressor disks

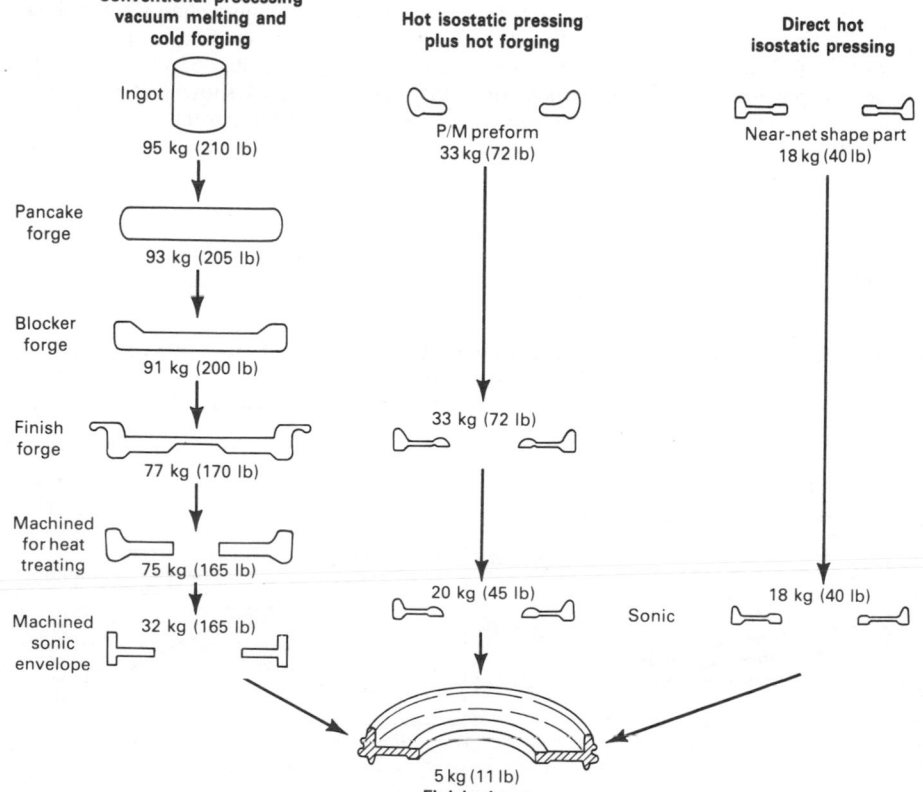

Fig. 3 Comparison of mechanical properties of T-700 engine hardware produced from René 95 by conventional cast and wrought processing and by hot isostatic pressing plus forging of P/M materials

(a) Ultimate tensile stength. (b) 0.2% yield strength. (c) 0.2% creep strength. *P*, Larson-Miller parameter; *T*, temperature (°F); *t*, time (h). ○ indicates T-700 requirement. Source: Ref 1

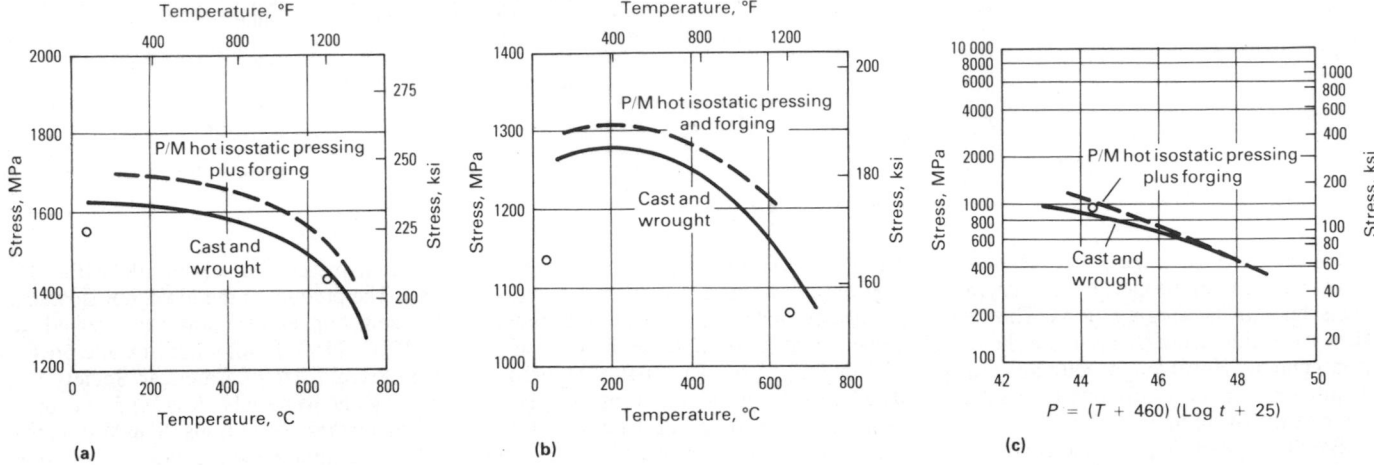

(a)   (b)   (c)

for F-101 engines also are produced by direct hot isostatic pressing. These components are large diameter rings approximately 12.7 mm ($^1/_2$ in.) thick, and 63.5 cm (25 in.) outside diameter by 55.8 cm (22 in.) inside diameter. All conventional means of production, such as cast and wrought or P/M techniques incorporating forging, cannot compete economically with direct hot isostatic pressing. Direct hot isostatic pressing to near-net shape offers at least a 50% reduction in cost and material utilization for this application.

Other applications for P/M René 95 include high-pressure turbine disks and forward outer seals for F-101 engines. These large components are produced from P/M billet that is consolidated by hot extrusion. The extruded billet is then cut and isothermally forged to near-net shape disks. A combination of extrusion and isothermal

forging enables a high degree of thermomechanical working of the material to produce a highly refined structure for these applications that is competitively priced with direct hot isostatic pressing. This process has also resulted in increased capabilities of ultrasonic inspection of these components.

Direct Hot Isostatic Pressing. Manufacturing processes used to produce as-hot isostatically pressed parts from argon-atomized René 95 powder generally consist of the following steps. A shaped container is prepared from low-carbon or stainless steel sheet that is carefully cleaned and filled with René 95 powder. The container is then vacuum outgassed and sealed to maintain a vacuum in the container. Hot isostatic pressing of the powder-filled container to full density is then accomplished. Heat treatment, ultrasonic inspec-

tion of the near-net shape part, and machining to final shape complete the manufacturing sequence.

First- and second-stage high-pressure turbine disks and cooling plates for T-700 engines, as well as F-101 engine high-pressure turbine blade retainers, are produced in this manner. The typical weight of an as-hot isostatically pressed T-700 engine disk is about 9 kg (20 lb). The shape used for ultrasonic inspection weighs about 6 kg (13.5 lb), and the finished part weighs about 3.2 kg (7.0 lb). Mechanical property data in Fig. 4 show that parts produced by direct hot isostatic pressing followed by heat treatment, inspection, and machining result in property levels in excess of T-700 design requirements. See the article "Hot Isostatic Pressing of Metal Powders" in this Volume for detailed information on the fundamentals of this

Fig. 4 Mechanical property comparisons between hot isostatically pressed plus forged hardware and direct hot isostatically pressed hardware of René 95 for T-700 application

(a) Ultimate tensile strength. (b) 0.2% yield strength. (c) 0.2% creep strength. *P*, Larson-Miller parameter; *T*, temperature (°F); *t*, time (h). ○ indicates T-700 requirement. Source: Ref 1

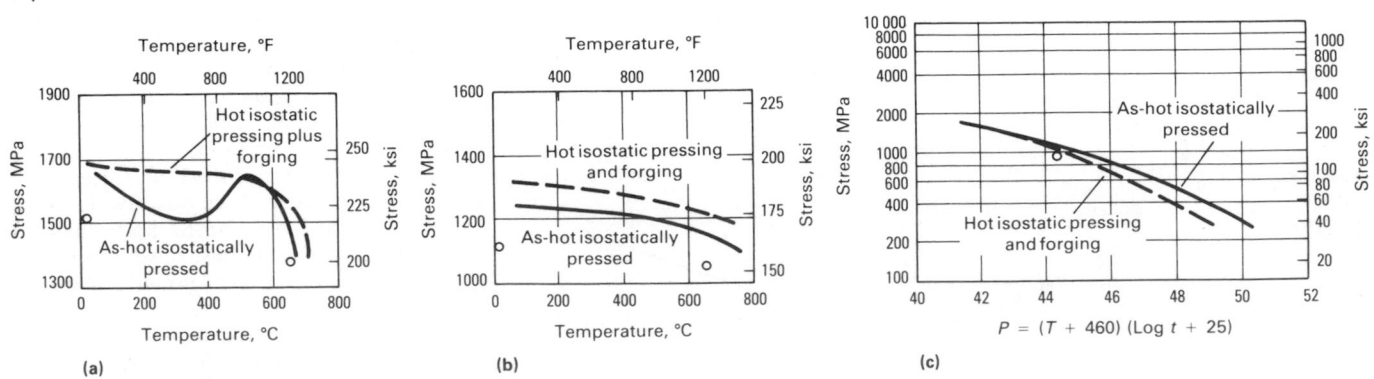

(a)   (b)   (c)

P/M consolidation method and additional information on hot isostatic pressing of nickel-based superalloys.

Extrusion Plus Isothermal Forging. To further improve fatigue properties and inspectability of René 95 components, thermomechanical processing techniques have been developed and used to produce large engine hardware. An effective technique of manufacturing thermomechanically processed hardware is to produce billets by extrusion, followed by isothermally forging the billet into component shape.

Manufacturing processes used to produce P/M René 95 extruded billet from argon-atomized powder typically consist of the following steps. A cylindrical container is prepared from low-carbon steel sheet and is carefully cleaned. The container is filled with René 95 powder, vacuum outgassed, and sealed. The evacuated container is then hot compacted to about 95% density. The compacted powder can is hot extruded into a billet, with a typical area reduction ranging from 5 to 7:1. The steel container is removed, and the billet is carefully inspected. It is then sectioned into forging multiples (mults) of the proper weight for isothermal forging.

Isothermal forging is accomplished in vacuum or inert atmosphere using TZM molybdenum dies that are preheated to the same temperature as forging mults (isothermal forging). Forging of the mults to near-net shape is accomplished by using a controlled strain rate that capitalizes the superplastic state of the powder product. Heat treating, cleaning, inspection, and machining of the forging to final shape completes the processing. A typical large disk forging weighs about 150 kg (330 lb), and finish disk weight is about 63.5 kg (140 lb).

Quality control testing for these critical René 95 P/M parts typically includes qualification of each batch of argon-atomized powder, ultrasonic inspection of billets and sonic shapes, and destructive testing of the first part and subsequent parts at preselected intervals. Mechanical testing consists of tensile, stress rupture, creep, low-cycle fatigue, and crack growth rate testing. See the article "P/M Forging" and the Appendix to the article "Hot Pressing" for detailed information on these P/M consolidation processes.

IN-100 Turbine Disks for F-100 Engines

Using isothermal forging techniques for producing near-net shapes, components such as disks, seals, and spacers for F-100 engines have been produced from P/M su-

peralloy stock (Ref 4, 5). Comparison of material requirements of conventional and isothermal forging highlights the benefit of near-net shape production. Table 2 gives weight comparisons for a variety of F-100 engine parts produced by conventional forging and isothermal forging. For the engine, a total savings of 106 kg (233 lb) of IN-100 alloy results by isothermal forging (see Table 1 for alloy composition). Figure 5 compares a conventionally forged disk, an isothermally forged disk, and a finish machined disk. The configuration of the isothermally forged disk is closer to that of the finished shape than the configuration of the conventionally forged disk. As indicated in Fig. 5, a weight savings of 15.4 kg (34 lb) results by isothermally forging the first-stage turbine disk, as opposed to conventionally press forging the disk.

Comparisons of mechanical property data of isothermally forged material and spec-

ification requirements are given in Table 3. Isothermal forging of billets prepared from P/M alloys results in improved creep, stress rupture, and low-cycle fatigue resistance properties. Additional data on properties of nickel-based and other P/M alloys can be found in the article "Mechanical Properties of P/M Materials" in this Volume.

Astroloy Turbine Disks in JT8D-17R Turbofan Engines (Ref 6)

A project of the NASA Materials for Advanced Turbine Engine (MATE) program resulted in the application of P/M low-carbon Astroloy (see Table 1 for alloy composition) for the high-pressure turbine disk in the JT8D-17R turbofan engine. Conventionally forged Waspaloy was replaced by this hot isostatically pressed and heat treated P/M product after successful

Table 2 Part weights of F-100 engine components

Part	Current forging weight kg	lb	Wrought P/M near-net shape forging weight kg	lb	Net weight savings kg	lb
First-stage turbine disk	53	117	37.6	83	15.4	34
Second-stage turbine disk	50.3	111	30.8	68	19.5	43
Third-stage turbine disk	34.9	77	24.9	55	10.0	22
Fourth-stage turbine disk	32.2	71	18.6	41	13.6	30
1-2 turbine rim spacer	47.6	105	29.9	66	17.7	39
Thirteenth-stage cone seal	29.5	65	(a)	(a)	29.5	65
Total savings					105.7	233

(a) Wrought P/M near-net shape process uses the center slug material from the 1-2 turbine rim spacer forging as the preform.
Source: Ref 4

Fig. 5 Weight savings realized by use of near-net shape forging capability (isothermal or hot die) versus conventional forging for first-stage turbine disk of F-100 engine
Source: Ref 4

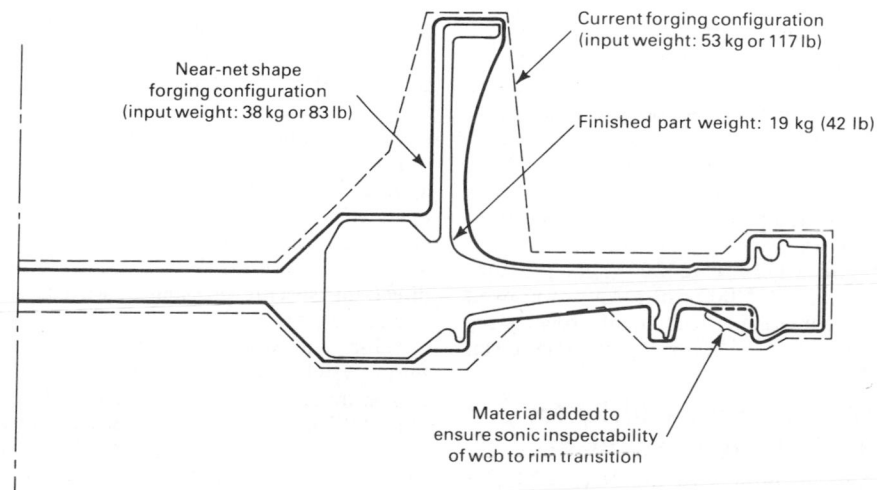

Table 3 Mechanical properties of wrought P/M IN-100

Material	Yield strength MPa	ksi	Tensile strength MPa	ksi	Elongation, %	Reduction in area, %
At room temperature						
P/M IN-100	1161	168.4	1620	235.0	22.1	26.1
PWA 1074	1124	163.0	1586	230.0	22.0	25.5
At 705 °C (1300 °F)						
P/M IN-100	1083	157.0	1241	180.0	18.8	23.0
PWA 1074	1089	158.0	1272	184.5	18.5	21.0

Material	Stress MPa	ksi	Hours to 0.1% strain	Hours to 0.2% strain
Average creep properties at 705 °C (1300 °F)				
P/M IN-100	552	80	184.8	245.6
PWA 1074	552	80	140.3	178.0

Material	Stress MPa	ksi	Hours to rupture	Elongation, %
Stress rupture properties at 730 °C (1350 °F)				
P/M IN-100	655	95	44.2	8.8
PWA 1074	655	95	34.4	...

Material	Average cyclic life	Material	Average cyclic life
Low-cycle fatigue (1% total strain)		**Bolt hole low-cycle fatigue(a)**	
At 540 °C (1000 °F)		**At 540 °C (1000 °F)**	
P/M IN-100	6 514	P/M IN-100	51 195
PWA 1074	5 912	PWA 1074	11 500
At 650 °C (1200 °F)		**At 650 °C (1200 °F)**	
P/M IN-100	9 159	P/M IN-100	17 870
PWA 1074	3 803	PWA 1074	7 500

Note: PWA 1074 is a Pratt & Whitney Aircraft Specification. (a) Maximum stress: 758.5 MPa (110 ksi); minimum stress: 37.9 MPa (5.5 ksi) at a rate of 10 cycles/min

Table 4 Mechanical-property requirements for hot isostatically pressed Astroloy P/M disks

At room temperature

Yield strength, MPa (ksi) 826 (120)
Ultimate strength, MPa (ksi) 1241 (180)
Elongation, % 15
Reduction in area, % 11

At 540 °C (1000 °F)

Yield strength, MPa (ksi) 758 (110)
Ultimate strength, MPa (ksi) 1103 (160)
Elongation, % 15
Reduction in area, % 18

At 730 °C (1350 °F)

Stress rupture (a), MPa (ksi) .... 552 (80 ksi)

At 705 °C (1300 °F)

Creep limit (b), MPa (ksi) .... 510 (74 ksi)

(a) Rupture occurred in 23 h, with 8% elongation. (b) 0.1% creep in 100 h
Source: Ref 6

engine testing. Material savings of 55 kg (121 lb) of superalloy and a cost savings of 20% were project goals. Near-net shape capability of the direct hot isostatic pressing process resulted in reproducing a machining envelope of 2.5 mm (0.1 in.).

Process steps included containerization of gas-atomized low-carbon Astroloy powder in shaped containers, hot isostatic pressing at 1215 °C (2220 °F) at a pressure of 105 MPa (15 000 psi) for 3 h, followed by heat treating. Mechanical property requirements of P/M Astroloy disks are given in Table 4.

MERL 76 Turbine Disks for Turbofan Engines (Ref 6)

Another project of the NASA-MATE program involved application of P/M MERL 76 for turbine disks in a high by-pass ratio turbofan engine (see Table 1 for alloy composition). The goals of this project included increased material utilization, reduced production costs, and an increase in disk rim temperature capability of 20 °C (30 °F). This increase in disk rim temperature allows an increase in engine operating temperature, which subsequently increases the fuel efficiency of the engine. A 272-kg (600-lb) disk was produced by direct hot isostatic pressing of MERL 76 powder to near-net shape.

Subsequent work, sponsored by the Air Force Wright Aeronautical Laboratories, demonstrated that the input weight of another MERL 76 turbine disk for the JT9D engine could be reduced from 360 kg (800 lb) to less than 270 kg (600 lb) using the ceramic mold process (Ref 7). A detailed discussion of the ceramic mold process is

contained in the article "P/M Lightweight Metals" in this Volume.

René 95 Parts in F-404 Turbofan Engines

Successful application of René 95 P/M superalloy parts in the T-700 engine prompted the use of such parts in other engines (Ref 8-10). The F-404 turbofan engine, produced for the U.S. Navy F/A-18 Hornet strike fighter and the Canadian CF-18 strike fighter, was selected for further implementation of powder metallurgy. Hot isostatically pressed René 95 P/M parts were installed in high-temperature sections of the engine. The shaft and disk of the back section of the compressor were made of hot isostatically pressed René 95. The turbine nozzle vanes were made of Inconel MA 754, an oxide-dispersion-strengthened alloy produced in powder form by mechanical alloying. René 95 P/M superalloy parts produced as near-net shapes by direct hot isostatic pressing were cost effective due to improved material usage and significant reduction in machining requirements.

Inconel MA754 Turbine Vanes

Inconel MA754 is a nickel-based superalloy produced in bar form from mechanically alloyed powder (see Table 1 for composition). The uniform dispersion of fine oxides required to produce the desired mechanical properties in this alloy can only be achieved with P/M techniques. Used as the dispersoid, yttria and the other powders used to produce Inconel MA754 are containerized and hot extruded to full density. The extruded bar is then hot rolled to final mill shape. Forging also may be used as a part-shaping method to produce this alloy. The term "hot" is used loosely, because recrystallization does not occur during rolling. Following rolling (or forging), the bar stock (or forging) is recrystallized by zone annealing to produce a crystallographically oriented microstructure. The large elongated grain structure that results from this heat treatment maximizes longitudinal elevated-temperature properties.

Inconel MA754 is used for high- and low-pressure turbine vanes in selected engines. This alloy is used in these applications instead of cast superalloys because of its excellent high-temperature creep and rupture properties, high melting point, and excellent environmental resistance. Detailed information on the processing, properties, and applications of oxide-strengthened P/M alloys can be found in the article "Dispersion-Strengthened Materials" in this Volume.

Stellite 31
Turbine Blade Dampers
in TF30-P100 Engines

Winner of a Metal Powder Industries Federation (MPIF) Award of Achievement in the 1982 P/M Part of the Year Design Competition, a P/M turbine blade damper made of Stellite 31 is being used in the TF30-P100 engine (Ref 11, 12). This engine powers the USAF F-111F airplane. Compacts were cold pressed in rigid tooling on a 89-kN (10-ton) hydraulic press and vacuum sintered to high density. The tensile properties of the part are:

Yield strength, MPa (ksi)	450 (65)
Tensile strength, MPa (ksi)	620 (90)
Total elongation, %	3

Source: Ref 11

A total of 68 of these 6-g (0.21-oz) parts are used in the first-stage rotor assembly of the jet engine. Powder processing proved to be more economical than precision casting, as minimal grinding was required to achieve the final dimensional tolerances. The near-net shape capability of P/M processing resulted in a significant cost advantage for this three-level part.

Gas Turbine
Combustor Liners

To improve jet engine efficiency, increased engine operating temperatures are required. For the combustor, this entailed switching from Hastelloy X (Ref 13). Oxide-dispersion-strengthened alloys prepared from mechanically alloyed powder offer a 170 °C (300 °F) temperature improvement through increased creep strength and cyclic oxidation resistance. Incoloy MA 956 (Fe-20Cr-4.5Al-0.5Y_2O_3) and Haynes Development Alloy 8077 (Ni-16Cr-4Al-1.3Y_2O_3) were selected for testing by NASA as part of the MATE program. Based on mechanical property results, Incoloy MA 956 was selected for combustor liner trials; both the new material and the new liner design were evaluated. Although not in production at this time, this application typifies the potential of P/M superalloys in turbine engines.

Inconel MA6000E for Small
Aircraft Gas Turbine Blades

The TFE731 small aircraft gas turbine engine was originally designed to use solid IN-100 cast turbine blades with an equiaxed grain structure. Substitution for this initial blade specification progressed from internally cooled IN-100 cast blades with an equiaxed grain structure to solid directionally solidified cast blades of MAR-M247. Part of the NASA-MATE program is focused on the use of solid turbine blades of Inconel MA6000E alloy (Ref 14). This alloy is a nickel-based superalloy with oxide-dispersion strengthening that is produced from mechanically alloyed powder by hot extrusion and hot rolling into bars. Zone annealing produces a directionally recrystallized microstructure with large elongated grains. Mechanical property testing indicates the capacity to increase operating temperatures by 80 °C (150 °F) at equivalent stress rupture levels of 138 MPa (20 ksi) for Inconel MA6000E, as compared to the directionally solidified MAR-M247 cast alloy.

Radial Wafer Turbine Blades

Wafer technology for producing internally cooled turbine blades and applications for rapidly solidified powder products are currently under development (Ref 15-17). Wafer technology involves etching patterns onto wafers, assembling the wafers systematically into a stack, and bonding the wafers such that the etched patterns form cooling passages. Bonding may be achieved by press diffusion bonding or gas pressure bonding. In its simplest form, two halves of a turbine blade are prepared, etched, and then press diffusion bonded together. Powder metallurgy techniques play an important part in this process, because rapid solidification technology can be used in conjunction with the enhanced cooling capability of wafer technology.

Prealloyed powder is produced by a suitable process, such as centrifugal atom-

Fig. 6 Schematic of turbine blade production by radial wafer technique
Source: Ref 17

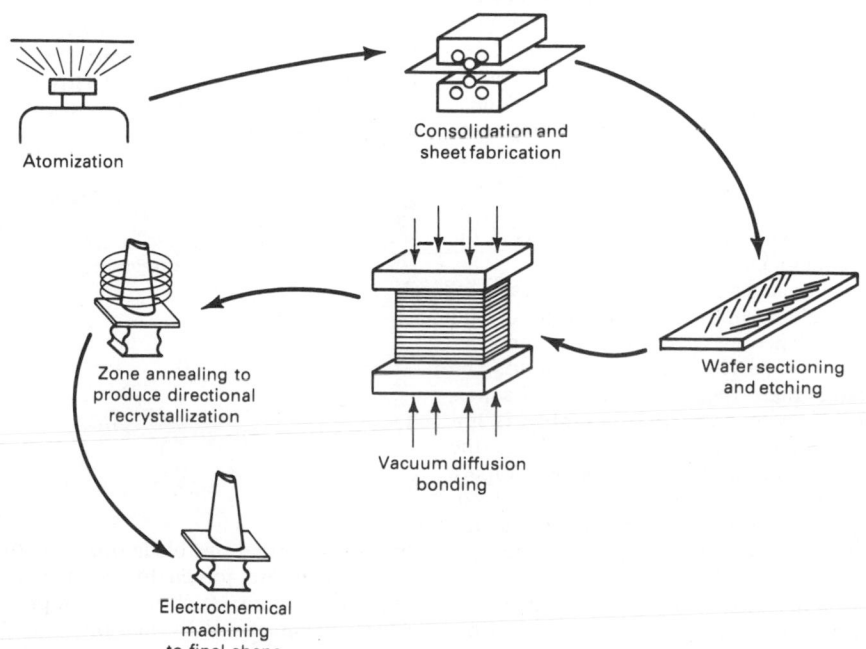

Atomization

Consolidation and sheet fabrication

Wafer sectioning and etching

Zone annealing to produce directional recrystallization

Vacuum diffusion bonding

Electrochemical machining to final shape

izization or inert gas atomization. Fully dense wafer stock must then be produced from this powder. Typically, powder is vacuum hot pressed to high density and then rolled, forged, or extruded. Deformation not only produces the proper wafer size, but it also guarantees that the stock is 100% of theoretical density. Wafer stock is sectioned and etched to produce the correct cooling passage pattern. Wafers are then stacked and diffusion bonded in vacuum to produce a monolithic shape with integral cooling passages. Next, the material is zone annealed to produce a directionally recrystallized microstructure consisting of large elongated grains. The final shape of the turbine blade relies on advanced machining processes, such as electrochemical machining. A schematic of turbine blade production by the radial wafer technique is shown in Fig. 6.

Mechanical property data indicate that a rapidly solidified alloy, RSP 185 (Ni-14.4Mo-6.7Al-6.1W-0.04C), offers an improvement in creep resistance of 83 °C (150 °F) over directionally solidified MAR-M200 plus hafnium, a precision cast alloy. Figure 7 illustrates this improvement in creep resistance. For the fundamentals of rapid solidification technology, see the article "Atomization" in this Volume.

Dual-Alloy/Dual-Property Engine Components

Powder metallurgy technology is currently under development to produce dual-alloy, dual-property turbine disks. By use of this technology, the high tensile and low-cycle fatigue property requirements of the bore can be attained with a fine-grained P/M product. Creep and stress rupture requirements of the rim and/or blades can be attained by special processing techniques, such as single-crystal casting, directionally solidified casting, fiber reinforcement, or mechanical alloying. The ceramic mold process is being developed to produce dual-property turbine wheels for small engines (Ref 18).

One concept uses a blade ring made by tack welding individual cast single-crystal blades. The ring is then sandwiched between two halves of a ceramic mold. The mold is then filled with prealloyed superalloy powder, and the assembly is hot isostatically pressed to consolidate the powder and bond it to the blades. Figure 8 shows turbine wheels made by this technique. Figure 9 shows the microstructure of the bond line between the powder and cast materials.

Another dual-property wheel technique involves the use of a preconsolidated hot isostatically pressed P/M hub and a direc-

Fig. 7 Creep comparison of rapidly solidified alloy RSP 185 with MAR-M200 plus hafnium
Source: Ref 15

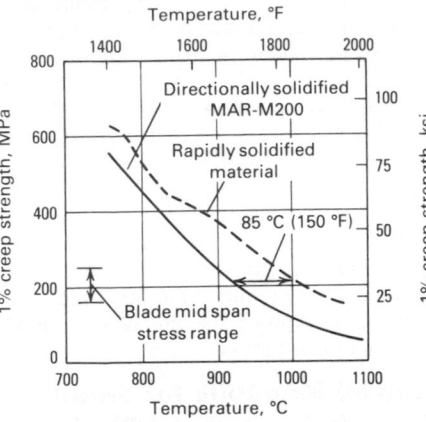

Fig. 8 Dual-property turbine wheel made from cast blades and a P/M hub
Source: Ref 18

Fig. 9 Microstructure at hot isostatically pressed bond line between cast alloy C103 (top) and P/M alloy PA-101 (bottom)
Source: Ref 18

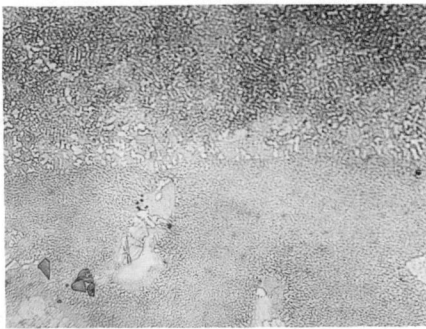

tionally solidified cast blade ring. The two components are joined by hot isostatic pressing. Figure 10 shows a completed wheel manufactured by this process. As shown in this figure, the coarse-grained

Fig. 10 Dual-property turbine wheel made by bonding (hot isostatically pressing) a fine-grained hub to a coarse-grained directionally solidified cast blade ring
Courtesy of Detroit Diesel Allison

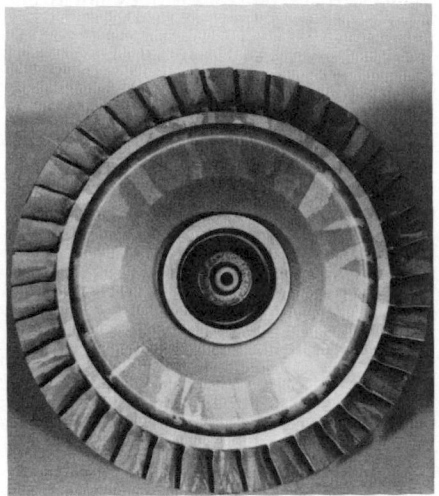

directionally solidified rim and the fine-grained P/M hub are easily visible. For discussion of other P/M dual-alloy techniques that have successfully demonstrated metal container shape technology, see Ref 19.

Aluminum P/M Applications

With the emergence of wrought P/M aluminum alloys that exhibit high strength, high corrosion resistance, and improved elevated-temperature properties compared to conventional aluminum alloys, aluminum powder metallurgy is experiencing emerging use in the aerospace industry (Ref 20-25). Experimental, or example, parts are prevalent that have resulted largely from research and development programs. Chemical compositions of several of these experimental alloys are given in Table 5. These compositions cannot be processed by conventional ingot metallurgy techniques. They require rapid solidification technology to produce metastable phases that result in higher strengthening. They also rely on the development of fine precipitates and dispersoids that cannot be achieved by ingot metallurgy techniques.

Structural members and engine components made of these alloys currently are being evaluated. One such application is a P/M alloy 7090 forging used in the main landing gear support structure of the Boeing 757 airplane. A photograph of the main

Table 5 Compositions of wrought aluminum P/M alloys

Alloy	Cu	Mg	Ce	Cr	Fe	Mn	Zn	Ni	Co	Li	C	O	Al
7090	1.0	2.5	···	···	···	···	8.0	···	1.5	···	···	0.35	rem
7091	1.5	2.5	···	···	···	···	6.5	···	0.4	···	···	0.35	rem
Al-Al$_4$C$_3$	···	···	···	···	···	···	···	···	···	···	3.0	···	rem
Al-Cu-C	1.0	···	···	···	···	···	···	···	···	···	1.5	···	rem
Al-Mg-C	···	2.0	···	···	···	···	···	···	···	···	1.5	···	rem
IN-9021	4.0	1.5	···	···	···	···	···	···	···	···	1.2	0.75	rem
IN-9051	···	4.0	···	···	···	···	···	···	···	···	0.7	0.6	rem
IN-9052	···	4.0	···	···	···	···	···	···	···	···	1.0	0.75	rem
Al-Fe-Co	···	···	···	···	3-8	···	···	···	2-7	···	···	···	rem
Al-Fe-Ce	···	···	3.5	···	8.0	···	···	···	···	···	···	···	rem
Al-Li-Cu	2-3.5	0.5	···	···	···	0.6	···	···	···	2.5-3.2	···	···	rem
Al-Mn-Co	···	···	···	···	···	3-8	···	···	1.5-6.5	···	···	···	rem
Al-Mn-Ni	···	···	···	···	···	3-8	···	1.6-7	···	···	···	···	rem
Al-Fe-Cr	···	···	···	1.5	8.5	···	···	···	···	···	···	···	rem
Al-Ni-Co	···	···	···	···	···	···	···	2-5	2-5	···	···	···	rem

Table 6 Typical tensile properties of die forgings of P/M alloys 7090-T7E71 and 7091-T7E69

Thickness: ≤76.2 mm (3.0 in.)

Alloy/test direction	Tensile strength MPa	ksi	Yield strength MPa	ksi	Elongation (4D), %
7090-T7E71					
Longitudinal	614	89	579	84	10
Transverse	579	84	545	79	4
7091-T7E69					
Longitudinal	579	84	531	77	13
Transverse	545	79	496	72	9

Fig. 11 Fuselage braces of Ti-6Al-6V-2Sn alloy P/M part produced by direct hot isostatic pressing using the ceramic mold process

Source: Ref 27

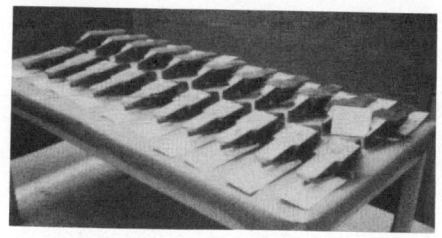

Fig. 12 Candidate engine parts of P/M titanium alloys made by direct hot isostatic pressing using the ceramic mold process

(a) Compressor rotor. (b) Compressor spool. (c) Fan disk. Source: Ref 28

 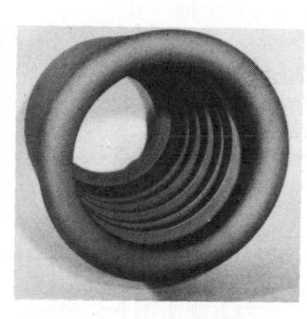 

(a) (b) (c)

landing gear link and the landing gear door actuator can be found in Fig. 15 of the article "P/M Lightweight Metals" in this Volume. Typical tensile properties of P/M aluminum alloy die forgings are given in Table 6.

Substitution of aluminum for titanium alloy members is an effective method of reducing overall weight and improving fuel efficiency in airplanes. Alloying additions of lithium, cobalt, zinc, iron, nickel, and molybdenum are under examination for improved stiffness-to-density ratios, as well as improved strength.

Engine applications under investigation include impellers in small turbine engines and compressor components. Both of these applications, which require improvements in elevated-temperature strength, are aimed at replacing titanium alloys with aluminum. Additional information on the properties and applications of high-strength, wrought P/M aluminum alloys can be found in the article "P/M Lightweight Metals" in this Volume.

Titanium P/M Applications

Many aerospace components produced from titanium alloys are forged to a rough shape and then machined to a final configuration. This often results in only 16% of the starting stock being used in the final part. Various P/M titanium processes are being evaluated to improve the efficiency of titanium part manufacture and material utilization.

One major approach to titanium powder metallurgy involves the use of spherical, prealloyed powder produced by the plasma rotating electrode process. In this process, high-quality alloy bar stock is rotated rapidly as the end of the bar is melted by a plasma arc. The rotating motion spins the powder off from the molten pool. This powder is ideally suited to compaction by hot isostatic pressing.

Hot isostatic pressing involves loading the powder into shaped metal or ceramic molds, evacuating and sealing the molds, and then hot isostatically pressing to 100% of theoretical density. Although there are no hot isostatically pressed near-net shape parts in current production, implementation of several parts is anticipated in the near future (Ref 26-28). Possible applications include structural aircraft components, such as the fuselage braces shown in Fig. 11, and high-performance engine components, such as those shown in Fig. 12.

Table 7 contains comparisons by weight of these P/M parts and parts produced by conventional forging. Improved material utilization by direct hot isostatic pressing to near-net shape is clearly demonstrated.

Extensive property evaluations have been conducted on test blocks and parts produced from plasma rotating electrode powder and hot isostatic pressing. Tables 8 and 9 give room-temperature tensile and fracture toughness data for P/M Ti-6Al-4V and P/M Ti-6Al-6V-2Sn, respectively. Both alloys exhibit strength, ductility, and

Table 7 Trial P/M titanium alloy airframe and engine parts and weight comparisons

Part	Forging billet kg	Forging billet lb	Hot isostatically pressed kg	Hot isostatically pressed lb	Final part kg	Final part lb
Boeing 747 walking beam	25	55	14	30	9.5	21
General Dynamics F-16 pivot shaft	67	148	24	53	14.5	32
General Electric TF34 compressor spool	66	147	29	65	6.8	15
McDonnell Aircraft F-15 dropout link	52	115	25.5	56	6.4	14
McDonnell Aircraft F-15 keel splice	3	6	0.4	0.8	0.2	0.4
Northrop F-18 arrestor hook	82	181	25	55	13	28
Pratt & Whitney Aircraft F-100 fan disk	54	120	29	65	12	27
Williams International F-107 compressor rotor	14.5	32	2.8	6.2	1.6	3.6

smooth and notched fatigue strength of P/M Ti-6Al-4V and wrought Ti-6Al-4V. In both instances, the P/M product falls within the property range of the wrought product.

Blended elemental alloy production offers another approach to titanium powder metallurgy, which involves elemental titanium and alloy powders. Titanium sponge fines and elemental and/or master alloy powders are blended, cold pressed to a green density of 88 to 91% of theoretical, and then vacuum sintered to a density of approximately 99% of theoretical. Advantages of this approach include:

- Cold compactability, which allows conventional P/M pressing techniques to be used

toughness comparable to cast and wrought material.

The hot isostatically pressed P/M products also exhibit excellent fatigue properties. Figures 13 and 14 compare the

Table 8 Room-temperature tensile and fracture toughness properties of hot isostatically pressed Ti-6Al-4V plasma rotating electrode processed powder

Material is in the vacuum mill annealed condition

Property	P/M material(a)	Specification minimum
Tensile strength, MPa (ksi)	951 (138)	896 (130)
0.2% offset yield strength, MPa (ksi)	896 (131)	827 (120)
Elongation, %	15	10
Reduction in area, %	39	20
Fracture toughness (K_{Ic}), MPa $\sqrt{m}$ (ksi $\sqrt{in.}$)	83 (76)	55 (50)

(a) Data developed on six test blocks representing two lots of material. Source: Ref 29

Table 9 Room-temperature tensile and fracture toughness properties of hot isostatically pressed Ti-6Al-6V-2Sn plasma rotating electrode processed powder

Material is vacuum annealed at 760 °C (1400 °F)

Property	P/M material
Tensile strength, MPa (ksi)	1055 (153)
Yield strength, MPa (ksi)	1006 (146)
Elongation, %	18
Reduction in area, %	36
Fracture toughness (K_{Ic}), MPa $\sqrt{m}$ (ksi $\sqrt{in.}$)	63 (57)

Source: Ref 30

Fig. 13 Room-temperature fatigue results for load-controlled smooth specimens made of 6.4-mm ($^1/_4$-in.) diam P/M engine mount supports and witness blocks

Both parts made using hot isostatically pressed Ti-6Al-4V powder. R = 0.1. K_T = 1.0. Source: Ref 31

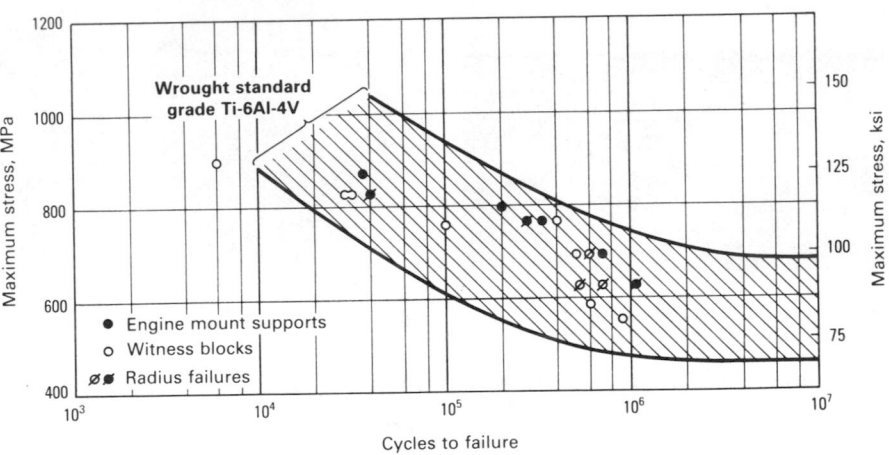

Fig. 14 Room-temperature fatigue results for notched specimens from P/M engine mount supports made of Ti-6Al-4V

R = 0.1. K_T = 3.0. Source: Ref 31

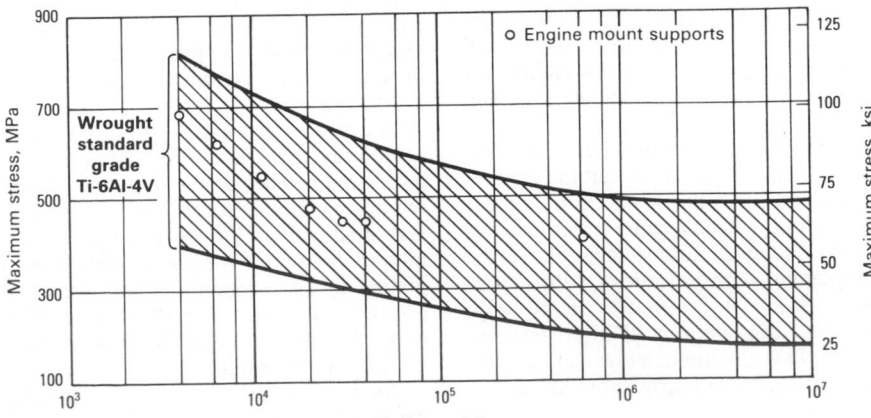

Fig. 15 Typical tensile properties of blended elemental P/M alloys
Shaded areas represent observed ranges. Source: Ref 31

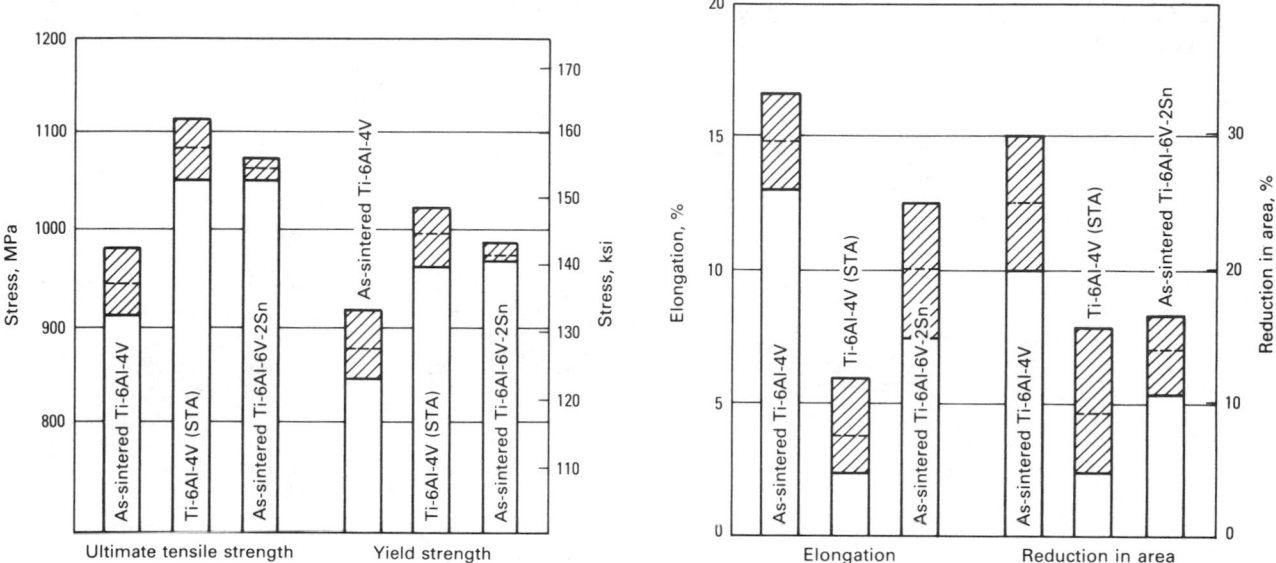

Fig. 16 Comparison of room-temperature smooth axial fatigue behavior of Ti-6Al-4V blended elemental and prealloyed P/M compacts with wrought annealed material

$R = 0.1$. Source: Ref 35

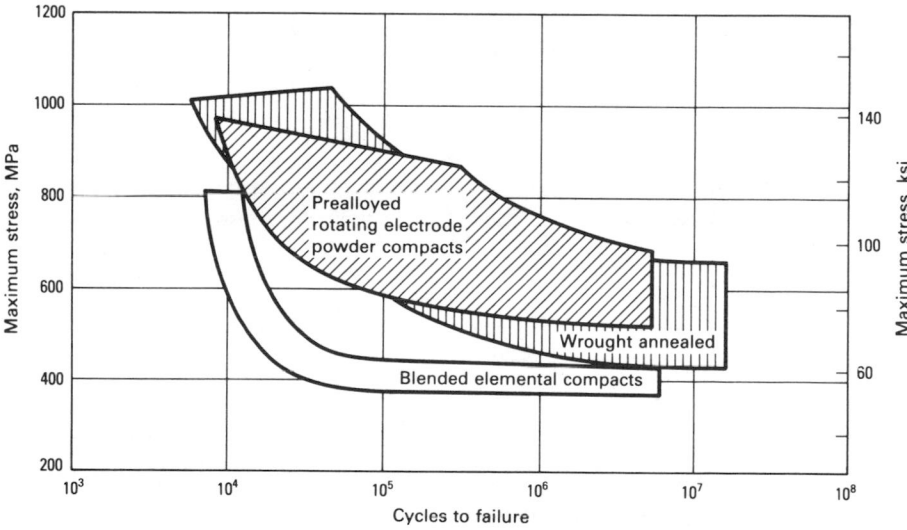

- Low powder costs (sponge fines are a by-product of the conventional titanium production process)
- Low compaction pressures due to use of elemental titanium, which has lower strength than titanium alloys

The disadvantages of the process include the presence of residual chloride from the sponge fines, which affects particle bonding and properties, and residual porosity, which is inherent to the press and sinter process. Using sponge fines of commercially pure titanium and an aluminum/vanadium-rich master alloy powder, a compressor stator link for the F-100 engine was produced by press and sinter technology (Ref 32-34). Material utilization for P/M processing of the link was 95%, and a projected cost saving of 71% of the wrought link was realized. The only finish machining required was drilling of four transverse holes. A photograph of this part is shown in Fig. 24 of the article "P/M Lightweight Metals" in this Volume.

Figure 15 illustrates tensile properties of several titanium alloys produced from blended elemental powders compacted by cold pressing and sintering. Strength levels of the pressed and sintered materials are comparable to wrought materials, but ductility is generally lower. Figure 16 compares the fatigue properties of blended elemental pressed and sintered compacts with hot isostatically pressed, prealloyed powder.

Another source of powder is the scrap from machining, stamping, or trimming of titanium (Ref 36). Alloy scrap is hydrided, which results in a brittle material at room temperature. This hydrided scrap can then be crushed and milled into powder form. Hydrogen is subsequently removed by heating the powder or compacted powder in a vacuum. Powders produced by the hydride technique can be consolidated by hot isostatic pressing or by vacuum hot pressing.

Detailed information on titanium P/M technology can be found in the articles "Production of Titanium Powder," "Ordnance Applications," and "P/M Lightweight Metals" in this Volume.

REFERENCES

1. Bartos, J.L., P/M Superalloys for Military Gas Turbine Applications, in *Powder Metallurgy in Defense Technology*, Vol 5, Metal Powder Industries Federation, Princeton, NJ, 1980, p 81-113
2. Rizzo, F.J., Lane, J., and Moll, J.H., Production of Near-Net Shape Superalloy Hardware, *Metal Powder Rep.*, Vol 38 (No. 1), 1983, p 7-13
3. Moll, J.H., Petersen, V.C., and Du-

lis, E.J., Powder Metallurgy Parts for Aerospace Applications, in *Powder Metallurgy: Applications, Advantages, and Limitations*, Klar, E., Ed., American Society for Metals, 1983, p 247-298

4. Walker, B.H. and Fuss, L.N., "Manufacturing Methods for Near-Net Shape Disks by Isothermal Forging," in Tri-Service Metals Manufacturing Technology Program Status Review —Addendum, Daytona Beach, 1979, p A-1 to A-20

5. Allen, M.M., "Iso-Forging of Powder Metallurgy Superalloys for Advanced Turbine Engine Applications," *Advanced Fabrication Techniques in Powder Metallurgy and Their Economic Implications*, AGARD-CP-200, 1976, p 5.1-5.15

6. Dreshfield, R.L. and Miner, R.V., Jr., "Application of Superalloy Powder Metallurgy for Aircraft Engines," NASA Technical Memorandum 81466, June 1980

7. Malley, D.R., Stulga, J.E., and Ondercin, R.J., Production of Near-Net Shapes by Hot Isostatic Pressing of Superalloy Powder, *Progress in Powder Metallurgy 1982*, Vol 38, American Powder Manufacturers Institute, Princeton, NJ, 1983

8. Hurst, A., Powder Metal in New Engine, *Metalworking News*, Sept 1, 1980

9. IncoMAP, Inco's New ODS Superalloys for the Eighties, *Metal Powder Rep.*, Vol 38 (No. 1), 1983, p 32-35

10. Thompson, F.A., "Forging of Mechanically Alloyed Materials," Frontiers of High Temperature Materials Conference, sponsored by INCO, Ltd., May 1981

11. Johnson, P.K., P/M Replaces Components Made by Investment Casting, Machining and Stamping, *Int. J. Powder Metall. Powder Technol.*, Vol 18 (No. 4), 1982, p 355-359

12. Church, F.L., Industrial, Consumer Products Dominate P/M Design Awards, *Modern Metals*, May 1982, p 11-20

13. Henricks, R.J., "Evaluation of Oxide Dispersion Strengthened Sheet Alloys for Aircraft Gas Turbine Combustor Applications," Frontiers of High Temperature Materials Conference, sponsored by INCO, Ltd., May 1981

14. Hoppin, G.S., III and Schweizer, F.A., "MA6000E For Small Aircraft Gas Turbine Blades," Frontiers of High Temperature Materials Conference, sponsored by INCO, Ltd., May 1981

15. George, D.G., Brown, B.T., and Cox, A.R., "The Application of Rapid Solidification Rate Superalloys to Radial Wafer Turbine Blades," AIAA Paper 79-1226, June 1979

16. Kear, B.H. and Thompson, E.R., Aircraft Gas Turbine Materials and Processes, *Science*, Vol 208, 1980, p 847-856

17. Anderson, R.E., Cox, A.R., Tillman, T.D., and Van Reuth, E.C., "Use of RSR Alloys for High Performance Turbine Airfoils," 2nd International Conference on Rapid Solidification Processing, Reston, VA, March 1980

18. Moll, J.H., Schwertz, J.H., and Chandhok, V.K., P/M Dual-Property Wheels for Small Engines, *Progress in Powder Metallurgy*, Vol 37, Metal Powder Industries Federation, Princeton, NJ, 1982

19. Kortovich, C.S. and Marder, J.M., "Development of Materials and Process Technology for Dual-Alloy Disks," NASA CR-165224, Oct 1981

20. Graham, R.H., Wrought Aluminum P/M Alloys, in *Powder Metallurgy: Applications, Advantages and Limitations*, Klar, E., Ed., American Society for Metals, 1983, p 235-246

21. Jangg, G. and Huppmann, W., The Production of Dispersion Strengthened Aluminum Parts by Powder Forging, in *Modern Developments in Powder Metallurgy*, Vol 13, Hausner, H.H., Antes, H.W., and Smith, G.D., Ed., Metal Powder Industries Federation, Princeton, NJ, 1981, p 419-428

22. Lewis, R.E., "Development of Advanced Aluminum Alloys from Rapidly Solidified Powders for Aerospace Structural Applications," Interim Report on DARPA Contract F33615-78-C-5203, March 1979

23. Adam, C.M., Bourdeau, R.G., and Broch, J.W., "Application of Rapidly Solidified Alloys," Quarterly Report on DARPA Contract F33615-76-C-5136, June 1979

24. Pickens, J.R., Aluminum Powder Metallurgy Technology for High-Strength Applications, *J. Mat. Sci.*, Vol 16, 1981, p 1437-1457

25. Benjamin, J.S. and Schelleng, R.D., Dispersion Strengthened Aluminum Alloy IN-9051 by Mechanical Alloying, *Met. Trans. A*, Vol 124, 1981, p 827

26. Moll, J.H., HIPing the High-Performance Alloys, *Mech. Eng.*, Nov 1981, p 56-61

27. Moll, J.H., Petersen, V.C., and Dulis, E.J., P/M Parts for Aerospace Applications, in *Powder Metallurgy: Applications, Advantages and Limitations*, Klar, E., Ed., American Society for Metals, 1983, p 247-298

28. Petersen, V.C., Chandhok, V.K., and Kelto, C.A., Hot Isostatic Pressing of Large Titanium Shapes, *Powder Metallurgy of Titanium Alloys*, AIME, 1980

29. Bruce, J.S. and Witt, R.H., "Progress on Hot Isostatic Pressing," 28th National SAMPE Symposium, Anaheim, CA, April 1983

30. Eloff, P.C. *et al.*, Properties of P/M Titanium Alloys Produced from Elemental Blends: The Current Status of Development, *Progress in Powder Metallurgy 1981*, Metal Powder Industries Federation, Princeton, NJ, 1982

31. Bruce, J.S., Sheinker, A.A., Bohlen, J.W., and Chanani, G.R., "Evaluation of Ti P/M Technology for Naval Aircraft Components, " 28th National SAMPE Symposium, Anaheim, CA, April 1983

32. Miller, J.A. and Brodi, G., "Consolidation of Blended Elemental Ti-6Al-4V Powder to Near-Net Shape," Tri Service Metals Manufacturing Technology Program Status Review, Addendum, Daytona Beach, Sept 1979, p B1-B13

33. Andersen, P.J., Alber, N.E., and Thellmann, E.L., Aerospace Applications for Pressed and Sintered P/M Titanium Alloys, in *Powder Metallurgy in Defense Technology*, Vol 5, Metal Powder Industries Federation, Princeton, NJ, 1980, p 23-32

34. Andersen, P.J. *et al.*, Fracture Behavior of Blended Elemental P/M Titanium Alloys, *Modern Developments in Powder Metallurgy*, Vol 13, Hausner, H.H., Antes, H.W., and Smith, G.D., Ed., Metal Powder Industries Federation, Princeton, NJ, 1981, p 537-549

35. Krishnamurthy, S., Vogt, R.G., Eylon, D., and Froes, F.H., "Developments in Titanium Powder Metallurgy," 1983 Annual Powder Metallurgy Conference, New Orleans, sponsored by MPIF and APMI, May 1983

36. Peebles, R.E., "Advanced Manufacturing Methods for High Quality, Low Cost Titanium Powder Production," General Electric Co. Aircraft Engine Business Group, AFWAL-TR-82-4113, Air Force Contract F-33615-76-C-5421, Nov 1982

Medical and Dental Applications

By Phillip J. Andersen
Manager, Metallurgical Research
Zimmer, Inc.

USE OF METAL POWDERS IN MEDICINE AND DENTISTRY began in 1896, when G.V. Black discovered several silver-tin dental amalgams. Currently, more than 160 million amalgam restorations are placed each year. Powder variables such as composition, shape, and size have a pronounced effect on the properties of these dental restorations.

Fig. 1 Total hip and knee replacements

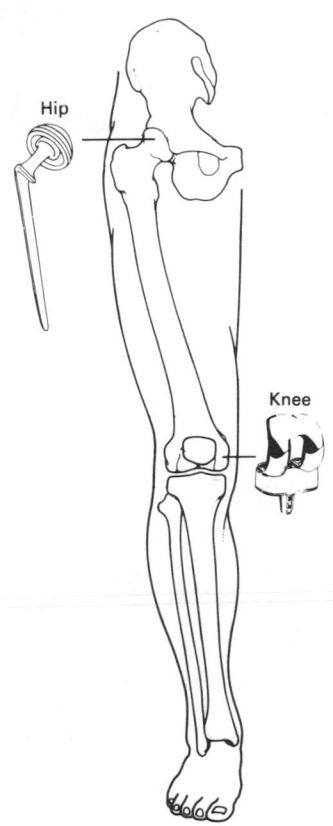

Powder metallurgy techniques are used to make orthopedic implants, as well. Fully dense implants are made by hot isostatic pressing of prealloyed powders to provide materials with excellent mechanical properties. Sintering techniques are used to bond large powder particles to the surfaces of other implants to provide a porous surface into which bone can grow, thus affixing the implant.

The orthopedic implants described in this article are used to replace diseased joints within the human body. Joint replacement may be necessary in patients with diseases such as rheumatoid arthritis, which may cause extreme pain and lack of mobility. Figure 1 shows hip and knee joints, which constitute the most common total joint replacements. Artificial finger, shoulder, elbow, and ankle joints also are used.

Because implant failure may necessitate reoperation, long-term reliability of these devices is paramount. Materials used must exhibit good mechanical properties, corrosion resistance, compatibility with the body, and superior friction and wear characteristics. Mechanical properties of interest include ultimate tensile and yield strengths, ductility, modulus of elasticity, and fatigue strength. Because these devices may be exposed to millions of load cycles, fatigue strength is especially important.

Long-term corrosion resistance is required, because the internal human environment is an oxygenated saline solution that contains proteins. Also, the implant material should not cause any adverse tissue reactions; consequently, materials inducing a topic response cannot be used. Friction and wear characteristics of articulating joints are of prime importance. Most implants have one surface made of ultra-high-molecular-weight polyethylene and the other surface made of highly polished metal. Suitable metallic materials that are available commercially include stainless steels such as American Society for Testing and Materials (ASTM) F 138 type 316, cobalt-chromium alloys (ASTM F 75 and F 799), and titanium-based alloys such as extra-low interstitial grade Ti-6Al-4V (ASTM F 136). Compositions of these materials are listed in Table 1.

Porous Orthopedic Implants

The major implants produced by hot isostatic pressing are total hip replacements made from a cobalt-chromium-molybdenum alloy that meets the composition requirements of ASTM F 799. Titanium-based alloy implants also may be made by similar techniques in the future. Use of P/M technology provides grain size refinement, improved material homogeneity, and near-net shape capability, which leads to high-quality, cost-effective implants.

Initially, cobalt-based implants were made only by lost wax investment techniques. Casting of these alloys results in extremely coarse grains (Fig. 2). The carbides present in as-cast materials also are coarse, and shrinkage porosity that is undetectable by x-ray inspection may be present. A representative cast microstructure is shown in Fig. 3.

Hot isostatic pressing of prealloyed powders produces a much finer microstructure. A typical cobalt-chromium-molybdenum structure produced by hot isostatic pressing is shown in Fig. 4. Because each powder particle is, in effect, a small ingot of the alloy being atomized,

Table 1 Orthopedic implant alloys

Element	Minimum	Maximum
Investment cast cobalt-chromium-molybdenum alloy (ASTM F 75)		
Chromium	27.0	30.0
Molybdenum	5.0	7.0
Nickel	···	1.0
Silicon	···	1.0
Manganese	···	1.0
Iron	···	0.75
Carbon	···	0.35
Cobalt		rem
Wrought extra-low interstitial Ti-6Al-4V (ASTM F 136)		
Aluminum	5.5	6.50
Vanadium	3.5	4.5
Iron	···	0.25
Oxygen	···	0.13
Carbon	···	0.08
Nitrogen	···	0.05
Hydrogen	···	0.0125
Titanium		rem
Wrought 316L stainless steel (ASTM F 138)		
Chromium	17.00	19.00
Nickel	12.00	14.00
Molybdenum	2.00	3.00
Manganese	···	2.00
Silicon	···	0.75
Copper	···	0.50
Nitrogen	···	0.10
Carbon	···	0.030
Phosphorus	···	0.025
Sulfur	···	0.010
Iron		rem
Thermomechanically processed cobalt-chromium-molybdenum alloy (ASTM F 799)		
Chromium	26.0	30.0
Molybdenum	5	7
Iron	···	1.5
Nickel	···	1.0
Silicon	···	1.0
Manganese	···	1.0
Carbon	···	0.35
Nitrogen	···	0.25
Cobalt		rem

the structural features of the material are of a much smaller scale. Carbides are extremely fine; porosity is eliminated; and grain size is greatly reduced. The influence of these changes on mechanical properties is discussed later in this article.

Powder metallurgy processes have the ability to produce near-net shapes. Because the cobalt-based alloys used in orthopedics are difficult to machine, near-net shape capability is thus desirable.

Powders. Requirements for powders used in implant production include compositional control, consistent tap densities (to ensure consistency of final part dimensions when working with fixed-mold cav-

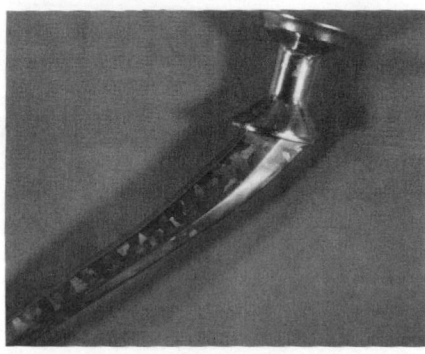

Fig. 2 Macroetched investment cast cobalt-chromium alloy hip stem exhibiting large grain size

Fig. 3 Microstructure of investment cast cobalt-chromium
The structure consists of a cobalt-chromium-molybdenum solid solution with large carbides. Macrograin size is ASTM 7.5.

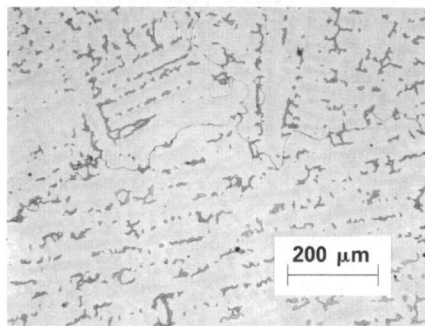

ity dimensions), and a high degree of powder cleanliness.

Currently, all P/M processed orthopedic implants have been made with powders produced by inert gas atomization (see the article "Atomization" in this Volume for more information on this process). Powder suppliers observe the same precautions and use the same handling procedures employed for aerospace applications. Typical inert-gas-atomized cobalt-chromium-molybdenum powders are shown in Fig. 5. Particles tend to be spherical in shape; smaller satellite particles attached to large particles are evident.

An alternate powder processing method is the plasma rotating electrode process (also described in the article "Atomization" in this Volume). Powder particles produced by this technique tend to be more spherical and smoother than inert-gas-atomized powder particles (Fig. 6). Use of powders produced by the plasma rotating electrode process for fabrication of actual implants has been considered. However, powder cost will ultimately determine the extent of this usage.

Fig. 4 Microstructure of a cobalt-chromium-molybdenum alloy produced by hot isostatically pressing prealloyed cobalt-chromium-molybdenum powder
Note the fine carbides and significant reduction in grain size as compared to Fig. 3. Macrograin size is ASTM 12-14

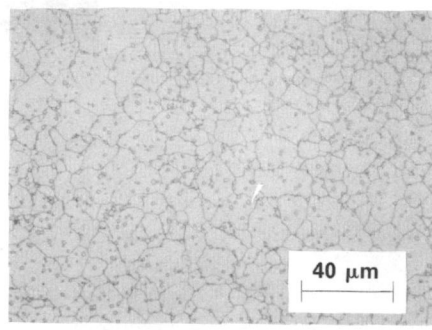

Fig. 5 Scanning electron micrograph of inert-gas-atomized powder used to produce dense implants
Magnification: 180×

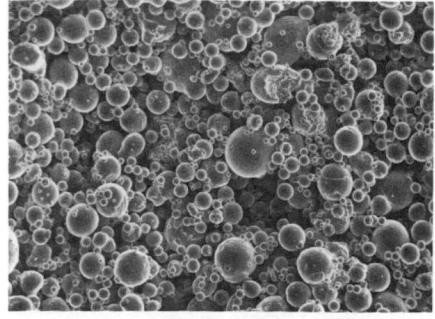

Fig. 6 Scanning electron micrograph of cobalt-chromium-molybdenum alloy powder made by the plasma rotating electrode process
Magnification: 20×

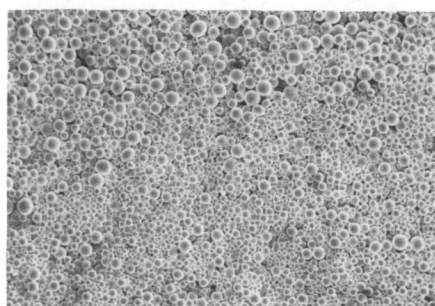

Production Processes. To date, all dense P/M hip stems have been made by hot isostatic pressing. Rapid omnidirectional compaction has been considered as an alternative to hot isostatic pressing. Rapid omnidirectional compaction uses

Fig. 7 Flowchart of the rapid omnidirectional compaction process

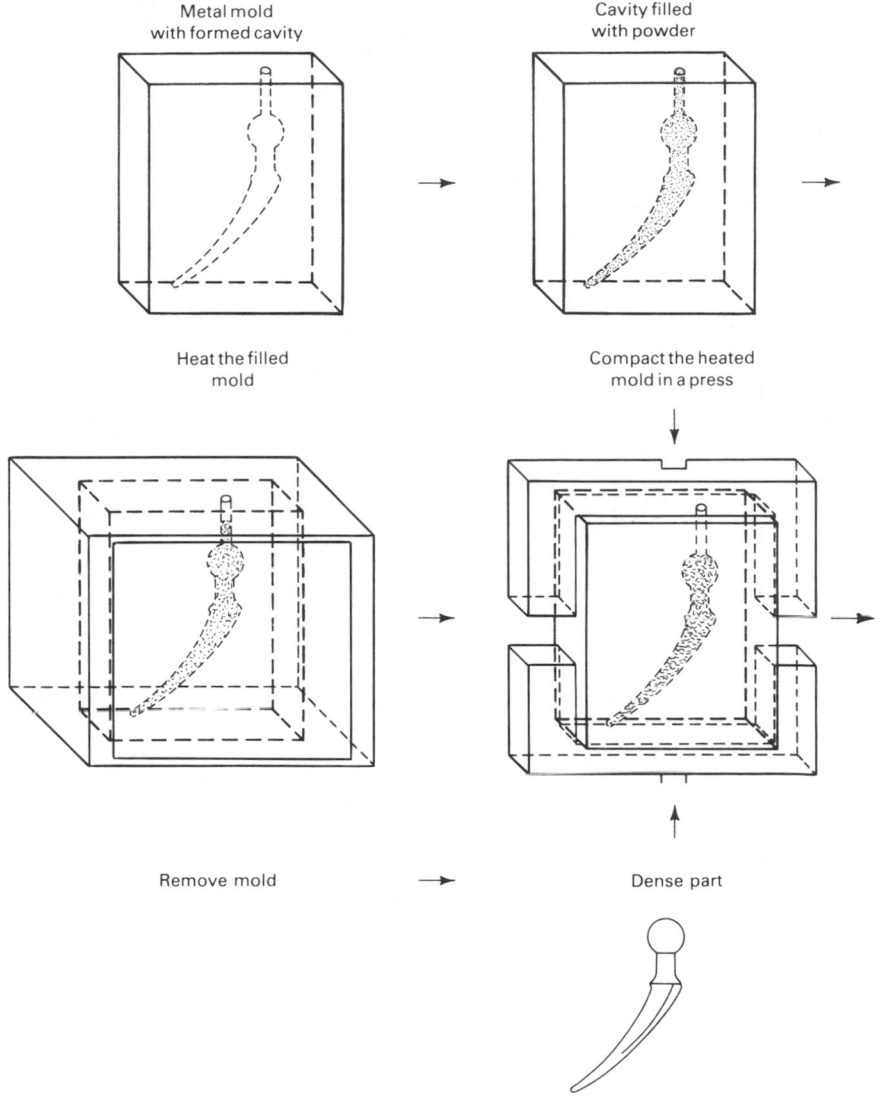

Metal mold with formed cavity

Cavity filled with powder

Heat the filled mold

Compact the heated mold in a press

Remove mold → Dense part

Fig. 8 Microstructure of cobalt-chromium-molybdenum alloy obtained by rapid omnidirectional compaction

As with hot isostatically pressed cobalt-chromium-molybdenum, the material exhibits a fine grain size and small carbides. Magnification: 250×

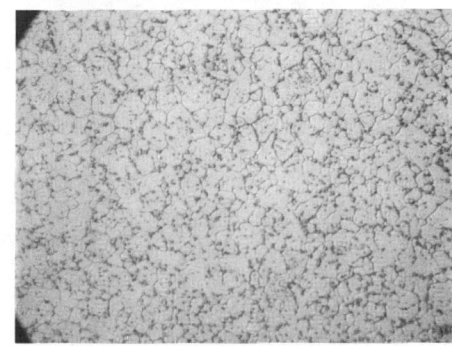

thick metal molds (which may be ferrous or nonferrous) that contain a preformed cavity. The molds are filled, sealed, heated, and compressed in a conventional press. Metal molds act in a quasi-isostatic fashion to compact the powder. This process is attractive to manufacturers with suitable press capacity that do not have equipment for hot isostatic pressing. Figure 7 is a flowchart of the process.

Initial applications of this technology to cobalt-based alloys have been successful. Typical microstructures (Fig. 8) obtained with rapid omnidirectional compaction are similar to those of materials produced by hot isostatic pressing. Consequently, mechanical properties of materials produced by rapid omnidirectional compaction are similar to those of the hot isostatically pressed material.

The process used for production of implants closely parallels hot isostatic pressing of nickel-based superalloys for aerospace applications (see the articles "Hot Isostatic Pressing of Metal Powders" and "Aerospace Applications" in this Volume for more information). Inert-gas-atomized powder, stored under inert gas, is fed into metal cans, which are then evacuated and sealed. Canned powder is processed by hot isostatic pressing. The cans are removed, and the preforms are forged to final size. Typical processing parameters are 1100 °C (2012 °F) and 100 MPa (14.5 ksi) for 1 h.

Mechanical Properties. Cobalt-based alloys produced by P/M techniques have much finer microstructures and greatly improved mechanical properties than cast cobalt-based alloys of identical composition. Figure 9 compares tensile properties

of ASTM F 75 cobalt-chromium-molybdenum alloy produced by casting, hot isostatic pressing, and rapid omnidirectional compaction. All static properties are significantly higher in P/M parts.

Figure 10 shows rotating beam fatigue properties of the same materials. The endurance strength of P/M parts is more than twice that of cast material. Fatigue strength is extremely important in medical devices because of the need for long component life; the enhanced fatigue properties of P/M materials provide significant advantages. For additional information on property data, see the article "Mechanical Properties of P/M Materials" in this Volume.

Corrosion Resistance. Corrosion behavior of ASTM F 75 cobalt-based alloys is determined primarily by matrix composition. Thus, the different microstructures exhibit similar corrosion resistance. Potentiostatic corrosion curves of cast and P/M cobalt-chromium-molybdenum alloys exhibit similar rest potential, passivation behavior, and pitting potential. The cobalt-chromium-molybdenum composition has demonstrated good corrosion resistance in the human body, regardless of whether it is investment cast or produced by P/M techniques. Reference 1 discusses corrosion resistance in some detail.

Porous Coatings for Implant Fixation by Bony Ingrowth

Traditionally, implants have been affixed by using polymethyl methacrylate as a cement. The surgical technique required to insert an implant involves preparing sized cavities within the bone, placing polymethyl methacrylate in a viscous, partially

Fig. 9 Tensile properties of cobalt-chromium-molybdenum alloy

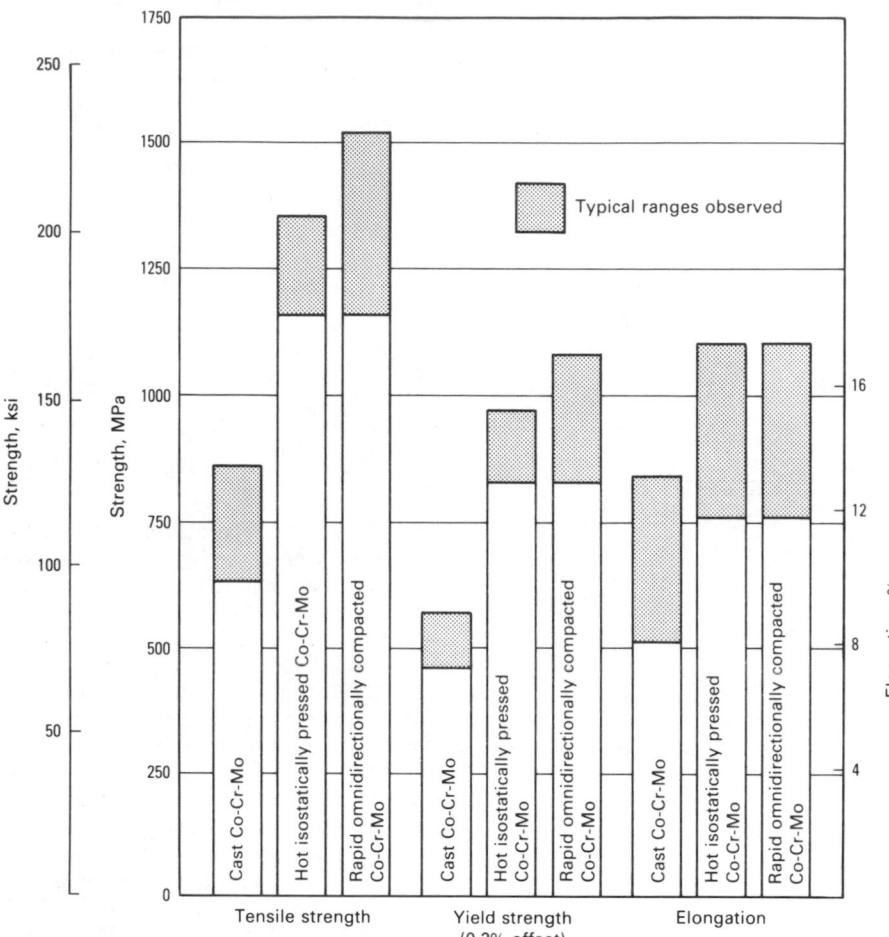

Fig. 10 10^7 cycle rotating beam fatigue endurance limit for cobalt-chromium-molybdenum alloy

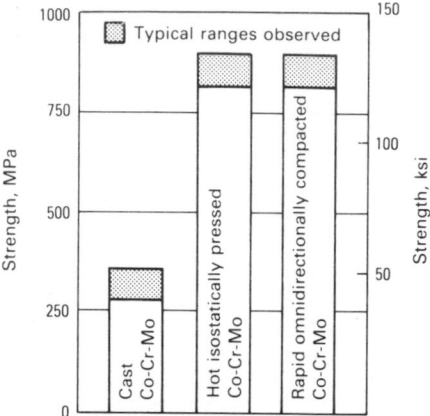

cured state within the prepared site, and inserting the implant. The polymethyl methacrylate then cures and fixes the implant in place. This fixation is primarily mechanical.

Cement penetrates into the cancellous bone and locks onto small surface irregularities on the implant. Shrinkage of the cement during curing also locks the cement onto the stem of the device. However, over time, the implant may become loose within the cement. In some cases, this loosening may either cause pain or increased stress on the implant and subsequent implant failure. Both of these effects may necessitate reoperation.

Currently, the use of roughened or porous coatings on implant surfaces to eliminate loosening of the implant within the cement is being studied. Porous-coated implants provide enhanced fixation through either cement interlocks with the porous structure or fixation via ingrowth of bone tissue.

For mineralized bone to grow into the porous coating, certain minimum pore sizes

must be achieved. If these minimums are not met, either soft tissue will form or ingrowth will not occur at all. These minimum levels reflect biological considerations. Research to determine the minimum interconnection size between pores has shown some variation, which is apparently due to material effects and experimental animal variation. In unloaded implants, the minimum interconnection between pores is approximately 50 μm. In loaded implants, such as artificial hips or knees, the minimum interconnection between pores needed to support healthy bone is approximately 100 to 150 μm. In porous coatings made from sintered powder particles, pore size is controlled by selection of powder size. For more information, see Ref 2-4.

Several techniques have been devised for applying porous coatings to implants. Cobalt-chromium or titanium powders can be applied using various binders and then sintered to the surface of cobalt-chromium or titanium alloy/titanium implants. Wires or fibers can be formed into porous pads

and diffusion bonded to the surface of titanium alloy implants. Void-metal composites have been made using fugitive pore formers. Plasma spraying techniques can be used to apply porous coatings. Porous polymeric coatings also have been applied to implants. For a discussion on fabrication of porous coatings, see Ref 5.

Conventional sintering techniques for bonding large powder particles to cobalt-chromium alloys typically involve sintering temperatures that are 90 to 95% of the melting point of the substrate. This elevated-temperature exposure has a significant effect on the microstructures and properties of fine-grained, high-strength implant materials such as hot isostatically pressed metal powders or wrought cobalt-chromium-molybdenum. Typically, these materials undergo significant grain growth, with resultant decreases in fatigue strength. Current cobalt-chromium devices use sintered powder coatings that are applied to coarse-grained cast implants; thus, these devices do not have the high fatigue strength of hot isostatically pressed cobalt-chromium-molybdenum metal powders. The porous layer may also create notch effects, which further reduce fatigue strength.

Some reduction in fatigue strength is tolerable in these implants, because good fixation by bony ingrowth results in load sharing between the implant and bone. Proper implant design prevents fatigue strength degradation. Additional design considerations include clinical requirements, such as insertion and removal.

Methods of reducing sintering temperatures by use of compositional modifications or by simultaneous application of pressure and temperature are being investigated. Porous titanium wire pads can be

Fig. 11 Porous cobalt-chromium-molybdenum coating produced by gravity sintering
(a) Scanning electron micrograph. (b) Metallographic cross section

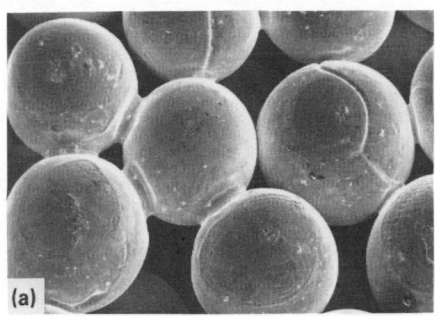
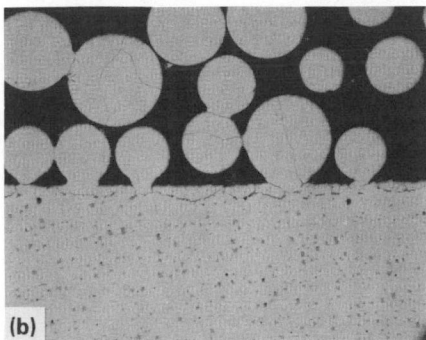

Fig. 12 Histological section of a porous coated canine femoral knee component after implantation of 9 months
Well-developed bone is observed within the pores. Source: Bobyn, J.D., *et al.*, Biologic Fixation and Bone Modeling, with an Unconstrained Canine Total Knee Prosthesis, *Clin. Orthopaedics Related Res.*, No. 166, 1982, p 307

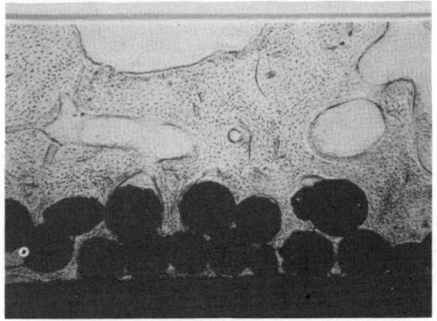

Fig. 13 Scanning electron micrograph of a lathe-cut dental amalgam powder
Magnification: 250×

diffusion bonded to titanium alloy implants at much lower temperatures than those needed for conventional gravity sintering.

Figure 11 shows scanning electron micrograph and cross-sectional views of sintered cobalt-chromium-molybdenum particles ($-20+45$ mesh). Note the necking between the particles. Particle-to-particle bonding is also evident in the cross-sectional view.

Figure 12 shows a section from a cobalt-chromium powder-coated canine femoral knee implant after an implantation of 9 months. The pores have been filled with well-developed bone. This type of fixation is superior to the mechanical fixation provided by the polymethyl methacrylate cement.

While porous-coated hip and knee implants have received the most attention, other implants also may benefit from porous coating technology. For example, porous coatings may be used to fix heart pacemaker electrodes in place. In this case, the implant would be affixed by ingrowth of soft tissue.

Dental Amalgams

Dental amalgams are formed by the reaction of liquid mercury with silver-based alloy powders. Powder compositions can be divided into two basic categories: low and high copper content. Low-copper alloys contain greater than 65 wt% Ag and 29 wt% Sn, with no more than 6 wt% Cu. High-copper alloys developed in the 1960's and 1970's have higher copper contents (up to approximately 30 wt%).

When silver-tin alloy powders make contact with mercury, silver and tin dissolve in the mercury, while mercury diffuses into the powders. Because solubility of both silver and tin in mercury is limited, precipitates are formed: Ag_2Hg_3,

known as γ_1, and a tin phase, which is approximately Sn_8Hg, known as γ_2. These reactions continue until the mercury is consumed. The final result is a composite of unreacted alloy powders surrounded by silver-mercury and tin-mercury intermetallics. Some porosity is also present in the final amalgam. Performance of the amalgam depends on the mercury-to-powder ratio, powder variables such as composition and morphology, and dental technique. High-copper alloy powders may be blends of silver-copper and silver-tin powders or a prealloyed silver, tin, and copper powder. The basic mechanisms of amalgam formation (interdiffusion and precipitate formation) are the same as with low-copper alloys.

In mixtures of binary silver-copper and silver-tin powders, Cu_6Sn_5 is formed around the silver-copper particles. This phase is formed by diffusion of dissolved tin to the surfaces of the silver-copper particles. This reaction eliminates most or all of the formation of the γ_2 (Sn_8Hg) phase, which is the weakest component in the amalgam, as well as the phase with the least resistance to corrosion in the mouth.

When a prealloyed silver-tin-copper composition is amalgamated with mercury, similar reactions occur, but the morphology of the Cu_6Sn_5 phase is different. In this case, a rod-like precipitate of Cu_6Sn_5 forms, which covers the particles. This rod-like morphology may improve bonding with the matrix and improve deformation resistance. The final amalgam consists of a matrix of γ_1 (Ag_2Hg_3) surrounding unconsumed alloy particles, which are covered with Cu_6Sn_5. Formation of the undesirable γ_2 is thus reduced or eliminated (Ref 6).

Powders. The first powders used to make dental restorations, referred to as lathe-cut powders, were made by machining cast ingots. This technique produces

dense, somewhat elongated chips (Fig. 13). Prior to machining, the ingots are homogenized by heat treatment below the incipient melting point of the alloy being processed. The phase distribution in the homogenized ingot may be altered by control of the cooling rate used at the end of the homogenization cycle.

Further processing may follow homogenization and machining to form the powder. For example, the machined powder may be ball milled for further size reduction. Powders may be washed with proprietary acid treatments to alter behavior during amalgamation. Powders also may be subjected to a stress-relief cycle to remove residual stresses resulting from cutting.

Atomized powders produced by conventional atomization techniques are somewhat irregular in shape (Fig. 14). Smaller particles tend to be spherical. Atomized powders also undergo heat treating to increase the grain size in the powder. This slows the rate of reaction with

Fig. 14 Scanning electron micrograph of an atomized dental amalgam powder

Magnification: 250×

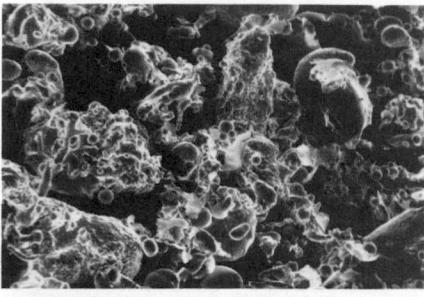

Fig. 15 Scanning electron micrograph of a mixture of lathe-cut and atomized dental amalgam powders

Note spherical atomized particles.
Magnification: 250×

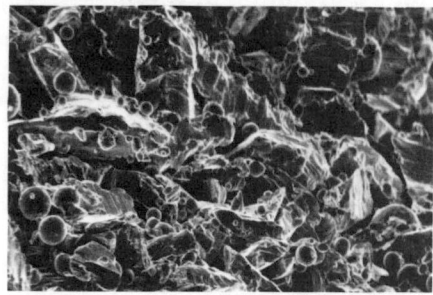

mercury. Atomized powders frequently are washed with acid. Mixtures of lathe-cut and atomized powders also are used (Fig. 15).

Mechanical Properties and Powder Variables. Important properties of dental amalgams include compressive and tensile strength, corrosion resistance, and dimensional stability. Powder variables such as particle shape, particle size distribution, surface area, and composition have a significant effect on some of these properties. Particle sizes of commercially available powders are approximately −325 mesh, with average sizes around 30 μm. These powders tend to provide more rapid hardening of the amalgam and a higher level of strength in the early stages than do powders with larger particles. However, it is possible to overdo this effect. Powders with a significant fraction of very fine powder (3 μm or less) have very high surface area and require increased amounts of mercury to produce a workable amalgam.

Particle shape has several effects on amalgam behavior. Spherical powders have less surface area for a given size than do lathe-cut particles, thus permitting the use

Table 2 Compressive strength of amalgams

Residual mercury, %	Compressive strength MPa	psi
52	57	8300
54	54	7800
56	43	6300
58	36	5200

Source: Swartz, M. L. and Phillips, R. W., Residual Mercury Content of Amalgam Restorations and Its Influence on Compressive Strength, *J. Dental Res.*, Vol 35 (No. 3), 1956, p 463

of less mercury (desirable from a mechanical properties viewpoint). Table 2 illustrates the relationship of compressive strength of a lathe-cut powder amalgam to mercury content. These results demonstrate that a significant decrease in compressive strength can occur with excessive mercury contents.

Amalgams made with spherical powders are also more plastic than those made with lathe-cut powders. When these powders are compacted by the dentist (condensation), higher strength is developed with low pressures. However, because of the increased plasticity, care must be taken to provide properly contoured restorations.

As mentioned previously, high-copper amalgams contain little or no γ_2 phase. Because γ_2 has lower strength, creep, and corrosion resistance than γ_1, the absence of γ_2 appears to be beneficial. Testing of low- and high-copper amalgams shows that the high-copper materials have higher initial and aged compressive strength and lower creep resistance.

Zinc is added to some alloys to act as a deoxidizer during melting. While this improves the ductility of the alloy, it also leads to excessive expansion of the amalgam if moisture is present while the amalgam is being mixed and condensed. This expansion is related to generation of hydrogen and begins 4 to 5 days after the amalgam is placed. Expansion may cause pain; use of low-zinc, high-copper alloy powders prevents this problem.

Porous Coatings for Implant Fixation Via Ingrowth

Porous coatings enjoy wide usage for dental implants. Several uncoated dental implants of various designs and materials have been produced. Basically, these devices are inserted in the mandible or maxilla without the use of polymethyl methacrylate. Implants generally are intended to support various dental appliances in the mouth. Figure 16 shows some blade implants designed for this purpose. The success of dental implants has been mixed.

Fig. 16 Uncoated dental implants

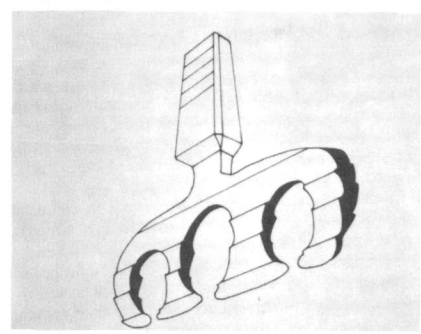

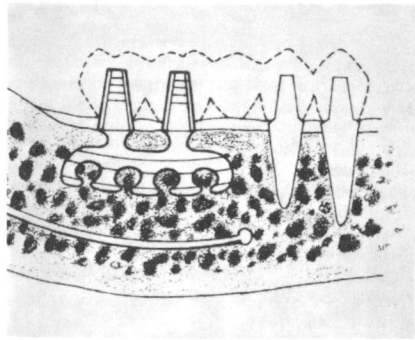

Fig. 17 Typical dental implants

(a) Cobalt-chromium-molybdenum powder coated dental implant. (b) Titanium wire-coated dental implant

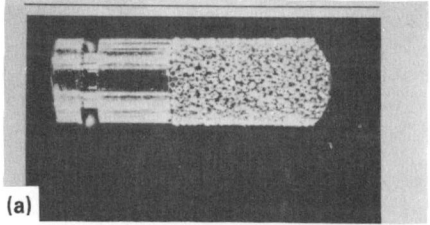

(a)

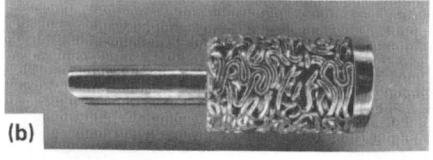

(b)

Recently, both powder and wire porous coatings sintered to metallic substrates have been used in an attempt to get bonding between the device and the bone.

As in orthopedics, titanium and cobalt-chromium-molybdenum alloys are the preferred materials. Examples of a cobalt-chromium-molybdenum powder-coated implant and a titanium wire-coated implant are shown in Fig. 17. The structure of these coatings is the same as those shown previously for orthopedic implants. Early clinical trials of the titanium wire-coated implant have shown success rates greater

Fig. 18 Injection molded stainless steel orthodontic brackets

Courtesy of Parmatech

than 90%. Thus, it appears that porous-coated dental implants have a promising future (Ref 7).

Dense Components. Typical dental devices are rather small for the known elevated-temperature consolidation processes such as hot isostatic pressing. However, injection molding has been used to make small stainless steel orthodontic brackets such as those shown in Fig. 18

(for further information, see the article "Injection Molding" in this Volume). This technology seems to be well suited for production of this type of component.

REFERENCES

1. Mears, D.C., *Materials and Orthopaedic Surgery*, Williams and Wilkins, Baltimore, MD, 1979
2. Klawitter, J.J. and Hulbert, S.F., Application of Porous Ceramics for the Attachment of Load Bearing Internal Orthopaedic Applications, *J. Biomed. Mater. Res. Symp.*, No. 2, 1971, p 161-229
3. Nilles, J.L. and Lapitsky, M., Biomechanical Investigations of Bone-Porous Carbon and Porous Metal Interfaces, *J. Biomed. Mater. Res. Symp.*, No. 4, 1973, p 63-84
4. Welsh, P., Pilliar, R.M., and McNab, I., Surgical Implants: The Role of Surface Porosity in Fixation to Bone and Acrylic, *J. Bone Joint Surg.*, Vol 53A, 1971, p 963-977
5. Hirschhorn, J.S. and Reynolds, J.T., Powder Metallurgy Fabrication of Cobalt Alloy Surgical Implant Materials, in *Research in Dental and Medical Materials*, E. Korostoff, Ed., Plenum Press, New York, 1969, p 137-151
6. Okabe, T., *et al.*, A Study of High Copper Dental Amalgams by Scanning Electron Microscopy, in *Microstructural Science*, Vol 7, Elsevier North Holland, 1979
7. Weiss, M.B. and Rostoker, W., Development of a New Endosseous Dental Implant. Part II: Human Studies, *J. Prosthetic Dentistry*, Vol 47 (No. 6), June 1982, p 633-645

SELECTED REFERENCES

- Phillips, R.W., *Skinner's Science of Dental Materials*, 8th ed., W.B. Saunders, Philadelphia, 1982
- Galante, J.O., *et al.*, Sintered Fiber Metal Composites as a Basis for Attachment of Implants to Bone, *J. Bone Joint Surg.*, Vol 53A, 1971, p 101-114
- Pilliar, R.M., Cameron, H.U., and McNab, I., Porous Surface Layered Prosthetic Devices, *Biomed. Eng.*, Vol 10, 1975, p 126-131

Nuclear Applications

POWDER METALLURGY techniques have been widely applied to material fabrication problems associated with nuclear power reactors. Features that make P/M processing attractive for nuclear applications include the ability to (1) control grain size, especially the production of fine-grained materials; (2) make alloys consisting of elements with large differences in density or melting point; and (3) make multiphase materials, particularly metal-ceramic types.

Of particular interest have been fissile materials, such as uranium, uranium dioxide, uranium carbide, and uranium nitride, that are used in nuclear fuel elements. During the 1950's, considerable work was conducted on the fabrication of dispersion-type nuclear fuel elements (Ref 1, 2). These consisted of the fissile compound, commonly uranium dioxide dispersed in a metallic matrix.

Selection of the matrix metal for dispersion-type nuclear fuel elements was based on minimum neutron capture cross section and adequate thermal conductivity and strength properties. Matrix materials chosen most frequently for these applications were stainless steel (300 series) and commercially pure aluminum.

The method of producing uranium dioxide-aluminum fuel elements involved the blending of −325 mesh atomized aluminum powder and uranium dioxide particles ranging from 45 to 105 μm. The mixture was subsequently pressed into compacts at a pressure of 350 MPa (50 ksi). The green compacts were then assembled into a "picture frame" configuration in which the uranium dioxide-containing compact was surrounded by aluminum on all four sides and on the top and bottom (Fig. 1).

The composite fuel plate was hot rolled at 590 °C (1095 °F) to a thickness of about 1.8 mm (0.07 in.) in six passes. The plates were then annealed and cold rolled to the desired final thickness and fabricated into a fuel assembly by brazing. Similar techniques were used in producing fuel elements with a stainless steel matrix (Ref 3).

Dispersion-type fuel elements are no longer used in commercial practice. Current commercial fuel elements, which consist of uranium dioxide pellets housed in a long, thin-walled tube (fuel clad), will be the focal point of this article.

Another P/M material used in nuclear applications is beryllium, which is an excellent moderator and solid reflector of thermal neutrons. A high scattering cross section (7.0 b) makes beryllium very effective in reducing neutron energies to a level consistent with efficient reactor operation. Beryllium also reflects or scatters leakage neutrons back into the reactor core.

Neutrons are thus conserved because beryllium has a low absorption cross section (0.010 b). Consequently, the probability of neutron capture is low. In reactors where conservation of weight and space is important, beryllium is particularly advantageous. Both neutron economy and flux density are enhanced by the use of beryllium in nuclear reactors.

Nuclear Fuel Elements

By Gerald R. Kilp
Advisory Engineer
Westinghouse Nuclear Fuel Division

URANIUM DIOXIDE, which is the dominant fuel form used in commercial reactors, has recently benefited from P/M technology, in that nuclear reactor fuel pellets are made using these techniques. This section will discuss the manufacture of fuel for light-water-moderated and cooled reactors, which include pressurized water reactors and boiling water reactors. These reactors comprise the majority of commercial reactors (Ref 4). The principles discussed in this section, however, also are applicable to some gas-cooled reactors and the Canadian heavy-water-moderated (CANDU) reactors, which also use uranium dioxide as a fuel.

Many of the uranium dioxide fuel elements used in the light-water-moderated plants consist of a long, thin-walled metal tube (fuel clad) filled with a column of cylindrical uranium dioxide fuel pellets, a pellet hold-down device (usually a spring), and two welded-on end plugs (Fig. 2). The upper end space (partially occupied by the spring) is called the plenum. This space allows for longitudinal expansion of the pellet stack and accommodates fission gases released from the fuel during operation.

The plenum and the diametral gap between the pellets and the clad normally contain helium to improve heat transfer between the fuel. In pressurized water reactors, internal helium pressure helps to minimize loss of the diametral gap by creep deformation at high operational pressures.

Almost all of the cladding used is made from dilute zirconium alloys, such as Zircaloy-2 (Zr-1.5Sn-0.12O-0.13Fe-0.10Cr-0.05Ni) or Zircaloy-4 (Zr-0.2-0.3Sn-0.2-0.3Fe-0.5C max). Fuel rods made from these alloys are susceptible to internal hydriding failure. Consequently, a prime requirement of the uranium dioxide fuel is low hydrogen content. In practice, this is equivalent to a low moisture content, because most hydrogen carried into the fuel rod is the result of adsorbed water on the fuel pellets.

Uranium in commercial fuel typically is enriched to about 1 to 5 wt% (expressed as metal) of the fissionable ^{235}U isotope. Natural uranium (uranium as found in nature) is about 0.7 wt% ^{235}U, although ore bodies with lower ^{235}U content have been used.

In addition to control of hydrogen (moisture) content, commercial fuel pellets must meet tight dimensional tolerances and must be strong and largely free of cracks and chips. Very high-purity starting materials are used, and the manufacturer must pay careful attention to minimizing contamination during fabrication. In particular, elements with high neutron absorption cross sections (boron, hafnium, and many of the rare earths) must be closely controlled.

Pellet dimensions for most fuel designs range between 7 and 15 mm (0.3 and 0.6 in.) in diameter and are less than 8 mm

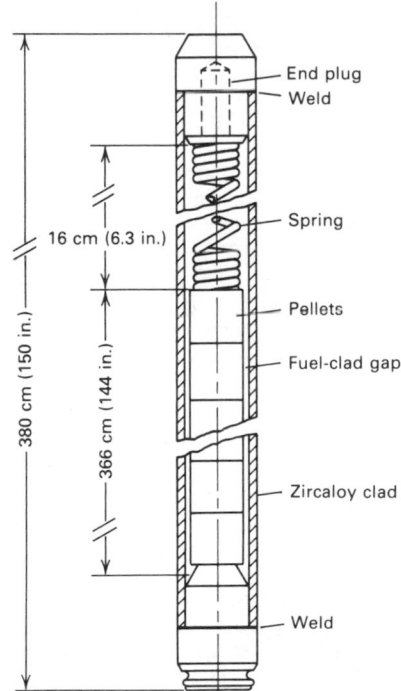

Fig. 2 Typical uranium dioxide fuel rod

(0.3 in.) in length. Length-to-diameter ranges typically range from 0.5 to 1.5. End dishes, chamfers, and pellet identification markings may be formed on the pellets as desired.

Uranium dioxide is the preferred fuel material, primarily due to its excellent chemical stability (especially in water re-actor coolants), high melting point (approximately 2800 °C, or 5070 °F), resistance to irradiation damage, and excellent compatibility with cladding materials (Zircaloy-2, Zircaloy-4, and type 304 stainless steel).

Fully dense uranium dioxide is not used primarily because of the practical problem of achieving fully dense materials with commercial P/M fabrication methods. Fuel densities are typically between 93 and 96% of theoretical density (10.96 g/cm^3).

Fuel Pellet Fabrication

Fuel pellet specifications require almost exactly stoichiometric uranium dioxide (oxygen-to-uranium ratios of $\approx$2.00). For example, American Society for Testing and Materials standard ASTM C 776 states that "the oxygen-to-uranium ratio shall be within the range of 1.99 to 2.02."

However, most uranium dioxide powders used to fabricate pellets contain considerable excess oxygen, which is removed during pellet sintering. Excess oxygen content in the powder can be quite high (oxygen-to-uranium ratios of 2.2 or higher are not uncommon) and still produce a satisfactory product. In fact, some excess oxygen is desirable in the powder, because it increases sinterability.

The oxygen content of production powders depends greatly on the type of process used to produce the powder. The average oxygen content of the powder press feed is also affected by any uranium oxide scrap that may be added. For instance, it is common practice to oxidize and recycle some of the uranium oxide scrap in the form of U_3O_8. The amount of scrap that can be recycled as U_3O_8 is normally limited to a few percent, because its use often leads to adverse effects on pellet strength and density.

Uranium dioxide fuel has been made by several conventional powder fabrication techniques (Ref 5), including extrusion. However, almost all commercial fuel is made by cold pressing and sintering, although universal procedures for commercial uranium dioxide pellet fabrication do not exist. The variability in uranium dioxide powders mandates differences in process details.

Generally, however, uranium dioxide pellet fabrication begins by production of virgin uranium powder. To this lot, several forms of uranium oxide scrap may be blended in, as well as various additives to serve as binders, lubricants, or pore size control agents. Dry and wet blending sequences can be used. Lubricants also may be added in later blending operations.

If the virgin powder has a relatively low bulk density, the blended powder may require a precompaction or "slugging" step, followed by granulation and screening. Precompaction improves powder flowability and ensures that the press feed is sufficiently dense to adequately fill the automatic press die cavities.

The powder is then pressed to a predetermined green density in a high-volume automatic press. The required density may vary considerably depending on the powder characteristics, but it usually ranges from 40 to 60% of theoretical density, which is around 10.96 g/cm^3. The required pressure, which also varies greatly, is commonly about 300 MPa (20 tsi).

After pressing, the green pellets are sintered to the required density and stoichiometry in large walking-beam or push-type furnaces that are heated by molybdenum heating elements. Reducing atmospheres are typically used; dry or wet hydrogen is preferred, although dissociated ammonia or hydrogen/inert gas mixtures can also be employed. Sintering times and temperatures also vary, but normally range from 3 to 8 h at 1550 to 1800 °C (2820 to 3270 °F).

Depending on the additive used, a separate low-temperature de-binding step may be employed. This is usually achieved, however, by simply controlling the rate of pellet temperature increase in the preheating section of the furnace. After sintering, the pellets are then cooled to ambient temperature before exposure to room atmosphere. Pellets are then centerless ground using water as a coolant, dried in air at a temperature slightly above 100 °C (212 °F), and subsequently loaded into fuel rods.

Although pellet fabrication processes may vary widely, the objectives of all pellet manufacturers are essentially the same—they must produce a sound pellet with a structure that is strong and resistant to moisture adsorption and that minimizes the fine pore content of the pellet. The number of fine pores (typically less than 2 μm) must be controlled in uranium dioxide pellets to prevent in-reactor densification. Densification refers to the in-reactor removal of fine pores (Ref 6, 7). The removal of the fine pores can lead to pellet shrinkage, which must be controlled. The means of controlling fine pore content depends greatly on the uranium dioxide powder properties.

With some highly active powders that are capable of sintering to 97 or 98% of theoretical density, in-reactor pore removal is not a problem, and a pore former may be used to ensure a large fraction of large pores if a lower density is desired.

The pore former (a material such as ammonium oxalate that thermally decomposes to gases) is added to the powder in small discrete particles that leave voids almost proportional to the particle size when the pellet is sintered. Less active powders can be heated to higher temperatures or for longer times, because fine pores can also be minimized by thermal processes.

Safety Considerations

Because enriched uranium is radioactive,[*] fissionable, and is classified as a heavy metal poison, it is subject to strict government regulations. Because it is fissionable, mass limit and/or geometry control are used during powder, pellet, and fuel rod fabrication and handling operations. This places restrictions on the size and shape of equipment and also affects storage of final and intermediate product forms. Due to its radioactivity and the heavy metal poisoning hazard, suitable precautions must be taken to guard against personnel contamination from airborne particulates or from direct contact. However, the radioactivity level is below the level that requires personnel shielding.

These health and safety considerations and the high cost of uranium naturally underscore the importance of scrap recovery and reprocessing. The fabricator is penalized twice for any wasted uranium—in the extra cost of the finished product and in the cost of scrap and waste disposal.

Nuclear Control Rods and Shielding

By William E. Ray
Manager, Product Development
Westinghouse Electric Corp.

DISPERSIONS OF PARTICLES of a neutron-absorbing material, such as europium oxide (Eu_2O_3) or boron carbide (B_4C), in aluminum, copper, stainless steel, or Zircaloy metal matrices have been in use for nuclear reactor shielding and reactivity control application for over 20 years (Ref 8, 9). Although the performances of

stainless steel-europium oxide and Zircaloy-boron carbide in particular have been adequate in military pressurized water reactor applications, these dispersions have not found significant control applications in current commercial nuclear power plants. The reasons are apparently economic; cylindrical boron carbide pellets canned in stainless steel tubing (cladding) are less costly to fabricate.

With the development of liquid sodium-cooled fast breeder reactors, stainless steel-europium oxide control rods were re-evaluated. The higher operating temperatures of this type of reactor had caused concern over the swelling performance of boron carbide pellets (Ref 10, 11). In the United States and the United Kingdom, selection of control rod material was resolved by testing under typical reactor conditions. Consequently, boron carbide pellets were selected again (Ref 12, 13). In France, West Germany, and Japan, both absorber materials are in use. In Russia, the BOR-60 fast breeder reactor has operated satisfactorily with one europium oxide control rod since 1972 (Ref 14). Europium oxide is canned in pellet form, much like the same shape used in boron carbide control rods for light-water power reactors.

Powder metallurgy dispersions are enjoying continued use in low-temperature (below 300 °C, or 570 °F) research reactors (Ref 8, 9). Compositions containing natural boron, boron enriched in the mass ten isotope, B_4C, Gd_2O_3, Eu_2O_3, and rare earth oxide mixtures are frequently used. The matrix metal is commercially pure aluminum, although stainless steel may be specified for high-temperature experiments. As in power reactor control rods, the composition of the protective cladding nominally matches that of the metallic matrix.

Although considerable research was conducted, especially before 1970, to develop P/M dispersions as reactor shielding materials, boron-containing P/M compositions have been replaced by cast alloys, as-cast or cast and wrought depleted uranium, lead, and special concretes. Recently, reduction in weight of the shipping casks used to transport radioactive material has renewed interest in P/M shielding. In particular, considerable work has been done to qualify copper-boron carbide for this application (Ref 15).

REFERENCES

1. Weber, C.E. and Hirsch, H.H., Dispersion Type Fuel Elements, Proceedings First International Conference on Peaceful Uses of Atomic Energy, Vol 9, 1953, p 196-202
2. Boyle, E.J. and Cunningham, J.E., MTR Type Fuel Elements, Proceedings First International Conference on Peaceful Uses of Atomic Energy, Vol 9, 1953, p 203-207
3. Lenel, F.V., *Powder Metallurgy—Principles and Applications,* Metal Powder Industries Federation, Princeton, NJ, 1980, p 490-491
4. Belle, J., "Uranium Dioxide: Properties and Nuclear Applications," U.S. Government Printing Office, Washington, DC, 1961
5. "Power Reactors in Member States," International Atomic Energy Agency, Vienna, Austria, Available from UNIPUB, New York, 1980
6. Chubb, W., Hott, A.C., Argall, B.M., and Kilp, G.R., The Influence of Fuel Microstructure on In-Pile Densification, *Nuclear Technol.,* Vol 26, 1975, p 486-495
7. Stehle, H. and Assman, H., The Dependence of In-Reactor UO_2 Densification on Temperature and Microstructure, *J. Nucl. Mater.,* Vol 52, 1974, p 303-308
8. Anderson, W.K., Therlacker, J.S., Ed., "Neutron Absorbers for Reactor Control," USAEC, Washington, DC, 1962
9. Ray, W.E., "Fabrication of Control Rods for Nuclear Reactors," ASM/USAEC, Rowman and Littlefield, New York, 1963
10. Pasto, A.E. *et al.,* "Eu_2O_3 as a Potential LMFBR Control Material," USAEC Report No. ORNL-TM-4226, Oak Ridge National Laboratory, 1973
11. Martin, M.M. *et al.,* "Description of FTR-Oriented Irradiation Test of Eu_2O_3 in EBR-II," USERDA Report No. ORNL-TM-4996, Oak Ridge National Laboratory, 1975
12. Shiely, W.F. and Dahl, R.E., B_4C Performance in Fast Reactors, *Trans. Am. Nucl. Soc.,* Vol 20, 1975, p 291
13. "A Compilation of B_4C Design Support Data for LMFBR Control Elements," USERDA Report No. HEDL-TME-75-19, Hanford Engineering Development Laboratory, 1975
14. Boilo, Y.V. *et al.,* "Development and Investigation of Properties of Disperse Boron-Containing Materials for Control Rods of Nuclear Reactor," Vestsi Akad. Navuk BSSR, 1978
15. "Development of Boron Carbide-Copper Cermets," SAND-78-2317, Sandia Laboratories, 1979

---

[*]Uranium-235 has a half-life of 7×10^8 years, and uranium-238 has a half-life of 4.5×10^9 years.

P/M Parts for Business Machines

By Frank D. Leone
Senior Manufacturing Engineer
Process Engineering
Pitney Bowes Inc.

POWDER METALLURGY plays an important role in the manufacture of business machines, such as copiers, word processors, high-speed printers, and mail-handling equipment. Second only to the automotive market, the greatest growth in P/M parts usage is anticipated in the business machine industry. The increasing ability of P/M technology to provide more complex shapes, higher mechanical and physical properties, and closer tolerances combined with cost savings makes P/M parts attractive to the business machine manufacturer.

This article presents a series of recent case studies that outline the material requirements, design features, process techniques, and cost considerations that make P/M components a viable part of the business machine industry. It should be noted that these examples, which deal with iron-based compositions, represent only a small portion of the potential applications. Aluminum P/M parts used in business machines are discussed in the article "P/M Lightweight Metals" in this Volume. Another important market for metal powders—the use of iron powders in copier machines—is discussed in the article "Copier Powders" in this Volume.

Machine Drive Gears

A flat-belt drive gear weighing 490 g (17.3 oz), which is compacted to a density of 7.45 g/cm³, is shown in Fig. 1. This type of gear is used in high-speed mail-handling and inserting machines. This part, consisting of electrolytic iron preblended with 0.35% C, replaced a two-piece gear assembly made of American Iron and Steel Institute (AISI) 1030 steel, thus resulting in a total cost savings of 66%.

Fig. 1 Iron-carbon P/M drive gear

(a) Finished machined, hardened, ground part. The hub diameter is machined oversized (+0.254 mm, or +0.010 in.) to allow for final grinding.
(b) P/M blank as-pressed

(a)　　　　　(b)

The powder was first compacted on a 5340-kN (600-ton) press to a density of 6.8 to 6.9 g/cm³ at a compaction pressure of 414 MPa (30 tsi). It was then presintered and re-pressed to a density of 7.45 g/cm³ at 965 MPa (70 tsi). After re-pressing, the part was sintered at 1120 °C (2050 °F) for 30 min.

As shown in Fig. 1, considerable machining (turning) of the hub was required to generate the final shape. In use, this gear rotates and a magnetic clutch rides on the machined hub. When the operator trips the cycle control on the machine, the clutch closes around the small hub diameter, and the load is transferred to the P/M gear. Pressing of a large hub ensures a uniform density in the cross-sectional area, but requires additional machining of the P/M blank.

After machining, the hub area is carbonitrided to a case depth of 0.127 mm (0.005 in.). Case hardness is 700 DPH (minimum), while core hardness is 550 DPH. To maintain a 0.025-mm (0.001-in.) diam tolerance between the hub and the magnetic clutch, the hub diameter is ground to final dimensions.

Typical properties for this P/M part are:

Tensile strength	414-586 MPa (60-85 ksi)
Yield strength	365-414 MPa (53-60 ksi)
Elongation	1-2%
Transverse-rupture strength	930 MPa (135 ksi)

High-Precision, High-Strength Gears

The high-precision, high-strength gears shown in Fig. 2 are made of 4600 alloy powder prealloyed with 0.25% Mn, 1.7 to 1.9% Ni, and 0.5 to 0.6% Mo and subsequently preblended with 0.75% Cu and 0.35 to 0.45% C. These gears are used in postage meters for counting applications.

The powder was compacted in a floating die press system to a density of 6.85 to 7.0 g/cm³. Sintering was conducted in an endothermic atmosphere at 1120 °C (2050 °F). After sintering, the parts were carbonitrided to provide a case hardness of 700 DPH for a 0.127-mm (0.005-in.) case depth and a core hardness of 400 DPH. After case hardening, the parts were deburred and oil impregnated.

The original method for manufacturing these gears consisted of a two-piece copper-brazed and gear-hobbed assembly of AISI 1030 steel. The larger (bottom) gear was first turned as a disk with a center hole. Gear teeth were then hobbed.

The smaller (top) gear was also turned and gear hobbed. The two parts were then

Fig. 2 P/M transfer gear made of high-strength low-alloy steel

(a) Original P/M processing technique, which required machining of flange section. (b) Modified P/M technique, which required no additional machining

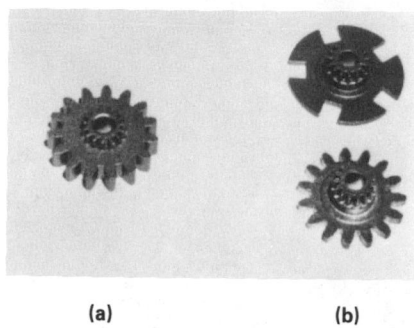

(a)   (b)

press fitted and copper brazed. At this point, inspection was required to reject and scrap any gears that shifted during brazing, thus losing the critical tooth-to-tooth relationship required between the small-diameter gear and the large-diameter gear. The rejection rate due to assembly errors ran as high as 50%.

Figure 2(a) shows the single-piece transfer gear manufactured using P/M techniques. The flange section was made thick enough so that the hub section could be machined as a secondary operation. The normal timing of the floating die press mechanism forced powder to transfer from the small gear teeth into the flange section, thus resulting in low density in the small (top) gear teeth. Consequently, tooling was modified to produce the gear configurations shown in Fig. 2(b), which required no machining of the gear teeth located on the hub section. Typical finished P/M properties are:

Ultimate tensile strength	586-793 MPa (85-115 ksi)
Yield strength	552-655 MPa (80-95 ksi)
Elongation	<1%
Transverse-rupture strength	1172-1379 MPa (170-200 ksi)

Geneva Pinions

Geneva pinion gears are used in counting devices. These parts were originally made by pressing gear blanks from electrolytic iron powder with 0.35% C added. Final shape was achieved by:

- Pressing to 6.8 to 6.9 g/cm³ density at 414 MPa (30 tsi)
- Presintering at 1000 °C (1830 °F) to just below the solution level of the carbon
- Re-pressing to 7.45 g/cm³ at 965 MPa (70 tsi) to increase density to near 95% of theoretical to ensure that parts could

Fig. 3 P/M Geneva pinions

(a) Original P/M design prior to milling. (b) Finished P/M pinions and mating parts. (c) Tooling design for the scallop Geneva section

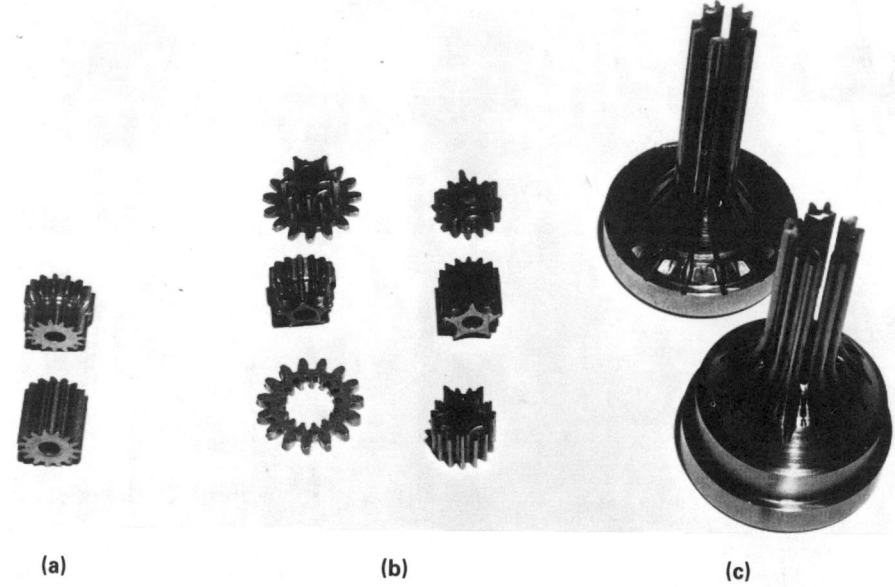

(a)   (b)   (c)

undergo carbonitriding to a 0.127-mm (0.005-in.) case
- Resintering at 1120 °C (2050 °F) for 30 min in endothermic atmosphere
- End milling of pinion teeth
- Carbonitriding to a case hardness of 700 DPH for a minimum 0.127-mm (0.005-in.) case and a core hardness of 400 DPH

The original P/M blanks prior to milling are shown in Fig. 3(a).

The critical requirement in this application is a high-strength, burr-free part that will facilitate smooth sliding motion between the Geneva pinion and the mating gear. The teeth of the two parts must be perfectly mated to ensure a free sliding motion. In addition to being costly, the milling operation in the original P/M process used to cut the pinion teeth generated burrs, which were difficult to remove without damaging the teeth. As a result, an alternative P/M technique was sought.

The solution to this problem was realized with a new tooling design (Fig. 3c), which enabled these parts to be processed without finish machining.

The tooling material was vacuum melted high-speed tool steel. In addition, a new high-strength steel composition replaced the previously used iron-graphite composition. This material consisted of a 4600 alloy powder prealloyed with 0.25% Mn, 1.7 to 1.9% Ni, and 0.5 to 0.6% Mo and subsequently preblended with 0.75% Cu and 0.35 to 0.45% C. The density of the finished part is 7.0 g/cm³.

With the new material and tooling,

Fig. 4 P/M miter gear made from an iron-carbon-copper alloy

(a) Top and bottom view of P/M blank. (b) Finished machine part. (c) Lower punch (top right) and upper punch (bottom right) used to mold gear

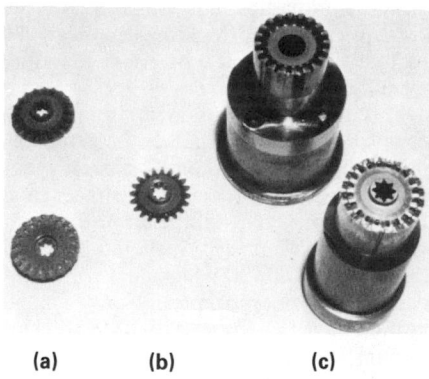

(a)   (b)   (c)

loading on the tooling was reduced 30%, which extended tool life 50%. Figure 3(b) shows the completed P/M parts with the mating gears.

Miter Gears

A typical miter gear used as the main drive gear of a postage meter is shown in Fig. 4. The material used for this application was reduced iron preblended with 0.86% C and 5% Cu, which was compacted to a density of 6.4 to 6.6 g/cm³. This part was originally made of a wrought medium-carbon steel that was gear hobbed.

Two critical parameters had to be met with this part—it had to pass a 10⁷ life cycle

Fig. 5 Sequence of machining steps to produce final dimensions of P/M miter gear

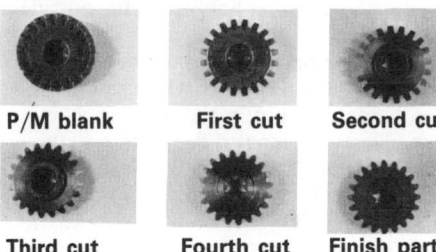

P/M blank   First cut   Second cut

Third cut   Fourth cut   Finish part

Fig. 6 P/M hammer guide assembly for a high-speed printer

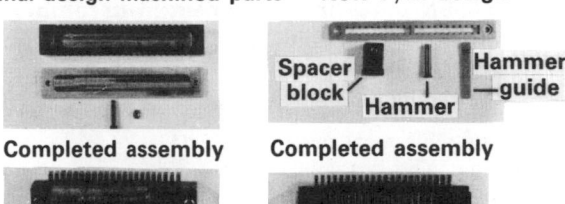

Original design machined parts   New P/M design

Spacer block   Hammer   Hammer guide

Completed assembly   Completed assembly

(a)   (b)

Fig. 7 P/M printing wheel for high-speed printing equipment
(a) P/M blank. (b) Rubber-molded finished part

(a)   (b)

Fig. 8 P/M gear insert (a) and molded plastic gear and P/M insert assembly (b)

(a)   (b)

test, and it had to possess uniform density across the tooth miter angle. Figure 4 shows the major tooling components and top and bottom views of the P/M blank.

These blanks have raised areas (projections) above the teeth and angled depressions that enable the part to achieve a uniform density in the miter angle. The larger tool in Fig. 4 (upper right) shows the lower punch that molds the miter angle and hub. The other punch (lower right in Fig. 4) is the upper or top punch, which has the density compensators on its face.

After compaction at 550 MPa (40 tsi), these parts were machined to final dimensions. Figure 5 shows the P/M blank (top left) with the raised and depressed surfaces and the first through fourth cut. All machining operations were done semiautomatically on a five-spindle chucking machine at a very low cost.

The gears, used as-sintered, meet the two criteria mentioned earlier. Typical as-sintered properties of these parts are:

Tensile strength	448-517 MPa (65-75 ksi)
Yield strength	379-448 MPa (55-65 ksi)
Elongation (ductility)	1%
Hardness	65 HRB min

Printer Hammer Guide Assembly

A hammer guide assembly for a high-speed electronic impulse printer is shown in Fig. 6. In the original design, which was made from 300 series stainless steel components, the main body was machined to shape, end milled to form a hollow inside pocket, then 26 slots were milled across the hollowed-out area to accommodate the impacting hammer guides. This setup required a cover to contain the hammers. The assembly was held together with two screws.

This extensive machining generated excessive burrs and scrap. The slots had to be completely burr free. In use, these hammers are actuated by electrical impulses and must move to the impact point in

milliseconds. Consequently, even the smallest burr could not be tolerated. Final deburring was done manually, which resulted in high costs.

Figure 6(b) shows the components in the P/M design. The hammer guide, which is used at the top and bottom of the assembly, is illustrated as well as spacer blocks, which are used at both ends of the assembly, and the P/M hammers. Each assembly used 22 hammers.

The hammer guides are made from a 4600 alloy powder prealloyed with 0.25% Mn, 1.7 to 1.9% Ni, and 0.5 to 0.6% Mo and then preblended with 0.75% Cu and 0.35 to 0.45% C. The parts are pressed to a density of 6.4 to 6.6 g/cm^3. The spacer blocks are also made from this material. The hammers are pressed to a density of 7.2 g/cm^3 and case hardened.

The major improvement in the P/M design is the perfect fit between the top and bottom hammer guides. At the top left corner of each guide is a 45° angle, which ensures that the guide is in alignment when placed in the assembly fixture. Next, the two spacers are installed, and the second guide is installed to form the cover. The parts are riveted together as an assembly. The slots in the hammer guide are formed by core rods that are assembly coded to prevent improper assembly in the press setup.

The guide slots are deburred by glass bead shot peening. After deburring, the guides are teflon coated (the coating is subsequently baked) to lessen the wear between the guide and hammers.

Printing Wheels

Another P/M part used in high-speed printing equipment is the printing wheel shown in Fig. 7, which is used in an ink printing mode in postage meter applications. The material used is reduced iron with 0.85% C and 5% Cu pressed to a density of 6.4 to 6.6 g/cm^3 at a compaction pressure of 550 MPa (40 tsi).

After sintering, the parts were oil impregnated and placed in an automatic

drilling center, which drills ten holes from the outside diameter of the ten paddles into each of the molded holes, to form a "T" channel in the part.

After drilling, the parts were vibratory deburred and given a steam treatment, which provided good wear resistance due to the hard iron oxides formed and good adhesion for subsequent rubber molding. Rubber injection molding forced high-pressure rubber into the "T" channel and formed a mechanical lock of the rubber-molded characters on the outside surfaces of each of the ten paddles.

P/M Gear Insert

The P/M gear insert shown in Fig. 8 illustrates how the strength and durability of P/M techniques can be combined with high-wearing and low-noise plastics. Figure 8(a) shows the compacted gear insert. This part is used in motor-driven mail-handling and inserting machines. The material is reduced iron with 0.65% C and 5% Cu, compacted to 6.4 to 6.6 g/cm³ at 550 MPa (40 tsi) and sintered. The plastic gear is injection molded over the P/M part.

A groove was machined on the hexagonal flange of the part as a secondary operation. This formed a double mechanical lock for plastic molding. The hexagonal shape prevented the outer plastic gear from turning from the driven P/M gear surface during operation. The groove machined into the hexagonal surface prevented lateral movement between the plastic gear and the P/M insert.

After the plastic gear was molded onto the P/M insert, the entire assembly was oil impregnated, which provided added lubrication to the P/M insert, as it had over 21% porosity remaining. The P/M material exhibited excellent strength and can be hardened for additional wear resistance if required. Typical properties for the assembly are:

P/M insert

Ultimate tensile strength	448-517 MPa (65-75 ksi)
Shear strength	379-448 MPa (55-65 ksi)
Elongation	6%
Compressive strength	724 MPa (105 ksi)
Apparent hardness	90 HRB

Plastic gear

Ultimate tensile strength	52 MPa (7.5 ksi)
Shear strength	55 MPa (8 ksi)
Compressive strength	31 MPa (4.5 ksi)
Density	1.5 g/cm³
Melting point	156 °C (339 °F)

Fig. 9 One-piece P/M drive gear (a) that replaced two-piece wrought assembly (b)

(a)   (b)

As an added advantage, the plastic is four times lighter than the P/M material and allows a significant weight reduction. Also, the P/M material provides excellent sound damping characteristics at the specified density, which is important in business machine applications and consumer products.

Drive Gears

This application offers a direct comparison of P/M materials and wrought counterparts. In the production example below, a single P/M part replaced a two-piece wrought assembly. Figure 9(a) shows the completed one-piece P/M part, which replaced the two-piece steel assembly (Fig. 9b). This part is used as a postage meter drive gear in high-speed applications.

The original design, which was made from AISI 1030 steel, was made by first machining the hub from bar stock. The flange was blanked from sheet stock. Next, the two parts were press fitted together and copper brazed. The inside diameter was then broached to form the keyway, and the gear teeth were gear hobbed on the outside diameter of the flange section. Following hobbing, the part was carburized to a case depth of 0.254 mm (0.010 in.) and induction hardened only in the gear tooth section. This was done to facilitate a flattening operation to remove warpage on the flange section. Gear flatness was within 0.152 mm (0.006 in.) from the keyway to the gear teeth.

The P/M replacement was designed as one piece. The material was a 4600 alloy powder prealloyed with 0.25% Mn, 1.7 to 1.9% Ni, and 0.5 to 0.6% Mo and subsequently preblended with 0.75 to 0.9% Cu and 0.35 to 0.45% C.

The part was compacted on a 1960-kN (220-ton) mechanical multilevel press at a compaction pressure of 830 MPa (60 tsi). It was then sintered at 1120 °C (2050 °F) for 30 min. The only secondary machining operation was a turning operation to complete a locking clip groove at the edge of the small hub diameter.

The part was then carburized and induction hardened in the tooth area. After hardening, it was oil impregnated, as the density was only 6.8 to 7.0 g/cm³. A porosity of 10% existed, which provided internal corrosion resistance and lubrication. A total savings of 56% was realized.

Flatness was held to within 0.127 mm (0.005 in.) after sintering and hardening. Consequently, flattening operations were not required.

P/M Parts for Farm, Lawn, and Garden Equipment

By John R. Howell
Metallurgist
Materials Technology
Deere & Co.

POWDER METALLURGY TECHNIQUES have become viable alternatives for the production of farm, lawn, and garden products that were previously made by the machining of sand cast iron or the shaping of wrought steel into parts. There are two major factors that have been primarily responsible for the initial and increasing use of sintered structural steel parts. The first factor is cost; P/M processing provides a means of producing parts at a lower cost. The second factor has been the change in gray iron foundries. Due to changing technology, environmental controls, and economics, many of these foundries have been closed. Those remaining are highly automated, and the small, precise castings made in the older foundries are either no longer available or have become too costly. Under these conditions, P/M processing offers considerable advantages over the more costly machined castings or wrought steel parts.

This article will examine a variety of P/M structural parts that are used in load-bearing, fluid-sealing, wear, and adjustment applications. At present, these parts are produced by the conventional press-plus-sinter process that may include infiltration, impregnation, re-pressing, and/or heat treatment as specified. The compacting presses used with secondary operations are predominantly mechanical, although several of the parts are compacted on hydraulic presses. Compacting pressures are in the 414 to 616 MPa (30 to 45 tsi) range. Sintering for the majority of the parts discussed in this article is accomplished in a continuous belt furnace operating at 1125 °C (2060 °F) in an endothermic atmosphere.

Farm Equipment

In the farm industry, design engineers are now considering powder metallurgy in the initial stages of design in order to determine what the design may contribute in terms of cost savings and improved function or reliability. Production volumes are relatively low when compared to the automotive industry. Consequently, materials savings and cost are paramount considerations. These low-volume criteria are especially important in applications that require hard tooling. Powder metallurgy processing must produce a lower cost part, as well as pay for the new tooling within a reasonable length of time.

In this section, five major pieces of farm equipment and the P/M components that are used in these applications are discussed. Included is discussion of the function of the equipment itself and the function of the P/M component as a part of the machine. In addition, details on the material and process specifications are given.

Tractors

Currently, two-and four-wheel drive tractors use a variety of P/M materials. Figure 1 shows a two-wheel drive row crop tractor with mechanical front-wheel drive assist. Powder metallurgy is one of many materials processes used in the manufacture of this piece of equipment. Several of the P/M parts used in this application are discussed below. Compositions and properties of these components are detailed in the tables that accompany the figures.

Power Steering Metering Pumps. The tractor in Fig. 1 uses fully hydrostatic steering; that is, there is no mechanical connection between the steering wheel and the front wheels. The operator steers the tractor by controlling a pressurized column of oil with the power steering metering pump. This pump activates the steering valve, which in turn directs the steering motor. It is the steering motor that actually steers the front wheels of the tractor.

The power steering metering pump consists of a cover, base, body, and two gears, all of which are made by P/M processes (see Fig. 2). Gray cast iron was considered for these parts, but was rejected because of the capital investment required for equipment to do additional machining. Minimal machining of the P/M parts is required and is limited to lapping and machining of a cross hole. The cover, base, and body must be leakproof to perform in this pressurized hydraulic environment. This is achieved through epoxy impregnation of these parts. Pressurized nitrogen leak testing is specified for the cover, base,

Fig. 1 Two-wheel drive row crop tractor

Fig. 2 P/M parts for power steering metering pump

	Cover	Base	Body	Gear	Gear
MPIF material	FC-0408-P	FC-0408-P	⋯	FC-0408-P	FC-0408-P
Carbon, %	0.60-1.00	0.60-1.00	0.8	0.60-1.00	0.60-1.00
Copper, %	2.00-6.00	2.00-6.00	2.0	2.00-6.00	2.00-6.00
Nickel, %	⋯	⋯	0.40-0.50	⋯	⋯
Manganese, %	⋯	⋯	0.25-0.35	⋯	⋯
Molybdenum, %	⋯	⋯	0.55-0.65	⋯	⋯
Iron, %	91.00(a)	91.00(a)	⋯	91.00(a)	91.00(a)
Density(a), g/cm³	6.4	6.4	6.4	6.4	6.4
Apparent hardness(a)	55 HRB	55 HRB	⋯	25 HRC	25 HRC
Particle hardness(a), HRC	⋯	⋯	55	55	55
Compressive strength, MPa (ksi)	⋯	⋯	515 (75)	515 (75)	515 (75)
Crush strength, kN (ton)	⋯	⋯	⋯	26.7 (3.0)	26.7 (3.0)
Impregnation	Epoxy	Epoxy	Epoxy	⋯	⋯
Finished weight kg (lb)	1.400 (3.10)	1.460 (3.20)	0.460 (1.00)	0.125 (0.28)	0.159 (0.35)
Leak testing	Must withstand 690 kPa (100 psi) nitrogen for 5 min without leaking				

(a) Minimum

Fig. 3 Rockshaft servo cam and rockshaft cam follower

	Cam	Follower
MPIF material	FC-0408-P	FC-0408-P
Carbon, %	0.60-1.00	0.60-1.00
Copper, %	2.00-6.00	2.00-6.00
Iron, %	91.00 min	91.00 min
Density, g/cm³	6.4 min	6.4 min
Copper infiltrate	Hole area to 7.2 g/cm³	⋯
Particle hardness, HRC	52 on cam face	52
Apparent hardness, HRB	80-100 in hole area	⋯
Heat treatment	Induction harden cam face	Quench and temper
Finished weight, kg (lb)	0.735 (1.620)	0.136 (0.300)

Fig. 4 Transmission oil pump gears

	Idler gear	Driven gear
MPIF material	FC-0408-P	FC-0408-P
Carbon, %	0.60-1.00	0.60-1.00
Copper, %	2.00-6.00	2.00-6.00
Iron, %	91.00 min	91.00 min
Density, g/cm³	6.35 min	6.35 min
Particle hardness, HRC	55	55
Heat treatment	Quench and temper	Quench and temper
Finished weight, kg (lb)	0.181 (0.40)	0.318 (0.70)

Fig. 5 Four-row cotton picker

and body, and crush testing is specified for the gears.

Rockshaft Servo Cam and Followers. A load-and-depth sensing three-point hitch for tillage implements can be used to control the ground engagement depth. Alternatively, the sensing device can be used to control the drawbar load, of the tractor shown in Fig. 1. A rockshaft servo cam and follower (Fig. 3) is used to perform this function. The original designs of these parts were made by P/M processes, although stamping and machining was considered. Powder metallurgy was chosen because it provided improved wear resistance and required minimal machining.

The cam is induction hardened on the active face to a particle hardness of 52 HRC minimum. A machining operation is required to drill and tap the cross hole. During testing, threads in this cross hole were found to promote brittle fracture. This problem was eliminated by copper infiltration in the threaded area to improve toughness. The follower is produced from the same material as the cam and is through hardened to 52 HRC minimum particle hardness.

Engine oil pump gears were converted to P/M components from machined steel. Transmission oil pump gears were originally designed as P/M parts. Powder

Fig. 6 Cross-sectional view of picker unit

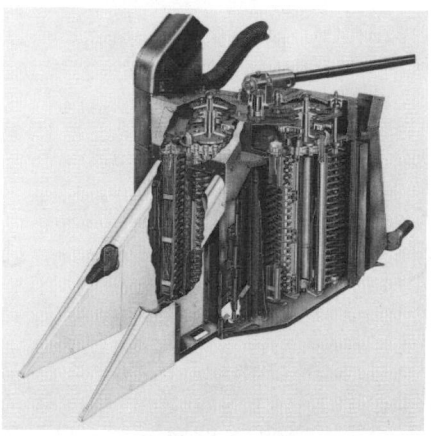

metallurgy processing eliminates machining and part handling. Two oil pump gears that comprise a transmission oil pump are shown in Fig. 4.

Cotton Pickers

Typically, cotton pickers use several different components to perform the picking operation. Although the annual production volume of cotton pickers may be low, the production volume of certain parts may be in the millions. Two of the P/M parts in current use on the cotton picker include a picker bar drive gear and a picker bar pivot stud. Figure 5 shows a four-row cotton picker that uses 96 of each of these parts per machine. Figure 6 provides a cutaway view of the picker unit in which these parts are located. These P/M components are discussed below. Compositions and properties of these parts can be found in the tables accompanying the figures.

Picker bar drive gears drive a bar that supports 20 picking spindles. Figure 7 illustrates this drive gear, which was originally designed as a P/M component. It currently is produced by re-pressing to a density of 7.2 g/cm³ and is through hard-

ened by quenching. The re-pressing operation ensures that all the gears perform as required. A statically loaded tooth strength test was implemented as a qualifier for part lots. Induction hardening is another process that has been considered for industrial use, because it permits a lower overall density (7.1 g/cm³); the compressive stresses introduced during the induction hardening operation compensate for the change in density.

Picker bar pivot studs, as shown in Fig. 8, support the lower end of the picker bar described above. This part is through hardened to an apparent hardness of 25 HRC minimum at a density of 7.0 g/cm³. Torsion testing is used to qualify lots of these parts.

Planters

Planters in current use perform several auxiliary functions in addition to planting seeds. These components apply liquid or dry fertilizer, herbicide, and insecticide. Figure 9 shows a typical planter in operation. Extensive design effort has been undertaken to ensure proper control of the amount of seed, fertilizer, herbicide, and insecticide that is deposited during the planting operation. Although many P/M parts are used in this type of agricultural equipment, several of the components used in the quantitative control aspects of this machinery are discussed below. Compositions and properties of these parts can be found in the tables that accompany the figures.

Floating Cams. The floating cam is used in a planter that uses the finger-pickup seed metering system. This mechanism, shown cut-away in Fig. 10, was designed originally for corn planting, but is also suitable for confectionery sunflower seeds. The pickup features 12 spring-loaded fingers that open and close around individual seeds. The P/M floating cam (Fig. 11) controls the action of the fingers. The cam must be resistant to wear, because the fin-

Fig. 8 Picker bar pivot stud

MPIF material	FN-0405-S
Carbon, %	0.3-0.6
Nickel, %	3.0-5.5
Copper, %	2.0 max
Iron, %	89.9 min
Density, (g/cm³)	7.0 min
Apparent hardness, HRC	25 min
Heat treatment	Quench and temper
Finished weight, kg (lb)	0.310 (0.68)

Fig. 9 Planter

gers that ride against it are case hardened and chromium plated. The cam is through hardened and tempered to an apparent hardness of 58 HRA minimum and an average hardness of 60 HRA to provide adequate wear resistance.

Dual Sprocket Assemblies. The dual sprocket assembly shown in Fig. 12 is made of two P/M sprockets that are staked together. This provides the flexibility to produce different gear ratios from pairs of

Fig. 7 Picker bar drive gear

	Version 1	Version 2
MPIF material	FN-0405-T	FL-4605-S
Carbon, %	0.3-0.6	0.35-0.65
Nickel, %	3.0-5.5	1.80-2.20
Molybdenum, %	···	0.30-0.70
Manganese, %	···	0.25-0.45
Copper, %	2.0 max	0.50 max
Iron, %	89.9-96.7	95.00 min
Density, g/cm³	7.2-7.6	7.1 min
Apparent hardness, HRC	40-52	35 min
Particle hardness, HRC	···	50 min
Heat treatment	Quench and temper	Induction harden
Case depth, mm (in.)	···	0.5 (0.02) min at root
Finished weight, kg (lb)	0.180 (0.40)	0.178 (0.39)

Fig. 10 Cross-sectional view of finger pickup mechanism

Fig. 11 Floating cam used in finger pickup mechanism

MPIF material	FN-0208-R
Carbon, %	0.60-1.00
Copper, %	1.90-2.00
Nickel, %	1.65-1.90
Molybdenum, %	0.50-0.65
Iron, %	rem
Density, (g/cm^3)	6.6 min
Apparent hardness, HRA	58 min, 60 avg
Heat treatment	Quench and temper
Finished weight, kg (lb)	0.225 (0.50)

single sprockets, rather than manufacturing a complete dual sprocket for each ratio required. The P/M staked assembly replaced a machined and welded one at a significant cost savings. The dual sprocket fits on a shaft that is fixed to the seed-metering device. A drive chain powers one of the dual sprockets to rotate the sprocket assembly and in turn drive the seed-metering device.

Sprocket Assemblies. The other sprocket on the assembly described above

Fig. 12 Dual sprocket assembly

	11-tooth sprocket	19-tooth sprocket
MPIF material	F-0005-S	F-0005-S
Carbon, %	0.26-0.60	0.26-0.60
Iron, %	97.40 min	97.40 min
Density, g/cm^3	6.9-7.3	6.9-7.3
Finished weight, kg (lb)	0.227 (0.50)	0.680 (1.50)
Additional requirements	Stake sprockets together	

Fig. 13 Sprocket assembly (sprocket plus retainer)

	Sprocket	Retainer
MPIF material	FC-0208-P	Sheet steel
Carbon, %	0.60-1.00	...
Copper, %	1.00-2.50	...
Iron, %	94.5 min	...
Density, (g/cm^3)	6.1-6.5	...
Finished weight, kg (lb)	0.091 (0.20)	...

Note: Assembled in green condition and brazed together during sintering

(Fig. 12) drives a sprocket assembly on the herbicide/insecticide meter of the planter through use of an additional chain. This sprocket assembly (Fig. 13) is produced by compacting the P/M sprocket and then brazing the sheet metal retainer to it during sintering.

Idler Arms. Figure 14 illustrates a planter component that is made from two P/M parts assembled in the green state and sintered together into a single piece. This

Fig. 14 Idler arm

	Arm	Hub
MPIF material	F-0008-P	FC-0408-P
Carbon, %	0.61-1.00	0.60-1.00
Copper, %	...	2.00-6.00
Iron, %	97.00 min	91.00 min
Density, g/cm^3	6.1-6.5	6.1-6.4
Heat treatment	Steam blue after sintering	
Finished weight, kg (lb)	0.132 (0.30)	0.132 (0.30)

Fig. 15 Feed roll and shutoff assembled in feedup

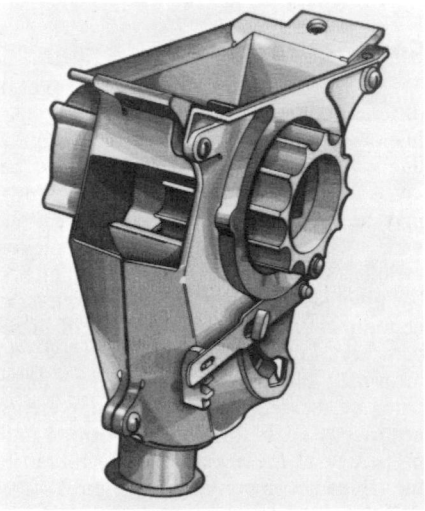

assembly is used to apply tension, through a nylon roller, to the chain that drives the dual sprocket assembly discussed above.

Grain Drills

Grain drills are used to plant small grains, soybeans, and grasses. Two assemblies in the grain drill that are formed using P/M techniques are the feedroll and shutoff, which operate in the fluted feed cup, and the planetary gear assembly, which drives the fertilizer-metering device. These components are discussed below. Compositions and properties of these parts can be found in the tables accompanying the figures.

Feed Rolls and Shutoffs. A fluted feed cup controls the amount of seed dispensed onto the grain drill. This volumetric device can be used to plant almost any type of seed. Figure 15 schematically illus-

Fig. 16 Feed roll and shutoff

| | Feed roll | | Shutoff |
	Flute	Stem	
MPIF material	F-0005-N	FC-0400-N	FC-0208-P
Carbon, %	0.26-0.60	0.30 max	0.60-1.00
Copper, %	···	2.00-6.00	2.00-6.00
Iron, %	97.40 min	90.70 min	91.00 min
Density, g/cm³	5.7-6.1	5.8-6.2	6.1-6.5
Apparent hardness, HRA	55 min(a)	···	···
Heat treatment	Carbonitride		Steam treat
Additional requirements	···	···	Oil impregnate
Finished weight, kg (lb)	0.455 (1.00)	0.455 (1.00)	0.255 (0.562)

(a) Tip and bottom of flute

Fig. 17 Planetary gear assembly

	Internal	Planet	Sun
MPIF material	FC-0208-P	FC-0208-P	FC-0208-P
Carbon, %	0.60-1.00	0.60-1.00	0.60-1.00
Copper, %	1.00-2.50	1.00-2.50	1.00-2.50
Iron, %	94.50 min	94.50 min	94.50 min
Density, g/cm³	6.1-6.5	6.1 min	6.1-6.5
Tensile strength(min), MPa (ksi)	345 (50.0)	345 (50.0)	345 (50.0)
Finished weight, kg (lb)	0.509 (1.12)	0.100 (0.220)	0.055 (0.120)

Fig. 18 Cross-sectional view of combine with corn head

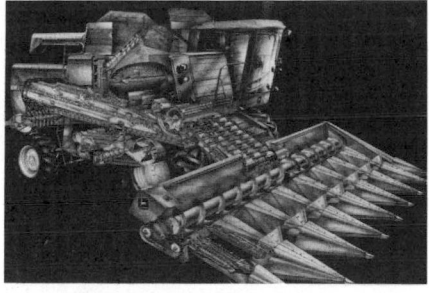

trates the feed roll and shutoff system assembled in the feedcup; these parts are shown in Fig. 16.

The feed roll rotates during the planting operation, while the shutoff remains stationary so that the amount of seed dispensed is controlled by the distance the shutoff extends into the feed cup. The feed roll is compacted as two P/M pieces, assembled in the green state, and sintered together. This assembly was originally cast in gray iron and was used in the as-cast condition. A chilled surface was provided on the flutes for wear resistance. Powder metallurgy was selected for these parts because it was difficult to obtain suitable high-quality gray iron at an acceptable cost. Carbonitriding of the P/M material to an apparent hardness of 55 HRA minimum has improved wear resistance.

Planetary gear assemblies used in the grain drill (Fig. 17) were originally designed as P/M components. Powder metallurgy has helped modify the mechanism of metering and distributing fertilizer during planting. Prior to the development of the P/M planetary gear assembly, fertilizer application rate was regulated by the opening and closing of a gate valve that controlled the gravity flow of the fertilizer. This method was subject to many problems that characteristically affect gravity flow—buildup of fertilizer on the gate, change of ground slope, and humidity, for example.

The planetary gear assembly is part of a system modification that provides positive feed of the fertilizer by using driven feed wheels. Application rate is adjusted by changing the speed of the feed shaft on which the feed wheels are mounted.

Combines

Combines typically are compact pieces of materials handling equipment (Fig. 18). They cut, convey, thresh, and temporarily store grain crops. Currently, combines use several P/M parts, including cams, gears, and manifolds. These applications are discussed below. Compositions and properties of these components are detailed in the tables that accompany the figures.

Inner and outer 45° cams, which are shown in Fig. 19, work together as torque-sensing devices that vary the tension of

the V-belt that drives a reversible, variable speed feeder house. They are carbonitrided to improve the wear resistance of the active cam faces and the inside and outside diameters. This is necessary because the outer cam also acts as a bushing for a pulley half. Cams are finish ground on the inside and outside diameters after heat treatment.

Auger bevel gears (Fig. 20) are used in pairs to transmit power from a drive shaft to an auger around a 90° corner. The augers extend almost the entire length of the combine and are used to move the threshed grain to the front of the cleaning systems.

Hydraulic Wheel Motor Manifolds. In combines powered by four-wheel drive, two of the wheels are powered by hydraulic cam lobe motors. Centrally located within the motor is a large P/M part that acts as a manifold (Fig. 21). This manifold weighs approximately 15 kg (32 lbs) and features 30 close-tolerance radial oil passages. Figure 22 shows this part in position within the hydraulic wheel motor.

The manifold was designed for P/M processing, because the machining required with other processes would be cost prohibitive. Using P/M technology, this is not an easy part to manufacture. The tooling is complex and features a dual upper punch motion, which forms the 30 radial oil passages. Two lower punches and a stepped core rod impart a taper to the manifold.

Fig. 19 Inner and outer 45° cams

	Inner	Outer
MPIF material	FC-0205-R	FC-0205-R
Carbon, %	0.3-0.6	0.3-0.6
Nickel, %	1.0-3.0	1.0-3.0
Copper, %	2.5 max	2.5 max
Iron, %	91.9-98.7	91.9-98.7
Heat treatment	Carbonitride	Carbonitride
Case depth, mm (in.)	0.51 (0.02) after grinding	0.51 (0.02) after grinding
Finished weight, kg (lb)	1.415 (3.12)	2.041 (4.50)

Fig. 20 Auger bevel gear

MPIF material	FX-2008-T
Carbon, %	0.61-1.00
Copper, %	15.00-25.00
Iron, %	69.00-84.39
Infiltrate	Copper
Density, g/cm^3	7.1-7.6
Heat treatment	Carbonitride
Particle hardness, HRC	58 min
Apparent hardness, HRB	95 min
Finished weight, kg (lb)	0.340 (0.75)

Fig. 21 Hydraulic wheel motor manifold

MPIF material	FX-2008-T
Carbon, %	0.61-1.00
Copper, %	15.00-25.00
Iron, %	69.00-84.39
Density, g/cm^3	7.1-7.6
Copper infiltrate	2.7-kg (6.0-lb) wafer
Finished weight, kg (lb)	12.973 (28.60)
Leak testing	Pressurize alternate ports with 415 kPa (60 psi) nitrogen, leakage must not exceed 75 L/min (19.8 gal/min)

Fig. 22 Exploded view of cam lobe hydraulic wheel motor

The manifold is compacted on a 8.9-MN (1000-ton) press and is copper infiltrated during sintering to provide a final density of 7.1 to 7.6 g/cm^3. Copper infiltration ensures pressure tightness and adds lubricity in operation.

Some machining is required before the manifold can be used. However, this is limited to turning the inside and outside diameters and drilling several holes. Because of the complexity of this part, cost savings are high, even at low volumes (5000 to 10 000 units).

Lawn and Garden Equipment

Lawn and garden equipment manufacturing is a competitive industry; consequently, cost-competitive processes such as powder metallurgy are ideal candidates for several applications. Lawn and garden equipment lends itself to the press-sinter method more readily than does farm equipment because its parts are of smaller size.

Ferrous parts typically are pressed at 400 MPa (30 tsi). The most frequently used presses for P/M compacting have compacting pressures around 5 MN (560 tons). Because of the pressure used and the size of the press, the typical part may have an allowable surface area of 130 cm (51.2 in.), such as a 14-cm (5.5-in.) diam gear with a 2.5-cm (1-in.) bore.

Fig. 23 Typical lawn and garden tractors

Fig. 24 Welded power take-off handle

MPIF material	F-0008-F
Carbon, %	0.6-1.0
Iron, %	97.0-99.1
Density, g/cm³	6.4-6.8
Finished weight, kg (lb)	0.055 (0.120)

Lawn and Garden Tractors

Lawn and garden tractors and riding mowers have power capacities ranging from 6 to 15 kW (8 to 20 hp) and sizes ranging from 135 to 410 kg (300 to 900 lb). Figure 23 shows several typical lawn and garden tractors. Some of these tractors feature aluminum engine blocks, hydrostatic drive, and a wide variety of attachments. Many components in lawn and garden tractors take advantage of the close-tolerance near-net shape capability and high performance-to-cost ratio offered by P/M technology. Several of the P/M components used in lawn and garden tractors are discussed below.

Welded Power Take-Off (PTO) Handle. Figure 24 shows a P/M cam used to actuate switches that are part of a neutral start system when the power take-off handle is engaged. The cam is a P/M part and features a straight sidewall, good tolerances, and the ability to perform multiple functions without additional machining. The cam is welded to the handle; to ensure problem-free welding, the part should not be dipped in or impregnated with oil prior to welding.

Power Take-Off Cam and Followers. Figure 25 shows the components used

to engage the power take off, which normally drives a mower deck. As the follower rotates through a 30° arc, the cam rides along a helix ramp that produces a 90° change in motion to axially move a shaft. The axial movement of the shaft engages and disengages the power take off. Design of this assembly capitalizes on these two P/M parts, thus offering a significant cost reduction. This P/M component replaced a riveted assembly, thus making manufacture of surrounding parts easier.

Idler pivots (Fig. 26) are part of a belt tensioning clutch on a lawn and garden tractor. Powder metallurgy was selected for this part so that some hardware could be integrated into the design. In this case, the use of two locating pins became unnecessary, because they were formed directly into the P/M version. These pins are locally infiltrated with copper to improve toughness. The part is then oil impregnated to improve corrosion resistance.

Mower hubs (Fig. 27) are used on small lawn and garden tractors and on some mowers. This hub is part of a design modification that replaced a welded hub/shaft assembly to reduce cost. The new design features a P/M hub that threads into a wrought steel shaft. The original material specified for the hub was Metal Powder Industries Federation MPIF FC-0208-R. Material specifications were subsequently changed to MPIF FN-0400-R to improve impact properties.

Front-Mounted Mower Neutral Arm. An example of a front-mounted mower is shown in Fig. 28. This mower is used on large estates, golf courses, and cemeteries, where its high maneuverability can be used to advantage. The neutral arm shown in Fig. 29 is a relatively simple P/M design that capitalizes on the close tolerances and shape flexibility offered by P/M technology. The three holes are unequally spaced and are assembled in one orientation. The notch in the end of the arm acts as a locator for assembly to ensure that the arm is placed in the correct orientation. The neutral arm is used in a neutral return linkage that controls the hydrostatic drive of the front-mounted mower.

Fig. 26 Idler pivot

MPIF material	FC-0208-R
Carbon, %	0.6-1.0
Copper, %	1.5-3.9
Iron, %	93.1-97.9
Density, g/cm³	6.4-6.8
Additional requirements	Copper infiltrate two 6.35-mm (0.25-in.) holes, oil impregnate
Finished weight, kg (lb)	0.164 (0.360)

Fig. 27 Mower hub

MPIF material	FN-0400-R
Carbon, %	0.15-0.3
Copper, %	2.0 max
Nickel, %	3.0-5.5
Iron, %	90.2 min
Density, g/cm³	6.4-6.8
Apparent hardness, HRB	30-50
Finished weight, kg (lb)	0.318 (0.70)

Fig. 28 Front-mount mower

Fig. 25 Power take-off cam and follower

	Cam	Follower
MPIF material	FC-0208-P	FC-0208-R
Carbon, %	0.6-1.0	0.6-1.0
Copper, %	1.5-3.9	1.5-3.9
Iron, %	93.1-97.9	93.1-97.9
Density, g/cm³	6.0-6.4	6.4-6.8
Additional requirements	Steam treat, oil dip	Steam treat, oil dip
Finished weight, kg (lb)	0.209 (0.460)	0.168 (0.370)

Fig. 29 Neutral arm

MPIF material	FN-0205-S
Carbon, %	0.3-0.6
Nickel, %	1.0-3.0
Copper, %	2.5 max
Iron, %	91.9-98.7
Density, g/cm^3	6.8-7.2
Additional requirements	Oil impregnate
Finished weight, kg (lb)	0.090 (0.20)

Fourth Reduction Gears. Lawn and garden tractors require gears that can withstand extremely high wear and heavy loading. For example, the fourth reduction gear shown in Fig. 30 is a drive gear in a six-speed transmission. A crush load of 4000 kg (8800 lb) per tooth is specified. High tooth density is required of the P/M part. To provide this feature, re-pressing the part to a density of 7.3 to 7.5 g/cm^3 would seem to be the solution. However, the part is too large to be compacted on a 4.45-MN (500-ton) press. To overcome this limitation, the part is compacted to three densities—6.4 to 6.6 g/cm^3 in the hub, 6.6 to 6.8 g/cm^3 in the inner flange, and 6.9 to 7.0 g/cm^3 in the tooth area.

Fig. 30 Fourth reduction gear

MPIF material	FN-0205
Carbon, %	0.5
Nickel, %	2
Molybdenum, %	0.5
Density, g/cm^3	
Tooth	7.3-7.5
Inner flange	6.6-6.8
Hub	6.4-6.6
Tooth apparent hardness, HRC	50-56
Tooth crush strength, kN (ton)	40 (4.5)
Finished weight, kg (lb)	1.042 (2.300)

After presintering, the gear is re-pressed only in the tooth area to a density of 7.3 to 7.5 g/cm^3. Carbon content is kept low to improve compressibility during re-pressing, after which the teeth are carburized.

Rototiller gear assemblies transmit power from the transmission to a chain that drives the ground tilling mechanism. The assembly shown in Fig. 31 consists of a gear and sprocket that are compacted and

Fig. 31 Rototiller gear assembly

	Gear	Sprocket
MPIF material	FN-0205-T	FN-0208-R
Carbon, %	0.5	0.6-0.9
Nickel, %	2.0	1.0-3.0
Copper, %	···	2.5 max
Molybdenum, %	0.5	···
Iron, %	···	91.6-98.4
Density, g/cm^3	7.2-7.4	6.6-6.8
Push out force, kN (ton)	67 (7.5)	
Finished weight, kg (lb)	0.485 (1.07)	
Additional requirements	Press together after full sinter, braze by local copper infiltration	

sintered individually. After sintering, the gear and sprocket are pressed together and brazed in place by local copper infiltration during an additional sinter. Machining operations are limited to honing of the bore and pressing in of the needle bearings.

Ordnance Applications

Introduction

By Saul Isserow
Metallurgist
Army Materials & Mechanics
Research Center

ORDNANCE APPLICATIONS of powder metallurgy are comprised of applications that resemble commercial applications and applications that necessitate rigorous performance requirements, which consequently limit production runs. Both of these types of applications are discussed in this article, with emphasis on the latter. Ordnance has many applications that require components similar to commercial counterparts. Such components are mass produced in sufficient quantities to justify the cost of equipment and tooling. The main goal in parts production is to achieve specified dimensions and properties at minimal cost, not necessarily with full densification. Such components usually are manufactured by conventional pressing and sintering processes.

For specialized applications, production requirements of ordnance components may be lower than conventional commercial production runs. Typically, parts producers usually require 20 000 parts, whereas ordnance systems may require only several thousand parts. Ordnance parts usually have special demands for premium performance, such as full density or compositional requirements. Performance requirements often override traditional commercial considerations and justify the premium cost of shorter production runs. Despite lower production requirements, the cost of the tooling can be amortized, thus reducing the cost of a P/M part below that of a cast or wrought counterpart through improved material utilization and reduced machining costs.

Low-volume production, with prime emphasis on material properties and quality, is not unique to ordnance applications. Aerospace and nuclear energy applications are similar examples in which performance is the overriding consideration. Special compositional requirements include stainless steels, cobalt-based alloys, and refractory metals. Low-volume specialty production may be performed in a captive facility established by a parts producer. Such an in-house arrangement permits close control over processing and properties and encourages interaction with the parts designer that can lead to tailoring of materials for specific requirements.

As another alternative, specialty producers are available who produce shorter runs, perhaps specializing in a class of materials or in certain proprietary processing methods. Choice of processing method may be determined by the low cost of tooling, such as cold isostatic pressing, for example. Such processing has found particular applications for titanium parts (Ref 1).

Tungsten-based penetrators constitute a unique P/M ordnance application (Ref 2). These mass-produced materials are produced only by P/M fabricating methods. These components exhibit high density combined with strength, toughness, and machinability. The fabrication sequence incorporates liquid-phase sintering, in which full densification of a cold pressed compact is achieved through the melting of a matrix alloy in a concentration of about 5 to 10 wt%. For tungsten carbide penetrators, the matrix is cobalt. Consequently, the materials and technology resemble those of carbide tools (hard metals).

For metallic tungsten, which has a higher density and ductility, the matrix is either a nickel-copper or a nickel-iron alloy. These materials resemble those used for shielding or counterweights (heavy alloy). With either matrix, the technology can produce a variety of geometries. For penetrators, a long cylindrical rod is the typical configuration. Similar technology is applicable for fragmentation devices.

Forging of P/M steel preforms has been used to produce an assortment of ordnance components, including machine gun parts, various gears, and guide fins for projectiles (Ref 3). These applications have advanced production technology, although these advancements have not necessarily been implemented in ordnance applications. Implementation of technology has been hampered by economic barriers; prior investment in conventional technology and the cost of qualifying new materials for high-performance weapons components have been prohibitive (Ref 4).

Work is underway to establish interchangeability of P/M parts with conventional counterparts. As in previous work on steels for ordnance applications, attention has focused on prealloyed 4600 steel powders. Process variables are being defined to eliminate residual oxides and porosity in order to realize the full capabilities of the alloy compositions, notably dynamic mechanical properties (Ref 5). Acceptance of P/M components thus will proceed from interchangeability to incorporation of such components into the initial design.

Aluminum powder metallurgy is cost effective for fuze parts (Ref 6), but as noted above for steel, adoption of the technology is limited by prior investment in conventional technology. With the development of high-strength wrought P/M alloys, such as X7090 and X7091, aerospace and military applications of aluminum P/M parts should increase. For additional information on these high-strength P/M alloys, see the article "P/M Lightweight Metals" in this Volume.

Injection molding is an attractive method for producing small ordnance parts. Irregular, intricate shapes are possible that can be justified economically for varied production volumes (Ref 7). For additional information on process parameters and potential applications, see the article "Injection Molding" in this Volume.

Use of P/M technology for ordnance applications should increase as designers become familiar with process benefits.

Eventually P/M hardware will be included in original design specifications instead of being limited to replacement components.

Titanium P/M Parts

By Edward L. Thellmann
Manager, Applied Materials
Technology
Gould, Inc.
Ocean Systems Division

BECAUSE OF the unique physical and mechanical properties of titanium, such as high strength-to-weight ratio and corrosion resistance, titanium and its alloys are ideal candidates for military and aerospace applications. However, its relatively high cost has limited wide usage. Consequently, the economics of P/M titanium processing has been a subject of extensive research. Much of this P/M research has centered around the production of commercially pure titanium and production of titanium alloys from blended elemental or prealloyed powders. Detailed information on these processing methods can be found in the article "Production of Titanium Powder" in this Volume.

Historical Development

Figure 1 (Ref 8) shows an early P/M titanium part that was fabricated from a P/M titanium alloy. This Navy sonar reflection plate was manufactured to specific density, tensile, and modulus requirements. Use of P/M techniques yielded a substantial cost reduction over machining this part from wrought stock.

Figure 2 (Ref 8) shows a P/M nose cone for missiles. Several hundred of these cold isostatically pressed and sintered parts were fabricated, with a significant economic advantage over wrought machine parts. All components passed proofing tests.

Another research project completed for the Navy involved development of a procedure to extrude P/M titanium blanks into 20-mm shell casings (Ref 8). Conventional cartridges corrode rapidly when stored in ready racks above deck, exposed to the elements. The initial production process was completed in seven stages (Fig. 3). Shell cases were produced and successfully test fired. However, a change in

Fig. 1 Sonar reflection plate
Courtesy of Imperial Clevite Technology Center

Fig. 2 Missile nose cones
Courtesy of Imperial Clevite Technology Center

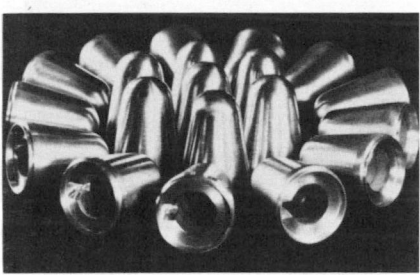

Fig. 3 20-mm cartridge cases
Courtesy of Imperial Clevite Technology Center

Fig. 4 Hollow bolts
Courtesy of Imperial Clevite Technology Center

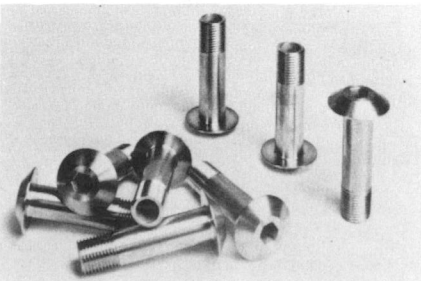

ordnance philosophy prevented further production of titanium blanks. Plastic and coated-brass cartridge cases eliminated the need for the corrosion-resistant titanium cartridges.

Figure 4 shows a hollow titanium bolt (Ref 8). Originally, this bolt was to be manufactured from a high-strength plastic for an aerospace application. The high-temperature environment, however, made use of plastic bolt materials impractical. Because design weight was based on use of plastic, the replacement material had to maintain the light weight, with no sacrifice of strength at elevated temperatures. These requirements were met by designing and fabricating a hollow bolt with a hex shape in the head and a round bore through the remainder of the bolt. This part was made by cold isostatic pressing and sintering. Finishing included a finish cut on the outer diameter and the addition of threads.

The excellent resistance of titanium to seawater corrosion makes it an ideal candidate for marine ordnance. Titanium P/M specimens were evaluated in seawater at Wrightsville Beach and in a marine atmosphere at Kure Beach, North Carolina (Ref 9). Titanium P/M test panels exhibited excellent resistance to corrosion after 15 years of exposure to seawater. Figure 5(a) shows a P/M test panel covered by marine organisms. Although P/M titanium is not resistant to biofouling in seawater, it is highly resistant to localized corrosion that can occur from the attachment of these fouling organisms. In Fig. 5(b), original machining marks on the test panels are visible after biofouling has been removed.

Another unique application of P/M titanium is its use in the production of incendiary materials. Historically, zirconium and Mischmetal have been used extensively in these applications. However, as indicated in Table 1, utilization of P/M titanium for incendiary applications offers many advantages. Titanium provides more Btus per pound or per cubic inch at a lower cost. Additional information on incendiary materials can be found in the article "Metal Powders for Fuel Propellants, Pyrotechnics, and Explosives" in this Volume.

Current Applications

Conventionally pressed and sintered P/M titanium represents a theoretical density of 94 to 96%. Further working or processing is required to achieve full density and to maximize properties, as shown by the forging preform in Fig. 6. This pro-

Fig. 5 Titanium P/M test panels for use in marine environments

(a) P/M panel covered by marine organisms. (b) P/M panel cleaned of organisms. Note original machining marks revealed after removal of biofouling. Courtesy of Imperial Clevite Technology Center

(a)

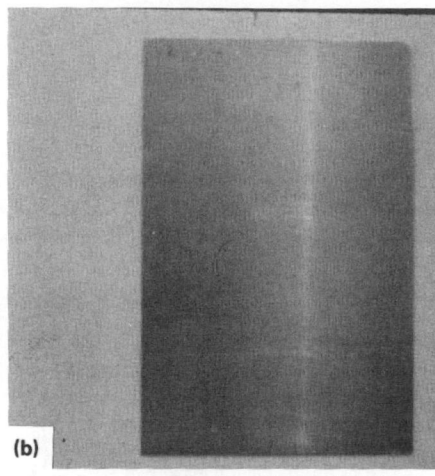

(b)

Table 1 Incendiary materials comparison

Material	kJ/kg	Heat content Btu/lb	GJ/m³	Btu/in.³	Density, g/cm³
Titanium	19 000	8200	85.8	1333	4.5
Zirconium	12 000	5150	77.6	1206	6.5
Mischmetal	7 000	3330	51.1	794	6.6

Fig. 6 Cold isostatically pressed compressor blade made from blended elemental Ti-6Al-4V powder

Part on left is the preform. Part on right is the finished forged product. Courtesy of Imperial Clevite Technology Center

totype compressor blade preform was cold isostatically pressed with blended elemental Ti-6Al-4V powder and sintered to approximately 95% density (Ref 10). The finished forged product was achieved with only one blow, resulting in minimal flash. Consequently, minimal scrap resulted. Conventional forging of this part would have required substantially more bar stock and several sets of breakdown tooling. A substantial improvement in economics was realized with P/M processing.

Currently, a P/M titanium alloy dome housing for the Sidewinder Missile is in commercial production (Ref 10). Previously, this component was machined from a billet blank weighing over 2.2 kg (5 lb). Final shape was machined to close tolerances and a thin wall of 0.635 mm (0.025 in.). Alternatively, isostatic pressing and vacuum sintering of this component resulted in a preformed blank weighing only 0.56 kg (1.25 lb) and significantly reduced the amount of finish machining necessary. Additionally, because final machining was conducted with a free-machining material, close tolerances were easily achieved. The isostatic pressed and sintered preform and the machined component are shown in Fig. 7.

In some cases, the cost of the P/M preform may exceed the cost of billet stock, but significant cost savings may be achieved through reduced machining cost. With the housing, however, the cost of preforming is only 60% of the cost of a wrought billet, with additional savings achieved in final machining.

An example of the flexibility offered by P/M processing is the missile wing configuration shown in Fig. 8 (Ref 10). This missile wing is currently under development and is fabricated by cold isostatic pressing into a preform of approximately 95% density, which is ultimately forged to full density.

A feasibility study of forging compressor stator vanes for auxiliary-powered turbines made from blended elemental titanium alloy powder metal preforms is currently being conducted (Ref 11). Preliminary cost studies show that cost savings of 40 or 50% are possible with this

Fig. 7 P/M titanium alloy dome housing for the Sidewinder Missile

Part on the left is the isostatically pressed and sintered preform. Part on the right is the finished machined component. Courtesy of Dynamet Technology

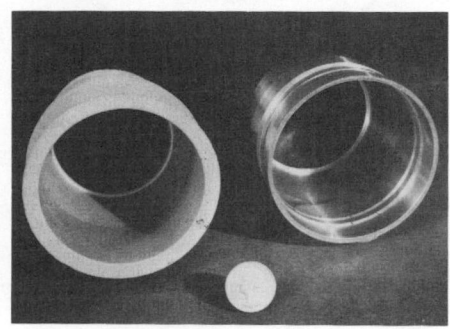

Fig. 8 Cold isostatically pressed plus forged titanium P/M missile wing

Courtesy of Dynamet Technology

Fig. 9 Compressor stator vane made from blended elemental titanium alloy powder

Courtesy of Garrett Turbine Engine Company, Division of Garrett Corp.

Fig. 10 Preform and machined lens housing
Courtesy of Imperial Clevite Technology Center

P/M approach. The forged compressor stator vane is shown in Fig. 9.

Figure 10 is a typical P/M titanium ordnance component. The preform and finished machine part are shown. This nonrotating part is a lens housing for the Maverick Missile. Powder metallurgy effectively reduced the "buy-to-fly ratio" from 15 to 1 to 3 to 1.* Material savings were approximately 80%. Additional savings resulted from the superior machinability of P/M titanium, even though this part was machined to very close tolerances (Ref 12).

Figure 11 shows commercially pure titanium fasteners produced by P/M fabricating methods (Ref 8). Acceptance of these fasteners has been exceptional because of significant cost savings realized by P/M processing. This P/M fastener has captured a large segment of the titanium hex nut market and is likely to find use in ordnance applications.

Major airframe and engine companies have parts in bench testing and in-flight testing to qualify P/M titanium hardware. Figure 12 shows a keel splice manufactured by conventional forging and hot isostatic pressing (Ref 12). Forging weight is approximately 2.1 kg (4.67 lb); the P/M near-net shape is approximately 0.45 kg (1 lb), with a final machine part weighing 0.18 kg (0.4 lb). See the article "Hot Isostatic Pressing of Metal Powders" in this Volume for additional information on near-net shape processing of titanium parts.

Several other components, such as disks, impellers, airframe braces, and a compressor stator connecting link for the F-100 jet engine, are also in production

*Buy-to-fly ratio indicates the number of kilograms (pounds) of purchased material for each kilogram (pound) of finished part.

Fig. 11 Commercially pure titanium hex nuts
Courtesy of Imperial Clevite Technology Center

Fig. 12 Titanium keel splice
(a) Conventional forging, 2.1 kg (4.67 lb). (b) Hot isostatically pressed near-net shape P/M part, 0.45 kg (1 lb). (c) Final finished P/M part, 0.18 kg (0.4 lb). Courtesy of Crucible Inc.

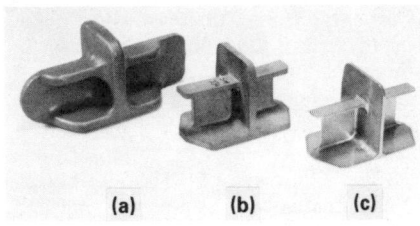

(a)　　(b)　　(c)

status or under investigation (Ref 13). These and other applications are discussed in the article "Aerospace Applications" and "P/M Lightweight Metals" in this Volume.

P/M Ferrous Materials

By Andrew Crowson
Metallurgist
U.S. Army Research & Development Center
Fire Control & Small Arms Weapons System Laboratory

ORDNANCE COMPONENTS made of ferrous materials are classified as ammunitions or weapons. A wide variety of ammunition parts are fabricated from P/M ferrous materials. The most common example is the sintered iron rotating band that was developed during World War II in Germany because of a shortage of copper.

Refinement of the process occurred in the 1960's with the adoption of a U.S. military specification (MIL-R-11073). Other examples of successful ammunition parts fabricated from ferrous P/M materials include ammunition cases, frangible ammunition, mortar bodies, and projectile shells. Conventional forming and forging technology is combined with P/M processing in many of these products to obtain acceptable physical and mechanical properties.

Weapon components also are successfully made from P/M ferrous materials. However, use has been restricted mainly to small arms weapons, specifically machine gun components. The accelerator for the 0.50-caliber M-85 machine gun is a typical ferrous P/M application. In the 1970's, this part was made by forging of P/M steel. Recently, fabrication of larger weapon components also has been investigated. Preliminary efforts have demonstrated that the breech block of a 175-mm howitzer can be fabricated by hot isostatic pressing.

Successful adaptation of P/M ferrous materials for ordnance applications is contingent on meeting the material property and performance standards required by the military and by overcoming established methods. Standards usually are based on physical and mechanical property requirements, such as density, surface finish, and dimensional tolerances, and on actual service performance.

Close dimensional tolerances are always a prerequisite for acceptance and require detailed dimensional and coordinate measurements of the part. Mechanical properties vary with type of application and service environment and are specified by standard American Society for Testing and Materials (ASTM) methods or specialized functional or service test procedures. In most cases, standards are based on a service life capability of two to three times the actual intended service.

Processing Variables

The application of P/M ferrous-based materials for ordnance parts generally includes the following: selection and production of powders, consolidation, presintering, sintering, and post-consolidation (forming).

Powders. Chemical compositions of powders that are suitable for ordnance applications are given in Table 2. Pyron 100 and Ancorsteel 101 are sponge iron powders produced by hydrogen reduction of select iron ores (see the article "Production of Iron Powder" in this Volume). They

Table 2 Typical ferrous-based powders for ordnance applications

Designation or tradename	Fe	Ni	Mo	Mn	Cu	Cr	P	S	Si	C
Pyron 100	97.5-98.5	...	...	0.45-0.65	...	...	0.012	0.005	...	0.015-0.022
Ancorsteel 101	98.8	...	...	...	...	...	...	0.01	...	0.20
Ancorsteel 1000	99.2	...	...	0.20	...	...	...	...	...	0.02
Ancorsteel 1000B	99.3	...	...	...	...	...	...	0.016	0.01	0.01
Ancorsteel 4600	97.2	1.77	0.48	0.23	0.05	0.05	<0.01	0.02	0.07	...

Fig. 13 Compactibility of prealloyed 4600 steel powder

ρ is the compact density; P_c is the compaction pressure; 2.38 and 0.238 are compaction equation constants.

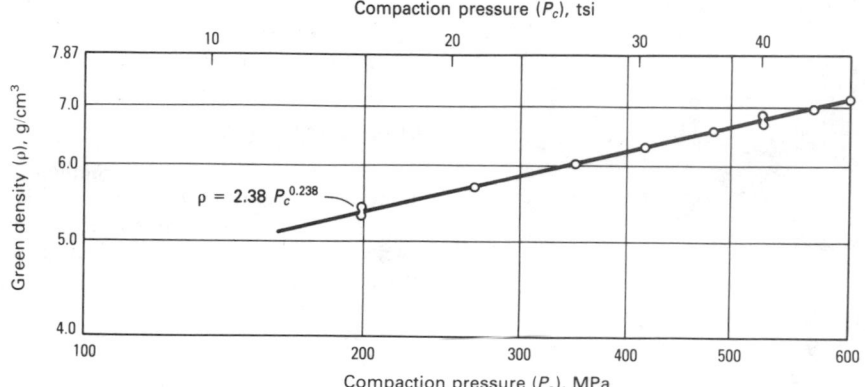

are used for fabricating rotating bands and ammunition cases. Ancorsteel 1000, 1000B, and 4600 are prealloyed powders designed for parts with densities over 6.7 g/cm³. These powders are produced by water atomization of molten metal (see the articles "Production of Low-Alloy Steel, Tool Steel, and Stainless Steel Powders" and "Atomization" in this Volume). Each particle has a homogeneous composition. Typical applications for these powders include frangible ammunition, mortar bodies, projectile shells, and weapon component parts.

Blending. Mixing of lubricant and graphite with ferrous or steel powders frequently is required before compaction. Standard lubricants such as stearates and waxes, in amounts up to $^1/_2$ to 1%, are commonly used (see the article "Lubrication of Metal Powders" in this Volume). Graphite in flake powder form is added to obtain the desired carbon content in the final product. The amount added is equal to the desired combined carbon content of the sintered preform plus an additional small amount to counteract losses caused by oxide content in the powder. These losses are due to the carbon-oxygen reduction reaction of the sintering process. Blending of constituents is accomplished by mixing in a blender for $^1/_2$ to 1 h. Longer mixing times result in segregation of the constituents (see the ar-

ticle "Blending and Premixing of Metal Powders" in this Volume).

Compaction of the powders is accomplished by using closed, confined die sets. Double-action or multiple-motion floating die sets are recommended to minimize density gradients in the compacted preform (see the articles "Shape Fundamentals: Rigid Tool Compaction" and "Production Presses and Tooling" in this Volume). Compacting pressures vary, depending on the desired preform density. When conventional P/M parts, such as iron rotating bands, are used for a particular application, typical densities range from 6.10 to 6.49 g/cm³. If additional consolidation techniques (such as forging or extrusion) are required after sintering, required preform density ranges from 6.30 to 7.00 g/cm³. A typical density-compaction pressure relationship is shown in Fig. 13 for 4600 powder. Similar relationships exist for other powders and can be calculated by:

$$\rho = aP_c^b$$

where ρ is the compact density; P_c is compaction pressure; and a and b are constants.

Presintering of the compacted preform completely removes (burns off) the admixed lubricant and imparts sufficient strength to the compact for handling. Usually, presintering is conducted in a dissociated ammonia atmosphere at temper-

atures of 430 to 650 °C (800 to 1200 °F) for 30 min.

Sintering. Conventional sintering generally is conducted in continuous or batch-type sintering furnaces. Preforms are heated in dissociated ammonia for 1 h at 1120 to 1150 °C (2050 to 2100 °F). In P/M forging applications, sintering is performed primarily in batch-type sintering furnaces at 1200 °C (2200 °F). Hydrogen and dissociated ammonia are both suitable atmospheres for this process. See the articles "Production Sintering Equipment," "Production Sintering Atmospheres," and "Production Sintering Practices for P/M Materials" in this Volume for additional information.

Forming. Additional reconsolidation operations are undertaken after sintering if full or near-full density is required. Typical forming operations include coining, extrusion, and hot forging. Coining and extrusion are used to cold form ammunition cases and mortar bodies from 101 and 1000 series steel powders (see Table 2 for compositions). Hot forging has been used to fabricate projectile shells and weapon components from 4600 series steel powders. See the article "P/M Forging" and the Appendix on hot extrusion in the article "Hot Pressing" for additional information.

Ordnance Applications

The following examples are typical P/M parts that have been used successfully in ordnance applications. The first example employs conventional P/M processing of iron powder. Production of mortar shell bodies incorporated cold working of the P/M preform to densify the porous material into a specific shape. The last two production examples utilize P/M hot forging to fabricate complex, load-bearing parts.

Iron Rotating Bands

The rotating band is an essential component of spin-stabilized ammunition. On an artillery projectile, the band performs the following operations during firing:

- Engages the spiral gun rifling and imparts a spinning motion to the projectile for improved flight stability
- Minimizes leakage of gaseous propellant products
- Engages the gun tube rifling cleanly, with minimum wear of the tube, without generating fragments or residues that may foul the gun tube
- Adheres to the projectile in flight

Fig. 14 Sintered iron 105-mm rotating band blank

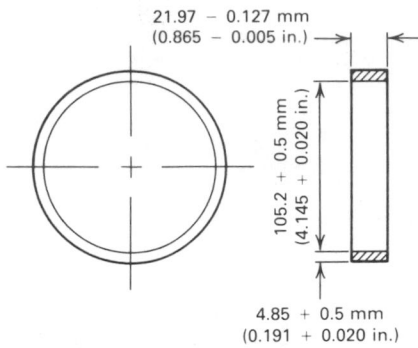

21.97 − 0.127 mm
(0.865 − 0.005 in.)

105.2 + 0.5 mm
(4.145 + 0.020 in.)

4.85 + 0.5 mm
(0.191 + 0.020 in.)

Table 3   Properties and wax contents specified in MIL-R-11073C for various classes of band blanks

Class	Density, g/cm³	Tensile strength(a) MPa	Tensile strength(a) ksi	Ductility(b), %	Wax content(a), wt%
1 ...	5.30-5.99	76	11	3	3.25
2 ...	5.60-5.89	90	13	4	2.50
3 ...	5.90-6.19	110	16	5	1.75
4 ...	6.20-6.49	141	20.5	6.5	1.00

(a) Minimum. (b) Minimum expansion of band circumference

Sintered iron blanks for rotating bands are simple in design and consist of a ring with a specified inside diameter, width, and wall thickness. The 105-mm band blank shown in Fig. 14 is typical of all such blanks, except for its size.

Fabrication of sintered iron rotating blanks begins by blending hydrogen-reduced iron powder, such as Pyron 100, with approximately 1% of standard admixed lubricant (Ref 14). The powder is then compacted, using a double-action floating die. Compacts are presintered in air at 430 to 650 °C (800 to 1200 °F) for 30 min to burn off lubricant. Sintering follows at 1120 to 1150 °C (2050 to 2100 °F) for 60 min in dissociated ammonia. Compacts are impregnated with molten microcrystalline wax by immersion for 1 to 5 h.

Properties and wax contents for the four density classes of band blanks covered by current military specification are shown in Table 3. Densities are determined using ASTM B 328, which includes checking segments cut across the width of the band blanks to ensure that a uniform density of ±0.2 g/cm³ is attained.

Tensile strengths are determined using a "saddle fixture" tension test, as shown in Fig. 15. Testing is performed at 0.635 cm/min (0.25 in./min). Tensile strength is calculated by dividing the load at fracture

Fig. 15 Split saddle tensile testing of 105-mm sintered iron rotating band

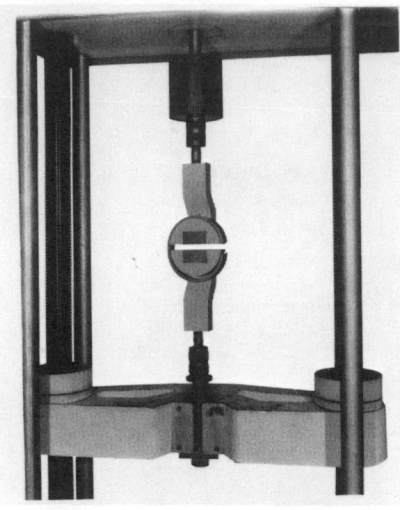

Fig. 16 Mandrel expansion testing of 105-mm sintered iron rotating band

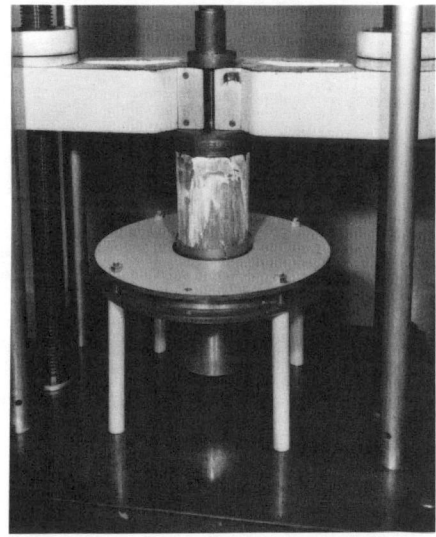

by twice the cross-sectional area of the band blank.

Ductility is determined using a "mandrel" expansion test, as shown in Fig. 16. A band is placed on a tapered mandrel and manually pushed downward until the band is firmly wedged on the taper of the mandrel. The diameter of the mandrel at the leading edge of the band is equal to the inside diameter of the band. The band is then pushed along the mandrel, without causing fracture, for a sufficient distance to determine compliance with the minimum amount of expansion specified for that particular density class.

Advantages. Sintered iron has definite advantages over copper-based alloys for rotating bands on artillery shells (Ref 15). Iron is particularly well suited for low-velocity howitzer shells because of its slower engraving pressure. Sintered iron is also suitable for high-velocity projectiles because of its higher melting point. At velocities exceeding 760 m/s (2500 ft/s), bands develop high surface temperatures from frictional heating; copper-based alloys tend to melt and coat the inside of the gun barrel. Additionally, the cost of sintered iron bands is appreciably less than the cost of copper alloy bands, and iron is more readily available for such applications.

Mortar Shell Bodies

A mortar shell body is a fragmenting munition used by the Army against "soft targets," such as personnel, jeeps, and trucks. For fragmenting munitions to be effective, the material must be able to withstand relatively low launching stresses and have a fracture pattern that will provide an optimum fragment size. In the past, the material used for 60- and 81-mm mortar bodies has been a low-alloy manganese steel, such as AISI 1340.

Fabrication. The typical processing sequence used for the production of 60-mm M-49 mortar bodies from sintered P/M preforms is shown in Fig. 17 (Ref 16). Atomized iron powder (Ancorsteel 1000) is blended with 0.4% graphite powder and 0.75% paraffin wax. The blended powder is compacted to 86% theoretical density using a double-action die pressing arrangement. Presintering at 540 °C (1000 °F) for 1 h follows to burn off lubricant. The preform is sintered for 1 h at 900 °C (1650 °F) in dissociated ammonia.

The compact is lubricated with a zinc phosphate soap coating. The preform undergoes a coining operation in a mechanical press to approximately 96% of theoretical density. It is re-sintered for 1 h at 1120 °C (2050 °F) in dissociated ammonia and subsequently lubricated with a zinc phosphate soap coating. The coined preform is backward extruded in a mechanical press to approximately 98% of theoretical density. Annealing for 1 h at 700 °C (1300 °F) in dissociated ammonia follows, with an additional lubrication with zinc phosphate soap coating.

Mortar shell body walls are forward extruded in a mechanical press, followed by lubrication with a zinc phosphate soap coating. The tail is tapered by cold extrusion in a mechanical press; a final lubrication with a zinc phosphate soap coating follows. Finally, the nose section of the

Fig. 17 Cold extrusion sequence of a P/M 60-mm M-49 mortar shell body

| Metal powder | P/M preform | Coined preform | Backward extrusion | Forward extrusion | Taper tail | Nose formed | Final machined shell |

Fig. 18 Accelerator for 0.50-caliber M-85 machine gun

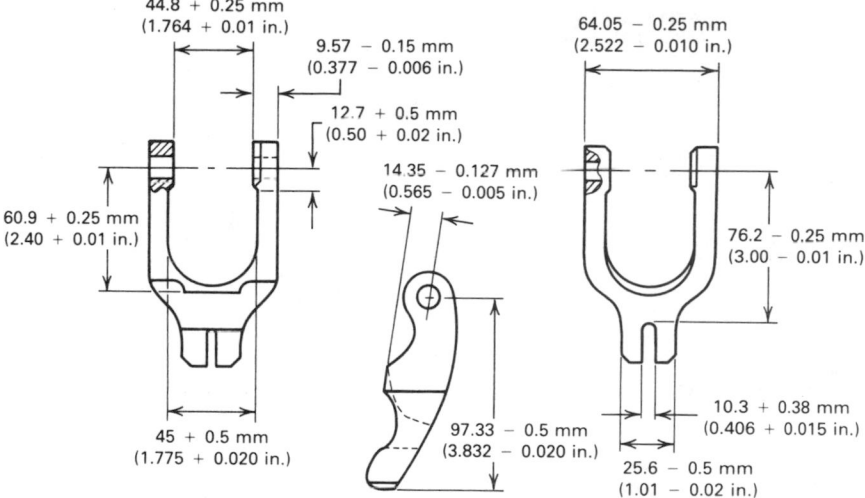

44.8 + 0.25 mm
(1.764 + 0.01 in.)

9.57 − 0.15 mm
(0.377 − 0.006 in.)

12.7 + 0.5 mm
(0.50 + 0.02 in.)

14.35 − 0.127 mm
(0.565 − 0.005 in.)

64.05 − 0.25 mm
(2.522 − 0.010 in.)

60.9 + 0.25 mm
(2.40 + 0.01 in.)

76.2 − 0.25 mm
(3.00 − 0.01 in.)

45 + 0.5 mm
(1.775 + 0.020 in.)

97.33 − 0.5 mm
(3.832 − 0.020 in.)

10.3 + 0.38 mm
(0.406 + 0.015 in.)

25.6 − 0.5 mm
(1.01 − 0.02 in.)

mortar body is formed by re-pressing in a mechanical press. Stress relief annealing at 430 °C (800 °F) for 1 h in dissociated ammonia completes processing. Machining to final configuration may be required.

Advantages. Use of P/M steels as preforms for conventional cold forming into shell bodies offers a number of advantages over standard wrought steel. Powder metallurgy steel preforms can be readily designed and processed to optimum configurations and weight with minimal scrap losses occurring during final machining.

Conventional methods of fabricating mortar shell bodies require excess starting material because of the large dimensional and weight variations among slugs that result from cutting and shearing of the bar stock. Total scrap losses between the original weight of the slug and that of the finished shell body can be as much as 15 to 20%. Inclusions, seams, and laps frequently are encountered with wrought material, which may cause fractures and defects in the formed shell bodies.

Improved control of metallurgical properties can be achieved with P/M steel preforms. The number of cold working operations required to form a mortar body may be reduced by incorporating a P/M preform that can be introduced at an intermediate stage of cold working. Composition of P/M materials can be adjusted easily by elemental powder additions to the iron powder.

Properties. Acceptance criteria for mortar shell bodies are based on minimum property levels (yield strength of 310 MPa, or 45 ksi, and elongation of 12%) and fragmentation characteristics. Mechanical properties are determined using R-5 tensile bars according to ASTM E 8, "Tension Testing of Metallic Materials." Fragmentation is based on pit fragmentation tests; following explosive fragmentation of a shell body, the fragments are recovered, counted, and weighed. From these data, the total number of fragments weighing 0.03 g (0.5 grains) and more and the average fragment mass in grains are

calculated and compared to the fragmentation behavior of a standard mortar body.

Accelerators for Machine Guns

Accelerators are high-performance weapon components used in the firing mechanisms of machine guns. A typical accelerator used in the 0.50-caliber M-85 machine gun is illustrated in Fig. 18. Firing rates and loads are such that the accelerator is subjected to a highly stressed, impact-loaded environment. Past material requirements specified medium-carbon alloyed steel (AISI 4340) in the forged and heat treated condition. With conventional processing, the material is worked through a series of dies that plastically form the metal into the desired shape (see Fig. 19). Excess stock, which is necessary to completely fill the die, remains in the form of flash and must be trimmed. Final forging subsequently requires an extensive amount of machining before the desired part is obtained.

Fabrication. A more cost-effective approach for fabricating the accelerator is the P/M steel forging process (Ref 17). In contrast to conventional methods, P/M forging technology utilizes a single, flashless forging operation (Fig. 19). Major cost advantages are incurred through more efficient material utilization and a reduction of machining operations.

Processing begins by blending 4600 prealloyed powder with 0.48% graphite and 0.50% zinc stearate. The blended powder is compacted in a double-action floating die to 85% of theoretical density. Presintering at 540 °C (1000 °F) in a reducing atmosphere follows to burn off lubricant.

Fig. 19 Comparison of conventional forging and P/M forging processes for producing 0.50-caliber M-85 machine gun accelerator

Conventional processing shown on top. P/M processing shown on bottom. See text for details.

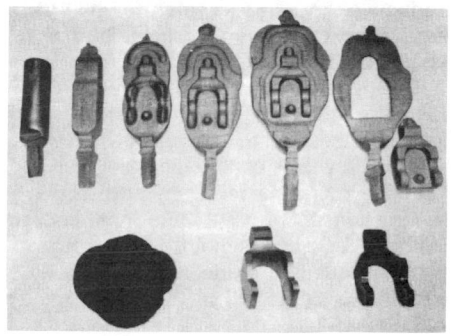

Fig. 20 Response of 4640 P/M steel forging to heat treatment

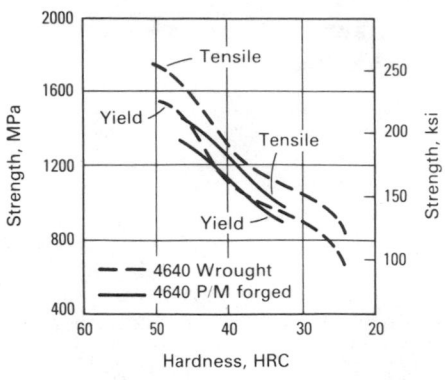

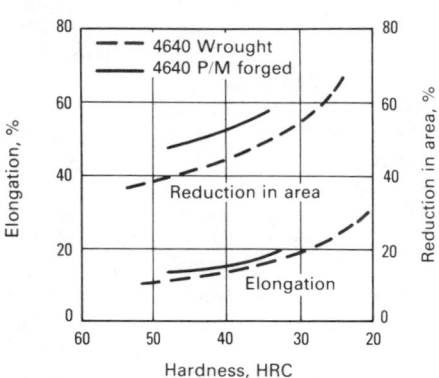

Fig. 21 Guided projectile fin requirements

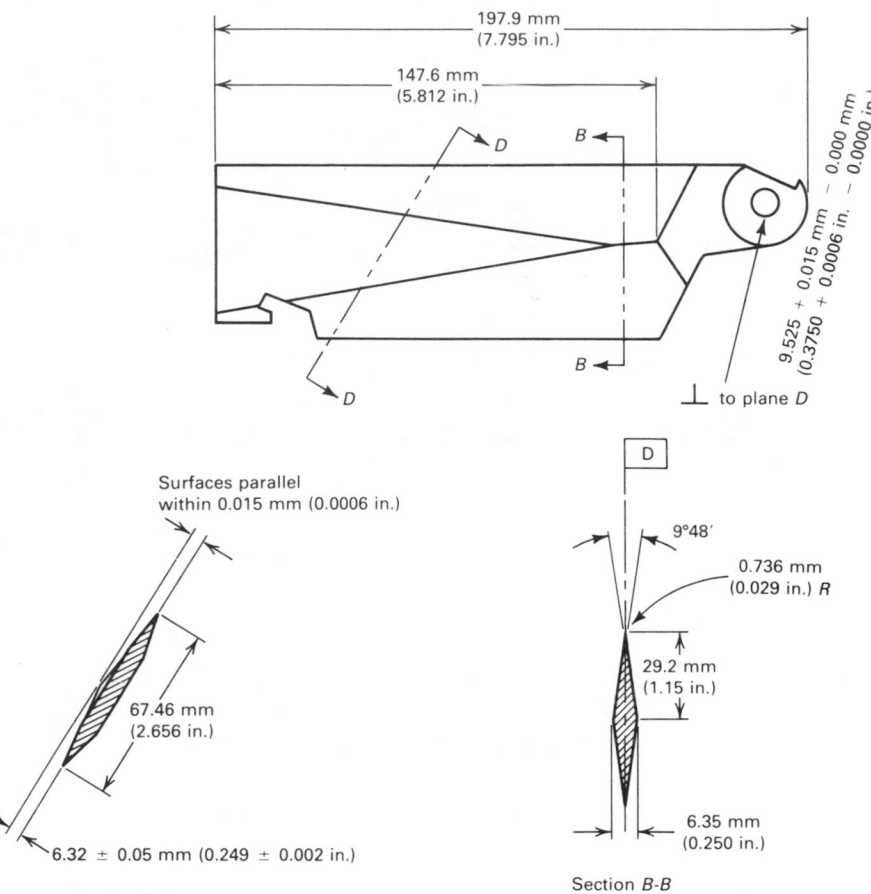

Section D-D

Section B-B

The compact is sintered for 1 h at 1200 °C (2200 °F) in a hydrogen atmosphere containing 1 vol% methane (Ref 18). The preform is coated with graphite prior to preheating. Preheating for 15 min at 1200 °C (2200 °F) follows in a hydrogen atmosphere containing 1 vol% methane. The preform is forged at 550 MPa (80 ksi) in a closed, confined die preheated to 200 °C (400 °F).

Finishing operations for forged accelerators include heat treating to 46 to 51 HRC, machining, chrome plating, and applying a final protective finish.

Properties. Military material acceptance requirements (MIL-F-45961) for producing the M-85 accelerator by P/M forging are contingent on achieving tensile properties comparable to wrought 4640 (AMS specification 6317B). Typical tensile values for 4640 P/M forgings as a function of heat treatment are shown in Fig. 20. Tensile values of wrought 4640 are included for comparison. In addition to meeting tensile property requirements, a minimum impact strength of 34 J (25 ft·lb) at 32 HRC and an oxide content of <300 ppm are also required in the final forged part (Ref 19).

Fig. 22 Isothermal transformation diagram for 4640 steel

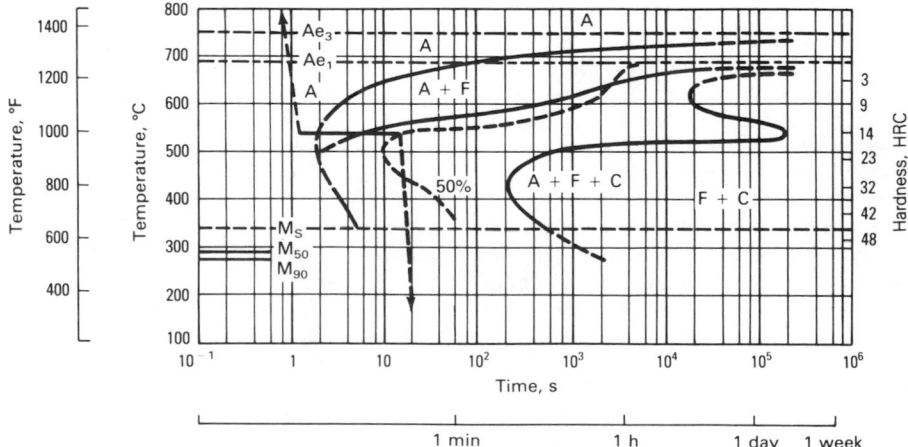

Guided Projectile Fins

The trailing fin for a Navy projectile is shown in Fig. 21 (Ref 20). The overall length is about 198 mm (7.8 in.), the airfoil width is 58 mm (2.3 in.), and the thickness varies from 9.14 mm (0.360 in.) in the attachment area to a tip thickness of 2.67 mm (0.105 in.), with an edge radius of 0.74 mm (0.029 in.). The airfoil is essentially a parallelogram cross section of tapering thickness with flats imposed. Extremely close tolerances, especially with regard to airfoil surfaces, must be maintained for the projectile to function satisfactorily in flight.

Fins were originally designed to be produced from 17-4 PH steel castings, with a minimum yield strength of 1170 MPa (170 ksi). Powder metallurgy steel forgings may be substituted as an alternative material, providing cost savings through more efficient material utilization.

Fabrication. Atomized steel powder (4600) is blended with 0.48% graphite and 0.75% zinc stearate. The blended powder is compacted in a closed, confined die to 80 to 85% of theoretical density. Presintering at 540 °C (1000 °F) follows to burn off lubricant. The compact is sintered for 1 h at 1200 °C (2200 °F) in a hydrogen atmosphere.

The preform is then coated with graphite prior to preheating for 15 min at 1200 °C (2200 °F) in a furnace containing a hydrogen plus 0.25 vol% methane atmosphere. Forging at 550 MPa (80 ksi) follows in a closed, confined die preheated to 200 °C (400 °F).

The preform is oil quenched and austenitized at 930 °C (1700 °F) for 30 min. Coining in a closed, confined die follows at 540 °C (1000 °F) for 10 s at 275 MPa (40 ksi) pressure, and the compact is oil quenched again. Preheating for 15 min at 500 °C (925 °F) in a hydrogen plus 0.25 vol% methane atmosphere follows. The compact is again coined in a closed, confined die for 1 min at 500 °C (925 °F) with a 275 MPa (40 ksi) pressure and is slowly cooled in silocel.

Coining after forging minimizes thermal and residual stresses. Substantial improvement in the parallelism and flatness of fins can be achieved by using this procedure instead of conventional oil quenching. A near flatness deviation of 0.061 mm (0.0024 in.) and parallelism of 0.081 mm (0.0032 in.) are possible with coining at 540 °C (1000 °F).

Properties. Specifications require tensile testing and a 100% inspection of all characteristics. Fins manufactured by the specified procedure have hardnesses ranging from 36.5 to 37.5 HRC and the following tensile properties:

Tensile strength	1132 MPa (164 150 psi)
Yield strength (0.2%)	1019 MPa (147 800 psi)
Elongation	11.5%
Reduction in area	37.1%

The 1019 MPa (147 800 psi) yield strength was below the required 1170 MPa (170 000 psi) for the finished fin. However, by coining at 650 °C (1200 °F) after austenitizing, the desired 1170 MPa (170 000 psi) minimum yield strength was obtained through increased transformation

time at the higher temperature (see Fig. 22).

The 100% inspection criteria combine a coordinate measurement and analysis program. Special fixtures are used to measure characteristics such as flatness, parallelism, and hole perpendicularity with respect to the center plane through the fins. Additionally, fins are inspected on a coordinate measuring machine and with a minicomputer. Output data are a function of Mach number, which indicates the performance capability of the particular fin being inspected.

P/M Parts for Military Vehicles

By Donald T. Ostberg
Materials Engineer
U.S. Army Tank-Automotive Command
and
James L. Chevalier
Chief, Armor & Components Function
U.S. Army Tank-Automotive Command

INCREASED INTEREST in P/M processing has recently focused on forging of P/M preforms to net shape and near-net shape high-performance gears. The technology is generic in nature and can be applied to wrought forgings. Fabricating technology is discussed in terms of P/M forging as a cost-effective method of gear production.

Implementation of P/M technology was a two-stage process. The first stage included the forging and gear evaluation of a 95.25-mm (3.75-in.) diam, 28-tooth spur gear with a small hub (Fig. 23). Forging of accessory gears and final drive input pinion gears for military combat vehicles constituted the second stage of process implementation.*

To eliminate costly trial-and-error methods of designing adequate preforms and preform dies, a computer-assisted design program was developed for AISI 4600 series powder. A die assembly that accepts sets of interchangeable inserts and

*The data presented in this section are part of an ongoing U.S. government project (contract DAAK30-80-C-0115) funded to TRW.

Fig. 23 Forged 95.25-mm (3.75-in.) 28-tooth spur gear made from 4600 series powder

punches was built so that a series of gear sizes under approximately 150 mm (6 in.) in diameter could be forged with the same die set and ejection system.

Test gears were suitable for testing on the NASA-Lewis gear test equipment. Preforms of AISI 4600 steel powder and graphite were die pressed at a load of 1334 kN (150 tons). Zinc stearate was used as a lubricant.

Control of several forging variables was critical to obtaining the desired final gear dimensions, including preform preheat temperature, die temperature, transfer time from the sintering furnace to the die, lubrication, and deformation loads. Sintered preforms were transferred from the dissociated ammonia atmosphere furnace at 1200 °C (2200 °F) to the forging die within 4 s. The part was forged with one blow in a 6225-kN (700-ton) capacity crank press. Die preheat temperature was maintained between 180 and 200 °C (355 and 385 °F). Deltaforge 31 lubricant, which is a graphite suspension in water, was sprayed on the top punch, bottom punch, and preform. Sintering for 30 min (minimum) serves as preheating for forging.

Sixty P/M gears were produced for evaluation: 20 oversized 4620 forgings carburized and ground to final tooth size, 20 net shape carburized 4620 forgings, and 20 net shape through-hardened 4640 forgings. In production, die operating temperature is controlled by cooling the dies. However, due to the limited number of gears forged in this production application, four cartridge heaters were embedded in the die and augmented by a gas burner.

NASA-Lewis Research Center reported B-10 lives of 13×10^6 cycles for the 4620 ground carburized tooth gears and 5×10^6

Fig. 24 Accessory gear for the M-1 Abrams tank AGT 1500 engine

cycles for the 4620 carburized net shape tooth gears. At the time of this writing, results were not yet available from NASA on the 4640 forgings. All tests were run at 10 000 rpm and 1710 MPa (248 ksi) maximum pitch line Hertz stress.

These test results show that P/M gears can be designed for high reliability in moderately to severely loaded environments. The P/M forging process produces components with properties equivalent to wrought material of the same composition. However, minor dimensional changes in the net shape forging cause some variations. For lightly or moderately loaded automotive grade gears, net shape forged P/M gears provide acceptable performance at minimum cost. For highly loaded or aircraft-quality gears, grinding may be necessary.

To aid implementation of P/M technology, actual production gears were forged as part of the second stage of the program. Number 6 accessory gears for the M-1 Abrams tank AGT 1500 engine and the final drive input pinion gear for the M-2/M-3 Fighting Vehicle were chosen as prototypes. These gears were designed to be interchangeable with existing gears and to have equivalent or superior properties.

The No. 6 accessory gear (Fig. 24) is a 114.2-mm (4.495-in.) diam, 61-tooth, 25-mm (1-in.) thick spur gear originally made from AMS 6414 (AISI 4340) and heat treated to a through hardness of 34 to 37 HRC. The P/M accessory gears, which were made from AISI 4640 steel and heat treated to a through hardness of 34 to 37 HRC, were sintered and forged to net shape with the same process used for the NASA test gears.

The final drive input pinion is a 154-mm (6.08-in.) diam, 19-tooth spur gear.

This size diameter is near the maximum for the existing die nest. The gear is 114 mm (4.50 in.) deep, including an integral shaft and an internal 1.4- to 1.8-mm spline. Current material requirements for AISI 4320H, AISI 4815 to AISI 4820H, and AISI 9310H require a case carburized layer with a hardness of 58 to 62 HRC and a case depth of 1.397 to 1.778 mm (0.055 to 0.070 in.). Core hardness is 28 to 42 HRC at the tooth center line.

The P/M final drive input pinion gears made from AISI 4620 were sintered and forged with the same process used for the other gears. These gears were forged to near-net shape and carburized to obtain a case hardness of 58 to 62 HRC. Instead of forging the spline to net shape, a smooth, undersize hole was forged. Machining was required only for the spline.

By forging sintered P/M preforms, components such as gears can be fabricated with properties equivalent to wrought material. The cost differential between the P/M gear and the conventionally fabricated gear depends on the shape of the component, dimensional tolerances, material, and operational requirements. If the component can be forged to net shape, expensive machining operations can be eliminated.

Kinetic Energy Penetrators

By Thomas W. Penrice
Vice President—Technology
Teledyne Firth Sterling

FABRICATION of high-density tungsten alloys into kinetic energy penetrators represents the major usage of powder metallurgy in ordnance applications. The high density of tungsten (19.3 g/cm^3) improves performance of kinetic energy penetrators. Tungsten alloys with densities ranging from 17.0 to 19.0 g/cm^3 are almost exclusively fabricated by P/M techniques; consequently, material selection dictates the use of P/M techniques, rather than competition between alternative fabrication processes.

Kinetic energy penetrators are slugs of metal accelerated to a high velocity by combustion of propellant in a cannon. Directed toward a target, the stored energy is used for penetration and incapacitation of the target. Within the limits imposed by mobility, ammunition handling, rate of fire,

and cost, a cannon with the largest practical caliber, or bore size and length, is the most effective. When used with an energetic propellant, large quantities of initial energy are available. In the 120-mm, 50-caliber antitank cannon, operating at a pressure of 690 MPa (100 ksi), 9.8 million J (7.2 million ft · lb) of energy accelerates a 7-kg (15-lb) projectile to a velocity of 1650 m/s (5400 ft/s).

A projectile that is equal in size to the full caliber loses energy rapidly in flight due to excessive aerodynamic drag. This loss can be minimized by discarding outer parts of the projectile "sabots" after launch. The remaining subprojectile, at a diameter approximately 25% of full caliber, has a cross section of less than 10% of the original. This arrangement not only reduces energy loss in flight, but also directs a high energy density at the target. Consequently, the subprojectile, which is essentially the penetrator, takes the form of a long rod with a length-to-diameter ratio of 10:1 to 20:1.

A lightweight nose cone is added to minimize further aerodynamic loss; rear fins stabilize flight trajectory, because spin stabilization is not effective for length-to-diameter ratios greater than about 4 to 1. Typically, at a range of 2000 m (1.24 miles), allowing for loss of mass of discarded parts and velocity falloff, only 55% of the original energy remains to attack the target. In this case, 5.4 million J (4 million ft · lb) impact on an area of 7 cm^2 (1.1 in.^{2}). Components for a 105-mm projectile are shown in Fig. 25.

Pure tungsten, with a theoretical density of 19.3 g/cm^3, is an ideal candidate for this application. Fabricated by traditional P/M techniques of cold pressing, direct resistance sintering at 2500 to 2800 °C (4500 to 5000 °F), and swaging to about 50% reduction in area, a density of 19.1 g/cm^3 can be achieved for 30-mm (1.18-in.) diam bars. However, tungsten has negligible ductility, is extremely difficult to machine, and is expensive to fabricate.

Fabrication Techniques

There are alternate ways to utilize tungsten for this application. Tungsten carbide bonded with 9 to 13 wt% of nickel or cobalt was originally used as a kinetic energy penetrator in World War II. Fabrication by cold-press liquid-phase sintering P/M techniques produces the following typical properties:

Density . . . . 14.2 to 14.4 g/cm^3
Transverse-rupture strength . . 2750 MPa (400 ksi)
Hardness . . . . 89.5 HRA

Fig. 25 Components of a 105-mm (4.13 in.) projectile

Courtesy of General Defense Corp., Flinchbaugh Division

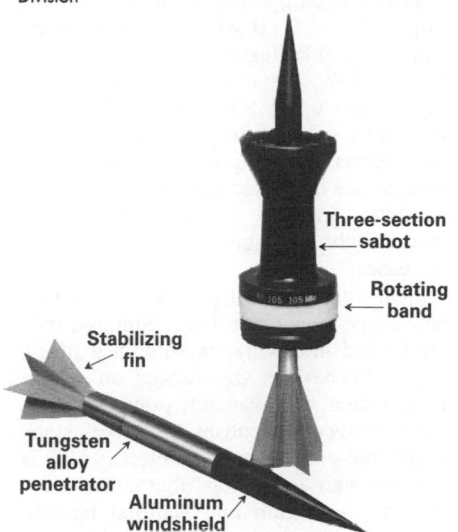

Three-section sabot

Rotating band

Stabilizing fin

Tungsten alloy penetrator

Aluminum windshield

Fig. 26 Components of a spin-stabilized medium-caliber projectile

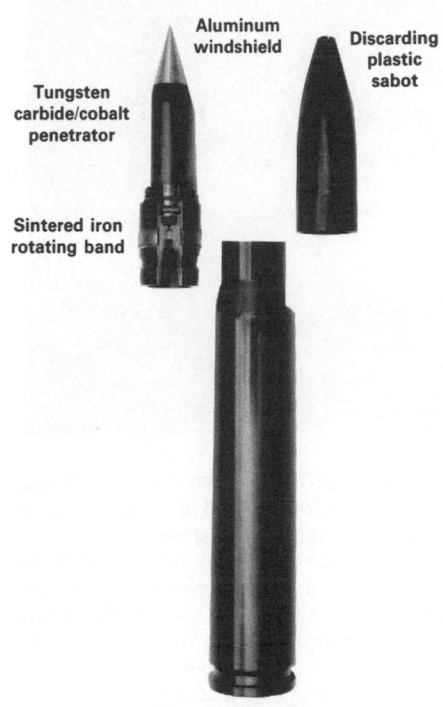

Aluminum windshield

Discarding plastic sabot

Tungsten carbide/cobalt penetrator

Sintered iron rotating band

20-mm armor-piercing discarding sabot ammunition

Fig. 27 Flowchart of the fabrication of tungsten alloy penetrators

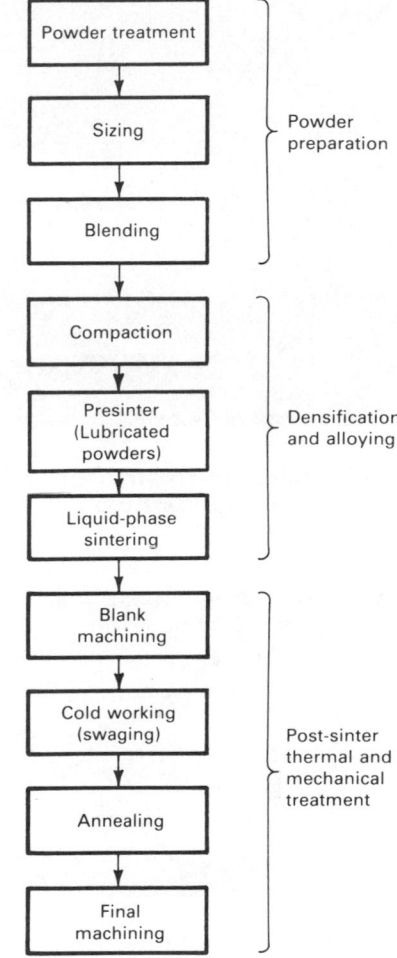

Powder treatment ⎫
Sizing ⎬ Powder preparation
Blending ⎭

Compaction ⎫
Presinter (Lubricated powders) ⎬ Densification and alloying
Liquid-phase sintering ⎭

Blank machining ⎫
Cold working (swaging) ⎬ Post-sinter thermal and mechanical treatment
Annealing
Final machining ⎭

Material properties are adequate to provide good performance against single-plate steel targets at near-normal impact angles. Tungsten carbide/cobalt hard metals are still used for spin stabilized small- to medium-caliber designs (5.62-mm through 20-mm, or 0.22-in. through 0.79-in. caliber), as shown in Fig. 26.

Currently, penetrators are made of tungsten-nickel-copper or tungsten-nickel-iron alloys. Alloys containing nickel-iron generally are preferred because of their superior mechanical properties. Alloys with tungsten contents of 90 to 97% can be fabricated to near-theoretical density (17.0 to 18.5 g/cm^3). The nickel-to-iron ratio of the alloy is not critical, but is commonly 7 to 3. A flowchart for fabrication of tungsten alloy penetrators is shown in Fig. 27.

Hydrogen-reduced tungsten powder, with particle sizes ranging from 3 to 5 μm, is mixed with nickel and iron powders (3 to 10 μm particle size) prepared from the metal carbonyl (see the articles "Production of Tungsten, Molybdenum, and Carbide Powders," "Production of Iron Powder," and "Production of Nickel and Nickel Alloy Powders" in this Volume). Ball mills or mixers are used to blend these materials. The introduction of work hardening to the powders or generation of new surfaces, common to ball or attrition milling, is not necessary to achieve good results in processing these alloys.

The metal powder mixtures develop reasonable green strength when compacted without binders or lubricants. Isostatic pressing at 200 MPa (30 ksi) pressure provides green strength that is adequate for handling and shaping prior to sintering. Binders, which decompose during presintering, may leave carbon residue that degrades mechanical properties of sintered material.

Sintering is almost exclusively performed in molybdenum resistance-type electric furnaces with a hydrogen or nitrogen-hydrogen atmosphere (Fig. 28). Densification occurs rapidly, but sintering continues to promote grain growth associated with good mechanical properties (Ref 21). The 3- to 5-μm tungsten grains enlarge to 50 to 90 μm.

The microstructure of a typical liquid-phase sintered tungsten alloy is shown in Fig. 29. Grain growth apparently occurs by preferential solution of one grain and precipitation onto another across a thin matrix film between adjacent particles. The energy level associated with the crystal orientation of the faces of the grains most likely determines which grain grows.

On cooling, tungsten retains only a trace of nickel and iron. The matrix, however, retains a substantial amount of tungsten in solution (up to 25 wt%), which may be controlled by the addition of other alloying elements, notably copper in nickel-iron alloys. Nickel-copper alloys are prone to void formation in the matrix on rapid cooling from the sintering temperature. Ductility of both alloy systems is affected by cooling rate. Slow cooling improves ductility.

Batch-type sintering furnaces that permit numerous combinations of heating ramp rate and temperature hold to achieve full densification and desired microstructure also allow control of cooling rate and, consequently, ductility. Batch-type furnaces, however, are expensive to operate and as a result are not well suited to sintering of long rod, high-energy density penetrators.

Post-Sintering Operations. To maximize ductility, it is necessary to reduce process dependence on the sintering cooling rate by using a secondary heat treatment. Phosphorus and sulfur impurities segregate at the tungsten matrix interface with concentration factors of 1500 and 4000, respectively (Ref 22). Concentration factors are segregation enrichment ra-

Fig. 28 Molybdenum resistance-type electric sintering furnace that uses a hydrogen atmosphere

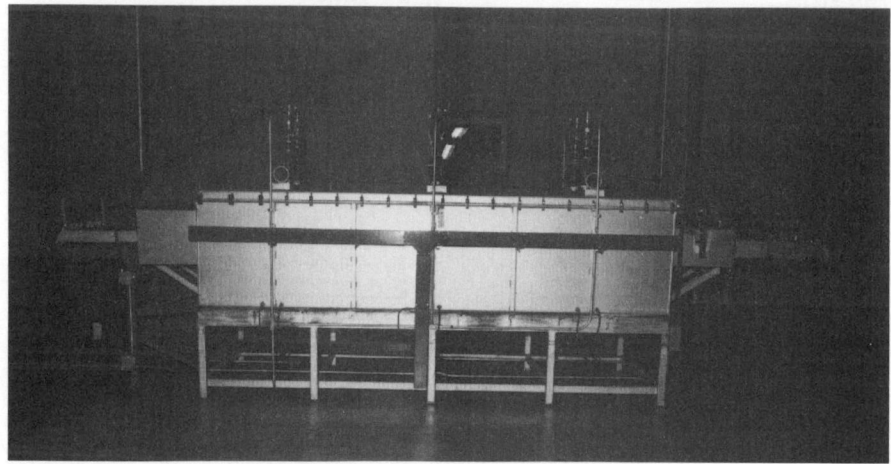

Fig. 29 Typical microstructure of a W-Ni-Cu-Fe liquid-phase sintered alloy

Magnification: 250×

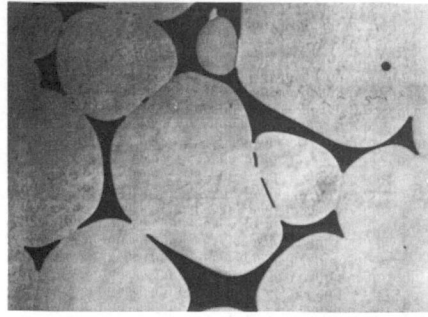

tios calculated from chemical content and the number of monolayers measured at grain boundaries. Phosphorus, which is uniformly distributed over the interface boundaries, adheres preferentially to the matrix or binder phase side of the interface. Consequently, phosphorus concentration is lower after solution annealing, whereas sulfur concentration is less uniformly distributed and is found equally on the tungsten and matrix sides of the interface. Furthermore, sulfur concentration is increased by a factor of two during solution annealing and quenching from 1350 °C (2460 °F), which reduces phosphorus concentration by a factor of about 4 to 1.

Solution annealing at 1000 to 1250 °C (1830 to 2280 °F) and quenching from 1100 to 1150 °C (2010 to 2100 °F) effectively improves ductility of sintered components. For 97.5 and 90.0 wt% tungsten alloys, elongation at failure in tension is typically 13 and 30%, respectively. After annealing, values up to 50% are possible. The effects of solution annealing on the fracture surface of a tungsten alloy penetrator material are illustrated in Fig. 30. With low ductility, fracture is intergranular and increasingly transgranular as ductility increases.

Heat treating usually is conducted in a batch-type vacuum furnace. Sintered rods are loaded into work baskets on a tilting rack. The baskets are stacked on a base plate fitting with vacuum pumping ports and thermocouple entries. The work stack is covered with an Inconel retort, which is covered with a "hot-top" that contains refractory insulation and Kanthal heating elements. The work area inside the retort and the hot-top are evacuated simultaneously, thereby preventing retort collapse during high-temperature processing.

Gas must be used to assist heat transfer during heating and cooling. Pumping of argon at 1000 °C (1830 °F) requires the use of coolers to protect the pumps. Cooling is accelerated by removing the hot-top after cooling to about 800 °C (1470 °F). A typical heating cycle for 8 metric tons (9 tons) of material takes about 32 h floor-to-floor time, including 10 h at 1050 °C (1920 °F).

Hardness is approximately 30 HRC, which remains almost unchanged by annealing. Strength in tension increases slightly, however, to about 900 MPa (130 ksi). This hardness may be considered too low to effectively penetrate hardened steel armor. These alloys are strain-rate sensitive, however, and at impact velocity they appear to have higher hardness and strength, with correspondingly reduced ductility.

Alloys may readily be cold worked—by rotary swaging, for example—to increase strength and hardness. Typical values are ultimate tensile strength of 1300 MPa (188 ksi) and hardness of 43 HRC, with a corresponding decrease in ductility after cold working. The range of properties available by varying alloy composition and cold working is shown in Fig. 31.

Cold working also may be achieved by upset forging; compression to more than 50% is generally possible without cracking. Further variation of properties is possible by aging after cold working. A 93% tungsten alloy penetrator cold worked to 25% reduction in area and aged at 500 °C

Fig. 30 Effect of solution annealing on the fracture surface of a 90W-7Ni-3Fe alloy

(a) Fracture surface showing intergranular fracture characteristic of low-ductility material. (b) Fracture surface after solution annealing to improve ductility with predominantly transgranular fracture. Magnification: 325×

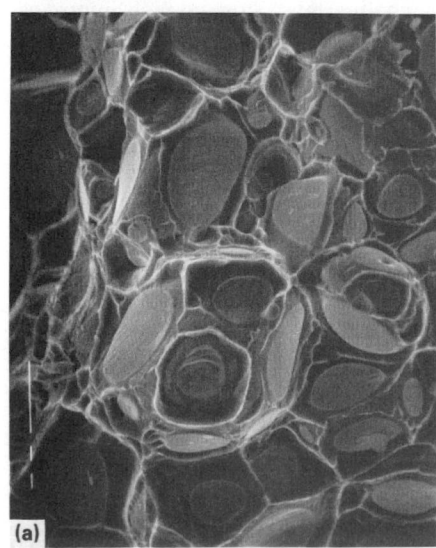

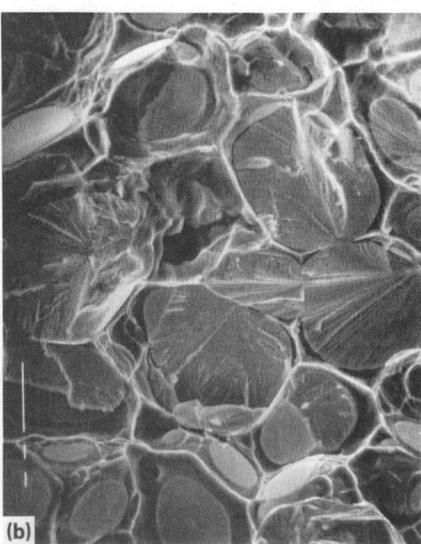

(a)

(b)

Fig. 31 Strength versus ductility by tungsten content for varying amounts of swaging cold work

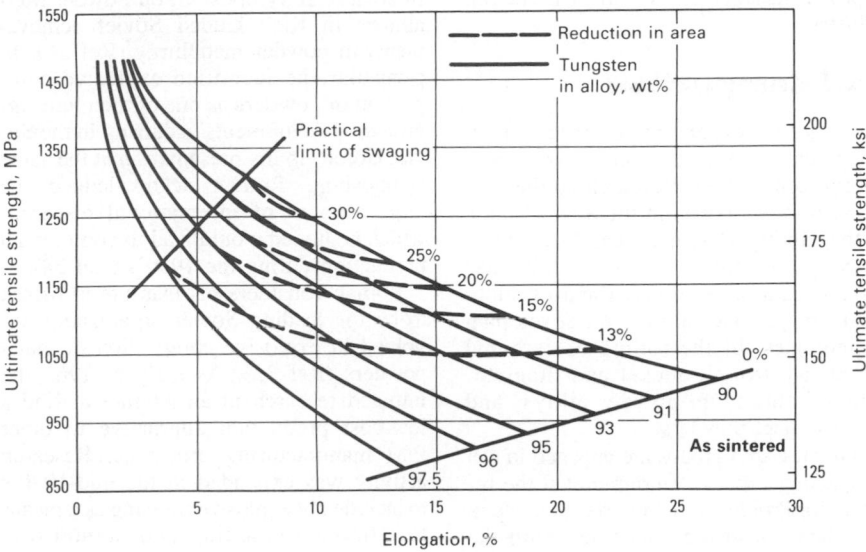

Fig. 32 Test firing of a fragmentation device incorporating zirconium pellets
Source: Teledyne Wah Chang Albany

(930 °F) for 1 h exhibits an ultimate tensile strength of 1500 MPa (218 ksi), when tested at a rate of 30 MPa/s (4.3 ksi/s), and a hardness of 47 HRC.

These alloys machine readily in all metallurgical conditions. Continuous chips form only for materials with extreme ductility. A class 2 cemented carbide is used to machine with or without coolant, typically at 150 m/min (500 sfm). Feed rate and depth of cut are limited only by surface finish, component rigidity, and available power. Collection of machined chips in a clean condition allows low-cost reprocessing to powder by comminution or by oxidation-reduction processes, rather than by more costly chemical recovery processes.

Fragmentation Devices

The same density, energy density, and velocity falloff requirements and guidelines outlined for penetrators also apply to fragmentation devices. Typical fragmentation devices include anti-aircraft shells exploded by a proximity fuse and bomblets released from a container delivered by missile or aircraft. Fragments may be 3- to 4-mm (0.12- to 0.16-in.) diam tungsten carbide or tungsten alloy spheres or fractured, prescored tungsten alloy hemispheres. The effectiveness of these munitions is doubled by the use of high-density tungsten alloys.

Inclusion of compacted zirconium sponge pellets makes air-to-ground bomblets effective incendiary devices. Figure 32 shows a test firing of such a device, in which burning zirconium fragments at the outside border of the explosion readily ignite fuel in the containers in the foreground.

Training ammunition for kinetic energy penetrators must simulate the flight characteristics of live ammunition, but should ideally self-destruct for safety reasons. A number of solutions for such ammunition have been developed. For example, a partially densified iron powder component is encased in a low-strength, thermally degradable plastic container. On light impact or after a few seconds of flight, the casing heats sufficiently to cause failure, and the iron powder compact disintegrates within the distance permitted by range safety requirements. Procurement quantities for such training ammunition generally exceed those for live ammunition.

Appendix: Soviet P/M Technology

William F. Marley Jr.
Metallurgist
U.S. Army Foreign Science and Technology Center

THE RAPID GROWTH in research, development, and production of powder metallurgy in the Soviet Union is readily evident. Although the Soviets credit themselves with the first significant developments in P/M processing during the early 1800's (the pressing and sintering of platinum powder into coins), it was not until about 1950 that they began to produce iron and nonferrous metals and alloys for industrial applications.

In a November 1978 speech before the Communist party, Leonid Brezhnev underscored the need to save metal. A newspaper account of Brezhnev's speech highlighted several well-known benefits of powder metallurgy, including reduced labor costs and the development of new materials with specific properties (Ref 23).

Powder metallurgy in the Soviet Union now receives high-level government support, which should allow the Soviets to keep abreast, if not move to the forefront, in certain aspects of the P/M state of the art. A significant statement from the Fundamental Directions for the Economic and Social Development of the USSR agency regarding planned progress through the year 1990 appeared as follows in Ref 24: "We plan to increase by a factor of three the production of metal powder in order to obtain products with enhanced durability, wear, corrosion resistance and also to reduce labor and metal consumption in making machines and mechanisms."

Since the advent of its commercial availability, powder metallurgy has developed rapidly into a major production technology that is used throughout Soviet industry. Many P/M technologies have been developed; however, those receiving the greatest research and development attention include roll compaction, explosive compaction, rapid solidification technology, and nickel and titanium P/M fabrication.

Roman, director of the Scientific Research Institute of Powder Metallurgy in

Fig. 33 Projected demand for metal powders in the Soviet Union

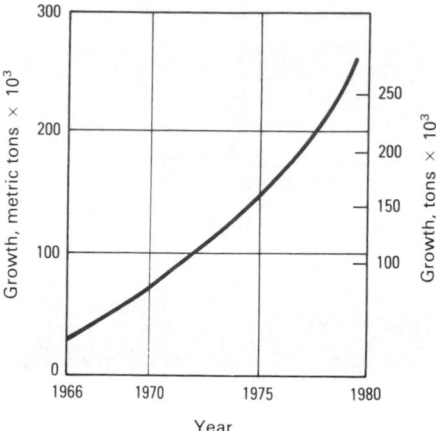

Minsk, has noted that in 1978 the State Committee on Science and Technology conceived a program for the development of powder metallurgy for the 11th Five Year Plan, which terminates in 1985 (Ref 25). This thrust can, in part, be attributed to Brezhnev's determination to develop a viable P/M industry.

Figure 33 illustrates the projected growth for metal powders through 1980. From these data, the Soviets apparently realized that they were not fulfilling their basic requirements for metal powders in the early 1970's and made the decision to accelerate powder production. Over the last 5 years, the Soviets have made an effort to upgrade their powder production capability through the purchase of Western technology. The West German P/M plant, reported to be the world's largest water atomization plant to produce iron and steel powders (capacity about 80 000 metric tons/year, or 88 000 tons), should eliminate powder deficiencies when operation begins in 1983 (Ref 26). Davy McKee Limited has built a high-speed tool steel powder plant that can produce 60-kg (130-lb) batches of tool steel powders (Ref 27).

The creation of a special P/M Problem Council under the State Committee for Science and Technology should contribute to improved management, particularly by accelerating applications of P/M components (Ref 28). Scientific problem councils are established to coordinate research and development programs related to key technical problems in a developing industry (Ref 29). Although the councils do not have administrative powers, their recommendations appear to carry considerable weight, largely because the membership of the councils is generally composed of highly respected scientists and technical personnel. Apparently, problem councils are created to effect better coordination and improve management within a particular industry.

Roll Compaction

A large and expanding research and development program is in progress regarding roll compaction. Research on the rolling of powders began in the Soviet Union at the Gorkiy Polytechnical Institute in 1948. Since that time, this facility has maintained a large research and development group, concentrating on developing technologies for the rolling of carbonyl nickel, electrolytic nickel and titanium, sintered aluminum powder alloys, and stainless steel powders.

Other research groups are engaged in roll compacting studies. Vinogradov at the Institute for Problems of Material Science in Kiev heads a group concerned with the theory and mechanisms of powder rolling (Ref 30, 31). Priorities include development of methods for calculating the density and thickness of strips for a given rolling condition and the development of the relationships between roll pressure and the density of the rolled products.

The Belorussian Polytechnical Institute in Minsk also is involved in powder rolling research. For over 22 years, it has been concerned primarily with the engineering aspects of roll compaction. One program has led to the development of technology for rolling carbide strip of any length with a thickness of 0.2 to 6 mm (0.008 to 0.25 in.). Use of this technology has reduced the cost of carbide tool tips and has reduced use of expensive materials by 50 to 67%. Tools made by this process are now being used at two production plants. The technology is being introduced at an aluminum plant for rolling aluminum powder into sheet (Ref 32). Also, work is being directed at the rolling of porous stainless steel sheet, as well as at the design of powder rolling equipment (Ref 33).

Another program at this institute deals with the rolling of dispersion-strengthened materials, such as nickel powder mixed with alumina (Ref 33). Recent applications of ultrasonic vibration during the rolling process increased dispersion of the oxide particles, resulting in improved elevated-temperature properties in the final product.

Explosive Compaction

The Soviets are actively advancing their capabilities in high-energy rate forming, particularly explosive compaction of both metal and ceramic powders. A paper presented by Fedovchenko at the May 1977 International Symposium on Powder Metallurgy in Kiev lauded Soviet achievements in powder metallurgy (Ref 34). In particular, he identified explosive compaction of powders as one of the more significant developments occurring in the Soviet Union. In his presentation at the same symposium, Roman acknowledged the large amount of international research, which he hoped would lead to commercial production during the 1980's (Ref 34).

Roman considers his institute in Minsk to be the leading Soviet organization involved in explosive compaction of metal powders (Ref 35). As early as 1963, he initiated research in an attempt to find a low-cost production alternative to other P/M manufacturing processes. Research activity was expanded in the mid-1970's to include hot explosive pressing as a means to eliminate cracking and nonuniform densities experienced under normal explosive compaction conditions. Roman is credited with establishing the Scientific Research Institute of Powder Metallurgy in 1972 and is now its director (Ref 36).

Roman appears to enjoy strong governmental support for the explosive compaction programs. According to Ref 11, an entire complex is to be built, consisting of a laboratory, an experimental facility for explosive compaction research, and a pilot plant with an annual production capacity of between 5000 and 10 000 metric tons (5500 and 11 000 tons) of P/M components. The complex will also have its own P/M design bureau.

The Soviet Academy of Sciences is actively supporting explosive compaction. About 1973, at the Institute of Metallurgy in Moscow, Shorshorov began studying the explosive compacting of iron, titanium, and nickel powders (Ref 37). Shorshorov is involved in composite materials research and may be investigating the use of explosive compacting as a production process for composites as well as for metallic powders.

Rapid Solidification Technology

One of the fastest growing metallurgical technologies—not only in the United States, but also in the Soviet Union—is the P/M processing of alloys formerly made by conventional melting and casting techniques. Following the lead of the United States, Soviet technologists presently are involved in a major effort to develop rapid solidification technology and incorporate it in P/M manufacturing capabilities. Although rapid solidification technology is useful for processing several alloy powder systems, Soviet research has focused on nickel and titanium alloy powder manufacturing (Ref 38).

Soviet interest in rapid solidification technology is manifested in the decision to assign its developmental and research coordination responsibility to the USSR Scientific Council (Ref 28). It has been reported by Belov that the USSR Ministry of Higher and Specialized Education has been asked to establish courses to train engineers in rapid solidification technology (Ref 39). These developments at high levels of the Soviet government underscore the emphasis being given to rapid solidification technology as a means to achieve higher strength values of structural materials.

Rapid solidification technology also is being investigated to produce steel powders, as well as powders from nickel alloys, cobalt alloys, aluminum alloys, titanium alloys, and intermetallic compounds. Linking rapid solidification technology with various consolidation techniques, such as explosive compaction, appears to offer the Soviets considerable latitude in utilizing advanced materials technology.

The Soviets are interested in aircraft gas turbine engine applications for rapidly solidified nickel powder (Ref 40, 41). Advantages of rapid solidification technology are well documented by Belov, including high structural homogeneity and physicomechanical properties that exceed those obtained by conventional ingot metallurgy methods (Ref 42).

Nickel Powder Metallurgy

Nickel powders have been produced in the Soviet Union since the late 1950's. However, the Soviets did not begin to develop a nickel P/M technology base until about 1975 (Ref 43). The Soviets now realize the importance of this technology and have greatly accelerated their research and development programs.

The Soviets understand the importance of powder quality and characteristics. They know that powders produced by atomization methods have higher levels of purity, low levels of segregation, uniformity of size, and improved properties due to faster cooling rates than powders produced by other means. Consequently, atomization methods are now receiving increased attention.

The Soviets have begun to report on isothermal forging of powdered nickel

alloys and are interested in achieving a capability in superplastic forming. One of the first publications confirming their interest appeared in 1974. Atomized ZhS6K powder (Ni-3.8Mo-5.3Al-2.8Ti-10.8Cr-4.9W-4.5Co) was used, which was probably produced by argon atomization (Ref 43).

The Soviets have become active developers of superplastic forming. Articles have been published on the mechanisms of plastic flow during sintering and forming operations. Several papers favor a grain boundary sliding mechanism to account for superplastic behavior. They have found that powders should have small but polycrystalline particles, indicating a preference for atomized or rotating electrode powders. This work supports an understanding of the mechanisms of superplasticity and would advance Soviet efforts in superplastic forming (Ref 44).

The Soviets have shown an interest in hot isostatic pressing of nickel alloy powders. Following the lead of the United States, they have initiated at least one hot isostatic pressing program. Engineers at the All Union Institute of Light Alloys in Moscow have been involved in nickel powder metallurgy since the mid-1970's; they recently have focused on designing steel cans for hot isostatic pressing of superalloy powder aircraft engine disks. Disks up to 60 cm (24 in.) diam were produced in plain carbon steel cans with 5- to 7-mm (0.2- to 0.3-in.) thick walls (Ref 45).

Readily evident at the 1981 Paris Air Show was the Soviet intention to introduce cost-effective P/M nickel-based alloys into jet engines and airframe applications (Ref 26). The Soviets are moving in the same direction as American engineers regarding the use of P/M nickel alloys for aircraft engine components.

Titanium Powder Metallurgy

The Soviets have been exceptionally active in titanium powder metallurgy. They are striving to develop specific compositions for alloys and have shown interest in hot isostatic pressing and isothermal forging (Ref 46). The Soviets also continue to maintain a strong titanium powder alloy development program.

Because P/M alloys possess greater homogeneity, the Soviets are apparently seeking alloys with higher solute contents and thereby are attempting to produce alloys with elevated-temperature strength and stability. According to Ref 47, P/M titanium alloys have not reached large-scale

production, but a major effort is underway to reach that goal.

The fact that the Soviets are aware of the potentialities of hot isostatic pressing and isothermal forging of titanium alloy preforms is confirmed by developments at several research and development facilities. Apparently, research has been underway since the early 1970's. Success of the programs is unknown, because only electrolytically produced powders have been used. The Soviets may be forced to use electrolytically produced powders because their atomized powders do not consistently maintain required levels of purity (Ref 43).

The main advantage gained from isothermal forging relates to the use of atomized powders with low oxygen contents produced at reasonably fast cooling rates in inert atmospheres. Consolidation of the powder into high-density preforms that possess characteristics conducive to reaching a superplastic state during isothermal forging and extrusion yields components with optimum properties at the lowest cost (Ref 43).

The Soviets have published little on atomized titanium powder. Dendrite arm spacing data generally are not reported, which would aid in identifying powder cooling rates and, consequently, gases employed for quenching the powder.

One aspect of titanium powder metallurgy worthy of discussion is the development of dispersion-strengthened titanium alloys. Dispersion-strengthened materials are not new, but recently have attracted worldwide attention for some high-temperature applications. Realizing several metallurgical advantages, the Soviets have established an effort to develop these materials by P/M techniques. The limited Soviet literature reveals a marked increase in both short-term and, to a much greater extent, long-term strength properties.

The Soviets have experienced a problem in controlling the amount of oxide dissolution during sintering; however, this may not be serious. In fact, according to Soviet data, elevated-temperature properties can be optimized by a balance of solid solution and dispersion strengthening. Oxides that have been investigated include magnesium oxide, aluminum oxide, titanium dioxide, and zirconium dioxide. Electrolytic titanium powder has been found to provide better properties than hydrided titanium powder (Ref 43).

According to published data, several oxide dispersion-strengthened titanium alloys have been investigated by the Soviets. Nominal properties include:

Tensile strength:
At room temperature .... 882-1117 MPa
(127-161 ksi)
At 550 °C (1020 °F) .... 400-630 MPa
(58-92 ksi)
At 450 °C (840 °F)
for 1 h .... 400 MPa (58 ksi)
Elongation (room temperature) 5-9%
Reduction in area (room
temperature) 2.5-14%

Optimum mechanical properties were associated with alumina and zirconia additions. Zirconium oxide favorably affected fatigue properties.

At least six titanium P/M alloys have been identified in Soviet literature. The following chemical compositions are representative of these alloys:

Composition	Soviet designation
Ti-3Al-5Cr-3Fe	IPM-6/2
Ti-4Al-1.5Sn	...
Ti-3Al-2V	IMP-7
Ti-1.5Al-2Mo	...
Ti-2Al-2Mo	2M2A
Ti-3Al-2Mo	...

Three Ti-Al-Mo alloys are being studied for manufacturing fastening nuts (Ref 48, 49). Another application for titanium P/M alloys in the Soviet Union is automobile internal combustion engines (Ref 50). Success also has been achieved with titanium P/M connecting rods. These components have been successfully tested in the Zaporozhets-969, a ½-metric ton utility vehicle similar to the M-422 Marine Jeep.

Another application for titanium P/M parts is filters. Titanium filters are used in aircraft engines to ensure reliable operation of the hydraulic and fuel systems. One of the more important applications for titanium P/M components is compressor blades and disks for gas turbine engines. According to Ref 24, titanium P/M alloy blades and disks are being successfully developed.

REFERENCES

1. Abkowitz, S., Isostatic Pressing of Complex Shapes from Titanium and Titanium Alloys, in *Powder Metallurgy of Titanium Alloys*, The Metallurgical Society of the American Institute of Mechanical Engineers, Warrendale, PA, 1980, p 291-302
2. Penrice, T.W., Developments in Materials for Use as Kinetic Energy Penetrators, *Powder Metall. Defense Technol.*, Vol 5, 1980, p 11-21
3. Lally, F.T. and Toth, I.J., "Forged P/M Alloy Steels," Technical Paper EM78-278, Society of Manufacturing Engineers, Dearborn, MI, 1978
4. Pease, L.F., III, "An Assessment of
Powder Metallurgy Today and Its Future Potential," Technical Paper 831042, Society of Automotive Engineers, 1983
5. Crowson, A. and Anderson, F.E., Properties of Powder Metallurgy Steel Forgings, *Prog. Powder Metall.*, Vol 33, 1977, p 47-65
6. Crowson, A., Fabrication of Aluminum P/M Fuze Parts, *Powder Metall. Defense Technol.*, Vol 5, 1980, p 37-50
7. Billiet, R., Plastic Metals from Fiction to Reality with Injection Molding P/M Materials, *Prog. Powder Metall.*, Vol 38, 1982, p 45-52
8. Thellmann, E.L., Titanium Powder Metallurgy in Ordnance, *Powder Metall. Defense Technol.*, Vol 4, 1978
9. Garriott, R.E. and Thellmann, E.L., "Titanium Powder Metallurgy a Commercial Reality," Proceedings of the 1976 International Powder Metallurgy Conference, Metal Powder Industries Federation/American Powder Manufacturers Institute, 1976
10. Abkowitz, S., "Isostatic Pressing of Complex Shapes from Titanium and Titanium Alloys," Conference proceedings, 109th Annual Meeting, American Institute of Mining, Metallurgical, and Petroleum Engineers, Las Vegas, Feb 1980
11. Hulewicz, P., private communication, September 2, 1983
12. Thellmann, E.L. and Lyndrup, J.F., "Focus on Titanium Powder Metallurgy Technology," American Institute of Aeronautics and Astronautics/Society of Automotive Engineers/American Society of Mechanical Engineers, 15th Joint Propulsion Conference, Las Vegas, June 1979
13. Andersen, P.J., Alber, N.E., and Thellmann, E.L., "Aerospace Applications for Pressed and Sintered P/M Titanium Alloys," Powder Metallurgy and Defense Technology Seminar, Metal Powder Industries Federation, Yuma, AZ, Nov 1979
14. Zaleski, F.I., "Fabrication and Testing of 105mm Sintered Iron Rotating Bands," International Automotive Engineering Congress, Society of Automotive Engineers, Detroit, MI, Jan 1969
15. McIntire, H.O., "Production of Sintered-Iron Rotating Band Blanks," Defense Metals Information Center, Battelle Memorial Institute, Columbus, OH
16. Buttleman, D.P., "The Effect of Working on the Properties and Fragmentation Behavior of P/M Steel Pre-
forms" Powder Metallurgy in Ordnance Seminar, Philadelphia, Oct 1975
17. Lally, F.T., Toth, J.J., and Di Benedetto, J., Forging of Steel Powder Products, in *Forging of Powder Metallurgy Preforms*, Hausner, H.H., Roll, K.H., and Johnson, P.K., Ed., Metal Powder Industries Federation, New York, 1973, p 104-130
18. Crowson, A., "Sintering Cycle Influence on P/M Steel Forgings' Dynamic Properties," ARSCD-TR-79008, U.S. Army Research and Development, Dover, NJ, Aug 1979
19. Crowson, A., Properties of Powder Metallurgy Steel Forgings, *Prog. Powder Metall.*, Vol 33, 1977, p 47-65
20. Jost, M.M., Precision Forging Process for P/M Fins, *Powder Metall. Defense Technol.*, Vol 5, 1979, p 115-128
21. Penrice, T.W., Developments in Materials for Use as Kinetic Energy Penetrators, *Powder Metall. Defense Technol.*, Vol 5, 1980, p 15
22. Lea, C., Muddle, B.C., and Edmonds, D.V., Segregation to Interphase Boundaries in Liquid-Phase Sintered Tungsten Alloys, *Metall. Trans.*, Vol 14A (No. 4), April 1983, p 667
23. Tikhonov, G.F., *Sbornik Statey Mezhvuzovskay Nauchnoy Konferentsii Gor'kovskikh Uchenykh*, Gorkiy, 1969, p 161-165
24. *Aviatsiya i Kosmonavtika*, No. 8, Aug 1982, p 36-37
25. *Kommunist Belorussii*, No. 8, Aug 1979, p 65-72
26. *Metal Powder Report*, No. 11, 1981, p 539-543
27. *P/M Technology Newsletter*, Vol 12 (No. 3), April 1983, p 1
28. Belov, A.F., Problem of New Processes for the Production and Processing of Metallic Materials, *Soviet Sci./USSR Acad. Sci. Bull.*, Vol 45 (No. 5), 1975, p 29-36
29. Marley, W.F., Soviet Powder Metallurgy Technology, *Powder Metall. Defense Technol.*, Vol 3, 1977, p 139-144
30. Vinogradov, G.A., The Rolling of Metal Powders, *Metallurgiya*, Moscow, 1969
31. Vinogradov, G.A. and Yu.I. Kovalenko, *Poroshkovaya Metallurgiya*, Vol 153 (No. 9), 1975, p 21-25
32. Lozhechnikov, Ye., *Sovetskaya Belorussiya*, No. 135, June 12, 1974, p 2
33. *Sovetskaya Belorussiya*, No. 230 (15315), Oct 4, 1981, p 2

34. International Symposium on Powder Metallurgy, Kiev, May 11-15, 1977, *Int. J. Powder Metall.*, Vol 14 (No. 1), 1978, p 63-64

35. *Kommunist Belorussii*, No. 8, Aug 79, p 65-72

36. *Sovetskaya Belorussiya*, No. 169, July 22, 1975, p 2

37. Atroshchenko, E.S. and Shorshorov, M. Kh., *Fizika i Khimiya Obrabotki Materialov*, No. 1, 1974, p 119-123

38. Miroshnichenko, I.S., Rapid Solidification, *Metallurgiya*, Moscow, 1982

39. *Sovetskaya Belorussiya*, No. 8, Jan 10, 1975, p 2

40. *Elonomicheskaya Gazeta*, No. 17, April 1981, p 7

41. *Sotsialisticheskaya Industriya*, No. 197 (3388), Aug 28, 1980, p 4

42. Belov, A.F., *Izv. Akad. Nauk SSSR Metallurgia*, No. 6, 1982, p 11-20

43. Belov, A.F., *Izvestiya Metally*, No. 2, 1974

44. Skorokhod, V.V., *Soviet Powder Metall. Metal Ceramics*, No. 5, May 1978, p 359-364

45. Garibov, G.S., Samarov, V.N., and Feigin, V.I., *Soviet Powder Metall. Metal Ceramics*, No. 2, Feb 1979, p 136-140

46. The Fourth International Titanium Conference, May 1980, Kyoto, Japan

47. Radovskiy, V.S. and Silyaev, A.F., Poroshkovaya Metallurgiya Zharoprochnykh Splavov i tugoplavkikh metallov (The Powder Metallurgy of Refractory Alloys and High-Melting Metals), *Metallurgiya*, Moscow, 1974

48. Avkhimovich, G.F. and Rakovskiy, V.S., *Poroshkovaya Metallurgiya*, No. 12 (144), Dec 1974, p 89-93

49. Ustinov, V.S., Powder Metallurgy of Titanium, *Metallurgiya*, 1973

50. Kovalev, Yu., *Pravda Ukrainy*, No. 110, May 1974, p 4

P/M Porous Parts

By James J. Snyder
Senior Metallurgical Engineer
Ocean Systems Division
Gould Inc.

POROUS PARTS, which are characterized by interconnected porosity, can be grouped into the following primary categories: filters; damping devices and flame arrestors; metering devices; storage reservoirs for liquids, including self-lubricating bearings; and battery elements (Ref 1, 2). Starting materials for these applications include glass, ceramics, cellulosic materials, and metallic materials such as metal powder, woven wire mesh, chopped or cut wire fibers, machined shavings, and continuous wire filaments. Sintered metal powder, however, is the most effective starting material in terms of meeting the various design and performance characteristics of porous parts—high strength, heat and corrosion resistance, durability, shock resistance, and controlled porosity and permeability.

Metal powders used for porous parts are selected according to the application. The most commonly used powders include bronze, stainless steel, nickel and nickel-based alloys, titanium, and aluminum. Materials used less frequently include the refractory metals (tungsten, molybdenum, and tantalum) and the noble metals (silver, gold, and platinum) (Ref 3, 4).

The various methods of producing these powders are discussed in separate articles in the Section of this Handbook "Production of Metal Powders." This article addresses the performance characteristics that are paramount in porous P/M part materials, the fabrication and properties of various porous parts, and applications of P/M porous parts.

Performance Characteristics*

The characteristics that are most important in porous mediums are:

*This section is taken from Ref 5 and 6.

- Adequate mechanical strength
- Retention of solid particles to a specified minimum size
- Fluid permeability, or the ability to pass a given volume of fluid of a specified viscosity through a unit area and thickness of a filter under a given pressure gradient
- Resistance to environmental attack

Mechanical strength governs the ratio of area to thickness of large filters, particularly under high-pressure conditions. Mechanical strength is also a reliable measure of interparticle bonding. Compared with nonmetallic filter materials, metallic filters have adequate strength, which prevents particles of the filter material from breaking away in service and entering the stream of the filtered material (liquid or gas). Additionally, the filter material must have sufficient ductility to prevent fracturing under mechanical or thermal shock.

According to Hoffman and Kapoor (Ref 6), the shear strength test is the preferred method for determining the mechanical strength of filter materials. A comparison of shear and tensile strengths of a stainless steel powder as a function of density is shown in Fig. 1. This curve indicates that coarser particle sizes (300 to 500 μm) reduce the tensile strength, while the shear strength is unaffected. This phenomenon may be due to the influence of notch stresses on tensile strength. For this reason, shear strength is preferred as an indicator of the behavior of filter materials.

Both shear and tensile strengths of a porous structure are inversely related to the percentage of porosity present in the material. Shear strength of type 316L powder as a function of porosity is shown in Fig. 2.

Solid Particle Retention and Permeability. Generally, the smaller the

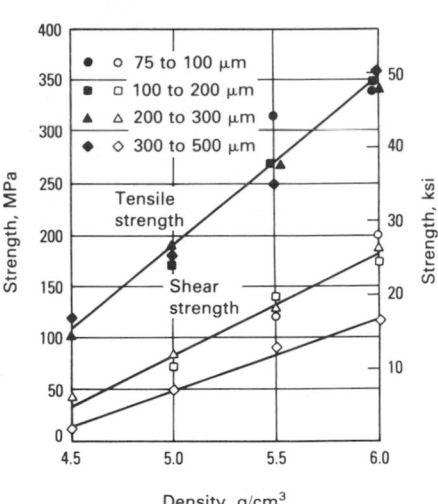

Fig. 1 Tensile and shear strength of type 316L-P powder as a function of density
Source: Ref 6

minimum size of particles retained in a filter, the lower is its permeability. To produce a filter that passes a sufficiently large volume of fluid in a given time while retaining particles to a minimum size, it is frequently necessary to design filters with large cross-sectional areas.

Retention of solid particles may be determined by passing a suspension of particles of known size, such as spheres of glass or plastic, through the filter and subsequently determining to which size these particles are retained by the filter. When suspensions of particles of a range of sizes are passed through a metallic filter of appreciable thickness, the filter retains not only all particles above a given size, but also many particles smaller than this size (Ref 7).

A widely used method for obtaining an estimate of particle retention is the bubble

Fig. 2 Shear strength of type 316L powder as a function of porosity
Source: Ref 6

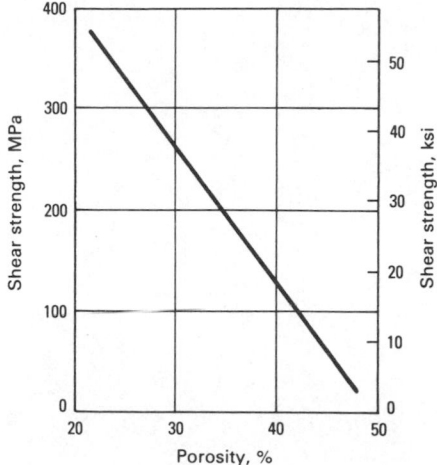

test pore size (Ref 7, 8) evaluation. A form of this test has been published as the International Standards Organization standard ISO 4003. The test sample is impregnated with and immersed in a test liquid. Air is introduced at the underside of the piece under gradually increasing pressure. The pressure at which bubbles are first emitted from the surface of the piece is a measure of the maximum pore size:

$$d = \frac{4\gamma}{p}$$

where d is the maximum pore diameter (m); γ is the surface tension of the test liquid (N/m); and p is the pressure difference across the test piece (Pa), equal to $p_g - p_i$ where p_g is the gas pressure, and p_i is equal to 9.81 ρh where ρ is the density of the liquid (kg/m^3), and h is the distance between the upper surface of the test piece and the level of the test liquid.

A comparison of the values of maximum pore diameter determined by this test and a test based on actual retention of particles shows that the bubble test gives values of maximum pore size considerably larger than the pore size determined by retention. For filters produced from uniform spherical particles, the ratio of bubble test diameter and particle retention diameter is approximately 3.1. For porous metal made from irregular particles, the ratio is approximately 5.

The fluid permeability of a permeable sintered metal material (Ref 7, 9) may be determined by a test method described in ISO 4022. The pressure drop and the volumetric flow rate are measured as a test fluid of known viscosity and density is passed through a test sample. When the flow of the fluid is strictly viscous, Darcy's law applies:

$$\frac{\Delta P}{e} = \frac{Q \cdot \eta}{A \psi_v}$$

where ΔP is the pressure drop (Pa); e is the thickness of the test piece (m); A is its cross-sectional area (m^2); η is the absolute dynamic viscosity of the test fluid (kg/m · s); Q is the volumetric flow rate of the fluid (m^3/s), equal to the mass flow rate of the fluid divided by its density (ρ); and ψ_v is the viscous permeability coefficient (m^2).

The flow of fluids through porous materials involves several simultaneous mechanisms. Additionally, viscous flow, inertia, and slip flow are important variables to monitor. Inertial flow gives an indication of the loss of energy due to the changes in the direction of the fluid in passing through tortuous porosity in combination with the onset to local turbulence. Forchheimer (Ref 10) has shown that the presence of inertia flow can be controlled by introducing an inertia permeability coefficient with units of length into the equation of fluid flow:

$$\frac{\Delta P}{e} = \frac{Q \cdot \eta}{A \cdot \psi_v} + \frac{Q^2 \cdot \rho}{A^2 \cdot \psi_i}$$

where ψ_v and ψ_i are the viscous and the inertia coefficients of permeability, respectively. By measuring the rate of flow of a given fluid at a series of pressure drops, both coefficients necessary to characterize the flow can be determined. In experiments with filters with three different pore sizes produced from 50 to 80 mesh, 100 to 200 mesh, and 200 to 325 mesh stainless steel powder, using air as the fluid with downstream atmospheric pressure and pressure drops up to 400 kPa (60 psi), German has shown that the Forchheimer equation gives an excellent reproduction of the experimental data (Ref 11).

The term "slip flow" refers to flow in which the mean free path of molecules between intermolecular collisions approaches the size of the pores in the filter. It is an important variable to monitor for gases at low pressures and high temperatures and for filters with very small pore sizes. Under these conditions, inertia flow is absent. Under slip flow conditions, the permeability coefficient is pressure dependent and, as Klinkenberg (Ref 12) has shown, may be written as:

$$\psi_s = \psi_v \left(1 + \frac{B}{P_{av}}\right)$$

where ψ_s and ψ_v are the slip and viscous permeability coefficients; P_{av} is the average between inlet and outlet pressure ($P_1 + P_2/2$); and B is the Klinkenberg factor, which is a constant for a given gas and porous material and has the dimensions of a pressure. By measuring over a range of different absolute pressures P_1 and P_2 and by plotting ψ_s versus $2/P_1 + P_2$, a straight line is obtained, whose slope is $B \cdot \psi_s$ and whose intercept is equal to the viscous permeability ψ_v.

German produced porous metal gas flow restrictors for mass spectrometer calibrations from −325 mesh water-atomized type 304L stainless steel powder by compacting, sintering, coining, and annealing (Ref 11). The flow rate of helium was measured as it passes through these restrictors with inlet pressures from 70 to 4000 kPa (10 to 600 psi) absolute and vacuum on the outlet side. Using the slip flow correction of Klinkenberg for the viscous permeability coefficient, German reproduced the flow pressure relationship for flows ranging from 10^{-10} to 10^{-7} m^3/s (10^{-4} to 10^{-1} cm^3/s).

Resistance to Environmental Attack. By proper selection of materials, corrosion resistance can be ensured. The most widely used metallic filter materials are bronze and stainless steel. In special applications, such as filtration of highly corrosive fluids, titanium, Monel, or Inconel are used because of the higher corrosion resistance of these materials.

Fabrication and Design

Depending on part size, configuration, material, and the degree of porosity required, most porous metal products are produced by one of the following processes.

Compacting and Sintering. Metal powder is pressed in a die at pressures that make the powder particles adhere at contact points resulting in sufficient strength so that the compact can be handled after ejection from the die. The "green" (unsintered) strength of the compact depends on the metal powder characteristics, such as composition, particle size, shape, and purity, and the forming pressure (see the article "Green Strength of Compacted Metal Powders" in this Volume for more information on green strength). After forming, green parts are then heated, or sintered, in a controlled atmosphere at a temperature that is below the melting temperature of the metal, but is sufficient to bond the particles, thus markedly increasing part strength. For more information on this process, see the article "Production Sintering Practices for P/M Materials" in this Volume.

Stainless steel, titanium, and certain bronze parts are frequently made by compacting and sintering. Advantages include high production rates, good permeability control, and good dimensional reproducibility (Ref 1).

Gravity Sintering. Gravity or "loose powder" sintering is used to make porous metal parts from powders that diffusion bond easily. Most production parts are made from bronze. The first step in producing bronze filters is to fill spherical bronze powder of a closely controlled particle size range into a mold. The mold is leveled and vibrated. The powder may be atomized from a liquid of 90% Cu and 10% Sn alloy or may be a spherical copper powder coated with a layer of fine tin powder that has been partially alloyed with copper.

The particle size range is adjusted according to the size of the particles the filter must remove and may be as coarse as 20 to 30 mesh or as fine as 80 to 125 mesh. Graphite molds and molds from stainless steel are widely used. Mold shape reproduces that of the filter to be sintered. The mold filled powder is sintered in a reducing atmosphere at a temperature near the solidus temperature of the alloy for approximately 30 min. The powder particles sinter together; the filter is dimensionally smaller than the mold in which particles are sintered. Porosities range from 40 to 50%, and the resulting product has adequate strength for a filter due to the strong bonds between the spherical powder particles.

When considering gravity sintering of a part, the design engineer must visualize the final shape as completely enclosed or surrounded by a mold, with the exception of a top access hole used to fill the cavity with powder. A formed shape that can be easily extracted from that mold after sintering is also a design requirement. Generally, a draft of 1° on the sides of the part is adequate for removal from the mold, although this varies with the depth of fill and material grade.

Part tolerances should be as liberal as possible. Although inside diameters tend to shrink in a predictable manner, material usually shrinks to the core during sintering. This causes the outside diameter to vary among production pieces, depending on size, shape, material, and density of the powder fill. Consequently, design tolerances of ±2% should be used. When required, closer tolerances can be produced by sizing. Overall part length should be specified to liberal dimensions as well, primarily to circumvent the need for secondary processing. Usually, a tolerance of

3% is adequate. Specific values vary with material grade or part shape (Ref 1).

Sheetmaking. Porous P/M sheet is available in stainless steel, bronze, nickel-based alloys, and titanium. Sheet is made by gravity sintering the powder in wide, shallow trays. Specified porosity is obtained by selecting the proper powder particle size. Powder metallurgy sheet is available in a variety of thicknesses ranging from 1.5 to 3 mm ($^1/_{16}$ to $^1/_8$ in.) and in dimensions ranging from 60 to 150 cm (24 to 60 in.). Sheet can be sheared, rolled, and welded into various configurations (Ref 1).

Isostatic compacting is a process in which pressure is applied uniformly to a deformable container that holds the metal powder to be compacted. This technique is particularly useful in the manufacture of parts with large length-to-diameter ratios. The isostatic compacting system generally includes a pressure vessel designed to contain a fluid under high pressure, a deformable container, and arbors (or cores) if tubes or special shapes are being made. An example of a cold isostatically pressed porous tube is shown in Fig. 3.

By proper selection of pressurizing fluid and containers, isostatic compaction can be used at elevated temperatures (hot isostatic pressing), although most porous parts are isostatically compacted at room temperature. After removal from the isostatic tooling, the green part is then sintered. Sintering is inherent in the hot isostatic pressing process, however; a subsequent sintering operation is usually not required. This process may be used with all conventional P/M materials (Ref 1). Detailed information on isostatic compaction can be found in the articles "Hot Isostatic Pressing" and "Cold Isostatic Pressing" in this Volume.

Metal Spraying. A porous metal structure can be manufactured by spraying molten metal onto a base surface. Control of porosity is achieved by monitoring spraying conditions or by spraying a soluble or a reactable material that may be removed (Ref 1).

Machining. Although some techniques for machining porous parts exist that do not result in pore closure, most machining operations such as turning, milling, boring, or grinding tend to close the surface pore structure by smearing action of the porous metal. As a result, flow permeability is lost, rendering the resultant porous metal structure ineffective as a flow-control device. If at all possible, therefore, design of the P/M tooling should be such that the porous part does not require

finishing of the working surfaces after sintering.

If machining or grinding is required to achieve a specified shape, dimension, or surface finish, sharp tooling with a slight negative rake angle should be used. Machined areas should be subsequently etched to remove cutting fluids and re-open smeared pores. Careful cleaning is also required to remove the etchant (Ref 1).

Materials and Properties

Bronze, stainless steel, nickel-based alloys, titanium, and aluminum are primarily used in P/M porous metal applications. A description of these materials and their properties follows.

Bronzes. The major advantage of P/M bronze materials over other porous metals is lower cost. Bronze filters usually are made by gravity sintering of spherical bronze powders. Two methods of producing bronze powders are used, the most common of which is the atomization of molten prealloyed bronze powder. For additional information on this procedure, see the article "Production of Copper Alloy Powders" in this Volume.

These powders typically contain 90 to 92% Cu and 8 to 10% Sn. Filters made from atomized bronze have sintered densities ranging from 5.0 to 5.2 g/cm^3. To produce filters with the highest permeability for a given maximum pore size, powder particles of a uniform particle size must be used.

Fig. 4 Filter characteristics of type 316L-F powder
(a) Viscous permeability coefficient. (b) Filter grade by glass bead test. Source: Ref 6

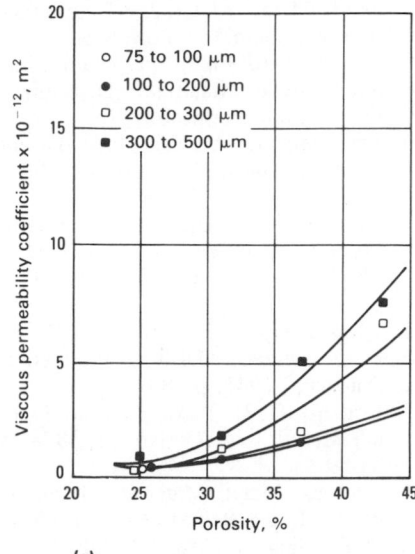

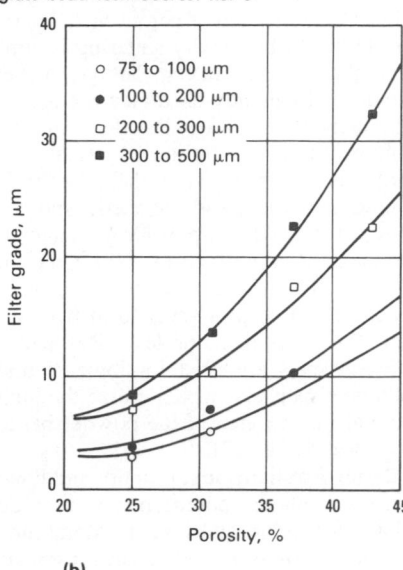

(a)

(b)

Fig. 5 Disk and three-element filters made of type 316L stainless steel

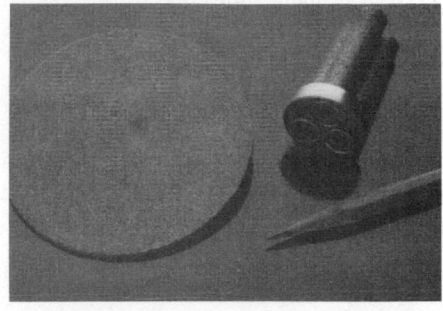

Fig. 6 Porous titanium disk filters
Courtesy of Gould, Inc.

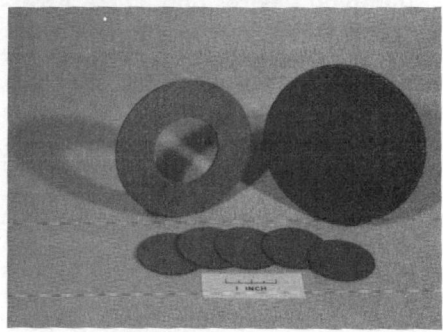

Filters made from tin-coated cut copper wire with tin contents ranging from 2.5 to 8% are also used to a lesser extent. Filters made from these materials have sintered densities ranging from 4.6 to 5.0 g/cm³.

Stainless Steels. The most common commercial composition of porous stainless steel parts is the American Iron and Steel Institute (AISI) type 316L (Fe–0.03C–16-18Cr–10-14Ni). From a manufacturing standpoint, however, any of the austenitic grades (300 series) of stainless steel may be used. These powders are produced by atomization (see the article "Production of Low-Alloy, Stainless Steel, and Tool Steel Powders" in this Volume).

Properties of filter materials composed of individual powders depend on the porosity and particle size of the given powder. Permeability is affected to a greater extent by variations in porosity than by variations in particle size. Figure 4 illustrates the viscous permeability coefficient and filter grade of type 316L-F powder as a function of porosity of several sieve fractions.

While bronze filters are produced by gravity sintering, this production method is not suitable for stainless steel. The methods used for making stainless steel filter materials depend on the final shape of the filter. For making sheet, loose powder mixed with resin is spread in a mold, lightly pressed at a temperature that cures the resin, and the resulting sheet is sintered. During sintering, the resin decomposes. The porous sheet is densified by repressing and is then resintered. It may be subsequently formed into hollow cylinders and seam welded. Another method for fabricating porous hollow cylinders is by cold isostatic pressing (see the article "Cold Isostatic Pressing" in this Volume). Typical examples of stainless steel filters are shown in Fig. 5.

Nickel-Based Alloys. For corrosion- or heat-resistant applications in which stainless steel has proved inadequate, nickel-based alloys are frequently used. Filter-grade powders usually are available in nickel, Monel, Inconel, and some Hastelloys. Before specifying these materials, the manufacturer should be consulted; a material substitution at the design stage may eliminate the need for a special powder.

Monel and Hastelloy usually are used in acidic waters where stainless steel suffers crevice corrosion. Inconel and Hastelloy are used in high-temperature and severe corrosion applications. Carbonyl nickel finds special application in battery and fuel cell plates and also is used as a diffusion barrier in uranium isotope separation.

Titanium. Porous titanium offers excellent corrosion resistance and is suitable for filtration applications in certain corrosive environments. Produced from powders made by the rotating electrode process, titanium porous materials are available in sheet form, with controlled porosity ranging from 3 to 40 μm and density ranging from 25 to 75% of theoretical. Porous titanium parts pressed to a density above 92% of theoretical lose the interlocking (interconnecting) porosity required for filtering. Sheet is available in widths ranging from 30 to 90 cm (1 to 3 ft). Porous titanium tube is available in lengths up to

60 cm (2 ft) and with diameters up to 12.7 cm (5 in.) (Ref 1). Several typical P/M titanium disk filters are shown in Fig. 6.

Aluminum is used less frequently in porous form than the materials discussed above. However, porous aluminum has a major weight advantage. An aluminum part is about one third the weight of a comparable bronze or stainless steel part. Other advantages include cost, ease of fabrication, and superior nonmagnetic properties. Specialized applications requiring aluminum porous parts include filters, metering orifices, shock absorbers, vacuum forming molds, and venting bodies (Ref 2).

Applications

Filters constitute one of the primary applications of porous metals. Compared to porous metal filters, fine screens are more fragile. Organic felt materials are more easily deformed, and paper filters are weaker. Felt and paper filters also have a much lower operating temperature range. Filtration separates solid contaminants from a liquid or gaseous stream.

Sintered bronze filters are capable of filtering air in pneumatic systems by separating water from air via differences in surface tension. Porous metal filters are used for automotive and oil-burner fuel filtration. Porous stainless steel and Monel are used to filter uranium compounds in nuclear applications and to filter liquid sodium or liquid sodium-potassium alloys in heat exchangers (Ref 2-4, 13).

A porous filter usually is used at the outlet of pressurized bottled gas to protect the valve, nozzle, and meter from contaminant particles. Porous metal filters are used in cryogenic, pharmaceutical, food, beverage, brewing, textile, chemical processing, and aerospace applications. In cryogenic applications, liquefied gases (oxygen, nitrogen, and helium) are filtered. In pharmaceutical use, antibiotic crystals can be separated from liquids. Fuel, hydraulic, and air filtration are performed in aerospace systems (Ref 13-15).

Damping Devices and Flame Arrestors. Porous metals are used for acoustic dampers and flame arrestors with gases and liquids. The flow resistance of porous metals is utilized in snubbers to dampen pressure surges and in pneumatic time delay devices and silencers. Cupronickel and bronze parts are used as flame arrestors. Porous metals also are used to dampen sound waves. As acoustic silencers and sound dampers, porous metals suppress noise in jet engines and submarine silent valves. The acoustic impedance property of porous metal parts improves frequency response in telephone components (Ref 4, 5, 14, 16).

Metering Devices and Distribution Manifolds. The controlled permeability properties of porous metal components are beneficial in liquid and gas flow control. Flow resistance properties of porous metals are utilized in metering devices and distribution manifolds of liquids and gases. The flow of gases is controlled through porous metal devices in air bearings, aerators, and in spargers for bubbling gases and fluidized beds. Fluidized beds are used to handle and convey bulk particulate material, such as gypsum. Stainless steel porous metal layers sandwiched to sheet metal are used in constructing fluid de-icers for aircraft applications. Liquid ethylene glycol-alcohol mixtures are exuded to the leading edges of wings and stabilizers (Ref 14, 15).

Storage Reservoirs for Liquids. One of the largest applications of porous metals is in self-lubricating bearings. Bronze is the most common material used for these bearings, in which the porous structure serves as an oil reservoir (see the article "P/M Bearings" in this Volume). Porous tungsten carbide, used in ball-point pens, eliminates wear, while the porosity provides a leak-free, controlled ink flow.

Batteries. Porous metals are used as electrodes for alkaline batteries and for fuel cells. The fabrication of porous nickel parts for electrodes by gravity sintering is similar to that of bronze filters. Carbonyl nickel powder is frequently used (see the article "Production of Nickel Powder" in this Volume for a description of metal carbonyl technology). After the powder is poured into the mold, leveled, and vibrated, the mold and powder are sintered for about 15 min at 700 to 1000 °C (1290 to 1830 °F). These electrodes are produced in thin strip or sheet form that is 1 mm (0.04 in.) thick or less. Refractory porous metals are used for ionizers and cathode dispensers, which utilize the high internal surface area of the powder-based strip (Ref 5, 14, 17).

Other Applications. In tooth and bone surgical implants, porous metals provide a base for tissue attachment. Membrane support for dialysis and reverse osmosis depend on the structural strength of a porous metal support. Boundary layer control of aerodynamic and hydrodynamic shapes represents another porous metal application (Ref 14, 18).

Porous metals are used for transpiration cooling in gas turbines and rocket engines. Transpiration-cooled nose tips for missiles are made of porous 316L stainless steel. Sintering conditions are adjusted to obtain the required gas permeability during manufacture (Ref 19).

REFERENCES

1. *Porous Metal Design Guidebook*, Porous Metals Council of P/M Industries Assoc., Metal Powder Industries Federation, Princeton, NJ, 1980
2. From Powder to Porous Metal Parts, *Design Engineering*, Porous Metals Council of P/M Industries Assoc., Metal Powder Industries Federation, Princeton, NJ, March, 1981
3. Johnson, W.R., Putting P/M Pores to Work, *Mach. Design*, 25 July, 1974
4. Miska, K.H., Porous Metals Control Sound, Liquids and Gases, *Mat. Eng.*, Feb 1972
5. Lenel, F.V., Powder Metallurgy—Principles and Applications, Chapt 15, "Porous Metals," Metal Powder Industries Federation, Princeton, NJ, 1980
6. Hoffman, G. and Kapoor, D., Properties of Stainless Steel P/M Filters, *Int. J. Powder Metall. Powder Technol.*, Vol 12 (No. 4), 1976
7. Morgan, V.T., Filter Elements by Powder Metallurgy, *Symp. Powder Metallurgy*, The Iron and Steel Institute, London, 1956, p 81-89
8. ISO standard 4003, "Permeable Sintered Metal Materials—Determination of Bubble Test Pore Size"
9. ISO standard 4022, "Permeable Sintered Metal Materials—Determination of Fluid Permeability"
10. Forchheimer, P., *Zeitschrift Verein duetscher Ingenieure*, Vol 45, 1901, p 1782
11. German, R.M., Gas Flow Physics in Porous Metals, *Int. J. Powder Metall. Powder Technol.*, Vol 15, 1979, p 23-30
12. Klinkenberg, L.J., American Petroleum Institute, "Drilling Production Practice," 1941, p 289
13. Rothero, G.R., Porous Media, *Powder Metall.*, Vol 21 (No. 2), 1978, p 85-89
14. "Porous Metal Materials Technology," Eng. Report 618, DYNA-PORE, Garden City, MI, July 1975
15. Nickolas, N. and Ray, R., Porous Stainless Steel—The Unique Filter Medium, *Modern Developments in Powder Metallurgy*, Vol 5, Metal Powder Industries Federation, Princeton, NJ, 1971, p 187-199
16. Andersen, P.J., Alber, N.E., and Thellmann, E.L., Aerospace Application for Pressed and Sintered P/M Titanium Parts, *Powder Metallurgy in Defense Technology*, Vol 5, Metal Powder Industries Federation, Princeton, NJ, 1980, p 23
17. U.S. Patent 3 308 530, 14 March 1967
18. Johnson, W.R. and German, R.M., Gas Flow Controlled by Porous P/M Media, *Modern Developments in Powder Metallurgy*, Vol 12, Metal Powder Industries Federation, Princeton, NJ, 1981, p 821-833
19. "Manufacturing Technology Note," U.S. Army Material Development and Readiness Command, Report No. AMMRC TR-80-30, May 1983

SELECTED REFERENCES

- "Introduction to Engineering Controlled Porosity Products," Mott Metallurgical Corp., Cat. No. 1000
- Zholobov, V.V., Zhevnovatyi, A.I., and Shenberg, G.F., Production of Porous Tube Preforms by the Method of Transverse Extrusion, *Sov. Powder Metall.*, July 1973
- Anderson, J.E., P/M Porous Media for Metering, Filtering and Acoustics, *Powder Metallurgy in Defense Technology*, Vol 27, Metal Powder Industries Federation, Princeton, NJ, 1972, p 89

P/M Friction Materials

By Seyfi Ozsever
Senior Project Engineer
Industrial Division
Raymark Corp.

FRICTION MATERIALS are the components of a mechanism that converts mechanical energy into heat upon sliding contact. The conversion product, heat, is absorbed or dissipated by the friction material. The coefficient of friction, an index of shearing force of the contacting parts, determines the degree of performance of the friction material (Ref 1). The required level of the coefficient of friction depends on the operating conditions and the end use of the product.

Metallic friction materials are used in heavy-duty applications such as aircraft brake linings and as clutch facings on tractors, heavy trucks, earth-moving equipment, and heavy presses. There are two principal types of applications or operating conditions for metallic friction materials: "wet" and "dry." Under wet conditions, the friction components, such as clutches in powershift and automatic transmissions, are immersed in oil. Dry operating conditions involve direct contact of friction components without oil, such as in aircraft brakes and standard clutches.

Figure 1 shows typical clutch and disc brake pad designs. They can be applied either wet or dry, depending on design requirements. Brake linings retard the relative movement between two surfaces (Ref 2). Heat, produced as a result of the contact, is dispersed throughout the lining. Clutches transmit the energy of a power source to another mechanism, which is brought to the speed of the power source.

Manufacturing of Sintered Friction Materials

Metallic friction materials are produced by compacting and sintering mixes of metal powders and friction-producing ceramic materials such as silicon dioxide or aluminum oxide. Friction materials consist of a dispersion of a friction-producing ingredient in a metallic matrix. Originally, a copper-tin alloy was used as the metallic matrix, in which copper powder, tin powder, the friction-producing ingredient in powder form, and other ingredients that modify the frictional behavior were mixed, compacted, and sintered. This matrix material is still used, but other compositions such as copper-zinc matrix materials have been developed, mainly for oil-immersed (wet) applications. The copper-zinc materials have the ability to maintain a strong, yet porous, matrix that retains oil. The more porous copper-zinc materials have higher

Fig. 1 Disc brake pad (a) and clutch plate (b)

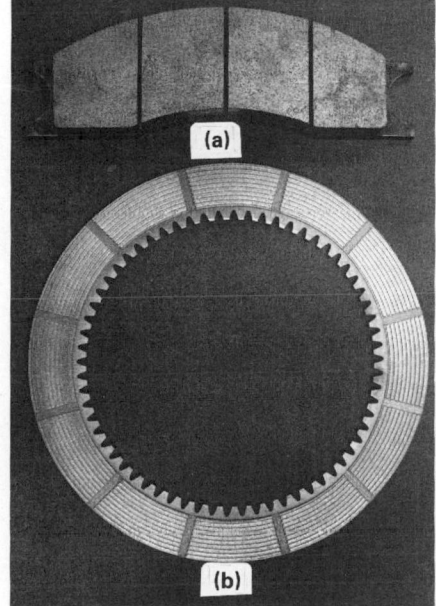

friction coefficients and higher energy absorption capacity than copper-tin materials.

Raw Material Blending. Sintered metal friction materials involve a wide variety of compositions. The choice of the composition depends on the nature of the application, such as the use and energy-power requirements. In general, metallic friction materials can be classified as either copper or iron based. Table 1 provides the range of compositions of metallic friction materials. Proportions of the components greatly affect physical properties of the materials. By varying the percentages of the individual components, different coefficients of friction may be achieved. Before mixing, powders generally are brush screened to break up agglomerated particles. Mixing usually is done in cone-type blenders. During mixing, small amounts of additives are introduced, mainly to prevent possible segregation of components by specific gravity. These additions are light-fraction oils that are volatilized easily during sintering.

Compacting. Most compacting of powders is done in hydraulic presses. Compacting pressure varies from 165 to 276 MPa (12 to 20 tsi) and is determined by the type of powder and the compacted density required. Before starting a production run, a series of density checks should be made. Low densities often cause handling problems, especially with thin cross sections. This can be overcome by increasing pressure. Surface parallelism is a major consideration and, in general, parts are kept in the 0.05 to 0.10 mm (0.002 to 0.004 in.) tolerance range.

Sintering of Clutch Plate Facings. Green compacts, placed on supporting steel backing plates, are stacked in sintering furnaces. Backing plates or cores generally are cleaned and copper plated to

Table 1   Nominal compositions of copper-based and iron-based friction materials

	Composition, %						
Premix	Copper	Iron	Lead	Tin	Zinc	Silicon dioxide	Graphite
Copper based 65-75		...	2-5	2-5	5-8	2-5	10-20
Iron based 10-15		50-60	2-4	2-4	...	8-10	10-15

achieve good bonding with the friction material. The type of sintering furnace used depends on the shape of the parts being produced. Bell-type furnaces are generally used for clutch discs because of the compression requirement. However, for disc brake pads and other odd shapes, the use of bell-type furnaces is not essential because compression is not required during sintering.

In bell furnaces, pressure is applied on the vertical stack of discs to reach the desired sintered density level and to prevent warpage or distortion. However, excess pressure causes high sintered densities and loss of low-melting-point metals from the friction material. Sintering is carried out at temperatures of 550 to 1000 °C (1020 to 1830 °F) in a protective atmosphere to prevent oxidation. Figure 2 shows the structure of a copper-based friction material, copper-plated layer, and steel backing plate after sintering at 650 °C (1200 °F) for 2 h.

Final Operations. Dimensional accuracy and additional design considerations are of major importance for applications. Parts are machined after sintering to meet dimensional specifications.

Grooving is required in the production of friction facing. Various types of grooves may be used, and each performs in a different manner. Two of the most frequently used types of grooves are combined in one friction facing (Ref 3), as shown in Fig. 1(b). The purpose for grooving varies from wet friction to dry friction materials, but grooves are very important in wet clutch systems. In wet applications, the heat energy is partly removed by the oil. Grooves allow cooling oil to flow across the surface of the friction faces, yet allow fast oil runoff during engagement to provide minimum oil thickness on the surfaces and maximize the coefficient of friction.

In dry systems, grooves serve somewhat different purposes than in wet systems. These include (1) prevention of the crushing effect of thermal expansion during high application temperatures, as in the disc brake pads of heavy-duty vehicles; (2) removal of operation debris; and (3) transfer of water and other liquids if the surface becomes wet.

Friction Applications

The coefficient of friction (μ) is the most important property in selection of a facing material. Friction may be expressed as static or dynamic values. The static coefficient of friction is the friction value of two surfaces at zero speed, while the dynamic coefficient of friction is measured at speeds greater than zero. The coefficient of friction is a function of conditions such as rubbing speed, pressure, and temperature. The relationship between rubbing speed and the coefficient of friction is shown in Fig. 3. With increasing rubbing speeds of the two surfaces, the coefficient of friction tends to drop. Applied pressure has a similar effect on the stability of the coefficient of friction. In wet friction applications, any increase in the temperature of the two rubbing surfaces above 150 °C (300 °F) will cause a substantial drop in the coefficient of friction. This temperature is commonly referred to as "breakdown temperature."

In wet applications, wear rates are usually low; therefore, thinner facings may be produced. Wear rate depends on factors such as temperature, number of engagements, mating or coupling plate surface finish, coupling plate, and facing material.

Dry friction applications are simple and lightweight compared to wet applications. Their major disadvantage is their erratic behavior in wet environments (Ref 2). For instance, on passenger cars, considerable fade is encountered when water enters the brake assembly. Figure 4 shows the effect of pressure and speed on the dynamic coefficient of friction of an automobile disc brake. Similar to wet friction materials, the drop in the dynamic coefficient of friction with pressure is noticeable.

Dry friction materials vary in composition from wet friction materials. In wet

Fig. 2 Structure of a sintered copper-based friction material

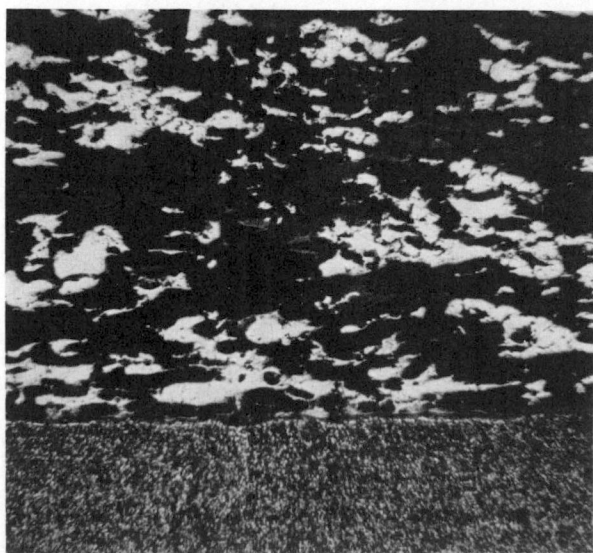

Copper-based friction material→

Copper-plated layer→

Steel backing→ plate

Fig. 3 Effect of unit pressure and rubbing speed on the coefficient of friction in a wet system

Solid line indicates dynamic coefficient of friction. Dotted line indicates static coefficient of friction.

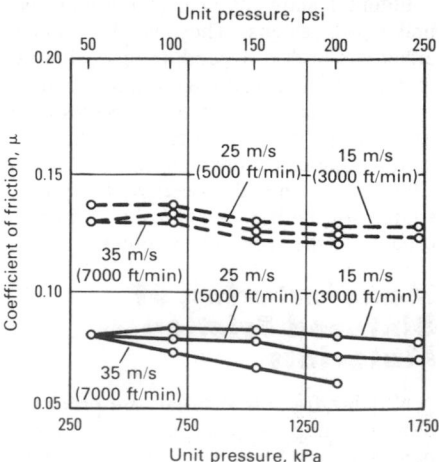

Fig. 4 Effect of pressure and speed on the coefficient of friction in an iron-based dry braking surface

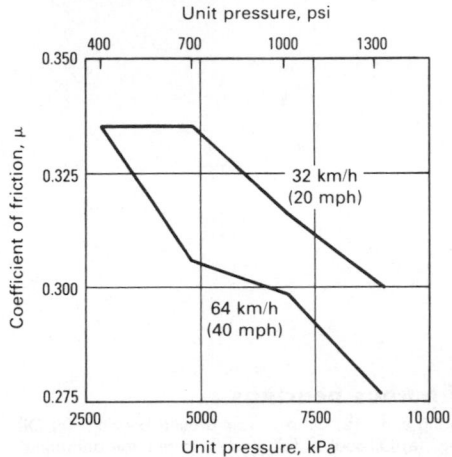

Unit pressure		Coefficient of friction
kPa	psi	

32 km/h (20 mph) line

2760	400	0.336
4825	700	0.336
6900	1000	0.317
8965	1300	0.291

64 km/h (40 mph) line

2760	400	0.336
4825	700	0.306
6900	1000	0.298
8965	1300	0.277

Source: Raymark Corporation

friction compounds, the matrix of the sintered material is copper, but in dry friction compounds the percentage of copper is reduced and iron is increased. The effect of temperature on dry friction material is shown in Fig. 5(a) and 5(b). Between 95 and 315 °C (200 and 600 °F), the drop in

Fig. 5 Effect of temperature on the coefficient of friction for a proprietary iron-based compound for brake usage

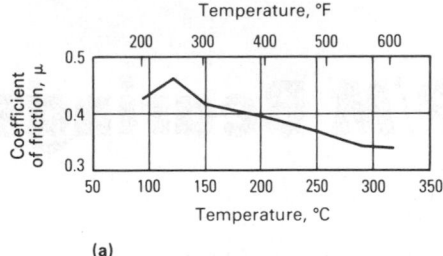

(a)

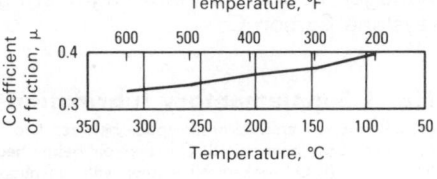

(b)

Temperature		Coefficient of friction
°C	°F	

Fig. 5(a)

93	200	0.425
120	250	0.460
150	300	0.415
180	350	0.405
205	400	0.395
230	450	0.380
260	500	0.360
290	550	0.350
315	600	0.345

Fig. 5(b)

315	600	0.330
260	500	0.340
205	400	0.360
150	300	0.370
93	200	0.400

Source: Raymark Corporation

the dynamic coefficient of friction is noticeable. After a certain level, the rate of decrease slows and eventually stabilizes.

In some materials, a slight increase in the coefficient of friction with increasing temperature is found after stabilization. This is mainly because of a change in surface morphology and the high-temperature effect of graphite in the facing material. In many cases, the drop in the friction level is recoverable, as shown in Fig. 5(b). As the temperature drops, the coefficient of friction increases. At 95 °C (200 °F), frictional recovery is almost complete. However, the level of recovery depends on the material used.

Clutch and brake applications include:

Dry friction materials

- Earth moving equipment
- Agricultural equipment
- Cranes and hoists
- Lift trucks
- Highway trucks (clutches)
- Aircraft (brakes)

Wet friction materials

- Earth moving equipment
- Agricultural equipment
- Military
- Lift trucks (clutches)

REFERENCES

1. Davies, R., *Friction & Wear*, Elsevier Publishing, Amsterdam, 1959, p 37-38
2. Jenson, W., Friction Materials, *Machine Design*, Jan 1972, p 108-113
3. Fish, R.L., Wet Friction Applications: Some Design Considerations, Technical Report, Raymark Corp., Stratford, CT

SELECTED REFERENCES

- Friction and Antifriction Materials, *Perspectives in Powder Metallurgy*, Vol 4, Plenum Press, New York, 1970
- Dixon, R.H.T. and Clayton, A., *Powder Metallurgy for Engineers*, Machinery Publishing Co., Ltd., London, 1971

P/M Bearings

By Norbert A. Arnold
Manager, Powdered Metal Engineering
Keystone Carbon Co.

SELF-LUBRICATING BEARINGS are one of the oldest industrial applications of porous P/M parts, dating back to the mid-1920's (Ref 1-3). They remain the highest volume part produced by the P/M industry. The major advantage of porous bearings is that porosity in the bearing acts as an oil reservoir. The pores are filled with a lubricant that comprises about 25 vol% of the material.

When the journal in an oil-impregnated self-lubricating bearing starts to turn, friction develops, the temperature rises, and oil is drawn out of the pores because of the greater coefficient of expansion of the oil compared with the metal and because of the hydrodynamic pressure differential in the oil film between the journal and the bearing. When rotation stops and the bearing cools, the oil is re-absorbed by capillary action.

For many self-lubricating bearings, lubricant contained in the pores of the bearing remains for the entire service life of the bearing. On some heavy-duty bearing applications, an oil reservoir that feeds additional oil through the bearing wall may be provided on the outside diameter of the bearing. Figure 1 shows typical examples of arrangements for supplementary lubrication of porous bearings. The most common shapes of self-lubricating bearings are shown in Fig. 2.

Every component that requires rotary motion relies on a bearing of some type. Since the introduction of porous bronze bearings, applications have increased; P/M self-lubricating bearings can be found in almost every part requiring rotary motion. Examples of the wide usage of P/M bearings include:

- *Automotive*: Heater motors, window lift motors, air conditioners, windshield wiper motors, power antenna motors,

Fig. 1 Supplementary lubrication of porous bearings

(a) Oil reservoir created in the space between two bearing ends. (b) Oil reservoir around bearing. (c) Oil reservoir above bearing. (d) Oil reservoir below bearing. (e) Oil-soaked felt washer to provide additional lubrication. (f) Oil-soaked felt washer with self-aligning bearing

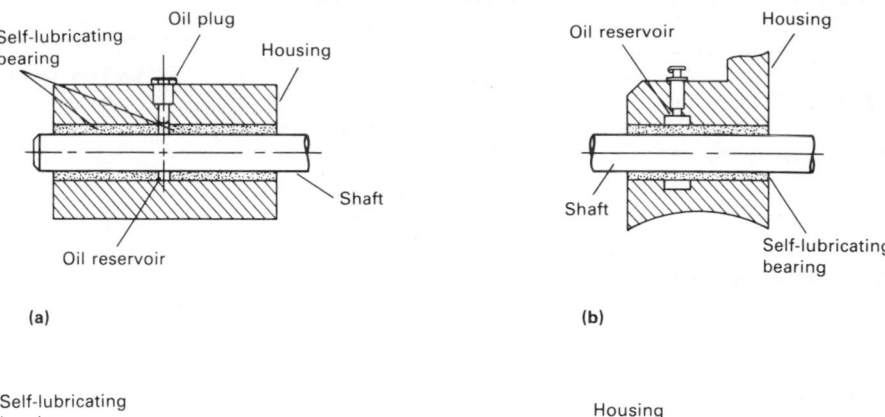

(a)

(b)

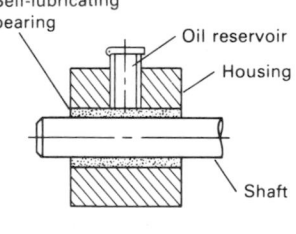

(c)

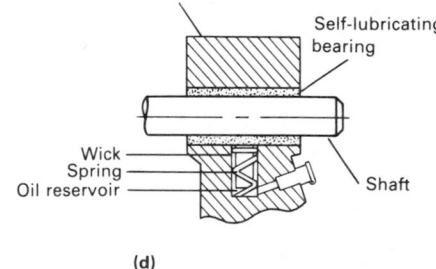

(d)

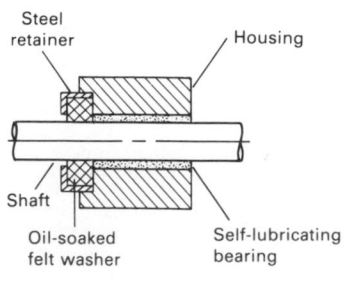

(e)

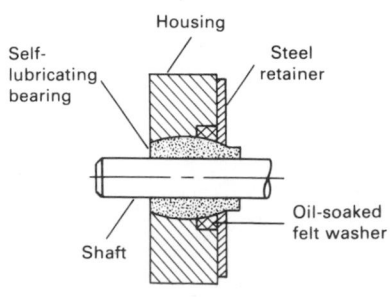

(f)

Fig. 2 Typical bronze self-lubricating bearings

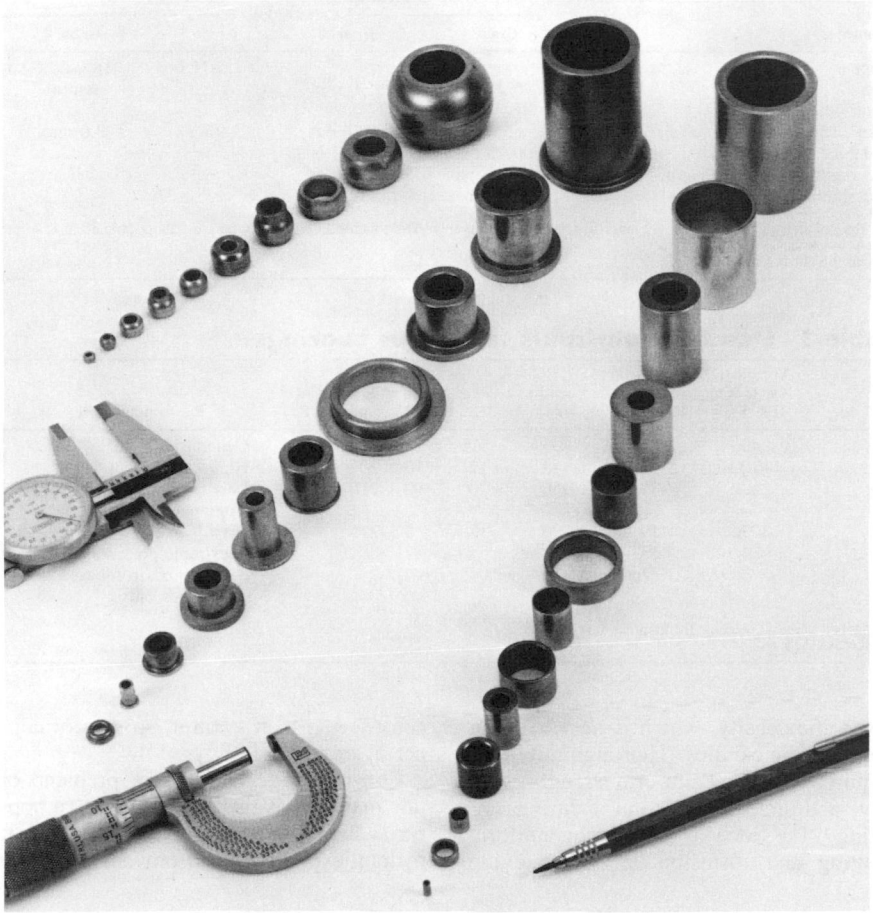

nal and most widely used P/M bearing material is 90%Cu-10%Sn bronze, with or without the addition of graphite (1% fine natural graphite is often added to enhance fabrication, as well as to improve bearing properties). The 90%Cu-10%Sn bronze material is superior in bearing performance to the iron-based and iron-bronze compositions, which are lower in cost and used in less severe applications.

Sintered bronze bearings, which are covered in the American Society for Testing and Materials (ASTM) standard B 438, are made primarily from elemental copper, tin, lead, and graphite powders. These bearings are available in two grades, plain bronze or leaded bronze, as shown in Table 1. Each grade is available in two classes: without graphite (class A) or with graphite (class B). Each grade and class, in turn, is available in four types, determined by density range:

Type	Density, g/cm³
I	5.8-6.2
II	6.4-6.8
III	6.8-7.2
IV	7.2-7.6

Source: ASTM B 438

A maximum density limit of 6.2 g/cm³ has been established on type I to ensure a minimum oil content of 27%. Satisfactory bearings also can be produced between types I and II. These bearings have slightly higher strength and slightly lower oil content. Oil content of sintered bronze bearings typically is:

Type	Oil content, vol%, min
I	27
II	19
III	12
IV	8

Source: ASTM B 438

Iron-based sintered bearings, which are covered in ASTM B 439, consist of four iron-based compositional grades (Table 2). Grades 1 and 2 are iron-carbon materials, while grades 3 and 4 are iron-copper alloys. Density and oil content of these grades are:

Grade	Density, g/cm³ min	max	Oil content, vol%
1 and 2	5.7	6.1	20
3 and 4	5.8	6.2	19

Source: ASTM B 439

trunk-closing motors, seat adjuster motors, and tape deck motors
- *Portable power tools*: Drills, reciprocating saws, jigsaws, and sanders
- *Home appliances*: Washers, dryers, refrigerators, blenders, mixers, food processors, fans, and clocks
- *Consumer electronics*: Phonographs, high-fidelity equipment, stereo equipment, tape players, and video recorders
- *Business machines*: Typewriters, computers, and copiers

- *Farm and lawn equipment*: Tractors, combines, cotton pickers, lawn mowers, string cutters, and chain saws
- *Marine equipment*: Outboard motors

Bearing Compositions

Porous self-lubricating bearings are divided into three groups: sintered bronze bearings, iron-based sintered bearings, and iron-bronze sintered bearings. The origi-

Iron-bronze sintered bearings, which are covered in ASTM B 612, are iron-copper-tin-graphite oil-impregnated bear-

Table 1 Chemical composition of self-lubricating sintered bronze bearings

Element	Composition, % Grade 1 Class A	Class B	Grade 2 Class A	Class B
Copper	87.5-90.5	87.5-90.5	82.6-88.5	82.6-88.5
Tin	9.5-10.5	9.5-10.5	9.5-10.5	9.5-10.5
Graphite	0.1 max	1.75 max	0.1 max	1.75 max
Lead	(a)	(a)	2.0-4.0	2.0-4.0
Iron	1.0 max	1.0 max	1.0 max	1.0 max
Total other elements by difference	0.5 max	0.5 max	1.0 max	1.0 max

(a) Included in other elements
Source: ASTM B 438

ings of one composition, commonly referred to as diluted bronze. The microstructure of a diluted P/M bronze alloy is shown in Fig. 3. Density of these bearings is 6.0 to 6.4 g/cm³, while the oil content is not less than 18%. Chemical composition requirements for these bearings are as follows:

Element	Composition, %
Copper + tin(a)	38-44
Total carbon	0.5-1.3
Combined carbon (on basis of iron)	0.5 max
Iron	53.5-59.5
Total other elements	1.25 max

(a) Bronze composition: copper 89.5-90.5%; tin 9.5-10.5%
Source: ASTM B 612

Blending and Mixing

Mixing or blending of fine elemental powders normally is done in a double-cone blender. For bronze bearings, mixtures of electrolytic, reduced, or atomized copper powders, atomized tin powders, and natural graphite powders are used. The particle size distribution of the copper powder is 100% −100 mesh and between 40 to 70% −325 mesh. The tin powder is 95% −325 mesh. Both reduced and atomized iron powders are used for iron-based bearings. The powders are combined with a fine lubricant such as stearic acid, zinc stearate, or Acrawax. The lubricant, which burns off during sintering, acts as a die release.

Blending of elemental and/or prealloyed powders is done for a predetermined time to produce a homogeneous mixture of all components and to supply a mix that has

Fig. 3 Microstructure of diluted bronze P/M bearing material
Magnification: 150×

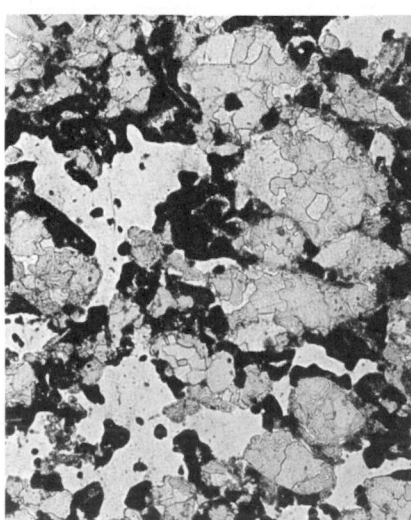

Table 2 Chemical composition of self-lubricating iron-based sintered bearings

Element	Composition, % Grade 1	Grade 2	Grade 3	Grade 4
Copper	...	...	7.0-11.0	18.0-22.0
Iron	96.25 min	95.9 min	rem(a)	rem(a)
Total other elements by difference	3.0 max	3.0 max	3.0 max	3.0 max
Combined carbon(b)	0.25 max	0.25-0.60	...	...
Silicon	0.3 max	0.3 max	...	...
Aluminum	0.2 max	0.2 max	...	...

(a) Total of iron plus copper is 97%, min. (b) On basis of iron only. The combined carbon may be a metallographic estimate of the carbon in the iron.
Source: ASTM B 439

Table 3 Standard lubricants for porous bearings

SAE No.	Viscosity(a) at 38 °C (100 °F), s	Viscosity(a) at 99 °C (210 °F), s	Viscosity index	Pour point °C	Pour point °F	Flash point °C	Flash point °F	Recommended use
30	400	60	105	−9.5	+15	240	465	General purpose
10	146-150	44	113	−12	+10	205	400	High speeds, low loads
(b)	72	37	150	−70	−90	210	410	Low temperature or wide temperature variation
90	1076	89	90	−18	0	215	415	High loads, low speed, extreme-pressure lubricant (iron-based bearings only)

(a) Saybolt universal seconds. (b) Proprietary grade
Source: ASTM B 612

a specific density, which is necessary for fixed fill die cavities. Thorough mixing is required to achieve uniform strength, density, and dimensional control during sintering. The idea is to obtain uniform melting and diffusion of low melting constituents such as tin in bronzes or copper in iron-based alloys.

Compacting. The mix or preblend of raw materials is then transferred to a hopper, which is attached to a mechanical or hydraulic press. Loose powder flows by

Fig. 4 Compaction of iron-based link arm bearing

Fig. 5 Compaction of spherical bronze bearing

Fig. 6 Alpha bronze microstructure in 90%Cu-10%Sn P/M bearing alloy
Magnification: (a) 150×. (b) 300×

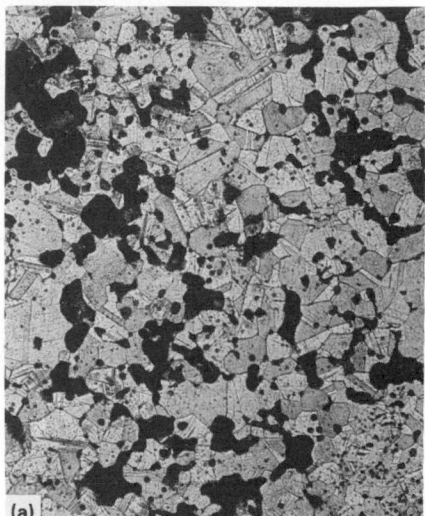

(a)

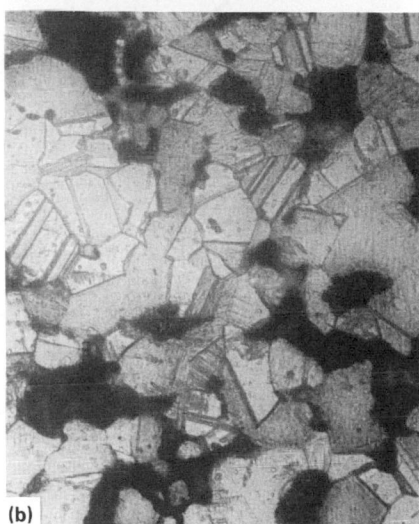

(b)

gravity (some vibration may be required) to the die cavity, where it is compacted to the desired shape and density. Compacting pressure for typical bearing compositions varies from 138 to 413 MPa (10 to 30 tsi). Depending on final dimensional accuracy, the green compact, or bearing, is molded 1 to 5% oversize to allow for material movement during sizing.

Tools used to press the loose powder are precision tooling. The die and core pin normally are made of tungsten carbide, while the upper and lower punches are made of high-grade tool steel. Figures 4 and 5 illustrate the compacting of P/M bearings.

Sintering transforms the green bearing into a metallurgically bonded alloy. The green bearing, an elemental mixture of copper and tin, is transformed into a true alpha bronze alloy at a temperature that is below the melting point of copper but above the melting point of tin. Under these liquid-phase conditions, sintering occurs without an appreciable change in the size of the bearing.

Sintering temperature for bronze may vary from 830 to 900 °C (1525 to 1650 °F), depending on the time at temperature (normally 3 to 8 min) and the desired sintered size. Suitable bronze bearings have microstructures that are essentially alpha bronze, as shown in Fig. 6. Diluted bronze bearings require sintering temperatures of 870 to 980 °C (1600 to 1800 °F), whereas iron-copper bearings should be sintered above the melting point of copper —normally in the range of 1105 to 1125

°C (2020 to 2060 °F). See the article "Production Sintering Practices for P/M Materials" in this Volume for additional information.

Sizing. Most precision bearings require a sizing operation after sintering. Sizing is done in highly polished, dimensionally accurate tools. Sizing pressures range from 206 to 551 MPa (15 to 40 tsi), depending on the amount of sizing required and the composition of the sintered part. Sizing is done primarily to improve dimensions; however, it cannot be done effectively without increasing density.

Typical sized inside diameter tolerances range from 0.0076 mm (0.0003 in.) for spherical bearings up to 16 mm (0.63 in.), with an overall length of up to 19 mm (0.8

Table 4 Permissible loads for sintered bronze self-lubricating bearings

| Shaft velocity(a) | | | Permissible loads, grades 1 and 2 | | | | | | | |
| | | | Type I | | Type II | | Type III | | Type IV | |
m/s	ft/min		MPa	psi	MPa	psi	MPa	psi	MPa	psi
Slow, intermittent	Slow, intermittent	. . . .	22	3200	28	4000	28	4000	28	4000
0.13	25	. . . .	14	2000	14	2000	14	2000	14	2000
0.26-0.51	50-100	. . . .	3.4	500	3.4	500	3.9	550	3.9	550
Over 0.51-0.77	Over 100-150	. . . .	2.2	365	2.2	325	2.5	365	2.5	365
Over 0.77-1.02	Over 150-200	. . . .	1.7	280	1.7	250	1.9	280	1.9	280

Note: With a shaft velocity of less than 0.255 m/s (50 ft/min) and a permissible load greater than 6.89 MPa (1000 psi), an extreme-pressure lubricant should be used. With good heat dissipation and heat removal techniques, higher PV ratings can be obtained.

(a) For shaft velocities over 1.02 m/s (200 ft/min), permissible loads may be calculated: $P = \dfrac{50\,000}{V} \left(P = \dfrac{1.75}{V} \right)$

Source: ASTM B 438

Fig. 7 Determination of the reduction of PV

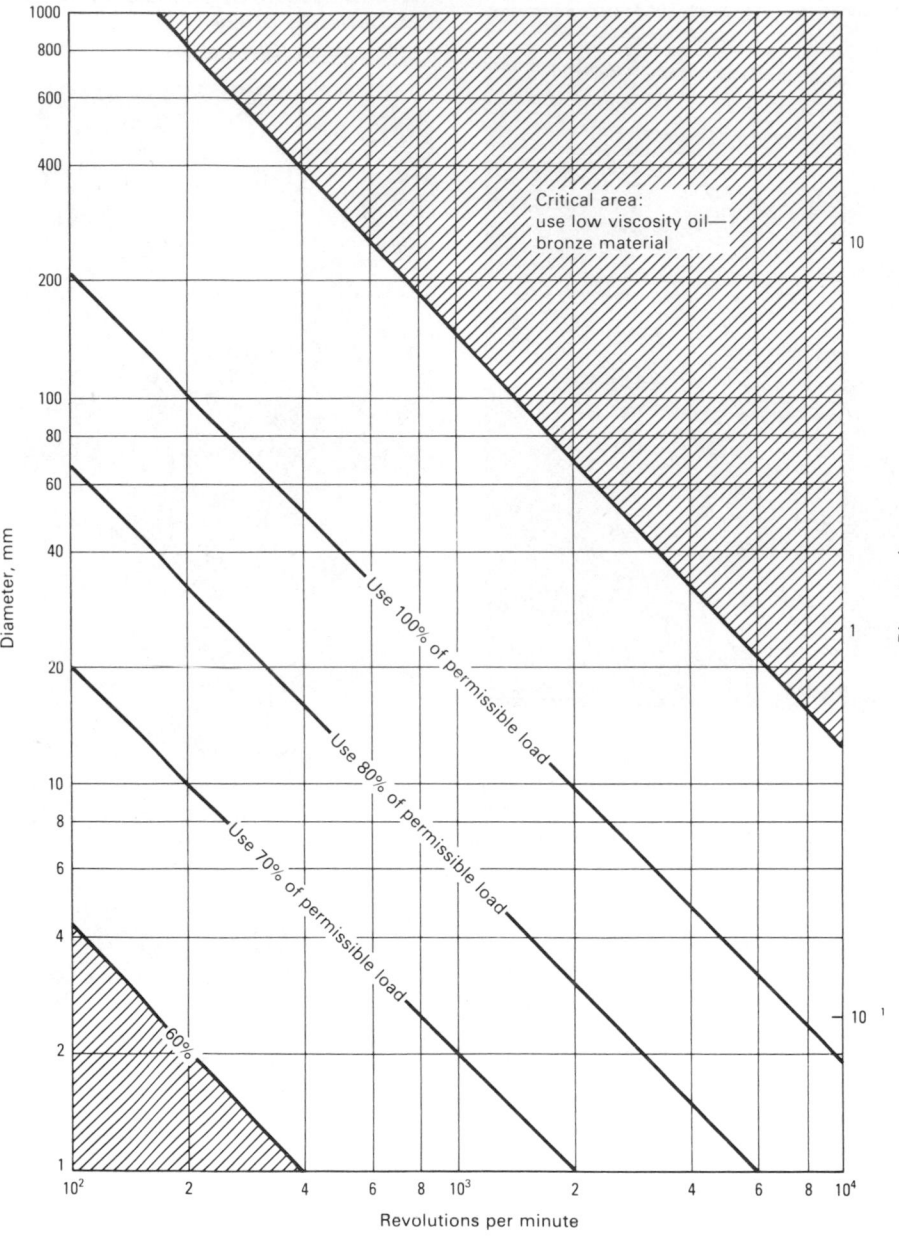

in.). Straight-wall bearing inside diameter tolerances normally are 0.03 mm (0.001 in.) per 25 mm (1 in.) of diameter for lengths up to 38 mm (1.5 in.).

Impregnation. Powder metal bearings appear solid, but are actually 20 to 25% porous. This characteristic—porosity—makes parts made by P/M processing ideal bearing materials. Porosity consists of thousands of small capillaries that are interconnected throughout the structure of the bearing and act as a built-in reservoir for the lubricant. These pores can be filled with a lubricant that is sufficient for the life of the unit for most applications. However, for longer bearing life, oil-saturated felt or wicking can surround the bearing.

The pores are impregnated with oil normally under vacuum. Bearings are placed in the tank or autoclave, the lid is closed, and the vacuum pump is turned on. Pressure is reduced to at least 709 torr (28 in. of mercury) before oil is permitted to flow over the bearings. Oil is introduced into the tank without shutting off the vacuum pump, and the tank, now filled with bearings and oil, is evacuated for another few minutes or until the pressure is again reduced to 709 torr (28 in. of mercury).

After obtaining the desired vacuum, the vacuum pump is shut off, a valve is opened, and the oil is forced out of the impregnating tank to the storage reservoir. A properly impregnated bearing contains sufficient oil to fill 90% of the available interconnecting porosity or approximately 20 vol% oil (standard density 6.4 to 6.8 g/cm^3 bearing) and $2^1/_2$ to 3 wt% oil.

The principal considerations in selecting a lubricant are the temperature of operation, bearing load or pressure between moving parts, operating speed, clearance between mating parts, and atmospheric conditions such as air temperature, dust, and moisture. In general, the most versatile type of lubricant for applications re-

Table 5   Permissible loads for iron-based sintered self-lubricating bearings

| Shaft velocity(a) | | | Permissible loads | | | |
| | | | Grades 1 and 2 | | Grades 3 and 4 | |
m/s	ft/min		MPa	psi	MPa	psi
Slow, intermittent	Slow, intermittent	....	25	3600	55	8000
0.13	25	....	12	1800	20	3000
0.26-0.51	50-100	....	3.1	450	4.8	700
Over 0.51-0.77	Over 100-150	....	2.1	300	2.8	400
Over 0.77-1.02	Over 150-200	....	1.6	225	2.1	300

(a) For shaft velocities over 1.02 m/s (200 ft/min), permissible load may be calculated using $P = \dfrac{50\,000}{V}$.

Source: ASTM B 439

Table 6   Permissible loads for iron-bronze sintered self-lubricating bearings

| Shaft velocity(a) | | Permissible load | |
m/s	ft/min	MPa	psi
Slow, intermittent	Slow, intermittent	28	4000
0.13	25	14	2000
0.26-0.51	50-100	2.8	400
Over 0.51-0.77	Over 100-150	2.1	300
Over 0.77-1.02	Over 150-200	1.4	200

(a) For shaft velocities over 1.02 m/s (200 ft/min), permissible load may be calculated using $P = \dfrac{40\,000}{V}$.

Source: ASTM B 612

Fig. 8 Determination of bearing length

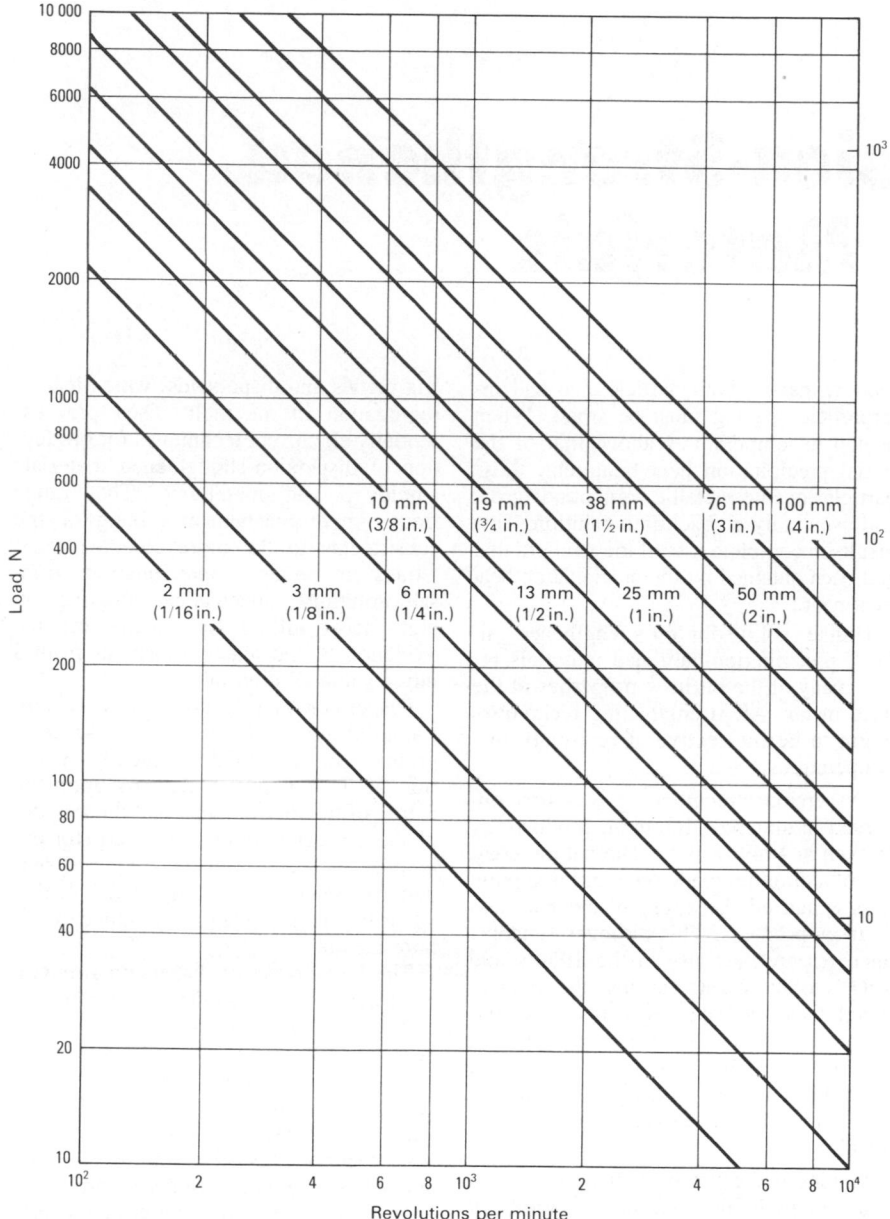

quiring self-lubricating bearings is a high-grade turbine oil. Several standard grades of oils for self-lubricating bearings, along with their properties, are listed in Table 3.

Load-Carrying Capacities

The PV formula should be used when determining the load-carrying capacity of self-lubricating bearing materials. The formula used to determine loads and speeds is:

$$PV = \frac{W}{Ld} \times \frac{\pi dn}{12} = \frac{\pi W n}{12L}$$

where P is the load (in megapascals or pounds per square inch) on the projected bearing area (bearing inside diameter × length), V is surface velocity of the shaft in metres per second (feet per minute), W is bearing load in kilograms (pounds), L is bearing length in millimetres (inches), d is the inside diameter of a bearing in millimetres (inches) (cancels out of formula), and n is shaft speed in revolutions per minute.

When the surface speed of the shaft is sufficient to maintain a satisfactory oil film, self-lubricating bearing materials have a permissible PV factor of 50 000. Thrust bearings are the exception, where a PV

factor of 10 000 is suggested. Supplementary lubrication increases the permissible PV. Adverse conditions, such as the presence of dirt or abrasive particles, misalignment of shaft, or an operating temperature in excess of 93 °C (200 °F), reduce the permissible PV factor.

When determining bearing size, the shaft diameter is established by the requirements of the application. The outside diameter of a bearing should be at least 25% larger than the inside diameter. If the speed of the shaft is known, Fig. 7 can be used to determine how much the PV should be reduced because of the surface speed of the shaft. Once this reduction is determined, Fig. 8 (based on 50 000 PV) can be used to determine what bearing length is required. The following examples illustrate how to determine the reduction of the PV and bearing length.

The bearing size for a 12.7-mm (0.5-in.) diam shaft operating at 1750 rpm and supporting a load of 445 N (100 lb) can be determined as follows. The outside diameter of the bearing is 125%, which equals a minimum diameter of 16 mm (0.63 in.). Figure 7 shows that a 12.7-mm (0.5-in.) diam shaft at 1750 rpm will not require a PV reduction, and Fig. 8 shows that the bearing should be 25 mm (1 in.) long.

Another bearing shaft with a 6-mm (0.25-in.) diam operating at 500 rpm supports a 222-N (50-lb) load. The outside diameter of the bearing is 125%, which equals a minimum of 7 mm (0.3 in.). Figure 7 shows that this surface speed is too slow and that a PV of less than 50 000 is required. According to Fig. 7, 74% of the load chart of Fig. 8 should be used. A 6-mm (0.25-in.) diam bearing 6 mm (0.25 in.) long will support a load of 423 N (95 lb). Consequently, 74% of 423 N (95 lb) is 313 N (70 lb), which is more than the required 222-N (50-lb) load, assuming that conditions do not require a larger safety factor. Permissible loads for sintered bronze, iron-based, and iron-bronze bearings are given in Tables 4, 5, and 6, respectively.

REFERENCES

1. U.S. Patent 1,556,658, Williams, H.M., General Motors Corp., 1925

2. U.S. Patents 1,642,347; 1,642,348; and 1,642,349, Williams, H. M. and Boegehold, A.L., General Motors Corp., 1927

3. U.S. Patent 1,607,389, Claus, C., Bound-Brook Oilless Bearing Co., 1926

Dispersion-Strengthened Materials

DISPERSION-STRENGTHENED MATERIALS, fiber-reinforced composites, ceramics, and cermets are being developed to meet the increasing demand for materials that perform well in severe environments. These materials are characterized by unusually high strengths at elevated temperatures. This article discusses the properties and applications of dispersion-strengthened materials.

Metals can be strengthened by strain hardening, solid-solution strengthening, and precipitation or age hardening. In strain hardening, the metal is subjected to cold work or plastic deformation below its recrystallization temperature. For most pure metals, the recrystallization temperature is between about one third to one half its melting point, measured on the absolute temperature scale. When a cold worked metal is heated to the recrystallization temperature, essentially all prior strengthening is lost.

Solid-solution strengthening is accomplished by adding other elements to the base metal. Atoms of the additive elements enter into the crystal lattice of the base metal to form a solid solution. These additive atoms inhibit slip of adjacent atom planes over one another, thus resisting plastic deformation. Solid-solution alloys lose much of their strength at relatively low temperatures, that is, at approximately half the solidus temperature on the absolute temperature scale. Additionally, important intrinsic properties of the base metals, such as electrical and thermal conductivities, are altered.

In age hardening, elements are added to the base metal to form a metastable solid solution; a subsequent precipitation heat treatment causes clusters or particles of intermetallic compounds to form in the base metal or matrix. These particles inhibit the slip of atom planes.

The most effective strengthening is obtained when these particles are extremely fine and distributed uniformly within the matrix; that is, both particle size and interparticle spacing must be small. When heated to temperatures above that of the initial precipitation heat treatment, these particles of intermetallic compounds grow and eventually go back into solution. This results in complete loss of the strength that had been obtained by the precipitation heat treatment.

Unlike solid-solution-strengthened alloys, precipitation-hardened materials retain many of the intrinsic properties of the base metal. All strengthening techniques begin to be ineffective at relatively low temperatures.

Dispersion strengthening was first observed in sintered aluminum products by Irmann in 1949 (Ref 1). The natural coating of aluminum oxide on aluminum powders enhanced the strength and hardness of these products. This discovery encouraged research activity in the 1950's and 1960's to investigate various other matrix-metal-dispersed phase systems. The matrix metals studied include copper, nickel, cobalt, iron, silver, platinum, gold, and lead; while the dispersed phases (dispersoids) studied include oxides, carbides, and nitrides.

Oxides make the best dispersoids because of their high hardness, stability at high temperatures, inertness or insolubility in the matrix metals, and relative availability in fine particulate form. Oxides of reactive elements such as aluminum, silicon, beryllium, magnesium, thorium, zirconium, and yttrium are preferable, because they are more stable at high temperatures than oxides of the more noble metals such as copper and nickel, which are more suitable as matrix metals.

Conventional melting and casting techniques are not useful for production of dispersion-strengthened materials because of the very high interfacial energy between the molten metal and oxide, which leads to flocculation, and because of the disparities in densities between the matrix metals and dispersoids, which lead to segregation in the melt. Therefore, essentially all current techniques for production of dispersion-strengthened materials employ powder metallurgy. They range from simple mechanical mixing of the constituents to the more complex techniques involving coprecipitation from salt solutions, mechanical alloying by high-energy milling, and selective internal oxidation of the reactive element from a dilute solid-solution alloy.

Dispersion quality and cost vary substantially among these methods, so compromises must be made frequently. Selection of the technique depends upon the nature of the matrix metal and the kinetics of the critical reactions involved (for example, coprecipitation and internal oxidation). Selection is complicated further when the matrix is an alloy rather than a pure metal.

The higher cost of dispersion-strengthened materials has limited their commercial usage. However, some important factors are helping to change this. Design engineers are learning to use dispersion-strengthened materials in cost-effective ways by redesigning components to reduce section sizes and eliminate secondary support systems. Material producers have been developing cost-effective processes and using more efficient equipment. More significant cost reductions are expected to result from economies of scale derived from the production of higher volumes.

Commercially available dispersion-strengthened materials, including copper, silver, platinum, and nickel-based alloys, are discussed in detail in the following sections. Dispersion-strengthened nickel was one of the first commercially available dispersion-strengthened materials, but has been surpassed by dispersion-strengthened nickel-based alloys. Dispersion-strengthened lead was made in developmental quantities, but is no longer in use.

Dispersion-Strengthened Copper

By Anil V. Nadkarni
Manager, Non-Ferrous Full Dense
Technology
SCM Metal Products
and
James E. Synk
Metallurgist
SCM Metal Products

COPPER is widely used in industry because of its high electrical and thermal conductivities, outstanding corrosion resistance, and ease of fabrication. Copper has relatively low yield strength and fatigue resistance in its pure form. A wide variety of copper alloys are commercially available that offer higher strength levels, but with lower electrical and thermal conductivities and varying degrees of corrosion resistance. Precipitation-hardened alloys offer high strength and high electrical and thermal conductivities. However, the strength and conductivity of these alloys are drastically reduced when they are subjected to prolonged heating at temperatures above that of the initial precipitation heat treatment, which is generally in the range of one third to one half the melting point of the copper matrix. Dispersion-strengthened copper overcomes some of the shortcomings of these copper alloys.

A series of dispersion-strengthened copper materials has been commercially available in the United States since 1973 and offers a unique combination of high strength and high electrical and thermal conductivities. These materials also have the ability to retain most of these properties after prolonged exposure to temperatures approaching the melting point of the copper matrix, thus extending considerably the useful temperature range of copper alloys. These characteristics also enable fabrication of parts by high-temperature joining processes, such as brazing, without losing strength. They can also be used in applications involving high operating temperatures.

The properties of these dispersion-strengthened copper materials arise from a fine, uniform dispersion of aluminum oxide particles in the copper matrix. These particles range in size from about 3 to 12 nm (30 to 120 Å), with interparticle spacing ranging from 50 to 100 nm (500 to 1000 Å). The aluminum oxide particles

are hard and thermally stable at high temperatures. They retain their original particle size and interparticle spacing even at temperatures approaching the melting point of copper. The quality of the dispersion depends greatly on the method of manufacture.

Manufacture of Dispersion-Strengthened Copper

Dispersion-strengthened copper can be made by any of the techniques mentioned in the introduction to this article. Internal oxidation produces the finest dispersoid particles and most uniform particle distribution. The copper matrix is well suited for internal oxidation, because of the high diffusion rate of oxygen in it.

For effective internal oxidation, oxygen must diffuse in the matrix several orders of magnitude faster than the solute element, such as aluminum. Because internal oxidation depends on diffusion of oxygen into the matrix, the reaction time is proportional to the square of the distance through which the oxygen must diffuse to complete the reaction.

To hold reaction times within practical limits, diffusion distance must be small. In wrought form, internal oxidation can only be practical in thin wire or strip, severely limiting the use of dispersion-strengthened materials. Powder metallurgy offers a unique solution to this problem, because powder particles can be internally oxidized rapidly and then consolidated into almost any shape.

The process involves melting a dilute solid-solution alloy of aluminum in copper and atomizing the melt by use of a high-pressure gas such as nitrogen. The resulting powder is blended with an oxidant, that consists primarily of fine copper oxide powder. The blend is heated to a high temperature; the copper oxide dissociates, and the oxygen produced diffuses into the particles of solid-solution copper-aluminum alloy. Because aluminum forms oxide more readily than copper, the aluminum in the alloy is preferentially oxidized to aluminum oxide. After complete oxidation of all the aluminum, excess oxygen in the powder is reduced by heating the powder in hydrogen or a dissociated ammonia atmosphere.

Full theoretical density is essential to realize the potential properties of dispersion-strengthened copper. The powder is fabricated into fully dense shapes by various techniques. Mill forms, such as rod and bar, are made by canning the powder in a suitable metal container (generally

copper) and hot extruding it to the desired size. Wire is made by cold drawing coils of rod. Strip is made either by rolling coils of extruded rectangular bar or by directly rolling powder with or without a metal container. Large shapes that cannot be made by hot extrusion are made by hot isostatic pressing of canned powder; alternatively, such shapes can be made by hot forging canned powder or partially dense compacted preforms.

Properties of the consolidated material depend on the amount of deformation introduced into the powder particles. Consequently, low deformation processes such as hot isostatic pressing and, to a lesser extent, hot forging develop materials with lower strengths and ductilities than those produced by extrusion.

Finished parts can be made from consolidated shapes by machining, brazing, and soldering. Fusion welding is not recommended, because it causes the aluminum oxide to segregate from the liquid copper matrix, resulting in loss of dispersion strengthening. However, flash welding, in which the liquid metal is squeezed out of the weld joint, and electron beam welding, in which a small heat-affected zone is created, has been used successfully. Solid-state welding (with multiple cold upsets in a closed die) has also been used with success to join smaller coils into a large coil for wire drawing.

Properties of Dispersion-Strengthened Copper

Dispersion-strengthened copper offers a unique combination of high strength and high electrical and thermal conductivities. More important, it retains a larger portion of these properties during and after exposure to elevated temperatures than any other copper alloy.

The properties of dispersion-strengthened copper can be modified to meet a wide range of design requirements by varying its aluminum oxide content and/or the amount of cold work. Figure 1 shows the ranges in tensile strength, elongation, hardness, and electrical conductivity obtained as a function of aluminum/aluminum oxide contents. These properties are typical for rod stock in the hot extruded condition. Cold work can be used to broaden the ranges in tensile strength, elongation, and hardness; its effect on electrical conductivity is minimal.

Two grades of dispersion-strengthened copper are commercially available. They are designated as C15760 and C15715.*

* Designation is pending approval by the Copper Development Association.

Fig. 1 Properties of dispersion-strengthened copper
Courtesy of SCM Metal Products

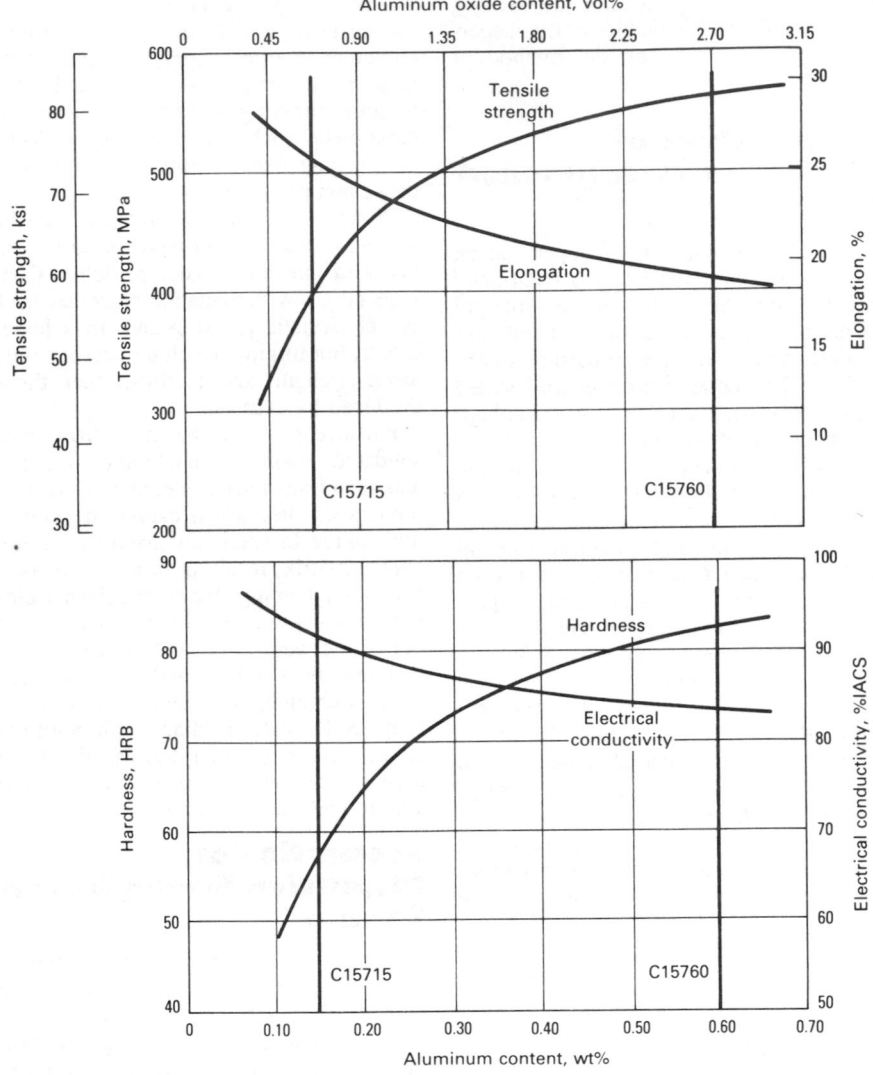

cause the matrix melts and the aluminum oxide separates from the melt. Density, modulus of elasticity, and coefficient of thermal expansion are similar to those of pure copper.

High electrical and thermal conductivities are particularly interesting to design engineers in the electrical and electronics industries. At room temperature, these range from 78 to 92% of those for pure copper. Coupled with the high strengths of these materials, they enhance increased current-carrying or heat-dissipating capabilities for a given section size and structural strength. Alternatively, they enable reduction of section sizes for component miniaturization without sacrificing structural strength or current- and heat-carrying capabilities. At elevated temperatures, the electrical and thermal conductivities of dispersion-strengthened coppers closely parallel those of pure copper (Fig. 2).

Room-temperature mechanical properties of dispersion-strengthened coppers in available mill forms are given in Tables 3 through 8, which provide a wide range of sizes, typified by various amounts of cold work by drawing and rolling, for example. Properties obtained after annealing the material at elevated temperatures are also shown.

Dispersion-strengthened copper has strength comparable to many steels and conductivity comparable to copper. It has a large capacity for being cold worked. For

Fig. 2 Electrical and thermal conductivities of dispersion-strengthened and pure copper at elevated temperatures
Courtesy of SCM Metal Products

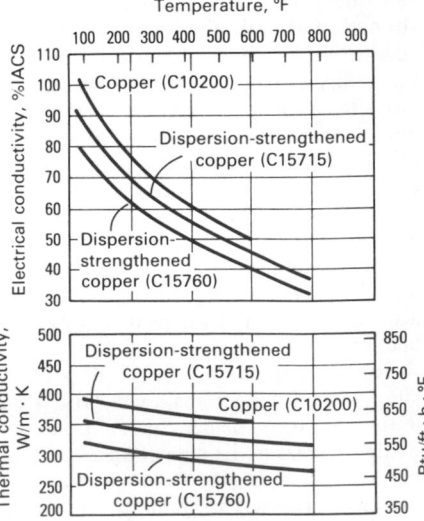

Nominal compositions of the two grades are given in Table 1 and are shown in Fig. 1 for varying aluminum oxide contents. Other grades can be produced easily to specified property requirements.

The free or reducible oxygen content of C15760 and C15715, generally about 0.02 to 0.05 wt%, is present in the form of dissolved oxygen and cuprous oxide. Alloys in this state are prone to hydrogen embrittlement at high temperatures. Oxygen-free compositions are available in both of these grades, in which the reducible oxygen is converted to nonreducible oxides. These grades are then immune to hydrogen embrittlement and must be specified for applications in which the components are likely to be subjected to reducing atmospheres during manufacture or use.

Table 1 Chemical compositions of dispersion-strengthened coppers

Grade	Copper wt%	Copper vol%	Aluminum oxide wt%	Aluminum oxide vol%
C15715	99.7	99.3	0.3	0.7
C15760	98.9	97.3	1.1	2.7

Physical Properties. Because dispersion-strengthened copper contains small amounts of aluminum oxide as discreet particles in an essentially pure copper matrix, its physical properties closely resemble those of pure copper. Table 2 gives physical properties of the two commercial dispersion-strengthened coppers and oxygen-free copper. The melting point is essentially the same as for copper, be-

Table 2 Physical properties of dispersion-strengthened and oxygen-free copper

Property	C15715	C15760	Oxygen-free copper
Melting point, °C (°F)	1083 (1981)	1083 (1981)	1083 (1981)
Density at 20 °C (68 °F), g/cm³ (lb/in.³)	8.84 (0.319)	8.81 (0.318)	8.94 (0.323)
Electrical resistivity at 20 °C (68 °F), nΩ · m (Ω · circular mil/ft)	18.6 (11.19)	22.1 (13.29)	17.1 (10.28)
Electrical conductivity at 20 °C (68 °F), 1/nΩ · m (% IACS)	0.054 (92)	0.045 (78)	0.058 (101)
Thermal conductivity at 20 °C (68 °F), W/m · K (Btu/ft · h · °F)	365 (211)	322 (186)	391 (226)
Coefficient of thermal expansion(a) at 20-1000 °C (68-1830 °F), m/m · °C (in./in. · °F)	16.6×10^{-6} (9.2×10^{-6})	16.6×10^{-6} (9.2×10^{-6})	17.7×10^{-6} (9.8×10^{-6})
Modulus of elasticity at 20 °C (68 °F), GPa (ksi)	115 (17×10^{3})	115 (17×10^{3})	115 (17×10^{3})

(a) Average value over the range of 20 to 1000 °C (68 to 1830 °F); variations within narrow segments of this range are negligible.
Source: SCM Metal Products

example, C15715 material can be drawn from an extruded size of 28.6 mm (1.1 in.) to a wire of 0.25 mm (0.01 in.) without intermediate annealing treatments. This is unlike most copper alloys and is due to the slow work hardening rate of dispersion-strengthened copper. Although both commercial grades of dispersion-strengthened copper can be cold worked readily, C15715 is generally recommended for applications requiring extensive cold work.

Dispersion-strengthened copper exhibits a high yield strength to ultimate tensile strength ratio and retains much of this strength even after annealing. This ratio is generally high for copper and its alloys in the highly cold worked condition, but drops drastically when the alloys are annealed. It also exhibits a high yield strength retention ratio, which may be defined as yield

Table 3 Room-temperature mechanical properties of C15715 rod and bar

Rod diam mm	Rod diam in.	Cold work, %	Condition annealing temperature(a) °C	Condition annealing temperature(a) °F	Ultimate tensile strength MPa	Ultimate tensile strength ksi	Yield strength, 0.2% MPa	Yield strength, 0.2% ksi	Elongation (16-mm, or 0.64-in. gage length), %	Reduction in area, %	Yield strength/ ultimate tensile strength, %	Yield strength retention ratio (annealed to drawn), %	Hardness, HRB
28.6	1.13	0	As extruded		393	57	324	47	27	68	82	...	57
			315	600	393	57	324	47	27	70	82	100	57
			650	1200	393	57	324	47	28	70	82	100	56
			980	1800	386	56	317	46	29	73	82	98	55
19.1	0.75	55	As drawn		428	62	407	59	24	67	95	...	63
			315	600	428	62	399	58	24	69	94	98	62
			650	1200	393	57	345	50	27	69	88	85	60
			980	1800	393	57	331	48	27	72	85	81	56
12.7	0.5	80	As drawn		455	66	434	63	21	66	95	...	67
			315	600	448	65	421	61	21	68	94	97	64
			650	1200	399	58	352	51	25	69	88	81	62
			980	1800	393	57	331	48	27	70	85	76	59
7.0	0.28	94	As drawn		496	72	469	68	19	65	95	...	68
			315	600	462	67	434	63	19	65	94	93	67
			650	1200	407	59	359	52	24	66	88	76	63
			980	1800	393	57	331	48	27	69	85	71	60

Note: Copper cladding is machined off prior to testing.
(a) Annealed in nitrogen for 1 h
Source: SCM Metal Products

Table 4 Room-temperature mechanical properties of C15715 wire

Wire diam mm	Wire diam in.	Cold work, %	Condition annealing temperature(a) °C	Condition annealing temperature(a) °F	Ultimate tensile strength MPa	Ultimate tensile strength ksi	Yield strength, 0.2% MPa	Yield strength, 0.2% ksi	Elongation (25-cm, or 10-in. gage length), %	Yield strength/ ultimate tensile strength, %	Yield strength retention ratio (annealed to, drawn), %
2.54	0.1	99.2	As drawn		496	72	469	68	2	95	...
			315	600	448	65	421	61	3	94	90
			650	1200	386	56	338	49	11	88	72
			980	1800	338	49	290	42	12	85	62
1.27	0.05	99.8	As drawn		524	76	496	72	2	95	...
			315	600	445	66	428	62	3	94	86
			650	1200	399	58	352	51	10	88	71
			980	1800	358	52	303	44	11	85	61
0.51	0.02	99.9	As drawn		600	87	572	83	2	95	...
			315	600	483	70	455	66	3	94	80
			650	1200	407	59	359	52	9	88	63
			980	1800	352	51	296	43	10	85	52
0.36	0.014	99.9	As drawn		607	88	579	84	1	95	...
			315	600	503	73	476	69	3	94	82
			650	1200	407	59	359	52	8	88	62
			980	1800	365	53	310	45	9	85	54

(a) Annealed in nitrogen for 1 h
Source: SCM Metal Products

Table 5 Room-temperature mechanical properties of C15715 strip

Strip thickness mm	in.	Cold work, %	Condition annealing temperature(a) °C	°F	Ultimate tensile strength MPa	ksi	Yield strength, 0.2% MPa	ksi	Elongation (50-mm, or 2-in. gage length), %	Yield strength/ ultimate tensile strength (annealed to rolled), %	Yield strength retention ratio (annealed to rolled), %
6.35	0.25	0	As extruded		379	55	317	46	6	82	...
			315	600	372	54	310	45	26	83	98
			650	1200	372	54	296	43	28	80	93
			980	1800	359	52	283	41	30	78	89
2.29	0.1	64	As rolled		483	70	455	66	7	95	...
			315	600	448	65	393	57	13	88	86
			650	1200	421	61	359	52	21	86	79
			980	1800	365	53	290	42	22	79	64
1.27	0.05	80	As rolled		503	73	476	69	6	95	...
			315	600	476	69	421	61	11	88	88
			650	1200	421	61	358	52	20	86	75
			980	1800	358	52	283	41	22	79	59
0.76	0.03	88	As rolled		510	74	483	70	6	95	...
			315	600	476	69	421	61	11	88	87
			650	1200	434	63	372	54	20	86	77
			980	1800	358	52	283	41	21	79	59
0.25	0.01	96	As rolled		607	88	579	84	4	95	...
			315	600	483	70	428	62	8	88	74
			650	1200	441	64	379	55	16	86	65
			980	1800	358	52	283	41	18	79	49

(a) Annealed in nitrogen for 1 h
Source: SCM Metal Products

strength in the annealed condition to yield strength in the work hardened condition. High values of this ratio at annealing temperatures close to the melting point of the copper matrix are indicative of its ability to resist weakening or softening.

Annealing enhances the ductility (or formability) of dispersion-strengthened copper. Due to its high yield strength retention ratio, dispersion-strengthened copper offers the highest yield strength for a given amount of ductility among the high-conductivity copper alloys.

Dispersion-strengthened copper has excellent fatigue resistance. It exhibits a high fatigue ratio (endurance limit/tensile strength). Fatigue properties of the two commercial dispersion-strengthened coppers are shown in Fig. 3. The tests were conducted at room temperature in a Krause cantilever bending-rotating beam mode at a frequency of 10 000 cpm. The C15715 rod used in the test underwent 94% cold work, while the C15760 rod underwent 14% cold work. At comparable levels of cold work, C15760 has significantly higher fatigue strength than C15715.

Mechanical Properties at Elevated Temperatures. Dispersion-strengthened copper has excellent strength at elevated temperatures. Figure 4 shows the 100-h stress rupture strengths of C15760 and C15715 at temperatures up to 870 °C (1600 °F). Other high-conductivity copper-based materials are shown in Fig. 4 for comparison. Ranging from pure copper on the low end to precipitation-hardened alloys on the high end, there is a sharp drop in stress

Fig. 3 Fatigue resistance of dispersion-strengthened copper

Tests conducted at room temperature in a Krause cantilever bending-rotating beam mode at a frequency of 10 000 cpm. C15760 underwent 14% cold work, and C15715 underwent 94% cold work prior to test. Courtesy of SCM Metal Products

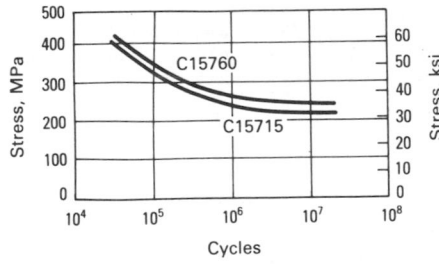

rupture strength in the 200 to 450 °C (400 to 850 °F) temperature range. Above about 400 °C (750 °F), the dispersion-strengthened coppers are superior to any of the other alloys. Above 600 °C (1100 °F), the dispersion-strengthened coppers have rupture strengths comparable or superior to some stainless steels. The log stress/log rupture life plots have extremely flat slopes for dispersion-strengthened coppers; therefore, the rupture strengths for lives longer than 1000 h are not much different from 100-h rupture strengths. For example, at 650 °C (1200 °F), a C15760 rod having 100-h rupture strength of 200 MPa (29 ksi) would have an extrapolated 1000-h rupture strength of 186 MPa (27 ksi), and a 10 000-h rupture strength of 172 MPa (25 ksi). Dispersion-strengthened copper has

excellent thermal stability at high temperatures, because the aluminum oxide particles retain their original particle size and distribution even after prolonged heating and do not allow recrystallization of the matrix.

Cold work significantly enhances the stress rupture properties of dispersion-strengthened copper; the higher the temperature, the more noticeable the enhancement (Table 9).

Applications of Dispersion-Strengthened Copper

Dispersion-strengthened copper has wide market acceptance in several applications, and design engineers are continually developing new applications. The major applications are described below.

Resistance Welding Electrodes. Dispersion-strengthened copper electrodes are used in Resistance Welder Manufacturers Association (RWMA) class II and some class III welding applications, mainly in automotive and appliance industries. They last four to ten times longer than conventional class II copper-chromium electrodes. Due to their slow mushrooming (electrode tip deformation) rates and nonsticking characteristics against galvanized steel, they minimize down times associated with electrode dressing and changing operations. This is particularly important in automatic press and robot welding applications in high-volume assembly lines.

Table 6 Bend test data for C15715 strip

Strip thickness mm	in.	Bend radius mm	in.	Condition annealing temperature(a) °C	°F	No. bends (b), longitudinal	No. bends (b), transverse	Minimum bend radius(c) Longitudinal mm	in.	Transverse mm	in.
2.29	0.09	2.29 (1T)	0.09	As rolled		1	0	...		...	
				315	600	3	2	...		...	
				650	1200	4	3	...		...	
				815	1500	5	4	...		...	
				980	1800	6	5	...		...	
		4.57 (2T)	0.18	As rolled		3	1	1.27	0.05	3.81	0.15
				315	600	6	4	0.76	0.03	2.29	0.09
				650	1200	9	7	0.51	0.02	1.27	0.05
				815	1500	10	8	0.25	0.01	0.76	0.03
				980	1800	12	9	0.25	0.01	0.76	0.03
		6.86 (3T)	0.27	As rolled		5	2	...		...	
				315	600	10	8	...		...	
				650	1200	12	10	...		...	
				815	1500	15	13	...		...	
				980	1800	18	17	...		...	
1.27	0.005	1.27 (1T)	0.05	As rolled		2	0	...		...	
				315	600	3	2	...		...	
				650	1200	5	3	...		...	
				815	1500	6	4	...		...	
				980	1800	7	5	...		...	
		2.54 (2T)	0.10	As rolled		2	0	0.76	0.03	3.81	0.15
				315	600	6	3	0.51	0.02	1.27	0.05
				650	1200	8	6	0.51	0.02	0.51	0.02
				815	1500	10	8	0.25	0.01	0.25	0.01
				980	1800	12	9	<0.25	<0.01	0.25	0.01
		3.81 (3T)	0.15	As rolled		4	1	...		...	
				315	600	8	6	...		...	
				650	1200	10	8	...		...	
				815	1500	15	12	...		...	
				980	1800	18	15	...		...	
0.76	0.03	0.76 (1T)	0.03	As rolled		1	0	...		...	
				315	600	3	2	...		...	
				650	1200	5	4	...		...	
				815	1500	8	6	...		...	
				980	1800	9	6	...		...	
		1.52 (2T)	0.06	As rolled		3	0	0.25	0.01	2.54	0.10
				315	600	5	3	0.25	0.01	0.51	0.02
				650	1200	9	8	<0.25	<0.01	<0.25	<0.01
				815	1500	11	10	<0.25	<0.01	<0.25	<0.01
				980	1800	12	10	<0.25	<0.01	<0.25	<0.01
		2.29 (3T)	0.09	As rolled		6	0	...		...	
				315	600	11	6	...		...	
				650	1200	13	12	...		...	
				815	1500	16	15	...		...	
				980	1800	18	15	...		...	
0.25	0.01	0.25 (1T)	0.01	As rolled		2	0	...		...	
				315	600	6	4	...		...	
				650	1200	7	6	...		...	
				815	1500	8	7	...		...	
				980	1800	9	8	...		...	
		0.51 (2T)	0.02	As rolled		3	0	<0.25	<0.01	0.76	0.03
				315	600	8	6	<0.25	<0.01	<0.25	<0.01
				650	1200	9	8	<0.25	<0.01	<0.25	<0.01
				815	1500	10	8	<0.25	<0.01	<0.25	<0.01
				980	1800	11	9	<0.25	<0.01	<0.25	<0.01
		0.76 (3T)	0.03	As rolled		4	1				
				315	600	10	7				
				650	1200	11	9				
				815	1500	15	13				
				980	1800	18	15				

(a) Annealed in nitrogen for 1 h. (b) One bend defined as a 90° fold over radius and back to original upright position without cracking. All bends identical. (c) Minimum bend radius is smallest radius that allows one bend. Source: SCM Metal Products

With increasing automation and wider usage of galvanized steel in automobiles, dispersion-strengthened copper electrodes are being used more frequently. The slow mushrooming rate also allows less frequent current step-ups during welding and leads to substantial energy savings.

Lead Wires. Dispersion-strengthened copper wire is used in leads for incandescent lamps. Its high-temperature strength retention capability facilitates pressing of glass stems without undue softening of the leads. This eliminates the need for expensive molybdenum support wires without sacrificing lead stiffness. Because of the superior strength of the leads, the diameter can be reduced to conserve materials. Thinner lead wires also minimize heat loss from the filament, enabling the lamp to provide higher light output at lower wattage.

Dispersion-strengthened copper wire also may be used in leads for discreet electronic components such as diodes. Advantages of dispersion-strengthened copper wire include its high-temperature strength retention during hermetic sealing and its stiffness, which enable multiple insertions in circuit boards.

Commutators. Dispersion-strengthened copper has been used in commutators for helicopter starter motors. High strength is required to combat the high deformation stresses associated with high rotational speeds. Dispersion-strengthened copper also can be used in commutators for submerged fuel pumps in automobiles equipped with fuel injection systems. In this application, resistance to both wear and corrosion in "sour" gasoline (stored gasoline that has generated hydrogen peroxide, which is corrosive to copper) is of particular interest.

Relay blades and contact supports are current-carrying arms that move between the fixed contact points to "make" or "break" an electrical circuit. In a limit switch, the force that moves the contact arm (or contact support) to the "make" position is provided by mechanical means, such as a moving object. In a relay, the contact arm (relay blade) is moved by electromagnetic action. In most cases, the spring force of the arm itself moves the contact back to the "break" position.

Generally, the relay blades and contact supports have silver contacts brazed or riveted to them at the point where they touch the fixed contacts. The strength retention capability of dispersion-strengthened copper after exposure to elevated temperatures allows brazing of contacts to the blade without appreciable strength loss. Because of its higher electrical conductivity, dispersion-strengthened copper has replaced conventional copper alloys such as phosphor bronze and beryllium copper in some relays. Such relays can carry a higher current than was possible with the prior relays.

Miscellaneous Applications. Other potential applications for dispersion-strengthened copper include continuous casting molds, side dam blocks for Hazelett casting machines, gas metal arc welding tips, seam welding wheels, high-current welding cables, microwave tube components, and electrical connectors. In

Table 7   Room-temperature mechanical properties of C15760 rod and bar

Rod diam		Cold work, %	Condition annealing temperature(a)		Ultimate tensile strength		Yield strength, 0.2%		Elongation (16-mm, or 0.64-in. gage length), %	Reduction in area, %	Yield strength/ ultimate tensile strength, %	Yield strength retention ratio (annealed to drawn), %	Hardness, HRB
mm	in.		°C	°F	MPa	ksi	MPa	ksi					
12.7	0.5 14		As drawn		572	83	545	79	16	56	95	...	83
			315	600	565	82	538	78	18	57	95	99	81
			650	1200	524	76	496	72	22	59	94	92	78
			980	1800	496	72	455	66	22	45	92	84	75
7.0	0.28 74		As drawn		621	90	600	87	14	50	97	...	86
			315	600	614	89	586	85	16	51	96	98	84
			650	1200	579	84	545	79	18	53	94	91	80
			980	1800	524	76	490	71	18	40	93	82	77

Note: Copper cladding is machined off prior to testing.
(a) Annealed in nitrogen for 1 h
Source: SCM Metal Products

Table 8   Room-temperature mechanical properties of C15760 wire

Wire diam		Cold work, %	Condition annealing temperature(a)		Ultimate tensile strength		Yield strength, 0.2%		Elongation (25-cm, or 10-in. gage length), %	Yield strength/ ultimate tensile strength, %	Yield strength retention ratio (annealed to drawn), %
mm	in.		°C	°F	MPa	ksi	MPa	ksi			
2.54	0.1 .... 96.6		As drawn		627	91	607	88	4	97	...
			315	600	579	84	558	81	4	96	92
			650	1200	552	80	517	75	7	94	85
			980	1800	496	72	462	67	7	93	76
1.27	0.05 .... 99.1		As drawn		655	95	634	92	3	97	...
			315	600	641	93	614	89	4	96	97
			650	1200	593	86	558	81	5	94	88
			980	1800	503	73	469	68	5	93	74
0.51	0.02 .... 99.9		As drawn		710	103	690	100	2	97	...
			315	600	648	94	621	90	3	96	90
			650	1200	621	90	586	85	4	94	85
			980	1800	531	77	496	72	4	93	72

(a) Annealed in nitrogen for 1 h
Source: SCM Metal Products

most applications, dispersion-strengthened copper must be incorporated in the design to demonstrate cost effectiveness and to exploit its full capabilities.

Dispersion-Strengthened Silver

By James E. Synk
Metallurgist
SCM Metal Products

DISPERSION-STRENGTHENED SILVER containing discreet particles of cadmium oxide in a silver matrix is the most common electical contact material. Its outstanding characteristics are good arc erosion resistance, low contact resistance, antiwelding characteristics, and thermal and electrical conductivities approaching those of fine silver. These materials are used for medium- to high-current-carrying contacts

Table 9   Effect of cold work on stress rupture properties of C15715

Cold work, %	Test temperature		100-h stress rupture strength	
	°C	°F	MPa	ksi
0	425	800	124	28
	650	1200	55	8
	870	1600	14	2
94	425	800	241	35
	650	1200	145	21
	870	1600	76	11

Source: SCM Metal Products

(Ref 2) that convey 30 to 1500 A and 120 to 600 V (Ref 3). These contacts are used in make-and-break switches, in which arcing reduces the contact life. For an overall discussion of electric contact materials, see Ref 4.

Silver/cadmium oxide materials differ from other dispersion-strengthened materials in several respects. In dispersion-strengthened superalloys and dispersion-strengthened copper, the dispersoids primarily strengthen the matrix; consequently, the dispersoids are hard, coherent with the matrix, and uniformly sized on an angstrom scale.

In silver/cadmium oxide composites, the dispersoid chiefly inhibits contact welding and erosion, so the dispersoid is soft and incoherent with the matrix, and varies widely in size on a micron scale. The matrix-strengthening effect of the cadmium oxide is a secondary benefit.

Cadmium oxide particles confer antiwelding characteristics by distributing the arc over the contact surface; when molten silver forms on the contact, the oxides agglomerate at the solid-liquid interface to embrittle any welds that form (Ref 5). Cadmium oxide also retards contact erosion in two ways. When an arc forms between two contacts, the cadmium oxide endothermically decomposes to cool the surrounding matrix and quench the arc. Secondly, the high cadmium oxide surface tension retains molten silver on the contact surface and prevents it from being blown away by the arc (Fig. 5).

Silver/cadmium oxide also differs from other dispersion-strengthened materials in oxide content. Typically, dispersion-strengthened copper alloys and superalloys contain up to about 1 wt% oxide dispersoid, but in commercially available

Fig. 4 Stress rupture properties of dispersion-strengthened copper

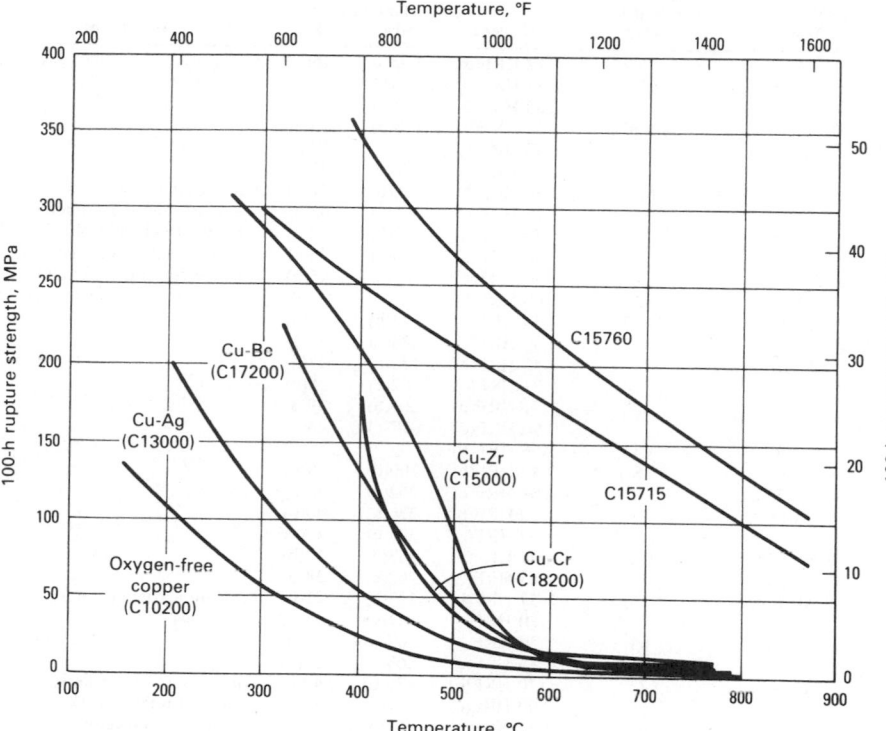

Fig. 5 Arc erosion as a function of cadmium oxide content in dispersion-strengthened silver
Source: Ref 6

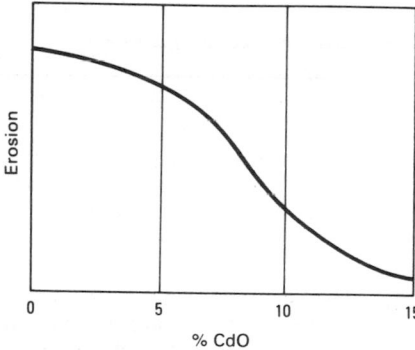

contact materials the dispersoid amount ranges from 2.5 to 25 wt%.

Manufacturing Methods and Properties

Manufacturing methods for commercially available silver/cadmium oxide materials employ powder or wrought ingot techniques. Either approach necessitates compromises among microstructure/performance, dimensional versatility, and cost. For example, the wrought ingot or internal oxidation approach offers the low-

est cost, but provides the least consistent microstructural/performance uniformity throughout the contact life.

These materials characteristically have low erosion rates, but tend to weld at high current. Various powder techniques ensure a more uniform microstructure/performance throughout the contact life, but at a higher cost.

Microstructural uniformity refers primarily to size and distribution of second-phase oxide particles. Powder techniques are more amenable to contacts with a thick cross section and high oxide content. Typically, P/M silver/cadmium oxide displays superior antiwelding properties but higher arc erosion (Ref 6). Contacts produced in this manner are receiving increasing use, because their antiwelding characteristics at high powder density are essential for device miniaturization (Ref 7). Table 10 lists properties and production techniques for commercial silver/cadmium oxide materials. Table 11 compares materials made by various production techniques.

Wrought Ingot Metallurgy: Conventional Internal Oxidation

The conventional internal oxidation process has been the most widely used

method of producing dispersion-strengthened silver and does not employ powder techniques. An alloy of silver-cadmium is cast and worked into fully dense wire, sheet, or rivets. The cadmium is internally oxidized *in situ* by heating the wrought product in air so that the oxygen diffuses inward through the noble silver. Oxidation time is proportional to the square of the oxidized thickness (Ref 8). This process is restricted to thin sections; oxidation of a thin strip may take a day or longer for completion at a typical temperature of 800 °C (1500 °F). This process is limited to low and medium oxide contents.

Microstructural variation through the thickness results from concentration gradients during diffusion. While the oxygen diffuses inward, cadmium diffuses outward. Consequently, the outer surfaces exhibit a fine oxide distribution, which coarsens with depth into a more lamellar structure and ultimately becomes an oxide-denuded zone at the cross section core. This depletion zone is characterized by a low oxide concentration of a relatively few fine oxide particles (Fig. 6 and 7).

The undesirable microstructural deviations lead to variations in contact performance throughout the contact life. The internally oxidized contacts generally offer lower arc erosion rates and faster arc extinguishing than P/M-produced contacts, but are also prone to weld at high currents. Welding tendency and contact arc erosion increase with wear toward the denuded zone (Fig. 8). This causes unreliable performance in applications involving significant wear, but may be unimportant for applications involving minimal wear.

Although it is difficult to completely eliminate structural variations, various alloying additions (Ref 9, 10) and processing controls substantially improve the microstructural homogeneity of these materials. Other techniques capitalize on the microstructural nonuniformity. For example, two alloy sheets may be welded together and separated after internal oxidation, and the silver-rich surface is brazed to a backing material.

P/M Oxide-Dispersion Techniques

Blending. The simplest, least expensive P/M approach to dispersing cadmium oxide with silver is to blend the powders together. Silver powder is commercially available in a 4- to 6-μm range, and cadmium oxide powder is commercially available in a submicron size. Unlike the

Table 10 Properties of silver/cadmium oxide composites for make-and-break contacts

Nominal composition	Manu-facturing method(a)	Density, g/cm³ Theoretical	Typical	Electrical conductivity, %IACS	Hardness	Tensile strength MPa	ksi	Applications
97.5Ag-2.5CdO	PSR	10.42	10.21	85	22 HRF(b)	110(b)	16(b)	
	PSE	10.42	10.42	95	37 HRF(b)	131(b)	19(b)	
					60 HRF(c)	172(c)	25(c)	
95Ag-5CdO	PSR	10.35	9.50-10.14	80-90	32 HRF(b)	110(b)	16(b)	
	PSE	10.35	10.35	92	40 HRF(b)	131(b)	19(b)	
					70 HRF(c)	172(c)	25(c)	Aircraft circuit breakers,
	IO	10.35	10.35	80	40 HRF(b)	186(b)	27(b)	aircraft relays, automotive
					75 HRF(c)	241(c)	35(c)	relays, truck controls,
	PPSE	10.35	10.35	85	70 HRF(b)	207(b)	30(b)	snap switches, contactors,
					90 HRF(c)	248(c)	36(c)	motor controllers, circuit
90Ag-10CdO	PSR	10.21	9.30-9.80	72-85	42 HRF(b)	103(b)	15(b)	breakers, governor relays
	PS+A	10.21	10.11	82	60 HRF(b)	...	...	
	PSE	10.21	10.21	84-87	46 HRF(b)	172(b)	25(b)	
					80 HRF(c)	228(c)	33(c)	
	IO	10.21	10.21	75	45 HRF(b)	186(b)	27(b)	
					81 HRF(c)	262(c)	38(c)	
	PPSE	10.21	10.21	82	71 HRF(b)	269(b)	39(b)	
					90 HRF(c)	317(c)	46(c)	
87Ag-13CdO	...	...	9.20	43	56 HRF(b)	...	...	
86.7Ag-13.3CdO	IO	10.11	10.11	68	48 HRF(b)	200(b)	29(b)	
					84 HRF(c)	262(c)	38(c)	
86.5Ag-13.5CdO	PPSE	10.11	10.11	75	70 HRF(b)	276(b)	40(b)	
					90 HRF(c)	324(c)	47(c)	
85Ag-15CdO	PSR	10.06	8.60-9.58	55-75	35 HRF(b)	83(b)	12(b)	
	PS+A	10.06	10.00	73	65 HRF(b)	262(b)	38(b)	
	PSE	10.06	9.90-10.06	55-75	57 HRF(b)	193(b)	28(b)	Pressure and temperature
					80 HRF(c)	241(c)	35(c)	controls
	IO	10.06	10.06	65	50 HRF(b)	207(b)	30(b)	
					85 HRF(c)	269(c)	39(c)	
	PPSE	10.06	10.06	72	70 HRF(b)	276(b)	40(b)	Aircraft circuit breakers,
					90 HRF(c)	331(c)	48(c)	aircraft relays, truck
83Ag-17CdO	IO	10.01	10.01	62	52 HRF(b)	214(b)	31(b)	controls, contactors,
					88 HRF(c)	276(c)	40(c)	circuit breakers, governor
	PPSE	10.01	10.01	70	70 HRF(b)	276(b)	40(b)	relays
					90 HRF(c)	352(c)	51(c)	
80Ag-20CdO	PS+A	9.93	9.84	63	70 HRF(b)	290(b)	42(b)	
	PPSE	9.93	9.93	68	70 HRF(b)	276(b)	40(b)	
					90 HRF(b)	345(c)	50(c)	
75-Ag-25CdO	PS+A	9.79	9.66	53	74 HRF(b)	317(b)	46(b)	
	PPSE	9.79	9.79	60	...	...	...	

(a) PSR, press/sinter/re-press; PS+A, press and sinter with sintering aid; PSE, press/sinter/extrude; IO, internal oxidation; PPSE, preoxidize/press/sinter/extrude. (b) Annealed. (c) Cold worked
Source: SCM Metal Products

Table 11 Comparison of silver-cadmium oxide materials made by different methods

Properties	Press, sinter, re-press	Press, sinter, extrude	Internal oxidation	Preoxidize, press, sinter without aid (extrude without aid)
Performance characteristics				
Resistance to arc erosion 3		2	1	1
Resistance to sticking and welding 1		1	2	2
Low contact resistance and temperature rise 1		1	1	1
Arc interruption 3		2	1	1
Resistance to corrosion 1		1	1	1
Material characteristics				
High mechanical properties 3		2	2	1
Resistance to annealing 3		2	2	1
Electrical and thermal conductivities 2		1	1	1
Flexibility of composition 2		2	2	1
Uniform cadmium oxide distribution 1		1	3	1

Note: 1 indicates that, under most conditions, this is the preferred material; 2 indicates that, under most conditions, the material is preferable to 3, but not as good as 1; 3 indicates that the material may be acceptable, but under typical operating conditions it is not as good as 1 or 2.
Source: Ref 4

conventional internal oxidation of strip, this approach maintains oxide uniformity in size and distribution throughout the contact cross section and also allows high oxide contents.

The primary disadvantage is that oxide particles are present only on the surface of the silver grains, not within them. However, simple blending imparts good anti-welding ability to the finished contact, and is in widespread commercial use.

Coprecipitation. A common method of producing mixtures of silver and cadmium oxide is coprecipitation of silver and cadmium compounds. The proper mixture of silver and cadmium oxide powders is dissolved in a nitric acid solution. Hydroxide or carbonate solutions are added to change the pH so that silver carbonate and cadmium carbonate coprecipitate.

These powders are calcined at 500 °C (930 °F) to yield a mixture of silver and cadmium oxide. Adequate washing con-

Fig. 6 90Ag-10CdO composite produced by conventional internal oxidation

Note oxide particle size variation from fine (on surface) to coarse (interior). Magnification: 100×. Courtesy of GTE Products Corp.

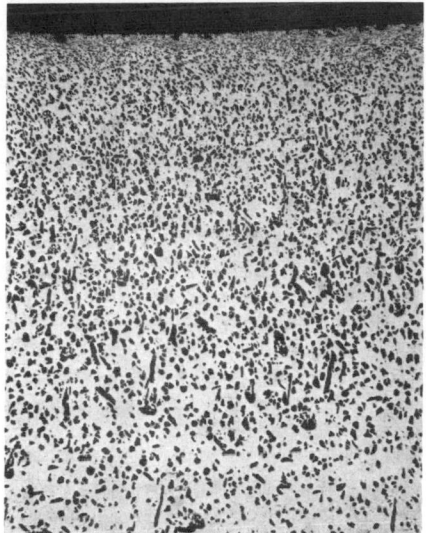

Fig. 7 85Ag-15CdO internally oxidized strip after roll bonding to fine silver backing

Light central area is oxide depletion zone. Magnification: 23×. Courtesy of Texas Instruments, Inc.

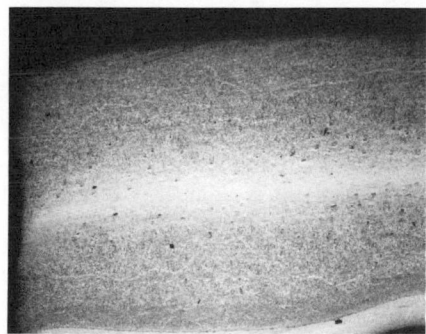

trols the content of deleterious alkali metal impurities to the 10- to 300-ppm range. Coprecipitation imparts oxide uniformity, long-range microstructural homogeneity, and an improved oxide distribution compared to simple blending. However, as in blending, the oxides are retained on the surface of the silver particles, not in their interior.

Preoxidation distributes the cadmium oxide particles within the silver particles and allows a wide selection of oxide content and particle sizes. It accomplishes this by forming a silver-cadmium alloy powder and internally oxidizing/sintering to a

Fig. 8 Effect of cadmium oxide dispersion in fully internally oxidized 85Ag-15CdO material

Source: Ref 6

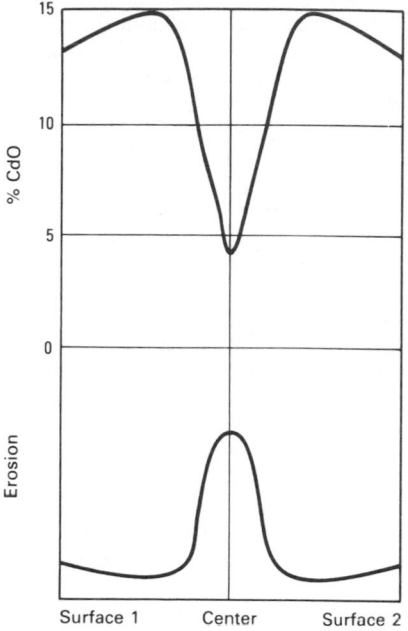

controlled oxide size. This technique also is known as the internally oxidized alloy powder process.

Commercially available cadmium oxide and silver powders are used in the internally oxidized alloy powder process (Ref 6). Cadmium oxide powder has a cubic morphology and submicron size, while silver powder has an irregular morphology in the range of 3 to 9 μm. These powders are blended in the desired proportions and heated in hydrogen, first at 200 °C (400 °F) and then at 400 °C (750 °F).

The low-temperature exposure reduces the cadmium oxide to cadmium metal; the high-temperature exposure melts the cadmium, which diffuses into the silver by liquid-phase and solid-state mechanisms.

The resulting powder is a loosely sintered, homogenized sponge of silver cadmium alloy, which is mechanically ground to porous 0.5-mm (0.02-in.) aggregates. These lumps are completely oxidized internally in air at 600 °C (1100 °F) within several minutes, about 1000 times faster than conventional internal oxidation.

The powder then contains a uniform distribution of very fine (0.05-μm) oxide particles, without undesirable oxide films in the grain boundaries. Subsequent thermal treatments coarsen the oxides to the desired 2-μm size. The high sintering temperature and low oxygen potential of the sintering atmosphere have the greatest effect on oxide growth.

Silver/cadmium oxide is also produced by oxidizing alloy shot. Silver and cadmium are melted, atomized into an alloy shot 2 to 3 mm (0.08 to 0.12 in.) in diameter, and internally oxidized within a few hours at elevated temperature. The silver/cadmium oxide pellets have a denuded core, but this structure is changed by extruding into 1- to 3-mm (0.04- to 0.12-in.) diam wire. Thus, the end product has a uniform structure.

P/M Consolidation Techniques

Press/Sinter/Re-Press. The conventional press/sinter/re-press consolidates a blend of silver and cadmium oxide powders. The compact is sintered and cold worked. As shown in Table 10, however, these materials typically possess only 95% of full calculated density. The same high-temperature cadmium oxide decomposition that prevents contact welding also inhibits densification during sintering. However, full density seems necessary for long contact life.

In addition, the re-pressing operation often initiates cracks that are damaging to ductility, fracture strength, and arc erosion in the finished part (Ref 11). Additions of cadmium oxide are limited to 10 wt%. The conventional press/sinter/re-press process is useful for large parts in low-performance applications.

Press/Sinter/Aid. An addition of 0.1 wt% lithium nitrate sintering agent to preoxidized silver/cadmium oxide develops densities exceeding 99% theoretical after sintering (Ref 10). This addition suppresses cadmium oxide dissociation and changes the oxides from a faceted to a more desirable rounded morphology through a series of complex reactions. Consequently, lithium-nitrate-bearing compacts have higher electrical conductivity and improved fracture strength, with twice the ductility and arc erosion resistance of powders processed without the sintering aid.

This sintering aid permits cadmium oxide contents up to 25 wt% and eliminates the need for additional cold work after sintering. Small germanium oxide additions further strengthen the P/M contacts. As a result, press/sinter/aid contacts are comparable to fully dense, high-performance contacts produced by any other method. The press/sinter/aid process adds a fine silver backing to the contact for future brazing by co-pressing a small amount of silver on the surface (Fig. 9).

Press/Sinter/Extrude. This process produces fully dense contacts from pow-

Fig. 9 Near fully dense 85Ag-15CdO with lithium nitrate sintering aid after sintering

Material was copressed with fine silver backing. Note uniform oxide distribution characteristic of P/M contacts. Magnification: 25×. Courtesy of GTE Products Corp.

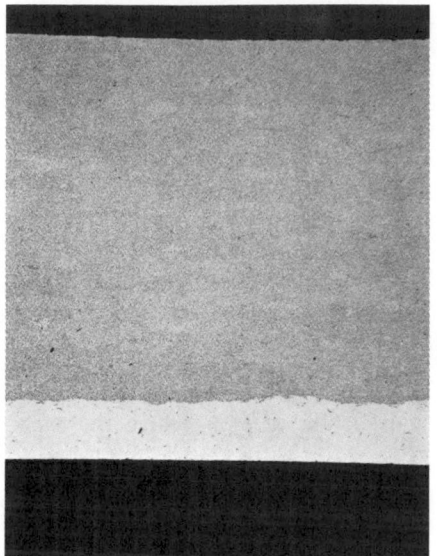

der blends, coprecipitated powder, or pre-oxidized powder. The powders are pressed into an ingot, sintered, extruded, and often worked further. Although this process ensures full density and a mechanically sound structure, it also requires another major processing steps to add the fine silver backing for future joining (Ref 11).

Fastening of Contacts to Base Materials

Before use in a switching device, silver/cadmium oxide contacts are mounted on a backing support material that conducts current and provides the mechanical strength to withstand contact loading. These backing, or base, materials should also be corrosion resistant and formable, brazeable or weldable, and low in cost (Ref 3).

Generally, the contact is mounted onto the base metal by brazing. This requires a thin silver layer, typically 10% of the contact thickness, on the back surface of the silver/cadmium oxide contact. On discreet pressed/sintered/re-pressed parts, the silver layer is co-pressed with the contact; on continuous strips, the layer can be solid-state bonded.

One commonly available configuration is called "top lay," in which one or more strips of contact material are brazed to a wider strip of base metal to provide a contact elevation of 0.5 to 3.2 mm (0.019 to 0.125 in.) from the base-metal surface. The contact surface is profile rolled and may be coined for precise dimensions (Ref 3).

Advantages of this technique are good electrical and thermal conductivities because of a uniform metallurgical bond, low cost by placing the precious metal only where it is needed, good strength from subsequent cold rolling, accurate dimensional control, and design versatility. Table 12 lists the base materials available by this process. Top lay is available in strip or fabricated parts.

Solid-state bonding joins silver/cadmium oxide to a base metal without brazing. Overlay is one such configuration, in which the contact material completely covers one or both sides of the base metal (Table 12). The precious metal strip or strips are rolled onto the base metal either at elevated or room temperature, followed by an elevated-temperature sinter anneal.

The precious and base metals are bonded at a heavy gauge and then rolled to finish size. This technique, like strip brazing, offers good strength, high thermal and electrical conductivities across the bimetal interface, accurate dimensional control, and design versatility. Simple mechanical fastening, as with rivets, also joins contacts to base metals.

Applications of Silver/Cadmium Oxide Contacts

Table 10 summarizes applications for silver/cadmium oxide contacts. All involve make-and-break switches, in which substantial current endangers welding or arc erosion of pure silver and high temperature threatens oxidation of less noble materials.

Silver/cadmium oxide is a prime contact material for applications involving a surge of current when the circuit is completed by the switch, where the possibility of contact welding is the greatest. A typical application is use in a tungsten heater or lamp, in which the initial temperature and electrical resistance of the relatively cool tungsten are low.

The contact material experiences a very high transient current until the temperature and electrical resistance increase to steady-state conditions and the current drops. Other applications with high initial currents include starting motors, solenoids, contactor coils, and capacitive-loading devices.

Silver/cadmium oxide is also a prime contact material for applications involving a heavy inrush of current at the point of breaking a contact, where the possibility of arc erosion is greatest. Examples include inductive circuits, such as solenoids and induction coils, that utilize a direct current that surges with a collapsing magnetic field.

An alternating current power factor of less than unity also loads the contacts similarly, because the voltage and current lag impair arc quenching (Ref 3). Arc quenching is difficult in direct current circuits, which are also an ideal application for silver/cadmium oxide contacts.

Table 12 Base metal selection for silver/cadmium oxide top lay and overlay

Base metal	Top lay	Overlay
Copper	X	X
Yellow brass	X	X
Red brass	X	X
Phosphor bronze	X	X
Nickel silver	X	X
Cupronickel	X	No
Beryllium copper	No	No
Nickel	X	X
Kovar	X	No
Monel	X	X
Copper-clad low-carbon steel	X	X
Low-carbon steel	X	X
Stainless steel	No	No

Source: Ref 3

Dispersion-Strengthened Platinum

PLATINUM is unique as a structural material in high-temperature oxidizing environments where the absence of chemical erosion is essential. However, the pure metal has very limited strength and creep resistance at high temperatures. Also, the high cost of pure platinum can be prohibitive.

Platinum is usually strengthened by alloying with rhodium, but the benefit of such solid-solution strengthening becomes less marked at very high temperatures; grain coarsening occurs and the platinum-rhodium alloy fails due to grain-boundary sliding.

Platinum-rhodium alloys cannot be used in certain applications, however. For example, it cannot be used in applications with optical glass, which discolors when contaminated with rhodium. Such applications are excellent candidates for the use of dispersion-strengthened platinum. Dispersion-strengthened platinum relies on the indirect strengthening resulting from the stable substructure and grain structure that can be attained in the presence of the oxide particles. The extent of the strengthening can be controlled by the amount of oxide addition.

Properties of Dispersion-Strengthened Platinum

Table 13 lists the physical properties of dispersion-strengthened platinum and a dispersion-strengthened platinum-rhodium alloy in comparison with pure platinum and platinum-rhodium alloys. Information on rupture strength, creep deformation, and work hardening follows.

Rupture Strength. Pure platinum has insufficient high-temperature strength for the most demanding applications. Consequently, platinum-rhodium alloys are generally used. However, platinum-rhodium alloys are still susceptible to grain coarsening and will fail after a period of time at high heat.

In contrast, the dispersion of oxide particles in dispersion-strengthened platinum gives a microstructural stability that imparts a high-temperature strength superior to all platinum-rhodium alloys. Tests on sheet material have shown that the 100-h rupture stress of dispersion-strengthened platinum is at least five times greater than that of pure platinum and at least twice that of 20% platinum-rhodium alloys (Fig. 10).

Figure 11 illustrates the stress-rupture time relationship for dispersion-strengthened platinum and dispersion-strengthened platinum-rhodium alloy in comparison with three commonly used platinum-rhodium alloys. The dispersion-strengthened materials can withstand higher pres-

Fig. 10 Comparison of rupture strength of dispersion-strengthened platinum with platinum and platinum-rhodium alloys

Sheet thickness: 1 mm (0.04 in.) in air. Source: Ref 12

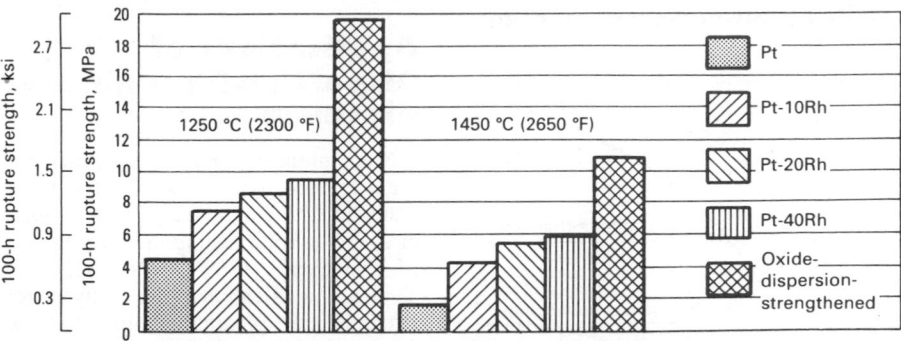

Fig. 11 Stress rupture properties of dispersion-strengthened 10% platinum-rhodium (A), dispersion-strengthened platinum (B), 40% platinum-rhodium (C), 20% platinum-rhodium (D), and 10% platinum-rhodium (E) alloys

Curves are based on tests made on sheet specimens 1.5 mm (0.06 in.) thick tested in air at 1400 °C (2550 °F). Source: Ref 13

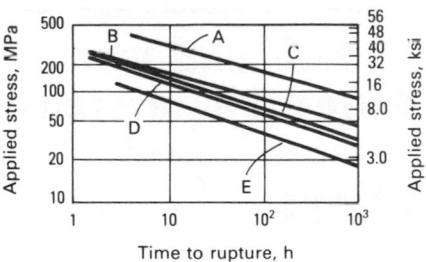

Fig. 12 Minimum creep curves for dispersion-strengthened 10% platinum-rhodium and platinum-rhodium alloys

Tested in sheet form in air at 1400 °C (2550 °F). Source: Ref 13

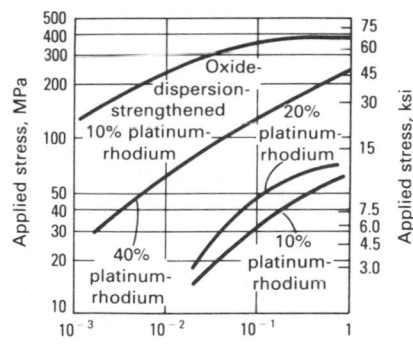

sures for longer periods of time before rupture occurs.

Creep Deformation and Work Hardening. Another benefit of disper-sion-strengthened platinum is improved resistance to creep deformation (Fig. 12). This is an important parameter, as there is an increased demand for larger platinum alloy components that must perform for extended periods without distortion. The minimum creep rate for dispersion-strengthened platinum sheet at 1450 °C (2650 °F) is two to three orders of magnitude lower than that of platinum-rhodium alloys under the same conditions. Figure 13 illustrates the effect of the introduction of a dispersoid into a platinum alloy on the work-hardening characteristics of that alloy.

Fabrication of Dispersion-Strengthened Platinum

Dispersion-strengthened platinum and platinum alloys are prepared using standard P/M techniques. The powder metal (platinum), the dispersoid or oxide start-

Table 13 Physical properties of platinum, platinum-rhodium alloys, and dispersion-strengthened materials

Property	Platinum	10% platinum-rhodium	20% platinum-rhodium	40% platinum rhodium	Dispersion-strenthened platinum	Dispersion-strengthened 10% platinum rhodium
Density at 20 °C (68 °F), g/cm³	21.45	20.0	18.8	16.8	21.28	19.8
Electrical resistivity at 20 °C (68 °F), μΩ·cm ...	10.6	18.4	20.0	17.5	10.8(b)	21.2
Tensile strength(a) at 20 °C (68 °F), MPa (ksi)	161.3 (23.4)	331.0 (48.0)	478.5 (69.4)	563.8 (81.8)	200 (29.0)	355.0 (51.3)
Elongation(a) at 20 °C (68 °F), % ..	40	35	33	30	40	30
Vickers hardness(a) at 20 °C (68 °F), HV	40	75	115	130	55	110

(a) Annealed. (b) At 0 °C (32 °F)
Source: Ref 12 and 13

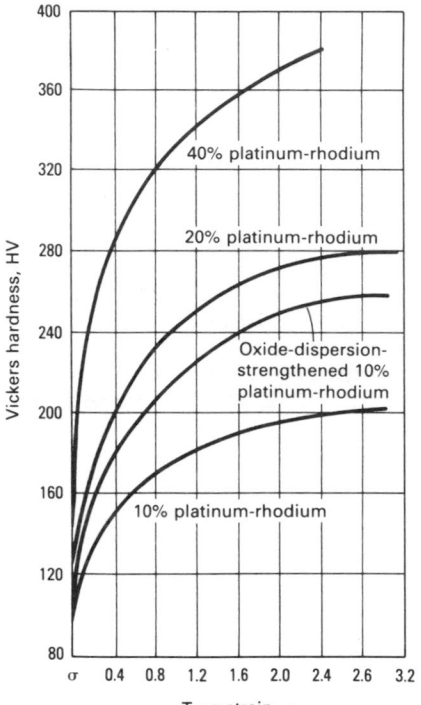

ing powder (thoria, yttria, zirconia, or zirconium chloride), and the master alloy (rhodium, if an alloy is being fabricated) are intimately mixed. At this point, any one of several compaction methods can be used; for example, the P/M mixture can be isostatically pressed or hot extruded.

Although dispersion-strengthened platinum possesses superior high-temperature strength, its hardness and ductility at room temperature are similar to those of pure platinum (Table 13). Therefore, dispersion-strengthened platinum can be readily worked by standard methods such as rolling, drawing, and spinning.

In large components, joining of dispersion-strengthened platinum may be necessary. However, the oxide dispersion is lost if the sections are heated above the melting point, as in fusion welding. If fusion welding must be employed, rhodium-rich filler metal is added to the weld pool. This usually provides sufficient solution hardening at the dispersion-strengthened platinum interface to make up for the local loss of the oxide dispersion. The component often is designed so that the fusion welds are made away from any critically stressed areas. Solid-state can also be per-

formed on oxide-dispersion-strengthened platinum because of the virtual absence of any stable oxide films on the heated metal surface.

Applications of Dispersion-Strengthened Platinum

Many of the components for which platinum is used require the material to operate under stress at very high temperatures, often approaching the melting point of the metal (1769 °C, or 3220 °F). Typical applications for dispersion-strengthened platinum alloys include furnace windings, heater tapes and ignition coils, glass processing equipment, and crucibles for holding molten minerals and ceramics. A particularly important industrial application is for bushings or rotors used in the manufacture of glass fibers. Typically, the equipment in these operations is expected to run continuously at temperatures above 1000 °C (1830 °F) for periods greater than 1 year.

Nickel- and Iron-Based Dispersion-Strengthened Alloys

By Paul S. Gilman
Senior Metallurgist
NOVAMET
Inco Alloy Products Co.
and
John S. Benjamin
General Manager, IncoMAP
Inco Alloy Products Co.

NICKEL- AND IRON-BASED SUPERALLOYS are produced by mechanical alloying, a relatively new development in process technology. The strength of metals at high temperatures can be increased by the addition of a fine dispersion of insoluble refractory oxides. Although there are many methods that can produce such dispersions in simple metal systems, these techniques are not applicable to the production of more highly alloyed materials, such as those required for gas turbine engines. For instance, conventional P/M techniques either do not produce an adequate dispersion or do not permit the use of reactive alloying elements

such as chromium and aluminum, which confer the required property characteristics, including corrosion resistance and intermediate temperature strength. Simple mixing techniques either do not disperse the oxides well enough, produce a powder so fine that it is pyrophoric, or introduce excessive contamination. Chemical reduction processes that are successful for simple dispersion-strengthened alloys—for example, dispersion-strengthened nickel or nickel-chromium—cannot reduce the oxides and salts of the highly reactive elements that are needed to produce complex superalloys.

Mechanical alloying is a dry, high-energy ball milling process for producing composite metallic powders with a controlled, fine microstructure. It is carried out in a highly agitated ball charge by the repeated cold welding and fracturing of a mixture of metal powders to which some nonmetal powders may be added. Its widest use has been in the production of dispersion-strengthened nickel- and iron-based superalloys for service at temperatures of 1000 °C (1830 °F) and above.

Unlike mechanical mixing processes, mechanical alloying produces a material whose internal homogeneity is independent of starting powder particle size. Thus, ultrafine dispersions (less than 1-μm interparticle spacing) can be obtained with relatively coarse initial powders (50- to 100-μm average diameter).

Although chemical processes have been used successfully to produce simple dispersion-strengthened alloys, mechanical alloying permits use of relatively reactive elements, such as aluminum and titanium. A full range of complex dispersion-strengthened nickel-, cobalt-, and iron-based superalloys can be prepared by this method.

Materials for Mechanically Alloyed Dispersion-Strengthened Superalloys

The raw materials used for mechanically alloyed dispersion-strengthened superalloys are widely available commercially pure powders that have particle sizes that vary from about 1 to 200 μm. These powders fall into the broad categories of pure metals, master alloys, and refractory compounds. The pure metals include nickel, chromium, iron, cobalt, tungsten, molybdenum, and niobium. The master alloys include nickel-based alloys with relatively large amounts of combinations of aluminum, titanium, zirconium, or hafnium.

These master alloys are relatively brittle when cast and easily comminuted to powder. In addition, because they consist of relatively exothermic intermetallic compounds, the thermodynamic activity of the reactive alloying elements, such as aluminum and titanium, is considerably reduced compared to that of the pure metals.

The oxygen contents of the commercially pure metal powders and the master alloys range from 0.05 to 0.2 wt%. The refractory compounds that can be added include carbides, nitrides, and oxides. For the production of dispersion-strengthened materials, such additions are limited to very stable oxides, such as yttria, alumina, or less frequently thoria. These oxides, which are prepared by calcination of oxalate precipitates, consist of crystallites of about 50 nm agglomerated into pseudomorphs of about 1 μm.

The only restriction on the mixture of powder particles for mechanical alloying (other than the particle size range mentioned above and the need to minimize excessive oxygen) is that at least 15 vol% of the mix should consist of a compressibly deformable metal powder. The function of this component, which can consist of any one or all of the pure metals, is to act as a host or binder for the other constituents during the process.

Other similar metals, such as copper, zinc, and magnesium, are suitably ductile but not normally added to superalloys. For dispersion-strengthened superalloys, the amount of refractory oxide added ranges from about 0.4 to 1.5 wt% (1 to 2.7 vol% for yttria).

Equipment for Mechanical Alloying

The machinery used for mechanical alloying consists of one of several types of high-energy ball mills (Fig. 14). These are selected on the basis of given processing times, ranging from hours to tens of hours. The types of ball mills employed include shaker mills, vibratory mills, stirred ball mills, centrifugal ball mills, and conventional ball mills with diameters greater than about 1 m. The restriction on conventional ball mills arises from the relatively low energy density of operation of smaller mills, which leads to excessive processing times.

Unlike the procedure used in ball milling for comminution, the ratio of balls to powder in mechanical alloying is relatively high. These ratios range from 6 to 1 by weight to as high as 30 to 1, but most commonly are in the range of 10 to 1 to 20 to 1. The balls themselves range from

Fig. 14 Schematic of high-energy ball mill
Source: Ref 14

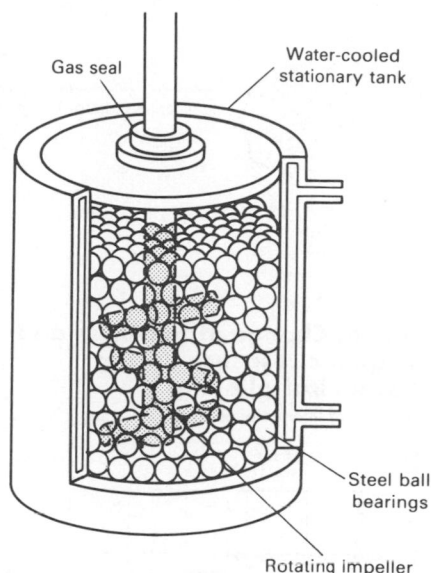

4 to 20 mm (0.16 to 0.8 in.) in diameter but are usually 8 to 10 mm (0.32 to 0.4 in.) in diameter and are made of a through-hardened steel, such as 52100. The environment within the grinding machine is controlled wherever practical; water cooling and atmosphere control are employed. The milling atmosphere consists of either nitrogen or argon with measured trace amounts of oxygen. Liquids can also be used.

Mechanical Alloying Process

A mixture of metal, master alloy, and nonmetal powders calculated to yield the composition of the desired alloy is charged into a suitable high-energy ball mill. The starting powder constituents for mechanical alloying and their deformation characteristics are illustrated schematically in Fig. 15. During each collision of the grinding balls, a multitude of powder particles is trapped along the line of centers (Fig. 16). Typically, around 1000 particles with an aggregate weight of 0.2 mg (7×10^{-6} oz) are trapped during each collision.

These particles undergo severe plastic deformation, which increases their surface-to-volume ratio and ruptures the surface films of adsorbed contaminants. It is estimated that the time of such a collision, during which the grinding balls are decelerated and transfer their energy to the plastically deforming metal, is on the order of 10^{-4} s. Where metal particles overlap, atomically clean metal interfaces are brought into intimate contact, forming cold welds and building up composite metal particles consisting of various combinations of the starting ingredients.

Examination of sectioned grinding balls withdrawn from an attritor during mechanical alloying indicates that powder is welded to the surfaces of the grinding balls. An irregular, knobby layer is built up with a maximum thickness of approximately 200 μm (Fig. 17). As larger aggregates are

Fig. 15 Representative constituents of starting powders used in mechanical alloying showing deformation characteristics
Courtesy of International Nickel Co., Inc.

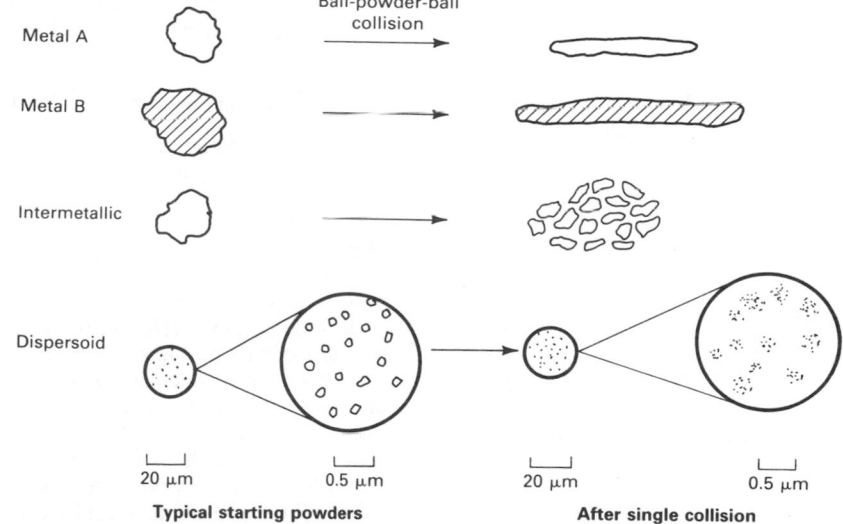

Fig. 16 Ball-powder-ball collision of powder mixture during mechanical alloying

(a) Cold welding. (b) Powder fracture. Courtesy of International Nickel Co., Inc.

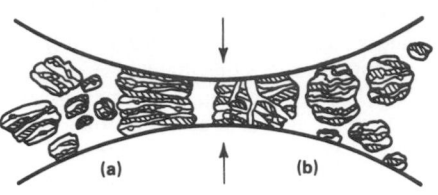

(a)   (b)

Fig. 17 Surface of sectioned nickel pellet showing welded composite layer

Magnification: 115×. Courtesy of International Nickel Co., Inc.

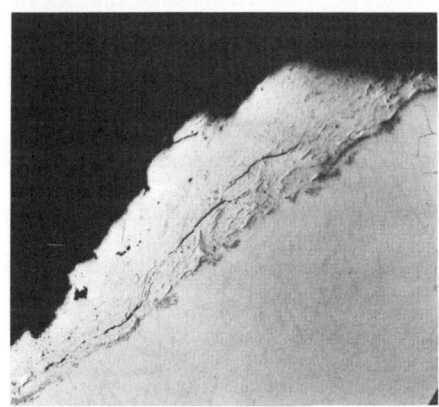

formed, each particle contains substantially all of the starting ingredients, including the refractory oxide, in the proportions of the master blend (Fig. 18).

With continued plastic deformation, the hardness of the powder increases, and fracturing and cold welding occur. A balance is achieved between the rate of weld-

Fig. 18 Early stage of mechanical alloying process

Particles are layered composites of starting constituents. Courtesy of International Nickel Co., Inc.

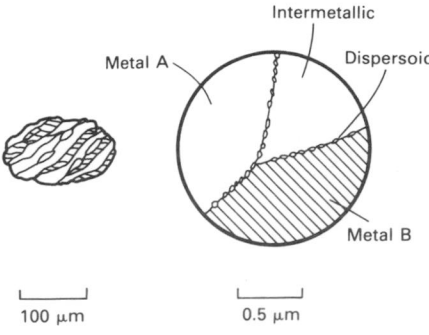

100 μm   0.5 μm

Fig. 19 Completion of mechanical alloying process

Each powder particle composition is equivalent to starting powder blend and contains uniform distribution of dispersoids. Courtesy of International Nickel Co., Inc.

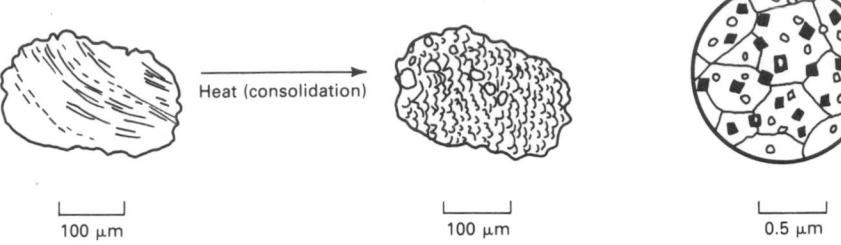

Heat (consolidation)

100 μm   100 μm   0.5 μm

Fig. 20 Change of hardness and magnetic response during mechanical alloying

Courtesy of International Nickel Co., Inc.

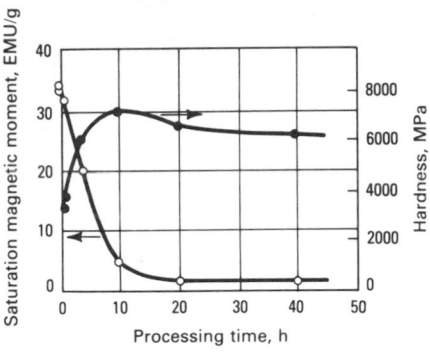

ing, which tends to increase the average composite particle size, and the rate of fracturing, which tends to decrease the average composite particle size. With continued processing, this leads to a steady-state particle size distribution of the composite metal particles. The constant interplay between welding and fracturing establishes a kneading action that continues to refine the internal structures of the composite metal powders (Fig. 19).

Powder Characteristics

A given sample of mechanically alloyed superalloy powder may contain particles ranging from 10 to 500 μm, with an average particle size between 50 and 200 μm. The internal structure of the powder is independent of particle size once the steady state is achieved.

Because of the severe plastic deformation that occurs during mechanical alloying, very high hardnesses are achieved in the powders. Hardness increases almost linearly during the initial stages of the process (Fig. 20), reaching a saturation value, after which time it is presumed that work softening balances further cold work.

It has also been noted by magnetic measurements (Fig. 20) that true alloying occurs, as evidenced by the loss of saturation magnetic moment, with increasing processing time in a high-nickel composition. This may be due to a combination of factors. First, the extreme amount of cold work leads to low activation energy and pipe diffusion. Second, the temperature is moderately elevated to approximately 150 °C (300 °F) during the process. Third, relatively short interdiffusion distances (less than 1 μm) exist toward the end of the process.

Consolidation, Thermomechanical Processing, and Properties

Figure 21 is a schematic showing the process path of thermomechanical processing from raw material to final product. Mechanically alloyed superalloy powders are consolidated either by extrusion or by hot isostatic pressing in hermetically sealed cans. This operation, which is carried out at temperatures between 900 and 1100 °C (1650 and 2000 °F), yields a monolithic body with an extremely fine grain size (less than 0.5 μm).

At appropriate temperatures and strain rates, the consolidated material exhibits superplastic hot deformation characteristics. Thermomechanical processing, which consists of hot working or, in some cases, hot and cold working, controls the fine grain size within a very narrow range, typically 0.1 to 0.4 μm, in order to prepare the alloy for a germinative grain growth process. Germinative grain growth or secondary recrystallization is carried out at temperatures between 1100 °C (2000 °F) and the solidus temperature of the alloy being processed.

Nickel-based superalloys containing more than approximately 2% Al contain equilibrium precipitates based on the in-

Fig. 21 Thermomechanical processing from raw material to final product
Source: Ref 14

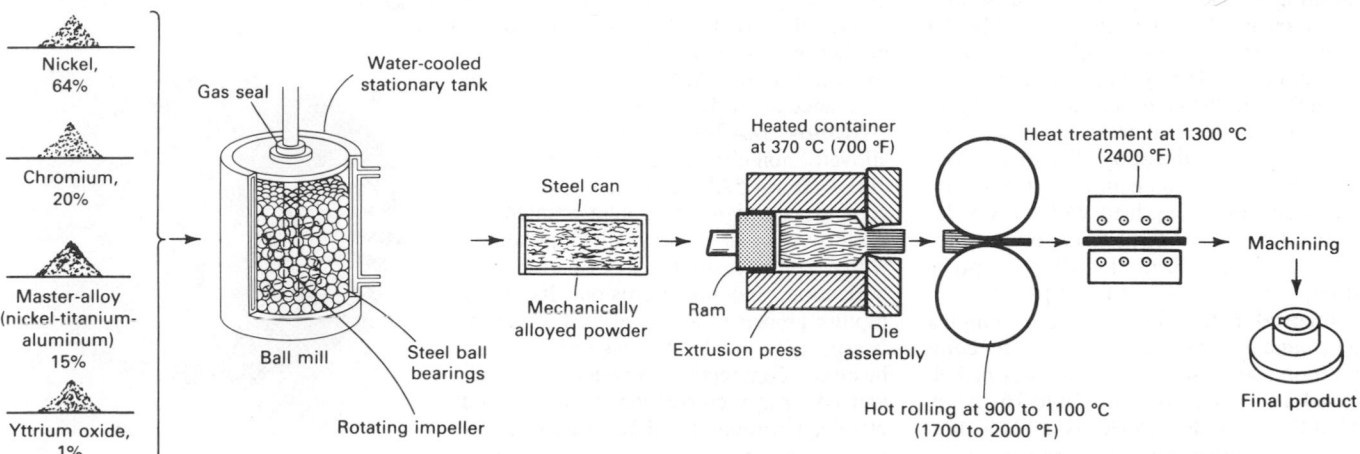

tcrmetallic compound Ni₃Al. Although it is not known whether these compounds form during the mechanical alloying process itself, they do form on heating of powder for consolidation. In the final alloy, these gamma prime precipitates contribute significantly to strength at intermediate temperatures (700 to 800 °C, or 1290 to 1470 °F). During heat treatment following critical hot working, these precipitates form grain boundaries and prevent grain growth.

For gamma-prime-containing dispersion-strengthed alloys, the germinative grain growth temperature must also exceed the gamma prime solvus temperature. The germinative grain growth process can be carried out isothermally or by the passage of a moving zone in the direction of maximum thermomechanical processing extension, that is, the principal working direction.

This step results in an increase in grain size and grain aspect ratio (ratio of maximum grain length to minimum grain dimension). Minimum grain dimensions of about 1 mm, with grain lengths in excess of 1 cm, are typical with conventional anneals. Zone annealing can lead to effectively infinite grain aspect ratios, with grain diameters of 1 cm and grain lengths in excess of 20 cm, and are limited only by the length of the workpiece.

Dispersion-Strengthened Nickel- and Iron-Based Alloys

Inconel MA754 is a dispersion-strengthened nickel-chromium alloy (20NiCr-1.0Fe-0.3Al-0.5Ti-0.6Y₂O₃) produced by mechanical alloying. Table 14

Table 14 Longitudinal and long transverse 100-h rupture stresses of Inconel MA754

	Rupture at:			
	760 °C (1400 °F)		1090 °C (2000 °F)	
Direction of testing	MPa	ksi	MPa	ksi
Longitudinal	214	31.1	102	14.9
Longitudinal transverse	172	25.0	41	6.0

Table 15 Physical properties of Inconel MA754

Approximate melting point (solidus)	1400 °C (2550 °F)
Modulus of elasticity at 20 °C (68 °F)	151 GPa (22 × 10⁶ psi)
Density	8.3 g/cm³

gives longitudinal and long transverse 100-h stresses, and Table 15 lists physical properties.

The stress rupture properties of Inconel alloy MA754 compared to other bar materials are shown in Fig. 22. The yttria dispersoid imparts exceptional high-temperature strength and creep resistance to the alloy.

The strength, high melting point, and microstructural stability of this alloy make it an attractive material for gas turbine components and other extreme service condition applications. With alloy MA754, the advantages of a wrought material can be obtained in service environments in which only cast alloys were used previously.

MA754 is thermomechanically processed to create a stable, recrystallized grain structure that is coarse and highly elongated in the direction of hot working (Fig.

Fig. 22 Stress rupture properties of Inconel MA754 compared to other bar materials

Temperature: 1100 °C (2000 °F). Courtesy of International Nickel Co., Inc.

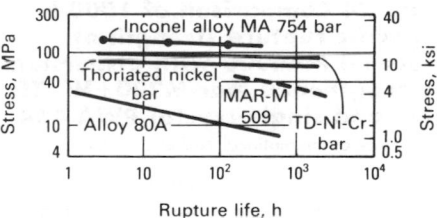

23). Grain length-to-width ratios may be as high as 10 to 1.

For applications requiring good resistance to thermal fatigue, such as gas turbine vanes, Inconel MA754 is given a strong texture. The majority of the grains are aligned so that their <100> axes are parallel to the principal working direction and along the length of the bar. Such texture results in a low modulus of elasticity (149 GPa or 21.6 × 10³ ksi) in the lon-

Fig. 23 Coarse, elongated grain structure of Inconel MA754

Courtesy of International Nickel Co., Inc.

gitudinal direction. The low modulus improves resistance to thermal fatigue by lowering stresses for given thermal strains.

The outstanding characteristic of MA754 is high strength, especially at elevated temperatures. The dispersion strengthening achieved through mechanical alloying remains effective at temperatures to the melting point of the alloy. The dispersion-strengthening mechanism is augmented by the high grain length-to-width ratio developed by thermomechanical processing. The 1095 °C (2000 °F) 100-h rupture strength is 102 MPa (14 800 psi).

Inconel MA6000 is a more complex mechanically alloyed superalloy. Its composition and those of other complex nickel-based alloys are given in Table 16. Alloy MA6000 was developed as a blade material for advanced gas turbines. It is strengthened by both an oxide and a precipitate dispersion for high creep and rupture properties at 1095 °C (2000 °F) as well as 760 °C (1400 °F) (Fig. 24).

Unlike cast alloys, MA6000 exhibits nearly flat rupture-life curves at high tem-

peratures. This results from oxide dispersion strengthening combined with high grain length-to-width ratios (typically greater than 10 to 1). Thus, rupture-life predictions are more reliable. Because of its composition, MA6000 has excellent resistance to oxidation and sulfidation.

Incoloy MA956 is a mechanically alloyed iron-chromium-aluminum alloy ($74Fe-20Cr-4.5Al-0.5Ti-0.5Y_2O_3$) that combines the high-temperature strength and stability of oxide dispersion strengthening with excellent resistance to oxidation, carburization, and hot corrosion. Its stress rupture properties are given in Fig. 25. It is especially suitable for gas turbine combustion chambers, components of advanced energy-conversion systems, and other applications involving rigorous service conditions.

Alloy MA956 sheet is thermomechanically processed to form a coarse, pancake-shaped grain structure that contributes to strength and toughness in all directions in the sheet plane. Alloy MA956 is also produced as bar and tubing. It can be fabricated into intricate components, such as the power station swirler shown in Fig. 26.

Advantages of Mechanical Alloying

The major advantage of mechanical alloying as applied to nickel- and iron-based high-temperature materials is increased load-bearing capability at high temperatures. Thus, at 1100 °C (2000 °F), one such alloy can withstand a stress of 126 MPa

Fig. 26 Power station swirler fabricated from Incoloy MA956
Courtesy of Central Electric Generating Board, United Kingdom

(18.3 ksi) for 1000 h, while an alloy of similar composition without an oxide dispersion can withstand a stress of only 45 MPa (6.5 ksi) for the same period of time.

Secondary advantages of mechanical alloying relate directly to its ability to add reactive alloying elements such as aluminum, titanium, and chromium. These can lead to relatively high strengths at lower temperatures through gamma prime precipitation hardening or to the development of combined strength and corrosion resistance.

REFERENCES

1. Irmann, R., SAP: Ein Neuer Werkstoff der Pulvermetallurgie aus Al, *Technische Rundschaw*, Vol 36, 1949, p 19-26
2. Kleis, J.D., Metallurgy of Contacts, *Proceedings of the Holm Seminar on Electrical Contact Phenomena*, 1970, p 1
3. "Electrical Contacts, Parts and Materials," Texas Instruments, Inc., Attleboro, MA
4. Electric-Contact Materials, in *Metals Handbook*, Vol 3, 9th ed., American Society for Metals, 1980
5. Freudinger, E., Electrical Contact Materials—Properties and Selection, *Electro-Technology*, June 1962
6. Hirsbrunner, H.G., Improved Sintered Silver Cadmium Oxide Material for Electrical Contacts, *Proceedings of the Holm Seminar on Electrical Contacts*, 1974
7. Pedder, D.J., Douglas, P., McCarthy, J.P., and Brugner, F.S., I.O.A.P.—A Novel Process for the Manufacture of Long Life Silver-Cadmium Oxide AC Contact Material, *Proceedings of the Holm Seminar on Electrical Contacts*, 1976

Fig. 24 Comparison of 1000-h specific rupture strength of Inconel MA6000 with dispersion-strengthened Mar-M200+Hf, TD-Ni, and single-crystal PWA 454
Courtesy of International Nickel Co., Inc.

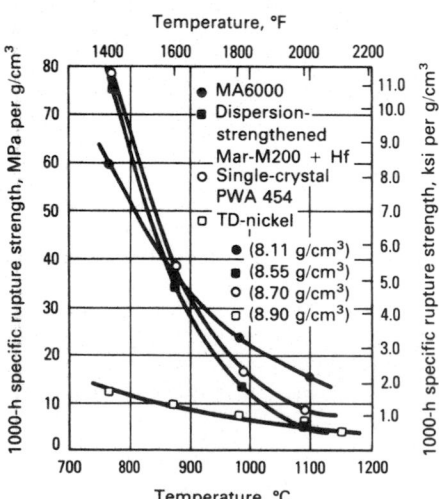

Fig. 25 Stress rupture properties of Incoloy MA956 sheet compared with other materials
Temperature: 1100 °C (2000 °F). Courtesy of International Nickel Co., Inc.

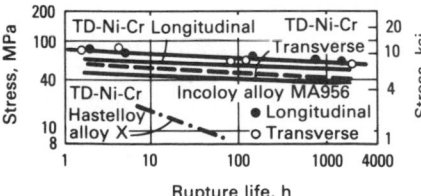

Table 16 Nominal compositions of Inconel MA6000 and nickel-based superalloys

Alloy	Al	Ti	Ta	V	Nb	Cr	Mo	W	Co	C	Zr	B	Y_2O_3
MA6000	4.5	2.5	2.0	…	…	15.0	2.0	4.0	…	0.05	0.15	0.01	1.1
Inconel 792	3.0	4.5	4.0	…	…	12.4	2.0	3.8	9.0	0.12	0.10	0.02	…
Inconel 738	3.4	3.4	1.75	…	0.9	16.0	1.75	2.6	8.5	0.17	0.10	0.01	…
Inconel 100	5.5	4.7	…	1.0	…	10.0	3.0	…	15.0	0.18	0.06	0.014	…
713C	5.9	0.6	…	…	2.0	12.0	4.5	…	…	0.05	0.10	0.01	…

(a) rem nickel

8. Rhines, F.N., Johnson, W.A., and Anderson, W.A., *Trans. Met. Soc. AIME*, Vol 147, 1942, p 205

9. Comey, K.R., Jr. and Happ, M.B., The Effect of Elemental Additives on the Contact Behavior of Silver Cadmium Oxide, Texas Instruments Inc., prepared for presentation at the Engineering Seminar on Electrical Contact Phenomena, Nov 1967

10. Shen, J.S., Zdanuk, E.J., and Krock, R.H., The Effect of Additives from Groups IA and IIA on the Morphology and Properties of Silver Cadmium Oxide Contacts, *Proceedings of the Holm Seminar on Electrical Contact Phenomena*, 1971

11. Kim, H.J., Development of Improved Ag-CdO Material by a Reactive Sintering Process, *Proceedings of the 10th International Conference on Electrical Contact Phenomena*, Budapest, Hungary, 1980

12. Stanley, R.G. and Wilson, F.G., ODS (Oxide Dispersion Strengthened) Platinum—A Unique High Temperature Material for the Most Demanding Applications, *Met. Powder Rep.*, Vol 37 (No. 4), 1982, p 175-178

13. ZGS 10% Rhodium-Platinum (Dispersion Strengthened Platinum-Base Alloy), *Alloy Digest*, March 1979, p 2

14. Benjamin, J.S., Mechanical Alloying, *Scientific Am.*, Vol 234 (No. 5), 1976, p 40-48

SELECTED REFERENCES

• Alexander, G.B., Iler, R.K., and West, S.F., "Metal Oxide-Metal Composition," U.S. Patent 2 972 529, 1961

• Benjamin, J.S., Dispersion Strengthened Superalloys by Mechanical Alloying, *Trans. AIME*, Vol 1, 1970, p 2943-2951

• Benjamin, J.S. and Volin, T.E., The Mechanism of Mechanical Alloying, *Met. Trans.*, Vol 5, 1974, p 1929-34

• Bevington, R.C. and Kim, H.J., Thermophysical Properties of Ag-CdO Composite Materials, *IEEE Trans. Components, Hybrids, Man. Tech.*, Vol CHMT-2 (No. 1), March 1979

• Coxe, C.D. and McDonald, A.S., "Process for Internally Oxidation Hardening Alloys, and Alloys and Structures Made Therefrom," U.S. Patent 3 184 835, 1965

• Crosby, R.L. and Desy, D.H., U.S. Bureau of Mines Report 7266, June 1969, p 17

• Gilman, P.S. and Benjamin, J.S., Mechanical Alloying, *Ann. Rev. Mat. Sci.*, Vol 13, 1983

• Grimwade, M.F. and Jackson, K., The Preparation and Properties of Copper, Nickel and Iron Containing a Dispersed Oxide Phase, *Powder Metall.*, Vol 10, 1962, p 13-33

• Hammer, G.H., Kaufman, D., and Clasing, M., Dispersion Strengthening of Platinum by Simultaneous Precipitation of Pt-ZrOCl$_2$ Powders, *Modern Developments in Powder Metallurgy*, Vol 14, H.H. Hausner, H.W. Antes, and G.D. Smith, Ed., Metal Powder Industries Federation, Princeton, NJ, 1981, p 205-214

• Meijering, J.L. and Druyvesteyr, M.J., Hardening of Metals by Internal Oxidation, *Phillips Research Report*, Vol 2, 1947, p 260-280

• Morse, J.P. and Benjamin, J.S., Mechanical Alloying, in *New Trends in Materials Processing*, American Society for Metals, 1976, p 165-99

• Preston, O. and Grant, N.J., Dispersion Strengthening of Copper by Internal Oxidation, *Trans. AIME*, Vol 221, Feb 1961, p 164-172

• Stamford, M.S., *Copper*, Vol 29, 1966, p 7-11

• Unpublished data, SCM Metal Products Research Laboratory, Cleveland

• Wilson, F.G. and DesForges, C.D., Preparations and Properties of Some Advanced Oxide Dispersion Strengthened Alloys for High Temperature Use, *Proceedings of the Behavior of High Temperature Alloys in Aggressive Environments Conf.*, Oct 1979, Petten, The Netherlands, 1980, p 1017-1030

• Zwilsky, K.M. and Grant, N.J., Dispersion Strengthening in Copper-Alumina System, *Trans. AIME*, Vol 221, April 1961, p 371-377

P/M Stainless Steels

By Ralph W. Stevenson
Project Manager/Materials Group
Energy Technology Division
Midland-Ross Corp.

STAINLESS STEEL P/M PARTS represent an important and growing segment of the P/M industry. The selection of stainless steel components made by P/M techniques begins by recognizing the need for a stainless steel alloy. Because the cost of stainless steel usually greatly exceeds that of iron or steel, the potential application should require at least one of the following:

- Corrosion resistance
- Oxidation resistance
- Good physical appearance
- Good mechanical properties (ductility and impact strength)
- Wear resistance
- Nonmagnetic characteristics (300 series)
- Hardenability (400 series)

The design engineer and parts manufacturer must identify the specific properties that are required for each application to obtain the optimum combination of P/M stainless steel alloy, density, and processing conditions.

All commercial compacting-grade stainless steel powders are produced by atomization. Most powders that are conventionally pressed and sintered are water atomized. This process produces a completely alloyed powder that is characterized by an irregular particle shape (Fig. 1). Gas (nitrogen) atomization is also used to produce spherical powder, which is consolidated by hot isostatic pressing or hot extrusion.

The chemical compositions of P/M stainless steel alloys are held to closer tolerances than the compositions of American Iron and Steel Institute (AISI) wrought stainless steels. Low carbon and high nickel contents in the austenitic grades contribute to good compressibility. The austenitic alloys are the most widely used grades, with type 316L accounting for about one third of all stainless steel powders. Chemical compositions, physical properties, and screen analyses of the most widely used P/M stainless steel powders are listed in Table 1.

Fig. 1 Scanning electron micrograph of water-atomized type 304L stainless steel (−100 mesh)

Magnification: 150×

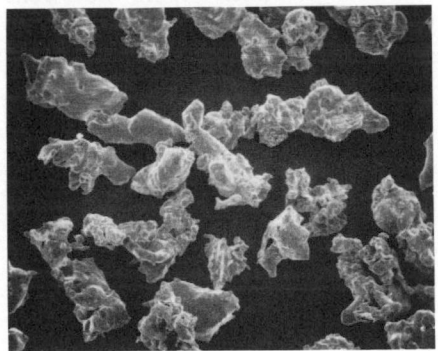

Processing Sequence

Design and processing of stainless steel P/M parts are subject to the same basic considerations as are other P/M materials. However, compared with low-alloy ferrous powders, stainless steel powders require higher compacting pressures and have lower green strength. Typical compaction characteristics of austenitic and martensitic grades are shown in Fig. 2. Compaction pressures ranging from 550 to 830 MPa (40 to 60 tsi) are common in commercial practice.

Table 1 Properties of stainless steel compacting-grade powders

	303L (a)	304L	316L	830 (b)	410L	434L
Chemical analysis, %						
Chromium	17.5	18.5	16.5	20.5	12.0	17.0
Nickel	12.5	11.5	13.5	30.0	...	...
Molybdenum	...	...	2.1	2.5	...	1.0
Manganese	0.2	0.2	0.2	0.2	0.5	0.2
Silicon	0.7	0.8	0.7	1.0	0.8	0.8
Carbon	0.02	0.02	0.02 (max)	0.02	0.02	0.02
Iron	rem	rem	rem	rem	rem	rem
Physical properties						
Apparent density, g/cm³	3.1	2.7	2.7	2.8	2.9	2.8
Flow rate, s/50 g	26	30	30	30	28	29
Screen analysis, %						
+100 mesh	1	1	1	1	1	1
−100+150 mesh	7	12	11	9	14	11
−150+200 mesh	13	20	18	14	20	17
−200+325 mesh	24	25	26	27	26	27
−325 mesh	55	42	44	49	39	44

Note: Type 830 is used for the manufacture of P/M parts where superior corrosion resistance is of primary consideration. Parts made from type 830 exhibit improved resistance to oxidizing media and sulfuric acid. This grade is not recommended for conditions involving unstable chlorides.
(a) 0.2% sulfur added for machinability. (b) 3.5% copper also present

Fig. 2 Compactibility of stainless steel powders

(a) Type 316L austenitic stainless steel. (b) Type 410L martensitic stainless steel

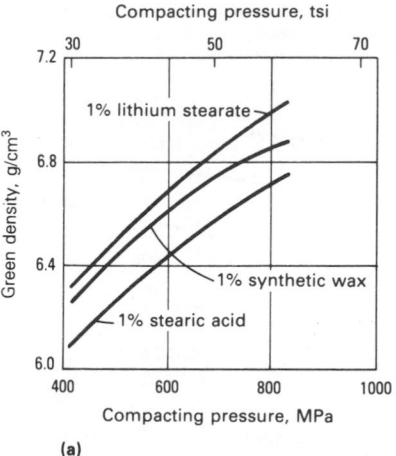

(a)

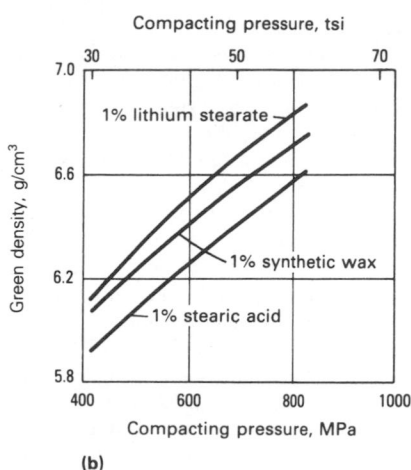

(b)

Fig. 3 Effect of lubricant and compacting pressure on green strength of stainless steel powder

(a) Type 316L austenitic stainless steel. (b) Type 410L martensitic stainless steel

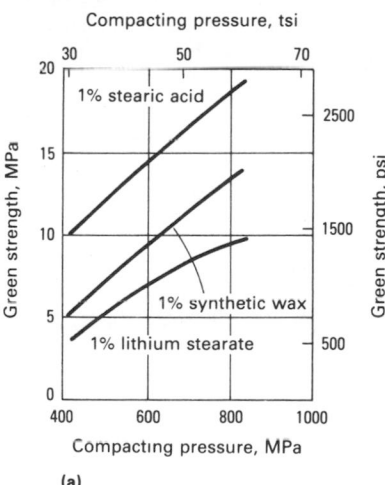

(a)

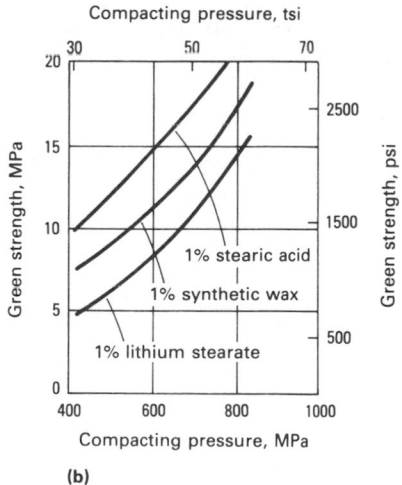

(b)

The green strength of stainless steel compacts, which is about half that of P/M iron, is influenced by compaction pressure and the type of lubricant. Lubricants that provide high green strength, such as stearic acid (see Fig. 3), generally cause lower compactibility (Fig. 2). Therefore, lubricant selection is an important factor in determining successful application and fabrication of P/M stainless steels.

In fabrication of P/M stainless steel parts, carbide tooling is required, because highly alloyed powders are inherently harder than pure metals such as iron or copper. In addition, because of the relatively low green strength of P/M stainless steel, complex parts require special care in punch ejection and handling.

Sintering is the most important step in processing stainless steel P/M parts. Because carbon lowers corrosion resistance, the lubricant must be removed to prevent carbon diffusion into the part. For many applications, stainless steels are sintered

at 1120 to 1150 °C (2050 to 2100 °F). However, when improved mechanical properties and corrosion resistance are required, sintering temperatures up to 1315 °C (2400 °F) and higher are used.

Mechanical properties of austenitic stainless steels depend on the atmosphere in which they are sintered. The most common commercial atmosphere for sintering stainless steel is dissociated ammonia (75 vol% hydrogen and 25 vol% nitrogen). Sintering in this atmosphere introduces as much as 0.3% nitrogen into the composition, which increases strength, but lowers ductility (Table 2). Typical tensile properties of type 316L stainless steel sintered in dissociated ammonia are given in Fig. 4.

Although considerable data are available regarding hydrogen-based sintering atmospheres for stainless steels, there is little commercial use of hydrogen because

Fig. 4 Tensile properties of standard type 316L stainless steel Metal Powder Industries Federation tensile bars

Bars pressed to densities indicated and sintered in −40 °C (−40 °F) dew point dissociated ammonia for 45 min after lubricant was burned off in air.

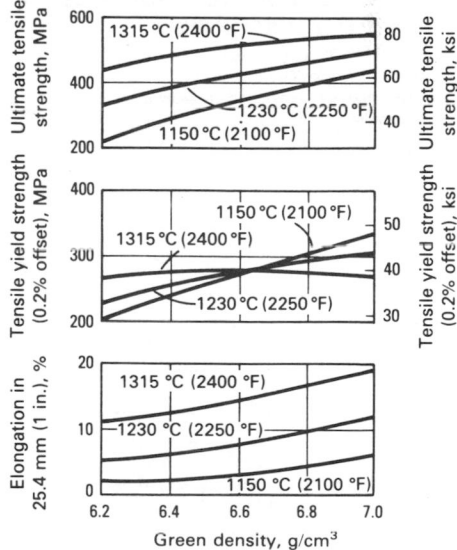

Table 2 Influence of sintering atmosphere on mechanical properties of type 316L pressed to 6.85 g/cm³

Property	Sintered in dissociated ammonia	Sintered in hydrogen
Ultimate tensile strength, MPa (psi)	365.4 (53 000)	288.2 (41 800 psi)
Yield strength, MPa (psi)	274.4 (39 800)	183.4 (26 600 psi)
Elongation in 25 mm (1 in.), %	7.0	10.9
Apparent hardness, HRB	67	47

Note: Compacts sintered for 30 min at 1120 °C (2050 °F)

of its high cost. The principal alternative to dissociated ammonia or nitrogen-based atmospheres is vacuum. For material sintered in vacuum using an argon backfilling gas, properties are similar to those achieved in hydrogen-based atmospheres. When nitrogen is used for backfilling during vacuum sintering, sintered properties are comparable to those achieved in dissociated ammonia.

The effect of sintered part density, sintering temperature, and atmosphere on the tensile strength of vacuum-sintered type 316L stainless steel is illustrated in Fig. 5. Strength increased as part density and sintering temperature increased when argon was used as the backfilling gas. In addition, the use of nitrogen as the backfilling gas almost has a doubling effect on strength, especially at lower densities. Therefore, sintered density is not the sole factor to be considered when designing a type 316L stainless steel P/M part in which strength is a critical factor. Both sintering temperature and atmosphere must be known to determine tensile strength.

The effect of sintering parameters on yield strength is also shown in Fig. 5. In

this case, the effect of sintering temperature is not as pronounced as for tensile strength. However, the strengthening effect of nitrogen in austenitic stainless steel is evident. Note also the flattening of yield strength above 6.8 g/cm^3 under the nitrogen sintering conditions.

Sintering conditions have the most significant effect on the ductility of type 316L stainless steel, as shown in Fig. 5. Although the influence of final part density is recognized as a controlling factor in determining ductility, the effect of sintering temperature has not been emphasized. In comparing the ductility curves in Fig. 5, ductility at 7.2 g/cm^3 for a 1120 °C (2050 °F) sinter cycle was less than for a 6.0 g/cm^3 part sintered at 1200 °C (2200 °F). Note also the dramatic rise in ductility for the 1290 °C (2350 °F) sinter cycle above the 7.0 g/cm^3 level. At these densities, ductility begins to approach that of wrought stock. The nitrogen backfill limits ductility, especially at lower densities. However, just as the yield strength flattened above 6.8 g/cm^3, ductility begins to rise rapidly above this density level and approaches ductility for the 1200 °C (2200 °F) argon cycle. At higher densities, nitrogen diffusion is limited to bulk diffusion rates and can no longer enter the alloy through the decreasing interconnected porosity network.

Stainless steels exhibit greater shrinkage during sintering than do P/M irons. The dimensional change for type 304L sintered in dissociated ammonia is shown in Fig. 6, in which it can be seen that shrinkage increases with increasing temperature. Sintering in either hydrogen or vacuum results in greater shrinkage. More detailed information on sintering of stainless steels can be found in the article "Production Practices for P/M Materials" in this Volume.

Applications

Stainless steel P/M parts are used in a wide variety of applications. The primary reason for selecting a stainless steel composition usually is its improved corrosion resistance. Because the corrosion behavior of P/M materials is not as well documented as it is for comparable wrought alloys, field testing is advised to properly assess the corrosion resistance of P/M stainless steels.

The corrosion resistance of sintered stainless steel depends on the processing method, the material, its porosity, and the type and severity of the environment to which the part is exposed. Generally, high-temperature sintering improves corrosion

Fig. 5 Tensile properties of standard type 316L stainless steel Metal Powder Industries Federation tensile bars

Bars sintered in vacuum for 2 h using argon and nitrogen as backfilling gases. A partial pressure of 400 μm mercury was used to prevent vaporization of chromium.

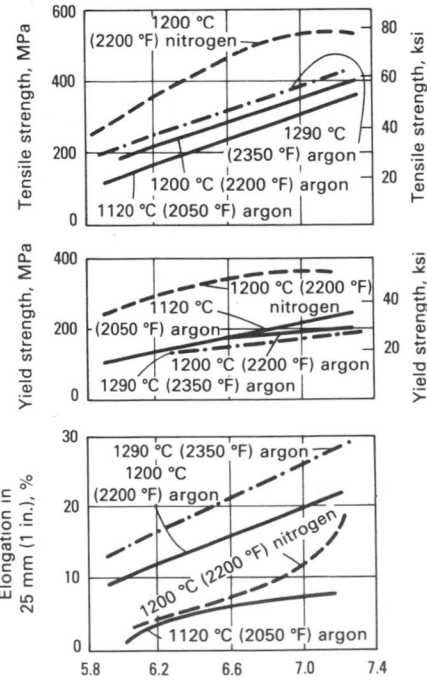

Fig. 6 Dimensional changes for type 304L stainless steel

Dimensional changes were determined on transverse-rupture bars sintered for 45 min in −40 °C (−40 °F) dew point dissociated ammonia and were calculated from die size.

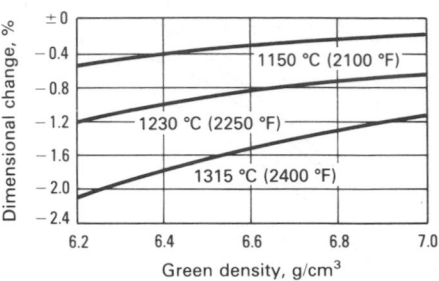

resistance, particularly in nitrogen-based atmospheres. The carbon content must be maintained as low as possible. If the furnace atmosphere must be decarburizing and reducing, a carbon level below 0.03% can be attained during sintering.

The following summary provides an overview of the various market segments for P/M stainless steel and the type of parts used.

Part	Alloy
Aerospace	
Seatback tray slides	316L
Galley latches	316L
Jet fuel refueling impellers	316L
Foam generators	316L
Agriculture	
Fungicide spray equipment	316L
Appliances	
Automatic dishwasher components	304L
Automatic washer components	304L
Garbage disposal components	410L
Pot handles	316L
Coffee filters	316LSi
Electric knives	316L
Blenders	303L
Can opener gears	410L
Automotive	
Rearview mirror mounts	316L, 434L
Brake components	434L
Seat belt locks	304L
Windshield wiper pinions	410L
Windshield wiper arms	316L
Manifold heat control valves	304L
Building and construction	
Plumbing fixtures	303L
Spacers and washers	316L
Sprinkler system nozzles	316L
Shower heads	316L
Window hardware	304L, 316L
Thermostats	410L

Part	Alloy

Chemical

Filters 304L-Si, 316L
High-corrosion-resistance filters 830
Cartridge assemblies 316L-Si

Electrical and electronic

Limit switches 410L
G-frame motor sleeves 303L
Rotary switches 316L
Magnetic clutches 410L, 440A
Battery nuts 830
Electrical testing probe jaws 316L

Hardware

Lock components 304L, 316L
Threaded fasteners 303L
Fasteners 316L
Quick-disconnect levers 303L, 316L

Industrial

Water and gas meter parts 316L
Filters, liquid and gas 316L-Si
Recording fuel meters 303L
Fuel flow meter devices 410L
Pipe flange clamps 316L
High polymer filtering 316L-Si

Jewelry

Coins, medals, medallions 316L
Watch cases 316L
Watch band parts 316L

Marine

Propeller thrust hubs 316L
Cam cleats 304L

Medical

Centrifugal drive couplings 316L
Dental equipment 304L
Hearing aids 316L
Anesthetic vaporizers 316L

Office equipment

Nonmagnetic card stops 316L
Dictating machine switches 316L
Computer knobs 316L

Recreation and leisure

Fishing rod guides 304L, 316L
Fishing rod gear ratchets 316L
Photographic equipment 316L
Soft drink vending machines 830, 316L
Travel trailer water pumps 316L

Porous Parts. One of the major applications for P/M stainless steels is in the fabrication of porous mediums. Porosity generally is an undesirable, although tolerable, characteristic of P/M parts for structural applications. However, for porous mediums, this porosity is an advantage. Type 316 stainless steel powder is used in the production of many parts with controlled interconnected porosity. Appli-

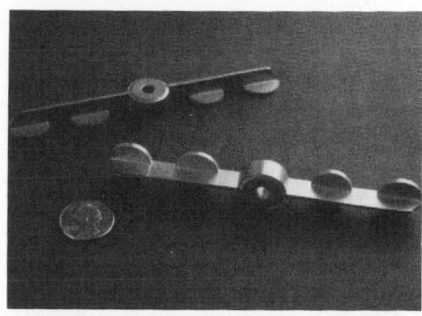

Fig. 7 Type 316L stainless steel ejector pad for refrigerator automatic icemaker

cations include liquid retention (as in bearings), filtering, metering of liquids or gases, and sound attenuation in telephones, microphones, and hearing aids. Detailed information on stainless steel porous mediums can be found in the article "P/M Porous Parts" in this Volume.

Appliances. Parts for appliances represent another sizable market for P/M stainless steels. Figure 7 shows a type 316L steel ejector pad that is used for a refrigerator automatic icemaker. The ejector pad is pressed to a density of 6.6 g/cm^3 and has a hardness of 65 HRB. After the part is pressed, it is sintered at 1260 °C (2300 °F) or higher in dissociated ammonia and coined or restruck to attain straightness.

The complex shape (five individual pads joined by a thin rail) requires precise part flatness and straightness. The bottom of the center pad must be parallel with the bottom of the other pads within 0.254 mm (0.010 in.) from center pad to end pad. It must be straight within 0.127 mm (0.005

Fig. 8 Garbage disposal part made from type 410L stainless powder

in.) maximum from the theoretical centerline through the pads over a part length of 140 mm (5.514 in.).

The improved straightness allows a tighter fit in the mating die casting, resulting in improved ice cube ejection. Barrel finishing to a smooth surface texture prevents ice cubes from sticking when freezing. Thermal passivation imparts corrosion protection.

Figure 8 shows a garbage disposal part made from type 410L stainless steel powder. This martensitic grade of stainless steel powder is intended for use in the manufacture of hard, wear-resistant P/M parts. Type 410L powder contains less than 0.03% C, but sufficient hardening is obtained by sintering in dissociated ammonia. When maximum hardness and wear resistance are required, carbon in the form of graphite can be added to the mix to give added hardenability. The effects of sintering atmosphere, graphite additions, and heat treatment on the strength and

Table 3 Properties of sintered type 410L stainless steel

Processing treatment	Graphite added, %	Sintering atmosphere(a)	Tempering temperature °C	Tempering temperature °F	Tensile strength MPa	Tensile strength ksi	Apparent hardness, HRB
As sintered and cooled in water-jacketed zone of furnace	0	Dissociated ammonia	· · ·	· · ·	724	105	102
	0.10	Dissociated ammonia	205	400	683	99	103
	0	Hydrogen	· · ·	· · ·	393	57	68
	0.10	Hydrogen	175	350	710	103	95
Reheated in dissociated ammonia and oil quenched from 950 °C (1750 °F)	0	Dissociated ammonia	205	400	627	91	106
	0.10	Dissociated ammonia	220	430	703	102	102
	0	Hydrogen	· · ·	· · ·	752	109	106
	0.10	Hydrogen	220	430	717	104	105
Reheated in hydrogen and oil quenched from 950 °C (1750 °F)	0	Dissociated ammonia	205	400	731	106	104
	0.10	Dissociated ammonia	205	400	745	108	104
	0	Hydrogen	205	400	641	93	95
	0.10	Hydrogen	220	430	800	116	101

(a) Sintered for 30 min at 1120 °C (2050 °F)

Fig. 9 Copier machine parts made from type 316L powder

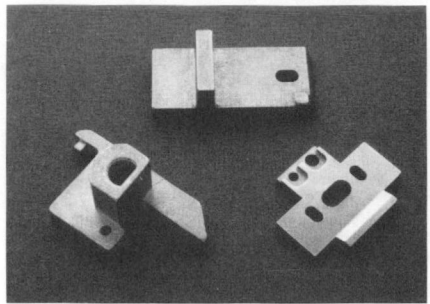

Fig. 10 P/M parts produced from stainless steel powder

(a) Specialty fasteners made from type 303L powder. (b) Sailboat cleats made from type 316L powder. (c) Hunting knife handles made from type 316L powder. (d) Splined coupling for pump made from type 304L powder

(a)

(b)

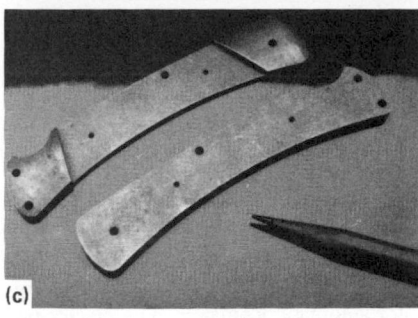
(c)

(d)

hardness of type 410L stainless steel are shown in Table 3. Additional information on stainless steel parts used in appliances can be found in the article "P/M Parts for Appliances" in this Volume.

Business Machines. Figure 9 shows a variety of P/M stainless steel parts used in copier machines. Over 91 metric tons (100 tons) of stainless steel powder are used annually in the manufacture of business machine parts. Use of stainless steel powder pressed to near-net shape eliminates the expensive machining operations required when using wrought bar stock. Additional information can be found in the article "P/M Parts for Business Machines" in

this Volume.

Other Applications. Stainless steel P/M parts also are used in the automotive,

hardware, and leisure products industries. Figure 10 illustrates a number of parts used in these applications.

P/M Copper-Based Alloys

By Ralph W. Stevenson
Project Manager/Materials Engineering
Energy Technology Division
Midland-Ross Corp.

THE USE OF COPPER in the P/M industry dates back to the 1920's, when porous bronze bearings became commercialized. These self-lubricating bearings, which are the oldest application of porous P/M parts, were developed independently in the research laboratories of General Motors Corp. (Ref 1, 2) and Bound-Brook Oilless Bearing Co. (Ref 3). Currently, these bearings account for the major portion of P/M copper and copper alloy applications. Other important applications for copper and copper-based P/M materials include friction materials, brushes, filters, structural parts, electrical parts, additives to iron powders (alloying as well as infiltration), paints, and pigments.

Annual P/M copper consumption (1982 data) totals about 16 000 metric tons (18 000 tons). The diversified and extensive applications of copper-based powder metals include (Ref 4):

Abrasive wheels

- Bonding (copper)

Agriculture

- Fungicides (copper)
- Lawn and garden equipment (bronze)
- Soil conditioning (copper)

Aerospace

- Brake linings (copper)
- Counterweights (copper-tungsten)
- Filters (bronze)

Automotive

- Brake bands and liners (copper, brass, copper-lead, copper-lead-tin)
- Bushings (bronze)
- Instruments (nickel silver)

Building and construction

- Conductive and nonsparking floors (copper)
- Decorative plastics (copper, bronze, brass)
- Domestic water filters (brass)
- Pipe joint compounds (copper)

Chemical

- Catalysts (copper)
- Filters (bronze)
- Valves and pumps (copper nickel)

Coatings

- Antifouling paints (copper)
- Conductive paints and plastics (copper, brass)
- Decorative paints (copper, brass, bronze)
- Lacquers (brass, bronze)
- Mechanical (peen) plating (copper, brass)
- Spray coating (copper, brass)
- Vacuum metallizing (copper)

Coins, medals, medallions (copper nickel, brass)

Electrical and electronic

- Brushes (copper)
- Brush holders (nickel silver)
- Contacts (copper)
- Heatsinks (copper, dispersion-strengthened copper)
- Printed circuits (copper)
- Semiconductor stud bases (copper, dispersion-strengthened copper)
- Telephone components (brass, bronze)
- Switch and contactor components (copper, brass)

Hardware

- Locks and keys (brass, bronze)
- Nuts (brass)

Industrial

- Balancing weights (copper-tungsten)
- Bearings and bushings (bronze, copper-lead, copper-lead-tin)
- Filters, liquid and gas (bronze)
- Flame arrestors (bronze)
- Control instruments (nickel silver)

Joining

- Brazing compounds (copper, bronze, brass)
- Resistance welding electrodes (copper, dispersion-strengthened copper)

Lubricants

- Antigalling pipe joint compounds (copper)
- Copper lubricants (copper)
- Plastic-filled metal (copper, bronze)

Machining

- Electrical discharge machining (copper)
- Electrochemical machining (copper)

Office equipment

- Business machines (brass)

Ordnance

- Armor-piercing cores (copper)
- Fuze parts (brass)
- Projectile rotating bands (copper, brass)

Personal products

- Cordless electric toothbrushes and razors (copper)
- Fingernail polish (copper)
- Photographic equipment (bronze, brass, nickel silver)
- Poker chips (brass, bronze, copper nickel)

Printing inks

- Metallic inks for offset, letterpress, gravure (copper, brass)

Radio and television

- Printed circuits (copper)

Railroads

- Brake linings (bronze, copper-lead, copper-lead-tin)
- Friction strips on pantographs (copper)

Self-lubricating parts

- Oil-filled (bronze)
- Plastic-filled (copper)

Ships

- Antifouling paint (copper)

Physical and mechanical properties of copper and copper alloy P/M parts are comparable with cast and wrought copper-based materials of similar composition on a density-related basis. Additionally, P/M processing permits flexibility in part design. For example, P/M copper parts may vary in density from the low density typical of self-lubricating bearings or filters to the near-theoretical density of wrought parts.

Copper powder containing more than 99% Cu is available commercially. Powder metallurgy parts produced from these powders have electrical conductivities of 80 to 95% IACS in the as-pressed and sintered condition. Further densification provides higher conductivities, approaching 100% IACS. Copper P/M parts provide an added benefit in the production of electronic and electrical components that require superior electrical or thermal conductivity.

Copper and copper alloy powders, which exhibit excellent ductility representative of wrought copper grades, enjoy wide usage in industrial applications. As discussed earlier, their primary application is in the production of self-lubricating bearings. This application illustrates the advantage of producing a part with controlled interconnected and surface-connected porosity through P/M techniques. A similar production method is used in the manufacture of metallic bronze filters. For additional information, see the articles "P/M Bearings" and "P/M Porous Parts" in this Volume.

Alloyed with tin, zinc, nickel, and/or other elements, copper powder is used for many structural parts and friction materials. Brasses, bronzes, and other copper alloys produced by P/M methods exhibit physical and mechanical properties comparable to their cast and wrought counterparts. Copper also is used as an alloying element in iron powder components to enhance mechanical properties and control dimensional changes during sintering. Copper addition is made by either blending or infiltration. For additional information, see the article "Infiltration" in this Volume.

Production of Copper Powders

The four major methods for producing unalloyed copper powders are atomization, electrolysis, hydrometallurgy, and oxide reduction (Ref 5). Brief descriptions of these processes follow. For additional information pertaining to the production and properties of copper powder, see the article "Production of Copper Powder" in this Volume.

Atomization begins with melting of the copper. The liquid metal flows through a sized orifice from a tundish. Liquid flow is impinged by a high-velocity stream of gas or liquid, such as water. This breaks the liquid stream into particles that solidify rapidly. Particle size and shape are greatly influenced by the atomizing medium, pressure, and flow rate.

After atomization, particles may be annealed in a reducing atmosphere to minimize surface oxide formed during atomization. Copper powder can be produced as either spherical or irregular in shape. Purity, depending on processing technique, exceeds 99%.

Electrolysis. Electrolytic copper powder is produced by electrodeposition of copper to obtain a nonadherent powder deposit, rather than a smooth, continuous, adherent layer. Formation of loose deposits is induced by a low copper ion concentration in the electrolyte, high acid concentration, and high cathode current density. The addition of colloids to the plating bath aids the formation of a uniform copper deposit. The starting material is a pure copper cathode.

After deposition, the powder is washed to remove all traces of the electrolyte and may be annealed in a reducing atmosphere. The reduced copper product is fed into a high-velocity impact mill to uniformly reduce the powder. The powder then is screened, classified, and blended to the desired particle size distribution. Powder morphology is dendritic. A wide range of powders with varying apparent densities and high green strengths can be obtained by electrolysis.

Hydrometallurgy is used to produce copper powder from copper cement, concentrates, or scrap copper. Copper is leached from these materials with sulfuric acid or ammoniacal solutions, and the resulting solution is separated from the residue by filtration. Copper is precipitated as powder. The wet powder is dried in a reducing atmosphere, then milled, classified, and blended to obtain the desired particle size distribution. The powder has fine particle size, with relatively low apparent density and green strength.

Oxide Reduction. With this method, also referred to as solid-state reduction, oxides (including mill scale) are ground to controlled particle sizes and then reduced in gas atmospheres containing carbon monoxide, hydrogen, and cracked natural gas at temperatures below the melting point of copper. Generally, powders produced by oxide reduction are porous and have intermediate apparent densities, with good strength.

Production of Alloy Powders

Nearly all alloy powders are produced by atomization. Preblended powders are mixtures of selected compositions, with or without lubricant, that form the desired alloy during sintering. Prealloyed powders are produced by atomization of the alloy material by the same methods used for the production of copper powder. Prealloyed powder also is produced by sintering a preblend and grinding the material to attain the desired powder characteristics.

Alloy powders are available in various compositions, including brasses ranging from 95Cu-5Zn to 60Cu-40Zn (as well as leaded grades of these alloys) and nickel silvers, tin bronzes, aluminum bronzes, and beryllium bronzes. See the article "Production of Copper Alloy Powders" in this Volume for additional information on the production of alloy powders.

Processing

Consolidation of copper powder usually is performed in a closed die, although other means, such as hot pressing, roll compaction, isostatic pressing, extrusion, and forging, can be used. When pressure is applied, a series of events occurs: (1) rearrangement of the particles with minimal deformation, (2) elastic compression of contacts between particles, (3) plastic deformation at the contacts to form increasingly larger areas, and (4) massive deformation of total powder mass.

Various materials require different compacting pressures. For example, pure

Table 1   Typical compacting pressure and compression ratios of various copper P/M parts

| P/M parts | Compacting pressure | | Compression ratio |
	MPa	tsi	
Brass parts	414-689	30-50	2.4-2.6 to 1
Bronze bearings	193-275	14-20	2.5-2.7 to 1
Copper-graphite brushes	345-414	25-30	2.0-3.0 to 1
Pure copper parts	206-248	15-18	2.6-2.8 to 1

Source: Ref 4

copper P/M parts are made with relatively low pressures. Compacting pressures of 205 to 250 MPa (15 to 18 tsi) are recommended for thin sections, although higher pressures are used for heavier sections. Recommended pressures and compression ratios for copper and copper alloy parts are summarized in Table 1.

Sintering is a process in which particles in a mass of powder are bonded by atomic or molecular attraction in a solid state by heating. During sintering, changes in the shapes of the pores and a reduction in volume reduce surface energy.

Sintering proceeds in three phases. Initially, neck growth between particles proceeds rapidly, but powder particles retain their identity. In the second phase—when most of the densification occurs—the structure is recrystallized, whereupon powder particles lose their separate identities and diffuse into each other. In the third phase, isolated pores begin to spheroidize, and densification progresses at a slower rate. Figure 1 shows the progressive densification of compacted copper powder as a function of time and temperature.

The rate of sintering has a pronounced effect on compact properties and can be modified by either physical or chemical treatments of the powder or compact, or by adding reactive gases in the sintering atmosphere. These treatments are known as activated sintering. Activation improves densification, strength, and electrical conductivity of the compacts. For a detailed account of the fundamental principles of sintering and enhanced densification during sintering, see the article "Physical Fundamentals of Consolidation" in this Volume.

Liquid-Phase Sintering. In this process, a mixture of two or more powders is sintered at a temperature below the melting point of the high-melting-point constituent, but above that of the low-melting-point constituent. The sequence of events in heating a mixed phase begins with a mixture of elemental powders, such as copper-tin or copper-iron premixes. When the liquid forms, it flows to wet the solid particles and begins acting on the solid phase. Melt penetration between the solid

particles can cause swelling of the compact. Along with melt penetration, there is rearrangement of particles. The combination of wetting, liquid flow, and rearrangement contributes to a rapid change in the volume of the compact. The distinct stages of liquid phase sintering, along with the driving forces and material transport mechanisms that are involved in this process, are discussed in detail in the article "Physical Fundamentals of Consolidation" in this Volume.

In the copper-tin system, the tin melts and alloys with copper to form a bronze, with accompanying expansion of the compact. In the iron-copper system, the copper melts, becoming saturated with iron, and the copper-iron alloy diffuses into the iron skeleton, causing expansion of the skeleton. Pores remain at the sites vacated by the copper.

In both the copper-tin and iron-copper systems, growth or shrinkage of compacts can be altered by the addition of carbon in the form of graphite. Thus, graphite can be used to control dimensional changes in these systems. In the copper-tin-carbon system, sintering is inhibited by mechanical separation of the constituents; as a result, expansion increases. In the iron-copper-carbon system, the amount of the liquid phase is increased by the formation of a ternary iron-copper-carbon eutectic that restricts expansion. See the article "Production Sintering Practices for P/M Materials" in this Volume for detailed information regarding sintering of copper-based materials.

Pure Copper P/M Parts

Pure copper P/M parts are used mainly in electrical and electronic applications. It is essential to use very pure copper powders ($\geq$99.95% purity) or to bring about the precipitation of soluble impurities during sintering. As little as 0.023% Fe in solid solution in copper lowers its conductivity to 86% of that of pure copper. Small amounts of iron mechanically mixed with the copper powder lower the conductivity much less, unless the iron dissolves in the copper during sintering. If high-purity

Fig. 1 Effect of sintering temperature and time on densification of copper powder compacts
Source: Ref 6

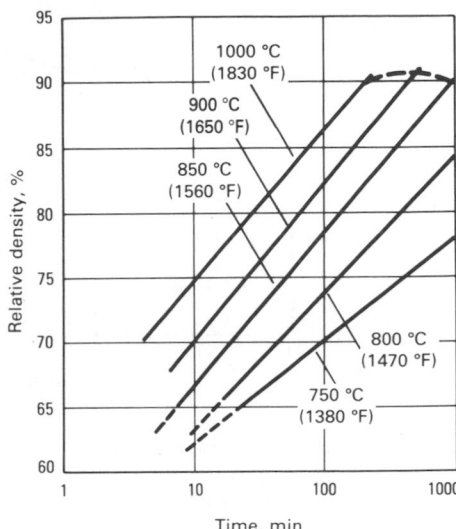

copper is used, or if soluble impurities are precipitated during sintering, it is possible to obtain the values of strength and conductivity shown in Fig. 2. Physical properties of pure copper are:

Melting point	1083 °C (1981 °F)
Density	8.94 g/cm^3 at 20 °C
	(0.323 lb/in.^{3} at 68 °F)
Coefficient of thermal	
expansion	17.0 $\times$ 10^6/°C at 20 to 100 °C
	(9.4 $\times$ 10^6/°F at 68 to 212 °F)
Thermal conductivity	398 W/m · K
	(226 Btu/ft · h · °F)
Electrical resistivity	1.71 $\mu\Omega$ · cm at 20 °C
	(10.3 Ω · cir mil/ft at 68 °F)
Electrical conductivity(a)	101% IACS
	at 20 °C (68 °F)
Specific heat	384.9 J/kg · K at 20 °C
	(0.092 Btu/lb · °F at 68 °F)
Modulus of elasticity	
(tension)	117 GPa (17 000 ksi)
Modulus of rigidity	44 GPa (6 400 ksi)

(a) Volume basis
Source: Ref 4

Conductivity is directly related to porosity; the greater the void content, the lower the conductivity. Electrical conductivity of as-pressed and sintered pure copper parts varies from 80 to 90% IACS. Full-density properties, as shown above, are reached or approached by compacting at moderate pressures of 205 to 250 MPa (15 to 18 tsi), sintering at temperatures 50 to 150 °C (90 to 270 °F) below the melting point of copper (1083 °C, or 1981 °F), followed by re-pressing, coining, or forging.

Typical applications of pure copper parts in which high electrical conductivity is re-

Fig. 2 Effect of density on electrical conductivity and tensile properties of P/M copper

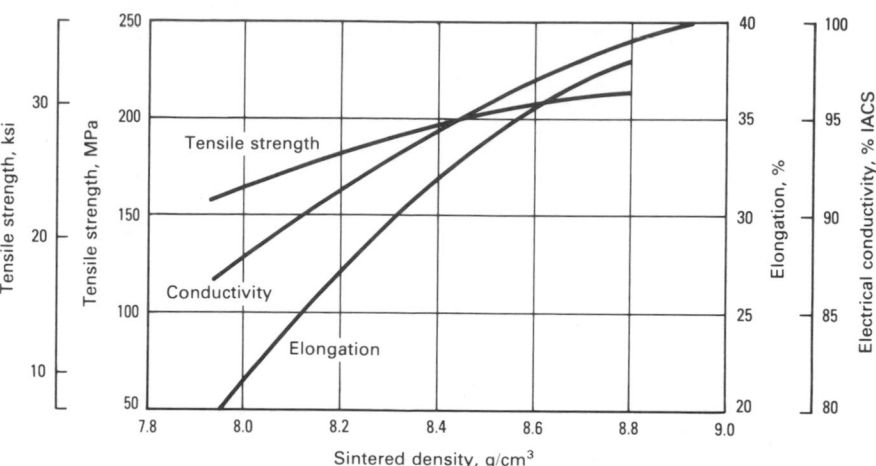

Fig. 3 Effect of density on the strength of copper-tin and copper-tin-graphite compacts
Source: Ref 7

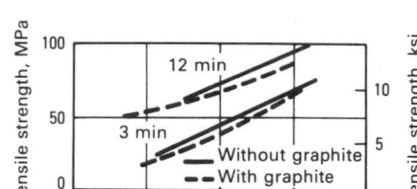

Fig. 4 Pressing characteristics of premixed and prealloyed 90Cu-10Sn powders
Source: Ref 8

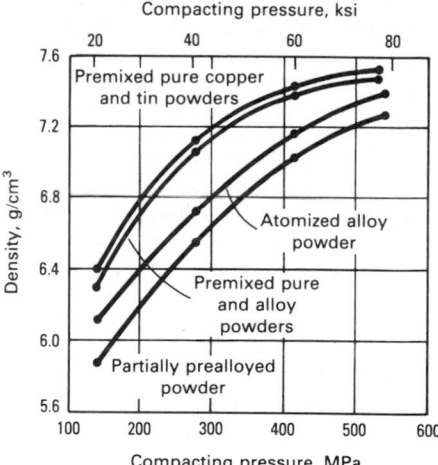

quired include commutator rings, contacts, shading coils, nose cones, and electrical twist-type plugs. Copper powders also are used in copper-graphite compositions that have low contact resistance, high current-carrying capacity, and high thermal conductivity. Typical applications include brushes for motors and generators and moving parts for rheostats, switches, and current-carrying washers.

Bronze P/M Parts

Most tin bronze P/M parts are produced from elemental premixes of tin and copper. Typical strength/density curves for 90Cu-10Sn bronzes with and without graphite additions are shown in Fig. 3. Sintering of these materials requires certain precautions in order to develop optimum properties and microstructures (see the articles "Production Sintering Practices for P/M Materials" and "P/M Bearings" in this Volume).

Some structural parts, however, requiring densities >7.0 g/cm³ are fabricated from prealloyed powder. Prealloyed powders have higher yield strengths and work-hardening rates than premixed powders. Therefore, pressing loads required to achieve green density are higher than pressures required for elemental powders. Differences in pressing characteristics of premixed and prealloyed powders are compared in Fig. 4.

Bearings. Self-lubricating porous bronze bearings continue to consume the major portion of the copper powder produced each year. These bearings are made by pressing elemental powder blends of copper and tin, followed by sintering at 800 to 875 °C (1500 to 1600 °F) for about 15 min. Typical compositions are Cu-10Sn and Cu-10Sn-0.5C. Pore volumes range from 25 to 35%.

Self-lubricating bearings are sold either dry or saturated with oil. Sizes range from 0.8 to 76 mm (1/32 to 3 in.) in diameter. Development of these bearings had a great impact on the home appliance industry. By eliminating periodic lubrication, the self-lubricating bearing ensured continuous operation of appliances and led to great expansion within the industry.

Typical applications of sintered bronze bearings include:

Automotive components

- Starters
- Generators
- Oil and water pumps
- Windshield wipers
- Hood and window raisers

Home appliances

- Dishwashers
- Clothes dryers
- Washing machines
- Sewing machines
- Vacuum cleaners
- Refrigerators
- Food mixers

Leisure products

- Phonographs
- Record changers
- Tape recorders

Industrial and business equipment

- Textile machines
- Packaging machines
- Electric fans
- Business machines

Detailed information concerning the manufacturing properties and applications of bronze bearing materials can be found in the article "P/M Bearings" in this Volume.

Filters. The ability to achieve close control of porosity and pore size is the main reason metal powders are used in filter applications. Most producers of nonferrous filters prefer atomized spherical powder of closely controlled particle size to permit the production of filters within the desired pore size range.

Tin bronze is the most widely used filter material, but nickel silver and copper-nickel-tin alloys may also be used. The effective pore size of P/M filters generally ranges from 5 to 125 μm. Porous P/M bronze filters can be obtained with tensile strengths ranging from 20 to 140 MPa (3 to 20 ksi) and appreciable ductility, up to 20% elongation. In addition, P/M bronze has the same corrosion resistance as cast bronze of the same composition and therefore can be used in a wide range of environments.

Table 2   Typical properties of sintered bronze structural parts

Density, g/cm³	Ultimate tensile strength MPa	ksi	Elongation in 25 mm (1 in.), %	Yield point in compression MPa	ksi	Hardness, HRH
6.4	93	13.5	1	76	11.0	45
6.8	110	16.0	2	103	15.0	55
7.2	138	20.0	3	138	20.0	65

Source: ASTM B 255

Powder metallurgy bronze filters are used to filter gases, oils, refrigerants, and chemical solutions. They have been used in fluid systems of space vehicles to remove particles as small as 1 μm. Bronze diaphragms can be used to separate air from liquids or mixtures of liquids that are not emulsified. Only liquids capable of wetting the pore surface can pass through the porous metal part.

Bronze filter materials can be used as flame arrestors on electrical equipment operating in flammable atmospheres, where the high thermal conductivity of the bronze prevents ignition. They also can be used as vent pipes on tanks containing flammable liquids. In these applications, heat is conducted away rapidly so that the ignition temperature is not reached. Additional information on the manufacture, properties, performance characteristics, and applications of P/M bronze filters can be found in the article "P/M Porous Parts" in this Volume.

Bronze structural parts generally are produced by methods similar to those used for self-lubricating bearings. The compositions of these materials, according to ASTM standard B 255, are:

Element	Composition, %
Copper	87.5-90.5
Tin	9.5-10.5
Carbon, max	1.75
Iron, max	1.0
Total other elements, max	0.5

Corresponding properties for these materials are given in Table 2. Powder metallurgy bronze parts for structural applications frequently are selected because of the corrosion and wear resistance of bronze. A one-way bearing clutch for automotive applications is shown in Fig. 5. The powder mix, which is a homogeneous blend of copper, tin, graphite, and dry lubricants, is compacted to a density of 6.8 g/cm³ and sintered at 840 °C (1550 °F) to produce a uniform alpha bronze structure, as shown in Fig. 6. The part is then repressed or sized to 7.0 g/cm³ density to maintain dimensional tolerances.

Bronze P/M components also are used in applications such as copiers, outboard motors, and paint-spraying equipment. A complex bronze sleeve used in the counterbalance mechanism for a copier machine is shown in Fig. 7. The bronze sleeve, made to comply with Metal Powder Industries Federation (MPIF) material designation CT-0010-S, is compacted to densities ranging from 6.8 to 7.2 g/cm³. It has an ultimate tensile stength of 124 MPa (18 ksi) and a compressive yield strength of 117 MPa (17 ksi). Density control in the part is particularly important because of the cone shape, bevel slot, and sleeve. Machining from bar stock and investment casting proved to be uneconomical when compared with P/M processing. After coining, the bronze sleeve is oil impregnated to provide self-lubrication in the assembly.

Brass and Nickel Silver P/M Parts

In contrast to bronze structural parts, parts made from brass, leaded brass, and nickel silver are produced from prealloyed atomized powder. Sintering of brass and nickel silver compacts normally is carried out in a dissociated ammonia atmosphere at temperatures ranging from 815 to 930 °C (1500 to 1700 °F), although endothermic and nitrogen-based atmospheres are also used. In all cases, to avoid excessive distortion and/or blistering of the compacts, sintering temperatures should not exceed the solidus temperature of the alloy. Composition and property requirements for these materials are covered in ASTM B 282, "Standard Specification for Sintered Brass Structural Parts," and ASTM B 458, "Standard Specification for Sintered Nickel Silver Structural Parts."

Commercial prealloyed brass powders are available in leaded and nonleaded compositions. Brass alloys include 90Cu-10Zn through 65Cu-35Zn; however, leaded versions of 80Cu-20Zn and 70Cu-30Zn frequently are used for the manufacture of sintered structural parts that may require secondary machining operations.

The only nickel silver powder used has a nominal base composition of 64Cu-18Ni-18Zn, which is modified by the addition of lead to improve machinability. Typical compositions and corresponding mechan-

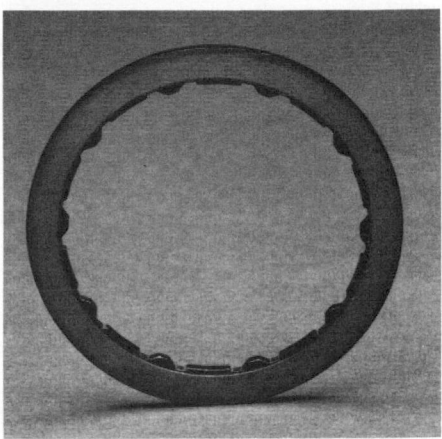

Fig. 5 Bronze one-way bearing overrunning clutch
Courtesy of Metal Powder Industries Federation

Fig. 6 Alpha bronze structure of sintered bronze clutch section
Magnification: 250×. Courtesy of Keystone Carbon Co.

Fig. 7 Bronze sleeve used in the counterbalance mechanism for a copier machine
Courtesy of Metal Powder Industries Federation

Table 3 Typical mechanical properties of brass and nickel silver P/M compacts pressed at 414 MPa (30 tsi)

Nominal composition	Sintered density, g/cm³	Tensile strength MPa	ksi	Elongation in 25 mm (1 in.), %	Hardness
Brass					
90Cu-10Zn 8.1		207	30	20	77 HRH
85Cu-15Zn 8.2		217	31.5	20	82 HRH
70Cu-30Zn 8.1		262	38	21	87 HRH
88.5Cu-10Zn-1.5Pb 8.4		207	30	25	76 HRH
80Cu-18.5Zn-1.5Pb 8.2		238	34.5	31	82 HRH
68.5Cu-30Zn-1.5Pb 7.7		239	34.6	29	71 HRH
Nickel silver					
64Cu-18Ni-18Zn 7.9		234	34	12	83 HRB
64Cu-18Ni-16.5Zn-1.5Pb 7.8		193	28	11	84 HRB

Source: Ref 4

Fig. 10 Leaded brass rack for a stereo three-dimensional microscope

Courtesy of Metal Powder Industries Federation

Fig. 8 Brass rack guide for rack-and-pinion steering column of an electric outboard motor

Courtesy of Metal Powder Industries Federation

Fig. 9 Leaded brass horn ring-adjusting nut for an industrial paint sprayer

Courtesy of Metal Powder Industries Federation

ical properties of brass and nickel silver P/M compacts are given in Table 3.

These alloys usually are blended with lubricants in amounts from 0.5 to 1.0 wt%. Lithium stearate is the preferred lubricant because of its cleansing and scavenging action. However, bilubricant systems are common, such as lithium stearate and zinc stearate, which minimize the surface staining attributed to excessive lithium stearate.

Next to bronze bearings, brasses and nickel silvers are the most widely used materials for structural P/M parts. Typical applications include hardware for latch bolts and cylinders for locks; shutter components for cameras; gears, cams, and actuator bars in timing assemblies and in small generator drive assemblies; and dec-

orative trim and medallions. Figures 8 through 11 illustrate typical applications of P/M parts that are indicative of the diversity of copper-based P/M technology.

A brass rack guide used in a battery-operated electric outboard trolling motor is shown in Fig. 8. This part, which operates in a rack and pinion steering column of the electric motor, is fabricated to a density range of 7.3 to 7.9 g/cm³. Originally designed as an extrusion, the rack guide required extensive machining. The selection of P/M technology eliminated machining, except for four cover attaching holes in the side walls of the rack. After compacting and sintering, the guide is coined for dimensional properties.

A complex leaded brass (MPIF material designation CZP-0220-U) horn ring-adjusting nut for an industrial paint sprayer is shown in Fig. 9. This P/M part replaced a brass forging and eliminated eight production operations, providing a manufacturing cost savings of 75%. It is fabricated to a density of 8.0 g/cm³ and has a tensile strength of 207 MPa (30 ksi) and a yield strength of 97 MPa (14 ksi). The complex shape required an innovative tooling design to obtain uniform density throughout the part. Powder metallurgy also provided closer tolerances and improved surface finish for chromium and cadmium plating.

A leaded brass (MPIF material designation CZP-0220-U) rack used for a stereo three-dimensional microscope is shown in Fig. 10. The rack is used to control the microscope focusing mechanism and is mounted to the slide unit. The rack is fabricated to a density of 8.0 g/cm³ and has a tensile strength of 193 MPa (28 ksi) and a yield strength of 90 MPa (13 ksi).

Tooling is critical for the rack, which required a thin top and bottom punch to form very fine teeth. Compacting and sizing tools were designed to allow for proper shrinkage to size the part without tool breakage. After sizing a 0.109-mm (0.0043-in.) projection of 0.685-mm (0.027-in.) diam wire over the top of the teeth, the dimensional tolerance must be held throughout the full length of the rack profile. The expense of machining the rack from brass bar stock was determined to be a considerably higher cost alternative.

A leaded brass reticle mount used in a zoomscope sight for sporting rifles is shown in Fig. 11. This P/M part replaced a reticle mount machined from brass bar stock. The machined mount required a turning operation, two milling and drilling operations, and special deburring. Powder

Fig. 11 Leaded brass reticle mount used in rifle zoomscope sight

Courtesy of Metal Powder Industries Federation

metallurgy offered more consistent parts and improved production rates.

The brass P/M part is compacted from leaded brass powder (MPIF material designation CZP-0228-T) and coined to a density of 7.2 to 7.8 g/cm³. Apparent hardness is 30 to 35 HRB. The part must accept black oxide finish and a solder joint that will not fail during fatigue testing. Critical dimensional requirements of this part were met using P/M technology.

Copper-Nickel P/M Parts

Copper-nickel P/M alloys containing 75Cu-25Ni and 90Cu-10Ni have been developed for coinage and corrosion-resistant applications. The 75Cu-25Ni alloy pressed at 772 MPa (112 ksi) has a green density of 89% of theoretical. After sintering at 1090 °C (2000 °F) in dissociated ammonia, elongation is 14%, and apparent hardness is 20 HRB. Re-pressing at 772 MPa (112 ksi) increases density to 95%.

This alloy has the color of stainless steel and can be burnished to a high luster. The 90Cu-10Ni alloy has a final density of 99.4% under similar pressing and sintering conditions. It has a bright bronze color and also can be burnished to a high luster.

In one method of producing coins, medals, and medallions, a mixture of 75Cu-25Ni powders with zinc stearate lubricant is compressed, sintered, coined, and re-sintered to produce blanks suitable for striking. These blanks are softer than rolled blanks, because they are produced from high purity materials. Therefore, they can be coined at relatively low pressures and achieve greater relief depth with decreased die wear.

In another procedure, organic binder is mixed with copper or copper-nickel pow-

Fig. 12 Applications of sintered metal friction materials

Source: Ref 9

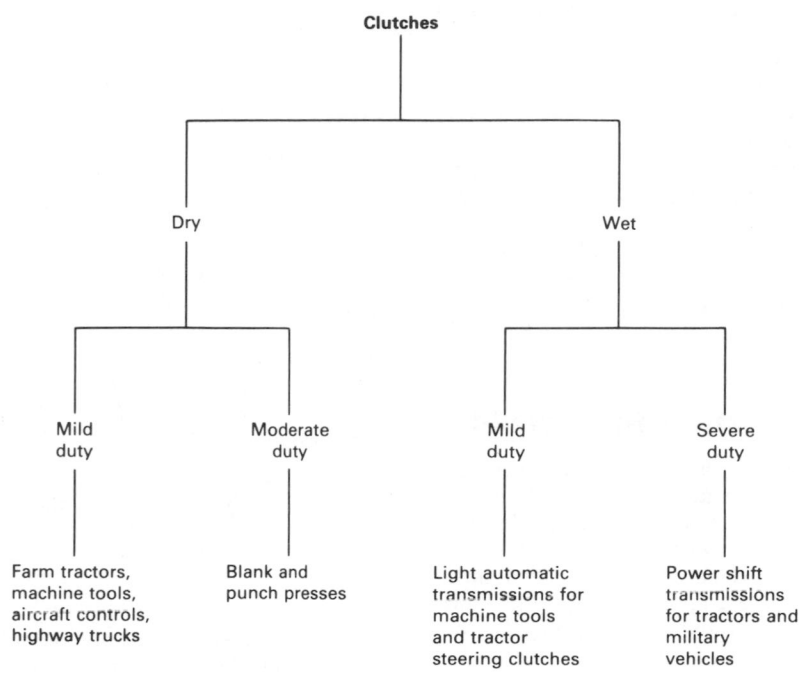

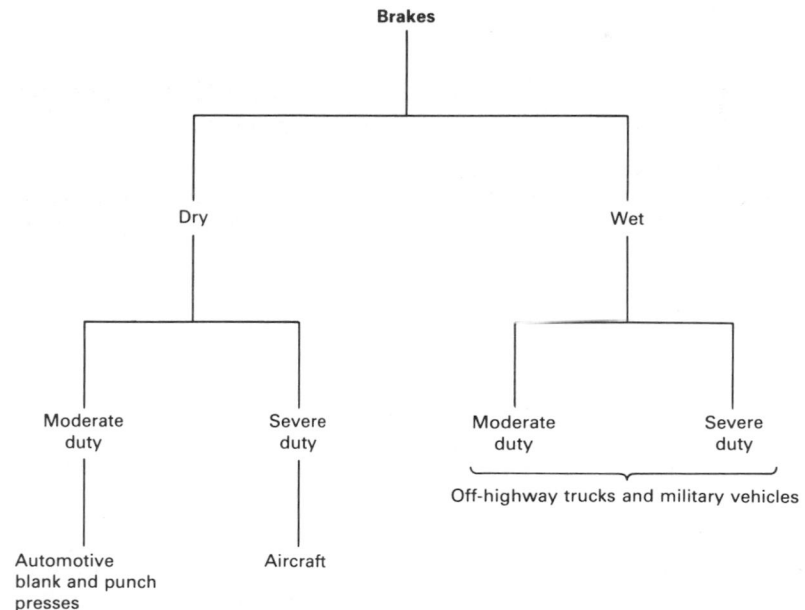

ders and rolled into "green" sheets. Individual copper and copper-nickel sheets are pressed together to form a laminate, and blanks are punched from it. Blanks are heated in hydrogen to remove the organic binder and sinter the material. The density of the "green" blanks is low (45% of theoretical), but coining increases density to 97%. After pressing, the blanks are annealed to improve ductility and coinability.

Copper-Lead P/M Parts

Copper and lead, which have limited solubilities in each other, are difficult to alloy by conventional ingot metallurgy. Copper-lead powder mixtures have excellent cold pressing properties; they can be compacted at pressures as low as 76 MPa (11 ksi) to densities as high as 80% and, after sintering, can be re-pressed at pressures as low as 152 MPa (22 ksi) to produce essentially nonporous bearings.

Steel-backed copper or copper-lead-tin P/M materials are used to replace solid bronze bearings. They are produced by spreading the powder in a predetermined thickness on a steel strip, sintering, rolling to theoretical density, re-sintering, and annealing. The end product has a residual porosity of about 0.25%. Blanks of suitable size are cut from the bimetallic strip, formed, and drilled with oil holes or machined to form suitable grooves. These materials include Cu-25Pb-0.5Sn, Cu-25Pb-3.5Sn, Cu-10Pb-10Sn, and Cu-50Pb-1.5Sn alloys.

P/M Friction Materials

A basic advantage of P/M processing is its ability to combine materials in powder form that are otherwise immiscible. This facilitates the production of friction materials, in which copper and other metal powders are combined with solid lubricants, oxides, and other compounds. Metallic friction materials can be operated at higher loads and temperatures than organic friction materials.

Powder metallurgy friction materials, which are used as clutches and brakes, may be classified in terms of dry/wet and mild/moderate/severe applications, as shown in Fig. 12. The majority of the clutch applications are for wet (oil) operation. Early P/M friction materials were copper-based materials. Currently, copper-based materials still are used in all applications, but iron-based compounds have been developed for moderate-to-severe-duty dry applications.

Copper-based materials are used mainly where semifluid friction occurs. For dry friction applications, these materials are suitable only where operating conditions are relatively mild (<350 °C, or 660 °F).

Typical processing techniques for P/M friction materials include hot pressing and sintering at high temperatures for long times, followed by re-pressing. Fine, reactive powders are preferred. The friction segments usually are brazed, welded, riveted, or mechanically fastened to supporting steel members or are pressure bonded directly to the assembly. Powder metallurgy friction materials are multiphase composites that contain between four and eight different materials. Typical compositions include 75Cu-6Pb-7Sn-5graphite-4molybdenum disulfide-3feldspar for dry clutches and brakes, and 74Cu-3.5Sn-2Sb-16graphite-4.5galena for wet

Fig. 13 Dispersion-strengthened copper lead wires in a light bulb
Source: Ref 9

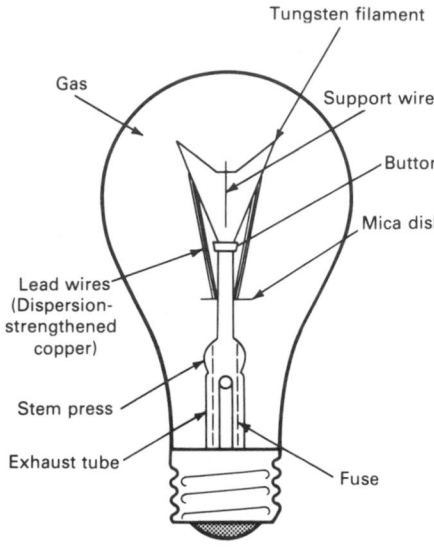

clutches and brakes. For additional information on the fabrication properties and applications of these materials, see the article "P/M Friction Materials" in this Volume.

Dispersion-Strengthened P/M Copper Materials

Copper P/M components can be strengthened by using finely dispersed particles of oxides such as alumina, titania, beryllia, thoria, or yttria in the matrix. Dispersoids can be added by mechanical mixing, internal oxidation, or co-precipitation.

The main benefit of dispersion strengthening is to produce a material that resists softening and grain growth at temperatures approaching the melting point of copper. Dispersion-strengthened materials are superior in structural stability to precipitation-hardenable alloys such as copper-chromium or copper-beryllium, because oxides do not dissolve at high temperatures—a characteristic of precipitation-hardenable alloys.

The major application for dispersion-strengthened copper is electrode caps for nonconsumable resistance welding. This application is discussed in the article "Electrical and Magnetic Applications" in this Volume. These materials also are used for the lead wires to the tungsten filaments in light bulbs, as shown in Fig. 13. Additional information can be found in the

article "Dispersion-Strengthened Materials" in this Volume.

Infiltrated Parts

Iron-based P/M parts can be infiltrated with copper or a copper alloy by placing slugs of the infiltrant (which has a lower melting point than the porous body) on the compact before sintering. On melting, the infiltrant is completely absorbed as a liquid into the pores by capillary action to produce a component with a composite structure. The resulting properties generally depend on the metals that make up the infiltrated part and the method and proportion in which they are combined.

Infiltration increases density, with a resulting improvement in mechanical properties. It improves corrosion resistance by closing pores and also improves machinability and brazeability. Tensile strengths ranging from 483 to 621 MPa (70 to 90 ksi) can be achieved in iron porous bodies infiltrated with 15 to 25% Cu. For a detailed description of this process, its applications, and properties of infiltrated parts, see the article "Infiltration" in this Volume.

REFERENCES

1. Williams, H.M., U.S. Patent No. 1 556 658, General Motors Corp., 1925
2. Williams, H.M. and Boegehold, A.L., U.S. Patent No. 1 642 347; 1 642 348; 1 642 349, General Motors Corp., 1927
3. Claus, C., U.S. Patent No. 1 607 389, Bound-Brook Oilless Bearing Co., 1926
4. Everhart, J.L., *Copper and Copper Alloy Powder Metallurgy Properties and Applications*, Copper Development Association, New York
5. Taubenblat, P.W., Importance of Copper in Powder Metallurgy, *Int. J. Powder Metall. Powder Technol.*, Vol 10 (No. 3), 1974, p 171
6. Cable, R.I. and Gupta, T.K., Intermediate Stage Sintering, in *Sintering and Related Phenomena*, Gordon and Breach, New York, 1967
7. Rowley, A.K.S., Wasser, E.C.C., and Nash, M.J., The Effect of Some Variables on the Structure and Mechanical Properties of Sintered Bronze, *Powder Metall. Int.*, Vol 4 (No. 2), 1971, p 71
8. Price, A. and Oakley, J., Factors in the Production of 90/10 Tin Bronze Compacts of Higher Density (7.49 g/cm³), *Powder Metall.*, Vol 8, 1965, p 201
9. Klar, E., Ed., *Powder Metallurgy—Applications, Advantages and Limitations*, American Society for Metals, 1983

P/M Lightweight Metals

LIGHTWEIGHT METALS are finding increased use in the P/M industry due to their unique physical and mechanical properties. Of the lightweight metals most prevalent in the P/M industry (aluminum, titanium, and beryllium), aluminum P/M parts enjoy the widest range of applications. Due to their high strength-to-weight ratio, corrosion resistance, and superior finishing properties, aluminum P/M parts are used in business machine, automotive, aerospace, and appliance applications.

Advances in P/M techniques that offer control of part shape and dimension with minimal scrap loss have significantly increased titanium P/M parts applications. In addition to being strong and lightweight, titanium is essentially chemically inert under a wide range of corrosive environments and exhibits excellent high-temperature capabilities at sustained temperatures up to 550 °C (1020 °F). Titanium P/M parts are finding increased usage in aerospace, chemical, and structural applications.

Practically all beryllium is commercially produced by P/M techniques. Beryllium is finding increasing use in the aerospace and nuclear fields. Initially, beryllium was used exclusively in nuclear applications because of its ability to slow neutrons to thermal velocities without reacting with them. The next major application was for inertial guidance systems, where a lightweight, dimensionally stable material was needed. Beryllium P/M parts are currently finding increased usage in the aerospace, x-ray, and optical industries.

This article examines the applications of lightweight metal P/M parts along with their processing and property data. More detailed information on P/M processing of these metals can be found in individual articles dealing with the production and sintering of aluminum, titanium, and beryllium powders.

Aluminum P/M Technology

By Ralph W. Stevenson
Project Manager/Materials Group
Energy Technology Division
Midland Ross Corp.

Conventionally pressed and sintered aluminum powder metal parts have been commercially available for many years. Aluminum P/M parts are competitive with many aluminum castings, extrusions, and screw machine products that require expensive and time-consuming finishing operations. In addition, aluminum P/M parts compete with other metal powder parts in applications where some of the attractive physical and mechanical properties of aluminum can be utilized.

Commercially available aluminum powder alloy compositions (Table 1) consist of blends of atomized aluminum powders mixed with powders of various alloying elements, such as zinc, copper, magnesium, and silicon. The most common heat treatable grades are comparable to the 2xxx and 6xxx series wrought aluminum alloys. Alloy 201AB is most similar to wrought alloy 2014. It develops high strength and offers moderate corrosion resistance. Alloy 601AB is similar to wrought alloy

Table 1 Compositions of typical aluminum P/M alloy powders

Grade	Cu	Mg	Si	Al	Lubricant
601AB	0.25	1.0	0.6	rem	1.5
201AB	4.4	0.5	0.8	rem	1.5
602AB	⋯	0.6	0.4	rem	1.5
601AC	0.25	1.0	0.6	rem	⋯
201AC	4.4	0.5	0.8	rem	⋯
202AB	4.0	⋯	⋯	rem	1.5

6061. This alloy offers high strength, good ductility, corrosion resistance, and can be specified for anodized parts. Alloy 601AC is the same as 601AB, but does not contain an admixed lubricant. It is used for isostatic and die wall lubricated compaction. When high conductivity is required, alloy 602AB often is used. Conductivity of 602AB ranges from 72.3×10^{-8} to $83.51 \times 10^{-8} \ \Omega \cdot m$ (42.0 to 48.5% IACS), depending on the type of heat treatment selected. Electrical and thermal conductivities of sintered aluminum alloys are discussed later in this article. Aluminum wrought P/M alloys, which are comparable to the 7xxx series, are discussed later in this article in the section on high-strength P/M aluminum alloys.

Material Properties

The benefits of using P/M techniques for lightweight metal applications are exemplified by aluminum. In using aluminum P/M parts for special applications, a variety of improved material characteristics are available.

Weight Reduction. Aluminum enjoys a considerable weight advantage over competing P/M materials, assuming strength requirements are not high. Aluminum P/M parts weigh one third as much as the same parts made from iron or copper powders. Lightweight aluminum parts reduce power requirements in drive motors and minimize vibration and noise in machinery that has unbalanced rotary or reciprocating motions. Aluminum P/M components are lower in cost than titanium or beryllium P/M components.

Corrosion Resistance. Aluminum P/M parts are used extensively in both structural and nonstructural applications because of corrosion resistance properties. Corrosion resistance can be further improved by the application of chem-

Fig. 1 Machining chips from a wrought aluminum alloy (right) and from a P/M aluminum alloy (left)

Courtesy of Aluminum Company of America

ical conversion coatings or anodizing treatments.

Machinability. Secondary finishing operations such as drilling, milling, turning, or grinding can be performed easily on aluminum P/M parts. Aluminum P/M alloys provide excellent chip characteristics; compared to wrought aluminum alloys, P/M chips are much smaller and are broken more easily with little or no stringer buildup, as can be seen in Fig. 1. This results in improved tool service life and higher machinability ratings.

Table 2 Electrical and thermal conductivity of sintered aluminum alloys, wrought aluminum, brass, bronze, and iron

Material	Temper	Electrical conductivity(a) at 20 °C (68 °F), % IACS	Thermal conductivity(b) at 20 °C (68 °F), cgs units
601AB	T4	38	0.36
	T6	41	0.38
	T61	44	0.41
201AB	T4	32	0.30
	T6	35	0.32
	T61	38	0.36
602AB	T4	44	0.41
	T6	47	0.44
	T61	49	0.45
6061 wrought aluminum	T4	40	0.37
	T6	43	0.40
Brass (35% Zn)	Hard	27	0.28
	Annealed	27	0.28
Bronze (5% Sn)	Hard	15	0.17
	Annealed	15	0.17
Iron (wrought plate)	Hot rolled	16	0.18

(a) Determined with FM-103 Magnatester. (b) Converted from electrical conductivity values

Electrical and Thermal Conductivity. Aluminum has higher electrical and thermal conductivities than most other metals. Table 2 compares the conductivities of sintered aluminum alloys with wrought aluminum, brass, bronze, and iron.

Mechanical Properties. Aluminum P/M parts can be produced with strength that equals or exceeds that of iron or copper P/M parts. Tensile strengths range from 110 to 345 MPa (16 to 50 ksi), depending on composition, density, sintering practice, heat treatment, and repressing procedures. Table 3 lists typical properties of four nitrogen-sintered P/M alloys. Table 4 compares the properties of several copper alloys with aluminum P/M alloys 601AB,

Table 3 Typical properties of nitrogen-sintered aluminum P/M alloys

Alloy	Compacting pressure MPa	tsi	%	Green density g/cm³	Green strength MPa	psi	%	Sintered density g/cm³	Temper	Tensile strength(a) MPa	ksi	Yield strength(a) MPa	ksi	Elongation, %	Hardness
601AB	96	7	85	2.29	3.1	450	91.1	2.45	T1	110	16	48	7	6	55-60 HRH
									T4	141	20.5	96	14	5	80-85 HRH
									T6	183	26.5	176	25.5	1	70-75 HRE
	165	12	90	2.42	6.55	950	93.7	2.52	T1	139	20.1	88	12.7	5	60-65 HRH
									T4	172	24.9	114	16.6	5	80-85 HRH
									T6	232	33.6	224	32.5	2	75-80 HRE
	345	25	95	2.55	10.4	1500	96.0	2.58	T1	145	21	94	13.7	6	65-70 HRH
									T4	176	25.6	117	17	6	85-90 HRH
									T6	238	34.5	230	33.4	2	80-85 HRE
602AB	165	12	90	2.42	6.55	950	93.0	2.55	T1	121	17.5	59	8.5	9	55-60 HRH
									T4	121	17.5	62	9	7	65-70 HRH
									T6	179	26	169	24.5	2	55-60 HRE
	345	25	95	2.55	10.4	1500	96.0	2.58	T1	131	19	62	9	9	55-60 HRH
									T4	134	19.5	65	9.5	10	70-75 HRH
									T6	186	27	172	25	3	65-70 HRE
201AB	110	8	85	2.36	4.2	600	91.0	2.53	T1	169	24.5	145	24	2	60-65 HRE
									T4	210	30.5	179	26	3	70-75 HRE
									T6	248	36	248	36	0	80-85 HRE
	180	13	90	2.50	8.3	1200	92.9	2.58	T1	201	29.2	170	24.6	3	70-75 HRE
									T4	245	35.6	205	29.8	3.5	75-80 HRE
									T6	323	46.8	322	46.7	0.5	85-90 HRE
	413	30	95	2.64	13.8	2000	97.0	2.70	T1	209	30.3	181	26.2	3	70-75 HRE
									T4	262	38	214	31	5	80-85 HRE
									T6	332	48.1	327	47.5	2	90-95 HRE
202AB: Compacts	180	13	90	2.49	5.4	780	92.4	2.56	T1	160	23.2	75	10.9	10	55-60 HRH
									T4	194	28.2	119	17.2	8	70-75 HRH
									T6	227	33	147	21.3	7.3	45-50 HRE
Cold formed parts (19% strain)	180	13	90	2.49	5.4	780	92.4	2.56	T2	238	33.9	216	31.4	2.3	80 HRE
									T4	236	34.3	148	21.5	8	70 HRE
									T6	274	39.8	173	25.1	8.7	85 HRE
									T8	280	40.6	250	36.2	3	87 HRE

(a) Tensile properties determined using powder metal flat tension bar (MPIF standard 10-63), sintered 15 min at 620 °C (1150 °F) in nitrogen

Table 4   Mechanical properties of typical P/M materials

Material	Density, g/cm³	Tensile strength MPa	ksi	Elongation, %
90-10 bronze	6.4-7.2	96-138	14-20	1-3
80-20 brass	7.2-8.0	138-255	20-37	10-21
Nickel silver	7.5-8.4	172-290	25-42	10-12
601AB	2.42-2.55	171-255(a)	24.9-37(a)	2-6
602AB	2.42-2.55	120-193(a)	17.5-28(a)	2-10
201AB	2.50-2.64	245-356(a)	35.6-51.7(a)	5½

(a) Heat treated

602AB, and 201AB. Thermal treatments, which greatly affect the properties of aluminum P/M alloys, include:

Temper	Description
O	Annealed 1 h at 413 °C (775 °F), furnace cooled at maximum rate of about 30 °C (50 °F) per hour to 260 °C (500 °F) or below
T1	Cooled from sintering temperature to 425 °C (800 °F) (601AB and 602AB) or 260 °C (500 °F) (201AB) in nitrogen, air cooled to room temperature
T4	Heat treated 30 min at 520 °C (970 °F) (601AB and 602AB) or 505 °C (940 °F) (201AB) in air, cold water quenched, and aged minimum of 4 days at room temperature
T6	Heat treated 30 min at 520 °C (970 °F) (601AB and 602AB) or 505 °C (940 °F) (201AB) in air, cold water quenched, and aged 18 h at 160 °C (320 °F)

Aluminum P/M Part Processing

Basic design details for aluminum P/M parts involve the same manufacturing operations, equipment, and tooling that are used for iron, copper, and other metal powder compositions.

Compacting. Aluminum P/M parts are compacted at low pressures and are adaptable to all types of compacting equipment. The pressure density curve, which compares the compacting characteristics of aluminum with other metal powders, indicates that aluminum is simpler to compact. Figure 2 shows the relative difference in compacting characteristics for aluminum and sponge iron or copper.

The lower compacting pressures required for aluminum permit wider use of existing presses. Depending on the press, a larger part often can be made by taking advantage of maximum press force. For example, a part with a 130-cm² (20-in.²) surface area and 50 mm (2 in.) deep is formed readily on a 4450-kN (500-ton) press. The same part in iron would require a 5340-kN (600-ton) press. In addition, because aluminum responds better to compacting and moves more readily in the die, more complex shapes having more precise and finer detail can be produced.

Fig. 2 Relationship of green density and compacting pressure

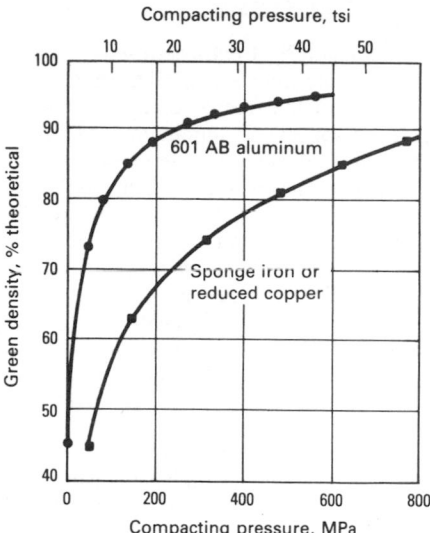

Sintering. Aluminum P/M parts can be sintered in a controlled, inert atmosphere or in vacuum. Sintering temperatures are based on alloy composition and generally range from 595 to 625 °C (1100 to 1160 °F). Sintering time varies from 10 to 30 min. Nitrogen, dissociated ammonia, hydrogen, argon, and vacuum have been used for sintering aluminum; however, nitrogen is preferred because it results in high as-sintered mechanical properties (Table 3). It is also economical in bulk quantities. If a protective atmosphere is used, a dew point of −40 °C (−40 °F) or below is recommended. This is equivalent to a moisture content of 120 mL/m³ (120 ppm) maximum.

Aluminum preforms can be sintered in batch furnaces or continuous radiant tube mesh or cast belt furnaces. Optimum dimensional control is best attained by maintaining furnace temperature at ±2.8 °C (±5 °F). Typical heating cycles for aluminum parts sintered in various furnaces are illustrated in Fig. 3, 4, and 5.

Mechanical properties are directly affected by thermal treatment. All compositions respond to solution heat treating,

Fig. 3 Heating cycle for batch furnace sintering of aluminum P/M parts

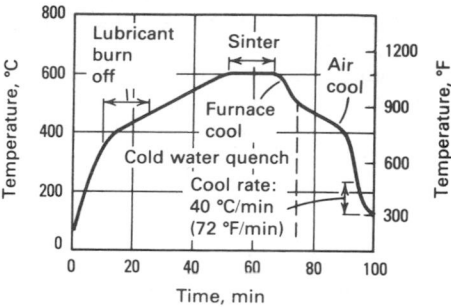

Fig. 4 Heating cycle for continuous furnace sintering of aluminum P/M parts

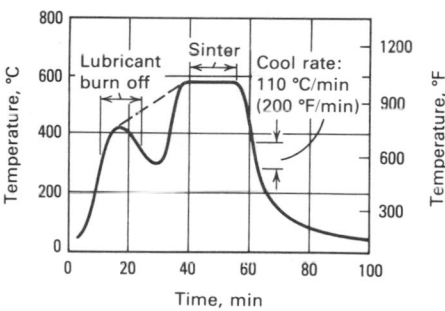

Fig. 5 Heating cycle for vacuum sintering of aluminum P/M parts

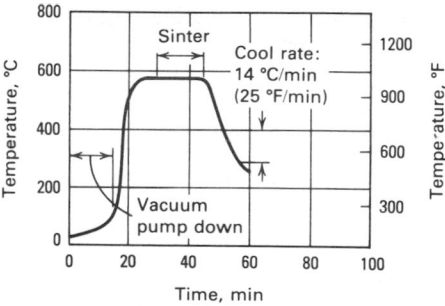

quenching, and aging in the same manner as conventional heat treatable alloys. More detailed information on sintering of aluminum can be found in the article "Production Sintering Practices for P/M Materials" in this Volume.

Forging of aluminum is a well-established technology. Wrought aluminum alloys have been forged into a variety of forms, from small gears to large aircraft structures, for many years. Aluminum lends itself to the forging of P/M preforms to produce structural parts.

In forging of aluminum preforms, the sintered aluminum part is coated with a graphite lubricant to permit proper metal flow during forging. The part is either hot or cold forged; hot forging at 300 to 450 °C (575 to 850 °F) is recommended for parts requiring critical die fill. Forging pressure usually does not exceed 345 MPa (50 ksi). Forging normally is performed in a confined die so that no flash is produced and only densification and lateral flow result from the forging step. Scrap loss is less than 10% compared to conventional forging, which approaches 50%. Forged aluminum P/M parts have densities of over 99.5% of theoretical density. Strengths are 40 to 60% higher than nonforged P/M parts and, in many ways, are similar to conventional forgings. Fatigue endurance limit is doubled over that of nonforged P/M parts.

Alloys 601AB, 602AB, 201AB, and 202AB are designed for forgings. Alloy 202AB is especially well suited for cold forging. All of the aluminum powder alloys respond to strain hardening and precipitation hardening, providing a wide range of properties. For example, hot forging of alloy 601AB-T4 at 425 °C (800 °F) followed by heat treatment gives ultimate tensile strengths of 221 to 262 MPa (32 to 38 ksi) and a yield strength of 138 MPa (20 ksi), with 6 to 16% elongation in 25 mm (1 in.).

Heat treated to the T6 condition, 601AB has ultimate tensile strengths of 303 to 345 MPa (44 to 50 ksi). Yield strength is 303 to 317 MPa (44 to 46 ksi), with up to 8% elongation. Forming pressure and percentage of reduction during forging influence final properties.

Ultimate tensile strengths of 358 to 400 MPa (52 to 58 ksi) and yield strengths of 255 to 262 MPa (37 to 38 ksi), with 8 to 18% elongation, are possible with 201AB heat treated to the T4 condition. When heat treated to the T6 condition, the tensile strength of 201AB increases from 393 to 434 MPa (57 to 63 ksi). Yield strength for this condition is 386 to 414 MPa (56 to 60 ksi), and elongation ranges from 0.5 to 8%.

Properties of cold formed aluminum P/M alloys are increased by a combination of strain hardening densification and improved interparticle bonding. Alloy 601AB achieves 257 MPa (37.3 ksi) tensile strength and 241 MPa (34.9 ksi) yield strength after forming to 28% upset. Properties for the T4 and T6 condition do not change notably between 3 and 28% upset. Alloy 602AB has moderate properties with good elongation. Strain hardening (28% upset) results in 221 MPa (32 ksi) tensile and 203 MPa (29.4 ksi) yield strength. The T6 temper parts achieve 255 MPa (37 ksi) tensile strength and 227 MPa (33 ksi) yield strength. Highest cold formed properties

are achieved by 201AB. In the as-formed condition, yield strength increases from 209 MPa (30.3 ksi) for 92.5% density to 281 MPa (40.7 ksi) for 96.8% density.

Alloy 202AB is best suited for cold forming. Treating to the T2 condition, or as-cold formed, increases the yield strength significantly. In the T8 condition, 202AB develops 280 MPa (40.6 ksi) tensile strength and 250 MPa (36.2 ksi) yield strength, with 3% elongation at the 19% upset level.

Aluminum P/M Applications

Aluminum P/M parts are used in an increasing number of applications. The business machine market currently uses the greatest variety of aluminum P/M parts. Other markets that indicate growth potential include automotive components, aerospace components, power tools, appliances, and structural parts. Due to their mechanical and physical properties, aluminum P/M alloys provide engineers with flexibility in material selection and design. These factors, coupled with the economic advantages of this technology, should continue to expand the market for aluminum P/M parts.

Business Machines. Parts designers are interested in manufacturing parameters only to the extent that they underscore process capabilities and design considerations. Cost-effectiveness and performance of competing processes and materials in a given application are of prime importance. The light weight of an aluminum P/M part is an important characteristic. Consequently, increasing the speed of the product and reducing power requirements and vibrations are continuing goals in the business machine industry.

Lighter weight components reduce inertia on startup and stopping. Moving parts such as drive belt pulleys, hubs, end caps, and connecting collars have proven to be excellent applications for aluminum P/M parts. Their corrosion resistance eliminates the need for costly plating operations or rust preventive oils. Aluminum P/M parts offer a cost-effective alternative to stainless steel parts in applications where the manufacturer cannot afford potential corrosion problems. Machinability of aluminum P/M parts is advantageous for parts that require extensive secondary work. Increased production rates and longer tool life result in substantial cost savings over iron P/M parts in such cases.

Many typical aluminum P/M parts are fabricated from alloy 201AB with a minimum density of 2.6 g/cm^3. Parts usually

Fig. 6 Aluminum P/M pulley in T1 condition for copier machine applications
Courtesy of New Industrial Techniques, Inc.

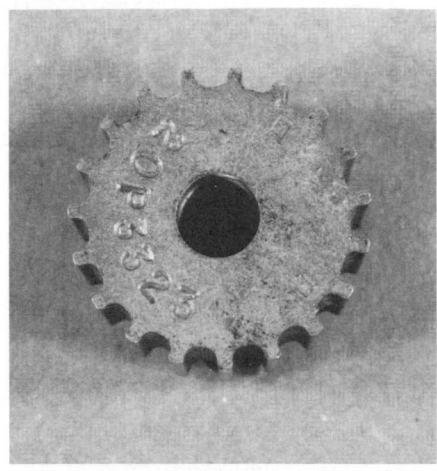

Fig. 7 Aluminum P/M pulley in T4 condition for copier machine applications
Courtesy of New Industrial Techniques, Inc.

Fig. 8 Aluminum P/M cap brush roll
Courtesy of New Industrial Techniques, Inc.

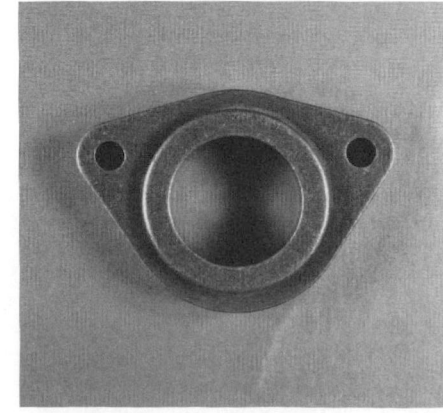

Fig. 9 Aluminum P/M pistons for automotive shock absorbers
Courtesy of Aluminum Company of America

Fig. 10 Aluminum P/M sensor housings for a truck antiskid device
Courtesy of Aluminum Company of America

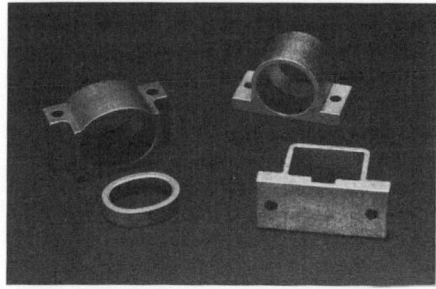

Fig. 11 Aluminum P/M sewing machine parts
Courtesy of Aluminum Company of America

are produced in the T1 and T4 conditions. T1 temper is cooled from the sintering temperature to 260 °C (500 °F) in nitrogen and air cooled to room temperature. In the T4 condition, parts are heat treated for 30 min at 505 °C (940 °F) in air, quenched in cold water, and aged for a minimum of 4 days at room temperature. Sintering temperature range is 565 to 650 °C (1050 to 1200 °F).

Aluminum P/M parts used in copier machine assemblies are shown in Fig. 6, 7, and 8. The pulley shown in Fig. 6 is pressed, sintered, sized, tumbled, and used in the T1 condition. The small gear shown in Fig. 7 is pressed, sintered, heat treated, sized, tumbled, and used in the T4 condition. The cap brush roll shown in Fig. 8 is pressed, sintered, heat treated, sized, tumbled, drilled and tapped, and used in the T4 condition.

Automotive Components. Materials substitution is a key concern in the automotive industry. The desire to use lightweight, corrosion-resistant materials should promote increased interest in aluminum P/M parts. Figure 9 shows an aluminum P/M shock absorber piston that offers 15% more strength than competing iron P/M parts. Figure 10 shows a sensor housing used for a truck antiskid braking system. This part, which formerly was made of 6061 wrought aluminum alloy, is now made of alloy 601AB.

Other Applications. Aluminum P/M parts also are being used for appliances. Figure 11 shows typical sewing machine parts. The lightweight aluminum parts reduce noise level and vibration during operation. Aluminum P/M parts also are used for thermostat control gears for countertop ovens. The thermal conductivity of aluminum (Table 2) makes it an ideal material for heatsinks (Fig. 12). Structural spacers, which were formerly made by conventional stamping, are now produced from P/M aluminum. These spacers, as shown in Fig. 13, are used on structural electric transmission towers and other similar applications where corrosion resistance is required.

High-Strength P/M Aluminum Alloys

High-strength parts fabricated from atomized aluminum, zinc, magnesium, copper, and cobalt alloy powders have superior strength, ductility, resistance to stress-corrosion cracking, and fracture toughness. Powder metallurgy methods yield fine metallurgical structures and compositions that cannot be produced by ingot metallurgy (I/M) methods.

Fig. 12 Aluminum P/M heatsink
Courtesy of Aluminum Company of America

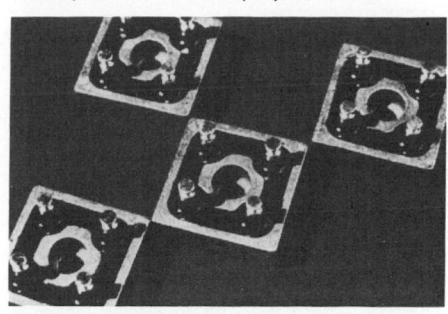

Fig. 13 Aluminum P/M structural spacers
Courtesy of Aluminum Company of America

The nominal chemical compositions for several high-strength aluminum P/M and I/M alloys are compared in Table 5. Alloys X7090 and X7091, the first high-strength P/M alloys used in commercial applications, are similar to the 7xxx series of conventionally processed wrought aluminum alloys, because the primary alloying element is zinc. The addition of cobalt in quantities generally higher than those contained in conventional I/M alloys (1.5 and 0.4% for X7090 and X7091, respectively) refines the grain structure. Rapid solidification during atomization ensures homogeneous distribution of alloying elements. Examples of transverse etch sections of die forgings of P/M X7090, P/M X7091, and I/M 7050 are compared in Fig. 14.

Table 5 Nominal chemical composition for several high-strength aluminum P/M and I/M alloys

Alloy	Composition, %						
	Cu	Mg	Cr	Zn	Co	O	Al
P/M alloys							
X7090	1.0	2.5	...	8.0	1.5	0.35	rem
X7091	1.5	2.5	...	6.5	0.4	0.35	rem
I/M alloys							
7050	2.3	2.25	...	6.2	...	...	rem
7075	1.6	2.5	0.23	5.6	...	...	rem
7175	1.6	2.5	0.23	5.6	...	...	rem
7178	2.0	2.8	0.26	6.8	...	...	rem

Fig. 14 Comparison of microstructures of die forgings

Transverse sections from flange across flash plane. Magnification: 88×. Keller's etch. (a) P/M X7090. (b) P/M X7091. (c) I/M 7050. Courtesy of Aluminum Company of America

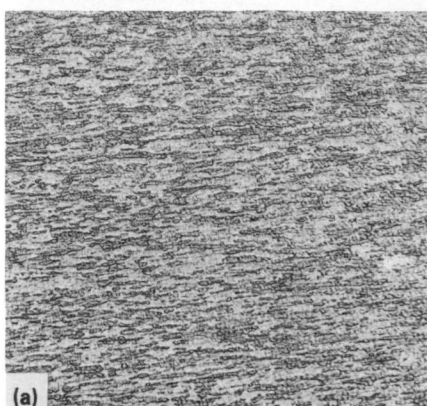

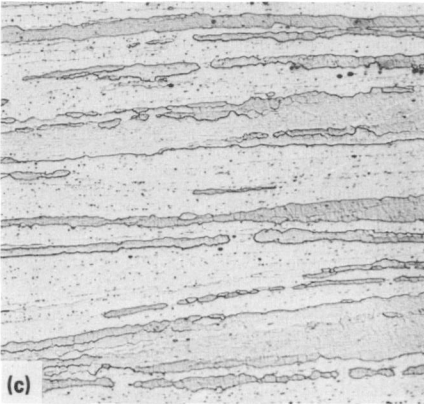

Processing Conditions. Wrought P/M alloys typically are processed in the following production sequence. Molten prealloyed aluminum is atomized and rapidly solidified. Rapid solidification refines the dendritic structure and reduces the size of constituent particles. Powder particles are compacted isostatically and encapsulated in an aluminum canister. The compact is heated, and gases are evacuated from the canister. The heated canister and compact are pressed into a billet with 100% density. Then the canister is machined from the surface of the billet and the billet is ready for subsequent fabrication, using conventional techniques such as forging or extruding. Larger billets are fabricated into sheet and plate products. More detailed information on rapid solidification techniques is available in the article "Atomization" in this Volume.

Available Product Forms. Aluminum alloys in forged and extruded form are particularly well suited for aerospace and ordnance parts that are subjected to unusual corrosion or mechanical stress conditions. Available product forms and tempers include:

Alloy-temper	Available product forms
X7090-T7E71	Extrusions and forgings
X7090-T7E75	Die forgings (hot water quenched)
X7090-T7E80	Hand and die forgings (compression stress relieved)
X7091-T7E69	Extrusions and non-stress relieved forgings
X7091-T7E70	Extrusions and non-stress forgings
X7091-T7E76	Die forgings (hot water quenched)
X7091-T7E77	Die forgings (hot water quenched)
X7091-T7E78	Hand and die forgings (compression stress relieved)
X7091-T7E79	Hand and die forgings (compression stress relieved)

Note: All extrusions stress relieved by stretching

Fig. 15 Aluminum alloy 7090 P/M aerospace forgings

Main landing gear link (left) and landing gear door actuator (right) for Boeing 757 aircraft. Courtesy of Aluminum Company of America

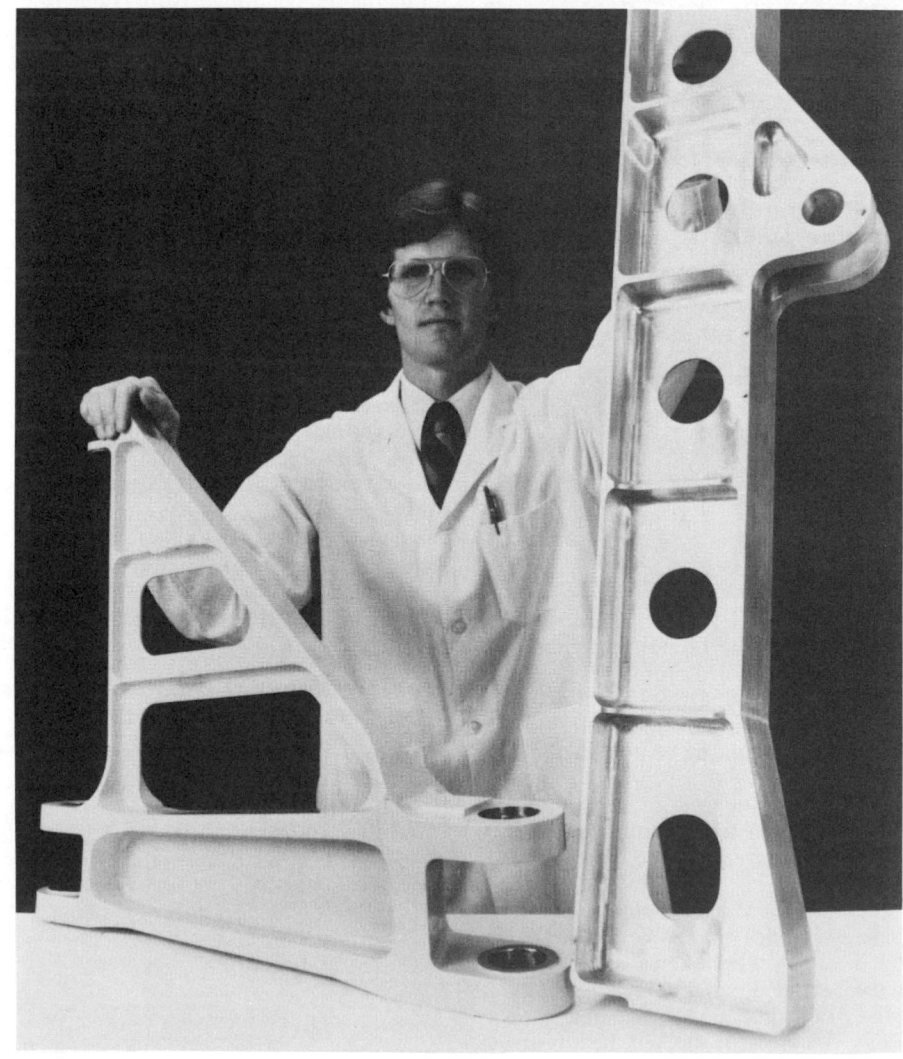

Fig. 16 Comparison of P/M and I/M aluminum alloys

(a) X7090 P/M alloy. Magnification: 500×. (b) 7178 I/M alloy. Magnification: 500×

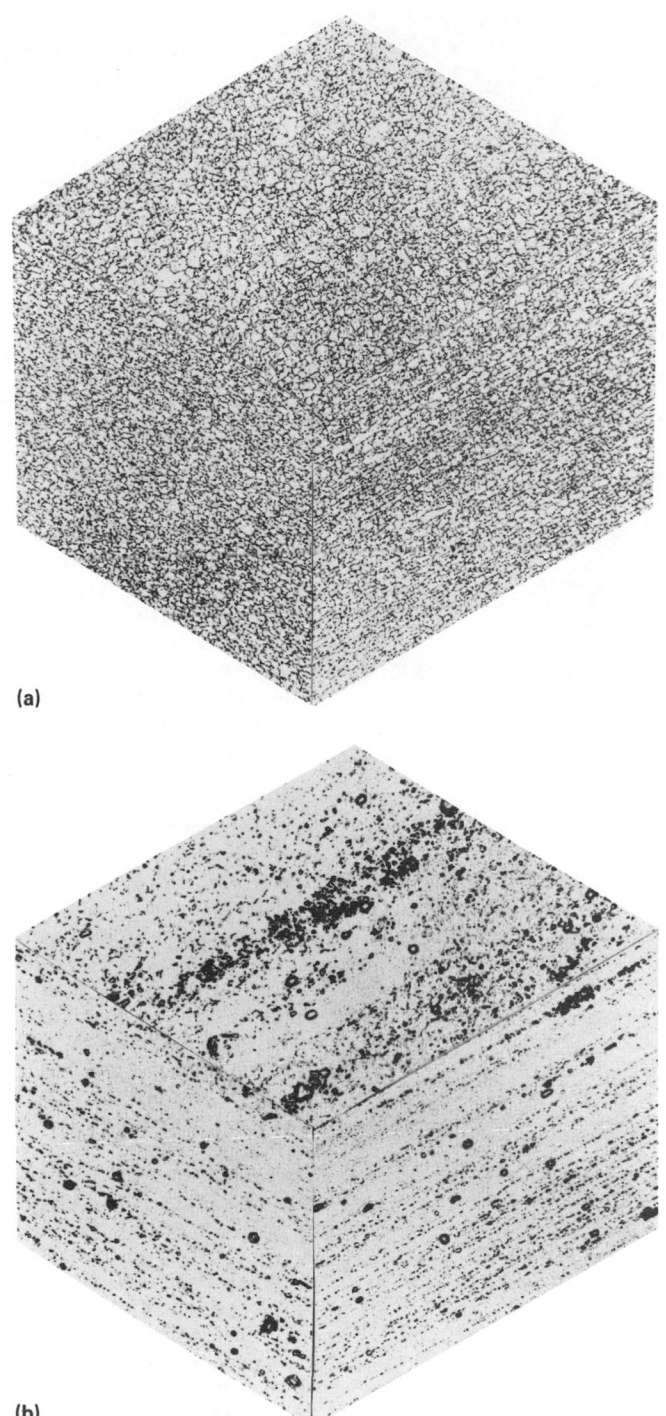

(a)

(b)

Fig. 17 Fracture toughness versus yield strength for high-strength aluminum alloy extruded shapes

Specimens were 6.4 to 38 mm (0.25 to 1.5 in.) thick.

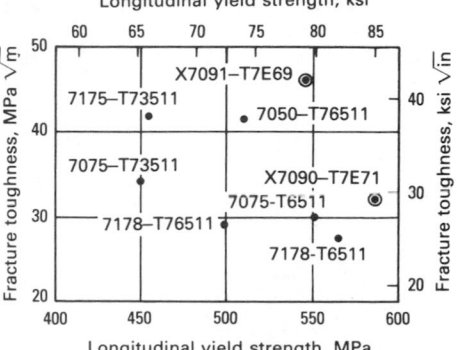

Table 6 Density and modulus of typical P/M and I/M alloys

Alloy	Density, g/cm³	Modulus of elasticity(a), MPa × 10³
P/M alloys		
X7090 .... 2.850		73.8
X7091 .... 2.823		72.4
I/M alloys		
7050 .... 2.823		71.7
7075 .... 2.796		71.7
7178 .... 2.823		71.7

(a) Average of tension and compression moduli

Figure 15 shows two 7090 P/M alloy forgings for a Boeing 757—the main landing gear support beam link and the landing gear door actuator support fitting. Longitudinal yield strength of the extruded P/M alloy 7090-T7E71 is approximately 15% higher than the I/M alloy 7050-T76511, while the yield strength of the P/M alloy 7091-T7E69 is 7% more than 7050-T76511. Use of these parts provides a 6.4-kg (14 lb) weight savings, which increases fuel economy.

In addition, these P/M alloys have excellent corrosion resistance to stress-corrosion cracking that is equivalent to 7050T73—the I/M alloy considered to have optimum corrosion properties. In exposure to a seacoast atmosphere for over a 5-year period, these P/M alloys in the T76 and T6 tempers also show no evidence of exfoliation. This improvement in corrosion resistance is due to the finer grain structure and the improved homogeneous distribution of alloying elements in the P/M alloys. Figure 16 illustrates the differences between P/M and I/M microstructures.

Fracture Toughness and Strength. Initial testing of the strength-toughness relationship of the P/M alloys indicates their superiority to most of the 7xxx series I/M alloys. The relationship between yield strength and fracture toughness of extruded shapes 6.4 to 38 mm (0.25 to 1.5 in.) thick in the longitudinal-transverse orientation is illustrated in Fig. 17. A comparison of density and modulus of elasticity between P/M and I/M alloys is given in Table 6.

Fig. 18 Typical forged and extruded P/M alloy shapes
Courtesy of Aluminum Company of America

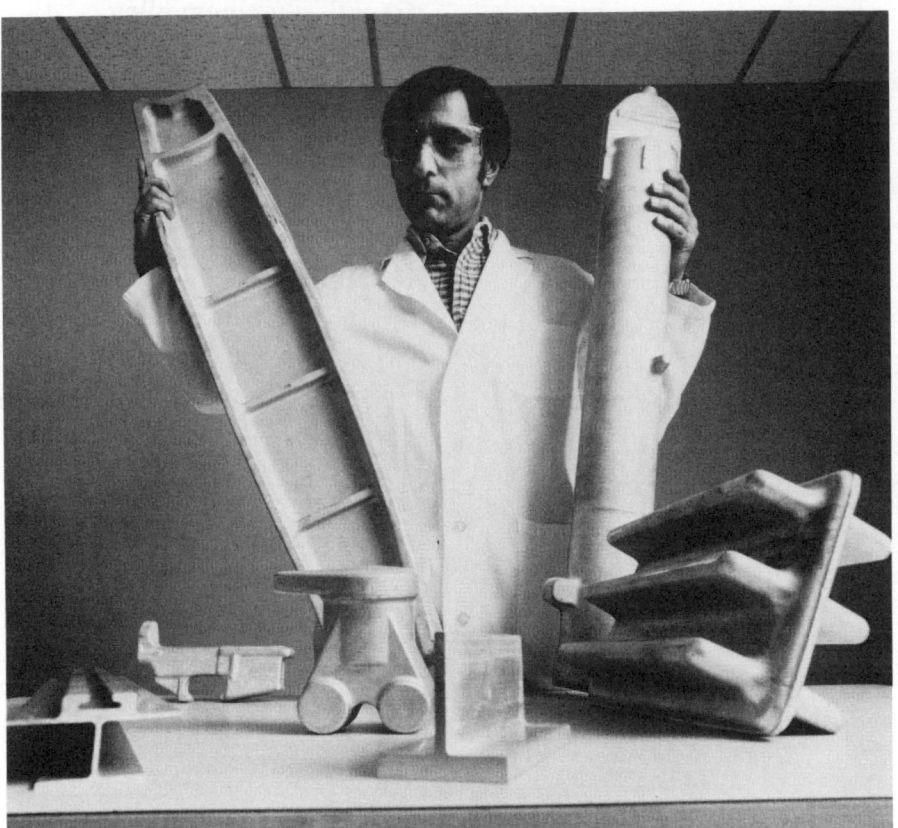

Table 7 Titanium content of military airframes

System	Early design, wt%	Final concept, wt%
C-5 (cargo)	24	3
B-1 (bomber)	42	22
F-15 (fighter)	50	34

Potential Applications. High strength combined with superior corrosion resistance of wrought P/M alloys render them candidates for aerospace and military applications. Upper wing skins from plate or extrusions, stiffeners, landing gear components (Fig. 15), helicopter rotors, and other major airframe primary load-carrying structural members are typical aerospace applications where high-strength P/M alloys may find use. Figure 18 shows typical forged and extruded shapes that have been produced.

A significant factor contributing to military interest in the development of P/M materials is the potential for improved hardware performance through (1) improved corrosion resistance at required strength level, (2) weight reduction permitted by higher strength, or (3) the po-

tential of maximizing the combination of strength and fracture toughness. Potential ordnance applications for high-strength wrought P/M alloys include rifle receiver forgings, mortar and artillery projectile fuses, cartridge cases, artillery shells, and rocket warheads.

Titanium P/M Technology

By F.H. Froes
Technical Area Manager
U.S. Air Force
AFWAL Materials Laboratory
and
D. Eylon
Metcut-Materials Research Group
and
Gerald Friedman
TRW Inc.
Materials and Manufacturing
Technology Center

Titanium and titanium alloys are the design choice for many aerospace, medical,

chemical processing, energy, marine, and industrial applications, because they offer low density, excellent corrosion resistance, and good mechanical properties at room and moderately elevated temperatures. The cost of titanium alloy components is relatively high because of the initial cost of the alloy, the cost of processing and forging, and the cost of machining (Ref 1, 2). Despite the fact that titanium is the best design choice for many systems from the standpoint of performance durability, less titanium is used in final production because of the high cost of the components (Ref 2). Some of the newest Air Force aircraft are prime examples, as shown in Table 7.

As a result, net shape or near-net shape technologies are one of the major avenues for reducing the cost of titanium alloy components. These processes include isothermal forging, casting, superplastic forming, and powder metallurgy (Ref 1). Titanium P/M fabrication offers the potential for true net shape capability (Ref 3-5), combined with mechanical properties that are equal to or exceed cast and wrought products. This is due to a lack of texture and segregations and the fine, uniform grain structure inherent to titanium P/M products (Ref 3, 4).

Titanium alloy P/M components are produced by two different processes: the blended elemental method and the prealloyed method. Basically, the blended elemental method is a cold press (or cold isostatic press) and sinter process that results in less than fully dense material. As a result, this method provides relatively lower cost products with good tensile strength, fracture toughness, and fatigue crack growth rates, but with somewhat lower fatigue strength. On the other hand, prealloyed powder compacts are fully dense with good mechanical properties, including fatigue strength. Blended elemental components are relatively inexpensive and may be suitable for noncritical applications. The prealloyed method is intended to produce high-performance components in complex shapes.

This section of this article reviews the consolidation, shapemaking capabilities, mechanical properties, and applications of blended elemental and prealloyed titanium alloy powders. Unless otherwise stated, all references in this section of this article are to the commonly used Ti-6Al-4V alloy.

Consolidation

Blended Elemental Powders. The basic methods of blended elemental compaction are cold isostatic pressing or press consolidation. These processing steps are followed by sintering. Consolidation may include hot isostatic pressing, which increases density (Ref 3, 4). Recently, additional treatments following hot isostatic pressing have been evaluated: (1) isothermal forging and (2) conversion to plate, sheet, or foil by hot rolling (Ref 6), both of which are discussed in the Section "Consolidation of Metal Powders" in this Volume.

Prealloyed Powders. A number of processes have been developed to improve the performance of the end product by removing contaminants or adjusting the final microstructure (Ref 4). These include jet classification, electrostatic separation, electrodynamic degassing (all of which are discussed in the article "Cleaning of Metal Powders" in this Volume), and strain energizing. Jet classification divides particles according to density and size, separating hollow particles, nonmetallic inclusions, and dense inclusions such as tungsten. It is moderately successful in achieving these goals. Electrostatic separation divides ceramic and metal particles according to discharge rates after charging in a high-voltage (25-kV) electrostatic field; some success has been achieved with this technique. Electrodynamic degassing is designed to eliminate gas inclusions and is applied mainly to superalloy powders. Strain energizing involves deformation by rolling of titanium powder particles, thereby storing substantial strain in the material. A subsequent anneal—which may be thermal exposure during compaction—recrystallizes the microstructure, producing a finer microstructure and thus providing some flexibility in microstructure and ultimately the property combination achievable.

Characterization of both the as-produced powder particles and the foreign particles which may be present at the loose powder stage can assist in quality control. It has been clearly demonstrated that both the basic microstructure and contaminants present influence mechanical properties, particularly fatigue. A method that is useful in separating out foreign particles for subsequent classification (but not as a cleaning method for titanium) is water elutriation, which distinguishes between particles on a density basis. A typical titanium powder particle microstructure is shown in Fig. 19.

Three consolidation processes currently are being investigated, all of which are capable of producing fully dense compacts (Ref 4). Hot isostatic pressing within a heated pressure vessel or autoclave is the most common procedure. By simulta-

Fig. 19 Martensitic microstructure of rotating electrode powder Ti-6Al-4V

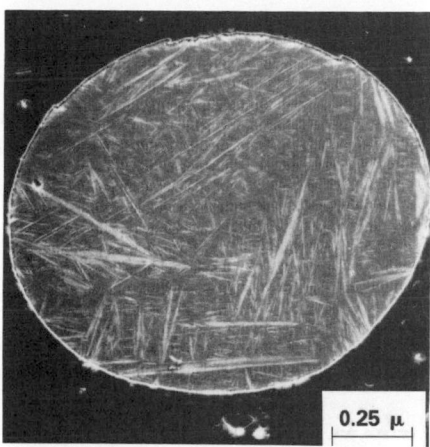

0.25 μ

neously applying temperature and pressure, full density in the part is attained by transmittal of the applied pressure. Press consolidation (Fig. 20) allows rapid (or lower temperature) compaction of powder inside a shaped evacuated can. Finally, powder can be compacted by vacuum hot pressing, in which powder is hot compacted in a forge press that is adapted to a vacuum system. Dies press the powder to 100% density in the required shape. Advantages of this technique include lower equipment investment than hot isostatic pressing units, ability to easily adapt existing equipment to the process, and shorter cycle times. Currently, the major disadvantage of vacuum hot pressing appears to be the lack of flexibility in shapes that are pressable and the size of parts that can be produced. All of these consolidation processes are discussed in the Section "Consolidation of Metal Powders" in this Volume.

Shapemaking

Blended Elemental Powders. Cold isostatic pressing with elastomeric molds can produce extremely complex shapes, such as the impeller shown in Fig. 21. Because the mold material is elastomeric (as opposed to hard punch and die tooling), lubricant is not needed, thus eliminating a presintering treatment to remove lubricant, as well as eliminating lubricant contamination. Figure 22 shows a bag and mandrel used for the production of a Ti-6Al-4V gyro base. Figure 23 depicts a thick-walled rubber bag and steel mandrel for hydrostatically pressing sponge fines to produce a cold pressed commercially pure titanium thin-walled splined cylinder

Fig. 20 Schematic of powder consolidation using the press consolidation process

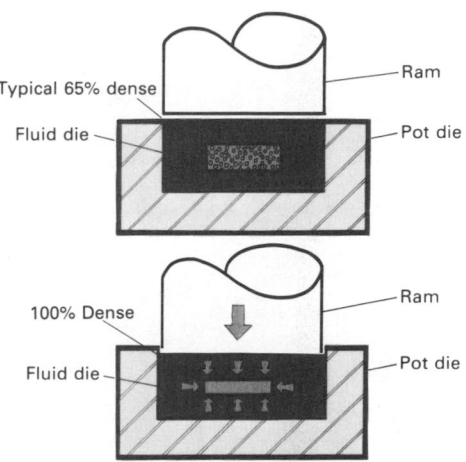

Fig. 21 Impeller made of Ti-6Al-4V elemental blend powder
Produced by cold isostatic pressing, using an elastomeric mold. Courtesy of Dynamet Technology Inc.

for a gyro application. However, part size is limited to a maximum diameter of 60 cm (24 in.); length is limited by the availability of cold isostatic pressing equipment. Dimensional tolerance has not as yet been determined by production data, but appears to be about ±0.02 mm/mm (±0.02 in./in.) length.

Press consolidation is not limited by size; presses up to 45 000 metric tons (50 000 tons) are available in the United States. Consequently, a part of almost 13 000 cm^2 (2000 in.^{2}) can be produced. Intricate shape capabilities, however, do not approach those possible by cold isostatic pressing. Production data are required to establish dimensional tolerance, but these tech-

Fig. 22 Elastomeric bag and mandrel for the production of cold isostatically pressed blended elemental Ti-6Al-4V gyro base
Courtesy of TRW, Inc.

Fig. 24 Connector link arm for the Pratt & Whitney F-100 engine
Produced from elemental blend Ti-6Al-4V powder by the elemental powder consolidation method. Courtesy of Imperial Clevite, Inc.

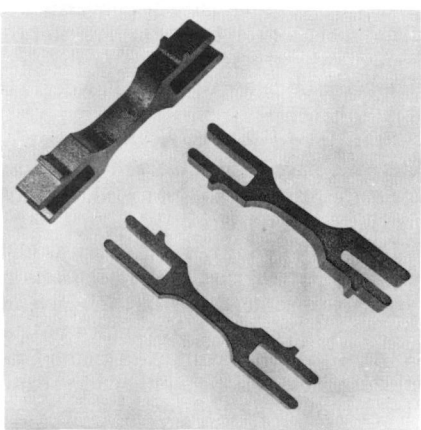

Fig. 23 Thick-walled rubber bag and steel mandrel for hydrostatically pressing sponge fines to produce cold pressed titanium thin-walled splined cylinder for a gyro
Courtesy of TRW, Inc.

Molds are prepared by the lost-wax process. Combining this process and P/M hot isostatic pressing offers the potential for producing low-cost net and near-net shapes for a variety of applications (Ref 5). Wax patterns are prepared in shapes that are intentionally larger than the final configuration. This is necessary because the wax pattern, which subsequently becomes the shell mold, must allow for shrinkage of about 35% as the powder is consolidated, in contrast to investment casting where the liquid metal shrinks about 7% during solidification.

Once the final compact configuration is known, the required wax pattern shape can be defined by completing calculations based on the packing density of the powder. For complex shapes, three or more steps are presently required to reach the desired configuration, as shown in Fig. 25. Fortunately, trial wax patterns can be fabricated inexpensively until starting shape dimensions are established. Waxes can be produced with injection tooling if an actual production run of parts is required. With either approach, wax patterns are used to form a ceramic shell (mold) with an inside shape precisely defined by the shape of the wax pattern. This is done by dipping the wax alternately in a refractory slurry and a granular stucco coating, followed by drying at room temperature. The wax is then melted from the molds, which are fired to produce a durable shaped container that is filled with the metal powder.

The main problem in adapting the ceramic mold method to titanium P/M processing is the selection of mold coatings that are chemically inert to the titanium, but that retain shape reproducibility dur-

niques appear capable of ±0.01 mm/mm (±0.01 in./in.) length. Figure 24 represents a typical part produced by the press consolidation process.

Prealloyed Powders. Production of complex shapes currently is achieved by three competing techniques: metal can, ceramic mold, or fluid die. The first two methods use hot isostatic pressing consolidation, while the fluid die process is adaptable to either hot isostatic pressing or hot press compaction. The metal can is shaped to the desired configuration by state-of-the-art sheet metal methods, such as brake bending, press forming, spinning, or superplastic forming. Carbon steel is the best suited container material, because it reacts minimally with titanium—forming titanium carbide, which inhibits further reaction. Fairly complex shapes can be produced by this technique.

Basically, the Colt-Crucible ceramic mold process relies on technology developed by the investment casting industry.

Fig. 25 Manufacturing sequence of ceramic mold process

(a) Fabricate mold. (b) Fill with powder. (c) Add secondary metal can with secondary pressing medium and evacuate. (d) Hot isostatically press. (e) Finished part

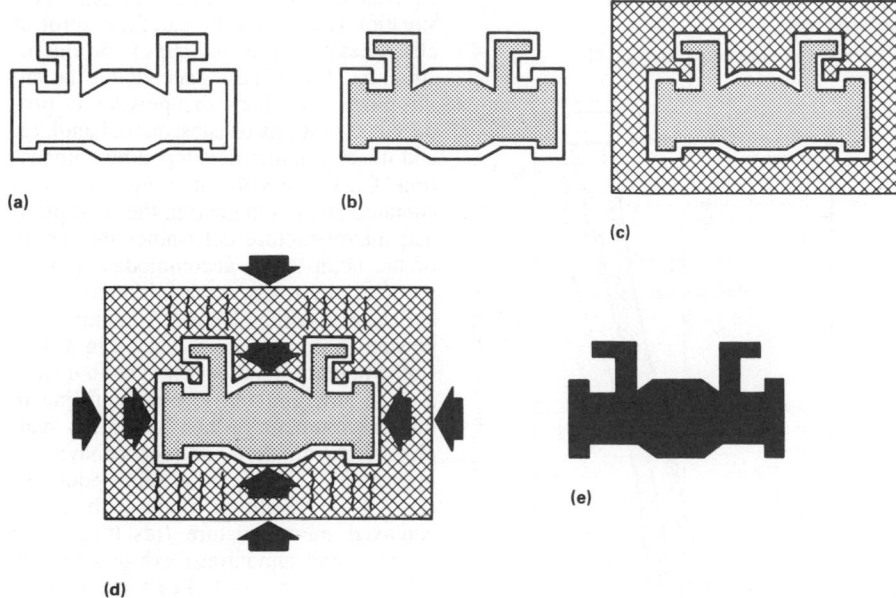

Fig. 27 Flowchart of the Colt-Crucible ceramic mold process

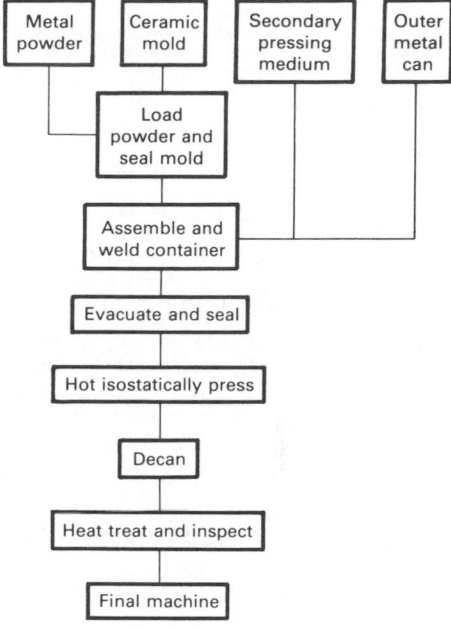

Fig. 26 Containerization system for Colt-Crucible ceramic mold process

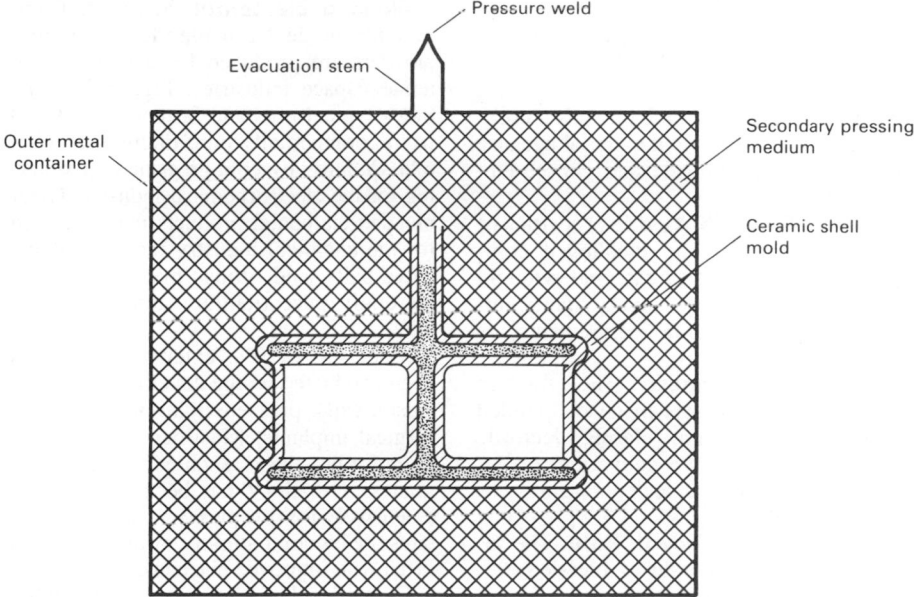

ing powder consolidation. Before titanium powder is loaded, the ceramic mold is cleaned to ensure internal integrity of the part. During powder loading, the mold is vibrated for maximum packing density. Powder loading is carried out under clean room conditions with filtered air under positive pressure to prevent contamination of the powder. The filled mold is sealed and placed in a steel container that is backfilled with a granular refractory material. A schematic of the containerization system in the ceramic mold process is shown in Fig. 26.

The auxiliary pressing medium transmits the pressure during hot isostatic pressing from the outer steel can walls to the titanium powder in the ceramic mold.

An evacuation stem is fitted to the top of the steel outer container, and the loaded steel canister is evacuated and outgassed prior to hot isostatic pressing of the assembly. Evacuation stems are subsequently sealed by hot pressure welding.

During consolidation, the contained assembly is heated under argon gas. For titanium P/M processing, the assembly is heated to 870 to 980 °C (1600 to 1800 °F), depending on alloy grade, at a pressure of 103 MPa (15 000 psi) for 2 to 8 h. Generally, titanium alloys are consolidated below the beta transus temperature. No mold wall friction is encountered during consolidation; therefore, production of intricate parts to near-net shapes, such as recessed angles, is achieved by joining wax patterns. Following hot isostatic pressing, the assembly is opened, and the compacted metal shape is removed and sandblasted to remove mold residue. The part is then prepared for heat treatment, inspection, chemical milling, and/or machining. A flowchart of the Colt-Crucible ceramic mold process is shown in Fig. 27.

The fluid die process practiced by Kelsey-Hayes (Ref 5) is an outgrowth of earlier work on glass containers. Dies are made from carbon steels, copper alloys, or ceramic materials and are of sufficient mass and dimension to behave as a viscous liquid under pressure at temperature. Dies are comprised of two halves, with inserts where necessary to simplify manufacture. The two halves are then welded

Table 8   Tensile properties of Ti-6Al-4V powder compacts

Material	Yield strength(a) MPa	ksi	Ultimate tensile strength MPa	ksi	Elongation, %
P/M compacts					
Blended elemental powder(b)	868	126	951	138	15
Rotating electrode powder(c)	923	134	985	143	17
Rotating electrode powder(d)	937	136	985	143	16
I/M wrought mill annealed					
Plate	923	134	978	142	17
MIL-T-9047(e)	827	120	896	130	10

(a) 0.2% offset. (b) Cold compacted and sintered. (c) Hot isostatically pressed. (d) Fluid die press consolidated. (e) Typical minimum properties

Table 9   Typical fracture toughness of Ti-6Al-4V P/M and I/M compacts

Material	Fracture toughness(a) ksi√in.	MPa √m
P/M		
Blended elemental powder(b)	55	61
Plasma rotating electrode powder compact	78	85
I/M		
Plate, mill annealed(c)	46	50
Forging, mill annealed	55	61

(a) K_{Ic} values. (b) 99.5% dense. (c) Longitudinal-transverse orientation

together to form a hermetic seal. Powder loading and consolidation follow. The fluid die process combines the ruggedness and fabricability of metal with the flow characteristics of glass to generate a replicating container capable of producing extremely complex shapes. Like the ceramic mold process, a number of shapemaking steps are required to reach the final part configuration. Generally, the latter two processes are capable of generating more complex and intricate shapes than the metal can method.

Mechanical Properties

Recently, the mechanical properties of both blended elemental and prealloyed powder compacts have been significantly upgraded (Ref 5). In fact, the mechanical properties of prealloyed compacts are equivalent to comparable I/M products, including initiation critical fatigue. As

Fig. 28 Comparison of fatigue crack growth rate of Ti-6Al-4V P/M compact with I/M material heat treated to various conditions

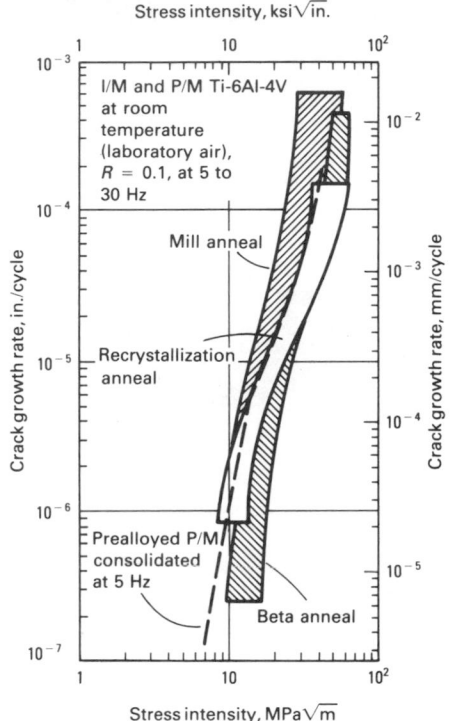

mentioned earlier, fatigue remains a problem in blended elemental material because of the residual chloride and consequent porosity.

Tensile Strength. Tensile properties of blended elemental and prealloyed compacts are equivalent to cast and wrought I/M products. Strength levels significantly exceed specification requirements, as shown in Table 8.

Fracture Toughness. The fracture toughness of compacts made of blended elemental and plasma rotating electrode powder frequently are equivalent or superior to I/M mill annealed forgings, as shown in Table 9.

Fatigue Crack Growth Rate. The fatigue crack growth rate of blended elemental and prealloyed compacts is equivalent to I/M material with the same microstructure (Fig. 28).

Fatigue. The fatigue behavior of titanium P/M compacts is compared to I/M products in Fig. 29. The blended elemental material is inferior to prealloyed compacts and I/M materials. This is caused by residual chlorides and consequent porosity; also, chemical heterogeneity may lead to areas of similarly aligned alpha plates. Blended elemental compacts, however,

compete well with many titanium alloy castings in fatigue strength. Prealloyed powder compacts exhibit fatigue behavior equivalent to that of I/M materials. This situation is achieved by careful control of cleanliness (powder handling) and microstructure. Cleanliness depends on the environment in which the powder is produced, conditions of subsequent handling, and microstructure developed by compaction. Cleanliness dictates the amount of contamination contained in the final product; microstructure determines the ability of the compact to accommodate foreign particles and resist crack initiation.

Plasma rotating electrode titanium alloy powders currently are produced in a dedicated titanium facility that is contained in a clean environment. Container filling in this environment further minimizes handling contamination. Fatigue behavior is optimized as the alpha becomes equiaxed (Fig. 30). Prealloyed plate with a fine equiaxed microstructure (resulting from working and annealing) exhibits fatigue behavior in excess of I/M cast and wrought levels (Ref 6).

Applications

Blended Elemental Powders. Components made from blended elemental powders are produced for aerospace and nonaerospace industries. Figure 31 illustrates typical nuts and fasteners produced from commercially pure titanium and Ti-6Al-4V alloy by cold pressing and sintering for the electrochemical industry. Other blended elemental products manufactured for the chemical industry include valves, valve balls, and fittings made from commercially pure titanium and Ti-6Al-4V alloy.

Since the 1950's, extensive work has been conducted on the development of hot isostatically pressed titanium alloys for surgical implants. Most early use of titanium for prostheses entailed the unalloyed grades. Ti-6Al-4V has been investigated extensively for this purpose, because it retains the biocompatibility of the unalloyed metal with improved mechanical properties.

Currently, Ti-6Al-4V is the only alpha-beta alloy in use for surgical implants. The grade of alloy used in the United States is covered by American Society for Testing and Materials standard ASTM F136 and is the extra-low-interstitial (ELI) grade in that it contains a low concentration of the interstitial elements nitrogen, hydrogen, and oxygen.

Partial and total joint replacements for the hip, knee, elbow, jaw, finger, and shoulder have been produced commer-

Fig. 29 Comparison of smooth axial fatigue behavior of Ti-6Al-4V blended elemental and prealloyed P/M compacts with wrought annealed material
Room temperature. $R = 0.1$

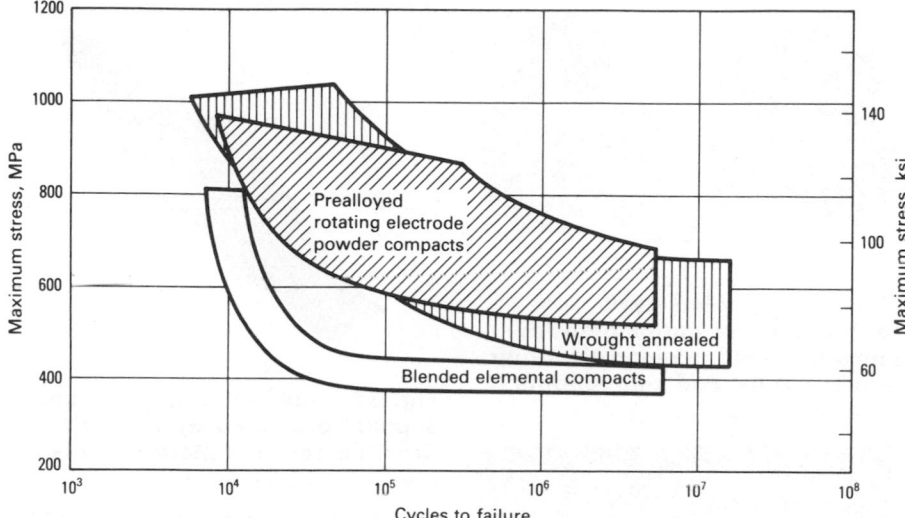

Fig. 30 Equiaxed microstructure of Ti-6Al-4V rotating electrode powder compact with good crack initiation resistance properties

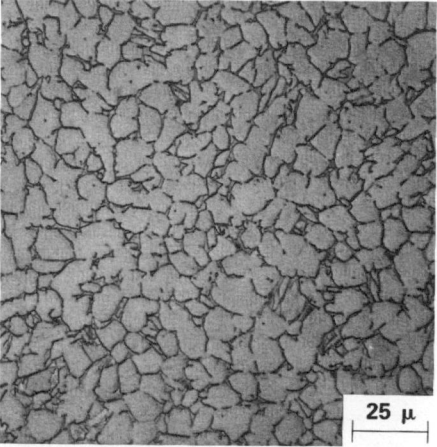

cially from both unalloyed titanium and Ti-6Al-4V. Figure 32 shows a hip prosthesis and knee joint implant made from P/M Ti-6Al-4V; Fig. 33 shows fasteners used for prosthetic fixation made from the same alloy.

For partial joint replacements in which the implant articulates with the cartilage, titanium has been found to have the same minimal wear exhibited by stainless steel and cobalt-chromium-molybdenum alloys.

In aerospace applications, blended elemental components are used for nonfatigue critical components such as missile housings or gyro gimbals. These parts typically are made of Ti-6Al-4V or Ti-6Al-6V-2Sn. The gyro base in Fig. 34 was consolidated by hydrostatic methods. Sponge fines were blended with an aluminum-vanadium master alloy to produce a mixture containing 90% Ti and 10% 60Al-40V. The fines were hydropressed with a bag and mandrel. The tooling objective for this part was to make the flat-sided hemispherical bowl at the top and the connecting interior passages hollow, with minimal or no machining in these areas. The base was cold isostatically pressed at 415 MPa (60 ksi) and sintered at 1200 °C (2200 °F) for 4 h. Hot isostatic pressing followed at 930 °C (1700 °F) at 105 MPa (15 ksi) for 4 h.

The Ti-6Al-4V gimbal ring in Fig. 35 was die pressed, sintered, and forged. The forging preform produced was a square-sided ring with flat and parallel top and bottom surfaces. Following sintering under the same conditions used for the gyro

base (Fig. 34), the compacts were coated with a lubricant, heated in air to 980 °C (1800 °F), and forged in a closed die between opposed punches. When the forgings cooled to room temperature, they were grit blasted and treated in a hot nitric-hydrofluoric acid solution to remove surface oxidation. The final assembled gyro containing five component parts produced by the blended elemental approach is shown in Fig. 36. Cost savings as high as 70% can be realized in the manufacture of complex parts using cold isostatic pressing, even when subsequent machining is required.

Fig. 31 Typical cold pressed titanium and Ti-6Al-4V nuts and parts produced by blended elemental compaction
Courtesy of Imperial Clevite, Inc.

Fig. 32 Hip prosthesis and knee joint implant made from P/M Ti-6Al-4V
Courtesy of Colt Industries

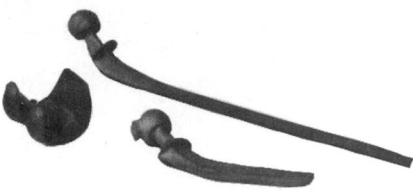

Fig. 33 Fasteners for external prosthetic fixation made of pressed and sintered Ti-6Al-4V
Courtesy of Imperial Clevite, Inc.

Fig. 34 Cold isostatically pressed and sintered Ti-6Al-4V gyro base
Courtesy of TRW, Inc.

Prealloyed Powders. The full density achieved by prealloying and the high level of mechanical properties and net shape capability of complex configurations (Ref 5)

Fig. 35 Die pressed, sintered, and forged Ti-6Al-4V gyro gimbal rings
Courtesy of TRW, Inc.

Fig. 36 Assembled gyro containing five major blended elemental cold pressed titanium and Ti-6Al-4V P/M components
Courtesy of TRW, Inc.

make this method suitable for aerospace components.

The majority of aerospace prealloyed titanium P/M parts currently are produced by the Colt-Crucible ceramic mold process (Ref 5). The Navy funded an early program to demonstrate the feasibility of producing a small airframe component (F-14 fuselage brace) by hot isostatic pressing consolidation (Fig. 37). This part had a high rib and deep pocket design, which normally requires significant machining.

An additional airframe part currently produced by the Colt-Crucible ceramic mold process is the F-18 engine mount support shown in Fig. 38. This part is basically a sheet with stiffeners, which involves extensive machining when produced from a forging. At the successful completion of the program, this part will be flight qualified and ready for produc-

Fig. 37 Ti-6Al-4V P/M F-14 fuselage brace produced by the Colt-Crucible ceramic mold process
Courtesy of Colt Industries

Fig. 38 Ti-6Al-4V engine mount support produced by the Colt-Crucible ceramic mold process
Courtesy of Colt Industries

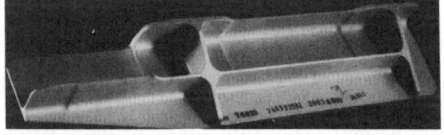

Fig. 39 P/M F-107 Cruise Missile engine radial compressor rotor produced by the Colt-Crucible ceramic mold process
Courtesy of Colt Industries

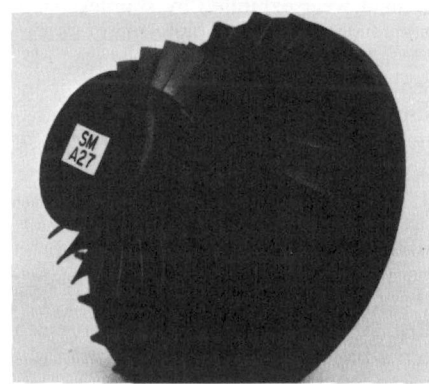

tion. Data generated from this program will become part of a large data bank (MIL-HDBK-5) and will form the basis for additional hot isostatically pressed Ti-6Al-4V data.

Recently, the Colt-Crucible ceramic mold process has advanced so that larger and more complex shapes can be made. These include the cruise missile engine F-107 radial compressor rotor, which is produced so closely to net shape that only a final chemical mill should be required (Fig. 39), thus eliminating expensive ma-

Table 10 Shape reproducibility of Ti-6Al-4V P/M impeller produced by the Colt-Crucible ceramic mold process

Dimension	Median mm	Median in.	Variation mm	Variation in.	Variation, %
Radius measured from centerline(OD)	88.9	3.500	±0.66	±0.026	±0.7
Radius measured from centerline(ID)	53.31	2.217	±0.63	±0.025	±1.1
Blade height on large impeller outer diameter	55.06	2.168	±0.38	±0.015	±0.7

Note: Values are based on 7 parts and 23 locations per part (see Fig. 21)

Fig. 40 Largest titanium P/M component produced by the Colt-Crucible ceramic mold process

F-14A Nacelle frame was fabricated by electron beam welding of four hot isostatically pressed sections at the locations indicated. Courtesy of Colt Industries

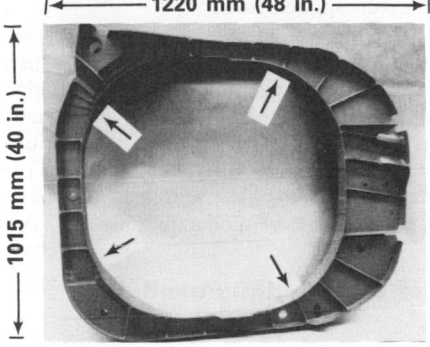

├──── 1220 mm (48 in.) ────┤

1015 mm (40 in.)

chining. The dimensional reproducibility of this method is demonstrated in Table 10. The generic nature of this configuration should lead to widespread P/M engine applications.

The largest P/M part currently produced by the Colt-Crucible ceramic mold process is the F-14A Nacelle frame, which is 100 by 120 cm (40 by 48 in.). This component consists of a high rib and deep pocket design that currently requires significant machining during fabrication. This part is to be fabricated by electron beam welding of four hot isostatically pressed powder parts into an oval-shaped F-14A frame section (Fig. 40), which increases the difficulty of producing this part.

As discussed previously, cost reduction is the major reason for using P/M processing instead of conventional alloy production. Because the bulk of the cost of titanium component fabrication is concentrated in forging and machining (Ref 1, 2), selection of appropriate parts can be made only after consideration of these factors, as well as material selection. Forging and machining costs are directly proportional to the size and complexity of the part. Material costs are higher as the buy-to-fly ratio increases, that is, the weight of in-coming mill product compared to the

weight of the final component. Generally, P/M processing is attractive for large, complex parts with a high buy-to-fly ratio when fabricated by conventional methods. However, because the largest autoclave currently available is 120 cm (48 in.) in diameter by 245 cm (96 in.) in height, an upper size limit exists, unless the process is modified (by subsequent welding, for example) for fabrication of large components (Fig. 40). Table 11 lists the current forging weight, P/M product weight, final part weight, and anticipated potential cost savings for various parts produced by the prealloyed technique. These estimates suggest that cost savings realized by P/M processing compared to forged parts could range between 20 and 50%, depending on the size and complexity of the part and production quantity. Higher volume runs result in higher savings. An additional advantage, which assumes even greater importance with material shortage, is that lead-time can be reduced by 50% or more for P/M parts over equivalent forged parts.

Beryllium P/M Technology

Beryllium, due to its unique mechanical and physical properties, is finding increased use in the nuclear, x-ray, aero-

space, and consumer fields. Interest in beryllium parts was stimulated with the advent of the nuclear energy program. Initial applications were as a moderator and a neutron source in nuclear weapons and reactors.

The next major application was for inertial guidance systems, where a lightweight and dimensionally stable material was needed. Extensive use of beryllium in inertial guidance systems has continued for many years. The ability of beryllium to efficiently absorb heat is a major reason for its use in re-entry systems. Beryllium heat shields have been used on Polaris and Poseidon missiles, as well as Mercury and Gemini capsules. Recently, beryllium has been selected as the heatsink material for brakes on C5A, F-14, B-1B, F-18, and S3A aircraft.

Beryllium is used as a structural material in both the Minuteman missile and the Agena. Both of these applications require superior strength, thermal diffusivity, and rigidity at elevated temperatures in conjunction with light weight.

Other applications include precision instruments; x-ray windows; satellite structures, including antennas, control rods, small rocket nozzles, mechanical fasteners, and high-speed rotating shafts; and consumer products, such as phonograph cartridge components. The wide acceptance of beryllium by the optical industry for use as support structures and mirror substrates is directly related to the density, stiffness, and strength properties of beryllium.

During the last 20 years, efforts have increased to overcome the problem of brittleness associated with beryllium. New grades of unalloyed beryllium with strength and ductility superior to those available in the 1960's have been developed. Breakthroughs in P/M technology have been achieved with the use of high-purity electrolytic source beryllium, the introduction of impact attritioning as a means of be-

Table 11 Typical titanium prealloyed P/M parts

Component	Forged billet kg	Forged billet lb	P/M part kg	P/M part lb	Final part kg	Final part lb	Potential cost savings, %
Colt-crucible ceramic mold process							
F-14 fuselage brace	2.8	6.2	1.1	2.5	0.77	1.7	50
F-18 engine mount support	7.7	17.0	2.5	5.5	0.5	1.1	20
F-18 arrestor hook support fitting	79.4	175.0	24.9	55.0	12.8	28.4	25
F-107 radial compressor impeller	14.5	32.0	2.8	6.2	1.6	3.6	40
F-14 Nacelle frame	142.8	315.0	82.1	181.0	24.1	53.2	50
Fluid die compaction							
AH64 radial compressor impeller	9.5	21.0	2.2	5.0	1.06	2.34	35

ryllium powder production, and the use of cold and hot isostatic pressing techniques for consolidation. Additional information on beryllium processing can be found in the articles "Production of Beryllium Powder" and "Production Sintering Practices for P/M Materials" in this Volume.

Physical Properties

Beryllium—the lightest structural metal, with a density of 1.8477 g/cm^3—has a combination of physical properties that is not found in any other metal or alloy. Beryllium has a high melting point (1283 °C, or 2341 °F), high modulus of elasticity (Young's modulus is 290 GPa, or 42.5 × 10^3 ksi), high specific heat (1.886 kJ/kg · K, or 0.46 Btu/lb · °F at room temperature), and a low coefficient of thermal expansion (average of 10.66 × 10^{-6}/°C in the temperature range of 2 to 38 °C, or 36 to 100 °F). The strength-to-weight ratio for wrought beryllium is higher than titanium. The following properties make beryllium one of the more desirable structural materials in the aerospace field:

Density	1.8477 g/cm^3
Melting point	1283 °C (2341 °F)
Modulus of elasticity	292 Gpa
	(40 to 44 × 10^6 psi)
Specific heat	1.886 kJ/kg · K
	(0.46 Btu/lb · °F)
Thermal conductivity	190 W/m · K
	(104 Btu · ft/ft^2 · h · °F)

Coefficient of thermal expansion	
Temperature range, °C	Linear expansion per °C × 10^{-6}
25-100	11.6
25-300	14.5
25-600	16.5
25-1000	18.4

The following physical characteristics also contribute to the versatility of beryllium and its wide usage in a variety of applications:

- A protective oxide surface film, similar to that of aluminum, is formed in air.
- Oxidation at elevated temperature is not excessive below about 815 °C (1500 °F).
- It conducts electricity (electrical conductivity: 40% IACS).
- It is nonmagnetic.
- It reflects light (optical reflectivity, 50%; ultraviolet reflectivity, 55%; infrared (10.6 μm) reflectivity, 98%).
- It transmits sound (velocity of sound in beryllium is 12 600 m/s, 41 300 ft/s, 2^1/$_2$ times that of steel).
- It is transparent to x-rays.

The low thermal neutron absorption cross section of beryllium (0.009 b) and high neutron scattering cross section (6.9 b)

make it useful as a moderator and a reflector for nuclear reactors. Beryllium is particularly well suited for reactor applications in which weight and space economy is important. Neutron economy and flux density are increased by the use of beryllium in nuclear reactors.

Mechanical Properties

The main factor limiting the extensive use of beryllium in lightweight structures has been its brittleness at temperatures below approximately 200 °C (390 °F). This brittleness can be attributed to the nature of interatomic bonding and the deformation behavior of the close-packed hexagonal crystal, which is anisotropic. Mechanical properties of polycrystalline beryllium greatly depend on grain orientation and are sensitive to grain size. Impurity content is also important because impurities may alter the deformation behavior of single crystals.

Since the late 1960's, efforts have been made to improve the ductility of polycrystalline beryllium by the control of metallurgical process parameters. As a result, the tensile elongation of high-purity powder source material is comparable to many other aerospace alloys. Table 12 summarizes tensile properties for a variety of beryllium product forms. Approaches to achieve ductile beryllium at room temperature have included control of preferred orientation, improved purity, reduction of inclusions (beryllium oxide content, for example), altering inclusion distribution, and reducing grain size.

Preferred Orientation. Hot upsetting or rolling beryllium tends to align the basal planes parallel to the plane of rolling. This alignment of basal planes improves the ductility in the plane of sheet or block, because longitudinal stresses are mostly sustained by prism planes. Stresses resolved across the basal planes also are lower for a given applied load.

Grain Size. Strength of beryllium is related directly to grain size (Fig. 41), which is primarily a function of the particle size and oxide content of the powder used. Historically, attempts to produce fine-grained, high-ductility beryllium by reducing the size of the source powder have met with only moderate success. Al-

Table 12 Typical tensile properties of commercially available beryllium

Material	Test orientation	Ultimate tensile strength MPa	ksi	Yield strength(a) MPa	ksi	Elongation(b), %
Block						
Normal-purity (hot-pressed)						
structural grade(c)	L	370	53	266	38	2.3
	T	390	56	273	39	3.6
Thermal or brake grade	L	294	42	196	28	2.7
	T	322	46	196	28	4.6
High-purity (isostatically pressed)	L	455	65	287	41	3.9
	T	455	65	287	41	4.4
High-oxide instrument grade						
(hot pressed)	L	476	68	406	58	1.5
	T	511	73	413	59	2.7
Fine grain size (isostatically pressed)	L	580	84	407	59	3.7
	T	587	85	407	59	4.2
Sheet (1 to 6.4 mm, or 0.040 to 0.250 in. thick)						
Normal-purity powder	...	531	77	372	54	16
Normal-purity ingot	...	352	51	172	25	7
Extrusions						
Normal-purity powder	...	655-690	95-100	345-518	50-75	8-13
High-purity powder	...	828-655	95-120	345-518	50-75	8-13
Forgings						
Normal-purity	...	483-600	70-87	435-600	63-87	0-4.5
Wire (0.05 to 0.64 mm, or 0.002 to 0.025 in. diameter)						
High-purity ingot	...	966	140	793	115	3

(a) 0.2% offset. (b) Elongation in 50 mm (2 in.), except 250 mm (10 in.) for wire. (c) Structural grade contains about 1.8% beryllium oxide and thermal or brake grade about 0.9%
Source: Ref 7

Fig. 41 Effect of grain size on the elevated temperature tensile properties of vacuum hot pressed beryllium

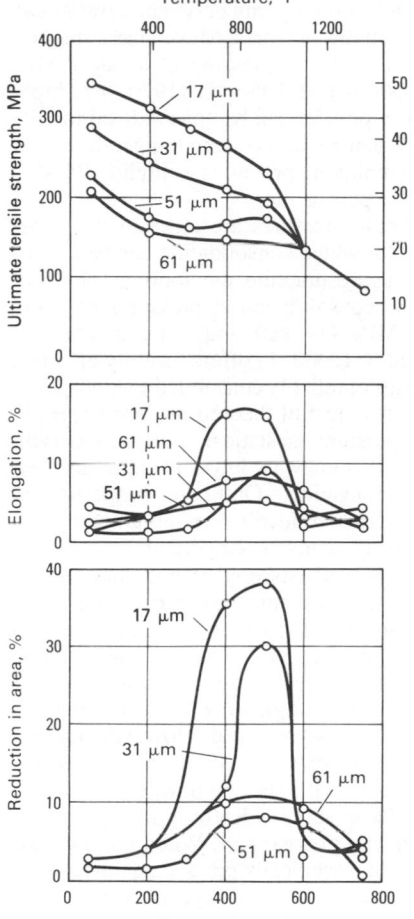

Fig. 42 Impact attrition mill

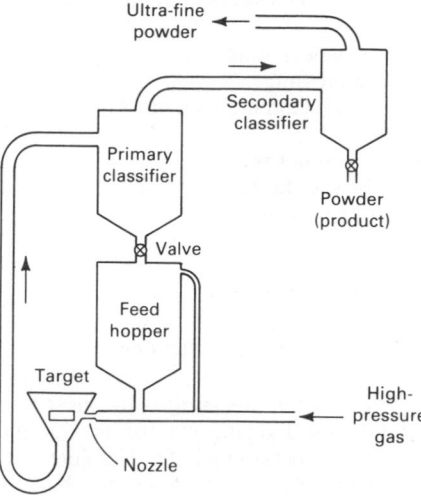

tents) are discarded. High-pressure gas is turned off when the feed hopper is empty. When the system returns to ambient pressure, the valve at the bottom of the primary classifier is opened to deposit the coarse particles into the feed hopper for the next cycle.

Impact attritioning has enabled consistent control of powder composition by reducing impurity contamination and oxidation of powder particles. The process also results in improved powder configuration and morphology, resulting in improved isotropy and cleaner microstructure of the consolidated product.

Improvement in Purity

Impurity elements that affect the behavior of beryllium include aluminum, iron, silicon, magnesium, copper, cobalt, nickel, oxygen, and carbon. Aluminum and iron can cause deleterious effects in beryllium. When present at the grain boundaries, aluminum causes reduced hot ductility in components for high-temperature applications.

Iron forms an intermetallic compound, $FeBe_{11}$, which results in precipitation hardening. Yield point phenomena observed in beryllium have been attributed to the pinning of dislocations by $FeBe_{11}$ precipitates. If the pinning of dislocations persists until the ultimate strength is reached, the metal fractures with little plastic deformation. However, moderate yield points that peak before the fracture stress may have little effect on the mechanical behavior of beryllium and may be of considerable advantage where elastic limit is a design factor. Aluminum and iron, when present together in beryllium, can form an $FeAlBe_4$ compound. The

though the grain size is reduced, conventional powder manufacturing techniques such as disk mill attritioning and ball milling increase oxide content rapidly as particle size of the powder decreases. This has an embrittling effect, which often nullifies the effect of reduced grain size.

The impact attrition system, which is illustrated in Fig. 42, currently is used for production of beryllium powder. In this system, coarse powder from the feed hopper (Fig. 42) is fed into the gas stream, and the gas-borne powder is carried through the nozzle, where the velocity of the gas and powder mixture increases. The powder impacts the beryllium target and is comminuted. Gas-borne powder is transported to the primary classifier, where coarse particles drop out, and fines are carried to the secondary classified. In this stage, the coarse particles (the product) drop out, and ultrafines (powders with high oxides and intermetallic contaminant con-

proper balancing of iron and aluminum content can eliminate the embrittling effects of aluminum without resorting to ultrahigh-purity materials.

Silicon acts as a sintering aid in the consolidation of beryllium powder. In practice, the hot pressing of beryllium powder with low silicon content (<100 ppm) to full density is difficult because of the pressure tolerable with graphite dies. The high-pressure capability of hot isostatic pressing often is used for the consolidation of low-silicon, high-purity electrolytic powders. Alternatively, the material is doped with silicon either by elemental additions at the vacuum melting stage or by adding trichlorosilane at the vacuum hot pressing stage.

When present at grain boundaries, silicon, along with other low-melting constituents such as aluminum and magnesium, also affects the morphology, size, and distribution of beryllium oxide. In general, these elements cause agglomeration and growth of beryllium oxide particles and also affect recrystallization, grain growth, and creep strength of beryllium. Copper, cobalt, and nickel exhibit extensive solubility in beryllium and are excellent alloying additions. Copper strengthens beryllium, without reducing its tensile elongation.

Among the nonmetallic impurities, only carbon and oxygen are present to any appreciable extent. Both have minimal solubility in beryllium. Carbon occurs as the cubic refractory compound beryllium carbide, which usually is found within the grains. This carbide accelerates the corrosion of beryllium, as it slowly hydrolyzes in the presence of moisture.

Oxygen forms stable beryllium oxide, which is primarily a grain boundary constituent. However, it also may be found in the matrix. Typical concentrations of beryllium oxide in beryllium range from 0.7% to about 4% in commercial products.

Beryllium oxide, because it is a major impurity constituent in beryllium, greatly affects the mechanical properties of beryllium. Figure 43 shows the effect of decreasing beryllium oxide content on the tensile elongation of high-purity powder beryllium. By electrorefining beryllium, impurities can be reduced. Metallic impurities also can be carefully controlled during the chemical processing of beryllium ores.

Powder Production and Consolidation

The production of beryllium powder begins with extraction of beryllium from ore

Fig. 43 Effect of beryllium oxide content on the tensile elongation of high-purity powder source beryllium

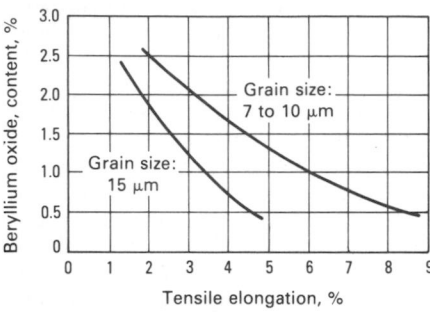

Table 13 Nominal compositions of nuclear and structural grades of beryllium powders

Chemical composition, %	Nuclear grade	Structural grade
Beryllium assay(a) 99.0		98.0-98.5
Beryllium oxide(b) 0.9		1.2-2.0
Aluminum(b) 0.075		0.14-0.16
Boron(b) 0.00020		...
Cadmium(b) 0.0002		...
Calcium(b) 0.01		...
Carbon(b) 0.10		0.15
Chromium(b) 0.01		...
Cobalt(b) 0.0005		...
Copper(b) 0.015		...
Iron(b) 0.075		0.15-0.18
Lead(b) 0.002		...
Lithium(b) 0.0003		...
Magnesium(b) 0.05		0.08
Manganese(b) 0.012		...
Molybdenum(b) 0.002		...
Nickel(b) 0.02		...
Nitrogen(b) 0.02		...
Silicon(b) 0.06		0.08
Silver(b) 0.0010		...
Other metal impurities(b)	...	0.04

(a) Minimum. (b) Maximum

by chemical, thermal, and mechanical processing. Extraction produces beryllium "pebble," a relatively impure product containing magnesium fluoride and other impurities. These pebbles are vacuum remelted and cast into ingots, which exhibit a coarse grain size, low ductility, and low resistance to fracture. The vacuum cast ingots are broken up into chips by lathe turning. Chips are subsequently comminuted by impact attritioning (Fig. 42) to a particle size suitable for P/M fabrication. Table 13 gives the nominal compositions for nuclear and structural grades of beryllium powders. Production of other grades of beryllium is possible by varying the powder particle size, chemical composition, and processing pressure and temperature. Beryllium is one of the few metals that is fabricated exclusively by P/M tech-

niques. Beryllium powders are consolidated to near-theoretical density by:

- Pressureless sintering
- Cold pressing
- High-vacuum/high-temperature sintering
- Vacuum hot pressing
- Cold isostatic pressing followed by vacuum hot pressing
- Cold isostatic pressing followed by pressureless sintering
- Cold isostatic pressing followed by hot isostatic pressing
- Direct hot isostatic pressing

Vacuum hot pressing is the most frequently used technique for producing commercial beryllium billet. Graphite dies are commonly used because they have a low coefficient of thermal expansion, which simplifies removal of the beryllium block after cooling. Graphite is available in sufficiently large sizes to produce vacuum hot pressings at more than 560 mm (22 in.) in diameter.

Pressings that are too large in diameter for graphite dies are produced in steel dies. Removal of the pressing is more difficult in steel dies compacting, and creep deformation of the die limits the allowable compacting pressure, which can result in reduced density of the pressing. The use of nickel-based superalloy dies has demonstrated significant improvements over steel. Filament-wound dies made of graphite fibers provide improved hoop strength and size capability, but are more costly.

Beryllium powder is vibratory loaded into the dies to achieve a density of about 55%. The die assembly is then placed in a vacuum hot pressing chamber. Vacuum, heat, and pressure are applied concurrently. Typical pressing temperatures range from 1000 to 1100 °C (1830 to 2000 °F), and pressure ranges from 3.5 to 14 MPa (500 to 2000 psi) with vacuum conditions below 500 torr, depending on powder characteristics and size of the compact. High purity and fine particle size powder require much higher pressures.

Isostatic pressing offers advantages over vacuum hot pressing in that higher pressures are available and the limitations imposed by graphite dies are eliminated. The 3.5 to 14 MPa (500 to 2000 psi) pressure normally considered the maximum for graphite dies is marginal for some grades of beryllium and inadequate for high-purity grades. Hot isostatic pressing is generally done at 103 MPa (15 ksi) at temperatures varying from 760 to 1040 °C (1400 to 1900 °F).

High-purity beryllium powders lack sufficient concentration of impurity elements such as aluminum and silicon that act as sintering aids in the vacuum hot pressing of normal-purity powders. Although normal-purity beryllium powder can be vacuum hot pressed to nearly theoretical density by pressing at about 7 MPa (1000 psi) at 1050 °C (1920 °F), high-purity powder can be consolidated to only 98% density at the same temperature only by employing pressures as high as 24 MPa (3500 psi) or more.

These pressures are close to the upper limit at which consolidation can be carried out using graphite die tooling. Hot isostatic consolidation at pressures of about 96 MPa (14 ksi) and a temperature of 1100 °C (2000 °F) offers the only effective means of reliably consolidating high-purity powders to full theoretical density. High-temperature isostatically pressed beryllium has somewhat lower strength and better elongation. Optimum properties of strength and ductility are obtained by iso-pressing at lower temperatures.

Improved isotropy of both microstructure and mechanical properties is a major advantage of isostatic pressing. Because pressure is applied from all sides, shrinkage is uniform in all directions, and directionality originating from the pressing operation is prevented. However, there is some preferential alignment of powder particles in the starting powder bed, which results in some anisotropy in the hot isostatically pressed part. When hollow parts are hot isostatically pressed against a mandrel (usually made of copper), there is more shrinkage in the thickness direction, because friction between the mandrel and the part impedes free movement in the axial direction. Thus, the basal planes become aligned in the circumferential direction.

Isostatic consolidation of high-purity powder produces improved physical and mechanical properties, including fine grain size, well-distributed oxide content, clean microstructure, good tensile and elongation properties, a smooth stress-strain curve that is free of yield point, a high degree of isotropy, good compression creep properties, good high-temperature properties, and no grain growth below 1260 °C (2300 °F).

Wrought Product Forms

Wrought forms of beryllium with optimum properties are produced from vacuum hot pressed P/M billets by conventional hot working techniques. Wrought products exhibit improved mechanical

Fig. 44 Beryllium thrust tube assembly for the Japanese CS-2 Satellite
Courtesy of Brush Wellman, Inc.

Fig. 45 Hughes Satellite, which incorporates beryllium structural members
Courtesy of Brush Wellman, Inc.

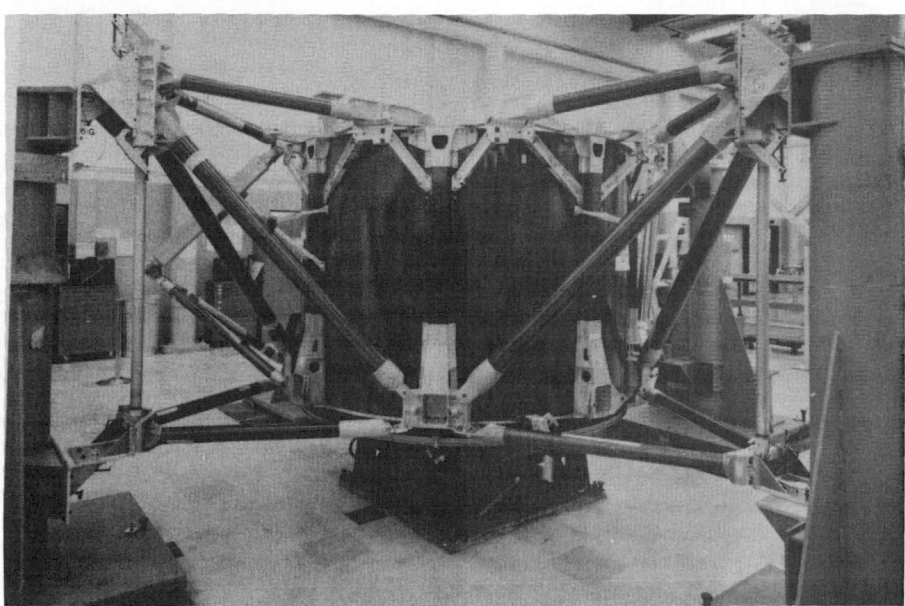

properties in the predominant direction of metal flow. Mechanical properties of wrought products depend greatly on crystallographic orientation. Biaxial or triaxial deformation procedures frequently are used to produce parts with balanced directional properties.

Sheet, plate, and foil are the most commonly produced wrought forms of beryllium. These flat rolled products are produced by encasing beryllium rolling blocks in steel jackets, followed by cross rolling at a moderate temperature. Biaxial deformation is accomplished by alternating the direction of the rolling pack through the rolling mill. This procedure produces flat rolled products with uniform biaxial properties. See Table 12 for typical tensile properties of beryllium sheet. The standard condition that sheet and plate are supplied in is hot rolled, stress relieved, ground, and pickled.

Beryllium foil is available in thicknesses from 0.025 to 0.5 mm (0.001 to 0.019 in.) and panel sizes up to 125 by 300 mm (5 by 12 in.). Beryllium foil is also available in disk form in diameters up to 200 mm (8 in.) for x-ray window applications.

Beryllium extrusions are available with consistent mechanical properties. Sections are supplied to dimensions that are within commercial tolerances. Mechanical properties of extrusions are superior to the mechanical properties of hot pressed block in the direction of metal flow. Directional properties are produced in varying degree, as a function of crystallographic orientation.

Input billets for extrusions usually are machined from hot pressed block. Vacuum cast ingots and unsintered powder can be extruded for custom applications. Extrusion billets are jacketed in low-carbon steel cans that have shaped nose plugs and are extruded through a steel die between

900 to 1065 °C (1650 to 1950 °F). Warm extrusions are produced at about 425 °C (800 °F) when the volume of the product is large enough to justify the special tooling. Steel jackets are chemically removed from the extrusions.

Rod, tubing, and structural shapes are also available. Rod is produced in sizes from 9.5 to 135 mm (0.375 to 5.250 in.) in diameter. Tubing is available from 6.4 mm (1/4 in.) outside diameter by 10 mm (0.040 in.) wall thickness up to 150 mm (6 in.) outside diameter by 3.20 mm (0.125 in.) wall thickness. A wide variety of structural and special shapes are also available. Current applications include structural shapes for aerospace applications, finned tubing for nuclear reactors, bar stock for forging and finish machining, fastener stock, rotating shafts, fuel element cladding, and draw stock for wire. Table 12 lists typical tensile properties of normal-purity and high-purity beryllium extrusions.

Forgings. The use of beryllium hot pressed billets is limited in some structural applications, primarily due to its relatively low ductility. Forging improves the strength and ductility of beryllium by raising the yield strength of a high-strength grade from 350 to 550 MPa (50 to 80 ksi), with improved ductility in selected directions. Forging also minimizes metal losses that occur in machining parts from solid hot pressed block.

Input for the forging operation usually is hot pressed billet or extruded material.

Powder that has been canned also can be used. Forging is performed, either canned or bare, in a die design that allows compressive forces to act on the surface of the billet, thus minimizing or eliminating forging tears. Back extrusion and upsetting operations also are used. Certain minimum deformation processes can be used successfully without external compressive restraint. Typical tensile properties for normal-purity beryllium forgings are given in Table 12.

Applications of Beryllium

Aerospace Applications. The low density and high modulus of elasticity of beryllium make it an ideal choice as a structural member in aerospace components. The development of higher ductility grades of beryllium in various wrought forms, including sheet metal, has increased the potential application of beryllium in the aerospace field. Figure 44 illustrates a beryllium thrust tube (cone) assembly for the Japanese CS-2 satellite. Tests to determine the structural load-carrying capacity indicated that this structure has carried uniform loads at and/or above yield strength limits. Post-test analysis and inspection of the beryllium assembly indicated that it was flight qualified; loads were increased to over 160% of the theoretical limit. Figure 45 illustrates a beryllium structural member in an aerospace component. Both the body and

Fig. 46 Beryllium optical bench for the McDonnell Douglas mast-mounted helicopter sight
Courtesy of Brush Wellman, Inc.

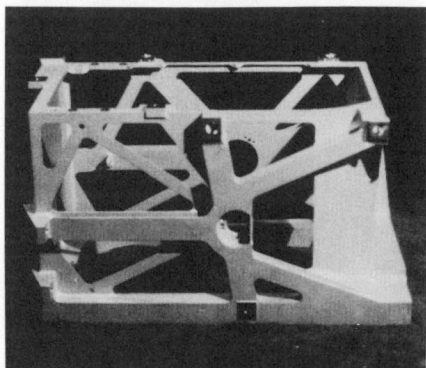

Fig. 47 Modifications being made to the Bell OH-58 helicopter
(a) Note the location of the mast-mounted sight, which incorporates the beryllium optical bench shown in Fig. 46. Source: Ref 8. (b) Helicopter in flight. Courtesy of Brush Wellman, Inc.

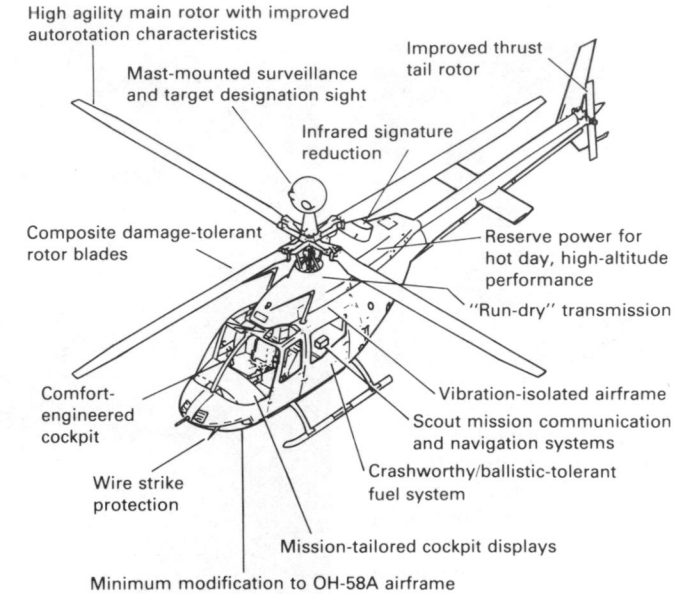

High agility main rotor with improved autorotation characteristics

Mast-mounted surveillance and target designation sight

Improved thrust tail rotor

Infrared signature reduction

Composite damage-tolerant rotor blades

Reserve power for hot day, high-altitude performance

"Run-dry" transmission

Comfort-engineered cockpit

Vibration-isolated airframe

Scout mission communication and navigation systems

Wire strike protection

Crashworthy/ballistic-tolerant fuel system

Mission-tailored cockpit displays

Minimum modification to OH-58A airframe

(a)

(b)

support struts in this satellite are made of beryllium.

The low density, high elastic modulus, and dimensional stability of beryllium are exploited in the optical bench for the McDonnell Douglas mast-mounted helicopter sight shown in Fig. 46. Laser, television, and infrared sensors are mounted to the bench, which must maintain precise alignment for proper operation of the target detector and designator system. Figure 47 shows the location of the mast-mounted sight in relation to the other modified components on the Bell OH-58 helicopter.

Another recent development is the Infrared Astronomy Satellite (IRAS), which was launched in January 1983. Its success is partly due to the precision and design of the beryllium mirror and support system shown in Fig. 48 and 49. The high infrared reflectivity, low weight, low inertia, and high stiffness of the mirror and support member are all primary design considerations for telescope optics. In addition, the mirror shown in Fig. 48 is cooled to liquid helium temperature (4.2 K) to eliminate infrared "noise" from the telescope. The high thermal conductivity of beryllium is therefore a decided advantage.

To date, the only beryllium alloy series that has shown promise as a candidate for critical stiffness members of spacecraft and military airplanes is the beryllium-aluminum alloy family. These alloys, which contain 24 to 36 wt% Al, have been under development since the early 1960's.

Lockalloy, the most popular beryllium-aluminum alloy, generally contains 38 wt% Al and is produced as extrusions and sheet that is hot rolled from extruded bar stock. A flowchart of the manufacture of Lockalloy sheet and plate is shown in Fig. 50. Aerospace structural parts are not, how-

Fig. 48 Beryllium mirror used in the Infrared Astronomy Satellite
Courtesy of Brush Wellman, Inc.

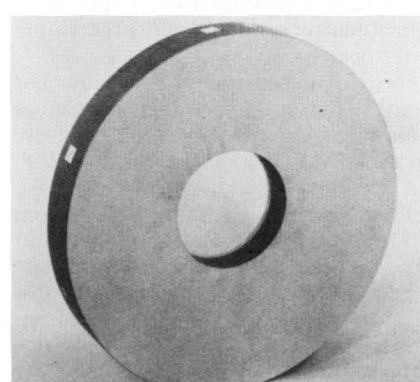

Fig. 49 Beryllium support system for the Infrared Astronomy Satellite telescope optics
Courtesy of Brush Wellman, Inc.

Fig. 50 Lockalloy sheet and plate manufacturing sequence

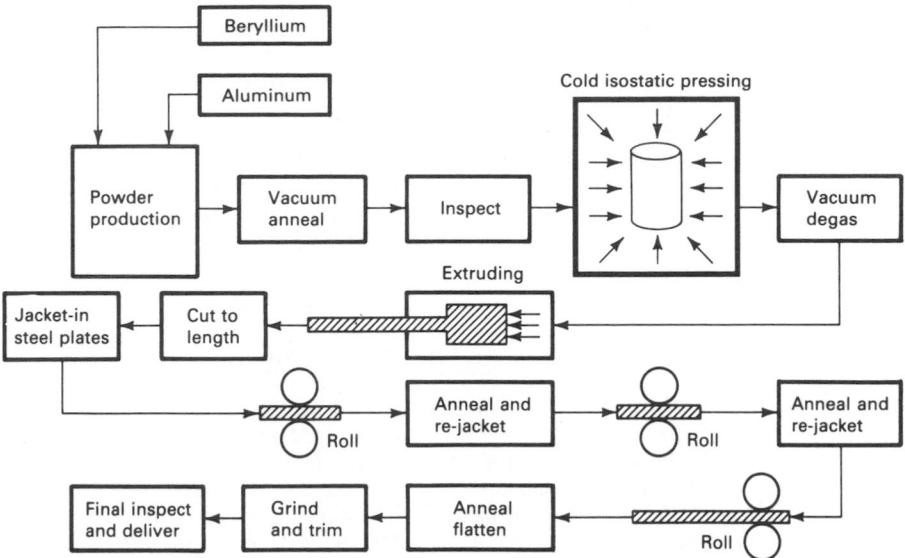

Fig. 51 Brazed beryllium x-ray window assembly
Courtesy of Brush Wellman, Inc.

Fig. 52 Beryllium Advanced Inertial Reference Sphere for the Intercontinental Ballistic Missile

Fig. 53 Beryllium gyroscope (center) and related components
Courtesy of Brush Wellman, Inc.

ever, currently being produced from these alloys.

X-Ray and Nuclear Applications. Because it has excellent transparency to soft x-rays and other radiation such as gamma rays, beryllium is used in x-ray tube windows. Because of its low atomic weight, beryllium passes x-rays 17 times better than an equivalent thickness of aluminum. Beryllium x-ray windows allow the use of long-wave x-rays that have greater intensity. Most of these require application disks that are thicker than 0.25 mm (0.01 in.) and are cut from sheet or machined from vacuum hot pressed blocks.

The stiffness of beryllium and its transparency to x-rays is combined in medical radiation detection devices such as computerized axial tomographic scanners and other related equipment with pressurized detection chambers. A stiff beryllium window resists the internal gas pressure with minimum deflection. A brazed x-ray window assembly is shown in Fig. 51. The capability of rolling beryllium to a thin foil permits the transparency of beryllium to electromagnetic radiation to be used to the greatest advantage.

The growth of the nuclear energy field has provided new applications for beryllium components. Beryllium continues to be used as a reflector material in research and fast reactors. Beryllium has also been used to some degree in power reactors, for moderators, reflectors, and fuel cladding material.

Instrument Applications. Beryllium is used in precision-machined components in inertial guidance systems. Since its initial use in the mid-1950's, beryllium has come to be accepted as the base line structural material for precision navigational instruments. It is now used in precision navigational aides for aircraft, spacecraft, missiles, ships, and submarines.

The Advanced Inertial Reference Sphere (AIRS) for the Peacekeeper Intercontinental Ballistic Missile (ICBM) is shown in Fig. 52. This beryllium component provides the Peacekeeper (also referred to as the MX missile) with improved accuracy over previous missile systems. The high modulus, high precision elastic limit, and dimensional stability under high G-forces make beryllium the material of choice for this application. Figure 53 shows a gyroscope and related components made of beryllium, which are used for a tactical aircraft guidance and inertial reference system.

High-Temperature Applications. Beryllium has the highest specific heat of all structural materials, good thermal conductivity, relatively high melting point, and resistance to oxidation up to about 800 °C (1470 °F). It is a highly efficient heatsink, heat shield, and aircraft and automobile brake material. Beryllium also is used for making rocket engine nozzles and switchgear components. Thermal-grade beryllium has been used successfully for load-carrying disc brakes in military cargo and fighter aircraft and for making thrust chambers of rocket engines.

Consumer Applications. High-frequency trackability of phonograph cartridges has been a major concern of the audio industry, particularly with the advent of digitally mastered and direct-to-disc recordings. To improve the accuracy of phonograph cartridges, a method of low-ering the effective tip mass had to be developed. Although all the moving parts of a stylus assembly contribute to its effective mass, the largest single contributor usually is the cantilever, or shank.

Beryllium, which has the highest stiffness-to-mass ratio of any metal, is theoretically an ideal candidate for this com-

Fig. 54 Cross sections of beryllium stylus shanks

(a) Solid rod, 250 μm (10 mil) in diameter. (b) Thick-walled tube, 300 μm (12 mil) in diameter, 50 μm (2 mil) wall thickness. (c) Microwall/Be, 450 μm (18 mil) in diameter, 12.5 μm (0.5 mil) wall thickness

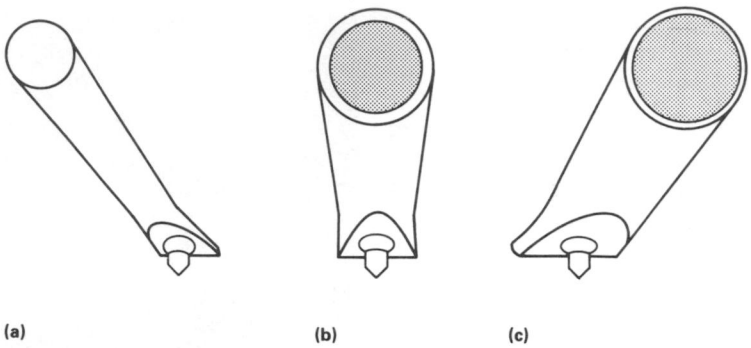

(a)   (b)   (c)

Fig. 55 V15 type V phonograph cartridge with thin-walled beryllium shank

Courtesy of Shure Bros., Inc.

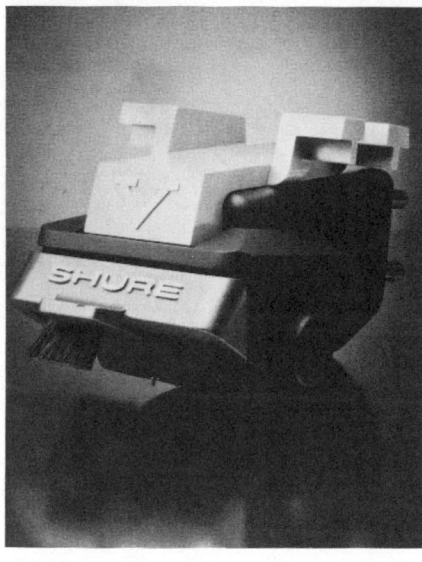

ponent. Until recently, however, difficulties in working the material prevented any performance advantage. While previous shank designs used beryllium, they were solid rods that were inferior to aluminum tubing.

This drawback, however, has been eliminated with the development of thin-walled beryllium tubing for the shank or cantilever material. The wall thickness of the tubing, which is made by rolling beryllium foil, is 12.5 μm (0.5 mil). As a result, the effective mass contribution of the beryllium shank is 0.096 mg, less than half of the 0.213 mg contribution of a telescoped aluminum shank. However, total stiffness has been increased. Stiffness is necessary to ensure that the cartridge transducer accurately follows the motion of the stylus tip. Linear density of the material determines its contribution to effective mass, an important factor in achieving high-frequency trackability.

Thin-walled beryllium tube has a clear advantage in both respects when compared to other geometries. Figure 54 shows

cross sections of typical beryllium shank geometries. Figure 55 is the improved V15 type V phonograph cartridge with a beryllium stylus shank. Table 14 compares the properties of typical beryllium shank geometries.

REFERENCES

1. Eylon, D., Field, M., Froes, F.H., and Eichelman, G.E., Manufacturing Cost-Affordable High Performance Components for Advanced Air Force Systems, *SAMPE Quart.*, Vol 12 (No. 3), 1981, p 19-25
2. Tupper, N.G., Elbaum, J.K., and Burte, H.M., Opportunities for Cost-Affordable Titanium Aerospace Structures, *J. Metals*, Vol 30, 1978, p 7-13
3. Froes, F.H., Eylon, D., Eichelman, G.E., and Burte, H.M., Developments in Titanium Powder Metallurgy, *J. Metals*, Vol 32 (No. 2), 1980, p 47-54
4. Froes, F.H. and Smugeresky, J.E., Ed., *Titanium Powder Metallurgy*, Proc. AIME Symposium, Las Vegas, Feb 1980
5. Parsons, L., Bruce, J., Lane, J., and Froes, F.H., Titanium Powder Metallurgy Comes of Age, *Met. Prog.*, 1984, to be published
6. Work in progress, Air Force Materials Laboratory, 1983
7. Floyd, D.R. and Lowe, J.N., Ed., *Beryllium Science and Technology*, Vol 2, Plenum Press, New York, 1979, p 321
8. *Aviation Week Space Technol.*, March 28, 1983, p 48

SELECTED REFERENCES

Aluminum

- Cebulak, W.S., Johnson, E.W., and Markus, H., High Strength Aluminum P/M Mill Products, *Int. J. Powder Metall. Powder Technol.*, Vol 12 (No. 4), Oct 1976
- Lyle, J.P. and Cebulak, W.S., Powder Metallurgy Approach for Control of Microstructure and Properties in High Strength Aluminum Alloys, *Metall. Trans. A*, Vol 6A, April 1975, p 685
- Bruns, M., "Aluminum P/M Processing Data for Specific Aluminum P/M Parts," personal communication, New Industrial Techniques, Inc., Coral Springs, FL, Handy & Harman Co., 1983
- "Wrought P/M Alloys—A Balance of Properties for Demanding Aerospace Applications," Alcoa Headquarters Office, Pittsburgh, A03-14363
- Roesel, R.E., Aluminum Powder Metallurgy—Production, Properties and

Table 14 Property comparisons of typical beryllium shank geometries

Property	Shank geometry			
	450 μm (18 mil) OD, 12.5 μm (½ mil) wall	350 μm (14 mil) OD, 25 μm (1 mil) wall	300 μm (12 mil) OD, 50 μm (2 mil) wall	250 μm (10 mil) diam rod
Relative linear density (proportional to effective mass)	0.35	0.52	0.80	1.00
Relative stiffness	2.13	1.74	1.64	1.00
Stiffness to density ratio (resonance frequency)	6.25	3.44	2.06	1.00

Potential, *Powder Metall. Int.,* Vol 5 (No. 2), 1973

- Aluminum Company of America, "The Properties and Performance of Aluminum P/M Parts," F38-13347
- Dudas, J.H. and Thompson, C.B., "Improved Sintering Procedures For Aluminum P/M Parts," Aluminum Company of America, F38-12964, 1970
- Buchovecky, K.E. and Rearick, M.R., Aluminum P/M Forgings, *Met. Prog.,* Feb 1982, reprinted F38-21403
- Dudas, J.H. and Brondyke, K.J., Aluminum P/M Parts—Their Properties and Performance, Technical Paper No. 70041, Society of Automotive Engineers, New York, F38-12903
- Cebulak, W.S., "Program to Develop High-Strength Aluminum Powder Metallurgy Products—Phase IVA—Second Quarterly Report," Alcoa Research Laboratories, Contract DAAA25-72-CO593 (Frankford Arsenal), 1973
- Cebulak, W.S. and Truax, D.J., "Program to Develop High-Strength Aluminum Powder Metallurgy Products—Phase III—Scale-Up A," Alcoa Research Laboratories, Contract DAAA25-70-CO358 (Frankford Arsenal), 1972
- Lyle, J.P., Jr., Cebulak, W.S., and Buchovecky, K.E., *Progress in Powder Metallurgy 1972,* Vol 28, Metal Powder Industries Federation, New York, 1972, p 93-113
- Lyle, J.P., Jr. and Cebulak, W.S., *Met. Eng. Quart.,* Vol 14 (No. 1), 1974, p 52-63
- Lyle, J.P., Jr. and Cebulak, W.S., *Powder Metallurgy for High Performance Applications,* Syracuse University Press, 1972, p 231-254
- Roberts, S.G., "Powder Fabrication of Aluminum Alloys," WADC Report No. 56-481, April 1957
- Roberts, S.G., "Research Study for Development of Aluminum Base Alloys by Powder Metallurgy Techniques," Summary Project Report No. MSPR61-69, Contract No. DA-04-200-507-ORD-886, 1961
- Towner, R.J., "Development of Aluminum Base Alloys," Annual Progress Report for the Period September 29, 1961, to September 30, 1962, Project No. 593-32-004, Contract No. DA-36-034-ORD-3559RD, 1962
- Haarr, A.P., "Development of Aluminum Base Alloys—Section III," Contract No. DA-36-034-ORD-3559RD, 1966
- Dean, W.A. and Spear, R.E., *Strengthening Mechanisms,* Syracuse University Press, 1966, p 268-270

Titanium

- Schwertz, J.H., Chandhok, V.K., Petersen, V.C., and Thompson, V.R., "Consolidation of Titanium Powder to Near-Net Shapes," AFML-TR-78-41, 1978
- Paez, C., "Built-Up Low-Cost Advanced Titanium Structures (BLATS)," Contract F33615-77-C-3109, 5th Quarterly Progress Report, Grumman, Feb 1979
- Roberts, P.R. and Loewenstein, P., "Titanium Alloy Powders Made by the Rotating Electrode Process," Nuclear Metals, Inc.
- Fleck, J.N. and Clark, L.P., Advancements in Titanium Powder Processing, *SAMPE Quart.,* Oct 1976, p 10-20
- Dulis, E.J., Chandhok, V.K., Froes, F.H., and Clark, L.P., "Manufacturing Procedures for the Production of Large Titanium PM Shapes: Current Status," 10th National Society for the Advancement of Material and Process Engineering Technical Conference, Kiamesha Lake, NY, Oct 1978
- Barker, C.S. and Nagy, P., "Application of HIP Ti-6Al-4V Powder to the Radial Compressor for the Cruise Missile Engine," Williams Research Corp.
- Chanani, G.R., Highberger, W.T., Kelto, C.A., and Petersen, V.C., "Application of Titanium Powder Metallurgy for Manufacture of a Large and Complex Naval Aircraft Component," Northrop Corp.
- Petersen, V.C. and Chandhok, V.K., "Manufacturing Process for the Hot Isostatic Pressing of Large Titanium PM Shapes," AFML-IR-184-7T
- Petersen, V.C., Chandhok, V.K., and Kelto, C.A., Hot Isostatic Pressing of Large Titanium Shapes, in *Powder Metallurgy of Titanium Alloys,* Froes, F.H. and Smugeresky, J.E., Ed., Conference Proceedings, American Institute of Mechanical Engineers 109th Meeting, Las Vegas, 1980
- Anderson, P.J., Svoyatytsky, V.M., Froes, F.H., and Mahajan, Y., Fracture Behavior of Blended Elemental P/M Titanium Alloys, in *Modern Developments in Powder Metallurgy,* Vol 13, Metal Powder Industries Federation, Princeton, NJ, 1981
- Moll, J.H., Petersen, V.C., and Dulis, E.J., Powder Metallurgy Parts for Aerospace Applications, in *Powder Metallurgy: Applications, Advantages and Limitations,* American Society for Metals, 1983
- Lowenstein, P., Specialty Powders by the Rotating Electrode Process, in

Progress in Powder Metallurgy, Vol 37, Metal Powder Industries Federation, Princeton, NJ, 1982

- Petersen, V.C., Chandhok, V.K., and Kelto, C.A., Hot Isostatic Pressing of Large Titanium Shapes, in *Powder Metallurgy of Titanium Alloys,* American Institute of Mechanical Engineers, New York, 1980
- Dulis, E.J., Moll, J.H., Chandhok, V.K., and Hebeisen, J.C., Progress in P/M Superalloy and Titanium for Aircraft Applications, in *The 1980's —Payoff Decade for Advanced Materials,* Vol 25, Society for the Advancement of Material and Process Engineering, 1980
- Dulis, E.J. and Moll, J.H., Properties of HIP Consolidated Superalloy and Titanium Alloys, *Rapid Solidification Processing,* Claitor's, Baton Rouge, 1978
- Friedman, G., The Powder Metallurgy of High-Strength Titanium Alloys, *Met. Prog.,* March 1975
- Barker, C.S. and Nagy, P., Application of HIP Titanium Ti-6Al-4V Powder to the Radial Compressor for the Cruise Missile Engine, *Powder Metallurgy of Titanium Alloys,* American Institute of Mechanical Engineers, New York, 1980
- Schwertz, J.H., Chandhok, V.K., Petersen, V.C., and Thompson, V.R., "Consolidation of Titanium Powder to Near-Net Shapes," AFML-TR-78-41, May 1978
- Chanani, G.R. *et al.,* Application of Titanium P/M for Manufacture of a Large and Complex Naval Aircraft Component, in *Powder Metallurgy of Titanium Alloys,* American Institute of Mechanical Engineers, New York, 1980
- Witt, R.H. and Magnuson, J., "Flight Qualification of Titanium F-14A Airframe Components Manufactured by Hot Isostatic Pressing," Final Report, Contract N00019-76-C-0143, Grumman Aerospace, June 1977
- Dawson, D.B. and Ulitchny, G., Characterization of the Tensile Properties of Hot Isostatically Pressed Ti-6Al-6V-2Sn, *Powder Metallurgy of Titanium Alloys,* American Institute of Mechanical Engineers, New York, 1980
- Mahajan, Y., Eylon, D., Bacon, R., and Froes, F.H., Microstructure Property Correlation on Cold Pressed and Sintered Elemental Ti-6Al-4V Powder Compacts, in *Powder Metallurgy of Titanium Alloys,* American Institute of Mechanical Engineers, New York, 1980
- Boyer, R.R., Magnuson, J.E., and Tripp, J.W., Characterization of Pressed and Sintered Ti-6Al-4V Powders, in

Powder Metallurgy of Titanium Alloys, American Institute of Mechanical Engineers, New York, 1980

- Eloff, P.C. *et al.*, Properties of P/M Titanium Alloys Produced from Elemental Blends: The Current Status of Development, in *Progress in Powder Metallurgy 1981*, Metal Powder Industries Federation, Princeton, NJ, 1982
- Witt, R.H. and Highberger, W.T., Hot-Isostatic Pressing of Near-Net Titanium Structural Parts, in *Powder Metallurgy of Titanium Alloys*, American Institute of Mechanical Engineers, New York, 1980
- Anderson, P.J. and Eloff, P.C., Development of Higher Performance Blended Elemental Powder Metallurgy Titanium Alloys, in *Powder Metallurgy of Titanium Alloys*, American Institute of Mechanical Engineers, New York, 1980

Beryllium

- Floyd, D.R. and Lowe, J.N., Ed., *Beryllium Science and Technology*, Plenum Press, New York, 1979
- The Metallurgy and Applications of Beryllium, *Trans. Ind. Inst. Met.*, Vol 35 (No. 2), April 1982
- Proceedings of the International Conference on Metallurgy of Beryllium, Oct 1961, Institute of Metals, Chapman and Hall, London, 1963
- Schetky, L.M. and Johnson, H.A., Ed., *Beryllium Technology*, Gordon and Breach Science, New York, 1966
- Conference Internationale sur la Metallurgie du Beryllium, Grenoble, Societe Francaise de Metallurgie Presses Universitaires de France, Paris, May 1965
- Proceedings of Beryllium Conference, March 1970, Arlington, VA, National Materials Advisory Board, Publication NMAB-272, National Academy of Sciences—National Academy of Engineering, Washington, DC, 1970
- Beryllium 1977, Fourth International Conference on Beryllium, The Metals Society, London, Oct 1977
- Saxton, H.J. and London, G.J., *Beryllium Science and Technology*, Webster, D. and London, G.J., Ed., Vol 1, Plenum Press, New York, 1979, p 115-144
- Pinto, N.P., Denny, J.P., and London, G.J., *Light Metals 1974*, Forberg, H., Ed., Vol 2, American Institute of Mechanical Engineers, New York, 1974, p 489-511
- Hanes, H.D., Beryllium Processing—The Foundation of Structural Powder Metallurgy, Beryllium 1977, Fourth International Conference on Beryllium, The Metals Society, London, Supplementary Paper, 1977
- Channon, S.L., Confidence in Beryllium, Beryllium 1977, Fourth International Conference on Beryllium, The Metals Society, London, Paper 48, 1977
- Fullerton-Batten, R.C. and Hawk, J.A., A Review of Present and Future Applications of Beryllium, Beryllium 1977, Fourth International Conference on Beryllium, The Metals Society, London, Paper 49, 1977
- Zenczak, S., Beryllium in Aircraft Brakes, Beryllium 1977, Fourth International Conference on Beryllium, The Metals Society, London, Paper 50, 1977
- Grant, L.A., Successful Application of Fabricated Beryllium Structures on Satellites, Beryllium 1977, Fourth International Conference on Beryllium, The Metals Society, London, Paper 58, 1977
- Hathaway, R.G.W., Fabrication Methods for Beryllium Space Craft Components, Beryllium 1977, Fourth International Conference on Beryllium, The Metals Society, London, Paper 60, 1977
- Terry, E.L. and Ely, W.E., Applications of Beryllium Brake Discs on the Lockheed C-5A Aircraft, Proceedings of Beryllium Conference, March 1970, Arlington, VA, National Materials Advisory Board, Publication NMAB-272, National Academy of Sciences—National Academy of Engineering, Washington, DC, 1970, p 171
- Hausner, H.H., Beryllium—Its Metallurgy and Properties, University of California Press, Berkeley, 1965
- White, D.W. and Burke, J.E., *The Metal Beryllium*, American Society for Metals, 1955, p 49-62
- Kaufmann, D.F., Lerine, E.D., Pickett, J.J., and Aronin, L.R., The Effect of Zone Refining on the Purity and Mechanical Properties of Beryllium Single Crystals, Atomic Energy Commission Conference 170, Gatlinburg, TN, 1963, p 69-89
- Grant, L.A. and Kamper, L.F., Beryllium Fabrication Techniques and Their Related Applications, 20th National Society for the Advancement of Material and Process Engineering, Symposium and Exhibition, San Diego, 1975

P/M High-Temperature Materials

By John B. Lambert
Vice President and Technical Director
Fansteel, Inc.
and
Robert E. Droegkamp
Manager of New Products
Metals Division
Fansteel, Inc.

THE REFRACTORY METALS—tungsten, molybdenum, niobium, tantalum, and rhenium—are distinguished by several common characteristics, including high density, high melting point, and superior resistance to wear and acid corrosion. Tungsten, for example, has a density over twice that of iron and a melting point of 3410 °C (6170 °F), the highest of any element. These metals have body-centered cubic crystal structures (with the exception of rhenium, whose crystal form is hexagonal). All are subject, however, to severe oxidation above 500 °C (930 °F) and must be protected for service by coatings or nonoxidizing atmospheres.

The refractory metals are extracted from ore concentrates, processed into intermediate chemicals, and then reduced to metal, which may be in the form of powder. The pure or alloyed powders are pressed, sintered, and subsequently worked. Typical applications for P/M refractory metals and their alloys include:

Electronics

- Lamp filaments
- X-ray targets
- Capacitors
- Electrical contacts
- Electrodes

Alloying (as an additive)

- High-strength low-alloy steel
- Tool steels

Alloying (heavy metal)

- Counterweights
- Projectiles

Nuclear power

- Radiation shielding
- Moderators
- Liquid metal handling

Aerospace

- High-temperature alloys

Chemicals

- Catalysts
- Corrosion-resistant equipment

Metal cutting; wear parts

- Cemented carbides

The mechanical and physical properties of refractory metals are compared in Table 1; their strengths vary, depending on product form and processing. Alloying or combining the metals in composites frequently provides properties that are superior to those of the base metal. Tungsten- and molybdenum-based alloys must be hot worked, whereas tantalum and niobium are easily cold worked. Rhenium is used primarily as an alloy additive. Table 2 compares the chemical corrosion resistance of the unalloyed refractory metals under a variety of conditions.

Tungsten

Although the most common application of tungsten is for cemented carbides for metal cutting, mining, and oil drilling (see the article "Cemented Carbides" in this Volume), tungsten's high density, high melting point, and low vapor pressure lend it to other applications as well. Figure 1 is a flowchart of the production sequence from tungsten ore to final products.

Tungsten Metal

Tungsten rod or wire has many uses, including light bulb filaments, electron emitters, welding electrodes, electrical contacts, and reinforcing filaments for metal-matrix composites. The rod is fabricated by isostatic pressing of powder and sintering, usually by resistance heating, in a hydrogen atmosphere. The rod is reduced to wire or other forms by hot swaging and is cut to length with diamond saws.

Non-sag wire, prepared from tungsten powder containing about 0.1% potassium aluminum silicate, is particularly useful for hot filaments. During sintering or annealing, the aluminum and silicon volatize, producing bubbles several angstroms in size. On subsequent working, these bubbles align and promote the formation of interlocking, elongated recrystallized grains (Fig. 2).

Additions of thoria or zirconia stabilize the grains and strengthen tungsten rod and wire. Because these compositions may contain up to 1 to 2% oxide, they are, strictly speaking, tungsten alloys. The thoria or zirconia is added as an aqueous nitrate salt solution to tungstic oxide (blue

Table 1 Typical mechanical and physical properties of pure refractory metals

Property	Tungsten	Rhenium	Metal Tantalum	Molybdenum	Niobium
Density, g/cm^3	19.3	21.04	16.6	10.22	8.57
Melting point, °C (°F)	3410 (6170)	3180 (5756)	2996 (5425)	2610 (4730)	2468 (4474)
Electrical conductivity, % IACS	31.0	9.3	13.9	34.0	13.2
Electrical resistivity, μΩ·cm	5.5	19.1	13.5	5.7	14.1
Thermal conductivity, (cal/°C) (cm/s)	0.397	0.17	0.13	0.35	0.125
W/m·K	166.105	71.128	54.392	146.44	52.30
Tensile strength, MPa (ksi):					
at room temperature	689-3445 (100-500)	1929.2 (280)	241.15-482.3 (35-70)	826.8-1378 (120-200)	206.7-413.4 (30-60)
at 500 °C (950 °F)	689-2067 (100-300)	923.26 (134)	172-310 (25-45)	241.2-447.9 (35-65)	137.8-275.6 (20-40)
at 1000 °C (1830 °F)	344.5-516.75 (50-75)	454.74 (66)	89.6-117.13 (13-17)	137.8-206.7 (20-30)	55.12-103.35 (8-15)
Modulus of elasticity, 10^5 MPa					
at room temperature	4.065 (59)	4.616 (67)	1.860 (27)	3.169 (46)	1.034 (15)
at 500 °C (930 °F)	3.789 (55)	3.790 (55)	1.723 (25)	2.825 (41)	0.896 (13)
at 1000 °C (1830 °F)	3.445 (50)	···	1.516 (22)	2.687 (39)	0.793 (11.5)

Table 2 Resistance of refractory metals to corrosive media

Media	Conditions	W	Mo	Ta	Nb
Liquid ammonia	Oxygen-free	■	■	▲	▲
Aqua regia	···	▲	□	■	▲
Chlorinated hydrocarbon	···			■	▲
Ferric chloride	···		□	■	■
Fuming sulfuric acid (oleum)	···	□	□	□	□
Hydrochloric acid (concentrated)	···	▲	▲	■	▲
Hydrofluoric acid	···	▲	▲	□	□
Nitric acid (concentrated)	···	▲	□	■	■
Nitric acid (fuming)	···	□	□	■	□
Phosphoric acid	<4 ppm Fluorine	▲	▲	■	▲
Potassium hydroxide (dilute)	<100 °C (210 °F)	▲	▲	▲	
Potassium hydroxide (fused)	···	□	□	□	□
Sodium hydroxide (dilute)	<100 °C (210 °F)	□	□	▲	
Sodium hydroxide (fused)	···	□	□	□	□
Sulfuric acid (concentrated)	To 160 °C (320 °F)	▲	▲	■	▲
	Above 160 °C (320 °F)		▲	▲	□
Liquid metals					
Gallium	To 450 °C (840 °F)	■	■	■	■
Lead	To 1000 °C (1830 °F)		■	■	▲
Lithium	To 1000 °C (1830 °F) (oxygen-free)		■	■	■
Mercury	To 600 °C (1110 °F)	■	■	■	■
Sodium	To 1000 °C (liquid and vapor)	■	■	■	■
Zinc	To 500 °C (930 °F)	■	▲	▲	
Gases					
Bromine	Wet		▲	■	■
	Dry, below 200 °C (390 °F)	■	■	■	
Carbon dioxide	To 1000 °C (1830 °F)	▲	▲	□	□
Chlorine	Wet	▲	▲	■	■
	Dry below 175 °C (345 °F)	▲	■	■	■
Fluorine	···	□	□	□	□
Hydrogen	To 1000 °C (1830 °F)	■	■	□	□
Nitrogen	To 500 °C (930 °F)	■	■	▲	▲
	>500 °C (930 °F)	■	■	□	□
Oxygen	To 500 °C (930 °F)	▲	▲	▲	▲
	>500 °C (930 °F)	□	□	□	□
Sulfur trioxide	···	□	□	□	□
Water vapor	<500 °C (930 °F)	▲	▲	▲	▲
	>500 °C (930 °F)	□	□	□	□

Key: ■ Satisfactory; ▲ variable, depending on temperature and concentration; □ not resistant

oxide) before reduction. The material is then dried and reduced in hydrogen.

Because thoriated tungsten is stronger than pure tungsten, higher sintering temperatures and longer sintering times are necessary. Swaging and drawing is similar to that performed on pure tungsten except that the rate of work hardening is greater, requiring more intermediate anneals.

Non-swag wire is better when distortion is a problem; however, thoria or zirconia dispersion-strengthened tungsten has superior hot strength and electron emission characteristics. Consequently, thoriated tungsten is widely used for inert gas welding electrodes.

Although silicide and noble metal coatings are effective oxidation-resistant coatings, protective atmospheres such as helium or argon are used to protect tungsten heating and incandescent elements and welding electrodes. Superior heat resistance also makes tungsten an ideal candidate for trays and radiation shields in vacuum or controlled-atmosphere furnaces.

Superior electrical conductivity, combined with resistance to wear and spark erosion, makes tungsten, tungsten-copper, and tungsten-silver useful for electrical contacts. Tungsten is also used in contacts for automotive ignition systems, broadcasting equipment, light switches, and voltage-control thyratron devices.

Tungsten Alloys

Tungsten is frequently used as an alloying element in tool steels and high-temperature alloys. Homogeneous powders of several of these alloys containing up to 30% W have been made by melting and atomization. Parts are then manufactured by pressing and sintering. Hot isostatic pressing has been employed to reduce or eliminate porosity.

The addition of 25% Re to tungsten provides greatly enhanced ductility. This alloy is usually prepared from powder blends and then pressed and sintered, hot compacted, or cast by arc or electron beam melting. The alloy can be rolled at 1000 °C (1830 °F) as much as 86% without cracking, but is difficult to work because of its strength at high temperatures. Alloys with 20 or 10% Re possess ductility that falls in the range between that of pure tungsten and the 25% alloy.

Other useful powders, prepared either by co-reproduction or by blending, are mixtures of tungsten and silver or tungsten and copper. The silver or copper additive improves the ductility, malleability, and thermal or heat conductivity of parts in electrical contact applications. Tungsten

Fig. 1 Flowchart of processing sequence from tungsten ore to finished products

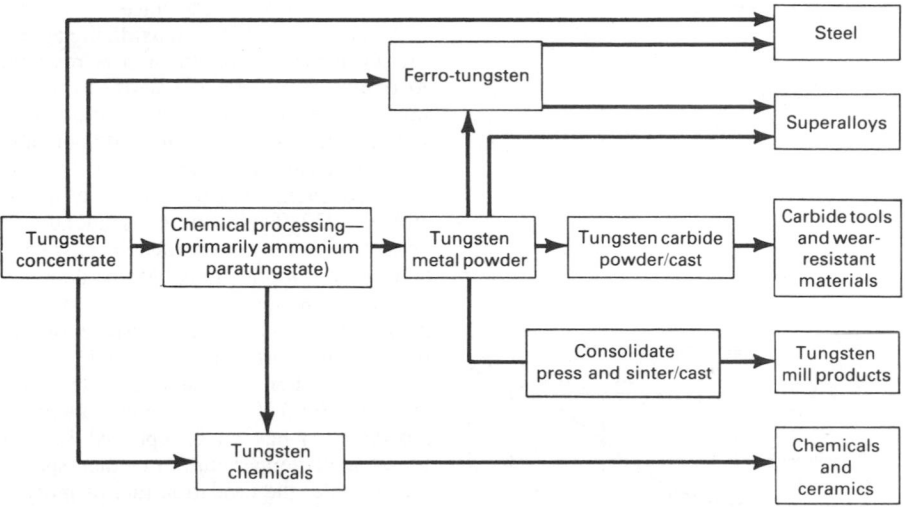

Fig. 2 Comparison of pure tungsten wire (top) and non-sag wire when heated to 100 °C (210 °F), 1800 °C (3270 °F), and 2300 °C (4170 °F) (left to right)

Magnification: 250×. Source: Ref 1

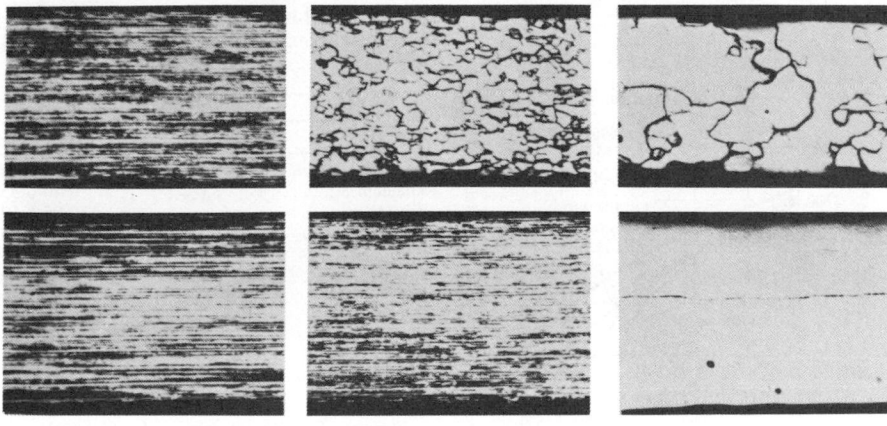

Fig. 3 Flowchart of processing sequence from molybdenum ore to finished products

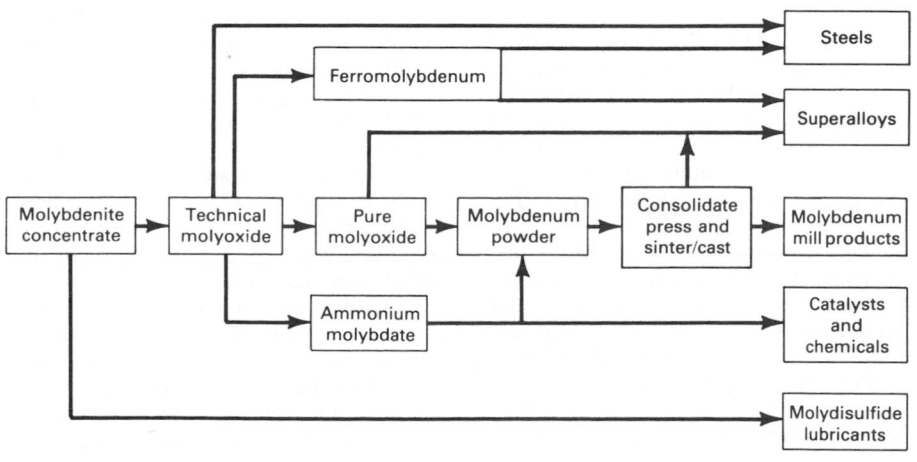

imparts high-temperature and spark crosion resistance to the composite.

Tungsten Heavy Metals

Tungsten heavy metals contain 90 to 97% W and the remainder either copper and nickel or iron and nickel plus additional elements. Densities are high, ranging from 16.8 to 18.5 g/cm^3. These compositions are easier to machine than pure tungsten. Tensile strength is several times higher than carbon steel and twice that of brass or cast iron, ranging from 760 to 970 MPa (110 to 140 ksi). Young's modulus for tungsten heavy metals is from 270 to 340 GPa (40 to 50 × 10^3 ksi).

Tungsten heavy metals have low thermal expansion coefficients (about half that of iron). They retain dimensional stability over a wide temperature range and exhibit excellent corrosion resistance. These materials can be fabricated economically by P/M techniques, including liquid phase sintering and infiltration. In the latter process, a porous tungsten powder compact is placed in contact with a second, lower melting point metal, which is then heated to its melting point. The liquid metal infiltrates the pores of the tungsten compact. Heavy metal parts are readily machined to close tolerances.

Typical applications include use in aircraft and helicopter counterweights, rudders, and elevator controls and in helicopter rotor blades. Tungsten heavy metals are used in radiation shielding, which requires the high density of these alloys to attenuate gamma and x-rays. Smaller shielding devices are possible with substantial weight savings. These alloys recently have enjoyed wide usage in sporting goods applications for inserts for golf club heads and to make well-balanced, accurate darts.

Molybdenum

More molybdenum is consumed annually than any other refractory metal. Molybdenum ingots, produced by melting of P/M electrodes, are extruded, rolled into sheet and rod, and subsequently drawn to other mill product shapes, such as wire and tubing. These materials can then be stamped into simple shapes. Molybdenum is also machined with ordinary tools and can be gas tungsten arc and electron beam welded, or brazed. A flowchart of the production process is shown in Fig. 3. Typical physical and mechanical properties are shown in Fig. 4.

Molybdenum has outstanding electrical and heat-conducting capabilities and relatively high tensile strength (Fig. 4a and b). Thermal conductivity is approximately

Fig. 4 Physical and mechanical properties of molybdenum compared to molybdenum alloys and other refractory metals

(a) Specific resistance. (b) Specific heat. (c) Thermal expansion. (d) Strength versus temperature

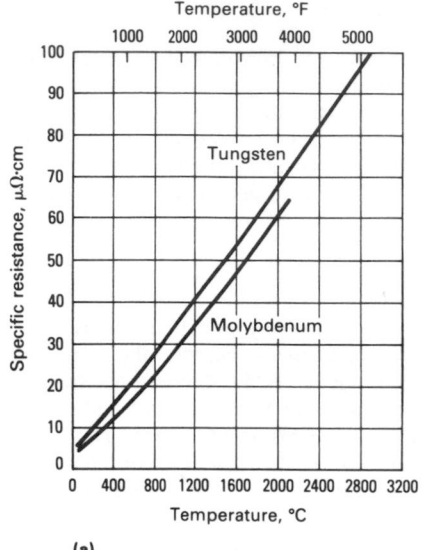

(a)

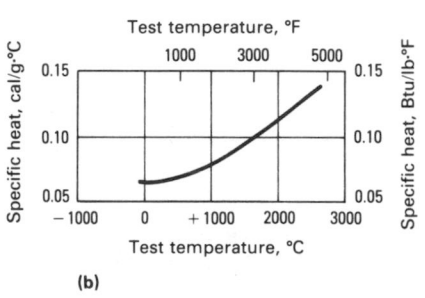

(b)

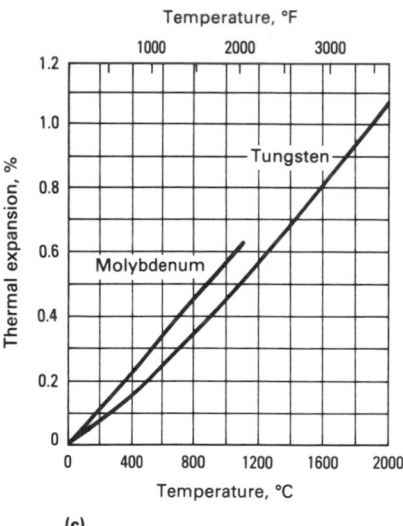

(c)

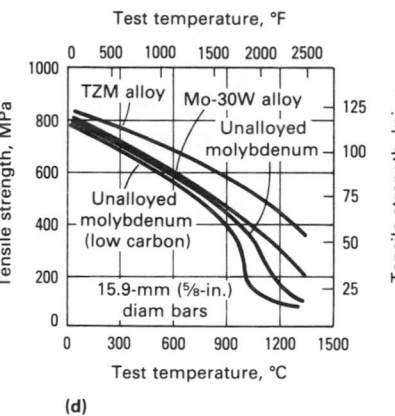

(d)

stainless steels, and nickel- or cobalt-based superalloys to increase hot strength, toughness, and corrosion resistance.

As shown in Table 2, is oxidizing agents are not present, molybdenum is resistant to corrosion by mineral acids, even hydrofluoric acid at room temperature. It is not subject to hydrogen embrittlement and does not form a hydride.

Molybdenum oxidizes rapidly at about 500 °C (930 °F) in air; consequently, a protective coating is needed in elevated-temperature applications. Many coatings involve formation of a thin layer of molybdenum silicide on the surface of the molybdenum part. The compound has outstanding oxidation resistance up to about 1650 °C (3000 °F). In vacuum, uncoated molybdenum has an exceptional service life at high temperature. In outer space, for example, the oxidation rate of molybdenum is insignificant.

Molybdenum Alloy TZM

The molybdenum alloy of greatest technological importance is the high-strength, high-temperature alloy TZM. The material is manufactured either by P/M or arc-cast processes. The composition of TZM is:

Element	%
Carbon	0.01-0.04
Titanium	0.40-0.55
Zirconium	0.06-0.12
Oxygen	<0.0025
Hydrogen	<0.0005
Nitrogen	<0.002
Iron	<0.010
Nickel	<0.002
Silicon	<0.008
Molybdenum	rem

Figure 5 shows typical strength-to-density ratios for TZM compared to other refractory materials. At about 1100 °C (2010 °F), its strength is about twice that of unalloyed molybdenum. The alloy is ideal for structural applications at 1000 °C (1830 °F) under conditions where unalloyed molybdenum normally is used.

TZM has a higher recrystallization temperature and higher strength and hardness at room and at elevated temperatures than unalloyed molybdenum. It also exhibits adequate ductility. Its superior mechanical properties are due to the dispersion of complex carbides in the molybdenum matrix. TZM is well suited to hot work applications because of its combination of high hot hardness, high thermal conductivity, low thermal expansion, and high resistance to heat checking compared to hot work steels. Major uses include:

50% higher than that of steel, iron, or nickel alloys. It consequently finds wide usage as heatsinks. Its electrical conductivity is the highest of all refractory metals, about one third that of copper, but higher than nickel, platinum, or mercury. The coefficient of thermal expansion of molybdenum plots almost linearly with temperature over a wide range (Fig. 4c). This characteristic, in combination with its heat-conducting capabilities, accounts for its use in bimetal thermocouples. Methods of doping molybdenum powder with potassium aluminosilicate to obtain a non-sag microstructure comparable to that of tungsten also have been developed.

The mechanical properties of unalloyed molybdenum depend greatly on the degree of work done below the recrystallization temperature. For optimum ductility, parts should be given at least 50% reduction in area. The metal exhibits superior strength and hardness at elevated temperatures (Fig. 4d). However, when hot strength is required, a molybdenum alloy rather than the element is the material of choice. Titanium and zirconium additions to molybdenum produce alloys with hot strength and recrystallization temperatures above those of unalloyed molybdenum.

The major use for molybdenum is as an alloying agent for alloy and tool steels,

- Die inserts for casting aluminum, magnesium, zinc, and iron
- Rocket nozzles
- Die bodies and punches for hot stamping
- Tools for metalworking (due to the high abrasion and chatter resistance of TZM)
- Heat shields for furnaces, structural parts, and heating elements

Figure 6 is a cross-sectional view of a cold seal pressure vessel constructed of TZM. In an attempt to improve the high-temperature strength of P/M TZM alloys, alloys have been developed in which titanium and zirconium carbide is replaced by hafnium carbide. Such an alloy may contain 1.0% Hf and 0.06% C. It has a tensile strength of 590 MPa (85 ksi) at 1315 °C (2400 °F), compared to 480 MPa (70 ksi) for TZM. Creep rate at 1205 °C (2200 °F) at a stress of 330 MPa (48 ksi) is 0.038%/h, compared to 0.05%/h for TZM.

Other Molybdenum Alloys

Alloys of molybdenum and rhenium are more ductile than pure molybdenum. An alloy with 35% Re can be rolled at room temperature to more than 95% reduction in thickness before cracking. For economic reasons, molybdenum-rhenium alloys are not widely used commerically. Alloys of molybdenum with 5 and 41% Re are used for thermocouple wires.

Rhenium

Among the elements, rhenium has the highest melting point, except for tungsten and carbon. Its density is exceeded only by osmium, iridium, and platinum. A ductile-to-brittle transition temperature does not exist in pure rhenium. Rhenium is the only refractory metal that does not form carbides.

Rhenium has a high electrical resistivity over a wide temperature range. Figure 7 compares electrical resistivity values for tungsten and rhenium. Rhenium typically exhibits higher resistivity values than tungsten. This characteristic, combined with a low vapor pressure, makes it ideally suited for filament applications; additionally, it maintains ductility and is not affected by the oxidation/reduction cycle experienced in these applications, as is tungsten.

One of the largest applications for rhenium is for mass spectrometer filaments. These are available in commercial (99.99%) and zone-refined (99.995%) purities.

Rhenium is not attacked by molten copper, silver, tin, or zinc. It dissolves readily in molten iron and nickel, but is stable

Fig. 5 Strength-to-density ratio for TZM compared to refractory alloys

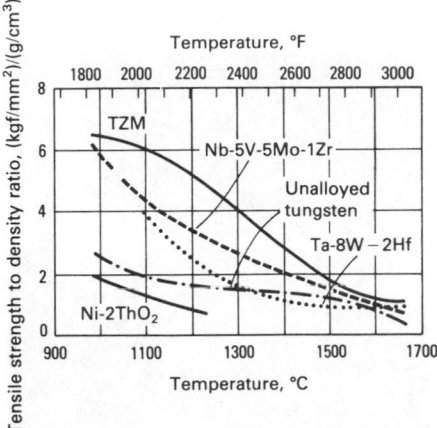

in the presence of aluminum. Rhenium has a significant hardening effect on platinum.

At elevated temperature, rhenium resists attack in hydrogen and inert atmospheres. It is resistant to hydrochloric acid and seawater corrosion and to the mechanical effects of electrical erosion.

Rhenium is widely used in combination with platinum in catalysts, particularly for selective hydrogenation and crude oil reforming. Rhenium catalysts display high resistance to poisons such as nitrogen, sulfur, and phosphorus.

Rhenium Alloys

Rhenium is a beneficial alloying addition with other refractory metals. Rhenium greatly enhances the ductility and tensile strength of refractory metals and their alloys (Fig. 8). This tendency is maintained on heating above the recrystallization temperature.

Rhenium alloys are used in nuclear reactors, semiconductors, electronic tube components, thermocouples, gyroscopes, miniature rockets, electrical contacts, thermionic converters, and other commercial and aerospace applications. Tungsten-rhenium alloys, applied by vapor deposition, are used to coat the surface of molybdenum targets in x-ray tube manufacture. Other rhenium alloys (with tungsten or molybdenum) are used for filaments, grid heaters, cathode cups, and ignitor wires in photoflash bulbs. Figure 9 shows thin-walled tubes of a rhenium alloy used in magnetrons as a cathode support.

Tantalum

World production of tantalum powder is about 1000 metric tons (1100 tons) annually. Most of the powder is earmarked

Fig. 6 Sectional view of TZM cold seal pressure vessel

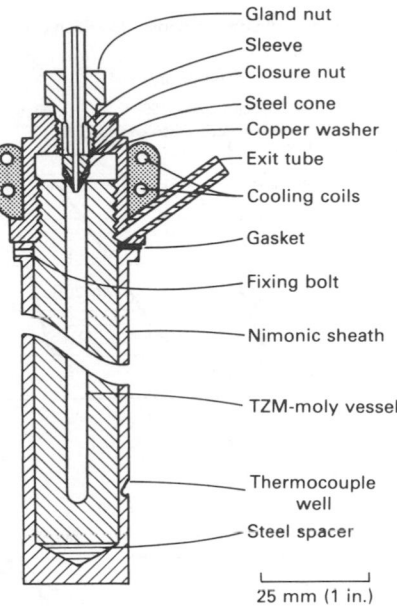

- Gland nut
- Sleeve
- Closure nut
- Steel cone
- Copper washer
- Exit tube
- Cooling coils
- Gasket
- Fixing bolt
- Nimonic sheath
- TZM-moly vessel
- Thermocouple well
- Steel spacer

25 mm (1 in.)

Fig. 7 Electrical resistivity versus temperature for rhenium, its alloys, and other refractory metals

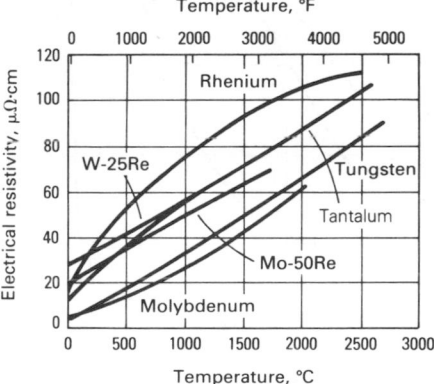

for production of electronic products, primarily capacitor anodes, foil, wire, and cases. The most common capacitor construction (Fig. 10) consists of a porous tantalum pellet with an embedded lead wire. When tantalum is anodized, an amorphous tantalum oxide film forms on the porous exterior and interior metal surface.

This oxide acts as the capacitor dielectric, and the capacitance is directly related to the total anode surface area. After anodization, capacitor manufacture is completed by impregnating the pellet with an electrolyte. For the solid tantalum capacitor, this is accomplished by dipping the anode in manganese nitrate solution and

Fig. 8 Effect of alloying additions of rhenium on ductility

(a) Rhenium-molybdenum alloys. (b) Rhenium-tungsten alloys

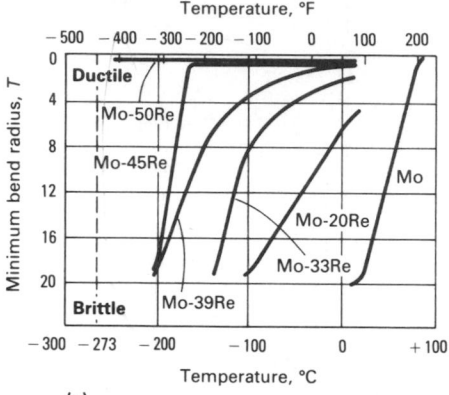

(a)

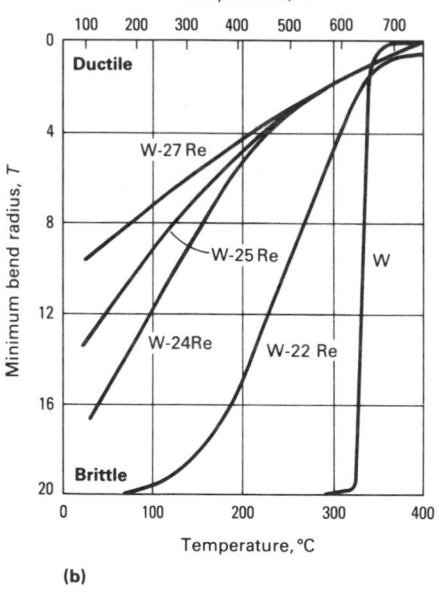

(b)

Fig. 9 Mo-48Re alloy seamless thin-walled tubes used in magnetrons as a cathode support

Courtesy of Rhenium Alloys, Inc.

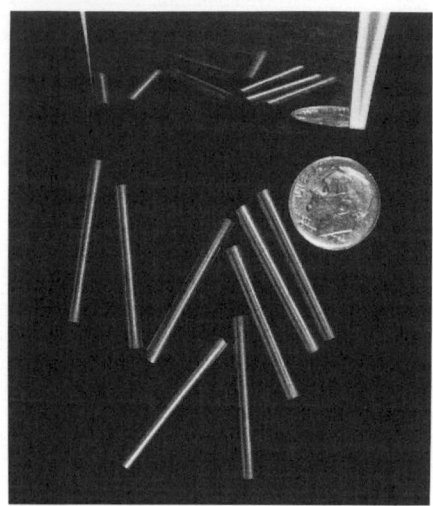

Fig. 10 Solid electrolyte metal case tantalum capacitor

Courtesy of Mallory Capacitor Co.

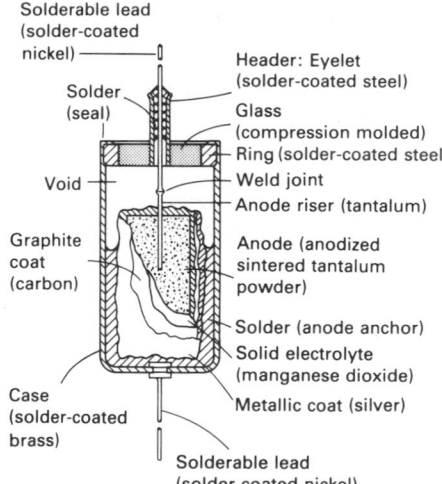

pyrolyzing repeatedly until the interval voids are filled with manganese oxide.

The capacitor is then coated with conductive films of carbon and silver to ensure electrical contact with the cathode lead. Subsequently, the entire unit is encased in a metal, plastic, or epoxy resin. Other types of capacitors use a wet sulfuric acid electrolyte, and the pellet is hermetically sealed in a silver or tantalum capsule or case.

Manufacture of the tantalum pellet begins by cold pressing the powder into a low-density pellet. The capacitance rating of the device determines the powder weight required. On the average, the weight per pellet is about 0.2 g (0.007 oz). The pellet is then sintered in high vacuum (1.33 × 10^{-4} Pa, or 10^{-4} torr). The sintering temperature is set low enough to retain as much

of the porous structure as possible, while sintering the "necks" of adjacent particles to form a continuous path of conducting metal.

Further surface purification of the metal also occurs during sintering by volatilization of the monoxides of tramp cations, the presence of which may otherwise cause defects in the anodic oxide film. Sintering temperatures usually range from 1450 to 2000 °C (2640 to 3630 °F). Selection of temperatures is determined by the type of powder and device required; there is a trade-off between loss of surface area and improved purity as temperature increases.

Additions of small amounts (10 to 100 ppm) of sintering inhibitors, such as phosphorus, are frequently employed. These agents allow finer powders to be used than would be possible in the absence of inhibitor. Alternatively, higher sintering temperatures can be used without loss of surface area, thereby aiding vaporization of undesirable impurities. Dielectric properties are thereby improved.

Sodium-reduced tantalum powders are used for 80 to 90% of capacitor applications. Electron beam melted, degassed-hydride powders account for the remainder. Sodium-reduced material is used in high-capacitance devices at lower voltage ratings (<50 V) because these powders can be produced in finer particle sizes, and the material has improved pressing characteristics. The latter permits anodes to be pressed at lower green density.

Consequently, sodium-reduced powders are more efficient, requiring less powder to achieve a given capacitance rating in a finished device. However, electron beam melting provides degassed-hydride powders with an additional purification step. Because of this higher purity, the anodic oxide film can be electrolytically formed, or stressed, to higher voltages without breakdown. Therefore, electron beam melted powders frequently are required for high-reliability capacitors rated for 50- to 125-V use. Figure 11 compares capacitance at several anode sintering temperatures for sodium-reduced and electron beam melted, degassed powders.

Although the dielectric properties of niobium oxide are inferior to those of tantalum oxide, niobium has been evaluated as a means of producing low-cost capacitors. Considerable niobium powder development work is still required before niobium capacitors will be offered commercially.

Tantalum Mill Products: Wire, Sheet, and Foil

Powder metallurgy mill products begin by cold isostatic pressing the powder into bars. The bars are then direct resistance sintered by clamping the ends firmly between water-cooled copper terminals in a vacuum furnace at 1.33 × 10^{-2} Pa (10^{-4} torr). Heating occurs by the passage of an electric current.

For tantalum, temperatures above 2300 °C (4170 °F) are required to achieve full density and adequate purification (removal of interstitial impurity elements). Sintered bars are then rolled or drawn at room temperature, with appropriate intermediate vacuum anneals.

Fig. 11 Capacitance of sodium-reduced and electron beam melted, degassed-hydride tantalum powder

For 30-min anode sintering temperature at anode green densities commonly used for each powder

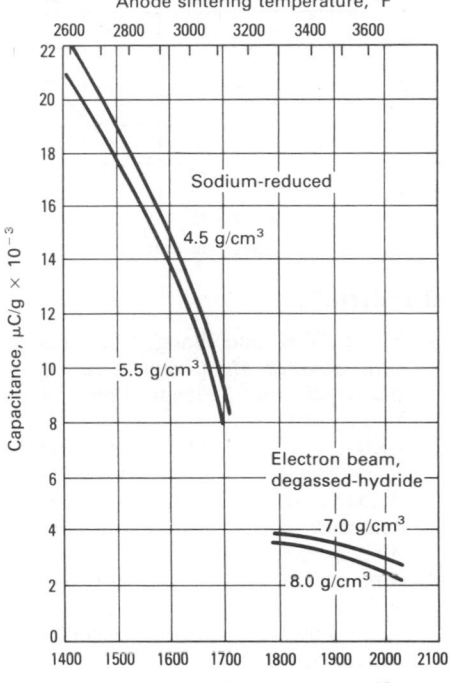

Table 3 Properties comparison of Ta-7.5%W sheet made by powder metallurgy and from ingot

Measured for 2.1-mm (0.083-in.) thick, annealed and recrystallized sheet

Process	Hardness, DPH	Yield strength, σ_0 (0.2% offset) MPa	ksi	Modulus of elasticity, $E \times 10^{-6}$ MPa	ksi	Modulus of resilience(a), U_R MPa	ksi
P/M	291	676	98	196	28.5	1172	170
Ingot	150	483	70	196	28.5	593	86

(a) $U_R = \dfrac{\sigma_0^2}{2E}$ (see Ref 2)

Powder metallurgy products differ from those produced from cast materials primarily because surface oxides are present in the starting powders. These oxides are present in the microstructure of the P/M material. Recrystallization is inhibited and finer grain metal is produced. The product is usually harder and has higher yield and tensile strength but somewhat less ductility.

Oxide microalloying additions, such as thoria or yttria, are sometimes intentionally added to control grain size and strength after recrystallization. Small quantities of silicon are also added, particularly to bars that are subsequently drawn to capacitor wire.

Silicon aids in the retention of ductility of the wire by preventing embrittlement of the tantalum lead-wire embedded in the capacitor pellet. During anode vacuum sintering, oxygen diffuses from the surrounding powder into the embedded wire, which has lower oxygen, and collects at grain boundaries, thus causing embrittlement. Silicon acts as a getter for the excess oxygen. There is evidence that the microstructure is modified with precipitated oxides in the form of globules rather than platelets. A portion of the silicon

vaporizes harmlessly as volatile silicon monoxide.

Tantalum Alloys

Tungsten and hafnium are the alloying additions commonly added to improve the strength properties of tantalum. Although tantalum powder may be an intermediate, most tantalum alloys are melted. As examples, Ta-2.5%W is used for the fabrication of corrosion-resistant heat exchangers, valves, and other chemical equipment. Ta-10%W, although less ductile, is a harder, stronger, and more wear-resistant alloy and is particularly suited for the fabrication of furnace hardware, missile parts, nozzles, and fasteners.

A Ta-7.5%W alloy, useful for springs, bellows, and leaf springs in severe corrosion environments, such as dry chlorine, bromine, or hydrochloric acid, is made by powder metallurgy. For springs, the modulus of resilience, U_R, is an important parameter. Resilience is defined as the "ability of a material to absorb energy when deformed elastically and retain it when unloaded" (Ref. 2). Because the yield strength of the P/M product is higher than ingot material of the same tungsten content, its modulus of resilience is also considerably higher, as shown in Table 3.

Niobium

The major use for niobium is as an alloying agent for austenitic and high-strength low-alloy steels. It is also an additive in several nickel- and cobalt-based superalloys. Niobium provides the alloys with many of its refractory properties without adding weight. Consequently, alloys containing niobium are widely used in aircraft turbines and aerospace rocket engines. Niobium is added to the alloy during melting as a master alloy such as ferroniobium, which contains 63 to 68% Nb.

In addition to its high melting point and acid corrosion resistance, niobium has a number of useful properties. Its affinity for oxygen at elevated temperature has led to

its use as a getter in radio and electronic equipment. For structural use at high temperature in air or oxidizing atmospheres, however, niobium must be coated (with a silicide-type coating, for example) for satisfactory performance. Because the metal forms a stable, anodic oxide film, platinized niobium electrodes are used for cathodic protection against galvanic seawater corrosion, for example, on off-shore structures such as oil-drilling rigs.

Niobium has the highest superconducting transition temperature, 9.2 K, of any element; Nb_3Ge has the highest of any compound known thus far, 23.6 K. Thus niobium and its alloys are being used as superconductors at cryogenic temperatures for transport of electrical power without resistance loss. In such applications, energy consumption can be reduced as much as 80%.

Niobium has lower density and a low thermal neutron cross section than other refractory metals. This property, together with its resistance to corrosive attack by liquid metal coolants, makes niobium and its alloys (for example, Nb-1%Zr) useful in nuclear reactors.

Niobium Alloys

There are presently no known commercial uses of P/M niobium or its alloys. However, with the increasing need to conserve critical materials and the economic promise of the near-net shape concept, the aerospace industry is evaluating the substitution of high-strength niobium P/M alloy parts for ones currently made by forging and machining. One example now in development is C-103 (Ref. 3).

C-103 alloy (Nb-10Hf-1Ti) is used for hot gas exhaust and attitude vector control nozzles in missile and space vehicle systems. Precise machining of forged parts is time consuming, difficult, and inefficient. Recently, alloy powders made by hydriding, crushing, and dehydriding of electron beam or arc melted ingots have been used. Experimental work is also being conducted on rotating electrode powders (Ref 4).

Table 4   Room-temperature and 1650 °C (3000 °F) tensile properties for hot isostatically pressed C-103 P/M alloy

Alloy	Processing temperature °C	°F	Test temperature °C	°F	Ultimate tensile strength MPa	ksi	0.2% yield strength MPa	ksi	Elon-ga-tion	Reduction in area, %
C-103 alloy	1350	2450	Room		516.4	74.9	364.7	52.9	30	56
	1350	2450	Room		506.8	73.5	357.8	51.9	25	49
	1350-1600	2450-2900	Room		426.8	61.9	302.7	43.9	50	67
	1350	2450	1650	3000	31.7	4.6	28.3	4.1	16	20
	1260-1600	2300-2900	Room		528.1	76.6	464.7	67.4	27	32
	1260-1600	2300-2900	Room		545.4	79.1	421.9	61.2	23	37
	1260-1600	2300-2900	Room		517.8	75.1	401.3	58.2	29	51
	1260-1600	2300-2900	1650	3000	17.9	2.6	15.9	2.3	48	19
	1260-1600	2300-2900	1650	3000	17.2	2.5	13.8	2.0	36	16
Recrystallized wrought	···	···	Room		413.7	60.0	31.03	45.0	40	···
Bar stock from ingot	···	···	1650	3000	34.5	5.0	29.0	4.2	70	···

Table 5   Creep properties of C-103 alloy at 1650 °C (3000 °F)

C-103 source	Initial stress MPa	ksi	Test duration, h	Creep ductility, %	Minimum creep rate × 10⁶, sec⁻¹
Hot isostatically pressed powder	12.40	1.80	8.4	>125	11.6
Wrought ingot	12.40	1.80	6.9	>120	10.0
Hot isostatically pressed powder	13.78	2.00	5.5	104	15.0
Wrought ingot	13.78	2.00	5.1	>100	19.2
Hot isostatically pressed powder	16.54	2.40	2.9	98	30.0
Wrought ingot	16.54	2.40	2.6	>113	30.0

Table 4 compares the tensile properties of hot isostatically pressed compacts of C-103 powder versus the same alloy made from ingot. Creep properties are compared in Table 5, and Fig. 12 shows the creep curve for the P/M alloy.

Fig. 12 Typical creep properties curve for C-103 alloy

Powder source C-103 alloy sample tested at 1650 °C (3000 °F) at an initial stress of 13.8 MPa (2 ksi)

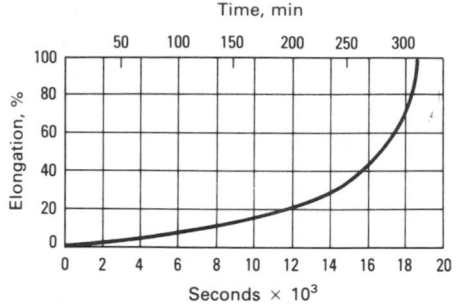

REFERENCES

1. Yih, S.W.H. and Wang, C.T., *Tungsten, Sources Metallurgy, Properties and Applications*, Plenum Press, New York, 1979
2. Dieter, G.E., Jr., *Mechanical Metallurgy*, McGraw-Hill, New York, 1961
3. Wadsworth, J., Roberts, C.A., and Rennhack, E.H., *J. Mat. Sci.*, Vol 17, 1982, p 2539-2546
4. Himmelblau, C.D., *et al.*, in *Progress in Powder Metallurgy 1983*, Vol 79, Metal Powder Industries Federation, Princeton, NJ, 1984

SELECTED REFERENCE

• Schussler, M., *Corrosion Data Survey on Tantalum*, Fansteel, Inc., North Chicago, 1972

Cemented Carbides

By Ralph W. Stevenson
Project Manager/Materials Group
Energy Technology Division
Midland-Ross Corp.

CEMENTED CARBIDES are a class of very hard, wear-resistant materials produced by P/M processing. Significant contributions toward the development of modern cemented carbides were made in Germany in the early 1920's by Karl Schröter, who used tungsten monocarbide (6.13% C) as the hard component in his first patent. Tungsten and carbon were stoichiometrically proportional. This carbide was mixed with up to 10% iron binder. The patent also described the compacting and sintering operations; sintering occurred at temperatures close to the melting point of the binder. In Schröter's second patent, the binder metal was cobalt (up to 20 wt%). This is still the most important composition in current cemented carbide production. In addition, the sintering procedure developed by Schröter is widely used to produce cemented carbides.

The first cemented carbides were developed in response to demands for a material sufficiently wear resistant for drawing dies used to produce tungsten wire. These materials were also suitable for machining, possessing higher hardness and wear resistance, including higher hardness at elevated temperatures, than high-speed tool steels. Currently, cemented carbides are used for metal cutting, mining, rock drilling and stone cutting, metal forming tools and structural components, wear parts, and abrasive grits.

This article examines the methods of powder production, consolidation, and sintering of cemented carbides, their physical properties, classification, and applications. Only unalloyed and alloyed grades of tungsten carbide/cobalt cemented carbides are discussed. Additional information on carbides such as chromium carbide, titanium carbide, tantalum carbide, and niobium carbide can be found in the article "Cermets" in this Volume.

Classification of Cemented Carbides

Tungsten Carbide/Cobalt Grades. These grades are often referred to as unalloyed grades, straight tungsten carbides, cast iron cutting grades, or edge wear-resistant grades of cemented carbides. They consist of fine angular particles of tungsten carbide bonded with metallic cobalt, although nickel can also be used as a binder material. Cobalt may vary from 3 to 13% for cutting tool grades and up to 30% for wear-resistant parts. The average size of the carbide particles varies from less than 1 to 8 μm. Increasing amounts of cobalt and increasing tungsten carbide grain size lower hardness, but increase mechanical shock resistance of these grades.

Grades containing only tungsten carbide in the carbide phase are used for machining cast iron, nonferrous metals, and nonmetallic materials, but generally not for machining steel. They also are used for nonmachining applications of cemented carbides. Additional information on these grades can be found in the sections of this article on selection and application of cemented carbides.

Alloyed Tungsten Carbides. These grades are often referred to as steel cutting or crater-resistant grades. These materials, in which cobalt is the binder phase and tungsten carbide is the major constituent of the carbide phase, also contain titanium carbide and/or tantalum carbide. The basic compositions of the grades contain 3 to 12% cobalt, 60 to 85% tungsten carbide, 4 to 25% titanium carbide, and up to 25% tantalum carbide.

These grades are used primarily for cutting steel, in which straight tungsten carbide grades are subject to cratering, or the erosion of the tool surface by the chip. This is caused primarily by a diffusion reaction between the tool and chip at high temperatures reached during machining. The extent of cratering depends on the material being machined and machining conditions. Grades of cemented carbides containing titanium carbide and/or tantalum carbide were developed in the early 1930's to combat cratering during machining of steel. Additional information on these grades can be found later in the section of this article on selection and application of cemented carbides.

Production of Tungsten Carbide

The first stage in the production process (Fig. 1) is the blending of powders and their subsequent mixing and milling. Mixtures may be composed of a blend of tungsten powder, titanium dioxide, tantalum oxide, and carbon, usually added in the form of lampblack (see the Section entitled "Production of Metal Powders" in this Volume for detailed information on the production of tungsten, titanium, tantalum, and carbide powders). Powder mixes are blended together in a rotating mixer or ball mill. After blending, the mixture of tungsten and carbon (94 parts by weight of tungsten and 6 parts by weight of carbon) is charged into an induction heated vacuum furnace and heated to approximately 1550 °C (2800 °F) to carburize the tungsten and carbon into the monocarbide tungsten carbide.

Alloyed grades of titanium carbide are produced by carburizing titanium and carbon at about 2200 °C (4000 °F), while tantalum carbide is produced by carburizing tantalum and carbon at about 1650 °C (3000 °F). Part of the tungsten carbide is then replaced by titanium carbide, tantalum carbide, or a combination of the two. The alloy addition (titanium or tantalum car-

Fig. 1 Flowchart of cemented carbide manufacturing process

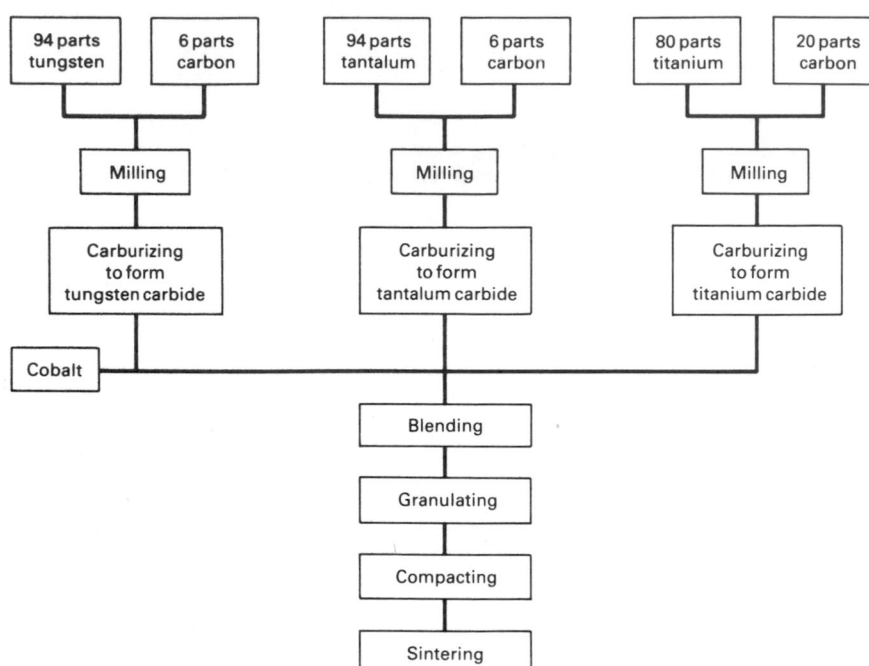

bide) may be introduced into the final product as single carbides or as solid solutions in combination with part of the tungsten carbide.

To provide the necessary strength, a binding agent (usually cobalt) is added to the tungsten carbide in fine powder form. These two materials are rod or ball milled for several hours to several days to obtain a uniform homogeneous product. Milling generally is conducted in an organic liquid such as hexane. With the exception of mixtures that are to be hot pressed, a lubricant (usually paraffin) is added to the mixture. After milling, the organic liquid is removed by one of several drying methods.

Compaction. The milled powder usually is consolidated by cold pressing. The dies employed are made to the shape of the desired end product. The size of the die must be greater than the final product size to allow for dimensional shrinkage that occurs during final sintering. These dies are expensive and usually are made with tungsten carbide liners. Therefore, a sufficient number of compacts is required to justify the expense involved in manufacturing special dies. Ordinarily, cold compacting pressures are about 200 MPa (15 tsi). Higher pressures have little effect on further densification of the compact and cause increased die wear. Lower pressures cause greater shrinkage during final sintering, with increased distortion and porosity.

An alternative method of powder compaction is the hot pressing of powders in graphite dies at the sintering temperature. After cooling, the part has attained full hardness. Because the graphite dies are expendable, this method generally is used only when the part to be produced is too large for cold pressing and sintering. See the article "Hot Pressing" in this Volume for additional information.

Another method normally used for large pieces is isostatic pressing. Powders are charged into a closed, flexible container that is suspended in a liquid in a closed pressure vessel. Pressure in the liquid is raised to the point where the powders become properly compacted. This system is advantageous for pressing large pieces, because the pressure acting on the powders operates equally from all directions, resulting in a compact of uniform pressed density. See the article "Cold Isostatic Pressing of Metal Powders" in this Volume for additional information.

Powders also can be compacted by extrusion methods, in which the cemented carbides are mixed with a plasticizer. This process is used for producing shapes that are difficult to form by conventional pressing and for applications in which the distortion during presintering to remove lubricant and during final sintering can be tolerated.

Presintering and Preforming. Presintering is a low-temperature heat treatment, carried out in hydrogen or a vacuum, to remove die lubricant and to impart sufficient strength to the compact to permit preforming or shaping operations if required. The preforming or shaping operation may be done with conventional mechanical shaping techniques, such as drilling, slicing, grinding, and turning. As in compacting, sintering shrinkage must be considered.

Sintering of tungsten carbide/cobalt compacts is performed with the cobalt binder in the liquid phase. The compact is heated in a hydrogen atmosphere or vacuum furnace to 1350 to 1600 °C (2500 to 2900 °F), depending on composition. Time and temperature must be carefully controlled to achieve optimum properties and geometry. In spite of the high pressures employed in the compacting of powders, the pore volume of the compact is very high. At the sintering temperature, densification to ultimate maximum theoretical density occurs, practically eliminating the pore volume. The American Society for Testing and Materials has issued a standard test method (ASTM B 276) for determining apparent porosity in cemented carbides.

The compact shrinks approximately 16% linearly, or 40% volumetrically. The exact amount of shrinkage depends on particle size of the powders and the composition of the grade. Control of size and shape is most important. This is particularly true for grades of cemented carbides with higher cobalt contents. Because cobalt is less dense than tungsten carbide, it occupies a greater portion of the volume than is indicated by its weight percentage.

Accordingly, higher cobalt-bearing grades have a disproportionately high percentage of the mass in the liquid phase, and extreme care is required to control the magnitude and direction of shrinkage. For detailed information on the mechanisms of liquid phase sintering, see the article "Physical Fundamentals of Consolidation" in this Volume. For specifics on sintering of various carbides, see the article "Production Sintering Practices for P/M Materials" in this Volume.

Properties of Tungsten Carbide/Cobalt Grades

The tungsten carbide/cobalt grades of cemented carbides are the most resistant to abrasive wear. These grades primarily are used for machining operations where the materials to be machined yield discontinuous chips or chips that do not cause tool wear by cratering. Typical microstructures of straight tungsten carbide/cobalt grades and their corresponding

Fig. 2(a) Effect of tungsten carbide grain size on hardness and abrasion resistance of fine-grained 94WC-6Co

Magnification: 1500×. Courtesy of Carmet Co.

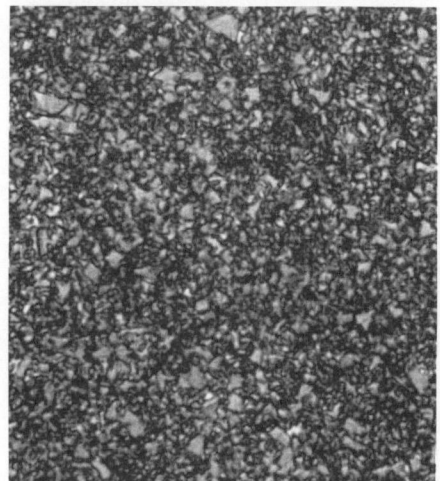

Hardness	92.2 ± 0.4 HRA
Density	14.90 g/cm³
Transverse rupture	2070 MPa (300 ksi)
Elastic modulus	640 GPa (93 × 10⁶ psi)
Compressive strength	4.6 GPa (675 ksi)
Abrasion resistance	36

Fig. 2(b) Effect of tungsten carbide grain size on hardness and abrasion resistance of coarse-grained 94WC-6Co

Magnification: 1500×. Courtesy of Carmet Co.

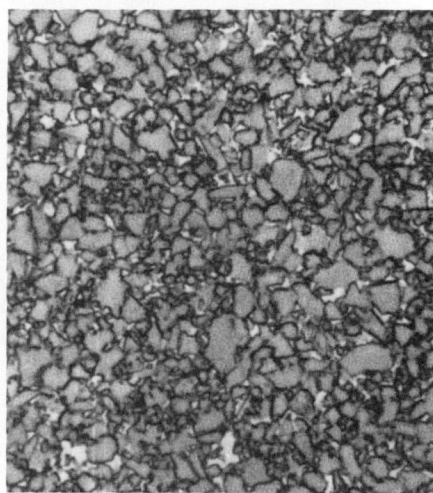

Hardness	91.0 ± 0.5 HRA
Density	14.90 g/cm³
Transverse rupture	2070 MPa (300 ksi)
Elastic modulus	630 GPa (92 × 10⁶ psi)
Compressive strength	4.4 GPa (645 ksi)
Abrasion resistance	16

Fig. 4 Typical hot hardness values for straight tungsten carbide grades and high-speed tool steels

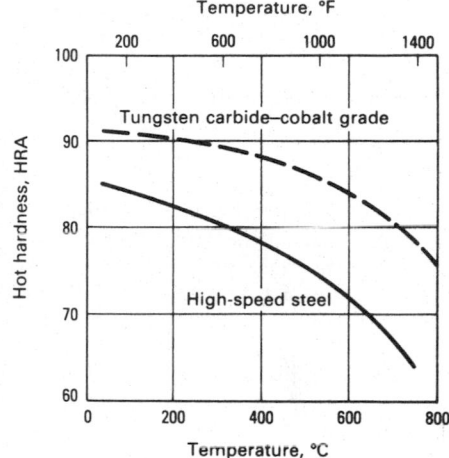

physical properties are shown in Fig. 2 and in the Appendix to this article.

Hardness. In the United States, hardness of cemented carbides generally is measured on the Rockwell A scale (diamond indenter, 60-kg load); however, Vickers hardness numbers are widely used in other countries. Apparent hardness is primarily influenced by cobalt content and tungsten carbide grain size. Figure 3 shows the relationship of Rockwell A hardness of grades with different cobalt contents versus tungsten carbide grain size. The effect of grain size in commercial practice can best be demonstrated by the fact that it is not unusual for a manufacturer to produce two grades of the same chemical composition, but with a considerable difference in hardness due to the grain size as shown in Fig. 2(a) and (b).

The American Society for Testing and Materials has issued a recommended practice (ASTM B 390) for evaluating apparent grain size and distribution of cemented carbides. See the Appendix to this article for hardness values of unalloyed and alloyed grades of cemented carbides.

Hot Hardness. The ability to retain hardness at high temperatures is a function of the tungsten carbide/cobalt structure. Carbide particle hardness is not appreciably affected by the temperatures reached

Fig. 3 Hardness of tungsten carbide/cobalt alloys with different cobalt contents as a function of tungsten carbide grain size

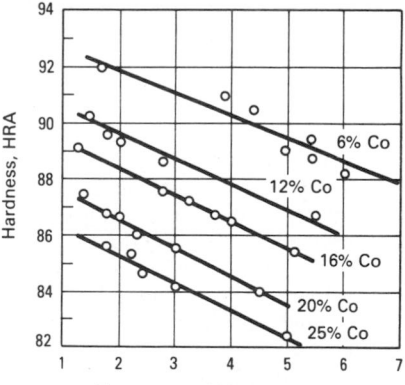

during normal machining operations. Critical hardness losses result when the cobalt binder absorbs sufficient heat to transform it into the plastic range where deformation occurs. Figure 4 compares typical hardness values of high-speed tool steels and straight tungsten carbide grades of cemented carbides.

Transverse-rupture strength is another important mechanical property of

tungsten carbide. Because of its extreme hardness, carbide does not respond well to the tensile test used to determine the strength of molten metal products. Instead, the transverse-rupture test on bar specimens is used to determine its strength and resistance to crack propagation.

A test method to determine this property of cemented carbides has been standardized by the American Society for Testing and Materials (ASTM B 406). As with hardness, transverse-rupture strength is influenced by cobalt content and grain size. As shown in Fig. 2(b), tungsten carbide particles (light gray, outlined) are uniformly dispersed in a continuously interlocking matrix of cobalt. Minimal coherence, therefore, exists between the particles of tungsten carbide, so that the strength of a particular grade is a function of the amount of binder phase present, or the thickness of the mean free path between carbide grains. Figure 5 shows the

Fig. 5 Transverse-rupture strength as a function of cobalt content

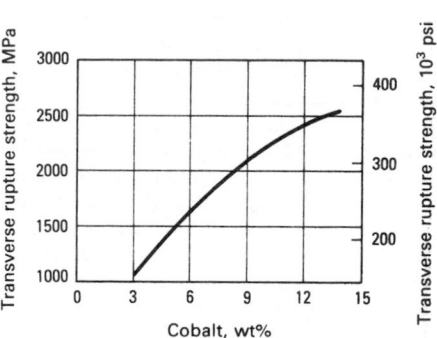

effect of cobalt content on transverse-rupture strength of unalloyed tungsten carbide grades. See the Appendix to this article for transverse-rupture strength values of unalloyed and alloyed grades of cemented carbides.

Properties of Alloyed Tungsten Carbide Grades

The alloyed grades of cemented carbides are used in the machining of steel and other materials, yielding strong, continuous chips that contribute to the formation of craters on the top face of a cutting tool. Typical microstructures and corresponding physical properties for various alloyed tungsten carbide grades of cemented carbides can be found in the Appendix to this article.

Tungsten Carbide/Titanium Carbide/Cobalt Grades. The most significant contribution of titanium carbide or a titanium carbide/tungsten carbide solution in a carbide cutting tool is the reduction of the tendency of chips to weld or adhere to the cutting edge. The weld strength of these alloyed grades is less than that of tungsten carbide/cobalt; consequently, adhesion does not occur, and the incidence of particle fracture from the cutting tool is reduced.

At high operating speed, temperatures developed at the tool/chip interface are high, and tool wear results from diffusion between work and tool. Additions of titanium carbide to the basic carbide matrix reduce this diffusion wear process. Also, this addition delays the formation of craters in the rake face of the tool, which is the most common tool failure in steel cutting operations.

Hot hardness also is improved with additions of titanium carbide. Figure 6 shows relative hardness for two typical grades of carbides. Both grades contain 9.0 wt% Co. The alloyed compact is somewhat harder at room temperature. As temperature increases, however, the spread between the hardness of the two grades increases.

As titanium carbide content increases, transverse-rupture strength is reduced, providing the cobalt content remains constant. Consequently, the transverse weakening effect of titanium carbide additions is offset with increasing amounts of cobalt to achieve adequate strength for the application for which the grade is designed. See the Appendix to this article for typical microstructures and physical properties of tungsten carbide/titanium carbide/cobalt grades of cemented carbides.

Fig. 6 Effect of titanium carbide on the hot hardness of tungsten carbide/cobalt alloys

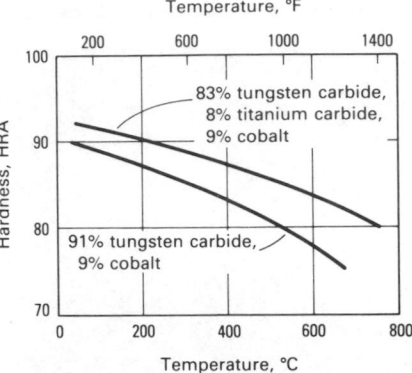

Tungsten Carbide/Titanium Carbide/Tantalum Carbide/Cobalt Grades. Many commercial-grade carbides contain titanium carbide and tantalum carbide alloy additions to the basic tungsten carbide/cobalt alloys for a given cobalt content.

As described above, addition of titanium carbide has an adverse effect on transverse-rupture strength, but additions of tantalum carbide to tungsten carbide/titanium carbide/cobalt alloys result in higher transverse-rupture strength at cutting temperature. This improvement is attributed to the ability of tantalum carbide to form pure solid solutions and to inhibit grain growth of the carbide phase. Additionally, the hot hardness of tungsten carbide/titanium carbide/tantalum carbide/cobalt grades is higher than that of corresponding tungsten carbide/titanium carbide/cobalt alloys. Also, hot compressive strength is significantly higher, thus reducing plastic deformation of the point of the tool in the cut.

Other important benefits can be derived from high alloy additions to tungsten carbide/cobalt alloys. As mentioned above, most steel cutting operations are performed at high temperatures. The temperature at which a carbide will weld or adhere to a steel chip may be increased by as much as 50%, if the carbide contains a high titanium carbide/tantalum carbide content. See the Appendix to this article for typical microstructures and physical properties of tungsten carbide/titanium carbide/tantalum carbide/cobalt grades of cemented carbides.

Tungsten Carbide/Tantalum Carbide/Cobalt Grades. In carbide cutting tool applications, tantalum carbide additions prolong tool life in cutting operations of materials such as steel, which yield long, continuous chips. Small additions of

tantalum carbide to tungsten carbide/cobalt alloys inhibit recrystallization of the carbide phase; therefore, the resulting alloys generally have a finer grain size and higher hardness than corresponding compositions that do not contain tantalum carbide. Such alloys also permit wider sintering ranges and are less sensitive to oversintering.

However, because tantalum carbide is a softer carbide, it reduces the hardness of the final product. Usually, additions of more than 12% are not effective in contributing to cutting tool performance in most applications. However, tantalum carbide does not have the reducing effect on transverse-rupture strength that titanium carbide has, and exhibits excellent resistance to thermal shock.

Consequently, tantalum carbide is very useful when added to tungsten carbide/cobalt alloys to produce grades that can be used for special applications. These applications include flash trimming on welded tubes, deep slotting work where heat generation is great and coolant application is difficult, and in machining of uranium where chip adhesion is a serious problem. See the Appendix to this article for typical microstructures and physical properties of tungsten carbide/tantalum carbide/cobalt grades of cemented carbides.

Composition has various effects on the properties of cemented carbides. Increasing cobalt content contributes to lower wear resistance, lower hot hardness, lower resistance to thermal deformation, and lower crater resistance. Higher cobalt content does, however, increase strength.

Generally, the higher the titanium carbide content, the higher the wear resistance, hot hardness, resistance to thermal deformation, and crater resistance exhibited by the carbide. Strength is lower, however. The higher the tantalum carbide content, the higher the hot hardness, resistance to thermal deformation, and crater resistance exhibited; however, wear resistance and strength are lower.

Selection of Cemented Carbides

As cutting tool materials, tungsten carbide materials may be separated into two types. First and most common is the simple two-phase type consisting of tungsten carbide with a cobalt binder (C-1 through C-4). If metal removal involves generation of a short, weak, discontinuous chip, as in the machining of cast iron, this type of material is most effective. These carbides exhibit extreme hardness and excellent resistance to abrasive wear. However, if the workpiece material yields a strong, continuously curling chip, such as with steel, the alloyed grades C-5 through C-8 must be used.

Tungsten carbide and cobalt usually are alloyed with additional carbide(s), usually titanium carbide and/or tantalum carbide. These alloying additions delay crater development and increase tool life. The classification system for carbides used in machining applications is shown in Table 1.

Straight tungsten carbide grades are most resistant to abrasive wear and have the greatest strength for a given hardness value. Two possible variables affect mechanical properties: grain size of the tungsten carbide particles and the percentage of cobalt. The following factors should be considered in grade selection:

- Finer grain size results in higher hardness.
- Lower cobalt content results in higher hardness.
- Higher hardness results in greater abrasive wear resistance.
- Lower cobalt content results in lower strength.
- Finer grain size results in lower strength.

Straight tungsten carbide grades are classified in the C-1 through C-4 range. They are used to machine most cast irons, 200 and 300 series stainless steels, most high-temperature alloys, nonferrous alloys, and nonmetals.

Alloyed tungsten carbide grades are used where the predominant wear factors are adhesion and diffusion due to high cutting forces and high heat. The following factors should be considered in grade selection:

- Higher titanium carbide content results in greater resistance to cratering.
- Higher titanium carbide content results in higher hot hardness.
- Higher titanium carbide content results in higher wear resistance.
- Higher titanium carbide content results in lower strength.
- Higher tantalum carbide content results in greater resistance to thermal deformation and cratering.
- Higher tantalum carbide content results in higher hot hardness.
- Higher tantalum carbide content results in lower strength.
- Higher tantalum carbide content results in lower wear resistance.

Alloyed tungsten carbide grades are classified in the C-5 through C-8 range. These grades are used to machine most carbon and alloy steels, 400 and 500 series stainless steels, and malleable and ductile irons.

Machining Applications

A recent development in cutting tool design is the indexable throw-away cemented carbide insert, which is mechanically held in a toolholder. Inserts are available in several thicknesses and a variety of sizes and shapes, as shown in Fig. 7. Round, square, triangle, and diamond shapes account for the greatest percentage of inserts produced. Many other shapes, including the parallelogram, hexagon, and pentagon, are used to meet specific machining requirements. Each shape has advantages and limitations dictated by operational and economical factors. The manufacturer should be consulted as to the type of insert needed for a specific application.

Several types of coated carbide insert grades are also available. The most widely used coating process is chemical vapor deposition using titanium tetrachloride and

Fig. 7 Typical indexable, throw-away cemented carbide inserts for machining applications

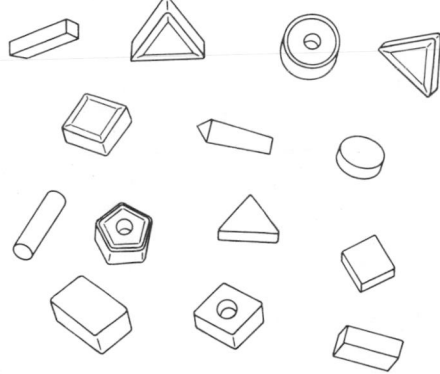

Table 1 Classification system for machining-grade cemented carbides

Grade	Application	Carbide characteristics
C-1	Roughing	Medium-high shock resistance; medium-low wear resistance
C-2	General purpose	Medium shock resistance; medium wear resistance
C-3	Finishing	Medium-low shock resistance; medium-high wear resistance
C-4	Precision finishing	Low shock resistance; high wear resistance
C-5	Roughing	Excellent resistance to cutting temperature; shock and cutting load; medium wear resistance
C-50	Roughing and heavy feeds	Excellent resistance to cutting temperature; shock and cutting load; medium wear resistance
C-6	General purpose	Medium-high shock resistance; medium wear resistance
C-7	Finishing	Medium shock resistance; medium wear resistance
C-70	Semi-finishing and finishing	High cutting temperature resistance; medium wear resistance
C-8	Precision finishing	Very high wear resistance; low shock resistance

Fig. 8 Cross section of vapor-deposited carbide insert

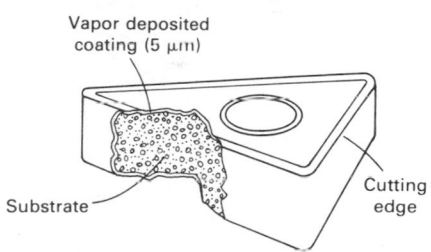

a hydrocarbon vapor with hydrogen as the carrier gas. The thin titanium carbide coating, which is approximately 5 μm thick, causes less friction between the chip and the insert cutting edge during cutting. Reduction of friction eliminates heat, increases resistance to edge wear, crater, and deformation, and greatly improves metal removal rates and tool life. In addition to titanium carbide, several other coatings have been developed, including titanium carbonitride, titanium nitride, hafnium carbide, hafnium nitride, and aluminum oxide. A cross section of a vapor-deposited carbide insert is shown in Fig. 8.

Cemented Carbides for Wear Applications

By John Campbell
Research Director
Carmet Co.

MANY APPLICATIONS exist for cemented carbides in noncutting areas. These may be generally categorized as wear parts. Properties that are paramount for wear part applications include abrasion resistance, impact strength, and corrosion resistance. Due to the physical property requirements, suitable materials for wear part applications include tungsten carbide and cobalt.

Tantalum carbide may be added in some applications for antigalling and corrosion resistance. Nickel may be substituted for cobalt in corrosion-resistant applications. Titanium carbide, widely used in metal cutting grades, is never used in wear applications, because it lowers strength and abrasion resistance.

Mechanical Properties

Abrasion resistance is directly related to hardness in the tungsten carbide/cobalt system. It generally is measured by Cemented Carbide Producers Association procedure CCPA-112. Abrasion resistance, as shown in Fig. 9, is a function of the cobalt binder metal content and the tungsten carbide grain size. The lower the binder content and finer the tungsten carbide grain size, the harder and more abrasion resistant is the final product. Small additions of tantalum carbide (less than 1%) do not affect abrasion resistance, because tantalum carbide is a softer carbide than tungsten carbide. Large additions of tantalum carbide, however, lower abrasion resistance. Nickel binder grades have lower abrasion resistance than comparable cobalt binder grades.

Impact strength, or shock resistance, is a function of cobalt binder content and tungsten carbide grain size. Higher cobalt levels and larger tungsten carbide grain

Fig. 9 Effect of tungsten carbide grain size and cobalt content on the relative abrasion resistance of cemented carbides

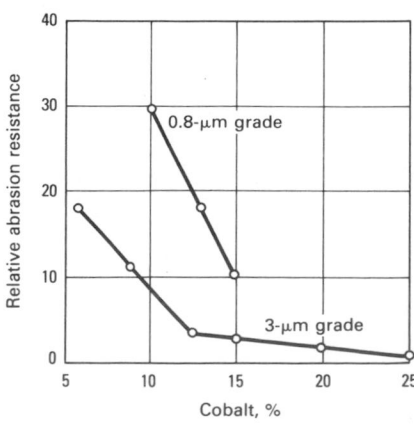

Fig. 10 Effect of cobalt content on the impact strength of cemented carbides

Average tungsten carbide grain size: 3 μm

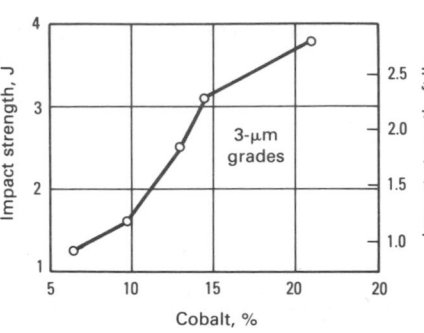

sizes increase impact strength (Fig. 10). Tantalum carbide additions do not affect impact strength, however. Conditions that increase impact strength also increase thermal shock resistance.

Corrosion Resistance. In corrosive attack of cemented carbides, the tungsten carbide component usually is not affected. The binder metal corrodes, exposing the angular tungsten carbide grains. A roughened surface results, which may cause tool wear or work material galling. To increase corrosion resistance, less binder metal should be exposed to attack. This can be accomplished by using a finer tungsten carbide grain size and lower binder metal content. Finer tungsten carbide grain size also yields more tungsten carbide surface area and results in a higher dissolved tungsten content in the binder metal, with improved corrosion resistance. Frequently, nickel binder grades exhibit higher corrosion resistance than comparable cobalt binder grades (Table 2). Small additions of chromium to both nickel and cobalt increase corrosion resistance, but cause embrittlement of the cemented carbide.

Applications

Cemented carbides used for wear applications may be divided into the following categories:

Light impact, high abrasion resistance applications

- Blast nozzles
- Spray nozzles
- Reamers
- Gun drills
- Wear pads
- Powder compaction dies
- Glass cutters
- Plastic extrusion dies
- Guide rings
- Circuit board drills
- Paper slitters

Medium impact, medium abrasion resistance applications

- Tape slitters
- Metal draw dies
- Aluminum extrusion dies
- Shear knives
- Guide rolls
- Snowplow blades

High impact, low abrasion resistance applications

- Cold header dies
- Stamping dies
- Bar mill rolls

Corrosion resistance applications

- Seal rings
- Nozzles
- Ball valves
- Metering chokes
- Aluminum can tooling

Specialty applications

- Nuclear components
- Nonmagnetic guidance gyros
- Nonmagnetic dies for magnets

Light Impact, High Abrasion Resistance Applications. A large number of applications capitalize on the abrasion resistance of cemented carbides. These include sliding friction, light uninterrupted cutting, or particle abrasion applications, which involve minimal mechanical shock. Generally only straight tungsten carbide/cobalt grades are used. High abrasion resistance grades typically have low cobalt

Table 2 Corrosion rates of cemented carbide test samples immersed for 48 h at 22 °C (72 °F)

Material	Tungsten carbide grain size, μm	10% nitric acid g/m²/h	10% nitric acid mg/in.²/h	10% hydrochloric acid g/m²/h	10% hydrochloric acid mg/in.²/h	10% sulfuric acid g/m²/h	10% sulfuric acid mg/in.²/h
94WC-6Co	1.5	2.45	1.58	0.09	0.06	0.19	0.12
85WC-15Co	3	18.85(a)	12.16(a)	0.14	0.09	1.4	0.91
85WC-15Co	0.8	19.82(a)	12.79(a)	0.12	0.08	0.68	0.44
94WC-6Ni	1.5	0.88	0.57	0.03	0.02	0.05	0.03

(a) Tungsten carbide surface layer spalled

contents (4 to 6%) and fine tungsten carbide particle sizes (1 to 2 μm).

High abrasion resistance grades, which are classified in the C-9 through C-12 range, have the lowest impact strengths (see Table 1 for classifications of machining-grade tungsten carbides).

Submicron grades with 6 to 10% cobalt also are used for these applications. Abrasion-resistant grades have the lowest thermal shock resistance. As such, care should be taken when brazing cemented carbides to other materials such as steel, which have significant differences in thermal expansion. The difference in thermal expansion rates can result in fracture during brazing.

Medium impact, medium abrasion resistance applications require a combination of abrasion resistance and impact strength. Generally only straight tungsten carbide/cobalt grades (C-13 classification) are used for these applications. Grade composition ranges from 9 to 13% Co, and tungsten carbide particle size ranges from 2 to 3.5 μm. Submicron grades with from 10 to 15% Co also are suitable for this usage. Tantalum carbide may be added to prevent galling or to improve elevated-temperature properties (Fig. 11). Due to the higher cobalt content and coarser tungsten carbide particle size, these grades have improved thermal shock properties.

High Impact, Low Abrasion Resistance Applications. In many high-impact applications, cemented carbides compete with tool steel. Suitable cemented carbide grades (C-14 classification) have cobalt contents ranging from 15%, with 3 to 4 μm tungsten carbide particle sizes for lamination dies, to 25% with 4 to 5 μm tungsten carbide particle sizes for cold header dies. Generally, submicron grades are not used in impact applications.

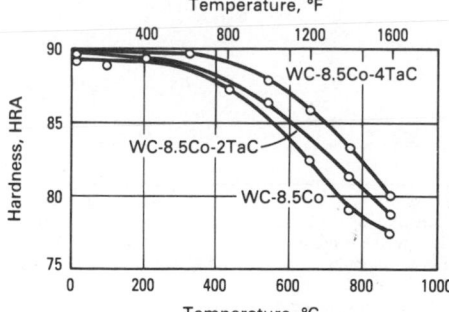

Fig. 11 Effect of tantalum carbide additions on hot hardness of tungsten carbide/cobalt cemented carbides

The 25% Co binder content is the highest available because of the difficulty maintaining shape during liquid-phase sintering. Tantalum carbide is used in some applications for galling resistance. High cobalt grades (18% and above) may be machined with aluminum oxide and silicon carbide wheels and cemented carbide tooling. Below 17 to 18% Co, diamond tools are required.

Corrosion Resistance Applications. Cemented carbides frequently are used for their combined corrosion and abrasion resistance. Typically, the binder metal corrodes. Finer tungsten carbide particle size and lower cobalt content improve corrosion resistance. Grades with binder metal contents as low as 1.5% are used.

Nickel binder grades with fine tungsten carbide grain sizes are finding increasing usage. Important corrosion resistance applications include the use of cemented carbides for valve stems and seats, chemical seal rings, and for aluminum can tooling for use with synthetic lubricants. Grades with small chromium additions are used for sour gas applications (natural gas with high hydrogen sulfide). Ultimately, the optimum cemented carbide composition for a corrosion application depends on the intended service environment.

Specialty Applications. Many wear part applications require the special properties of cemented carbide in addition to abrasion resistance and impact strength. These include nuclear and nonmagnetic applications. In parts for the nuclear industry (seal rings, wear pads, and hardfacing), nickel binder grades are suitable, because nickel has a much shorter radioactive half-life than cobalt.

For nonmagnetic applications (guidance gyros, dies for ceramic magnets, and tape player heads), nickel binder grades with high dissolved tungsten or chromium contents are used. Cobalt binder grades are ferromagnetic. Another prominent application for cemented carbides is gage blocks. A chromium carbide/nickel grade typically is used. This material has a low coefficient of thermal expansion, is corrosion resistant, nonmagnetic, and abrasion resistant, and exhibits high reflectivity when lapped.

Failure Mechanisms

In any wear part application, the predominant failure mechanism can be a guide to proper grade selection. Major failure modes and their causes and solutions are discussed below.

Breakage and chipping can occur due to lack of sufficient impact strength. If this occurs, the carbide grade selected is too hard. This problem may also occur with improper punch-to-die clearance. The use of cemented carbides with higher cobalt content and coarser tungsten grain size lessens the possibility of breakage and chipping.

Excessive abrasive wear can occur if the cemented carbide grade is too soft. Grades with lower cobalt content and finer tungsten carbide should be used to prevent or minimize abrasive wear.

Thermal Deformation. Edges of cemented carbide tooling can become rounded or deformed through softening of the cobalt due to excessive heat buildup. Grades with tantalum carbide additions, lower cobalt content, and finer tungsten carbide grain size should be used to prevent or minimize thermal deformation.

Corrosion. Attack of cobalt binder metal due to incompatibility between carbide and lubricant or water results in excessive wear and subsequent edge breakage. If a change in lubricant does not abate the problem, grades with lower cobalt content and finer tungsten carbide grain size should be used. The use of nickel binders also may lessen the possibility of corrosion.

Galling. Adherence of work material to the carbide die or punch may be caused by excessive heat buildup. Frequently this tendency may be eliminated by using a smoother grind or lubricant. Grades with additions of tantalum carbide, lower cobalt content, and finer tungsten carbide grain size should be used to minimize or prevent galling.

Porosity. Because cemented carbides tend to be brittle, they are sensitive to surface flaws. Porosity can greatly lower strength, resulting in suitable sites for corrosion and metal galling. Hot isostatic pressing can ensure porosity-free cemented carbides and is widely used on wear parts. Typical hot isostatic pressing conditions for cemented carbide are 1300 °C (2370 °F) at a gas pressure of 15 000 psi (argon). Typically, the cobalt binder metal is liquid at hot isostatic pressing temperatures, and argon gas pressure induces cobalt and tungsten carbide to flow and close any existing pores, resulting in minimized porosity.

Appendix:
Microstructures and Physical Properties of Cemented Carbides

Microstructure and physical properties of 98.5WC-1.5Co
Magnification: 1500×. Courtesy of Carmet Co.

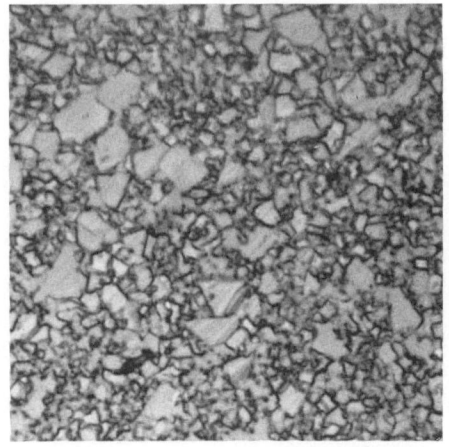

Hardness	93.0 ± 0.4 HRA
Density	15.55 g/cm^3
Transverse rupture	1380 MPa (200 ksi)
Compressive strength	5.8 GPa (850 ksi)
Abrasion resistance	76

Microstructure and physical properties of 95.6WC-4.4Co
Magnification: 1500×. Courtesy of Carmet Co.

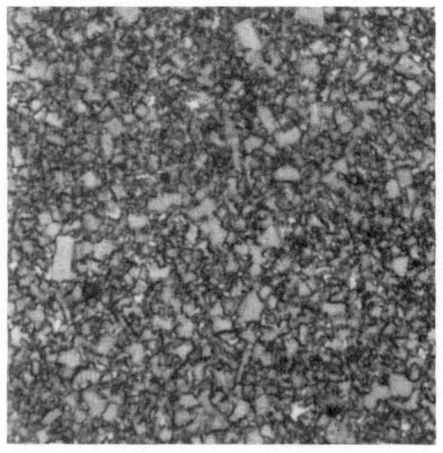

Hardness	92.4 ± 0.2 HRA
Density	15.10 g/cm^3
Transverse rupture	1790 MPa (260 ksi)
Elastic modulus	670 GPa (97 × 10^6 psi)
Compressive strength	5.2 GPa (750 ksi)
Abrasion resistance	56

Microstructure and physical properties of 94WC-6Co
Magnification: 1500×. Courtesy of Carmet Co.

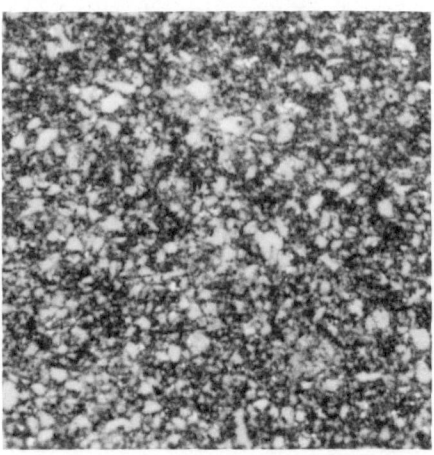

Hardness	93.0 ± 0.3 HRA
Density	14.90 g/cm^3
Transverse rupture	2240 MPa (325 ksi)
Elastic modulus	675 GPa (98 × 10^6 psi)
Compressive strength	4.9 GPa (720 ksi)
Abrasion resistance	65

Microstructure and physical properties of 91WC-9Co
Magnification: 1500×. Courtesy of Carmet Co.

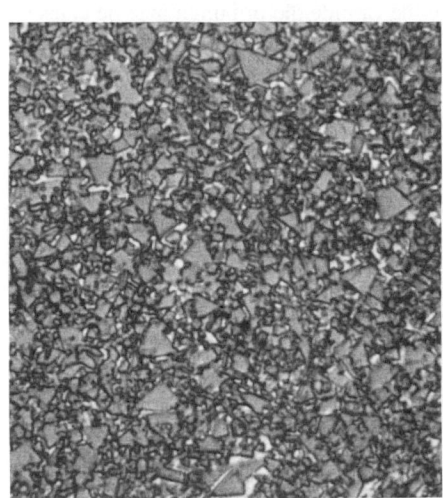

Hardness	90.0 ± 0.5 HRA
Density	14.60 g/cm^3
Transverse rupture	2760 MPa (400 ksi)
Elastic modulus	615 GPa (89 × 10^6 psi)
Compressive strength	4.2 GPa (610 ksi)
Abrasion resistance	11

Microstructure and physical properties of 90WC-10Co
Magnification: 1500×. Courtesy of Carmet Co.

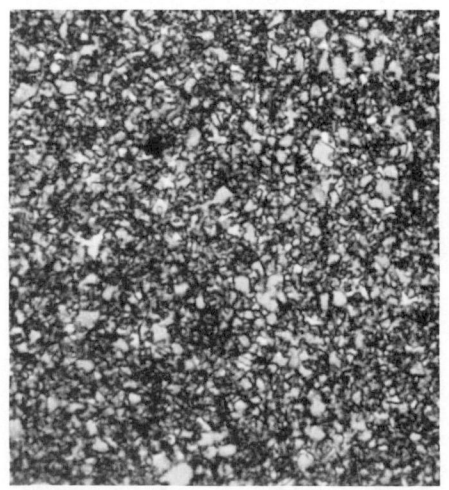

Hardness	92.0 ± 0.5 HRA
Density	14.53 g/cm^3
Transverse rupture	2755 MPa (400 ksi)
Elastic modulus	590 GPa (86 × 10^6 psi)
Compressive strength	4480 MPa (650 ksi)
Abrasion resistance	30

Microstructure and physical properties of 87WC-13Co
Magnification: 1500×. Courtesy of Carmet Co.

Hardness	88.5 ± 0.5 HRA
Density	14.15 g/cm^3
Transverse rupture	3100 MPa (450 ksi)
Elastic modulus	545 GPa (79 × 10^6 psi)
Compressive strength	3.8 GPa (550 ksi)
Abrasion resistance	4.2

Microstructure and physical properties of 86WC-14Co

Magnification: 1500×. Courtesy of Carmet Co.

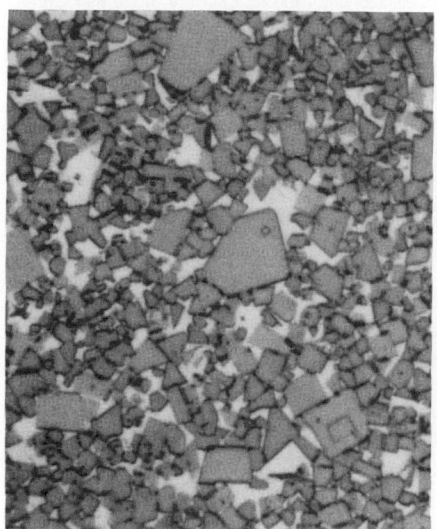

Hardness	88.0 ± 0.5 HRA
Density	14.10 g/cm³
Transverse rupture	3275 MPa (475 ksi)
Elastic modulus	530 GPa (77 × 10⁶ psi)
Compressive strength	3.7 GPa (535 ksi)
Abrasion resistance	4

Microstructure and physical properties of 85WC-15Co

Magnification: 1500×. Courtesy of Carmet Co.

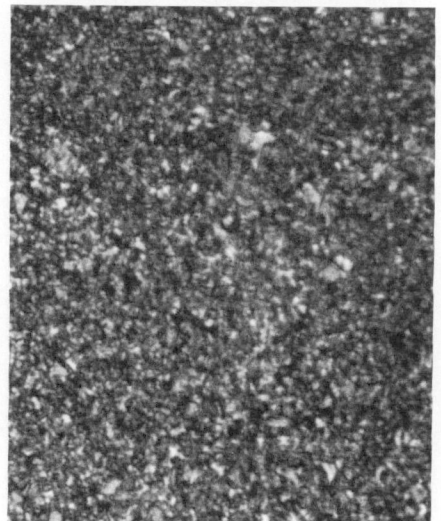

Hardness	89.7 ± 0.5 HRA
Density	14.00 g/cm³
Transverse rupture	3445 MPa (500 ksi)
Compressive strength	4.0 GPa (580 ksi)
Abrasion resistance	10

Microstructure and physical properties of 85WC-15Co

Magnification: 1500×. Courtesy of Carmet Co.

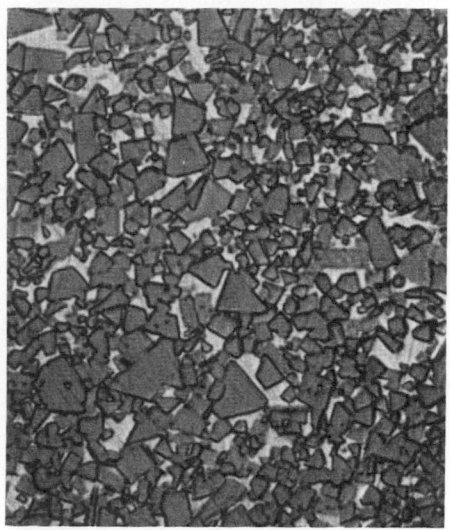

Hardness	87.5 ± 0.5 HRA
Density	14.00 g/cm³
Transverse rupture	3310 MPa (480 ksi)
Elastic modulus	525 GPa (76 × 10⁶ psi)
Compressive strength	3.6 GPa (525 ksi)
Abrasion resistance	3.5

Microstructure and physical properties of 82.4WC-8.3TiC-3.3TaC-6Co

Magnification: 1500×. Courtesy of Carmet Co.

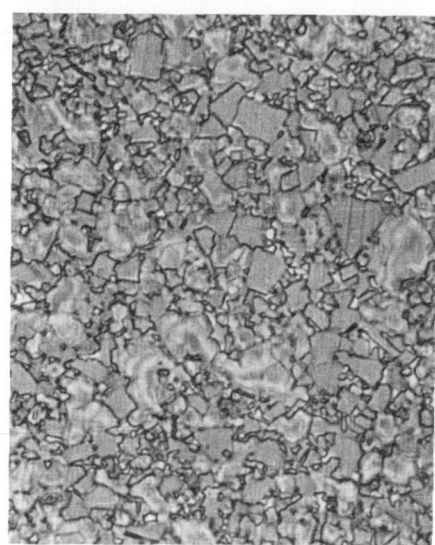

Hardness	92.0 ± 0.5 HRA
Density	12.75 g/cm³
Transverse rupture	1725 MPa (250 ksi)
Elastic modulus	550 GPa (80 × 10⁶ psi)
Compressive strength	3.9 GPa (575 ksi)
Abrasion resistance	15

Microstructure and physical properties of 80WC-5TiC-5TaC-10Co

Magnification: 1500×. Courtesy of Carmet Co.

Hardness	90.5 ± 0.5 HRA
Density	13.10 g/cm³
Transverse rupture	2275 MPa (330 ksi)
Elastic modulus	515 GPa (75 × 10⁶ psi)
Compressive strength	3.3 GPa (475 ksi)
Abrasion resistance	10

Microstructure and physical properties of 78.4WC-8.3TiC-3.3TaC-10Co

Magnification: 1500×. Courtesy of Carmet Co.

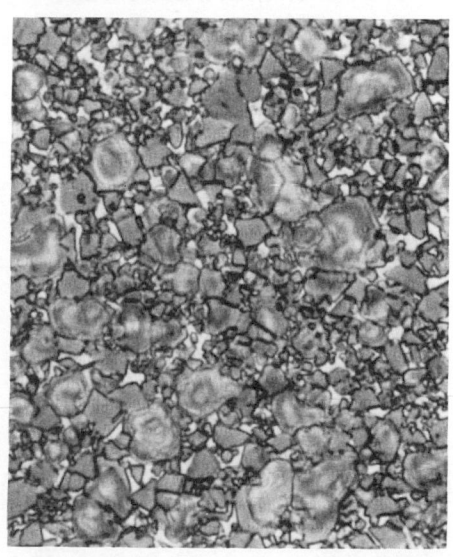

Hardness	90.8 ± 0.5 HRA
Density	12.50 g/cm³
Transverse rupture	2070 MPa (300 ksi)
Elastic modulus	515 GPa (75 × 10⁶ psi)
Compressive strength	3.4 GPa (500 ksi)
Abrasion resistance	10

Microstructure and physical properties of 75.75WC-6.25TiC-10.0TaC-8Co

Magnification: 1500×. Courtesy of Carmet Co.

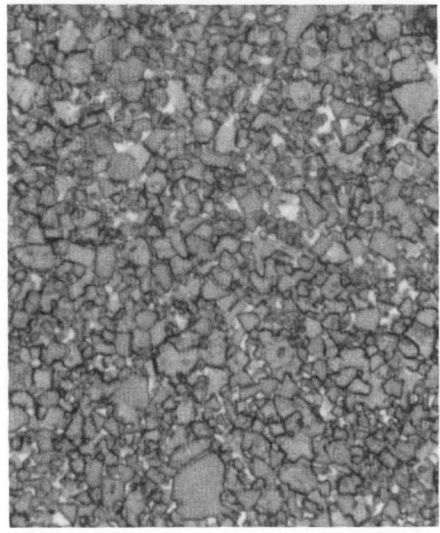

Hardness	91.2 ± 0.5 HRA
Density	12.90 g/cm³
Transverse rupture	1895 MPa (275 ksi)
Elastic modulus	495 GPa (72 × 10⁶ psi)
Compressive strength	3.8 GPa (550 ksi)
Abrasion resistance	10

Microstructure and physical properties of 72WC-8TiC-12TaC-8Co

Magnification: 1500×. Courtesy of Carmet Co.

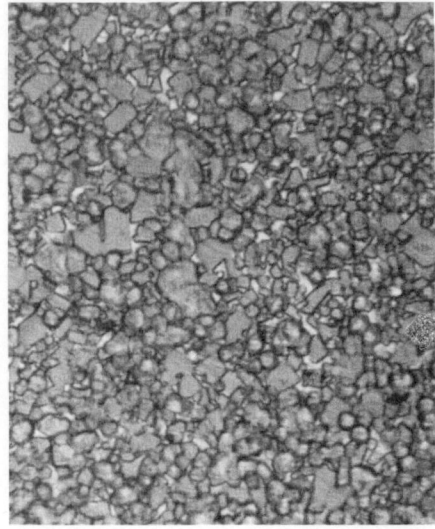

Hardness	92.0 ± 0.5 HRA
Density	12.55 g/cm³
Transverse rupture	1725 MPa (250 ksi)
Elastic modulus	550 GPa (80 × 10⁶ psi)
Compressive strength	3.6 GPa (525 ksi)
Abrasion resistance	9

Microstructure and physical properties of 66.5WC-14.3TiC-10.7TaC-8.5Co

Magnification: 1500×. Courtesy of Carmet Co.

Hardness	91.3-92.3 HRA
Density	11.00 g/cm³
Transverse rupture	1515 MPa (220 ksi)

Microstructure and physical properties of 91WC-3TaC-6Co

Magnification: 1500×. Courtesy of Carmet Co.

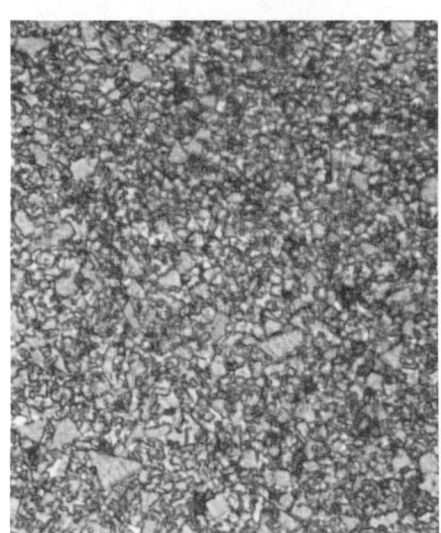

Hardness	92.2 ± 0.4 HRA
Density	14.90 g/cm³
Transverse rupture	2070 MPa (300 ksi)
Compressive strength	4.4 GPa (650 ksi)
Abrasion resistance	34

Microstructure and physical properties of 71.5WC-5TaC-23.5Co

Magnification: 1500×. Courtesy of Carmet Co.

Hardness	83.0-83.6 HRA
Density	13.18 g/cm³
Transverse rupture	2755 MPa (400 ksi)

Microstructure and physical properties of 73TiC-12.5Mo₂C-14.5Ni

Magnification: 1500×. Courtesy of Carmet Co.

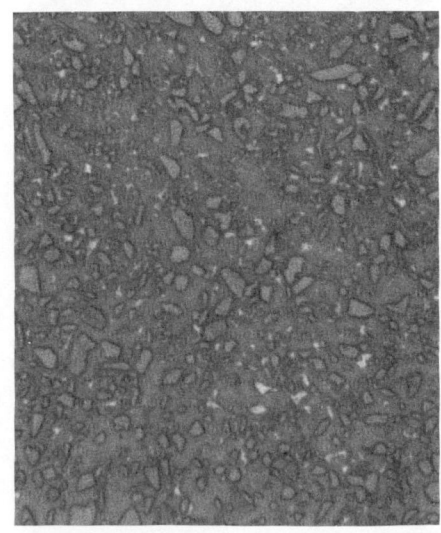

Hardness	92.5-93.0 HRA
Density	6.10 g/cm³
Transverse rupture	1550 MPa (225 ksi)

Microstructure and physical properties of 62TiC-10WC-18Ni-10Mo

Magnification: 1500×. Courtesy of Carmet Co.

Hardness .... 91.0 ± 0.5 HRA
Density ....6.25 g/cm^3
Transverse rupture .... 1725 MPa (250 ksi)
Compressive strength ....3.4 GPa (500 ksi)
Abrasion resistance 7

Microstructure and physical properties of 85Cr$_3$C$_2$-15Ni

Magnification: 1500×. Courtesy of Carmet Co.

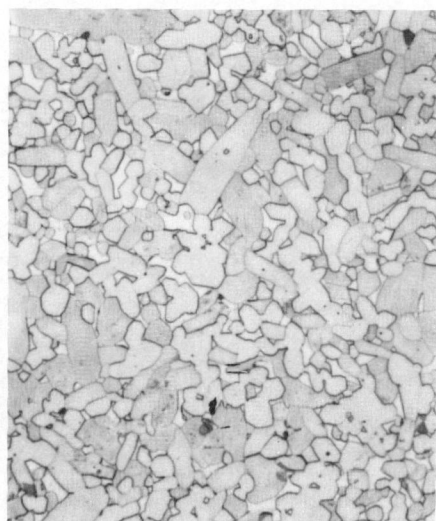

Hardness .... 88.5 ± 1.0 HRA
Density .... 7.0 ± 0.10 g/cm^3
Transverse rupture .... 690 MPa (100 ksi)
Elastic modulus .... 345 GPa (50 × 10^6 psi)
Compressive strength ....3.6 GPa (520 ksi)
Abrasion resistance 1.8

Microstructure and physical properties of 90WC-10Ni

Magnification: 1500×. Courtesy of Carmet Co.

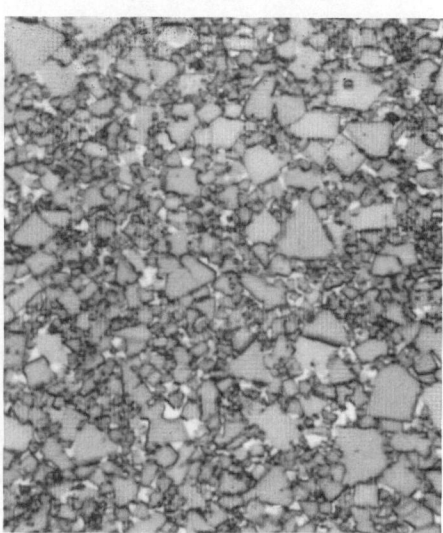

Hardness .... 89.5 ± 0.5 HRA
Density ....14.65 g/cm^3
Transverse rupture .... 2070 MPa (300 ksi)
Elastic modulus .... 615 GPa (89 × 10^6 psi)
Compressive strength ....4.1 GPa (600 ksi)
Abrasion resistance 9

P/M Tool Steels

By Ralph W. Stevenson
Project Manager/Materials Group
Energy Technology Division
Midland-Ross Corp.

HIGH-SPEED TOOL STEELS are used extensively for cutting and forming tools, as well as for wear parts, because of superior mechanical properties and abrasion resistance. However, wrought tool steels exhibit a strong tendency to segregate during solidification in a mold, resulting in a coarse, nonuniform microstructure with variations in chemical composition. This tendency limits toughness, creates heat treating problems, and decreases grindability.

Powder metallurgy processing is being used increasingly to overcome these problems. Advantages of P/M high-speed steels include superior strength and toughness. Powder metallurgy produces a finer grain structure and prevents the carbide clusters (banding) that subsequently limit strength in wrought high-speed steels.

Improved size stability in heat treating is also achieved in P/M parts, because distortion or cracking due to segregation is minimized. This characteristic also provides more uniform hardness during heat treatment.

Improved grindability facilitates toolmaking and resharpening and decreases damage due to burning. Longer tool life results in increased productivity because of reduced downtime for tool-changing. Productivity is frequently increased by the higher speeds and/or feed rates achieved with tools made from these steels. In addition, it is now possible to make highly alloyed grades (for example, ASP 60). Previously, this type of material could not be produced satisfactorily using conventional means.

Since the early 1970's, several P/M techniques for producing high-speed tool steels have been developed, including controlled spray deposition, rapid omnidirectional compaction, consolidation at atmospheric pressure (CAP process), the STAMP process, and injection molding.

These processes are discussed in the Section of this Volume entitled "Special and Developing Consolidation Processes."

This article will discuss procedures for producing an alloy powder by inert-gas atomization, followed by compaction by isostatic pressing. These processes include the Anti-Segregation process (ASP), developed in Sweden by Stora Kopparberg and ASEA, and the Crucible Particle Metallurgy (CPM) process, developed in the United States by Crucible Inc. The FULDENS process, which uses water-atomized powders compacted in a vacuum sintered environment, is also discussed. It was developed in the United States by Consolidated Metallurgical Industries, Inc.

For additional data concerning the classification, composition, heat treatment, and properties of conventionally processed and P/M processed tool steel materials, see the article "Tool Steels" in Volume 3 of the 9th edition of *Metals Handbook*.

The Anti-Segregation Process

The Anti-Segregation process, also referred to as the ASEA-STORA process, is used to produce high-speed steels by powder metallurgy. In this process, an alloy steel melt is atomized in inert gas into a powder. The spherical powder is then poured into cylindrical sheet steel capsules (cans), which are vibrated to pack the particles as tightly as possible. A cover is welded onto the capsule after air has evacuated. The capsule and its contents are cold isostatically pressed at 400 MPa (58 000 psi).

The capsule is then hot isostatically pressed at 100 MPa (14 500 psi) at 1150 °C (2100 °F) to full density. After compaction, the steel is worked conventionally by forging and rolling to the desired

dimensions. Figure 1 compares the processing of conventional (wrought) tool steels with that of ASP tool steels.

This processing results in a fine-grained material with uniform distribution of small carbides. The homogeneous material, completely free from segregation, always has an identical structure, regardless of bar size and alloy content. Figure 2 compares the microstructures of conventional high-speed steel and ASP steel.

Properties of ASP Steels*

The primary benefits of ASP techniques include improved toughness and yield strength due to uniform carbide distribution and the absence of metallurgical defects. Improved grindability due to the small carbide size and improved dimensional stability in heat treatment due to the absence of segregation are also benefits. Additionally, the content of alloying elements and subsequently wear resistance can be increased, without sacrificing toughness and grindability.

Currently, ASP steel is available in three grades: ASP 23, 30, and 60 (ASP 60 can only be made by the powder metallurgy process). Compositions and recommended applications of these grades are given in Table 1. Additional information on applications of ASP steels can be found later in this article.

Wear Resistance. In most tooling applications, wear is caused by a number of factors. Wear resistance is generally a function of the hardness of the tool and the specific alloy content or type of carbide. The higher hardness that is possible with P/M steels, plus higher carbon and vanadium content (see Table 1), directly improve wear resistance.

Toughness of a tool or high-speed steel usually is defined as a combination of

*Source: Ref 1

Fig. 1 Schematic comparison of conventionally (wrought) processed tool steel and ASP tool steel

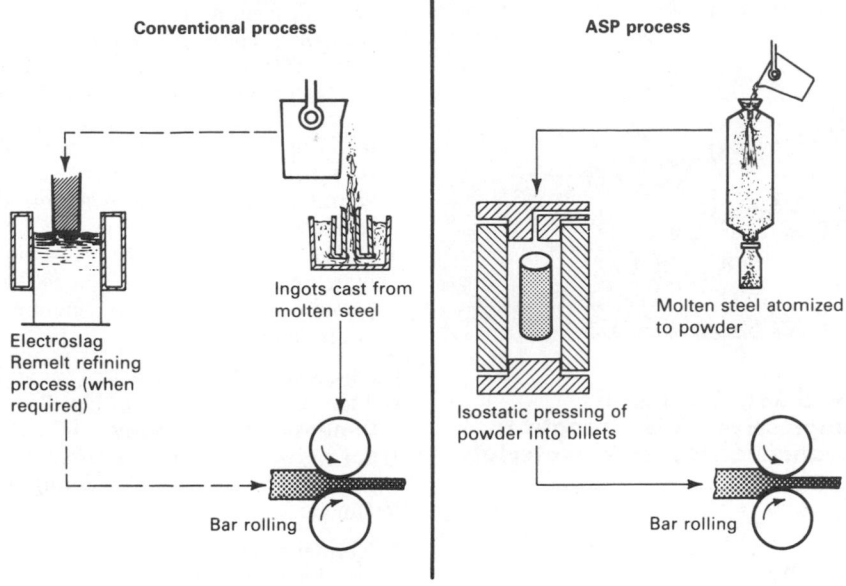

Fig. 2 Comparison of microstructures of conventional and P/M steel

(a) Conventional high-speed steel microstructure showing carbide segregations. (b) Microstructure of ASP steel showing small, uniformly distributed carbide particles, regardless of size. Courtesy of Uddeholm Corp.

Table 1 ASP steel grades, compositions, and applications

ASP grade	Composition, %						Recommended applications
	C	Cr	Mo	W	Va	Co	
23	1.27	4.2	5.0	6.4	3.1	···	For ordinary applications of most cutting tools where hot hardness is not of primary concern. Also for tools used in cold working applications
30	1.27	4.2	5.0	6.4	3.1	8.5	For cutting tool applications where hot hardness is important. Suitable for cutting most stainless steels and superalloys, and for cutting at higher speeds. Also for cold work tools where wear resistance is critical
60	2.30	4.0	7.0	6.5	6.5	10.5	For cutting tools where wear resistance and hot hardness are critical. Particularly suitable for extra tough applications—cutting titanium high hardness materials, and iron forgings

strength and ductility, or as resistance to breaking or chipping. A tool that deforms from lack of strength is useless; a tool that lacks adequate ductility will exhibit brittle failure. Toughness of a tool steel refers to its ability to be unyielding.

The importance of toughness of high-speed steel is illustrated in Fig. 3. A cutting edge may suffer from repeated microchipping. As shown in Fig. 3, the ASP 23 cutting edge shows minimal wear. The M2 cutting edge, however, shows microchipping under the same service conditions. Microchipping blunts the cutting edge, increases stress, and accelerates other wear factors.

One method of measuring toughness of high-speed steel is bend testing. Bend yield strength, ultimate bend strength, and deflection are measured on 5-mm (0.2-in.) diam test bars on which a load is exerted. These tests provide information about the properties of a material after heat treatment. The values, based on laboratory measurements, are confirmed by shop experience.

As shown in Fig. 4, toughness and hardness can be controlled by varying the hardening temperature. A low hardening temperature produces good toughness. Raising the hardening temperature increases hardness, but lowers toughness.

Compressive yield strength of punches and dies is an important factor to consider when selecting material for cold forming tools. For example, the degree of strength required for a punch is influenced by workpiece shape, composition, and hardness of the metal being extruded, as well as by production requirements. Figure 5 illustrates that compressive strength increases proportionally to hardness. The presence of retained austenite in the material reduces yield strength.

In actual practice, useful yield strength is limited by the requirements of toughness. Therefore, a more uniform microstructure, as shown in Fig. 2, yields better toughness. It subsequently allows higher hardness.

Grindability of ASP steel is superior to that of conventional high-speed steel of the same chemical composition. This is due to the small carbide size and the uniform distribution of carbides, regardless of bar size. Figure 6 compares the grindability of several tool steels. These data are based on laboratory measurements, but results are confirmed by shop experience.

Heat Treatment of ASP Steels

Only with proper heat treatment can optimum mechanical properties of tools and dies be obtained. Improper heat treatment may result in a tool with greatly reduced productivity or even an unusable tool.

Heat treatment consists of the following stages:

Fig. 3 Comparison of cutting edge wear of conventional and P/M tool steels

(a) Cutting edge of tool made of conventional AISI M2 material, showing severe microchipping. (b) Cutting edge of tool made of ASP 23 material, showing no microchipping under same conditions as (a). Courtesy of Uddeholm Corp.

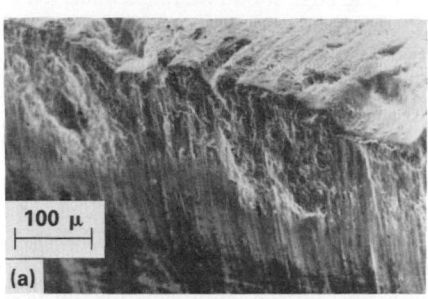

Fig. 4 Bend test results to determine toughness of ASP steels

(a) Bend strength of a test bar of ASP 23 steel after hardening and tempering to 560 °C (1040 °F) (three times for 1 h). (b) Bend strength of a test bar of ASP 30 steel with a 5-mm (0.2-in.) diam after hardening and tempering to 560 °C (1040 °F) (three times for 1 h). (c) Bend strength of a test bar of ASP 60 steel with a 5-mm (0.2-in.) diam after hardening and tempering at 560 °C (1040 °F) (three times for 1 h). Hardness values may vary ±1%; ultimate bend strength may vary ±10%; bend yield strength may vary ±5%.

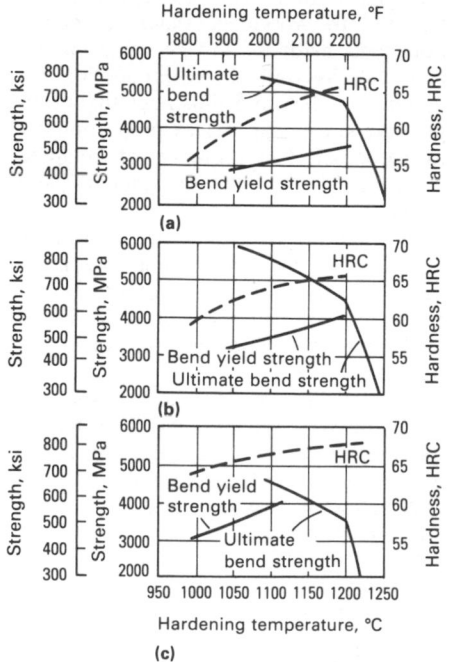

- Preheating
- Austenitizing
- Quenching
- Tempering

The heat treatment procedure for ASP steels is essentially the same as for wrought high-speed steels. Optimum heat treating tem-

Fig. 5 Relationship of compressive yield strength to hardness of tool steel materials

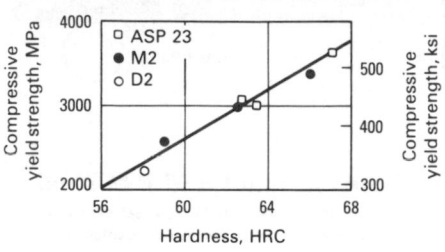

Fig. 6 Grindability of P/M and conventional tool steel materials

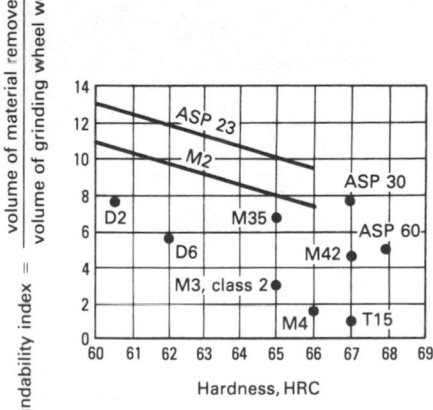

peratures may vary however, even if chemical compositions are identical.

The following procedures should be used to heat treat ASP steels:

- *Soft anneal:* Heat to 850 to 900 °C (1560 to 1650 °F). Slow cool 10 °C/h (18 °F/h) to 700 °C (1290 °F). Hardness values are 260 HB maximum for ASP 23, 300 HB for ASP 30, and 340 HB for ASP 60.

- *Stress relieve:* At 600 to 700 °C (1110 to 1290 °F). Hold for approximately 2 h at temperature. Slow cool to 500 °C (930 °F) in furnace.

- *Harden:* Preheat in two steps, first at 450 to 500 °C (840 to 930 °F) and then at 850 to 900 °C (1560 to 1650 °F). Austenitize at 1050 to 1180 °C (1920 to 2175 °F) and quench, preferably in a neutral salt bath. Cool to hand warmth. See Table 2 for recommended temperatures.

- *Temper:* At 560 °C (1040 °F) or higher three times for at least 1 h at full temperature. Cool to room temperature between draws.

Hardnesses of ASP steel after hardening and tempering are shown in Fig. 7.

Dimensional Stability in Heat Treatment. Three types of distortion are experienced metallurgically during heat treatment:

- Normal volume change due to phase transformations in the steel
- Variations in volume change in different parts of the tool due to the segregation in the steel
- Distortion due to residual stress caused by machining or nonuniform heating and cooling during heat treatment

ASP grades, however, differ significantly from conventionally manufactured high-speed steels. Generally, ASP grades exhibit less volume change than conventional high-speed steels, but higher volume change than cold work steels. Dimensional changes are more uniform in all directions. Because ASP steels are completely segregation free, variations in dimensional change are smaller. As a result, dimensional change occurring during hardening can be predicted more accurately. Conventionally processed high-speed steels go out-of-round in a four-sided pattern. The extent of distortion during heat treatment depends on the type and degree of segregation. In P/M steels, anisotrophy

Table 2 Austenitizing temperatures of ASP 23 steel

Hardness(a), HRC	Temperature(b) °C	Temperature(b) °F	Salt Bath(c), min/in.	Other furnace(d), min
58	1010	1850	15	30
60	1050	1925	12	25
62	1090	2000	10	20
64	1135	2075	8	15
66	1175	2150	6	10

(a) After triple draw at 560 °C (1040 °F); hardness values may vary by ±1%. (b) If vacuum or muffle furnace is used, increase hardening temp. by 14 °C (25 °F) above the recommended temperature. (c) Total immersion time in minutes per inch after preheating. (d) Holding time (min) after tool has reached full temperature

Fig. 7 Hardness of ASP steels after hardening and tempering a 25-mm (1-in.) diam specimen three times for 1 h

(a) ASP 23, oil cooled. (b) ASP 30, oil cooled. (c) ASP 60, cooled in step bath

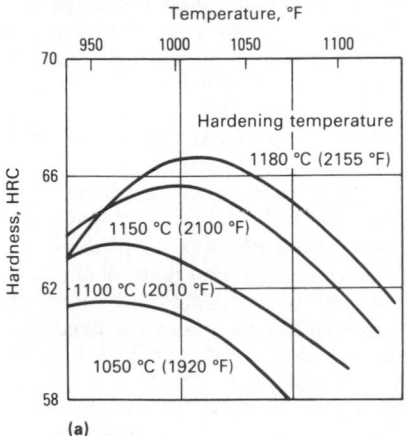

(a)

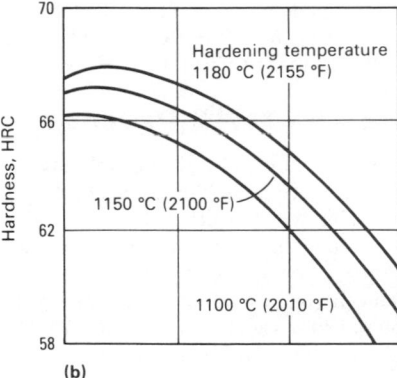

(b)

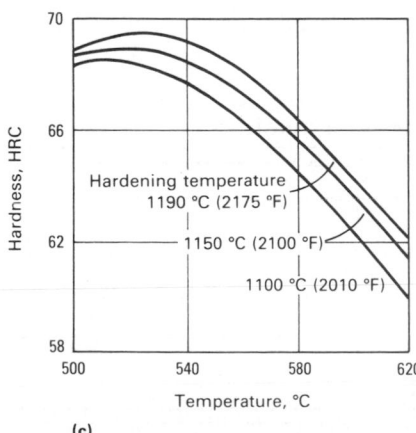

(c)

is smaller, and "out-of-roundness" occurs in a close circular pattern. Figure 8 shows typical results of measuring 100-mm (4-in.) diam disks after hardening and tempering. With ASP steels, cracking

Fig. 8 Out-of-roundness measurements on test disks after hardening and tempering

Test disks machined from 100-mm (4-in.) diam bars

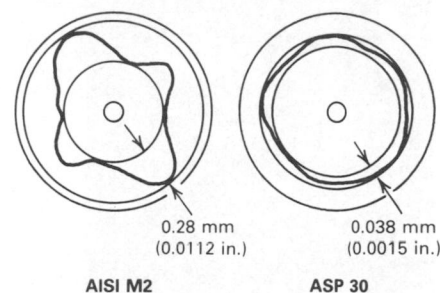

0.28 mm (0.0112 in.) AISI M2

0.038 mm (0.0015 in.) ASP 30

and variation of hardness are minimized, because of their fine-grained, uniform structure.

The same precautions must be taken to control distortion due to residual stresses during heat treating. Mechanical stresses from soft machining can be eliminated by stress relieving prior to finished machining and heat treating.

Crucible Particle Metallurgy Process

Another P/M method to produce tool steel materials is the Crucible Particle Metallurgy process. The process consists of induction melting and atomizing, screening, and containerizing prealloyed particles, followed by hot isostatic pressing to full density. See Fig. 9 for a schematic of the process elements. The desired chemical composition is melted, and the molten stream is poured into an atomizing chamber where high-pressure gas jets disperse it into spheroidal droplets that are rapidly quenched to ambient temperature. Powder is removed from the atomizing chamber, dried, and screened to obtain the desired size fraction. It is then poured into cylindrical steel cans that are evacuated and sealed. Cans are subsequently heated to a specific temperature and hot isostatically compacted to achieve a fully dense product. Compacts are processed to the desired billet and bar sizes by conventional rolling and forging (Ref 2).

As stated earlier, the most detrimental tendency of conventionally produced high-alloy tool steels is the high degree of alloy and carbide segregation that occurs during ingot solidification. This segregation not only reduces hot workability and fabrication of these alloys, but also results in reduced mechanical properties and tool performance. An increase in the carbon and alloy content results in increased segre-

Fig. 9 Schematic of CPM processing

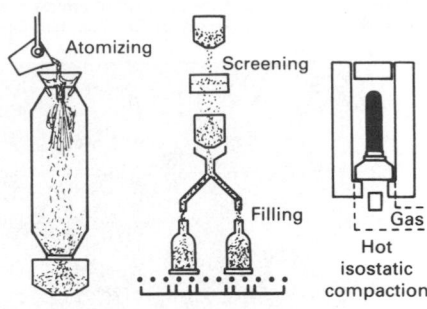

gation and low product yield after hot working.

The CPM process was developed to minimize alloy segregation in standard high-alloy tool steel grades. Additionally, the CPM process is used to produce more highly alloyed grades than can be made by conventional practices.

Alloy T15 (Fe-1.55C-4Cr-5.0V-12.25W-5Co) demonstrates the CPM process advantages. This high-speed tool steel is one of the most wear- and heat-resistant grades of the standard American Iron and Steel Institute (AISI) tool steel materials. However, T15 usage has been limited by conventional production; this highly alloyed, carbide-rich steel is difficult to produce. The CPM process makes it possible to produce such difficult compositions.

The main carbide size distributions in CPM and conventionally produced T15 have been measured. Most carbides in CPM high-speed steel are less than about 3 μm, whereas those in the conventional product cover the entire size range to approximately 34 μm, with a median size of 6 μm. Microstructures of CPM and conventionally processed T15 are compared in Fig. 10.

CPM Alloy Development

Chemical compositions of a variety of CPM tool steels are given in Table 3. These materials are difficult or impossible to produce by conventional methods. Alloy CPM 10V (Table 3) is made by producing a melt composition H-11. In this melt, additional vanadium and carbon are dissolved before it is atomized. The atomized powder is hot isostatically pressed, hot worked, and heat treated. After heat treatment, the microstructure of the material consists of small, spherical, uniformly distributed carbides in a matrix of a tough air-hardening medium-alloy steel, as shown in Fig. 11.

This alloy has a unique combination of wear resistance and toughness at temper-

Fig. 10 Microstructures of CPM T15 (left) and conventional T15 high-speed steel

Carbide segregation and its detrimental effects are eliminated with the CPM process, regardless of the size of the products. Courtesy of Crucible Inc., Specialty Metals Division

 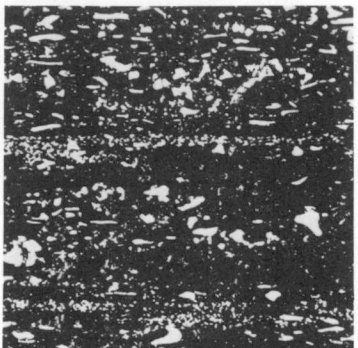

Fig. 11 Photomicrograph of CPM 10V alloy

Note small, spherical, uniformly distributed carbides in a matrix of tough air-hardening medium-alloy steel. Courtesy of Crucible Inc., Specialty Metals Division

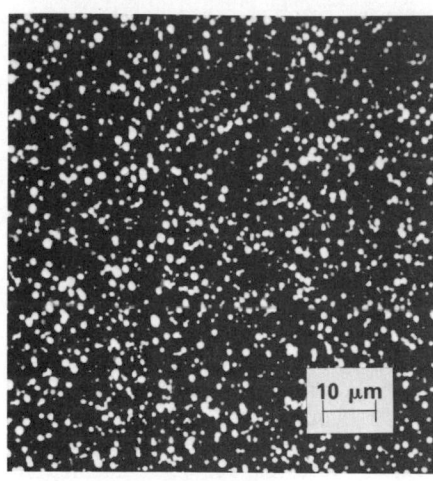

temperature in high-speed steels, thereby permitting the use of high austenitizing temperatures to achieve greater solutioning of alloying elements. Cobalt also enhances the secondary hardening reaction, which results in a 1 to 2 Rockwell C hardness advantage in the fully heat treated condition. It also enhances hot hardness and temper resistance, thus allowing a tool to retain a sharp cutting edge at higher machining speeds that generate heat. Despite the advantages of cobalt additions, the high cost and lack of availability of cobalt have necessitated the development of cobalt-free alternatives.

During initial alloy development, prime objectives were to formulate a composition with satisfactory room- and elevated-temperature hardness, as well as temper resistance equal to that of conventional and CPM T15. If these objectives were successfully met, then impact and bend fracture strength, wear resistance, and tool performance properties could be compared.

The result of this alloy development program was CPM Rex 25 (see Table 3 for chemical composition), which was processed to bar products to obtain property requirements comparable to cobalt-rich tool steels. The results of these comparisons are shown in Tables 4 through 6.

Tables 4 and 5 show the results of temper resistance and hot hardness comparisons for specimens that were heat treated to full hardness. These results show that CPM Rex 25 and CPM T15 are equivalent in both temper resistance and hot hardness. Table 6 compares the Charpy-notch impact and bend fracture strength values obtained for CPM Rex 25 in three heat treated conditions with those obtained for CPM and conventional T15. Both the Charpy-notch impact and the bend fracture strength of CPM Rex 25 are slightly lower than those of CPM T15, but are notably higher than those of conventional T15 and were considered acceptable for this alloy. Table 7 shows the results of laboratory lathe tool tests in single-point cutting on H13 and titanium alloy. The overall performance of CPM Rex 25 was comparable to that of CPM T15.

The FULDENS Process

Another method for the consolidation of P/M tool steels is the FULDENS process. This process differs from the others discussed in this article in that it uses water-atomized powders compacted either mechanically or by cold isostatic pressing and sintered in a vacuum to full density. For a discussion of mechanical properties of water-atomized powders that have been compacted and sintered, see the article "Mechanical Properties of P/M Materials" in this Volume. The FULDENS process allows close-tolerance, complex shapes to

atures up to 480 °C (900 °F). It has been successfully used for P/M compacting dies and in many other tool and die applications. It has frequently replaced more expensive cemented carbide tooling.

Alloy CPM Rex 76 is a cobalt-rich high-speed steel with exceptional hot hardness and wear resistance and greatly increased tool life in difficult cutting operations. The high alloy content—32.5% compared to 27.8% for T15 and 25% for M42—renders this alloy unforgeable if produced by conventional processing.

Two prominent cutting tool grades used in machining difficult-to-machine superalloys and titanium alloys used by the aircraft industry are T15 and M42, which contain 5 and 8% Co, respectively (see Table 2). Cobalt increases the solidus

Table 3 Compositions of CPM processed tool steels

Alloy	Composition, %							
	C	Cr	V	W	Mn	Mo	Co	Fe
CPM 10V	2.40	5.3	9.8	0.3	0.5	1.3	⋯	rem
CPM Rex 76	1.50	3.75	3.0	10.0	⋯	5.25	9.0	rem
CPM Rex 42	1.10	3.75	1.10	1.5	⋯	9.5	8.0	rem
CPM Rex 25	1.80	4.0	5.0	12.5	⋯	6.5	⋯	rem
CPM Rex 20	1.30	3.75	2.0	6.25	⋯	10.5	⋯	rem
CPM T15	1.55	4.0	5.0	12.25	⋯	⋯	5.0	rem

Table 4 Temper resistance of CPM alloys

Alloy grade	Hardness, HRC				
	As heat treated at 1230 °C (2250 °F) + 550 °C (1025 °F) 3 times/2 h	As heat treated + 595 °C (1100 °F)/ 2 h	As heat treated + 595 °C (1100 °F)/ 2 + 2 + 4 h	As heat treated + 650 °C (1200 °F)/ 2 h	As heat treated + 650 °C (1200 °F)/ 2 + 2 + 4 h
CPM Rex 25	68	66.5	65.5	62.5	55
CPM RexT15	67	65.5	64.5	61	54.5

Source: Ref 2

Table 5 Hot hardness of CPM alloys

Alloy grade	Room temperature before test	540 °C (1000 °F)	595 °C (1100 °F)	650 °C (1200 °F)	Room temperature after test
CPM Rex 25	67.8	57.8	54.0	48.4	64.3
CPM Rex T15	66.7	57.1	54.2	48.2	63.4

Source: Ref 2

Table 6 Charpy-notch impact and bend fracture strengths of CPM alloys and conventional alloy

Alloy	Austenitizing temperature(a) °C	°F	Hardness, HRC	Charpy-notch impact strength J	ft·lb	Bend fracture strength MPa	ksi
CPM Rex 25	1230	2250	68	13	9.5	3654	531
CPM Rex T15	1230	2250	67	19	14	4674	678
Conventional T15	1230	2250	66	6	4	2150	312

(a) 4-min soak in salt bath and oil quenched. Tempered at 550 °C (1025 °F) three times for 2 h.
Source: Ref 2

Table 7 Lathe tool test results on CPM alloys

Alloy grade	Austenitizing temperature °C	°F	Hardness, HRC	Intermittent cut on H13 steel at 33 HRC	Continuous cut on H13 steel at 33 HRC	Continuous cut on Ti-6Al-6V-2Sn at 33 HRC
CPM Rex 25	1230	2250	68	18.5	16	17
CPM Rex T15	1230	2250	67	33.5	15	17

Test conditions

Speed, m/s (sfm)	0.25 (50)	0.3 (60)	0.18 (35)
Feed, mm/rev (in./rev)	0.10 (0.004)	0.16 (0.0065)	0.24 (0.0095)
Depth of cut, mm (in.)	1.57 (0.062)	1.57 (0.062)	1.57 (0.062)
Coolant	None	None	None

Source: Ref 2

be made repeatedly with mechanical properties and performance characteristics equal to or better than those of equivalent parts made by conventional machining, with considerable material and labor savings.

Processing Steps. Figure 12 is a flowchart for the FULDENS process. The powders, usually water atomized, are specially prepared both as to composition and particle shape and size distribution. The powder is then annealed and pressed into green compacts either by conventional mechanical pressing or by cold isostatic pressing. When part geometry allows, the part is compacted by filling a closed die with annealed powder and compacting with pressure ranging from 414 to 690 MPa (30 to 50 tsi). Cold isostatic pressing is more suitable for parts of lower volume, higher complexity, and more liberal tolerance; mechanical pressing is more suitable for parts of higher volume, lower complexity, and closer tolerance. Examples of isostatically pressed and mechanically pressed parts are shown in Fig. 13. For more information on cold isostatic pressing, see the article "Cold Isostatic Pressing of Metal Powders" in this Volume.

The compact is then sintered in a specially built vacuum sintering furnace. After sintering, it emerges from the furnace fully dense. For more information on vacuum sintering, see the article "Production Sintering Practices for P/M Materials" in this Volume.

Advantages. Hardenability, elongation, and impact strength of materials drop significantly if small amounts of porosity (even 1 or 2%) are present. This has often limited the applicability of powder metallurgy in demanding applications. Because the FULDENS process provides a finished part of nearly 100% density, improved materials properties and product performance can be expected.

In the FULDENS process, only the net weight of materials in the preform is actually used. In conventional manufacturing, much material is wasted in the form of chips. With higher alloy high-speed steels that contain cobalt, molybdenum, and tungsten (materials in short supply), savings in scrap reduction by using P/M processing can be substantial. Figure 14 illustrates the savings in materials. The

FULDENS process also eliminates much pre-heat treat machining labor.

Applications. Components produced by the FULDENS process should be considered for applications in which the following conditions exist:

- Low net to gross weight ratios
- Relatively high strength to ductility relationships (and sensitivity to impact strength)
- High wear environments
- High temperature environments
- High Hertz stresses

Applications in which fully dense P/M parts are currently in use include screw machine tooling, gear cutting tools, high-speed steel indexable inserts, and forming tools. A significant potential application is diesel engine parts.

Applications of P/M Tool Steels

Milling. Milling cutters, such as those shown in Fig. 15, are emerging as a major application of P/M tool steels. Stock removal rates generally can be increased by raising the cutting speed and/or feed rate. In general, the feed per cutter tooth is increased in roughing operations, and the cutting speed is increased for finishing operations.

The performance of conventionally processed and P/M end mills (Fig. 16) in milling Ti-6Al-4V have been evaluated.* In these tests, ASP 30 and ASP 60 were compared to M42. The cutting conditions used for this evaluation are given in Fig. 17, which shows tool life versus cutting speed. Both feed per tooth (0.203 mm or 0.008 in.) and cutting speeds (>45.7 m/min or >150 ft/min) are higher than those used in production practice for machining aircraft parts. At a constant metal removal rate that corresponds to a cutting speed of 53.3 m/min (175 ft/min), ASP 60 and ASP 30 lasted eight times and 4.5 times longer, respectively, than the M42 end mill.

Other materials machined by P/M milling cutters include tough hardened steels such as 4140, heat-resistant high-alloy steels such as 49K (22Cr-4Ni-9Mn) used for tooling for superplastic forming of titanium, austenitic stainless steels such as AISI 316, and nickel-based superalloys such as Nimonic 80.

Hole Machining. Reamers, taps, and drills (Fig. 18) are also made from high-strength P/M tool steels. In one applica-

*Tests were conducted by Metcut Research, Cincinnati, Ohio.

Fig. 12 Schematic of FULDENS process
Source: Ref 3

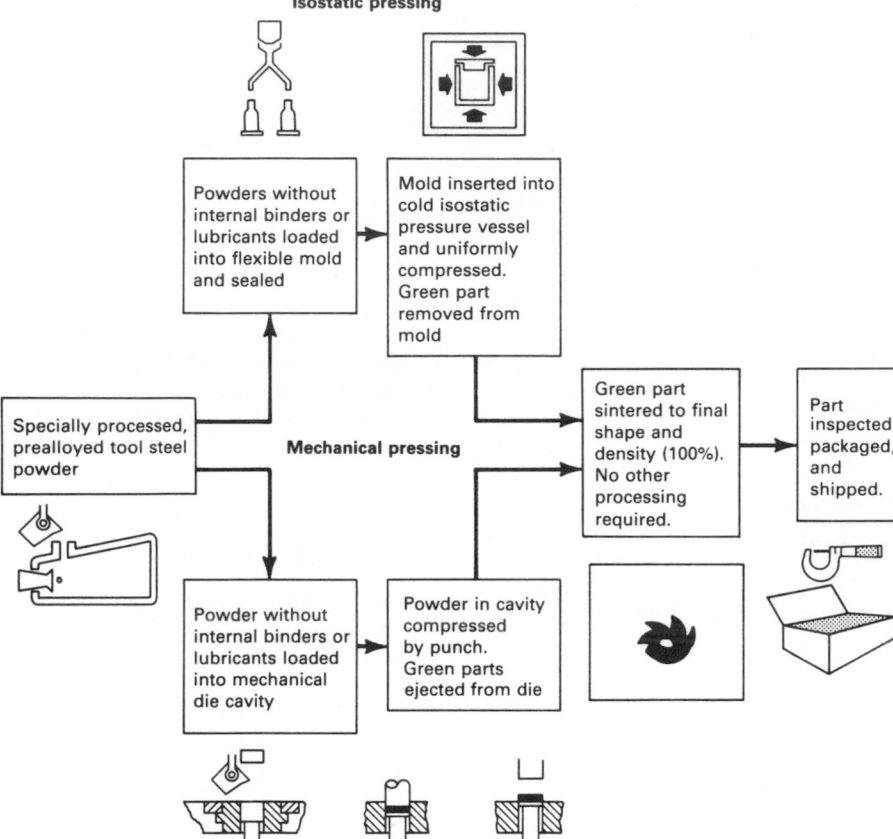

Isostatic pressing

Powders without internal binders or lubricants loaded into flexible mold and sealed

Mold inserted into cold isostatic pressure vessel and uniformly compressed. Green part removed from mold

Specially processed, prealloyed tool steel powder

Mechanical pressing

Green part sintered to final shape and density (100%). No other processing required.

Part inspected, packaged, and shipped.

Powder without internal binders or lubricants loaded into mechanical die cavity

Powder in cavity compressed by punch. Green parts ejected from die

Fig. 14 Solid tap blank and FULDENS cold isostatically pressed finished part

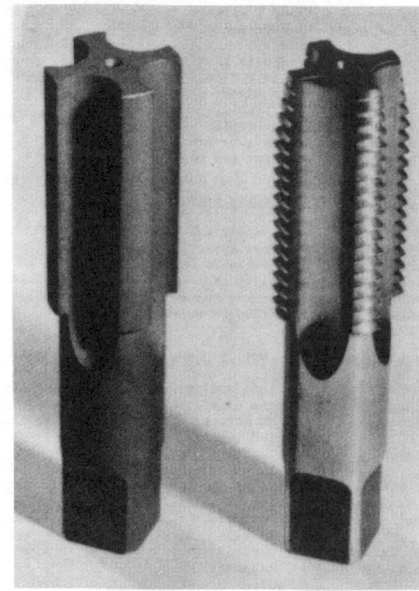

tion, tool life of four-flute reamers made of M44 and ASP 30 was compared. The operation consisted of reaming 16-mm ($^5/_8$-in.) diam holes to a depth of 15 mm (0.591 in.) in oil drill heads made from AISI 8620 steel heat treated to a hardness of 37 to 43 HRC.

Fig. 13 Examples of parts manufactured by the FULDENS process
Note the complexity of shapes attainable by this process. (a) Using cold isostatic pressing. (b) Using mechanical pressing

(a)

(b)

Fig. 15 Typical milling cutters made from P/M tool steels
Courtesy of Uddeholm Corp.

Fig. 16 End mills made from P/M tool steels
Courtesy of Uddeholm Corp.

Fig. 17 End mill test on Ti-6A1-4V
Hardness: 321 HB

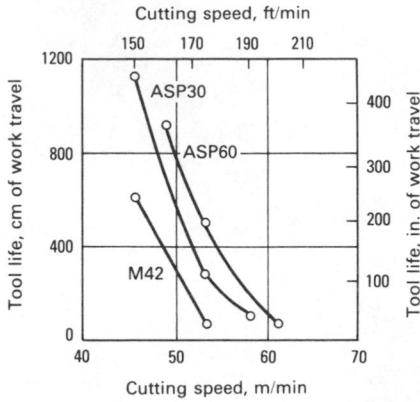

Cutter 25-mm (1-in.)
                                    diam end mills
Feed .... 0.203 mm/tooth
                                    (0.008 in./tooth)
Radial depth of cut .... 6.35 mm (0.250 in.)
Axial depth of cut .... 25.4 mm (1.000 in.)
Cutting fluid .... Soluble oil (1:20)
Tool life end point ..... 0.5 mm (0.020 in.) wear

Fig. 18 Reamers, taps, and drills made from P/M tool steels
Courtesy of Uddeholm Corp.

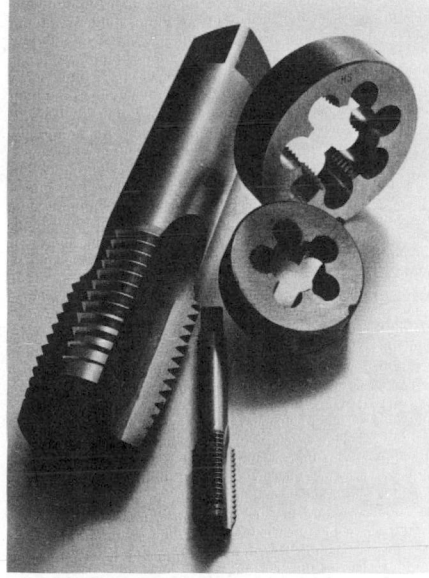

About 0.40 mm (0.016 in.) of stock was removed per side, while maintaining the hole diameter to a tolerance of ±0.013 to 0.038 mm (±0.0005 to 0.0015 in.). Cutting speed was increased from 124 to 189 rev/min, and the feed rate was maintained at 0.040 mm/rev (0.0016 in./rev); consequently, machining time per hole was reduced from 3 min for M44 reamers to 2

min for ASP 30 tools. Simultaneously, tool life of the four-flute reamers was increased from 60 to 180 holes, and regrinding was easier.

Broaching tools constitute another major application for high-speed P/M tool steels, because tool life is often improved when broaching difficult-to-cut materials such as case-hardened steels and superal-

loys. One application required broaching six ball tracks that are used in front wheel drive automobiles in constant-velocity joint hubs made of a case-hardening steel. Figure 19 shows the joint hubs and broaching tool used in this application.

In this broaching application, surface finish and form tolerance requirements are high, because subsequent machining is not performed on the ball tracks. In broaching tools 18 mm (0.709 in.) in diameter and 185 mm (7.283 in.) in length made from high-carbon M35 steel (similar to M41 in chemical composition), the total number of hubs machined per tool was 5600. With a P/M tool, 20 000 parts were produced. This improvement was possible because the M35 tools experienced severe flank wear and the presence of a large built-up edge, which produced poor surface finishes.

Large broaching tools, such as those shown in Fig. 20, are also being made from high-speed P/M tool steels. One application for these tools is the broaching of involute splines in bores of truck transmission gear blanks. Bores up to 300 mm (12 in.) in diameter by 1380 mm ($54^1/_4$ in.) long have been cut using such tools.

Gear Manufacturing. Gear hobs (Fig. 21) made from P/M tool steels can also provide substantial cost reductions by increasing machining rates. One application calls for hobbing of rear axle gears for heavy-duty trucks and tractor differentials manufactured at the Volvo plant in Lindesberg, Sweden. Hobs made of conventionally processed AISI M35 (65 HRC) and ASP 30 (67 HRC) were compared. Test parameters for both materials were:

- *Hob dimensions:* 152-mm (6-in.) diam by 50-mm (2-in.) diam by 205 mm (8 in.) long
- *Work material:* Case-hardening steel (Fe-0.2C-1Cr-1.2Ni-0.12Mo); hardness: 160 to 180 HB
- *Cutting speed:* 70 m/min (230 ft/min)
- *Spindle speed:* 150 rpm
- *Roughing:* Feed: 4.24 mm (0.167 in.); depth of cut: 15.0 mm (0.591 in.)
- *Finishing:* Feed: 5.92 mm (0.233 in.); depth of cut: 0.80 mm (0.032 in.)
- *Coolant:* Cutting oil
- *Number of parts per resharpening:* 20

Production results show that the flank wear land on the hobs made of ASP 30 (0.44 mm or 0.017 in.) is much less than those made of M35 (0.72 mm or 0.028 in.). Chipping of the edges is infrequent on the ASP 30 hobs, while the M35 hobs are frequently subject to such damage. ASP 30 is now the standard grade for hobs used by Volvo.

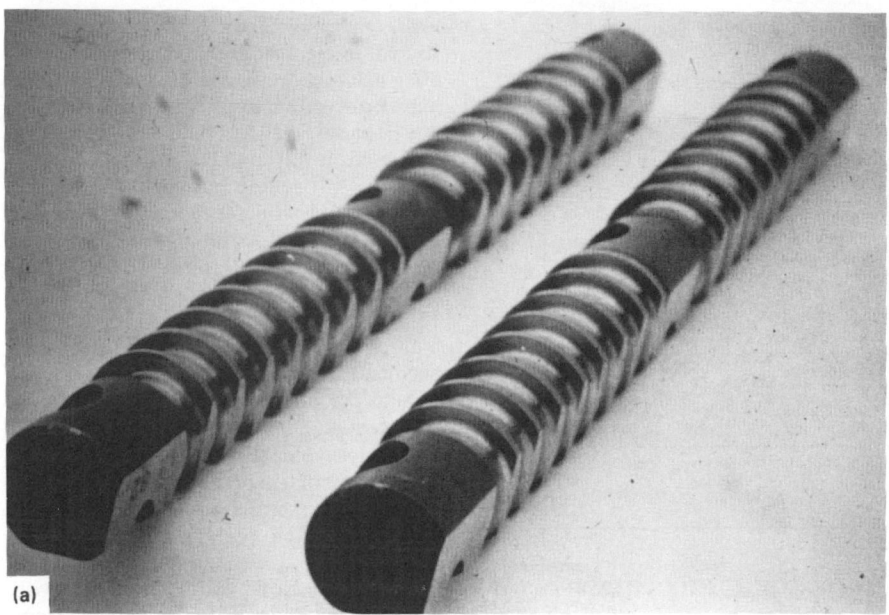

(a)

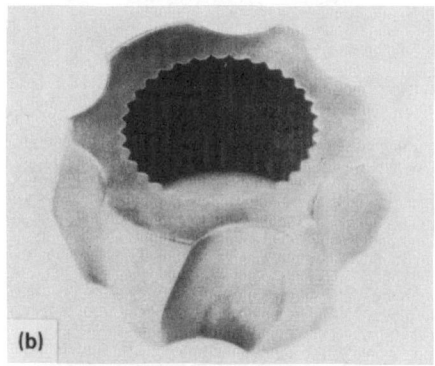

(b)

Fig. 19 Broaching application
(a) Tool used for producing ball tracks on joint hub. (b) ASP 30 tools produced 20 000 parts compared to 5600 parts by tools made from conventional high-speed steel. Courtesy of Uddeholm Corp.

Fig. 20 Large broaching tools made from P/M tool steel used for broaching involute splines in bores of truck transmission gear blanks.
Courtesy of Uddeholm Corp.

Fig. 21 Gear hobs made from P/M tool steels
Courtesy of Uddeholm Corp.

Other applications of P/M high-speed steels include their use for punches and dies for conventional and fineblanking tooling; cold heading, forming, compacting, and extruding tools; and thread rolling dies. Superior wear resistance and toughness, plus excellent grindability, make them ideal candidates for such applications.

Figure 22 shows several applications where P/M tool steels have been used for hole punching. These include:

- Punching of fine slots in brass parts for clocks
- Punching of 3-mm (0.118-in.) holes in high-carbon steel (0.7% C) sheet (240 HB)
- Punching of holes in chain links made of 0.55% carbon steel with a thickness of 2 mm (0.079 in.) and a hardness of 290 HB

Figure 23 shows a variety of parts that

have been blanked, formed, extruded, or pressed using P/M tools. Typical applications include:

- Blanking of watch cases for wrist watches made from 18-8 stainless steel
- Compacting of iron powder for screen covers for television picture tubes
- Fineblanking of components for automobile safety belts

Fig. 22 Parts with holes punched by P/M tool steels
Courtesy of Uddeholm Corp.

Fig. 23 Parts produced with P/M tool steel tooling
Courtesy of Uddeholm Corp.

Fig. 24 Tool bits made from P/M tool steels
Courtesy of Uddeholm Corp.

- Backward cold extrusion of 40-mm (1.575-in.) diam aluminum tubes with narrow thickness tolerances

Figure 24 shows tool bit designs that have been produced using P/M tool steels. Typical applications include:

- Machining of turbine blades made from superalloys
- Turning of hardened steels, such as En 24 (0.4C-19Cr-3.5Ni)

REFERENCES

1. Siegwarth, O., "Higher Productivity with ASP Tooling Material," Society of Manufacturing Engineers Technical Paper MF 81-137, 1981, p 1-22
2. Dax, F.R., Haswell, W.T., and Stasko, W., Cobalt-Free CPM High Speed Steels, *Processing and Properties of High Speed Tool Steels*, The Metallurgical Society of AIME, 1980, p 148-158
3. Podob, M.T. and Harvey, R.P., Advantages and Applications of CMI's FULDENS Process, *Processing and Properties of High Speed Tool Steels*, The Metallurgical Society of AIME, 1980, p 181-195

SELECTED REFERENCES

- Dulis, E.J. and Neumeyer, T.A., Particle-Metallurgy of High-Speed Tool Steel, in *Materials for Metal Cutting*, The Iron and Steel Institute, London, Publication 126, 1970, p 112-118
- Kasak, A. and Dulis, E.J., Powder-Metallurgy Tool Steels, *Powder Metall.*, Vol 21 (No. 2), 1978, p 114-123
- Kasak, A., Steven, G., and Neumeyer, T.A., "High-Speed Tool Steels by Particle Metallurgy," SAE Paper 720182, Society of Automotive Engineers, Warrendale, PA, 1972, p 2-5
- Berry, J.T., "High Performance High Hardness High Speed Steels," Climax Molybdenum Co., Greenwich, CT, 1970
- Roberts, G.A., Hamaker, J.C., Jr., and Johnson, A.R., *Tool Steels*, American Society for Metals, 1962, p 710-713
- Cehlin, Bengt-Ake, "Improving Productivity With High Strength P/M High Speed Steel Cutting Tools," SME Technical Paper MR82-948, presented at Increasing Productivity With Advanced Machining Concepts Clinic, Los Angeles, 1982
- Henderer, W.E. and von Turkovich, B.F., "The Influence of Heat and Surface Treatments on the Performance of M1 HSS Taps," ASM Symposium on Hole Making Operations, Boston, May 1977
- Hellman, P. and Wisell, H., "Effect of Structure on Toughness and Grindability of High Speed Steels," Colloquium on High Speed Steels, Saint-Etienne, France, Nov 1975
- Söderberg, S., Hogmark, S., Haag, H., and Wisell, H., "Wear Resistance of High Speed Steel Milling Tools," Uppsala University, Institute of Technology, Report No. 821 OR
- Hellman, P. *et al.*, The ASEA-STORA-Process, *Modern Developments in Powder Metallurgy*, Vol 4, Plenum Press, 1970, p 573-582
- Alvelid, B. and Wisell, H., Wear of High Speed Steel in Orthogonal Milling, *Scand. J. Metall.*, Vol 9, 1980, p 59-67
- Hellman, P., Wear Mechanism and Cutting Performance of Conventional and High-Strength P/M High-Speed Steels, *Powder Metall.*, Vol 25 (No. 2), 1982
- Carlson, E.A., Hansen, J.E., and Lynn, J.C., Characteristics of Full-Density P/M Tool Steel and Stainless Steel Parts, in *Modern Developments in Powder Metallurgy*, Vol 13, Metal Powder Industries Federation, Princeton, NJ, 1980
- "Tool Steels Today Newsletter," AISI Committee on Tool Steel Producers, Washington, D.C., Jan 1976
- Kasak, A., Stevens, G., and Neumeyer, T.A., Colt Industries, "High-Speed Tool Steels by Particle Metallurgy," SAE Paper No. 720182, Society of Automotive Engineers, Detroit, 1972
- *Metals Handbook*, Vol 3, 9th ed., American Society for Metals, 1980
- "Crucible CPM Rex—High Speed Steel for Superior Cutting Tools," Colt Industries, Pittsburgh, PA
- Huppmann, W.J. and Beiss, P., Sintering of P/M Tool Steels to Full Density, *Sintermetallwerk Krebsoge GMBH*, presented at Fully Dense P/M Materials for High Performance Applications, and MPIF Short Course, New Orleans, Feb 1982

Amorphous Powder Metals

By Ranjan Ray
President
Marko Materials, Inc.

AMORPHOUS METALLIC SOLIDS are produced by rapid quenching of metal from a melt. They are characterized by metallic bonding of atoms and the absence of three-dimensional, long-range crystallographic periodicity. Amorphous metals with noncrystalline structures produced by continuous cooling from the melt are called metallic glasses. They lack the grain boundaries, dislocation faults, and segregated characteristics typical of crystalline materials.

Since the synthesis of the first metallic glass alloy based on gold-silicon in 1960, many new amorphous compositions based on iron, nickel, cobalt, copper, palladium, platinum, titanium, and zirconium have been formulated. Development of engineering applications for these new materials has generated increased interest in them. This article will emphasize the fabrication, properties, and applications of metallic glasses based on iron and nickel. Data on other compositions can be found in Ref 1.

Metallic Glasses

Metallic glasses exhibit excellent mechanical strength, outstanding soft magnetic properties, and superior corrosion resistance (Ref 1, 2). Rapid solidification techniques based on melt-spinning are well developed, with the capability to fabricate vast quantities of metallic glasses, primarily as thin ribbons or tapes. In melt-spinning, liquid metal is brought in contact with a rapidly rotating substrate of high thermal conductivity (usually copper), whereby the melt is rapidly solidified as a thin ribbon approximately 40 μm (1.6 mils) thick. The most commonly produced compositions contain 80 at.% iron, nickel, and/or cobalt, with additions of carbon,

boron, silicon, and phosphorus totaling up to 20 at.%. Iron-based metallic glasses are expected to be suitable substitutes for conventional iron-silicon and iron-nickel alloys for power transformers and electronic devices.

Disadvantages of Metallic Glass Alloys. Metallic glasses undergo transformation by crystallization (devitrification) into very brittle phases at relatively low temperatures of 350 to 500 °C (660 to 930 °F). This phenomenon makes them unsuitable for high-temperature applications. Also, because of the need for extremely fast heat extraction during processing, metallic glass alloys are normally fabricated only as thin bodies, such as wires, ribbons, and sheets. Their application in bulk products is not feasible.

Microcrystalline Alloys Produced From Metallic Glasses

Recent research has led to synthesis of new types of metallic glasses. These alloys contain critical combinations of transition metals such as iron, nickel, and cobalt with metalloid additions (for example, boron, carbon, and silicon). These optimized compositions permit devitrified materials to attain improved ductility with heat treating at high temperatures. Most metallic glasses have eutectic/hypereutectic compositions containing approximately 20 at.% (total) boron, phosphorus, carbon, and silicon. The new heat treatable alloys contain metalloids in the range of 5 to 13 at.% (total), as shown in Table 1.

Table 1 Selected groups of heat-treatable bulk microcrystalline alloys made from metallic glasses

Alloy system, at.%	Examples
$R_{30-75} R'_{10-30} Cr_{0-30} M_{0-15} B_{5-12} (P,C,Si)_{0-2.5}$	
R is one of the group consisting of iron, nickel, and cobalt; R' is one or two elements of the group consisting of iron, nickel, and cobalt other than R; and M is an element of the group consisting of molybdenum, tungsten, niobium, and tantalum; the sum of chromium, R', and M must be at least 12%, and the boron content is 80% or more of the combined metalloid content (boron, phosphorus, carbon, and silicon).	$Fe_{50} Cr_{25} Ni_{16} Mo_5 B_{10}$ $Fe_{45} Cr_{20} Ni_{15} Mo_{12} B_8$ $Ni_{40} Cr_{25} Co_{10} Fe_{10} Mo_5 B_{10}$ $Co_{40} Cr_{30} Ni_{10} Fe_{10} B_{10}$ $Co_{55} Ni_{10} Fe_{15} W_6 Mo_6 B_8$
$Fe_{30-80} Cr_{0-40} (Co,Ni)_{0-2} (Mo,W)_{0-20} B_{5-12} (P,C,Si)_{0-2.5}$	
The sum of chromium, cobalt, nickel, molybdenum, and/or tungsten cannot be less than 10%, and when the content of molybdenum and/or tungsten is less than 10%, chromium content must be equal to or greater than 8%; maximum metalloid content should not exceed about 12%.	$Fe_{71} Cr_{15} Mo_4 B_{10}$ $Fe_{63} Cr_{22} Ni_3 Mo_2 B_8 C_2$ $Fe_{75} Cr_8 Mo_5 W_2 B_{10}$ $Fe_{70} Cr_{10} Ni_3 Mo_7 B_8 C_1 Si_1$
$Ni_{48-75} Cr_{0-20} Mo_{10-30} B_{5-12}$	
Molybdenum is greater than 20%, and chromium must be equal to or less than 15%.	$Ni_{55} Cr_{15} Mo_{20} B_{10}$ $Ni_{65} Mo_{25} B_{10}$ $Ni_{60} Mo_{30} B_{10}$ $Ni_{57} Cr_{10} Mo_{25} B_8$

Metalloid-lean metallic glasses, when heat treated sufficiently above the crystallization temperature (T_c), are transformed into alloys containing an aggregate of microcrystalline phases. Microcrystalline alloys have a tensile strength generally above 1240 MPa (180 ksi), good ductility, and good thermal stability up to about 700 °C (1290 °F).

Microstructure. The superior mechanical properties of the new microcrystalline alloys is due to their microstructures. The isotropic glassy phase, which has a critically balanced composition after heat treating at temperatures sufficiently above T_c (preferably between 0.6 and 0.95 of the solidus temperature on the Celsius scale), transforms to yield a unique microstructure consisting of an aggregate of microcrystalline phases. Metallographic studies by optical transmission and scanning electron microscopy reveal that devitrified alloys exhibiting high tensile strength values consist of extremely fine homogeneous microstructures (Fig. 1). The matrix based on iron, nickel, and/or cobalt typically is constituted of ultrafine grains of 0.2 to 0.3 μm (8 to 12 μin.) that are uniformly dispersed among 0.1 to 0.2 μm (4 to 8 μin.) intermetallic phases (such as borides).

Microcrystalline alloys can be classified as new generic alloys based on carefully selected and controlled compositional, microstructural, and processing principles. They have evolved from a series of hypoeutectic metallic glass alloys by reducing metalloid content to significantly low levels. A low boron content is critical in achieving desirable microstructures of devitrified alloys, with an optimum content (approximately 20 to 30 vol%) of finely dispersed intermetallic phases. A high boron content will lead to formation of a continuous network of brittle boride phase in the devitrified alloy; such microstructures lead to extreme brittleness.

Metallic Glass Powders. One method of fabricating bulk microcrystalline alloys involves hot isostatic pressing and/or hot extruding of tape into rolls of metallic glass ribbons, with rolls packed in a sealed can under vacuum. Alternatively, metallic glasses are processed as powders, followed by subsequent P/M consolidation to bulk microcrystalline alloys.

Several methods of producing rapidly solidified powders have been developed. Generally, atomization is used, followed by rapid cooling. These processes have limitations; for example, a wide range of liquid droplet sizes generally is produced. Because of their differing sizes, these droplets cool at different rates, which results in a nonuniform product. Larger

Fig. 1 Typical microstructures of microcrystalline alloys devitrified from metallic glass phase

(a) Optical micrograph. Alloy: $Ni_{40}Cr_{25}Co_{10}Fe_{10}Mo_5B_{10}$ (at.%). Etchant: Marbles reagent. Magnification: 1000×. (b) Transmission electron micrograph. Alloy: $Ni_{40}Cr_{25}Co_{10}Fe_{10}Mo_5B_{10}$ (at.%). Dark regions represent the intermetallic phases; white regions represent matrix grains consisting of microtwins. Magnification: 36 000×. (c) Scanning electron micrograph. Alloy: $Fe_{50}Cr_{20}Ni_{15}Mo_5B_{10}$ (at.%). Etchant: Marbles reagent. White phase is boride. Magnification: 7000×. (d) Transmission electron micrograph. Alloy: $Fe_{45}Cr_{20}Ni_{15}Mo_{12}B_8$. Dark regions represent the intermetallic phase; white regions represent matrix grains consisting of microtwins. Magnification: 40 000×.

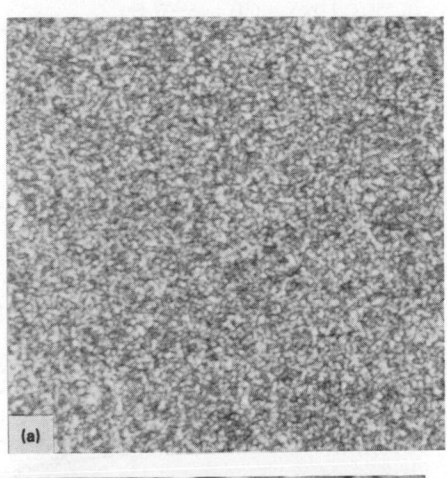

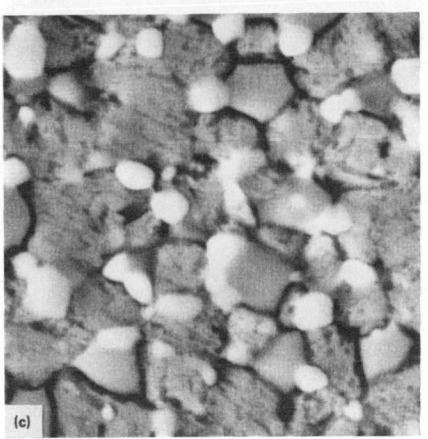
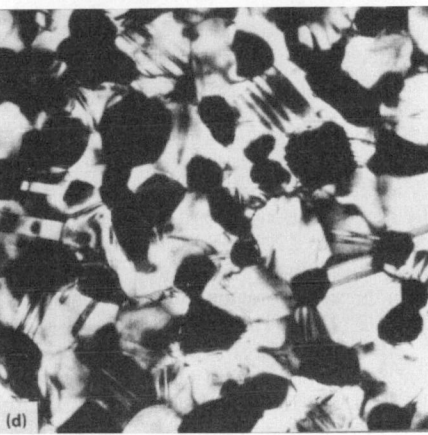

droplets also may cool at rates that are lower than the rate required to form metallic glasses (10^6 °C/s, or 1.8×10^6 °F/s). Screening to separate only the finest particles provides a low yield and is a less economical process.

Metalloid-lean metallic glasses form brittle ribbons upon melt-spinning, in striking contrast to metalloid-rich metallic glass ribbons, which have high tensile strength and excellent bend ductility. The brittle ribbons of low-metalloid glassy alloys are pulverized readily into powders by hammer milling. Efficient high-volume production of such metallic glass powders involves on-line pulverization of many ribbons spun simultaneously at 30 m/s (6000 ft/min) off the chilled rotating substrate. The quench rate of the liquid is more uniform in the melt-spinning process.

In powders made from pulverized ribbons, particles of all sizes undergo essentially the same quenching process, creating a highly uniform consolidated product. The melt-spinning and pulverization procedure is used to provide a high yield (approximately 95%) of a relatively fine powder (100 U.S. mesh, or 150 μm) suitable for subsequent P/M handling and processing.

Metallic glass powder particles have a platelet-shaped morphology, with relatively smooth edges and surfaces, resulting from fracture of hard and brittle ribbons (Fig. 2).

Microcrystalline Alloys

Iron-based microcrystalline alloys with atomic weight compositions that are

Fig. 2 Scanning electron photograph of metallic glass powder prepared by pulverization of ribbons
Magnification: 170×

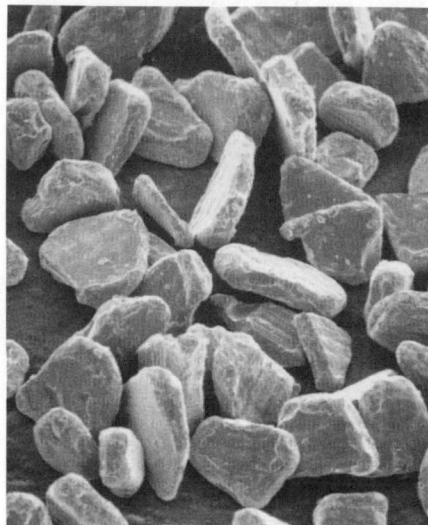

Table 2   Hot hardness of iron-based microcrystalline alloys
After 30 min at temperature

Alloy	Room temperature	Hot hardness, HRC, at: 315 °C (600 °F)	540 °C (1000 °F)	650 °C (1200 °F)
$Fe_{70}Cr_{18}Mo_2B_9Si_1$ 44		43	43	42.5
AISI H21 steel (0.3%C-0.4%V-3.25%Cr-9.5%W-Fe) 49.5		45	34.5	19
AISI H26 steel (0.55%C-1.0%V-4.0%Cr-18%W-Fe) 46.5		42	32.5	20

Fig. 3 Average hot hardness of microcrystalline nickel-molybdenum-iron-boron alloys and commercial hot work tool alloys

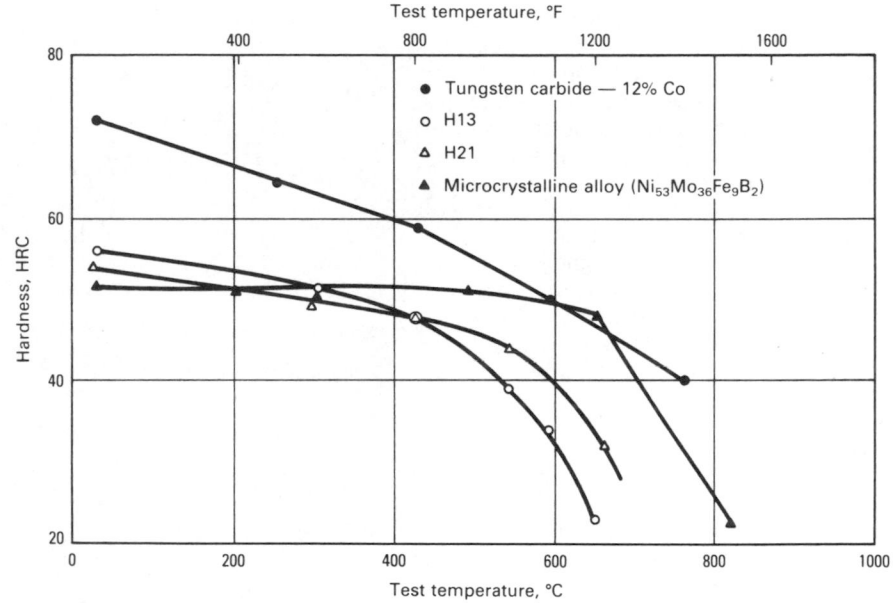

similar to those of $Fe_{70}Cr_{13}Ni_6Mo_1B_9Si_1$ and $Fe_{62.5}Cr_{16}Mo_{11.5}B_{10}$ possess ultimate tensile strengths of 1380 to 1585 MPa (200 to 230 ksi). Alloys containing chromium and refractory metals such as molybdenum and tungsten are characterized by high hot strength/hardness at 540 to 650 °C (1000 to 1200 °F), excellent oxidation resistance, and good Charpy V-notch impact energy at elevated temperatures up to 540 °C (1000 °F).

Alloys for Tool Applications. Based on early laboratory studies, microcrystalline alloys have the properties required for hot work tool applications. Table 2 shows the superior hot hardness of an iron-based microcrystalline alloy, compared to commercial tungsten-bearing hot work tool steels. The mechanical properties of a nickel-molybdenum-based microcrystalline alloy having the composition $Ni_{53}Mo_{36}Fe_9B_2$ are:

Ultimate tensile strength, MPa (ksi)	1419 to 1835 (205 to 265)
Yield strength at 0.2% offset, MPa (ksi) ...	1073 to 1419 (155 to 205)
Transverse-rupture strength, MPa (ksi)	3358 (485)
Unnotched impact strength, J (ft · lb)	68 (50)
Charpy V-notched impact strength, J (ft · lb)	4.75 (3.5)
Coefficient of thermal expansion, μm/m · K (μin./in. · °F)	8.98 (5)
Density, g/cm³ (lb/in.³)	9.13 (0.33)

This alloy has been used successfully for extrusion dies at 850 °C (1530 °F) in the primary reduction of hot billets of copper and brass to redraw bars and shapes. The principal advantages of these dies, compared to dies of high-cobalt steel alloys, are long service life and good surface finish retention. These properties reduce grinding and finishing costs prior to subsequent redrawing operations. Hot hardness values of nickel-molybdenum-based microcrystalline alloys, compared to commercial hot work die materials, are shown in Fig. 3.

Microcrystalline iron-based alloys containing small additions of carbon can be heat treated to change the hardness and ductility of cutting tools in the same manner that the hardness and ductility of steel may be changed with heat treatment. Hardness values in excess of 68 HRC can be obtained. Such high hardness is attributed to their unique microstructures, which consist of ultrafine boride particles in an ultrafine-grained martensitic matrix.

These alloys are devitrified during hot consolidation to bulk forms from the corresponding glassy state (powders). After suitable heat treatment (such as annealing), they can be machined to any desired shape, such as cutting tools or wear-resistant parts. Machined parts are subsequently heat treated for specific properties such as high hardness and toughness. The extremely homogeneous, fine microstructure of these alloys enhances the grindability of tools made from them.

Aluminum Die Casting Mold Inserts. Die casting molds for aluminum alloys usually are machined from H13 steel, which has good heat resistance. The life of an H13 mold in the critical gate area where the hot die casting alloy enters the mold is approximately 75 000 to 125 000 cycles (or "shots"). Die casting molds typically fail by erosion and thermal fatigue. Cyclic thermal stresses cause heat cracks and fissures in the mold surface. Microcrystalline nickel-molybdenum alloy ($Ni_{53}Mo_{36}Fe_9B_2$) has demonstrated a service life of over 250 000 shots.

Microcrystalline alloys crystallized from an isotropic amorphous phase are structurally homogeneous and isotropic. Mi-

Table 3 Aqueous corrosion rates of a microcrystalline alloy ($Ni_{52.7}Cr_{20.3}Mo_{22.8}Fe_{2.9}C_{0.8}B_{0.5}$)

		Test temperature		Test time,	Corrosion rate	
Solution	Concentration, wt%	°C	°F	h	µm/yr	mils/yr
Hydrochloric acid	10	24	75	48	12.7	0.5
	10	80	175	24	81.3	3.2
Nitric acid	10	80	175	24	157.5	6.2
Sulfuric acid	10	80	175	48	25.4	1.0
Phosphoric acid	40	80	175	48	2.5	0.1
Formic acid	20	80	175	24	53.3	2.1
Ferric chloride	10	24	75	24	101.6	4.0

Table 4 Wear and corrosion resistance of arc plasma spray coatings of metallic glass alloys

	Corrosion rate					
	1% HCl at 50 °C (120 °F)		10% H_2SO_4 at 50 °C (120 °F)		Abrasive wear volume loss, mm^3	
Alloy composition, wt%	µm/yr	mils/yr	µm/yr	mils/yr	ASTM G 65(a)	ASTM D 2714(b)
$Ni_{53}Mo_{36}Fe_9B_2$	330	13	...	...	7-10	0.05
$Ni_{54}Mo_{44}B_2$	102	4	500	20	9-10	0.045

(a) Sand/rubber wheel. (b) Falex/block and ring

crocrystals are not organized into large crystalline matrices with anisotropic characteristics that can lead to nonuniform stresses. With uniform distribution of thermal stresses, cracks are less likely to occur during thermal expansion of the insert surface. Microcrystalline nickel-molybdenum alloy has a low coefficient of thermal expansion, which minimizes cyclic strain and fatigue during die casting.

Metal-Bonded Abrasive Wheels. Metallic glass powders are used as metal binders for diamond, silicon carbide, and other abrasive materials in oil well and mine drilling bits, masonry saws, and grinding wheels. Typically, nickel-molybdenum-iron-boron and nickel-molybdenum-boron alloys as metallic glass powders are mixed with the abrasive material and are then hot pressed at 950 °C (1740 °F) and 34 to 48 MPa (5 to 7.5 ksi) into a 97% dense bulk shape. During hot pressing, the internal structure of the powder changes from metallic glass to microcrystalline at temperatures of about 550 °C (1020 °F). The ma-

trix of the consolidated composite product is hard and wear resistant, with the capability to maintain substantial hardness at elevated temperatures.

Corrosion-Resistant Microcrystalline Alloys. Laboratory studies indicate that nickel-based microcrystalline alloys containing chromium and molybdenum generally possess excellent corrosion resistance in different acidic and alkaline solutions. Superior corrosion resistance is attributed to excellent chemical homogeneity and uniformity of structures. Table 3 lists aqueous corrosion rates for various acidic solutions of a microcrystalline alloy with a nominal composition $Ni_{52.7}Cr_{20.3}Mo_{22.7}Fe_{2.9}C_{0.8}B_{0.5}$ prepared by hot extrusion of metallic glass powder.

Corrosion-Resistant Plasma Spray Hardfacing. Metallic glass powders are used as corrosion-resistant arc plasma hardface coatings. They are applied with a wide range of plasma spray equipment to achieve uniformly dense, well-bonded amorphous coatings of high hardness.

Amorphous coatings do not crack or flake from the substrate when they are deformed. Table 4 gives wear- and corrosion-resistance rates of two nickel-molybdenum-boron metallic glass alloy plasma sprayed coatings.

REFERENCES

1. Gilman, J.J. and Leamy, H.J., Ed., *Metallic Glasses,* American Society for Metals, 1978
2. Gilman, J.J., Ferrous Metallic Glasses, *Met. Prog.,* July 1979, p 44-47

SELECTED REFERENCES

- Cantor, B., Ed., *Rapidly Quenched Metals III, 3rd International Conference,* Vol 1 & 2, The Metals Society, London, 1978
- Glickstein, M.R., Patterson, R.J., and Shockey, N.E., Rapid Solidification Effects of Micron-Size Droplets, proceedings of the Second International Conference on Rapid Solidification Processing, Reston, VA, March 1980, p 46
- Grant, N.J., A Review of Various Atomization Processes, proceedings of the Second International Conference on Rapid Solidification Processing, Reston, VA, March 1980, p 230-240
- Kushnik, J. and Sawheny, D.L., Recent Applications of Ultra Rapid Solidification Technology (URST) Microcrystalline and Amorphous Powder Materials, proceedings of the Third International Conference on Rapid Solidification Processing, National Bureau of Standards, Gaithersburg, MD, Dec 1982
- Ray, R., Bulk Microcrystalline Alloys From Metallic Glasses, *Met. Prog.,* June 1982
- Ray, R., Bulk Microcrystalline Alloys From Metallic Glass Powder, *Int. J. Powder Metall. Powder Technol.,* Vol 18 (No. 3), 1982, p 209

Cermets

By Claus G. Goetzel
Consultant, Lecturer
Department of Materials Science
and Engineering
Stanford University

CERMET is an acronym that is used worldwide to designate "a heterogeneous combination of metal(s) or alloy(s) with one or more ceramic phases" (Ref 1). An American Society for Testing and Materials (ASTM) Study Committee on cermets described the material as "a heterogeneous combination of metal(s) or alloy(s) with one or more ceramic phases in which the latter constitutes approximately 15 to 85% by volume and in which there is relatively little solubility between metallic and ceramic phases at the preparation temperature" (Ref 2). A good definition of the term ceramic can be found in the *Ceramic Glossary* (Ref 3): "Any of a class of inorganic, nometallic products which are subject to a high temperature during manufacture or use. Typically, but not exclusively, a ceramic is a metallic oxide, boride, carbide, or a mixture or compound of such materials; that is, they include anions that play important roles in atomic structures and properties." With particular reference to cermets, this definition of the ceramic component could be broadened to include nitrides, carbonitrides, and silicides.

In contrast to composite laminates, the combination of metal and nonmetal in cermets occurs on a microscale. The nonmetallic phase is usually not fibrous, but consists of more or less equiaxed fine grains that are well dispersed in and bonded to the metal matrix. If either the ceramic or the metallic component is predominantly fibrous, the material should be designated as a fiber composite. The bond between the nonmetallic phase and the metal matrix makes important contributions to the cermet; it is strongly affected by the phase relations, solubilities, and wetting properties that exist in the relationship between the ceramic and metallic components.

The size of the ceramic component varies, depending on the system and application. It may be as coarse as 50 to 100 μm, as in some types of uranium dioxide (UO_2) based cermets for nuclear reactor fuel elements, or as fine as 1 to 2 μm, as in the micrograin type of cemented carbides. If the ceramic component is even finer and is present in small amounts, the material may be considered to fall outside the accepted definition for cermets (see the article "Dispersion Strengthened Materials" in this Volume).

The basic concept of combining metal and ceramic on an intimate scale is incorporation of the desirable qualities of either species and suppression of the undesirable properties. When cermets were conceived, about 35 years ago (Ref 2 and 4), expectations were that the refractoriness, strength, and corrosion resistance of the ceramic phase could be mated advantageously on a proportional basis with the high ductility and thermal conductivity of the metallic phase and that some superior new materials would become available for a multitude of high-temperature applications.

Unfortunately, these goals were not fulfilled, despite a major development effort in the United States and Europe during the 1950's. The degree of ductility and toughness imparted by the metallic binder phase remains inadequate for most critical applications, such as turbojet and stationary gas turbine blades or nozzle vanes. In other areas, however, cermets have proven their value as engineering materials, notably in some types of nuclear fuel elements. Uranium dioxide based cermets, as well as uranium carbide (UC) based cermets, offer potential for advanced fuel elements. Cermets based on zirconium boride (ZrB_2) combined with silicon carbide (SiC), or others containing aluminum oxide (Al_2O_3), silicon dioxide (SiO_2), titanium carbide (TiC), and graphite or diamond, possess unique properties. Several are used commercially in applications that range from high-speed cutting or hot machining tools to ultrahigh-temperature components in chemical rockets to use in grinding wheels and furnace fixtures.

The metallic binder phase may consist of a variety of elements, such as nickel, cobalt, iron, chromium, molybdenum, and tungsten, alone or in combination, and also other metals, such as stainless steel, bronze, or superalloys. The volume fraction depends entirely upon the intended properties and end use and may range, as previously cited, from 15 to 85%.

Classification

Four main groups of cermets exist: those based on oxides, those based on carbides, those based on borides, and those that contain carbon as graphite or diamond.

Oxide-based cermets constitute a category composed primarily of UO_2 or thorium dioxide (ThO_2) for a major fission component in nuclear reactor fuel elements, Al_2O_3 or other highly refractory oxides for components in liquid-metal manipulation (for example, pouring spouts) and general furnace parts, and SiO_2 for a minor constituent in friction elements. Combinations of Al_2O_3 with TiC are suitable for hot machining tools.

Boride-based cermets constitute a class that has a boride of one of the transition metals as the dominant phase. These cermets provide excellent high-temperature corrosion resistance against attack by active metals in the molten or vapor state; for example, a combination of ZrB_2 and

SiC is resistant against erosion from the propulsion gases of chemical rockets.

Carbide-based cermets constitute the bulk of the cermets. This group includes the broad field of cemented carbide cutting tools and wear parts based on tungsten carbide (WC), the so-called bonded TiC materials for tools and special form parts, the TiC cermets for high-temperature components in propulsion systems, and chromium carbide (Cr_3C_2) based materials for corrosion resistance applications.

Carbon-containing cermets are a group comprised of materials that contain graphite in different proportions for electrical contacts or as a minor constituent to provide some lubrication in friction elements. Also included are diamond particles within metal matrices that are used in special tools.

Fabrication Techniques

The methods used for powder preparation, forming, firing or sintering, and post-treatments of cermets generally are similar to conventional ceramic and powder metallurgy processing techniques. Figure 1 is a flow chart of the different powder metallurgy techniques applicable to cermets, or at least to their metallic constituents. Table 1 summarizes the relative characteristics of the major forming methods that are practiced in producing oxide-based and most other types of cermets. The principal processes are cold pressing and sintering, hot pressing and pressure sintering, slip casting, and infiltration. Both cold and hot pressing include static uniaxial and isostatic multiaxial compaction.

Cold pressing and sintering is the most often used production method for cemented carbides and many other types of cermets. Powders of the ceramic and metal are milled together to a particle size that is generally finer than 325 mesh. In this operation, which is conducted in ball mills, the components are thoroughly blended together, and where metallic binders are part of the system, the ceramic particles become coated with a film of metal.

The powder mixtures are compacted at pressures of 35 to 100 MPa (5 to 14.5 ksi). The predominant method is pressing dry wax-lubricated powder from two sides in hardened steel dies. For long rods or tubes of uniform cross section, these dies are used for extrusion of a paste in which the powder particles are embedded in a suitable organic binder or wax. To form complex shapes, the dry powder is placed in a pliable mold and compacted from all sides by hydrostatic pressure inside a sealed, reinforced steel cylinder.

Fig. 1 Powder metallurgy production methods for cermets
Source: Ref 5

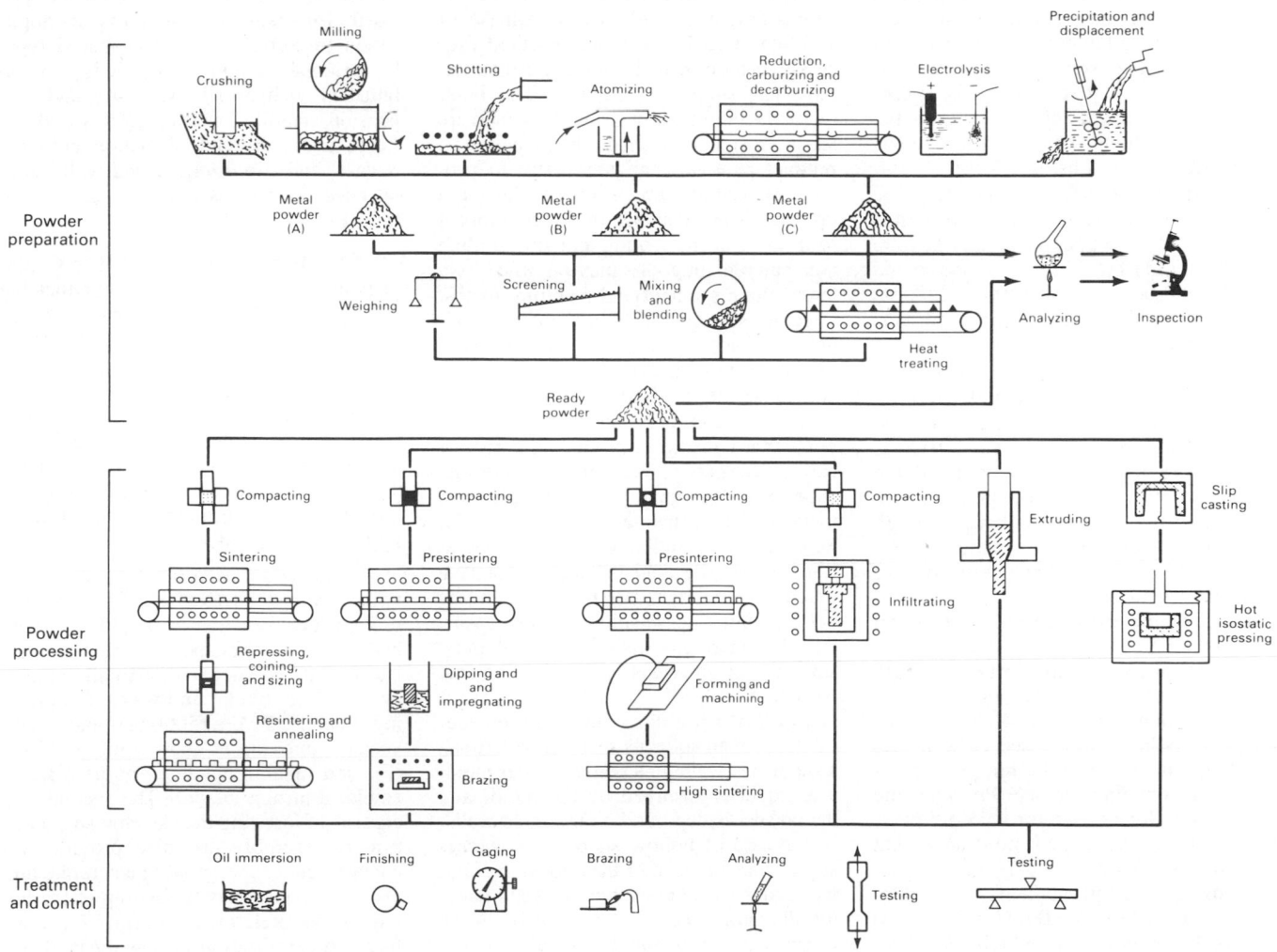

Table 1 Cermet forming techniques

Technique	Size capability	Shape capability	Mold or die requirement	Production Rate	Production Labor cost
Slip casting .... Limited by green strength		Intricate, undercuts	Wood or metal patterns, plaster molds	Low	High
Static cold pressing .... Limited by press capacity		Straight sided, no undercuts	Hardened steel or carbide dies	High	Low to moderate
Cold hydrostatic pressing ..... Limited by pressure vessel capacity		Intricate, limited dimensional accuracy	Rubber or plastic bags	Medium	Low
Static hot pressing .... Limited by mold strength		Straight sided, no details	Graphite or aluminum oxide molds	Low	Moderate to high
Hot isostatic pressing .... Limited by pressure vessel capacity		Intricate, limited dimensional accuracy	High-temperature-resistant sheet metal containers	Medium	Low to moderate
Extrusion .... Long lengths, limited cross sections		Straight dimensions	Alloy steel or ceramic abrasion-resistant dies	High	Low
Infiltration .... Limited by melting point difference, liquid metal infiltration rate		Intricate, requires machining green compacts	Hardened steel or carbide dies, ceramic molds	Low	Moderate
Plasma spraying .... Limited in thickness by deposition rate		Intricate, controlled by substrate	Expendable patterns, no molds	Medium to high	Low

Source: Ref 6

Sintering converts the green compacts into solid, strong bodies. The operation generally requires a controlled atmosphere such as dry hydrogen, dissociated ammonia, or a vacuum and a temperature high enough to bond and weld the particles together. Sintering temperatures depend entirely on the ceramic-metal systems involved and on whether sintering is to take place entirely in the solid state or in the presence of a liquid phase, which generally yields higher densities. Typical temperatures range from 850 to 1050 °C (1560 to 1920 °F) for products that contain a bronze, silver, or copper metal matrix; 1300 to 1500 °C (2370 to 2730 °F) for cemented carbides and borides; and 1700 to 2200 °C (3100 to 4000 °F), or even higher, for certain ceramic oxide-based cermets.

For applications in which intricate contours require fine machining and grinding, as in many cemented carbide parts and tools, presintering is performed at 1000 to 1100 °C (1830 to 2010 °F) to bond the metallic contact points and give enough strength to the body to withstand rough machining. Allowance is made for the substantial shrinkage that occurs during subsequent sintering.

Hot pressing and pressure sintering is a method for production of cermets in which the pressure and temperature are applied simultaneously. The powder mixtures are either compacted directly in the hot press mold or prepressed cold in dies and then transferred to the hot press tools or pressure sintering furnace. Pressures are considerably lower than for the cold press method. They may range from dead weight loads up to 3 MPa (500 psi) for pressure sintering (of friction elements, for example) or from 10 to 35 MPa (1500 to 5000 psi) for hot pressing, in which the lower

end of the range applies to liquid-phase systems.

Sintering temperatures are reached by induction or resistance heating of the mold or of the powder compact directly. In the former case, the mold material consists of graphite. This is the more practical process, because usually no controlled atmosphere supply is required. The latter method requires ceramic molds, which are sensitive to thermal shock, break easily on product removal, and are costly to produce to accurate dimensions of the mold opening. The advantage of direct compact heating—that the tooling and surrounding area can remain cool—may be offset by a temperature gradient and resulting microstructure segregation effects in the product. For most systems with readily oxidizing metal matrices, a controlled atmosphere is required.

Hot isostatic pressing is increasingly popular as a means of producing boride or carbide cermets of very high density and superior surface quality; for example, cermets used in cemented carbide rolls for Sendzimir-type mills. Often, this operation follows the cold press and sintering procedure, in which case the free-standing body requires no sheathing, and pressurized hot inert gas inside a heated pressure vessel provides protection against oxidation.

Slip casting utilizes fine-grain powder mixtures in an aqueous suspension to form a slip that is cast into a porous plaster mold. The liquid is absorbed by the mold, and the powder is deposited on the mold walls. In the case of hollow shapes, the excess slip is drained off after the deposit reaches the required wall thickness; for solid parts, the slip must remain and dry slowly. When completely dried and hardened, the de-

posit has shrunk sufficiently to be readily stripped from the plaster mold. The part is then further densified and strengthened by sintering. The slip casting process is used widely in the ceramic industry, primarily for small, intricate parts. Its application to cermets (the oxide-based type, for example) requires knowledge of parameters such as slip viscosity and suspension stability, wetting agents and deflocculants, slip-mold interaction and mold release, and slip drainage and wall build-up rates, as well as casting crack formation (Ref 7).

Infiltration is a process that is similar to liquid-phase sintering, except that the solid phase is first formed into a porous skeleton body, and the liquid-metal phase is introduced during sintering from the outside and allowed to penetrate the pore system. Excessive shrinkage associated with *in situ* liquid-phase sintering is avoided, and dimensional stability of the product is obtained, except for about 1% growth due to a thin surface film formed by the liquid metal.

The procedure used for TiC cermets involves two steps (Ref 8). First, an approximately 60% dense carbide skeleton body of near-net shape is formed by mixing the TiC powder with a small percentage of nickel binder and wax, cold pressing at about 35 MPa (5000 psi) into a slab, vacuum sintering at about 1300 °C (2370 °F), and machining the contour (for example, a turbine blade). The second step consists of inserting the skeleton shape into a mold assembly that also provides for containment of the metal in a ceramic tundish on top and gravity feeding of the liquid to the skeleton at preferred contact faces. An infiltration arrangement of this type is shown in the article "Infiltration" in this Volume.

The mold assembly is made of graphite, and its cavity is lined with a refractory ceramic in powder form that interfaces with the TiC skeleton. The ceramic liner is chosen so that it does not react with the TiC up to infiltration temperature and shrinks at a controlled rate, permitting formation of a uniform gap all around. The mold assembly is heated in a vacuum furnace to about 1400 to 1500 °C (2550 to 2730 °F); that is, well above the melting temperature of the infiltrating alloy, such as 80%Ni-20%Cr or 70%Co-24%Cr-6%Mo. During infiltration, the liquid metal first fills the gap between the liner and skeleton exterior by capillary forces and then penetrates the interior of the porous TiC part. After furnace cooling, the fully infiltrated product is readily extracted by fragmenting the sintered ceramic liner without degrading the graphite mold assembly that can be reused.

By varying the density of the TiC skeleton through special die filling and multiple-step pressing, graded cermet parts can be produced. For instance, a turbine blade can be made that has a high TiC concentration and, therefore, high strength at the center of the foil and in the transition to the root. The turbine blade also has a metal-rich jacket around the foil and especially at the mechanical shock-sensitive blade edges, as well as around the serrated root needed for blade attachment to the turbine disk.

Bonding and Microstructure

The physiochemical aspects of the bond between the dissimilar phases and the size of the ceramic grains embedded in the metallic matrix are vitally important to the properties and performance of cermets. Humenik and Whalen (Ref 9) have discussed the fundamentals involved in the bonding mechanism, particularly the TiC-metal systems.

Bonding. In general, due to the basic difference in the nature of the ceramic and metallic components in cermets, none of the known solid-state bonds applies singly. Instead, combinations are formed between ionic bonds, covalent bonds, and metallic bonds. The first type predominates in cermets based on oxide ceramics, and little force beyond simple adhesion exists to hold onto a metallic phase. The second bond type applies chiefly to systems involving silicon and carbon, such as graphite, diamond, and SiC. Again, the strength of this bond is rather limited. The

cermet gains substantially in its mechanical cohesiveness only where metallic bonding combines with the covalent type of bonding. The metal-carbide and metal-boride systems are examples of this combination.

Solubility. The high bond strength between metal and ceramic in cemented carbides and borides is enhanced further by mutual or partial solubility. During sintering, the active surfaces of the carbide or boride particles are dissolved in the liquid phase, and the carbon, boron, and transition metal atoms are reprecipitated on the solid particles during cooling. Depending on the system, traces to several percent of these elements remain dissolved in the binder phase.

Even in other cermet types, this partial solubility mechanism is beneficial, as it generates a metalloid interface. Metals are know to bond more readily to metalloids, such as silicides and borides, than oxides, an example of which is the Cr-Al_2O_3 system. In this system, a surface layer of chromium oxide (Cr_2O_3) on the chromium particles forms a solid solution with Al_2O_3 during sintering in a closely controlled, mildly oxidizing atmosphere, and the result is greatly enhanced bonding (Ref 10, 11). Similar metal-metalloid-oxide transition-type bond enhancement is obtainable with an intermediate layer of copper oxide (CuO) in the Cu-Al_2O_3 systems, or a layer of titanium nitride (TiN) in the nickel-magnesium oxide (Ni-MgO) system (Ref 9).

Wetting. Another important aspect of the bonding mechanism is the wettability of the solid phase by the liquid-metal component. This is controlled by the surface energies of the system during liquid-phase sintering (Ref 9). As shown in Fig. 2, the wetting ability is indicated by the contact angle (θ) that is formed by a liquid drop resting on a solid substrate. Relationships of the surface force vectors are given by the equation:

$$\sigma_{SV} - \sigma_{SL} = \sigma_{LV} \cos \theta$$

Fig. 2 Surface forces acting at the point of intersection of a liquid resting on a solid

See text for explanation of symbols. Source: Ref 9

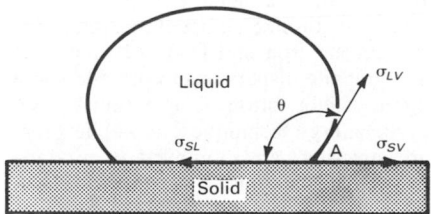

where σ_{SV}, σ_{SL}, and σ_{LV} are the surface energies of the solid-vapor interface, solid-liquid interface, and liquid-vapor interface, respectively. The contact angle is a parameter that can be measured with precision. In metal-ceramic oxide systems, the surface energy of the liquid-vapor interface of the metal is greater than the surface energy of the solid-vapor interface of the oxide, and the contact angle is much larger than 90°. Consequently, during liquid-phase sintering or infiltration in neutral atmosphere, the liquid metal is not retained in the pores of the solid, but tends to sweat out. If the contact angle is less than 90°, however, the liquid-metal phase is retained in the pore system of the ceramic; as the contact angle approaches zero, the bond becomes stronger. This is the case with the cemented carbides that have cobalt binders.

Microstructure. The nature of the bond in cermets is very closely related to the microstructure. This is especially significant for the carbide cermets, which have properties that are greatly affected by variables such as shape, size, and dispersion of the carbide grains, amount of carbide grains in the metal matrix, composition and structure of the matrix, and, of course, the degree of bonding of the two phases (Ref 12). Although these variables act in conjunction with one another, they can be singled out for specific effects.

A very fine carbide grain size tends to yield high strength and hardness, but a somewhat coarser size of about 2.2 μm tends to increase fracture toughness in cemented carbides (Ref 13). Sharp corners prevail in WC grains and cause only minimal harm to the strength of low-binder cemented carbide grades; in cermets with higher metal content, however, they affect localized stress raisers in the ductile metal matrix.

A good dispersion of the carbide grains in the metal matrix provides isolation of the grains, which limits the tendency of a crack initiated in one grain to propagate to others that are coalesced with the first. The same reasoning applies to any secondary hard phases that may form during sintering and bridge the original grains. Chromium carbide and nickel aluminide (Ni_3Al) are examples of reaction products from the matrix that may deposit on the carbide grains of TiC cermets, thereby contributing to the continuity of the brittle hard-phase structure (Ref 11).

The volume fraction of carbide in the cermet has a major influence on the mechanical and physical properties of the end product. Due to the mutual solubility and enhanced bond between the disparate phases, the rule of mixture does not strictly apply to the TiC systems. Generally, strength and hardness increase, while the coefficient of thermal expansion decreases with carbide content. Ductility improves as the metal matrix becomes the continuous phase and as it increases sufficiently in volume to avoid a triaxial state of stress.

The composition of the matrix is important in several respects. High ductility and toughness are essential to relieve the stresses caused by the hard phase and to provide a modicum of safety against catastrophic failure in service. Alloy selection must entail consideration of the matrix as a possible source for the brittle reaction products that may coalesce with the carbide grains into a continuous hard phase. Other contributory effects by the metal matrix are in the areas of oxidation and corrosion resistance, machinability, and weldability.

Good bonding between the carbide and metallic phase is essential, because the bond must translate the stresses from one individual phase to the other. Therefore, any gaps between carbide particle surface and matrix are detrimental. The example of bond improvement of Ni-TiC by the addition of molybdenum is striking. The contact angle of liquid nickel on the TiC in hydrogen is 17°, but the angle changes to almost zero with the molybdenum addition. As seen from the micrographs of Fig. 3, the molybdenum-free composition suffers excessive carbide grain growth, while the Ni-Mo-TiC displays a fine-grained, well-dispersed carbide phase in a continuous metal matrix (Ref 9).

Oxide Cermets

This class of material combines oxide-ceramic and metallic constituents on a microscopic scale. Thus, it fits the term cermet in the true sense of the word. More than most other mechanically mixed combinations of interstitial compounds and metallic phases, oxide cermets are negatively affected by poor thermal shock resistance and inadequate fracture toughness that limit their usefulness in a great many situations involving high temperatures and dynamic stresses. Some of these materials, however, possess excellent resistance to oxidation or corrosion at high temperatures, and others exhibit unique physical properties, such as nuclear fission. Generally, oxide cermets can be produced to withstand high-temperature stresses greater than those tolerated by most nonmetallic oxide ceramics.

About a half dozen different oxide ceramic-metal-type cermets have been

Fig. 3 Microstructure of titanium carbide cermets

Sintered 1 h in vacuum at 1400 °C (2550 °F) on graphite. Magnification: 1000×. (a) 50 wt% TiC and 50 wt% Ni. (b) 50 wt% TiC, 37.5 wt% Ni, and 12.5 wt% Mo. Source: Ref 9

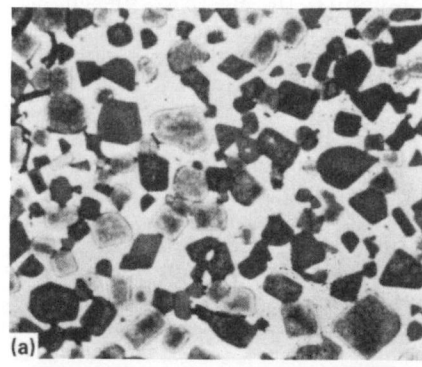

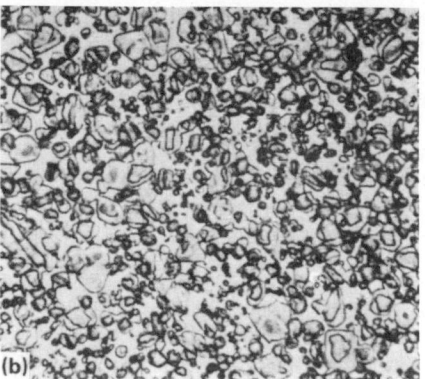

developed; several are used industrially. Generally, they differ from the oxide dispersoid-type alloys (see the article "Dispersion Strengthened Materials" in this Volume) by having a much larger particle size and, usually, by occupying a greater proportion of the ceramic phase.

Silicon Oxide-Containing Cermets. The classical combination of ceramic and metal can be found in the metallic friction materials, in which the ceramic produces the hard phase. Industrial machinery clutches and heavy-duty brakes, including those for airplanes, are the major fields of application. The ceramic phase is a relatively coarse granular (for example, 200 mesh) SiO_2 to which Al_2O_3 sometimes is added; it amounts to about 2 to 7 vol% of the material. The metallic matrix consists of brass or bronze compositions and also may contain iron and lead. All materials have graphite dispersions to provide some degree of lubrication. Conventional powder metallurgy techniques, as well as pressure sintering, are employed to produce the friction materials in the form of disks that fit into special attachment cups or

plates and strips that are bonded directly to the structural steel support.

Aluminum Oxide-Based Cermets. In this type of cermet, the ceramic is the dominant phase and the metal serves only as a binder. Aluminum oxide-based cermets are used in cutting tool bits for very high-speed machining with light chip removal (Ref 14). The oxide is milled to great fineness (usually only 1 to 3 μm), then mixed and milled together with nickel powder. Because the binder phase rarely exceeds 5 to 10 vol%, the cermet is very brittle after pressing and sintering, and press lubricants and organic binders are required to facilitate handling. Sintering is carried out in dry hydrogen, dry nitrogen, or preferably, in vacuum at temperatures of about 1450 to 1550 °C (2640 to 2820 °F). Finishing is a delicate operation.

A different type of aluminum oxide cermet is used for high-temperature, heat-resistant applications, such as the furnace components, jet flame holders, pouring spouts, flame protection rods, and thermocouple protection tubes and seals shown in Fig. 4 (Ref 6). The metallic component of this cermet is either chromium or chromium-molybdenum, in which case a small percentage of titanium dioxide (TiO_2) is added to the ceramic phase. The volume ratio of metal to ceramic is about 65 to 35, and that of chromium to molybdenum is about 4 to 1. The latter is conveniently charged as a crushed coarse-grained master alloy, and the powder mixture is ground to about 10-μm particle size. Consolidation is achieved by slip casting, cold pressing, or hydrostatic pressing, followed by high-temperature sintering at 1560 to 1700 °C (2840 to 3090 °F). The furnace atmosphere is high-purity hydrogen that contains controlled amounts of water vapor to cause surface oxidation of the chromium particles. The chromium oxide diffuses into the alumina, forming a solid solution at the contact areas that results in strong bonds between the grains (Ref 15). Some typical properties of these cermets are listed in Table 2, and Fig. 5 shows the effect of temperature on transverse-rupture, tensile, and stress-rupture strength.

The chromium content has a significant bearing on the creep resistance of these cermets in the 1380 to 1530 °C (2515 to 2785 °F) temperature range (Ref 16). With up to about 25 vol% Cr, the Al_2O_3 forms a coherent matrix and the chromium occurs mainly as a statistically distributed phase. For higher chromium concentrations, a network of the metal forms that is fairly continuous at 50 vol%. Consequently, the dominance of the creep

Fig. 4 Aluminum oxide-chromium cermet products

(a) Furnace muffle and tube. (b) Flow control pin. (c) Flame protection rod. (d) Thermocouple protection tubes. (e) T-shaped pouring spout. (f) Mechanical seals. (g) Jet flame holders. Source: Ref 6

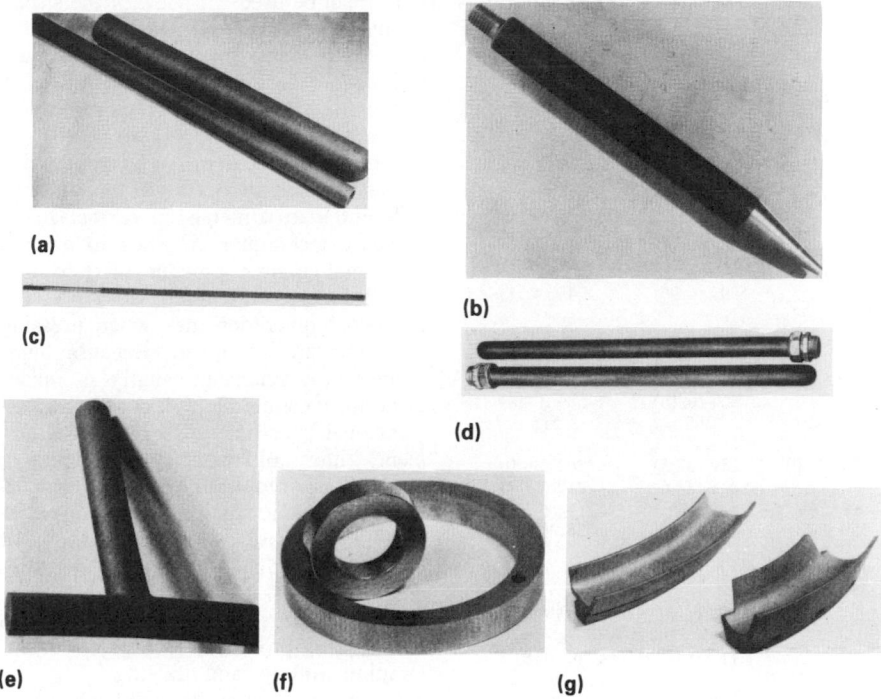

(a)

(c)

(b)

(d)

(e)   (f)   (g)

strength of the Al_2O_3 is lost as the formation of the metallic network becomes complete.

Many other metals have been mated with Al_2O_3 on an experimental basis (Ref 17, 18, 19) with the objective of developing high-temperature, serviceable cermet materials that have acceptable engineering properties. Metals used in these studies include nickel, cobalt, iron, molybdenum, tungsten, copper, and silver, and the main effort was directed at a better understanding of the bonding mechanism. None of these combinations, however, has achieved commercial realization.

Magnesium Oxide-Containing Cermets. Chromium also has been used as the metallic phase in magnesia-based cermets (Ref 15). Results of experiments with various metal-ceramic ratios have been reported in literature, with the magnesium oxide (MgO) fraction ranging from 50 vol% (Ref 17) to as low as 6 vol% (Ref 20). Although in none of these could a combination of properties be obtained that is superior to the Al_2O_3-Cr cermets, in the MgO-Cr system an intermediate reaction product (a $MgO \cdot Cr_2O_3$ spinel) was also observed between the ceramic and metal phases (Ref 15).

The 6% MgO-containing material is extrudable and exhibits 10% elongation and more at room temperature after sintering the extruded powder mixture (Ref 20). Yield and tensile strength are about 200 and 350 MPa (30 and 50 ksi) up to 600 °C (1100 °F), but taper off at higher temperatures. These strength properties can be maintained at the higher level up to about 1000 °C (1800 °F) by alloying the chromium with a small amount—for example, 1%—of niobium; however, this degrades the ductility. A measurable degree of room temperature elongation could also be observed in 30 vol% Cr material that was made by hydrostatic compression of a coarse powder mixture and sintering at 1600 °C (2900 °F). Unfortunately, the ductility in this highly refractory material is rapidly lost due to nitride formation, if the cermet is heated in air above 1100 to 1200 °C (2000 to 2200 °F).

Nickel, iron, and cobalt and alloys of these with chromium are other metals that have been investigated in MgO-based cermets (Ref 17, 19). The MgO-Co cermets, in particular, exhibt interesting mechanical as well as electrical properties over a wide composition range (Ref 19). Stress-rupture strength at 850 °C (1560 °F) for 100 h peaks at 80 MPa (11.2 ksi) for a cermet with 50 wt% Co, which despite its almost continuous metal phase (approximately 30 vol%) is an insulator. There is

no abrupt change in strength as the material transits from high to low electrical resistance.

Beryllium Oxide-Based Cermets. According to Ryshkewitch (Ref 21), beryllia cermets bonded with tungsten possess better thermal shock resistance and also soften at a higher temperature than most of the chromium-alumina materials. They have been used successfully as crucibles, and at one time, they were proposed for rocket nozzle throat inserts. Ryshkewitch (Ref 21) also has proposed the use of combinations of beryllia with beryllium metal up to 50 vol% for high-temperature thermal insulators and nose cones for re-entry bodies, despite the disadvantages of extreme brittleness and toxicity.

Zirconium Oxide-Based Cermets. Zirconia is another ceramic that can be bonded with metal to give useful refractory products. Even when combined with only small amounts of metal, such as 5 to 15 at.% Ti, strong and thermal shock-resistant materials suitable for crucibles to melt rare and reactive metals can be produced (Ref 22, 23). If the oxide is combined with molybdenum, the resulting cermet exhibits excellent corrosion resistance against molten steel, in addition to high-temperature strength and limited sensitivity to thermal shock (Ref 24), especially when the metal content is approximately 50 vol%. Thermocouple sheaths for temperature measurements of metallic melts, extrusion dies used for forming nonferrous metals, and wear-resistant parts made from these cermets with somewhat higher ceramic content, such as 60 vol% (Ref 24), are some of the applications cited.

Thorium Oxide-Containing Cermets. Cronin (Ref 25, 26) discusses metal-ceramic materials in which finely divided thoria is combined with molybdenum or tungsten to form a number of products used in the electronics industry. The principal P/M operations include screening the oxide and metal powders through 325-mesh screens, weighing, dry blending, compacting, sintering in a reducing atmosphere at 2000 °C (3630 °F) for the molybdenum material and somewhat higher in the case of tungsten, and finish machining to specified size and tolerances.

These products take the form of cylinders and sleeves in high-power pulse magnetrons that can deliver up to several million watts, or they have simple disk shapes for evacuated electron beam tubes (klystrons), traveling wave tubes, and special-purpose guns. In some high-voltage operations, the thermionic emission cathodes operate over a wider temperature range

Table 2 Composition and properties of aluminum oxide-containing cermets

Property	Al$_2$O$_3$-Cr	Cr-Al$_2$O$_3$	Cr-Al$_2$O$_3$	Cr-Mo-Al$_2$O$_3$-TiO$_2$
Composition, wt%				
Chromium	30	72	77	59
Molybdenum	...	...	...	20
Al$_2$O$_3$	70	28	23	19
TiO$_2$	...	...	...	2
Density, g/cm^3 (lb/in.^{3})	4.7 (0.17)	5.9 (0.21)	5.9 (0.21)	6.0 (0.22)
Electrical resistivity, μΩ·cm				
at 25 °C (75 °F)	...	...	87	65
Mean coefficient of thermal expansion, μm/m/°C (μin./in./°F)				
At 25-800 °C (75-1470 °F)	8.65 (4.81)	8.64 (4.80)	...	...
At 25-1000 °C (75-1830 °F)	...	...	8.93 (4.96)	8.48 (4.71)
At 25-1315 °C (75-2400 °F)	9.45 (5.25)	10.35 (5.75)	...	...
Thermal conductivity, W/m·K (Btu·in./ft^2·h·°F)				
at 260 °C (500 °F) avg	...	...	50.2 (348)	...
Specific heat, J/kg·K (Btu/lb·°F)	...	...	669 (0.16)	586 (0.14)
Hardness, HV	1100	...	365	515
Modulus of elasticity, GPa (10^6 psi)				
At 25 °C (75 °F)	361 (52.3)	324 (47.0)	259 (37.5)	262 (38.0)
At 1000 °C (1830 °F)	...	...	225 (32.6)	217 (31.4)
Transverse-rupture strength, MPa (ksi) at 25 °C (75 °F)	380 (55)	550 (80)	310 (45)	385 (56)
Tensile strength, MPa (ksi) at 25 °C (75 °F)	240 (35)	270 (39)	145 (21)	...
Compressive strength, MPa (ksi) at 25 °C (75 °F)	2210 (320)	...	760 (110)	1655 (240)
Shear modulus, GPa (10^6 psi) at 25 °C (75 °F)	...	...	117 (17)	103 (15)
Shear strength, MPa (ksi) at 25 °C (75 °F)	...	...	276 (40)	...
Poisson's ratio in flexure, at 25 °C (75 °F)	...	...	0.20-0.22	0.25-0.27
Impact resistance, microcharpy, unnotched, J (in.·lb) at 25 °C (75 °F)	1.35 (<12)	1.35 (<12)	...	...
Thermal shock resistance, max temp °C (°F)	1315 (2400)	1040 (1900)	...	...
Oxidation resistance, long time, max temp °C (°F)	1500 (2730)	1200 (2200)	1200 (2200)	1200 (2200)

Source: Ref 6, 15

(1000 to 1700 °C or 1830 to 3100 °F), but the range is narrowed to about 1300 to 1500 °C (2370 to 2730 °F) for cathodes in the average tubes. Because the ThO$_2$ is present in the refractory metal as a well-dispersed, fine-particulate minor phase that rarely exceeds 4 to 5 vol%, the materials may be referred to as a dispersion-type alloy, rather than a cermet, which is the usual reference to it in literature (Ref 25, 26).

Uranium Oxide-Containing Cermets. These cermets are used in the fuel elements of nuclear reactor cores. They consist of a dispersion of the fissionable UO$_2$ in a sintered matrix of aluminum (Ref 27), stainless steel, or tungsten (Ref 28). Compared to plain oxide fuel, these cermets have better retention of fission products and an increased thermal conductivity that inhibits melting at high operating temperatures. Usually, the ceramic component does not exceed 35 vol% of the cermet to ensure a continuous and coher-

ent metal matrix and to limit the radiation damage caused by the fissionable UO$_2$. The cermet is contained inside structural stainless steel supports, such as casings or frames.

Powder metallurgy fabrication details have been given by Loewenstein and co-workers (Ref 29). The UO$_2$ may vary in purity, depending on its processing, and should be of stoichiometric composition. The ceramic particles are fairly coarse and must be strong enough to withstand subsequent working without fracturing. Typical particle sizes are 44 μm minimum (+325 mesh) for bicrystals, 35 to 44 μm for monocrystals, and 40- to 50-μm diam for agglomerates of very fine (0.1- to 1-μm) irregular crystals. These UO$_2$ powders are fired at 1600 to 1700 °C (2900 to 3100 °F) in hydrogen to increase particle size, strength, and density (of the agglomerates). Small additions of TiO$_2$ increase the sintering rate. Procedures used con-

ventionally for mixing oxide and metal powder must be modified—glovebox or other environmentally controlled operations must be used—in view of the strong radioactivity of the UO$_2$. Where large differences in density between the ceramic and metal exist, tumbling is inadequate to prevent segregation, and ball milling is required.

Table 3 lists the density and some physical and nuclear properties of the UO$_2$ and different matrix metals in cermet fuels. Standard techniques are used to process the metal-ceramic powder mixture into consolidated cermet fuel. Cold pressing achieves higher densities when pressing lubricants are employed. Because high-temperature sintering usually is inadequate for meeting the high density and dimensional specifications of the fuel element, either cold working and sizing or machining is required. Some oxide particle fragmentation and waste is unavoidable in this case. Alternate methods of consolidation that yield higher densities and minimize fragmentation, but increase production cost, include hot pressing and hot working processes such as extrusion, swaging, rolling, and drawing.

Cermets containing 50 vol% each of UO$_2$ and tungsten have been fabricated, using methods such as high-energy compaction, isostatic or vacuum hot pressing, powder rolling, or coextrusion, into fuel elements for gas-cooled reactor cores that have coolant temperatures of 1500 °C (2730 °F) and higher (Ref 28).

Carbide Cermets

Another main group of cermets is carbide cermets, which consists of tungsten, chromium, and titanium carbide-based cermets.

Tungsten Carbide-Based Cermets. In the broadest sense, the industrially important tungsten carbides and tungsten-based multicarbides can be considered to belong to the family of cermets. For a discussion of the production practices involved and applications for these materials, see the articles "Production of Refractory Metal and Carbide Powders," "Production Sintering Practices for P/M Materials," and "Cemented Carbides" in this Volume.

A class of special tungsten carbide cermets is related to the so-called micrograin carbides. The WC grains are finer than 2 μm, and their growth is inhibited by an inert separate phase that is dispersed through the binder metal, such as "Baxtron" DBW and DBA, produced by DuPont (Ref 30), or "Carmet-300," man-

Fig. 5 Effect of temperature on strength properties of aluminum oxide-chromium cermets

(a) Transverse-rupture strength. (b) Tensile strength. (c) Stress-to-rupture strength. Composition: —— 30%Cr-70%Al_2O_3; —— 72%Cr-28%Al_2O_3; ----- 77%Cr-23%Al_2O_3; —·— 59%Cr-20%Mo-19%Al_2O_3-2%TiO_2. Source: Ref 6, 15

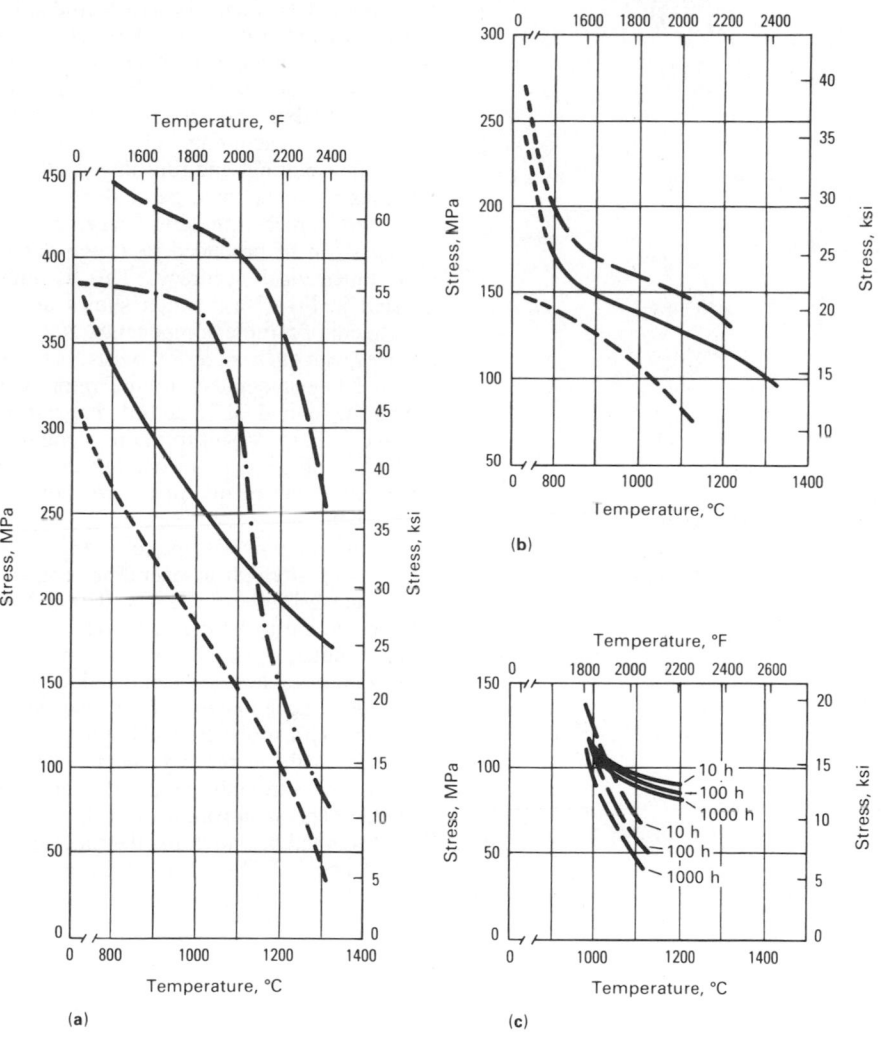

(a)　(b)　(c)

Table 3　Properties of uranium dioxide and candidate matrix metals in nuclear reactor fuel cermets

Material	Thermal absorption cross section, barns/atom	Density		Melting point	
		g/cm³	lb/in.³	°C	°F
Uranium dioxide					
(0.53 uranium volume ratio)	0.002	10.96	0.397	2500	4530
Beryllium	0.01	1.848	0.067	1277	2330
Magnesium	0.06	1.738	0.063	650	1202
Zirconium	0.18	6.489	0.235	1852	3366
Aluminum	0.23	2.699	0.098	660	1220
Niobium	1.1	8.57	0.31	2468	4474
Molybdenum	2.4	10.22	0.37	2610	4730
Iron	2.4	7.87	0.285	1537	2799
Stainless steel, type 304	2.9	~7.90	~0.286	~1400	~2550
Chromium	3.1	7.19	0.26	1875	3407
Nickel	4.5	8.90	0.322	1453	2647

Source: Ref 29

ufactured by Allegheny Ludlum (Ref 31). In another type, "Ramet I," made by Fansteel (Ref 32), unique cutting properties (positive rake) are realized by associating less than 1% Cr_3C_2 particles of 0.01- to 0.05-μm size with the 0.2- to 1.5-μm WC grains.

Chromium Carbide-Based Cermets. Cermets that contain chromium carbide (CrC) as the major constituent possess some unique properties that make them useful for certain applications in the tool and chemical industries (Ref 33, 34). This class of material is essentially a cemented chromium carbide of the Cr_3C_2 modification that is bonded with nickel or a nickel-tungsten alloy. Cr_3C_2 powder is produced by reacting Cr_2O_3 with carbon at about 1600 °C (2900 °F). Minor additions of carbides of lower chromium content are added to control the carbon balance and keep the free-carbon content low. Standard cemented carbide manufacturing practice is applied to the production of these cermets. Some interesting properties and specific applications of these CrC-based cermets are:

- Very low density, which makes the material useful in applications such as the making of valve balls in oil well valves
- Relatively high coefficient of thermal expansion, which permits direct brazing to steel, provided that boron-containing fluxes are used
- Bright and durable surfaces of high reflectivity, which permit finishing to optical flatness. These properties, together with the thermal expansion characteristics, make the material suitable for gage blocks, micrometer tips, and other measuring tools.
- Virtually nonmagnetic nature, which eases measuring tasks, in spite of the nickel binder
- Excellent wear and corrosion resistance—for example, against salt water attack up to 85 °C (185 °F)—which makes the cermet suitable as a bearing and seal material or for fishing rod guide rings
- Outstanding high-temperature erosion resistance at least to 1000 °C (1830 °F), which makes the material ideally suitable for dies used in the automatic hot extrusion of brass products held in the 1000 to 1100 °C (1830 to 2000 °F) temperature range

Additional areas of application for the chromium carbide cermets include high-temperature bearings and seals, various valve components, nozzles, guides, dies operating at elevated temperatures, and a multitude of gaging components. Typical

Table 4   Properties of typical chromium carbide-based cermets

Property	Type A	Type B
Composition, wt%		
Cr_3C_2	83	88
Nickel	15	12
Tungsten	2	
Density, g/cm^3 ($lb/in.^3$)	7.0 (0.253)	6.9 (0.250)
Electrical resistivity, $\mu\Omega\cdot$cm at 25 °C (75 °F)	84	70
Electrical conductivity,		
Mhos(cm), at 25 °C (75 °F)	0.012	...
%IACS	2.1	...
Thermal conductivity, W/m·K		
(Btu·in./ft^2·h·°F) at 50 °C (120 °F)	10.88 (75.5)	12.55 (87.1)
Mean coefficient of thermal expansion,		
μm/m/°C (μin./in./°F) at 25-595 °C (75-1100 °F)	10.71 (5.95)	11.10 (6.17)
Hardness, HRA at 25 °C (75 °F)	88.3	...
Hardness, HV		
At 25 °C (75 °F).............................	...	1300
At 800 °C (1470 °F)	...	900
Modulus of elasticity in compression, GPa		
(10^6 psi) at 25 °C (75 °F)	345 (50)	333 (48)
Elastic limit in compression, MPa		
(ksi) at 25 °C (75 °F)	900 (130)	...
Compressive strength, MPa		
(ksi) at 25 °C (75 °F)	3450 (500)	3725 (540)
Transverse-rupture strength, MPa		
(ksi) at 25 °C (75 °F)	780 (113)	735 (107)
Ductility in compression, % at 25 °C (75 °F)	1.1	...
Poisson's ratio	0.28	...
Impact resistance, Izod, unnotched,		
J (ft·lb) at 25 °C (75 °F)	0.158 (1.4)	...
Resistance to oxidation		
Short time, max temp, °C (°F)	1100 (2000)	...
Long time, max temp, °C (°F)	1000 (1830)	...
Resistance to corrosion, 10-24 h immersion,		
wt loss, g/m^2/day		
50% H_2SO_4	3.1	...
35% HNO_3	1.9	...
50% NAOH	0.1	...
5% lactic acid	0.15	...
$1N$ solution H_2SO_4	...	5
$1N$ solution HNO_3	...	>10

Source: Ref 33, 34

physical, mechanical, and chemical properties of representative CrC cermets are given in Table 4. Figure 6 shows the effect of temperature on thermal conductivity, the expansion coefficient, and transverse-rupture strength (Ref 33, 34).

Titanium Carbide-Based Cermets. Next to the cemented WC materials, this class of cermets has received, by far, the widest attention within the family of metal-ceramic materials. With the advancement of high-thrust turbojet engines for military aircraft and the advent of commercial jet transports shortly thereafter, the need for better materials for certain critical stationary and, in particular, moving parts in the power plants created the need for a major effort in utilization of the TiC cermets.

The desire to combine, on a microscale, the high strength and relatively good oxidation resistance at elevated temperatures, plus a low specific gravity of the ceramic with a metallic alloy phase that imparts good resistance to mechanical and thermal shock, was the underlying moti-

vation for the large development effort that was conducted in the United States and Europe throughout the 1950's. In these TiC cermets, the metallic phase was varied over a broad range; that is, from about 30 to 72 wt%. The principal alloys were Ni-Mo, Ni-Mo-Al, Ni-Cr, and Ni-Co-Cr types (Ref 35, 36, 37). Some more complex alloys similar to commercial superalloys were also used to bond the TiC. Where high-temperature oxidation resistance of the binder metal was inadequate, such as in Ni-Mo, this property could be enhanced in the cermet by complexing the ceramic phase through prealloying a small amount of niobium, tantalum, or titanium, and carbon in the solid solution (Nb, Ta, Ti)C with the TiC (Ref 36).

Titanium carbide powder is produced industrially by a reaction that uses TiO_2 and carbon powders as starting materials. Carburization occurs through the gas phase (carbon monoxide) at a reaction temperature of 1600 to 1700 °C (2900 to 3100 °F). The reaction is terminated only when

the free-carbon content of the product is below 0.8%. High-quality powder has only 0.1 to 0.2% free carbon and a minimum of 80.0% Ti.

Two manufacturing routes have been used to produce these cermets—conventional cemented carbide production practices and infiltration. The first, still in use today, employs uniaxial or isostatic compaction of mixtures of the ceramic and metal powders, presintering, shape generation by machining, high-temperature vacuum sintering, and finishing. With fastidious manipulations, particularly during the early processing steps, very complex shapes can be produced to a high degree of dimensional accuracy. This is illustrated in Fig. 7(a), which shows an assortment of turbine components. Not only blades with different airfoil twists, but even completely integrated turbine rotor units with the individual blade foils hogged out of the disk, can result from this technique. The individual steps in making the complete rotor out of one piece are shown in Fig. 7(b).

If the high stress-rupture strength and low creep strength at operating temperatures in the 1000 to 1100 °C (1830 to 2000 °F) range are to be derived from the TiC phase alone, the binder content must be low (comparable to the cemented tungsten carbides). Such materials have insufficient toughness, a property that improves only slightly with increasing metal content, while the strength properties are degraded. This is apparent from the data in Tables 5 and 6, which list the mechanical properties of TiC cermet grades, and also from Fig. 8 and 9, in which the effect of temperature on strength and toughness is plotted.

The brittle nature of the cemented TiC cermets affected their performance as gas turbine blades. In engine tests reported by Deutsch (Ref 38), blade failures occurred in the base of the fir tree roots, due to insufficient notch toughness; at the tips of the airfoil, due to low resistance to impact by hard carbon particles; and across the airfoil, due to an incapacity to accommodate dynamic interference with protrusions or bows of the stationary shroud by some degree of plastic deformation. Low fatigue resistance near the tip of the deeply scalloped foils was another shortcoming of the material.

Considerable progress was made toward production of an acceptable TiC cermet for these critical turbojet engine applications by increasing the amounts of cermet material and simultaneously strengthening the binder phase through alloying. The composition was varied in ac-

Fig. 6 Effect of temperature on thermal properties and strength of chromium carbide cermets

(a) Thermal conductivity ambient of 83%Cr₃C₂-15%Ni-2%W. (b) Mean coefficient of thermal expansion from ambient to temperature indicated on scale for 83%Cr₃C₂-15%Ni-2%W. (c) Transverse-rupture strength of 88%Cr₃C₂-12%Ni. Source: Ref 33, 34

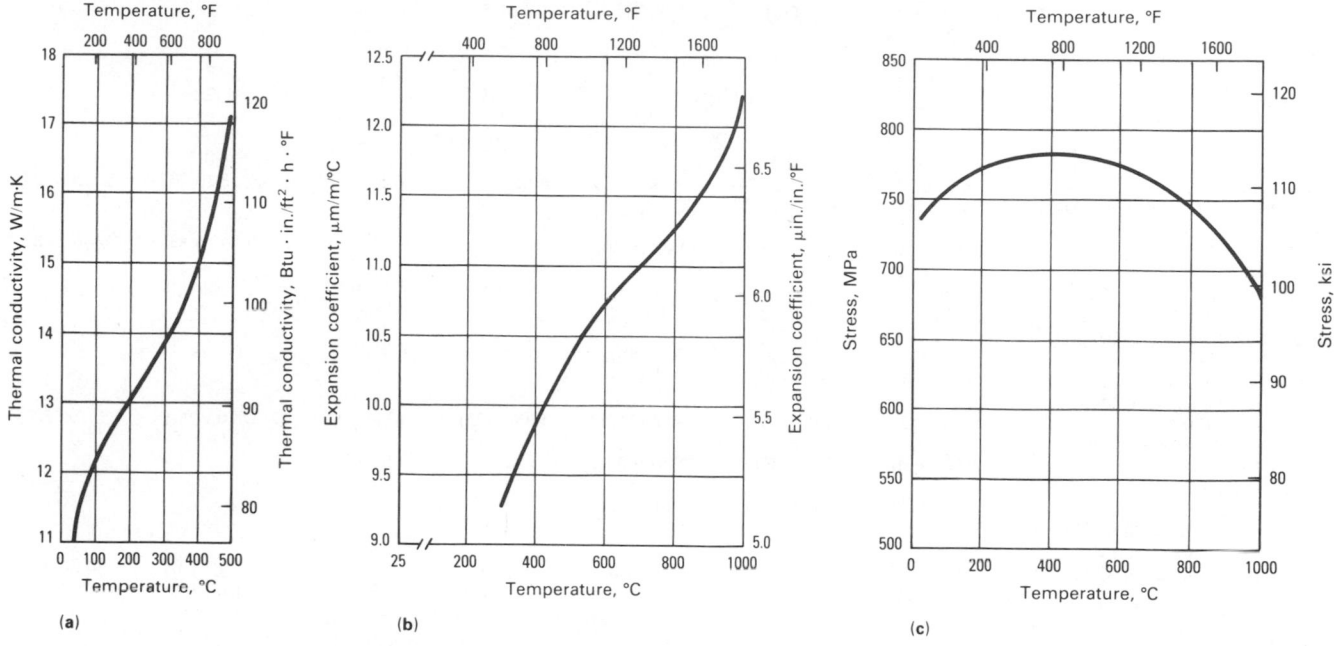

(a)   (b)   (c)

Fig. 7 Titanium carbide cermet gas turbine components

(a) Top, blades of different shape and stages of finish; bottom right, cermet blades attached to separate disk; bottom left, one-piece rotor with integrated blades. Reproduced from Pfaffinger, K., Hartmetalle auf Titankarbidbasis für Hochtemperaturanwendungen, *Planseeberichte für Pulvermetallurgie*, Vol 3 (No. 1), Feb 1955, p 17. (b) Fabrication of one-piece rotor, showing (clockwise from bottom left) a presintered disk, the disk after machining the blades, the uniformly shrunk rotor after high temperature sintering, and the final product after finish grinding. Reproduced from Kieffer, R. and Kölbl, F., Über die Entwicklung und Eigenschaften warm- und zunderfester Hartlegierungen auf Titankarbidbasis mit Nickel-Kobalt-Chrom-Bindern, *Planseeberichte für Pulvermetallurgie*, Vol 1 (No. 1), Oct 1952, p 34

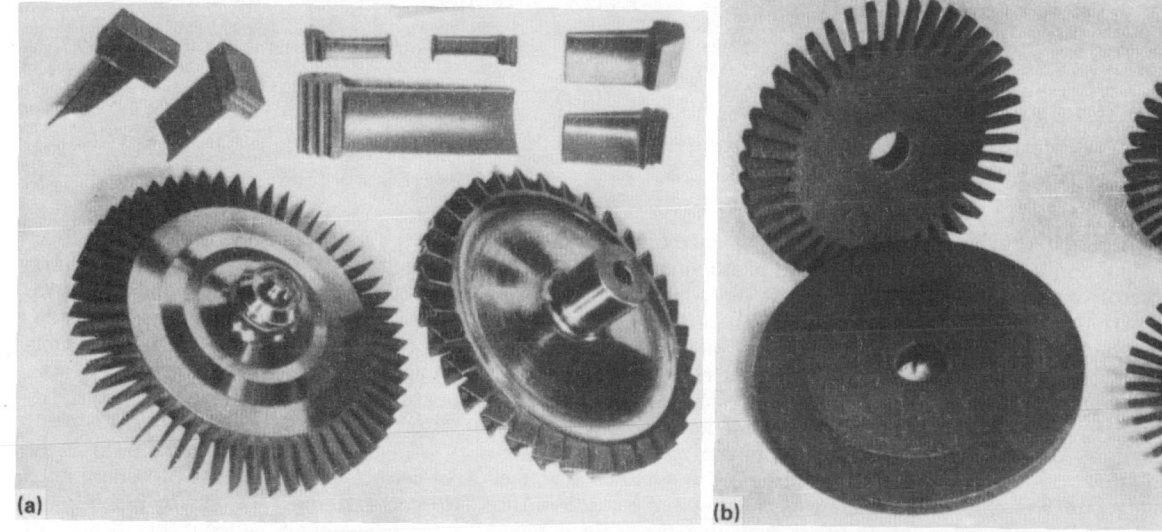

(a)   (b)

cordance with the strength and ductility requirements for different parts of the blades, using the infiltration process for making graded products with alloy accumulations at the root and at the edges and tip of the airfoil (Ref 8, 12, 39, 40). This approach is discussed in the article "Infiltration" in this Volume. The improvements in ductility and toughness, without sacrifice of high-temperature, stress-rupture properties, is shown in Fig. 9 and Tables 5 and 6, which compare infiltrated and sintered TiC cermet grades (Ref 12, 35, 36, 37).

The development of serviceable engine components made of infiltrated carbide cermets was stopped short of reaching its goal because of simultaneous develop-

Table 5 Composition and properties of titanium carbide-based cermets

Property	Liquid-phase sintered					Infiltrated	
Composition, wt%							
Titanium carbide	70(a)	60(a)	50(a)	61.5	44.5	47	50
Nickel	25	32	42.5	22	50	42	5
Cobalt	...	...	...	7.5	...	...	30
Iron	...	...	...	...	...	3	...
Chromium	...	2.5	...	7.5	5.5	7.5	12
Molybdenum	5	3	7.5	1.5	...	...	...
Tungsten	...	...	...	...	3	...	...
Aluminum	...	2.5	...	...	...	0.5	...
Density, g/cm^3 (lb/in.^{3})	6.0 (0.217)	6.2 (0.224)	6.6 (0.239)	6.05 (0.219)	6.55 (0.237)	6.55 (0.237)	6.5 (0.235)
Mean coefficient of thermal expansion, μm/m/°C (μin./in./°F) at 25-985 °C (75-1800 °F)	9.54 (5.30)	10.08 (5.60)	10.08 (5.60)	...	...	9.97 (5.54)	9.79 (5.44)
Hardness, HRA							
At 20 °C (70 °F)	89	87	84	87.5	79	80	83
At 760 °C (1400 °F)	74	76	69	...	...	...	...
Modulus of elasticity, GPa (10^6 psi)							
At 20 °C (70 °F)	395 (57)	345 (50)	345 (50)	...	...	...	...
At 870 °C (1600 °F)	330 (48)	...	275 (40)	...	...	...	...
Transverse-rupture strength, MPa (ksi), at 20 °C (70 °F)	1275 (185)	1268 (184)	1462 (212)	1206 (175)	1344 (195)	1310 (190)	1379 (200)
Compressive yield strength, MPa (ksi), at 20 °C (70 °F)	2100 (450)	3180 (461)	2930 (425)	...	...	...	...
Tensile strength, MPa (ksi), at 20 °C (70 °F)	772 (112)	717 (104)	868 (126)	...	...	...	...
Tensile elongation, % at 20 °C (70 °F)	0.21	...	0.28	...	...	...	...
Impact resistance, microcharpy, unnotched, J (ft·lb) at 20 °C (70 °F)	0.17-0.32 (1.5-2.8)		1.80 (16)	0.25-0.34 (2.2-3.0)	0.66-0.79 (5.8-7.0)	1.19-1.64 (10.5-14.5)	0.56-0.71 (5.0-6.3)

(a) Contains 6% of (Nb, Ta, Ti) C solid solution
Source: Ref 12, 35-37

Table 6 Impact resistance of titanium carbide-based cermets

| Material | Impact resistance at test temperature(a) of: | | | | | | | |
| | 20 °C (70 °F) | | 870 °C (1600 °F) | | 980 °C (1800 °F) | | 1095 °C (2000 °F) | |
	J	in.·lb	J	in.·lb	J	in.·lb	J	in.·lb
Ni-cemented Ti(NbTa)C	0.46-0.58	4.1-5.1	0.59-0.70	5.2-6.2	0.89-0.96	7.9-8.5	0.37-0.50	3.3-4.4
Ni-Mo cemented Ti(NbTa)C	0.17-0.32	1.5-2.8	0.25-0.36	2.2-3.2	0.28-0.34	2.5-3.0	0.19-0.26	1.7-2.3
Ni-cemented Ti(Cr)C	0.20-0.34	1.8-3.0	0.56-0.60	5.0-5.3	0.60-0.68	5.3-6.0	0.33-0.42	2.9-3.7
Ni-Cr-Al infiltrated TiC	0.63-0.94	5.8-8.3	0.96-1.36	8.5-12.0	0.97-1.19	8.6-10.5	0.50-0.67	4.4-5.9
Ni-Cr infiltrated TiC	1.19-1.64	10.5-14.5	1.36-2.20	12.0-19.5	1.98-2.66	17.5-23.5	0.64-0.82	5.7-7.3
Co-Cr-Mo infiltrated TiC	0.62-0.85	5.5-7.5	0.82-0.99	7.3-8.8	0.94-1.10	8.3-9.7	0.89-1.10	7.9-9.7
Co-Cr-W infiltrated TiC	0.56-0.71	5.0-6.3	0.75-0.89	6.6-7.9	0.89-1.12	7.9-9.9	1.07-1.36	9.5-12.0
Superalloy X-40	4.52-5.64	40-50	5.64	50	5.64	50	...	...
Superalloy X-40, aged	3.39-4.52	30-40	5.64	50	5.64	50	...	...

(a) Measured optically. Microcharpy test specimens, 1 cm^2 (0.16 in.^{2}) by 40 mm (1^1/$_2$ in.), were torch-heated to temperature, kept for 1 min to attain temperature equilibrium prior to impact testing.
Source: Ref 12

ments in industrial vacuum metallurgy that produced advanced precipitation-strengthened superalloys capable of operating reliably and for sustained periods at increasingly higher gas temperatures. Titanium carbide cermets are still in use, however, in several less sensitive applications, especially those in which the low density or high-temperature oxidation resistance, which is better than that of cobalt-cemented WC, is advantageous. Examples are furnace parts of all sorts, seals and bearings operating at elevated temperatures, sliding contacts, and especially, wear-resistant parts and metal-working tools.

Titanium carbide cermets with steel as the binder phase have several unique characteristics. The steel-bonded carbide is produced in a sequence of operations, including liquid-phase sintering in vacuum, that are similar and comparable in cost to those used for cemented tungsten carbide (Ref 41). This sequence combines some of the superb wear-resistance and cutting properties of the carbide with the heat treatability of tool and high-speed steels (Ref 42, 43). In the annealed condition, the material is soft enough to be machined with ordinary single-point carbide or diamond tools, replacing the more cumbersome and costly machining that follows presintering and grinding after final sintering.

After shaping, heat treating, such as austenitizing followed by oil quenching, causes martensite to form in the binder phase and results in hardness levels that closely approach those of commercial cemented WC. Although the ferrous binder may amount to as much as 60 vol% of the total composition, the high hardness is obtained through alloying of small amounts of chromium and molybdenum in the steel matrix that contains the homogeneous TiC dispersion.

An additional characteristic of this type of cermet is its capability of being joined to steel backings, not only by brazing, but also by welding, without any danger of cracking during cooling or heat treatment (Ref 42). This is the result of a favorable coefficient of thermal expansion that matches that of steel more closely than do tungsten carbide tool materials. The early martensitic, steel-bonded TiC materials have more recently been supplemented by

Fig. 8 Effect of temperature on some mechanical properties of sintered titanium carbide cermets

(a) Compressive yield strength. (b) Ultimate tensile strength. (c) Tensile elongation. Composition: —— 70%TiC-25%Ni-5%Mo; —·— 60%TiC-32%Ni-3%Mo-2.5%Cr-2.5%Al; ---- 50%TiC-42.5%Ni-7.5%Mo. In each, the titanium carbide contains 6% of (Nb, Ta, Ti)C solid solution. Source Ref 35, 36

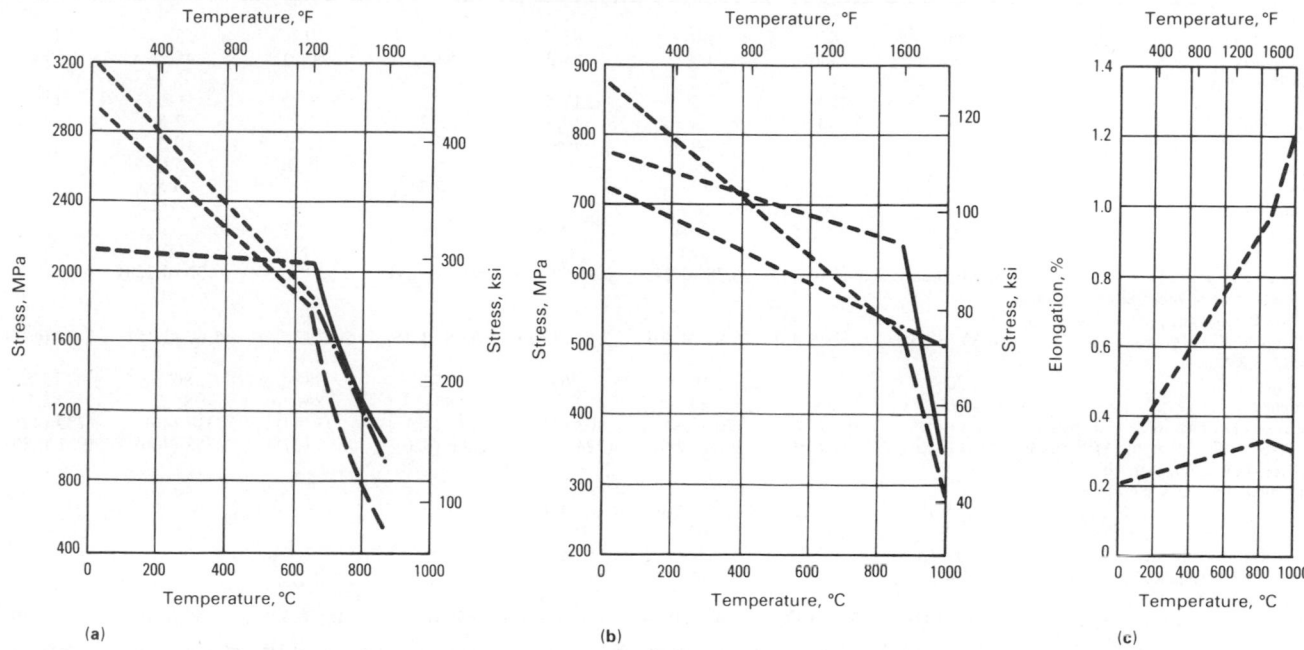

(a)   (b)   (c)

Fig. 9 Impact and stress-rupture properties of sintered and infiltrated titanium carbide cermets

(a) Effect of temperature on impact resistance. (b) Effect of temperature on stress-to-rupture strength. (c) Effect of time at 985 °C (1800 °F) on stress-to-rupture elongation. Composition: —·— 70%TiC-25%Ni-5%Mo, sintered; —··— 50%TiC-42.5%Ni-7.5%Mo, sintered; ----- 61.5%TiC-22%Ni-7.5%Co-7.5%Cr-1.5%Mo, sintered; ––––– 47%TiC-42%Ni-7.5%Cr-3%Fe-0.5%Al, infiltrated; —— 50%TiC-30%Co-12%Cr-5%Ni-3%W, infiltrated. In each, the titanium carbide contains 6% of (Nb, Ta, Ti)C solid solution. Source: Ref 12, 35-37

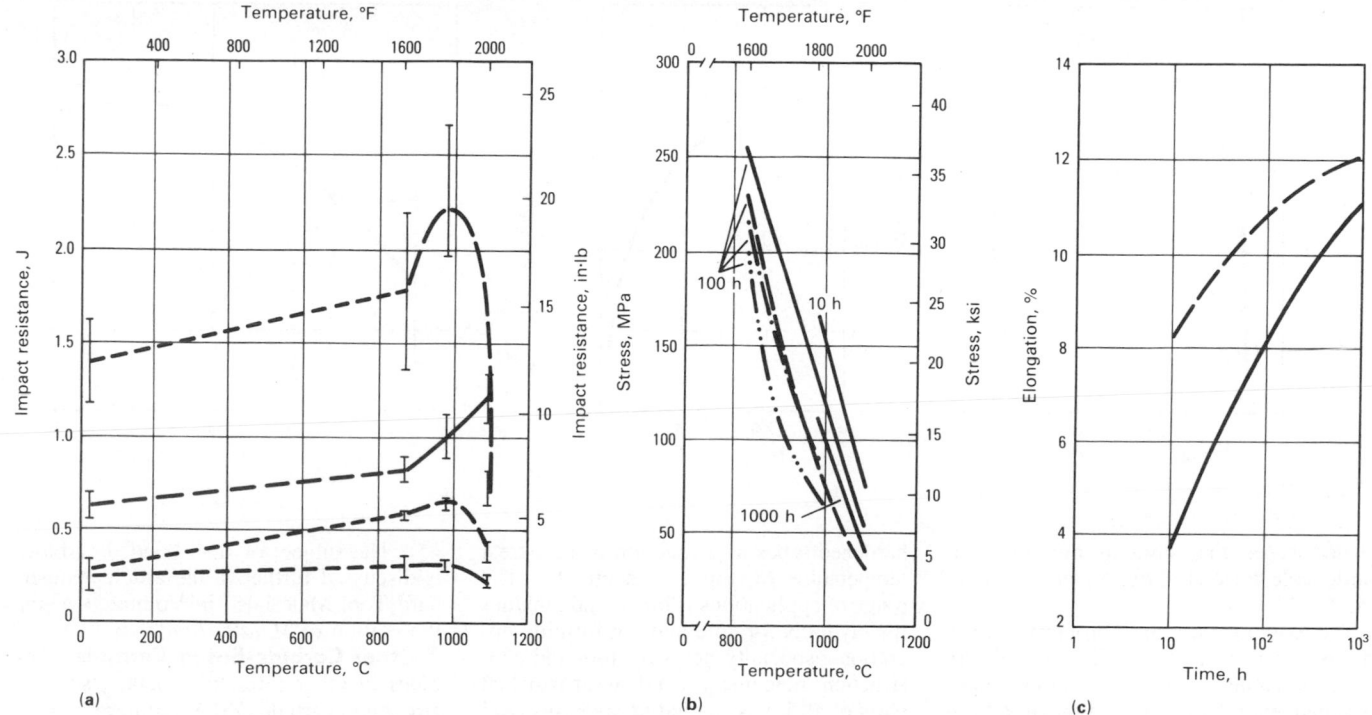

(a)   (b)   (c)

Table 7 Properties of quench and precipitation-hardenable titanium carbide-steel cermets

Property	Cermet grades							
	A	B	C	D	E	F	G	H
Composition, vol%								
TiC	45	45	40	45	45	50	55	45
Steel matrix	55(a)	55(b)	60(c)	55(d)	55(e)	50(e)	45(e)	55(f)
Composition, steel matrix, wt%								
Chromium	3	10	5	17.5	...	...	...	13
Molybdenum	3	3	4	0.5	3.2	2.9	2.5	2
Nickel	...	...	0.5	...	12	11.2	10.5	8
Cobalt	...	...	...	...	5.7	5.3	5.0	...
Titanium	...	...	...	...	0.7	0.6	0.5	...
Aluminum	...	...	...	...	...	...	...	1
Carbon	0.6	0.85	0.4	0.75	...	...	...	...
Iron	(g)	(g)	(g)	(g)	(g)	(g)	(g)	(g)
Density, g/cm³ (lb/in.³)	6.59 (0.238)	6.45 (0.233)	6.78 (0.245)	6.45 (0.233)	6.67 (0.241)	6.53 (0.236)	6.33 (0.229)	6.53 (0.236)
Mean coefficient of thermal expansion, μm/m/°C (μin./in./°F) at 20-93 °C (70-200 °F)	6.39 (3.55)	6.22 (3.46)	8.82 (4.90)	5.54 (3.08)	6.05 (3.36)	5.87 (3.26)	5.67 (3.15)	7.74 (4.30)
Hardness, HRA								
Annealed	70	73	69	76	75.5	78	80	75.5
Hardened	86.5	86	84	85.5	83	85	85.5	82.5
Modulus of elasticity, GPa (10⁶ psi)	303 (44)	303 (44)	268 (39)	303 (44)	296 (43)	303 (44)	317 (46)	289 (42)
Transverse-rupture strength, MPa (ksi)	2068 (300)	2137 (310)	2068 (300)	1724 (250)	2206 (320)	2585 (375)	2758 (400)	1931 (280)
Impact resistance, microcharpy, unnotched, for cross section of:								
1 cm², J	5.55	4.20	8.70	4.38	...	...	...	...
1 in.², in.·lb	317	240	497	250	...	...	...	...
0.16 in.², in.·lb	49.15	37.2	77.0	38.75	...	...	...	...
Thermal shock resistance, quench cycles	4	1	15	2	...	...	...	...
Maximum operating temperature, °C (°F)	205 (400)	540 (1000)	540 (1000)	430 (805)	455 (850)	455 (850)	455 (850)	455 (850)

(a) Medium-alloy steel. (b) High-chromium tool steel. (c) Hot working tool steel. (d) Martensitic stainless steel. (e) Precipitation-hardenable martensitic steel. (f) Precipitation-hardenable martensitic stainless steel. (g) Balance
Source: Ref 44, 45

Fig. 10 Room temperature hardness of heat treated titanium carbide cermets with ferrous metal binder
(a) Effect of austenitizing temperature on quench-hardenable material. (b) Effect of tempering temperature on quench-hardenable material. (c) Effect of aging time at temperature on precipitation-hardenable material. Precipitation temperatures of: —— 480 °C (900 °F) and —— 540 °C (1000 °F). Source: Ref 45, 46

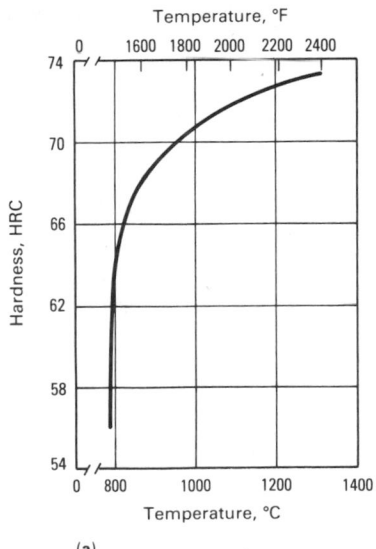

(a)

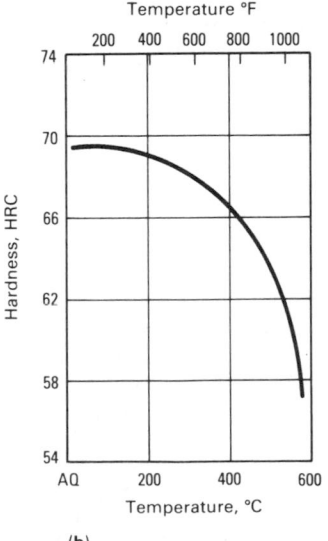

(b)

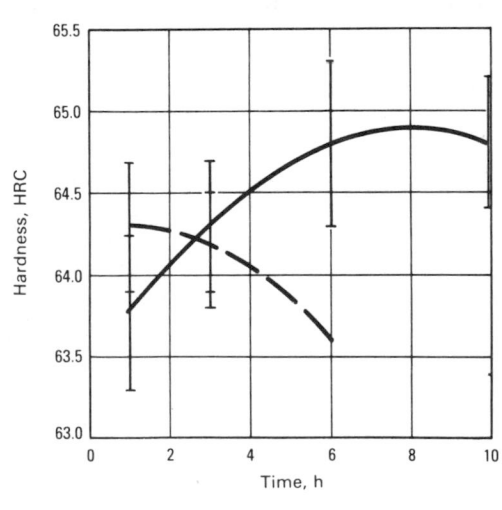

(c)

related types that contain precipitation-hardenable ferrous alloys as binders (Ref 44, 45).

The composition and properties of a number of heat treatable TiC cermets are listed in Table 7. Figure 10 shows hardness in quench-hardened and precipitation-hardened types as a function of heat treat temperature or time, respectively. The range of applications is broad and includes mainly tools for cold and hot forming operations, specialty items for tool and construction machinery, and wear-resistant parts of all kinds, including overlays (Ref 45). The subject of steel-bonded carbides is discussed further in the article "Superhard Tool Materials" in Volume 3 of the 9th edition of *Metals Handbook*.

Other Carbide-Based Cermets. Carbides of other refractory metals, such as zirconium carbide (ZrC), hafnium carbide

(HfC), tantalum carbide (TaC), and niobium carbide (NbC), have been produced experimentally and investigated for high-temperature applications. They have the highest melting points of all compounds known (Ref 47). Hafnium carbide melts at 3890 °C (7030 °F), TaC at 3800 °C (6870 °F), ZrC at 3530 °C (6380 °F), and NbC at 3500 °C (6330 °F). All of these carbides exhibit poor oxidation resistance at high temperatures and are extremely brittle. Cementing these carbides with ductile binder metals does not improve these properties sufficiently to make the resulting cermets competitive with the industrial carbides available for high-temperature structural and tool applications (Ref 48). The use of small amounts of tantalum and niobium carbides as additions to titanium carbide to produce cermets of enhanced high-temperature oxidation resistance has been cited before.

Uranium carbide cermets are of some interest in nuclear reactor technology (Ref 29). Since carbon has a low neutron cross section, UC would be desirable as a fuel element because of neutron economy. The compound has higher thermal conductivity than UO_2, a high melting point (2300 °C or 4170 °F), and creep resistance to 1000 °C (1830 °F). The major disadvantages of UC are brittleness, poor thermal shock resistance, and susceptibility to corrosion in aqueous environments at elevated temperatures. Binder matrix metals chosen for their low thermal absorption cross section, such as beryllium, zirconium, niobium, molybdenum, or iron, do not alleviate the disadvantages of the UC ceramic phase. Hence, this type of cermet has not found industrial applications beyond experimental reactor technology in the United States and abroad.

The carbides of the metalloids boron and silicon, B_4C and SiC, are of considerable industrial significance and enjoy such diverse applications as superhard tools and electrical resistor heating elements. These compounds are processed and used without metallic binder phases and hence are considered to fall outside the material classification for cermets.

Boride Cermets

Because metal borides generally are more refractory than TiC, cermets based on borides are of interest in situations that require a material of extreme heat and corrosion resistance, such as applications in contact with reactive hot gases or molten metals. The diborides of the transition metals hafnium, tantalum, zirconium, and titanium have extremely high melting

points, descending in the order given from 3250 to 2800 °C (5880 to 5070 °F). Molybdenum boride (MoB) and chromium boride (CrB) melt considerably lower (at 2180 and 1550 °C or 3960 and 2820 °F, respectively). The oxidation resistance of the transition metal diborides above a temperature of 1100 °C (2000 °F) is considerably better than that of TiC and roughly follows the descending order of the melting point (Ref 49). The oxidation resistance and strength properties at high temperatures can be further enhanced by reacting the boride crystals with small amounts of other thermally stable compounds, such as SiC or molybdenum disilicide ($MoSi_2$), prior to processing the powder into solid bodies.

Because these metal borides have relatively high thermal conductivity and high-temperature stability, they do not depend on a supportive metallic binder matrix for thermal shock resistance and strength as do the Ni-Cr alloys in TiC cermets. The boride phases alone in their highly purified state are extremely hard and abrasive, however; their consolidated bodies pose problems in fabrication to useful products, as well as in service, especially in environments involving dynamic gas or liquid-metal flow. This shortcoming can be alleviated in some instances by a metallic binder phase. For thermodynamic reasons, this binder phase generally is limited to 2 to 5 at.% up to a maximum of 10 at.%.

The principal candidate metals for cementing the boride grains are iron, nickel, cobalt, chromium, molybdenum, tungsten, and boron, or some of their alloys. Low-melting eutectics in the systems boron-iron (1161 °C or 2122 °F), boron-cobalt (1102 °C or 2015 °F), and especially boron-nickel (990 °C or 1814 °F) restrict the amount of the respective binder metals to a small percentage. The effectiveness of iron and cobalt, in particular as binder for titanium diboride (TiB_2) and zirconium diboride (ZrB_2), is further diminished by the formation of very brittle intermetallic compounds; whereas chromium and boron, singly or combined, produce tougher, higher melting eutectics with these borides (Ref 50). Addition of up to 5 wt% B and 10 wt% Mo or W can be successfully used as binder (of ZrB_2, for example) without forming low-melting phases (Ref 51, 52).

The transition metal borides are produced as pure crystals by processes such as solid-state reaction of metal boron, reaction of the metal or its oxide with boron carbide, reduction of boron and metal oxides with carbon or reactive metals, or

fused-salt electrolysis. Mixtures of these borides and binder metal powder are processed into cermet products by ceramic or powder metallurgy techniques, such as hydrostatic pressing or slip casting followed by vacuum sintering, or by hot uniaxial or isostatic pressing. The high costs of producing the borides and handling the brittle products with the necessary care have limited applications to those cases in which the unusual properties are an essential requirement. Table 8 lists the physical and mechanical properties of the commercially available metal borides and their cermets.

Zirconium Boride-Based Cermets. A comprehensive study of the properties of this transition metal boride has been reported by Kaufman and Clougherty (Ref 55). The very high melting point and good high-temperature mechanical properties, as well as a noticeable reduction in brittleness with rising temperature, make it one of the few borides that have gained commercial status. Steinitz (Ref 51) has shown the beneficial effect of 2 to 5 wt% B binder on the high-temperature properties of ZrB_2 that render the material suitable for extremely high-temperature applications, including high-performance burner, rocket, and jet reaction systems.

Kaufman and Clougherty (Ref 56, 57) have shown that the oxidation resistance of ZrB_2 can be further enhanced by reacting it with up to 15% SiC and that the consolidated cermet bodies can successfully withstand oxidizing environments in the 1900 to 2500 °C (3450 to 4530 °F) temperature range. These materials have been the object of an extensive investigation for nozzle throat inserts for liquid propellant rockets (Ref 58). Probably the most outstanding characteristic of ZrB_2 is its high-temperature corrosion resistance and non-wetting property in comparison to molten aluminum, brass, zinc, and lead. As a result, applications for this cermet have opened up in systems handling molten metals. Typical examples include impellers and bearings in pumps for liquid-die casting alloys, spray nozzles for atomizing metal powders, and furnace parts that come in contact with molten reactive metals or vapors. Figure 11 shows an assortment of parts made of boron-bonded ZrB_2 cermets.

Titanium Boride-Based Cermets. The physical and mechanical properties at low and high temperatures of TiB_2 do not vary greatly from those of ZrB_2 (Ref 55). As single boride or in solid solution with chromium boride (CrB_2), TiB_2 is considered by some to be the most promising of the transition metal borides (Ref 50). Applications include evaporation vessels for

Table 8 Properties of metal borides and boride-based cermets

Property	TiB₂(a)	ZrB₂(b)	ZrB₂-B(c)	Cermet grades CrB(d)	CrB-Ni(e)	CrB-Cr-Mo(f)	Mo₂NiB₂(g)
Melting point or range, °C (°F)	2980 (5400)	3040 (5500)	2955-3010 (5350-5450)	2050 (3720)	1650-1760 (3000-3200)	1930-1980 (3500-3600)	1430 (2600)
Density, g/cm³ (lb/in.³)	4.5 (0.163)	6.1 (0.221)	4.97-5.27 (0.180-0.191)	6.15 (0.222)	6.16-6.27 (0.223-0.225)	6.77-7.27 (0.245-0.263)	8.40 (0.305)
Electrical resistivity, μΩ·cm at 25 °C (75 °F)	15.3	16	17-23	20	38-58	37-54	66-71
Mean coefficient of thermal expansion, μm/m/°C (μin./in./°F)	6.39 (3.55)(h)	7.50 (4.17)(h)	5.76 (3.20)(k)	...	9.81 (5.45)(j)	9.90 (5.50)(j)	...
Thermal conductivity, W/m·K (Btu·in./ft²·h·°F) at 200 °C (500 °F)	25.9 (180)	23.0 (160)	...	...	...	...	...
Hardness							
HK	3370	2300	...	2140	...	...	...
HRA	...	...	88-90	...	75-86	77-88	88-90
Modulus of elasticity, GPa (10⁶ psi)	365 (53)	441 (64)	...	...	...	...	...
Transverse-rupture strength, MPa (ksi)							
At 20 °C (70 °F)	130 (19)	200 (29)	...	...	813 (118)	...	≤690 (≤100)
At 980 °C (1800 °F)	...	...	434 (63)	...	550-950 (80-138)	620-965 (90-140)	...
Tensile strength, MPa (ksi)	127 (18.4)	196 (28.5)	...	...	...	...	...
Stress-to-rupture strength, MPa (ksi) at 980 °C (1800 °F) (m)	...	...	128 (18.5)	...	83-137 (12-20)	96-103 (14-15)	≤82 (≤12)

(a) 100 wt% TiB₂. (b) 100 wt% ZrB₂. (c) 95 wt% ZrB₂, 5 wt% B. (d) 100 wt% CrB. (e) 85 wt% CrB, 15 wt% Ni. (f) 80 wt% CrB, 16 wt% Cr, 4 wt% Mo. (g) 100 wt% Mo₂NiB₂. (h) At 20-760 °C (70-1400 °F). (j) At 20-980 °C (70-1800 °F). (k) At 20-1205 °C (70-2200 °F). (m) 100 h
Source: Ref 53-55

reactive metals, electrodes for aluminum refining, and in general, parts that are exposed to molten zinc and brass. The addition of TiB₂ to TiC in a composite structure has been successfully used for cutting tools, and complex cermets of TiC-TiB₂ with a Co-Si alloy binder (Ref 60, 61) or of TiB₂-MoSi with a graphite binder (Ref 58) have been experimented with for use as nozzles.

Chromium Boride-Based Cermets. Of the generally excellent corrosion- and oxidation resistant boride materials, CrB was one of the first borides to be investigated for its high-temperature potential.

Sindeband (Ref 62) successfully bonded the compound with cobalt, nickel, nickel-chromium, and nickel-copper. A composition containing as much as 15 wt% Ni can be hot pressed without exuding much liquid phase. The cermet is oxidation resistant up to 950 °C (1740 °F) and has a high hot hardness and transverse-rupture strength of about 890 MPa (130 ksi). The microstructure of this material is shown in Fig. 12. These properties were greatly improved by using Cr₂B crystals and cementing them with up to 10 wt% of an 80%Cr-20%Mo alloy (Ref 50, 63). These cermets have good stress-to-rupture prop-

erties and sufficient mechanical shock resistance to be candidate materials for steam and gas turbine blades, valve seats and inserts for internal combustion engines, and exhaust nozzles and tubes for jet engines. Although compositions much higher in nickel (for example, cermets containing the compound chromium-nickel boride, Cr₂NiB₄) suffer from the low-melting eutectics, this is used as an advantage for wear- and erosion-resistant overlay coatings and hardfacing applications.

Molybdenum Boride-Based Cermets. The molybdenum borides MoB and Mo₂B have less thermal stability than the previously discussed metal borides, but their electrical properties, hardness, and wear resistance are very good, and when

Fig. 11 Assortment of zirconium boride cermet parts
Included are crucibles, thermocouple protection tubes, and vaporizing vessels. Source: Ref 59

Fig. 12 Microstructure of 85%CrB-15%Ni cermet
Magnification: 375×. Source: Ref 62

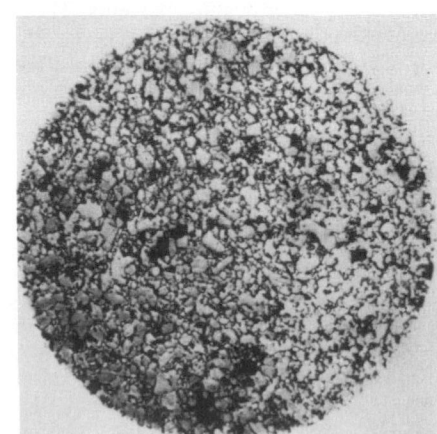

cemented with nickel, these cermets have excellent corrosion resistance—for example, to dilute sulfuric acid (Ref 53). Nickel-bonded molybdenum boride exhibits interesting behavior in two areas: If the composition corresponds to the compound molybdenum-nickel boride (Mo_2NiB_2), if the cermet contains Mo_2B in addition to Mo_2NiB_2, or if a low-melting, intermetallic binder containing chromium boride and nickel is used, cutting tool materials result that are comparable to commercial WC tool tips for machining brass, aluminum, and cast iron (Ref 64, 65). Secondly, the Mo_2NiB_2-type composition has thermal expansion characteristics that closely match those of the refractory metals and a favorable melting temperature that makes it ideally suitable as a high-temperature braze for molybdenum and tungsten, without risk of excessive grain growth or embrittlement of the primary metal structure (Ref 53, 59). When used in rod form with shielded arc welding equipment, this cermet is suitable for brazing electronic components in applications such as vacuum tubes and magnetrons.

Other Refractory Cermets

The nitrides, carbonitrides, and silicides of certain transition metals have gained importance for specific uses in operations involving high temperatures. The main mode of application for these refractory cermets, however, is in the form of coatings, such as TiN and TiC-TiN in various ratios for high-speed cutting tools or $MoSi_2$ for surface protection of molybdenum against high-temperature oxidation. In a very few cases, these compounds are used as solids, either in the pure state or cemented with a lower melting metallic phase.

Carbonitride- and Nitride-Based Cermets. Kieffer, Ettmayer, and Freudhofmeier (Ref 66) have found TiN and titanium carbonitrides to be suitable as hard phase for tool materials. The best binder is an alloy of 70%Ni-30%Mo, and optimum hardness in the 1000 to 2000 HV range is obtained with 10 wt% binder. The hardness increases progressively with the TiC component of the solid solution. The same trend prevails for the hardness of a cermet containing 14 wt% binder in which the values increase from about 1400 to 1900 HV for the straight cemented TiC composition. Transverse-rupture strength does not follow any trend; the best values reach about 1300 MPa (188 ksi) for a 10 wt% binder composition with a 72-to-18 TiN-TiC ratio and a 14 wt% binder material with a 69-to-17 TiN-TiC ratio. This compares with 1070 and 1275 MPa (155 ksi and 185 ksi), respectively, for the straight TiC cermets with 10 and 14 wt% binder. The hardness of TiN alone cemented with 10% of the 70%Ni-30%Mo alloy has a level of about 1050 HV and a transverse-rupture strength of about 785 MPa (115 ksi).

Combinations of nitrides and borides, with or without metallic binder, can also be fabricated into tools. Holtz and Parikh (Ref 67) have hot pressed a mixture of 60 wt% tantalum nitride (TaN) and 40 wt% ZrB_2 into tool bits that have performed very well at very high cutting speeds.

Nitride products based on the metalloids boron and silicon, like their carbide counterparts, have gained some significant commercial uses since their early development in the 1950's and 1970's, respectively. The normal hexagonal crystal lattice of boron nitride (BN) can be converted to a cubic crystal form by reacting boron powder with nitrogen at 1650 °C (3000 °F) minimum temperature while simultaneously applying pressure in excess of 7000 MPa (1000 ksi) with the aid of special press tools adopted from synthetic diamond manufacture. The product is extremely hard and is considered to be one of the best electrical insulators known, especially at high temperatures up to about two thirds of its melting point—in the vicinity of 2730 °C (4950 °F) (Ref 23, 68).

Cermets exhibiting excellent cutting performance have been achieved by bonding carefully graded particles of the superhard cubic boron nitride with cobalt or similar hard metal binders. Hot pressing is the preferred method of powder consolidation, and tool bits made in this manner outperform tungsten carbide tips by a factor of two-to-one and better (Ref 69).

The nitride of silicon and its combination with different oxides, notably Al_2O_3 (the "Sialons"), as well as the different silicon ceramics based on SiC, belong to the increasingly important new class of refractory materials known as structural ceramics. Additives of these cermets are nonmetallic and serve mainly to control the sintering mechanism. They do not contribute to a strengthening of the hard particle structure in the sense of a metallic binder and, in fact, cause a weakening of the grain-boundary network at high temperature in many systems. Therefore, these silicon ceramics are considered to lie outside the material classification for cermets.

Silicide-Based Cermets. Only isolated instances of commerical use are known for the metallic silicides. This is due chiefly to the extreme brittleness of these compounds and their concomitant problems of fabrication into solid objects. Because of its outstanding high-temperature oxidation resistance in conjunction with favorable coefficients of thermal expansion and electrical resistance, $MoSi_2$ is an important material for heating elements. Poor resistance to mechanical and thermal shock is the major deficiency of $MoSi_2$ and limits the applications of this material to simple cylindrical or rectangular shapes. Additions of metallic elements to remedy this handicap have been only partially successful, and cermets of $MoSi_2$ with nickel, cobalt, and platinum binder metals are still too brittle for fabrication into complex shapes (Ref 70). High-temperature bearings have been made experimentally by infiltrating molten silver into hard matrices containing $MoSi_2$, tungsten disilicide (WSi_2), or vanadium disilicide (VSi_2) and have shown good antifriction behavior against steels at elevated temperatures (Ref 71).

Graphite- and Diamond-Containing Cermets. Materials that contain a combination of carbon in the form of graphite or diamond with metals constitute a border region for cermets and are usually not designated as such. Because the carbon and metallic components are most often intimately mixed and uniformly distributed in the microstructure, they are considered pertinent to this discussion.

Graphite-metal combinations for electrical contact applications fall basically into two types of materials. For metallic brushes used in motors and generators, the metallic phase consists of copper or bronze; in the case of sliding contacts involving relatively low rubbing speeds and light contact pressures, it is silver. In brushes, the graphite particle content may spread over a wider range—from 5 to 70 wt%. A typical binary composition contains 70% Cu and 30% graphite. To improve wear and bearing properties, many brushes also contain up to 10% Sn and/or Pb and up to 12% Zn (Ref 72). The graphite content in the silver contact composition generally ranges between 2 and 50 wt%.

Graphite-containing metallic friction materials for brake linings and clutch facings have a predominantly metallic matrix to utilize a high thermal conductivity. This property permits rapid energy absorption, making this type of material suitable for service under a more severe wear and temperature environment than that which is possible for organic, resin-bonded asbestos friction elements. The most important contribution of a cermet-type lining material in aircraft brakes probably has been an increased energy capacity without additional weight or larger size of the brake unit (Ref 73). The friction coefficient of these cermets is tailored to the requirements of the particular application, principally by varying the ratio of a friction-producing ceramic to the graphite, which acts as a solid lubricant. The metallic matrix phase is essentially a bearing alloy containing 60 to 75 wt% Cu and 5 to 10% each of tin, lead, zinc, and/or iron. Graphite content falls within the 5 to 10% range, and the ceramic, mainly SiO_2 with the possibility of some Al_2O_3 additions, amounts to 2 to 7% (Ref 74).

Cermets composed of diamond varying in size from coarse splinters to fine dust inside a metal matrix are used for grinding, lapping, sawing, cutting, dressing, and truing tools. The size of the diamond is important for the efficiency of the tool; although finish improves as the grain or grit size becomes finer, the cutting speed is slower. For dressing tools, 5 to 35 diamond splinters are embedded per carat with a size of approximately 1 to 2.5 mm (0.04 to 0.1 in.). For rough grinding, the grit size is in the range of 0.15 to 0.5 mm (0.006 to 0.02 in.); for fine polishing, it falls between 0.05 and 0.15 mm (0.002 and 0.006 in.). Even finer diamond powder is used in combination with WC for specialized applications such as polishing plane surfaces of hard metal tools or finishing the rolls for Sendzimir-type mills. Typical compositions of these tools contain 12 to 16 wt% diamond dust embedded in a WC matrix cemented with 13% Co (Ref 75).

Other metallic bonding substances are based on copper, iron, nickel, molybdenum, or tungsten. Examples for copper matrices are bronzes with 10 to 20% Sn or 2 to 4% Be, which may be strengthened by precipitation hardening, and a 47%Cu-47%Ag-6%Co alloy. Bonding metals suitable for somewhat higher temperature service include iron-nickel, iron-nickel-chromium, and iron-tin-antimony-lead alloys, Permalloy, and nickel alloys containing 2 to 8% Be. Refractory metal-based matrices are alloys of the molybdenum-copper, molybdenum-cobalt, or tungsten-nickel-copper types and tungsten-nickel-iron heavy alloys (Ref 75). In general, the bond materials must be selected with consideration of lowest possible processing temperatures to avoid the danger of a transformation of the diamond to graphite.

REFERENCES

1. Report of Task Group B on Cermets, ASTM Committee C-21, American Society for Testing and Materials, Philadelphia, 1955
2. Tinklepaugh, J.R. and Crandall, W.B., *Cermets*, Reinhold, New York, 1960, Ch. 1
3. Van Schoick, E.C., Ed., *Ceramic Glossary*, The Ceramic Society, Columbus, OH, 1963
4. Kieffer, R. and Benesovsky, F., *Hartmetalle*, Springer, New York, 1965, p 437-489
5. Goetzel, C.G., *Treatise on Powder Metallurgy*, Vol 1, Interscience, New York, 1949, p 4
6. Marshall, C.L., Chromium-Alumina Base Cermets, *Cermets*, Reinhold, New York, 1960, p 109-118
7. Weber, B.C., Slip Casting of Cermets, *Cermets*, Reinhold, New York, 1960, p 58-73
8. Goetzel, C.G., Infiltration Process, *Cermets*, Reinhold, New York, 1960, p 73-81
9. Humenik, M., Jr. and Whalen, T.J., Physiochemical Aspects of Cermets, *Cermets*, Reinhold, New York, 1960, p 6-49
10. Blackburn, A.R. and Shevlin, T.S., Fundamental Study and Equipment for Sintering and Testing Cermet Bodies: V, Fabrication, Testing and Properties of 30 Chromium-70 Alumina Cermets, *Journal of the American Ceramic Society*, Vol 34 (No. 11), 1951, p 327-331
11. Hauck, C.A., Donley, J.C., and Shevlin, T.S., "Fundamental Study and Equipment for Sintering and Testing of Cermet Bodies," U.S. Air Force Report WADC-TR-173, March 1956
12. Goetzel, C.G., Titanium Carbide-Metal Infiltrated Cermets, *Cermets*, Reinhold, New York, 1960, p 130-146
13. Chermant, J.L. and Osterstock, F., Fracture Toughness and Fracture of WC-Co Composites, *Journal of Materials Science*, Vol 11, 1976, p 1939-1951
14. Hatschek, R.L., Take a New Look at Ceramics/Cermets, Special Report 733, *American Machinist*, Vol 125 (No. 5), 1981, p 165-176
15. Shevlin, T.S., Oxide-Base Cermets, *Cermets*, Reinhold, New York, 1960, p 97-109
16. Engelhardt, G. and Thümmler, F., Creep Deformation of Al_2O_3-Cr Cermets with Cr-Content up to 50 vol%, *Modern Developments in Powder Metallurgy*, Vol 8, Metal Powder Industries Federation, Princeton, NJ, 1974, p 605-626
17. White. A.E.S., *et al.*, Metal-Ceramic Bodies, *Symposium on Powder Metallurgy 1954*, Special Report No. 58, The Iron and Steel Institute, London, 1956, p 311-314
18. Baxter, J.R. and Roberts, A.L., Development of Metal-Ceramics from Metal Oxide Systems, *Symposium on Powder Metallurgy 1954*, Special Report No. 58, The Iron and Steel Institute, London, 1956, p 315-324
19. Harris, G.T. and Child, H.C., The Rupture Strength of Some Metal-Bonded Refractory Oxides, *Symposium on Powder Metallurgy 1954*, Special Report No. 58, The Iron and Steel Institute, London, 1956, p 325-330
20. Watkins, R.V., Reed, G.C., and Schalliol, W.L., Hot-Extruded Chromium Composite Powder, *Progress in Powder Metallurgy*, Vol 20, Metal Powder Industries Federation, New York, 1964, p 149-158
21. Ryshkewitch, E., Oxide-Metal Compound Ceramics, *Metals for the Space Age*, Plansee Proceedings 1964, Springer, New York, 1965, p 823-830
22. Weber, B.C. and Schwartz, M.A., Metal-Modified Oxides, *Cermets*, Reinhold, New York, 1960, p 119-121
23. Weber, B.C. and Schwartz, M.A., Container Materials for Melting Reactive Metals, *Cermets*, Reinhold, New York, 1960, p 154-158
24. Heitzinger, F., Molybdenum + Zirconia, A New Metalceramic Material for New Applications, *Modern Developments in Powder Metallurgy*, Vol 8, Metal Powder Industries Federation, Princeton, NJ, 1974, p 371-390
25. Cronin, L.J., Refractory Cermets, *American Ceramic Society Bulletin*, Vol 30, 1951, p 234-238
26. Cronin, L.J., Electronic Refractory Cermets, *Cermets*, Reinhold, New York, 1960, p 158-166
27. Weber, C.E. and Hirsch, H.H., "Dispersion Type Fuel Elements," *Proceedings First International Conference on Peaceful Uses of Atomic Energy*, Vol 9, 1953, p 196-202
28. Holden, A.N., *Dispersion Fuel Elements*, Gordon and Breach, New York, 1967, p 80-91, 152-167
29. Loewenstein, P., Corzine, P.D., and Wong, J., *Nuclear Reactor Fuel Elements*, Interscience, New York, 1962, p 393-394, 396-398
30. Sandford, J.E., The New Angles in Cutting, *Iron Age*, Vol 203 (No. 25), 1969, p 69-76
31. Dispersion of Micro-Particles in Tungsten Carbide Produces Superior Tool Materials, *Industrial Heating*, Vol 36 (No. 11), 1969, p 2162, 2164
32. Hummer, J., "Hard, Sintered, Cobalt-Bonded Tungsten Carbide Bodies Containing Chromium Carbide," U.S. Patent 3,480,410 (applied 1968, issued 1969)
33. Hinnüber, J. and Rüdiger, O., Chromium Carbide in Hard-Metal Alloys, *Symposium on Powder Metallurgy 1954*, Special Report No. 58, The Iron and Steel Institute, London, 1956, p 305-310

34. Pozzo, R.F. and West, J.V., Chromium Carbide Applications, *Cermets*, Reinhold, New York, 1960, p 150-153

35. Wambold, J. and Redmond, J.C., Recent Developments in Sintered Titanium Carbide Compositions, *High Temperature Materials*, John Wiley, New York, 1959, p 125-139

36. Wambold, J., Properties of Titanium Carbide-Metal Compositions, *Cermets*, Reinhold, New York, 1960, p 122-129

37. Havekotte, W.L., Titanium Carbide-Base Cermets, *Sintered High-Temperature and Corrosion-Resistant Materials*, Plansee Proceedings 1955, Pergamon Press, London, 1956, p 111-129

38. Deutsch, G.C., The Use of Cermets as Gas-Turbine Blading, *High Temperature Materials*, John Wiley, New York, 1959, p 190-204

39. Lavendel, H.W. and Goetzel, C.G., "A Study of Graded Cermet Components for High Temperature Turbine Applications," U.S. Air Force Report WADC-TR-57-135, ASTIA Document AD 131031, May 1957

40. Lavendel, H.W. and Goetzel, C.G., Recent Advances in Infiltrated Titanium Carbides, *High Temperature Materials*, John Wiley, New York, 1959, p 140-154

41. Epner, M. and Gregory, E., Titanium Carbide-Steel Cermets, *Cermets*, Reinhold, New York, 1960, p 146-149

42. Ellis, J.L., A Machinable, Heat Treatable, and Weldable Cemented Carbide for Tooling Purposes, *Tool Engineer*, Vol 38 (No. 4), 1957, p 103-105

43. Tarkan, S.E. and Mal, M.K., Hardening Steel Bonded Carbides, *Metal Progress*, Vol 105 (No. 5), 1974, p 99, 100, 102

44. Miska, K.H., Refractory Metal Carbides Fight Wear, Heat and Corrosion, *Materials Engineering*, Vol 82 (No. 6), 1975, p 68-70

45. Robisch, T.J., Mal, M.K., and Tarkan, S.E., Steel Bonded Titanium Carbides Effectively Used as Hardenable Wear Resistant Overlays, *Industrial Heating*, Vol 49 (No. 5), 1982, p 18-20

46. Ellis, J.L., Gregory, E., and Epner, M., Heat Treatable Steel-Bonded Carbides—New Construction Materials for Tools and Wear Resistant Components, *Progress in Powder Metallurgy*, Vol 16, Metal Powder Industries Federation, New York, 1960, p 76-83

47. Hammer, R.L., Manly, W.D., and Bridges, W.H., *Carbides and Cermets, Reactor Handbook*, 2nd ed., Vol 1, C.R. Tipton, Jr., Ed., Interscience, New York, 1960, p 508

48. Kieffer, R. and Benesovsky, F., *Hartmetalle*, Springer, New York, 1965, p 469-470

49. Kaufman, L., Clougherty, E.C., and Berkowitz-Mattuck, J.B., Oxidation Characteristics of Hafnium and Zirconium Diboride, *AIME Transactions*, Vol 239 (No. 4), American Institute of Mining, Metallurgical and Petroleum Engineers, 1967, p 458-466

50. Kieffer, R. and Benesovsky, F., *Hartmetalle*, Springer, New York, 1965, p 475-479

51. Steinitz, R., Borides—Part B: Fabrication, Properties and Applications, *Modern Materials*, Vol 2, Academic Press, New York, 1960, p 191-224

52. Halcombe, C.E., Jr., "Slip Casting of Zirconium Diboride," U.S. Atomic Energy Commission Report Y-1819, 28 Feb 1972

53. Everhart, J.L., New Refractory Hard Metals, *Materials and Methods*, Vol 40 (No. 2), Aug 1954, p 90-92

54. Latva, J.D., Selection and Fabrication of Ceramics and Intermetallics, *Metal Progress*, Vol 82 (No. 4), Oct 1962, p 139-144, 180, 186

55. Kaufman, L. and Clougherty, E.V., Investigation of Boride Compounds for High Temperature Applications, *Metals for the Space Age*, Plansee Proceedings 1964, Springer, New York, 1965, p 722-758

56. Kaufman, L. and Clougherty, E.V., "Investigation of Boride Compounds for Very High Temperature Applications," U.S. Air Force Materials Laboratory Report RTD-TDR-63-4096, Part 1, Dec 1963

57. Clougherty, E.V., Pober, R.L., and Kaufman, L., Synthesis of Oxidation Resistant Metal Diboride Composites, *AIME Transactions*, Vol 242 (No. 6), American Institute of Mining, Metallurgical and Petroleum Engineers, 1968, p 1077-1082

58. Clougherty, E.V., et al., "Research and Development of Refractory Oxidation Resistant Diborides," U.S. Air Force Materials Laboratory Report AFSC-ML-TR-68-190, Part 1, Oct 1968; Part 2, Vol 1-7, Nov 1969 —June 1970; Part 3, May 1970

59. Blum, A. and Ivanick, W., Recent Developments in the Application of Transition Metal Borides, *Powder Metallurgy Bulletin*, Vol 7 (No. 3-6), April 1956

60. Greenhouse, H.M., Stoops, R.F., and Shevlin, T.S., A New Carbide-Base Cermet Containing TiC, TiB$_2$ and CoSi, *Journal of the American Ceramic Society*, Vol 37 (No. 5), 1954, p 203-206

61. Montgomery, E.T., et al., "Preliminary Microscopic Studies of Cermets at High Temperature," U.S. Air Force Report WADC-TR-54-33, Part 1, April 1955; Part 2, Feb 1956

62. Sindeband, S.J., Properties of Chromium Boride and Sintered Chromium Boride, *AIME Metals Transactions*, Vol 185, American Institute of Mining, Metallurgical and Petroleum Engineers, Feb 1949, p 198-202

63. Binder, I. and Moskowitz, D., "Cemented Borides," Office of Technical Services, U.S. Department of Commerce, PB 121346, 1954-1955

64. Steinitz, R. and Binder, I., New Ternary Boride Compounds, *Powder Metallurgy Bulletin*, Vol 6 (No. 4), Feb 1953, p 123-125

65. Binder, I. and Roth, A., An Evaluation of Molybdenum Borides as Cutting Tools, *Powder Metallurgy Bulletin*, Vol 6 (No. 5), May 1953, p 154-162

66. Kieffer, R., Ettmayer, P., and Freudhofmeier, M., About Nitrides and Carbonitrides and Nitride-Based Cemented Hard Alloys, *Modern Developments in Powder Metallurgy*, Vol 5, Plenum Press, New York, 1971, p 201-214

67. Holtz, F.C. and Parikh, N.M., Developments in Cutting Tool Materials, *Engineers' Digest*, Vol 28 (No. 1), 1967, p 73, 75, 99

68. Borazon—Man Made Material Is Hard as Diamond, *Materials and Methods*, Vol 45 (No. 5), 1957, p 194, 196

69. Pipkin, N.J., Roberts, D.C., and Wilson, W.I., Amborite—A Remarkable New Cutting Material from De Beers, *Industrial Diamond Review*, June 1980, p 203-206

70. Kieffer, R. and Benesovsky, F., *Hartmetalle*, Springer, New York, 1965, p 487-489

71. Baskey, R.H., An Investigation of Seal Materials for High Temperature Applications, *Transactions of American Society of Lubricating Engineers*, Vol 3 (No. 1), 1960, p 116-123

72. Lenel, F.V., *Powder Metallurgy*, Metal Powder Industries Federation, Princeton, NJ, 1980, p 556

73. Heron, R.H., Friction Materials—A New Field for Ceramics and Cermets, *Ceramic Bulletin*, Vol 34 (No. 12), 1955, p 295-298

74. Lenel, F.V., *Powder Metallurgy*, Metal Powder Industries Federation, Princeton, NJ, 1980, p 485

75. Goetzel, C.G., *Treatise on Powder Metallurgy*, Vol 2, Interscience, New York, 1950, p 171-174

Metal Powders Used for Filler Materials*

By John Mikurak
Technical Manager of Welding Products
Sandvik Steel Co.
and
Michael S. Sierdzinski
Research Engineer
Alloy Rods, Inc.

METAL POWDERS USED IN ELECTRODE COATINGS include a wide range of ferroalloys and pure metals. Ferroalloys of boron, manganese, silicon, chromium, niobium, molybdenum, and vanadium and metal powders such as nickel, manganese, chromium, and iron are used in various types of electrodes. When selecting a metal powder for electrode coatings, the material should be reduced to the desired particle size without creating excessive industrial hazard, and the finely ground material should not react with liquid alkaline silicate during mixing or fabrication (extrusion) of the electrodes. Iron meets the above requirements and is widely used. According to the Metal Powder Industries Federation, 10% of the iron powder produced in North America is used by the welding industry.

High deposition rates of filler metal can be obtained by using iron powder in welding electrodes, flux-cored wire, and joint fill for submerged arc welding. In general, iron powder intended for use in welding must have few impurities (in particular, little sulfur and phosphorus) and uniform chemical and physical properties. Also, cost of the iron powder must be comparable to that of the core wire for low-alloy filler metals. Reduced and water-atomized iron powders satisfy these requirements and are used almost exclusively in welding.

Reduced (sponge) and atomized powder differ in particle shape and structure, and have different properties because of differences in their method of manufacture. This article examines the types of iron powders used in welding applications, their production methods, the effect of type and quality of iron powder on the production of coated electrodes, and the subsequent performance characteristics of these electrodes. Additional detailed information on selection and classification of the electrodes discussed in this article can be found in Ref 1.

Functions of Electrode Coatings

There are four general functions controlled or influenced by electrode coatings: welding parameters, slag property control, weld deposit quality, metal transfer, and compatibility with extrusion equipment. Success in fulfilling these functions depends on the proper balance of the constituents of the electrode coating.

Control of Welding Parameters. The electrode coating determines the type of welding current and the arc stability. For example, core wire used for an E6010 direct current electrode may be the same as that used in an E7024 alternating current/direct current electrode, but the coatings are different. Arc voltage and optimum current are functions of the type and formulations of the coating. Therefore, electrodes of the same classification produced by different manufacturers may require minor current adjustments for optimum

welding in field applications. To prevent side-arcing, the coating should be a good insulator.

Slag Property Control. Control of the many properties of a molten slag is very difficult. Wetting characteristics of the slag help determine the contour of the deposit and the degree of slag coverage. Expansion of the slag affects markedly the slag removal. Fluidity or viscosity of the slag alters the out-of-position characteristics of the finished electrode. Freezing point of the slag affects bead contour and is a controlling factor in determining the acceptance of the electrode for vertical and overhead positions. All these interrelated properties are controlled by the electrode coating composition. Only extensive laboratory testing can produce a coating with the optimum balance of all these properties.

Weld Deposit Quality. The electrode coating controls the metallurgical structure and quality of the weld deposit. Service requirements demand that every weld deposit must be free from porosity and slag inclusions. The coating must contain a sufficient quantity of deoxidizing material to degas the molten metal and to ensure that all resulting oxide and silicate inclusions float off in the slag.

Compatibility With Extrusion Equipment. Modern high-speed extrusion equipment places rigid demands on electrode coating ingredients. A proper balance of particle size among the various ingredients is essential for ease of extru-

*Originally published as "The Significance of Iron Powder as Additive Material for Welding Practice," Technical Bulletin, Hoeganaes Corp., Riverton, NJ, 1975. Reprinted with permission.

sion, ability to withstand mechanical damage, and a pore structure that permits immediate, continuous drying.

Manufacturers of ceramics are able to extrude nonplastic materials, such as beach sand, by reducing the particle size of the individual grains to less than 1 μm. Based on this experience, electrode manufacturers might be prone to use only extremely fine materials to obtain maximum plasticity and extrudability. However, if such fine materials were used, alloy losses would be prohibitive, even if the coatings could be dried in modern ovens. On the other hand, coarse-grain materials, such as unground beach sand, could be dried readily in continuous ovens if they could be extruded. Therefore, the electrode manufacturer must carefully select coating ingredients to capitalize on good extrusion and drying characteristics.

Electrode Coating Constituents

Coating materials are divided into four classifications: binders and extrusion aids, arc stabilizers, fluxes and slag modifiers, and alloying materials. These classifications are based only on the primary end use in a coating. Many materials might fall into two or more classifications if their overall effects are considered. For example, sodium and potassium silicate may be equally effective as binders, but potassium silicate is preferred where additional arc stability is desired. Titanium dioxide and titania-bearing materials frequently are used to improve arc stability. However, such additions may markedly alter the properties of the resultant slag.

Binders and extrusion aids, commonly used by the electrode industry, include liquid sodium and potassium silicates, celluloses, clays, bentonites, talc, and mica. The alkaline silicates and celluloses provide dry strength, green strength, or assist in controlling the drying characteristics. Other inorganic materials such as talc, mica, clay, or bentonite may be used to increase plasticity, that is, slippage under extrusion pressure.

Arc stabilizers are predominantly titanium- or potassium-bearing materials. Titanium dioxide, rutile, potassium titanate, ilmenite, and potassium carbonate or hydroxide are the most common. Lithium-bearing materials have arc-stabilizing properties, but their cost is somewhat higher.

Fluxes and Slag Modifiers. Almost all of the more common minerals and fluxes used by the ceramics industry have been employed as additions to electrode coating compositions. In commercial practice, calcium carbonate, fluorspar, silica flour, alumina, feldspar, manganese oxides, iron oxides, wallastonite, and zirconium-bearing materials appear to be the most widely used.

Alloying Materials. A wide variety of ferroalloys and metal powders are commonly used in electrode coatings. In addition to providing the necessary alloying contents, materials such as ferromanganese and ferrosilicon serve as deoxidizers.

Table 1 gives compositions of electrode coatings for classes of electrodes used in shielded metal arc welding of low-carbon steel and lists the primary and secondary functions of the coating constituents. The

thickness of the coatings varies from 10 to 55% of the total diameter of the coated electrode, depending on the type of coating.

Types of Iron Powder for Welding

A variety of methods are used to produce iron powder for welding applications. A brief summary of these production methods follows. Detailed information on these processes can be found in the article "Production of Iron Powder" in this Volume.

Reduced Iron. The oldest iron powder production method is the Swedish Sponge Iron process, in which pure magnetite (Fe_3O_4) is reduced by a carbonaceous material. A photomicrograph of a sponge iron powder is shown in Fig. 1. Table 2 lists compositions and properties of reduced iron welding-grade powders.

Fine (−100 mesh) sponge iron powder having an apparent density of 2.50 g/cm^3 is recommended for use in small-diameter basic electrodes and flux-cored wire. This powder has a large specific surface due to the microporous particles and irregular grain shape. This contributes to increased electrical conductivity, good arc stability, and smooth arc action of the welding electrode.

Sponge iron powder having an apparent density of 2.40 g/cm^3 and a particle size range of −40+200 mesh has found wide application in production of the low-hydrogen electrode. It is also used in rutile and acid electrodes with low to medium deposition efficiency. Sponge iron powder of higher apparent density (2.80 g/cm^3) normally is used for electrodes having a

Table 1 Functions and composition ranges of coating constituents on low-carbon steel arc welding electrodes

Coating constituent	Function of constituent Primary	Secondary	E6010, E6011	E6012, E6013	E6020, E6022	E6027	E7014	E7016	E7018, E7048	E7024	E7028
Cellulose	Shielding gas	Slag modifier	25-40	2-12	1-5	0-5	2-6	...	...	1-5	...
Calcium carbonate	Shielding gas	Slag modifier	...	0-5	0-5	0-5	0-5	15-30	15-30	0-5	0-5
Fluorspar	Slag modifier	Fluxing agent	...	...	...	...	...	15-30	15-30	...	5-10
Dolomite	Fluxing agent	...	...	...	...	...	...	...	...	...	5-10
Titanium dioxide (rutile)	Slag modifier	Arc stabilizer	10-20	30-55	0-5	0-5	20-35	15-30	0-5	20-35	10-20
Potassium titanate	Arc stabilizer	Slag former	(a)	(a)	...	...	...	...	0-5	...	0-5
Feldspar	Slag modifier	Stabilizer	...	0-20	5-20	0-5	0-5	0-5	0-5	...	0-5
Mica	Extrusion aid	Stabilizer	...	0-15	0-10	...	0-5	...	...	0-5	...
Clay	Extrusion aid	Slag former	...	0-10	0-5	0-5	0-5	...	...	...	...
Silica	Slag modifier	...	...	...	...	5-20	...	...	...	...	...
Manganese oxide	Slag modifier	Alloying	...	...	0-20	0-15	...	...	...	...	...
Iron oxide	Slag modifier	...	...	...	15-45	5-20	...	...	...	...	...
Iron powder	Deposition rate	Contact welding	...	...	...	40-55	25-40	...	25-40	40-55	40-55
Ferrosilicon	Deoxidizer	Alloying	...	...	0-5	0-10	0-5	5-10	5-10	0-5	2-6
Ferromanganese	Alloying	Deoxidizer	5-10	5-10	5-20	5-15	5-10	2-6	2-6	5-10	2-6
Sodium silicate	Binder	Fluxing agent	20-30	5-10	5-15	5-10	0-10	0-5	0-5	0-10	0-5
Potassium silicate	Arc stabilizer	Binder	(a)	5-15(a)	0-5	0-5	5-10	5-10	5-10	0-10	0-5

(a) Used in E6011 and E6013 electrodes to permit welding with alternating current

Fig. 1 Sponge iron powder (−100 mesh) produced by reduction of iron ore
Magnification: 500×

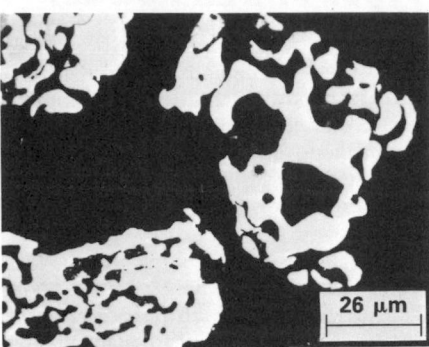

Fig. 2 Water-atomized iron powder
(a) Scanning electron micrograph of particles. Magnification: 2000×. (b) Polished cross section of particles. Magnification: 150×

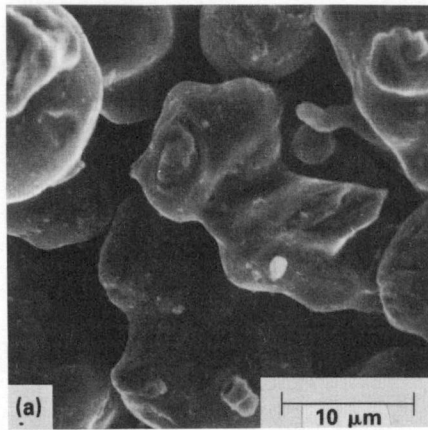

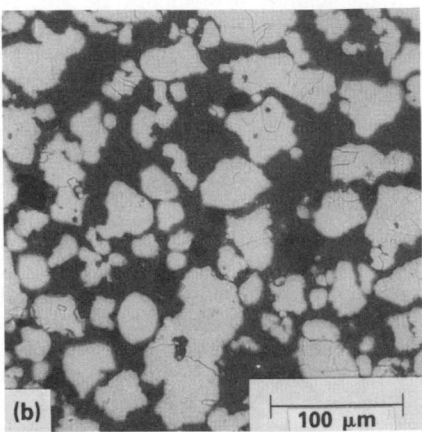

medium deposition efficiency of 120 to 170%.

The deposition efficiency (yield) of an electrode is calculated as follows:

$$(A \div B) \times 100 = R$$

In this equation, A is the deposited weld metal, B is the weight of melted core wire, and R is the deposition efficiency (yield).

Atomized Iron. Water-atomized iron powder, as shown in Fig. 2, is also used in electrode coatings. Properties and compositions of these powders are given in Table 2. The low carbon content and iron oxide content produce a smooth arc and a slag that is easy to remove. This type of powder, which has an apparent density of 3.20 g/cm³, is designed for production of low-hydrogen and titania-type electrodes using 50% iron powder in the coating.

High-density (≥3.50 g/cm³) atomized iron powder has relatively smooth parti-

cles that help reduce manufacturing costs and facilitate the use of high-deposition-efficiency electrodes. Low-hydrogen electrodes (Type E7028) can be made with coatings containing 65% iron powder. Such electrodes have deposition efficiencies to 225%. This type of iron powder also can be used for E6027, E6024, E7024, and gravity feed electrodes having deposition efficiencies from 190 to 210%. Classifications of coated electrodes are listed in Tables 3 and 4.

In the manufacture of electrodes, the higher density, relatively smooth particles require less silicate, allow higher extrusion pressure, and have faster extrusion rates, better brushing, and less swelling in baking. The resulting high-strength coating resists chipping and cracking.

Use of high-density iron powder elec-

trodes saves time and costs through high deposition rates, longer bead lengths, better penetration, no undercutting, and slag that is self-removing. Powder characteristics promote smooth arc action and better recovery of metallics, thus providing uniform physical properties.

Alternative Production Methods. Welding-grade iron powder can also be produced by methods that combine reduction and atomization, such as the Domfer process and the Quebec Metal Powder process. Properties and compositions of these powders can be found in the article "Production of Iron Powder" in this Volume.

Methods of Improving Welding Electrode Production

Currently, welding electrodes are produced by a continuous high-speed production system. Consistent physical properties of electrode coating components help to attain shorter throughput times. Extrusion problems tend to multiply as coating thickness and the amount of iron powder added to the coating are increased.

Extrusion pressure is influenced greatly by the physical properties of iron powder and the amount of silicate added to the electrode coating. Figure 3 shows results from extrusion tests on a rutile high-yield electrode. By using 65% of iron powder having a 3.50 g/cm³ apparent density at a coating-to-core-wire ratio of 2 to 1 and 170 g silicate/kg (75 g silicate/lb) of dry coating material, a welding efficiency yield of 185% was obtained from this electrode. A pressure range of 35 to 45 MPa (5000 to 6500 psi) has provided the best conditions. The ideal extrusion pressure of 39

Table 2 Typical properties of reduced and water-atomized iron welding powders

Property	1(a)	2(a)	Powder 3(a)	4(b)	5(b)
Chemical composition, %					
Iron, total	98.20	98.0	98.0	98.0	98.0
Hydrogen loss	0.40	0.45	0.45	0.50	0.50
Carbon	0.05	0.05	0.05	0.05	0.05
Sulfur	0.01	0.015	0.015	0.025	0.025
Phosporus	0.01	0.01	0.01	0.01	0.01
Sieve analysis, U.S. standard, %					
+40	...	Trace	Trace	2	2
−40+80	1	54	55	61	62
−80+200	48	44	43	32	33
−200	...	2	2	5	3
−200+325	24	...	...	...	...
−325	27	...	...	...	...
Apparent density, g/cm³	2.5	2.4	2.8	3.2	3.65

(a) Reduced iron powder. (b) Water-atomized iron powder

Table 3   Electrode classification of iron powder, titania, and low-hydrogen electrodes

Iron powder content, %	AWS classification	Type of covering	Capable of producing satisfactory welds in positions shown(a)	Type of current(b)

E60 series: minimum tensile strength of deposited metal in as-welded condition, 410 MPa (60 ksi) or higher

Iron powder content, %	AWS classification	Type of covering	Positions	Type of current
10 max	E6010	High cellulose, sodium	F, V, OH, H	DCEP
10 max	E6011	High cellulose, potassium	F, V, OH, H	ac or DCEP
10 max	E6012	High titania, sodium	F, V, OH, H	ac or DCEN
10 max	E6013	High titania, potassium	F, V, OH, H	ac, DCEN, or DCEP
25	E6020	High iron oxide	H-fillets	ac or DCEN
			F	ac, DCEN, or DCEP
30-45	E6027	Iron powder, iron oxide	H-fillets	ac or DCEN
			F	ac, DCEN, or DCEP

E70 series: minimum tensile strength of deposited metal in as-welded condition, 480 MPa (70 ksi) or higher

U.S.	European	AWS	Type of covering	Positions	Type of current
18	25	E7014	Iron powder, titania	F, V, OH, H	ac, DCEN, or DCEP
25-35	44	E7018	Iron powder, low hydrogen	F, V, OH, H	ac or DCEP
45-50	65	E7024	Iron powder, titania	H-fillets, F	ac, DCEN, or DCEP
50	65	E7028	Iron powder, low hydrogen	H-fillets, F	ac or DCEP

(a) Abbreviations indicate these welding positions: F, flat; H, horizontal; H-fillets, horizontal fillets; V, vertical; OH, overhead.
(b) DCEN, direct current electrode negative (straight polarity); DCEP, direct current electrode positive (reverse polarity)

MPa (5600 psi) was arrived at by testing the following additions of silicate:

Iron powder apparent density, g/cm³	Amount of silicate added, g
3.50	150
2.85	185
2.40	215

The slopes of the curves in Fig. 3 show that comparatively small changes in quantity of added silicate or in the physical properties of the iron powder make the extrusion pressure fall beyond the best extrusion conditions. Pressures that are too high would cause longitudinal cracks immediately after extrusion. Transversal cracks might appear after the baking. Conversely, an extrusion pressure that is too low produces a coating that is too soft. In such a case, the electrode may easily be deformed during handling and transportation.

For those high-yield electrodes investigated, it was found that when silicate addition was unchanged, a powder with smooth surface structure gave a comparatively lower extrusion pressure. When adding large portions of iron powder, rough, irregularly shaped grains might augment critically the internal friction within the coating material, thereby increasing the extrusion pressure. Thus, use of a smoother, heavier powder made possible the use of similar additions of silicate or other plasticizing ingredients. This is desirable technically and economically, especially in the production of this type of high-yield electrode.

Fig. 3 Effect of silicate content on extrusion pressure

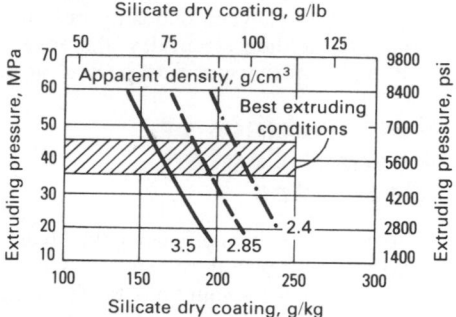

Basic electrodes are somewhat of an exception from the above observation. Cracks in the coating may appear after the electrode is baked, or they may be due to the coating composition or the relatively high temperature applied. The coating can be reinforced through large silicate additions, thus counteracting the tendency to form contraction cracks during baking. It has also been observed that light, fine iron powder allows an increased addition of silicate without rendering the coating too soft after extrusion. With basic electrodes, improved arc stability is desirable; one way to attain this is by increasing the addition

Table 4   Classification of low-hydrogen, high-strength electrodes

Iron powder content, %	AWS classification	Type of covering	Capable of producing satisfactory welds in positions shown(a)	Type of current(b)

E70 series: minimum tensile strength of deposited metal in stress-relieved condition, 480 MPa (70 ksi)

Iron powder content, %	AWS classification	Type of covering	Positions	Type of current
25-35	E7018-X	Iron powder, low hydrogen	F, V, OH, H	ac or DCEP
35	E7020-X	High iron oxide	H-fillets	ac or DCEN
			F	ac, DCEN, or DCEP
20-45	E7027-X	Iron powder, iron oxide	H-fillets	ac or DCEN
			F	ac, DCEN or DCEP

E80 series: minimum tensile strength of deposited metal, 550 MPa (80 ksi)

30	E8018-X	Iron powder, low hydrogen	F, V, OH, H	ac or DCEP

E90 series: minimum tensile strength of deposited metal, 620 MPa (90 ksi)

30	E9018-X	Iron powder, low hydrogen	F, V, OH, H	ac or DCEP

E100 series: minimum tensile strength of deposited metal, 690 MPa (100 ksi)

25	E10018-X	Iron powder, low hydrogen	F, V, OH, H	ac or DCEP

E110 series: minimum tensile strength of deposited metal, 760 MPa (110 ksi)

20	E11018-X	Iron powder, low hydrogen	F, V, OH, H	ac or DCEP

E120 series: minimum tensile strength of deposited metal, 830 MPa (120 ksi)

15-20	E12018-X	Iron powder, low hydrogen	F, V, OH, H	ac or DCEP

(a) Abbreviations indicate these welding positions: F, flat; H, horizontal; H-fillets, horizontal fillets; V, vertical; OH, overhead.
(b) DCEN, direct current electrode negative (straight polarity); DCEP, direct current electrode positive (reverse polarity)

of potassium silicate. On the other hand, such an increase may make the coating more hygroscopic, which is a disadvantage. Therefore, when choosing raw materials for welding electrodes, the advantages and disadvantages must be evaluated.

Performance of Powder-Coated Electrodes

Both the deposition efficiency (yield) and performance of welding electrodes may be influenced by the amount of silicate added and the apparent density of the iron powder used in the coating. The tests on which Fig. 4 through 7 are based were carried out with a high-yield rutile electrode. Normally, such an electrode would have a yield of 180 to 185%, provided that 61% of high-density iron powder and approximately 150 g silicate/kg (70 g silicate/lb) of dry coating are added. This electrode is characterized by very good welding performance when operating on horizontal fillet welds.

Figure 4 illustrates yield as a function of the apparent density of the iron powder

Fig. 4 Deposition efficiency (yield) of a coated electrode as a function of apparent density of iron powder addition

Electrodes extruded at 39 MPa (5600 psi).

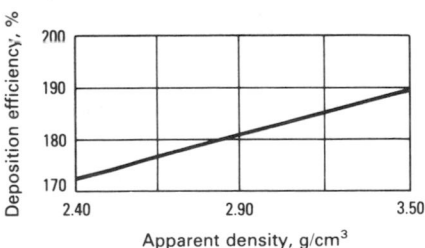

Fig. 5 Weight of a coated electrode (after baking) as a function of the apparent density of the iron powder addition

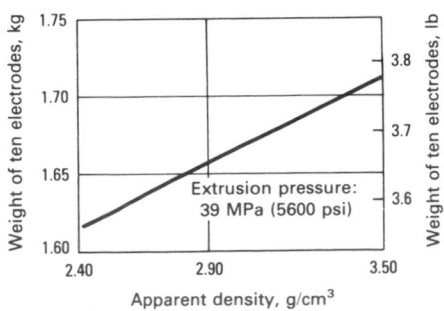

Fig. 6 Effect of silicate content and extrusion pressure on deposition efficiency of coated electrodes

Electrodes extruded with high-density (3.5 g/cm³) iron powder.

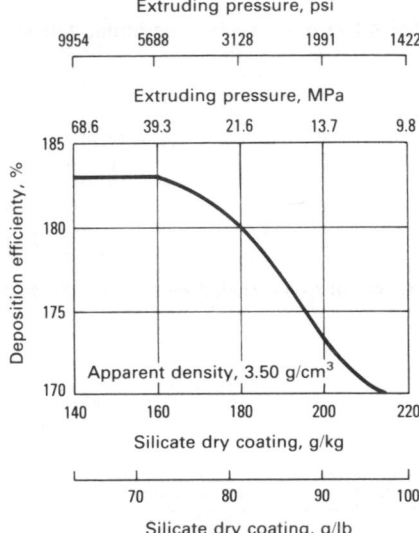

Fig. 7 Amount of weld metal deposited (melting rate) as a function of welding amperage for extruded coated electrodes

Electrodes contain iron powder of varying densities, as indicated in the figure.

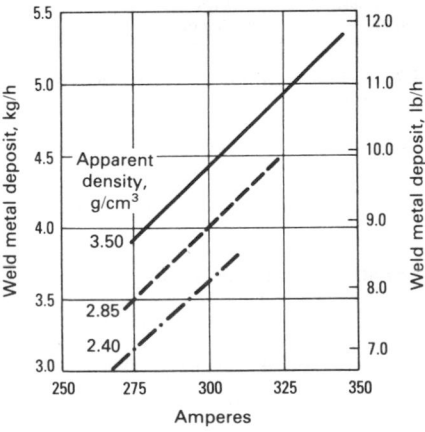

used for electrodes extruded at 39 MPa (5600 psi). Approximately 15% lower yield was obtained from the electrode made with iron powder that had an apparent density of 2.4 g/cm³, as compared to an electrode made with iron powder that had an apparent density of 3.5 g/cm³.

Figure 5 illustrates data that explain the above observation. The actual weight of

the electrode after baking varies with the apparent density of the powder. The nature of the curve is the same as that in Fig. 4, indicating the oxidation losses are independent of powder particle shape and apparent density. Furthermore, coating thickness, as well as the weight of the iron powder addition, is directly proportional to the apparent density. However, this only applies when the amounts of carbon and iron oxide in the iron powders are fairly equal. When using iron powder with very large amounts of both carbon and iron oxide, the deposition efficiency (yield) may drop considerably because of increased spatter losses.

Figure 6 shows the deposition efficiency (yield) of electrodes extruded with high-density iron powder (3.5 g/cm³) as varying amounts of silicate were added. The deposition efficiency decreases sharply when the amount of silicate is increased. The correct amount of silicate addition with respect to extruding pressure would be 160 g/kg (72 g/lb) of dry mix. With such an addition, the electrode will give a yield of 175%.

Figure 7 shows the melting rate calculated as weight (kilograms or pounds) of weld deposit per hour as a function of the welding amperage applied on various electrodes. The electrodes were extruded by using iron powder with varying apparent densities. Faster welding speeds may be achieved when using high-density iron powders. The amperage considered maximally applicable in normal welding practice is higher for electrodes made with high-density iron powders. Thus, the differences between various iron powders are more pronounced under practical conditions, such as welding of horizontal V-fillet welds or horizontal fillet welds.

The following conclusions can be drawn from the examples in Fig. 4 through 7:

- The amount of silicate added for a specific type of electrode and a specific extrusion pressure depends on the apparent density, particle shape, and sieve analysis of the powder.
- Deposition efficiency and melting rate rise when the silicate content is reduced and the apparent densities of the iron powder are increased. Therefore, only high-density iron powder welding grades are recommended for high-yield electrodes that have deposition efficiencies greater than 160%.
- Narrow specification ranges for physical properties, particle shape, apparent density, and sieve analysis of the iron powder are necessary for reproducible production of welding electrodes.

High-Deposition Submerged Arc Welding With Iron Powder Joint Fill

Iron powder is important in submerged arc welding. Steel fabrication plants require higher deposition rates from the submerged arc welding process to improve the economics of welding heavy plates and structural shapes. This need becomes quite apparent when conventional, multipass, automatic submerged arc welding procedures are used for welding heavier joints with depths ranging from 13 to 50 mm ($^1/_2$ to 2 in.). These conventional procedures require slag cleaning after each pass; a relatively large number of passes maintain weld soundness and the mechanical properties required by structural welding codes.

To improve welding economics while maintaining quality, experimental welds were made using an iron powder additive in the weld joint. This welding method used the automatic submerged arc welding process with a single-pass welding procedure. Based on later production experience, it became apparent that a tandem arc approach offered advantages over a single arc. The following describes a method of submerged arc welding using one or two arcs with a powdered iron additive (joint fill) for use in high-deposition, single-pass welding.

Submerged Arc Welding Process. The submerged arc process (Ref 2) is illustrated schematically in Fig. 8. The iron powder in a tandem arc welding setup completely fills the joint ahead of the lead arc and flux supply. The lead arc parameters are established for complete penetration through the joint fill into the backing strip. The flux supply covers the welding area as in conventional submerged arc welding. A relatively high current with reverse polarity (electrode is positive) is applied to the first electrode. The trail electrode supplies the additional deposited metal required for smooth top bead reinforcement and fusion. The second flux supply covers the trail arc area. The trail electrode is supplied with direct-current, straight-polarity power for maximum deposition rate. The molten flux following the trail arc solidifies into slag, which protects the cooling weld deposit from oxidation.

Iron powder joint fill has been applied with submerged arc welding, and the process has been flexible in terms of providing high productivity and deep penetration on single-pass welds. Iron powder joint fill can be used with single or tandem arc setups for butt or fillet welds. The tandem arc setup provides greater ease of adjustment to achieve adequate penetration and desirable weld reinforcement on the top bead. Other welding processes have been used in a few instances, for example, using the semiautomatic flux cored arc welding process with a small amount of iron powder to help control penetration on poorly fitted joints.

The submerged arc process has been applied successfully to structural steel ranging in thickness from 9.5 to 75 mm ($^3/_8$ to 3 in.). Weld lengths typically range from 0.6 to over 12 m (2 to over 40 ft).

High-current, tandem submerged arc welding requires heavy-duty equipment. Figure 9 illustrates a gantry fixture equipped with twin tandem heads, which typically are used for heavy structural steel fabrication. In actual production jobs, iron

Fig. 8 Schematic of submerged arc welding process

When using the iron powder joint fill method, the iron powder is applied ahead of the lead arc and flux supply.

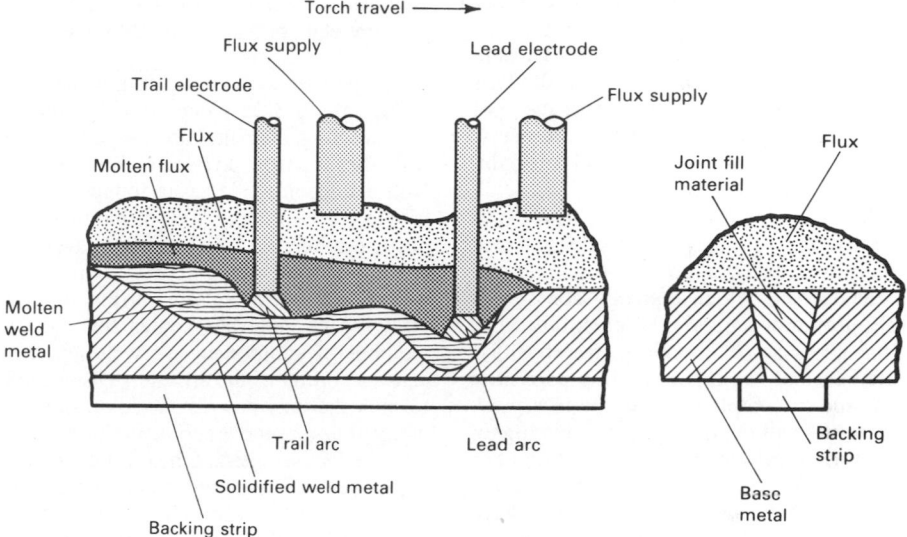

Fig. 9 Tandem arc equipment suitable for high-current groove or fillet joints

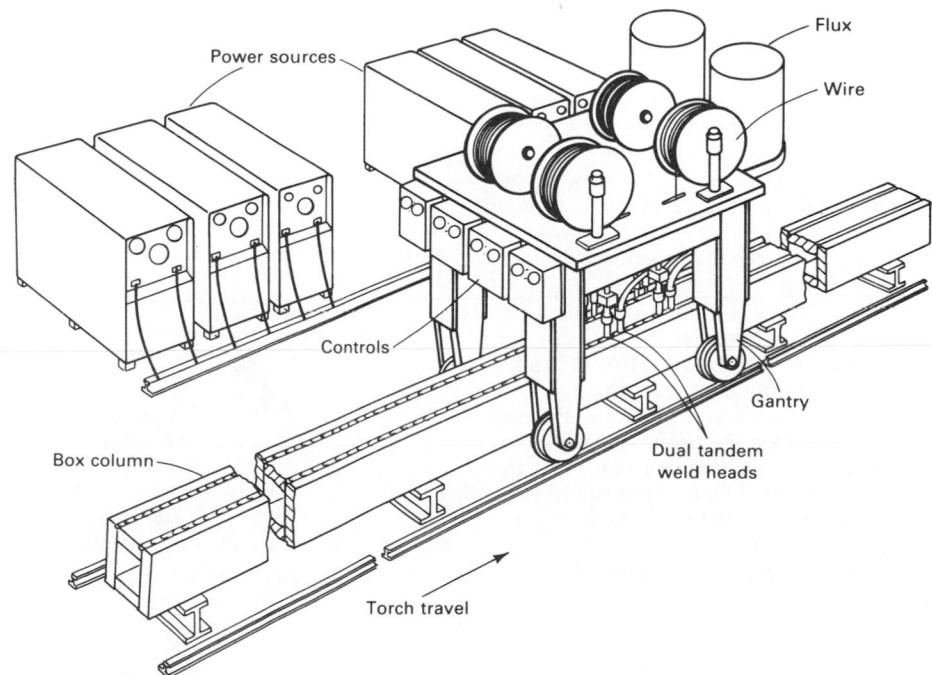

powder has been added either by hand or by an automatic device similar to that used for dispensing flux. Because a weld joint level full of iron powder usually is desired, a simple leveling device on the bottom end of a flux hopper is used where automatic application is required. This system is custom designed to fit the specific welding application requirements. Quantity requirements depend on joint thickness and weld length.

Productivity is increased with submerged arc welding for two principal reasons. First, the iron powder fuses into the weld deposit, which typically consists of 40% iron powder and 60% electrode wire. This gives a 67% increase in deposition rate. Second, a higher welding current is required to completely fuse the iron powder to the full depth required. This higher current typically increases deposition rates 100% over conventional currents. Total rates of 45 to 85 kg/h (100 to 190 lb/h) are usually realized in production. Other productivity increases may result because welding generally is done in a single pass, eliminating slag removal between passes. In some cases, iron powder added in small quantities helps to maintain uniform penetration and fusion in joints with varying fitup.

Filler materials for the submerged arc process consist of flux, electrode wire, and the iron powder joint fill. Successful practice of this welding method requires that all components be formulated for compatibility under typical user conditions. Actual test results with filler materials meet requirements of American Welding Society Specification D1.1-72, Welding Procedure Qualification, and have been approved by the American Bureau of Shipping. Each fabricator is required to qualify his own procedures to whatever specification is required in the contract.

Iron Powder Type. The iron powder joint fill used with this method of welding was developed to have a uniform particle size and homogeneous chemical analysis. It is produced by atomization, which involves breaking a stream of molten iron into very small particles by means of fluid jets aimed at the stream. Prior to atomizing, the impurity content of the iron powder can be controlled by refining the molten iron through the proper choice of raw materials and melting practice. The iron powder produced by this process has very low porosity, and the particles are roughly spherical. A specific powder composition can be formulated by changing the metal-

lic content of the melt. Iron powder produced in this way has been found to be free from segregation, because each particle has a chemical analysis similar to that of every other particle made from the same melt.

Iron powder with an apparent density of about 3.00 to 3.25 g/cm^3 provides excellent welding stability and adequate fill contribution. Iron powder of this density with a full joint filled was found to contribute approximately 40% to the total deposit, resulting in a 67% increase over the amount of weld metal deposited from the electrode alone at a given current. Figure 10 compares deposition rates for various methods of submerged arc welding. Because a portion of the arc energy is needed to melt the iron powder, welds produced using this method generally show a reduced effective welding heat input at similar currents. This reduced heat input contributes to better mechanical properties, especially toughness in the heat-affected zone, other factors being equal. The most effective particle sizing for iron powder appears to be approximately $-70+325$ mesh. Particles of this size are sufficiently small to fill relatively narrow grooves, yet

Fig. 10 Deposition rates for various methods of submerged arc welding

Lower values in each range are for larger electrode with reverse polarity; higher values are for electrodes with straight polarity.

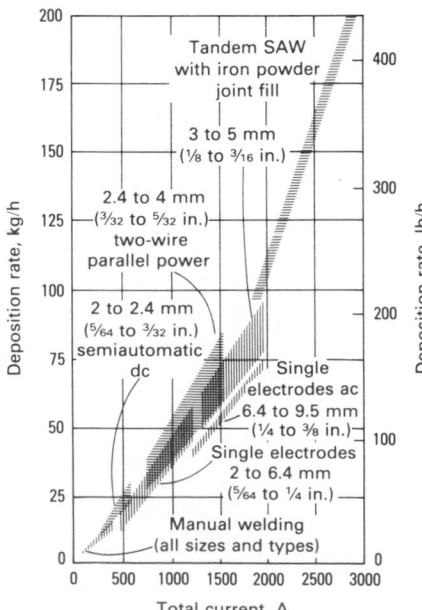

are large enough for practical handling. The shape and density of the iron particles contribute to natural, uniform packing when the particles are allowed to fall into the joint freely. No tamping or packing is recommended or required with this powder.

Desirable weld properties for the iron powder are a tensile strength of 480 MPa (70 ksi) and a yield strength of 345 MPa (50 ksi). These properties correspond to those of steels that are widely used in buildings, bridges, heavy machinery, and earthmoving equipment.

Submerged arc welding with iron powder joint fill can be applied to a variety of joints where welding is done essentially in the flat position. Joints may be designed for fillet welds, square butt welds, single- or double-bevel groove welds, and single- or double-V groove welds.

Process Advantages. Submerged arc welding with iron powder joint fill has several advantages. A 67% increase in deposition rate has been measured when iron powder is fused into the weld deposit. The deposit consists of 60% melted wire and 40% melted iron powder. An additional 100% increase in deposition rate results when iron powder is applied through use of higher welding currents. A total of 85 kg/h (190 lb/h) has been measured on heavy joints.

A reduction in weld passes is usually possible. One-pass welding can be done on most applications. Higher deposition efficiencies have been measured. Typically, 67% has been achieved, while 40 to 50% is common for conventional submerged arc welding. A greater tolerance to joint gap variation may be obtained with special procedures using iron powder.

Distortion and gross heat input can be reduced substantially. A ten-pass weld had four times more lateral angular distortion than a single-pass weld with iron powder. Mechanical property tests of deposits consistently have exceeded welding qualification test requirements of the American Welding Society, American Bureau of Shipping, and American Society of Mechanical Engineers.

REFERENCES

1. *Metals Handbook*, 9th ed., Vol 6, American Society for Metals, 1983
2. Troyer, W. and Mikurak, J., High Deposition Submerged-Arc Welding with Iron Powder Joint-Fill, *Weld. J.*, Vol 53 (No. 8), Aug 1974, p 494-504

Metal Powders Used for Hardfacing

By Kishor M. Kulkarni
Operations Manager
Powders and P/M Parts
Wear Technology Division
Cabot Corp.
and
Vidhu Anand
Engineering Associate
Cabot Corp.

HARDFACING is the application of hard, wear-resistant material to the surface of a component by welding, thermal spraying, or a similar process for the main purpose of reducing wear. The control of wear is growing in importance because wear adversely affects the satisfactory performance of equipment and components, causing poor efficiency or costly downtime. Hardfacing is used on components of original equipment, as well as for repair and rebuilding of worn components. Hardfacing filler (consumable) material is available in different forms, such as powder, solid welding rods, and tube rods.

This article discusses hardfacing processes and applications that use powder. However, the use of metal powders in the fabrication of consumable electrodes for hardfacing will not be discussed. In addition, applications in which corrosion or oxidation resistance is of prime importance and surface modification techniques such as flame hardening, nitriding, and ion implantation are not covered in this article. For detailed information on hardfacing operations for wear control and weld overlay operations for corrosion control, see the articles "Hardfacing" and "Weld Overlays" in Volume 6 of the 9th edition of *Metals Handbook*.

Hardfacing of a selected alloy onto a base metal (substrate) provides a protective surface layer that resists wear or loss of material by adhesion, galling, abrasion,

impingement erosion, cavitation erosion, fretting, or corrosion. Adhesive wear is caused by the sliding action between two metallic components in which abrasives are not intended to be present. At low load, wear is mild because of the thin oxide film on the components. At some transition load, metallic bonds form between the surface asperities of mating materials, and wear becomes severe.

The term "galling" is used to describe conditions in which adhesive wear is severe enough to cause seizure of the moving components. Abrasive wear is caused by the cutting action of sliding abrasives between mating surfaces that suffer scratches or gouges. Impingement erosion refers to the type of material loss that is caused by the cutting action of moving particles. Cavitation erosion is caused by shock waves that result from collapse of air bubbles on the metal surface in turbulent fluid flow. Fretting is a combination of mild adhesive (oxidative) and abrasive wear caused by very small amplitude vibrations at mechanical connections.

In severe abrasive wear service, such as rock crushing and pulverizing, pounds of material are rapidly worn away. In control valves, however, a few thousandths of an inch of wear is intolerable. Wear and corrosion are encountered in knives used in the food processing industry and in valves and pumps handling corrosive fluids. Diesel engine valve seats are subjected to severe conditions involving erosion, fatigue,

and hot corrosion. Commercially available hardfacing materials thus help to solve these industrial problems.

The choice of filler material requires careful consideration of service conditions, available equipment, hardfacing processes, equipment and process cost, hardfacing operation, and equipment downtime. Technical assistance from the powder supplier and testing of samples under operating conditions are strongly recommended.

Many of these material systems are used as powders because of their specific advantages. A wide range of compositions is possible through P/M processing; some of these cannot be made by conventional techniques. Use of powders allows uniformity of composition and microstructural features. Powder processes minimize heat input to the base metal and hence dilution of the deposit. Furthermore, in many applications, P/M techniques provide more effective coating of intricate surfaces.

General Powder Characteristics

Gas and water atomization are the most widely used methods for producing hardfacing powders, although milling and crushing operations are used to produce tungsten carbide powders, some grades of nickel-based powders, and other friable materials. These methods provide broad ranges of particle size and size distribu-

Fig. 1 Powder size ranges for different hardfacing processes

At least 85% of the powder by weight is in the size range shown.

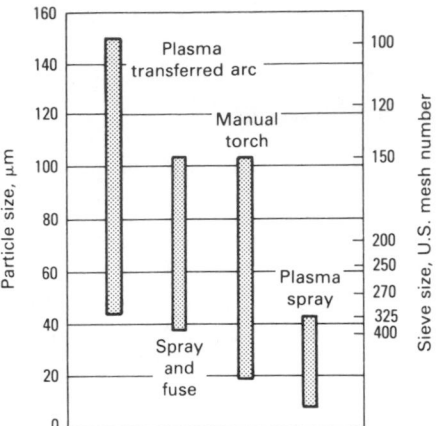

Fig. 2 Scanning electron micrograph of oxide-free cobalt-based self-fluxing alloy

Magnifications: (a) 125×. (b) 300×. (c) 650×. (d) 1300×. Courtesy of Hoeganaes Corp.

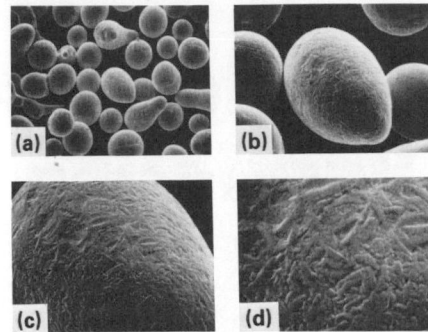

Fig. 3 Gas-atomized cobalt-based alloy powder particles with attached satellites

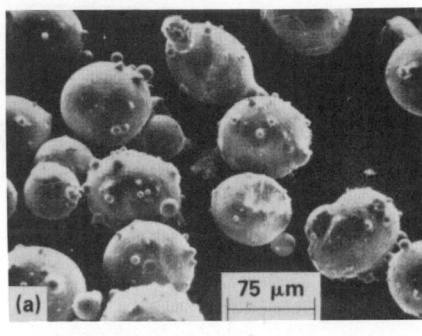

tion; therefore, screening (sieving) and air separation are used to ensure a narrow distribution for specific grades of powder.

Figure 1 shows typical particle size ranges of different grades of powder used for various hardfacing processes. Approximately 85 wt% of any grade of powder falls within the ranges shown. For the same process, however, equipment made by different manufacturers may require different particle size ranges.

The plasma spray operation requires fine powders (−325 mesh) because of the high jet speed and short dwell time of the particles in the heating zone (even though the zone temperatures are high). However, special dispensing devices are required to prevent powder flow problems. In some applications (plasma transferred arc), use of excessive fines is damaging to the deposit because fines tend to have higher surface oxides and burn-off tendencies, thus causing smoking.

Type of application also governs powder size requirements. For instance, thin overlays typically use fine powders. Therefore, the user must consider the process, equipment, and the final application before selecting a given powder size.

Powder shape and cleanliness are also important considerations in any hardfacing operation. For most applications, the deposit must be nearly free of porosity or entrapped fluxes. The use of spherical powders, as shown in Fig. 2, is beneficial to the service and integrity of the coatings. Spherical powder flows better than angular powder because of minimal interparticle friction; this is important when using gravity powder-feed devices.

Larger atomized particles sometimes have smaller particles, or satellites, at-

tached. If excessive, these satellites can cause flow problems. If loosely attached, they may break up during spraying, resulting in oxidation and smoking. Gas-atomized powders generally are spherical in shape and prone to satellite formation (Fig. 3). Water-atomized powder may be irregular or spherical in shape, depending on the alloy used and the quenching rate. Surface oxidation may be a potential problem with water-atomized powder (particularly in cobalt-based alloys). Self-fluxing nickel-based powders are routinely made by this process, because oxygen pickup is minimal and the process is more economical.

Apparent density and flow rate also depend on the characteristics mentioned above. Irregular particle shape, numerous satellites, or a narrow size distribution lowers apparent density and slows the flow rate of the powder. Equipment with gravity powder-feeding devices requires powders that flow faster. In any specific application, compatibility between the equipment and operating conditions and reproducibility of these powder properties are considerations.

Deposit hardness depends on the powder composition, as well as on the conditions used for depositing. Hardness range frequently is reported in relation to powder characteristics. The bulk hardness of a deposit, as measured for example on the Rockwell C scale, depends on the deposit matrix hardness, as well as on the type and quantity of hard constituents, such as carbides and borides, or phases such as martensite or Laves phase intermetallics.

Hardness is also influenced by the degree of dilution. For a specific base metal and powder alloy combination, hardness normally decreases with increased dilution of the deposit by the base metal. Dilution of 5% means that the deposit contains 5%

base metal and 95% hardfacing powder alloy. Dilution increases as the base metal receives more heat input and more of base metal melts.

The plasma transferred arc process provides more dilution than the manual torch process, whereas plasma spray provides essentially no dilution. More dilution can make the deposit tougher and more resistant to cracking.

In practice, hardfacing processes and techniques should be selected to control dilution to less than 20%. In a multilayer deposit, the first layer experiences the largest degree of dilution and is the softest. Generally, deposit hardness can be used to verify powder quality only if the testing method is rigidly standardized; even then, hardness may differ from the hardness of the deposit on the components to be hardfaced.

Powder Composition

Chemical composition of hardfacing powders is the most important influence on the properties of the deposit. The main criteria in selecting an alloy for a given hardfacing application are its response to the type of wear encountered, the base metal to be hardfaced, and material cost. Other influential properties are impact strength, corrosion, oxidation, and thermal behavior.

Generally, impact resistance of hardfacing alloys decreases as the carbide content increases. Consequently, in applications in which a combination of impact and abrasion resistance is desired, a compromise between the two must be made. In applications in which impact resistance is important, austenitic manganese steels are used to build up worn parts.

Frequently, wear is accompanied by aqueous corrosion from acids or alkalis, as in the chemical processing and petroleum industries or in flue gas scrubbers.

Few iron-based hardfacing alloys possess the necessary corrosion resistance in such aqueous mediums. As a result, nickel- or cobalt-based surfacing alloys generally are recommended when a combination of corrosion resistance and wear resistance is required. For example, knives used to cut tomatoes in a food processing plant will last many times longer if their edges are made of a cobalt-based alloy rather than tool steel.

Oxidation and hot corrosion resistance of iron-based alloys is also generally poor. Typically, boride-containing nickel-based alloys do not contain sufficient chromium in the matrix to resist oxidation. Hence, Laves phase or carbide-containing nickel- or cobalt-based alloys typically are recommended for applications in which wear resistance combined with oxidation or hot corrosion resistance is required.

The ability of an alloy to retain strength at elevated temperatures is important for wear applications such as hot forging dies or valves for service at 870 °C (1600 °F), as well as service in coal gasification/liquefaction applications. Iron-based alloys with martensitic structures lose hardness at elevated temperatures. Generally, the high-temperature strength retention of a hardfacing alloy increases with its tungsten or molybdenum content. In applications requiring elevated-temperature strength and wear resistance, cobalt-based alloys or Laves phase alloys are recommended.

Wear-resistant materials typically are classified by type and alloy content, which largely determines their properties. These materials have structures consisting of hard phases (carbides, borides, or Laves phases) in a matrix that is hardened to a level that depends on the desired alloy properties. Table 1 provides a comprehensive guide to selection of surfacing alloy systems. Powder alloys may be custom made for various applications by using the same composition, but varying the sieve size or distribution. Because American Welding Society (AWS) designations do not exist for many of the hardfacing alloys discussed in this article, trade names will be used. It should be noted that many of these compositions are available from various manufacturers under different names.

Cobalt-Based Powders

The two types of commercially available cobalt-based hardfacing alloys are carbide-containing alloys and alloys containing Laves phase, which are described later in this article. Carbide-containing cobalt-based alloys have been widely used since the early 1900's, when a cobalt-based alloy with the nominal composition of Co-28Cr-4W-1.1C was first developed. Typically, this alloy, referred to in the hardfacing industry as Alloy No. 6, is a eutectic alloy containing 16 to 17 vol% M_7C_3-type carbides in a cobalt-based matrix alloyed with chromium and tungsten.

The lightest colored phase in the microstructure is the matrix, which has a face-centered cubic (fcc) crystal structure. The dark phase is M_7C_3; the exact composition is $(Cr_{0.85}, Co_{0.14}, W_{0.01})_7C_3$, which has a hexagonal crystal structure. The carbide phase in Alloy No. 6 forms between matrix dendrites. These carbides are often referred to as eutectic carbides. Generally, resistance to abrasive wear is imparted by the presence of carbides, whereas resistance to corrosion and/or elevated-temperature hardness retention is imparted by the matrix. Galling and metal-to-metal wear properties are similarly determined by the matrix alloy. The hardness of Alloy 6, which is designated by AWS as RCoCrA, is 35 to 43 HRC.

Nominal compositions and average deposit hardness data for a number of alloys in the Co-Cr-W-C system are given in Table 2. These alloys retain high hardness at elevated temperatures primarily from carbide and solid-solution strengthening rather than precipitation hardening. They contain up to 30% Cr, up to 15% W, and up to 3.0% C. Some of these alloys also contain combinations of nickel, silicon, molybdenum, boron, and manganese.

Alloys No. 1 (AWS RCoCrC) and 12 (AWS RCoCrB) are higher hardness modifications of Alloy No. 6 because of additional carbon and tungsten contents. For example, Alloy No. 1 contains primary M_7C_3, eutectic M_7C_3 plus matrix, and secondary M_6C. Alloy No. 12 is similar to No. 6, except it has more M_6C and thus contains interdendritic eutectic carbide (tungsten and chromium rich). As shown in Table 3, the weight percent of carbides decreases, as does abrasion/erosion resistance, when comparing these alloys. Specific abrasive and adhesive wear, hot hardness, and corrosion rates for some of these alloy designations are given in Tables 4 and 5.

Cobalt-based alloys, when used in overlay applications, are prone to cracking during cooling. Typically, this problem can be overcome by preheating the base metal prior to application. These alloys are deposited readily by plasma transferred arc, which is a low-dilution, high-deposition-rate process.

Cobalt-based hardfacing powders are unsurpassed for use in combined wear-resistant, corrosion-resistant, and high-temperature applications. Cost of these materials is affected by the prevailing price of cobalt and tungsten and tends to be higher than nickel- and iron-based alloy powders. Several alloys with lower cobalt content, or no cobalt, have been developed as low-cost replacements for cobalt-based grades. However, these substitutes have not found broad commercial acceptance.

Nickel-Based Powders

Commercially available nickel-based hardfacing alloys can be divided into three groups: boride-containing alloys, carbide-containing alloys, and Laves phase-containing alloys, which are described later in this article. Based on the Ni-Cr-B-Si system, these alloys are finding increased usage for wear-resistant surfacing applications, particularly in oil pumping and in glass, ceramic, cement, plastics, and steel production. The oil pumping and glass industries are the major users of nickel-based powders.

Additions of boron and silicon to nickel lower the melting point so that an oxy-acetylene torch can be easily used. Nickel and boron form a $Ni-Ni_3B$ eutectic at 1092 °C (1998 °F). Silicon additions further lower the eutectic. However, excessive additions of boron and silicon lower plasticity. Because silicon forms a substitutional solid solution with nickel and because the solubility of boron in nickel is small (0.1 wt%), the abrasion resistance of these alloys is a function of the amount of hard borides present. Most of these alloys contain nickel boride (Ni_3B), chromium borides (CrB, Cr_5B, and Cr_2B), and other complex borides as hard phases.

Typical alloy compositions are listed in Table 2. Deposit hardness in these alloys is as high as 60 HRC, depending on the chromium, boron, and silicon contents. Usually, the boron content varies from 1.5 to 3.5%, while chromium varies from 0 to 17%. Alloys containing large amounts of boron, such as Ni-14Cr-4Si-3.4B-0.75C, are extremely abrasion resistant, but have poor impact toughness.

The Ni-Cr-B-Si alloys are self fluxing. The low-melting eutectic combines with the surface oxides on powders to form borosilicates, which promote wetting of the substrate. Some manufacturers add copper and molybdenum to improve corrosion and pitting resistance. However, hot tearing has been reported due to copper additions. These alloy powders have good abrasive and metal-to-metal (adhesive) wear resistance, although hot hardness and corro-

Table 1 Selection guidelines for hardfacing alloy systems

Surfacing material type	Composition	Hardness		Maximum service temperature		Resistance	
		Hot	Cold	°C	°F	Abrasion	Impact
Iron-based, low-alloy	0.3-1.0 C, 1.0-7.0 W, 3.0-5.0 Cr, 4.0-9.0 Mo, rem Fe	47 HRC at 595 °C (1100 °F) 30 HRC at 650 °C (1200 °F)	55-60 HRC (30 HRC, annealed)	595	1110	Superior to low-carbon steel	Moderate; increases with temperature
Iron-manganese	0.5-0.9 C, 11.0-16.0 Mn, 2.75-6.0 Ni, or 0.5-1.5 Mo, rem Fe	⋯	170-230 HB (workhardens to 450-550 HB)	260-315	500-600	Moderate	Excellent
Iron-chromium	3.0-5.0 C, 4.0-8.0 Mn, 26.0-32.0 Cr, 2.0 Mo, rem Fe	47 HRC at 480 °C (900 °F) 5 HRC at 650 °C (1200 °F)	51-62 HRC	425-480(b)	800-900(b)	Excellent for low-stress abrasion	Good for light impact
Cobalt-chromium	0.7-3.0 C, 2 Mn, 3-14 W, 3 Ni, 25-33 Co, 5 Fe, rem Cr	20-39 HRC at 650 °C (1200 °F)	23-58 HRC	980	1800	Excellent in low-stress; wears 5-20% as fast as carbon steel	Fair; deteriorates as carbon increases
Nickel-chromium	0.3-1.0 C, 1.0-1.5 Co, 8.0-18.0 Cr, 1.25-5.5 Fe, 2.0-4.5 B, 1.25-5.5 Si, rem Ni	19-42 HRC at 540 °C (1000 °F)	24-62 HRC	955	1750	Excellent in low-stress; increases with carbon content	Good for light impact; decreases as carbon increases
Copper-based	0.25-5.0 Fe, 0.01-15.0 Al, 0.02 Zn, 0.04-4.0 Si, 9.0 Sn, rem Cu	⋯	70-320 HB	205	400	Poor	CuAl: good; decreases with increased Al content, CuSi: good. CuSn: poor. CuZn: very poor
Tungsten-carbide	60% carbide granules, 40% steel tube	41-47 HRC 650 °C (1200 °F)	90-95 HRA (granules) 30-60 HRC (bonding matrix)	540	1000	Excellent	Good under light impact

(a) Ratings are relative between material types. (b) Softens but retains other properties up to 980 °C (1800 °F)

Table 2 Chemical compositions and hardnesses of typical cobalt- and nickel-based hardfacing alloys

Alloy	Composition, wt%										Hardness	
	B	C	Co	Cr	Fe	Mo	Mn	Ni	Si	W	HRC	Diamond pyramid
Cobalt-based alloys												
Alloy No. 1	⋯	2.5	rem	30	3 max	1 max	1 max	3 max	1	12.5	51-58	550-685
Alloy No. 6	⋯	1.1	rem	28	3 max	1 max	1 max	3 max	1	4	39-43	380-425
Alloy No. 12	⋯	1.4	rem	29	3 max	1 max	1 max	3 max	1.4	8	47-51	480-550
Alloy No. 21	⋯	0.25	rem	27	2 max	5.5	1 max	2.8	2 max	⋯	28-32	285-320
Alloy No. SF1(a)	2.2	1.3	rem	19	3	⋯	1 max	13	3	13	54-58	600-685
Alloy No. SF6(a)	1.7	0.7	rem	19	3	⋯	1 max	13.5	2.3	7.5	43-46	425-460
Alloy No. SF12(a)	1.8	1	rem	19	3	⋯	1 max	13	2.8	9	48-50	490-540
Alloy No. SF20(a)	3	1.3	rem	19	2 max	⋯	1 max	13	2.8	15	60-65	740-900
Nickel-based alloys												
Alloy No. 22	1.5	0.2 max	⋯	⋯	1 max	⋯	⋯	rem	2.8	⋯	19-24	200-250
Alloy No. 35	1.6	0.3	⋯	3.7	1.2	⋯	⋯	rem	3.1	⋯	32-37	320-365
Alloy No. 40	1.7	0.35	⋯	7.5	1.5	⋯	⋯	rem	3.5	⋯	35-42	350-415
Alloy No. 50	2.4	0.45	⋯	11	3	⋯	⋯	rem	4	⋯	49-52	525-575
Alloy No. 60	3.5	0.8	⋯	15.5	4	⋯	⋯	rem	4.3	⋯	59-62	710-790

(a) Spray-and-fuse powders. Source: Cabot Corp.

to wear(a) Corrosion	Adhesion	Mechanical properties	Machinability	Microstructure	Heat treatment	Applications
Fair to poor	Good	Good compressive strength; takes a high polish for low coefficient of friction	Good; anneal, machine, then retreat to harden	Martensite, austenite, carbides	Surfacing: preheat base metal to 150 °C (300 °F)	Where hardness and toughness are required to 595 °C (1100 °F); cutting tools, shear blades, reamers, dies, ingot tongs
Low (similar to carbon steel)	Excellent	Low yield point deformation	Poor; grind if required	Austenite saturated with carbon; iron-manganese carbides at grain boundaries in thin, flat, brittle plates	Heat to reaustenitize; water quench	High-impact service; rail tracks, frogs, switches; use on carbon or manganese steel base
Resists oxidation except in moist air; poor in liquid service	Excellent	Low coefficient of friction; high strength	Difficult to grind	Hard carbides in austenite matrix	None	Agricultural machinery, coke shutes, brick-making equipment
Excellent for salt, stream, and some acids up to 980 °C (1800 °F)	Excellent for low-stress; poor for high-stress; increases with carbon content	Excellent metal-to-metal; takes a high polish, low coefficient of friction	Poor	Chromium, tungsten, cobalt carbides	None	High-temperature, high-corrosion applications; pump shafts, engine valves
Excellent for salt, steam, some acids and alkalis	Excellent for low-stress	Excellent metal-to-metal wear	Poor; finish by grinding	Chromium carbides, chromium borides, in eutectic matrix	Preheat to 315-425 °C (600-800 °F) to prevent cracking; slow cool	Seal rings, cement pump screws, valves, cams
Oxide coating, good protection from atmosphere; good for many acids, bases, and saltwater	Good	Elastic limit 35 to 1170 MPa (5-170 ksi)	Good	See composition	None	Gears, sheaves, cams, bearings, wear plates, dies for forming stainless steel, carbon steel, aluminum
Similar to steel; low resistance in oxidizing mediums	Poor	Yield same as high-carbon steel	Unmachinable; finish with diamond grinding wheel	W and WC granules in a ferritic matrix	Carburizing and hardening may harden base metal	Cutting teeth and edge holding surfaces of rock drill bits; quarrying, digging, earthmoving equipment

sion resistance are slightly inferior to those of cobalt-based alloys (Tables 4 and 5). The most common use of Ni-Cr-B-Si powders is in spray-and-fuse and manual torch applications in the glass industry. They are also used in the plasma transferred arc process.

The use of carbide-containing nickel-based alloys has been limited. The most popular and widely used alloys in this group are included in the Ni-Cr-Mo-C system. A typical alloy in this system is Hastelloy* Alloy C, which has a nominal composition of Ni-17Cr-17Mo-0.12C and a hardness level of 95 HRB. This alloy, which has good inherent corrosion resistance, is normally deposited by the plasma spray process.

Carbide-containing nickel-based alloys of the Ni-Cr-Mo-Co-Fe-W-C system are gaining popularity as low-cost alternatives to cobalt-based alloys. A typical alloy in

*Hastelloy and Haynes are registered trademarks of Cabot Corp.

this system is Haynes* Alloy 716, which has a nominal composition of Ni-11Co-26Cr-29Fe-3.5W-3Mo-1.1C-0.5B and a hardness of 32 HRC. These alloys, depending on precise composition, contain M_7C_3- or M_6C-type carbides. They are normally deposited by the plasma transferred arc process.

Carbides

Carbide hardfacing powders are used for severe abrasion and cutting applications. Tungsten carbide is the most widely used carbide in these types of powders. Comminuted tungsten carbide usually has a mixture of WC and W_2C carbides. (Tungsten carbide/cobalt powder contains WC and M_6C carbides.) Fine carbides are used when smooth surfaces and improved impact resistance are desired.

Larger particles yield a rougher surface, improved abrasion resistance, but lower toughness, which often results in fracture and subsequent loss of carbides. Some

Table 3 Carbide percentage and hardness of cobalt-based alloys

Alloy	Carbides, wt%	Hardness, HRC
Alloy No. 1	28-29	48-55
Alloy No. 12	16-17	43-50
Alloy No. 6	12-13	35-43
Alloy No. 21	<5	28-32

Note: As the weight percent of carbides decreases, so does abrasion/erosion resistance. Source: Cabot Corp.

commercial grades also contain percentages of cobalt and nickel to improve aqueous corrosion or high-temperature oxidation resistance of the deposit. However, oxidation of tungsten carbide particles limits the upper service temperature to 600 °C (1100 °F). These powders are used primarily in plasma spray and manual torch processes, with small amounts used in plasma transferred arc and spray-and-fuse powders. Table 6 lists the hardnesses of various carbides and select materials.

Table 4 Wear data for cobalt- and nickel-based alloys

Alloy	Average hot hardness (DHP), kg/mm²			Adhesive wear volume loss(a), mm³		Average abrasive wear(b)		Galling test(c) (self mated) at 4080 kg (9000 lb)
	At 540 °C (1000 °F)	At 650 °C (1200 °F)	At 760 °C (1400 °F)	At 68.2-kg (150-lb) load	At 95.5-kg (210-lb) load	Volume loss, mm³	Wear coefficient(d)	Degree of damage, mm³
Cobalt-based alloys								
Alloy No. 1 465	390	230	0.6	0.7	46	15×10^{-4}	0.63	
Alloy No. 6 275	260	185	2.6	9.5	66	13×10^{-4}	1.88	
Alloy No. 12 325	285	245	2.4	7.6	53	11×10^{-4}	...	
Alloy No. 21 145	135	115	5.2	10.3	70	8×10^{-4}	No damage	
Nickel-based alloys								
Alloy No. 60 440	250	115	0.2	0.2	11	3×10^{-4}	13.7	
Alloy No. 50	...	...	0.3	0.3	12	3×10^{-4}	...	
Alloy No. 40 270	195	80	0.19	0.22	21	5×10^{-4}	...	

(a) Average of two or more tests in air against a case hardened SAE 4520 steel ring (63 HRC). (b) Tested for 2000 rpm at a load of 13.6 kg (30 lb) using a 230-mm (9-in.) diam rubber wheel and dry sand. (c) Pin on block test; reciprocating motion through 120°. Surface is analyzed via profilometry. (d) Wear coefficient (K) is calculated from $V = KPL/h$, where V = wear volume in mm³; P = load in kg; L = sliding distance in mm; and h = diamond pyramid hardness. Source: Cabot Corp.

Table 5 Corrosion rates for typical cobalt- and nickel-based alloys

Five 24-h test periods; determined in laboratory tests; samples should be tested under actual plant conditions.

Alloy	Corrosion rates for:			
	30% formic acid at 66 °C (150 °F)	30% boiling acetic acid	5% sulfuric acid at 66 °C (150 °F)	65% nitric acid at 66 °C (150 °F)
Cobalt-based alloys				
Alloy No. 1 .... E	G	S	G	
Alloy No. 6 .... E	E	E	U	
Alloy No. 12 .... E	G	E	E	
Alloy No. 21 .... E	E	E	E	
Nickel-based alloys				
Alloy No. 60 .... G	U	U	U	
Alloy No. 50 .... S	U	U	U	
Alloy No. 40 .... S	U	U	U	

Note: E (excellent) 0.13 mm/yr (5 mils/yr); G (good) 0.13 to 0.51 mm/yr (5 to 20 mils/yr); S (satisfactory) greater than 0.51 to 1.27 mm/yr (20 to 50 mils/yr); U (unsatisfactory) greater than 1.27 mm/yr (50 mils/yr). Source: Cabot Corp.

Table 6 Approximate hardness of carbides and select materials

Material	DPH	Hardness HK	Mohs
Diamond	...	8000	10
SiC 3200	2750	9.2	
W₂C 3000	2550	+9	
VC 2800		+9	
TiC 2800	2750	+9	
Cr₃C₂ 2700			
Alumina	2100	9	
WC 2400	1980	+9	
Cr₇C₃ 2100	...	...	
Cr₂₃C₆ 1650	...	...	
Mo₂C 1570	...	8	
Zircon	1340	...	
Fe₃C 1300	Cementite	...	
Quartz 1000	800	7	
Lime	560	...	
Glass	500-600	...	

Iron-Based Powders

Iron-based hardfacing alloys are more widely used than cobalt- and/or nickel-based hardfacing alloys and constitute the largest volume use of hardfacing alloys. Although iron-based powders are most frequently used as filler materials in consumable electrode hardfacing processes, they are also used in hardfacing techniques that use powder as the consumable. Iron-based hardfacing alloys offer low cost and a broad range of desirable properties.

Most equipment that undergoes severe wear, such as crushing and grinding equipment and earthmoving equipment, is usually very large, rugged, and often subject to contamination. Parts subjected to wear usually require downtime for repair. For this reason, they are usually hardfaced with the lowest cost and most readily available materials. As a result, literally hundreds of iron-based hardfacing alloys are in use today.

Due to the large number of alloys involved, iron-based hardfacing alloys are best classified by their suitability for different types of wear and their general microstructures rather than by chemical composition. Most iron-based hardfacing alloys can be divided into:

- Pearlitic steels
- Austenitic steels
- Martensitic steels
- High-alloy irons

Properties of several typical iron-based hardfacing alloys of major types are listed in Table 7 as a general guide for selection of hardfacing materials.

Pearlitic steels are essentially low-carbon steels with minor adjustments in composition to achieve weldability. These alloys contain low carbon (0.25%) and low amounts of other alloying elements, resulting in a pearlitic structure. Pearlitic steels are useful as buildup overlays, primarily to rebuild machinery parts back to size. Examples include shafts, rollers, and other parts in heavy machinery subjected to rolling, sliding, or impact loading. Typically, this group of alloys has high impact resistance and low hardness (in the range 25 to 35 HRC), as well as excellent weldability.

Austenitic Steels. Austenite in this group of alloys is usually stabilized by manganese additions. Austenitic iron-based hardfacing alloys essentially are modeled after Hadfield steels. Most commercially available alloys in this category can be broadly subdivided into low-chromium and high-chromium alloys.

Low-chromium alloys usually contain up to 4% Cr and 12 to 15% Mn and some nickel or molybdenum. Low-chromium austenitic steels generally are used to re-

Table 7 Composition and hardness of select iron-based hardfacing alloys

Nominal composition	Nominal hardness DPH	HRC	Unlubricated sliding wear volume(a), mm³	Abrasive wear volume(b), mm³	Density, g/cm³
Pearlitic steels					
Fe-2Cr-1Mn-0.2C 318		32	0.5	55	7.75
Fe-1.7Cr-1.8Mn-0.1C 372		38	0.6	67	7.47
Austenitic steels					
Fe-14Mn-2Ni-2.5Cr-0.6C . . . . 188 HRB		88 HRB	0.4	86	7.75
Fe-15Cr-15Mn-1.5Ni-0.2C 230		18	0.3	113	7.75
Martensitic steels					
Fe-5.4Cr-3Mn-0.4C 544		52	0.4	54	7.47
Fe-12Cr-2Mn-0.3C 577		54	0.3	60	7.47
High-alloy irons					
Fe-16Cr-4C 595		55	0.3	13	7.47
Fe-30Cr-4.6C 560		53	0.2	15	7.19
Fe-36Cr-5.7C 633		57	0.1	12	7.47

(a) Wear measured from tests conducted on Dow-Corning LFW-1 against 4620 steel ring at 80 rpm for 2000 revolutions varying the applied loads. (b) Wear measured from dry sand rubber wheel abrasion tests. Tested for 2000 revolutions at a load of 14 kg (30 lb) using a 23-cm (9-in.) diam rubber wheel and AFS test sand.

build machinery parts subjected to high impact (impact crusher or shovel lips). Low-chromium austenitic steels are not recommended for joining manganese steel parts due to the possibility of cracking. Welding to plain carbon steels should be carefully executed. Martensite sometimes forms in zones that are low in manganese content, thereby embrittling the interdiffusion zone.

High-chromium austenitic steels, which may normally contain 12 to 17% Cr in addition to about 15% Mn, were developed to preclude diffusion zone embrittlement. Furthermore, the as-deposited hardness of high-chromium steels is higher (16 to 20 HRC) than that of low-chromium steels (86 to 88 HRB). High-chromium austenitic steels typically are used for rebuilding manganese steel and carbon steel parts subjected to high metal-to-metal pounding, such as railroad frogs and steel mill wobblers, as well as for joining manganese steels.

Martensitic Steels. Alloys in this category are designed to form martensite on normal air cooling of the weld deposit. As a result, these steels are often termed "self-hardening" or "air hardening," and they resemble tool steels with hardnesses in the range of 40 to 45 HRC. The carbon content of the martensitic steels ranges up to 0.5%. Other elements such as molybdenum, tungsten, nickel (up to 3%), and chromium (up to 15%) are added to increase hardenability and strength and to promote martensite formation. Manganese and silicon usually are added to aid weldability.

The major hardfacing applications for martensitic steels include unlubricated metal-to-metal rolling or sliding parts such as undercarriage parts of tractors. The impact resistance of martensitic steels is inferior to that of pearlitic or austenitic alloys, but there is a compensating increase in hardness and resistance to abrasive wear.

High-Alloy Irons. This group of alloys is referred to as irons because of their similarities to cast irons. These alloys contain large amounts of chromium and/or molybdenum carbides in what is essentially a martensitic matrix. The carbon content ranges from 2 to 6%, providing large amounts of carbides. Accordingly, resistance to abrasion is improved considerably over the low-carbon iron-based alloys. Impact resistance and toughness of these alloys, however, are correspondingly lower.

High-alloy iron hardfacing deposits usually are limited to one or two passes in thickness due to an extremely high susceptibility to form tension cracks. In some alloys containing 4 to 8% Mn or up to 5% Ni, the austenitic phase is stabilized in the matrix, which reduces the tendency to crack on cooling. Most high-alloy irons have extremely high compressive strength, but very low tensile strength. The hardness range for high-alloy irons is typically 52 to 62 HRC.

Special Compositions

Each manufacturer produces many different powder compositions, some of which are unique. Several examples of powders with unique properties are discussed below.

Laves Phase Alloys. Laves phase is a type of topologically close-packed intermetallic compound. Historically, the presence of Laves phase has been avoided in most alloys due to its detrimental effect on mechanical properties. However, in the early 1960's, the usefulness of Laves phase in resisting metal-to-metal wear was discovered; subsequently, alloys containing Laves phase have become commercially available.

The deposits have smooth surfaces that can be finished to a 2-μm (78-μin.) surface finish. Alloys are particularly well suited to applications in which adequate lubrication is a problem. Laves phase materials maintain mechanical properties over a wide range of temperatures. Their main uses are for plasma spray hardfacing operations. An important application is for plasma spray coating of gas turbine blades.

There are two different Laves phase-containing cobalt-based alloys commercially available for hardfacing applications: Co-28Mo-8Cr-2Si and Co-28Mo-17Cr-3Si. These alloys contain at least 50 vol% of Laves phase bound in a cobalt-based matrix that is alloyed with chromium and molybdenum.

Laves phase has a hexagonal structure similar to M_7C_3 carbides, but a hardness value between 1000 and 1200 DPH, which is less than that of carbides. Consequently, the Laves phase-containing alloys are less abrasive to mating materials than carbide-containing alloys in metal-to-metal wear situations.

Only one Laves phase-containing nickel-based alloy is commercially available—Ni-32Mo-13Cr-3Si. This alloy, like most nickel-based alloys, is difficult to weld using the oxyacetylene process, but can be readily welded using gas tungsten arc welding or the plasma transferred arc process. It can also be applied using the plasma or detonation gun spray technique. Although it has excellent metal-to-metal wear resistance and moderate abrasive wear resistance, it exhibits poor impact resistance.

Nickel-Aluminum Powders. Protective coatings based on nickel-aluminum composite powders are deposited by plasma spray techniques. Such composite powders are inherently heterogeneous, in that each particle consists of a nickel core and an aluminum coating. The main advantage of using nickel-aluminum powders is their superior bond strength, which results

Fig. 4 Spray-and-fuse torch in which the powder is conveyed by gravity into the flame and conveyed to the workpiece by the force of the combustion gases
(a) Schematic. (b) Typical torches. Courtesy of Hoeganaes Corp.

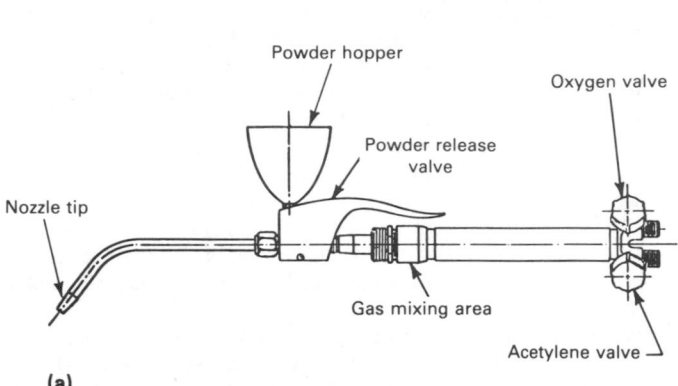

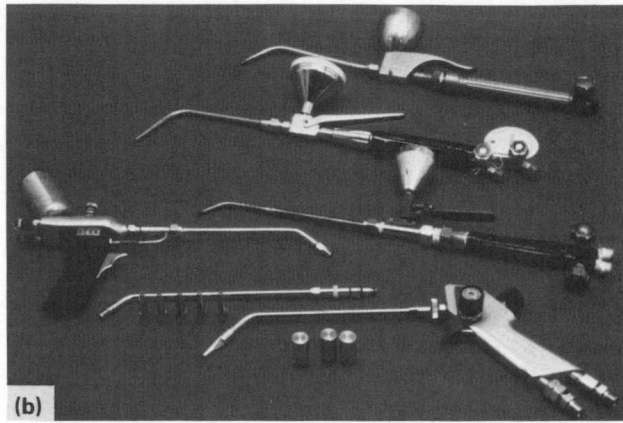

Fig. 5 Schematic of spray-and-fuse hardfacing process

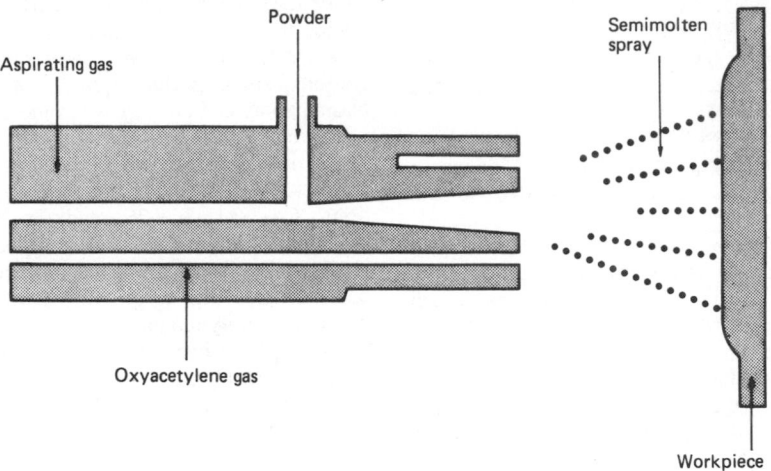

from exothermic reactions that occur within or on the surface of the powders as they pass through the plasma spray flame. Some diffusion also results in the substrate (in the absence of heat), thus further improving the overall coating bond strength. These composite powders are also referred to as self-bonding, one-step, or exothermic hardfacing powders. See the article "Production of Composite Powders" in this Volume for additional information.

Ceramics are being used increasingly as hardfacing powders. In addition to resistance to wear, they can have other desirable properties such as insulation and chemical resistance. The processes used for depositing ceramics include plasma spray and detonation gun techniques. The material may be a single oxide, mixed oxides, or other type of ceramic. Examples of some materials in this category are alumina (Al_2O_3) and chromium oxide (Cr_2O_3). Ceramic coatings have high hard-nesses, ranging up to 1200 DPH, and very little porosity. In addition, these coatings can be ground and lapped to obtain a very smooth surface finish.

High-Temperature Powders. A relatively new type of hardfacing powder is that based on the (Fe, Ni, Co)-Cr-Al-Y system. These yttrium-containing powders are used for selective coating of gas turbine blades by the plasma spray process. The coating is highly resistant to high-temperature oxidation and corrosion and is claimed to be superior to coatings made by the pack cementation, diffusion coating technique.

Hardfacing Processes, Equipment, and Applications

For satisfactory performance of the coating, selection of the proper hardfacing process is equally as important as selec-tion of the proper hardfacing material. Process selection requires careful consideration of many factors, such as required coating thickness, metallurgical integrity of the coating, size and shape of the component to be hardfaced, compatibility with the base metal, welder skill, cost of filler material, and equipment downtime.

The term "hardfacing" is sometimes used to describe true welding processes, in which both the filler material and a small amount of the base metal are molten and are bonded metallurgically. Such processes are preferred for applications that require dense and relatively thick coatings with excellent bonding between the hardfacing material and the base metal. Thermal spray processes are used for applications that require a thin, hard coating applied with minimum work distortion and where a purely mechanical bond between the hardfacing material and the base metal is acceptable. Unless otherwise stated, the term "hardfacing" is used in this section to include both the welding and thermal spray types of processes.

As discussed earlier, the properties of the hardfacing material are affected by dilution, which in turn varies with the hardfacing process. The success of a hardfacing operation depends on factors such as surface preparation of the base metal, preheating conditions, and specific process parameters.

Spray-and-Fuse Process

The spray-and-fuse process is a two-step process in which powdered coating material is deposited by conventional thermal spraying, usually using either a combustion gun (or torch) or a plasma spray gun, and subsequently fused using either a heating torch or a furnace. Coatings usu-

Fig. 6 Spray-and-fuse sequence for hardfacing a tubular part
(a) Grit-blasted, mounted tube. (b) Flame spraying. (c) Fusing operation. (d) Finished (fused) part. Courtesy of Hoeganaes Corp.

ally are made of nickel or cobalt self-flux-ing alloys to which hard particles, such as tungsten carbide, may be added for in-creased wear resistance. Deposits ranging from 0.51 to 2.5 mm (0.020 to 0.100 in.) thick can be made by building up several layers at a rate of 0.13 to 0.25 mm (0.005 to 0.010 in.) per pass.

Deposit thickness is controlled by the traverse speed of the torch, speed of ro-tation (when done between centers on cy-lindrical parts), powder flow, and the number of layers applied. Deposit rates of 3.5 to 5.5 kg/h (8 to 12 lb/h) can be achieved. Spray-and-fuse hardfacing is particularly effective on small cylindrical parts, as well as on irregular shapes. Cy-lindrical parts can be sprayed and fused between centers on a converted lathe. Work travels at about 0.8 m/s (150 sfm).

Several types of combustion powder spraying systems are used for the initial deposition step, differing primarily in the method used to feed the powder into the flame. Compressed air can be used as a propellant for the powder, which is in-jected directly into the flame through a central orifice in the nozzle. Another sys-tem feeds the powder into aspirating gas

positioned ahead of the nozzle prior to ig-nition. As shown in Fig. 4, gravity feed can be used to feed the powder directly into the flame. Powder is then conveyed to the workpiece by the force of the com-bustion gases.

Acetylene combined with oxygen typi-cally is used as the fuel gas, although other fuel gases such as propane and MAPP gas are sometimes used. Powder may also be encapsulated in plastic tubing for use with wire-type spray equipment. A typical flame spray schematic is shown in Fig. 5. Al-ternatively, a plasma spray torch rather than a combustion gun may be used. In all set-ups, the powdered alloy becomes molten or semimolten as it passes through or into the combustion flame or plasma effluent, thus facilitating the required initial bond prior to fusing.

Fusion is usually accomplished in a sep-arate operation using an oxyacetylene torch with a multi-tip flame designed to spread the heat over a wide area. Alternatively, fusion may be accomplished by induction heating or by heat treatment in a vacuum or reducing atmosphere furnace. As-sprayed coatings are stressed; conse-quently, fusing should immediately follow

spraying to eliminate the possibility of cracking of the coating.

In the spray-and-fuse process, the com-ponent, which usually is cleaned by grit blasting, is heated uniformly to approxi-mately 345 °C (650 °F). The torch is then placed over the area where fusion is to ini-tiate, and the temperature of this area is raised to about 760 °C (1400 °F). Move-ment of the torch is then stopped, and the heat is concentrated at the starting point.

The temperature is raised at this posi-tion until the surface exhibits a shiny mol-ten layer (glazing). The torch is then moved along the coating to fuse it uniformly, as indicated by glazing; heating should not be excessive, or loss of shape or running may result. Fusing should not be inter-rupted, and the entire coating should be fused at once. The as-deposited powder coating is bonded mechanically, and upon fusing, a high-quality metallurgical bond also results between the deposit and the base metal. The above sequence is de-picted in Fig. 6, which shows a mounted tube being hardfaced by the spray-and-fuse process.

The spray-and-fuse process can be used to coat a wide range of materials. How-

Fig. 7 Microstructure of self-fluxing nickel-based alloy (top) deposited by the spray-and-fuse process on a mild steel substrate.
Courtesy of Cabot Corp.

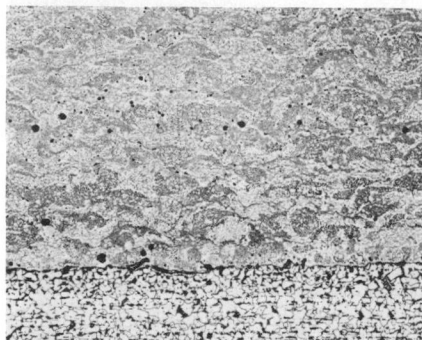

Fig. 8 Defective casting being repaired by the spray-and-fuse hardfacing process
(a) Part being repaired. (b) Repaired casting. Courtesy of Hoeganaes Corp.

(a)

(b)

ever, some materials, such as tool steels, austenitic stainless steels, and steels with carbon contents higher than 0.25%, require preheating and careful cooling after fusing to prevent cracking of the deposit. Use of type 303 stainless steel or other free-machining metals containing sulfur, manganese, phosphorus, lead, or selenium can result in porous overlays due to the emanation of gas during fusion, and these materials usually are not considered for this process. Titanium- or aluminum-bearing alloys (type 321 stainless steel) are also not considered because of the highly stable oxide layers on these base metals, which interfere with proper bonding of the overlay to the base metal during the fusion operation.

Self-fluxing nickel-based powders are the most common hardfacing materials used with the spray-and-fuse process. As shown in Table 2, these powders are available for a wide range of deposit hardnesses. The lower hardness grades find extensive use in the glass mold industry. The higher hardness grades are used extensively on pump components, particularly in the oil industry.

The quality and properties of the deposit depend not only on the hardfacing powder composition, but also on the particle size distribution and the care exercised in making and fusing the deposit. Figure 7 shows a typical microstructure of a self-fluxing nickel-based alloy deposit. Porosity up to 5% is common in the majority of hardfacing applications. The spray-and-fuse process is also used with cobalt-based powders (Table 2) and with carbide-containing composite powders. Composite powders are more difficult to use, and their deposits exhibit more porosity.

Spray-and-fuse powders generally are applied to cylindrical parts such as pump shafts, packing gland sleeves, and pistons. Other typical applications include ammonia applicators, cams, fan blades, glass molds, pump sleeves and pump plungers, valve seats and stems, bushings, sprockets, wire drawing capstan rings, overlays on dics and molds, and casting repairs, as shown in Fig. 8.

Plasma Spray Process

In the plasma spray process, a plasma gas stream heats and propels powder particles onto the workpiece surface. Because plasma temperatures are extremely high, the plasma spray process can be used to apply refractory coatings that cannot be applied by the spray-and-fuse process.

Although plasma spray coatings have inherent porosity and are principally mechanically bonded to the substrate, they generally have higher density and better adhesion than is achieved with single-step flame spraying. Powders can be applied manually (Fig. 9) or with mechanized equipment at rates ranging from less than 0.45 kg/h (1 lb/h) to over 14 kg/h (30 lb/h), depending on power of the equipment and type of powder used. In most applications, coating thickness is in the range of 0.13 to 0.4 mm (0.005 to 0.015 in.); depending on the application, much thicker coatings, such as thermal barrier

Fig. 9 Manual application of a hardfacing coating by the plasma spray process
Courtesy of Cabot Corp.

coatings, can be applied. Usually, the workpiece is maintained below 150 °C (300 °F), resulting in little or no distortion of the component and no dilution of the coating by the substrate. Very smooth deposits can be obtained, and finishing may not be required.

Figure 10 schematically depicts a typical plasma spray gun, showing alternative powder inlet positions. A gas, usually argon or nitrogen (or a mixture of these) with hydrogen or helium, flows through a water-cooled copper anode, which serves as a constricting nozzle. An arc is maintained internally from an axial rear tungsten electrode.

Fig. 10 Schematic of plasma spray hardfacing process

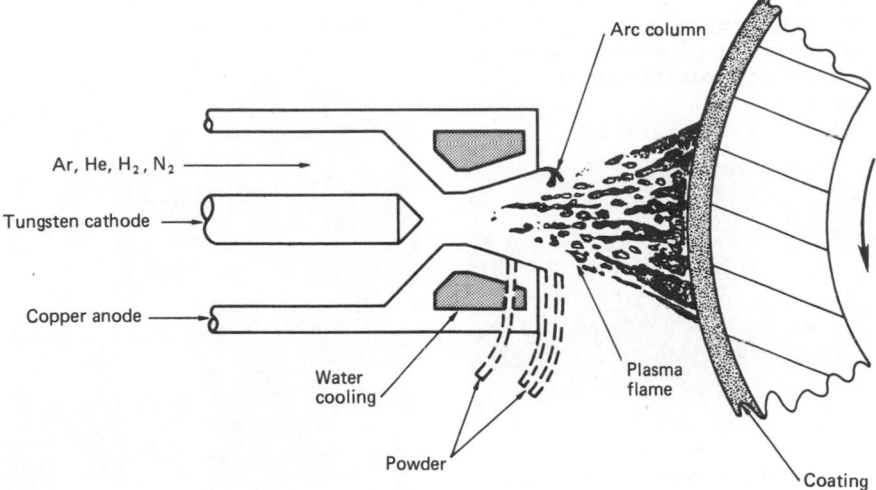

Fig. 11 Microstructure of a plasma spray deposited Laves phase nickel-based alloy

Magnification: (a) 60×. (b) 300×. Courtesy of Cabot Corp.

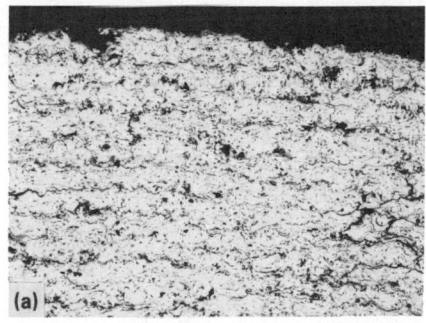 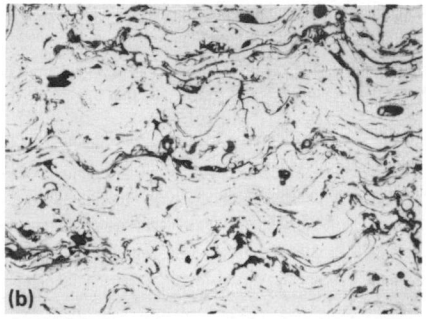

Powder introduced into the plasma stream is heated to a molten or semimolten state and propelled at high velocity onto the substrate. Plasma spray deposits with some high-velocity (>305 m/s, or 1000 ft/s) torches can have bond strengths in excess of 70 MPa (10 ksi). The powder is fed at a precisely regulated rate into an inert carrier gas stream and is usually introduced into the plasma stream either in the diverging portion of the nozzle or just beyond the exit.

Usually, the carrier gas is the same composition as the primary plasma gas (argon and/or nitrogen). The choice of powder-feed rate is important. If the powder is fed faster than it can be properly heated, deposition efficiency decreases rapidly, and the coating contains trapped, unmelted powder. In contrast, if the feed rate is too low, operation costs significantly increase.

Materials for the plasma spray process must be in powder form and must be sized appropriately to achieve melting. For the highest coating density, powders are sized between about 44 μm (325 mesh) and 10 μm. The lower limit ensures free flow. Coarser powder may be sized up to 100 μm, with typical distribution falling between 88 μm (170 mesh) and 44 μm (325 mesh). Powder shape ranges from spherical to acicular; spherical is best because of reduced surface area and consistent flow.

Any material that melts without subliming and is available as a properly sized powder has potential as a plasma spray process coating. Plasma spray coatings may be elemental (aluminum, molybdenum, nickel, or chromium), alloyed (nickel-, iron-, or cobalt-based), compounds (Al_2O_3, Cr_2O_3, or Cr_3C_2), composites (nickel-aluminum, cobalt-bonded tungsten carbide TiO_2, or clad Al_2O_3), or mechanical blends. A microstructure of a plasma-sprayed nickel-based Laves phase alloy is shown in Fig. 11.

In one process variation, the plasma gun and workpiece are enclosed in a vacuum chamber, and the entire operation is carried out in an inert atmosphere at low pressure (50 torr). Advantages of this low-

Fig. 12 Schematic of the plasma transferred arc hardfacing process

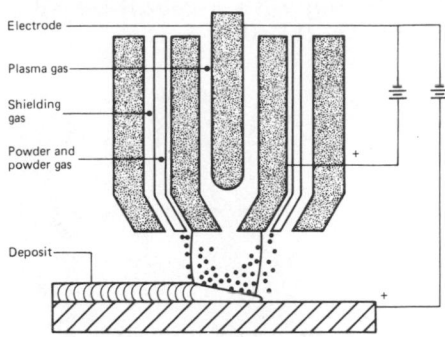

pressure plasma technique include improved bonding and density of the deposit, improved control over coating thickness (even with an irregular work surface), and higher deposit efficiency. However, facility cost is higher, and workpiece size is limited by the vacuum chamber size. This modification is suitable for coating turbine airfoils, blade tips, and shroud segments.

Very fine powder (−325 mesh) generally is used for low-pressure plasma spraying to achieve maximum density. If high-purity coatings are also desired, then the powder must be manufactured under inert cover and atomized in an inert gas to preserve purity.

Suitable applications for plasma spraying include those requiring a thin, smooth, wear-resistant coating, with low heat buildup in the workpiece. This process is used for hardfacing seal rings of aircraft jet turbine engines and for other turbine components. The coating provides a thermal barrier, resists high-temperature erosion, and forms a tight seal between rotating parts. Other applications include bottling machine parts, mixer and feeder parts, shafts and sleeves, wear rings, arc disks, and thread guides.

Plasma Transferred Arc Process

Plasma transferred arc surfacing, as shown schematically in Fig. 12, is a welding process in which the powder is introduced into a combined arc/plasma stream to form a molten pool on the workpiece. The resulting deposit is homogeneous and dense, with excellent metallurgical bonding with the base metal (workpiece). This process uses a constricted arc. The alloy is carried from a powder feeder to the plasma torch in a stream of argon gas, which forms the plasma and is directed away from the torch into the arc effluent, where it is melted and fusion bonded to

Fig. 13 Automotive diesel valves before (right) and after (left) plasma transferred arc hardfacing with a cobalt-based alloy (Co-28Cr-4W-1.1C)
Courtesy of Cabot Corp.

Fig. 14 Microstructure of a cobalt-based (Co-28Cr-4W-1.1C) alloy (top) deposited by plasma transferred arc hardfacing on a martensitic stainless steel substrate
Magnification: 75×. Courtesy of Cabot Corp.

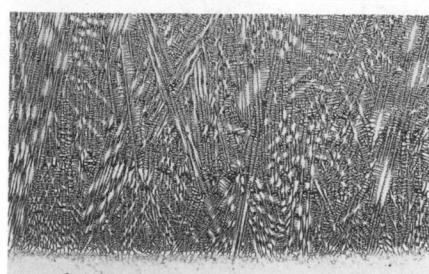

Fig. 15 Pump sleeve being hardfaced on a plasma transferred arc machine
Courtesy of Cabot Corp.

Fig. 16 Extruder screw being hardfaced on a plasma transferred arc machine
Courtesy of Cabot Corp.

the base metal. A direct current power source connecting the tungsten electrode and the workpiece provides energy for the transferred arc.

A second direct current power source connecting the tungsten electrode and the arc-constricting orifice supports a nontransferred arc. This nontransferred arc supplements the heat of the transferred arc and serves as a pilot arc to initiate the transferred arc. Argon is passed through a gas diffuser to provide a blanket of shielding gas in and around the arc zone.

Deposits ranging from 0.64 to 3.2 mm (0.025 to 0.125 in.) thick can be produced rapidly in a single pass. Penetration into the base metal can be controlled closely to very low levels. However, a dilution of 5 to 20% in the deposit is typical. Powder used for plasma transferred arc surfacing typically ranges from about 150 to 45 μm in size. Coarser powders can be used, because the transferred arc can easily melt the powder. Use of particles finer than 45 μm must be limited to prevent smoking and oxide inclusions in the deposit.

The plasma transferred arc process is well suited to automation operation and is therefore a frequent choice for high-volume applications such as valves (Fig. 13). The exhaust valves of internal combustion engines are subjected not only to mechanical degradation (due to impact and sliding stresses) but also to hot corrosion from the by-products of combustion. In diesel engines, hot corrosion is particularly detrimental, especially if low-grade (residual) fuels are used.

To combat hot corrosion and mechanical wear of valve seat faces that must maintain integrity for efficient engine operation, these components are commonly coated with cobalt-based alloys. The microstructure of a plasma transferred arc deposited cobalt-based alloy (Co-28Cr-4W-1.1C) is shown in Fig. 14.

Another application of the process is shown in Fig. 15, in which a pump sleeve made of type 316 stainless steel is hardfaced with a Co-30Cr-12.5W-3Ni-3Fe-2.5C alloy. Preheating was achieved at 93 °C (200 °F). Argon was used as the plasma and shielding gas, and helium was used as the carrier gas. The transferred arc current was 180 A. Deposit thickness of 3.8 mm (0.15 in.) was obtained as a single layer. Figure 16 shows an extruder screw being hardfaced by the plasma transferred arc system.

Manual Powder Torch Welding

Also termed "powder welding," this is a true welding process in which powder application and fusion occur in a single operation using a special oxyacetylene torch. Powder is fed from a small hopper mounted on a gas welding torch into the fuel gas supply and conveyed through the flame to the workpiece surface. By following general oxyacetylene surfacing practice, smooth, thin, dense deposits are obtained. Deposit thickness is controlled by powder flow rate and movement of the torch. A dilution of 1 to 5% in the deposit occurs from this process.

Deposits ranging from 0.76 to 3.2 mm (0.030 to 0.125 in.) in thickness can be applied at a rate of 1.3 mm (0.050 in.) per pass. Sweating of the base metal during surfacing generates a fusion-type bond with the hardfacing alloy. Ease of application and material recovery are enhanced be-

cause of close control of the powder flow into the welding flame.

Oxyacetylene torch flame adjustment is essential to produce good powder deposition. The amount of excess acetylene required for hardfacing varies depending on alloy composition. The amount of excess acetylene is conveniently measured in terms

Fig. 17 Manual powder torch welding application
(a) Spare tips. (b) Deposition process in progress. Courtesy of Cabot Corp.

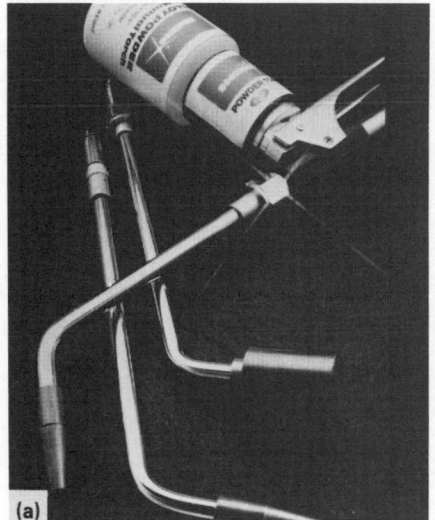

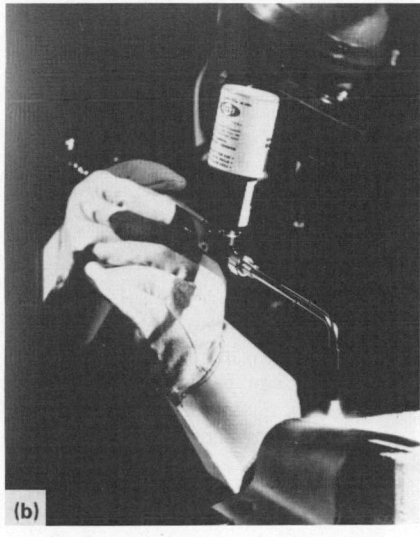

(a)

(b)

Fig. 18 Microstructure of a manual powder torch deposit of a Ni-2.8Si-1.5B-1Fe-0.2C alloy (top) on a mild steel substrate
Magnification: 100×. Courtesy of Cabot Corp.

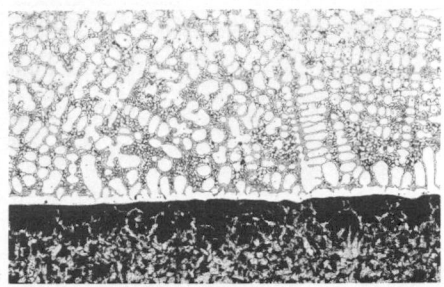

of the flame geometry as the ratio between the length of the oxyacetylene feather and the inner cone. The length of the inner cone, indicated by the value X, is the distance from the torch tip to the extreme end of the inner cone.

The length of the oxyacetylene feather is measured from the torch tip to the extreme end of the intermediate zone, generally expressed in multiples of the length of the inner cone. A neutral flame is thus classified as 1×. Cobalt-based powders are usually applied using 2× to 3× flame. The oxyacetylene feather is two to three times the length of the inner cone.

Tungsten carbide-containing composite powders should be applied using a 1.5× flame. Nickel-based powders are welded with a neutral or slightly oxidizing flame. Because flame characteristics change somewhat when powder is being sprayed, the powder feed handle should be depressed and final flame adjustment made while the powder is being sprayed. Alloys commonly welded with this process contain boron and silicon, which slag off any oxide skin to provide a clean weld pool.

The manual torch process is well suited for small repairs or for hardfacing of small areas requiring a wide range of hardnesses (about 20 to 60 HRC). Typical applications of manual torch welding include glass dies and molds, brick molding equipment, components for cement industry, cast iron parts for foundries, feed screws, hammer mill hammers, and pump parts. Figures 17 and 18 depict hardfacing with a manual torch and the microstructure of a nickel-based (Ni-2.8Si-1.5B-1Fe-0.2C) alloy, respectively.

Specialized Hardfacing Techniques

The hardfacing processes described previously in this article are well established. However, other proprietary techniques offer special advantages in select applications. This section reviews some of the specialized techniques that are based on significantly different principles than the methods discussed up to this point.

Detonation Gun. The detonation gun spray process is markedly different from other flame spraying processes and was initially developed for the deposition of hard, wear-resistant materials such as oxides and carbides. The extremely high particle velocities achieved in the detonation gun result in coatings with higher density, greater internal strength, and superior bond strength than can be achieved with conventional plasma spraying or single-step flame spraying.

This process, which is depicted schematically in Fig. 19, uses a controlled series of explosions of a mixture of oxygen and acetylene to blast the powder onto the surface of the workpiece. The powder to be sprayed is introduced into the long chamber in the mixture of oxygen and acetylene. When the gas mixture is ignited, the denotation products rapidly heat and accelerate the powder particles. Temperatures of 3300 °C (6000 °F) are reached inside the gun.

The process is capable of spraying a wide range of both metallic and nonmetallic powders with thicknesses of up to 0.5 mm (0.020 in.). The coating produced is similar to, but denser than, a plasma spray coating. It has an excellent bond strength and can be finished to an excellent surface finish. The workpiece can be processed below 150 °C (300 °F) in a carbon dioxide spray cooling system. This process has found wide acceptance in aerospace and oil exploration industries.

Jet Kote* Surfacing System. This is a high-pressure combustion thermal spray process, in which the spray gun burns a fuel (either MAPP gas or propane) with a high volume of oxygen within a combustion chamber. Using a nitrogen carrier gas, the powder is injected into the combustion products and is rapidly heated and accelerated to impact at high speed against the work surface to be coated. Unlike the detonation gun process, the Jet Kote system is based on continuous combustion within the gun.

Systems with high operating pressure provide a smooth, dense coating that is comparable to coatings produced by the detonation gun process. Figure 20 shows the microstructure of a Jet Kote-deposited tungsten carbide/cobalt coating. The process can also use both metallic and ceramic powders comparable to those used in plasma spraying. One of the main advantages of the Jet Kote system is low capital equipment cost compared to plasma spraying. Typical applications include those in which high-quality, thermal spray coatings are required, such as in the oil exploration industry.

Bulkwelding Process. The bulkwelding process, also referred to as joint welding, is an "add-on" to conventional submerged arc welding. It utilizes the existing heat produced by a conventional submerged arc weld to melt powders ahead of the melt pool to increase overall deposition rates. The process uses a powder composition that matches the feed wire composition.

*Jet Kote is a registered trademark of Cabot Corp.

Fig. 19 Schematic of detonation gun hardfacing process

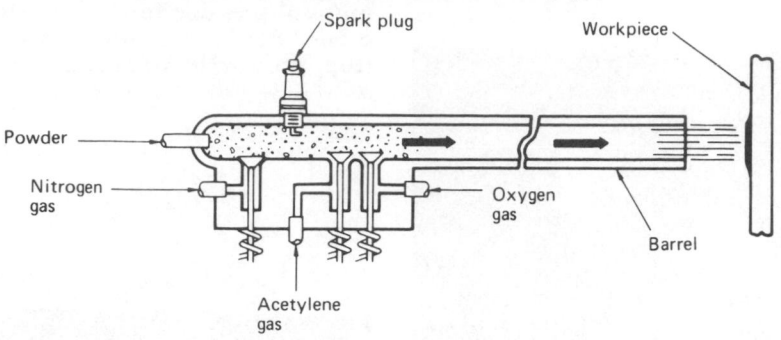

Fig. 20 Microstructure of tungsten carbide-cobalt deposit made by the Jet Kote process
Magnification: 100×. Courtesy of Cabot Corp.

Fig. 21 Schematic of the bulkwelding process using metal powder joint fill

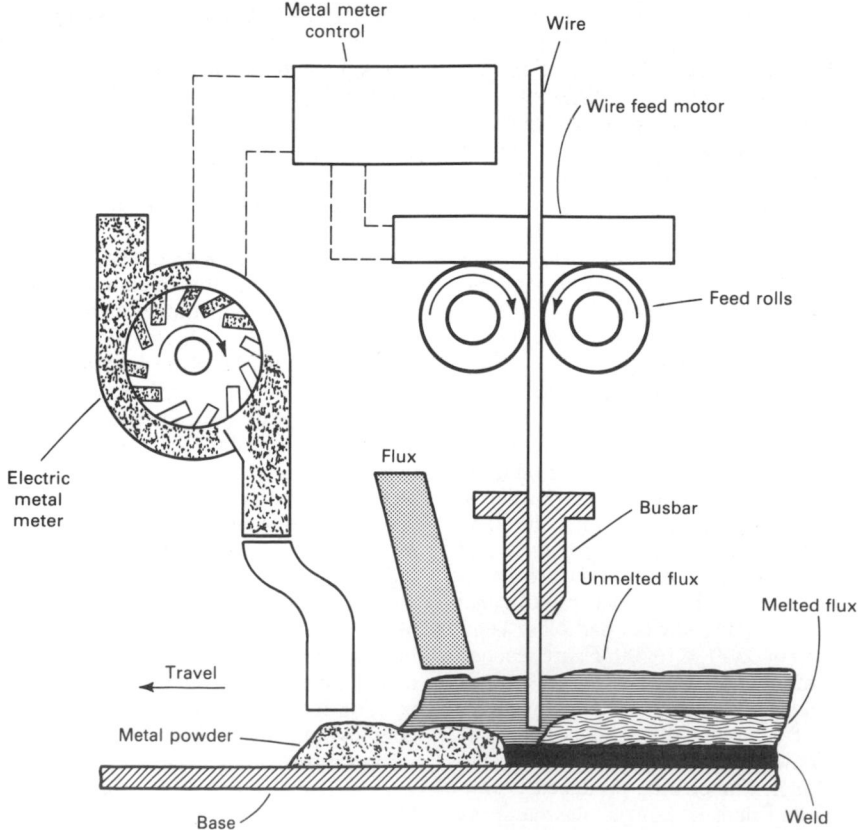

The quantity of powder used is governed by the application. The powder metal to wire ratio can be as high as 2 to 1. However, for most applications, this ratio is 1 to 1. A schematic of the process is shown in Fig. 21.

The following advantages are associated with this process:

- Deposition rates with bulkwelding can be increased two or three times over single-wire submerged arc welding without increasing heat input. The deposition rate depends on the amount of powder metal used in relation to the wire.

- Flux consumption per pound of deposited metal is reduced considerably.
- Less distortion is evident due to lower heat input.
- Accurate powder metering prevents spreading and other related problems (unmelted powder and rough edges) and improves overall deposition efficiency.
- A reduction in weld passes is possible. One-pass welding can be done on most applications.

The process is used for shipbuilding, building fabrication in which partial- or full-penetration weld and fillet welds are required, heavy machinery, bridges, and vessels. Most powders used are iron based. However, some cobalt-based powders are also used.

Conforma Clad Process. This process uses a flexible cloth made of polytetrafluoroethylene and a hardfacing powder of desired composition. The main benefit derived from this technique is that the cloth can be cut into any required size and placed on the surface to be coated, regardless of the profile of the workpiece. During subsequent furnace brazing, polytetrafluoroethylene is vaporized, and the hardfacing powder forms the deposit layer. The powder particles can be irregular or spherical in shape and can vary in size from 1 to 150 µm. Typical powder compositions are similar to the nickel-based alloys 40, 50, and 60 and cobalt-based alloy SF6 in Table 2. In addition, composites of these nickel-based alloys with tungsten carbide or chromium carbide are also produced.

Deposits produced by this process are highly dense and wear resistant. The primary advantage is the ability of the Conforma Clad process to build up thick coatings (up to 2.5 mm or 0.1 in. thick) on selected areas of parts. Typical applications of the process are for agricultural equipment and valves.

SELECTED REFERENCES

- Antony, K.C., Wear-Resistant Cobalt-Base Alloys, *J. Metals,* Vol 35 (No. 2), Feb 1983, p 52-60
- Knotek, O. and Lugscheider, E., Brazing Filler Metals Based on Reacting Ni-Cr-B-Si Alloys, *Welding J. (Welding Res. Suppl.),* Oct 1976, p 314S-318S
- Mayer, C.A., How to Select Hard-Surfacing Materials, *Welding Des. Fabrication,* Vol 55, Oct 1982, p 61-67
- Moskowitz, L.N. and Klar, E., Structure-Property Relationships in Flame Sprayed Nickel-Base Powder Coatings, *Proc. 1980 Int. Powder Metall. Conf.,* June 1980, Metal Powder Industries Federation, Princeton, NJ

Metal Powders Used for Brazing and Soldering

By Robert L. Peaslee
Vice President
Wall Colmonoy Corp.

METAL POWDERS are used extensively as filler metals in the brazing and soldering industries. These powders offer a convenient method of applying filler metal to parts, although alternative filler metal forms (wire, rod, clad sheet, foil, preforms and plating) also are used. For brazing, filler metals have a liquidus temperature above 450 °C (840 °F) and below the solidus of the base metal. For soldering, filler metals (solders) have a liquidus temperature below 450 °C (840 °F) and below the solidus of the base metal.

In both brazing and soldering, on heating to a suitable temperature with a suitable flux or protective atmosphere, the filler metal is distributed between closely fitted surfaces of the joint by capillary action. For detailed information on the various types of brazing and soldering operations, selection of filler materials, flux composition and selection, and protective atmospheres, see Volume 6 of the ninth edition of *Metals Handbook*.

Powder Types

Atomized Powders. Most powders currently are produced by atomization, because it offers a convenient and cost-effective means of producing large quantities of powder. Atomization consists of melting the metal or alloy in an induction melting furnace and then pouring it into a crucible, which has a hole in its bottom and is fitted with a gas nozzle at the bottom of the hole. In most atomizing processes, a high-pressure stream of fluid (usually water or gas) impinges on the molten metal stream, thus breaking it into droplets.

Materials such as solders can be readily air atomized, while most silver-based, nickel-based, and other alloys are atomized with a high-pressure argon or nitrogen gas. For additional information on the atomization process, see the article "Atomization" in this Volume.

Powders are subsequently screened to the desired particle size distribution. Most brazing filler metals are sieved to one of four standard sieve analyses as follows:

Sieve size	Size analysis, %
100 mesh	
−60	100 min
−100	95 min
140 C mesh	
+100	Trace
+140	10 max
−325	20 max
140 F mesh	
+100	Trace
+140	10 max
−325	55 max
325 mesh	
+200	Trace
+325	10 max
−325	90 min

Source: Ref 1

Brazing filler metal atomized powders are available in the following types:

- Nickel alloys
- Cobalt alloys
- Silver alloys
- Gold alloys
- Copper-phosphorus alloys
- Copper
- Aluminum-silicon alloys

These powders are generally screened to American Society for Testing and Materials standard ASTM B 214 ("Standard Method for Sieve Analysis of Granular Metal Powders"), using sieves with wire size to ASTM E 11 ("Standard Specification for Wire-Cloth Sieves for Testing Purposes"). Table 1 lists the compositions of various brazing powders, as well as their respective solidus, liquidus, and brazing temperatures.

Fine prealloyed solder powders (−325 mesh) are used to make soldering pastes and creams, in which the solder alloy is suspended in the fluxing material. The ratio of solder to flux is 80 to 90 wt%. A standard specification for solder metal can be found in ASTM B 32. Table 2 lists the compositions of some typical solder alloy powders used in soldering pastes and creams, as well as their respective solidus, liquidus, and soldering temperatures.

Ball Milled Powders. Historically, nickel filler metals have been produced by ball milling. In this process, alloy shot is ball milled into fine powders. Currently, atomization techniques are used to produce these powders.

Mechanically Filed Powders. Early filler metal powders made of silver and gold filler metals were mechanically filed, using standard commercial filing machines. These powders, however, were not as readily usable in paste form. The long, thin filings do not flow through a thin nozzle when in paste form; consequently, mechanical filing, while still available, is decreasing in usage and has been converted mostly to atomized powder manufacture.

Oxide Powders. Most paste-type copper filler metals are made of red copper oxide that has been milled to a fine pow-

Table 1 Compositions and properties of typical brazing alloy powders

Nickel alloys

Filler metal	Cr	B	Si	Fe	C	P	S	Al	Ti	Mn	Cu	Zr	Ni	Other elements total	Solidus temperature °C	Solidus temperature °F	Liquidus temperature °C	Liquidus temperature °F	Brazing temperature range °C	Brazing temperature range °F
BNi-1	13.0-15.0	2.75-3.50	4.0-5.0	4.0-5.0	0.6-0.9	0.02	0.02	0.05	0.05	...	...	0.05	rem	0.50	977	1790	1038	1900	1066-1204	1950-2200
BNi-1a	13.0-15.0	2.75-3.50	4.0-5.0	4.0-5.0	0.06	0.02	0.02	0.05	0.05	...	...	0.05	rem	0.50	977	1790	1077	1970	1077-1204	1970-2200
BNi-2	6.0-8.0	2.75-3.50	4.0-5.0	2.5-3.5	0.06	0.02	0.02	0.05	0.05	...	...	0.05	rem	0.50	971	1780	999	1830	1010-1177	1850-2150
BNi-3	...	2.75-3.50	4.0-5.0	0.5	0.06	0.02	0.02	0.05	0.05	...	...	0.05	rem	0.50	982	1800	1038	1900	1010-1177	1850-2150
BNi-4	...	1.5-2.2	3.0-4.0	1.5	0.06	0.02	0.02	0.05	0.05	...	...	0.05	rem	0.50	982	1800	1066	1950	1010-1177	1850-2150
BNi-5	18.5-19.5	0.03	9.75-10.50	...	0.10	0.02	0.02	0.05	0.05	...	...	0.05	rem	0.50	1079	1975	1135	2075	1149-1204	2100-2200
BNi-6	...	...	...	...	0.10	10.0-12.0	0.02	0.05	0.05	0.04	...	0.05	rem	0.50	877	1610	877	1610	927-1093	1700-2000
BNi-7	13.0-15.0	0.01	0.10	0.2	0.08	9.7-10.5	0.02	0.05	0.05	...	...	0.05	rem	0.50	888	1630	888	1630	927-1093	1700-2000
BNi-8	...	...	6.0-8.0	...	0.10	0.02	0.02	0.05	0.05	21.5-24.5	4.0-5.0	0.05	rem	0.50	982	1800	1010	1850	1010-1093	1850-2000

Cobalt alloys

Filler metal	Cr	Ni	Si	W	Fe	C	B	P	S	Al	Ti	Zr	Co	Other elements total	Solidus temperature °C	Solidus temperature °F	Liquidus temperature °C	Liquidus temperature °F	Brazing temperature range °C	Brazing temperature range °F
BCo-1	18.0-20.0	16.0-18.0	7.5-8.5	3.5-4.5	1.0	0.35-0.45	0.7-0.9	0.02	0.02	0.05	0.05	0.05	rem	0.50	1121	2050	1149	2100	1149-1232	2100-2250

Silver alloys

Filler metal	Ag	Cu	Zn	Cd	Ni	Sn	Li	Mn	Other elements total	Solidus temperature °C	Solidus temperature °F	Liquidus temperature °C	Liquidus temperature °F	Brazing temperature range °C	Brazing temperature range °F
BAg-1	44.0-46.0	14.0-16.0	14.0-18.0	23.0-25.0	...	...	...	...	0.15	607	1125	618	1145	618-760	1145-1400
BAg-1a	49.0-51.0	14.5-16.5	14.5-18.5	17.0-19.0	...	...	...	...	0.15	627	1160	635	1175	635-760	1175-1400
BAg-2	34.0-36.0	25.0-27.0	19.0-23.0	17.0-19.0	...	...	...	...	0.15	607	1125	702	1295	702-843	1295-1550
BAg-2a	29.0-31.0	26.0-28.0	21.0-25.0	19.0-21.0	...	...	...	...	0.15	607	1125	710	1310	710-843	1310-1550
BAg-3	49.0-51.0	14.5-16.5	13.5-17.5	15.0-17.0	2.5-3.5	...	...	...	0.15	632	1170	688	1270	688-815	1270-1500
BAg-4	39.0-41.0	29.0-31.0	26.0-30.0	...	1.5-2.5	...	...	...	0.15	671	1240	779	1435	779-899	1435-1650
BAg-5	44.0-46.0	29.0-31.0	23.0-27.0	...	...	...	...	...	0.15	677	1250	743	1370	743-843	1370-1550
BAg-6	49.0-51.0	33.0-35.0	14.0-18.0	...	...	...	...	...	0.15	688	1270	774	1425	774-871	1425-1600
BAg-7	55.0-57.0	21.0-23.0	15.0-19.0	...	...	4.5-5.5	...	...	0.15	618	1145	652	1205	652-760	1205-1400
BAg-8	71.0-73.0	rem	...	...	...	...	...	...	0.15	779	1435	779	1435	779-899	1435-1650
BAg-9	64.0-66.0	rem	13.0-17.0	...	...	...	...	...	0.15	671	1240	718	1325	718-843	1325-1550
BAg-10	69.0-71.0	rem	8.0-12.0	...	...	...	...	...	0.15	691	1275	738	1360	738-843	1360-1550
BAg-13	53.0-55.0	rem	4.0-6.0	...	0.5-1.5	...	...	...	0.15	718	1325	857	1575	857-968	1575-1775
BAg-13a	55.0-57.0	rem	...	...	1.5-2.5	...	...	...	0.15	771	1420	893	1640	871-982	1600-1800
BAg-18	59.0-61.0	rem	...	...	...	9.5-10.5	...	...	0.15	602	1115	718	1325	718-843	1325-1550
BAg-20	29.0-31.0	37.0-39.0	30.0-34.0	...	...	...	...	...	0.15	677	1250	766	1410	766-871	1410-1600
BAg-21	62.0-64.0	27.5-29.5	...	...	2.0-3.0	5.0-7.0	...	...	0.15	691	1275	802	1475	802-899	1475-1650
BAg-22	48.0-50.0	15.0-17.0	21.0-25.0	...	4.0-5.0	...	...	7.0-8.0	0.15	680	1260	699	1290	699-830	1290-1525
BAg-24	49.0-51.0	19.0-21.0	26.0-30.0	...	1.5-2.5	...	...	...	0.15	660	1220	707	1305	750-843	1305-1550
BAg-25	19.0-21.0	39.0-41.0	33.0-37.0	...	...	...	...	4.5-5.5	0.15	738	1360	790	1455	790-845	1455-1555
BAg-26	24.0-26.0	37.0-39.0	31.0-35.0	...	1.5-2.5	...	...	1.5-2.5	0.15	707	1305	802	1475	800-870	1475-1600
BAg-27	24.0-26.0	34.0-36.0	24.5-28.5	12.5-14.5	...	...	...	...	0.15	607	1125	746	1375	745-860	1375-1575
BAg-28	39.0-41.0	29.0-31.0	26.0-30.0	...	...	1.5-2.5	...	...	0.15	649	1200	710	1310	710-843	1310-1550

(continued)

Table 1 (continued)

Gold alloys

Filler metal	Composition(a), % Au	Cu	Pd	Ni	Other elements total	Solidus temperature °C	°F	Liquidus temperature °C	°F	Brazing temperature range °C	°F
BAu-1	37.0-38.0	Bal	...	...	0.15	991	1815	1016	1860	1016-1093	1860-2000
BAu-2	79.5-80.5	Bal	...	...	0.15	891	1635	891	1635	891-1010	1635-1850
BAu-3	34.5-35.5	Bal	...	2.5-3.5	0.15	974	1785	1029	1885	1029-1091	1885-1995
BAu-4	81.5-82.5	...	...	Bal	0.15	949	1740	949	1740	949-1004	1740-1840
BAu-5	29.5-30.5	...	33.5-34.5	35.5-36.5	0.15	1135	2075	1166	2130	1166-1232	2130-2250
BAu-6	69.5-70.5	...	7.5-8.5	21.5-22.5	0.15	1007	1845	1046	1915	1046-1121	1915-2050

Copper-phosphorus alloys

Filler metal	Composition(a), % P	Ag	Cu	Other elements total	Solidus temperature °C	°F	Liquidus temperature °C	°F	Brazing temperature range °C	°F
BCuP-2	7.0-7.5	...	Bal	0.15	710	1310	793	1460	732-843	1350-1550
BCuP-3	5.8-6.2	4.8-5.2	Bal	0.15	643	1190	813	1495	718-816	1325-1500
BCuP-4	7.0-7.5	5.8-6.2	Bal	0.15	643	1190	718	1325	691-788	1275-1450
BCuP-5	4.8-5.2	14.5-15.5	Bal	0.15	643	1190	802	1475	704-816	1300-1500
BCuP-6	6.8-7.2	1.8-2.2	Bal	0.15	643	1190	788	1450	732-816	1350-1500
BCuP-7	6.5-7.0	4.8-5.2	Bal	0.15	643	1190	771	1420	704-816	1300-1500

Copper

Filler metal	Composition(a), % Cu	Zn	Sn	Fe	Mn	Ni	P	Pb	Al	Si	Other elements each	total	Solidus temperature °C	°F	Liquidus temperature °C	°F	Brazing temperature range °C	°F
BCu-1a	99.0 min	...	...	...	...	...	...	...	...	...	...	0.30	1083	1981	1083	1981	1093-1149	2000-2100
BCu-2	86.5 min	...	...	...	...	...	...	...	...	...	...	0.50	1083	1981	1083	1981	1093-1149	2000-2100

Aluminum-silicon alloys

Filler metal	Composition(a), % Si	Cu	Fe	Zn	Mg	Mn	Al	Other elements each	total	Solidus temperature °C	°F	Liquidus temperature °C	°F	Brazing temperature range °C	°F
BAlSi-4	11.0-13.0	0.30	0.8	0.20	0.10	0.15	Bal	0.05	0.15	577	1070	582	1080	582-604	1080-1120

(a) Single values are maximum percentages, unless otherwise indicated.
Source: AWS A5.8, "Specification for Brazing Filler Metals"

Table 2 Compositions and properties of typical soldering alloy powders

ASTM designation	Composition, %									Solidus temperature		Liquidus temperature		Soldering range	
	Tin	Silver	Antimony	Bismuth(a)	Copper(a)	Iron(a)	Aluminum(a)	Arsenic(a)	Lead	°C	°F	°C	°F	°C	°F
10 B	8-12	···	0.2-0.5	0.25	0.08	0.02	0.005	0.02	rem	268	514	299	570	329-369	624-696
20 B	18-22	···	0.2-0.5	0.25	0.08	0.02	0.005	0.02	rem	183	361	277	531	307-347	585-657
30 B	28-32	···	0.2-0.5	0.25	0.08	0.02	0.005	0.02	rem	183	361	255	491	285-325	545-617
40 B	38-42	···	0.2-0.5	0.25	0.08	0.02	0.005	0.02	rem	183	361	238	460	268-308	514-586
50 B	48-52	···	0.2-0.5	0.25	0.08	0.02	0.005	0.03	rem	183	361	216	421	213-282	415-540
60 B	58-62	···	0.2-0.5	0.25	0.08	0.02	0.005	0.03	rem	183	361	190	374	220-282	428-540
63 B	61-65	···	0.2-0.5	0.25	0.08	0.02	0.005	0.03	rem	183	361	183	361	212-253	414-487
62 P	61.5-67.5	1.75-2.25	0.3-0.5	0.25	0.08	0.02	0.005	0.005	rem	179	355	179	355	209-249	409-481
96.5 TS	95-96	3.3-3.7	0.2-0.5	0.25	0.08	0.02	0.005	0.005	···	221	430	221	430	251-291	484-556

Note: Other solder alloys are also available in paste form. Typical compositions, impurity levels, and designations are given in ASTM B 32.
(a) Maximum level

der. This powder is then mixed with a vehicle that provides a convenient paste for the filler metal application. Because copper oxide dissociates readily, it makes a good filler metal for furnace brazing of carbon steels in strongly reducing atmospheres such as combusted city gas (nitrogen/carbon monoxide/hydrogen) or prepared mixed atmospheres (nitrogen-hydrogen).

Preforms From Metal Powders. Preforms are used because they are readily applied with automated equipment. Preforms are easily produced from wire and sheet. Consequently, powders are generally not used to make preforms other than stampings from transfer tape materials, which are plastic-bonded sheets with an adhesive on one side. Solder powders generally are not used for making preforms.

Flux-Powder Mixtures. Brazing fluxes used in flux-powder applications are generally proprietary. These fluxes usually are composed of various mixtures of some of the following: borates of sodium, potassium, and lithium; fused borax; fluoroborates of potassium or sodium; fluorides of sodium, potassium, lithium; chlorides of sodium, potassium, and lithium; boric acid (frequently calcined); alkalis such as potassium hydroxide and sodium hydrides; wetting agents; and water (Ref 2).

Depending on the desired reactivity of the brazing flux and its melting characteristics, these materials are used in various quantities. Some fluxes are suitable for direct manufacture of flux-powder pastes, while other flux-powder pastes are made using a long-chain viscous organic vehicle that may exhibit thixotropic properties.

Solder fluxes are categorized as corrosive, intermediate, and noncorrosive fluxes. These fluxes are mixed with fine solder powder to form solder creams and pastes. Corrosive fluxes contain one or more of the following components:

- Zinc chloride
- Ammonium chloride
- Stannous chloride
- Sodium or potassium chloride
- Lithium chloride
- Aluminum chloride
- Sodium or potassium fluoride
- Boron trifluoride
- Hydrochloric acid
- Hydrofluoric acid
- Orthophosphoric acid
- Fluoboric acid

Intermediate-type powder/paste fluxes contain one or more of the organic acids (lactic acid, citric acid, stearic acid, and oxalic acid, for example) or organic hydrohalides (glutamic acid hydrochloride, aniline hydrochloride, and hydrazine hydrochloride, for example). Amines and other compounds such as urea, glycerol and diethylene diamine, or triamine also are used. For flux-powder pastes, one or more of these fluxing agents is added to the powder and is mixed with a long-chain viscous organic vehicle that may exhibit thixotropic properties.

Noncorrosive flux-powder pastes contain rosin as the main constituent, which is active at soldering temperatures but inactive at room temperature. Occasionally, additives are used to increase the reactivity of the rosin.

Paste Mixtures

Brazing paste mixtures usually are prepared for furnace brazing in a protective atmosphere. Consequently, they do not require flux additions. For special-purpose applications, a small amount of flux occasionally may be added.

Copper oxide powder pastes are extremely fine; consequently, vehicles such as ethylene glycol and mineral oil are used to form the paste mixture. On heating in the presence of copper or copper oxide, these vehicles chemically crack, leaving a residue of carbon. At the brazing temperature, carbon reacts with the water in the higher dew point (−1 to 26 °C, or 30 to 80 °F) furnace brazing atmosphere to produce carbon monoxide gas, thus eliminating the carbon from the surface of the part. If the furnace atmosphere is too dry, the carbon remains on the surface of the part.

Paste mixtures of nickel, silver, and gold filler metals use a long-chain organic vehicle that may exhibit thixotropic properties. These vehicles or binders are vaporized on heating and are forced out of the furnace by the gas atmosphere. If sufficient moisture exists in the furnace atmosphere—approximately −1 to 26 °C (30 to 80 °F) dew point—the vehicle reacts with the moisture to produce hydrogen and carbon monoxide. In a high-vacuum furnace (approximately 10^{-3} torr and below), the vehicle vaporizes when heated and is forced out into the cold portions of the furnace and usually is pumped out of the system.

Application of Filler Metal

Filler metal can be easily applied with the use of automated brazing machines by using powder in a paste mixture of either flux and/or binder, as required by the particular type of equipment. If the parts are to be brazed in an automated machine with air-gas torch heating, the use of flux is required to provide a suitable braze and to protect parts from oxidation, while allowing brazing or soldering filler metal to wet and flow over the part. When powder is to be used in automated equipment or in furnaces with controlled atmospheres, flux is not required. A binder is used to hold the powder to the work; when heated, it is gassed off and pumped out of the system.

Brazing and soldering powders, pastes, and creams are applied in different manners. One method consists of the use of an inexpensive plastic bottle that allows manual application of the filler metal during brazing or soldering furnace operations. The operator puts the parts on the furnace belt and then drops a quantity of filler metal in the proper location at the joint. More sophisticated types of application equipment apply the filler metal pneumatically or volumetrically to the parts to be brazed or soldered on an automated brazing or soldering machine. Powder and flux can be added either separately or as a single material on the brazed joint, as the parts move through the machine.

Solders generally are supplied in paste or cream form, which can be applied by screening, brushing, extruding, or rolling onto the substrate surfaces. Solder pastes and creams, which are used widely in microelectronic component soldering, are often applied using a syringe. Other methods of application of brazing filler metal include spraying, dipping, applying with a syringe, or applying dry powder and wetting with a binder.

Mechanical Properties of Brazed and Soldered Joints

The mechanical properties of joints made with powdered filler metals have properties comparable to those of joints made from solid rod, wire, and foil, provided that powders are clean and free of oxides and that the protective flux or atmosphere is maintained at the proper activity to prevent oxidation of the specific metal or alloy. Solder pastes and creams containing fluxes protect these powders from oxidation.

Brazing filler metals that contain chromium, silicon, boron, and other highly oxidizing elements experience some oxidation on heating to the brazing temperature. When the furnace atmosphere is not properly controlled, the large surface area of the brazing powder particles may pick up enough oxygen to encapsulate each particle or may remove most of the boron, thus causing poor brazing. This constitutes inadequate quality control and is easily avoidable.

Powder particle size may affect the flow of the brazing filler metal when it contains elements with stable oxides such as chromium, silicon, boron, aluminum, or titanium. The high surface area of -325 mesh powders, which contain elements with oxides that are difficult to dissociate, may pick up sufficient oxygen from the atmosphere to reduce the flow of the filler metal. The degree of oxygen pickup depends on the specific element present and the partial pressure of oxygen present in the brazing atmosphere during the heating portion of the cycle.

For example, if Ni-19Cr-10Si filler metal (BNi-5) is heated in a marginal atmosphere on type 304 stainless steel base metal, the base metal may emerge bright and clean; however, as the brazing temperature is approached, the chromium and silicon in the powdered filler metal will oxidize to some extent. The degree of oxidation depends on the partial pressure of oxygen in the atmosphere surrounding the powder. In a high-quality atmosphere, the degree of oxidation is so minimal that there is no deleterious effect on the melting and flow characteristics of the brazing filler metal.

As the partial pressure of oxygen in a brazing atmosphere increases, more surface oxidation of the powder occurs; smaller particles may be completely oxidized and thus will not melt and flow when the brazing temperature is reached. Larger particles may not be completely oxidized and thus will melt and flow; however, oxides and residual unmelted powders remain at the point of application. Unmelted powder results from the loss of silicon through oxidation.

Because high-surface-area brazing filler metal powders (-325 mesh) are more susceptible to oxidation, the majority of brazing operations utilize 140 F mesh powders. It should be noted, however, that if the brazing atmosphere is strictly controlled to maintain a low partial pressure of oxygen, -325 mesh powders that contain one or more of the elements having stable oxides can be used. For a lower quality atmosphere (higher partial pressure of oxygen), a coarser powder such as 140 C mesh is used.

As a result, brazing filler metal powder that is of a given mesh size and that contains one or more readily oxidizable elements is a good indicator of atmosphere quality. Every time the melting and flow characteristics of one of these filler metals is tested, the quality of the brazing atmosphere is being tested to an even greater extent. For this reason, extreme care must be exercised when interpreting the results of melting and flow tests. With proper control of the furnace brazing cycle (including atmosphere), each type of filler metal, regardless of its form (wire, sheet, or powder) will have the same mechanical properties in the joint.

REFERENCES

1. AWS Specification A5.8, "Brazing Filler Metal," American Welding Society, Miami, 1981
2. Fluxes and Atmospheres, in *Brazing Manual*, 3rd ed., American Welding Society, Miami, 1976

Metal Powders Used for Flame Cutting

METAL POWDER CUTTING is a technique that supplements an oxyfuel torch with a stream of iron or blended iron-aluminum powder to facilitate flame cutting of difficult-to-cut materials. The powdered material propagates and accelerates the oxidation reaction, as well as the melting and spalling action of the materials to be cut. Metal powder cutting was developed after World War II to provide an effective method of cutting stainless steels. These highly alloyed materials resist conventional oxyfuel cutting because a refractory oxide layer forms when stainless steel is heated in the presence of oxygen.

This article examines the uses of metal powders for cutting, lancing, and scarfing operations. Emphasis is placed on the types of powders commercially available and powder production methods, as well as the equipment used to transport the powder to the cutting surface, and the industrial applications of metal powder cutting.

Process Fundamentals

In all oxyfuel cutting processes, cutting is accomplished through a thermochemical reaction in which preheated metal is cut, or removed, by rapid oxidation in a stream of pure oxygen. During metal powder cutting, finely divided iron powder is carried by dry compressed air or nitrogen into the cutting oxygen stream along the line of the cut. Combustion of this powder increases the temperature of the heated zone; consequently, the refractory oxides are melted and fluxed by the action of the powder particles.

The extreme heat created by the burning of the powder enables cutting without preheating of the metal. As the cut progresses, the heat and fluxing action of the burning iron powder enable the cutting oxygen stream to oxidize the base metal continuously. For additional information on the fundamentals of oxyfuel cutting processes, the chemistry of cutting, and

the selection and properties of fuel gases, see the article "Thermal Cutting" in Volume 6 of the 9th edition of *Metals Handbook*.

Powder Selection and Production

Many metallic and nonmetallic powders have been used for cutting operations. The ideal powdered solid fuel, whether metallic or nonmetallic:

- Exhibits nonhygroscopic action (does not readily absorb moisture)
- Liberates considerable heat during combustion
- Generates combustion products that flux refractory oxides
- Generates nonirritating and nontoxic combustion products
- Provides good process economy

Nonmetallic powders, such as sodium bicarbonate, remove refractory oxides by forming liquid slags that have a fluxing effect. Such powders, however, give off little or no heat. Consequently, cutting speeds are significantly slower than for metallic powders.

The most suitable metallic powders for cutting applications are iron and iron-aluminum blends. For metal powder cutting of stainless steel, iron powder can be used. For cutting oxidation-resistant materials other than stainless steels, blends containing iron and atomized $-100+325$ mesh aluminum powder are preferable.

Iron powder used for metal cutting applications is produced by water atomization, or by a combination of granulation of molten iron by high-pressure water jets and chemical reduction. According to the Metal Powder Industries Federation, 2% of all iron powder produced in North America is used for cutting applications. Table 1 provides data on iron cutting powder produced by the Quebec Metal Powder (QMP) process. Cutting-grade iron powder of similar density and composi-

tion also is produced by the Domfer process. Detailed information on the production and properties of QMP- and Domfer-processed iron powders can be found in the article "Production of Iron Powder" in this Volume.

Aluminum powder is sometimes added to iron powder to form blends used for cutting and lancing concrete, firebrick, difficult-to-cut nonferrous metals, and refractory-laden scrap, such as slag deposits on ladle pouring spouts. Table 2 provides data on cutting-grade aluminum powders that are produced by air atomization of molten aluminum. Detailed information on the production and properties of atomized aluminum powder can be found in the article "Production of Aluminum Powder" in this Volume.

Powder blends consisting of iron and 10 to 40 wt% Al are used to cut a variety of materials. Although preblended compositions consisting of iron and 15 wt% Al are available, iron and aluminum powders can be custom blended for specific applications, as shown in Table 3.

To blend powders properly, components should be poured simultaneously into a dry, nitrogen-filled container and tumbled for about 5 min. Additional blending occurs as the iron-aluminum powder is screened in the powder dispenser. Slight nonuniformities in the mixture have little effect on cutting performance.

Equipment

Metal powder cutting processes require a means of introducing and burning powder in the cutting oxygen stream after the powder is conveyed to the cutting torch. A typical powder cutting equipment setup includes the following components to accomplish this:

- Dispenser to supply powder during cutting
- Source of dry compressed air or nitrogen to convey powder from the dis-

Table 1 Typical properties of iron powder used for metal powder cutting

Screen analysis

U.S. mesh	Opening, μm	%
On 70	212	Trace
−70+100	150	5
−100+140	106	28
−140+200	75	23
−200+325	45	24
−325	...	20

Chemical analysis, %

Iron	98
Carbon	0.20 max
Manganese	0.10 max
Hydrogen loss	1.0 max
Silica acid insolubles	0.30 max

Physical properties

Apparent density	2.9 g/cm³
Flow rate	26 s/50 g

Table 2 Typical properties of atomized aluminum powder for metal powder cutting and lancing

Screen analysis

U.S. mesh	%
On 100	Trace
−100+200	52-67
−200+325	31-42
−325	0-4

Chemical analysis(a), %

Aluminum	99.7
Iron	0.18-0.25 max
Silicon	0.12-0.15 max
Other metallics, each	0.01-0.03 max
Other metallics, total	0.15 max

Physical properties

Apparent density	1.1 g/cm³
Tap density	1.4 g/cm³
Surface area	0.10-0.20 m²/g

(a) Chemical analysis excludes 0.4% aluminum oxide, which exists on the surface of the particles.

Table 3 Iron-aluminum blends used for cutting and lancing difficult-to-cut materials

Nominal composition	Applications
Fe–10-15Al	Refractory scrap, thin nonferrous sections up to 25 mm (1 in.) thick
Fe–25Al	Aluminum, brass, bronze
Fe–30Al	Nickel, Monel, Inconel, Hastelloy, and concrete
Fe–40Al	Copper, brass, and bronze heavy sections 15 cm (6 in.) thick and up

Fig. 1 Pneumatic powder dispenser for use in metal powder cutting

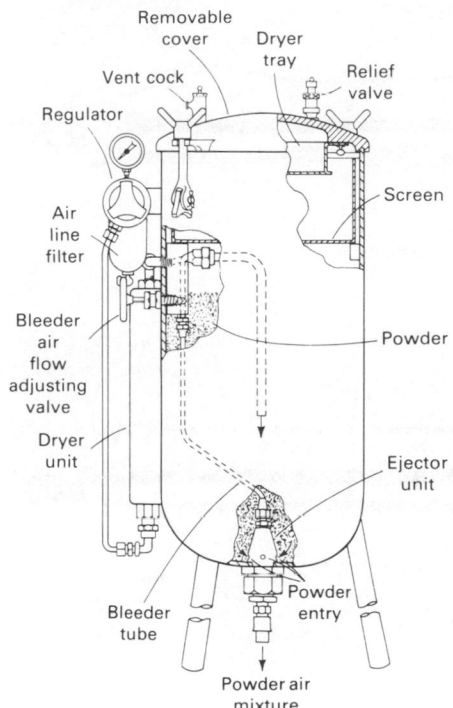

penser to the cutting oxygen stream
- Powder nozzle or attachment to introduce the powder into the cutting oxygen stream
- Cutting torch (blowpipe) to control the gases required to burn the oxygen/powder mixture

Powder Dispensers

Powder dispensers may be of two general types: pneumatic and vibratory. Generally, all dispensers consist of a compressed air/nitrogen filter and regulator, a powder hopper, and a means of expelling the powder from the dispenser.

Pneumatic Dispensers. During operation of a pneumatic dispenser (Fig. 1), powder flow to the ejector unit is the result of gravitational force and the pressure differential between the discharge end of the air line in the ejector and the surface of the powder when the torch powder valve is opened. Compressed air (or nitrogen) passing through the ejector draws powder from the ejector baffle plate into the gas stream, forming an air-powder (or nitrogen-powder) mixture which then passes through the ejector outlet tube to the torch or lance. The ejector adjusting screw controls the amount of powder falling on the baffle plate; the regulator and

the air throttling valve control the compressed air or nitrogen supplied to the ejector and the hopper. By manipulating the ejector adjusting screw, the regulator pressure, and the air throttling valve, the proper compressed air-powder (or nitrogen-powder) mixture can be obtained. This powder-conveying gas mixture is carried through a hose (6.4 mm or ¼ in. max diam by 7.5 m or 25 ft long) connected to the cutting torch.

A removable cover that is fitted with a pressure relief valve allows the hopper to be filled. A drying tray and a removable screen for eliminating over-sized powder particles are also fitted into the top of the hopper.

Vibratory dispensers allow a quantity of powder to be dispensed from a hopper that is governed by a vibrator. Powder flow can be controlled by adjusting the amplitude of vibration.

Vibratory dispensers are generally used where uniform and accurate powder flow is required. Typical applications include precision cutting of materials, such as stainless steel, as well as production of high-quality, sharp top edges on cuts in carbon steels.

Powder-Cutting Apparatus

Attachments for metal powder cutting are designed for use with conventional oxyfuel cutting torches. Attachments are available in three types: single-tube, multijet, and dual-tube. In addition to powder cutting attachments, specially designed powder cutting torches are available.

Single-tube attachments, which are clamped to a conventional machine or hand-held oxyfuel cutting torch as shown in Fig. 2(a), externally discharge a single, uniform stream of powder into the cutting oxygen stream at an angle of about 25°.

Multijet attachments, as shown in Fig. 2(b), are equipped with an adapter that fits over the standard cutting nozzle. The adapter (Fig. 2c) completely encircles the nozzle and feeds powder from a ring of ports through the preheat flames and into the cutting oxygen stream.

Dual-tube attachments are used for high-capacity machine cutting torch applications in which the size and cost of a multijet attachment are prohibitive. Two powder nozzles are used to feed powder into the high-velocity cutting oxygen stream.

Powder cutting torches are designed for manual powder cutting. Figure 3 shows a typical powder cutting torch. Depressing the cutting lever of the torches initiates the flow of both cutting oxygen and powder. The nozzles used in these torches are made

Fig. 2 Powder cutting attachments

(a) Single-tube attachment. (b) Multijet attachment. (c) Enlargement of powder nozzle adapter. Courtesy of Union Carbide Corp., Linde Division

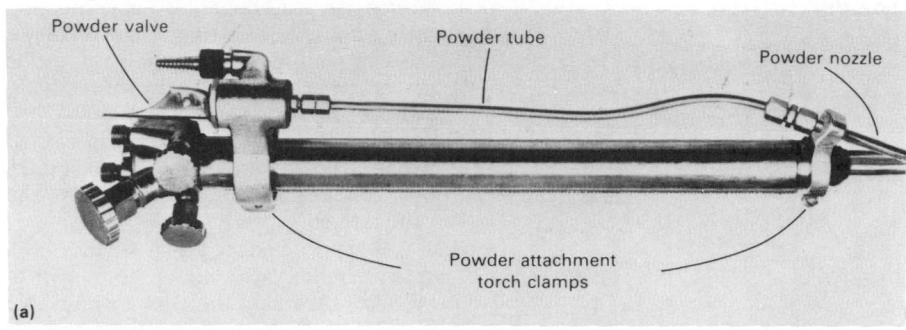

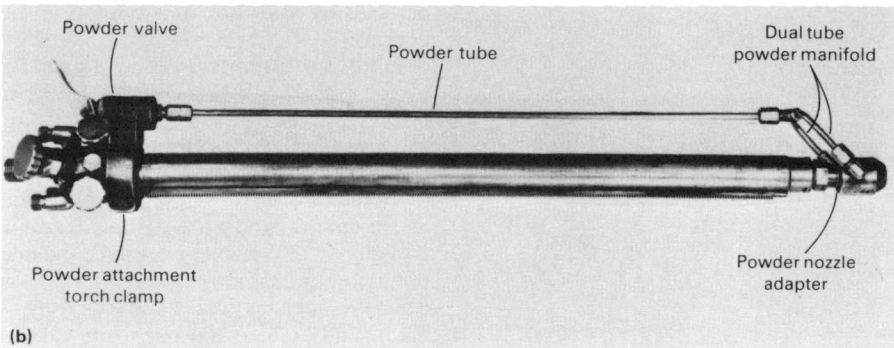

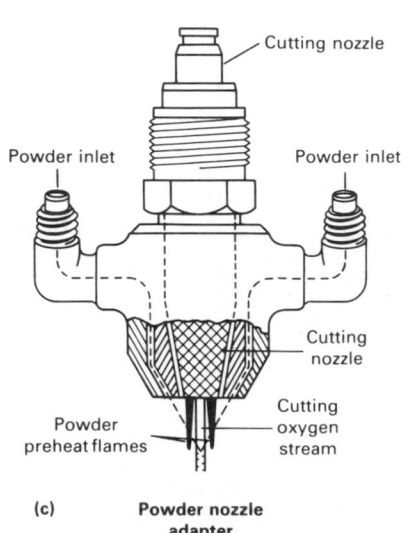

Table 4 Typical oxyfuel gas powder cutting conditions for 18-8 austenitic stainless steel

| Metal thickness | | Cutting oxygen orifice diameter | | Cutting oxygen pressure | | Cutting speed | | Gas consumption | | | | Powder flow | |
| | | | | | | | | Oxygen | | Acetylene | | | |
mm	in.	mm	in.	kPa	psi	mm/s	in./min	L/min	ft³/h	L/min	ft³/h	g/min	oz./min
13	½	1.02	0.040	344	50	5.9	14	59	125	7	15	113	4
25	1	1.52	0.060	344	50	5.1	12	106	225	11	23	113	4
51	2	1.52	0.060	344	50	4.2	10	142	300	11	23	113	4
76	3	2.03	0.080	344	50	3.8	9	260	550	15	32	142	5
102	4	2.54	0.100	344	50	3.4	8	319	675	18	38	170	6
127	5	3.05	0.120	414	60	3.0	7	378	800	21	45	198	7
152	6	3.56	0.140	414	60	2.5	6	425	900	30	63	227	8
203	8	3.56	0.140	483	70	1.7	4	472	1000	30	63	227	8
254	10	4.06	0.160	517	75	1.5	3.5	520	1100	35	75	227	8
305	12	4.06	0.160	517	75	1.3	3	566	1200	35	75	227	8

Table 5 Typical powder scarfing conditions for 18-8 stainless steel

| Cutting oxygen orifice diameter | | Oxygen pressure | | Acetylene pressure | | Powder flow | |
mm	in.	kPa	psi	kPa	psi	kg/h	lb/h
0.5	0.020	345-552	50-80	55-83	8-12	7-14	15-30
1.02	0.040	448-517	65-75	55-83	8-12	14-34	30-75

Note: Distance of nozzle from work: 64 mm (2½ in.); impingement angle: 30° (approximate); lateral angle: 50 to 10° (approximate); speed: 41 to 71 mm/s (96 to 168 in./min)

Applications

Metal powder cutting is frequently used in cutting, scarfing, and lancing of oxidation-resistant materials. This type of from chrome-plated pure copper inserted into an abrasion-resistant sleeve.

metal cutting was originally developed for cutting stainless steels. The alloying elements (chromium and nickel) that impart desirable properties to stainless steels have made these steels difficult to process by conventional oxyfuel cutting. Chromium is the most detrimental; when oxyfuel cutting is applied to stainless steels, chro-mium, which has a high affinity for oxygen at elevated temperatures, immediately forms the highly refractory chromium oxide on the faces of the kerf and prevents further oxidation. These refractory oxides are melted and fluxed by the combustion of the powder particles during metal powder cutting. Table 4 gives typical operating conditions for cutting of 18-8 stainless steel.

Nonferrous metals with the following thicknesses are suitable candidates for powder cutting:

| Material | Thickness | |
	cm	in.
Aluminum	25	10
Nickel	15	6
Brass and bronze	25	10
Copper	15	6
Hastelloy	16.5	6.5
Inconel and Monel	25	10

Powder Scarfing. In steel mills, powder metal cutting is used to scarf metals with alloy contents that are too high for oxyfuel scarfing. Typical applications include powder scarfing of large, bloomed ingots and small, unrolled ingots, slabs, and billets. Conditioning of these product forms by powder scarfing is less expensive and less time consuming than mechanical methods such as grinding. Table 5 provides typical operating conditions for scarfing of 18-8 stainless steel.

Fig. 3 Metal powder cutting torch
Courtesy of Union Carbide Corp., Linde Division

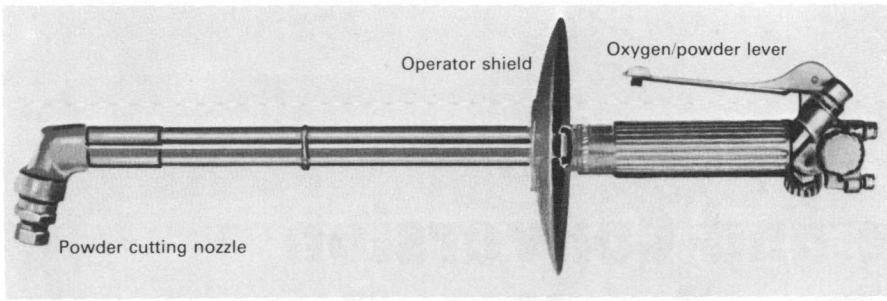

Operator shield

Oxygen/powder lever

Powder cutting nozzle

Powder lancing is a piercing process in which the energy of the powder reaction is applied to oxygen lancing, thus permitting rapid, effective piercing of many materials that are difficult to pierce with a standard oxygen lance. These include iron and steel containing inclusions, reinforced concrete, firebrick, cinder block, aluminum billets, and sand and metal incrustations inside large castings.

Typical powder lancing applications include:

- Removal of blast furnace bosh plates
- Removal of large masses of iron (salamanders) that are deposited at the base of a blast furnace
- Cleaning of furnace linings
- Furnace tapping to remove slag
- Cleaning of soaking pits
- Removal of ladle skulls
- Piercing holes in reinforced concrete walls and floors

The efficiency of the powder lance in piercing reinforced concrete is preferred to the time-consuming, costly alternative of drilling with pneumatic tools. Figure 4 shows a 20-metric ton (22-ton), 46-cm (18-in.) thick section of reinforced concrete being removed by powder lancing.

Fig. 4 Removal of reinforced concrete by powder lancing
Courtesy of Union Carbide Corp., Linde Division

Appendix: Metric and Conversion Data for Powder Metallurgy

This Appendix is intended as a guide for expressing weights and measures in the Système International d'Unités (SI) for use in the powder metallurgy industry. The purpose of SI units, developed and maintained by the General Conference of Weights and Measures, is to provide a basis for world-wide standardization of units and measure. For more information on metric conversions, the reader should consult the following references:

- Standard for Metric Practice, ASTM E 380, American Society for Testing and Materials, 1916 Race Street, Philadelphia, PA 19103
- Metric Practice, ANSI/IEEE 268-1982, American National Standards Institute, 1430 Broadway, New York, NY 10018
- SI Units and Conversion Factors for the Steel Industry, 1978, American Iron and Steel Institute, 1000 16th Street, NW Washington, DC 20036
- SP 330, The International System of Units, National Bureau of Standards Order from Superintendent of Documents, U.S. Government Printing Office, Washington, DC 20402
- Metric Editorial Guide, 3rd ed., American National Metric Council, 1625 Massachusetts Ave., NW Washington, DC 20036
- ASME Orientation and Guide for Use of SI (Metric) Units, ASME Guide SI 1 (9th edition, 1982), The American Society of Mechanical Engineers, 345 East 47th Street, New York, NY 10017

Base, supplementary, and derived SI units

Measure	Unit	SI symbol formula
Base units		
Length	metre	m
Mass	kilogram	kg
Time	second	s
Electric current	ampere	A
Thermodynamic temperature	kelvin	K
Amount of substance	mole	mol
Luminous intensity	candela	cd
Supplementary units		
Plane angle	radian	rad
Solid angle	steradian	sr
Derived units		
Frequency	hertz	Hz
Force	newton	N
Pressure, stress	pascal	Pa
Energy, work, quantity of heat	joule	J
Power, radiant flux	watt	W
Quantity of electricity, electric charge	coulomb	C
Electric potential, potential difference, electromotive force	volt	V
Capacitance	farad	F
Electric resistance	ohm	Ω
Conductance	siemens	S
Magnetic flux	weber	Wb
Magnetic flux density	tesla	T
Inductance	henry	H
Luminous flux	lumen	lm
Illuminance	lux	lx
Activity (of radio-nuclides)	becquerel	Bq
Absorbed dose	gray	Gy

SI prefixes—names and symbols

Exponential expression	Multiplication factor	Prefix	SI symbol
10^{18}	1 000 000 000 000 000 000	exa	E
10^{15}	1 000 000 000 000 000	peta	P
10^{12}	1 000 000 000 000	tera	T
10^{9}	1 000 000 000	giga	G
10^{6}	1 000 000	mega	M
10^{3}	1 000	kilo	K
10^{2}	100	hecto(a)	h
10^{1}	10	deka(a)	da
10^{0}	1	BASE UNIT	
10^{-1}	0.1	deci(a)	d
10^{-2}	0.01	centi(a)	c
10^{-3}	0.001	milli	m
10^{-6}	0.000 001	micro	μ
10^{-9}	0.000 000 001	nano	n
10^{-12}	0.000 000 000 001	pico	ρ
10^{-15}	0.000 000 000 000 001	femto	f
10^{-18}	0.000 000 000 000 000 001	atto	a

(a) Nonpreferred. Prefixes should be selected in steps of 10^3 so that the resultant number before the prefix is between 0.1 and 1000. These prefixes should not be used for units of linear measurement, but may be used for higher order units. For example, the linear measurement, decimeter, is nonpreferred, but square decimeter is acceptable.

U.S. standard sieve sizes

Sieve number	Nominal size, μm	Nominal size, in.
No. 25	710	0.0278
No. 30	600	0.0234
No. 35	500	0.0197
No. 40	425	0.0165
No. 45	355	0.0139
No. 50	300	0.0117
No. 60	250	0.0098
No. 70	212	0.0083
No. 80	180	0.0070
No. 100	150	0.0059
No. 120	125	0.0049
No. 140	106	0.0041
No. 170	90	0.0035
No. 200	75	0.0029
No. 230	63	0.0025
No. 270	53	0.0021
No. 325	45	0.0017
No. 400	38	0.0015

Conversion factors

To convert from	to	multiply by
Angle		
degree	rad	1.745 329 E − 02
Area		
in.^{2}	mm^2	6.451 600 E + 02
in.^{2}	cm^2	6.451 600 E + 00
in.^{2}	m^2	6.451 600 E − 04
ft^2	m^2	9.290 304 E − 02
Bending moment or torque		
lbf · in.	N · m	1.129 848 E − 01
lbf · ft	N · m	1.355 818 E + 00
kgf · m	N · m	9.806 650 E + 00
ozf · in.	N · m	7.061 552 E − 03
Bending moment or torque per unit length		
lbf · in./in.	N · m/m	4.448 222 E + 00
lbf · ft/in.	N · m/m	5.337 866 E + 01
Current density		
A/in.^{2}	A/mm^2	1.550 003 E − 03
A/ft^2	A/m^2	1.076 400 E + 01
Electricity and magnetism		
gauss	T	1.000 000 E − 04
maxwell	μWb	1.000 000 E − 02
mho	S	1.000 000 E + 00
Oersted	A/m	7.957 700 E + 01
Ω · cm	Ω · m	1.000 000 E − 02
Ω circular-mil/ft	μΩ · m	1.662 426 E − 03
Electrode force		
lbf	N	4.448 222 E + 00
kgf	N	9.806 650 E + 00
Energy (impact, other)		
ft · lbf	J	1.355 818 E + 00
Btu (thermochemical)	J	1.054 350 E + 03
cal (thermochemical)	J	4.184 000 E + 00
kW · h	J	3.600 000 E + 06
W · h	J	3.600 000 E + 03
Flow rate		
ft^3/h	L/min	4.719 475 E − 01
ft^3/min	L/min	2.831 000 E + 01
gal/h	L/min	6.309 020 E − 02
gal/min	L/min	3.785 412 E + 00
Force		
lbf	N	4.448 222 E + 00
tonf	kN	8.896 443 E + 00
kgf	N	9.806 650 E + 00
Force per unit length		
lbf/ft	N/m	1.459 390 E + 01
lbf/in.	N/m	1.751 268 E + 02
Fracture toughness		
ksi$\sqrt{\text{in.}}$	MPa$\sqrt{\text{m}}$	1.098 800 E + 00

To convert from	to	multiply by
Heat content		
Btu/lb	kJ/kg	2.326 000 E + 00
cal/g	kJ/kg	4.186 800 E + 00
Heat input		
J/in.	J/m	3.937 008 E + 01
kJ/in.	kJ/m	3.937 008 E + 01
Length		
Å	nm	1.000 000 E − 01
μin.	μm	2.540 000 E − 02
mil	μm	2.540 000 E + 01
in.	mm	2.540 000 E + 01
in.	cm	2.540 000 E + 00
ft	m	3.048 000 E − 01
yd	m	9.144 000 E − 01
mile	km	1.609 300 E + 00
Mass		
oz	kg	2.834 952 E − 02
lb	kg	4.535 924 E − 01
ton (short, 2000 lb)	kg	9.071 847 E + 02
ton (short, 2000 lb)	kg × 10^3(a)	9.071 847 E − 01
ton (long, 2240 lb)	kg	1.016 047 E + 03
Mass per unit area		
oz/in.^{2}	kg/m^2	4.395 000 E + 01
oz/ft^2	kg/m^2	3.051 517 E − 01
oz/yd^2	kg/m^2	3.390 575 E − 02
lb/ft^2	kg/m^2	4.882 428 E + 00
Mass per unit length		
lb/ft	kg/m	1.488 164 E + 00
lb/in.	kg/m	1.785 797 E + 01
Mass per unit time		
lb/h	kg/s	1.259 979 E − 04
lb/min	kg/s	7.559 873 E − 03
lb/s	kg/s	4.535 924 E − 01
Mass per unit volume (includes density)		
g/cm^3	kg/m^3	1.000 000 E + 03
lb/ft^3	g/cm^3	1.601 846 E − 02
lb/ft^3	kg/m^3	1.601 846 E + 01
lb/in.^{3}	g/cm^3	2.767 990 E + 01
lb/in.^{3}	kg/m^3	2.767 990 E + 04
Power		
Btu/s	kW	1.055 056 E + 00
Btu/min	kW	1.758 426 E − 02
Btu/h	W	2.928 751 E − 01
erg/s	W	1.000 000 E − 07
ft · lbf/s	W	1.355 818 E + 00
ft · lbf/min	W	2.259 697 E − 02
ft · lbf/h	W	3.766 161 E − 04
hp (550 ft · lbf/s)	kW	7.456 999 E − 01
hp (electric)	kW	7.460 000 E − 01
Power density		
W/in.^{2}	W/m^2	1.550 003 E + 03
Pressure (fluid)		
atm (standard)	Pa	1.013 250 E + 05
bar	Pa	1.000 000 E + 05

To convert from	to	multiply by
in.Hg (32 °F)	Pa	3.386 380 E + 03
in.Hg (60 °F)	Pa	3.376 850 E + 03
lbf/in.^{2} (psi)	Pa	6.894 757 E + 03
torr (mmHg, 0 °C)	Pa	1.333 220 E + 02
Specific heat		
Btu/lb · °F	J/kg · K	4.186 800 E + 03
cal/g · °C	J/kg · K	4.186 800 E + 03
Stress (force per unit area)		
tonf/in.^{2} (tsi)	MPa	1.378 951 E + 01
kgf/mm^2	MPa	9.806 650 E + 00
ksi	MPa	6.894 757 E + 00
lbf/in.^{2} (psi)	MPa	6.894 757 E − 03
Temperature		
°F	°C	5/9 · (°F − 32)
°R	°K	5/9
Temperature interval		
°F	°C	5/9
Thermal conductivity		
Btu · in./s · ft^2 · °F	W/m · K	5.192 204 E + 02
Btu/ft · h · °F	W/m · K	1.730 735 E + 00
Btu · in./ft^2 · h · °F	W/m · K	1.442 279 E − 01
cal/cm · s · °C	W/m · K	4.184 000 E + 02
Thermal expansion		
in./in. · °C	m/m · K	1.000 000 E + 00
in./in. · °F	m/m · K	1.800 000 E + 00
Velocity		
ft/h	m/s	8.466 667 E − 05
ft/min	m/s	5.080 000 E − 03
ft/s	m/s	3.048 000 E − 01
in./s	m/s	2.540 000 E − 02
km/h	m/s	2.777 778 E − 01
mph	km/h	1.609 344 E + 00
Velocity of rotation		
rev/min (rpm)	rad/s	1.047 164 E − 01
rev/s	rad/s	6.283 185 E + 00
Viscosity		
poise	Pa · s	1.000 000 E + 01
strokes	m^2/s	1.000 000 E − 04
ft^2/s	m^2/s	9.290 304 E − 02
in.^{2}/s	mm^2/s	6.451 600 E + 02
Volume		
in.^{3}	m^3	1.638 706 E − 05
ft^3	m^3	2.831 685 E − 02
fluid oz	m^3	2.957 353 E − 05
gal (U.S. liquid)	m^3	3.785 412 E − 03
Volume per unit time		
ft^3/min	m^3/s	4.719 474 E − 04
ft^3/s	m^3/s	2.831 685 E − 02
in.^{3}/min	m^3/s	2.731 177 E − 07
Wavelength		
Å	nm	1.000 000 E − 01

(a) kg × 10^3 = 1 metric ton

Abbreviations and Symbols

a crystal lattice length along *a* axis

A ampere

Å angstrom

ac alternating current

Ac₁ temperature at which austenite begins to form upon heating

Ac₃ temperature at which transformation of ferrite to austenite is completed upon heating

Ac$_{cm}$ in hypereutectoid steel, temperature at which cementite completes solution in austenite

ACGIH American Conference of Governmental and Industrial Hygienists

Ae$_{cm}$, Ae₁, Ae₃ equilibrium transformation temperatures in steel

AES Auger electron spectroscopy

AFWAL Air Force Wright Aeronautical Laboratories

AIRS Advanced Inertial Reference Sphere

AISI American Iron and Steel Institute

AMS Aerospace Material Specification (of SAE)

ANSI American National Standards Institute, Inc.

AOAC Association of Official Analytical Chemists

API American Petroleum Institute

APMI American Powder Metallurgy Institute

APT ammonium paratungstate

Ar₁ temperature at which transformation to ferrite or to ferrite plus cementite is completed upon cooling

Ar₃ temperature at which transformation of austenite to ferrite begins upon cooling

Ar$_{cm}$ temperature at which cementite begins to precipitate from austenite on cooling

ASM American Society for Metals

ASME American Society of Mechanical Engineers

ASP Anti-segregation process or ASEA-STORA process

ASTM American Society for Testing and Materials

at.% atomic percent

atm atmosphere (pressure)

AWG American wire gage

AWS American Welding Society

B magnetic induction

B$_r$ magnetic remanence

b barn (unit of nuclear cross-section)

bal balance

bcc body-centered cubic

BET Brunauver-Emmett-Teller

Btu British thermal unit

cal calorie

CAP consolidation by atmospheric pressure

CCPA Cemented Carbide Producers Association

CCT continuous cooling transformation

CDA Copper Development Association

CE carbon equivalent

CIP cold isostatic pressing

cm centimetre

CP commercial purity

CPM Crucible Particle Metallurgy

CPU central processor unit

CVN Charpy V-notch

dB decibel

DBTT ductile brittle transition temperature

dc direct current

DCEN direct current electrode negative

DCEP direct current electrode positive

diam diameter

D$_o$ diffusion coefficient

d$_m$ mass median particle diameter

DPH diamond pyramid hardness

DRI direct reduced iron

DS dispersion strengthened

E modulus of elasticity

e natural logarithm base, 2.71828...

EBW electron beam welding

EDXA energy dispersive x-ray analysis

EIA Electronics Industries Association

ELI extra-low interstitial

ENS electrostatic nonmetallic separator

EPA Environmental Protection Agency

Eq equation

ESCA electron spectroscopy for chemical analysis

eV electron volt

F farad

fcc face-centered cubic

FDA Food and Drug Administration

Fig. figure

FIOR fluid iron ore reduction

FN ferrite number

ft foot

F-test significance test for variances (powder sampling)

G gauss

G shear modulus

g gram

gf gram force

GJ gigajoule

GMAW gas metal arc welding

GPa gigapascal

GTAW gas tungsten arc welding

H magnetic force

h hour

HAZ heat-affected zone

HB Brinell hardness

H_c critical magnetic field

hcp hexagonal close-packed

HF hardenability factor

HIB high iron briquettes

HIP hot isostatic pressing

HK Knoop hardness

hp horsepower

HRA Rockwell "A" Hardness

HRB Rockwell "B" Hardness

HRC Rockwell "C" Hardness

HRE Rockwell "E" Hardness

HRF Rockwell "F" Hardness

HRH Rockwell "H" Hardness

HSLA high-strength low-alloy

HV Vickers hardness

Hz hertz

IACS International Annealed Copper Standard

IARC International Agency for Research on Cancer

ICBM Intercontinental Ballistic Missile

ID inside diameter

I/M ingot metallurgy

in. inch

IOAP internally oxidized alloy powders

IRAS infrared astronomy satellite

ISO International Organization for Standardization

I_{sp} specific impulse

J joule

JIS Japanese Industrial Standard

K Kelvin

K strength constant

kg kilogram

kHz kilohertz

K_{Ic} plane-strain fracture toughness

K_{Iscc} threshhold stress intensity for stress corrosion cracking

kN kilonewton

kPa kilopascal

ksi 1000 pounds per square inch

kV kilovolt

kW kilowatt

L litre, longitudinal (direction)

lb pound

LEL lower explosive limit

ln natural logarithm (base e)

log common logarithm (base 10)

LOR loss on reduction

LPG liquified petroleum gas

LT long transverse (direction)

m metre

MAPP or MPS methyacetylene propadiene stabilized

max maximum

MC or MeC metal carbide

Me metal

M_f temperature at which martensite formation finishes during cooling

mg milligram

MIL military

min minimum, minute

mL millilitre

MM Mischmetal

mm millimetre

MN meganewton

MO or MeO metal oxide

mol mole

MPa megapascal

mPa millipascal

MPIF Metal Powder Industries Federation

M_s temperature at which martensite starts to form from austenite upon cooling

MW molecular weight

N newton

NBS National Bureau of Standards

NDTT nil ductility transition temperature

NEMA National Electrical Manufacturers Association

NFPA National Fire Protection Association

nm nanometre

No. number

OD outside diameter

ODS oxide dispersion strengthened

Oe oersted

OF oxygen free

ONIA Office National Industrial d'Azote

OSHA Occupational Safety and Health Administration

oz ounce

p page

Pa pascal

P_c compaction pressure

PCB printed circuit board

PEL permissible exposure limits

pH negative logarithm of hydrogen-ion activity

PH precipitation hardenable

P/M powder metallurgy

ppb parts per billion

ppm parts per million

pps pulses per second

Pr Prandtl number

PREP plasma rotating electrode process

PROM programmable read-only memory

psi pounds per square inch

psia pounds per square inch (absolute)

psig pounds per square inch (gage)

PTO power take off

R gas constant

RA reduction in area

rad radian

RAM random access memory

RBV relative bioavailability

RDX cyclotrimethylenetrinitramine

RE rare earth

Re Reynolds number

Ref reference

rem remainder

REP rotating electrode process

RF radio frequency

rms root mean square

ROC rapid omnidirectional compaction

rpm revolutions per minute

RWMA Resistance Welder Manufacturers Association

S siemens

s second

SAE Society of Automotive Engineers

SCR silicon controlled rectifier

SEM scanning electron microscopy

sfm surface feet per minute

SIMS secondary ion mass spectroscopy

ST short transverse (direction)

STEL short term exposure limit

T tesla, or transverse (direction)

t time

T_c transition temperature from normal to superconducting state, or crystallization temperature

TD thorium dioxide dispersion strengthened

TEM transmission electron microscopy

TLV threshold limit values

T_m melting temperature

TNT 2,4,6-trinitrotoluene

TRS transverse rupture strength

tsi tons per square inch

t-test test of the significance of means (powder sampling)

TTT time-temperature transformation

UHF ultra-high frequency

UNS Unified Numbering System (ASTM-SAE)

USDA United States Department of Agriculture

UTS ultimate tensile strength

V volt

VHF very high frequency

vol volume

vol% volume percent

W watt

Wb weber

wt% weight percent

XCLB x-ray compositional line broadening

XPS x-ray photoelectron spectroscopy

yr year

YS yield strength

° degree, angular measure

°C degree Celsius (centigrade)

°F degree Fahrenheit

⇌ direction of reaction

÷ divided by

= equals

> greater than

≥ greater than or equal to

∫ integral sign

∞ infinity

< less than

≤ less than or equal to

± maximum deviation

μin. micro-inch

μm micron

− minus, negative ion charge, through mesh

× multiplied by, diameters (magnification)

· multiplied by

/ per

% percent

+ plus, in addition to, including, positive ion charge, on mesh

√ surface roughness

√ square root

~ similar to, approximately

α angle

Δ change in quantity, application of heat in a chemical reaction

ε strain

μ magnetic permeability, coefficient of friction

ν Poisson's ratio

π pi (3.141592)

ρ density

Σ summation of

σ stress

σ_g geometric standard deviation

τ_{sol} solidification time

τ_{sph} spheroidization time

Ω ohm

INDEX